COMPREHENSIVE MEDICINAL CHEMISTRY II

COMPREHENSIVE MEDICINAL CHEMISTRY II

Editors-in-Chief

Dr John B Taylor
Former Senior Vice-President for Drug Discovery, Rhône-Poulenc Rorer, Worldwide, UK

Professor David J Triggle
State University of New York, Buffalo, NY, USA

Volume 2

STRATEGY AND DRUG RESEARCH

Volume Editor

Dr Walter H Moos
SRI International, Menlo Park, CA, USA
University of California–San Francisco, San Francisco, CA, USA

ELSEVIER

AMSTERDAM BOSTON HEIDELBERG LONDON NEW YORK OXFORD
PARIS SAN DIEGO SAN FRANCISCO SINGAPORE SYDNEY TOKYO

Elsevier Ltd.
The Boulevard, Langford Lane, Kidlington, Oxford OX5 1GB, UK

First edition 2007

Copyright © 2007 Elsevier Ltd. All rights reserved

1.05 PERSONALIZED MEDICINE © 2007, D Gurwitz
2.12 HOW AND WHY TO APPLY THE LATEST TECHNOLOGY © 2007, A W Czarnik
3.40 CHEMOGENOMICS © 2007, H Kubinyi
4.12 DOCKING AND SCORING © 2007, P F W Stouten

The following articles are US Government works in the public domain and not subject to copyright:
1.08 NATURAL PRODUCT SOURCES OF DRUGS: PLANTS, MICROBES, MARINE ORGANISMS, AND ANIMALS
6.07 ADDICTION

No part of this publication may be reproduced, stored in a retrieval system
or transmitted in any form or by any means electronic, mechanical, photocopying,
recording or otherwise without the prior written permission of the Publisher

Permissions may be sought directly from Elsevier's Science & Technology Rights
Department in Oxford, UK: phone (+44) (0) 1865 843830; fax (+44) (0) 1865 853333;
email: permissions@elsevier.com. Alternatively, you can submit your request online by
visiting the Elsevier web site at http://elsevier.com/locate/permissions, and selecting
Obtaining permission to use Elsevier material

Notice
No responsibility is assumed by the publisher for any injury and/or damage to persons
or property as a matter of products liability, negligence or otherwise, or from any use
or operation of any methods, products, instructions or ideas contained in the material
herein. Because of rapid advances in the medical sciences, in particular, independent
verification of diagnoses and drug dosages should be made

British Library Cataloguing in Publication Data
A catalogue record for this book is available from the British Library

Library of Congress Catalog Number: 2006936669

ISBN-13: 978-0-08-044513-7
ISBN-10: 0-08-044513-6

For information on all Elsevier publications
visit our website at books.elsevier.com

Printed and bound in Spain

06 07 08 09 10 10 9 8 7 6 5 4 3 2 1

Working together to grow
libraries in developing countries

www.elsevier.com | www.bookaid.org | www.sabre.org

ELSEVIER   BOOK AID International   Sabre Foundation

Disclaimers

Both the Publisher and the Editors wish to make it clear that the views and opinions expressed in this book are strictly those of the Authors. To the extent permissible under applicable laws, neither the Publisher nor the Editors assume any responsibility for any loss or injury and/or damage to persons or property as a result of any actual or alleged libellous statements, infringement of intellectual property or privacy rights, whether resulting from negligence or otherwise.

Knowledge and best practice in this field are constantly changing. As new research and experience broaden our knowledge, changes in practice, treatment and drug therapy may become necessary or appropriate. Readers are advised to check the most current information provided (i) on procedures featured or (ii) by the manufacturer of each product to be administered, to verify the recommended dose or formula, the method and duration of administration, and contraindications. It is the responsibility of the practitioner, relying on their own experience and knowledge of the patient, to make diagnoses, to determine dosages and the best treatment for each individual patient, and to take all appropriate safety precautions. To the fullest extent of the law, neither the Publisher, nor Editors, nor Authors assume any liability for any injury and/or damage to persons or property arising out or related to any use of the material contained in this book.

Contents

Contents of all Volumes	xi
Preface	xix
Preface to Volume 2	xxi
Editors-in-Chief	xxiii
Editor of Volume 2	xxiv
Contributors to Volume 2	xxv

Introduction

2.01 The Intersection of Strategy and Drug Research   1
W H MOOS, *SRI International, Menlo Park, CA, USA and University of California–San Francisco, San Francisco, CA, USA*

2.02 An Academic Perspective   85
D J TRIGGLE, *State University of New York, Buffalo, NY, USA*

2.03 An Industry Perspective   99
W WIERENGA, *Neurocrine Biosciences, Inc., San Diego, CA, USA*

Organizational Aspects and Strategies for Drug Discovery and Development

2.04 Project Management   137
Y NAKAGAWA, *Novartis, Emeryville, CA, USA* and L LEHMAN, *Gilead Sciences, Inc., Foster City, CA, USA*

2.05 The Role of the Chemical Development, Quality, and Regulatory Affairs Teams in Turning a Potent Agent into a Registered Product   159
S A MUNK, *Ash Stevens Inc., Detroit, MI, USA*

2.06 Drug Development   173
P PREZIOSI, *Catholic University of the Sacred Heart, Rome, Italy*

2.07 In-House or Out-Source   203
R D CONNELL, *Pfizer Global Research and Development, Groton, CT, USA*

2.08 Pharma versus Biotech: Contracts, Collaborations, and Licensing   225
D CAVALLA, *Arachnova Ltd, Cambridge, UK*

2.09	Managing Scientists, Leadership Strategies in Science A M SAPIENZA, *Simmons College, Boston, MA, USA*	239
2.10	Innovation (Fighting against the Current) R ROOT-BERNSTEIN, *Michigan State University, East Lansing, MI, USA*	253
2.11	Enabling Technologies in Drug Discovery: The Technical and Cultural Integration of the New with the Old M WILLIAMS, *Feinberg School of Medicine, Northwestern University, Chicago, IL, USA*	265
2.12	How and Why to Apply the Latest Technology A W CZARNIK, *University of Nevada, Reno, NV, USA* and H-Y MEI, *Neural Intervention Technologies, Inc., Ann Arbor, MI, USA*	289
2.13	How and When to Apply Absorption, Distribution, Metabolism, Excretion, and Toxicity R J ZIMMERMAN, *Zimmerman Consulting, Orinda, CA, USA*	559
2.14	Peptide and Protein Drugs: Issues and Solutions J J NESTOR, *TheraPei Pharmaceuticals, Inc., San Diego, CA, USA*	573
2.15	Peptidomimetic and Nonpeptide Drug Discovery: Receptor, Protease, and Signal Transduction Therapeutic Targets T K SAWYER, *ARIAD Pharmaceuticals, Cambridge, MA, USA*	603
2.16	Bioisosterism C G WERMUTH, P CIAPETTI, B GIETHLEN, and P BAZZINI, *Prestwick Chemical, Illkirch, France*	649
2.17	Chiral Drug Discovery and Development – From Concept Stage to Market Launch H-J FEDERSEL, *AstraZeneca, Södertälje, Sweden*	713
2.18	Promiscuous Ligands S L McGOVERN, *M.D. Anderson Cancer Center, Houston, TX, USA*	737

Targets

2.19	Diversity versus Focus in Choosing Targets and Therapeutic Areas D A GIEGEL, *Celgene Corporation, San Diego, CA, USA*, and A J LEWIS, *Novocell, Inc., Irvine, CA, USA*, and P WORLAND, *Celgene Corporation, San Diego, CA, USA*	753
2.20	G Protein-Coupled Receptors W J THOMSEN and D P BEHAN, *Arena Pharmaceuticals, Inc., San Diego, CA, USA*	771
2.21	Ion Channels – Voltage Gated J G McGIVERN, *Amgen Inc., Thousand Oaks, CA, USA* and J F WORLEY III, *Vertex Pharmaceuticals Inc., San Diego, CA, USA*	827
2.22	Ion Channels – Ligand Gated C A BRIGGS and M GOPALAKRISHNAN, *Abbott Laboratories, Abbott Park, IL, USA*	877
2.23	Phosphodiesterases D P ROTELLA, *Wyeth Research, Princeton, NJ, USA*	919
2.24	Protein Kinases and Protein Phosphatases in Signal Transduction Pathways T K SAWYER, *ARIAD Pharmaceuticals, Cambridge, MA, USA*	959
2.25	Nuclear Hormone Receptors N T ZAVERI and B J MURPHY, *SRI International, Menlo Park, CA, USA*	993

2.26 Nucleic Acids (Deoxyribonucleic Acid and Ribonucleic Acid) 1037
E E SWAYZE, R H GRIFFEY, and C F BENNETT, *ISIS Pharmaceuticals, Carlsbad, CA, USA*

2.27 Redox Enzymes 1053
J A DYKENS, *EyeCyte Therapeutics, Encinitas, CA, USA*

List of Abbreviations 1089

List of Symbols 1103

Subject Index 1111

Contents of all Volumes

Volume 1   Global Perspective

Historical Perspective and Outlook
1.01   Reflections of a Medicinal Chemist: Formative Years through Thirty-Seven Years Service in the Pharmaceutical Industry
1.02   Drug Discovery: Historical Perspective, Current Status, and Outlook
1.03   Major Drug Introductions

The Impact of New Genomic Technologies
1.04   Epigenetics
1.05   Personalized Medicine
1.06   Gene Therapy

Sources of New Drugs
1.07   Overview of Sources of New Drugs
1.08   Natural Product Sources of Drugs: Plants, Microbes, Marine Organisms, and Animals
1.09   Traditional Medicines
1.10   Biological Macromolecules
1.11   Nutraceuticals

Animal Experimentation
1.12   Alternatives to Animal Testing

The Role of Industry
1.13   The Role of Small- or Medium-Sized Enterprises in Drug Discovery
1.14   The Role of the Pharmaceutical Industry
1.15   The Role of the Venture Capitalist
1.16   Industry–Academic Relationships

Drug Discovery: Revolution, Decline?
1.17   Is the Biotechnology Revolution a Myth?
1.18   The Apparent Declining Efficiency of Drug Discovery

Healthcare in the Social Context
1.19   How Much is Enough or Too Little: Assessing Healthcare Demand in Developed Countries
1.20   Health Demands in Developing Countries
1.21   Orphan Drugs and Generics

Ethical Issues
1.22   Bioethical Issues in Medicinal Chemistry and Drug Treatment
1.23   Ethical Issues and Challenges Facing the Pharmaceutical Industry

Funding and Regulation of Research
1.24   The Role of Government in Health Research
1.25   Postmarketing Surveillance

Intellectual Property
1.26   Intellectual Property Rights and Patents
Subject Index

Volume 2   Strategy and Drug Research

Introduction
2.01   The Intersection of Strategy and Drug Research
2.02   An Academic Perspective
2.03   An Industry Perspective

Organizational Aspects and Strategies for Drug Discovery and Development
2.04   Project Management
2.05   The Role of the Chemical Development, Quality, and Regulatory Affairs Teams in Turning a Potent Agent into a Registered Product
2.06   Drug Development
2.07   In-House or Out-Source
2.08   Pharma versus Biotech: Contracts, Collaborations, and Licensing
2.09   Managing Scientists, Leadership Strategies in Science
2.10   Innovation (Fighting against the Current)
2.11   Enabling Technologies in Drug Discovery: The Technical and Cultural Integration of the New with the Old
2.12   How and Why to Apply the Latest Technology
2.13   How and When to Apply Absorption, Distribution, Metabolism, Excretion, and Toxicity
2.14   Peptide and Protein Drugs: Issues and Solutions
2.15   Peptidomimetic and Nonpeptide Drug Discovery: Receptor, Protease, and Signal Transduction Therapeutic Targets
2.16   Bioisosterism
2.17   Chiral Drug Discovery and Development – From Concept Stage to Market Launch
2.18   Promiscuous Ligands

Targets
2.19   Diversity versus Focus in Choosing Targets and Therapeutic Areas
2.20   G Protein-Coupled Receptors
2.21   Ion Channels – Voltage Gated
2.22   Ion Channels – Ligand Gated
2.23   Phosphodiesterases
2.24   Protein Kinases and Protein Phosphatases in Signal Transduction Pathways
2.25   Nuclear Hormone Receptors
2.26   Nucleic Acids (Deoxyribonucleic Acid and Ribonucleic Acid)
2.27   Redox Enzymes
List of Abbreviations
List of Symbols
Subject Index

Volume 3   Drug Discovery Technologies

Target Search
3.01   Genomics
3.02   Proteomics
3.03   Pharmacogenomics
3.04   Biomarkers
3.05   Microarrays
3.06   Recombinant Deoxyribonucleic Acid and Protein Expression
3.07   Chemical Biology

Target Validation
3.08 Genetically Engineered Animals
3.09 Small Interfering Ribonucleic Acids
3.10 Signaling Chains
3.11 Orthogonal Ligand–Receptor Pairs

Informatics and Databases
3.12 Chemoinformatics
3.13 Chemical Information Systems and Databases
3.14 Bioactivity Databases
3.15 Bioinformatics
3.16 Gene and Protein Sequence Databases
3.17 The Research Collaboratory for Structural Bioinformatics Protein Data Bank
3.18 The Cambridge Crystallographic Database

Structural Biology
3.19 Protein Production for Three-Dimensional Structural Analysis
3.20 Protein Crystallization
3.21 Protein Crystallography
3.22 Bio-Nuclear Magnetic Resonance
3.23 Protein Three-Dimensional Structure Validation
3.24 Problems of Protein Three-Dimensional Structures
3.25 Structural Genomics

Screening
3.26 Compound Storage and Management
3.27 Optical Assays in Drug Discovery
3.28 Fluorescence Screening Assays
3.29 Cell-Based Screening Assays
3.30 Small Animal Test Systems for Screening
3.31 Imaging
3.32 High-Throughput and High-Content Screening

Chemical Technologies
3.33 Combinatorial Chemistry
3.34 Solution Phase Parallel Chemistry
3.35 Polymer-Supported Reagents and Scavengers in Synthesis
3.36 Microwave-Assisted Chemistry
3.37 High-Throughput Purification

Lead Search and Optimization
3.38 Protein Crystallography in Drug Discovery
3.39 Nuclear Magnetic Resonance in Drug Discovery
3.40 Chemogenomics
3.41 Fragment-Based Approaches
3.42 Dynamic Ligand Assembly
Subject Index

Volume 4 Computer-Assisted Drug Design

Introduction to Computer-Assisted Drug Design
4.01 Introduction to the Volume and Overview of Computer-Assisted Drug Design in the Drug Discovery Process
4.02 Introduction to Computer-Assisted Drug Design – Overview and Perspective for the Future
4.03 Quantitative Structure–Activity Relationship – A Historical Perspective and the Future
4.04 Structure-Based Drug Design – A Historical Perspective and the Future

Core Concepts and Methods – Ligand-Based
4.05　Ligand-Based Approaches: Core Molecular Modeling
4.06　Pharmacophore Modeling: 1 – Methods
4.07　Predictive Quantitative Structure–Activity Relationship Modeling
4.08　Compound Selection Using Measures of Similarity and Dissimilarity

Core Concepts and Methods – Target Structure-Based
4.09　Structural, Energetic, and Dynamic Aspects of Ligand–Receptor Interactions
4.10　Comparative Modeling of Drug Target Proteins
4.11　Characterization of Protein-Binding Sites and Ligands Using Molecular Interaction Fields
4.12　Docking and Scoring
4.13　De Novo Design

Core Methods and Applications – Ligand and Structure-Based
4.14　Library Design: Ligand and Structure-Based Principles for Parallel and Combinatorial Libraries
4.15　Library Design: Reactant and Product-Based Approaches
4.16　Quantum Mechanical Calculations in Medicinal Chemistry: Relevant Method or a Quantum Leap Too Far?

Applications to Drug Discovery – Lead Discovery
4.17　Chemogenomics in Drug Discovery – The Druggable Genome and Target Class Properties
4.18　Lead Discovery and the Concepts of Complexity and Lead-Likeness in the Evolution of Drug Candidates
4.19　Virtual Screening
4.20　Screening Library Selection and High-Throughput Screening Analysis/Triage

Applications to Drug Discovery – Ligand-Based Lead Optimization
4.21　Pharmacophore Modeling: 2 – Applications
4.22　Topological Quantitative Structure–Activity Relationship Applications: Structure Information Representation in Drug Discovery
4.23　Three-Dimensional Quantitative Structure–Activity Relationship: The State of the Art

Applications to Drug Discovery – Target Structure-Based
4.24　Structure-Based Drug Design – The Use of Protein Structure in Drug Discovery
4.25　Applications of Molecular Dynamics Simulations in Drug Design
4.26　Seven Transmembrane G Protein-Coupled Receptors: Insights for Drug Design from Structure and Modeling
4.27　Ion Channels: Insights for Drug Design from Structure and Modeling
4.28　Nuclear Hormone Receptors: Insights for Drug Design from Structure and Modeling
4.29　Enzymes: Insights for Drug Design from Structure

New Directions
4.30　Multiobjective/Multicriteria Optimization and Decision Support in Drug Discovery
4.31　New Applications for Structure-Based Drug Design
4.32　Biological Fingerprints

Subject Index

Volume 5 ADME-Tox Approaches

Introduction
5.01　The Why and How of Absorption, Distribution, Metabolism, Excretion, and Toxicity Research

Biological and In Vivo Aspects of Absorption, Distribution, Metabolism, Excretion, and Toxicity
5.02　Clinical Pharmacokinetic Criteria for Drug Research
5.03　In Vivo Absorption, Distribution, Metabolism, and Excretion Studies in Discovery and Development
5.04　The Biology and Function of Transporters
5.05　Principles of Drug Metabolism 1: Redox Reactions

5.06	Principles of Drug Metabolism 2: Hydrolysis and Conjugation Reactions
5.07	Principles of Drug Metabolism 3: Enzymes and Tissues
5.08	Mechanisms of Toxification and Detoxification which Challenge Drug Candidates and Drugs
5.09	Immunotoxicology

Biological In Vitro Tools in Absorption, Distribution, Metabolism, Excretion, and Toxicity

5.10	In Vitro Studies of Drug Metabolism
5.11	Passive Permeability and Active Transport Models for the Prediction of Oral Absorption
5.12	Biological In Vitro Models for Absorption by Nonoral Routes
5.13	In Vitro Models for Examining and Predicting Brain Uptake of Drugs
5.14	In Vitro Models for Plasma Binding and Tissue Storage
5.15	Progress in Bioanalytics and Automation Robotics for Absorption, Distribution, Metabolism, and Excretion Screening

Physicochemical tools in Absorption, Distribution, Metabolism, Excretion, and Toxicity

5.16	Ionization Constants and Ionization Profiles
5.17	Dissolution and Solubility
5.18	Lipophilicity, Polarity, and Hydrophobicity
5.19	Artificial Membrane Technologies to Assess Transfer and Permeation of Drugs in Drug Discovery
5.20	Chemical Stability
5.21	Solid-State Physicochemistry

In Silico Tools in Absorption, Distribution, Metabolism, Excretion, and Toxicity

5.22	Use of Molecular Descriptors for Absorption, Distribution, Metabolism, and Excretion Predictions
5.23	Electrotopological State Indices to Assess Molecular and Absorption, Distribution, Metabolism, Excretion, and Toxicity Properties
5.24	Molecular Fields to Assess Recognition Forces and Property Spaces
5.25	In Silico Prediction of Ionization
5.26	In Silico Predictions of Solubility
5.27	Rule-Based Systems to Predict Lipophilicity
5.28	In Silico Models to Predict Oral Absorption
5.29	In Silico Prediction of Oral Bioavailability
5.30	In Silico Models to Predict Passage through the Skin and Other Barriers
5.31	In Silico Models to Predict Brain Uptake
5.32	In Silico Models for Interactions with Transporters
5.33	Comprehensive Expert Systems to Predict Drug Metabolism
5.34	Molecular Modeling and Quantitative Structure–Activity Relationship of Substrates and Inhibitors of Drug Metabolism Enzymes
5.35	Modeling and Simulation of Pharmacokinetic Aspects of Cytochrome P450-Based Metabolic Drug–Drug Interactions
5.36	In Silico Prediction of Plasma and Tissue Protein Binding
5.37	Physiologically-Based Models to Predict Human Pharmacokinetic Parameters
5.38	Mechanism-Based Pharmacokinetic–Pharmacodynamic Modeling for the Prediction of In Vivo Drug Concentration–Effect Relationships – Application in Drug Candidate Selection and Lead Optimization
5.39	Computational Models to Predict Toxicity
5.40	In Silico Models to Predict QT Prolongation
5.41	The Adaptive In Combo Strategy

Enabling Absorption, Distribution, Metabolism, Excretion, and Toxicity Strategies and Technologies in Early Development

5.42	The Biopharmaceutics Classification System
5.43	Metabonomics

5.44 Prodrug Objectives and Design
5.45 Drug–Polymer Conjugates
List of Abbreviations
List of Symbols
Subject Index

Volume 6 Therapeutic Areas I: Central Nervous System, Pain, Metabolic Syndrome, Urology, Gastrointestinal and Cardiovascular

Central Nervous System
6.01 Central Nervous System Drugs Overview
6.02 Schizophrenia
6.03 Affective Disorders: Depression and Bipolar Disorders
6.04 Anxiety
6.05 Attention Deficit Hyperactivity Disorder
6.06 Sleep
6.07 Addiction
6.08 Neurodegeneration
6.09 Neuromuscular/Autoimmune Disorders
6.10 Stroke/Traumatic Brain and Spinal Cord Injuries
6.11 Epilepsy
6.12 Ophthalmic Agents

Pain
6.13 Pain Overview
6.14 Acute and Neuropathic Pain
6.15 Local and Adjunct Anesthesia
6.16 Migraine

Obesity/Metabolic Disorders/Syndrome X
6.17 Obesity/Metabolic Syndrome Overview
6.18 Obesity/Disorders of Energy
6.19 Diabetes/Syndrome X
6.20 Atherosclerosis/Lipoprotein/Cholesterol Metabolism
6.21 Bone, Mineral, Connective Tissue Metabolism
6.22 Hormone Replacement

Urogenital
6.23 Urogenital Diseases/Disorders, Sexual Dysfunction and Reproductive Medicine: Overview
6.24 Incontinence (Benign Prostatic Hyperplasia/Prostate Dysfunction)
6.25 Renal Dysfunction in Hypertension and Obesity

Gastrointestinal
6.26 Gastrointestinal Overview
6.27 Gastric and Mucosal Ulceration
6.28 Inflammatory Bowel Disease
6.29 Irritable Bowel Syndrome
6.30 Emesis/Prokinetic Agents

Cardiovascular
6.31 Cardiovascular Overview
6.32 Hypertension
6.33 Antiarrhythmics
6.34 Thrombolytics
Subject Index

Volume 7 Therapeutic Areas II: Cancer, Infectious Diseases, Inflammation & Immunology and Dermatology

Anti Cancer
7.01 Cancer Biology
7.02 Principles of Chemotherapy and Pharmacology
7.03 Antimetabolites
7.04 Microtubule Targeting Agents
7.05 Deoxyribonucleic Acid Topoisomerase Inhibitors
7.06 Alkylating and Platinum Antitumor Compounds
7.07 Endocrine Modulating Agents
7.08 Kinase Inhibitors for Cancer
7.09 Recent Development in Novel Anticancer Therapies

Anti Viral
7.10 Viruses and Viral Diseases
7.11 Deoxyribonucleic Acid Viruses: Antivirals for Herpesviruses and Hepatitis B Virus
7.12 Ribonucleic Acid Viruses: Antivirals for Human Immunodeficiency Virus
7.13 Ribonucleic Acid Viruses: Antivirals for Influenza A and B, Hepatitis C Virus, and Respiratory Syncytial Virus

Anti Fungal
7.14 Fungi and Fungal Disease
7.15 Major Antifungal Drugs

Anti Bacterials
7.16 Bacteriology, Major Pathogens, and Diseases
7.17 β-Lactam Antibiotics
7.18 Macrolide Antibiotics
7.19 Quinolone Antibacterial Agents
7.20 The Antibiotic and Nonantibiotic Tetracyclines
7.21 Aminoglycosides Antibiotics
7.22 Anti-Gram Positive Agents of Natural Product Origins
7.23 Oxazolidinone Antibiotics
7.24 Antimycobacterium Agents
7.25 Impact of Genomics-Emerging Targets for Antibacterial Therapy

Drugs for Parasitic Infections
7.26 Overview of Parasitic Infections
7.27 Advances in the Discovery of New Antimalarials
7.28 Antiprotozoal Agents (African Trypanosomiasis, Chagas Disease, and Leishmaniasis)

I and I Diseases
7.29 Recent Advances in Inflammatory and Immunological Diseases: Focus on Arthritis Therapy
7.30 Asthma and Chronic Obstructive Pulmonary Disease
7.31 Treatment of Transplantation Rejection and Multiple Sclerosis

Dermatology
7.32 Overview of Dermatological Diseases
7.33 Advances in the Discovery of Acne and Rosacea Treatments
7.34 New Treatments for Psoriasis and Atopic Dermatitis
Subject Index

Volume 8 Case Histories and Cumulative Subject Index

Personal Essays
8.01 Introduction
8.02 Reflections on Medicinal Chemistry Since the 1950s

8.03	Medicinal Chemistry as a Scientific Discipline in Industry and Academia: Some Personal Reflections
8.04	Some Aspects of Medicinal Chemistry at the Schering-Plough Research Institute

Case Histories

8.05	Viread
8.06	Hepsera
8.07	Ezetimibe
8.08	Tamoxifen
8.09	Raloxifene
8.10	Duloxetine
8.11	Carvedilol
8.12	Modafinil, A Unique Wake-Promoting Drug: A Serendipitous Discovery in Search of a Mechanism of Action
8.13	Zyvox
8.14	Copaxone
8.15	Ritonavir and Lopinavir/Ritonavir
8.16	Fosamax
8.17	Omeprazole
8.18	Calcium Channel α_2–δ Ligands: Gabapentin and Pregabalin

Cumulative Subject Index

Preface

The first edition of *Comprehensive Medicinal Chemistry* was published in 1990 and was intended to present an integrated and comprehensive overview of the then rapidly developing science of medicinal chemistry from its origins in organic chemistry. In the last two decades, the field has grown to embrace not only all the sophisticated synthetic and technological advances in organic chemistry but also major advances in the biological sciences. The mapping of the human genome has resulted in the provision of a multitude of new biological targets for the medicinal chemist with the prospect of more rational drug design (CADD). In addition, the development of sophisticated in silico technologies for structure–property relationships (ADMET) enables a much better understanding of the fate of potential new drugs in the body with the subsequent development of better new medicines.

It was our ambitious aim for this second edition, published 16 years after the first edition, to provide both scientists and research managers in all relevant fields with a comprehensive treatise covering all aspects of current medicinal chemistry, a science that has been transformed in the twenty-first century. The second edition is a complete reference source, published in eight volumes, encompassing all aspects of modern drug discovery from its mechanistic basis, through the underlying general principles and exemplified with comprehensive therapeutic applications. The broad scope and coverage of *Comprehensive Medicinal Chemistry II* would not have been possible without our panel of authoritative Volume Editors whose international recognition in their respective fields has been of paramount importance in the enlistment of the world-class scientists who have provided their individual 'state of the science' contributions. Their collective contributions have been invaluable.

Volume 1 (edited by Peter D Kennewell) overviews the general socioeconomic and political factors influencing modern R&D in both the developed and developing worlds. Volume 2 (edited by Walter H Moos) addresses the various strategic and organizational aspects of modern R&D. Volume 3 (edited by Hugo Kubinyi) critically reviews the multitude of modern technologies that underpin current discovery and development activities. Volume 4 (edited by Jonathan S Mason) highlights the historical progress, current status, and future potential in the field of computer-assisted drug design (CADD). Volume 5 (edited by Bernard Testa and Han van de Waterbeemd) reviews the fate of drugs in the body (ADMET), including the most recent progress in the application of 'in silico' tools. Volume 6 (edited by Michael Williams) and Volume 7 (edited by Jacob J Plattner and Manoj C Desai) cover the pivotal roles undertaken by the medicinal chemist and pharmacologist in integrating all the preceding scientific input into the design and synthesis of viable new medicines. Volume 8 (edited by John B Taylor and David J Triggle) illustrates the evolution of modern medicinal chemistry with a selection of personal accounts by eminent scientists describing their lifetime experiences in the field, together with some illustrative case histories of successful drug discovery and development.

We believe that this major work will serve as the single most authoritative reference source for all aspects of medicinal chemistry for the next decade and it is intended to maintain its ongoing value by systematic electronic upgrades. We hope that the material provided here will serve to fulfill the words of Antoine de Saint-Exupery (1900–44) and allow future generations of medicinal chemists to discover the future.

'As for the future, your task is not to foresee it but to enable it'
Citadelle (1948)

John B Taylor and David J Triggle

Preface to Volume 2

This volume addresses the pharmaceutical (and biotechnology) research and development (R&D) enterprise from the perspective of strategy, directed to medicinal chemists in particular. Thus, we cover drug research from numerous complementary viewpoints to elucidate how to think about many interesting and important topics in this field.

A volume on strategy is new to *Comprehensive Medicinal Chemistry*, not having been part of the first edition. This volume will be of interest to anyone who follows the broad biotechnology and pharmaceutical sector, whether technical or lay person. Different vantage points and more details from strategic and technical perspectives on the topics discussed herein can be found elsewhere in the series, particularly in Volumes 1 and 8 (covering a global perspective and case histories, respectively). More details from scientific viewpoints can be found in Volume 3 (technologies), Volume 4 (computer-aided drug design), Volume 5 (pharmacokinetics and related approaches), Volumes 6 and 7 (therapeutic areas), and in the case histories in Volume 8.

It is the editor's hope that the present volume will provide a general background to this exciting area as well as a suitable entry point to other chapters in this major reference work and to leading references in the general literature. Readers will find references to classic articles, reviews, primary literature, and also the latest newspaper and Internet citations. Scattered throughout are examples, anecdotes and curiosities, and brief but representative case studies.

What is strategy in drug research? A strategy is a plan of action designed to achieve a specific objective – in very simple terms, a means to an end. The ultimate goal of drug research is to advance a product candidate progressively through later stages of R&D toward commercialization. Once that goal has been achieved, with patients being treated successfully with the new drug, research does not stop. Rather, additional safety data are collected, new therapeutic indications are pursued, alternative formulations and dosing regimens are tested, new approaches to disease are advanced, and much more, in a seemingly never-ending cycle. At times, this process may seem to be little more than trial and error, albeit sophisticated, on a large and expensive scale. Of course, it is much more than that, as you will see.

Strategy is even more important in the pharmaceutical and biotechnology industries, since it takes years to develop a drug – a decade or two, costing a billion dollars or more – and with much attrition along the way. Where academia and industry intersect, the strengths and weaknesses of the old and the new, the crossroads where tactics and strategy interweave, and how market forces operate, are all covered in this volume. The basics of management and leadership, project management, transitions from research to development, decisions about what to conduct in-house or what to outsource, what to originate internally and what to in-license, are all discussed at a practical level. The challenges and opportunities in managing scientists, leadership and both entrepreneurs and intrapreneurs, as well as fostering innovation, are all considered. We also compare and contrast pharmaceutical and biotechnology companies, and review how and why to apply the latest technologies and approaches, whether it be peptides or peptidomimetics, bioisosteres or chiral drugs, or the next advance to captivate the industry. And, finally, we discuss decisions companies make regarding therapeutic areas, drug targets (including background information thereto), and diversity versus focus.

Health care is very critical to all of us at one or other stage of our lives. It is therefore a great privilege and pleasure to work in the arena of pharmaceutical and biotechnology R&D. If this volume helps in some small way to teach the state of the art so that more lives around the world can be saved and improved, then we will have achieved our goal. With that humble mission in mind, please read on.

Walter H Moos

Editors-in-Chief

John B Taylor, DSc, was formerly Senior Vice President for Drug Discovery at Rhône-Poulenc Rorer. He obtained his BSc in chemistry from the University of Nottingham in 1956 and his PhD in organic chemistry at the Imperial College of Science and Technology with Nobel Laureate Professor Sir Derek Barton in 1962. He subsequently undertook postdoctoral research fellowships at the Research Institute for Medicine and Chemistry in Cambridge (US) with Sir Derek and at the University of Liverpool (UK), before entering the pharmaceutical industry.

During his career in the pharmaceutical industry Dr Taylor spent more than 30 years covering all aspects of research and development in an international environment. From 1970 to 1985 he held a number of positions in the Hoechst Roussel organization, ultimately as research director for Roussel Uclaf (France). In 1985 he joined Rhône-Poulenc Rorer holding various management positions in the research groups worldwide before becoming Senior Vice President for Drug Discovery in Rhône-Poulenc Rorer.

Dr Taylor is the co-author of two books on medicinal chemistry and has more than 50 publications and patents in medicinal chemistry. He was joint executive editor for the first edition of Comprehensive Medicinal Chemistry, a visiting professor for medicinal chemistry at the City University (London) from 1974 to 1984 and was awarded a DSc in medicinal chemistry from the University of London in 1991.

David J Triggle, PhD, is the University Professor and a Distinguished Professor in the School of Pharmacy and Pharmaceutical Sciences at the State University of New York at Buffalo. Professor Triggle received his education in the UK with a BSc degree in chemistry at the University of Southampton and a PhD degree in chemistry at the University of Hull working with Professor Norman Chapman. Following postdoctoral fellowships at the University of Ottawa (Canada) with Bernard Belleau and the University of London (UK) with Peter de la Mare he assumed a position in the School of Pharmacy at the University at Buffalo. He served as Chairman of the Department of Biochemical Pharmacology from 1971 to 1985 and as Dean of the School of Pharmacy from 1985 to 1995. From 1996 to 2001 he served as Dean of the Graduate School and from 1999 to 2001 was also the University Provost. He is currently the University Professor, in which capacity he teaches bioethics and science policy, and is President of the Center for Inquiry Institute, a secular think tank located in Amherst, New York.

Professor Triggle is the author of three books dealing with the autonomic nervous system and drug–receptor interactions, the editor of a further dozen books, some 280 papers, some 150 chapters and reviews, and has presented over 1000 invited lectures worldwide. The Institute for Scientific Information lists him as one of the 100 most highly cited scientists in the field of pharmacology. His principal research interests have been in the areas of drug–receptor interactions, the chemical pharmacology of drugs active at ion channels, and issues of graduate education and scientific research policy.

Editor of Volume 2

Walter H Moos, PhD, has 25 years of experience in the pharmaceutical and biotechnology industries. He is currently Vice President of the Bioscience Division at SRI International, one of the world's largest nonprofit research institutes, founded as the Stanford Research Institute in 1946. For a number of years, Dr Moos served as Chairman and CEO of MitoKor (now Migenix). Previously, he was a corporate officer at Chiron (now Novartis), last holding the position of Vice President of R&D, Technologies Division. Dr Moos' earlier position at Warner-Lambert/Parke-Davis (now Pfizer) was as Vice President, Neuroscience and Biological Chemistry. Dr Moos has served on a number of boards of directors, including Alnis, Anterion, Axiom, the Biotechnology Industry Organization, CMPS, the Critical Path Institute, the Keystone Symposia, Migenix, Mimotopes, Oncologic, Onyx, and Rigel. He is a co-founding member of the CEO Council, Red Abbey Venture Partners. Dr Moos has edited several books, helped to found multiple journals, and has a total of 140 manuscripts and issued patents to his name. In addition, he has held adjunct faculty positions at the University of Michigan, Ann Arbor, and the University of California, San Francisco. He has also served on academic and related advisory committees, including the US National Committee at the National Academy of Sciences. Dr Moos holds an AB from Harvard University and received his PhD in chemistry from the University of California, Berkeley.

Contributors to Volume 2

P Bazzini
Prestwick Chemical, Illkirch, France

D P Behan
Arena Pharmaceuticals, Inc., San Diego CA, USA

C F Bennett
ISIS Pharmaceuticals, Carlsbad, CA, USA

C A Briggs
Abbott Laboratories, Abbott Park, IL, USA

D Cavalla
Arachnova Ltd, Cambridge, UK

P Ciapetti
Prestwick Chemical, Illkirch, France

R D Connell
Pfizer Global Research and Development, Groton, CT, USA

A W Czarnik
University of Nevada, Reno, NV, USA

J A Dykens
EyeCyte Therapeutics, Encinitas, CA, USA

H-J Federsel
AstraZeneca, Södertälje, Sweden

D A Giegel
Celgene Corporation, San Diego, CA, USA

B Giethlen
Prestwick Chemical, Illkirch, France

M Gopalakrishnan
Abbott Laboratories, Abbott Park, IL, USA

R H Griffey
ISIS Pharmaceuticals, Carlsbad, CA, USA

L Lehman
Gilead Sciences, Inc., Foster City, CA, USA

A J Lewis
Novocell, Inc., Irvine, CA, USA

J G McGivern
Amgen Inc., Thousand Oaks, CA, USA

S L McGovern
M.D. Anderson Cancer Center, Houston, TX, USA

H-Y Mei
Neural Intervention Technologies, Inc., Ann Arbor, MI, USA

W H Moos
SRI International, Menlo Park, CA, USA and University of California–San Francisco, San Francisco, CA, USA

S A Munk
Ash Stevens Inc., Detroit, MI, USA

B J Murphy
SRI International, Menlo Park, CA, USA

Y Nakagawa
Novartis, Emeryville, CA, USA

J J Nestor
TheraPei Pharmaceuticals, Inc., San Diego, CA, USA

P Preziosi
Catholic University of the Sacred Heart, Rome, Italy

R Root-Bernstein
Michigan State University, East Lansing, MI, USA

D P Rotella
Wyeth Research, Princeton, NJ, USA

A M Sapienza
Simmons College, Boston, MA, USA

T K Sawyer
ARIAD Pharmaceuticals, Cambridge, MA, USA

E E Swayze
ISIS Pharmaceuticals, Carlsbad, CA, USA

W J Thomsen
Arena Pharmaceuticals, Inc., San Diego, CA, USA

D J Triggle
State University of New York, Buffalo, NY, USA

C G Wermuth
Prestwick Chemical, Illkirch, France

W Wierenga
Neurocrine Biosciences, Inc., San Diego, CA, USA

M Williams
Feinberg School of Medicine, Northwestern University, Chicago, IL, USA

P Worland
Celgene Corporation, San Diego, CA, USA

J F Worley III
Vertex Pharmaceuticals Inc., San Diego, CA, USA

N T Zaveri
SRI International, Menlo Park, CA, USA

R J Zimmerman
Zimmerman Consulting, Orinda, CA, USA

2.01 The Intersection of Strategy and Drug Research

W H Moos, SRI International, Menlo Park, CA, USA and University of California–San Francisco, San Francisco, CA, USA

© 2007 Elsevier Ltd. All Rights Reserved.

2.01.1	**Overview**	2
2.01.2	**Organizational Aspects**	4
2.01.2.1	Introduction	4
2.01.2.2	Major Issues Including Attrition	11
2.01.2.2.1	New product development tool kit for the critical path	11
2.01.2.3	Project Management Including Teams	12
2.01.2.3.1	Overall project assessment scheme	14
2.01.2.3.2	Target validation	15
2.01.2.3.3	Selected decision points and transitions from discovery to new drug application	16
2.01.2.4	Recruitment and Management of Human Resources	16
2.01.2.4.1	Guidelines for medicinal chemists early in their careers	17
2.01.2.5	Multiple Sites Including Outsourcing	19
2.01.2.6	Contrasting Biotech with Pharma and Innovation	20
2.01.2.7	Business Development Including Aspects of Portfolio Management	25
2.01.2.8	Drug Development	26
2.01.3	**Drug Discovery**	31
2.01.3.1	Introduction	31
2.01.3.2	Technologies	32
2.01.3.3	Chemical Biology and Chemical Genomics	44
2.01.3.4	From Small Molecules to Macromolecules	47
2.01.3.5	Computational Approaches	48
2.01.3.6	Evolution of Preclinical Research over Several Decades	52
2.01.3.7	Transition from Research to Development, Including Absorption, Distribution, Metabolism, Excretion, and Toxicity	54
2.01.3.8	The Perspective of Successful Drug Hunters	54
2.01.4	**Targets**	56
2.01.4.1	Transporters as an Example of an Important Class of Targets	56
2.01.4.1.1	Current nomenclature	58
2.01.4.1.2	Gene/molecular biology classification and protein structure	59
2.01.4.1.3	Physiological function	60
2.01.4.1.4	Prototypical pharmacology and therapeutics	62
2.01.4.1.5	Genetic diseases	66
2.01.4.1.6	New directions	67
2.01.4.2	General Discussion of Targets	68
2.01.4.3	Project Selection	73
2.01.4.4	Other Target Classes	73
References		**76**

2.01.1 Overview

In this overview, readers will find a preview of what follows. We will start with one of the fundamental issues facing the industry today. As may be evident from even a cursory reading of the daily newspaper, drug research is a complicated, time-consuming, and expensive process. It is the proverbial high-risk, high-payoff industry. Alarmingly, the capitalized, out-of-pocket, pre-approval costs for new drug development have increased by a factor of 3–4 over estimates at the end of the 1980s and nearly six times that of the 1970s. Yet, even with increased spending, industry productivity has dropped. Given these financial and other issues, the sustainability of the industry has been called into question, with the need for each major pharmaceutical company to produce a blockbuster at least once a decade to remain healthy financially, and in the face of continuing safety issues and withdrawals of major marketed drugs. Can the drug discovery process be redirected to provide a stronger pharmacoeconomic future?

If ever there were an industry in need of a better, faster, cheaper strategy, the pharmaceutical industry would be it. Despite this somber picture, the 1990s were quite prosperous for many pharmaceutical companies, and many biotechnology and pharmaceutical companies are doing exceedingly well today financially and otherwise. Venture capital funded biotechnology companies represent an increasing proportion of new drugs entering the marketplace. Indeed, the term biotech is often a misnomer, since many biotech companies focus on small molecule drugs rather than on protein biologics.

Safety has been a major issue in the new millennium, with failures of selective cyclooxygenase inhibitors such as rofecoxib (Vioxx) commanding attention. Even without safety issues, drugs face continued hurdles once marketed. For example, despite demonstrated clinical significance in treating Alzheimer's disease, certain regulatory bodies have decided that the benefits of cholinesterase inhibitors such as donepezil (Aricept) are not sufficient to justify their costs.

How does one take into account today's major issues – including attrition, productivity, efficiency, success rates, costs, and shots on goal – and make things better? The answer is not clear, but most would agree that the expected acceleration in drug discovery and development with the advent of new technologies such as combinatorial chemistry and genomics has, unfortunately, been slow to materialize. In reality, it may simply take more time for the results of recent advances to become more tangible.

Pharmaceutical R&D can be organized in various ways, taking into account line versus matrix considerations, departments versus teams, decision points, interfaces, and transitions. Before embarking on a new project, and as part of periodic project reviews, many groups evaluate the project from a number of perspectives. Decision making of this sort is important in almost any profession, albeit less complicated in many other industries. Successfully managing the interface of drug discovery and drug development requires a clear target product profile that team members and management have agreed upon. With each successive stage of drug discovery and development, new groups must interface to advance the project. For example, as a drug moves into early development, scale-up is usually required, in order to prepare enough material under controlled conditions to proceed with toxicology and other studies.

Internal research must be coordinated within and without, including with contract organizations, to which work may be outsourced. Whether a company is a giant pharmaceutical corporation or a small biotechnology firm, it may be resident at more than one site. Even the largest companies will contract out certain work. Small companies with only a few people may contract out most or all of their laboratory studies. Preclinical development and clinical and regulatory tasks are commonly conducted by contract research organizations specializing in such areas. Perhaps contrary to expectations, basic research is also frequently outsourced. Most of the large biotech-pharma alliances are essentially outsourcing arrangements, with the biotech company contributing new technologies and approaches while big pharma provides the funding and later stage resources required to take a product through preclinical and clinical development to marketing and sales. Though the First World has dominated these arenas historically, the Pacific Rim has positioned itself to capitalize on many of these opportunities in the future, with substantial growth in biotechnology and pharmaceutical R&D. Southeast Asia, including China, India, Singapore, and Taiwan, as well as other countries, represents growing competition for the future, and offshoring of high technology jobs from the US and Europe has become a topic of discussion and concern.

What about the people who drive this industry? How does one recruit and manage the human resources required to be successful in medicinal chemistry and pharmaceutical R&D? There are, as one might expect, many factors to consider. Research laboratory management requires paying attention to both the forest and the trees – the big picture as well as the details. It requires finding ways to stay informed without micromanaging. Managing well requires many leadership skills, including an appreciation for and a knack for leveraging different styles, the ability to communicate effectively on many different levels, ways of addressing conflict and change constructively, comfort (or if not comfort then a productive disharmony) with the corporate culture, and an understanding of motivation.

Companies go through several stages as they develop, moving from start-up through focus, integration, optimization, and finally reaching maturity. Management practices must evolve during these progressions, or crises and even business failures will result. Thus, whereas the management profile in early business stages is often vision-driven and planning is ad hoc, later-stage companies require more structured processes and formal planning. Similarly, business goals evolve from being flexible and entrepreneurial to being more consistent, integrated, and measurable.

Innovation is often discussed but rarely practiced, and even more rarely studied with rigor. In fact, many people fight innovation as it moves through the several stages of a revolution. In the end, most new science needs to be hyped to have a chance. New technologies are always greeted with skepticism, many people have vested interests in competing old approaches, and people do not like change in general. Moreover, the timeframe to develop practical applications of new technologies is always longer than expected.

The latter part of the twentieth century witnessed the birth of numerous technologies that advanced the state of the art in drug research. Technological advances often spawned a new cadre of start-up ventures. That trend was quite evident in the combinatorial chemistry field in the 1980s and 1990s. However, as is common with these waves of change, many of the original companies were acquired, failed, or changed directions, becoming unrecognizable or untraceable over a few short years.

Many technologies critical to today's medicinal chemistry (and pharmaceutical R&D in general) were developed starting in the 1980s. These include combinatorial chemistry, high-throughput screening, genomics, and proteomics. The development of combinatorial chemistry is a particularly interesting case study in the biotechnology and pharmaceutical industries. One might trace its formal beginnings to the development of solid-phase peptide synthesis by Merrifield in the early 1960s, though other developments including natural products chemistry may also be considered seminal events. Nanotechnology (or nanobiotechnology) is the latest platform to employ the skills of the medicinal chemist. The concept of targeted therapeutic nanoparticles is being tested in a number of academic and industrial laboratories around the world. Much of this work is being done in the cancer field.

While the beginnings of quantitative structure–activity relationships (QSARs) in medicinal chemistry have roots that date back to the 1940s and 1950s, drawing heavily on linear free energy relationships and physical organic chemistry, methods advanced remarkably once computers became commonplace. Still, sometimes the best techniques – back of the envelope calculations – are the simplest. Advances in computing have allowed the development of molecular modeling and other tools for structure-based design, sometimes referred to as computer-aided drug design. While this can be a very technical discipline, practical guides exist for those who are interested in learning more or even becoming practitioners. The thousands of three-dimensional structures of macromolecules that have been solved provide a good basis for drug design.

Chemical space, nearly infinite, is a continuum, with pockets of biologically relevant properties. Various strategies have been used to calculate this n-dimensional physicochemical property space. Evaluating the diversity of a given set of compounds from first principles may utilize one of a number of different approaches, such as clustering, receptor site interaction simulation, similarity, and other techniques.

What other considerations should be mentioned? Even if one had a large and diverse library of drug-like molecules, it is conceivable that some of the biological targets of interest would not be druggable – that is, it might not be possible to modulate some targets through interactions with orally available small molecules. Others have tried to devise new ways of screening libraries virtually – by computer – before proceeding to laboratory work.

In the old days, chemists, biochemists, and pharmacologists worked together in pharmaceutical companies to identify promising new drug leads by a slow, iterative process, often driven principally by data in animals. What has happened to the old ways with the advent of novel technologies and new scientific disciplines? For one, molecular biologists became a major part of drug discovery teams in the 1980s and 1990s, developing molecular assays to prioritize compounds for animal testing. The basics have not changed, but many new technologies have been added to the medicinal chemist's armamentarium, including combinatorial chemistry.

While there have been major advances in many of the early stages of drug discovery, a bottleneck remains in preclinical development – in absorption, distribution, metabolism, excretion, and toxicity (ADMET) – both in vitro at a high-throughput level, or with high content assays, and in vivo in animals for screening and for formal drug development tasks. Addressing this part of drug development requires different thought processes and procedural rigors because of regulations. Various methods have been developed to predict or optimize drug absorption and pharmacokinetics.

Successful drug hunters persevere, have a whatever-it-takes mindset, and occasionally use tips and tricks that others miss. The key can be as simple as determining when it is time to move on – proceeding to the next stage of evaluation even if the molecule is not ideal, since more questions will be answered by going forward than have been answered to date. Drug hunting can require fighting against the current, for example, working on dirty drugs or polypharmacy when others are focused on very selective drugs.

Drug discovery targets fall into various classes and families. Many are receptors or enzymes. One analysis of drug targets concluded that there were more than 400 in total, based on drugs that had reached the marketplace, including enzymes, ion channels, receptors, and other targets. The largest percentage, about one-third, were targets associated with synaptic and neuroeffector junctions and the central nervous system.

Some have tried to assess the genome from the standpoint of druggability. They estimate that only about 3000 druggable targets exist in the human genome. G protein-coupled receptors and kinases figure prominently in this analysis. Considering only the currently marketed drugs, G protein-coupled receptors account for almost one-third of this group, with ion channels a distant second. Another evaluation suggests that there are no more than 1500 exploitable drug targets based on the intersection of druggable targets and genes linked to disease.

Selecting the right target is more an art than a science. The process to select a target requires consideration of corporate, business, and R&D strategy, as well as portfolio balance, in addition to scientific and medical arguments for or against. It must also consider the interplay of stakeholders, consumers, and the drug firm, with attendant factors, including those that are technical, legal, economic, demographic, political, social, ecological, and infrastructural. Unfortunately, it is frequently the case that marketing projections used to prioritize drug opportunities are not anywhere close to actual sales.

There is an old saying: The difficult we do immediately; the impossible takes a little longer. In some ways, this is the challenge in front of us. The health and well being of the world rests on our success in tackling these challenges and opportunities. Wish us luck!

2.01.2 Organizational Aspects

2.01.2.1 Introduction

Drug research is a complicated, time consuming, and expensive process[1,2] – high risk with a potentially high payoff (**Table 1**). PhRMA (Pharmaceutical Research and Manufacturers of America) estimates that it takes an experimental drug an average of 10–15 years to advance from early-stage laboratory research to US patients. (For purposes of illustration, the pharmaceutical R&D process leading to commercialization will be outlined here as if the intended path is with the US Food and Drug Administration (FDA)). During these dozen or so years, thousands of compounds are evaluated in discovery and preclinical testing, just five enter human clinical trials, and only one is approved by the FDA.[3] (Company and R&D statistics can be found in various industry compilations, such as those prepared by PAREXEL.[4])

The capitalized, out-of-pocket, pre-approval costs for new drug development have recently been estimated to exceed $800 million (MM), an increase of 3–4 times over estimates at the end of the 1980s and 6 times that of the 1970s (**Figure 1**). Thus, it should be no surprise that major pharmaceutical companies invested more than $30 billion in R&D in 2002, a number that has doubled every 5 years since 1970 (**Figure 2**).[5] Increased spending is also evident at the National Institutes of Health (NIH). It should also be no surprise that drug prices can be high, given the astronomical costs to develop each drug, the long timelines, and the major risks of failure. Nonetheless, the debate over drug pricing rages on in an effort to contain healthcare costs, including arguments that patients are double charged for drugs – once via tax dollars and a second time when they need a drug.[6–9] This debate began in the 1950s, and continues to this day.[10]

Even with increased spending, the number of regulatory applications for new therapeutics has dropped (**Figure 3**). Success rates for approval of investigational drugs, in particular, have been studied extensively. If one looks at the period from 1963 to 1999, the number of investigational new drug applications (INDs) for new chemical entities (NCEs) has not changed much, but the duration of the clinical and approval phases has gone up by more than 50%, from just over 4 years to almost 7 years. The time from synthesis to approval of NCEs went up from 8.1 years in 1963–1969 to 14.2 years in 1990–1999. In the latter period, cancer drugs took the longest, at 16.2 years, while anti-infectives took 12.3 years. Interestingly, during the 1980s these studies showed a reduction in clinical approval rates for self-originated drugs, but an increase in clinical approval rates for acquired compounds, suggesting that companies found it more productive to develop other companies' products than their own.[11,12] Overall, despite the passage of several legislative acts in recent decades (e.g., Prescription Drug User Fee Act of 1992 (PDUFA) and FDA Modernization Act of 1997 (FDAMA)) that could have speeded the development of new drugs, overall approval times for new therapeutics in the US are little changed, though there have been apparent temporary improvements from time to time at one stage or another in drug development.[13,14]

Add to these concerns the projection that the compound annual growth rate of large pharmaceutical companies may not support their valuations. For example, predicted growth rates for 2003–2008 range from 1.1% to 11.3% among the

Table 1 Pharmaceutical R&D and approval process, timeline, and costs

| | Preclinical | Clinical | | | | Postmarketing | | |
	Discovery & early development	IND	Phase I	Phase II	Phase III	NDA	FDA	Phase IV
Years	6.5		1.5	2	3.5		1.5	
Test subjects	In vitro, in vivo, animal studies		Healthy volunteers	Patient volunteers	Patient volunteers		Review process	Surveillance
			20–100 subjects	100–500 patients	1000–5000 patients			Up to millions of patients
Purpose	Bioactivity, safety, pharmaceutical properties		Safety & dose	Efficacy & side effects	Efficacy, adverse reactions, longer term use		Approval	Other studies per FDA
Success rate	≥5000 compounds evaluated		5 drugs enter clinical trials				1 drug approved	
Cost	$335 million		—	$467 million				$95 million

IND, Investigational New Drug Application; NDA, New Drug Application; FDA, Food and Drug Administration.

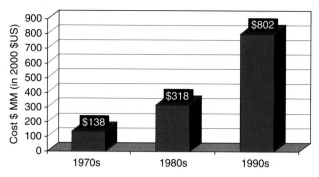

Figure 1 Average capitalized costs of new drug R&D.

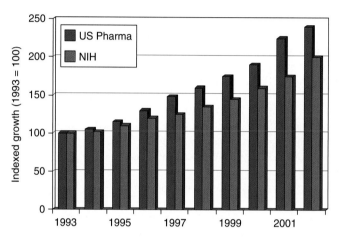

Figure 2 Rising costs of R&D in the US.

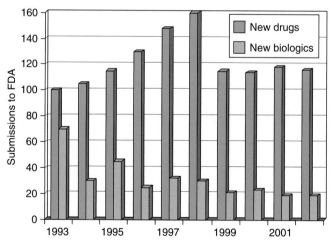

Figure 3 Decreases in new drug and biologic submissions in the US.

top 13 pharmaceutical companies. At the other end of the scale, the sometimes unrealistic wishes of company founders may not support the valuations of small biotech companies.[15–16]

Given these financial and other issues, the sustainability of the industry has been called into question, with the need for each major pharmaceutical company to produce a blockbuster at least once a decade to remain healthy financially, and in the face of continuing safety issues and withdrawals of major marketed drugs. Can the drug discovery process be redirected to provide a stronger pharmacoeconomic future? Major drugs continue to be withdrawn from the market for safety reasons, such as astemizole (Hismanal), cisapride (Propulsid), fenfluramine (Pondimin), and terfenadine (Seldane) for cardiovascular problems, phenylpropanolamine for stroke risk, and troglitazone (Rezulin) for liver toxicity.[17] Pharmaceutical and biotechnology companies have never faced so many challenges, including financial, technical, regulatory, social, and ethical issues.[18,19] If ever there were an industry in need of a better, faster, cheaper strategy, the pharmaceutical industry would be it. Some, however, have questioned whether it will ever get better without radical changes, yet major changes are hard to enact in such a heavily regulated industry.[20]

Despite this somber picture, the 1990s were quite prosperous for many pharmaceutical companies. A look at more than 40 drugs that have changed the world has been published, from aspirin to sildenafil (Viagra) and more.[21] The 1990s were dominated by antiulcer drugs such as ranitidine (Zantac) and omeprazole (Prilosec). Various case histories can be found on these subjects.[22–24] Given that treatment options for high levels of gastric acidity were rather limited only 30 years ago, the gastrointestinal field has made great strides forward, first with histamine H_2 blockers, and later with proton pump inhibitors. This field has seen the gamut: the rise of billion dollar products, prescription (R_x) to over-the-counter (OTC) switches, and the development and commercialization of single enantiomers (e.g., esomeprazole (Nexium) from omeprazole). Of recent great interest has been the so-called magic bullet for certain cancers, namely, imatinib (Glivec or Gleevec). The history of imatinib has also been summarized as a case history, in both journal article and book form. It is a story that started around 1990, with initial leads identified in a protein kinase C (PKC) screen, continuing to 1992, when imatinib was first synthesized, to 2001, when imatinib was approved by the FDA to treat chronic myelogenous leukemia (CML).[25,26] Imatinib has been a great success story for Novartis and its employees, target-based therapies and kinase inhibitors, cancer research and cancer patients, and drug development in general, but it still took 9 years from first synthesis to regulatory approval. Moreover, the drug industry had been working on PKC inhibitors for many years prior to the discovery of imatinib.

The US is the largest market for pharmaceuticals in the world. Based on an analysis in mid-2004, it represented nearly $230 billion in sales, which was 46% of global sales. Japan was second largest, at 11.1%. Germany, France, Italy, Spain, and the UK taken together accounted for 20.8% of global sales. Antiulcer drugs have for a long time occupied the top spot, as noted above, with antidepressants not far behind, but no longer. Agents that reduce cholesterol and triglycerides now hold first place with nearly $30 billion in global sales. The recent market leader has been atorvastatin (Lipitor), sold for hypercholesterolemia. However, possibly as many as 20 recent top drugs will lose their patent protection in the period between 2000 and 2010, and thus their sales will likely fall precipitously as generic competition grows and new drugs are introduced. In the face of this blockbuster mania, it is interesting to note that while some companies have grown largely on their own (e.g., Roche), others have grown through mergers. An example is Pfizer, which today encompasses Agouron, Gödecke, Jouvenal, Monsanto, Pharmacia, Searle, Upjohn, and Warner-Lambert/Parke-Davis, among others.[27]

Though many reviews focus on sales as the index worth measuring when it comes to top drugs, another measure is the total number of prescriptions. **Table 2** shows a comparison for 2003. Returning to sales, the numbers are staggering – each of the top 48 drugs in the US sold $1 billion or more in 2003. Moreover, every drug in the top 200 sold $200 MM or more in the US in 2003.

Table 2 Top drugs in 2003 by number of prescriptions and sales in the US[28]

Top drugs by number of prescriptions	Top drugs by sales
Hydrocodone with acetaminophen	Lipitor ($6.3 billion)
Lipitor	Zocor
Synthroid	Prevacid
Atenolol	Procrit
Zithromax	Nexium
Amoxicillin	Zyprexa
Furosemide	Zoloft
Hydrochlorothiazide	Celebrex
Norvasc	Epogen
Lisinopril	Neurontin

Venture capital (VC) funded biotechnology companies account for an increasing proportion of new drugs entering the marketplace. However, 'biotech' is often a misnomer, since many biotech companies focus on small molecule drugs rather than on protein biologics. In recent years, almost 25% of new drugs, both new molecular entities (NMEs or NCEs) and new biological entitites (NBEs), came from the biotech industry.[29] The Hatch-Waxman Act benefited drug development significantly in the US, especially for big pharma, providing for patent term extensions, among other things.[30] It has also led to many patent challenges, which could in the long run be detrimental to all concerned.[31] The 1980 Bayh-Dole act was even more important, especially for the academic sector and for the creation of a strong biotechnology industry. It put patents issued based on government-funded research in the hands of the universities and other institutions where the work was conducted. From the standpoint of technology transfer and the creation of new companies, products, and personal wealth, this regulatory act has been a major success. It may be the prime reason for continued US dominance of biotechnology and related areas. However, not everyone has been pleased with the results, since material transfer agreements (MTAs) that keep rights in check can slow and hinder research, and some have complained about a growing corporate for-profit mentality among university administrators and researchers.[32]

The Cohen and Boyer patent on recombinant deoxyribonucleic acid (DNA) techniques signaled a turning point for pharmaceutical research and the creation of the biotechnology industry. Granted in 1980, the processes represented by this patent underpinned much of the industry's early growth and the influx of major sums of venture capital in the two decades leading up to the twenty-first century.[33] Thus, it should be no shock how important patent protection is to this industry. Patentability is determined based on novelty, inventive steps, enablement, and support.[34] Of course, patents provide legal monopolies over new therapies, which has led to concerns about monopolistic drug pricing, as alluded to above.

Biotechnology companies have had a number of banner years in recent times. The year 2000 is the record holder for monies raised in initial public offerings (IPOs), with over $8 billion. In 2004, another good year, more than $2 billion was raised. VC investments were in excess of $5 billion in 2004, even higher than in 2000.[35]

This is indeed an industry that counts time in decades and even centuries. Consider for illustration the case of nonsteroidal anti-inflammatory drugs (NSAIDs). It has been known for centuries that the bark of willow trees and other plant substances have medicinal effects. On hydrolysis, the active antipyretic ingredient in willow bark, salicin, yields salicyclic alcohol. Based on this discovery, salicylates including aspirin came into use as pure drugs in the mid to late 1800s. Many years later, in the early 1960s, the project that produced piroxicam (Feldene) was initiated, building on the knowledge of other NSAIDs, including aspirin. Piroxicam was first launched by Pfizer in 1980, 18 years after the project began, and well over 100 years since the discoveries relating to salicin had been made. Aspirin and other NSAIDs are now known to exert many of their actions (analgesic, anti-inflammatory, antipyretic) through the inhibition of prostaglandin biosynthesis, this in turn being effected through the inhibition of two forms of cyclooxygenase, COX-1 and COX-2. Based on the differential expression patterns of cyclooxygenases, a number of companies sought COX-2–selective agents to minimize the gastric toxicity that is so problematic with chronic NSAID use. This work ultimately yielded COX-2–selective agents such as celecoxib (Celebrex), which became billion dollar drugs before it was realized that marketed COX-2 inhibitors increased cardiovascular risk.[36,37] The chemical structures of representative NSAIDs are shown in **Figure 4**.

Indeed, the fate of COX-2 inhibitors is under debate at the time of this writing. The recall of rofecoxib (Vioxx), a billion dollar product for inflammatory pain, led to an immediate collapse of Merck stock. Whereas Merck had been the darling of the industry for many years, and was widely respected for high-quality research, its star fell almost overnight. Will Merck be able to survive alone, or will it be forced to merge with another entity to weather the storm, given adverse market forces and public perceptions?[38]

Is a drop in new drug application (NDA) approval times the reason for an increase in drug withdrawals from the marketplace, as with COX-2 inhibitors? Though some have suggested this to be the case, it is not clear. However, it is a fact that NDA approval times have dropped considerably in the last 20 years, from 1986, when median approval times for all NDAs were approximately 33 months, to 2003, when priority reviews were accomplished in under 7 months and standard reviews in about 23 months.[39] COX-2 inhibitors are not the only therapeutics to come under fire in recent years. Deaths of multiple sclerosis patients following therapy with natalizumab (Tysabri, formerly antegren, approved in 2004) has led to further scrutiny of the FDA early approval process. Other drugs that have received accelerated approval include indinavir (Crixivan, human immunodeficiency virus/acquired immune deficiency syndrome (HIV/AIDS), 1996), Gleevec (imatinib, leukemia, 2001), gefitinib (Iressa, lung cancer, 2003), and bortezomib (Velcade, multiple myeloma, 2003).[40]

The failure of major COX-2 inhibitors postmarketing should be a wake-up call for the importance of monitoring the safety of new drugs[41] – and importantly for personalized medicine – diagnostics, prognostics, biomarkers, surrogate markers, pharmacogenetics, and pharmacogenomics. Whether suitable application of the latest technologies could have

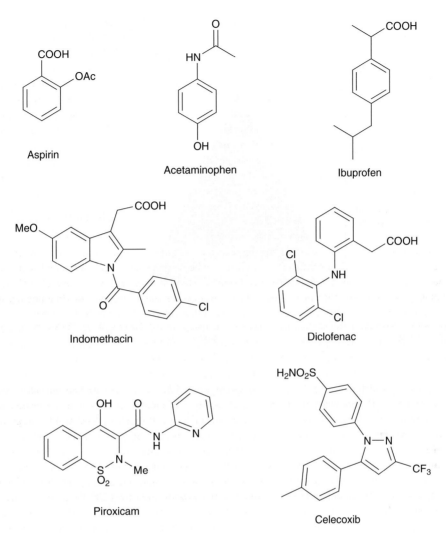

Figure 4 Representative NSAID chemical structures.

prevented such a fiasco is unknown, but the future will (or at least should) demand application of the tools of personalized medicine. To say this in a different way to drive the point home, the variation of drug responses between individuals and populations can be significant, quite literally meaning a difference of life or death.[42–44]

Even without safety issues, drugs face continued hurdles once marketed. For example, despite demonstrated clinical significance in treating Alzheimer's disease (AD), certain regulatory bodies have decided that the benefits of cholinesterase inhibitors such as donepezil (Aricept), rivastigmine (Exelon), and galantamine (Reminyl) are not sufficient to justify their costs.[45]

While the basic research that underpins drug discovery may start years or even many decades earlier, the discovery and preclinical testing phase of pharmaceutical research is generally regarded as lasting 6–7 years. In this phase, a large number of compounds are studied in vitro and in vivo to evaluate their biological effects, safety, and formulations. Roughly 1 in 1000–10 000 (or even millions!) of these compounds advance to filing of an IND with the FDA (**Table 1** and **Figure 5**). An IND typically describes the discovery and preclinical testing of the drug candidate, the clinical plans, and various chemical and biological data, including manufacturing details, presumed mechanism of action, and toxicology. In addition, institutional review boards (IRBs) must review and approve all clinical trials at each clinical site. If the FDA does not disapprove an IND or put it on hold within 30 days, human testing may begin.

Clinical trials include three phases of testing prior to filing of an NDA with the FDA (**Table 1**). In Phase I, which takes about 1.5 years to complete, up to 100 healthy human volunteers are studied to determine safety and dosage

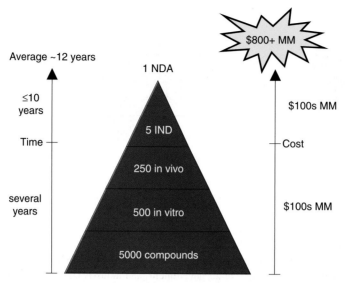

Figure 5 The drug development pyramid – long timelines, high costs, heavy attrition. IND, Investigational New Drug Application; NDA, New Drug Application. (Reprinted with permission from Moos, W. H. Introduction: Combinatorial Chemistry Approaches the Next Millennium. In *A Practical Guide to Combinatorial Chemistry*; Czarnik, A. W., DeWitt, S. H., Eds.; American Chemical Society: Washington, DC, 1997, pp 1–16. Copyright 1997 American Chemical Society.)

parameters. In Phase II, which may take 2 years to complete, up to 500 patients are studied to evaluate drug efficacy and side effects. In Phase III, thousands of patients are studied to confirm efficacy and monitor adverse reactions with longer term use. Approximately one in five of these compounds progress through an NDA to FDA approval. An NDA describes all of the information that a company has amassed, and may amount to more than 100 000 pages of reports. (Fortunately, FDA filings are increasingly electronic, rather than hard copy.)

Of course, there are exceptions. For instance, in some diseases, such as cancer, in which the drugs may be very toxic, initial human testing is conducted in patients rather than normal individuals, and preclinical testing may be abbreviated.[46] In orphan diseases, defined in the US to be diseases with fewer than 200 000 patients, smaller numbers of patients may be studied.[47]

Congress first signed the Orphan Drug Act into law in 1983, recognizing that there are many life-threatening or very serious diseases and conditions that have not received attention because of their small market size. Developing drugs for these indications would not be profitable without financial and other incentives. This group includes such diseases as amyotrophic lateral sclerosis (ALS; Lou Gehrig's disease), Huntington's disease, and muscular dystrophy, which affect small numbers of individuals. Incentives are now in place in the US, Europe, and elsewhere for orphan diseases, providing facilitated interactions with regulatory bodies, grants and other financial incentives, and marketing exclusivity for a period of years, when certain conditions are met.

Once a drug is approved, and physicians begin prescribing it, the FDA may require further testing. This postmarketing surveillance period is sometimes referred to as Phase IV (**Table 1**). During this period, companies must continue to submit regular reports to the FDA describing adverse reactions and other relevant records.

Given all of these circumstances, many of which may be incomprehensible to lay persons, how does the public view the pharmaceutical industry? Medicinal chemists and their employers should be pleased to hear that 78% of adults say that ethical drugs make a big difference, and 91% believe that drug companies make significant contributions to society through R&D.[48] Despite this positive view, one can find a number of unflattering (and some would say unfair) assessments of the industry and its players.[49]

And what about the media? Though the media serves an important role in society, sometimes it works at cross purposes with pharmaceutical and biotechnology R&D, in part by promoting sensationalism and fear mongering at its worst. Both reporters and scientists can be at fault when this relationship and its checks and balances fail.[50–52] It is hoped that this chapter, and the other chapters in this volume and series, will in some small way aid in making the relationship between the media and scientists practicing this art one of honesty and mutual understanding. May the next scientific breakthrough be viewed with a new perspective of comprehension and reported in a balanced manner.

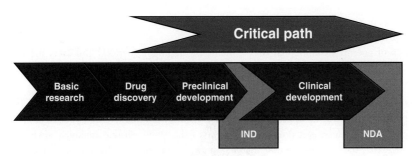

Figure 6 The FDA critical path for medical product development.

2.01.2.2 Major Issues Including Attrition

How does one take into account today's major issues – including attrition, productivity, efficiency, success rates, costs, and shots on goal – and make things better? The acceleration expected in drug discovery and development with the advent of new technologies such as combinatorial chemistry and genomics has, unfortunately, been slow to materialize. This delay has led to much discussion in academic, government, and industry circles, with various proposals resulting. For example, the FDA has analyzed the so-called pipeline problem and published a white paper on its assessment of the challenges and opportunities in the critical path to new medical products.[53] During the last decade, both the total budget of NIH and US pharmaceutical R&D spending have more than doubled (**Figure 2**). Despite these increases in spending, the total number of NCEs and NBE license applications (BLAs) submitted to the FDA has dropped by roughly a factor of two (**Figure 3**). While most stages and phases of drug discovery and development appear to have slowed and become more expensive, it is in what FDA calls the critical path (**Figure 6**) – from preclinical development to product launch – that the changes are most evident. The critical path investment required for one successful product launch has increased by more than $500 MM since 1995. A new product development tool kit has been proposed to address the pipeline problem, including modernized scientific and technical methods in several areas. Areas of focus in the new product development tool kit include animal models, computer-based predictive models, biomarkers for efficacy and safety, and clinical evaluation techniques, including the following.[53]

2.01.2.2.1 New product development tool kit for the critical path
Efficacy:

- Quantitative metrics and surrogate markers correlating mechanism of action with efficacy
- Imaging technologies
- Pharmacogenomic strategies for optimal patient selection

Safety:

- Proteomic and toxicogenomic approaches
- Predictive in silico toxicology
- Accurate assessment of cardiovascular risk, including QTc interval prolongation

What causes attrition? The average clinical success rate, from first-in-man studies to registration, was only 11% from 1991 to 2000 among the 10 largest drug companies. In the clinic, a lack of efficacy or safety concerns each account for about 30% of the failures. While pharmacokinetics and bioavailability have been cited as the reasons for around 40% of attrition in prior years, more recent estimates show that these factors account for less than 10% of the failures. In recent years, cardiovascular drugs have had the highest success rate (20%), whereas oncology and women's health drugs have had the worst success rates (no more than about 5%). Some authors have proposed that one can reduce attrition by paying attention to various factors, such as the following[54]:

The following factors may influence attrition:

- strong evidence supporting mechanism of action and links to specific disease pathways;
- elimination of leads with mechanism-based toxicity using modern tools such as gene knockouts;
- identification of biomarkers for dosing and efficacy;

- early proof-of-concept studies in humans;
- more appropriate animal models such as transgenics; and
- earlier resolution of potential commercial issues, including alignment of R&D and sales and marketing and more accurate forecasts of changes in medical best practices and competitive products.

It may be possible to achieve the desired output by brute force. The larger the number of drugs in development, the greater the chance that drugs will be approved – this is a quantity argument that ignores quality in calculating the potential for success. Nonetheless, is it possible that some of the largest drug companies, such as Pfizer, will in fact become large enough that they will have enough shots on goal to be successful almost regardless of the odds.

2.01.2.3 Project Management Including Teams

Pharmaceutical R&D can be organized in various ways, taking into account line versus matrix considerations, departments versus teams, decision points, interfaces, and transitions. Some companies have developed strong line management-based departments, such as chemistry, biological sciences, pharmacokinetics and drug metabolism, and toxicology. In these organizations, projects may be driven by individuals within a department, who on their own find collaborators in other departments to move projects forward (department-driven projects, **Figure 7**). In others, matrix teams are in control, though departments based on scientific discipline provide a home for team members (matrix-driven projects, **Figure 8**). Team leaders in a matrix organization report directly to the head of research, as well as reporting to their department heads. A hybrid of these two approaches places all of the necessary resources for each project under one person's control, regardless of scientific discipline (self-contained projects, **Figure 9**). In these organizations, there is no scientific discipline-based department. The project is the department, with the necessary chemistry, biology, and other scientific disciplines all self-contained. The choice of organizational structure is based more on personal preference and experience than on clear evidence that one system works better than another. In the matrix-driven project model,

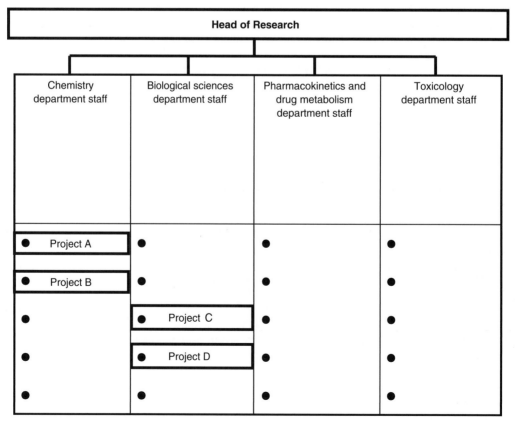

Figure 7 Department-driven projects.

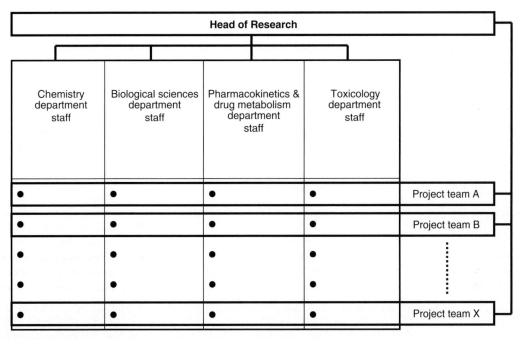

Figure 8 Matrix-driven projects.

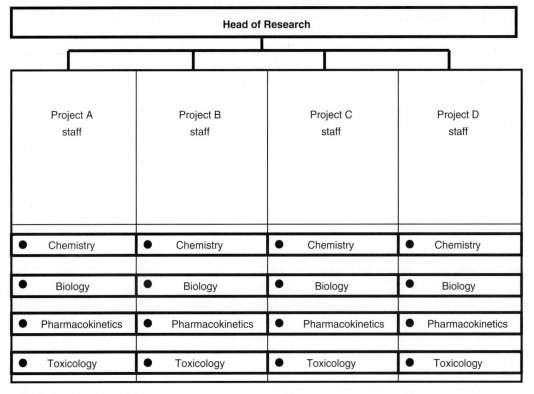

Figure 9 Self-contained projects.

one may have the best of all worlds, but significant coordination is required, necessary resources may not always be available, and leaders may have more than one boss. In self-contained projects, all of the resources necessary to be successful are under project control, but the quality of the science may suffer because there is no larger home for each scientific discipline.

Table 3 Project team members are selected to represent different departmental mixes at different stages of development

Scientific members	Other team members
Biochemistry	Business development
Chemical development	Contracts
Clinical development	Intellectual property
Core facilities	Project management
Drug metabolism	Regulatory
Medicinal chemistry	Sales and marketing
Molecular biology	
Pharmacokinetics	
Pharmacology	
Toxicology	

Project team members are drawn from various departments to address the needs of a project at different stages (Table 3). Thus, medicinal chemistry and discovery biology are critical in the early stages of the project, whereas chemical development, clinical and regulatory, are more important as one enters the clinic. A project team might range in size from a few people to two or three dozen members.

Project team chairpersons are responsible for the success of their projects. The chair provides overall leadership for the project from a scientific and management perspective. The chair should push the project forward to ensure that state-of-the-art (SOTA) technologies are being applied, group cohesiveness is maintained, and progress is being made toward team goals. Ultimately, the team leader is responsible for communicating the project's successes and failures to management. Team vice- or co-chairs assist the chairs in all aspects of the project. The vice-chair assumes responsibility for the project in the chair's absence. In addition, the vice-chair is responsible for recording the team's progress, or lack thereof, and disseminating this information as appropriate.

Before embarking on a new project, and as part of periodic project reviews, many groups evaluate the project from a number of perspectives. Some companies refer to this as a SWOT analysis – Strengths, Weaknesses, Opportunities, and Threats. The questions asked might include the following:

- What is the overall objective, and over what time period?
- Is this an area of major unmet medical need?
- Is the market large enough to warrant the time and cost required to develop a new drug?
- How are you and your competitors approaching the problem, and how will you win this race or competition?
- What are the clinical, regulatory, and other hurdles that you need to surmount?
- What constitutes proof-of-concept, and how will you know when you reach your objective?
- What are the resources you need to be successful?

Decision making of this sort is important in almost any profession, albeit less complicated in many other industries. Managing the interface of drug discovery and drug development successfully requires a clear target product profile that team members and management have agreed upon.[55] Some groups start by constructing a hypothetical package insert for the product they would like to develop.

2.01.2.3.1 Overall project assessment scheme
Executive summary:

- Synopsis of R&D program
- Objectives, including timeframes

Medical need:

- Disease incidence/prevalence
- Demographics/geographical distribution

Strategies to meet objectives:

- Current industry/medical dogmas
- What is your primary approach?
- What constitutes proof-of-concept?
- Rationale for choosing these approaches
- What are the alternatives?

Competitive assessment:

- Competitive technologies and approaches
- Status of top products already marketed or in clinical trials

Market assessment:

- Overall business/commercial potential
- Potential impact of disease trends
- Potential impact of new drugs reaching market within 5–10 years

Regulatory position:

- Pharmacogenomic opportunity?
- Nonclinical R&D issues (pre-IND)
- Clinical R&D issues (pre-NDA)

Resources:

- Team chairs and champions
- Resources, including staffing plans and other major expenses
- Major consultants

Proposed goals/milestones:

- Project history
- Plans for next 12 months
- What does success look like 5 years hence?

One of the basic decisions to consider at the earliest stage of drug discovery is which scientific or therapeutic area a given team of people should focus on. Once a group is operational, targets need to be validated, and decision points and transitions need to be considered.

2.01.2.3.2 Target validation

- Are the targets relevant to the disease?
- What is the desired profile of a drug acting against a given target, and how should one screen for this activity?
- Once hits have been identified, and improved leads have been prepared, when are the leads good enough to proceed to exploratory development, such as rising-dose, range-finding toxicology?

2.01.2.3.3 Selected decision points and transitions from discovery to new drug application

- Therapeutic area choice
- Target validation
- From screening hits to optimized leads
- Entry into exploratory preclinical development
- Advancement to full-scale preclinical development
- IND preparation and submission
- Conduct and review of clinical Phases I–III
- NDA submission

With each successive stage of drug discovery and development, new groups must interface to advance the project. As a drug moves into early development, scale-up is usually required, in order to prepare enough material under controlled conditions to proceed with toxicology and other studies

Despite these never ending complications, a very large number of drugs have made it to the finish line, being marketed in many countries around the world. Several books provide details on these drugs and their properties from the perspective of medicinal chemistry or from other pharmacological and medical perspectives.[56–59]

2.01.2.4 Recruitment and Management of Human Resources

How does one recruit and manage the human resources required to be successful in medicinal chemistry and pharmaceutical R&D? There are, as one might expect, many factors to consider:

- How many people are needed?
- What mix of talents and personalities will be most functional?
- Who is the right boss for various occasions?
- How does one provide for career growth and training?
- What about other factors that fall under categories such as morale, management, and leadership?

Research laboratory management requires paying attention to both the forest and the trees – the big picture as well as the details.[60] It requires finding ways to stay informed without micromanaging.[61] Managing well requires many leadership skills, including an appreciation for and a knack for leveraging different styles, the ability to communicate effectively on many different levels, ways of addressing conflict and change constructively, comfort (or if not comfort, then a productive disharmony) with the corporate culture, and an understanding of motivation.[62] A terrific scientist and friend once described to me what makes for a successful scientist in Biotech or Pharma (paraphrased below)[63]:

> Often the scientists who are most successful in industry have an in-depth knowledge of some specific area within their discipline, and, to the extent that they are vocal promoters of such a subdiscipline, could be viewed as prima donnas. The major question for such an individual is not whether this type of attachment should be readily abandoned, but rather whether other approaches are viable, respectable, and supportable as practiced by colleagues with different expertise. The scientist who believes that s/he has the only solution to any problem does not fit well into a biotech or pharma setting because s/he sends others clear signals that they have no value except to serve as acolytes to his/her vision and technical approach. The true prima donna is so self-centered and intolerant of others' ideas that s/he causes too much collateral damage to justify his/her limited success. The super subspecialist who values others' approaches is, on the other hand, a potentially fine colleague who can make substantial contributions to a multidisciplinary organization. For the manager of such a person, the challenge is to prevent that individual from becoming isolated and evolving a program that does not employ alternative practices. Though one could focus on examples of failures, this is perhaps less productive than highlighting successes. Successes can arise from proper contextual use of subspecialty expertise, such as combinatorial chemistry as one avenue of drug discovery, while separately evaluating specialty libraries related to biological target families, and also using scaffolds derived from specialty libraries to enable new combinatorial approaches. As another example, rational drug design approaches, which are usually employed only in a convergent fashion – to advance from lead to preclinical candidate – can also be employed in a divergent fashion to suggest novel scaffolds for novel combinatorial programs.

While the importance of interviewing can be lost on busy technical people in companies both large and small, hiring new researchers who have the ability to make both intellectual and emotional contributions to an organization (especially small organizations) can mean the difference between self-destruction and success.[64] How do you judge a candidate for a job? Can and will s/he do the job? What about compatibility with the current team? What are the candidate's values, work ethic, motivations, hot buttons, etc.? How do you make sure to avoid legal pitfalls during the interview, possibly related to age, race, religion, sex, sexual persuasion, or marital status? If the job candidate had a problem at their prior employer, will that problem repeat itself at your place of work? There are some desirable characteristics to look for when recruiting[65]:

- role awareness and loyalty;
- high and focused energy levels;
- inner motivation and good reasons to want the job;
- emotional maturity and compatibility;
- fire in the belly (but not a rebel without a cause); and
- drive to complete tasks that have been started.

As one progresses up the management ladder, there are many factors that may influence successful transition from technical jobs to management, as outlined below[66]:

- treating time as a most precious asset;
- dealing well with aggravations;
- selling yourself versus the competition;
- mentorship (for you and for your staff);
- leveraging strengths;
- understanding and compensating for one's weaknesses (possibly more important than having great strengths);
- teaming and collaboration;
- confidentiality and corporate obligations; and
- the art and science of negotiation.

After reading a memo on completed staff work in a military journal, an early head of a major pharmaceutical company summarized his thoughts as follows.[67] He certainly offers something to consider:

1. It is your job to advise your chief what he ought to do, not to ask him what you ought to do.
2. Your views should be placed before him in finished form so he can make them his views simply by signing his name (…'or by making very minor corrections').
3. …[A] draft needn't be neat but ought to be complete.

Setting expectations can be important from many perspectives – to maximize performance and to minimize frustrations, amongst others. In this context, what does one tell aspiring medicinal chemists just starting out in their careers? Perhaps they need to know, at a very basic level, their missions, the keys to success, and their responsibilities, as outlined below – points that might be only rarely taught to those who end up practicing medicinal chemistry, if at all.[69] The exact origins of this listing of overview and responsibilities are unclear, but it is something that the author has used and developed in various forms over the years.

2.01.2.4.1 Guidelines for medicinal chemists early in their careers
2.01.2.4.1.1 Overview

2.01.2.4.1.1.1 Mission of a medicinal chemist   To use all techniques and knowledge available to the modern organic chemist, combined with a sound understanding and application of medicinal principles, in order to discover innovative, safe, and efficacious new drug candidates in a timely and efficient manner.

2.01.2.4.1.1.2 Keys to success   Highly intelligent, skilled, and motivated scientists whose contributions, individually and as part of a team, both in ideas and in the synthesis of compounds, are absolutely essential to novel drug discovery and development, to the company, and to the ultimate beneficiaries of new medicines.

2.01.2.4.1.2   Responsibilities

2.01.2.4.1.2.1 Literature work   Appropriate literature searches should be conducted to evaluate possible synthetic routes and to determine whether target compounds or analogs are known. Knowledge of current literature is important. Independent reading of relevant scientific literature during off hours is essential for optimum performance.

2.01.2.4.1.2.2 Synthesis of known compounds   You should know if the compounds you are making are known. It makes little sense to reinvent the wheel – in the long run you will get farther by consulting the literature first. If a compound is known but not fully characterized, then you should complete the characterization. Where physical data have been reported, you must compare some but not necessarily all data. Application of appropriate and thorough chemical expertise is essential. After appropriate literature work has been done, new compounds should be prepared by the best possible route. Preparation involves setting up and carrying out reactions, monitoring reaction processes, working up reactions, and purifying and isolating the desired products. Full chemical characterization ideally implies at least the following: nuclear magnetic resonance (NMR) spectroscopy and other spectra to ascertain structure, mass spectrometry and/or elemental analysis to determine composition, and chromatography to show the degree of purity. Where desirable, other data should be obtained. Careful interpretation of these spectral and analytical data is expected.

2.01.2.4.1.2.3 Submission of compounds for biological evaluation   Appropriately characterized target compounds should be submitted for biological evaluation as directed, with inclusion of as much relevant information as possible on submission forms (which are increasingly electronic). The forms may include information regarding solubility, project classification, synthetic routes, and references.

2.01.2.4.1.2.4 Maintaining accurate and well-documented experimental records   Research notebooks should be kept orderly and up-to-date, with cross-references wherever necessary. Experimentals should include starting reagents, a description of the experiment (setup through isolation), references where appropriate, cross-references to spectra, analyses, and the like, and an overall assessment of mass balance and yield.

2.01.2.4.1.2.5 Invention records   Significant ideas for new target structures, modifications to existing leads, or new methods of compound use should be recorded in a full and timely manner. To ensure the maximum value of invention records, new ideas should be investigated, developed, and exploited with diligence.

2.01.2.4.1.2.6 Safe laboratory conduct   General safe laboratory conduct is essential. This includes maintaining clean and organized work areas, and proper disposal of wastes. Safety glasses must be worn at all times in the laboratory, and no food or drinking material should be present in the laboratory. Water awareness is expected – to reduce the chance of floods – with hoses clamped and unused water lines turned off.

2.01.2.4.1.2.7 Oral and written reports   Reports should be completed in a neat, concise, and timely manner whenever requested. Clear reports at various meetings summarizing recent personal work are expected. Again, preparation of invention records, patents, and manuscripts should be done in a timely fashion.

2.01.2.4.1.2.8 Careful treatment of laboratory equipment   Equipment should be treated with respect. This includes maintaining instruments, keeping equipment clean, and not abusing equipment. For example, vacuum pump oil should be changed regularly, glassware breakage should be kept to a minimum, stirrers and hot plates should be wiped off periodically, cracked or frayed electrical cords should be replaced, and glassware should be cleaned thoroughly and soon after use.

2.01.2.4.1.2.9 Independence and initiative   A certain degree of independence should be demonstrated in the completion of assigned tasks. Similarly, initiative should be taken to design and implement synthetic routes to unknown compounds. Routine purifications, isolations, and characterizations should be carried out without intervention. Supplies of important intermediates should be monitored, and when low, further supplies should be synthesized or purchased. An appropriate range of alternatives should be considered prior to making decisions.

2.01.2.4.1.2.10 Personal interactions Personal interactions should be handled professionally. Interpersonal problems should be confronted in a constructive and positive manner. Diplomacy, consideration, and the ability to compromise (when appropriate) are important traits.

2.01.2.4.1.2.11 Monitoring biological activity The activity of compounds should be monitored carefully. A good understanding of the biomedical underpinnings of the project is necessary for optimum performance. New compounds should be suggested for synthesis based upon sound medicinal reasoning. Where appropriate, computer-assisted methods should be employed. Knowledge of compound numbers, names, and structure–activity relationships (SARs) is expected.

2.01.2.4.1.2.12 Awareness of goals An awareness of corporate, department, group, and personal goals is important.

2.01.2.4.1.2.13 Timeliness The importance of developing, implementing, and maintaining effective plans should not be underestimated. Deadlines should be adhered to; even better, complete the task ahead of schedule. Potential problems should be anticipated. The status of tasks and assignments should be reviewed periodically to ensure successful completion.

2.01.2.4.1.2.14 Management Assigned personnel should be managed effectively. This requires the use of appropriate leadership styles for the circumstances at hand, and administering relevant policies, practices, and procedures.

Additional words to the wise for thought processes, decision making, and related topics have been summarized elsewhere.[70] It is important to develop, maintain, and where possible record, the institutional memory of organizations. This tacit knowledge can be a significant strategic advantage – or a troublesome stumbling block.[71]

Scientists may not be as focused on money as people in other professions, but money can still be a factor in medicinal chemistry, as in any other profession – for recruitment, retention, and job satisfaction. Though the media and the general public often complain about the compensation paid to pharmaceutical and biotechnology executives, is their compensation truly out of line with leaders in other professions? (This is always a hotly debated topic.) In recent years, salaries for biotech executives have trended upward. The salary range generally depends on company size, from small companies with market capitalization below $50 MM with executives paid $275 000 annually, to large companies with billion dollar market capitalization with executives paid $500 000 annually. A bonus based on performance is often granted on top of the salary, as are stock options (see below).[72] As one comparison, consider professional baseball, with average salaries greater than $2 MM, versus biotechnology chief executive officers (CEOs), with average salaries (base plus bonus) of $310 000. Successful mortgage and real estate brokers, as well as those involved in various segments of the financial industry, are routinely compensated well above biotechnology company CEOs. At the top end, while the CEO of Pfizer made almost $17 MM in 2004, including stock-related compensation, and the CEO of Genentech made $23 MM, many professional athletes, radio, television, and movie stars, and real estate moguls made the same or more. Thus, if one sets aside the debacles of Enron, ImClone, Tyco, WorldCom, and a few others, healthcare and other corporate executives are in fact underpaid versus other professions, especially given the importance of healthcare.[73–80]

2.01.2.5 Multiple Sites Including Outsourcing

Internal research must be coordinated within and without, including with contract organizations, to which work may be outsourced. Whether a company is a giant pharmaceutical corporation or a small biotechnology firm, it may be resident at more than one site. Large multinational companies often have R&D sites in several countries. For example, Warner-Lambert's Parke-Davis Pharmaceutical Research Division (now part of Pfizer) had R&D sites in France, Germany, the UK, and the US (California, Michigan, and New Jersey) at various points in time. Even small biotechnology companies may have sites in more than one country. The small, privately held biotechnology company MitoKor (now part of Migenix, based in British Columbia, Canada) at one time had sites in Massachusetts and California in the US and also in Australia.

With size and multiple sites, though, come difficulties in managing a large and far-flung empire. The concept of making each site more or less self-contained is one that some companies have gravitated toward, with what GlaxoSmithKline calls Centers of Excellence (as in Centers of Excellence in Drug Discovery (CEDD)). This approach makes a given site more like a biotechnology company, with all of the resources necessary to take drug candidates into early clinical development, and ideally through human proof-of-concept.[81–83] Johnson and Johnson (J&J) claims to take this approach as well, with various acquisitions historically operating rather independently (e.g., ALZA, Centocor, and Janssen). Warner-Lambert also followed this approach, with its UK site focusing on neuroscience, and its California site focusing on structure-based design in areas such as antivirals and anticancer agents.

Having multiple sites can lead to we/us versus they/them issues, lack of coordination, duplication of resources, and many more challenges. If the sites are in distant time zones, there may never be a convenient time to talk, though faxes, emails, and cell phones help in this regard. Some companies have chosen to organize different sites around different therapeutic areas. This structure provides focus and critical mass, which makes sense from a business standpoint. However, it can limit the potentially profitable follow-up of scientific discoveries that lead one outside the site's therapeutic areas, a limitation that makes no sense from the perspective of scientific innovation because it is often the unexpected results that open the door to major opportunities. Unfortunately, the organization of research facilities is only one of many areas where the viewpoints of business and science may differ. Clearly, to apply the principles of portfolio management properly and most successfully, one must consider the elements of both business and science.[84]

Even the largest companies will contract out certain work. Small companies with only a few people may contract out most or all of their laboratory studies. Preclinical development and clinical and regulatory tasks are commonly conducted by contract research organizations (CROs) that specialize in such areas, such as Covance, MDS, PAREXEL, Southern Research Institute, SRI International (formerly the Stanford Research Institute), and others that have preclinical or clinical capabilities. In recent years, both chemical discovery (especially chemical libraries) and chemical development have found outlets in chemistry CROs, including Albany Molecular, ArQule, Array, Ash Stevens, and Discovery Partners, to name a few.

Decisions to outsource R&D may be based on practical business needs, practical scientific limitations of one's organization, or even personal relationships. However, there are often a number of advantages to using CROs. They provide expertise and resources (people, contacts, laboratory space, and equipment) in areas that some firms may not have, and there is no commitment to continue funding the staff or the work beyond a certain point. For small companies, the alternative, namely, recruiting and making a long-term commitment to the necessary staff, outfitting a laboratory, and getting operations up and running, usually would take too long and would commit too much capital to be worthwhile. For a large company, CROs provide additional resources that extend what a firm might be able to accomplish during periods of peak activity.

Even basic research is frequently outsourced. Most of the large biotech-pharma alliances are essentially outsourcing arrangements, with the biotech company contributing new technologies and approaches while big pharma provides the funding and later stage resources required to take a product through preclinical and clinical development to marketing and sales. The need for and the value of such deals have been established since the 1980s through numerous partnerships, though disappointment is not uncommon. One could argue that big pharma should outsource even more of its research, focusing internally on its stronger capabilities in development and commercialization. Indeed, breakthrough products discovered and developed internally at major pharmaceutical companies have become a rarity, even though R&D expenditures at these same companies have roughly doubled every 5 years since 1970. Biotechnology companies, on the other hand, appear to be playing an increasingly important role in generating new therapeutic product candidates – and they are doing this with far less investment than their larger competitors. The factors driving this evolution range from financial to management to scientific, an understanding of which provides a more profitable future. Pharmaceutical and biotechnology companies that place themselves at the center of a cluster of relationships with partners (see **Figure 10**) may significantly outpace their more internally focused competitors, by increasing the productivity of their overall R&D efforts, at lower cost, and with less risk. Thus, in the view of some, "the twentieth century… encompassed the rise and decline of major pharmaceutical discovery research laboratories."[85]

The Pacific Rim has experienced substantial growth in biotechnology and pharmaceutical R&D. Southeast Asia, including China, India, and Singapore, as well as others, represents growing competition for the future, and offshoring of high technology jobs from the US and Europe has become a topic of discussion and concern. Companies such as Bridge Pharmaceuticals have been set up to take advantage of offshore CROs and other opportunities in mainland China, Taiwan, and elsewhere. Government-supported expansion has been especially evident in Singapore since the 1980s, with its Economic Development Board (EDB) and, more recently, the Agency for Science, Technology, and Research (A*STAR). Various companies, large and small, have established significant research efforts in Singapore as a result, including Novartis and MerLion, to name just two.[86–87]

2.01.2.6 Contrasting Biotech with Pharma and Innovation

Companies go through several stages as they develop, moving from start-up through focus, integration, optimization, and finally reaching maturity. Management practices must evolve during these progressions, or crises and even business

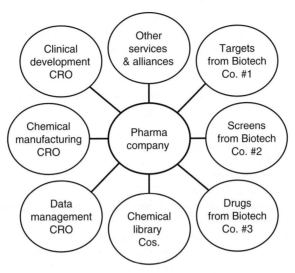

Figure 10 A pharma-biotech cluster model.

failures will result. Thus, whereas the management profile in early business stages is often vision-driven and planning is ad hoc, later-stage companies require more structured processes and formal planning. Similarly, business goals evolve from being flexible and entrepreneurial to being more consistent, integrated, and measurable.[88]

Biotech can be contrasted with pharma and innovation in the following ways:

- entrepreneurs versus intrapreneurs;
- planning versus chaos;
- strengths and limitations of each model and individual;
- new frontiers versus me-too products;
- effect of research environment; and
- champions.

Innovation is often discussed but too rarely practiced (and even more rarely studied with rigor, though a few examples can be found in the recent literature[89]). In fact, many people fight innovation as it moves through the several stages of a revolution. The stages of a scientific revolution have been summarized in various ways for centuries. For example, the German philosopher, Arthur Schopenhauer, who lived from 1788 to 1860, has been quoted as follows. "All truth passes through three stages. First, it is ridiculed. Second, it is violently opposed. Third, it is accepted as being self-evident."[90] Others have elaborated on the general theme that Schopenhauer stated so clearly and simply. In the first stage of a revolution, your friends tell you that you are wrong, dead wrong. You are not even close to the truth. Some of your friends stop being friends as you continue to pursue your disruptive idea. In the second stage, a few of your colleagues begin to admit that there may be some validity to your hypothesis. Don't let this lull you into a sense of security, however, because they continue to deride you, discount the value of your new approach, and may even attack you in public. Even worse, they may undermine you and your hypothesis behind your back. In the third stage, the soundness of your concept is at last recognized by a growing majority in your field, but most people still spend their time finding holes in your theorem. Finally, even the opinion leaders – the power brokers of your field – agree that there is significance in your work, and they eventually announce that it was in fact their idea in the first place.[91]

In the end, new science usually needs to be hyped to have a chance. New technologies are always greeted with skepticism, many people have vested interests in competing old approaches, and people do not like change in general. Moreover, the timeframe to develop practical applications of new technologies is always longer than expected. However, for those who persevere with a truly valuable new technology, proof-of-concept (PoC) or proof-of-principle (PoP) comes when the eureka moment was so long ago that most people cannot remember the original experiments or who performed them, successful applications are numerous, the new science is fully integrated into everyday R&D, and the approach is referred to by abbreviation (and many can't remember what the letters in the abbreviation stand for). Finding the right balance – to be cutting edge but not so novel that one's work is too risky – is a comfort space that can be difficult to find, as many unique enterprises, such as those pioneering mitochondrial medicine, have observed.[92]

Recognizing the innovative members of your organization, and creating an environment to get the most out of them, is easier said than done.[93] Some might even say that creativity has been held hostage by corporate control systems and capitalism, among other forces.[94] There are many obstacles and inducements to the process of discovery, but there are a series of research strategies that may help.[95] In the end, there is no one right way to conduct research, and unusual paths can lead to interesting but unexpected results.[96]

Thus, according to Root-Bernstein, there are ten research strategies to facilitate discovery:

- do the unthinkable;
- get as much noise as possible out of the system;
- run controls over and over and over again;
- make analogies from your work and extrapolate;
- test experimental variables over a very wide range of conditions;
- determine the specific criteria you are looking for and then try everything you can think of until you generate the desired results;
- generate theories of existing processes in order to predict new ones;
- work simultaneously on several problems in the hope that one will cross-fertilize another;
- reevaluate an old discovery from the modern perspective; and
- turn problems on their heads and reevaluate.

The promise and disappointments of innovative science of recent decades are manifold. Advances in analytical techniques and separations technologies, especially NMR spectroscopy, mass spectrometry (MS), and high-performance gas and liquid chromatography (GC and HPLC), allowed the SOTA in the pharmaceutical industry to move forward dramatically in the 1960s and 1970s. These advances were followed in the 1980s by the adoption as routine tools of such techniques as computer-aided molecular modeling (structure-based design, SBD) and molecular biology, including the polymerase chain reaction (PCR). In the 1990s, combinatorial chemistry, high-throughput screening (HTS), and genomics took center stage. Several technologies are vying for the spotlight as we start the new millennium, including proteomics and ribonucleic acid (RNA) interference (RNAi). While many of these techniques are now commonplace, the jury is still out on the routineness and value of antisense and gene therapy approaches (**Table 4**). Will they become widely accepted, or are they destined to be niche plays? In this writer's opinion, it is likely to be the former, but acceptance will clearly take more time. Some examples of pharmaceutically relevant scientific advances of recent decades are given in **Table 5**.

Technological advances often spawn a new cadre of start-up ventures. That trend was quite true of the combinatorial chemistry field in the 1980s and 1990s. However, as is common with these waves of change, many of the original companies were acquired, failed or changed directions, becoming unrecognizable or untraceable over a few

Table 4 Which technologies have made it, and which have not (yet)

Widely accepted
- High-performance liquid chromatography (HPLC)
- Nuclear magnetic resonance (NMR) spectroscopy
- Mass spectroscopy
- Polymerase chain reaction (PCR)
- Personal computers

Widely adopted
- Structure-based design (SBD)
- Combinatorial chemistry
- High-throughput screening (HTS)
- Genomics

Jury still out
- Antisense
- Gene therapy
- Pharmacogenomics
- Proteomics

Table 5 Examples of pharmaceutically relevant scientific advances of recent decades

1970s
- NMR spectroscopy and MS
- HPLC

1980s
- Computer-aided molecular modeling
- Molecular biology

1990s
- Combinatorial chemistry and HTS
- Genomics

2000s
- Proteomics
- Ribonucleic acid interference (RNAi)

Table 6 A selection of yesterday's pioneers and today's public independents in the field of combinatorial chemistry

A selection of yesterday's pioneers	*Some of today's public independents*
Affymax	Albany Molecular
Arris	ArQule
CombiChem	Array
Diversomer	Pharmacopeia
Genesis	
Irori	
Mimotopes	
Molecumetics	
NexaGen	
Parnassus	
Protos	
Selectide	
Sphinx	
Terrapin	
Trega	

short years. (They are, at least, hard to remember.) Some of today's largest independent public biotechnology companies didn't exist in the early years of combinatorial chemistry's debut (**Table 6**). The path of acquisition for some companies has been long and tortured, ending with their going out of business (**Table 7**).

The mantra in many of these new fields has some or all of the elements of doing things better, faster, and cheaper. The paradigm shifts of recent decades have often led to subtle improvements, which are possible only with extra spending, and therefore are certainly not cheaper (**Table 8**).

The timeframe for the development and adoption of new technologies is typically several decades. Using combinatorial chemistry as an illustrative case, one can trace its origins to the 1960s and early solid-phase peptide chemistry. If one accepts this starting point, then it took almost 40 years for combinatorial chemistry to become commonplace, and its full impact may not be evident until 2010–2020, given the long timelines of the pharmaceutical industry. Given that Mother Nature had a head start eons ago, it took even longer (**Figure 11**).

Innovative science takes time to develop, usually decades (**Table 9**).

Table 7 New technologies and start-up companies may take tortured business routes, being acquired, failing, changing directions, or even going out of business[a]

Commonwealth Serum Laboratories > CoSelCo Mimotopes > Chiron > Houghton Joint Venture > MitoKor > Fisher Scientific > PharmAus/EpiChem
Stauffer > Unilever > Cheseborough-Ponds > Imperial Chemical Industries > Zeneca > Oxford/Cambridge Combinatorial > Millennium > Signature > Out of business
CombiChem > Dupont > Bristol Myers Squibb > Deltagen > Out of business
Arris > AxyS > Celera > Out of business
Protos > Chiron > Novartis
Parnassus > Kosan

Note: The symbol '>' denotes that the previous company became or was acquired by the company after the symbol.
[a] These examples are taken from the field of combinatorial chemistry.

Table 8 Are new technologies better, faster, and cheaper?

Better
- Often yes
- However, old science is still extremely valuable and new science simply adds to it

Faster
- Often yes
- However, different bottlenecks become rate limiting and speed is still restricted

Cheaper
- Often additive to existing expenses
- Rarely cheaper until many years later

Date	Event
5 billion BC	Earth's formation & first molecular & biodiversity
5 million BC	First human beings
500 BC	Time of Hippocrates, the father of modern medicine
1960 AD	First publications on solid-phase *peptide* synthesis
1975 AD	First articles on solid-phase *organic* synthesis (nonpeptide)
1985 AD	First publications on rapid parallel synthesis
1985 AD	First companies devoted to combinatorial chemistry
1990 AD	New drugs approved after 2000 were first studied
2000 AD	Combinatorial chemistry became mainstream

Figure 11 Putting time into perspective: the full timeline for the development of combinatorial chemistry. (Reprinted with permission from Moos, W. H. Introduction: Combinatorial Chemistry Approaches the Next Millennium. In *A Practical Guide to Combinatorial Chemistry*; Czarnik, A. W., DeWitt, S. H., Eds.; American Chemical Society: Washington, DC, 1997, pp 1–16. Copyright 1997 American Chemical Society.)

Table 9 The development of combinatorial chemistry

1960s
Solid-phase peptide synthesis, e.g., Merrifield at Rockefeller University

1970s
Solid-phase nonpeptide synthesis, e.g., Rapoport at UC Berkeley

1980s
Combinatorial libraries, e.g., Mimotopes in Australia

1990s
Marriage of high-throughput chemistry and biology, e.g., Chiron in California

2.01.2.7 Business Development Including Aspects of Portfolio Management

Before one places major resources behind a project, a formal evaluation is required in many organizations. This evaluation may include an assessment of medical need, strategic approach, the competition, regulatory hurdles, market potential, timelines, and resources required, as discussed earlier in this chapter. Many questions arise in the process. For example: How many projects should one work on, in which therapeutic or technological areas? Are you a mile wide and an inch deep, or a mile deep and an inch wide? How should one evaluate internal products versus projects and products that have been acquired or partnered? Where does intellectual property come into play? As in many situations, a fair amount of judgment is required in this area, and colleagues, investors, and other stakeholders may find each other at odds.

One of the most important strategies to balance risk, gain access to cutting edge technologies, and to increase success rate is for larger companies to partner with smaller ones. These alliances are in principle beneficial to both partners. The financial terms of such an arrangement are critical, though probably more so to the smaller partner. Typical deal terms include up-front payments, license fees, equity purchases, staffing costs (full-time equivalents (FTEs) or equivalent full-time staff (EFTS)), and success-based milestone payments and finally royalties after commercialization. There are many factors to analyze when looking at deal terms, as outlined in **Table 10**.[97]

Because most therapeutic projects fail, it can be better for companies to place multiple bets. That is, taking into account odds of success and failure, a chance to receive 10% each on 10 projects is more likely to yield a payout than 100% on one project. On the other hand, if a company has only one project, it will probably need to keep that project to itself to have a chance of success.

There is a phenomenon sometimes referred to as a U-shaped value curve (which is really an incremental return-on-investment [iROI] curve; **Figure 12**). At the earliest stages of a project, such as when a new molecular target has been

Table 10 Deal terms in alliances between pharmaceutical companies and biotechnology firms

	R&D tools	In vitro data	In vivo data	Clinical safety	Clinical efficacy	NDA filed
	Patents, targets, libraries, screens	Lead series	Optimized lead	Phase I	Phase II–III	Approved
Years of work	1+	2+	3+	5+	6++	8+++
Deal royalties[a]	0–6%	2–8%	4–10%	6–12%	8–14%	10–20%
Deal value created ($MM)[a]	$0–3 each	$1+	$5+	$20+	$70+	
Private valuation ($MM)	$0–5 each	$5–30	$10–60	$40–120	$100++	
Public valuation ($MM)	$1–3 each	$10–20	$20–30	$30–50	$150++	$400++

[a] Clinical success milestones (IND, Phases I–III, NDA) typically yield at least $8–12 MM in addition.

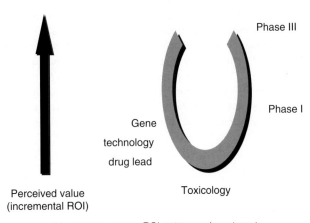

Figure 12 U-shaped project value and investment curve. ROI, return on investment.

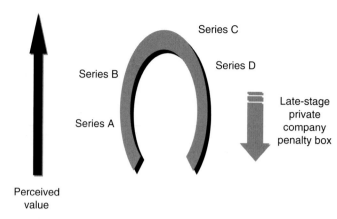

Figure 13 Inverted U-shaped value curve for later stage private companies.

identified, the project is often valued very highly. It is new, exciting, and little bad is known about the project. Then, as the project progresses and one learns about limitations or frank problems with the approach, its apparent value (and iROI) may decrease, often hitting a low point in preclinical development, when toxicities begin to appear. Once the drug enters clinical trials, the value (and iROI) increases again, with Phase I/II being a crossover point for deal making. On the other hand, a project may not be partnerable until it reaches a later stage of development, so despite a lower iROI for additional studies, it is critical to spend the money to advance the compound to later stages of development.

There is another phenomenon that produces a U-shaped value curve, but this time it is an inverted U (**Figure 13**). Later stage private companies that have not yet gone public begin to lose value, ending up in a penalty box for not graduating with their class. Such companies may end up being forced to do down rounds (i.e., reducing the price of their stock, sometimes quite significantly), to merge with other companies, or to go out of business. There may be simply too much baggage from the company's earlier years, too many tired investors, and an over-shopped story, to stand a reasonable chance of pulling out of a downward spiral.

For many early-stage technology companies, the transition to product development is difficult if not impossible. Even in the genomics space, where rich partnering deals and public offerings propelled the industry for several years, companies were forced to forward integrate to clinical stage therapeutic products, often by merger or acquisition, in order to protect their value and viability.[98] Millennium is a good example, having acquired LeukoSite and COR Therapeutics to get where it is today with products in the clinic or on the market. The standard model, where pharmaceutical companies have been strongly in control of the product outcomes of partnerships with smaller biotechnology firms, is evolving to one where biotechnology companies retain much more control over their targets and resulting drug candidates.[99]

At the core of a project's value is the intellectual property that surrounds it. Intellectual property includes patents, trademarks (TMs) or service marks (SMs), and copyrights.[100] Other confidential information such as manufacturing processes might be maintained as trade secrets. For R&D, patents are usually of most interest, as they give the inventors or their assignees the ability to exclude others. Once filed, patents generally have a 20-year lifetime. Patents start with an invention, recorded as an invention record in many companies, followed by filing of a provisional or full patent application. Patents are usually filed in more than one country, and are valid only in those countries where they issue. To be patentable, an invention must be novel, among other things. If someone tries to make, use, or sell an invention in a country where you hold a patent, they are said to be infringing, and various legal remedies exist. Once a product composition is commercialized, TMs or SMs become more interesting. It is the ability to have a legal monopoly on a given product or its use for a period of time that makes it worthwhile for companies to invest in novel research. Interestingly, certain large companies worry more about freedom-to-operate than being able to exclude others from R&D. They assume that they can out-muscle anyone who invades their territory, provided they have freedom to operate.

2.01.2.8 Drug Development

Drug development, whether chemical or biological, is a complicated but well-documented and heavily regimented process.[101–102] There are many interesting aspects to this stage of R&D, and chemistry has a major impact at many points along the path toward commercialization. Consider, for example, the switch to predominantly one enantiomer that has occurred in the last decade or so. Curiously, as recently as the late 1980s, industry information sources were still uncertain as to which drugs were single isomers and which were not.[103]

The representative stages of drug development in which chemistry has significant impact are as follows:

- chemical development;
- process research;
- chirality;
- formulations (especially novel approaches such as liposomes and other nanoparticles, nanosuspensions, etc.[104,105]);
- clinical stages and novel strategies and approaches thereto; and
- commercialization and cost of goods.

Practical aspects of drug development include determining the best salt form and formulation, characterization of the active pharmaceutical ingredient (API), and defining purity requirements. Salt forms can influence bioavailability, manufacturing, purification, solubility, stability, and even toxicity.[106] Salts are often chosen empirically, and systematic approaches to choosing salts have been proposed.[107] Hydrochlorides have been the most common salts historically, representing >40% of the FDA-approved commercially marketed drug salts in 1974. Other common anionic salts include acetates, bromides, chlorides, citrates, hydrobromides, iodides, maleates, mesylates, phosphates, sulfates, and tartrates (Table 11). Calcium, potassium, and sodium are common metallic cation salts.[108] For injectable products, one needs to be wary of potential toxicity resulting from co-solvents and excipients. Certain intravenous solutions may, for example, cause hemolysis.[109–112]

Purity is also critical. One might target 95–98% purity for early preclinical studies, with no single impurity at levels greater than 1%. Any impurity present at a level of 1% or higher should be characterized both chemically and biologically. (Why is this important? Consider the case where a 1% impurity is 100 times more potent than the main product – its effects thus being equal to that of the main product despite much lower levels.) As development progresses, full-scale toxicology and clinical studies should generally use material that is at least 98% pure, with no single impurity at levels greater than 0.5%.

As development progresses, one must adhere to Good Laboratory and Good Manufacturing Practices (GLP and GMP). Ultimately, a drug manufacturing process can be codified as a Drug Master File (DMF), allowing reference to its contents and general approvals for subsequent use by others.

Thus, there are many challenges in the transformation of research laboratory methods to process R&D at scales suitable for manufacturing and under conditions that meet regulatory guidelines. The low profile of process R&D in past years has become a much higher profile with more complicated molecules entering development (e.g., discodermolide). Ignoring total quantities that might be required, it is clear that dealing with amoxicillin, with an API cost of about $40/kg, is less of an issue than dealing with paclitaxel, at $200 000/kg, and drugs such as discodermolide would be off the charts (see below).[113]

Novartis successfully carried out a 39-step synthesis to provide tens of grams of discodermolide, a potent tumor cell growth inhibitor, for clinical testing. While the compound subsequently ran into trouble, the Novartis achievement

Table 11 Common salts found in marketed drugs

Acetate
Bromide
Calcium
Chloride
Citrate
Hydrobromide
Hydrochloride
Iodide
Maleate
Mesylate
Phosphate
Potassium
Sodium
Sulfate
Tartrate

represents a synthetic tour de force.[114] Discodermolide is a polyketide lactone containing 13 stereogenic centers, a tetra-substituted lactone, two multisubstituted alkenes, a carbamate, and a terminal diene. It was originally isolated in small quantities from a Caribbean sponge, *Discodermia dissoluta*, and exhibits potent immunosuppressive and cytotoxic effects. The activity of discodermolide is typical of certain natural products that affect microtubules and the mitotic spindle, including cancer drugs such as epothilone and paclitaxel (Taxol) – see **Figure 14**. To prepare enough discodermolide for clinical trials, Novartis resorted to chemical synthesis using fragments produced by fermentation, since biological production of the molecule as a whole had not been successful by the time scale-up was required. With an overall yield of up to 10%, and at least 35 total synthetic steps, 60 g of material were prepared.[115]

Prior to the initiation of each phase of human clinical trials, animal toxicity studies that support the planned doses must be completed (**Table 12**). These animal studies are typically done in two species, rodents and nonrodents, with the intent of providing adequate safety data to support uninterrupted human clinical trials. Among various exams that are conducted, observations are made regarding the animals' eating and other behaviors, body weights and metabolism, liver and kidney function, and blood parameters.[116] A number of studies have evaluated from a regulatory perspective the proper conduct and duration of toxicology studies, including carcinogenicity testing, and considerations for specific therapeutic areas, including respiratory drugs and anticancer agents.[117–120] This preclinical R&D process, taken (1) from idea to target discovery and validation, (2) through drug discovery, pharmacokinetics (PK), absorption, distribution,

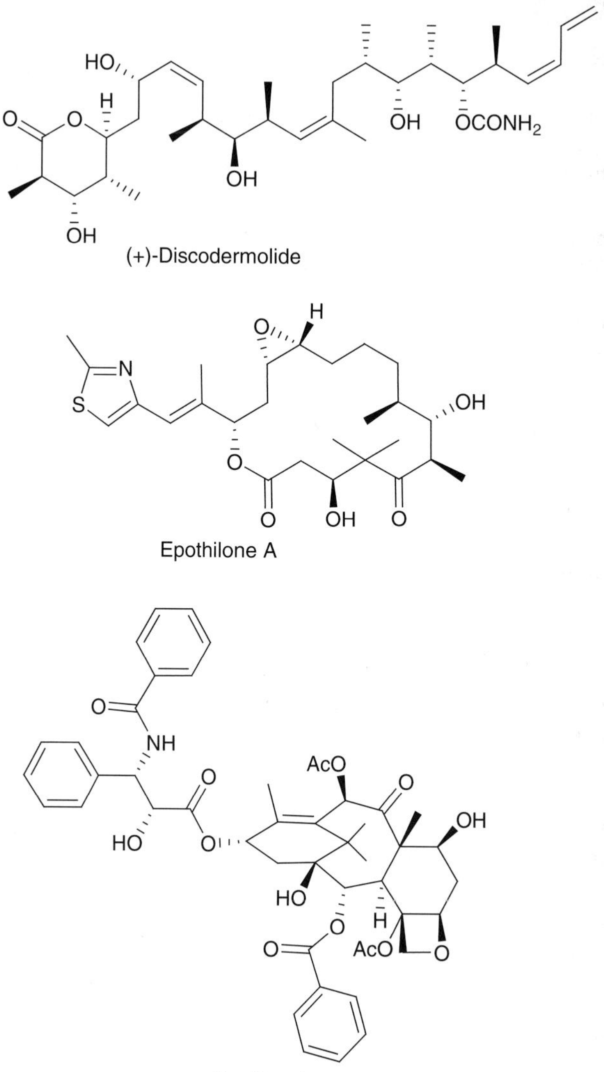

Figure 14 The complicated chemical structures of the drugs discodermolide, epothilone, and paclitaxel.

Table 12 Subacute or chronic toxicology studies required to support human clinical testing of orally administered drugs

Clinical trial duration	Duration of repeated dose toxicology studies	
	Rodents	Nonrodents
Single dose	⩽4 weeks	2 weeks
⩽2 weeks	⩽4 weeks	⩽4 weeks
⩽1 month	⩽3 months	⩽4 weeks
⩽3 months	⩽6 months	⩽3 months
⩽6 months	6 months	⩽12 months
>6 months	6 months	⩽12 months

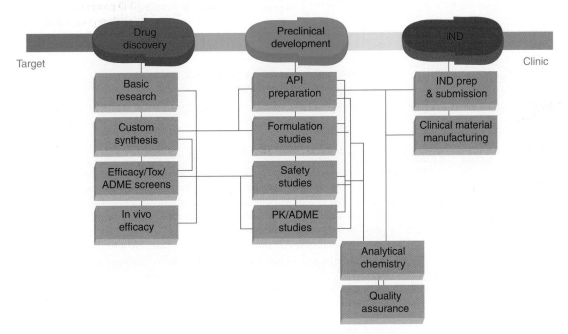

Figure 15 The process from idea to IND.

metabolism, and excretion (ADME), and toxicity (ADMET), as well as formulation, among other steps, (3) to filing of an IND, requires the interplay of many different groups and disciplines (**Figure 15**).

If suitable data have been obtained to support human clinical testing and all regulatory hurdles have been cleared, the drug enters clinical pharmacology studies – first in man (FIM) or Phase I. FIM may be conducted in normal volunteers or in patients, depending on the disease target, with rising dose PK and tolerability (safety), and occasionally with biomarker endpoints.

To accelerate the entry of new drugs into clinical trials, as well as to select the best lead for advancement, microdosing or Phase 0 studies have been devised to determine human metabolism data on very small quantities of material. Accelerator mass spectrometry (AMS) has been used to provide pharmacokinetic data, and positron emission tomography (PET) has been used to generate pharmacodynamics information.[121]

There are various types of clinical trial design, each with its own set of strengths and weaknesses (**Figure 16**). Some of the common terms used to describe clinical studies are given in **Table 13**. For example, a nonrandomized intervention study using historical controls ensures that every patient receives therapy (no one receives placebo). It can be easier to recruit patients into these studies and less expensive to run the trial. However, the lack of a concurrent control can lead to incorrect conclusions, particularly when the current patient population differs from the historical population or when disease management has evolved to a new standard of care. While different diseases and special circumstances may sometimes dictate otherwise, for most Phase III trials the gold standard is a double-blind,

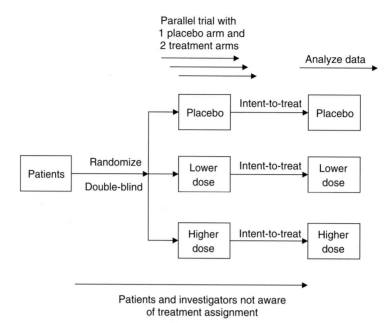

Figure 16 Schematic of a parallel, randomized, double-blind, placebo-controlled clinical trial using an intent-to-treat analysis.

Table 13 Some of the common terms used to describe clinical studies

Cohort	Groups exposed to the therapy of interest are followed for a period of time and then compared with unexposed groups using outcome measures specified up front
Crossover	Subjects to be treated are divided into at least two treatment groups. Each group receives all treatments sequentially after random selection of the first treatment, usually with a washout period between treatments
Cross-sectional	Treatment exposures and outcome measures are determined at a single time point
Double-blind	Neither investigators nor subjects know whether subjects have been assigned to drug or placebo
Intent-to-treat analysis	All randomized patients are analyzed in the groups to which they were originally assigned, regardless of compliance – even if they fail to adhere to the protocol or withdraw from the study
Meta-analysis	Statistical analysis of a combination of separate clinical trials of the same drug
Multicenter	Studies conducted at more than one site
Open-label	Nonblinded study – both investigators and subjects know if subjects have been assigned to drug or placebo
Parallel-arm	Subjects receive one or two different treatments throughout the study
Prospective	Looking forward from a given starting point to assess drug effects
Randomized	Subjects allocated to treatment groups randomly

randomized, parallel-arm study, with the aim of demonstrating statistically significant superiority over placebo. Such a design reduces potential biases from affecting trial outcomes and is less affected by issues such as missing data.[122] One alternative to a superiority trial is a noninferiority or equivalency trial, with the endpoint being either effectiveness or safety.[123] Occasionally, trials will be stopped in midstream for interim analyses. The pros and cons of such an approach must be weighed carefully at the outset, and statistical parameters must be carefully considered.[124]

For any selected design, one must guard against false-positives (type I or α errors) or false-negatives (type II or β errors). To avoid false-positives in analyzing clinical data, a significant outcome is usually considered to be only those results that reach statistical significance at the level of $p = 0.05$ or better. To avoid false-negatives, the sample size (number of patients) is adjusted (powered) based on the targeted difference between treated and control arms so that the study is likely to show statistical significance.[125] Biomarkers or surrogate endpoints are being used increasingly to help guide clinical trials. Thus, while the definitive endpoints in renal transplantation are several, one of the definitive endpoints is acute rejection, and creatinine serves as a straightforward surrogate endpoint. In prostate cancer, the definitive endpoints are mortality and disease progression, but prostate-specific antigen (PSA) levels have been used as a surrogate endpoint (though not without controversy).[126] In targeted cancer therapies, a number of changes have been made to the standard clinical protocols. For instance, instead of looking for a maximum tolerated dose and dose-limiting toxicities in Phase I, the desired outcome might instead be the optimal biological dose with no toxicity.[127,128] Ultimately, substantial evidence of efficacy must be demonstrated to gain approval of an NCE, which may require multiple adequate and well-controlled clinical trials, using prospectively determined assessment variables for major outcomes and both a global clinical improvement rating and a specific quantitative assessment related to the disease that is accepted to be valid and reliable.[129]

Ultimately, the path to drug approval is not always a straight line – far from it in many cases. Tacrine, an aminoacridine, was the first drug approved by the FDA to treat Alzheimer's disease, in 1993, just a few years after gaining notoriety in a controversial 1986 publication on studies in Alzheimer's patients.[130] Yet, it was originally identified as an antibacterial agent in 1945.[131]

2.01.3 Drug Discovery

2.01.3.1 Introduction

The latter part of the twentieth century witnessed the birth of numerous technologies that have advanced the SOTA in drug research.[132] The search for new approaches was in part propelled by the prosperous 1990s. As discussed earlier in this chapter, the 1990s were a period dominated by antiulcer products, with ranitidine surpassing the first billion dollar drug, cimetidine,[133,134] and holding the top spot in the first part of the decade, after which it was unseated by an antiulcer agent acting by a different mechanism, omeprazole.[135] In the 1990s, the number of billion dollar drugs rose from six to 35. During this period, worldwide sales of the top drug increased from $2 billion to almost $6 billion, and by 1999 the 100th largest selling drug raked in almost $500 MM. However, although antiulcer drugs dominated the top spots, there were more cardiovascular drugs in the top 10 in the 1990s than any other drug class. While no company had multiple top 10 drugs in all of the 1990s, Merck came closest, missing only 1990.[136] (And yet, what will happen with Merck after the rofecoxib debacle?)

There are multiple hit identification strategies (**Figure 17**), and new drugs come from many sources, both rational and brute force.[137] The exploitation of drug prototypes has been reviewed extensively.[138] Some have argued that starting with an old drug is the best way to find a new drug. This has been called selective optimization of side activities (SOSA). Thus, one starts with a diverse collection of marketed drugs in order to find and then optimize hits against new pharmacological targets. Because the starting points have already proved useful in man, the hits are guaranteed to be drug-like. A number of successful SOSA examples have been reported starting from sulfonamides, calcium channel blockers, beta blockers, antidepressants, neuroleptics, and so on.[139] Despite the power of combinatorial chemistry and HTS, natural products still provide unique structural diversity. Natural molecules have intrinsic advantages, of course, having evolved alongside biological systems. Mother nature had eons of a head start versus medicinal chemists (**Figure 11**). Thus, it might have been predictable that a large percentage, almost 30%, of the NCEs approved by the FDA between 1981 and 2002 were natural products or compounds derived from natural products. Natural products (or derivatives) with important therapeutic uses today include the antibiotic vancomycin, the anticholinergic scopolamine, and the opiate morphine, to name just a few.[140] However, in recent years, natural products have gone out of favor in some circles, in part because they have been less compatible with other new technologies, such as HTS.[141]

New drugs come from many sources including:

- focused synthesis of new chemical structures;
- modification of old drugs (new analogs);
- new uses for old drugs (repositioning or repurposing)[143];
- endogenous substances found in humans;
- natural products from nonhuman sources; and
- random screening.

Though the field of combinatorial chemistry is still evolving, it has already made a mark on the industry. It has asked lofty questions such as: How high is up? That is, the vastness of chemical space is not easy to fathom, and a vanishingly small proportion of this space has been studied by medicinal chemists. The challenge of harnessing this space, defining

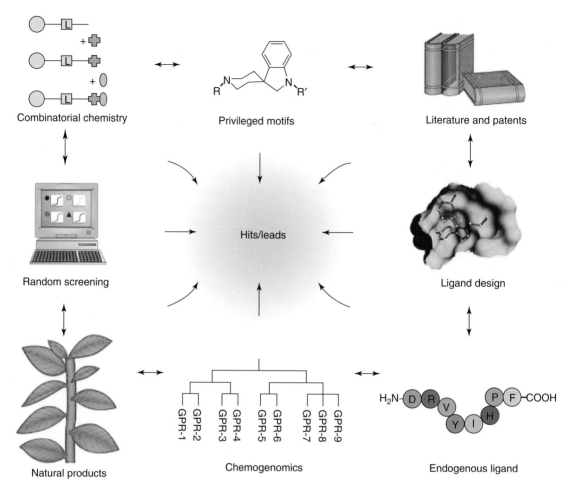

Figure 17 Hit identification strategies. The most commonly applied hit identification strategies today range from knowledge-based approaches, which use literature- and patent-derived molecular entities, endogenous ligands or biostructural information, to the purely serendipity-based 'brute force' methods such as combinatorial chemistry and high-throughput screening. The amalgamation of both extremes is anticipated to deliver more high-content chemical leads in a shorter period of time. (Reprinted with permission from Bleicher, K. H.; Böhm, H.-J.; Müller, K.; Alanine, A. I. *Nat. Rev. Drug Disc.* **2003**, *2*, 369–378. Copyright 2003 Nature Publishing Group.[142])

what parts are biologically relevant in order to create improved pharmaceuticals, is being tackled by many scientists around the world.[144] The rapidity of change in this new field has manifested itself in various ways. Totally new ways of thinking about and visualizing the problem were required, such as flower plots (**Figure 18**). As another example, during the preparation of one recent book on combinatorial chemistry, nearly 60 chapter authors had changed jobs, moved, joined new companies, or been involved in a merger or acquisition.[145]

The pharmaceutical paradigm *c.* 2000 is a marriage of the old and the new, in many cases driven by advances in chemical and biological technologies. Whereas the old pharmaceutical model was low-tech trial and error, taking 10–20 years and $800 MM or more to reach the marketplace, the new model is high tech and information rich (**Figure 19**). Drug discovery is a subset of this complicated paradigm. It combines targets, HTS, compound libraries, and informatics to produce leads that are evaluated iteratively for biodisposition parameters, safety, and efficacy (**Figure 20**).

2.01.3.2 Technologies

Many technologies critical to today's medicinal chemistry were developed starting in the 1980s.[146] These include combinatorial chemistry, HTS, genomics, and proteomics. The development of combinatorial chemistry is a particularly interesting case study in the biotechnology and pharmaceutical industries. One might trace its formal beginnings to the

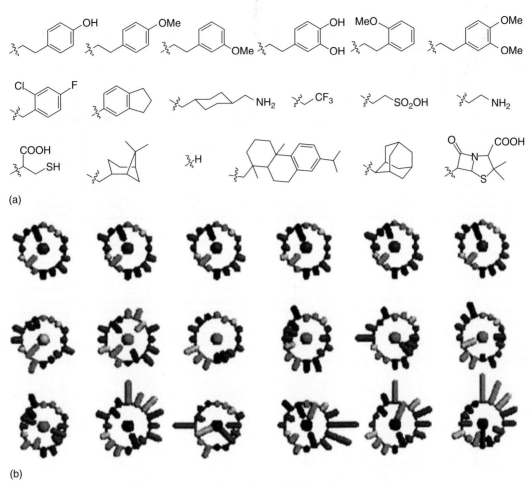

Figure 18 Flower plots as a way of analyzing molecular diversity. Structures and 'flower plots' of 18 side chains from a biased NSG peptoid combinatorial library based on the tyramine submonomer. (a) The top row side chains are from tyramine and its five closest available analogs. The 12 side chains in the lower rows were chosen by D-optimal design from a pool of 721 amines. (b) Corresponding flower plots each represent all 16 properties for a single side chain. Petals for positive values point outward, and negative petals point toward the center. The radius is 3 standard deviations, and the center has been colored by similarity to tyramine. (Reprinted with permission from Martin, E. J.; Blaney, J. M.; Siani, M. A.; Spellmeyer, D. C.; Wong, A. K.; Moos, W. H. J. Med. Chem. **1995**, 38, 1431–1436. Copyright 1995 American Chemical Society.)

development of solid-phase peptide synthesis by Merrifield in the early 1960s,[147] though other developments including natural products chemistry may also be considered seminal events.

Perspectives on the application of combinatorial technologies to drug discovery have been published by various groups,[148–151] and several books have appeared on the subject.[152–155] As reviewed by Furka, three main methods were developed starting in the late 1980s. These included mixed reactant, portioning-mixing, and light-directed synthesis methods.[156] The early work by Geysen and co-workers,[157] Houghten and co-workers,[158] and the group at Chiron (Geysen, Moos, Rutter, Santi, and the former Mimotopes and Protos) is noteworthy. Personal perspectives on the early days of this field have been reported by Lebl. Much of the early work was performed on libraries of peptides and nucleotides (e.g., containing phosphodiester, phosphorothioate, methylphosphonate, phosphoramidate, guanidine, or peptide nucleic acids [PNAs], antisense oligodeoxynucleotides[159]). While each group contributed significantly to this revolution in synthesis, it was probably the Parke-Davis group that first showed how broadly organic syntheses could be conducted on solid-phase supports. Their work commanded attention because it focused on heterocycles, the bread and butter of big pharma, rather than the peptides and nucleotides that biotech companies emphasized. Thus, the fledgling field of combinatorial chemistry finally got the pharmaceutical industry's attention.[160] Academic groups also joined the fray, preparing libraries of benzodiazepines using the multipin method.[161]

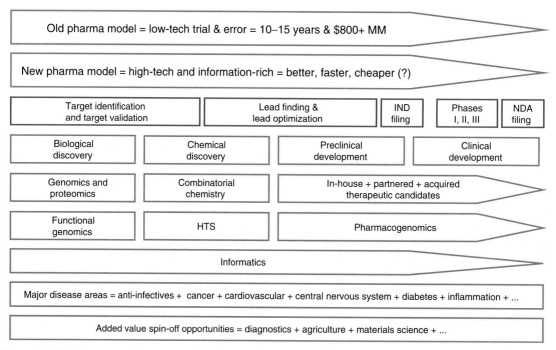

Figure 19 Today's elaborate biotechnology-driven pharmaceutical paradigm is a marriage of the old and the new.

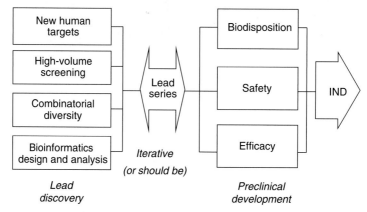

Figure 20 The iterative process of drug discovery and preclinical development illustrated as a flow chart. (Reprinted with permission from Moos, W. H. Introduction: Combinatorial Chemistry Approaches the Next Millennium. In *A Practical Guide to Combinatorial Chemistry*; Czarnik, A. W., DeWitt, S. H., Eds.; American Chemical Society: Washington, DC, 1997, pp 1–16. Copyright 1997 American Chemical Society.)

The use of natural and unnatural nucleic acids as tools to study biological processes has blossomed in recent years. Whether antisense, small interfering RNAs (siRNAs, part of the RNAi revolution), or other oligonucleic probes, these readily prepared and highly specific molecules are not only useful as tools for research, but also have utility as diagnostic probes and therapeutic agents. Current and emerging approaches with nucleic acids and analogs include[162]:

- triplex-forming oligonucleotides;
- ribozymes that alter DNA sequences;
- siRNAs;
- messenger RNA (mRNA) cleavage and repair using ribozymes or deoxyribozymes;
- riboswitches;

- RNA and DNA aptamers; and
- nucleoside analog therapeutics.

While their potential applications are broad, finding a magic bullet antisense therapeutic has been slower than some might have expected, requiring easy synthesis in bulk, in vivo stability, and the requisite cellular efficacy, half-life, safety, and selectivity.[163] The medicinal chemistry of this class of compounds had already been extensively studied in the early 1990s (**Figure 21**).[164] However, economically viable large-scale synthetic processes for antisense compounds took a long while to develop.[165] Fomivirsen (Vitravene) was the first antisense product to gain clearance for marketing. It is used to treat cytomegalovirus (CMV) retinitis in AIDS patients, and was approved approximately 20 years after oligonucleotide therapeutics were first proposed.[166]

The task of converting bioactive peptides into stable, orally available small molecule drugs has been the focus of much work. A new understanding of proteases in the 1980s served as a catalyst for a large number of companies and laboratories to tackle proteases as targets and peptides as drugs (**Table 14**).[167] This work was facilitated by growing datasets of structural information on which molecular modeling could be performed. Improved technologies for the synthesis of peptides and peptidomimetics also aided this work, often in the areas of cardiovascular and antiviral research.

Perspectives on how to bridge the gap between peptides and peptidomimetics and peptoids were addressed regularly starting in the 1980s, at a time when many other advances were being made in the understanding of endogenous

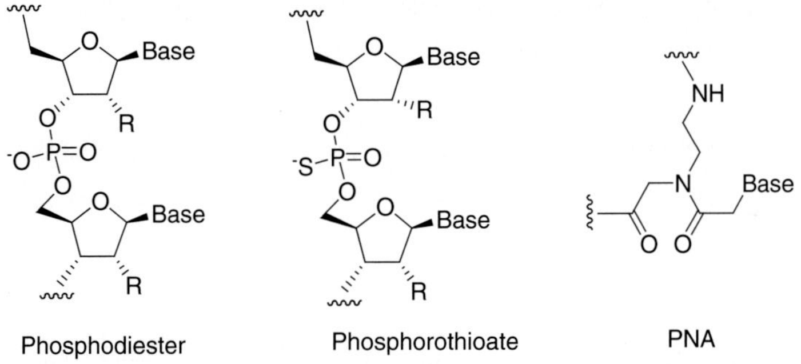

Figure 21 Phosphodiester and other antisense backbones.

Table 14 Protease target classes

Active site	Examples
Aspartic	Pepsin
	Renin
	Cathepsins D and E
Cysteine	Papain
	Cathepsins L, B, and H
Metallo	Collagenases
	Thermolysin
Serine	Chymotrypsin
	Elastase
	Trypsin

peptides, especially neuropeptides.[168,169] Numerous obstacles stood in the way of making the leap from peptides to peptidomimetics, however. For example, because food and proteins are digested and absorbed in the intestine, peptides and peptide-based drugs are attacked by numerous enzymes. Progress has been made, however, in a number of areas. Some of the early work focused on angiotensin-converting enzyme (ACE) inhibitors, renin inhibitors, and cholecystokinin (CCK) antagonists. CCK regulates the secretion of pancreatic enzymes, contractility of the gall bladder, and gut motility. It is commonly found in the gut (and the brain) as an octapeptide (8-mer). Screening of fermentation broths for CCK antagonists led to a breakthrough with the discovery of asperlicin, the initial lead around which a number of orally active analogs were developed. Of course, Mother Nature had already devised her own peptidomimetic stories, including that of the naturally occurring morphine and the later discovered endogenous opiate peptide enkephalins, originally termed morphinomimetics. See **Figure 22** for two pairs of endogenous substances and their mimics. Drug delivery approaches have also been tried, including novel formulations, penetration enhancers, and intranasal delivery.[170]

Common obstacles in the development of peptidomimetic drugs have included:

- poor intestinal absorption;
- alternative routes of administration, other than intravenous (IV), also lead to poor blood levels;
- rapid metabolism, including hydrolysis (proteolysis) at peptide amide bonds; and
- rapid excretion.

Though the term peptoid was first used more broadly, it found regular use to describe a family of N-substituted glycine (NSG) oligomers (**Figure 23**). Subsequently, a number of novel biopolymer systems were proposed or reported. This includes β-peptide oligomers, oligocarbamates, oligosulfones, and vinylogous oligopeptides – far from a complete list (**Figures 24** and **25**[171–172]).[173–175]

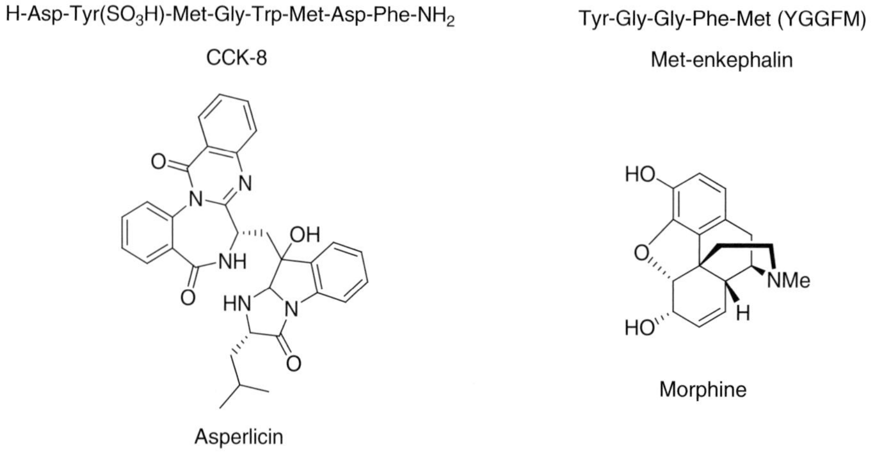

Figure 22 The structures of some biologically active endogenous peptides and their corresponding peptidomimetics.

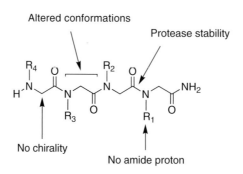

Figure 23 NSG peptoid backbone.

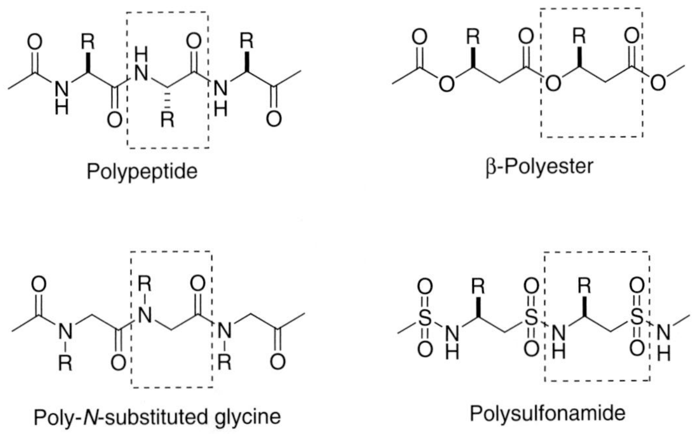

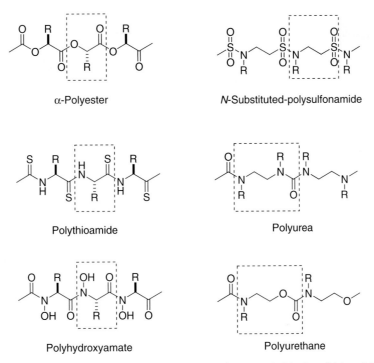

Figure 24 Various biopolymer systems that have been proposed or reported to be of interest in drug discovery or combinatorial chemistry, including peptides and NSG peptoids.

Figure 25 Additional biopolymer systems that have been proposed or reported to be of interest in drug discovery or combinatorial chemistry.

How were these new backbones developed? Oligo-NSG-peptoids were originally conceived as a new class of diverse peptide-like molecules able to be prepared using combinatorial methods. Initial studies showed NSG peptoids to be resistant to proteolysis and to have activity as protease inhibitors and as binders to nucleic acid structures.[176–178] The development of a breakthrough submonomer method of synthesis made the preparation of diverse peptoid libraries straightforward using automated and robotic synthesizers.[179] Further work showed that carefully designed peptoid libraries allowed the discovery of a wide variety of bioactive molecules through mix-and-split and deconvolution strategies, yielding compounds with nanomolar potency at G protein-coupled receptors (GPCRs), including adrenergic and opiate receptors,[180–181] and also uncovering novel antibiotics[182–183] (see **Figures 26** and **27**). The easy preparation of small oligopeptides led to the study of larger polypeptides. Polypeptoids with diverse side chains have been synthesized dozens of residues in length, and some have been shown to have stable secondary structure despite the achirality of the peptoid backbone. For example, peptoid 5-mers show properties similar to peptide α-helices,[184] and

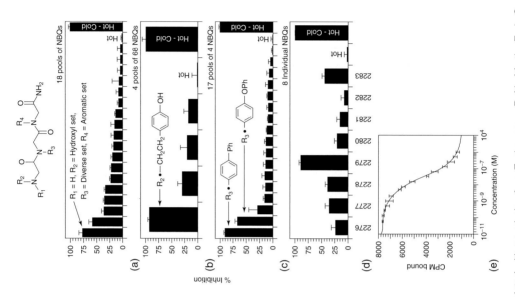

Figure 26 Deconvolution strategy employed in the discovery of NSG peptoid ligands for GPCRs. (Reprinted with permission from Zuckermann, R. N.; Martin, E. J.; Spellmeyer, D. C.; Stauber, G. B.; Shoemaker, K. R.; Kerr, J. M.; Figliozzi, G. M.; Goff, D. A.; Siani, M. A. *J. Med. Chem.* **1994**, *37*, 2678–2685. Copyright 1994 American Chemical Society.)

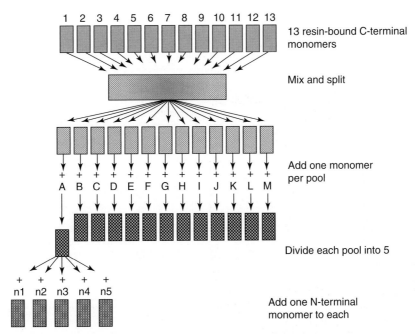

Figure 27 Mix-and-split combinatorial synthesis used in the discovery of NSG peptoid antibiotics. A one-mix library. A total of 13 amines were attached through the C-terminus to Rink type resin. The amine resin was then mixed and the population containing 13 different resins was split into 13 tubes. To each mixed resin-containing tube a defined second position amine was added. Each of the 13 tubes was then split into five tubes so that five different defined N-terminal amines could be added to each pool. This protocol results in 65 pools of 13 compounds, in which the N-terminal and middle positions are defined. (Reprinted from Ng, S.; Goodson, B.; Ehrhardt, A.; Moos, W. H.; Siani, M.; Winter, J. *Bioorg. Med. Chem.* **1999**, *7*, 1781–1785, with permission from Elsevier.)

libraries of 15-mers have uncovered molecules that show helical behavior.[185] Ligation is possible to create dimers of oligopeptides, allowing the synthesis of even larger polypeptides.[186] Cationic peptoid oligomers and conjugates have been demonstrated to facilitate gene transfer.[187]

The shift from screening fully characterized single pure compounds, synthesized in solution, to assaying mixtures of large numbers of compounds, often prepared using solid-phase supports, required a sea change in thinking. Given the entrenched views of the pharmaceutical industry at that time, it is perhaps not surprising that much of this work was pioneered in small entrepreneurial companies and at academic laboratories or research institutes. Even for the latter, moving beyond natural building blocks was a difficult early step in combinatorial chemistry (**Figure 28**). The game of building blocks and numbers that combinatorial chemistry became in the 1990s has evolved into a higher level game in the twenty-first century, though the numbers were quite impressive even in the beginning (**Table 15**).[188]

In the early days of combinatorial chemistry, there was perhaps more emphasis on quantity than quality – large numbers of compounds were prepared, but arguably these libraries were not very interesting from a small molecule drug discovery perspective. The general concept was a good one – more data, less effort, more successful choices (**Figure 29**). As the field developed, and researchers developed combinatorial methods to prepare libraries of more complicated and more drug-like molecules, and as people became more sophisticated at analyzing molecular diversity parameters, there was a clear shift toward library quality.[189] Today, libraries of more complex molecules can be prepared using one or more of a variety of solid-phase or solution methods (**Figure 30**).[190]

With new libraries of compounds, often mixtures of compounds, new HTS methods had to be developed (**Figure 31**). Making HTS more efficient has been the goal of many groups. Doing so requires development of the right assays, selection of the right compound sets, and methods to filter out unwanted hits including promiscuous hitters. Assays may be homogeneous, using techniques such as fluorescence, or separation-based, using techniques such as filter-binding. A hybrid method for selecting the right compounds has been referred to as rapid elimination of swill (REOS), which uses simple counting schemes to weed out molecules that are too large or too reactive, among other features.[191]

What if one could put all known drug targets and all known drug candidates in a test tube, allow the highest affinity pairs to self-assemble, and then separate the bound pairs? While not possible today at such an extreme, this approach, known as affinity selection, has been demonstrated on a much smaller scale. Thus, compound libraries have been

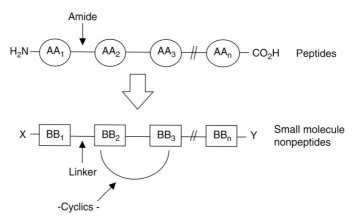

Figure 28 The early days of combinatorial chemistry: From amino acid building blocks to nonpeptides assembled like beads on a string (AA, amino acid; BB, building block). (Reprinted with permission from Moos, W. H. Introduction: Combinatorial Chemistry Approaches the Next Millennium. In *A Practical Guide to Combinatorial Chemistry*; Czarnik, A. W., DeWitt, S. H., Eds.; American Chemical Society: Washington, DC, 1997, pp 1–16. Copyright 1997 American Chemical Society.)

Table 15 The combinatorial chemistry game of numbers

Oligomer size	Number and type of building block at each position	Total number of compounds possible in library
Oligopeptides		
2-mer	20 Standard amino acids	$20^2 = 400$
		$20^3 = 8000$
3-mer		$20^4 = 160\,000$
4-mer		$20^5 = 3\,200\,000$
5-mer		
Oligonucleotides		
2-mer	4 Standard nucleic acids	$4^2 = 16$
3-mer		$4^3 = 64$
4-mer		$4^4 = 256$
5-mer		$4^5 = 1024$
Oligopeptoids or other unnatural oligos		
2-mer	10 building blocks	$10^2 = 100$
	100 building blocks	$100^2 = 10\,000$
	1000 building blocks	$1000^2 = 1\,000\,000$
3-mer	10 building blocks	$10^3 = 1000$
	100 building blocks	$100^3 = 1\,000\,000$
	1000 building blocks	$1000^3 = 1\,000\,000\,000$

Reprinted with permission from Moos, W. H. Introduction: Combinatorial Chemistry Approaches the Next Millennium. In *A Practical Guide to Combinatorial Chemistry*; Czarnik, A. W., De Witt, S. H., Eds.; American Chemical Society: Washington, DC, 1997, pp 1–16. Copyright 1997 American Chemical Society.

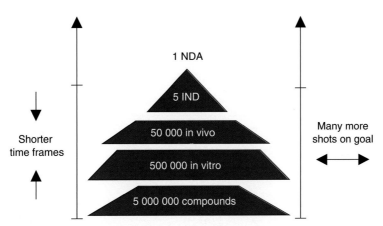

Figure 29 In theory, combinatorial chemistry promised more data, less effort, and more successful choices. (Reprinted with permission from Moos, W. H. Introduction: Combinatorial Chemistry Approaches the Next Millennium. In *A Practical Guide to Combinatorial Chemistry*; Czarnik, A. W., DeWitt, S. H., Eds.; American Chemical Society: Washington, DC, 1997, pp 1–16. Copyright 1997 American Chemical Society.)

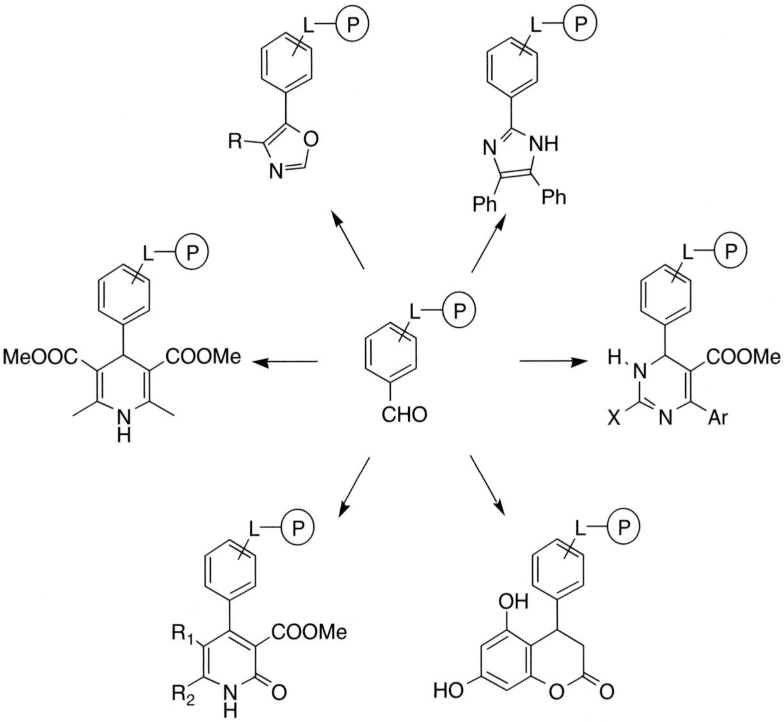

Figure 30 Evolution from amino acid-like combinatorial building blocks to true drug-like scaffolds took time, but eventually diverse drug libraries were being prepared using readily available common intermediates. (Reprinted with permission from Moos, W. H. Introduction: Combinatorial Chemistry Approaches the Next Millennium. In *A Practical Guide to Combinatorial Chemistry*; Czarnik, A. W., DeWitt, S. H., Eds.; American Chemical Society: Washington, DC, 1997, pp 1–16. Copyright 1997 American Chemical Society.)

mixed with purified proteins, size-exclusion chromatography has been used to separate bound from free, and the resulting receptors with bound ligands have been separated and the ligands then detected by mass spectrometry (a process called affinity selection mass spectrometry, ASMS); see **Figure 32**. The method has been used successfully to study ligands for both receptors and enzymes.[192–196]

Technologies may also take the form of cellular compartments or drug classes. Thus, companies such as Cytokinetics have focused on cytokinetic pathways, Idun on cell death and caspases, Isis on antisense, MitoKor on mitochondria,

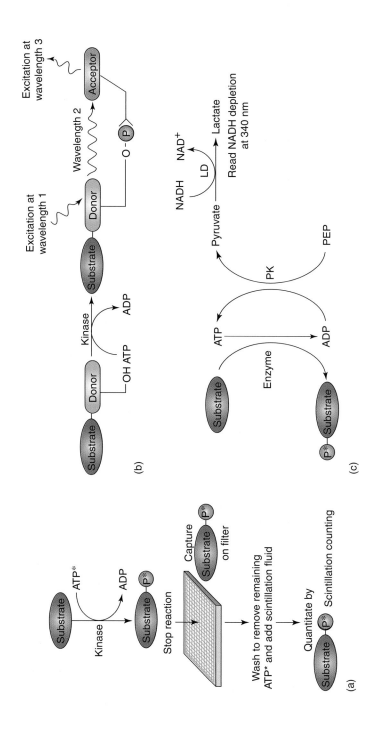

Figure 31 Several examples of screening assays used in HTS paradigms. This figure outlines the principles behind three assay types used for screening. The examples illustrate their application to kinase assays, but in all cases the methods are applicable to other target types. A standard separation-based assay is depicted in (a), in which radiolabeled ATP transfers a labeled phosphate moiety from ATP to substrate. The substrate is subsequently captured on a membrane, and separated from the ATP substrate by filtration. The signal is measured by addition of scintillate and quantitated by counting. (b) An outline of the general principles of FRET, a homogeneous assay format. In this case, a labeled substrate is phosphorylated allowing it to bind to a second molecule labeled with an acceptor group. The proximity of the two groups allows energy transfer between the donor and acceptor. This causes a shift in wavelength for the signal emitted from the assay and detection of the product without separation from the substrate. (c) Another homogeneous assay format in which the conversion of ATP to ADP is coupled to the conversion of NADH to NAD using two enzymes, pyruvate kinase (PK) and lactate dehydrogenase (LD). This decreases the A340 of the assay, which is used to monitor the progress of the reaction. (Reprinted with permission from Walters, W. P.; Namchuk, M. *Nat. Rev. Drug Disc.* **2003**, *2*, 259–266. Copyright 2003 Nature Publishing Group.)

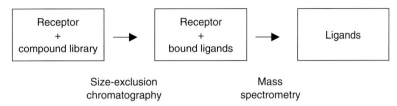

Figure 32 Affinity-selection mass spectrometry as a means of screening compound libraries to identify agents that bind to targets of interest, such as receptors.

X-Ceptor on nuclear hormone receptors, and so on. Such foci can lead to fascinating developments; consider mitochondria, for example. Mitochondria are cytoplasmic organelles involved in calcium regulation, cell death, energy production, free radicals, and metabolism. They are cellular powerhouses, functioning as the batteries, generators, and engines of cells. Almost all cellular activity involves mitochondria, so they are essential to human life. In many ways, mitochondria function like a cell within a cell, and they house proteins derived from two genomes, the nuclear genome, inherited from both the mother and the father, and a small mitochondrial genome, inherited in almost all cases only from the mother. Whereas the mitochondrial genome has only 37 genes, nuclear DNA encodes thousands of mitochondrial proteins. Mitochondria play a critical role in human health and disease, and mitochondrial dysfunction has been linked to more than 75 diseases, including major diseases of aging.[197–199] Because of the central role of mitochondria in so many cellular events, many drug discovery groups ultimately work on this organelle directly or indirectly.

Which provides a segue to genomics... Subsequent to the combinatorial chemistry revolution, genomics took hold. When applied to drug discovery and development, genomics generally refers to the study of genes associated with or implicated in human diseases, and the set of technologies that enables such work.[200] Since the start of the Human Genome Project in 1988 and its completion in draft form in 2001, and for the foreseeable future, genomics has commanded and will command the attention of countless researchers. Though a number of initial estimates predicted around 100 000 genes in the human genome, many were surprised when this number had to be rounded down to something closer to 20 000–30 000 genes.

Exploring genome space requires bringing together geneticists, computer scientists, biochemists, cell biologists, structural biologists, and physiologists in order to understand gene function. Medicinal chemists can help to validate these potential new drug targets and their function by discovering ligands that prove the point.[201]

Completing the human proteome, being more challenging, has lagged developments in genomics. Yet a map of all human proteins was discussed as early as 1981,[202] and a human protein index was considered by the US Congress long before the Human Genome Project was conceived. However, because of its much greater complexity – probably at least two orders of magnitude more complex than the human genome – human proteomics is in many ways still in its infancy.[203] Proteomics was advertised as the next genomics, but as suggested above, while the human genome may have only 20 000–30 000 genes, the total human proteome is no doubt millions of proteins when one takes into account posttranslational modifications and other diversity enhancing changes. At a very basic level, proteomics researchers must sequence peptides, typically via mass spectrometry. At a more complicated level, and critical to its use in personalized medicine, proteomics must be extrapolated from the bench to the bedside,[204] applying proteomics to early diagnosis, to disease-tailored therapeutic targets, and finally to personalized medicine. Taking cancer as the example, full exploitation of proteomics requires, ideally, knowledge of or access to protein signaling pathways, tumor-host interactions, mass spectrometry, laser capture microdissection, and microarrays. To break down the problem into smaller parts, certain groups have chosen to focus on organellar proteomics, including mitochondria, in order to establish a beach head in this area.[205–207] Others have directed their attention to epigenetic phenomena, such as DNA methylation.[208]

Interestingly, whereas drug discovery scientists are used to working in the range of millimolar (10^{-3}) to femtomolar (10^{-15}) concentrations, genomics and proteomics have added to our vocabulary because they deal with concentrations in the attomolar (10^{-18}), zeptomolar (10^{-21}), and even yoctomolar (10^{-24}) ranges. Given the deluge of information flowing from both genomics and proteomics, bioinformatics has become a new focal point for both education and R&D.[209]

What is the potential impact of genomics and related technologies on the pharmaceutical and biotechnology industries? It is expected to be quite broad, including small molecule drugs, gene and protein therapies, and diagnostic tool kits. It is not without its challenges, however, including questions about data overload, who owns the data, who pays for what, ethical objections to certain uses, and so on.[210]

Potential genomics opportunities in healthcare include:

- gene therapy;
- protein therapy;
- small molecule drugs; and
- diagnostics.

2.01.3.3 Chemical Biology and Chemical Genomics

Chemical biology and chemical genetics are at their core all about interfaces and boundaries with other scientific disciplines, including biochemistry, pharmacology, and genomics. It is often said that the action in science is at the interfaces, and this has never been more true of chemistry, biology, and physics, with chemistry at the center of the action, of course (**Figure 33**).[211] Whereas NMR revolutionized organic chemistry in the 1960s and 1970s, and became a common in vivo diagnostic imaging methodology in the 1980s and 1990s, its recent use to view chemical processes and protein conformations within living cells may herald yet another major step forward for biochemical and pharmacological applications of medicinal chemistry post-2000.[212]

Phrases with the word 'chemical' followed by biology, genetics, or genomics have cropped up in recent years, particularly from the work of Schreiber.[213–217] One may think about applying medicinal chemistry to controlling pathways and systems at various levels, including genetic manipulations (**Figure 34**). Such applications have been accomplished with cell-cycle machinery, for example.[218] Powerful manipulations of cellular pathways can be effected by bridging different protein subunits with small molecules. Such manipulations have been accomplished through chemical inducers of dimerization in systems such as the immunophilins with drugs based on cyclosporine and FK-506 (**Figure 35**).[219] Chemistry-to-gene assays have also been established in model organisms such as *Caenorhabditis elegans*,[220] and reporter-gene based mouse models are being developed,[221] all of which may make chemical genomics more useful in the future.

Combinatorial biosynthesis has been exploited for many years through natural products. More recently, a number of academic and industrial groups have managed to harness some of these processes to produce new collections of related molecules. Progress is particularly evident in the production of classes of polyketides and nonribosomal peptides that have proven useful in pharmaceutical and agricultural products. Examples include antibiotics such as erythromycin, penicillin, rifamycin, and vancomycin, as well as anticancer drugs, cholesterol-lowering agents, and other important classes of pharmaceuticals. Now it is possible to exploit controlled biological manipulations to effect Claisen-like condensations, cyclizations, decarboxylations, dehydrations, epimerizations, reductions, and other chemistries in order to produce new analogs that would otherwise be very difficult to prepare synthetically (**Figure 36**).[222–223]

Both hybrid enzymes and catalytic antibodies have been used to synthesize or modify complex molecules.[224] Applying the power of the immune system by using antibodies to facilitate transformations such as metallation, pericyclic, and redox reactions has allowed the synthesis of molecules that might be hard to prepare in other ways.[225]

Synthetic biology is another variation on the theme. One group of synthetic biologists uses unnatural molecules to mimic or create artificial life. Another group studies natural systems and then applies these functions to unnatural systems. Through these approaches, diagnostics have been created, as have a variety of devices.[226] For example, the standard 20 amino acids are used as building blocks by essentially all organisms to generate the peptides and proteins of

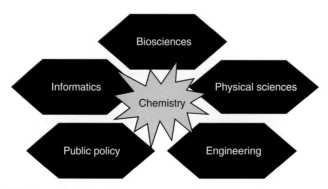

Figure 33 Leveraging the interfaces of science.

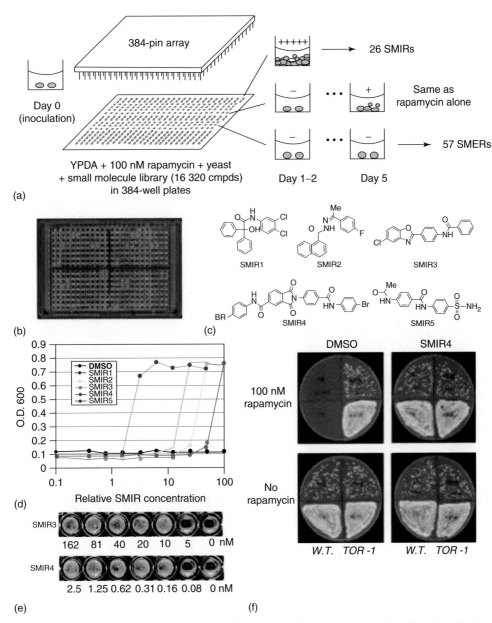

Figure 34 A chemical genetic screen for small molecules that modulate rapamycin's antiproliferative effect in yeast. (a) Schematics of the screen. Compounds were transferred from library plates to assay plates (containing growth medium, rapamycin, and yeast cells) by using 384-pin arrays. SMER, small-molecule enhancers of rapamycin. (b) Retest of SMIRs in a 384-well plate. White wells indicate compound-induced yeast growth in the presence of rapamycin; black (transparent) wells indicate no growth. (c) Chemical structures of the fast-acting SMIRs (yeast growth indentifiable on day 1, same as the 'no rapamycin' control). (d) Dose–response curves for SMIRs in wild-type (rapamycin-sensitive EGY48) cells inoculated in YPDA containing 100 nM rapamycin. (e) Minimal concentrations of SMIR3 and SMIR4 required for yeast growth in YPDA containing 20 nM rapamycin. (f) SMIR4 treatment and the *TOR1-1* (S1972R) mutation both confer rapamycin resistance. Cells were plated at two different densities on the upper versus lower halves of the plates (1:1000). (Reprinted with permission from *Proc. Natl. Acad. Sci. USA* **2004**, *101*, 16594–16599. Copyright 2004 National Academy Sciences, USA.)

life. Additional diversity is generated through posttranslational modifications. However, it is only in the last few years that scientists have devised ways to add novel amino acids and to expand the genetic code directly. Using unique codons, over 30 novel amino acids have been incorporated through genetic manipulations, including azido, fluorescent, sugar, keto, metallo, polyethyleneglycol (PEG), and unsaturated alkyl moieties.[227,228] For example, one can incorporate alkyne residues into proteins in *Escherichia coli* bacteria.[229] These changes may ultimately allow the routine production

Cyclosporin A

FK-506

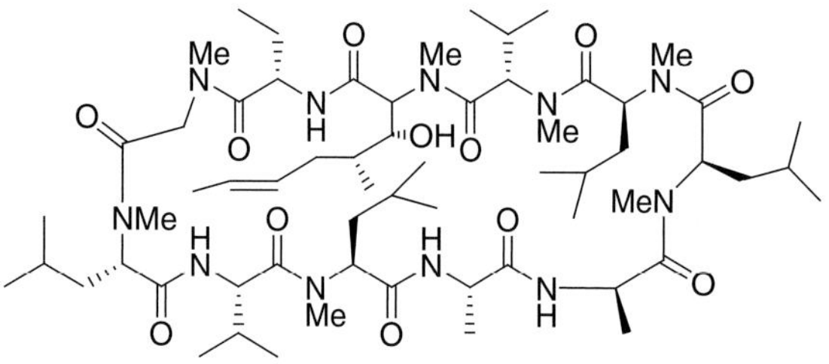

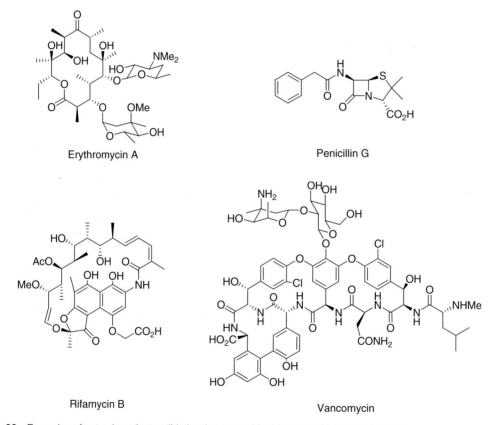

Figure 35 Structures of representative immunophilin ligands that inhibit the cell cycle.

Erythromycin A

Penicillin G

Rifamycin B

Vancomycin

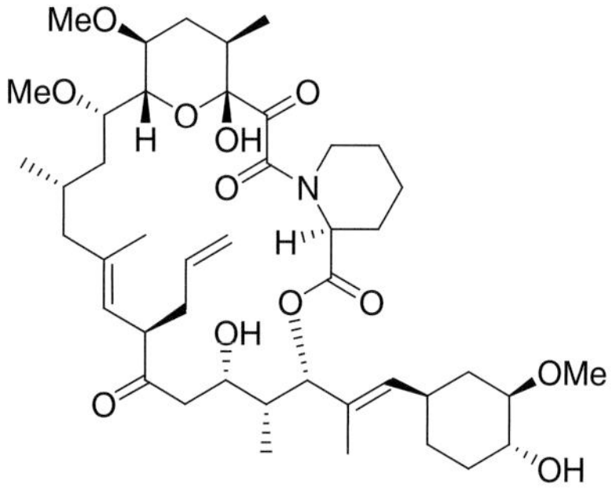

Figure 36 Examples of natural product antibiotics that are polyketides or nonribosomal peptides.

of new and improved biomolecules and even new life forms.[230] This approach also provides new routes for labeling proteins in a site-specific manner for NMR studies.[231] It has further been used to modify viral combinatorial peptide systems such as phage display.[232]

Overcoming certain obstacles in stem cell biology, such as large-scale production, will be required for broader use of stem cells for healthcare purposes. Chemistry can play a role in advancing this field too, for example, through the use of small molecules that regulate the fate of stem cells. This includes simple solvents such as dimethyl sulfoxide (DMSO), steroids such as dexamethasone, and anticancer drugs such as imatinib.[233]

Nanotechnology (or nanobiotechnology) is the latest platform to employ the skills of the medicinal chemist. Heterocyclic peptide nanotubes have been prepared.[234] The concept of targeted therapeutic nanoparticles is being tested in a number of academic and industrial laboratories around the world. Much of this work is being done in the cancer field. Thus, a particle of perhaps 10 nm diameter can be derivatized on its surface with a ligand that preferentially targets cancer cells; the particle itself is sized to allow enhanced penetration into cancerous tissues, and inside the particle is a payload of a therapeutic agent to kill the cancer cells. These nanoparticles are approximately the size of an antibody, and may provide a platform that allows one to detect, image, and treat all at once.[235–236]

2.01.3.4 From Small Molecules to Macromolecules

Small molecule drugs, typically less than 500 Da in molecular weight, have been the mainstay of pharmaceutical R&D for the last century. Earlier sections of this chapter discuss peptides, peptoids, peptidomimetics, and the use of new technologies to advance the SOTA in drug discovery. From these sections, it should be evident that the promise and reality of applying biotechnology to pharmaceutical applications is a relatively recent phenomenon.[237–238] What about proteins and other biological macromolecules – drugs that weigh tens to hundreds of thousands of daltons? The routine chemical synthesis of peptides was limited to relatively short peptides of up to about 50 amino acids until the 1990s. However, methods such as chemical ligation now allow the synthesis of proteins of 100 amino acids or longer, with the chemical synthesis of the 99-residue HIV-1 protease being a major event in this field (**Figure 37**). These new methods provide access to hitherto unreachable analogs and in some cases higher purity than is available using biological means of production.[239] For both short and long sequences, various protecting groups and solid phases have been employed to prepare peptide analogs, including Rink and Wang resins.[240–242]

It was 1975 when Milstein and Köhler discovered how to isolate monoclonal antibodies (MAbs) using hybridomas. This discovery helped to start the UK biotech industry and led to the first MAb diagnostic product in 8 years and the first FDA-approved MAb therapeutic product in 11 years. By the end of 2004, 18 MAb products were approved for therapeutic use in the US, and half this number in Europe. Many of these MAbs have been approved for cancer therapy, and some have achieved billions of dollars in sales. Drug discovery has benefited from advances in the use of MAbs as research tools, and it is conceivable that selective targeting strategies will require substantial help from medicinal chemistry as the field progresses.[243–244]

RNAi was discovered in the 1990s, and is finding increasing applications in drug discovery and as therapeutics. Specific inhibition of mRNA reduces protein levels in a cell, sometimes essentially silencing a given transcript ($>75\%$ knockdown). By using siRNAs, one can determine the relevance of certain proteins to disease pathology, thus aiding studies to define the druggable human genome.[245–246]

We have only scratched the surface in the potential tools, applications, and products that will result from these arenas. Blocking protein:protein interactions continues to be a challenge for small molecules, but new methodologies are cropping up regularly, including the targeting of susceptible domains.[247]

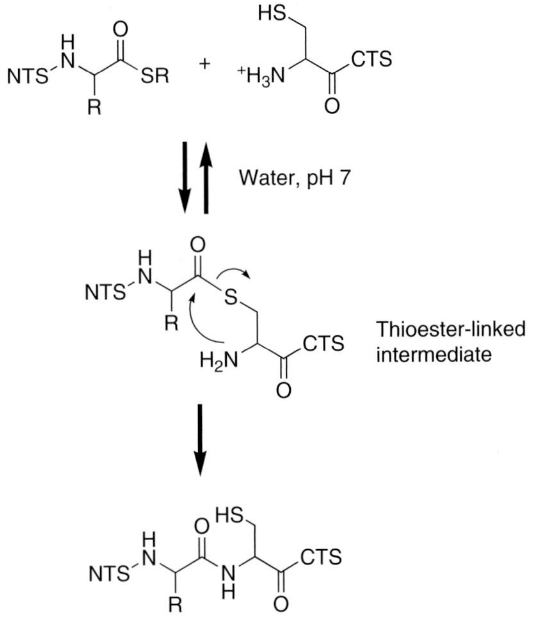

Figure 37 Chemical ligation method for the synthesis of large proteins from smaller peptides. NTS, N-terminal segment; CTS, C-terminal segment.

2.01.3.5 Computational Approaches

The structural requirements for receptor binding that lead to agonism can be exceedingly tight. In the case of the muscarinic cholinergic agonist, arecoline, the addition of a single methyl group reduces affinity and/or agonism by a factor of 10 or more (**Figure 38**).[248] However, most interactions are more forgiving than this example, and thus amenable to a wide variety of structure-based design and other computational approaches, including quantitative structure–activity relationships (QSAR).

While the beginnings of QSAR in medicinal chemistry have roots that date back to the 1940s and 1950s,[249] drawing heavily on linear free energy relationships and physical organic chemistry,[250] methods advanced remarkably once computers became commonplace. Sometimes the best techniques are the simplest. For example, a study of various central nervous system (CNS)-active drugs yielded a common phenethylamine pharmacophore.[251] This pharmacophore is evident in opiate analogs[252] and many other classes of CNS drugs. Operational schemes such as the Topliss Tree can be powerful yet simple 'back-of-the-envelope' approaches to deciding which analogs to synthesize in a medicinal chemistry program (**Figure 39**).[253] Manual methods for Hansch-type applications have also been proposed,[254] and the contributions of various functional groups to binding interactions have been calculated.[255]

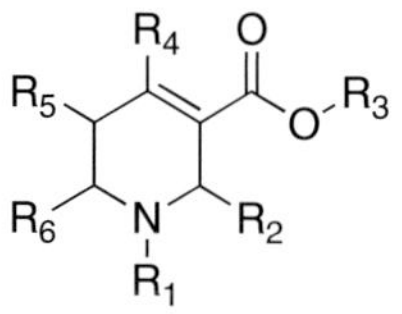

Analog	R_1	R_2	R_3	R_4	R_5	R_6	IC_{50}	Relative agonism[a]
1	Me	H	H	H	H	H	>100	–
2	H	H	Me	H	H	H	50	39
3 (arecoline)	Me	H	Me	H	H	H	10	100
4	Me_2+	H	Me	H	H	H	50	93
5	Et	H	Me	H	H	H	350	8
6	Me	H	Et	H	H	H	10	10
7	Me	Me	Me	H	H	H	>100	–
8	Me	H	Me	Me	H	H	1800	1
9	Me	H	Me	H	Me	H	>100	–
10	Me	H	Me	H	H	Me	>100	–

[a] Arecoline set arbitrarily at 100.

Figure 38 Tight binding requirements for muscarinic agonism among arecoline analogs.

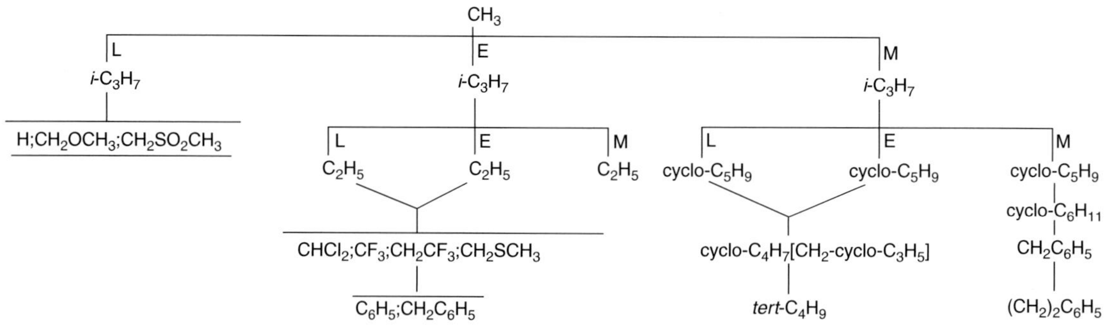

Figure 39 Topliss operational scheme for deciding which analogs should be synthesized in a medicinal chemistry program. M, more active; E, equiactive; L, less active. Descending lines indicate sequence. Square brackets indicate alternatives. (Reprinted with permission from Topliss, J. G. *J. Med. Chem.* **1972**, *15*, 1006–1011. Copyright 1972 American Chemical Society.)

Though QSAR analyses are not as common as they once were, they can be very helpful in understanding and guiding analog programs.[256] However, one must be careful not to analyze too many variables in QSAR studies, as the risk of a chance correlation increases with the number of variables considered.[257–258] Occasionally, almost perfect correlations have been found between biological activity and parameters such as lipophilicity (log P).[259] Lipophilicity was once determined laboriously by partitioning experiments. It was facilitated greatly when HPLC techniques became available to estimate octanol–water partition coefficients.[260] Today, most medicinal chemists use calculated log P (ClogP) values.[261] After these many years, the field of QSAR has generated considerable data of utility in drug discovery. For example, the Hansch group has entered nearly 12 000 equations into its database, representing over 40 years of work. This work has been summarized as a collection of human chemical–biological interactions, including taste, odor, metabolism, absorption, excretion, and toxicity, among others.[262]

Advances in computing have allowed the development of molecular modeling and other tools for SBD, sometimes referred to as computer-aided drug design (CADD).[263] While this can be a very technical discipline, practical guides exist for those who are interested in learning more or even becoming practitioners.[264–265] The thousands of three-dimensional structures of macromolecules that have been solved provide a good basis for drug design. This approach has been used to develop novel inhibitors in many systems, some of the first being inhibitors of matrix metalloproteases, HIV protease, and renin.[266] Despite shrinking budgets in many institutions, structural biology has been picking up the pace lately. More than 600 unique structures have been solved and entered into the Protein Data Bank (PDB),[267] though most proteins cluster into just four classes of structures.[268] While current SBD approaches can be very sophisticated, the field began with physical models, which were followed by hand-drawn models and small molecule x-ray crystal structures, which led to various conformational and charge-based modeling techniques, and ultimately many other strategies were employed to yield technical insights via molecular modeling (see **Figure 40**).

A more recent approach, fragment-based lead discovery, has been used successfully by several groups, often utilizing crystallography or NMR techniques, together with molecular modeling. In these methods, small molecular fragments are allowed to interact with targets of interest, and different binding elements are then linked together or tethered to increase the potency and specificity of the binding interaction.[270–272] **Figure 41** shows an early example of the results obtained by using an NMR-based approach on an FK-binding protein (FKBP) target. In this and other SBD-related areas, in some cases the approaches have evolved from traditional lock and key systems to induced fit models and the stabilization of conformational ensembles.[273] Even membrane-bound targets are being studied at a structural level using solid-state NMR.[274]

With these and other computational technologies in place, the stage was set for applying computers to thinking about and calculating the diversity space that combinatorial chemistry could in theory tackle.[275–278] Thus, combinatorial chemistry was described as a molecular diversity space odyssey as we approached 2001.[279] Consistent with the space odyssey theme, others have made analogies between mapping biological activity space and exploring uncharted territories amongst the stars.[280] For most intents and purposes, chemical diversity space is infinite. However, there are subsets of this vast space that represent preferred characteristics of drug classes and families of molecules that interact with gene superfamilies, such as the GPCRs. There are also subsets that one might consider to contain drug-like or druggable chemical series.[281] (The term 'druggable' may also refer to drug targets, namely, those targets that can be readily attacked with small organic molecules.) At this point in time, many approaches to determining criteria for drug-likeness have been proposed.[282–283] Moving beyond drug-likeness to the design of libraries with specific properties, such as CNS activity, has been the subject of some studies as well,[284] though other principles such as blood–brain barrier (BBB) transport are also important in such an analysis.[285] With any screening program, one needs to watch out for false positives. For instance, it turns out that some types of molecule are promiscuous, perhaps resulting from aggregation phenomena.[286]

Chemical space, whether infinite or not, is a continuum, with pockets of biologically relevant properties, and various strategies have been used to calculate this n-dimensional physicochemical property space (**Figure 42**).[287] Evaluating the diversity of a given set of compounds from first principles may utilize one of a number of different approaches, such as clustering, receptor site interaction simulation, similarity, and other techniques.[288–290] The vastness of this space is exemplified by the following: there may be a larger number of drug-like compounds that could in theory be synthesized than there are particles in the universe.

What other considerations should be mentioned? Even if one had a large and diverse library of drug-like molecules, it is conceivable that the biological target of interest would not be druggable – that is, it may not be possible to modulate some targets through interactions with orally available small molecules.[291] Others have tried to devise new ways of screening libraries virtually – by computer – before proceeding to laboratory work. While much heralded in the 1980s, this SBD technique has yet to live up to its full potential.[292–293] In the face of these many powerful new technologies, it can be easy to forget the power of simple thought paradigms that can be exploited on the back of an

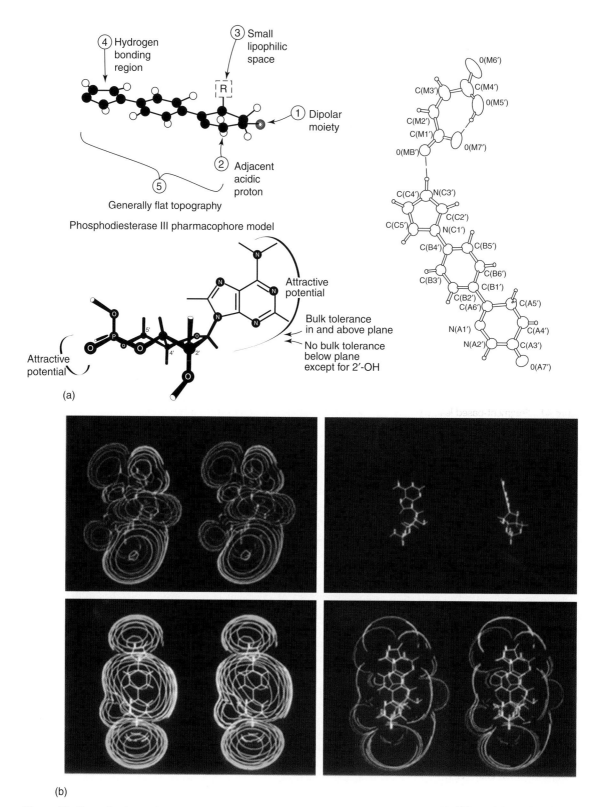

Figure 40 Example of an early structure-based design approach, with phosphodiesterase inhibitors studied as potential new drugs for congestive heart failure: (a) five-point model for cAMP PDE inhibition and inotropic effects, use of x-ray crystal structures to better understand small molecule structures, and revised five-point model based on cAMP; (b) comparing molecular frameworks and electrostatic potentials of PDE inhibitor (imazodan) and cAMP. (Reprinted with permission from Moos, W. H.; Humblet, C. C.; Sircar, I.; Rithner, C.; Weishaar, R. E.; Bristol, J. A.; McPhail, A. T. *J. Med. Chem.* **1987**, *30*, 1963–1972. Copyright 1987 American Chemical Society.[269])

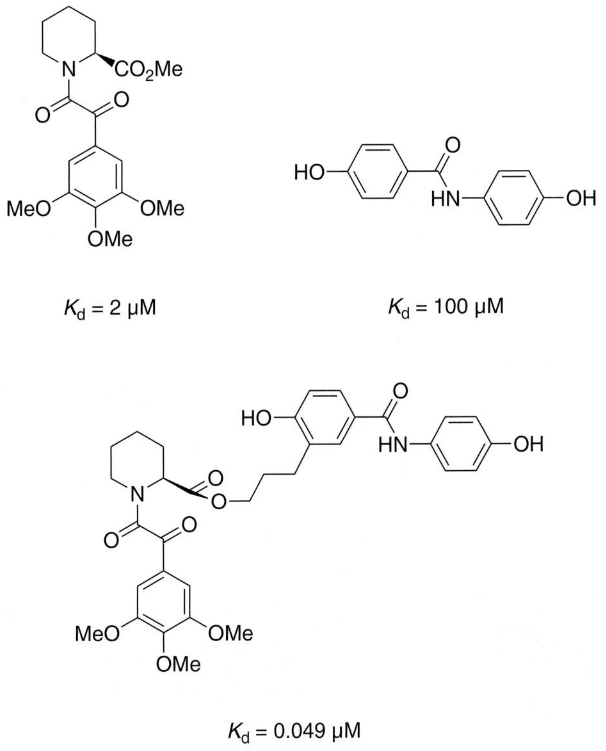

Figure 41 Fragment-based lead discovery of high-affinity ligands for proteins using SAR by NMR.

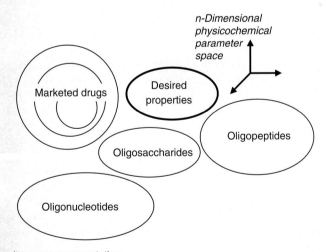

Figure 42 Chemical diversity space representation.

envelope. Lipinski's rule of five fits this rubric.[294] The rule of five predicts poor absorption or permeation to be more likely when certain thresholds are surpassed.

Some common characteristics of drug-like molecules are[295]:

- molecular weight ~250–500;
- oral bioavailability ~1–100%;
- lipophilicity (as calculated by ClogP) ~1–5;
- elimination half-life ~1–5 h;

- plasma protein binding up to >90%;
- soluble in water and alcohol; and
- one or more rings and a basic nitrogen.

In the midst of all of this new diversity that is being synthesized and calculated, a recent review of all oral drugs approved from 1937 to 1997 showed that, despite small trends in physical properties, there has been little substantial change over the years, except for molecular weight. Molecular weight increased from generally below 300 to often above 400 during this period. Only 2% of these drugs have molecular weights greater than 500 and more than three hydrogen-bond donors. Fewer than 5% have more than four hydrogen-bond donors.[296] Another study showed that the average molecular weight of oral drugs decreases as they progress through clinical trials, converging on the average molecular weight of marketed drugs.[297]

2.01.3.6 Evolution of Preclinical Research over Several Decades

In the old days, chemists, biochemists, and pharmacologists worked together in pharmaceutical companies to identify promising new drug leads by a slow, iterative process, often driven principally by data in animals. What has happened to the old ways in the face of SOTA technologies and new scientific disciplines? For one, molecular biologists became a major part of drug discovery teams in the 1980s and 1990s, developing molecular assays to prioritize compounds for animal testing. The basics have not changed,[298] but many new technologies have been added to the medicinal chemist's armamentarium, including combinatorial chemistry.

Solid-phase chemistry in the pharmaceutical industry was very rare, except in areas such as peptide synthesis, until the 1990s. Yet the field took off, and by 1997 well over 150 different types of solid-phase-based chemical products had been described,[299] and more than 80 different compound libraries had been prepared.[300] Only a few years later, hundreds of compound libraries had been reported.[301]

Opinions vary on the success of new drug discovery technologies, such as combinatorial chemistry and HTS.[302] However, it is hard to contest that combinatorial chemistry has led a revolution in medicinal chemistry – the world of synthetic chemistry has been changed forever – and it is getting better with each new generation of approaches, including that of the latest natural products-based, diversity-oriented libraries.[303] While the costs of what has been called big dumb science are not insignificant, applying these new processes in a focused and rational manner has begun to show real benefits in the time it takes and the quality of the leads that are progressing into the clinic.[304] Genomics is being applied increasingly to identifying molecules that affect targets of interest. For example, a genome-wide screen in yeast has been used to identify small molecules that inhibit the activity of kinases.[305] Some argue that microarrays and other molecular profiling techniques will revolutionize the way we develop drugs (**Figure 43**).[306,307]

A case study that displays some of the points of interest in the never-a-dull-moment evolution of preclinical research over several decades can be found in aging research and CNS diseases such as AD.[308] The number of centenarians per million people in the US increased from 15 in 1950 to 263 in 2000.[309] Though the US government declared the 1990s the Decade of the Brain, progress has been slow given the complexity of this organ. Nonetheless, at various points in time, roughly 10% of the drugs in development globally targeted CNS indications. The CNS space is an interesting example, highlighting some of the many issues that face drug hunters today in studies of the brain and other physiological systems. Whether the issues faced in this field and others will trigger a return to the fundamentals seems doubtful, even if such a change would address the current ailments of the industry – the seemingly poor productivity, the 30 years that it takes to move from scientific discovery to marketed products, megamergers that underperform, infatuation with new technologies long before their time, major failures postmarketing, and so on. There are multiple opinions on this subject, as one might expect.[310–314] A list of the major classes of CNS drugs is given in **Table 16**.

The representative issues facing CNS drug hunters today are:

- the complexity of the brain and our limited understanding;
- incomplete information regarding the pathophysiology of major diseases;
- relegation of behavioral pharmacology to a secondary role versus molecular neuroscience, starting in the late 1980s; and
- biases against funding what some consider to be more difficult, costly, and time-consuming hurdles in CNS clinical study (biases that are not always correct).

Possibly no other area has seen such major advances in the late twentieth century as AD. This curious ailment of old age was first reported by Alois Alzheimer in 1907.[315,316] The field of AD and related disorders has given rise to many classifications of age-related cognitive decline, from age-associated memory impairment (AAMI) to the more recent mild

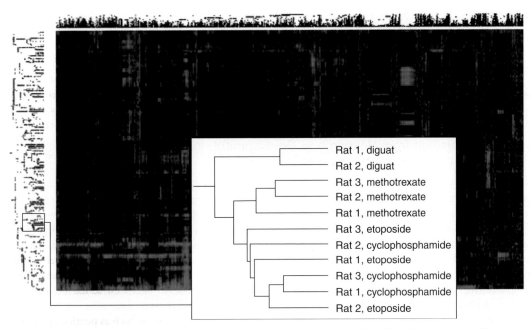

Figure 43 Microarray data showing changes in gene expression for a series of 48 different hepatotoxins. (Reprinted with permission from Ulrich, R.; Friend, S. H. *Nat. Rev. Drug Disc.* **2002**, *1*, 84–88. Copyright 2002 Nature Publishing Group.)

Table 16 Major classes of CNS drugs

- Anxiolytics
- Antipsychotics
- Anticonvulsants
- Antidepressants
- Analgesics
- Hypnotics and sedatives
- Stroke therapies
- Substance abuse medications
- Anti-Parkinsonian agents
- Alzheimer's disease therapeutics

cognitive impairment (MCI).[317,318] About 10% of Americans in a community setting over age 65 have probable AD, ranging from a few per cent in the 65–74 age group, to nearly half of all subjects over 85. Given the high percentages of AD in older persons, the public health impact of this disease will continue to rise with increasing longevity.[319]

This is a field full of dogmas, where some people equate β-amyloid with AD, and researchers take to their pathophysiological dogmas like religions, whether it be the β-amyloid Baptists, the τ protein Taoists, or the mitochondrial dysfunction Mitochondriacs. It is also a field that ignores studies that don't fit the latest schema. Thus, even in the face of numerous studies reporting positive data in one or more measures, large, politically charged, government-funded studies carry more weight, biasing medical practice in directions that may not be correct. Witness the dramatic decline in estrogen use for dementia or cognitive decline (and for menopause symptoms) after the results of the Women's Health Initiative Memory Study (WHIMS) were published.[320] The conclusions of WHIMS authors are not, however, fully accepted by all estrogen researchers.[321] Notwithstanding the WHIMS results, finding new drugs from old drugs (described earlier as SOSA) is an approach that has borne fruit in the estrogen field. Here, less feminizing or nonfeminizing estrogens are finding utility as neuroprotectants, possibly via mitochondrial mechanisms.[322–323] These include components of estrogen therapies that have been administered to humans for more than 60 years, such as the less hormonal (less feminizing) 17α-estradiol, an isomer of the most potent natural feminizing hormone, 17β-estradiol.[324] The activity of compounds like these has been reported in transgenic models of AD.[325]

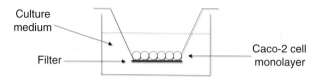

Figure 44 Assessing permeability of drugs using Caco-2 cell monolayers and filter-well setups.

2.01.3.7 Transition from Research to Development, Including Absorption, Distribution, Metabolism, Excretion, and Toxicity

While there have been major advances in many of the early stages of drug discovery, a bottleneck remains in preclinical development – in ADMET – both in vitro at an HTS level, or with high content assays (HCAs), and in vivo for screening and for formal development. Addressing this part of drug development requires different thought processes and procedural rigors because of regulations. Various methods have been developed to predict or optimize drug absorption and pharmacokinetics.[326–328] Properties that influence oral bioavailability have been studied extensively.[329] One popular system uses human colon epithelial (Caco-2) cell lines. In this system, permeability is typically assessed using monolayers of Caco-2 cells on filter wells (**Figure 44**).[330,331] Physicochemical measurements and plasma protein binding data have been used to predict volume of distribution values.[332] Molecular modeling has been used to predict cytochrome P450 (CYP)-mediated drug metabolism.[333] Pharmacokinetics has become more integrated into the drug discovery process over time.[334]

Many groups have worked to reduce attrition rates via ADME profiling earlier in the drug discovery process. These efforts include studying a wide variety of parameters.[335] Similarly, the chemistry of metabolism and toxicology as it relates to safety and toxicity profiling, prediction, and analysis has been reviewed. Modern techniques and approaches include the following:

- solubility and solubilization;
- permeability and active transporters;
- ionization or dissociation constants;
- lipophilicity;
- chemical integrity and stability;
- metabolic stability and clearance;
- CYP inhibition and other metabolism-related drug interactions;
- safety and toxicity threshold concepts from the perspective of adverse effects in general, carcinogenicity in particular, and individual human variation[336];
- mechanistic insights into idiosyncratic toxicities, including reactive metabolites, immune-mediated responses, low levels of infections accompanying disease, and mitochondrial damage[337];
- in silico predictive toxicology and ADME, including PK and phenotyping, using molecular modeling, expert systems, and data-driven systems, which of course still need to be validated and accepted by regulatory bodies[338–340];
- prediction of drug interactions, focusing on CYP, kinetics, genetic variability, and computational methods[341,342];
- designing against metabolic activation and analytical methods for metabomics (also referred to as metabonomics or metabolomics), such as has been used in the design of leukotriene antagonists and PDE inhibitors[343–344];
- drug delivery and BBB models, including in silico methods, endothelial cell monolayers, co-cultures of endothelial cells and glia, dynamic flow-based systems, and other models of permeability including multidrug resistance[345]; and
- model organisms like the zebrafish and cell signaling networks, including developmental toxicology, toxicogenomics, proteomics, xenobiotic metabolism, and toxicity of various systems, such as cardiovascular, CNS, and the immune system, and mapping drug and gene function relationships.[346,347]

2.01.3.8 The Perspective of Successful Drug Hunters

Annual Reports in Medicinal Chemistry is a good, concise, annual summary of the latest success of the industry's drug hunters.[348] Successful drug hunters persevere, have a whatever-it-takes mindset, and occasionally use tips and tricks that others miss. The key can be as simple as determining when it is time to move on – proceeding to the next stage of evaluation even if the molecule is not ideal, since more questions will be answered by going forward than have been answered to date. Success can require fighting against the current, such as working on dirty drugs or polypharmacy when others are focused on very selective drugs. Multicomponent therapeutics is a recent example of the latter,[349] though Janssen apparently espoused this approach decades ago.

The following are a few thoughts that might go through a drug hunter's mind:

- research and development can be very different, requiring different temperaments;
- breaking the rules creatively in research (innovative chaos) may lead to the next breakthrough;
- following the rules exactly – competence, pure and simple – may be the only viable path through highly regulated phases of development;
- key experiments should be run as soon as possible to get to proof-of-concept;
- from a portfolio and media standpoint, a drug's development in the clinic should not be halted until you have a suitable replacement ready to go;
- if possible, go/no-go decisions on safety should be made as early as possible, but decisions on efficacy should be delayed until Phase III; and
- a decision should not be revisited unless there is new information.

The push for personalized medicine became more forceful in the 1990s with major advances in genomics and proteomics, and personalization will definitely affect the drug hunters' game in this century. However, the regular use of pharmacogenomics to guide clinical trials and choice of therapy remains a distant goal, given the complexities of the human genome, first sequenced in 2001.[350–354] Nonetheless, marrying diagnostic advances with the latest therapeutics to enable the next generation of personalized therapies could be just over the horizon. Once personalization is achieved regularly as the norm, industry-wide economic effects can be expected. Will large companies be at a disadvantage to smaller ones if personalized medicine decreases the market size of each new therapeutic to only those individuals who will benefit most? At the time of this writing, we can only speculate.

Some of the thorny issues faced by drug hunters at the more than 1000 biotech companies that exist today can be summarized as follows[355]:

- rising costs in R&D;
- a weak economy following the 'froth' of 2000;
- capital formation difficulties;
- Prescription Drug User Fee Act of 1992 (PDUFA);
- FDA heads, staff, and appropriations (turnover, vacancies, and insufficient funds);
- stem cell research and gene therapy;
- medical records confidentiality;
- generic biotechnology products;
- animal rights activism;
- pricing and controls, including Medicare and reimportation of drugs; and
- intellectual property and technology transfer issues.

With all of these issues, what's a drug hunter to do? A few additional comments from experienced drug hunters follow regarding the yin and yang of pharmaceutical research and development[356–359]; though not obviously related, some of the same thought patterns can be helpful in running a company[360]:

- Decisions, once made, must be acted upon promptly and communicated well.
- A decision should be made when you have enough information to make a good one, or if you don't have time to delay; once it is made, never look back (e.g., because someone or something may be gaining on you!).
- What are the 'key' experiments? Many of the killer questions can be asked up front, but are never answered until it is too late – when the issues are about to kill the project.
- With analogy to football, research is often a passing game, lots of razzle-dazzle excitement, where you want to fail early and often, whereas development is the ground game, with success often the result of grit, determination, and sheer competence.
- Never waver in development.

- Don't avoid risk, especially in pharmaceutical research, where over 90% of everything you do will fail – try something new rather than following the pack.
- Many companies have done well with me-too products, so taking a risk is not always the best move from a development or financial perspective.

2.01.4 Targets

At the end of this volume are a series of chapters that introduce strategy and drug research in the context of several important classes of drug targets. While the present volume does not endeavor to present an all encompassing review of major drug targets, it sets the stage for a more detailed consideration of therapeutics and their targets in later volumes (see in particular Volumes 6 and 7 in this reference work – for example, chapters that cover drugs for attention deficit hyperactivity disorder (ADHD), anxiety, cancer, depression, infectious diseases, and neurodegeneration). In considering how to evaluate drug targets from the perspective of medicinal chemistry, one is generally less interested in the information that might be found in molecular biology or pharmacology texts. Rather, one should be more interested in information concerning gene classification, protein structure, localization, physiological function, second messengers (if any), prototypical pharmacology and therapeutic or other products, association with diseases, genetic defects, and future directions – all with the intent of serving as a sound and understandable introduction to someone entering a new area. Thus, it is hoped that this brief introduction will provide the chemist or other interested readers with sufficient background to understand and interpret more comprehensive and detailed experimental and theoretical works.

It is with the above in mind that we now introduce target strategy and drug research. This introduction will be even less complete than those chapters that conclude this volume, and the subsequent volumes in this reference work, but such a layered approach, at greater and greater levels of detail, should serve to lead everyone, from beginner to expert, to their most appropriate entry points into the literature.

We will begin with transporters, a popular set of CNS targets, especially in the last couple of decades, and also a broad class of proteins with a significant role to play in both drug discovery and preclinical development, including ADMET properties (see Volume 5). A brief discussion of other targets will follow.

2.01.4.1 Transporters as an Example of an Important Class of Targets

Transporters, pumps, and ion channels have evolved to move ions, proteins, and other molecules across membrane barriers.[461] Transporters are membrane proteins that actively or passively control the influx and efflux of organic and inorganic molecules in all organisms. This flux may be the primary site of action for certain drugs, or it may modulate the efficacy or toxicity of others. Transporters also play a role in drug absorption and disposition. Unlike channels, which are typically on or off and may form pores in membranes, transporters participate more intimately in the movement of drugs, ions, and other substrates across membranes, often requiring a conformational change to perform their function. As such, their transfer rates are usually much slower than that of channels. Transporters may also be called pumps, reuptake sites, and translocases, among other names.

Drug transporters are expressed widely – in brain, intestine, kidney, and liver, for example – and play a key role in both drug action and in ADMET, as introduced above. Given such a central role in cellular physiology, it should be no great revelation that a significant percentage of human genes have been surmised to encode transporters or related proteins.[361] In the initial sequencing of the human genome, more than 500 genes were identified as putative transporters,[353] though this number may be an underestimate because many genes whose molecular functions were not known at the time may ultimately be found to encode transporter functions. Approximately 4% of marketed small-molecule drugs act via transporter targets, which is on a par with drugs working via nuclear hormone receptors. Except for the dominance of GPCRs and enzymes, which respectively account for 30% and 47% of the targets through which marketed small molecules act, only ion channels (7%) are more common than transporters in this respect.[362]

Serotonin (5-hydroxytryptamine, 5HT) reuptake (serotonin transporter, SERT) inhibitors used in treating depression, such as fluoxetine (Prozac, **Figure 45**), represent a common class of drugs acting at transporter targets. Fluoxetine heralded a breakthrough in treating depression when it was approved by the FDA in 1987, some 13 years after it was first described in the scientific literature as a selective serotonin (re)uptake inhibitor (SSRI). SSRIs subsequently became one of the most widely used drugs for CNS disorders. Particularly worth reading are a case history of fluoxetine and a recent review on neurotransmitter transporters and their impact on the development of psychopharmacology.[363,364] Since the original discoveries were made, fluoxetine and other SSRIs have also been shown to have utility in treating not just depression but also anxiety, obsessive compulsive disorder, and other diseases.[365] In recent years, several transporter

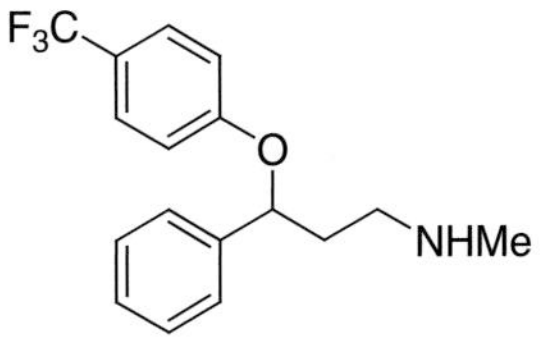

Figure 45 Chemical structure of fluoxetine, a prototypical selective serotonin reuptake inhibitor.

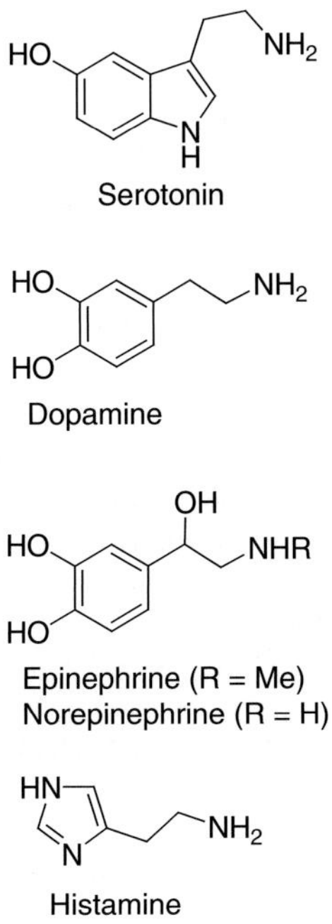

Figure 46 Chemical structures of selected small-molecule monoamine neurotransmitters.

inhibitors achieved a billion dollars or more in sales, among these being citalopram (Celexa, Forest), fluoxetine (Prozac, Lilly), paroxetine (Paxil, GlaxoSmithKline), sertraline (Zoloft, Pfizer), and venlafaxine (Effexor, Wyeth).[366]

The dopamine transporter (DAT) – a Na^+ and Cl^--dependent neuronal transmembrane protein – was first cloned in the early 1990s, and is involved in locomotor control, including functions lost in Parkinson's disease. DAT is also involved in reward systems, and thus in addiction to drugs such as amphetamine and cocaine, and in ADHD and Tourette's syndrome, among other illnesses.[367–371] Indeed, the actions of many small molecule neurotransmitters containing a basic amine are modulated through transporter sites, including those of adrenaline and noradrenaline (epinephrine or norepinephrine), dopamine, histamine, and serotonin (**Figure 46**). That is, transporters serve to modulate synaptic neurotransmitter levels through reuptake into nerve terminals, and once inside they are taken up into vesicles via different transporters (**Figure 47**). These effects lend themselves to being studied via in vivo imaging techniques, as outlined later.

Multidrug and drug-specific transporters are involved in the efflux of anticancer and antibiotic chemotherapeutics, which leads to significant resistance in cancer cells and a variety of pathogenic organisms, including bacteria, fungi, and parasites.[372] This is often referred to as multidrug resistance (MDR), with P-glycoprotein (P-gp) being a common MDR site in cancer cells. P-gp first generated interest when it was discovered in the late 1980s to be overexpressed in cancer cells. Some drugs, including verapamil (**Figure 48**), a calcium channel blocker, have been found to inhibit MDRs.[373] While such inhibitors might eventually prove useful in dealing with resistance in cancer and infectious diseases, success to date has been limited.

Transporters have also been exploited in drug delivery. The use of Pluronic block copolymers to modulate P-gp-mediated BBB drug efflux has been studied. These polymers have been demonstrated to enhance BBB penetration.[374] The choline transporter has also been proposed as a BBB vector for drug delivery.[375] Other drug delivery examples are described below.

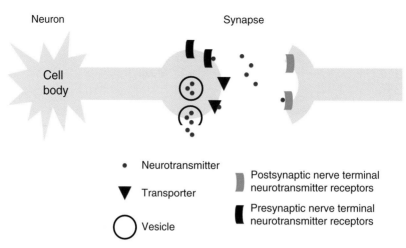

Figure 47 Neuronal synapse containing neurotransmitter receptors and transporter sites.

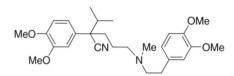

Figure 48 Chemical structure of verapamil, an MDR inhibitor and a calcium channel blocker.

Table 17 Websites containing information on transporters, nomenclature, and related data

Information	Website
Nomenclature of membrane transport proteins	http://www.chem.qmul.ac.uk/iubmb/mtp/
ABC transporters	http://nutrigene.4t.com/humanabc.htm
SLC transporters	http://www.bioparadigms.org
Transporter genes in the human genome	http://www.gene.ucl.ac.uk/nomenclature/
Pharmacogenetics of transporters	http://www.pharmgkb.org
	http://www.pharmacogenetics.ucsf.edu
Drug transporters	http://www.TP-search.jp
Drug interactions with neurotransmitter receptors	http://pdsp.cwru.edu/pdsp.asp
Drug and drug target information	http://redpoll.pharmacy.ualberta.ca/drugbank/

The transporter arena has proved so important that biotechnology companies have been founded to exploit the effects of transporters on drug absorption. Consider, for example, XenoPort.

2.01.4.1.1 Current nomenclature

Several superfamilies of transporters have been identified. Most notably, these include ATP-binding cassette (ABC) and solute carrier (SLC) transporters (**Table 17**).

ABC transporters are also known as the major facilitator superfamily (MFS). There are seven subfamilies of MFS (denoted A through G). Many ABC transporters work to actively pump substrates across membranes using energy derived from ATP hydrolysis. The ABC transporters, including MDR and P-gp proteins, are of interest not only in studying drugs used in cancer and infectious disease but also in studying the transmembrane conductance regulator in cystic fibrosis (CFTR, an ABC transporter that functions like a chloride channel[462]), the sulfonyl urea receptor (SUR)

Chloroquine antimalarial   **Tolbutamide (SUR ligand used to treat diabetes)**

Figure 49 Chemical structures of chloroquine and tolbutamide.

ion channels in diabetes. (Note that KATP channels are made up of Kir6, part of the Kir channel family, and SUR, which has been called an ion channel but is an ABC transporter family member.[463]) P-gps have been compared across several species.[464] Though it is true that well-studied transporters such as MDR play an important role in drug resistance, new transporters with potential relevance to drug resistance are still being discovered, such as Pfcrt, which is responsible for resistance to chloroquine (**Figure 49**) in malaria, and the hexose transporter, which may be a worthy new drug target in malaria.[376–378]

Neurotransmitter transporters are part of the SLC transporter family, and include biogenic amine transporters of two types, the first comprising DAT, the norepinephrine transporter (NET), and SERT, and separately the vesicular monoamine transporters (VMATs), a distinct proton-dependent gene family (**Table 18**). Excitatory and inhibitory amino acid neurotransmitters also have transporters. Many neurotransmitter transporters retain the predicted 12 transmembrane spanning domains of this broad family, and range from about 500–800 amino acids in length.[379]

2.01.4.1.2 Gene/molecular biology classification and protein structure

The ABC and SLC superfamilies of transporters are multi-membrane-spanning proteins.

We will consider for structural illustration one family of SLCs, namely the SLC22 transporter family, which includes transporters for organic anions (OATs) and organic cations (OCTs), as well as zwitterions and cations (OCTNs).

Examples of SLC22 substrates include:

- α-ketoglutarate;
- cAMP;
- cGMP;
- choline;
- L-carnitine;
- prostaglandins; and
- uric acid salts.

These generally polyspecific transporters contain 12 predicted transmembrane α helices and one large extracellular loop between the first and second transmembrane domains (**Figure 50**). SLC22 transporters function as co-transporters (as with Na$^+$ and L-carnitine), exchangers, and uniporters. From the cloning of the first SLC22 in 1994, the rat OCT1, a number of SLC antiporters, symporters, and uniporters have been cloned. SLC22 genes have been found on a number of chromosomes, including 1, 3, 5, 6, 11, and 14.[380] Related proteins include organic anion transporting polypeptide transporters (OATPs), which are involved in drug disposition.[381]

The gene organization and other features of the human DAT in neuropsychiatric disorders have been reviewed.[382] DAT functional groups have been analyzed in detail at the level of amino acid topography. From these studies it has been ascertained that drug–DAT binding occurs at multiple distinct sites. An atlas of these functional groups has been published.[383] Structure–function studies of the DAT have shown that it can operate in normal and reverse or efflux modes.[384] Promiscuous inhibitors have been reported that affect DAT, NET, and SERT to significant but varying degrees, which has provided insight into designing more selective monoamine neurotransmitter transporter inhibitors.[385]

Because transporters are membrane proteins, which are difficult to crystallize, bacterial proteins that can be expressed at high levels have been most amenable to x-ray crystallography. Successful structures to date include MsbA, an *E. coli* ABC transporter that is homologous to MDR,[386] and LacY, an *E. coli* MFS lactose permease.[387] LacY is a proton symporter, co-transporting protons and oligosaccharides, and is encoded by a structural gene in the *lac* operon of *E. coli*. As with other members of the MFS family, LacY causes the accumulation of (in this case) oligosaccharides, which is driven uphill energetically using energy released by the downhill electrochemical proton translocation gradient. The

Table 18 Neurotransmitter transporters

Neurotransmitter transporters (T)	Subtypes	Location	Selected inhibitors	Primary substrates
Monoamine transporters (12)[a]				
Dopamine (DA)	DAT	Plasma membrane	Mazindol	Dopamine
Norepinephrine (NE)	NET	Plasma membrane	Imipramine	Norepinephrine
Serotonin (SER)	SERT	Plasma membrane	Fluoxetine	Serotonin
Vesicular monoamine (VM)	VMAT-1	Vesicular	Tetrahydrobenazine, reserpine	Monoamine neurotransmitters
	VMAT-2	Vesicular	Tetrahydrobenazine, reserpine	Monoamine neurotransmitters
Amino acid transporters (6, 8, 10, or 12)[a]				
Excitatory amino acid (EAA)	EAAT1 (GLAST)	Plasma membrane	L-*trans*-2,4-Pyrrolidine-2,3-dicarboxylic acid	Glutamate, aspartate
	EAAT2 (GLT-1)	Plasma membrane	L-*trans*-2,4-Pyrrolidine-2,3-dicarboxylic acid	Glutamate, aspartate
	EAAT3 (EAAC1)	Plasma membrane	L-*trans*-2,4-Pyrrolidine-2,3-dicarboxylic acid	Glutamate, aspartate
	EAAT4	Plasma membrane	L-*trans*-2,4-Pyrrolidine-2,3-dicarboxylic acid	Glutamate, aspartate
	EAAT5	Plasma membrane	L-*trans*-2,4-Pyrrolidine-2,3-dicarboxylic acid	Glutamate, aspartate
γ-Amino-butyric acid (GABA)	GAT-1 (mGAT-1)	Plasma membrane	Nipecotic acid, guvacine	GABA
	GAT-2 (mGAT-3)	Plasma membrane	Nipecotic acid, guvacine	GABA, β-alanine
	GAT-3 (mGAT-4)	Plasma membrane	Nipecotic acid	GABA, β-alanine
	BGT-1 (mGAT-2)	Plasma membrane	EGYT-3886	GABA, betaine
	VGAT (VIAAT)	Vesicular	Nipecotic acid, vigabatrin	GABA
Glycine	GlyT-1	Plasma membrane	Sarcosine	Glycine
	GlyT-2	Plasma membrane	ALX 1393	Glycine

[a] Putative number of transmembrane domains.

mechanism of transporters such as LacY, and their hand-off of protons and substrates, may involve inward-facing and outward-facing conformations (**Figure 51**). Computational models have been developed to identify potential P-gp inhibitors and substrates.[465,466]

2.01.4.1.3 Physiological function

As one might expect from a very broad and important class of proteins, transporters play a wide variety of roles in physiology. A sampling of the roles they play can be found below.

Transporters act as reuptake sites for neurotransmitters, and the pathophysiology of major depressive disorders is believed to involve serotoninergic systems, including SERTs. SERTs, for example, are reduced in depression, and various genetic studies implicate these transporters in the disease.[388] Long-term exposure to reuptake inhibitors causes downregulation, as one might expect. However, this effect is reversible in 48 h.[389] There is also a remodeling that appears to be necessary for therapeutic utility, as antidepressants such as SSRIs take from 2 weeks to 4 weeks to achieve their full effects.

Adenine nucleotide transporters (ANTs), also called translocases or translocators, are reversible transporters of ADP and ATP, and have been implicated in mitochondrial apoptotic processes (programmed cell death). ANTs are believed

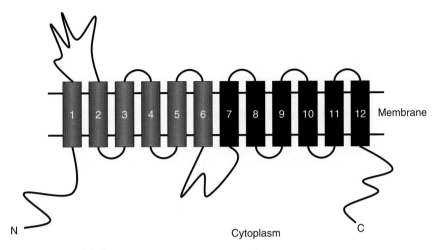

Figure 50 General structure of SLCs.

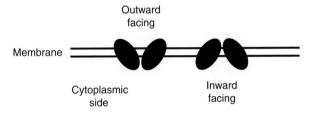

Primary active transport
Uses the energy from ATP to transport substrates uphill (against the electrochemical gradient)

Secondary active transport
Antiporters transport substrates in the opposite direction to the ion providing the uphill driving force

Symporters transport substrates in the same direction as the ion providing the uphill driving force

Passive transport
Substrates diffuse passively or in a facilitated manner downhill energetically

Figure 51 Different conformations of ABC transporters that may explain how these proteins transport their substrates.

to participate in a complex of proteins that govern mitochondrial membrane permeabilization, a key assemblage in a number of apoptotic pathways. Mitochondrial dysfunction has been implicated in a wide variety of disorders, ranging as far afield as AD and viral infections.[390] Mitochondria, with their inner and outer membranes, also use transporters to move proteins around, such as the translocases of the outer membrane (TOMs) and translocases of the inner membrane (TIMs).[391]

The potential for drug–drug metabolism–transporter interactions has been an increasing worry for the industry and for regulatory bodies because of major drug development failures and safety issues with marketed or late-stage drugs in recent years. This includes interactions between antihistamines and antibiotics, for example. Conferences have been held and articles published to debate how best to deal with these issues.[392,393]

SLC OATs are critical to the renal elimination of many charged drugs, toxins, and xenobiotics. This role in clearance and detoxification is essential for survival. Moreover, drug–drug interactions may occur and nephrotoxicity may also result from interactions with these transporters.[394]

Polyamine transporters (PATs) have been identified at elevated levels in tumor cells. There appears to be broad structural tolerance among PATs. The molecular requirements for selective polyamine delivery in cells containing PATs have been defined.[395]

Cholestasis is the syndrome that results from the impairment of bile formation, which may occur because of autoimmune, genetic, or metabolic disorders. Bile secretion depends on a variety of membrane transport systems, including organic and ion transporters, such as OATP, MDR1, and others. Defects in these transporters are associated with various liver diseases in humans.[396]

Deficiencies in HLA molecules occur when transporters associated with antigen processing (TAPs) are defective.[397]

2.01.4.1.4 Prototypical pharmacology and therapeutics
2.01.4.1.4.1 Therapeutics

The activity of drugs against efflux transporters can be assessed readily by using one or more in vitro and in vivo systems. This includes accumulation, efflux, transport, ATP hydrolase or synthase (ATPase) assays in vitro, and behavioral, transgenic, and mutant models in vivo.[398] Thus there are many ways to uncover and study drugs that may modulate transporter function. A few examples of therapeutics and their mechanistic classes are given below.

Again, one of the most successful classes of transporter inhibitors are the SSRIs. SSRIs are more selective for neurotransmitter transporters than other classes of antidepressants, but the tricyclic class of drugs also interacts with the transporters (**Figures 52** and **53**). These various antidepressant classes bind to various neurotransmitter receptors as well, which explains many of the side effects (**Table 19**). There are a number of thorough reviews of the therapeutic potential of neurotransmitter transporter inhibitors, both in this reference work (*see* 5.01 The Why and How of Absorption, Distribution,

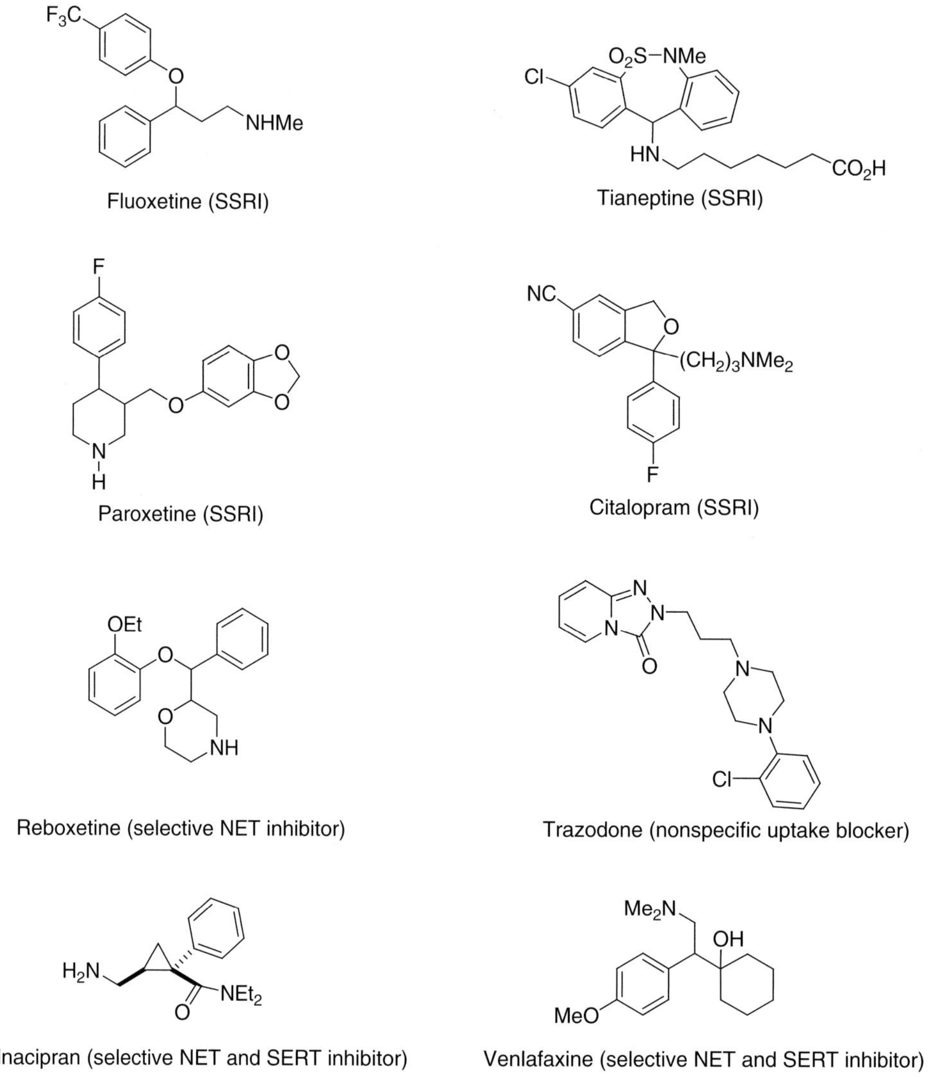

Figure 52 Chemical structures of representative monoamine neurotransmitter transporter inhibitors.

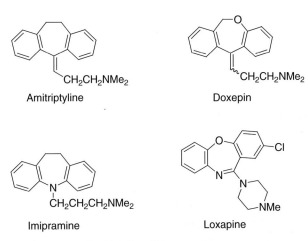

Figure 53 Chemical structures of representative tricyclic antidepressants.

Metabolism, Excretion, and Toxicity Research; 5.04 The Biology and Function of Transporters; 5.32 In Silico Models for Interactions with Transporters; 6.01 Central Nervous System Drugs Overview; 6.03 Affective Disorders: Depression and Bipolar Disorders; 6.05 Attention Deficit Hyperactivity Disorder; 7.01 Cancer Biology; 7.02 Principles of Chemotherapy and Pharmacology; 7.03 Antimetabolites; 7.09 Recent Development in Novel Anticancer Therapies) and elsewhere.[399,400]

Neurotransmitter transporter sites are the targets of drugs for ADHD, anxiety disorders, and depression, among other diseases. This includes a wide range of structural classes, perhaps most notably SSRIs such as fluoxetine. The SSRIs have also been an interesting area for exploration of chiral substitutions, since drugs such as citalopram and fluoxetine are racemic mixtures, which raises questions around ADME, efficacy, and safety.

SSRIs in some cases have seen expanded use and have challenged anxiety drugs such as benzodiazepines. With fewer side effects, including lesser liability of dependence and withdrawal effects, transporter inhibitors have as a result been able to carve out a worthwhile niche beyond depression. Transporter inhibitors are also finding use in veterinary indications; for example, to treat depressed or nervous pets.

DAT is a key interaction site for cocaine (**Figure 54**), and may be a critical pathway leading to its liability for abuse.[401] Indeed, novel DAT inhibitors have been studied as potential therapies for drug abuse.[402]

Drugs such as fenfluramine and 3,4-methylenedioxymethamphetamine (MDMA, ecstasy) cause the neuronal release of serotonin by acting as substrates for SERT proteins (**Figure 55**). Agents such as these affect appetite, but can lead to neurotoxicity, primary pulmonary hypertension, and valvular heart disease. Their potential in treating psychiatric disorders, such as addiction, depression, and premenstrual syndrome, is predicated on eliminating the side effects.[403]

Potent and selective inhibitors of the type 2 glycine transporter (GlyT-2) have been described (**Figure 56**). As glycine is one of the major inhibitory neurotransmitters in vertebrates, it has been proposed that selective inhibitors of GlyTs could have novel effects as analgesics, anesthetics, or muscle relaxants.[404]

One of the largest selling drugs in history, the antiulcer omeprazole (**Figure 57**), inhibits the secretion of gastric acid. Omeprazole blocks the monovalent inorganic cation transporter activity of H^+/K^+-ATPase, also known as the proton pump.[405,405a]

Sodium-co-dependent bile acid transporter inhibitors (**Figure 58**) have been studied as potential therapeutics to reduce levels of low-density lipoproteins in atherosclerosis, to prevent coronary heart disease.[406]

Peptide inhibitors of P-gp have been developed based on the structure of the P-gp transmembrane domains in the hope of discovering agents that reduce drug resistance (**Figure 59**). These peptides sensitize resistant cancer cells to doxorubicin.[407]

Thiazide diuretics (**Figure 60**) block the Na^+, Cl^- transporter. Diuretics are used to treat patients with edema, including those with cirrhosis, congestive heart failure, nephrotic syndrome, and renal insufficiency.[408]

2.01.4.1.4.2 Imaging

Despite therapeutic successes, imaging studies of SERT and studies of post-mortem brain tissue have not to date yielded unequivocal conclusions about the role of these transporters in depression and suicide.[409] The use of and challenges in exploiting imaging to monitor disease progression in neurodegenerative diseases such as Parkinson's disease have been reviewed, with a focus on disease-related and drug-induced changes in DAT expression and single

Table 19 Classes of antidepressant neurotransmitter transporter inhibitors and their indications, side effects, and potency[a]

Inhibitor classes	Prototypical inhibitors	Primary indications	Representative side effects	Transporter inhibition (relative order of magnitude potency (10^{-x}))			Receptor binding (relative order of magnitude potency (10^{-x}))[b]						
				DAT	NET	SERT	α_1	α_2	D_2	H_1	M	$5HT_{1A}$	$5HT_2$
Tricyclics	Amitriptyline, desipramine, imipramine, nortriptyline	Anxiety, atypical pain syndromes, depression	Blurred vision, cardiac arrhythmias (prolonged QRS or QT intervals), constipation, dry mouth, orthostatic hypotension, sedation, seizures, sexual dysfunction, urinary retention, weight gain	6	8	9	8	6	6	9	8	7	8
Reuptake blockers (transporter inhibitors)	Fluoxetine, fluvoxamine, paroxetine, sertraline, venlafaxine	Anxiety, bulimia nervosa, depression, OCD, panic disorders, post-traumatic stress syndrome, social phobias	Sexual dysfunction	6	7	10	6	5	–	6	6	5	7

[a] Potency: comparing amitriptyline and fluoxetine.
[b] Receptors: α_1, α_2, adrenergic; D_2, dopamine; H_1, histamine; M, muscarinic cholinergic; $5HT_{1A}$, $5HT_2$, serotonin.

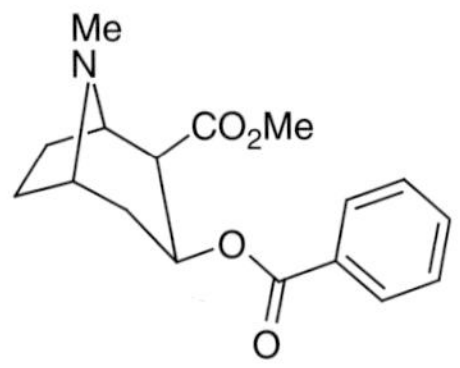

Figure 54 Chemical structure of a major drug of abuse, cocaine.

Figure 55 Chemical structures of selected serotonin releasers.

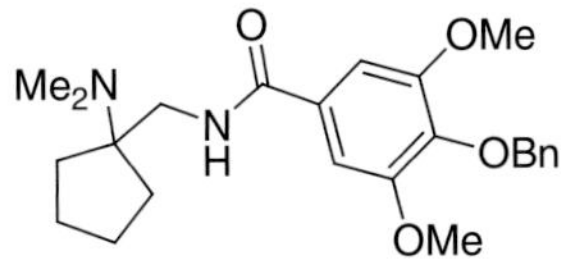

Figure 56 Chemical structure of a GlyT-2 inhibitor.

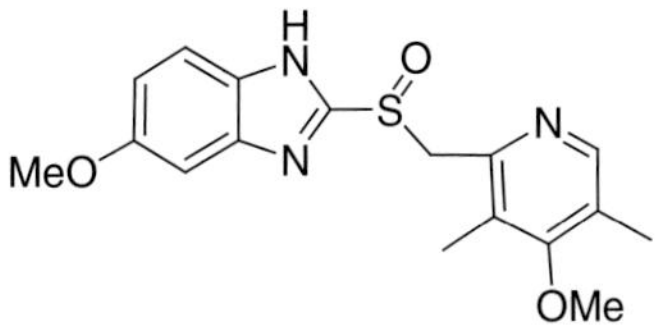

Figure 57 Chemical structure of the proton pump inhibitor omeprazole.

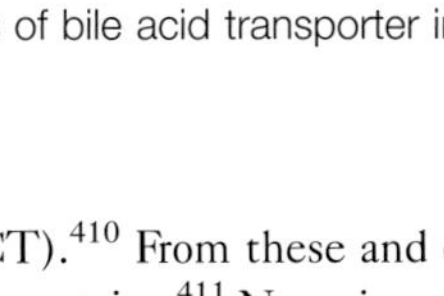

Figure 58 Generic chemical structure of a class of bile acid transporter inhibitors.

photon emission computed tomography (SPECT).[410] From these and other studies, the turnover rate of DAT protein has been shown to be similar to that of receptor proteins.[411] Neuroimaging data suggest that levodopa may modify DAT after chronic use. This is important because levodopa is widely used to reduce the symptoms of Parkinson's disease.[412]

Positron emission tomography (PET) often uses ^{18}F-fluorodeoxyglucose (FDG) to monitor cellular glucose consumption in patients. FDG is taken up into cancer and other cells via glucose transporters, and its uptake provides information on the consumption or metabolism of glucose, and on glycolysis. Other radiotracers are used in PET, including suitably labeled thymidine and fluorothymidine, which are taken up via nucleoside transporters. In the latter cases, information is generated on DNA synthesis and the proliferation of tumor cells.[413]

Fragment of the P-gp transmembrane sequence
KAITANISIGAAFLLIYASYALAFWYGTTLVLSGE

LIYASYALAFWYGTTLVLSGEGSDD
Micromolar peptide inhibitor

Figure 59 Peptide inhibitors of P-gp.

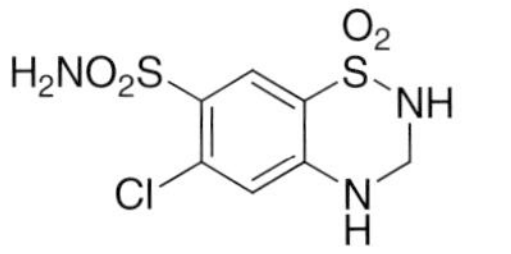

Figure 60 Chemical structure of hydrochlorothiazide, a diuretic.

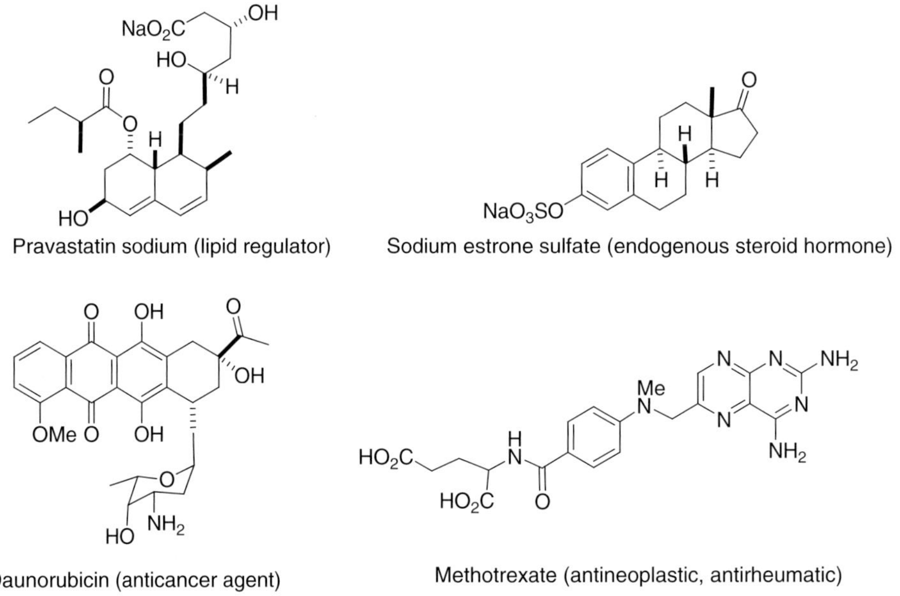

Pravastatin sodium (lipid regulator)

Sodium estrone sulfate (endogenous steroid hormone)

Daunorubicin (anticancer agent)

Methotrexate (antineoplastic, antirheumatic)

Figure 61 Chemical structures of additional endogenous and xenobiotic substances whose ADMET properties are modulated by transporters.

2.01.4.1.4.3 Drug delivery

Transporters such as the SLCs have been studied as drug delivery targets as well as therapeutic targets and targets for ADMET and drug resistance (**Figure 61**). For example, the lipid regulator pravastatin distributes selectively to the liver via SLCs, which reduces side effects in other parts of the body. P-gps can restrict the entry of a variety of endogenous substrates (e.g., estrone sulfate) or xenobiotics (e.g., daunorubicin and methotrexate) into the brain or other tissues.[414] Prodrugs have also been designed to take advantage of transporters involved in transporting amino acids, carboxylic acids, nucleosides and nucleoside bases, and peptides, among others.[415]

2.01.4.1.5 Genetic diseases

The high variability in psychopharmacologic drug responses makes pharmacogenetics in areas such as the study of drugs acting at serotonin or other catecholamine transporters complex. Nonetheless, pharmacogenetic studies on a number of drugs have yielded interesting results on phenotypes and response rates, including those with SSRIs such as fluoxetine and citalopram.[416] Indeed, drug development strategies for behavioral disorders in neurodegenerative diseases have

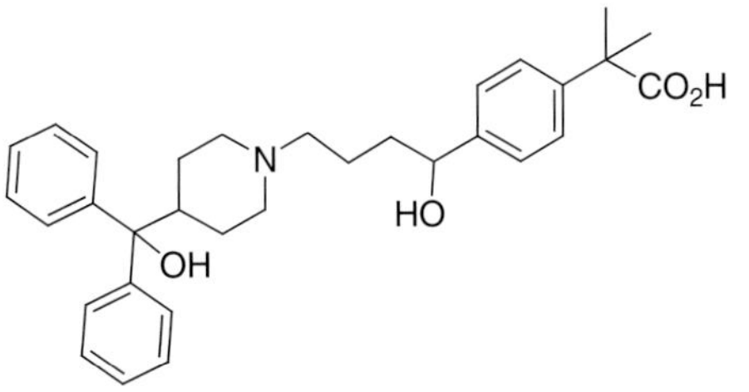

Figure 62 Chemical structure of fexofenadine, an antihistamine.

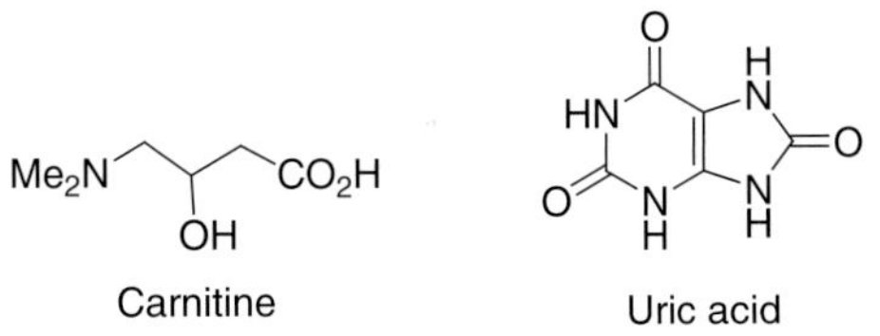

Figure 63 Chemical structures of carnitine and uric acid.

been discussed in the context of gene polymorphisms associated with several neuropsychiatric diseases or symptoms. For example, a 44 base pair insertion in *SERT* has been associated with psychosis and aggression.[417]

The primary site of action of SSRIs is now well known to be *SERT*.[418] As cited earlier, because patients with depression and related illnesses respond so variably to SSRIs, a number of studies have been undertaken to determine whether there is a genotypic explanation for the large differences in clinical response. Current information is not conclusive as a diagnostic tool, but *SERTPR s/s* and *STin2 10/12* genotypes in Caucasians and Asians, respectively, may be correlated with less favorable responses.[419,420] Meta-analyses of population- and family-based studies of *SERT* gene polymorphisms and bipolar disorder have concluded that there is a small but detectable effect of the transporter on the odds ratio of the disease.[421] Other studies have concluded that variants in the long promoter region of *SERT* (5-*HTTLPR*) are associated with transporter efficiency and cytokine and kinase gene expression relevant to treatment outcome, and with lithium treatment outcome in prophylaxis against mood disorders.[422,423] A broad review of the genes, genetics, and pharmacogenetics of *SERTs* has appeared.[424] Genetic or acquired deficiencies in another neurotransmitter transporter, NET, have been associated with orthostatic intolerance and tachycardia.[425]

Though much of the historical work on variations in drug metabolism has focused on CYP enzymes, more recent studies have shown that ABC transporters such as P-gps lead to altered distribution and bioavailability parameters, owing to functional genetic polymorphisms.[426] Because such a wide range of molecules are transported by P-gps, the effects of *MDR1* polymorphisms on disease risk, drug disposition, and drug efficacy have been studied extensively.[427] Single nucleotide polymorphisms (SNPs) in *ABCB1* have also been shown to affect fexofenadine (**Figure 62**) concentrations in patients.[428]

One in three epilepsy patients have drug-resistant epilepsy. It has been reported that specific ABC transporter polymorphisms (i.e., *ABCB1*, which is *MDR1* or *P-gp 170*), which lead to increased expression, may be either causal or linked with a causal variant of drug-resistant epilepsy.[429] In addition, mutations in the *ABCA3* gene cause surfactant deficiencies in newborns, which is fatal.[430] Defects in the SLC22 class of transporters causes systemic deficiencies (e.g., in carnitine or uric acid (**Figure 63**)) and changes in drug absorption and excretion.[380]

Statins and the effects of SNPs on hepatic uptake transporters have been studied, as mentioned above. However, how these SNPs contribute to drug disposition and therapeutic efficacy and safety indices remains to be clarified.[431]

2.01.4.1.6 New directions

The current state of the art in antidepressant therapy dates back in origin to the 1940s and 1950s, with the discovery of serotonin as 'enteramine,' and the finding that the antitubercular iproniazid improved the mood of sanitarium patients through a mechanism ultimately defined as inhibition of monoamine oxidase (MAO) (**Figure 64**). These serendipitous findings helped dramatically to deepen our understanding of monoamines, neurotransmitters, and CNS diseases such as depression.[432]

The history of this field has been summarized by several authors. While at most 50–70% of clinically depressed individuals find solace in monoamine therapies and other psychiatric approaches, many patients are poorly served. Thus, newer approaches that are non-monoamine in mechanism are being sought, with limited success to date,

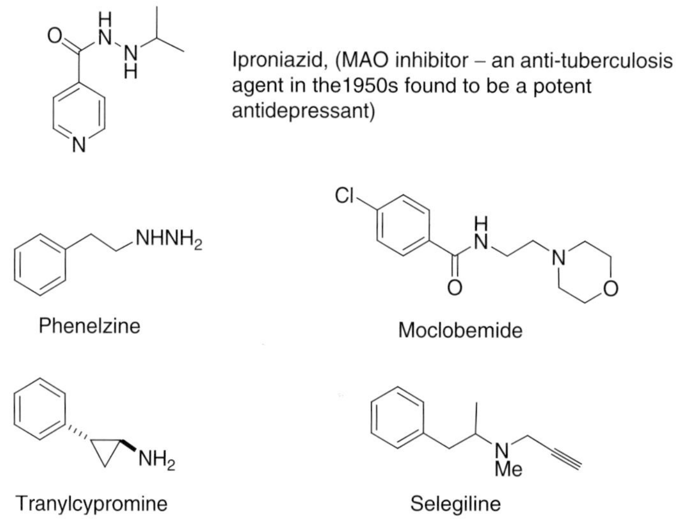

Figure 64 Chemical structures of iproniazid and other MAO inhibitors.

unfortunately. If one considers the first wave of antidepressant drug development as starting with classic pharmaceutical research and resulting in the tricyclic antidepressants, and the second wave starting with clinical observations and yielding SSRIs such as fluoxetine, then the current or third wave would start with pathophysiological, pharmacogenomic, or etiological models and will hopefully yield a new generation of therapies that extend beyond monoamine therapeutics. Representative new approaches are listed below[433,434]:

- circadian rhythm gene products;
- corticotropin-releasing factor;
- glucocorticoids;
- glutamatergic agents;
- hypothalamic feeding peptides;
- neurokinins;
- neurotrophic factors;
- phosphodiesterase inhibitors; and
- vasopressin receptor antagonists.

Finally for this discussion, with the worldwide increase in illicit drug use (e.g., more than 15 million US citizens over the age of 12 years are estimated to be illicit drug users), the need for addiction therapies continues to grow. Targeting monoamine transporters for such therapies has been the focus of a number of laboratories in recent years.[435]

2.01.4.2 General Discussion of Targets

Having considered transporters as drug targets, we would be remiss if we did not mention other major classes, such as G protein coupled receptors, ion channels, and so on, since drug discovery targets fall into various classes and families (many are receptors or enzymes). One analysis of drug targets concluded that there were 417 in total, based on drugs that had reached the marketplace, including enzymes, ion channels, receptors, and other targets. The largest percentage, about one-third, were targets associated with synaptic and neuroeffector junctions and the CNS.[436]

Some have tried to assess the genome from the standpoint of druggability (**Figure 65**). They estimate that only about 3000 druggable targets exist in the human genome. GPCRs and kinases figure prominently in this analysis. Considering only the currently marketed drugs, GPCRs account for almost one-third of this group (**Table 20**), with ion channels a distant second (7%). Another evaluation suggests that there are no more than 1500 exploitable drug targets based on the intersection of druggable targets and genes linked to disease.[437] Hopefully this is a vast underestimate.

The major drug discovery target classes are as follows:

- GPCRs;
- kinases and phosphatases;

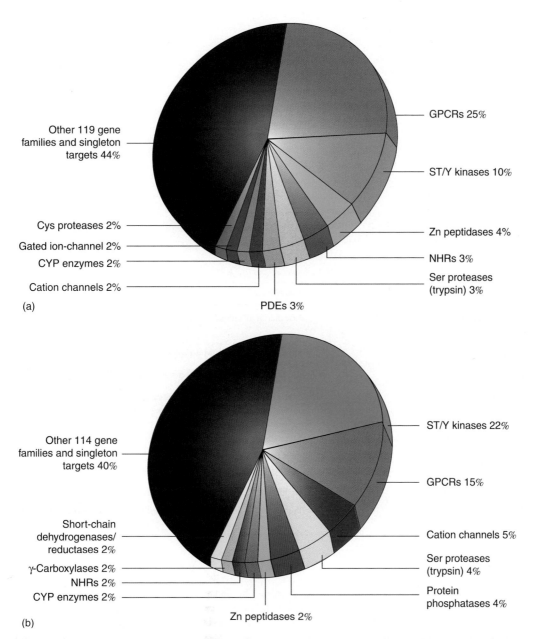

Figure 65 Drug discovery target classes and families, with reference to the druggable genome. Gene family distribution of (a) the molecular targets of current rule-of-five-compliant experimental and marketed drugs, and (b) the druggable genome. Serine/threonine and tyrosine protein kinases are grouped as one gene family (ST/Y kinases), as are class 1 and class 2 G protein-coupled receptors (GPCRs). CYP, cytochrome P450; Cys, cysteine; NHR, nuclear hormone receptor; PDE, phosphodiesterase; Zn, zinc. (Reprinted with permission from Hopkins, A. L.; Groom, C. R. *Nat. Rev. Drug Disc.* **2002**, *1*, 727–730. Copyright 2002 Nature Publishing Group.)

- ligand- and voltage-gated ion channels;
- nuclear hormone receptors;
- nucleic acids (DNA and RNA);
- phosphodiesterases;
- proteases;
- redox enzymes;
- second messenger metabolizing enzymes; and
- transporters (as described above).

Table 20 Selected examples of GPCRs

Adenosine	Interleukin
Adrenergic	Leukotriene
Angiotensin	Melanocortin
Bradykinin	Muscarinic
Calcitonin	Opiate
Cannabinoid	Prostanoid
Cholecystokinin	Thromboxane
Dopamine	Serotonin
Endothelin	Somatostatin
γ-Amino-butyric acid	Tachykinin
Galanin	Thrombin
Histamine	Vasopressin

Table 21 Key features of drug discovery targets to consider

Key features	*Description*
Current Nomenclature and Molecular Biology Classification	International unions, societies, etc.
Protein Structure	Protein sequences and 3-dimensional structure
Physiological Function	Role of target family in mediating physiological functions
Second Messengers	Role of biochemical intermediates (cAMP, inositol phosphates, etc.) in mediating physiology and their targets and metabolizing enzymes (cascades and pathways)
Prototypical Pharmacology	Major drugs and classes of drugs associated with target subtypes and the function of agonists, antagonists, and inverse agonists
Prototypical Therapeutics	Major therapeutic applications
Genetic Links to Diseases	Diseases associated with target and mutations
Future Directions	Problems and limitations associated with target, and possible new directions

Selecting the right target is more an art than a science. The process to select a target requires consideration of corporate, business, and R&D strategy, as well as portfolio balance. It must also consider the interplay of stakeholders, consumers, and the drug firm, with attendant technical, legal, economic, demographic, political, social, ecological, and infrastructure factors, in addition to scientific and medical arguments. Unfortunately, it is often the case that marketing projections used to prioritize drug opportunities are not anywhere close to actual sales – examples include tamoxifen (Nolvadex), captopril (Capoten), cimetidine (Tagamet), fluoxetine (Prozac), and atorvastatin. Thus, the research director must evaluate – and discount – all these inputs prior to making the judgment call.[438]

Various strategies have been proposed to improve decision making in preclinical studies, whether it be hit-to-lead generation or predicting how a compound will behave in humans from the standpoint of pharmacokinetics, metabolism, efficacy, and safety. Other strategies include decision gates regarding target validation, proof-of-concept, and druggability. Given that about 45% of currently marketed drugs target receptors, and 28% target enzymes, clearly these target classes are druggable. Regardless of the target, many approaches end up in the same place, focused on reducing attrition, optimizing multiple properties simultaneously, moving from hit to lead, and arguing about quality versus quantity in chemical libraries.[439] Some of the features to consider when evaluating drug discovery targets can be found in **Table 21**.

GPCRs, also known as 7TMs, 7-spanners, or serpentine receptors, are widely considered the most important class of drug targets, at least historically. However, kinases are gaining ground. Kinases are today implicated directly or indirectly in more than 400 human diseases.

Diseases with inflammatory or proliferative responses in which kinases may play a significant role include

- arthritis,
- asthma,
- cancer,
- cardiovascular disorders,
- neurological disorders, and
- psoriasis.

Kinase activity can be modulated by blocking adenosine triphosphate (ATP) binding, disrupting protein–protein interactions, or downregulating gene expression. These approaches have yielded several small molecule drugs and antibodies on the market or in development (**Figures 66** and **67**). The first major breakthrough in this area was a Novartis drug, imatinib, a protein-tyrosine kinase inhibitor that blocks ATP binding, discussed earlier in the chapter. Imatinib was first marketed in 2001 for chronic myeloid leukemia (CML), and crossed the billion dollar sales threshold in 2003. Unfortunately, clinical resistance has been observed in the targeted oncogene, the result of mutations. Other examples of kinase inhibitors developed for cancer include gefitinib and erlotinib (Tarceva). To date, clinical studies have focused on cancer – colorectal, nonsmall cell lung, pancreatic, renal – though arthritis, AD, diabetes, diabetic neuropathy, heart disease, inflammation, and Parkinson's disease are also under investigation.[440,441]

One of the givens with a new class of targets is that the early therapeutic indications may not be where the target ultimately proves useful. Consider the case of the endothelins (ET-1, ET-2, and ET-3) and their receptors (ET$_A$, ET$_B$, and ET$_C$). The first ET receptor ligand was identified in 1988 as a very potent, vascular endothelium-derived vasoconstrictor peptide, 21 residues in length.[443] Given the vasoactivity of this substance, many companies originally targeted cardiovascular disease, though the area was believed to represent a major opportunity for many diseases. The wide range of targets is not surprising given that ET receptors are widely distributed in mammalian tissues, and elevated levels of ET-1 have been observed in atherosclerosis, congestive heart failure, hypertension, myocardial infarction, pulmonary hypertension, and renal failure. However, the direction of R&D in this area has changed

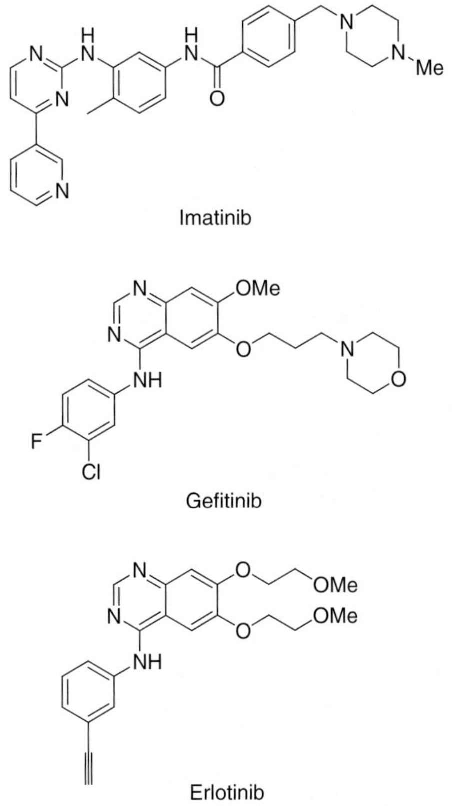

Figure 66 Chemical structures of the protein kinase inhibitors imatinib, gefitinib, and erlotinib.

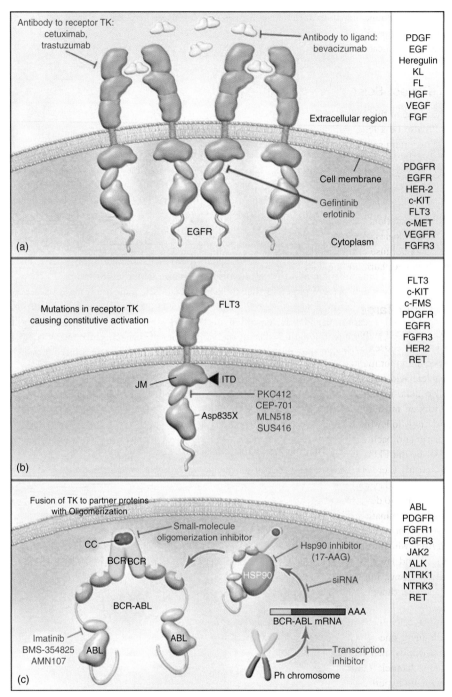

Figure 67 Mechanisms of tyrosine kinase function and drugs acting at various sites. (a) Overexpression of receptor TK or ligand; (b) mutations in receptor TK causing constitutive activation; and (c) fusion of TK to partner proteins with oligomerization. TK, tyrosine kinase; PDGF(R), platelet-derived growth factor (receptor); VEGF(R), vascular endothelial growth factor (receptor); EGF(R), epidermal growth factor (receptor); FGF(R), fibroblast growth factor (receptor); KL, KIT ligand; FL, FLT3 ligand; HGF(R), hepatocyte growth factor (receptor); HER-2, named after its gene *HER2/neu*, also known as *ErbB2* (named based on its similarity to avian erythroblastosis oncogene B), is the protein product of a protooncogene, originally named after human epidermal growth factor receptor, to which it is structurally related; RET, receptor tyrosine kinase; FLT, fms-like tyrosine kinase (fms, Feline Mcdonough Sarcoma); c-KIT, a member of the PDGFR family, c-KIT is a tyrosine kinase receptor that dimerizes following ligand binding and is autophosphorylated on intracellular tyrosine residues; c-FMS, a tyrosine kinase; c-MET, a tyrosine kinase; JM, juxtamembrane; ITD, internal tandem duplications; BCR, breakpoint cluster region; ABL, Abelson; HSP, heat shock protein; CC, coiled coil; JAK, Janus kinase; ALK, anaplastic lymphoma kinase; NTRK, a tyrosine kinase; ASP, aspartic acid; 17-AAG, 17-allylamino-17-demethoxygeldanamycin; Ph, Philadelphia. (Reprinted with permission from Krause, D. S.; Van Etten, R. A. *N. Engl. J. Med.* **2005**, *353*, 172–187. Copyright 2005 Massachusetts Medical Society. All rights reserved.[442])

significantly. Speedel announced in 2005 that its oral ET receptor antagonist (SPP301) successfully completed Phase II clinical trials in diabetic nephropathy,[444] and Abbott is seeking FDA approval for an ET receptor antagonist (atrasentan (Xinlay)) in oncology indications.[445]

2.01.4.3 Project Selection

It is beyond the scope of this introduction to go into detail on all of the techniques that researchers use to validate new targets. It would also be a long discussion, at times philosophical, to cover reasons to diversify one's research versus focusing efforts in certain areas. Small biotech companies face such decisions regularly. For example, a company named 7TM focuses on GPCRs, and as mentioned earlier, MitoKor focused on mitochondria. Other companies focus on therapeutic areas – Kosan and Telik being largely cancer based, for example, though each has its own technological focus as well. Still others focus on technologies or specific approaches, such as SGX (formerly known as Structural GenomiX) and its fragment and crystal SBD strategies. These examples notwithstanding, project selection – including the identification, validation, and prioritization of genes and targets emanating from genomics initiatives – ultimately draws on many technologies. While an enormous amount of data has been generated, we have barely started our journey to exploit these data because of the laborious and time-consuming process of validating new drug targets (**Table 22**)[446] and the many challenges surrounding data integration (**Figure 68**).[447]

2.01.4.4 Other Target Classes

In 1997, Drews and Ryser summarized what they termed classic drug targets.[448] These targets of marketed drugs were broken down into 11 classes comprising about 500 targets (**Table 23**). Receptor targets dominated these numbers, being twice as common for marketed drugs as enzyme targets, followed by factors and hormones, channels or unknown targets, and nucleic acid targets. Central nervous system (CNS) and related targets were the most common.

How have we done in various therapeutic and target areas? First, let us consider cancer. Modern chemotherapy started with nitrogen mustards in the 1940s. The NCI began systematic screening for new anticancer drugs in the 1950s. Chemotherapy following surgical removal of cancers was demonstrated to improve cure rates in the 1960s. Major drugs like cisplatin and paclitaxel were approved in the 1970s and 1990s. The tyrosine kinase inhibitor, imatinib, was approved in 2001 by the FDA, representing a new target-based paradigm in the field. Clearly a lot of progress has been made – but also clearly there is a long way to go to achieve the National Cancer Institute's goals.[449] If you are diagnosed with a serious cancer today, and a cure is discovered for this cancer today, you will probably still die of this cancer because it takes so long to develop new drugs, even for cancer. The war on cancer is thus far from won, but the NCI has stated as its goal to eliminate suffering and death from cancer by 2015.[450]

Moving beyond cancer, new targets in the 1980s included phosphodiesterases (PDE). Who could have predicted that PDE inhibitors, which were studied extensively in the 1980s for heart failure (e.g., inhibitors of type 3 cAMP PDE, like amrinone),[451] and later as antidepressants (e.g., rolipram),[452] would ultimately yield billion dollar products for erectile dysfunction (e.g., inhibitors of type 5 cGMP PDE, such as sildenafil (Viagra)).[453,454] This shift in emphasis raises other questions, though. A society that shifts its attention to "desires rather than diseases" is a society gone off track.[455]

The history of calcium antagonists provides an interesting case study in the development of a whole new therapeutic paradigm, one which continues to evolve as new compounds, new targets, and new medical uses are developed. Compounds such as verapamil, nifedipine, and diltiazem showed how widely divergent chemicals could all target similar ion channels (**Figure 69**).[456] Ion channels continue to be of great interest as drug targets, beyond traditional therapy for hypertension, anesthesia, anxiety, epilepsy, and insomnia, as new opportunities are uncovered for this well-established therapeutic class in pain, autoimmune disease, cancer, and other areas.[457]

Table 22 Functional genomics technologies leading to new drug targets

Gene-family mining and comparative genomics	Identity and similarity to known gene families
RNA profiling	High-density expression arrays
Proteomics	Gel electrophoresis and mass spectrometry
Oligonucleotides	Knockdowns, RNAi, microRNAs
Systematic analyses	Analysis of gene function in cell culture, cDNAs
Model organisms	Drosophila, zebrafish

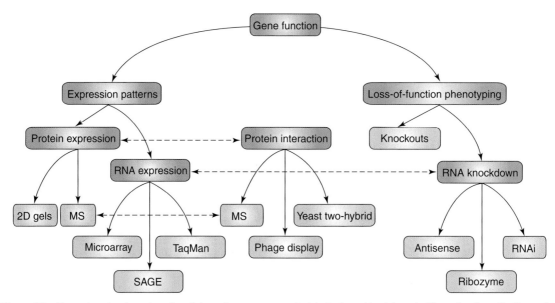

Figure 68 Examples of various domains of drug discovery research data that must be integrated for optimal results. Examples of scientific data domains (by no means exhaustive) relating to gene function, arranged in a taxonomy to indicate their natural nesting from general themes down to specific data-generating technologies (in pink). Such a hierarchy could continue upwards to culminate in an amalgamation of the major domains of biology, chemistry, and medicine. Integration of heterogeneous domains can occur vertically from the bottom up in an initiative to provide comprehensive data access or consolidated data collections; it can also occur horizontally between domains, such as RNA expression and protein interaction, in specific integrative 'experiments.' The challenges of integration actually begin with ostensibly homogeneous data, such as various readouts of expression, and even in combining results of several experiments based on the same technology. The dashed arrows connect domains dealing with the same biological entities (proteins and RNA) or with instances of the same method (mass spectrometry), each of which should be related by means of common data standards for naming and formatting so as to ease integration. MS, mass spectrometry; SAGE, serial analysis of gene expression; RNAi, RNA interference. (Reprinted with permission from Searls, D. B. *Nat. Rev. Drug Disc.* **2005**, *4*, 45–58. Copyright 2005 Nature Publishing Group.)

Table 23 Therapeutic targets of marketed drugs

Number	Therapeutic or functional area
137	Synaptic, neuroeffector, or CNS
55	Hormones and their antagonists
50	Inflammation
47	Renal and cardiovascular
39	Neoplastic diseases
38	Blood and bloodforming organs
19	Infectious disease
16	Immunomodulatory
15	Gastrointestinal functions
10	Vitamins
8	Uterine motility

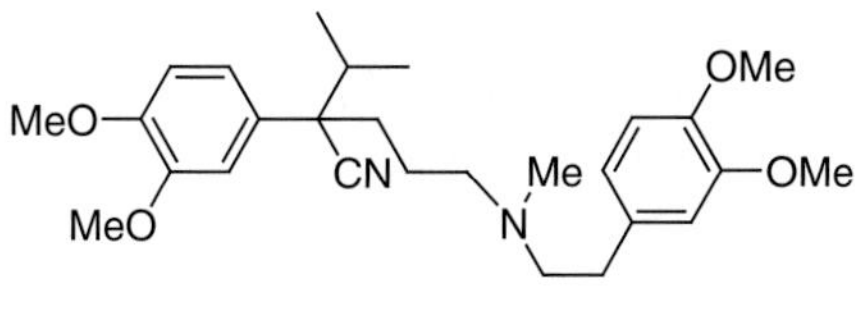

Verapamil

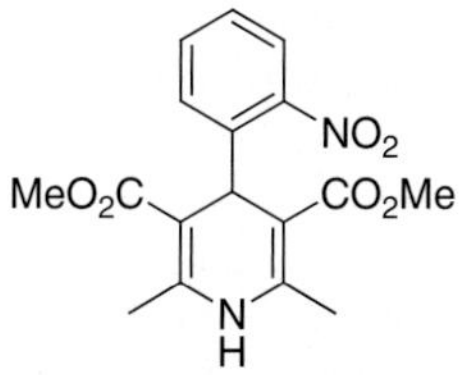

Nifedipine

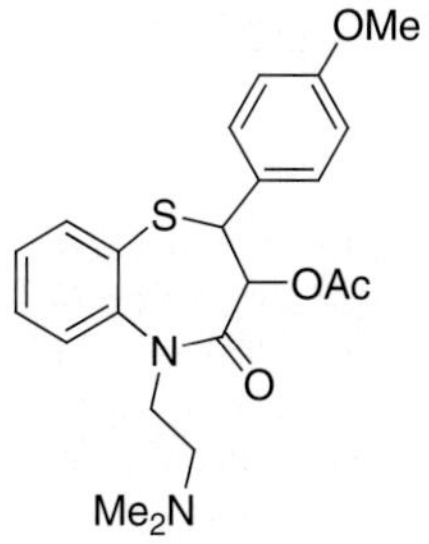

Diltiazem

Figure 69 The diverse nature of calcium channel blocker chemical structures.

Table 24 Properties of drug candidates to consider when contemplating full-scale development

Chemistry	*Pharmacology*	*Pharmacokinetics and drug metabolism*	*Preclinical safety*
Novel chemical structure	Validated target	Measurable plasma levels	Low pharmacologic risk (cardiovascular, CNS, gastrointestinal, renal, respiratory)
Strong patent position	Mechanism of action linked to pathophysiology of disease	High oral bioavailability	Negative in mutagenicity tests
Inexpensive synthesis possible on large scale	Potent	Suitable half-life	High therapeutic index based on toxicology studies
Isomers separated (and compared)	Selective versus other targets	Low first pass metabolism	
High purity	Reproducible efficacy in predictive animal model	Limited induction of hepatic enzymes	
Salt form studied and selected	Clear dose response	Limited interaction with other drugs	
	Suitable duration of action		
	Active after repeated dosing (minimal tolerance)		
	Advantages versus competitive agents		

While many companies have exited research on agents for infectious disease, new biodefense initiatives are bringing such work back into vogue.[458] This renewed interest may yield drug candidates of use both for terrorist attacks and for the more traditional community-acquired infections, bacterial, fungal, or viral, including flu pandemics and the like.

Though many of the new technologies and ways of thinking about molecular targets, druggability, and ADMET are indeed worthwhile advances, one must not lose sight of the principles of classical medicinal chemistry. Indeed, the belief that pharmacokinetics is the major reason for drug development failures is not borne out by the data. Whereas pharmacokinetics was the reason for 39% of drug development failures in a survey of large UK companies from 1964 to 1985, this number drops to 7% if one excludes anti-infectives from the list. Thus, decades of medicinal chemistry experience must be merged with the new technologies to achieve the desired levels of success.[459] And even the simplest screening methods can be useful, such as using silica gel plates with targets such as cholinesterase.[460] In the end, a variety of properties must be evaluated the old-fashioned way before advancing a drug into development, whether the target is a good one or a bad one. In closing, **Table 24** serves as a reminder of what properties matter in such decisions, including target validation, mechanism of action, dose response, bioavailability, and other pharmaceutical properties.

References

1. Smith, C. G. *The Process of New Drug Discovery and Development*; CRC Press: Boca Raton, FL, 1992.
2. Silverman, R. B. *The Organic Chemistry of Drug Design and Drug Action*, 2nd ed.; Academic Press: San Diego, CA, 2004.
3. PhRMA. Medicines in Development for Neurologic Disorders, 2006. See www.phrma.org (accessed July 2006).
4. Mathieu, M. P., Ed. *PAREXEL's Pharmaceutical R&D Statistical Sourcebook 2004/2005*, 10th ed.; PAREXEL: Waltham, MA, 2004.
5. DiMasi, J. A.; Hansen, R. W.; Grabowski, H. G. *J. Health Econ.* **2003**, *22*, 151–185.
6. Vagelos, P. R. *Science* **1991**, *252*, 1080–1084.
7. Gregson, N.; Sparrowhawk, K.; Mauskopf, J.; Paul, J. *Nat. Rev. Drug Disc.* **2005**, *4*, 121–130.
8. Scherer, F. M. *N. Engl. J. Med.* **2004**, *351*, 927–932.
9. Goozner, M. *The $800 Million Pill*; University of California Press: Berkeley, CA, 2004.
10. Scherer, F. M. *N. Engl. J. Med.* **2004**, *351*, 927–932.
11. DiMasi, J. A. *Clin. Pharmacol. Ther.* **2001**, *69*, 286–296.
12. DiMasi, J. A. *Clin. Pharmacol. Ther.* **2001**, *69*, 297–307.
13. Reichert, J. M. *Nat. Rev. Drug Disc.* **2003**, *2*, 695–702.
14. Berndt, E. R.; Gottschalk, A. H. B.; Philipson, T. J.; Strobeck, M. W. *Nat. Rev. Drug Disc.* **2005**, *4*, 545–554.
15. Frantz, S. *Nat. Rev. Drug Disc.* **2005**, *4*, 93–94.
16. Frei, P.; Leleux, B. *Nat. Biotechnol.* **2004**, *22*, 1049–1051.
17. Preziosi, P. *Nat. Rev. Drug Disc.* **2004**, *3*, 521–526.
18. Moos, W. H.; Feldbaum, C. B. *Drug Dev. Res.* **2002**, *57*, 45–50.
19. Aronson, D.; Best, S. G.; Werner, M. J.; Moos, W. H. *Drug Dev. Res.* **2003**, *63*, 89–92.
20. Bains, W. *Drug Disc. World* **2004**, *5*, 9–18.
21. Daemmrich, A. A.; Bowden, M. E. *Chem. Eng. News* **2005**, *83*, 28–136.
22. Bardhan, K. D.; Nayyar, A. K.; Royston, C. *Dig. Liver Dis.* **2003**, *35*, 529–536.
23. Olbe, L.; Carlsson, E.; Lindberg, P. *Nat. Rev. Drug Disc.* **2003**, *2*, 132–139.
24. Berkowitz, B. A.; Sachs, G. *Mol. Interventions* **2002**, *2*, 6–11.
25. Capdeville, R.; Buchdunger, E.; Zimmermann, J.; Matter, A. *Nat. Rev. Drug Disc.* **2002**, *1*, 493–502.
26. Vasella, D.; Slater, R. *Magic Cancer Bullet. How a Tiny Orange Pill is Rewriting Medical History*; Harper Collins: New York, 2003.
27. Class, S. *Chem. Eng. News* **2004**, *82*, 18–29.
28. See http://www.rxlist.com (accessed July 2006).
29. Kneller, R. *Nat. Biotechnol.* **2005**, *23*, 529–530.
30. Mossinghoff, G. J. *Food Drug Law J.* **1999**, *54*, 187–194.
31. Glass, G. *Nat. Rev. Drug Disc.* **2004**, *3*, 1057–1062.
32. Kennedy, D. *Science* **2005**, *307*, 1375.
33. Hughes, S. S. *Isis* **2001**, *92*, 541–575.
34. Webber, P. M. *Nat. Rev. Drug Disc.* **2003**, *2*, 823–830.
35. Lawrence, S. *Nat. Biotechnol.* **2005**, *23*, 164.
36. Hardman, J. G.; Limbird, L. E.; Gilman, A. G.; *Goodman and Gilman's The Pharmacological Basis of Therapeutics*, 10th ed.; McGraw-Hill: New York, 2001.
37. Lombardino, J. G.; Lowe, J. A. *Nat. Rev. Drug Disc.* **2004**, *3*, 853–862.
38. Oberholzer-Gee, F.; Inamdar, S. N. *N. Engl. J. Med.* **2004**, *251*, 2147–2149.
39. Okie, S. *N. Engl. J. Med.* **2005**, *352*, 1063–1065.
40. Mathews, A. W.; Hechinger, J. *Wall Street Journal*, March **2005**, *1*, B1.
41. Breckenridge, A.; Woods, K.; Raine, J. *Nat. Rev. Drug Disc.* **2005**, *4*, 541–543.
42. Schena, M.; Greene, W. *Drug Disc. World* **2005**, *6*, 9–18.
43. Naylor, S. *Drug Disc. World* **2005**, *6*, 21–30.
44. Daar, A. S.; Singer, P. A. *Nat. Rev. Genetics* **2005**, *6*, 241–246.
45. Boseley, S.; Gould, M. *The Guardian*, March 1, 2005 http://www.guardian.co.uk/1427534,00.html (accessed July 2006).
46. DeGeorge, J. J.; Ahn, C.-H.; Andrews, P. A.; Brower, M. E.; Giorgio, D. W.; Goheer, M. A.; Lee-Ham, D. Y.; McGuinn, W. D.; Schmidt, W.; Sun, C. J. et al. *Cancer Chemother. Pharmacol.* **1998**, *41*, 173–185.

47. See http://www.fda.gov/orphan/oda.htm (accessed July 2006).
48. Kaufman, M. *Washington Post*, 26 Feb, 2005, A03.
49. Angell, M. *The Truth About the Drug Companies. How They Deceive Us and What to Do About It*; Random House: New York, 2004.
50. Campion, E. W. *N. Engl. J. Med.* **2004**, *351*, 2436–2437.
51. Moos, W. H.; Weisbach, J. A. *Pharm. News* **2000**, 7, 6–7.
52. Schwitzer, G.; Mudur, G.; Henry, D.; Wilson, A.; Goozner, M.; Simbra, M.; Sweet, M.; Baverstock, K. A. *PLoS Med.* **2005**, *2*, 576–582.
53. FDA White Paper *Innovation or Stagnation Challenge and Opportunity on the Critical Path to New Medical Products*; US Department of Health and Human Services: Washington, DC, 2004, pp 1–31.
54. Kola, I.; Landis, J. *Nat. Rev. Drug Disc.* **2004**, *3*, 711–715.
55. Kennedy, T. *Drug Disc. Technols.* **1997**, *2*, 436–444.
56. Williams, D. A.; Lemke, T. L.; Foye, W. O. *Foye's Principles of Medicinal Chemistry*, 5th ed.; Lippincott Williams and Wilkins: Philadelphia, PA, 2002.
57. Abraham, D. J., Ed. *Burger's Medicinal Chemistry and Drug Discovery*, 6th ed.; Wiley-Interscience: Hoboken, NJ, 2003.
58. Sweetman, S. C., Ed. *Martindale: The Complete Drug Reference*, 34th ed.; Pharmaceutical Press: London, UK, 2004.
59. *Physician's Desk Reference*, 53rd ed.; Medical Economics Co.: Montvale, NJ, 1999.
60. White, V. P. *Handbook of Research Laboratory Management*; ISI Press: Philadelphia, PA, 1988.
61. Sapienza, A. M.; Lombardino, J. G. *Drug Dev. Res.* **2005**, *64*, 99–104.
62. Sapienza, A. M. *Managing Scientists: Leadership Strategies in Biomedical Research and Development*; Wiley-Liss: New York, 1995.
63. Lewis, R. A. Personal communication, 2005.
64. Simmerman, S. B. *Drug Dev. Res.* **2002**, *57*, 103–105.
65. Wareham, J. *CHEMTECH* **1982**, 396–402.
66. Dougherty, D. E. *From Technical Professional to Corporate Manager*; Wiley-Interscience: New York, 1984.
67. Boyer, F. *Thoughts on Business Policies*; Smith Kline and French Laboratories: Philadelphia, PA, 1975.
69. Ganellin, C. R.; Mitscher, L. A.; Topliss, J. G. *Med. Res. Rev.* **1998**, *18*, 121–137.
70. Moos, W. H. *Pharm. News* **1999**, *6*, 37.
71. Sapienza, A. M.; Lombardino, J. G. *Drug Dev. Res.* **2002**, *57*, 51–57.
72. Piskora, B. *The Scientist* **2004**, *18*.
73. Brenner, L. *Parade*, 13 March, 2005, pp 4–10.
74. Brenner, L. *Parade*, 14 March, 2004, pp 4–10.
75. Ovide, S. *Wall Street Journal*, 9 March, 2005.
76. Piskora, B. *The Scientist* **2004**, *18*, 48.
77. Lewis, M. *Moneyball: The Art of Winning an Unfair Game*; W. W. Norton and Company: New York, 2003.
78. Anonymous. Time: Fall Style Design Supplement, 2004, p 16.
79. Brumfiel, G. *Nature* **2004**, *430*, 957.
80. Caporizzo, A. W., Johnson, D. G., Ohrnberger, J. M., Vicidomino, J., Pearce, B., Buckley, S., Holodnak, B. *Compensation and Entrepreneurship Report in Life Sciences*, Wilmer Cutler Pickering Hale and Dorr, Ernst and Young, J. Robert Scott: Boston, MA, 2004.
81. See http://www.drugresearcher.com/news/news-ng.asp?id=11048-gsk-sets-up (accessed Sept 2006).
82. Warner, S. *The Scientist* **2005**, *19*, 36.
83. Mullin, R. *Chem. Eng. News* **2005**, *83*, 13.
84. Drews, J.; Ryers, S. *Drug Disc. Today* **1997**, *2*, 365–372.
85. Weisbach, J. A.; Moos, W. H. *Drug Dev. Res.* **1995**, *34*, 243–259.
86. Entzeroth, M. *Drug Dev. Res.* **2004**, *62*, 293–294.
87. Buss, A. D.; Butler, M. S. *Drug Dev. Res.* **2004**, *62*, 362–370.
88. Mardis, W. E.; Aibel, J.B. Biotechnology Management Practices 2003; Mardis Aibel & Associates: New York, 2003; Vol. 1.
89. Cohen, F. J. *Nat. Rev. Drug Disc.* **2005**, *4*, 78–84.
90. See http://www.quotationspage.com/quotes/Arthur_Schopenhauer/ (accessed July 2006).
91. Moos, W. H.; Davis, R. E. *Pharm. News* **1998**, *5*, 38–39.
92. Howell, N. *Drug Dev. Res.* **2002**, *57*, 75–82.
93. Root-Bernstein, R. S. *CHEMTECH* **1994**, May, 15–20.
94. Cuatrecasas, P. Corporate America: Creativity Held Hostage. In *Creative Actions in Organizations: Ivory Tower Visions and Real World Voices*; Ford, C. M., Gioia, D. A.. Eds.; Sage: Thousand Oaks, CA, 1995, pp 201–205.
95. Root-Bernstein, R. S. *Res. Technol. Management* **1989**, *32*, 36–41.
96. Root-Bernstein, R.; Dillon, P. F. *Drug Dev. Res.* **2002**, *57*, 58–74.
97. Moos, W. H.; O'Connell, S. K.; Sanders, T. G.; Deane, R. et al. Unpublished work (terms culled from the personal experience of several colleagues since the 1980s).
98. Fersko, R. S.; Garner, E. S. *Drug Dev. Res.* **2002**, *57*, 83–96.
99. Cunningham, B. C. *Drug Dev. Res.* **2002**, *57*, 97–102.
100. Hofer, M. A.; Leonardo, M. S.; Sorell, P. B. *J. Biolaw Bus.* **2001**, Special Suppl, 85–108.
101. Mathieu, M., Ed. *New Drug Development: A Regulatory Overview*, 3rd ed.; PAREXEL: Waltham, MA, 1994.
102. Mathieu, M., Ed. *Biologics Development: A Regulatory Overview*; PAREXEL: Waltham, MA, 1993.
103. Drayer, D. *Clin. Pharmacol. Ther.* **1987**, *42*, 364.
104. Torchilin, V. P. *Nat. Rev. Drug Disc.* **2005**, *4*, 145–160.
105. Rabinow, B. E. *Nat. Rev. Drug Disc.* **2004**, *3*, 785–796.
106. Stahl, P. H.; Wermuth, C. G. *Handbook of Pharmaceutical Salts: Properties, Selection, and Use*; Wiley-VCH: Weinheim, Germany, 2002.
107. Gould, P. L. *Int. J. Pharmaceutics* **1986**, *33*, 201–217.
108. Berge, S. M.; Bighley, L. D.; Monkhouse, D. C. *J. Pharm. Sci.* **1977**, *66*, 1–19.
109. Fu, R. C.; Lidgate, D. M.; Whatley, J. L. *J. Parenteral Sci. Technol.* **1987**, *41*, 164–168.
110. Fort, F. L.; Heyman, I. A.; Kesterson, J. W. *J. Parenteral Sci. Technol.* **1984**, *38*, 82–87.
111. Reed, K. W.; Yalkowsky, S. H. *J. Parenteral Sci. Technol.* **1987**, *41*, 37–39.
112. Spiegel, A. J.; Noseworthy, M. M. *J. Pharm. Sci.* **1963**, *52*, 917–927.

113. Federsel, H.-J. *Nat. Rev. Drug Disc.* **2003**, *2*, 654–664.
114. Freemantle, M. *Chem. Eng. News* **2004**, *82*, 33–35.
115. Mickel, S. J. *Curr. Opin. Drug Disc. Dev.* **2004**, 7, 869–881.
116. Dixit, R. *Am. Pharm. Outsourcing* **2004**, *March/April*, 1–8.
117. DeGeorge, J. J.; Meyers, L. L.; Takahashi, M.; Contrera, J. F. *Toxicol. Sci.* **1999**, *49*, 143–155.
118. Contrera, J. F.; Jacobs, A. C.; DeGeorge, J. J. *Regul. Toxicol. Pharmacol.* **1997**, *25*, 130–145.
119. DeGeorge, J. J.; Ahn, C. H.; Andrews, P. A.; Brower, M. E.; Choi, Y. S.; Chun, M. Y.; Du, T.; Lee-Ham, D. Y.; McGuinn, W. D.; Pei, L. et al. *Regul. Toxicol. Pharmacol.* **1997**, *25*, 189–193.
120. Clark, D. L.; Andrews, P. A.; Smith, D. D.; DeGeorge, J. J.; Justice, R. L.; Beitz, J. G. *Clin. Cancer Res.* **1999**, *5*, 1161–1167.
121. Lappin, G.; Garner, R. C. *Nat. Rev. Drug Disc.* **2003**, *2*, 233–240.
122. Morrow, G. R.; Ballatori, E.; Groshen, S.; Oliver, I. *Support Care Cancer* **1998**, *6*, 261–265.
123. Landow, L. *Anesthesiology* **2000**, *92*, 1814–1820.
124. Whitehead, J. *Nat. Rev. Drug Disc.* **2004**, *3*, 973–977.
125. Nottage, M.; Siu, L. L. *J. Clin. Oncol.* **2002**, *20*, 42s–46s.
126. Lachenbruch, P. A.; Rosenberg, A. S.; Bonvini, E.; Cavaillé-Coll, M. W.; Colvin, R. B. *Am. J. Transplant.* **2004**, *4*, 451–457.
127. Fox, E.; Curt, G. A.; Balis, F. M. *Oncologist* **2002**, *7*, 401–409.
128. Deplanque, G.; Harris, A. L. *Eur. J. Cancer* **2000**, *36*, 1713–1724.
129. Leber, P. D. Clinical Trial Issues: The Example of Dementia. In *Cognitive Disorders: Pathophysiology and Treatment*; Thal, L. J., Moos, W. H., Gamzu, E. R., Eds.; Marcel Dekker: New York, 1992, pp 355–378.
130. Summers, W. K.; Majovski, L. V.; Marsh, G. M.; Tachiki, K.; Kling, A. *N. Engl. J. Med.* **1986**, *315*, 1241–1245.
131. Albert, A.; Gledhill, W. *J. Soc. Chem. Industry* **1945**, *64*, 169–172.
132. Clark, C. R.; Moos, W. H. *Drug Discovery Technologies*; Ellis Horwood: Chichester, UK, 1990.
133. Brimblecombe, R. W.; Duncan, W. A.; Durant, G. J.; Emmett, J. C.; Ganellin, C. R.; Leslie, G. B.; Parsons, M. E. *Gastroenterology* **1978**, *74*, 339–347.
134. Duncan, W. A.; Parsons, M. E. *Gastroenterology* **1980**, *78*, 620–625.
135. Olbe, L.; Carlsson, E.; Lindberg, P. *Nat. Rev. Drug Disc.* **2003**, *2*, 132–139.
136. Moos, W. H.; Ghosh, S. S. *Pharm. News* **2000**, *7*, 56–59.
137. Spilker, B.; Cuatrecasas, P. *Inside the Drug Industry*; Prous: Barcelona, Spain, 1990.
138. Sneader, W. *Drug Prototypes and Their Exploitation*; John Wiley: New York, 1996.
139. Wermuth, C. G. *J. Med. Chem.* **2004**, *47*, 1303–1314.
140. Clardy, J.; Walsh, C. *Nature* **2004**, *432*, 829–837.
141. Koehn, F. E.; Carter, G. T. *Nat. Rev. Drug Disc.* **2005**, *4*, 206–220.
142. Bleicher, K. H.; Böhm, H.-J.; Müller, K.; Alanine, A. I. *Nat. Rev. Drug Disc.* **2003**, *2*, 369–378.
143. Ashburn, T. T.; Thor, K. B. *Nat. Rev. Drug Disc.* **2004**, *3*, 673–683.
144. Dobson, C. M. *Nature* **2004**, *432*, 824–828.
145. Gordon, E. M.; Kerwin, J. F., Jr.; personal communication, 1998.
146. Mei, H.-Y.; Czarnik, A. W.; *Integrated Drug Discovery Technologies*; Marcel Dekker: New York, 2002.
147. Merrifield, R. B. *J. Am. Chem. Soc.* **1963**, *85*, 2149–2154.
148. Moos, W. H. *Drug Dev. Res.* **1994**, *33*, 63.
149. Hogan, J. C., Jr. *Nature* **1996**, *384*, 17–19.
150. Gallop, M. A.; Barrett, R. W.; Dower, W. J.; Fodor, S. P. A.; Gordon, E. M. *J. Med. Chem.* **1994**, *37*, 1233–1251.
151. Gallop, M. A.; Barrett, R. W.; Dower, W. J.; Fodor, S. P. A.; Gordon, E. M. *J. Med. Chem.* **1994**, *37*, 1385–1401.
152. Czarnik, A. W.; DeWitt, S. H.; *A Practical Guide to Combinatorial Chemistry*; American Chemical Society: Washington, DC, 1997.
153. Gordon, E. M.; Kerwin, J. F.; *Combinatorial Chemistry and Molecular Diversity in Drug Discovery*; Wiley-Liss: New York, 1998.
154. Moos, W. H.; Pavia, M. R.; Kay, B. K.; Ellington, A. D., Eds., *Annual Reports in Combinatorial Chemistry and Molecular Diversity*; ESCOM: Leiden, the Netherlands, 1997; Vol. 1.
155. Pavia, M. R.; Moos, W. H., Eds., *Annual Reports in Combinatorial Chemistry and Molecular Diversity*; Kluwer: Dordrecht, the Netherlands, 1999; Vol. 2.
156. Furka, A. *Drug Dev. Res.* **1995**, *36*, 1–12.
157. Geysen, H. M.; Meloen, R. H.; Barteling, S. J. *Proc. Natl. Acad. Sci. USA* **1984**, *81*, 3998–4002.
158. Houghten, R. A. *Proc. Natl. Acad. Sci. USA* **1985**, *82*, 5131–5135.
159. Matteucci, M. D.; Wagner, R. W. *Nature* **1996**, *384*, 20–22.
160. Lebl, M. *J. Comb. Chem.* **1999**, *1*, 3–24.
161. Bunin, B. A.; Plunkett, M. J.; Ellman, J. A. *Proc. Natl. Acad. Sci. USA* **1994**, *91*, 4708–4712.
162. Breaker, R. R. *Nature* **2004**, *432*, 838–845.
163. Stein, C. A.; Cheng, Y.-C. *Science* **1993**, *261*, 1004–1012.
164. Cook, P. D. *Anti-Cancer Drug Des.* **1991**, *6*, 585–607.
165. Geiser, T. *Ann. NY Acad. Sci.* **1990**, *616*, 173–183.
166. Crooke, S. T. *Annu. Rev. Pharmacol. Toxicol.* **1992**, *32*, 329–376.
167. Kay, J.; Dunn, B. M. *Biochim. Biophys. Acta* **1990**, *1048*, 1–18.
168. Farmer, P. S.; Ariëns, E. J. *Trends Pharmacol. Sci.* **1982**, *3*, 362–365.
169. Hökfelt, T. *Neuron* **1991**, *7*, 867–879.
170. Plattner, J. J.; Norbeck, D. W. Obstacles to Drug Development from Peptide Leads. In *Drug Discovery Technologies*; Clark, C. R., Moos, W. H., Eds.; Ellis Horwood: Chichester, UK, 1990, pp 92–126.
171. Simon, R. J.; Bartlett, P. A.; Santi, D. V. Modified Peptide and Peptide Libraries with Protease Resistance, Derivatives Thereof and Methods of Producing and Screening Such. U.S. Patent 5,965,695, Oct 12, 1999.
172. Simon, R. J.; Bartlett, P. A.; Santi, D. V. Modified Peptide and Peptide Libraries with Protease Resistance, Derivatives Thereof and Methods of Producing and Screening Such. U.S. Patent 6,075,121, June 13, 2000.
173. Appella, D. H.; Christianson, L. A.; Klein, D. A.; Powell, D. R.; Huang, X.; Barchi, J. J.; Gellman, S. H. *Nature* **1997**, *387*, 381–384.
174. Moran, E. J.; Wilson, T. E.; Cho, C. Y.; Cherry, S. R.; Schultz, P. G. *Biopolymers (Peptide Science)* **1995**, *37*, 213–219.

175. Hagihara, M.; Anthony, N. J.; Stout, T. J.; Clardy, J.; Schreiber, S. L. *J. Am. Chem. Soc.* **1992**, *114*, 6568–6570.
176. Simon, R. J.; Kania, R. S.; Zuckermann, R. N.; Huebner, V. D.; Jewell, D. A.; Banville, S.; Ng, S.; Wang, L.; Rosenberg, S.; Marlowe, C. K. et al. *Proc. Natl. Acad. Sci. USA* **1992**, *89*, 9367–9371.
177. Miller, S. M.; Simon, R. J.; Ng, S.; Zuckermann, R. N.; Kerr, J. M.; Moos, W. H. *Bioorg. Med. Chem. Lett.* **1994**, *4*, 2657–2662.
178. Miller, S. M.; Simon, R. J.; Ng, S.; Zuckermann, R. N.; Kerr, J. M.; Moos, W. H. *Drug Dev. Res.* **1995**, *35*, 20–32.
179. Zuckermann, R. N.; Kerr, J. M.; Kent, S. B. H.; Moos, W. H. *J. Am. Chem. Soc.* **1992**, *114*, 10646–10647.
180. Zuckermann, R. N.; Martin, E. J.; Spellmeyer, D. C.; Stauber, G. B.; Shoemaker, K. R.; Kerr, J. M.; Figliozzi, G. M.; Goff, D. A.; Siani, M. A.; Simon, R. J. et al., *J. Med. Chem.* **1994**, *37*, 2678–2685.
181. Gibbons, J. A.; Hancock, A. A.; Vitt, C. R.; Knepper, S.; Buckner, S. A.; Brune, M. E.; Milicic, I.; Kerwin, J. F.; Richter, L. S.; Taylor, E. W. et al. *J. Pharmacol. Exp. Ther.* **1996**, *277*, 885–899.
182. Goodson, B.; Ehrhardt, A.; Ng, S.; Nuss, J.; Johnson, K.; Giedlin, M.; Yamamoto, R.; Moos, W. H.; Krebber, A.; Ladner, M. et al. *Antimicrob. Agents Chemother.* **1999**, *43*, 1429–1434.
183. Ng, S.; Goodson, B.; Ehrhardt, A.; Moos, W. H.; Siani, M.; Winter, J. *Bioorg. Med. Chem.* **1999**, *7*, 1781–1785.
184. Kirshenbaum, K.; Barron, A. E.; Goldsmith, R. A.; Armand, P.; Bradley, E. K.; Truong, K. T. V.; Dill, K. A.; Cohen, F. E.; Zuckermann, R. N. *Proc. Natl. Acad. Sci. USA* **1998**, *95*, 4303–4308.
185. Burkoth, T. S.; Beausoleil, E.; Kaur, S.; Tang, D.; Cohen, F. E.; Zuckermann, R. N. *Chem. Biol.* **2002**, *9*, 647–654.
186. Horn, T.; Lee, B.-C.; Dill, K. A.; Zuckermann, R. N. *Bioconj. Chem.* **2004**, *15*, 428–435.
187. Lobo, B. A.; Vetro, J. A.; Suich, D. M.; Zuckermann, R. N.; Middaugh, C. R. *J. Pharm. Sci.* **2003**, *92*, 1905–1918.
188. Geysen, H. M.; Schoenen, F.; Wagner, D.; Wagner, R. *Nat. Rev. Drug Disc.* **2003**, *2*, 222–230.
189. MacDonald, A. A.; Nickell, D. G.; DeWitt, S. H. *Pharm. News* **1996**, *3*, 19–21.
190. Schreiber, S. L. *Science* **2000**, *287*, 1964–1969.
191. Walters, W. P.; Namchuk, M. *Nat. Rev. Drug Disc.* **2003**, *2*, 259–266.
192. Kaur, S.; McGuire, L.; Tang, D.; Dollinger, G.; Huebner, V. *J. Prot. Chem.* **1997**, *16*, 505–511.
193. Blom, K. F.; Larsen, B. S.; McEwen, C. N. *J. Comb. Chem.* **1999**, *1*, 82–90.
194. Huyer, G.; Kelly, J.; Moffat, J.; Zamboni, R.; Zongchao, J.; Gresser, M. J.; Ramachandran, C. *Anal. Biochem.* **1998**, *258*, 19–30.
195. Chu, Y.-H.; Kirby, D. P.; Karger, B. L. *J. Am. Chem. Soc.* **1995**, *117*, 5419–5420.
196. Cancilla, M. T.; Leavell, M. D.; Chow, J.; Leary, J. A. *Proc. Natl. Acad. Sci. USA* **2000**, *97*, 12008–12013.
197. Scheffler, I. E. *Mitochondria*; Wiley-Liss: New York, 1999.
198. Beal, M. F.; Howell, N.; Bódis-Wollner, I. *Mitochondria and Free Radicals in Neurodegenerative Diseases*; Wiley-Liss: New York, 1997.
199. Dykens, J. A.; Davis, R. E.; Moos, W. H. *Drug Dev. Res.* **1999**, *46*, 2–13.
200. Kennedy, G. C. *Drug Dev. Res.* **1997**, *41*, 112–119.
201. Vukumirovic, O. G.; Tilghman, S. M. *Nature* **2000**, *405*, 820–822.
202. Clark, B. F. *Nature* **1981**, *292*, 491–492.
203. Kenyon, G. L.; DeMarini, D. M.; Fuchs, E.; Galas, D. J.; Kirsch, J. F.; Leyh, T. S.; Moos, W. H.; Petsko, G. A.; Ringe, D.; Rubin, G. M. et al. *Mol. Cell. Proteomics* **2002**, *1*, 763–780.
204. Weinshilboum, R.; Wang, L. *Nat. Rev. Drug Disc.* **2004**, *3*, 739–748.
205. Steen, H.; Mann, M. *Nat. Rev. Mol. Cell Biol.* **2004**, *5*, 699–711.
206. Petricoin, E. F.; Zoon, K. C.; Kohn, E. C.; Barrett, J. C.; Liotta, L. A. *Nat. Rev. Drug Disc.* **2002**, *1*, 683–695.
207. Taylor, S. W.; Fahy, E.; Zhang, B.; Glenn, G. M.; Warnock, D. E.; Wiley, S.; Murphy, A. N.; Gaucher, S. P.; Capaldi, R. A.; Gibson, B. W. et al. *Nat. Biotechnol.* **2003**, *21*, 281–286.
208. Robertson, K. D. *Nat. Rev. Genet.* **2005**, *6*, 597–610.
209. Kingsbury, D. T. *Drug Dev. Res.* **1997**, *41*, 120–128.
210. Wallis, K.; Richmond, M.; Patchett, T. *The Impact of Genomics on the Pharmaceutical Industry: A Pharmaceutical Group White Paper*; PricewaterhouseCoopers: London, UK, 1998.
211. Martin, A. B.; Schultz, P. G. *Trends Cell Biol.* **1999**, *9*, M24–M28.
212. Serber, Z.; Keatinge-Clay, A. T.; Ledwidge, R.; Kelly, A. E.; Miller, S. M.; Dotsch, V. *J. Am. Chem. Soc.* **2001**, *123*, 2446–2447.
213. Schreiber, S. L. *Bioorg. Med. Chem.* **1998**, *6*, 1127–1152.
214. Owens, J. *Drug Disc. Today* **2004**, *9*, 299–303.
215. Strausberg, R. L.; Schreiber, S. L. *Science* **2003**, *300*, 294–295.
216. Blackwell, H. E.; Peréz, L.; Stavenger, R. A.; Tallarico, J. A.; Eatough, E. C.; Foley, M. A.; Schreiber, S. L. *Chem. Biol.* **2001**, *8*, 1167–1182.
217. Clemons, P. A.; Koehler, A. N.; Wagner, B. K.; Sprigings, T. G.; Spring, D. R.; King, R. W.; Schreiber, S. L.; Foley, M. A. *Chem. Biol.* **2001**, *8*, 1183–1195.
218. Hung, D. T.; Jamison, T. F.; Schreiber, S. L. *Chem. Biol.* **1996**, *3*, 623–639.
219. Crabtree, G. R.; Schreiber, S. L. *Trends Biochem. Sci.* **1996**, *21*, 418–422.
220. Jones, A. K.; Buckingham, S. D.; Sattelle, D. B. *Nat. Rev. Drug Disc.* **2005**, *4*, 321–330.
221. Maggi, A.; Ciana, P. *Nat. Rev. Drug Disc.* **2005**, *4*, 249–255.
222. Cane, D. E.; Walsh, C. T.; Khosla, C. *Science* **1998**, *282*, 63–68.
223. Pfeifer, B. A.; Khosla, C. *Microbiol. Mol. Biol. Rev.* **2001**, *65*, 106–118.
224. Schultz, P. G. *Science* **1988**, *240*, 426–433.
225. Schultz, P. G.; Lerner, R. A. *Science* **1995**, *269*, 1835–1842.
226. Benner, S. A.; Sismour, A. M. *Nat. Rev. Genet.* **2005**, *6*, 533–543.
227. Cropp, T. A.; Schultz, P. G. *Trends Genet.* **2004**, *20*, 625–630.
228. Deiters, A.; Cropp, T. A.; Summerer, D.; Mukherji, M.; Schultz, P. G. *Bioorg. Med. Chem. Lett.* **2004**, *14*, 5743–5745.
229. Deiters, A.; Schultz, P. G. *Bioorg. Med. Chem. Lett.* **2005**, *15*, 1521–1524.
230. Wang, L.; Schultz, P. G. *Angew. Chem. Int. Ed.* **2005**, *44*, 34–66.
231. Deiters, A.; Geierstanger, B. H.; Schultz, P. G. *ChemBioChem* **2005**, *6*, 55–58.
232. Tian, F.; Tsao, M.-L.; Schultz, P. G. *J. Am. Chem. Soc.* **2004**, *126*, 15962–15963.
233. Ding, S.; Schultz, P. G. *Nat. Biotechnol.* **2004**, *22*, 833–840.
234. Horne, W. S.; Stout, C. D.; Ghadiri, M. R. *J. Am. Chem. Soc.* **2003**, *125*, 9372–9376.
235. Moos, W. H.; Barry, S. *Drug Dev. Res.* **2006**, *67*, 1–3.

236. Ferrari, M. *Nat. Rev. Cancer* **2005**, *5*, 161–171.
237. Moos, W. H.; DiRita, V. J.; Oxender, D. L. *Curr. Opin. Biotechnol.* **1993**, *4*, 711–713.
238. Maulik, S.; Patel, S. D. *Molecular Biotechnology: Therapeutic Applications and Strategies*; Wiley-Liss: New York, 1997.
239. Dawson, P. E.; Kent, S. B. H. *Annu. Rev. Biochem.* **2000**, *69*, 923–960.
240. Perlow, D. S.; Erb, J. M.; Gould, N. P.; Tung, R. D.; Freidinger, R. M.; Williams, P. D.; Veber, D. F. *J. Org. Chem.* **1992**, *57*, 4394–4400.
241. Rink, H. *Tetrahedron Lett.* **1987**, *28*, 3787–3790.
242. Wang, S.-S. *J. Am. Chem. Soc.* **1973**, *95*, 1328–1333.
243. Greener, M. *The Scientist* **2005**, *19*, 14–16.
244. Stacy, K. M. *The Scientist* **2005**, *19*, 17–19.
245. Bortone, K.; Michiels, F.; Vandeghinste, N.; Tomme, P.; van Es, H. *Drug Disc. World* **2004**, *5*, 20–27.
246. Miller, C. P. *Drug Disc. World* **2005**, *6*, 41–46.
247. Dev, K. K. *Nat. Rev. Drug Disc.* **2004**, *3*, 1047–1056.
248. Moos, W. H.; Bergmeier, S. C.; Coughenour, L. L.; Davis, R. E.; Hershenson, F. M.; Kester, J. A.; McKee, J. S.; Marriott, J. G.; Schwarz, R. D.; Tecle, H. et al. *J. Pharm. Sci.* **1992**, *81*, 1015–1019.
249. Hansch, C. *Acc. Chem. Res.* **1969**, *2*, 232–239.
250. Shorter, J. *Correlation Analysis in Organic Chemistry: An Introduction to Linear Free-energy Relationships*; Oxford University Press: Oxford, UK, 1973.
251. Lloyd, E. J.; Andrews, P. R. *J. Med. Chem.* **1986**, *29*, 453–462.
252. Moos, W. H.; Gless, R. D.; Rapoport, H. *J. Org. Chem.* **1981**, *46*, 5064–5074.
253. Topliss, J. G. *J. Med. Chem.* **1972**, *15*, 1006–1011.
254. Topliss, J. G. *J. Med. Chem.* **1977**, *20*, 463–469.
255. Andrews, P. R.; Craik, D. J.; Martin, J. L. *J. Med. Chem.* **1984**, *27*, 1648–1657.
256. Martin, Y. C. *Quantitative Drug Design*; Marcel Dekker: New York, 1978.
257. Topliss, J. G.; Costello, R. J. *J. Med. Chem.* **1972**, *15*, 1066–1068.
258. Topliss, J. G.; Edwards, R. P. *J. Med. Chem.* **1979**, *22*, 1238–1244.
259. Moos, W. H.; Szotek, D. S.; Bruns, R. F. *J. Med. Chem.* **1985**, *28*, 1383–1384.
260. Haky, J. E.; Young, A. M. *J. Liq. Chromatogr.* **1984**, *7*, 675–689.
261. See http://www.daylight.com/dayhtml/doc/clogp/ (accessed July 2006).
262. Verma, R. P.; Kurup, A.; Mekapati, S. B.; Hansch, C. *Bioorg. Med. Chem.* **2005**, *13*, 933–948.
263. Sauer, W. H. B. *Drug Disc. World* **2005**, *6*, 65–68.
264. Perun, T. J.; Propst, C. L.; *Computer-Aided Drug Design: Methods and Applications*; Marcel Dekker: New York, 1989.
265. Clark, T. *A Handbook of Computational Chemistry*; Wiley-Interscience: New York, 1985.
266. Blundell, T. L. *Nature* **1996**, *384*, 23–26.
267. See http://www.rcsb.org/pdb/ (accessed July 2006).
268. Service, R. *Science* **2005**, *307*, 1554–1557.
269. Moos, W. H.; Humblet, C. C.; Sircar, I.; Rithner, C.; Weishaar, R. E.; Bristol, J. A.; McPhail, A. T. *J. Med. Chem.* **1987**, *30*, 1963–1972.
270. Rees, D. C.; Congreve, M.; Murray, C. W.; Carr, R. *Nat. Rev. Drug Disc.* **2004**, *3*, 660–672.
271. Shuker, S. B.; Hajduk, P. J.; Meadows, R. P.; Fesik, S. W. *Science* **1996**, *274*, 1531–1534.
272. Erlanson, D. A.; Braisted, A. C.; Raphael, D. R.; Randal, M.; Stroud, R. M.; Gordon, E. M.; Wells, J. A. *Proc. Natl. Acad. Sci. USA* **2000**, *97*, 9367–9372.
273. Bursavich, M. G.; Rich, D. H. *J. Med. Chem.* **2002**, *45*, 541–558.
274. Watts, A. *Nat. Rev. Drug Disc.* **2005**, *4*, 555–568.
275. Geysen, H. M.; Houghten, R. A.; Kauffman, S.; Lebl, M.; Moos, W. H.; Pavia, M. R.; Szostak, J. W. *Mol. Diversity* **1995**, *1* (Editorial), 1–3.
276. Martin, E. J.; Blaney, J. M.; Siani, M. A.; Spellmeyer, D. C.; Wong, A. K.; Moos, W. H. *J. Med. Chem.* **1995**, *38*, 1431–1436.
277. Moos, W. H.; Okajima, N. *Kagaku (in Japanese)* **1996**, *51*, 472–476.
278. Moos, W. H.; Banville, S. C.; Blaney, J. M.; Bradley, E. K.; Braeckman, R. A.; Bray, A. M.; Brown, E. G.; Desai, M. C.; Dollinger, G. D.; Doyle, M. V. et al. In *Medicinal Chemistry: Today and Tomorrow*; Yamazaki, M., Eds.; Blackwell Science: Oxford UK, 1997, pp 137–142.
279. Moos, W. H. *Pharm. News* **1996**, *3*, 23–26.
280. Stockwell, B. R. *Nature* **2004**, *432*, 846–854.
281. Lipinski, C.; Hopkins, A. *Nature* **2004**, *432*, 855–861.
282. Muegge, I.; Heald, S. L.; Brittelli, D. *J. Med. Chem.* **2001**, *44*, 1841–1846.
283. Ajay; Walters, W. P.; Murcko, M. A. *J. Med. Chem.* **1998**, *41*, 3314–3324.
284. Ajay; Bemis, G. W.; Murcko, M. A. *J. Med. Chem.* **1999**, *42*, 4942–4951.
285. Pardridge, W. M. *J. Neurochem.* **1998**, *70*, 1781–1792.
286. McGovern, S. L.; Caselli, E.; Grigorieff, N.; Shoichet, B. K. *J. Med. Chem.* **2002**, *45*, 1712–1722.
287. Muegge, I. *Med. Res. Rev.* **2003**, *23*, 302–321.
288. Willett, P.; Winterman, V.; Bawden, D. *J. Chem. Inf. Comput. Sci.* **1986**, *26*, 109–118.
289. Parks, C. A.; Crippen, G. M.; Topliss, J. G. *J. Comput.-Aided Mol. Des.* **1998**, *12*, 441–449.
290. Bone, R. G. A.; Villar, H. O. *J. Comp. Chem.* **1997**, *18*, 86–107.
291. Lipinski, C.; Hopkins, A. *Nature* **2004**, *432*, 855–861.
292. Shoichet, B. K. *Nature* **2004**, *432*, 862–865.
293. Kitchen, D. B.; Decornez, H.; Furr, J. R.; Bajorath, J. *Nat. Rev. Drug Disc.* **2004**, *3*, 935–949.
294. Lipinski, C. A.; Lombardo, F.; Dominy, B. W.; Feeney, P. J. *Adv. Drug Deliv. Rev.* **2001**, *46*, 3–26.
295. This includes elements of Lipinski's rules as well as input from personal communications with W. C. Ripka, J. G. Topliss, and others.
296. Proudfoot, J. R. *Bioorg. Med. Chem. Lett.* **2005**, *15*, 1087–1090.
297. Wenlock, M. C.; Austin, R. P.; Barton, P.; Davis, A. M.; Leeson, P. D. *J. Med. Chem.* **2003**, *46*, 1250–1256.
298. Krogsgaard-Larsen, P.; Liljefors, T.; Madsen, U.; *A Textbook of Drug Design and Development*; Harwood: Amsterdam, the Netherlands, 1996.
299. James, I. W. *Annu. Rep. Comb. Chem. Mol. Div.* **1999**, *2*, 129–161.
300. Dolle, R. E. *Annu. Rep. Comb. Chem. Mol. Div.* **1999**, *2*, 93–127.
301. Dolle, R. E. *J. Comb. Chem.* **2004**, *6*, 623–679.

302. Mullin, R. *Chem. Eng. News July* **2004**, *July 26*, 23–32.
303. Borman, S. *Chem. Eng. News Oct* **2004**, *Oct 4*, 32–40.
304. Beeley, N.; Berger, A. *Br. Med. J.* **2000**, *321*, 581–582.
305. Luesch, H.; Wu, T. Y. H.; Ren, P.; Gray, N. S.; Schultz, P. G.; Supek, F. *Chem. Biol.* **2005**, *12*, 55–63.
306. Ulrich, R.; Friend, S. H. *Nat. Rev. Drug Disc.* **2002**, *1*, 84–88.
307. Stoughton, R. B.; Friend, S. H. *Nat. Rev. Drug Disc.* **2005**, *4*, 345–350.
308. Williams, M.; Coyle, J. T.; Shaikh, S.; Decker, M. W. *Annu. Rep. Med. Chem.* **2001**, *36*, 1–10.
309. Pfizer. *Annual Report*, 2004, p. 14.
310. Williams, M. *Curr. Opin. Ther. Patents* **1991**, *1*, 693–723.
311. Williams, M. *Curr. Opin. Invest. Drugs* **2004**, *5*, 29–33.
312. Williams, M. *Curr. Opin. Invest. Drugs* **2005**, *6*, 17–20.
313. Duyk, G. *Science* **2003**, *302*, 603–605.
314. Milne, G. M. *Annu. Rep. Med. Chem.* **2003**, *38*, 383–396.
315. Wilkins, R. H.; Brody, I. A. *Arch. Neurol.* **1969**, *21*, 109.
316. Jarvik, L.; Greenson, H. *Alzheimer Dis. Assoc. Disord.* **1987**, *1*, 7–8.
317. Crook, T.; Bartus, R. T.; Ferris, S. H.; Whitehouse, P.; Cohen, G. D.; Gershon, S. *Dev Neuropsychol.* **1986**, *2*, 261–276.
318. Petersen, R. C. *Nat. Rev. Drug Disc.* **2003**, *2*, 646–653.
319. Evans, D. A.; Funkenstein, H. H.; Albert, M. A.; Scherr, P. A.; Cook, N. R.; Chown, M. J.; Hebert, L. E.; Hennekens, C. H.; Taylor, J. O. *J. Am. Med. Assoc.* **1989**, *262*, 2551–2556.
320. Shumaker, S. A.; Legault, C.; Kuller, L.; Rapp, S. R.; Thal, L.; Lane, D. S.; Fillit, H.; Stefanick, M. L.; Hendrix, S. L.; Lewis, C. E. et al. *JAMA* **2004**, *291*, 2947–2958.
321. Brinton, R. D. *Curr. Opin. CNS Drugs* **2004**, *18*, 405–422.
322. Dykens, J. A.; Simpkins, J. W.; Wang, J.; Gordon, K. *Exp. Gerontol.* **2003**, *38*, 101–107.
323. Howell, N.; Taylor, S. W.; Fahy, E.; Murphy, A.; Ghosh, S. S. *TARGETS* **2003**, *2*, 208–216.
324. Dey, M.; Lyttle, C. R.; Pickar, J. H. *Maturitas* **2000**, *34*, S25–S33.
325. Levin-Allerhand, J. A.; Lominska, C. E.; Wang, J.; Smith, J. D. *J. Alzheimer's Dis.* **2002**, *4*, 449–457.
326. Van de Waterbeemd, H.; Smith, D. A.; Beaumont, K.; Walker, D. K. *J. Med. Chem.* **2001**, *44*, 1313–1333.
327. Egan, W. J.; Merz, K. M.; Baldwin, J. J. *J. Med. Chem.* **2000**, *43*, 3867–3877.
328. Yoshida, F.; Topliss, J. G. *J. Med. Chem.* **2000**, *43*, 2575–2585.
329. Veber, D. F.; Johnson, S. R.; Cheng, H.-Y.; Smith, B. R.; Ward, K. W.; Kopple, K. D. *J. Med. Chem.* **2002**, *45*, 2615–2623.
330. Rubas, W.; Jezyk, N.; Grass, G. M. *Pharm. Res.* **1993**, *10*, 113–118.
331. Audus, K. L.; Bartel, R. L.; Hidalgo, I. J.; Borchardt, R. T. *Pharm. Res.* **1990**, *7*, 435–451.
332. Lombardo, F.; Obach, R. S.; Shalaeva, M. Y.; Gao, F. *J. Med. Chem.* **2002**, *45*, 2867–2876.
333. de Groot, M. J.; Ackland, M. J.; Horne, V. A.; Alex, A. A.; Jones, B. C. *J. Med. Chem.* **1999**, *42*, 4062–4070.
334. Welling, P. G.; Tse, F. L. S. *Pharmacokinetics*; Marcel Dekker: New York, 1988.
335. Wang, J.; Urban, L. *Drug Disc. World* **2004**, *5*, 73–86.
336. Johnson, D.; Smith, D. A.; Park, B. K. *Curr. Opin. Drug Disc. Dev.* **2005**, *8*, 24–26.
337. Waring, J. F.; Anderson, M. G. *Curr. Opin. Drug Disc. Dev.* **2005**, *8*, 59–65.
338. Helma, C. *Curr. Opin. Drug Disc. Dev.* **2005**, *8*, 27–31.
339. Votano, J. R. *Curr. Opin. Drug Disc. Dev.* **2005**, *8*, 32–37.
340. Williams, J. A.; Bauman, J.; Cai, H.; Conlon, K.; Hansel, S.; Hurst, S.; Sadagopan, N.; Tugnait, M.; Zhang, L.; Sahi, J. *Curr. Opin. Drug Disc. Dev.* **2005**, *8*, 78–88.
341. Hutzler, J. M.; Messing, D. M.; Wienkers, L. C. *Curr. Opin. Drug Disc. Dev.* **2005**, *8*, 51–58.
342. Shou, M. *Curr. Opin. Drug Disc. Dev.* **2005**, *8*, 66–77.
343. Evans, D. C.; Baillie, T. A. *Curr. Opin. Drug Disc. Dev.* **2005**, *8*, 44–50.
344. Pelczer, I. *Curr. Opin. Drug Disc. Dev.* **2005**, *8*, 127–133.
345. Cucullo, L.; Aumayr, B.; Rapp, E.; Janigro, D. *Curr. Opin. Drug Disc. Dev.* **2005**, *8*, 89–99.
346. Parng, C. *Curr. Opin. Drug Disc. Dev.* **2005**, *8*, 100–106.
347. Berg, E. L.; Hytopoulos, E.; Plavec, I.; Kunkel, E. J. *Curr. Opin. Drug Disc. Dev.* **2005**, *8*, 107–114.
348. Doherty, A. M. *Annu. Rep. Med. Chem.* **2004**, *39*, 1–433.
349. Keith, C. T.; Borisy, A. A.; Stockwell, B. R. *Nat. Rev. Drug Disc.* **2005**, *4*, 71–78.
350. Gurwitz, D. *Drug Dev. Res.* **2004**, *62*, 71–75.
351. Roses, A. D. *Drug Dev. Res.* **2004**, *62*, 79–80.
352. Hakonarsson, H.; Stefansson, K. *Drug Dev. Res.* **2004**, *62*, 86–96.
353. Venter, J. C.; Adams, M. D.; Myers, E. W.; Li, P. W.; Mural, R. J.; Sutton, G. G.; Smith, H. O.; Yandell, M.; Evans, C. A.; Holt, R. A. et al. *Science* **2001**, *291*, 1304–1351.
354. Lander, E. S.; Linton, L. M.; Birren, B.; Nusbaum, C.; Zody, M. C.; Baldwin, J.; Devon, K.; Dewar, K.; Doyle, M.; FitzHugh, W. et al. *Nature* **2001**, *409*, 860–921.
355. Moos, W. H.; Feldbaum, C. B. *Drug Dev. Res.* **2002**, *57*, 45–50.
356. Moos, W. H. *Pharm. News* **1998**, *5*, 45.
357. Moos, W. H. *Pharm. News* **1999**, *6*, 40.
358. Moos, W. H. *Pharm. News* **1999**, *6*, 44–45.
359. Moos, W. H. *Pharm. News* **1999**, *6*, 46–47.
360. Moos, W. H. A Biotech CEO's Perspective. In *Insiders Guide to Venture Capital*; Fichera, D., Ed.; Prima Communications: Roseville, CA, 2001, pp 383–391.
361. Giacomini, K.M.; Sugiyama, Y. Membrane Transporters and Drug Response. In *Goodman and Gilman's The Pharmacological Basis of Therapeutics*, 11th ed.; Brunton, L.L., Lazo, J.S., Parker, K.L., Eds.; McGraw-Hill: New York, 2006, Chapter 2, pp 41–70.
362. Hopkins, A. L.; Groom, C. R. *Nat. Rev. Drug Disc.* **2002**, *1*, 727–730.
363. Wong, D. T.; Perry, K. W.; Bymaster, F. P. *Nat. Rev. Drug Disc.* **2005**, *4*, 764–774.
364. Iversen, L. *Br. J. Pharmacol.* **2006**, *147*, S82–S88.

365. Baldessarini, R.J. Drug Therapy of Depression and Anxiety Disorders. In *Goodman and Gilman's The Pharmacological Basis of Therapeutics*, 11th ed.; Brunton, L.L., Lazo, J.S., Parker, K.L., Eds.; McGraw-Hill: New York, 2006; Chapter 17, pp 429–459.
366. Anon. *MedAdNews* **2004**, May.
367. Uhl, G. R. *Movement Disorders* **2003**, *18*, S71–S80.
368. Uhl, G.; Lin, Z.; Metzger, T.; Dar, D. E. *Methods Enzymol.* **1998**, *296*, 456–465.
369. Dutta, A. K.; Zhang, S.; Kolhatkar, R.; Reith, M. E. A. *Eur. J. Pharmacol.* **2003**, *479*, 93–106.
370. Bannon, M. J. *Toxicol. Appl. Pharmacol.* **2005**, *204*, 355–360.
371. Madras, B. K.; Miller, G. M.; Fischman, A. J. *Biol. Psychiatr.* **2005**, *57*, 1397–1409.
372. Calabrese, D.; Bille, J.; Sanglard, D. *Microbiology* **2000**, *146*, 2743–2754.
373. Mizuno, N.; Niwa, T.; Yotsumoto, Y.; Sugiyama, Y. *Pharmacol. Rev.* **2003**, *55*, 425–461.
374. Kabanov, A. V.; Batrakova, E. V.; Miller, D. W. *Adv. Drug Deliv. Rev.* **2003**, *55*, 151–164.
375. Allen, D. D.; Lockman, P. R. *Life Sci.* **2003**, *73*, 1609–1615.
376. Howard, E. M.; Zhang, H.; Roepe, P. D. *J. Membrane Biol.* **2002**, *190*, 1–8.
377. Peel, S. A. *Drug Resistance Updates* **2001**, *4*, 66–74.
378. Joët, T.; Krishna, S. *Acta Tropica* **2004**, *89*, 371–374.
379. Watling, K. J., Ed. *The Sigma-RBI Handbook of Receptor Classification and Signal Transduction*, 4th ed.; Sigma-RBI: Natick, MA, 2001.
380. Koepsell, H.; Endou, H. *Pflüg. Arch. – Eur. J. Phy.* **2004**, *447*, 666–676.
381. Kim, R. B. *Eur. J. Clin. Invest.* **2003**, *33*, 1–5.
382. Bannon, M. J.; Michelhaugh, S. K.; Wang, J.; Sacchetti, P. *Eur. Neuropsychopharmacol.* **2001**, *11*, 449–455.
383. Volz, T. J.; Schenk, J. O. *Synapse* **2005**, *58*, 72–94.
384. Chen, N.; Reith, M. E. A. *Eur. J. Pharmacol.* **2000**, *405*, 329–339.
385. Greiner, E.; Boos, T. L.; Prisinzano, T. E.; De Martino, M. G.; Zeglis, B.; Dersch, C. M.; Marcus, J.; Partilla, J. S.; Rothman, R. B.; Jacobson, A. E. et al. *J. Med. Chem.* **2006**, *49*, 1766–1772.
386. Shilling, R. A.; Balakrishnan, L.; Shahi, S.; Venter, H.; van Veen, H. W. *Int. J. Antimicrobial Agents* **2003**, *22*, 200–204.
387. Abramson, J.; Smirnova, I.; Kasho, V.; Verner, G.; Kaback, H. R.; Iwata, S. *Science* **2003**, *301*, 610–615.
388. Neumeister, A.; Young, T.; Stastny, J. *Psychopharmacology* **2004**, *174*, 512–524.
389. Horschitz, S.; Hummerich, R.; Schloss, P. *Biochem. Soc. Trans.* **2001**, *29*, 728–732.
390. Boya, P.; Roques, B.; Kroemer, G. *EMBO J.* **2001**, *20*, 4325–4331.
391. Koehler, C. M. *FEBS Lett.* **2000**, *476*, 27–31.
392. Tucker, G. T.; Houston, J. B.; Huang, S.-M. *Pharm. Res.* **2001**, *18*, 1071–1080.
393. Benet, L. Z.; Cummins, C. L.; Wu, C. Y. *Curr. Drug Metab.* **2003**, *4*, 393–398.
394. You, G. *Med. Res. Rev.* **2002**, *22*, 602–616.
395. Wang, C.; Delcros, J.-G.; Cannon, L.; Konate, F.; Carias, H.; Biggerstaff, J.; Gardner, R. A.; Phanstiel, O., IV *J. Med. Chem.* **2003**, *46*, 5129–5138.
396. Trauner, M.; Meier, P. J.; Boyer, J. L. *N. Engl. J. Med.* **1998**, *339*, 1217–1227.
397. Klein, J.; Sato, A. *N. Engl. J. Med.* **2000**, *343*, 782–786.
398. Zhang, Y.; Bachmeier, C.; Miller, D. W. *Adv. Drug Deliv. Rev.* **2003**, *55*, 31–51.
399. Iversen, L. L.; Glennon, R. A. Antidepressants. In *Burger's Medicinal Chemistry*; 6th ed., Vol. 6, *Nervous System Agents*; Abraham, D. J., Ed.; John Wiley: New York, 2003; Chapter 8, pp 483–524.
400. Currie, K. S. Antianxiety Agents. In *Burger's Medicinal Chemistry*, 6th ed., Vol. 6, *Nervous System Agents*; Abraham, D. J., Ed., John Wiley: New York, 2003, Chapter 9, pp 525–597.
401. Howell, L. L.; Wilcox, K. M. *J. Pharmacol. Exp. Ther.* **2001**, *298*, 1–6.
402. Newman, A. H.; Kulkarni, S. *Med. Res. Rev.* **2002**, *22*, 429–464.
403. Rothman, R. B.; Baumann, M. M. *Pharmacol. Ther.* **2002**, *95*, 73–88.
404. Caulfield, W. L.; Collie, I. T.; Dickins, R. S.; Epemolu, O.; McGuire, R.; Hill, D. R.; McVey, G.; Morphy, R.; Rankovic, Z.; Sundaram, H. *J. Med. Chem.* **2001**, *44*, 2679–2682.
405. Hoogerwerf, W. A.; Pasricha, P. J. Pharmacotherapy of Gastric Acidity, Peptic Ulcers, and Gastroesophageal Reflux Disease. In *Goodman and Gilman's the Pharmacological Basis of Therapeutics*; 11th ed.; Brunton, L. L., Lazo, J. S., Parker, K. L., Eds.; McGraw-Hill, New York, 2006; Chapter 36, pp 967–981.
405a. See http://redpoll.pharmacy.ualberta.ca/drugbank/ (accessed July 2006).
406. Tremont, S. J.; Lee, L. F.; Huang, H.-C.; Keller, B. T.; Banerjee, S. C.; Both, S. R.; Carpenter, A. J.; Wang, C.-C.; Garland, D. J.; Huang, W. et al. *J. Med. Chem.* **2005**, *48*, 5837–5852.
407. Tarasova, N. I.; Seth, R.; Tarasov, S. G.; Kosakowska-Cholody, T.; Hrycyna, C. A.; Gottesman, M. M.; Michejda, C. J. *J. Med. Chem.* **2005**, *48*, 3768–3775.
408. Brater, D. C. *N. Engl. J. Med.* **1998**, *339*, 387–395.
409. Stockmeier, C. A. *J. Psychiatr. Res.* **2003**, *37*, 357–373.
410. Winogrodzka, A.; Booij, J.; Wolters, E. C. *Parkinsonism Related Disord.* **2005**, *11*, 475–484.
411. Kuhar, M. J. *Life Sci.* **1998**, *62*, 1573–1575.
412. Parkinson Study Group. *N. Engl. J. Med.* **2004**, *351*, 2498–2508.
413. Juweid, M. E.; Cheson, B. D. *N. Engl. J. Med.* **2006**, *354*, 496–507.
414. Sai, Y.; Tsuji, A. *Drug Disc. Today* **2004**, *9*, 712–720.
415. Majumdar, S.; Duvvuri, S.; Mitra, A. K. *Adv. Drug Deliv. Rev.* **2004**, *56*, 1437–1452.
416. Lesch, K. P.; Gutknecht, L. *Progr. Neuro-Psychopharmacol. Biol. Psychiatr.* **2005**, *29*, 1062–1073.
417. Cummings, J. L.; Zhong, K. *Nat. Rev. Drug Disc.* **2006**, *5*, 64–74.
418. Murphy, D. L.; Li, Q.; Engel, S.; Wicherns, C.; Andrews, A.; Lesch, K.-P.; Uhl, G. *Brain Res. Bull.* **2001**, *56*, 487–494.
419. Smits, K. M.; Smits, L. J. M.; Schouten, J. S. A. G.; Stelma, F. F.; Nelemans, P.; Prins, M. H. *Mol. Psychiatr.* **2004**, *9*, 433–441.
420. Serretti, A.; Benedetti, F.; Zanardi, R.; Smeraldi, E. *Progr. Neuro-psychopharmacol. Biol. Psychiatr.* **2005**, *29*, 1074–1084.
421. Cho, H. J.; Meira-Lima, I.; Cordeiro, Q.; Michelon, L.; Sham, P.; Vallada, H.; Collier, D. A. *Mol. Psychiatr.* **2005**, *10*, 771–781.
422. Rausch, J. L. *Progr. Neuro-psychopharmacol. Biol. Psychiatr.* **2005**, *29*, 1046–1061.
423. Serretti, A.; Lilli, R.; Mandelli, L.; Lorenzi, C.; Smeraldi, E. *Pharmacogenomics J.* **2001**, *1*, 71–77.

424. Murphy, D. L.; Lerner, A.; Rudnick, G.; Lesch, K.-P. *Mol. Interventions* **2004**, *4*, 109–123.
425. Shannon, J. R.; Flattem, N. L.; Jordan, J.; Jacob, G.; Black, B. K.; Biaggioni, I.; Blakely, R. D.; Robertson, D. *N. Engl. J. Med.* **2000**, *342*, 541–549.
426. Brinkmann, U.; Eichelbaum, M. *Pharmacogenomics J.* **2001**, *1*, 59–64.
427. Schwab, M.; Eichelbaum, M.; Fromm, M. F. *Annu. Rev. Pharmacol. Toxicol.* **2003**, *43*, 285–307.
428. Evans, W. E.; McLeod, H. L. *N. Engl. J. Med.* **2003**, *348*, 538–549.
429. Siddiqui, A.; Kerb, R.; Weale, M. E.; Brinkmann, U.; Smith, A.; Goldstein, D. B.; Wood, N. W.; Sisodiya, S. M. *N. Engl. J. Med.* **2003**, *348*, 1442–1448.
430. Shulenin, S.; Nogee, L. M.; Annilo, T.; Wert, S. E.; Whitsett, J. A.; Dean, M. *N. Engl. J. Med.* **2004**, *350*, 1296–1303.
431. Kim, R. B. *Clin. Pharmacol. Ther.* **2004**, *75*, 381–385.
432. Owens, M. J.; Nemeroff, C. B. *Depression Anxiety* **1998**, *8*, 5–12.
433. Berton, O.; Nestler, E. J. *Nat. Rev. Neurosci.* **2006**, *7*, 137–150.
434. Wong, M.-L.; Licinio, J. *Nat. Rev. Drug Disc.* **2004**, *3*, 136–151.
435. Carroll, F. I. *J. Med. Chem.* **2003**, *46*, 1775–1794.
436. Drews, J. *Nat. Biotechnol.* **1996**, *14*, 1516–1518.
437. Hopkins, A. L.; Groom, C. R. *Nat. Rev. Drug Disc.* **2002**, *1*, 727–730.
438. Knowles, J.; Gromo, G. *Nat. Rev. Drug Disc.* **2003**, *2*, 63–69.
439. Pritchard, J. F.; Jurima-Romet, M.; Reimer, M. L. J.; Mortimer, E.; Rolfe, B.; Cayeri, M. N. *Nat. Rev. Drug Disc.* **2003**, *2*, 542–553.
440. Cohen, P.; Goedert, M. *Nat. Rev. Drug Disc.* **2004**, *3*, 479–487.
441. Melnikova, I.; Golden, J. *Nat. Rev. Drug Disc.* **2004**, *3*, 993–994.
442. Krause, D. S.; Van Etten, R. A. *N. Engl. J. Med.* **2005**, *353*, 172–187.
443. Yanagisawa, M.; Kurihara, H.; Kimura, S.; Tomobe, Y.; Kobayashi, M.; Mitsui, Y.; Yazaki, Y.; Goto, K.; Masaki, T. *Nature* **1988**, *332*, 411–415.
444. See http://www.speedel.com/section/7/subsections/4?form_link=1120715251 (accessed July 2006).
445. See http://www.psa-rising.com/med/research/atrasentan.htm (accessed July 2006).
446. Kramer, R.; Cohen, D. *Nat. Rev. Drug Disc.* **2004**, *4*, 965–972.
447. Searls, D. B. *Nat. Rev. Drug Disc.* **2005**, *4*, 45–58.
448. Drews, J.; Ryser, S. *Nat. Biotechnol.* **1997**, *15* (Special Issue).
449. Chabner, B. A.; Roberts, T. G. *Nat. Rev. Cancer* **2005**, *5*, 65–72.
450. See http://www.cancer.gov/newscenter/benchmarks-vol3-issue2 (accessed July 2006).
451. Bristol, J. A.; Sircar, I.; Moos, W. H.; Evans, D. B.; Weishaar, R. E. *J. Med. Chem.* **1984**, *27*, 1099–1101.
452. O'Donnell, J. M.; Zhang, H.-T. *Trends Pharmacol. Sci.* **2004**, *25*, 158–163.
453. Terrett, N. K.; Bell, A. S.; Brown, D.; Ellis, P. *Bioorg. Med. Chem. Lett.* **1996**, *6*, 1819–1824.
454. Yu, G.; Mason, H.; Wu, X.; Wang, J.; Chong, S.; Beyer, B.; Henwood, A.; Pongrac, R.; Seliger, L.; He, B. et al. *J. Med. Chem.* **2003**, *46*, 457–460.
455. Triggle, D. J. *Drug Dev. Res.* **2005**, *64*, 90–98.
456. Triggle, D. J. *Drug News Perspect.* **1991**, *4*, 579–588.
457. Southan, A.; James, I. F.; Cronk, D. *Drug Disc. World* **2005**, *6*, 17–23.
458. Burnett, J. C.; Henchal, E. A.; Schmaljohn, A. L.; Bavari, S. *Nat. Rev. Drug Disc.* **2005**, *4*, 281–297.
459. Kubinyi, H. *Nat. Rev. Drug Disc.* **2003**, *2*, 665–668.
460. Kiely, J. S.; Moos, W. H.; Pavia, M. R.; Schwarz, R. D.; Woodard, G. L. *Anal. Biochem.* **1991**, *196*, 439–442.
461. Ashcroft, F. M. *Nature* **2006**, *440*, 440–447.
462. Gadsby, D. C.; Vergani, P.; Csana'dy, L. *Nature* **2006**, *440*, 477–489.
463. Nichols, C. G. *Nature* **2006**, *440*, 470–476.
464. Xia, C. Q.; Xiao, G.; Liu, N.; Pimprale, S.; Fox, L.; Patten, C. J.; Crespi, C. L.; Miwa, G.; Gan, L.-S. *Mol. Pharmaceutices* **2006**, *3*, 78–86.
465. Crivori, P.; Reinach, B.; Pezzetta, D.; Poggesi, I. *Mol. Pharmaceutics* **2006**, *3*, 33–44.
466. Raub, T. J. *Mol. Pharmaceutics* **2006**, *3*, 3–25.

Biography

Walter H Moos has more than 20 years of experience in the pharmaceutical and biotechnology industries. He is currently Vice President and Head of the Biosciences Division at SRI International. For a number of years Dr Moos served as Chairman and Chief Executive Officer of MitoKor, until it merged with Micrologix to create MIGENIX. Previously, he was employed by Chiron (now Novartis) as a corporate officer, last holding the position of Vice President of R&D – Technologies Division. Dr Moos held earlier positions at Warner-Lambert/Parke-Davis (now Pfizer), last serving as Vice President, Neuroscience and Biological Chemistry. Dr Moos has served on the boards of directors of Alnis, Anterion, Axiom, the Biotechnology Industry Organization, CMPS, MIGENIX, Mimotopes, Oncologic, Onyx, Rigel, and the Keystone Symposia. He is a co-founding member of the CEO Council, Red Abbey Venture Partners. Dr Moos has edited several books, helped to found multiple journals, and has a total of more than 140 manuscripts and issued patents to his name. In addition, he has held adjunct faculty positions at the University of Michigan – Ann Arbor, and since 1992 has been an adjunct professor at the University of California – San Francisco. He has also served on academic and related advisory committees, including the US National Committee at the National Academy of Sciences. Dr Moos holds an AB from Harvard University and received his PhD in chemistry from the University of California – Berkeley.

2.02 An Academic Perspective

D J Triggle, State University of New York, Buffalo, NY, USA

© 2007 Elsevier Ltd. All Rights Reserved.

2.02.1	Changing Faces, Places, and Directions	85
2.02.2	Academia in a State of Flux	89
2.02.3	The Ethos of Science	91
2.02.4	The University in a Market Environment	93
2.02.5	Globalization and Science in the Rich and Poor Worlds	94
2.02.6	Implications for Academic Medicinal Chemistry	94
2.02.7	Conclusions	94
References		95

"The achievements of chemical synthesis are firmly bound in our attempts to break the shackles of disease and poverty".

Roald Hoffman, 1993
Nobel Laureate Chemistry

2.02.1 Changing Faces, Places, and Directions

A number of recent reviews have discussed at some length the nature of medicinal chemistry education in both Europe and the USA and have also examined the issues of the educational characteristics necessary for medicinal chemists practicing in the pharmaceutical industry.[1–4] These reviews constitute a useful comparative resource to which reference should be made for details of curriculum recommendations and related matters. Some of these findings and recommendations are presented in **Tables 1** and **2**. Additionally, the Carnegie Foundation for the Advancement of Teaching has published valuable essays by Breslow and Kwiram on doctoral education in chemistry.[5,6] Extensive discussion on the future of graduate education, albeit solely from a US perspective, may be found at 'Re-envisioning the PhD,'[7] which includes links to academic studies of many issues around doctoral education in both the sciences and humanities. This website contains the reports of many studies from which may be drawn some general conclusions, amongst the most important being that the programs are too long, too specialized, and too focused on preparing their students for solely academic careers, which careers will not be available for the overwhelming majority of graduates. A specific report on graduate education in the pharmaceutical sciences was commissioned by the American Association of Colleges of Pharmacy and published in 1999.[8] Indeed, the output of only a very small number of graduate programs is sufficient to fill new faculty recruiting demand and this creates a very competitive and recruiter-advantaged situation. The most recent study, for example, of faculty recruitment into Departments of Chemistry (and other departments from Colleges of Arts and Sciences) reveals that over 50% of the faculty at the top 25 American research universities obtained their PhD degrees from the top 10 universities (the rankings are those provided annually by *US News and World Report*)[9].

Finally, and reflecting the well-established observations that increased higher education correlates with higher lifetime income, the Royal Society of Chemistry in conjunction with the Institute of Physics commissioned PriceWaterhouseCoopers to re-examine this with particular reference to graduates of physics and chemistry programs in the UK. Their report found that graduates of these programs will earn on average 30% more than the corresponding high-school graduates.[10] This study was taken in light of the unpopularity of chemistry programs in the UK that has led to the closure of a number of university departments.[11] This situation is not unique to the UK. The USA also has serious concerns about its education achievements in general, with overall concerns for literacy, science, and mathematics. Recent reports indicate that not only does the USA rank poorly in science on an international comparative scale, but that there are widespread inequalities within populations[12] and the USA has increasingly depended on non-USA students, particularly science graduates from China and India, to fill its graduate programs in science and engineering.[13–15] In 2003, non-US students, for example, obtained approximately 40% of science doctorates and almost 60% of engineering doctorates. At the National Institutes of Health some 50% of the doctorate staff and fellows are non-US and

Table 1 Expertise sought in the recruitment of new medicinal chemists in the pharmaceutical industry

	Percent response				
	Germany	Italy	Japan	UK	USA
Knowledge area					
Organic chemistry	83	78	48	100	88
Biochemistry	17	22	17	11	27
Pharmacology	17	11	17	0	6
Molecular biology	8	0	9	0	3
Computer modelling	0	22	9	6	0
Physical organic chemistry	0	11	0	22	15
Medicinal chemistry					
Expertise					
Synthetic organic chemistry	58	56	57	83	61
Experimental skills	8	0	9	83	6
Computer experience	8	0	17	6	3
Biochemistry	17	0	0	0	0
Pharmacology or biology	17	0	0	0	9
Qualifications					
Postdoctoral	58	11	9	29	55
PhD	42	56	43	88	6
MS (or equivalent)		33	43		
Graduate			13	71	

Data are taken from Busse, W. D.; Ganellin, C. R.; Mitscher, L. A. *Eur. J. Med. Chem.* **1996**, *31*, 747–760. Reference should be made to this article for a discussion of the limitations of the questionnaire employed.

the 2000 US Census indicates that almost 40% of doctoral-level employees are foreign-born. The Business Round Table arguing for a new scientific education initiative has discussed these issues in the United States.[16]

And the Western world continues to attract scientifically and medically qualified individuals to its shores, despite the impoverishing effects that this may have on the countries of origin.[17,18] Whether these demographics reflect actual true shortages of scientifically qualified individuals or simply market-driven efforts to dampen wage and salary expectations and the cost of education remain in question.[19–24] However, many initiatives are under way with mechanisms designed to make chemistry a more attractive study area.[25] Whether these will be productive is questionable, at least in the USA, where new post-Scopes fundamentalist views of science now appear to be endorsed by both the public at large and their political leaders.[26,27]

In this chapter, however, a rather broader view will be taken in an effort to place medicinal chemistry within the context of ongoing changes in the underlying sciences, in the roles and structures of universities, in the changing social role of science, and of the impact of globalization on science in both the rich and poor worlds.

Science is not a static entity. Consider physics. At the beginning of the twentieth century everything was in its place in Newton's clockwork universe. And there was no reason to suspect that this clockwork precision would not accommodate what little else there was to discover. Yet in a few short years it was all overturned and physics had to start again. That revolution was a catalytic event for other changes in the twentieth century. Biology transformed itself in a

Table 2 What additional courses in medicinal chemistry should be offered as options to chemistry students

Subject	Percent response				
	Germany	Italy	Japan	UK	USA
Modeling or QSAR	50	33	35	40	36
Pharmacology or physiology	58	33	22	40	33
Molecular biology	25	11	26	27	30
Enzymology	17	0	9	33	24
Biochemistry or ADMET	17	22	13	40	21
Drug design	0	0	9	0	18
Case histories	0	0	0	33	6
Drug action	0	22	9	40	9
Biological chemistry	17	11	9	7	9
Toxicology	25	0	0	0	3
Microbiology	8	11	4	0	3
Cell biology	0	0	13	13	0

Data are taken from Busse W. D.; Ganellin, C. R.; Mitscher, L. A. *Eur. J. Med. Chem.* **1996**, *31*, 747–760. Reference should be made to this article for a discussion of the limitations of the questionnaire employed.

few short years from an essentially descriptive discipline to one largely molecular in nature, in significant part because of the entry of physicists into the field. And that biological revolution in now spreading to other disciplines, notably chemistry.

The underlying paradigms of molecular biology – 'diversity,' 'evolution,' 'self-organization,' and 'replication' – are changing the face of chemistry and of medicinal chemistry in particular. Ehrlich's magic bullet may in the future be, not the laboriously handcrafted molecule of the 1900s, but rather a self-synthesizing, reproducing, evolving, and targeting molecular machine or bioreactor designed to synthesize on demand the appropriate molecule(s) for cellular repair or destruction. These changes will continue to impact both the definition and the content of medicinal chemistry.[28–31]

Originally a discipline largely grounded in and practicing synthetic organic chemistry and guided in its direction by in vivo and some in vitro pharmacological data, medicinal chemistry has expanded in the past three decades to embrace not merely the dramatically expanded synthetic and technologic advances in organic chemistry, but also to include expanded analyses of structure–activity relationships through detailed knowledge of target structure, and increasingly of structure–property relationships and to be guided in direction by genomics-driven technologies that provide biological data directly from defined human targets. Thus, the medicinal chemist is increasingly dealing with pharmacokinetic, metabolic, genomic, and toxicological inputs all of which also guide the directions of organic synthesis. The influence of these inputs is clearly seen in **Figures 1** and **2** that demonstrate respectively the old and new pathways of drug discovery.

As synthetic chemistry continues to be further influenced by molecular biology through, for example, template-guided processes there will be increasing need for definitions of molecular biology to expand even further its disciplinary horizons.[32–35] The IUPAC definition[3] of medicinal chemistry of 1998:

"Medicinal chemistry is a chemistry-based discipline, also involving aspects of biological, medical, and pharmaceutical sciences. It is concerned with the invention, discovery, design, identification, and preparation of biologically active compounds, the study of their metabolism, the interpretation of their mode of action at the molecular level and the construction of structure–activity relationships."

will continue to develop and to expand to include a greater biologic input. In this context, the words of George S. Hammond[90] in his 1968 Norris lecture to the American Chemical Society assume particular significance:

"The most fundamental and lasting objective of synthesis is not production of new compounds, but production of properties."

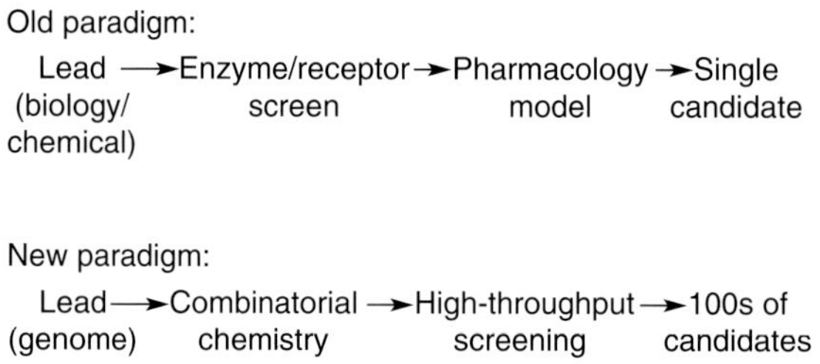

Figure 1 The two pathways of drug discovery that compare the classical (top) and the genomics-driven (bottom) routes.

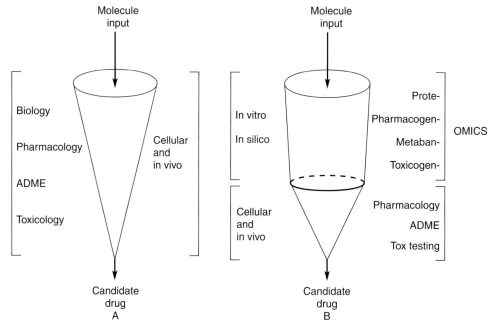

Figure 2 A comparison of the pathways of drug discovery indicating in pathway B the role of high-throughput and high-information inputs to the process subsequent to molecular input compared to the classical low-throughput pathway A. The implications for medicinal chemistry are the increasing importance of genomics-based disciplines to the total drug discovery process and the necessity for the medicinal chemist to factor these considerations into drug and molecular design.

These changes in disciplinary content and focus mean that the old walls and barriers that have traditionally separated disciplines are eroding rapidly. Biology is no longer confined to Departments of Biology and chemistry is practiced in academic departments from Anthropology to Zoology. These changes impact the teaching and practice of medicinal chemistry and related pharmaceutical disciplines. As Schools of Pharmacy, long the dominant home of Departments of Medicinal Chemistry, focus increasingly on the clinical aspects of their teaching and research so will their basic science disciplines be increasingly found in other departments. Thus, many Departments of Chemistry with their increasing biological emphasis also have strong programs in medicinal chemistry. Similarly, drug delivery is increasingly seen an integral component of bioengineering programs housed in Schools of Engineering, rather than a major activity of Schools of Pharmacy.

Such changes create tension in the faculty, but they are to be welcomed as part of the inevitable maturation of scientific disciplines and as further indications of the progressive integration of scientific knowledge. Nowhere is this to be seen more prominently than in the rapidly emerging discipline of systems biology, a discipline intended to provide through the integrated application of chemical, biological, and physical disciplines a complete understanding and mapping of biological pathways and function.[32,33] Both the practice and teaching of medicinal chemistry will be influenced significantly by this emerging discipline. The medicinal chemist of the future will thus be an individual armed with a toolbox of discipline-derived technologies that she will simultaneously use, replace, and replenish as necessary. It is, in fact, one of the great ironies of our twentieth-century reductionist approach to science that we have arrived at a stage where integration of knowledge is an even greater priority than the mere accumulation of new knowledge. The twentyfirst-century will be heavily dependent on the interdisciplinary sciences.[34]

> "The challenge for the university will be how to accommodate the learning necessary to achieve such a skill set within an environment that has a very strong tradition of silo-based disciplinary organization and that is also increasingly assuming some of the properties and characteristics of the market economy."

The traditional expectation from the pharmaceutical industry has been that the principal training for a medicinal chemist should be in synthetic organic chemistry with the anticipation that biological and related knowledge will be added later through experience or through specific training.[2–4] It remains true today that the medicinal chemist must have a thorough grounding in synthetic organic chemistry – 'how to put molecules together and take them apart again' – and to be able to correlate structure and biological activity. However, there have been very significant changes in the drug discovery process during the past two or three decades. Prior to this period the chemist collaborated with the pharmacologist who typically worked with in vivo or in vitro tissue preparations. Considerations of pharmacokinetic and toxicological issues came late in the discovery process and were not of primary concern to the chemist. In the new drug discovery programs there is much more emphasis on biological considerations and the nature and character of the drug target, which may well have been defined from genomic considerations. Additionally, biological assays are often primarily cell-based with cloned receptors, enzymes or channels and ADME (absorption, distribution, metabolism, and excretion) considerations are now part of the molecular design process. Finally, the increasing sophistication of structure-based approaches and the use of in silico technologies necessitate at least a familiarity by the chemist of these computational techniques.[35] It is thus possible that the current view of a medicinal chemist as primarily a synthetic organic chemist with grafted-on biological knowledge will be replaced by a broader definition and even the creation of subspecialities of medicinal chemist.[36] How this will be achieved is a matter of some current interest.

The university will not, of course, be the sole player in the teaching of any discipline or collection of disciplines in this learning environment. Thus, for medicinal chemistry there are a number of national and international summer schools and similar learning environments designed to provide at least the rudiments of medicinal chemistry, typically to newly trained synthetic organic chemists.[37] A broadly representative program from one of these summer schools is presented in **Table 3**. Valuable as these programs are the major responsibility for education must nonetheless fall on the university. On possible solution will be the creation of 'certificate' programs that are created jointly by universities and the pharmaceutical industry and that offer validated knowledge in a variety of medicinal chemistry subdisciplines – molecular modeling, in silico chemistry, ADME, etc.

Similarly, it is plausible to consider some further sharing of the drug discovery enterprise where academia has typically been involved only in the very earliest stages. For example, the Molecular Libraries Initiative of the National Institutes of Health Roadmap for Medical Research will generate a library of small-molecule chemical probes that will be screened through the development of new assay formats and procedures and that will permit predictive ADME/Toxicology.[38,39] The data from this will be available in a public database, PubChem, and should provide a new stimulus for academic involvement in the early phases of drug discovery.

2.02.2 Academia in a State of Flux

> "Today we are at a crossroads
> One road leads to hopelessness and utter despair
> The other leads to total extinction
> Let us pray that we have the wisdom to choose wisely."
>
> Woody Allen, Address to the Graduates

The preceding discussion, albeit focused primarily on medicinal chemistry, must however be seen as part of a larger set of changes that are affecting the academic environment of the twentyfirst-century. A better understanding of the challenges facing medicinal chemistry may be gained by some consideration of the changing faces and directions of the academy – an organization in a considerable state of flux. In this broader context it should be noted that in this pace of

Table 3 Curriculum for Drew University Residential School in Medicinal Chemistry 2005

Day One
  1. Strategic issues in drug discovery
  2. Lead discovery and modification (1)
  3. Lead discovery and modification (2)
  4. Drugs affecting ion channels
  5. Drug transporters
  6. Enzyme inhibition

Day Two
  1. Receptor binding assays
  2. Plasma protein binding
  3. Receptor structure and agonist–antagonist assays
  4. Drug delivery
  5. High throughput screening
  6. Patents

Day Three
  1. Bacterial genomics and small-molecule discovery
  2. QSAR in drug research
  3. Chemical diversity
  4. Structure-based drug design
  5. Seminar I: Molecular modeling^a
  6. Seminar II: Pharmacokinetics and ADME^a

Day Four
  1. Toxicology
  2. Cell-based assays
  3. Seminar III: Metabolic inactivation pathways^a

Day Five
  1. Case histories: I
  2. Case histories: II
  3. Case histories: III

^{a} Small group sessions.

change there is nothing particularly unique to the discipline of medicinal chemistry. Michael Gibbons, a former director of a science policy unit in the UK has argued that much of today's leading science does not consist of the patient process of putting one brick on top of another, but rather of solving complex problems that cross many disciplines. He sees the university in an environment that he calls "a socially distributed knowledge production system" in which the university is but one of many generators of knowledge.[40] Thus, in this viewpoint the idea of a university as a tidy institution living harmoniously within its own self-defined boundaries may well be dead. Rather, we may need to see the university as an open structure within a cloud of continuously forming and dissolving relationships. This is a pattern that is already common to individual faculty who are long used to being accused of having greater loyalties to their disciplines than to their own institutions.

Universities always appear to be in a state of crisis or at least flux, whether it be the salary of the football coach, the nature of core programs and their lack of funding or, and more simply, some new sex scandal focused on the administration. This apparent universal state of crisis is puzzling since universities are, in fact, our oldest and arguably amongst our most successful institutions. The first true university was founded in Bologna in the eleventh-century and those of Paris, Oxford, and Padua a century later. But universities are really much older and can trace their origin to the library of Alexandria and the Greek protocols of scientific inquiry. These ancient foundations, and thousands like them, continue to grow and prosper. But they are changing – the students are no longer monks, the teachers no longer priests, and the curriculum has long been stripped of Greek, Latin, logic, and rhetoric and even of much philosophy. Over a century ago Cardinal John Henry Newman provided a definition of the university based on his belief that it is necessary to separate the pursuit of truth from mankind's 'necessary cares.' Newman's university would be: "the high protecting power of all knowledge and science, of fact and principle, of inquiry and discovery, of experiment and speculation."[41,42] Newman had recognized that which Francis Bacon had already announced some 400 years ago that knowledge is power: "knowledge and human power are synonymous, since the ignorance of the cause frustrates the effect, for nature is only subdued by submission."[43]

The reality is, however, that universities have always been nonstatic entities, but that the pace and nature of change are today both quantitatively larger and qualitatively more distinct that in previous centuries. And these changes are not universally welcomed.[44–46] The factors generating change include:

1. Exponential increases in scientific knowledge.
2. Fragmentation of scientific disciplines.
3. Increasing cost of scientific research and education.
4. Increasing ethical issues, particularly around biology-based disciplines.
5. Rapid changes in learning technologies.
6. Increased demand – demographics and economic market.
7. Increased general competition.
8. New learning organizations – virtual universities.
9. Proprietary competition – corporations buying or creating universities.
10. Rapidly growing role for education as credentialing.
11. Education as a commodity.
12. Diminishing public funding as demand increases.
13. Expectations of education as an immediate economic benefit.
14. Worldwide competition to Western dominance of knowledge.
15. Privatization of knowledge as part of a neoconservative 'free market' economy.

The challenge for universities thus appears to be twofold. First, a dramatic increase in the rate of generation of scientific knowledge and the associated specialization of knowledge and second and increased public and political expectation of universal higher education and of attendant direct economic returns from university-based, publicly funded research. For universities the questions are thus whether to remain true to Newman's concept and serve as the guardian and generator of knowledge for the public good and how to accommodate to the increasing expectation that knowledge should be treated as a commodity of rapidly translatable economic benefit. But the challenge may be more apparent than real. Although the faculty, particularly those from the humanities, may well dislike the term "The Knowledge Factory," used by *The Economist*,[47] applied to their institutions, universities have, at least until comparatively recently, nonetheless navigated quite successfully the competing demands of generating the intellectual commons and of contributing to economic development. This dual role is eminently in line with the concept of the public funding of research as defined by Vannevar Bush in the USA in the immediate post-World War II environment[48]:

"The Government should accept new responsibilities for promoting the flow of new scientific knowledge and development of scientific talent in our youth. These responsibilities are the proper concern of the Government, for they vitally affect our jobs, our health, our security."

Although the model of scientific support outlined by Vannevar Bush has often been interpreted as a linear description of scientific research that distinguishes 'basic' and 'applied' research as separate extremes, it was noted by Donald Stokes in *Pasteur's Quadrant*[49] that the distinction is in reality far less clear and that much scientific research falls into the category of 'use-inspired basic research' (**Figure 3**). Much biomedical research, including medicinal chemistry falls into this category.

2.02.3 The Ethos of Science

"Nature is often hidden, sometimes overcome, seldom extinguished."

Francis Bacon, 1625

The traditional role of academic science, and of university function in general, has been to operate as a 'gift' economy to which members of this economy contribute intellectual and creative gifts. The rewards in this economy are to those who provide the most and the most valuable gifts and are in the form of academic honors, distinctions, and recognition.[50–52] This gift economy contrasts with the commodity economy of the market place whereby money constitutes both the price of entry and the recognition of participation. The gift economy model of academic science is entirely consistent with the ethos of science as defined by Robert Merton some 50 years ago.[50,53,54] Merton defined this ethos based on the values of the free and open exchange of knowledge, the unrestricted pursuit of this knowledge

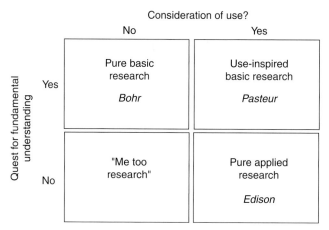

Figure 3 The quadrant model of scientific research. (Reproduced with permission from Stokes, D. E.; *Pasteur's Quadrant: Basic Science and Technological Innovation*; Brookings Institution Press: Washington, DC. 1997.)

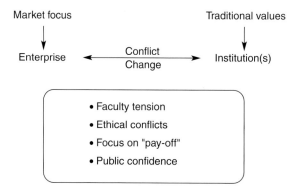

Figure 4 The conflict between the traditional and the new 'market' focus of universities.

independent of self-interest, and an acceptance that science is a product of nature and not of politics, religion, or culture. Merton defined four norms that characterized this view of science – "universalism", "communalism," "disinterestedness," and "organized skepticism." These norms define the intellectual commons of science – a freely generated and freely available pool of "certified knowledge." It is this environment that has characterized the role of the university and that has stood society well: this role may be changing and it imposes stress on the traditional structure and values of the academy (**Figure 4**).

From the specific perspective of medicinal chemistry Ralph Hirschmann, who has been long and sequentially associated with both the pharmaceutical industry (Merck Research Laboratories) and academia (University of Pennsylvania) the situation is clear (*see* 1.01 Reflections of a Medicinal Chemist: Formative Years through Thirty-Seven Years Service in the Pharmaceutical Industry):

> "Medicinal chemistry has always stood with one foot in the academic world and the other in the industrial world ... The roles of these two sectors need to be profoundly different if their combined efforts are to benefit the patient. It is the responsibility and function of academe to undertake fundamental research, and it is the task of industry to perform the developmental research required if the knowledge generated by the former is to lead to medicines and vaccines..."

Similarly, Sir James Black has written[55]:

> "I think that you should understand that at no point have I ever professionally made claims about what I thought that work would achieve, for the simple reason I've never thought that it was going to achieve anything other than answer a question. The challenge was to make a molecule which would have the properties of allowing you to answer a physiological question."

Finally, and to reinforce the previous statements, it is worthy of some note that between 1993 and 2002 only 14 new chemical or biological entities had their origins in academic laboratories out of the total of 355 new introductions (*see* 1.13 The Role of Small- or Medium-Sized Enterprises in Drug Discovery).

2.02.4 The University in a Market Environment

> "There is big pressure on universities now to go into places that are determined politically by the funding councils and I think that this very bad for discovery. There is too much pressure to go into applied areas."
>
> Robin Ganellin, 2004[56]

There is an increasing awareness that universities are becoming increasingly involved in a variety of commercial activities and that this may change their traditional role of contributors to and keepers of the intellectual commons. To be sure this concern is not new and was expressed almost a century ago by Thorstein Veblen in his volume *The Higher Learning in America*.[57] And the American land-grant university has always had a significant role in fostering the practical arts.[58] However, there is a concern well expressed by a number of academic authors that the increasing commercialism may change significantly both the roles and the prestige of universities.[59–62] This has been well expressed by Slaughter and Leslie[62]:

> "Increased global competition interacted with national and state/provincial spending priorities so that less money was available from government, when measured as a share of higher education revenue or as constant dollars per student. In the 1980's globalization accelerated the movement of faculty and universities towards the market…began to undercut the tacit contract between professors and society because the market put as much emphasis on the bottom line as on client welfare. The raison d'etre for special treatment of universities, the training ground for professionals, was undermined."

The commercialization of universities is not a uniquely American phenomenon, although the USA has certainly led the way. Increasingly, European and other universities are also following suit, driven by arguments that competitive scientific wealth in the global environment can be extracted from university research by commercializing and focusing its research.

Whether such approaches will be as successful as their promoters argue is yet to be determined. The successes of the past that led to the creation of the high technology hubs of San Francisco, Boston, San Diego, and elsewhere are due largely to events that had occurred prior to the Bayh–Dole legislation of 1980 and, in any event, were due only in part to the intellectual achievements of the university.[61,63,64] The temptation in universities to become increasingly large research institutions may, in fact, diminish their intrinsic creativity. Even in a Viagra-fueled world bigger is not always better, as was recognized by Theodore Schultz as early as 1980[65]:

> "We have become enamored of large research institutes and large programmes…but they are difficult to administer and they are subject to what is known in economics as diminishing returns to scale."

According to Jonathan Cole, former Provost of Columbia University, size brings a corollary set of problems. The increased dependence on industry, government, and business brings increased public exposure and hence the universities are increasingly susceptible to political and public manipulation and control.[66] This is not, of course, new. American universities have seen this before during the era of Senator McCarthy and their leaders did not stand up to be counted then either.[67] And prior to that both German and Russian science and their citizens paid an enormous price for the politicization of science by Hitler and Stalin, respectively.

The reflections of Derek Bok, former President of Harvard University, over a two-decade period are of some relevance here:

> "Alliances with industry will provide not only an alternative to federal largesse, but also revitalize intellectual life in universities, as corporations inject real world problems."[86]
>
> "The mounting requirements of the state might prove less onerous than the demands made by the market."[87]
>
> "I worry that commercialization may be changing the nature of academic institutions in ways which we will come to regret. By trying so hard to capture more money for their work, universities may compromise values that are essential to the continued confidence of the general public."[59]

But of course Adam Smith knew this almost 250 years ago[88]:

> "People of the same trade seldom meet together, even for merriment and diversion, but the conversation ends in a conspiracy against the public."

2.02.5 Globalization and Science in the Rich and Poor Worlds

"For unto whomsoever much is given of him much shall be required."

Luke 12: 48

Since 1945 a principal goal of the USA has been to maintain world leadership in the sciences. There is little doubt but that the USA has been enormously successful in science and technology as measured by a variety of indices including Nobel (and similar) prizes received, citations to publications, papers, and patents published, and the launching of technology-based industries, notably biotechnology. However, precise comparisons of scientific wealth in nations is a difficult task, and when citation comparisons are made with reference to both population and GDP, the smaller European countries (Scandinavia, the Netherlands, and Switzerland) all perform more strongly.[68,69] In significant part, the success of these nations is due to the role of universities. A comparison of the top 500 world universities published by the Shanghai Institute of Education and based on a number of criteria, including Nobel prizes, articles published in *Science* and *Nature*, numbers of publications and numbers of faculty members, shows that in the top 100 institutions there are 58 US and 31 European universities.[70] However, regardless of these differences it is striking that a mere 31 of the world's 193 countries produce 97.5% of the most cited papers in science.[69] These figures represent, of course, only one component of a grossly unequal world.[71]

More recently, the USA has announced a goal of unchallengeable world supremacy in both military and economic fields.[72] This goal is neither desirable nor attainable and is, in fact, delusional. It is most unlikely that in a globalized world one nation can maintain such supremacy. Indeed, there are already significant observations that the distribution of scientific wealth is shifting between nations and that the USA is losing its claimed dominance in the sciences.[73] In a comparison of science citations China and India are rising rapidly and the citation gap between the USA and the EU is closing.[74–76] Indeed, some comparisons suggest that in some fields the EU is ahead of the USA.[77] These changes are part of a global power shift in the making that may well end with a transfer of economic power from the West to the East, or at least a significant reduction of the existing dominance by the West.[78–82] According to Raghunath Mashelkar, Director-General of the Council of Scientific and Industrial Research of India:

"If India plays its cards right it can become by 2020 the world's number-one knowledge production center."[80]

2.02.6 Implications for Academic Medicinal Chemistry

"As for the future, your task is not to foresee but to enable it."

Antoine de Saint-Exupéry[89]

Outside of changes in our knowledge base of medicinal chemistry there are two important implications of the preceding discussions for the practice and achievements of medicinal chemistry in the academic environment. First, the changing nature of university-based science in general with its increasing emphasis on patenting, inventions, and rapid commercialization. As a component of the shift (or reduction) of economic and scientific dominance by the West it may be anticipated that the existing distribution of medicinal chemistry expertise will change at both the academic and industrial levels. Associated with this shift the compulsion by TRIPS (Trade Related aspects of Intellectual Property RightS) for noncompliant countries to shift from copying to creativity in the pharmaceutical (and other) fields will increase the need for medicinal chemistry development at both academic and industrial levels. There is significant evidence that this is already occurring through both outsourcing from the USA and the increase in drug discovery chemistry in both China and India.[78] Although major pharmaceutical companies have long outsourced components of drug discovery chemistry to US-based companies such as Millennium, Pharmacopeia, Albany Molecular, ArQule, and Discovery Partners, the shift is now to non-US-based companies, notably in India.[79–85] This dual stimulus of outsourcing and TRIPS will expand both the need and dimensions of medicinal chemistry education in India and elsewhere in Asia and it is to be anticipated that over the next one to two decades the teaching and practice of medicinal chemistry will increasingly be focused and delivered in Asia.

2.02.7 Conclusions

Medicinal chemistry is a discipline firmly grounded in organic chemistry. Traditionally, in fact, a subdiscipline of synthetic organic chemistry, medicinal chemistry has increasingly expanded at both the physical and biological interfaces of organic chemistry. This expansion represents a challenge both to definition as well as to the practice of teaching and research within an academic environment. Departments of medicinal chemistry have long been associated with schools and colleges of pharmacy, but this association is changing and medicinal chemistry, together with allied pharmaceutical science disciplines, is now increasingly associated with other departmental structures, notably chemistry. Additionally, with the globalization of science non-US and non-European countries are increasingly seen as significant academic and industrial science players and Asia will likely be a major player in both academic and industrial medicinal chemistry by the second and third decades of the twenty-first century.

References

1. Ganellin, C. R; Mitscher, L. A.; Topliss, J. G. *Annu. Rep. Med. Chem.* **1995**, *30*, 329–337.
2. Ganellin, C. R.; Mitscher, L. A.; Clement, C; Kobayashi, T.-H.; Kyburz, E.; Lafont, O.; Marcinal, A.; Tarzia, G.; Topliss, J. G. *Eur. J. Med. Chem.* **2000**, *35*, 163–174.
3. Busse, W. D.; Ganellin, C. R.; Mitscher, L. A. *Eur. J. Med. Chem.* **1996**, *31*, 747–760.
4. Mitscher, L. A.; Topliss, J. G. *Med. Res. Rev.* **1998**, *18*, 121–137.
5. Kwiram, A. L. *Reflections on Doctoral Education in Chemistry*; The Carnegie Foundation for the Advancement of Teaching, 2003.
6. Breslow, R. *The Doctorate in Chemistry*; The Carnegie Foundation for the Advancement of Teaching. 2003.
7. Re-envisioning the PhD http://www.grad.washington.edu (accessed April 2006).
8. Triggle, D. J; Miller, K. W. *Am. J. Pharm. Ed.* **1999**, *63*, 218–248.
9. Wu, S. *Academe*, **2005**, July/Aug. http://www.aaup.org/publications/Academe/2005/05ja/05jawu.htm (accessed April 2006).
10. The Royal Society of Chemistry and The Institute for Physics. *The Economic Benefits of Higher Education Qualifications*; Price Waterhouse Coopers: London, 2005.
11. Connor, S. *The Independent (UK)*, Dec 22, 2004. www.independent.co.uk (accessed April 2006).
12. National Center for Educational Statistics. *NAEP 2004 Trends in Academic Progress: Three Decades of Student Performance in Reading and Mathematics*. http://nces.ed.gov (accessed April 2006).
13. Brown, H. Challenges and trends in international student admissions. Presented at the Council for Graduate Schools, Mar 9, 2005, http://www.cgs.net (accessed April 2006).
14. National Academy of Sciences USA. *Implications of International Graduate Students and Postdoctoral Fellows in the United States*; National Academies Press: Washington, DC, 2005.
15. Oliver, J. National Science Foundation: Directorate for Social, Behavioral, and Economic Sciences – InfoBrief, Aug 2005. http://www.nsf.gov/publications/orderpub.jsp (accessed April 2006).
16. The Business Round Table. *Tapping America's Potential: The Education for Innovation Initiative*. http://www.businessroundtable.org (accessed April 2006).
17. Martineau, T.; Decker, K.; Bundred, P. *Health Pol.* **2004**, *70*, 1–10.
18. Elliott, L. *The Guardian (UK)*, Dec 19, 2004, www.guardian.co.uk (accessed April 2006).
19. Triggle, D. J.; Miller, K. W. *Am. J. Pharm. Ed.* **2002**, *66*, 287–294.
20. Broad, W. J. *New York Times*, May 5, 2004, www.nytimes.com (accessed April 2006).
21. Monastersky, R. *Chronicle Higher Education*, July 9, 2004, http://www.chronicle.com (accessed April 2006).
22. Teitelbaum, M. *The Public Interest*, Fall **2003**, pp 40–53. www.researchcaucus.com; www.thepublicinterest.com/current/article2.html. (accessed April 2006).
23. Mervis, J. *Science* **2003**, *300*, 1070–1074.
24. Kennedy, D.; Austin, J.; Urquhart, K.; Taylor, C. *Science* **2004**, *303*, 1105.
25. Lougheed, T. *New Scientist*, Aug 27, 2005, pp 48–49.
26. Mooney, C. *The Republican War on Science*; Basic Books: New York, 2005.
27. Triggle, D. J. *Drug Dev. Res.* **2005**, *63*, 112–120.
28. Triggle, D. J. *Annu. Rep. Med. Chem.* **1993**, *28*, 343–350.
29. Triggle, D. J. *Drug Dev. Res.* **2003**, *59*, 269–291.
30. Erhardt, P. W. *Pure Appl. Chem.* **2002**, *74*, 703–785.
31. Wess, G.; Urman, M.; Sickenberger, B. *Angew. Chem. Int. Ed. Engl.* **2001**, *40*, 3341–3350.
32. Davidov, E. J.; Holland, J. M.; Marple, E. W.; Naylor, S. *Drug Disc. Today* **2003**, *8*, 175-183.
33. Henry, C. M. *Chem. Eng. News* **2003**, May 19, 45–56.
34. Winnacker, E. L. *Curr. Opin. Biotechnol.* **2003**, *14*, 328–331.
35. Lombardino, J. G.; Lowe, J. A., III. *Nat. Rev. Drug Disc.* **2004**, *3*, 853–862.
36. Russo, E. *Nature* **2003**, *424*, 594–596.
37. Residential School on Medicinal Chemistry, Drew University, Madison, NJ, USA, http://www.depts.drew.edu/resmed; Medicinal Chemistry Residential School, Royal Society of Chemistry, University of Nottingham, UK, http://www.rsc.org/pdf/education/MedChem2005.pdf; Leiden/Amsterdam Center for Drug Research, Noordwijkerhout, The Netherlands, http://www.medchem.leidenuniv.nl/; Swiss Course on Medicinal Chemistry, New Swiss Chemical Society, Leysin, Switzerland, http://www.pharma.ethz.ch; University of California at San Diego Extension Medicinal Chemistry, USA, http://bioscience.ucsd.edu/MedChemSS.html; American Chemical Society Short Courses, http://www.acs.org (accessed April 2006).
38. Austin, C. P.; Brady, L. S.; Insel, T. R.; Collins, F. S. *Science* **2004**, *306*, 1138–1139.
39. *Genome Technol.*, Jan/Feb 2005, pp 27–35.
40. Nowotny, H.; Scott, P.; Gibbons, M. *Rethinking Science. Knowledge and the Public in an Age of Uncertainty*; Polity Press: London, 2001.
41. Newman, J. H. *The Idea of a University*; Yale University Press: New Haven, CT, 1996 (Originally published in 1899).
42. Pelikan, J. *The Idea of The University: A Reexamination*; Yale University Press: New Haven, CT, 1992.
43. Bacon, F. *Novum Organum*; 1620.
44. Readings, B. *The University in Ruins*; Harvard University Press: Cambridge, MA, 1996.
45. Triggle, D. J. *Am. J. Pharm. Ed.* **1998**, *4*, 207–217.

46. Kirp, D. L. *Shakespeare, Einstein and the Bottom Line: The Marketing of Higher Education*; Harvard University Press: Cambridge, MA, 2003.
47. *The Economist*, Oct 4, 1997, pp 1–20. http://economist.com (accessed April 2006).
48. V. Bush, *Science: The Endless Frontier*, A report to the President, July **1945**. www.nsf.gov/od/lpa/nsf50/vbush/1945.htm (accessed April 2006).
49. Stokes, D. E. *Pasteur's Quadrant: Basic Science and Technological Innovation*; Brookings Institute Press: Washington, DC, 1997.
50. Merton, R. K. *Philos. Sci.* **1938**, *5*, 321–337.
51. Merton, R. K. *Am. Sociol. Rev.* **1957**, *22*, 635–659.
52. Merton, R. K. *The Sociology of Science*; University of Chicago Press: Chicago, IL, 1973, pp 267–278.
53. Baird, D. Scientific Instrument-Making, Epistemology and the Conflict between the Gift and Commodity economies. http://scholar.lib.vt.edu/ejournals/spt/v2n3n4/baird.html. Quoted by J. Kovac, Gifts and Commodities Inchemistry, http://www.hyle.org/journal/issues/7/kovac.htm
54. Menninger, J. R. *Chronicle Higher Education*, Sep 14, 2001. http://chronicle.com (accessed April 2006).
55. Black, J. Daydreaming Molecules. In *Passionate Minds: The Inner World of Scientists*; Wolpert, L., Richards, A., Eds.; Oxford University Press: Oxford, 1997, pp 124–129.
56. Ganellin, C. R. *Drug Disc. Today* **2005**, *9*, 158–160.
57. Veblen, T. *The Higher Learning in America: a Memorandum on the Conduct of Universities by Businessmen*; Transaction Publishers: New York, 1918.
58. The Morrill Act of 1862. http://www.ourdocuments.gov/doc.php?doc=33 (accessed April 2006).
59. Bok, D. *Universities in the Market Place*; Princeton University Press: Princeton, NJ, 2003.
60. Krimsky, S. *Science in the Private Interest*; Rowan and Littlefield: Oxford, UK, 2003.
61. Washburn, J. *University Inc.: The Corporate Corruption of Higher Education*; Basic Books: New York, 2005.
62. Slaughter, S.; Leslie, L. L. *Academic Capitalism: Politics, Policies and the Entrepreneurial University*; Johns Hopkins University Press: Baltimore, MD, 1997.
63. Triggle, D. J. *Drug Dev. Res.* **2005**, *63*, 139–149.
64. Mowery, D. C.; Nelson, R. R.; Sampat, B. N.; Ziedonios, A. A. *Res. Pol.* **2001**, *30*, 99–119.
65. Schultz, T. W. *Minerva* **1980**, *18*, 644–651.
66. Cole, J. R. *Daedalus* **2005**, *Spring*, 5–17.
67. Lamont, C. *Freedom Is as Freedom Does: Civil Liberties in America*; Continuum Press: New York, 1957.
68. May, R. M. *Science* **1997**, *275*, 793–796.
69. King, D. A. *Nature* **2004**, *430*, 311–316.
70. Shanghai Institute of Technology, 2004. http://ed.sjtu.edu.cn/ranking.htm (accessed April 2006).
71. Sachs, J. *The End of Poverty*; Penguin Press: New York, 2005.
72. *Rebuilding America's Defenses*. Report of the Project for the New American Century, Washington, DC, 2000.
73. Broad, W. J. *New York Times*, May 3, 2004. www.nytimes.com (accessed April 2006).
74. Shelton, R. D.; Holdridge, G. M. *Scientometrics* **2004**, *60*, 353–363.
75. ScienceWatch. U.S. falls in physical science output. http://www.sciencewatch.com/sept-oct99/sw_sept-oct99_page1.htm (accessed April 2006).
76. ScienceWatch. U.S. slides in world share continues as European Union, Asia Pacific advance. http://www.sciencewatch.com/july-aug2005/sw_july-aug2005_page1.htm (accessed April 2006).
77. Hoge, J. F., Jr. *Foreign Affairs* **2004**, *Jul/Aug* 2–7.
78. Cookson, C. *Financial Times*, June 9, 2005, p 15. www.ft.com (accessed April 2006).
79. Friedman, T. L. *New York Times*, June 3, 2005. www.nytimes.com (accessed April 2006).
80. Mashelkar, R. A. *Science* **2005**, *307*, 1415–1417.
81. Harris, B. *Naturejobs* **2005**, Feb 23, p 902. www.nature.com (accessed April 2006).
82. *The Economist*, Mar 5, 2005, pp 3–16. www.economist.com (accessed April 2006).
83. *Good News India*, Nov 2000, www.goodnewsindia.com (accessed April 2006).
84. Siva Sankar, Y. Interview with Dr. R. A. Mashelkar: India can be a Biotech Superpower in the 21st century. Nov **2000**. http://www.rediff.com/money/2000/nov/10inter.htm (accessed April 2006).
85. Varawalla, N. *PRA Int.*, Autumn **2004**, www.praintl.com (accessed April 2006).
86. Bok, D. *Beyond the Ivory Tower*; Harvard University Press: Cambridge, MA, 1982.
87. Bok, D. *Universities and the Future of America*; Harvard University Press: Cambridge, MA, 1990.
88. Smith, A. *The Wealth of Nations*; 1776.
89. Saint-Exupéry, A. de. *The Wisdom of the Sands*; 1950.
90. Department of Chemistry. www.chem.utah.edu (accessed April 2006).

Biography

David J Triggle, PhD, is the University Professor and a Distinguished Professor in the School of Pharmacy and Pharmaceutical Sciences at the State University of New York at Buffalo. Professor Triggle received his education in the UK with a PhD degree in chemistry at the University of Hull. Following postdoctoral fellowships at the University of Ottawa (Canada) and the University of London (UK) he assumed a position in the School of Pharmacy at the University at Buffalo. He served as Chairman of the Department of Biochemical Pharmacology from 1971 to 1985 and as Dean of the School of Pharmacy from 1985 to 1995. From 1996 to 2001 he served as Dean of the Graduate School and from 1999 to 2001 was also the University Provost. He is currently the University Professor, in which capacity he teaches bioethics and science policy, and is President of the Center for Inquiry Institute, a secular think tank located in Amherst, New York.

Professor Triggle is the author of three books dealing with the autonomic nervous system and drug–receptor interactions, the editor of a further dozen books, some 280 papers, some 150 chapters and reviews, and has presented over 1000 invited lectures worldwide. The Institute for Scientific Information lists him as one of the 100 most highly cited scientists in the field of pharmacology. His principal research interests have been in the areas of drug–receptor interactions, the chemical pharmacology of drugs active at ion channels, and issues of graduate education and scientific research policy.

2.03 An Industry Perspective

W Wierenga, Neurocrine Biosciences, Inc., San Diego, CA, USA

© 2007 Elsevier Ltd. All Rights Reserved.

2.03.1	**Business Model**	**99**
2.03.1.1	Pharmaceuticals	100
2.03.1.2	Specialty Pharma	102
2.03.1.3	Diversified Model	102
2.03.1.4	Biotech	102
2.03.2	**Pharmaceutical R&D**	**104**
2.03.2.1	How to Organize for Drug Discovery and Drug Development	106
2.03.2.2	Size (Biotech)	110
2.03.2.3	Philosophy	111
2.03.2.3.1	Target versus disease	112
2.03.2.3.2	New molecule entity versus life cycle management	115
2.03.2.3.3	Chemical versus biological	115
2.03.2.3.4	Best in class versus first in class	118
2.03.2.3.5	Decision making versus organizational alignment	120
2.03.2.3.6	Fail fast versus follow the science	121
2.03.2.4	Culture	121
2.03.2.5	Partnering	123
2.03.3	**R&D Productivity**	**123**
2.03.3.1	Drug Discovery	124
2.03.3.2	Drug Development	126
2.03.3.3	Infrastructure	130
2.03.3.3.1	Intellectual property	130
2.03.3.3.2	Information technology	131
2.03.3.3.3	Regulatory	131
2.03.3.4	Technology	132
2.03.4	**Conclusion**	**133**
	References	**134**

2.03.1 Business Model

The landscape of what is typically referred to as the life sciences pharmaceutical sector is usually divided into four categories by the investment community and Wall Street. The first two categories are the pure play pharmaceuticals businesses and the specialty pharma businesses. A third category is the diversified businesses which typically includes consumer products, chemical or agriculture as part of a pharmaceutical business. The last category is biotechnology. Until about 1980 only three of these four categories coexisted, and, in fact during the twentieth century these categories underwent substantial changes in definitions. The biotechnology sector arose with the advent of recombinant DNA methods in the 1970s, generating, for the first time, the opportunity to make proteins in cell culture in the laboratory in a designed fashion. Of course, prior to that time there were protein-based drugs, such as insulin and growth factors, but these were derived through laborious extraction processes from natural sources. From today's vantage point this landscape of four business models is now well established and likely to extend for some time into future decades. Nonetheless, through the ever-changing consolidation and rebirth of new companies within the overall industry there will undoubtedly be significant changes in populating these categories.

2.03.1.1 Pharmaceuticals

Inherent in the processes of discovering and developing drugs are the fundamental aspects of understanding diseases, their cross-species penetrance, and the capability of modeling these diseases in the laboratory. The utilization of this knowledge predictably led to the development of drugs for both the treatment as well as the prevention of disease or disease progression. In addition, there is knowledge generated to identify the disease in the first instance (diagnostics) and evaluate the progression of the disease in the patient. To the extent that these diseases are similar between animals and humans, the exploitation of this knowledge leads to the development of veterinary products as well. Over the last century the business model for pharmaceuticals has frequently included the development and commercialization of products as treatments for disease, vaccines for the prevention of disease, diagnostics for the detection of disease, and utilization of all three of these in animal models of disease. While the focus of this chapter is not on veterinary medicines per se, a significant business driver in veterinary medicine has been the utilization of drugs in food-producing animals; specifically, the use of antibiotics and hormones for the efficient growth of food-producing animals and the prevention of concomitant disease has revolutionized the animal-products industry. However, over time the business of veterinary medicine has developed more and more in parallel with human medicine, including the treatment of cancers and chronic inflammatory diseases such as arthritis in animals. In addition, the reimbursement side of this business has also begun developing parallels with human medicine with the advent of insurance for payers for treatment of animal diseases. Therefore, we can reasonably expect the pharmaceutical companies will continue to be interested in veterinary medicine as well.

All of the pharmaceutical companies have been involved in therapeutics, and their growth has been driven by success in therapeutic drugs. Some of the companies have also been interested in prevention: vaccines. While vaccines have had tremendous impact on human and animal health in disease prevention, interest was not universal within the industry due, in part, to the special requirements of manufacturing vaccines. In addition, while the return on investment was never as high as therapeutics, it dropped significantly with perceived safety issues and increased litigation in the 1970s. A resurgence of interest has occurred, due to new biotechnology-based approaches, reproducible methods of manufacturing, and increased demand, both on medical need and government support. Biotech companies and large pharmaceutical companies have been coparticipants in this reemergence of interest in vaccines.

In a somewhat analogous fashion, diagnostics has only been part of a subset of the pharmaceutical companies' business interests. The margins have been lower than therapeutics and the product cycle shorter. However, with advances in biotechnologies such as polymerase chain reaction (PCR) and, more recently, genomics and proteomics, there is a resurgence of interest in diagnostics. Furthermore, there is the expectation that many of tomorrow's drugs will be co-marketed with a diagnostic.

From today's vantage point we view the pharmaceutical business as a very complex business that is both capital and time intensive. It has very high thresholds for entry because of the time and capital commitments coupled with the high knowledge-base requirements. The term fully integrated pharmaceutical company (FIPCO) is used routinely today to describe this model. The top 20 pharma companies today are all FIPCOs and possess all the requisite infrastructure for basic research, drug discovery, drug development, and manufacturing for commercialization for their products as well as marketing and sales forces for distribution and sales of their products. The most obvious value driver, and a significant barrier to entry, is the extraordinary resources required for the discovery and development of drugs. **Table 1** shows the largest pharmaceutical companies in 2004 and their aggregate R&D spend of nearly US$40 billion. This translated into about US$250 billion in revenues for those companies in 2004. The global pharmaceutical sales for 2004 was US$550 billion.

However, this was not always so. It was not until the 1950s and 1960s that the concept of a fully integrated organization really materialized. R&D spend at that time was not double-digit percentages of sales, as it is today, and the key value drivers were ability to manufacture and penetrate the European, North America, and Japanese marketplaces. This required manufacturing as well as sales in a number of countries and represented the internationalization of the business. The larger pharmaceutical companies of that time outsourced very few of their services and often times included internal services for maintenance, employee activities, and transportation. It is important to recognize however, that the FIPCO model was one that evolved over a number of prior decades. This is perhaps most readily understood by exemplification; the consolidation of a number of companies that led to today's GlaxoSmithKline.

The parentage of GlaxoSmithKline can be traced back to Beecham's Pills in 1842, John K. Smith & Company founded in 1841, Burroughs Wellcome & Company begun in 1880, and McLeans Ltd. founded in 1919.[1,2] Three of these names have already disappeared from the current GlaxoSmithKline moniker but they were very substantial contributors to today's present GlaxoSmithKline corporation. Beecham's Pills was a laxative business that grew into a

Table 1 Aggregate R&D expenditures of major phamaceutical companies, 2004

Company	R&D spend (US$ billion)	Percentage of sales
Pfizer	7.7	15
Sanfi-Aventis	5.2	16
GlaxoSmithKline	4.8	16
Novartis	4.0	14
AstraZeneca (AZ)	3.8	18
Roche	3.8	16
Lilly	2.6	19
Bristol-Myers Squibb (BMS)	2.4	11
Wyeth	2.4	14
Amgen	1.9	20

diversified pharmaceutical manufacturer in the mid-twentieth century and ultimately became a major international pharmaceutical producer of antibiotics in the late twentieth century. In 1989 Beecham merged with the then Smith Kline Beckman. The Beckman name was dropped and Smith Kline Beecham became the name of the new entity. The name Glaxo has its roots in milk powder production back in the late nineteenth century established, originally by Joseph Nathan in Wellington, New Zealand. Glaxo's move into prominence was relatively recent in the timetable of pharmaceutical company maturation and was evident with its blockbuster drug, Zantac, which evolved in the 1970s and 1980s. However, it was the merger with Wellcome in 1995, generating GlaxoWellcome, that really provided a diversified pharmaceutical business. Burroughs Wellcome, founded in London by two American pharmacists, Henry Wellcome and Silas Burroughs in 1880, ultimately straddled the Atlantic between the USA at Research Triangle Park and Wellcome in London, majority owned by the Wellcome Foundation in London. They had substantial competencies in chemotherapies and were one of the early entrants into the production of interferon. In 2000, after a couple of attempts, Glaxo Wellcome merged with Smith Kline (having dropped Beecham from the name) and is now the third largest pharmaceutical company. However, even this brief litany does not do justice to the evolution of becoming a FIPCO, for, within the background of the Beecham history, there is the SE Massengill Company, perhaps (unfortunately) best known for its elixir of sulfanilamide, which included the toxic cosolvent diethyleneglycol. Beecham also acquired McLeans Ltd. (McLeans toothpaste) and Glaxo acquired Meyer Laboratories in 1978.

The big pharma model has been remarkably robust and continues today with more than a half a decade of validation. The model has been the basis for significant earnings flow and growth, probably unsurpassed by any other industry. Undoubtedly the driver for this has been the development of new treatments that have not only extended life but have significantly improved the quality of life of millions of people. Nonetheless, the history of pharma is littered with examples where a successful product itself is not sufficient for sustaining the business model. Examples include ibuprofen from Boots, naproxen from Syntex, and sodium chromoglycate from Fisons. However, even companies with a portfolio of products have ultimately failed to sustain their independence and identity, even though they were FIPCOs. Many of these companies like Squibb, Roussel, Robbins, Miles, Lederle, Kabi, Boehringer Mannheim, Upjohn, and others were acquired with the belief that size and scale are the ultimate drivers to sustaining growth. Complicating this picture, however, is that there has been a strong financial driver for mergers and acquisitions, rather than just size/scale responding to a changing environment. In addition to size and scale there has been the belief that capturing cost synergies is important. Cost synergies have materialized; unfortunately, virtually all pharmaceutical companies that have participated in mergers and acquisition activities have failed to increase market share at the end of the day. It was only a decade ago that the largest pharma company had revenues representing only 4% of the total market. Today, Pfizer has 13% of the global market share, by virtue of its recent acquisitions of Parke Davis/Warner Lambert and Pharmacia. However, 13% share is less than the additive shares of the three companies prior to merger. Further growth of Pfizer is unlikely to occur by more consolidation, if only for antitrust reasons. In large part because of this there have been alternative models for life sciences and pharmaceuticals and these have included diversified, specialty pharma, and biotech.

2.03.1.2 Specialty Pharma

Specialty pharma has arisen as a more prominent sector in the overall pharmaceutical landscape over the last 10–15 years. This is primarily due to the move of big pharma into blockbuster driven sales and marketing with increasing emphasis on primary care sales forces and direct to consumer advertising. Therefore companies with a focus on specialty products have arisen by in-licensing products and investing more focused sales efforts on already marketed products. This sector currently enjoys significant value assessment because of the absence of significant R&D costs together with lower sales costs associated with specialty product opportunities. In parallel there has been a growing resurgence and interest in generic products. During the mid-1990s generic products were plateauing, after a significant rise in the 1980s, and big pharma were selling off their generics businesses to specialty companies focusing on that sector. However, in the first decade of this century the number of large products opportunities that have become generic has mushroomed, and this has precipitated a resurgence in interest in not only the specialty pharma but also big pharma acquiring these products and competing in this landscape. In parallel, some specialty pharma have also moved further back into the value creation paradigm and in-licensed early staged product development opportunities to carry them forward to commercialization. They are back integrating therefore into ultimately becoming FIPCOs. Similarly, several companies in the service sector, that is contract research organizations (CROs), particularly those focusing on clinical services, have also in-licensed development stage compounds to ultimately attempt to commercialize them.

2.03.1.3 Diversified Model

As noted earlier, a number of today's pharmaceutical companies had their roots in a strong chemical business. More recently a number of companies with strength in that area have attempted to enter into the pharmaceutical arena. Companies such as Monsanto, Dupont, Kodak, and Proctor & Gamble, to name a few, have attempted to add pharmaceuticals to their diversified business interests. One of the rationales offered was to balance a cyclic business that has significant peaks and valleys such as pharmaceuticals, with more stable businesses, thus modulating the overall swings in a company's fortunes. However, there is usually only modest synergy, if any at all, between the various businesses. There are different competencies in research, the development areas, and ultimately sales and marketing, and this basically results in multiple companies within the corporation contributing to the bottom line of a corporation in an independent fashion. Companies that have recognized this have been able to manage this diversification. Johnson & Johnson is one such example. The key to success seems to be allowing those various companies to operate with autonomy. However, companies that have attempted to integrate these have rarely been successful and ultimately have lead to deconsolidation. Examples of this include Warner Lambert, Bristol-Myers Squibb (BMS), American Home and, most recently, Bayer.

2.03.1.4 Biotech

Biotechnology began in the 1970s at the commercial level with the formation of Cetus (1971), Biogen (1976), Chiron (1977), and Genex (1977). Shortly thereafter one of the most well-known biotech companies, Genentech, began and this was followed by a substantial increase in biotech start-ups securing funding through venture capital and ultimately proceeding to become public companies. The exclusive focus of these biotechnology companies was the production of protein therapeutics with recombinant DNA technology. Some of these proteins were replacements for already marketed protein therapeutics; many were new. The former category includes such products as insulin, growth hormone, and interferon and the latter category includes such products as tissue plasminogen activator (TPA), erythropoietin (EPO), and granulocyte colony stimulating factor (G-CSF). In the 1980s the breadth of therapeutic opportunities increased particularly as it began to include the development of antibodies as therapeutic proteins. However, the incubation phase for success with antibodies was longer than anticipated, and really did not materialize until the mid-1990s. Even with biologicals as products, it is still a decade of time and investment to the first product (**Figure 1**).

In the intervening time a new type of biotech company came into being. These new companies were start-ups that focused on small molecules or developing tools for the discovery of small molecules. In essence, these start-ups competed directly in the landscape of big pharma. These neopharma of course had limited funding, staffing, and experience, but were offering new targets, faster, smarter ways of discovering drugs and an entrepreneurial enthusiasm for risk-taking in a very risk-tolerant environment.

Unlike their former biotechnology/biological cousins in the biotech sector, the investment in the small molecule type drug was relatively modest until one proceeded into later stage clinical evaluation. The challenge for developing biologicals was the manufacturing of these biologicals and the investment in costly facilities early in the development

Company	Founded	IPO	Years to IPO	First product launch	Product	Years to first product	First full-year profit	Years to profit from founding	Years to profit from IPO
Amgen	1980	1983	3	1989	Epogen	9	1986	6	3
Biogen	1978	1983	5	1989	Interferon alpha	11	1989	11	6
Cephalon	1987	1991	4	1999	Provigil	12	2001	14	10
Chiron	1981	1983	2	1990	HCV diagnostics	9	1990	9	7
Genentech	1976	1980	4	1982	Protropin somatrem	6	1979	3	NA
Genzyme	1981	1986	5	1988	Clindamycin	7	1991	10	5
Gilead	1987	1992	5	1996	Vistide	9	2002	15	10
Idec	1986	1991	5	1997	Rituxan	11	1998	12	7
MedImmune	1988	1991	3	1991	CytoGam	3	1998	10	7

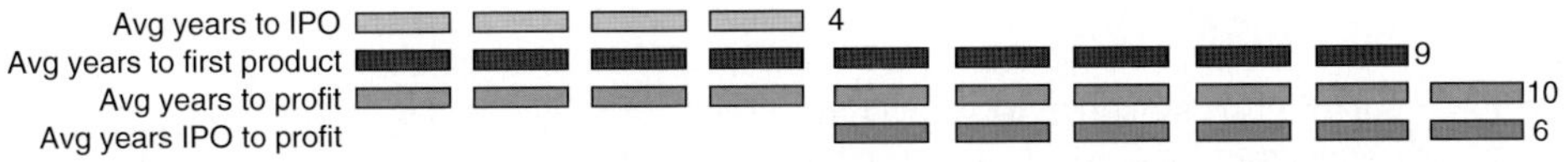

Avg years to IPO 4
Avg years to first product 9
Avg years to profit 10
Avg years IPO to profit 6

Figure 1 Selected profitable biotech companies and the time it took for them to sell an initial public offering (IPO), launch their first product, and turn the corner on profitability. Genentech was able to reach and sustain a profit prior to launching its first product through collaborative revenue. Chiron became profitable based on revenues from the sale of extrahepatic hepatitis C virus (HCV) diagnostics through its joint venture with Ortho. (Reproduced with permission from *BioCentury publications, Inc.*, September 3, 2002, page A4 of 21.)

cycle of the product. This pioneering work in biologicals manufacturing was primarily unique to biotechnology. However, the capability to scale up small molecule drugs, and manufacturing them, of course, had generations of experience in big pharma. It was natural then that the neopharma, small molecule biotech companies would look to partner relatively early on with big pharma to exploit their respective strengths and weaknesses. This paradigm of corporate partnering was in fact happening on a modest scale in the 1980s, but grew to be very substantial in the 1990s and, in fact, represented a validation of the technology of the small biotech company.

Another significant opportunity arose in the mid-1990s in biotech and it revolved principally around two major initiatives. The first initiative was the sequencing of the human genome and the genomes of number of species. The second was the much more diversified but nonetheless significant investment in automation, miniaturization, and informatics. The combination of these two waves of technology development and information development led to a significant paradigm change in the discovery of drugs. This paradigm change was the large-scale adoption of an already evolving philosophy of molecular targets for drugs, irrespective of whether the drugs are biologicals or small molecules. With the initiatives in these two domains being primarily in biotech, this also represented a significant partnering opportunity for big pharma, with their realization that they needed to access this information and these technologies. The scale of this was unprecedented and required in many ways a substantial increase in the investment in R&D for drug discovery and drug development. Furthermore, this changing paradigm likely contributed to the reduction in productivity in the development of new drugs during the 1990s by big pharma and, even to the current time. During this time biotechnology matured and companies such as Genentech and Amgen achieved market capitalizations equivalent to big pharma with even more enviable price/earnings (P/E) ratios. Biotechnology demonstrated not only that proteins could be drugs but that they could have a substantial impact on the treatment of diseases heretofore not treated with small molecules. Often these biologicals were at a higher cost, but the cost–benefit ratio was acceptable. So what was once a chemistry-driven industry has now become appropriately a blend of chemistry and biologicals with a growing emphasis on the molecular understanding of the basis of diseases. Success depended on the integration of the biological and chemical sciences to discover drugs. In 2005 there are at least 500 publicly traded biotechnology companies in the world and they are providing an increasing number of new discoveries and drug development for the overall industry. Eleven of the 80 blockbuster drugs in 2004 originated from biotech companies.

Within the last several years a new type of start-up company has arisen which focuses exclusively on drug development. These have been termed NRDOs (no research, development only). Given the history of biotechnology, it is hard to call these biotech companies, as they are almost the antithesis of biotech. These companies, like specialty pharma, minimize investment in research and find their drugs by in-licensing. They secure presently undervalued assets, perform clinical studies, add value, and then partner with commercial organizations, big pharma, or ultimately build their own sales organizations. Venture capital, at this time, appears to favor this business model as it shortens the time to a product-driven revenue stream over the more traditional biotech model, and presumably reduces risk, although that remains to be seen. In a similar vein is a growing interest in repurposing old drugs and/or resurrecting old drug candidates that never quite made it.

Lastly, the contribution of CROs and the increasing importance of outsourcing have contributed dramatically to this changing landscape of biotechnology relative to big pharma. Nearly every element of the competencies and activities needed for drug discovery, development, and commercialization can be outsourced. This outsourcing now extends globally and is an extremely competitive environment. To this end, small drug discovery companies can contract out toxicology studies, formulation work, clinical trials, regulatory services, and ultimately manufacturing. Many recent drug development companies are virtual in the sense that they have a small group of management that oversees the development of drugs they have in-licensed, completely dependent on CROs. Indeed even big pharma contract out a substantial amount of their clinical trials and, to varying degrees, manufacturing and information management.

2.03.2 Pharmaceutical R&D

Pharmaceutical R&D is most often described with a picture. This picture, an example of which is shown in **Figure 2**, depicts a sequence of events, over time, which are key milestones in the drug discovery and development process. In addition, with each of these key milestones there are a series of competencies or capabilities that are required to achieve these particular milestones. Typically these key milestones or accomplishments start with the identification of a drug candidate based on laboratory generated knowledge, followed by the preclinical evaluation of the drug candidate for appropriate safety in animals. Subsequently, a package of information is assembled, typically in the form of an investigative new drug application (IND), that includes a description of the drug, a process of how it is made, the metabolic profile of the drug, the safety profile and the pharmacokinetic performance of the drug in animals, and of course, its mechanism of action, in vivo pharmacology, and intended treatment of patients with diseases appropriate to the drug's mechanism of action.

The next stage of evaluation is clinical studies in human subjects, typically divided into Phase I, Phase II, and Phase III. These primary phases of clinical evaluation represent increasingly extensive exposure to greater number of subjects and patients over time. Phase I is the initial exposure of the drug, typically in volunteers, in a single dose followed by a multiple-dose format. Key questions to be addressed include the safety evaluation of the drug as well as the pharmacokinetics of the drug. Phase II typically represents the evaluation of the drug, usually at several doses, within a patient population for which the drug was originally intended. Often times there are several Phase II studies which explore the full dosage range of the drug in patients together with different lengths of time of treatment. The end result of these studies is to establish the appropriate efficacious dose or dose range for the larger Phase III clinical trials. These studies involve a much greater number of patients over a more extensive treatment period, followed by a safety evaluation on longer-term treatment. (During the 1990s, a series of conferences were held to develop harmonized guidelines between regulators in Europe, Japan, and the USA for use by drug developers. These International Conference on Harmonization (ICH) guidelines stipulate that at least 100 patients are required for

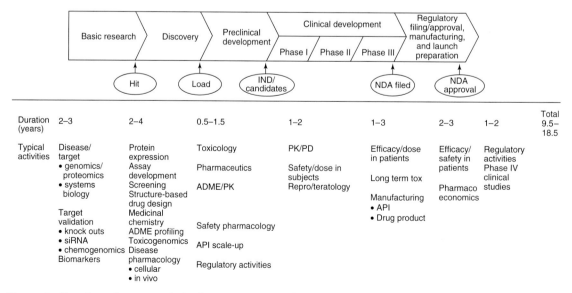

Figure 2 The drug discovery and development process. ADME, absorption, distribution, metabolism, excretion; PD, pharmacodynamics; PK, pharmacokinetics.

longer-term safety evaluation.) Ultimately, the package of information from the clinical studies, together with much more extensive toxicology studies including up to 2 years evaluation in animals, plus extensive manufacturing data on the active ingredient as well as the final drug product, are assembled into a new drug application (NDA) and submitted to the regulatory authorities. Typically the review period for these NDAs is on the order of 1–2 years and the drug is then approved for commercial distribution. The overall process from project initiation to approval usually extends beyond 10 years. The duration of the various phases are shown as ranges in **Figure 2**. The activities between lead development candidate and IND are where regulations begin to impact and this regulation continues through approval and beyond. Therefore, the duration of these later phases or stages is more consistent, since many of the activities are common to most drugs. There are exceptions where regulators, such as the Food and Drug Administration (FDA), have allowed for abbreviated toxicology studies in animals or accelerated approval based on a single pivotal clinical study.

Drug discovery and drug development can be described as a process but also as a complex and interdependent array of multidisciplinary competencies. Throughout the history of drug discovery and development an essential competency has been the ability to synthesize organic compounds. While historically some drugs were secured from natural sources, and a few still are today, most drugs are synthesized in a laboratory and then scaled up in much larger reactors and equipment in manufacturing facilities. Therefore, primary prerequisites for drug discovery are the competencies of organic chemistry synthesis and structure determination. These expertises are required at the beginnings of the process of drug discovery and are also required throughout the drug development path to commercialization. However, the primary need is in the earlier stages of drug discovery and drug development. Other competencies, principally in the domain of biology, are required in the earlier stages of drug discovery and are critical to dissecting pathways to understand the etiology of diseases. Similarly, these biological expertises are important for establishing laboratory surrogates for disease pathways, and for generating tools for understanding mechanism of action for drug candidates. These scientific disciplines and expertise are presented in **Table 2**. Some of the more recent additions to the scientific disciplines and expertises required for an integrated R&D enterprise include computational and informatic technologies, as well as the competencies of genomics and proteomics to better translate laboratory information to the clinic regarding the drug and the disease. This expertise frequently partners with pharmacokinetics to understand exposure response relationships and establish what are termed PK/PD correlations, that is, pharmacokinetics/pharmacodynamics. (PD are surrogate markers that measure the impact of the drug on the patient's disease or on the patient per se (e.g., safety).) Another recent trend has been to establish better models in the laboratory for preclinical predictability of potential clinical findings for safety and ADME (absorption, distribution, metabolism, excretion). For example, laboratory tools and assays to predict potential liabilities due to cytochrome P450 metabolism or inhibition, QTc prolongation, certain liver toxicities, and substrates for unique transporters are now routine in today's modern drug discovery laboratories. The goal for these kinds of activities is to minimize the failure rate that occurs when the drug enters human clinical study, either for pharmacokinetic reasons, safety reasons, or efficacy reasons.

The pharmaceutical value paradigm is depicted in **Table 3** and brings together some of these competencies by phase or stage of development and commercialization of a drug. While some of the competencies bridge a number of these areas of expertise in drug discovery, development, manufacturing, marketing and sales, nonetheless there are important differentiating characteristics between some of these stages of the pharmaceutical value paradigm. Many of the activities in drug development, manufacturing and commercialization are readily scalable, and in today's environment, readily outsourced. On the other hand, many of the activities in drug discovery are not readily scalable and require critical, state-of-the-art expertise, but often on a modest scale. Given the complexity of the underlying sciences, many of these activities are collaborative in nature and are often interfaced with academic investigators or scientists in research institutes funded by various government agencies or foundations. Similarly, while the later stages of drug development are capital intensive, the earlier stages of drug development, and indeed drug discovery, are 'human capital' intensive. Typically the proportion of the R&D budget that is allocated to drug discovery is on the order of 25%, while 75% is used for drug development. The more significant contributors to the increasing costs of drug development are increasing regulatory requirements, increasing size and breadth of clinical trials, as well as their inherent cost, and also the scale-up of the drug entity under development to prepare for ultimate commercial scale. In the context of the overall pharmaceutical value paradigm, drug discovery and development represent about 20% of the overall cost, with another 10% incurred for manufacturing, 10% for marketing, and, ultimately, 25% for sales and distribution.

Overall the drug discovery and development process is complex, characterized by many more failures than successes, and is time and capital intensive. It is ultimately characterized by long cycle times and extraordinary efforts at gathering data from diverse sources and assembling usable knowledge into a dossier for a new drug registration.

Table 2 Scientific disciplines and expertises required in the integrated R&D enterprise

Biology	
	Molecular biology
	Protein expression/purification
	Electrophysiology
	Systems biology
	Cell biology
	Microbiology
	Genomics
	Computational biochemistry
	Pharmacology
	Proteomics
	Bioinformatics
Chemistry	
	Synthetic organic
	Computational
	Process (scale-up)
	Medicinal
	Spectroscopy
	Engineering
	Analytical
	Crystallography
	Radiolabeling
Pharmaceutics	
	Formulation/excipients
	Encapsulation
	Analytical
	Tableting
	Engineering
	Injectables
Pharmacokinetics/metabolism	
	Pharmacokinetics
	Analytical
Medical regulatory	
	Clinical pharmacology
	Biostatistics
	Clinical operations
QA/QC, IT	

2.03.2.1 How to Organize for Drug Discovery and Drug Development

As previously noted, the processes of drug discovery and development are complex, highly interdependent, and multidimensional. This complexity has challenged various organizational models over the decades and no single model has evolved as the unifying solution. It is not only the multidimensionality and complexity that challenges organizational models, but also the value drivers that are sought by the company engaged in the business of developing drugs. For example, if one of the value drivers in the organization is scientific excellence in various disciplines and technologies contributing to drug discovery and development, this value is orthogonal to higher-level, multidimensional knowledge across a variety of scientific disciplines and/or disease categories. Furthermore, many drugs are useful in treating more than one disease, and these diseases, while perhaps interrelated, are typically organized in different therapeutic areas. And, lastly, there is the important dimension of project/program evaluation and prioritization. Drug discovery or development programs are categorized by stage of development as well as importance and urgency of development. An organizational model for drug discovery and development therefore, must be able to prioritize across drug development and discovery programs which span stages of development, disease and therapeutic areas, and

Table 3 Pharmaceutical value paradigm

	Drug discovery	Drug development	Manufacturing	Marketing	Sales
Functions	Target validation	Formulations	Pilot production	Research	Recruitment
	Protein generation	Tox	Large scale API	Medical education	Training
	Screening	PK	Chemical/biological	Promotion	Development
	Lead identification	Clinical studies	Good manufacturing practice	PR	Compensation
	Medicinal chemistry	ADME	Drug product	Pricing	
	In vitro pharmacology	Regulatory	Packaging		
	Mode of action studies		Supply chain		
	Off target profiling				
	Pharmacology				
	ADME/PK				
Characteristics	Not readily scalable	——————Scalable——————			
	Modest capital (human)		——Capital intensive——		
	Collaborative/limited outsourcing		——Readily outsourced——		
	Networks/pathways		——Parallel processes——		
Percent of R&D spend	25%	75%			
Percent of cost	5%	15%	10%	10%	25%

An Industry Perspective   107

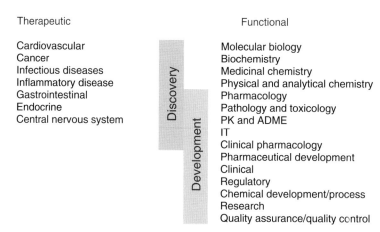

Figure 3 R&D organizational categories.

current and future business interests, and which account for the reality of the annual budget and the competitive landscape for that particular program.

With these attributes in mind, most pharmaceutical organizations are either structured around therapeutic areas or functional disciplines. **Figure 3** depicts the typical functional disciplines and therapeutic categories of R&D. The categories listed are not meant to be comprehensive but represent the most commonly chosen categories, both therapeutically and functionally. In addition to this listing of therapeutic categories there are also specific diseases within these categories, and some diseases that are not readily captured or segregated into a therapeutic category. Additional complicating factors include diseases that span therapeutic categories. There are many examples of these complications, but to name just a few: is multiple sclerosis, an autoimmune disease, listed under central nervous system (CNS) or under inflammatory or immunupathological diseases? is type 1 diabetes, another autoimmune disorder, listed under endocrine diseases? is inflammatory bowel disease an inflammatory disease or listed in the gastrointestinal category? A third complicating dimension to therapeutic categories is the recognition that a single category can encompass quite a diverse group of diseases and, similarly, a diverse group of drug discovery targets. For example, CNS disorders can include psychiatric diseases such as anxiety, depression, schizophrenia, bipolar disorder, and neurological disorders including epilepsy and insomnia, as well as neurodegenerative diseases, such as Alzheimer's and Parkinson's diseases. It also can include neurotrauma, such as stroke, traumatic brain injury, or subarachnoid hemorrhage. The evaluation of each one of these diseases and disease subcategories is actually quite different in the clinic as well as in the laboratory. Therefore, research into these disease subcategories can have many more unique than common features in a therapeutic area. Each of these disease categories can independently require a substantial number of scientists and clinicians to appropriately discover and develop drugs to treat these disorders.

The alternative R&D organizational option is organization aligned by function. This type of structure attempts to utilize the critical path value and state of the art competency attributes in those functional areas to assure scientific excellence in the quality of the work that is ultimately executed in the laboratory or the clinic. It also facilitates the flexible reallocation of staff to various projects, based on functional needs, with sufficient resources to meet the need of that particular project. These needs can vary frequently from project to project as well as by stage of development.

Not unlike the therapeutic categories, the functional categories also have significant diversity within them. For example, in molecular biology, there are different skill sets for producing proteins in various cell-based systems, particularly between prokaryotic and eukaryotic systems. Physical and analytical chemistry encompasses a diverse array of knowledge and expertise in a variety of instruments including sophisticated techniques such as x-ray crystallography, nuclear magnetic resonance (NMR), and mass spectrometry, as well as specialization in various separation technologies and the detection/identification of minute amounts of samples. Indeed there is even some overlap with biological analysis between the analytical sciences of chemistry and the analytical sciences of molecular biology (gene chips and proteomic chips). A function such as chemical development and chemistry process research involves a cadre of chemical engineers and synthetic chemists involved in novel processes research, as well as synthetic chemists expert in large-scale production. It also includes groups of analytical chemists and experts in environmental assessment for manufacturing and disposal.

Some of these functional areas are common to both drug discovery and development. Typically, information technologies (IT) are required for all phases of drug discovery and development. In addition, PK and ADME,

Table 4 Key factors for organizing R&D

Attributes	*Considerations*
Scientific/technical competencies/depth	Program funding versus department funding
Disease knowledge	Annual budget
Therapeutic area knowledge	Priorities
Drug discovery knowledge	Speed versus breadth/depth of program
Drug development knowledge	
Commercial/market opportunities	
Competitive/analysis (includes IP)	

toxicology, and pharmacology, both laboratory and clinical, as well as pharmaceutical sciences and product development, are integral to both drug discovery and development. Therefore, any functional-based organization with alignment to drug discovery versus drug development needs to take into consideration the functional overlaps. **Table 4** summarizes the key attributes and considerations in organizing R&D.

The R&D organizational model most often utilized by smaller organizations is the functional model. The most compelling factors for this choice are sufficiency and critical mass of these various functions together with the flexibility to reallocate resources to priority programs. In addition, most small organizations usually focus on one or two therapeutic categories, or perhaps more accurately, several different diseases. Therefore, for all practical purposes, they represent a therapeutic category that has functional organizational structure. At the other end of the size spectrum are the large pharmaceutical R&D organizations involving several thousand scientists. The majority of these organizations organize by therapeutic category to capitalize on knowledge of diseases and increased ability to interface with their commercial counterpart organizations, as most commercial organizations are aligned by product and therapeutic category. In addition, this provides for a more focused, team-based approach on particular products in a specialized therapeutic category. Lastly, these organizations are large enough to justify a critical mass of most of these functional areas within a therapeutic category.

There are challenges and deficiencies with this model, however, in the large organizations. One of the challenges is dealing with the stage of development of the project. For example, should a therapeutic category of cardiovascular diseases encompass, as an organizational model, both drug discovery and development? The benefits of collaboration and coordination between clinical and preclinical activities within a therapeutic category spanning both drug discovery and development are maximized. However, the effective management of an organization that spans a 10-year time horizon per project within a therapeutic division is a significant undertaking and challenge. The head of the therapeutic area has to have significant experience and expertise in a number of the diverse functional areas contributing to the ultimate success of their drug discovery and development programs. In addition, the senior management of such an organization has to make priority decisions between different therapeutic areas, as opposed to projects, and translate these into budget and staffing. Furthermore, the present-day realities of the significant global investment in late stage clinical studies, international regulatory review, and manufacturing of product present extraordinary challenges to the therapeutic-based organizational structure, since these cannot economically be replicated in each therapeutic organization. Similarly, there are some technical subspecialties that also cannot be replicated in each therapeutic category. Lastly, therapeutic-area-based organizations are notorious for not sharing resources or drug discovery or drug development programs between therapeutic areas. So the concept of free-standing therapeutic divisions within an R&D organization that span the stages of drug discovery all the way to commercialization is extremely difficult to justify.

One of the larger global pharmaceutical companies, GlaxoSmithKline, has chosen a hybrid model that incorporates some of the elements of both therapeutic categories and functional categories. Several years ago, they established centers of excellence in drug discovery (CEDD) that were organized by therapeutic category. These divisions typically were staffed with anywhere from 300 to 500 scientists and had responsibility for the stages of late drug discovery through clinical proof of concept. This required these CEDDs to staff in areas of medicinal chemistry, analytical chemistry, pharmacology, PK and ADME, clinical pharmacology, and some elements of toxicology and pharmaceutical sciences. Furthermore, they had the biological functions within their organization unique to their therapeutic category. Nonetheless, they were able to tap into (indeed, required in most instances) global functional organizations that included the basic drug discovery technologies as well as global Phases II and III clinical trials, regulatory and chemical development, pharmaceutical development, and ultimately manufacturing. In addition, GlaxoSmithKline established two external CEDDs through biotech partnerships. This type of organization, therefore, attempts to optimize for the strengths of therapeutic focus and knowledge expertise, and at the same time provide access to those competencies in drug discovery technologies, and drug development that are scalable, typically global in nature, and not unique to any particular therapeutic category or disease-based drug discovery/development. The challenges to this type of hybrid model are the fact that there are now several handoffs in the process from the earliest stages of concept and target validation through to NDA approval. In addition, there are multiple interfaces within the various global drug discovery technology functions and drug development functions. There still exists the challenge of developing a drug that may have applications across therapeutic areas, and lastly, there is the challenge to senior management regarding decision making in therapeutic areas for funding staffing and the relative priorities of areas versus projects.

Another model that has been explored to varying degrees is the business unit. The business unit model evolves from the goal of enhanced coordination and integration of drug development with the commercial organization. This model maximizes input from marketing, marketing research, and sales together with reimbursement policies into the decisions of which drugs to develop and how to develop them. This is basically an R&D organizational model that has the maximal commercial input and responsibility. This type of model has not been utilized very often in modern pharmaceutical R&D organizations. However, as the organizations are getting larger, the appeal of this type of model grows, primarily for specialty sales and marketing. In fact, the specialty pharma business represents an example of this type of model, particularly as specialty pharma is now back integrating by licensing in drugs in earlier stages of drug development rather than only acquiring drugs that are on the market. The challenges for this type of model are inherent in the success or failure of the drugs in late stage development. There are still quantifiable failure rates for drugs in Phases II and III trials and even, for that matter, for drugs recently launched. With the success of the business unit model hinging on perhaps just a couple of products, there could be significant swings in funding, as well as staffing this type of organizational structure. To be successful, this model would have to encompass a significant portfolio of development projects and launched drugs.

Three other very important elements in organizing for drug discovery and development are the practical issues of size and location, together with a centralized or decentralized philosophy of managing the overall organization. Clearly as size increases and extends beyond a core location, centralized decision making and management tends to yield to some elements of decentralized management. Global pharmaceutical R&D enterprises have at least five major R&D centers that are often spread between Europe, the USA, and Japan. It would not be surprising to see a geographic expansion over the next 10–15 years into China and India. The size of the individual R&D site appears to range anywhere from 300 or 400 to several thousand. Behaviorists note that when an organization grows beyond the ability to have daily or weekly contact with people that you know, there is a risk of losing the elements of efficient teamwork and goal-oriented team execution. Growing beyond the size of about 300 or so brings the risk of losing those advantages and of course also requires significant additional investment in other forms of communication to make them as facile as possible. On the other hand, sites that are staffed with fewer than 300 risk the ability to rapidly execute on programs because of deficiencies in resident expertise. Clearly for large sites there is an economy of scale in terms of infrastructure and services and not unimportantly other business interests such as recruiting base and cost of living.

A key element for some R&D organizations is proximity to other centers of excellence, most typically, academic institutions and research institutes. In fact, most pharmaceutical organizations are located close to major academic centers with international reputations. The concept is a simple one, that is, to maximize interaction with academic investigators which are often a source of new ideas, collaborative research, and clinical investigations. In addition, such locations provide an important catchment basin for recruiting the best scientists to an R&D organization. With recruitment being on an international scale, locations close to the top universities have enhanced ability to recruit internationally as well as providing a variety of employment options for dual-career families. One of the challenges however, is these areas tend to be higher cost of living. Companies like Lilly and Pfizer, for example, have chosen to put several of their R&D centers in areas that are not close to academic centers. Both of these R&D organizations tend to have more centralized decision making in management and their R&D centers exceed several thousand people per site.

2.03.2.2 Size (Biotech)

Over their 30-year history biotechnology companies have evolved from small start-ups to companies such as Amgen and Genentech that today rival the market caps of big pharma and also compare favorably with the size of the R&D organizations in big pharma. However, the vast majority of the biotech companies are still relatively modest in size, averaging fewer than 500 in total staff in R&D. The issue of size has already been alluded to relative to both the stages of drug discovery versus drug development, as well as the critical competencies and expertise that are required to

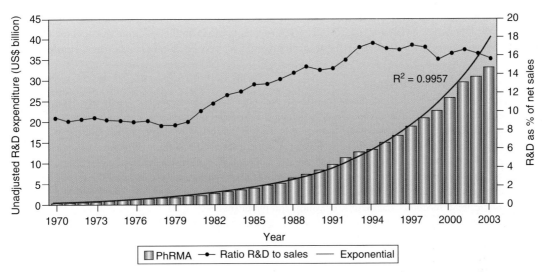

Figure 4 Pharmaceutical industry R&D spending 1970–2003. Pink columns represent unadjusted annual R&D expenditures of Pharmaceutical Research Manufacturers of American (PhRMA) member companies as reported in their annual surveys. Spending is as reported, unadjusted for gross domestic product. An exponential curve (red) has been fitted to these data along with the square of the correlation coefficient. The blue curve represents annual R&D expenditure as a percentage of net sales. (Reproduced with permission from Cohen, F. J. et al. Nat. Rev. Drug Disc. **2005**, 4, p 81 © Nature Publishing Group; http://nature.com/.)

effectively perform drug discovery and development. It has been demonstrated many times that the process of drug discovery can often involve only a relatively modest team of scientists who are focusing on the translation of basic research into the use of a particular drug candidate to treat a specific disease. Frequently there is also an important link to academic investigators who were involved in the basic concepts to be validated or who provided collaborative insights into a novel model of a particular disease. Size is not a prerequisite for the initial adaptation of the science or even the optimization of the initial discovery to find a drug candidate. Indeed, even a modest-sized organization can effectively utilize CROs to evaluate an initial drug candidate in small clinical trials. It is axiomatic then that size is not a competitive advantage in the early stages of drug discovery but, in fact, could lead to lack of efficiencies and productivity.

Alternatively, those functions that require a considerable capital investment and are readily scaled are competitive advantages if available on a significant size, primarily in areas of drug development. Nonetheless scale has been a critical factor in some areas that involve drug discovery. The sequencing of the human genome is one such example. Additional examples include high-throughput screening (HTS), the genotyping of many human diseases, and the human proteome initiative. None of these would be possible without substantial coordination of many collaborative sites so that the overall effort can be appropriately scaled. Furthermore, systems biology is a new endeavor, which requires scalable technologies to interrogate, in parallel, networks and pathways at the signaling level, in cellular functional output, and ultimately at the anatomic and species level. The output will contribute to drug discovery, but is not the result of hypothesis driven research, the latter of which has historically been the basis for most of drug discovery. This is paradigm shift.

The past 10 years has provided some insights into the lack of correlation with R&D investment, size, and productivity. Productivity, as measured by new molecular entities being launched, has been flat over the last 10 years while R&D investment, together with size of R&D organizations and big pharma, has gone up dramatically. **Figure 4** describes the overall productivity as a function of R&D spend over the last several decades. It is evident that the trend of the last 10 years is a continuation of a trend from the two previous decades. In 2001 the major pharmaceutical companies were credited with discovering one-quarter of the new molecule entities (NMEs) approved by the FDA that year, but were responsible for two-thirds of the global R&D spending.[3,4] Furthermore, in 2005 the in-licensing and partnering activities in big pharma are as significant as their internal R&D efforts. Most of the in-licensing and partnering is done with biotech companies. This provides further evidence that size and productivity are inversely correlated.

2.03.2.3 Philosophy

The various approaches to drug discovery and development can have a significant impact on the ultimate success of the endeavor. The choices described below are presented as either/or but the reader should be cognizant that these represent ends of a spectrum in every case, and that while choices need to be made, the choices are frequently blends of these various approaches. As in most cases involving options, the position on the spectrum that is chosen has various advantages and disadvantages. These are, at best, qualitative and are very often context and time dependent.

2.03.2.3.1 Target versus disease

A drug is approved by the health regulatory authorities in each country for the treatment of a disease. The labeling and patient package insert for the drug provides information on the dose and duration of administration to provide relief of symptoms associated with the disease or the reduction in disease burden. Most diseases have multiple etiologies and these etiologies may actually take on more or less prominent roles as the disease progresses or regresses. As the molecular basis for the etiologies has unfolded over the last half century, drug discovery has focused on the designing of drugs that would interact with the key molecular targets in an etiologic pathway of the disease. These molecular targets are, of course, primarily proteins or glycoproteins and, for drug discovery purposes, are subdivided into cell surface receptors, intracellular receptors and factors, enzymes, and ion channels.

Small-molecule drugs, much like the organism's own metabolome, interact with these molecular targets in a selective way and usually at unique binding sites. This selectivity in small molecule to macromolecule recognition defines a unique three-dimensional fit. By definition, this fit or recognition is unique to the molecular target as well as the drug or drug class. By extension, therefore, it is difficult to design a drug that can interact with more than one molecular target because of the unique definition of the binding elements. This binding leads to a functional outcome, which is either inhibition of the function of the molecular target or the enhancement of the function of the molecular target. Common terminology includes inhibitors, antagonists, agonists partial agonists, and modulators. The binding of proteins or other macromolecules as drugs to molecular targets follows similar rules, although the binding is usually more complicated, involving a number of points of recognition, some of which may be noncontiguous. This usually translates into greater specificity in macromolecule-to-macromolecule interaction. One caveat is that it is possible to design drugs that recognize members of a class of molecular targets if the binding site/recognition element is common to the members; this is termed 'pan-specific.' In totality, this describes then the mechanism of action of the drug. Given the relative specificity of drugs for molecular targets and the multiple etiologies of diseases, the logical outcome is that more than one drug is used to treat a disease. Indeed the practice of combination therapy is now common, particularly for most chronic and degenerative diseases.

An analysis of the number and type of molecular targets employed by the drug discovery industry, prior to the human genome initiative of the 1990s,[5–7] identified about 400 targets. The two major categories of targets were, not surprisingly, enzymes and cell-surface receptors. A more recent analysis disclosed a breakdown of marketed small-molecule drugs by class, showing nearly half of the marketed drugs targeting enzymes and about one-third of the marketed drugs targeting G protein-coupled receptors (GPCRs).[8] By definition these are druggable targets **Figure 5**.

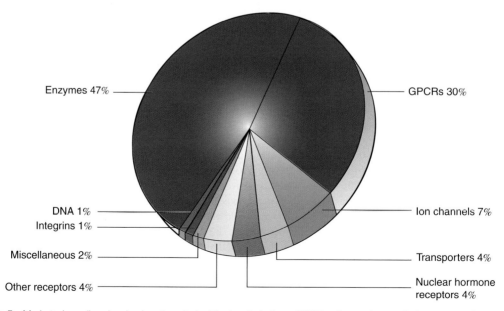

Figure 5 Marketed small-molecule drug targets by biochemical class. GPCRs, G protein-coupled receptors. (Reproduced with permission from Hopkins, A. L.; Groom, C. R. *Nat. Rev. Drug Disc.* **2002**, *1*, p 730 © Nature Publishing Group; http://www.nature.com/.)

An Industry Perspective   113

The human genome program ultimately determined that there were about 30 000 unique genes in the human genome, suggesting that there would be a significant new drug discovery landscape of new molecular targets. Indeed, estimates ranged as high as 10 000 new molecular targets based on either the number of disease related genes[5] or the number of ligand binding domains.[6] However, a more recent estimate suggests that the drug targets available for small-molecule drugs is probably closer to 1 500.[9] **Figure 6** presents two different ways of discriminating the druggable genome. **Figure 6a** represents a distribution of molecular targets using a combination of marketed drugs and drugs in development. An extension of this to the druggable genome in **Figure 6b** describes the available families of drug targets and their relative percentages, based on the sequencing of the human genome. An analysis of the distribution of the gene family populations shows that there are no large, undiscovered protein families, even

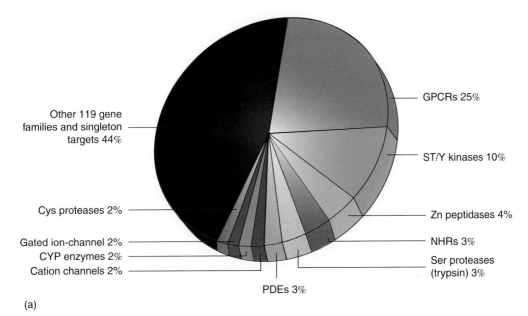

(a)

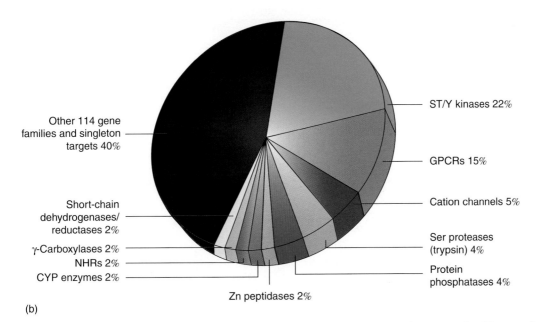

(b)

Figure 6 Drug-target families. (a) Gene-family distribution of the molecular targets of current rule-of-five-compliant experimental and marketed drugs; (b) gene-family distribution of the druggable genome. CYP, cytochrome P450; NHRs, nuclear hormone receptors; PDEs, phosphodiesterases; ST/Y, serine/threonine tyrosine. (Reproduced with permission from Hopkins, A. L.; Groom, C. R. *Nat. Rev. Drug Disc.* **2002**, *1*, pp 728, 730 © Nature Publishing Group; http://www.nature.com/.)

though there is still a significant percentage of proteins expressed by the genome which are functionally unclassified. Therefore, our most current analytics of drug target families would suggest that while the number of families might grow, they will be populated by relatively few numbers of targets and that, for all practical purposes, the druggable genome may expand threefold over the next decade or two. Concomitant with the expansion of the number of druggable molecular targets has been the expansion of the number of diseases. It had been estimated that there were about 100 diseases for which we had drug therapy available prior to the beginning of the human genome project.[10] **Table 5** represents a compilation of diseases newly treated in the last decade and a half that expands this disease treatment universe by at least a third. In addition, there are about 500 new drugs that have come to market over that same time frame that not only provide treatments for these diseases but also give better treatments for previously treated diseases.

A cataloging of the distribution of known molecular targets by therapeutic areas provides some correlation and guidance to the interrelationships of targets and diseases. **Table 6** shows that there are some significant differences in molecular target types depending on therapeutic area. For example, CNS diseases are dominated by GPCRs, whereas infectious diseases and cancer are dominated by enzymes as drug discovery targets. The option to orient drug discovery toward molecular targets versus diseases has a focus on a particular class of targets at one end of the spectrum and a focus on a specific disease or disease category at the other end of the spectrum. The advantages of this focus are obvious and can represent competitive advantages in drug discovery. A focus on a disease category provides in-depth knowledge of that disease category, both at the laboratory as well as at the clinical and marketing levels. However, that specific disease category may require multiple families of drug discovery targets requiring a depth of knowledge unique to these individual families of targets. Alternatively, a focus on a specific class or family of targets may well require a breadth of therapeutic area or disease target applications to fully exploit that drug discovery effort. Furthermore, some of these families of targets represent a significant degree of heterogeneity. For example, enzymes represents a diverse group of proteases, kinases, and phosphatases. GPCRs have enjoyed radically different degrees of success in yielding to drug targeting. Structure-based drug design (SBDD) can much more effectively be applied to enzymes as targets as well as some nuclear hormones and nuclear receptors, whereas integral membrane proteins are much more refractory to SBDD approaches. HTS approaches typically measure only binding as an endpoint and secondary, lower throughput/high content assays are required for determining function. Therefore, the investment in expertise in cellular pharmacology can be different with certain classes of enzymes versus GCPRs versus nuclear receptors. Ion channels

Table 5 Diseases newly treated in last 15 years

Infectious diseases
- HIV/AIDS
- Hepatitis B, C
- Respiratory syncytial virus (RSV)

Inflammatory diseases
- Chronic obstructive pulmonary disease (COPD)
- Inflammatory bowel disease (IBD)
- Irritable bowel syndrome (IBS)

CNS diseases
- Alzheimer's
- Neuropathic pain
- Multiple sclerosis (MS)
- Migraine
- Attention deficit hyperactivity disorder (ADHD)
- Restless leg syndrome (RLS)

Cardiovascular diseases
- Coronary artery disease (CAD)
- Restenosis
- Acute coronary syndrome

Blood
- Leukopenia
- Hematopoiesis

Cancer
- Certain lymphomas and leukemias
- Head and Neck cancers
- Glioblastoma
- Pancreatic cancer
- Gastrointestinal stromal tumor (GIST)

Endocrine
- Endometriosis
- Obesity
- Osteoporosis
- Benign prostatic hypertrophy (BPH)

Muscular/skeletal
- Erectile dysfunction
- Overactive bladder

Other
- Cystic fibrosis
- Gaucher's disease
- Paget's disease
- Emesis

Table 6 Distribution of known molecular targets by therapeutic areas

Therapeutic area	Receptors	Nuclear receptors/factors	Enzymes	Ion channels
Central nervous system	115		12	8
Inflammation/immunomodulation[a]	30	8	20	1
Cardiovascular	17	1	17	12
Gastrointestinal	11		2	1
Endocrine	30	18	16	3
Blood	8	30	4	
Infectious diseases[a]	1		50	1
Cancer[a]	7	9	20	

[a] Excluding DNA/RNA as target.

represent a relatively unique expertise requiring electrophysiology and other tools for analyzing ion flux and ligand versus voltage gating functional output. In conclusion, it is important to evaluate the multiple overlap of molecular target classes with diseases and therapeutic areas to maximize for success in drug discovery and development.

2.03.2.3.2 New molecule entity versus life cycle management

A continuing issue in managing R&D is the conflicting choices of investing in NME research versus investing in life cycle management. As projects achieve the stage of commercialization, there is an inevitable continuing investment in these projects to further understand the safety of the product postmarketing, to explore broader indications of the product, and, at times, to explore new formulations and dosing schedules for the product. This work, of course, requires resources that frequently are in the domain of R&D and therefore compete with investment in NMEs at earlier stages of evaluation. Another component of this competition for overall budget spend is a Phase IV clinical trials program for a recently launched product. Phase IV studies are important, and involve evaluation of the product in patients with comorbidities related to the target disease for the product. However, usually the funding and sponsorship of these Phase IV studies is under the commercial organization of a pharmaceutical company and only indirectly competes for R&D resources.

The investment in life cycle management for a recently launched product is usually quite significant, particularly for products that are seeking label expansion to include longer-term morbidity and mortality measurements. The data to support this usually requires large clinical trials. The market forecast and justification for this investment is more readily quantified and much less discounted than a market forecast for a new molecular entity that is in Phase I or Phase II clinical studies. By definition this forecast carries more credibility and provides more assurances on return on investment. However, companies grow the top line by adding new products and this can only happen if R&D organizations are investing sufficiently in NME research. If investment in NME research is coming at the expense of investment in life cycle management, then the organization is likely to suffer the longer-term consequences of these shorter-term views or perspectives. The bottom line is that success is expensive, investment cannot stop at the door of commercialization, and life cycle management is a reality. It is imperative that the organization takes a long-term view and sufficiently invests in NME research and development together with life cycle management. The consequence of underfunding NME research is that the company increases its risk of losing independence within a short time frame.

2.03.2.3.3 Chemical versus biological

The history of the pharmaceutical organization is replete with examples of drug discovery that were dominated by chemistry and medicinal chemistry. As chemists achieved a better understanding of what properties were important for a chemical to be a therapeutic, they also understood the abilities to morph a chemotype or structural motif or chemical core (e.g., benzodiazepine) into a drug that could be useful for other disease pathways. An example of this is the history of sulfanilamides (**Figure 7**).[5] The transformation of one of the initial sulfanilamides as an antibiotic to a hypoglycemic, then to a diuretic, and finally to an antihypertensive over a 30-year time span illustrates the medicinal chemistry learnings of these concepts.

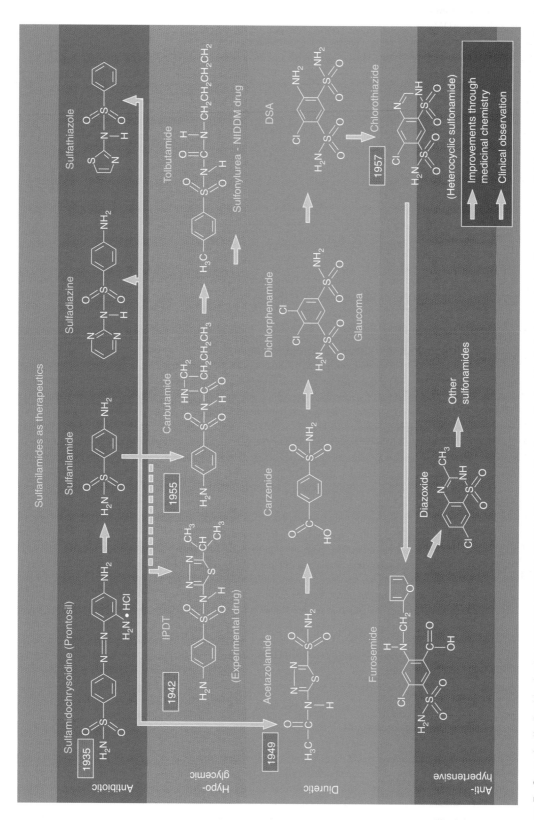

Figure 7 Sons of sulfanilamide. A schematic representation of drugs that originated from sulfanilamide. A single chemical motif gave rise to antibiotics, hypoglycemic agents, diuretics, and antihypertensive drugs. (Reprinted (abstracted/ excerpted) with permission from Drews, J. Science **2000**, 287, 1960–1964. Copyright 2000 AAAS.)

Terminology for small-molecule drug properties includes such terms as promiscuous ligands, 'rule of five,' bioisosterism, chirality, ADME, biopharmaceutical profiling, and off-target profiling.[11] Today's drug discovery armamentarium is basically multidimensional profiling wherein the chemical is optimized for structure against a number of factors at a very early stage prior to advanced in vivo evaluation. The biggest challenge for small molecules as drugs is, and still remains, safety evaluation. The ability of small molecules to interact with a number of macromolecules in the in vivo milieu can lead to safety issues. In addition, safety issues can arise from extended pharmacology, from extended half-life, variable absorption, or metabolic induction. Our laboratory models for predicting human safety are the least predictive of all of our tools in drug discovery. Most small drugs fail for animal toxicity or human adverse effects. Improvements in this dimension will undoubtedly improve the productivity of developing small molecules as drugs. The merger of these predictable tools with medicinal chemistry will have a dramatic impact.

Biologicals (i.e., proteins) have been used as therapeutics for a number of decades but have been limited by source. The advent of recombinant DNA technology in the 1970s revolutionized our abilities to make proteins in the laboratory and has led to the introduction of a number of proteins onto the marketplace for treatment of various diseases. The challenge in the development of proteins, unlike small molecules, is not from the toxicology or safety perspective, but more from effective drug delivery and manufacturing perspectives. Since the production of biologicals is defined at the regulatory level as 'the process is the product,' the process must be well defined when clinical trials are initiated. Since there can be significant changes in the process as it is scaled up, the material that is produced for the key Phase II and certainly pivotal Phase III trials, of necessity, has to be at the manufacturing scale. To accomplish this, the R&D enterprise needs to invest in large-scale facilities for biologicals manufactured early in the process of drug development. This is a significant capital investment and an equally significant investment in time and talent. The delivery of macromolecules as drugs limits their breadth of application. Oral administration is yet to be commercialized in spite of many clever approaches, so parenteral administration is the only route. Inhalation and intranasal approaches are nearing commercialization. A related limitation with biologics as drugs is the difficulty of using a macromolecule for intracellular drug targets.

A third, not inconsequential, issue for developing biologicals is the potential to generate an antibody response in humans that binds and/or neutralizes the biological. Therefore, significant research has gone into engineering proteins and antibodies to minimize an immunological response in humans and optimize for overall efficacy. On the plus side, one of the advantages of developing biologicals is that the antibody response in animal safety studies precludes any meaningful safety findings and therefore longer-term animal toxicology studies are not required. One of the other reasons for biologicals requiring less toxicology is that many are endogenous substances, and thus already well tolerated by patients at physiological levels.

With these relative advantages and disadvantages in mind, most large pharmaceutical companies today are involved in developing both small-molecule drugs and biologicals. Amgen, originally a biologics-only biotech company, has expanded its original biological platform toward small-molecule drug discovery and development. Genentech, on the other hand, has remained consistently with biologicals in drug discovery and development. Most small biotech companies are doing either small-molecule or biological development, not both. Big pharma was late getting into biologicals; however, over the last 7–8 years there has been a considerable increase in commitment (**Figure 8**).

There have been significant attempts to develop other large molecules as drugs that are not proteins. These include both DNA and RNA based oligonucleotide drugs. Thus far, development of these types of drugs has had very limited

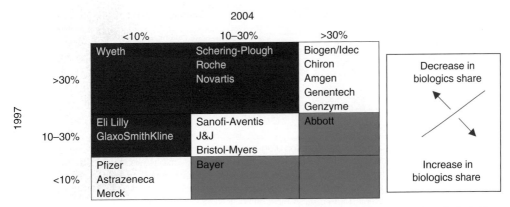

Figure 8 Share of biologicals as a percentage of total late-stage product pipeline. Late-stage products include Phases II and III and preregistration trials. (Reproduced with permission from Windhover's In Vivo.)

success. The primary difficulty has been delivering these macromolecules to the target organ or target cell type. Achieving sufficient PK, both in the whole organism and also at the cellular level, has been extremely difficult and this has only been one of a number of challenges in developing oligonucleotides as drugs. The initial challenge was simply the ability to synthesize and scale these up. In addition, many of the oligonucleotides presented unique toxicology in animals as well as in humans. These very substantive hurdles have now been largely overcome. However, efficacy has still eluded the pioneers in this area of new drug discovery and development. Lastly, gene therapy has struggled to yield a successful therapeutic, in spite of over a decade of significant time and investment. Delivery of genes in various formats with various vectors has been validated in animals as well as in humans. However, local regional delivery versus systemic exposure remains a challenge and safety continues to define the most significant hurdle yet to be overcome in gene therapy approaches.

2.03.2.3.4 Best in class versus first in class

The debate on best in class versus first in class can be pictured as a pendulum that swings in one direction and then the opposite direction, historically over a 10-year period. The 1980s were characterized by best in class strategies relative to first in class or so-called innovative drugs. For example, better nonsteroidal anti-inflammatory drugs (NSAIDs) were being registered along with better beta blockers, more potent corticosteroids, and H_2 antagonists. Of course angiotensin-converting enzyme (ACE) inhibitors and statins were first introduced in the 1980s with follow-ons coming quickly behind. The 1990s, on the other hand, were characterized by biologicals coming to market as first in class and most R&D centers espousing first in class philosophies. In addition to the advent of biologicals, this swing in philosophy was driven largely by the anticipation of many new molecular targets coming out of the human genome project, as well as the belief that pricing pressures would be continually increasing due to generic competition and only first in class drugs could command premium pricing (**Figure 9**).

Several retrospective analyses of the relative merits and value of first in class versus best in class have concluded that, at least for blockbuster drugs, best in class has greater value creation[12]. The analysis segregates first in class to those drugs targeting unprecedented molecular targets, that is, molecular targets that had not yet been validated with drugs, selective for these targets, in Phase II clinical efficacy studies. The best in class drugs, of course, were defined by precedented approaches and were further subdivided into fast followers, differentiators, and latecomers. This analysis showed that drugs targeting precedented approaches (**Figure 9**), not only had a higher value creation but also had

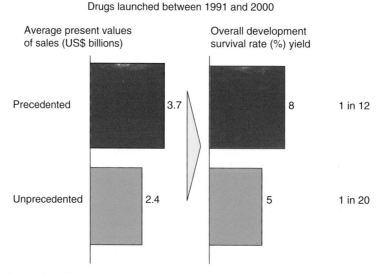

Figure 9 Analysis of precedented versus unprecedented drug targets. Drugs targeting precedented approaches create more value with a lower risk of failure (precedented approaches are defined as those which have established clinical effectiveness; as proxy, we use 40 months prior to market entry to designate that status, as Phase III and registration take ~40 months and clinical effectiveness is typically proven after smaller Phase IIa trials). Present values were calculated using sales from 1990 to 2000, adjusted for inflation, and estimated future sales discounted at 10% back to year 2000 dollars; overall development survival rate is the product of phase-specific survival rates from preclinical, Phase I, Phase II, and Phase III + Reg trials. (*Source*: McKinsey analysis; PBJ Publications Pharmaprojects; FDA/CDER; Company reports; analyst reports.)

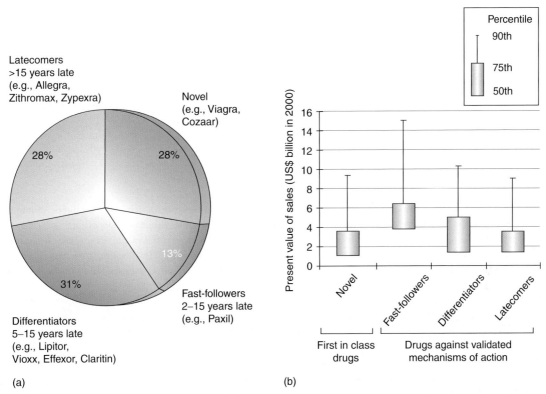

Figure 10 Most blockbusters had established mechanisms of action, and novelty does not always generate value. (a) The 32 blockbusters launched by the top 15 pharmaceutical companies during 1991–2000 and compared their mechanisms of action with presently marketed products. (This analysis was based on multiple data and information sources, including Elsevier Science, PJB Publications Pharmaprojects and the Evaluate Pharma Database). (b) The 189 drugs lunched by 15 top Pharmaceutical companies in the period of 1991–2000. Present values were calculated using actual sales, adjusted for inflation, and estimated future sales were discounted at 10% back to year 2000 US dollars; all present value figures showed the same trends as data comparing the sum of the first 4 years of sales. Shown are the ranges in present value of sales from the 90th (top 10%), 75th, and 50th (median) percentiles. (This analysis was based on the data and information from the following sources: Elsevier Science; Evaluate Pharma Databased; PBJ Publications Pharmaprojects; FDA/CDER; company reports; analyst reports, IMS Health.) (Reproduced with permission from Booth, B.; Zemmel, R. *Nat. Rev. Drug Disc.* **2003**, *2*, 838–841© Nature Publishing Group; http://www.nature.com.)

overall higher development survival rate. The breakdown of the best in class versus novel or first in class drugs is shown in **Figure 10a**. The first in class drugs represent only about 30% of the total number of blockbuster drugs launched during that decade. Similarly, **Figure 10b** depicts the relative value creation of the first in class drugs versus best in class, again, secondarily cataloged by fast-followers, differentiators, and latecomers. The largest number of drugs in the best in class were in the latecomer category.

In addition to value creation and overall survival through the development process, other factors to consider are development cost and time, stages at which the respective programs are being derisked, and the competitive landscape. With these considerations, the advantage should usually favor first in class. The competitive landscape should favor first in class. With development cost and time, there may well be a more abbreviated development path to commercialization, assuming it represents breakthrough therapy. Best in class drugs, on the other hand, claim the advantage of a precedented development path to commercialization and a higher likelihood of survival through the clinical phases of development. However, a critical attribute for best in class is differentiation. Differentiation will usually not be evident until the completion of Phase III clinical trials. While improved efficacy is a key differentiator, as important, if not more important, are safety, the breadth of the label claims, and convenience of administration. Safety is much more fully defined in Phase III clinical trials and ultimately in postmarketing analyses. Efficacy and label claims are differentiated also in Phase III clinical trials (**Table 7**).

It would be prudent for R&D management to assess carefully opportunities of best in class and first in class in the therapeutic areas of interest and within the context of molecular targets of interest. Appropriate risk/reward analysis is

Table 7 The relative advantages of best in class versus first in class drugs

Best in class	First in class
Higher value creation	Speed of development
Precedented development and commercialization	Costs of development
Higher survival rate	Physician/consumer interest
	Differentiation
	Novelty

relevant to either category (**Table 7**). First in class, as higher risk programs against unprecedented targets, should not be justified based on niche market opportunities. Similarly, with best in class differentiation occurring late in the development scenario, research on precedented targets requires an equally substantial market potential. Of course, if the aggregated industry worked only on best in class, there would ultimately be no new novel drug categories. Careful consideration should be given to an analysis of both of these categories as part of an overall R&D philosophy.

2.03.2.3.5 Decision making versus organizational alignment

The differentiation of decision making versus organizational alignment is an increasingly important factor as R&D organizations increase in size. Basically, these two options represent the ends of a spectrum defined by speed. Rapid decision making leads to rapid execution. Of course, if the decision is wrong, it can also lead to rapid failure. On the other hand, organizational alignment is achieved by consensus building, which is generally a slow process, and therefore can lead to a significant delay in execution. A further counterpoint is that the execution phase may be more effective and of higher quality, because of the increase in inherent ownership in the process due to the consensus approach employed to achieve the final decision. So, in addition to speed, there is also quality of execution that is an important factor in decision making versus organizational alignment.

The advantage for small organizations is they can achieve both effective and rapid decision making in concert with organizational alignment. However, as the majority of drug discovery and development programs are in larger organizations, and as noted, organizations that continue to increase in size, the ability to effect rapid decision making with organizational alignment is significantly compromised in larger R&D enterprises.

A large organization can have rapid and effective decision making using a top–down approach. This is an effective management option as is evidenced by its continual utilization in most military forces and in many business models including pharmaceuticals. However, in a diverse, highly technical environment, such as pharmaceutical R&D, it is imperative and essential to have involvement of the technical community in both the planning as well as the execution of the programs. Recognition of this complexity and valuing collective wisdom are important drivers for establishing the right balance of effective decision making and organizational alignment. The decision making process should be clear and transparent to the organization. Transparency is challenging in an environment where decisions are complicated and highly debatable, and where failure is the more frequent outcome. Decisions should be timely and clearly communicated. As importantly, there should be appropriate, periodic monitoring of the execution phase of the program to deal with focus and completion, as well as, mid-course corrections. Assuming high-quality competencies in an organization, the execution of the program is most effective if there is ownership and buy-in on the part of the stakeholders. This can be achieved, indeed highly valued, by soliciting proposals for programs or program changes, establishing peer review of the relative merits of the proposals, and then appropriately resourcing the agreed programs. Ultimately senior management must make these decisions, as R&D cannot be effectively accomplished in a loose federation of individual investigators. High-quality science must be valued not in its own right but for its applications to new medicine discovery and development. Fraud, or incomplete or inaccurate information, cannot be tolerated, for in the end, human safety is jeopardized.

The pace of change is increasing and the competitive environment is expanding both in depth as well as in breadth or globally. Nonetheless, improvements in drug discovery and development over the past several decades have been incremental and will likely continue to be incremental. The time from concept to commercialization will continue to be a decade or more in the foreseeable future. Therefore, an R&D organization must be willing and able to act quickly, but not prematurely. The decisions are complex and multifactorial and the urge to decide in the face of public or investor pressure should not override thoughtful and deliberative decision making.

2.03.2.3.6 Fail fast versus follow the science

The mantra 'fail early, fail fast' rose to prominence in the 1990s, particularly as new molecular targets for drug discovery were highly valued. The rationale for this perspective is quite obvious. Many new targets were being established and evaluated using HTS technologies. Differentiation of 'on target' versus 'off target' binding was readily available. Furthermore, as new predictive laboratory screens and tools for ADME were becoming available, the researcher could make decisions about lead compound quality much earlier in the process than was typically done in the past. In addition, with new targets being the focus, the failure rate for lack of efficacy in the laboratory and/or the clinic seemed to be increasing and of course, the cost of failure much later in the process is higher. Therefore, the logic is to take forward into development only those drug candidates that pass all the prescriptive and preclinical tests and demonstrate acceptable animal safety prior to Phase I evaluation in human subjects. By extension, this philosophy can be extended to early clinical evaluation, as drugs that exhibit a flaw in their early clinical profile can still be taken forward all the way through Phase III clinical studies. Therefore, the 'fail early, fail fast' philosophy exhorts the decision maker to evaluate critically the early clinical findings and discontinue development as soon as feasible. In fact, the philosophy encourages the R&D organization to celebrate program terminations as effective, critical decision making and saving the organization money downstream.

Most researchers in drug discovery and development would agree that the most difficult decision is when to terminate a program or project. Examples abound of drugs that made it to market after being 'killed' several times during their development. In addition, drugs can be deprioritized at various time points due to limited resources and the constraints of the annual budget. However, the discovery and development of a drug, while a tortuous path, with oftentimes several branch points, nonetheless is a path that develops data and new knowledge. This knowledge needs to be evaluated within the context of the significant limitations of our tools to predict human efficacy and safety in various disease categories. These tools are still quite rudimentary, particularly our ability to predict human safety based on animal toxicology studies. Development of new knowledge along the development pathway of a drug is another form of hypothesis driven research. 'Follow the science' means, simply, do appropriate experiments, assess the data, and understand the implications of the conclusions. Early, key milestones in drug discovery include the establishment of preclinical efficacy data on a drug candidate with acceptable biopharmaceutical properties, including the ability to scale up effectively the active ingredient. Toxicological evaluation in animals is not a black-and-white analysis, but in fact, a subjective one, which must be taken within the context of the targeted disease and patient population. The FDA, as one regulatory body, has encouraged sponsors to dialogue with them prior to initial clinical evaluation of a drug candidate. The purpose of this overture is to bring more collective wisdom to bear on the options to evaluate the drug candidate in humans, including both an acceptable safety threshold to proceed, as well as, the best path forward. Subsequent key milestones include Phase I human PK/PD and safety followed by Phase II efficacy (exposure response) in patients. Each one of these milestones generates critical data to extend the working hypotheses, modify them, or disprove them.

The drug discovery/drug development industry is diligently trying to improve the predictabilities of its preclinical tools for human efficacy and safety. However, the decision matrix does not yet readily yield to a box checking exercise with quantitative thresholds for go/no-go decisions. Hypotheses should be critically evaluated, 'the science should be followed,' and success milestones should be celebrated.

2.03.2.4 Culture

Culture is a word that has many connotations. A dictionary might define culture in the first instance as "a particular form of civilization, especially the beliefs, customs, arts and institutions of a society at a given time." The culture of an R&D organization has some parallels to this definition of a civilization or a form of a civilization. However, it is still difficult to describe it fully. But culture certainly includes important attributes such as views and values, practices, geography, and size. Some of these attributes have already been discussed in this chapter, and therefore, the focus of this brief section on culture will be on four additional elements important to drug discovery and development.

Organizations can typically be described as collaborative or insular; again, two ends of the spectrum. To some degree, competitive is another relevant adjective and can be in contradistinction to collaborative, but competitive can also be an integral part of collaborative. Fundamentally, the question is: "What are the advantages or disadvantages of being a collaborative organization?" Viewed in the larger context, 'collaborative' must include both intramural interdepartmental and interlaboratory collaboration as well as extramural collaboration with groups and institutions external to the R&D organization. Given the large number of interfaces in the overall process of drug discovery and development, together with a similarly large number of interdependencies between the various laboratories and disciplines, collaboration has to be the norm rather than the exception for effective drug discovery and development. All aspects of collaboration should be evident, including communication, in its largest sense, joint planning and execution, shared goals and responsibilities, data sharing, and collaborative analyses. When organizations have planning silos or departmental or decision making silos, both the efficiency and effectiveness of drug discovery and development are compromised. There does need to be an appropriate balance of 'need to know' versus 'nice to know' information access and information analyses. An individual scientist or small group of scientists are responsible for the detailed planning and execution of a particular experiment or study and has responsibility for the integrity of the data. They are, after all, the individuals with the most knowledge to understand and interpret the data. Nonetheless, in the highly interdependent, regulated, and resource-intensive environment of drug discovery and development, all experiments and studies are part of bigger programs with long time-lines. If poorly executed or poorly timed, they can have a serious impact on the value creation for a new drug candidate.

The opportunities for an R&D organization to collaborate with the scientific community in the external environment are multiple. Indeed every R&D organization has collaborations with academic scientists and research institutes on an ongoing basis. However, the philosophies in managing these relationships can vary. One end of the collaborative spectrum is a contractual obligation to the external party to perform certain tasks and, when completed, to deliver the final report. The other end of the spectrum is an agreement for activities that require resources at both institutions, wherein the activities are interdependent, much like an internal program within an R&D organization. Indeed in some cases, three-way collaborations exist. These kinds of collaborations minimize the we/they attitude that can happen with arm's length collaboration and can truly lead to synergies from the joint investment of the scientists and technologies of the two organizations. The collaborations do require careful management, project champions in both institutions, and alignment of strategy and tactics. Since these collaborations often involve a significant culture difference and/or size difference, sensitivities to these differences must be acknowledged up front and their unique features understood throughout the term of the collaboration.

Some R&D organizations have set up competitive collaborations, both internal and external. The rationale is that two approaches to discovering a drug candidate of a certain type are better than one and that competition brings out the best efforts of both parties, since ultimately a winner is defined. However, these strategies clearly establish silos within the organization or silos between two institutions and significantly diminish the collaborative attitudes within the culture of the organization.

Investing in a research program or a drug candidate development program is a long-term investment. Frequently difficult, seemingly intractable challenges arise during these programs that can halt progress, or at a minimum, diminish momentum. It is human nature that ownership defines loyalty. Individual scientists or teams truly invested in a program will exert the extra effort to find ways around problems that arise and to exercise the perseverance necessary for the longevity of the program to ultimately achieve regulatory approval and commercialization. Program champions are prerequisites for real ownership and investment. In the absence of a program or product champion, there is a loss of ownership because ownership and leadership are dispersed amongst committee members. When this is coupled with multiple committee responsibilities for team members and/or frequent turnover of leadership roles for programs, passion is diluted and focus is lost. Concern with product champion roles is that the champion is less than objective in their analysis of the value of the assets and the chance of technical success. These complicated programs, however, cannot be left to autopilot and therefore, a responsible R&D organization should appropriately balance the advantages of product/program champions with periodic objective review of the overall portfolio of programs. This review should address resource adjustments or program terminations to reflect the overall relative chances of success within the business context.

Any R&D organization has a number of programs at any one point in time to manage and resource. Its culture is defined by the objectiveness and transparency of this portfolio management, as well as by the factors that are valued in managing the portfolio. Some organizations discount considerably the value of market forecast in defining various phases of drug discovery and development process, whereas other organizations put considerable emphasis on market assessments for programs, from initiation through to ultimate regulatory approval. Should the absence of a current market preclude the initiation of a drug discovery program? Should the assessment of a limited commercial opportunity define the progression of a drug candidate into Phase I clinical studies? How important is the chance of technical success in defining program initiation or program progression through various stages of discovery and development? Is medical need a surrogate for market assessment? Should a program starting in drug discovery have a well-defined path laid out to ultimate regulatory approval at inception? The answers to these questions define to a large degree the philosophy of a portfolio management for the R&D culture. The realities are that unmet medical needs exist and new markets are created by addressing these needs with new drugs. Good primary and secondary market research can make reasonable estimates about market potential when there is no market. But the challenge remains to define market potential or market opportunity 10 years in advance of when the question is addressed. The medical need as well as reimbursement opportunities can change dramatically in that time frame. The typical scenario is for market research to impact in a more significant way as programs become more mature and closer to regulatory approval. However, the medical need and the marketing opportunities can and should play a role in even the earliest stages of drug discovery programs. Market research can help define a potential product profile, refine the competitive landscape, and provide one of the several discriminating factors for program prioritization.

The final factor for designing culture in an R&D organization is entrepreneurism. An entrepreneurial environment is one characterized by individual initiatives and exploratory endeavors, frequently involving high risk. An entrepreneurial environment is further characterized by decentralized management and decision making. Alternatively, the more regimented structured environment is characterized by centralized management and decision making. However, appropriate entrepreneurism is not just high-risk activity, but it should also be a program with a credible business plan and opportunity, where rewards can be articulated and quantified, and where failure is linked directly to continued employment. An entrepreneurial environment attracts scientists who are willing to invest themselves in programs with limited validation but with high potential. It is an environment that is likely to be a transforming environment. Nonetheless, to be successful and transforming, the R&D organization needs to assess and manage the risk, provide appropriate freedom to execute programs and demand responsibility for success and failure.

2.03.2.5 Partnering

Partnering between biotech and big pharma has become an industry in its own right. In 2004 Pfizer claimed that it had over 300 active R&D partnerships with biotechnology companies. These partnerships span the spectrum of drug discovery and development including novel technology access, drug discovery partnerships, and preclinical and early clinical phase deal making for products or projects, as well as later stage partnering/licensing arrangements. The change in landscape is not just in the number of deals being struck, but in the fact that the combination of these partnerships as well as traditional licensing now means that half of the products launched by pharmaceutical companies are products from licensing or partnering activities. The management of these interfaces is a critical function within companies and perceived as a competitive advantage.[13] In spite of the shared reward inherent in the nature of these partnerships, there is still a good return on investment for big pharma as well as biotech, considering the terms of the agreements and business needs and drivers of the companies involved in the partnering. The degree to which big pharma companies partner is dependent on their R&D philosophy but it also dependent on their nearer-term business needs. For example, during the 1990s Merck effected a modest number of partnerships compared to the other top five pharma companies. However, in the first decade of the twenty-first century, Merck has dramatically increased the number of partnerships and in-licensing deals reflecting the anticipated and significant reduction in revenues with products going off patent mid-decade.

2.03.3 R&D Productivity

R&D productivity has declined over the last 20 years, as measured by investment in R&D spend versus new molecular entitities approved and launched (**Figure 4**). Over the last three decades the group at Tufts University Center for the Study of Drug Development in Boston has studied the costs of drug development. Their most recent estimates[14] are based on a number of randomly selected drugs developed by 10 pharmaceutical companies, between the years 1983 and 1994. The reported development costs that were assessed run through the year 2000, and are therefore based on year 2000 dollars. Their calculated out-of-pocket costs per new drug was about US$400 million. Capitalizing these costs to market approval with appropriate discount rates yields an US$800 million cost in year 2000 currency. These data, together with earlier studies, suggest an annual rate of increase of drug development costs running at 7.4% above inflation. Therefore, today's cost for the successful development and commercialization of a new molecular entity is well over US$1 billion.

Many factors have contributed to the increasing costs and time required to discover and develop drugs, some outside of the landscape of R&D. Over the last two decades five critical pieces of legislation have affected the discovery and development of new therapeutics. The Patent and Trademark Amendments of 1980 (Bayh–Dole Act) and the Drug Price Competition and Patent Term Restoration Act of 1984 (Hatch-Waxman Act) have fostered translational research and extended patent lifetime as a function of development time extensions, but have also significantly boosted generic competition. The Orphan Drug Act of 1983 attempted to incentivize for drug discovery and development of treatments for rare diseases. In aggregate, these three pieces of legislation have stimulated continuing investment in R&D for drug discovery and development. During the 1990s the Prescription Drug User Fee Act of 1992 (PDUFA) and the FDA Modernization Act of 1997 (FDAMA) addressed the continuing elongation of drug development and review times. These two pieces of legislation required user fees for the filing of NDAs and then utilized these user fees to increase the number of reviewers at the FDA. Overall this has reduced review times modestly over the last decade, compared to the previous decade, and stabilized the time spent in the clinical phases of drug development.[15] However, the scope and size of clinical trials has continued to increase concomitantly with increasing costs of clinical studies.

In spite of the decrease in R&D productivity and increasing costs of drug development, the pharmaceutical and biotech industries have continued to grow and deliver new medicines to patients. These industries have been able to do this because they have essentially enhanced the overall value per drug by targeting commercially attractive, unmet needs that can be premium priced, and by expanding established market arenas with improved therapeutics. But the industry can only grow, and indeed survive, by addressing two other fundamental areas of productivity. It must increase efficiency by reducing costs or time or both, and it must decrease the cost of failure by terminating drug candidates sooner or improving the probability of success. The area for greatest impact is the area of decreasing cost of failure. Increasing efficiency will be incremental at best, since reducing time per program is unlikely. Similarly, reducing costs is very unlikely. How do we improve the probability of success and decrease the cost of failure? – focus on drug discovery.

2.03.3.1 Drug Discovery

A survey by Accenture of 15 pharmaceutical companies between 1997 and 2000 showed that the average company utilized 250 full-time employees (FTEs; 1 FTE = 1 year equivalent), or approximately US$70 million, for each new drug that enters into formal drug development.[16] Furthermore, while times had decreased from target identification to development candidate from 1997 to the year 2000, nonetheless, the average time was still around 5 years. Considering that a top five pharmaceutical company will have to launch six new molecular entitities per year in the 2010 year time frame, simply to achieve annual growth targets in the 5% range, the time and financial investment alone for multiple drug discovery programs funded in parallel with the historical attrition rates is simply not feasible.

Key factors in understanding productivity in drug discovery evolve from an analysis of bottlenecks in the overall process (**Figure 2**). In concert with overall attrition rates, there are a variety of attrition rates that are cited for the various stages of drug discovery. **Table 8** is a compilation of data from personal experience in the industry together with a number of different sources over the last 10 years, therefore it must be viewed as an approximation and several years out of date. The overall success rate of 4%, starting with molecular target identification and validation up to IND, underscores the high failure rate of a process that historically has taken 5 years to proceed from target identification to IND. Clearly, only modest improvements in several of these steps can have a dramatic impact on the overall success rate. Therefore, there has been significant time and attention paid to these various stages in reducing the failure rate and, concomitantly, reducing the time. Many R&D organizations today have shaved at least 1 year off this 5-year drug discovery cycle and some have reduced it even further. Researchers are improving the quality of their compound libraries in several dimensions: numbers of compounds, compound diversity, compound stability, compound characteristics (molecular weight, logP, H bonds, polar surface area, etc.).[17] HTS has grown up during the late 1990s into efficient, automated systems by the year 2005 with the capacity to do 384 well-plate screening routinely, with good informatics support, and the ability to cherry-pick samples for primary and secondary screening. Additional investment is being made in high content cell-based screens to evaluate the activity of the early hits for validation and prioritization. Increasing quality is going into the hit to lead process in drug discovery.[18] This is the stage that significant additional investment will be required to optimize the early leads, and this stage is still one of the bottlenecks in modern drug discovery. At this stage, medicinal chemistry is invested in the process and

Table 8 Success rates for stages of drug discovery

Stage	Success rate (%)
Target identification/validation	50
Screen	25
Hit	30
Lead compound	30
Clinical candidate	70
Investigational new drug	50

multidimensional structure–activity relationships (SARs) are explored to find the best compound. Such a compound must meet the clinical candidate profile of potency, selectivity, PK/bioavailability, metabolic stability, as well as key safety thresholds for cytochrome P450 inhibition, QTc liabilities, and genotoxicity. This stage of drug discovery can often take 2 years or more. It is highly iterative and is the most resource-intensive stage of the drug discovery program. It is because of these attributes that significant improvements are sought to reduce the number of compounds made to achieve a clinical candidate and reduce the time required to achieve a clinical candidate. The interested reader is referred to subsequent chapters in this volume which deal with these various activities in much more detail.

Another bottleneck in the drug discovery process is preclinical safety evaluation. A comprehensive perspective on this has to include the pharmacology of the drug in animal models of the disease being studied, together with the pharmacology and toxicology of the drug in rodents and nonrodents in the traditional preclinical safety evaluation. These studies require considerable time and resources to execute, and are often done in a serial fashion. Furthermore, these studies require significant amount of the active ingredient as well as significant support from analytical and pharmaceutical sciences. As a substantial amount of these activities are regulated by the FDA and come under ICH guidelines, there is little opportunity for modification. Nonetheless, there are strategies under investigation for predictive toxicology assays including toxicogenomics. Performing pharmacology and animal safety in a more parallel fashion can shorten time-lines.

There are more publications today describing ways to improve drug discovery productivity than there are actual initiatives to improve productivity. Not to be left out, the author offers the "Top 10 ways to improve drug discovery productivity" in **Table 9**. Some of these suggestions are more strategic in nature, such as the choice of drug target (precedented versus unprecedented), investment in biologicals as drug targets, and the philosophy of drug targeting selectivity versus specificity. Furthermore, SBDD or computational techniques are important considerations in improving productivity. Recognizing that these techniques are not yet universally applicable to all drug targets, they are however growing in throughput and breadth of application. Lastly, these suggestions include consideration for the earlier prioritization of various screens for drug candidate liability assessment within the overall screening and lead to optimization paradigm.

A very important consideration for improving R&D productivity and impacting both drug discovery and drug development attrition rates is the strategy of parallel development and backup/second generation candidates. To evaluate these options properly, the boundaries need to encompass advance lead in drug discovery through at least proof of concept in early Phase II clinical studies. Furthermore, an analysis has to look at a number of programs together with quantifying the number of compounds within a single program. Similarly, the cycle time for the compounds in serial versus parallel mode development requires specification. It is noted that this analysis is relevant to every medium to large-size organization since it requires no new technologies or unique skill sets.

Keelin and Reeds[19] have used a simulation approach to evaluate serial versus parallel development and utilized four categories of parallel, from light to maximum parallel. **Figure 11** describes the input format, which includes phases of development through to launch, numbers of compounds, phase success probability as well as terminal state probability. In addition, the simulation assumes full dynamic downstream decision making, including if and when a target is known to have failed, all development stops, and if and when a lead compound fails, backups are advanced according to the parameters of the strategy. In addition, cost, time, and failure probabilities are phase adjustable by program and a net present value (NPV) analysis of cost and value is presented for each scenario. **Figure 12** is one representation of the output of the simulation and shows that parallel strategies are up to 18 months faster to market, and, similarly, result in higher peak sales. Interestingly, significant advantages over serial development are noted with only moderate and evaluate at dose versus efficacy. Phase III clinical trials typically involve a thousand or several thousand patients and

Table 9 Top 10 ways to improve drug discovery productivity

- Increase degree of validation of drug targets
- Include biologics and small molecules
- Utilize biomarkers
- Minimize metabolism liabilities early
- Titrate 'off-target' selectivity appropriate to the disease(s) of interest and use more than one screen for 'on-target'
- Eliminate compounds with mechanism-based toxicity
- Prioritize PK/ADME analyses
- Maximize SBDD
- Employ 'rule of five' or similar guidance and early pharmaceutics profiling
- Integrate computational informatics and algorithm for quantitative structure–function relationships and virtual ligand screening (VLS)

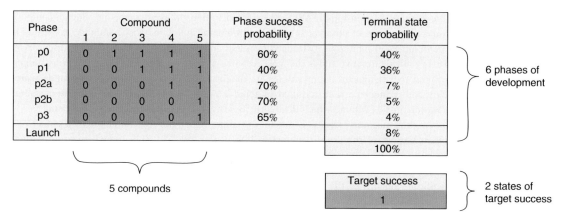

Figure 11 A simulation approach enabled considering the thousands of possible scenarios for each strategy. There are $6^5 \times 2 = 15\,552$ scenarios. We sample these by simulation according to the assigned probabilities (green cells potentially changed in each simulation).

reaffirm efficacy at a specific dose, but more importantly evaluate safety in a larger population of patients. The design of the Phase III or pivotal trials is usually reviewed with the regulatory authorities and forms the basis for establishing the important dosing and administration information of the drug for regulatory approval.

During the time periods in which drug development projects were analyzed, 1985–95, success rates were generally 1 in 9 or 10 entering Phase I achieving regulatory approval and launch. This has gone up somewhat over the last 10 years to 1 in 7 or 8. Attrition is most significant in Phase II where historically it has been typically greater than 50%. Phase I attrition is less, with Phase III realizing the lowest attrition rate. Clearly, the biggest impact of R&D productivity would be to reduce attrition rates in clinical evaluation. To that end, most of the focus has been on improving the drug candidate profile prior to entering the clinic, although certainly improvements in clinical trial design, particularly proof of concept in early Phase II, has been an area of further scrutiny and concentration. Another factor has been the focus on the correlation of Phase II endpoints with the pivotal Phase III trials. It is not uncommon that the Phase II or proof of concept studies can utilize surrogate endpoints that are different than the registration driven endpoints for Phase III required by the regulatory authorities. Examples of this include clinical studies of new cancer treatments, the evaluation of multiple sclerosis, endometriosis, benign prostatic hypertrophy, and others.

Other important factors affecting attrition rates in drug development include the target selection for the drug candidate under study and also the therapeutic/disease category. In addition, there appears to be a different success rate of biologicals versus small molecules, as well, in Phases II and III. **Figure 13** portrays data from one analysis of success rates on late stage parallel activities. Moderate parallel is defined 'as many as two backups pursued at any one time' with the first started simultaneously, while the second lags by one quarter. In contrast, maximum parallel involves four backups pursued through Phase IIa with no lag in starting backups after the lead. Therefore, moderate parallel is a resource moderate scenario.

2.03.3.2 Drug Development

The scrutiny of R&D productivity began with drug development. The logic was that drug development was the most significant investment in time and in funding in the overall drug discovery and development process. Secondly, drug development was more predictable and the activities were common across most drug development programs. In addition, it was in drug development that significant increases in time, cost, and regulation were most evident, and, lastly, failure, particularly in late stage drug development, was not only readily visible but also very costly. For example, data from the PhRMA trade association show that in the 10-year period of 1991–2001 R&D spend went from US$12 billion dollars to US$30 billion, nearly tripling. In addition, while research/drug discovery spending went from US$7 billion to US$12 billion, development spend went from US$5 billion in 1991 to US$19 billion in 2001. And while there was a fourfold increase over this 10-year time period in terms of total development spend, there was a fivefold increase in clinical trial costs. The aggregate spend for R&D for 2004 was over US$50 billion. Obviously, costs continue to rise, and continue to be disproportionate to the number of NME approvals.

Success rates for drugs in various stages of development is typically measured by success in the clinical phases of drug development: Phases I–III. To briefly reiterate, Phase I clinical studies are typically of short duration, involve fewer than 100 subjects, and evaluate dose (concentration) versus safety. More recently, Phase I studies have been used

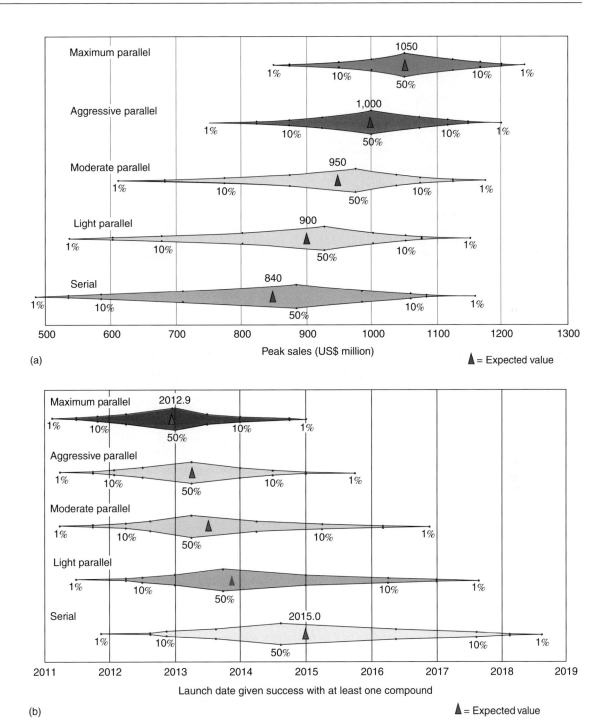

Figure 12 (a) Parallel strategies: higher peak sales. (b) Parallel strategies: faster to market. (Reproduced with permission from Keelin, T.; Reeds, M. *Points fo Light,* **2002** *10*, 17–18.)

to evaluate PK and PD as well. Phase II clinical studies typically involve 700 patients new molecular entities under study in 1997. This shows that Phase II clinical success is twice as high for precedented or best in class targets versus unprecedented or first in class targets. Furthermore, biological targets show a significant higher success rate than small-molecule or chemical drugs against unprecedented targets. The overall trend is evident as well in Phase III success rates but, the differences are not as dramatic.[20] Therefore, the portfolio choices of types of targets, as well as small-molecule versus biological can affect the overall attrition rate in R&D productivity.

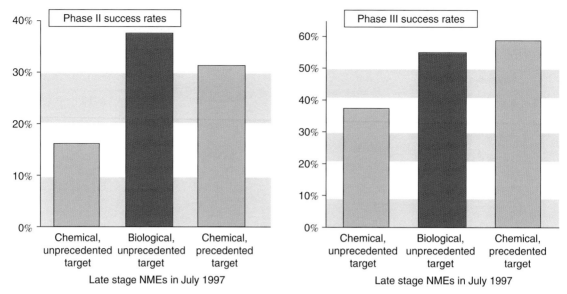

Figure 13 Success rates for compounds targeting novel mechanisms of action from Phase II that reach preregistration. (Reproduced with permission from Windhover's In Vivo.)

Similarly, choices of therapeutic areas can affect the attrition rates. For example, while the development of cancer drugs has enjoyed lower regulatory hurdles, both clinically and toxicologically, than other therapeutic categories, nonetheless, it has historically yielded lower success rates. Those therapeutic areas that have the best correlation of preclinical efficacy to clinical efficacy include antibiotics, certain cardiovascular products such as antihypertensives and lipid-lowering agents, and certain categories of the treatment of pain, particularly inflammatory-associated pain. In contrast, preclinical evaluation in animal models of cancer, many CNS disorders, immunopathologies, antivirals, and degenerative cardiovascular disorders have poor correlation with clinical evaluation. **Table 10** provides a perspective on various disease categories and the success rates in these categories.

Lastly, even within a therapeutic category, there can be significant differences in success rates depending on the types of drug targets that are chosen. Cancer, by definition, probably represents the highest diversity of types of molecular targets. In large part, because of the diversity of types of cancer and the large phenotypic and genotypic differences in the degree of progression of the stages of cancer, multiple molecular targets can change significantly in their validity as targets by tumor type and overtime. Historically, cancer treatment has been principally cancer chemotherapy where the goal was to utilize cytotoxic drugs to reduce or eliminate the cancer burden. These usually were not cancer or histiotype-specific treatments and acted on common molecular machinery common to different cancer types as well as normal cells. More recently, targeted therapies have arisen which focus on genes or their protein products unique to a type of cancer, often expressed on the surface of the tumor cell. However, there have also been targeted therapies going after unique signaling dysfunction within cancers or the disregulation of hormonal control within certain cancer types. These have included not only small-molecule approaches but also immunotherapy involving antibody treatment. These more recently introduced targeted therapies have also changed the paradigm of cancer treatment to one involving more chronic therapy and reducing, but not always completely eliminating, the tumor burden. Layered on this complexity is the frequent utilization of combination therapy attempting to enhance overall response by attacking more than one pathway as well as dealing with the frequent resistance development that occurs with many cancer treatments. In conclusion, many factors then impact the success rate in drug development relative to the clinical evaluation of the drugs including the predictability of the preclinical data, the type of molecular target, the therapeutic and disease category, and, lastly, the nature of the treatment.

Again, there are significant differences in acute versus chronic therapy, both in the speed with which one can secure an answer of success or failure, but also in the duration of the clinical studies. Historically, the mix has favored chronic therapies over acute treatments. However, more recently, big pharma have attempted to broaden their perspective in reinvesting in acute therapies and specialty care opportunities. In one respect the upsurge in drug discovery in cancer represents specialty care. Other examples in specialty disease areas, include sepsis, acute congestive heart failure, and severe, hospitalized infections. Finally, broadening the overall drug modality footprint to include biologicals is another component of maximizing success and return on investment.

Table 10 Therapeutic area analysis

Therapeutic area category	Diseases	Degree of difficulty (1–5)[a]	
		Drug discovery	Drug development
Central nervous system			
Neurology	Stroke, traumatic head injury (THI), subarachnoid hemorrhage (SAH)	2	5
	Epilepsy	2	2
Neurodegenerative	Alzheimer's, Parkinson's, Huntington's, multiple sclerosis	3	4
Psychiatric	Generalized anxiety disorder, major depressive disorder (MDD), bipolar, schizophrenia	2	3
Pain	Nociceptive, inflammatory	1	2
	Neuropathic	3	3
Cardiovascular			
Coronary artery disease	Hypertension	1	1/3
	Dyslipidemias, atherosclerosis	2	2/3
	Restenonis	3	3
Congestive heart failure	Acute, chronic	3	3
	Anticoagulants	3	2
Metabolic			
Diabetes	Type 1	5	4
	Type 2	2	2
Obesity	Energy, anorexigenic	4	3
Reproductive	Fertility	2	4
	Endometriosis, fibroids, benign prostatic hypertrophy	3	3
Inflammation			
Pulmonary	Asthma, chronic obstructive pulmonary disease	3	3
Arthritis	Rheumatoid, osteoarthritis	3	3
Gastrointestinal	Inflammatory bowel disease, irritable bowel syndrome	4	4
Gastrointestinal			
Antisecretory/antidiarrheal		3	2
Antiemetic		3	2
Overactive bladder		3	2
Gastroesophageal reflux disease		2	2
Infectious diseases			
Bacterial		1/3	1/3
Viral		2	2
Fungal		2	2
Cancer			
Hematological	Leukemias, lymphomas	2	2
Solid tumors	Colon, breast, lung, etc.	3	3

[a] Ranked 1–5, with 1 = low and 5 = high; includes assessment of technical success and duration/cost; split values reflect a subset value due to unique hurdles (e.g., 1/3 for infectious diseases for drug development due to 1 = low difficulty in clinical studies but 3 = moderate difficulty due to recent FDA stipulation of percent confidence intervals for achieving equivalence to standard therapy).

There are other components of drug development that can be improved to impact overall R&D productivity. The two most important ones are long-term toxicology in animals and drug scale-up and manufacture. The potential liabilities here are primarily addressed by timing. For biologicals, it is important to determine the scalability of the product early on before significant downstream investment. For long-term toxicology studies, particularly for unprecedented targets and first in class drugs, it is important to initiate the studies sooner in the paradigm to minimize overall investment in the program before this key finding can be established. A recent example is the requirement by the FDA to complete 2-year carcenogenicity studies in drugs for the treatment of diabetes, that act on peroxisome proliferator-activated receptor gamma (PPAR-γ), before longer-term clinical trials can be initiated. A final factor to consider is discontinuation for commercial reasons. Here again, addressing the commercial fit of the drug development portfolio should be done early on in the drug discovery or early stage drug development process. Alternatively, and exemplified most commonly in biotech companies, if there is no commercial fit, the product can be partnered or out licensed at Phase II, a stage in which significant asset value has been determined and the overall program reasonably derisked.

Efficiencies have been an area of focus as well for drug development. Efficiencies have focused primarily on the bottlenecks in clinical trials. One of the challenges in early phase clinical evaluation is the cost and time in establishing the PK of the drug candidate, or PK/PD. The typical scenario requires considerable investment in drug scale-up (usually Good Manufacturing Practice), 2–4-week toxicology in two species, bioanalytical development, and preparing regulatory submissions. The FDA has recently promoted 'phase 0' studies, which allow for microdosing and/or short-term, low doses that might exhibit pharmacology in human subjects, to address PK and PK/PD with a more limited preclinical package. Other bottlenecks have been enrollment, clinical trial execution, and dealing with the tremendous data burden generated by clinical trials, particularly large clinical trials. Significant improvement has been realized in patient recruitment and enrollment in clinical studies. Establishing large databases of patients, extending clinical trials on a global basis, advertising, earlier clinical site qualification, and other techniques have been used to address this bottleneck. The data challenge with clinical trial execution has primarily been addressed by trying to minimize paper and make the overall data acquisition as well as data management electronic. Electronic data capture at sites is now becoming a reality and real-time data acquisition has improved considerably. Undoubtedly, there are additional efficiencies that need to be realized but progress is being made. Lastly, the coordination of clinical trial supplies using interactive voice recognition services and just in time inventories has enhanced the efficiencies of clinical trial execution as well.

2.03.3.3 Infrastructure

Often overlooked in R&D productivity assessments is the role of intellectual property (IP), information technologies (IT), and regulatory affairs. These are critical areas of competency and expertise in drug discovery and development in that they provide the framework for describing and protecting the assets: the drug candidates. Expertise in IP, IT, and regulatory is required at a number of the stages of the drug discovery and development paradigm and therefore, the integration of these competencies is important in enhancing R&D productivity.

2.03.3.3.1 Intellectual property

Patents are essential to the pharmaceutical business model. Since the time from concept to commercialization is frequently more than a decade, some degree of exclusivity of the product is required to justify the extraordinary time, expense, and risk associated with drug discovery and development. Patents have therefore been part of drug discovery and development for many years. Novelty in therapeutics that has been patented has been primarily of two types: composition of matter and methods of use. The most rigorous and defensible patenting of proprietary work has been the composition of matter since it unequivocally describes the drug and its closely related analogs. Furthermore, this type of patent also describes how to make and use the novel entities. In recent times, a challenge to composition of matter patents has been the legalities of patenting genes, gene sequence, vectors with genes as therapies, and the like. Generally, the US Patent & Trademark Office (PTO) has been favorable while the European countries patent offices have not.

Alternatively, methods of use patents have been utilized to describe novel uses of prior art and afford a measure of exclusivity for the innovator. Over the past two decades there has been a significant increase in methods of use patents as it applies to drug discovery tools. These patents have typically described new screening methods and/or key components of screens, such as the molecular target of interest, as a way to identify potential drug candidates. While these types of patents have been more difficult to enforce than typical composition of matter patents, these patents are considered particularly important to small companies business models as this IP represents a potential revenue stream. There is inherent conflict with these patents and the freedom to practice in the domain of 'research purposes only.' The FDA has allowed companies that are preparing new submissions for drugs for regulatory approval to utilize the patented art of other companies to generate appropriate safety and efficacy information, for example, as comparators to their own drug candidate. This represents a second source of inherent conflict between method of use patents regarding drug discovery tools and the drug discovery and development paradigm. Ultimately, these conflicts will be resolved.

The protection of the novel art in drug discovery and development is an ongoing function within R&D, from the genesis of drug discovery through to the commercialization of the product. Patent attorneys with a science and/or medical background are essential for the execution of appropriate patent applications. They must be involved at the conception of a program since the initial drug discovery target could be novel in its own right, as well as, methods of using said target to discover new drugs. Furthermore, the composition of matter of the initial lead compounds should be described as soon as possible. The timing of a patent application takes on additional importance with the change in the American Inventors Protection Act of 1999, which states that any patent filed in 2000 and beyond will be published within 18 months of its application date. Prior to that time point a patent was not published until it was issued. Therefore, the publication of the application within 18 months discloses the intents of the innovating party sooner than has historically been the case. The timing of filing is critical both from the point of view of length of the exclusivity period as well as the protection of art in a highly competitive area of research.

As a program progresses in development, it is not uncommon that new uses for a drug are identified and explored. These also involve additional patent activity. New routes that are uncovered to produce and scale up the active ingredient also merit patent protection. Furthermore, novel forms of the active ingredient such as polymorphs, novel formulations, and/or unique delivery systems should be patented. Another dimension to patent activity is the breadth of filing patent applications. Global pharmaceutical companies usually file in many countries, countries that they have identified as potential commercial markets for their future product. This requires interfacing with the patent offices of many different countries, dealing with the languages unique to those countries as well as the downstream issuance fees. Today many countries, including now China and India, subscribe to the World Trade Organization (WTO) and General Agreement on Tariffs and Trade (GATT). This serves to harmonize some of the regulations on patenting among the many countries.

Patents are, of course, particularly important for small companies in that the value of the company is based on both the drugs in the company's portfolio that under development but also the validity of the exclusivity surrounding these drug candidates. Maintaining that exclusivity is not just a matter of the filing and issuance of patents, but also dealing with patent interferences that arise or lawsuits involving other parties that may be utilizing the innovator company's proprietary inventions. This investment in expertise in patent law within a pharmaceutical company or biotech company is no longer unique to those environments. Most academic institutions and government laboratories now are very aggressive in patent applications and the enforcement thereof. This has become a growing revenue stream for those institutions and furthermore, often represents a basis for spinout companies with tech transfer of the intellectual property discovered at these institutions.

2.03.3.3.2 Information technology

Several decades ago IT was relegated to specialists in computer science siloed in an organization cloistered around a main frame. With the advent of the internet, desktop computers, distributed computing networks, high-speed communications including wireless, IT are now an integral part of day-to-day activities for the drug discoverer and the drug developer. In fact, IT represents the most significant revolution over the last several decades in the life sciences and particularly in drug discovery and development. IT provides the opportunity to generate and analyze extraordinary amounts of data with high efficiency while providing the tools for the individual scientist or scientific teams to perform efficient multidimensional analyses of drug candidates. It does challenge the drug discoverer or developer to extract and utilize knowledge from all of these data. Indeed knowledge management is probably our biggest challenge.

The R&D organization virtually runs on its IT backbone. IT typically represents about 10–15% of the R&D budget in most organizations and includes not only computer scientists but experts in chemoinformatics, bioinformatics, and statistics, and computational scientists involved in simulation and modeling. The colocalization of these scientists with the chemists, biologists, and those scientists involved in drug development is a competitive advantage. Close working relationships are essential between the IT specialist and the scientist utilizing the applications so that effective and efficient research comes from the overall investment. The challenge exists to maintain critical mass and competency in IT centers, such as data management, computational chemistry, and infrastructure while maintaining an appropriate balance between the investment in these core competencies and the distributed nature of IT.

2.03.3.3.3 Regulatory

Regulatory affairs typically refers to that group of scientists who formulate the strategy for interacting with the regulatory authorities in various countries as well as the tactics of securing responses to questions dealing with submissions and maintaining communication postregistration. Many of the scientists who populate regulatory affairs groups were originally involved in drug discovery or development. They need to understand the processes involved in drug discovery and development to accurately represent the science to the regulatory authorities. They are also involved in the compilation of the information in the form of investigational new drug submissions or new drug applications for final registration. With the evolving landscape of regulatory guidances, they also need to remain state-of-the-art in what the regulatory agencies are thinking and saying.

Organizationally they become involved when preclinical toxicology studies initiate on a drug candidate, as this represents usually the first event in generating data under regulatory guidelines (Good Laboratory Practice (GLP)). From this point on regulatory affairs is involved with nearly every aspect of drug development through to commercialization. They are also involved in postcommercialization activities. For example, activities that involve an expansion of information to modify a current label on a marketed product, submissions for new indications, or new formulations and postmarketing surveillance of safety (pharmacovigilence) are part of regulatory. Regulatory affairs also plays a role in reviewing advertising and other communications that are used to describe the new product during and after an NDA.

2.03.3.4 Technology

In addition to IT, drug discovery and development has been dramatically impacted by new technologies. These technologies have been primarily advances in automation and miniaturization, and secondly in the abilities to sequence DNA in an extraordinarily high-throughput and high-fidelity fashion. These technologies are going to be elaborated in much more depth in subsequent chapters in this volume. As an overview, the areas of principle focus in new technologies are combinatorial chemistry, HTS, and genomics.

Combinatorial chemistry had its genesis with the early efforts in the late 1980s and early 1990s to synthesize rapidly peptides and subsequently small organic molecules in a combinatorial fashion. Varying several different appendages to a core synthetic element allowed a rapid parallel expansion of a library of compounds within this structural core. These resulting combinatorial libraries provided for an explosion of compound inventories. In turn, these libraries provided for the increased chemical diversity as well as increased design for HTS. This technique had its advent in the mid-1990s and incorporated engineering and robotics to screen rapidly for actives or hits in a multiwell format, such as 96, 384, or 1536 well-plates. HTS generated extraordinary amounts of data about binding of compounds in these libraries to a molecular target of interest and therefore served as a starting point for a new drug discovery program. These two technologies have now been incorporated into every drug discovery organization in one fashion or another. They are capital-intensive investments, not only in the hardware, but also in the human expertise required to utilize these technologies, and lastly the IT support for them. Even though it has been a decade since they came into use, it is not yet clear whether they have truly increased R&D productivity.

Genomics describes the multiple technologies involved in high-throughput sequencing and analysis of sequences to provide a gene map for humans and a number of other species. This has been a global effort on a massive scale and has resulted in the complete sequencing of multiple genomes in less than a decade. It has been estimated that over $15 billion in private capital alone was invested in the human genome project. The vision is that this information will provide for many new molecular targets for drug discovery, and will provide for a molecular understanding of diseases and therefore allow for better diagnostics and personalized medicine. The realization of this vision is however still sometime in the future. There is an extraordinary amount of biology that still needs to be done to understand the function of the gene in the first instance. Then follows the exploiting of this functional information in many dimensions including drug discovery and linkage to human diseases. Proteomics is a technology that has paralleled genomics in its development and describes the analysis of the proteins that are expressed from genes that are under active transcription. Proteomic analysis is now a ongoing technology and business opportunity, allowing for a better understanding of disease pathways, the discovery of new diagnostics, and the establishment of new druggable molecular targets for drug discovery and development. While there may be only a thousand to several thousand new molecular targets that have come out of the genomics efforts, posttranslational modification and compartmentalization of proteins within cell substructures means that there are multiple proteins that come from a single gene. Therefore it is expected that additional definition will be forthcoming about the true number of new druggable molecular targets based on proteomic analyses. Systems biology is the latest new descriptor of frontier efforts evolving from genomics and proteomics. It is the effort to understand the pathways and networks interrelated within the cell for a more global understanding of function. One area of genomics promise and application has been pharmacogenomics. In brief, this is the evaluation of genotypes or genotypic profile for a particular disease in patients using a variety of different techniques for genome location analysis and gene sequencing. Investigators, and many companies, in this new field are studying how variations in sequences contribute to phenotypic traits such as diseases, disease subpopulations, and/or drug responses. For example, one approach is analysis of millions of single nucleotide polymorphisms (SNPs) in a population that has specific traits or a disease. The applications of this information include molecular targets for drug discovery, diagnostics, biomarkers for drug development, subsetting patient populations that may be more responsive to a particular therapy (personalized medicine), or repurposing known drugs. The potential is extremely exciting, but the promise continues to be well ahead of the deliverables.

It is premature to conclude whether genomics and its related 'omics' technologies have increased or decreased R&D productivity. However, they are essential components of modern-day drug discovery and development and as such, these obligatory tools are being used in the overall drug discovery and development paradigm. Drug discoverers will follow the science whether it is hypothesis-driven or a systems approach. Ignoring the science will be at their peril.

The growing power of reimbursement policies at the federal level in many countries in the pharmaceutical marketplace is having a significant impact on the current business models for big pharma as well as specialty and biotech. For many years, countries in Europe as well as Japan have had prices determined at the federal level and therefore have historically been markets with lower margins relative to the USA. In addition, the Japanese marketplace, which is characterized by higher prices on market entry, has the obligatory, biannual reduction in the price of a pharmaceutical after market entry. Recently, various countries in Europe have introduced the concepts of capitation of total healthcare budgets, reference price listing, and parallel imports between countries. All of these have served to reduce margins even more over the last 5–10 years in these countries for new pharmaceuticals. Furthermore, in the USA at the government and payer level, we are witnessing pressure to cover the uninsured, and pressure for a Medicare drug benefit which has now recently been enacted into law and will be put in place in 2006. As payer economics are suffering and Medicare bankruptcy is looming in the next 10–15 years, and with healthcare costs accounting for about 20% of the federal budget, there is a growing pressure to shift the burden to consumers and employers. These are seen with increasing co-pays and employers requiring employees to bear a greater share of the healthcare burden. We are seeing that employers are bearing about 40% of all costs versus 20% in 1990 and healthcare is the second largest payroll cost, representing between 6–8% of the total operating cost of employers. And, lastly, as the population is aging simultaneously with the uninsured population increasing, there will be greater pressures put on healthcare costs in the future.

With all of this there is also growing competition in major drug classes. Exclusivity periods are declining significantly in the last two decades and the cliff is even steeper for branded drugs when they lose patent, witnessing losses of 90% plus in less than a single quarter. While each one of these elements is not new, they are all trending in a direction that will result in a collision in the not too distant future. If that collision occurs, something will have to yield. Either society will have to be willing to pay more for medical treatment relative to other costs in society or the cost of discovering and developing a drug will have to be somehow be slashed or, lastly, the prices for drugs will have to be significantly reduced. If either of the latter two scenarios materializes, there is a high likelihood that there will be a significant diminution of new drugs becoming available to treat diseases in the next quarter century.

2.03.4 Conclusion

Drug discovery in the pharmaceutical industry has evolved over the last 100 years from serendipitous findings with rapid commercialization to a highly structured, regulated, and complex network of laboratory and clinical activities requiring more than 10 years and US$1 billion from concept to commercialization. The 'industry' has similarly evolved from a diverse array of companies, most of modest size, with significant proportions of nonpharmaceutical revenues, to a more consolidated landscape today with the largest pharma companies securing more than 10% market share and focusing exclusively on pharmaceuticals. In the last three decades the biotechnology sector, starting from a new technology in designing biologicals as drugs, has matured to an array of over 500 small to large companies heavily invested in the discovery and development of all types of new pharmaceuticals, diagnostics, and products for agriculture.

Revolutionary advances in science, particularly at the interfaces of chemistry, biology, engineering, and IT have led to increasing complexities and costs in the development of new drugs but have also led to new ways to design drugs, new drugs for diseases previously untreated, and much better understanding of the etiology of diseases. The challenge in today's pharmaceutical research environment is the appropriate integration of the requisite multidisciplinary functions at the various stages of drug discovery and development to effectively find, develop, and introduce new drugs. This challenge is exacerbated by the consolidation of the industry resulting in R&D organizations of extraordinary size and literally spanning the globe. The most efficient and effective model for drug discovery and development is the functional model within a therapeutic area, sized to enjoy critical mass of key functions and outsource other functions, yet compact enough for complete colocation of functional competencies. Size can be a strategic advantage in some drug development functions but is often an impediment to productive drug discovery.

Biotechnology has revolutionized drug discovery and development in several ways. Firstly, it represents the advent of designed biologicals as drugs. Secondly, with venture capital financing, the biotech industry has substantially solved the challenge of starting new companies in a business that is risky and capital intensive. Thirdly, it epitomizes the new biology and new technologies. Lastly, it defines entrepreneurism, a cultural asset largely ignored in larger pharmaceutical companies.

Philosophy matters! Technical competencies and motivated scientists are essential, but not sufficient, for successful drug discovery and development. Important philosophical choices regarding target versus disease, chemical versus biological, best in class versus first in class, and fail fast versus follow the science can significantly affect the success of a pharmaceutical or biotech enterprise. Related issues such as the relative resourcing of life cycle management of drugs on or about to enter the market versus development of new drug candidates are critical decisions for the longer-term building of the business. These choices cannot be generically prescribed, but must be evaluated and appropriately titrated to meet the needs of the organization at any defined state of maturation.

Culture is an often under-appreciated competitive advantage. Those R&D organizations that recognize the values of collaboration, foster an entrepreneurial environment, and balance product champions with good portfolio management usually thrive and exhibit the attribute of excellent execution. Lastly, partnering has become another essential password for growth and success, from the smallest to the largest companies and extending well beyond the private sector to academic and government laboratories.

In spite of the decrease in R&D productivity and increasing costs of drug development, the pharmaceutical and biotech industries have continued to grow and deliver new medicines to patients. A major reason for this is they have enhanced the overall value per drug by targeting commercially attractive unmet needs that can be premium priced, and by expanding established market arenas with improved therapeutics. But the industry can only grow, and indeed survive, by addressing two other fundamental areas of productivity. It must increase efficiency by reducing costs or time or both, and it must decrease the cost of failure by terminating drug candidates sooner or improving the probability of success. The area for greatest impact is the area of decreasing cost of failure. Increasing efficiency will be incremental at best, since reducing time per program is unlikely. Similarly, reducing costs is very unlikely. While significant improvements and efficiencies have been realized in drug development, the focus is on drug discovery. The future will belong to those organizations that develop the tools to dramatically improve success rates in clinical evaluation.

References

1. Glaxo Smith Kline timeline. www.gsk.com (accessed April 2006).
2. Lesney, M. S. *Modern Drug Disc.* **2004**, *January*, 25–26.
3. Center for Medicines Research International (CMR). International Pharmaceutical R and D Expenditure and Sales 2001, Data Report I. In *Pharmaceutical Investment and Output Survey*; CMR: London, UK, 2001.
4. Sherman, B.; Ross, P. *Acumen J. Sci.* **2004**, *1*, 46–51.
5. Drews, J. *Science* **2000**, *287*, 1960–1964.
6. Drews, J. In *Human Disease from Genetic Causes to Biochemical Effects*; Drews, J., Ryser, S., Eds.; Blackwell: Berlin, 1997, pp 5–9.
7. Drews, J.; Ryser, S. *Nat. Biotechnol.* **1996**, *14*, 15.
8. Bailey, D.; Zanders, E.; Dean, P. *Nat. Biotechnol.* **2001**, *19*, 207–209.
9. Hopkins, A. L.; Groom, C. R. *Nat. Rev. Drug Disc.* **2002**, *1*, 727–730.
10. Drews, J. *Nat. Biotechnol.* **1996**, *14*, 1516–1518.
11. Lombardino, J. G.; Lowe, J. A., III. *Nat. Rev. Drug Disc.* **2004**, *3*, 853–861.
12. Booth, B.; Zemmel, R. *Nat. Rev. Drug Disc.* **2003**, *2*, 838–841.
13. Minkel, J. R. *Drug Disc. Mag.* **2004**, *November*, 32–36.
14. DiMasi, J. A.; Hansen, R. W.; Grabowski, H. *J. Health Econ.* **2003**, *22*, 151–185.
15. Reichert, J. M. *Nat. Rev. Drug Disc.* **2003**, *2*, 695–702.
16. Banerjee, P. K.; Myers, S. D.; Baker, A.; Andersen, M.; Bicket, S. *High Performance Drug Discovery*; Accenture: Boston, MA, 2001.
17. Kola, I.; Landis, J. *Nat. Rev. Drug Disc.* **2004**, *3*, 711–715.
18. Gillespie, P.; Goodnow, R. A., Jr. *Annu. Rep. Med. Chem.* **2004**, *39*, 293–304.
19. Keelin, T.; Reeds, M. *Points of Light*, **2005**, *10*, 17–18.
20. Aghazadeh, B.; Boschwitz, J.; Beever, C.; Arnold, C. *In Vivo* **2005**, *January*, 59–64.

Biography

Wendell Wierenga, PhD is Executive Vice President, Research and Development at Neurocrine Biosciences in San Diego, CA. He joined Neurocrine in September 2003 and is responsible for all aspects of R&D including discovery research as well as preclinical and clinical development, manufacturing, and regulatory affairs. From August 2000 to August 2003, Dr Wierenga was Chief Executive Officer of Syrrx, Inc., a private biotech company specializing in high-throughput, structure-based drug design. Prior to joining Syrrx, from March 1997 to July 2000, he was Senior Vice President of Worldwide Pharmaceutical Sciences, Technologies and Development at Parke-Davis/Warner Lambert (now Pfizer), where he was responsible for worldwide drug development, including toxicology, pharmacokinetics/drug metabolism, chemical development, pharmaceutics, clinical supplies, information systems, and technology acquisition. From 1991 to 1997 he was Senior Vice President of Drug Discovery and Preclinical Development at Parke-Davis. Prior to Parke-Davis, Dr Wierenga was at Upjohn Pharmaceuticals for 16 years, where his last position was Executive Director of Discovery Research. Dr Wierenga led/participated in the research and development of more than 50 INDs, over 10 NDAs, and over 10 marketed products, including Lipitor and Neurontin. Dr Wierenga earned his BA from Hope College in Holland, MI, his PhD in Chemistry from Stanford University, and an American Cancer Society Postdoctoral Fellowship at Stanford. Dr Wierenga is on the SAB of Almirall and Agilent. In addition, he has been on the boards of directors of four biotech companies and is currently on the boards of Ciphergen, Onyx, and XenoPort.

2.04 Project Management

Y Nakagawa, Novartis, Emeryville, CA, USA
L Lehman, Gilead Sciences, Inc., Foster City, CA, USA

© 2007 Elsevier Ltd. All Rights Reserved.

2.04.1	**Introduction**	**138**
2.04.2	**Pipeline Phases and Appropriate Project Teams**	**138**
2.04.2.1	Project Phases: Activities and Goals Define Team Composition	138
2.04.2.1.1	Target identification and validation	138
2.04.2.1.2	Target development and high-throughput screening (HTS)	139
2.04.2.1.3	Hit-to-lead	139
2.04.2.1.4	Lead optimization	140
2.04.2.1.5	Predevelopment	141
2.04.2.2	Team Responsibilities	141
2.04.2.3	Metrics and Goals	142
2.04.2.3.1	Project goals	142
2.04.2.3.2	Individual and project functional area goals	142
2.04.2.4	Development Candidate Criteria	142
2.04.3	**Project Team Leadership**	**142**
2.04.3.1	Effective Leadership	142
2.04.3.2	Interpersonal Skills of Effective Leaders	144
2.04.3.3	Project Leader Responsibilities	144
2.04.3.4	Transitioning Project Leadership	144
2.04.4	**Effective Project Teams**	**145**
2.04.4.1	Productive Project Teams	145
2.04.4.2	Functional Departments versus Project Teams	145
2.04.4.3	High-Performing Project Teams	146
2.04.4.4	Rewarding Teamwork	146
2.04.5	**Matrix-Driven Organizations**	**146**
2.04.6	**Research Project Oversight and Decision Points**	**147**
2.04.6.1	Management Teams	147
2.04.6.2	Reviews and Feedback	148
2.04.6.2.1	Written project reports	148
2.04.6.2.2	Formal project reviews	148
2.04.6.2.3	Ad hoc meetings with project teams	148
2.04.6.2.4	Team member feedback	149
2.04.6.3	Project Decision Points	149
2.04.6.4	Project Management	149
2.04.6.4.1	Research project management	149
2.04.7	**Management of Collaborative Projects**	**151**
2.04.7.1	Collaboration Initiation	151
2.04.7.2	Logistical Considerations	151
2.04.7.3	Collaboration Oversight	152
2.04.8	**Transitioning Projects into Development**	**152**
2.04.8.1	Development Candidate Nomination Document	153
2.04.8.2	Transitioning of Project Team Membership	155
2.04.9	**Project Management Tools**	**156**
2.04.10	**Conclusion**	**156**
	References	**157**

2.04.1 Introduction

One of the most important assets to a company is its research project pipeline, and to ensure that the best decisions are made with respect to both advancing and terminating projects, proper oversight and management of the project portfolio are essential. Successful companies are able to translate good science into attractive drug candidates, and ultimately, a therapeutic drug. They are able to make good decisions as quickly as possible based on solid information and data. Successful project teams should be able to provide this information to the key decision-making committees in a succinct and timely manner, and skilled project management support should ensure that project teams are able to deliver on these needs. By making the right decisions, the company increases its probability of success by providing the appropriate resources (which are usually limited) to those projects that are the most promising and have the best chance of advancing into the clinic, and ultimately to the market. This chapter will address several important aspects of research project management: from project teams, leaders and managers, to organizational oversight, collaboration management, and transitioning of projects from research into development. The approaches described in this chapter are commonly utilized in many companies; however, there are other approaches that can be and are used in other organizations, and that may be better suited for their particular research environment and needs.

2.04.2 Pipeline Phases and Appropriate Project Teams

Development of a small-molecule drug is becoming an exceedingly expensive and complex process,[1] and requires the participation of numerous functional areas. The creation of a project team should enable better strategic decisions and operational planning, and cross-disciplinary project teams are essential to ensure that key aspects of candidate advancement are appropriately addressed. A high-performing team will identify and anticipate key issues and make appropriate decisions quickly, thereby keeping critical timelines on track.

Project team composition will evolve over the lifetime of the project, comprising mainly research functions in its early stages, and later including nonclinical, clinical, regulatory, and process development functions as needed. Commercial, business development, and intellectual property (IP) input are also important, and the extent of analysis and input will again be dependent upon the stage of the project. Similarly, project leadership may evolve over time as the key focus and goals of a project change as the project advances through the pipeline. Project leadership will be addressed in the next section.

Different companies have differing philosophies regarding the role and/or need for project management support, especially prior to entry of the project into development. An effective project manager will ensure that there is good communication and sharing of information between team members, that project timelines and budgets are adhered to, and working closely with the project team leader, that the project goals are met, and that senior management is informed of any key deviations from the original plan and/or strategy. If an experienced project team is in place, the need for project management support may be less critical, as the team members will understand the importance of clearly defined goals and open communication, the issues and activities which need to be addressed and implemented, and presumably will have satisfactory organizational skills. An experienced and well-organized research project leader can often serve in the dual role of a Project Leader/Manager. However, once a project enters into development, formal project management support is usually required, as project tracking, resource management, and communication to appropriate management teams become even more critical. Further aspects of project management will be discussed in Section 2.04.6.4.

To address appropriate project team composition at various stages of the pipeline, it is first necessary to describe the general activities and goals for each phase of a project (**Figure 1**).

2.04.2.1 Project Phases: Activities and Goals Define Team Composition

2.04.2.1.1 Target identification and validation

Depending on the company strategy and therapeutic area, target identification and validation may or may not be part of the drug-development pipeline. Due to the potential risks of working on clinically yet-to-be-validated targets,

Figure 1 Typical research pipeline stages.

many companies elect to focus their drug discovery efforts on known, validated targets. If a company is interested in identifying novel targets, or in further validating putative targets, this stage is critical as significant resources will eventually be placed on the target of interest. The availability of tool or reference compounds and existing structural data are considerations that can impact the attractiveness of a target. A poorly validated target can result in wasted time and resources, and therefore strict criteria must be established and met before a target is accepted into the pipeline.[2]

New literature can dramatically affect the validity of a target (both positively or negatively); therefore much diligence must be placed on following the scientific literature at this point of the pipeline. In addition, the project team must be aware of the IP status of the target; freedom to operate considerations surrounding the target (including assay reagents and technologies) are important factors in deciding whether to progress a target of interest. The project team at this initial stage is usually small and comprised primarily of research biologists and bioinformaticists.[3] However, input from clinical and commercial groups should also be solicited at this point for a reality check on the downstream development feasibility and commercial attractiveness of the target/indication.

In addition to the target validation biology efforts, protein biochemistry resources may be needed for reagent generation for future activities, including primary and secondary screening, structural chemistry efforts, etc.

2.04.2.1.2 Target development and high-throughput screening (HTS)

Once a target is approved by the appropriate oversight committee, it is necessary for assay development scientists to develop a robust, high-throughput screen for screening the company's compound collection. If only a subset of compounds will be screened based on the target, a lower throughput screen will be acceptable. Many new technologies are being used for high- and medium-throughput screening, and may be addressed in Chapter 2.12, pp 90–95. Hits from the screening effort need to be confirmed for activity and identity. In addition, nascent structure–activity relationships (SARs) from a compound series are often a requirement for advancing a compound or scaffold into the next phase. During this phase, robust secondary assays as well as selectivity screens should be developed. Medicinal chemistry and chemical informatics resources often become involved from this stage to help evaluate the attractiveness of the chemical starting points, as well as to help evaluate all the biological data available on the compounds of interest.

2.04.2.1.3 Hit-to-lead

Once a project has advanced past the screening phase and identified confirmed hits with a discernible SAR, a project team is usually created to take the project to the next phase. It is at the hit-to-lead stage where the project requires dedicated participation from different scientific disciplines. Typically, the biology champion of the project and a medicinal chemist will team to lead the effort to identify a few attractive chemical lead series based primarily on in vitro data. It is important for the project team to develop a target product profile (TPP) by this stage, which will define the properties of the ideal drug candidate.[4] Input from downstream functions such as clinical and commercial groups will be necessary to develop a useful TPP. The TPP defines the required efficacy and side-effect (safety) profile for the drug, the dose regimen and how it will be administered, in which patient populations and for what indication (**Table 1**). It should target an acceptable cost of goods based on anticipated competitor products. Ideally, it will define the performance requirements of the drug that will enable the commercial organization to assess the market impact upon introduction of the product, and estimate the commercial return.

The project team will also need to develop a screening cascade, which clearly delineates the critical path and criteria that need to be met to advance a compound further down the pipeline. The proposed screening cascade will need to be approved by the Research Management Team, since this allows them to estimate the resources that will be required for moving the project (as well as other projects in the portfolio) forward. Frequent and open communication between project teams and research management is critical from this point on, since significant resources (both people and external dollars) start being deployed at this stage, and any new result could change the strategy of the project. Both the TPP and the screening cascade should be dynamic and flexible; they should be able to reflect changes in new biology and competitive landscape, and frequently be revisited to ensure that they serve to drive decision-making and advance the most promising compounds (**Figure 2**).

In addition to the in vitro primary and secondary assay and selectivity data, it may be appropriate at this stage to better characterize the biochemical mechanism of activity, for example, reversible versus irreversible inhibition, competitive versus noncompetitive or allosteric inhibition. The physicochemical properties of the leads should also be assessed early on. This would include analytical chemistry measurements of compound solubility, stability, and preliminary assessment of potential metabolic liabilities, among others.

Table 1 Example of a target product profile (TPP)

Profile parameter	Target claim
Indication	
In vitro potency/selectivity profile and MOA	
drug metabolism and pharmacokinetics (DMPK)	
Preclinical efficacy and biomarker analysis	
Safety	
Clinical efficacy	
Dose regimen, route	

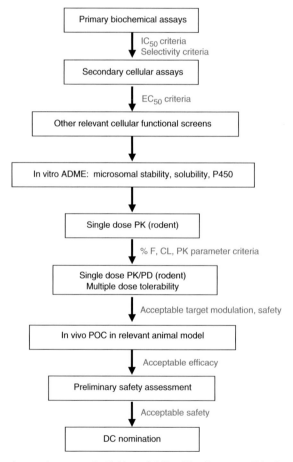

Figure 2 Example of a general screening cascade. F, bioavailability; CL, clearance; PK, pharmacokinetics.

Ideally at this stage, alternative scaffolds will also be identified, in case the lead series should fall out later in the pipeline due to unfavorable physicochemical or biological properties.

2.04.2.1.4 Lead optimization

Once attractive leads that meet all the defined in vitro criteria have been identified, the compounds are optimized based on their in vivo properties. This occurs during the lead optimization phase of a project, and therefore nonclinical functions of pharmacology, pharmacokinetics (PK), and to some extent, toxicology are integrated into the project team. Companies are increasingly realizing that concomitant optimization of absorption, distribution, metabolism, and excretion (ADME) and safety parameters relatively early in the drug discovery process (lead selection and optimization phases) can help to improve the probability of success, so that fewer candidates fail in development for reasons that could have been predicted in research.[5,6] Medicinal chemistry resources continue to be critical, not only for generation of new compounds to test, but for initial compound scale-up for the in vivo studies. Projects may also require evidence of target modulation (pharmacokinetics/pharmacodynamics (PK/PD)) using translational medicine approaches to demonstrate that the compounds are exerting their pharmacologic effect via the desired mechanism/target.

Appropriate animal models must be available to assess the activity of the candidates. Attention should be given to ensure that the animal model is relevant, that is, that the observed disease is driven by the targeted mechanism. The clinical development organization should also be consulted to select the relevant testing models as well as clinical dose regimens. The team must revisit its criteria for advancement (efficacy, PK, safety), and at the end of this phase, select a small number (1–3) of promising candidates to advance into the predevelopment phase.

2.04.2.1.5 Predevelopment

The goal of this phase of the drug-discovery pipeline is to identify the Development Candidate (DC) from the 1 to 3 advanced preclinical candidates. Further in vivo evaluation to determine efficacy, optimal dose scheduling, and safety will be conducted. PK studies will need to be conducted in nonrodent as well as rodent species, as will preliminary safety studies. Further benchmarking against competitor compounds should also be performed to assess the relative activity and safety of the possible drug candidates. These studies may provide insights regarding areas of anticipated superiority of the internal candidate. The project team composition will expand to include more involved participation from functional areas which are also important in investigational new drug-enabling and other development activities. In particular:

- Chemical development representation to ensure compound scale-up procedures are feasible.[7]
- Formulation development to perform preformulation work, salt selection, polymorph evaluation, etc.
- Toxicology support since more extensive safety assessment will be needed prior to candidate selection.
- Clinical: preliminary clinical strategy will be required to design appropriate safety studies (with some guidance from Regulatory on any considerations that could impact the clinical development of the candidate).

2.04.2.2 Team Responsibilities

From the previous section, it is apparent that as a project advances through the drug discovery pipeline, the size of the project team will increase due to the increasing number of activities that need to occur. Meetings comprised of large groups of people are rarely effective; therefore a smaller Core Team should be defined. The Core Team is usually comprised of more senior representatives from the key functional areas of the project team. This working group (usually 4–6 people) is responsible for meeting on a regular basis to discuss new project developments, make appropriate decisions, and then to communicate those decisions and action items back to their functional areas. To keep the entire project team informed and engaged, it may be appropriate to hold monthly or quarterly project team meetings where the whole team is invited to hear the latest updates, accomplishments, and key issues the project was facing (mechanisms for sharing project information/data are addressed in Section 2.04.9). These meetings should be primarily informational in nature; however, there may be times when it is appropriate to include a larger group to discuss data and reach decisions. Functional areas may hold department meetings on a regular basis where all department scientists attend, and at these meetings key issues or developments regarding a particular project may be presented and discussed. It is usually within this forum that younger team members receive exposure to real project-related problems, and how they are addressed. In addition, open project reviews serve to educate and expose less-experienced researchers to real-life drug discovery issues (and solutions).

The project team is also responsible for keeping senior management updated on the status of the project. It is usually the project leader and/or project manager who will communicate both project achievements as well as issues to the management team. It is particularly important that delays or challenges facing the team are brought forward quickly, as they could impact allocation of resources within the organization. If a team does bring forward issues to the management team, they should also be prepared to recommend solutions to the challenge at hand.

2.04.2.3 Metrics and Goals

2.04.2.3.1 Project goals

To keep the team on track, it is imperative to establish project goals and metrics with associated timelines that have been approved by senior management. At the outset of a project, it will be recognized that the timing is a best guess estimate based on previous experience and knowledge. However, as the project progresses, definition of tighter delivery dates should become possible, and it becomes the challenge of the leader, manager, and team to meet the goals and timelines. Nomination and acceptance of drug candidates into development are often highly visible (and easy to measure) metrics.

Examples of key milestones/goals for a project team include:

- Advancement to next project phase
- Identification of advanced preclinical candidate(s)
- Completion of candidate scale-up for toxicology studies
- Completion of the first safety study
- Nomination of DC.

2.04.2.3.2 Individual and project functional area goals

From the high-level team goals, functional area goals for the project can be set. It is critical that the project leader and functional area heads agree upon the functional area goals for each project in the portfolio to eliminate the possibility of conflicting functional versus project metrics. One way to accomplish this is to establish written goals for each team member, which are signed off by both the project leader and functional area head. In this manner, the functional area head is better able to assess all the activities across the entire R&D portfolio for which his/her group is responsible, and can ensure that there are adequate resources available to provide the agreed-upon deliverables. Without this understanding, there is a risk that conflicting priorities (due to limited resources to support both project and functional goals) will result in delays to project progress. This possible conflict is addressed further in Section 2.04.5 on matrix organization.

Senior management will review the organization's portfolio of projects and should ensure that the approved timelines and delivery dates for each project are compatible with available resources – both research and anticipated downstream development resources. Development Candidate nomination dates should be agreed upon by both research and development management teams, because it is important that once approved, the compound advances immediately into investigational new drug-enabling studies. Based on such consideration of resources, if two projects project a DC nomination at similar times and adequate resources are not available to advance both at the same time, a decision may be made to delay one DC nomination date to provide the team more time to identify and nominate the best possible candidate. However, any delay in an anticipated DC nomination may result in idle capacity in the development organization, which was planning to receive the candidate. In understanding the dependence of resource allocation and budgeting upon the meeting of project goals and timelines, the importance of proper project team oversight and management becomes clear.

2.04.2.4 Development Candidate Criteria

In addition to defining the goals for the project team, it is important to define the criteria by which a DC will be selected (**Table 2**). As do the timelines, as the project matures, the criteria may become more stringent and better defined. Depending on the nature of the project and the therapeutic indication, the criteria for selectivity, therapeutic index, compound half-life, metabolic liabilities, and other parameters will vary. Each project will have its unique set of criteria, but listed below are parameters that are commonly considered in selecting a DC. Other considerations involved in transitioning projects into development are further addressed in Section 2.04.8.

2.04.3 Project Team Leadership

2.04.3.1 Effective Leadership

Effective leadership is as important to the success of a project team as it is to the overall success of a pharmaceutical or biotechnology company. There does appear to be a paradox, however, between empowered groups of coworkers who

Table 2 Development candidate criteria considerations

Parameter	Required information
In vitro potency	• K_i or EC_{50} • Selectivity against relevant off-targets and related targets • Other testing funnel parameters (e.g., MDR cell lines)
Chemistry/physicochemical properties	• Efficient, cost-effective route for API production at kg scale (including analytical methods for qualification of API) • Chiral centers • Crystallinity • Solubility • Stability • Acceptable preclinical formulation
Drug metabolism	• P450 isozymes (3A4, 1A2, 2C9, 2C19, 2D6), lack of inhibition or activation • Microsomal stability in relevant species (human, rodent, dog) • Plasma stability • Hepatocyte stability • Protein and red blood cell (RBC) binding • Other possible metabolic liabilities • Drug–drug interactions
Pharmacokinetics	• Pharmacokinetics in two species (rodent and nonrodent) • % F (>30%) • $T_{1/2}$ (supports desired dosing regimen) • C_{max}, AUC • V_d, tissue distribution • Clearance • Dose linearity • Permeability (Caco-2), transporter (P-gp) studies if relevant • Major routes of elimination
Efficacy	• Efficacy in two animal models, if available • Evidence of target modulation, if applicable
Safety	• hERG (negative) (*human ether-a-go-go related* gene) • Ames mutagenicity (negative) • Micronucleus (negative) • Mouse lymphoma assay • 7-day pilot studies in rodent and nonrodent • PanLabs/other screening tox panel (no significant interactions)

collaborate to achieve team goals and the notion of having a leader who, in traditional terms, directs a team. Project teams need strong leaders who help to drive the effort to define goals and who will motivate, facilitate, and coordinate individual efforts of team members. Project leadership is best embodied in a leader of colleagues rather than a leader of subordinates (supervisor/employee), typical of most functional departments. A project leader must have the scientific experience necessary to lead their team. An effective leader will jointly set goals, use talent effectively, and question and develop each team member, appreciating their value and their contributions. A project leader may also coordinate meetings and agendas and may be responsible for writing meeting summaries and team communications (in the absence of a dedicated project manager). The shaping of project reviews and the updating of senior management, as appropriate, are also the responsibility of a project leader.

Previous leadership experience can be invaluable. Having experienced the drug discovery or preclinical development pathway before and knowing the next steps often can prevent backtracking and can shorten the development pathway. For example, a team might have planned to perform the study that would support recommending a DC, but the leader realized that that particular study would have to be repeated under Good Laboratory Practice (GLP) conditions for an investigational new drug application. It might be more efficient to take the extra time to qualify the active pharmaceutical ingredient (API), rather than repeat the study, saving cost and time. However, sometimes the less experienced can be successful team leaders if they have effective interpersonal skills and are diligently listening and learning from those around them.

2.04.3.2 Interpersonal Skills of Effective Leaders

Interpersonal skills are among the keys to effective leadership. Good communication, in particular, is important. A leader must listen to the team and facilitate the listening between various functional team members. Issues should be addressed and brought to the forefront, not ignored in hopes that they will vanish. It is important to identify gatekeeping issues as soon as possible since they impact the team cross-functionally. No team member should force a path forward over the objections of other team members. For instance, a chemist might want to move a molecule forward because of potency. However, lack of solubility might preclude good bioavailability or limit the options for formulations. A good leader will negotiate with the team on activities and their priorities, maintaining particular focus on activities that lie at the interface of different groups. An effective leader delegates activities, yet checks in to ensure that timelines on deliverables are not shipping. This could have an impact on other team members' activities as well as delay the project team's effort to achieve the goals communicated to senior management. If timelines are slipping, the leader should ensure that the entire team is aware of the changes so that they can adjust their activities accordingly. Lastly, an effective leader will motivate a team to work for the greater good and to transcend the individual functional goals. An example of this might be when the project team leader convinces the core team member that it is more important to have his/her team focus on the mechanism of action of the compound, in support of the project, than to achieve the biology department's goal of setting up a new target for screening.

2.04.3.3 Project Leader Responsibilities

A leader must be willing to assume responsibility for the success or failure of the project. The failure of a project will not be evaluated as a bad outcome if the project was well executed but the science did not support the theory. Terminating a project early saves the company time and money and permits resources to be deployed on more promising therapeutic targets.

A leader can help to develop and motivate a team by providing structure and expectations. Structure takes many forms; regular meetings with consensus agendas, enforcement of teamwork and frank discussions, regular challenging of the team's direction as well as the direction of the individual functional teams. Effective leadership includes challenging the team in a productive manner to think strategically as its information base grows and changes.

It is important that the project team leader, as well as the team, is aware of relevant external events. Awareness of the competitive intelligence and maintaining a liaison with the commercial organization are critical. The leader should keep the team up to date on external and competitive advances and ask the team and the commercial group how this information can impact or change the current TPP of the project team. For instance, if the competitor has identified an effective agent that was dosed as a once daily dose regimen, the team may need to adjust the product profile from twice-daily dosing and focus on other attributes such as no cytochrome P450 (CYP) inhibition to avoid drug–drug interactions, which were a liability of the competitor.

A project leader's role is to orchestrate the moving parts of a project while regularly communicating on a high level with senior management. S/he has the responsibility to keep management informed of progress and issues as they arise. Leaders ensure adequate resourcing to achieve the team's objectives and advocate with senior research management for those resources. Resources include not only full time equivalents, but also equipment, contracts, and collaborations that can facilitate achievement of team goals. The leader must understand the needs and issues at such a detailed level as to offer sufficient justification to senior management.

2.04.3.4 Transitioning Project Leadership

As a team transitions through drug discovery to preclinical development, there may be a need to transition leadership as well. Some leaders are able and willing to lead projects on into early clinical development while some leaders' strengths and interests are better suited to particular stages of a project. Sometimes it is best for the team to transition to leadership that can better lead the cross-functional team through a critical point. An example might be a team whose most significant challenge is the optimization of pharmacokinetic parameters of the lead molecules, but the team leader's strengths focus on the biological mechanism of action. The team needs the leadership of someone who is comfortable driving the team to focus on the drug metabolism and pharmacokinetics (DMPK) experiments and on feedback to medicinal chemists. Additionally, a project leader might not be/feel capable of leading a team in preclinical development and would rather move back onto an earlier staged project team. Different leadership skills are needed at different times in the drug discovery pipeline and research management should be gauging leadership effectiveness at all times, adjusting it as necessary. If a team is not achieving its goals, is it because of less than effective leadership? If so, a change should be made. Although teams do not always embrace change and adjust as quickly as they should, management should make changes swiftly, reassuring the team that these changes are in the best interest of the team's success. At the earlier stages in the pipeline (hit-to-lead, lead optimization), a co-leadership model (e.g., a biologist pairing with a chemist) is an option which can lead to favorable results.

2.04.4 Effective Project Teams

The definition of a project team is a group of individuals with a common vision, set of performance goals, and an agreed-upon approach for which they hold themselves mutually accountable.[8] High-performing project teams share some basic features: the project goals are clear and the team has a 'can-do' approach to their work. Project teams must have a sense of urgency and the skills to achieve dependable results.

2.04.4.1 Productive Project Teams

Productive project teams generally exhibit the following characteristics:

- *Strategic thinking*: Action plans are well-conceived and logical, and work plans are adjusted to accommodate various outcomes.
- *Interdependence*: Team members solicit ideas and input from their various colleagues while defining the goals and the implementation of the work plan. A high-performing project team member welcomes the input of his/her colleagues, recognizing that it might help the project team make better decisions, thereby reducing wasted time and effort.
- *Project commitment*: Team members are committed to the success of the project – achievement of the team's objectives being the highest priority of each member. Their individual success is tied to the success of the project.
- *Pragmatic innovation*: Effective teams develop innovative but practical solutions to problems and work through team issues in a pragmatic manner.
- *Diversity*: Diverse skill sets, views and ways of thinking breed novel solutions to unanticipated issues.
- *Thoughtful responsiveness*: Effective teams do not overreact to new problems – they make a clear assessment and develop a solution.
- *Effective communication*: Project team members clearly communicate progress, issues, and obstacles among themselves and within discovery research.
- *Clarity*: Effective teams clearly understand the team goals and each member's responsibilities.
- *Commitment to the team*: Team members are committed to each other's success since successful teams are comprised of successful individuals.
- *Authority and responsibility*: Teams have the authority to make decisions and the responsibility to execute plans to achieve the goals. They are focused and follow through on activities.
- *Empowerment*: Quality teams have the power to make decisions and have clear decision-making processes.
- *Passion*: Team members feel urgency and enthusiasm for the accomplishment of the team's goals. Individuals care about the outcome and the results.
- *Reliability*: Project team members are able to rely on each other. Since project team members are interdependent, the inability of a team member to deliver affects other functions of the team and the team's overall success.
- *Flexibility*: Science can be unpredictable at times. Good teams have the flexibility to deal with unanticipated outcomes.
- *Honesty and trustworthiness*: Team members are honest with each other, trust each other, and operate with integrity.
- *Supportiveness*: The team supports each member's efforts and members are mutually appreciative of each other's efforts. They are good listeners, and they question and challenge each other to achieve.
- *Enjoyment*: Effective teams are composed of talented individuals who enjoy their own work as well as being part of a team.

These are some of the characteristics of effective teams. There is a tendency to call a collection of people who work together a team, but it is the above characteristics that distinguish such a group from a highly productive team.

2.04.4.2 Functional Departments versus Project Teams

Functional departments are usually comprised of researchers with common scientific background and training, and goals for the department are usually focused on improving technologies and efficiencies of the department, as well as advancing the science in specific areas for which the group may be known. It is critical for department members to keep an update on the scientific literature and competitive landscape to ensure that they remain competitive and are aware of developments that could impact both the department and the projects that the department supports. Within a given functional department, researchers' projects may not be related and people may work independently of each other, but they will have the same supervisor and will be rewarded based on their individual contributions toward meeting the functional area metrics.

Productive teams, on the other hand, have clear common goals and objectives and are rewarded primarily based on the team's performance. Highly effective teams meet to solve problems, resolve issues, and to make decisions. The project team members are interdependent on each other, are skilled at communicating and juggling tasks, and therefore put out enhanced performance, quality decisions and a greater work product than could be achieved if each individual worked alone.

2.04.4.3 High-Performing Project Teams

High-performing project teams utilize team meetings effectively to communicate, coordinate, and cooperate on team activities. Successful meetings are thoroughly planned well ahead of time. Effective leaders will seek input on the agenda and communicate the purpose of every meeting. Occasionally, special meetings will be necessary to deal with unpredicted results or issues. Agendas should be distributed in advance of meetings so that members can be prepared to discuss issues and share new data. During the project meetings, the team will review progress made toward commitments and goals and articulate action items that evolve from each discussion. Brief meeting minutes or summaries should be written after each meeting to capture the discussions, conclusions, and action items. Each team member thereby knows who is responsible for each action item.

2.04.4.4 Rewarding Teamwork

It is critical that the success of teams be rewarded as opposed to individual efforts.[9] The culture of drug discovery must truly recognize and compensate teamwork. Line managers in pharmaceutical companies are often responsible for the evaluation of individuals. They must seek input from project leaders and other team members. This information can be invaluable in recognizing an individual's contributions and, in providing constructive feedback, to enhancing the individual's value to the team and the effectiveness of his or her teamwork.

Project leaders should provide to the reviewing line manager an assessment of each of their team members. Information provided should include:

- Does the team member fully represent their functional area, providing relevant, high-quality data in a timely manner?
- Does the team member participate in the intellectual development and creative expansion of the project?
- Does the team member always act in the best interest of the team, maintaining good working relationships with other team members and project team leaders, and supporting the project decisions?

Similarly, research senior management must provide an assessment of the project leader's performance. Information provided should include:

- *Leadership skills*: Does the project leader articulate project goals and inspire commitment from the team? Does the leader possess the energy, scientific insight, and the personal commitment necessary to lead the team?
- *Communication*: Does the leader foster an environment in which the team communicates honestly and openly and is information shared through regular team meetings? Does the project leader effectively communicate project issues to senior research management and the scientific advisory board as required?
- *Teamwork*: Does the leader promote collaboration to achieve team goals?
- *Operations*: Does the leader develop project plans and deliverables, have a command of the project issues and make critical decisions? Each project team member and project leader must know that their individual performance and resulting compensation is closely tied to the effectiveness of their performance on the team.

2.04.5 Matrix-Driven Organizations

Matrix-driven organizations exist in various forms across a wide range of R&D organizations.[10] They are the center of a continuum of management possibilities between a purely functional (line) organization and a purely product-based organization.[11,12] If the environment of a company is relatively simple (as defined by considering of the following four factors: complexity, diversity, rate of change, and uncertainty), a more traditional hierarchical management structure may be preferred. However, if the environment of a company is complex, as is the nature of most biopharmaceutical or pharmaceutical companies, a matrix organization is probably more appropriate.[13] The formation of cross-functional project teams by definition involves a matrix organization since it involves bringing two or more usually distinct functional areas together to focus on the common project goals. Team members will therefore have dual lines of authority, responsibility, and accountability that violate the traditional 'one boss' principle involved in line management.

One of the primary advantages associated with the formation of formal project teams is the creation of cross-functional (horizontal) communication linkages. Improved communication between functional groups is critical for project success, and therefore this type of team structure emphasizes and values development of communication skills within a team. Improved information flow within project teams allows information to more quickly permeate within an organization, and therefore can lead to improved decision-making and response time.[14,15] In addition, in a matrix organization, individuals have the opportunity to participate on a variety of projects with different individuals from across the organization, allowing for sharing of ideas, knowledge, perspectives, and experiences. Thus, being a member of a project team should result in increasing the individual's breadth and awareness of other disciplines and activities within the organization, ultimately contributing to his or her career development.

One of the primary disadvantages of a matrix- or project-driven organization results from the creation of dual/multiple authority and influence.[16] The most common authority conflicts are those between functional and project leaders over project priorities. For example, functional group responsibilities require and encourage individuals to remain current and aware of new technical and scientific developments within their area of expertise, and this could be in conflict with performing more routine studies that the project may require. It is therefore very important that there is agreement between the project and functional leaders regarding the expectations and priorities for each project (as described in Section 2.04.2.3). It is also critical that the reporting structure for the project teams is clearly defined: the project teams are accountable to the Research Management Team, and the functional heads, as members of the management team must agree upon the priorities of the project versus their functional metrics.

Cross-functional teams force individuals to take more personal initiative in defining their roles and goals, negotiating conflict, assuming responsibility, and making decisions. This increased level of autonomy and responsibility can lead to increased stress among team members; however, if managed properly, the matrix organization can result in a more productive and satisfying work environment.

2.04.6 Research Project Oversight and Decision Points

2.04.6.1 Management Teams

As mentioned in previous sections, proper project oversight is critical to monitor progress and to ensure that projects are resourced appropriately for success, or conversely, to terminate projects and make resources available to other projects if the situation (e.g., scientific data, budget considerations) warrants this. Typically, the Research Management Team will comprised of the Head of Research along with the heads of key functional areas. This group is ultimately responsible for ensuring that the research organization meets the agreed-upon annual metrics. Project teams should report to the Research Management Team to avoid the possibility of conflicting interests or agendas between project goals and functional area goals. The management team must act consistently and transparently in managing department priorities and resources in order to achieve the organization's goals.

A typical Research Management Team may be comprised of:

- Head of Research
- Head of Biology
- Head of Bioassays
- Head of Chemistry
- Head of Nonclinical (pharmacology, PK, toxicology)
- Head of Research Project Management/Project Mentor (if applicable).

Ad hoc participation by representatives from human resources, operations, finance, legal, and business development is often necessary to address the range of issues the management team is responsible for. Typical responsibilities of the management team include:

- Oversight of research (project) portfolio – ensure scientific rigor of projects
- Definition and oversight of research metrics
- Development and oversight of research budget
- Identification of business development opportunities (in- and out-licensing)
- Career development opportunities
- Staffing considerations (hires and terminations).

Effective decision-making by the Research Management Team should minimize risk and maximize benefit to the research organization.

2.04.6.2 Reviews and Feedback

For all the reasons that have been previously stated, it should be apparent that management teams must be updated regularly on the status of the projects in the research portfolio. There are several mechanisms by which this can occur.

2.04.6.2.1 Written project reports

Formal written reports (using a defined format or template) can provide useful information which is readily available when needed for reference. For projects, either bimonthly or quarterly updates are probably appropriate. Each project should address:

- Progress toward project goals (e.g., on track or delayed)
- Key highlights/achievements
- Issues identified and possible solutions
- Budget
- Next milestones, next steps.

Writing a project report often forces the project leader to stop and assess the status of the project, to review the progress as well as to identify key issues, and to see how well the team performed relative to what was projected in the last report. Many project leaders start out with the view that they have more important things to do other than writing a project report/update; however, they often eventually appreciate that taking the time to critically analyze the project progress, issues and planning the next steps is a valuable exercise and helps them become a more effective leader.

2.04.6.2.2 Formal project reviews

Management teams can use open forum, formal project reviews as a mechanism for receiving updates on a project, as well as a means to educate their research organization. It is also an opportunity to include the development organization in open discussions, allowing personnel to learn more about projects that are in the late research pipeline. By inviting the research organization to regular project reviews, a large audience can be exposed to real drug discovery projects – real issues that are faced, questions that should be considered, projects that are making good progress, and projects that are floundering. Over time, the audience should be able to evaluate which projects are running well and why, and conversely, which projects are not running well and why. Management should be pleased if when making the difficult decision to terminate a project, the organization understands and agrees with the decision. Researchers will also hopefully apply lessons learned from the presentations and discussions to their own projects and areas of research.

2.04.6.2.3 Ad hoc meetings with project teams

In addition to regular project updates (either reports or presentations), the Research Management Team must be able to meet with project teams on an ad hoc basis – either at the request of the team, or when the management team is concerned about certain issues or needs clarification.

It is important that the teams receive feedback from their efforts to update management on the status of their project – both from a technical perspective and to know how effectively their message was communicated. This feedback should be delivered to the team in a timely manner – ideally within 48 h of the presentation or receipt of the report, but certainly within a week. Teams will be much more motivated to deliver a quality product if they know management is going to take the time to read and review the material and offer thoughtful, useful, and timely feedback. This is probably one of the most important jobs of the management team – to oversee the projects in the research pipeline, educate project teams, and make adjustments where necessary.

2.04.6.2.4 Team member feedback

With matrix-driven project teams, it is important that each team member be evaluated for his/her performance in advancing both functional and project goals. Contributions toward achieving functional goals are usually easier to assess by the direct supervisor, since presumably there is a direct reporting relationship of a team member and functional supervisor. Contributions to the team may be more difficult for the direct supervisor to assess, as the supervisor will not usually be a member of the team, and therefore must rely upon both the project leader and manager for formal feedback on team performance. This is an important consideration, since a team member risks not having an accurate and fair performance evaluation on team performance, especially if most of their time is spent on project-based activities.

2.04.6.3 Project Decision Points

Based on information provided, it is the responsibility of the management team to make decisions on prioritization and resourcing of projects. Typical project decision points include:

- Acceptance of a target for further target development/HTS activities (deployment of assay development, protein biochemistry resources)
- Acceptance of preliminary hits into hit-to-lead activities (e.g., deployment of medicinal chemistry resources)
- Advancement of project into lead optimization (deployment of preclinical resources)
- Identification of 1–3 advanced preclinical candidates for predevelopment activities (deployment of further preclinical resources, process/chemical development)
- Acceptance of DC (transition project from research into development and IND-enabling studies).

2.04.6.4 Project Management

Given the importance of team performance in achieving project (and organizational) goals, it is not surprising that many companies elect to have project managers working in concert with the project lead to ensure coordination and oversight of key team activities. A project manager is usually assigned to a project once it advances to a phase where numerous functional areas are involved (e.g., lead optimization/predevelopment stage), or if there is an external collaboration that needs to be carefully managed (*see* Section 2.04.7). The vast majority of companies will deploy formal project management support once a compound officially enters development (investigational new drug-enabling activities), due to the numerous activities with critical timelines and regulatory requirements that need to be managed.

2.04.6.4.1 Research project management

Depending on the level of experience of the project teams, management can elect to provide project management support during the later phases in research. If project management support is not provided, the organizational aspects of running a project will fall upon the project leader and other members of the core team. Contrasting roles and responsibilities for project leaders and managers are presented in **Table 3**, with the caveat that it is also possible for one person to serve both functions. There are traits in common that make a person an effective project leader and/or manager. It has been observed that to be an effective leader, one has to know when to manage; to be an effective manager, one has to know when to lead.

Many of the operational activities of a project (meeting scheduling, developing timelines and budgets, collaboration management, etc.) are not technical in nature, therefore research team members often prefer to have someone on the team whose responsibility is to attend to these important, but nontechnical activities. It is extremely useful, however, for the research project manager to have a scientific background so that she can understand key project issues, and help the team to develop and implement solutions. For a less-experienced team, if the project manager has drug development experience, she can work with the project leader to define the project goals while the team works to achieve them. An experienced project manager will also look to the future and anticipate needs and identify potential issues while the team members are dealing with more immediate activities. If truly all the team needs is advanced organizational and administrative support, the role is probably best filled by a project coordinator, whose scientific

Table 3 Roles and responsibilities of project leaders and managers

Project leader	*Project manager*
• **Key communicator: project status, progress and results/issues** – Updates and alerts key stakeholders/management and oversight committees, partners as appropriate – Spokesperson/representative of project team internally and externally – Project champion – Surfaces alignment issues between project needs and functional area support	• **Owns overall project plan** – Works with team to develop project plan and timeline based on agreed-upon project goals – Integrates functional plans and deliverables – Maintains accurate budget/schedule – Forecasts/anticipates possible gaps or events which could impact budget or timeline – Ensures functional area deliverables occur on time and with quality

• **Leads/monitors technical execution and strategy for project team**
– With team and management, define clear goals and deliverables and timelines
– Sets scientific strategy for achieving goals
– May have key functional role in project plan
– Ensures technical soundness of project implementation and results (data)
	• **Manage project processes**
	– Project planning and tracking (Gantt charts, liaison with functional areas)
	– Budget planning, tracking, and forecasting (interfacing with finance)
	– Team effectiveness
	  ○ Meeting planning and facilitation
	  ○ Team communication
	  ○ Meeting follow-up/action items
	  ○ Identification of key issues/gaps and possible solutions

• **Lead quality decision-making**
– Accountable for quality of decisions
– Ensures functional integration and alignment to maximize probability of success (PoS)
– Holds team members and functional heads accountable for planning and execution of agreed-upon deliverables
	• **Manage support interfaces**
	– Ensure communication, alignment, and integration with
	  ○ Business development
	  ○ Finance
	  ○ External collaborators
	  ○ IP
	  ○ Noncore team functional areas

experience/expertise would not be as advanced as that of a research project manager. This is a suitable role for an associate level researcher who is looking to move out of a laboratory setting into a more administrative role, but still has interest in being involved in drug discovery. The risk in limiting a research project manager to these activities is the lack of a challenge and the potential resentment of having to perform administrative duties at the expense of keeping up with his or her area of technical or scientific expertise. This can be offset, of course, if there are other challenges associated with the position, such as those described in next paragraph. Depending on the experience of the project manager and the working relationship with the project leader, there may even be times when it is appropriate for the project manager to act on behalf of the project leader. The attractive feature of the project manager position is that it is the central point of the project, and she must know what is happening on all aspects of the project, not just in one particular functional area.

The project manager will often act as the liaison between functional areas, and should be viewed as an 'objective' member of the team. His/her job is to ensure that the project is on track, and to identify where interactions can be improved for the benefit of the project. This is possible since the project manager is not a functional area member, and therefore should not be perceived as having any sort of hidden agenda. An effective project manager will know who are the key contacts in each of the relevant functions and establish good working relationships with them. The project manager should at all times be viewing the project from the '30 000 foot' vantage point, rather that on single components of the project, to assess how the various pieces of the project are (or are not) coming together in support of the team's goals. The working relationship between the project leader and project manager can be compared with that of the Chief Executive Officer (CEO) and Chief Operating Officer (COO) of a company. The leader defines the (technical) strategy and goals for the project, and the manager works from the operational side to ensure that the strategy is executed and the goals are met on time and within the allocated budget. **Table 4** provides other examples of the complementarity and potential overlap of the two roles.

Table 4 Duality (and potential overlap) of project leader/project manager responsibilities

Activity	Leader	Manager
Creating the project plan	Establishes direction and strategy, scientific rationale, and project goals	Defines task and develops budget
Developing the team and network to implement the plan	Aligns and directs team with technical expertise and experience	Directs with definition of roles, structures, resources required, and timelines
Execution of the project plan	Motivates, leads, and inspires; communicates across and up	Controls by tracking activities and communicating across and down
Achieving the goals	Produces quality product (target, DC), future opportunities for team members	Produces order and predictability; reduces unforeseen obstacles/delays

Where the project management function resides organizationally varies within different companies. Many companies do not offer formal project management support in research, and individual team members (in particular, the project leader) assume the planning and operational responsibilities. In some organizations, the research project management group resides within research, whereas some companies deploy project management support for the research from out of the development project management department. In the transitioning of a candidate between research and development, project leadership will often change hands and the project management support can provide a continuum during this process.

As mentioned earlier in this section, if project management resides within research, the head of the function will often be a member of the management team, as she can inform the group of project-related issues that may not be readily apparent to individual functional unit heads. In smaller research organizations, rather than having a project management group, a senior level experienced individual (Head of Discovery Team Management, Project Mentor) can function as a resource to all project teams, providing drug discovery/development as well as project management type of advice. This position would have responsibility for the overall performance of project teams, function as team mentor for all teams, integrate with business development and collaborators as appropriate, and create/initiate new project teams once approved by the Research Management Team.

2.04.7 Management of Collaborative Projects

2.04.7.1 Collaboration Initiation

If management makes a decision to collaborate on a project with an external partner, it is critical at the outset that each party's roles and responsibilities are clearly defined, as well as the goals of the collaboration. Typically, this means that the relevant participants meet prior to the execution of the agreement to work out such details, so that once the agreement is executed, people are ready to start immediately working toward the project goals and milestones. Optimally, at the initiation of a collaboration, a clearly delineated research plan will have already been drafted. This plan should detail the project goals and milestones, activities/responsibilities of both parties with target delivery dates, number of full time equivalents in various disciplines, and other anticipated project expenses. At the kick-off meeting, this plan should be further vetted with team members, discussed and formalized, and there should be agreement between the team/project leaders and senior management that the deliverables, resource allocations, and time frames are reasonable.

It will either be the responsibility of the project leader or the project manager to ensure that communication between the parties is frequent and productive. A collaboration will only be successful if the two parties trust and respect each other, and therefore it is often beneficial to ensure that scientists have a chance to meet face to face and interact with each other soon after the collaboration starts, if not before. A sign of a good collaboration is when scientists from each organization feel comfortable contacting and communicating with their colleagues whenever necessary – not just at the scheduled joint meetings. The sooner the team can transition from a 'we versus them' mentality to a 'we are one team' mentality, the more likely it is that a productive and successful collaboration will result.

2.04.7.2 Logistical Considerations

It is important to include chemical informatics and information technology representatives during the initial planning stages of a collaboration, as it will be critical to have easy and secure exchange of information (e-mail, structures, and data)

between the two parties. Details regarding protection/isolation of company databases, project data, samples, etc. must be addressed. There are several 'virtual office' software packages available which allow secure communication both within and between organizations (Flypaper, Groove).[17] This type of software allows team members from both parties to share files, manage meetings, track activities – basically to function as if they were working at one location.

Whatever a project leader or manager can do to encourage open communication between colleagues will ultimately benefit the collaboration. The timing of key discussions and decisions cannot always be coordinated with scheduled joint team meetings, therefore it must be made clear to team members that any material discussions be formally documented and communicated to the project leaders. Given the advances in teleconference capabilities (secure shared databases and e-mail systems, sharing of slides, etc.) – joint project team teleconference meetings can be effective and scheduled on a regular basis. However, one cannot underestimate the value of face-to-face meetings between colleagues as a way to instill the commitment, camaraderie, and trust needed for project success, especially at the outset of the collaboration.

If a project is collaborative in nature, a project manager will often be assigned, as it is recognized that additional organizational and logistical activities are usually involved, and are critical to the success of the project. As trivial as they might seem, details regarding joint teleconference calls and meetings, project-specific confidential databases, routine communication, travel, etc., can significantly impact the quality of the collaboration and should not be underestimated.

2.04.7.3 Collaboration Oversight

Collaboration oversight is typically performed by a Joint Management Team (JMT), which is comprised of an equal number of key representatives from both parties (usually 3–5 from each). Assignment of a Chair for the committee can vary; some collaborations have an assigned member who acts as Chair for all JMT meetings, while others may elect to have the Chair alternate depending on which party is hosting the meeting. The composition of a typical JMT would include from each company:

- Vice Presidents of key functional areas (Biology, Chemistry)
- Project leader
- Core team member/area experts
- Project or alliance manager (or business development representative)
- IP or other legal representative (if appropriate)

The JMT should meet on a regular basis for the duration of the term of the collaboration (usually quarterly or bi-annually). The responsibilities of the joint Research Management Team include:

- Approval of and monitoring of the research plan
- Approval of any modifications to the research plan (including budget)
- Approval of any changes to DC selection criteria

It is critical that the details of dispute resolution are clearly described in the collaboration agreement. If the JMT is not able to reach an agreement on a particular issue, the issue will be addressed by the management representatives identified in the agreement.

By leveraging the capabilities of the two parties, the probability of success for the project is increased. A well-run collaboration will identify and call on the strengths of each organization, but will also recognize that there may be times when resources are limited. Under these circumstances, to minimize delays in the project timeline, a partner may be requested to perform activities normally assumed by the other party.

2.04.8 Transitioning Projects into Development

As a drug candidate meets the criteria of the TPP, the project team must obtain the support of research management, and therefore should schedule a project review with their management. This review should be an oral presentation by the core project team members on the attributes of the molecule which support the team's nomination to move the candidate into clinical development. This review should promote discussion and critical review of the data. If management does not agree with the team's recommendation, they must provide critical feedback to the team on experiments that should be performed, or additional issues that need to be addressed before they can endorse the team's recommendation.

2.04.8.1 Development Candidate Nomination Document

If the Research Management Team supports the nomination of the candidate into development, they will usually need to justify the nomination to a larger strategic body, such as a Development Committee or Strategic Review Committee within the company. It is customary and recommended that the research project team assemble into one document all the information and data generated on the individual DC. This information can be distributed as preread material before the formal discussion and agreement by the larger strategic decision-making body. The document could be called a Development Candidate Nomination (DCN) and should contain as much of the following information as is available:

Development Candidate Nomination
Table of Contents

I. Executive Summary
II. Introduction
    a. Rationale
    b. Key Distinguishing Product Features
    c. Competition
    d. IP Position
III. Product Profile
    a. Draft Profile
IV. Chemistry
    a. Name
    b. Chemical Structure
    c. Molecular Formula
    d. Molecular Weight
    e. Formula Weight
    f. Physical Appearance
    e. Melting Point
    h. Solubility
V. Pharmacology
    a. Mechanism of Action Statement
    b. In Vitro
    c. In Vivo
VI. Drug Metabolism and Pharmacokinetics
    a. In Vitro Profiling
    b. In Vitro Metabolism
    c. In Vivo Profiling
    d. Drug Interaction Profile
VII. Toxicology
    a. General Toxicology
    b. Safety Pharmacology
    c. Genetic Toxicology
VIII. Estimated Human Dose
IX. Pharmaceutical Development:
    a. Synthetic Route
    b. Physicochemical Properties and Salts
    c. Process (Scale-up) Assessment
    d. Estimated Costs of Goods
    e. Analytical Methods
    f. Formulation
X. Major Issues, Assumptions, Constraints
XI. Team Recommendation
XII. IND Timeline and Resources
XIII. Project Team Members

Executive Summary: An executive summary should be developed after the rest of the DCN is assembled. It should include the top-line information that supports the advancement of the compound into development, including: therapeutic area, key data, predicted human dose, summary of synthetic process, IP position, and rationale.

Introduction: The introduction provides the background information on the unmet medical need, the market opportunity and the current standard of care and its limitations. It also provide, the project background, origins of leads, and collaborators if applicable.

Product Profile:

a. *Rationale*: Provide epidemiology and prevalence of the disease which should support the subsequent elaboration of the unmet medical need and limitations of current therapies. Describe the current market and market potential with appropriate citations. This section can be developed with the assistance of the commercial development team. Justify the advancement of the candidate into your company's marketing and commercial strategy. Describe the targeted product attributes and outline the TPP which the team developed early in its life time with the input of the commercial strategy team.

b. *Key Distinguishing Product Features*: Describe how the product would work, how it could be rationally combined with other molecules within your portfolio or the market or could displace the current standard of care.

c. *Competition*: Summarize the status of the competition, describing the individual compounds, the stage of development (Phase I, Phase II, etc.), the mechanism of action, the companies and publicly known status on efficacy.

d. *IP Position*: Have your patent group make a statement on the patentability of the proposed DC. If patents have not been issued, provide an update on the status of all applications.

Chemistry: Include the IUPAC name of the molecule along with the structure and in-house company name. Summarize the molecular formula, the molecular weight, and formula weight. Describe the physical appearance, melting point, and solubility at various pH values.

Pharmacology:

a. *Mechanism of Action*: Describe the mechanism of action and the studies carried out to support the mechanism.

b. *In Vitro Activity*: Summarize all in vitro activity of the molecule in enzyme or cell culture systems. Include selectivity versus other cell lines and related targets and cytotoxicity observed. Provide results on experiments that describe the impact of serum shift on efficacy, if relevant.

c. *In Vivo Activity*: Describe all in vivo efficacy experiments, doses utilized, and results as compared with a known standard.

Drug Metabolism and Pharmacokinetics: Describe the bioanalytical methods developed and applied for the quantification of the drug candidate in the plasma of preclinical species.

a. *In Vitro Profiling*: Describe the relevant in vitro disposition properties of the drug candidate studies in a variety of tissues from both animal and human origin. These in vitro systems describe aspects of the absorption, metabolic stability, and distribution properties and can be predictive of the in vivo pharmacokinetic properties in preclinical species. Property studies should include: (1) kinetic solubility; (2) absorption in Caco-2 permeability systems; (3) metabolic stability in human and nonhuman hepatic microsomes that can lead to a prediction of in vivo clearance rates; and (4) protein binding in the plasma of human and all nonhuman systems tested in vivo.

b. *In Vitro Metabolism*: Describe the in vitro metabolism studies on the drug candidate which were utilized to identify the major metabolites and the proposed metabolic pathway. The in vitro data should suggest that the drug candidate would have adequate metabolic stability and free fraction such that it would be reasonable to expect sufficient active pharmacological concentrations in human subjects utilizing the desired route of administration.

c. *In Vivo Profiling*: Describe the in vivo disposition of the drug candidate in preclinical species, including the clearance, volume of distribution, half-life, and percent bioavailability. Discuss whether the in vivo pharmacokinetic profile data is consistent with and predictable from the in vitro data. Include results of dose proportionality studies and the impact on C_{max}, T_{max}, AUC (area under the curve) and percent bioavailability, as well as the results of salt forms, if applicable, on the fraction of the dose absorbed.

d. *Drug Interaction Profile*: Describe the CYP inhibition potential against the most prevalent human CYPs. High IC_{50}s relative to the potency at the drug target might suggest the potential for drug–drug interactions is unlikely. The potential for CYP induction should also be examined.

Toxicity: Describe the results of a general 7-day toxicity study in rat or dog. Include clinical observations, effects on body weights and food consumption, hematology and clinical chemistry. Confirm exposure on the first and last day of the study. At the end of the study, describe the macroscopic observations and organ weights. Determine the no observed adverse effect level (NOAEL) for the compound. Safety pharmacology should be investigated in a panel of receptor or enzyme assays to determine if some off-target pharmacology would be predicted. Genetic toxicity potential should be determined in the Ames screening assay.

Estimated Human Dose: Based on the in vitro and in vivo PK correlation presented for the preclinical species, determine the estimated human pharmacokinetic parameters for the drug candidate as determined from the in vitro data from human test systems. A one-compartment model to simulate pharmacokinetics of the drug enables the calculation of total plasma concentrations required to exceed the trough target EC_{50} or IC_{50}. Estimated dosing regimen, quantity and frequency, is important for future preclinical and clinical studies.

Pharmaceutical Development:

a. *Synthetic Route*: Provide the synthetic route for the synthesis of the drug candidate from commercially available starting materials.
b. *Process (scale-up) Assessment for Active Pharmaceutical Ingredient (API)*: Identify issues which will need to be addressed for the further development of the synthetic process prior to scale-up.
c. *Estimated Cost of Goods for Active Pharmaceutical Ingredients*: Estimate the full time equivalents and time for process development and for the GLP and GMP campaigns. The cost estimates can then be provided from these estimates.
d. *Analytical Methods*: Outline the specification and methods for raw materials, intermediates and in-process testing which will need to be developed. Estimate the analytical support necessary for process scale-up and pilot plant manufacturing. In addition, specifications and methods for drug substance and drug product need to be developed and batch analyzed for toxicology and clinical studies. Estimate the resources required for the development and release of materials.
e. *Physicochemical Properties of the Free Base and Salts, if Appropriate*: Provide pH–solubility profiles and melting point and heat of fusion as determined by differential scanning calorimetry.
f. *Formulation Assessment of the Pharmaceutical Drug Product*: Outline the activities and resources necessary for preformulation and formulation efforts.

Major Issues, Assumptions, and Constraints: List unresolved major issues and assumptions included in the document that might impact the decision to advance the drug candidate.

Recommendation: The recommendation is usually fairly straightforward. The team is recommending the initiation of IND-enabling activities. The IND timelines with supporting assumptions are outlined in the next section.

IND Timeline and Resources: Include a Gantt Chart and list of assumptions.

Project Team: List project team members and their role on the team, so that individuals can be contacted for additional questions or clarifications. Phone numbers, office location, and other relevant contact information (administrative assistants, e-mail address, etc.) should also be included as appropriate.

References: List references utilized.

Appendix: Additional supporting material if necessary.

Once the document has been generated, it will serve not only as a resource to support the formal declaration as a DC, but also as a reference for the new project team down the clinical development path.

2.04.8.2 Transitioning of Project Team Membership

As mentioned previously, project leadership or project team composition may change as projects mature through the drug-discovery process; through target identification, assay development, high-throughput screening, lead identification, and lead optimization. The most dramatic changes in project team composition occur, however, as projects transition from research into development. At this time, a project team becomes focused on the advancement of a specific drug candidate. It is very important that during this process, the project is handed off in a manner to ensure that critical information is transferred and team efficiency is not lost. It is often recommended to keep a few key individuals from the research project team on the new development project team.

As a research project team realizes that they are close to identifying a DC that meets the desired TPP, it is important to begin to include representatives from process chemistry, clinical research, and development project management in the research project team meetings. The first objective is to familiarize the development organization with the issues and possible timing of the recommendation of a DC. The second objective is to begin to solicit input from development colleagues; input on decisions and on the required information for the DCN document.

The preclinical and subsequent clinical development of a product requires a more structured and well-defined development process than that occurring in drug discovery. Regulatory requirements by the Food and Drug Administration (FDA) and European Medicines Agency (EMEA) agency require that a drug's sponsor must submit data demonstrating that the drug candidate is safe for use in initial Phase I clinical trials. A preclinical team must undertake GLP studies to support the safety of administering the drug candidate to humans. Depending on the therapeutic use, the requirements for preclinical testing may vary. After developing a preclinical plan, it is recommended that a pre-IND meeting or a conference call with the FDA be scheduled to ensure that the team is performing the necessary experiments to support human dosing. Minimally, the FDA will require information on (1) the pharmacological profile of the drug; (2) the acute toxicity of the drug in at least two preclinical species (one nonrodent); and (3) subacute toxicity studies to support the proposed frequency and duration of dosing in Phase I human trials.[18] The data for the FDA must comply with GLP regulations. The regulatory agencies utilize the GLP standards to ensure the quality of the preclinical data.

2.04.9 Project Management Tools

There has been considerable pressure on pharmaceutical research departments over the past decades to become more efficient due to the increased costs of pharmaceutical research and development. As such, the industry has adopted many automated tools to improve productivity. Many of these improvements are early stage research tools for research to help identify new biological targets, synthesize compound libraries, and screen the libraries against the new targets. More recently, there has also been a push to increase the effectiveness of the later stages of drug discovery. There has been an emphasis on integrating these processes to achieve more success, more efficiently with information management and simulation tools. Many companies agree, however, that discovery project management tools have not advanced significantly over the years. Data generated by project teams should be easily accessed by team members on company databases that should be updated nightly. Project team documents meant to be communicated between project teams and research management can be saved into folders that are automatically sent to both parties. These documents might include project team meeting summaries, project reviews or action items generated by research management. The creation of a research information management site on the company intranet provides a common area for all project team information to be stored and accessed by parties with the required access privileges. This type of site would typically be maintained by a Chemical Informatics group, and it is the responsibilities of the team members (or the project manager) to ensure that project information remains update.

A commonly used project management tool is Microsoft Project, which can be used to track and integrate team activities, especially interdependencies and critical path activities. Probably one of the most useful capabilities is the ability to generate a Gantt chart for the project, which serves to clearly represent a timeline of key activities to the team as well as management. The Gantt chart shows the duration for each activity, dependencies, and key decision and delivery dates. It also can immediately recalculate the timeline if one critical path activity is delayed. Although this and other more sophisticated tools are available for development project teams (e.g., OPX2 by Planisware),[19] research project teams often do not use these types of tools until there are clear activities associated with a single candidate (e.g., at the predevelopment stage), or when the team is contemplating recommending a DC and are estimating IND filing dates as part of the DCN documentation.

2.04.10 Conclusion

The discovery and development of quality drug candidates is a complex, cross-functional process. Highly effective teams and project leaders teams can achieve their goals by working cooperatively in a matrix-type organization. Successful project teams have clearly defined goals and objectives and are able to communicate effectively among themselves and with research management. It is important for project leaders and functional area heads to agree upon the project goals and priorities to avoid conflict between line and project metrics. Effective project management can help drive the definition of the project goals, ensure that the project remains on track and is resourced appropriately, as well as anticipate possible issues which could impact project timelines or budgets. Research management is responsible for oversight of the project pipeline, ranging from project resourcing and providing technical advice and mentoring, to portfolio management and making go/no go decisions on projects themselves. Research management should empower project team leaders and their teams, internal or collaborative, as much as possible to successfully plan, monitor, and execute on team objectives. By instilling the discipline of good project/research management and leadership, companies can improve their probability of success for delivering successful drug candidates into clinical development, as well as creating a stimulating and productive environment in which their scientists can work.

References

1. DiMasi, J.; Hansen, R.; Grabowski, H. *J. Health Econ.* **2003**, *22*, 151–185.
2. Hopkins, A. L.; Groom, C. R. *Nat. Rev. Drug Disc.* **2002**, *1*, 727–730.
3. Ma, P.; Zemmel, R. *Nat. Rev. Drug Disc.* **2002**, *1*, 571–572.
4. Kennedy, T. *Drug Disc. Tech.* **1997**, *2*, 436–444.
5. Biller, S. A.; Custer, L.; Dickinson, K. E.; Durham, S. K.; Gavai, A. V.; Hamann, L. G.; Josephs, J. L.; Moulin, F.; Pearl, G. M.; Flint, O. P. et al. The Challenge of Quality in Candidate Optimization. In *Pharmaceutical Profiling in Drug Discovery for Lead Selection*, Borchardt, R. T.; Kerns, E. H.; Lipinski, C. A.; Thakker, D. R.; Wang, B., Eds.; Biotechnology – Pharmaceutical Aspects Series; American Association of Pharmaceutical Scientists Press: Arlington, VA, 2003, pp 413–429.
6. Pritchard, J. F.; Jurima-Romei, M.; Reimer, M. L. J.; Mortimer, E.; Rolfe, B.; Cayen, M. N. *Nat. Rev. Drug Disc.* **2003**, *2*, 542–553.
7. Bernstein, D. F.; Manrell, M. R. *Drug Info. J.* **2000**, *34*, 909–917.
8. Lundy, J. L. *Teams: Together Each Achieves More Success*; Dartnell: Illinois, 1992; Chapter 21, pp 137–149.
9. Thompson, L. L. *Making the Team*, 2nd ed., Pearson Education, Inc.: New Jersey, 2004; Chapter 3, pp 42–68.
10. Ford, R. C., Randolph, W. A. *J. Mgmt.* **1992**, June.
11. Katz, R.; Allen, T. J. *Acad. Mgmt. J.* **1985**, *28*, 67–87.
12. Keller, R. T. *Acad. Mgmt. J.* **1986**, *29*, 715–726.
13. Ford, R. C.; Armandi, B. R.; Heaton, C. P. *Organization Theory: An Integrative Approach*; Harper & Row Publishers: New York, 1988.
14. Larson, E. W.; Gobeli, D. H. *Cal. Mgmt. Rev.* **1987**, *29*, 37–43.
15. Randolph, W. A.; Posner, B. Z. *Getting the Job Done: Managing Project Teams and Task Forces for Success*; Prentice-Hall: Englewood Cliffs, NJ, 1992.
16. Barker, J.; Tjosvold, D.; Andrews, R. I. *J. Mgmt. Stud.* **1988**, *25*, 167–178.
17. Visit Flypaper: http://www.flypaper.com and Groove Networks: http://www.groove.net for more details and information (accessed Aug 2006).
18. Smith, C. G. *The Process of New Drug Discovery and Development*, CRC: Boca Raton, FL, 1992; Chapter 12, pp 103–114.
19. Visit Planisware website for more information: http://www.Planisware.com (accessed Aug 2006).

Biographies

Yumi Nakagawa is a Senior Director of Research at Novartis in Emeryville, CA. She received her BS in Chemistry from the University of California, Berkeley, in 1979, and her PhD from the University of California, Los Angeles. Dr Nakagawa was also an NIH Postdoctoral Research Fellow at UC Berkeley from 1984 to 1987. Prior to joining Chiron in 1996, Dr Nakagawa was a Senior Research chemist at Chevron, where she worked in agricultural chemistry and later, in catalyst research. She is the inventor/co-inventor on 31 US patents. At Chiron, Dr Nakagawa managed several large technology transfer and target collaborations, and eventually led the Research Project Management group in Chiron Technologies. In 2003 she became head of Chiron BioPharmaceuticals Research Operations, and later worked in the Office of the Chief Scientific Officer until the acquisition by Novartis in spring 2006.

Lori Lehman is currently Senior Director of Research at Gilead Sciences, Inc. in Foster City, CA. She received her BS in Chemistry and MS in Organic Chemistry from Bucknell University, Lewisburg, PA, in 1979 and her PhD in Organic Chemistry from Duke University, Durham, NC, in 1983. Prior to joining Gilead, Dr Lehman held positions as Senior Scientist at Schering-Plough working in the Departments of Allergy and Inflammation and Cardiovascular Disease; Senior Director of Chemistry at Amylin Pharmaceuticals working in Diabetes; and as Vice President of Research at RiboGene, Inc. working in Infectious Diseases. During her career, Dr Lehman led several research and development projects and managed collaborative relationships with large and small pharmaceutical companies and academic research groups.

2.05 The Role of the Chemical Development, Quality, and Regulatory Affairs Teams in Turning a Potent Agent into a Registered Product

S A Munk, Ash Stevens Inc., Detroit, MI, USA

© 2007 Elsevier Ltd. All Rights Reserved.

2.05.1	**Introduction**	**159**
2.05.2	**Development Chemistry Tasks**	**160**
2.05.2.1	Regulatory Tasks for Drug Approval	161
2.05.2.1.1	Drug substance physical and chemical properties	161
2.05.2.2	Statistical Evaluation of Process Variables	164
2.05.2.3	Cleaning of Equipment and Facilities	165
2.05.2.4	Analytical Methods	166
2.05.2.5	Analytical Method Validation	166
2.05.2.6	Impurities	167
2.05.2.7	Reference Standard of the Drug Substance	167
2.05.2.7.1	Stability of the active pharmaceutical ingredient	167
2.05.2.7.2	Packaging	168
2.05.2.8	Reprocess and Rework of Materials	169
2.05.3	**Development Chemistry Timeline**	**169**
References		**171**

2.05.1 Introduction

Organic chemistry is a science practiced by two groups in the pharmaceutical industry who have radically different aims. The first group consists of medicinal chemists who use the tools of organic synthesis to construct unique structures with the aim of identifying interesting biological activity. The only concern of this group is the architecture of the molecule to achieve a predetermined pharmacologic purpose. This group of scientists is relatively unconcerned with the practicality of the synthetic route selected as medicinal chemists typically prepare materials on the gram scale. As a compound moves from discovery through early development and ultimately into the manufacturing arena, the amounts of drug substance required to support the studies and postapproval marketing ranges from kilos to as much as hundreds of metric tons of material. It is at the development stage that the second group of organic chemists becomes involved. In sharp contrast to the medicinal chemist's world, the concern of the chemical development team is a manufacturing route that can be carried out reproducibly and economically while being executed by chemical operators at a large scale.

A 'drug product' consists of a pharmacologically active agent: a 'drug substance' or 'active pharmaceutical ingredient' (API) that has been compounded with appropriate excipients into a system that can effectively deliver that agent to its site of action within the body of a patient. Approval to market a drug product in a timely fashion can mean substantial dollars to the sponsor of a new drug application. If one assumes that the new drug product is a blockbuster, having a market value of one billion dollars per year, every month that an application's approval is delayed costs the sponsor almost one hundred million dollars! This has a substantial effect on the ultimate economic value of the product to the enterprise. It is money that can never be recovered, as the drug patent's lifetime is finite and not related to the time required to answer questions from a regulatory agency.

Regulatory approval to market a drug product from the US Food and Drug Administration (FDA) and other Ministries of Health can be obtained upon successful completion of two important review processes. The first component is the scientific package of toxicology and clinical data that support the safety and efficacy of the drug for its proposed indication. The chemical development team has no control over that aspect of the approval process. This is an inherent property of the molecule and the quality of biological and clinical evaluation of the molecule. The second component is the scientific review of the chemistry, manufacturing, and controls sections of regulatory package assembled for submission to licensing agencies. This is a property of regulatory compliance. The regulatory approval process is mapped out in a set of regulations promulgated by the FDA as well as other regulatory bodies throughout the world. This chapter deals principally with small-molecule drug substances regulated under Title 21 of the US Code of Federal Regulations (21 CFR), the regulations that resulted from the Food, Drug and Cosmetic Act of 1938 which has been updated from time to time.[1] Regulation of these small molecule drug substances specifically falls under the Center for Drug Evaluation and Research (CDER) of the FDA in the USA. The USA, Europe, and Japan under the auspices of the World Health Organization (WHO) determined that a consistent set of regulatory requirements and format would be to everyone's advantage. They formed the International Conference on Harmonization (ICH) to address this issue. The result of those conferences was to publish a set of principles and issue a set of guidance documents,[2] collectively referred to as the current Good Manufacturing Practices or cGMPs, which govern the manufacture of drug substances. It is compliance with these regulations that the chemical development team controls and where that team can generate substantial value for its organization. Chemical development is an inherently costly process requiring an investment of many millions of dollars. While it is difficult to compress the overall regulatory approval cycle time, a properly documented, carefully executed development package will generate few questions from regulatory agencies. While almost all chemistry regulatory questions can be answered, the process of clarification and resolution of outstanding issues involves a delay in approval to market as well as an additional investment. It is this aspect of the development process that will be the focus of this chapter.

In addition to the chemical development team's role in securing approval to market a drug, the team has a number of other ongoing material supply requirements that must be met during the development cycle. These responsibilities include supplying materials to support ongoing toxicology studies and formulation development. These studies do not require that the API be manufactured following the requirements of cGMP. In order to ensure that the material fairly represents material that will ultimately be used in clinical trials and subsequently marketed; the development team should provide material that is comparable to that anticipated to be generated from the manufacturing process as well as good documentation of the synthesis methodology employed. The material should have a comparable profile of impurities and be the same particle size and polymorph to ensure that dissolution properties do not change, thus changing availability of the drug to the living system. Materials will also be needed to support ongoing clinical trials in people. Federal law mandates that all materials destined for human use must be manufactured following cGMP requirements.

A term that is often misused is GLP or Good Laboratory Practice. These regulations are not a less stringent version of the GMP requirements; rather these regulations mandate how the analytical laboratory used to evaluate toxicology samples is managed.[3]

All of the regulatory requirements surrounding the drug product and the API have been mandated to ensure that the pharmaceutical industry's products are safe, pure, and effective. Additionally, these requirements ensure that the manufacturing processes are designed to be consistent from batch to batch and ensure that patients receive the same pharmacologic benefit with each dose of medicine. Defining a chemical process that is consistent is a critical component of process development. Many of the current requirements have been instituted to minimize the opportunity for human error.

2.05.2 Development Chemistry Tasks

A well-established principle in the drug-manufacturing arena is that the quality and purity of the drug substance cannot be ensured simply by analytical evaluation of the API at the end of the manufacturing process. Rather, the purity of the drug substance depends on maintaining proper control of the materials used and proper control of the synthetic process used for the manufacture of the material. Therefore, the task of the chemical development group can be simply stated as the identification of a robust, economically viable, and, most importantly, consistent process. A robust process is required to allow chemical operators to execute the process in large, fixed equipment. The process must be consistent to assure that the patient receives the same product with each dose (vide supra). An economical process ensures that the product is viable in the marketplace.

Economics is playing an increasingly prominent role in the pharmaceutical industry and occasionally development of pharmacologically interesting classes of molecules is terminated for reasons of high cost of goods. For example, in the 1980s many major pharmaceutical companies had research efforts to identify compounds to manage hypertension by inhibiting the enzyme renin. Today, there are no renin inhibitors marketed as antihypertensive drugs, even though many drug candidates of that class of agent were shown to reduce blood pressure in humans. The high cost of goods was a concern in many of those programs. Interestingly, the drug design concepts elucidated in the renin area provided the foundations for many of the currently marketed human immunodeficiency virus (HIV) protease inhibitors.

The first tasks for the team are definition of the synthetic route, reagents, and process parameters. Chemists operating in a medicinal chemistry laboratory have many technical options available to execute a synthesis. Reaction volumes, management of heat, and mass flow are not concerns in a medicinal chemistry setting. These issues become prominent in a plant setting. There are a number of excellent monographs covering these topics in detail. Anderson presents a particularly detailed text covering the chemical operational topics that must be considered.[4] Additionally, some chemical steps that can be executed in a matter of hours with laboratory equipment can require days in a plant setting. In the discovery setting, one has easy access to temperatures as low as $-100\,°C$ and a wide variety of chromatographic systems to purify the most challenging reaction mixtures. In the plant setting, one would prefer to employ less technically challenging, costly unit operations. That is not to imply that technically difficult unit operations are precluded. Atorvastatin calcium (Lipitor) is manufactured using low-temperature chemistry.[5] During scale-up work toward SDZ-NKT343, Prashad *et al.* used chromatography early in the program but successfully eliminated that type of purification upon further process development.[6] Such operations are costly and, in general, worth avoiding. They must, however, be evaluated on a case-by-case basis.

In addition to the design of an efficient process, the development team must be concerned about reagents selected for production. Certain reagent classes are easy to use in research but pose substantial hazards for production materials destined for consumption by humans. For example, lead- or mercury-containing reagents might prove useful in a discovery laboratory but should be avoided in a process setting.

Medicinal chemists have easy access to a very wide variety of solvents, including chloroform and diethyl ether. When used at production scale, chloroform is a solvent that has serious environmental and operator exposure liabilities. Diethyl ether is a potential fire hazard and can form peroxides rapidly and thus should be avoided in a plant setting. These types of solvent are generally not permitted in a drug-manufacturing process. Regulatory agencies have categorized solvents as class 1, 2, or 3 and have imposed residual solvent limits based on potential hazards to human health (**Table 1**). Class 1 solvents should be avoided whenever possible. In some instances, they cannot be avoided, for example if benzene is required as a core-starting subunit in a drug synthesis or if the specific reactions fail when conducted with class 2 or class 3 solvents. If they must be used, the residual solvent limits required by the FDA are generally well under 10 PPM. Class 2 solvents have general residual solvent limits in the range of 100–1000 PPM while class 3 solvents have a residual solvent limit of up to 5000 PPM under certain circumstances without any additional justification. Higher limits are permitted with appropriate justification if the process cannot achieve lower levels of the residual solvent. Toxicology studies must be completed to support the safety of the material, and general GMP guidelines must be followed. There are exceptions to these generalities. Methanol has a limit of 3000 PPM, although it is listed as a class 2 solvent. Specific limits for individual solvents can be found in the FDA *Guidance*.[7]

The tasks that the chemical development team must execute are described below and that discussion is roughly patterned after the current requirements suggested by the ICH guidelines for a common technical document (CTD), summarized in M4Q.[8] This and related FDA guidance documents provides the framework for assembly of regulatory documents in a format that permits effective review by both US and international agencies of the application for approval to market the drug.

2.05.2.1 Regulatory Tasks for Drug Approval

The major sections of the CTD that concern the chemical development team are those covering 'drug substance.' Within that major section are a number of sections and subsections that describe all elements of the chemical process, analytical methods, characterization of materials, and requirements of cleaning processes.

2.05.2.1.1 Drug substance physical and chemical properties

The first section of a CTD covers nomenclature, chemical structure, and physicochemical properties of the drug substance. One is required to provide the International Union of Pure and Applied Chemistry (IUPAC) name of the agent and the US Adopted Name (USAN) as well as any common names for the material. A standard graphical representation of the structure is presented along with general physicochemical information about the material. The properties described should include spectroscopic properties, for example ultraviolet/visible spectrophotometry (UV-Vis), melting or degradation range, and solubility properties. The rotation for optically active substances should be provided as well as pK_a's if appropriate. The morphology of a solid drug substance must be evaluated for APIs destined for solid dosage forms of drug products. In the case of solution drug products (oral liquids or those destined for

Table 1 Solvent classification for pharmaceutical products

Class 1 solvents	Class 2 solvents	Class 3 solvents
Benzene	Acetonitrile	Acetic acid
Carbon tetrachloride	Chlorobenzene	Acetone
1,2-Dichloroethane	Chloroform	Anisole
1,1-Dichloroethene	Cyclohexane	1-Butanol
1,1,1-Trichloroethane	1,2-Dichloroethene	2-Butanol
	Dichloromethane	Butyl acetate
	1,2-Dimethoxyethane	tert-Butylmethyl ether
	N,N-Dimethylacetamide	Cumene
	N,N-Dimethylformamide	Dimethyl sulfoxide
	1,4-Dioxane	Ethanol
	2-Ethoxyethanol	Ethyl acetate
	Ethylene glycol	Ethyl ether
	Formamide	Ethyl formate
	Hexane	Formic acid
	Methanol	Heptane
	2-Methoxyethanol	Isobutyl acetate
	Methyl butyl ketone	Isopropyl acetate
	Methylcyclohexane	Methyl acetate
	N-Methylpyrrolidone	3-Methyl-1-butanol
	Nitromethane	Methylethyl ketone
	Pyridine	Methylisobutyl ketone
	Sulfolane	2-Methyl-1-propanol
	Tetrahydrofuran	Pentane
	Tetralin	1-Pentanol
	Toluene	1-Propanol
	1,1,2-Trichloroethene	2-Propanol
	Xylene	Propyl acetate

injection), morphology is not a concern. As the API is already in solution, the crystalline form of the solid is irrelevant unless it is a concern for dissolution during the manufacturing of the formulated product. The evaluation should define the crystalline form using optical microscopy and provide information concerning polymorphs – different crystal lattice forms of the same chemical agent. Understanding the polymorphic form of the drug and controlling that precise arrangement of molecules within the crystal during production and storage of the bulk active ingredient aid in an understanding and definition of the rate of dissolution of the drug in the body.[9] The rate of dissolution has a substantial effect on the rate at which the drug is delivered to the patient. In addition to the role the polymorph of a drug has on the patient, polymorphs are patentable as unique compositions of matter. Thus there are economic reasons to consider polymorphism in addition to safety and efficacy concerns.

In addition to the polymorph of an API having an impact on bioavailability, particle size can also have a tremendous impact on the efficacy of a drug in a human. Different particle sizes dissolve at differing rates, causing the drug to be delivered to its site of action at different rates. During process development, consideration should be given to the appropriate crystallization conditions as well as the need to modify the bulk particle size.[10] The latter can be accomplished through various milling, micronization, or sieving techniques. As with morphology, particle size is of particular concern for drugs that are to be administered as solids. It is much less important for drugs that are administered orally in solution or parenterally.

The names and addresses of the manufacturer(s) of the drug substance and all subcontractors involved in the work should be provided. Approval to manufacture a given product also mandates the site(s) of manufacture and the specific types of processing equipment that can be used. While these points are at the discretion of the applicant, they must be included in regulatory filings. The sites and type of equipment that are employed can be changed, but most such changes require revalidation of the process, updates to regulatory filings, and approval from regulatory agencies before product incorporating the changes can be sold.

The next section details the manufacturing steps for the drug substance. A flow diagram for the process should be provided along with a written description of the synthetic route. The specific type of equipment used is presented in this section. Examples might include a description of the type of reaction vessels, for example a glass-lined vessel; and isolation equipment utilized, for example, isolation via centrifugation. A critical component of the section is the description of the in-process controls (IPCs) used in the process to assess progress of a reaction or quality of an isolated intermediate. Typically, each step of the process is monitored to assure completion of the step without buildup of untoward impurities or degraded material. Ideal IPCs to monitor reaction progress are information-rich, reproducible, accurate, and easily executed by a chemical operator in a plant setting. A typical IPC to monitor reaction progress might include the conduct of a chromatographic test to assess progress of the reaction. This might be a thin-layer chromatography (TLC), high-performance liquid chromatography (HPLC), or gas chromatography (GC) analysis. Visual tests can often be effective tools. For example, a color change might be an indication of reaction completeness (presence of the blue color of Na^0 in liquid ammonia to signify excess sodium in ammonia in a sodium–ammonia reduction). Often spectroscopic methods such as UV or infrared spectroscopy are employed. Occasionally pH can be used as an indication of reaction progress.

The specifications for the quality attributes and manufacturers of the key starting materials should be included in the manufacturing section together with the analytical methods used to evaluate them. In addition to key starting materials, the manufacturing section should provide a discussion of the specifications for reagents, solvents, and ancillary materials.

This section of the CTD also presents the rationale for the definition of the starting materials for the process. The team preparing the filing would generally like to define the starting materials as late in the synthetic sequence as possible. Materials used in the production of an API and which are incorporated as a significant structural fragment into the structure of the API are commonly defined as 'API starting materials.' They are often separated from the API by several synthetic steps.

An API starting material should be an article of commerce or it should be well described in the chemical literature. It can be a material purchased from one or more suppliers under contract or commercial agreement, or alternatively, an API starting material might be produced in-house.

The approach of defining starting material as an advanced intermediate in the scheme affords maximum flexibility in sourcing those materials. It further allows changes to be incorporated in the manufacture of that compound without regulatory impact provided that testing demonstrates that no new impurities are generated. In general, it is advisable to define an API starting material and ensure that the position that has been taken is agreeable to the FDA and other regulatory bodies through discussions of the definition with those agencies prior to filing documents for regulatory approval. This approach avoids costly surprises late in the registration process. GMP requirements do not apply to steps prior to the defined API starting material.

Starting materials, reagents, and intermediates, final APIs, as well as packaging materials and processing aids all require well-defined specifications and appropriate analytical evaluation to control the attributes of the materials used to produce drug substances. These specifications and tests should be information-rich and should provide data that assure that the materials introduced into the process are of suitable quality to afford API that meets the clinical requirements. In general, specifications for late-stage intermediates are more critical than for those of early-stage materials or common reagents. The intent is to provide appropriate characterization of the material to determine suitability of the material for the next step in the process.

Setting specifications for the final API requires careful thought. At the stage of toxicology studies, material should be of the minimum level of purity acceptable for evaluation and contain all likely impurities that might be encountered as the process is scaled. The reason for this is that, as a process is scaled, equipment is changed from laboratory glassware to fixed equipment in a plant. With unit operations in a plant often requiring longer times than typically found in the laboratory, impurities can build up and purification can become an issue. In a medicinal chemistry laboratory, there are many easy-to-manage techniques available to purify an agent. Flash chromatography is a very common laboratory technique, as an example. As noted, these operations become more challenging in the plant setting. Additionally, many plant operations are conducted with technical-grade solvents rather than American Chemical Society (ACS) materials. An ACS-certified material indicates that the reagent meets all of the specifications and requirements found in *Reagent Chemicals*, published and periodically revised by the ACS while 'technical-grade' materials meet less stringent requirements and, as a consequence, are less expensive. Decisions about solvent and reagent requirements should be considered at an early stage of the program. Typically, for toxicology studies, the API should be in the range of 96–98% pure. As the process is refined and more is learned about the material, the purity should be increased to a level above 98% purity. While it might be straightforward to prepare the first few hundred grams at a level of purity of greater than 99% pure, this can prove very limiting, as larger amounts are needed. Bridging toxicology studies are required if new impurities are introduced into the API. Such studies are quite costly, in terms of both time and money.

Critical parameters should be determined by experiment and scientific judgment and typically should be based on knowledge derived from research, scale-up batches, or manufacturing experiences. Prior to validation, the appropriate range for each critical process parameter that will be used during routine manufacturing should be defined.

2.05.2.2 Statistical Evaluation of Process Variables

In the 1920s the British geneticist and statistician Sir Ronald Fisher developed the initial concepts of statistical design of experiments, also referred to as design of experiments or DoE.[11] These concepts provide a formal, rigorous framework to rationally design a set of experiments to relate the quantitative effect of variation of specific parameters with the resulting measured response(s) analyzed using statistical methods. The original concepts were put forth to meet the needs of agricultural experimentation with which he and his colleagues were involved. The basic idea was that there is a simple underlying geometric structure to such experimentation. Fisher and his colleagues further developed the statistical methods and analyses that became the standard procedures of the discipline.

Many chemists are unfamiliar with the techniques. Workers at DuPont were instrumental in moving these concepts into the chemical industry and taught these techniques to many outside organizations, beginning in the 1970s.[12] DoE has been shown to be quite useful in the identification of appropriate ranges for critical parameters in a chemical process. The method allows multidimensional space to be probed and considers the effect of variation of one parameter on other parameters in the process. It is also very efficient, using a statistically based mathematical model to generate the interaction of a very large number of parameter interactions through variation of multiple parameters in a single experiment.

Fortunately, there are a number of commercial, off-the-shelf software products available so that the practicing process chemist does not need to become an expert statistician to use these effective techniques. Many in the pharmaceutical industry have used Fusion Pro, developed by S-Matrix, with success. This software is commercially available and parameters such as temperature can be monitored directly by computer. Parameters that might be considered include reaction temperature, mixing rate, cooling rate, and ratio of reagents utilized in the process. While the DoE approach appears complicated at first, that approach proves to afford much more information with fewer experiments than a traditional linear approach where a single variable is explored in a single experiment. When unexpected results are obtained, it is quite probable that some parameters that were not considered critical to the manufacturing process in the initial experimental design are indeed critical.

Parameters that have been determined to be critical process parameters should be controlled and monitored during process validation studies. Process validation should confirm that the impurity profile for each individual lot of API is within the limits set in the specifications and is comparable to, or purer than, the profile of material used for toxicological and pivotal clinical studies.

A written set of documents detailing the strategy planned for process validation should be in place and followed for validating the manufacturing processes for the API. The purpose of the study is to ensure that the manufacturing process that has been developed is reliable and provides product that is consistent and homogeneous, and meets the predetermined set of specifications. Among the attributes of the API that should be considered are purity, qualitative and quantitative impurity profiles, physical characteristics, including particle size, density, and polymorphic form, moisture and solvent content, homogeneity, and microbial load. For drug substances destined for injection, bacterial endotoxin or pyrogen testing is also required.

The protocol should define the scope and purpose of process validation. The number of process validation batches to be prepared should be addressed. The number of complete batches to be prepared under the protocol is typically three consecutive successful runs. Given the high value of many APIs, it is acceptable to have executed only one run at the time of the FDA preapproval inspection (PAI) with a commitment to complete the subsequent batches successfully prior to launch. This is referred to as 'concurrent validation.' Analytical test methods that will be used are listed. Often, a number of additional analytical tests are performed during validation that might subsequently be eliminated. An example of an analytical test method that might be used during validation but not during routine product release could be an x-ray analysis to examine polymorphic form. These data are collected under protocol, and if the results obtained during validation are consistent, subsequent x-ray analysis of the product would not be required. A detailed discussion of the chemical transformations, unit operations, and a process flow diagram should be included. The description of the unit operations should include all major processing equipment to be used as well as appropriate qualification of the equipment. Equipment qualification can be directly incorporated into the validation protocol or by reference to prior reports. Based on the parametric studies that were executed, the critical process parameters and operating ranges must be described. The protocol should include plans for product sampling. This includes a description of sampling points, frequency of removing samples, quantities of each sample required, and procedures for collecting samples.

The protocol needs to define criteria by which the executed validation will be deemed acceptable. Deviations from the plan must also be documented. The plan should also describe what measures will be taken in the event of a failure. It is wise to execute at least one full-scale batch using the proposed batch production records prior to initiating validation. This is often referred to as a demonstration batch. This affords confidence that the production procedures, test methods, and specifications have been appropriately defined. Validation should be conducted prior to the commercial distribution of an API and when there have been significant changes to the manufacturing procedure.

2.05.2.3 Cleaning of Equipment and Facilities

This is an appropriate juncture to discuss cleaning procedures.[13] These procedures involve equipment-cleaning as well as facility-cleaning. There are three types of contaminants that can adulterate an API: (1) physical contamination (particulates); (2) biological contamination (bacteria, yeast, and mold; also called 'bioburden'); and (3) chemical contamination (cross-contamination).

The physical facility, including the heating, ventilation, and air conditioning (HVAC) system and associated particulate filtration should be of an appropriate design and be able to be cleaned. The facility should also be monitored to demonstrate that it is reasonably free of microbial contamination. The API itself is also evaluated to insure that it has an appropriately controlled level of bioburden. Subpart C of 21 CFR part 211 details the requirements for facility design.

Cleaning of multi-use manufacturing equipment requires detailed and validated cleaning procedures and analytical methods. The potential limits for product carryover into subsequent batches should be considered when setting limits for the appropriate level of cleaning of the equipment. With the exception of penicillin analogs and other extremely sensitizing agents, which should be manufactured in dedicated facilities, the regulatory agencies offer little firm guidance on an appropriate level of decontamination of facilities and equipment that must occur. Limits that are set for carryover should be 'logical,' 'practical,' 'achievable,' and 'verifiable' according to regulatory agencies. Often, an appropriate approach to setting a limit for permissible amount of carryover is based upon pharmacologic data for a particular material. The limit is set to allow a maximum carryover of 1/1000 of the usual therapeutic dose to be present in a dose of the subsequent product based on the assumption that such a low dose would have minimal effect on a patient.

If the process train, or set of equipment used to produce a material, is to be used to produce the same product multiple times in that process train, a 'minor cleaning' procedure between batches is required. A 'minor cleaning' is also appropriate if the process train is used to conduct the next step in a chemical sequence to produce a product. The cleaning process used in this case should be effective for the removal of process residues. For the purposes of minor cleaning, however, a visual inspection demonstrating that there is no material remaining in the process train is all that is required to insure that the cleaning process has been effective.

Process equipment in a chemical plant is seldom dedicated, as such equipment costs millions of dollars to design and install. Typically, therefore, process equipment will be used for a variety of different products and 'major cleaning' processes must be developed and validated by executing three successful, successive cleaning operations. This includes analytical verification that the process was effective. The procedure might involve aqueous conditions, including a detergent, which is itself a potential chemical contaminant. If detergents are used, their absence after cleaning must be verified using an appropriate and validated analytical method. Organic solvents and physical methods are excellent choices for cleaning. Process equipment for the pharmaceutical industry is designed to have minimal dead zones where materials can collect; however, there are still some areas where residue can remain. These areas include seams or valves. The cleaning program should include an evaluation of those difficult-to-clean areas. The delineation of hard-to-clean sites can be predicted based on prior experience; however a formal challenge is preferable. Often, equipment will be coated with a nontoxic, surrogate material which is easy to visualize using either visible or UV light. Riboflavin is an example of such a surrogate. The cleaning procedure is executed and the equipment is inspected to assess the effectiveness of the cleaning procedure. Most of the agent will have been removed, but it is not unusual for material to be observed in valves, seals, and other difficult-to-clean areas.

Once a piece of process equipment has been cleaned, the cleanliness must be verified. Equipment can be swabbed; a direct sampling of the surface of the processing equipment, to demonstrate the validity of the cleaning process. It can also be rinsed to evaluate the effectiveness of a given cleaning procedure. The swabs and rinses can then be evaluated by analytical techniques that are preferably specific to the given product to be removed from the equipment. Appropriate analytical methods include HPLC or a UV-Vis analysis. Under certain conditions, an analysis of total organic carbon might be appropriate; however, that type of an analysis is not material-specific and is predicated on the materials to be evaluated being soluble in water. The methods for these analyses must be validated (vide infra). As part of the cleaning program, swab and rinse transfer efficiencies and material solubility must be evaluated. Typically, small coupons of the surface being cleaned (Teflon, glass, or stainless steel, for example) can be evaluated in the laboratory to determine the efficiency of the transfer of the analyte to the swab and subsequently from the swab to the analytical medium.

2.05.2.4 Analytical Methods

The analytical techniques used to characterize fully the API are very important topics to address. While simple spectroscopic techniques and elemental analyses are sufficient in a medicinal chemistry laboratory, regulatory agencies require a much more elaborate data set to register a drug. A typical set of analytical methods for full characterization of the material might include IR, nuclear magnetic resonance (NMR) spectroscopy (both ^{1}H and ^{13}C), elemental analysis, UV-Vis, mass spectrometry (MS), optical rotation, differential scanning calorimetry (DSC), HPLC, GC, residue on ignition/sulfated ash, thermogravimetric analysis (TGA), as well as an analysis of moisture in the product. x-ray powder diffraction (XRPD) for determining polymorphic form and a determination of particle size, often using a laser light-scattering method, of the solid drug substance must also be described. Again, morphology questions are irrelevant issues for APIs destined for a solution formulation but critical issues for solid-dosage forms of drug products.

2.05.2.5 Analytical Method Validation

The analytical group within the chemical development team has the responsibility to demonstrate that the analytical methods developed for a particular drug substance are validated.[14] The methods employed must be shown to be accurate, precise, linear, sensitive, specific for a given material, and reproducible. Analytical methods are progressively optimized throughout the development cycle. For the purpose of the investigational new drug application (IND), a method that has been developed at a rudimentary level is adequate, but the method must be scientifically sound, consistent with ICH Q7A. As with the synthetic process, optimization continues throughout the development cycle. All analytical methods should be fully optimized and validation completed prior to initiation of phase III trials. By completing these activities in advance, the chemical development team minimizes the potential for erroneous results in clinical trials, when many patients are involved and many dollars are at stake.

Prior to each analysis of a drug substance, there must be an evaluation of the analytical system suitability. This is an established procedure that ensures that the equipment, analytical procedure, and samples to be analyzed are working properly at the time of analysis. System suitability parameters should be defined and included as an integral section of the analytical procedure.

As with all validation activities in the pharmaceutical industry, a validation protocol should be developed prospectively after a thorough set of experiments has been executed to define appropriate conditions for the analysis. That protocol details the specifics of the acceptance of successful execution of the validation.

The process of validation of an analytical method generally demonstrate properties of:

- specificity: ability to measure desired product in a complex mixture;
- accuracy: agreement between measured and real value;
- linearity: proportionality of measured value to concentration;
- precision: agreement between a series of measurements;

- range: concentration interval where method is precise, accurate, and linear;
- robustness: reproducibility under normal but variable laboratory conditions;
- limits of detection: lowest amount of material that can be detected;
- limits of quantitation: lowest amount of material that can be measured.

2.05.2.6 Impurities

A detailed discussion of impurities found in the API is also required. This discussion should include a listing of impurities that are introduced from the starting materials, reagents, and solvents, as well as those arising as a consequence of the mechanistic pathway of the reactions utilized in the process. Forced degradation studies of the product are useful guides in this endeavor. In addition to providing access to potential impurities that can be used as standards of the impurities found in the drug substance, forced degradation experiments demonstrate that the stability studies that must be conducted are indeed stability-indicating. By demonstrating that impurities can arise and that they can be detected with the analytical methods used, confidence is gained that those methods are stability-indicating and can be validly used in the stability studies. Modes of forced degradation include subjecting the material to high temperatures, light, or oxidative conditions. The material should also be subjected to hydrolytic conditions employing either acidic or basic conditions. Reference standards of impurities can be isolated using preparative HPLC. Typically, the structure of all impurities present in the product at greater than 0.1% should be elucidated. HPLC, MS, and NMR analyses prove to be a primary set of techniques used for this purpose.

For critical impurities, one often needs authentic materials. One approach to obtain the standards is independent synthesis of the suspected impurities. This can be a time-consuming endeavor. Alternatives that can be employed include using preparative HPLC to isolate the desired impurities from the drug substance. Often, the residual solvents used for precipitation or crystallization of a material that contains the impurities (that supernatant is also called a 'mother liquor') or other operations conducted during the synthesis of the API are rich sources of such materials.

2.05.2.7 Reference Standard of the Drug Substance

The primary reference standard for a drug substance is a sample of the drug substance that is generally the highest-purity material that can be prepared. An independent synthesis route is often used to prepare this standard. It may also be synthesized using the manufacturing route with additional purification steps. If an alternative synthetic strategy was utilized, it must be described in detail. If the usual route of synthesis is employed, the material should be purified by additional crystallizations or by chromatographic purification and these should also be detailed in the CTD. The primary reference standard must be shown to be the desired chemical structure by a very detailed chemical analysis. This involves analytical testing above and beyond the requirements for normal testing and release of material. The reference standard is maintained as the benchmark against which all working standards are qualified.

Working standards of authentic materials are prepared from routine manufacturing batches of material. The purity of working standards is determined by direct comparison with the primary reference standard. This material is used as the standard for testing and release of routine production batches of drug substance. As with the primary reference standard, additional analyses are required to establish the attributes of this material versus routine release.

2.05.2.7.1 Stability of the active pharmaceutical ingredient

Formal, documented studies to provide experimental data to justify all holding times for intermediates as well as to support the proposed API expiration or requalification date must be conducted.[15] Analytical methods should be validated and shown to be stability-indicating prior to initiating stability studies for regulatory filings to ensure that the results of the studies are meaningful. The APIs used in stability studies should be packaged in a comparable fashion to the manner that the bulk API will be stored (vide infra). While three representative batches of material must be evaluated in the stability study, the batches do not have to be prepared at full manufacturing scale. Rather, pilot batches produced using the same synthetic procedures and process can be used to provide material to support a regulatory filing. Small-scale batches can be utilized to manage cash flow as manufacturing at full scale is quite costly. Companies often do not want to manufacture a large number of full-scale batches until approval appears imminent. A commitment is made at the time of filing to subject the first three full manufacturing batches to ongoing stability studies as well as an additional batch per year thereafter. Many products are stable under ambient conditions and typical stability studies are conducted under the conditions presented in **Table 2**.

Some APIs, however, are unstable under ambient conditions and lower temperature storage conditions are required. **Table 3** presents standard storage conditions for materials to be stored under refrigerated conditions and **Table 4**

Table 2 Typical conditions for ambient stability studies complying with ICH guidelines

Study	Storage conditions	Minimum time period covered with data at time of submission
Long-term	25 °C ± 2 °C/60% RH ± 5% RH or 30 °C ± 2 °C/65% RH ± 5% RH	12 months
Intermediate	30 °C ± 2 °C/65% RH ± 5% RH	6 months
Accelerated	40 °C ± 2 °C/75% RH ± 5% RH	6 months

RH, relative humidity.

Table 3 Typical conditions for refrigerated stability studies complying with ICH guidelines

Study	Storage conditions	Minimum time period covered with data at time of submission
Long-term	5 °C ± 3 °C	12 months
Accelerated	25 °C ± 2 °C/60% RH ± 5% RH	6 months

RH, relative humidity.

Table 4 Typical conditions for freezer storage stability studies complying with ICH guidelines

Study	Storage conditions	Minimum time period covered with data at time of submission
Long-term	−20 °C ± 5 °C	12 months
Accelerated	25 °C ± 2 °C/60% RH ± 5% RH	6 months

details conditions used for standard freezers. If one requires still lower temperatures, those must be dealt with on a case-by-case basis. Stability studies include 'accelerated conditions' to provide an early indication of product instability. Additionally, one should consider the effect of potential temperature excursions that might arise during shipping and design specific experiments to demonstrate that the material is stable during its travels. Stability studies are only required to be long enough to support proposed length of API storage. Typically, 2 years of stability data are adequate; however, for very-low-volume products, stability studies might be extended for up to 5 years.

The frequency of analysis during a stability study should establish the stability profile of the drug substance. For drug substances with a labeled stability of greater than 12 months, analysis of long-term storage samples should be every 3 months during the first year, every 6 months during the second year, and annually thereafter through the proposed length of the study. Evaluation of all attributes for the assigned specification is not necessary at all time points. Rather, a subset of stability-indicating tests is conducted at intermediate time points and the full set of assays is only executed at the initial time point, typically the original release date, and at the end of the study. Intermediate assays might include identity by IR, purity using HPLC, and moisture analysis. Attributes such as microbial load and residue upon ignition are not expected to change during storage and are thus only evaluated at the beginning and the end of the study.

For intermediate and accelerated storage conditions, a minimum of three time points, including the initial and final time points (e.g., 0, 3, and 6 months) for a 6-month study are appropriate. If the material appears to be decomposing, the intermediate and accelerated studies should be extended to define the limits of stability of the API.

2.05.2.7.2 Packaging

Packaging of the API is a detail often taken for granted in a discovery laboratory. Typically, in a research laboratory, one has easy access to a variety of small vials. In a plant setting, however, the quantities stored are substantially larger. While glass bottles might prove acceptable for hundreds of grams or even a small number of kilos of material, generally packaging options for bulk drug substances include plastic or lined drums, or large totes. The materials of construction of packaging material must be compatible with the materials being stored. As noted, the same material of construction, closure system, and configuration should be used for stability studies as will be used for storage and shipping of the material. As an example, small plastic bags sealed with twist ties in fiber drums might be utilized for the stability study if the packaging configuration for commercial materials is to be plastic bags sealed with twist ties inserted into fiber drums.

2.05.2.8 Reprocess and Rework of Materials

Reprocessing and reworking are not issues that often arise during the development cycle but can become very serious issues postapproval. On presumably rare occasions, a batch of an intermediate or a final API fails to meet the required specifications. Regulatory agencies recognize that there is a tremendous amount of value in each batch and those agencies do permit reprocessing and reworking operations to be executed. These operations are distinct both operationally and from a regulatory perspective.

Reprocessing is defined as resubjecting the material to identical conditions detailed in process description. Reprocessing might involve the addition of fresh reagents or solvents to unreacted material if an in-process test demonstrates that a reaction is incomplete. It could also be the repetition of a purification operation such as a crystallization step or a distillation that are part of the established manufacturing process.

There is a risk for formation of by-products during a reprocessing operation as the material could be subjected to additional warming or oxygen during the operations. Any reprocessing should involve an evaluation by the process, manufacturing, and quality teams to be certain that the quality of the material is not adversely impacted and that the potential formation of by-products is considered.

If reprocessing is required for many batches, the process might in reality be out of control. If the reprocessing always affords material that meets the specification, the reprocessing operation should be incorporated into the standard manufacturing process.

Subjecting out-of-specification material to a process that is not part of the established manufacturing procedure is defined as a reworking process rather than a reprocessing operation. For example, a rework might include a crystallization from a solvent not typically found in the original manufacturing process. Prior to reworking materials, a formal investigation led by the quality team into the origins of the nonconformance should be conducted and the issues causing the problem should be identified. Batches that have been reworked should be tested to demonstrate that the material has the same analytical attributes as material arising from the original process. In particular, a very detailed analysis of the impurity profile of the reworked material is required to demonstrate that no new impurities were introduced and that the material is of the same quality as the material produced by the original, validated process. Stability testing will be required for materials produced through a new, rework process. The new process will also require formal validation. In this case, concurrent validation of the process can occur during the actual execution of the process. As with all validation activities, the generation of a protocol is required, describing the rework procedure and defining the expected results. If there is only one batch to be reworked, a report can be written and the batch released once the material has been shown to be acceptable.

2.05.3 Development Chemistry Timeline

As can be seen from the discussion of the tasks that the chemical development team must execute prior to achieving regulatory approval, there is a tremendous amount of work required. The entire development process, however, takes some years and the tasks can occur in a programmed fashion, which also affords effective management of cash flow.

Figure 1 graphically depicts the critical tasks for the chemical development team and the phases of development at which the tasks should be executed. The timeline from biological concept to approved product is often 10 years or more. The process development team's involvement typically begins at the stage when a novel chemical agent is elevated to the status of a drug candidate. At that time, larger quantities of materials are required for toxicological evaluation. The amount of material required at this point is typically in the range of hundreds of grams to many kilos. The amount of material required is determined by the potency of the candidate, the proposed therapeutic indication, and dosing schedule of the drug.

While the analytical characterization of the material tends to be rudimentary at the stage of toxicology studies, it is very important to understand enough about the molecule and its properties such that the data generated during toxicology studies are meaningful. As such, analytical methods, in particular, HPLC methods used for an assessment of the purity of the drug substance, should be linear, precise, reproducible, and have an established limit of detection. The methods are likely to be refined throughout the development cycle. Therefore, while a fully validated analytical method is not needed at this time, the data must be justified from a scientific perspective. It is also necessary to understand the morphology of the product at this point if the product is to be administered from a solid delivery system.

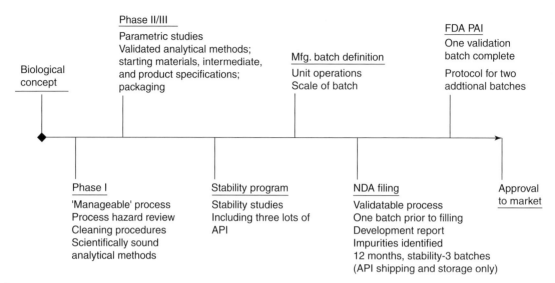

Figure 1 Development chemistry task timeline.

When the compound has been shown to have an appropriate safety profile in animals, human evaluation is set to begin. As one begins human evaluation of the product, a process that is manageable at a multi-kilogram scale is often required. As the scale moves into the range of kilograms, it is an appropriate time to consider in detail the safety of the chemical operations. Under certain conditions, the Occupational Safety and Health Administration (OSHA) mandates that a formal process safety management (PSM) program be implemented, as large quantities of chemicals are used.[16] The regulations provide specific guidelines as to when such a PSM evaluation is required and the rules are dependent on very specific amounts of materials used in a process. For example, a PSM evaluation is required when 100 lb or more of propargyl bromide is to be used, while such analysis for hydrogen chloride is not required until a threshold limit of 5000 lb is to be used in a process. Even when a formal PSM is not required, many elements of the program should be considered as an integral part of good business practices. In particular, an initial process hazard analysis (PHA) should be conducted by a team including representatives from engineering, operations, and individuals who have hands-on experience in executing the process prior to preparing material on the kilo scale. This analysis is a systematic evaluation to identify and analyze the significance of potential hazards associated with the proposed chemical process. Typical factors considered in a PHA are the potential for a fire, an explosion, and various chemical releases. Issues discussed should include an evaluation of the equipment to be used in the process, instrumentation, utilities, and the potential for human error. This analysis also specifically addresses potential failure modes in a proposed synthetic process.

From the perspective of cGMP compliance requirements, at the stage of phase I clinical trials, cleaning procedures must be considered and protocols should be developed; however, the methods need not be validated. As described above, the analytical methods should be sound; however, the final set of cleaning methods and assays has probably not been defined at this stage of the program.

As the clinical data for phase I are analyzed and the candidate is elevated into phase II clinical trials, the process must be refined and should substantially resemble the process proposed for use in the manufacturing setting. Parametric studies should be initiated to defend the proposed ranges for temperature, mode and time for reagent addition, stirring rates, holding times, and any other process parameters that have been deemed critical from the studies by process chemistry and engineering.

As phase III clinical trials are initiated, the process and analytical methods should be locked and preparation for the assembly of the new drug application should be underway. Analytical methods should be validated and formal stability studies should be underway using three separate lots of API. The formal development report covering all aspects of the chemical process and impurities should be assembled. This report will be evaluated by regulatory agencies during the inspection process.

Once the new drug application has been filed, the FDA or foreign regulatory agencies, depending on where regulatory filings have occurred, will typically conduct a PAI of the facilities. At the time of the inspection, regulatory agencies expect that the first batch of API at full scale, manufactured according to the validation protocol, will have been complete and that the material will have been released. The validation protocol must include plans to complete at least three batches according to that protocol. Regulatory agencies understand the expense of executing multiple batches. Some firms elect to postpone execution of the remaining validation batches until approval appears imminent. More conservative companies will execute all three batches prior to an inspection to assure themselves that no untoward events occur.

As the new drug application review process occurs, the process is typically formally transferred from the process development team into the manufacturing department. Collectively, those teams with input from those responsible for evaluating the size of the market and thus demand for the product postapproval must ensure that enough material is available to support a successful product launch.

While changes to the process, methods, type of processing equipment, and site of manufacture can be initiated at any point, it is important to remember that these changes require a detailed understanding of the regulatory process. It is wise not to initiate such changes during the new drug application approval process. Depending on the scope of the changes involved, new validation studies will be required, as will completion of additional stability testing with the API that resulted from incorporation of those changes. The material incorporating any changes will need to shown to be equivalent to the original API. Agencies understand product lifecycle management and allow such changes; however, adequate time must be allowed to obtain proper regulatory approval.

In summary, there are many tasks that must be executed for a successful approval of a new drug application. The tasks are logical and follow a scientific progression. Attention to the details and well-executed science will ensure success in the approval process.

References

1. *Drug Substance; Chemistry, Manufacturing, and Controls Information (Draft Guidance)*, US Department of Health and Human Services, Food and Drug Administration, Center for Drug Evaluation and Research, 2004.
2. ICH Guidance Documents. URL http://www.ich.org/cache/compo/2762541.html GMP requirements are specifically covered in Q7A: *Good Manufacturing Practice Guide for Active Pharmaceutical Ingredients*.
3. *Good Laboratory Practice Regulations*, 21 CFR Part 58.
4. Anderson, N. G. *Practical Process Research and Development*; Academic Press: New York, 2000.
5. Winslow, R. *The Birth of a Blockbuster: Lipitor's Route Out of the Lab*; The Wall Street Journal, January 24, 2000.
6. Prashad, M.; Prasad, K.; Repic, O.; Blacklock, T. J.; Prikoszovich, W. *Org. Proc. Res. Dev.* **1999**, *3*, 409–415.
7. *Guidance for Industry Q3C. Tables and List*, US Department of Health and Human Services; Food and Drug Administration; Center for Drug Evaluation and Research, Center for Biologics Evaluation and Research, 2003.
8. *ICH Harmonized Tripartite Guideline: The Common Technical Document for the Registration of Pharmaceuticals for Human Use: Quality, M4Q*, 2002.
9. Datta, S.; Grant, D. J. W. *Nat. Rev. Drug Disc.* **2004**, *3*, 42–57.
10. Fujiwara, M.; Nagy, Z. K.; Chew, J. W.; Braatz, R. D. *J. Proc. Controls* **2005**, *15*, 493–504.
11. Fisher, R. A. *Design of Experiments*, 7th ed.; Oliver and Boyd: Edinburgh, 1960.
12. *DuPont Applied Technology Division* The, *Strategy of Experimentation*, E. I. du Pont de Nemours: Wilmington, DE, 1974.
13. Bismuth, G.; Neumann, S. *Cleaning Validation, A Practical Approach*; Interpharm Press: Englewood, CO, 2000.
14. *ICH Guidance for Industry Q2A: Text on Validation of Analytical Procedures*, 1994 and *ICH Guidance for Industry Q2B: Validation of Analytical Procedures*, 1996.
15. *ICH Guidance for Industry Q1A(R2): Stability Testing of New Drug Substances and Products*, 2003.
16. *Process Safety Management of Highly Hazardous Substances*, PSM, 29 CFR Subpart H (1910.119).

Biography

Stephen A Munk, PhD, is the President and CEO of Ash Stevens Inc. (ASI), an organization serving as an API development and manufacturing contractor operating in the state of Michigan. Dr Munk is a member of its board of directors. He has experience in drug discovery, development, and manufacturing, both as a scientist and as a manager.

Prior to joining ASI in 1997, he worked at Allergan, Inc. in the area of drug discovery, focusing on adrenergic agents. Ash Stevens Inc. received nine FDA approvals to manufacture APIs under his leadership through 2005. These include bortezomib, clofarabine and 5-azacitidine during the period of 2003–04. Dr Munk's work has been summarized in a number of publications and patents covering topics in medicinal chemistry and process development. He received his PhD in organic chemistry from the University of California at Berkeley under Henry Rapoport's guidance and he was an American Cancer Society Postdoctoral Fellow, conducting studies in medicinal chemistry and molecular biology, at Purdue University in Dale Boger's laboratory. Dr Munk also serves as an adjunct Associate Professor at Wayne State University. His activities at Wayne State are in the Departments of Chemistry and Medicinal Chemistry.

Comprehensive Medicinal Chemistry II
ISBN (set): 0-08-044513-6

ISBN (Volume 2) 0-08-044515-2; pp. 159–172

2.06 Drug Development

P Preziosi, Catholic University of the Sacred Heart, Rome, Italy

© 2007 Elsevier Ltd. All Rights Reserved.

2.06.1	**Introduction**	**173**
2.06.2	**Discovery and Development of a New Chemical Entity**	**175**
2.06.2.1	Hits and Leads	175
2.06.2.2	Principles of New Chemical Entity Testing Prior to Human Trials	177
2.06.3	**Clinical Testing**	**180**
2.06.3.1	Models and Techniques Used for Clinical Drug Testing	181
2.06.3.2	Statistical Analysis	184
2.06.3.3	Biomarkers, Real and Surrogate Endpoints	185
2.06.3.4	Evaluating a Protocol for a Clinical Trial of a New Chemical Entity	186
2.06.3.5	The Phases of Clinical Testing	187
2.06.3.5.1	Phase 0	188
2.06.3.5.2	Clinical phase I	188
2.06.3.5.3	Clinical phase II	188
2.06.3.5.4	Clinical phase III	189
2.06.3.5.5	Clinical phase IV	189
2.06.4	**Modern Technologies of Interest for the Process of Drug Development**	**192**
2.06.4.1	Imaging Technology	192
2.06.4.1.1	Computed tomography	192
2.06.4.1.2	Magnetic resonance imaging or nuclear magnetic resonance	192
2.06.4.1.3	γ-Scintigraphy	192
2.06.4.1.4	Positron-emission tomography	193
2.06.4.2	Nuclear Magnetic Resonance Spectroscopy, Mass Spectrometry, and Human Tissue Histology Technologies	194
2.06.4.3	The 'Omic' Technologies in Drug Development	194
2.06.4.3.1	Proteomics	196
2.06.4.3.2	Transcriptomics	196
2.06.4.3.3	Pharmacogenomics	196
2.06.4.3.4	Metabolomics	197
2.06.4.3.5	Metabonomics	197
2.06.4.3.6	Toxicogenomics	197
2.06.4.3.7	Usability and criticism	197
2.06.5	**Time and Cost Requirements for Development of a New Chemical Entity and Factors Contributing to Project Failure**	**199**
2.06.6	**Regulatory Problems**	**200**
	References	**200**

2.06.1 Introduction

Over the last 50 years, many important new drugs and new classes of drugs have been developed, which have undoubtedly contributed to the battle against serious illnesses, prolonging the average life expectancy and improving quality of life. Nonetheless, there are still several areas of medicine in which new pharmaceutical weapons are needed:

- viral infections (e.g., respiratory tract infections, viral hepatitis, acquired immune deficiency syndrome (AIDS), etc.)
- control of neoplastic growth

- degenerative diseases of the central nervous system (CNS) (e.g., Alzheimer's and Parkinson's disease, multiple sclerosis)
- muscular dystrophies
- infections caused by bacteria resistant to antimicrobials, which are increasing in frequency
- malignant vomiting syndrome during pregnancy (drugs without adverse effects on the pregnancy or embryo/fetus)
- nonaddictive pain control
- rheumatoid arthritis
- control of the side effects of glucocorticoid therapy

Over the past decade, the search for new drugs has become a frantic race to identify biological targets, hits, and leads, and to develop them as rapidly as possible.[1–5]

The identification of a new chemical entity (NCE), i.e., a medication containing an active ingredient that has not been previously approved for marketing in any form, as a 'lead' requires important decisions. An initial series of preclinical studies will be necessary to define the lead's fundamental characteristics, confirm its validity, and evaluate the prospect of its further development. If the decision is made to proceed, the molecule will be evaluated in preclinical pharmacotoxicological studies, which are a prerequisite for requesting authorization to proceed with additional studies in humans (claimed investigational exemption for a new drug: IND). Development of the NCE then proceeds through phases I, II, and III, and continues, after the new drug application (NDA) has been approved by the US Food and Drug Administration (FDA), with the postmarketing surveillance (PMS) of phase IV (Figure 1).[6]

The pages that follow will provide a detailed analysis of the R&D process and highlight some of the possibilities offered by new technology for accelerating this process and hopefully reducing its costs.

It must be stressed at this point that R&D of an NCE requires a substantial investment of private capital, with the objective of obtaining realistic returns within an acceptable period of time. According to stock market analysts, a drug that can be considered a 'blockbuster' should generate sales exceeding $1 billion a year,[5] while costs for the realization of an innovative NCE currently range from $800 million to $1 billion (see below). The identification of the target, selection of the lead, and every subsequent step in the development process are thus tied to the producer's strategy – corporate strategy, business strategy, R&D strategy, and portfolio.[5]

There are three basic types of R&D projects, each with its own costs and research commitments:

1. Innovative research: a project that will lead to the development of a drug with greater efficacy (e.g., the statins, which were more effective than earlier lipid-lowering drugs) or greater efficiency (in terms of a more convenient route of administration, less frequent dosing, and/or better tolerability than earlier drugs used to treat the same condition).
2. Innovative improvement (also referred to as 'enlightened opportunism'[7]): these projects produce a drug with higher potency, less frequent dosing, and/or greater selectivity than those currently used for the same purpose (e.g., the selective β_1-blocker, atenolol, versus nonselective $\beta_1\beta_2$-blockers like propranolol).
3. Me-too projects (or 'unenlightened opportunism'[7]): these efforts involve justifiable improvement of a drug, a new NCE that can be patented, but no substantial originality.

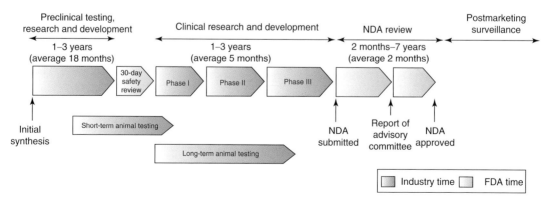

Figure 1 The drug approval process in the US. (Reproduced from Dickson, M.; Gagnon, J. P. *Nat. Rev. Drug Disc.* **2004**, *3*, 417–429 © Nature Publishing Group.)

2.06.2 Discovery and Development of a New Chemical Entity

2.06.2.1 Hits and Leads

Traditionally, drugs were discovered by testing compounds synthesized in time-consuming multistep processes against a battery of in vivo biological screening tests.[8] Today's environment calls for rapid screening, hit identification, and hit-to-lead development.[9] The identification of synthetic or naturally occurring molecules as potential drug leads generally involves automated testing of large collections (libraries) of compounds for activity as inhibitors or activators of a specific biological target, such as a cell receptor or enzyme.[9]

Identification of the target is a fundamental step in the R&D of an NCE. Until a few years ago, the target was considered the 'receptor for the drug molecule,' but the term has now been expanded.[10] A recent analysis of current therapeutic targets indicates that, while almost half of drugs presently on the market are directed toward cell membrane receptors (48%), close to one-third (28%) act on enzymes, and 11% target hormones or hormonal factors; the rest are directed against ion channels (5%), nuclear receptors (2%), nucleic acids (2%), or unknown targets (<7%).[11] Disease-linked proteins are common targets.[5]

The identification of a 'hit' and the rapid transformation of this molecule into a series of high-content 'leads' are key activities in the R&D process[11] (**Figure 2**). According to Bleicher *et al.*,[11] a 'hit' is a primary active compound with nonpromiscuous binding behavior that exceeds a certain threshold value in a given assay. After the hit has been subjected to an identity and purity evaluation, an authentic sample is obtained (or resynthesized), and its activity is confirmed in a multipoint determination. This validated hit is a molecule capable of producing physiologically or pathologically significant modification in a biological substrate. A lead, according to the same authors, is "a prototypical chemical structure, or series of structures, that demonstrates activity and selectivity in a pharmacological or biochemically relevant screen. This forms the basis for a focused medicinal chemistry effort for lead optimization and development with the goal of identifying a clinical candidate. A distinct lead series has a unique core structure and the ability to be patented separately."

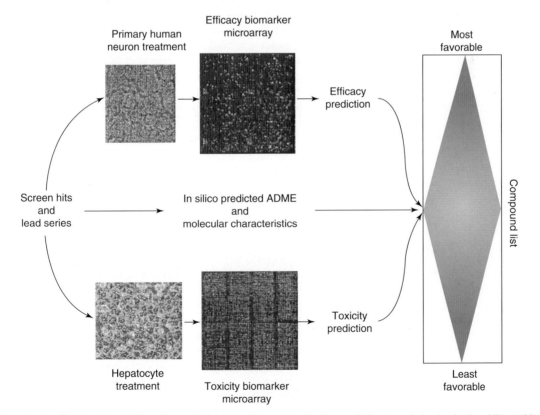

Figure 2 Application of predictive efficacy, toxicogenomics, and molecular prediction to early lead selection. Hits and lead series molecules' characteristics might be evaluated using in vitro assays of efficacy and toxicity biomarker expression, combined with in silico predictions of ADME and drug-like qualities, to prioritize compounds for further development. (Reproduced from Lewin, D. A.; Weiner, M. P. *Drug Disc. Today* **2004**, *9*, 976–983, with permission from Elsevier.)

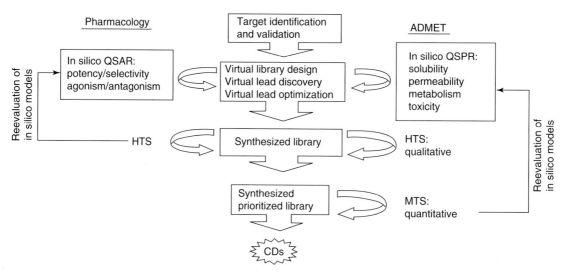

Figure 3 Knowledge-based, parallel drug discovery setting. Computational prioritization of a virtual library is followed by chemical synthesis. In silico and in vitro pharmacological and ADMET screening are performed simultaneously. ADMET, absorption, distribution, metabolism, excretion, and toxicology; CDs, candidate drugs; MTS, medium throughput screening; QSPR, quantitative structure–permeability relationships. (Reproduced with permission from Bergström, C. A. S. *Basic Clin. Pharmacol. Toxicol.* **2005**, *96*, 157 © Blackwell Publishing.)

The hit identification strategies most commonly used today range from knowledge-based approaches, which utilize literature- and patent-derived molecular entities, endogenous ligands, or biostructural information, to methods based on 'pure serendipity' or 'brute force,' such as combinatorial chemistry and high-throughput screening (HTS).[11] The latter method involves robotized analysis of enormous libraries of natural and synthetic compounds with the aim of identifying substances potentially capable of interacting with, e.g., binding, activating, or inhibiting, specific biological targets, such as surface receptors or enzymes.[9,13] Combinatorial chemistry allows the synthesis of large series of closely related chemicals using the same chemical reaction and appropriate reagents. These series, or libraries, can then be subjected to HTS, based largely on assays of binding to target biomolecules, to identify hits, around which more highly focused series are designed and synthesized in a subsequent round.[13] The amalgamation of both extremes is expected to deliver more high-content chemical leads in a shorter period of time[11,13] (**Figure 3**).

Once a target has been identified, its 'druggability' must be assessed, that is, its potential to be significantly modified by a small ligand molecule with physicochemical and absorption, distribution, metabolism, and excretion (ADME) characteristics that will allow it to be developed into a candidate drug (i.e., a lead) with the desired therapeutic effects.[11]

As a rule, the molecular and structural properties of the lead are subjected to relatively minor modifications to produce the launched candidate drug.[11] In this process, knowledge of the pharmacophore is also important. Of the various definitions provided for this term,[9,13,15] the most complete seems to be that furnished by Yamashida and Hashida:[13] "the steric and electronic features that are necessary to ensure optimal interactions with a specific biological target structure and to trigger (or block) its biological response."

The protein nature of the target must also be considered. Most targets belong to one of the principal classes of enzymes and proteins noted above, i.e., cell membrane and nuclear receptors, protein kinases, proteases, and protein phosphatases, although structural homologs of these proteins (with > 30% sequence identity) can also be considered as targets.[16] Knowledge of the three-dimensional structure of the target protein or its homolog (especially membrane proteins, which are the most common drug targets) is essential to understand the drug's mechanisms of action and also for rational drug design. It can be determined by means of x-ray crystallography, NMR spectroscopy, or cryoelectron microscopy.[10] In this context, it is important to consider the process of computationally placing a virtual molecular structure into a binding site located on a biological macromolecule (docking) or rigidly relaxing the respective structure and then ranking (scoring) the complementarity of the fit.[11] The process begins with the application of docking algorithms that pose small molecules in the active site. The calculations do not succeed if they do not differentiate correct poses from incorrect ones and if 'true' ligands cannot be identified. The pose score is often a rough measure of the fit of a ligand into this site; the rank score is generally more complex and may serve as a substitute for binding assays.

The in silico approach for predicting the metabolism of an NCE includes quantitative structure–activity relationship (QSAR) and three-dimensional QSAR studies, protein and pharmacophore models, and predictive database.[8,16] There have recently been a number of QSAR studies focusing on drugs' ability to cross Caco-2, Madin–Darby canine kidney (MDKC) cells, parallel artificial membrane permeability assay (PAMPA), which are widely used as a model of intestinal absorption.[13,17] Filters can be installed on computers to allow ADME analysis of computationally designed drug-like molecules before they are actually synthesized. In this manner, synthesis can be reserved for those that have been predicted in silico to have acceptable potency and developability.[17] For this type of ADME analysis, there is a need for more rapid high-throughput (HT) technology. Those that are currently available do not come within the two orders of magnitude routinely achieved for these analyses in the industry, e.g., in ligand receptor assays.[18] In addition, none of the in vitro systems for ADME analysis has been validated thus far, although some validation is included in the long-term plans of the European Centre for Validation of Alternative Medicines (ECVAM).[18] More information on ADME can be found in Ekins et al.[19] and in the fifth volume of Comprehensive Medicinal Chemistry II.

Gene expression microarray analysis can be valuable as the basis for cell-based HTS to identify compounds that alter gene expression patterns in a particular way.[20] Gene expression profiling in in vitro cell systems, animal models, and clinical settings can all lead to the identification of additional targets that feed back into earlier stages of the drug discovery process.[20]

Before embarking on the development of a lead, the manufacturer must be reasonably certain that the molecule can be synthesized or otherwise produced (for biological products), in accordance with good manufacturing practices (GMP) at a reasonable cost and that it can be formulated in an appropriate manner for administration to animals and ultimately humans. The drug's stability and measurability in biological matrices must also be ascertained in a preliminary manner.

2.06.2.2 Principles of New Chemical Entity Testing Prior to Human Trials

As the Nobel Prize winner, Sir James Black,[21] has pointed out, the *non*selectivity of an NCE may not be detectable until it reaches the development stage involving testing in intact animals. Too much combinatorial chemistry might well come to be seen as a risk factor to the corporate health. The process of preclinical drug discovery and development passes through a number of phases, ranking from the earliest activities when a therapeutic target is identified, through to animal studies preparing for the handover to clinical development and the first dose in humans.[22]

According to Greaves et al.,[23] preclinical testing of a lead includes the following phases:

- primary pharmacology
- secondary pharmacology
- toxicology
- kinetics and toxicokinetics

Primary pharmacological studies involve a variety of experimental methods and focus on various organs and systems[24] depending on the type of lead and its clinical orientation. Depending on the specific characteristics of the NCE, they should include a phase of investigation at the cellular and molecular levels of its in vitro receptor binding (to ascertain its affinity and selectivity); its effects on enzymes that play key roles in the function of the neurons of the autonomic, peripheral, and central nervous systems (e.g., acetylcholinesterase, tyrosine hydroxylase, dopamine-3′ hydroxylase, monoaminoxidase); its effects on the hepatic cytochrome P450 system (complete with in silico ADME modeling, when possible); assays of receptor agonism or antagonism in cultured cells; effects on neoplastic cells, microbial or fungal strains, isolated tissues, blood vessels, heart, lungs, ileum of rats or guinea pigs, diverse types of tumor cells, and antimicrobial effects. In vitro approaches can be used to study the concentration of free active drug at the level of the receptor. Studies have shown that a concentration capable of producing a 75% rate of receptor occupation is predictive of therapeutic concentrations for many G protein-coupled receptors,[28] although the specific rate required varies depending on the target disease (see later discussion of the in vitro therapeutic coefficient).

These activities must be supplemented by in vivo studies of the drug's activity in animals with experimentally induced disease, e.g., models of inflammation, neoplastic disease, infections, diabetes, atherosclerosis, depression, and dementia, and, when possible, in animals with natural or spontaneous pathology or those subjected to gene knockout, knockdown, or knockin procedures.

Secondary pharmacological studies include assessments of safety, which should focus on the cardiovascular, respiratory, immune, and central and peripheral nervous systems. The animal species most suitable for these types of studies are listed in **Table 1**.

Table 1 Animal species suitable for evaluating the various pharmacotoxicological aspects of an NCE

Organ or system affected	Species used for testing
Central nervous system	Dogs and monkeys (better than rats, especially at high doses)[a]
Sensory organs	Rats, dogs, monkeys
Cardiovascular system	Dog (conscious or anesthetized), minipig
Gastrointestinal system	Rodents, dogs[b] better than monkeys (e.g., monkeys are more resistant to vomiting)
Kidneys	Dogs and rodents (concordance is weak)
Endocrine system and metabolism	Rodents, dogs
Bone marrow	All species: concordance is good as long as the dose is calculated in terms of mg/m^2 of body surface
Coagulation	Rabbit
Skin	Pig, rabbit
Site of administration	Rabbit, pig (serum creatine phosphokinase activity, histopathology)

Adapted from various authors.[22,24–26]

[a] Mice and rats are preferable for evaluating certain CNS effects (degree of sedation, muscle relaxation, motor activity, which is correlated with dizziness in humans, stimulation).
[b] The gastrointestinal tract of the dog is physiologically very similar to that of humans in terms of motility patterns, gastric emptying, and pH, particularly in the fasting state.[13]

Toxicology studies must be performed according to good laboratory practice (GLP)[25] and evaluate the effects of single or repeated doses of the drug on the general condition of the animal (it is important to recall that, despite careful checks, 3–5% of the animals are unhealthy at the beginning of any study[27]), on blood chemistry parameters, blood cell morphology, and autopsy findings. These studies will be followed by in vitro tests that investigate genotoxic endpoints,[26] in vivo mutagenic potential,[24] effects on reproduction, and carcinogenesis, as summarized in **Table 2**.[24]

These toxicological studies identify the no-effect dose and the maximum tolerated dose for single-dose administration and for repeated treatment, and the therapeutic coefficient. With reference to the duration of treatment shown in **Table 2**, a month should be sufficient to ascertain most toxic phenomena.[23]

An interesting in vitro approach to evaluation of the therapeutic coefficient involves simultaneous assessment, based on measurement of appropriate parameters, of the concentration of the NCE that is sufficient to produce a therapeutic effect (e.g., 75% occupation of the target receptor), the concentration of free drug in the plasma, and the safety threshold for the parameter being considered. For instance, in studies of terfenadine, terolidine, and cisapride, close correlations have emerged between the free drug concentration in plasma, lengthening of the QT segment in dogs and humans (a very important parameter of cardiac toxicity), and the concentrations associated with in vitro blockade of K^+ human Ether-a-gogo Related Gene (hERG) channels[28] (**Figure 4**). The dependence of QT prolongation on free plasma concentrations lent support to the application of a 30-fold safety multiple between therapeutic activity and concentrations causing QT prolongation.[28]

Drugs known to affect the QT interval in humans have also been shown to affect heart rate in zebrafish embryos.[29,30] Angiogenesis inhibitors induce specific vascular defects in the patterning of embryonic vasculature.[29,30] There are marked species differences in the response to peroxisome proliferators: mice and rats are highly responsive, while humans respond poorly or not at all.[31]

Studies of kinetics and pharmacokinetics should verify the presence and the destiny of the drug in organic liquids (maximum plasma concentration attained (C_{max}), half-life of drug ($t_{1/2}$), area under the plasma concentration–time curve (AUC), apparent volume of distribution (V_d), clearance) of rats, dogs, and monkeys at different times after single and multiple administrations.[23]

For small molecules intended for oral administration, several key properties are also required (Lipinski's rule of five): molecular mass <500 Da; fewer than five donor hydrogen bonds; fewer than 10 acceptor hydrogen bonds; and an octanol–water partition coefficient (an indicator of the NCE's ability to cross biological membranes) less than 5.[9]

All in all, even if studies in the various models are carried out simultaneously, the testing discussed above will require 2–4 years, after which the producer will be able to present an adequately documented application for an IND. For this reason, it is common practice to carry out preliminary explorative pharmacotoxicology studies without strict

Table 2 Safety tests

Type of test	Approach	Comment
Acute toxicity	Acute dose that is lethal in approximately 50% of animals and the maximum tolerated dose. Usually two species, two routes, single dose.	Compare with therapeutic dose.
Subacute toxicity	Three doses, two species. 4 weeks to 3 months may be necessary prior to clinical trial. The longer the duration of expected clinical use, the longer the subacute test.	Clinical chemistry, physiologic signs, autopsy studies, hematology, histology, electron microscopy studies. Identify target organs of toxicity.
Chronic toxicity	Rodent and nonrodent species. 6 months or longer. Required when drug is intended to be used in humans for prolonged periods. Usually run concurrently with clinical trial.	Goals of subacute and chronic tests are to show which organs are susceptible to drug toxicity. Tests as noted above for subacute. 3 dose levels plus controls.
Effect on reproductive performance	Effects on animal mating behavior, reproduction, parturition, progeny, birth defects, postnatal development.	Examines fertility, teratology, perinatal and postnatal effects, lactation.
Carcinogenic potential	Two years, two species. Required when drug is intended to be used in humans for prolonged periods	Hematology, histology, autopsy studies. Tests in transgenic mice for shorter periods may be permitted as one species.
Mutagenic potential	Effects on genetic stability and mutations in bacteria (Ames test) or mammalian cells in culture; dominant lethal test and clastogenicity in mice.	Increasing interest in this potential problem.
Investigative toxicology	Determines sequence and mechanism of toxic action. Discover the genes, proteins, pathways involved. Develop new methods for assessing toxicity.	May allow rational and earlier design and identification of safer drugs. Possibly run at higher compound throughout.

Reproduced with permission from Berkowitz, B. A. Basic and Clinical Evaluation of New Drugs, In *Basic and Clinical Pharmacology*, 9th ed.; Katzung, B. G.; Ed.; Lange Medical/McGraw-Hill: New York, 2004, pp 67–74 © The McGraw-Hill Companies, Inc.

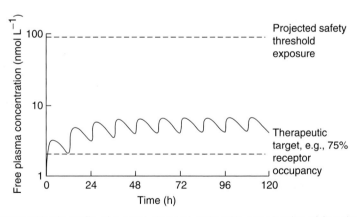

Figure 4 Modeled pharmacokinetic profile of a novel chemical entity with consideration of free drug exposure relative to projected efficacy levels (based on, e.g., receptor occupancy) and safety threshold (based on, e.g., level at which QT prolongation is anticipated). (Reproduced with permission from Walker, D. K. *Br. J. Clin. Pharmacol.* **2004**, *58*, 601 © Blackwell Publishing.)

observance of the GLPs. These can provide information on the minimum effective and maximum tolerated doses for single- and multiple-dose administration (from 14 days to 1 month, via two routes of administration and in two species, preferably the rat and the dog), the in vitro mutagenic potential of the drug in bacterial and mammalian cells,[23] and fundamental ADME properties. A final focus, especially for larger-molecule drugs, is the molecule's possible effects on the immune system. New technology (such as surface plasma resonance and cell-based bioassays) is available to determine whether antibodies have formed or if immune system biomarkers have changed during toxicology testing.[25]

Conventional approaches of experimental pharmacology, together with repeat-dose toxicity studies of up to 1 month's duration, can predict adverse events in first-dose-to-human studies with a reasonable level of accuracy, i.e., more than

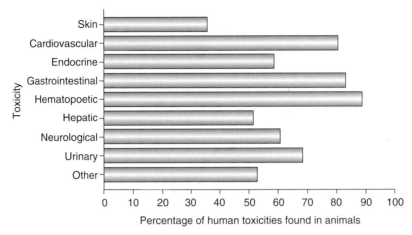

Figure 5 Percentage concordance between animal and human toxicities, grouped by organ. Similar to data on anticancer drugs, correlation is better for toxicities in the gastrointestinal tract, and hematopoetic and cardiovascular systems. (Reproduced with permission from Greaves, P.; Williams, A.; Eve, M. *Nat. Rev. Drug Disc.* **2004**, *3*, 226 © Nature Publishing Group.)

90% of the toxic effects that can be detected in animal models.[23] When there is no prior knowledge of organ toxicity, it is advisable to evaluate the effects of repeat administration of a high dose of the drug for at least a few days, a period that is generally sufficient to permit the development, expression, and identification of organ pathology.[22] **Figure 5** shows the rates of concordance between toxic effects produced by drugs in animals and humans.[23]

The toxicological tests outlined above cannot exclude the risk of milder toxic effects, such as nausea, headache, distortion of the sense of taste, vertigo, or dry mouth.[23,32,33] They are also no guarantee against idiosyncratic reactions. Unlike the effects that can be predicted based on the results of toxicological testing, idiosyncratic reactions are infrequent adverse responses to a drug characterized by the absence of dose dependence and by variable periods of latency following drug administration despite extensive and essentially negative preclinical testing and large clinical trials (troglitazone,[23] trovafloxacin,[31] bromfenac,[23] fialuridine[24]) (*see* Section 2.06.4.3.6). They can nonetheless be quite serious and in some cases fatal. They are often manifested by severe hepatotoxicity, which some investigators attribute (e.g. fialuridine) to mitochondrial toxicity.[23,31] In light of the need for first-dose-to-human studies, integrated primary and secondary pharmacology data and repeat-dose toxicity information across the experimental models used are required to provide meaningful data to justify the experiment, to manage diverse events, and to avoid serious organ toxicity.[23]

The first dose to humans may be one-fifth to one-tenth of the dose per kilogram that has proved to be effective in 50% of the animals in specific models of the target disease or condition and at least 10–20 times lower than the dose that produces no effect in dogs and one other animal species when administered repeatedly for at least 1 month.

2.06.3 Clinical Testing

A clinical trial is essentially a prospective controlled trial that includes human subjects and is designed to evaluate the efficacy of a therapeutic intervention, diagnostic procedure, or medical/surgical device.[34] All clinical trials must be conducted according to the ethical principles contained in the Helsinki Declaration, which is periodically updated. (See versions drafted in Tokyo, Venice, Hong Kong, South Africa, Edinburgh, and a note of clarification added in Washington, 2002.)[35] The study protocol is evaluated by the institutional review boards (IRB) and ethical committees, whose establishment and duties are regulated by the Council of International Organizations of Medical Sciences (CIOMS)/World Health Organization/WHO)[36] and, in the US, by the FDA.[37] The investigator must also ensure that the trial will adhere to the basic principles of GLP outlined in the code cited above.

Evaluation of the NCE's efficacy and safety in healthy and diseased humans currently involves phase I trials, which are normally conducted on healthy human subjects, and those of phases II, III, and IV, which are conducted on patients affected by the disease that the NCE is expected to influence.[24,38–41] Specific experimental methods and techniques have been developed to reduce the risk of problems related to the following factors: (1) the clinical history of the disease; (2) the presence of other risk factors; and (3) placebo and nocebo effects, i.e., positive and negative effects experienced by human subjects that are due exclusively to the administration of a substance identified as a drug. These effects, which can occur in a substantial percentage of trial participants, are manifested by subjectively reported improvement in a symptom (placebo effects) or adverse effects such as nausea, abdominal pain, and headache (nocebo effects) with no objectively detectable correlates.

The clinical history of the disease should allow one to determine: (1) whether or not its course is sufficiently stable; (2) its type is sufficiently uniform over a period of at least several months; (3) the absence of sudden improvement or worsening; and (4) the nature and (if possible) the timeframe of its outcome.

Potential risk factors include the presence of other diseases, particularly those affecting the liver, lungs, or thyroid, and lifestyle factors, such as smoking, alcoholism, or drug use. The patients treated with the NCE and those receiving the control treatment must be similar in terms of diet and other types of therapy. The agent used to treat the control group can be:

1. the currently available drug considered most effective for treatment of the target disease, e.g., for evaluation of a new antiarrhythmic drug for the control of atrial fibrillation, the control group might be treated with amiodarone, flecainide, propafenone, or quinidine, as the experimenter sees fit. As a rule, the control treatment and the one being evaluated in the trial are administered at fixed doses for the duration of the study
2. a placebo, i.e., an inert compound with no pharmacological action, which has been formulated in such a way that it appears identical to the test drug. In some cases, placebos are designed to produce some of the same adverse effects predicted for the test drug (e.g., dry mouth) since the absence of these effects in a double-blind study could reveal to the experimenter the group origin of control subjects.

The inclusion of placebo treatment in clinical trials is designed to eliminate the risk of biases, on the part of the subject or the researcher or both, in favor of the NCE under conditions of variable disease history. Responses to a placebo are based on the effects of suggestion (administration of any form of therapy is the best way to lead the patient to believe he or she is improving) and those of spontaneous improvement. Many patients seek treatment when they are experiencing symptoms, and many symptoms, such as pain, anxiety, depression, insomnia, diarrhea, and constipation, can regress or subside spontaneously if they are not being caused by a disease that is progressive. Therefore, in any trial designed to assess the effect of a treatment (pharmacological or nonpharmacological) on a symptom, it is appropriate, whenever possible (see Section 2.06.3.1), to include a group of patients who will be treated under double-blind conditions with a placebo.

2.06.3.1 Models and Techniques Used for Clinical Drug Testing

In clinical trials, especially those of phase II, III, or IV, the problems outlined above can be avoided to some extent by:

1. inter- or intraindividual cross-over study designs[24,46]
2. use of uncontrolled or parallel multiple-arm study design
3. random assignment of trial enrollees to the NCE and control treatment arms[43–45]

Trials conducted according to an interindividual cross-over design begin with a 'run-in' period, in which the patient is allowed to stabilize and vital parameters are evaluated. Thereafter, each randomized group of patients is divided into arms, which are treated with the NCE or the control agent for a predetermined period during which efficacy parameters are periodically assessed. This interval is followed by a second 'washout period,' during which all treatment is suspended, and the treatment arms are then reversed: the arm treated with the NCE now receives the control treatment, and vice versa. This second treatment period is identical to the first in terms of duration and patient evaluation. All treatments are administered under blind or double-blind conditions. (In the former design, the patients, but not the investigators, are unaware of which drug they are receiving; in the latter, the investigators are also blinded.) At the end of the study, all data collected in the two cross-over groups will be divided into two pools: (1) those recorded during treatment with the NCE; and (2) those recorded during the control treatment.

In trials conducted according to an intraindividual cross-over design, in which three treatments are being compared – the NCE (A), the standard treatment (B), and a placebo (C) – the randomized patients are divided into three groups. Patients in group I are treated with drug B, then with drug C, and finally with drug A. In the other two groups, the order of the treatments is different (group II: C–A–B; group III: A–B–C). Each treatment is administered for an identical period of time (e.g., 1 week), during which prescribed evaluations are carried out, and the three treatment periods are separated by a week-long washout period. At the end of the trial, data from the three groups are pooled, and values recorded during each of the three treatments are compared and subjected to statistical analysis.[24]

The methods described above are useful for testing NCEs designed to treat diseases characterized by slow evolution or long periods of stability (e.g., stable angina, type 2 diabetes, moderate hypertension, asthma without crises, rheumatoid arthritis, osteoarthritis).

For drugs that will be used to treat short-lasting conditions (e.g., infectious disease, acute cystitis, seasonal rhinitis), the trial must be conducted in an uncontrolled manner (blind or double-blind) with no form of cross-over. In uncontrolled trials,[41] all eligible patients are treated with the NCE, and its effects are evaluated against the normal course of the disease, which is considered to be well known, and if possible based on the evolution of suitable biomarkers (see below). These trials are unsuitable for drugs used to treat diseases with variable outcomes (e.g., mortality in community-acquired pneumonia,[47] AIDS,[48] etc.). They should only be used when: (1) the disease is very likely to be fatal (e.g., coagulopathy caused by Ebola virus infection); (2) the efficacy of the NCE is dramatically evident during the initial observations; (3) the adverse effects are acceptable in light of the expected benefit; (4) there are no alternative treatments for use as a control; and (5) there are plausible biological and pharmacological bases for the NCE's efficacy.

The use of nonrandomized controlled trials with parallel groups is also to be discouraged. In these studies, the NCE and the control treatment are administered to two groups of eligible patients, and their responses are comparatively analyzed. However, treatment assignment is not random, but based on other criteria (e.g., assignment to one arm on odd or even days), which are predictable and thus subject to influence by the investigator's preferences. Consequently, the two treatment groups may not be truly comparable.

In another type of nonrandomized controlled trial, all enrolled patients are given the test drug, and the results are compared with those documented in a historical control group consisting of patients treated in the past with a traditional form of therapy. The control data in these studies are collected retrospectively, usually from hospital charts. One of the major drawbacks of this trial design is that the group treated with the NCE has certain advantages over the historical control group in terms of more modern (and presumably more effective) forms of diagnosis, assistance, and therapy (e.g., computed tomography (CT), magnetic resonance imaging (MRI) or γ-stratigraphy for evaluation of tumor dimensions versus conventional imaging for the historical controls).

In randomized clinical trials (RCTs), the NCE and control drug (or placebo) are assigned to eligible patients in a random manner, and the groups are treated according to a parallel design (or, less commonly, another type of design). Randomization enhances the comparability of the groups.[44] Each subject is assigned a numerical code; this measure allows researchers to identify a single subject (for any type of problem related to that individual) without interrupting the entire trial. The trial is conducted under double-blind (or, less commonly, single-blind) conditions, with a cross-over design when possible, depending on the characteristics of the disease being treated. The indices and parameters to be evaluated are identified before the study begins, as are the times for their evaluation. The data collected in the groups are subjected to comparative statistical analysis (described in greater detail below).

The discussion above refers to trials with two or three arms, but there may be more than three. This approach is used, for example, when multiple doses of the NCE are being assessed or when the NCE's effects are being compared with those of two or more standard treatments. In these cases, extreme caution is essential to avoid treatment mix-ups, particularly in double-blind studies in which the drugs are identified exclusively by a numerical code. Placebo treatment of the control group can only be justified when all of the following are true: (1) there are currently no effective standard means for treating or preventing the disease being studied; (2) the treated and control groups will be exposed to the same risk; and (3) a controlled trial is absolutely necessary to determine whether or not an NCE is actually effective.

In phase II and III, an alternative method for evaluating the effects of an NCE is the sequential trial approach.[38,49,50] In the conventional trials used for pharmacological testing during phases II and III, the number of subjects to be treated in each arm – 10, 20, 30, 100, or more – is established before the study begins, whereas in nonrandomized controlled studies, the analysis extends to all eligible subjects enrolled within predetermined intervals of time (for example, on odd or even days of the month or every other month).

In sequential trials, the comparison is made in matched pairs of subjects with disease of the same severity, one treated with the NCE, treatment A, the other with conventional therapy, treatment B. An alternative approach involves consecutive administration of the two treatments in the same subject, who thus serves as his/her own control. (In this case, as in cross-over studies, the patient's disease must be fairly stable.) Based on a predetermined criterion which is evaluated in terms of pre-established endpoints (as hard as possible), a 'preference' is assigned to one of the two treatments, i.e., it is rated as better than the other treatment. Pairs in which no preference can be assigned are excluded from the analysis. Each preference is plotted on a graph, like the one shown in **Figure 6**, producing a line that extends diagonally upward (if treatment A is truly better) or downward (if the opposite is true). When the number of preferences assigned to one of the treatments exceeds the pre-established statistical limits in the graph, that treatment

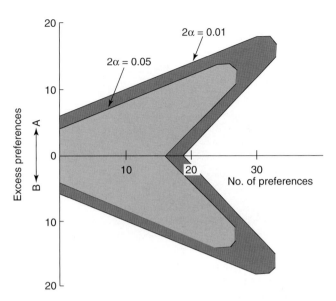

Figure 6 Boundary diagram for a sequential trial. Here $1 - \beta = 0.95$, but $\theta_1 = 0.85$. Two boundary diagrams are shown, one for $2\alpha = 0.05$, the other for $2\alpha = 0.01$. Values of α and β at the boundaries are approximate. See text for explanation. (Reproduced with permission from Goldstein, A.; Aronow, L.; Kalman, S. M. *Principles of Drug Action. The Basis of Pharmacology*, 2nd ed.; John Wiley: Chichester, 1974, p 793.)

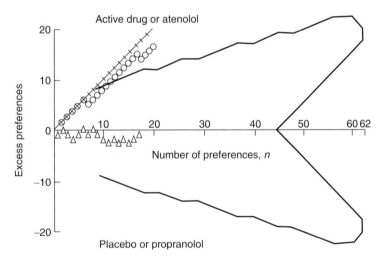

Figure 7 Sequential analysis of a trial of the effects of atenolol (x) and propranolol (O) versus placebo in hyperthyroid patients using an objective assessment of effects, including heart rate. Both drugs are significantly better than placebo ($P = 0,05$) but a comparison of atenolol and propanolol (Δ) fails to distinguish between them in efficacy. (Reproduced with permission from Rogers, H. J.; Spector, R. J.; Troance, J. R. *A Textbook of Clinical Pharmacology*; Hodder & Strughton: London, 1981, p 169 by permission of Hodder & Stoughton Ltd.)

is declared to be significantly superior to the other (**Figures 7** and **8**). If this does not occur, the null hypothesis is declared to be true: the two treatments are not significantly different.

Sequential trials offer certain advantages. They can generally be conducted with fewer patients than conventional trials, and they are considered to be more ethical for studies involving severe or potentially lethal diseases because conclusions regarding the drug's efficacy can be reached more rapidly. Moreover, they are less costly, and they allow organizers to identify a single well-defined time point for the end of the study. On the other hand, it can be difficult to find sufficiently well-matched pairs of subjects, individuals who, during the same period of time, are suffering from identical forms of the same disease, and this can lead to errors of the type associated with nonrandomized controlled trials.

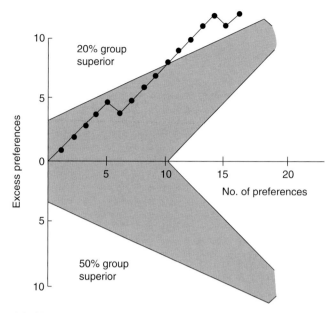

Figure 8 Result of sequential trial: myocardial infarction or death in two treatment groups. Patients with angina pectoris or a previous myocardial infarction were maintained on the anticoagulant phenindione at two dosage levels. The higher dosage reduced the prothrombin levels to 20% of normal, on the average; at the lower dosage, prothrombin levels were 50% of normal. The infarction rate in the 20% group was significantly lower than in the 50% group ($P < 0.01$). If a patient developed a myocardial infarction or had a cardiovascular death, a preference was given to the group to which the other patient in the pair belonged. Here $2\alpha = 0.05$, $1 - \beta = 0.95$, $\theta = 0.90$.

2.06.3.2 Statistical Analysis

In clinical trials designed to establish the efficacy or identify the adverse effects of an NCE, selection of the statistical methods that will be used to evaluate the significance of differences must be based on the type of scale used for recording the results. An interval scale is a simple numerical scale. Blood pressure values, for example, might be expressed in millimeters of mercury, heart rate as beats per minute, body weight in grams or kilograms, or body temperature in Celsius or Fahrenheit degrees. When a nominal scale is used, items are assigned to qualitative groups or categories. Pain or anxiety, for example, may be classified as severe, moderate, minimal, or absent. No quantitative information is conveyed and no ordering of the items is implied. Nominal scales are particularly useful for trials involving neuropsychiatric disease. The third approach is ordinal. In this case, the responses to different drugs are originally recorded using an interval or nominal scale. The results are then compared and re-elaborated in an ordered manner, e.g., in producing a given therapeutic effect, drug A is superior to drug B, which in turn is superior to drug C.

Results measured on an interval scale can be compared after the means and standard errors have been defined for the values recorded in each group, and the significance of differences that emerge can be assessed with the Student t-test. When there are multiple variables that can influence the response to the drug, analysis of variance can be used. Confidence limits must also be calculated.

The statistical significance of results measured on a nominal scale can be evaluated with the chi square test (χ^2), while those measured on an ordinal scale should be assessed with a nonparametric test, e.g., the Wilcoxon or other ranking methods. Multivariate techniques of analysis of variance, covariance, and dispersion are particularly useful because they minimize differences related to other factors, e.g., interindividual variability, time-related differences, highlighting in this manner the differences actually related to the treatments. In most cases, however, these types of analysis require the assistance of a biostatistician.

In trials designed to compare the efficacies of two drugs, e.g., an NCE and the standard treatment or placebo, the analysis of statistical significance will indicate the frequency with which a given difference will occur purely by chance. In other words, if the null hypothesis (i.e., the absence of difference between the drugs) is indeed true, and if the experiment is repeated 100 times, how many times will a difference of this magnitude be observed? 1/100 (0.01), 5/100 (0.05), 10/100 (0.1). The level of statistical significance P (the lower the frequency, the greater the significance of the difference) has been set at less than 0.05 ($P<0.05$). For tests such as controlled clinical trials (including those that are randomized), the significance may be influenced by numerous variables. In general, a low level of significance, e.g., $P<0.05$, is acceptable when the trial involves an NCE with a high safety index that is active against diseases associated with high mortality and morbidity rates. In contrast, higher levels, e.g., $P<0.01–0.001$, are required when the NCE has substantial adverse effects and is being used for treatment of diseases that are relatively benign. According to a glossary of terms used in therapeutics and diagnosis that is useful for statistical analyses[51] the term relative risk reduction (RRR) refers to "the proportional reduction in rates of bad events between experimental (experimental event rate, EER) and control (control event rate, CER) patients in a trial, calculated as (EER − CER)/CER and accompanied by a 95% confidence interval (CI). In contrast, the RRI (relative risk increase) refers to an increased rate of bad events in the experimental drug group, compared with controls. It is calculated in the same manner used for the RRI. The RRI is also used for assessing the effect of risk factors for a disease." The CI is an index of the uncertainty surrounding a measurement. It is usually reported as the 95% CI, i.e., the range of values within which we can be 95% sure that the true value for the whole population lies.

Intention-to-treat and interim analyses are important parts of drug trials.[40,49] In the former, the groups are analyzed based on the manner in which the patients were randomized rather than on the treatment they received during the study. In an intention-to-treat analysis, the results are included for all eligible patients, regardless of their adherence to the protocol. Those who withdraw before treatment is begun are not considered. The rationale for this type of analysis is that, if a treatment is effective in 80% of the patients who tolerate it, but it has to be suspended in 70% of the cases treated due to the development of serious side effects, the treatment can be said to be effective in only 24% of those treated, i.e., 80% × 30% of those for whom the treatment was prescribed.

Interim analyses are those that are conducted to determine whether or not a trial should be ended before its planned completion date. From a methodological point of view, this approach is incorrect: the data should only be analyzed at the end of the trial because the sample required to satisfy the aims of the study with the chosen study design has already been established in the protocol. Nevertheless, interim analyses may be justified by ethical, scientific, practical, or economic considerations. They might be indicated, for instance, in a long-term study based on mortality or other strong endpoints, e.g., a trial designed to test the effects of an angiotensin-converting enzyme inhibitor or beta-blocker on survival in patients with heart failure. If the interim analysis demonstrates that the test drug is clearly more or less effective than the control, it would be unethical to continue the trial, administering a treatment known to be inferior or ineffective to patients with disease of this severity.

2.06.3.3 Biomarkers, Real and Surrogate Endpoints

Evaluation of biomarkers, surrogate endpoints, and clinical endpoints is very important in clinical trials. A biomarker is, in simple terms, any observable physiological parameter. A valid biomarker is one that is measured in an analytical test system within a well-established scientific framework or body of evidence that elucidates the physiologic, toxicologic, pharmacological, or clinical significance of the results[12]; it is evaluated as an indicator of normal biological processes, pathogenic processes, or pharmacological responses to a therapeutic intervention.[52] Use of a reliable and specific biomarker as an early predictor of long-term efficacy and toxicity can reduce the time, costs, and size requirements for clinical trials. Its specificity is an essential feature, and insufficient specificity can be due to several factors. The failure of a biomarker can be due to any of the following[53]: (1) the marker is not in the pathophysiological pathway of the disease (i.e., several pathways are present); (2) the biomarker is also in other pathophysiological pathways (i.e., several such markers are present); (3) the biomarker is not in the pathway affected by the therapeutic intervention; (4) the intervention acts through different and unknown pathways; and (5) the biomarker does not correlate well with traditional clinical endpoints, being more sensitive than the latter. In a metanalysis of patients with colorectal carcinomas in which tumor mass was used as a biomarker, correlation between the response of the tumor (reduction in mass) and the clinical endpoint, survival, was only 38%. In this case (number 2 in the list above), other factors influence the correlation between tumor response and survival.[53]

Surrogate endpoints are biomarkers used as substitutes for clinical endpoints.[52,54–56] Clinical investigators use epidemiologic, therapeutic, pathophysiologic, and other scientific evidence to select a surrogate endpoint that can be expected to predict clinical benefit, harm, or lack of benefit.[52] As defined by Temple,[56] a surrogate endpoint in a clinical trial is "a laboratory measurement or a physical sign used as substitute for a clinically meaningful endpoint that measures directly how a patient feels, functions, or survives. Changes induced by a therapy in a surrogate endpoint are expected to reflect changes in a clinically meaningful endpoint." Surrogate markers must be well validated before they can be used to provide documented evidence of the activity of an NCE for submission to regulatory authorities.[53]

In effect, a surrogate is a test or set of tests that indicates the progression or regression of a specific human pathology. Surrogate markers functioning as endpoints can be classified as follows:[13]

- type 0 – those associated with the natural history of the disease
- type I – those that indicate a known response to therapeutic intervention
- type II – those that reflect a clinical outcome

Surrogate endpoints are used to obtain acceptably reliable evidence that the NCE is actually influencing the target disease, and to do so more rapidly and more economically than would be possible with the real endpoint. Surrogates are obviously not necessary if the true clinical endpoint is rapidly attained and easy and economical to measure.

Toxic changes affecting soft tissues may be evaluated by means of correlating changes in gene expression in more readily accessible cells, e.g., peripheral blood mononuclear cells (PBMCs), buccal mucosal epithelial cells,[31] or accessible tumor cells,[57] which can be used as surrogate biomarkers. Numerous groups have used bleomycin sensitivity of peripheral blood lymphocytes as a surrogate for identifying individuals with increased sensitivity to environmental mutagens. Those with increased sensitivity to bleomycin have an increased risk of developing lung cancer.[57]

A recent review showed that surrogate endpoints, in particular tumor size as measured by MRI, were used in well over half of all oncology trials (53 of 71 trials).[53] Their use has recently been criticized as an unacceptable means for accelerating approval of cancer drugs.[58]

Clinical (real, true) endpoints are characteristics or variables that reflect how a patient feels or functions or how long a patient survives.[52,55]

2.06.3.4 Evaluating a Protocol for a Clinical Trial of a New Chemical Entity

In the protocol, or plan, for a clinical trial, the title of the study and its objectives must be stated in clear and simple terms. The scientific rationale for the study should be fully explained and documented with acceptable references. The problem being analyzed and its current importance should both be clearly defined. The project must be described clearly to give the reader an overall idea of what is being planned. Caution should be exercised in the statement of objectives, both primary and secondary: multiple objectives and/or those that are not unclear should be avoided. It is also important to specify any potential conflicts of interest, i.e., all material and nonmaterial benefits that might be derived from the study by the proponent or the institution conducting the research. The pathology being treated must also be clearly defined, and methods used for diagnosis and prognostication should be specific and reproducible.

In addition to the study objective(s), the protocol must define the primary and secondary endpoints that will be measured. To provide an overall view of the research project, the following points must be precisely described:

- experimental design and methods (randomization, placebo or active drug controls, double-blinding, etc.)
- the site at which the trial will be conducted, existing structures, relation to structures providing emergency treatment or intensive care
- the coordinating center for multicenter trials
- the duration of the trial
- the methods to be used for analysis of data (intention-to-treat, statistical elaboration)
- the statistical power of the protocol (to ensure that the study has sufficient statistical power, the number of subjects to be enrolled needs to be established with the aid of a statistician).

Attention should also be given to the facilities available in the structure where the study will be conducted. Rapid access to an intensive care unit is particularly important for adequate treatment of any serious adverse event that might occur during the study. The person or person in charge of the research, and in particular the team leaders, must be identified by name and function. In defining the criteria for recruiting trial participants, the researcher must consider all variables, known and unknown, that could influence the results. The number and type (patients or healthy volunteers) of subjects to be enrolled must be specified, along with their characteristics (age, sex, etc.), their possible relations to the researcher (as employees, students, etc.) and the method of enrollment. Each individual being considered for enrollment must receive information on the study itself, and those enrolled must provide written consent to all study procedures. (This aspect is explained in greater detail in the section on the phases of clinical testing, below.) If the study is going to involve multiple centers in more than one country, it is essential to specify the number of subjects to be enrolled in each center. The inclusion and exclusion criteria should also be spelled out clearly and in detail, along with the treatments and dosage regimens (doses and administration frequencies) that will be used, the methods to be used for assigning participants to study arms and for evaluating the results, i.e., the types of testing to be performed, divided into invasive and noninvasive categories, and the frequency at which they will be carried out. Procedures might also be classified as 'minor' (collection of urine or stool samples, nasal or pharyngeal swabs, or moderate volumes of blood for nondiagnostic purposes) and 'major' (physically invasive involving biopsy, removal of body tissues, exposure to radiation). Use of procedures of the latter type should be carefully evaluated by the IRB or the ethics committee. The types and frequency of follow-up evaluations should also be specified.

It is also important to evaluate the information that will be given to potential trial participants in order to obtain their informed consent. (The information must be provided in a form that is clear and easy to understand. Alarmism should be avoided, but the candidate must be adequately informed on all operations or maneuvers planned in connection with the study and on all possible adverse effects of the NCE and of the control treatment.)

The description of methods used for recording and reporting results requires particular care. The protocol must specify the type(s) of measurements that will be made, when and how they will be carried out, and the procedures that will be adopted to safeguard the trial from biases on the part of participants and/or study personnel. It should provide for groups and assessment methods based on considerations of sex, age, renal and hepatic function, and the possibility of genetic polymorphisms. The possibility of a pregnancy and the risk of genetic damage to sperm (e.g., caused by cytotoxic chemotherapy) must always be considered, even if the subjects being enrolled are males.

The protocol should also specify other drugs that will be administered in response to particular situations that might arise during the trial. For example, in a study involving an asthma drug or one used to treat chronic obstructive pulmonary disease, an acute episode of severe bronchospasm might be treated with a short-acting bronchodilator, or in a trial focusing on a new drug for control of mild to moderate hypertension, nifedipine or some other rapid-acting antihypertensive might be identified as the treatment for critical blood pressure elevations.

It is also advisable to draft a letter for the participant's personal physician, particularly for phase II and III trials. The letter should provide a brief explanation of the objective and nature of the study and instructions for managing the patient during his/her participation in the study (signs and symptoms to look for, drugs – prescription and over-the-counter – and herbal preparations to be avoided).

Ethical evaluation of a trial protocol should focus on the appropriateness of the scientific method, the possible use of a placebo, and the information provided for patients during the enrollment procedure. It will also examine the potential benefits to the patient in relation to the risks, invasiveness, and secondary effects of the study treatment, and the inclusion of safeguards of the participant's right to withdraw voluntarily from the study without penalties or loss of benefits.

There have been trials in which the informed-consent procedure was criticized on ethical grounds. For example, in the study comparing foscarnet and placebo for the treatment of cytomegalovirus retinitis in AIDS patients, another active antiviral agent could have been used as a control, but the information was not given to enrolled patients. A similar situation occurred during testing of recombinant tissue plasminogen activator for treatment of myocardial infarction: streptokinase could have been used instead of placebo as the control treatment. Other examples include the use of sham therapy in controls enrolled in trials to evaluate the use of endoscopic sclerotherapy in the treatment of cirrhosis, as well as clinical trials of dubious validity that have been conducted in Third-World populations.[60]

2.06.3.5 The Phases of Clinical Testing

Clinical testing of an NCE is carried out in four phases: I, II, III, and IV. With the exception of the first phase and an aspect of phase I known as phase 0, all have therapeutic significance. In effect, all research conducted on diseased humans to evaluate the effects of a drug that can potentially improve that disease must be considered therapeutic in scope, in contrast to the scientific research that offers no possible benefits to the individual patient despite its value to humanity as a whole. In any case, the necessity for scientific research can never justify the violation of an individual's rights.

Assessment of the effects of an NCE on humans, whether healthy or otherwise, raises a number of issues, the most important of which are:

- how to guarantee the safety of trial participants
- how to ensure that trial participants are aware of all the implications of treatment with the NCE (informed consent)
- the existence of factors that may be used to exert pressure on a candidate for enrollment, such as:
- possible retribution
- indemnity for damages other than the insurance protection provided by the sponsor
- subjects considered to be particularly vulnerable to indirect psychological influences (children, handicapped individuals, psychiatric patients, etc.).

2.06.3.5.1 Phase 0

The appropriateness of the term phase 0 has been discussed at length. It actually refers to an early phase of phase I testing that involves the administration of ^{14}C-labeled microdoses of the NCE, i.e., doses decidedly lower than the first dose used in healthy volunteers.[11,61,62] Levels of the drug in the blood and other biological fluids are then determined by means of mass spectrometry (MS), a method that is both reliable and highly sensitive. With appropriate modifications, this approach could allow considerable reductions in the number of laboratory animals needed for pharmacokinetic studies.[62] Criticism has been raised regarding the possible interference of P glycoprotein (P-gp) with microdoses of drugs like α_1-antagonists, β-antagonists, or neurokinin-2 (NK_2) antagonists, which could destroy the linearity of the results. When the microdose method was used to investigate a new phosphodiesterase V inhibitor, a 14-fold difference in bioavailability was noted between doses of 10 and 800 mg.[28]

2.06.3.5.2 Clinical phase I

These are open studies, in most cases nonrandomized, that are performed in a small number (25–50) of human participants (normal volunteers or patients). Their scope is to obtain information about the pharmacodynamic (dose–effect range for vital parameters) and pharmacokinetic (ADME in humans) properties of the NCE. They are useful in identifying minimal effective and maximal tolerated dosages and in establishing a safe dosage range.

The exact number of subjects needed for a phase I trial varies depending on the NCE, its similarity to existing drugs, and on the initial testing results. For evaluation of an NCE that appears to offer therapeutic benefits for a serious and potentially fatal illness, such as neoplastic disease or AIDS, phase I testing may be conducted on patients suffering from the targeted disease instead of healthy volunteers. In this manner, the risk of potentially significant toxic effects (which might be unacceptable for a healthy subject) is offset by the possibility that the enrolled patient might benefit from a therapeutic effect. Trials of this type naturally have to be evaluated with care by IRBs or ethical committees.

Initially, an NCE is administered orally (when possible) at the lowest possible dose (*see* Section 2.06.2.2) to small groups of subjects, generally 1–5. After an appropriate washout period (generally, a couple of weeks), the same subjects are treated with a higher dose, and the process continues until the maximum tolerated dose is reached. This approach is known as definite dose finding. If repeated administration of the NCE is being evaluated, a dose midway between the lowest effective and highest tolerated doses is given for a period of time that is proportionally equivalent to the duration of multiple-dose toxicity studies in animals. The results are compared with the subject's pretreatment condition, as far as subject aspects are concerned, and with baseline values of preselected biomarkers. It goes without saying that the measurement methods and evaluation criteria used must remain constant for the duration of the trial.

In order to provide informed written consent to enrollment in a phase I trial, a potential participant must be given the following information:

- a new and potentially useful drug has been discovered
- the nature, expected duration, and scope of administration of the study drug
- methods and means used for the administration
- methods and means used for collection of data (in particular, invasive methods)
- certain risks are possible (adverse effects associated with use of the NCE: describe them)
- other unexpected risks are also possible (list possibilities).

The information sheet for prospective participants, along with the protocol itself, must be reviewed and approved by IRB or the Ethics Committee (for further information on this subject, *see* Section 2.06.3.4.).

Subjects under 18 years of age, elderly persons (at least in the first phase), smokers, women in their reproductive years, and subjects who have taken part in other trials in the last few months are excluded from phase I trials, as are those individuals who might be subject to psychological pressure (students, employees of the company that manufactures the NCE, members of the armed forces, etc.). Phase I trials are generally conducted in research centers specialized in clinical pharmacology. Attempts to predict human pharmacokinetics and pharmacodynamics with a mechanistic approach have yielded variable results: some predictions compared well with clinical results, but others did not.[63]

2.06.3.5.3 Clinical phase II

Clinical phase II trials are performed in a small number of patients with a disease or condition of interest and focus on the safety, efficacy, and appropriate dose range of an NCE in comparison with a standard treatment (positive control) or placebo preparation for treatment of the targeted indication. All types of study design may be used: open, randomized, blinded or double blind.

They normally include phase IIa, which is usually an open trial conducted on 25–50 subjects, and phase IIb, which includes 80–250 patients. The NCE is compared with an active drug or placebo to determine whether it is therapeutically active against the target pathology or pathologies. They will be conducted as dose–response studies using the doses considered safe for repeated administration based on the results of phase I studies. Surrogate endpoints are used.[64,65] These trials should furnish a precise indication of the short-term safety and efficacy of the NCE, satisfying the proof-of-concept requirement for subsequent clinical testing of the product. They must be completed with additional studies on the metabolism and elimination of the NCE in healthy subjects as well as those with the disease of interest. Two-stage[66] designs were proposed that are optimal in the sense that the expected sample size is minimized if the regimen has low activity subject to constraint upon the size of the type 1 and 2 errors. Two-stage designs that minimize the maximum sample size are also determined.[59] Clinical trial simulation in late-stage drug development could predict the outcome of phase III trials using phase II data.[64]

In order to provide informed written consent to enrollment in a clinical phase II, a potential participant must be given the following information:

- a trial is being planned to test a new and potentially useful drug
- for an accurate evaluation of the drug, some participants will be treated with the drug and others will not
- the beneficial health effects that are expected to be observed in subjects treated with the new drug
- the type of treatment that will be administered to the group of participants who do not receive the test drug (adequate standard treatment, which must be specified, or placebo)
- the nature and duration of NCE administration
- methods and means used for the administration
- methods and means used for collection of data (in particular, invasive methods)
- possible undesirable effects and risks
- other unexpected risks and undesirable effects (list possibilities)
- adverse effects that may be associated with use of the NCE (and the control treatment)
- the existence of alternative therapies that might be available for the patient.

2.06.3.5.4 Clinical phase III

These are large trials performed to confirm the safety, efficacy, and optimal dosage range of an NCE in comparison with placebo or with a standard treatment for the targeted indication (*see* Section 2.06.3.1[86]). Clinical drug trials in the twenty-first century are primarily multicenter and multinational and are conducted at various sites.[34] The statistical power of the study must be adequate (>1000–2000 patients), and the trials must be conducted according to a randomized controlled double-blind design with endpoints that are exclusively clinical. The objective is to evaluate the short-term efficacy and safety of the NCE, ideally in different age groups, and to outline a final therapeutic profile for the NCE (indications, dosing, route(s) of administration, side effects, possible interactions with other drugs, and contraindications): in short, an initial risk–benefit assessment. NDA approval of an application for an NCE requires that the drug's efficacy be proven in at least two well-designed randomized double-blind placebo-controlled clinical trials.[6]

Depending on the number of subjects enrolled, a clinical trial should be able to detect adverse effects that occur with a frequency of 1:1000–1:1500.[38,67,68] Rare adverse effects occurring with a frequency of 1:10 000 or less are generally the result of idiosyncratic reactions, and some are quite dangerous. Since the number of subjects enrolled in clinical trials rarely exceeds 3000, these effects are rarely detected unless by pure chance. This issue is discussed in greater detail in the section on phase IV postmarketing surveillance (PMS), below. In order to provide informed written consent to enrollment in a phase III RCT, a potential participant should be given the same information reported above for clinical phase II trials.

Clinical trials involving pediatric patients are associated with specific problems related to informed consent: this may be provided by the adults responsible for the children, or by children themselves after reviewing information sheets specially formulated for comprehension by children of different age groups. For these reasons, phase I clinical trials are generally conducted on adults, if possible, and those of phases II–III begin in adolescents (13–18 years of age) and proceed backward to younger age groups (infants up to 2 years of age and newborns a few weeks old).

2.06.3.5.5 Clinical phase IV

These large-scale long-term postmarketing studies[34] are conducted to identify the morbidity, mortality, and adverse events associated with an NCE. In some cases, they also reveal new indications for the drug. Phase IV testing can be considered an indepth analysis in a normal clinical setting of the new drug's effects under approved usage conditions (indications and dosage). These trials can include all patient types for which the drug has been approved. It serves to confirm the results of phase II and III trials. All types of study design can be used. The new agent can be compared with other drugs or with formulations of the new agent that have not yet been approved. Its effects can be compared in different age groups, populations, or specific population subgroups, or at different approved doses. The number of patients that must be enrolled varies. Phase IV trials do not include testing aimed at identifying new indications for the drug, which can only be authorized after a new phase III trial.

PMS[67,68] focuses on those potential adverse effects of the NCE that are likely to emerge only when the drug is used in large populations under the conditions of normal clinical practice. In effect, the latter conditions are substantially different from those that characterized phase II and III testing, in terms of dosage; and patient factors like the presence of associated disease, which may be complex; the use of other drugs, including over-the-counter products and herbal medicines; interactions with foods; and a large number of other variables.

For various reasons, premarketing studies of an NCE in diseased subjects (phases II and III) are an imperfect simulation of the conditions under which the drug will actually be used in clinical practice. First of all, the number of patients who receive the drug is limited.[67,68] As noted above, these trials rarely include more than 3000 participants,[67] a population that is large enough to detect only those high-frequency adverse effects that must be excluded before marketing. In the second place, even when phase II and III testing is lengthy, the period of administration is usually shorter than it is in normal clinical practice (e.g., antihypertensives or drugs used for controlling diabetes have to be taken for life). And, finally, the subjects receiving the NCE are generally carefully selected, and their behavior during therapy is monitored to ensure compliance with the NCE treatment protocol and exclude the use of other drugs (e.g., those used on a pro re nata (prn) basis).

In contrast, PMS is designed to detect and characterize the adverse effects (but also unexpected therapeutic effects) that are liable to occur during widespread use of the drug under common clinical conditions. The objective is to identify special risks or benefits related to the presence of associated disease, sometimes highly complex, with use of other drugs, including over-the-counter products and alternative medicines (homeopathic or herbal preparations), and exposure to and interaction with a variety of foods, and a large number of other variables (e.g., ethnic origins).

PMS may also reveal novel properties of an NCE that were not recognized in the previous phases of clinical testing. What appear to be 'side effects' may some day be developed into therapeutic effects for the treatment of completely different types of illness. Some notable examples include:

- the hypoglycemic and diuretic effects of some sulfonamides
- the antidepressant effects of the tuberculosis drug, iproniazide, or the neuroleptic agent, imipramine
- the antiarrhythmic effects of the antiepileptic drug, phenytoin
- the hypotensive properties of propranolol given to control arrhythmias.

New controlled phase III trials must obviously be conducted for any new indication that might emerge during the PMS phase.

During the PMS, various means are used to ensure that the regulatory authorities are promptly informed of any adverse effect of a newly marketed drug that may have escaped detection during phase II and III trials. Reports can be filed by physicians, pharmacists, other types of healthcare workers, caregivers, and even patients themselves. For certain NCEs, authorization is granted with the temporary restriction that the drug be prescribed exclusively within a hospital setting, a measure that favors early detection of adverse events (possibly by the hospital's pharmacology or pharmacy services).

At this point, it will be helpful to review the characteristics of unwanted or unfavorable actions, known as adverse drug reactions (ADRs). ADRs include all involuntary harmful responses observed during administration of a drug at doses normally used for its intended purpose (prophylaxis, diagnosis, therapy). A fundamental requirement is the demonstration of a causal relation between use of the drug and the unexpected clinical event. This 'imputability' is often very difficult to establish, and one can never exercise too much caution. Two types of ADR are commonly distinguished: type A, which refers to those events characterized by dose dependence, predictability, high morbidity rates, and relatively low mortality rates, and type B, which are bizarre, unpredictable events that are unrelated to dose and characterized by limited morbidity but high mortality. These are the so-called idiosyncratic reactions mentioned earlier in Sections 2.06.2.2 and 2.06.3.5.4. They are the ones that have the greatest impact on the general population because, while they are extremely rare (occurring at frequencies $> 1:10\,000$ and in some cases $> 1:100\,000$), their severe and often fatal nature can be extremely frightening and will create serious doubts in the general public regarding the drug's safety.[62] Type B reactions are the ones most commonly detected during PMS. Because of their higher frequency, dose dependence, and other characteristics, type A ADRs are more likely to be identified in premarketing surveillance or in large phase II or phase III trials involving 2000–3000 participants.

Establishing causality can be difficult for several reasons:

1. the effect may occur after a long period of latency, sometimes lasting years. An example is the hormone diethylstilbestrol which was given to pregnant women. Some 18–20 years later, the daughters born to the treated women presented an increased rate of genital cancer
2. the effect occurs infrequently
3. the effect resembles a form of spontaneous pathology that is fairly common
4. it is associated with ambiguous symptoms
5. difficulties in differentiating a drug-induced syndrome from one or more concomitant diseases
6. the possibility that the event is due to interaction with other drugs being used by the patient.

PMS utilizes methods used in epidemiologic studies: the descriptive approach (description of an effect related to the drug's toxicity or to its efficacy) and the analytical approach (to determine when the relation between the drug and adverse event is causal or not).

The historical prototype of the descriptive approach, still valid today, is the letter to the editor in a medical journal, but its important is now outweighed by spontaneous notifications filed with the regulatory authorities by physicians, pharmacists, and the other groups listed above. Letters to the editor were responsible for establishing the relationships between thalidomide and congenital malformations, chloramphenicol and aplastic anemia, phenylbutazone and leukopenia, aminofumarate and primary pulmonary hypertension, practolol and the oculomucocutaneous syndrome, and α-methyldopa and hemolytic anemia. Spontaneous notifications led to the recognition of important ADRs, including the liver damage caused by amiodarone, halothane- and α-methyldopa-induced forms of hepatitis, the neurologic changes caused by nalidixic acid, the hepatic lesions associated with ibufenac, thromboembolic disease caused by oral contraceptives, the extrapyramidal effects of metoclopramide, the adverse cardiovascular effects of nonsteroidal antiinflammatory drugs (including the new cyclooxygenase-2 inhibitors), and the photosensitizing effects of protryptiline. Data are available from 1982 through 2002.[62]

The spontaneous notification approach failed to identify the early cases of practolol-induced oculomucocutaneous syndrome. It also failed to identify estolate (and not other salts of the antibiotic) as the specific cause of erythromycin-induced cholestatic hepatitis.

In the analytical approach, random and nonrandom sampling is used:

1. experimental – This involves the promotion of controlled clinical trials similar to the RCTs of phases II and III without randomization. These studies are obviously ambiguous in many respects, and they also raise ethical questions that are extremely difficult to resolve, e.g., informed consent to take part in a study focusing on an adverse effect that is already known to occur
2. nonexperimental – These are cohort studies in which the incidence of a potential adverse reaction is compared in representative samples of the population, who are or are not exposed to the agent being investigated (in this case, the NCE to be tested), or case-control studies, in which the number of drug-treated subjects is determined in representative samples of individuals with a pathology that may be related to an ADR (e.g., aplastic anemia, pseudomembranous colitis, endometrial cancer).

These studies never involve the deliberate administration of the drug to patients by investigators: instead they are based on observations of patients who are being treated with the drug for therapeutic purposes by their own physicians.

In both cases, prospective and retrospective assessments are also possible. In the former, the possibility of adverse effects is evaluated in patients taking the drug for an approved indication on the orders of their physician. In retrospective analysis, the possible relation between the adverse event and drug use is evaluated after the event occurs. Excluding the effects of so-called confounding factors, a methodologically correct observational study can be a valid tool. Metanalyses can also be used to detect rare adverse effects.

As previously noted, the statistical weight of the study plays a determining role in the interpretation of the results that emerge. In most cases, all of the unacceptable high-frequency (type A) adverse effects of an NCE will be identified before the drug is marketed, during the course of phase II and III trials,[68] or in any case years after the hit has become a lead and advanced into the phase of preclinical pharmacotoxicological testing.

Clinical testing of an NCE in a population of 1000–3000 subjects guarantees a confidence level of 95%. That is, the adverse reaction will occur at a frequency no higher than 1:500–1:1000.

Bizarre type B reactions generally occur in 1 out of every 10 000–60 000 persons (e.g., the aplastic anemia caused by chloramphenicol), but some are even rarer ($\geq$1:100 000). Thirty thousand patients have to be treated to detect an adverse reaction with a frequency of 1:10 000.[67] To attribute an event of rarer type to a drug with an acceptable level of certainty, the drug would have to be tested in about 2 million subjects. In addition, ADRs may be very difficult to differentiate from spontaneous forms of pathology.[69]

Between 1982 and 2002, approximately 30 drugs were withdrawn from the world market due to unacceptable adverse reactions detected 4–290 months (and in one case, several years) after the drug had been marketed.[62]

2.06.4 Modern Technologies of Interest for the Process of Drug Development

The modern technologies of interest for the process of drug development include:

- imaging technologies
- NMR spectroscopy, MS, and human tissue histology technologies
- the 'omic technologies.'

2.06.4.1 Imaging Technology

CT, MRI, positron-emission tomography (PET), a dual photon nuclear imaging technique, single-photon emission computed tomography (SPECT), and other types of imaging technology exploit the interaction with tissues of various forms of energy to visualize the body in a noninvasive manner.[70] Among other things, they can be used to determine whether or not a putative drug is reaching its target, to evaluate the target's expression and function (up- and downregulation, activation or inactivation), and ultimately whether an NCE has a disease-modifying effect.

2.06.4.1.1 Computed tomography

When x-rays pass through different types of tissue, they are differentially deflected or absorbed. In CT, an x-ray-emitting source is rotated around the subject or the structure of interest, and the x-rays transmitted are measured from different angles to produce a three-dimensional image.

2.06.4.1.2 Magnetic resonance imaging or nuclear magnetic resonance

MRI is a powerful diagnostic method that uses radiofrequency waves in the presence of a magnetic field to extract information from certain atomic nuclei (^{1}H in particular). It furnishes more parameters than CT: proton density, T1 (the longitudinal relaxation time), T2 (the traverse relaxation time), flow, tomographic differentiation, and susceptibility.[53,70] It is used primarily for anatomical studies but also provides information on the physicochemical state of tissues, flow, diffusion, movement, and, more recently, molecular targets. MRI was used to measure tumor dimensions as a surrogate endpoint in 53 out of 71 studies of oncological drugs.[53] It has also been used successfully for the diagnosis, prognostic evaluation, and treatment of multiple sclerosis, and MRI findings are thus considered an important surrogate endpoint in the assessment of NCEs designed to treat this disease. Contrast-enhanced MRI has been identified as a promising technique for the evaluation of the glycosaminoglycan content of articular cartilage, which is abnormally high in areas where the cartilage surface is damaged.[52] Whereas clinical scoring showed no significant differences between etanercept over methotrexate (the standard therapy at the time) in the treatment of erosive bone lesions of rheumatoid arthritis,[53] the imaging-based erosion score showed statistically significant differences[53] and that of interferon-β treatments over relapsing-remitting multiple sclerosis, the NCE PT787 in the treatment of colon cancer.[53] With the technique known as functional MRI (fMRI), physiological parameters can be measured in live subjects.[70] Molecular imaging may be used to visualize specific molecules and targets (distribution and binding of an NCE) or the target itself (for example, receptor expression and modulation of downstream targets).[70] Pien et al.[53] have tabulated several hundred imaging biomarkers on their website.[90]

2.06.4.1.3 γ-Scintigraphy

γ-Scintigraphy is a noninvasive method that uses standardized radiolabeling techniques to investigate the presence of a test substance in the body. A computerized image of the emitted γ-rays reveals the presence of the substance in the body and distinguishes different concentration levels.[25]

2.06.4.1.4 Positron-emission tomography

PET is a tomographic imaging method[53,70] that identifies the positron emission associated with the breakdown of nuclides. It can be used to measure modifications in blood flow associated with cerebral activities based on the positrons emitted by selectively radioactive substances, such as ^{18}F-deoxyglucose (^{18}FdG),[25] a labeled form of glucose that emits fluoride protons and is used for metabolic studies that involve glucose uptake.[53] ^{18}FdG PET has been used for early identification of responders and nonresponders in trials evaluating antitumoral NCEs like imatinib or bevacizumab. In patients with CT-documented responses, partial or lasting, ^{18}FdG PET reveals reductions in glucose metabolism at the level of the tumor days or even weeks before decreases in the tumor mass can be appreciated on CT. It has also been used in studies of second-line therapy or surgical treatment for tumors of the lung or colon, again identifying nonresponders and allowing them to avoid useless treatment. Concordance has been observed between PET-documented reductions in glucose metabolism and both progression and survival times in patients receiving chemotherapy for nonsmall cell lung cancer (NSCLC), confirming the therapeutic value of the treatment. Research on patients with esophageal cancer has confirmed the value of this early metabolic study of tumor mass over the method based on CT study of tumor vascularization: the PET-based approach displayed 93% sensitivity and 95% specificity.[53]

PET has also proved to be useful in pharmacokinetic studies. The concentrations of ^{18}F fluconazole observed with this technique compare favorably with those required to inhibit the in vitro growth of fungal pathogens, and this observation is important for establishing dosages.[53] Concordance has also been observed between doses, blood levels, and receptor occupation. In some cases, PET studies have revealed no substantial differences in receptor binding after administration of high versus low doses of an NCE, justifying decisions to abandon further development of the molecule. [^{18}F]-labeled substance P antagonist receptor quantifier, a radioactive tracer peptide that binds NK_1 receptors, was recently used to determine the active dose to be used in treatment with the new antiemetic drug, aprepitant, which is an NK_1-antagonist. The study demonstrated that maximal receptor occupation could be achieved with an intermediate dose of the drug, compared with the maximum dose that had been tested, eliminating the need for clinical assessment of the maximum dose that had been tested.[52]

PET can also be useful for following microdoses of a drug during phase 0 studies of pharmacokinetics, metabolism, and excretion in particular (see Section 2.06.3.5.1).

The use of microdose studies for this purpose is not universally accepted: critics cite the activity of P-g, which facilitates extrusion of the drug, and the variability of the results obtained with different drugs. SPECT is a nuclear imaging technique in which radioactive tracers generating single photons of a specific energy are injected into subjects to produce imaging.

The imaging techniques discussed above can also be combined: MRI + CT + PET for evaluation of tumor angiogenesis, fMRI + CT for situations involving ischemic stroke, and MRI + PET or SPECT for neurodegenerative conditions like Alzheimer's disease.

Application of imaging technologies in drug discovery and development include[52]:

- neuroreceptor mapping (with PET or SPECT tracers) to investigate receptor occupation by an NCE or, in clinical trials, specific transmitters involved in CNS system[25] pathology
- structural imaging to evaluate morphological modifications and their consequences (CT, MRI)
- metabolic mapping (with ^{18}FdG NMR spectroscopy, molecular MRI) for neuroanatomic studies and assessment of the CNS effect of a drug
- functional mapping (e.g. by PET with ^{18}FdG or fMRI) to evaluate drug–disease interactions.

Nuclear techniques, PET imaging in particular, can now be used on small rodents, and they are routinely used in canine and primate models. Small-animal micro-CT, microultrasound, and micro-PET detectors may allow identical protocol designs for in vivo assessment across the species, enabling truly comparative pharmacology, directing proof-of-concept studies, and reducing risk in decision making. For example, animals can be studied in small PET chambers to obtain data on receptor occupation relative to blood levels that can be achieved in vivo in patients or animal models.

Fluorescence and bioluminescence imaging (BLI) techniques are widely used in drug R&D because of their low cost, versatility, and HT capacity. A reconstruction method based on fluorescence-mediated tomography (FMT) that was developed for in vivo imaging of fluorescent probes produces images of deep structures mathematically that are reconstructed by solving diffusion equations, based on the assumption that photons have been scattered many times.[70] Fluorescence reflectance imaging (FRI) is a simple method of image acquisition that is similar to fluorescence microscopy except that different optics allow imaging acquisition of whole animals. It is best suited for imaging of surface tumors or those that have been surgically exposed.[70] The use of imaging endpoints instead of time-consuming dissection and histological and histochemical studies can significantly reduce the workload required for tissue analysis, and this can greatly accelerate the evaluation of a candidate drug.[70] Image activators are caged near-infrared fluorochromes (NIRF), paramagnetic agents that change spin lattice relaxivity on activation or superparamagnetic sensors.

2.06.4.2 Nuclear Magnetic Resonance Spectroscopy, Mass Spectrometry, and Human Tissue Histology Technologies

NMR spectroscopy utilizes signals that are weaker than those produced by ^{1}H, e.g., ^{31}P, ^{13}C, or ^{15}N. It can provide information on specific compounds present in tissues. NMR spectra of biofluids serve two distinct but closely related purposes: they provide quantitative metabolic fingerprints as well as information on metabolite (biomarker) structure. NMR spectroscopy can be used to detect metabolites and biomarkers of toxicity or safety in biofluids or tissues[62,71] and to identify the type of bonds formed between drugs/toxic agents and living material. Reliable analysis of plasma or cerebrospinal fluid (CSF) can be obtained with a sample of ~5 μL, which means that even small animals, like mice, can be subjected to multiple assays.[27] NMR spectroscopy is being used with increasing frequency to investigate the biochemical changes induced in a tissue as a result of a toxic insult. High-resolution magic angle spinning (HRMAS) can be used directly on tissues, but it requires much more work and cannot be automated.[27] MS-NMR on intact tissues allows direct correlation of tissue indices and biological modifications. Future technologies include ^{13}C NMR spectroscopy, superconducting NMR probe technology (cryoprobes), and relaxivity on activation or supermagnetic sensors. More information on the direct applicational aspects of these methods can be found in Section 2.06.4.3.5.

Technologies such as liquid chromatography coupled with tandem mass spectrometric detection (LC/MS/MS) or liquid chromatography coupled with NMR spectroscopy (LC/NMR), liquid chromatography coupled with NMR and MS (LC-NMR-MS), ultrahigh resolution and sensitivity of Fourier transform ion cyclotron resonance (FT-ICR-MS) have dramatically improved the sensitivity of pharmacokinetic and toxicokinetic assays: detection limits are now in the low picogram per milliliter ranges. Assays can now be easily performed on minute blood samples (10–50 μL) from animals used in pharmacological and toxicological testing.[25,62] More detailed characterization of metabolites can be obtained with hybrid quadrupole-time-of-flight (Q-TOF) LC/MS/MS systems.[25] Advances in NMR, MS, and high-performance liquid chromatography (HPLC) technology have resulted in hyphenated LC-NMR and now LC-NMR-MS.[9] There are also several new technologies that are still in the developmental stages, such as surface-coated protein chips coupled with MS detection (SELDI).[62,72] The need to analyze large numbers of well-characterized human tissues constitutes a major bottleneck in drug discovery and development. Up to 1000 assays (including immunohistochemistry, fluorescence in situ hybridization (FISH) or ribonucleic acid (RNA) hybridation) can be performed with tissue microarrays.[73]

2.06.4.3 The 'Omic' Technologies in Drug Development

The human genome contains somewhere up to 32 000[74] and 35 000 genes[2,29] and numerous RNA splice variants.[29] Around 15–20% of these genes have been identified and their functions established; of the remaining 80–85%, presumed functions based on sequence analysis have been assigned to fewer than half. These genes can synthesize over 100 000 different proteins, approximately 5000–10 000 of which have been identified as potential drug targets[4] and 600–1500 as 'druggable' potential targets.[3] Fewer than 1000 gene products have been identified to date as potential targets for therapeutics.[74] The genes responsible for almost all monogenically inherited human diseases have already been identified. Gene expression is a complex process involving coordination of dynamic events, which are subject to regulation at multiple levels: the transcriptional level (transcription initiation, elongation, and termination), the posttranscriptional level (RNA translocation, RNA splicing, RNA stability), the translational level (translation initiation, elongation, and termination), and the posttranslational level (protein splicing, translocation, stability, and covalent modifications). Whether or not messenger (m)RNA modulation results in an immediately measurable change in the steady-state level of the encoded protein will depend on the protein turnover rate and other factors.[12] Gene expression can be assessed by measuring the quantity of the endproduct, i.e., protein, or the mRNA template used to synthesize the protein.[20] In general, an array experiment includes the following processes: (1) fabrication of the array; (2) RNA isolation and labeling; (3) application of the labeled sample to the array and measurement of hybridization; and (4) data analysis and interpretation.[20]

Gene expression microarray analysis can be used at all stages of the process of drug discovery, including target identification and validation, mechanism of action studies, and the identification of pharmacodynamic endpoints.[20] The advent of genetics and genomics has led to the discovery of a high level of conservation across the species in terms of deoxyribonucleic acid (DNA) sequences, proteins, gene functions, and signals.[27] Conservation of biochemical and

Table 3 Pharmacokinetic consequences of CYP2D6 polymorphism

Pharmacokinetic parameter	Consequences for the PM relative to EM
Bioavailability	2–5 fold
Systemic exposure	
C_{max}	2–6 fold
AUC	2–5 fold
Half-life	2–6 fold
Metabolic clearance	0.1–0.5 fold

Reproduced with permission from Shah, R. R. *Br. J. Clin. Pharmacol.* **2004**, *58*, 452 © Blackwell Publishing.

signaling pathways during the evolution of species makes it possible to use animal models as surrogates for human patients. The knowledge gained through the study of these animals can be applied to similar processes, pathways, and mechanisms that have been conserved in the human species.[29] Studies of the zebra fish, the fruit fly, *Drosophila melanogaster*, and the nematode, *Caenorhabditis elegans*, have played and continue to play key roles in resolving numerous human health problems.[12,20,29]

Genetic polymorphism refers to the differences in DNA sequences among individuals, groups, or populations.[34] In some cases, the alteration involves only one nucleotide (A, T, C, or G) in the genome sequence. These single-nucleotide polymorphisms (SNP) are basepair changes observed with a frequency of at least 1% within a given population.[74–76] They can occur in noncoding, coding, or regulatory regions of DNA.[74] They are abundant within the human genome, stable, and detectable with HT techniques like polymerase chain reaction and fluorescence. At present, testing is available to identify over 1.5 million different SNPs in the human genome.[74] The most common functional changes (e.g., traits referable to diseases) are manifested as SNPs. SNPs can have a considerable impact on responses to drugs. An example of a genetic factor that can determine interindividual and interpopulation variability in drug responses is the polymorphic expression of numerous P450 enzymes and glucuronyltransferase.[18] Many drug-metabolizing enzymes are polymorphic as a result of SNPs, gene deletions, or gene duplications. The ones that have been most widely studied are CYP2D6[77] (Table 3) and CYP2C19.

SNPs are responsible for interindividual variations in a drug's efficacy: for example, losartan, an angiotensin type 1 (AT1) angiotensin II receptor antagonist, is less effective in individuals with reduced CYP2C19 activity. This difference is not seen with another AT1 antagonist, irbesartan, which is not metabolized by the CYP2C19 coenzyme. The anticoagulant effects of warfarin are also related to CYP2C19 polymorphism.[28] The presence of specific mutations in the promoter region of the gene for serotonin is predictive of the response to a selective serotonin reuptake inhibitor.[77] Other examples of response variability have emerged during clinical trials of several new drugs. For example, the anticancer monoclonal antibody, trastuzumab (Herceptin; Genentech/Roche), has proved to be active only in women who carry multiple copies of the ERBB2 gene (also known as HER2/*neu*),[25] and the FDA has approved the use of histochemical studies (Dako Herceptest) and FISH (Vysis Patuvision) for the selection of patients to be treated with this drug.[74] Potentially fatal idiosyncratic hypersensitivity reactions to abacavir have been related to two different genes on chromosome 6.[62,74] During clinical trials with tranilast, hyperbilirubinemia was documented in roughly 10% of the patients treated with this drug. This evidence, which is suggestive of toxic liver damage, in all probability would have blocked further drug testing if an association had not been discovered between the cases of hyperbilirubinemia and the uridine diphosphoglucose (UDG)-glucuronyl transferase 1 (*UTG1*) gene. Instead, within the time course of a phase III trial, the manufacturer was able to attribute the 'adverse reaction' to a genetic risk factor associated with a clinically benign form of Gilbert syndrome.[62,76] In IND and NDA applications presented to the FDA, pharmacogenetic/pharmacogenomic tests included in the early phase of development have shown that certain adverse effects are related to drug metabolism by 350P enzyme isoforms.[78] The outcome of drug therapy can be influenced by polymorphisms of genes within various functional categories, including those involved in drug transport (for example, glycoprotein P_1 ($ABCB_1$), which affects plasma levels of digoxin), drug metabolism (e.g. genes encoding thiopurine S-methyltransferases can influence the risk of thiopurine toxicity), and expression of target receptors (for example, the gene encoding the β_2-adrenoceptor (β_2AR), a G protein-coupled receptor that mediates the actions of catecholamines).[75] In asthmatics, a clear correlation has been demonstrated between the outcome of treatment with β_2-agonist albuterol and specific haplotypes of the *ADRβ2* gene encoding β_2AR.[52]

In 50% of cases the pharmacogenetic polymorphism is not significantly associated with drug response therapy.[34]

Despite our increased knowledge of the human genome and that of other species, the fundamental nature of many disease processes has remained obscure, and this has stimulated the development of methods for correlating gene expression with phenotypic results.

The so-called 'omic' technologies include:

- proteomics
- transcriptomics
- (pharmaco)genomics
- metabolomics
- metabonomics
- toxicogenomics.

These technologies should be able to reveal modifications produced in vitro or in vivo by drugs and toxic agents at the level of the gene itself (for example, toxicogenomics) and/or in the expression of proteins (proteomics). These pharmacological/toxicological fingerprints of an NCE's action on living material are useful in terms of predictive toxicology, identification of toxicological mechanisms (a major part of NCE risk assessment), and the preclinical and clinical development of NCEs. This approach could eliminate the need for costly histological, electron microscopy, and biochemical studies of living tissues, which are currently used to detect subtle forms of structural damage.[62]

An entire volume in this book has been dedicated to the technologies. The present discussion will be limited to the aspects that are most important to the field of drug development. Genomics and structure–activity relationships (SARs) can lead to more rapid selection of new drug targets and new drugs and more rapid identification of mechanisms underlying toxicity.[23]

2.06.4.3.1 Proteomics

The proteome is defined as the expressed protein complement of a cell, organ, or organism and it includes all isoform and posttranslational variants.[74] According to Kramer and Cohen,[29] proteomics is used to determine differential protein expression, posttranslational modifications, and alternative splicing and processed products. Two-dimensional gel electrophoresis is commonly used to fractionate proteins, which are separated first on a polyacrylamide gel according to isoelectric point (first dimension) and then at a 90° angle on the basis of molecular mass (second dimension). The technique is often used to separate the numerous proteins from a cell or tissue, which are subsequently analyzed by MS to identify those that are differentially expressed or modified proteins. Control of proteome can be affected at the translational, posttranslational, and functional level.[74] A recent example of this innovative strategy was used to study the anticancer drug bortezomid. In animal models, a single dose of this compound was associated with 80% inhibition of proteosome activity in PBMCs and no irreversible toxicity. The inhibitory effect was no longer detectable 72 h after drug administration.[79]

2.06.4.3.2 Transcriptomics

Transcriptome analysis provides a measure of RNA abundance, and it is usually limited to isolated tissues or peripheral blood lymphocytes.

2.06.4.3.3 Pharmacogenomics

Pharmacogenomics uses biological indicators – DNA, RNA, proteins – to predict the efficacy of a drug[80] and the likelihood of an adverse event in individual patients.[28,81] It is a relatively new field that includes the study of genetic variations (polymorphisms) between individuals and how these variations influence responses to therapeutic drugs.[18,75,82–84] Although 'pharmacogenomics' and the older term 'pharmacogenetics' are often used interchangeably, the former term is broader in scope.[34] Pharmacogenomic research is proceeding along two main pathways: (1) the identification of new genetic targets associated with various diseases; and (2) the identification of specific genetic polymorphisms linked to the responsiveness to particular drugs.[28]

In short, in pharmacogenomics the static aspect of pharmacogenetics becomes dynamic.[74] In effect, the genome is static whereas the proteome is in a state of continuous flux in response to internal and external stimuli. The proteome is the expressed protein complement of a cell, an organ, or an organism, and it includes all isoforms and posttranslational variants. A cell contains a single genome and multiple proteomes. The proteome can be modulated at the translational, posttranslational and functional levels.

2.06.4.3.4 Metabolomics
Metabolomics is used to study the metabolic regulation in single cell types.

2.06.4.3.5 Metabonomics
Metabonomics is the study of endogenous metabolites, systemic biochemical profiles, and data on regulation of organic function based on analysis of biofluids and tissues.[74] It can provide a 'snapshot' of the small molecules (amino acids, organic acids, sugars) present at a given time in a cell, fluid, or tissue. In MS analysis of samples, the peaks are compared with those contained in a standard library to identify and name known biochemicals and to catalogue unknown ones.[12,27,74] NMR spectroscopy is an equally valuable tool in metabonomics due to its ability to furnish results rapidly and with reduced workloads.[74] Metabolic profiles of biological fluids (plasma, CSF, urine) reflect normal variations as well as the physiological impact on organs or organ systems of toxic insults and diseases. NMR spectroscopic analysis of biofluids has identified new metabolic indicators of organ-specific toxicity in rodents, and this explorative role is one in which the new technology excels.[85] For example, the combined alterations of a series of urinary parameters, trimethylamine-N-oxide, NN-dimethylglycine, dimethylamine, and succinate, have been shown to reflect damage to the renal papilla, a condition for which no reliable marker was previously available[27,85] (**Figure 9**).

MS-NMR on intact tissues allow direct correlation of tissue indicators and biological modifications. Metabonomic criteria can also be used to evaluate the efficacy of an NCE or other therapeutic agent. It may allow identification of the organ targeted by a toxic effect and the underlying biochemical mechanisms and define biochemical indicators of the onset, progression, and regression of the toxic lesions. HRMAS NMR spectroscopy can investigate biochemical changes in tissues directly. Unfortunately, this approach is rather labor-intensive and currently not amenable to automation.[27] It is possible to foresee future situations in which metabonomics, gene expression, and proteomic data are integrated using multivariate approaches to provide a unified vision of an organism under various types of stress, based on indicators furnished by all three platforms. Q-TOF, LC/MS/MS systems are also popular for the characterization of metabolite profiles. The configuration of a Q-TOF results in an instrument capable of high sensitivity, mass resolution, and mass accuracy in a variety of scan modes.[25]

2.06.4.3.6 Toxicogenomics
Toxicogenomics is the application of microarray analysis toward toxicology. It aims to identify targets and mechanisms. An in vitro toxicogenomic assay will only find practical applications in drug discovery if gene expression patterns or markers can be identified that show close correlation with the potential drug toxicity.[20,31] Because changes in gene expression precede changes at the histological or anatomopathological level, transcriptome analysis has the potential to identify compounds with toxic liability early in the drug discovery process.[20,31] This task, however, has proven to be considerably more difficult than anticipated. In most cases, it is highly unlikely that a given toxic effect will be strongly correlated with an expression change involving any single gene. The results showed that, based on gene expression analysis, liver slices, followed by primary hepatocytes in culture, were the most similar to intact rat livers.[31]

Using gene expression profiling, trovafloxacin, a quinolone responsible for severe liver toxicity in a small number of patients not observed in preclinical studies, can clearly be distinguished from other quinolones; overall treatment with trovafloxacin resulted in far more gene expression changes than treatment with other quinolones.[31]

2.06.4.3.7 Usability and criticism
The significance of genotyping is different in the various phases of clinical testing[34]:

- Phase I: identify polymorphisms correlated with phenotypic elements (e.g., pharmacokinetic and pharmacodynamic properties, excretion, and serum levels)
- Phase II: genotype–phenotype correlations. Associate specific polymorphisms with efficacy differences using the candidate gene approach. Data on molecular profiles associated with the response to an NCE
- Phase III: use the results obtained in phases I and II to design an optimal large-scale phase III study. Test candidate genes for efficacy and metabolism. Can be useful for carrying out large-scale genotyping studies to discover new pharmacogenomic indicators. Identify responders versus nonresponders to certain drugs of to the NCE itself
- Phase IV: the studies conducted in this phase might be useful: to identify rare adverse effects and the relations between these events and specific populations; for marketing considerations (that is, to determine whether a diagnostic test is capable of distinguishing the drug in question from its competitors and whether the market for a given drug would justify the development of diagnostic tests).

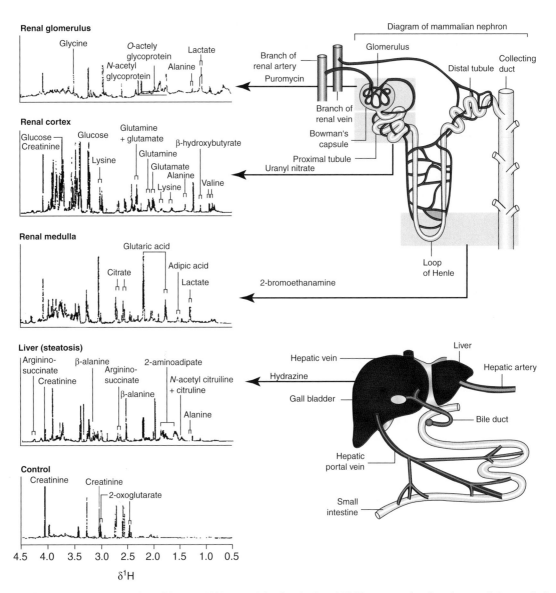

Figure 9 Metabonomic detection of liver and kidney toxicity. Stack plot of NMR spectra showing characteristic metabolic fingerprints of tissue-specific toxicity produced by different site-selective xenobiotics given in single doses to rats in relation to an untreated control. Each 600-MHz spectrum represents one time point after dosing for each toxic compound. The xenobiotics affect specific regions within the organs (depicted in the figure by shaded boxes); puromycin affects the renal glomeruli; uranyl nitrate affects the lower regions of the proximal tubules; 2-bromoethanamine affects the renal medulla, including the loop of Henle and the collecting ducts; and hydrazine affects the hepatic parenchymal cells. (Reproduced with permission from Nicholson, J. K.; Connelly, J.; Lindon, J. C.; Holmes, E. *Nat. Rev. Drug Disc.* **2002**, *1*, 153–161 © Nature Publishing Group.)

Pharmacogenomic studies could also reduce costs in preclinical and clinical phase III testing by allowing the early elimination of certain drugs and identifying a more selective target population for others. The selection of drug therapy based on individual genetic profiles can lead to a clinically important decrease in adverse outcomes.[80] On the other hand, however, individual subjects or groups excluded from trials based on genotype data may be deprived of the benefits associated with trial participation. Pharmacogenomics thus provides a new setting for reassessing the ethics of research on human subjects in terms of the tests performed for recruitment and the eligibility of various groups.

Ideally, a variety of drugs would be available, each designed for the genotype of a given individual, but it seems more likely that a patient will simply be classified as a responder or nonresponder with respect to a given treatment for a given disease or condition.

It is important to recall that environmental factors are also important. Many common diseases are polygenic and multifactorial and subject to the influence of gene–gene and gene–environment interactions. In addition, roughly 50% of all polymorphisms have no effect on the dose response to drugs.[34]

As for the actual possibilities for using the omic technologies in drug development, it is important to recall that, at present, there is no universal consensus within the scientific community regarding the best design for genomic tests. It was hoped that HT genomic studies would clarify associations between ADRs and SNPs, but thus far no methods have been established for analyzing the data from clinical trials currently present in the literature.[34]

Basal gene expression profiling of cancer cells can classify and predict long-term outcome, but these studies have yet to predict the initial response to therapy.[20] The immediate benefits promised by the genomic revolution remain to be seen[73] and the true value of toxicogenomics in the field of drug discovery and development has yet to be demonstrated.[31]

In most cases, drug responses depend on a complex series of factors, including multiple genes and environmental factors.[75,82,83] Response variability can also be the result of other mechanisms, such as epigenetic factors that do not modify DNA sequences. Mechanisms of this type might explain the resistance to alkylating nitrosourea compounds that has been observed in certain cases of glioma.[84]

In light of these considerations, cautious optimism can be justified, but there is still an enormous amount of work to be done in the validation and use of the omic technologies in drug development. The journey, in short, is just beginning.[74]

2.06.5 Time and Cost Requirements for Development of a New Chemical Entity and Factors Contributing to Project Failure

Drug development is risky and time-consuming, and it requires substantial investments in terms of capital (~$500 million), human resources, research skill, and technological expertise.[6] Costs are high, even in the initial phases. Determining whether or not a small molecule can be developed currently requires about 6 months and $500 000.[25] Long-term animal testing requires approximately $3–7 million, and a detailed breakdown of these costs has been published.[62] The development of new drugs requires nearly $1 billion and 12 years to bring the average NCE to commercialization.[53] $200 million might be attributed to compound failure.[12] Recently, an NCE targeting the peroxisome proliferator activated receptor γ family ragaglitazar (NN622), a dual acting insulin sensitizer, was found to be positive in the carcinogenic bioassay study (urinary bladder tumors were observed in mice and rats), and the drug had to be dropped during phase III trials, after many millions of dollars had been spent on its development.[25] The costs of fundamental clinical testing for an NCE are even higher (in millions of dollars): 15.2 for phase I, 15.7 for phase II, and 27.0 for phase III.[6] Compounds that are abandoned during the testing process (some as late as phase III) represent an enormous loss, amounting on the average to roughly $200 million.[12] Since the mid-1960s, the process of drug approval has been modified to significantly improve the safety and the efficacy of new drugs for use in general practice. However, one of the consequences of these changes has been an increase in the time and costs associated with placing a new drug on the market.[6] It has been shown that small improvements in clinical trial outcomes and decision making translate into hundreds of millions of dollars of development cost-saving and a faster time to market.[53]

The mean interval of time between synthesis of a compound and approval of the NDA was 7.9 years in 1960: by 1990, it had risen to 12.8 years. Estimates for 2004 dropped to between 3.2 and 8.5 years,[6] but intervals of 10–20 years are not uncommon. The increasing time requirements are the result of the growing complexity of clinical tests, the demand for increasingly rigid testing protocols, administrative aspects of testing, and the indispensable inclusion in the study of particular population subgroups, such as the elderly.[6] The duration of clinical testing ranges from 53 to 86 months for NCEs belonging to major drug categories: antiinfective agents: 74 months; antineoplastic drugs: 116 months; cardiovascular drugs: 103 months; endocrine agents: 115 months; and immunological drugs: 100 months.[86] A 25% reduction in the time needed for the clinical phases of drug development would decrease the capitalized cost of NCE development by 16%.[5]

Approximately one out of every 12 leads completes the process and becomes an NDA[59] (1:5000[25]). Between 1983 and 2001, the overall final clinical success rate for all investigational drugs was 21.5%.[6] The main causes of project failure in the advanced phases of NCE development are pharmacokinetic issues (39%[27]–40%[8]) and animal toxicity (11%[8]). Late evaluation of safety and efficacy, low therapeutic indexes (the ratio of the maximum tolerated dose to the therapeutic dose per kilogram based on repeated treatment in two animal species), and the time, size, and costs of clinical studies are other major reasons for compound attrition.[12] The latter variables can be reduced by the use of reliable and specific biomarkers that can be used as early predictors of efficacy and long-term toxicity.[12] Minor improvements in clinical trial outcomes and decision making have been shown to translate into major savings in terms of cost (hundreds of millions of dollars) and time to market.

The increasing number of potential hits generated each year has not resulted in a corresponding increase in the number of NCEs that reach the clinical trial stage, and this is due in part to inadequacies in the investigation of the candidate drug's pharmacokinetic properties.[14] The advisability of requiring early pharmacokinetic studies in humans has recently been underlined by some authors.[22]

In any case, due to the high cost and specificity of the drug development process, most manufacturers adopt a step-by-step approach, with well-defined points for evaluating the results obtained and deciding whether or not to proceed. A balance must be struck between investments and the length of intermediate periods between testing phases, and a continuous evaluation of the project time is needed to make decisions and ensure a reasonable possibility of success as early as the lead stage of the NCE.

2.06.6 Regulatory Problems

As noted at the beginning of this chapter in the US, the FDA reviews and acts on NDAs as well as applications for INDs. In Europe, the European Medicines Agency (EMEA) authorizes the registration of innovative drugs, and it is currently the only body authorized to evaluate through the so-called centralized procedure biotechnical products and, more recently, cancer agents.[87–89] An NCE that has been approved by the EMEA can be marketed in all member states of the EU. There are also national regulatory authorities in each state, which can authorize phase I testing and marketing within their respective national territories of drugs belonging to categories other than those indicated above or drugs that are already being marketed in other European countries.

In the US, the time required for evaluation of a NDA and its approval by the FDA decreased in the late 1990s but rose again during the period 2000–2002, and this latest increase has been associated with a decrease in the time devoted to clinical development and an increase in the number of applications. Between 1980 and 2001, the FDA approved 504 small molecules, 40 recombinant proteins, and 10 monoclonal antibodies.[86]

In 1993 with the passage of the Food and Drug Administration Modernization Act (FDAMA),[6,86] the approval process was accelerated for drugs designed to treat serious or life-threatening diseases or conditions for which no treatment is currently available. These 'fast-track' products are eligible for priority review.

Fast-track approval can also be granted when efficacy data are based on surrogate endpoints, i.e., clinical endpoints other than survival or irreversible morbidity, but only when the NCE provides significantly greater benefits than those offered by existing treatments. In these cases, approval is sometimes issued with restrictions that ensure that the drug will be used safely. In addition, for all NDAs approved based on surrogate endpoint studies, the FDA requires well-controlled postmarketing studies to verify the drug's efficacy in the target population.

The FDA will review and act on 90% of standard original applications within 3 months of receipt and on 90% of priority original applications within 6 months.[86] The FDA has also pledged to assist industry in the preparation of high-quality applications in an attempt to reduce the need for multiple review cycles.

In efficacy assessment of an NCE, parameters measured with imaging techniques (CT, MRI, PET, etc.) can be used as surrogate endpoints or even as biomarkers if they are reasonably accurate predictors of clinical outcome. This principle was established in the FDA Modernization Act of 1997, which explicitly authorizes the FDA to approve drugs for the treatment of serious or life-threatening conditions on the basis of studies using endpoints of this type.

In a recently issued guidance, the FDA has attempted to clarify its policy on the use of pharmacogenomic data in the drug application review process.[86,91] The Administration recognizes that, at present, most gene expression profiling data are of an explanatory or research nature and would therefore not be required for submission.[31]

References

1. Fitzgerald, J. D. *Drug Dev. Res.* **1998**, *43*, 143–148.
2. International Human Genome Sequencing Consortium. *Nature* **2001**, *409*, 860–921.
3. Hopkins, A. L.; Groom, C. R. *Nat. Rev. Drug Disc.* **2002**, *1*, 727–730.
4. Drews, J. *Science* **2000**, *287*, 1960–1964.
5. Knowles, J.; Gromo, G. *Nat. Rev. Drug Disc.* **2003**, *2*, 63–69.
6. Dickson, M.; Gagnon, J. P. *Nat. Rev. Drug Disc.* **2004**, *3*, 417–429.
7. Maxwell, R. A. *Drug Dev. Res.* **1984**, *4*, 375–389.
8. van de Waterbeemd, H.; Gifford, E. *Nat. Rev. Drug Disc.* **2003**, *2*, 192–204.
9. Koehn, F. E.; Carter, G. T. *Nat. Rev. Drug Disc.* **2005**, *4*, 206–220.
10. Dahl, S. G.; Sylte, I. *Basic Clin. Pharmacol. Toxicol.* **2005**, *96*, 151–155.
11. Bleicher, K. H.; Böhm, H. J.; Müller, K.; Alanine, A. I. *Nat. Rev. Drug Disc.* **2003**, *2*, 369–378.
12. Lewin, D. A.; Weiner, M. P. *Drug Disc. Today* **2004**, *9*, 976–983.
13. Yamashita, F.; Hashida, M. *Drug Metab. Pharmacokinet.* **2004**, *19*, 327–338.
14. Bergström, C. A. S. *Basic Clin. Pharmacol. Toxicol.* **2005**, *96*, 156–161.

15. Kitchen, D. B.; Decornez, H.; Furr, J. R.; Bajorath, J. *Nat. Rev. Drug Disc.* **2004**, *3*, 935–949.
16. Egner, U.; Krätzschmar, J.; Kreft, B.; Pohlenz, H. D.; Schneider, M. *ChemBioChem* **2005**, *6*, 468–479.
17. Van de Waterbeemd, H. *Basic Clin. Pharmacol. Toxicol.* **2005**, *96*, 162–166.
18. Pelkonen, O.; Turpeinen, M.; Uusitato, J.; Rautio, A.; Raunio, H. *Basic Clin. Pharmacol. Toxicol.* **2005**, *96*, 167–175.
19. Ekins, S.; Waller, C. L.; Swaan, P. W.; Cruciani, G.; Wrighton, S. A.; Wikel, J. H. *J. Pharmacol. Toxicol. Methods* **2001**, *44*, 251–272.
20. Clarke, P. A.; te Poele, R.; Workman, P. *Eur. J. Cancer* **2004**, *40*, 2560–2591.
21. Black, J. *Pharm. Policy Law* **1999**, *1*, 85–92.
22. Parrot, N.; Jones, H.; Paquereau, N.; Lavè, T. *Basic Clin. Pharmacol. Toxicol.* **2005**, *96*, 193–199.
23. Greaves, P.; Williams, A.; Eve, M. *Nat. Rev. Drug Disc.* **2004**, *3*, 226–236.
24. Berkowitz, B. A. Basic and Clinical Evaluation of New Drugs. In *Basic and Clinical Pharmacology*, 9th ed.; Katzung, B. G., Ed.; Lange Medical/McGraw-Hill: New York, 2004, pp 67–74.
25. Pritchard, J. F.; Jurima-Romet, M.; Reimer, M. L. J.; Mortimer, E.; Rolfe, B.; Cyen, M. N. *Nat. Rev. Drug Disc.* **2003**, *2*, 542–553.
26. Bernauer, U.; Oberemm, A.; Madle, S.; Gunert-Remy, U. *Basic Clin. Pharmacol. Toxicol.* **2005**, *96*, 176–181.
27. Griffin, J. L.; Bollard, M. E. *Curr. Drug Metab.* **2004**, *5*, 389–398.
28. Walker, D. K. *Br. J. Clin. Pharmacol.* **2004**, *58*, 601–608.
29. Kramer, R.; Cohen, D. *Nat. Rev. Drug Disc.* **2004**, *3*, 965–972.
30. Austen, M.; Dohrmann, C. *Drug Disc. Today* **2005**, *10*, 275–282.
31. Yang, Y.; Blomme, E. A. G.; Waring, J. F. *Chem. Biol. Interact.* **2004**, *150*, 71–85.
32. Fletcher, A. P. *J. R. Soc. Med.* **1978**, *71*, 693–696.
33. Olson, H.; Betton, G.; Robinson, D.; Thomas, K.; Monro, A.; Kolaja, G.; Lilly, P.; Sanders, J.; Sipes, G.; Bracken, W. et al. *Regul. Toxicol. Pharmacol.* **2000**, *32*, 56–67.
34. Issa, A. M. *Nat. Rev. Drug Disc.* **2002**, *1*, 300–308.
35. World Medical Association. Declaration of Helsinki. http://www.wma.net/e/policy/pdf/17c.pdf.(accessed Aug 2006).
36. World Health Organization. *Operational Guidelines for Ethics Committees That Reviews Biomedical Research*; World Health Organization: Geneva, 2000.
37. US Food and Drug Administration, *Code of Federal Administration (CFR)*, Rockville, revised April 2001; subchapter D, part 312, 'Responsibilities of Sponsors and Investigators'; part 50, 'Protection of Human Subjects;' and part 56, 'Institutional Review Board'. www.access.gpo.gov/cgi-bin/cfrassemble.cgi?title = 200121.
38. Goldstein, A.; Aronow, L.; Kalman, S. M. *Principles of Drug Action. The Basis of Pharmacology*, 2nd ed.; Wiley: Chichester, 1974, pp 779–801.
39. Popock, S. J. *Clinical Trials. A Practical Approach*; Wiley: Chichester, 1983.
40. Pitt, B.; Desmond, J.; Popock, S. *Clinical Trials in Cardiology*; W. B. Saunders: London, 1997.
41. Friedman, L. M.; Furberg, C. D.; DeMets, D. L. *Fundamentals of Clinical Trials*; Wright: Boston, 1996.
42. Byar, D. P.; Schoenfeld, D. A.; Green, S. B.; Amato, D. A.; Davis, R.; De Gruttola, V.; Finkelstein, D. M.; Gatsonis, C.; Gelber, R. D.; Lagakos, S. et al. *N. Engl. J. Med.* **1999**, *323*, 1343–1348.
43. Sibbald, B.; Roland, M. *Br. Med. J.* **1998**, *316*, 201.
44. Roberts, C.; Sibbald, B. *Br. Med. J.* **1998**, *316*, 1898.
45. Torgerson, D. J.; Sibbald, B. *Br. Med. J.* **1998**, *316*, 360.
46. Sibbald, B.; Roberts, C. *Br. Med. J.* **1998**, *316*, 1719.
47. Fine, M. J.; Auble, T. E.; Yealy, D. M.; Hanusa, B. H.; Weissfeld, L. A.; Singer, E. D.; Coley, C. M.; Marrie, T. J.; Kapoor, W. N. *N. Engl. J. Med.* **1997**, *336*, 243–250.
48. Mellors, J. W.; Munoz, A.; Giorgi, J. V.; Margolick, J. B.; Tassoni, C. J.; Gupta, P.; Kinsley, L. A.; Todd, J. A.; Saah, A. J.; Detels, R. et al. *Ann. Intern. Med.* **1997**, *126*, 946–954.
49. Rogers, H. J.; Spector, R. J.; Troance, J. R. *A Textbook of Clinical Pharmacology*; Hodder & Stoughton: London, 1981, pp 166–170.
50. Turner, P.; Richens, A. *Clinical Pharmacology*, 2nd ed.; Livingstone: Edinburgh, 1975, pp 1–8.
51. Evidence Based Medicine, a journal of the *Br. Med. J.* publishing group: *Glossary*. www.evidence-basedmedicine.com (accessed Aug 2006).
52. Frank, R.; Hargreaves, R. *Nat. Rev. Drug Disc.* **2003**, *2*, 566–580.
53. Pien, H. H.; Fishman, A. J.; Thrall, J. H.; Sorensen, A. G. *Drug Disc. Today* **2005**, *10*, 259–266.
54. Fleming, R. T.; DeMets, D. L. *Ann. Intern. Med.* **1996**, *125*, 605–613.
55. Epstein, A. E.; Hallstrom, A. P.; Rogers, W. J.; Liebson, P. R.; Seals, A. A.; Anderson, J. L.; Cohen, J. D.; Capone, R. J.; Wise, D. G. *JAMA* **1993**, *270*, 2451–2455.
56. Temple, R. J. A Regulatory Authority's Opinion about Surrogate Endpoints. In *Clinical Measurement in Drug Evaluation*; Nimmo, W. S., Tucker, G. T., Eds.; Wiley: New York, 1955, pp 3–22.
57. Sausville, E. A.; Holbeck, S. L. *Eur. J. Cancer* **2004**, *40*, 2544–2549.
58. Garattini, S.; Bertelé, S. *Br. Med. J.* **2002**, *325*, 269–271.
59. Girard, P. *Basic Clin. Pharmacol. Toxicol.* **2005**, *96*, 228–234.
60. Barry, M. *N. Engl. J. Med.* **1988**, *319*, 1083–1085.
61. Stoughton, R. B.; Friend, S. H. *Nat. Rev. Drug Disc.* **2005**, *4*, 345–350.
62. Preziosi, P. *Nat. Rev. Drug Disc.* **2004**, *3*, 521–526.
63. Meno-Tetang, G. M. L.; Lowe, P. *J. Basic Clin. Pharmacol. Toxicol.* **2005**, *96*, 182–192.
64. Simon, R. *Controlled Clin. Trials* **1989**, *10*, 1–10.
65. The Italian Study Group for the Di Bella Multitherapy trials. *Br. Med. J.* **1999**, *318*, 224–228.
66. De Ridder, F. *Basic Clin. Pharmacol. Toxicol.* **2005**, *96*, 235–241.
67. Lewis, L. A. *Trends Pharmacol. Sci.* **1981**, *2*, 93–94.
68. Freeman, J. They Test New Drugs, Don't They? In *Drug Safety: A Shared Responsibility*; International Drug Surveillance Department (IDSD), Glaxo Group, Eds.; Churchill Livingstone: Edinburgh, 1991, pp 13–26.
69. Berkowitz, B. A. Basic and Clinical Evaluation of New Drugs. In *Basic and Clinical Pharmacology*, 7th ed., Katzung, B. G., Ed.; Appleton and Lange Medical: Stanford, CT, 1998; pp 62–72.
70. Rudin, M.; Weissleder, R. *Nat. Rev. Drug Disc.* **2003**, *2*, 123–131.
71. Holmes, E. *Toxicol. Lett.* **2003**, *144*, S4.
72. Thompson, D. C. *Toxicol. Lett.* **2003**, *144*, S4.
73. Sauter, G.; Simon, R.; Hillan, K. *Nat. Rev. Drug Disc.* **2003**, *2*, 962–972.

74. Bilello, J. A. *Curr. Mol. Med.* **2005**, *5*, 39–52.
75. Watters, J. W.; McLeod, H. L. *Trends Pharmacol. Sci.* **2003**, *24*, 55–58.
76. Roses, A. D. *Nat. Rev. Drug Disc.* **2002**, *1*, 541–549.
77. Shah, R. R. *Br. J. Clin. Pharmacol.* **2004**, *58*, 452–469.
78. 2002 Workshop on Pharmacogenetics/Pharmacogenomics in drug development and regulatory decision – making. www.fda.gov/CDER/Calendar/default.htm (accessed Aug 2006).
79. Sausville, E. A. *J. Chemother.* **2004**, *16*, 16–18.
80. Phillips, K. A.; Veenstra, D. L.; Oren, E.; Lee, J. K.; Sadee, W. *JAMA* **2001**, *286*, 2270–2279.
81. Norton, R. M. *Drug Disc. Today* **2001**, *6*, 180–185.
82. Evans, W. E.; McLeod, H. L. *N. Engl. J. Med.* **2003**, *348*, 538–549.
83. Weinshilboum, R. *N. Engl. J. Med.* **2003**, *348*, 529–537.
84. Weinstein, J. N. *N. Engl. J. Med.* **2000**, *9*, 1408–1409.
85. Nicholson, J. K.; Connelly, J.; Lindon, J. C.; Holmes, E. *Nat. Rev. Drug Disc.* **2002**, *1*, 153–161.
86. Reichert, J. M. *Nat. Rev. Drug Disc.* **2003**, *2*, 695–702.
87. De Andres-Trelles, F. *J. Chemother.* **2004**, *16*, 19–21.
88. Directive 2004/27/EC of the European Parliament and of the Council of 31 March 2004 amending Directive 2001/83/EC on the Community code relating to medicinal products for human use. http://pharmacos.eudra.org/F2/review/doc/final.publ/Dir.2004.27.2004030.EN.pdf.
89. Regulation (EC) no 726/2004 of the European Parliament and of the Council of 31 March 2004 laying down Community procedures for the authorisation and supervision of medicinal products for human and veterinary use and establishing a European Medicines Agency. http://pharmacos.eudra.org/F2/review/doc/final.publ/Reg.2004.726.20040430.EN.pdf.
90. http://biomarkers.org (accessed Aug 2006).
91. http://www.fda.gov/cder/guidance (accessed Aug 2006).

Biography

Paolo Preziosi is Professor and Chairman of the Institute of Pharmacology, Catholic University, School of Medicine, Rome. He received his MD degree from Federico II Naples University and completed his training at the local Institute of Pharmacology, followed by a research fellowship at the Department of Pharmacology of Ghent University, Belgium, under the mentorship of Nobel Prize winner Corneille Heymans. He has authored or co-authored over 160 peer-reviewed papers in international journals in various pharmacological areas, mainly related to the pharmacology of the pituitary–adrenal axis, hypophyseal hormone release, mechanisms of endocrine and immune-related toxicity by anticancer drugs, and pharmacodynamics of various classes of drugs, and more than 300 abstracts. He holds many positions and offices in the pharmacological field: he was formerly President of the International Union of Toxicology and the Italian Societies of Pharmacology, Toxicology and Chemotherapy; Member of the Board of International Union of Pharmacology, of the European Society of Toxicology, of the Scientific Committee for Medical Products and Medical Devices, European Commission, Brussels, and the Scientific Committee for evaluation of medicinal drugs, Ministry of Health, Italy; and he is a fellow of the Royal College of Pathologists, the Czechoslovakian Academy of Medicine J Purkinje, the Academia Europaea, the French Académie Nationale de Pharmacie. He is also a member of the British Society of Pharmacology, the Society of Toxicology (US), the Société Française de Pharmacologie, and the Société belge de physiologie et de pharmacologie fondamentales et cliniques. He is honorary member of the International Society of Chemotherapy, of the European Society of Toxicology, and of Italian Societies of Pharmacology and of Toxicology. He is on the editorial boards of several journals.

2.07 In-House or Out-Source

R D Connell, Pfizer Global Research and Development, Groton, CT, USA

© 2007 Elsevier Ltd. All Rights Reserved.

2.07.1	**Introduction**	**204**
2.07.1.1	Context for Chemistry Out-Sourcing	204
2.07.1.2	The Stages of Drug Discovery	205
2.07.1.3	Common Sourcing Terms	205
2.07.1.3.1	Domestic in-house	205
2.07.1.3.2	Domestic out-sourcing	205
2.07.1.3.3	Offshore out-sourcing	205
2.07.1.3.4	Captive offshoring	206
2.07.1.3.5	Domestic providers	206
2.07.1.3.6	Offshore providers	206
2.07.1.3.7	Hybrid vendors	206
2.07.1.3.8	University-based providers	206
2.07.2	**Why Out-Source Discovery Chemistry?**	**207**
2.07.2.1	Expand or Improve Capabilities	207
2.07.2.2	Gain Access to a Larger or More Diversified Talent Pool	207
2.07.2.3	Staffing Flexibility	207
2.07.2.4	Reduce Labor Cost	208
2.07.2.5	Minimize Risks Associated with Acquiring Assets and Financial Flexibility	208
2.07.2.6	Increase Value-Added Activities of Internal Staff	208
2.07.2.7	Improve Performance	208
2.07.2.8	Commercial Considerations	209
2.07.3	**Developing a Discovery Chemistry Sourcing Strategy**	**209**
2.07.3.1	Defining the Primary Chemistry Activities	209
2.07.3.1.1	Scale-up or compound resynthesis	211
2.07.3.1.2	Analog production (singletons)	211
2.07.3.1.3	Parallel chemistry-enabled synthesis	212
2.07.3.2	Determining the Sourceability of a Chemistry Activity	212
2.07.3.2.1	The maturity of the process	213
2.07.3.2.2	Control of the process outcome	213
2.07.3.2.3	Providing a competitive advantage	213
2.07.3.2.4	Potential to generate intellectual property	214
2.07.3.2.5	Activity interdependency	214
2.07.3.2.6	Existence of external capability	214
2.07.3.2.7	Current company or industry practice	214
2.07.3.3	Determining the Offshore Ability of an Activity	215
2.07.3.3.1	Communication requirements of the activity	215
2.07.3.3.2	Import/export considerations	215
2.07.3.3.3	Intellectual property risk	215
2.07.3.3.4	Existence of external capability	216
2.07.3.4	Clustering of Activities for Further Analysis	216
2.07.4	**Assessing Performance**	**217**
2.07.4.1	Analog Production	218
2.07.4.2	Scale-Up Activities	218
2.07.4.3	Parallel Chemistry	218
2.07.4.4	General Caution on Metrics	218

2.07.5	**Risks**		**219**
2.07.5.1	Intellectual Property Loss		219
2.07.5.2	Staff Morale		219
2.07.5.3	Competition for Quality Service Providers		219
2.07.5.4	Introducing Organizational Complexity		220
2.07.5.5	Communication		220
2.07.5.6	Publicity		220
2.07.5.7	Erosion of the Cost Differential		220
2.07.6	**Conclusion**		**220**
References			**221**

2.07.1 Introduction

In his book on strategic out-sourcing, Maurice Greaver frames the issue of out sourcing in a clear and compelling context:

> Out sourcing is not a secret formula for success, not a magic wand for vaporizing problems, and not a management fad. It is just a management tool – not 'the tool' but a newer tool, which is not well understood. In a capable executives toolbox, it coexists with tools such as benchmarking, reengineering, restructuring, and activity-based cost management.[1]

To many chemistry managers and heads of discovery research, the discussion of chemistry out-sourcing takes on a much less dispassionate feel than the view expressed by Greaver. Many in the pharmaceutical industry view discovery chemistry as a 'core competency.' For this reason, out-sourcing this highly iterative, cycle-time critical activity with significant potential to generate intellectual property (IP) appears counter-intuitive at first blush. There are advocates on both sides of the equation. On one hand, the success and growth of the out sourcing market related to pharmaceutical R&D has caused some to ask whether all of pharmaceutical R&D can be out-sourced.[2] Others assert that certain activities (like the lead optimization phase of drug discovery) do not fit into the compartmentalized component model of out-sourcing.[3] Others who take a broader perspective on the business of out-sourcing business process functions now suggest that out-sourcing will lose luster within large organizations.[4]

This chapter is directed to stakeholders of chemistry sourcing in order to help place this type of sourcing in perspective. It is directed to the managers of R&D organizations who are considering what activities are appropriate for sourcing and the risks to consider when out sourcing all or a portion of their chemistry activities. It is also directed at contract research organizations (CROs) participating in chemistry services. By offering a perspective on the perceived risks and benefits from a big pharma perspective, it may be possible to better consider what services, activities and performance criteria are considered important to your customers. Lastly, this chapter may also provide bench chemists in industry with a business perspective of the chemistry landscape that may help them better reflect on what they or their management may view as value-added activities within the chemistry discipline.

2.07.1.1 Context for Chemistry Out-Sourcing

Given the well-documented productivity gap of big pharma,[5,6] R&D organizations face the business imperative to increase the output of clinical candidates relative to the amount of resources invested. The initial focus was on in licensing of advanced clinical candidates. However the competition in this area has produced a dearth of advanced clinical candidates, resulting in companies seeking to license compounds that are substantially less advanced in their clinical development. It remains to be seen if this early stage in licensing will indeed achieve its goal of reducing attrition and increasing productivity.

Sourcing of all varieties (out-sourcing and in-sourcing) has always been a component of the R&D process that has helped reduce the cost associated with drug development through efficient management of internal capacity and workflow.[7] In discovery, companies have routinely established strategic alliances with biotechnology companies to gain access to a technology without buying the physical assets, the equipment, or the company in order to do so. Estimates vary, but some suggest that by 2010, more than 40% of R&D is projected to be out-sourced to more specialized firms in order to efficiently maintain a strong and vital pipeline for new blockbuster drugs.[8]

In the chemistry business area, early deals and alliances tended to focus on access to natural products: an area that many companies lacked a scientific expertise in. However in the late 1990s, the desire to grow small-molecule, 'druglike,' compound collections led discovery organizations to sign deals with companies that produced compounds of a parallel chemistry origin. The smaller companies making these deals tended to offer a platform where parallel chemistry-enabled compounds could be produced in large numbers and at a lower cost than the traditional methods used inside many major R&D organizations. Over time, as the ability to prepare parallel chemistry enabled compounds grew, many small companies shifted their business model to licensing access to their small-molecule collections that were billed as more 'druggable' relative to previously offered compound collections.

As the 'out-sourcing' of parallel chemistry-enabled chemistry was increasing, the merger and acquisition within big pharma led to the growth of small CROs that focused on the type of chemistry activities critical to drug discovery that were not suitable to parallel chemistry-enabled technologies. Many of these companies were started in the countries where the large R&D organizations housed their discovery centers such as the US, EU, and UK. More recently, expatriates who had worked in large pharmaceutical organizations have started to found companies in China and India, a trend seen in a number of industries.[9] Clark and Newton have provided an excellent review of this history and the partnerships that were formed during this period.[10]

2.07.1.2 The Stages of Drug Discovery

Before delving into the sourcing analysis, it is worthwhile to review the stages of drug discovery that relate to the quality and value of the lead compounds or lead matter. It is lead matter that is systematically modified in the drug discovery process that ultimately produces clinical candidates. This discovery process is often broken out into three distinct stages:

1. Hit follow up. This is the stage that follows the screening of the corporate compound file against an enzyme or target protein. The compounds that are identified at this stage are validated for their identity and potency and related analogs are prepared and tested to confirm a trend in activity.
2. Lead development. At this stage, the lead series are optimized for in vitro and in vivo activity. Initial screens for selectivity, safety, and drug metabolism are introduced to ensure the lead is not just potent but is becoming druglike.
3. Candidate seeking. At this stage, the compounds are optimized for their suitability as a clinical candidate. Leads that are evaluated at this stage are assessed more extensively for their pharmacological profile, safety, and drug metabolism properties. Pharmaceutical science plays a significant role at this stage and the drug attributes such as solubility, salt form, synthetic route and cost of goods come under increased scrutiny.

A lesser-known element of the discovery process is the high level of attrition associated with the progression of lead matter from hit follow-up to clinical development. An article by George Milne of Pfizer in 2003 indicated that of a hundred programs or targets that begin the discovery process, only one progresses to the market (see **Figure 1**).[11]

2.07.1.3 Common Sourcing Terms

Sourcing can be characterized into four main categories depending on the relationship of the service provider to the client.

2.07.1.3.1 Domestic in-house
A model where the activity is conducted internally, within the current footprint of the R&D organization and performed within the same country or region from which the activity is sponsored.[12]

2.07.1.3.2 Domestic out-sourcing
A model where the activity is conducted externally using third-party resources within the same country/region as the client R&D organization.[13]

2.07.1.3.3 Offshore out-sourcing
A model where the activity is conducted externally using third-party resources in a different country or region from that of the sponsor location. Often referred to as offshoring, this type of relationship has often been associated with clients in the US or European Union (EU) and providers countries such as the UK,[14] India,[15] China,[16] and Russia.[17]

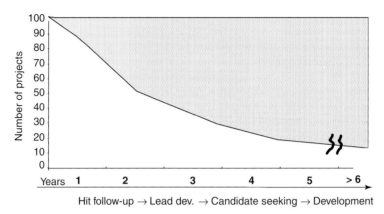

Figure 1 Historical attrition of discovery programs as a function of stage. © Pfizer Inc. 2005.

2.07.1.3.4 Captive offshoring

In this model, the activity is conducted within the R&D organization located in a different country or region from that of the larger, more established US or European research centers. The facilities are typically located in a far-shore venues. Examples of this operating model include the Roche[18] operations in Shanghai, China, and the GSK,[19] Novartis[20] or AstraZeneca operations in India. Lilly,[21] and more recently Wyeth[74] have established slightly different organizational models of a captive offshore facility. In these build-to-own (BTO) models, the companies have a dedicated set of chemists and support staff working in a facility that has the corporate logo on it, but is maintained and managed by an external service organization. Robert Ruffolo, R&D President at Wyeth recently described this arrangement at a presentation in 2005[22]:

> "As we speak we're opening up medicinal chemistry centers. They'll have the Wyeth name on it [but] it will be a building ... with no Wyeth employees in it, except one managing it. According to this report, the centers will be located in a number of different countries though no specific countries were named at the time."

A related set of terms is emerging to describe the variety of service providers who operate in the discovery chemistry space.

2.07.1.3.5 Domestic providers

These CROs are colocalized in the countries of their clients. Many tend to be founded or led by faculty of nearby universities or were founded by ex-pharmaceutical alumni who started their own companies.[23]

2.07.1.3.6 Offshore providers

This refers to any service providers who are on a different continent that the client. For US companies, this often refers to companies in the EU, Russia, India, or the Asian rim. Scientists or managers formerly from large, US- or UK-based pharmaceutical companies have founded a surprising number of these companies. Within this group of providers, one can further break providers down into near-shore providers and far-shore providers. While the descriptors of 'near' and 'far' shore do refer to geographical distance from the client R&D organization, of greater relevance is that 'far' shore venues typically are in developing countries with lower cost bases than near-shore venues.

2.07.1.3.7 Hybrid vendors

This is an emerging type of business model where a CRO will have a presence in both low- and high-cost sourcing markets. More commonly it will be headquartered in a high-cost market but there are instances where it could be based in a low-cost market and have a presence in a high-cost market. This formation of a hybrid organization would usually be driven by the need to obtain greater breadth of service/capability than a single country model. For example, some US-based CROs have announced either a captive offshore facility in Asia,[24] while several Chinese-[25] or Russian-based[26] companies have established US facilities. For clients who are either unwilling or unable to tap into low-cost markets directly, these hybrid arrangements may represent a convenient way to access this labor pool.

2.07.1.3.8 University-based providers

This is a different kind of hybrid model where the CRO leases and uses laboratory space from a university. In some models, a professor at the university may be on the payroll of the CRO and postdoctoral scientists or graduate students may be employees of the CRO. Many small start-up companies use this approach to build a client list and identify key talent prior to building their own facilities. While reports have emerged citing the success of this model,[27] the paucity of companies operating with this kind of model may reflect its niche or transitory role in the larger sourcing industry.

2.07.2 Why Out-Source Discovery Chemistry?

Although chemistry is central to the drug discovery process, there are several reasons to consider sourcing all or a portion of selected chemistry activities. Some of these points are listed below. Additional aspects related to the topic of out sourcing drug discovery have been previously described in the literature.[28]

2.07.2.1 Expand or Improve Capabilities

A given service provider could offer unique synthetic expertise (e.g., experience with heterocyclic chemistry, chiral transformations, or resolutions[29]), capabilities (to handle highly corrosive or explosive transformations involving fluorine, nitro-containing compounds, hydrazine, azide, diazomethane, or peroxide chemistries), high potency compounds such as beta-lactam or cytotoxic chemistries, etc.,[30] or technologies (massive parallel synthesis capabilities,[31] high pressure chemistry,[32] high temperature reactions) that are not available internally. The need for these competencies tends to be episodic in an R&D organization. However, when they are needed, time and access to knowledgeable staff are often critical factors in the decision to procure these services externally. While the cost avoidance associated with the purchase of assets is an independent driver to sourcing (see below), this aspect is especially important in the areas of chemistry described above where the need for this expertise is not anticipated to be routine.

2.07.2.2 Gain Access to a Larger or More Diversified Talent Pool

The sourcing of activities to regions outside of ones typical hiring market provides a firm with access to a larger talent pool. If the sourcing partner is located in a region outside of where the client operates, access to a diversified talent pool may expose the client's staff to different ways of thinking or approaches to problem solving. Countries with a strong reputation for developing talented scientists in chemistry include Russia,[33] China, and India.[34,35] In addition, many Western-educated scientists from these regions are now returning to their home countries to start or join local chemistry service providers and pharmaceutical or biotech companies. The cultural or industrial experience that these expatriates provide can further enhance the efficacy of communication between the CRO and the pharmaceutical clients they work with. Of course, an external collaborator does not necessarily need to be from another country to provide a diversity of thought. Many domestic providers who have been trained or have worked in a different corporate culture can also provide a diverse perspective that enhances the decision-making within an organization.

2.07.2.3 Staffing Flexibility

There are times in an R&D organization where the demand for synthetic chemistry will exceed capacity. Faced with this prospect, the company can invest for the long term and grow the size of its permanent staff by starting or leveraging a recruiting, hiring, and training program.[36] However if the long-term demand for chemistry capacity is uncertain, or if the need for an immediate infusion of trained chemists is high, having access to quality out-sourcing capabilities will allow the organization to resource its immediate needs while it weighs the decision to grow its internal organization for the future.

Similar to the rise and fall of short- and long-term staffing needs, there are times when the financial resources of an organization can fluctuate. During times of financial surplus, out-sourcing chemistry projects that are important but not urgent can be achieved when they might otherwise not have been addressed. This is especially useful for the sourcing of monomers and templates (M&Ts: see below). Conversely, during times of financial shortfalls or distress (bankruptcy, attrition of clinical candidates or marketed drugs), the ability to leverage the contractual flexibility of CRO staff may reduce the need to cut back permanent staff within the R&D organization.

Lastly, there is an efficiency component of dynamic staffing that is sometimes overlooked in terms of flexibility. The orientation of CRO chemists is to simply provide whatever target compounds are requested, regardless from which programs they originate. In contrast, the decision to move internal chemists from one program to another can have implications related to a sense of program completion or anxiety about missing the key compound that could turn the program around. If internal program staffs are moved frequently from one program to another, there may be aspects of career development that need to be factored into the decision. Such aspects are not a factor for the typical CRO chemists, particularly contract chemists who provide more synthetic capabilities rather than medicinal chemistry. This is not to devalue the tenacity or instincts that internal chemists routinely demonstrate. In fact, this tenacity and a knowledge of the program details has also rescued many a clinical candidate from program termination.[37] However with regard to advantages to the flexibility of sourcing, this particular human aspect is a point worth noting.

2.07.2.4 Reduce Labor Cost

Aside from technology-assisted areas of chemistry, synthetic chemistry continues to be a people-intensive activity. In labor-intensive efforts such as analog production, the ability for a CRO to produce compounds at a lower cost per chemist will lower the overall cost of compound production. The differences in labor rates are difficult to determine with great precision. Kathleen M. Schulz of BMGT said that PhD chemist salaries in India and China are 20–30% of US salaries.[38] An article by McGoy noted it costs roughly $60 000 a year to hire a full-time equivalent (FTE) chemist in China or India while rates in the US range from $200 000 to $250 000.[39] For this reason, companies continue to leverage their global reach not only to gain access to intellectual capital from around the world, but also to leverage the arbitrage of labor costs that presents itself globally in selected areas of synthetic chemistry.

When looking at labor rates, it is important to assess the fully loaded costs of a CRO chemist relative to the internal cost. For example, it is worth determining if the lower labor costs are offset or enhanced by the cost of reagents, capital, and/or fees associated with customs or other import/export transactions. It is also important determine if the quality and productivity are similar to that provided by the client's internal chemists. This latter aspect is discussed in the metrics portion of the chapter.

2.07.2.5 Minimize Risks Associated with Acquiring Assets and Financial Flexibility

For large and small companies alike, the ability to sample a technology via an out-sourcing arrangement provides a significant advantage. An organization may de-risk the investment by acquiring access to assets that have been field tested by a service organization on the corporations' projects. This deferral may result in the organization buying the same technology or the next generation of that technology, thereby reducing the cost of upgrading the technology had it been purchased directly. The aforementioned example with parallel chemistry is such a case where this type of 'renting' prior to buying can help to inform the decision whether to purchase the technology internally.

In addition, the ability to out-source functions or tasks may free an organization to invest in areas that maximize cash flow or profits, or provide financial flexibility. For large pharmaceutical companies, stock value is predominately driven by growth versus assets creating incentives to utilize out-sourcing to minimize capital investments. This capital avoidance may also be pertinent to small companies that are just reaching the point where chemistry resources are needed. Small companies avoid financing assets that will not yield positive cash flow such as buildings, associated equipment (rotary evaporators, computers, etc.) and the supporting analytical instrumentation (nuclear magnetic resonance (NMR) instruments, mass spectrum devices, spectrophotometers, etc.), needed to enable a drug discovery organization.

2.07.2.6 Increase Value-Added Activities of Internal Staff

Discovery chemists take on a variety of tasks associated with drug discovery. While some are highly innovative activities which result in generating IP for the organization, others are repetitive, low-value tasks (compound resynthesis, preparation of custom monomers for a parallel chemistry library, etc.) that, while necessary, could be delegated to others or taken off-line from other, higher valued or time-critical activities where the organization should be focusing its attention. As Steven M. Hutchins, director of out-sourcing at Merck and Co., was quoted as saying, "We want our medicinal chemists concentrating on projects and not spending a lot of time producing intermediates and reagents."[40] From a strategic perspective, aligning high-salaried staff with highly valued activities and moving low-value work to lower-cost providers makes good business sense.

2.07.2.7 Improve Performance

An article by Chrai highlighted that a nonfinancial benefit to out-sourcing is that it tends to promote competition.[41] In his analysis, he asserts that the competition among suppliers ensures the availability of higher-quality goods and services. An article by Tharp suggested that a careful assessment of performance versus cost metrics revealed a 30–80% cost savings by out-sourcing activities that were run both internally and externally.[42] Easton asserts that CROs are expected to perform at higher levels of productivity than our own (internal) function in order to generate profits on what for the client may be an overhead function.[43] He also cites inferior travel policies and flatter management for contributing to this enhanced performance of sourcing organizations. This presumably results in a higher percentage of the CRO staff working at the bench, which should translate to a higher output per employee than what may be observed in traditional Western-styled drug organizations. Others note that contract staff are positioned to make more efficient use of their time for the client since they do not have to be involved with personnel issues.[44]

Prahalad notes that because remote development and delivery demands clear documentation, the process capabilities of both the customer and the vendor improve as a result of a sourcing relationship.[45] Most often, for sourcing relationships to work, the customers have to get their legacy processes cleaned up. Many firms have found that out-sourcing helps in better documentation of internal processes.

2.07.2.8 Commercial Considerations

Large pharmaceutical companies typically invest significantly in research. At the 2004, American Chemical Society/Pharma Leaders Meeting in Cambridge, MA, it was noted that the large potential market size and growing demand of customers in these parts of the world, coupled with the potential cost efficiencies they present, make movement of certain R&D functions an attractive option for companies.[46] James Shaw, president for China of Eli Lilly & Co., noted that the $10 billion pharmaceutical market in China is growing at double-digit rates and is likely to be the world's third largest, after the USA and Japan, within a decade.[18] "China is important today," Shaw says. "But in the future it will be critical." In addition, countries such as India are considering 10-year tax holidays to pharmaceutical companies involved in R&D in India.[48]

2.07.3 Developing a Discovery Chemistry Sourcing Strategy

For both the small biotech and the mature, fully integrated pharmaceutical organization, it is essential to establish a process to determine which activities lend themselves to being sourced. The first step in this process is to disaggregate the unit of work being assessed into discrete activities. Once the list of target activities is established, it is important to assess the activity against a set of criteria supporting the dimensions of activity ownership or control.[49] This step will help assess the suitability of the activity to be done by others outside of the R&D organization or whether it is better done inside the confines of the R&D organization.

If an activity is assessed as being sourceable, the next step is to determine the appropriateness of offshoring the activity. This methodology requires additional steps including assigning weights and scores to the criteria per activity, calculating the fully loaded cost differential for each activity, and assessing the external marketplace per activity. However for the purpose of this chapter, the application of this sourcing assessment will focus only on the initial three steps.

2.07.3.1 Defining the Primary Chemistry Activities

Many chemistry departments tend to be divided into the major subtheses of chemistry such as analytical chemistry, computational chemistry (which may include various components of information technology (IT) support), or synthetic chemistry (often used interchangeable with medicinal chemistry). Pharmaceutical sciences tends to be a development activity, but some companies have this as a preclinical function. For small companies that presently lack these large department functions, these departments units may be an appropriate sized bundle of activity to assess the suitability of procuring these services rather than building them.

For large pharmaceutical organizations that may have these functions already in place, it is difficult to do an analysis on such large departments that may have multiple subactivities with differing activity or control criteria. Moreover, not all companies have the same activities in the same organizational lines. For these reasons, it is important to look at the subactivities or skills that make up an activity when initiating a sourcing assessment. When one starts to get to this level of detail, it is possible to tabulate a series of subspecialization or subactivities within larger units or departments and generate hierarchies such as that shown in **Figure 2**.

A review of a hypothetical mapping process will identify several types of activities. Some, like combustion analysis, are stand-alone activities that few companies today run internally. Others activities, like the design of an improved molecule for a lead optimization program, involve related activities such as data visualization, target or analog design, and perhaps even route design skills. For such activities, it may be beneficial to cluster all three activities into the broad category 'Lead design activity' though separate skills and competencies may be involved. Lastly there will be some activities that are similar in methodology but different in application when viewed across the organization. An example may be route design (in the synthetic chemistry unit) and process chemistry in the pharmaceutical science area. There may or may not be benefits to assessing this process as a unit rather than as two separate sourcing assessments.

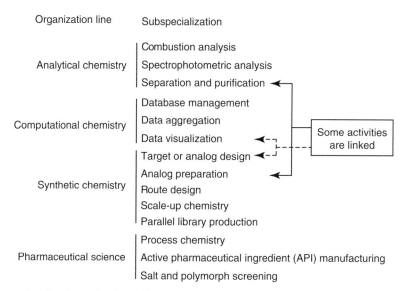

Figure 2 Deaggregating chemistry-related activities. © Pfizer Inc. 2005.

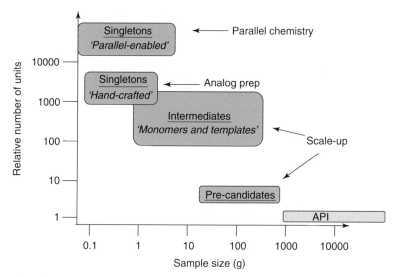

Figure 3 Subcategories of synthetic chemistry. © Pfizer Inc. 2005.

When a target activity is broad and complex, is composed of distinctly different work streams, and offers an attractive opportunity to evaluate available external market offering (e.g., structure–activity relationship (SAR) analysis), it may be helpful to break down the activity into smaller subcategories. Subcategories are activity-centric and can be determined by an understanding of where resources are focused (staff/budget) and as such are typically either labor-intensive activities (large staff count) or specific technology-dominated areas (low staff count but technology intensive). For example, synthetic chemistry can be thought of as the art and science of making molecules. However, depending on the amount of material prepared, the preparation may involve some unique skills or capabilities. Alternatively, if the number of compounds produced per unit time is anticipated to be large for one type of activity and sparse for another, this could become a distinguishing characteristic to create subcategories of activities. As **Figure 3** indicates, one way to visualize certain subactivities of synthetic chemistry is to array them in a grid according to number of compounds produced versus the quantity of material typically desired. For the purposes of this analysis, the major activities in this space include parallel chemistry, analog preparation, and scale-up of compounds. While vendors exist for the preparation of metabolites,[50] or reference standards,[51] this represents a small percentage of the discovery chemistry business and will not be expanded upon in this chapter.

The process of identifying subactivities can be either introspective as defined above or may be more opportunistic based on benchmarking the activities that peer organizations have out sourced or for which CROs have offered services. With an eye toward this latter approach, it is worth reviewing three activities in particular that tend to attract a lot of discussion when companies identify areas to review for their sourcing potential.

2.07.3.1.1 Scale-up or compound resynthesis

The out-sourcing of active pharmaceutical ingredients (API) is a well-established area of pharmaceutical out-sourcing.[52,53] This activity is distinct from discovery chemistry sourcing in that it often centers on a well-characterized compound that has entered clinical development. It is sourced in larger quantities (kilogram to kiloton) than the scale-up typically encountered in discovery (typically kilogram or less).

In the discovery space, the step prior to clinical candidate identification often produces a small set of compounds that may themselves be named as clinical candidates. These pre-candidates are often identified late in a program, and tend to be needed in multigram quantities for more advanced, in vivo and safety profiling. The number of compounds identified as pre-candidates are usually small (fewer than five) and the amount required for profiling may vary depending on how extensive a set of studies will be required prior to nomination. Some key qualitative aspects to out sourcing pre-candidates include:

- An established protocol. Since these compounds have been prepared already, there will be a procedure for how to make these compounds. Depending on how robust the original synthetic procedure is, a reputable CRO should be able to reproduce the original procedure (base case) or improve upon the yield or process (best case).
- Time is a critical factor. While the protocol to make the compound may not be optimal, it may be important to have the compound made by the known, existing route than invest time and resources in optimizing the route or seeking to improve the yield. Often the resynthesis of compound in large quantities is on the critical path for in vivo profiling, and the scheduling of these studies with either internal or external collaborators requires adherence to strict timelines. Delays in delivering pre-candidates from a CRO could significantly jeopardize the nomination of a candidate and may irreparably harm the reputation of a CRO with regard to future business from within the client's organization.

An equally important, though more common, scale-up activity in the discovery sourcing area is the production of monomers and templates (M&Ts). M&Ts do not tend to have significant pharmacological activity as fragments; rather they tend to be the building blocks for more advanced leads or pre-candidates. While some M&Ts are available commercially, companies tend to source these fragments for reasons of cost or convenience. The time-criticality of these agents can vary depending on their use. In programs where analogs for an active discovery program are produced from a template, it may be very important to have a template produced in a timely fashion. In contrast, there are some M&Ts that are used as part of a proprietary corporate inventory for use in library production or as a repository for future use. These fragments may be less time-critical and their sourcing may be driven more by cost and quality rather than speed to delivery to the R&D organization. Some key qualitative aspects to out sourcing M&Ts include

- Their IP potential. Depending on the novelty, M&Ts may or may not be novel or have appeared previously in patents or publications. Hence there may be significant IP associated with their efficient synthesis, despite their lack of pharmacological activity and building block use in discovery. To this end, it is important to select CROs that have a rigorous IP protection policy in place to ensure that 'IP slippage' does not occur.[54]
- Cost may be a significant driver. In the area where M&Ts are used to stock a corporate repository, the cycle time between requests for synthesis and compound arrival may be less of a factor than the cost per gram. While many of these fragments will have a protocol for their production, an opportunistic CRO with an experienced process chemistry team may seek to optimize the route in order to deliver the desired compound within specifications at a lower cost. This may position the CRO for favorable consideration when it comes to the re-supply of these fragments in the future.

2.07.3.1.2 Analog production (singletons)

Analog synthesis that is not technology-enabled relates to the preparation of compounds in the traditional 'hand-crafted' ways using bench chemistry. In contrast to library production, the production of individual compounds or 'singletons' is often reserved for compounds that are produced on scale or compounds that do not lend themselves to parallel chemistry production. For the purposes of these discussions, the analog production stage represents the first time a compound is made for testing. As such, there are several key aspects to out sourcing analog production including:

- IP protection. As mentioned above, many of these compounds are made early in the discovery phase when the size and scope of the chemical space explored is small and the pharmacology and drug metabolism of these compounds are not yet optimal. For this reason, companies may choose to defer filing patent applications on the initial compounds until a clearer SAR has emerged. Given that the out-sourcing of this activity may have novel, unpatented compounds prepared at a CRO for the first time, the contract language and due diligence related to externally sourcing this activity is exceptionally important.
- Problem-solving is key. As noted previously, the external sourcing of novel analog will necessitate that the CRO prepare target compounds without an established route. For this reason, a key skill for CRO chemists to demonstrate is the ability to problem-solve in the area of route design. Having noted that, the ability and willingness of the chemists at the contracting R&D organization to research and provide as much information as possible to the CRO chemists remains the primary determinant of the success of out-sourcing analog production.

2.07.3.1.3 Parallel chemistry-enabled synthesis

As mentioned previously, parallel chemistry represents one of the most actively out-sourced areas of discovery chemistry. The heavy investment required for new technologies between 1995 and 1999 made pharmaceutical companies wary of buying or building platform technologies resulting in a preference for fee-for-service or leasing-type deals.[55] As the technology associated with parallel chemistry became more established, the capital costs were reduced and multiple companies able to offer this service started to appear. This resulted in two phenomena that had an impact on the sourcing of parallel chemistry:

- Commoditization of the market. The ready access to parallel chemistry-enabled technologies allowed both R&D organizations as well as second-generation CROs to start parallel chemistry teams in their organizations. Over time, as the chemistry on plates became more established and robust synthetic protocols were growing in number, the market for parallel chemistry started to fractionate into the high-volume, commodity-type library providers versus companies that could claim some element of added value to their process or products. This bifurcation of an activity is not unique to parallel chemistry and has been reported for other 'craft' processes that have become standardized.[56]
- Migration up the value chain. While the commoditization of the market led to some consolidation,[57] several companies that started as library-based organizations have either migrated their science platforms,[58] or merged to jump-start their migration to an integrated drug company.[59] The risks of sourcing parallel chemistry to companies with their own internal drug discovery engine are discussed later in the chapter.

As a result of the standardization and reduced cost of parallel chemistry-enabling technologies, it is now common that most companies doing analog production sourcing have the capabilities to do parallel chemistry-enabled work as well.

These three activities (scale-up, analog preperation, parallel chemistry) are highlighted because of the maturity of the marketplace associated with sourcing of these activities. For CROs looking to provide synthetic chemistry services to a client, having a service platform that encompasses most if not all three of these activities positions them favorably for medicinal chemistry out-sourcing where a diversity of subactivities are often required.

For companies seeking to build a sourcing strategy, it is important to assemble a level of scholarship on the activities being assessed. Specifically, it is important to understand the sourceability and offshore ability of an activity.

2.07.3.2 Determining the Sourceability of a Chemistry Activity

Given the central role of discovery chemistry activities to drug discovery, the decision to source a particular chemistry activity will depend on a number of factors. Some of the most important factors to consider are found below. These factors may be weighted differently for each activity and the weightings may differ from one organization to another. To some extent, this begins the first step in an R&D organization process toward assessing what its core competencies are.

2.07.3.2.1 The maturity of the process

There are many activities in discovery chemistry that are so routine and well credentialed from a protocol standpoint that they lend themselves to being done externally. The analysis of carbon, hydrogen, and nitrogen content for a compound is a classic example of a mature process, and few pharmaceutical companies perform this analysis in-house. More recent examples include the resynthesis of compounds with established protocols (e.g., M&Ts), pre-candidates, and even parallel chemistry projects where established protocols for coupling and purification are well established. In these types of 'mature' processes, there is reduced uncertainty as to the ability to deliver the desired outcome without compromising quality. In general, the prototypical mature process is one with robust documentation that can be handed to a third party with sufficient detail that they can read and executed against by the CRO staff. Ideally, it should not require a lot of judgement/interpretation.

2.07.3.2.2 Control of the process outcome

An activity, no matter how discretely defined, may still have a several substages to it. Some may include upfront planning and design that influence the success of the outcome, others are simply tactical and executional in nature. To illustrate this point, one could break down the chemistry-related steps of lead optimization into three stages with varying degrees of influence on the process outcome:

1. Lead design. The first step in the process involves an upfront, design phase where trends in biology and drug metabolism data are identified. Based on these, a hypothesis is generated directing what the next target to be made should be. Computational tools that aid in the prioritization of one lead target over several may aid this design phase. Clearly the overall success of this activity (i.e., a more valuable lead) will be highly dependent on the outcome of this first step of the process.
2. Route design and target synthesis. The design of the lead has some element of synthetic do-ability factored in. Nevertheless, the route design phase may represent a mid-level activity with regard to the impact on the outcome of the process (i.e., a more valuable lead). For lead matter requiring new routes to access the target, there is still a significant potential to influence the process outcome. Therefore the quality of the staff and quality of the idea may still impact the ability to achieve the outcome. Both this and the lead design stages rely on a significant amount of experience and judgement to execute in processes that are less mature.
3. Target resynthesis. In a situation where the prototype has been made and found to be an improvement over the original lead, there may be a need to prepare more material. To the extent the original synthesis is substantially suitable for scale-up, the activity of preparing more material could be described as executional or tactical in nature.

This example serves to demonstrate how an activity (lead optimization) has several subactivities, between which there are various levels of control for the outcome. From a risk perspective, the target resynthesis activity has little suspense (hopefully) regarding the outcome, which is different from the target design stage that involves more planning and design.

2.07.3.2.3 Providing a competitive advantage

When assessing an activity, it is equally important to determine if internally executing the activity gives the R&D organization a competitive advantage. Can the R&D organization deliver the outcome of that activity faster, more cheaply, or in higher quality if the activity were to be introduced or retained in the organization compared to external resources? An illustrative example may be the ability of two different companies to convert the same lead for a kinase program to a clinical candidate. While both companies may have chemists who could design and synthesize compounds, perhaps one company has a competitive advantage in this arena by virtue of their staff (experienced in the lead matter of this gene family and/or a track record of converting kinase leads to candidates), their screening strategy (knowing what counter screens to run and how to weigh the relative importance of the data from these screens), their technology (access to data visualization tools, structure-based drug design (SBDD), computational chemistry tools), their understanding of safety science (knowing the prior art around safety related findings and the relative importance of some data over others to drug attrition) plus other key skills, assets, trade secrets, or experience within their organization. While no segment of the process may be uniquely advantageous to the organization relative to the industry, the ability to leverage these tools effectively and in concert may be the competitive advantage. Having said that, a competitive advantage can be realized even if the skills and technologies are physically separate from each other (i.e., acquired through strategic alliances with other service providers) if the integration provides a value greater than the individual parts.

2.07.3.2.4 Potential to generate intellectual property

While IP can be generated at each stage of drug discovery, some activities have more potential to generate IP than others and some IP has more commercial value than others. For example, the resynthesis of a compound that is in the public domain with no pharmacological activity and no commercial potential for the R&D organization offers little in the way of an IP risk. Hence it is unlikely to be an area the organization will look to patent and also an area where there is reduced concern about exposing information.

Consider the production of novel analogs, and the situation becomes less clear. The value of the IP surrounding the compounds increases if the analogs enter clinical development. However, given the attrition curves seen in **Figure 1**, it is difficult to determine which programs, and therefore which lead matter, will advance to the clinic. However, the same graph would suggest that novel analog work done in the candidate-seeking stage is more likely to generate candidate-centric IP than the same work done in the hit follow-up stage where lead matter attrition is still relatively high.

Of the activities that are IP-generating, special attention and care should be made around activities that are associated with higher than average commercial value. For example, process chemistry innovations and research on salt or polymorph forms for a drug candidates undergoing clinical development may lead to IP or trade secrets that are more valuable from a commercial perspective than other applications in a candidates patent estate.[60]

2.07.3.2.5 Activity interdependency

Drug discovery is a highly interactive process that involves a variety of chemists, biologists, safety, and drug metabolism scientists to optimize lead matter and identify clinical candidates. It is therefore quite likely that some of these activities are linked to other discovery functions and are therefore cycle time critical. In addition, the same activity might have different time criticalities depending on the stage in which the activity takes place. For example, in the lead development stage, it is likely that the ability to design new target compounds will depend on the speed at which biological data can be generated. If there is a significant delay in generating data for new analogs, there will either be a slowdown in analog production or the analogs produced may not be fully informed with the new data. In the hit follow-up stage, internal chemistry resources may be directed toward the more advanced programs while lead matter in an exploratory program languishes with a waiting assay but no new analog production activity. This is a huge frustration for both the biologists and the chemists in a discovery organization. Under these conditions, analog production is rate limiting, and the interdependency of data may be less cycle time critical.

In addition to cycle time considerations, it is important to consider the complexity of integrating data or samples into the internal work stream for activities that are highly interdependent. For data streams, it may be possible to have the CRO format data in a way that lends itself more readily to circulation to the team and/or archiving in corporate repositories. For this, automation of the data uploading becomes a key enabler. Sample logistics becomes a key consideration, and it is important to think through the steps needed to have the external samples get to the place they need to be (materials management, project team chemists, biologists who test the samples, etc.). Regardless of the activity, thinking through the supporting logistics and cycle time factors associated with integrating an external provider's data or material into the clients operating unit will be an important aspect of assessing an activities interdependency.

2.07.3.2.6 Existence of external capability

The ability to source an activity relies on the existence of a quality pool of suppliers. Ideally one wants both breadth (so that one has options and leverage) and depth (expertise) of the external market. In the three main areas of synthetic chemistry (scale-up, analog production, and parallel chemistry) there appears to be an ample supply of quality providers, too great to list in this chapter. Increasingly, a market is emerging for a number of highly specialized and eclectic activities associated with drug development such as protein crystallography[61] and salt form screens.[62]

2.07.3.2.7 Current company or industry practice

For a decision-maker who is considering out-sourcing, it may be helpful to know if companies in related situations have out-sourced the activity of interest. To the extent the prior experience in sourcing was successful in the company the barrier to additional or related sourcing will be lower. Where the prior art (internal or industry-wide) contains negative experiences, understanding the root causes of the failure will either confirm the decision not to source an activity, or provide knowledge to inform the present decision to source. This decision to source an activity in the face of a negative experience assumes there are lessons to be learned and steps to be taken that will lead to a more positive outcome. In cases where only industry experience is available, it is important to evaluate as broad a data set as possible to determine which data, if any, are most informative for the situation under consideration. Having compelling examples (either internal or industry-wide) is useful for building confidence with internal stakeholders who will be needed to ensure the success of the relationship.

2.07.3.3 Determining the Offshore Ability of an Activity

After an activity has been determined to be sourceable, it is important to determine if the activity is best suited for domestic or offshore sourcing. As the sourcing industry continues to become more global, it is important to carefully consider the risk and cost benefit of domestic sourcing versus near-shore and far-shore sourcing relationships. On one extreme, domestic sourcing for US-centric companies offers the prospects of tapping into a mature and experienced CRO market that provides a shared time zone, language, business culture, and the convenience of domestic shipping and receiving of compounds. On the other extreme, sourcing from far-shore CROs in India, China, and other Asian countries offer a US-centered company the potential to tap into a highly educated workforce with a significantly reduced labor cost. If cost and experience were the only factors, the decision would be relatively straightforward.

Many of the same criteria used to assess the activity/ownership criteria for an activity (activity interdependency, external capability, etc.) can be reassessed with an eye toward the offshoreability of an activity. However when considering the offshore question, some additional criteria need to be considered (see below). As with the out-sourceability of an activity, the decision to offshore a particular activity may be weighted differently for each activity and the weightings may vary from one organization to another.

2.07.3.3.1 Communication requirements of the activity

Several articles have espoused the importance of communication to the success of pharmaceutical out sourcing.[63–65] The advent of e-mail, telecommunications, and overnight delivery of packages has enabled a number of collaborations to take place at a distance. It is therefore important to assess if the target activity require frequent, person-to-person contact to achieve the desired quality or speed for a successful outcome.

While all out-sourcing relationships benefit from excellent communication loops, some activities require less oversight or interaction than other activities. For example the resynthesis of compounds with established, robust protocols requires little oversight. However the synthesis of novel analogs with uncharted synthetic routes may benefit from more intensive feedback loops where options can be discussed in a more timely fashion when problems occur. Additionally, in project teams where data is altering the priority list on a regular basis, having the ability to reprioritize rapidly is a plus.

E-mail and telecoms can be leveraged for many of the small and acute adjustments that need to be made on the fly. However the need or desire to meet face to face on a regular basis will influence the decision to move an activity offshore verses near-shore. This personal contact is important for new relationships with established vendors but is especially critical for relationships with start-up CROs who may not have worked with big pharma before and may therefore be unfamiliar with some of the drivers the client weighs heavily in a relationship. Thus, it is important to consider how frequently the client and CRO staff will be required to meet face to face to effect a productive collaboration and to factor any additional management and oversight costs into the calculation for offshoring an activity.

2.07.3.3.2 Import/export considerations

Unlike information technology, finance, or business process out-sourcing, the transactions associated with chemistry sourcing involve the movement of material from the CRO to the R&D client. To the extent the R&D projects are cycle time critical, the introduction of import/export issues to the equation may increase the risk of time delays to the discovery process. If these risks can be managed or are not critical to the particular activity being sourced, these factors can be weighted less than other factors.

Depending on the countries involved and the nature of the material being shipped, there may also be additional costs associated with the movement of compounds across borders. While these are likely to be a small portion of the overall cost of sourcing, they still need to be factored into to the fully loaded cost of sourcing compounds.

2.07.3.3.3 Intellectual property risk

In addition to the previous IP assessment of the intrinsic activity, it is important to assess the IP environment of the region where an activity is being considered for out-sourcing. A key development of IP protection took place in 1998 when the World Trade Organization came to agreement on the Trade-Related Aspects of Intellectual Property (TRIPS). Countries had until January 2005 to comply with the terms of TRIPS, principally by allowing product patents rather than just process patents under the current regime. Understanding where a country stands with regards to TRIPS and reviewing any country laws related to inventorship that may be different from the client country is an important step to de-risking the IP risk associated with offshoring.

2.07.3.3.4 Existence of external capability

The last 2 years has seen an explosion in the number and size of offshore providers for selected chemistry activities. There is a mature market for cost-driven activities such as the synthesis of M&Ts or related scale-up activities. However there continues to be significant growth in the more technically challenging areas such as analog production and parallel chemistry. When the decision to assess the offshore potential is made, it is important to rapidly get past the general assessment of a country or the description of a company by its own documentation. An excellent way to assess a CRO's potential is by an on-site visit of qualified scientists to determine the effective external capacity for a specific activity.

2.07.3.4 Clustering of Activities for Further Analysis

With an assessment of the activities for the ownership/control potential as well as their offshore potential, it is possible to array specific activities into one of four quadrants (see **Figure 4**).

Quadrant 1 These activities are considered to have low out-sourcing potential and a low offshore potential. These activities may be at or near the core competencies of an R&D organization. If they were to be resourced by staff outside of the R&D organization, it would likely be via the in-sourcing model.

Quadrant 2 These activities are considered to have better out-sourcing potential but a low offshore potential. These activities meet the criteria that they lend themselves to being done outside of the R&D organization thought principally with domestic providers.

Quadrant 3 These activities have high out-sourcing potential and have the potential to be carried out on offshore locations. Given the cost saving potential associated with far-shore vendors, activities that fall into this quadrant should be looked at with more granularity if they are presently being sourced domestically or not being sourced in part or at all.

Quadrant 4 These activities are considered to have low out-sourcing potential but have the potential to be carried out on offshore locations. However, it may be the activity/ownership assessment suggest these activities are more suitable for work done in a captive offshoring setup.

When a series of activities are assessed against their out sourceability and offshore ability criteria, and scored on a scale of 1–5, it should be possible to generate a grid such as that shown in **Figure 5** where activities are distributed across the four quadrants. Depending on the activity, some activities may straddle the boundary between quadrants

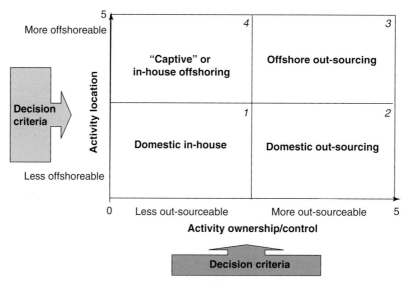

Figure 4 Clustering of activities by quadrants. © Pfizer Inc. 2005.

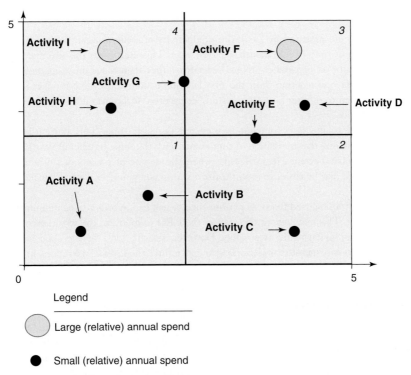

Figure 5 Simulated array of activities. © Pfizer Inc. 2005.

(i.e., activities E and G). In these situations, it may be worth assessing if the activity has a range of subscenarios associated with it. For example, if 'scale-up' were the initial activity, its composite score may be composed of the score for the scale-up of pre-candidates and the score for the resynthesis of M&Ts. Rerunning the analysis with the two smaller activities may provide outcomes that are more informative and actionable. As noted previously, it may be that the organization has activities being sourced domestically (quadrant 2) that this analysis suggests fall into quadrant 3 and its offshore potential. Alternatively, it may be that an activity is being sourced in the appropriate quadrant (i.e., quadrant 2 or quadrant 3), but only at a small percentage of the spend for that activity. It may be worth probing the activity in more detail to assess if there is a higher threshold percentage that could be realized and the effect a change in the percentage would have on the outcome of the activity.

To take this analysis further, it may be worth focusing on those activities that reflect the highest annual spend for the R&D organization. For example, while both activity F and activity D both fall in the quadrant 3, the higher spend on activity F suggest it may be worthwhile to start assessing this activity more closely with regard to its out-sourcing potential. Similarly, when considering the activities in quadrant 4, consider whether the spend is significant enough to warrant a captive offshore strategy, either alone or in combination of other smaller or interrelated activities.

Note that this is an oversimplification to serve as an example for the process. In reality, an organization may look at related activities in quadrants 1 and 4 together in assessing captive sourcing for any one line of business. However it is unlikely that any one activity in the discovery space would be of sufficient size or scope for an organization to buy or build a captive offshore facility. The decision to build a facility is most likely addressed as a part of a broader strategic discussion that includes commercial/market considerations as opposed to just R&D execution.

2.07.4 Assessing Performance

When analyzing sourcing relationships, many groups indicate that the lack of quantifiable metrics or performance standards is the main reason organizations become dissatisfied with out-sourcing. As with any service activity, it is important to determine measurements or 'metrics' of success. Depending on the activity, these metrics may center on cost, quality, and/or speed.

2.07.4.1 Analog Production

In an area where compounds are prepared by hand, the productivity of a chemist can be measured by assessing how many reactions are run per unit time (week, month, quarter, etc.). This metric has the advantage that it assesses the ability of the chemist to multitask or parallel-process reactions rather than focus on assignments in a slower, serial manner (i.e., not starting a second reaction until the first reaction is complete). The number obtained may be less meaningful if care is not taken to differentiate 'successfully completed' reactions rather than just reactions that were initiated but never completed.

An alternative method is to measure compounds prepared per unit time. This bottom-line metric is a good assessment to use when comparing the production of compounds with the same number of synthetic steps or the same degree of difficulty. This metric becomes less effective when the degree of precedence differs for one assignment versus another. For example, it may be easier to resynthesize a known compound or a related analog than to prepare the first prototype for a series.

Should either of these metrics be used to assess performance, it will be important to communicate this upfront prior to the relationship beginning. This may inspire the staff at the CRO to innovate the way they organize their work in order to drive the assessment as favorably as possibly. However, as with most metrics, too heavy an emphasis on throughput as a singular or primary measure of performance may reduce the level of risk-taking in an analog program or discourage colleagues who make be positioned to identify improvements to the process chemistry along the way.

2.07.4.2 Scale-Up Activities

There are a number of metrics associated with the sourcing of multigram amounts of material. One important metrics to consider is timely delivery of material. This metric can be assessed as the percentage of deliveries that arrive prior to or on the agreed-upon delivery date. Another performance standard for scale-up activities in discovery chemistry relate to compliance to the requested amount. High marks are given to a provider that provides the agreed-upon amount of material (or more). Conversely, if a vendor establishes a track record of underdelivering the desired amount, it may be an indication of poor raw-material planning or synthetic execution. An interesting article by Tharp and colleagues describes a performance metric for scale-up activities in which the overall performance of a sourcing activity is defined as the multiple of the two metrics (delivery date and amount of material).[42]

2.07.4.3 Parallel Chemistry

Like analog preparation metrics, the number of compounds produced per unit time is also an important metric for parallel chemistry. Since parallel chemistry is so heavily enabled by the particular technology platform used to prepare compounds, the number of chemistry transformations that can be provided by a CRO and their platform becomes an important metric. Alternatively, the CRO may be assessed on the number of new protocols they produce and use to enable chemistries. Lastly, the large technology capability that is associated with industrial-scale analog production may offer cost advantages to one CRO relative to another for the same type of compound libraries. Hence cost per compound produced may be a strong differentiator in this area, especially with chemistries that are deemed straightforward and are well precedented by protocols.

2.07.4.4 General Caution on Metrics

As with measuring the performance of internal staff, there is seldom one metric that will capture the essence of an outsourcing activity. For this reason, several metrics are often assessed in parallel. By gathering and assessing multiple metrics, it may be possible to identify trends that inform further questions about the process. Over time, one to two metrics may emerge as especially informative for an activity, or that certain metrics need to be refined to make them more informative. For all types of activities, it is important to factor in other qualitative aspects such as problem-solving, vendor–client interactions, or route optimization tasks that may not be obvious when assessments are based only quantitative metrics.

There is a cost associated with gathering metrics. Regardless of who captures the data (the R&D organization or the CRO) it takes time from somebody's day to measure and analyze the data. If the R&D organization does this, this adds to the cost of oversight. If the CRO does this, this may increase the cost of the collaboration or decrease the productivity of the laboratory chemists if they are spending less time in the laboratory on metric-related tasks. For this reason, it is important to balance the justifiable desire to gather information against the cost associated with compiling and analyzing this information.

Lastly the maturity of the sourcing market and the need to track metrics for success will lead organizations to take a closer look at the comparison between internal and external performance. This is difficult to assess when sourcing new capabilities but more straightforward when an organization has exposure to CROs that provide additional capacity. To this end, it will be important to ensure that sensible data sets are used for these comparisons as the service providers may have a more focused mission than an internal scientists. The internal staff may have a broader set of commitments to the organization than the specific activity being measured (*see* Section 2.07.5.2).

2.07.5 Risks

While external sourcing provides the prospects of significant advantages to the R&D organization, there are also a number of risks associated with such relationships that need to be factored into the decision to source an activity.

2.07.5.1 Intellectual Property Loss

As noted previously, the most important risk associated with any collaboration, partnership, or sourcing relationship centers on the potential for IP loss. As Cavella noted in an excellent review article, the product from pharmaceutical development is essentially knowledge and as long as companies intend to keep corporate secrecy at or near their top priority, the incentive to out-source will not be strong enough.[66] While the contract language that governs the confidentiality of a relationship should provide sufficient legal and contractual protection, the inadvertent or unintended loss of IP from the client to the CRO and the CRO to other customers has to be prevented. The term 'IP slippage' has been coined to describe this type of IP loss, which differs from espionage, or the actions of a rogue employee.[54]

The magnitude of the loss due to IP slippage will vary. One can lose IP if the patentable materials were disclosed through the CRO to the public domain before the client files a patent application (regardless of whether the disclosure was inadvertent or not). Alternatively, one can lose competitive advantage if trade secrets (not patentable or not intended to be patented) were taken advantage of by the CRO or appeared in the public domain. Lastly, one can lose competitive advantage or future IP if trade secret or patentable materials provided to the CRO were used as a source of information to advance competitors' programs or the CRO's internal programs. IP slippage is difficult to detect, and is easier to prevent with thorough site audits and a rigorous assessment of the IP protection policies of a potential CRO prior to engaging in a relationship. It is also worth monitoring during a relationship as business models,[68] collaboration partners, and senior leadership at the CRO can change over time.[69]

2.07.5.2 Staff Morale

The sourcing of activities previously done by internal staff within an R&D organization can generate a lot of anxiety among the personnel. Unlike business process out-sourcing where an entire activity is out-sourced, many chemistry activities are only partially sourced and the line between what can be sourced and what is being sourced can appear vague to the internal staff. For this reason, it is important to discuss the drivers for sourcing within an organization and make it clear what activities are being sourced and the reasons behind those tasks' selection for sourcing. If this is done in an organization that actively encourages and fosters staff and skill development, there will be less tension and conflict arising from the sourcing experience. However for this strategy to succeed, the staff needs to be willing and able to move to higher valued activities that are less commoditized or sourceable.

2.07.5.3 Competition for Quality Service Providers

When a company sources an activity for the first time, it is difficult to determine which service providers provide the best value for the dollar in the regions of interest. A CRO may be initially selected for a longer-term contract based on a presentation they made, a project they bid on, or word of mouth. If the CRO delivers value, and does the same with all their clients, there will invariably be other clients who will be looking to hire or expand their utilization of the CROs services. This competition for the quality service providers may translate to higher rates, less capacity to grow with the client organization, or a change in the CRO staff that manage the relationship.[70] For this reason, it may be worth considering what steps an organization is willing to take to retain or grow the relationship with quality vendors. For example, these steps could involve longer contract terms or identifying specific individuals from the CRO to remain on the team working for the client. It may also involve growing the collaboration in business areas related to the primary activity resulting in a more value-added product for both parties involved.

2.07.5.4 Introducing Organizational Complexity

The sourcing of an activity that is presently run within an organization provides unique challenges when compared to sourcing an activity that is new to an organization. For example, when a client procures an activity that introduces a new capability to their organization, the client may be much more forgiving about the disruptions that the CRO introduces to work streams, data flows, and team dynamics. However, when an organization procures additional capacity of an activity that is presently done internally, there may be a focus more on the disruption than the value (different data format, migrating data through firewalls, different sized vials, etc.). When this happens, it is usually because the teams do not see the service provider as bringing something special to the table, particularly since they have been procured simply for more capacity. The fact that the service provider is not internal requires that additional steps need to be taken to bring the same type of services, data, and material into the workflow of the project team. For this reason, it is important to anticipate the disruptions that a service provider may bring to an organization and manage the expectations around change that this relationship will impose on the organization.

2.07.5.5 Communication

In addition to managing expectations internally, there is a need to prepare for the unique challenges that working with remote service providers will bring. Discovery organizations tend to underestimate how much informal communication takes place in the matrixed environment between management and staff, members of a department or discipline, and members of project teams. For this reason, when a company procures additional capabilities with a service provider, there is often a period of adjustment where assumptions and unvoiced expectations lead to misunderstandings, unsatisfactory workflow, mistrust, or acrimony. It is therefore important to 'sweat the details' when forming or establishing a relationship and consider the value of launch meetings and face-to-face visits when starting a working relationship. Establishing a pattern post launch to discuss progress and encouraging informal or ad hoc discussions over the phone or e-mail will ensure mechanisms are in place as teams start working together. Lastly, it is important to recognize that all newly formed teams go through a forming–storming–norming–performing process of evolution.[71] By having project teams and managers discuss these stages regularly with their service provider, it is possible to move the sourcing relationships more rapidly into a high-performance mode by creating a culture where both sides can lay issues open for discussion.

2.07.5.6 Publicity

There is an unusual tension regarding the publicity associated with working with service providers. The CRO may view their ability to issue a press release on a newly signed collaboration a key value driver for their organization. From their perspective, the announcement of a deal serves to validate their business model in the industry and may drive other clients to their door. From the client's perspective, this may not be viewed as a positive development. Given the negative view that out-sourcing has in some circles, clients may prefer to avoid press releases on new sourcing collaborations, especially when they involve offshore vendors.[72] As noted above, a client that has found a productive CRO may prefer to keep the CRO in relative anonymity in order to avoid increasing competition for the staff at the service provider. To avoid miscues and misunderstandings, it is advised that both parties discuss the level of disclosure they are comfortable with prior to deal signing. This will ensure there are no misunderstandings or surprises after the deal is signed.

2.07.5.7 Erosion of the Cost Differential

While differences in labor costs can be a significant driver in the decision to source an activity, some question how long this delta in pricing will continue. A recent report in the *New York Times* notes annual raises of 10% or more are now the norm in India's $17.2-billion-a-year out-sourcing industry.[73] As the competition for English-speaking scientific talent drives up salary costs, it may also increase staff turnover which may increase the aforementioned risk of IP loss or slippage.

2.07.6 Conclusion

Success in chemistry sourcing will ultimately depend on how the management prepares the organization for its introduction and how it is utilized within an R&D organization by the staff.

Chemistry sourcing is not a new phenomenon and there are fundamental, compelling reasons that companies have utilized sourcing to complement their internal activities (expand or improve capabilities, reduce cost associated with acquiring capital or assets, etc.). What is relatively new is how companies are looking to apply out-sourcing to activities it has traditionally done in-house such as analog production and scale-up activities pre-API, activities long deemed core competencies within large pharmaceutical organizations.

Companies that will look to out-source discovery chemistry would benefit from doing an internal review of the activities it undertakes in-house and objectively assess the suitability of these activities for external sourcing. By establishing assessment criteria other than cost, such as IP risk, cycle times, and quality, a company may be able to identify tasks that lend themselves (all or in part) to be more efficiently sourced.

Once an activity is identified, it will be important to consider what process changes need to be undertaken to fully enable the sourced task to succeed. For example, do the IT systems need to be modified to enable data to be readily absorbed into the client's data network? Is there a strategy for sample logistics? Do the staff need to be retrained or educated on what changes in their roles and skill sets will be needed to enable sourcing to work for their project teams or departments? If an organization does not think through the process changes required to enable effective sourcing within their organization, they will be destined to be disappointed with the outcome.

References

1. Greaver, M. F., II. *Strategic Out Sourcing*; American Management Association: New York, 1999.
2. Crossley, R. *Drug Disc. Today* **2004**, *9*, 15.
3. Cavalla, D. *Drug Disc. Today* **2004**, *9*, 635–636.
4. Landis, K.; Mishra, S.; Porrello, K. *Calling a Change in the Outsourcing Market*; Deloitte Consulting: New York, NY, 2005.
5. Does Lack of Launches Spell End of Expansion? *Scrip Mag.* **2005**, *3025*, 24.
6. A New Dawn for Dealmakers. *Scrip Mag.* **2005**, *3025*, 38.
7. Whiting, R. Getting A Dose of Cost Cutting: Big Pharma Must Respond to Concerns about Rising Health-Care Costs and the Drug-Approval Process. *Information Week* **2005**, Sept 26.
8. Pharmaceutical and Biotechnology Companies Adopt Outsourcing Practices to Combat Rising Costs. *PR Newswire* **2005**, Nov 24.
9. Balfour, F.; Roberts, D. Stealing Managers from the Big Boys. *Business Week* **2005**, Sept 26.
10. Clark, D. E.; Newton, C. G. *Drug Disc. Today* **2004**, *9*, 492–500.
11. Milne, G. M. *Annu. Rep. Med. Chem.* **2003**, *38*, 383–396.
12. For examples of "in-sourcing" in the Chemistry space, see: Gura, T. *Science* **2004**, *303*, 303–305.
13. For recent deals that have been announced in this area, see: Seattle Signs Up Albany Molecular to Produce SGN-35. *Scrip Mag.* **2005**, May 10, 6.
14. For recent deals that have been announced in this area, see: Roche and Evotec OAI Extend Medchem Collaboration. *Scrip Mag.* **2005**, *3046*, 11. Almirall and Evotec Sign Library Synthesis Deal. *Scrip Mag.* **2005**, *3076*, 11.
15. Biocon Ties Up with Novartis for Contract Research. *Business Line* **2004**, Sept 2.
16. WuXi PharmaTech Ranks in Top Ten of Deloitte Tech Fast 500 Asia Pacific Winners. *Xinhua-PRNewswire* **2004**, Dec 14.
17. For recent deals that have been announced in this area, see: Trimeris and ChemBridge Research Laboratories Sign Antiviral Drug Discovery and Development Agreement. *Business Wire* **2005**, June 14.
18. Einhorn, B.; Magnusson, P.; Barrett, A.; Capell, K. Go East, Big Pharma. *Business Week* **2004**, Dec 13, *3912*, 28.
19. Mukherjee, W. GSK Bio Targets India for R&D. *The Economic Times* **2005**, Jan 10.
20. Novartis Buys the India Story, Looks for Partners Here. *The Times of India* **2004**, Nov 26.
21. Yidong, G. *Science* **2005**, *309*, 735.
22. Wyeth Shifting R&D Funds to Early-Stage Compound Research and Licensing. *Pink Sheet* **2005**, Oct 3, 19.
23. Companies that trace their founders to big Pharma include (a) Argenta which can be traced back to Aventis' Dagenham Research Centre (see http://www.argentadiscovery.com (accessed April 2006)), (b) Biofocus which had staff from the Wellcome Research Laboratories in the UK (see http://www.biofocus.com (accessed April 2006)), (c) Kalexsyn which was founded by ex-Upjohn/ Pharmacia chemists following the merger with Pfizer in 2003 (http://www.kalexsyn.com (accessed April 2006)).
24. Rouhi, A. M. *Chem. Eng. News* **2004**, *82*, 48–50; McCoy, M. *Chem. Eng. News* **2004**, *82*, 11; AMRI Builds in Singapore: Albany Molecular Research, Inc. Announces Preliminary Fourth Quarter and Full Year 2004 Results. *Business Wire* **2005**, Mar 4.
25. Several US/Chinese hybrid organization exist including Astatech (http://www.astatech.com.cn (accessed April 2006)) and Pharmaron (blou@pharmaron.com).
26. Several US/Russian hybrid organization exist including Chembridge (http://www.chembridge.com (accessed April 2006)) and Asinex (http://www.asinex.com (accessed April 2006)). These and several other Russian hybrids tend to be located in Moscow, which has emerged as a vibrant chemistry city for outsourcing.
27. Yarnell, A. *Chem. Eng. News* **2004**, *82*, 34.
28. Mander, T.; Turner, R. *J. Biomol. Screen.* **2000**, *5*, 113–118.
29. Mullin, R. *Chem. Eng. News* **2005**, *83*, 43.
30. Watkins, K. J. *Chem. Eng. News* **2001**, *79*, 17–19.
31. McCoy, M. *Chem. Eng. News* **2004**, *82*, 45–46.
32. Thayer, A. M. *Chem. Eng. News* **2005**, *83*, 54–61.
33. Stephens, T., State Department program brings delegation of scientists from former Soviet Union to UCSC. *UC Santa Cruz Currents Online*, **2004**, June 14 (http://currents.ucsc.edu/03-04/06-14/scientists.html (accessed April 2006)).
34. Gilman, V.; Schulz, W. G. *Chem. Eng. News* **2004**, *82*, 67–70.
35. Research & Technology Executive Council. *Out Sourcing Chemistry R&D to Low Cost Countries*; City, 2004.
36. Given the boom/bust cycles of scientific recruiting, companies are looking to a model of virtual recruitment where scientific recruiting itself is out-sourced. Mills, E. *Nat. Biotechnol.* **2002**, *20*, 853.

37. Clader, J. W. *J. Med. Chem.* **2004**, *47*, 1–9.
38. Wu, M. L. *Chem. Eng. News* **2004**, *82*, 40.
39. McCoy, M. *Chem. Eng. News* **2005**, *83*, 14–18.
40. McCoy, M. *Chem. Eng. News* **2003**, *81*, 21.
41. Chrai, S. S. *Am. Pharm. Outsourcing* **2002**, *3*, 6–13.
42. Tharp, G.; Eckrich, T. *Am. Pharm. Outsourcing* **2002**, *3*, 26–33.
43. Easton, J. C. *Am. Pharm. Outsourcing* **2005**, *6*, 28–35.
44. Panayotatos, N. *Nat. Biotechnol.* **2003**, *21*, 131.
45. Prahalad, C. K. The Art of Outsourcing. *The Wall Street Journal* **2005**, June 8.
46. Laranag-Mutlu, T. *Chem. Eng. News* **2004**, *82*, 69–70.
48. Dogra, S. A Pep for Indian R&D. *Express Pharma Pulse* **2005**, Aug 4.
49. Calvert, C. In *Creating an Outsourcing Strategy: The Make vs Buy Decision*, 9th Annual Outsourcing of Pharmaceutical Chemistry, Preclinical Development and Contract Manufacturing Conference, Jersey City, NJ, July 18–19, 2005; Strategic Research Institute: Jersey City, NJ, **2005**.
50. Dove, A. *Nat. Biotechnol.* **2004**, *22*, 953–957.
51. For companies who provide reference samples, see http://www.chromadex.com (accessed April 2006) and http://www.psd-solutia.com/cms/carbogen-amcis (accessed April 2006).
52. Harris, A. R. *Am. Pharm. Outsourcing* **2005**, *6*, 16–20.
53. Klein, A. *Am. Pharm. Outsourcing* **2003**, *4*, 8–16.
54. Connell, R. D. *Am. Pharm. Outsourcing* **2006**, 7, 51–54.
55. Crossley, R. *Drug Disc. Today* **2002**, 7, 756–757.
56. Davenport, T. H. *Harv. Bus. Rev.* **2005**, *83*, 100–108.
57. For an assessment of the Medichem acquisition by deCODE, see McCoy, M. *Chem. Eng. News* **2002**, *80*, 11–15.
58. Infinity Announces Small Molecule Collaboration With Novartis. *PR Newswire* Jan 6, 2005. Infinity Pharmaceuticals (http://www.ipi.com (accessed April 2006)) is an example of a company that has migrated from library-based collaborations to a company developing pharmaceutical compounds against known targets and targets which were previously thought to be "non-druggable."
59. Arqule Purchases Cyclis for $25M. *Boston Business Journal* **2003**, July 17. The article notes that Arqule is "a chemistry services company that is trying to make the transition to a drug company" (http://www.bizjournals.com/boston/stories/2003/07/14/daily32.html (accessed April 2006)).
60. For a case study on the relative importance of certain patents within a marketed drugs patent estate, see Connell, R. D. *Expert Opin. Ther. Patents* **2004**, *14*, 1763–1771.
61. For companies providing protein crystallography services, see Activesite (http://www.active-sight.com (accessed April 2006)).
62. For companies that provide salt form screening for API, visit Cardinal Health (http://www.cardinal.com (accessed April 2006)) or Avantium (info@avantium.com).
63. Zaret, H. E. *Am. Pharm. Outsourcing* **2002**, *3*, 30–34.
64. Wu, Z.-P. *Am. Pharm. Outsourcing* **2005**, *6*, 36–41.
65. Dunnington, D. *J. Biomol. Screen.* **2000**, *5*, 119–122.
66. Cavalla, D. *Drug Disc. Today* **2003**, *8*, 267–274.
68. McCoy, M. *Chem. Eng. News* **2005**, *83*, 15.
69. McCoy, M. *Chem. Eng. News* **2005**, *83*, 29–30.
70. The demand for contract workers has provided them with a significant negotiating position in areas where scientists re inhigh demand. See reference 12.
71. Tuckman, B. W. *Psychol. Bull.* **1965**, *63*, 384–399.
72. For an analysis of the rising, negative sentiment in the media, see reference 4.
73. Rai, S. Outsourcers Struggling to Keep Workers in the Fold. *The New York Times* Nov 12, 2005, p 13.
74. Wyeth signs outsourcing deal with Indian CRO. *Scrip Mag.* **2006**, *3124*, 12.

Biography

Rick D Connell was born in Lowell, MA and received his BS in chemistry in 1984 from Merrimack College in North Andover, MA. He was accepted into the PhD program at the University of Notre Dame and worked under the direction of Prof Paul Helquist. During that period, he received a John G Bergquist Fellowship from the American Scandinavian

Foundation and spent 6 months at the Royal Institute of Technology in Stockholm, Sweden where he carried out research in the area of organo-palladium catalysis under the direction of Prof Björn Åkermark. He received his PhD in 1989 and worked as a National Institutes of Health postdoctoral fellow at Harvard under the direction of Prof E J Corey.

In 1990, he began his industrial career at Bayer pharmaceuticals in West Haven, CT. He worked as a medicinal chemist in the field of immunology until 1993 when he transferred to Wuppertal, Germany to do research in the cardiovascular (lipid lowering) division at Bayer's main pharmaceutical campus. He returned to West Haven in 1995 and subsequently was asked to lead the Medicinal Chemistry group focused on diabetes and obesity research. In 1999, Rick accepted a position at Pfizer's Groton, CT campus as the head of cancer Medicinal Chemistry. He spent a year on secondment at the Pfizer site in Sandwich, UK where he was asked to be head of the Medicinal Chemistry group for sexual health before returning to Groton in 2003. Shortly after his return to Groton, he was promoted to Executive Director of Discovery operations at Pfizer's Groton campus with responsibility for the antibacterial, immunology, and cancer (AIC) therapeutic area zones.

In July of 2005, Rick was asked to establish and lead a Chemistry sourcing group within the Discovery Chemistry Discipline. Centered in Groton, CT, this organization has responsibility for setting and implementing an integrated, worldwide chemistry sourcing strategy for the seven Pfizer R&D sites located in the USA, UK, and Japan. This organization is responsible for managing the network of sourcing-related collaborations around the globe as well as compound logistic activities within the Pfizer organization.

Rick is an inventor on over 35 patents and patent applications. He is a section editor for the journal *Expert Opinion on Therapeutic Patents* and serves on the Editorial Advisory Panel for the journal *Therapy*. Rick lives in East Lyme, CT with his wife Nancy, and their two children: Ricky (age 15) and Kelly (age 13).

Comprehensive Medicinal Chemistry II
ISBN (set): 0-08-044513-6

ISBN (Volume 2) 0-08-044515-2; pp. 203–223

2.08 Pharma versus Biotech: Contracts, Collaborations, and Licensing

D Cavalla, Arachnova Ltd, Cambridge, UK

© 2007 Elsevier Ltd. All Rights Reserved.

2.08.1	Definitions	225
2.08.2	Service Companies	226
2.08.3	Advantages and Disadvantages of Outsourcing	227
2.08.4	Virtual Companies and Extended Enterprises	229
2.08.5	Comparison of Contracts, Collaborations, and Licensing	229
2.08.6	Trends in Biotechnology Sector Companies: Products versus Technology Platforms	231
2.08.7	Types, Stage, and Value of Alliances with Large Pharmaceutical Companies	232
2.08.8	The Growth of Licensing	234
2.08.9	Changing Role of Contract Research Organizations	236
2.08.10	Conclusion	237
	References	237

2.08.1 Definitions

Implicit in the title for this chapter is an understanding of the meaning of the term 'biotechnology.' The word grew from the early focus of a first generation of new alternatives to traditional pharmaceutical companies, upon biologically produced therapeutics rather than chemically manufactured ones. But pharmaceutical R&D on biologicals is not confined to what most people understand to be the biotechnology sector (for example, drotrecogin alfa (Xigris) is a recombinant protein for sepsis from Lilly). Nor is this sector confined to R&D on biotechnological approaches to new drugs. Many companies in the so-called biotechnology sector have little or nothing to do with the discovery and development of biological medicines, preferring to follow traditional small-molecule approaches to new therapeutics.

Another metric that may be used to define a biotechnology company is the notion that it is a start-up, or a small–medium enterprise (SME), usually loss-making and backed by private investors (typically venture capital). This is also clearly unsatisfactory since the first 'biotechs' like Genentech or Amgen are now large organizations, making substantial profits. In common with a growing minority of biotechnology companies, and like their large pharmaceutical cousins, their shares are traded on the public stock markets. A crude attempt at drawing out the different strategies and characteristics of different 'biotechnology' companies is set out in **Table 1**, from which it is clear that companies are often grouped into this sector without any clear common element. Like 'Modern Art,' it is a term that is difficult to define, but much easier to recognize.

Jurgen Drews, a well-known commentator on the strategy of pharmaceutical R&D, has suggested that the "first tier of biotechnology companies is likely to become the most effective segment of the drug industry."[1] Many such companies, in themselves quite large entities, have a mixed portfolio of developments, some biological and some small molecules.

Overall, the biotechnology sector comprises a significant portion of the pharmaceutical R&D spend. As shown in **Figure 1**, the biotechnology sector spends about half as much as large pharmaceutical sector on R&D.[2] The biotechnology sector spends about 60% of its revenue on R&D compared to about 15–20% for the pharmaceutical sector (as composed of members of the US PhRMA).

An alternative way to define sectors in pharmaceutical R&D is not by what they do, but by the way in which they do it. A discussion of the strategic organization of new medicines discovery, particularly in comparing the attributes of the large pharmaceutical versus the biotechnological approaches, cannot be complete without bringing another factor into the discussion. That is whether a company is involved in the provision of services or in the conduct of its own R&D programs.

Table 1 Categorization of biotechnology companies by various factors

Company	Public or private	Discovery or development focus	Biological or small molecule focus	Product or technology based
Microscience	Private	Development	Biological	Product
Speedel	Private	Development	Small molecule	Product
AERES Biomedical	Private	Discovery	Biological	Technology
Astex Molecules	Private	Discovery	Small molecule	Technology
Genentech	Public	Both	Biological	Product
Millennium	Public	Both	Both	Product
Cambridge Antibody Technology	Public	Both	Biological	Product
Sepracor	Public	Development	Small molecule	Product
Abgenix	Public	Discovery	Biological	Technology
Vertex	Public	Discovery	Small molecule	Both

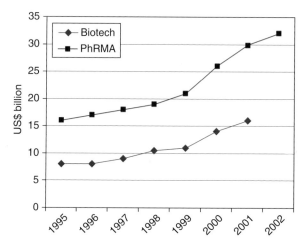

Figure 1 Biotech versus PhRMA: R&D spend in US$ Billion. (Data from Kenley.[2])

2.08.2 Service Companies

Some of the companies mentioned in **Table 1** are increasingly involved in technology service provision, whereas others are becoming more focused on developing their own projects, at least up until proof of concept clinical trials. Biotechnology sector companies that provide services have large comparators in the contract research organization (CRO) sector.

Historically, contracting out or outsourcing of pharmaceutical R&D began with chemical scale-up and bulk manufacturing activities, which had been part of the operation of the more mature industrial chemical industry for many years and became available for the younger pharmaceutical industry. The capital cost of building and maintaining chemical plant required that it should be fully utilized in order to maintain profitability; its use by a number of clients has obvious cost-saving elements. On the other hand, the requirement for biological studies changed in 1955, when the tragedy of thalidomide revealed the importance of adverse toxicology and transformed the public policy surrounding drug safety, and was furthered by the introduction in 1975 of Good Laboratory Practice (GLP). Since these two events, the pharmaceutical industry has accepted outsourcing preclinical safety studies as an integral part of its overall strategy. This can be similarly applied to the clinical work that is performed to supply proof of safety and efficacy of new medicines.

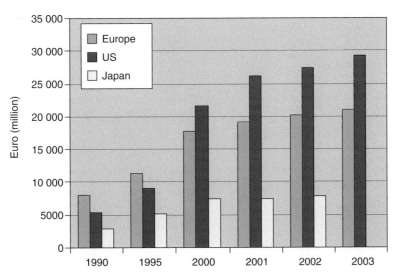

Figure 2 Pharmaceutical R&D expenditure in Europe, the United States, and Japan (Euro million), 1990–2003. (Data from Barnett International.[4])

In the past two decades, the concept of the CRO has grown considerably, with some of the larger companies such as Quintiles or Covance having filed more New Drug Applications (NDAs) with the Food and Drug Administration (FDA) than many pharmaceutical companies. In a recent report, outsourcing in the pharmaceutical industry was estimated at about 30% of overall R&D spend, and rising.[3] Given that pharmaceutical R&D was approximately US$50 billion for the 2003,[4] this amounts to some US$15 billion of expenditure annually (**Figure 2**). Growth of external R&D is estimated to be 50% higher than internal R&D for the foreseeable future and overall outsourcing market is expected to grow to perhaps 40–50% of pharmaceutical R&D. In some areas such as outsourcing of chemistry-related functions the figure has recently been rising at a compound annual rate of 40–50%.

In the past few years globalization, through 'offshoring' has added to the growth of contractually outsourced R&D. The two countries representing the bulk of this activity are India and China; in 2004, the Indian outsourcing market was estimated at US$120–470 million, growing at a compound annual rate of 75%. It is divided roughly equally into nonclinical and clinical research, and is able to offer substantially reduced costs, in particular much cheaper full-time employee (FTE) rates. In India, there is a large talent pool, with particular expertise in chemistry, and since January 2005, a new patent law has enabled the grant and validity of pharmaceutical product patents; there is also an increasingly widespread adherence to high quality standards (albeit from a low base), and of course the wide use of the English language makes India an easy place for scientists to communicate internationally. There are two other aspects that favor the conduct of clinical research in India, namely the existence of 16 000 hospitals and 171 medical colleges, and a large pool of naive patients, enabling fast clinical trial recruitment. Offshoring represents a significant commercial challenge to the service organizations based in the US, Europe, and Japan. Presently, the price advantage of outsourcing to India and China is offset by the better quality generally available in the higher-cost countries.

2.08.3 Advantages and Disadvantages of Outsourcing

Much has already been written on this subject,[5] and it is not the intention of this review to repeat the detailed points here. The main advantages and disadvantages are shown in **Table 2**. Over the years many of the more commonplace problems have been ironed out, and the enhanced ability to communicate by telephone and e-mail has greatly facilitated distal outsourced relationships.

The advantages of outsourcing are particularly appealing for most product-focused biotechnology sector companies, which do not have the resources to put in place internal organizations for pharmaceutical development, and as a result the proportion of externally contracted work is generally higher in this sector. An example of the way in which multiple providers can be brought together by a small biotechnology company is provided by Ionix Pharmaceuticals (Cambridge, UK), a specialist pain therapeutics company. The company has extensive discovery relationships with a variety of chemical and formulation companies, including EvotecOAI AG (Oxford, UK) and Tripos, Inc. (Cornwall, UK) for medicinal chemistry, West Pharmaceuticals Services, Inc. (Lionville, PA) for intranasal formulation, and Xenome Ltd

Table 2 Advantages and disadvantages of outsourced relationships

Advantages	*Disadvantages*
Reduction in fixed costs (traded against increased variable costs)	Loss of control
Better allocation of resources in a project with variable demand	Difficulty of coordination and management
Access to specific technology, expertise, or skills either not present internally or cheaper than internal alternative or quicker than internal alternative	Increase in variable costs, traded against reduced fixed costs
Flexibility in disengagement from unsuccessful research; greater objectivity in making that decision	Differing cultures of external party
Better management of risk	Time taken to agree contracts
	Difficulties in agreeing ownership/split of intellectual property rights
	Instability in case external party becomes financially insolvent, merges or is acquired

(Brisbane, Australia) for peptide chemistry. Ionix is developing intranasal buprenorphine, either alone or in combination with other drugs, for the treatment of acute and chronic pain, and has used its contractual network in order rapidly to advance its product development plans.

Management of outsourcing is a much more complex process than that of internal R&D. Unlike the intramural alternative, it starts with the identification and selection of a suitable partner for the required work, followed by the negotiation of a contract: with more than 1000 companies in the business of offering contract pharmaceutical services, this can be a lengthy and complex process. Although certain of the larger CROs are offered a wide menu of services, the risk for the buyer is that the quality and value for money is not always equally high across all areas. Beyond the choice of partner, the success of outsourcing depends critically on the quality of the project management, because problems need to be anticipated before they occur, and can be difficult to rectify.

There is some evidence that, done correctly, the outsourced approach does offer improved efficiency across the industry as well as substantial improvements in time to achieve certain goals. Protodigm, as a subsidiary of Roche and through comparisons with its parent, documented a 38% advantage in the time taken to advance a cancer project through phase I/II relative to the internal alternative. In 2000 the company separated from its parent, Roche, and changed its name; Fulcrum Pharma Developments now operates as a contractual resource to provide customized development programmes that lead to early proof of concept studies, for client pharmaceutical companies. It does so using a customized network of contract research organisations assembled specifically for each project. Similar comparisons carried out by Barnett International Benchmarking Group[6] have established reductions of 31–41% during phases II and III, and an overall time saving of at least 2 years on conventional intramural drug development times using the outsourced alternative (**Figure 3**).

Relative to other business sectors, pharmaceuticals has some features which particularly suits it to an outsourced approach. One is the multiplicity of scientific disciplines and skills necessary to advance a project through research and development: not all are likely to be present inside one organization. Another is the fact that specialized providers exist to offer expertise in all sections of pharmaceutical R&D, as well as sales and marketing. Third, the product from a pharmaceutical development is essentially knowledge, either in terms of intellectual property in the narrow sense of a patent, or in the broader sense of an information package for regulatory submission. In an era of electronic communication the mobility of such packages across international boundaries makes pharmaceutical outsourcing a global enterprise.

There are other features of the industry that mitigate against outsourcing, notably the very complexity of the development process, which make project management a difficult task in the absence of adequate competence in a wide range of disciplines (this argument is countered by the observation that in other business sectors complicated projects such as large-scale construction is conducted through a framework of outsourcing). Second, for historical reasons, the pharmaceutical industry has been late in adopting outsourced forms of operation. It is a relatively new industry, heavily regulated; and most ethical pharmaceuticals are protected from free market competition by patent coverage. Corporate secrecy is sometimes seen as a priority that conflicts with outsourcing in certain sensitive areas, where intellectual property is capable of being generated.

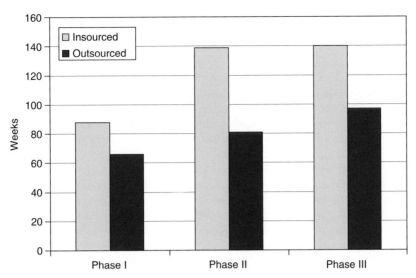

Figure 3 Reductions in pharmaceutical development cycle times using internal and outsourced resources. (Redrawn from Welcome to Kendler.[6])

The counterpart to the rise of outsourcing is the increasing emphasis placed on management of project and of external relationships, to the point that some companies focus on this as a core competence.

2.08.4 Virtual Companies and Extended Enterprises

The virtual pharmaceutical company is one that concentrates on project management of extramural operations, and may operate within a part of the overall R&D process (for instance, from the end of preclinical development to the completion of proof of concept clinical trials). The management and direction of the project(s) are retained in house, and the operations are carried on externally by a network of suppliers, with whom the virtual company has strong contractual relationships.

The concept is not unique to pharmaceuticals: Nike, for instance, legally does only marketing and R&D. It manages its suppliers as part of a corporation, putting its own personnel in production sites, visiting plants regularly, running codevelopment teams, and sharing information extensively. Its strength relies substantially on the tight relationships it has with its suppliers.

The management of pharmaceutical discovery as opposed to development poses two additional challenges. One is that substantial detailed scientific expertise across multiple disciplines is necessary, and may be hard to find in a small virtual company. The other is that in research, changes in direction and iterative methods of working are essential, and this is difficult to manage via contractual relationships. A successful case of totally outsourced discovery from identification of a lead molecule and continuation successfully through to phase II has been reported from Napp Pharmaceuticals.[7] This is not a widely replicated model and the difficulties should not be underestimated.

In pharmaceuticals, in addition to contractual relationships, knowledge and technology are vital parts of the process which are also amenable to transfer between organizations. The term "extended enterprise" (EE) has been introduced to cover this idea: its accepted definition is a "dynamic, networked organization."[8] The phrase has both a broader meaning and applies to a wider range of businesses than 'virtual.' Companies with their own internal focus may necessarily couple that with external elements to accomplish certain goals. The methods used to establish and manage these relationships are the same regardless of size and strategy.

The scope of relationships possible in pharmaceutical extended enterprises encompasses three forms: contracts, collaborations, and licenses. As will become clear, all these forms are used by both large pharmaceutical companies and biotechnology companies in their operations.

2.08.5 Comparison of Contracts, Collaborations, and Licensing

The three modes of link employed in pharmaceutical extended enterprises, namely collaboration, contracts, and licensing, are typically found at different stages of discovery, development, and marketing, and between different players in the overall process. **Figure 4** represents some of the forms of interaction between the main components of the pharmaceutical R&D supply chain. In addition to the various genres of company mentioned above, this figure also includes the academic sector as a source of technology which is licensed or forms the basis of a collaboration with a product-based biotechnology or large pharmaceutical company. For simplicity the role of large pharma has been represented in development and marketing, whereas in reality such companies have large investments in pharmaceutical discovery too, and indeed the primary purpose of collaborations with platform biotechnology companies is to support these efforts.

A summary of the place of contracts, collaborations, and licensing in the pharmaceutical value chain is outlined in **Table 3**. Whereas contracts focus on the provision of services, collaborations allow for technology or expertise as well as services, and licenses rarely incorporate any element related to services, and concentrate on technology or knowledge transfer. While contractual relationships allow for better risk management and more efficient resource allocation at an operational level, collaborations and licenses permit trading in intellectual property in a global marketplace, and association of the companies and institutions most capable of inventing new products and technology with those most capable of commercializing and mass marketing.

Collaborations between large pharmaceutical companies and other parties are particularly appropriate for the discovery phase because management of external discovery is more difficult than development. Sufficient expertise and project direction needs to be left with the external collaborator to enable them to deal with upsets and failures, to solve

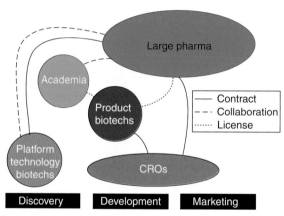

Figure 4 Examples of contractual, collaborative, and licensing arrangements between different components of the pharmaceutical R&D supply chain.

Table 3 Summary of differentiating characteristics of contracts, collaboration, and licensing

	Contracts	*Collaborations*	*Licenses*
Requirement	Access to resources not available or not sufficient internally	Access to technology for improved research productivity	Access to products (or occasionally technology) for enhancing development pipeline
Optimum stage of establishment	Development, and to a lesser but increasing extent, discovery	Discovery	Development
Reliance on external technology	Low	Medium	High
Aim	Completion of defined piece of (developmental) work	Integration of external technology into intramural discovery	Commercialization of external technology/product
Payments	Based on work done	Based on time allocated plus possible access fee, and success-related milestones and low royalties	Based on value of product development, plus success-related milestones and royalties

certain problems independently, and to apply their particular expertise in the most appropriate way. As they are earlier in the R&D process, collaborations are inherently more risky than later-stage development projects, and a component of the monetary exchange is often success-related. This may take the form of milestones and low rates of sales-based royalty as the project develops.

Licenses can take a wide variety of forms, but most often give the right for a company to develop, manufacture, and/or sell a product in a certain territory. They are thus oriented around a product rather than work or technology. Typically, licenses take the form of success-related payments (milestones) and royalties on sales. However, as discussed below, large pharmaceutical companies have sought more often to use in-licensing to bolster weak developmental pipelines, competition among licensees has increased and deals have grown in complexity. It is now not uncommon for biotechnology company licensees to demand (and get) an increasingly large share of the upside, including equity participation, codevelopment, and options to market in certain territories or well-defined indications.

2.08.6 Trends in Biotechnology Sector Companies: Products versus Technology Platforms

One of the key messages from **Figure 4** is the divergence between the product-based and the platform technology-based biotechnology sector company. During the early 1990s, the first generation of new technology in genomics, combinatorial chemistry, and high-throughput screening underpinned a dramatic change in the way in which pharmaceutical research was carried out. Mechanization would turn a cottage industry to mass production. The development of this new technology outside the main companies provoked a rush to collaborate with new, small, high-tech start-up companies, and substantially changed the attitudes towards outsourcing of pharmaceutical discovery.

With time, two things have happened to dampen the initial fervor. First, the rate of innovation in discovery technology has slowed, the technology has become more widespread, and much has been incorporated within large pharmaceutical companies, where large capital investments have been made. The bottleneck in R&D productivity has moved beyond discovery to early preclinical development.

Second, the promise of the triumvirate of new discovery technologies has largely not been realized despite the investment. That promise was to increase the productivity of pharmaceutical R&D and dramatically increase the number of new medicines reaching the market.[9] That promise has not been realized: in order to be able to deliver 10% year-on-year increases in sales, and 15% growth in earnings per year, the largest companies such as Pfizer (which has annual sales of US$50 billion) require a sales increment roughly the equivalent of that of the total sales of a company like Amgen, the biggest in the biotechnology sector.

As a result in the decreased value of discovery collaborations, there has been a concomitant trend towards them becoming contractual in nature. This trend was foreseen as early as 1997, and by the end of the decade there was a widespread belief that demand for drug discovery collaborations would decline and dramatic action was needed.[10] The sentiment has been reflected in the investment community, which was largely caught out in the perimillennial technology rush. Today, investors in the sector are much more risk-averse and focused on products. Such investment criteria are ultimately important drivers for biotechnology sector strategy.

The changed climate quickly became manifest. In December 2001, Pfizer signed a deal for chemistry services with ArQule involving contractual payments of US$117.5 million over 4–5 years and additionally an equity purchase arrangement. Unlike previous arrangements between large pharma and a combinatorial chemistry company, the deal did not involve milestone or royalty payments. ArQule would provide expanded lead generation capabilities using its automated solution phase chemistry and would transfer on a nonexclusive basis its proprietary library design and informatics platform. ArQule and Pfizer would work together to improve this platform to enhance the process of lead generation for early drug discovery. ArQule might also collaborate with Pfizer on lead optimization during the alliance.

Some technology platform biotechnology sector companies have acceded to the trend, and have become discovery contract organizations; others have invested the profits received in the era of large collaborative technology-based deals back into a new corporate strategy based on product development.

Two salient examples of these alternatives are Millennium and Lion Biosciences. Millennium Pharmaceuticals, Inc., historically a provider of genomic technology and the recipient of a huge 5-year collaborative deal with Bayer Healthcare AG, first announced a change of strategy in 2000 with the acquisition of Cambridge Combinatorial Chemistry. In 2002 this was followed by a further merger with Cor Therapeutics, whereby Millennium assumed rights to Integrilin (eptifibatide) for acute myocardial infarction. In 2003 the company announced a restructuring and loss of 600 jobs, in order to finance the launch of its first product Velcade (bortezomib) for advanced stage multiple myeloma. Millennium aims to reach profitability by late 2006.

In contrast, Lion Biosciences also started its internal discovery strategy (iD3) in 2000 and acquired Trega Biosciences, followed a year later by the construction of a chemistry R&D centre to support its internal research program. The cost of this transformation proved impossible to sustain, however, and in 2002 Lion abandoned iD3, announcing a corporate restructuring in 2003 with the loss of 86 jobs. Currently Lion's business model is that of a typical software company. Income is generated from the sale of software licenses, maintenance, and support, and from offering services in chemoinformatics and bioinformatics. It has been unable to achieve its goal to break even as a purely service company, and is currently not forecast to do so.

These examples demonstrate similar initial strategies based on a recognition of the transient appeal of discovery technology, both initiated in 2000. The difficulty of realizing the ambition and the substantial investment required for such a change led to one company abandoning the strategy and reverting to the service provider model, while the other achieved its aim, after the expenditure of hundreds of millions of dollars. The profitability of the service model is not assured, although many discovery contract organizations are performing much better than Lion.

Despite these trends, there remain certain technologies, normally in the developmental field, for which biotechnology companies can retain long-term value for the projects they work on.

One area where examples can be found is in bilateral agreements concerning controlled release formulation technology: for instance, SkyePharma offers such technology in the context of drug development support including formulation, clinical, regulatory, and manufacturing skills. The company worked with GlaxoSmithKline to develop a controlled release version of the antidepressant Paxil/Seroxat (paroxetine). The work was performed under a contract between the parties which also stipulated that SkyePharma receives low single-digit royalty payments on the sales of the product, clearly demonstrating the long-term value sharing element of the relationship.

Another example is in technology to enhance form and formulation. For instance, TransForm Pharmaceuticals was founded in 2001 around physical chemistry and novel high-throughput platform technologies, supported by a data informatics platform. It uses these capabilities to optimize drug form and/or formulations, and increase the clinical and commercial value of pharmaceutical products, across the entire spectrum of pharmaceutical development from early stage to life cycle management. In research and development, TransForm is working with partners such as Alza Corporation and Lilly to help them make better candidate selection decisions and reduce attrition and development time and cost. In the collaboration with Lilly, TransForm worked on several discovery and early developmental stage compounds. TransForm's work on the developmental stage compounds could generate various success-based payments including milestones and royalties. In some cases where TransForm's technology had, for instance, reinvigorated a compound whose development was otherwise stalled could give rise to a partial 'ownership' position on the drug. For later stage and marketed products, TransForm set up a collaboration with partners such as Johnson & Johnson to enhance the life cycle management of their products through improved bioavailability, broad intellectual property protection, and investigations into new indications. In 2005, this latter collaboration led to the wholesale acquisition of TransForm Pharmaceuticals for the headline figure of US$230 million.

2.08.7 Types, Stage, and Value of Alliances with Large Pharmaceutical Companies

An analysis of pharmaceutical collaborations and alliances (which involve one of the top 20 pharmaceutical companies), and categorization by phase of alliance has been published for the period 1988–2002 (**Figure 5**).[11] The phase of alliance relates to the stage of development of the technology or compound at the point of entering into the alliance. This analysis demonstrates that in the period 1998–2002, there has been a roughly constant percentage (54–68%) of alliances related to projects in discovery or lead stages, 10–19% in preclinical development, phase I or phase II, and 20–29% in phase III, NDA filing, or approved. The preponderance of alliances in the discovery and lead phases of drug discovery has persisted for this entire period, although the focus of the alliance has perhaps changed.

More interestingly, an analysis of the type of deal, whether it relates to a diagnostic, drug delivery technology, or to a therapeutic focus, reveals an informative trend over the period 1988–2002 (**Figure 6**). The dominance of therapeutics as the foundation for alliances has grown from 75% of all deals done to 86% in the latest period 2000–02.

A second interesting point relates to the value of the deals done. As shown in **Figure 7**, the greatest value is being devoted to deals that are done at late stages in development, particularly in the latest period 2000–02. The improved terms have an important beneficial effect on the commercial outlook for product-based biotechnology companies, and as described in more detail below, reflect a changing dynamic in licensing in particular. The current environment also tends to favor such licensees hanging on to products until later in development when the added value is better rewarded.

Pharma versus Biotech: Contracts, Collaborations, and Licensing 233

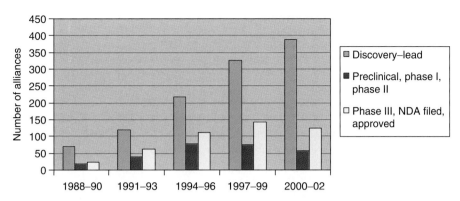

Figure 5 Alliances by stage involving a top 20 pharmaceutical company over the period 1988–2002. (*Source*: Recombinant Capital http://www.recap.com.)

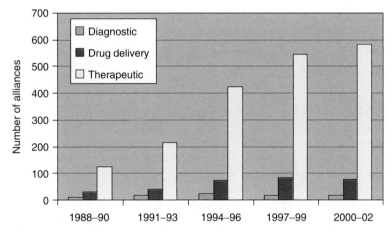

Figure 6 Alliances by type involving a top 20 pharmaceutical company over the period 1988–2002.

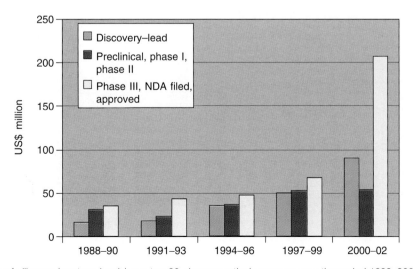

Figure 7 Value of alliances by stage involving a top 20 pharmaceutical company over the period 1988–2002.

2.08.8 The Growth of Licensing

License opportunities, particularly from biotechnology sector companies for mid- to late-stage developmental compounds, are increasingly important to address deficits in large pharmaceutical companies' R&D pipelines. While, since 1995, the number of active compounds in early to mid-development has increased, there has not been a concomitant increases in numbers of compounds in late development (**Figure 8**).[12] Most large companies have now set up multidisciplinary teams composed of licensing executives and scientific expert assessors to scout for suitable opportunities. They evaluate thousands of opportunities each year, through a series of scientific and commercial filters; it is typical for approximately 1% of these opportunities to succeed to a completed license, thus emphasizing that a substantial effort is made to bring in as many as possible for consideration but only a very low proportion are ultimately accepted.

There is intense competition among the major companies for the most attractive in-licensing opportunities: the factors that lead to positive assessments (mainly scientific and commercial in nature) are likely to pick similar winners. In 2003, for instance, there were fewer than a dozen high-quality phase III products of commercial significance available for worldwide licensing, and demand for these opportunities significantly outstripped supply. In 2002, seven deals were worth headline figures of more than US$250 million, and four exceeded US$450 million in value. Large companies are promoting themselves to smaller companies as attractive licensing partners. For companies not in the largest tier, this is likely to include factors other than marketing capability. Roche, for instance, emphasizes its ability to do deals more quickly than others: 4.5 months as opposed to the industry average of 9–10 months. Such factors can have a substantial impact on development time, period of marketing exclusivity, and extent of competition when the product is introduced.

Another strategy for large pharmaceutical companies to enhance their ability to gain access to attractive licence deals with biotechnology companies is to enter into broad corporate alliances which offer options on such opportunities following a period of collaborative development. A good example of this is the arrangement between Theravance and GlaxoSmithKline.

In 2004, GlaxoSmithKline and Theravance entered into a broad-based collaboration to develop and commercialize novel therapeutic agents across a range of therapeutic areas, including bacterial infections, urinary incontinence, and respiratory and gastrointestinal disorders. Under the terms of the agreement, Theravance would receive US$129 million, much of which significant amount related to an equity investment by GlaxoSmithKline in Theravance. In return, GlaxoSmithKline would also receive an exclusive option to license potential new compounds from all current and future programs on a worldwide basis, through to August 2007. Upon accepting a new program, GlaxoSmithKline would be responsible for all development, manufacturing, and commercialization activities. Depending on the success of these new programs, Theravance would receive clinical, regulatory, and commercial milestone payments and

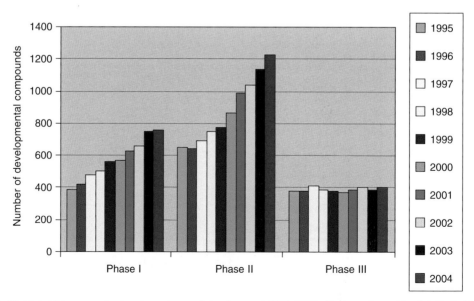

Figure 8 Number of compounds at various stages of development 1995–2004. (Redrawn from Lloyd, I. *Scrip Mag.* **2005**, *February*, 24–25.)

significant royalties on sales of medicines. Theravance retained its operational independence to continue discovering and developing new medicines outside of the agreement.

Later in 2005, GlaxoSmithKline exercised its option to license one of Theravance's programs in the respiratory field, led by an inhalable muscarinic antagonist/beta-2 adrenoceptor agonist developed to treat chronic obstructive pulmonary disease (COPD) and asthma. Using the principles of multivalent drug design, Theravance had discovered a series of long-acting inhaled bronchodilators that are bifunctional; one molecule functions as both a muscarinic receptor antagonist and a beta-2 receptor agonist, thereby producing a potential medicine with greater efficacy and equal or better tolerability than single-mechanism bronchodilators. The lead compound was under preclinical investigation for its potential in COPD and asthma at the time of the deal with GlaxoSmithKline. Following take-up of the option, GlaxoSmithKline would fund all future development, manufacturing, and commercialization activities for product candidates in this program.

The background to this deal was that the value of GlaxoSmithKline's respiratory franchise was expected to drop from US$8 billion in 2005 to US$5.7 billion by 2010. GlaxoSmithKline's leading respiratory product, Seretide/Advair, was expected to show a revenue decline over 5 years from US$4.9 billion in 2005 to US$4.2 billion in 2010, accounting for 74% of franchise sales at that point. As a result of these expectations, GlaxoSmithKline was looking to expand its pipeline in this area, and reduce its dependence on Seretide/Advair, which had been the victim of an adverse patent ruling in 2004 in the UK. Programs from the emerging biopharmaceutical world offer a potentially vital means of gaining new compounds, which would benefit from having GlaxoSmithKline's marketing and production muscle behind them.

The proportion of marketed products that derive from in-licensing has been the subject of some study.[13] Based on the position in 2003, Bristol-Myers Squibb derives the highest proportion of revenues from in-licenses, at 64%, followed by Schering Plough, at 55% (**Figure 9**). But the greatest overall revenues are found in Pfizer's portfolio, for which US$17.5 billion of revenue comes from in-licensed products, out of a total of nearly US$50 billion. Products licensed in by the top 20 large pharmaceutical companies over the past 15 years have generated over 30% of the revenues in the cardiovascular, musculoskeletal/pain, and anti-infective areas. These figures underpin the importance of externally derived products to large pharmaceutical companies' revenues, and that the expectations of these companies in seeking to license developmental projects are based on the proven experience of the past, not just hope.

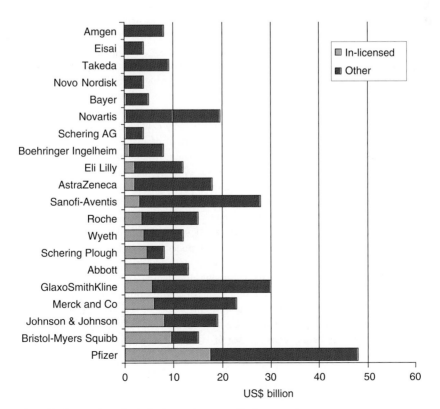

Figure 9 Proportion of in-licensed products for large pharmaceutical companies.

Two final points from this analysis: **Figure 9** includes one company that may not be typically regarded as in the large pharmaceutical sector, namely Amgen, the world's largest biotechnology sector company. Although devoid of significant revenue from significant in-licensing activities among its present marketed portfolio, this is unlikely to remain the case for much longer. Large biotechnology sector companies in general with an existing product range are seeking in-licenses just like their 'large pharmaceutical sector' cousins. Biogen-Idec, for instance, has set itself a target of deriving 50% of its pipeline from in-licensed products by 2010. In 2004, the company completed a deal with Vernalis giving it access to a phase I developmental product for Parkinson's disease; the product was an adenosine A2A antagonist.

Lastly, as a result of the increased in-license activity from large pharmaceutical companies, there has been a reassessment of some of the internal projects leading to them being assigned for out-licensing. These projects are seen as commanding too low a priority for their development within a reasonable time-frame. If the problems associated with the project are relatively easily solved, or if for instance they have insufficient market potential, they could be attractive opportunities for smaller firms to develop. GlaxoSmithKline has been particularly active in seeking licensees for such programs, often to small biotechnology companies in return for an equity stake in the company.

2.08.9 Changing Role of Contract Research Organizations

As well as biotechnology sector companies, potential licensees for projects at this stage from large pharmaceutical companies include some of the larger CROs. For instance, PPD has entered into a number of deals involving licensing opportunities at the late preclinical or phase I stage, which it then developed through clinical proof of principle to phase II or phase III, and sought a late stage development and/or marketing partner. Quintiles has set up a special division called PharmaBio specifically to pursue this strategy.

An example of a project which has been partially developed in this way includes dapoxetine for premature ejaculation. Dapoxetine is a selective serotonin reuptake inhibitor, structurally related to fluoxetine, which was initially being developed by Eli Lilly for the potential treatment of depression, for which it was in phase I trials; however, no development regarding this indication has been reported after 1995. PPD GenuPro, a subsidiary of PPD, which was established from a collaboration between Eli Lilly and PPD, gained the rights to develop dapoxetine in December 1999. Dapoxetine was evaluated in a phase II trial in the USA for the treatment of premature ejaculation. Results from the 155-patient trial indicated a significant increase in ejaculatory latency in dapoxetine recipients, relative to placebo. Dapoxetine moved into phase III development for the treatment of premature ejaculation in July 2003. In December 2000, ALZA Corporation, a wholly owned subsidiary of Johnson & Johnson, licensed exclusive rights to dapoxetine from GenuPro for genitourinary therapies, including premature ejaculation. Under the terms of the agreement, ALZA was to develop and commercialize dapoxetine and be responsible for manufacturing, clinical, regulatory, and sales and marketing costs resulting from the license. In exchange, GenuPro was to receive an undisclosed upfront payment. In addition to royalties on net sales, the terms of the agreement include milestone payments based on product approval and meeting certain sales levels. In December 2004, ALZA Corporation submitted a new drug application to the US Food and Drug Administration (FDA) for dapoxetine hydrochloride in the treatment of premature ejaculation. Very clearly, PPD used a series of license deals interspersed with financing of the development of a compound through some critical stages of pharmaceutical development, thereby adding value which it was to realise in the later of the license arrangements.

PharmaBio, Quintiles' development group, offers innovative partnering solutions to pharmaceutical and biotech companies. For instance, it has participated in various investments, either directly or indirectly, into biotechnology sector companies (e.g., Innapharma Inc., a New Jersey company engaged in the development of a series of 'small chain' peptide-based products to treat central nervous system disorders, such as major depression). It also is active in product partnering, in which PharmaBio acquires product rights or products from pharmaceutical companies to help them develop and manage product portfolios, utilizing the parent company Quintiles' development and commercialization capabilities. Finally, PharmaBio can also participate in various risk-sharing agreements for product development, in which a share in the end product value is returned for a financial contribution to product development.

These new ways in which CROs work emphasize the changing nature of the contract sector, and the desire of such companies to seek commercial advantage by offering a broader range of development alternatives which deflect the usual comparisons with competitors based solely on cost of service. These types of arrangements are currently only offered by the very largest CROs. With time, they may become more important as value drivers for the companies offering these services relative to the sector as a whole. These services may also be important differentiating factors for companies based in Europe, USA, and Japan when set against their competitors in India and China. As described above, offshoring is fast becoming a significant factor in contractual outsourcing and forcing the companies with higher cost bases to consider greater added value offerings.

2.08.10 Conclusion

Outsourced relationships are becoming increasingly common for pharmaceutical innovation. It is now very rare for a single company or organization to be solely responsible for the entire research and development and commercialization process. It is probable that some form of research collaboration (e.g., with an academic institution) will have played a part in the discovery of the product and/or parts of the development process will have been conducted under contract by a CRO. While contractual relationships operate on a fee-for-service basis, collaborations and licensing agreements normally include staged payments based on success. This reflects the fact that these types of relationship include the transfer of knowledge or intellectual property.

Presently many new medicines that originate in the biotechnology sector are licensed to large companies for later development and commercialization, and form a key source of growth, currently comprising roughly 30% of revenue and growing over 50% faster than internally developed products. A typical deal may involve a biotechnology sector licensor and a large pharmaceutical company licensee in relation to a project that has proven clinical potential in phase II trials. While the 'plain vanilla' license terms typically involve an upfront payment, milestones, and royalties on sales, increasing competition for the most attractive licensable products has permitted additional terms related to equity participation, codevelopment, and licensor options to market in certain territories or well-defined indications. The partners can often come from outside the biotechnology and large pharmaceutical sectors, and some CROs are becoming increasingly interested in leveraging their developmental and commercial services for a risk-based gain. Equally, it is certainly not unknown for middle and large biotechnology companies to comprise the licensee in licensing transactions.

Some commentators have proposed that the biotechnology sector will increasingly become the 'most effective segment of the drug industry.' The companies in the biotechnology sector are incredibly diverse in the kinds of discovery projects they work upon, the kinds of product they develop – whether large biomolecules or small conventional therapeutics, and the size of company – many being privately owned while a few have a public shareholder base. In addition, over the past 5 years an increasing gulf has opened between the high-risk, high-gain companies aiming to produce therapeutic products, and low-risk, low-gain companies operating in the service sector. As sentiment has cooled towards drug discovery technology, some have attempted to make the transition to product-focused entities, but not all have succeeded. The service providers in the biotechnology sector have to operate in an increasingly fierce competitive environment particularly given the growing importance of India and China as sources for similar services.

References

1. Drews, J. *Drug Disc. Today* **2003**, *8*, 411–420.
2. Kenley, R., http://www.ispe.org/newengland/presentations/0104_kenley.pdf (accessed April 2006).
3. Deutsche Banc Alex. Brown, Equity Research, Pharmaceutical Outsourcing 1999, London.
4. Barnett International. http://www.barnettinternational.com/RSC_ImgUploads/PRD04%20Sample%20Pages.pdf (accessed April 2006).
5. Cavalla, D.; Flack, J.; Jennings, R. *Modern Strategy for Preclinical Pharmaceutical R&D: Towards the Virtual Research Centre*; John Wiley: Chichester, UK, 1997.
6. Welcome to Kendler. http://www.kendle.com/investor.ppt (accessed April 2006).
7. Cavalla, D.; Gale, D. D. *Drug News Perspect.* **1997**, *10*, 470–476.
8. Deloitte. http://www.deloitte.com/dtt/cda/doc/content/LifeSci(1)(3).pdf (accessed April 2006).
9. Arlington, S. *Pharma 2005 An Industrial Revolution in R&D*; PriceWaterhouseCoopers: London, 1998.
10. Cavalla, D. *Drug News Perspect.* **1997**, *10*, 197.
11. Edwards, M.; van Brunt, J. *Signals Mag.* **1999**.
12. Lloyd, I. *Scrip Mag.* **2005**, February, 24–25.
13. Benyon, K; Plieth, J. *Scrip Mag.* **2005**, February, 38–40.

Biography

David Cavalla, following his studies at Cambridge University, UK and postdoctoral research in Washington, USA, joined Glaxo Group Research, where he was involved with the research projects that led to the marketed drugs Zofran and Imigran. Subsequently, as Head of Biosciences at Napp Research Centre, he initiated the PDE4 inhibitor project and directed the research strategy for asthma. This work, from discovery to clinic, was the first example of a successful research project carried out using an extended enterprise strategy, completed with the use of externally contracted resources coordinated by a small team of multidisciplinary project managers.

David is the author of Modern Strategy for Preclinical R&D – Towards the Virtual Research Company, published by John Wiley. He has been closely involved with the UK Society for Medicines Research, for which he is past-Chairman, and interested in the multidisciplinary aspects of drug discovery and development.

As a founder of Arachnova, he has been a pioneer in the strategy of drug repurposing as a means to substantially improve R&D times and costs, with minimized risks in the discovery of novel medicines.

2.09 Managing Scientists, Leadership Strategies in Science

A M Sapienza, Simmons College, Boston, MA, USA

© 2007 Elsevier Ltd. All Rights Reserved.

2.09.1	**Context**	**239**
2.09.1.1	Why (and How) Leadership Matters	239
2.09.1.2	Scientific Training and Leading Scientists	240
2.09.1.3	Scientists' Own Views on Good and Bad Leadership	241
2.09.1.3.1	The effective leader	241
2.09.1.3.2	The ineffective leader	241
2.09.1.3.3	Climate in the laboratory	242
2.09.2	**Scientists' Most Difficult Problems**	**242**
2.09.2.1	Becoming a Leader	243
2.09.2.2	Dealing with Conflict	243
2.09.2.3	Motivating People	243
2.09.2.4	Communicating Effectively	243
2.09.3	**The Twenty-First Century Scientific Workforce: Issues of Diversity and Inequity**	**243**
2.09.3.1	The Issue of Diversity	244
2.09.3.2	The Issue of Inequity	244
2.09.4	**Leading People**	**245**
2.09.4.1	Motivating	245
2.09.4.1.1	Reasonable working situation	246
2.09.4.1.2	Competent people trained appropriately for their job	246
2.09.4.1.3	Assurance of link between effort and outcomes	246
2.09.4.1.4	Equity and fairness	246
2.09.4.1.5	Not micromanaging	247
2.09.4.2	Communicating and Listening	247
2.09.4.3	Resolving Conflict	248
2.09.4.3.1	Sources of potential conflict	248
2.09.4.3.2	Resolving conflict	248
2.09.5	**Managing R&D**	**249**
2.09.5.1	Structure	249
2.09.5.2	Size	250
2.09.5.3	Systems	250
2.09.6	**Conclusion**	**250**
References		**251**

2.09.1 Context

2.09.1.1 Why (and How) Leadership Matters

Does leadership matter? In a series focusing on the science and technology of medicinal chemistry, a chapter on leadership of R&D is at least an anomaly. Of course, most scientists would respond that leadership – to be defined shortly – does indeed matter. The crucial issues are: why might leadership matter; and, how?

Many books on creativity and innovation,[1,2] as well as investigations of corporate innovations[3] and biographies and autobiographies of innovators,[4,5] concur that leadership matters for reasons described in a study of US National Aeronautics and Space Administration (NASA) scientists:

> [It] is no longer enough to be excellent in [one's] scientific discipline ... A research leader needs to get work done with and through other people ... Time, money, morale, and quality of product are only a few of the elements that are at risk when ineffective leaders are at the helm.[6]

Leadership also matters because of its indirect effect on industries and nations. The US National Science Foundation's annual reviews of the status and role of science, engineering, and technology demonstrate that global economies both benefit from and depend on high-technology industries and services. The latter are defined by "their high R&D spending and performance ... [that produce] innovations that spill over into other economic sectors."[7]

When the output of research is high-quality innovation, institutions investing in R&D enjoy positive economic returns. More importantly, "[r]eturns to society overall are estimated to be even higher. Society often gains more from successful scientific advancements than does the organization conducting the research."[7] It is not too much of an exaggeration, or simplification, to state that effectively led science contributes to social and economic welfare.

If social and economic welfare, as well as 'time, money, morale, and quality of product,' provides an answer as to why leadership matters, the more intriguing question remains. How does leadership affect the creativity and caliber of science?

The purpose of this chapter is to provide a brief overview of how leadership matters in terms of people and the organization of scientific work. Before continuing further, however, let me provide the definitions of leading, managing, and an effective leader that will be assumed throughout the chapter[8]:

- Leading: being an exemplar and inspiration, as well as directing people in their work, in decision making, and in problem solving
- Managing: administering the R&D organization (structure, systems, etc.)
- An effective leader: one who is capable of developing and maintaining an enthusiastic, energetic, and creative group of scientists and of managing R&D successfully.

2.09.1.2 Scientific Training and Leading Scientists

If the leader of R&D 'needs to get work done with and through other people,' it will be instructive to reflect on general scientific education and training as preparation for that role.

Individuals with a terminal degree in science (as well as medicine, engineering, and so on) can be described as narrowly trained professionals (i.e., specialists).[9] Narrowly trained professionals typically undergo a lengthy and particularly designed education and socialization.[10] During this time, their progress and rewards are based on individual intellectual accomplishments, such as solving technical problems, crafting hypotheses to explain new phenomena, devising bench experiments to test hypotheses, carrying out suitable experiments, and interpreting the results.[11] They may work in groups and sometimes conduct experiments with colleagues, but they are recognized for their own ideas and receive individual grades and individual degrees. Working cooperatively or collaboratively is typically neither required nor rewarded in graduate school or postdoctoral training.

When scientists graduate, they may work for a time as an individual contributor. However, whether or not they remain on a technical ladder, nearly all must take on the challenge of supervising people in their laboratory. When they do, they discover that the rules have changed. Although highly trained in the cognate sciences and competent in their specialty, they are likely to be ill prepared to deal with the behavioral aspects of their job such as working in and with teams and leading other people:

> ...While we are truly experts in the technical aspects of our job, we oftentimes have difficulty excelling at our non-technical responsibilities. This can lead to varied success managing people and projects, promoting our programs and initiatives, and ascending to high-level management or leadership positions within our organization.[12]

Moreover, by the very paradigms of their disciplines, scientists are predisposed to be skeptical of the sciences they perceive as "soft." Although even quantitative subjects like finance and accounting may be questioned as being hard sciences, behavioral and managerial theories are often of highly suspect credibility. "The idea of management as a professional skill with its own disciplines can be a tough sell. New training has much to overcome in the way of old prejudices."[13]

2.09.1.3 Scientists' Own Views on Good and Bad Leadership

Surprisingly, there is little information on how scientists themselves define effective and ineffective leadership (there are biographies and autobiographies of scientists and certainly anecdotal reports[14]). To address the gap, a colleague and I began to survey expert panels of scientists in 1996, because we were interested in scientists' experience of leadership – both leading and being led.[15]

Since then, seven panels consisting of 228 individuals have completed the survey. They include scientists from the USA, Europe, and Asia and are primarily PhDs (about 80%), with the remainder MDs (or MD/PhD) and MS (very few have only a master's degree). Most are working in the life sciences and have degrees in chemistry, biology, or medicine, although a number have doctorates in engineering, mathematics, or physics. They are about equally distributed between academia and government scientific enterprises and industry (large pharmaceutical and small biotechnology firms). Because this ongoing study is predominantly exploratory as well as descriptive, the sample is not meant to be representative of all scientists but rather to provide a window into what it may feel like to lead and to be led in R&D.

Scientists are asked three open-ended questions, either verbally or written on a short survey. The numbers of individuals and the responses that could be coded separately, per question, are as follows:

1. Describe the best example of scientific leadership you have encountered and explain why this person was effective (191 scientists generated 338 responses).
2. Describe the worst example of scientific leadership you have encountered and explain why this person was ineffective (164 scientists generated 219 responses).
3. Describe the most difficult of the typical problems you encounter in your scientific position, (200 scientists generated 301 responses).

2.09.1.3.1 The effective leader

If you were asked the same question, what would you describe as the most important attribute of an effective leader of R&D?

When I have posed this in a discussion, often a scientist will respond "intelligence, technical skills." In fact, qualities of what I have termed being a nice person are most frequently listed. That attribute is followed by skills in management; then, intellectual accomplishment; and finally, being a good role model. To date, those four classes account for nearly 80% of answers to the question as to why the leader was effective. An effective leader is described as:

- Caring, compassionate, supportive, and enthusiastic (27% of all responses)
- Managerialy skilled in communicating, listening, resolving conflict, and organizing (23% of responses)
- Technically accomplished to lead a scientific effort (15% of responses)
- Good role model, mentor, and coach (11% of responses).

Scientists also list diplomacy, fairness, patience, and having a sense of humor as important qualities in a leader.

2.09.1.3.2 The ineffective leader

Although the first question generated the largest number of responses per person, the second question generated the most intense feelings, as reflected in scientists' language. More than half of the responses to this question describe the worst example of scientific leadership as involving a boss who:

- Is abusive, publicly humiliates subordinates, provides only negative feedback (19% of responses)
- Is selfish, exploitive, dictatorial, disrespectful of others' contributions (19% of responses)
- Cannot deal with conflict (14% of responses).

Other attributes of the ineffective leader are being disorganized, having unrealistic expectations, taking prolonged absences from the laboratory, and being dishonest.

Does good science emerge from these conditions? As the next section illustrates, there is a sizable difference in the climate created by an effective leader and that created by an ineffective leader. Scientists' own responses attest to the impact of leadership on morale and, thence, on the caliber of results.

2.09.1.3.3 Climate in the laboratory

The best leaders are described as "scientifically very competent, and compassionate and caring deeply for collaborators and subordinates." One respondent noted: "[the best leader is] caring but assertive. Good working rapport as well as friendship in the lab. Overall feeling of appreciation for the work done." The effective leader "not only criticizes but also praises. A lot of people tell you when you've done something wrong. Very few people tell you when you've done something right."

In a detailed example, one scientist described his former boss as:

> ... a great scientific leader and manager. He held regular group meetings, included everyone in the discussions, took risks scientifically and in management, and was not afraid to speak up. He kept everyone focused and was a real 'cheerleader' when it came to motivating us, keeping us a very focused and excited research team. He gave us a certain amount of independence and expected us to plan our work thoroughly. He also spent a lot of time in the lab, talking with us individually about the work. Our team was VERY productive! [respondent's capitalization and punctuation].

Scientists describe a very positive climate in the laboratory produced by leaders who are "highly enthusiastic and support others' unorthodox ways of thinking." Such leaders create an atmosphere in which professional growth and scientific innovation seem just to happen without struggle. The effective leader "can get the best out of each person"; ensures that each individual "feels a part of what is happening and wants to do a good job"; and has "the ability to inspire and make everyone enthusiastic about the research." These leaders generate a "fun and productive atmosphere in which each person can thrive in his/her own individual way"; they support a "stimulating environment," they "encourage ingenuity," and are able to "appreciate innovative/novel/different ideas."

In contrast, the climate created by an ineffective leader is at best emotionally stressful and at worst pernicious. Respondents write that they are yelled at publicly, berated, nagged continuously, belittled. One scientist described "lab meetings [as] notorious for being forums for public denigration. [X] is abusive in meetings and often bluffs his way through things he knows little about." Numerous scientists attest to the fact that ineffective leaders cannot deal with conflict. Instead, they "avoid conflicts and let problems fester"; they "look the other way" or they "hide from conflict." One scientist wrote that the director "uses the technique of avoidance and, when problems are arising, simply never shows up in the lab." Another gave an example of a situation in which the principal investigator "delayed dealing with interpersonal problems until they grew out of hand – then asked a post-doc to handle the issues."

A powerfully negative climate – in other words, poor morale – can be created by an ineffective leader. Moreover, not one respondent has noted, in all the descriptions of ineffective leaders, that scientists are nevertheless productive. In fact, the opposite is described:

> ... I often find not only in my experiences but observing others that negative motivation doesn't work. It makes me much less productive.
> ... There is much waste of human and financial resources in science from ineffective leadership.
> ... Having had both extremes – great and horrible – as leaders, I'm aware of the productivity associated with a good leader and the lack of productivity associated with a bad leader.
> ... The most ineffective leader I have had used intimidation to squeeze results from people. Fear is a great motivator, but resentment kills trust, certainty, and creativity.

2.09.2 Scientists' Most Difficult Problems

Given the earlier description of scientific education and training, it should be clear that being trained as a scientist does not mean that one is therefore competent to lead scientists, as a survey respondent admitted:

> I would like to become an effective manager and as far away as possible from being an ineffective one. There is a difference in management, and time, and self-motivation that leading others requires and that being a bench scientist did not.

To date, four classes of problems account for almost three-fourths of all answers to the third question ('of the typical problems you encounter, describe the most difficult'):

- Becoming a leader (e.g., being authoritative, staying focused, balancing scientific efforts with management responsibilities, delegating) (26% of responses)
- Dealing with conflict (21% of responses)

- Motivating people, generating enthusiasm (15% of responses)
- Communicating effectively; especially, providing feedback (11% of responses).

Other difficulties include 'not being taken seriously as a leader,' 'lack of respect and support from people in authority,' and 'being undermined by colleagues, mentors, even secretaries.' Each of the four major problems is described below.

2.09.2.1 Becoming a Leader

Moving from a position as colleague and peer to being a leader with some authority over a group brings with it an array of problems. Scientists report that they are sometimes overwhelmed by administrative responsibilities ("NON-SCIENCE activities," in the exact words and capitalization of one respondent), including "space conflicts and limited reagents." Laboratory administration requires that they "solve equipment and material problems," "deal with parking," and "chase after borrowed equipment that was not returned." These enlarged responsibilities result in the pressure of "too much work, too little time, and too few hands," perhaps because (as one scientist stated) of the difficulty of "saying 'No'." Becoming a leader may also highlight a scientist's lack of "confidence to delegate."

The most challenging issues, as implied by the earlier citations from the survey, involve dealing with people, not only those in one's laboratory. A principal investigator discovered that he also had to manage "difficult – arrogant and abrasive – people in other labs with which we must deal on a regular basis. I struggle with getting my point across, without causing a bigger dispute."

2.09.2.2 Dealing with Conflict

Being skilled in resolving the inevitable conflicts that arise when people work together is an important attribute of the effective leader. With regard to their own most difficult problems, respondents write that they struggle with how to resolve disagreements that range from 'which music is played in the lab to which experiments have higher priority.' They struggle to 'keep people from sniping at each other'; and they struggle with how to handle 'jealousy,' 'moodiness,' and 'one bad apple who poisons the atmosphere.'

Conflict that is ignored and avoided invariably entangles more individuals than were first involved. They may not realize it, but scientists and technicians are drawn to take sides, which escalates the issues. People who are involved in a conflict even to a limited degree find that more and more of their energies go to the conflict situation rather than to the science, to the detriment of the science.

2.09.2.3 Motivating People

In their new role, scientists discover how hard it can be to generate 'enthusiasm equal (or at least closer) to my own.' The effective leader of science has been described, above, as a "'cheerleader' when it came to motivating..." Motivating people does require a great deal of time praising, supporting, cajoling, encouraging. The effective leader spends "a lot of time in the lab, talking with [people] individually about their work." Thus, motivating people and communicating effectively are interdependent leadership challenges.

2.09.2.4 Communicating Effectively

Newly promoted supervisors in any organization face a common problem in giving feedback that will not be felt as 'personal attacks.' One scientist described her difficulty in "being able to convince people that they are going in a (likely) wrong direction in a way that will leave no resentment behind."

Communicating effectively allows a leader to critique without "sounding confrontational" or "hurting [people's] feelings." Because R&D almost always requires multidisciplinary team efforts, because it involves complex processes and careful management of the flow of information, effective communication is critical both to motivation and to "keeping all team players focused on the critical path." Communicating effectively is in fact a foundation skill for leading effectively.

2.09.3 The Twenty-First Century Scientific Workforce: Issues of Diversity and Inequity

Leading R&D is challenging for a number of reasons. First, leadership really does matter – it can have an impact on social and economic welfare, as well as on cost, quality, and novelty of the output from R&D. Second, scientific education and training, to some degree, are counterproductive to the leadership skills of getting "work done with and through other people." Third, and the topic of this section, the workforce that a leader of science faces will be diverse in terms of national origin, age, and education yet not diverse in terms of gender. Both issues present leadership challenges.

2.09.3.1 The Issue of Diversity

Let me start by describing a hypothetical average laboratory of about a dozen people in the USA: a number of different national cultures will be represented – several scientists will be foreign-born, predominantly from Asia. Nearly half will be between 35 and 49 years of age; one or two will be younger than 29 years and perhaps one will be 60 years or over. Several will be women. About half will have a doctorate and about half a bachelor's degree, with several distinct disciplines represented.[16]

A major challenge raised by diversity is communication. Given the fact that our hypothetical group involves people trained in languages other than English, there are likely to be linguistic misunderstandings. In addition, though, national "[c]ultural differences in style, expectations, and work attitudes can create misunderstandings that impede the flow of information and the development of science."[17] Moreover, language barriers exist between the disciplines that are represented in the laboratory. Each discipline represents a group of people who share a common meaning for their language, which may be different from that shared by people trained in other disciplines.

Although laboratories in relatively homogeneous countries may have a smaller proportion of international scientists, their R&D workforce will also include people from a number of different countries and with different disciplines.[18] The challenge of communication is inherent in the modern science organization and is directly related to the degree of diversity. Paradoxically, and often underappreciated by leaders of R&D, diversity is linked to the quality of thinking that can occur in the laboratory. In other words, diversity presents a sizable communication challenge to leaders of science, yet sufficient diversity is crucial to the caliber of output from R&D.

An early study of group problem solving showed that, among other qualities, there must be sufficient difference in (or diversity of) approaches, views, perspectives, and so on. With too little diversity, the scope of information search and the quality of the processing of the information by the group was degraded.[19] In short, the caliber of thinking was compromised.[20] Put another way, "science is fundamentally a way of thinking, and people from other countries think differently," as do people of different races and gender.[17] Assuming requisite intellectual competencies of those involved, the caliber of science in a laboratory is dependent on the inclusion of people of different ages, from different countries, from different racial and ethnic groups, from different disciplines, and from both genders. The latter characteristic, gender, brings us to the issue of inequity.

2.09.3.2 The Issue of Inequity

In all industrialized nations, there is a marked under representation of women scientists. For example, according to two news reports:

> The [US] National Academy of Sciences (NAS) announced 72 new members..., nearly 25 percent of them female, representing the largest proportion of women ever elected.... The new members boost the total number of women in the Academy to about 160, or approximately 8% of the 1,922 active members.[21]
>
> A record nine women are among the 42 new fellows elected by the U.K.'s Royal Society this year. Women now make up 4.4% of the Royal Society's total fellowship of 1290.[22]

The president of Harvard University said recently that there is a lack of "women's representation in tenured positions in science and engineering at top universities and research institutions" that is not restricted to the USA.[23] An earlier study of the academic career outcomes of men and women postdocs (Project Access) confirmed both the fact of under representation and the fact of pay and advancement inequities in academia. Of scientists who received their PhD before 1978, women were only half as likely as men to become full professor. Men published more articles, although articles published by women were cited more frequently. Career obstacles for women were "small in themselves in effect, but large in numbers ... [so that] a small set of misfortunes or disadvantages throughout the career accumulated in the same direction, so as to deflect the women in one direction."[24,25]

Apparently, female and male scientists can be judged differently, as illustrated by a 1997 study reported in *Nature* of the Swedish peer review system for postdoctoral fellowship applications. The authors found that "peer reviewers cannot judge scientific merit independent of gender." Regression analyses of factors influencing the judgment of competence showed that:

> [Female] applicants started from a basic competence level of 2.09 competence points... and were given an extra 0.0033 competence point by the reviewers for every impact point they had accumulated. Independent of scientific productivity, however, male applicants received an extra 0.21 points for competence. So, for a female scientist to be awarded the same competence scores as a male colleague, she needed to exceed his scientific productivity by 64 impact points....[26]

The same under representation and inequity hold true in industry. In Europe, "private industry is funding 56% of all European scientific research ... [yet] by now, only 15% of the industrial researchers are women."[27] The gender pay gap also remains sizable in the European private sector, with women holding an advanced degree earning between 60% and 83% of what their male counterparts earn.[28]

A study of women in industrial research found that, in most Organization of Economic Co-operation and Development (OECD) countries, "women scientists and engineers working in the industrial sector are underrepresented and are also more likely to leave technical occupations, as well as the labor force, than women working in other sectors."[28] That situation presents, for Europe, a "crisis" in terms of the shortfall of researchers more generally.[29]

Even when women are present, they may not be heard. So-called "male stereotyping and preconceptions" can influence scientific communication:

> The commonplace presence of women in the laboratory, and the occasional inclusion of women in management positions, masks one residual problem: scientific competence is still judged by male communication styles. Objective, unemotional, assertive. This stereotypic male style seems to be the very essence of research.[30]

Responses from the expert panels reflect the reality many women scientists face:

- Being ignored ("My advisor ignores me when experiments don't work; refuses to have productive discussions; takes experiments away from me if I don't succeed; and is generally not supportive.")
- Being excluded from important committees (One respondent wrote that her most difficult problem was "exclusion from the committees that contribute to the way the department is run.")
- Not being taken seriously ("If I am too soft or quiet, then I am not taken seriously as a leader.").

The question of why women may receive inequitable treatment is beyond the scope of this chapter (the European Women in Industry report cites barriers to recruitment, lack of career opportunities, and gender pay gap and gender stereotypes, among others[8]). What is within the scope of this chapter is the result. When women are underrepresented in the R&D organization, the level or degree of diversity among scientists may be suboptimal for creative thinking:

> [Women] have different qualities than men, and therefore more women in research increases diversity, changes modes of communication and brings something new to the innovation process and thus improving the sector's competitiveness.[28]

In summary, good leadership matters. Good leaders understand that they may lack people skills, despite their scientific education and training. Good leaders appreciate the issues of diversity and inequity in the modern scientific workforce and the challenges they bring.

With this as background, in the following section I describe three positively stated people skills that are critical to leading R&D effectively: the skill of motivating, the skill of communicating and listening, and the skill of resolving conflict. The fourth people skill must be stated in the negative: not micromanaging.

2.09.4 Leading People

2.09.4.1 Motivating

It should be very clear that leadership affects motivation – how people feel about their scientific work, how eager they are (or are not) to come to the laboratory in the morning, how enthusiastically (or not) they approach difficult scientific and technical problems.

The word "motivation" comes from the Latin movere, meaning "to move" – not in the sense of picking up an object and carrying it to a different place but in the sense of a person's being moved to action. This is also the root of the word, "motive." Motives are defined as "relatively stable dispositions to strive for [in other words, be moved to action towards] certain classes of goals."[31] A laboratory with motivated people is one in which there is high morale, or a positive climate. These terms are essentially synonymous. Studies in organizational behavior have demonstrated that being motivated depends among other things on a number of conditions being met. Each is quite straightforward and, one must admit, rather obvious. Yet, my experience suggests that they are not always achieved.

2.09.4.1.1 Reasonable working situation

Safety in the laboratory must be ensured; space must be at least adequate and decently appointed; the required equipment must be available to do the job, and so forth. Although these obviously influence morale, a reasonable working situation can be jeopardized by, for instance, commonly imposed budget constraints. Whether temporary or permanent, draconian approaches to containing or cutting the R&D budget can result in scientists' using outdated equipment and/or sharing critical equipment so that project timelines are affected. They can also result in the outsourcing of tasks to less interested third parties. And, budget constraints always produce what one respondent called "increasing administrative time sinks: paperwork, forms, and hurdles." The point of this example is not that limited space, or equipment problems, or even budget constraints are inherently bad or unavoidable. The point is that, if a problematic working situation is not addressed by the leader, it will be demotivating. (Of course, safety problems are inherently wrong as well as demotivating.)

2.09.4.1.2 Competent people trained appropriately for their job

We can assume that people hired into an R&D organization possess the scientific and technical skills required for a particular position, because human resource systems and personnel have matched technical competencies of recruits with technical demands of jobs. Jobs, however, "demand" human skills, and some require even more human or people skills than technical. Hence, we can also assume that some scientists in a leadership position are not competent for that role, as illustrated by respondents' descriptions of leaders who are "capricious and insulting," "intimidating and confrontational," "given to public tirades," and so on.

Even if competent for their role – at least insofar as they do not exhibit the above egregious behaviors – scientists in leadership positions are rarely trained appropriately for 'getting work done with and through other people.' Ineffective leaders, for example, are also described by respondents as individuals who "rarely praise," who "do not take the responsibility for dealing with unpleasant issues," or who "do not make the effort to listen to people in their lab." Such scientists are clearly uneducated and untrained in what the role of leader requires. Ironically, poorly trained (or incompetent) leaders are also likely to be poorly motivated leaders, with poorly motivated scientists and technicians working for them.

2.09.4.1.3 Assurance of link between effort and outcomes

People must believe that their effort will lead to the desired work goal; they must believe that achieving this goal will lead to certain outcomes (e.g., personal recognition by their scientific peers); and they must value those outcomes. Some examples of leader behaviors (from the expert panels) that produced demotivating conditions, because one of the above links was absent or damaged, include:

- Asking people 'for the impossible – way impossible'
- Expecting 'too much from people and setting unrealistic deadlines'
- Pitting 'individuals against each other' in unachievable contests
- Doing a 'flip-flop' so often on project requirements that the goal literally disappeared.

There is also an important relationship between inequity, as discussed earlier in Section 2.09.3.1, and this condition for motivation. Believing that goal achievement will lead to a desired outcome, such as recognition by one's peers, may be difficult for women scientists. If "a female scientist to be awarded the same competence scores as a male colleague [needs] to exceed his scientific productivity by 64 impact points,"[26] then believing that she will be recognized by her peers may be hard to sustain by a woman scientist in the modern science organization.

2.09.4.1.4 Equity and fairness

This condition, which stipulates that people must be treated and paid fairly in the organization, as compared with similar organizations, is related as well to the above discussion of inequity. If women are not only underrepresented in the scientific workforce but also underpaid in comparison with men, then their morale will of necessity be affected.

2.09.4.1.5 Not micromanaging

This last condition must be stated as a required absence rather than a presence (e.g., 'equity and fairness'). Micromanagement is demotivating and very clearly influences innovation. From studies of scientists at the Center for Creative Leadership, a number of factors were identified as antithetical to creativity[32]:

- Constrained choice
- Overemphasis on tangible rewards
- Evaluation
- Competition
- Perceived apathy by leaders toward the scientists' work
- Unclear goals
- Insufficient resources
- Overemphasis on the status quo
- Time pressures.

Several of these factors characterize the situation that can be provoked by micromanagement: constrained choice, evaluation, overemphasis on the status quo, and time pressures. Another manifestation of micromanaging behavior is the leader who attempts to stay on top of everything. A leader who tries to know every detail reduces morale by reducing scientists' autonomy – and stifles creativity.[11]

2.09.4.2 Communicating and Listening

An article in *The Scientist* a few years ago stated: "You can never do enough training around your overall communication skills."[34] Communication, which comes from the Latin communicare, means to share or impart.[35] Note that what most of us take for granted as communicating is in fact the second aspect (to impart) – of primacy is sharing, which requires listening.

Communicating and listening figure prominently in the expert panel responses describing the most effective leader. Such individuals are articulate and direct. They seek feedback from their intended audience throughout the process of imparting information (in person or electronically), to prevent misunderstandings. They recognize the language barriers that may hinder communication in a group and act as translators and facilitators. People working for them 'catch' their enthusiasm, because such leaders communicate – share – their ideals and passion for the science.

Effective leaders also listen well, which means they:

- 'put aside bad moods when interacting with others'
- 'take time to listen'
- 'look at me directly when speaking and listening'
- 'are never too busy to discuss results.'

The most-repeated adjective associated with 'listening well' is open-minded. Effective leaders are described by the expert panel respondents as open to others, open to ideas, open to different views, open to alternatives. That characteristic of open-mindedness is the prerequisite for what is termed active listening, which describes both a mental stance (i.e., deliberate emotional state) and a set of skills.[36] The mental stance needed to listen well is being nonevaluative. If we judge either the person who is speaking or what we hear, then we are not listening well.

Being nonevaluative is necessary but not sufficient; additional skills required to listen well include:

- Paraphrasing manifest content
- Reflecting the implications
- Reflecting underlying feelings
- Inviting further contributions
- Using nonverbal listening responses (e.g., nodding, consistent eye contact, and so on).

Communicating and listening well are also crucial for the times when leaders must deal with the inevitable conflicts that arise in R&D, as described next.

2.09.4.3 Resolving Conflict

Conflict – resolving or not resolving – appears in each set of responses to the expert panel survey questions. Effective leaders can resolve conflict. Ineffective leaders cannot resolve conflict. Resolving conflict is the second most frequently mentioned difficult problem that scientists face in their role as leader. Because this is such an important topic, I first identify sources from whence conflict is likely to arise and then review approaches for resolving actual conflict.

2.09.4.3.1 Sources of potential conflict

Major sources of differences in R&D represent sources of potential conflict. These sources include individual differences, group differences, and power differences.

Differences among individuals are a likely source of conflict, because individuals may perceive the same issue differently and behave consistently with their perceptions. What we call personality affects our perception, as do age, education, race, gender, ethnicity, experience, religion, and so on.

Differences among groups are a likely source of conflict, because a group of people who have been working together for some time will share common experience and common understanding. Like personality differences, group differences result in a particular way of looking at the world and behaving in response to that perception. A pioneering study (in nonscience firms) revealed that:

> [Group] differences include orientations toward the particular goals of the department, emphasis on interpersonal skills, and time perspectives. Departments, therefore, vary not only in the specific tasks they perform but also in the underlying behavior and outlooks of their members.[37]

The final major source of differences is power; it is, perhaps, also the most obvious:

> The power variable is a relational one; power is meaningless unless it is exercised … Power relationships entail mutual dependency.[37]

The greater the dependence of one person (or one part of the organization) on another, the greater the ability of the more powerful to influence the less powerful. Power derives from having resources that another wants and being in control of their allocation, especially when there are no alternatives. If there are alternatives (whether they are being used or not), the power of the agent holding the resources is diminished. Generally, what the powerful agent in an organization controls is access. For example, senior faculty in the university control access to key information, critical relationships, authorship order, as well as to space, equipment, and funds. The director controls access to general support as well as key information and critical relationships. These differences (or asymmetries) in control of access between the powerful and the dependent agents in science organizations are potential sources of conflict.

2.09.4.3.2 Resolving conflict

Management texts generally describe five approaches to dealing with conflict, of which the most common is avoidance.[38] In a very few situations, avoidance may be appropriate. For instance, if an individual is in serious emotional distress, then the leader needs to help him/her find assistance rather than to deal with the conflict at that moment. But, the problem with avoidance is simple: 'Avoiding a conflict neither effectively resolves it nor eliminates it. Eventually, the conflict has to be faced.'

A second approach is smoothing, which means that the leader minimizes the differences between individuals or groups and emphasizes the commonalities (smoothes the bumps, so to speak). Smoothing can be effective, if the parties generally get along well or if the issue is minor. 'But, if differences between groups [or individuals] are serious, smoothing – like avoidance – is at best a short-run solution.'

A third approach is to compromise, by asking each party to give up something or share a resource equally. Again, although such an approach may be effective, it remains problematic. 'With compromise, there is no distinct winner or loser, and the decision reached is probably not ideal for either group.'

A fourth approach is forcing, or using authority to command a particular decision. In some time-critical or crisis situations, forcing may be appropriate. When time permits, however, the real source of the conflict must be addressed:

> [Forcing] usually works in the short run. As with avoidance, smoothing, and compromise, however, it doesn't focus on the cause of the conflict but rather on its results. If the causes remain, conflict will probably recur.

What is an effective approach to resolving conflict? Confrontation has been found to be an important technique, because it addresses two of the three sources of conflict, individual and group differences:

> The confrontation method of problem solving seeks to reduce tensions through face-to-face meetings of the conflicting [parties]. The purpose of the meetings is to identify conflicts and resolve them ... For conflicts resulting from misunderstandings or language barriers, the confrontation method has proved effective.

The word, "confrontation," is derived from the old French term for sharing a common frontier.[35] To be skilled in effective confrontation is to be skilled in finding the common frontier between people or groups. As should be clear, communication and listening skills are paramount in the process of confrontation.

Although effective in resolving conflicts caused by individual or group differences, confrontation is less effective for conflicts caused by power differences. Instead, leaders should consider the superordinate goals technique, which involves

> developing a common set of goals and objectives that can't be attained without the cooperation of the [parties] involved. In fact, they are unattainable by one [party] singly and supersede all other goals of any of the individual[s] ... involved in the conflict.[38]

The effectiveness of this technique is based on a willingness to discuss the issues by the more powerful person. The reality may be that a power asymmetry represents too high a gradient for meaningful communication between parties, no matter how skilled the leader in communicating and listening. Although not a conflict resolution technique, banding together in the face of power asymmetries and trying to make a general problem (e.g., verbal abuse by a boss) more visible can sometimes achieve the desired results.

2.09.5 Managing R&D

Let me repeat my definition of an effective leader, from Section 2.09.1: an individual who is capable of developing and maintaining an enthusiastic, energetic, and creative group of scientists and of managing R&D successfully. Managing R&D comprises all the 'NONSCIENCE activities' required to produce results, such as designing the organization of work (structure), determining the right number of people in work groups (size), and setting up formal processes that support work (systems). Each of the latter activities, and its impact on the caliber of science, is briefly addressed in this section.

2.09.5.1 Structure

Structure, or how work is organized, refers to the pattern by which people relate to each other and communicate with each other. (Structure should not be assumed to be the same as the organization chart, which depicts formally established lines of authority.) My purpose in this discussion is not to describe all the concrete ways in which work can be organized but rather the two basic patterns by which work can be organized. One is vertical (superior to subordinate, hierarchical relationship and communication) based on the parent–child model. The other pattern is lateral (equal to equal, horizontal relationship and communication) based on the sibling (peer) model. Each pattern is more effective in certain situations.[39]

Vertical structure is more effective in a stable environment and for work well defined by rules and established procedures. For this type of work (e.g., some late-stage development projects), a vertical structure (superior–subordinate, hierarchical relating and communicating) will be more effective. Lateral structure is more effective in an environment of rapid change and high uncertainty and for work ill defined by few rules and established procedures (e.g., some discovery projects). One reason for the effectiveness of lateral structure under such conditions is that people relating to each other as peers are more apt to collaborate informally and to communicate openly. Peer relationships and open communication foster the intellectual challenge that supports good and novel science.

Cognitive scientists describe two types of problems that correspond to the above definitions of types of work. The first is a well-defined problem, which has "an unambiguous problem state, a well articulated goal state, a limited number of potential operators, and clear criteria for when the solution has been found. Ill-defined problems lack one or several of these characteristics."[40] Thus, we can also say that the vertical structure will be better for solving well-defined problems and the lateral for solving ill-defined problems.

2.09.5.2 Size

Size follows logically from structure. The number of people who can be organized in a vertical structure will be larger than the number who can be organized in a lateral structure. Lateral structure, by its nature, constrains the size of a work group. Scientists and technicians can only relate as peers with those whom they know (by definition, a limited number of people). They can only communicate directly with a limited number of people. Estimates of optimum size range from 8 to 12 for a work team up to 200 in the total work organization, such as a discovery unit:

> Size [of an organization] is again the potential killer [of innovation]. The amount and level of communication required to sustain a consistently far-reaching, leading-edge, collectively identified community ... is very high and needs to have great fidelity.[41]

Even in late-stage technology development, problems may arise that can only be solved by work defined by few rules and established procedures (i.e., ill-defined problems). Under these conditions, the problem solving unit should be organized as a lateral structure. A small group of people on a task force, or a small team from vertical functions, should relate to and communicate with each other as peers, collaborating and challenging each other's ideas. These groups should be informal, responsibility should be broadly delegated, and the pattern of relating and communicating should be horizontal.

Let me emphasize that vertical structure is commonly found in R&D and is (frankly) the easier structure to design and manage. Work is divided by department, responsibility is clearly ordered, and there is a visible 'top' and 'bottom.' But, as was noted by one observer of R&D organizations: "Parents are not creative. They are defenders of the paradigm. They use words like 'ought' and 'must'."[41] If the science will benefit from open challenge and debate, then vertical structures will be less effective. Unfortunately, when faced with qualities of the lateral structure, attributes of creative people, and the need to meet cost and/or time constraints, leaders' response may be to put in tighter controls and more explicit rules, and to expand the hierarchy. Such responses prevent a 'collectively identified community' from ever developing.

2.09.5.3 Systems

The caliber of science can also be influenced by formal systems, such as recruitment, performance appraisal and reward, decision making and approval, and information systems. For example, recruitment systems may be unwittingly designed to find people who 'fit'; thus, they may screen out those whose personal and intellectual styles are different from the majority. The consequences are likely to be lack of diversity, conformity and uniformity, people who are unlikely to challenge the status quo, and suboptimal creativity:

> ... the power of 'people who are like me' to be directly interpreted as 'people who are best for our organisation' is enormous. Its converse: 'If I do not like them as people, it cannot be good for me to have to work with them,' is a fundamental error of any manager in a creative field.[41]

Like recruitment systems, performance appraisal and reward systems may unintentionally promote mediocrity, by (for example) rewarding risk aversion. Leaders must be willing to reward people who use good judgement and take risks, even if they are not successful. Effective leaders understand the difference between disappointing results and poor judgement.

With regard to decision making and approval systems, I have observed that leaders of R&D have often designed, or accepted the design of, decision making and approval systems that are cumbersome, slow, and indicate lack of trust in the intelligence of those who use them. In fact, very few formal hurdles should be necessary.

Finally, information systems can provide numerous opportunities to enhance the caliber of science. Collaboration can be supported by electronically linking scientists both within and outside the institution. And, creativity can be enhanced by ensuring wide distribution of provocative material, such as work-in-progress within the organization, and by providing easy access to useful external data banks.

2.09.6 Conclusion

Even this brief overview of leadership issues in R&D would be remiss, if no mention were made of organizational culture. At the end of the day, a leader is also responsible for discerning what are the fundamental values held by people in the organization and how those values either support or hinder good science:

> The means by which creativity and innovation can be influenced [at the organizational level] are many and various but hard to measure. This is cultural.... The organization's culture must reward and encourage behaviour which challenges it, which burns with the desire for change.[41]

In some organizations, the qualities required for creative thinking and novel science are not valued, or they are given only lip service. There is little true collaboration across disciplines; little intellectual challenge from those who might bring a radically different way of thinking; little candid communication (under the guise of valuing civility); and little genuine risk taking (under the guise of valuing quality). Good science emerges when culture – values and norms – encourages and supports the right climate (i.e., morale) and right behaviors.

I do not know anything about the cultures of the institutions in which the expert panel respondents work, but it is a reasonable hypothesis that some norms support (or, at least, do not dissuade) the egregious behaviors of ineffective leaders. It is also a reasonable hypothesis that some values support the behaviors of effective leaders. Effective and ineffective leadership does not exist in a vacuum. If leaders' behaviors persist over time, they have to be supported by some larger context. That context is culture.

References

1. Katz, R., Ed. *The Human Side of Managing Technological Innovation*; Oxford University Press: New York, 1997.
2. Afuah, A. *Innovation Management*; Oxford University Press: New York, 1998.
3. Whatmore, J. *Releasing Creativity: How Leaders Develop Creative Potential in Their Teams*; Kogan Page: London, 1999.
4. Bliss, M. *The Discovery of Insulin*; University of Chicago Press: Chicago, IL, 1982.
5. Baenninger, A.; Costa e Silva, J. A.; Hindmarch, I.; Moeller, I. J.; Rickels, K. *Good Chemistry*; McGraw-Hill: New York, 2004.
6. Day, S. Dissertation, Submitted in partial requirement for the PhD degree, University of Maryland, College Park, MD, 2002.
7. National Science Foundation. *Science and Engineering Indicators*; Washington, DC, 2001.
8. Sapienza, A. M. *Managing Scientists: Leadership Strategies in Scientific Research*, 2nd ed.; Wiley-Liss: New York, 2004.
9. Sapienza, A. M.; Stork, D. Presented at the 39th Annual Meeting of the Eastern Academy of Management, New Haven, CT, May 2002.
10. Turner, B. S. *Medical Power and Social Knowledge*; Sage: London, 1987.
11. Sapienza, A. M.; Lombardino, J. *Drug Dev. Res.* **2005**, *64*, 99–104.
12. Elizer, R. M. *Inst. Transport. Eng. J.* **2000**, *70*, 22–26.
13. Brickley, P. *The Scientist* **2001**, *15*.
14. See, for example, articles in *The Scientist*.
15. The initial workshops (1996 and 1997) were led by Carl M. Cohen and sponsored by the National Science Foundation during the American Society for Cell Biology Annual Meetings.
16. National Science Foundation. *Science and Engineering Indicators 2002*; Washington, DC, 2002. Data are based on 1999 surveys.
17. Park, P. *The Scientist* **2001**, *15*.
18. This observation is based on the author's experience in Europe and Japan.
19. Schroder, H. M.; Driver, M.; Streufert, S. *Human Information Processing*; Holt, Rinehart & Winston: New York, 1967.
20. Cox, T. H.; Blake, S. *Acad. Mgmt. Exec.* **1991**, *5*, 45–56.
21. Hitt, E. *The Scientist* **2003**, *5*.
22. Bhattacharjee, J. *Science* **2003**, *300*, 1217.
23. Summers, L. H. 2005. http://www.president.harvard.edu/ (accessed April 2006).
24. Finn, R. *The Scientist* **1995**, *9*, 3–9.
25. Sonnert, G.; Holton, G. *Gender Differences in Scientific Careers*; Rutgers University Press: New Brunswick, NJ, 1996.
26. Wenneras, C; Wold, A. *Nature* **1997**, *387*, 341–343.
27. *Women in Industrial Research*, Report of the International Conference, Berlin, Germany, October 10–11, 2003.
28. *Women in Industrial Research: Analysis of Statistical Data and Good Practices of Companies*. European Commission report EUR 20814, 2003.
29. EU faces researcher shortfall of 700,000 by 2010, expert group says. Pharma People/PharmaBiz.com, Brussels, 4 May 2004. http://www.pharmabiz.com/article/detnews.asp?articleid=21689&sectionid=17.
30. Barker, K. *The Scientist* **2002**, *16*.
31. Winter, D. G. In *Motivation and Personality: Handbook of Thematic Content Analysis*; Smith, C. P., Ed.; Cambridge University Press: Cambridge, UK, 1992.
32. Amabile, T. M. In *Theories of Creativity*; Runco, M. A., Albert, R. S., Eds.; Sage: Newbury Park, CA, 1990.
34. *The Scientist* **2002**, *16*.
35. *Webster's Third New International Unabridged Dictionary*; IDG Books: Foster City, CA, 2001.
36. Osland, J. S.; Kolb, D. A.; Rubin, I. M. *Organizational Behavior: An Experiential Approach*, 7th ed.; Prentice-Hall: New York, 2001.
37. Hall, R. H., Ed. *Organizations*, 8th ed.; Prentice Hall: New York, 2002. (Reference is to a 1967 study by P. Lawrence and J. Lorsch).
38. Gibson, J. L.; Ivancevich, J. M.; Donnelly, J. H. *Organizations: Behavior, Structure, Processes*, 10th ed.; Irwin McGraw-Hill: Boston, MA, 2000.
39. Gerwin, D. In *Handbook of Organizational Design*; Nystrom, P. C., Starbuck, W., Eds.; Oxford University Press: Oxford, UK, 1981.
40. Brun, H.; Sierla, S. *Soc. Stud. Sci.* **2006**, under review.
41. Thorne, P. *Organizing Genius*; Blackwell: Oxford, UK, 1992.

Biography

Alice M Sapienza, DBA received her bachelor degree in chemistry *magna cum laude* and was awarded US Public Health Service fellowships for doctoral studies in biochemistry to Johns Hopkins and the College of Physicians and Surgeons at Columbia University. After working for a number of years in industry, she received her MBA from Harvard Business School. Between then and entering the doctoral program at the School, she was a general manager in a Harvard teaching hospital. During her doctoral studies, she was awarded the Hawthorne Fellowship as outstanding doctoral student in Organizational Behavior, as well as the School's Division of Research Fellowship for her dissertation on organizational culture.

Now Professor of Healthcare Administration at Simmons College, Dr Sapienza's graduate teaching focuses on organizational strategy in the healthcare environment; her executive teaching focuses on leadership and organizational development. In her academic research, Dr Sapienza has studied the management of research and development and strategic planning in high-technology firms. In addition to articles, case studies, and book chapters, she has written *Managing Scientists: Leadership Strategies in Scientific Research* (Wiley, 2004, 2nd edn), *Forscher Managen* (VCH, 1997), and *Creating Technology Strategies* (Wiley, 1997) and is co-author of *Leading Biotechnology Alliances: Right from the Start* (Wiley, 2001) and *Successful Strategies in Pharmaceutical R&D* (VCH, 1996).

Dr Sapienza has an active consulting practice with high-technology organizations, particularly multinational pharmaceutical and biotechnology companies, working with scientists and executives on problems ranging from culture and organization development to competitive intelligence, strategy determination, and project management. Her public sector clients have included the US Veterans' Administration, Centre for Medicines Research (UK), Birmingham National Health Service (UK), and US National Aeronautics and Space Administration (NASA).

2.10 Innovation (Fighting against the Current)

R Root-Bernstein, Michigan State University, East Lansing, MI, USA

© 2007 Elsevier Ltd. All Rights Reserved.

2.10.1	Swimming against the Current	253
2.10.2	Failing Innovation	254
2.10.3	Innovation: Problem Generation as the Route to Novel Solutions	255
2.10.4	Individual Strategies to Foster Innovation	257
2.10.5	Innovators are Unusual, Inherently Difficult People	259
2.10.6	Managerial Strategies to Foster Innovation	260
2.10.7	Corporate Strategies to Foster Innovations	261
2.10.8	Conclusions	263
	References	263

2.10.1 Swimming against the Current

Sir James Black is one of the founders of modern pharmacology.[1] While working at ICI (1958–64) and Smith, Kline & French (SK&F) (1964–72), he pioneered mechanism-driven research, in the process inventing two novel classes of drugs: adrenergic beta blockers, which are used to treat heart disease and hypertension; and histamine antagonists, which are used to treat stomach ulcers. His inventions of propranolol and cimetidine earned him a Nobel Prize (1988), so one might expect his research to be a case study in how industry correctly identified a bright individual with promising ideas and allowed him to pursue them successfully. In fact, Black's story is a case history of how an innovator fought against the current of scientific opinion, overcame pharmaceutical dogma, and battled corporate distrust to create two of the greatest successes of the century.

Like most innovators, Black was not trained in the field to which he contributed. He was a physician who turned to physiology because of his disgust with the nonscientific basis of most medical care. His research was directed not at drug development, but at solving basic scientific questions and his initial contacts with ICI were not for employment, but for basic research support. In a recent interview, he stated that:

> I think you should understand that at no point have I ever professionally made claims about what I thought my work would achieve, for the simple reason I've never thought that it was going to achieve anything other than answer a question ... The challenge was to make a molecule which would have the properties of allowing you to answer a physiological question. Now, if the answer had come out a different way in each case, there would have been no drugs ...[2]

The fact that Black's beta blockers and histamine antagonists have medicinal value was thus serendipitous. Such basic research projects are not the sort that most pharmaceutical or biotech companies would even consider today and not ones that ICI and SK&F much liked at the time.

Black's second handicap was that he was neither a team player nor socially adept. One colleague described him as "a very vexing man ... He wants you to work in a particular way and he won't suffer fools."[3] Another says that Black is very "disorderly."[3] Black agrees, describing himself as congenitally and combatively skeptical, impatient, loud, and argumentative:

> The only difference between me and some other colleagues would be temerity. I think I ask questions of chemists and physiologists which are really quite preposterous. And I don't seem to mind being thought stupid; I don't seem to care what people think about my questions. So I think in so far as what I've done has led to things, they have all involved turning something around the other way. I think my brain automatically does this. It always is challenging the accepted view of things. So if someone says to me, 'The speed of light is constant', then I'll say, 'Well, what would happen if it wasn't?'[2]

Needless to say, many colleagues at ICI and SK&F balked at his hectoring and a considerable number quit rather than continue working with him.[3] Black notes laconically that his bosses "tolerated" him[3] and, "fought many battles on my behalf to keep the initially controversial programme[s] going."[4]

On the other hand, Black lived what he preaches. His invention of beta blockers began with the question of what causes angina. The standard dogma at the time maintained that angina resulted from the heart getting insufficient blood due to blocked coronary arteries. In consequence, the pharmaceutical industry was searching for ways to increase the amount of oxygen reaching the heart, which in practice meant increasing the heart rate using adrenalin (epinephrine) and its agonists. The problem with increasing the heart rate was that the heart itself then used more energy and needed even more oxygen, simply exacerbating the problem. Black, typically, turned the problem on its head. "Let us see if we can decrease how much blood the heart actually needs,"[2,4] he proposed: block the adrenergic compounds that increase heart rate and blood pressure and see if that relieves the heart. The problem was, however, complicated by the fact that a chemical cousin of epinephrine (adrenalin in British terminology), norepinephrine (noradrenalin), was released by sympathetic nerves innervating the heart, and this raised blood pressure, increasing oxygen delivery to heart tissue, an outcome Black desired. Black therefore did not want to block norepinephrine's actions, especially as previous studies had shown that blocking norepinephrine release exacerbated angina. Black, in short, needed to find a new class of compounds that would selectively block epinephrine activity (which was mediated by beta receptors) but not norepinephrine activity (which was medicated by alpha receptors).[2,4]

Black, essentially ignorant of the art of chemical design and synthesis, but a quick learner and aided by the young medicinal chemist, John Stephensen,[4] began the long process of finding a compound that would selectively block epinephrine's beta receptors without interfering with norepinephrine's alpha receptors. The eventual result was propranalol. But even with the compound in hand, the discovery process had only begun, because it was not until the clinical pharmacology was under way that Brian Pritchard of University College, London, discovered that beta blockers had antihypertensive effects in addition to relieving angina.[4]

With a major success in hand, one might have expected ICI to have rewarded Black with the freedom to investigate another fundamental physiological problem in hopes of an equally rewarding result. The opposite occurred. Black proposed to follow up his beta blocker success by addressing a similar problem regarding histamine production related to stomach ulcers.[3] ICI administrators balked, believing that lightning never strikes twice in the same place. Black, holding in his head a novel approach to studying the entire field of monoamine receptor regulation, refused to be bridled, and moved to SK&F where former colleagues created a place for him. There the same science-centered, antidogmatic approach to exploring histamine regulation produced the same corporate doubts and difficulties with management and subordinates, but, within a few years, yielded cimetidine.[2-4] Thus, in answering his second basic science question, Black created his second blockbuster drug and, in turn created his second unforeseen market. By that time, however, Black had found himself unable to work within the SK&F corporate structure, and had already left the company for academia. The man who had created two of the most profitable classes of drugs in pharmaceutical history could not work within the industry.[2-4]

2.10.2 Failing Innovation

If Black's saga were unique, we could dismiss it as an aberration that could be ignored. Unfortunately, Black's experiences in beginning with a basic research question, developing novel approaches to drug design, having to overcome scientific dogma, needing protection from managerial interference, and eventually leaving the industry, are all too typical, as I shall demonstrate below. These aspects of the pharmaceutical innovation process raise some hard questions for the industry. For example, how can research leading to blockbuster drugs and novel technologies be fostered if the process is often serendipitous and must employ difficult-to-manage, controversial individuals whose research may not even be focused on the development of drugs? Conversely, how can the independent, self-motivated scientists most likely to create real breakthroughs survive (or better, thrive) within cut-throat corporate environments in which acquiescence to corporate goals through teamwork and concensus are often required to obtain access to research resources? Make no mistake about it: the climate that grudgingly permitted Black the independence to perform his miracles is quickly disappearing to be replaced by top-down management practices too often driven by marketing executives who know nothing about research. Pharmaceutical innovation is the victim.

A dramatic decline in pharmaceutical innovation is undeniable. For the past two decades, the pharmaceutical and biotechnology industries have undergone a series of corporate mergers that have resulted in ever-larger multinational conglomerates.[5,6] Two of the most commonly cited reasons for these mergers are to increase the efficiency of marketing and research divisions. Corporations with strong sales forces but weak product pipelines will often merge with those having strong sets of products but lacking the means to market them effectively. While such a strategy looks promising at first blush, and may be so in the short term, it often results in centralizing and streamlining research, and enslaving it to market demands, with long-term adverse consequences.

It is an uncomfortable fact that, in the last 20 years, over half of the revenues accruing to the five largest pharmaceutical companies have come from drugs in-licensed from smaller companies, despite the fact that these five corporations invest as much in research as the rest of the industry combined.[7] Even as the amount of money spent on R&D has increased exponentially within the industry as a whole, from around $5 billion per year in 1980 to about $35 billion in 2005, the numbers of new chemical entities, new drug applications, and FDA-approved drugs have declined continuously for more than a decade.[5–8] The cost per new chemical entity launched has increased over the same time period from $44 million per NCE to almost $900 million.[7] The vast majority of approved drugs begin their lives in the smaller, less affluent, but intellectually more vibrant, cauldrons of start-up companies and smaller firms.[4–6] For many reasons, some of which will be discussed below, these small companies lose their innovativeness following mergers. Companies made ever larger by mergers require ever more blockbuster drugs to support their increased overhead, but fail to develop such drugs, requiring further rounds of mergers that become ever less effective at providing the necessary product pipeline. The innovators who support the industry find their intellectual freedom constrained by their new managers, undercutting their value to their new owners. As mergers continue, more and more innovators are leaving the large conglomerates, taking their ideas and techniques with them. Many of these entrepreneurs are successfully developing their drugs in new start-up companies that are able to salvage approaches the larger pharmaceutical firms have discarded or refused to support.[9]

In sum, the large pharmaceutical conglomerates are creating their own competitors through their failure to value innovation within their own corporations. The resulting situation is paradoxical: the largest companies with the greatest resources and the most flexibility take the fewest risks and reap the fewest discoveries, while the smallest companies, constrained by inadequate resources and lack of financial stability, take the greatest risks and harvest the most breakthroughs. As Theodore Shultz concluded in 1980 in a highly prescient analysis of industrial research innovation:

> We have become enamoured of large research institutes and large programmes. [But] they are difficult to administer and they are subject to what is known in economics as diminishing returns to scale.[10]

The problem to be addressed in this chapter is how the pharmaceutical industry can best reverse this trend toward diminishing returns to scale by better fostering innovation.

2.10.3 Innovation: Problem Generation as the Route to Novel Solutions

On the pages that follow, I will be using a very specific definition of innovation. Innovation is the process by which novel, often unanticipated, modes of solving practical problems come into being. Innovation is not be confused with discovering (which is the search for basic principles or understanding of nature), or inventing (the goal of which is a new technology), or engineering (which is the application of existing technologies to solve specific, well-defined problems). Innovation combines discovery, invention, and engineering and is arguably the most difficult of all to achieve. Thus, I consider Black's invention of beta blockers to be an innovation because it created a new paradigm for discovering and designing drugs that he, and other people, were able to apply fruitfully to other classes of problems. Propranolol itself is merely an invention – an exemplar of the innovation. This distinction is extremely important, because too much literature concerning pharmaceutical innovation pretends that every new drug, or every drug that treats a new class of diseases, is automatically an 'innovation' whether it involves novel methods of production, new theories of design or action, or is merely the application of well-understood and widely practiced techniques.

Intrinsic in the definition of innovation adopted here are two important points. One is that innovations, unlike inventions, cannot be planned. One of the major failings of pharmaceutical managers is to strive to produce a particular solution to a well-defined problem. The objective, all too often, is a product that will treat a specific indication and compete in a well-characterized market. Such top-down, end-driven research can only result in 'me-too' drugs and only rarely, usually as a result of failure or serendipity, in the kind of unexpected surprises that open up new fields. This leads to the second important point: the assumption behind most pharmaceutical research is that innovation consists of finding novel solutions to problems. In fact, the basis for all innovation is found in the choice and definition of the problem that is to be attacked.[11,12] As Jack Kilby, one of the inventors of the integrated circuit, or 'chip,' has written:

> The definition of the problem becomes a major part of the innovation. A lot of solutions fail because they're solving the wrong problem, and nobody realizes that until the patent is filed and they've built the thing.[13]

As a general rule, all innovations begin with the recognition of a novel problem or the recasting of a well-known problem into new terms. Innovators are therefore people who pose new sorts of problems, which in turn create possibilities that no one has previously imagined. There can be no innovation without such problem invention. The freedom to choose a solvable problem, and to define the problem in the way most likely to result in a solution, is therefore the most critical aspect of fostering innovation.

Examples of inventing new problems or recasting old ones are legion in the history of pharmaceutical innovation. While most cardiologists posed the problem of angina in terms of the heart needing more oxygen, Black recast it as how to let the heart get along with the oxygen that was available. By recasting the problem, he was led to develop a new class of drugs with actions that no one else was even considering.

Kary Mullis, the Nobel Prize winner who invented the polymerase chain reaction (PCR) concept, tackled the problem of artificially replicating a genetic sequence by first recasting an existing problem.[14] The Cetus Corporation, for which Mullis worked at the time (about 1983–84), was trying to develop highly sensitive and accurate tests for the mutations in the beta-hemoglobin gene that are responsible for sickle-cell anemia. The standard approach was to create an artificial primer complementary to the open reading frame of the beta-globin gene and then to add DNA polymerase in the presence of adequate DNA bases to synthesize a complementary strand that could be analyzed for its sequence. Unfortunately, this approach was not sensitive enough. So Mullis's first step in recasting the problem was to wonder what would happen if he doubled the amount of information he acquired by making two primers. One primer would be against the mutated beta-globin gene and the other against its (untranslated) complementary strand. In the presence of the polymerase, short polynucleotides mimicking both the gene and its complementary strand would be synthesized and analyzed, each sequence providing confirmation of the information gained from the other: twice as much information, twice the sensitivity. Not long after this thought occurred to Mullis, he was

> suddenly jolted by a realization: the strands of DNA in the target, and the extended oligonucleotides [synthesized from the primers], would have the same base sequences. In effect the mock reaction would have doubled the number of DNA targets in the sample![15]

Primed by previous work he had performed as an avocation on recursive functions in computer programs (such as loops and fractals), Mullis immediately realized that what he could do once, he could do as many times as he liked to make a theoretically unlimited number of copies of any gene sequence:

> Eureka! By then, I'd forgotten all about the problem of diagnosis. I realized that this is a process that I can use anywhere I want to amplify a little section of DNA ...[15]

After doing a literature search, aided by Cetus's librarian, Mullis discovered that no one had previously considered the possibility of gene amplification. It was a new problem posing new possibilities that created a new market.

The invention of an early form of combinatorial chemistry at Parke-Davis Pharmaceutical Research Division by Walter Moos's Bioorganic Group occurred in a similar way. The group was considering how best to invest several million dollars in peptide arrays to be synthesized by Chiron Corporation, a pioneer in peptide array synthesis. I had been hired as a consultant to the group and, not being a medicinal chemist, asked naively how the results of these arrays would be used. It was patiently explained to me that the peptide leads would be painstakingly converted into small molecules using a series of strategies well known to the industry. This process of finding a peptide lead and converting it seemed wasteful to me, so even more naively, I asked the group why they did not just make combinatorial arrays of non-peptide-organic compounds based on the same strategy by which peptide arrays are made, thereby saving themselves an expensive and time-consuming step. Once the question was posed, the group instantly realized that analogous combinatorial chemistry using small organic building-blocks could be done. It required only a few more weeks to produce a prototype set of compounds and a few months to produce a proof of concept. In retrospect, it is probable that every medicinal and organic chemist had the knowledge to have invented combinatorial chemistry, but none had asked the right question.[16]

Failures are another major source of fruitful problems. Leo Sternbach's invention of the benzodiazepine antianxiety drugs, such as Valium, resulted from his early experience as a dye chemist. While working as a young assistant in a chemistry laboratory at the University of Krakow, Sternbach spent several years trying to develop azo dyes, or their intermediates. During his researches, he came across the 4,5-benzo-[hept-1,2,6-oxadiazines], which he was able to produce in large, pure yields, but which turned out to be unsuitable as dyes. Sternbach did not publish his data and the class of compounds was almost entirely ignored by chemists for another generation. Then, some 20 years later,

Sternbach realized that these compounds had the right chemical properties to be biologically active, stable, and to permit novel chemical modifications that would provide a wide range of potential psychotherapeutic drug candidates.[17]

A combination of basic research and clinical failure also led to the identification of sildenafil citrate (Viagra), a phosphodiesterase-5 inhibitor that increases nitric oxide (NO) production, by Pfizer. The discovery of NO and its function as a smooth muscle relaxant came about, like Black's discoveries, from investigators asking basic science questions without regard for their potential applications. Robert Furchgott, Louis Ignarro and Ferid Murad, working independently, each became interested in the role that cyclic GMP plays in smooth muscle contraction and relaxation. Furchgott,[18] originally trained as a physical chemist, provided the impetus to the research by discovering an endothelium dependent relaxation factor (EDRF) in aortic smooth muscle, which led to Murad and Ignarro's work. Murad,[19] a physician with an additional degree in pharmacology, showed that compounds such as nitroglycerin released NO, which in turn activates cyclic GMP. Ignarro,[20] a pharmacologist, then demonstrated that NO itself relaxes vascular smooth muscle.

Notably, several major pharmaceutical companies had opportunities to support and capitalize on these men's work and did not. Ignarro's work was begun at Geigy Pharmaceuticals, where he reports having had unusual "freedom to pursue basic research on biochemical pharmacology" in addition to directed drug development research.[20] This freedom disappeared around 1972 after Geigy merged with Ciba and Ignarro left the company for academia shortly thereafter. Similarly, Murad's work on NO and its mechanisms of action, which began at Stanford University, was developed while he was a Vice President at Abbott Laboratories, where he became the company's senior scientist. Citing marketing pressures to develop drugs for which there was inadequate scientific basis, Murad left to become the founder, President and CEO of a new biotech company, Molecular Geriatrics Corporation and with him went his NO research.[19] The discovery of Viagra thus had to occur by serendipity.

Around 1990, Pfizer scientists developing new classes of heart disease drugs discovered that chemical compounds of the pyrazolopyrmidinone class acted as phosphodiesterase-5 blockers, antagonizing cyclic GMP breakdown, and thereby maximizing NO effects. Sildenafil citrate was chosen from this class of compounds for clinical testing. Rather than exploring all the possible uses of the compound, a market-driven approach was adopted, resulting in its testing as an antihypertensive and angina medication in 1994. It failed. One unauthenticated story has it that patients refused to give back the pills they had been given when the clinical trial was terminated. The patients' odd response alerted the trial directors, Nicholas Terrett and Peter Ellis, that the drug must have some unexpected benefit and further inquiry revealed that men taking the drug had improved sexual performance. Another story has it that inquiries about side effects revealed enhanced erections. Pfizer researchers have steadfastly refused to comment on the actual discovery process.[21] Whatever the case, the unexpected observation that sildenafil had aphrodisiac effects led Pfizer to reinvestigate the compound and apply for approval to test the drug to treat impotence, a novel indication for which it, and its competitors, have been very successful.

In short, one of the major obstacles to innovation is the belief that the role of the scientist is to solve existing problems posed by the marketing department, managers, or competitors using well-proven techniques. Quite the contrary, innovation can only occur when dogmas are questioned, questions are asked in new and different ways, and failures are used as clues to new problems that require new kinds of solutions.

2.10.4 Individual Strategies to Foster Innovation

While every innovation comes about in a unique way, the process is always similar and I have found that creative scientists share many research strategies that are foreign to the average scientist and unknown to most managers (**Table 1**). Black, for example, has said in his typically understated way that "I'm always just following something [puzzling], and always aiming for something which I ought to solve fairly quickly by asking simple questions."[2] This brief sentence contains three strategies shared by most innovators that are critical for developing breakthrough projects: (1) innovations are usually rooted in basic, rather than applied, research; (2) they involve problems that can be addressed quickly, easily, and cheaply; and (3) they concern simple, often antidogmatic, questions that can be clearly stated in testable ways.

More often than not, the questions that lead to major pharmaceutical innovations concern fundamental or basic science problems rather than applied or market-targeted problems. An analysis by Thompson shows that most blockbuster drugs developed in recent decades have originated in university laboratories funded by federally sponsored research programs such as the National Institutes of Health.[24] A more general review of biomedical innovations by Comroe and Dripps[25] has revealed similarly that over 60% of new therapies resulted from research that was not directed at creating a therapy or product. Examples of pharmaceutical innovations beginning with basic research questions are common, include Black's work on beta blockers and antihistamines as well as the discovery of NO, which eventually led

Table 1 Some strategies for innovating[11,12,16,22,23]

1. Pay attention to anomalies, failures, and outliers.
2. Turn every practice and theory on its head.
3. Extrapolate, iterate, and recurse.
4. Think simple.
5. Court serendipity by running myriad controls.
6. Court serendipity by trying to solve 'impossible problems.'
7. Analogize: what worked for one problem may work for another.
8. Work on multiple problems simultaneously and look for intersections.
9. Look back: recreate old discoveries with new techniques.
10. Look back: apply old, simple techniques to new problems.
11. Work on the problem that needs to be solved *after* your current problem.
12. When nothing else works, try the unthinkable.

to the invention of Viagra, Cialis, and other impotence drugs. Francis Colpaert's description of the path he and his colleagues took at Janssen Pharmaceutica to the creation of the antipsychotic risperidone also depended on the freedom to investigate basic science problems:

> It occurred after a sinuous, unpredictable route to which previous drug discoveries (for example, those of haloperidol and fentanyl) gave impetus, through novel technologies (for example DD [the animal-based Drug Discrimination test invented at Janssen]) and through fundamental research (for example, in receptor theory and [the development of a novel theory of] opioid tolerance), which in some instances took decades for the scientific community to at least tolerate, if not endorse ... Other important factors in the development of risperidone were a resourceful medicinal chemistry group, and a climate of exceptional intellectual freedom and genuine concern with medical progress, in which boundaries between the theoretical and the applied, between the academic and the industrial, and between the hypothesis and the treatment were blurred ... It was fuelled by the desire to treat and understand, inspired by intense fundamental research.[26]

Black's second strategy is that the problem to be tackled be addressable quickly, cheaply, and simply. This criterion differentiates breakthrough research from developmental research. One of the best examples of such qualitative research comes from Albert Szent-Gyorgyi's discovery of vitamin C. In order to screen many compounds for antioxidant activity very quickly, he developed a simple color reaction that took only minutes to perform. This test allowed him to narrow his research focus to a specific compound very rapidly.[27] As Nobel Laureate G. P. Thompson has argued,[28] the object of such breakthrough research is not to produce publication-quality quantitative data but to create quick-and-dirty qualitative techniques that can rapidly and extremely inexpensively test the potential of a wide range of possible research targets.

Black's third strategy is that the problem should pose the possibility of turning a field upside down. Innovations (as opposed to inventions) rarely emerge from incremental modifications of existing technologies. New classes of drugs, whether they are the first antibiotics, the first beta blockers, the first impotence treatments, or novel technologies such as PCR, combinatorial chemistry, or genetically engineered proteins, result when people stop trying to improve current practice incrementally and look instead for a completely new way to address the problem. As C. Robin Ganellin, one of Black's colleagues at SK&F (who also left for academia) has said, "The desire to play safe kills research."[29]

Thus, innovating often requires questioning daily practice. Students of medicinal chemistry are taught to use the latest techniques to address the most modern problems, yet many discoveries are made by adapting age-old techniques to new problems or by addressing old problems with new techniques. Szent-Gyorgyi wrote that one of his favorite techniques for making discoveries was to reproduce a 50- or 100-year-old discovery and to look at it with modern eyes.[27] It is therefore noteworthy that Edward Shorter has recently suggested that pharmacologists look back, rather then forward, in their search for new drugs:

> The idea is not to recycle the golden oldies, but to determine the receptor profile of forgotten drugs of proven efficacy, and to determine what patentable compounds today might have a similar receptor profile. Since the introduction of modern psychopharmacology [for example] with the first chlorpromazine trials in 1952, thousands of compounds have been synthesized. Many of these have shown efficacy in open-label trials or anecdotally, only to be cast aside.[30]

Similarly, old techniques can also be resurrected and combined with new problems. When I was consulting for Chiron Corporation around 1995, I realized that the drug discovery team had large libraries of compounds called 'peptoids' (*N*-alkylated glycine trimers) that they had combinatorially synthesized and tested for specific indications, but which now lay unused in storage. In my constant efforts to foster serendipitous discoveries, I approached various members of the team with the idea of screening these compound libraries for unintended and unexpected activities. In order to make such screening programs worthwhile, screens had to be (in accordance with Black's criteria) extremely rapid, inexpensive, and simple, and they had to hold out the possibility of yielding very surprising, useful results. One outstanding problem fit the bill, and that was the inexorable emergence of antibiotic-resistant bacteria and the lack of new classes of antibiotics to treat them. Combining this major contemporary problem with one of the oldest drug discovery methods, I therefore suggested that all existing compound libraries be screened against antibiotic-resistant strains of bacteria using the same technique used by Alexander Fleming in the 1920s to discover penicillin: plate the bacteria on agar and then treat with mixtures from the compound libraries.[23] If the bacteria died, then one could assume that something in the mixture of chemicals was lethal to them. The mixtures could then be deconvoluted to single compounds that could be tested individually for their antimicrobial activity. The project was so simple that it was begun as an unfunded voluntary effort in the evenings and on weekends by virologist Jill Winter and several co-workers, including Mike Siani, a computer expert with minimal laboratory experience.[31] Within a few months, a novel class of drugs with potent, broad-spectrum antibiotic properties, including rapid activity against antibiotic-resistant strains of *Staphylococcus aureus* and *Escherichia coli*, had been isolated. Within 6 months, the project had moved from an unfunded exploratory exercise to a formally recognized and funded focus of the preclinical research team. Chiron's CHIR29498 became the first of what may prove to be a broad class of peptoid-based antibiotic compounds.[32,33]

2.10.5 Innovators are Unusual, Inherently Difficult People

Innovators are, by their personal natures and in consequence of what they are attempting to achieve, difficult to manage.[34,35] One cannot expect individuals who pose novel, often antidogmatic problems, and thereby attempt to change the daily practice and thinking of their colleagues and managers, to be comfortable team players. On the contrary, they are much more likely to be independent, demanding, overly confident, egotistical, self-motivated, difficult people. They are also likely to be trained outside of medicinal or organic chemistry, or even pharmacology, making them doubly suspect. Black, Furchgott, Mullis, Winter are all examples. Any company that wants to foster innovation has to be willing to work with such unusual people (**Table 2**).

Working with innovators is never easy. Sternbach, the inventor of Librium, Valium, and the other benzodiazapines, was so independent that he often caused conflicts at the Roche laboratories, Nutley, NJ:

> In general, [he] was very critical of his superiors, especially those trained as chemists, and extremely demanding toward his subordinates. Critical, demanding, and implacable, occasionally unfair as well, Sternbach puts it in a nutshell: 'Those who were above me were not my favorites.'[17]

Kary Mullis of PCR has been described in similar terms by colleagues at Cetus: "His irreverence, verging on belligerence, and his increasingly trenchant criticisms of current procedures produced clashes with [colleagues]."[15] His boss later described Mullis as being "abrasive and combative and often times his comments would be counterproductive in meetings where people have to try and work together. Mullis had a grudge against his critics

Table 2 Some characteristics of innovators[11,13,34–42]

---
1. Unusually and broadly trained.
2. Outspoken, critical, and independent.
3. Actively engaged in creative avocations.
4. Physically active and energetic.
5. Self-motivated.
6. Able to work very intensively.
7. Unwilling to work on 'useless' projects.
8. High standards and expectations of selves and others.
9. Competitive.
10. Constantly exploring new fields and technologies.

and they had a grudge against him."[15] The grudges escalated to fist fights and at one point Mullis threatened a co-worker with a gun. Mullis even flouted the Human Relations regulations of the company by having serial affairs with co-workers at the laboratory. In retrospect, it is astounding that he was not fired.

The members of the Bioorganic Group who invented an early form of combinatorial chemistry at Parke-Davis were not as outrageous as Mullis, but several were generally considered to be difficult to work with when they were assigned to Walter Moos. Mike Pavia was known as a brash, outspoken chemist who would never have succeeded as a diplomat. Sheila Hobbs DeWitt was a fiercely independent chemist who, like Sternbach, expected the same superlative results from every colleague and subordinate that she expected from herself. John Kiely, the third senior chemist in the group, was skeptical almost to a fault and so critical that he frequently antagonized colleagues with his blunt assessments of their perceived failings. The extraordinary thing is that this group of 'difficult' individuals quickly formed a high-functioning team once they found a common problem to which they could all contribute their unique skills.[16] Unfortunately, Parke-Davis was not ready for the innovation they produced, and every member of the Bioorganic Group left Parke-Davis within a couple of years of their innovation, as it became clear that the company had neither the intention nor the capability to implement combinatorial chemistry itself. In a classic case of a company creating its own competitors, each Bioorganic Group member became a successful senior executive at a different biotechnology corporation that valued their innovative spirit.

What it comes down to is that people who are capable of the best science know that they are capable of the best science. They have vision and they have the drive to make their visions reality. They are not, therefore, the type of people who take orders. They do not respect managers who lack the scientific credentials to make the decisions that need to be made daily in the laboratory. They disdain those who cannot see as they do how a small breakthrough signifies the potential of a much bigger one. They trust in their own ability to be pioneers, hacking their way into the uncharted waters of medicine with only their intuition, experience, and courage as guides. Black has summed up the problem for the pharmaceutical industry in these words:

> People like me … are … bad news for industry because I was constantly rebelling. I was constantly challenging the assumptions they made. I challenged their practices. I was awful … It wasn't that I set out to give them a hard time. It was simply that being there put me in an environment which wasn't conducive to what I wanted to do … I was always sort of unhappy.[2]

Discontent is part of the innovator's life since the object of the innovator is to shake up existing modes of working and thinking. Innovators are themselves discontent because they see how things could be done better, faster, cheaper, more effectively. Innovators make everyone else discontent because innovation forces people out of their comfortable routines. Companies that truly want to foster innovation must be willing to put up with such problems. Any manager who thinks that it will be easy is fooling him- or herself.

2.10.6 Managerial Strategies to Foster Innovation

Despite the poor record that many pharmaceutical companies have in managing innovation, it is indisputable that some unusually enlightened corporate managers have played major roles in bringing every major pharmaceutical innovation from its inception in the mind of the pioneering inventor to its proof of concept and thence to its actual production. They tend to evaluate projects and people in unusual ways that ignore industry standards (**Table 3**).

One way to minimize problems is to isolate innovators as much as possible from the rest of the corporation. When ICI's managers refused to believe that James Black could follow up his invention of beta blockers with a similarly conceived approach to developing histamine blockers to treat stomach ulcers, Black began looking for other employment and quickly convinced a former colleague, Edward Paget, that what he had really developed was not a drug, but a novel approach to drug discovery. Paget, a manager at Smith, Kline & French Laboratories (based in Philadelphia) hired Black to be chief of pharmacology at the British Welwyn Research Institute and gave another open-minded colleague, William Duncan, oversight of Black's project. Black promised an antiulcer compound within 6 months and Paget let him loose to completely restructure Welwyn's research organization, carefully hiding from his bosses at SK&F in Philadelphia the revolutionary nature of Black's new approach to the problem. And then, despite Black's failure to provide such a compound within 6 months, or even six times that, Paget and Duncan continued to defend and protect Black's project, "isolat[ing] the scientists from the political and financial hardships of a company undergoing major reorganization and a new-product famine."[3] In the end, the problem was not the drug design program, which in retrospect had, in fact, yielded active drugs within the 6 months that Black had projected, but rather the pharmacological test that the company had adopted for testing the compounds. The standard test was incapable of

Table 3 Some criteria for evaluating innovative projects[1,2,3,11,13,22,23,28]

1. Is the problem being addressed fundamental?
2. Will a solution alter the way professionals think and act?
3. Is the concept or experiment simple?
4. Do clear-cut *qualitative* tests exist?
5. Is it inexpensive to test?
6. Can it be tested quickly and easily?
7. Is it testable using existing equipment and personnel?
8. Are there clear criteria for success or failure?
9. Might it create a *new* market?
10. Can it be used as a model for solving similar types of problems?

demonstrating antihistamine activity. When one of Black's employees secretly designed a new, more appropriate test, the efficacy of the compounds immediately became apparent, and SK&F, which was slowly going under, staked its hopes on the new class of drugs that resulted – and succeeded.[3]

Similarly, PCR would have been many years, if not decades, slower to be developed had not the management team at Cetus Corporation, especially Tom White, not coaxed, cajoled, guided, and perhaps even threatened Kerry Mullis into translating his idea into a proof of concept. Not only did White have to enlist reluctant colleagues to help Mullis test the concept despite severe questions as to its viability, but White also had to protect Mullis from administrators and colleagues who wanted him fired for his outrageous behavior. Everyone agrees it would have been much easier to ignore Mullis's hair-brained idea and simply fire him.[15]

What each of these managers did was what General Electric's most successful director of research (and Nobel Laureate) Irving Langmuir did: create the conditions under which innovations may flourish. And the most important condition is that of freedom for the investigator. As Langmuir wrote:

> You can't plan discoveries. But you can plan work that will lead to discoveries. You can organize a laboratory so as to increase the probabilities that useful things will happen there. And, in so doing, keep the flexibility, keep the freedom … We know from our own experience that in true freedom we can do things that could never be done through planning.[43]

Langmuir's advice is echoed by fellow Nobel Laureate Baruch Blumberg, also a scientific director of several laboratories. Blumberg has written that

> flexibility is essential and the scientists themselves have the responsibility of deciding the appropriate directions, within guidelines. The role of management is to supply funds and to not get in the way of the creativity of the scientists, while, when required, providing them with appropriate direction.[44]

I would go further and argue that the manager's responsibility is to create conditions conducive to constant change by protecting innovators both from the complacency and routine practiced by the average scientist and from the all-too-often blinkered visions of upper management.[45] Innovations are like babies: they require time to mature before they are asked to compete in the adult world. Managers of innovation must be like parents, guiding innate tendencies in fruitful directions and finding the best opportunities for their offspring to succeed.

2.10.7 Corporate Strategies to Foster Innovations

Given the obstacles to innovation that inevitably exist within most large companies, it is worthwhile asking whether innovations must simply be licensed from biotech firms and universities where the freedom to explore is still possible. This is certainly one plausible model, and one that is being increasingly exploited. It is not, however, optimal. As Genallin has argued:

> There is big pressure on universities now to go into areas that are determined politically by the funding councils and I think this is very bad for discovery. There is too much pressure to go into applied areas. One should allow people a lot more freedom to speculate and make their own discoveries. For that, you need to remove some of the pressure that is on people.[29]

In addition, universities are increasingly spawning their own start-up companies that have every incentive to develop innovations as independently as possible in order to nurture their value.

Fortunately, other models exist. One was pioneered by Eli Lilly to produce the first genetically engineered drug (insulin). Lilly, one of the oldest manufacturers of insulin, realized early in the biotechnology revolution that insulin would eventually be produced by some company using recombinant DNA techniques. They wanted to be the first. Unfortunately, they quickly encountered a corporate culture that made it virtually impossible to hire the rare individuals who had appropriate recombinant DNA experience. (Sheila Hobbs DeWitt encountered similar problems in developing combinatorial chemistry at Diversomer Technologies, Inc., a failed Parke-Davis spin-off company, when the chemist-dominated Parke-Davis culture would not allow hiring of the robotics and computer experts necessary to the project.[16]) Lilly eventually partnered with Genentech, which supplied the needed biotechnology expertise.[46] Such partnering is much more likely to produce innovations than simple in-licensing, which can only occur after the innovation has been made, but still leaves the larger company without the in-house expertise needed to make additional innovations. In addition, the Lilly model can be applied only to products such as insulin that already have well-defined and lucrative markets. In this context, it is worth remembering that neither Lilly nor any other pharmaceutical company discovered insulin. That distinction went to a surgeon without academic credentials (Frederick Banting) working with a doctoral student in physiology (Charles Best) in a borrowed laboratory without funding or adequate equipment. Pioneers work in undeveloped wildernesses.

The trick, then, is to create an environment that mimics "outside the pale" inside the corporation (**Table 4**). A model for such an environment not only exists, but has been working with great efficiency and effectiveness in the aerospace (Lockheed), computer (IBM, Bell Labs) and manufacturing (3M) industries, and it goes by the now trademarked named 'Skunk Works.'[47–49] The purpose of a skunk works is to take proven innovators and give them a rich research environment completely protected from corporate planners with the simple expectation that they do whatever research they believe will most likely yield a fundamental, dogma-shifting innovation. Middle managers guide this process but do not determine the problems addressed, or how they are addressed. The key to making a skunk works work is to hire the right people, and this selection is not as difficult as it may at first appear. Studies have shown that there are three types of scientists who are identifiable by their very different patterns of research.[11,39–41] There are those who have average or below average productivity (in terms of producing high-impact papers and patents) all of which remain in the field in which they were originally trained, and often focus on the topic of their dissertation research. These are competent, noncreative individuals. Among those who are very creative (again in terms of high-impact papers and patents), there are two groups. There are those who spend a few years exploring various possible areas of research and strike upon one that attracts their full and undivided attention for most of the rest of their careers. Such individuals will produce at most one major innovation during their lives because they know only one area of science. Examples are Jonas Salk and his polio vaccine or Kary Mullis and PCR. Such individuals are too focused and too narrow to make another important contribution in another area of science. The people who are most appropriate for a skunk works are those in the second group of highly innovative individuals. This group is characterized by performing research in multiple areas sometimes serially, but most often concomitantly, and by making at least one major breakthrough by the time they are in their early forties. Such individuals are very likely to continue to be innovative for the rest of their careers because they are constantly exploring new problems, acquiring new skills, and changing fields. People like Black and Sternbach are good models. Such people must be rewarded not by being promoted into managerial positions, but by being given increasing freedom and resources to do what they do best: innovate!

Table 4 Some guidelines for managing innovation

1. Provide freedom for investigators.
2. Keep innovation groups small.
3. Let the groups themselves define clear, fundamental, and compelling problems.
4. Let the groups direct their own research.
5. Don't hire people whose judgement you don't trust.
6. Choose independent, unusually trained, self-motivated individuals.
7. Generate multiple targets and approaches.
8. Use tight time and fiscal restraints to force groups to choose best targets.
9. Simultaneously provide long-term security for group.
10. Don't promote innovators! Reward them with more opportunities to innovate.

The key pitfall to avoid from a corporate level is the allure of efficiency. Efficiency is an oxymoron in the context of research. Consider a concrete example. More and more pharmaceutical houses are turning to massively robotized, high-throughput systems for creating and testing compounds. There is no evidence that this approach is working any better than having small numbers of chemists make individual compounds. The reason for this is quite simple. Machines can be designed to perform only those functions that have previously been codified in practice. Since innovations are, by their very nature, unanticipated and unpredictable surprises, and since the discovery process itself has not yet been codified, the use of machine codifiable procedures is a guarantee that nothing fundamentally new can be discovered. Robots are only good at optimizing approaches we already use, not for inventing new approaches, or new problems. If discovery technologies are a focus of a company's innovation strategy, then the focus must be on creating novel technologies, such as combinatorial chemistry, PCR, or recombinant DNA.[50]

2.10.8 Conclusions

In sum, top-down, highly organized, efficient research organizations are the enemies of innovation. Innovative organizations focus on small, flexible, independent groups that generate and solve their own problems, bootstrapping their success stage by stage, beginning with simple, cheap, quick qualitative proofs-of-concept and building progressively toward demonstrations of effectiveness. The people who succeed in such groups cannot work in highly structured environments or for people less creative than themselves. They are intrinsically difficult to manage and impossible to direct. Exceptional managers can, however, create conditions that will bring out the best in such individualists by creating appropriate challenges and setting firm limits. In addition, innovators can employ many strategies to increase their chances of successfully navigating the innovation process. Among the most successful is identifying inexpensive, quickly addressable, simple problems that have the potential to overturn existing dogmas and practices. Turn problems on their head; redefine them; find rate-limiting steps; try lots of things in search of serendipitous, and surprising, results. Assume competitors will solve the problem first and ask what problems their solution will create that no one has even begun to think about. Create the conditions that foster innovation (freedom, independence, persistence). Plan for the unexpected. Be flexible. Innovation is not something that just happens. Innovation is a way of living and working.

References

1. Warne, P.; Page, C. *Drug News Perspect.* **2003**, *16*, 177–182.
2. Wolpert, L.; Richards, A.; *Passionate Minds: The Inner World of Scientists*; Oxford University Press: Oxford, UK, 1997, pp 124–129.
3. Nayak, P. R.; Ketteringham, J. M. *Breakthroughs*; Rawson Associates: New York, 1986, pp 102–129.
4. Black, J. W. Autobiography. http://nobelprize.org/medicine/laureates/1988/black-autobio.html (accessed April 2006)
5. Harris, G. *New York Times*, 2003, Oct 5, *sec 3*, pp 1 and 8.
6. Class, S. *Chem. Eng. News* **2004**, *82*, 18–29.
7. Cunningham, B. C. *Drug Dev. Res.* **2002**, *57*, 97–102.
8. Service, R. F. *Science* **2004**, *303*, 1796–1797.
9. McCoy, M.; Mullin, R. *Chem. Eng. News* **2005**, *83*, 33–44.
10. Schultz, T. W. *Minerva* **1980**, *18*, 644–651.
11. Root-Bernstein, R. S. *Discovering*; Harvard University Press: Cambridge, MA, 1989.
12. Root-Bernstein, R. S.; Dillon, P. F. *Drug Dev. Res.* **2002**, *57*, 58–74.
13. Reid, T. R. *The Chip: How Two Americans Invented the Microchip and Launched a Revolution*. Random House: New York, 2001, pp 65–66.
14. Mullis, K. *Dancing Naked in the Mind Field*; Vintage: New York, 2000.
15. Rabinow, P. *Making PCR*. University of Chicago Press: Chicago, IL, 1996, pp 95–98.
16. Root-Bernstein, R. S. In *Scientific Research Effectiveness: The Organisational Dimension*; Hurley, J., Ed.; Kluwer: Dordrecht, The Netherlands, 2003, pp 165–196.
17. Baenninger, A.; Costa de Silva, J. A.; Hindmarch, I.; Mueller, H.-J.; Rickels, K. *Good Chemistry: The Life and Legacy of Valium Inventor Leo Sternbach*; McGraw Hill: New York, 2004, pp 49–50.
18. Furchgott, R. F. Autobiography. http://nobelprize.org/medicine/laureates/ (accessed April 2006).
19. Murad, F. Autobiography. http://nobelprize.org/medicine/laureates/ (accessed April 2006).
20. Ignarro, L. J. Autobiography. http://nobelprize.org/medicine/laureates/ (accessed April 2006).
21. Viagra, the patenting of an aphrodisiac. http://inventors.about.com/library/weekly/aa013099.htm (accessed April 2006).
22. Root-Bernstein, R. S. *Res. Tech. Mgmt.* **1989**, *32*, 36–41.
23. Root-Bernstein, R. S. *Chemtech* **1994**, *24*, 15–23.
24. Thompson, R. B. *FASEB J.* **2001**, *15*, 1671–1676.
25. Comroe, J. H., Jr.; Dripps, R. D. In *Biomedical Innovation*; Roberts, E. B., Levy, R. I., Finkelstein, S. N., Moskowitz, J., Sondik, E. J., Eds.; MIT Press: Cambridge, MA, 1981, pp 101–122.
26. Colpaert, F. C. *Nature Rev. Drug Disc.* **2003**, *2*, 315–320.
27. Szent-Gyorgyi, A. In *Current Aspects of Chemical Energetics*; Kaplan, N. O., Kennedy, E. P., Eds.; Academic Press: New York, 1966, pp 63–76.
28. Thompson, G. P. *The Inspiration of Science*. Oxford University Press: Oxford, UK 1961, pp 128–132.

29. Ganellin, C. R. *Drug Disc. Today* **2004**, *9*, 158–160.
30. Shorter, E. *Nat. Rev. Drug Disc.* **2002**, *1*, 1003–1006.
31. Ng, S.; Goodson, B.; Ehrhardt, A.; Moos, W. H.; Siani, M.; Winter, J. *Bioorg. Med. Chem.* **1999**, *7*, 1781–1785.
32. Goodson, B.; Ehrhardt, A.; Ng, S.; Nuss, J.; Johnson, K.; Gidelin, M.; Yamamoto, R.; Moos, W. H.; Krebber, A.; Ladner, M.; Giacona, M. B.; Vitt, C.; Winter, J. *Antimicrob. Agents Chemother.* **1999**, *43*, 1429–1434.
33. Humet, M.; Carbonell, T.; Masip, I.; Sanchez-Baeza, F.; Mora, P.; Canton, E.; Gobernado, M.; Abad, C.; Perez-Paya, E.; Messeguer, A. *J. Comb. Chem.* **2003**, *5*, 597–605.
34. Raelin, J. A. *The Clash of Cultures. Managers and Professionals*; Harvard Business School Press: Boston, MA, 1986.
35. Spiro, H. *Drug Ther.* 1990, *March*, 27–30.
36. Zimmerman, H. *Scientific Elite: Nobel Laureates in the United States*; Free Press: New York, 1977.
37. Eiduson, B. T. *Scientists: Their Psychological World*; Basic Books: New York, 1962.
38. Mansfield, R. S.; Busse, T. V. *The Psychology of Creativity and Discovery: Scientists and Their Work*; Nelson-Hall: Chicago, IL, 1981.
39. Root-Bernstein, R. S. *Res. Tech. Mgmt* **1989**, *32*, 43–50.
40. Root-Bernstein, R. S.; Bernstein, M.; Garnier, H. *Creativity Res. J.* **1993**, *6*, 243–329.
41. Root-Bernstein, R. S.; Bernstein, M.; Garnier, H. *Creativity Res. J.* **1995**, *8*, 115–137.
42. Root-Bernstein, R. S.; Root-Bernstein, M. M. In *Creativity: From Potential to Realization*; Sternberg, R. J., Grigorenko, E. L., Singer, J. L., Eds.; American Psychological Association: Washington, DC, 2004, pp 127–151.
43. Halacy, D. S., Jr. *Science and Serendipity: Great Discoveries by Accident*; Scribners: Philadelphia, PA, 1967, pp 12–14.
44. Blumberg, B. S. In *Scientific Research Effectivenes: The Organisational Dimension*; Hurley, J., Ed.; Kluwer: Dordrecht, The Netherlands, 2003, pp 37–48.
45. Hollingsworth, R. In *Essays on the History of Rockefeller University*; Stapleton, D., Ed.; Rockefeller University Press: New York, 2002, pp 17–63.
46. Johnson, I. S. *Nature Rev. Drug Disc.* **2003**, *2*, 747–750.
47. Rich, B.; Janos, L. *Skunk Works. Little*; Brown: New York, 1996.
48. The Graphing Calculator Story. http://www.pacifict.com/.
49. Building a Better Skunk Works. http://www.fastcompany.com/magazine/92/ibm.html (accessed April 06)
50. Gershell, L. J.; Atkins, J. H. *Nat. Rev. Drug Disc.* **2003**, *2*, 321–327.

Biography

Robert Root-Bernstein earned his AB in Biochemistry (1975) and PhD in the History of Science (1980) at Princeton University. As a Post-Doctoral Fellow in Theories in Biology at the Salk Institute for Biological Studies he was awarded one of the first MacArthur Fellowships (1981). This award enabled him to explore two areas that have remained central to his research: (1) the nature of scientific creativity; and (2) the roles of molecular complementarity in the evolution of physiological systems and as a guide to drug design. He has consulted for a number of pharmaceutical and biotechnology companies, including Parke-Davis, Chiron, and MitoKor. He is the author of several patents, over 100 peer-reviewed papers, 20 book chapters, and four books: *Discovering: Inventing and Solving Problems at the Frontiers of Science* (Harvard University Press, 1989), *Rethinking AIDS* (Free Press, 1993), and (with Michele Root-Bernstein) *Honey, Mud, Maggots and Other Medical Marvels* (Houghton Mifflin, 1997) and *Sparks of Genius* (Houghton Mifflin, 1999). He is currently a Professor of Physiology and member of the Lyman Briggs School at Michigan State University.

2.11 Enabling Technologies in Drug Discovery: The Technical and Cultural Integration of the New with the Old

M Williams, Feinberg School of Medicine, Northwestern University, Chicago, IL, USA

© 2007 Elsevier Ltd. All Rights Reserved.

2.11.1	**Introduction**	**265**
2.11.2	**The Drug Discovery Process**	**266**
2.11.2.1	Phases of Drug Discovery	267
2.11.2.1.1	The empirical/physiological phase (1885–1948)	267
2.11.2.1.2	The biochemical phase (1948–1987)	267
2.11.2.1.3	The biotechnology phase (1987–2001)	267
2.11.2.1.4	The genomic phase (2001–present)	269
2.11.3	**Facets of Drug Discovery**	**269**
2.11.3.1	Societal Expectations for the Pharmaceutical Industry	269
2.11.3.1.1	Drug costs	270
2.11.3.1.2	Consumer advocacy groups	270
2.11.3.1.3	Governmental initiatives in drug discovery	271
2.11.3.2	Drug Discovery Strategies, Culture, and Organizational Effectiveness	271
2.11.3.3	Scientific Expertise, Competence, and Career Expectations	272
2.11.3.3.1	Data focus	272
2.11.3.3.2	The PowerPoint generation and its consequences	273
2.11.3.4	Disease Complexity and Novel Targets	274
2.11.3.4.1	Novel targets	274
2.11.3.4.2	Old target opportunities and metrics	274
2.11.3.4.3	Systems biology and pharmacology	276
2.11.3.4.4	Re-engineering/re-positioning known drugs	276
2.11.3.4.5	Animal models	277
2.11.3.4.6	Intellectual property	277
2.11.3.5	Filling the Pipeline	278
2.11.4	**Technological Impact and Integration**	**279**
2.11.5	**Future Challenges**	**282**
2.11.5.1	Cultural	282
2.11.5.2	Operational	283
2.11.5.3	Scientific	284
References		**285**

2.11.1 Introduction

Driven by the widespread availability of the personal computer and the subsequent internet revolution, the past two decades have seen spectacular advances in the sophistication of the technology platforms used to synthesize/identify and characterize new chemical entities (NCEs) as potential drugs.[1] As a result, more data has been generated regarding the physicochemical and biological properties of the many NCEs synthesized in the past 20 years than was ever known for the relatively smaller number of NCEs that had existed as discrete samples in the prior 80 years or so. The ability to interrogate this data[2] (assuming the existence of a database architecture at a facile, user-friendly level) to retrospectively understand NCE action(s) can provide critical information to prospectively design new NCEs against defined biological targets and thus represents a powerful tool in the drug discovery process. Such knowledge has not, however, apparently aided in the end game, namely that of the practical application of the advances in biomedical knowledge to the treatment of human diseases in the form of new drugs, the introduction of which has decreased over the past decade.[3]

Following publication of draft maps of the human genome in 2001,[4,5] there were various predictions[6–8] that new drugs, acting at hitherto unknown, disease-associated targets, having improved efficacy and fewer side effects, based on their origins in, and use of, genomic information (especially for clinical trial patient selection), were but a few years away from human use. Five years later, it appears unlikely that genome-based medicine will have a major impact on healthcare in the foreseeable future,[9,10] a result of the dramatic oversimplification of the challenges involved in reducing one of the most complex data sets known to practice.[11] Similarly, proteomics, in itself an infinitely more complex science than that of genomics due to the multiple possibilities resident in the posttranslational protein modification process,[12] requires the dissection of primary tissue and cellular proteomes into numerous subproteomes.[13–15] Nonetheless proteomics rapidly replaced genomics when it became apparent that the latter was a far greater challenge than had been anticipated and that it would be easier (for a time) to publish in a new area than productively work out issues with an existing one. The receptorome,[16] the interactome[17] and the epigenome[18] are the more recent 'omes' predicted to change the practice of human healthcare. The interactome, derived from initial drafts of the human haplotype map (HapMap), is a collection of binary protein–protein interactions, even more complex than either the genome or the proteome. The epigenome involves environmental influences on gene expression and reflects the interface between nature and nurture.

Unfortunately with each new 'ome,' the initial anticipation for a wealth of new information has inevitably given way to skepticism, an unfortunate example of the Hype Cycle[19] concept especially as initial genomic and proteomic data sets have repeatedly failed replication both within, and between, data sets and disease states.

In both genomics and proteomics, the critical challenge of target validation[15] has been repeatedly trivialized to the level of an enthusiastic naivety,[20] lacking in apparent logic, appropriate insight, and necessary rigor. As an example, the $5HT_6$ receptor has been known for well over a decade, being implicated as a potential target in CNS disorders that include schizophrenia and Alzheimer's disease (AD). Several programs have been initiated in the industry to find $5HT_6$-selective ligands and at least one antagonist, SG518, is being evaluated in the clinic for use in cognitive impairment. The $5HT_6$ receptor is, however, a long way from being validated in terms of representing the site of action of a bona fide drug with discrete therapeutic opportunity. Activities directed toward target validation are thus more correctly characterized as *target confidence* building and are far more complex than had been — or even are — envisaged.[15,21]

Several books[22,23] and a variety of articles[3,24–32] have highlighted the lack of productivity in drug discovery — the "unrelenting attrition statistics"[3] — since 1996. While there are inevitably divergent viewpoints, all parties agree that productivity, as measured either in terms of Investigative New Drug Applications (INDs) or New Drug Applications (NDAs), is inconsistent with the investments being made in drug R&D and in the future viability of the pharmaceutical industry. Opinions differ as to the root causes of the productivity shortfall with additional opinions of what needs to be done to 'save' drug discovery, including the ever evolving role of the biotech industry in being the "main driver for new drugs" and "center[s] for drug innovation."[26] Few of these 'rescue' recipes appear to have a current basis in reality.

The reasons cited for the productivity decrease include: (1) the intrusion of business mores into the science of drug discovery[26,28]; (2) the excessive funding of marketing to the detriment of research; (3) more stringent regulatory hurdles for drug approval[29]; (4) the intrinsic challenge of the disease states being currently addressed (stroke, AD, septic shock, cancers, etc.)[24,25] that are focused on novel, often 'unvalidated' disease targets/hypotheses, these being consequently more difficult to test in the clinic with unproven/ill-defined clinical trial endpoints that confound efficacy outcomes; that all the low-hanging fruit in drug discovery have been picked[25]; (5) that novel technologies/scientific concepts take considerable time to become productive[25]; and (6) that the patenting of gene-related materials has had a negative impact on research progress.

New technologies and additional resources are thus failing to address the current shortfall in the drug discovery enterprise and Milne has predicted that in continuing on the 'current' path of drug discovery, the risk of continued failure is 'arguably absolute'[3] and has advocated "investment in quality thinking and experimentation...to improve the outcome" with "extra months spent at the front end [saving] time" reflecting a higher priority in strategizing the use of new technologies and new paradigms and how these will be used tactically in the context of a hypothesis-driven drug discovery effort.

2.11.2 The Drug Discovery Process

In his seminal monograph on drug discovery, Sneader[33] unsurprisingly equated effective drugs with the conquest of human disease, an activity that dates back to the traditions of Greek, Egyptian, Arabic, and ayurvedic medicine where: (1) products derived from natural sources evolved on a trial-and-error basis (those individuals using toxic drugs rarely having the opportunity to complain) as safe and effective drugs for human use; and (2) the 'astrologer-priest-physician' prescribed the use of a medicine to be filled by a local apothecary with limited, if any, production standards. Issues with the latter led to the emergence of the pharmaceutical industry with its initial role being in a quality control capacity. The natural product pharmacopoeia still includes drugs like morphine and aspirin that remain widely used, albeit it in different formulations and delivery systems, highlighting one path to success in the drug discovery process that has largely fallen out of favor.[34,35]

The two key disciplines that have driven successful drug discovery to date are medicinal chemistry[3] and pharmacology.[30] Working closely together in an iterative manner with ancillary disciplines like physical and computational chemistry, high-throughput screening (HTS), and molecular biology to name but four of a potential 30 or more, medicinal chemistry and pharmacology have used the conceptual 'lock and key' hypothesis, now over a century old, as the framework in which to identify new drugs (**Figure 1a**). It has been noted that the chemistry contribution to successful outcomes in drug discovery is greater than 85%,[3] if not absolute[21] almost exclusively in the identification of drug-like molecules. However, experience and insight into the biological evaluation of these NCEs can be considered of equal value.

2.11.2.1 Phases of Drug Discovery

The human use of natural product-based drugs for therapeutic purposes dates back to two centuries BC. While the first pharmaceutical company, Takeda, was founded in Japan in the early seventeenth century, the foundations of modern day drug discovery in Germany were based on the systematic physiological studies of Betrand, Kekulé's dream of the structure of benzene, and market-driven changes in the German dye industry in the mid-to-late nineteenth century.

Drug discovery can be loosely divided into four phases: the empirical/physiological phase, the biochemical phase, the biotechnology phase, and the genomic phase.

2.11.2.1.1 The empirical/physiological phase (1885–1948)
This involved the identification of NCEs from natural product sources with a long history of human use, or derived from the products of the aniline chemical industry, which were evaluated for their antimicrobial activity,[35] and in bioassays or in animals to assess phenotypic changes and other outcomes thought to be reflective of the targeted disease state.[36] During this period, the 'lock and key' theory of drug action, the 'magic bullet,' was conceptualized as the basis of pharmacology.[37] Following from this "the first task of the young science of pharmacology was to purge the pharmacopoeia by casting out useless drugs," the latter being loosely defined (those without a mechanism of action[38]).

2.11.2.1.2 The biochemical phase (1948–1987)
This originated in the seminal work of Alquist (1948) on adrenoceptor classification with the previous empirical, bioassay-based research approach being complimented by more precise in vitro biochemical approaches at the cellular and tissue levels to study both the mechanism of action of new chemical entities (NCEs) thought to have therapeutic potential and the molecular causes of disease. This approach, now reincarnated as 'chemical genomics' or 'chemigenomics'[39] used NCEs to characterize/perturb drug targets (receptors, enzymes, etc.) and their subtypes and were used in turn to characterize NCEs in vitro for their target efficacy and selectivity to prioritize their advancement into more complex in vivo disease models. An important facet of the biochemical era was that it was used in conjunction with, and as an extension of, the empirical approach rather than replacing the former in a wholesale and exclusionary manner.

2.11.2.1.3 The biotechnology phase (1987–2001)
This emerged from the biotechnology revolution of the 1980s and encompassed four major themes: (1) the use of molecular biology to create a novel scientific approach termed *synthetic biology*[40] that focused on examining drug targets/disease processes using target cloning, recombinant protein expression, and target mutations to create cell lines and animals[41]; (2) the application of high-throughput, industrial-scale technology-driven approaches to NCE characterization (HTS)[42]; (3) *computational-based/bioinformatics* approaches critical to the storage, access, and real-time interrogation of NCE-based data sets that numbered in the many millions[2]; and (4) a redefinition of the culture of biomedical research, which is discussed in detail below. In addition to its somewhat reductionistic approach[32,43] the biotechnology era also became both exclusionary and revisionary in that the hype attendant on its introduction and revolutionary premise necessitated the dismissal of previous approaches to drug discovery as archaic and irrational. A direct result of this was a de-emphasis of the role of pharmacology and a movement away from classical animal disease models,[30,36] the latter being replaced with elegant transgenic mouse models,[41] flies and zebrafish, and sophisticated in vitro systems.[43]

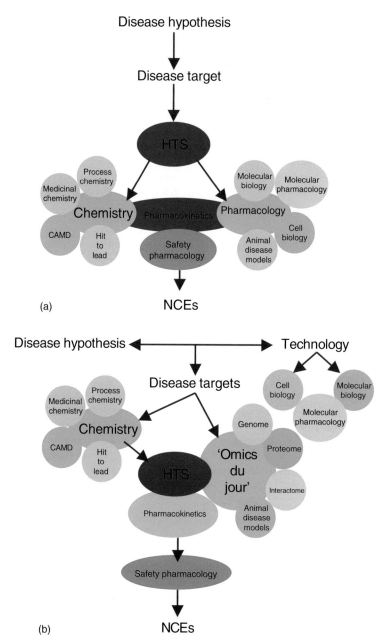

Figure 1 Cartoon representation of key elements of chemo-/pharmaco-centric approach as compared to the technology-driven approach. (a) Key elements of the hypothesis-driven approach. A hypothesis is formulated based on current knowledge from both preclinical and clinical studies and targets identified. An integrated chemo-/pharmaco-centric plan is put in place to find compounds and biological assay systems to gain proof of concept. Once this has been determined, an integrated, iterative, hierarchical flow chart is put in place to identify, test, and optimize NCEs for efficacy, side effect liability, and drug-like properties. (b) Key elements of the technology-driven approach to drug discovery. Disease hypotheses and targets can be linked with technology platforms around a rubric like that of systems biology to iteratively evaluate the three in order to identify NCEs that can then be optimized by a traditional chemistry driven campaign. The approach is heavily influenced by evolving trends on 'omics' technologies, represented here by the 'Omics du jour' kaleidoscope, an ever-changing technology resource. Astute readers will note that the more complex the 'ome,' the further away it is from NCE generation. The path to clinical NCEs is less integrated than that for the hypothesis-driven approach, and for those involved in technology evaluation it will come as no surprise that IND-enabling activities required by the FDA are sometimes planned to occur late in Phase II.

2.11.2.1.4 The genomic phase (2001–present)
This phase has been conceptually driven by the mapping of the human genome in 2001[4,5] and an expectation, thus far generally unrealized, that disease-associated genomic targets, their proteomic progeny, and related interactomic 'pathways' would provide the means to more rapidly and accurately identify new generations of drugs that would be highly specific in their disease-related, beneficial effects and hence more efficacious and freer of side effects than existing drugs.

2.11.3 Facets of Drug Discovery

In two thought-provoking articles on biomedical research,[24,25] the late David Horrobin questioned the rationale of the present approach to drug discovery, not only in terms of core technologies, but also the basic culture as well as some obvious disconnects that had entered the realm of cognitive dissidence. Initially, critiquing the failure of technologies like HTS and combinatorial chemistry to deliver compounds, and the faulty premise that pharmacogenomics would reduce the risk and cost in clinical trials,[24] Horrobin also questioned whether modern-day drug discovery had any basis in reality[25] noting that both the pharmaceutical industry and biomedical research in general were 'failing' in their attempts to find solutions to human disease states. In assessing the possibilities for this failure, he considered: (1) a lack of investment by the pharmaceutical industry; (2) that all the easy problems (e.g., the proverbial 'low-hanging fruit') had been solved with only the difficult ones remaining; and (3) that the new technologies needed additional time and investment to bear fruit. Horrobin concluded that none of these were causal, rather biomedical and pharmaceutical science had "taken a wrong turn in their relationship to human disease," equating modern-day biomedical research with the world of Castila in Hesse's *The Glass Bead Game*. In this novel, popular in the 1970s, the brightest and best scholars once highly educated were recruited to a "magical state within a state, the isolated world of Castila" where they played the "Glass Bead Game," an intellectual game that Horrobin described as "beautifully refined and internally self-consistent," but designed to "make...no real contribution to real world issues." To Horrobin, current biomedical research, using "heavily funded and heavily hyped techniques," e.g., cell culture and 'unvalidated' animal models, represented the biomedical equivalent of the "Glass Bead Game" and resulted from a reductionistic approach to biomedical research. Elements of Horrobin's views have been echoed to varying degrees in a slew of recent papers,[1,3,21,26,28,30,44] especially those focusing on the newly perceived value of systems biology in 'salvaging' drug discovery.[32,43]

While it is a relatively facile process to address the challenges of productivity in drug discovery in terms of the metrics, e.g., numbers of compounds, screens and scientists, and address these by adding more of each, it is far more difficult to focus on issues related to organizational and individual culture, competence, and drug discovery experience. Such concerns have been raised by Black[45] who in focusing on the evolving culture of drug discovery, noted a lack of focused commitment in the present research environment, of a trend of giving up on difficult problems "when the breaks are not coming," and transferring energies from unsolved problems to new ones – the peripatetic, and unproductive, path of least resistance. Interestingly, a recent survey of drug discovery scientists[9] who while acknowledging that drug discovery was an experimental science, noted that "in many cases, people did not want to perform the experiments." This was an interesting viewpoint, given that Wong *et al.*,[46] in their case history of the discovery and development of the SSRI fluoxetine, had noted that "it is essential that ideas are steadfastly championed by passionate believers to achieve the final goal."

Not unexpectedly, nearly all aspects of the drug R&D process have changed markedly over the past decade and a half, driven to a very major extent by the biotech revolution, and the consequent mingling of science with business. Four key themes deserve special comment: (1) societal expectations for the pharmaceutical industry; (2) drug discovery strategies, culture, and organizational effectiveness; (3) scientific expertise, experience, and expectations; and (4) disease complexity and novel targets.

2.11.3.1 Societal Expectations for the Pharmaceutical Industry

The pharmaceutical industry and its biotechnology sister are 'for profit,' technology-based, innovation-driven enterprises whose products are life saving/enhancing drugs. Society looks to the pharmaceutical industry for these drugs to treat life-threatening conditions, e.g., infections, AIDS, cancer, depression, heart disease, etc., and to improve the quality of life, e.g., better analgesics, birth control pills, drugs to help prevent obesity and erectile and bladder dysfunction, etc. These expectations should also be consistent with those of the industry's shareholders to provide an adequate return on investment in order to provide a sustainable path forward in terms of corporate productivity and viability given the 10–15 years it takes for a drug to reach the marketplace.

The pharmaceutical industry is thus celebrated in terms of its contribution to the gross national product (especially in the US) and for its historical ability to effectively apply the findings of basic biomedical research to discover useful drugs. At the same time the industry is vilified for the cost of its products and for its marketing strategies,[22,23,28] which have been described as "undermining the cause of public health,"[47] a viewpoint not helped by the slew of lawsuits against Merck over the cardiovascular liabilities of the cyclooxygenase-2 (COX-2) inhibitor Vioxx[48] that have made the industry second only to the tobacco industry in the level of public distrust and legal opportunism. The repercussions of these lawsuits are profound and will have a long-lasting impact on how the industry conducts clinical trials, how the latter are reviewed by regulatory authorities, how these outcomes are published in key, peer reviewed medical journals (e.g., *JAMA*, *Lancet*), and how the industry then markets approved drugs.[28,48]

2.11.3.1.1 Drug costs

The well-known cost of discovering a marketable drug ($800 million), the NCE attrition rate in preclinical and clinical phases (80–90%), global regulatory hurdles, the time to approval and government as a third party payer[49] make the pharmaceutical industry unique as compared to other technology-based industries, e.g., telecommunications, computer software, and the oil industry (to which the pharmaceutical industry is frequently compared in terms of the analogy of drilling dry wells).

The global pharmaceutical industry has traditionally invested heavily in research and development for new generations of drugs and has depended on both innovation and serendipity for success. There is however considerable skepticism as to the true cost of drugs, a consumer commodity that unlike clothing, books, music, alcohol etc., the patient has little objective choice in using and often does not personally select, the latter being the purview of the prescribing physician. A public survey by Kaiser in the US equated rising health care costs with the high profits made by drug companies.[112] Drug costs in the US, and the perception that these are driven to a major degree by marketing budgets,[22,23] have resulted in both states and individuals importing discount prescription drugs from Canada and Europe (some of dubious origin and quality), where prices are 40% lower.[50] This situation has been described as an economic 'free-rider'[51] where the major cost of drug development is underwritten by the US consumer[112] and where, if there were no US market, "no one would have to pay for drug development… [a situation involving]…economic illiterates…. who believe in magic".[51] This view has been disputed by a study where pharmaceutical R&D spending was compared to the Gross National Product (GNP) in countries where pharmaceutical research was conducted.[50] Switzerland spent 0.55% of its GNP on pharmaceutical R&D, the UK, 0.32%, and the US, 0.24%. The authors then argued that in Canada, where 0.08% of the GNP was spent on pharmaceutical R&D, income from domestic sales was 10 times greater than R&D costs. Given that pharmaceutical R&D effort in Canada is, at best, modest and is technically conducted by subsidized subsidiaries of foreign (US and UK) owned companies, this argument appears naïve if not untenable. If drug prices in the US were the same as those in Europe, Canada, and Japan where federal agencies rather than the marketplace set drug prices, it is highly questionable whether European, Canadian, and Japanese pharmaceutical companies could afford R&D without the US market to recoup costs. The analogy is similar to that of Boeing and Airbus who would certainly not build planes without customers outside their immediate domestic markets. Indeed the majority of customers for the Airbus 380, the world's largest commercial aircraft are from the Middle and Far East. There is also a major trend by European pharmaceutical companies to move their research HQs to the US to more effectively compete such that the industry, while worldwide, is becoming increasingly 'US-centric'.

It then becomes debatable whether the global pharmaceutical industry would exist in its present form and whether the public could have reason to expect new medications for diseases like cancer and AD where there is considerable unmet medical need. Furthermore, the proposition that reducing drug prices will either decrease healthcare costs[49] or aid in increasing the flow of innovative new drugs is in direct contrast to data showing that a dollar spent on newer drugs reduces nondrug expenditures, especially those involving hospital costs, sevenfold, e.g., $7.[52] Even though drugs represent less than 15% of the annual individual cost of healthcare, this 15% is typically a direct payment from the consumer and is thus the most apparent. Effectively taxing the pharmaceutical industry as a means to address the monumental economic challenges of providing healthcare to an ever-increasing number of patients that, due to the drugs produced by the industry, is living longer, is a solution that defies logic especially in the face of societal norms where paying $25 for a book or compact disk, the cost of which to produce is in the 25¢ range, is deemed a fitting recognition of intellectual property rights, albeit for products that are used at the exclusive choice of the consumer.

2.11.3.1.2 Consumer advocacy groups

Consumer advocacy groups have also, to some extent, changed the way in which drug companies allocate their resources. The most visible of these was the AIDS group ACT-UP who were instrumental not only in advocating with drug companies to accelerate the development of first-generation AIDS medications, but also in helping educate patients regarding their effective use. The wisdom of special interest groups in becoming involved in the science underpinning healthcare issues is, however, of questionable benefit.[53]

There have been major political inroads into the funding of stem cell research, notably by the States of California and New Jersey, the UK and South Korea, as a means to find improved treatments for neurodegenerative and degenerative disorders, edging toward the search for longevity that was the topic of Huxley's *After Many a Summer Dies the Swan*. These initiatives, however, have ignored previously documented 'side effects' of fetal cell transplants where aberrant sprouting of transplanted dopaminergic neurons led to Parkinsonian patients developing uncontrollable and irreversible spontaneous movements that proved to be worse than the actual disease.[54] Considerable caution is therefore required in conveying perceptions of the immediacy of the application of cell replacement therapies given the woefully limited knowledge of the control of cell differentiation, regeneration, and functional recovery.[55] The expectation that stem cell therapies, like the genome, represent the wave of the future for medicine and that religious, rather than scientific, concerns have impeded the widespread acceptance of stem cell technology has led to a less than rigorous scientific assessment of progress, that has been equated with the Piltdown Man Hoax of 1912. This confusion of scientific logic with religious opinion and politics has led to the recent fraud in human embryonic stem cell cloning in South Korea perpetrated by Hwang Woo Suk.[56] The circumstances surrounding this case reflect how science can be subverted by politics. The chairperson of the US President's Council on Bioethics, Leon Kass, had noted[53] that "… American scientists and the American media have been complicit in the fraud, because of their zeal in politics of stem-cell and cloning research and their hostility to the Bush funding policy. Concerted efforts have been…to hype therapeutic cloning, including irresponsible promises of cures around the corner… The need to support these wild claims and the desire to embarrass cloning opponents led to the accelerated publication of Dr Hwang's findings… We even made him exhibit A for the false claim that our moral scruples are causing American science to fall behind".

In trying "to pursue profits too earnestly.[in]…pharmaceutical research…[that]…. defeat[s] its own objectives"[57] the industry is perceived as losing focus on its strengths and the inherent value of innovative R&D, a view that falls in line with the various outside perceptions.[22,23,48]

2.11.3.1.3 Governmental initiatives in drug discovery

In light of the shortfall of new drugs from the established industry, the US government has also begun to focus on initiatives in drug discovery. These are the US National Institutes of Health's (NIH) 'NIH Roadmap Initiatives'[58] that identified three major strategic themes in current biomedical research that required prioritization (which, by the way, will in many cases simply duplicate resources that first existed in big pharma, were later duplicated and advanced by biotech, and now will go through another generation of duplication and advancement in academia – is this an effective use of tax dollars?): (1) New Pathways to Discovery; (2) Research Teams of the Future; and (3) Re-engineering the Clinical Research Enterprise, in essence, moving the NIH into front stage of the drug discovery process.[59] Additional facets of the NIH Roadmap were efforts in diversity-oriented synthesis (DOS) supported by NIH-funded centers for Chemical Methodologies and Library Development (CMLDs)[21] and efforts to seed translational medicine initiatives[28] via Clinical and Translational Science Awards (CTSAs) at academic medical centers.[112] This had been reinforced by the FDA's 'Critical Path Initiative' (CPI), a white paper[60] that identified the need to "strengthen… and rebuild… the disciplines of physiology, pharmacology and clinical pharmacology." An on-line update on the CPI in March 2006[112] identified new biomarkers, the streamlining of clinical trials, increasing bioinformatics and improving manufacturing processes as key to improve productivity in drug discovery. It has been further suggested[21] that the NIH also establish an institute of clinical practice to conduct clinical trials where existing medicines are compared (both to one another and NCEs), to generate best-practice guidelines for physicians to prescribe drugs, and to look for new uses for old drugs, the latter being a strategy that is already gaining currency.[16] The funding for this new institute was suggested as a puzzling 1% surtax on what the author already had concluded were drugs that were too expensive.[22] Given concerns related to the NIH's ability to productively focus on the women's health care issues over a 15-year time period[113] and the questionable returns on investment in the NCI's landmark search for new cancer drugs, it is questionable whether the NIH Roadmap Initiatives will have much impact on the drug discovery process. What is certain, however, is that funding for the basic research that has been the hallmark of NIH's success will be reduced and the number of acronyms will increase.

2.11.3.2 Drug Discovery Strategies, Culture, and Organizational Effectiveness

There are many approaches, real and perceived, to success in the drug discovery process: those that have succeeded based on historical perspective; those that have conceptual promise but are unproven and; those that result from frequent reorganizations to address failure.[26] Key elements vary from the concept of the 'drug hunter,' individuals like Paul Janssen, James Black, Pat Humphrey, George Hitchings, Trudy Elion, etc., who effectively led research teams to success based on their passion, insights, and dedication, to organizations that with the help of management consultants continually re-engineer the 'process' to be more effective, the latter a corporate metrics-based approach that leads to a paralysis in decision making.

Historically, drug discovery has been populated with individuals with an overwhelming passion for their research, scientists who would change jobs rather than give up on an idea for a drug they thought was worthwhile, totally consistent with "ideas [being] steadfastly championed by passionate believers to achieve the final goal."[46] In effectively executing a drug discovery paradigm it is important to recognize it as a distinct scientific discipline with goal-oriented challenges, day-to-day activities that can be extremely tedious and repetitive that require both matrix and hierarchical, time-based, goal-oriented and integrated management,[61–63] together with individuals with the experience to make a difference in execution and outcome. This is a marked difference from how research is conducted in government and academia.

2.11.3.3 Scientific Expertise, Competence, and Career Expectations

In conducting R&D, it is assumed that a drug discovery organization has reached some level of strategic and tactical vision in its R&D operations that is hopefully underpinned by an objective, data-driven decision-making process, an integrated in-licensing function, and adequate resources. There is also an implicit assumption, recently evolved, that the full-time employees (FTEs), whether chemists or biologists, excluding the context of overt discipline specialization (chemist versus biologist), are reliably interchangeable elements with basic training, competencies, experience, and team-oriented motivation in common. In identifying a need for "rigorous data standards and quality" to improve productivity, Milne[3] has obliquely raised the issue that perhaps this latter assumption is incorrect. In the biological realm, the recent trend toward an overtly reductionistic approach has resulted in a change in the way that experiments are planned, conducted (if at all),[9] interpreted, and used as a basis for additional research initiatives. The overt focus on technology platforms leads to a perceived loss of individuality, scientific commitment, and changes cultural and ethical standards.[26] The current ephemeral approach to science and the lack of adherence to the Law of Mass Action have been extensively discussed.[30] These reflect a failure to replicate experiments, the arbitrary choice of NCE and drug concentrations used to perturb biological systems, the absence of dose/concentration response curves, the lack of characterization of NCE effects using the corresponding agonist or antagonist, and the lack of a null hypothesis approach (the use of negative controls), all of which were second nature to biologists trained in the 1970s. These are causes for concern, as is the increasing trend to ignore or even accept as 'noise,' failed experiments that in themselves can often have a tale to tell. In today's research environment, Sir Alexander Fleming's penicillin spores would probably have been consigned to the trash.

2.11.3.3.1 Data focus

In considering causes for this situation, biomedical science in the last 15 years has focused on reductionism (a powerful tool in an appropriate context), megaprojects (genomics, etc.), intense efforts to commercialize early stage science (often before its time), and data generation. In the pharmaceutical industry, scientists have been increasingly rewarded for producing data, and less for scholarship in terms of publications, peer reviewing activities, introspection, and investment in quality thinking,[3] the focus being on quantity rather than quality. The incredible flow of biological data (with the *Journal of Biological Chemistry* publishing over 30 000 pages a year) and the focus on producing ever more data has decreased the time and ability to think. Firestein and Pisetsky, in developing guidelines for the publication of microarray experiments, noted that microarray-based experimentation has made "hypotheses… superfluous to the pursuit of research and that specific questions need no longer be asked" leading to "the illusion that all…information can be downloaded"[64] They recommended that microarray approaches be: replicated (!); subjected to statistical analysis(!); adequately controlled; and confirmed via nonarray techniques, all standards routinely used in classical pharmacology. This does not, however, devalue the premise of hypothesis-seeking as contrasted to hypothesis-testing experiments[65] provided the former are viewed as leading to the latter.

An additional facet of the current scientific training process is that scientists are trained less and less to think (or act) in an independent manner with a trend toward seeking external approval (condonment) of their activities, abnegating their personal responsibility for standards that should be second nature as the result of their training. If an '*n*' value of 1 on a bioassay or making 1 mg of an NCE for biological testing is approved/accepted by an outside entity – supervisor, organization, grant review committee, etc. (the devil being in the details and the outcome, the path of least resistance) – then common sense and intrinsic values go by the wayside. With such externalization of scientific approval, it is no surprise that the "brain turns off"[1] and researchers "behave like lemmings in the fog, running behind every new concept or method whether it is validated or not ... [relying] ... on artificial in vitro systems hoping that the information from bits and pieces holds true for the whole system"[1]. The passion for research diminishes and innovation and common sense become scarce commodities. Extrinsic (salary, stock options, travel budgets) then replaces intrinsic (scientific inquisitiveness, passion) motivation leading to confusion in goals and rewards. Certainly, many biotechs have valued revenue above bona fide scientific progress.

Data using a single, millimolar concentration of a known drug interacting at a novel target has been used as evidence that the existing targets where these drugs were known to act at nanomolar concentrations were incorrect and that the newly cloned receptors were the 'real' targets for their efficacious actions. The "triumphant pervasiveness" of molecular biology[66] that emerged in 1992 has led to a qualitative, rather than quantitative, science, the latter based on the use of the Law of Mass Action to generate IC_{50}, EC_{50}, and pA_2 values.

More recently, and again in the context of external scientific approval, scientists in industry are willing to change their therapeutic areas of choice and their scientific disciplines in exchange for job security resulting in "a greater [scientific] uniformity."[26] With complacency replacing the passion of discovery, it becomes just a job and it is then all too easy to give up on a topic when problems emerge.[45] Similarly, in academia, federal funding is dictated largely by what is perceived to be in scientific vogue. While the federal government via the US Food and Drug Admistration (FDA) laments the limited availability of pharmacologists[60] and the NIH is now re-investing in their training,[67] the lack of seasoned pharmacologists, especially in the in vivo area, is a direct consequence of funding decisions made by the NIH in the 1980s and 1990s.[30,68]

2.11.3.3.2 The PowerPoint generation and its consequences

There has also been a considerable focus on simplifying science for investors and management staff lacking a scientific background. Microsoft's PowerPoint has now become the norm for scientific discussions even in the absence of a presentation, heavy on conclusions but light on data and controversy. This has further encouraged superficial thinking, for if the data is too complicated to be reduced to the context of a PowerPoint slide, then surely it lacks the import to warrant discussion.

The ability to simplify science was at one time viewed as a gift, with the communicator, in full knowledge of the limitations of the data, being able to communicate with an audience in a direct and engaging manner. This contrasted with the gifted scientific 'nerd,' popular in books and movies, who was so engrossed with minutiae that the audience could not comprehend what he or she was talking about. The value placed on facile communication soon progressed from the eloquence and insight of a George Poste to the scientific superficiality and self-serving obfuscation of a Sam Waksal.[69] While useful in communicating with investors, the currency of facile communication has de-emphasized the focus on bona fide scientific issues and created a new level of scientific management — that of the *scientific poseur* or spokesperson — good on memorable, if nonsensical, sound bites, short on science, and thus eminently qualified to communicate the complexities of science to nonscientists and to lead science in a technology-oriented environment where context and complexity remain unappreciated. Those that had continued to worry over scientific issues soon became naysayers with divergent opinions that were considered to be overtly conservative (nihilistic) with such individuals being avoided, and their input discounted. And scientific standards inevitably suffered. One apocryphal story is that of the biotech CEO or CSO, relentlessly pressured by investors to advance an NCE to Phase I trials despite limited scientific merit. Despite protests that the NCE is not ready, the CEO/CSO is told that if the NCE is not advanced, job security will be an issue and that the CEO/CSO is being too cautious. Acquiescing to the investors, the CEO/CSO advances the NCE to the clinic where it fails, being written up in *BioWorld*; the CEO/CSO is then accused by the same investors who have seen the value of their investment plummet, of not being firm enough in upholding his/her scientific standards. The CEO/CSO in this 'damned if you do, damned if you don't' scenario is then fired.

Additional stereotypes have emerged or multiplied over the past decade and include the *technocrat*, certainly not a new phenomenon, but one that has become more pervasive and populated given the amount of funding available for technology acquisition. The technocrat has always been less interested in solving problems than in finding new ways to ask questions (often the same one), incrementally building on technology platforms that had yet to be proven either useful or productive. An example of this is the multiplication of formats for HTS with ever more NCEs being assayed in shorter time periods. The shift from the 48 assay tubes in racks used in the 1970s to the 96-well microtiter plates represented a major advance in resource utilization. Today 384-well microtiter plates are deemed as becoming inadequate with 1536 approaching the norm and with rare (and unreplicated) experiments focusing on 'well' -type formats in excess of 8000. As a result, the endpoint in screening, the ability to evaluate a total chemical library, can be done in far less time such that new bottlenecks have been created with an overwhelming amount of data to archive and sift through.

The continuing investments in technology and the need to provide evidence to investors and upper management that these were being used in an exciting and productive manner, the focus more on the former than the latter, led to multiple boards of directors and senior management teams being introduced to 3D molecular docking models on Silicon Graphics workstations that were simple, direct, and easy to grasp, albeit representing only a very small part of the drug discovery equation. While important in its own right and in an appropriate context, computer-assisted molecular design (CAMD) was perhaps not as uniquely self-contained a technology as nonscientists were repeatedly led to believe. It then became a difficult task to easily explain slow progress and failures when the audience had been convinced that the key to success was only an image on a Silicon Graphic terminal and a mouse click away.

From the *technocrat*, it became a short step to the *hype addict*,[70] usually an individual in management, driven by a need (or desperation) to acquire the latest technology, believing in all the things that the newly proposed technology may be able to do at face value, despite a long stream of previously acquired technologies that failed to live up to promises, in the hope and expectation that the next one might be the right one.

2.11.3.4 Disease Complexity and Novel Targets

2.11.3.4.1 Novel targets

The implicit opportunity in the human genome map was that of new, disease-associated targets. Of additional importance was the expectation that these newer targets would be discovered in an 'agnostic' manner, with no preconceived ideas to fit into an existing hypothesis. New targets would develop new hypotheses, correct older ones, and lead to new drugs. Instead, disease association has been tainted or subverted with existing hypotheses. The protein product of the novel schizophrenia-associated gene *g72* was used as a bait in yeast two-hybrid experiments. Among some 3000 prey, the enzyme D-amino acid oxidase (DAAO) was identified.[71] The activity of DAAO is normally associated with bacterial detoxification in the kidney, making it, at first blush, an unlikely pathway candidate for a psychiatric disorder like schizophrenia. However, in earlier work it had been shown that the unnatural D-amino acid, D-serine, formed in mammalian brain by the enzyme serine racemase, was a potent allosteric modulator of the N-methyl-D-aspartate (NMDA) receptor and that D-serine, when given with traditional antipsychotic medications, had positive effects on symptomatology in schizophrenics[72] (*see* 6.02 Schizophrenia).

While DAAO is still the subject of active research, its identification was not based on an 'agnostic' null hypothesis but rather on the heuristic relationship of the enzyme to the glutamate hypothesis of schizophrenia, interpreting 'new' science in the context of old. Whether a more novel, paradigm-shifting approach to schizophrenia may have existed in the other 2999 or so *g72* protein-associated pull downs or its partner gene, *g90*, remains to be seen.

2.11.3.4.2 Old target opportunities and metrics

In considering the apparently urgent need for new targets and their full exploitation, it is worthwhile considering existing targets of known structural homology and function. In addition to the diversity of 7TM receptors and their signaling pathways and ancillary proteins, there are more than 500 members of the human kinome (protein kinases), over 1300 distinct phosphatases, 18 members of the histone deacetylase (HDAC) family, and, at a minimum, 14 members of the phosphodiesterase family, all of which, based on: (1) the identification of important drugs/NCEs acting at these targets; and (2) their key roles in cell function, may represent viable drug targets.

Three other well-known receptor families have also been the focus of considerable research efforts in academia and the pharmaceutical industry. The first, the histamine 7TM receptor family, has been an active research area since the 1930s and has yielded many valuable drugs active at H_1 and H_2 receptor subtypes including the antagonists mepyramine, diphenhydramine, astemizole, cimetidine, and ranitidine.[33] Current research efforts are focused on the two newer, recently cloned members of the histamine receptor family, H_3 and H_4, for psychiatric and hematopoetic diseases, respectively.[73] Thus, this now 75-year-old research area, proven to be a fertile source for drugs, is still filled with untapped opportunity using a rich portfolio of existing NCEs in conjunction with molecular biology probes on which to base new chemistry efforts.

Acetylcholine (ACh) produces its pharmacological effects via the activation of two classes of cholinergic receptor, the ligand-gated nicotinic receptor family ($\alpha 7$, $\alpha 4 \beta 2$, etc.)[74] and the 7TM muscarinic receptor family (m1 – m5),[75] both of which have been seminal in the history of receptorology. Prompted by studies on acetylcholinesterase inhibitors that began in the mid-nineteenth century[33] and long before molecular biology was a facile and widely used technique, the nicotinic receptor from *Torpedo electroplax* was the first to be obtained in any quantity. Coupled with the availability of natural products like nicotine, dihydro-β-erythroidine, and methyllycaconitine, this provided the biological and chemical tools to study receptor perturbation in the 1950s, even with the relatively crude tools then available. However, it was not until the early 1990s that the culmination of research efforts at several academic laboratories (Changeux, Lindstrom, and Patrick) and at SIBIA, R.J. Reynolds/Targacept, and Abbott provided the context for bona fide drug discovery efforts in the nicotinic area, the ability to search for NCEs that had the positive attributes of nicotine on CNS function without many of its receptor subtype-selective side effects. From the late 1950s until the 1990s, the nicotinic ion channel remained essentially unexploited, especially in terms of medicinal chemistry, and it has taken another 15 years, from 1990 until 2005, for the first novel nicotinic drug, Pfizer's varenicline, to be approved for use in smoking cessation. And there is obviously still more to be done with unexploited uses of novel nicotinic receptor ligands in pain, cognition, attention deficit hyperactivity disorder, depression, anxiety, and schizophrenia.[74] The muscarinic receptor was the first 7TM/GPCR to be cloned in 1989 and was also the first to be freely available to all investigators from the NIH. While many ligands have been made to emulate classical ligands like atropine, pilocarpine, pirenzepine, etc., the majority, e.g., xanomeline, despite promising preclinical efficacy data, have been confounded by side effect liabilities. Darifenacin is a selective M_2 antagonist for the treatment of urinary retention.

Another class of receptors that are rich in complexity and function yet lack a diversity of drug-like NCEs and priority therapeutic targets are those for adenosine and ATP, the P1 and P2 receptor families, numbering 4 and approximately 15, respectively. Some 30 years of research in industry and academia, much of the latter focusing on NCEs as in vitro research tools rather than as leads for optimization, have resulted in only one NCE reaching late-stage clinical trials, the A_{2A} antagonist, istradefylline, which by acting as an indirect dopamine antagonist represents the next generation of treatment for Parkinson's disease.

From these examples, it may be argued that each drug target can take a minimum of 20 years of full-time, undiluted research effort to result in NCEs that have the potential to be drugs. The development of novel, receptor subtype NCEs and their subsequent use to perturb biological systems inevitably leads to new information that can then be used in an iterative manner to design newer NCEs to optimize their efficacy, therapeutic index, and drug-like characteristics. With this paradigm, one may then ask the question as to whether the obsessive search for new targets is truly a vision of the future or a reluctance to do the hard work required to properly exploit an existing drug target. A modern-day drug discovery scientist can spend his or her career always looking for something new without actually planning to achieve anything of substance.

An interesting exercise would be to take the industry metric of the FTEs required to support a discovery project at critical mass, currently considered in the range of 24–35, and multiply this by 15 as an average expectation of target life span for success. This results in a range of person years of between 360 and 525 and this number can then be used as: (1) a denominator for existing discovery scientists in the worldwide pharma/biotech industries; and (2) a multiplier for a conservatively useful 5000 new targets present in the human genome.

In the former instance, it can easily be imagined that the number of existing drug discovery scientists will be rapidly exhausted long before there were any need to delve into the genome for new targets. For the latter, the aggregate 1.8–2.6 million person years required to exploit the genome rapidly approaches the science fiction realm of the 3×10^{62} drug-like molecules that could possibly exist (W. Michne, personal communication). The translational medicine paradigm[28,112] is key to improving this situation although it is probable that effectively extending the NCE research paradigm into Phase IIa will be more successful than the concept of CTSAs.

2.11.3.4.2.1 Target centric versus multiple targets

While the current approaches to drug discovery have an almost exclusive focus on finding NCEs that are selectively active at discrete molecular targets, a 'target-centric' approach, it is well known that many existing drugs have multiple targets that contribute to their efficacy and also their side effect liabilities.[76,77] Imatinib (Gleevec), while originally characterized as a potent and selective *Bcr-Abl* kinase inhibitor, has since been found to be active at PDGF and KIT kinases.[78]

The most cited of these multitarget drugs is the atypical antipsychotic clozapine.[76] This compound has superior efficacy in schizophrenics, an absence of extrapyramidal symptoms, albeit with several limiting side effects, the most important being that of potentially fatal agranulocytosis (*see* 6.02 Schizophrenia). The latter may be due to the interaction of clozapine with the histamine H_4 receptor while efficacy has been ascribed to: (1) a multitude of receptors in the CNS including dopamine D_2 and D_4, $5HT_{2A}$, $5HT_{2C}$, $5HT_6$, etc.; and (2) the activity of its major metabolite, desmethylclozapine, as an allosteric modulator/agonist at muscarinic receptors.[79] The search for an improved version of clozapine lacking the liability for agranulocytosis has been ongoing for nearly 35 years on an empirical, chemically driven, 'trial and error' basis, since, despite a wealth of data on putative targets for clozapine and other antipsychotics, it is still unknown what mix of these activities, including partial receptor agonism, contributes to the superiority of clozapine. Similarly, many kinase inhibitors have multiple effects, from the ATP-mimetic, staurosporine, which interacts promiscuously with multiple kinases,[80] to compounds that while still not totally selective, can inhibit drug-resistant mutants of ABL, KIT, and epidermal growth factor (EGF) kinases.[78] More recently, there have been deliberate efforts to design NCEs with multiple molecular actions.[81]

2.11.3.4.3 Systems biology and pharmacology

A logical extension of the multiplicity of putative molecular targets for existing drugs has been the evaluation of NCEs for their phenotypic effects in complex systems to derive a 'system response profile.'[32] While being presented as a totally novel approach to drug discovery under the rubric of systems biology, a means for "rescuing drug discovery," [32,43] pharmacologists have used similar approaches to study NCEs in tissues and whole animal models for nearly a century,[30,36] one notable example being the classical Irwin test,[82] now some 40 years old, that is used to study the effects of NCEs on CNS function in terms of the behavioral phenotype they produce in rodents.

The systems biology approach is argued as a more cost effective and rapid way to do drug discovery and, as a consequence, increase productivity. Butcher[43] has naively proposed that since an NCE can enter human trials with established preclinical safety and, in some instances, efficacy in the absence of a known mechanism of action, the drug discovery paradigm should move from a 'target-centric' approach, based on the challenging and often futile task of target validation, to one of compound validation. Once shown to be efficacious in human trials, the mechanism(s) of action of compounds with a defined system response profile could then be determined, apparently for a fraction of the cost of a target-centric-based approach.

For industry researchers actually involved in such challenges, this is easier said than done. In the absence of a defined mechanism, it is exceedingly difficult, if not impossible, to establish a structure–activity relationship (SAR) and identify second-generation compounds that have improved efficacy and/or safety. While various mechanisms have been proposed to underlie the efficacy of the antiepileptic, valproic acid, as a mood stabilizer and as a treatment for AIDS, none have been entirely convincing such that second-generation NCEs have been designed on a structurally empirical basis leading to compounds like ABT-769[83] that have an improved safety profile as compared to valproic acid but still lack a mechanism of action (see 6.11 Epilepsy).

While systems biology is certainly a logical approach in helping to decrease NCE attrition rates, it, like molecular biology, is being currently used in an exclusionary manner rather than in conjunction with the proven tools of medicinal chemistry and pharmacology. The concept has been extended to that of 'systems pharmacology',[32] and has been further refined 'via multiplexing' using transgenic animals.[16] Despite the varied nomenclature, it appears very much a recapitulation of what was being done in drug discovery in the 1970s, albeit with more sophisticated technologies and improved data analysis and storage systems.

2.11.3.4.4 Re-engineering/re-positioning known drugs

A key part of the NIH Pathway Initiative has been the Molecular Library Initiative that has facilitated the setting up of screening centers at the NIH, National Institutes of Mental Health (NIMH), Harvard, Rockefeller University, and Johns Hopkins University to evaluate NCEs against targets that are either of limited interest or not available to pharmaceutical companies.[16] This initiative involved the setting up of large chemical libraries (currently including 500 000 compounds with projections to grow to 3 million by 2009) to provide academic laboratories with the opportunity to screen NCEs. This initiative also provides defined compound sets like that of 1040 FDA-approved drugs. From the latter, the β-lactam antibiotic, ceftriaxone, was found to upregulate levels of the glutamate transporter GLT1 in vitro,[84] functionally decreasing extracellular levels of this excitotoxic neurotransmitter and thus acting as a novel neuroprotectant. While these effects of ceftriaxone were confirmed in mouse models, they contrasted to the lack of efficacy of the antibiotic in human amyotrophic lateral sclerosis (ALS) trials, the latter of which have been criticized based on design and powering.[16] This represents a newly emerging trend where the clinical trial outcomes for NCEs that have failed efficacy trials are discounted in light of 'interesting' and novel preclinical data. A similar situation exists for the propargylamine GAPDH ligand, TCH346, which, having failed in the clinic for the treatment of Parkinson's disease, has been reassessed in terms of its anti-apopotic properties in cell and animal models and still apparently represents a viable molecular entity for clinical trials.[85]

Systems biology has provided intriguing evidence of the potential for 5HT receptor antagonists to prevent viral entry into cells and for the potential use of nonsteroidal anti-inflammatory drugs (NSAIDs), statins, lipid-lowering agents, and PPARγ inhibitors in AD, although, at least for the NSAIDs, the data has been inconclusive in terms of prospective evidence for their efficacy in retarding AD progression despite the high positive retrospective efficacy of indomethacin.[86]

The repositioning (or 'repurposing') of existing drugs[87] has also been extended to the evaluation of drug combinations (e.g., NCEs active at multiple targets) to assess compound synergy in vitro using classical pharmacological approaches in conjunction with highly sophisticated, computerized data analysis techniques, e.g., Loewe and Bliss independence modeling.[88] While preclinical data on drug combinations have provided interesting uses for approved drugs, perhaps acting at unknown, off-target sites, the evolving FDA's combination drug policy may be a hurdle in advancing these as NCEs.

2.11.3.4.5 Animal models

The animal models routinely used to characterize NCEs as part of the hierarchical process, while not infallible, are key in providing both evidence of putative efficacy to advance NCEs to the clinical testing stage and to assess NCE absorption, distribution, metabolism, and excretion (ADME) profiles to determine the drug-like characteristics of an NCE and to help in predicting doses for use in clinical trials. Few animal disease models recapitulate the human situation leading to considerable debate on their utility.[25,89] Rodent models of chronic pain include: spinal nerve ligation, an acute and precise surgical insult; streptozotocin-induced models of diabetic neuropathy; and various cytotoxic treatments to model cancer therapy-induced nerve damage (see 6.14 Acute and Neuropathic Pain). The majority of these models are typically studied for periods of 1–4 months, a chronic time period in an animal but very different from the years or even decades that are thought to precede the development of chronic pain in humans – often with idiopathic causality. These longer term changes in humans may result in more subtle, more complex, and longer lasting synaptic/neuronal changes than those occurring in the shorter lived animal model – confounding the transition of NCEs from animals to humans. Animal models of psychiatric diseases like schizophrenia also have limited relationship to the human disease. A classical assay to assess NCEs for antipsychotic potential, rat catalepsy, is more a model of dopamine receptor blockade than a model of psychosis with the prepulse inhibition (PPI) model more accurately reflecting the occurrence of similar phenomena in humans[90] (see 6.02 Schizophrenia). For antidepressants, iterations of the classical behavioral despair model,[91] now nearly 30 years old, remain state-of-the-art despite early inroads in assessing behavioral phenotypes in transgenic mice.[92] Transgenically modified mice, fruit flies, and zebra fish have been used as surrogate models of human disease states with an apparent remarkable degree of success, albeit via a retrospective analysis.[41]

An issue with transgenic models is whether the gene alteration, especially in the embryo, results in compensatory developmental changes that eliminate the molecular lesion due to systems redundancy. For instance, while there are clearly defined pathways for pain transmission, these can be highly redundant as are the systems involved in responses to septic shock. Removal of a single molecular component of these systems may have no phenotype in surviving animals or they may be lethal. Given that many cellular systems active in the process of development are quiescent in adults, the embryonic ablation of a key protein in a pathway may indeed be lethal at the wrong stage of development. Conditional knockouts may provide very different phenotypes in adult animals that may help in determining the role of a target in a given disease state. Similarly, other systems may malfunction as the result of gene removal or insertion and in this context there is emerging evidence that ADME properties in transgenic animals may be different from their wild-type controls.

Animal models are also widely used to assess side effect potential[93] (from NCE effects on blood pressure and sedation to overt toxicity) and can be used to determine the therapeutic index – a ratio of the efficacious dose/plasma level of an NCE to the dose/plasma level producing robust side effects and to provide first approximations of compound dosing for clinical trials.

2.11.3.4.6 Intellectual property

Prior to the biotech revolution, much of the intellectual property (IP) from biomedical research conducted in academic and federal laboratories was freely available in the form of publications and presentations at scientific meetings. Such information could then be used by the pharmaceutical industry to drive their drug discovery efforts. With the advent of the biotech revolution, driven to a major degree by the Bayh–Dole Act,[94] intellectual property became a highly viable currency for academia, making technology transfer operations in universities akin to venture capital (VC) firms, in some instances resulting in a 'gold digger mentality.'[95] This led to many inventions being excluded from industry use without a license, which was often exclusive in that only one party could use the invention. This latter situation was designed to avoid a 'tragedy of the commons'-type situation[96] in that the previous system of patenting and disseminating new findings had left more than 95% of federally funded inventions 'gathering dust' as companies would not invest to commercialize without owning exclusive title. The Bayh–Dole Act, now 25 years old, transferred the title to the IP of discoveries made with federal research grants to the universities where such discoveries were made in order to facilitate their commercialization, and has been credited[94] with helping create the biotechnology industry. In 1979, the year before Bayh–Dole, American universities received 264 patents. By 2004, universities were filing over 10 000 new patent applications and receiving approximately $1.4 billion per year in revenues from license fees.[95] While many universities and their in-house inventors became wealthy from federally funded research, from 1992 through 2003 pharmaceutical companies engaged in approximately 500 patent suits, "more than the number filed in the computer hardware, aerospace, defense, and chemical industries combined" – "a giant, hidden 'drug tax'... [that]..shows up in higher drug costs."[94]

The ease of filing and obtaining patents in the US, the number of patent applications having tripled since 1980 with the number of patents issued increasing fourfold,[97] has led to concerns, from the FTC's "questionable patents are a significant competitive concern and can harm innovation,"[97] to Heller and Eisenber's "anticommons" where "proliferation of [IP] rights upstream may be stifling life-saving innovation downstream"[97] to the *Economist's*, "there is ample evidence that scientific research is being delayed, deterred or abandoned due to the presence of patents and proprietary technologies."[95] In the anticommons argument, fragmented and overlapping patents are viewed as impeding product development, certainly the case some years ago when the total royalty rates on a diagnostic assay based on the use of several innovative technologies that had been negotiated at a 2–5% rate on individual technologies were found to total 47%, a level at which commercialization was not economically feasible. The impact of licensing on drug discovery research has thus been profound.

On the positive side, the ability to license patents exclusively has incentivized companies to work in an area. However, on the negative side, some biotech companies have taken good ideas from academia and created extensive and highly expensive-to-maintain patent portfolios ostensibly to commercialize the protected technology but, lacking in-house expertise and experience in drug discovery, have failed to add value to their assets, thus restricting access to promising technology – "coddled and kept out of the rain"[94] – in essence, squandering valuable scientific opportunities that a more experienced organization could have made into a value proposition. In the 1990s, a now long since acquired biotech company held a patent, colloquially known in the industry as '629,' that apparently covered the signaling ability of every drug target system, both artificial and natural. Thus, in addition to covering luciferin–luciferase reporter systems, this patent claimed adenylyl cyclase-linked systems in native cell lines, as well as the behavioral phenotypes of animals following NCE administration. In fact, just about any experiment run in biomedical research. This company aggressively attempted to enforce this patent forcing another promising biotech out of business and gaining collaborations with certain big pharmas. When patent 629 was finally challenged in court, it was disallowed. Had this not occurred, nearly every experiment being run in biological systems throughout the world today would require paying this unnamed company royalties.

The ability to patent the gene-based elements of cell function, receptors, biochemical pathways, etc., has led to hotly disputed patents on the knowledge of life itself rather than true inventions involving composition of matter or trade secrets. The negative impact of this patenting strategies is now being actively debated[98,99] with the NIH advocating the free[94] or nonexclusive[95] licensing of research tools, a position supported by the National Research Council.

Access to patented research tools has traditionally been more expensive in industry than academia (even though those same tools in academia were used to found companies or generated revenue for the inventors via licensing deals) or unavailable when the patent holder has licensed exclusive rights to a third party. The cost of these license fees and of accompanying collaboration milestones has increased the costs sunk in research. To the in-house costs of running a collaborative research program must then be added licensing (and legal) fees, milestones that can average from $1–5 million before an NCE is tested in humans to $30–120 million or even more when an NCE is approved via an NDA. These costs are all 'at risk' in that the sponsoring partner must pay them before a penny in revenue from the NCE occurs. To the projected $800 million in cost for bringing a drug to market based on historical numbers must then be added another 15–25% in financial costs, which do not include the cost of money over the 8- to 12-year period.

2.11.3.5 Filling the Pipeline

The inverse relationship between the considerable investments in drug discovery, both financial and technological, to the current IND and NDA output[3,44] has necessitated that drug companies and biotechs focus considerable efforts (often with less scientific expertise) on opportunistic compound in-licensing activities to sustain their product pipelines. As part of an integrated, global strategy that includes both consideration and inclusion of the company's R&D efforts, in-licensing can be a highly productive endeavor. Several major pharma companies have deliberately licensed up to 50% of their current product portfolio from outside sources or via company acquisitions, e.g., Pfizer and Pharmacia, Amgen with Immunex and Abgenix, Abbott and Knoll, to both reduce risk and build depth in their pipeline portfolio. Similarly, several of the biotechs that have survived from the boom of the 1990s have achieved sustainable success with in-licensed compounds, often in therapeutic areas distinct from that in which the company was initially founded, the initial scientific concept of the company having proven faulty.

The in-licensing paradigm has been clearly successful when "diamond[s] in the rough"[26] have been identified, e.g., the novel wake promoting agent, modafinil (*see* 8.12 Modafinil, A Unique Wake-Promoting Drug: A Serendipitous Discovery in Search of a Mechanism of Action).

The continuation of in-licensing activities at their current feverish rate creates a vicious cycle that needs to be placed in the context of the many and considerable in-licensing failures that are forgotten or ignored. Biotechs continue to search for compounds with overlooked and unexploited commercial potential in the archives of major pharmaceutical companies, an activity in which they often compete directly with the latter as the big pharmas similarly compete with one another to identify such opportunities in one another's 'shelved' portfolios. However, the number of Phase II/Phase III failures reported on an increasingly regular basis in *BioWorld* reinforces the fact that many of these 'opportunities' were indeed the 'coals' that the originating company had already passed judgment on rather than the elusive 'diamonds' that the licensee had anticipated.

One unfortunate outcome from in-licensing activities is that drug discovery and development organizations, both big and small, begin to question the value of their internal research groups, arguing that the monies spent on the unpredictable and long-term 'black hole' of research would be better used in finding more near-term, external 'diamonds,' overlooking the fact that these 'diamonds:' (1) had to originate from a research-based organization; (2) are in increasingly short supply and/or of increasing questionable quality; and (3) because of industry consolidation and downsizing and the bursting of the biotech bubble, are not being replenished. As a result, an opportunistic in-licensing approach becomes ineffective unless it is part of a longer-term 'strategy' to sustain company growth and stability. Indeed, as many in-licensing activities are purely opportunistic, any corporate strategic plan can be subject to almost instant change in direction via the next acquisition, making the long-term vision required for the 10–12 year time frame of the R&D process a major challenge unless proactively and consistently managed.[61,62,100] Also, the opportunities in in-licensing activities are often driven by the 'grass being greener' elsewhere in terms of compound quality and opportunity (and led at one point to the author's suggestion that each drug company should put a select number of their second and third tier compounds into a 'basket' to provide the in-licensing functions, a portfolio of compounds that could be used as the self-perpetuating basis for deal making (you buy my compound, I'll license yours) while R&D were left to create long-term value without the irritation of supporting ephemeral initiatives from the business development function) and effectively reduces the considerable investments being made in internal technology to enhance the drug discovery process to that of a prospecting expedition. This can result in a factious dynamic with the internal R&D staff who, in the process of due diligence, can transition an ill-conceived 'billion dollar' opportunity to the status of ill-conceived and poorly executed science, and are then perceived as territorial. An additional facet of the 'grass being greener' is that many companies intending to out-license compounds have second thoughts as other companies find value in their portfolio, and decide it is better to keep an asset than give it to a competitor who may show it had potential, an example of the "100% of nothing being better than 50% of something" school of drug R&D. As an anecdote, the author was at one time involved in due diligence on an in-licensing opportunity that had tremendous commercial potential but for which there were a number of serious and unresolved side effect issues. The result was that the opportunity went to another company and resulted in the head of marketing at the author's company decrying the lack of vision in research, the proverbial NIH ('Not-Invented-Here') syndrome, for this compound that apparently had revenues of $200 million in its first 2 months on the market. Within the following 6 months, the compound was withdrawn from the market following major safety issues and the acquiring company became immersed in litigation in the tens of billions of dollars. When confronted with the fact that the objective research evaluation had saved the author's company from a similar fate, the head of marketing refused to comment.

2.11.4 Technological Impact and Integration

The imperative in building new biotech companies has been more often driven by financial opportunism than by any innate belief that the scientific concepts on which such companies were based would lead to new drugs and improved healthcare. Despite a technology revolution that promised safer, more efficacious drugs in a more rapid and consistent timeframe, it still takes from 8 to 14 years from the time an NCE is discovered to market introduction.

The revisionist approach inherent in the biotech revolution has resulted in drug discovery being considered a predictable commodity with drug targets, irrespective of their intrinsic complexity and dynamic nature, being totally interchangeable. Similarly, as already noted, the individuals conducting the research have become viewed as interchangeable such that a chemist can function as a biologist (despite marked differences in training and culture) and a neuroscientist could easily work as a cell biologist in anti-infective drug discovery. This has reduced all aspects of drug research to a production line-like metric, discounting the fact that insight based on experience, innovation, and serendipity are unpredictable and highly necessary for success.

While the reductionistic approach of molecular biology has led to major advances in the understanding of cell function, the subtle nuances of the native target are frequently absent when human DNA is expressed in a cell line. 7TM receptors require key accessory proteins, e.g., β-arrestins and phosphatases, that are either missing in transfected cell systems or atypically linked due to receptor/target overexpression,[101] such that reports of new signaling pathways for known receptors in transfected cell lines should be viewed with considerable caution. Receptors also undergo dimerization,[102] trafficking, and internalization and may also have tissue- and disease-specific differences in their signaling pathways.[103] For protein kinases, the phosphorylation state of the substrate, which can be a function of the disease state, the endogenous ATP concentration, and the nature of the natural protein substrate (which is frequently unknown), all alter inhibitor activity as do disease-related constitutive forms of the enzyme.[104,105] Finally, a molecular networks approach[106] to synaptic protein interactions (the synaptome?) can help delineate functional interactions for drug targets and their cognate signaling pathways in a highly complex, albeit physiologically rigorous manner.

Similarly, the interaction of NCEs with their target can be far more complex than the simple classification of either agonists or antagonists/activators or inhibitors. The demonstrated existence of partial agonists, allosteric modulators, neutral and inverse antagonists, and inverse agonists in multiple systems re-emphasizes the critical need for concentration/dose assessment of NCE activity rather than single-dose SAR.[37,107] Extensions of such concepts include constitutive activity of 7TM receptors, the activity of receptor systems that occurs independently of the presence of the ligand, collateral efficacy, and permissive antagonism[103] and redefine both the ligand and the nature of its interaction with its target(s).

In operating almost exclusively at the cellular level, or integrating findings from a cell-based approach with a transgenic mouse that often recapitulates the test tube, the reductionistic approach has limited the ability to interpret data related to NCE efficacy, target selectivity, side effect liability, and ADME in increasingly more complex systems. The revisionist biotech agenda de-emphasized the integrative, hierarchical approach inherent in classical pharmacology, where NCE assessment progressed from a relatively simple tissue homogenate, sequentially to the whole cell, tissue and animal levels, with data being reviewed and integrated at each step to provide a more thorough understanding of NCE actions at the molecular, tissue, and whole animal levels (**Figure 1a**; **Table 1**) before advancing these to clinical trials. The challenge then became to take the newly synthesized NCE and get it into humans as rapidly as possible without bothering with all these 'old fashioned' constraints (and often ignoring not just the time-tested approaches but also the data already in the literature) (*see* Section 2.11.3.4.4).

The perceived importance and ease of cellular studies led to these being increasingly funded, not as additional tools to enhance the biomedical research endeavor, but to the exclusion of more traditional integrative research approaches. In 1994, 11% of NIH pharmacology fellowships involved whole animal research,[67] resulting in a de-emphasis in the teaching of classical pharmacology, and with the aging of the existing population of pharmacologists, a dearth of expertise and experience.[36]

As the topic of drug discovery now extends beyond the pharmaceutical and biotechnology industries into the NIH and academia, it is clearly time that a more concerted effort be placed on integrating all the available disciplines relevant to drug discovery, both old and new, in an effort to improve productivity. Rather than continue the poseur/hype addict/technocrat approach with an exclusionary, revisionist focus on technologies to enable drug discovery, it is a critical necessity to take what worked in the past, e.g., pharmacology and medicinal chemistry, and revitalize it, while redirecting new technologies in strategies for drug discovery that have been proven to work (**Figure 1**). The author has addressed the need for a pharmacologically based integrative approach to address the shortfall in productivity in the context of an *iPharm* concept based on Apple's highly successful iPod.[30] In identifying a need in the marketplace, a design team at Apple took existing, off-the-shelf hardware and software to create a technologically sophisticated, well-designed, and functional portable digital storage device for music, photographs, movies, and books, the utility of which (and demand for) has far exceeded the sum of its parts. Thus, appropriate, cost-effective technology was successfully used to reach a previously defined endpoint rather than Apple searching for a project to justify the existence of their technology. The *iPharm* concept may provide a truly integrative framework to understanding compound action at all levels rather than a 'systems biology' approach that seeks to re-invent pharmacology while discounting the value implicit in knowing the mechanism of action of a drug, a 'modern-day back-to-the basics approach.'[108]

Table 1 The pharmaceutical industry then and now: A comparison of organizational and individual cultures in drug discovery

		Drug Discovery 1976	*Drug Discovery 2006*
Organizational environment		Golden age of drug discovery	Empty pipelines, perception of poor prospects for new drugs, generic competition
		Merck 'most admired company'	Distrust of industry • Vioxx law suits • Conflicts of interest in clinical trials • Excessive emphasis on marketing

		First $billion/year drug–cimetidine	Mutiple $billion/year drugs
			    Statins–$12 billion/year
			    SSRIs–$7 billion/year

		Major US, European, and Japanese pharmas	Multiple mergers to form mega pharmas
		Multiple local/family boutique pharmas	*Pfizer* = Warner Lambert + Pharmacia + Monsanto + Upjohn + Sugen + Goedecke + Agouron + Vicuron + Idun + AngioSyn + Esperion + Rinat
		Local markets	*Sanofi-Aventis* = Sanofi-Synthelabo + Aventis + Hoechst + Roussel Uclaf + Hoechst Marion Roussel + Marion + Richardson Merrill + Merrill Dow + Dow Pharmaceuticals + Rhone Poulenc Rorer + Rorer, etc.
		No biotech industry	Biotechnology industry firmly established – 4000 companies worldwide
			Emergence of Indian and Chinese pharmaceutical companies
		European/US pharmaceutical R&D	Consolidation of global pharma R&D in the US
		Research – science based	Research–market driven
Scientific approach		Hypothesis driven	Hypothesis seeking
		Using technologies to answer questions	Numbers and technology driven – using technology to find questions
		Integrated chemo-/pharmaco-centric molecular target-based approach	Reductionistic, exclusionary approach
		Evolving hierarchical complexity and integration in NCE evaluation in natural systems: • membrane • cell • tissue • animal	
			Pharmacology by any other name
			Chemigenomics and emergence of systems biology
		DNA revolution evolving–no cloning, no expression	Multiple 'omics' sciences – genomics, proteomics, interactomics, epigenomics
		Traditional medicinal chemistry	
		Facile interface between industry, academia, and government	
		New scientific findings freely published and presented at meetings	
		HTS – 200 compounds a week	HTS – 1 million + compounds a week
		Important new data initially published in peer-reviewed journals with information to ensure facile replication	New data 'published' in *Wall Street Journal, New York Times* or *BioWorld* – no peer review, often never published, and when published, key details for replication withheld for IP purposes
			Structures and patented 'know how' omitted from papers

continued

Table 1 Continued

	Drug Discovery 1976	*Drug Discovery 2006*
Individual motivation and focus	Discipline/disease specialization	Technology orientation
	Passion for science	Passion for science tempered by concern for stock options
	Inquisitive	Directed
	Intrinsically (scientifically) motivated	Trend toward extrinsic (stock options, $$) motivations
	Motivated by drug discovery as an end	
	High level of career stability	
	George Merck ethos	Business influenced
	Card-carrying members of the scientific community	Scientific standards slipping[26,30]
	Publications	
	Peer reviewing	

2.11.5 Future Challenges

The concerns regarding the pharmaceutical industry related to lack of productivity, to drug costs, and to issues with the drug approval and marketing processes[22,23,28] may be viewed either as the end of the golden era of drug discovery or a bona fide opportunity to remold the research enterprise based on the richness of information, technology, and experience that are now available. To do this requires a renewed focus on process management and best practices together with integration of all activities involved in drug discovery at the individual, organizational, cultural, and technological levels. The productivity gap can only be addressed if the drug discovery paradigm, as it now exists in big pharma, undergoes significant change.[3] The consequences of not adopting a 'new' paradigm will continue to negatively impact the vitality and even the existence of big pharma in its present form and continue the productivity woes and inefficiencies.[21] Thus, the risk of continued failure with existing approaches to drug discovery is 'arguably absolute.'[3]

2.11.5.1 Cultural

As noted, biologists are being trained at a different level than they were 30 years ago. It may also be argued that current training in the chemical sciences tends to oversimplify the complexity and inherent variability of biological systems. The need to re-instill inquisitiveness and independence of thought is a key issue that extends to society as a whole but is particularly acute in the area of drug discovery where these qualities have proven to be critical to success. To do this will take time and requires leadership that can provide the challenge and mentorship to allow scientists to expand to their full potential by letting them invest time in quality thinking before the experiments begin.[3] It is all too easy to find a library and set up HTS screens. Diminishing the visibility, impact, and presence of both the scientific poseur and hype addict within the research environment will be a major help in a return to reality.

A research organization cannot tolerate anything other than the best science; it must insist on timely decision making driven by data and encourage risk taking and the hiring of passionate people. It should have a high level of tolerance for individuality and should encourage its scientists to take risks on a daily basis without consequence.[63] As another anecdote, the author had considerable troubles with his management when celebrating with a project team the demise of the project just 2 weeks before the IND candidate was to go into humans. A class effect, presumably related to the target, had emerged, compromising the safety of the therapeutic approach. The risk of going forward to the clinic was deemed unacceptable. In light of the new data, the project team recommended that the project be terminated, a clear decision that was as worthy of acknowledgment as moving an IND forward. To be successful, a research organization must tolerate such failures.

The continued dependence of pharmaceutical management on input from outside thought leaders from academia and management consultants to help guide and organize drug R&D, respectively, in essence second guessing their own employees, is somewhat akin to BMW asking Michael Schumacher to design its production lines or the paints industry looking to Leonardo da Vinci to spend time away from his code, advising its chemists on how to develop the most durable paint composition for the Golden Gate Bridge. A major opportunity that has resulted from the widespread mergers and downsizing in major pharma over the past decade is that individuals (drug hunters) with many years of practical experience, "precious knowledge gained in past projects,"[1] are now available as consultants, providing unprecedented insights and wisdom into the fundamentals of the process.

In a somewhat tongue-in-cheek anecdote, Lowe[31] has noted that the "highest correlation with eventual success.. [in drug discovery is that]... for your drug candidate to be a winner, someone has to have tried to kill it" – preferably at the VP or higher level and even more preferably numerous times. While amusing, this concept underlines two facets of modern day drug discovery. The first is that the process is highly dynamic, with many twists and turns in the process of advancing an NCE, with many opinions, new data sets, and decision points requires an imperative to operate in an objective, data driven mode. The second is that if management perceives an individual as being on the wrong track in a drug discovery project then that individual is inherently right from a scientific perspective. Unfortunately, for every Jim Black or Pat Humphrey, there are many well-intentioned, yet misguided scientists who fail to take an objective viewpoint as to what their data is telling them, a situation made worse by the 'turn off the brain' phenomenon.[1] The more they are told they are mistaken and the more their project is killed, the more zealous they become, perhaps providing another reason for the lack of productivity in the drug discovery process.

Drug discovery at its basic level reflects the interaction between medicinal chemists and pharmacologists to objectively test disease-related hypotheses in order to identify NCEs that may in time become drug candidates and eventually drugs. This is a highly risk-oriented enterprise inasmuch as the tools required to validate new hypotheses result from the activities being conducted, making the process extremely iterative and far from linear. Certain aspects, like HTS and NCE synthesis are less risky in that the activities are better defined and easier to execute. However, in the biology arena, predictability is a rare commodity and a biologist needs to remain highly attuned to discrepant, yet potentially informative, data to ensure that the path being followed is logical and that the right experiments are being planned and executed. With a reductionistic approach, unexpected data inevitably results in a dogmatic, often negative response, that concludes the data is wrong. However, the ability to objectively evaluate the data and place it in context can aid in designing future experiments and even result in breakthroughs. Some 25 years ago, in the days before ADME was a routine part of the early-stage drug discovery process, the author worked on a peptide receptor project. None of the NCEs worked in the selected in vivo model, which led to considerable consternation and questioning of the 'druggability' of the target. After a significant campaign to measure plasma levels of the peptidic NCEs, it was found that the NCEs had a maximal half-life of 2 min. The in vivo paradigm involved measuring the NCE effects 15 min after administration.

2.11.5.2 Operational

A research organization should maintain a high level of consistency. Thus, standards applied to the evaluation of external research opportunities should be similarly applied to internal activities[100] and the best opportunity to provide a drug candidate for the organization should take priority over any internally invested project that objectively cannot achieve its goals either due to scientific challenges or timing. All research projects should be objectively benchmarked against the pharmaceutical standard of care, if one exists, in a disease area to provide both the clinical and marketing organizations with information on the potential of an NCE. To this end, it is important that an optimal profile for the targeted NCE, a 'quasi-package insert', be developed early on and used to guide project activities.

The major theme of this chapter had been the reintroduction of the integrative discipline of pharmacology into the drug discovery process. This requires positing a hypothesis-related question and then using all available technologies, chemical, molecular, in vivo and clinical, to provide answers with a view to understanding the impact of NCEs on biological systems at increasing levels of hierarchical complexity. As argued, this should avoid exclusionary revisionist and political agendas, including those that dismiss any inherent value in animal models to those that argue against doing an experiment because the outcomes may provide contrary data. In this latter context, it is an absolute imperative that experiments be well planned with a clear idea of possible outcomes and plans to act on these using a decision tree-like approach. The existence of clear go/no-go decision points should aid in the design of these experiments. Project team activities should involve a critical path flowchart and avoid extraneous recapitulations of known activity via parallel experimentation. If the potential outcomes of a hypothesis-driven experiment in the context of the project flow chart will not provide data for making a decision, then the experiment should not be done. All research should be conducted at a peer-reviewed level so that the quality of internal and regulatory documents maintains the highest standards achievable. Quantitative rather than qualitative data with full replication based on a null hypothesis-based approach focused on "systematic and comprehensive approaches to understanding the mechanism of action of drugs,"[109] is sine qua non. An operational perspective is provided in **Figure 1**, where the current fragmented and reductionistic technology-driven approach is compared to the more integrated 'pharmacocentric' approach.

A major hurdle to the rapid assessment of an NCE as a potential drug is its transition from animals to humans. Once safety has been established for an NCE, it should enter clinical trials as soon as possible to obtain a proof of concept. Examples of NCEs that showed half-lives of under an hour in animal models and were later found to have half-lives in excess of 10 h in humans are legendary with the inevitable question "which species predicts human?" In this context, much has been written on the topic of translational medicine, the ability to advance an NCE to human trials in an exploratory fashion.[28] For this approach to work effectively it requires clinical oversight with an appreciation of basic biological research. Too often, 'pilot clinical studies,' designed to get some idea of efficacy or proof of concept, end up being underfunded and underpowered and can rapidly transition to pivotal decision-making trials, making their execution a risk verging on the suicidal.

2.11.5.3 Scientific

There is a somewhat schizoid dynamic in the drug discovery process, that of the ability to maintain the scientific ethos in a 'for-profit' environment. In the 1970s, joining industry was judged to be passing to the 'dark side' of science, sacrificing one's scientific standards in return for money. While Drews[26] has suggested that "good science" in industry has been replaced by "marketing dogma," drug discovery research does require a well-managed process with clear goals, the resources to execute, and a forum for objective, data-driven decision making. To many drug discovery scientists, any form of management is anathema to their level of working comfort, raising concerns regarding the stifling of their creativity and innovation while to others, there is a comfort level in being directed with others making decisions and taking responsibility. As the lifeblood of research is success in delivering drugs, the ideal individual is the motivated person with an appreciation of order and a well-grounded sense of scientific self to help guide the organization in doing things in a consistent and scientifically credible manner. Within a well-organized, well-led research organization there is room for both a focused, time-mandated research effort and the opportunity to be innovative either concomitantly or sequentially. There should be little room in an effective research organization for individuals who have anointed themselves as the 'thinkers,' a quasi-elite, incapable of taking an idea to a tactical level without the help of more pragmatic individuals.

Today's industry is overly focused on the new, from technology to targets. The outcome has been an increase in the cost of research with a corresponding decrease in the tangible output of new drug candidates. Nonetheless, drugs are an integral part of twenty-first century healthcare that immeasurably reduces the overall costs of healthcare. History may in time reflect that COX-2 inhibitors and sildenafil were business-driven aberrations in an otherwise illustrious record of providing vaccines, statins, and other drugs that have improved both longevity and quality of life. However, with the obsessive focus on new targets and the ever-changing kaleidoscope of 'omes' (**Figure 1b**), it is important to remember that many well-known targets (histamine receptors, cholinergic receptors, ATP receptors, kinases, PI3/AKT, etc.) have yet to be exploited to their full therapeutic potential. This will need a focused commitment and for each target will most likely take a 20-year time frame for success. The time spent in looking for and negotiating the rights to new targets and increasing comfort in their value is a high-risk endeavor that detracts from this effort, and must be considered in decisions regarding resource prioritization.

In a review of successful business enterprises in 1995, Jennings and Grossman[110] identified the key to business longevity under the rubric of the "Tony Bennett Factor." Mr Bennett, an established and successful entertainer (and accomplished painter) has a career that is now over half a century old and still viable (despite leaving his heart in San Francisco), which is based on Mr Bennett focusing on doing what he was doing in the 1950s, that is, entertaining; in essence, understanding his business and staying with it. In asking the question, "what business are we in?," each of the companies surveyed by Jennings and Grossman "knew their strengths, developed strong market presences based on these strengths and never forgot their roots." It may indeed be argued that today's pharmaceutical industry has lost its way and has, as suggested by David Horrobin, "taken a wrong turn in [its] relationship to human disease."[26] With research-based drug companies becoming involved in managed healthcare pharmacies and then divesting them and with Eastman Kodak and Proctor and Gamble getting into (and exiting) ethical pharmaceuticals, this does beg the question of 'what business?'

For productivity to reemerge requires that the pharmaceutical and biotech industries refocus on their core competency, that of innovation in applied biomedical research directed at ameliorating human disease. The industry must also recognize and commit to what business it is in and also recognize that the 'old ways' of carrying out drug discovery (**Figure 1a; Table 1**) have not been, and perhaps never will be, replaced by technology-driven in vitro and in silico approaches, nor will the passion of the drug hunter be replaced with either the golden-tongued sound bites of the scientific poseur or the ephemeral knowledge of the management consultant.

While the path forward in drug discovery has been viewed as "how we organize to implement new paradigms"[3], it is important that the key elements that worked in the past are not overlooked. Abou-Gharbia has noted that "if you don't have capable chemistry, you won't have successful drug discovery"[21] (and one might add, drug-like NCEs) while Reynolds has commented that "It's the people who carry out drug R&D and their ability to exercise good judgment at many stages of the process that have perhaps always been and will continue to be the key in making drug discovery more efficient."[21] Finally, in the context of the Tony Bennett principle is George Merck's seminal comment in an interview with Time magazine in 1952 that "We [Merck] try never to forget that medicine is for the people. It is not for profits. The profits follow, and if we have remembered that, they will never fail to appear. The better we remember that, the larger they have been." A return to Mr. Merck's sentiments with the pharma and biotech industries focusing on reinforcing their commitment to research driven by objective, innovative, and intellectually commited science and to supporting scientists rather than technologies will do much to reverse the current paucity in drug discovery.

References

1. Kubinyi, H. *Nat. Rev. Drug Disc.* **2003**, *2*, 665–668.
2. Lutz, M.; Kenakin, P. *Quantitative Molecular Pharmacology and Informatics in Drug Discovery*; Wiley: Chichester, UK, 1999.
3. Milne, G. M., Jr. *Annu. Rep. Med. Chem.* **2003**, *38*, 383–396.
4. Lander, E. S.; Linton, L. M.; Birren, B.; Nusbaum, C.; Zody, M. C.; Baldwin, J.; Devon, K.; Dewar, K.; Doyle, M.; FitzHugh, W. et al. International Human Genome Sequencing Consortium. *Nature* **2001**, *409*, 860–921.
5. Venter, C.; Adams, M. D.; Myers, E. W.; Li, P. W.; Mural, R. J.; Sutton, G. G.; Smith, H. O.; Yandell, M.; Evans, C. A.; Holt, R. A. et al. *Science* **2002**, *291*, 1304–1351.
6. Collins, F. S.; Green, E. D.; Guttmacher, A. E.; Guyer, M. S. *Nature* **2003**, *422*, 835–847.
7. Kramer, R.; Cohen, D. *Nat. Rev. Drug Disc.* **2004**, *3*, 965–972.
8. Stoughton, R. B.; Friend, S. H. *Nat. Rev. Drug Disc.* **2005**, *4*, 345–350.
9. Carney, S. *Drug Disc. Today* **2005**, *10*, 1025–1029.
10. Weatherall, D. *Royal Society, London*, 2005. http://www.royalsoc.ac.uk/ (accessed Aug 2006).
11. Jones, S. *Milbank Memorial Fund Report*, June, 2000.
12. Figeys, D. In *Industrial Proteomics*; Figeys, D., Ed.; Wiley-Interscience: Hoboken, NJ, 2005, pp 1–62.
13. Huber, L. A. *Nat. Rev. Mol. Cell Biol.* **2003**, *4*, 74–80.
14. Jeffery, D. A.; Bogyo, M. *Curr. Opin. Biotechnol.* **2003**, *14*, 87–95.
15. Kopec, K.; Bozyczko-Coyne, D. B.; Williams, M. *Biochem. Pharmacol.* **2005**, *69*, 1133–1139.
16. O'Connor, K. A.; Roth, B. L. *Nat. Rev. Drug Disc.* **2005**, *4*, 1005–1014.
17. Rual, J.-F.; Venkatesan, K.; Hao, T. *Nature* **2005**, *437*, 1173–1179.
18. Jones, P. A.; Martienssen, R. *Cancer Res.* **2005**, *65*, 11241–11246.
19. Linden, A.; Fenn, J. *Research ID R-20-197*; Gartner: Stamford, CT, 2003.
20. Wilmut, R., Eds., *All The Words. The Complete Monty Python's Flying Circus*; Pantheon: New York, 1989; Vol. 2, pp 63–64.
21. Borman, S. *Chem. Eng. News*, June 19, 2006, pp 56/78.
22. Goozner, M. *The $800 Million Pill: The Truth Behind the Cost of New Drugs*; University of California Press: Berkeley, CA, 2004.
23. Avorn, J. *Powerful Medicines: The Benefits, Risk and Costs of Prescription Drugs*; Knopf: New York, 2004.
24. Horrobin, D. F. *Nat. Biotechnol.* **2001**, *19*, 1099–1100.
25. Horrobin, D. F. *Nat. Rev. Drug Disc.* **2003**, *2*, 151–154.
26. Drews, J. *Drug Disc. Today* **2003**, *8*, 411–420.
27. Williams, M. *Curr. Opin. Investig. Drugs* **2004**, *5*, 1–3.
28. FitzGerald, G. A. *Nat. Rev. Drug Disc.* **2005**, *4*, 815–818.
29. Schmid, E. F.; Smith, D. A. *Drug Disc. Today* **2005**, *10*, 1031–1039.
30. Williams, M. *Biochem. Pharmacol.* **2005**, *70*, 1707–1716.
31. Lowe, D. B. *Contract Pharma*, October 2005, 28/30. http://www.contractpharma.com (accessed July 2006).
32. Van der Greef, J.; McBurney, R. N. *Nat. Rev. Drug Disc.* **2005**, *4*, 961–967.
33. Sneader, W. *Drug Discovery: The Evolution of Modern Medicines*; Wiley: Chichester, UK, 1985.
34. Triggle, D. *Curr. Protocol Pharmacol.* **1998**, 9.1.
35. Monaghan, R. L.; Barrett, J. F. *Biochem. Pharmacol.* **2006**, *71*, 901–909.
36. In Vivo Pharmacology Training Group. *Trends Pharmacol. Sci.* **2002**, *23*, 13–18.
37. Kenakin, T. A. *Pharmacology Primer. Theory, Application and Methods*; Elsevier/Academic Press: San Diego, CA, 2004.
38. Gaddum, J. H. *Nature* **1954**, *173*, 14–15.
39. Stockwell, B. R. *Nat. Rev. Genet.* **2000**, *1*, 116–125.
40. Brent, R. *Nat. Biotechnol.* **2004**, *22*, 1211–1214.
41. Zambrowicz, B. P.; Sands, A. T. *Nat. Rev. Drug Disc.* **2003**, *2*, 38–51.
42. Posner, B. A. *Curr. Opin. Drug Disc. Dev.* **2005**, *8*, 487–494.
43. Butcher, E. C. *Nat. Rev. Drug Disc.* **2005**, *4*, 461–467.
44. Kola, I.; Landis, J. *Nat. Rev. Drug Disc.* **2004**, *3*, 711–716.
45. Black, J. *J. Med. Chem.* **2005**, *48*, 1687–1688.
46. Wong, D. T.; Perry, K. W.; Bymaster, F. P. *Nat. Rev. Drug Disc.* **2005**, *4*, 764–774.
47. Economist. *Economist*, Nov 4, 2004.
48. Topol, E. J. *N. Engl. J. Med.* **2005**, *351*, 1707–1709.
49. Pipes, S. C. *Miracle Cure*; Pacific Research Institute: San Francisco, CA, 2004.

50. Light, D. W.; Lexchin, J. *Br. Med. J.* **2005**, *331*, 958–960.
51. G. Colvin, *Fortune*, Nov 1, 2004, p 82.
52. Lichtenberg, F. R. NBER Working Paper # W8996, 2002 http://ssrn.com/abstract = 315993 (accessed April 2006).
53. Stephens, B. *The Wall Street Journal*, Jan 7–8, 2006, p A6.
54. Ranalli, P., 1999 http://www.nrlc.org/Baby_Parts/ranalli.html (accessed April 2006).
55. Bjorklund, A.; Lindvall, O. *Nat. Neurosci.* **2000**, *3*, 537–544.
56. Normile, D.; Vogel, G.; Holden, C. *Science* **2005**, *310*, 1886–1887.
57. Kay, J., 1998. http://www.johnkay.com/society/133 (accessed April 2006).
58. Zerhouni, E. *Science* **2003**, *302*, 63–72.
59. Couzin, J. *Science* **2003**, *302*, 218–221.
60. FDA. Innovation/Stagnation. Challenge and Opportunity on the Critical Path to New Medical Products; FDA: Bethesda, MD, April, 2004 http://www.fda.gov/oc/initiatives/criticalpath/whitepaper.html (accessed April 2006).
61. Jensen, I.; Jorgenson, S.; Sapienza, A. *Drug Dev. Res.* **1998**, *35*, 1–6.
62. Sams-Dodd, F. *Drug Disc. Today* **2005**, *10*, 1049–1056.
63. Sapienza, A. *R&D Mgmt.* **2005**, *35*, 473–482.
64. Firestein, G. S.; Pisetsky, D. S. *Arthritis Rheum.* **2002**, *46*, 859–861.
65. Lazo, J. S.; Ducruet, A. P.; Koldamova, R. P. *Mol. Pharmacol.* **2003**, *64*, 199–201.
66. Maddox, J. *Nature* **1992**, *335*, 201.
67. Preusch, P. C. *Mol. Interventions* **2004**, *4*, 72–73.
68. Jobe, P. C.; Adams-Curtis, L. E.; Burks, T. F.; Fuller, R. W.; Peck, C. C.; Ruffolo, R. R.; Snead, O. C., III; Woosley, R. L. *Physiologist* **1994**, *37*, 79–84.
69. Prud'homme, A. *The Cell Game*; HarperCollins: New York, 2004.
70. Williams, M. *Curr. Opin. Investig. Drugs* **2006**, *7*, 1–4.
71. Chumakov, I.; Blumenfeld, M.; Guerassimenko, O.; Cavarec, L.; Palicio, M.; Abderrahim, H.; Bougueleret, L.; Barry, C.; Tanaka, H.; La Rosa, P. et al. *Proc. Natl. Acad. Sci. USA* **2002**, *99*, 13675–13680.
72. Tsai, G.; Coyle, J. T. *Annu. Rev. Pharmacol. Toxicol.* **2002**, *42*, 165–179.
73. Leurs, R.; Bakker, R. A.; Timmerman, H.; de Esch, I. J. P. *Nat. Rev. Drug Disc.* **2005**, *4*, 107–120.
74. Lloyd, G. L.; Williams, M. *J. Pharmacol. Exp. Ther.* **2000**, *292*, 461–467.
75. Caulfield, M. P.; Birdsall, N. J. M. *Pharmacol. Rev.* **1998**, *50*, 279–290.
76. Roth, B. L.; Sheffler, D. J.; Kroeze, W. K. *Nat. Rev. Drug Disc.* **2004**, *3*, 353–359.
77. Fritz, S. *Nature* **2005**, *437*, 942–943.
78. Carter, T. A.; Wodicka, L. M.; Shah, N. P.; Velasco, A. M.; Fabian, M. A.; Treiber, D. K.; Milanov, Z. V.; Atteridge, C. E.; Biggs, W. H., III; Edeen, P. T. et al. *Proc. Natl. Acad. Sci. USA* **2005**, *102*, 11011–11016.
79. Weiner, D. M.; Meltzer, H. Y.; Veinbergs, I.; Donohue, E. M.; Spalding, T. A.; Smith, T. T.; Mohell, N.; Harvey, S. C.; Lameh, J.; Nash, N. et al. *Psychopharmacology* **2004**, *177*, 207–216.
80. Fabian, M. A.; Biggs, W. H., III; Treiber, D. K.; Atteridge, C. E.; Azimiora, M. D.; Benedetti, M. G.; Carter, T. A.; Ciceri, P.; Edeen, P. T.; Floyd, M. et al. *Nat. Biotechnol.* **2005**, *23*, 329–336.
81. Morphy, R.; Rankovic, Z. *J. Med. Chem.* **2005**, *48*, 6523–6543.
82. Irwin, S. *Psychopharmacologia* **1968**, *13*, 222–257.
83. Giardina, W. J.; Dart, M. J.; Harris, R. R.; Bitner, R. S.; Radek, R. J.; Fox, G. B.; Chemburkar, S. R.; Marsh, K. C.; Waring, J. F.; Hui, J. Y. et al. *Epilepsia* **2005**, *46*, 1349–1361.
84. Rothstein, J. D.; Patel, S.; Regan, M. R.; Haenggeli, C.; Huang, Y. H.; Bergles, D. E.; Jin, L.; Hoberg, M. D.; Vidensky, S.; Chung, D. S. et al. *Nature* **2005**, *433*, 73–77.
85. Hara, M. R.; Thomas, B.; Cascio, M. B.; Bae, B.-L.; Hester, L. D.; Dawson, V. L.; Dawson, T. M.; Sawa, A.; Snyder, S. H. *Proc. Natl. Acad. Sci. USA* **2006**, *103*, 3887–3889.
86. McGeer, P. L.; McGeer, E.; Rogers, J.; Sibley, J. *Lancet* **1990**, *335*, 1037.
87. Ashburn, T. T.; Thor, K. N. *Nat. Rev. Drug Disc.* **2004**, *3*, 673–683.
88. Keith, C. T.; Borisy, A. A.; Stockwell, B. R. *Nat. Rev. Drug Disc.* **2005**, *4*, 71–78.
89. Spedding, M.; Jay, T.; Costa e Silva, J.; Perret, L. *Nat. Rev. Drug Disc.* **2005**, *4*, 467–478.
90. Powell, S. B.; Risbrough, V. B.; Geyer, M. A. *Clin. Neurosci. Res.* **2003**, *3*, 289–296.
91. Porsolt, R. D.; Anion, G.; Blavet, N.; Jalfre, M. *Eur. J. Pharmacol.* **1977**, *47*, 379–391.
92. Cryan, J. F.; Holmes, A. *Nat. Rev. Drug Disc.* **2005**, *4*, 775–790.
93. Porsolt, R.; Williams, M. 2005. http://www.xpharm.com/citation?Article_ID = 119961 (accessed March 2006).
94. Leaf, C. *Fortune*, Sept 19, 2005, pp 250–268.
95. *Economist*, Dec 20, 2005, p 109.
96. Heller, M. A.; Eisenberg, R. S. *Science* **1998**, *280*, 698–701.
97. Surowiecki, J. *New Yorker*, Dec 26, 2005, p 50.
98. Kesselheim, A. S.; Avorn, J. *JAMA.* **2005**, *293*, 850–854.
99. Paradise, J.; Andrews, L.; Holbrook, T. *Science* **2005**, *307*, 1566–1567.
100. Bussey, P.; Pisani, J.; Bonduelle, Y. In *Industrialization of Drug Discery: from Target Selection through Lead Optimization*; Handen, J. S., Ed.; CRC Press/Taylor and Francis: Boca Raton, FL, 2005, pp 191–218.
101. Kenakin, T. *Pharmacol. Rev.* **1996**, *48*, 413–463.
102. Angers, S.; Salahpour, A.; Bouvier, M. *Annu. Rev. Pharmacol. Toxicol.* **2002**, *42*, 409–435.
103. Maudsley, S.; Martin, B.; Luttrell, L. M. *J. Pharmacol. Exp. Ther.* **2005**, *314*, 485–494.
104. Settleman, J. *Drug Disc. Today: Disease Mechanisms* **2005**, *2*, 139–144.
105. Vieth, M.; Sutherland, J. J.; Robertson, D. H.; Campbell, R. M. *Drug Disc. Today* **2005**, *10*, 839–846.
106. Pocklington, A. J.; Cumiskey, M.; Armstrong, J. D.; Grant, S. G. N. *Mol. Syst. Biol.* **2006**, *2*, 2006.0023.
107. Kenakin, T. *Nat. Rev. Drug Disc.* **2005**, *4*, 919–927.
108. Handen, J. S. In *Industrialization of Drug Discovery: from Target Selection through Lead Optimization*; Handen, J. S., Ed.; CRC Press/Taylor and Francis: Boca Raton, FL, 2005, pp 1–12.
109. Duyk, G. *Science* **2003**, *302*, 603–605.

110. Jennings, M. M.; Grossman, L. *The Wall Street Journal*, June 26, 1995, p A12.
111. Mullin, R. *Chem. Eng. News*, June 19, 2006, pp 30/55.
112. FDA: Critical path opportunities report. March 2006. http://www.fda.gov/oc/initiatives/criticalpath/reports/opp_report.pdf (accessed June 2006).
113. Parker-Pope, T. *The Wall Street Journal*, February 28, 2006, pp A1/A13.

Biography

Michael Williams received his PhD in 1974 from the Institute of Psychiatry and his Doctor of Science degree in Pharmacology (1987) both from the University of London. Dr Williams has worked for 30 years in the US-based pharmaceutical industry at Merck, Sharp, and Dohme Research Laboratories, Nova Pharmaceutical, CIBA-Geigy, and Abbott Laboratories. He retired from the latter in 2000 and served as a consultant with various biotechnology/pharmaceutical companies in the US and Europe. In 2003 he joined Cephalon, Inc. in West Chester, where he is Vice President of Worldwide Discovery Research. He has published some 300 articles, book chapters, and reviews and is Adjunct Professor in the Department of Molecular Pharmacology and Biological Chemistry at the Feinberg School of Medicine, Northwestern University, Chicago, IL from which vantage point he publishes his personal viewpoints on the drug discovery process.

2.12 How and Why to Apply the Latest Technology*

A W Czarnik, University of Nevada, Reno, NV, USA
H-Y Mei, Neural Intervention Technologies, Inc., Ann Arbor, MI, USA

© 2007 A W Czarnik. Published by Elsevier Ltd. All Rights Reserved.

2.12.1	**Introduction**	**295**
2.12.2	**Target Identification and Validation: Coupling Genotype to Phenotype**	**304**
2.12.2.1	Introduction	304
2.12.2.2	Target Identification/Validation	305
2.12.2.2.1	Genomics	305
2.12.2.2.2	Proteomics	314
2.12.2.2.3	Pharmacogenetics and pharmacogenomics	315
2.12.2.2.4	Bioinformatics/structural informatics/chemi-informatics	317
2.12.2.3	Conclusion	318
2.12.3	**Functional Genomics**	**319**
2.12.3.1	Introduction	319
2.12.3.2	Proteins: Structure and Function	319
2.12.3.2.1	Molecular evolution	320
2.12.3.2.2	Sequence and protein families	323
2.12.3.3	Molecular Pathways	324
2.12.3.3.1	Signal transduction	324
2.12.3.3.2	Protein–protein interaction	325
2.12.3.3.3	Tertiary structure and convergence	325
2.12.3.3.4	Gene expression and protein kinetics	326
2.12.3.3.5	Autoregulation of gene expression	326
2.12.3.4	Genomics Techniques	326
2.12.3.4.1	High-throughput screening	328
2.12.3.4.2	Microarrays	328
2.12.3.4.3	Single-nucleotide polymorphism detection	328
2.12.3.4.4	Deletion detection	329
2.12.3.4.5	Differential display by hybridization	329
2.12.3.4.6	Expression monitoring	329
2.12.3.4.7	Finding targets for drug development	329
2.12.3.5	Overview	330
2.12.4	**Integrated Proteomics Technologies**	**330**
2.12.4.1	Introduction	330
2.12.4.2	Gel Electrophoresis	331
2.12.4.2.1	Two-dimensional gel electrophoresis	331
2.12.4.2.2	Staining methods in two-dimensional electrophoresis	333
2.12.4.2.3	Imaging and image analysis	335
2.12.4.2.4	Protein spot excision	336
2.12.4.3	Protein Identification	336

*This chapter is adapted with permission from Mei, H.-Y. and Czarnik, A. W., Eds., Discovery Technologies; Marcel Dekker: New York, Basel, 2002. The contributors to this book were: Amon, L.; Beugelsdijk, T. J.; Blackledge, J. A.; Bradley, M.; Burke, J. W.; Cai, H.; Coassin, P. J.; Combs, A. P.; Czarnik, A. W.; Devlin, J. P.; Du, P.; Fromount, C.; Giuliano, K. A.; Haque, T. S.; Howard, B.; Kapur, R.; Keifer, P. A.; Kilby, G. W.; Kniaz, D. A.; Lebl, M.; Lepley, R. A.; Li, A. P.; Loo, J. A.; Ogorzalek Loo, R. R.; Macri, J.; Mathis, G.; McMaster, G.; Mei, H.-Y.; Morris, T. S.; Niles, W. D.; Nolan, J. P.; O' Hagan, D.; Olson, K. R.; Pomel, V.; Rapundalo, S. T.; Rohlff, C.; Rufenach, C.; Scheel, A.; Sterrer, S.; Stevenson, T. I.; Thompson, L. A.; Turner, R.; Upham, L. V.; Vasudevan, C.; Wang, J.; White, P. S.; Wierenga, W.; and Woo, E. S.

2.12.4.3.1	General concepts	336
2.12.4.3.2	Automated proteolytic digestion	337
2.12.4.3.3	Peptide mapping by matrix-assisted laser desorption/ionization mass spectrometry	339
2.12.4.3.4	High-performance liquid chromatography-mass spectrometry and tandem mass spectrometry	340
2.12.4.3.5	Computer-based sequence searching strategies	341
2.12.4.3.6	Other methods used for proteomics research	346
2.12.4.4	Conclusions	348
2.12.5	**Where Science Meets Silicon: Microfabrication Techniques and Their Scientific Applications**	**348**
2.12.5.1	Introduction	348
2.12.5.2	Traditional Methods of Microfabrication	349
2.12.5.2.1	Lithography and photolithography	349
2.12.5.2.2	Newer techniques and materials	353
2.12.5.3	System Integration Issues	355
2.12.5.3.1	Interfaces	355
2.12.5.3.2	Fluidics and electronic integration	355
2.12.5.3.3	Integral detection systems	356
2.12.5.4	Applications of Microtechnologies	356
2.12.5.4.1	Bead-based fiber-optic arrays	356
2.12.5.4.2	Deoxyribonucleic acid arrays	357
2.12.5.4.3	Electronically enhanced hybridization	357
2.12.5.4.4	Microfluidic devices	359
2.12.5.5	Nanotechnology	361
2.12.5.6	Overview	362
2.12.6	**Single-Nucleotide Polymorphism Scoring for Drug Discovery Applications**	**362**
2.12.6.1	Introduction	362
2.12.6.1.1	Single-nucleotide polymorphism applications	362
2.12.6.1.2	Single-nucleotide polymorphism discovery and scoring	362
2.12.6.2	General Considerations	363
2.12.6.2.1	Sample preparation	363
2.12.6.2.2	Sample analysis	363
2.12.6.2.3	Automation and multiplexing	363
2.12.6.3	Single-Nucleotide Polymorphism-Scoring Chemistries	364
2.12.6.3.1	Sequencing	364
2.12.6.3.2	Hybridization	364
2.12.6.3.3	Allele-specific chain extension	364
2.12.6.3.4	Single-base extension	365
2.12.6.3.5	Oligonucleotide ligation	366
2.12.6.3.6	Nuclease-based assays	366
2.12.6.3.7	Overview	366
2.12.6.4	Platforms	367
2.12.6.4.1	Electrophoresis	367
2.12.6.4.2	Microplate-based assays	367
2.12.6.4.3	Mass spectrometry	367
2.12.6.4.4	Flat microarrays	368
2.12.6.4.5	Soluble arrays	368
2.12.6.4.6	Overview	368
2.12.6.5	Conclusions and Prospects	369
2.12.7	**Protein Display Chips**	**370**
2.12.7.1	Introduction to Protein Chip Technologies	370

2.12.7.2	Surface Plasmon Resonance-Based Technology: How Optical Biosensors Revolutionized Protein–Protein Interaction Studies	371
2.12.7.2.1	Optical biosensor technology: technical background and development	371
2.12.7.2.2	Types of surface plasmon resonance applications and experimental considerations	372
2.12.7.2.3	Surface plasmon resonance in the drug development process: conclusion and outlook	374
2.12.7.3	Time-of-Flight Mass Spectrometry-Based Technologies	374
2.12.7.3.1	Matrix-assisted desorption techniques: technical background and development history	374
2.12.7.3.2	Surface-enhanced laser desorption/ionization: protein display chips for mass spectrometry	376
2.12.7.3.3	Surface-enhanced laser desorption/ionization applications: general principles and experimental considerations	376
2.12.7.3.4	Surface-enhanced laser desorption/ionization in the drug development process: a bright future	379
2.12.7.3.5	Biomolecular interaction analysis/mass spectrometry: connecting a popular surface plasmon resonance platform to matrix-assisted laser desorption/ionization-time-of-flight	380
2.12.7.4	Conclusion: Protein Display Chips in Drug Discovery – the Future is Here	380
2.12.8	**Integrated Proteomics Technologies and In Vivo Validation of Molecular Targets**	**380**
2.12.8.1	Proteomics in Molecular Medicine	380
2.12.8.1.1	Proteome terminology	381
2.12.8.1.2	Genome versus proteome	381
2.12.8.1.3	Changes in post-translational modifications associated with pathogenesis	381
2.12.8.1.4	New drug candidates directed against molecular targets modulating protein post-translational modification	383
2.12.8.1.5	Protein expression mapping	384
2.12.8.1.6	Cellular proteomics	385
2.12.8.1.7	Proteomic analysis of body fluids	386
2.12.8.1.8	Analyses of protein complexes	386
2.12.8.1.9	Combined mRNA/protein expression analysis for pathway mapping and candidate target selection	387
2.12.8.2	In Vivo Validation of Molecular Targets	388
2.12.8.2.1	Disease model relevance	388
2.12.8.2.2	Animal models of Alzheimer's disease	388
2.12.8.2.3	Animal models of cancer	388
2.12.8.2.4	Knockout and mutagenesis models, and tissue-specific inducible gene in vivo models	389
2.12.8.2.5	Gene-related functional proteomics	389
2.12.8.2.6	Pharmacoproteomics	390
2.12.8.3	Proteomic Technology in the Molecular Characterization of Novel Therapeutic Targets: Future Perspectives	390
2.12.9	**High-Throughput Screening as a Discovery Resource**	**391**
2.12.9.1	Background	391
2.12.9.2	Where We Are	393
2.12.9.3	Test Substance Supply	394
2.12.9.3.1	Compound library development	394
2.12.9.3.2	High-throughput organic synthesis applications	395
2.12.9.3.3	Natural products as a discovery resource	397
2.12.9.3.4	Structure-based design	398
2.12.9.4	Bioassay Development and Implementation	399

2.12.9.4.1	Fluorescence detection systems	400
2.12.9.4.2	Comparative functional genomics	401
2.12.9.4.3	Absorption, distribution, metabolism, and excretion, and toxicological profiling	401
2.12.9.5	Informatics	402
2.12.9.6	Management and Personnel Issues	403
2.12.9.7	Projections	403
2.12.10	**Fluorescence Correlation Spectroscopy and FCS-Related Confocal Fluorimetric Methods (FCS$^+$plus): Multiple Read-Out Options for Miniaturized Screening**	**404**
2.12.10.1	Introduction: A Rationale for New Read-Out Methods in Drug Discovery	404
2.12.10.2	Single-Molecule Confocal Fluorescence Detection Technology	404
2.12.10.2.1	FCS$^+$plus: multiparameter fluorescence read-out technology	405
2.12.10.2.2	Advantages of using FCS$^+$plus for miniaturized high-throughput screening	406
2.12.10.2.3	Case studies 1: using FCS$^+$plus multiparameter read-outs	407
2.12.10.2.4	Case studies 2: flexibility of multiple read-out modes	408
2.12.10.3	Overview	411
2.12.11	**Homogeneous Time-Resolved Fluorescence**	**411**
2.12.11.1	Introduction	411
2.12.11.2	Unique Properties of Homogeneous Time-Resolved Fluorescence Chemistry	412
2.12.11.2.1	Specific requirements of homogeneous assays	412
2.12.11.2.2	Lanthanide cryptates: a new type of fluorescent label	412
2.12.11.2.3	Homogeneous time-resolved fluorescence signal and measurement	414
2.12.11.3	Applications of Homogeneous Time-Resolved Fluorescence for High-Throughput Screening	416
2.12.11.3.1	Immunoassays	417
2.12.11.3.2	Enzyme assays	418
2.12.11.3.3	Receptor-binding assays	419
2.12.11.3.4	Protein–protein interactions	421
2.12.11.3.5	Nucleic acid hybridizations	422
2.12.11.4	Homogeneous Time-Resolved Fluorescence Assay Optimization	423
2.12.11.4.1	Buffer selection and stability	423
2.12.11.4.2	Choice of label for each assay component	424
2.12.11.4.3	Suggested assay controls	425
2.12.11.5	Overview	425
2.12.12	**Screening Lead Compounds in the Postgenomic Era: An Integrated Approach to Knowledge Building from Living Cells**	**425**
2.12.12.1	Introduction	425
2.12.12.2	Genomics and Proteomics as the Foundation for Cell-Based Knowledge	425
2.12.12.3	Creation of New Biological Knowledge	426
2.12.12.4	High-Content Screening of Target Activity in Living Cells	426
2.12.12.5	High-Content Screening of Drug-Induced Microtubule Cytoskeleton Reorganization	428
2.12.12.6	Specialized Fluorescent Reagents that Find Dual Use in High-Content Screening and High-Throughput Screening	428
2.12.12.7	Coupling High-Throughput Screening and High-Content Screening with Microarrays of Living Cells: The Cellchip System	431
2.12.12.8	The Cellomics Knowledgebase – A New Paradigm for Bioinformatics: Creation of Knowledge from High-Content Screening	433
2.12.12.9	Prospectus	436

2.12.13	**Miniaturization Technologies for High-Throughput Biology**	**436**
2.12.13.1	Introduction	436
2.12.13.1.1	Key integrated miniaturization technologies	437
2.12.13.1.2	Why is miniaturization critical to ultrahigh-throughput screening development?	437
2.12.13.2	The Screening Process	437
2.12.13.2.1	Assay development	437
2.12.13.2.2	Compound distribution	438
2.12.13.2.3	Loading the assay	438
2.12.13.2.4	Incubation	439
2.12.13.2.5	Detection	439
2.12.13.2.6	Results database	440
2.12.13.3	Plate Design and Performance	440
2.12.13.4	Microfluidic Technology	441
2.12.13.4.1	Sample distribution robot system	441
2.12.13.5	Solenoid-Based Dispenser Technology	444
2.12.13.5.1	The reagent distribution robot	444
2.12.13.6	Fluorescence Detection	445
2.12.13.7	High-Throughput Functional Biology	447
2.12.13.8	Overview	447
2.12.14	**Data Management for High-Throughput Screening**	**448**
2.12.14.1	Introduction	448
2.12.14.1.1	Background	448
2.12.14.1.2	Issues in high-throughput screening data management	449
2.12.14.2	Acquisition of Data	449
2.12.14.2.1	Integration with automation equipment	449
2.12.14.2.2	Integration with other information systems	450
2.12.14.2.3	Well types and plate layouts	451
2.12.14.2.4	Calculation, reduction, and normalization of data	451
2.12.14.3	Validation of Experimental Results	452
2.12.14.3.1	Detection of experimental error	452
2.12.14.3.2	False positives versus false negatives	453
2.12.14.3.3	Sources of error	453
2.12.14.4	Decision Support	453
2.12.14.4.1	Finding systematic error	453
2.12.14.4.2	Locating hits quickly	454
2.12.14.4.3	Integration of other data	454
2.12.14.4.4	Graphical display versus tabular display	454
2.12.14.4.5	Reporting	455
2.12.14.5	Data Analysis and Mining	455
2.12.14.5.1	Cross-assay reports	455
2.12.14.5.2	Visualization of data	455
2.12.14.6	Logistics	458
2.12.14.6.1	Plate handling	458
2.12.14.6.2	Special cases	459
2.12.14.7	Future Technologies	459
2.12.14.7.1	Higher densities and ultrahigh-throughput screening	459
2.12.14.7.2	Wider use of high-throughput screening techniques	459
2.12.14.7.3	High-throughput screening kinetics and high-content screening	460
2.12.14.8	Overview	460
2.12.15	**Combinatorial Chemistry: The History and the Basics**	**460**
2.12.15.1	Definition	460
2.12.15.2	History	461

2.12.15.3	Small Organic Molecules	462
2.12.15.4	Synthetic Techniques	462
2.12.15.5	Philosophy and Criteria	463
2.12.16	**New Synthetic Methodologies**	**463**
2.12.16.1	Introduction	463
2.12.16.1.1	Strategies for combinatorial library syntheses	464
2.12.16.2	Solid Phase Synthesis Methodologies	464
2.12.16.2.1	Introduction to polymer-supported synthesis	464
2.12.16.2.2	Synthetic transformations on solid supports	466
2.12.16.2.3	Resin-to-resin transfer reactions	474
2.12.16.3	Complex Multistep Synthesis on Solid Supports	475
2.12.16.3.1	Oligomers – natural and unnatural	475
2.12.16.3.2	Heterocycle/pharmacophore synthesis	479
2.12.16.3.3	Natural product total syntheses on a solid support	480
2.12.16.4	Solution Phase Synthesis Methodologies	482
2.12.16.4.1	Solution phase polymer-supported synthesis	482
2.12.16.4.2	Solution phase reactions involving scavenging resins	485
2.12.16.4.3	Solution phase reactions involving resin-bound reagents	488
2.12.16.4.4	Multistep synthesis with resin-bound scavengers and reagents	491
2.12.16.5	Conclusion	492
2.12.17	**Supports for Solid Phase Synthesis**	**493**
2.12.17.1	Resin Beads	493
2.12.17.1.1	Polystyrene gel-based resins	493
2.12.17.1.2	Macroporous resins	495
2.12.17.1.3	Polyethyleneglycol-containing resins	495
2.12.17.2	Synthesis on Surfaces	499
2.12.17.2.1	Pins and crowns	499
2.12.17.2.2	Sheets	502
2.12.17.2.3	Glass	503
2.12.17.3	Overview	506
2.12.18	**The Nuclear Magnetic Resonance 'Toolkit' for Compound Characterization**	**506**
2.12.18.1	Introduction	506
2.12.18.2	Basic Nuclear Magnetic Resonance Tools	506
2.12.18.2.1	One-, two-, three-, and multidimensional nuclear magnetic resonance	506
2.12.18.2.2	High-field nuclear magnetic resonance	508
2.12.18.2.3	Broadband decoupling	508
2.12.18.2.4	Spin locks	508
2.12.18.2.5	Shaped pulses	509
2.12.18.2.6	Indirect detection	510
2.12.18.2.7	Pulsed-field gradients	511
2.12.18.3	Specialized Nuclear Magnetic Resonance Tools	512
2.12.18.3.1	Biomolecular structure elucidation	512
2.12.18.3.2	Water (solvent) suppression	514
2.12.18.3.3	Hardware developments	515
2.12.18.3.4	Nuclear magnetic resonance of solid phase synthesis resins	519
2.12.18.3.5	Flow nuclear magnetic resonance	521
2.12.18.3.6	Mixture analysis: diffusion-ordered and relaxation-ordered experiments	526
2.12.18.3.7	Other probes: superconducting probes, supercooled probes, and microcoils	529
2.12.18.3.8	Combination experiments	529
2.12.18.3.9	Software	529
2.12.18.3.10	Quantification	529
2.12.18.3.11	Automation	530

	2.12.18.3.12	Research nuclear magnetic resonance versus analytical nuclear magnetic resonance	530
	2.12.18.4	Overview	530
	2.12.19	**Materials Management**	**531**
	2.12.19.1	Introduction	531
	2.12.19.2	Materials Management: A Definition	531
	2.12.19.3	Role of Materials Management in Drug Discovery	531
	2.12.19.4	Job Functions Involved in Materials Management	533
	2.12.19.5	Materials Management Work Processes	533
	2.12.19.5.1	Reagent sourcing	533
	2.12.19.5.2	Reagent receiving and tracking	534
	2.12.19.5.3	New compound registration and sample submission	534
	2.12.19.5.4	Master plate preparation and distribution for high-throughput screening	535
	2.12.19.5.5	Follow-up screening	535
	2.12.19.6	Materials Management Technologies	535
	2.12.19.6.1	Chemical structure-based reagent selection tools and products	535
	2.12.19.6.2	ERP systems, supplier catalogs, and electronic commerce	536
	2.12.19.6.3	Robotic weighing and liquid-handling workstations	536
	2.12.19.6.4	Materials management systems	536
	2.12.19.6.5	Automated sample storage and retrieval systems	537
	2.12.19.7	Future Directions	537
	References		**537**

2.12.1 Introduction

If one were to start the process of drug discovery de novo, one would probably not begin; it is an incredibly daunting task. If you look at what one is up against in terms of discovering and developing a drug, the challenges are almost overwhelming. There is a veritable physiological labyrinth for determining drug efficacy and safety. What does a pharmacological agent, whether it is a large or small molecule, have to go through to be of some benefit to an individual with some pathologic state? It must survive gastric pH and gastrointestinal enzymes if it is an orally administered drug. Absorption is a key issue: it has to get to the systemic circulation. There is also the problem of enterohepatic recycling. There are barriers – lymphatic barriers, endothelial barriers, blood–brain barriers, and blood–retina barriers – blocking the movement of the molecule to the appropriate site of action or pathology. All of these are very important challenges for a potential new drug. Many of these barriers do not have good laboratory models at the present time. And this does not even take into account what the body likes to do with xenobiotics; namely, to metabolize, conjugate, excrete, and eliminate them. Eventually, some of the drug must get to the right organ, cell type, and molecular target (enzyme, receptor) to effect a beneficial response. In so doing, it must have sufficient selectivity relative to potentially deleterious effects on normal tissue to yield an acceptable benefit/risk ratio. This physiological labyrinth often takes more than a decade of research time for a single drug to traverse, while, in parallel, many more drugs fail. There are yet other prerequisites for a drug. It has to be stable, of course. It has to have a decent shelf life, but not only as the bulk material on a large scale under a variety of conditions; the formulation must also be stable and compatible with human life. There has to be an economical source, because, in the end, if one cannot produce the drug in a reasonably cost-effective manner, it is not going to be available to the marketplace. Lastly, one must have the necessary analytical tools for determining the various physical and chemical properties of the molecule and its presence in various biological fluids. All these tools need to be made available for one eventually to have something called a drug.

How did we discover drugs? **Table 1** divides drug discovery into five categories. The two that are highlighted in bold – modifying the structure of known drugs and screening inventories of natural products (primarily) – represent the history of drug discovery in the 1950s, 1960s, and 1970s. The others – proteins as therapeutics, modifying the structure of natural substrates, and structure-based, computer-aided drug design – emerged in the past two decades or so, and are really beginning to make an impact only today. We will review each of these categories of approaches to drug discovery.

Table 2 shows a few examples of drugs created by the modification of known drugs. There are certainly many more, but these are a few examples from the Parke-Davis research laboratories of building on past leads and establishing

Table 1 How did we discover drugs?

- Modify structure of known drugs
- Screen inventories of natural products in laboratory models of diseases
- Proteins as therapeutics/vaccines
- Modify structure of natural substrate of enzyme or receptor
- Structure-based design computer-assisted drug design (CADD)

Table 2 Examples of drugs created by modifying the structure of known drugs

- ACE inhibitors
- Quinolone antibiotics
- Cephalosporin antibiotics
- κ-agonist analgesics
- HMG-CoA reductase inhibitors

Table 3 Screening: extensive history of natural products and microbial secondary metabolites

- Antibiotics
  - Macrolides
  - β-Lactams
  - Aminoglycosides
  - Tetracyclines

- Antifungals
  - Amphotericin
  - Griseofulvin

- Atherosclerosis
  - Mevinolin

- Hypertension/angina
  - Digoxin

- Anticancers
  - Vinca alkaloids
  - Anthracyclines
  - Etoposide
  - Mitomycin
  - Taxol

- Neurologics
  - Atropine
  - Cocaine

- Immunosuppressives
  - Cyclosporine
  - FK 506

improvements in a molecule based on a previously known drug: angiotensin-converting enzyme (ACE) inhibitors, quinolone antibiotics, cephalosporins, κ agonists, and 3-hydroxy-3-methylglutaryl coenzyme A (HMG-CoA) reductase inhibitors. In some cases, these examples were very successfully marketed; other cases were not commercially successful but resulted in the creation of effective drugs. Nonetheless, the modification of known drugs is a successful paradigm for drug discovery, and represents the vast majority of the research investment in the 1960s and 1970s in drug discovery.

In many ways, this approach was predated by natural products. Many of our early drugs came from plant sources or, later on, from *Streptomyces* or other microorganisms. **Table 3** is a nonexhaustive but representative list of natural products that were sources of drugs, discovered using a screening paradigm. In some cases, the knowledge of 'traditional medicines' and folklore contributed significantly to the nonrandom sourcing for the eventual isolation and purification of the natural product. These secondary metabolites are still having a significant impact on medical therapy today. There are anti-infective products (antibiotics and antifungals), as well as products for the treatment of

Table 4 Screening: a more recent history of the chemical library

Receptor agonists	*Enzymes*

• G protein-coupled receptors
  ○ Substance P (NK-1)
  ○ Oxytocin and vasopressin
  ○ Bombesin/gastrin-releasing peptide
  ○ Bradykinin
  ○ Cholecystokinin
  ○ Neuropeptide Y
  ○ Adrenergic/serotinergic/dopaminergic
• Chemokine/lymphokine receptors
  ○ MCP-1
  ○ CCR-5
  ○ IL-8
• Gultamatergic receptors
  ○ AMPA
  ○ NMDA
  ○ MGlu
• Nuclear receptors
  ○ ER, PR
  ○ PPAR/RXR-RAR
• Integrins
  ○ gp llb/llla
	• Asp proteases
	  ○ Renin
	  ○ HIV-1
	• Ser proteases
	  ○ Thrombin
	  ○ Factor Xla
	• Matrix metalloproteases
	  ○ MMP 1–13
	• Kinases
	  ○ Cyclin dependent
	  ○ Tyrosine
	  ○ Ser/Thr
	• Phosphodiesterases
	• Polymerases/reverse transcriptases
	• Topoisomerases
	• Ion channels
	  ○ Calcium (L, N)
	  ○ Sodium
	  ○ Potassium/ATP

atherosclerosis. Mevinolin (also known as lovastatin) is a paradigm for very important treatments of dyslipidemias and atherosclerosis (i.e., HMG-CoA reductase inhibitors). There are also treatments for hypertension and angina. There is an expanding group of anticancer drugs, many of them based on original plant- or microorganism-derived secondary metabolites that exhibited cytotoxic activity in a number of tumor cell assays.

Finally, there are neurologics and immunosuppressives. The category of drugs comprising natural products and derivatives is extensive.

This sourcing of drugs from natural products has changed over the past 10–15 years, with increasing numbers of drugs coming from chemical libraries rather than from natural products. The reason for this is that over the last four decades, pharmaceutical companies and chemical companies have established significant inventories of small molecules. These inventories were not being exploited as potential sources of drugs because of the absence of broad-based, automated screening methods and biologically relevant assays. This changed dramatically in the last decade of the twentieth century. Today, one can come up with an extensive list of potential drug candidates arising from the screening of chemical libraries (**Table 4**), and this has become the basis for significant new technologies in the area of high-throughput screening (HTS) and combinatorial chemistry. So, how did one discover drugs in the past? To recapitulate, in large part, one did it by modifying the structure of known drugs or screening inventories of natural products.

With the advent of molecular biology in the late 1970s, proteins as therapeutics came onto the scene. While the list of protein-based drugs is not yet long, it is growing, and it represents a very important drug category (**Table 5**). These drugs are important not only from the point of view of medical therapy but also because of their economic impact. This particular list of 16 proteins represents currently over US $7 billion in sales annually. These proteins are having a dramatic impact on our industry, and they are having a dramatic impact on patient therapy. The fourth approach to drug discovery, the modification of natural substrates, has had less success in yielding new drugs. **Table 6** lists examples taken from research efforts at Parke-Davis. These include adenosine deaminase inhibitors, renin inhibitors, angiotensin receptor antagonists (a very large area of investment for many pharmaceutical research laboratories over the past 15 years), and cholecystokinin B (CCK-B) receptor antagonists. While this is another valid approach for drug discovery, one subset of this approach – peptidomimetics as drugs – continues to be characterized by poor pharmacokinetics and delivery-related challenges.

There was an interesting statement in the May 1991 issue of *Business Week*: "The old way – screening thousands of chemicals in a hit-or-miss search – is inefficient and wastes time. That is why it can now cost more than $200 million to bring one drug to market." One can debate that statement, but it is still the case that "Boger and others like him are carrying the flag for a new wave of research and development, often called rational drug design." While rational drug

Table 5 Biologicals (rDNA based)

Drug	Indication
Human growth hormone	Growth
tPA	Thrombolysis
Interferon-α/β	Cancer, viral diseases, multiple sclerosis
Insulin	Diabetes
Erythropoietin	Anemia
G-CSF and GM-CSF	Leukopenia
Recombinant hepatitis B vaccine	Hepatitis B
IL-2	Cancer
Recombinant hemophilic factors	Hemophilia
PDGF gel	Diabetic foot ulcers
α-Glucocerebrosidase	Gaucher's disease
TNF receptor/binding protein	Rheumatoid arthritis
Ab-EGF receptor	Cancer
DNase	Cystic fibrosis
Ab-gpIIb/IIIa	Thromboembolism/unstable angina
Ab-CD20 B cells	Lymphoma

Table 6 Examples of drugs created by modifying natural substrates

- Adenosine deaminase inhibitors
- Renin inhibitors
- Angiotensin receptor antagonists
- CCK-B receptor antagonists

design includes the approach of modifying natural substrates, it is most often thought of as structure-based and computer-aided drug design. This approach to drug discovery started in the early 1980s, and is now a significant skill set in many pharmaceutical companies.

The factor that drives this approach is specificity. The driving force for many in drug discovery is to find agents that are more specific for a particular molecular target because that should enhance efficacy, reduce side effects, and yield an agent that is unique to a particular pathology. However, the task is very daunting, as one can see by going down the list of molecular targets in **Table 7** and thinking about the implications. For instance, we are good at finding specific inhibitors of enzymes, but there are multiple classes, and even within classes there are isozymes. Similarly, with receptors, there are isoforms. Receptors are found not only in the plasma membrane but in the cytoplasm and nucleus as well. Ion channels have multiple subunits. The complications go on and on. There is, in each one of these vasopressin categories of potential molecular targets, a multiplicity of targets with, in many cases, a lot of similarity between those targets in terms of binding sites for the putative antagonist or modulator. There can be active sites and allosteric binding sites. Also, one must not forget other important carriers or modifiers to drugs in the biological milieu: serum proteins, cytochrome P-450 modification, glucuronidation, and sulfation, to name but a few. These all represent important factors in this aspect of specificity, and they must be taken into account in the area of rational drug design.

How, then, does one achieve specificity? Specificity is fundamentally molecular information. One builds in specificity by understanding the 'rules of recognition' between one molecule and another through the use of hydrogen bonds, ionic interactions, hydrophobic interactions, and, of course, the overall three-dimensional structure. If one analyzes our success in drug discovery so far, one can derive a 'specificity scale' (**Table 8**) relating the complexity of molecular targets to the amount of information one can incorporate into a potential inhibitor or binding agent. We have

Table 7 Specificity, The Holy Grail

- Enzymes: multiple classes, isozymes
- Receptors: multiple classes, isoforms
- Ion channels: multiple subunits and classes
- Intracellular transducers: kinases, phosphatases, G proteins, lipases, prenylases, ion-binding proteins
- Transcription factors
- DNA: nucleus, mitochondria, coding/noncoding, binding proteins, repair enzymes
- RNA: cytosolic (rRNA), nuclear (tRNA, mRNA)
- Serum proteins, cytochrome P-450s, glucuronidation/sulfation

Table 8 Specificity equals information

- Molecular information
  - Hydrogen bonds
  - Ionic bonds
  - Hydrophobic interactions (van der Waals' forces)
  - Three-dimensional structure

- Specificity Scale
  | Multimeric complexes (e.g., TFs)
  | Nucleic acids
  | Antibodies
  | Dimeric/domain recognition (e.g., signaling proteins, receptors)
  ↓ Enzymes

Table 9 Approaches to enhance drug specificitjy

- Proteins (IFN, IL, CSF, EPO, HGH, insulin)
- Peptides (ANP, LHRH)
- Antibodies
- Antisense
- Ribozymes
- Structure-based drug design

been most successful with enzymes in terms of specificity. On the other hand, transcription complexes, including transcription factors, activators, and repressors, often seven to nine components along with DNA in the milieu, represent a much more daunting challenge. They may very well epitomize the most complicated state so far for targets of drug discovery. In the middle tier are nucleic acids, antibodies, dimeric domain recognition kinds of targets, signaling proteins, and receptors. If one reviews the past decade of drug discovery progress, it parallels the specificity scale.

The challenge, then, is to build specificity into our drugs through the use of molecular information. Using proteins as drugs is one way of enhancing specificity (**Table 9**). By definition, you can build a lot of information into a large molecule, even if it has repeating units of amide bonds or, with an oligonucleotide, repeating units of phosphodiester bonds. There are related approaches to enhance specificity that have yielded to drug discovery in the last decade, such as antisense and ribozymes targeting RNA or DNA in the nucleus or the cytosol, having exquisite specificity for a particular sequence. Thus, structure-based drug design is increasing in importance as an approach to drug discovery by enhancing drug specificity.

However, one should not forget that nature has already shown us that small molecules can have sufficient information built into them to achieve specificity as drugs. Consider steroids, leukotrienes, nucleosides, mononucleotides, catecholamines, excitatory amino acids, or even something as simple as acetylcholine (**Table 10**). These are small-molecule messengers acting extra- or intracellularly, or both, and represent drugs with legitimate specificity in terms of their molecular targets. One does not necessarily need a molecular weight of 3000 in a drug to achieve appropriate selectivity or specificity. This provides continuing impetus that small molecules can, through various approaches, yield the degree of selectivity necessary to achieve the goal of being a drug for a particular disease. In spite of the efficacy of some of these newer approaches, however, one cannot forget screening because screening has

Table 10 Small-molecule messengers

- Steroids
- Leukotrienes
- Mononucleotides (cAMP, ADP)
- Catecholamines
- Acetylcholine
- Excitatory Amino Acids

Table 11 Promiscuous scaffolds or chemotypes

- Steroids
- Tricyclics
- Benzodiazepines
- Phenylpiperidines
- β-Hydroxyalkylamines
- Sugars
- Macrolides
- Cyclopeptides

advanced in speed, quality, and in diversity of sources and targets. Over the past several years, the industry has moved from high throughput to ultrahigh throughput. Today, one can set up a screen and go through a library of 100 000 compounds in a day, even with a cell-based screen. Rapid, automated, high-volume techniques and instrumentation, as well as novel sources for diversity, biological as well as chemical, and the generation of molecular targets through biotechnology, have revolutionized screening as an important paradigm in drug discovery.

In a parallel fashion, another translational technology for enhancing screening as a drug discovery tool is the explosion in diversity of screening sources, namely through combinatorial libraries. Many libraries have been generated, including those for peptides, monoclonal antibodies, oligosaccharides, and oligonucleotides. Over the past several years, an additional focus on small-molecule combinatorial chemistry has generated multiple additions of low molecular weight 'organic' compound libraries for HTS. Rapid generation of libraries is becoming routine. Indeed, we are finding certain chemotypes recurring as frequent 'hits' (**Table 11**).

The power of this technology has been brought to bear, together with modification of natural substrate or structure-based drug design in this decade. If one looks back to 1990, one can find hardly any small molecules that bound to G protein-coupled receptors (GPCRs) and demonstrated appropriate agonist or antagonist activity. Within the past decade, this field has exploded, in large part because of these technologies. For example, some early work from the GlaxoSmithKline laboratories used ligand-based modification to generate cyclopeptides based on the natural ligand vasopressin, and GlaxoSmithKline scientists were able to generate nanomolar level antagonists targeting the V_1 receptor.[1105] Concomitantly, scientists at Otsuka Pharmaceuticals in Japan were actually screening a chemical library for V_1 antagonists, and came up with a compound with lower molecular weight than the cyclopeptide on which the GlaxoSmithKline scientists had been working.[1106] (Scientists at Yamanouchi had a parallel effort.) This represented an early example of a small molecule derived via screening that could act as an appropriate antagonist with reasonable selectivity, in comparison with the rather complicated natural ligand vasopressin or the cyclopeptides from GlaxoSmithKline.

In a similar fashion, discoveries were made regarding the related receptor oxytocin (OT). It had been shown that meperidine bound the OT receptor extremely weakly, but by modifying the structure, scientists generated a small molecule, where the natural ligand was a cyclopeptide for the OT receptor and, again, exhibited quite high selectivity.[1107] A natural product that had high selectivity for the OT receptor was found by screening natural product libraries in parallel research efforts at Merck.[1108]

The relatively huge investment in angiotensin (AT) receptor antagonists really began from an initial disclosure of a Takeda Pharmaceutical patent on a small molecule that was based on a two-amino-acid component structure of angiotensin. From this came a plethora of AT1 receptor antagonists from many laboratories. At Parke-Davis a different receptor was being screened, and it was found that a structurally unrelated low molecular weight compound was quite specific for the AT2 receptor. This turned out to be an interesting lead in terms of understanding the pharmacology of the AT2 receptor relative to AT1. Again, the screening of small molecules in combination with structure-based design approaches generated interesting potential drug candidates.

Some laboratories (e.g., Parke-Davis) took a reductionist approach to looking for low molecular weight compounds that would act as agonists or antagonists at the CCK receptors. This approach led to the conclusion that Trp and Phe were the two key recognition sites in a portion of the natural ligand for binding to the CCK-B receptor. Parke-Davis scientists were able to generate a 2 nM level antagonist at the CCK-B receptor; this agent eventually went into clinical trials (**Figure 1**).[1109,1110] This agent has a molecular weight of about 600, which is, of course, much smaller than the natural ligand itself. This is an approach exemplifying structure-based ligand modification to generate a drug candidate. In a parallel fashion, drug discovery efforts generated low molecular weight (benzodiazepine-like) CCK-B antagonists at several other pharmaceutical companies. These molecules came out of a screening approach based on finding a natural product, asperlicin, that exhibited high binding to the CCK-B receptor. So, a combination of approaches yielded potential drugs that were evaluated in the clinic against several disease states.

Structure-based, computer-aided drug design, the last category of the five approaches, is an area of increasing investment for many groups in drug discovery. **Table 12** lists four examples from the Parke-Davis laboratories. These approaches illustrate the increasing importance of structure-based drug design in the area of drug discovery. The HIV-1 aspartyl protease and its dimeric structure has really become the first major success story of this approach. We learned much in the early 1980s about inhibitors of a related aspartyl protease, renin, and applied this science to the first generation of inhibitors of HIV-1 protease. The industry now has a number of inhibitors that are approved, on the market, and being used, representing a significant advance in the treatment of HIV-positive individuals. Many share a key element, which is a hydroxyl group that is involved in the binding site to water and aspartic acid in the recognition site of the dimeric aspartyl protease. They are elongated peptide-like structures, and represent a challenge to achieving acceptable bioavailability and drug–drug interactions.

Second-generation agents (**Figure 2**) are under clinical evaluation, and represent smaller versions of the first-generation molecules, or even radically modified or divergent forms. The drug PNU 140690 was disclosed from Pharmacia and Upjohn, and has little structural relationship to the other inhibitors. It does have a close structural relationship to a series discovered through parallel research in the Parke-Davis laboratories. These compounds represent the combination of structure-based drug design with an iterative cycle of modeling, synthesis, testing, and crystallography, to come up with improvements in specificity, selectivity, and bioavailability.

In early 1992, through screening of the compound library at Parke-Davis, a particular pyrone, shown in **Figure 3**, was found to be a very modest, but nonetheless reproducible, inhibitor of HIV-1 protease. It was selective for HIV protease over other aspartyl proteases and related families. An iterative design cycle was begun, consisting of the structure

	CCK-B binding (K_i)
Asp-Tyr(SO$_3$H)-Met-Gly-Trp-Met-Asp-Phe (CCK 26-33)	3 nM
Trp-Met-Asp-Phe (CCK 30–33)	3 nM
Boc-Trp-Phe	70 μM
Boc-αMe-Trp-Phe	6 μM
CI-988	2 nM

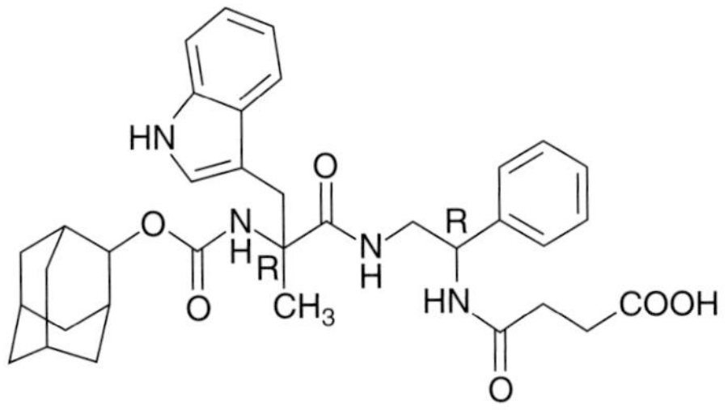

Figure 1 Peptidomimetic cholecystokinin antagonists.

Table 12 Examples of structure-based, computer-aided drug design

- HIV-1 protease inhibitors
- Thymidylate synthetase inhibitors
- M$_1$ selective muscarinic agonists
- Purine nucleoside phosphorylase inhibitors

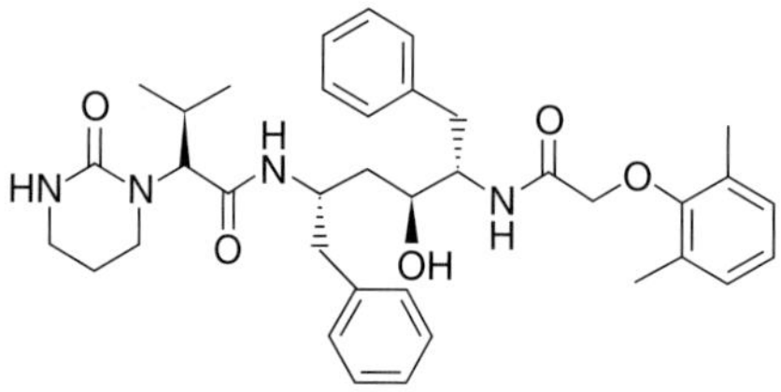

ABT 378

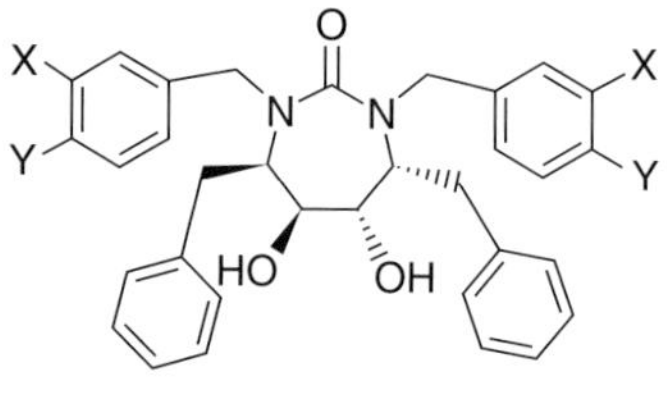

DMP323: X=H, Y=CH_2OH
DMP450: X=NH_2, Y=H

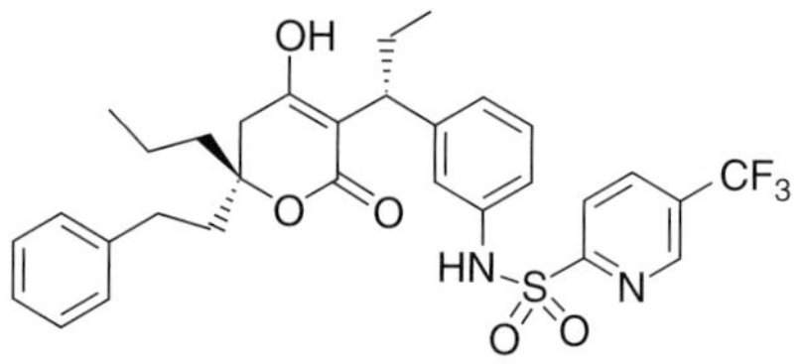

Tipranavir
PNU140690

Figure 2 Second-generation HIV-1 protease inhibitors.

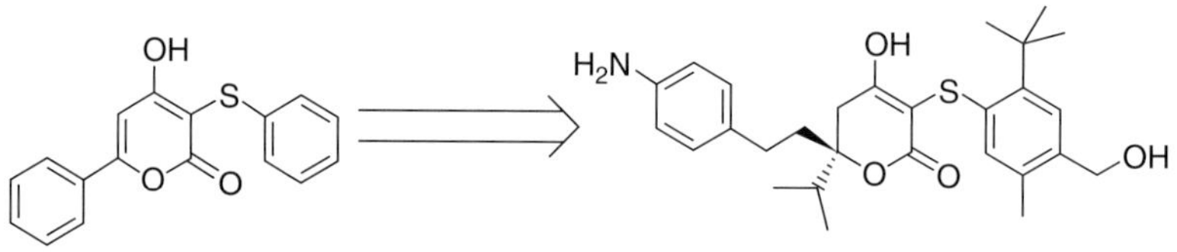

Mass screen hit
K_i = 11 700 nM

PD0178390
K_i = 0.028 nM
>410 000 fold increase

Figure 3 Third-generation, nonpeptidic HIV protease inhibitors.

determination (using x-ray crystallography, co-crystallization, and soaking experiments with additional inhibitors), modeling, additional chemistry to modify the structure based on what had been learned from modeling, and looking at the requisite biochemistry and enzymology of the agents. This cycle was repeated to determine the appropriate binding of these molecules in the three-dimensional structure using structure-based drug design. Parke-Davis scientists were able to alter this relatively simple pyrone with a series of modifications, principally in the P2 and P2-prime sites, which very rapidly led to nanomolar level inhibitors and represented legitimate advances in nonpeptide-based HIV protease inhibitors. This has now been extended to an interesting molecule whose IC_{50} is 5 nM. It has a different structure, but nonetheless binds in the active site, does not use water within the binding site, and is quite different from the first- and second-generation inhibitors.[1111] This demonstrates the power of structure-based drug design, coming originally from a screening approach, to yield a drug candidate lead.

In the muscarinic area, quite a different approach was taken. Parke-Davis scientists had determined that there was a three-point pharmacophore based on a series of agonists that to bind to the M_1 receptor; however, the pharmacophore analysis had gotten bogged down at that point. As this had been modeled within the GPCR model, where it was binding was considered, and the other muscarinic receptors (M_2 or M_5) were looked at in terms of differences in the transmembrane region. It was hypothesized that, not unlike retinal binding to rhodopsin, if this very small agonist was binding in the transmembrane region, this simple pharmacophore could be elongated (much like retinal is extended) to generate something that would have greater specificity. In fact, the basic pharmacophore was able to be extended to produce a molecule that has much greater specificity for the central versus the peripheral muscarinic receptors. This compound is now under investigation in clinical studies targeting the treatment of patients with Alzheimer's disease (AD). Thus, this original idea of molecular modeling and pharmacophore analysis is in fact a validated approach to drug discovery.

How will we discover drugs in the future? While I believe that all five categories will be used (see **Table 9**), modification of known drugs (validated chemotypes) will probably be of decreasing interest over time. The integration

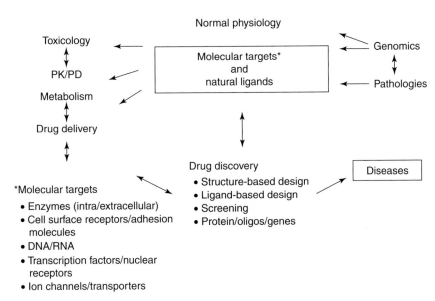

Figure 4 New drug discovery paradigm.

of the other four approaches will become more important in the future application of drug discovery. The targets are manifold. **Table 4** is a highly simplistic representation of the source of molecular targets that we currently spend so much time identifying and finding antagonists or agonists for. They can be extracellular, membrane-based receptors, nuclear receptors, or cytosolic signaling receptors. There is a growing list of targets, such as serine/threonine kinases, tyrosine kinases, and phosphatases, as well as a significant interest in the cell cycle and understanding the regulatory enzymes involved in the cycle. Probably the newest frontier is transcription factors, where many are working on ways of finding agents that would regulate gene expression and affect some particular pathological state.

The technological foci for drug discovery, shown in **Figure 4**, are organized for a specific reason. The technologies of HTS, and biomolecular structure determination, molecular modeling, and combinatorial chemistry are having a dramatic impact on drug discovery today, and will continue to become even more important in the future. However, they are very dependent on finding the molecular targets. A veritable avalanche of new targets will be coming via genomics and proteomics. Concomitantly, there is the critical aspect of managing the plethora of data that emerge as a result of these data-rich approaches, and understanding from those data which are the truths that are important in the drug discovery process.

Genomics is a simple word, but it envelops many components that are more than simply the genetic map, or the physical map, or even the gene sequence. The advances made in these three areas during the last decade has been striking, and much more rapid than any of us would have imagined. In many ways, gene sequencing, as well as the maps, will probably be well in hand for a number of genomes in the next couple of years. However, gene function and gene regulation represent a formidable challenge for drug discovery and the molecular sciences. Genetic functional analysis has a number of tools already available, and there will undoubtedly be more to come. Transgenics, knockouts, and gene replacement are very powerful technologies in our understanding of gene function. Antisense is already available, and of course the two-hybrid technique is being exploited in many laboratories investigating gene function. Synteny, differential display, and single-nucleotide polymorphism (SNP) analyses are additional tools. Nonetheless, this is a bottleneck, and improvements are needed before we can move forward from sequence to function and understand regulation. The challenges in this field will include sequencing, informatics, multiple species, and the fact that it is not only the natural state that we are interested in but the pathological state as well. We need to understand function and mutations relevant to the pathological state. The output is the genetic footprint. Disease phenotype is what we are interested in. It has implications for diagnostics as well as for drug discovery, and it has implications down the road for preventive medicine and gene therapy.

Proteomics is also an important area. In fact, this is here and now, not in the future. Two-dimensional gel electrophoresis, together with matrix-assisted laser desorption/ionization mass spectrometry and image analysis, is used to determine the output of genomics. This is an area of intense investment in many laboratories that must be included in the bioinformatics database that is being generated. This database will be used to help determine which targets are the appropriate ones for drug discovery or for diagnostics.

Lastly, there are some important components for which tools are slowly evolving. We need to optimize a drug candidate not only for selectivity but for bioavailability, toxicity, target organ selectivity, stability, and scalability. These are all legitimate and important components. We have drug class issues in drug discovery and development that act as guides for us, but nonetheless there are going to be discoveries unique to particular drug candidates. Small molecules continue to present difficulties with toxicity and bioavailability. Proteins have associated cost and delivery issues. Peptides have stability and bioavailability issues. Oligonucleotides often have the same class issues as proteins and peptides. Gene therapy certainly has to face the safety, delivery, and duration-of-effect issues.

What, then, are the future issues for drug discovery and development, given all of these technological foci? Determining the importance of the molecular targets is one such issue. For the next few years, we will continue to operate with a pretty tenuous linkage of molecular target to disease. In many cases, our molecular targets are hypothesized to be linked to a particular disease; we do not really determine this until we find inhibitors and take them to the clinic to see if they work. Also, the diseases we are facing today and tomorrow are more challenging than those that confronted us yesterday and the day before. Many are chronic diseases with multiple etiologic factors, and will probably require combination therapy – another complication to drug development. Of course, we will have novel toxicities. Finally, as we move further into the area of gene regulation, nuclear targets will represent additional complexities. The opportunities and the challenges will yield new drugs, combination therapies, and patient-specific treatments during the next decade.

2.12.2 Target Identification and Validation: Coupling Genotype to Phenotype

2.12.2.1 Introduction

There are three basic components necessary for drug discovery and lead identification: (1) targets, (2) screens, and (3) compounds. Each of these plays a pivotal role in the drug discovery process and, ultimately, the success of clinical trials. During the past 50 years there has been a paradigm shift in drug discovery (**Figure 5**), closely bound to the development of technology.[1] Initially, companies focused on the effects of putative therapeutics on whole-animal models to study pharmacology and efficacy. Later, tissue and cell cultures led to a refinement in pharmacological studies and in the number of screenings performed. More recent progress has involved advanced purification and molecular biology technologies, which in the past 10 years has led to a wholesale move toward selecting molecular targets to screen. Most recently, there has been an explosion of genomics, proteomics, bioinformatics, and information technologies, which will significantly increase the potential to identify and validate novel disease genes and their corresponding pathways, representing enormous value to the pharmaceutical industry. Of the approximately 80 000–150 000 human genes, there may be as many as 10 000 potential 'drugable' targets, that is, amenable to low-molecular-weight molecules (enzymes, receptors, ion channels, hormone receptors, etc.) for treatment of the most common multifactorial diseases.[2]

In addition to these novel disease targets, there are a significant number of partially validated targets associated with major human diseases, such as nuclear receptors and GPCRs.[3–6] These targets, combined with genomic technologies (including genetics, bioinformatics, data mining, and structural informatics) and compounds, offer potential for very selective drugs of the future. The best example of these are the so-called selective estrogen receptor modulators (SERMs), where a compound can be agonist in one cell type and an antagonist in another. In addition, one needs to consider that multifactorial diseases themselves have overlap (e.g., diabetes and obesity) and that single genes

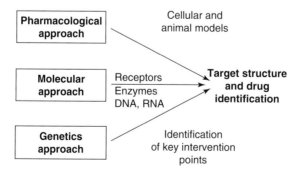

Figure 5 There has been a paradigm shift in the drug discovery process during the past 50 years. The pharmacological approach has used whole-animal and cellular models to discover drugs. Later, molecular targets were employed to screen for drugs. Most recently, genetics and genomics opened up a whole new approach to discover disease genes and pathways, creating enormous potential value to the pharmaceutical industry.

(e.g., chloride transporter) can be involved in the etiology of multiple diseases (e.g., cystic fibrosis, asthma, pancreatitis, and male infertility).

The process of 'gene to drug' will accelerate with the ever-increasing volumes of new genetic information, with increased biological understanding of how the pathways are connected, and with the technologies that make it possible for each step of the process to be completed more quickly and efficiently. The purpose of this chapter is to demonstrate to the reader which technologies can be used to identify gene targets and validate them to speed up the quality of drug discovery by providing an increased number of better validated targets.

2.12.2.2 Target Identification/Validation

2.12.2.2.1 Genomics

Thomas Roderick[7] coined the term 'genomics' in 1986. Genomics describes the study of a genome by molecular means distinct from traditional genetic approaches. The term is derived from 'genome,' a fusion of the words 'gene' and 'chromosome,' and is used to describe the complete collection of genes possessed by an organism.[8]

Genomics as currently practiced can be categorized as follows:

(1) structural genomics;
(2) functional genomics; and
(3) pharmacogenomics.

Genomics technologies, which are already more or less impacting drug discovery identification and validation, are as follows:

- positional cloning and association genetics;
- genome sequencing (humans, animals, and pathogens);
- expressed sequence tag (EST) sequencing;
- derivation of full-length cDNAs and their sequences (gene families);
- microarrays of oligonucleotides and cDNAs (chips);
- subtractive hybridization;
- differential display;
- antisense;
- in situ hybridization;
- serial amplification of gene expression (SAGE);
- Taq-Man/quantitative polymerase chain reaction (PCR);
- protein–protein interactions using yeast 2 hybrid techniques;
- reporter genes (e.g., green fluorescent proteins, GFPs);
- transgenic/knockout animals (especially mice);
- model organisms (e.g., yeast, *Caenorhabditis elegans*, *Drosophila*, zebrafish, mouse, and rat);
- two-dimensional gel electrophoresis; and
- databases (bioinformatics, structural informatics).

Presently, this collection of genomics technologies for drug discovery is used in variations or combinations of two approaches:

(1) the 'top-down' approach, which uses positional cloning and association genetics to identify disease-related genes; and
(2) the 'bottom-up' approach, which predominately focuses on the random sequencing of cDNA and an inference of the function of the protein product based on sequence similarity to well-characterized genes and proteins.

In the future, emerging technologies will further impact drug discovery target identification and validation. These are:

- integrated databases (structure to function);
- single-nucleotide polymorphism (SNP) identification/scoring in high-throughput mode;
- second-generation solid phase DNA arrays;
- single-cell PCR;

- laser capture;
- protein–protein and protein–compound interactions;
- in-solution RNA profiling technologies (liquid arrays);
- phage display libraries (antibodies);
- genetic suppressor elements (GSEs);
- zinc finger proteins;
- conditional knockout mice (Cre/lox);
- inducible promoters in vitro and in vivo (e.g., tetracycline on/off systems);
- gene trapping;
- biochips;
- microfluidics;
- in vitro and in vivo imaging;
- fiber optics;
- structures of membrane-bound proteins (e.g., GPCRs); and
- RNA as targets for infectious diseases

The approach of this chapter is to first describe the existing technologies and their impact on target identification and validation and then to turn to many of the emerging technologies to predict their potential impact on drug discovery.

2.12.2.2.1.1 Genetic mapping and positional cloning

One of the objectives of the Human Genome Project was to advance genetic mapping from the restriction fragment length polymorphisms used initially to microsatellite markers.[9] The accent now is on biallelic marker systems using SNPs, which are very prevalent within the genome.[10–12] This has allowed the cloning of many single genes causing human diseases inherited in a simple Mendelian fashion. Some of the most important genes controlling hypertension have been found by studying the molecular genetics of blood pressure variation in this way.[13,14] Recently, the ATP-binding cassette transporter 1 (ABC-1) was positionally cloned using Tangiers patients.[15] The ABC-1 transporter is involved in cholesterol efflux, and is linked to low levels of high-density lipoprotein and coronary heart disease. Furthermore, the implication of potassium channels in long-QT syndrome and epilepsy by their positional cloning[16] simply reinforces the power of genetics for biochemical understanding and for providing a genetic validation of a biological hypothesis.[17,18] It also illustrates that the comprehensive study of a gene family (e.g., potassium or sodium channels) by using data mining and genomics approaches in concert with genetic mapping is a paradigm for the study of more complex traits. It may also be a rapid way to find new 'genetically validated targets' for drug intervention.[19] Unfortunately, until now the unraveling of the biochemical pathology of common diseases in humans based on an understanding of complex, polygenic disease traits by the cloning of relevant genes has been somewhat disappointing. A number of linkages have been made for several diseases such as diabetes, obesity, asthma, and cognition, but very few of the genes have been uncovered.[20–22]

Positional cloning has also been used to isolate mouse genes, which cause Mendelian diseases in this species. Several of these genes have been instructive regarding similar human syndromes.[23] One of the best studied is leptin and the leptin receptor for obesity derived from analysis of ob/db mice.[24] The mouse has certainly become one of the most useful models of polygenic models of human disorders. Often, the phenotypic characteristics are strikingly similar to those for the analogous human disorder.[25] Generally, the predicted amino acid sequence homology ranges from approximately 80% to near-identity. Therefore, in some cases it is likely that drugs that interact with these gene targets using the mouse as the model will have similar physiological effects in humans, although the downstream signaling (e.g., the interleukin-4 receptor) may not be identical in the mouse and human. However, the mouse can be 'humanized' (i.e., the mouse gene exchanged for the human in a knockout/knock-in approach).[25]

In addition, there has been conserved organization of the gene order and relative position in the genomes of human and mouse for large chromosomal regions. This is known as synteny, and the ability to jump between human and mouse genetic information based on synteny is of great advantage in gene identification and localization.[26] The fundamental approach is to use conserved synteny to determine if the same genomic region is functionally involved in a disease phenotype in both human populations and the mouse.[27] If the same chromosomal region is involved, then it is generally easier to identify the gene in the mouse than in humans, and one can use mutation detection/analysis to determine if the same gene is involved in human polygenic disease. "Although several textbooks and other reference works give a correct definition, the term synteny nowadays is often used to refer to gene loci in different organisms located on a chromosomal region of common evolutionary ancestry,"[28] as just described. Originally, the term 'synteny' (or 'syntenic') referred to gene loci on the same chromosome regardless of whether or not they are genetically linked by classical linkage analysis. However, molecular biologists inappropriately use the original definition of synteny and its etymological derivation, especially as this term is still needed to refer to genes located on the same chromosome. "Correct terms exist: 'paralogous' for genes that arose from a common ancestor gene within one species and 'orthologous' for the same gene in different species."[28] To make a long story short, the process described above as synteny should be described as orthologous. Nonetheless, the approach is very powerful. Gene mapping of complex disorders has resulted in the identification of a number of cases that are clearly orthologous between humans and the mouse.[28,29] Furthermore, the similarity in gene structure between the mouse and humans is very high. This homology extends from the exon–intron organization of many genes to the predicted amino acid sequence of the gene products.[20]

One of the best examples of the power of genetics has been provided by studies on Alzheimer's disease (AD). All of the genes known to be involved unequivocally were found by positional cloning or by association genetics (i.e., the amyloid precursor protein (APP), presenilin (PS1 and PS2), and ApoE4 genes).[30] There is unlikely to be a pharmaceutical or biotechnology company investigating AD, which does not have a drug discovery program, based on one or other of these genes. It is also worth mentioning that access to transgenic animals is still an important component of the drug discovery process (see Section 2.12.8.2). In AD, for example, various transgenic mice strains have been used to examine amyloid deposition. Genetic crosses have shown that ApoE4 and transforming growth factor β1 influence amyloid deposition, and that the presenilins act synergistically with APP in the development of the pathology. Some of these animals may become models for the disease, with utility for finding compounds that modify the pathology of plaque formation.[31] None of this would have been possible without having the genes in hand. The lesson is simple: genetics is the key to mechanism; mechanism is the key to therapeutic discovery; and therefore genetics is the key to therapeutic discovery.

2.12.2.2.1.2 Genome sequencing

At present, the greatest impact of genomic research has come from DNA sequencing projects.[32] One of the first applications of genomics to drug discovery was through the development of EST sequencing and the creation of large gene sequence databases.[20] This approach to generating large-scale DNA sequence information was carried out by Craig Venter at TIGR and by Human Genome Sciences. These initiatives and the development of commercial EST sequence databases and data-mining tools by companies such as Human Genome Sciences and Incyte have had a rapid effect on drug discovery, by giving pharmaceutical companies access to potentially new targets related to previously known ones. The best (and most obvious) examples are targets such as GPCRs, steroid hormone receptors, ion channels, proteases, and enzymes. The value from this approach is considerable, although it is not always clear what disease a relative of a known target might be useful for without further validation. Any antagonist or agonist of such a target can, however, be used in animals to rationalize or validate the protein as a target for intervention in a particular disease. A reasonable example would be the development of a novel serotonin receptor of active small molecules working via a new receptor.[33] Evidence of the anticipated impact on drug discovery of such sequence databases is clear from the rapid take-up of subscriptions to the Incyte (ESTs) and Celera (genomic) databases by many of the major pharmaceutical companies.

Nowhere has comprehensive genome sequencing had more effect than in bacterial genetics. The sequences of well over 20 different bacterial genomes are now available.[34–36] The information derived from the comparisons of these sequences within and between microorganisms and with eukaryotic sequences is truly changing microbiology. The greatest benefit from these approaches comes from being able to examine homologies and gene organization across whole genomes at once. For target identification this has obvious benefits, because if the potential target is conserved among target organisms but not in the host, then it may be a reasonable target. This is not new thinking; the development of dihydrofolate reductase inhibitors and some antifungal drugs followed exactly the same rationale. In the bacterial studies, procedures such as 'signature-tagged mutagenesis' and in vivo expression methods are tying together genomic and biological data to reveal genes permissive for virulence and pathogenicity.[37,38] Recently, the smallest of all genomes, that of *Mycoplasma genitalium*, which has only 517 genes, was sequenced at Celera. The big surprise was that only 300 of the 517 genes are essential for the life of the organism, and 103 of these 300 essential genes are of unknown function.[39] This tells us what is in store for the future when we consider that the human genome is estimated to have some 80 000–150 000 genes. Still, this approach will be valid for higher eukaryotes too, once the genomes of the mouse and humans have been fully sequenced and assembled.[40]

Genomics is made possible through the large-scale DNA sequencing efforts of many public and private organizations, including the Human Genome Project, which has provided major impetus to the discovery of the genome and information technologies.[9] This international program to determine the complete DNA sequence (3000 million bases) was first completed in draft form in 2001 through the combined efforts of a very large number of researchers.[180,1112]

It is fair to say that early estimates were, in many respects, roughly on target, though conservative. (It was originally projected that by 2002 a comprehensive working draft (90% of the genome) should be expected, and that by 2003 the entire sequence would be completed.[41]) Recently, chromosome 22 has been completely sequenced as part of the Human Genome Project. Chromosome 22 is especially rich in genes. While sequencing the 33 million base pair chromosome, Dunham et al.[42] identified 679 genes, 55% of which were previously unknown. Approximately, 35 diseases have been linked to mutations in chromosome 22. These include immune system diseases, congenital heart disease, and schizophrenia, among others.

The purpose of structural genomics is to discover, map, and sequence genetic elements. Annotation of the sequence with the gene structures is achieved by a combination of computational analysis (predictive and homology-based[1090]) and experimental confirmation by cDNA sequencing. Recently, a website entitled the 'Genome Channel'[1091] has offered a road atlas for finding genes by zooming in from chromosomes to annotated DNA sequences. Furthermore, the site links the user to sequencing centers for 24 organisms, including humans, mice, and *Escherichia coli*, to name a few. Detecting homologies between newly defined gene products and proteins of known function helps to postulate biochemical functions for them, which can then be tested. Establishing the association of specific genes with disease phenotypes by mutation screening, particularly for monogenic/single disorders, and provides further assistance in defining the functions of some gene products, as well as helping to establish the cause of the disease. As our knowledge of gene sequences and sequence variation (*see* Section 2.12.2.3) in populations increases, we will pinpoint more and more of the genes and proteins that are important in common, complex diseases.[41] In addition, by comparing corresponding genomic sequences (comparative genomics) in different species (man, mouse, chicken, zebrafish, *Drosophila*, *C. elegans*, yeast, *E. coli*) regions that have been highly conserved during evolution can be identified, many of which reflect conserved functions, such as gene regulation. These approaches promise to greatly accelerate our interpretation of the human genome sequence.[43] In the future, it will be possible to understand how specific sequences regulate the expression of genes in the genome.

2.12.2.2.1.3 Functional genomics

Complete sequencing of the human genome is really only the beginning. As the geneticist Eric Lander has characterized, "Molecular and cell biologists are still only approaching a phase of development of their discipline reached by chemists 100 years ago, when the periodic table was first described."[44] We still need to link the genetic makeup of an organism (its genotype) to its form and function (its phenotype) and how the environment effects them. To that end, whereas structural genomics seeks to discover, map, and sequence these genetic elements, functional genomics is the discipline that seeks to assign function to genetic elements. A more detailed understanding of the function of the human genome will be achieved as we identify sequences that control gene expression. In recent years, our knowledge of gene sequence has increased massively, principally due to large-scale cDNA and genome sequencing programs.[41] The availability of this information resource has fueled efforts to develop ways of analyzing gene expression systematically, and as a result there are a range of approaches available that allow parallel analysis of a large number of genes. These tools can provide a comprehensive view of the genes expressed in samples of tissue and even individual cells, and in so doing will advance our understanding of biochemical pathways and the functional roles of novel genes. Most importantly, genome sequences will provide the foundation for a new era of experimental and computational biology, providing the essential resources for future study.

Genes are sections of genomic DNA that encode proteins. They are copied into messenger RNA (transcribed), and it is this mRNA that carries the genetic code from the nucleus to the cytoplasm to direct protein synthesis. It is thought that there are 80 000–150 000 genes in the genome of humans and other mammals. In each cell, only a portion of these genes is active at any one time, thought to be in the region of 10 000–15 000. The expression of these genes instructs the protein synthetic machinery to produce a specific set of proteins that are required for the cell to perform its normal functional role. Certain genes are expressed in all cells all of the time, and encode so-called housekeeping proteins, whereas others are expressed only in certain cell types and/or at certain times. It is these latter proteins that give a cell its unique structural and functional characteristics and, ultimately, make one cell different from another. However, the complement of genes expressed by a cell is not a fixed entity, and there are many genes whose expression can be induced or reduced as required This provides the flexibility in biological systems to respond and adapt to different stimuli, whether they are part of the normal development and homeostatic processes, or a response to injury, disease, or drug treatment. As the transcription status of a biological system reflects its physiological status, the ability to study the complement of genes expressing its 'signature' and the abundance of their mRNAs in a tissue or cell will ultimately provide a powerful insight into their biochemistry and function.[45]

However, there are certain inherent difficulties in the study of gene expression. Unlike DNA, which is essentially the same in all cells of an individual, there can be enormous variation in the abundance and distribution of a particular mRNA species between cells. Some genes are highly expressed, that is, their mRNA is abundant (>1000 copies per cell), whereas other genes are weakly expressed, with their transcripts present at only a few copies per cell. In addition, because most tissues are composed of distinct cell populations, an mRNA, which is only present in one of those cell types, perhaps already at low levels, becomes even rarer when the RNA is extracted from that tissue as it is diluted in the RNA derived from nonexpressing cells. It is also becoming increasingly apparent that for many genes the transcripts can exist in different forms, so-called alternative splice variants. These splice variants allow an even greater diversity in the complement of proteins that can be generated from the genetic code, but adds another level of complexity to the analysis of gene expression. Furthermore, these mutations of splice sites can lead to reduction of a biological function (e.g., sodium channels[19]).

Knowledge of the sequence of a novel gene alone usually provides few clues as to the functional role of the protein that it encodes. However, sequence information can be used for further characterization of the gene, and is a good starting point in defining its expression. If the expression of a gene is limited to certain tissues or cell types or is changed during disease, then we can begin to postulate and focus in on those sites. Likewise, if the expression of other genes maps to the same cells, we can begin to understand which protein complex or biochemical pathway the gene product might interact with.[45]

The other major goal of RNA expression profiling has always been the identification of genes that are expressed at different levels between one system or experimental paradigm and another. Knowledge of these genes not only sheds light on the biochemical events underlying the change but in some cases also provides a list of potentially interesting genes, such as which genes are expressed in malignant but not in normal cells. For these reasons, expression profiling will help in our understanding gene function and the biology of complex systems, as well as in many aspects of the drug discovery process.[45]

The past several years have witnessed the development of a plethora of new methodologies to study gene expression:

- subtractive hybridization[46];
- subtractive PCR[47,1092];
- differential display[48];
- in situ PCR[49,50];
- single-cell PCR[51,52];
- SAGE[53,1093];
- DNA chips[54,1094]; and
- microarrays.[55,1095]

These all can provide insights into the complement of genes expressed in a particular system. However, what really makes the difference and ultimately determines how widely they are used in the future is the detail of how they work and quality of data they yield. We also need to distinguish what is important in (1) the discovery of novel genes (e.g., subtractive hybridization, subtractive PCR, and SAGE), (2) large-volume RNA profiling of known genes (e.g., PCR, DNA chips, and microarrays), and (3) the expression of genes occurring in a subcellular region of a tissue (in situ hybridization and in situ PCR).

The main issues in expression profiling are therefore as follows:

- Novel versus known genes – is the objective to discover novel genes or to evaluate known genes in a high-throughput mode?
- Localization within a tissue – can the technology distinguishes expression at the cellular level within a tissue?
- Sensitivity – can the method detect low-abundance sequences and how much starting material (mRNA) is required?
- Specificity – is the assay highly specific for the transcript of interest?
- Quantification – how accurately can the technique measure mRNA abundance?
- Reproducibility – how robust is the methodology?
- Coverage – how many transcripts can be analyzed at once and does the approach have the ability to detect previously uncharacterized genes?
- Redundancy – how often is each transcript sampled? (Some systems have the potential to analyze the same mRNA a number of times, therefore increasing the complexity of the data.)
- False positives – how often do things appear to be differentially expressed but turn out not to be?
- Scale of analysis – is the approach amenable to HTS data output and does the assay give results that are easy to interpret?
- Cost – how much does it cost for the initial investment and cost/gene/assay?

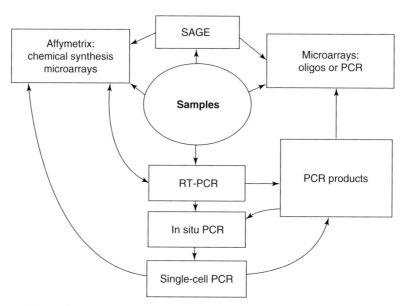

Figure 6 Numerous RNA-profiling technologies are available, which are interdependent. Depending on the question posed, the experimenter may choose to use a specific combination of technologies. SAGE provides the best approach to discover and quantify novel and known genes. However, the throughput is lower than that of Affymetrix DNA chips or microarrays. Affymetrix chips are convenient to use, but cost more than microarrays and are less flexible for following up on genes of interest in a high-throughput mode. PCR microarrays are more sensitive than oligomicroarrays; however, PCR microarrays cannot detect splice variants such as oligomicroarrays.

Depending on the question the experimenter is asking, one or a combination of these technologies will be needed. This is outlined in **Figure 6**, which illustrates that any one RNA sample may require use of one of these technologies (e.g., SAGE) to detect novel and known genes at low throughput and thereafter require the use of DNA chips or arrays for high-volume throughput and reproducibility testing. In situ hybridization, in situ PCR, and/or single-cell PCR would test the most important genes identified and validated by chips and arrays.

As mentioned above, though the methodologies have been widely used in the research environment, not all are applicable for high-throughput, routine analysis of gene expression. However, one approach above all others shows most promise in this respect, and is now driving the field of expression profiling: the use of DNA chips and microarrays. Although there are some important differences between DNA chips and microarrays, both work by a similar mechanism (i.e., the hybridization of complex mixtures of DNA or RNA to complementary DNA probes immobilized on a solid surface). DNA chips are presently available through one company, Affymetrix who pioneered this approach.[54] Affymetrix makes microchips of overlapping oligonucleotides ('oligos') available, representing sequences from thousands of genes. These chips can also be used to measure sequence variation and for the detection of SNPs. One of the perceived disadvantages of the Affymetrix format is its lack of flexibility.

Other chips are available from Synteni and elsewhere that comprise cDNAs or oligos[56] covalently attached to a solid phase. Oligos have an advantage over cDNAs in that oligos can detect splice variants, which can make very interesting drug targets. These chips can be interrogated by using fluorescently labeled hybridization probes, and the results displayed by color integration. By far the most impressive use of such chips has been carried out in Pat Brown's laboratory at Stanford University. Two elegant experiments have been done. The first compared the pattern of yeast gene expression during sporulation,[55] and the second looked at the expression of a subset of the genes expressed in fibroblasts after serum starvation and activation. The take-home messages from these beautiful experiments are as follows: the reproducibility of the biology of the system is crucial for consistent results; (2) the need for sophisticated pattern matching is absolutely essential for data interpretation.[57,58] The function of an unknown gene based on its pattern of expression under different conditions is compared with a known data set. From the experiments reported above, the functions of many 'unknown' yeast and human genes were suggested, based on their expression characteristics relative to known sets such as cell cycle genes or genes involved in DNA replication. Now that the *C. elegans* and *Drosophila* genomes are completely sequenced,[59] it will not be long before worm and fly gene chips are available for the same kind of experiments. In fact, the first signs are here; White *et al.*[60] published an elegant paper on the microarray analysis of *Drosophila* development during metamorphosis, where both known and novel pathways were assigned to metamorphosis.

So what role will RNA profiling have in the future of drug discovery? High-throughput RNA expression profiling is still in its infancy, and clearly much remains to be done to apply it for maximum impact on drug discovery. We have yet to identify all human genes and, to an even greater extent, those of model organisms (the mouse and rat). There is also considerable room for improvement on the basic technologies and methodology. Judging by the number of recent small start-up companies and new emerging industries (e.g., Motorola, HP, Corning, and IBM), one of the most active areas of genomics is the use of microarrays for the large-scale measurement of gene expression.[61] In addition, second-generation technologies (e.g., Corning, Third Wave Technologies, Luminex, Curagen, and QuantumDot) are underway. These second-generation technologies are expected to (1) design better surfaces, (2) produce more sensitive dyes, and (3) develop totally new approaches such as the in-solution assays (liquid arrays), e.g., 'Invader' technology,[1096] or by exploiting electronics by designing digital micromirrors.[62]

The ability to examine the expression of all or a large number of genes at once will provide new insights into the biology of disease, such as in the identification of the differences between normal and cancerous cells,[63] as well as gene expression during aging[64] or during HIV infection.[65] Undoubtedly, there must be gene products that play a crucial role in metastasis that could be targeted by drug treatment. However, working out which those are, which cancers they are active in, and their specificity to tumor cells should now be possible. Golub *et al.*[63] have demonstrated that molecular classification of cancer types could be predicted by gene expression profiling. Furthermore, this information can be integrated into the genetics of cancer and histopathology building a three-dimensional model.[66] Companies such as LifeSpan provide disease and normal tissue banks of more than 1 million samples or a 'customized approach' to genomics. In addition, Velculescu *et al.*[67] released approximately 3.5 million SAGE transcript tags from normal and diseased tissues.

Indeed, expression profiling is a powerful research tool for analyzing the differences between tumor types and for distinguishing between tumors, which up to now have been classified based on their morphological characteristics and presence of a few markers. The recently developed technology 'laser capture' should also help to better define tumor types in vivo in combination with chips[68] and how they relate to mouse xenografts, not to forget just defining differences or reliable markers in tumor cell lines for in vivo studies. Another example is the case of genetic disorders that have a complex multigenic basis but that can be mapped back to so-called quantitative trait loci. It has been recently demonstrated how microarrays can be used alongside classical genetic approaches to identify the genes underlying the defect.[69]

Genome-wide arrays also have more direct application in drug discovery. It is possible to use them to profile the effect of various drugs or compounds on gene expression so as to assess their potential efficacy or side effects.[70] It is also possible to compare the profile of genes expressed in a deletion mutant compared to wild type. If the deletion mutant has a 'desirable' phenotype it may be possible to phenocopy the deletion by finding a chemical inhibitor of the protein product or the deleted gene using whole-scale gene expression screening. In addition, the expression of the gene can be induced or repressed in vitro or in vivo, and then its responsiveness characterized by RNA profiling. Such examples are (1) tetracycline on/off,[71] (2) GSEs,[72,73] (3) zinc finger proteins,[74] (4) gene trapping,[8] and (5) correcting the mutation using chimeric RNA–DNA oligonucleotides.[75–83]

Still, one of the major problems with the simultaneous analysis of the expression of thousands of genes is the shear weight of data generated (*see* Section 2.12.3.4). Not only is it essential to link the result from each DNA probe back to the parent sequence; it is also necessary to decide which result is significant, and then generate a list of 'interesting genes.' If, say, 300 genes change in a given experimental paradigm, due to experimental error, which ones are interesting and worthy of follow-up and what do they tell us about what is really happening in the system under investigation? How also can one store and integrate this enormous amount of data over years of experimentation and even compare data generated by different laboratories? These problems will undoubtedly remain for sometime; however, too much data is better than not enough. While clearly there is much to do on many fronts, it is almost certainly true to say that expression profiling will help revolutionize the way in which we perform biological investigations in the future. RNA profiling, together with the many other genomic technologies, promises to have a big impact on the process of drug discovery.[45]

Recently, three new approaches to validate target genes have been developed:

- GSEs;
- zinc finger proteins; and
- gene trapping.

GSE (retroviral based) is a unique technology to rapidly discover and validate novel pharmaceutical targets, as well as design therapeutic agents that regulate the function of these targets in specific disease processes.[72,73] Presently, there are four start-up biotechnology companies dominating the field: Arcaris (formerly Ventana), Rigel Pharmaceuticals, Genetica, and PPD. GSE technology is based on the rapid creation of high-titer helper-free recombinant retroviruses capable of infecting nearly any higher eukaryotic cell type.[84] Novel in its conception and application, this technology facilitates both target and drug discovery in diseases previously resistant to standard approaches. Unlike traditional small-molecule screening approaches, which can underrepresent a given library, the GSE technology delivers highly complex 'expression libraries' into target cells where each cell contains unique information. This information is decoded to synthesize a single peptide or antisense RNA, which interacts with potential targets within the cell. By examining the specific physiological effect of the introduced molecule, the peptide or antisense RNA capable of changing cellular physiology in a desired manner is obtained, and the essential link to identifying a target is made.[85–87] To date these studies have been primarily in oncology and virology.[88,89] Other fields of immediate potential use include antimicrobials, neurosciences, immunology and transplantation, and cardiovascular and metabolic diseases.

Zinc finger proteins are DNA-binding proteins that mediate the expression, replication, modification, and repair of genes. Pabo and Pavletich[74] solved and published the first crystal structure of a zinc finger protein bound to its cognate DNA sequence – perhaps the seminal publication in the field of zinc finger protein rational design. The discovery of zinc finger DNA-binding proteins and the rules by which they recognize their cognate genetic sequences (up to 18 base pairs) has made possible the rational design of novel transcription factors that can recognize any gene or DNA sequence. The start-up biotechnology company Sangamo[1097] has combined these protein–DNA recognition rules with powerful selection methods to allow the rapid generation of proteins that recognize and bind to target DNA sequences. Sangamo can rationally design zinc finger proteins that selectively up- or down-regulate the expression of target genes and that in vitro or in vivo.

Traditional gene-trapping approaches, in which genes are randomly disrupted with DNA elements inserted throughout the genome, have been used to generate large numbers of mutant organisms for genetic analysis. Recent modifications of gene-trapping methods and their increased use in mammalian systems are likely to result in a wealth of new information on gene function. Various trapping strategies allow genes to be segregated based on criteria such as the specific subcellular location of an encoded protein, the tissue expression profile, or responsiveness to specific stimuli. Genome-wide gene-trapping strategies, which integrate gene discovery and expression profiling, can be applied in a massively parallel format to produce living assays for drug discovery.[8] Gene trapping was originally described in bacteria.[90,91] Since then it has been used in many other organisms, including plants,[92] *C. elegans*,[93] *Drosophila*,[94] mouse embryonic stem cells,[95] zebrafish,[96] and yeast.[97]

Gene trapping provides another approach to help validate targets for modern drug discovery efforts, including bioassays to sort through the large number of potentially active compounds. The need for an integrated technology platform to discover genes, pathways, and corresponding drug candidates drives the development of novel approaches, termed 'gene-to-screen genomics.' Gene trapping/tagging strategies may provide such a link to drug discovery.[98]

2.12.2.2.1.4 Model organisms

"Connecting genotype to phenotype is not always straightforward."[99] Thus, the power of less complex organisms for understanding the function of unknown human genes derives from the essential homology of many genes from different organisms.[20] Examples are transcription factors, neurotransmitters and their receptors, growth factors, signaling molecules, apoptosis factors, GPCRs, and membrane trafficking and secretion factors.[100,101] During the past several years, a great deal of work, much of it in yeast, has identified a network of proteins, constituting 'checkpoints' of cell cycle regulation. One has been linked to the hereditary disease Li–Fraumeni syndrome, which leaves patients prone to developing multiple forms of cancer.[102] Yeast genetic screening has also been used to identify mammalian nonreceptor modulators of G-protein signaling.[103] Yeast screening has resulted in the identification of novel biochemical pathways based on a biochemical genomics approach to identify genes by the activity of their products.[104] Bikker *et al.*[4] took the global approach by developing an array of 6144 individual yeast strains, each containing a different yeast open reading frame (ORF) fused to a reporter gene/purification tool, glutatione *S*-transferase (GST), as originally described by Simonsen and Lodish.[105] The strains were grown in defined pools, GST–ORFs purified, and activities of each strain identified after deconvoluting. This approach yielded three novel enzyme activities.

The nematode *C. elegans* is a principal organism for the analysis of the development and function of the nervous system. This is because it is especially amenable to molecular genetic analysis.[106] Its complete cell lineage and nervous system connectivity have been mapped.[107] *C. elegans* is the first animal for which the complete genome has been sequenced.[59] Since the *C. elegans* genome was sequenced, an enormous amount of novel functions have been discovered and linked to more complex organisms, including humans.[106,107–111] For example, presenilins and APP of *C. elegans* are particularly attractive given the ease of making animals lacking the function of a single gene either genetically or by using RNA interference.[108] Subsequently, suppressor mutations are selected in genes in the same pathway and their DNA sequence is determined. One of the most impressive examples of this approach is in studies of apoptosis, where the major components of apoptosis pathway are conserved in *C. elegans* and mammalian cells.[112] Another example is the identification that p66^{shc} expands life by enhancing resistance to environmental stresses such as ultraviolet light and reactive oxygen species.[113] p66^{shc} knockout mice show the same phenotype as the worms.[114] Taking a global two-hybrid approach, Walhout et al.[115] have functionally annotated 100 uncharacterized gene products starting with 27 proteins of known function. The most striking set of interactions involves the Rb tumor suppressor protein complex, which regulates gene expression during the cell cycle. The screens yielded 10 interacting proteins, comprising three known interactors and seven known to interact with other proteins of the complex. This publication showed that by having the total sequence of *C. elegans*, it is now possible on a genome-wide scale to map the interacting proteins and enhance our understanding of molecular mechanisms both in this organism and in humans. The model organism *Drosophila* has also provided insights into mammalian signaling pathways and their importance in development,[116] and will continue to, especially since its genome has been sequenced.[117] Both yeast and *C. elegans* are used for drug screening, both at the expression level using chips and at the organism level, and by creating strains designed (by phenotype and genotype) for screening compounds against a particular biochemical step or pathway.[20,118] Andretic et al.[119] demonstrated that cocaine sensitization could be demonstrated in '*Drosophila*'. Cocaine sensitization was absent in all but the mutant fly strains missing the *timeless* gene. This implication of a subset of the circadian clock genes in drug responsiveness echoes recent suggestions that at least some of these genes may act in more places and in more functions than just the brain's clock.[120]

Over the past 40 years, the mouse has certainly become one of the most useful models of human genetic disease. Multiple single-gene and polygenic models of human disorders have been characterized in the mouse. Often the phenotypic characteristics are strikingly similar to that for the analogous human disorder.[121–126] The mouse has many advantages over other model organisms for these kinds of studies:

- The relatively short generation time facilitates breeding approaches for genetic linkage studies and accelerates the development of genetic models.[25]
- The genetic understanding and characterization of the genome in the mouse is second only to humans in depth and breadth among mammalian species. The ability to genetically map in the mouse[1098] and then clone single-gene disorders rapidly with emerging technologies will allow the economical identification of the polygenes controlling quantitative traits.[27,127,128] There are numerous examples of single-gene mutations in the mouse having virtually identical phenotypes in humans, and cloning of the mouse gene has led to cloning of the human homolog.[129]
- The sequence/structural homology between humans and the mouse is quite good. The mouse can be used in parallel with human studies for gene discovery and target identification. This parallel approach can be used for the identification of disease genes (and putative targets or pathways involved in pathogenesis) as well for as phenotypic characterization. As described above (see Section 2.12.2.2.1.3), the fundamental approach is to use conserved 'synteny' (or orthology) to determine if the same genomic region is functionally involved in a disease phenotype in both human populations and the mouse. If the same chromosomal region is involved, then it is generally easier to identify the gene in the mouse than in humans, and one can use mutation detection/analysis to determine if the same gene is involved in human polygenic disease.
- The ability to make transgenic and gene-targeted deletions or substitutions both classical and conditional is now a mature technology for the mouse.[121,122,130,131]
- The ability to phenotype the mouse is becoming more precise and achievable despite the small size of the animal (e.g., magnetic resonance imaging[132–134]). The ability to phenotype is of utmost importance. This is why obesity was one of the first therapeutic indications tackled. Many obesity genes have been identified in the mouse during recent years.[135–138] Therefore, the mouse is emerging as one of the premier models for disease gene discovery and utility in the drug discovery process.

The premise of genomics in drug discovery and validation is that genes found to be associated with disease are prevalidated. If mutations in these genes affect the disease risk or phenotype, then we know that these genes are likely to be important in the disease process.[25] However, the identification of an association between a gene mutation or sequence variant and a disease trait does not necessarily imply that the gene, or its protein product, is a 'drugable' target. In addition, one must not forget the role of the environment even in well-controlled genetic backgrounds. Crabbe et al.[139] have demonstrated that subtle environmental differences between laboratories could have a significant effect on behavioral measures in inbred and mutant mouse strains, not to mention nongemomic transmission across generations of maternal behavior and stress responses found recently in rats.[140] Thus, these genes may simply be important players in the disease process, and an understanding of their role may serve to provide access to the pathways involved in disease and, ultimately, the appropriate targets for therapeutic intervention. Still, the utility of the mouse extends beyond being just a gene discovery tool to provide prevalidated targets. It can also be used for the development of animal models, and the testing of compounds in specifically constructed transgenic and knockout strains to further define the target and pathway of a therapeutic compound.[25]

2.12.2.2.2 Proteomics

The proteome of an organism is the complete set of proteins that it can produce. Proteomics is the study of the proteome of cells, tissues, or organisms, including the interaction of the proteome with the environment, where the proteomic signature is the subset of proteins whose alteration in expression is characteristic of a response to a defined condition or genetic change.[141] Besides genomics there is also a great interest in working at the genome level with polypeptides or proteomics. In fact, proteomics is one of the most important 'postgenomic' approaches to understanding gene function. Given that the mRNA concentration in any cell and the amount of cognate protein are not always correlated (owing to the differential kinetics of both types of molecule), there may be a good deal of new information to be found by studying proteins in this way. It is especially interesting because post-transcriptional regulation of gene expression is a common phenomenon in higher organisms.[142]

Presently, proteomics primarily uses two-dimensional gel electrophoresis to profile gene expression and cell responses at the protein level. Protein spots are identified by matrix-assisted laser desorption/ionization (MALDI) and tandem mass spectrometry techniques, using proteomics for target identification, mechanism of action studies, and for comparing compounds based on the similarity of proteomic signatures they elicit. Probably the most advanced organization taking the global approach to proteomics, especially in higher eukaryotes, is UCB, which, along with Incyte, has launched an integrated proteomics database that includes information about how mRNA expression levels correlated to protein expression levels. Large Scale Biology has focused on establishing databases of hundreds of compounds, effects on rat liver, and Proteome has established excellent yeast (*Candida albicans*) and *C. elegans* protein databases. To date, the bottleneck with the two-dimensional gel approach has not been running the gels but rather sensitivity, the quantitative detection of spots, as well as the time needed to 'gel gaze,' (i.e., to image the gel so as to identify differences from gel to gel). To that end, Genomic Solutions[1099] has assembled and developed a complete suite of technologies and systems for protein characterization-based electrophoretic separation, which includes a fairly sophisticated and proven imaging technology that permits the identification, quantification, and comparison of two-dimensional separations. Other companies, such as Amersham/NycoMed, are developing second-generation dyes and stains to increase the sensitivity of the detection of the spots in the two-dimensional gels. Ruedi Abersold at the University of Washington is developing an approach whereby he is miniaturizing classical electrophoresis technologies in a way that is compatible with automation.[143]

As described in a recent *Nature* survey[143] on the prospects of proteomics, the now well-established two-dimensional gel approach has many limitations. Still, the development of the more advanced technologies (e.g., protein chip, antibody, and protein array) that may deliver fast and parallel quantitative analyses of protein distributions has a way to go (e.g., see below – Phylos, Ciphergen, CAT, Morphosys, etc.).

One such technology that is further advanced comes from Ciphergen Biosystems. The ProteinChip system that Ciphergen uses is the patented SELDI (surface-enhanced laser desorption/ionization) ProteinChip technology, to rapidly perform the separation, detection, and analysis of proteins at the femtomole level directly from biological samples. The ProteinChip system can replace and complement a wide range of traditional analytical methods, which not only are more time-consuming but require specialized scientific expertise. ProteinChip arrays allow the researcher to affinity capture minute quantities of proteins via specific surface chemistries. Each aluminum chip contains eight individual chemically treated spots for sample application; this setup facilitates simultaneous analysis of multiple samples. A colored hydrophobic coating retains samples on the spots and simultaneously allows for quick identification of chip type. Typically, a few microliters of sample applied on the ProteinChip array yields sufficient protein for analysis with the ProteinChip Reader. Designed with proprietary technology, the Reader takes advantage of modern time-of-flight (TOF) mass spectrometry to determine the precise molecular weight of multiple proteins from a native biological sample. To enhance the appearance and facilitate interpretation of the protein mass data collected, ProteinChip software offers various presentation formats or 'data views.' Recently, Garvin et al.[144] has employed MALDI-TOF to detect mutations in gene products by tagging a PCR fragment after in vitro transcription and translation. The process can be multiplexed, and is amenable to automation, providing an efficient high-throughput means for mutation discovery and genetic profiling.

To bring functional genomics to the protein arena, Phylos has pioneered PROfusion technology, whereby proteins are covalently tagged to the mRNA.[145] Using this technology, both synthetic and natural libraries (representing the repertoire of proteins naturally expressed in a given cell type or tissue source) have been constructed. Due to the in vitro nature of library construction, Phylos libraries are the largest described to date, up to 1014 in size. From a library of such molecules, one can select for a protein function of choice, with the benefit that the genetic material is linked to the protein for subsequent PCR amplification, enrichment, and, ultimately, identification.[146] The key to the Phylos technology is in the development of a puromycin-containing DNA linker that is ligated to RNA prior to translation. Puromycin is an antibiotic that mimics the aminoacyl end of tRNA, and acts as a translation inhibitor by entering the ribosomal A site and accepting the nascent peptide, a process catalyzed by the peptidyltransferase activity of the ribosome. To accomplish this, puromycin is chemically protected and attached to controlled-pore glass for automated synthesis of a DNA linker consisting of 5′-dA27-dC-dC-P (P is puromycin). The PROfusion molecule represents a very unique entity with characteristics that can be utilized for the production of protein chips. The presence of the RNA tail offers a convenient appendage for anchoring the antibody mimic onto a microarray similar to a DNA chip. Since the technical aspects of chemical attachment of nucleic acids to a chip surface have been well worked out, one can take advantage of this reagent and convert an existing DNA chip to a protein chip through the hybridization of a pool of PROfusion molecules. Because the PROfusion molecule has a unique genetic tag associated with it, specific capture probes can be designed and a specific DNA chip produced for the pool of molecules. Upon hybridization of the PROfusion molecule to the DNA chip, the surface now displays the covalently attached peptide or protein portion of the PROfusion molecule.[145] Furthermore, by using a drug or protein physically anchored either to a solid support or to a chemical ligand (e.g., biotin), the PROfusion cellular library is passed over the target and the interacting proteins eluted and amplified for the next round. Such a methodology brings with it a number of significant advantages. Direct selection of interacting proteins from a cellular PROfusion library through rounds of selection and amplification allows for the identification of both low-abundance and low-affinity targets.[145]

Other approaches to identifying interactions of chemical entities with target peptide libraries are described by Kay et al.[147] These libraries can be generated on pins, beads, or in solution, expressed in bacteria attached to phage or synthesized in vitro off of polysomes. However, the complexities only range from 108 to 1010. A method of quantitative analysis of complex protein mixtures using isotope-coded affinity tags has been reported by Gygi et al.[148] The method is based on a class of new chemical regents, termed isotope-coded affinity tags, and tandem mass spectrometry. The first successful attempt using this technology was comparing yeast using various sources of carbon to grow on. The future will show if this technology can be used to quantitatively compare global protein expression in other cell types and tissues.

Another protein chip approach has been demonstrated by Bieri et al.[149] by which they were able to develop a novel assay of GPCR activity that employs immobilized receptors on a solid surface and in a functional form. To achieve this, the authors used a specific labeling procedure to biotinylate the carbohydrate located in the glycosylated extracellular domain region of the rhodopsin receptor, a G protein-coupled receptor responsible for dim light detection in animals. This in turn reacts with the surface on the chip using lithography. Rhodopsin interacts with a specific Ga subunit, transducin, to activate cyclic GMP phosphodiesterase, and thus can activate the immobilized rhodopsin on the sensor chip with the flash of light, which is monitored by receptor activation through the release of transducin from the chip surface.

Monoclonal antibodies raised against purified native protein have typically been used as the optimal reagent for many research purposes. However, creation of this type of reagent is both costly and time-consuming. A more rapid and cost-effective alternative, but one that is less attractive, is the use of polyclonal antibodies to the target protein or peptides derived from the target. These reagents are generally inferior to monoclonal antibodies with respect to affinity of binding, specificity of reactivity, and utility. An attractive alternative to these approaches is phage display antibody technology (e.g., marketed by Cambridge Antibody Technologies, Morphosys, Dyax, and UbiSys). Phage display antibodies are at least equal, and in most cases superior, to monoclonal and polyclonal antibody approaches with regard to affinity of interaction, selectivity of reactivity, utility in a variety of applications, speed of antibody generation, and numbers of antibodies that can be generated. Moreover, single-chain variable regions with affinities and selectivity equal to monoclonal antibodies can be generated in a period of weeks (as opposed to months). As a result, it is possible to generate much larger numbers of antibodies in the same period of time. It has become clear that there is a critical need for potent and selective antibodies for target identification and validation.

2.12.2.2.3 Pharmacogenetics and pharmacogenomics

The intimate connection between drug discovery and genetic information has only become widely recognized in recent years, primarily due to the fact that the two disciplines were perceived to have nothing in common. The discovery of

SNPs has united the two fields.[150,151] SNPs are the predominant basis of genetic variability in the human population. SNPs will provide the genetic markers for the next era in human genetics.[41] Particular advantages of using SNPs over other types of genetic marker include the relative ease of automating SNP typing assays for cheap, robust, large-scale genotyping, and the abundance of SNPs throughout the genome. Comparisons of the DNA sequence between two individuals reveals, on average, one nucleotide difference for every 1000 base pairs.[152] This frequency increases as more samples are included in the comparison.[153,154] The majority of changes are single-base substitutions (i.e., SNPs), although deletions and insertions of one or more bases are also observed Heritable sequence variations arise as a result of a copying error, or damage and misrepair of DNA that results in alteration of the genomic sequence in germ cells. As a result, the changes are passed on in subsequent generations, and the two or more alternatives at a particular position, or locus, become established in the population. Further diversity arises as a result of recombination between homologous segments of chromosomes during meiosis.[41,155]

In the early days of the genomics revolution, when a comprehensive working draft of the human genome was yet to be completed, and when 'The Human Pharmaco SNP Consortium' was still in the beginning stages of identifying hundreds of thousands of common sequence variants (SNPs), no one could accurately predict the outcome. In addition, private companies were assembling SNP databases (e.g., Celera, Curagen, Genset, Orchid, Phenogenex, etc.) in partnership with larger pharmaceutical companies. In the interim, pharmaceutical companies limited their analysis of SNPs to selected candidate genes (pharmacogenetics) to identify associations between polymorphism and drug.[41] Today, though many of the original companies who pioneered the area have faded away, these data are used regularly by academia and the entire healthcare industry in support of various R&D tasks.

The goals for pharmacogenetics/pharmacogenomics are to associate human sequence polymorphisms with drug metabolism, adverse events, and therapeutic efficacy. Defining such associations will help decrease drug development costs, optimize selection of clinical trial participants, and increase patient benefit. Retrospective analysis and rescue of nonconclusive studies as well as determination of the genetic causality of disease are also possible. In addition, the ascertainment of DNA samples from clinical trial patients will facilitate the discovery and validation of targets using linkage disequilibrium and association methods. Candidate genes identified by positional cloning in the mouse can be followed up by using the appropriate patient's DNA to confirm the importance of the gene(s) in disease.[156]

Pharmacogenetics covers a new aspect of drug development, which will apply to discovery, development, and marketed programs. To date, benchmarking-type information is not available. However, it is expected to play an important role in the near future for many pharmaceutical companies. To associate polymorphism with drug effects, DNA samples are obtained from both drug- and placebo-treated patients. To associate polymorphism with disease risk (i.e., target discovery/validation), DNA from patients affected by a specific disease should be compared with DNA from patients unaffected by that disease. The latter group may be affected with other disease states; hence, patient DNA samples from one trial may serve as controls in another, provided that they have been obtained from a population of comparable ethnic composition. In effect, supplementary DNA collection will not be necessary to effect properly controlled pharmacogenetic and target validation studies. Moreover, power simulations suggest that sample sizes obtained during standard clinical trials are sufficient to detect genetic association, given that the genetic effect is of sufficient magnitude and the allele is not rare.[157] Drugs can be more accurately profiled for drug–drug interactions by looking at the metabolism of the compound by the known enzymes and variants thereof. If a drug is found to have limited efficacy but virtually no (or very mild) side effects, then it may also be possible to subtype the patients by genotype so as to distinguish responders from nonresponders. Similarly, if a compound has a potentially severe but very rare side effect (but good efficacy), then genotyping may be used to identify patients in whom a side effect is more likely.

Polymorphisms in coding, regulatory, intronic, untranslated, and other regions can all cause variation in gene activity. Cytochrome P-450 2D6 gene is an example of a coding polymorphism that modulates the activity of the gene and thereby influences a drug's metabolism rate. Finally, there is growing activity in profiling compounds for toxicity by using SNP and expression chip 'signatures.'[158] Good examples of genotyping currently being done in clinical trials include measuring ApoE4 genotypes for AD and measuring polymorphism in various cytochrome P-450 genes. As the general population to be treated is genetically diverse, the statistical considerations of these approaches are complex. However, recent examples show that in outbred populations, linkage disequilibrium can be regularly observed at distances of 25 kb from a causative polymorphism. Hence, a useful genome-wide SNP map will be composed of 300 000 evenly distributed SNPs localized to specific points in the genome.[159] Given the sample sizes available in clinical trials, the rare allele in a biallelic SNP must be present in a population at a frequency greater than 15%. The 'Holy Grail' is undoubtedly a genome-wide 'SNP' map by which correlation of SNP haplotypes with drug–person phenotypes will be made. From a commercial perspective, there will be a trade-off between marketing a drug in a subset of the population and prescribing it for everyone with the disease. Diagnostics will go hand in hand with pharmacogenetic/pharmacogenomic testing.[20] Presently, there are numerous start-up companies developing technologies to score SNPs in a very high-throughput mode to meet the needs of pharmacogenetics and, in the future, pharmacogenomics. These technologies include mass spectrometry and DNA arrays to fiber-optic arrays.[159] For more information, see the websites of the following companies: Third Wave Technologies, Kiva Genetics, Hexagen, CuraGen, GeneTrace, Sequenom, Orchid, PE, RapidGene/Chiroscience, Affymetrix, Nanogen, Illumina, Renaissance Pharmaceuticals, Varigenics, Celera, PPGx, and Sangamo, among others. It is safe to say that profiling humans by genotyping will become part of drug development in the future; it is simply a question of when.[20]

2.12.2.2.2.4 Bioinformatics/structural informatics/chemi-informatics

Bioinformatics encompasses the acquisition, storage, and analysis of information obtained from biological systems. The information is derived primarily from genomics (genomic and EST DNA sequences, RNA profiling, and proteomics) and HTS.[160] Structural informatics refers to the acquisition, storage, and analysis of structural genomics data, especially that from whole genomes (i.e., genomes that have been totally sequenced, such as those of microbes, yeast, and *C. elegans*).[161,162] "High level protein expression systems, robotic crystallization, cryogenic crystal handling, x-ray area detectors, high field NMR [nuclear magnetic resonance] spectrometers, tunable synchrotron radiation sources and high performance computing have together catapulted structural biology from an esoteric niche to biological mainstream."[162] The next step is to integrate the bioinformatics and structural informatics data. There are both private (Inpharmatica) and public (Structural Research Consortium[1100]) initiatives underway. As described above in the introduction, defining SERMs by using the three-dimensional structure of the estrogen receptor, combining expression profiling (chips) to specific cell types (both in vitro and in vivo), and linking the data to chemi-informatics will be most fruitful for drug discovery in the future.

Many pharmaceutical and biotechnology organizations have expressed the desire to create a means for tying together structural genomics and chemical information. A GenChem Database would map the gene sequence (actually, the protein sequence) to the chemical structure in a multidimensional matrix. Two dimensions would be the sequence and the chemical compounds, where each dimension could be examined for similar or dissimilar sequences or chemistry (technology for both exists). The additional dimensions are the different ways to fill in the two-dimensional chemistry versus sequence matrix. A simple method that has some predictability is to score sequences against chemicals for binding ability. You would then find, for example, that proteases cluster by sequence similarity and by inhibitor similarity. Another way to map the sequence-to-chemical matrix in terms of whether the chemical compound upregulates or downregulates the gene expression.

Researchers are now populating the multidimensional matrix with different types of data (ligand-binding affinities, expression changes, etc.). In addition, so-called digital organisms are already on the horizon. These are computer programs that self-replicate, mutate, and adapt by natural selection.[163] The availability of these comprehensive data sets will fundamentally change biomedical research and healthcare practices. Bioinformatics, structural informatics, and chemi-informatics seek to lead the industry into this new era by bringing genomic information to bear on drug discovery, development, and therapy (**Figure 7**).[1114] However, in the 'postgenomics' world, everyone has realized that bottlenecks exist in the measurement and analysis of data. Participants at the Genome V session held on October 6–10, 1999[164] discussed strategies for systematic information gathering on genes and proteins, how to improve pattern recognition algorithms, and ways of analyzing large data sets efficiently without the need for supercomputers. Many meetings such as this one were held to push genomics to the next stage of development.

Genomic studies have shown that many genes in mammals are present in four copies relative to metazoan genomes.[165] Individual members of these gene families have, in many cases, evolved to take on new functions. Nonetheless, there are numerous cases of functional redundancy in mammalian genomes. Therefore, drugs that target single proteins may induce the activity compensatory proteins that cancel the effect of the drug. Using sequence similarity algorithms, bioinformatics may identify targets likely to have functionally redundant gene relatives. Reports of multifunctional proteins have become increasingly common; for example, signal transduction pathways have proteins that act on multiple substrates. Drugs that inhibit multifunctional proteins may induce unwanted effects. Gene expression (RNA profiling) and protein expression (proteomics) tools reveal patterns that characterize the response of a cell or tissue to compounds.[160] When correlated to efficacy, toxicity, and biological mechanism, these patterns become powerful tools for generating testable hypotheses. In the future, when the databases are large and robust enough, such patterns will be diagnostic and will not require further experimental validation.

A common problem in large, multidimensional data sets is variable data quality. It is seldom possible to apply weighting factors in genomic data sets, and therefore early genomic data sets will be noisy. Key in identifying important

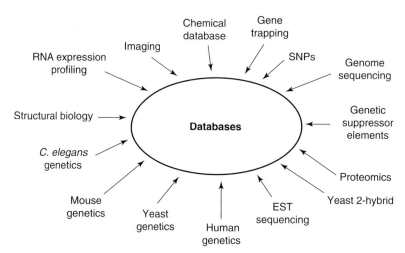

Figure 7 To capitalize on genomics, the most important task is the integration of databases. The plea is for better visualization and data-mining tools as well as better interfaces for databases.

signals in this background are larger experiments (increased sampling frequencies) and the creation of novel algorithms tuned to identify biologically significant patterns. A typical array experiment generates thousands of data points and creates serious challenges for storing and processing data. Informatics can include both 'tools' and 'analyzers.' Tools include software that operates arraying devices and performs image analysis of data from the readers, databases to hold and link information, and software to link data from individual clones to web databases. The DeRisi Lab has made available software for operating custom-built arrayers.[1101]

The quality of image analysis programs is crucial for the accurate interpretation of signals for slides (chips) and filters. Yidong Chen (NHGRI) has developed a software image analysis package for slides and filters called 'deArray' that is available but not supported. Mark Boguski and colleagues have developed software to analyze and link databases such as Entrez and UniGene.[58] Software packages for commercial arrayers and chips are also available:

- Synteni (Gem Tools);
- Genomic Solutions (Visage Suite); and
- Silicon Genetics/Affymetrix (GeneSpring).

Additional development of algorithms for pattern recognition and cluster analysis of complex genomic data sets is key to the successful completion and integration of the complex data sets.[164]

Most EST databases and all gene expression technologies are tuned to detect genes expressed at moderate-to-high levels (100 transcripts per cell or better). Genomic sequence, subtracted normalized cDNA libraries, and gene-finding sequence analysis algorithms will all help to identify genes expressed at low levels. Tools for highly parallel measurement of mRNA levels will require significant improvements before gene expression changes for rare transcripts can be reliably measured. Similar gains in technology will be required to approach comprehensive sampling of the proteome and molecular identification of its constituents. Presuming adequate quality control, genomic databases (sequence, polymorphism, gene expression, and proteomic) all increase in value as more data are deposited. To maximize the investment in genomics and bioinformatics, data must be stored in a series of central databases. These databases must be made available for browsing/mining to scientists and those interested in multiple therapeutic areas. Moreover, links between databases will facilitate the integration of information from multiple types of experiments and data. A current common theme in the pharmaceutical industry is the importance of information technology as a key support factor in drug discovery and development.

2.12.2.3 Conclusion

As the reader can see, there are two major approaches using numerous technologies to identify and validate targets: 'top-down' and 'bottom-up.' The choice of approach depends on the knowledge of the disease indication at hand, and in some cases both approaches may apply. Irrelevant of the approach or approaches, key is information technologies. The biggest concern of pharmaceutical companies in terms of drug discovery in the postgenomic era is data integration, particularly as it relates to target validation. Their plea is for better visualization and data-mining tools as well as better interfaces for databases.[166] At present, there is no quick way to validate a target. Indeed, there is no universal definition of what a validated target is. All of the data being generated on a genome-wide scale must be captured and integrated with other data and with information derived by classical, hypothesis-driven biological experimentation to provide the picture from which completely new targets will emerge. This is a daunting bioinformatics challenge.[167] However, one thing is clear: target validation has become a term that is often used but not rigorously applied. The evolution of truly validated targets is a much slower process than target identification. At present, even large drug discovery organizations in the biggest pharmaceutical companies are not capable of producing more than a handful of well-validated targets each year.[2] Thus, the technologies described in this section must be applied such that the number of disease gene targets identified is rigorously validated to improve the quality of drug discovery while providing an increased number of better validated targets because even the best validated target does not guarantee lead compounds for development. Targets are only truly validated when a successful drug (not compound), working through that mechanism, has been found.

2.12.3 Functional Genomics

2.12.3.1 Introduction

A well-known paradigm in biology is that the function of a biological entity is closely linked to its physical structure. Based on this principle, predictions of biological function have been made by analyzing the structure of the organic molecules on which cells are constructed, such as DNA, RNA, and proteins. Since proteins provide the raw materials by which the superstructures of cells are constructed, it is reasonable to conclude that the amount, kind, and state of proteins found within a cell can be used to predict how a cell or organism will respond to chemicals, drugs, and the environment. Due to the hierarchical nature of genetic information (i.e., DNA to RNA to protein), it has been possible to approximate the structure and function of proteins by studying the information contained within the primary sequence and abundance of DNA and RNA, respectively. The massively parallel research methods used to determine the sequence and abundance of DNA and RNA are collectively called genomics. Functional genomics is the study of the DNA and RNA component of gene expression, coupled with biological and biochemical experimental data, to approximate the structure and function of proteins.[168]

Since it is with proteins that drugs primarily interact, investigating the properties of proteins may lead to innovations in the treatment of human disease.[169] Nevertheless, due to the public resources now available through the efforts of the Human Genome Project, it is the study of the DNA and RNA component of gene expression that is most immediately available. For this reason, genomic methodologies allowing for the parallel characterization of DNA and RNA are now available and routinely practiced. Functional genomics now serves as a tool by which predictions of cellular function and potential drug–cell interaction can be made. Moreover, given the complexity of cellular signaling and regulatory pathways, numerous simultaneous measurements of the interdependent changes occurring during biological processes can be used to predict how drugs will affect these pathways.[170] This section will highlight the primary considerations given to the use of functional genomics as an approach to drug discovery. These are (1) the functional characterization of proteins, (2) the molecular pathways by which proteins provide complex cellular function, and (3) the use of genomic technologies to determine function and find drug targets.

2.12.3.2 Proteins: Structure and Function

All proteins rely on a linear arrangement of 20 amino acids to create the complicated structures that serve a cell or organism with such properties as locomotion, exchange of raw materials, and procreation. Although there are as many as 10.24 trillion combinations of these 20 amino acids for every 10 residues, based on the proteins sequenced to date many of the same sequences are observed. This observation suggests that very few protein sequences translate into proteins that can perform some function. In the hope of better understanding how certain protein sequences are selected over others we must explore the way in which proteins originate and improve the reproductive strength of the cell or organism – its evolutionary history.

The term 'molecular evolution' suggests that, just as species are selected based on their reproductive advantage, proteins are selected based on their ability to improve the health and well-being of a cell at the molecular level. The origin of all proteins starts with the production of an open reading frame through genetic alterations introduced by such events as point mutations, chromosomal rearrangements, and transposable elements. Therefore, it is the environment that provides the physical and functional constraint by which proteins are selected through the evolving genomes of the world. By relying on the form and fit of proteins to the molecules of the cellular microenvironment, such as nucleic acids, ions, and other proteins, a subset of all possible protein sequences have been selected and modified to produce the diversified adaptations observed in nature.[171–173] Moreover, it is this connectivity between sequence and function that provides biologists with the primer for unraveling the cellular role of proteins with unknown function ('orphans'). By observing the conservation of sequence elements, the profile of protein expression, and the effects of mutations, a crude approximation of the function of an orphan protein can be ascertained.[174–179]

2.12.3.2.1 Molecular evolution

Since it is the function of the protein that keeps it from being lost during evolutionary selection, it is reasonable to conclude that any segment of a protein that has not been altered in sequence over great evolutionary distances must be performing some role in the health of the organism. The most abundant and extensively characterized class of proteins, first observed in prokaryotes, appears to be selected to perform a series of functions related to the maintenance of the cellular microenvironment (i.e., housekeeping genes). These consist mostly of proteins involved in energy metabolism, the biosynthesis of required amino and nucleic acids, transcription/translation, and the replication of the genome of the cell (**Figure 8**).[180]

These fundamental requirements of all cells provide the backbone on which subsequent protein adaptations are made during evolution. As a consequence of speciation, many proteins are composites of functional units acquired during the fusion or rearrangement of ancestral genes. These small segments of protein sequence are called domains or modules.[176,181,182] These modules are made up of independent segments (motifs), which provide the module with important contact and energetic attributes, allowing the module to perform a task. It has been suggested that during molecular evolution the exons containing these functional units are partially responsible for the shuffling of modules to create alternative composite proteins.[175,176,182,183] Many environmental changes occur over small periods of evolutionary time, and therefore require adaptations on behalf of the organism to survive. Due to this immediate need in the production of proteins required to compensate for the environmental change, it is easy to see how the coordinated rearrangement of functional domains within a genome would arise. By using bioinformatic sequence alignment tools it has been possible to group these modules into classes, and demonstrate the conservation and expansion of shared modules during the course of evolution (**Table 13**).

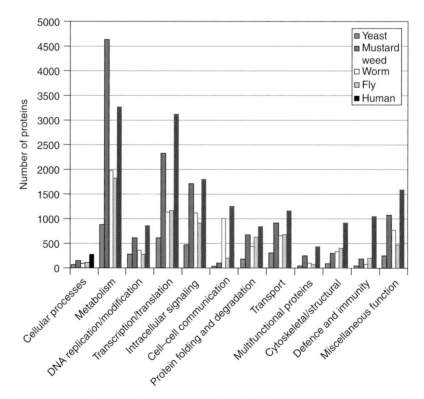

Figure 8 Functional categories in eukaryotic proteomes. The classification categories were derived from functional classification systems, including the top-level biological function category of the Gene Ontology Project. Initial sequencing and analysis of the human genome.[180]

Table 13 Domain sharing and order conservation within humans and between humans and other eukaryotes

Human versus	Domain sharing						Identical domain arrangements				
	No. of proteins sharing domains	No. of cases					Total No. of domains in a protein	No. of Identical arrangements/No. of human proteins[a]			
		No. of domain types						No. of domain types			
		1	2	3	>3			2	3	>3	
Human	2	214	194	73	61		1	—	—	—	
	3	147	88	25	18		2	141/56	—	—	
	4	123	38	17	5		3	57/208	21/62	—	
	5	67	17	5	3		4	53/168	18/63	4/10	
	6	56	19	5	0		5	44/173	11/27	5/16	
	>6	377	79	20	5		>5	150/605	66/172	34/78	
Fly	1	143	129	32	23		1	—	—	—	
	2	134	65	14	12		2	119/337	—	—	
	3	97	47	11	5		3	35/98	10/18	—	
	4	83	19	7	0		4	28/65	10/24	1/1	
	5	51	9	2	2		5	25/74	8/17	5/13	
	>5	359	65	14	2		>5	58/137	11/19	12/16	
Worm	1	136	92	27	9		1	—	—	—	
	2	124	56	11	12		2	89/307	—	—	
	3	94	38	9	7		3	28/118	10/24	—	
	4	84	17	5	2		4	16/39	6/20	0/0	
	5	46	8	2	2		5	16/60	3/8	3/8	
	>5	355	61	11	1		>5	43/118	8/16	9/13	

continued

Table 13 Continued

Human versus	Domain sharing	Identical domain arrangements									
	No. of proteins sharing domains	No. of cases				Total No. of domains in a protein	No. of Identical arrangements/No. of human proteins				
		No. of domain types					No. of domain types[a]			
		1	2	3	>3		2	3	>3
Yeast	1	135	51	8	2	1	—	—	—
	2	91	27	5	0	2	51/199	—	—
	3	64	18	2	0	3	9/20	4/12	—
	4	58	5	0	0	4	4/7	3/3	0/0
	5	41	3	0	0	5	3/6	1/2	1/1
	>5	260	24	4	1	>5	36/16	1/3	0/0
Fly	1	75	24	4	1	1	—	—	—
Worm	2	78	16	3	0	2	26/145	—	—
and	3	49	12	1	0	3	4/10	1/3	—
yeast	4	48	3	0	0	4	3/5	0/0	0/0
	5	33	2	0	0	5	3/6	0/0	1/1
	>5	249	21	3	1	>5	7/18	0/0	0/0

[a] Number of unique domain arrangements/number of human proteins in which these arrangements are found. The second number is larger than the first because many proteins may share the same arrangement. For example, in the case 4/10, there are four unique arrangements of three-domain proteins with two domain types. In other words, the arrangement A-B-A has three domains but only two domain types, A and B, which have been conserved among humans, flies, worms, and yeasts. In humans there are 10 such proteins, hence the 4/10 designation.

Ancestral proteins of prokaryotic origin have given rise to eukaryotic proteins with new functions, such as the suspected transition of the bacterial FtsZ to the eukaryotic tubulin protein,[171] as well as immunological proteins restricted to multicellular organisms. It is the molecular evolution of these ancestral genes that have given rise to a multitude of proteins that have diverse function yet can be categorized into families on the basis of their relatedness.

In addition, through genetic and biochemical experimentation of proteins with similar motifs and domains, the functions of these sequences are being identified and will provide us with even more information about proteins with undefined function. By the association of sequence and biochemical/biological data, the suggestion of biological function for proteins is possible.

2.12.3.2.2 Sequence and protein families

As stated previously, the proteins selected by molecular evolution are a subset of all possible proteins, and can be categorized into families (**Table 14**). It is the sequence similarity of DNA, as well as the protein, that serves as the measure of evolutionary distance, and perhaps suggests altered or new protein function.

Through the use of bioinformatic tools, as well as the extensive amount of sequence data available in public databases, it is now becoming a common practice to investigate the relationships between the protein sequence of a gene and to classify it in a specific protein family. Comparing the protein and DNA sequence of novel proteins with other proteins of known function and similar structure, we are able to predict biochemical or biological properties based on previously collected data on these family members.[174–176,181–184] An important caveat and difficulty with this process is that many proteins in higher organisms have evolved through the fusion of the functional domains of existing proteins, and are a composite of more than one protein ancestral family (**Figure 9**).

The homeodomain family of transcription factors serves as the best example of this modular design. Since many transcription factors rely on a DNA-binding domain as well as an additional domain that allows for the activation of the transcription factor, multiple protein sequences will align and cluster under these separate domains. Therefore, insight into the biological function of these new proteins from its family association can be misleading. Nevertheless, these bioinformatic techniques are expected to provide a wealth of detailed information to the biologist searching for clues as

Table 14 The largest protein families[a]

Family	Source	Modules in Swiss Prot	Found where?
C2H2 zinc fingers	PF00096	1826	Eukaryotes, archaea
Immunoglobulin module	F000471	351	Animals
Protein (Ser/Thr/Tyr) kinases	PF00069	928	All kingdoms
EGF-like domain	PF00008	854	Animals
EF-hand (Ca binding)	PF00036	790	Animals
Globins	PF00042	699	Eukaryotes, bacteria
G protein-coupled receptor–rhodopsin	PF00001	597	Animals
Fibronectin type III	PF00041	514	Eukaryotes, bacteria
Chymotrypsins	PR00722	464	Eukaryotes, bacteria
Homeodomain	PF00046	453	Eukaryotes
ABC cassette	PF00005	373	All kingdoms
Sushi domain	PF00084	343	Animals
RNA-binding domain	PF00076	331	Eukaryotes
Ankyrin repeat	PF00023	330	Eukaryotes
RuBisCo large subunit	PF00016	319	Plants, bacteria
Low-density lipoprotein receptor A	PF00057	309	Animals

[a]The sources for these numbers of modules are Pfam (PE) or Prints (PR).

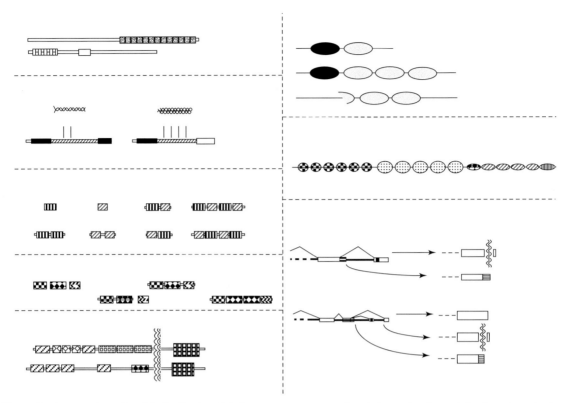

Figure 9 A comparison of protein families in which the sequence conservation during molecular evolution has given rise to proteins of similar yet distinct function.

to how the orphan protein of interest is involved with the model under investigation.[185] **Table 14** is a short list of the largest protein families and where they can be found.[181]

As a consequence, the DNA sequence similarity of most proteins is highly conserved between protein families as well as species in regions of functional importance. The most common observation is that the portion of the protein that is responsible for the interaction of that protein with other molecules has at least a weak similarity in protein sequence and is the focus of family categorization.

2.12.3.3 Molecular Pathways

Through large-scale analysis of human genes and the proteins that they produce, a detailed map of the human body will provide a blueprint of molecular pathways, which will describe how the complicated interactions of proteins support our existence. A blueprint of this interdependent network is vital to the development of drugs that will be used to influence the quality of life through pharmaceutical intervention. The final response to the environmental signal is largely dependent on the abundance and state of the proteins in the pathway. The lack of expression or the inactivation of the downstream proteins can lead to a breakdown in the communication of the signal or the diversion of the signal toward an alternative pathway.

Drug discovery is based on determining which compounds are best suited for the systemic, or tissue-specific, manipulation of biological function for the long-term chemical compensation of patients with a weak genetic background or the temporary modulation of life-sustaining systems threatened as a result of injury.[168–170,186] Functional genomics promises to provide researchers with detailed feedback on the effectiveness, as well as negative consequences, of this intervention at the molecular level. In order to produce drugs that manipulate biological function but do not produce side effects that prohibit its use, it will become commonplace to incorporate large-scale genomic methods that reveal the shortcomings of the drug being tested.

2.12.3.3.1 Signal transduction

The foundation of most drug discovery projects relies on the ability of cells within a tissue to be influenced by the local environment. This response is due to the presence of proteins on the surface of the cell, like the platelet-derived growth factor (PDGF) receptor, that sense the environmental changes and transduce this signal into the cell and alter its biochemical state. By altering the behavior of even a subset of cells within a tissue, the function of whole organs can be altered, leading to the adjustment of life-sustaining parameters necessary to maintain higher order functions. Signal transduction is made possible by networks of interacting proteins working in concert to achieve changes in cellular function. The state of the cell is defined by parameters such as the complement of proteins being synthesized and processed, their enzymatic activity, and their location within the cell, as well as changes in the equilibrium of ions across intracellular and extracellular membranes. The state of the cell determines how it will respond to the environment.[187] For instance, will it secrete specific humoral factors, such as insulin, or contract upon neuronal stimulus like muscle cells? The presence of receptors that are inserted into the cell membrane allow for the interaction of signaling molecules, called ligands, on the outside of the cell to influence the function of cells by altering the state of protein networks within the cell.

2.12.3.3.2 Protein–protein interaction

The cellular signaling mechanism responds to perceived signals from receptors through networks of proteins that come in direct contact with one another. This protein–protein interaction serves as one criteria for protein domain conservation during molecular evolution,[172,173] and can be recognized by the massively parallel collection of sequence and expression data offered by genomic technologies. Sequence alignment of protein family members across many species, in conjunction with biologically relevant experimentation, provides important insight into the function of these conserved sequence elements, and therefore must be carefully considered during the assignment of function of any one protein.

2.12.3.3.3 Tertiary structure and convergence

In some cases the functions of two proteins separated by great evolutionary distances converge. In these cases, it is common to have very little similarity in protein sequence yet similar function (**Figure 10**).[173] By viewing the tertiary structure of two proteins that share a similar function, general similarities can be seen. The tools for comparing the three-dimensional conformation of proteins in order to understand function are still in their infancy. Therefore, it is necessary to rely on the information contained within the primary sequence of the two proteins to investigate the rules that control the ability of that protein to perform its function. Determining what structural characteristics allow proteins to interact with the rest of its molecular pathway will make it possible to more accurately predict the consequences of their individual contribution to that pathway. For example, it is well known that defined mutations within the homeodomain of certain transcription factors will specifically alter the sequence specificity on which the transcription factor binds, required for the induction or repression of gene transcription. Moreover, a better understanding of their structural characteristics would allow for the logical design of small molecules to modulate specific components of that pathway, potentially translating into drugs with higher fidelity and fewer side effects, and increasing the number of targets on which drugs are designed.

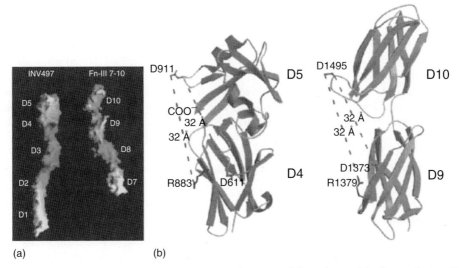

Figure 10 (a) Note the structural similarities between the invasion extracellular region and the fibronectin type III domain. (b) A ribbon diagram illustrating the tertiary structure with only 20% sequence identity.

2.12.3.3.4 Gene expression and protein kinetics

Proteins function at the molecular level, and are governed by the laws of probability and molecular kinetics. If the concentration of a protein is increased beyond some critical value, the probability that that protein will come into contact with its correct partner is increased. Yet, as a consequence of its increased rate of interaction with all molecules in the cell, interactions with inappropriate branches of the intracellular regulatory network become common. This altered protein function is similar to the altered specificity of drugs that have been given at high doses to achieve an effect but cause unwanted side effects. By monitoring the normal concentration of regulatory proteins in the cell, it may be possible to recognize patterns of expression that are associated with disease, as well as define the correct abundance ranges of proteins to achieve stability and avoidance from a pathological state. **Figure 11** represents data collected through the use of microarray technology in which mRNA from patients with the premature aging disorder progeria is compared with that of 'normals' of differing age groups. These data provide an example of similarities in the abundance of proteins expressed within these two groups, and suggests that these proteins are somehow involved in progeria.

Techniques that provide information on the abundance of proteins in cells will be crucial to understanding how the many molecular pathways coordinate the maintenance of the intracellular environment. It is also well known that cells of multicellular organisms do not express all available proteins in every cell. By expressing specific proteins at a particular abundance, alternative cell types are available to the organism. To understand biological function, it is necessary to determine which protein complements are required for any one specific cellular contribution to the organism.

Another possibility for the use of this kind of information is that it may make it possible to treat one cell type so that it may carry out the function of another. By modifying the abundance of a subset of proteins required to compensate for the loss of that cell type, pharmaceutical intervention may allow for the compensation of diseases in which cells of a particular type have been destroyed. Potential targets for such treatment are found in diabetic patients where the loss of insulin-producing pancreatic islet cells has been effected through an autoimmune response, as well as dopamine-producing cells of the central nervous system involved in Parkinson's disease. The ability to control cell identity is dependent on being able to determine which signaling molecules are necessary to achieve changes in protein expression, thereby mimicking the cell that has been destroyed. For this, the expression profile of that cell must serve as the defining characteristic to be achieved by pharmaceutical intervention.

2.12.3.3.5 Autoregulation of gene expression

The state of a cell is regulated by the activity as well as the complement and abundance of proteins. Therefore, it is common to have immediate changes in cellular function without any change in the expression of proteins. It is this immediate early response that allows for quick adaptation and compensation to environmental conditions. The presence of phosphate on proteins that alter their specificity or enzymatic rates is a prime example of changes of state that lead to rapid cellular responses, as is subcellular localization. To use genomic techniques for the characterization of novel drugs that are designed to influence these sorts of immediate early responses, it is necessary to identify changes in protein activity. This is classically performed through the addition of radioactive phosphate that identifies proteins that have incorporated or lost these molecular markers. The purpose of this experimental approach is to ascertain if a regulatory pathway has been altered. Since genomic approaches rely on the measurement of mRNA and only approximate protein abundance, immediate early responses are not easily recognized. Nevertheless, the induction of early response mechanisms can lead to secondary effects on the abundance of proteins through transcriptional modulation. Transcription factors that regulate their own expression have been observed, and it is this autoregulation that allows the inference of changes in regulatory pathways by levels of protein expression.[188] It is the change in gene expression observed at times later than the initial treatment of cells that can be used to better understand the way in which a drug has influenced the state of the cells. Functional genomics is a tool, and all of the limitations must be recognized when inferring functions onto proteins, cells, and drugs based on changes in gene expression. Nevertheless, the power of large-scale experimentation may provide the information necessary for the informed selection of efficacious drugs.

2.12.3.4 Genomics Techniques

Genomic methodologies, like microarrays, have been recently implemented to supplant techniques like Northern and Southern blot analyses because, in general, they provide the same type of information with regard to the kinds and abundance of proteins in cells.[189–192] Northern blot analysis serves as a measure of the steady state abundance of mRNA for a transcript of interest, and this is used to approximate the abundance of a protein, since for most genes the

Middle age

Cell cycle control proteins

Acc.#	FoldΔ	Gene name
X13293	-4.8	B-myb
U74612	-3.3	Hepatocyte nuclear factor-3/fork head homolog 11 A (HFH-11A)
Z36714	-12.5	Cyclin F
X51688	-5.4	Cyclin A
M25753	-2.9	Cyclin B
U01038	-2.8	pLK
U05340	-2.9	p55CDC

Chromosomal processing and assembly

Acc.#	FoldΔ	Gene name
U30872	-2.6	Mitosin (CENP-F)
X67155	-3.5	Mitotic kinesin-like protein-1
U37426	-3.6	Kinesin-like spindle protein (HKSP)
X14850	-2.2	Histone (H2A.X)
U14518	-3.3	Centromere protein-A (CENP-A)
U63743	-2.8	Mitotic centromere-associated kinesin
X13546	-2.0	Non-histone chromosomal protein HMG-17

Protein processing

Acc.#	FoldΔ	Gene name
U73379	-2.3	Cyclin-selective ubiquitin carrier protein

Old age

Cell cycle control proteins

Acc.#	FoldΔ	Gene name
X13293	-4.9	B-myb
U74612	-9.0	Hepatocyte nuclear factor-3/fork head homolog 11A (HFH-11A)
Z36714	-13.1	Cyclin F
X51688	-5.8	Cyclin A
M25753	-5.2	Cyclin B
U01038	-3.0	pLK
U05340	-4.3	p55CDC
S78187	-7.6	CDC25B
U56816	-8.7	Kinase Myt1 (Myt1)
X54941	-2.6	Ckshsl Cksl protein homolog
M30448	-2.2	Casein kinase II beta subunit
U37022	-4.0	Cyclin-dependent kinase 4 (CDK4)
U49844	-3.0	FRAP-related protein (ATR, ATM)
X74008	-3.0	Protein phosphatase 1γ

Chromosomal processing and assembly

Acc.#	FoldΔ	Gene name
U30872	-2.6	Mitosin (CENP-F)
X67155	-3.5	Mitotic kinesin-like protein-1
U37426	-3.6	Kinesin-like spindle protein (HKSP)
X14850	-2.2	Histone (H2A.X)
U14518	-3.3	Centromere protein-A (CENP-A)
U63743	-2.8	Mitotic centromere-associated kinesin
X13546	-2.0	Non-histone chromosomal protein HMG-17
M97856	-3.0	Histone binding protein
D38076	-2.9	RanBP1(ran-binding protein 1)
X62534	-4.3	HMG-2
Y08612	-4.1	Nup88 protein
L43631	-2.5	Scaffold attachment factor (SAF-B)
U33286	-3.6	Chromosomal segregation gene homolog CAS
D26361	-4.3	KIAA0042 (centromere protein-E)
M37583	-3.0	Histone (H2A.Z)
U72342	-2.3	Platelet activating factor acetylhydrolase (45 KDa subunit LIS1)
D43948	-2.6	KIA00097 (36% similar to yeast suppressor of tubulin STU2)

Protein processing

Acc.#	FoldΔ	Gene name
U73379	-3.5	Cyclin-selective ubiquitin carrier protein
AB003102	-2.7	26S proteasome subunit p44.5
AB003103	-3.7	Proteasome subunit p55
D00760	-2.6	Proteasome subunit HC3
D00762	-3.2	Proteasome subunit HC8
D11094	-3.2	MSS1 (26S proteasome subunit)
D78275	-2.4	Proteasome subunit p42

Progeria

Cell cycle control proteins

Acc.#	FoldΔ	Gene name
X13293	-7.0	B-myb
U74612	-8.7	Hepatocyte nuclear factor-3/fork head homolog 11 A (HFH-11A)
Z36714	-8.7	Cyclin F
S78187	-3.4	CDC25B
U56816	-11.3	Kinase Myt1 (Myt1)
X54941	-4.5	Ckshsl Cksl protein homologue
M30448	-2.0	Casein kinase II beta subunit
L08246	-2.2	Myeloid cell differentiation protein (MCLI)

Chromosomal processing and assembly

Acc.#	FoldΔ	Gene name
X14850	-6.1	Histone (H2A.X)
U14518	-4.3	Centromere protein-A (CENP-A)
U63743	-4.8	Mitotic centromere-associated kinesin
X13546	-3.0	Non-histone chromosomal protein HMG 17
M97856	-2.9	Histone binding protein
D38076	-2.7	RanBP1(ran-binding protein 1)
X62534	-4.4	HMG-2
Y08612	-3.5	Nup88 protein
L43631	-3.2	Scaffold attachment factor (SAF-B)
U35451	-2.5	Heterochromatin protein p25

Figure 11 Data collected through the use of microarray technology in which mRNA from patients with the premature aging disorder progeria are compared with 'normals' of differing age groups.

mRNA abundance in the cell is at least proportional to the amount of the protein. The primary difference between genomic and blotting techniques is that many simultaneous approximations of the protein complement of a cell or tissue can be obtained in a fraction of the time. Furthermore, simultaneous measurement of experimental conditions reduces the many false assumptions made from experimental data accumulated over great periods of time. This section will survey and explain a few genomic techniques so that the relevance of the techniques may become obvious.

2.12.3.4.1 High-throughput screening

HTS relates to the parallel or serial processing of many individual samples in an attempt to obtain a large data set by which subsets will be identified through selection criteria of many forms. Like looking for the proverbial needle in a hay stack, the elementary question raised by HTS is whether or not a particular 'straw' matches your selection criteria. Unlike the more traditional one-straw-at-a-time approach of a sequential search, HTS is characterized by performing massively parallel searches. That is, it tries to look at large numbers of 'straws' simultaneously, with each one assessed for the same criteria. As simplistic as it may seem, it is the most direct method by which one evaluates a sample and decide whether it is involved with one's model or not, and the rate at which one proceeds depends greatly on how quickly and accurately the screening takes place. Microarrays are glass slides on which DNA of many different types is placed in an array format. This substrate serves as a methodological platform to identify and measure the level of expressed genes. By placing either of these substrates in direct contact with labeled cDNA, a fluorescent copy of the expressed genes of the cell, a measure of gene expression can be made. The question of whether gene A is on or gene B is off can be answered in parallel. If the question is when gene A gets turned on, a million samples may have to run in series before the answer is 'yes,' therefore making HTS a necessity.

2.12.3.4.2 Microarrays

Microarrays and other solid substrates can augment the progression of genetic analysis. The early to mid-1960s saw the beginning of a genetic revolution, which led to discoveries that until then had not previously been thought possible. Regularly, such journals as *Cell*, *Science*, and *Nature* were publishing articles pertaining to the cloning and characterization of novel genes. As technology progressed, it became clear that the regulation of the newly found genes was complicated and interlaced with environmental influences, as well as intercellular communication pathways. To make any headway into the analysis of such a complicated network of gene regulation, a fast and effective protocol for looking at the expression of these genes was critical. Microarray technologies were developed to help researchers investigate the complex nature of gene expression, and continue to be utilized on an enormous scale. To better understand the full scope of microarray and other technologies, it is helpful to be introduced to the applications in which the genomic tools are used.

2.12.3.4.3 Single-nucleotide polymorphism detection

Polymorphism is the property of taking on many forms, and it is used to describe the natural changes in the DNA sequence of a gene relative to other alleles. Microarrays and new techniques can quickly and reliably screen for single-base-pair changes in the DNA code for an entire gene (polymorphism) by using the sequence by hybridization method.[192] Sequencing through the hybridization of a target molecule to a known array of oligonucleotides has been performed successfully, and will be used extensively for the monitoring of changes related to disease.[192] With the help of genomic technologies, or by knowing the sequence of a gene associated with a disease, the diagnoses of patients with a higher degree of confidence is predicted to mature.[169,170,186,189] Since the sequence of a gene can identify phenotypic characteristics that are associated with many metabolic processes, the technique may also make it possible to use this information in deciding the proper course of treatment for patients when the ailment is known; this is called pharmacogenomics. It is well known that patients with varying genetic backgrounds respond to treatments of disease to varying degrees. Through the sequencing of particular genes related to drug-metabolizing enzymes and genes as yet unknown, we may obtain a rationale for drug therapy and design.[189]

SNP detection will be very similar to the sequencing of a whole gene, but there will only be representatives of known polymorphisms. An example of this kind of approach to diagnostics was presented by Smith and his co-workers in 1994. They used PCR to produce a defined probe, hybridized this probe to a matrix of known oligonucleotides representing exon 4 (the coding component of messenger RNA) of the human tyrosinase gene, and then scored and assessed the presence of polymorphisms. Another example of this kind of approach was performed by Klinger and her co-workers, whereby a complex, defined PCR probe was used to assess the presence of polymorphisms in the genes involved in cystic fibrosis, beta-thalassemia, sickle cell anemia, Tay–Sachs disease, Gaucher's disease, Canavan's disease, Fanconi's anemia, and breast cancer.

2.12.3.4.4 Deletion detection
Microarray and other analyses may also be performed to characterize a deletion, insertion, or mutation in genes that have been correlated to disease, such as the p53 gene to cancer. If it is suspected that a gene is mutated, clinicians will isolate DNA from the patient, use this DNA to create a probe, and then hybridize this probe to an array that can identify mutations in that gene.

2.12.3.4.5 Differential display by hybridization
Use of microarrays and other substrates for hybridization with cDNA produced from cells or tissue will make possible the identification of regulatory genes involved in the control of human disease.[189] There are models for cellular differentiation and diseases in which cDNA probes can be obtained for the purpose of hybridization to microarrays and others. The genes identified by differential hybridization to these substrates can be used as candidates in the characterization of human disease and, once identified, confirm the involvement of these candidate genes in disease. If this approach is successful, these disease genes can be used as a target for the production of drugs that influence the expression of the genes. An example of this approach is the treatment of prostate cancer cells with a chemopreventive agent to determine its involvement in prostate cancer. The observation that makes this example important is that the cells were cultured under two different conditions but, of the 5000 genes that were screened, all but a few were differentially expressed. This suggests that the change induced by the treatment may be minor and understandable. These sorts of experiments, in conjunction with additional supporting data, will provide candidates that can be screened for their usefulness as drug targets.

One of the most important questions asked by today's research scientists is, what is different between two cell types (e.g., disease versus normal, or treated versus untreated)? The only methods available for this kind of query are differential display, representational difference analysis, subtractive hybridization, serial analysis of gene expression, and differential hybridization to cDNA libraries. An exceptional example of this approach was performed by Jordan, who implemented microarrays in his hunt for genes that define the existence of three cell types in the murine thymus. By creating a 3' complex probe and hybridizing this to a microarray, he found novel genes that were differentially expressed between the three cell types. This methodology can be implemented for other models of disease as well, and may lead researchers to rational therapies for the devastating medical disorders that plague our society.

2.12.3.4.6 Expression monitoring
Expression monitoring is another approach that takes advantage of the large databases of sequenced cDNAs. By designing arrays that contain representatives of the most important genes that control the cell cycle and involve cellular regulation, an evaluation of cellular and tissue state is possible. The power of new bioinformatics software will improve the usefulness of this method, and such software is frequently used. Another possible application that has been discussed is the diagnosis of disease by association of a pattern of hybridization with that disease, as compared with positive and negative controls. In other words, does a tissue resemble a specific expression pattern as a consequence of disease?

With the recent advances in bioinformatics technology and the development of microarrays and tools that contain the number of probes necessary for the analysis of gene expression, it is now possible to confront the daunting task of human genetic analysis. Moreover, some pathological conditions caused by environmental insults instigate a cascade of changes in gene expression that can have drastic outcomes, as experienced by patients with the autoimmune disorder rheumatoid arthritis. Given the complexity of cellular signaling and regulatory pathways in a cell, it has been impossible to address these numerous interdependent changes occurring during biological process, be it pathological or normal. This allows physicians and scientists to address the needs of the cell, clearly a substantial contribution to the reduction of human morbidity.

2.12.3.4.7 Finding targets for drug development
As mentioned previously, most drugs are small organic molecules that are ingested and introduced into the circulatory system, whereupon they bind to receptors on the surface of cells whose function is to be influenced. There are approximately 30 000–35 000 genes in the human genome, producing an even greater number of proteins through alternative splicing and differences in post-translational modification. Unfortunately, only a handful of these proteins serve as targets for pharmaceutical intervention.[169] To increase the range of influence available to clinicians through drug treatment, it will be necessary to broaden the scope of protein targets used during drug development. Yet, before drugs can be specified for the treatment of clinical symptoms, a clear picture of the physiological and molecular pathways involved with the disease or pathological process must be assessed. To understand the function of any system,

all the components of that system as well as their interactions must be known. It is not enough merely to catalog all of the proteins expressed in a cell type. The interrelationships of expressed proteins must be determined through careful planning of experimental models that have defined measurable physiological or molecular changes. The extent to which the model has been characterized can synergistically increase the amount and value of the conclusions drawn from the individual experiment. Like a jigsaw puzzle, the next piece of data must be based on the existing framework. In order to increase the rate of discovery, experimental data must be gathered in parallel. This has been made possible through the advent of molecular techniques such as that associated with microarray technology. The probability of success in any experimental fishing expedition is the size and quality of the net used. With microarrays, the expression of approximately 10 000 genes can be monitored from as little as 5 μg of total mRNA. Since there are many experimental models that have distinct physiological or molecular changes associated with a defined cell type, a wealth of data can be acquired in parallel. Nevertheless, the quality of the data rests in the hands of the individuals who are responsible for the sample handling and tracking. As there is so much data to be analyzed, a small error in sample tracking can lead to devastating errors in the conclusions drawn from experiments. Sample tracking is a crucial component of the data collection and review process. In this case, the adage 'garbage in, garbage out' rings true. If the acquisition and storage of highly parallel data can be performed with some degree of confidence, then the sky is the limit. Similarities and differences, or the correlation of expressed genes with changes in responses of cells to their environment, will expand our understanding of the human machine at rates not thought of 10 years ago.

2.12.3.5 Overview

Functional genomics is a term used to describe the assembly of data to better understand the form and function of the cell. Much is already known about the mechanisms that govern the interaction of a cell with its environment, and it is this base of knowledge that will act as the primer for solving the puzzles of biology. DNA, RNA, and proteins act together to maintain the integrity of the cellular environment, and how they interact with each other is the information accessible through many newly developed high-throughput technologies. By applying high-throughput technologies to the information already known about how cells respond to their environment (functional genomics), the development of efficacious drugs with few side effects through intelligent design will be possible.

2.12.4 Integrated Proteomics Technologies

2.12.4.1 Introduction

Proteomics is a powerful approach for integration into drug discovery because it allows the examination of the cellular target of drugs, namely proteins. Understanding how drugs affect protein expression is a key goal of proteomics in a drug discovery program. Mapping of proteomes, the protein complements to genomes, from tissues and organisms has been used for the development of high-throughput screens, for the validation and forwarding of new protein targets, for the development of structure–activity relationships and for exploring mechanisms of action or toxicology of compounds, and for the identification of protein biomarkers in disease. For example, Arnot and co-workers describe an integrated proteomics approach to identify proteins differing in expression levels in phenylephrine-induced hypertrophy of cardiac muscle cells (myocytes).[193] Once a proteome is established, and expressed proteins are linked to their respective genes, then the proteome becomes a powerful means to examine global changes in protein levels and expression under changing environmental conditions. The proteome becomes a reference for future comparison across cell types and species. It is expected that proteomics will lead to important new insights into disease mechanisms and improved drug discovery strategies to produce novel therapeutics.

The application of proteomics to the study of biochemical pathways and the identification of potentially important gene products as targets for drug discovery is well established in the literature.[194–196] Celis et al. have explored the possibility of using proteome expression profiles of fresh bladder tumors to search for protein markers that may form the basis for diagnosis, prognosis, and treatment.[197] Ultimately, the goal in these studies is to identify signaling pathways and components that are affected at various stages of bladder cancer progression and that may provide novel leads in drug discovery. The explosive growth of the proteomics arena is illustrated by the rapid increase in the number of publications in the past few years (**Figure 12**). Studying individual proteins in the context of other cellular proteins is complementary to the information gathered from a genomics-based approach.

In order to increase the capacity and throughput of the information flow derived from a proteomics-based research program, several recent technologies must be highly integrated to yield a complete and efficient proteomics-based methodology.[198,199] The traditional approach involves the separation of the highly complex protein mixture, with

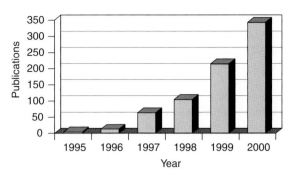

Figure 12 'Proteomics' literature publications as a function of publication year. The literature search was accomplished through the SciFinder program (American Chemical Society).

two-dimensional electrophoresis (2DE) – the most popular method because of its high sample capacity and separation efficiency. Associated with 2DE are the necessary means to analyze and store the gel images for protein correlation and data tracking. A major component of the proteomics assembly is bioinformatics, including methods that correlate changes observed in the gel patterns (attributed to changes in the environment) and software that identifies proteins from genomic and protein sequences.

A proteomics-based approach has been practiced for many years, to monitor differential protein expression. However, at the time the term 'proteomics' was coined,[200] the technology presented by new mass spectrometry (MS) methods was developed to make the identification of proteins separated by polyacrylamide gel electrophoresis (PAGE) more amenable, and it has greatly expanded the range of applications to which proteomics can contribute. MS provides a robust method for protein identification (provided the genome or protein sequences are known and available) and is being applied in many laboratories worldwide. The method can be automated, thereby greatly increasing the throughput of the analysis if many proteins need to be identified. The virtues of MS have played an important role in honing the burgeoning field of proteomics, and MS will continue to support these escalating efforts. This section serves to outline the major technologies associated with a proteomics-based approach to drug discovery.

2.12.4.2 Gel Electrophoresis

2.12.4.2.1 Two-dimensional gel electrophoresis

Since its inception, 2DE has provided researchers with a powerful tool that is capable of resolving a complex protein mixture into its individual components.[201] As implied by its descriptive nomenclature, 2DE systems separate proteins on the basis of distinct properties in each of two dimensions. In each dimension, the chosen property of the individual proteins within the mixture defines the mobility of individual proteins within an electrical field. As defined in this section, the term 2DE describes a technique that separates proteins in the first electrophoretic dimension on the basis of their charge and in the second electrophoretic dimension on the basis of their molecular mass. While 2DE separations of protein mixtures can theoretically be carried out in free aqueous solvent systems, the resultant separations are of little practical value, due principally to thermal convection and diffusion effects. Thermal effects caused by joule heating during the separation, and the effects of diffusion following the separation will serve to severely limit protein separation and, ultimately, protein resolution. To minimize these effects, 2DE is carried out in a polyacrylamide support medium. The utility of polymerizing acrylamide monomers into an electrophoretic support matrix capable of reproducibly separating complex protein mixtures was recognized long before the advent of 2DE systems.[202,203] While either electrophoretic dimension alone can independently resolve 100–200 individual proteins, the combined separation properties resident in 2DE systems offer investigators the ability to resolve up to 10 000 individual proteins.[204] The route to this level of understanding of how 2DE systems can facilitate our understanding of the proteome has been a scientifically arduous evolution.

Protein separation in the first dimension is based on the isoelectric point (pI) of individual proteins. Isoelectric focusing (IEF) depends on the formation of a continuous pH gradient through which a protein migrates under the influence of an electrical field. The net charge on the protein varies with the pH but approaches zero at its pI. At a pH equivalent to its pI, movement of the protein in the electrical field ceases, and the protein is 'isoelectrically focused.' In practical terms, IEF is achieved using either carrier ampholyte gels or immobilized pH gradient (IPG) gels. Carrier ampholytes are oligoamino and oligocarboxylic acid derivatives with molecular weights ranging from 600 Da to 900 Da.

By blending many carrier ampholyte species together, highly reproducible continuous pH gradients between pH 3 and 10 can be formed in a polyacrylamide support matrix. Typically, the first-dimension pH gradient and polyacrylamide support matrix are cast in a tube gel format that conforms to the physical measurements of the second-dimension gel system. Carrier ampholyte IEF tube gels possess several advantages over IPG strip gels. First, they are easily prepared and do not require specialized gradient-forming instrumentation. Second, carrier ampholyte IEF gels can be experimentally tailored to generate linear or nonlinear pH gradients in broad or very narrow ranges.[205] Disadvantages of carrier ampholyte use include (1) lot-to-lot variation in the ampholytes themselves, attributable to complicated organic synthetic methods, and (2) the physical chemical effect termed 'cathodic drift' that prevents a truly stable pH equilibrium state from being obtained in the first-dimension gel system. The tangible effect of cathodic drift is a progressive deterioration of the basic side of the pH gradient that can be compensated for by addressing factors such as the time and power requirements.[206]

IPG gel chemistry is based on Pharmacia Immobiline acrylamido buffers.[207] These chemical entities are bifunctional. The general chemical structure $CH_2=CHCONH\ R$ defines one end of the molecule, designated R, as the titrant moiety where a weak carboxyl or amino group provides an ionizable buffer capacity that defines the range of the continuous pH gradient. At the other end of the molecule, the $CH_2=CH-$ moiety provides an acrylic double bond that copolymerizes into the polyacrylamide support matrix. It is the polymerization of the Immobiline buffers into the polyacrylamide support matrix that provides an IPG. Because the Immobiline molecules themselves are rather simple chemical entities, the discrete molecules can be produced in a very reproducible manner with minimal lot-to-lot variation. This eliminates a principal concern associated with carrier ampholytes. IPG gradient gels also offer the prospect of increased sample loading over the carrier ampholytes. Carrier ampholyte tube gels can routinely accommodate up to 100 μg of protein, and in certain instances several hundred micrograms of protein. In comparison, IPG gels offer higher protein-loading potentials that can accommodate up to and in certain instances more than 1000 μg of protein.[208] Another benefit associated with IPG gels is the virtual elimination of cathodic drift, which permits basic proteins to be resolved.[209] While both carrier ampholyte and IPG gels offer similar pI resolution, IPG gels have been viewed as providing less reliable positional focusing for discrete protein species than carrier ampholyte gels.[205] Although casting procedures for IPG gels require specialized gradient-forming instrumentation, the increased commercial availability of IPG gels from several manufacturers minimizes this potential disadvantage.

Sample preparation presents a problem that must be considered in the context of the first-dimension IEF chemistry.[210] The first-dimension IEF chemistries that produce the desired continuous pH gradient depend on protein charge to effect zonal concentration of discrete protein species. Consequently, the introduction of charged species, especially detergents, with the sample must be minimized. This constraint makes sample preparation for 2DE analysis a daunting task. To maximize protein representation with the sample and simultaneously retain a fixed linear relationship with respect to the original material being sampled necessitates that (1) the starting material be handled appropriately to minimize modification to the protein due to degradation and spurious post-translational covalent modification, (2) the disruption and solubilization of proteins that participate in intermolecular interactions be ensured, (3) highly charged species such as nucleic acids are removed, (4) intra- and intermolecular disulfide bonds are disrupted, and (5) transmembrane proteins are solubilized and removed. To accomplish this task, basic 2DE sample preparation buffers employ high concentrations of a chaotropic agent such as urea, employ nonionic or zwitterionic detergents, use nucleases, and depend on reductants such as dithiothreitol in the presence of alkylating agents to disrupt disulfide bonds. Consideration must also be given to post-first-dimension buffer exchanges that facilitate the egress of proteins from the first-dimension gel and permit efficient penetration of the second-dimension gel matrix. An excellent compendium of recent procedures that address protein solubilization concerns associated with specific organisms and specific protein classes has been provided in a text edited by Link.[211]

Protein separation in the second dimension is based on the molecular size of individual proteins. Many variations exist as to the composition of the second dimension. Specific applications have been designed that optimize protein separation under denaturing or nondenaturing conditions, reducing or nonreducing conditions, linear or nonlinear polyacrylamide support matrixes, and numerous buffer compositions. The most frequently used second-dimension buffer systems are based on the pioneering work of Laemmli.[212] In the Laemmli system, the polyacrylamide support matrix is used to sieve proteins that have been subject to denaturation in the presence of sodium dodecyl sulfate (SDS). This crucial step is based on the premise that SDS binds uniformly to denatured macromolecules with a fixed stoichiometry, and in so doing forces them to assume a prolate ellipsoid shape. Theoretically, all protein species assume the same shape and migrate through the polyacrylamide support matrix with rates that are dependent on only their hydrodynamic radius. Consequently, the position to which a protein species migrates during a fixed time is correlated to its molecular size. This fact offers investigators the opportunity to move to dimensionally larger gel formats to increase protein loading and subsequently resolution. A larger 2DE gel system can accommodate more protein. This increases the likelihood that low-abundance proteins will be detected and that these can be separated from other proteins in the gel. An alternative approach to this problem is present in so-called zoom gels. Zoom gel systems run the same sample on narrow first-dimension pH gradients and separate on as large a second-dimension gel as possible. The broadest pH range and most extensive molecular weight range can then be reassembled into a single composite image by visual means or attempted using specialized software applications. In practice, this is very difficult to accomplish.

2.12.4.2.2 Staining methods in two-dimensional electrophoresis

As is the case for most technologies in which there are numerous acceptable methods, there is no single method that is universally used to detect proteins separated by 2DE. Various stains are widely available to visualize proteins.[213] A number of parameters need to be evaluated before a stain or dye can be chosen for a particular application. The weight of each parameter in the decision-making process is a function of the goals of the study. The sensitivity, dynamic range, and compatibility with analytical techniques associated with protein identification, as well as the cost, are all important considerations that must be addressed in order to select the most appropriate visualization agent.

The needs of a study are critical determinants when choosing a staining approach. Many factors require consideration prior to initiating 2DE, including the need to quantify protein abundance or changes in protein expression levels, the amount of protein sample available, the number of samples involved, and the need for protein identification. Most proteomic-based studies are relatively large undertakings, requiring substantial time and effort. The use of methods to visualize and detect proteins of interest consistently and to a required degree of sensitivity will ensure the success of those projects.

2.12.4.2.2.1 Coomassie stains

The Coomassie blue stains have been the most widely used for proteins separated by 2DE. Two forms of the Coomassie blue stain are available, namely Coomassie blue G-250 and Coomassie blue R-250. First used by Meyer and Lambert to stain salivary proteins, Coomassie blue stains can be used to detect as little as 200 ng of protein mm^{-2} on a 2D gel.[214] The mechanism of binding between the stain and the protein occurs through several different interactions, including the binding between basic amino acids and the acid dyes, as well as by hydrogen bonding, van der Waals attraction, and hydrophobic interactions between protein and dye.[214] A common Coomassie blue staining protocol utilizes 0.1% Coomassie blue R-250 dissolved in water–methanol–glacial acetic acid (5:4:1). The duration of gel staining increases with the thickness of the gel and decreases as a function of temperature. Gels that are 1.5 mm thick are generally stained overnight at room temperature. The quickest way to remove excess stain from the gel is by destaining in a solution containing 30% methanol–10% acetic acid. An alternative method dissolves the stain in trichloroacetic acid (TCA). The relative insolubility of Coomassie stain in TCA results in the preferential formation of protein–dye complexes. The result is the rapid staining of a 2D gel with little destaining required.[214]

The major disadvantages with the use of Coomassie stains are the inability to accurately quantify proteins and the relative insensitivity of the stain. Despite these drawbacks, Coomassie blue stains have a number of advantages that will ensure their continued utility for visualizing proteins separated by 2DE. There are many instances in which a high degree of sensitivity is not required. The proteins of interest may be in high abundance, either naturally or through manipulation of the sample preparation using affinity chromatography. In addition, the inherent pH stability of the IPG strips allows the application of milligram quantities of protein to the first dimension,[215] dramatically increasing the abundance of protein spots present in the SDS–PAGE gel. Most importantly, unlike silver stain, proteins stained with Coomassie blue can easily be processed with analytical procedures used in protein identification. Finally, Coomassie stains are inexpensive and reusable, making them attractive to laboratories with low budgets.

2.12.4.2.2.2 Silver staining

The use of silver stain to visualize electrophoretically separated proteins was proposed in 1979 by Switzer and co-workers.[216] With a sensitivity approximately 100 times that of Coomassie blue, silver stain offers the ability to detect as little as 1 ng of protein. Numerous studies have focused on optimizing the various steps in the silver staining protocol.[217] The majority of methods conducted have utilized either silver diammine (alkaline methods) or silver nitrate (acidic methods) as the silvering agent. The acidic methods can be further categorized into lengthy procedures, which require greater than 5 h, and the less sensitive rapid methods, whereby staining can be completed in less time. A comparative study conducted by Rabilloud determined protocols utilizing silver diammine as well as glutaraldehyde as an enhancing agent (discussed below) to be the most sensitive staining methods.[218] Although somewhat laborious, the silver staining of 2DE gels is a relatively straightforward procedure. The initial step of the silver stain procedure involves the immersion of the gels in a fixative solution. This solution (generally containing acetic acid and either methanol or ethanol) functions to precipitate the proteins within the gel matrix as well as remove interfering substances such as detergents, reducing agents, and buffer constituents (e.g., Tris).[217] The second step is aimed at enhancing the subsequent image formation. Enhancement can be achieved through increasing the silver binding (referred to as amplification) to the proteins through the use of aromatic sulfonates or dyes such as Coomassie blue. The process of minimizing the formation of a background image through the use of oxidizing agents (permanganate or dichromate) is known as contrastization, and is usually used only in the acidic methods. A third major enhancement technique (sensitization) involves increasing the rate at which silver reduction occurs on the proteins. Sensitization can be accomplished using sulfiding agents (thiourea) and/or reducing agents (e.g., dithiothreitol or DTT). Many of the existing staining protocols contain a combination of enhancement techniques to improve the subsequent image formation. Following silver impregnation, image development is accomplished through the reduction of Ag^+ to metallic Ag. Development solutions used with silver nitrate generally contain formaldehyde, carbonate, and thiosulfate, while those methods utilizing silver diammine are composed of formaldehyde and citric acid.[219]

Despite the high sensitivity of silver stain, there are a number of major drawbacks and limitations associated with this technique. The routine use of silver stain is relatively expensive due to the high price of reagents as well as the cost of disposal of hazardous waste. In addition, poor images may result from high background noise, often resulting from the use of poor-quality water. A study investigating the contributing factors of background silver staining also suggested the involvement of a redox initiator system such as ammonium persulfate and an amine (e.g., N,N,N',N'-tetramethylethylenediamine).[219] There is also a significant protein-to-protein variability relative to the extent of silver deposited on the protein. Some proteins do not stain at all, or are detected as negatively stained in contrast to a dark background.[220] The polypeptide sequence and the degree of glycosylation function to influence the intensity and color of the resulting stained proteins. Silver stain is not exclusive for proteins, since DNA as well as lipopolysaccharides may also be stained. Other limitations of silver staining include poor linearity to protein concentration[221] and relative incompatibility with subsequent analytical techniques. A study by Gharahdaghi and co-workers demonstrated an improvement in sensitivity when MS was used to identify proteins from silver-stained gels destained with Farmer's reducer prior to enzymatic digestion and analysis.[222] The ability to routinely identify silver-stained proteins using MS analysis would be a significant advancement in the field of proteomics.

2.12.4.2.2.3 Fluorescent stains

One of the most promising new developments in the area of visualizing proteins on 2DE gels has been the recent commercial availability of reliable fluorescent stains. A widely used fluorescent stain is the Sypro line, available from Molecular Probes. Stains such as Sypro ruby have a number of distinct advantages over both Coomassie and silver stains. Several of these advantages are a function of the binding that occurs between the proteins and the stain. Rather than binding to specific functional groups or portions of the polypeptide backbone, the Sypro stains actually bind to the SDS molecules coating the protein.[223] This type of interaction minimizes the protein-to-protein signal variation and allows quantitative comparison between proteins. In addition, the noncovalent binding of the stain to protein–SDS complexes does not mask antigenic sites, permitting Western blot analysis when the gels are processed using a nonfixative staining procedure.[223] The dynamic range of the fluorescent stains is 5–10 times higher than either Coomassie or silver stains, which allows for accurate determination of protein expression levels. Equally impressive is the fact that many of the new fluorescent stains have sensitivities equal to or greater than silver stain. In addition, the protocol for fluorescent staining is very simple, requiring only a brief incubation (30 min) in a fixative solution followed by incubation in the dye from 90 min to overnight. The fluorescently labeled proteins can be visualized using an ultraviolet transilluminator, a blue-light transilluminator, or a laser scanning instrument with documentation being achieved through the use of black-and-white print film, charge-coupled device (CCD) camera, or laser scanning instrument. Several of the newer stains, such as Sypro ruby, are completely compatible with MS and microsequencing.[223] Despite all of the positive attributes of fluorescent stains, the major drawback is the cost of the stain, which can be prohibitive to most laboratories running large-format 2D gels.

Fluorescent dyes such as the cyanine dyes have also been used to detect proteins separated by 2DE.[224] The major advantage of fluorescent dyes is the ability to detect protein differences between samples using a single gel. The technique employed is referred to as difference gel electrophoresis (DIGE). DIGE involves labeling the samples of interest with different dyes, and combining the samples into a single sample that is then subjected to 2DE. The fluorescence images obtained from the different dyes are then superimposed to detect protein differences between samples. The use of a single gel eliminates the gel-to-gel variability associated with 2DE as well as the difficulties associated with matching spots between gels.[224] The success of DIGE is based on identical proteins labeled with different dyes having the same electrophoretic mobility. Similar to the case of fluorescent stains, the high cost associated with fluorescent dyes is the major drawback, along with the fact that many are not readily available as consumables.

2.12.4.2.3 Imaging and image analysis

Rigorous attention to detail from sample preparation to image analysis is necessary if the 2DE project is designed to generate valid comparative and numerical data. While precisely determining how much protein is loaded on the first-dimension gel determines the validity of subsequent image analysis, the type of 2DE image is in large part determined by the nature of the sample itself. Autoradiographic images derived from isotopically labeled samples can be collected on film or phosphor imaging systems. Difficulties associated with film response to weak beta-emitting isotopes and a modest linear dynamic range of 300:1 have led most investigators to employ phosphor imaging devices. Phosphor imaging systems have linear dynamic ranges that span up to five orders of magnitude and are as much as 250 times more intrinsically sensitive to radiolabeled proteins than film.[225] Other advantages of phosphor imaging systems include fast imaging times, acquisition of digitized data ready for computational analysis, and the chemical development of films is not required. A significant drawback to phosphor imaging systems is their cost and the requirement that different screens be used to measure specific radioisotopes. A particularly attractive use of phosphor imaging technology is the application of double-label analysis. Experimental samples can be prepared with two different radiolabels. Phosphor images can be acquired that track two different biochemical processes within the same sample and from the same gel by imposing selective shielding between the source and the target following the initial image. This approach circumvents problems associated with intergel reproducibility. Nonradiolabeled samples can be imaged using chromogenic or chemiluminescent immunological methods[226,227] or conventional stains or dyes. Current methods for 2DE image acquisition depend primarily on CCD camera systems and document scanning devices.[228,229] CCD camera systems employ CCD devices in lieu of conventional film to acquire the image. A characteristic emission spectrum and stability of the light source as well as a uniform diffusion of the light across the 2DE gel surface are necessary to ensure that optimized, consistent images are obtained over time. The CCD chip image map can be readily converted to digital form and downloaded to a computer for subsequent analysis. Document scanners acquire 2DE image information through an array of photodetectors that are moved across a 2DE gel illuminated by a white or filtered light source. For 2DE gel applications it is important that the scanner obtain and record a uniform image map across the length and width of the gel. Document-scanning features that change the gain settings to optimize text contrast during image acquisition should be disabled to ensure that a uniform image is obtained for subsequent analysis. During image acquisition, any changes in the relationship between the image CCD device or scanner must be noted. A gray scale step tablet should be used initially, to calibrate and to recalibrate the system following any changes.

All steps of the 2DE process culminate in the generation of an image and establish its analytical value. The central tenet of 2DE image analysis (i.e., to provide a comparative means to detect and measure changes) has changed little since the QUEST system was designed and developed by Garrels *et al.* about three decades ago.[230] The specific intent for the QUEST system was to develop a system for quantitative 2DE that was exemplary in gel electrophoresis, image processing, and data management. Although advances in computer computational speed and power have made more complex analyses possible by more sophisticated algorithms in less time, these principal goals remain. Currently, several 2DE image analysis software systems are available commercially; among them are Melanie II and PDQUEST (BioRad Laboratories), BioImage (Genomic SolutionsI), and Phoretix 2D Advanced (Phoretix International). While each software system possesses unique features, the basic approach to 2DE gel analysis is quite similar. The process of image analysis begins with spot picking and image editing. Spot-picking algorithms are effective but they frequently miss spots, fail to resolve multiple spots, or identify artifacts as spots. These occurrences must be visually identified and corrected by the investigator. Once edited, reference spots are identified that appear consistently and are well resolved on each gel in a project. These are used to register the individual gels across a project. Matching individual spots across all gels in a project follows the gel registration process. This process is arduous and time-consuming. For each hour spent producing a 2DE gel, 4–6 h may be spent in image editing and analysis. Once the imaged spots have been matched, statistical processes can be applied that detect and define significant differences between spots across respective groups within a project. At this point, data management and data visualization tools are required.

Data management and data visualization are critical concerns in the evolution of a 2DE project. Although all investigators confront these issues when a 2DE project is completed, an integrated approach that considers data management and visualization during the planning stages of a 2DE project offers the best opportunity to optimize the process and achieve intended project goals. A clear understanding of the specific aim of the project at its onset can define several important strategic concerns. How will the 2DE data be analyzed and what level of validation is needed? Is visual examination of the 2DE gel data sufficient or are statistical methods required? Is the experimental design appropriate and are sample numbers adequate to ensure that the experimental outcomes can be determined without ambiguity? What form of raw data must be harvested from the 2DE images and by what analytical method? Does this project stand alone or will data be integrated into a larger ongoing project? Will data be added to a database? If so, what annotation will be necessary to ensure continuity with existing data? Will important spots be excised from the gel for subsequent identification? In simple terms, the outcome of a 2DE project is determined by careful planning. If visual inspection of the 2DE gel data is sufficient to achieve project-specific aims, then 2DE images must be evaluated in a consistent manner and the results cross-checked and agreed on by several observers. If statistical methods are to be used, then the type of raw data that will be used in the analysis determines the output of the image analysis system. The statistical analysis can be performed at a single level with an analysis that employs *t* tests or can be multilevel, with each successive level dependent on the previous analysis. The type of assay that is used is an important determinant of whether the final 2DE data represent a completed project or become a covariant in a larger model that includes data from other sources. Preliminary statistical analysis of the spot data can be accomplished using a straightforward *t* test procedure based on individual spot intensity values or integrated spot intensity values. Mapping statistically changed spot matches onto the 2DE gel images can provide a visual signature that is characteristic of a treatment, organism, or pathologic process. Signature patterns can be visualized on gels or by more complex statistical methods that cluster data and create strong visual links between related and disparate statistical data. There are many alternative approaches to the visualization of complex computational data sets.[231] In the absence of established standards, investigators should be guided by a desire to present data graphically with simplicity, clarity, precision, and efficiency.[232]

2.12.4.2.4 Protein spot excision

The excision of protein spots from a 2DE gel can fulfill a tactical objective or be a component of a more strategic objective. Fundamentally, spots are excised from 2DE gels to provide a source of material that can be used to identify the protein(s) resident in the spot. From a tactical perspective, a 2DE project can represent a completed experimental objective. In this context, the identification of important spots derived from analytical means applied completes the project. In contrast, a 2DE project can be a component of a much larger ongoing project. In this circumstance, exhaustive protein spot excision that accounts for all spots in a 2DE gel may be used to establish a repository of information that can be stored and drawn on at a later date. Regardless of the intent, protein spot excision requires that spots be precisely removed, and that protein(s) present in the spot be abundant with respect to the detection limit of the identification technology and be stained in a manner that does not interfere with identification technology. Several approaches to protein spot removal should be considered. With respect to protein spots, the 2DE gel is quite variable with respect to spot density. Consequently, picking methods that remove defined sections of the 2DE gel in a grid pattern are inefficient in some regions of the gel and prone to cross-contamination from multiple spots in others. Conversely, a picking method that punches specific spots performs equally across the gel but requires considerable investigator skill and time. An alternative is the application of a spot excision robot to the process. This is feasible only if large numbers of protein spots are to be excised from many gels over time due to the fact that robotic systems are costly and have only recently become commercially available (e.g., Genomic Solutions), suggesting that performance and reliability remain largely unknown. Many wide-bore needles can be beveled flat and used as spot excision devices for the price of an automated robotic system. In regions of high density, great accuracy must be used to remove only the spot intended without contaminating the target spot or affecting the ability to excise proximal spots. To accomplish this manually or through automation requires that the excision device be precisely centered on the spot and have a diameter that does not exceed the diameter of the spot. The flip side of this is that in instances where a large spot needs to be excised in its entirety, a larger punching device or multiple punches must be taken. Given the present state of robotic systems, this is much easier to accomplish manually. Another factor should be considered in the sampling of low-abundance protein spots. First, staining is not necessarily linearly correlated to spot intensity. If multiple spots are to be pooled to increase the apparent abundance of a protein, a point of diminishing returns may be reached where excision from multiple 2DE gels fails to increase the protein extracted from the gel above the detection limit of the detection device. In this instance, preparative sample loading on a fresh 2DE gel is often the better solution. Keratin contamination is a major concern during spot excision. Keratin proteins from many sources must be assumed to be present in the laboratory environment unless specific and heroic measures are taken to ensure and guarantee its absence. Precautions against keratin contamination must be taken at all steps of the 2DE gel process to avoid contamination of the sample, gel chemistry components, and the surface of the gel. During spot excision, the investigator should wear protective clothing, gloves, a hair net, and a mask. If a robotic system is used, careful maintenance, cleaning, and use of an enclosure system will greatly reduce potential keratin contamination.

2.12.4.3 Protein Identification

2.12.4.3.1 General concepts

Proteins are separated first by high-resolution 2D-PAGE and then stained. At this point, to identify an individual or set of protein spots, several options can be considered by the researcher, depending on the availability of techniques. For protein spots that appear to be relatively abundant (e.g., >1 pmol), traditional protein characterization methods may be employed. Specifically, methods such as amino acid analysis and Edman sequencing can be used to provide necessary protein identification information. With 2DE, approximate molecular weight and isoelectric point characteristics are provided. Augmented with information on amino acid composition and/or amino-terminal sequence, accurate identification can be achieved.[198,233,234]

However, most of the emphasis the last few years has focused on employing MS-based methods. The sensitivity gains of using MS allows for the identification of proteins below the 1 pmol level and in many cases in the femtomole range. More confident protein identifications can be made with the mass accuracy of the MS molecular mass measurement. For peptide fragments less than 3000 Da, a mass accuracy of better than 100 ppm can be obtained. For example, for a peptide of molecular mass 1500 Da, the accuracy of the MS measurement can be ± 0.075 Da (50 ppm). Moreover, MS is highly amenable to automation, as steps from sample preparation, to sample injection, to data acquisition, and to data interpretation can be performed unattended. Automation is an important concept as capacities increase. It can reduce sources of sample contamination as well as sample-handling errors. A general approach employing MS for protein identification is shown in **Figure 13**. Great strides have been made toward the automation of these procedures, and many more improvements will become evident in the near future. The rapid increase in proteomic sample throughput arising from constant improvements in the automation of sample preparation and data acquisition have resulted in ongoing generation of an overwhelming amount of data that must be mined for significant results. This has necessitated the parallel development of increasingly efficient computer-based data-searching strategies, which can rapidly provide protein identifications from experimentally derived polypeptide molecular masses and sequences. With these clear advantages of using an MS-based approach, proteomics research has embraced MS with enthusiastic support.

2.12.4.3.2 Automated proteolytic digestion

For the application of MS for protein identification, the protein bands/spots from a 2D gel are excised and are exposed to a highly specific enzymatic cleavage reagent (e.g., trypsin cleaves on the C-terminal side of arginine and lysine residues). The resulting tryptic fragments are extracted from the gel slice, and are then subjected to MS methods. One of the major barriers to high throughput in the proteomic approach to protein identification is the 'in-gel' proteolytic digestion and subsequent extraction of the proteolytic peptides from the gel. Common protocols for this process are often long and labor-intensive. An example of an in-gel digest and extraction protocol[235] is as follows:

(1) Washing and dehydration of the gel:

   (a) Add 100 μL of a 1:1 mixture of 100 mM NH_4HCO_3/CH_3CN to each tube containing a gel piece and let stand for 15 min.
   (b) Remove the supernatant and dehydrate with 25 μL of CH_3CN for 10 min.
   (c) Remove the supernatant and dry by vacuum centrifugation for 5 min.

(2) Reduction and alkylation of the protein:

   (a) Add 10 μL of 10 mM dithiothreitol in 100 mM NH_4HCO_3 and let stand for 60 min at 56 °C.
   (b) Allow to cool, then add 10 μL of 100 mM fresh iodoacetic acid in 100 mM NH_4HCO_3 and let stand for 45 min at 45 °C in the dark.

(3) Washing and dehydration of the gel:

   (a) Add 50 μL of 100 mM NH_4HCO_3 and let stand for 10 min, then remove the supernatant.
   (b) Add 25 μL of CH_3CN and let stand for 10 min, then remove the supernatant.
   (c) Repeat steps a and b.
   (d) Dry by vacuum centrifugation for 5 min.

(4) Trypsin digestion:

   (a) Swell the gel disk in 15 μL of 50 mM NH_4HCO_3 containing 10 ng μL^{-1} of TPCK-modified trypsin in an ice bath for 45 min.
   (b) Add 15 μL 50 mM NH_4HCO_3 and digest overnight at 37 °C in the dark.

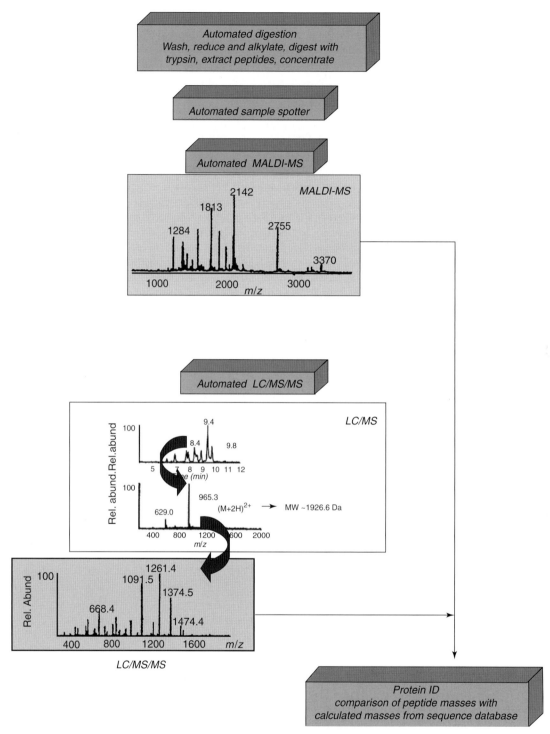

Figure 13 A general scheme for protein identification utilizing an MS-based approach.

(5) Extraction of tryptic peptides from the gel:

(a) Extract with 30 μL of 50% CH$_3$CN containing 1% trifluoroacetic acid (TFA) three times for 10 min each.
(b) Combine the supernatant for each of the three extractions and dry by vacuum centrifugation.
(c) Resuspend peptides in 5 μL of 30% CH$_3$CN containing 1% TFA and store at −80 °C.

Using the above protocol, a scientist can comfortably digest and extract 25 samples in less than 36 h, which limits the throughput to approximately 100 in-gel digests and extractions per week. (In our experience, significant human error can be introduced over time while manually processing more than 25 samples per day.) Reducing the digest time from overnight to 2–3 h can improve the throughput, but not significantly. However, a 2D gel of a cell lysate can contain literally thousands of unique proteins, seen as stained spots. Even if the spots are excised from the gel manually, hundreds of samples requiring digestion and extraction can be generated per day. Setting aside the fact that in a proteomics project it is rare that each experiment would require every single spot from the gel to be identified, it is obvious that many more samples can be generated per week per scientist than digested and extracted manually.

A welcome alternative to the lengthy and laborious manual digest and extraction process are several commercially available automated digest and extraction robots.[236–238] The most useful of these robotic systems have the following features in common: some form of computer control, temperature control, options for allowing digestion with multiple enzymes, and the ability to digest gel pieces and extract the subsequent proteolytic peptides in a 96-well plate format, allowing parallel sample processing of 96 samples at a time. This will increase the throughput to approximately 400 samples per week per robot using the above protocol, and potentially far more using a shorter digestion period.

Commercial units from Genomic Solutions (ProGest) and AbiMed (Digest-Pro) incorporate a particularly ingenious mechanism for overcoming possible sample contamination issues arising from the liquid-handling syringe system. The syringe never touches the solutions in the wells. Instead, it pressurizes the wells such that the liquid flows from two small holes laser drilled into the bottom sides of each of the wells in the 96-well PCR plate. The robot slides the reaction PCR 96-well plate over either a waste position for adding and discarding reaction reagents or a sample collection position for the extraction of proteolytic peptides. This methodology also decreases the opportunities for sample loss to adsorptive surfaces, although this remains a significant issue for low-level samples.

2.12.4.3.3 Peptide mapping by matrix-assisted laser desorption/ionization mass spectrometry

A mass spectrum of the resulting digest products produces a 'peptide map' or a 'peptide fingerprint'; the measured masses can be compared with theoretical peptide maps derived from database sequences for identification. There are a few choices of mass analysis that can be selected from this point, depending on the available instrumentation and other factors. The resulting peptide fragments can be subjected to MALDI[239] and/or electrospray ionization (ESI)-MS[240] analysis.

MALDI-MS is a tool that has rapidly grown in popularity in the area of bioanalytical MS. This is due to recent improvements in time-of-flight (TOF) technology through enhancements in resolution and, subsequently, mass accuracy, which have increased the usefulness of MALDI-MS data. Currently, the TOF analyzer is the most common system for MALDI. Improvements such as the reflectron analyzer and time-lag focusing (delayed extraction) have greatly improved the quality of MALDI-MS data.[241] Automation has also played an important role in the increased use of MALDI-MS as a tool for proteome analysis by allowing automated acquisition of data, followed by fully automated database searching of protein sequence databases such as SWISS-PROT and NCBI.[194,242,243]

To better understand the reasons for the popularity of MALDI-MS, one must first understand the simplicity and speed of the approach. The peptide analyte of interest is co-crystallized on the MALDI target plate with an appropriate matrix (i.e., 4-hydroxy-α-cyanocinnamic acid or 3,5-dimethoxy-4-hydroxycinnamic acid (sinapinic acid)), which are small, highly conjugated organic molecules that strongly absorb energy at 337 nm. In most MALDI devices, 337 nm irradiation is provided by a nitrogen laser. Although other lasers operating at different wavelengths can be used, size, cost, and ease of operation have made the nitrogen laser the most popular choice. The target plate is then inserted into the high-vacuum region of the source, and the sample is irradiated with a laser pulse. The matrix absorbs the laser energy, and transfers energy to the analyte molecule. The molecules are desorbed and ionized during this stage of the process. The ions are then accelerated under constant kinetic energy (V) down the flight tube of the TOF instrument by using acceleration potentials up to 30 kV. This acceleration potential imparts the ions with nearly the same kinetic energy, and they each obtain a characteristic flight time (t) based on their mass-to-charge (m/z) ratio and total flight distance (L):

$$t = \sqrt{[(m/z)(1/2v)]}$$

As a result, ions of different masses are separated as they travel down the field-free region of the flight tube of the mass spectrometer (lighter ions travel faster than larger ions), then strike the detector and are registered by the data system, which converts the flight times to masses.

A MALDI-TOF mass spectrometer consists of six major sections:

(1) Laser system: a device capable of supplying an appropriate wavelength of energy sufficient to desorb and ionize matrix and analyte from the target surface.

(2) Source: a section of the instrument that contains the target plate that holds the matrix and analyte of interest. Ions generated in the source region are accelerated down the flight tube.
(3) Flight tube: a field-free region of the mass spectrometer where separation of the ions based on their characteristic mass-to-charge ratio occurs.
(4) Detector: a device that detects and measures ions.
(5) Data system: a computer that converts the detector output into an easily interpreted form (mass spectrum) and allows storage of the spectral data.
(6) Vacuum system: components necessary to evacuate the source and free-flight region, creating pressures in the 10^{-6}–10^{-7} torr range.

To obtain the best mass accuracy possible, MALDI-MS instrumentation equipped with a reflectron and time-lag focusing is commonly employed.[241] A reflectron is used to compensate for the initial energy spread that the ions may have following desorption off the sample target plate. Ions of the same mass-to-charge ratio, which have slightly different energy, have different final velocities, and therefore arrive at the detector at slightly different times, resulting in loss of resolution. To compensate for this effect, an ion mirror or reflection is used to focus the ions by creating a flight distance gradient: ions with higher kinetic energy penetrate more deeply and thus travel a longer total distance than ions of lower energy. In addition, time-lag focusing (commonly referred to as delayed extraction) allows the initial energy spread to be partially focused prior to accelerating the ions. The combination of time-lag focusing with an ion mirror provides for higher order energy focusing, resulting in significantly enhanced mass-resolving power and improved mass accuracy. In many cases, this combination can increase MALDI-TOF mass accuracy to better than 20 ppm for peptides of molecular mass 500–3000 Da.

The large number of samples generated by a proteomics approach requires a capability for very high throughput that can be achieved through automation of the entire process. Most commercial MALDI-MS instruments have capabilities for automated data acquisition. However, the amount of interactive control feedback can vary greatly, and this feature should be considered carefully when evaluating this type of analysis. The quality of the data and the speed at which it is obtained can be greatly affected by laser energy and the homogeneity of the analyte spot. To minimize the effects caused by inhomogeneous analyte crystals and searching for 'the sweet spot' across a sample spot, an automated MALDI sample preparation robot can be used. Currently, there are several automated sample preparation robots available commercially, including ones from Micromass, Bruker, and PerSeptive Biosystems.

The MALDI sample spotter supplied by Genomic Solutions is capable of spotting the target configurations of various instrument manufacturers' systems. In addition, the system can be programmed with sample preparation protocols to remove gel contaminants (i.e., salts, ammonium bicarbonate, SDS, etc.) that could interfere with or degrade spectral quality. For example, the use of pipette tips packed with reversed-phase media (e.g., C18 ZipTips from Millipore) effectively desalts and concentrates peptides for MALDI-MS analysis.

A small aliquot of the digest solution can be directly analyzed by MALDI-MS to obtain a peptide map. The resulting sequence coverage (relative to the entire protein sequence) displayed from the total number of tryptic peptides observed in the MALDI mass spectrum can be quite high (i.e., greater than 80% of the sequence), although it can vary considerably depending on the protein, sample amount, and so forth. The measured molecular weights of the peptide fragments along with the specificity of the enzyme employed can be searched and compared against protein sequence databases using a number of computer searching routines available on the internet. An example of such an exercise is depicted in **Figure 14** and is discussed further in Section 2.12.4.3.5.

2.12.4.3.4 High-performance liquid chromatography-mass spectrometry and tandem mass spectrometry

An approach for peptide mapping similar to MALDI-MS involves the use of ESI-MS. ESI is a solution-based ionization method. Analyte solutions flowing in the presence of a high electric field produce submicrometer-sized droplets. As the droplets travel toward the mass spectrometer orifice at atmospheric pressure, they evaporate and eject charged analyte ions. These ions are sampled by the mass spectrometer for subsequent mass measurement. A peptide map can be obtained by the direct analysis of the peptide mixture by ESI-MS. A major advantage of ESI over other analytical approaches is its ease of coupling to separation methodologies such as high-performance liquid chromatography (HPLC). Thus, alternatively, to reduce the complexity of the mixture, the peptides can be separated by reversed-phase HPLC with subsequent mass measurement by on-line ESI-MS (see **Figure 13**). The measured masses can be similarly compared with sequence databases. An example of an LC/MS application is shown in **Figure 15**.

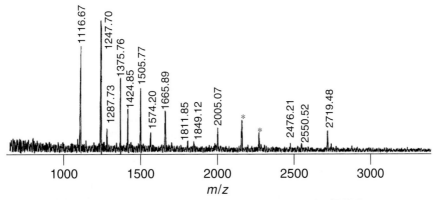

Figure 14 MALDI-MS peptide map of *Escherichia coli* protein, Cell Division Inhibitor MinD (MIND_ECOLI, accession #P18197), separated by 2DE. The molecular masses labeled on the spectrum were compared with the sequence of the protein for identification ('*' denotes trypsin autolysis peaks). The portions of the protein sequence for the observed tryptic peptides are highlighted on the sequence.

To provide further confirmation of the identification, if a tandem mass spectrometer is available, peptide ions can be dissociated in the mass spectrometer to provide direct-sequence information. Peptide fragmentation typically occurs along the polypeptide backbone to produce products termed 'y-type' and 'b-type' fragment ions (in which the y ions contain the C-terminal portion of the peptide and the b ions contain the N-terminal portion). These product ions from an MS/MS spectrum can be compared with available sequences using powerful software tools as well. In many examples, laboratories may use nanoelectrospray with MS/MS to examine the unseparated digest mixtures with a high degree of success.[194]

One of the most common tandem mass spectrometers used for proteomics applications is the quadrupole ion trap mass spectrometer.[193,243–245] An ion trap is a ion storage device that utilizes radiofrequency voltage across a ring electrode to contain ions.[246] As the radiofrequency amplitude increases, ions of increasing mass become unstable in the trap, and are ejected toward a detector. The ion trap mass spectrometer has become a workhorse instrument for proteomics because of its ease of use and because of its high efficiency for generating sequence information. Examples of MS/MS spectra derived from tryptic peptides are shown in **Figure 16**. For the examples shown, the MS/MS spectra were readily interpreted via computer programs.

For a single sample, LC/MS/MS analysis included two discrete steps: (1) LC/MS peptide mapping to identify peptide ions from the digestion mixture and to deduce their molecular weights, and (2) LC/MS/MS of the previously detected peptides to obtain sequence information for protein identification. An improvement in efficiency and throughput of the overall method can be obtained by performing LC/MS/MS in the data-dependent mode. As full-scan mass spectra are acquired continuously in the LC/MS mode, any ion detected with a signal intensity above a predefined threshold will trigger the mass spectrometer to switch to the MS/MS mode. Thus, the ion trap mass spectrometer switches back and forth between the MS mode (molecular mass information) and the MS/MS mode (sequence information) in a single LC run. This feature was implemented to generate the spectra shown in **Figure 16**. The data-dependent scanning capability, combined with an autosampler device, can dramatically increase the capacity and throughput for protein identification.

2.12.4.3.5 Computer-based sequence searching strategies

The concept of database searching with mass spectral data is not a new one. It has been a routine practice for the interpretation of electron ionization mass spectra for many years. However, there is a fundamental difference between database searching of electron ionization spectra and proteomic spectra. In the former, an experimentally acquired

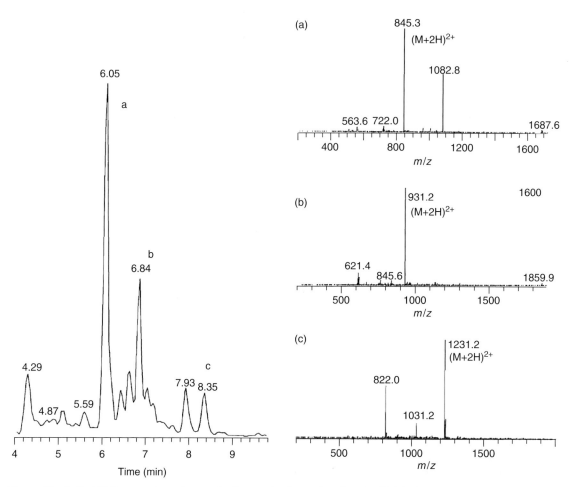

Figure 15 (Left): HPLC mass chromatogram for the tryptic digest of ovalbumin. (Right): the resulting mass spectra at the indicated retention times. Approximately, 0.25 μg of protein was loaded onto an SDS-PAGE gel, and the protein was stained with zinc imidazole. An ESI ion trap mass spectrometer (LCQ, Finnigan MAT, San Jose, CA) coupled on-line with capillary HPLC (MicroPro Syringe Pumping System, Eldex Laboratories, Napa, CA) in conjunction with a capillary peptide trapping precolumn was used for peptide analysis.

spectrum is compared with a large collection of spectra acquired from known, authentic standards. Fundamental features, such as the ions present and their relative intensities, are compared, to establish a match and a 'quality of fit.' This type of computer-based searching is typically referred to as 'library searching.' This is in contrast to the strategy employed with proteomic data, where a list of experimentally determined masses is compared with lists of computer-generated theoretical masses prepared from a database of protein primary sequences. With the current exponential growth in the generation of genomic data, these databases are expanding every day.

There are typically three types of search strategies employed: searching with peptide fingerprint data, searching with sequence data, and searching with raw MS/MS data. These strategies will be discussed below. While the strategy employed often depends on the type of data available, there is also a logical progression, based on the cost of each strategy in terms of time and effort. While many of the computer-based search engines allow for the purchase of an on-site license for use with proprietary data, most are available on the internet for on-line access and use with public domain databases (**Table 15**). One limiting factor that must be considered for all of the following approaches is that they can only identify proteins that have been identified and reside in an available database or are highly homologous to one that resides in the database.

2.12.4.3.5.1 Searching with peptide fingerprints

MALDI peptide fingerprinting is probably the most rapid technique with which to identify an unknown protein. Computer software is used to theoretically 'digest' the individual protein components of a protein database using the same protease, generating a list of theoretical peptide masses that would be derived for each entry. The experimentally

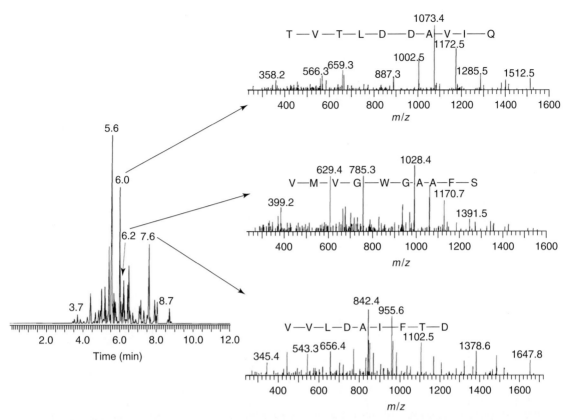

Figure 16 HPLC-MS/MS data-dependent scan results for a tryptic digest of yeast enolase. The ion trap mass spectrometer automatically acquires MS/MS data for peptide peaks above a defined abundance threshold. The HPLC mass chromatogram is shown on the left, and representative MS/MS spectra are shown on the right. From top to bottom, the observed tryptic peptides of enolase are TAGIQIVADDLTVTNPK (residues 312–328), AAQDS-FAAGWGVMVSHR (residues 358–374), and SGETEDT-FIADLV-VGLR (residues 375–391).

Table 15 Some representative internet resources for protein identification from mass spectrometry data

Program	Website
BLAST	http://www2.ebi.ac.uk/blastall/
Mascot	http://www.matrixscience.com/cgi/index.pl?page=/home.html
MOWSE	http://srs.hgmp.mrc.ac.uk/cgi-bin/mowse
PeptideSearch	http://www.mann.embl-heidelberg.de/Services/PeptideSearch/PeptideSearchIntro.html
ProteinProspector	http://prospector.ucsf.edu/
Prowl	http://www.proteometrics.com/

determined peptide masses are then compared with the theoretical ones in order to determine the identity of the unknown (as discussed in Section 2.12.4.3.6 and depicted in **Figure 17**).

The majority of the available search engines allow one to define certain experimental parameters to optimize a particular search. Among the more typical are: the minimum number of peptides to be matched, allowable mass error, monoisotopic versus average mass data, the mass range of the starting protein, and the type of protease used for digestion. In addition, most permit the inclusion of information about potential protein modification, such as N- and C-terminal modification, carboxymethylation, and oxidized methionines. However, most protein databases contain primary sequence information only, and any shift in mass incorporated into the primary sequence as a result of post-translational modification will result in an experimental mass that is in disagreement with the theoretical mass. In fact,

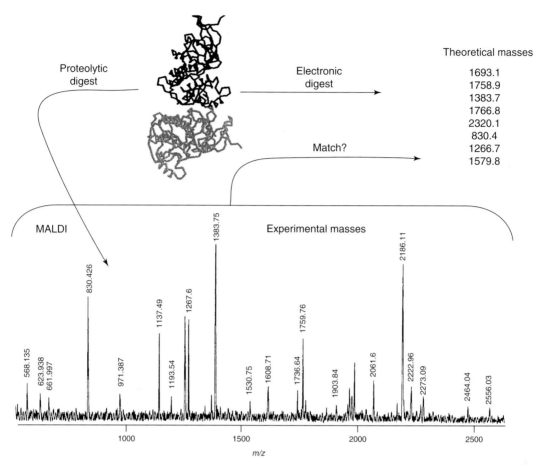

Figure 17 Principles of protein identification by MALDI-MS peptide fingerprinting.

this is one of the greatest shortcomings of this particular method. Modifications such as glycation and phosphorylation can result in missed identifications. Furthermore, a single amino acid substitution can shift the mass of a peptide to such a degree that even a protein with a great deal of homology with another in the database cannot be identified. Interestingly, one software package, ProteinProspector, has a feature within its MS-Fit subprogram that allows one to search peptide fingerprint data in the 'homology mode.' In this mode, the software considers possible mutations of individual amino acid residues in an effort to transform a theoretical peptide mass to match an experimental one. This can be an extremely powerful tool for the identification of a protein that is not in the database yet is significantly homologous to one that is in the database.

A number of factors affect the utility of peptide fingerprinting. The greater the experimental mass accuracy, the narrower you can set your search tolerances, thereby increasing your confidence in the match and decreasing the number of false-positive responses.[247] A common practice used to increase mass accuracy in peptide fingerprinting is to employ an autolysis fragment from the proteolytic enzyme as an internal standard to calibrate a MALDI mass spectrum. Nevertheless, in terms of sample consumption, throughput, and the need for intensive data interpretation, peptide fingerprinting is always a logical first approach for protein identification.

Peptide fingerprinting is also amenable to the identification of proteins in complex mixtures. Peptides generated from the digest of a protein mixture will simply return two or more results that are a good fit. As long as the peptides assigned for multiple proteins do not overlap identification is reasonably straightforward. Conversely, peptides that are 'left over' in a peptide fingerprint after the identification of one component can be resubmitted for the possible identification of another component.

2.12.4.3.5.2 Searching with sequence information

Computer-based searching of databases with sequence information is the oldest and probably the most straightforward of the three strategies. Typically, an experimentally determined partial amino acid sequence is compared with the sequences of all of the proteins listed in a database, and a list of those proteins containing the same partial sequence is generated. Virtually all of the web-based search applications are able to search with sequence information. One advantage of this approach is that typical search engines, such as BLAST, allow for increasing levels of ambiguity in the submitted experimental sequence, facilitating the identification of homologous known sequences.

Protein identification via searching with sequence information employs similar strategies irrespective of how the sequence information is obtained. Automated Edman degradation is one of the traditional methods for obtaining sequence information. However, several drawbacks make this approach somewhat less attractive. The collection of data is time-consuming, typically requiring 12–24 h. The subsequent interpretation of the data to extract an amino acid sequence is still largely done by hand, making it a difficult process to automate. And information is limited to the N terminus of the protein; if interior information is desired, the protein must be proteolytically digested, and the individual peptides separated and collected for off-line analysis.

The ability to rapidly generate amino acid sequence information via MS/MS experiments, whether it is with a triple-quadrupole instrument, an ion trap, or via postsource decay with MALDI-TOF, has revolutionized the practice of generating sequence information. Many software packages can use a combination of sequence information and mass spectral information, such as the molecular weight of the individual peptide under investigation. With an accurate value of the parent ion mass of a peptide, and even partial sequence information, it is possible to get a strong match for an unknown. A disadvantage of this approach is that it requires manual interpretation by the operator, making it more difficult to automate, and can require two separate experiments to get both pieces of information.

Some software packages employ a combination of predictive MS/MS data and sequence information. An example is the so-called error-tolerant software that is part of the PeptideSearch package (available from the EMBL website). This strategy is employed when a certain amount of sequence information is available from an MS/MS spectrum, but not the complete sequence. The starting mass of the partial sequence, followed by the sequence itself and then the ending mass, are submitted for database searching. The partial sequence must be manually derived from the experimental MS/MS data, which requires operator input and is not amenable to automation. However, a weak spectrum in which only 4–5 amino acid residues can be deduced typically does not yield enough information to unequivocally identify a protein, or may return a deluge of poor matches. By incorporating the ion mass at which the partial sequence starts and stops, as well as the mass of the peptide itself, PeptideSearch is often able to generate a strong match from marginal data. Potential candidate proteins in the database must contain the partial sequence, must generate a theoretical proteolytic peptide of the correct mass, and must contain the partial sequence positioned appropriately within that theoretical peptide, based on the starting and ending masses of the partial sequence. If a candidate protein meets all of these requirements, then a strong match can be argued for even a very short experimentally derived sequence.

2.12.4.3.5.3 Searching with raw tandem mass spectrometry data

Current mass spectral technology permits the generation of MS/MS data at an unprecedented rate. Prior to the generation of powerful computer-based database-searching strategies, the largest bottleneck in protein identification was the manual interpretation of this MS/MS data to extract the sequence information. Today, many computer-based search strategies that employ MS/MS data require no operator interpretation at all. Analogous to the approach described for peptide fingerprinting, these programs take the individual protein entries in a database and electronically 'digest' them, to generate a list of theoretical peptides for each protein. However, in the use of MS/MS data, these theoretical peptides are further manipulated to generate a second level of lists that contain theoretical fragment ion masses that would be generated in the MS/MS experiment for each theoretical peptide (**Figure 18**). Therefore, these programs simply compare the list of experimentally determined fragment ion masses from the MS/MS experiment of the peptide of interest with the theoretical fragment ion masses generated by the computer program. Again, as with the peptide fingerprint strategy, the operator inputs a list of masses, and typically has a choice of a number of experimental parameters that can be used to tailor the search as appropriate. This is a very processor-intensive function, and, due to the size of current databases, is only possible on a routine basis due to the explosive increase in desktop computing power.

The recent advent of data-dependent scanning functions on an increasing number of mass spectrometers has permitted the unattended acquisition of MS/MS data. Another example of a raw MS/MS data-searching program that takes particular advantage of this ability is SEQUEST. The SEQUEST software processes the input from a data-dependent LC/MS chromatogram, automatically strips out all of the MS/MS information for each individual peak, and submits it for database searching using the strategy discussed above. The appeal of this approach is that each peak is treated as a separate data file, making it especially useful for the on-line separation and identification of individual components in a protein mixture. No user interpretation of MS/MS spectra is involved.

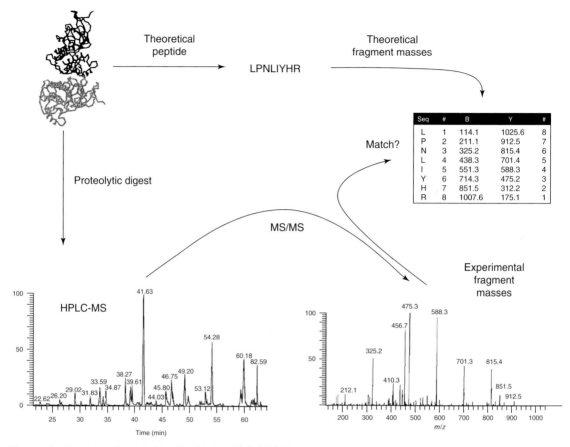

Figure 18 Principles of protein identification by HPLC-MS/MS peptide sequencing.

Several different strategies are applicable in the computer-based searching of sequence databases for the identification of unknown proteins. The choice of strategy employed is often dictated by the format of the data available. However, one must always use caution and examine the results with a critical eye. It is the responsibility of the investigator to examine the list of resultant matches with respect to quality of fit. Often this is aided by examining the individual MS/MS spectra and ensuring that most of the abundant ions are being used for identification, not simply those that are in the 'grass.'

The capacity to perform these searches using web-based programs provides direct access to the rapidly expanding collection of public domain protein databases that are also on the internet. Furthermore, most of the programs take great advantage of the ability of HTML to weave various sources of information together through hyperlinks, so that when a strong candidate for a protein identification is found, a wealth of additional information is only a mouse click away.

2.12.4.3.6 Other methods used for proteomics research

Seeking to avoid the difficulties associated with extracting proteins embedded in gel matrices, other approaches to proteome analysis dispense with polyacrylamide gel separations and rely on multidimensional chromatography. One such approach has paired size exclusion chromatography with HPLC for intact protein analysis, offering convenient preparative capabilities by directing column effluent to a fraction collector.[248] For identification purposes, ESI-MS mass analysis offers an intact molecular weight accurate to about $\pm 0.1\%$, and fractions can also be analyzed at significantly lower throughput by Edman degradation for N-terminal sequence determination, enzymatic digestion followed by successive stages of MS and MS/MS for higher confidence identification and/or characterization, or by other assays. While promising, the comprehensive chromatographic method is currently limited by dynamic range (ability to see low-level proteins present at only a few copies per cell in the presence of abundant proteins present at tens of thousands or more copies per cell). Moreover, identification on the basis of molecular weight alone is risky, particularly in complex systems presenting many post-translational modifications; the availability of chromatographic retention times does not enhance the confidence of proposed identifications significantly. Finally, membrane proteins and other

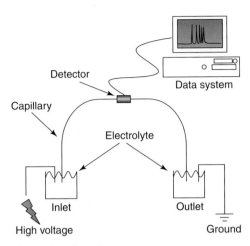

Figure 19 Schematic of a capillary electrophoresis system.

hard-to-solubilize proteins may not be recovered by this methodology, and tend to be harder to handle as intact proteins versus tryptic peptides.

Capillary electrophoresis (CE) is a very high-resolution technique similar to gel electrophoresis, except that the separations take place in a capillary filled with an electrolyte. The term 'CE' encompasses all of the electrophoretic separation modes that can take place in a capillary. A schematic of a CE system is shown in **Figure 19**. In the simplest terms, separations are achieved in CE based on differences in the electrophoretic mobilities of charged species placed in an electric field. However, one other phenomenon present in CE also contributes to the movement of analytes in an applied potential field, that being the electro-osmotic flow. Electro-osmotic flow describes the movement of fluid in a capillary under the influence of an applied electric field, and is brought about by the ionization of silanol groups on the inner surface of the capillary when in contact with the electrolyte. An electrical double layer is formed when hydrated cations from the electrolyte associate with the negatively charged ionized silanol groups. When a potential is applied, the hydrated cations migrate to the cathode, creating a net flow in the same direction. The velocity of the electro-osmotic flow can be directly affected by changes in the field strength, electrolyte pH, and viscosity of the electrolyte solvent system used. The contribution of electro-osmotic flow to the movement of charged species in a capillary can most easily be determined experimentally by observing the migration time to the detector of a neutral marker species. The flow characteristics of electro-osmotic flow are plug-like, in comparison with the laminar flow achieved in LC. As a result, far more theoretical plates can be generated in CE as compared with comparable column lengths in LC. There are any number of excellent texts available that cover in-depth the theories associated with CE.[249,250]

As described for HPLC, by far the most widely applied ionization method used in interfacing CE to MS for the analysis of biomolecules is ESI. An uninterrupted electrical contact is essential for both the continued operation of CE and for the generation of the electrospray when interfacing CE with ESI-MS. Several interfaces have been developed to achieve this electrical contact. The three most widely applicable interfaces are liquid junction, sheath liquid, and sheathless interfaces.

Although detection limits down to attomole ranges have been reported, CE is generally recognized as having a very low concentration limit of detection (CLOD). To achieve the best resolution and peak shape, it is necessary to inject only very small volumes (low nanoliters) of sample, which forces the use of highly concentrated samples initially. Several groups have developed various preconcentration techniques to attempt to overcome this CLOD.[251] All of these techniques involve trapping or preconcentrating the samples on some type of C18 stationary phase or hydrophobic membrane in the case of tryptic digest mixtures.

Capillary IEF (CIEF) has been combined with on-line ESI Fourier transform ion cyclotron resonance (ESI-FTICR) MS to examine desalted, intact *E. coli* proteins.[252] The methodology's promises of simultaneous ppm mass measurement accuracy, high sensitivity, and ultrahigh MS resolution would be particularly attractive for intact protein analyses from proteomics samples. Protein pI in combination with molecular weight can provide a useful means for proposing protein identifications. However, the CIEF conditions employed for on-line MS have so far employed native or near-native separation conditions, yielding pI values that are not predictable from sequence alone and that do not necessarily correlate with the denatured pI values. MS/MS dissociation techniques compatible with FTICR MS, such as infrared multiphoton dissociation or sustained off-resonance irradiation, may provide structural information to yield higher confidence protein identifications.

The most developed, automated nongel methodology for proteome analyses was demonstrated by the analysis of *E. coli* periplasmic proteins, partially fractionated by using strong anion exchange chromatography.[253] Each fraction was digested with trypsin and then analyzed by using microcolumn LC/ESI/MS. The MS/MS spectra were used to search the *E. coli* sequence database, from which a total of 80 proteins were identified. The procedure limits the amount of sample handling, and by manipulating the proteins as mixtures, the higher abundance proteins act as carriers for lower abundance proteins, further reducing losses. However, the presence of a single highly abundant protein can potentially suppress the acquisition of MS/MS spectra for lower abundance peptides present in the mixture (a dynamic range limitation). Also, because the procedure is a superficial sampling of the peptides present, there is the possibility that a protein may be present but no MS/MS data for peptides from that protein are acquired. For relatively simple mixtures this approach greatly increases the speed and efficiency of analysis.

Relative quantification can be difficult to achieve with all of the above methods. One trend, designed to deliver quantitative information, has been to isotopically label samples reflecting different conditions.[254] Based on the isotope ratios of tryptic peptides (or appropriately sized intact proteins when Fourier transform MS is employed[255]), one can attempt to determine whether a protein is upregulated or downregulated. Clearly this methodology can only be employed when it is possible to isotopically label cells (e.g., bacterial cell cultures).

A newer approach to the delivery of quantitative information on all types of samples relies on alkylating cysteine residues with unique tags (e.g., isotopically encoded) that incorporate biotin functionalities.[256] Cysteine-containing tryptic peptides can be withdrawn selectively from mixtures for subsequent LC/MS/MS analysis, yielding both identification and quantification. The approach should also reduce limitations on dynamic range because only cysteine-containing proteins would be loaded onto the LC column, albeit at the cost of losing all information about noncysteine-containing proteins.

2.12.4.4 Conclusions

Much of the technological advances to support proteomics research were developed during the past decade. However, it is obvious to those active in the field that many more developments will be unveiled in the near future. With improvements in sensitivity, throughput, and ease of use, proteomics will continue to flourish as a promising biomedical research endeavor. The tools used to generate the data will be better crafted, and the way in which the data generated from a proteomics approach are used to further the goals of drug discovery will become more sophisticated.

2.12.5 Where Science Meets Silicon: Microfabrication Techniques and Their Scientific Applications

2.12.5.1 Introduction

The historical trend in chemical and biological instrumentation and the development of analysis protocols over several decades has been to work with increasingly smaller sample volumes and quantities of analytes and reagents. This trend has been greatly facilitated by the development of very sensitive measurement technologies, such as mass spectrometry and laser-induced fluorescence. Additional benefits have included reduced reagent costs as well as lower environmental impact.

Microtechnologies merely continue this historical trend, but with a new twist – the jump to an entirely new technology platform. Moreover, new scientific problems can now be studied that take advantage of phenomena unique to micrometer scale domains, such as greatly increased surface to volume ratios. As an example, chemical reactions that are too energetic at a macro-scale and that require synthetic detours through less reactive intermediates can often be conducted directly on a micro-scale, where heat and mass transfer can be controlled much more efficiently. New intermediate products, never seen before, may also result.

The construction of microdevices is based on many of the fabrication technologies employed in the semiconductor industry. However, only recently has its use in microfluidic applications become a reality. A seminal paper that anticipated the use of silicon as a structural material for micromechanical systems is the one written by Petersen.[257] He suggested and illustrated that microstructures with mechanical functionality were indeed possible with silicon as the construction material. This led gradually to the construction of active microfluidic components, such as valves, pumps, motors, and separators on chips.

Much early research focused on the use of different micromechanical systems and devices, such as pumps and chemical functionalities, for reaction, separation, and detection. As these became more robust and more routine, increased complexity resulted in packaging of entire chemical protocols on a microdevice. On-board detection or interfaces to detection systems are also appearing in research laboratories and the marketplace.[258]

Microdevices are highly suited to field or portable use. We will see increased usage of these devices in medical care and environmental monitoring. Coupled to network systems, these devices can serve as a distributed laboratory. The economies of scale realized in the production of these systems will result in single-use or disposable systems.

This section will cover some of the traditional fabrication methods and discuss trends in new materials and techniques. Some discussion will focus on systems packaging and integration techniques and issues. This will be followed by descriptions of a selection of commercial systems that are emerging. It is difficult to predict where these technologies will lead. However, it is safe to assume that they will have a widespread and fundamental impact on many areas of scientific endeavor. These areas include materials research, environmental monitoring, healthcare, drug discovery, and national security.

The microtechnology field literature is rich.[259–261] It is not the purpose of this section to restate this material but rather to provide a quick overview from the perspective of the use of these technologies in the chemical and biological laboratory.

Microtechnologies in the semiconductor industry cover a size regime of roughly 0.1–100 µm. For example, microprocessors are currently being fabricated with 0.18 µm feature sizes that make possible not only faster performance but also the packaging of greatly increased functionality on the same size chip. This is the same scale that we will use to describe microtechnologies for chemical and biological applications.

Another new emerging field – nanotechnology – covers the atomic and molecular domains from a few nanometers to about 100 nm. Nanotechnology is defined as the creation of useful materials, devices, and systems through the control of matter on the nanometer length scale and the exploitation of novel properties and phenomena developed at that scale. As such, nano- and microregimes overlap and, indeed, there will be synergies between the two. For example, we will see nanoscale sensors as detection components in microscale laboratory devices. A brief discussion of nanotechnology and its applications will also be given.

2.12.5.2 Traditional Methods of Microfabrication

The term 'microfabrication' broadly refers to all techniques for fabricating devices and systems on the micrometer scale. As such, it borrows heavily from the processes and techniques pioneered in the semiconductor industry. To fabricate such small devices, processes that add, subtract, modify, and pattern materials are heavily used. The term 'micromachining' can also be used to describe technologies to make devices other than electronic and semiconductor circuits. Similarly, these technologies can be characterized as additive (i.e., deposition of materials, metals, etc.) or subtractive (i.e., removal of materials through mechanical means, etching, laser processes, etc.). Micromachining processes can further be described as either surface or bulk. Surface techniques act on layers of materials above a base or substrate whereas bulk techniques also involve removal or modification of the substrate.

2.12.5.2.1 Lithography and photolithography

Lithography is the most basic method for the construction of microscale devices, and is an old technique first developed in the early nineteenth century. Used originally and still widely in use today in the printing industry, lithography is also the mainstay of modern semiconductor manufacturing.

Lithography is basically a technique or suite of techniques for transferring copies of a master pattern onto the surface of a solid material. The most widely used form of lithography in the construction of microdevices is photolithography, where the master pattern (i.e., a mask) is transferred onto a photosensitive solid material using an illuminated mask. Transfer of the mask pattern can be accomplished through either shadow printing, whereby the mask is held close to (10–20 µm) or in contact with the surface of the photosensitive resist layer, or by projection printing, whereby the pattern is projected through an optical lens system onto the target layer. The latter can take advantage of magnifying or reducing optics to make very small features such as those seen in very large scale integrated devices.

2.12.5.2.1.1 Bulk micromachining

The exposed resist material is then removed or developed through various etching processes. The exposed portions of the resist layer are removed if positive resist materials are used and, conversely, unexposed portions of the resist are removed if negative resist materials are used. The basic positive resist process is illustrated in **Figure 20**.

In bulk micromachining, the substrate is removed wherever it is not protected by the resist material during the etching process. Etching can be either isotropic or anisotropic. In isotropic etching, removal of material occurs in all directions at the same or approximately the same rate. Isotropic etching tends to produce structures with rounded

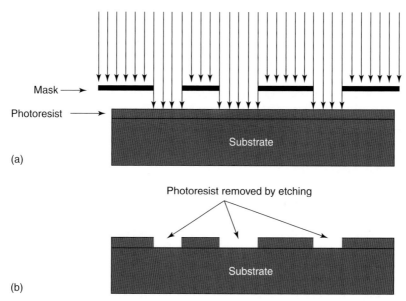

Figure 20 (a) The basic photolithographic process is illustrated wherein a photosensitive resist material on a substrate is exposed through a patterned mask. (b) After exposure the resist material is removed through a developing step. Positive photoresist material is shown in this example.

features. Anisotropic etching, on the other hand, tends to proceed along a preferred direction, and results in more sharply defined features. While these terms describe the extremes of potential etching behavior, it is rare that either process operates to the exclusion of the other, and the resulting structures are generally along a continuum somewhere between the two extremes.

2.12.5.2.1.1.1 Isotropic etching   Isotropic etching removes material in all directions at the same or nearly the same rates. Deviations from true isotropic behavior are generally attributed to differences in chemical mass transport of etchants to target surfaces. Agitation of the material during etching greatly improves true isotropic performance. **Figure 21** shows the individual steps involved in chemical etching, namely diffusion from the bulk etchant solution and adsorption to the surface of the substrate where chemical reaction occurs. This is followed by desorption of the reaction products and, finally, diffusion of the reaction products back into the bulk solution.

Material removal is often accomplished through etching in acidic media. A mixture of hydrofluoric acid (HF)/nitric acid (HNO_3)/water and acetic acid is commonly used.

The overall reaction[262] is given by

$$Si + HNO_3 + 6HF \rightleftharpoons H_2SiF_6 + HNO_2 + H_2 + H_2O$$

Mechanistically, this reaction can be broken down into several sequential half-reactions. Local oxidation of silicon takes place in nitric acid. The anodic reactions are also described as the injection of 'holes' into the silicon, and are given by

$$Si \rightleftharpoons Si^{2+} + 2e^-$$

$$Si + 2H^+ \rightleftharpoons Si^{2+} + H_2$$

The corresponding cathodic reaction is the attachment of OH^- to Si^{2+} to form $Si(OH)_2$, and is given by

$$2H_2O + 2e^- \rightleftharpoons 2OH^- + H_2$$

followed by

$$Si^{2+} + 2OH^- \rightleftharpoons Si(OH)_2$$

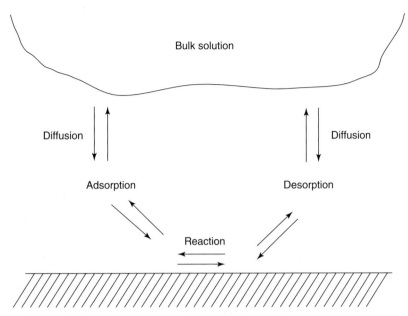

Figure 21 Chemical etching involves five steps: diffusion of reactants to the surface where they are adsorbed, reaction with the surface, and desorption and diffusion of reaction products into the bulk solution.

Table 16 Isotropic HNA-based silicon etching solutions

Nitric acid	HF	Acetic acid	Comment
100	1	99	Removes shallow implants
91	9		Polishing etch at 5 µm min^{-1} with stirring
75	25		Polishing etch at 20 µm min^{-1} with stirring
66	34		Polishing etch at 50 µm min^{-1} with stirring
15	2	5	Planar etch
6	1	1	Slow polishing etch
5	3	3	Nonselective polishing
3	1	1	Selective etch of n$^+$ or p$^+$ with added NaNO$_2$ or H$_2$O$_2$

and, finally:

$$Si(OH)_2 \rightleftharpoons SiO_2 + H_2$$

This is followed by the dissolution of the SiO$_2$ by HF:

$$SiO_2 + 6HF \rightleftharpoons H_2SiF_6 + 2H_2O$$

At high HF and low HNO$_3$ concentrations, the etch rate is controlled by the HNO$_3$ concentration and the rate of oxidation of silicon. Conversely, at low HF and high HNO$_3$ concentrations, the etch rate is controlled by the ability of HF to remove SiO$_2$. The addition of acetic acid extends the oxidizing power of nitric acid over a wider range of dilution. The dielectric constant of acetic acid of approximately 6, compared with a value of 81 for water, results in less dissociation of the nitric acid. The etching reaction is diffusion controlled, and the concentration determines the etch rate, so that mixing is important for uniform etching behavior. **Table 16** summarizes some typical isotropic etching solution concentrations and their effects.

Table 17 Typical etch rates and conditions for anisotropic etchants

Formulation	Temp (°C)	Etch rate ($\mu m\ min^{-1}$)	$\langle 100 \rangle / \langle 111 \rangle$ etch ratio	Mask film
KOH, water, isopropanol	85	1.4	400:1	SiO_2 (1.4 nm min^{-1})
				Si_3N_4 (negligible)
KOH, water	65	0.25–1.0		SiO_2 (0.7 nm min^{-1})
				Si_3N_4 (negligible)
TMAH, water	90	3.0	10	SiO_2 (0.2 nm min^{-1})

Although isotropic etching agents are simple to use, there are several drawbacks. It is difficult to mask with high precision with a masking agent such as SiO_2 because the masking agent is also etched to a measurable degree. Moreover, the etch rate is sensitive to temperature and stirring conditions. For this reason, anisotropic etching agents were developed.

2.12.5.2.1.1.2 Anisotropic etching The discovery of anisotropic etchants in the 1960s addressed the lack of geometrical control presented by isotropic bulk etching agents. These agents preferentially etch along a particular crystalline direction with a discrimination or anisotropy ratio (AR) that can exceed 500:1. The anisotropy ratio is defined as

$$AR = (hkl)_1\ \text{etch rate} / (hkl)_2\ \text{etch rate}$$

AR is defined as approximately 1 for isotropic etchants, and can be as high as 400/1 for the $\langle 110 \rangle / \langle 111 \rangle$ crystal orientations using potassium hydroxide.

A wide variety of etching solutions are available for anisotropic etching. Two popular ones are KOH and tetramethylammonium hydroxide (TMAH). **Table 17** gives some of the properties of these popular anisotropic etchants. Others include the hydroxides of other alkali metals (e.g., NaOH, CsOH, and RbOH), ammonium hydroxide, ethylenediamine pyrocatechol (EDP), hydrazine, and amine gallates. For each, a key feature is the fact that the $\langle 111 \rangle$ crystal directions are attacked at rates at least 10 times, and typically 20–100 times, lower than other crystalline directions. Anisotropic etching can also be further controlled by the introduction of dopants or by electrochemical modulation.

2.12.5.2.1.1.3 Etch stopping It is often desirable to stop the etching process at precisely defined places. Highly doped p^+ regions, generally obtained with gaseous or solid boron diffusion into the silicon substrate, greatly attenuate the etching rate. It is believed that doping leads to a more readily oxidizable surface. These oxides are not readily soluble in KOH or EDP etchants.

Other attenuation methods include electrochemical modulation. This involves holding the silicon positive to a platinum counter-electrode. A very smooth silicon surface can result (referred to as electropolishing). Photon-pumped electrochemical etching is also a popular modulation technique, especially for high-aspect-ratio features.

2.12.5.2.1.2 Surface micromachining

Surface micromachining refers to those processes that act on layers above a substrate. **Figure 22** illustrates the principle. As in bulk micromachining, both additive and subtractive processes can occur with surface micromachining. A problem unique to surface techniques is sticking, whereby supposedly freestanding surface micromachined structures adhere to the substrate after the last rinse.

The primary cause of sticking is the capillary force of water that pulls the structure against the wafer. Ethanol can be added to the rinse water to reduce surface tension. Ethanol and t-butyl alcohol followed by freeze drying can also be used as can the addition of antisticking surfaces such as fluorocarbon film.

Many structures have been made with surface techniques. These include digital micromirror devices (DMDs) or arrays (DMMAs), which are used in digital light projectors and have now also seen application in the construction of oligonucleotide hybridization arrays. Other common surface micromachined devices include valves[263] and nozzles.[264]

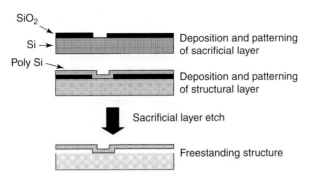

Figure 22 Surface micromachining operates on layers deposited on the surface of a substrate.

2.12.5.2.1.3 Metal electrodeposition

Electroplating is a very useful process for producing additive metal structures above a substrate. A large number of metals can be electroplated, including copper, gold, silver, nickel, and platinum. These are generally reduced from a salt solution, typically a sulfate or cyanide. Either DC or AC current can be used, with the latter generally resulting in much finer grain growth.

As an alternate to electricity, some metals can be deposited using chemical reducing agents. These are the so-called electroless processes, and can be used for gold and nickel. Chemical reducing agents include potassium borohydride and dimethylamine borane for gold and NaH_2PO_2 for nickel. Aspect ratios for structures generated by electroless plating can reach 10:1.

2.12.5.2.2 Newer techniques and materials

Many techniques have evolved and been developed in the last three decades. A few of these are profiled below. This list is far from complete, and there are also many variations on the ones described below. However, these techniques are commonly used, and therefore some familiarity will be useful.

2.12.5.2.2.1 Plasma and deep reactive ion etching

There are several dry-etching processes that include the use of reactive gas plasmas and ions. External energy in the form of radiofrequency power drives the chemical reactions. In a plasma etcher, the wafer is one of two electrodes, and is grounded. The powered electrode is the same size as the wafer. Plasmas are generally low temperature, and operate in the 150–250 °C range; however, some operate down to room temperature. In DRIE, the powered electrode is the wafer with the other electrode being much larger. Very-high-AR devices (40:1) can be obtained with these techniques. Reactant gases generating ions in the radiofrequency field generally have unusually high incident energies normal to the substrate. Some common etch gases and mixtures are $CClF_3$ plus Cl_2, $CHCl_3$ plus Cl_2, SF_6, NF_3, CCl_4, CF_4 plus H_2, and C_2ClF_5.

2.12.5.2.2.2 X-ray LIGA

LIGA is an acronym for the German 'Lithographie, Galvanoformung, Abformung,' a succession of lithographic, electroplating, and molding processes. LIGA can generate high-aspect-ratio devices in excess of 100:1. Developed by Ehrfeld,[265] LIGA is a template-guided microfabrication process that uses an extremely well-collimated synchrotron x-ray radiation source to transfer a pattern onto a substrate.

Figure 23 shows the basic steps of the LIGA process. A primary substrate is coated with a conductive top layer. Polymethyl methacrylate (PMMA) is then applied typically through a series of multiple spin casts or as a commercially prepared thin sheet solvent bonded to the substrate. The PMMA is then exposed through a mask opaque to x-rays and then developed, resulting in high-aspect structures. Metal is then electroplated onto the PMMA mold, and the resulting assembly is diamond lapped and polished to the desired height. The remaining PMMA is then developed away, and a chemical etch removes the conducting top layer, thus freeing the part.

One drawback of LIGA is the need for a synchrotron radiation source, which restricts access to the technique dramatically, even though there are many such light sources available worldwide. Many devices have been constructed using the LIGA process. These include microvalves,[266] micropumps,[267] fluidic amplifiers,[268] and membrane filters.[269] Some of these devices, while manufactured by the LIGA technique, have subsequently proven inferior to or more expensive than those made by other technologies.

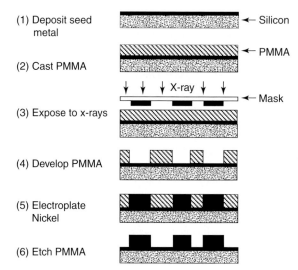

Figure 23 The LIGA process. Steps 1–5 are conventional lithography. Nickel is electroplated in step 6, after which the wafer is lapped to the desired thickness. Step 6 shows the remaining PMMA removed. The structures can then be freed from the seed metal base layer.

2.12.5.2.2.3 Ultraviolet LIGA

To overcome the limitation of synchrotron light source requirement, thick ultraviolet-sensitive resists, such as polyimides, Hoechst AZ-4000, and Epon SU-8, have been developed as alternatives to the x-ray-sensitive PMMA. SU-8 is an epoxy-based, transparent negative photoresist material that can be built up into 200 μm-thick layers. AZ-4000 is a positive resist material that can be built up into layers 15–80 μm thick with multiple spin coats. These materials and the technique certainly are more accessible than LIGA. Aspect ratios of 5:1 to 15:1 can be achieved with these materials.

2.12.5.2.2.4 Polymer replication

The fabrication of microdevices with conventional plastic molding technologies, such as reaction injection and thermoplastic molding and hot embossing, should not be overlooked. These continue to be by far the cheapest alternatives to making microstructures in materials. The entire CD industry is dependent on these technologies, so that micrometer-sized features in plastic are readily achievable. Many of the microfluidic devices currently appearing on the market are made by this technique, and are designed to be inexpensive and disposable.

2.12.5.2.2.5 Laser ablation

Powerful laser beams have sufficient energy to ablate silicon through thermal evaporation. The 1.06 μm Nd:YAG and carbon dioxide laser systems are used routinely for drilling silicon. Lasers can be used for subtractive as well as additive processes, and have been used for annealing, deposition, and welding. In laser beam cutting, coherent light replaces electrons as the cutting tool. Removal rates tend to be slow. Lasers have also been used to assist in chemical etching to photodisassociate the reactive etchant.

2.12.5.2.2.6 Ion beam milling

Ion beam milling is a thermomechanical drilling technique wherein the drill bit is a stream of ions. Typically, a liquid metal, such as gallium, is focused to a submicrometer diameter. Like laser ablation, ion beam milling is a slow serial technique. Atomic force microscope tools are made by ion beam milling, but the technique is generally too expensive to use in a mass production mode.

2.12.5.2.2.7 Ultraprecision machining

Single-crystal diamond tools have made ultrahigh-precision mechanical machining possible. A machining accuracy in the low-submicrometer range is possible. Throughput is extremely slow, and there are stringent environmental requirements on the machine. This technique will thus find limited use in the mass fabrication of microscale laboratory devices.

2.12.5.2.2.8 Porous silicon as a sensor material

Porous silicon was first discovered[270] during the electropolishing of silicon in 1956. At low current densities, partial dissolution of silicon results during etching with hydrofluoric acid, with the formation of hydrogen gas:

$$Si + 2F^- + 2H^+ \{ReversReact\} \ SiF_2$$

$$SiF_2 + 2HF \{ReversReact\} \ Si4_2 + H_2$$

A very porous silicon structure results that can have aspect ratios up to 250. Silicon can be made either microporous or macroporous. The pores follow the crystallographic orientation, and vary in size from 2 nm to 10 µm. Porous silicon oxidizes readily, and etches at a very high rate. A prepattern is generally defined by a photolithographic transfer. The etch pits formed on the exposed area of the mask serve as nucleation centers for macropores.

Porous silicon has been used in a number of applications, including electrochemical reference electrodes, high-surface-area gas sensors, humidity sensors, and sacrificial layers in micromachining. Recent research has focused on using porous silicon as a substrate for chemical sensors.[271] The silicon pillars can be derivatized using well-known chemistries, and linked to chemical recognition molecules (i.e., antibodies or oligonucleotides). When immersed in a solution of target molecules and exposed to light, an interference pattern results that is extremely sensitive to concentration. These porous silicon sensors can serve either as standalone instruments or as the sensing element in an integrated fluidic chip.

2.12.5.3 System Integration Issues

Integration covers a whole spectrum of issues from the semiconductor and fluidic levels to the control, information display and treatment, and user interface levels. Integration must address electronic interfaces, mechanical and optical interfaces, and materials and packaging issues, as well as the chemical compatibility and functionality.

Many single components, such as valves, pumps, separation columns, and optical and electrochemical detectors, have been made in the last three decades, and have been proven to work individually. The focus of much current work is to show how these components can be assembled into a multiple-function device. Activities such as sample preparation, detection, and data presentation are increasingly being collocated.

The choice of materials is wide. Materials must be chosen not only for ease and expense of fabrication but also for chemical and biochemical compatibility. **Table 18** lists some common choices, along with their advantages and disadvantages.

2.12.5.3.1 Interfaces

Fluid interconnects do not exist for microfluidic devices, and those that have been used vary widely from pipette tips and glue to gas chromatographic fittings and microfabricated fittings.[272,273] In general, the 'macro-to-micro' interface issues have not been well resolved. Various technologies, including electrokinetic samples made by Caliper, allow for the sampling of multiwell microtiter plates.

2.12.5.3.2 Fluidics and electronic integration

Monolithic integration places all components on the same device. Typically, actuators and sensing elements are made in the same process. Little or no assembly is required, and monolithic systems are highly reliable and low in cost. The

Table 18 Microchip systems material selection guide

Material	Advantages	Disadvantages
Silicon	IC compatible, large technology base, bulk etching, surface micromachining, and dry or wet processing; can include electronics	Chemically reactive surface
Glass	Insulating, transparent, good chemical compatibility, bulk etching (like silicon), long laboratory experience	Little commercial technology base
Plastics	Low cost, used widely in commercial diagnostics systems, replication, using hot embossing on a nanometer scale	Untried commercially for line resolution, hydrophobic chemicals absorbed

Source: S Verpoorte and J Harrison (short course on microfluidics taught at LabAutomation2000, Palm, Springs, CA, January 22–23, 2000).

limitations include incompatibilities with process steps, low overall yield due to many points of failure, and high costs for small quantities. Several layers, all with their own processing requirements, can be sandwiched together to make a final assembly in a collective system. The benefits of collective integration include being able to segregate chemical and processing step incompatibilities. Higher individual step yields are countered by the final assembly requirements, and are often highly variable. As in collective integration, hybrid integration results in a sandwich construction. The layers of silicon, glass, or plastic can be either fluidic, controlled, or electronic, and are assembled and bonded together.

2.12.5.3.3 Integral detection systems

Sample measurement in chips presents new challenges to detection systems. The most popular modalities are fluorescence, absorbance, luminescence, and chemiluminescence. Other methods include the use of mass sensors, such as mass spectrometry and acoustic wave devices; electrochemical sensors, such as potentiometric, amperometric, and conductivity; and, finally, thermal sensors, such as bolometers and thermopiles. All are handicapped to an extent by the small sample sizes encountered on chips, and therefore must be very sensitive or coupled to chemistries that exhibit amplification.

2.12.5.4 Applications of Microtechnologies

2.12.5.4.1 Bead-based fiber-optic arrays

The ability to etch silicon materials predictably has given rise to an interesting technology based on fiber optics.[274–276] Optical fibers are made from two kinds of glass or plastic: a core and a cladding. The core has a slightly higher refractive index than the cladding, permitting the fiber to transmit light over long distances through total internal reflection. Today's high-bandwidth telecommunications technology is based on optical fibers.

The fibers can be made with 3–7 μm outside diameters. One thousand of these fibers can be bundled together into a 1 mm-diameter bundle. The ends are then etched with hydrofluoric acid, to produce micrometer- or femtoliter-sized pits. **Figure 24** shows an image of these wells taken with an atomic force microscope.

It is possible to insert individual microspheres into these wells. The beads are sized so that they self-assemble, one to a well, when the fiber bundles are dipped into a 20% solids slurry. One milliliter of such a slurry can contain 100 million beads. Each individual bead can be optically encoded with fluorescent dyes[277] so that it reports a unique optical 'bar code' when interrogated with a laser. Moreover, it is possible to construct an array of 384 of these fiber bundles on a 4.5 mm pitch to fit into a standard 384-well microtiter plate. This increases the amount of information gathered in a single experiment to more than 2 million data points. This technology has been commercialized by Illumina, Inc.[278]

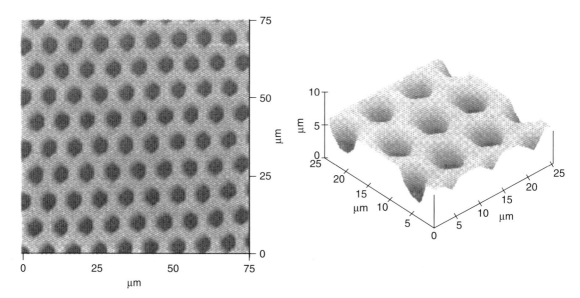

Figure 24 An atomic force microscope image of the wells generated by etching the inner core of each optical fiber in an optical fiber bundle. The capacity of each well is of the order of a few femtoliters. (Reproduced by courtesy of the D R Walt group, Tufts University).

Illumina's proprietary Oligator technology complements its BeadArray technology. The Oligator synthesizes in parallel many different short segments of DNA to meet the requirements of large-scale genomics applications. With unique segments of DNA attached to each bead, the BeadArray technology can be used for single-nucleotide polymorphism (SNP) genotyping.

2.12.5.4.2 Deoxyribonucleic acid arrays

Much has been written about oligonucleotide, DNA or cDNA, and RNA microarrays. This technology was first commercialized by Affymetrix, Incyte, and Hyseq. It relies on surface micromachining techniques wherein the 'layers' are successively photolithographically masked in light-directed oligonucleotide synthesis steps. For example, the Affymetrix technique works by shining light through a photolithographic mask onto a light-sensitive surface. The surface reactions are controlled so as to carry out the standard phosphoramidite oliognucleotide chemistry. Various cycles of protection and deprotection are required to build up a typical 20-mer oligo at specific sites.[279]

Some of the limitations of this approach to making arrays are that masks are not cheap or quick to make and are specific to the pattern of oligonucleotides that make up the array. A new approach, based on surface micromachining semiconductor technology from Texas Instruments,[280] uses a DMD. A DMD consists of individually addressable mirrors on a dense 600×800 array in a $10\,mm \times 14\,mm$ area. There are nearly half a million addressable locations available for light-directed synthesis of oligonucleotides. DMD technology is in common use today in digital light projectors. A research group at Wisconsin[281] has demonstrated the use of DMD in the construction of oligoarrays. Light from an arc lamp directed by digitally controlled micromirrors cleaves photolabile protecting groups off the ends of DNA molecules. The illuminated areas then couple to the next base to be added while the dark areas remain unreactive.

The ability to spatially control selective chemistries is the basis of much array-based technology. Even higher densities can be obtained with arrays of femtoliter microwells made with standard polymer-embossing techniques. Whitesides et al.[282] have described soft lithographic techniques to construct planar arrays of approximately 3 fL wells with 10^7 wells cm^{-2}. Basically, an array of posts formed in a photoresist on a silicon wafer is used as a 'master' to form the arrays of microwells. The master is generated by rapid prototyping or standard photolithography with a chrome mask. The arrays of wells are fabricated by molding an elastomeric polymer against the master. The wells have aspect ratios ranging from 1:1 to 1.5:1. The authors describe a technique for filling the wells that takes advantage of the difference in the interfacial free energies of the substrate and the liquid of interest and controlling the topology of the surface. Liquid is allowed to drain off an array of microwells either by gravity or by pulling the array from a bulk solution.

One application of DNA array technology is that of Hyseq. The HyChip system uses arrays of a complete set of DNA probes in conjunction with a target-specific cocktail of labeled probes to identify differences between a reference and test samples. Since the chip contains all possible probes, the HyChip system has the distinct advantage of being able to sequence DNA from any source. Other available DNA chip technologies require prior knowledge of the gene sequence and mutations prior to chip design, and require no less than one chip for each gene being tested. The HyChip system is therefore universal, and requires only a single chip for a wide variety of targets.

The arrays produced by the companies above tend to be very high density, with thousands if not millions of uniquely addressable locations. There is also a very large market for lower density arrays produced on coated surfaces by mechanical systems such as robots and Cartesian spotting devices. **Table 19** lists some of the leading manufacturers of microarray spotting systems, slide suppliers, scanners, image-processing software, and data analysis software.

2.12.5.4.3 Electronically enhanced hybridization

DNA sequencing by hybridization[283,284] and other oligonucleotide hybridization protocols rely on a rather time-consuming hybridization step. Although there are many variables to control for a successful hybdridization, migration of the DNA to the fixed complementary probe tends to be slow. Often, this step requires hours to overnight reaction times. Nanogen has commercialized a chip that electronically drives the target DNA to the probes, greatly speeding up this process.[285] After hybridization, the chip is washed, and detection occurs, reducing a typical several hour experiment to less than 30 min.

Nanogen's technology allows small sequences of DNA capture probes to be electronically placed at, or 'addressed' to, specific sites on the microchip. A test sample can then be analyzed for the presence of target DNA molecules by determining which of the DNA capture probes on the array bind, or hybridize, with complementary DNA in the test sample. In contrast to nonelectronic or passive hybridization with conventional arrays on paper or glass 'chips,' the use of electronically mediated active hybridization to move and concentrate target DNA molecules accelerates

Table 19 Selected companies providing microarray technologies, surfaces, scanners, imaging software, and data analysis tools

Company/institution	Website
Beecher Instruments	http://www.beecherinstruments.com
BioRobotics	http://www.biorobotics.com
Cartesian Technologies	http://www.cartesiantech.com
Engineering Services	http://www.esit.com/
Genetic Microsystems	http://www.geneticmicro.com
Genetix	http://www.genetix.co.uk
Gene Machines	http://www.genemachines.com
Intelligent Automation Systems	http://www.ias.com
Packard	http://www.packardinst.com
Amersham Pharmacia Biotech	http://www.apbiotech.com
Corning CoStar	http://www.cmt.corning.com
Surmodics	http://www.surmodics.com
Telechem	http://www.arrayit.com
Axon	http://www.axon.com
GSI Luminomoics	http://www.gsilminomics.com
Genomic Solution	http://www.genomicsolutions.com
Molecular Dynamics	http://www.mdyn.com
Virtek	http://www.virtek.com
BioDiscovery	http://www.biodiscovery.com/
Imaging Research	http://imaging.brocku.ca/arrayvision.html
National Human Genome Research Institute	http://www.nhgri.nih.gov/DIR/LCG/15K/img_analysis.html
Stanford University	http://rana.stanford.edu/software/
The Institute for Genomic Research	http://www.tigr.org/softlab/
Silicon Gentics	http://www.signetics.com
Spotfire	http://www.spotfire.com

hybridization so that hybridization may occur in minutes rather than the hours required for passive hybridization techniques. In addition to DNA applications, this technology can be applied to a number of other analyses, including antigen–antibody, enzyme–substrate, cell–receptor, isolation of cancer cells from peripheral blood cells,[286] and cell separation techniques.

2.12.5.4.3.1 Electronic addressing

Electronic addressing is the placement of charged molecules at specific test sites. Since DNA has a strong negative charge, it can be electronically moved to an area of positive charge. A test site or a row of test sites on the microchip is electronically activated with a positive charge. A solution of DNA probes is introduced onto the microchip. The negatively charged probes rapidly move to the positively charged sites, where they concentrate and are chemically bound to that site. The microchip is then washed, and another solution of distinct DNA probes can be added. Site by site, row by row, an array of specifically bound DNA probes can be assembled or addressed on the microchip. With the ability to electronically address capture probes to specific sites, the system allows end-users to build custom arrays through the placement of specific capture probes on a microchip. In contrast to current technologies, these microchip arrays can be addressed in a matter of minutes at a minimal cost, providing research professionals with a powerful and versatile tool to process and analyze molecular information.

2.12.5.4.3.2 Electronic concentration and hybridization
Following electronic addressing, electronics are used to move and concentrate target molecules to one or more test sites on the microchip. The electronic concentration of sample DNA at each test site promotes rapid hybridization of sample DNA with complementary capture probes. In contrast to the passive hybridization process, the electronic concentration process has the distinct advantage of significantly accelerating the rate of hybridization. To remove any unbound or nonspecifically bound DNA from each site, the polarity or charge of the site is reversed to negative, thereby forcing any unbound or nonspecifically bound DNA back into solution, away from the capture probes. In addition, since the test molecules are electronically concentrated over the test site, a lower concentration of target DNA molecules is required, thus reducing the time and labor otherwise required for pretest sample preparation.

2.12.5.4.3.3 Electronic stringency control
Electronic stringency control is the reversal of electrical potential to quickly and easily remove unbound and nonspecifically bound DNA as part of the hybridization process. Electronic stringency provides quality control for the hybridization process and ensures that any bound pairs of DNA are truly complementary. The precision, control, and accuracy through the use of the controlled delivery of current in the electronic stringency process permits the detection of single-point mutations, single-base-pair mismatches, or other genetic mutations, which may have significant implications in a number of diagnostic and research areas. Electronic stringency is achieved without the cumbersome processing and handling otherwise required to achieve the same results through conventional methods. In contrast to passive arrays, Nanogen's technology can accommodate both short and long single-stranded fragments of DNA. The use of longer probes increases the certainty that the DNA that hybridizes with the capture probe is the correct target. Nanogen's electronic stringency control reduces the required number of probes and, therefore, test sites on the microchip, relative to conventional DNA arrays. In contrast, traditional passive hybridization processes are difficult to control and require more replicants of every possible base pair match so that correct matches can be positively identified.

2.12.5.4.4 Microfluidic devices

There are many university research laboratories and institutes across the world working in the area of microfluidic devices of all kinds. Much early work was done by such groups as Ramsey's at Oak Ridge National Laboratory,[287–289] which, in addition to the groups of de Rooij,[290] Manz,[291] and Harrison,[292,293] for example, continue to be very active.

Microfluidic technologies have over the past decade and a half matured sufficiently to become the basis of many companies. Many of these companies also perform fundamental research while providing their technologies to customers either as products or in early technology access partnerships. Still other companies use microfluidics technology in a specialized application, and market the value added in a specialized field of endeavor, such as environmental monitoring, patient diagnostics, genetic testing, or drug discovery. For example, complete instrument systems based on microfluidic separations are emerging, many integrated with detection.

As will be seen from the selected examples discussed below, many products and services are being developed based on microarray and microfluidic components. **Table 20** lists some of the companies, to which the readers are referred for additional information.

2.12.5.4.4.1 Microfluidic separations
Caliper Technologies was founded on the technologies originating in many of these research laboratories. In the late 1990s, Caliper introduced its first two LabChip systems – a personal laboratory system for the life sciences market and a HTS system to aid pharmaceutical companies in discovering new drugs. Co-developed with Agilent, the first laboratory system consists of the Agilent 2100 bioanalyzer, a desktop instrument designed to perform a wide range of scientific experiments using a menu of different LabChip kits. Each kit contains chips and reagents designed for specific applications. The LabChip system brings the benefits of miniaturized, integrated, and automated experimentation to the researcher's desktop. Agilent launched this product in 1999 for applications including DNA and RNA analysis. The analyzer works with a series of chips optimized for DNA and RNA analysis (e.g., the DNA 7500 LabChip assay kit provides size and concentration information for DNA fragments ranging in size from 100 to 7500 base pairs).

Table 20 Selected companies and institutes

Company/institution	Website	Comment
Oak Ridge National Laboratory	http://www.ornl.gov	Large microfluidic development effort
Caliper Technologies	http://caliperls.com	One of the early companies in microfluidic systems, DNA applications
Monogram Biosciences[a]	http://www.aclara.com/	Microfluidic systems, DNA applications
Illumina	http://www.illumina.com/	Fiber-optic, self-assembling bead arrays, very high-throughput assay technology
Hyseq[b]	http://www.hyseq.com/	Developer of sequencing by hybridization (SBH)
Nanogen	http://www.nanogen.com	Electronically driven hybridization
Orchid Biosciences	http://www.orchidbio.com	Early microfluidic company focused on chip-based synthesis and SNP scoring
Micronics	http://www.micronics.net/	Licensor of University of Washington microfluidic technologies, microcytometry
Gyros Microlab AB	http://www.gyrosmicro.com/	Lab-on-a-CD disk

[a]ACLARA Biosciences merged with Monogram Biosciences.
[b]This company now trades under the name Nuvelo, and its direction has changed.

Caliper's LabChip HTS system utilizes chips that draw nanoliter volumes of reagents from microplates for analysis. The system performs a wide range of experiments using a menu of different chips. Caliper currently offers chips to perform drug screening for several classes of targets and anticipates offering more chips, currently in development, to increase both capability and throughput.

Using a similar approach, Monogram Biosciences has demonstrated rapid, high-resolution electrophoretic separations in plastic chips.[294–296] DNA fragment sizing is required in numerous applications and represents an attractive target market wherein Monogram Biosciences provides ready-to-use disposable plastic chips, prefilled with a proprietary gel. Monogram Biosciences produces plastic chips in polymers, thereby enabling the mass production of low-cost disposable chips. They use two main steps in the production of plastic chips: (1) the formation of microstructures in a base layer and (2) sealing of the base layer with a cover layer. To form the base layer, microstructure patterns are replicated from a micromachined master (or submaster) onto a polymeric substrate. Monogram Biosciences has developed several different replication technologies that span a range of capabilities and economies of scale, enabling the company to select the most appropriate technology for any application under development.

2.12.5.4.4.2 Microfluidic synthesis

Traditional means of transferring fluids with pipettes are severely challenged much above 384 wells in a microtiter plate. The microfluidic technology developed by Orchid Biocomputer[297] provides a viable route to achieve much higher levels of densification, unachievable by traditional systems. Orchid is developing modular chip-based systems with 96, 384, 1536, and 12,288 (8 × 1536) reactor arrays that accommodate 100–80 nL sample volumes.

The unique features of the Orchid Biocomputer chips include the ability to process hundreds of reactions in parallel through the use of precise fluidic delivery methods. These chips use hydrostatic pressure pulses with nonmechanical microvalves fabricated in the multilayer collective device to transfer fluids vertically and horizontally into 700 nL microreaction wells. The highly complex three-dimensional architecture of these chips enables the broadest range of capabilities of any chip in the industry. With this enclosed system there is no risk of evaporation or reagent degradation. One important application of Orchid Biocomputer's microfluidic technology is in parallel chemical processing, as embodied in the Chemtel chip.

2.12.5.4.4.3 Chip-based flow cytometry for medical diagnostics

A multidisciplinary team was formed at the University of Washington School of Medicine's departments of bioengineering, mechanical engineering, electrical engineering, and laboratory medicine. The developments resulting from this multidisciplinary research have been licensed from the University of Washington to Micronics.[298]

Micronics is further developing microfluidics-based systems for application to clinical laboratory diagnostics and analytical and process control chemical determinations. In both instances, the small size and autocalibrating characteristics of the microfluidics technology lends itself to application at the point of the sample. This avoids the transportation of sample to centralized laboratories – a step that results in both delays and degradation of the sample – and it provides answers at the point where decisions can be made immediately.

A major area of Micronics's technology is in microfluidic-based cell cytometry. In operation, biological cells from a sample, such as blood, pass in single file through a channel on which is focused a laser beam. Light-scattering measurements are taken at multiple angles, and these multiparameter scatter measurements provide a 'fingerprint' for the various types of cells. This technology, known as flow cytometry, is not new. However, Micronics is implementing the technology on a microfluidic scale using new miniaturized optical technology developed for CD readers. Micronics microcytometer technology has demonstrated the capacity to count and classify platelets, red blood cells, and various white cell populations by means of laminate-based microfluidic flow channels and light-scattering optics. Additional light-scattering and data analysis channels will be used to extend the capabilities of the microcytometer toward a complete blood cell assay, including a five-part white cell differential.

2.12.5.4.4.4 Microfluidics in a rotating CD

Gyros Microlabs has integrated a microfluidics system based on a simple modular design, consisting of a spinner unit, a flow-through noncontact liquid microdispenser, a CD with proprietary applications and chemistries, detector(s), and software.

Many complex steps, such as sample preparation, fluid handling, sample processing, and analysis, can be carried out seamlessly within the confines of a single CD. The CD is made up of intricate molded microchannels in plastic, creating interconnected networks of fluid reservoirs and pathways. No pumps are required because spinning the CD moves the liquids around. What makes the technology versatile is the ability to rapidly and easily create application-specific CDs, where the applications are embedded in the intricate microstructures on the CD.

One application explored is high-throughout SNP scoring on a CD in combination with solid phase pyrosequencing.[299] The centripetal force in the CD device allows for parallel processing without complex tubing connections. One of their current designs integrates the whole process for SNP analysis, including sample preparation, achieving more than 100 000 SNPs per day.[300]

Gyros has also developed a flow-through, noncontact, piezoelectric dispenser allowing the deposition of a precise amount of liquid into each microstructure while the CD is spinning at 3000 rpm, thus managing many operations simultaneously without any evaporation problems. The flow-through principle facilitates the use of thousands of different samples or reagents. Gyros is developing a full range of dispenser variants including arrays, which will revolutionize the way small volumes of liquids can be handled.

2.12.5.5 Nanotechnology

Nanotechnology[301] is a very broad emerging discipline; it describes the utilization and construction of structures with at least one characteristic dimension measured in nanometers. Such materials and systems can be deterministically designed to exhibit novel and significantly improved physical, chemical, and biological properties, phenomena, and processes because of their size. Since they are intermediate between individual atoms and bulk materials, their physical attributes are often also markedly different. These physical properties are not necessarily predictable extrapolations from larger scales. Currently known nanostructures include carbon nanotubes, proteins, and DNA.

In biology, recent insights indicate that DNA sequencing can be made many orders of magnitude more efficient with nanotechnology. It can also provide new formulations and routes for drug discovery, enormously broadening the therapeutic potential of drugs. For example, drugs or genes bound to nanoparticles can be administered into the bloodstream and delivered directly to cells. Furthermore, given the inherent nanoscale of receptors, ion channels, and other functional components of living cells, nanoparticles may offer a new way to study these components.

Nanosensors with selectivities approaching those of antigen–antibody or hybridization reactions will greatly influence chemistry and instrumentation development. These sensors would be naturally complementary to microfluidic systems, and we should expect to see these systems integrated in the future.

Scientists are just now beginning to understand how to create nanostructures by design. A more complete understanding will lead to advances in many industries, including material science, manufacturing, computer technology, medicine and health, aeronautics, the environment, and energy, to name but a few. The total societal impact of nanotechnology is expected to be greater than the combined influences that the silicon integrated circuit, medical imaging, computer-aided engineering, and manufactured polymers have had. Significant improvements in performance and changes of manufacturing paradigms will lead to several industrial revolutions in this century. Nanotechnology will change the nature of almost every man-made object. The major question now is, how soon will this revolution arrive?

2.12.5.6 Overview

The above examples are merely illustrative and not intended to be a comprehensive coverage of all the available systems of current research. This field is changing very rapidly; new products, services based on microfluidic systems, and even new companies are constantly emerging. Most of the above products were made by combinations of the fabrications technologies described in Section 2.12.5.1. They also illustrate how fundamental developments made in various research groups are migrating to the private sector and, eventually, to the marketplace.

These devices are the tip of the iceberg. They are indicative of the state of the technology and also serve as a prelude to products to come.

2.12.6 Single-Nucleotide Polymorphism Scoring for Drug Discovery Applications

2.12.6.1 Introduction

An important byproduct of the Human Genome Project is an appreciation of the nature and degree of individual genetic variation.[302] While small insertion and deletions and variation in the length of repetitive DNA elements are common, by far the most frequently occurring type of genetic variation is the SNP. An SNP is a nucleotide position in genomic DNA where the nucleotide base varies within a population.[303] In humans, it is estimated that between any two genomes, 1 in 1000 bases will be an SNP. When a large population consisting of many genomes is considered, as many as 1 in 100 bases will be polymorphic. Thus, in the human genome, which contains some 3 billion base pairs, there are expected to be more than 3 million SNPs. Many of these SNPs will have biomedical uses ranging from the identification of potential drug targets to the diagnosis of disease susceptibilities and targeting of therapies. The analysis of millions of SNPs in millions of patients presents a daunting challenge, with new types of considerations and requiring the development of new approaches. In this section, we will briefly cover some of the uses of SNPs and describe some of the assay chemistries and analysis platforms that will allow SNPs to be used on a large scale for these emerging pharmaceutical applications.

2.12.6.1.1 Single-nucleotide polymorphism applications

SNPs have several potential applications relevant to drug discovery and development. First, SNPs located in the protein coding or regulatory regions of genes can affect protein function or expression. In some cases, such alterations can be identified as the cause of a disease, thereby pointing to a potential drug target as well as having diagnostic uses. While in some diseases a single mutation can be identified as the cause of a disease, more often disease is expected to be the result of an interacting combination of differences in a number of genes.[304] Case–control association studies that focus on protein-altering SNPs within a set of candidate genes can reveal the involvement of specific genes in such complex diseases.[303] In a set of related applications, SNPs can be used as mapping markers to identify genes that may be associated with a particular disease. In these scenarios, hundreds of thousands of SNPs are scored in affected and control populations consisting of hundreds to thousands of individuals. While the experimental design and statistical considerations for large-scale association or linkage disequilibrium studies are still the subject of debate (discussed and reviewed by Terwilliger and Weiss[305]), it is hoped that this approach may help address the difficult problem of diseases with complex genetic susceptibilities.

Similarly, amino acid-altering SNPs can affect a patient's response to a drug. Variation among individuals with respect to drug targets as well as drug metabolism can result in important differences in drug efficacy and toxicity.[306,307] As for complex diseases, drug responses will most often have a polygenic character, and patient responses will be determined by an individual variation in the relevant genes. The understanding of pharmacogenetic variation will not only be important for diagnostic uses but for the design of clinical trials for new drugs as well. Ensuring that known drug target and metabolism genotypes are represented in a clinical trial population should make it possible to more precisely predict drug efficacy and safety in the general population. In the broad vision of pharmacogenomics, disease and susceptibilities as well as highly individualized therapies would be ascribed on the basis of a patient's genotype, thereby ensuring safe and optimal treatment for every patient.

2.12.6.1.2 Single-nucleotide polymorphism discovery and scoring

A prerequisite for the applications described above is the identification and mapping of SNPs throughout the genome. Most of this task will be accomplished in the course of various large- and small-scale sequencing efforts.[308–311] There are currently active efforts in both the public and private sectors to discover and map SNPs on a genome-wide scale.[312,313] There have been several reviews on the methods and considerations that are relevant for SNP discovery.[302,314] While large-scale sequencing is the source of the majority of newly discovered SNPs, there are also efforts to develop methods of 'scanning' genomic DNA for variation, with sequencing being used to confirm and identify SNPs. Although most SNP discovery methods can be configured for subsequent scoring of a SNP in unknown samples, scanning methods do not, in general, scale well for high-throughput genotyping applications. Scoring of large numbers of known SNPs is more efficient using one or more methods designed to interrogate an SNP directly.[315] In this section, we will consider the methods and analysis platforms available for large-scale SNP scoring in a drug discovery environment.

2.12.6.2 General Considerations

Before discussing in detail the chemistries and platforms used to score SNPs, it is useful to consider the important characteristics of large-scale SNP analysis for drug discovery applications, because the requirements in this setting are very different from those of a smaller scale research setting. As discussed above, though estimates vary for applications related to gene discovery, on the order of 100 000 SNP markers will have to be screened in thousands to tens of thousands of individuals. Thus, any single application will likely require the scoring of millions of SNPs. This scale of analysis involves particular consideration of issues of throughput and cost that will influence the choice of assay chemistry and analysis platform. Two issues that are especially important in this regard are the amenability of methods to automation and the multiplexed analysis of many SNPs in each sample.

2.12.6.2.1 Sample preparation

In most cases, the starting material for an SNP-scoring project will be purified genomic DNA, which is then subjected to polymerase chain reaction (PCR) to amplify the regions containing the SNPs. PCR amplification is a major cost for most SNP-scoring methods, and although there is much interest in developing 'PCR-less' assays, as yet the combination of sensitivity and specificity of PCR has not been surpassed. Fortunately, PCR reactions are readily set up and performed in microwell plates with commercially available laboratory automation equipment. Additional cost and time savings can be obtained by multiplexing the PCR step, although the technical challenges to performing many PCR reactions in one tube are still significant enough to keep this approach from being routine on a large scale.

Once the DNA template is prepared, some assay chemistry is performed, generally consisting of a hybridization or binding procedure, often coupled to an enzymatic reaction, that allows identification of the nucleotide base located at the SNP position. Such chemistries usually require incubations ranging from minutes to hours, but these need not be rate-limiting if they can be performed in parallel in microplates. Specialized enzymes, fluorescent nucleotides, or other reagents are often required, and represent another major cost for SNP-scoring assays. Again, multiplexed sample processing of many SNPs in a single tube can dramatically decrease the cost per SNP scored, as well as increase the throughput.

2.12.6.2.2 Sample analysis

If upstream sample-processing steps can be conducted in parallel, sample analysis is often the rate-limiting step in SNP-scoring applications. The ideal method would permit analysis of the products of SNP-scoring reactions directly, without the need for further purification, including highly multiplexed analysis of many SNPs per sample, as well as the analysis of many samples simultaneously and/or a high serial sample analysis rate. The method should also be integrated with commercial laboratory automation instrumentation commonly used in a drug discovery environment. While no single analysis platform meets all of these criteria, a few meet several of them, and continued instrument development will likely improve performance in these respects.

2.12.6.2.3 Automation and multiplexing

Especially important features for large-scale applications are multiplex and automation capabilities. Microplate-based assays are particularly well suited to automation, and many commercial robotic systems can perform large-scale sample preparation and analysis using microplates. Automation permits the processing of large numbers of samples, and microplate assays can be configured to use small volumes (a few microliters), providing a reduction in assay reagent costs. An even bigger reduction in assay costs can be obtained by performing the multiplexed scoring of many SNPs in a single sample. For some SNP-scoring chemistries the expensive reagents (enzymes and/or labeled nucleotides) are in excess, so that additional sites can be scored using the same pool of these reagents. This results in a decrease in the cost per SNP scored that is directly proportional to the level of multiplexing.

While microplates are well suited to the parallel processing of many samples, it is difficult to multiplex solution-based SNP-scoring assays beyond a few sites. Electrophoresis-based methods can provide a modest degree of multiplexing combined with parallel sample analysis, and automation capabilities are improving. Microarray-based analysis methods are especially well suited for highly multiplexed assays, enabling the analysis of many more SNPs in a single sample than microplate-based assays. Flat microarrays (DNA chips) facilitate the multiplexed scoring of 1000 or more SNPs from a single sample, while soluble microarrays (DNA microspheres) can supports the analysis of dozens to a hundred SNPs simultaneously, and potentially many more. Sample preparation and analysis for DNA chips can be automated using vendor-supplied equipment, while multiplexed microsphere-based assays are compatible with parallel sample preparation in microplates and conventional laboratory automation equipment.

2.12.6.3 Single-Nucleotide Polymorphism-Scoring Chemistries

The objective of an SNP-scoring assay is to determine the nucleotide base at specific sites in the genome. A variety of approaches have been developed to score SNPs, some of which exploit differences in the physical properties of the DNA and oligonucleotide probes, while others rely on highly specialized enzymes to analyze DNA sequences. These methods vary in their ease of use, flexibility, scalability, and instrumentation requirements, and these factors will affect the choice of approaches employed in a HTS environment.

2.12.6.3.1 Sequencing

Conceptually, the simplest approach to scoring SNPs is probably direct sequencing, since sequencing is used to discover and confirm SNPs initially. However, in practice, direct sequencing is a fairly inefficient way to score SNPs. While the speed and throughput of sequencing instruments continues to improve, with capillary-based sequencers capable of generating 96 sequencing reads of 400–500 bases each in 2–3 h, such read lengths will, in general, contain only one to a few SNPs on average. Considering this throughput rate (96 SNPs every 2–3 h per instrument) and sample preparation costs (PCR amplification or cloning followed by one or more sequencing reactions for each target), direct sequencing is not suitable for most large-scale SNP-scoring applications. In some highly variable regions of the genome – certain human leukocyte antigen (HLA) genes, for example – a sequencing read may contain dozens of potential SNP sites. In such cases, sequencing can be quite efficient, providing multiplexed SNP scoring in one reaction. However, such cases are rare exceptions, and for most applications, sequencing is not a viable high-throughput SNP-scoring method.

2.12.6.3.2 Hybridization

SNPs can often be detected by measuring the hybridization properties of an oligonucleotide probe to a template of interest. The melting temperature for hybridization of an oligonucleotide of modest length (15–20-mer) can differ by several degrees between a fully complementary template and a template that has a one-base difference (i.e., an SNP). By carefully designing oligonucleotide probes and choosing hybridization conditions that allow the probe to bind to the fully complementary template, but not to a template containing an SNP, it is possible to use hybridization to distinguish a single-base variant (**Figure 25a**).

In practice, this approach generally requires careful optimization of hybridization conditions because probe melting temperatures are very sensitive to sequence context as well as to the nature and position of the mismatched base. This difficulty in choosing optimum hybridization conditions (buffer and temperature) can be eliminated by continuous monitoring of probe hybridization as the temperature is increased, allowing the melting temperature for matched and mismatched probes to be determined directly in a process known as dynamic allele-specific hybridization. Alternatively, the use of highly parallel microarrays enables an SNP-containing template to be interrogated by many different hybridization probes for each SNP,[316] providing a redundancy of analysis that effectively increases the signal-to-noise ratio. PNA, which are oligonucleotide analogs with a structure that make hybridization very dependent on correct base pairing (and thus very sensitive to mismatches),[317,318] can improve hybridization-based assays. Locked nucleic acids, another oligonucleotide analog, are likely to be similarly useful.

2.12.6.3.3 Allele-specific chain extension

The combination of a hybridization probe with a sequence-sensitive enzymatic step can offer very robust and accurate SNP-scoring chemistry. Allele-specific chain extension (ASCE) involves the design of a primer whose 3′ terminal base will anneal to the site of the SNP.[319,320] If the base at the SNP site on the template DNA is complementary to the

```
                                      Probe B
                                 5'-GCGCGCTTCATGCA
                        Probe A
                   5'-GCGCGCTGCATGCA
        3'-ACGTACGTACGTACGTCGCGCGCGCGAcGTACGTACGTACGTACGTACGTACGTACGT
                                Template
(a)
```

```
                                       Primer B
                              5'-TGCATGCAGCGCGCGCGCTT
                           Primer A
              5'-TGCATGCAGCGCGCGCGCTGCATGCATGCATGCATGCATGCATGCATGCA
        3'-ACGTACGTACGTACGTCGCGCGCGCGAcGTACGTACGTACGTACGTACGTACGTACGT
                                Template
(b)
```

```
                                                      dT-*
                           Primer
                   5'-TGCATGCAGCGCGCGCGCTG-*
        3'-ACGTACGTACGTACGTCGCGCGCGCGAcGTACGTACGTACGTACGTACGTACGTACGT
                                Template
(c)
```

```
                                            Probe B
                                     TCATGCATGCATGCATG-3'
                    Primer                  Probe A
          5'-TGCATCGAGCGCGCGCT           GCATGCATGCATGCATG-3'
        3'-ACGTACGTACGTACGTCGCGCGCGCGAcGTACGTACGTACGTACGTACGTACGTACGT
                                Template
(d)
```

Figure 25 SNP scoring chemistries. (a) Hybridization-based methods to interrogate a variable site (lower case) in a DNA template involve the design of probes that, under the appropriate incubation conditions, will anneal only to one specific allele. (b) Allele-specific chain extension employs primers that anneal at the SNP-containing site such that only the primer annealed perfectly to a specific allele (primer A) is extended, while a primer specific for another allele (primer B) is not. (c) In single-base extension, a primer is designed to anneal immediately adjacent to the SNP-containing site, and a DNA polymerase is allowed to extend the primer one nucleotide using labeled dideoxynucleotides. The identity of the incorporated dideoxynucleotide reveals the base at the SNP site on the template DNA. (d) The oligonucleotide ligation assay uses two types of oligonucleotides. The first (primer) anneals to the template DNA immediately adjacent to the SNP-containing site. The second type (probe) anneals on the other side of the SNP such that the terminal base will pair with the SNP. If the terminal base is complementary to the base at the SNP site, the first (primer) and second (probe A) oligos will be covalently joined by DNA ligase. If the terminal base of the second oligo (probe B) is not complementary to the SNP base, ligation will not occur.

3' base of the allele-specific primer, then DNA polymerase will efficiently extend the primer (**Figure 25b**). If the SNP base is not complementary, polymerase extension of the primer is inefficient, and under the appropriate conditions no extension will occur.

Originally implemented as a PCR-based assay, allele-specific PCR (AS-PCR) used gel electrophoresis to detect the PCR product as a positive indicator of the presence of the SNP of interest.[319,320] AS-PCR is widely used in diagnostic assays for several disease-associated mutations[321] as well as for HLA typing.[322] In addition, by using two allele-specific primers, AS-PCR can be used to distinguish whether two SNPs are present on the same chromosome.[323] AS-PCR assays can also be multiplexed to some extent, enabling several SNPs to be assayed in a single sample.[324] More recently, ASCE has been adapted to detection platforms more compatible with large-scale analysis, including microarrays.[325,326]

2.12.6.3.4 Single-base extension

The ability of DNA polymerase to discriminate correctly paired bases is the principle underlying conventional DNA sequencing as well as ASCE. Another approach employing the sensitive sequence discrimination of DNA polymerases is SBE of a primer, also known as minisequencing or genetic bit analysis. In SBE (**Figure 25c**), an oligonucleotide probe is designed to anneal immediately adjacent to the site of interest on the template DNA. In the presence of dideoxynucleoside triphosphates, DNA polymerase extends the annealed primer by one base.[327,328] This is essentially the same dye terminator chemistry widely used in conventional sequencing, and, as long as the SBE primer is specific, can be very accurate with an excellent signal-to-noise ratio. Most often, the dideoxynucleotide is labeled. Usually the label is a fluorescent dye, but in some configurations biotinylated dideoxynucleotides can be used and detected with a labeled avidin in a secondary labeling step. If fluorescent dideoxynucleotides are used, all four bases can be scored in a single reaction if each of the four dideoxynucleotides is tagged with a different fluorophore. Thermal cycling of the SBE reaction using a temperature-stable polymerase can provide signal amplification, although, at present, amplification of the genomic template by PCR is still required.

2.12.6.3.5 Oligonucleotide ligation

Another method that takes advantage of enzyme discrimination of correctly paired bases is the oligonucleotide ligation assay (OLA). Two types of oligonucleotide probes are designed: one that anneals immediately adjacent to the SNP site on the template DNA and one that anneals immediately adjacent to the other side of the SNP (**Figure 25d**) such that the terminal base will pair with the SNP. If the terminal base of the second primer is complementary to the SNP base, then a DNA ligase enzyme will covalently join the two primers. If the terminal base is not complementary, ligation will be very inefficient. As for SBE, OLA exploits highly specialized enzymes to discriminate correctly matched bases, and can be very accurate. In addition, because it uses two primers to interrogate the SNP, OLA can also be configured to detect small insertions or deletion mutations.

Often, one primer is immobilized on a solid substrate[329] and the second primer is labeled, either directly with a fluorophore or with a biotin for subsequent detection with labeled avidin. OLA is generally performed on PCR-amplified template DNA, but a variant, the ligase chain reaction, uses genomic DNA as the template for the ligation reaction, the products of which are then amplified.[330] Another variant suitable for use with genomic DNA template is 'padlock probe' ligation, which generates a circular oligonucleotide product, which can then be amplified many hundred-fold using rolling-circle amplification.[331,332]

2.12.6.3.6 Nuclease-based assays

Several SNP-scoring methods based on the nuclease cleavage of oligonucleotide probes have been demonstrated. A relatively straightforward assay targets SNPs that alter the recognition site for a restriction endonuclease, resulting in a change in the electrophoretic migration of the template molecule. More general approaches that can be configured as homogeneous assays are Taq-Man[333] and Invader,[334] both based on hybridization combined with an enzymatic 5'-nuclease activity.

In Taq-Man, the 5'-nuclease activity of Taq DNA polymerase cleaves a hybridization probe that specifically anneals to the template DNA during PCR. The Taq-Man hybridization probe is labeled with fluorescent donor and acceptor molecules that exhibit fluorescence resonance energy transfer (FRET) that is released upon cleavage. The Taq-Man probes must be carefully designed to hybridize only to a perfectly matched template and not to a template with a mismatched base. The Invader assay uses the 5'-nuclease activity of a thermostable flap endonuclease enzyme, to cleave a 5' 'flap' displaced by an invading probe. Probe cleavage produces a short oligonucleotide bearing a label that can be detected. Alternatively, the short cleavage product can serve as an Invader probe for a second stage of probe cleavage. This results in an amplification of the signal under isothermal conditions, and detection of signals from genomic DNA without PCR has been reported.[334] Both the Taq-Man and Invader assays are gaining popularity for the analysis of single-point mutations. Highly multiplexed analysis of many SNPs simultaneously, such as is desired for large-scale applications, has not yet been demonstrated for these assays.

2.12.6.3.7 Overview

For large-scale SNP scoring, the SBE chemistry is probably the most robust, flexible, and easily multiplexed chemistry. For SBE, one primer is designed for each SNP, which are then labeled by extension with dideoxynucleotides. The extension of primers with labeled dideoxynucleotides by DNA polymerase is essentially the same chemistry used in conventional sequencing, and is generally very accurate. Allele-specific chain extension requires a different primer for each allele, but also benefits from the specificity of DNA polymerase. Ligation assays can be very accurate as well, but they require a primer and a different probe to score each SNP. On the other hand, ligation assays are more readily configured to score small insertions or deletions than SBE. Ligation, ASCE, and SBE assays are not well suited to score SNPs in highly variable regions of the genome, as variability in primer binding sites makes primer design more difficult. In these rare instances, conventional sequencing may be the most effective method of assessing the genotype.

2.12.6.4 Platforms

Just as there are a variety of assay chemistries for scoring SNPs, there are several platforms on which samples can be analyzed. Although all chemistries are not compatible with every platform, most can be analyzed by a variety of instruments. Key considerations for large-scale SNP scoring in a drug discovery environment are compatibility with highly parallel and automated sample preparation and the ability to perform multiplexed scoring of many SNPs in each sample.

2.12.6.4.1 Electrophoresis

Being a standard technique in molecular genetics laboratories, gel electrophoresis is widely used for SNP scoring. Most often, mobility through the gel or sieving matrix is used to identify a particular DNA fragment by its size. In addition, fluorescence detection of one or more probes can be used to identify primers or incorporated nucleotides. These two features form the basis of automated DNA sequencing, but also have applications in SNP scoring. In allele-specific PCR, agarose gel electrophoresis is often used to identify amplified products based on size.[324] For SBE assays, different-sized primers, each interrogating a different SNP, can be resolved by electrophoresis, enabling multiplexed analysis of several sites simultaneously.[335,336]

The limitations of electrophoresis as an SNP-scoring platform stem from fairly low throughput. The current state of the art in electrophoresis for sequencing applications, capillary electrophoresis instruments with 96-sample capacity, can perform an analysis every 2–3 h. Assuming a size-based primer multiplex to analyze 10 SNPs simultaneously, one could score roughly 4000 SNPs per workday. New-generation instruments with 384-capillary capacity could quadruple that throughput.

2.12.6.4.2 Microplate-based assays

Many SNP-scoring assays have been adapted to a microwell plate format, enabling highly parallel and automated sample processing and analysis. This platform is compatible with commercial high-throughput sample handling and measurement instruments that are widely used in the pharmaceutical industry. SNP-scoring assays in microplates can be configured as heterogeneous (requiring wash steps to remove excess reagents) or homogeneous (no wash steps required), with fluorescence or absorbance detection.

For the heterogeneous approaches, the microwell is often used as a solid support for oligonucleotide probes. For example, OLA and SBE assays have been configured with the SNP primer immobilized on the microplate bottom.[329,337] PCR-amplified template, reporter oligos or nucleotides, and enzyme (ligase or polymerase) are added, and the enzymatic reaction causes the immobilized primer to be labeled. Excess reagents are then washed away, and the signal remaining in the well is measured.

For homogeneous assays, the microwell serves as a cuvette in which fluorescence is monitored. The key to the most homogeneous assays is the use of FRET, which results in a change in the fluorescence spectra of a sample when two fluorescent dyes are brought into proximity. In addition to the example of the Taq-Man assay described earlier, FRET-based approaches have been configured for hybridization using molecular beacons,[338] OLA,[339] and SBE.[340]

Fluorescence detection provides sensitive detection for the microwell format, and the use of FRET provides essentially real-time scoring of the SNP. An entire microwell plate can be measured in less than a minute, enabling high sample throughput, and, in the case of 384 or higher densities, the reaction volumes can be quite small (on the order of 10 μL). A disadvantage of microplate-based detection is the limited capacity for the multiplexed detection of many SNPs per sample. The use of differently colored probes can enable simultaneous detection of multiple signals, but because currently available fluorophores have relatively broad emission spectra, it will be very difficult to configure homogeneous solution assays with a multiplex capacity of more than three or four SNPs per sample.

2.12.6.4.3 Mass spectrometry

There is a significant effort aimed at adapting mass spectrometry to genomic applications, including SNP scoring. The advantages of mass spectrometry as an analytical tool stem from its capacity to produce very rapid and precise mass measurements. This allows many assays to be configured without the use of exogenous labels. For instance, SBE reactions can be analyzed by measuring the mass of a dideoxynucleotide added to a primer. Because each of the four dideoxynucleotides has a distinct mass, it is possible to identify which nucleotide(s) was added to the primer.[341] In addition, by using several primers of different masses (lengths) to interrogate several SNPs, it is possible to perform a multiplexed analysis on each sample.[342]

Current limitations of the mass spectrometry-based SNP scoring include a marked dependence on the size of DNA molecule being analyzed. Single-base resolution decreases for larger oligonucleotides, especially between nucleotides of similar mass such as A and T. The use of mass-tagged nucleotides (i.e., isotopically labeled nucleotides) can increase this resolution,[342–344] but at added expense. Another limitation is the requirement for highly processed and purified samples. Current mass spectrometry-based genotyping protocols call for the purification of samples before analysis. This requirement increases the cost and time of the procedure, although these steps can be automated and performed in parallel. It is very likely that some of these limitations will be overcome, given the current high level of interest in adapting mass spectrometry to genomic applications.

2.12.6.4.4 Flat microarrays

Perhaps one of the best publicized biological technologies of the last decade is the use of microarrays to perform large-scale genomic analysis.[345,346] These microarrays, typically constructed on glass using photolithographic or robotic printing or spotting processes, can contain hundreds to hundreds of thousands of different DNA molecules on a surface of just a few square centimeters or less. Assays are generally configured using fluorescent reporter molecules, and arrays are read using scanning methods such as confocal microscopy.

For SNP-scoring applications, DNA microarrays have generally been used in conjunction with hybridization assays.[316] In these assays, the sensitivity of hybridization to the local sequence context of an SNP is overcome by interrogating each SNP with multiple hybridization probes, providing a redundancy to the analysis that reduces ambiguous SNP scoring. Chips designed to interrogate hundreds of SNPs simultaneously have been developed,[316] and it should be possible to make chips with even higher SNP-scoring capacity. In addition to hybridization-based SNP scoring, the SBE chemistry has been adapted to flat microarrays.[328] Currently, flat DNA microarray technology is expensive and not very flexible. Acquisition of proprietary or specialized array manufacturing technology and dedicated array analysis instrumentation requires a large initial investment. However, for large-scale SNP-scoring applications, the investment may be justified.

2.12.6.4.5 Soluble arrays

An approach that has not yet generated the publicity of the DNA chip but will probably have a major impact in a variety of areas is the use of soluble arrays and flow cytometry. Soluble arrays are composed of microspheres that are dyed with different amounts of one or more fluorophores such that many distinct subpopulations of microspheres can be identified on the basis of their fluorescence intensity using flow cytometry. Conceptually, the soluble array approach is similar to that of flat microarrays, with different levels of fluorescence intensity replacing x–y positions on a surface. Currently, soluble microsphere arrays of 100 elements are commercially available (Luminex),[347] but this number could in principle be expanded to thousands with the use of additional dyes to create multidimensional arrays.

The analysis of soluble arrays by flow cytometry has a number of advantages over flat surface microarrays. First, oligonucleotides are readily immobilized on microspheres using well-known benchtop chemistry. Second, because each array element is a distinct population of microspheres, array preparation consists of the combining of individual microspheres into a mixture using a pipette. Reconfiguration of the array requires only the modification of individual microsphere populations, followed by the preparation of a new mixture. The use of universal oligonucleotide address tags[348,349] to capture tagged primers after a solution-phase SNP-scoring reaction further increases the flexibility of the array by allowing assays to be redesigned. Third, because the array is in a liquid phase, conventional liquid-handling hardware can be used for highly parallel sample preparation in microplates. In addition, improved sample handling capabilities for flow cytometry[350] should boost throughput by more than a factor of 10 from the current 1–2 samples per minute. Fourth, because flow cytometry can discriminate between free and particle-bound probe, under most conditions no wash or purification step is required. Finally, because flow cytometry is capable of measuring tens of thousands of particles per second, soluble arrays can be analyzed in just a few seconds.

A variety of SNP-scoring chemistries are compatible with the use of microspheres as solid supports and fluorescence detection by flow cytometry. Hybridization-based assays have been used to detect PCR products[351] as well as to interrogate SNPs by multiplexed competitive hybridization.[352] In addition, the OLA and SBE chemistries have been adapted to microspheres.[349,353,354] While soluble array technology does not yet have the parallel analysis capacity of the flat microarrays, advantages in serial sample throughput, ease of automation, and flexibility will make it attractive for many applications.

2.12.6.4.6 Overview

The key feature of a large-scale SNP-scoring platform is the ability to analyze highly multiplexed assays with high serial sample throughput in an automated manner (**Table 21**). Microplates are an excellent platform for automated parallel sample processing, but cannot offer a high level of multiplexed analysis. Flat DNA microarrays offer very highly

Table 21 Characteristics of SNP scoring platforms

Platform	Hardware costs (US$)	Throughput	Automation compatibility	Multiplex capacity
Gels	Low–high	Low–moderate	Low–moderate	Moderate
Microplates	Moderate–high	High	High	Low
Mass spectrometry	High	Moderate–high	Moderate–high	Moderate
Flat arrays/scanner	High	Low	Moderate	High
Microspheres/flow cytometry	Moderate–high	Moderate–high	Moderate–high	Moderate–high
Low	<20 000	<1 sample min^{-1}	Custom high-throughput screening	<5 SNPs per sample
Moderate	20 000–100 000	1–10 samples min^{-1}	Custom commercial hybrid	5–50 SNPs per sample
High	>100 000	>10 samples min^{-1}	Commercial high-throughput screening	>50 SNPs per sample

multiplexed analysis (thousands of SNPs), but a low sample analysis throughput. Soluble arrays have a high multiplex capacity (dozens to a hundred SNPs), with a fairly high serial sample throughput. Microplate-based sample processing combined with a highly multiplexed array-based measurement probably represents the most efficient combination of current technologies.

2.12.6.5 Conclusions and Prospects

Many strategies are being employed to score SNPs. Large-scale SNP scoring such as is required for drug discovery applications requires special consideration of the issues of throughput and cost. Especially important for optimizing throughput and cost are the ability to automate sample preparation and analysis and the ability to perform multiplexed scoring of many SNPs in a single sample. In this section, we have focused on those methods currently available that could be configured for large-scale applications. However, with the intense interest this area of research is generating, improvements in all aspects of sample preparation and analysis are to be expected. In closing, we would like to point to a few areas of likely progress.

Most SNP-scoring assays involve PCR amplification of the DNA template. PCR not only improves assay sensitivity but provides added specificity for the resolution of specific SNP sites against similar sites in gene homologs or pseudogenes. For the efficient highly parallel analysis of multiple SNPs, such as is possible with microarrays, the PCR amplification step must be highly multiplexed. Multiplexed PCR is a challenging problem, requiring the design and pooling of PCR primers that will specifically amplify many sequences under the same conditions yet will not interact with each other to produce primer artifacts. Because this is a universally important problem, with applications beyond SNP scoring, it is hoped that automated informatics and computational tools will be developed to supplement the largely empirical approach to multiplex PCR currently employed.

Applications of DNA microarray technologies will continue to be developed, providing highly parallel multiplexed analysis. Flat surface microarrays should prove to be much more useful for SNP scoring when combined with SBE and universal capture tags. The parallel analysis capabilities of soluble microsphere arrays are expected to increase as improved dye chemistry enables the production of multidimensional arrays, providing the parallel analysis throughput of flat microarrays with conventional liquid handling automation. Newer formats, such as microarrays configured on the ends of optical fibers (*see* Section 2.12.7), may also prove to have unique advantages as detection platforms for SNP-scoring applications. Finally, the push of micro- and nanofabrication toward integrated lab-on-a-chip approaches will likely have an impact on reducing costs, especially if these can provide a highly multiplexed analysis. An understanding of human genetic variation, especially in the form of SNPs, offers enormous potential for enhancing and improving the process of drug discovery and development. The realization of this potential depends on the ability to score rapidly and efficiently large numbers of SNPs in large numbers of samples. Technologies that combine automated sample preparation with highly multiplexed analysis will play a central role in meeting these needs.

2.12.7 Protein Display Chips

2.12.7.1 Introduction to Protein Chip Technologies

The genome-sequencing revolution has presented protein biochemists with blessings and challenges. At the DNA level, high-throughput technologies enabled the rapid discovery progress that the gene discovery community has enjoyed. Innovative, chip-based technologies have provided molecular biologists with a completely new set of tools. At the same time, the incredible amount of information from the various genomic analysis efforts has added additional layers of complexity to the way a lot of biological and medical questions are addressed. The simultaneous genetic analysis of many species and tissue types has broadened our view and quickened our understanding of key physiological processes on a genetic level. As will be discussed in this section, technology platforms originally introduced for nucleic acid analysis would eventually also have a significant impact on protein biochemistry in the hands of drug discovery researchers.

The rapid developments in the DNA and RNA analysis field have led the way to smaller sample volumes and rapidly increasing sample throughput. Arraying technologies, as well as microfluidic and robotic systems, have revolutionized industrial molecular biology. However, the fast and furious pace in that area has widened the gap in the understanding of proteins and protein–protein interactions, which has not progressed at the same speed. The reasons for this are both of a philosophical and a practical nature. In the minds of the molecular biology community, proteins have been, for a long time, gene expression products. The umbrella of this term very conveniently shields the molecular complexity and diversity with which proteins present us. This molecular diversity is the main practical reason why the study of proteins is much harder to press into a multiplexed high-throughput format. The beauty of nucleic acid biochemistry lies in the simple variation of a known theme, the repetitiveness of a very limited number of building blocks, with the triplet-based sequence as the brilliant key to the language of life. This key encodes the bountiful variety of the protein world, where things become much less predictable. Already, the 20 amino acids as components with very different chemical characteristics present a considerable analytical challenge when one considers the combinatorial possibilities. In addition to that, proteins can undergo post-translational modifications with other classes of molecules, such as lipids or carbohydrates, which can alter the behavior and characteristics of the entire molecule. Yet from the earliest studies in protein biochemistry, especially in the area of enzymology, the need for efficient as well as meaningful protein analysis has been clear, especially in two aspects: proteins as antagonist targets and, lately, also as therapeutics themselves in the fight against disease. The increasing pace of discovery in cell biology and molecular biology discovery research has made that need all the more pressing. Protein development efforts drive large programs in the pharmaceutical industry. The focus of these programs has undergone many changes over the years. Enzymology clearly made its mark on drug discovery research early on. The advent of recombinant bacterial and eukaryotic protein expression systems now allows the production of large amounts of functional proteins for small-molecule antagonist screening, structural characterization, and functional studies.

Tagging recombinant proteins further facilitated upstream optimization as well as downstream processing and analytical work. Tagging systems represented the first step toward an integrated assay platform for proteins that went beyond polyacrylamide gels. At the same time, advances in the fields of immunology and signal transduction presented new challenges. The functionality of a protein in these areas could not be measured through a catalytic reaction that was monitored as a color change or fluorescence enhancement. Instead, the function of a protein was based on its proper interaction with another protein or other class of effector molecule. This called for the development of new types of assays that allowed real-time monitoring of binding kinetics and the determination of binding specificity in a physiological setting. Also, the genome sequencing efforts led to the discovery of whole families of novel genes with unknown functions. This presented an additional need for interaction studies that would allow the identification of binding or reaction partners. HTS groups started to expand into the areas of rapid biological assays for proteins with unknown activities. Aside from these exploratory and discovery applications, functional assays have become increasingly important at downstream junctures of the drug development process. To catch up with the speed of gene discovery, protein development was in need of sensitive, multiplexed assays that still would have the flexibility to accommodate the diverse needs of the protein world. As it turned out, the technology toolbox that revolutionized the nucleic acid world also changed the way we look at proteins. This section will attempt to cover the advances in the area of chip-based protein analysis technologies over the past decade and a half, and highlight the most important implications of these technologies for the drug discovery process.

2.12.7.2 Surface Plasmon Resonance-Based Technology: How Optical Biosensors Revolutionized Protein–Protein Interaction Studies

2.12.7.2.1 Optical biosensor technology: technical background and development

Development in the field of optical biosensors received significant impetus from the need of improvement in the area of laboratory-based immunodiagnostics. Optical techniques seemed an obvious choice for this area, since refractive index is one of the few physical parameters that vary upon the formation of an immune complex.[355] While interest in diffraction phenomena and the curious absorption anomalies of metallized optical diffraction gratings in particular dates back to the turn of the last century, more systematic theoretical work in the field did not start until the 1950s, with the major advances published again much later.[356–359]

Optical biosensor technology relies on a quantum mechanical detection phenomenon called the evanescent field. It is used to measure changes in the refractive index that occur within a few hundred nanometers of a sensor surface. These refractive index changes are caused either by the binding of a molecule to a second molecule that has been immobilized on the surface or by the subsequent dissociation of this complex.[360] There are several ways to create an evanescent field.[361–363] SPR, which is the technology platform of the Biacore instrument, has established itself as the most commonly used over the past 15 years. In SPR, polarized light is shone onto a glass prism that is in contact with a thin gold–glass surface. Light is reflected at all angles off the gold–glass interface, but only at the critical angle does the light excite the metal surface electrons (plasmons) at the metal–solution interface. This creates the evanescent field, and causes a dip in the intensity of the reflected light. The position of this response is sensitive to changes in the refractive index as well as the thickness of the layer in the vicinity of the metal surface.[364] Resonance is described by an arbitrary scale of resonance units (RU). A response is defined as the RU difference compared with a baseline that is normally established at the beginning of an experimental cycle with the immobilization of the capture protein or molecule. **Figure 26** shows the setup of a typical SPR detection system, consisting of a detector, the sensor chip, and an integrated microfluidics system, which allows for the continuous flow of buffer, sample, and reagents.

In a typical experiment, a capture molecule is first immobilized covalently on the sensor surface. Subsequently, the residual binding sites are blocked. After a wash with a buffer of low ionic strength, the interaction sample is passed over the sensor, followed again by buffer. **Figure 27** depicts a typical SPR sensogram. Response is measured with time, and is proportional to the mass of the adsorbed molecule.[360,365] As ligand adsorption on the immobilized capture molecule occurs, the adsorption profile allows the determination of the association rate constant, k_{on}. After passing of the sample

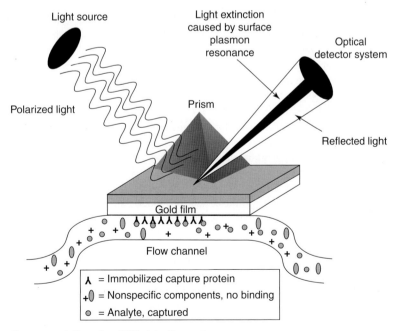

Figure 26 Schematic representation of an SPR detection system.

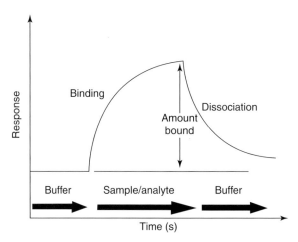

Figure 27 Typical SPR sensogram.

plug, dissociation of the complex follows, as buffer passes over the sensor. The logarithmic decay of the dissociation reaction then allows the calculation of the dissociation rate constant, k_{off}. Earlier studies exploring the potential of SPR in the field of biomolecular interactions employed simple adsorption to the metal surface for the immobilization of the capture component. This strategy comes with certain limitations. Some proteins may denature when absorbed directly to the metal; and hapten-type ligands are very difficult to immobilize on metal. Furthermore, the performance of the sensor depends on the even and reproducible distribution as well as correct orientation of the adsorbed molecules. With a directly immobilized ligand there can be undesired interaction of solutes with the metal as well as exchange processes between the ligand and solute. To overcome these limitations, a carboxylated dextran surface coupled to the gold film was developed, which evolved into the standard surface medium for the covalent immobilization of capture molecules in the Biacore system, the CM5 sensor chip. The concept of this surface technology was first described by Johnsson et al.[364] Their hydrogel-modified surface was designed to minimize nonspecific adsorption of ligands at physiological ionic strength and at the same time facilitate accessibility of the ligand in the subsequent interaction study. Since the Biacore system has grown into a generally accepted and commercially available analysis platform, the associated surface chemistry has been refined and diversified to accommodate the constantly growing number of applications. Today's gold–dextran surfaces basically have five different layers: the glass basis layer, the gold film, a linker layer, the dextran layer, and a specific layer. In the original surface design, the covalent coupling of the biological capture molecules occurred directly by nucleophilic displacement of N-hydroxysuccinimide (NHS) esters on the carboxymethyldextran chains.[366] Modern CM5 surfaces come preactivated for four different covalent coupling strategies: amine, ligand thiol, surface thiol, or aldehyde. In addition to the covalent binding, hydrophobic surfaces are now available for the capture of membrane complexes, and nickel–chelate surfaces allow the selective binding of His-tagged proteins.

The popularity of biotinylation as an efficient tag for the capture of both proteins and nucleic acids via streptavidin has prompted the development of streptavidin precoated surfaces.

On the instrumentation front, development has yielded a higher degree of sophistication and sensitivity, addressing some of the earlier problems associated with reproducibility and difficulty in detecting low-affinity interactions. At the same time, a greater level of automation has standardized and facilitated the use of the technology, opening it to the nonspecialized user. High-end systems allow the most elaborate kinetic analyses of biological interactions, whereas more simply designed yet fully automated systems allow rapid and reliable screening of a large number of samples.

2.12.7.2.2 Types of surface plasmon resonance applications and experimental considerations

Since its original development in the area of immunodiagnostics, optical sensor technology has expanded into a wide variety of biological and medical areas of interest. An obvious choice for signal transduction research, SPR has become a standard tool for the analysis of receptor–ligand interactions.[367,368] The field of chaperonins also discovered the technology early on.[369–371] Beyond mere protein–protein interactions, SPR has found applications in small-molecule screening[372] as well as protein–carbohydrate[373] and protein–nucleic acid interactions.[374–376] Even viral particle binding to cell surface receptors has been analyzed by SPR.[377] At downstream junctures of the drug development process, specialized optical biosensors are used for bioprocess monitoring and product analysis.[378] In spite of the great diversity of molecules analyzed, SPR studies mainly fall into two categories as far as experimental goals and design are concerned. Screening applications look for the relatively simple answer of an approximate kinetic ranking, basically to see if an analyte is specifically binding to a given ligand or not. Functional studies look at the mechanism of a certain biological interaction by determining quantitative rate constants. Clearly, a functional study with the aim of retrieving key mechanistic data requires a different level of insight and experimental premeditation than a simple 'quick and dirty' screen. Yet it is crucial to realize that the mimicking of a biological interaction event on a two-dimensional platform is not a trivial endeavor per se. Failure to understand the key biochemical parameters, caveats, and limitations of the system will inevitably lead to the failure of even the simplest of studies.

The continuously advancing simplification and automation of commercially available SPR systems has made life easier for the inexperienced user, and it has become more and more tempting to forget about the cumbersome underlying theory. Increasing simplicity and user friendliness have allowed an explosion in SPR-related references in the scientific literature. Reviewed over time, the literature reveals a typical trend for an evolving technology platform. Most of the earlier papers in the field were devoted predominantly to experimental designs, technological details, and, most of all, the analysis of kinetic data.[379–381] More recent papers are overwhelmingly application driven, and for the most part have moved the more theoretical aspects into the background.[382,383] This development clearly comes at a price, for it appears that less and less effort is spent on both careful experimental design and the critical evaluation of the answers provided, leading to a situation where more data actually provide less value but increased confusion instead.[384] Therefore, it seems worthwhile to revisit the key factors that control both the experimental as well as the data analysis parts of SPR experiments.

Clearly, when a reaction that usually occurs in solution is forced into a two-dimensional setup, the limitations of mass transfer will have to be considered. At the same time, the immobilization of one binding partner on the surface can actually help mimic a biological situation where the capture molecule is immobilized in a molecular superstructure such as a membrane. While no systematic efforts have gone into the comparative analysis of SPR and solution binding, several researchers have revisited their SPR-derived affinities using other techniques, such as isothermal titration calorimetry, for confirmation. They have generally found good agreement, provided that the experimental design was chosen wisely.[385–387]

Obviously, background understanding of the biological system to be analyzed will help in the design of the study. If the ligand or the analyte is a novel protein of unknown function, the data obtained in the experiment will have to be qualified with the possibility in mind that the binding partners may not have been presented in the correct orientation or configuration for an interaction to take place. When coupling the ligand to the surface in a covalent fashion (e.g., via the primary amines), this basically constitutes a random binding event that may hamper the formation of the correct capture configuration to receive the analyte. Tumor necrosis factor and similar ligands, for example, bind their receptors in a trimeric fashion: three receptor molecules have to form a homologous trimer to receive an analyte trimer. The addition of tags can help to overcome this problem: the use of Fc fusions, for example, allows capture on the surface via a protein A bridge that orients all capture molecules in the same way. Also, Fc tags even without the help of protein A, force their fusion partner into homologous dimers with other molecules. While this can help to increase the percentage of correctly oriented capture ligands on the sensor surface, it also has to be considered as an additional source of artifacts. Some tags, such as Fc tags, are quite bulky, and can lead to considerable structural changes in the proteins they are attached to. They may destabilize them to the point of breakdown or simply lead them into an inactive configuration. If the fusion breaks down, the empty Fc will still bind to other molecules and produce inactive multimers. Molecule size does matter for another reason in this context: it should be considered in the decision which partner of the binding experiment to immobilize covalently: the bulkier antibody, receptor, fusion protein for example, or the small peptide analyte. In many situations, this decision is made out of necessity rather than choice, as, for example, when one binding partner is available in a purified form but the other needs to be fished out of conditioned cell culture supernatant.

The purity and homogeneity of the ligand and analyte play a key role in the kinetics of the experiment. If contaminants are present that interfere with binding or act as competitors, this will undoubtedly at least complicate the situation, and at most render kinetic data useless. Contaminants can also obstruct the formation of specific multimers necessary for capture. Especially if the stoichiometry of the reaction is unknown, it is important to analyze a wide concentration range of both the ligand and analyte. Problems of mass transfer have been extensively studied from a theoretical point of view, and the reader is referred to the literature for a comprehensive treatise on the phenomenon.[388–390] In a high-density binding situation the crowding of molecules in the association state will push the balance from the desired multimerization to unwanted aggregation. It has been observed that, in general, valid kinetic measurements are best obtained with low levels of ligand immobilization.[391] If the analyte concentration chosen is too high, the mass transfer limitations hamper its access to and from the surface. The valency of the interaction also needs consideration, since multivalency leads to avidity effects, even in a low-density association. This has been extensively studied using the example of whole antibodies that are at least bivalent.[392]

Since SPR analysis represents the in situ monitoring of an interaction with one binding partner in flow, the time component of the setup naturally plays an important role, and the optimization of flow rates deserves some consideration. As a general rule, a higher flow rate helps to minimize mass transfer limitations.[393] In a typical Biacore experiment, the flow rate is kept constant, and the sample is delivered to the surface in a small sample plug (typically in a volume between 50 and 200 µL) at a low analyte concentration, preceded and followed by low ionic strength buffer. In this way, a discreet amount of analyte is delivered to the chip while the continuous flow necessary for kinetic analysis is maintained. This configuration is useful for first-order or pseudo-first-order kinetics and the determination of affinity constants. However, because of the finite volume of the sample plug and the limited data collection time associated with it, some binding reactions cannot be measured to equilibrium. To do this, the sample has to be recirculated over the surface or delivered in the buffer at a constant concentration.[394,395]

The scope of this section precludes in-depth discussion of kinetics and data analysis methods that apply to SPR applications, and the reader is referred to the more specialized literature for that area. However, it seems appropriate to at least touch on the most important parameters involved in deriving reliable kinetics and a sound statistical analysis of experimental data. In a very insightful 1994 review, O'Shanessey critiqued SPR studies described in the literature thus far.[396] His article shows that inappropriate interpretation of raw experimental data and incorrect mechanistic assumptions for the kinetics of a reaction are as much a source of error and failure as are faulty experimental design and execution. The single biggest error observed in documented SPR studies to this day is the application of simplistic kinetics and statistical models to reactions that in reality are very complex. This has in part been driven by the goal to create user-friendly software analysis platforms for commercial automated SPR systems. In a lot of cases, to keep things simple, attempts are made to fit data to the simplest 1:1 interaction model. This works out if the data actually fit this situation, but, as described earlier, the reality is often a lot more complex. Then the derived numbers become apparent rate constants that bear no resemblance to the reality of the molecular interaction. At the same time, use of the wrong statistical fits can be equally detrimental, and the user should invest some effort in the evaluation of linear, nonlinear, and global fitting methods.

2.12.7.2.3 Surface plasmon resonance in the drug development process: conclusion and outlook

The usefulness of SPR as a label-free monitoring technique for biological interactions at the many junctures of drug discovery and development is obvious, and the annual number of publications in the field bears substantial testimony to this. With commercially successful and widely used platforms such as that from Biacore, the innovative technology has now crossed into the field of established methodology, attracting more and more nonspecialized users. Both an increase in sample throughput and advanced automation permit the efficient handling of large numbers of samples. Currently, the more automated systems at the same time apply simpler analysis methods and are designed for 'quick and dirty' answers. For these answers to be of any value downstream, SPR systems need to be integrated with complementary analysis tools for the independent confirmation of positives. Eventually, hits from screens also have to be revisited with a more sophisticated SPR tool for more meaningful mechanistic studies. As the body of experience with the technology continues to grow, and systems at the same time become more and more sophisticated electronically, eventually even high-throughput systems will allow more elaborate answers – that is, provided the right questions are asked.

2.12.7.3 Time-of-Flight Mass Spectrometry-Based Technologies

2.12.7.3.1 Matrix-assisted desorption techniques: technical background and development history

Before we attempt to discuss the two currently available true protein chip technologies for mass spectrometry, surface-enhanced laser desorption/ionization (SELDI) and biomolecular interaction analysis/mass spectrometry (BIA/MS), we need to take a step back. The development of mass spectrometry from an analytical chemistry technique into a more user-friendly biochemistry and biology tool deserves some attention. The technology was established mainly for purity analysis and molecular weight confirmation of organic chemicals and peptides with molecular weights under 1000 Da. Fragmentation-related mass limitations initially prevented the analysis of larger molecules. Several desorption techniques, such as field desorption, secondary ion mass spectrometry, fast atom bombardment, plasma desorption, and, finally, laser desorption/ionization, extended the molecular weight range of analytes above 100 000 Da.[397] This obviously made a big difference for the usefulness of mass spectrometry in the area of protein biochemistry. The technique of MALDI was simultaneously and independently discovered in two laboratories in 1987 and reported at international conferences that same year. Results from both groups were first published in 1988.[398,399] The technology has since evolved into a standard technique in the area of protein characterization by mass spectrometry. With MALDI it became possible to very accurately and sensitively analyze large, intact biomolecules, and even derive information about their quaternary and tertiary structure. Commercially available MALDI platforms have become user friendly as well as much more powerful over the years. Mass accuracy in the 50–100 ppm peptide range is the expected norm. Eventually, the field of mass spectrometry started to open to nonexpert users, though real biology applications remained limited.

To put this technique into perspective, it is important to understand the key factors that make a MALDI-TOF experiment work. Very simply speaking, in a TOF mass spectrometry experiment the protein of interest is turned into a molecular ion. Subsequently, the travel time or TOF of this molecular ion through a vacuum tube toward a charged detector is measured. For singly charged ions, the TOF is proportional to the square root of the molecular weight of the analyte. There are different strategies to turn an uncharged molecule into a molecular ion. In the case of MALDI, an organic compound, the so-called matrix, in combination with an energy source, usually an ultraviolet laser, is used. A better term for 'matrix' would actually be 'energy-absorbing molecules' (EAMs), since that is exactly the purpose of these compounds. After the protein of interest has been mixed with an EAM solution and applied to the instrument probe in a tiny drop, EAMs and protein co-crystallize in a dry film. When the laser hits this film, the EAMs absorb the energy, expand into the gas phase, and carry the analyte molecules with them. At the same time, a charge transfer between the EAMs and the analyte leads to the creation of intact molecular ions. Usually the EAMs are present in vast excess over the analyte. The high matrix/sample ratio reduces the association of sample molecules and at the same time provides protonated and free radical products for the ionization of the molecules of interest. As the sample molecules acquire one or multiple charges, they are propelled down the vacuum tube toward a charged detector.

The matrices used in MALDI are mostly organic acids. They are prepared as saturated solutions in a solvent system typically consisting of an organic and a strongly acidic inorganic component. The choice of the matrix and solvent system very much depends on the nature of the molecule analyzed and is a crucial determinant for the success or failure of a MALDI experiment. Sinapinic acid and α-cyano-4-hydroxycinnaminic acid are among the most popular matrix compounds for general protein and peptide analysis. The preparation in 0.5% trifluoroacetic acid and 50% acetonitrile is a good starting point for a first experiment. The purity of the matrix component as well as of the solvent chemicals can critically influence the outcome of the experiment. Accordingly, the choice of the highest grade material is well advised.

Figure 28 shows a spectrum derived in a typical MALDI experiment. The x axis represents mass/charge (m/z), and the y axis the percentage of signal intensity. The two peaks visible represent the singly and doubly charged entities of the same molecule. The singly charged entity is detected at the molecular mass plus 1 (H+), the so-called monoisotopic mass, the doubly charged entity shows at apparently half the molecular mass. In addition to the analyte, matrix molecules are ionized as well, and show up at the very low end of the mass spectrum. This has to be kept in mind when attempting to detect very small molecules with molecular masses between 0 and 300 Da. When analyzing larger molecules that have hydrophobic qualities, matrix molecules can adhere to the analyte and create artifactual shoulders in the detected peaks. Fortunately, the generally high accuracy of mass detection allows distinction between the different molecular species and substraction of matrix-related peaks.

Outside of finding the best matrix and solvent composition for a successful experiment with a particular protein, the purity of the protein and the buffer system used have a crucial influence on the outcome of the MALDI analysis.

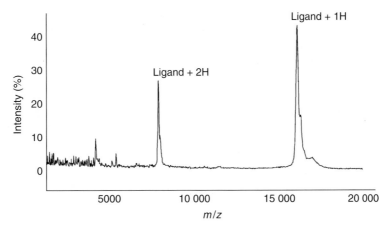

Figure 28 Typical MALDI spectrum of a purified protein.

Certain ions, such as Na$^+$, cause a phenomenon called ion suppression that usually results in tremendous chemical noise and inhibited ionization of the protein of interest. Due to this, Tris-based buffers as well as other buffers with a high content of sodium ions are unsuitable. Detergents can be very problematic, too. Very few can be considered compatible with the technique. Even under ideal conditions, the protein forms a three-dimensional crystal with the large excess of matrix on the instrument probe. To obtain the optimal signal, it is often necessary to search wider areas of the probe surface. This limits the quantitative value of the data.

The overall biggest limitations of the MALDI technique are the restrictions on the quality and nature of the sample that can be analyzed. Unless the protein sample is relatively pure and soluble in ionization-compatible buffers, analysis is difficult and requires great skill and understanding in the area of sample preparation. This clearly limits the usefulness of the technique in the area of biological screening, and has driven the development of matrix-assisted technologies that address these limitations.

The development has mainly gone in three directions: use of micropipette tip clean-up methods for standard MALDI applications, the direct modification of mass spectrometry probes for specific analyte capture and clean-up (SELDI), and the use of Biacore chips as the MALDI platform (BIA/MS).[400]

2.12.7.3.2 Surface-enhanced laser desorption/ionization: protein display chips for mass spectrometry

The SELDI technique was developed by Hutchens and Yip at Baylor College of Medicine, and the first experiments were published in 1993.[401] With a background of protein purification and characterization from very dilute biological samples like tear fluid and urine, Hutchens and Yip had found the available MALDI techniques unsatisfactory and limited for the sensitive characterization of biological samples. They began experimenting with surface modifications of MALDI probes that would selectively tether biomolecules, allow for in situ clean-up and concentration, and then release the molecules again in the laser desorption analysis.[402,403] Soon they moved on to chemistries that would allow covalent molecule linkage, similar to the Biacore chips: that is, immobilize one binding partner covalently, capture a ligand from biological samples, wash away contaminants, and desorb the ligand.[404] As Hutchens' group changed from an academic setting into industry, their new partner, Ciphergen Biosystems, obtained comprehensive patent coverage for the chemical modification of mass spectrometry surfaces. SELDI and ProteinChip Arrays have become registered trademarks. Since then, the technique has been refined and, again similar to Biacore, turned into an integrated analysis platform that consists of a chip reader (a linear delayed-extraction TOF device), the arrays, and an analysis software package.

Being a lot younger than SPR technology, SELDI today is where the Biacore was probably in 1994. The technology is a few years past its commercial introduction, the first publications are starting to appear outside of the development community,[405,406] and users are starting to explore the potential and limitations of this new analysis strategy. Similar to the SPR situation, SELDI has brought a new analysis strategy into the hands of medical and biological researchers, for the most part users with no background knowledge in mass spectrometry. This has somewhat slowed the general acceptance of the technology, since initially a lot of inexperienced users had to discover the laws and limitations of MALDI by trial and error. Core laboratories and established mass spectrometry facilities initially avoided the technology widely, mostly due to the low accuracy of the first-generation mass reader. Ciphergen Biosystems has so far refused to license the surface technology to other instrument manufacturers, with the vision in mind to keep the hardware, chemware, and software under one roof. This has most certainly slowed the spread and development of the technology. Ciphergen Biosystems targeted the biotechnology and pharmaceutical industries as initial customers, and project collaborators rather than academia. This has led to a large body of unpublished proprietary research, very different from SPR, which stimulated significant interest in the academic community right away that proliferated into a greater number of publications from the start.

2.12.7.3.3 Surface-enhanced laser desorption/ionization applications: general principles and experimental considerations

Unlike SPR, SELDI quickly came along with a variety of different surfaces, mainly of two categories: preactivated surfaces for the covalent binding of biomolecules via primary amines, and chromatographic surfaces for the selective capture of proteins via charge, hydrophobicity, or metal–chelate interaction.

Preactivated surfaces support applications quite familiar to SPR users: the study of biomolecular interactions. While SPR looks at the affinity as expressed in on and off rates, SELDI takes a snapshot of everything that the capture molecule has pulled from a biological sample. Retention can be influenced by the chosen conditions for binding and washing. Molecules bound via the covalently attached capture protein are desorbed in the ionization process.

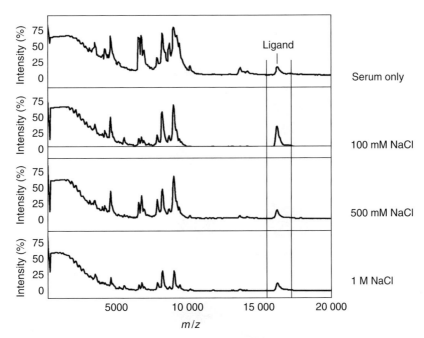

Figure 29 SELDI affinity study of a ligand captured via a receptor from human serum. Effect of different stringency modulators on nonspecific background binding.

Figure 29 shows a SELDI experiment where a ligand of interest is captured via a receptor from a background of 100% serum. As expected for this medium, a significant amount of nonspecific binding by serum proteins is detected. The experiment shows how the addition of stringency in binding reduces this background but does not completely eliminate it. However, complete background elimination in SELDI is not necessary. If the molecular weight of the ligand is known, competition binders are easily distinguished, as the mass spectrometer has a resolution superior to other molecular weight detection methods, generally much better than 0.1%. In an important difference from SPR, SELDI identifies the interacting molecules by molecular weight. This is important in a situation where the ligand of interest is captured from a complex biological background. In this situation, competition binding can lead to confusing results in SPR analysis. SPR and SELDI are very much complementary techniques in this application area: SPR allows the determination of affinities, and SELDI contributes protein identification capabilities. SELDI is limited by similar constraints as SPR; in both approaches a biological interaction is forced onto a two-dimensional platform. Different from SPR, though, the binding does not occur in a flow situation but rather in a tiny, stationary drop on the chip. With the limitations of sensitivity in a very small surface area, the high density of capture molecules is key. Since kinetic considerations do not apply to the same extent in this analysis strategy, the main concern with high molecule density has to be steric hindrance.

From a preparation point of view, SELDI experiments with preactivated chips share some similarity with enzyme-linked immunosorbent assay (ELISA) techniques. **Figure 30** shows a schematic outline of the process. The capture molecule is usually applied to each spot in a volume of less than $2\,\mu L$. For a protein, the concentration of the applied solution should be between 0.5 and $1\,mg\,mL^{-1}$. Because of the primary amine coupling, Tris and azide-containing buffers are incompatible. After binding of the capture molecule, residual sites are blocked with ethanolamine. To remove noncovalently attached capture molecules remaining on the surface, pH-cycle washes in phosphate and sodium acetate buffers with high salt are recommended. Right before ligand application the chip is washed with phosphate-buffered saline. The ligand is applied in a volume between 1 and $10\,\mu L$, depending on the concentration. After the ligand binding, the spots on the chip are washed with physiological buffers of appropriate stringency. The latter can be adjusted with salt or detergent. Due to the incompatibility of detergents and sodium salts with the desorption technique, a final wash in carbonate buffer or HEPES is necessary for good detection results. A conventional MALDI matrix mix is finally applied to the semi-dry sample for co-crystallization. Once dry, the chip is read in the instrument.

The even spread of molecules on the surface, forced by the specific capturing, and the layering of matrix on top of the biomolecules, dramatically enhance the quality of analysis compared with a conventional MALDI probe. More importantly, the opportunity to wash the surface without removing the protein of interest allows the use of previously

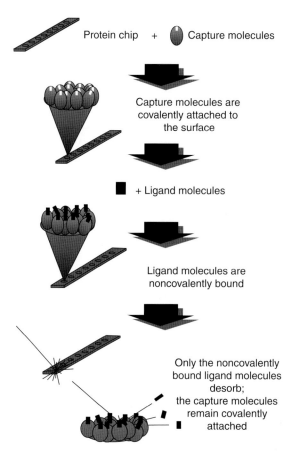

Figure 30 Flow chart of a typical SELDI experiment using preactivated arrays.

MALDI-incompatible buffers and detergents in all steps of the process prior to ionization. This is a significant breakthrough that opened the door to the successful analysis of complex biological samples. On-chip sample clean-up is probably the biggest benefit that the world of mass spectrometry sample preparation has seen in a long time, especially when considering that at the same time a biological affinity interaction is kept intact on a mass spectrometry probe. A carefully designed SELDI binding experiment explores a wide range of ligand concentrations and applies the same data collection parameters across the different spots of the same chip. In this case, quantitative correlation between the signal intensity and the actual concentration of the ligand can be quite impressive. The study of a wide range of ligand concentrations as well as the challenge of specificity in a competition binding experiment are as important as in the SPR technique to assure quality data.

The application type just described finds use at many junctions of the drug development process. In the discovery area, the method is a tool to obtain information about unknown ligands to known capture molecules. With known molecules such as purified antibodies, it is useful to confirm binding activity. In the area of pharmacokinetics, it allows the monitoring of molecules in patient serum: if the molecule of interest is a peptide with a molecular weight under 10 000, the mass accuracy will even allow reading of the degradation sequence from the molecule, providing it desorbs with decent resolution. In the current absence of post- or in-source decay capabilities for the mass reader, the degradation sequence has to be generated by on-probe carboxypeptidase digestions.

In direct comparison with ELISA methods, SELDI cannot quite compete with the levels of sensitivity, but the protein identification capability eliminates false positives by revealing cross-reactivity, if present. This can be as important as mere sensitivity, especially when dealing will molecule families that have a high level of conservation in structure and sequence, leading to immunological cross-reactivity. In one aspect SELDI is more limited than SPR: not all biomolecules are easily ionized and detected. In SPR the mode of detection is more or less independent of the molecular characteristics of the analyte; hence, the broad applicability to proteins, nucleic acids, even carbohydrates. The last two classes of molecules present special problems for matrix-assisted desorption techniques. They trap a lot of salt in their molecular backbone that causes ion suppression. This also affects the detection of proteins with carbohydrate modifications. Glycoproteins ionize very poorly, which affects the mass accuracy of detection, as well as the sensitivity of the assay. SELDI has greatly enhanced the number of ionizable proteins by the simple fact that on-chip washes are very powerful in removing incompatible chemicals and ions, even from very difficult molecules. However, some molecules just will not fly, as mass spectrometrists say. Still, the use of DNA or carbohydrates as capture molecules is possible in SELDI, as long as the ligand is easily detected. In SELDI, as in SPR, the choice of the binding partner to immobilize is very important but is often driven by necessity, as discussed earlier. If the capture molecule is small, a higher density can be achieved on the chip, and hindrance does not become a problem as quickly as with larger molecules. If the captured target ligand is very large, it is more difficult to ionize, and the assay loses sensitivity in the detection step. The random binding to the chip also presents the familiar problem of not having all the capture molecules correctly oriented to receive the ligand. In the case of antibodies and Fc-fusion proteins, protein A coupling is helpful despite the fact that it increases the level of chemical and biological noise in detection. In a protein A sandwich assay, both the capture protein and the ligand will be ionized; only the protein A stays attached to the chip. The quality of the protein A used for capture can critically influence sensitivity and noise levels in this type of experiment. In a protein–protein binding experiment, the goal generally is detection of a single binding entity, sometimes multimers of it, or at the most a heteromeric complex.

A completely different set of data is obtained when protein chips are used for profiling applications. These applications look at the protein inventory of a biological sample in a given molecular weight range. Protein-profiling applications are very interesting in the area of disease research. A comparison of lysates from normal versus diseased cells can reveal the expression of important marker proteins. For cancer research this approach has found a faithful following. Crude cell lysates are spotted on protein chips with different chromatographic surfaces, allowing a multidimensional binding picture based on different types of interaction. Control lysates are read against lysates of interest under the same data collection conditions, and subsequent peak subtraction allows the identification of differences. Once a peak of interest has been detected, this protein can be singled out for further analysis. This is made possible by the application of more stringent conditions to the surface. Depending on the type of interaction, this can be achieved in gradient washes of pH, salt, or organic solvent, followed by an ionization-compatible wash. Once a peak is singled out sufficiently, the possibility of on-chip digestion with proteolytic enzymes and subsequent analysis of the peptide patterns can yield important identification information. At the current stage of the technology, protein identification applications are still somewhat limited by the mass accuracy of the protein chip reader, and it is very important to use careful calibration techniques with known peptides of comparable molecular weight to obtain reliable peptide maps. Even with a mass accuracy in the millidalton range, the proteomic researcher is cautioned that the protein chest of nature contains enough redundancy and sequence conservation to make wrong identification calls based on peptide masses. It is therefore necessary to have powerful fragmentation and sequencing capabilities combined with a front end of protein chips to obtain high-resolution data and novel protein information directly from biological samples. Developments in this area are ongoing and already provide exciting glimpses into a very bright future for proteomics by SELDI mass spectrometry.

Aside from the described discovery profiling, chromatographic protein chips are very useful tools in the day-to-day routines of the industrial protein chemistry laboratory. They can be used for expression monitoring and purification screening, as well as a simple identification tool at the end of a high-throughput purification process based on the same type of affinity chromatography.

2.12.7.3.4 Surface-enhanced laser desorption/ionization in the drug development process: a bright future

In general, protein chips for mass spectrometry have a very bright future in the drug discovery laboratory. The versatility of SELDI and the speed of the analysis will eventually lead to very elaborate multiplexed systems that provide an incredible amount of information in a very short time. In discovery as well as process development, production, and quality control, the mass accuracy of mass spectrometry joined with a platform for chemical or biological affinity for very small sample volumes will prove to be a very powerful combination. The technology still needs time to mature. As learned from the Biacore example, a crucial step will be taken when the technology acquires a user base large enough to speed up the elimination of problems with experimental as well as instrument and platform design. It is safe to say that the future of protein biochemistry will be revolutionized by label-free chip technologies, and a few years from now protein gels may have become a thing of the past.

2.12.7.3.5 Biomolecular interaction analysis/mass spectrometry: connecting a popular surface plasmon resonance platform to matrix-assisted laser desorption/ionization-time-of-flight

We have already discussed some of the striking similarities that SPR and SELDI share in terms of surface activation and specific capturing of biological molecules on analysis chips. Hence, it is not surprising that researchers have looked at the possibilities of directly connecting a Biacore platform with a MALDI-TOF system. Two facts have driven the development in this area: the wide acceptance of the Biacore platform in industrial and academic laboratories, and the comprehensive patent coverage in the hands of Ciphergen Biosystems for the chemical modification of mass spectrometry probes. BIA/MS circumvents the issue of surface modification specifically for mass spectrometry, and allows users to retrieve additional information from an SPR experiment. Two experimental strategies are mainly used in BIA/MS. In the first technique, the captured molecule of interest is eluted from the Biacore chip with a MALDI-friendly solution (typically a small volume of formic or trifluoroacetic acid) and subsequently applied to the MALDI probe. This has been successful for elution from metal–chelate chips.[407] In an alternative strategy, the Biacore chip is directly used as a MALDI probe by the application of matrix after conclusion of the SPR cycle.[408]

Compared with SELDI, the BIA/MS strategy has several disadvantages. As has been discussed earlier, to obtain good results in SPR, it is important to maintain a low molecule density on the chip as well as a low analyte concentration. This challenges the detection capabilities of the best MALDI systems, especially when molecules with poor ionization qualities are analyzed. Furthermore, the Biacore chip was not designed as a mass spectrometry probe in the first place, and thus has several intrinsic problems when used as a probe.

Overall, BIA/MS is not a straightforward and easily adapted experimental system. So far, the literature in this area has mainly come from two groups.[409,410] BIA/MS advocates stress MALDI sensitivity as well as the tolerance of the technique for difficult buffer systems, but the examples demonstrated in the literature so far mainly focus on examples from *E. coli* lysate, where protein abundance and available sample preparation options are generally not limiting factors.[408] It remains to be seen how well the technique will fare with difficult biological backgrounds such as membrane systems. In those areas the opportunity for higher molecule density and sample clean-up on the SELDI chip will most like prove more advantageous.

2.12.7.4 Conclusion: Protein Display Chips in Drug Discovery – the Future is Here

This section shows that protein display chips are still very much a developing field, and there are far fewer players than for the respective technology for nucleic acids. Nevertheless, as the genomics field has come of age, development in the protein area has picked up, profiting from the great leaps that have been made in nanoscale analysis for DNA and RNA. Still, for proteins the most exciting discoveries are probably yet to come. Accommodating the incredible diversity of proteins in discovery, development, analysis, and validation is an enormous task, requiring not one or two but at least a handful of technology platforms. These in turn must be integrated to render a multidimensional analysis picture. The overlap between techniques such as SPR and SELDI as well as the limitations of both already point us in this direction. The techniques and platforms discussed give us a glimpse of the future that has finally arrived in protein biochemistry.

2.12.8 Integrated Proteomics Technologies and In Vivo Validation of Molecular Targets

2.12.8.1 Proteomics in Molecular Medicine

The identification and selection of a therapeutic target can determine the success or failure of a drug discovery program. This decision is influenced by the underlying biases of the drug discovery research strategy, which can be a reflection of the applied technologies. Since the vast majority of drug targets are proteins, the identification and characterization of a novel molecular target will be greatly facilitated by carrying out the analysis at the protein level, thus focusing "on the actual biological effector molecules"[411] of the gene. This section will examine the role of proteomics in drug discovery with a particular focus on preclinical target evaluation. Until we can truly rely on experimental medicine for drug development, disease models will play a pivotal role in molecular medicine for the safety and efficacy evaluation of a new therapeutic. To succeed with this strategy, the relevance of these models must be considered. Proteomics may facilitate this assessment through a comparison of the human disease phenotype to the one obtained in an animal model through a comprehensive analysis of protein expression maps (PEMs) of the relevant body compartments such as bodily fluids and the affected organ tissues.

2.12.8.1.1 Proteome terminology

The term 'proteomics' is twinned with another new term, the 'proteome,' which originated in 1995,[412] to describe the complete protein expression profile in a given tissue, cell, or biological system at a given time. Proteomics represents an extension of the earlier established concepts of the genome and genomics. The human genome delineates the entire human genetic information contained in two copies within the nucleus of each cell of the human body, with the exception of mature platelets. "Functional genomics is the attachment of information about function to knowledge of DNA sequence,"[413] giving an indication of whether a gene is transcribed but not necessarily translated in a disease- or tissue-specific manner. More specifically, functional proteomics describes the biochemical and biological characterization of proteins. These efforts are aimed at identifying structurally and functionally significant sequence motifs in the primary protein sequence with respect to the subcellular location of their activities and to relate these to disease- or tissue-specific changes. Pharmacogenomics is aimed at identifying genetic polymorphic variations in the human population that are relevant to the disease state or the ability to metabolize a drug. Recently, a lot of emphasis in the area of pharmacogenomics has been directed at the identification of the majority of single-nucleotide polymorphisms (SNPs) involved in human disease and drug response. SNPs are single-nucleotide variations between the DNA sequence of individuals that can be substitutions, insertions, or deletions. Information on these variations may allow the clinician of the future to stratify a patient population in clinical trials based on individuals' predicted drug sensitivity–response profile.[414] First applications are emerging, such as chip-based mutational analysis of blood, urine, stool, and breast effusions.[415] However, it remains to be determined what impact the presence of genetic polymorphisms will have on the application of animal models in preclinical development.

Although pharmacoproteomics aims at patient stratification, it is based on protein expression polymorphisms in body fluids and tissues, which could be indicative of susceptibility to disease and responsiveness to drugs in pharmacoproteomic pharmacology or toxicology studies. Longitudinal serum samples from drug-treated animals should identify individual protein markers or clusters thereof that are dose related and correlate with the emergence and severity of toxicity. Structural genomics could be considered an extension of functional genomics using structural information obtained by x-ray crystallography to infer the folding and three-dimensional structure of related protein family members. It can be envisioned that the discipline of structural proteomics could apply the information available on the tertiary structure of a protein or motif to establish its biological significance by systematically mapping out all its interacting protein partners. This information could suggest their role in the pathogenesis of a particular disease and assess its dependence on post-translational modifications (PTMs). We will illustrate some applications for structural proteomics in Section 2.12.8.1.8 on the proteomic analysis of protein complexes. Definitions of these terms are given in **Table 22**.

2.12.8.1.2 Genome versus proteome

Until recently, biological and biochemical analyses were carried out on single molecules. The dramatic advances in genomic sequencing and mRNA-based analyses of gene expression have generated important descriptive information. The field is now quickly evolving from its primary objectives of information collection, organization, and mining to functional and structural genomics, whereby one uses DNA-based technologies to make inferences about the structure and behavior of an organism mainly through large-scale analysis of gene expression. The Human Genome Project, together with the availability of an increasing number of methods for gene expression, has led pharmaceutical companies to review their strategies to discover new drugs. However, it is becoming apparent that purely gene-based expression analysis is not sufficient for the target discovery and validation process. There may not necessarily be a tight temporal correlation between gene and actual protein expression.[416] Differences can arise from the different stabilities and turnover of mRNA and proteins, post-transcriptional mRNA splicing yielding various protein products, and PTMs. Proteomics can take these variables into consideration, and contribute additional information through expression analysis of subcellular fraction and protein complexes (**Figure 31**).[417]

2.12.8.1.3 Changes in post-translational modifications associated with pathogenesis

An understanding of the control of cellular processes and of the molecular basis of inter- and intracellular signaling can only emerge when the protein and the protein complexes involved in these processes can be elucidated together with the dynamics of their PTMs and associated activities. It is largely through an understanding of such protein-based mechanisms that real insight into the molecular bases of complex phenotypic diseases will emerge, leading to more relevant molecular targets for drug discovery. Ample evidence has distinctly validated numerous novel enzymes modulating protein PTMs as targets in many diseases, including cancers, diabetes, and sepsis, as well as certain cardiovascular and neuronal abnormalities.[419–425] The rational design of modulators of these new molecular targets has

Table 22 Definition of terms used in genomic and proteomic research

Term	Definition
Genome	Total genetic information possessed by an individual
Genomics	Characterization and sequencing of the genome
Functional genomics	Systematic analysis of gene activity in healthy and diseased tissues
Structural genomics	Systematic analysis of the three-dimensional structure of a protein based on homology to a protein with a known structure
Pharmacogenomics	Stratification of patients by their genetic susceptibility to disease and responsiveness to drugs
Phenotype	Observable properties of an organism produced by the genotype (in conjunction with the environment)[a]
	Genetically inherited appearance or behavior of organism[b]
Proteome	Total protein profile of a cell or tissue at a given time
Proteomics	Systematic analysis of a protein expression map in a particular tissue or body fluid
Functional proteomics	Systematic analysis of protein expression and post-translational modification of proteins in various cellular compartments of healthy and diseased tissues
Structural proteomics	A proteomics approach for structure determination using native protein array technology
Pharmacoproteomics	Patient stratification based on protein expression polymorphisms in body fluids indicative of susceptibility to disease and responsiveness to drugs

[a] From Devlin.[511]
[b] From Forum on Data Management Technologies in Biological Screening.[512]

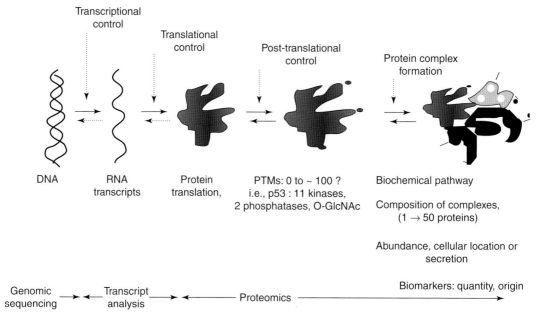

Figure 31 Genome to proteome. Gene expression can be regulated by the rate of transcription, translation, and further post-translational regulation of protein expression and activity are known.[418] Each protein may undergo various levels of PTM. In some instances, more than 10 serine-, threonine-, and tyrosine-phosphorylated and serine- or threonine O-linked-glycosylated amino acids can be detected on a single protein. Additional PTMs include asparagine-linked glycosylation, farnesylation, and palmitoylation, all of which can affect the activity, stability, and location of a particular protein (see Section 2.12.8.1.3). There are an increasing number of examples where enzymes carrying out these PTMs become targets for therapeutic intervention (see Section 2.12.8.1.4)[417].

yielded multiple drug candidates, promising more selective therapies. In cancer research, for example, several pharmacological approaches have focused on the enzymatic activity of enzyme-linked cell surface receptor signaling cascades such as the epidermal growth factor (EGF) receptor family signaling pathway.

2.12.8.1.4 New drug candidates directed against molecular targets modulating protein post-translational modification

As illustrated by the EGF receptor family signaling pathway, tumorigenesis via the erbB receptor signaling cascade has a proven biological significance in experimental tumor metastasis systems.[426] This provides ample opportunities for novel therapeutic modalities at multiple sites within the pathway such as the EGF receptor kinase antagonists[427] and trastuzumab (herceptin), an anti-HER2 monoclonal antibody blocking the activity of HER2 (ErbB-2), the second member of the EGF receptor family.[428] The importance of membrane anchoring via acetylation of certain signal proteins, such as Ras, is exemplified by the farnesyltransferase inhibitors, a putative cancer therapy with a better therapeutic index than standard chemotherapy. Originally designed to target tumors with mutant *ras* genes, farnesyltransferase inhibitors also affect growth of tumors with no known *ras* mutations.[429] Several protein kinases, including the immediate downstream effector of Ras, Raf kinase, and protein kinase C, are targeted clinically via the antisense approach.[430,431] Inhibition of the tumor growth-promoting protein kinase MEK1 (MAPK kinase) leads to arrest at the G_1 phase of the cell cycle. In addition to blocking cancer cell growth, MEK1 inhibitors reverse some of the transformation phenotype, such as resumed flattened morphology and loss of the ability to grow in the absence of a substrate.[432] A detailed understanding of such a pathway in the context of the disease may enable us to select the right drug directed to a particular step the pathway or a combination thereof for each subcategory of the disease.

Initial emphasis in the development of small-molecule protein kinase inhibitors focused on modulating catalytic activity directly. However, homology among the catalytic ATP-binding site of hundreds of protein kinases active in a mammalian cell at any given time severely limits the degree of selectivity attainable. A strategy to overcome this has been illustrated by MEK1 kinase inhibitors that appear to act through an allosteric mechanism of inhibition. This novel class of MEK1 inhibitors displays very good toxicity profiles, suggestive of a high degree of selectivity.[432] However, the MAPK signal pathway plays an important role in normal cell signaling for the immune response and neuronal functions, mandating thorough validation in vitro and in vivo.[433] Such validation studies increase the knowledge about the behavior of the small-molecule inhibitor in a complex biological system, and may lead to label extension for completely different indications. For example, the MEK1 kinase inhibitors are being evaluated for their ability to reduce postischemic brain injury.[434]

A similar approach of allosteric inhibition of intraprotein binding led to the development of small-molecule inhibitors that block the interactions between the tumor suppressor protein p53 and its inhibitor MDM2. Initial studies on the regulation of p53 focused largely on phosphorylation, the complexity of which is emphasized by at least nine kinases, two phosphatases,[435] and cytosolic O-linked glycosylation, modulating p53 location, stability, and activity.[436] p53 activity is further regulated by the proto-oncoprotein MDM2, which impedes its transcriptional function and targets p53 oligomers only for ubiquitin-dependent degradation.[437] Small-molecule inhibitors blocking this protein–protein interaction led to the release of active p53 to activate programmed cell death (**Figure 32**). These two examples emphasize the importance of studying protein complexes, as will be discussed in Section 2.12.8.1.8, since they may yield an immediate new target for therapeutic intervention.

The small-molecule MDM2 inhibitor mimics the activity of the endogenous cell cycle regulatory protein $p19^{ARF}$, which also binds MDM2 and stabilizes p53. The diverse expression of the $p19^{ARF}/p16^{INK4a}$ gene exemplifies the complexity of gene expression at another level of regulation. The second $p19^{ARF}/p16^{INK4a}$ gene product originates from an alternative splicing product, which is translated into $p16^{INK4a}$. $p16^{INK4a}$ has a peptide sequence different from $p19^{ARF}$ because the use of an alternative start exon causes a frameshift, and the mRNA is translated in a different reading frame. The alternative splicing occurs in a tissue-specific manner (**Figure 32**). A third splice form, $p12^{INK4a}$, was discovered to be of functional significance in pancreatic cancer.[440] Originally studied as a tumor suppressor in cancer, $p16^{INK4a}$ is now being evaluated as a novel therapeutic agent for the management of rheumatoid arthritis.[441] The physiological and therapeutic importance of distinct splice variant isoforms of a protein has already been recognized for seven transmembrane G protein-coupled receptors.[442] These splice variants arising from exon skipping, intron retention, alternative start exons, or splicing will also affect ligand-binding profiles, G-protein coupling, and receptor desensitization in a tissue-specific manner.[442]

In conclusion, the discovery and characterization of molecular targets resulting from abnormal cellular function through abnormalities in protein expression and their PTMs cannot be achieved through descriptive genetic analysis. Exploring how the protein products of individual genes interact dynamically to control cellular processes, as well as how

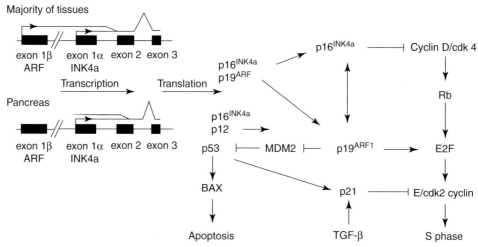

Figure 32 Gene sequence does not predict protein functions; tissue-specific transcription and translation of the p16^{INK4a}/p19^{ARF} gene. The tumor suppressor p16$^{INK4a/ARF}$ gene locus on human chromosome 9p21 is a site of frequent deletion in many cancers, including leukemias, gliomas, and non-small cell lung cancers.[513,514,515,515a,515b] Three splicing forms of the p16$^{INK4a/ARF}$ gene yield three distinct proteins: p16^{INK4a}, p19^{ARF}, and pl2. p16^{INK4a} blocks the cyclin D/cyclin-dependent kinase 4 complex, preventing phosphorylation and inactivation of Rb. Active Rb will inhibit the transcription factor E2F and prevent gene expression of cell cycle machinery protein for S phase progression, causing arrest in the G$_1$ phase of the cell cycle. The second splice form p19^{ARF} uses a different start exon, and yields a different peptide sequence due to a resulting frameshift. p19^{ARF} prevents MDM2 from binding p53. The third protein product of the pl6$^{INK4a/ARF}$ gene, p12, composed of the INK4a exon la and a novel intron-derived C terminus, also inhibits the cell cycle without interacting with the Rb pathway.[438] MDM2 blocks p53 tumor suppressor function by blocking its transcriptional activity and targeting it for ubiquitin-dependent degradation. MDM2 also binds a close homolog of p53, p73α, inhibiting its transcriptional activity without inducing its degradation.[439]

these protein interactions and activities are controlled by PTMs, will lead to greater insight into the control of the complex and dynamic processes of protein PTM that are central to cellular behavior and cellular responses. Understanding the control of subcellular protein expression and PTMs will lead to our understanding with greater precision the molecular basis of abnormalities in cellular control processes and protein networks, leading ultimately to more specific and more relevant molecular targets for disease control.[443] This implies a central role for the proteomic analysis of molecular targets in the evaluation process in vitro and in vivo. Although proteins reflect the physiology of normal and diseased tissue more closely than transcript analysis, the technology requirements for the reproducibility, sensitivity, and throughput of two-dimensional gel electrophoresis hampered the fast and direct access to this information. Significant advances in the technology combined with new approaches to disease models and access to clinically relevant tissues may provide us with a molecular fingerprint of diseases at the protein level, and identify further targets for therapeutic intervention.

2.12.8.1.5 Protein expression mapping

Protein expression mapping aims to define global protein expression profiles in tissues, cells, or body fluids. A protein expression profile may be representative of a given time point in an experiment or a disease state and become the basis for a comparative analysis. The most common implementation of proteomics is based on 2DE of proteins in a complex mixture and their subsequent individual isolation, identification, and analysis from within the gel by mass spectrometry (MS). If multiple isoforms of a protein exist within the sample due to PTMs and splice variants as described above, they may be identified because PTMs affect the charge and molecular weight of protein isoforms altering their mobility in the two-dimensional (2D) gel.

In the proteomic process, proteins are initially extracted from tissues, whole cells, or cell organelles using extraction procedures well documented in the literature.[444] The proteins are solubilized in 2D sample buffers and run on 2D polyacrylamide gels by separating them according to their charge (pI) in the first dimension using isoelectric focusing. They are then separated by size, using sodium dodecyl sulfate–polyacrylamide gel electrophoresis (SDS-PAGE), in the second dimension. Most commonly, proteins within the gel are visualized by Coommassie blue or silver staining. The proteins referred to as features in the raw primary image represent the proteome. At UCB, the image is scanned through fluorescent detection and digital imaging to yield a PEM. An electronic database containing all PEMs (now in digitized form) can be constructed. Each protein feature in each PEM is assigned a molecular cluster index (MCI). Each MCI has attributes of pI, relative molecular mass, quantity in each sample, frequency in each sample set, and a linkage to the feature ID in each gel for subsequent analysis of any proteins in any archived gel. By this process, all protein features in a sample are linked to MCIs, and are then available for quantitative analysis. A comparison of the proteomes generated from normal and disease samples is summarized in the proteograph. The PC-Rosetta software is capable of rapidly identifying those proteins that are unique in PEMs of control and diseased samples, or those that show increased or decreased expression. It provides information on its pI and molecular weight, and in large sample sets, statistical data on differentially expressed proteins are computed. Several other commercial companies also provide basic image analysis tools, such as Melanie (BioRad Laboratories[1102]) and Phoretix 2D (Nonlinear Dynamics[1103]).

Many modifications have been made at UCB to improve the handling, throughput, sensitivity, and reproducibility of 2DE. For example, proprietary attachment chemistry causes covalent attachment of the gel to the back plate. This greatly enhances reproducibility by avoiding warping and other distortions of the gel. Fluorescent dyes that bind noncovalently to SDS-coated proteins facilitates the detection of protein spots in the gel over a greater linear range than densitometric methods while not interfering during the MS analysis.[445]

Zoom gels can be used to further expand the resolution of a gel or to analyze a particular area of the gel containing a protein of interest with greater accuracy. For this purpose, proteins are separated in the first dimension in three isoelectric focusing (IEF) steps (pH 4–6, 5–7, and 6–8) rather than standard IEF (pH 3–10). Through this process the number of features resolved can be increased considerably. For example, the number of features resolved from a cell extract of Huh7 human hepatoma cells can be increased from approximately 2000 features to more than 5000 features.

Selected features are isolated from the gels and chemically or enzymatically fragmented in the gel slice[446,447] or after electrotransfer onto a suitable membrane for MS analysis.[448–451] The resulting peptides can be separated by on-line capillary high-performance liquid chromatography for subsequent MS analysis through matrix-assisted laser desorption ionization MS,[452,453] tandem MS (MS/MS), or electrospray ionization MS/MS, or as described by others.[454,455] Protein identification is based on searching the resulting peptide fragmentation spectra against genomic and EST databases using either the automated SEQUEST program or proprietary algorithms. Thus, protein identification has moved from the traditional laboratory N-terminal sequencing method to a highly dynamic fast MS-based method that correlates the MS data of peptides with information contained in sequence databases.[456]

As pointed out by Haynes et al.,[456] there are a number of alternative approaches to proteome analysis currently under development. There is a considerable interest in developing a proteome analysis strategy, which bypasses 2DE altogether, because it is a relatively slow and tedious process, and because of perceived difficulties in extracting proteins from the gel matrix for analysis. However, 2DE as a starting point for proteome analysis has many advantages compared to other techniques available today. The most significant strengths of the 2DE-MS approach include the relatively uniform behavior of proteins in gels, the ability to quantify spots, and the high resolution and simultaneous display of hundreds to thousands of proteins within a reasonable time frame.

In addition, a comparative analysis carried out with digitized PEMs, described above, enables us to compare hundreds of samples/PEMs obtained from 'normal' and 'disease' samples, each containing thousands of proteins. In concurrence with a statistical analysis of the fold changes for each protein, a disease-specific protein marker can be selected out of several hundred thousand proteins queried.

An alternative 2DE independent approach for a comparative global proteome analysis[457] is based on an isotope-coded affinity tag (ICAT) method combined with direct MS/MS analysis. The approach relies on an isotopically coded linker coupled to an affinity tag specific for sulfhydryl groups in the side chains of cysteinyl residues in the reduced protein sample. Two versions of ICAT, an isotopically light and a heavy form, are used to label protein mixtures from state A and state B of the sample. The samples are mixed and digested for direct MS analysis, as described above. The relative signal intensities for identical peptide sequences can be quantified by being traced back to the original sample based on the respective ICAT. This allows for simultaneous identification and quantification of components of protein mixtures in a single automated operation from a fairly limited amount of sample such as laser capture microdissection (LCM) (see below).

2.12.8.1.6 Cellular proteomics

To understand the molecular basis of a disease as the difference between normal and diseased tissues and organs, it is necessary to interpret the complex interactions between many different cell types at the protein level. In addition, it may be necessary to isolate and characterize abnormal cells within or adjacent to unaffected areas of the diseased tissue or organ. LCM technology[458] has made this difficult task achievable with a minimal effect on the gene and protein expression profile present in vivo. LCM allows for a precise identification and isolation of specific cells and their application to reverse transcriptase PCR,[458] cDNA microarrays,[459] enzyme activity,[458] and proteomic analysis.[460] Cells in frozen tissue slices are first identified through a conventional staining method of choice, before they are selected through a microscope. A solid-state near-infrared laser beam is then focused on the target cells, causing their selective adhesion to a transfer film. The film is lifted away from the surrounding tissue, and placed directly into DNA, RNA, enzyme assay, or protein lysis buffer. Since Emmert-Buck et al. published their original findings in 1996, LCM has become routine for the identification of tumor-specific mRNA/gene expression analysis in a variety of cancers. The technology has progressed in the meantime from applications in conjunction with cDNA microarrays to profile gene expression of adjacent neuronal subtypes.[459] At the same time, Banks et al.[460] described the first proteomic analysis of LCM-purified normal and malignant human cervix and renal cortical tissues. These technological advances provide us with an opportunity to map out complex interaction of many cell types throughout human anatomy in both the normal and diseased states at an unprecedented level of complexity and detail. In oncology research, such valuable information can give an understanding of the alterations taking place in the cells surrounding the diseased tissue, such as stromal and endothelial cells, as well as the cancer cell itself. These additional insights provide us with an opportunity to broaden our therapeutic approaches, as illustrated by novel cancer therapeutics targeted against tumors and/or surrounding activated vascular endothelial cells.[461,462] But more importantly for the purpose of this review, we can now assess how closely our chosen animal models mimic the in vivo situation in the patient. For example, in tumor angiogenesis and metastases, researchers are faced with the dilemma that xenograft tumors (human tumors implanted underneath the skin of mice with faulty immune systems) do not behave like naturally occurring tumors in humans,[461,463] as will be discussed further in Section 2.12.8.2.3. In such a case, a proteomic comparison of LCM-purified stromal and endothelial cells from a human biopsy and the mouse model may be revealing, and identify critical differences in the two protein expression profiles that may be linked to the deficiencies of the animal model.

2.12.8.1.7 Proteomic analysis of body fluids

Body fluids represent a central link between various body compartments to exchange metabolic 'fuels' and dispose of metabolic waste products and toxins. The complex mixtures of proteins present in systemic fluids such as serum or plasma are secreted from many organs, such as the liver, pancreas, lungs, and kidneys, and are central to energetic drug metabolism and catabolism and homeostasis. For example, many steroids and peptide growth factor-binding proteins constantly present in the serum are responsible for a tight regulation of these potent circulating cell regulatory molecules. Many alterations in the physiological or disease states of an animal or human change the levels of the proteins in body fluids. Information about the magnitude of the change of identified proteins and clusters thereof may be of diagnostic, prognostic, or therapeutic significance. Proteomics body fluid analyses have been carried out for plasma, serum, cerebrospinal fluid (CSF), urine,[464] joint fluid,[465] tears,[466] pancreatic fluid,[467,468] amniotic fluid,[469] ascitic fluid,[470] pleural fluid,[471] cyst fluid,[472] sweat,[473] milk,[474] and seminal fluid.[475]

The objectives of these studies were to discover and characterize disease-specific proteins in body fluids by building databases of high-resolution PEMs from the relevant body fluids obtained from individuals. Interrogation of these databases may reveal proteins whose expression is significantly increased in a disease-related manner for subsequent molecular characterization. Routine application of this approach for a rapid and reliable identification of disease-specific protein markers had been severely limited by the interference by high-abundance proteins such as albumin, haptoglobin, IgG, and transferrin. For example, without removal of these high-abundance proteins, no more than 300–500 individual features can be separated in CSF. UCB has refined the use of proteomics for these type of analyses through immunoaffinity-based depletion methods, in which these four highly abundant proteins are removed from the sample to be analyzed. After such an enrichment, many additional features previously masked by the high-abundance proteins become visible. Now, up to approximately 1500 features of a CSF sample can be separated through 2DE. As with the analysis of cell extracts, zoom gels may be used to further enhance the resolution and sensitivity of the separation. This approach can be extended to all known body fluids and the removal of any known interfering high-abundance proteins prior to a proteomic analysis. PEMs obtained from human biofluids may serve as a reference point for the selection of the appropriate animal model. A comparison of body fluid PEMs between randomly mutagenized mice and human disease samples may establish a link between the disease and a single genetic abnormality.

2.12.8.1.8 Analyses of protein complexes

Once a putative therapeutic target protein has been identified, the need to assign a biological function and establish the relevant pathway for such a protein drives our attempts for cataloging binding partners. As discussed in Section 2.12.8.1.3, changes in cellular location and protein binding partners in response to stimuli-driven PTMs may suggest

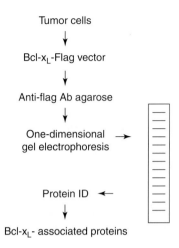

Figure 33 Example of a proteomic analysis of protein complexes as described previously.[411]

protein function. For example, proteomic mapping of purified protein complexes has led to the discovery of novel signal proteins such as FLICE,[476] IKK,[477] and components of the spliceosome complex.[478] For this purpose, the cell lysate is enriched for protein partners of the protein complex either with a tagged version of the protein such as FLAG[479] or GST[480] or through well-established immunoaffinity-based technologies. Again, it is 2DE that will assist in a reliable and quick identification of those proteins or isoforms thereof that are changed within the complex in the disease only, guiding the investigator toward the disturbances in a given pathway pertinent to the disease. In the case of the Ras pathway, multiple alterations cause uncontrolled proliferation and the loss of the ability to carry out programmed cell death (apoptosis), thereby promoting carcinogenesis. The proto-oncogene Raf kinase prevents apoptosis by phosphorylation-dependent activation of the Bcl-2 protein, a mechanism downregulated by the anticancer agent taxol in ovarian cancer.[481] The exact mechanism by which Bcl-2 and BclXL (a related protein) prevent cell death is not known. However, it has been suggested that alterations in the mitochondrial membrane permeability through a Bcl-2/BclxL Bnip3L (Bcl-2-interacting protein-3-like) protein heterocomplex may be a prerequisite for apoptotic cell death.[482] In a classic proteomic mapping exercise summarized in **Figure 33**, Gygi et al.[411] were able to identify 23 BclXL-associated proteins, many of which have already a proven association with apoptosis.

One can imagine how this process can be continued iteratively until an endpoint for a particular pathway has been reached or sufficient novel putative targets have been selected. The throughput of this technique may be further accelerated in some instances by applying technologies such as ICAT-MS/MS analysis (described above), and further improvements in MS now facilitate characterization of even complex mixtures of proteins.[457]

2.12.8.1.9 Combined mRNA/protein expression analysis for pathway mapping and candidate target selection

Although mRNA expression analysis may have no direct utility in the analysis of protein complexes, combined with proteomics it may yield useful information for additional target validation experiments. Effective integration of nucleic acid- and protein-based technologies may accelerate the selection of the best candidate target. Whereas the target discovery process described above is largely protein-driven, knockout technologies and overexpression strategies rely on molecular biology. In a case where a protein is downregulated in the disease, one can envision the use of inducible gene expression vectors as a quick means to restore protein expression and normal cell function (i.e., the reintroduction of a tumor suppressor such as p53). More commonly, an overexpressed disease-specific protein becomes the target for a therapeutic inhibitor strategy. If access to pre-existing small-molecule inhibitors are not available or cannot be generated easily, the gene can be readily ablated through antisense strategy. An antisense oligonucleotide can be synthesized or an antisense cDNA can be expressed in the cell in an inducible gene expression vector. In some cases, ribozyme constructs (small DNA fragments with intrinsic endonuclease activity that binds the target mRNA through a small DNA sequence complementary to a coding region in the target mRNA) are more useful. However, the effectiveness of these strategies is again monitored most accurately at the protein level. In particular, for the antisense strategy it is important to determine that the antisense DNA selectively blocks the expression of the target protein only, and does not result in nonspecific effects on protein expression. In the case of cell surface protein targets, quickly advancing phage display technologies now promise functional inhibitory antibodies within months. With these tools, one can now establish if interfering at different levels in the pathway will affect the protein expression profile differently, and which protein expression profile resembles that of the normal state most closely. This approach is facilitated by the fact that, in most cases, whole clusters of proteins rather than a single protein will be changed. By using new bioinformatics tools, clusters of proteins can be used in a specific PEM to describe a phenotype, thus increasing the predictive value of the analysis.

2.12.8.2 In Vivo Validation of Molecular Targets

In this new millennium, rapid advances in molecular medicine may provide us with a molecular fingerprint of the patient and disease phenotype that could represent the molecular and genetic bases for a preclinical animal model to evaluate novel candidate therapeutic targets. The preclinical evaluation of candidate targets requires complex animal models that will represent as many of the aspects of the drug discovery process as possible in order to identify the successful therapy. At the stage where small-molecule inhibitor leads have been identified, the relevance of the disease model becomes even more important, to investigate the mode of action, efficacy, and safety of these leads. To illustrate this, we will continue to use examples of oncology and central nervous system (CNS) research, with particular emphasis on AD.

2.12.8.2.1 Disease model relevance

The assessment of lead compounds in the most suitable animal model is of great importance because many agents that look highly promising in vitro or in animal models fail because of insurmountable toxicity problems or lack of efficacy in humans. Knockout and transgenic technologies have made the creation of mouse models of genetic disorders almost routine. As we will discuss below, specific gene mutations and conditional knockouts are now achievable in a tissue and/or time-dependent manner. An analysis of the complexity of most human disease necessitates a global comparison of the protein phenotype, well suited to a gene-related proteomic analysis (*see* Section 2.12.8.2.5).

2.12.8.2.2 Animal models of Alzheimer's disease

AD is a progressive neurodegenerative disorder that first presents clinically with memory impairment followed by continuous deterioration of cognitive function. The pathology is characterized by the presence of neurofibrillary tangles (NFTs) within critical neurons and neuritic plaques, causing synaptic and neuronal death. NFTs are largely composed of hyperphosphorylated tau, a structural microtubule-associated protein. The plaques contain deposits of the amyloid-β protein (Aβ), a cleavage product of the β-amyloid precursor protein (APP). Of the two isoforms of Aβ present in brain tissue and CSF, $A\beta_{42}$ aggregates more readily into plaques than $A\beta_{40}$.[483,484] In fact, the APP gene was the first gene identified causing familial AD. Mutations in APP affect its cleavage and the generation of Aβ (for a review, see Selkoe[485]). In addition, mutations in the presenilin genes *PS1* and *PS2* are also observed in familial AD, and patients with *PS1* mutations display increased $A\beta_{42}$ deposition as well.[486,487] The formation of amyloid plaques has been recognized as an early, necessary step in the pathogenesis of AD.[485] Many transgenic mice have been used to model the biochemical and neuropathological effects of the known genes implicated in familial AD. Animals overproducing human familial AD APP in neurons produce AD-like amyloid deposition,[488,489] with a morphology and regional distribution similar to that of humans.[490] While some of these transgenic mice strains show increased Aβ levels, amyloid, and correlative memory deficits,[491] NFTs or severe neuronal death were not observed in any of these models.[490] In addition, in some models, cognitive impairments appear prior to plaque formation.[492] Synaptic alterations in these mice have not yet been identified. It appears likely that sporadic forms of AD as well as familial AD, as modeled by familial AD mutants, originate from multiple causes. As a result, the complex phentotypic changes in many interacting cell types in critical areas of the brain must be evaluated comprehensively. This analysis should be carried out through a global analysis of the biological effector molecules (i.e., the protein expression of normal and diseased CSF and CNS tissue). This may be achieved through a systematic correlation of each gene mutation to complex changes in protein expression pattern through applications such as gene-related functional proteomics (*see* Section 2.12.8.2.5) supported by pharmacoproteomics (*see* Section 2.12.8.2.6) in transgenic animal models and human tissue with clinical aspects of the phenotype.

2.12.8.2.3 Animal models of cancer

An increasing number of studies in the analysis of signal transduction pathway of human cancers reveal that many regulatory proteins generate stimulatory and inhibitory signals at the same time. For example, in early stages of cancer, transforming growth factor β (TGF-β) can inhibit cancer growth via induction p21 (see **Figure 32**). However, in cancer cells where the Ras signal cascade is activated, TGF-β promotes invasiveness and tumor metastases.[462] Attempts to study cancer in an in vivo model such as the human xenograft are further complicated by the fact that human tumor-derived factors may interact differently with mouse host than human host machinery. For example, cancer-associated fibroblasts now have a recognized role in the tumorigenesis of transformed epithelial cells.[462,493] While this opens exciting new avenues for therapeutic intervention, it also points to further difficulties for the reliable use of human xenograft models in rodents. This might explain why some metastatic human tumors do not spread to other tissues in mice.[461] In fact, the US National Cancer Institute (NCI) has encountered serious challenges in its efforts to establish a meaningful screening system for candidate anticancer drugs. For example, the human xenograft models employed for the NCI in vivo screen show very limited predictive information for the clinical setting. When the NCI tested 12 anticancer agents used in patients against xenograft models, more than 60% of the tumors did not show a response.[463] In other cases, tumors showed opposite behaviors in vitro and in vivo. Tumors with a deletion in the p21 gene (see **Figure 32**) respond to radiation therapy in vivo, whereas the same tumor is equally resistant to radiation in an in vitro clonogenic assay.[494] In addition to similar findings in other models, this suggests that the efficacy of currently available chemotherapy may be linked to genetic mutations such as those associated with checkpoint or cell cycle-related genes.[495] This has lead many investigators to a molecular characterization of new targets and away from classical drug-screening efforts in models with poor predictability. A comprehensive summary of animal models of cancer is provided by DePinho and Jacks.[496]

2.12.8.2.4 Knockout and mutagenesis models, and tissue-specific inducible gene in vivo models

The examples discussed in Sections 2.12.8.2.1 and 2.12.8.2.2 signify the importance of gene-targeting technologies in mouse models of human disease (see the literature for reviews[497–500]). Recent advances in creating conditional and tissue-specific knockouts enable us to study the disease in a more relevant temporal and organ-specific context. For example, until recently it was not possible to study the *BRAC1* gene responsible for about 50% of heredity breast cancer in animal models, since homozygous Brca1 null mice would die early in embryogenesis. Conditional knockouts were generated with the Cre-loxP approach (for reviews, see Plück[501] and Stricklett *et al.*[502]) in mice expressing Cre recombinase with a mammary epithelial cell-specific promoter allowed for an analysis of a selective inactivation of the *BRCA1* gene in mammary glands.[503] These mice develop mammary tumors, and have abnormally developed mammary glands that would form the basis of a better understanding of *BRCA1* in tumorigenesis. Similar advances have been made for CNS models using the tetracyline-regulated system. Vectors have been reengineered for CNS-specific gene expression, and led to the development of an inducible gene expression system specific to several brain regions, including the CA1 region of the hippocampus, cerebellum, and striatum.[504] It is now conceivable that in the near future tedious gene knockout technology may be replaced by new tissue-specific gene-targeting systems that use rapid retroviral vectors in genetically engineered mice expressing the receptor for the retrovirus in a particular organ or tissue.[505] These approaches or a combination thereof may be particularly useful in conjunction with proteomics. For example, the viral receptor responsible for organ-selective infection by the retrovirus could also provide the molecular basis for immunobased cell purification procedure or a fluorescent tag to be recognized during LCM/proteomics. Such systems allow for a rapid and reproducible expression of many single genes or a combination thereof as well as selective purification of those cells only for a detailed molecular characterization of the disease at the protein level.

2.12.8.2.5 Gene-related functional proteomics

The definition of the phenotype of an organism was recently given as 'the integral of fulfilled gene functions and environmental effects' wherein 'the protein of a gene offers all the molecular structures and properties needed to fulfill the functions of a gene.'[506] Considering that any phenotype may be the result of more than one gene, it is important to determine the extent by which each gene contributes to the phenotype in order to determine the function of a particular gene. New approaches in gene-related functional proteomics (reviewed by Klose[506]) may allow us to gain new insights into the relationship between the genotype of an organism and the phenotype of its proteins. A disease such as AD defined by its clinical symptoms may be described by the phenotype at the morphological, physiological, biochemical, and molecular levels, and enable us to subdivide a disease into mechanistic subclasses of individual steps in the pathogenesis according to the proteins affected. This could facilitate the evaluation of many putative therapeutic targets specific to different aspects/pathways/stages of the disease.

In order to evaluate the role of a single gene product/protein in a disease, we need to consider existing protein polymorphisms, such as PTM-specific isoforms. Some of the origins of these protein polymorphisms may be revealed by mouse genetics. Tracing the origin of PTM variant protein through genetic techniques in animal models can lead to the definition of the origins of a protein at the genetic level, as discussed above for body fluid analysis. Furthermore, a 2DE comparison of 10 000 protein features obtained from the brains of various genetically distant mouse strains revealed variations in more than 1000 protein features. Identification of the subset of protein isoforms unique to a particular mouse strain correlated with genetic linkage studies, and gene mapping may reveal some of the responsible relevant genes.[506] For example, the tau polymorphism results in more than 100 protein spots in a 2DE gel of human brain proteins caused by various levels of phosphorylation and multiple splice variants.[507]

In addition, 2DE may identify PTM variants originating from gene mutations causing amino acid substitutions in flavin adenine dinucleotide from various mutations in APP, PS1, or PS2, all causing increased cerebral production and deposition of Aβ peptide in amyloid plaques (reviewed by Selkoe[508]). A protein polymorphism characterized at the molecular and genetic level will reveal more detailed information about its origins and significance in the diseased tissue. Studies of its relevance in normal tissues and other organs not affected by the disease or its occurrence at a particular age of the animal can be compared between the animal model and that observed in the patient. Contrary to the above example, where multiple proteins result in the same phenotype (e.g., increased amyloid deposition), a single protein affects the phenotype of several other proteins simultaneously, and a similar analysis can be applied.

In conclusion, it is important to define the origins of a detected protein polymorphism at the molecular and genetic level, in particular, for any multifactorial disease. A phenotypic analysis originating at the protein level rather than the gene level seems more likely to identify the primary defective protein linked to the disease as well as other proteins affected by the defective proteins. At this stage it is appropriate to identify the gene linked directly to the defective proteins.

2.12.8.2.6 Pharmacoproteomics

The application of preclinical drug development may be summarized as one aspect of pharmacoproteomics. Within preclinical development, proteomics can be used beyond the evaluation of a model, and confirm the modes of action of candidate drugs, once the appropriate model has been selected. For the majority of cancers, diagnostic, prognostic, or therapeutic proteins identified in serum may have great potential, since blood samples are readily available for proteomic analysis. In the CNS, the CSF bathes the brain almost completely. Therefore, it seems likely to contain many important secreted proteins and neuropeptides mediating the interaction of different cell populations, and reveal insights into disease-related changes when CSF from patients with neurodegenerative disease is compared with that of normal patients. Protein expression changes within these body fluids underlying specific disease states can be compared with the animal model counterparts. The degree of similarity may guide the investigator to the most relevant model available for the in vivo evaluation of candidate drugs.

At this stage in development, proteomics can be applied to identify serum protein markers associated with overt toxicity and efficacy or lack thereof. Such markers may be predictive of toxicity in a particular organ or indicative of efficacious therapy. This approach offers greater sensitivity than conventional toxicology, and enables candidate drugs to be screened for toxicity at the earliest possible stage of development. Thus, pharmacoproteomics may allow us to refine a multifactorial disease into subcategories more suitable for a particular therapy or less likely to bring about undesirable side effects. These findings may be applied during clinical development to enhance the successful selection of lead candidates for clinical trials (**Figure 34**).

2.12.8.3 Proteomic Technology in the Molecular Characterization of Novel Therapeutic Targets: Future Perspectives

Most human disease, as defined in clinical pathology, should be regarded as multifactorial, and its complexity must be understood at the molecular level. It has been recognized that all complex diseases, such as cancer, must be

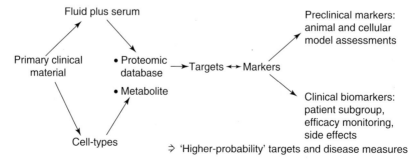

Figure 34 Emerging strategy for proteomic applications in drug discovery.

characterized in terms of the host organism.[509] Disease models will continue to play a significant role in modern drug discovery. Rapid genomic and proteomic discovery technologies reveal great insight into the disease phenotype at the gene and protein level. This increasing knowledge can be applied to a more mechanistically based and target-driven discovery process. In preclinical development, more emphasis will be based on a refined molecular characterization of the target rather than classical efficacy assessments in animal models. Proteins represent the main biological effector molecules of any given cell, and appear most suitable for the target evaluation process. Classical proteomic approaches offer the unique opportunity to identify disease-specific changes in the protein expression profile in a complex system comprising up to 10 000 proteins. This strategy will generate an increasing number of disease-specific protein markers that can be utilized for the evaluation of preclinical models.

The following important questions can now be addressed:

- Does cluster analysis of multiple protein changes reveal similar changes between normal and disease in humans and in the selected model?
- Are these changes reversed in the validation studies with candidate target protein inhibitors?
- Are these changes unique to a particular step in the pathway in a cell type-specific manner indicative of selectivity/specificity?
- Do small-molecule inhibitors modulate protein expression profiles of human disease and correlate with those observed in the validation studies?
- Do clinical lead compounds affect the same cluster of proteins in the animal model as in samples obtained in clinical studies from humans?

Interfaces with other technologies, such as protein chip arrays,[510] will quickly expand to a more versatile platform with the ability to evaluate biological systems of great complexity at great speed. An integrated proteomics tool can assign up to several hundred disease-specific proteins in body fluids and purified authentic tissues with statistical confidence. This information becomes valuable in a comparative analysis between the preclinical model and the clinical situation, which can feed back into a refined experimental design (**Figure 34**). Rapid advances in genetically engineered animal models should enable researchers to tailor design models in such a way that complex gene–protein interactions can be studied in a temporal and spatial context reflective of the clinical situation and with easy access to proteomic analysis.

2.12.9 High-Throughput Screening as a Discovery Resource

2.12.9.1 Background

Over the past two decades, we have witnessed the most dramatic change in drug discovery since the launch of antibiotic research in the 1940s. HTS evolved from a part-time, low-priority activity in the late 1980s to become the core of discovery operations in most pharmaceutical companies before the turn of the century. That development not only created unprecedented opportunities in the identification of new bioactive molecules; it also established a unique working relationship and dependence among different scientific and support disciplines that had never been achieved. Discovery became an integrated multidisciplinary activity.

Prior to 1990, new drug discovery was largely a departmental effort based on the contributions of innovative chemists and biologists. Biotechnology had made its entry, but the impact on 'small-molecule' discovery was, at best, vague. In most major companies, considerable emphasis was still based on analog synthesis, with breakthroughs dependent on the serendipitous finding of novel activities in new structural classes devised by chemists or on the skilled observations of biologists in the recognition of unusual activities in conventional bioassays. The standard therapeutic targets, which encompassed the commercial gamut of therapeutic significance, were believed to be well represented in clearly defined pharmacological models, and presumably supported by comprehensive structure–activity relationship (SAR) bases in medicinal chemistry. The situation was stagnant and breakthroughs became less common. The number of new-development compounds in a company's pipelines were dwindling, and costs increasing. 'Me-too' chemistry was also losing its appeal as markets became saturated and shares in those markets shrank to single digits. Management became concerned, and reached out for new opportunities.

Concurrent with the above were the dramatic developments in genetic engineering and molecular biology, along with the emergence of economical access to biological reagents not previously imagined. The ability to explore specific macromolecular interaction in rapid biochemical and cellular screens became an economic reality. Targets and protein ligands were soon identified (e.g., the interleukins) that promised new forms of therapy and the hope of useful intervention with small molecules. Whole-animal or tissue models, which had been the mainstay of primary screening for decades, slowly disappeared from the discovery armamentarium, and were replaced by biochemical and cellular assays that reflected mechanistic responses rather than whole-organism behavior. Screening became specific and economical.

With new biomolecular assays available and the ability to screen larger numbers of test compounds than the conventional pharmacological models, management turned to the chemist, with the anticipation of satisfying the need for the compounds required as screening candidates. Here lay the first problem. Designing a synthesis program that would ultimately yield a new molecule that interferes at a biological target is an impossible task when there is no chemical precedence to guide the initial probes. 'Rational' drug design had been touted as the answer, but despite two notable successes (cimetidine and captopril), it was clearly limited in application by strict demands of support data at the molecular and biological levels. The real answer was to resort to the almost 'unscientific' task of screening all possible structural types in the new assay with the hope that some chemical lead (a 'hit') would surface. It was the only solution, but the corporate infrastructure was not prepared.

This 'random screening' approach was not well received by industrial scientists, who were accustomed to more aesthetic research activities. Nevertheless, it was reluctantly accepted as a part-time activity of a few biochemistry and pharmacology laboratories. With the absence of a central focus, budgets were small and progress was slow. The available instrumentation was based on a variety of liquid handlers redesigned from diagnostic applications. HTS robotics was in its infancy. Information systems were primarily developed in-house, with the exception of established core software for data (e.g., Oracle) and chemical structure management (e.g., MDL's Chembase); again, the market was not fully developed, and incentives were few.

In 1990, assay throughput was at best 100 compounds per week, but that level was sufficient to handle the limited compound collections available. Compound archives in most companies were large, but highly focused and low in diversity. The organic chemist reluctantly took on a new responsibility in the discovery process: manual selection of a few thousand representatives from the corporate compound archives that would reflect the structural scope available. The many gaps in diversity in these small collections were filled by the purchase of compounds from conventional chemical suppliers, occasionally by direct purchase from academic chemists worldwide, and from the new breed of 'compound brokers' that had emerged.

A reluctance to share experiences in early HTS development was also a problem. In 1992, the first conference dedicated to HTS technologies was held at SRI International (Menlo Park, CA).[511] The purpose of that meeting was to bring together scientists involved or interested in HTS, with the intention of stimulating dialog in the implementation of this new technology, and discussing common problems in organization, technical integration, data management, and personnel issues. Finding the audience for this event was difficult at that time because screening had no internal management focus and the individuals involved were distributed throughout the departmental structure of most companies. Nevertheless, 140 participants gathered to discuss screening challenges and solutions. Open discussion periods provided enough catalysis to stimulate dialog and to make the conference a grand success. A by-product of the conference was the launch of the first publication on screening technologies, *Screening Forum*, which provided information and technology reports on HTS topics.[512]

So it began. HTS, now synonymous with 'grunt work,' was a part of the weekly schedule of R&D and accepted as such. Enthusiasm was initially low but, fortunately, a new breed of scientist emerged who recognized the challenge and the potential rewards of screening and was able to convince management of the immense potential if the technology were provided sufficient opportunity and resources. Such scientists were quick to take up the challenge of penetrating interdepartmental barriers and blending complex technologies, current instrumentation, and the available data management tools into functional and productive discovery machines. While instrumentation designed for HTS had yet to be developed, innovative adaptation of available instrumentation (e.g., from Beckman, Tecan, Packard, and Wallac) initially proved satisfactory. Notable was Tomtec's 96-well plate duplicator (Quadra) as an important asset at this early stage. The drive and foresight of these pioneers was the catalyst necessary to demonstrate the productivity of HTS and its importance as a discovery process.

As with any new technology, the standardization of tools became a critical matter. Fortunately, the 96-well microplate had already been proven in diagnostic applications, and quickly became the standard screening vessel. Its origin can be traced to 1960[512]; however, it was not until the mid-1970s, when KenWalls at the US Centers for Disease Control and Prevention started using it for enzyme-linked immunosorbent assay testing, that its benefit in automated assays became apparent. Robotic systems were developed to accommodate the microplate, and integrated test systems established. It was a natural transition to apply these systems to the demands of the new and now lucrative market of screening.

Targets and assays were primarily developed in-house, since they represented the company's proprietary interests and technologies; however, novel and more sensitive detection systems evolved externally that provided useful tools in optimal protocol development. New microplate designs[513,514,515,515a,515b] were similarly created in the support industries that saved time and reagents in assay development and operation. Packaged data management systems evolved from companies such as Molecular Design (now MDL Information Systems) and Tripos, which formed the core of chemical structure management. The SD file became the standard file format for structure display and storage. Compound libraries could be created by the purchase of small samples from a growing list of brokers who gleaned their collections from academic laboratories and 'chemical factories' throughout the world. However, the cost of establishing the all-important structurally diverse compound library was formidable and a serious bottleneck for start-up companies.

In 1993, MicroSource Discovery Systems[512] pioneered the provision of test compounds in microplate format. This approach provided efficiency, economy, and diversity in a single product with full data support. It was based on the realization that even with a modest level of assay miniaturization, a few milligrams of a test compound were sufficient to support years of screening. Purchases of the standard 50 mg sample were unnecessary, and the cost of acquisition of large libraries was cut by as much as 90%. Diversity was drawn from world leaders in chemical synthesis, and samples were provided in a format that avoided handling, documentation, and storage. Today, MicroSource Discovery Systems has evolved into a leading provider of natural products and drug standards in microplate formats.

By 1994, HTS had become an integral part of the drug discovery process. Large companies reorganized their internal structure to accommodate the multidisciplinary character of this new technology. The hierarchical distinction between HTS and conventional 'research' was resolved by the recognition of the distinct differences in the tools and goals of these activities and their equivalent and complementary contributions to the development of new drugs. Screening operations became centralized and new positions created. Topics such as assay miniaturization, integrated automation, compound acquisition, and data management became key activities in discovery operations. New companies sprang up that were based solely on new assay targets and the ability to screen efficiently. A myriad of support industries were also created to provide tools and resources that facilitated screening and improved return. Test compounds, automation systems, and even cloned target receptors became available from outsource groups.

In 1994, the second forum on HTS technologies was held in Princeton, NJ. It was a sellout in attendance, with almost 400 participants squeezed into a small conference space. The meeting was a tremendous success, and launched HTS as a formidable tool in discovery. That conference also provided the seed for the beginning of the Society for Biomolecular Screening, which today is the foremost source of information on HTS technologies.[517]

2.12.9.2 Where We Are

HTS, as applied in the pharmaceutical industry today, refers to the integrated technologies that permit the rapid evaluation of millions of compounds annually in scores of bioassays in search for new therapeutic agents. There are three objectives: de novo discovery, hit development, and the preliminary assessment of the metabolism and toxicity of lead candidates. The first two are widely applied and complementary; the last is new to HTS, but is rapidly becoming an important facet of drug development.

De novo discovery addresses the search for compounds that interact with a new biochemical or cellular target for which there is no precedence in chemical structure. The identification of both ligands and antagonists for most orphan receptors is an excellent example of such applications. The discovery of drugs against diseases that cannot be satisfactorily treated with existing therapies or for which no treatment is available is another. De novo HTS requires the use of large compound collections with as broad a structural diversity as possible and bioassays that accentuate the specific character of each target. The goal is to find one or more 'hits' that can provide a chemical focus for further screening. Such hits need not be at the activity level anticipated for a development candidate; structural information about a weak inhibitor is much better than no information at all. De novo HTS generally addresses a broad spectrum of targets.

Unlike de novo HTS, hit development begins with some level of chemical intuition. Such information may have been gleaned from de novo programs (see above), from historical data on the character of ligands or antagonists, or from computational analysis of the structure of the target receptor. This is the area where the use of high-throughput organic synthesis (HTOS) has provided the greatest benefit and has resulted in important advances in optimization of a lead and dramatic reduction in the time required for lead development. These benefits are discussed in greater detail below. Bioassays employed in developmental HTS focus on the overall activity profile anticipated for the test substance.

Preliminary assessment of the metabolism and toxicity of new drug candidates is a critical part of drug development. The requirements for such assessments are heavily regulated and, with the companion characteristics of absorption and distribution, constitute key elements in the success of any new drug application and its subsequent introduction. These processes are critically dependent on in vivo studies that are costly and time-consuming. Failure of a compound in any of these aspects usually results in its removal from development; the investment made to that point is essentially lost. Recent developments in the miniaturization of metabolic and toxicological assessment procedures permit high-throughput assessment of lead candidates at a preselection stage. Such in vitro information has yet to be accepted by regulatory bodies; however, it does provide an important preview of the performance of a lead. The use of such data in the selection of candidates for development can significantly improve compound survival rates and shorten development time.

The primary components of any HTS program can be roughly divided into three areas: test substance supply, bioassay development and implementation, and informatics (data management). How these are organized is another dimension that segregates throughput – low from high, high from ultrahigh, and so on. The result is the same: discovery. Instrumentation and systems integration have become integral parts of all aspects of HTS, and need not be addressed separately.

2.12.9.3 Test Substance Supply

A structurally diverse test substance supply is the key to discovery. It is the critical resource (i.e., the sole factor that determines the success level for every assay). Careful attention to the quality of the test substance resource is an essential aspect of HTS management. Essentially, a well-managed, structurally diverse test substance supply holds a company's equity in new drug discovery. This is true for any screening program, low and high capacity, in both small and large corporations.

A large variety of sources for test substances are available for HTS programs. Accessing any of these in a naïve 'numbers-only' approach can be disastrous. Heavy dependence on commercially available combinatorial libraries is not advisable. Structural diversity alone dictates the number of test substances required. A small but carefully selected collection of 10 000 compounds may be a much better resource for discovery than a mass-produced library with millions of components. The design and maintenance of an appropriate test substance resource is an important facet of corporate survival.

2.12.9.3.1 Compound library development

The demand for large and complex chemical libraries as test resources in HTS programs continues to grow to meet the dramatic increase in the throughput potential of today's HTS systems. However, it is noteworthy that a small, low-capacity screen can still yield important advances. This principle is manifested by the early screening successes in the potent PAF inhibitor WEB-2086 and the HIV reverse transcriptase inhibitor nevirapine. Both of these were discovered in libraries of less than 10 000 compounds without any indication of structural guides, certainly not 'pharmacophores.'

Despite the above comments, numbers are important if the compounds selected for inclusion in the corporate HTS pool have reasonable structural diversity; essentially, the higher the number of diverse test compounds, the better the chances of success. Today's million-compound libraries have simply improved the odds.

Compound collections that are truly diverse in skeletal and functional array are especially important in achieving a meaningful probe into the unknown three-dimensional space of a new receptor or enzyme. The classical approach in de novo discovery has been to draw on the structural riches of a well-designed and stocked compound collection that represents not only the scope of in-house synthesis but also the diversity available from external suppliers: direct chemical sources, brokers, and microplated compound libraries. The structural richness of the numerous pure natural products and their derivatives that are commercially available[512,521] must also be included since these displays cannot be mimicked by current HTOS.

Claims that high-throughput solid phase and solution phase chemical synthesis is a reliable source for the generation of limitless diversity are unfounded. This position is softened when such libraries are multitemplate based and constitute millions of representatives. Companies such as ArQule and Pharmacopeia offer such access, but cost again becomes significant. The unquestioned strengths of these sources are in the rapid development of series with some chemical insight. This aspect is addressed below.

Compound management and acquisition has become an important facet of HTS structure and function. It requires a continuous awareness of what is in hand, what has been depleted, and what is necessary to enhance the chemical and topological character of the corporate collection. The responsibility lies not only in filling gaps in the functional parameters and their display within the collection, but also in avoiding those substances that are inherently reactive with biological systems. Certain groups that are chemically reactive can give meaningless false positives that waste time and money and interrupt the discovery process. Acid chlorides, anhydrides, active esters, and the like are among these. Many halogenated compounds also fit into the category of nonspecific alkylating agents – α-haloketones, halomethyl aromatics, 2-halopyridines, and other reactive heterocycles are common constituents in commercial collections. There are other groups that are cytotoxic and unwelcome components; most organometallics and nitroso, alkyl nitro, and nitroso groups, are among these. Hydrazines, hydrazides, and similar structures should not be removed entirely, since these structural features have found their way into several important drugs; however, their ease of preparation makes them a more than necessary component in available collections. The same applies to aryl nitro representatives. Instability is another concern: t-butyl esters, acetals, aminals, and enol ethers are some examples. Other groups that have been the subject of extensive analog development (e.g., adamantane and per-haloalkanes) should also be carefully filtered. Redundancy is another problem. Analog representation is necessary to efficiently represent a new compound class; however, excessive representation can become an unnecessary expense and distraction.

It is nevertheless important to be sensitive to the rigors of implementing some of these rules. Some published guidelines recommend the removal of compounds that are potential substrates for Michael addition (e.g., α,β-unsaturated ketones). Such restriction is important if the group is especially activated, but that is frequently not the case – progesterone and a myriad of similar molecules are potential Michael addition candidates. Other restrictions, such as molecular size, the number of rotatable bonds, and chemical functional density, are fine on paper but destructive if applied generically in library design. The end product can lead to a library that is pretty in presentation but sterile in discovery! These matters become serious concerns when the tailoring of the corporate compound collection is assigned to individuals insensitive to the potential of discovery in complex systems that violate one or more of the above guidelines. There is no substitute (yet) for the practiced eye of a medicinal chemist.

Software packages are available[518,518a–e] that claim to provide assessments of diversity and recommendations for its enhancement. Very large libraries must depend on these or other computational techniques in such assessments, but caution must be exercised as to the extent that such packages are used. The utility of many of these programs is strained when the assessment attempts to include complex molecular arrangements. Diversity assessment is especially important in small collections (about 10 000 compounds); however, it is best in these instances to resort to manual review.

2.12.9.3.2 High-throughput organic synthesis applications

HTOS (also termed combinatorial chemistry) has had a dramatic impact on both the organization and productivity of industrial medicinal chemistry. It was first applied in new drug discovery in the late 1980s in the form of combinatorial peptide coupling. Solid phase synthesis of peptides, which had been developed by Merrifield (Rockefeller University) 20 years earlier, was used to create large polypeptides by affixing one end to a solid support (e.g., a resin bead) and adding the individual units sequentially by standard peptide-coupling techniques. Subsequent chemical cleavage from the resin yielded the target polypeptide. Application of these techniques to the simultaneous synthesis of families of related peptides was pioneered by Geysen (Glaxo) in the mid-1980s. The concept was further applied to new drug discovery in the synthesis of large libraries (20 000 to more than 1 million) of peptides fixed to beads or silicon chips and the assessment of the binding of these fixed libraries to labeled soluble targets (Selectide, Affymax, and others). Research managers were attracted to these systems by the remarkable numbers of test substances and the simplicity of the detection systems. Unfortunately, the impact of fixed supports in the binding of biomolecules and the limitations of peptides as structural leads in development were underrated. Soluble peptide libraries (e.g., from Houghton Pharmaceuticals) were similarly offered as discovery tools, but again applications were limited, and interest and corporate support waned.

Despite the shortcomings of large peptide libraries discussed above in de novo drug discovery, the use of peptide HTOS technologies in the identification of their direct roles in receptor responses per se or as surrogates for ligand definition in orphan receptor research brings forth a new and important role for these HTS approaches (see below).

The slump in the popularity of peptide-based HTOS in the early 1990s quickly turned around as the concept was applied to the synthesis of nonpeptide libraries using diverse synthetic techniques. Pavia (then at Sphinx) was an important pioneer in the development of these technologies and their implementation in microplate format. His innovation and that of his associates facilitated the introduction and adaptation of HTOS into HTS applications and encouraged the automation/robotics industries to develop instrumentation that would meet the needs of industrial chemistry. It revived HTOS as a discovery technology with little apparent interruption, but once more the enthusiasm was unchecked, and promises of satisfying the needs of discovery programs in structural diversity became unrealistic (see **Figure 54**).

Early HTOS also launched support industries that could take a chemical lead or hunch and expand it into large libraries that encompass relevant structural diversity. Companies such as ArQule and Pharmacopeia offered efficient application of HTOS in the exploitation of solution and solid phase syntheses in the development of libraries that addressed specific client targets. These two successful corporations have, in their years of operation, amassed libraries including millions of components. Notwithstanding the above-mentioned limitations of HTOS in the generation of structural diversity, these massive libraries can offer a first-pass opportunity in discovery simply on the basis of the huge library populations and the diversity of the templates.

Reactions performed on solid supports have a number of process advantages over classical solution methods. The ability to drive reactions to completion with large excesses of reagents is one such advantage; the ease with which the product is cleaned at each step is another. The reaction kinetics is often found to be different between solid phase and solution syntheses, and is often favorable to the solid phase approach. But there are shortcomings. Many reactions are impossible or at least exceedingly difficult to perform on solid phase due to the sensitivity of the linker or the resin itself. The inherent need to base such syntheses on common platforms also produces redundancy in the library generated. The common linkage site on each molecule further dampens structural diversity. Multiple platforms and nonfunctional linkage residues offer partial solutions to these problems. Nevertheless, today's HTOS compound libraries sorely need to be supplemented by selected products of classical synthesis and pure natural products to attain a meaningful three-dimensional, functional display for a de novo discovery program. This aspect is best illustrated by assessing how a hypothetical HTOS program would be designed if the goal were to rediscover (from scratch) representative nonplanar drugs, such as glucocorticoids, lipid-lowering agents (e.g., mevalonin), and taxol. Would today's solid phase or even solution phase synthesis collections achieve such a goal?

There are new directives in process that will provide some solutions to these problems. For example, the introduction of naturally derived, three-dimensional platforms can dramatically broaden the scope of HTOS. Variables will increase dramatically, and thereby create another dimension in numbers as libraries of millions of compounds become commonplace. MicroBotanica[515] has identified proprietary natural templates as the core of unique libraries and cheminformatics tools. These templates offer three-dimensional access not available to conventional laboratory-based synthons. ArQule, a leader in the progressive development of diverse and directed HTOS libraries, has made significant advances in improving the efficiency of HTOS in both solid and solution phase modes.[519,519a]

Simultaneously, HTOS was applied to bringing a HTS hit or otherwise-identified bioactive entity to lead status and its rapid development into a clinical candidate. In most cases, HTOS has replaced the drudgery of 'one-at-a-time' analog synthesis in the generation of SAR data and dramatically reduced development time. Essentially, all possible analogs are prepared rather than a small population selected by medicinal chemistry on the basis of immediate relevance and economics. HTOS libraries provide the full SAR analysis, with little reference to what would formerly be perceived as redundant. Such a 'bulk' approach leads to a more definitive analysis of SAR and a better selection of candidates. This application is clearly the most important contribution that high-throughput technologies have made to medicinal chemistry and perhaps to drug discovery in general. Its impact is best appreciated in consideration of development time and on-market patent life (**Figure 35**). Today, an annual market return of US $1 billion is not unusual for a new therapeutic entity. That translates into US $83 million for every month that the new drug is on the market and under patent protection. The development time for a new drug can easily consume half of its patent life. HTOS can reduce the development time and dramatically increase market returns.

Figure 35 illustrates the time-line of drug development from discovery to introduction. It does not reflect the additional savings that HTOS provides in hit-to-lead generation. That aspect can provide an additional year of benefit in earlier recognition of the lead. While that generally does not impact patent life (a subsequent event), it does give the advantage of 'first on the market' in a highly competitive field. The illustration (**Figure 35**) also does not take into consideration the emerging benefits of high-throughput absorption, distribution, metabolism, and excretion (ADME) and toxicity profiling, discussed below.

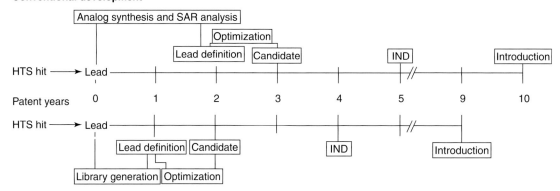

Figure 35 Impact of HTOS on new drug development (IND, investigational new drug).

2.12.9.3.3 Natural products as a discovery resource

The unique structural diversity inherent in natural products continues to be an important HTS resource. As with HTOS and other technologies, trends have had a disturbing influence on the adoption of natural products as a primary discovery resource (**Figure 36**). Notable is the stability of microbials as a discovery resource, but plants have had a sine wave history, and continue to be susceptible to technological pressures and prejudices.

Microbials are by far the most widely used natural resource in industrial screening programs. They have been the mainstay in HTS applications in several major companies, with little fluctuation in their use over time. The reason is threefold: economics, reproducibility, and a continued history of new biomolecular discovery. Once a company has established and staffed a microbiological facility on-site, costs are manageable and predictable.

The use of microorganisms over the past six decades in antimicrobial research and development has established a sound technological basis for the culture, preservation, and manipulation of bacteria and fungi. Clear projections of scope and cost can be made for their use at the semi-microscale in discovery and the macroscale in commercial processes. Microbials are accountable for most of the antibiotics in therapeutic use today, and, in addition, have served as biochemical factories for chemical modifications that cannot be reasonably achieved otherwise. In the past two decades, we have seen the successful use of microbials in broader pharmacological screening with remarkable success – ivermectin, FK-506, rapamycin, and mevalonin, to name only a few. They can also provide unique chemical substrates for further manipulation in the chemist's laboratory, and in this application constitute the basis for important semisynthetic variations in antibiotics and other products. The shortcomings of microbials are (1) the limited biochemical scope compared with plant resources, and (2) the critical need for dereplication at an early stage in screening. Novelty in new structural types is less common in microbial screening.

Marine organisms offer a remarkable variety of chemical variation. Many such natural products are not available from other natural sources, and certainly not through laboratory synthesis. Unfortunately, access to this resource is very expensive, limited to initial structural leads, and rarely amenable to commercial development. Even recollection for research purposes is limited by cost and environmental concerns.

Plants, although entrenched in the early history of drug development, have not been used as a discovery resource to the extent that microbials have in recent times. In the 1950s and early 1960s there was a flurry of tissue- and animal-based screening based on plant extracts (**Figure 36**). Important discoveries in the alkaloid classes (e.g., reserpine) created an inordinate level of optimism, and many companies established extensive plant-based screening programs in in vivo systems. Unfortunately, the narrow focus on alkaloids in both collection (selected plant families) and chemistry (nitrogen-containing constituents) led to a great deal of redundancy, so that by the mid-1960s many of these expensive industrial research programs were closed as discovery attended to the more lucrative area of analog synthesis.

The revival in natural plant products' research in the 1970s, on the heels of the then-fashionable, marine-based research, was cut short by the introduction of rational drug design through computational modeling. The high level of hype associated with this area was short-lived (**Figure 36**), but soon replaced by a similar optimism with peptide-based combinatorial chemistry (see above). Plant natural products as a classical (and expensive) program again lost favor.

This persistent quick-change attitude in research management reflected its frustration with conventional discovery resources, its determination to be at the cutting edge of new technologies and among the first to harvest the benefits. Happily, stability is returning, as HTOS finds its place in drug development, and the significance of assuring structural diversity in test substance supply is addressed in practical terms. Plant-derived natural products are beginning to resurface and take their rational place as a source of immense structural diversity.

The lack of adoption of plant sources as a test resource is somewhat paradoxical. The plant genome is much larger than other natural sources (bacteria, 1000 genes; fungi, 10 000 genes; plants, more than 100 000 genes), and therein

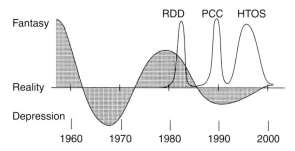

Figure 36 The rise and fall of technological hype in the use of plant natural products and sequential influence of early attempts in rational drug design (RDD), peptide combinatorial chemistry (PCC), and HTOS in HTS programs.

offers a broader biochemical network for the generation of unique chemical structures. Unfortunately, until recently, the technologies and protocols of collection, processing, recollection, and scale-up had changed little in 50 years. Each sample collected individually demanded shipping, storage, and handling expenses; the attention of a chemist and considerable laboratory-based preparation time prior to assay were also required. Few attempts had been made to adapt these classical collection and preparative techniques to modern screening programs and budgets. As a result, plant-based research commanded a much higher cost per test sample than microbials, and it is this cost, and the requirement for a dedicated research team, that turned program directors away from the use of plants as a significant HTS resource.

Companies that have continued to use plants over the past two decades have done so at low throughput, with the evaluation of only a few hundred species per year. In 1993, MicroBotanica addressed the economics and efficiency of plant collection, with the goal of making this unique source available to screening programs at a throughput level that permits broad evaluation in modern HTS systems and a cost that is compatible with HTS budgets. It established a joint program with three groups in Peru: Perubotanica srl, Universidad Nacional de la Amazonia Peruana, and Conservacion de la Naturaleza Amazonica del Peru. The goal was biodiscovery, and the collection, classification, documentation, and development of the plant resources in the Amazon Basin with strict compliance with the terms of the Convention on Biological Diversity.[520,521]

The technical approach is based on the position that the initial collection of test materials for HTS should be only what is required for the first pass and the initial confirmation of a hit (i.e., about 1 mg of a crude extract). Larger scale collection and extraction must be implemented only on the identification of hits. Such a 'microcollection' program dramatically reduced costs and increased the number of plants sampled. It also permitted extraction in the field, and thereby eliminated on-site specimen processing as well as laboratory handling time and expenses. This consortium can also respond immediately to a hit with sufficient material for confirmation and secondary profiling, and simultaneously assure recollection and the receipt of larger scale extract preparation in a timely manner – usually 2–3 weeks.

Another issue of special concern in the implementation of a plant-based natural product program is the sensitivity of many HTS assays to interference by polyphenols and tannins through nonspecific binding. Such substances are common constituents in plants, and their removal prior to bioassay is frequently required. This issue has also been addressed by MicroBotanica through the development and application of proprietary liquid–liquid partitioning amenable to automation or the more conventional use of polyamide binding. Today, MicroBotanica boasts ready access to samples of more than 30 000 extracts from approximately 13 000 plant specimens and the ability to follow through from recollection to isolation on any hit. MicroBotanica's participation in the Amazon consortium also makes it a unique source for the collection of and research on any plant endogenous to the Amazon Basin.[512]

2.12.9.3.4 Structure-based design

The use of the structure of the target as a basis for computational design is not a new concept. Two decades ago it ran rampant through the pharmaceutical industry (**Figure 36**). Research management, heady with the developments in molecular modeling software and x-ray crystallography that were current, was impressed with the purported ability to design new drugs on the basis of target structure alone. Misinformation, misinterpretation, and bad timing took its toll. Structural analyses were weak and presumptive; computational power was overestimated, and projections totally unrealistic. Fortunately, only a few companies made major commitments in this area, but those that did lost a great deal in program deletions, time, and credibility. Today, the de novo design of an active molecule for a new biomolecular target is still fanciful. Finding structural precedence in small-molecule interaction is the key. Enter HTS and the revival of structure-based drug design.

Today, structure-based drug design (SBDD) encompasses the integrated use of computational design software, HTOS, and HTS. In principle, SBDD is based on the sequence illustrated in **Figure 37**.

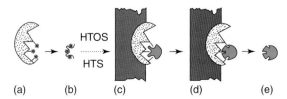

Figure 37 (a) Structure-based drug design. (b) Definition of target dimensions. (b, c) Computational analysis of structural features for small-molecule interaction. (c, d) HTOS generation of relevant compound libraries and identification of first hit through HTS. (e) Optimization of the hit to one or more leads.

Information based on x-ray crystallographic data, nuclear magnetic resonance, and other analytical techniques facilitates the definition of the molecular dimensions and functionality of the target and its active site (**Figure 37a**). Computational techniques are applied to the identification of the optimal small-molecule characteristics for interaction at the target site. This information is then used to screen real or virtual libraries for compounds suitable as first probes. A test library is then generated, acquired, or synthesized through HTOS, and screened in the relevant assay(s). If hits are not realized, the process is repeated by retooling the target definition or anticipated small-molecule characteristics. Once significant data are obtained (**Figure 37c**), larger libraries are prepared, and the molecule is optimized to lead status. Conventional medicinal chemistry usually takes over at this point, to provide the optimum candidate(s) for development. The system works. Of course, it still has both the requirement of a considerable knowledge base for the target and the risk associated with projections of biomolecular interaction to therapeutic significance. Nevertheless, it is reassuring to know that such a simple but elegant approach can be successfully implemented, provided all the tools are present and realistically evaluated.

SBDD is also applied to assess the significance of the myriad of therapeutic targets anticipated from comparative analyses in functional genomics (*see* Section 2.12.9.4.2). While estimates of new and therapeutically relevant receptors are in the hundreds, finding those in the sea of information is a Herculean task. The relevance of these targets, the identification of natural and artificial ligands, and the discovery of small molecules that can modulate their effects are goals in the practical application of this immense information base. Information gleaned from these studies will find application in drug discovery, diagnostics, toxicology, and other areas yet to be fathomed.

A recent and dramatic example of the efficient application of SBDD was in the identification of inhibitors of the nuclear factor of activated T cells (NFAT) in the search for new modulators of the immune response that do not have the limitations of cyclosporine or FK-506. Hogan and collaborators[522] applied HTOS in the form of peptide libraries of more than a billion units to identify optimal amino acid composition at seven sites of a hexadecapeptide sequence associated with calcineurin binding. This is an area that is by definition peptide in content and well suited to peptide HTOS. Transfer of the information gleaned into a useful immune-suppressive drug is yet to be achieved, but the ground is laid and open to innovation.

2.12.9.4 Bioassay Development and Implementation

The identification of a cellular or molecular target for drug discovery and the development and validation of a relevant bioassay are critical early milestones in a HTS program. There must be a persuasive reason to believe that the cellular or molecular (enzyme, receptor, protein–protein, or protein–DNA interaction) target is involved in a pivotal step in the targeted disease process. That link – the molecular target to the disease process – requires exhaustive analysis. Today, with the cost of screening at US $5–10 per well (compound and reagents) at the 96-well plate level, the assessment of a 50 000-member library will cost as much as US $500 000 before personnel costs and overhead are considered. Many of us can recount pre-HTS times when certain animal or biochemical assays formed the core of a discovery program only to prove of little or no value in the identification of a clinically active agent. Relevance must be carefully balanced against cost (**Figure 38**).

Once a target has been selected, the next step is to develop a bioassay that is appropriate and compliant with the parameters applied in the corporate HTS system. Typically, such assays have to be robust and avoid the use of temperamental or unstable reagents. Each assay must be amenable to miniaturization to at least the 96-well microplate level, but preferably beyond. Miniaturization to 384 wells is commonplace, but further reduction is encouraged, with

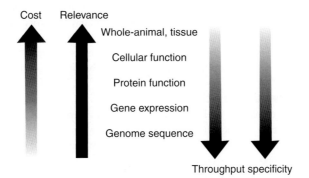

Figure 38 Relationship of bioassay targets to relevance and HTS compatibility.

the potential of lower cost and simultaneous higher throughput and data analysis. Cost savings are in attention time, and the amount of compound and reagents used. By the simple shift from 96 to 384 wells the practical reaction volume and, correspondingly, the required reagents and compound are reduced by 75%. The latter is of special significance since many compounds, especially pure natural products, are in limited supply, and the conservation of a supply is often critical. Shifts to 1536 plates or high-density formats have corresponding benefits, although sample handling becomes a problem. Moreover, some detection systems are unreliable beyond 384.

Assays and even therapeutic targets have a defined lifetime. Frequently, even the most comprehensive and structurally diverse compound collection will not yield a hit in some assays. Such events generally are attributed to the inability of a small molecule to interfere in a complex biological process. Protein–protein interactions are often difficult to disrupt or block. At some point, usually the exhaustion of the current compound library, the decision is made to remove the assay for re-evaluation. Occasionally this may result in a minor variation in the reagents or protocol, or may lead to the elimination of the assay or even the base target. The number of compounds screened through any bioassay may range from 10 000 to millions, depending on the culture of the organization, the extent and diversity of the chemical resources, and a judgment as to the importance of the target.

Assays are generally established to permit a return of 0.05% or less hits. High-density systems with large numbers and low diversity in the test substances used may be a log or two lower. The frequency with which assays are removed from a HTS program on the basis of 'no yield' can be reduced by critical analysis of the assay and the dose level prior to introduction into the HTS program. This is especially true in de novo applications where there is no drug reference to guide appropriate test concentrations and time. Nevertheless, the effects of a broad group of bioactive compounds on the assay response can be a useful directive. Fortunately, compound collections such as GenPlus (960 bioactive compounds) and NatProd (720 natural products) are available from MicroSource Discovery Systems[512] in microplate format. Such collections can be valuable tools in defining the sensitivity and specificity of any new assay. These are preplated at standard concentrations (mg mL^{-1} or 10 mM dimethyl sulfoxide), and can be easily blended into an assay validation protocol.

The living cell is the preferred format in discovery programs, offering a test response at the 'organism' level while still allowing automation and high throughput. Cells can be engineered to house complex receptor systems tailored to represent a myriad of therapeutic targets. Detection modes are colorimetric, fluorescent, chemiluminescent, and radioisotopic systems. While regulatory and disposal problems continue to encourage the displacement of radioisotopes by other detection modes, innovations in simple and sensitive homogeneous systems, such as the SPA developments by Amersham,[523] maintain a stable market.

The validation of a lead prior to further development involves the testing of the lead for efficacy in an animal model. The biological connection of a target to a disease, and the utility of such a target in drug discovery and development, is speculative without whole-animal data. Once a lead has been identified, it is essential to establish a 'proof of concept' through in vivo evaluation. The selection of such a model is a process that should be concurrent with the development of the HTS assay. Such models may already be established, but it is likely that, with the emergence of novel genomics-based targets, they will require creative pharmacology and critical validation before confidence is established.

Activity in a whole animal is still not definitive for activity in humans, but the absence of activity draws attention to the need to reconsider the utility of the target. It is best to have such revelations early in a discovery program. The time-line and milestones applied in the collaboration of Axys and Merck in the identification and development of a small-molecule inhibitor of cathepsin K as a therapeutic modality in osteoporosis is a good illustration of this approach.[524,524a]

All of these technologies are not only complementary but feed back on one another with dramatic amplification of discovery potential. This can be of significant benefit as new activities in new indications are uncovered; however, there is a serious danger of misdirection. Incorrect information derived from a missed false-positive, incorrect analysis of an assay output or poor assay design can expand into unproductive and expensive voids. Any new development in HTS in design, discovery, or data analysis must be carefully assessed before it becomes an integral part of the process.

2.12.9.4.1 Fluorescence detection systems

Fluorescence technologies are dominant in HTS. Numerous well-characterized fluorescence tags suitable for differentially labeling specific cell components and protein probes are now commercially available.[525] Many quantitative techniques have been developed to harness the unique properties of fluorescence; several of these are represented in the following chapters in this book.

An excellent example of such technology is provided in Section 2.12.10, discussing the development and advantages of functional screening based on high-resolution fluorescence imaging of multiple targets in intact cells. Section 2.12.15

discusses the application of multiple fluorescence read-out methods in the development of single-molecule detection technologies that enable a variety of parameters to be evaluated in a single sample.

Fluorescence resonance energy transfer (FRET) is an ultrahigh-throughput technology developed by Aurora, one of the most progressive biotechnology companies in the HTS arena. FRET uses a coumarin-linked phospholipid asymmetrically bound to the plasma membrane as the donor and negatively charged oxonol acceptors that partition across the plasma membrane as a function of the transmembrane electric field. In Section 2.12.16, the use of FRET in cell-based and reporter–receptor binding, protein–protein interactions, and other applications is discussed. Aurora's platform also includes ultrahigh-throughput screening technologies and discovery modules, which allows for computer-directed delivery and bioassay of more than 100 000 selected compounds per day in 96- or 384-well plates.

Homogeneous time-resolved fluorescence, as well as its chemical basis and application in a broad scope of cellular and biomolecular assays, is described in Section 2.12.17.

2.12.9.4.2 Comparative functional genomics

The identification of genome-based targets and representative bioassays is a more recent facet that already has spawned opportunities that were unimagined a decade ago. The use of human functional genomic data as a basis for new drug discovery is taking a commanding lead in industrial HTS programs. The technologies used in this area are not new, but the huge amount of data that has become available is, and their comparative analysis in normal and disease states constitutes one of the most rapidly growing and promising areas in the identification of new therapeutic targets.

Comparative functional genomics focuses on the identification of human genes associated with disease response in the whole organism and their role in the proliferation or suppression of the disease state. Such analysis has yielded hundreds of new targets that have no precedence in therapy or medicinal chemistry; many more are anticipated. The definition of the significance of these 'orphan' receptors relies on classical strategies of genetic engineering and disease gene modeling. These orphan receptors require validation before becoming an integral part of a discovery program; the identification of associated ligands is a first step.

Needless to say, these technologies have produced considerable hype in the projection of their impact on health management. Such hype is seen both in the marketplace and in industrial research planning. Fortunately, we are probably at the peak of the hype curve, and look forward to tangible contributions in new target modeling and ultimately in therapeutic benefit.

Proteomics, an essential supportive technology in functional genomic applications, addresses the characterization of the physicochemical properties of a protein expressed by a newly discovered disease-relevant gene and its production in large amounts through transfected cell cultures. Such probes have also been conducted in the HTS mode. These proteins, if beneficial, may lead to therapeutic agents in their own right, but for our purpose they can serve as tools for de novo assay development. Antibodies to such proteins provide insight into their function and location, and may also serve in therapy or assay development as specific antagonists.

2.12.9.4.3 Absorption, distribution, metabolism, and excretion, and toxicological profiling

ADME and toxicological studies are critical parts of any drug development program, and essential for compliance with regulatory guidelines. Historically they were conducted only on drug candidates that had survived the rigors of chemical optimization, process development, and pharmacological profiling. The reason for this segregation was simply that such studies invariably involved whole-animal models and therefore were time-consuming and expensive. It was not economically sound to expend such resources on candidates that were not firmly committed to development by other selection criteria. Unfortunately, when an ADME problem was finally detected, it was at late investigational new drug preparation or even in the clinic. Such events created serious disruption of the development process, and often resulted in closure of the project and a lost opportunity.

The parallel development of several candidates from the same compound class has been a standard procedure to avoid project termination in the event of the emergence of an untoward effect. The considerable cost imparted by this approach was justified as being necessary for project survival.

Today, the situation is changing rapidly and dramatically. ADME and toxicology technologies have evolved to permit the use of rapid and less expensive methods that have made the early assessment of drug candidates very attractive to the pharmaceutical industry. Major companies are shifting ADME assessment to become an integral part of the candidate selection process. Costs are still substantial but justified.

The goal is to move the assessment of drug metabolism and toxicity up in the discovery/development process (**Figure 39**). Metabolic profiles of a large group of compounds that are considered for development can provide important information at the preselection level, save considerable time, and significantly reduce the cost of new drug

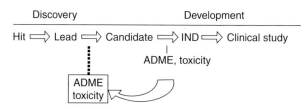

Figure 39 Current shift in the emphasis of ADME assessment in the drug development process.

development. Structural characteristics of test compounds that impart resistance to rapid metabolism are often independent of the biotarget pharmacophore and yield a different SAR analysis. Early ADME assessment can uncover such compounds before they are discarded on the basis of activity alone.

The principal source of drug metabolism is the cytochrome P-450 enzymes, which constitute a superfamily of monooxygenases (requiring NADPH). They are found primarily in the endoplasmic reticulum of the liver, although a prominent member is also found in the gut. In vitro ADME assays can be based on human liver microsomes or hepatocytes, which incorporate the full complement of cytochrome P-450 and other microsomal and cytosolic enzymes, or individual enzymes. Most cytochrome P-450 enzymes have been identified, and many are available through recombinant techniques; their monoclonal antibodies are also commercially available.[525] Information can be obtained through in vitro systems about the patterns of metabolism, which cytochrome P-450 enzymes are involved, metabolite hepatotoxicity, and drug–drug interactions. Such analyses are amenable to HTS. Metabolite isolation and identification is a HTS challenge that can build on the technology already in place in HTOS and combinatorial technologies.

HTS applications in ADME and toxicology are discussed in Section 2.12.19. The ease of isolation of human hepatocytes and their importance in metabolic and toxicological studies are underscored. These hepatocytes are stable and can be cryopreserved, thereby providing a regular supply of hepatocytes for screening without dependence on frequent access to fresh human liver.

Gentest[528] has described a microplate-based fluorimetric approach to rapid screening of test compounds for their susceptibility to and effects on cytochrome P-450 enzymes. The technology is efficient (100 compounds in 4 h) but limited in scope.

If a drug has an inhibitory effect on one enzyme and enhances the effect of another, its blood level in diverse ethnic populations can be dramatically different and create risks of ineffective dose or toxicity. It is likely that more stringent assessment requirements will be forthcoming for drugs that demonstrate inhibitory or enhancing activities on metabolic enzymes. HTS again can provide the tools.

2.12.9.5 Informatics

The analysis and efficient use of the massive amount of data generated in HTS and related areas has created the new discipline of bioinformatics, which has opened doors to new modes of discovery. Bioinformatics combines software, hardware, and database architecture to facilitate the storage, retrieval, and analysis of HTS data and to cross-reference such data with archival information on the basis of structure, physiochemical characteristics, biological profile, and therapeutic application. Such analyses enhance the efficiency of lead detection but also can provide unanticipated 'finds' in areas not previously considered; serendipity continues to play an important role.

The complex network of biological and physicochemical interactions of test compounds with receptors, endogenous proteins, metabolites, and other cellular constituents provides a 'pedigree' of behavior that constitutes the equity of the corporate screening resource. This pedigree also includes its chemical character, which incorporates chemical functionality and topological display. It is also important to add any data that are associated with compound-handling characteristics (solubility, instability, etc.), chemical character (reactivity, chelation potential, etc.), as well as the potential for interference in bioassay detection systems (color, fluorescence, etc.).

It is not enough to record that a compound is active or inactive in one or more assays. Any unusual observation with a test compound must be stored within its pedigree and not lost in the archives of a project. It is important to tabulate these findings with assay conditions and activity levels, although the simple 'yes/no' record for many of today's high-density assays limits such input. Much of this information is not readily available for large collections, but can be accumulated in time as aspects of these and other features surface. ADME information is also an important aspect of such an information base, but, today, that is especially limited. Storage and handling of these data and their integration into analytical and decision-making protocols yields a valuable information pool. This is the basis of HTS informatics.

Variation in the nature of assay protocols and performance must be addressed in the design of the information-handling system. Protocol compliance and experimental error – mechanical, biological, or chemical interference – are important factors in defining the integrity of the data retrieved. False positives and negatives and the relationship of their incidence to assay or compound character are other aspects that can improve data quality and future performance.

Today, the term 'data mining' brings forth the realization that the data retrieved in a HTS run have a value that goes beyond the initial assay intention. Such information can lead to pattern analysis that will impact future inquiries. Negative and positive data hold equal value and constitute valuable equity in a company's HTS operations. This aspect is of special importance as we enter into the realm of unknown receptors and ligands.

2.12.9.6 Management and Personnel Issues

Centralization of HTS activities was an early organizational change in R&D. It was essential on the basis of coordination of robotics, assay development and integration, data retrieval and management, and, most importantly, personnel development. This section began by acknowledging the arrival in the early 1990s of a new breed of scientist who recognized the challenge and the potential rewards of screening and the important catalyst that such individuals provided in the rapid growth and success of HTS. That breed has evolved into a new professional elite that forms the core of discovery operations in many companies.

The specific goals of HTS have made it a separate and important corporate entity. This has been a natural process as technologies and information output have become exclusive to the discipline. Personnel have also become highly specialized, with attention and skill directed to aspects not previously part of conventional discovery operations.

Even the equipment employed for HTS in major companies has introduced new dimensions, requiring huge space allocations and large staff commitments. Witness the means employed for the storage, retrieval, and dispensing of dry compounds and solution samples. Such systems significantly improve HTS efficiency and accuracy, but can require thousands of square meters for operation. Notable is the Haystack system from Automation Partnership, which has become the standard system for most large HTS operations. Haystack can automatically handle millions of samples as well as providing scheduling and tracking support. A more recent addition is the HomeBase system, which extends that capacity to include microplate management up to densities of 1536 wells with dispensing potential from 0.5 µL to 250 µL. Such equipment and associated robotics support constitute a 'production level' operation. This physical feature and the corresponding emphasis on massive data management and analysis separate the operation from the mainflow of research activity. While such separation is inevitable, there remains the need to ensure that feedback is given to the HTS team on the development of any hits and leads uncovered.

In small and large corporations, the integration of multidisciplinary technologies and personnel into a productive and innovative HTS team is as important as the anticipated advances in throughput and discovery. The latter follows automatically from a well-integrated, self-supportive team. Excellent reviews and commentaries by others on these all-important 'people' aspects have been published elsewhere.[530,531]

2.12.9.7 Projections

Throughput has been increased with higher density microplates, free-form systems, chips, and other innovative devices that take advantage of advances in microfluidics and the associated micronization of sample transfer, assays, and readouts. These advances also significantly reduce operational costs.

Microfabrication and microfluidic technologies are in a growth phase and creating a significant impact on innovation in screening design. They are a lucrative outsource activity, and likely to remain so as the scope of these technologies continues to be explored. Application in miniaturization in drug discovery and genomics are two areas that impact our interests. Monogram BioSciences[532] employs plastic and glass microfluidic array chips in sample processing and analysis at picoliter volumes. Caliper Technologies[533] applies similar microchip technologies in facilitating liquid handling and biochemical processes. Their 'lab-on-a-chip' technology uses nanoliter volumes of reagents in enzyme, receptor binding, and functional cell-based assays as well as systems that address toxicological and pharmacokinetic parameters.

Aurora developed the NanoWell assay plate, which boasts 3456 miniaturized wells suitable for fluorescent assay applications. The assay volume used in these plates is 100 times smaller than the conventional 96-well format. This reduces the cost per test and the amount of compound needed. The required microfluidic technologies have also been developed by Aurora for compound transfer (<1 nL) at rates of up to 10 000 wells per hour; fluorescence detectors capable of handling more than 500 000 assays per day are also in place.

How small can we go? Biosensors based on molecular recognition force measurements between individual molecules using atomic force microscopy (AFM) have been described.[534] These include interactions between individual ligand–receptor, antibody–antigen, and DNA–DNA molecules. Using similar single-molecule measurements, discrimination between two chiral molecules has also been achieved at the single-molecule level by chemical derivatization of the scanning probe tip in AFM.[534] These assay technologies are far from high-throughput, but offer considerable promise for such at the monomolecular level.

We have overcome the demands of sample throughput up to the point where meeting the demands of compound supply and diversity is a strain. Robotics, detection systems, sample density, and data management have increased to a level that was not conceivable a few years earlier. Unfortunately, assay systems that provide relevance as well as throughput are few and far between. All too often, relevance is set aside as a matter for secondary evaluation rather than the primary screen. Such an approach is fine, provided there is an appropriate second screening stage to bring both numbers and relevance to acceptable levels. The challenge now is in the development of systems that allow significant flexibility for assay enhancement in therapeutic relevance and the increase of information gleaned from each. Increase in the latter is important, but not at the expense of the former.

2.12.10 Fluorescence Correlation Spectroscopy and FCS-Related Confocal Fluorimetric Methods (FCS$^+$plus): Multiple Read-Out Options for Miniaturized Screening

2.12.10.1 Introduction: A Rationale for New Read-Out Methods in Drug Discovery

The process of drug discovery has been in a state of rapid change over the past decade and a half. The advent of genomics and combinatorial chemistry has lead to an increased reliance on efficient and effective techniques for screening large numbers of chemical compounds against an increasing number of potential pharmaceutical targets. The discipline of high-throughput screening (HTS) has been at the center of a revolution in drug discovery. Not only has HTS emerged as an important tool in the early stage of drug discovery, it is also the stage for the interaction of profound biological and chemical diversity with new technologies in the area of automation, robotics, and bioinformatics.[536,537] The way in which assays are performed for screening is changing significantly. Whereas once assay volumes of 100–150 µL were common for such screening techniques as enzyme-linked immunosorbent assay (ELISA) and radioligand binding assays, modern miniaturized HTS favors homogeneous 'add-and-read' assay formats with volumes as low as 1 µL per well. Miniaturization will become the rule throughout the drug discovery process as compounds will be needed for more assays while being synthesized in smaller quantities. The process of HTS is also seeing increased application of industrial standard automation and robotics technology to improve efficiency. In parallel developments, novel detection technologies and assay strategies are being implemented to improve the effectiveness of the process. Therefore, novel detection technologies will be tested in an environment that requires the maximum determination of biological information in the shortest amount of time with a minimum of reagent and compound, and with maximal precision.

The technology described in this section has emerged from physics laboratories and the analysis of rapid chemical reactions to find application in drug discovery. Use of the confocal optical systems described herein facilitates the analysis of biological interactions at the molecular level. Changes in the molecular environment of a fluorescently labeled biological molecule result in the change in signal used by FCS and related technologies. The changes in molecular environment typically occur as a result of binding or biological processing, and therefore provide an ideal basis for screening assays.[538]

The challenges with regard to modern assay techniques and the related technologies are large. These include the need to maintain sensitivity as volumes, and therefore quantities of reagents and compounds, decrease as much as 100-fold; to obtain as much information as possible from a single measurement; to help eliminate the disturbances often encountered in compound screening, such as compound autofluorescence and turbidity; and to be compatible with homogeneous assay formats. The following describe how FCS$^+$plus meets most of these challenges.

2.12.10.2 Single-Molecule Confocal Fluorescence Detection Technology

Fluorescence-based confocal detection technologies allow molecular interactions to be studied at the single-molecule level. Combining laser spectroscopy with confocal microscopic optical systems, the laser beam can be highly focused in such a way that only molecules in a volume of 1 fL are hit by the light – a volume equal to roughly the size of a bacterial cell.[539] With the laser illuminating such minute parts of the sample, even volumes of 1 µL or less are sufficient for sample testing. On the basis of this technology, new methods have been developed in recent years that allow the molecular properties of fluorescent biomolecules to be studied in 1 µL sample volumes, enabling a wide range of solution-based and cellular assays to be established for HTS.

The ability to make full use of the fluorescent molecule and all of its properties should be the benchmark of any fluorescence-based assay technology. While most assay strategies make use of only a single fluorescent property, the broad applicability of a screening system requires that flexibility of assay design be extended to the read-out technology by including a variety of detection modes. While most scientists may associate the use of fluorescence in biological assay systems solely with the measurement of fluorescence intensity, the measurement of additional fluorescent properties, such as lifetime, polarization, fluorescence energy transfer, and quenching, can yield a wealth of information from a single measurement. This ability to collect multiple data points per measurement not only provides an internal control but also contributes to screening efficiency by enabling rapid multiparameter evaluation of compound–target interactions.

2.12.10.2.1 FCS$^+$plus: multiparameter fluorescence read-out technology

FCS is used to determine the translational diffusion of fluorescent molecules.[540] Each fluorescent molecule that diffuses through the illuminated confocal focus of the laser gives rise to bursts of fluorescent light quanta. The length of each photon burst corresponds to the time the molecule spends in the confocal focus. The photons emitted in each burst are recorded in a time-resolved manner by a highly sensitive single-photon detection device. The detection of diffusion events makes possible the determination of a diffusion coefficient. Upon binding of a fluorescently labeled ligand to its receptor, the molecular weight, and therefore the diffusion coefficient, changes. Thus, the diffusion coefficient serves as a parameter to distinguish between free and bound ligand. The confocal optics eliminate interference from background signals, and allow homogeneous assays to be carried out.

Since translational diffusion relies on significant changes in molecular weight upon molecular interaction, new methods were developed at Evotec that use a variety of fluorescence parameters as the read-out. These new methods, collectively called FCS$^+$plus, evaluate fluorescence signals from single molecules on the basis of changes in fluorescence brightness, fluorescence polarization, fluorescence lifetime, or fluorescence spectral shift, by fluorescence energy transfer, or by confocal imaging (Table 23). Brightness analysis is a unique read-out method that allows one to determine concentrations and specific brightness values of individual fluorescent species within a sample.[541] As a measure of fluorescence brightness, the number of photon counts per defined time interval in the confocal volume is detected. Changes of molecular brightness during a binding event can be due to two mechanisms: (1) if one of the two partners is labeled with a fluorescent dye, quenching of fluorescence may occur upon binding; and (2) amplification of

Table 23 Read-out modes offered by FCS$^+$ plus

Method	Principle
Translational diffusion (fluorescence correlation spectroscopy, FCS)	Translational diffusion properties of molecules are dependent on the molecular weight. Diffusion coefficients of fluorescently labeled molecules therefore change upon interaction, and enable distinction between free and bound states of the molecule
Fluorescence brightness	The number of photon counts per defined time interval in the confocal volume serves as a measure of the molecular fluorescence brightness. Changes in fluorescence brightness of a fluorescent molecule upon binding are monitored
Fluorescence polarization	The fluorescence polarization of a molecule is directly proportional to its molecular volume. Changes in the molecular volume due to binding or dissociation of two molecules, conformational changes, or degradation can be detected as changes in polarization values of the fluorescently labeled molecule
Fluorescence lifetime	Fluorescence lifetime describes the average time that a fluorescent molecule remains in the excited state. The lifetime of the fluorescent signal is dependent on the molecular environment of the fluorescent tag, allowing monitoring of molecular interactions
Fluorescence energy transfer	The emitted light of a fluorophore serves as the energy source to excite a second fluorophore. The energy transfer between a donor fluorophore on molecule A to an acceptor fluorophore on molecule B depends on proximity, and can therefore serve as a measure of the interaction of A with B
Spectral shift	The excitation and/or emission wavelength of a fluorescent tag is dependent on the molecular environment. Changes of the spectral properties of a fluorescent molecule induced by molecular binding serve as a read-out for binding
Confocal imaging	Fluorescence changes on the cell surface or within cells caused by biological reactions are visualized by confocal imaging techniques combined with two-dimensional scanning

Table 24 Comparison of Evotec's technology and other fluorescence methodologies

Feature	Conventional fluorescence methods				Technologies at Evotec	
	FI	FP	FRET	HTRF	FCS	FCS$^+$ plus[a]
Homogeneous	0	+	+	+	+	+
Mass independent	+	−	+	+	−	+
Signal independent of assay volume	−	−	−	+	+	+
Single-molecule sensitivity	−	−	−	−	+	+
Insensitive to autofluorescence	−	−	0	+	0	+
No significant inner filter effects	−	−	−	0	+	+
Multiplexing	−	−	−	−	+	+
Components to be labeled	1 or 2	1	2	2	1	1 or 2

FI, fluorescence intensity (total); FP, fluorescence polarization; FRET, fluorescence resonance energy transfer; HTRF, homogeneous time-resolved fluorescence.
+, advantage always featured; 0, not always featured; −, disadvantage/not a feature.
[a] FCS$^+$ plus comprises the detection of fluorescence brightness, polarization (FP), molecular diffusion, and lifetime. In addition, assay systems based on FI and FRET can be applied with FCS$^+$ plus.

the molecular brightness takes place when both partners are labeled or a particle offers multiple binding sites for the fluorescent ligand (e.g., receptor-bearing vesicles or beads/bacteria with one binding partner immobilized on the surface). This technique has wide applications, since it can be used to study the interactions of proteins with similar molecular weight and binding of ligands to membrane receptors (see below).

Changes in any of the molecular parameters described in **Table 23** can be used as a read-out to characterize molecular interactions. In most cases, the same optical and electronic configurations are utilized, merely employing a different algorithm for analysis. This means that all fluorescent parameters noted above can be monitored using the same detection unit. Some of the fluorescence parameters described in **Table 23** can even be monitored simultaneously in a single measurement (multiplexing), resulting in exceptional data quality regarding reproducibility and statistics. A comparison of Evotec's detection technology with other commonly used fluorescence methodologies is shown in **Table 24**.

2.12.10.2.2 Advantages of using FCS$^+$ plus for miniaturized high-throughput screening

The major advantages of FCS$^+$ plus technologies over other detection technologies are summarized in the following sections.

2.12.10.2.2.1 Inherent miniaturization

All macroscopic fluorescence methods, whether based on intensity, polarization, or lifetime detection, measure by averaging all signals across an optical collection volume, which is usually a significant portion of the sample well. For such ensemble measurements, lowering the assay volume results in a lower number of fluorescence signals contributing to the ensemble measurement. In most cases, this results in reduced assay performance, with the signal decreasing relative to the background as the assay volume decreases toward the 1 μL range. With FCS$^+$ plus, fluorescence parameters are measured from individual molecules in a detection volume of 1 fL regardless of sample volume. The signal-to-background ratio is effectively independent of the sample volume. Using FCS$^+$ plus as a detection technology, miniaturized screening is possible in assay volumes of 1 μL and lower without loss of signal. Miniaturization applied to large-scale screening has a considerable effect on the costs: to run a screen with 100 000 compounds, the reagent savings of 100 μL versus 1 μL assay volumes can make a difference of US $1 million versus US $10 000, assuming average costs for reagents include standard purified proteins. The cost savings will increase all the more if precious reagents are needed.

2.12.10.2.2.2 Homogeneous assay format

Since the bound and unbound state of a fluorescent biomolecule can be distinguished by different fluorescent parameters (see **Table 23**), no physical separation of bound and unbound ligand is required. The elimination of washing and separation steps make such 'add-and-read' assay formats easily amenable to automation, and rapid to perform.

2.12.10.2.2.3 Increased safety of reagents
Fluorescence detection technologies avoid the use of hazardous radiochemicals and the production of large-scale radioactive waste.

2.12.10.2.2.4 Elimination of background effects
Due to the short pathlength of the confocal optical configuration, background effects and signal reduction caused by turbidity and ink-like solutions can be substantially reduced in comparison with other methods. In addition, other disturbing background effects, such as those from light-scattering or autofluorescent compounds, can be eliminated by registering only the signal of the fluorophore used for labeling.

2.12.10.2.2.5 Single-component labeling
In most cases, fluorescent labeling of one binding partner with standard dyes is sufficient using FCS$^+$plus. FCS$^+$plus is therefore more widely applicable than FRET or homogeneous time-resolved fluorescence (HTRF), where both partners have to be labeled.

2.12.10.2.2.6 Multiple read-out modes
Since FCS$^+$plus encompasses a variety of read-out modes (see **Table 23**), it offers unique flexibility: the most suitable read-out mode can be selected for each assay system. FCS$^+$plus is applicable to the study of protein–nucleic acid, protein–peptide, and protein–protein interactions, enzymatic reactions, the interactions of ligands with membrane fractions or live cells, the detection of secretory products, and intracellular events such as reporter gene activity or translocation events. Some of the read-out modes described in **Table 23** can even be applied in parallel (multiplexing).

2.12.10.2.2.7 Multiple read-out parameters
FCS$^+$plus provides an intrinsically information-rich output for each measurement: each read-out mode allows the detection of several read-out parameters, such as the total concentration of fluorescent species as well as absolute concentrations of bound and free ligand. These additional parameters provide valuable intrawell controls and help to eliminate false-positives results.

2.12.10.2.2.8 Throughput
FCS$^+$plus measurements are fast (typical read-out times are 1–2 s per well), allowing for high throughput. However, high-performance and HTS in a miniaturized format can only be carried out by combining the FCS$^+$plus technology in an automated HTS system with microfluidics and robotics technology working with the same level of precision. Using the EVOscreen system, a fully automated ultrahigh-performance screening platform, a throughput of up to 100 000 compounds per day can be achieved for most FCS$^+$plus assays.

2.12.10.2.3 Case studies 1: using FCS$^+$plus multiparameter read-outs
We have chosen a number of different biological systems as case studies that demonstrate the unique potential of the FCS$^+$plus technology in meeting the demands of modern drug discovery described in the introduction.

2.12.10.2.3.1 Effect of assay miniaturization
Since FCS$^+$plus uses highly focused confocal optics, the volume illuminated by the laser beam is as small as a bacterial cell. The signal-to-background ratio is therefore effectively independent of the sample volume. This is demonstrated in **Figure 40**, where the DNA-binding properties of topoisomerase were studied. Since the molecular weight of fluorescently labeled oligonucleotides increases by approximately a factor of 17 upon binding to topoisomerase, this event can be monitored using FCS. As shown in **Figure 40**, the performance of the assay in a miniaturized 1 µL format is identical to that in a 'large'-scale format of 20 µL with respect to both statistics and the observed protein–DNA affinity.

This demonstrates that in comparison with other methods currently used in industrial HTS, such as ELISA, fluorescence polarization, scintillation proximity assay (SPA), and HTRF, assay miniaturization down to a 1 µL volume is achieved without compromising assay performance.

Besides the need for a sensitive detection technology, screening in a miniaturized format requires significant expertise with liquid handling of nanoliter volumes. In addition, evaporation and adsorption effects must be overcome. Typically, target characterization and assay development are initially carried out in volumes of 20 µL. Subsequently, the assay volume is reduced to 1 µL for HTS adaptation and screening.

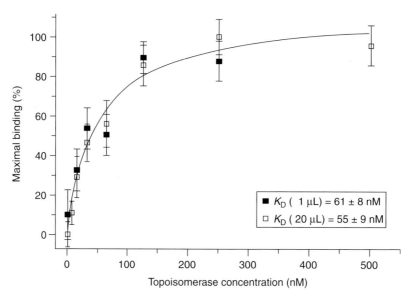

Figure 40 FCS$^+$plus allows assay miniaturization to 1 μL formats. The interaction of topoisomerase with a fluorescently labeled oligonucleotide was monitored using FCS in 1 μL and 20 μL assay volumes. Maximal binding was normalized to 100%.

2.12.10.2.3.2 Multiparameter read-out

The success of HTS strategies in drug discovery depends on the reliable determination of the compound activity, thus distinguishing between promising potential drug candidates and useless false positives (where inactive compounds score active in the assay). Elimination of false positives is a drain on time and resources because it requires retesting. While the use of fluorescence is usually associated solely with the measurement of fluorescence intensity, FCS$^+$plus allows the determination of several fluorescence parameters in a single measurement. Therefore, FCS$^+$plus delivers more information on a compound, allowing for efficient elimination of false positives in the primary screen.

To demonstrate FCS$^+$plus multiparameter read-outs, a model system based on biotin–streptavidin was used. Binding of fluorescently labeled biotin (molecular weight 1 kDa) to streptavidin (molecular weight 60 kDa) can easily be monitored by FCS. This allows one to determine the diffusion rates of free and bound biotin, and the distribution of fluorescent biotin between these two species. A test run was carried out in a miniaturized format in a HTS mode (1 μL sample volume, 2 s read time per well) using unlabeled biotin as a competitor. A sample of 40 wells from a single row of a high-density plate is shown in **Figure 41**. Three fluorescence parameters are obtained from a single measurement within each well: (1) the fluorescence count rate (average fluorescence intensity in the sample), (2) the particle number (a measure of the total number of fluorescent biotin molecules present in the confocal volume), and (3) the ratio of bound to total biotin (ratio of streptavidin-bound biotin relative to total biotin).

Other detection systems based on radioactivity or fluorescence typically deliver a single read-out parameter per well, usually the amount of complex formed. In this case, well 6 would score as a hit. The additional read-out parameters obtained by FCS$^+$plus show that both the count rate and the concentration of biotin (particle number) in this well are increased. This can be attributed to the failure of a dispenser to deliver the appropriate amount of streptavidin to this well. The resulting lower volume yields an increase in the count rate and the particle number (the final concentration of biotin is higher due to the lower volume), whereas the lower amount of streptavidin results in a decrease in the amount of complex formed. Thus, with the additional information obtained from the FCS$^+$plus read-out parameters, well 6 is discarded as a false positive.

This example shows how the simultaneous analysis of multiple fluorescence parameters helps to significantly improve the precision of primary screening by the efficient elimination of false-positive results. Using the EVOscreen platform, the screening database automatically eliminates false positives on-line during the primary screen if control parameters within a well, such as the total count rate or the particle number, deviate from a previously defined range.

2.12.10.2.4 Case studies 2: flexibility of multiple read-out modes

Ideally, a detection technology must allow maximal flexibility in assay design since assay types developed and screened in pharmaceutical companies vary enormously. Since FCS$^+$plus encompasses a variety of different read-out modes, for

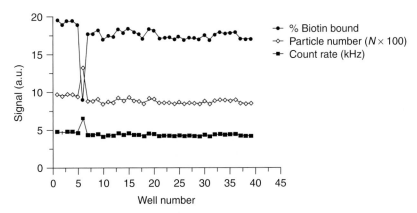

Figure 41 FCS$^+$plus offers multiparameter read-outs. Binding of fluorescently labeled biotin to streptavidin was studied with FCS on EVOscreen. Shown is a segment from a nanocarrier (total assay volume of 1 μL). The data was fitted using a two-component fitting procedure, assuming the presence of a fast-diffusing component (free fluorescent biotin) and a slowly diffusing component (fluorescent biotin–streptavidin complex). The fit procedure yields the percentage of fluorescent biotin present in the two fluorescent components ('% biotin bound' and '% biotin free'), the total fluorescence intensity (count rate), and the total concentration of fluorescent biotin (particle number).

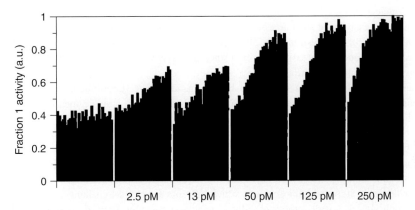

Figure 42 Enzyme kinetics in 1 μL assay volumes. Src kinase activity was determined on a nanocarrier (1 μL assay volume) using FCS. The production of phosphorylated peptide was monitored for 30 min using increasing enzyme concentrations.

each specific assay system the most suitable read-out mode can be selected. In the following sections, we have chosen several important target classes/assay systems in order to demonstrate the degree of flexibility and performance offered by FCS$^+$plus.

2.12.10.2.4.1 Enzyme assays

Kinases are important therapeutic targets in drug discovery programs. We have selected p60^{c-src} as an example of tyrosine kinases that are involved in controlling important cellular functions such as mitogenesis. Kinase activity is usually monitored by quantifying the amount of phosphorylated peptide formed using an anti-phosphotyrosine antibody. Since the molecular weight of a phosphopeptide increases significantly upon antibody binding, enzymatic activity can easily be determined with FCS using a fluorescently labeled peptide as the substrate.

Other detection methods, such as fluorescence polarization and HTRF, are also applicable. In comparison with these methods, FCS offers significant advantages. Using FCS, assays can be designed not only to determine the endpoint activity of a given kinase but also in a kinetic mode where enzyme activity is monitored over time.

Thus, during the assay development phase, new kinases are characterized by determining kinetic constants such as k_{on} and k_{off}. During assay adaptation to a miniaturized HTS format, suitable assay conditions are investigated, as shown in **Figure 42**. In this experiment, different amounts of enzyme were used, and the amount of product formed was monitored over time in a 1 μL format. After successful completion of this stage, the assay was run in a HTS mode on the EVOscreen platform (1 μL per assay, endpoint mode) in order to identify compounds that inhibit kinase activity. Thus, FCS$^+$plus can be used for target characterization, for assay development, for miniaturized, fully automated HTS, and for hit profiling, yielding maximum information using a minimum of assay components and compounds. Using the EVOscreen system, a throughput of up to 100 000 compounds is achieved within 24 h for this type of assay.

2.12.10.2.4.2 G protein-coupled receptor assays

GPCRs belong to another important target class because they are involved in a variety of diseases such as asthma, AIDS, and neurodegenerative and cardiovascular diseases.[542] However, technologies currently used have significant disadvantages as they typically involve the use of radioactively labeled ligands. This means exposure to hazardous radioactivity, limited shelf lives of labeled components, and the production of radioactive waste on a large scale. Also, radioactive assays are still carried out in the standard 96-well format using volumes of 100–150 μL, and are not easily amenable to miniaturization beyond the 384-well format (30–50 μL per well).

These assays involve the use of membrane fractions prepared from receptor-expressing cells; therefore, standard FCS cannot be applied because of the very slow diffusion rate of membrane vesicles. To overcome this restriction, a new read-out method was developed, based on brightness analysis. This method complements FCS since it measures the brightness of fluorescent species and not diffusion times; as a result, it is entirely mass-independent. It allows one to study the interaction of proteins with similar molecular weight but is also applicable to monitor interactions of soluble ligands with receptor-bearing vesicles. The principle of brightness analysis in this case is multiple fluorescently labeled ligands binding to a receptor-bearing membrane vesicle. Bound ligand can be distinguished from unbound because a membrane vesicle with many fluorescent ligand molecules bound is significantly brighter than a single fluorescent ligand molecule.

Figure 43 shows an example of such an assay using the chemokine receptor CXCR2 and its ligand, fluorescently labeled interleukin 8 (IL-8). The results from 500 wells during a screening run using the EVOscreen platform are displayed. The competitor added was easily identified since the assay yields a nice screening window with a signal-to-background ratio of approximately eightfold.

Using radioactive methods, only the amount of bound ligand can be determined. In contrast, with brightness analysis, the concentration of both bound and free ligand is obtained from a single measurement, thus providing a valuable internal control.

2.12.10.2.4.3 Cellular assays

A further challenge to modern HTS is the use of living cells. Cellular assays provide a more physiological approach, and put functional data into a biological context. However, the requirements for miniaturized HTS on cellular assays are high: assays must be single step and homogeneous, which presents a technical challenge.

FCS$^+$ plus is applicable to a wide variety of cellular assay systems. One example is shown in **Figure 44**, where binding of a fluorescent ligand to cell surface receptors was monitored with live cells. In this experiment, the activity of

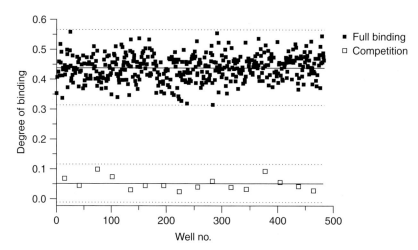

Figure 43 Test screen of the chemokine receptor CXCR2 in a miniaturized format. A sample of 500 wells from a nanocarrier is shown from a fully automated test screen using EVOscreen. 100 nL of 10% dimethyl sulfoxide with (□) or without (■) competitor were added from 96-well plates to nanocarriers. Ligand (100 nL) and membranes (800 nL) were added by dispensers, and the assay was analyzed by brightness analysis. The solid line represents the mean of each sample, and the dotted line represents the mean ± three standard deviations.

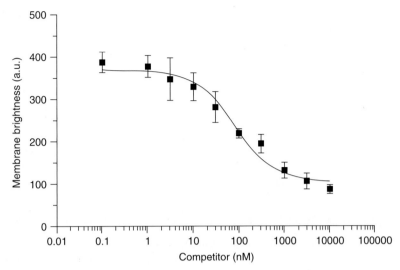

Figure 44 Studying ligand–receptor interactions using live cells. Live cells expressing the chemokine receptor CXCR2 were incubated with fluorescently labeled IL-8 in the presence of increasing concentrations of a low molecular weight compound analyzed by brightness analysis. The membrane brightness was determined by FCS$^+$plus.

a chemical compound was analyzed for competing with IL-8 binding to CXCR2 receptors expressed on the cell surface. In this case, a variation of the brightness analysis read-out described above was used to study the binding event. Here, the cells were placed at the bottom of a well, and fluorescently labeled ligand added. Cell-bound fluorescence was quantified by scanning with the confocal volume through the cells.

In addition to cell surface binding, secretory products from cells in the surrounding medium as well as gene induction using reporter systems can be measured by FCS$^+$plus. Due to the small confocal detection volume, subcellular resolution is obtained with FCS$^+$plus by using confocal imaging techniques. This makes assay systems accessible to HTS where the distribution of fluorescence across a cell changes while the total fluorescence remains unaltered (e.g., during translocation events).

The combination of sensitivity and miniaturization capabilities of FCS$^+$plus offers a further major advantage: primary cells can be used in HTS because restrictions resulting from the availability of primary tissue are overcome by the very small numbers of cells needed for each assay (1000–2000 cells per well).

2.12.10.3 Overview

FCS and related confocal fluorescence-based detection technologies provide a valuable platform for assay development, HTS, and subsequent optimization of assays. The technology is applicable to a wide variety of target classes and assay techniques, allowing great flexibility in assay design. Screening applications as varied as the kinetic study of compounds acting on tyrosine kinases, the binding characteristics of compounds acting on GPCR systems, and the study of binding to and secretion from living cells have been accomplished with this technology.

FCS$^+$plus technologies overcome the use of hazardous radioactive probes, are amenable to miniaturization, can be used in a kinetic or endpoint mode, and deliver internal control values for each sample measured. Because of the inherently small detection volume, the techniques are independent of sample size, suffering no loss of sensitivity in low volume assays, and therefore are ideally suited for assay miniaturization. The multiple parameters read in a single measurement improve the precision of each measurement taken. The potential to impact all areas of drug discovery is yet to be fully exploited, but FCS$^+$plus is a technology platform that we feel will contribute significantly to improvement of the drug discovery process.

2.12.11 Homogeneous Time-Resolved Fluorescence

2.12.11.1 Introduction

The evolution of high-throughput screening (HTS) has created the need for more sensitive, rapid, and easily automated assays. Combinatorial chemistry, expanding compound libraries, increasing numbers of drug targets, and increasing pressure for the discovery of new chemical entities in the pharmaceutical industry have fueled the expansion of HTS methods.[543] The advent of homogeneous, 'mix-and-measure' assays led to a significant leap in the throughput of typical HTS. Homogeneous assays eliminate the need to separate bound from free label, which also reduces waste and error. Furthermore, homogeneous assays are easier to automate since they require only addition steps. Scintillation proximity assay (SPA) (Amersham International) was the first mix-and-measure assay developed. With the TopCount scintillation and luminescence microplate counter (Packard Instrument Company) and the MicroBeta microplate scintillation counter (EG&G Wallac), SPAs made HTS a reality.[544] However, SPAs still require a radiolabeled binding partner, long counting times for accurate measurements, and correction for the quench and color interference effects of biological compounds. In addition, SPAs involve a scintillating solid phase bead, which settles out of solution.[545] The need for a nonradioisotopic, truly homogeneous alternative led to the development of HTRF, an in-solution, homogeneous, nonradioisotopic method. This section describes the theory and application of HTRF – a sensitive, robust, homogeneous, fluorescence method for HTS.

2.12.11.2 Unique Properties of Homogeneous Time-Resolved Fluorescence Chemistry

2.12.11.2.1 Specific requirements of homogeneous assays

Based on the specific requirements of homogeneous assays, a new class of long-lived fluorescent tracers was utilized to create this novel assay method. The following sections describe the unique challenges of creating a fluorescence-based, homogeneous method in general, and the unique solution provided by HTRF chemistry.

2.12.11.2.1.1 Unique tracers

Assaying a specific effect on biological targets without the luxury of a separation step requires unique tracers. The natural fluorescence of proteins and other compounds in biological samples and media creates limitations for the use of conventional fluorescent labels. Upon laser excitation, conventional fluorophores shift to a longer wavelength and are measured immediately, based on the change in wavelength.[546] Most background signals are prompt in nature, and also dissipate within 50 µs after excitation, making them difficult to separate from specific signals. HTRF utilizes the rare-earth lanthanide ion Eu^{3+}, which exhibits a signal as long lived as several hundred microseconds, permitting time-resolved measurement and elimination of prompt background fluorescence.[547]

2.12.11.2.1.2 Signal modulation

Another requirement of homogeneous assays is a method of modulating the signal between the affected and unaffected assay targets. Making the distinction between the bound and unbound target molecules by a method other than separation is necessary. HTRF technology utilizes a carefully selected pair of fluorescent molecules to generate signals specific to the bound and unbound states. The 'donor' molecule is europium cryptate, referred to as (Eu)K. When excited, (Eu)K transfers energy to an 'acceptor' molecule, a modified (stabilized) allophycocyanin called XL665. A long-lived signal at a specific wavelength is generated only when a binding event between donor and acceptor molecules occurs. For example, when the target molecules are unbound as a result of the presence of an inhibitor, the specific signal is not generated.[548] Such signal modulation is required for a truly homogeneous assay.

2.12.11.2.1.3 Resistance to biological media

Since no separation step is involved in a homogeneous assay, biological assay components and media remain in the presence of the target molecules of interest during measurement. The signal of a homogeneous assay should be resistant to unrelated, nonspecific effects of biological assay components, such as media and natural products, if involved. HTRF technology is measured by a patented 'ratiometric' method, described later in detail, which eliminates or corrects for nonspecific interference from biological assay components.[549]

2.12.11.2.2 Lanthanide cryptates: a new type of fluorescent label

2.12.11.2.2.1 Chelates as fluorescent labels

The rare-earth lanthanides europium, terbium, dysprosium, and samarium are naturally occurring fluorophores, and their chelates have a number of applications in immunological and biological assays.[550] In lanthanide chelates, the lanthanide is the fluorescent label. It is held by chelation that permits conjugation with biological components. Since lanthanide chelates are not fluorescent when conjugated to biological components, a dissociative enhancement step is required to free the lanthanide ion from the conjugated chelate, to make a new and different ion complex that can generate measurable fluorescence, as in the DELFIA chemistry (EG&G Wallac). The exception is the LANCE

(EG&G Wallac) chemistry, which has demonstrated a measurable signal in homogeneous form, but with a substantially diminished signal-to-noise ratio, relative to the DELFIA heterogeneous method.[550] Lanthanide chelates are subject to potential dissociation of the ion, which can undermine the integrity of the label, increasing the background and nonspecific binding contributions. In addition, lanthanide chelates are subject to inactivation by EDTA, and require separation and washing steps for best results.[551] These limitations of chelates are overcome by the use of the novel family of lanthanide cryptates as fluorescent labels, for which Professor Jean Marie Lehn was awarded the 1987 Nobel Prize in Chemistry (shared with J. Pederson and D. Cram).

2.12.11.2.2.2 Cryptates as fluorescent labels

Lanthanide cryptates are formed by the inclusion of a lanthanide ion in the cavity of a macropolycyclic ligand containing 2,2′-bipyridine groups as light absorbers (**Figure 45**). The cryptate can undergo intramolecular energy transfer when the cavitated species is Eu^{3+}. The cage-like structure of the cryptate protects the central ion, making it stable in biological media. Well-known heterobifunctional reagents can be used to conjugate the diamine derivative of the europium trisbipyridine cryptate, (Eu)K, to biological assay components, such as proteins, peptides, receptors, nucleic acids, and antibodies, without loss of reactivity. Even upon conjugation to such components, the photophysical properties of (Eu)K are conserved, making it particularly useful for homogeneous assays as a fluorescent label. No dissociation enhancement step is required, and cryptates are stable and kinetically very inert, as the activation energy required to reach the transition state of dissociation is significantly higher for cryptates than for chelates.[552] Excitation of (Eu)K at 337 nm yields an emission spectrum with a strong peak at 620 nm, as shown in **Figure 45**.[553]

2.12.11.2.2.3 Use of XL665 as an acceptor

The best choice for an acceptor molecule for the (Eu)K donor will complement the cryptate fluorescence energy spectrum, accept the transfer of energy efficiently, be stable in common assay media, and lend itself to appropriate chemistry for conjugation with biological components. For HTRF assays, the best choice for the acceptor molecule is one of the main constituents of the phycobilisomes of red algae. These protein–pigment complexes absorb light and channel it to the photosystem of the cell. In the phycobilisome, the last component that transfers light is allophycocyanin (APC), a phycobiliprotein of 105 kDa with an absorption band ranging from 600 to 660 nm and an emission maximum at 660 nm. APC, when modified by cross-linking, emits at 665 nm, and is called XL665 (**Figure 46**).

XL665 has a high molar absorptivity at the cryptate emission wavelength, which enables very efficient energy transfer of about 75% for a donor–acceptor distance of 7.5 nm. Spectral emission is high where the cryptate signal is insignificant, making it particularly complementary (**Figure 47**). The quantum yield of energy is high, at about 70%. XL665 is stable and not quenched by the presence of biological media.[553]

2.12.11.2.2.4 Signal amplification of the cryptate fluorescence

Forster theory (1948) defines the nonradiative energy transfer that occurs between fluorescence resonance energy transfer (FRET) pairs, such as the (Eu)K donor molecules and the XL665 acceptor molecules of HTRF. The efficiency of transfer is a function of the distance $(1/d^6)$ between the donor and acceptor pairs.[554] **Figure 48** illustrates the energy

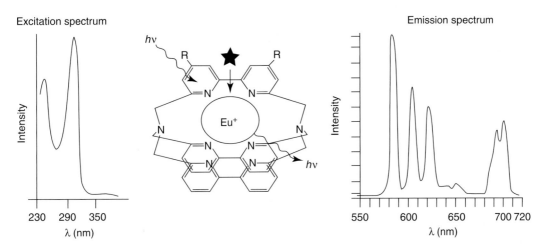

Figure 45 Europium trisbipyridine cryptate has an excitation peak at 337 nm and an emission peak at 620 nm.

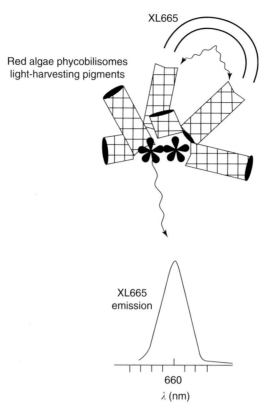

Figure 46 XL665, cross-linked allophycocyanin has a strong emission peak at 665 nm.

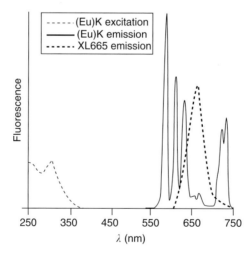

Figure 47 The complementary spectra of (Eu)K and XL665.

transfer between (Eu)K and XL665 in HTRF. (Eu)K excited separately has a spectrum that allows only about 60% of the signal to be measured. Paired with XL665, the signal is amplified by the transfer of energy to the XL665 spectrum, which provides a spectrum that allows one to measure essentially all emitted energy, at the shifted wavelength. Therefore, by being paired with XL665, the (Eu)K signal is effectively amplified.[555]

2.12.11.2.3 Homogeneous time-resolved fluorescence signal and measurement

The specific FRET pair of (Eu)K and XL665 exhibits both temporal and spectral emission characteristics that provide opportunities for measurement in unique ways. The Discovery HTRF microplate analyzer (Packard Instrument

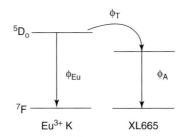

Figure 48 Energy transfer and amplification between (Eu)K and XL665. When the fluorescence is measured at the emission wavelength of the XL665, an amplification is obtained when $(\varphi_A)(\varphi_T) > \varphi_{Eu}$, where φ_A, and φ_{Eu} are the quantum yields of XL665 and the europium ion, respectively, and φ_T is the transfer efficiency.

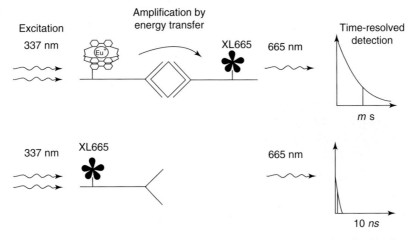

Figure 49 Principle of time-resolved detection of HTRF. The time delay of 50 μs allows short-lived background fluorescence and the signal from free XL665 to dissipate.

Company) was developed and optimized specifically for the detection of HTRF assays for HTS. Since no separation occurs between bound and free assay components, the signals from each individual fluorophore may be distinguished from the signal of the bound pair based on both time and wavelength. The Discovery analyzer takes into consideration the following characteristics of HTRF chemistry to make the most efficient and useful measurement.

2.12.11.2.3.1 Time-resolved measurement

Background and nonspecific fluorescence from microplates, media, and biological components has a short lifetime of emission. Unbound or isolated XL665 also emits a short-lived signal when excited at 337 nm. Unbound or isolated (Eu)K emits a long-lived signal at 620 nm, easily separated by optical discrimination. Together, the (Eu)K–XL665 pair emits a long-lived signal at 665 nm (**Figure 49**).

The Discovery analyzer excites each sample with a nitrogen laser excitation pulse at 337 nm. Measurements are taken after a 50 μs delay, allowing any short-lived fluorescence to dissipate before the emission light is collected. Time-resolved fluorescence measurement has been incorporated into a number of fluorometers, such as the Victor (EG&G Wallac) and the Analyst (LJL Biosystems), although the excitation sources vary.

2.12.11.2.3.2 Dual-wavelength measurement

Although time-resolved measurement is a feature now copied in and common to various vendors' instruments, simultaneous dual-wavelength detection is unique to the sophisticated Discovery instrument. **Figure 50** illustrates the optical design of the Discovery analyzer used to make measurements of HTRF assays at 620 nm and 665 nm, simultaneously. Together, time-resolved measurement and simultaneous dual-wavelength detection distinguish the bound from free fluorophores. In addition, the extent to which binding has occurred can be measured independently of interference effects from media, biological components, and physical assay complications such as turbidity. **Figure 51** shows the detection of an HTRF assay by the Discovery analyzer. During the course of 1 s per well, excitation by the

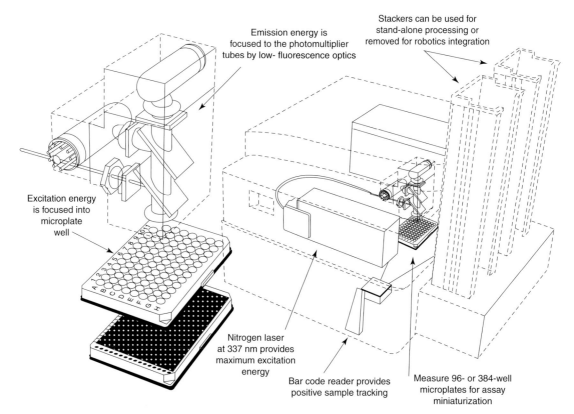

Figure 50 Optical configuration of the Discovery HTRF Microplate Analyzer. A nitrogen laser provides the excitation energy of 337 nm. Simultaneous dual-wavelength detection provides measurements at 620 nm and 665 nm after a time delay.

nitrogen laser at 337 nm elicits a long-lived signal at 620 nm from free (Eu)K and a long-lived signal at 665 nm from the (Eu)K bound to XL665 by biomolecular interaction.[548] The ratio of the two simultaneous measurements is a specific measure of the extent to which the labels are bound in the assay. Inhibitors of binding or enzyme effects on the binding partners can be measured directly without separation of individual assay components or biological media.

2.12.11.2.3.3 Data reduction

Assay data can be presented in terms of the ratio of the 665/620 nm measurements, R, to correct for any interference from media or absorption of excitation or emission wavelengths by the media. ΔR is the change in ratio that occurs with respect to the blank or negative sample. It is used to express the specific signal in the assay. ΔF is an instrument-independent measurement that shows the signal-to-noise ratio of the assay.[556] Both measurements utilize the patented ratio method of detection, which eliminates interference from test compounds, or turbidity of test samples.

$$\Delta R = (\text{ratio of the sample} - \text{ratio of the negative sample}) \times 10000$$

$$\Delta F = (\Delta R/\text{ratio of the negative sample}) \times 100$$

The Discovery analyzer can provide raw 665 nm counts, raw 620 nm counts, or ratios. This patented technique is unique to HTRF measurement, and is proprietary to this detection technology.

The following sections describe applications of HTRF chemistry for HTS.

2.12.11.3 Applications of Homogeneous Time-Resolved Fluorescence for High-Throughput Screening

HTRF assays can be constructed using several formats. The direct-assay format requires that the binding partners of interest be directly conjugated to the (Eu)K and XL665 fluorophores. The advantages of this format are that the assay

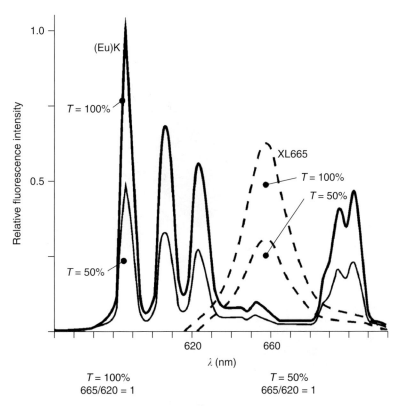

Figure 51 Ratio measurement of 620 nm and 665 nm wavelengths. The long-lived signals, 620 nm emission from free (Eu)K and the 665 nm emission from the bound (Eu)K and XL665 pair, provide a method for correcting for color or turbidity in the reaction.

optimization depends only on the particular interaction of the binding partners, with no consideration of antibody affinities. A disadvantage is that a substantial quantity of the binding partners is required because some material is lost in purification of direct conjugates. In addition, the integrity of the binding interaction may be affected by direct conjugation. The more commonly used indirect method utilizes a high-affinity pair, such as streptavidin/biotin, or antibodies specific to the binding partners to construct the assay. The semidirect method involves one directly labeled binding partner and another indirectly labeled partner.[557]

HTRF assays are typically run in a 200 μL volume in a 96-well plate or 70 μL volume in a 384-well plate. The Discovery analyzer measures plates in either format. HTRF assay technology is not appropriate for intracellular assays, but is particularly suitable for kinetic studies due to the fast read times and the ability to measure each sample repeatedly over time. Most widely performed assays include immunoassays, enzyme assays, receptor-binding assays with purified receptors or membrane fragments, protein–protein interactions, and nucleic acid hybridizations. The following are examples of HTRF assays and results as measured on the Discovery instrument.

2.12.11.3.1 Immunoassays

Immunoassays can be created using HTRF, as it was first developed for clinical applications and sold under the name of TRACE chemistry by CIS Bio International. These are conducted as sandwich assays in which monoclonal antibodies raised against the target are conjugated directly with (Eu)K and XL665. The following example compares the HTRF assay with a commonly used enzyme-linked immunosorbent assay (ELISA) method.

The ELISA version of this cytokine assay requires the first antibody to be bound to the well of a microplate. The second antibody to the same cytokine is linked to the reporter enzyme.[557] This assay requires three separate wash steps, four reagent additions, and four incubations, and can be used to screen a maximum of 10 000 assay points per day. The HTRF assay, as shown in **Figure 52**, requires no separation steps, one reagent addition, and one incubation at room temperature, and can be used to screen up to 10 000 assay points in 2 h. Samples contained red blood cells up to 5% by volume. **Figure 53** shows the comparison of the optical density of the ELISA assay at 405 nm with the ratio

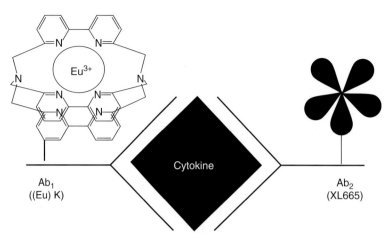

Figure 52 HTRF immunoassay configuration. Polyclonal antibodies are conjugated directly with (Eu)K and XL665. The presence of cytokine is measured by the binding of both antibodies.

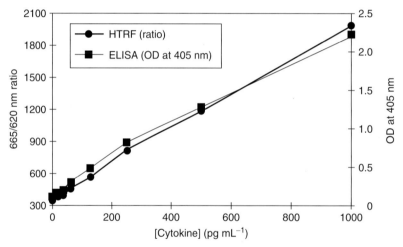

Figure 53 Comparison of ELISA and HTRF methods for the detection of cytokines. Both are capable of picogram levels of detection, but the HTRF assay has fewer steps.[558]

measurement of the HTRF assay. Both methods are capable of detecting picogram quantities of cytokine in complex samples; however, HTRF is considerably easier to perform for HTS purposes. An HTRF prolactin immunoassay has also been described previously.[553]

2.12.11.3.2 Enzyme assays

Enzyme-mediated reactions that cleave, synthesize, or modify compounds represent a significant group of targets for HTS. The HTRF assays can be constructed such that enzyme activity can be measured by the absence or presence of high signal. Protein tyrosine kinases play a critical role in the cellular signal transduction pathways, with significant implications for mechanisms of allograft rejection, allergic response, and autoimmune diseases. Hundreds of tyrosine kinases have been identified, and many are still to be discovered through genomic research.[559,560] The following is an example of an HTRF tyrosine kinase assay.

Biotinylated polypeptide substrates are incubated with kinase enzymes to catalyze the phosphorylation reaction. Eu(K) conjugated with PY20 antibody recognizes those peptides that are successfully phosphorylated. Streptavidin conjugated to XL665 binds to the biotinylated substrate, whether or not it is successfully phosphorylated. Those peptides that are phosphorylated effectively bring together the HTRF pair such that a high signal is achieved upon excitation at 337 nm with the nitrogen laser of the Discovery analyzer (**Figure 54**).

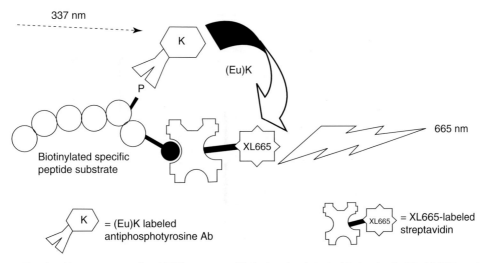

Figure 54 Tyrosine kinase assay using HTRF reagents. Biotinylated substrate binds streptavidin–XL665, and antiphosphotyrosine–(Eu)K recognizes those peptides that are phosphorylated.

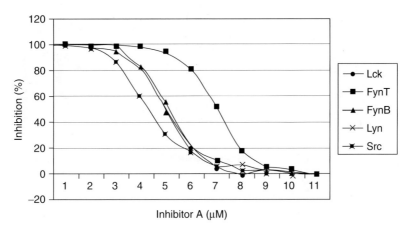

Figure 55 Titration of kinase inhibitor for the Src-1 family of enzymes.

A number of known inhibitors were titrated against five Src-family tyrosine kinases tested using this HTRF format. The IC_{50} values obtained from the HTRF assay were very similar to those achieved using a comparable radiometric SPA (Amersham International).[561] **Figure 55** shows the inhibition curves as measured by HTRF.

Sensitivity is a critical issue in HTS when the cost of reagents must be considered and the availability of enzyme and sample materials is limited. In order to determine the sensitivity of the HTRF chemistry for kinase activity, varying concentrations of Lck enzyme were added to reaction mixtures containing peptide and ATP. As time points up to 80 min, reactions were stopped and measured in the Discovery instrument. As little as 20 pM Lck enzyme gave significant results with a signal-to-noise background of 10:1 after 40 min (**Figure 56**). This result demonstrates that the HTRF assay method provided about two orders of magnitude more sensitivity than the SPA method, making it the first nonradioisotopic alternative capable of yielding results that could be achieved previously only with the use of radioisotopes. The high sensitivity of this HTRF method saves both enzyme and library samples, and enables screening of weaker inhibitors not easily measured by less sensitive methods such as SPA.

Additional examples of the application of HTRF specifically to kinase assays have been published.[562,563] Other HTRF enzyme assays that have been published include a viral protease assay[564] and a ubiquitination assay.[565]

2.12.11.3.3 Receptor-binding assays

Biomolecular interactions between nuclear receptors and their coactivators are associated with transcriptional regulation, and are therefore seen also as an important target for HTS in drug discovery. Ligand-binding promotes the

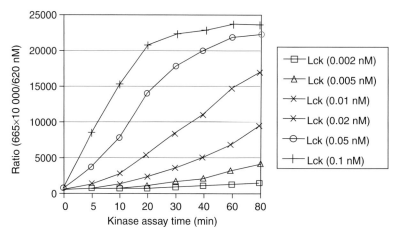

Figure 56 Time course of the Lck kinase assay. The concentration of the enzyme varies from 0.002 nM to 0.1 nM.

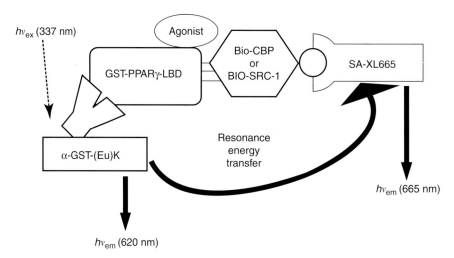

Figure 57 HTRF strategy for the detection and quantification of the interaction between a receptor and a coactivator.

association of nuclear receptors with nuclear proteins, such as CREB-binding protein (CBP) and steroid receptor coactivator 1 (SRC-1), which are believed to function as coactivators of transcriptional activity.[566] A description of how HTRF methods were used to characterize and screen for potential agonists to these ligand-dependent interactions follows.

Nuclear receptors contain a well-conserved, 200–300-mer amino acid ligand-binding domain (LBD) in which a number of functionalities are encoded, including ligand and coactivator binding and transactivation. The LBDs of peroxisome proliferator-activated receptor γ (PPARγ) were used as a model in this study. PPARγ–LBD was expressed and purified as a glutathione S-transferase (GST) fusion protein from E. coli strain DH5. Both coactivators CBP_{1-453} (amino acids 1–453) and $SRC-1_{568-780}$ (amino acids 568–780) were biotinylated.[566]

In this assay, (Eu)K is covalently bound to anti-GST antibody to create (Eu)K-∞-GST for labeling the nuclear receptor. The secondary fluorophore, XL665, is covalently bound to streptavidin (streptavidin–XL665) to label the biotinylated coactivators. Both streptavidin–XL665 and (Eu)K-∞-GST are available as 'generic' or 'off-the-shelf' reagents from Packard Instrument Company. **Figure 57** illustrates the indirect-assay format used to create the HTRF assay of nuclear receptors and coactivators. Agonist-induced interaction between the nuclear receptors and their coactivators brings the two fluorophores in proximity, and results in a high signal.

Known agonists have been shown previously to activate ligands for PPARγ in mammalian cell-based assays.[566] Using the HTRF assay approach described above, increasing concentrations of thiazolidinediones (TZDs) were added to reaction mixtures. **Figure 58** shows that TZDs induced a dose-dependent specific interaction between PPARγ–LBD

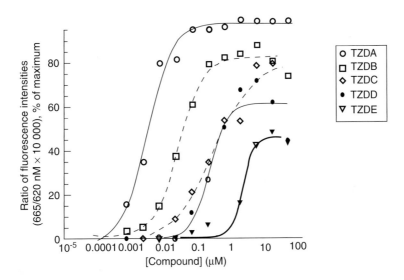

Figure 58 Effects of TZDs on PPARγ–CBP interactions using the HTRF approach.

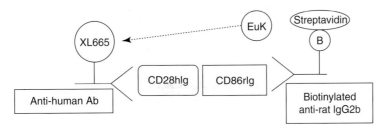

Figure 59 HTRF assay configuration for the detection of small-molecule inhibitors of the CD28/CD86 interaction.

and the nuclear receptor-binding domain of CBP with similar EC_{50} values to those obtained in transactivation studies. This assay indicates that HTRF assay methods can provide a powerful alternative to cell-based assays for screening large compound libraries for potential agonist drug candidates.[567]

2.12.11.3.4 Protein–protein interactions

A number of HTRF protein–protein interactions have been described previously. The JUN:FOS assay was done by directly labeling the FOS protein with (Eu)K and using biotinylated JUN to bind with the streptavidin–XL665.[568] Now, antispecies or antiepitope tag antibodies conjugated with (Eu)K are available and used more often than direct labeling and streptavidin–XL665.[556] The following example is an HTRF protein–protein assay using the indirect method.

Ligation of CD28 and CD86 leads to signals that are required for the production of interleukin-2, a process that is implicated in the regulation of T cell anergy and programmed cell death.[569–572] An HTRF assay was constructed to screen for small-molecule antagonists of this interaction, which could be considered possible drug targets (**Figure 59**). The acceptor, XL665, was covalently bound to antihuman antibody that recognizes the Fc region. CD28 was expressed as a fusion protein to the human immunoglobulin (Ig) domain. The other binding partner, CD86, was expressed as a fusion protein with a rat Ig domain, recognized by biotinylated sheep anti-rat antibody. Streptavidin–(Eu)K then binds to complete the HTRF reaction.[573]

Dose–response curves were created with blocking antibodies that prove that the binding of CD28/CD86 proteins causes the signal generation (**Figure 60**). Since the affinity of interaction between CD28 and CD80 was shown to be relatively low (200 nM to 4 µM) by different methods,[550,551] ELISA methods would result in dissociation of the bound complex during washing, and SPA methods would require high levels of radioactivity. The HTRF method uses readily available generic reagents, was shown to be resistant to color quench effects, and requires few steps, thereby minimizing the errors associated with pipetting. In addition, this modular approach makes development of other related assays for specificity testing easier than direct-labeling approach.[573]

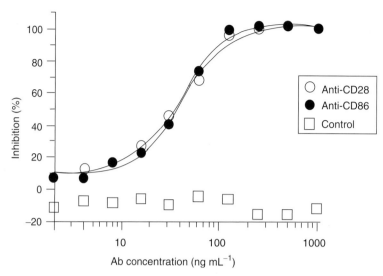

Figure 60 Dose–response curves illustrating the inhibition of binding by blocking antibodies.

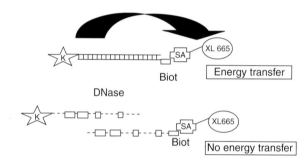

Figure 61 HTRF assay for the detection of nuclease activity.

Another published HTRF assay for measuring protein–protein interactions is given by Farrar *et al.* in their paper entitled "Stoichiometry of a ligand-gated ion channel determined by fluorescence energy transfer."[574] HTRF was used to elucidate the stoichiometry of subunits within an oligomeric cell surface receptor in a way that could be applied generally to other multisubunit cell surface proteins.

2.12.11.3.5 Nucleic acid hybridizations

DNA hybridization and related enzyme assays using nucleic acid-binding partners can be conducted in solution using HTRF. A simple measure of hybridization was conducted using one oligonucleotide directly labeled with (Eu)K, while the complementary strand was biotinylated and labeled with streptavidin–XL665. The extent to which the complementary strands are bound can be measured by the HTRF signal that results from the proximity of the two fluorophores.[553] The following assays were more recently developed to measure nucleic acid-related enzyme activity.

2.12.11.3.5.1 DNA nuclease assay

A 21 bp oligonucleotide was directly labeled with (Eu)K on the 3′ end, and biotinylated on the 5′ end. Streptavidin–XL665 was added to complete the FRET reaction (**Figure 61**). Increasing amounts of DNase enzyme were added, to show the specific signal decrease as activity increases. This assay may be used to screen for DNase inhibitors[575] (**Figure 62**).

2.12.11.3.5.2 Reverse transcriptase assay

Inhibitors of reverse transcriptase activity remain a major target for viral research in drug discovery. The following HTRF assay configuration enables one to screen for increases of inhibition of enzyme activity in high throughput. Oligonucleotide primers labeled directly with (Eu)K are added to RNA templates (**Figure 63**). Addition of reverse

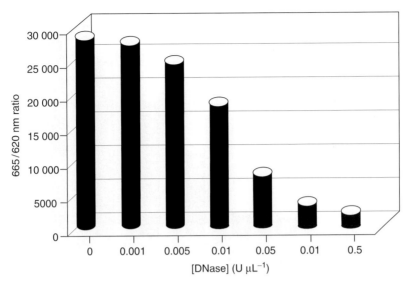

Figure 62 Increasing amounts of DNase enzyme decreases the HTRF signal.

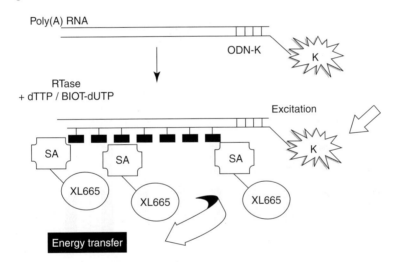

Figure 63 HTRF reverse transcriptase assay configuration.

transcriptase and biotinylated dUTP to nucleotide building blocks creates a biotinylated double-stranded molecule labeled with (Eu)K, to which streptavidin–XL665 is added to complete the reaction. Complete reverse transcriptase activity results in a high HTRF signal. Inhibition would decrease the signal to the extent that the enzyme is blocked. Increasing amounts of reverse transcriptase enzyme results in higher activity, as the HTRF fluorophores are brought into proximity while the double-stranded molecules form[572,576] (**Figure 64**).

2.12.11.4 Homogeneous Time-Resolved Fluorescence Assay Optimization

The following sections are included to provide an insight into how one might best develop and optimize an HTRF assay. HTRF assays are normally performed with only a few steps, mainly additions, so there are a smaller number of considerations when designing an HTRF assay than there may be with a more complicated method. The main considerations for optimizing an HTRF assay are buffer selection, whether label (Eu)K or XL665 should be used for each assay component, and what controls should be run.

2.12.11.4.1 Buffer selection and stability

Homogeneous assays by nature involve no washing step, so that all buffer and assay components are present during the ultimate measurement of activity or binding. As a result, the following work has been done to help determine how

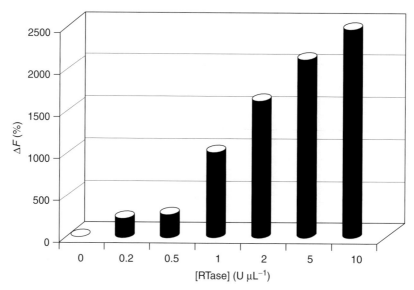

Figure 64 Increasing amounts of reverse transcriptase enzyme increase the HTRF signal.

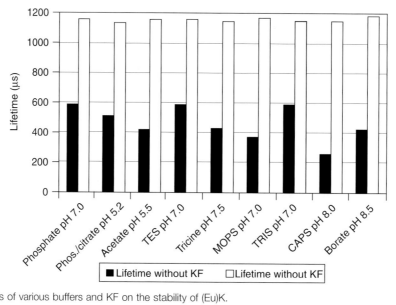

Figure 65 Effects of various buffers and KF on the stability of (Eu)K.

HTRF assays, specifically the (Eu)K component, are likely to perform in various buffers. (Eu)K was added to 200 μL samples of the most commonly used buffers, with and without 50 mM KF. KF provides a fluoride ion, which has been shown to form ion pairs with Eu^{3+}. The fluorescence lifetime was measured to determine if KF had an effect on the stability of the signal of the (Eu)K component as measured at its emission peak of 620 nm. **Figure 65** shows that, in all cases, the lifetime of the signal from the (Eu)K label was stabilized and, in effect, equalized by the presence of fluoride ions, illustrating the protective effect of the F^- ion.

2.12.11.4.2 Choice of label for each assay component

(Eu)K and XL665 are both available conjugated with streptavidin, biotin, antispecies, and antiepitope antibodies, as well as a number of other biological components. However, the following recommendations apply. Due to the difference in sizes of the fluorophores, it is better to use the (Eu)K label on the smaller assay component and XL665 on the larger assay component. In terms of the smaller component, if it can be biotinylated, it is likely acceptable to label it with (Eu)K. XL665 should be used on the component that is in excess in the assay, and (Eu)K should be used on that component that is most sensitive to change.

2.12.11.4.3 Suggested assay controls

For the purposes of assay development and analysis of screening results, the following controls should be run for each assay. A buffer blank should be created that includes buffer only at the volume of the assay. An (Eu)K blank should be included that contains buffer and (Eu)K at the concentration used in the assay. This control enables one to do an (Eu)K background subtract (or 'K blank') to subtract out the small peak that exists at the 665 nm wavelength. Finally, as always, a negative control should be created that includes total assay conditions with something to make it negative (e.g., completely inhibited or inactivated). Using ratio and ΔF calculations and the above controls helps to attain the most reproducible and precise data from the HTRF assay.

2.12.11.5 Overview

HTRF chemistry was developed to help overcome some of the obstacles to HTS in general. It is homogeneous, so that separation steps are eliminated. It is fluorescent in nature but has the time-resolved, simultaneous, dual-wavelength detection that enables one to achieve a high fluorescent signal with extremely low backgrounds. HTRF is best measured on the Discovery HTRF microplate analyzer, because it was designed specifically and optimized for the particular wavelengths of HTRF. However, a number of detection systems are now available for measuring TRF and HTRF applications. The HTRF reagent list is growing; thus, fewer assays require customized conjugation with (Eu)K and XL665. Many of the applications described utilize off-the-shelf reagents that are readily available and require no license fee. As a result, this tool is widely accepted in the literature as one of the few truly homogeneous methods available to meet the HTS demands of drug discovery today.

2.12.12 Screening Lead Compounds in the Postgenomic Era: An Integrated Approach to Knowledge Building from Living Cells

2.12.12.1 Introduction

The new face of drug discovery is focused on the living cell, with its myriad ionic, metabolic, macromolecular, and organellar networks as the ultimate target of drug activity. Wilson essentially set the stage for a cell-centric approach to drug discovery in his introduction to a textbook that was one of the earliest attempts to combine cytology, physiology, embryology, genetics, biochemistry, and biophysics into the newly coined discipline of cellular biology.[577] Among other insights, he stressed the importance of measurements made on living cells in the presence of "changed conditions in the physical or chemical environment." Since then, there have been countless reports on the characterization of isolated components of living cells. There have also been reports where components from single cells have been characterized. One early approach to single-cell biochemistry, which appeared around the same time that Watson and Crick described the structure of cellular DNA, involved the extraction of nucleic acids from single neurons, their hydrolysis, electrophoresis on copper fibers, and analysis with ultraviolet light.[578] Although knowledge of the components of the chemistry of life soared during the latter part of the twentieth century, resulting in the modern disciplines of genomics and proteomics, there remains the great challenge of discovering the functional integration and regulation of these living components in time and space within the cell. We envision that the integration of genomics and proteomics with a new knowledge base built from temporal and spatial data on the chemical and molecular interrelationships of cellular components will provide an extremely rich platform for drug discovery.

2.12.12.2 Genomics and Proteomics as the Foundation for Cell-Based Knowledge

Whereas other sections in this chapter specifically expand on the role of genomics and proteomics in modern drug discovery, we view the genome and proteome as important building blocks for the construction of new cell-based knowledge. The genomics era, begun two decades ago, has realized a one-dimensional description of human and other organismal genes, each a potential new drug target. Even those genes coding for molecules other than functional proteins are theoretically new drug targets.

The level of DNA organization in the genome is surprisingly simple. There are an estimated 100 000 genes in the human genome and they are encoded within about 3 billion base pairs of nucleotides. The precise sequence of genes within the context of these billions of units, which are themselves organized into larger structures such as chromosomes, has been the primary goal of the human genome project. The enormous impact of genomics on drug discovery can be attributed first to advances in automated sequencing instrumentation and reagents. A successive and arguably more important phase of genomics is the association of gene expression to normal and disease physiology. This phase has been empowered by technologies and platforms to measure message level changes in response to drug treatment and new informatics strategies to organize DNA sequence information into logical, searchable, and meaningful databases. Proteomic analysis has become the logical extension of genomics and its associated databases. The daunting task of defining the protein complement expressed by a genome within cells, tissues, and organisms can be rationalized with several arguments.[579] These include the much less than perfect correlation between mRNA and protein expression levels; post-translational modifications of proteins that affect intracellular localization, activity, or both; and proteome dynamics that reflect the physiological state of the cell, tissue, or organism. Nevertheless, the proteome of a single native organism has yet to be completely described, let alone a mapping of drug activity overlaid onto one. Although proteomic databases are continually growing, both the methodology to measure proteomic changes and the approaches to extract, analyze, and characterize proteomic data have yet to be attacked as systematically as the genomic platform development has been. It remains unclear if current sample preparation methods and two-dimensional protein electrophoresis, the highest-resolution approach available for proteome mapping, will be perfected to the point where an entire proteome can be mapped,[580] especially as sample sizes diminish toward the single-cell level. Several factors make proteome analysis inherently more complicated than genomic analysis. For example, many genes in the human genome code for multiple proteins depending on mRNA processing; proteins can be post-translationally modified; and proteins may have several distinct functions. Overlaid on the complexity of individual protein molecules are the temporal and spatial interactions between proteins and other molecules. At any one time, there is of the order of tens of thousands of different proteins arranged within a cell. Each protein may have up to hundreds of thousands of copies. Thus, out of the relative simplicity of the genome comes an astronomical number of molecular and biochemical reactions, nearly all of which are mediated by proteins, occurring in time and space within a living cell. As we move into the postgenomic era, the complementarity between genomics and proteomics will become apparent, and the connections between them will undoubtedly be exploited. However, neither genomics nor proteomics, nor their simple combination, will provide the data necessary to interconnect molecular events in living cells in time and space, especially the network of events that targeted drugs inevitably interrupt or modulate.

2.12.12.3 Creation of New Biological Knowledge

A new perspective on drug discovery is cell-centric rather than focused on isolated genes or proteins. Because the cell is the smallest unit of life that can live independently, it must be highly organized to perform life functions. This organization begins at the one-dimensional level encoded by its DNA. The intrinsic information of cellular structure and function is encoded by the simple language of the genome. The translation of genomic language into the proteins and other macromolecules that participate in every chemical reaction occurring within a living cell represents an even higher level of organization than the blueprint held within the DNA. However, cells are not composed of a random collection of these macromolecules, which includes the entire proteome. Additional orchestration of the highly organized interactions of ions, metabolites, and organelles makes cells living entities. The knowledge base of the living cell is built by connecting layers of these interactions into the pathways and networks that govern all aspects of cellular life. Cellomics has extended cell-based knowledge to include cellular responses to drugs, including pharmacokinetic effects as well as pharmacodynamic effects on multiple cellular targets, pathways, and networks under the heading of 'PharmacoCellomics.' All deep biological information will be captured, defined, organized, and searchable in the Cellomics Knowledgebase of cellular knowledge. Just as automated genome analysis and bioinformatics tools are pushing the genomics era to a conclusion, automated cell analysis systems and cellular knowledge will be key to the era of the cell. Coupling of cellular knowledge to the drug discovery process will be essential for faster and more effective drug discovery. Such a process, composed of automated whole-cell-based detection systems, fluorescent reagents, and informatics and bioinformatics, is referred to as high-content screening (HCS).

2.12.12.4 High-Content Screening of Target Activity in Living Cells

The wealth of information obtainable from genomic and proteomic databases has led to at least one bottleneck in the early drug discovery process – target validation. Here, potential targets are evaluated for suitability within new or established drug screens.[581] Once validated, targets are incorporated into primary drug screens that are typically high capacity and high throughput. Large compound libraries can be screened in a relatively short time, thus producing a plethora of 'hits,' that is, combinations of compounds and target activity that meet certain minimum, usually crude, criteria. Due to recent emergent technologies for primary screening, increasingly larger numbers of hits have created another bottleneck at candidate optimization, where hits are qualified to leads through structure–activity relationship investigations, cytotoxicity, and secondary screening assays.[581] Thus, both bottlenecks have arisen as a consequence of advances in technology, especially miniaturization and automation, at their respective preceding steps in the drug discovery process. To break these bottlenecks, Cellomics has developed HCS to automate steps downstream of bottlenecks and to obtain deeper biological information from cells. Cellomics's goal is to reduce the high failure rate of lead compounds by providing high biological content information at earlier stages in the drug discovery process. Such HCS information elaborates both temporal and spatial measurements of single or multiple target activities within cell populations at the level of single cells. Although the details about HCS reagents, automated imaging analysis systems, and the bioinformatics support have been described elsewhere,[581] some of the advantages of HCS are listed here (Table 25). It is important to stress that HCS is a platform for knowledge building; its advantages are derived from the unique combination of its component technologies.

HCS starts with live cells. Following compound treatment of live cells, two different HCS approaches can be taken. In one approach, cells are fixed and, in many cases, fluorescently labeled, prior to imaging analysis. The advantages of fixed-endpoint HCS include the flexibility to analyze samples without time constraints and the ability to optimize and augment analysis because fixed samples can be repeatedly processed and reanalyzed. Several single- and multiparameter assays have been designed using fixed-endpoint assays. These assays, designed and validated by Cellomics and several pharmaceutical companies, include transcription factor activation and translocation,[581,582] microtubule reorganization,[583] G protein-coupled receptor internalization,[584] and apoptosis.[581] A second approach to HCS is the live cell kinetic assay. In this mode, temporal as well as high-content spatial data on target activities are collected and analyzed simultaneously.

Designed specifically for HCS, the ArrayScan VTI system, developed by Cellomics, detects multiple fluorescence channels over the visible spectrum within living or fixed cells prepared with corresponding fluorophore-labeled targets. Importantly, the ArrayScan VTI system is distinguished from standard fluorescence microscopy systems by its fully automated acquisition, processing, and analysis of cell images with the capability of arraying cells in stacks of

Table 25 Advantages of HCS in the early drug discovery process

Cell-centric physiological context
- Target activities are measured within multiple living cell types
- Entry of drugs into specific cellular compartments is measurable
- Screening can be accomplished within mixed cell types
- No need to isolate and purify targets
- Effects of compounds on cell structure, communication, and development can be measured
- Cellular functions (e.g., division, endocytosis, and motility) become quantifiable drug targets
- Signal detection is done within volumes of the order of a picoliter
- The entire screening process can be miniaturized (e.g., the CellChip platform)

Multiparametric approach to screening
- Drug specificity is assessed by simultaneously measuring competing targets in the same cell
- Measures of toxicity are accompanied by target activity values
- Measurement of activities downstream of target activity is possible
- Unanticipated side effects of compounds can be measured

Automation of complex tasks
- Sample preparation uses optimized fluorescent reagent kits (e.g., HitKit existing automation technology)
- Imaging algorithms extract target activity values concurrent with data collection
- Automated kinetic assays that include automated liquid handling are possible

Generation of new biological knowledge
- New data management technology transforms enormous amounts of raw data into meaningful information
- Couple genomics, proteomics, and compound library databases with HCS data using the linking tools provided by the Cellomics Knowledgebase
- Discover complex cellular pathway interrelationships and the effect lead compounds have on them using the data-mining tools in the Cellomics Knowledgebase

high-density microplates. The ArrayScan VTI system has a total magnification range from approximately 10 × to 400 ×. This gives optical resolution ranges from less than 1 μm for subcellular spatial measurements to hundreds or thousands of micrometers for field-based measurements. During acquisition, fields of cells are automatically identified, focused upon, exposed, and imaged using a cooled charge-coupled device camera. A high-capacity plate stacker enables 'walk-away' operation of the system, and a live-cell environmental chamber maintains optimal temperature and atmospheric conditions for kinetic screening. Details of the unique multiparameter HCS capabilities of the ArrayScan VTI system are presented elsewhere.[581,1104]

The Cellomics KineticScan HCS Reader was designed to automate large-scale HCS of live cells, specifically facilitating drug-induced kinetic studies, by virtue of a 30-plate incubated plate stacker, on-board fluidics, and plate handling. Additional distinguishing features include eight excitation and eight emission channels, a continuous-focusing system, a fluorescence polarization capability, and a higher HCS throughput capacity. Currently, no alternative automated HCS platform exists for live cells. The KineticScan HCS Reader can be used as a stand-alone screening workstation or as a module within an integrated drug discovery platform comprising both high-throughput and high-content screening. Both fixed-endpoint and live-cell kinetic HCS assays provide a cellular context that in vitro assays of target activity do not match.

2.12.12.5 High-Content Screening of Drug-Induced Microtubule Cytoskeleton Reorganization

As detailed above, HCS assays fall into two general classes: fixed endpoint and kinetic. Here, we demonstrate both assay types in the quantification of microtubule targeted compounds. In the first assay, cells are treated with drugs, incubated for various times, fixed, and labeled to visualize microtubules. The labeling procedure involves immunofluorescence to visualize cytoplasmic microtubules. **Figure 66** shows the results of an assay where mouse fibroblasts have been treated with several known microtubule-disrupting drugs. The effects of these standard drugs are being used in the development of imaging algorithms to automatically classify and quantify the effects of potential lead compounds on the microtubule cytoskeleton.

Apart from immunofluorescence-based reagents, live-cell kinetic HCS assays necessitate the use of molecular-based fluorophores, most notable among them is the green fluorescent protein (GFP) from jellyfish[583] and other sources.[585] Protein chimeras consisting of mutant GFP molecules and a protein that interacts specifically with intracellular microtubules have been constructed, and at least one has been reported.[583] We present here a chimera that has been transfected into cells and acts to provide live measurements of microtubule dynamics in living cells. **Figure 67** shows cells transfected with this fluorescent chimera, which converts the cells into HCS reagents, and the effects of drugs on the microtubule cytoskeleton in these engineered sensor cells. In these cells, with this reagent, the microtubule-disrupting drugs had just as profound an effect on microtubules as they did in the fixed-endpoint assay described above. Although not shown here, the kinetics of microtubule disruption were measured over periods ranging from seconds to hours using these chimeric HCS reagents. Nevertheless, the data shown here represent well the cytoskeletal reorganization and microtubule dissolution that these drugs induce in living cells. Imaging algorithms similar to those developed for the fixed-endpoint assay of microtubule reorganization are being developed for the live-cell assay, but they contain a kinetics component as an added dimension. Therefore, the reagents described above for the fixed-endpoint and kinetic assays can be used either alone or in conjunction with other reagents to form the basis of multiparameter assays, a powerful application of HCS.[581]

2.12.12.6 Specialized Fluorescent Reagents that Find Dual Use in High-Content Screening and High-Throughput Screening

In many cases, high-content screens involve intracellular spatial measurements (e.g., translocation of molecules between cellular compartments), yet there is a class of screens where the fluorescent reagents can be engineered to yield high-content data from a high-throughput read-out. That is, using the example of intracellular translocation as an illustration, a fluorescent reagent can be designed not only to translocate between cellular compartments in response to some molecular activity but to alter its fluorescence spectral properties upon translocation. Thus, a measurement of the spectral change of the reagent, rather than a high-resolution measurement of spatial distribution, is used as a reporter of an intracellular translocation. Such dual-use reagents that report both their dynamic distribution and activity within living cells are often designed from macromolecules and are termed fluorescent protein biosensors.[586,587] Careful design of HCS reagents therefore makes it possible to build powerful platforms where high-throughput and high-content screens can be coupled.

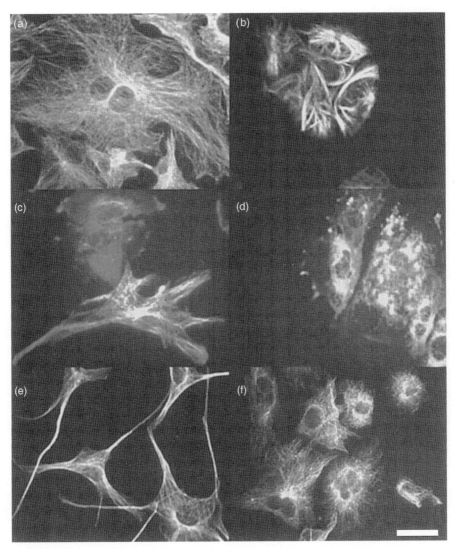

Figure 66 A fixed-endpoint, high-content assay of microtubule disruption. Mouse fibroblasts were treated with drugs followed by the visualization of the microtubule cytoskeleton using immunofluorescence techniques. The cells were treated as follows: (a) no treatment, (b) paclitaxel, (c) curacin A, (d) nocodazole, (e) staurosporine, and (f) colchicine. Each drug had a distinct effect on cell morphology and microtubule organization. Scale bar = 20 μm.

An example of a relatively simple fluorescent reagent that finds dual use in high-content as well as high-throughput assays is JC-1, a fluorescent dye that partitions preferentially into live mitochondria and acts as a reporter of membrane potential within the organelle.[588,589] JC-1 monomers (green fluorescence) are present in regions of mitochondria that exhibit relatively low membrane potential, whereas JC-1 aggregates (orange fluorescence) partition preferentially into regions of relatively high membrane potential. **Figure 68** shows an orange fluorescent image of JC-1 (Molecular Probes) in porcine epithelial cells. At the magnification shown, it was possible to resolve numerous mitochondria in a population of living cells. Furthermore, a heterogeneous distribution of JC-1 aggregates was measured in a single mitochondrion. Thus, a high-content measurement of JC-1 monomers and aggregates in living cell mitochondria can be used not only to assess the effects of lead compounds on mitochondrial physiology but also to identify the mitochondrial compartment as a fiduciary marker for the co-localization of other fluorescent reagents in the same cell.

JC-1 also finds use as a fluorescent reagent for high-throughput live-cell screening. The fluorescence signals emanating from all the cells can be measured simultaneously for both the monomeric and aggregated forms of JC-1 using, for example, a fluorescence microplate reader. Unlike the ArrayScan VTI system, a microplate reader provides no cellular spatial information, but rapidly assesses average cellular mitochondrial potential by measuring the ratio of green

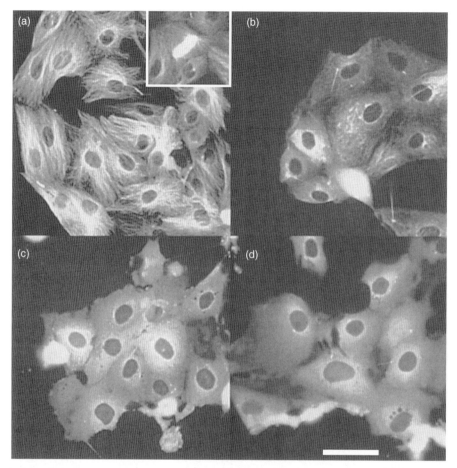

Figure 67 A live-cell reagent for kinetic high-content screening of microtubule disruption. Porcine epithelial cells were transfected with a plasmid encoding a chimera composed of a protein that readily incorporates into cellular microtubules and a mutant GFP. (a) Cells showing the distribution of microtubules in untreated cells, with the inset depicting a brightly labeled spindle in a mitotic cell. The other panels show the effects of microtubule disrupting drugs that include (b) vinblastine, (c) curacin A, and (d) colchicine.

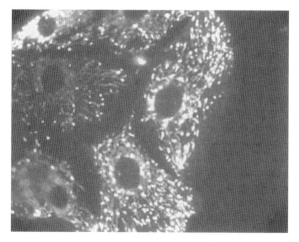

Figure 68 Porcine epithelial cells labeled with JC-1. LLCPK cells were labeled in culture with JC-1, and this micrograph depicts the intracellular distribution of the fluorescent probe. Once added to the extracellular medium, JC-1 rapidly crosses the plasma membrane and partitions specifically into mitochondrial membranes, where it reports the electrical potential.

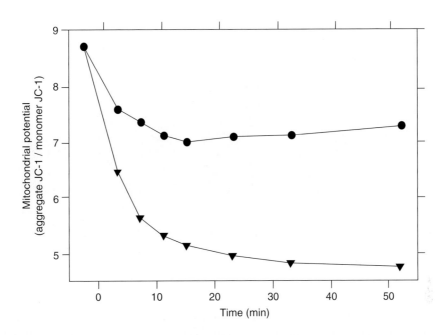

Figure 69 A prototype high-throughput assay for drugs that alter the mitochondrial potential. LLCPK cells, labeled with JC-1, were either left untreated (circles) or treated with 10 nM valinomycin (triangles) at time = 0. Kinetic measurement of the JC-1 fluorescence ratio is used to quantify the loss of mitochondrial membrane potential induced by the ionophore valinomycin. A high-throughput assay using this approach would simultaneously measure mitochondrial potential changes in multiple wells of a microplate.

and orange fluorescence from an entire monolayer of labeled cells. **Figure 69** shows an example of living cell mitochondrial potential measurements made with a fluorescence microplate reader. The kinetic data are presented as a fluorescence ratio (JC-1 aggregate/JC-1 monomer) that is independent of dye concentration but proportional to the mitochondrial potential averaged across an entire well. The kinetic trace for untreated cells exhibits a small decrease over the first minutes, which is likely due to the cooling of the cells within the microplate reader as well as the change in cell medium pH that occurs with the transfer of the microplate from an incubator (5% CO_2) to atmospheric conditions. Nevertheless, a significant decrease in mitochondrial potential was measured upon the addition of an inhibitor of oxidative energy production, valinomycin, to the living cells. Thus, a high-throughput assay of mitochondrial potential can be used to qualify promising lead compounds for more sophisticated high-content assays that may include spatially resolved mitochondrial potential measurements as well as macromolecular translocations[581,582] and organellar functional morphology changes within a population of the same cells.[584] Dual-use fluorescent reagents therefore will find utility in many distinct aspects of the drug discovery process, and become reagents that are even more powerful when used to couple high-throughput and high-content assays on the same drug discovery platform.

2.12.12.7 Coupling High-Throughput Screening and High-Content Screening with Microarrays of Living Cells: The Cellchip System

Miniaturization of assays is one of the major forces driving improved productivity in early drug discovery.[590,591] New drug development technologies and the explosion in genetic data as a result of the Human Genome Project have driven the development of miniaturized test beds or 'biochips' for genetic analysis based on microarrays of nucleic acid sequences. These biochips encompass a very diverse array of technologies and applications. Imprinted with several hundred different nucleic acid sequences, biochips use electro-optical means of addressing applications ranging from expression analysis to genotyping and mutation screening.[592,593] Biochips integrated into sample processing, metering, measuring, mixing, and sorting systems have laid the foundation for miniaturized 'sample-to-answer' systems. The first generation of biochips is focused on surface-anchored oligonucleotides and oligopeptides whose information content is dictated by the four-character nucleotide language of the DNA or the 20-character language of amino acids. Furthermore, the single-stranded DNA oligonucleotides are limited in their information content and applications, as they do not provide information on DNA–protein and other interactions.[594]

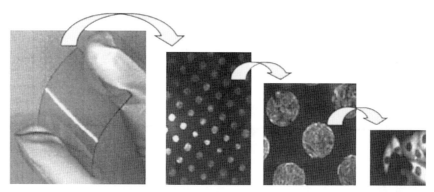

Figure 70 Colonies of cells microarrayed on polymer– and glass–solid phase supports. Each domain is fluidically and optically isolated. This enables massive parallelization of assays on a single chip. The production of a library of chips, where the library encompasses multiple cell types and multiple biochemical pathways, enables high sample throughput coupled with high information output of the pharmacological profile of compound libraries.

We believe that miniaturized cell arrays constitute the next logical frontier in early drug discovery. The multidimensional language of cellular physiology is only partly described by the combinatorial integration of nucleotide and amino acid languages. When high-content screens of living cells arrayed on miniaturized solid supports are used to dissect this complex language, they provide deep biological information on target distribution and activity in space and time. The Cellomics CellChip system is a combinatorial of surface chemistries, cell and tissue types, and HCS fluorescent reagents. On the chip, single or multiple engineered cell types are microarrayed in predetermined spatial addresses (**Figure 70**) on an optically clear polymer– or glass–solid phase support substrate.[590] The 'footprint' of each cellular domain can be adjusted to accommodate either a single cell or a colony of cells. The cell adhesive domains can be populated either with a single cell type or with multiple cell types by adhesive cell sorting from a mixed-cell population, according to selective adhesive interactions with particular cell-specific ligands coupled to the individual domains.[581]

HCS assays designed for microplates have been successfully transferred to the CellChip system. One of these assays – the cytoplasm-to-nucleus translocation of intracellular molecules – is a class of cell-based screens that tests the ability of candidate drug compounds to induce or inhibit transport of transcription factors from the cytoplasm to the nucleus. Sensor cells arrayed on CellChips are treated with a combination of chemical entities. The assays can be run as fixed-endpoint or live-cell kinetic assays. For example, in a fixed-endpoint assay, an array of sensor cells is treated with a chemical fixative and labeled with a fluorescent nucleic acid probe and an antibody against the transcription factor or stress-associated protein labeled with another fluorescent conjugate. The test consists of measuring the fluorescence from the antibody in the nucleus (the nucleus being defined by the nucleic acid probe), versus the cytoplasm defined by the cell domain outside of the nucleus. Proprietary algorithms facilitate quantification of the kinetics and the amount of transcription factor or stress protein translocation into the nucleus over time.[582] Using a polymer-based CellChip system, the activation-induced translocation of NF-κB, a transcription factor involved in cell stress molecular pathways, in response to tumor necrosis factor-α (TNF-α) has been quantified. Appropriate cellular domains on the CellChip platform were dosed with TNF-α; the cells were fixed, permeabilized, and labeled for the NF-κB p65 and nuclear domains (NF-κB Activation HitKit – Cellomics). As seen in **Figure 71**, there was a redistribution of the transcription factor to the nucleus because of stimulation. There was up to a fourfold increase in both the normalized nuclear intensity and the normalized ratio between nuclear and cytoplasmic intensities post-translocation. This increase is equivalent to the results obtained using the microplate platform.[582]

The advantages offered by miniaturization of a drug discovery platform include (1) a combined HTS and HCS platform with a single-pass read of the HTS data from all 'wells' prior to HCS 'drill-down', (2) higher throughput, (3) reduced processing time, (4) an increased number of tests run in a massively parallel format on one substrate, (5) smaller reagent volumes, (6) conservation of new chemical entities, and (7) reduced waste. The integration of such advantages translates into a dramatic reduction of cost and the acceleration of productivity in candidate compound testing. For example, migration of assays from 96-well microplates to 1536-well microplates reduces the reagent cost by 100-fold. Migration from the 96-well microplate format to microarrayed cells on chips will further reduce the volume of reagents and the cost of plate handling. For example, a chip with microarrayed cells in a density format of 100 wells cm^{-2} will reduce the reagent cost by 500-fold from the 96-well format. Furthermore, developing chips microarrayed with

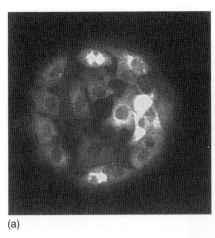 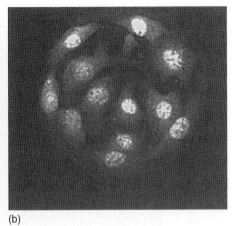

(a) (b)

Figure 71 Mammalian cells contained within the CellChip system employed in a high-content screen for transcription factor activation. As a result of stimulation with a proinflammatory factor such as TNF-α, there is a spatial redistribution of NF-κB. There is a fourfold increase in normalized nuclear intensity post-stimulation. (a) Unstimulated cells on the polymer-based CellChip platform. (b) TNF-α stimulated cells on the polymer-based CellChip platform.

tissue-specific cells will have a tremendous impact on screening the potency, specificity, toxicity, and efficacy of test compounds against a 'tissue-like' ensemble, leading to higher predictive relevance of the live-cell data. Developing multiple cell-based test beds microarrayed in addressable biochips will facilitate the use of these 'microscale tissues' as powerful indicators and predictors of the in vivo performance of the lead compound or toxin using HCS.

The evolution of a miniaturized cell-based drug discovery and identification platform, the CellChip system,[581,590] a variant of the generic biochip, will march in tandem with other miniaturization technologies aimed at shrinking benchtop instruments into their hand-held miniaturized versions.[595] The integration of HTS and HCS onto a single platform (**Figure 72**) will meet the high-sample-throughput and high-information-output needs of the pharmaceutical industry. This combined platform will reduce the data capture, processing, and analysis times, and provide a complete cell-based screening platform. Developing technologies to enable arrays of multiple cells on glass or plastic chips, with each cell carrying its own reagents in the form of single or multiple fluorescence reagents, including fluorescent protein biosensors,[587] adds multidimensional power to a complete drug-screening platform. Furthermore, reagent and assay technology developments made on platforms based on the present HCS technology or standard microplates will migrate directly to the CellChip system.

The future of biochips in early drug discovery is bright because enabling HTS technologies, such as biochip platforms for screening, have captured the attention of the biopharmaceutical market. The early biochips were designed for applications driven by genomics. 'Gene chips,' such as those manufactured by Affymetrix, have captured a significant portion of a billion dollar HTS market since the completion of the Human Genome Project. Close on the heels of the 'gene chips' are 'protein chips' that exploit the proteome as a tool for drug discovery. The CellChip system will build on present and future genomic and proteomic databases to provide multidimensional measures of cellular physiology. Furthermore, the integration of high sample throughput with high information output on the CellChip system will yield faster and more efficient winnowing of 'leads' from 'hits,' thereby optimizing the selection of the compounds early in the process of drug discovery.

2.12.12.8 The Cellomics Knowledgebase – A New Paradigm for Bioinformatics: Creation of Knowledge from High-Content Screening

Genomics, proteomics, and the knowledge base built from cellular information are critically important to the drug discovery process. The Cellomics Knowledgebase is the logical extension of genomics and proteomics because it defines the organization and analysis of cellular information based on new knowledge. The Cellomics Knowledgebase will contain all the knowledge related to cell structure and function. Pharmaceutical researchers will interact with this knowledge base in a biologically intuitive way to explore cellular mechanisms to be able to select optimal targets for drug discovery and to screen potential lead compounds more effectively.

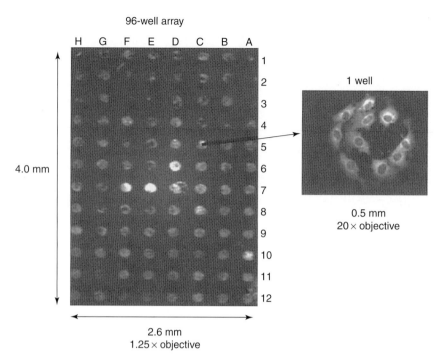

Figure 72 Combining HTS and HCS on the same platform. The massive parallelization achievable with miniaturization is shown in a simple simulation on this surface-modified plastic platform. The HTS is simulated here to detect 'hits' on the miniaturized chip platform. Lack of fluorescence signals in wells E3, F3, A3, and B4, for example, indicates 'non-hits.' HCS measurements are then made only on the 'hit' wells, to gain more in-depth information to produce more 'highly qualified hits.' Further depth and breadth of information can be obtained by arraying multiple organ-specific cells on a single chip and fluidically addressing each domain with a reagent of choice.

The Cellomics Knowledgebase captures the complex temporal and spatial interplay of all the components that compose the living cell. For example, the process of signal transduction relies on a highly coordinated network of intracellular ions, metabolites, macromolecules, and organelles. Most signal transduction processes are therefore a series of specific physical–chemical interactions. In this aspect, the functioning of a cell is conceptually similar to that of a computer neuronetwork. Computer neuronetworks are composed of simple subunits that perform rudimentary mathematical operations. However, there are many such units and, most importantly, these units are highly connected, using the output from one unit as the input for another. Cellular signal transduction offers a parallel strategy to the neuronetwork approach; the basic units of signal transduction are proteins and other cellular components. Each component can participate in one or more cellular functions while involved in a complicated interaction with one or more other cellular components in time and space. Therefore, the design of the Cellomics Knowledgebase reflects the functional significance of the interactions among the cellular constituents. **Figure 73** illustrates cellular interactions by showing a simplified fragment of the model for the Cellomics Knowledgebase. This fragment captures the core information of cellular pathways. This core is composed of a record of interactions of cellular constituents and the logical connections among these interactions. The cellular constituents are abstracted into the 'functional unit' class in the model. A functional unit can be either a 'component' or an 'assembly.' A component represents a single protein, metabolite, or other cellular constituent together with external stimuli. An assembly is composed of functional units, meaning that it can be made up of individual components or other assemblies. Functional units, assemblies, and components record the interactions of cellular constituents. The logical connections of these interactions are recorded in the 'transformation' class. A transformation can have many functional units acting as inputs, outputs, and effectors. For example, in the simple transformation depicted in **Figure 74**, the chemical conversion of a substrate into a product is catalyzed by an enzyme, a motif that is repeated for many of the hundreds of millions of chemical reactions that are occurring simultaneously in a living cell. In this model, for the transformation from the substrate to the product, the substrate acts as the input, and the output is a chemically altered product. In this case, an enzyme acts as an effector. If we drill down on any of the components shown in the scheme, we find that each may be a complex assembly made up of more components that maintain their own temporal, spatial, and chemical properties within the living cell. For

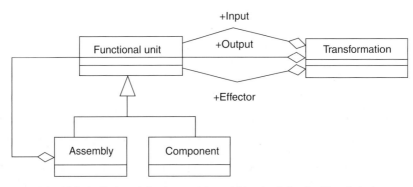

Figure 73 Fragment of the UML (unified modeling language) model for the Cellomics Knowledgebase.

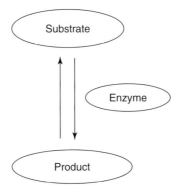

Figure 74 A transformation.

example, if the enzyme chosen in the model as an effector is pyruvate dehydrogenase, the enzyme that couples two major oxidative metabolic pathways, we find that it is itself a complex macromolecular assembly. The pyruvate dehydrogenase complex comprises multiple copies of three distinct enzymes that require five coenzymes for activity.[596] The pyruvate dehydrogenase complex is precisely assembled to optimize the oxidation of pyruvate to acetyl coenzyme A and CO_2. In eukaryotes, the enzyme complex is further localized into a higher-order structure, the mitochondrion, which in turn exhibits a temporal and spatial distribution of activities or transformations within the cell. Therefore, constituents of the transformation class exist as collections of cellular components in time and space from the continuum of ions, metabolites, macromolecules, and organelles.

Having captured cellular interactions, the design of the Cellomics Knowledgebase must also take into account the unique features of cell biological information, such as:

- The flow of information from genomics to proteomics to the living cell. Any insight into the function of cell biology will require seamless integration of information across this continuum.
- Cell biological information is often incomplete or in an unconfirmed state. For example, we need to be able to record the fact that a protein X interacts with an experimentally isolated assembly Y, although Y is not characterized well and the components of Y may be unknown. The database must be flexible enough to be rapidly and accurately updated as unknown components become characterized entities.

How might cellular knowledge be best visualized and probed by the user? One common way to represent cellular knowledge is through pathway diagrams. There are many well-defined metabolic and signaling pathways, but, in isolation, they have lost their physiological context. In reality, pathways often intersect with other pathways to form networks of molecular interactions. Within the Cellomics Knowledgebase, a pathway is a collection of transformations. This collection contains the same functional units and transformations as defined in a well-known pathway, and it can be generated de novo from a database query. Hence, the term 'molecular wiring diagrams', instead of 'pathway diagrams', is used by Cellomics to describe these connections.

The Cellomics Knowledgebase will also provide a new generation of data-mining capabilities. For example, the Cellomics Knowledgebase will provide functions for the analysis of complex cellular response patterns to drug candidates. Data-mining tools will make possible automated comparison of these patterns with the patterns of successful and unsuccessful leads, to guide the user further down the drug discovery process.

In summary, the Cellomics Knowledgebase is a knowledge management tool for cell biology and cell-based drug discovery. In the drug discovery process, knowledge is generated from 'data' and 'information.' Data are the measured values from cellular events that have been acquired, such as from HCS. Information is the result of putting data into a relevant context. For example, an HCS value becomes information when related to a particular assay with specific parameters such as cell type, reagent or reagents used, and the structure and concentration of the drug used in the assay. Knowledge therefore is the understanding of the biological meaning of information. With this understanding, the acquisition of data can be modified, and new information extracted to further improve the knowledge base. Knowledge is synthesized by forming connections between disparate pieces of information. Thus, the Cellomics Knowledgebase strives to provide tools for the analysis and visualization of these connections, such as the connection between screening results, compound structure, and biological context. This connection, and the knowledge built from it, will enable researchers from diverse backgrounds to interpret and utilize new and established information, thus unifying multiple screening projects from multiple laboratories.

The construction of complete cellular knowledge will be achieved with deliberate steps of discovery. Being able to incorporate these incremental discoveries into the context of the existing knowledge through the Cellomics Knowledgebase will become critical to a systematic understanding of cell biological functions. We believe that the pharmaceutical industry will be a primary beneficiary of this new level of understanding.

2.12.12.9 Prospectus

An epoch is rapidly approaching in which the sequence of many organismal genomes, including that of humans, will be deciphered and in which a genomics knowledge base will be built on the foundation of the voluminous data. Nevertheless, we envision this period in the near future as a prologue to the era of the cell where life science discovery, especially in the pharmaceutical industry, will be cell-centric rather than focused on isolated genes and proteins. New thinking will be required to interpret the exceedingly complex web of molecular processes that compose life at the cellular level. How, then, will we begin to unravel the integration of living components in time and space within cells? We predict that an amalgamation of molecular biological, biochemical, and cell biological disciplines will be tapped to supplant cells into being reporters of their own molecular activities. Furthermore, a novel approach to informatic and bioinformatic knowledge building integrated with a miniaturized platform for cellular data acquisition will become a major force in defining and shaping the molecular circuitry of the cell.

Several issues will be resolved during the evolution of the integrated technologies that will be required to define and structure cellular knowledge. First, new generations of luminescent reagents, including fluorescent protein biosensors, will be necessary to simultaneously measure multiple molecular processes in time and space within living cells. Innovative engineering of cells and the surfaces with which they interact will ensure that a wide range of cell types (adherent and nonadherent, mixtures of cell types, etc.) will be amenable to HCS. A fully automated system for extracting cellular data from images will speed the analysis of multiple cellular pathways and provide for computation concurrent with data acquisition. Finally, a fresh approach to informatics and bioinformatics will combine published data of cellular chemistry with HCS data into an unrivaled knowledge base of the multidimensional processes involved in the molecular basis of life. Therefore, the key to unlocking the door to this new era will be the intuitive integration of technologies required to decipher the language of the cell.

2.12.13 Miniaturization Technologies for High-Throughput Biology

2.12.13.1 Introduction

A strategy is summarized for the integration of cellular biology with instrument systems to miniaturize fluid delivery, assay containment, and detection for HTS and biological research. Specific components of the process include the handling of microliter scale volumes of liquids (microfluidics), fluorescence detection, sample handling, and bioassay technology amenable to high-density formats. The components enable ultrahigh-throughput (e.g., 100 000 samples per day) biological assays. These systems provide the increased productivity required for determining gene functions in a whole-cell context. Miniaturization of these technologies is essential to achieve the goal of 100 000 samples per day.

2.12.13.1.1 Key integrated miniaturization technologies
The key miniaturization technologies for high-throughput biological assays are:

- Small-volume assay plates (e.g., Aurora NanoWell plates).
- Sample handling and liquid dispensing:
  - piezo-based aspiration and dispensation;
  - solenoid valve-based aspiration and dispensation; and
  - reagent dispensation – cells, dyes, media, etc.
- Fluorescence detection.

This section discusses these technologies, using the procedures and instrumentation of Aurora as examples.

2.12.13.1.2 Why is miniaturization critical to ultrahigh-throughput screening development?
Miniaturization of the assay reactions enables conservation of compounds, reagents, and biologicals, and makes facile the scale-up for parallel (simultaneous) screening of a large number of compounds. Ultimately, miniaturization of the assay reaction is a critical and essential aspect of achieving high-throughput sample processing. Materials handling, robotic speed, material cost, and other logistical aspects would ultimately prevent optimal deployment of a high-throughput (>100 000 samples per day) system without assay miniaturization and increased reaction well density in the assay plate. Successful miniaturization requires optimization of the assay, the cell density, the expression level of the active read-out protein, and the fluorigenic substrate concentration inside the cells to provide sufficient signal intensity to facilitate rapid measurement of the activity of the cellular process under study. Typical assays are staged with positive and negative controls run in separate assay wells such that endpoint readings of fluorescence provide meaningful data, and the determination of kinetics is unnecessary. Kinetic or rate measurements are possible with the instrumentation, but typically require additional time-based liquid additions or optical measurements in order to establish analyte concentration.

2.12.13.2 The Screening Process

2.12.13.2.1 Assay development
Aurora is developing assays for cellular physiological processes based on fluorescence resonance energy transfer (FRET). In FRET, excitation energy is transferred from one fluorophore (the donor) to another fluorophore (the acceptor) when the two molecules are in close proximity (<10 nm) and in the optimal relative orientation, and the emission spectrum of the donor overlaps the absorption spectrum of the acceptor. Aurora has developed a variety of energy transfer fluorophore pairs that undergo efficient energy transfer when the donor is excited. Some of these FRET pairs are covalently bonded together, and this bond is broken to detect molecular change. Other FRET pairs are spatially associated but not covalently linked. The FRET rate constant is quantifiably diminished after the critical association between the donor and acceptor fluorophores is disrupted. FRET can also be effected by the bringing together of a donor and an acceptor. Either method provides a clear read-out of the molecular event. Aurora has commercially developed three different types of FRET-based fluorophores: Green fluorescent protein (GFP) and its variants, a gene expression reporter that is intrinsically fluorescent; β-lactamase, an enzyme-based reporter that cleaves a FRET-based substrate; and the voltage potential dyes, which make possible the detection of voltage potential across a cell membrane. The gene expression reporters can be introduced into the cell by transfection of the cell, with a plasmid encoding its sequence. Regulatory elements incorporated upstream from the enzyme-coding sequence can be specified to enable linking expression of the reporter with cellular physiological processes and signal transduction cascades. Some of the fluorophore pairs under development by Aurora are as follows:

- Pairs of mutated GFPs with overlapping fluorescence spectra. The proteins are covalently linked by an amino acid sequence tailored to be the substrate for specific proteases, phosphatases, and associative binding mechanisms.
- Coumarin and fluorescein dyes linked by a cephalosporin moiety that is a substrate for bacterial β-lactamase. The combined dual-fluorophore compound is synthesized in a cell membrane-permeant acetoxy ester form that can be passively loaded into cells and activated into the fluorescent form by endogenous esterases. With this compound, cells are transfected with an expression plasmid encoding β-lactamase under control by transcription factor elements that are gene regulatory elements for the physiological signal transduction cascade that is of interest to the investigator.

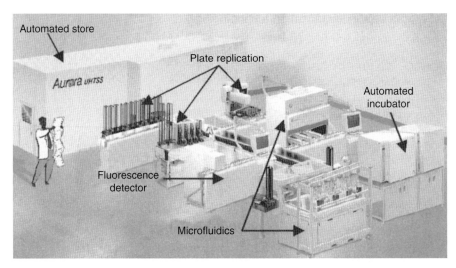

Figure 75 The Aurora UHTSS. The compounds are stored in industry-standard 96- and 384-well storage plates in the store. Compounds are screened for effects on physiological processes (probed in the assay cell lines) by automated selection from the store and conveyance of the compound plate to replication stations, where compounds are prepared for assay. Assays are staged by combining the compounds with assay reagents and cells in miniaturized reaction vessels at the microfluidic stations, incubating the reagents, and measuring fluorescence of donor and acceptor fluorophores at the detector.

- Coumarin dye linked to the headgroup of phosphatidylethanolamine and bisoxonol cyanine dye is a lipophilic fluorophore pair that partitions to lipid bilayer membranes of cells from the aqueous phase. Since the distribution of the charged oxonol is determined by the membrane potential of the cell, this probe pair is suited to assay membrane ion channels.

Assays are developed by transfection of cells with the desired expression plasmid. Once a cell type is successfully transfected and the transfected cells are selected and propagated as a clonal cell line, the miniaturized assay is ready for staging. Assays are constructed by the addition of the cells and the fluorigenic substrates in combination with either activators or inhibitors of the physiological process under study, together with the chemical compound under test in a reaction well. Fluorescence intensities of the donor and acceptor are measured under donor excitation. Aurora has constructed the Ultra-High-Throughput Screening System (UHTSS) as an industrial platform that automates the performance of miniaturized assays in a replicated format that can screen a large number of chemical compounds in a single pass through the platform (**Figure 75**). The UHTSS has been developed to enable reliable miniaturization of the construction and measurement of FRET-based cellular assays.

2.12.13.2.2 Compound distribution

Chemical compound libraries are screened by introducing the compound in cellular physiological assays, to determine whether the compound exerts a pharmacological effect and is of physiological interest. Library screening demands considerable precision of automation to handle the vast number of compounds in libraries, and accurate liquid handling to preserve a stock of potentially valuable material for a large number of assays. A database is used to maintain the information on each compound, such as its location in the storage facility, its well location in a bar-coded storage microplate, its stock concentration, and assay information such as the details of cellular assay experiments. Key automated steps are the retrieval of storage plates and redistribution into assay plates. High-throughput compound distribution is achieved with automated liquid handling, to enable a large number of parallel assays to be run on a large number of compounds at the same time. This requires that liquids be distributed rationally by the automation system into arrays of reaction wells that provide for a scientifically significant screen to ensure that each chemical compound is tested in a physiological assay and to decrease the possibilities of false-positive or false-negative responses that necessitate retesting. Miniaturization demands that the very small quantities of material be delivered accurately and precisely to individual small reaction volumes.

2.12.13.2.3 Loading the assay

To facilitate the construction of parallel assays for large numbers of test compounds at a time, special high-density reaction well formats have been created that enable scaling of biological assays to the reduced volume required. To construct assays in small volumes requires the confluence of accurate and precise fluid handling, the automated positioning of fluid dispensers, and assay well design, to avoid a series of potential pitfalls inherent in miniaturization. Thus, assays are constructed by bringing reaction wells that are formatted to evolving industry-standard designs under arrays of multiplexed fluid dispensers from which are ejected the constituents of the reaction. These include the engineered cells developed to serve as chemical and physiological platforms for the reaction, the cell culture medium, the buffers and any particular reagent requirements, the cell-permeant dual fluorophore, the traditional agonist (or antagonist) used to stimulate the transduction pathway, and the chemical compound under test. Each reaction requires a particular concentration of reagent or compound, and the delivery of each chemical may be constrained by particular physical properties such as solubility. Thus, loading an assay to a total final volume of 1–2 µL requires combining assay constituents that are dispensed in volumes that may range from 0.1 to 1 µL (i.e., a range of 10 000-fold). This range of volumetric liquid delivery requires dispenser modalities that are adapted and suitable for specific ranges of volumes such as 0.1–10 µL or 0.1–10 nL. Each range of dispensing is performed by a single modular platform that can be replicated for the different reagents or to improve speed of delivery. Typically, at these volume ranges, noncontact dispensation methods have inherent advantages over contact-based methodologies. Contact-based dispensation can be particularly effective in a 'touch-off ' mode with larger volumes or in dry plates, but becomes cumbersome for small volumes.

2.12.13.2.4 Incubation

With cellular physiological assays, biochemical action occurs predominantly inside the cells, even with cascades that are triggered at the cell surface, such as agonist binding to a receptor. Thus, once the assay is loaded into the reaction well, the cells are incubated to enable:

(1) The generation of a steady state level of the expressed read-out enzyme. This is determined by the activation (or inhibition) of the signal transduction cascade under study by the added agonist (antagonist) and the concomitant effect of the test compound (if any) on the steady state level of activation of the pathway. The steady state level of the read-out enzyme is a direct indicator of the level of pathway activation.

(2) The membrane-permeable dual-fluorophore compound (chemical name CCF2/CCF4) permeating the cells and becoming fluorescent by the action of endogenous esterases on the acetoxy esters that facilitate membrane permeation.

This requires that the cells be incubated to achieve physiological steady state after assay loading. Assays are typically constructed in several steps:

(1) the cells and medium or buffer are loaded, and the cells are allowed to attach to the reaction well bottom in one step;

(2) the physiological activators and test compound are loaded in other steps; and

(3) the reagents necessary for the fluorescent read-out are loaded.

Each step may be performed at a different temperature. The challenge for miniaturization in large formats is that both the high-density format array of (miniaturized) reaction wells and the incubator must be optimized for the maintenance of physiological conditions. This includes the maintenance of the atmospheric conditions of relative humidity and temperature, as well as buffer capacity (percentage of CO_2), at levels propitious for cell survival and for the avoidance of evaporation over the periods needed (up to 24 h) to stage the assay. In addition, this places demands on the plate material to be favorable for the growth of cells, i.e., no toxic constituents that interfere with cell survival, no materials that contribute to the fluorescence background, and high thermal conductivity to enable uniform warmth.

2.12.13.2.5 Detection

Miniaturized assays impose stringent requirements on the sensitivity and gain of the detection system. All Aurora-based assays are FRET based, and so it is necessary to measure both the donor and acceptor emission intensities in well-separated wavebands with high sensitivity. With miniaturized high-density format assays, the detector measures the fluorescence of each color and in each reaction well of the high-density format plate. Miniaturization forces the detector to provide the capacity to accurately measure the fluorescence signals arising from small numbers of cells (1000–10 000) within each small well. Thus, each cell must contain a sufficient number of fluorophores to contribute a detectable signal. Additional demands on sensitivity arise from heterogeneous responses in the population of cells within each well. Thus, the fluorescence detectors need to be relatively noise free and stable in order to accurately measure the fluorescence using a smaller number of spatial and temporal samples of each well. For high throughput, the detectors need high sensitivity and fast response to enable accurate measurement with a minimum of dwell or transit time over each well.

2.12.13.2.6 Results database

To discover new therapeutic lead candidates in the compound library, the effect of each test compound on the physiological processes under study in the cell-based assays needs tabulation in the compound database. The initial step is the rapid conversion of the raw output of the detectors into a format suitable for storage in a database and statistical analysis. Database records for each compound are linked to other databases that include spectroscopy and other data to provide the compendium of data for each compound. High-throughput, miniaturized, high-density format screening generates potentially a large number of individual data records that include (1) the test compound concentration, (2) the physiological processes tested, (3) the agonist identity and concentration for each assay, (4) the presence of modulators in addition to the test compound, and (5) the different physical and chemical conditions that may be used during an assay. Furthermore, in addition to these sets of data, the outcomes of various analytical procedures, such as EC_{50}, IC_{50}, inhibitor concentrations, competitive and noncompetitive K_m and V_{max}, and other types of analysis (Schild, Dixon, nonparametric indices of significance, etc.) are recorded. The limitation imposed by miniaturization is that an extremely large data set is produced for each compound. To fully understand the efficacy and potential effects of each compound, and to rapidly identify leads, large data reduction and visualization tools are used to enable the effect of a compound to be compared against other test compounds and standard agonists or modulators and across physiological assays.

2.12.13.3 Plate Design and Performance

Two criteria are important for the design of a miniaturized assay plate. The plate must provide a high-density array of identical reaction vessels to which reagents and cells can be readily added during assay construction. In addition, it must not contribute distortion to the fluorescence intensity signals being read. The high-density well array plate is constructed to be as nonreactive as possible and to provide an inert physical platform that physically stages the assay and the detection of the assay read-out. Aurora has designed and manufactured specialty NanoWell assay plates that consist of a 48 × 72 well array that is scaled from the industry standard 96-well (8 × 12) format. The plate has the same footprint as 96- and 384-well plates, and is fitted with a custom-designed lid that prevents evaporation and contamination. Each reaction well in the NanoWell plate is capable of holding 2.5 µL of liquid. The scaled format of wells makes possible automated positioning to be effected by subdividing the ranges used in 96-well format motor placement systems.

The bottom of the NanoWell plate (**Figure 76**) consists of a single <0.2 mm thin sheet of clear cyclo-olefin copolymer (COC) that provides both a chemically inert and an optically clear ($n = 1.65$) smooth surface. The plate can be brought down very close to the focusing optics of the fluorescence detector, so that bottom reading provides an unobstructed high numerical aperture sampling of the fluorescence emanating from each well. The circular well bottom is 0.95 mm in diameter, the diameter of the well at the top is 1.3 mm, and the well depth is 3 mm. The interstitial material between the wells consists of black dye mixed with COC polymer. The opacity minimizes the contamination of fluorescence measurements of each well by skew light originating from adjacent wells. The NanoWell plate bottom is optically planar to within 0.25 mm across its length, so that sphericity and astigmatism are reduced. Circular wells, in contrast to square-shaped wells, provide for the uniform distribution of cells across the bottom of the wells.

Chemical inertness results in the surface at the bottom of each well being difficult to derivatize with typical covalent reagents. This has both advantages and drawbacks. The advantage is that the surface composition is always known. In addition, the material is essentially nontoxic to cells introduced into the wells. The drawback is that cells need to interact with the surface in some normal physiological manner for survival (i.e., expression of metabolic phenotype, or entrance into a particular phase of the growth cycle). Chemical inertness thus requires some level of derivatization to facilitate cell–substrate adhesion (**Figure 77**). The well bottoms are easily treated to create an artificial extracellular matrix (poly-L-lysine, collagen, gelatin, fibro-/vitronectins, etc.). Nonpolarity of the well material favors adsorption of hydrophobic reagents, potentially reducing their bioavailability. This reduction in bioavailability is mitigated by indusion of extracellular serum factors that provide carrier capabilities. In general, bioavailability scales as the ratio of hydrophobic surface area to aqueous volume, which is approximately 50-fold greater for a well in the 96-well plate compared with a well in a NanoWell plate.

Figure 76 The Aurora NanoWell assay plate. The plate is constructed to have the same total-size 'footprint' as industry-standard microplates. The plate contains an array of 3456 reaction wells, each with a maximum volume of 2.7 μL. The reaction well arrangement comprised a 1.5 mm center-to-center distance and is scaled as a 6 × 6 well subdivision of the 9 mm center-to-center distance between the wells in the industry-standard 96-well microplate. The plate thickness is 3 mm. The plate is mounted in a custom-designed caddy. A close-fitting special lid covers the top of the plate, and effectively seals each well against evaporation.

Figure 77 Phase contrast micrograph of CHO cells grown in a single well of a NanoWell plate. The micrograph was obtained 6 h after seeding 2 μL of 106 cells mL^{-1}. The well bottom was coated with poly-L-lysine.

The inert plastic material also provides a very high electrical resistance. This effectively isolates each well from stray electrical currents created in ungrounded robotic positioning platforms. This also makes the plate susceptible to accumulation of static charge, which ultimately is not deleterious to the assay cells but may affect liquid dispensing.

2.12.13.4 Microfluidic Technology

2.12.13.4.1 Sample distribution robot system

The Piezo Sample Distribution Robot (PSDR) is designed for delivery to assays of liquid volumes in the subnanoliter range. The conceptual framework within which the need for subnanoliter dispensing was developed is that the individual compounds in the chemical library were synthesized in limited quantity and that replication of the synthetic run would not be efficacious if no useful leads were generated by the run. Hence, it is important to maximize the information derived from the limited quantity of each compound that is produced. The PSDR was designed to enable miniaturization of assays to submicroliter volumes by enabling the delivery of materials present in only limited quantity, such as the constituents of chemical compound libraries, or those that are prepared in solutions with relatively high concentrations. The PSDR dispenses by piezo-actuated compression of a liquid–vapor interface. This results in the ejection of a single drop of liquid from the interface with a drop volume of 300–700 pL. The PSDR enables the construction of microliter scale volumes by adding negligible quantities of concentrates.

To appreciate the value of the PSDR, it is useful to determine the number of molecules in a 2 μL reaction volume for various concentrations and then to calculate the stock concentration required if these molecules are delivered in a single bolus of liquid. For example, a 1 nM final concentration in 2 µL requires 1.2×10^9 molecules or 2 fmol. To contain this number of molecules in a single drop of 500 pL volume, the concentration of substance in the stock solution must be 4 µM. It is noteworthy that many compounds, such as small peptides and heterocyclic hydrocarbons, have aqueous solubility limits below this value. For a 1 µM final concentration, the situation is even more difficult in that the concentration of material in a single delivered drop must be 4 mM. The PSDR enables the delivery of these small amounts of compounds in extremely small volumes. This permits the use of stock concentrations within the solubility limit, and overcomes one of the significant limitations to assay miniaturization.

Each piezo-actuated compression dispenser (**Figure 78**) consists of a 1 mm diameter vertical glass microcapillary with a small (<100 µm diameter) orifice at the bottom end. The top portion of the microcapillary is connected to a hydraulic system that allows liquid to be moved from the system into the dispenser microcapillary during washes or into the microcapillary through the orifice when external liquid is aspirated prior to dispensing. About midway up, the glass tube is surrounded by an annular radially polled piezoelectric ceramic element. When a voltage pulse is applied between the inner and outer faces of the piezoelectric annulus, the element increases in volume and radially compresses the liquid-filled glass tube. At the liquid–vapor interface that spans the discharge orifice of the glass tube, the compression creates a transient capillary wave. If the voltage pulse is of sufficient amplitude, the capillary wave possesses enough energy to move liquid from the microcapillary to a wave crest to create a forming drop. Because the formation of a capillary wave requires a fixed energy, each generated drop has fixed energy. Thus, each drop generated has the same volume and is ejected with the same velocity. One square-wave electrical pulse to the piezoelectric element produces one drop. Since the amount of piezo-induced compression is contingent on the amplitude of the voltage pulse, the volume of a single drop is controlled by the pulse height. The number of pulses controls the total volume delivered to a well.

The PSDR consists of four parallel heads, each consisting of 96 dispensers arrayed in the industry-standard 12×8 format. In operation, the dispensers are filled with chemical compounds to be delivered to assay plates by removing a 96- or 384-well master compound plate from the store and transporting it to the PSDR. The plate is robotically placed under a 96-dispenser head and raised, so that the microcapillary tips are submerged in the compound solutions. The pressure control system is used to aspirate several microliters of test compound into each dispenser. The master compound plate is then lowered and transported back to the store, and a 3456-well NanoWell plate, in which an assay is being staged, is brought up to the head. Each dispenser is filled with a single compound from a chemical library. A total of 36 wells can be addressed, to create a formatted experimental grid in which the compound concentration is varied (by the number of drops delivered) and in which the experimental conditions are covered (e.g., a test compound without agonist or modulator). The test compound is then dispensed to one or more of the 36 wells that can be addressed by each dispenser. Each dispenser contains more than enough material to deliver more than 20 nL of test compound to each of the addressable 36 wells. Dispensing is uniform across the 96 dispensers in each head

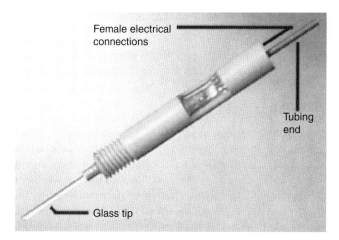

Figure 78 The piezo-actuated liquid dispenser. The glass tip is a microcapillary with an opening at the bottom end that serves as a discharge orifice. The glass microcapillary extends all the way through the dispenser, and at the top is connected with the liquid-filled pressure control system. The microcapillary is surrounded by an annular-shaped, radially polled piezoelectric ceramic material in the portion covered by the housing. The electrical stimulating leads to the piezoelectric material are exposed to the external electrical control system at the connections.

(Figure 79). The number of drops elicited in each visit of a dispenser tip to a well is controllable. Delivery to each well is reliably varied (Figure 80).

The spatial accuracy of dispensing can be affected by static charge that attaches to the nonconductive plate material surrounding each well. The charge has the effect of attracting the small drops of ejected material because the liquid is fairly conductive and the liquid–vapor surface is a site of dielectric discontinuity. The effect of this static potential is to deflect the discharged drop toward the site of accumulated charge. This is avoided by bringing the plate to within 0.25 mm of the tip bottoms. In addition, the static charge can be neutralized by grounding the robotic element chassis and by deionization. These treatments decrease the plate potential difference down to about $20\,\mathrm{V\,cm^{-2}}$. At this level, no spatial errors in dispensing are made when the dispenser tips are 0.5 mm above the well rims.

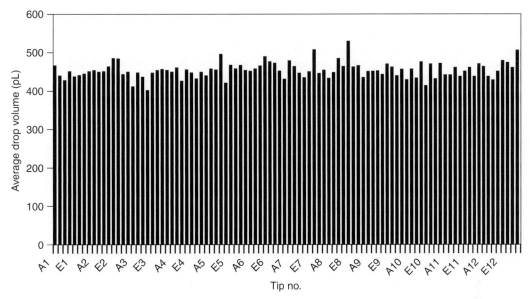

Figure 79 Dispensing accuracy. A few microliters of 5 mM stock fluorescein were aspirated into each of 96 dispenser tips in a head, and then 20 drops were delivered by each tip to a 96-well plate filled with 0.2 mL sodium borate buffer, pH 9.2. The height of each bar denotes the average volume of each drop. The average drop volume for the population of dispensers is 472 ± 46 pL.

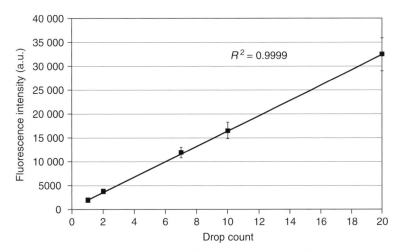

Figure 80 Linearity. In these experiments, three separate buffer-filled plates were used for each drop count. Each point shows the fluorescence of the borate buffer-diluted fluorescein averaged over all 96 dispenser tips in a head, and for all three plates. This confirms that each request for a delivered drop results in the subsequent dispensing of a drop. The implementation of the PSDR enables drop requests to range from 1 to 256 drops. Bars are $\pm$ standard deviation.

2.12.13.5 Solenoid-Based Dispenser Technology

2.12.13.5.1 The reagent distribution robot

2.12.13.5.1.1 Solenoid-controlled liquid dispensing

For dispensing volumes that are larger than the subnanoliter volumes dispensed by the piezo-actuated dispensers, more traditional fluid-handling technologies have been miniaturized to deliver the microliter-sized liquid volumes. These dispensers, termed reagent dispensers, utilize a solenoid-controlled valve to enact and disengage physical blockade of a hydraulic pathway in order to regulate liquid flow (**Figure 81**). The dispensing tube (vertically oriented) is divided into two parts by a solenoid-controlled valve. At the lower end of the tube below the valve is the discharge orifice with a specially fabricated nozzle (<1 mm diameter). Above the valve, the entire hydraulic system is maintained at a fixed positive pressure of about 10 psi (0.7 bar) when the pathway is filled with liquid. The solenoid valve consists of a metal core wrapped by an insulated wire that is, in turn, surrounded over a portion by an annular conductor that is clamped at a fixed current (detent). When current flows through the wire wrapped around the core, the newly generated magnetic field forces the solenoid core into alignment with the field established around the annular conductor (active). The solenoid is mechanically fixed to obstruct the flow path in the detent field, so that when the solenoid is activated, the flow path toward the discharge orifice becomes contiguous with the pressurized hydraulic system, and liquid moves out through the orifice. Because of the time course of the valve movement, the activating and deactivating electrical currents can only be incremented on a millisecond time-scale. This confines the linear range of volumes to 0.1–5 μL, which can be achieved by varying the duration over which the valve remains open. At short times, volume precision is limited by the inertia of the volume that must be moved in order to obtain any liquid dispensing into the target well, and at long times, the volume moved is sufficient to cause a drop in system pressure. Nonetheless, this range of volume is all that is required for dispensing reagent volumes to the miniaturized reaction wells (**Figure 82**).

2.12.13.5.1.2 High-throughput liquid dispensing to multiple reaction wells

The reagent dispensers are used to deliver the major components of the assays to the reaction wells. These constituents typically consist of the dispersed assay cells in cell growth media, fluorigenic precursor substrates, and agonists and other physiological modulators. The reagent dispensers are arrayed linearly in a head assembly that enables reagent delivery to each well in a single row of 48 wells in a high-density NanoWell assay plate at a single positioning of the plate relative to the dispenser. Because each dispenser solenoid is individually addressable, discharge through each outlet can be individually controlled, and each well could, in theory, receive entirely different reaction constituents.

In practice, each linear array of 48 dispensers is fed by a common supply line at fixed pressure that, in turn, is connected to a single large-volume bottle of one reagent (see **Figure 81**). Assays are constructed using multiple linear arrays, each supplied with a different reagent. An assay is staged by sequential visits of each well to the dispensing position under each orifice serving a column of wells. Thus, each linear array dispenses a reagent to each of the 48 wells in a column at a time. The plate is moved under the dispenser head 72 times, to position each of the 72 wells in a row under the orifice servicing the row. The plate is then moved to the next linear array of dispensers, to receive the next reagent. In this way, reactions can be mixed directly in each well over a short time course. This avoids the necessity of mixing reagents together in the supply volumes, with possible deleterious effects due to reaction, oxidation, and so forth.

In general, solenoid-based dispensers provide the major improvement to dispensing technology – over mechanical (plunger)-based delivery – with respect to the construction of microliter scale volume assays. Solenoid-actuated valves

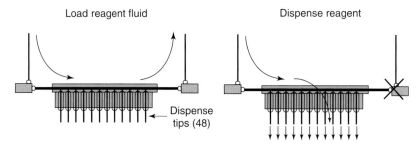

Figure 81 (a) Filling and (b) dispensing from the linear array. The common hydraulic line is connected to the supply volume of liquid to be dispensed on the left, while the right line goes to a disposal container. The common hydraulic line is subjected to a constant hydrostatic pressure. With both left and right valves open, the common line is filled with fluid to be dispensed. The outlet valve is closed to equalize pressure across the common line. The solenoid valves are then opened to enable dispensing. The volume is controlled by the duration for which the orifice valves are open.

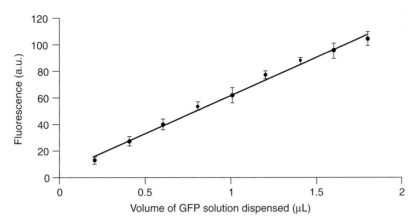

Figure 82 Linearity of dispensation of microliter volumes by solenoid-controlled dispensers. The supply line to a linear array of 48 orifices was filled with 105 mL^{-1} GFP-expressing CHO cells. A single NanoWell plate was passed under the orifice array, and eight sequential columns of 48 wells each received the same volume. The volume was set by adjustment of the duration for which the solenoid-gated valve was activated. The dispensed volume is directly proportional to the valve open time for volumes between 0.2 and 2.0 µL.

are useful in regulating small-diameter (0.1–1 mm) flow pathways due to the length scale over which relatively small electromagnets can produce forces sufficient to rapidly move the dead weight of the occlusion. Because the small outlets can be used in arrays that match the formats of high-density reaction well arrays, the solenoid-based valves are one of the most important technologies enabling miniaturization, and are used to deliver the bulk of the reagents to the assay wells.

2.12.13.6 Fluorescence Detection

The miniaturized assay technologies under development by Aurora all utilize measurement of FRET fluorescence. This requires measurement of the emission intensities of two fluorophores nearly simultaneously (or at least in close temporal sequence) in two reasonably well-separated spectral wavebands. This multiple-wavelength emission quantification is thus a subset of the broader domain of multicolor measurement. Because only two fluorophores are needed in a FRET measurement, the selection of a single waveband for the measurement of each dye that is relatively uncontaminated by fluorescence from the other probe (i.e., avoidance of 'spillover') obviates the requirement for continuous spectrometry. Instead, miniaturization creates challenges at the level of the optical interface between the sample and light collection, to enable the measurement of light from as few fluorophores in as few cells as possible. The NanoWell assay plate provides a useful cellular reaction platform that brings the cells to a distance from the optics that is essentially fixed. This enables the maximization of numerical aperture in such a way that the depth of field includes the entire layer of cells settled on the bottom of the well. Thus, the observation optics can be brought very close to the well bottom, and, moreover, each well is read with the same numerical aperture optics.

The major challenges of miniaturization are several in fluorescence detection:

- delivery of sufficient excitation light to the donor fluorophore to produce significant acceptor sensitization, so that changes in donor and acceptor emission are meaningful;
- collection of sufficient light from the sample from each fluorophore for measurement;
- separation of the light into the relevant wavebands; and
- reproducibly replicating the excitation emission optics at each assay well.

Aurora fluorescence measurement exploits the high internal reflectivity of fused-silica waveguides to accomplish delivery of light to and from the assay well bottom (**Figure 83**). An objective lens is interposed between the well bottom and the front of the light guide. For excitation of the donor, light from a high-intensity arc discharge lamp is directed through heat suppression and bandpass filters, to select the proper wavelengths. This filtered light is then focused by a condenser onto one end of a resin-clad fiber bundle that is brought to a lens positioned very close to the bottom of the well. Excitation light can either be focused on the cells at the bottom of the well (confocality) or allowed to remain nonconvergent (by using the condenser to focus the excitation beam at the back-focal plane of the objective

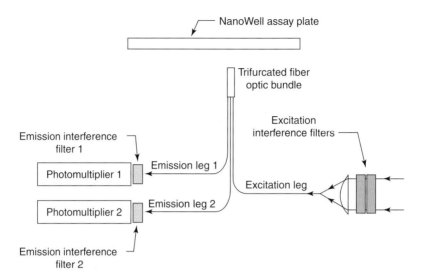

Figure 83 Schematic of the optical pathway in the NanoPlate fluorescence plate reader. The assay plate is scanned by the optical reader head by moving the plate. Excitation light is delivered to each well, and emission light is collected from each well by an optical reader head and a trifurcated fiber-optic bundle. The emission filter stacks are used to separate the emission light into the two wavebands originating primarily from the donor and acceptor fluorophore probes. With two photomultiplier tubes, one for each emission waveband, both donor and acceptor fluorescence intensities are measured simultaneously.

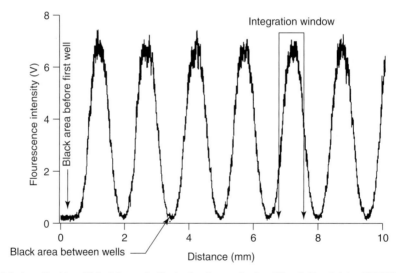

Figure 84 Raw data from the NanoWell plate reader illustrating the continuity of the digitized data. All 3456 wells are scanned in less than 2 min. The trace shows the digitized output of one photomultiplier tube during a scan of six wells in a single row. The raw data are processed to corrected fluorescence data originating from each well by averaging the fluorescence intensities in the integration window. The integration window is determined by the locations of local minimum and maximum fluorescence intensities and their values. The illustrated scan was obtained with 100 nM fluorescein in borate buffer pH 9.2 through the green emission channel.

lens). Confocal excitation delivers more light to fewer fluorophores. A duplicate paired set of light guides is used to collect the emitted light focused by the objective lens. One set leads to a stack of interference filters that isolates the light emitted by the donor fluorophores, and the other set leads to a stack that isolates the acceptor emission. Each stack sits atop the photocathode faceplate of a photomultiplier tube. Thus, the emissions from both fluorophores are obtained simultaneously.

In order to maximize the throughput of the miniaturized, high-density format system, the NanoWell plate is scanned relative to the optical fluorescence detector (**Figure 84**). In the Aurora instrument, the plate is moved by high-resolution motorized screws, and the optical system is fixed in place. This results in each well being represented as a set of discrete integrated intensities obtained for the scan. The interstitial material between the wells attenuates the fluorescence, so that the raw output of each photomultiplier resembles a periodic wave. Nearest-neighbor analysis is used to isolate the intensity peaks corresponding to the well centers in each scan of a row of wells. The well array format is used to isolate wells containing no fluorescent material. Furthermore, statistical analysis enables the exclusion of the rare, extremely intense yet spatially small fluorescent objects that are dust particles. Data acquisition must keep up with the rate of plate scanning and the amount of data that the two photomultipliers are capable of generating. In any case, fluorescence measurement requires a very small fraction of the time needed to construct and measure miniaturized cell-based physiological reactions in high density.

2.12.13.7 High-Throughput Functional Biology

Ultrahigh-throughput screening utilizing microliter volume assays is made possible with the types of instrumentation described for the manipulation of replicated small liquid volumes and for simultaneous two-color measurements repeated many times. This is demonstrated using a standard cell-based assay that is amenable for testing signal transduction pathways mediated by GPCRs. In this assay, Jurkat cells are transfected with an expression vector in which a gene encoding β-lactamase is under the control of a set of nuclear factor of activated T cells (NFAT) nuclear transcription activation sites. $β_2$-adrenergic receptor activation by isoproterenol or muscarinic activation by carbachol results in the heterotrimeric G-protein-dependent activation of phospholipase C. The resulting inositol triphosphate-activated release of calcium ions from intracellular stores activates calcineurin, a phosphatase that activates the NFAT transcription regulator, and promotes expression of β-lactamase. The activity of β-lactamase is determined by the use of the cephalosporin-linked coumarin-fluorescein dye pair (CCF2). Thus, the expression level of β-lactamase in these transfected Jurkat cells provides a direct indication of the extent of $β_2$-adrenergic activation. The enzyme activity is directly indicated by the amount of coumarin fluorescence, resulting from CCF2 cleavage and dequenching of the coumarin fluorescence by the disruption of FRET with fluorescein, relative to the amount of fluorescein fluorescence, which is sensitized by FRET with the directly excited coumarin. Distinguishing receptor activation in these cells is unambiguous (**Figure 85**).

To construct the assays shown in **Figure 86**, the NanoWell assay plate well was seeded with 2000 detached (spinner culture) cells in serum-free growth medium using a single array of the solenoid-based reagent-dispensing robot (RDR). The plate was then incubated at 37 °C for 6 h to allow the cells to attach to the well bottom. The carbachol was added after a second pass through the RDR, and, after a 2 h incubation, the CCF2 fluorigenic reagent was added. Fluorescence was measured after 4 h in CCF2 at room temperature. The receptor activation is clearly revealed as an inversion in the amount of blue fluorescence relative to green. In quantitative studies, the donor–acceptor fluorescence intensity ratio depends on the agonist concentration, indicating the pharmacological validity of the FRET sensitization ratio.

This relatively simple fluorescent assay for receptor activation consists of a significant number of operations, such as cell seeding in a medium, the addition of agonist, the addition of fluorigenic reagent, and several incubations at different temperatures. The steps needed to construct the assay must be performed within the constraints of time and physical condition. Nonetheless, the construction and measurement of the assay consist of relatively simple steps – fluid addition, incubation, and fluorescence reading – that are amenable to automation. The precision of the automation enables the assay to be miniaturized.

Further utility of the miniaturized assay is provided by a different set of transformed Jurkat cells that are activated by carbachol, and provide a convenient cell-based assay for screening chemical compounds for possible effects on signal transduction processes mediated by muscarinic receptors and heterotrimeric G-proteins (in addition to NFAT and NF-κB-activated gene transcription).

2.12.13.8 Overview

The technologies of liquid handling, assay construction, and fluorescence detection provide a set of operational modules that can be configured to perform a wide variety of assays. The key features are that they enable cell-based fluorescence measurements, and so are adapted to screening for effects on potentially intricate physiological pathways in intact living cells. In addition, the liquid handling can be parceled so that a particular instrument is specialized for a range of volumes that are consistent with the physical methodologies used to obtain control of the dispensed volume. Solenoid-based hydrostatic delivery is required for the range and precision of control of microliter-volume dispensing. At the subnanoliter level, more transient pressure control mechanisms are necessary, and so piezoelectric technology is useful. These technologies are fully amenable to the design of new assays for new physiological pathways and for the determination of gene product functions.

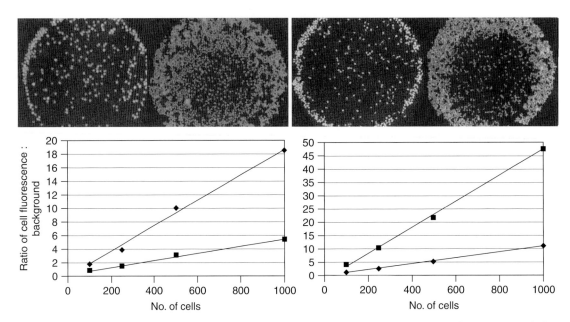

Figure 85 Isoproterenol stimulation of Jurkat cells transfected with NF-AT–β-lactamase vector. The upper panel shows fluorescence micrographs of individual wells seeded with Jurkat cells after 4 h incubation in the dual-FRET fluorophore CCF2. The pair of panels of cells on the left shows cells that were not treated with isoproterenol. The fluorescence originates by direct excitation of the coumarin and sensitization of fluorescein fluorescence by FRET in the intact CCF2 molecule. The leftmost panel was obtained by seeding 250 cells in the well, whereas the adjacent panel to the right shows 1000 cells. The plots show the quantitative relation between coumarin and fluorescein fluorescence intensities as a function of cell density. In unstimulated cells, green fluorescence predominates. The pair of panels on the right shows Jurkat cells after stimulation with isoproterenol. The expressed β-lactamase has cleaved the dual-FRET fluorophore, relieving the quenching of the coumarin fluorescence.

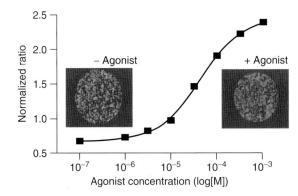

Figure 86 Dose–response curve for Jurkat cells transfected with a β-lactamase reporter gene and stimulated with carbachol. The emission intensity ratio at 460–520 nm was normalized by the sensitization ratio obtained at 10^{-5} M agonist concentration.

2.12.14 Data Management for High-Throughput Screening

2.12.14.1 Introduction

2.12.14.1.1 Background

Data management is an integral requirement of any HTS effort. Ultimately, the purpose of any screening program is to acquire information that can be turned into knowledge in the drug discovery process. In order to achieve this goal, the results of all tests, along with all appropriate contextual information, must be reliably captured by a data management system and made available to researchers throughout an organization. The tracking of sample locations, test conditions, the capture of raw results, and the transformation of raw results into calculated and meaningful information requires robust and flexible information systems with effective user interfaces, as well as links to inventory and other information systems. This section will highlight some of the issues involved in developing and maintaining an informatics system to support HTS.

2.12.14.1.2 Issues in high-throughput screening data management
2.12.14.1.2.1 Large volumes of data
The introduction of automation to biological screening has resulted in an increase in the volume of data of many orders of magnitude. In the early 1990s a large pharmaceutical research organization might screen tens of thousands of samples in a year. By the late 1990s some large research organizations were screening as many as 22 million samples in a single year. Clearly, in order to achieve this type of volumes, informatics systems must be a part of the overall strategy, both to aid with the logistics of HTS and to support decision-making.

2.12.14.1.2.2 Addition of automation
The defining feature of HTS is the application of automated systems to biological testing, greatly increasing the number of samples that can be tested in any time period. The increase in the number of samples being tested has a number of complicating side effects. First, since the goal of increasing throughput is to gain knowledge regarding the interactions of molecules and targets, results data must be collected, analyzed, and made accessible to researchers. Information processes that are possible by hand with small volumes of data require automated systems when large volumes of data are involved. Second, the use of automation introduces the possibility of mechanical error. Finding the error and isolating the results is a major issue for HTS informatics systems.

2.12.14.1.2.3 Variability inherent in biological experiments
Early in the days of HTS, many comparisons were made between the application of automation to biological screening and that in other industries, such as manufacturing. Many presentations and papers of the time talked about the 'industrialization of research.' While certain aspects of the process, particularly inventory and logistics, did become more industrialized, the predictions underestimated the rapid changes in the research process, as well as the variability inherent in biological systems. The nature of discovery research dictates rapid change. By comparison, the basic principles of double-entry accounting have not changed since the 1500 s. While there have been, and will continue to be, changes in the details, such changes pale in comparison to those in scientific research, where even many of the leading scientists today cannot tell you where the process will be going 5 years from now.

2.12.14.2 Acquisition of Data

In almost all cases, the endpoint of a HTS experiment is a change in the state of the contents of a well that may be detected by an automated device, typically referred to as a reader or a plate reader. The change in state may be an increase in fluorescence, radioactivity, or absorption. The plate reader will measure the state of the sample at a set time or over a period of time. In some experiments, the detector takes multiple readings. In almost all cases, the output of the reader is a series of raw results, which must be either normalized or reduced, and combined with other information, in order to produce meaningful results.

2.12.14.2.1 Integration with automation equipment
Plate readers most often produce one or more text files, which contain the raw results of the experiment. These results are often expressed in terms of counts per minute, optical density, or fluorescence. These results can be extremely difficult for scientists to interpret in their raw form for several reasons. An example of such a file is shown in **Figure 87**. There are several things that we can see regarding this file. First, we can see that besides the raw result, there is other potentially valuable information contained in the header section, such as the filters used and the temperature. Next, we see the results in a 'data block.' These are the raw results, but we can quickly surmise that reviewing the raw results would not be an efficient way to analyze the experiment. For one thing, there is no indication of what was in each well that was tested. While many readers today can accept the input from a bar code and will include the bar code plate identifier in the reader file, there is no indication of what sample was in the plate, what concentration was used for the test, or other information required. In addition, to make sensible decisions on the basis of these data, the scientist must perform other analyses, such as stability of the controls, and overall level of activity for the assay. While looking at a single plate of raw results might give an indication of the answer, one would have to keep a fair amount of information in one's head. For example: Where were the controls for this experiment located? What ranges of numbers indicate

Read time: 7/29/98 11:19
Plate ID: SampPlat101
Barcode:
Method ID: Galanin TRF screen
Comment: tic tac stds type 1 green red plate 1
Max cps: 1.78E+07 cps
Min counts: 236 counts
Microplate format: Packard White 96
Detection mode: F
Excitation side: Top
Excitation filter: 1 485-20 (Fluorescein)
Excitation polarizer filter: o
Attenuator mode: m
Emission side: Top
Emission filter: 1 535-25 (Fluorescein)
Emission polarizer filer: o
Z height: 4 mm Numeric
Conversion method: Comparator
Integration time: 100000 us
Total integration time: 100000 us
Readings per well: 1
Time between readings: 100 ms
Shake time: 0 s
Temperature: 23.8 °C
Instrument tag: Application's Lab
Serial number: AN0065
Data: Intensity
Units: cps

	1	2	3	4	5	6	7	8	9	10	11	12
A	15416754	16174624	17774608	16883962	15993355	16617787	15854868	15358044	17155580	16851530	17464322	49469
B	16749729	16732580	16919286	16919286	16883962	15901794	16649150	15505432	16764194	16706653	17259128	22175
C	15592631	16930910	16683497	16999124	16030343	15769842	16038291	15064792	16340762	15772406	16925004	18100
D	16147708	16193517	16715106	16051556	16285916	15757040	16482484	16706653	16925004	16810660	16580875	24924
E	15211197	16174624	21607	16883962	15993355	16617787	15854868	15358044	17155580	16851530	17464322	22933
F	16672032	16732580	16919286	16649150	15505432	15901794	16649150	15505432	16764194	16706653	17259128	65863
G	14889129	16930910	16683497	16999124	16030343	15769842	16038291	15064792	16340762	15772406	16925004	21038
H	16158464	16193517	16715106	16051556	16285916	15757040	16482484	15769842	17259128	16810660	16580875	72213

Figure 87 Raw output of an LJL Analyst plate reader. Automated plate readers can produce a great deal of information; however, the information is not always formatted in a way that is straightforward or easy for data systems to capture. Most modern plate readers can be programmed to produce data in desired formats. The 8 × 12 block of numbers in the center contains the raw results for this plate. Other information, such as filters used and temperature, may be included. Typically, the information system parses the information out of this type of file, processes the raw results, and stores all the information in appropriate fields in the database. Note also that ancillary information, such as sample ID or well type, is not included in this file.

activity versus inactivity? Are the results seen on this plate in line with what should be expected from this type of assay? While it may be possible for a scientist to manually analyze one plate's worth of data, imagine trying to do this for 1000 plates run in a single day. It may be possible, but it certainly would not be a good use of the researcher's time.

2.12.14.2.2 Integration with other information systems

In order to present the data to the scientist in a format that can aid in efficient decision-making, the raw results in the reader file must be parsed and brought into an information system, where the results will be integrated with information such as the sample tested, the kind of well (sample, control, reference, or other), and other conditions, then calculated and presented to the scientist so that decisions can be made. This integration can be challenging for a number of reasons:

- The plate read by the detection device is almost never the inventory plate. It is usually a copy made at the time of the test. Therefore, the information system must associate the ID of the assay plate with the inventory plate that was the source.
- The layout of the reader file can vary dramatically from one reader to the next. In fact, most readers have several programmable modes that allow them to create files with a number of different formats. Information systems must properly parse the information.
- The layout and configuration of the test plate is different for different experiments. In fact, sometimes within a single experiment multiple plate layouts may be used. The information system must be able to associate the correct layout with each plate.
- Finally, when making decisions, the scientist will frequently want to see results from other tests for the same samples. This means that the system should have the capability to query other databases.

2.12.14.2.3 Well types and plate layouts

One key piece of information that is not captured in either inventory systems or by the reader file is the plate layout. The plate layout describes the contents of each well in a plate. Plate layouts typically include information such as sample concentration, and the location of control wells or standards. While this is typical information, plate layouts can include any variable that the scientist has under his or her control, such as pH or salinity. In some cases, the scientist will even place different targets (enzymes, receptors, or cell lines) in the wells. This information is crucial for a meaningful calculation and interpretation of the results. Most HTS information systems have specialized input screens to support the definition of plate layouts. The variety of plate layouts used for different types of experiments requires that any general-purpose information system be extremely flexible.

2.12.14.2.4 Calculation, reduction, and normalization of data

Once the raw results are brought into the data system, some form of calculation is needed to make the data meaningful to the scientist. This is important because the raw numbers coming out of the reader file can vary widely based on the type of detection used as well as the specifics of the assay. Using just the raw numbers, it would be almost impossible for the scientist to quickly determine the quality of the assay run or to determine if any samples displayed unusual activity. Even within a single assay, results coming from plates taken at different times may vary substantially. So, unless the raw results are somehow normalized, the scientist can be easily misled. The most common calculated results are as follows:

- percent inhibition;
- percent of control;
- IC_{50} or EC_{50};
- K_i; and
- ratios.

2.12.14.2.4.1 Percent inhibition and percent of control

Percent inhibition and percent of control are perhaps the two most commonly used calculations in the HTS environment. In both cases, the calculations show the activity of the sample(s) being studied relative to the activity shown by a known set of controls or standards after accounting for systematic noise as expressed in the low controls. The percent inhibition calculation is

$$1 - (\text{sample} - \text{average of low controls})/(\text{average of high controls} - \text{average of low controls})$$

The percent of control calculation is

$$(\text{sample} - \text{average of low controls})/(\text{average of high controls} - \text{average of low controls})$$

Note that these two calculations are the inverse of the other. In many cases, researchers will calculate both and use the one that is most familiar to them. For the remainder of this chapter, we will use only percent inhibition, but we will mean both.

2.12.14.2.4.1.1 Variations   In many cases, experimentalists will attempt to reduce experimental error by using two or more wells to test each sample. The replicates are averaged, and then the averages compared. If there is a high deviation within the sample responses (e.g., if one replicate shows an inhibition of 80%, and the other replicate shows an inhibition of 5%), we have a good indication that one of them is wrong. Sometimes the replicates are on the same plate, and sometimes the entire plate is replicated. Again, the information system needs to be able to tell where the samples are, find the replicate, and perform the appropriate calculation.

In the most common application of percent inhibition tests, control wells are placed on each plate. However, in many cases experimentalists will choose to place the control wells on a designated control, reference, or quality control plate. The controls from this plate are then used to calculate results of samples on other plates.

Percent inhibition tests represent a good first-pass indication of the activity of a sample. When dealing with a large number of samples, where you are looking for 10-fold or better differences in activity level, these tests give you a fast, reasonable indication of which samples are worth following up. However, there are several shortcomings to these measurements. First, they are single-point measurements, so we only learn the activity level of a compound at a certain concentration. It may be impossible to make reasonable comparisons regarding the relative activity of different samples

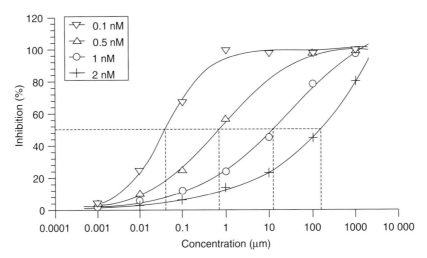

Figure 88 A set of fitted dose–response curves. The dashed line shows the IC$_{50}$ concentration (i.e., the concentration level where 50% inhibition is observed). Furthest to the left is the sample with the lowest IC$_{50}$, and in most cases would be indicative of the most potent sample.

if they were tested at different concentrations. In addition, these measurements give no indication of the nature of the activity. To obtain this type of information, other types of tests have to be run. The most common of these tests are the dose-response and the kinetics tests, described in the following sections.

2.12.14.2.4.1.2 Dose–response tests All dose–response tests measure the activity of a sample over a range of concentrations. The results of these tests are usually computed using one of many nonlinear curve fitting algorithms to find a measure called either the IC$_{50}$ or the EC$_{50}$. In both cases, these measures tell us the concentration at which the compound shows an activity level of 50%. This measure is superior to percent inhibition for several reasons. First, it allows reasonable comparisons between different compounds. Second, it is a much finer measure of activity because it also shows information about the nature of the activity. For example: How rapid is the response to changes in concentration? What is the concentration level at which activity maximizes? At what concentration is there no activity? (See **Figure 88**.)

2.12.14.2.4.1.3 Kinetics experiments Kinetics experiments measure the rate of change in activity of a sample over time. The usual measure of activity for a kinetics experiment is K_i, the maximum rate of change. A wide variety of models are used to calculate K_i. One of the most popular models is the Michealis–Menten model, which typically takes the form

$$v = V[A]/(K_{mA} + [A])$$

where V and K_{mA} are constants at a given temperature and a given enzyme concentration. A deeper explanation of kinetics and other calculations used in biochemical screening is published by the Nomenclature Committee of the International Union of Biochemistry.[596a]

2.12.14.2.4.1.4 Ratios A large number of experiments require that several results be compared. For example, in some tests we are interested in both protein growth and cell survival. In other tests, the exact level of activity is not considered as important as the change in activity from a baseline. In these and several other types of test, the results are typically reported as a ratio.

2.12.14.3 Validation of Experimental Results

2.12.14.3.1 Detection of experimental error

In any experiment, there is a chance of error. The addition of automation and the increased number of samples tested at any given time increase the probability of error. So, before the scientist even begins to look for meaning in the results of an experiment, he or she must determine if the results are useful at all, or if experimental error means that the results are invalid.

2.12.14.3.2 False positives versus false negatives

There are two types of experimental error that are generally to be accounted for: false positives and false negatives. False positives look like hits but are not, whereas false negatives are hits that are never revealed. In many cases, scientists care very little about these errors if they are random occurrences, impacting a small number of samples. The assumptions are that false positives will be found with relatively inexpensive follow-up tests, and that there is enough overlap in the chemical space that for any false negative there will be other similar compounds tested that will yield the correct answers. In short, given the nature of biological systems, the assumption is that there will be some level of random error, and most scientists will deal with that. The bigger concern for most HTS scientists is the detection of systematic error, which may invalidate the entire experiment or a large portion of the experimental results. These errors can be very expensive and difficult to recover from. Typically, most scientists depend on the use of controls, or known positives and negatives, to avoid experimental errors.

2.12.14.3.3 Sources of error

2.12.14.3.3.1 Automation equipment failure

Equipment failure is one of the most common sources of experimental error in the automated environment. Typical automation failures include improper dispensing of reagents, physical mishandling of plates, and the failure of incubators and detection devices. Computer systems and software that control the automation equipment are also subject to failure. It should be noted that each increase in throughput and/or plate density comes with a period of increased automation error. This is to be expected, as many of the issues become more problematic, and tolerances become tighter. As an example, a typical well in a 96-well plate can contain 200 μL. On a 1536-well plate, a typical well contains 10 μL. A variation of 1 μL will be 0.5% of a 96-well plate but 10% of a 1536-well plate.

2.12.14.3.3.2 Stability of reagents or targets

Many HTS tests involve the use of radioactive labels, which are subject to decay. Most involve the use of enzymes, receptors, proteins, or live cells, all of which are subject to degradation or contamination. Needless to say, if an enzyme breaks down, or if the cells die before the test, the results of the test are invalid.

2.12.14.3.3.3 Environmental issues

Most biological tests are designed to be performed within some range of conditions. Failure to maintain these conditions may result in invalid test results. As an example, in low-humidity environments, evaporation can be an extreme problem. If the liquid in a well evaporates before the test is completed, clearly the results will not be valid.

2.12.14.4 Decision Support

2.12.14.4.1 Finding systematic error

Given the possibility of some kind of systematic error and the dramatic effect that such errors can have on the results of a test, it makes sense for the scientist to validate that the experiment ran properly before searching for active compounds. The major methods of detecting systematic error are (1) the use of controls and standards and (2) the search for patterns within the data.

2.12.14.4.1.1 Use of controls and references

The most common way to detect systematic error is by the use of control wells on each plate. Control wells contain samples with a known activity or no sample at all. At the conclusion of the test, the scientist's first action will typically be to check the controls. Did they run as expected? How tightly did the results cluster? Was there wide variation in the results over time? Was there an acceptable difference in activity between the high and low controls (the signal-to-noise ratio)?

2.12.14.4.1.2 Search for patterns

The results of most HTS experiments should be random, with a small number of hits randomly distributed over the library tested. If the results indicate that one area (such as a specific row, column, or well location) had an unusual level of activity, it is usually indicative of some kind of systematic error. As with almost anything else in HTS, there are exceptions to this. For example, if screening a combinatorial library, where all members of a row contain a common substructure, increased activity for an entire row would not be uncommon, and might be exactly what you are looking for.

2.12.14.4.2 Locating hits quickly

In most HTS environments, the difference between a hit and other samples is somewhat dramatic, often of the order of 10-fold. Under these circumstances, it is fairly easy to locate the hits and move forward. There are a number of methods for locating hits, ranging from graphical display of the data to simple database queries, such as 'find the top $x\%$ in terms of activity.' Many organizations use standard deviation (σ), looking for samples that are 2σ more active than average. In many cases, locating the most active samples and producing a list of the identifiers is all that is required. This list is sent to a sample room (or back to the automation equipment) for replating and confirmation or secondary tests.

2.12.14.4.2.1 Evaluating a single run or an entire campaign

Often, the HTS laboratory will run an organization's entire library of compounds through a screen over a period of several weeks to several months. This is often referred to as a campaign. Many organizations will wait until the entire campaign is completed before evaluating samples. They do this because they have found that their libraries have sections (sublibraries) that are more or less active than the norm. If one of these sublibraries is run early, the threshold for activity may be set improperly. Therefore, it is more effective to wait until the entire library has been tested to make threshold evaluations.

2.12.14.4.3 Integration of other data

Sometimes it is desirable to consider information other than the results. Most often this information is used to narrow down the number of hits and to eliminate hits that are not worthy of further study. Some of the key information used to eliminate hits includes the following:

- Results from other tests. If a compound is found to be active against a number of other targets, it is likely to have undesirable side effects.
- Physicochemical data. The most common filtering method is the Lipinski 'rule of five.' This set of rules looks for compounds that are likely to have reasonable absorption characteristics and eliminates those that do not.

At this point, some people ask, 'Why run those compounds in the first place?' If, after all, they are just going to be eliminated after the primary screen, does it not make sense to remove them in the first place? There are two reasons why these compounds are typically run in primary screens. First, many companies maintain their entire libraries in 96-well plates. It is easier and less expensive to simply test everything than to pick out certain compounds. Second, in some cases these may be the only active compounds in a screen. While none of them will be drugs, the information gathered will give medicinal chemists a starting point in terms of how to develop compounds further.

2.12.14.4.4 Graphical display versus tabular display

There are many fine books on the subject of human interaction with and comprehension of data, and a detailed discussion of the subject is beyond the scope of this section. For now, we will simply say that both tabular and graphical displays of data have their place, and most systems provide a combination of displays. We will discuss some of the common graphical displays used, as well as some of the basic requirements for effective tabular displays.

2.12.14.4.4.1 Tabular display

A tabular display is basically a spreadsheet, showing the data in columns and rows. The most common spreadsheet in use today is Microsoft Excel. Most systems specifically designed for HTS data management try to offer an 'Excel-like' spreadsheet capability. Commercial examples of these systems include CSAP from Accelrys, Inc., and Assay Explorer from Elsevier MDL Information Systems. A very popular HTS package, ActivityBase from IDBS uses Excel as the front end. Tabular displays are useful in that they give scientists a detailed and precise view of specific results. Typically, tabular displays will need to sort and filter the results so that a meaningful subset can be displayed. It is unrealistic to assume that any scientist can find meaningful data in a tabular display of 100 000 records; however, scientists can raise a query such as 'Show me the 50 most active compounds from this test,' and a tabular report can be useful. Tabular reports could also be useful for statistical summaries of a large experiment.

2.12.14.4.4.2 Graphical display

Properly designed graphical displays allow scientists to quickly analyze large amounts of data, as well as to spot trends and outliers. Some of the typical graphs used in HTS include scatter charts and control charts. Scientists would generally request that any graphical display be interactive so that it can be used as a query device.

2.12.14.4.5 Reporting

In addition to display of the data, most organizations will want a hard copy of at least a summary of the results. Often the hard copy contains statistical data regarding the experiment and a summary of interesting results. These reports are usually stored in a laboratory notebook. Again, it is unlikely that anyone would want to create a hard-copy report showing the results of hundreds of thousands of compounds, 99% of which are inactive.

2.12.14.5 Data Analysis and Mining

Up to now, we have been discussing display and reporting of the results for a single experiment. The key decisions made at the single-experiment level include the following: (1) Did this experiment run as expected? (2) Were there any hits, or interesting compounds? While these are interesting and useful questions to ask, other questions can require even deeper analysis, and place an increased strain on information systems. For example: (1) Of the compounds that were active in this test, which were active (or not active) in other tests? (2) Are there structural components of the hits that would account for the activity? (3) Do any of the hits have physicochemical properties that would indicate that bioavailability, solubility, or potential toxic effects?

2.12.14.5.1 Cross-assay reports
2.12.14.5.1.1 Pivoted data

In most data management systems, results data are stored in some variant of a long, skinny table. This means that a single column in the table can hold multiple result types and that other columns are required to describe the results. An example of a simplified long, skinny table is shown in **Table 26**. Long, skinny tables make for efficient and flexible data storage because they do not require new columns or other modifications to define new variables. However, they do not present the data in a manner that is intuitive to the scientist. Most scientists prefer to see data in a short and wide format, in which each variable associated with a compound is represented as a column. An example of a short, wide table is given in **Table 27**. Note that each test is now a column, so the scientist can quickly answer questions such as 'Which compounds were active in this test, but not in others?' or, for the compounds of interest, 'Are there tests that were not run?' Converting data representation from tall/skinny, as it is stored, to short/wide is known as pivoting the data. The data pivot can take place in many different ways, and many biological information management systems have a data-pivoting function built in.

In **Table 27**, note how easy it is to tell which samples have been run in which tests, and which samples were active in one test but inactive in other tests. Unfortunately, this type of table makes for very inefficient storage, especially if there are sparsely populated columns. Also, this type of table is also very inflexible, since the addition of a new test requires the addition of new columns.

2.12.14.5.1.2 Classical structure activity

One of the goals of pharmaceutical research has been to find the relationship between small-molecule chemical structure and biological activity. One simple, but widely used, method of analysis is the structure–activity relation report. This report shows a chemical structure and pieces of information describing the activity on each row. The addition of quantitative measures of physicochemical properties (quantitative structure–activity relationships) is the province of molecular modelers. Structure–activity relationship reports are very useful, particularly for small numbers of compounds; however, the amount of data rapidly becoming available in most large life science companies will completely overwhelm the cognitive ability of most scientists to view and analyze the data. As we move into higher volumes of data, data reduction and visualization techniques become more important.

2.12.14.5.2 Visualization of data

When working with large volumes of data, humans tend to respond to visual cues much more rapidly than textual displays of the data. Several data visualization tools in use today allow users to simultaneously view multiple attributes. It has been established that by using a combination of size, shape, color, position, and texture, it is possible to perceive up to eight dimensions simultaneously. A popular data mining tool is Spotfire Pro, which allows the user to graphically reduce the data set to the results of interest; for example, displaying the results of multiple tests as dimensions, and the physicochemical properties of the samples as colors, shapes, and sizes.

Table 26 A simplified tall skinny table

Sample number	Test number	Result
SLA-0019	Test 4	80
SLA-0011	Test 7	71
SLA-0015	Test 9	93
SLA-0022	Test 5	28
SLA-0014	Test 8	95
SLA-0013	Test 3	37
SLA-0002	Test 1	49
SLA-0006	Test 5	90
SLA-0013	Test 6	39
SLA-0021	Test 4	80
SLA-0011	Test 2	37
SLA-0004	Test 2	63
SLA-0002	Test 3	38
SLA-0001	Test 8	45
SLA-0001	Test 9	0
SLA-0010	Test 9	52
SLA-0012	Test 8	66
SLA-0016	Test 7	69
SLA-0023	Test 2	55
SLA-0003	Test 5	58
SLA-0023	Test 8	95
SLA-0009	Test 8	3
SLA-0010	Test 6	46
SLA-0021	Test 7	28
SLA-0007	Test 6	97
SLA-0005	Test 9	48
SLA-0017	Test 1	95
SLA-0023	Test 6	91
SLA-0019	Test 3	2
SLA-0001	Test 4	0
SLA-0011	Test 3	4
SLA-0006	Test 4	17
SLA-0003	Test 8	84
SLA-0020	Test 5	52
SLA-0003	Test 2	92
SLA-0021	Test 9	82
SLA-0006	Test 9	78

Table 26 Continued

Sample number	Test number	Result
SLA-0013	Test 1	78
SLA-0012	Test 5	99
SLA-0001	Test 3	87
SLA-0009	Test 2	43
SLA-0013	Test 9	100
SLA-0015	Test 7	66
SLA-0010	Test 1	62
SLA-0020	Test 9	8
SLA-0001	Test 2	73
SLA-0017	Test 5	2
SLA-0002	Test 8	21
SLA-0012	Test 9	27
SLA-0021	Test 3	40
SLA-0012	Test 7	54
SLA-0022	Test 9	52
SLA-0022	Test 4	23
SLA-0015	Test 5	63
SLA-0008	Test 6	99
SLA-0011	Test 5	73
SLA-0010	Test 8	55
SLA-0008	Test 3	61
SLA-0006	Test 3	55
SLA-0014	Test 2	86
SLA-0019	Test 1	74
SLA-0003	Test 1	35
SLA-0020	Test 1	10
SLA-0018	Test 3	79
SLA-0008	Test 9	61
SLA-0019	Test 6	23
SLA-0020	Test 8	50
SLA-0009	Test 5	1
"	"	"
"	"	"
"	"	"
SLA-0005	Test 1	22
SLA-0004	Test 1	81

Table 27 A simplified short wide table

	Test 1	Test 2	Test 3	Test 4	Test 5	Test 6	Test 7	Test 8	Test 9
SLA-0017	95	39	68	48	2	62	12	22	24
SLA-0012	91	54		24	99	5	54	66	27
SLA-0011	90	37	4	32		68	71	50	73
SLA-0022	89	37	58	23	28	38		92	52
SLA-0001	87	73	87	0	71	25	23	45	0
SLA-0004	81	63	76	42	95	12	94	31	39
SLA-0008	78	52	61	25	32	99	73	95	61
SLA-0013	78	51	37	84	25	39	38	71	100
SLA-0019	74	2	2	80	43	23	99	99	100
SLA-0010	62	4	37	76	5	46	85	55	52
SLA-0009	51	43	59	68	1	71	10	3	65
SLA-0002	49	62	38	66	3	73	30	21	64
SLA-0021	35	66	40	80	63	88	28	54	82
SLA-0003	35	92	76	17	58	23	18	84	76
SLA-0006	25	38	55	17	90	2	99	11	78
SLA-0005	22	35	22	13	87	42	34	61	48
SLA-0015	20	82	4	31	63	39	66	46	93
SLA-0018	16	22	79	3	97	7	91	91	68
SLA-0014	15	86	86	4	62	85	100	95	89
SLA-0023	11	55	18	42	95	91	33	95	24
SLA-0007	11	14	26	90	83	97	92	58	100
SLA-0020	10	18	67	77	52	15	64	50	8
SLA-0016	5	42	22	7	1	91	69	48	92

2.12.14.6 Logistics

2.12.14.6.1 Plate handling

HTS is almost entirely performed in microplates, each containing between 96 and 3456 wells. During the past several years, the 384-well plate has supplanted the 96-well plate as the primary HTS vehicle. Tracking the contents of each well in each plate and associating the results with the proper sample is of crucial importance for a HTS system, but several factors must be overcome.

The detected plate (the assay plate) is rarely the plate stored in the inventory. Often, the process of detection itself is destructive. Whether detection is destructive or not, the addition of the biological reagents would make the assay plate unusable for future tests. The usual process has the automation equipment copy a plate taken from the inventory, to create the new assay plate. Reagents are then added to the assay plate, and the detection is performed. At some point, the data management system must be told what is in the assay plate or how to associate the assay plate with the information in an inventory plate. This can be done in several different ways. Two examples follow:

(1) The liquid-handling device, which actually makes the assay plate, keeps a running list of the assay plates that it makes as well as the inventory plates from which it makes each one. In this case, the data management system will have to look up the inventory plate for each assay plate and the get the appropriate inventory information.

(2) When making an assay plate, the liquid handler applies a bar code identifier that includes the inventory plate ID plus an extension. When the data system reads the plate ID (usually in the reader file), it needs to parse out the inventory plate ID and retrieve appropriate inventory information.

2.12.14.6.2 Special cases
2.12.14.6.2.1 Consolidation
An increasing percentage of HTS experiments are being performed on 384-well plates. However, most storage is still done using 96-well plates. During the process, four 96-well plates are typically combined into a single 384-well plate. So, as the sample is transferred from each of the 96-well plates to a different location on the 384-well plate, the information system must be informed about the new location of the sample.

2.12.14.6.2.2 Format changes
Often, samples are purchased from outside suppliers or are stored in the inventory in a format that is not compatible with the format used for the test. For example, many suppliers provide compounds in 96-well plates with all the wells filled; however, most tests require that one or two columns of wells be left blank for controls. In these cases, once again, samples must be moved to different locations on the new plates, and the information system must keep track of which samples are moved where.

2.12.14.6.2.3 Pooling/deconvolution
Some laboratories, in an effort to increase throughput, intentionally mix 8–12 samples per well. When a well displays activity, the samples contained in that well are tested individually to find the 'hot' sample. Assuming that 1% or less of all samples are active, this will result in a dramatic decrease in the number of wells tested, thus increasing throughput and decreasing costs. A special case of pooling is called orthogonal pooling. In this case, each sample is put into two wells, but the combination of samples is unique in each well, and any two wells will only have one sample in common. Therefore, when any two wells are active, the common sample can be located and retested for confirmation.

2.12.14.7 Future Technologies

Predicting the future in any case is a perilous thing to do. In the case of a rapidly changing field, such as HTS, it is downright foolhardy, but that will not keep us from trying. Most of the predictions we are making are just extrapolations of technologies under way. There is always the possibility that a major technological innovation will change the course of screening dramatic changes in HTS processes, or perhaps even obviating the need for HTS. Even in the absence of such dramatic discontinuities, we can see several new technologies that will have an impact on the practice of screening.

2.12.14.7.1 Higher densities and ultrahigh-throughput screening
First, the drive toward higher and higher densities will continue. The economics of HTS make a compelling case. Increases in speed of throughput are a secondary issue to reduction in the cost of reagents used. We fully expect the HTS standard to move from 96-well plates to 384, on to 1536, and possibly as high as 3456. The effect that this will have on informatics systems is fairly obvious. First, as noted earlier, smaller wells require much tighter tolerances; hence, the need for the capability to detect systematic errors will become even more acute. In addition, as we move to higher densities, new visualization tools will become a requirement, as many of the visualizations designed for 96- and 384-well plates will simply not be usable for 3456-well plates. Finally, the need for users to be able to quickly locate information for single samples will become more acute. Higher densities and faster instrumentation together are allowing throughput to increase dramatically. The increased throughput levels, referred to as ultrahigh-throughput screening, will put additional strain on information systems, forcing the systems to process more information at a higher rate.

2.12.14.7.2 Wider use of high-throughput screening techniques
One of the effects of the adoption of HTS throughout the pharmaceutical industry has been to move the bottleneck in the discovery pipeline downstream. Essentially, this means that HTS laboratories are producing hits at a faster rate than the therapeutic areas and development can absorb them. One strategy for dealing with this new bottleneck is to adapt HTS-style techniques for metabolism, toxicity, and other secondary and development assays. This change will have a dramatic effect on information systems, as the data produced by these tests are usually far more complex than those produced in classical HTS assays. For one thing, these tests can often have more than one endpoint. For example, some toxicity tests measure the effects of exposing a specific type of cell to different samples. The scientist may want

to record the percentage of cells that died (or survived), the reproduction rate, the growth rate, and if any specific types of proteins were expressed. Some tests go as far as measuring specific types of damage to the cell. The challenge to the information system is to be able to collect all of the relevant information so as to allow the scientist to declare all of the appropriate variables for the experiment and record the appropriate contextual information (e.g., how was the test run?).

2.12.14.7.3 High-throughput screening kinetics and high-content screening

Until recently, the goal of all HTS programs was simply to increase throughput, the sheer number of samples processed, and hence the volume of data produced. Many HTS scientists believe that this goal has been largely accomplished and that further investment in increasing throughput will bring diminishing returns. The next goal is to increase the quality of the data produced. Essentially, this means that the goal is no longer to simply test more samples but rather to learn more about each sample tested. Some organizations call this 'high-content screening' (HCS). HCS involves tests where multiple endpoints are achieved for each sample. In addition, many kinetics experiments (where multiple readings are taken for each sample over a course of time) have been adapted for automation. Now, instead of a single reading for a well, there can be up to 100 readings for each well. There is a great debate concerning what to do with this raw data. Should it be stored at all? Can it be archived? Should it be available for on-line queries? Most organizations have (to date) decided that it is impractical to attempt to store all the raw data on-line but that it can be archived in off-line storage, and the reduced data (such as K_i) can be stored on-line.

2.12.14.8 Overview

While it may not be possible to predict what HTS will look like in the future, several points are clear:

- The volume of data will continue to increase. Whether it is from higher throughput or higher content, or both, more data will be available to scientists. The challenge to the information system designer, then, is to create tools that allow the scientist to store all the relevant data and then retrieve it in such a way that information can be extracted quickly. The goal is to help the scientist find 'actionable' data, or data that cause the scientist to make a decision, such as 'this sample is worthy of further study' or 'the results of this test should be rejected.'
- As the volume of data increases, data storage systems will come under increasing pressure. Currently, most major pharmaceutical companies store all chemical or biological data in Oracle databases from Oracle Corporation. Oracle is migrating technology to object orientation, which can provide increased performance. It will be up to the developers of chemical and biological information management systems to utilize advances in the underlying core technologies as they become available.
- Data-mining tools are being developed for industries ranging from retail to financial services. While the volume of data being produced by HTS and related technologies calls for data-mining and statistical solutions, and there is a great deal of work going on in this area, it is unclear at the time of writing whether general-purpose data-mining tools can be adapted for drug discovery.
- Science will continue to evolve, and information systems must also continue to evolve to support the science.

2.12.15 Combinatorial Chemistry: The History and the Basics

2.12.15.1 Definition

What is combinatorial chemistry? There have been several opinions, some formulated very sharply, but most expressing what combinatorial chemistry is not. At the end of the conference 'Combinatorial Chemistry 2000' in London, a discussion session was organized to answer this question and make sure that our understanding of the term reflects the fact that at least some operations in the synthesis of the group of chemical compounds is performed in combinatorial fashion. Unfortunately, when it came to the public vote, the scientists in the audience voted for a much broader definition of combinatorial chemistry. The majority expressed the opinion that combinatorial chemistry is defined by the design process – that compounds designed by the combination of building blocks (and synthesized by whatever means) are the subject of combinatorial chemistry. In the literature, the term 'combinatorial chemistry' is used very often; however, the definition is found rarely. Seneci[597] says that "combinatorial chemistry refers to the synthetic chemical process that generates a set or sets (combinatorial libraries) of compounds in simultaneous rather than a sequential manner." The *Journal of Combinatorial Chemistry* defines combinatorial chemistry as "a field in which new chemical substances – ranging from pure compounds to complex mixtures – are synthesized and screened in a search for useful properties."

But do we really need to define combinatorial chemistry before discussing the history of this branch of science? Must we have consensus about the term 'combinatorial chemistry' before we start a new journal with the term in its title? (There are already several journals with the term on their covers.) Apparently not, and the precise definition is probably not as important as the fact that the novel techniques are being widely accepted and applied as needed for a variety of projects, starting from finding new drug candidates and ending in the discovery of new inorganic materials.

2.12.15.2 History

Maybe the best introduction to combinatorial chemistry is through its brief history. In 1959, the young chemist Bruce Merrifield had the idea that it would be extremely beneficial to modify the sometimes unpredictable behavior of growing peptide chain intermediates by attaching the chain to the polymeric matrix, the properties of which would be very uniform from step to step.[598–601] His invention of solid phase synthesis, for which he was awarded the Nobel Prize,[600] changed the field of peptide synthesis dramatically. The synthesis of oligonucleotides followed immediately[602]; however, the solid phase synthesis of organic molecules was pursued basically only in the laboratory of Professor Leznoff.[603,604] Even though solid phase synthesis was more or less accepted in the chemical community, it took another 20 years before the new ways of thinking about the generation of a multitude of compounds for biological screening brought combinatorial chemistry to life. Pressure from biologists motivated the development of combinatorial chemistry. Chemists could not keep up with the demand for new chemical entities. Big pharmaceutical companies started to screen their entire collections of chemical compounds against new targets, and the rate at which these collections grew seemed unsatisfactory. Ronald Frank in Germany,[605] Richard Houghten in California,[606] and Mario Geysen in Australia[607] devised ways to make hundreds of peptides or oligonucleotides simultaneously by segmenting the synthetic substrate–solid support. Frank used cellulose paper as the support for the synthesis of oligonucleotides. Cutting the circles of the paper and reshuffling the labeled circles for each cycle of the coupling was a very simple way to generate hundreds of oligos. Houghten enclosed classical polystyrene beaded resin in polypropylene mesh bags, later called 'tea-bags' or 'T-bags,' and used them for parallel synthesis of hundreds of peptides. The principle was the same: combine the bags intended for coupling the same amino acid and re-sort the bags after each cycle of coupling. Geysen used functionalized polypropylene pins arranged in the fixed grid. Each pin was then immersed in a solution of activated amino acid pipetted into the individual wells of microtiter plate. Pins were not re-sorted after each step, but the common steps of the synthesis (washing and deprotection) were done by the introduction of the pins into the bath containing appropriate solvent. These techniques cleared the way for the arrival of real combinatorial techniques applied to general organic chemistry and not only to the specific arena of peptides and oligonucleotides.

For biologists and biochemists, working with mixtures was absolutely natural – well, it was natural also for natural products chemists – however, organic chemists were (and still are) horrified when mixtures were mentioned. Therefore, the development of specific binders selected from the astronomically complex mixtures of RNA by selective binding and amplification of selected molecules by PCR was accepted enthusiastically, and papers describing it were published in *Science*[608] and *Nature*,[609,610] (Larry Gold and his colleagues were adventurous enough to build a company around this technology – NeXstar, subsequently merged with Gilead.) Relatively fast acceptance was given to the techniques generating specific peptides on the surface of the phage, panning for the binding sequences and amplification of the phage,[611,612] described by Smith. Again, the approach was basically biological. However, earlier attempts to publish papers describing the use of synthetic peptide mixtures for determination of epitopes in *Nature* were unsuccessful; the world was not ready for chemical mixtures. Geysen's seminal paper was eventually published in *Molecular Immunology*,[613] and did not find a large audience. In this paper, the mixture of amino acids was used for the coupling at the defined positions, thus generating large mixtures of peptides. Mixtures showing significant binding were 'deconvoluted' in several steps to define the relevant binding peptide sequence at the end.

The pioneer in development of the methods for creating the equal mixtures (of peptides) was Arpad Furka in Hungary. His method of 'portioning–mixing' was invented in 1982, and presented as posters in 1988 and 1989.[614,615] The method was not noticed until 1991, when it was reinvented and published in *Nature* by two independent groups, Lam *et al.* in Arizona (the 'split-and-mix' method)[616] and Houghten *et al.* in California (the 'divide–couple–recombine' method).[617] The technology of deconvolution of mixtures was the basis of the formation of Houghten Pharmaceuticals, later renamed Trega Biosciences and now known as LION Biosciences. Finding the active molecule requires synthesis of the second (and third, and fourth, etc.) generation mixtures of lower complexity based on the activity evaluation of the most active mixture from the first round of screening. An alternative method is positional scanning, in which mixtures of the same complexity with defined building blocks in all positions of the sequence are screened, and the importance of individual blocks is ascertained. The combinations of all 'important' residues are then assembled in the possible 'candidate sequences,' which are then tested individually.[618] The use of mixture-based libraries has been reviewed.[619]

Portioning–mixing (split-and-mix or divide–couple–recombine) is a simple but powerful method that not only allows the generation of equimolar mixtures of compounds but is also the basis of one-bead-one-compound technology for the screening of individual compounds (as recognized by Lam[616,620,621]). In this modification, the synthetic compounds are not cleaved from the resinous bead, and binding is evaluated by assay performed directly on the bead. The structure of a compound residing on a positively reacting bead is then established by direct methods or by reading 'the code' associated with that particular bead. The one-bead-one-compound technique can be modified for the release of the compound to solution,[622] or to semisolid media,[623] to allow for the use of assays not compatible with solid phase limitations. Again, this technology jump started the first combinatorial chemistry company, Selectide (now part of Aventis).

2.12.15.3 Small Organic Molecules

Libraries of peptides and oligonucleotides were relatively easy to handle both in the mixture and in the individual one-bead-one-compound format. The determination of the structure of peptides and/or oligonucleotides is made relatively easy by sequencing requiring picomolar or even lower amounts of material. At the same time, synthetic methodologies for their synthesis are well developed. However, good candidates for new successful drugs are being sought between 'small organic molecules.' Libraries containing nonoligomeric organic compounds were obviously the next step in the development of combinatorial chemistry. Jonathan Ellman recognized this need, and developed a method for the solid phase parallel synthesis of benzodiazepines.[624] His publication, together with published results from Parke-Davis[625] and Mimotopes,[626,627] started a flood of communications about the application of solid phase synthesis to the preparation of enormous numbers of different categories of organic compounds, with the major focus on heterocyclic molecules. (Numerous compilations of solid phase syntheses were published; e.g., see the examples in the reference list,[628–631] and a dynamic database of all relevant publications is available on the internet.[1105]).

Transformation of one-bead-one-compound libraries to the arena of small organic molecules requires methods allowing simple and unequivocal determination of the structure from the individual bead containing picomolar amounts of analyzable material. This problem was addressed by the inclusion of 'tagging' into the synthetic scheme.[632–635] The structure of the relevant molecule is determined by reading the 'tag.' A most elegant method for tagging was developed by Clark Still.[633] Again, as a rule in this field, the result was the formation of a new company, Pharmacopeia. In this method, the tagging of the organic molecule is achieved by a relatively small set of halogenated ethers attached to the bead as a defined mixture in each step of the synthesis, forming digital code (each molecule of the tagging substance is either present (1) or absent (0)), evaluated after the detachment from the bead by gas chromatography. It did not take long before the combinatorial techniques were applied to material science.[636–640] These libraries are produced usually in a spatially addressable form and were used to find new supraconductive, photoluminescent, or magnetoresistive materials.

2.12.15.4 Synthetic Techniques

Although the pressure to produce more compounds was visibly coming from pharmaceutical companies, some of the new techniques were developed at academic institutions. Big companies still did not embrace the new techniques, possibly due to the fact that they are quite simple and inexpensive to implement. Pharmaceutical companies do not want simple solutions; they would rather invest in enormous automation projects. Managers are judged by the budget they are able to invest, and a big room full of robotic synthesizers definitely looks impressive. Another major factor is the 'visibility' of the compound produced. The production of 100 nmol of the compound (about 50 µg of an average organic compound), which can make 100 mL of 1 µM solution (enough for 1000 biological assays), is unacceptable – simply because it is not 'visible.' Companies usually require 5–50 mg of the compound (more than enough for 1 million assays) just to 'have it on the shelf.' And techniques providing 100 nmol are definitely cheaper and require less automation than techniques needed to make milligram quantities of the compound.

A very elegant technique for synthesizing numerous organic compounds in parallel was introduced by Irori. This company was based on the idea that it is possible to label individual polymeric beads with a readable radiofrequency tag, which will be built during the split-and-mix synthesis of any type of molecule. The technique of 'Microkans' – small containers made from polymeric mesh material inside which are beads used for solid phase synthesis together with a radiofrequency tag[641,642] – is used in numerous laboratories.[643] The most recent incarnation of this technique (based on the original principle of the 'tea-bag' synthesis of Houghten[606]) is the labeling of small disks containing 2–10 mg of synthetic substrate, called 'NanoKans,' by a two-dimensional bar code on a small ceramic chip.[644]

On the other hand, thousands of compounds can be synthesized relatively inexpensively in polypropylene microtiter plates using either 'surface suction'[645] or 'tilted centrifugation.'[646] However, nothing can be more economical and versatile for the synthesis of up to couple of hundred compounds than disposable polypropylene syringes equipped with polypropylene frits, as introduced by Krchnak.[647] A syringe is charged with the solid support of choice, and all steps of the synthesis are performed by aspirating appropriate reagents using needles and (if needed) septum-closed bottles. The operation of syringes can be simplified by the use of domino blocks.[648]

2.12.15.5 Philosophy and Criteria

The different approaches to the synthesis of libraries illustrate the different philosophies of laboratories and companies. The same difference in thinking can be found in the value given to the purity of prepared compounds. Different companies apply different criteria. However, the only really important information that the chemist should be able to provide to the biologist is whether he or she can guarantee the preparation of the same sample tomorrow or a year from now. Does he or she have the stable, well-rehearsed protocol and reliable source of starting materials? If the biological activity is found in the impure sample, the likelihood that the active component of the mixture can be found after isolation of all components is pretty high. The probability that the activity is higher than observed in the mixture is also high. Occasionally, the active species might not be the one that was targeted but, rather, the side product of unexpected (and hopefully novel) structure.

We could go on discussing combinatorial chemistry, but because this text is intended to be an introduction to the history of the fields, we will stop here and refer readers to the published literature. The histories and personal recollections of the pioneers in this field were compiled in the inaugural issue of *Journal of Combinatorial Chemistry*,[649] and a similar format was used for a history of solid-supported synthesis.[650,651] In addition to books on the subject of combinatorial chemistry and solid phase synthesis,[652–670] we recommend attendance at the biannual symposia on solid phase synthesis and combinatorial techniques,[671] organized by Roger Epton. Reading of review articles[628–631,672–684] is also helpful. We also direct readers to the internet website compiling all papers published in this exciting and rapidly growing field.[1105]

2.12.16 New Synthetic Methodologies

The synthesis of libraries of diverse drug-like molecules is dependent on the availability of reliable and general synthetic methods. This section is intended to give the reader an overview of the latest developments in solid phase and solution phase syntheses directed to constructing libraries of compounds for biological screening. The brevity of this review with respect to some of the more established synthetic methods is by no means meant to diminish their importance, but rather to maintain the focus of this section on the most recent synthetic methodology advances. The reader is encouraged to consult the extensive literature cited in this section for complete details of these new synthetic methods.

2.12.16.1 Introduction

The development of high-throughput screening (HTS) technology in the late 1980s and early 1990s and the dramatic increases in the number of biological targets available from the Human Genome Project has fueled the desire of pharmaceutical companies for larger numbers of compounds to screen against these new biological targets. The solid phase synthesis of benzodiazepines was first demonstrated in 1977 by Camps Diez and co-workers.[685] Around this same time, several other organic chemists, including Leznoff and Frechet, also synthesized small molecules on solid supports.[686–688] However, little general notice was taken of these accomplishments until the early 1990s, when Ellman and Hobbs-Dewitt published their syntheses of benzodiazepines, and solid phase synthesis was recognized as an ideal method for the construction of large compound libraries.[689,690] Multiple parallel synthetic techniques first described by Geysen (using multipin arrays[691]) and Houghten (using 'tea-bag' methods[692]) and combinatorial synthesis by the split-and-mix method initially described by Furka *et al.* made possible the synthesis of vast numbers of diverse compounds for biological screening.[693,694] The rapid medicinal chemistry optimization of lead compounds discovered from these collections by analogous parallel synthesis methods promised to shorten the time required to advance these compounds to clinical trials. Although in 1992 it was clear that solid phase synthesis methods could significantly enhance the drug discovery process, the lack of robust solid phase synthetic methodologies severely limited its utility. The race was then begun to discover new synthetic methods, invent new robotics and instrumentation, develop computer-assisted library design and analysis software, and integrate these tools into the drug discovery process. The plethora of publications in combinatorial organic synthesis and the establishment of journals dedicated to combinatorial chemistry (e.g., the *Journal of Combinatorial Chemistry*, *Molecular Diversity*, and *Combinatorial Chemistry and High Throughput Screening*) demonstrates the commitment of the synthetic organic community to these endeavors.[695–723]

2.12.16.1.1 Strategies for combinatorial library syntheses

The efficient combinatorial synthesis of quality compound libraries can be reduced to a few basic principles: (1) compound libraries must be amenable to high-throughput synthesis; (2) the scope of chemistry must be sufficiently broad; (3) the building blocks must be readily available; (4) library purities must be excellent (>85% on average); and (5) the yield must be adequate (>25% on average). The ideal library synthesis thus consists of short chemical sequences composed of highly optimized synthetic reactions to ensure adequate yields and high purities of the library members. A variety of building blocks used in the synthesis would be available from commercial sources or one- to two-step syntheses, allowing for the rapid synthesis of large compound libraries.

Solid phase synthesis is one of the most powerful methods for construction of large compound libraries. It is amenable to 'split-and-pool' methods for extremely large (>10^4–10^7 members), single-compound-per-bead libraries, mixture libraries, or discrete compound syntheses. The ability to drive reactions to completion using excess reagents and simple removal of impurities from the crude reactions by washing the resin permits the multistep synthesis of complex scaffolds in excellent purities and yields. Recent advances in solution phase scavenging and solid-supported reagents are also proving to be effective in multistep synthesis of libraries. A comparison of the advantages and disadvantages of solid phase versus solution phase synthesis described by Coffin in an article by Baldino is provided in **Table 28** (refer to the article for clarification of the 'issues' discussed in the table). The obvious synergies of these two methods are apparent from the table. A proficient combinatorial chemistry laboratory is capable of utilizing either solution phase or solid phase high-throughput methods, and the decision of which method to use is based on a variety of factors. These factors include the type of chemistry to be performed, the size and scale of the library, and the intended use of the library (i.e., a lead discovery library versus a lead optimization or a structure–activity relationship library).

2.12.16.2 Solid Phase Synthesis Methodologies

Numerous advances in solid phase synthesis methodologies have been published in the literature in the past decade. In fact, most solution phase synthetic transformations have now been performed in some form on solid supports. This section presents brief descriptions of many of the most recent solid phase synthesis advances, and attempts to explain the importance of these methods for the rapid synthesis of compound libraries. The reader can expect to gain from this section a good overview of the current state of the art and where the field of solid phase synthesis is headed at the beginning of the twenty-first century.

2.12.16.2.1 Introduction to polymer-supported synthesis
2.12.16.2.1.1 Pros and cons of synthesis on a support

The concept of polymer-supported synthesis was popularized by Bruce Merrifield, with the publication in 1963 of his seminal paper describing the solid phase synthesis of peptides.[724] In this work, Merrifield covalently attached an amino acid to an insoluble polymer bead made from polystyrene cross-linked with 1–2% divinylbenzene (**Figure 89**). When placed in a reaction solution, the polystyrene beads swell and allow solvent to penetrate the bead, permitting a dissolved reagent access to the linked amino acid. Because the beads do not dissolve, they can be isolated from the solution by simply filtering them on a fritted funnel. The power of this method is the trivial isolation of the polymer (and thus the polymer-bound compound) from a reaction solution by filtration after the addition of each new reagent. This method completely eliminates the need for the isolation and purification of the intermediate products, each of which usually requires an aqueous extraction and chromatographic purification. This procedure is particularly useful in the synthesis of peptides, where the addition of each amino acid in solution requires two chemical steps (deprotection and coupling), as well as a separate time-consuming purification after the addition of each new amino acid (for more details, see **Figure 134** and Section 2.12.16.3.1.1 on peptide synthesis on a solid support). Using the solid phase procedure an entire peptide can be assembled while attached to the polymer bead, and only a single purification step at the end of the synthesis is necessary. Peptides of unprecedented length and purity were synthesized in a much shorter time using the Merrifield procedure than was previously possible using traditional solution phase peptide synthesis methods. The success of

Table 28 Advantages and disadvantages of solution phase versus solid phase parallel synthesis[a]

Issues	Solution phase	Solid phase
Range of accessible reactions	+ +	−
Production of congeneric sets of compounds in structure–activity relationship ordered arrays	+ +	−
Use of in-process controls	+	−
Effort required to 'combinatorialize' a synthetic reaction or scheme	+	−
Linker attachment sites	NA	−
Larger-scale resynthesis of bioactive library members	+ +	−
Choice of solvents	+	−
Operating temperatures	+	−
Heterogeneous reagents	+	−
Scavenger resins	+	−
Cost of reagents and materials	+	−
Abundance of literature precedents	+ +	−
Location-based sample identification	+ +	−
Tagging-based sample identification	−	+
Capital investment required	−	+
Maintaining inert conditions	−	+
Mass action reagent excess	−	+ +
Library transformations with no change in diversity	−	+
Protecting groups	−	+
Multistep synthesis	−	+
Use of bifunctional reagents	−	+
Access to split-and-pool amplification	−	+ +

[a] See the original article by Baldino[723] for a detailed discussion.

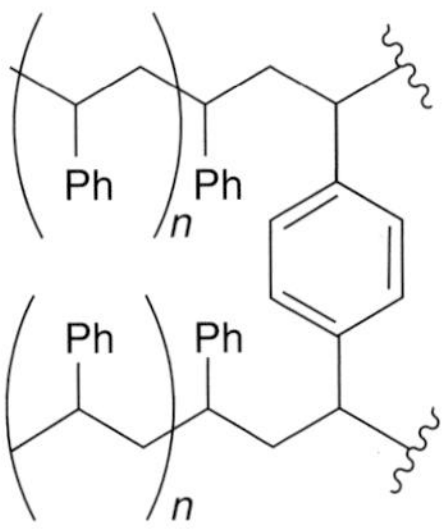

Figure 89 Structure of cross-linked polystyrene.

polymer-bound peptide synthesis also resulted in the development of similar technology for the construction of other biopolymers, including oligonucleotides, which are now routinely constructed in a completely automated fashion.

The concept of polymer-supported synthesis for molecules other than peptides was investigated in the late 1960s and early 1970s by a number of investigators who helped to define the scope and limitations of polymer-bound synthesis of organic molecules.[686,687] There are a number of advantages to polymer-bound synthesis. As mentioned above, the main advantage of an insoluble polymer support is that the insoluble polymer may be isolated from a reaction solution by filtration, making conventional purification of the desired compound unnecessary. This simplified purification procedure allows the use of a large excess of reagents to drive a reaction to completion while minimizing the time needed to purify and isolate the desired product from the complex reaction solution. Linkage of a molecule to a polymer support also facilitates encoding; an identifying tag of some kind can be linked to the polymer bead containing a particular molecule. Systems have been developed to allow the screening of millions of compounds followed by the identification of active structures by decoding tags attached to the polymer bead during the synthesis of the particular compound.[725] Although solid phase organic synthesis remains one of the most powerful techniques for parallel synthesis, it has drawbacks. Disadvantages of this method may include diminished reaction rates for certain reactions, formation of support-bound impurities that are released along with the desired molecule during cleavage, and the necessity of developing a method for linkage of the desired compound to the support. The analysis of resin-bound intermediates can also be complicated, and additional time is often required to develop robust and versatile chemistry for multistep syntheses on a solid support. A variety of tools have been developed or adapted to aid in the analysis of compounds on a solid support, including solid-state nuclear magnetic resonance (NMR), Fourier transform infrared and mass spectroscopies, as a number of color tests for the presence or absence of certain functional groups.[726–730] New technologies have been designed to try to retain the benefits of solid phase chemistry while minimizing the drawbacks (see Section 2.12.16.4.1 on solution phase polymer-supported synthesis).

2.12.16.2.2 Synthetic transformations on solid supports

2.12.16.2.2.1 Carbon–heteroatom coupling reactions on solid supports

Carbon–heteroatom coupling reactions are the most widely used synthetic transformations for the construction of compound libraries. N-Acylation, N-alkylation, N-arylation, O-acylation, O-alkylation, and O-arylation are just a few examples of these ubiquitous reactions performed on solid supports (**Scheme 1**). Nearly all reported multistep solid phase syntheses incorporate one or more of these reactions. In general, these reactions tend to be easily optimized, occur at or near room temperature, and are not particularly sensitive to air or moisture. These mild reaction conditions simplify tremendously the automation of the library synthesis, since reagents and reactions do not need to be cooled, heated, or kept under an inert atmosphere. Recent engineering advances in instrument design do permit the high-throughput synthesis of compound libraries under these more demanding conditions. Thousands of the reagents used in these syntheses are available from commercial sources. Diverse or focused sets of reagents can therefore be readily purchased for incorporation into diversity-oriented or target-directed compound libraries.

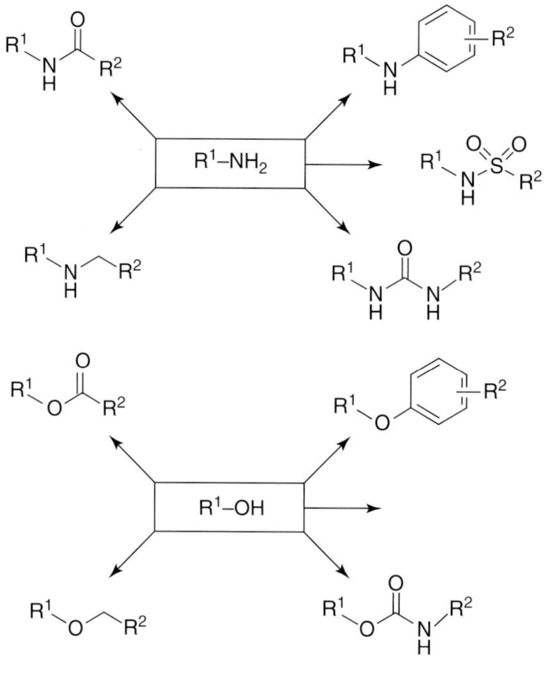

Scheme 1

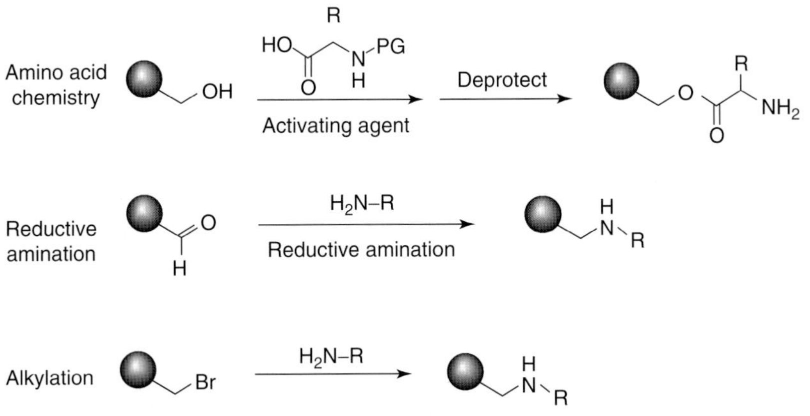

Figure 90 Examples of resin-bound amines.

2.12.16.2.2.1.1 *N-Acylation, N-alkylation, and N-arylation* Many synthetic methods for the derivatization of resin-bound amines have been optimized for the construction of diverse compound libraries. The ease of amine incorporation and subsequent derivatization, along with the large and diverse set of amine building blocks that are commercially available, has driven the exploitation of this functionality. A wide variety of resin-bound amines are thus readily available, since they can be incorporated onto the support in high yields and purities via standard amino acid chemistry, reductive amination of aldehyde-functionalized resins, or amine displacement of resin-bound halides, to name just a few synthetic methods (**Figure 90**).

Derivatization of resin-bound amines can be accomplished by many different synthetic transformations, thus allowing the generation of diverse product libraries. Several examples are given in **Scheme 1**. *N*-Acylation is a particularly useful diversity forming reaction, since thousands of building blocks (acids, acid chlorides, sulfonyl chlorides, isocyanates, chloroformates, carbamoyl chlorides, and others) are commercially available, and each can be performed in high yield (**Scheme 1**). *N*-Alkylation of resin-bound amines is another useful reaction sequence, since it affords basic secondary or tertiary amine products with dramatically different chemical and physical properties. The aldehyde, alkyl halide, or alcohol building blocks necessary for the synthesis of these *N*-alkylamine compounds (via reductive amination or *N*-alkylation, respectively) are also readily available, allowing large, diverse libraries of these compounds to be synthesized. A general method for the *N*-arylation of resin-bound amines is a relatively new addition to the optimized solid phase reactions.[731–734] A complete description of these transformations can be found in Section 2.12.16.2.2.3 on transition metal-mediated coupling reactions.

2.12.16.2.2.1.2 *O-Acylation, O-alkylation, and O-arylation* The carbon–oxygen bond is a ubiquitous motif found in many drugs. Therefore, solid phase synthetic methods for carbon–oxygen bond formation (*O*-acylation, *O*-alkylation, and *O*-arylation) have been utilized many times in compound library synthesis (**Scheme 1**). The robust nature of this chemistry with respect to the ease of synthesis and variety of building blocks that undergo efficient chemical reaction make these methods very attractive for diverse compound library construction. While the resultant esters can be somewhat unstable to proteases under physiological conditions, the ether and carbamate linkages are typically stable.

The hydroxyl functional group also offers a handle for linking a scaffold to a suitable polymer support. This technique has been a useful synthetic strategy for those libraries where a free hydroxyl group is desired in all members of the library. The solid phase syntheses of hydroxyethylamine aspartyl protease inhibitor libraries and prostaglandin libraries are excellent examples of this strategy (**Figure 91**).[735–737] The hydroxyl group is an integral part of the aspartyl protease inhibitor and prostaglandin pharmacophores. The linkage at the conserved hydroxyl group not only binds the core to the resin for subsequent derivatization but also protects the hydroxyl from undesired reactions during the library synthesis. The free hydroxyl group is only revealed upon cleavage of the final products from the solid support.

2.12.16.2.2.2 Carbon–carbon bond-forming reactions on solid supports
The biopolymer syntheses described in Section 2.12.16.3.1 involve the formation of a carbon–heteroatom bond (an amide bond for peptides, a glycosidic ether linkage for oligosaccharides, and a phosphate ester bond for

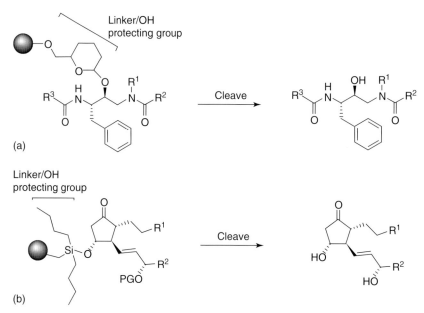

Figure 91 Use of a support/linker as a hydroxyl protecting group in the synthesis of (a) hydroxyethylamines and (b) prostaglandins.

oligonucleotides). While the ability to form these bonds is crucial to the synthesis of biopolymers, an expanded set of reactions is required to allow chemists to access more complex organic structures. A key tool for synthetic organic chemists is the ability to synthesize carbon–carbon bonds. There are numerous methods available for the generation of carbon–carbon bonds in normal solution chemistry (which in most cases translate well into soluble polymer chemistry). However, these methods must often be modified for reactions in the presence of a solid support, where factors such as polymer solubility (or 'swelling' ability), reagent solubility, and compatibility of any linker or previously existing functionality must be taken into consideration. Extensive effort has gone into developing conditions for a variety of carbon–carbon bond-forming reactions on solid supports, a number of which are now available to the chemist for use in the parallel synthesis of small organic molecules.[738] Several examples of these reactions are shown below. Numerous other carbon–carbon bond-forming reactions have been described, including metal-mediated coupling reactions (such as the Stille, Suzuki, and Heck reactions)[711] and multiple-component condensations (e.g., the Passerini, Ugi, and Biginelli reactions).[739] Many of these other reactions are described briefly elsewhere in this section, and in-depth discussions can be found in several recent reviews of solid phase synthesis.[698,701,710,711,713,738,740] The Grignard reaction involves the attack of a carbon nucleophile (for Grignard reactions, a carbon–magnesium halide salt) on an electrophilic carbon such as an imine, Weinreb amide, or ketone, as shown in the examples in **Figure 92**.[741] While the examples above show classical magnesium bromide Grignard salts acting as nucleophiles, similar reactions have been demonstrated on supports using organolithium and organocuprate reagents. (Cuprates are traditionally used to introduce a carbon nucleophile to an α,β-unsaturated carbonyl group at the β position of the double bond.) When performing reactions with strong nucleophiles/bases such as Grignard reagents, consideration must be given to factors such as the other functional groups present on the support, whether the linker can withstand reaction conditions involving strong nucleophiles, and so on. There exist in the literature numerous examples of Grignard reagents being successfully applied in a library format to a support-bound reactant.

The Wittig and Horner–Emmons reactions (**Figure 93**) are often used to install carbon–carbon double bonds into a target molecule, and have been readily applied to support-bound synthesis. The Baylis–Hillman reaction (**Figure 93**) involves the coupling of an aldehyde to an electron-poor alkene. Unlike most other carbon–carbon bond-forming reactions, it does not require an inert atmosphere, and can usually be conducted at room temperature. These mild reaction conditions make the Baylis–Hillman reaction well suited for combinatorial and parallel synthetic applications.[742] In an enolate alkylation, the enolate is generated by the reaction of a carbonyl-containing compound with a strong base (shown in **Figure 93**). The enolate is then reacted with an electrophile, resulting in the formation of the carbon–carbon bond. There have been various reports of enolates being generated and alkylated on a support to

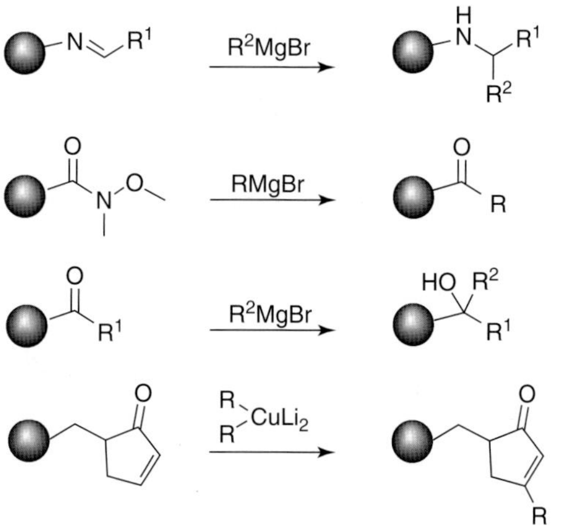

Figure 92 Carbon–carbon bond formation on a solid support: carbon nucleophiles (Grignard and cuprate additions).

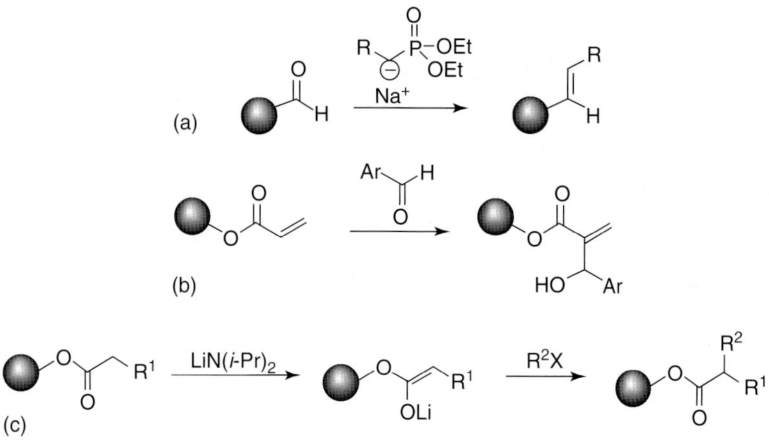

Figure 93 Carbon–carbon bond formation on a solid support: (a) Horner–Emmons reaction, (b) Baylis-Hillman reaction, and (c) enolate formation and reaction.

install a carbon–carbon bond adjacent to a carbonyl group. Chiral auxiliaries have been attached to a support adjacent to the reacting center to influence the stereochemistry of the chiral center that is formed, with diastereoselectivites that were found to be comparable to those obtained via the analogous solution phase reaction.

2.12.16.2.2.3 Transition metal-mediated coupling reactions

Transition metal-mediated coupling reactions have been utilized extensively in solid phase syntheses.[711] Palladium-mediated Heck, Suzuki, and Stille reactions are particularly useful for diversifying compound libraries due to their generality and efficiency of coupling (**Figure 94**). These biaryl- and arylalkene-forming reactions have been effected in high yield with either the metallated arene/alkene or the arylhalide bound to the solid support.

Transition metal-mediated olefin metathesis has been utilized in a number of solid phase syntheses for the construction of rings of various sizes, including several examples of macrocycles that would be inaccessible by other solid phase synthetic methods (**Figure 95**).[743–745] The generality of the method has made olefin metathesis a useful tool for the synthesis of various macrocycles. The resulting cyclic scaffolds are preferred over their linear counterparts since they often bind more tightly to proteins, due in part to their reduced entropy of binding. Medicinal chemists also glean additional structural information from leads discovered from these compound libraries, since their mode of binding can be predicted with greater precision due to decreased conformational flexibility of the molecules.

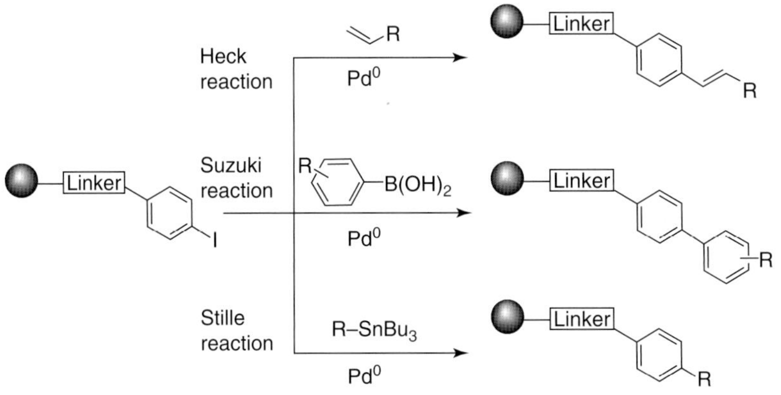

Figure 94 Support-bound Heck, Suzuki, and Stille reactions.

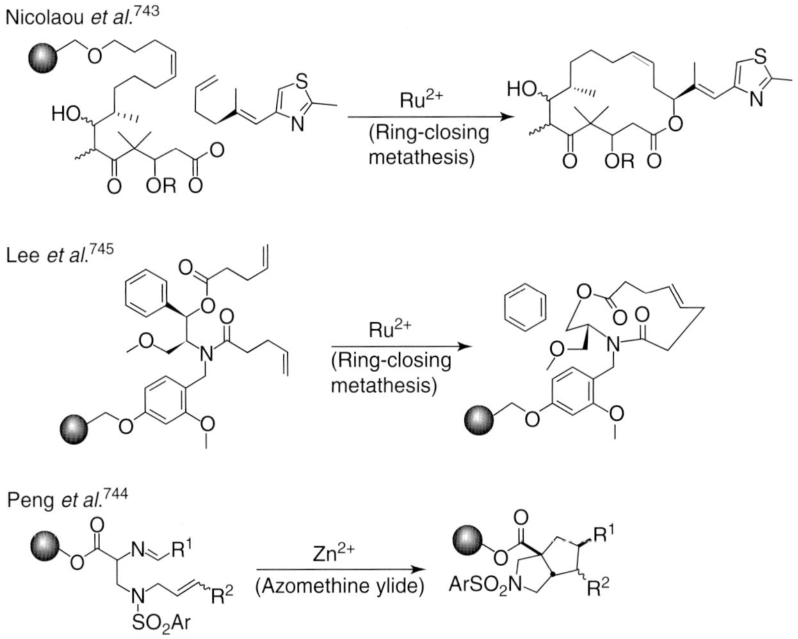

Figure 95 Three examples of transition metal-catalyzed ring-closing reactions on solid support.

Advances in palladium-mediated and copper-mediated N-arylation reactions have been demonstrated on solid supports (**Figure 96**).[731–734] These synthetic methods are particularly useful due to the omnipresent nature of the N-aryl bond in biologically active compounds. N-Arylation of support-bound amines by copper-acetate-mediated coupling of boronic acids afford N-arylated amines, while suitably substituted arylhalides on solid supports can be N-arylated with a variety of primary and secondary amines using palladium catalysis. These complementary reactions provide new synthetic avenues for the generation of diverse structures not previously available to the combinatorial chemist.

2.12.16.2.2.4 Intramolecular cyclization–resin-cleavage strategies

When conducting syntheses on a support, whether that support is a soluble polymer or an insoluble ('solid') phase, it is desirable to minimize the number of reaction steps required to obtain the final products. The time and material saved in such combinations of steps (e.g., by not having to wash a library on a solid support between reactions) is often worth the price of a small decrease in yield or purity. The two steps of cyclization to form a ring and cleavage of the product from the support have been combined in a number of reaction sequences. In the situation where cyclization directly

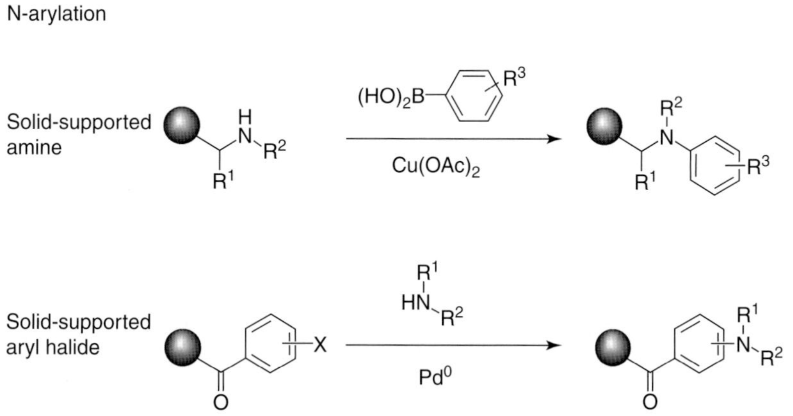

Figure 96 Two examples of N-arylation chemistry on a solid support.

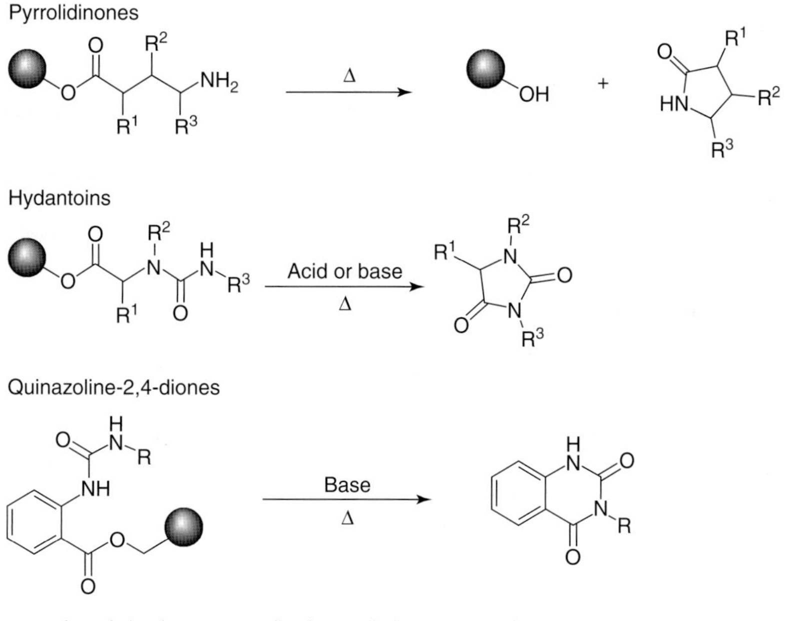

Figure 97 Three examples of simultaneous cyclization and cleavage reactions.

results in cleavage from the support, product purity is excellent. Typically, only the desired completed sequence is capable of cleavage, whereas any incomplete sequences remain on the support. When the cyclization and cleavage reactions occur sequentially (such as in the cyclopentapiperidinone case shown below, where acidic cleavage and cyclization are not necessarily simultaneous), and the cyclization does not proceed to 100% completion, both the cyclized and uncyclized material are released into solution. Both cyclization to release product and 'one-pot' cleavage followed by cyclization of product have been successfully applied to a number of small-molecule syntheses on a polymeric support, as shown in **Figures 97** and **98**.

Three examples of syntheses where cleavage from the support occurs as a direct result of cyclization (such that only the desired cyclized material is isolated) are shown in **Figure 97**.[746,747] In all three cases, only the cyclized product is obtained in high purity, though in varying yields, after the cyclization/cleavage step. All impurities are either washed from the resin prior to the final step or remain attached to the solid phase.

Two examples where cleavage and cyclization are not necessarily linked events are shown in **Figure 98**. The reaction conditions result in cleavage of material from the support, whether or not cyclization occurs. In the dihydrobenzodiazepine-2-one case, the product ester is generated after reaction with sodium methoxide at the terminal carboxylic acid/ester, while cyclization takes place via amine attack at the unsaturated carbon adjacent to

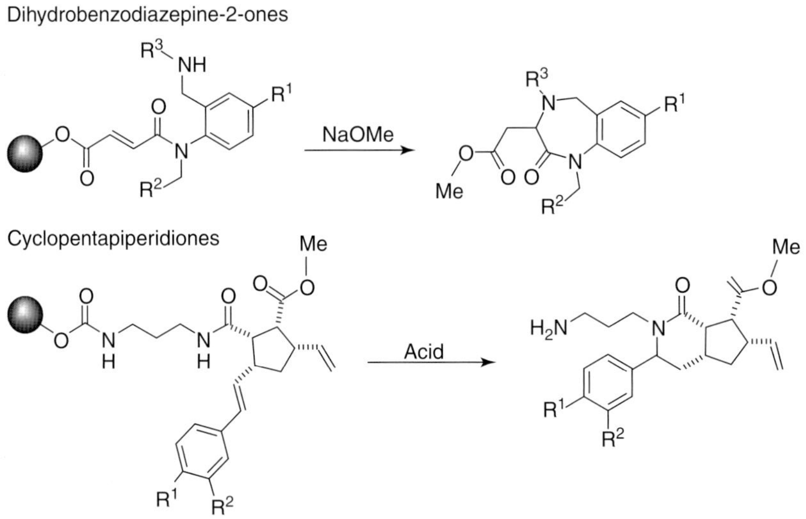

Figure 98 Two examples of cleavage from a support, followed by nonsimultaneous cyclization.

the amide.[748] In the cyclopentapiperidinone case, trifluoroacetic acid cleaves the free amine from the resin, then subsequently catalyzes the observed cyclization.[749] In these two cases, if cyclization could not occur, the molecules are still cleaved from support, and hence an additional impurity (the uncyclized material) may be present in the final product.

2.12.16.2.2.5 Cycloadditions

Carbocyclic and heterocyclic cores are frequently used to display pharmaceutically interesting functional groups. The cyclic core allows for the specific orientation of functional group side chains, often resulting in improved binding affinity or specificity versus more flexible linear cores. One method for obtaining cyclic and heterocyclic cores that has been exploited to a great extent in parallel and combinatorial synthesis is the cycloaddition reaction. Cycloadditions are very attractive to the chemist synthesizing a compound library, since in one step multiple bonds can be formed, introducing a carbocycle or heterocycle while concurrently installing one or more side chains in specific positions on the cyclic core. Strategies have also been developed whereby cyclization to provide the desired molecule and cleavage from the solid support occur simultaneously (see the previous section on intramolecular cyclization/cleavage strategies). A number of different support-bound cycloadditions have been reported, usually resulting in the formation of four-, five-, and six-membered rings with varying displays of side chains around the cores.[701,713,720,738,750] Several examples of cycloadditions on solid support are shown in **Figure 99**.

2.12.16.2.2.6 Multiple-component reactions on solid supports

Multiple-component reactions simultaneously condense three or more components to generate new complex molecules in a single synthetic transformation (**Figure 100**). Relatively few such synthetic transformations are known when compared with bimolecular reactions. Even so, these unique reactions are particularly desirable for generating libraries of complex molecules, since only a single step need be optimized and performed to bring together as many as five different components. These solid phase syntheses thus require substantially less time to optimize reaction conditions and are easier to perform in a high-throughput manner for large-compound library construction. Many of these reactions are also tolerant of a variety of functionalities on the individual components, and thus allow the combinatorial chemist to synthesize structurally diverse compound libraries. Although multiple-component reactions offer substantial efficiencies in the time and effort required to construct single-compound-per-well libraries, they do not allow for 'split-and-pool' synthesis strategies due to the inherent nature of the multiple-component condensation process.

The first multiple-component reaction to be performed both in solution and on a solid support was a three-component Passerini reaction, affording a library of azinomycin analogs (**Figure 101**).[705,739,751–753] A variety of multiple-component reactions have since been effected where one of the components is bound to the solid support, including the Ugi reaction, the Mannich reaction, the Biginelli reaction, the Grieco three-component reaction,

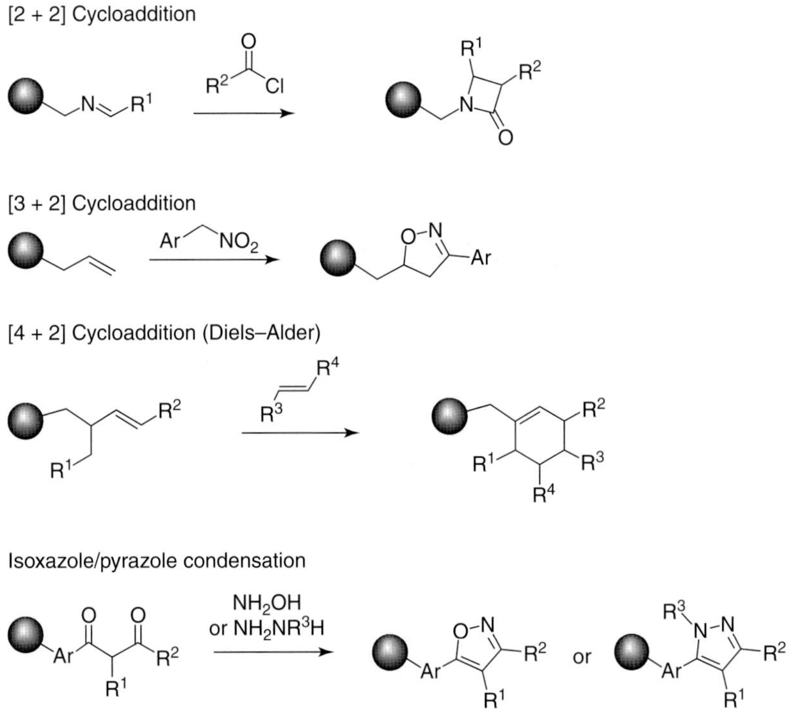

Figure 99 Carbon–carbon bond formation: cycloaddition and heterocyclic condensation reactions.

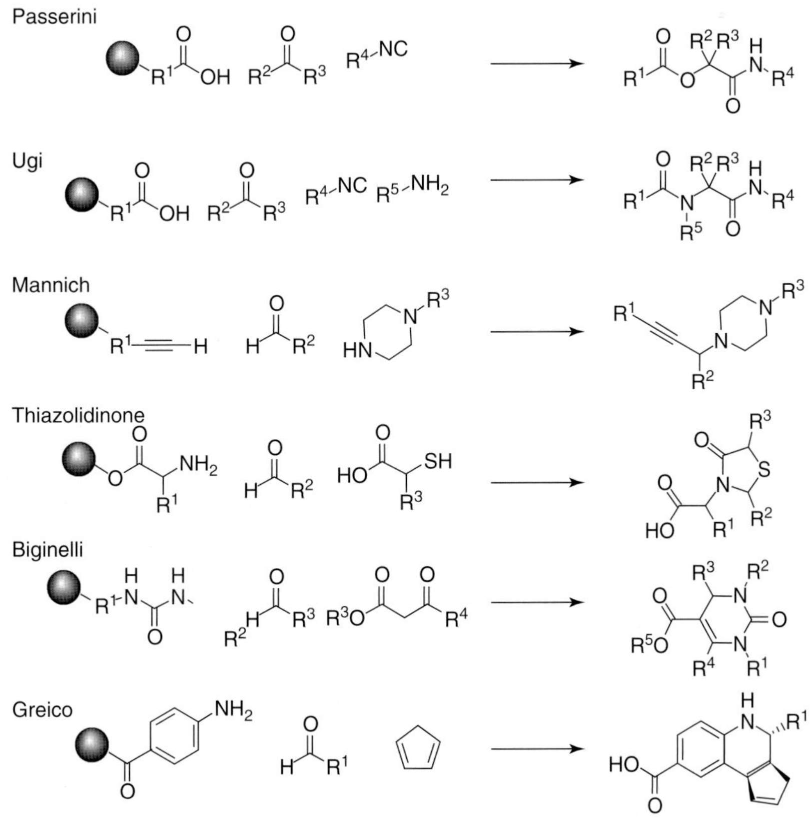

Figure 100 Representative multiple-component reactions performed on solid supports.

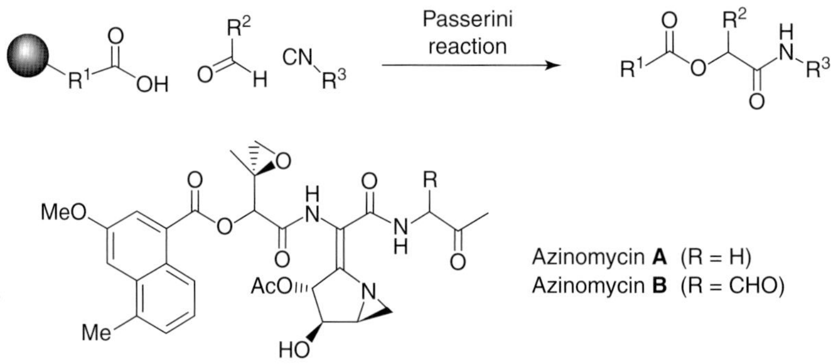

Figure 101 Passerini's three-component condensation to afford an azinomycin library.

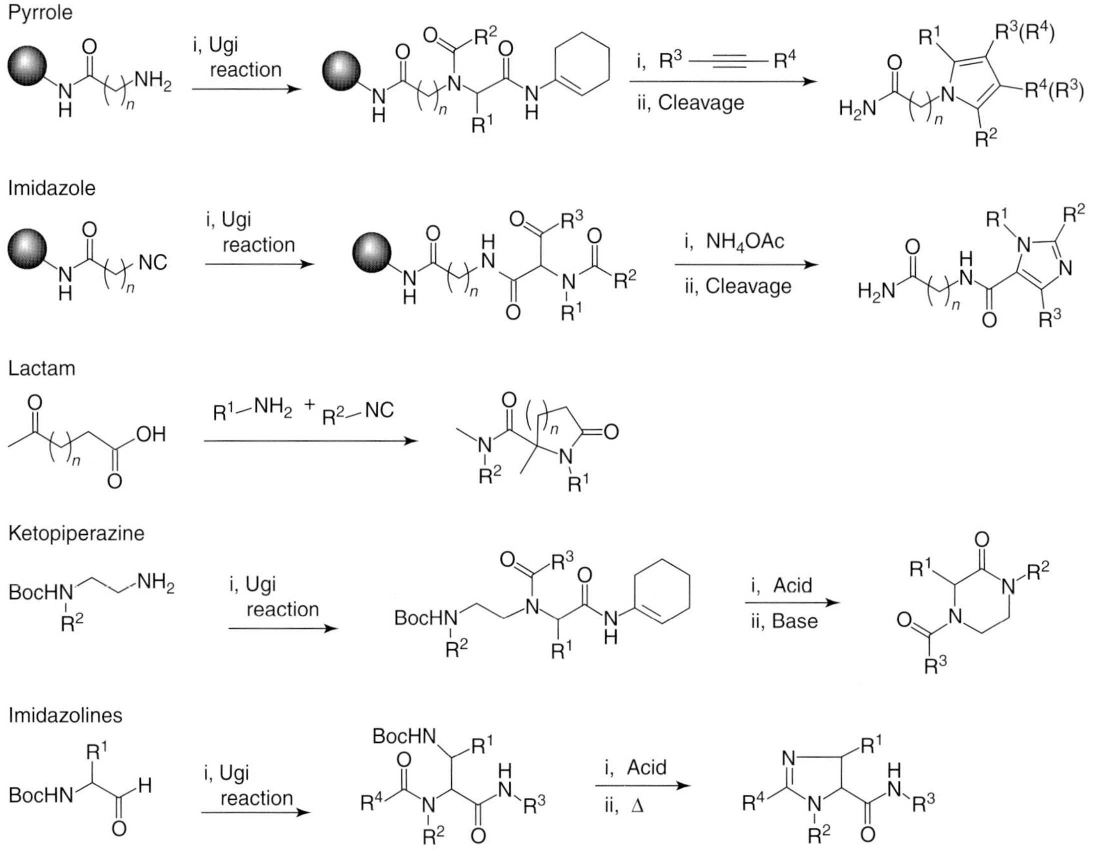

Figure 102 Examples of multiple-component reactions for both solid phase and solution phase synthesis of heterocycles.

and a thiazolidinone synthesis, to name but a few.[739,754] Recent advances in the use of multiple-component reactions have focused on the subsequent synthetic manipulation of the generic multiple-component scaffold to reduce the peptide-like nature of the products. The solid phase or solution phase synthesis of lactams,[755] pyrroles,[756] imidazoles,[757,758] imidazolines,[759] and ketopiperazines[760] from Ugi products are a few examples of such multiple-component reactions and subsequent synthetic elaboration to afford libraries of heterocycles (**Figure 102**).

2.12.16.2.3 Resin-to-resin transfer reactions

A synthetic method that would enable a chemist to synthesize a novel reagent on one resin and subsequently transfer it to another derivatized resin (i.e., resin-to-resin transfer) would be a useful tool for the convergent construction of diverse compound libraries. A report by Scialdone *et al.* demonstrates such a transformation, where an isocyanate was

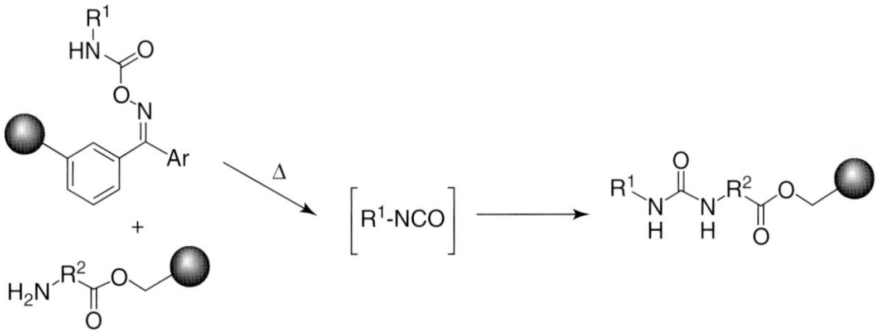

Scheme 2

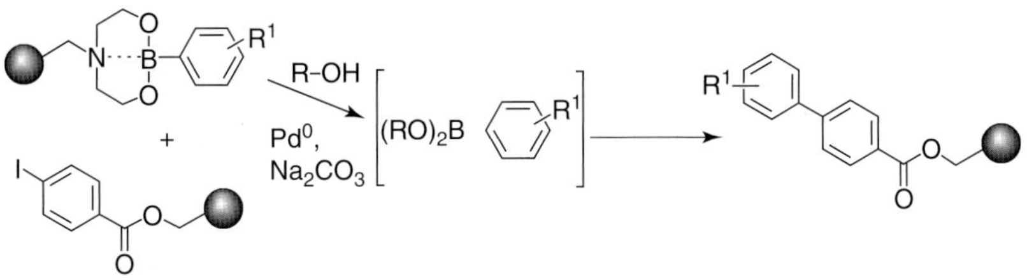

Scheme 3

generated from one resin, then diffused into and reacted with another appropriately functionalized resin to generate novel ureas (**Scheme 2**).[761]

Another example of resin-to-resin transfer, used in a Suzuki reaction, has been reported. In this case, the support-bound boronic ester is released from one resin via mild hydrolysis, and reacted with a resin-bound aryl iodide to yield the biaryl product on the support (**Scheme 3**).[762] The ability to synthesize increasingly complex building blocks by solid phase methods and subsequently utilize them as building blocks for a new solid phase synthesis is an important step toward the construction of structurally complex compound libraries.

2.12.16.3 Complex Multistep Synthesis on Solid Supports

2.12.16.3.1 Oligomers – natural and unnatural

Many of the synthetic techniques used for the synthesis of libraries of small organic molecules were originally developed for the chemical synthesis of biopolymers. Extensive research has gone into the required activating protocols and the wide array of protecting groups needed to synthesize large (20–100 steps) biopolymers in high yields and purities. The three major classes of biopolymers, peptides, oligosaccharides, and oligonucleotides, have all been chemically synthesized on supports, and each of these classes is discussed below.

2.12.16.3.1.1 Peptide synthesis

The chemical synthesis of peptides was performed by Merrifield in 1963, and represents the original application of the solid phase synthesis of organic molecules.[763] Whereas very large peptides and proteins are often synthesized by biochemical methods (e.g., overexpression in *E. coli*[764]), synthesis on a solid support is frequently used when peptides (typically lower than 50 amino acids) or peptides containing unnatural amino acids are desired (**Figure 103**). (Peptides containing unnatural amino acids can be expressed using Schultz's modified tRNA technique.[765] However, for short peptides or in cases where a significant amount of peptide is desired, chemical synthesis is still preferred.) While there have been a number of different routes developed for support-bound peptide synthesis, two techniques are used most frequently. One of these is the 9-fluorenylmethoxycarbonyl (Fmoc)-based approach, using amino acids that are protected at the backbone nitrogen with the base-labile Fmoc group and at side chain functional groups with acid-labile protecting groups. In the Fmoc–amino acid approach, side chain nitrogens are protected using *t*-butyloxycarbonyl (t-Boc) groups. Amino acids are coupled to the amino terminus of the growing peptide on a solid support via the use of an activating agent to boost the reactivity of the reacting carboxylate.[766,767] After each coupling step, the amino terminus is prepared for the next reaction by removal of the Fmoc protecting group. When completed, the peptide is

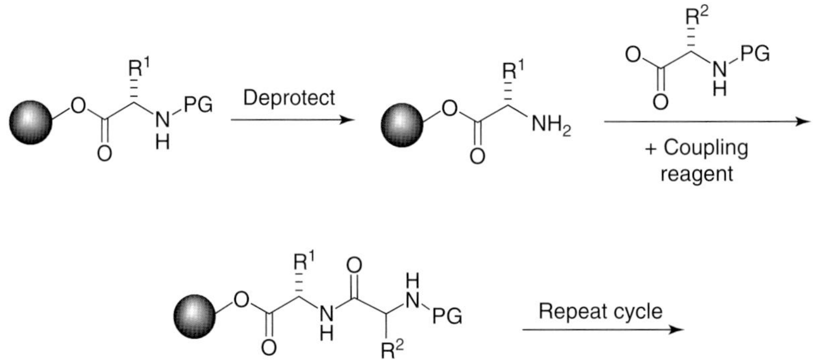

Figure 103 Solid phase peptide synthesis. Side chains are represented by R groups, and PG indicates a protecting group.

removed from the support by a strong acid cleavage step, which simultaneously removes all of the acid-labile side chain protecting groups, resulting in the desired peptide. Coupling conditions and reagents have been refined to provide high efficiency while minimizing racemization of the amino acid stereocenters.

Peptide libraries numbering in excess of 1 million compounds have been generated using standard Fmoc-based chemistry in a split-and-mix library synthesis format. For example, a 6.25 million member mixture-based library of tetrapeptides synthesized using Fmoc-based chemistry was used to identify several 0.4–2.0 nM agonists of the mu opioid receptor.[768] Synthesis of peptides on a solid support via the Fmoc-based strategy is easily automated, and a variety of peptide synthesizers that apply this chemistry have been developed. Many of the concepts and reagents that have been applied to Fmoc–amino acid peptide synthesis are used in combinatorial organic small-molecule synthesis as well, including various resins and linkers, coupling reagents, and protecting groups. In particular, a wide variety of the coupling reagents used to make amide and ester linkages in peptides have found extensive application in the generation of small-molecule libraries.

The second frequently applied support-bound peptide synthesis strategy is the t-Boc-protected amino acid approach, where the acid-labile t-Boc protecting group protects peptide backbone amines during coupling.[769] In this approach, side chains are protected using benzyl groups. These protecting groups are stable to the mildly acidic conditions used during deprotection of the peptide terminus. The support-bound t-Boc-protected amino terminus is deprotected by exposure to acid (typically trifluoroacetic acid in dichloromethane), and the next residue is added to the peptide via use of an activating agent, as described above. The completed peptide is cleaved from the support by a very strong acid, usually hydrofluoric acid or trifluoromethane sulfonic acid. A variation of the t-Boc–amino acid strategy is widely applied for solution phase synthesis of small peptides (i.e., free of polymeric support). The t-Boc–amino acid peptide synthesis technique is often used in industrial settings. However, it is applied less frequently than the Fmoc–amino acid strategy in academic or research settings. This most likely is due to the special equipment and extraordinary care that must be used with the t-Boc–amino acid strategy as a result of the extremely harsh cleavage conditions required to cleave the finished peptide.

2.12.16.3.1.2 Oligonucleotide synthesis

The synthesis of oligonucleotides on a solid support has been developed to such an extent that it is now readily automated; the synthetic steps are typically high yielding, and extremely long oligonucleotides can be synthesized. The protection and reactivity of only four different nucleotides had been considered in the optimization of DNA synthesis.[770] Molecular biology techniques, such as PCR, have increased the need for chemically synthesized oligonucleotides (e.g., to act as primer sequences in PCR). The demand for methods to chemically synthesize oligonucleotides is especially strong when synthetic, or 'unnatural,' nucleotides are to be incorporated, when milligram quantities of a oligonucleotide are needed, or when a library of relatively short oligonucleotides (less than 30 nucleotides) is desired. The technique most commonly used for oligonucleotide synthesis on a support is the phosphoramidite coupling method, a general outline of which is shown in **Figure 104**. In this technique a nucleotide is coupled to the free hydroxyl of a support-bound nucleotide, forming a phosphite triester. The phosphite is then oxidized to a phosphate triester, the protecting group (PG^2) is removed from the primary hydroxyl of the terminal nucleotide, and the coupling process is repeated. This method has been successfully applied in automated oligonucleotide synthesizers due to the high yields and purity of products typically observed for each step of the sequence.

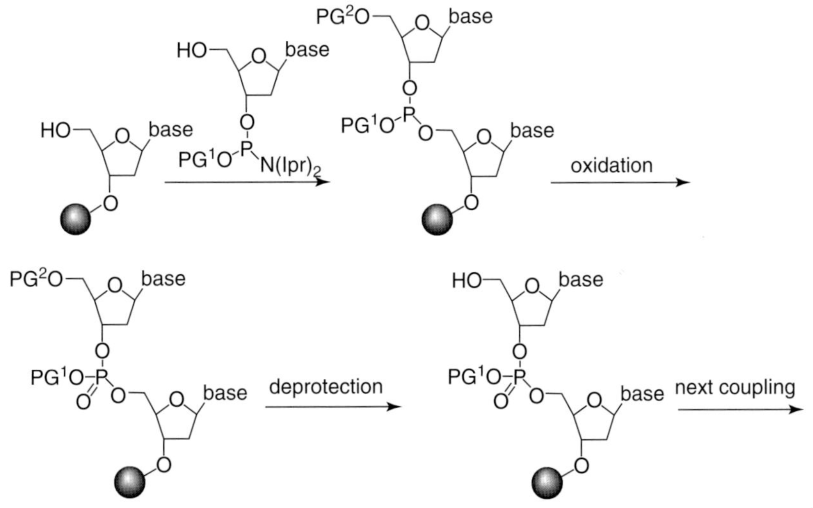

Figure 104 Support-bound oligonucleotide synthesis. PG^1 and PG^2 indicate two different protecting groups.

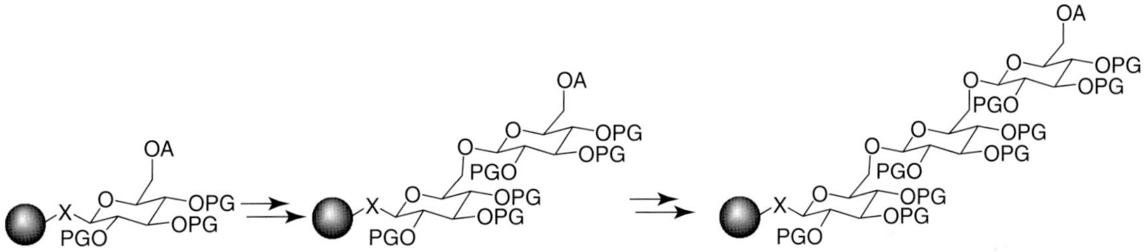

Figure 105 Representation of oligosaccharide synthesis using a support-bound glycosyl acceptor. PG indicates protecting groups, and OA indicates the glycosyl acceptor in the elongation reaction.

2.12.16.3.1.3 Oligosaccharide synthesis

Oligosaccharides were one of the last biopolymers to succumb to automated solid phase or solution phase chemical synthesis. They present a special challenge to chemical synthesis, either on a support or in solution, due to both the number of hydroxyl groups of similar reactivity that must be protected and the selective deprotection of a specific hydroxyl group that is required for glycosidic bond formation (**Figure 105**). Added to these issues is the requirement that glycosidic bond formation proceeds in a stereospecific manner, ideally with accessibility to either stereoisomer. The synthetically demanding requirements of obtaining regiospecific and stereospecific bond formation at each coupling step balances the advantages of support-bound synthesis (the ease of purification of intermediates and the ability to drive reactions to completion using excess reagents). While the oligosaccharide area of biopolymer synthesis is not as fully developed as the areas of peptides or oligonucleotides, many significant advances have been made in recent years.[771–773]

The parallel synthesis of oligosaccharides has been accomplished using both soluble and insoluble polymers. Glycoside elongation has been accomplished by either chemical or enzymatic bond formation. Glycopeptides have also been synthesized on a support, where the peptide is elongated after glycosylation; then the saccharide is attached to a peptide side chain and elongated, or, alternatively, the completed oligosaccharide is attached to the completed support-bound peptide.[772,774,775] Glycopeptides fall into two classes; O-linked glycopeptides and N-linked glycopeptides (examples of which are shown in **Figure 106**).

Oligosaccharides have been synthesized on a polymeric support using two approaches: either the glycosyl donor is on the support and the acceptor is in solution, or the glycosyl acceptor is on the support and the donor is in solution. An example of a support-bound glycosyl donor strategy applied to the synthesis of a β-linked oligosaccharide is shown in **Figure 107**.[776]

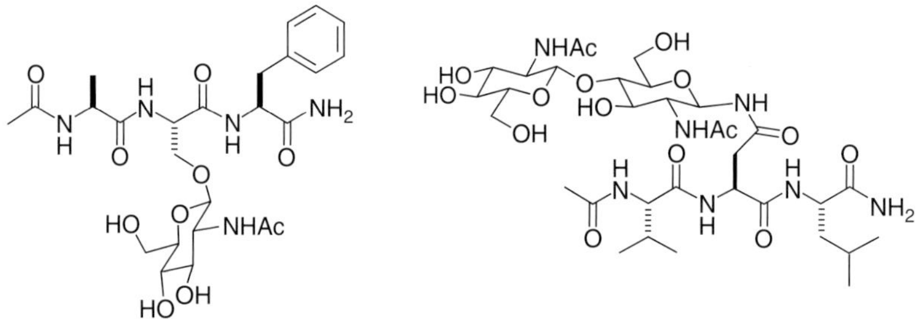

Figure 106 Examples of (a) O-linked and (b) N-linked glycopeptides.

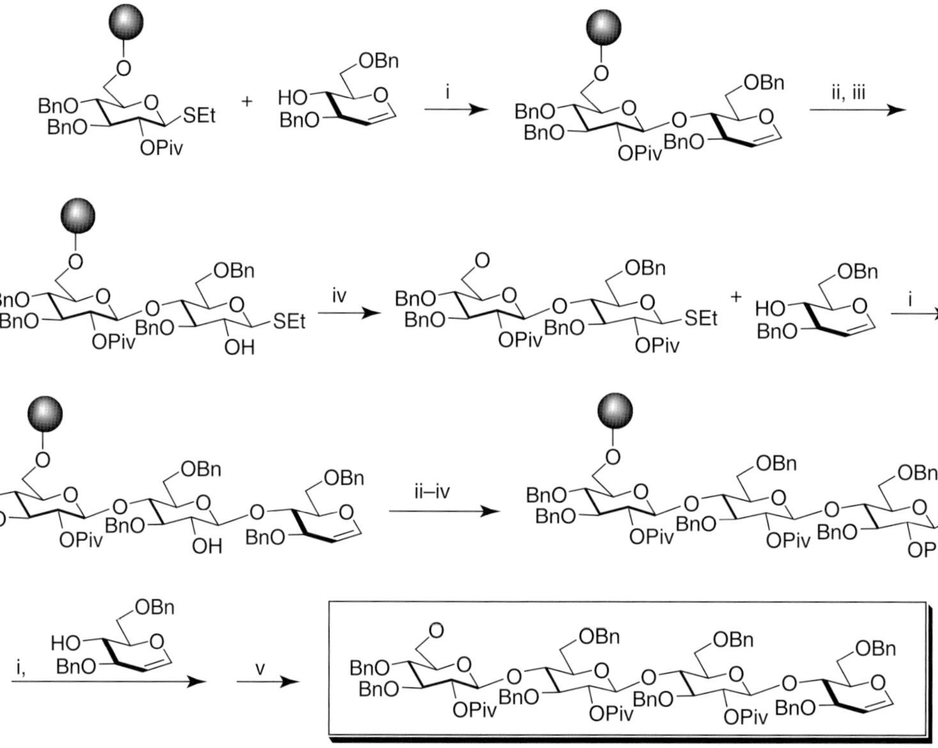

Figure 107 Tetrasaccharide synthesis using a support-bound glycosyl donor, as described by Zheng et al.[776] Bn indicates a benzyl protecting group, and Piv indicates a pivaloyl protecting group.

2.12.16.3.1.4 Synthesis of hybrid oligomers

The advances in techniques for the chemical synthesis of the three major classes of biopolymers discussed above have allowed scientists to explore hybrids of the biopolymers. As mentioned above, glycopeptides have been synthesized on a solid phase, with the oligosaccharide segment being elongated after coupling to a support-bound peptide. A related approach, where the completed oligosaccharide is coupled to the support-bound peptide, has also been reported. Likewise, peptide–DNA hybrids, or peptide nucleic acids (PNAs), have been synthesized.[777,778] In a PNA, the phosphate–sugar backbone of the oligonucleotide has been replaced by a peptide amide backbone (**Figure 108**). PNAs have received considerable attention as potential antisense and antigene drugs, as well as tools for probing DNA structure and helix stability.[779,780]

2.12.16.3.1.5 Unnatural oligomers

The proper orientation of pharmacologically important side chains around a central core is fundamental to medicinal chemistry. Nature accomplishes this orientation in peptides and proteins by displaying amino acid side chains around

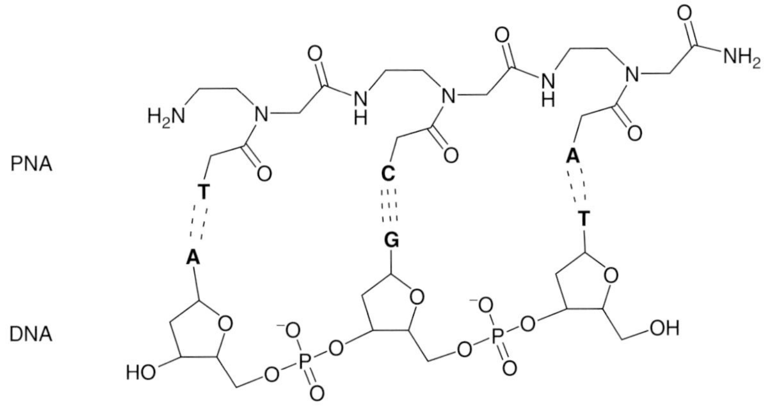

Figure 108 Base pairing of a segment of a DNA molecule with a PNA molecule, where C, G, A, and T represent the corresponding nucleic acids.

the peptide backbone. Libraries of peptides are often used to identify the proper side chains and orientation for binding to a receptor or enzyme active site. However, peptides (amino acid oligomers) have significant pharmacological drawbacks that typically make them unsuitable as orally bioavailable drugs. A number of unnatural oligomers have been reported in recent years, many of which have been synthesized on a solid phase or soluble polymer support. In their attempts to mimic the display of side chains by a normal peptide backbone, chemists have developed these various oligomers, while endeavoring to overcome some of the liabilities of peptides (susceptibility to proteolysis, poor bioavailability, etc.). Several examples of unnatural oligomers are shown in **Figure 109**, along with a natural peptide backbone for comparison.[781–789] This figure is not meant to represent an inclusive set, as several other unnatural oligomers (e.g., the PNAs mentioned in the previous section) have also been synthesized.

2.12.16.3.2 Heterocycle/pharmacophore synthesis

As the number of different chemical transformations available to chemists has expanded, the complexity of molecules that may be synthesized in parallel has likewise increased. Chemists have taken advantage of the new tools available to them to synthesize molecules of significant complexity. Several interesting examples of solid phase small-molecule synthesis are shown in **Figures 110–115**. Relatively early yet significant examples of small molecule combinatorial synthesis are the 1,4-benzodiazepine libraries, synthesized on a solid phase in the early 1990s.[689,790,791] Benzodiazepines are known to have a wide range of biological activities. The 1,4-benzodiazepine scaffold represents a versatile pharmacophore (or core structure), of which multiple diverse elements may be displayed. The synthesis shown in **Figure 110** allows for the incorporation of four diverse elements (represented as R^1–R^4) around the central benzodiazepine core. Another example of the synthesis of an important pharmacophore on a solid phase is that of the quinolones (**Figure 111**).[792] Quinolones compose a class of broad-spectrum antibiotics discovered relatively recently. In their demonstration of quinolone solid phase synthesis, MacDonald and co-workers included the known antibiotic ciprofloxacin in order to highlight the ability of their quinolone library to generate pharmaceutically relevant compounds.

Four recent pharmacophore syntheses on supports are shown in **Figures 112–115**. In all four cases, at least three diversity elements may be displayed around a central core. The bicyclic guanidines derive their diversity elements from amino acids; an acylated dipeptide is synthesized on a support, the amide carbonyl groups are reduced, and the resulting triamine is then cyclized (**Figure 112**).[793] The benzopiperazinones (**Figure 113**) also derive side chain diversity (at two positions) from amino acids, and a postcyclization alkylation on support introduces functionality at the R^4 position.[794] The oxopiperazine core is assembled via a tandem N-acyliminium ion cyclization–nucleophilic addition, resulting in the construction of both rings in a single step (**Figure 114**).[795] Finally, the prostaglandin synthesis shown in **Figure 115** is an excellent example of the application of a variety of solid phase reactions, including enantioselective carbon–carbon bond formation, and oxidation of an alcohol to a ketone.[737,796]

The Schreiber group has developed 'complexity-generating' reactions to access diverse compounds. One example of this approach is shown in **Figure 116**, where the 7-5-5-7 ring system was generated in six steps on a support. The synthesis involves two key complexity-generating steps: an Ugi reaction (followed by an intramolecular Diels–Alder

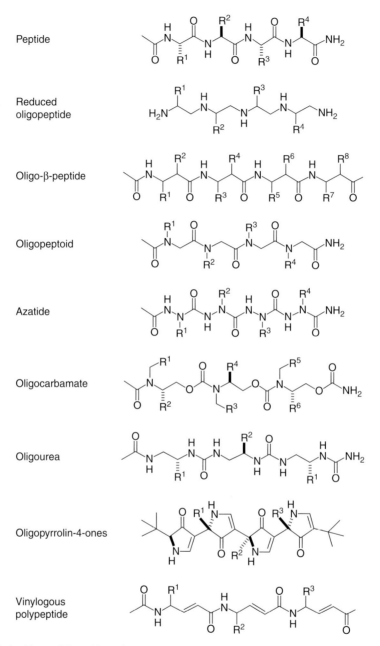

Figure 109 Peptide backbone, followed by various unnatural oligomer structures, indicating display of side chains (R groups).

reaction) and a ring-closing metathesis. The final product contains five stereocenters, and was isolated as a single stereoisomer. The authors report that the optimization of the support-bound reactions is underway, with the goal of generating a library via split-pool synthesis.

2.12.16.3.3 Natural product total syntheses on a solid support

The total synthesis of complex natural products offers formidable synthetic challenges even without the added difficulties involved with the optimization of solid phase synthesis.[797] These challenges have been met by several different research laboratories with the synthesis of such natural products as epothilones A and B,[743] the fumiquinazoline alkaloids,[798] and analogs of indolactam V[799] (**Figure 117**). The degree of difficulty of these

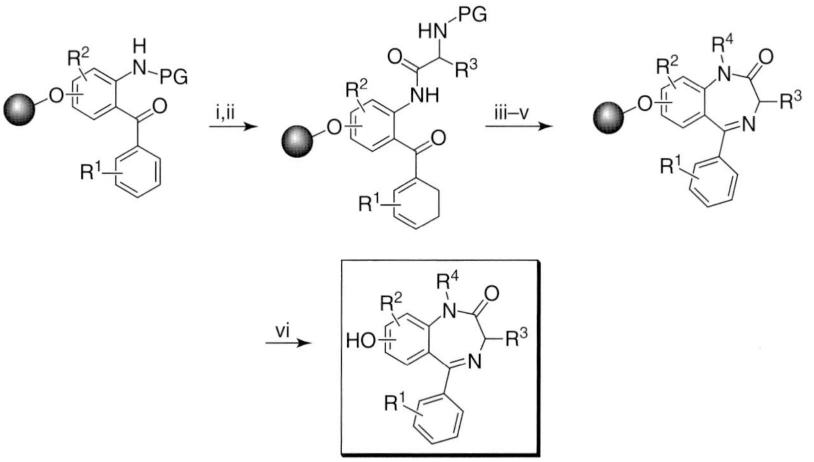

Figure 110 Solid phase synthesis of 1,4-benzodiazepines.

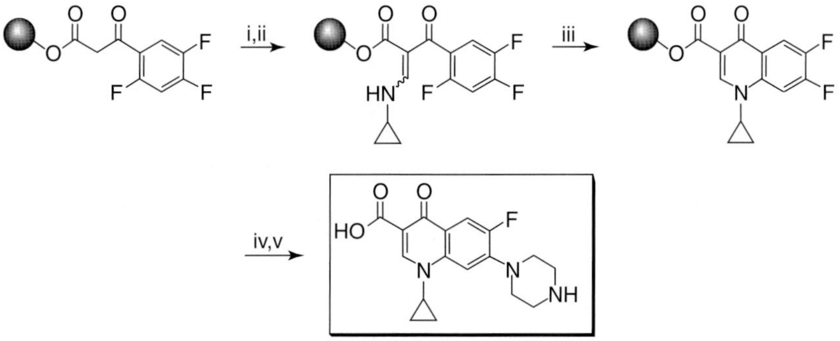

Figure 111 Solid phase synthesis of quinolones (ciprofloxacin is shown in the box as the final product).

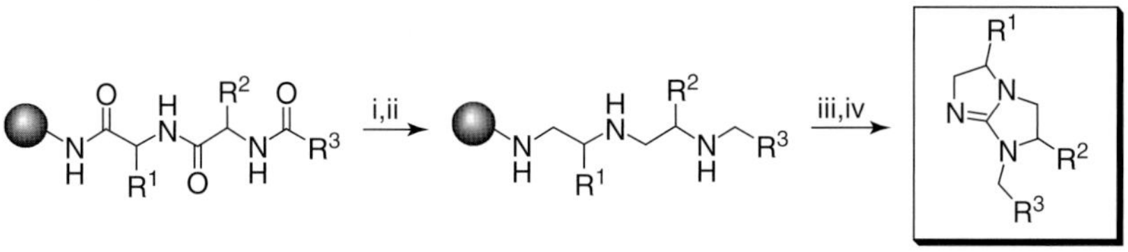

Figure 112 Solid phase synthesis of bicyclic guanidines.

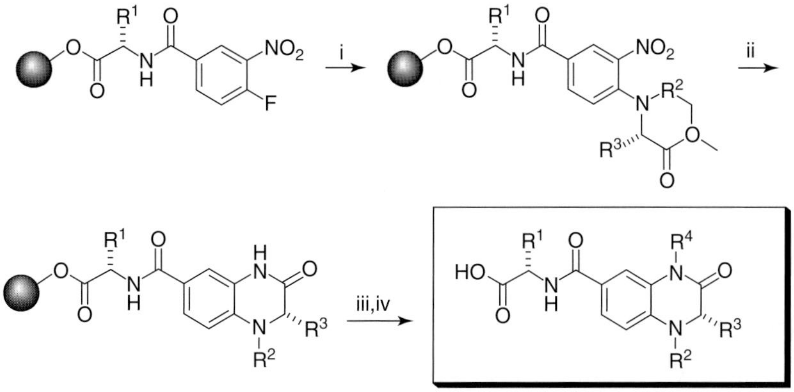

Figure 113 Solid phase synthesis of benzopiperazinones.

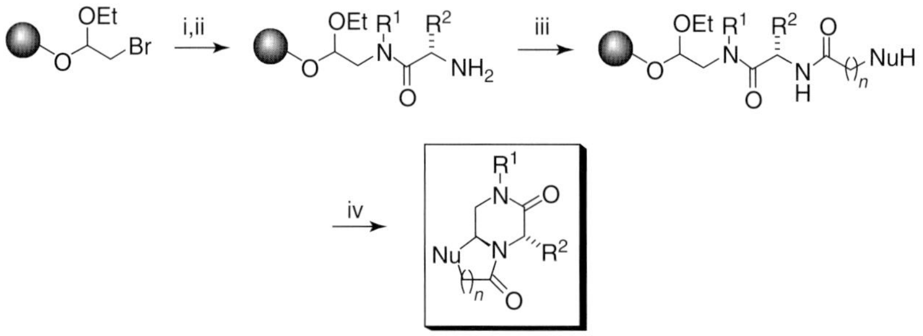

Figure 114 Solid phase synthesis of oxapiperazines.

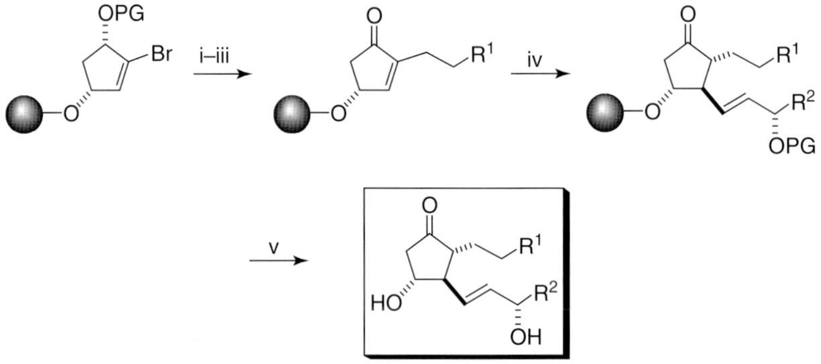

Figure 115 Solid phase synthesis of prostaglandins.

syntheses and proficiency with which they have been achieved are evidence that solid-supported synthesis has matured into an applicable method for the generation of complex molecules.

Ganesan's solid phase synthesis of (+)-glyantrytpine and several close analogs (fumiquinazoline alkaloids) from readily available starting materials demonstrates the potential of this technology to rapidly access natural product structures.[798] The synthesis began with Fmoc-protected L-tryptophan on a solid support. Subsequent elongation to a linear tripeptide was followed by a cyclative cleavage from the support, resulting in the desired fumiquinazoline alkaloids. The versatility of this approach was demonstrated by the synthesis of 17 unnatural fumiquinazoline analogs, where L-alanine, L-leucine, and L-phenylalanine were used in place of L-tryptophan. The final products were obtained in moderate to good yields and high purity.

Nicolaou's total synthesis of epothilones A and B reveals the potential synergies between solution and solid phase synthesis.[743] Complex subunits were synthesized in solution and then combined on the solid support to achieve a highly convergent method for the rapid construction of the uncyclized scaffold (**Scheme 4**). A strategic olefin linkage of the acyclic precursor to the solid support allowed for simultaneous macrocyclization and cleavage of the reaction product from the resin via ring-closing metathesis. A subsequent epoxidation in solution, followed by the resolution of diastereomers, led to the desired natural product. This type of hybrid solution phase and solid phase approach is particularly attractive to the medicinal chemist, since it allows for the rapid generation of analogs of these biologically active natural products.

Schreiber and co-workers have described the synthesis of a library of 'natural product-like' compounds. The library contained over 2 million compounds, and was derived from two cores that vary in the nature of the R group in the template (**Figure 118**).[800] Several other groups are also looking at cores related to natural products in order to access structural classes that were previously inaccessible by parallel synthesis.[744,801] This application of solid phase synthesis begins to bridge the gap between the relatively simple small molecules traditionally made via combinatorial methods and very complicated natural products, as chemists seek more sophisticated cores for their libraries.

2.12.16.4 Solution Phase Synthesis Methodologies

2.12.16.4.1 Solution phase polymer-supported synthesis

While the majority of reports of parallel and combinatorial synthesis still involve the attachment of reactants (either reagent or substrate) to an insoluble support, there have been numerous publications describing parallel synthesis using

Linker
i, Lutidine
ii, Piperidine deprotection
MeOH/THF
KHMDS,
CH2Cl2, 40 °C, 48 h
ii, HF-py

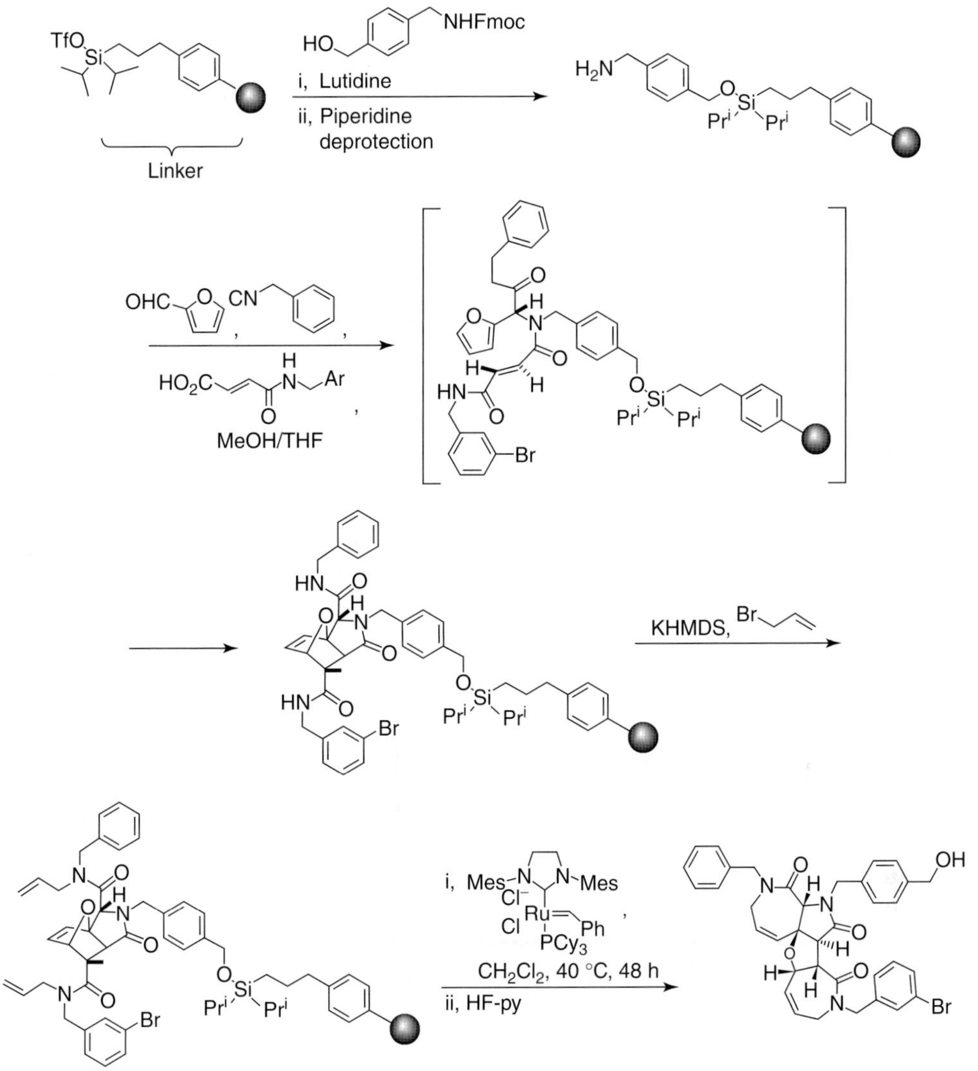

Figure 116 Solid phase synthesis using 'complexity-generating' reactions to produce a 7-5-5-7 ring system.

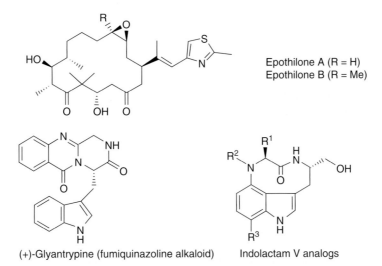

Epothilone A (R = H)
Epothilone B (R = Me)
(+)-Glyantrypine (fumiquinazoline alkaloid)
Indolactam V analogs

Figure 117 Natural products and natural product analogs synthesized on solid supports.

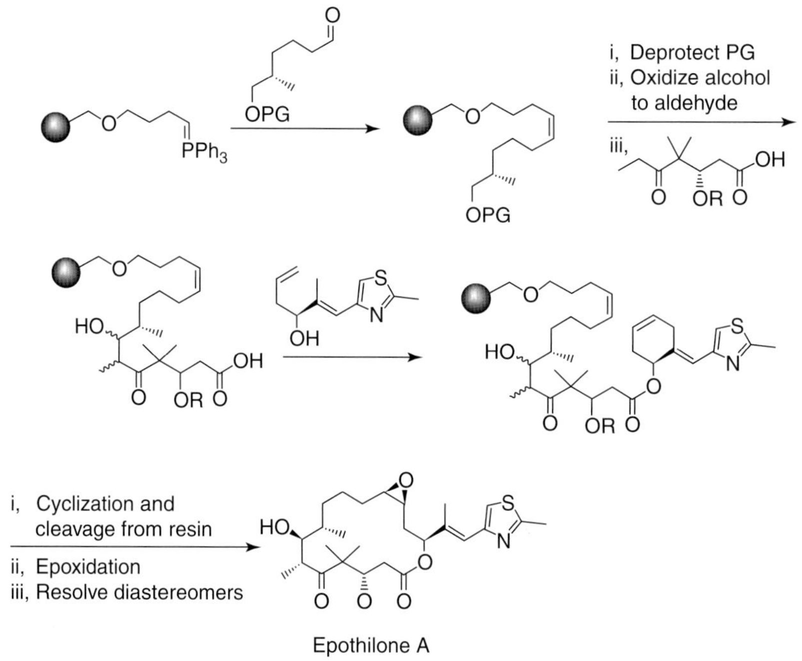

Scheme 4

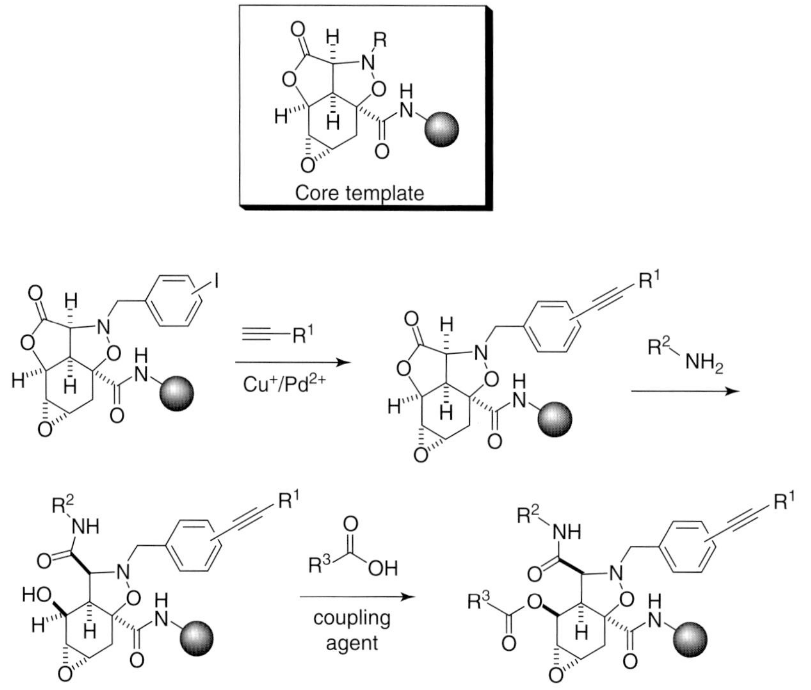

Figure 118 Core template (boxed) and representative synthesis of the 'natural-product-like' library from the benzyl-derived core and alkyne, amine, and carboxylic acid building blocks. For the core template, R represents either benzyl or phenyl side chains.

soluble polymers.[704,721,722] The major advantage that the use of a soluble polymer holds over a 'classical' insoluble polymer (e.g., polystyrene beads) is that reactions occur entirely in the solution phase, so that little or no modification of a traditional non-polymer-supported synthesis is necessary. A second advantage is that the reaction kinetics are often faster with the soluble polymer than with a two-phase system. Upon completion of the reaction, the desired product

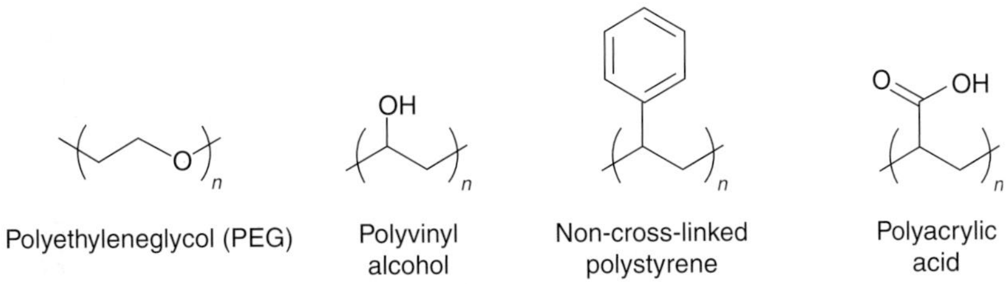

Figure 119 Examples of soluble polymer supports.

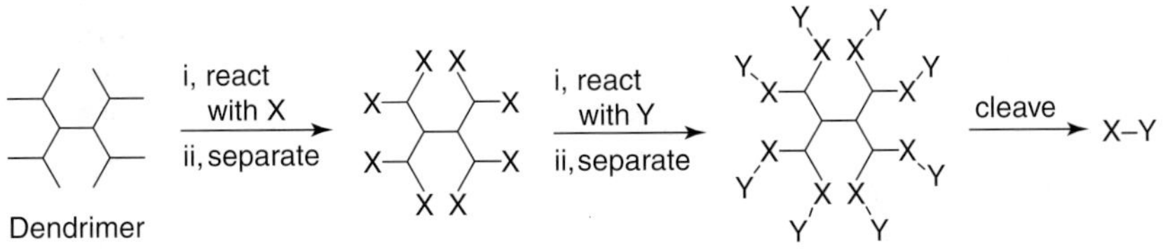

Figure 120 Dendrimer-supported synthesis.

(attached to the soluble polymer) is isolated by precipitation of the polymer, by extraction (e.g., into a fluorous solvent when using a fluorous polymer), or by means of separation based on molecular weight (such as equilibrium dialysis or size exclusion chromatography). Some examples of soluble polymers include polyethyleneglycol (PEG), polyvinyl alcohol, non-cross-linked polystyrene, and polyacrylic acid (**Figure 119**). Since the polymer-bound component can be separated from the other reagents at the end of the reaction, excess reagents may still be used to drive reactions to completion, affording high yields and purities of the products.

When using soluble polymer supports, such as PEG, consideration must be given to the size and solubility properties of the polymer to be used, as well as to the size of the desired product molecule. As the size of the attached molecule increases, its solubility properties can often moderate or dominate the solubility properties of the polymer, potentially resulting in loss of product due to incomplete crystallization or extraction. A number of different polymers have been applied to solution phase parallel synthesis, selected for their solubilizing characteristics and compatibility with reaction chemistry.

Chemistry has been developed to take advantage of the unique properties of highly fluorinated hydrocarbons. It has been recognized that solvents comprising these fluorinated hydrocarbons are immiscible with water and with many organic solvents, resulting in a 'fluorous' phase.[802] Compounds attached to highly fluorinated 'tags' may be reacted in solution in pure organic solvents or solvents containing a fluorous co-solvent, then extracted into a fluorous phase upon completion of the reaction. Likewise, reagents may be attached to highly fluorinated tags and extracted into a fluorous phase and away from the product. To selectively bind and separate the fluorous-tagged material from the reaction mixture, a third option is to combine a fluorinated solid phase (such as silica) with a fluorinated, bonded phase. Techniques involving the use of fluorinated tags have the advantages of allowing reactions to be run in solution, followed by straightforward separation of the product or reactant (depending on which contained the fluorous tag). No translation of solution phase procedures is necessary. However, such fluorous tagging has yet to find widespread use in parallel and combinatorial chemistry, due in part to the poor commercial availability of several of the required reagents.

Dendrimers have been suggested for use as soluble polymer supports for combinatorial libraries (**Figure 120**).[803] Separation from reactants is accomplished by methods such as size exclusion chromatography or ultrafiltration, where the larger dendrimer is isolated from the smaller reactants. Dendrimers typically have considerably higher loading (more molecules of product per gram of support) than insoluble polymers, and again there is no need to translate solution phase procedures for use with dendrimer supports. While it would appear that there are several advantages for the use of dendrimers over insoluble polymer supports, there have been relatively few reports of dendrimer-supported parallel or combinatorial syntheses. Only a few dendrimers are commercially available, and the methods of purification required between steps (size exclusion chromatography or ultrafiltration) have not yet been developed in a readily accessible high-throughput format.

2.12.16.4.2 Solution phase reactions involving scavenging resins

It was first recognized by Kaldor and co-workers that solid-supported nucleophiles and electrophiles could be used for rapid purification of compound libraries synthesized in a standard solution phase reaction.[702,804] The advantages of this technique include standard solution phase kinetics and easy analysis of reactions using conventional methods, and it also obviates the need for the development of linker chemistry. The largest incentive to pursuing this methodology is that the development time for the synthesis of new compounds can be much shorter than the time necessary to develop a new solid phase synthesis. This reduction in time is critical, particularly when medicinal chemists are developing methods for the exploration of structure–activity relationships in the intensely competitive pharmaceutical industry. Sequestering agents allow the most straightforward type of parallel synthesis; standard reactions are carried out in solution in parallel, and the replacement of the normally time-consuming work-up by simply mixing the reaction solutions with an appropriate scavenger on a support and filtering to provide the pure products eliminates purification.

There are several potential drawbacks to solution phase synthesis involving either scavenging resins or resin-bound reagents (discussed in Section 2.12.16.4.3). For example, it is not possible to design split-and-pool syntheses, as the reaction products are free in solution and not associated with any tag or bead to separate. This can limit the size of the library (number of compounds) produced. In addition, while several examples of longer syntheses involving scavenging or resin-bound reagents are shown below, reaction sequences using these reagents are typically very short, of the order of one to three steps. This is in contrast to syntheses on solid supports, which are often five or six steps long, and occasionally much longer. (For examples of longer sequences, see the preceding sections on complex multistep synthesis on solid supports.)

Support-bound scavengers have been developed for a variety of common reagents. There are two common varieties of scavengers: ionic reagents that remove impurities by forming salts, and reaction-based scavengers that react chemoselectively with an undesired excess reagent. Ionic reagents were the first such resins to be reported.[804] Amines can be removed by forming salts with support-bound sulfonic or carboxylic acids (sulfonic acid resins such as Dowex are common), and acids can be removed with support-bound amines, such as high-loading trisamine resins. Acid-functionalized resins are also commonly employed to scavenge excess metals and metal alkoxides from organometallic reactions, including Grignard and alkyllithium additions.[805] These reactions are relatively selective, as the unreacted polar reagents are usually removed from uncharged products. Although conceptually simple, these procedures are extremely powerful, and are used routinely in parallel synthesis.

Chemists have also used ion exchange resins to 'trap' products instead of impurities. In this inverse application of scavenger resins (often termed 'catch and release' purification), the charged product is captured by an appropriate ion exchange resin. The reaction impurities are then washed away, and the pure product is released from the ion exchange resin by elution with an appropriate buffer (**Figure 121**). This technique has also been applied in cases where the product is selectively reacted with a resin-bound reagent, to form a covalent link to the support. The impurities are then washed away, and the product is released via a second reaction or cleavage step.

Most reactivity-based scavengers are functionalized polystyrene resins. Nucleophilic impurities such as amines or thiols can effectively be removed by the use of electrophilic resins. Some examples of these reagents include polymer-bound isocyanates, sulfonyl chlorides, anhydrides, aldehydes, and chloroformates. Excess electrophiles, such as acid chlorides, isocyanates, aldehydes, alkyl halides, α,β-unsaturated carbonyl compounds, and sulfonyl chlorides, can be sequestered using polymer-bound nucleophiles such as amines and thiols.[806–808] Examples of several different classes

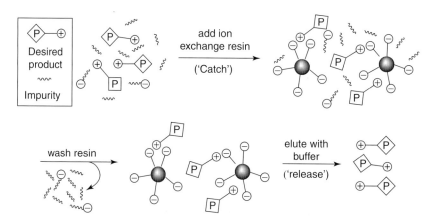

Figure 121 Outline of 'catch and release' purification: a cationic product is captured on an anionic ion exchange resin, the neutral and anionic impurities are washed away, and the pure product is eluted from the resin.

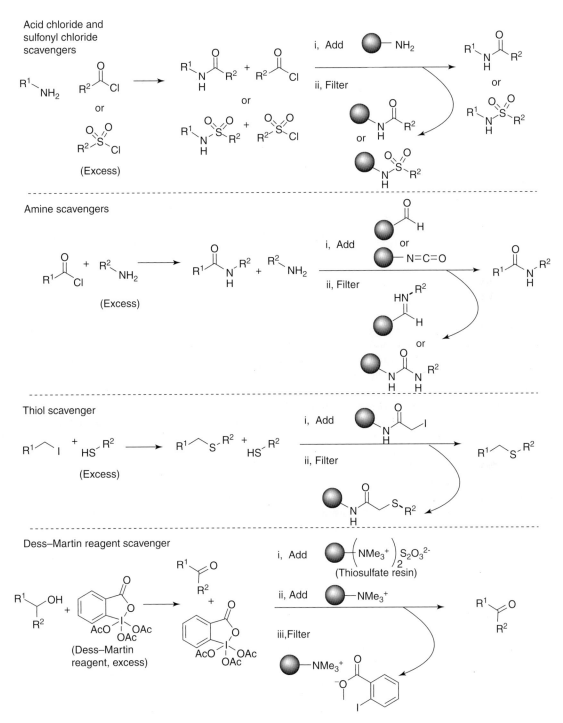

Figure 122 Reactivity-based scavenger resins.

of reactivity-based scavengers are shown (being used for purification of products) in **Figure 122**.[806–810] The large number of different reagents available for these purification techniques allows 'tuning' of the reactivity of a scavenger. For example, a thiol can be removed from a solution containing an amine by use of an α-iodoamide resin as an electrophile. This reagent will selectively react with the more nucleophilic thiol over the desired amine-containing product.[809] The loading levels of polymer-bound scavengers are important; the swelling of resins requires the use of large solvent volumes, such that higher resin loading levels of scavenger are desirable to minimize dilution of the

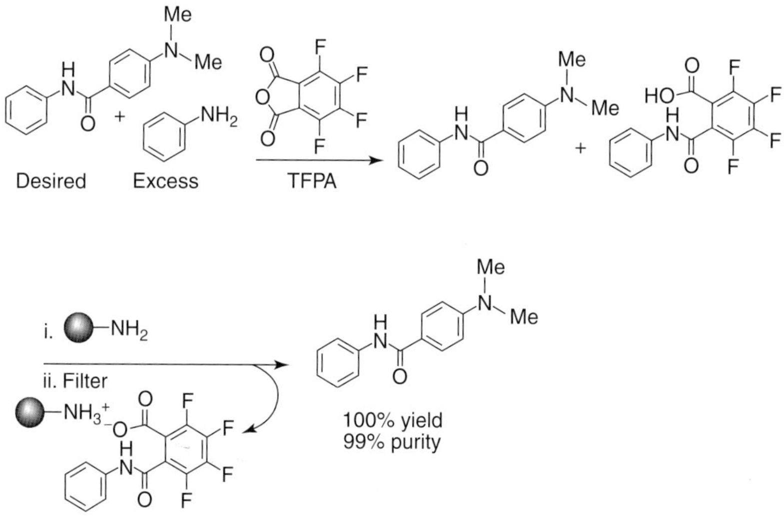

Scheme 5

reaction mixture and the amount of resin required. The following section highlights some more advanced scavengers that have been developed to remove specific impurities and demonstrate the power of this technique.

Parlow and co-workers have developed a scavenging method for the versatile Dess–Martin oxidation procedure.[810] Reaction of a primary or secondary alcohol with an excess of the Dess–Martin reagent in solution produces an aldehyde or ketone. The excess reagent is usually removed in solution phase reactions by reduction of the unreacted reagent with sodium thiosulfate followed by aqueous extraction. Parlow and co-workers have developed a thiosulfate resin by simply washing commercial chloride-form Amberlyst A-26 ion exchange resin with thiosulfate (**Figure 122**). Treating a crude reaction solution from a Dess–Martin oxidation with the thiosulfate resin reduces the unreacted Dess–Martin reagent to o-iodobenzoic acid, which can then be efficiently sequestered by any of a number of basic resins. This chemistry was demonstrated on a number of alcohols, to produce the corresponding aldehydes or ketones in good yields and excellent purities (>99% in almost all cases as measured by gas chromatography/mass spectrometry and ^{1}H NMR).

Ion exchange sequestering is an efficient procedure to remove excess impurities from reaction mixtures; however, in the pharmaceutical industry the use of sulfonic acid resins is limited by the presence of basic amines on a large number of desired products. Parlow and co-workers have devised a strategy to make possible the sequestration of a reactive amine from a solution containing a desired basic but nonnucleophilic amine.[811] By adding tetrafluorophthalic anhydride to such a reaction mixture, the excess reactive amine is consumed, leading to the formation of the phthalamide with a pendant carboxylic acid (**Scheme 5**). Both the resulting carboxylic acid and any excess anhydride can be sequestered using a basic resin, leaving only the desired basic amine in the reaction solution. This procedure is effective for removing relatively poor nucleophiles, such as anilines, that do not react completely with most support-bound electrophiles.

In another example of material design, Kirkland and co-workers prepared functionalized porous silica microspheres for use as scavengers in combinatorial chemistry.[812] The silica microspheres are more than 99.995% pure, and have a uniform particle size and shape that provide superior mass transfer properties. The microsphere beads are easy to handle, do not swell, and are efficient scavengers in both polar and nonpolar solvents. The microspheres can be functionalized with the same functional groups as polystyrene beads, and therefore can be used in many of the same situations as conventional scavengers. These silica-based materials have been evaluated for the ability to sequester a number of acids, acid chlorides, and amines from typical reaction solutions.

2.12.16.4.3 Solution phase reactions involving resin-bound reagents

A complementary approach to removing excess reagents from a reaction solution using a support-bound sequestering agent is to immobilize the reagent on an appropriate solid phase. The addition of an excess of this solid-supported reagent to a suitable substrate in solution provides a product that can be purified by simple filtering off of the support-bound reagent. This approach is conceptually relatively old but has been recently revitalized, as is evident by the vast number of support-bound reagent classes that have been developed for solution phase parallel synthesis. There are now commercial resins available that deliver support-bound equivalents of many types of reagents, including acids, bases

How and Why to Apply the Latest Technology   489

(tertiary amine, 4-dimethylaminopyridine, phosphazine, etc.), acylating agents (see below), coupling agents (carbodiimides), reducing agents (borohydride, cyanoborohydride, silylhydrides), oxidizing agents (permanganate, chromium, ruthenium, osmium), catalysts (palladium, asymmetrical catalysts), a wide variety of protecting groups, and many other reagent classes. A discussion of all of the reagents currently available is beyond the scope of this section. The interested reader is directed to the review articles mentioned in the introduction to the solution phase chemistry (*see* Section 2.12.16.4.1).[702,717,807] Several classes of these reagents are described in detail below.

Amide formation is a reaction of fundamental importance in medicinal chemistry; consequently, many techniques to form amides using polymer-assisted solution phase (PASP) chemistry have been developed. A convenient method is to utilize a support-bound alcohol that is functionalized using either a carboxylic acid and coupling agent or an acid chloride to provide a support-bound activated ester. These reagents can then be stored, and are typically stable for months. Simply stirring an excess of the support-bound activated ester with an amine provides (after a filtration step to remove the support) a high yield of a desired amide. Adamczyk and co-workers have developed an *N*-hydroxysuccinimide-functionalized resin for this purpose, utilizing a Michael addition of a support-bound thiol to *N*-hydroxymaleimide to provide the functionalized resins **1** (**Figure 150**).[813]

A resin similar to that discussed above has been reported by Salvino and co-workers whereby the reaction of aminomethyl resin with 4-carboxytetrafluorophenol provides a support-bound pentafluorophenyl ester equivalent (**2, Figure 123**).[814] Loading of the resin with a carboxylic acid or sulfonyl chloride provides a support-bound activated ester. This tetrafluorophenyl alcohol resin has the advantage that solid phase ^{19}F NMR can be utilized to quantitatively determine the loading level of the polymer-bound active ester reagents.[815] Other support-bound reagents for transformations of this type have been previously reported, including support-bound 1-hydroxybenzotriazole (HOBt) equivalents, a polymer-bound *o*-nitrophenol equivalent (**3, Figure 123**),[816] and the Kaiser oxime resin.[817] Chang and Schultz have analyzed resins **1**, the *o*-nitrophenyl ester resin, the Kaiser oxime resin, and commercial HOBt resin for their relative abilities to act as support-bound acylating agents.[818] By acylating all four resins with a carboxylic acid containing a dye and monitoring the rate of amide formation with benzylamine, the relative order of reactivity of the resins was determined to be as follows (P denotes the resin): P-HOBt ≫ P-*N*-hydroxysuccinimide > P-*o*-nitrophenyl ≫ P-oxime.

Several groups have developed 'safety catch' linkers whereby the linking unit is stable to a variety of acidic and basic conditions. The term 'safety catch' arises from the fact that the resins normally must be activated prior to release of the product (so the linker is 'safe' until intentionally modified to provide an activated ester equivalent). Kenner and Ellman have reported sulfonamide-based linkers that are stable under most reaction conditions.[819–821] Upon chemical activation, the linker serves as an activating group, and the product can be released by reaction with amines or alcohols to produce amides or esters (**Figure 124**). Scialdone and co-workers have published several applications of

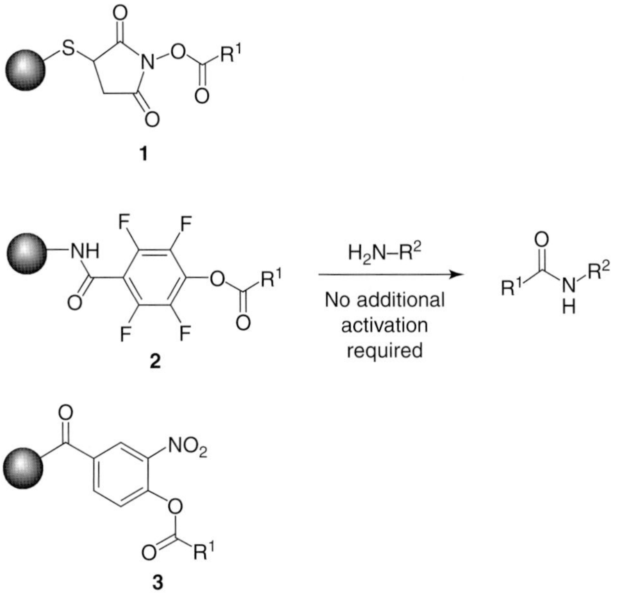

Figure 123 Support-bound activated esters that can be directly reacted with an amine nucleophile.

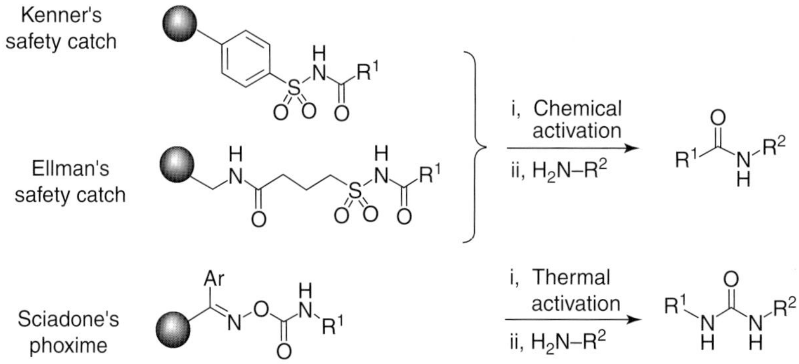

Figure 124 Support-bound reactants requiring activation prior to reaction with an amine nucleophile.

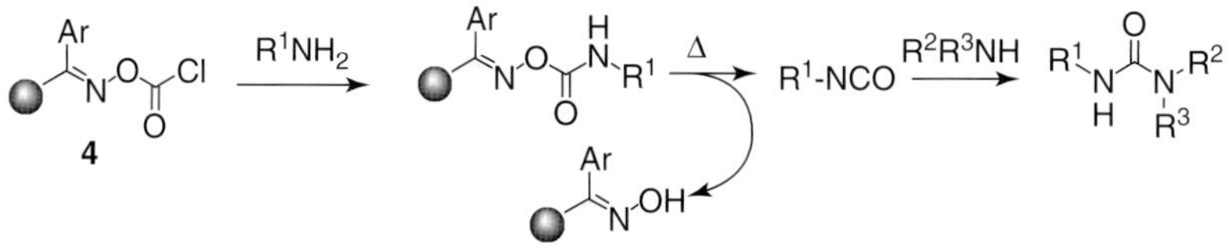

Scheme 6

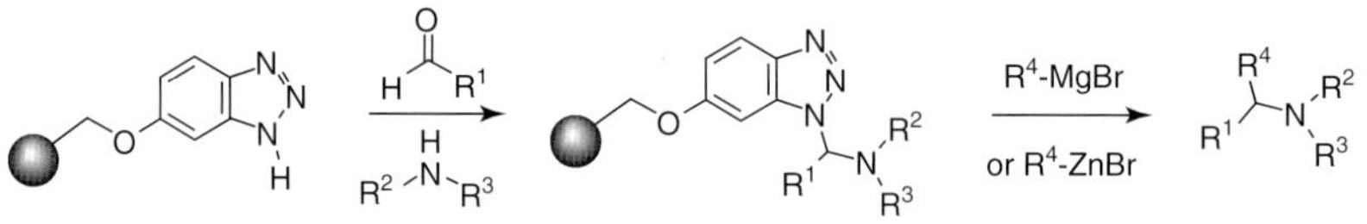

Scheme 7

phosgenated oxime (phoxime) resin (**4**, **Scheme 6**).[822] The phoxime resin also serves as a reactive equivalent, requiring thermal rather than chemical activation. This polymer-bound reagent traps amines as support-bound urethanes that are thermally labile and can be cleaved by thermolysis to provide isocyanates. The released isocyanate reacts with amines present in the cleavage solution producing highly substituted ureas (**Scheme 6**) (see also the previous section on resin-to-resin transfer).

A number of laboratories have investigated the synthesis and use of polymer-bound ^1H-benzotriazole as both a support-bound reagent and a traceless linker for library synthesis. It is fitting that the Katritzky group published the first report of such a reagent for the synthesis of amine libraries, as that same group pioneered the use of this reagent in traditional solution phase chemistry.[823] Treatment of any of the benzotriazole resins with an aldehyde and an amine under conditions suitable for formation of the Schiff base produces the polymer-bound benzotriazole–Schiff base adduct **6** (**Scheme 7**). Cleavage is then effected by substitution with a suitable nucleophile, such as a hydride, Grignard, or Reformatsky reagent, to produce a new, substituted amine (**7**) in good yield. Given the utility of benzotriazole in the synthesis of substituted heterocycles, it is certain that polymer-supported benzotriazole equivalents will be useful for many different heterocycle syntheses in the future.

Cycloaddition reactions utilizing support-bound substrates are popular due to the large increase in complexity of the final products afforded by these reactions. Smith has utilized a commercially available alkyldiethylsilane-functionalized polymer (PS-DES resin) activated as the triflate to prepare resin-bound silyl enol ethers of enones (**Scheme 8**).[824] These resin-bound intermediates then provide access to a variety of cycloaddition products. Smith has investigated the Diels–Alder chemistry of these intermediates, and Porco and co-workers[825] have demonstrated the utility of these precursors in providing functionalized esters from Claisen rearrangement chemistry. Given the utility of trimethylsilyl triflate and related silyl triflate catalysts in a variety of traditional solution phase transformations, such as glycosylation, PS-DES triflate is likely to be a useful reagent for many transformations in library synthesis.

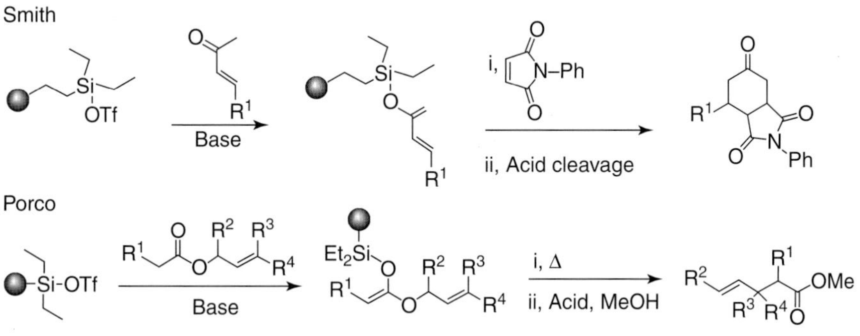

Scheme 8

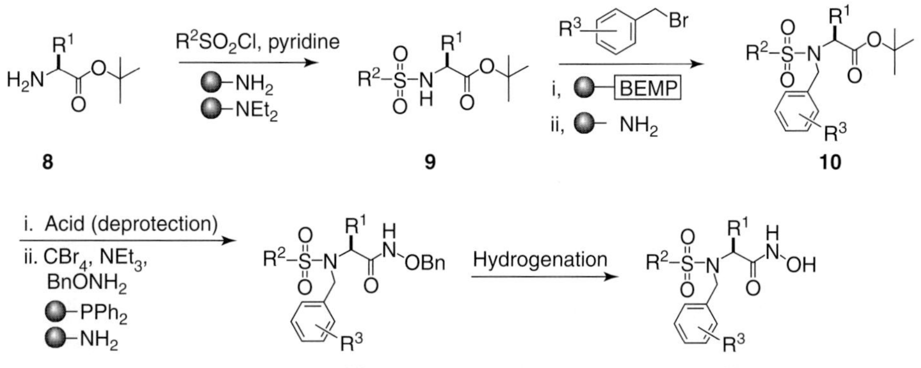

Scheme 9

2.12.16.4.4 Multistep synthesis with resin-bound scavengers and reagents

The development of a large number of strategies for PASP chemistry has allowed the synthesis of increasingly complex molecules using only fast PASP purification. This section details some of these advances, which provide a strong base for the future of the field.

One important synthetic target of the Ley group has been the hydroxamate inhibitors of the matrix metalloproteinases.[826] Inhibitors similar to the known inhibitor CGS-27023A (**Scheme 9**, **12** ($R^1 = Pr^i$, $R^2 =$ 4-MeO-phenyl, $R^3 =$ 3-pyridyl)) were prepared from amino acid *t*-butyl esters. The synthesis involved the use of support-bound amines to scavenge excess sulfonyl chloride and alkyl halide reagents, the use of a support-bound base in an alkylation, and the use of support-bound triphenylphosphine to generate an acid bromide in solution. Condensation with aromatic sulfonyl chlorides followed by work-up with P-NH$_2$ and desalting with ion exchange resin provides the sulfonamides **9**. Alkylation of the sulfonamide with P-BEMP (a resin-bound base) and a benzyl halide yields compounds **10** after removal of excess halide with P-NH$_2$. The *t*-butyl ester is cleaved with acid, and the hydroxamic acid functionality is introduced by preparing the acid bromide, followed by condensation with *O*-benzylhydroxylamine hydrochloride in the presence of base. Hydrogenation provides the final products **12** (27 examples). No chromatography was necessary, and the products were isolated in high (90–98%) overall purity and variable (39–100%) yields.

The Ley group has extended these concepts to include the multistep synthesis of natural products as well as the solution phase preparation of libraries. The synthesis of racemic oxomaritidine and epimaritidine (**Scheme 10**) was accomplished starting with 3,4-dimethoxybenzyl alcohol.[827] All six steps of the reaction sequence involve the application of a resin-bound reagent. Oxidation with polymer-supported perruthenate to the aldehyde followed by reductive amination with amine **15** using P-borohydride (a resin-bound reducing agent) gave the secondary amine **16**, which was then trifluoroacetylated using P-DMAP. Oxidative spirocyclization to **18** was achieved using polymer-bound (diacetoxyiodo)benzene. After removal of the trifluoroacetate protecting group with P-carbonate, the resulting amine

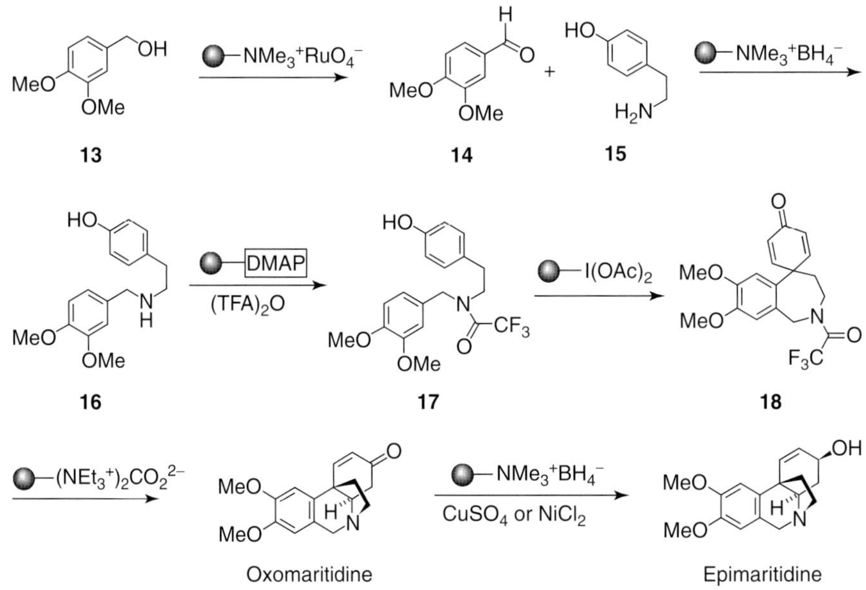

Scheme 10

spontaneously undergoes 1,4-addition to provide racemic crystalline oxomaritidine. Reduction with polymer-supported borohydride provides epimaritidine. Both natural products were prepared in excellent overall yield and purity.

A final example is the multistep synthesis of the analgesic natural product epibatidine, an alkaloid isolated from the Ecuadoran poison dart frog *Epipedobates tricolor*.[828] The 10-step synthesis involves support-bound reagents being used in oxidations, reductions, basic catalysis, and catch-and-release purification of the final product. The sequence begins with reduction of the acid chloride **19** with support-bound borohydride followed by oxidation with polymer-bound perruthenate to afford the aldehyde **20** (**Scheme 11**). A Henry reaction catalyzed by basic resin is followed by elimination to the styrene **22**. A Diels–Alder reaction with the diene **23** occurs smoothly in a sealed tube to yield the cyclohexanone **24**, which is then reduced with support-bound borohydride to the alcohol. Mesylation with P-DMAP affords the mesylate. Reduction of the nitro group without concomitant dehalogenation proved to be difficult; however, the modification of polymer-bound borohydride with nickel(II) chloride provided an impressive solution to this problem, and the desired product (**26**) was isolated in 95% yield as a 7:1 mixture of diastereomers. The critical transannular cyclization of **26** to *endo*-epibatidine occurred in 71% yield upon treatment with P-BEMP, and the unreacted *cis* isomer of **26** could be removed by reaction of the remaining mesylate using P-NH$_2$ (a nucleophilic support-bound amine). Finally, the natural product was prepared by isomerization to the desired *exo* isomer, followed by catch and release purification using a polymer-bound sulfonic acid. The natural product **27** was released from the support by eluting with ammonia in methanol, and was isolated as a 3:1 *exo:endo* mixture in more than 90% purity by ^{1}H NMR and liquid chromatography/mass spectrometry analysis.

2.12.16.5 Conclusion

Since Furka's and Geysen's seminal work on peptide library synthesis in the late 1980s, combinatorial chemistry has matured into its own discipline. Today, combinatorial chemistry encompasses an extremely broad range of technologies and methodologies that have brought many seemingly unrelated scientific fields together. The meshing of high-throughput synthesis, computational design, robotics, and informatics has provided a framework for modern research. The combinatorial chemistry laboratory is now highly automated, allowing hundreds of reactions to be performed in parallel. This new paradigm for discovery is transforming the way scientists think about setting up experiments. The basis of the scientific method is being retooled for the twenty-first century chemists, who are equipped to solve problems by asking not one question at a time but many. The development of new synthetic methods to create the desired libraries of molecules for the study of their specific function is critical, and will continue to determine one's ultimate success. Recent advancements in the field of combinatorial organic synthesis on solid phase and in solution phase are providing powerful tools to the chemist seeking to apply these new principles. The continued development of novel methodologies and technologies in combinatorial chemistry promises a new and bright future for science in general.

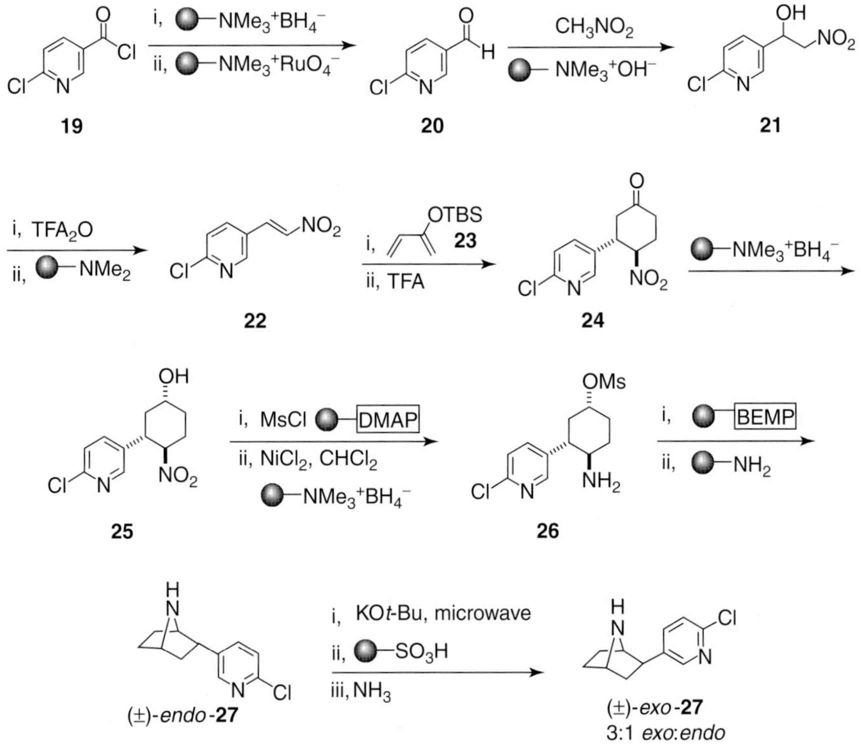

Scheme 11

2.12.17 Supports for Solid Phase Synthesis

Since the introduction by Merrifield of support-based peptide synthesis,[829] insoluble polymer supports have been widely used for numerous synthetic processes to accelerate synthesis and product purification.[830,831,831a] The original small-molecule solid phase efforts by Leznoff,[832] and more recent efforts, demonstrated the synthetic advantages of solid phase methods and have led to solid phase chemistry now playing a substantial role within most pharmaceutical companies, in part due to the ease of automation as well as the inherent advantages of solid phase chemistry. The increasing familiarity of synthetic chemists with solid phase synthesis has led research groups to investigate a wide variety of supports in combinatorial synthesis.[833,834] The main purpose of this section is to review the different types of supports used in solid phase synthesis, a crucial issue if any synthesis is ever to be successful and an area often ignored in a solid phase campaign.

The recent activity in solid phase organic chemistry has focused almost entirely on gel-type resins due to their prior impact in solid phase peptide synthesis. However, the more diverse chemistry that must be achieved nowadays has persuaded many groups and companies to investigate a wider range of solid supports compatible with different reaction conditions. Automation within pharmaceutical companies is also an issue in combinatorial chemistry, in so far as solid phase supports must be developed to facilitate automation and handling issues.

2.12.17.1 Resin Beads

2.12.17.1.1 Polystyrene gel-based resins

In practice, gel-based spherical resin beads of 50–200 μm diameter are commonly used in solid phase organic chemistry (SPOC). These can be easily synthesized or bought from a range of commercial suppliers. The size, shape, and uniformity are important features, and the uniformity of cross-linked polymer particles is vital for reproducible SPOC. Robust spherical beads with a uniform size are preferred for reproducible chemistry, with irregularly shaped particles being much more sensitive to mechanical destruction, often falling apart in the course of the chemistry. Bead size is also important, since reaction rates are inversely proportional to bead size. The technique of suspension polymerization is almost universally used for resin synthesis, and provides regular particles in a highly reproducible manner. Typically,

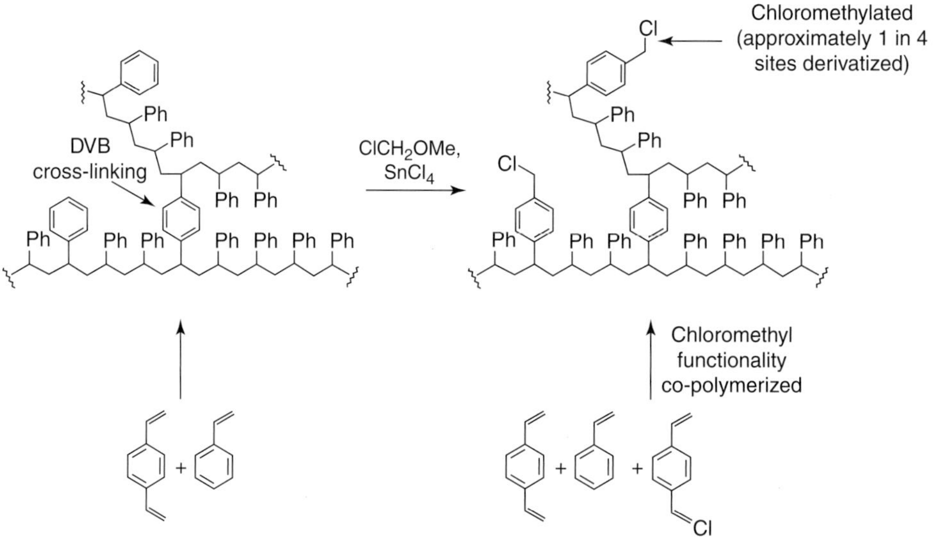

Figure 125 Synthesis of DVB-cross-linked PS resins and derivatization.

styrene and divinylbenzene (DVB) mixtures are dispersed as spherical droplets in an excess of an immiscible phase (water) containing the polymerization initiator. The aqueous phase generally contains a low level of some dissolved suspension stabilizer, a surface-active species, which prevents the organic monomer droplets from conglomerating. In the course of the polymerization, these droplets are converted to hard, glassy beads. Then the resin particles can be collected by filtration; unreacted monomers, initiator, and other by-products removed by solvent extraction; and the beads vacuum dried and sieved. The introduction of functionality onto a polymer support is usually achieved by direct chloromethylation, to give the classic Merrifield resin. However, an alternative approach is to use a co-monomer that already bears the desired functionality in the free polymerization mixture or some precursor that can subsequently be transformed. This second strategy can be useful in producing a structurally well-defined resin network with control of functional group ratios (**Figure 125**). So-called Merrifield resin and related polystyrene (PS) based analogs (e.g., aminomethyl resin) are the most widely used resins in SPOC today. They are characterized as being a gel-type polymer typically containing 1–2% of the cross-linking agent DVB, which provides the links between the linear PS chains, with the whole polymer solvated and with reactions taking place throughout the polymer network, controlled by diffusion into the bead.

The percentage of DVB used in bead synthesis is very important for the mechanical strength and insolubility of the support, but should not prevent swelling of the resin beads when immersed in an organic solvent. Indeed, swelling is an essential feature of gel resins. At low cross-linkings, sufficient swelling occurs in so-called good solvents to allow the diffusion of reagents within the PS network. Merrifield resin swells in solvents with a solubility parameter similar to the polymer (e.g., toluene), with swelling inversely related to the ratio of the DVB content. On the other hand, Merrifield resins are not compatible with highly polar solvents such as water and methanol. The swelling and shrinking process always occurs from the outside to the inside of the resin network; thus, very low levels of DVB (<1%) give mechanically weak networks that can be damaged by heating, handling, or solvent shock. On the other hand, a highly cross-linked resin network may not swell even in a 'good swelling' solvent, but clearly will offer much greater mechanically stability. If a PS network is fully swollen in a so-called good solvent and then shrunk in a bad solvent (e.g., during resin-washing steps), mechanical shock takes place and the bead may disintegrate. This effect (osmotic shock)[834] and the degree of cross-linking must thus be considered in any solid phase synthesis where many cycles of swelling and deswelling may be involved to reduce mechanical damage to the beads. Besides osmotic shock, PS supports have limitations when used with highly electrophilic reagents and at excessive temperatures (200 °C). Impurities from commercially available Merrifield resin can arise from incomplete polymerization and/or from trapped solvents, which contaminate the final products or decrease the actual loading. It should also be borne in mind that very high levels of initiator (sometimes 5%) are used in bead synthesis, which, depending on the initiator used, will undoubtedly have an influence on bead structure and chemistry. Although various methods have been investigated[835] to overcome resin-based impurities, such as prewashing of the resin beads, the main method appears to be based on final library purification, as being the best way to readily remove impurities from library compounds, although perhaps not ideal (the advent of automated HT purification means that low levels of impurities can be readily removed from the cleaved compounds at the end of the synthesis).

2.12.17.1.2 Macroporous resins

Macroporous resins have been used for many years in ion exchange applications, as polymeric adsorbents and for reverse-phase chromatography purifications. Macroporous resins are defined as a class of resins having a permanent well-developed porous structure even in the dry state (if they survive in this form). These resins are typically prepared by suspension polymerization of styrene–DVB mixtures containing a porogen or diluent (an organic solvent in general) at a well-defined ratio, providing resin beads with a defined pore structure. Removal of the porogen at the end of polymerization provides a heterogeneous PS matrix with some areas completely impenetrable and others free of polymer. Macroporous resins are characterized by a hard, rough surface having a defined total surface area. Unlike Merrifield resins, these materials do not need to swell to allow access of reagents through the PS network, as they possess a permanent porous structure that can be accessed by, essentially, all solvents; even solvents such as water can penetrate the macroporous PS–DVB matrix. When a solvent penetrates a macroporous resin, the polymer matrix tends to swell to some extent, and often rapidly, because the permanent holes provide rapid access through the network. However, due to the nature of the beads, swelling takes place in the pores and little swelling of the beads takes place. Due to the nature and speed of diffusion, macroporous resins show much better resistance to osmotic shock during swelling and deswelling processes.

Comparisons between macroporous resins and a 2% cross-linked gel-based Merrifield resin have been undertaken.[836] Resin washing efficiency is, not unexpectedly, much more efficient with macroporous resins due to the rigid pores speeding up the migration of material from the interior of the polymer to the bulk solvent. The kinetics for esterification between standard Merrifield resin and three macroporous resins from Rohm & Haas are shown in **Table 29**. This seems to show that a large number of sites are more accessible in the gel-based resins, while reaction rates are comparable. However, since reaction rates are bead size dependent for the gel-based resins, it is hard to draw any real conclusions from these data.

2.12.17.1.3 Polyethyleneglycol-containing resins

The development of polymer supports having both polar and nonpolar features has been of great interest for peptide synthesis. Although Merrifield resin has been widely used in solid phase peptide synthesis, the rate of incorporation of particular amino acid residues was found to decrease with increasing chain length. This decrease in yield was thought to be due to the unfavorable conformation adopted by the growing peptide chain and peptide aggregation. Thus, to improve the physicochemical compatibility of PS supports, polar hydrophilic acrylamide-based copolymers were introduced as an attractive alternative to all PS–DVB supports, most notably by Atherton and Sheppard.[837] However, an alternative approach, which leads to the second most popular support type in solid phase chemistry, after the PS gel-based resins, is the modification of PS resin by the addition of polyethyleneglycol (PEG) to generate PS-PEG-based materials. These hydrophilic solid supports were developed in theory to remove the reaction sites from the bulk 'rigid' PS matrix, to give a more solution-like environment as well as to broaden the solvent compatibilities of the support. However, another driving force was the need to prepare a solid support for continuous solid-flow solid phase peptide synthesis, which did not undergo dramatic volume changes during synthesis and arose at least in part from the early soluble PEG-based chemistries in the 1970s and 1980s. Various popular PEG-containing synthetic polymers as solid supports are summarized in **Table 30**.

Table 29 Kinetics for esterification between standard Merrifield resin and three macroporous resins

Resin (%) (48 h = 100%)	Loading (mmol g^{-1})			Relative loadings	
	$t = 4\,h$	$t = 20\,h$	$t = 48\,h$	$t = 4\,h$	$t = 20\,h$
XAD 16	0.09	0.46	0.62	14	74
XAD 2010	0.16	0.54	0.61	26	88
XAD 1180	0.14	0.47	0.55	25	85
PS-CM (1%)	0.23	0.71	1.14	20	62

Table 30 Some polyethyleneglycol-containing resins

Support	Description
Grafted PS-PEG	
PS PEG[838]	Preformed PEG covalently attached onto the PS. Chemistry takes place on the end of the PEG
TentaGel[839]	PEG polymerized onto 1% cross-linked hydroxymethyl-poly(styrene-*co*-divinylbenzene) resin beads
ArgoGel[840]	PEG polymerized onto a malonate-derived 1,3-dihydroxy-2-methylpropane poly(styrene-*co*-divinylbenzene) resin beads
NovaGel[841]	Methyl-PEG coupled onto 25% of resin sites. Chemistry takes place on the PS, not on the end of the PEG
Copolymers	
PEGA[842]	Poly(dimethylacrylamide-*co*-bis-acrylamido-PEG)-*co*-monoacrylamido PEG
TTEGDA-PS[843]	Poly(styrene-*co*-tetra(ethylene glycol)diacrylate)
CLEAR[844]	Poly(trimethylolpropane ethoxylate triacrylate-*co*-allylamine)

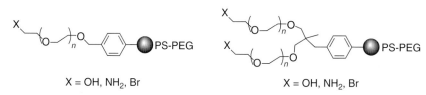

Figure 126 The two main PS-PEG supports used in solid phase synthesis.

2.12.17.1.3.1 Polystyrene-polyethyleneglycol supports

A preliminary approach to the synthesis of PS-PEG supports relied on covalent anchoring via amide linkages of defined PEGs onto amino-functionalized PS resins.[838] Later, a new version of PS-PEG was introduced by Bayer and Rapp,[839,840] obtained by ethylene oxide grafting, to give the material now known as TentaGel (**Figure 126**). Optimized TentaGel-grafted resins generally have PEG chains of about 3 kDa ($n = 68$ units, 70–80% by weight). PS-PEG beads display uniform swelling properties in a variety of solvents, from medium- to high-polar media, ranging from toluene to water.

The solution-like environment has an impact on reaction kinetics, allowing rapid access of reagents through the swollen resin to active sites. Rapp[841] has demonstrated that the kinetics of certain reactions on PS-PEG beads were similar to those in solution. However, TentaGel resin is far from being the perfect solid support due to (1) a low loading capacity, (2) Lewis acid complexation onto the PEG chain, and (3) instability of the PEG chain to acidic media, generating PEG contamination after trifluoroacetic acid cleavage. To overcome these problems, new families of PS-PEG resins with higher loading capacities, good acid stabilities, and low linear PEG impurities have been developed.[842] Thus, assuming that the acid instability of TentaGel resin was due to the benzylic ether PS–graft linkage, improved stability was obtained by using a longer linkage, while enhanced loading capacity has been obtained via bifurcation of functional groups prior to ethylene oxide grafting (**Figure 126**) and reduced PEG graftings.

It is worth noticing that at the end of the polymerization material must be contaminated by linear or cyclic PEG, and this removal may be the source of repeated claims of contamination. A series of graft copolymers have been elaborated by varying the ethylene oxide/initiator ratio, providing well-defined graft lengths with predictable loadings (0.3–0.55 mmol g^{-1}). Some swelling properties of PS and PS-PEG resins are summarized in **Table 31**.

2.12.17.1.3.2 Cross-linked ethoxylate acrylate resins

CLEAR resins (cross-linked ethoxylate acrylate resins), introduced by Barany,[843] are a unique family of supports that are highly cross-linked polymers (>95% by weight of cross-linker) yet show excellent swelling properties, contrary to the general expectation of the time. These supports are prepared by radical polymerization of the tribranched cross-linker trimethylolpropaneethoxylate-triacrylate **29**, with various amino-functionalized monomers, such as allylamine

Table 31 Swelling properties of PS and PS-PEG (mL g^{-1})

Resin	Solvent				
	Water	Tetrahydro Furan	Dichloromethane	Dimethylformamide	MeOH
PS-NH$_2$[a]	8.0	8.9	8.0	2.8	2.1
TentaGel-NH$_2$[b]	3.9	5.5	4.0	3.0	3.0
ArgoGel-NH$_2$[c]	6.4	8.6	5.0	4.9	4.0

[a] Obtained from Bachem.
[b] Obtained from Rapp Polymer.
[c] Obtained from Argonaut Technologies.

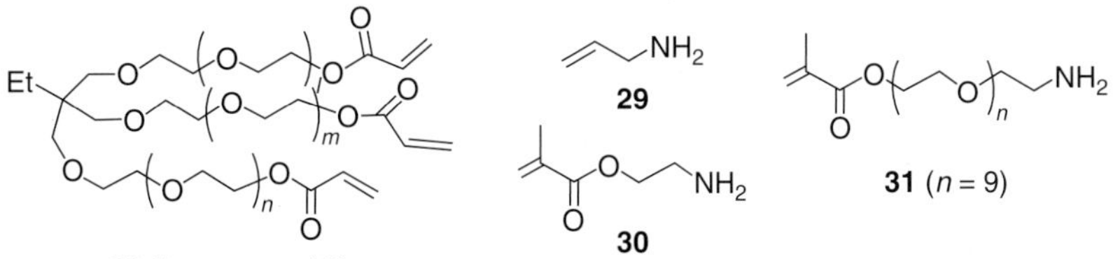

Figure 127 Monomers used in the highly cross-linked yet highly swelling CLEAR resins.

(**30**), or 2-aminoethyl methacrylate (**31**) (**Figure 127**). The CLEAR supports differ from other polymers in that they are synthesized from a branched cross-linker used in a high molar ratio. The fact that the amino functionality can be introduced into the bead in the polymerization process ensures that the required substitution level can be obtained.

A number of different formulations were tested and prepared by bulk suspension to obtain the optimal material. These supports swell in a wide range of hydrophobic and hydrophilic solvents, such as water. Although these materials showed great chemical and mechanical stability in acidic and weakly basic conditions, the resins will dissolve in the presence of ammonia or aqueous bases, due to the ammonolysis of the three ester linkages in the cross-linker.

2.12.17.1.3.3 Tetraethylene glycol diacrylate cross-linked polystyrene support (TTEGDA–PS)

A number of groups[844,844a,845] have developed resins based on varying the cross-linker to change the properties of the resin but maintaining the excellent chemical nature of the PS support. Pillai[844,844a] investigated the use of polyacrylamide and PS supports cross-linked with N,N-methylenebisacrylamide (NNMBA), TTEGDA, and DVB. It was shown that the polymer derived from TTEGDA-polyacrylamide had a significant increase in reaction kinetics when compared with NNMBA–polyacrylamide and DVB–polyacrylamide supports. Gel phase reaction kinetics with polyacrylamide-DVB supports showed that the first 90% reaction sites were homogeneously distributed throughout the bead and had equal reactivity, while the remaining 10% were less available to external reagents. The major problems with polyacrylamide supports is the lack of mechanical stability when compared with PS, and chemical reactivity when used in an organic chemistry sense, although there is an optimal hydrophobic–hydrophilic balance. TTEGDA–PS provides these two essential features in a single matrix with a hydrophobic PS backbone and a flexible hydrophilic cross-linker,[844,844a] thus being insoluble but highly solvated. PS supports with 4% TTEGDA (**Figure 128**) show good mechanical properties and good swelling characteristics, rendering this support suitable for long-chain peptide synthesis.

As mentioned earlier, the swelling properties of a solid support are a crucial feature in solid phase synthesis, and a measure of the scope of a new support. The difference between 1% DVB–PS and 4% TTEGDA–PS was investigated. As expected, TTEGDA-cross-linked PS had enhanced swelling properties in polar solvents compared with the DVB–PS support (**Table 32**), which renders maximum accessibility of reagents throughout the hydrophobic network.

2.12.17.1.3.4 Polyethyleneglycol-poly-(N,N-dimethylacrylamide) (PEGA) copolymer

Polar acrylamide supports were introduced successfully by Sheppard[837] and co-workers as an alternative to PS resin, and used as a versatile cross-linked support for peptide syntheses. Meldal[846,846a] introduced PEGA, a similar resin that

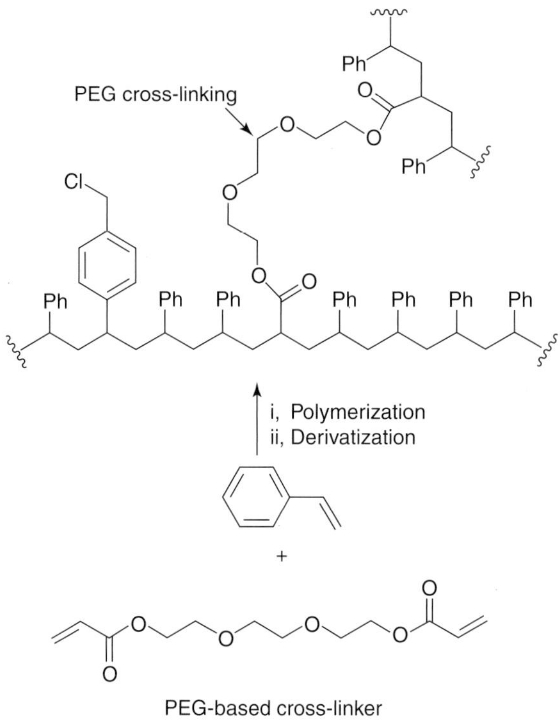

Figure 128 Hybrid PS- and PEG-based resins via PEG-based cross-linkers.

Table 32 Comparison of DVB–PS and TTEGD–PS swelling properties (mL g^{-1})

Solvent	1% DVB–PS	4% TTEGDA–PS
Chloroform	4.3	6.5
Tetrahydrofuran	5.2	8.7
Toluene	4.7	6.9
Pyridine	4.2	6.0
Dioxane	3.5	8.8
Dichloromethane	4.3	8.8
Dimethylformamide	2.8	5.5
Methanol	1.8	2.2

contains a PEG component. PEGA is superior to many other existing supports for peptide synthesis in terms of the solution-like nature of the structure, although it is associated with handling problems, and lacks chemical inertness. The PEG chain is the major constituent, providing a highly flexible and biocompatible matrix. The polymer backbone was designed to allow macromolecules, such as enzymes, to penetrate the polymer network, thereby facilitating solid phase enzymatic reactions.

Typically, preparation of PEGA resins is carried out by an inverse suspension polymerization reaction in a mixture of n-heptane and carbon tetrachloride (6:4 v/v) due to the water solubility of the monomers (**Figure 129**). Meldal and co-workers anticipated that PEGA 1900 ($n = 45$) had pores large enough to allow enzymes up to 50 kDa to penetrate the PEGA network.[847,847a] In order to identify new enzyme inhibitors for matrix metalloproteinase 9, which in its proform has a molecular weight of 92 kDa, longer cross-linked PEGs were used (4000–8000 Da, $n = 89$, 135, and 180) to increase the

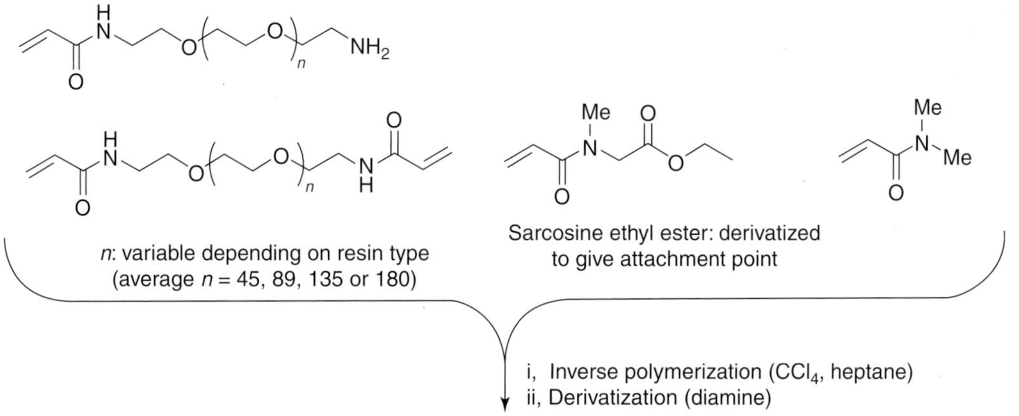

Figure 129 Monomers used for PEGA resin synthesis.

apparent 'pore size' of these supports. Bis-amino-PEGs were used to form partially or bis-N-acryloylated PEG as a macromonomer. These resins have high swelling properties, and are suited for enzyme reactions, despite initially low loading capacities (0.08–0.13 mmol g^{-1}). The introduction of acryloyl sarcosin with bis-acrylamido-PEG-1900, -4000, -6000, and -8000 and dimethylacrylamide resulted in PEGA copolymer-type resins with a much higher loading capacity.

2.12.17.1.3.5 Enhancing bead loading: dendrimers as powerful enhancers of functionality

One of the most powerful tools available to the combinatorial chemist is the process of split and mix in terms of the number of compounds that can be made and the economics of synthesis, with resin beads being used as the ultimate microreactor in synthesis. However, this method results in tiny amounts of compound being available, typically less than 1 nmol per bead. This might be enough to identify an active bead, but it is not enough to conduct numerous biological assays or to validate the compound as a lead. More importantly, it is insufficient to determine the compound's structure using conventional techniques such as ^{1}H NMR. Such a low loading requires a tagging system to identify the compound, and necessitates resynthesis on a larger scale either in solution or on the solid phase for rescreening. However, though elegant, the introduction of a tag during the library synthesis is very time-consuming. One solution to this problem is to increase the size of the beads. A larger bead will obviously bear more sites, although size is limited by bead stability, susceptibility to fracturing, and bead synthesis, as well as by poor reaction kinetics. Another approach is to multiply the functionalities on a bead by dendrimerization, using the hyperbranched nature of the dendrimer to amplify loading. Solid phase lysine-based dendrimers were published by Tam[848] as a means of generating a high density of peptide functionality for antigen presentation (the multiple antigen presentation system), but resins with low loading were deliberately used to avoid possible steric problems. However, the development of solid phase polyamidoamine (PAMAM) dendrimers has provided to be a powerful tool in increasing the loading of single beads.[849] The development of highly symmetrical dendrimers allows an extremely high loading to be obtained on the resin beads.

The efficiency of synthesis for the PAMAM dendrimers (Michael addition onto methyl acetate and ester displacement with a diamine) is achieved by using a large excess of reagent compared with the loading of the resin as in any normal solid phase synthesis, and produces dendrimers cleaner than those from the solution approach. The multiplication of the terminal amino functionalities makes possible the analysis of compounds cleaved from single beads.[850] Using a Tris-based monomer,[851] a loading of 230 nmol per bead was obtained for a second-generation dendrimer, which becomes a very acceptable loading for using the beads in a split-and-mix manner (**Figure 130**). The NMR spectrum and high-performance liquid chromatography (HPLC) data of material from one bead are shown. This is the Suzuki product formed between 4-methylphenylboronic acid and resin-attached 4-iodobenzoic acid. Using this method, dendrimers have been synthesized on a range of other solid supports, either to be used as such or to modify the surface properties of the support.

2.12.17.2 Synthesis on Surfaces

2.12.17.2.1 Pins and crowns

The limitations of linear 'solitary' peptide synthesis were first overcome by Houghten,[852] with the use of the 'tea-bag' methodology, and Geysen, with the 'pin' methodology.[853,853a] The pin method was a revolutionary approach to peptide synthesis and moved away from traditional PS-based resins, allowing the synthesis of peptides on the surface of

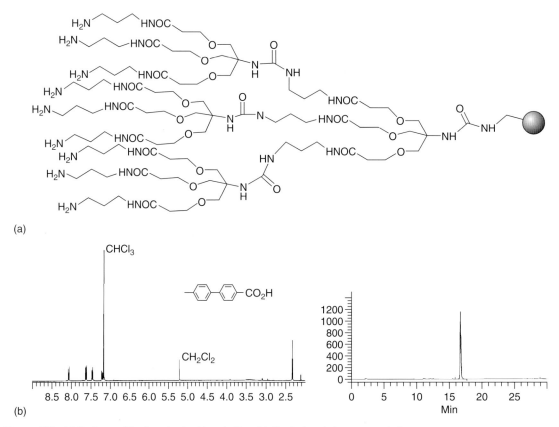

Figure 130 (a) Resin amplification via dendrimerization. (b) Single-bead cleavage analysis.

polyethylene supports (rod-like materials termed 'pins' (4 mm diameter × 40 mm)) in a multiple parallel manner. These derivatized materials were prepared by radiation grafting, in which the pins were placed in a γ-radiation source in the presence of an aqueous solution of acrylic acid. Radiation (initially 10^6 rad) activated the surface by radical formation to a depth of several micrometers, which reacted with the acrylic acid to give rise to a polyacrylic acid derivatized pin (**Figure 131**). It was this (immobilized) graft that could then be used as a support for peptide synthesis, initially by coupling to a suitably protected lysine residue (Boc-Lys-OMe). Due to the discrete and handleable nature of the pins, they could be arranged in an array format akin to the microtiter plates used for biological screening, with synthesis taking place by dipping the tips of the pins into the appropriate solutions. Although the whole pin was derivatized, only the tip was actually used synthetically. The major problems were thus logistical, due to compound numbers, rather than chemical. This basic procedure allowed the synthesis in 1984 of 208 overlapping peptides covering the entire peptide sequence of the coat protein VP1 of foot-and-mouth disease virus. This was accomplished using *t*-butyloxycarbonyl (t-Boc) chemistry, with the peptide directly attached to the pin as shown in **Figure 131**. Two assumptions made by the authors in this original work were that an initial loading of at least 1 nmol would be sufficient for even a low-efficiency synthesis to allow antibody binding, and that high purity was not necessary.

Developments and improvements on this original method took place at the end of the 1980s and in the 1990s, with three main driving forces: (1) the development of linkers to allow cleavage from the pins;[854,854a] (2) an improvement in their loading;[855] and (3) alterations in the grafting polymer to alter the physical and accessibility properties of the graft[856] (**Figure 132**). Thus, in 1990 and 1991, derivatization of the pin by acrylic acid was followed by saturation coupling with *t*-Boc-1,6-diaminohexane, and coupling to the most accessible sites with a limited amount of Fmoc-β-Ala-OH followed by capping of the remaining sites, to give a loading of 10–100 nmol per pin.[857] Linkers could then be added to the derivatized pins. The first approach was to use diketopiperazine (DKP) formation as a driving force for peptide release, allowing a 'safety catch'-based cleavage from the solid support. The second method used hydroxymethylbenzoic acid as a linker, which allowed the peptide to be subsequently cleaved from the pin with ammonia vapor.[858] Both methods allowed the release and analysis of peptides by HPLC and mass spectrometry.

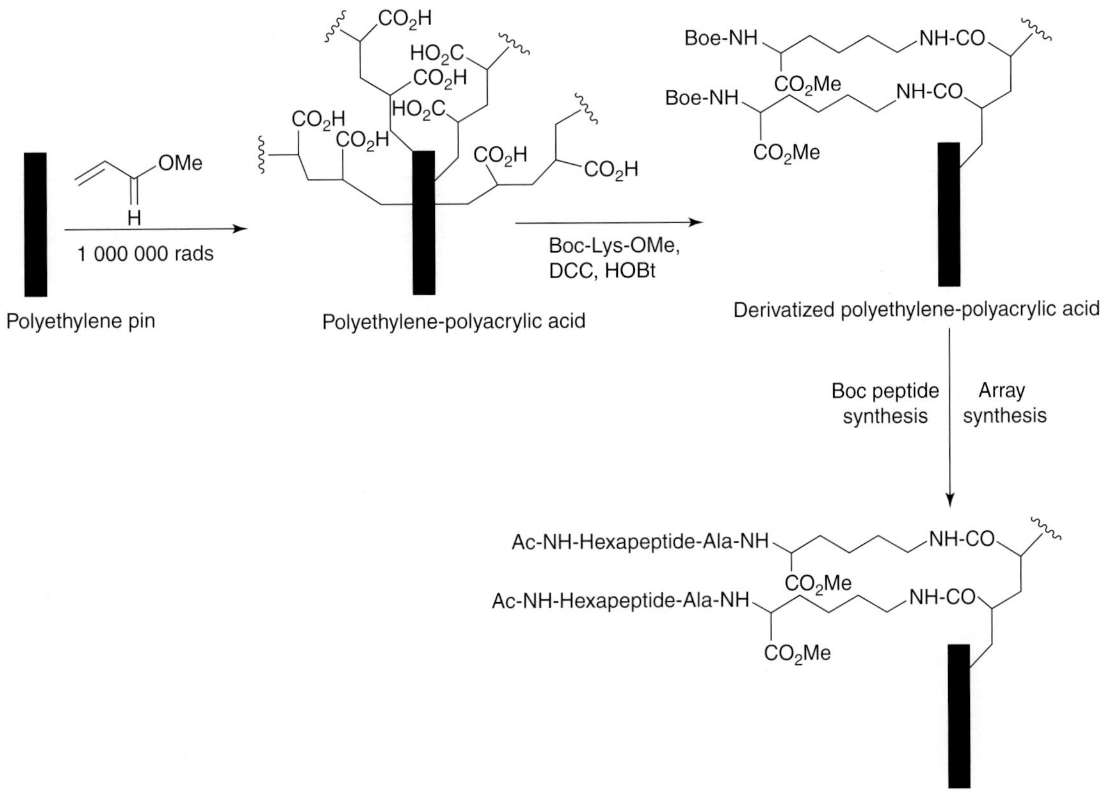

How and Why to Apply the Latest Technology

Figure 131 Initial method used for pin derivatization.

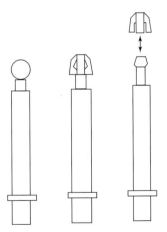

Figure 132 Radiation-grafted materials: Mimotopes crowns and removable pin heads.

Subsequent developments led to the removable crown (although confusingly also called pins or detachable pinheads in the literature) (**Figure 132**).[859] Crowns were prepared by injection molding of granular polyethylene. They were 5.5 mm high and 5.3 mm in diameter, but, importantly, had a surface area of 1.3 cm^2, much improved over that of the pins. The second improvement involved the use of alternative monomers (**Figure 133**). Variations in the vinylic monomer allowed alterations in the properties of the graft. Thus, changing from acrylic acid to hydroxyethyl methacrylate allowed much higher grafts weights to be obtained (the graft weight increased from 1% to 15%).[859] Thus, these materials could be esterified under normal dicyclohexylcarbodiimide/4-N,N-dimethylaminopyridine (DCC/DMAP) coupling conditions with Fmoc-β-Ala-OH and controlled reaction timings, allowing only the most reactive/accessible sites to be derivatized before the capping of all remaining sites. This procedure allowed crowns with loadings of 1.0–2.2 μmol per crown

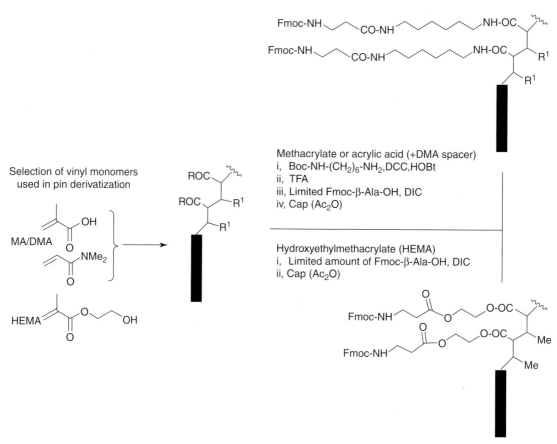

Figure 133 Second-generation pin/crown derivatization and loading.

to be prepared for solid phase peptide synthesis. Later papers used a variety of monomers, including methacrylic acid and dimethylacrylamide,[858] while PS grafts also became available. The crowns/pins thus became available with aminomethyl-PS functionalities, thereby allowing a wide range of linkers, including the acid-labile Wang and Rink type of linkers, to be introduced onto these supports via the aminomethyl group, thus permitting the straightforward cleavage, analysis, and screening of micromolar quantities of compound. The ease of handling these small discrete crowns, which could be prepared in a variety of different colors as an inbuilt coding mechanism, and the broad range of loadings available (1–40 μmol), obviously made them very user-friendly to the synthetic chemist for multiple parallel synthesis.

The potential of these crowns was demonstrated in a series of papers from Mimotopes[860,860a] and Ellman,[861,861a–d] including reductive aminations and Mitsunobu chemistries, and a large series of β-turn mimetics (using pins with a loading of 5.9 μmol) and 1680 1,4-benzodiazepines were prepared. Numerous groups have now used these materials in either an MPS- or a transponder-encoded defined split-and-mix synthesis. An addition to this area has been reported by Irori.[862] Radiation grafting was carried out as in the case of Mimotopes crowns, although in this case it was done on so-called microtubes (small plastic tubes) made of either polypropylene or a more chemically resistant fluoropolymer. However, they have not found widespread use.

2.12.17.2.2 Sheets

Sheet-type materials have been used for many years in the area of oligonucleotide and peptide synthesis. Early examples include the use of radiation-grafted polyethylene sheet, akin to Geysens pins, polypropylene membranes coated with cross-linked polyehydroxypropylacrylate, and derivatized glass or paper. Planar surfaces have a big advantage over beads used in the combinatorial manner in that the structure of the compound can be directly deduced by its location on the support. The first of these methods used low-density polypropylene sheet that was γ-irradiated while immersed in a solution of styrene, a method commonly used at the time for the radiation grafting of surfaces.[863] Graft weights of some 440% were obtained with the PS attached to the surface having a molecular weight of some $6 \times 10^6 \, \text{g mol}^{-1}$.

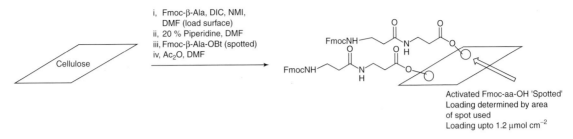

Figure 134 Spot array synthesis on paper.

Aminomethylation of the PS graft allowed linker attachment and peptide synthesis in good yield, with the sheet having a loading of about 29 µmol cm^{-2}. However, as far as we are aware, no more has been reported on the use of this material. At the same time, another approach was reported, although in this case the potential applications suggested were epitope mapping, affinity purification, and diagnostic testing as well as continuous-flow synthesis. The membrane used in this case was polypropylene coated with cross-linked hydroxypropylacrylate, which had a pore size of 0.2 µm.[864,864a] Unfortunately, no details of the coating process or of the material loading were given, but four peptides were prepared on this membrane with good purities. More recently, amino-functionalized polypropylene sheets have been used in DNA array synthesis, but the nature of the derivatization was not given.[865,865a]

In 1990, Frank[865] introduced the concept of spot array synthesis of peptides, the basic premise being the suitability of cellulose paper as a support for standard Fmoc/Bu^t peptide synthesis (**Figure 134**). Numerous peptides could be rapidly prepared by manual spotting of reagents, with up to 100 peptides prepared on an area of 4 cm^2, although 96 peptides based around a microtiter type of format became a standard format. Peptides were either screened on the paper (e.g., when looking for continuous epitopes), or released by DKP formation, analogous to the Mimotopes method, for solution screening, analysis, and characterization. Synthesis and screening of peptides was subsequently carried out on Inimobilon-AV from Millipore in an identical spot manner.

A new material and concept were developed by Kobylecki[866] in which 1% PS resin beads were sandwiched between two layers of polypropylene sheet, giving a loading of approximately 5–10 nmol cm^{-2}. This material was used in a defined split-and-mix synthesis, using three sheets of material. Thus, in the first step a different amino acid was added to each sheet. The material was then cut into three strips, and one strip from each sheet appended together. A different amino acid was added to each bundle of strips. The strips were lined up again as in the original three sheets, and cut into squares, with nine squares per sheet. The squares from each row were joined together, and again functionalized with another amino acid. Thus, this method uses the spatial nature of the material to define the synthetic sequence. In this case, it was used to prepare 27 (3^3) peptides.

Other sheet-like materials have been used in peptide/DNA synthesis.[867] Thus, amino-functionalized sintered polyethylene (Porex X-4920, nominal pore size 30 µm) has either been used directly for DNA synthesis (0.07 mmol g^{-1} loading) or derivatized with a chloromethylated styrene-based colloid having a diameter of 0.46 µm, to increase the loading to 9–12 µmol g^{-1} in an approach similar to that of the dendrimer methods described above. Another approach was described by Luo,[868,868a] in which porous polyethylene disks were oxidized to generate carboxylic acids on the surface that were subsequently derivatized with a diamine followed by carboxymethyl dextran. The acid sites were then loaded with diamines ready for synthesis. Thus, the polyethylene surface was, in essence, coated with derivatized dextran, making it highly hydrophilic and suitable for biological screening. However, accessibility to proteins might be very limited on these materials, depending on the pore size of the polyethylene used. However, these last two methods are anecdotal.

2.12.17.2.3 Glass

In 1991, an extension of 'spot' peptide synthesis was reported, with the introduction by Affymax of peptides and oligonucleotides synthesis on glass.[869,869a] In this process, light-based protecting groups were used for synthesis in conjunction with photolithographic masks, similar to those used in chip manufacture. The basic principle for photolithographic chemistry relies on a functionalized solid surface (glass/silica) that is treated with a photocleavable-protected linker or spacer to coat the surface (**Figure 135**). A laser or photolithographic mask allows the light-induced deprotection of a well-defined (size and position) portion of the slide (**Figure 136**). The disadvantage of this method is that it requires relatively complex instrumentation and the development of a new array of synthesis tools.

In the reported cases, a binary masking strategy was used, although later the permutational 'stripe' masking strategy was used to increase the library size. Using this method, peptide arrays of 1024 peptides could be prepared in an area of

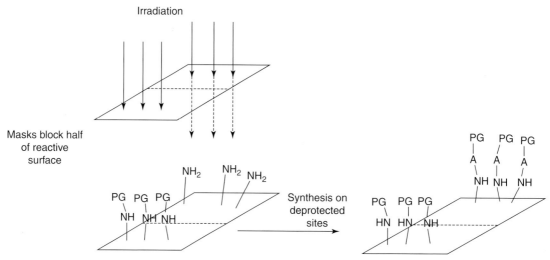

Figure 135 The concept of synthesis on glass using light-based protecting groups.

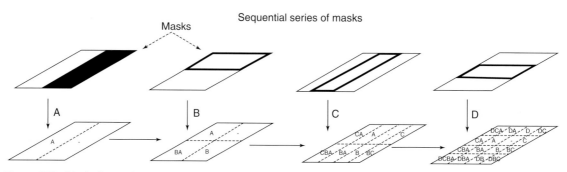

Figure 136 Masks for use in array synthesis.

1.28 cm², which could then be screened directly (each peptide site being 400 nm²). In a similar way, oligonucleotides could be prepared. Thus, the synthesis of short oligonucleotide probes and hybridization onto glass was published in 1994 by Fodor and co-workers, who demonstrated the synthesis of a 1.28 × 1.28 cm array of 256 nucleotides, and hybridized it toward the sequence 5′-GCGTAGGC-fluorescein.[870] The sequence specificity of the target hybridization was demonstrated by the fabrication of complementary probes differing by only one base, with the targets hybridizing almost totally specifically to its complementary sequence. However, a concern with all this chemistry, especially for oligonucleotide synthesis, is the efficiency of protecting-group removal.

In 1996, McGall[871,871a] published the preparation of DNA arrays (using standard solid phase oligonucleotide synthesis) using polymeric photoresist film as a physical barrier to mask selected regions of the surface; exposure to light reveals the surface, and allows the coupling of the first nucleoside. However, the harsh conditions required for deprotection could damage the library, and a bilayer process was developed. An inert polymeric film was introduced prior to the photoresist coat, with the underlayer functioning as both a barrier to the chemical deprotection step and as a protective layer to insulate the substrate surface from the photoresist chemistry and processing conditions. This was a direct application borrowed from the microelectronics industry, allowing, it was claimed, more precise deprotection of the surface, thereby permitting the creation of denser arrays. An array of 256 decanucleotides was synthesized and tested. For additional information, Wallraff and Hinsberg[872] published an overview of the lithographic imaging techniques. Southern[873] also reported using a physical masking approach to prepare DNA arrays, with the chemistry taking place on activated glass (**Figure 137**), with areas being masked or separated from each other by physical 'gaskets,' in this case silicon rubber. The advantage with this method is that well-developed chemistry could be used, the limitations being the number of positions available for synthesis, which is limited by the accuracy of masking. In 1994, the striped method was replaced by the regular shape of a diamond or disk mask in which linear translation of the

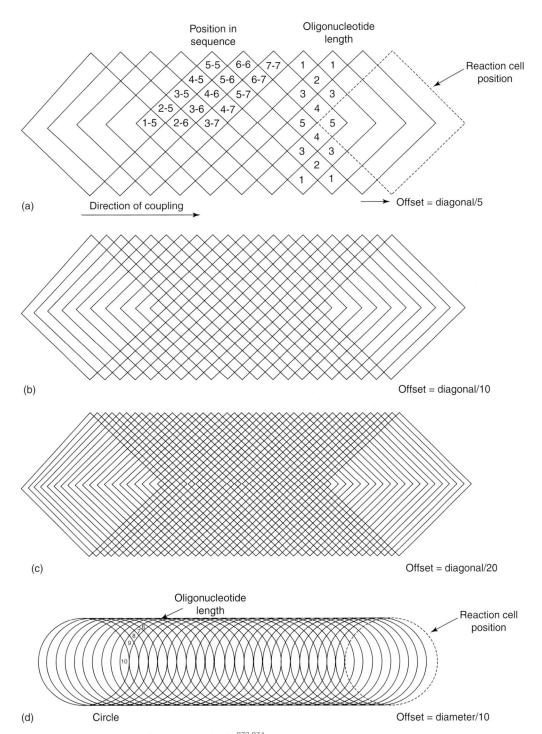

Figure 137 Physical masks for oligo array synthesis.[873,874]

mask allows the generation of a well-defined library,[874] since a small translation of the mask allows defined parts of the slide to be deprotected and reacted.

A library of decamers was synthesized, as an application of the technique, and used to study the hybridization behavior of the corresponding synthetic deca-nucleotide in solution. However, this approach is much less practical and much more cumbersome than the light-based masking strategy.

2.12.17.3 Overview

In conclusion, a wide range of materials have been used and continue to be used for solid phase synthesis. A number of these are very specific in application, such as synthesis on glass for DNA array synthesis. However, supports such as PS resins are undoubtedly the workhorse of the industry, with others, such as the pins and PS-PEGs, having a much narrower role. There is clearly much scope for improved resin design and the generation of new supports for solid phase synthesis.

2.12.18 The Nuclear Magnetic Resonance 'Toolkit' for Compound Characterization

2.12.18.1 Introduction

This section discusses the application of various NMR spectroscopy techniques to the process of drug discovery. To make it easier to understand how drug discovery incorporates NMR data, all drug discovery-related activities have been categorized into one of the following six disciplines: (1) natural products research; (2) medicinal chemistry; (3) rational drug design (protein structure and modeling); (4) metabolism; (5) combinatorial chemistry; and (6) drug production and quality control. Although this classification process may initially appear to be somewhat artificial, its use makes it easier to understand the applications and advantages of each NMR technique. However, readers should recognize that most drug discovery programs involve a blend of these six disciplines, and therefore usually benefit from using a combination of different NMR techniques.

Most advances in NMR technology have occurred when someone tried to address a particular limitation in one of these six disciplines. However, once a new tool is developed, it can sometimes solve a different problem in a different discipline or create a whole new field of NMR. An example of the former is the use of magic-angle spinning (MAS) first for solid state NMR, then for small-volume solution state NMR for natural products, and then for solid phase synthesis (SPS) resin NMR for combinatorial chemistry. An example of the last technique is the development of indirect detection, which helped spawn the use of isotopically labeled proteins and triple-resonance NMR for determining the structures of biomolecules in rational drug design programs.

In addition to the simple migration of NMR techniques from one discipline to another, two or more NMR techniques can often be synergistically combined. A good example of this is the combination technique of high-performance liquid chromatography–NMR (HPLC–NMR). Its current implementation involves the use of indirect detection, shaped pulses, broadband decoupling, pulsed-field gradient (PFG) sequences, PFG shimming, automation, and small-sample probe design.

The purpose of this section is to examine some of these NMR tools and to see how they have influenced drug discovery programs. First, we will see how they were initially used to solve a particular problem. Then we will see how some have been used to solve problems in other disciplines. Finally, we will see how others have been synergistically combined to develop much more powerful tools.

Section 2.12.18.2 covers the basic tools in NMR, which include multidimensional NMR, high-field NMR, broadband decoupling, spin locks, shaped pulses, indirect detection, and PFGs. Section 2.12.18.3 introduces some additional applications of NMR and the more specialized tools in use, most of which are based on the technologies introduced in Section 2.12.18.2. These include protein structure determinations and the multiple-channel indirect-detection experiments they spawned (Section 2.12.18.3.1), water suppression techniques (Section 2.12.18.3.2), and hardware and probe developments (Section 2.12.18.3.3). The probe developments that are covered include small-volume probes and high-resolution (HR)-MAS probes in Section 2.12.18.3.3.1, as well as flow probes and cryogenic probes, which are sufficiently important to be treated separately (see Sections 2.12.18.3.5 and 2.12.18.3.7, respectively). Detailed information about combinatorial chemistry applications can be found in the discussions on solid phase resin analysis using MAS (Section 2.12.18.3.4) and the analysis of libraries stored in titer plates using flow NMR (see Section 2.12.18.3.5.2). Lastly, there are individual discussions on analyzing mixtures, combined NMR techniques, quantification by NMR, and the automation of NMR in Section 2.12.18.3.11.

2.12.18.2 Basic Nuclear Magnetic Resonance Tools

2.12.18.2.1 One-, two-, three-, and multidimensional nuclear magnetic resonance

One of the first tools to be developed was multidimensional NMR. The first NMR spectra to be observed contained only one frequency axis, and so were termed 1D (even though they actually contain two dimensions of NMR information: frequency and amplitude). 1D spectroscopy was the only real option for the early generations of continuous-wave (CW) spectrometers, but with the advent of pulsed Fourier transform NMR (FT-NMR) spectrometers some 20 years later, it was soon recognized that additional dimensions of NMR information could be obtained. Because some 1D NMR spectra had a crowded frequency axis (especially large molecules such as proteins) or coincidental or overlapping chemical shifts (especially poorly functionalized molecules such as some steroids), methods to simplify these spectra were needed.

In the early 1970s, the first papers on 2D NMR spectroscopy began to appear.[875] Because 2D NMR allowed the NMR information to be spread in two dimensions, each of which could encode different kinds of information, most kinds of spectral overlap could be resolved. As an example, one dimension could contain chemical shift data while the other dimension could encode information about spin–spin couplings.

Different pulse sequences were then developed to select all kinds of information. The 2D correlations could arise from homonuclear or heteronuclear couplings, through-bond (scalar) or through-space (dipolar) couplings, or single-quantum or multiple-quantum couplings. They could also be designed either to detect the presence of or to actually measure the magnitude of a given kind of coupling. Experiments exploiting all of the various combinations of these possibilities were eventually developed, and remain in use today. These include experiments that generate 2D data to measure the presence of homonuclear scalar coupling (homonuclear correlation spectroscopy (COSY) and total correlation spectroscopy (TOCSY)); the magnitude of homonuclear scalar couplings (homonuclear 2D J spectroscopy (HOM2DJ)); the presence of heteronuclear scalar couplings (heteronuclear correlation spectroscopy (HETCOR), heteronuclear multiple-quantum correlation (HMQC), and heteronuclear multiple-bond correlation (HMBC)); the magnitude of heteronuclear scalar couplings (heteronuclear 2D J spectroscopy (HET2DJ)); the presence of homonuclear dipolar coupling (nuclear Overhauser effect (NOE) spectroscopy (NOESY)); the presence of heteronuclear dipolar coupling (heteronuclear NOE spectroscopy (HOESY)); and double-quantum, double-quantum-filtered, and multiple-quantum-filtered homonuclear scalar coupling correlation spectroscopy (DQ-COSY, DQF-COSY, and MQF-COSY, respectively). Many good books and reviews about 2D NMR can be found.[876–878]

It is useful to understand that a 2D pulse sequence generates data that contain both a direct (F_2) and an indirect (F_1) dimension. The direct dimension, sometimes called the 'real-time' dimension, consists of a series of one-dimensional experiments; each spectrum is completely acquired during one intact time interval, usually within a few minutes. However, the indirect dimension is generated by a 't_1' delay in the pulse sequence that is incremented many hundreds of times during a 2D experiment. A direct-dimension spectrum is acquired for each of these several hundred delay times, then the data are subjected to Fourier transformation a second time – along this indirect dimension – to produce the 2D data set. Each spectrum along the indirect dimension is reconstructed from hundreds of direct-dimension spectra, and each data point in an indirect-dimension spectrum arises from a different direct-dimension spectrum. This means that each indirect-dimension spectrum is composed of data points acquired over many minutes (or hours or days). Any time-dependent instabilities (in the spectrometer, the environment, the sample, the electrical power, etc.) can cause a corruption of data points, and therefore cause noise; since this noise runs parallel to the t_1 axis, it is called 't_1 noise.' We will see later how t_1 noise, which is ubiquitous, can be troublesome enough that some NMR techniques, such as PFG, were developed to suppress it.

After a wide range of 2D NMR pulse sequences were developed, it was recognized that they could be effectively combined to create 3D, 4D, and even higher 'nD' spectroscopies.[879,880] Each additional dimension is created by appending to the pulse sequence certain pulse sequence elements (that contain an incrementing delay) that generate the desired effect. A 3D pulse sequence has one direct dimension and two incrementing delays, whereas a 4D data set has three incrementing delays. Each dimension of an nD data set usually encodes for different information, so 3D HETCOR-COSY would have three distinct chemical shift axes (one ^{13}C and two ^{1}H) and three distinct faces. (If the data within the 3D cube are projected onto each of the three distinct faces, they would display 2D HETCOR, 2D COSY, and 2D HETCOR-COSY data, respectively.) These hyphenated experiments, combined with the power of 3D and 4D NMR, are proving to be very useful for elucidating the structures of large molecules, particularly for large (150–300 residue) proteins.[881]

Since a 2D data set contains three dimensions of information (two frequency axes and a signal amplitude), a contour plot presentation is usually used to represent the data on paper. A 3D data set is a cube in which each data point contains amplitude information. Since all this information is difficult to plot simultaneously, 3D data are usually plotted as a series of 2D planes (contour plots). 4D (and higher) experiments are easy to acquire (aside from the massive data storage requirements), but plotting and examining the data becomes increasingly difficult; as such, 3D spectroscopy is often the upper limit for routine NMR. We will see later how multidimensional NMR can be combined with multinuclear (triple-resonance) NMR experiments.

Although multidimensional NMR can provide more information, it also has some disadvantages. Its experimental time is longer; it usually requires more samples; set up of the experiment and processing and interpretation of the data require more expertise; it requires more powerful computer hardware and software; and it requires additional hard disk space. This means that a user should normally run the lowest dimensionality experiment that answers the question at hand. It is usually found in practice that the six disciplines of drug discovery use these multidimensional NMR techniques differently. Medicinal chemists, combinatorial chemists, drug production and quality control groups, and, to a lesser extent, metabolism groups usually use 1D NMR data to get most of their structural information. Medicinal chemists usually need only to confirm a suspected structure, and rarely use 2D NMR. Drug production and quality control groups will use 2D techniques if they encounter a tough enough problem. Combinatorial chemists might like to have 2D NMR data, but they do not have the time to acquire it, whereas metabolism chemists would often like to get 2D NMR data but do not have enough samples. Natural products chemists routinely depend on 2D NMR techniques to unravel their total unknowns. Rational drug design groups use 2D and 3D (and sometimes 4D) NMR techniques routinely.

2.12.18.2.2 High-field nuclear magnetic resonance

Multidimensional NMR was developed to simplify complex NMR spectra. Another way to simplify a complex or overlapped NMR spectrum is to place the samples in higher field-strength magnets to generate more dispersion along each axis. This helps resolve overlapping signals and simplifies second-order multiplets, while at the same time increasing NMR sensitivity. These benefits have driven magnet technologies to improve to the point where magnets having proton resonance frequencies of 800 MHz (18.8 T) are now routinely available. The disadvantage of these very high-field magnets is that they are much more expensive. (They primarily cost more to purchase, have higher installation and operational costs, especially for cryogens, and require bigger rooms with more sophisticated environmental controls.) These additional costs must be weighed against the time that can be saved by the improved performance, and both factors are balanced against the difficulties of the problems encountered. In practice, field strengths of 14.1 T (600 MHz) and above are routinely used in rational drug design programs, and occasionally in metabolism programs (due to their limited sample sizes). Natural products groups, and some drug production and quality control groups, usually use 400–600 MHz systems, while combinatorial chemistry, and especially medicinal chemistry groups, tend to use 300–500 MHz systems.

Improvements in magnet and shim coil technologies are also allowing the homogeneity of the magnetic field to be improved. This has allowed line shape to be cut in half over the last 15 years, and has improved the quality of solvent suppression in the NMR spectra of samples dissolved in fully protonated solvents.

2.12.18.2.3 Broadband decoupling

Another way to simplify spectra is to use heteronuclear decoupling.[882] This both collapses a coupled multiplet into one single frequency, and (through other mechanisms) produces taller resonances. Decoupling can also increase the sensitivity of certain X nuclei (e.g., ^{13}C) through nuclear Overhauser enhancements. In some triple-resonance experiments (discussed later), heteronuclear decoupling improves sensitivity by eliminating some undesirable relaxation pathways.

Although decoupling is generally desirable, the first available methods caused too much sample heating to be of routine use. More efficient decoupling schemes were eventually developed that produced a wider bandwidth of decoupling for a given amount of power. Starting from CW (unmodulated) decoupling, wider decoupling bandwidths were obtained with the development of noise – WALTZ, MLEV, XY32, TYCKO, GARP, and DIPSI decouplings (not necessarily in this order).[882,883] These modulation schemes allowed decoupling to become quite routine and allowed new techniques, such as indirect detection (discussed below), to become practical. As knowledge of shaped pulses improved (discussed below), ever more efficient decoupling schemes such as MPF1-10,[884] WURST,[885] and STUD[886] were developed; the last two by using advanced pulse shapes called adiabatic pulses.[887] No decoupling scheme is perfect for all applications or even for all field strengths. Each scheme provides a different balance of performance in terms of bandwidth per unit power, minimum linewidth, sideband intensity, complexity of the waveform (what hardware is needed to drive it), and tolerance of mis-set calibrations ('robustness'). The advantage of all these developments is that there is now a large set of different decoupling tools that are available to be used as needed.

2.12.18.2.4 Spin locks

Better theories and understanding of spin physics resulted from advances in broadband decoupling, and this allowed better spin locks to be developed. Spin locks were initially designed for experiments on solid state samples, but in the 1980s it was shown that spin locks could also be used for solution state experiments. Bothner-By and co-workers used a CW spin lock to develop the CAMELSPIN experiment (later known as rotating-frame Overhauser enhancement spectroscopy (ROESY)) to get dipolar coupling (through-space) information.[888] Other groups subsequently reported the use of both lower radiofrequency (RF) field strengths[889] and pulsed spin locks[890] to improve the quality of the ROESY data. Braunschweiler and Ernst used a spin lock to develop the homonuclear Hartmann–Hahn experiment (HOHAHA; later known as TOCSY), which provides a total scalar-coupling correlation map.[891] Bax and Davis showed that the MLEV-16 modulation scheme, used for broadband decoupling (discussed above), improved the quality of TOCSY data.[892] Kupce and co-workers built on their experience with adiabatic pulses to improve decoupling schemes to develop a spin lock using adiabatic pulses.[893] As discussed later, these spin locks are proving useful in obtaining TOCSY data on MAS samples. This is helping combinatorial chemists acquire better data on SPS resins. The hardware required to run spin-locked experiments is now routinely available, and these experiments usually provide better and more complete data than alternative experiments such as COSY.

The typical use of a spin lock is as a mixing scheme to allow the spins to exchange information. This is how they are used in TOCSY and ROESY experiments. However, spin locks have also been used to destroy unwanted magnetization. For this application, the pulse sequence is usually written so that the spin lock only affects resonances at a specific (selective) frequency.[894] This allows them to be used to perform solvent suppression experiments for samples dissolved in protonated solvents (e.g., H_2O). Non-frequency-selective spin locks have also been used to destroy unwanted magnetization.[895] This use of a spin lock as a general purge pulse is a viable alternative to the use of homospoil pulses,[896] and is in routine use.

The spin physics of the decoupling and spin lock sequences are related, and any given modulation scheme can often serve both purposes. In general, however, the sequences are usually optimized for only one of the two applications. Hence, MLEV-16, MLEV-17, and DIPSI are most commonly used as spin lock sequences, whereas WALTZ, GARP, WURST, and STUD are most commonly used as decoupling schemes.[882]

2.12.18.2.5 Shaped pulses

In addition to new decoupling and spin lock schemes, shaped RF pulses have also been developed. They are called 'shaped pulses' because all of these pulses change amplitude, phase, or frequency (or some combination) as a function of time.[897] In doing so, the pulse can be designed to be more frequency selective, more broadbanded in frequency, more tolerant of mis-set calibrations, or capable of multifrequency or off-resonance excitation. Different applications require different balances of these characteristics.

If a pulse is made more frequency selective, it can be used in a variety of new applications. One application is 'dimensionality reduction,' in which a selective 1D experiment is used in place of a 2D experiment, or a selective 2D experiment is used in place of a 3D experiment. Selective experiments let you obtain a selected bit of NMR information faster and with a higher signal-to-noise ratio – but only if you already know what you are looking for. This can be useful for studying smaller amounts of samples (or more dilute samples), as is often required for natural products. Examples include the use of 1D COSY,[898] 1D TOCSY,[899] 1D NOE,[900] DPFGSE-NOE (also sometimes erroneously called GOESY),[901,902] 1D HMQC (also called SELINCOR),[903] 1D HMBC (also called SIMBA),[904] and 1D INADEQUATE.[905] These sequences are less useful if you have many questions to be answered or do not know what you are looking for. Then, it is usually advantageous to go ahead and run the complete 2D or 3D experiment and analyze it as completely as is required to solve the problem at hand.

Another use of selective pulses is to perform frequency-selective solvent suppression. Presaturation is the simplest example of the use of a frequency-selective pulse for solvent suppression (although the pulse used in presaturation is typically not shaped). Examples of frequency-selective shaped pulses that are used in solvent suppression include the S and SS pulses,[906] Node-1,[907] and WET.[908] Note that although many commonly used water suppression sequences, such as 1331[909] and WATERGATE,[910] are frequency selective, they achieve that selectivity by using a train of nonshaped pulses, in the so-called DANTE technique,[911] instead of using shaped pulses. (The DANTE technique permits the use of simpler hardware but results in a less accurate frequency profile.)

Historically, the primary application of solvent suppression was to obtain spectra of biomolecules (proteins, RNA, etc.) dissolved in $H_2O:D_2O$ mixtures. However, more recent work, especially on direct-injection NMR (discussed below), indicates that it is now practical to obtain ^1H NMR spectra of organic samples dissolved in fully protonated nonaqueous solvents. This is being used especially in combinatorial chemistry to obtain ^1H NMR spectra on samples dissolved in DMSO-h_6, $CHCl_3$, and aqueous solutions of CH_3CN and CH_3OH.

Frequency-selective pulses can also be used for narrowband selective decoupling. Such pulses can be placed in evolution periods to simplify spectra.[912]

Not all shaped pulses are frequency selective (narrow bandwidth). In recent years, there have been many fine examples of the use of both band-selective and broadband-shaped pulses. Sometimes, these are used as single RF pulses, but more typically they have been used in decoupling schemes designed to hit either a selected, controlled, moderately wide region, or a very broadbanded region. A well-known example of band-selective pulses is the use of pulses[913] or decoupling schemes[914,915] to treat carbonyl-carbon resonances differently from aliphatic-carbon resonances in triple-resonance experiments (discussed below). Here, band-selective pulses are pulsing or decoupling regions of the spectrum that are from 1 kHz to 4 kHz in width. Applications using adiabatic pulses that are region selective have also been described.[916,917]

The use of shaped pulses that affect very wide (broadband) regions of a spectrum is also becoming more popular. Most of these examples exploit adiabatic pulses.[887] There are three advantages to using adiabatic pulses. First, they are uniquely capable of creating a good excitation profile over a wide frequency region. Second, they deposit less power into a sample (and hence cause less sample heating) for a given bandwidth than most other pulses. Third, they are quite tolerant of mis-set calibrations. This feature was originally exploited in applications that used surface coils – to compensate for the characteristically poor B_1 homogeneity of these coils – but some groups use this feature to simply make routine solution state NMR spectroscopy more robust and less sensitive to operator calibration errors. Adiabatic RF pulses are being used in high-field ^{1}H–^{13}C correlation experiments (e.g., HSQC) to achieve wider and more uniform inversion over the wide ^{13}C spectral width.[885,918] Adiabatic pulses are also being heavily used to create more efficient and wider bandwidth decoupling schemes (as discussed above in Section 2.12.18.2.3). Both of these applications are being used to facilitate high-field NMR spectroscopy of biomolecules.

One of the final tools in the shaped-pulse toolbox is the shifted laminar pulse (SLP),[919] which is sometimes (imprecisely) called a phase-ramped pulse. SLPs deliver their effects off-resonance without changing the transmitter frequency. (It is undesirable to change the frequency of a frequency synthesizer during a pulse sequence because this affects the phase of the transmitter.) SLPs that affect only one off-resonance frequency change phase linearly as a function of time; hence, the alternative name 'phase-ramped pulses.' SLPs that affect two (or more) off-resonance frequencies are both phase and amplitude modulated due to the interaction of the two different phase ramps. SLPs can be created to affect any number of frequencies (as long as enough power is available). They can also be created using any pulse shape or duration (or combination of shapes and durations). This means, for example, that an SLP can be made to hit three different frequencies, each with a different pulse width, and each with a different pulse shape (or phase) if desired. We will see later (in the discussion on LC–NMR – Section 2.12.18.3.5) how SLPs are being used to obtain multiple-frequency solvent suppression using only one RF channel.

2.12.18.2.6 Indirect detection

The development of indirect detection has probably had a bigger impact on all stages of drug discovery than any other NMR technique. Indirect detection is one of two methods for acquiring heteronuclear correlation data. The conventional method for acquiring 2D heteronuclear correlation data – called 'direct detection' and exemplified by the HETCOR experiment described above – measures a series of X nucleus spectra (typically ^{13}C) and extracts the frequencies of the coupled nuclei (typically ^{1}H) by using a Fourier transform in the t_1 dimension. The concept of indirect detection, first published in 1979,[920] inverts the process by acquiring a series of ^{1}H spectra, then Fourier-transforming the t_1 dimension of this data set to (indirectly) obtain the X (^{13}C) frequencies.[876,877] The major benefit of using indirect detection is improved sensitivity – up to 30-fold or more, depending on the nuclei involved.

In hindsight, indirect detection seems like an obvious development, but because the instrumental requirements of running indirect detection are more demanding (two-scan cancellation, broadband X nucleus decoupling, indirect-detection probes designed for water suppression), the method did not become truly routine until more than a decade later. They do not completely replace X-detected correlation experiments (which have superior X nucleus resolution), but these ^{1}H-detected experiments are now the default for most users. When sensitivity is an issue, the indirect detection version of an experiment is usually used.

There are three classic indirect-detection experiments: HMQC, HMBC, and HSQC. HMQC and HSQC detect one-bond H–X couplings, whereas HMBC detects long-range H–X scalar couplings. HMQC uses a multiple-quantum coherence pathway, while HSQC uses single-quantum coherence. HMQC was technically easier to run in the early days, but because the B_1 homogeneity of most NMR probes has improved since then, HSQC is now a better tool for most applications. Virtually all modern biomolecular experiments, and many natural product experiments, currently use HSQC-style (rather than HMQC-style) coherence pathways.[921,922] HMBC is the primary tool for de novo structure elucidations of small molecules, particularly for natural products.[923–927]

A big advantage of indirect detection is its increased sensitivity. A big disadvantage is its cancellation noise. Because only the ^{13}C satellites of a ^{1}H resonance are of interest in a ^{1}H{^{13}C} HSQC, the rest of the ^{1}H signal (98.9%) must be eliminated. In conventional pulse sequences this is accomplished by using a two-step phase cycle in which the ^{13}C satellites are cycled differently from the central ^{1}H–^{12}C resonance. Unfortunately, many spectrometer installations are not stable enough to provide clean cancellation (this is heavily dependent on the local environment), and the poorly canceled central-resonance signal ends up creating excessive t_1 noise in the spectrum. This is most visible in the study of unlabeled compounds, that is, those that have natural-abundance levels of ^{13}C (or ^{15}N). It becomes even more problematical in long-range HMBC-style experiments. Here, the t_1 noise is often greater than the signals of interest, and more experiment time is spent signal averaging to raise the signals of interest above the t_1 noise than to raise them above the thermal noise.

2.12.18.2.7 Pulsed-field gradients

In 1973, Paul Lauterbur discovered that linear gradients of B_0 (the magnetic field) could be used in NMR experiments to generate spatial information about what was located in the magnet.[928] This ultimately gave rise to the technique called magnetic resonance imaging, now used in many hospitals and clinics. As imaging technology developed, it became possible to generate pulses of linear B_0 gradients, of different strengths, in a controlled fashion. Barker and Freeman were the first to demonstrate that these pulsed magnetic field gradients could be used for coherence selection to acquire 2D NMR data.[929] This capability was expanded and popularized by Hurd and co-workers in the early 1990s, who showed that PFGs could be used in a number of different NMR experiments.[930–932] Thus, the PFG revolution was born.

PFGs can be used in several ways.[932,933] The original application was for imaging, and while this is usually used for clinical purposes, imaging techniques are also being used in drug discovery, primarily to study the metabolism of drugs.[934] PFGs are also useful in automating the shimming of high-resolution NMR samples.[935,936] Both ^{1}H and ^{2}H gradient shimming are in routine use for research samples as well as for samples being analyzed by routine automation. PFG techniques also allow old experiments to be run in different ways, and allow entirely new pulse sequences to be developed (discussed below).

PFGs embody experiments with a number of desirable attributes. One is that they allow some experiments to be acquired without phase cycling. This allows the data to be acquired much faster, as long as the quantity of sample is sufficient. This is especially useful for indirect detection, as well as for COSY experiments. PFGs also allow data to be acquired with less t_1 noise. This is useful for experiments in which the quality of the data acquired with conventional methods is significantly degraded by the presence of t_1 noise. A classic example of this is indirect detection. Conventional (phase-cycled) indirect-detection experiments use a two-step phase cycle to cancel the large signals from protons not bound to ^{13}C (or ^{15}N) nuclei, which is 98.9% of the signal in unlabeled samples (discussed above). Any imperfections in the cancellation leave residual signals that cause the t_1 noise. In HMBC in particular, because it detects the long-range multiple-bond correlations, which are often small, this t_1 noise may be greater than some of the desirable signals.[937,938] Since the PFG versions of HMBC select the desired signals (coherences) within a single scan, and hence contain no t_1 noise induced by the two-step cancellation technique, better-quality data can be obtained, and usually in less time. This is especially useful for elucidating the structures of natural products. Because the structures are usually unknown and the sample sizes are often quite small, every improvement in the quality of HMBC data is welcomed.

PFG experiments can also suppress unwanted solvent resonances, often easily and without any extra setup effort by the operator. In addition, the quality of this suppression is much less dependent on the quality of the NMR lineshape than with other experiments. (Presaturation of the water resonance in an H_2O sample is a classic example in which the NMR spectral quality depends heavily on how well the lineshape of the water resonance can be shimmed to have a narrow base. The corresponding PFG experiments typically do not need such exacting shimming.) This attribute is important because of the large number of NMR studies performed on biomolecules dissolved in $H_2O:D_2O$ mixtures.[939]

2.12.18.2.7.1 Pulsed-field gradient variations of existing experiments

The many beneficial attributes of using PFG experiments (listed above) have made PFG experiments quite popular. PFG versions of the indirect-detection experiments[940–942] are the most heavily used, but the PFG versions of many COSY-style experiments[943,944] are also popular. PFG-HMBC is a very striking example, and one that is very popular in natural products programs.[937,945] Alternatives such as GHSMBC,[946] ADEQUATE,[947,948] EXSIDE,[949] psge-HMBC,[945] and ACCORD-HMBC[950] have also been developed. Most people would benefit by using PFG-COSY more routinely, as opposed to a non-PFG (phase-cycled) COSY.[951] PFG-HSQC[952] has become a standard experiment in biomolecular NMR (for rational drug design programs). PFG versions of the selective experiments 1D TOCSY and 1D NOESY experiments have also been developed.[953]

2.12.18.2.7.2 New experiments made possible by pulsed-field gradient

PFGs have also allowed new pulse sequences to be developed. The WATERGATE spin-echo sequence was developed for biomolecular water suppression.[910] A newer nonecho solvent suppression technique called WET,[908] which combines PFG, shaped pulses, and SLP pulses, is now the standard for use in LC–NMR[954] – a technique that is becoming a popular tool for metabolism studies.[955] PFG-based WET suppression is also being used in the NMR analysis of combinatorial libraries dissolved in nondeuterated solvents.[956] Automated ^{2}H (and ^{1}H) gradient shimming, using gradients from both PFG coils as well as room temperature shim coils, is becoming routine for the automated walk-up NMR spectrometers used in medicinal chemistry programs.[936] Diffusion experiments,[957] which rely heavily on PFG, are being used to study ligand binding and evaluate library mixtures for compounds that bind to receptors in rational drug design programs.[958,959] Other PFG-based diffusion experiments are being used for solvent suppression.[960]

2.12.18.3 Specialized Nuclear Magnetic Resonance Tools

2.12.18.3.1 Biomolecular structure elucidation

Rational drug design programs are built on the idea that if the complete 3D structure of a receptor could be determined, one could custom design a small molecule to bind to its receptor site. The challenge in this process is to determine the structure of the receptor (a protein). While 3D protein structures can be determined using x-ray crystallography, it became apparent in the 1980s that this could also be done by using NMR spectroscopy, and NMR allows the solution-state structure of the protein to be determined.

Once the primary structure of a protein is known, the secondary and tertiary structure of the protein is determined with NMR by first making chemical shift assignments and then by measuring NOEs and coupling constants.[961] A ^{1}H–^{1}H NOE provides information about through-space distances. (If two protons show an NOE, and if they are located in different parts of the peptide chain, then this provides an intrastrand structural constraint that indicates that these protons must reside close to each other in the final 3D structure of the protein.) Proton homonuclear or heteronuclear coupling constants provide information about the torsion angles of the bonds in the peptide backbone. Either of these methods can provide a 3D structure if enough structural constraints can be determined.

Structure determinations can be very difficult for proteins whose molecular weights are greater than 10 kDa (even if the primary structure is obtained by other means). First, the chemical shift assignments become more difficult as the spectral complexity increases, and, second, the faster relaxation of larger molecules produces broader linewidths that make it harder to obtain coupling-constant information.

Biomolecular NMR became more important and grew significantly when three things happened. First, methods were developed for incorporating NMR-active isotopic labels (namely ^{13}C and ^{15}N) into proteins. Second, indirect-detection techniques were developed that facilitated the chemical shift assignments of these labeled proteins.[962,963] Third, 3D and 4D NMR techniques were developed to simplify and sort the vast amount of information needed to determine the structure of these large molecules.[964] These three developments greatly simplified the interpretation of protein NMR data, and made the determination of protein structures both more routine and more powerful. Although the techniques for acquiring indirect-detection NMR data on small molecules were well developed by the late 1980s, the requirements (and possibilities) of performing these experiments on proteins are a bit different. Much of this was worked out in the early 1990s, and the field of multiple-resonance protein NMR blossomed.[880,913,961,965] The resulting experiments (described below) are certainly powerful, and initially required new NMR hardware, but conceptually one can think of them as simply being composed of combinations of the fundamental techniques that had already been developed (described above).

2.12.18.3.1.1 Multiple radio frequency channel nuclear magnetic resonance

Since isotopically labeled proteins typically contain three kinds of NMR-active nuclei – ^{1}H, ^{13}C, and ^{15}N – the one-step heteronuclear coherence transfers normally used in heteronuclear correlation experiments can be concatenated into a multiple-step transfer. An example of a multiple-step transfer would be magnetization transfer from ^{1}H to ^{13}C, then to ^{15}N, where it evolves, back to ^{13}C, and then to ^{1}H for detection. Indirect-detection experiments allow any X nucleus to be examined, simply depending on where in the pulse sequence the t_1 evolution time is placed. 3D and 4D experiments allow multiple evolution times to be used. Proteins exhibit a range of homonuclear and heteronuclear scalar coupling constants (J), and the $1/2J$ delays in a pulse sequence can be finely tuned so as to select between several different scalar coupling pathways in the molecule. These capabilities, combined with the ability to select either complete or partial isotopic labeling within a protein, allow experiments to be designed that can correlate almost any two kinds of spins. In practice, at least one of these spins is usually a proton bound to an amide-bonded nitrogen. The identity of the second or third spin depends on the structural data needed. It may even be one of several

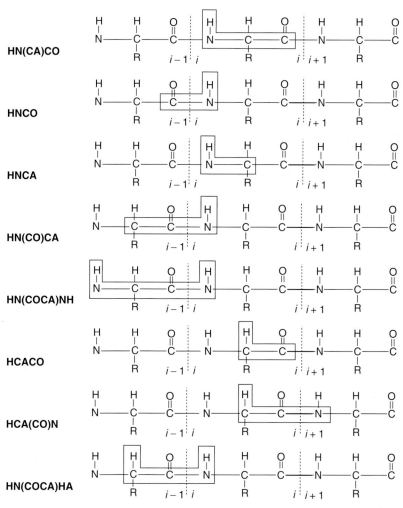

Figure 138 Internuclear correlations that can be detected on an isotopically labeled protein backbone using the listed triple-resonance experiments. Boxes with a solid outline designate the nuclei involved in each correlation experiment. Dashed vertical lines indicate the location of the peptide bonds.

structurally different carbon nuclei, since NMR can easily distinguish carbonyl carbons from all other carbon resonances quite easily (using frequency-selective pulses). **Figure 138** shows the intraresidue correlations that can be made using the HN(CA)CO experiment, as well as the interresidue correlations (to the previous adjacent carbonyl carbon) that can be made using the HNCO experiment. In HN(CA)CO, the magnetization starts on the NH proton, is transferred to the directly bonded ^{15}N (where there is an evolution time), is then transferred to the C-α ^{13}C (no evolution), and finally to the carbonyl ^{13}C, where it has its second evolution. It then reverses its path (CO to CA to N to H) for detection of ^{1}H. The parentheses around the CA in HN(CA)CO indicates that the C-α nuclei are used in the transfer, but there is no evolution time and they are not actually detected. To make correlations to the C-α carbons (instead of the carbonyl carbons), the corresponding experiments would be HNCA (for intraresidue assignments) and HN(CO)CA (for inter-residue assignments to the previous C-α carbon). In a similar manner, experiments such as HN(COCA)NH, HCACO, HCA(CO)N, and HN(COCA)HA, as well as many more, can be constructed (**Figure 138**).

As proteins (and other biomolecules) are large, their T_2 relaxation times are usually much shorter than those of small molecules. This causes their NMR resonances to have broader linewidths. Because most methods for determining accurate scalar coupling constants measure the separation of antiphase signals, and since this becomes less accurate as the linewidths increase, these experiments become less useful as the molecular weight of the compound increases. As a consequence, DQFCOSY experiments are less useful for large proteins, and so the determination of the 3D structure via torsion angle measurements becomes harder. Experiments have been designed in an attempt to overcome this limitation, principally through the use of combined hetero- and homonuclear scalar couplings.[966,967]

2.12.18.3.1.2 Newer biomolecular experiments: transverse relaxation-optimized spectroscopy, hydrogen bond J couplings, and dipolar couplings

Because large biomolecules have a faster T_2 relaxation than smaller molecules, their NMR signals will sometimes decay before they get through some pulse sequences. There are several schemes for addressing this problem. First, HSQC experiments can be used instead of HMQC experiments, because transverse magnetization in a single-quantum state (HSQC) relaxes even more slowly than transverse magnetization in a multiple-quantum state (HMQC).[921] Second, some pulse sequences now use a variety of heteronuclear and homonuclear decoupling schemes during evolution times to reduce T_2 decay rates. Third, some groups have resorted to perdeuteration of those large proteins that relax rapidly; this reduces the decay rate of signals in the transverse plane.[968,969] Fourth, the introduction of TROSY, which manipulates spins in a way that maintains those signals having longer relaxation times, is also allowing much larger proteins to be studied by NMR.[970,971]

Another tool for structural elucidation became available with the discovery that NMR allows observation of correlations due to scalar couplings across hydrogen bonds. This indicates that hydrogen bonds can have a measurable amount of covalent character, so the phenomenon has been called 'hydrogen bond J coupling.' Among the correlations observed are one-bond $^{1h}J_{HN}$ correlations,[972] two-bond $^{2h}J_{NN}$ correlations,[973] and three-bond $^{3h}J_{C'N}$ correlations,[974,975] all of which have been explained theoretically.[976] Much like a long-range NOE correlation, these correlations add structural constraints to 3D structures (especially for nucleic acids).

There is a lot of interest in using NMR to study ligand–receptor binding. This is normally done by looking for intermolecular NOEs between the ligand and the receptor, but because the NMR spectrum of a protein–ligand complex can be complicated, small NOEs can be hard to detect. The problem is simplified if isotopically labeled ligands are mixed with unlabeled proteins and an isotope-filtered NMR experiment used to selectively observe the labeled resonances.[977] As more complex ligands and receptors are studied, techniques to do the opposite – selectively observe the resonances of just the unlabeled molecules – were developed (in part through the use of adiabatic pulses).[978]

The final trend in biomolecular NMR is to use the internuclear dipolar couplings of oriented molecules to determine structures.[979] When molecules are in solution and tumbling freely, these couplings are averaged to zero; however, if the molecules are aligned with the magnetic field even partially, these couplings will be nonzero and measurable. If one measures the one-bond ^{15}N–^1H and ^{13}C–^1H dipolar couplings and knows the internuclear distances, then the orientation of the vectors of these bonds (with respect to the magnetic susceptibility tensor of the molecule) can be determined. This gives direct information about the torsion angles of the bonds, and hence information about the structure of the biomolecule. This information also complements that which can be obtained from analyses of NOEs and J couplings.

Because many molecules naturally exhibit only weak alignments with magnetic fields, a recent trend has been to use different kinds of liquid crystals and bicelles not only to create more alignment but to allow control over the degree of alignment.[980] Once aligned, measurement of the changes in the one-bond coupling constants (typically $^1J_{NH}$) or the ^{15}N chemical shifts, measured at a variety of magnetic field strengths, gives the dipolar couplings.[981] A more recent alternative is to measure the differences in chemical shifts and coupling constants between data acquired with and without MAS (using solution state samples and a nanoprobe; discussed below). The advantage of this technique is that only one set of measurements need to be made.

2.12.18.3.2 Water (solvent) suppression

In all of these biomolecular experiments, signals from the protons bound to the amide nitrogen are critical. Unfortunately, these amide protons also readily exchange with water, so the samples cannot be dissolved in D_2O. To keep the samples in a predominantly H_2O environment yet still supply a small amount of D_2O for a ^2H lock, they are usually dissolved in 90:10 or 95:5 $H_2O:D_2O$. Because that makes the H_2O signal 100 000 times larger than that of a single proton in a 1 mM protein, new techniques had to be developed for acquiring NMR data on these kinds of samples.

There are essentially three ways to deal with the problem. One is to suppress the water signal, the second is to not excite it in the first place, and the third is just to live with it. None of these strategies are perfect, and a wide range of techniques have been developed in an attempt to exploit at least the first two strategies. Many good reviews of water suppression techniques exist elsewhere.[982,983] It is of interest, however, to note two things. First, virtually all of the common water suppression techniques are combinations of shaped pulses, PFGs, and indirect-detection methods (all discussed above). Second, most techniques are designed only for water suppression, not for the more general concept of solvent suppression. Suppressing the NMR resonances of organic solvents is more difficult than water suppression because the ^1H signals contain ^{13}C satellites in addition to the central resonance. General solvent suppression is becoming important in several flow-NMR techniques, and will be discussed later.

2.12.18.3.3 Hardware developments

There are many examples of how developments in NMR techniques have required improvements in the hardware and software as well. The ever-higher magnetic fields used to simplify complex spectra have required improvements in superconducting magnet materials and magnet designs. FT-NMR required the development of pulsed RF, whereas 2D NMR required the development of RF hardware and pulse sequence control software that was capable of more flexible phase cycling and power level control. Indirect detection required the design of both more stable RF and of magnet antivibration hardware to improve the signal cancellation possible via RF phase cycling. Spin lock sequences (e.g., TOCSY) required phase-coherent RF to be developed. The complex pulse sequences of biomolecular multiple-resonance NMR required multiple RF channels (as many as five), each capable of delivering not only more but a wider variety of pulses per sequence. For example, any one RF channel may be called on to deliver high-power uniform-excitation broadband pulses, region-selective pulses, broadband or region-selective decoupling, broadband or region-selective spin locks, or highly frequency-selective pulses. These RF requirements have driven the development of faster and more flexible RF control, including more powerful pulse-shaping hardware and software. In addition, because PFG pulses of a variety of amplitudes are also freely intermixed with RF pulses, stronger gradients with ever-faster recoveries are always being sought.

2.12.18.3.3.1 Probe developments

In addition to hardware developments in the spectrometer console and magnet, advanced NMR experiments have also driven improvements in probe designs. The emphasis on water (solvent) suppression has created an emphasis on lineshape specifications. NMR lineshape specifications – defined as the width of the NMR peak at 0.55% and 0.11% of the height of the peak (0.55%/0.11%) – are now half as wide as they were a decade ago. Proton NMR linewidths of 4/6 Hz have been obtained. Narrower lineshape specifications do not guarantee good water suppression, but are often considered to be an essential element. The better specifications arise from improvements in probe designs, magnet designs, shim set designs, and installation procedures. Although NMR resolution (the width of the NMR peak at 50% of the height of the peak) for spinning samples has not improved, most high-field NMR spectrometers are now capable of – and specified to have – much better nonspin resolution than in the past. Since most nD NMR experiments, and all water suppression experiments, are run nonspinning, this is also now considered an important specification.

The second-most important probe development pertains to the ever-increasing probe sensitivities. The ^{1}H sensitivity specifications of modern probes are severalfold better than they were a decade ago. Instrument time is always precious, and a 3.2-fold increase in the signal-to-noise ratio can reduce the total time of an NMR experiment by an order of magnitude.

The emphasis on multiple-pulse NMR experiments (such as HSQC and the triple/multiple-resonance experiments) has driven improvements in three other parameters of probe design: shorter pulse widths, better RF homogeneities, and improved 'salt tolerance'. The first parameter is important because spectral widths in higher field NMR – as measured in hertz – are wider. This requires the RF pulses to become shorter (in time) to maintain the same excitation bandwidth (as measured in ppm). One way to get shorter pulses is to use higher power RF amplifiers, but as magnetic fields continue to increase, not only must the corresponding probes be redesigned to handle the ever-increasing power, but unfortunately the components that can handle these increased voltages are also usually larger. The space inside a triple-resonance PFG probe is already crowded, so the demands of increased power handling create conflicts that cause 800 MHz probes to be more difficult to make than 500 MHz probes (assuming the probe diameter remains the same). This is also why the first few probes for a new magnetic field strength often do not produce as much signal-to-noise improvement as would be predicted by theory.

The second parameter – better RF homogeneities – arises because every probe produces RF pulses that are spatially nonuniform to different extents, so that different parts of the sample receive RF irradiation equivalent to slightly different pulse widths. The net effect of this is that every pulse in an NMR experiment can degrade the resulting sensitivity, with the degradation directly proportional to the nonuniformity (inhomogeneity) of the pulse. Five 90° pulses in a sequence may generate only 80–95% of the sensitivity of a single 90° pulse. The specification is often quoted as the ratio of signal intensities for a 450° versus a 90° pulse (the 450°/90° ratio) or, on better probes, as the 810°/90° ratio. The homogeneities of both the observe (inner) and decoupler (outer) coils must be taken into account for multiple-channel experiments.

The third probe design factor – better salt tolerance – is driven by the need to study biomolecular samples dissolved in aqueous ionic buffers. As the salt concentration increases and the sample becomes more conductive, two things happen. The first is that the NMR sensitivity drops. A sample dissolved in a 0.5 M buffer may have only 90% of the sensitivity of the same sample dissolved in plain water. This means that a probe optimized for high sensitivity for organic samples (e.g., ethylbenzene) is not necessarily optimized for samples dissolved in aqueous buffers. With salty samples, the signal-to-noise ratio will decrease as the filling factor of the probe increases, as the probe volume increases, as the Q of the probe increases, and as the salt concentration increases. Large-volume (8 or 10 mm) ^{1}H probes, or superconducting probes, are particularly prone to this problem, and care must be taken when using them to study aqueous samples.

The second problem with acquiring NMR data on conductive samples is that they heat up as power is applied to the sample. For a given amount of power, the more conductive the sample is, the hotter it gets. Although this is a minor but measurable problem when running spin lock experiments (e.g., TOCSY), it is a significant problem with indirect-detection experiments because the X nucleus decoupling may use a relatively high power to increase its bandwidth. The resulting temperature increases change the chemical shifts of the solute (actually, the frequency of water changes with temperature, and this moves the ^{2}H lock frequency of D$_2$O) and can also change lineshapes and cause sample degradation. This has required probes to have better variable-temperature performance, partly to maintain more stable temperatures and partly to dissipate any heat generated within the sample by RF heating.

2.12.18.3.3.2 Probes for larger and smaller volume samples

Some samples do not allow acceptable signal-to-noise NMR data to be easily acquired with conventional NMR techniques. For solution state samples, there are two ways to resolve this problem. One, if the sample solubility is the limiting factor, use a probe whose RF coil can hold a larger volume of the solution. Two, if the sample quantity is the limiting factor, dissolve it in a smaller volume of solvent and place it in a more efficient small-volume probe. Samples dissolved in a smaller volume of solvent and run in a smaller-volume probe can exhibit higher NMR sensitivities and fewer interfering signals arising from excess solvent or solvent impurities.

The default probe geometry for solution state NMR is normally a '5 mm' probe (i.e., a probe with an RF coil optimized for 5 mm (OD) sample tubes; **Figure 139a**). If higher NMR sensitivity is needed and enough sample exists, probes that handle 8 and 10 mm sample tubes can be used. (Probes that use sample tubes larger than 10 mm in diameter were more common in the past but are rarely used anymore.) X observe probes of 10 mm diameter have been available for many years, whereas 8 mm and 10 mm diameter ^{1}H observe probes (including indirect-detection probes) have been available only for the last few years. Although data can be acquired on samples in 5 mm tubes placed in larger diameter (8 or 10 mm) probes (**Figure 139b**) (even though this decreases the net NMR sensitivity), the converse is not true; that is, 10 mm sample tubes cannot be placed in 5 mm probes.

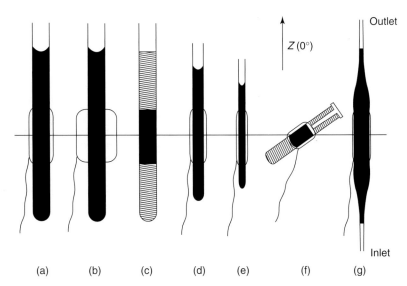

Figure 139 Representations of the various sample tube geometries available for acquiring solution state NMR data. Black areas represent the sample solution. Outlines of the RF coils for each probe geometry are shown to illustrate the active volume of the RF coils. Shown from left to right in order are: (a) a sample in a standard 5 mm tube and the RF coil from a 5 mm probe; (b) a 5 mm tube in a 10 mm probe; (c) a 5 mm tube with susceptibility-matched restrictive inserts; (d) a 3 mm microprobe sample; (e) a 1.7 mm submicroprobe sample; (f) a nanoprobe sample; (g) a flow probe. The nanoprobe tube has a 40 μL total (and active) volume; the flow probe has a 60 μL active volume. All tubes are drawn to the same scale.

Larger-diameter sample tubes are useful if samples have limited solubility and if plenty of solute exists. This is classically encountered when acquiring X-observe NMR spectra on inorganic complexes of limited solubility, but some groups studying biomolecules (i.e., those doing rational drug design) often use dilute solutions to minimize problems caused by aggregation, and sometimes they try to compensate for the reduced signal-to-noise ratio by using larger sample tubes. For the opposite situation, in which the amount of solute is limited but the solubility is not, better NMR sensitivity can be obtained by concentrating the sample. This is a common scenario in small-molecule pharmaceutical applications, especially when studying natural products or metabolites.

2.12.18.3.3.2.1 Microprobes and microcells   Once the sample has been concentrated into a smaller volume of solution, there are several options for obtaining NMR data. One option is to use a 5 mm- diameter sample tube with 'susceptibility plugs' (or certain types of microcells) – of which a variety of types are available – to reduce the depth (length) of the sample (**Figure 139c**). This gives higher sensitivity than **Figure 139a**, but the lineshape usually suffers a bit, and sample preparation can be significantly more difficult, largely because all air bubbles must be removed from the sample. A second option is to use a smaller-diameter sample tube in a probe with a smaller-diameter RF coil (**Figure 139d**). This also gives higher sensitivity than **Figure 139a**, and the lineshape is as good or often better. The increased sensitivity (in **Figure 139d** as compared with **Figure 139a**) is due in part to the ability to place a larger fraction of the sample volume in the active region of the RF coil without lineshape degradations. The better lineshape is due to the use of a smaller cross-section of the magnet. The sample volume depends on the diameter of the RF coil of the probe (which itself ranges from 2.5 to 3 mm in diameter), but typically ranges from about 80 to 150 μL. Probes like this are now usually referred to as 'microprobes.'[927,984,985]

Probes designed for even smaller volumes now exist. The most conventional option is to use an even smaller-diameter sample tube, typically with a diameter of 1.0 mm or 1.7 mm, in a probe with a matched-diameter RF coil (**Figure 139e**). Probes like this were first made available in the 1970s,[986] and might be making a comeback under the current moniker of 'submicroprobe.'[987] Higher NMR sensitivities per amount of sample can often be achieved with submicroprobes, but the small size and fragility of the sample tubes has made sample handling difficult.

2.12.18.3.3.2.2 Nanoprobes   A totally different solution to the problem of obtaining NMR spectra on small-volume solution state samples was introduced in the early 1990s. It uses a hybrid of MAS and traditional high-resolution probe technology. The resulting product, marketed by Varian as the NanoProbe, has had a large impact on pharmaceutical research, and is covered in detail in two review articles.[988,989]

All previous sample geometries for solution state samples were approximations of an infinite cylinder (the sample is long relative to the RF coil), because this is a reliable and convenient way to obtain good lineshapes. This geometry moves all discontinuities of magnetic susceptibility (e.g., liquid–air and liquid–glass interfaces) far away from the active region of the RF coil. The closer the discontinuities (interfaces) are to the RF coil, the more NMR line broadening they can introduce. Because perfectly cylindrical interfaces do not cause lineshape problems, the sample tubes themselves are made to be cylindrical. (Theory says that sample geometries that are spherical should also give good lineshapes; however, in practice the lineshape is usually substandard, presumably due to imperfections in the sphere such as the hole used to introduce the sample.)

Unfortunately, this 'infinite-cylinder approximation' sample geometry means that only a fraction of the sample can actually contribute to the NMR sensitivity. The available percentage that can be utilized depends on the diameter of the RF coil, but in a 5 mm probe the RF coil obtains most of its signal from the central 35% of the sample, and even a 3 mm probe uses only about 50% (**Figure 139**). (For any given probe style, the percentage claimed can vary among different vendors and generations of probes, and for a 5 mm probe may even be as high as 45% or 50%; however, higher percentages always cause a performance trade-off such as a worse lineshape or a reduced RF homogeneity.) Susceptibility plugs can constrict the entire sample to within the active region of the RF coil, but as the percentage approaches 100% the lineshape usually degrades. (This is because the plugs can never be a perfect match to the magnetic susceptibility of the sample.) Sample handling also becomes more difficult when susceptibility plugs are used.

Nanoprobes, on the other hand, work on an entirely different principle.[988–990] To maximize sensitivity, 100% of the solution state sample is placed the RF coil of the probe (**Figure 139f**). Normally this would result in severe broadening of the NMR lines because of the various liquid–air and liquid–glass interfaces (magnetic susceptibility discontinuities) close to the RF coil. To eliminate these magnetic susceptibility line broadenings and regain a solution-state-quality lineshape, the samples are spun about the magic angle (54.7° relative to the z axis).[991,992] This also allows samples that are smaller than the volume of the RF coil (40 μL in the case of a standard nanoprobe) to be properly studied by NMR. Smaller samples just leave a larger air bubble in the nanoprobe sample cell, and MAS completely eliminates this line broadening.

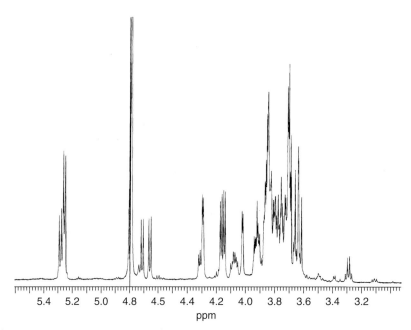

Figure 140 A ^{1}H NMR spectrum acquired using a ^{1}H nanoprobe. Several micrograms of the unknown oligosaccharide were dissolved in an H$_2$O:D$_2$O mixture. The sample was spun about the magic angle at 1.8 kHz; the data were acquired in 1024 scans using presaturation to saturate the water resonance (at 4.8 ppm). The narrow water resonance is characteristic of nanoprobe spectra, regardless of the sample volume. (Sample courtesy of A. Manzi, University of California, San Diego.)

Good NMR data have been obtained on samples dissolved in less than 2 µL of solvent; linewidths and lineshapes are good, water suppression is outstanding, and even a stable ^{2}H lock can be maintained on this volume of solvent.

Nanoprobes were initially designed to allow NMR spectra to be obtained on small-volume (40 µL) solution state samples.[988,989] This use is illustrated by five kinds of sample-limited applications. First, nanoprobes capable of ^{1}H-only detection have solved the structures of several unknowns using the data from a variety of 1D and 2D ^{1}H NMR experiments (**Figure 140**).[993,994] Second, ^{13}C-observe nanoprobes have used conventional 1D ^{13}C{^{1}H} NMR data to solve the structures of unknowns.[995,996] Third, ^{13}C-observe nanoprobes have generated high-sensitivity INADEQUATE data that determined the complete carbon skeleton of unknowns (in a more rigorous manner than an indirect-detection method such as HMBC could).[997–999] Fourth, nanoprobes have been used to acquire solvent-suppressed ^{1}H NMR spectra on samples dissolved in 90:10 H$_2$O:D$_2$O to solve structures using 1D and 2D NMR (homonuclear) techniques.[1000–1003] Fifth, nanoprobes capable of generating indirect-detection and PFG data are now available as well.[1004] These kinds of applications are primarily of benefit to natural product and metabolism chemists, both of which typically suffer from limited amounts of sample, but there is also an unrecognized potential to rational drug design programs (especially for the removal of dipolar couplings in large oriented molecules).

The advantages of nanoprobes are that they are the highest sensitivity per nucleus NMR probes commercially available, they shim equally easily for 40 and 4 µL samples, and have the unique capability of producing lineshapes for samples in H$_2$O that are very narrow at the base (which facilitates water suppression) (**Figure 140**). However, acquiring solution state data while using MAS does have some experimental limitations. Because the sample is spinning during the entire experiment, spinning sidebands may be present in all spectra (even multidimensional spectra). The sample spinning also diminishes the signal stability, and thus causes additional t_1 noise. This is a problem for experiments that utilize phase-cycled cancellation of large signals (e.g., phase-cycled indirect detection), although the availability of nanoprobes capable of running PFG experiments addresses this problem.

MAS is not a new invention. It has been in use since 1958 for narrowing linewidths in solid state samples. It was never used before for fully solution state samples for two reasons. First, MAS was initially designed to remove the much larger line broadenings that occur only in solid state samples (i.e., those that arise from interactions such as dipolar coupling and chemical shift anisotropy). The capability of MAS to eliminate the much smaller line broadenings that arise from magnetic susceptibility discontinuities was not considered nearly as important, although its utility was recognized early on by several groups.[988,989,1005,1006] Second, to obtain high-quality solution state lineshapes (<1 Hz, 10 Hz, and 20 Hz at the 50%, 0.55%, and 0.11% levels, respectively), NMR probes also need to be built using materials and designs that are susceptibility matched. While this has long been a criteria in the design of vertical-spinning solution state probes,[1007] it was never considered important in the design of solid state MAS probes until recently.[1008] The nanoprobe was the first MAS probe designed to handle solution state samples, and as such, it was the first MAS probe to incorporate this magnetic susceptibility-matched probe design technology. (Solid state samples often have ^{1}H linewidths of 100 Hz or more, so that an additional line broadening of 5–20 Hz coming from the probe design was considered insignificant. This additional line broadening is large, however, for solution state samples, whose natural linewidths can be well under 1 Hz.) Conventional MAS probes are built to emphasize very high-speed spinning and the ability to handle high RF powers. These characteristics are needed for solid state NMR but are unnecessary for (and often conflict with the requirements of) solution state NMR.

After the (Varian) NanoProbe demonstrated the virtues of using MAS on solution-state samples (and on a variety of semisolid samples, as discussed below), a variety of high-resolution MAS (HR-MAS) probes became available from several NMR probe vendors (including Bruker, Doty Scientific, and JEOL). The different probes exhibit a variety of performances caused by different balances of the contrasting needs of solution state versus solid state NMR. The biggest differences among the various HR-MAS probe designs are in the lineshape specifications and the filling factors (the sensitivity efficiencies). Less emphasized – but still important – differences include the maximum spin rate, the integrity of the sample container against leakage, the cost of the sample container, the cost and ease of automation, and the ease of changing between MAS and non-MAS probes. The reliability of sample spinning is usually not an issue with liquid (or semisolid) samples. There is one very big difference between MAS probes and microprobes (or other non-MAS probes). Non-MAS probes (those that orient samples along the vertical z axis of the magnet) use cylindrical (or spherical) sample geometries to eliminate line broadenings arising from those magnetic susceptibility discontinuities surrounding the sample, but this has no effect on magnetic susceptibility discontinuities within the sample itself. Only MAS can eliminate line broadenings that arise from these internal magnetic susceptibility discontinuities (those within the sample). This is a crucial and critical difference. A homogeneous (filterable) solution state sample has a uniform magnetic susceptibility only as long as there are no air bubbles in the sample. If any heterogeneity is introduced anywhere near the RF coil, either accidentally or purposefully – and this can be an air bubble, flocculent material, any amount of solid or liquid precipitate, or any form of emulsion, suspension, or slurry – the sample becomes physically heterogeneous to the NMR, and only MAS can be used to regenerate narrow lineshapes. (The author has encountered a surprising number of situations in which spectroscopists thought their samples were physically homogeneous – because they were visually clear and produced reasonable NMR spectra – until it was found that the spectra acquired under HR-MAS were measurably different.) This means that cylindrical (or spherical) microcells must be filled completely, with no air bubbles, or the NMR resonances will be broadened, whereas an MAS microcell will generate the same lineshape regardless of how completely the microcell is filled. Note that MAS does not necessarily guarantee that all resonances will be narrow: although MAS removes some line broadening effects, it does not affect other parameters (such as rapid T_1 or T_2 relaxation) that can broaden linewidths.

Probes are a very important part of an NMR spectrometer, and as such there are other styles of probes yet to be discussed. In particular, flow probes will be discussed in Section 2.12.18.3.5, and cryoprobes will be discussed in Section 2.12.18.3.7. However, the next section covers an important application of HR-MAS NMR in more detail. This attention is justified because an understanding of the physics involved in sample composition, sample preparation, and experimental parameters is necessary if optimal HR-MAS spectra are to be acquired. This contrasts with most other kinds of NMR spectroscopy, in which sample preparation, data acquisition, and data interpretation are usually more straightforward.

2.12.18.3.4 Nuclear magnetic resonance of solid phase synthesis resins

Techniques for obtaining NMR data on samples still bound to insoluble SPS resins have been a topic of considerable importance during the last few years, especially to combinatorial chemists.[988,989] The typical chemist wants to obtain a solution state-style spectrum, which normally means obtaining spectra with narrow lineshapes. This requires a consideration of magnetic susceptibilities, molecular motions, and how the three main types of NMR probes behave for heterogeneous SPS resin slurry samples.[990,1009]

The first thing a solid phase resin chemist needs to do to obtain narrow NMR linewidths is to swell the resin to a slurry with an excess of solvent. This increases the mobility of the bound substrates, and mobile nuclei typically exhibit narrower resonances than do solid state samples. (Solid state samples exhibit broader resonances than solution state samples because of several parameters, including faster relaxation, chemical shift anisotropy, sample heterogeneity, and homonuclear dipolar couplings.) Although NMR data can be obtained on SPS resins in the solid state by using traditional solid state tools such as cross-polarization MAS, this is of more interest to the fields of material science and polymer chemistry, and is not useful to organic or combinatorial chemists.

2.12.18.3.4.1 The observed nucleus

The second parameter to consider is which nucleus will be observed (^{1}H, ^{13}C, etc.). Different nuclei are affected differently by the sample heterogeneity of an SPS resin slurry. A slurry of SPS resin is a physically heterogeneous mixture that contains regions of free solvent, bound solvent, cross-linked polymer, bound samples, and possibly even long tethers, and each of these regions may possess its own unique magnetic susceptibility. This mix of magnetic susceptibility discontinuities causes an NMR line broadening for nearby nuclei that scales with the frequency of NMR resonance. This means that ^{1}H spectra suffer four times as much magnetic susceptibility line broadening as ^{13}C spectra because protons resonate at four times the frequency of ^{13}C nuclei. (For example, if ^{13}C nuclei resonate at 100 MHz, ^{1}H nuclei will resonate at 400 MHz.) This also means that resin spectra obtained on a 600 MHz NMR spectrometer will have linewidths twice as broad as those obtained on a 300 MHz system. Although this doubling of the linewidth may be completely offset by the corresponding doubling of the chemical shift dispersion (the number of hertz per ppm also doubles when going from 300 to 600 MHz), this is only true if all the line broadenings observed in a given spectrum arise from magnetic susceptibility discontinuities. This is certainly never the case for the resonances of the less mobile backbone of the cross-linked polymer, but it is a good approximation for the signals arising from the nuclei bound to the surface of the resin.

2.12.18.3.4.2 Choice of the resin and solvent

The third and fourth parameters that control the NMR linewidths in an SPS resin slurry are the choice of the resin used and the solvent used to swell the resin.[1009,1010] The more mobility the nuclei of interest have, the narrower their resonances will be. Many resins used today for solid phase organic synthesis contain long 'tethers' that allow bound solutes to be located a significant distance away from the more rigid cross-linked polymer portion of the resin bead. Since the tethers are usually flexible straight chains, the bound solute enjoys considerable freedom of motion. This is desirable for organic synthesis because it allows ready diffusion of reagents to the solute, and it is desirable for NMR because the additional freedom of motion decreases the efficiency of T_2 relaxation, and hence produces narrower NMR resonances. This means that the NMR spectra of resins with either short or no tethers (like the original Merrifield resins) contain NMR linewidths that are broad compared with the linewidths obtained on resins with long tethers (e.g., Tentagel or Argogel resins).

Then one must consider the solvent. Solvents that swell and properly solvate the resin (at least those portions of the resin that are around the bound solute) are more likely to produce narrow NMR linewidths. If parts of the resin are not properly solvated, the reduced molecular motion will increase the NMR linewidths.

In summary, this means that to acquire an NMR spectrum with narrow linewidths, you need both a proper resin and a proper solvent; a poor choice of either will produce a low-quality NMR spectrum (i.e., one with only broad NMR resonances).

2.12.18.3.4.3 Choice of nuclear magnetic resonance probe

The final parameter is the choice of which NMR probe is to be used: a conventional solution-state probe, an MAS probe, or a nanoprobe or HR-MAS probe. The importance of this parameter depends on the previous four parameters. Conventional solution state probes can generate high-resolution spectra for homogeneous liquids (because they use magnetic susceptibility-matched probe design technology, as discussed above) but they cannot eliminate magnetic susceptibility line broadening caused by heterogeneous SPS resin slurries. Conventional MAS probes can remove magnetic susceptibility line broadening caused by heterogeneous SPS resin slurries, but, because they do not use magnetic susceptibility-matched probe design technology, the probe induces additional line broadening of its own (about 3–30 Hz, depending on the resonance frequency). This is one reason why conventional MAS probes are inappropriate for acquiring NMR data on homogeneous liquids. Nanoprobes (and, to a lesser extent, other HR-MAS probes) use both MAS and magnetic susceptibility-matched probe design technology, and so they can acquire high-quality narrow-linewidth NMR spectra on both homogeneous liquids and heterogeneous resin slurries.

Does this mean a nanoprobe is required for all SPS resin spectra? No, it depends on what you need to do. A conventional solution state probe produces acceptable ^{13}C NMR data; this technique is common enough to have acquired the name of 'gel phase NMR,'[989,1011,1012] but ^{1}H NMR spectra acquired in this way are usually considered to exhibit unacceptably broad linewidths. A conventional MAS probe produces acceptable ^{13}C NMR data but poor-quality ^{1}H NMR data (although the data are better than they would be if a conventional solution state probe was used). A nanoprobe produces the best ^{1}H NMR data, and also produces the highest-resolution ^{13}C NMR spectra, but often the additional resolution in the ^{13}C spectrum is not needed, and data from either of the other two probes would have been just fine.[990] Most people agree that a nanoprobe (or other HR-MAS probe) is the only way to

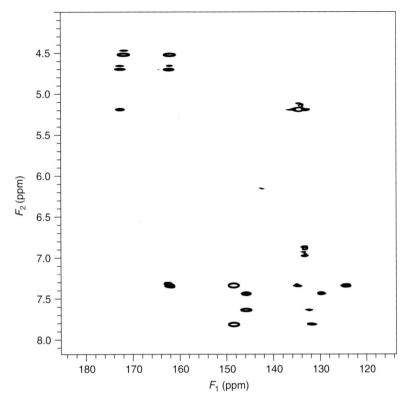

Figure 141 A multiple-bond ^{1}H–^{13}C PFG HMBC spectrum of a SPS resin acquired using a PFG indirect-detection nanoprobe. The sample consisted of 0.5 mg of Fmoc-Asp(OtBu)-NovaSyn TGA (Calbiochem) slurried in 20 μL of CD$_2$Cl$_2$. The absolute-value HR-MAS data set was acquired in 90 min using 32 scans for each of the 140 increments. The MAS spin rate was 2.1 kHz.

generate ^{1}H NMR data on SPS resins (**Figure 141**).[988,989,1013,1014] Nanoprobes were the first probes capable of acquiring high-resolution ^{1}H NMR spectra of compounds still bound to SPS resins,[1015] and they are responsible for starting the HR-MAS revolution. Nanoprobes are still the only probes capable of acquiring data on single 100 μm beads.[1016]

Note that if you are using a resin with no tether or have chosen a poor solvent, your spectra will contain broad lines regardless of which probe you use. (A nanoprobe may allow the resonances to be narrower by 50 Hz, but if the lines are already 200 Hz wide it will not help much.) All of these conclusions can also be extended to acquiring NMR data on any other kind of semisolid, slurry, emulsion, or membrane-bound compound; they are all examples of physically heterogeneous samples that contain enough localized mobility to allow NMR resonances to be narrow if the data are acquired properly. Biological tissues are another kind of semisolids; methods for using NMR to study them are an area of considerable interest to the pharmaceutical community,[1004] and investigations into the use of MAS and HR-MAS to acquire NMR spectra of tissues and intact cells are in progress.[1017–1019] This capability has the potential to have a big impact on both medical diagnostics and drug efficacy screening.

One final note about high-resolution MAS NMR: not all solution state experiments translate directly into the realm of MAS. It has been found that spin-locked experiments such as 2D TOCSY and ROESY do not always behave the same way as they do on conventional solution-state probes.[1020] The RF inhomogeneities of solenoidal coils, combined with sample spinning during the experiment, cause destructive interferences during the spin-locks that lead to signal intensities, which are a function of the sample spinning rate. The solution is to use either different spin locks[893] or different sample geometries (unpublished results).

2.12.18.3.5 Flow nuclear magnetic resonance

Some groups in the pharmaceutical industry are trying entirely different approaches to introducing samples into the NMR spectrometer. All of these techniques can be grouped into a category called 'flow NMR,' and they all use 'flow probes' (see **Figure 139g**), even though the samples may or may not be flowing at the time of NMR acquisition. The first example of this approach is HPLC–NMR, more commonly called just LC–NMR. More recently, direct-injection

NMR (DI-NMR) and flow injection analysis NMR (FIA-NMR) have been developed as ways to acquire NMR data without the use of the traditional precision-glass sample tubes. By interfacing robotic autosamplers and liquid handlers to NMR spectrometers, samples in disposable vials and 96-well microtiter plates are now routinely being analyzed by NMR.

2.12.18.3.5.1 Liquid chromatography–nuclear magnetic resonance

The traditional way to separate a complex mixture and examine its individual components is to perform a chromatographic separation off-line, collect the individual fractions, evaporate them to dryness (to remove the mobile phase), redissolve them in a deuterated solvent, and examine them by conventional NMR using microcells and microprobes if needed. This off-line technique has its place but is inappropriate if the solutes are volatile, unstable, or air sensitive. LC–NMR offers a way to perform an immediate analysis after an on-line separation, and as a technique it is becoming both powerful and more popular.

Although LC–NMR was first developed in 1978, for the first 10–15 years it was regarded more as an academic curiosity rather than a robust analytical tool. Early reviews[1021,1022] and later reviews[1023,1024] of LC–NMR exist, but the technique has evolved rapidly as flow probes have become more common, probe sensitivities have increased, and as techniques for working with nondeuterated solvents have been developed.[908] LC–NMR has now become almost a routine technique for metabolism groups,[955,1025] and is proving useful for drug production and quality control groups,[1026,1027] combinatorial chemistry groups,[1028] and groups doing natural products research.[1029–1031]

The unique advantage of LC–NMR over all other NMR techniques lies in its ability to separate components within a sample in situ. To many spectroscopists, this may seem either obvious or nonessential (by thinking that the separation could always be performed off-line), but it has been shown that there are some analyses that do not lend themselves to off-line separations.[1032–1036] Whenever the analyzed component is unstable to light, air, time, or the environment, LC–NMR will probably be the preferred NMR technique.

The hardware for LC–NMR consists of an HPLC system connected to an NMR flow probe. The experiments can be run either in an on-flow mode, in which the mobile phase moves continuously (useful for preliminary or survey data acquisition), or in a stopped-flow mode, in which peaks of interest are stopped in the NMR flow probe for as long as needed for NMR data acquisition (useful for careful examination of individual components). In the stopped-flow mode, either 1D or 2D NMR data can be acquired, although the limited amounts of sample usually tolerated by HPLC columns sometimes makes the acquisition of extensive 2D heteronuclear correlation data difficult. The stopped-flow mode can employ any one of three different kinds of sample handling. First, the samples may be analyzed directly as they elute from the chromatography column, one chromatographic peak at a time. This is often the first, or default, mode of operation. Second, the LC pump may be programmed to 'time slice' through a chromatographic peak, stopping every few seconds to acquire a new spectrum. This is useful for resolving multiple components (by NMR) within a peak that are not fully resolved chromatographically or for verifying the purity of a chromatographic peak. (Alternatively, on-flow acquisition with a very slow flow rate has been used for similar purposes.[1037]) The third method is to collect the chromatographic peaks of interest into loops of tubing (off-line) and then flush the intact fractions into the NMR flow probe one at a time as needed. A variation of this technique is to trap the eluted peaks onto another chromatographic column, to allow concentration of the solute, and then re-elute them with a stronger solvent into the flow probe as a more concentrated slug.[1038]

There are several aspects of acquiring LC–NMR data that are challenging. The first is that all mobile phases in LC–NMR are mixtures, and the solvents are rarely fully deuterated, so that usually several solvent resonances need to be suppressed. Additionally, the resonances of the organic solvents contain ^{13}C satellites that need to be suppressed. Also, when the samples are flowing through the probe, solvent suppression sequences such as presaturation take too much time and do not work well. All of these problems were solved when the WET solvent suppression experiment was developed.[906] WET uses a combination of shaped-pulse selective excitation, multifrequency SLP pulses, PFG, and indirect-detection ^{13}C decoupling during the shaped pulses to quickly and efficiently suppress multiple resonances using only a simple two-channel spectrometer.

Another challenge in acquiring LC–NMR data is that, by definition, the solvent composition changes during the experiment. In reversed-phase LC–NMR, one of the cosolvents is virtually always water, and unfortunately the chemical shift of water changes as the solvent composition changes. (It has been observed that this is also true, to a much lesser extent, for the resonances of several other solvents and modifiers.) In addition, if the mobile phase is fully protonated (nondeuterated) there will be no ^2H lock to keep the frequencies constant. (Also, D_2O serves as a poor lock signal because its chemical shift changes with solvent composition.) All of this means that the frequencies of solvent suppression are constantly changing. To allow this to be compensated for, and to allow the frequencies to be

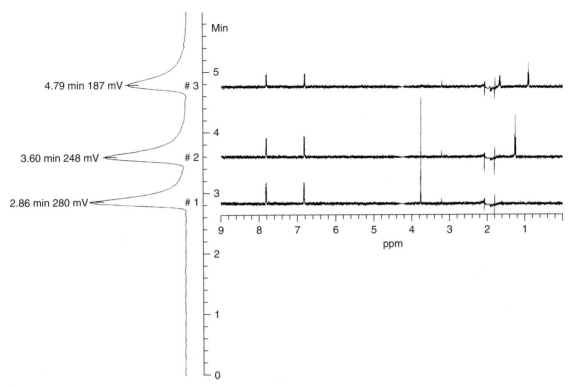

Figure 142 An on–flow LC–NMR spectrum. The three substituted benzoates were separated using a $CH_3CN:D_2O$ mobile phase. The ^1H chemical shift axis runs horizontally; the chromatographic time axis runs vertically. On the left is an 'NMRgram' showing the summed ^1H signal intensity between 7.5 and 8 ppm; this display is analogous to a chromatogram. On the top is the 1D spectrum (the 'trace') of the methyl benzoate LC peak that eluted at 2.9 min. The data were acquired while the sample flowed continuously through the probe by using two-frequency WET solvent suppression (at 1.95 and 4.2 ppm); suppression frequencies were automatically determined using the SCOUT scan experiment. The small signals flanking the CH_3CN resonance are traces of its ^{13}C satellites.

automatically optimized, the SCOUT scan technique was developed.[906] This technique takes a single-scan, small-tip-angle, nonsuppressed ^1H spectrum, moves the transmitter to the constant resonance (serving as a ^1H lock), measures where the other peaks moved to, and creates an SLP that suppresses these peaks. This whole process takes only a few seconds. It can be used in an interleaved fashion during an on-flow solvent gradient run or as a precursor to the signal-averaged stopped-flow data acquisitions. We will see later how it forms an integral part of the DI-NMR and FIA-NMR techniques.

The presentation of the resulting LC–NMR data depends on the type of experiment being run. On-flow LC–NMR data are usually displayed as a contour map, like a conventional 2D data set, although the Fourier transform is only applied along one axis (the F_2 axis) to give a frequency versus elution time plot. The 1D data plotted along the 'pseudo-t_1' axis may either be the LC detection output or a projection of the NMR data (**Figure 142**). Stopped-flow LC–NMR data, on the other hand, are usually presented as a series of individual 1D spectra either one spectrum per page or as a stacked plot (**Figure 143**).

LC–NMR is usually considered a very powerful (although usually not a fast) technique. Its power can be exploited by combining it with mass spectrometry, to produce LC–NMR–MS. Usually a fraction of the chromatographic effluent prior to the NMR flow cell is diverted to the mass spectrometer. LC–NMR–MS is considered by some to be the ultimate tool for pharmaceutical analysis, and although it is still in its early days, it is proving its worth in one-injection analyses of compounds that are either sample limited or proving to be tough structural problems.[954,955,1039,1040]

2.12.18.3.5.2 Flow injection analysis and direct-injection nuclear magnetic resonance

FIA-NMR and DI-NMR are essentially sample changers that exploit the speed and robustness of flow NMR. Neither technique uses chromatography, so they are not designed to analyze mixtures. DI-NMR is well suited for the analysis of

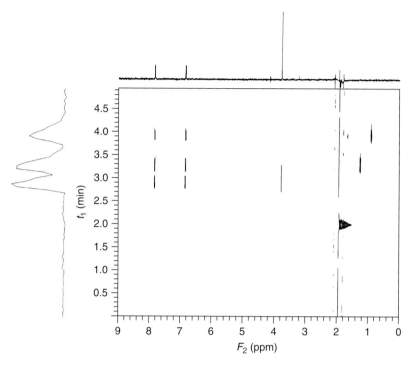

Figure 143 Stopped-flow LC–NMR data. The three substituted benzoates were separated using a $CH_3CN:D_2O$ mobile phase. The ^1H chemical shift axis runs horizontally; the chromatographic time axis runs vertically. On the left is the HPLC chromatogram; the corresponding ^1H NMR spectra are aligned with each peak. This is Varian's default (and automatic) output display for stopped-flow LC–NMR data. The data were acquired by using two-frequency WET solvent suppression (at 1.95 and 4.2 ppm) and a two-frequency DSP notch filters; suppression and filter frequencies were automatically determined using the SCOUT scan experiment. Each ^1H NMR spectrum was acquired in seven scans on a 25 μg on-column sample injection.

combinatorial chemistry library samples, whereas FIA-NMR appears to be more useful as a tool for repetitive quality control style analyses.

The hardware for FIA-NMR is similar to that for LC–NMR except that there is no chromatography column, so it is sometimes referred to as 'columnless LC–NMR.'[988,989] No additional detectors (other than the NMR spectrometer) are needed, nor is a pump capable of running solvent gradient methods. FIA-NMR uses the mobile phase as a hydraulic push solvent, like a conveyer belt, to carry the injected plug of sample from the injector port to the NMR flow cell. After the pump stops, the spectrometer acquires the SCOUT scan, analyzes it, and acquires the signal-averaged data (again using WET solvent suppression). When finished, a start signal is sent to the solvent pump, so that it can flush the old sample from the NMR flow cell and bring in the next sample. In classic FIA-NMR, as in LC–NMR, the sample always flows in one direction, and enters and exits the NMR flow cell through different ports. The flow cell is always full of solvent.

DI-NMR, which was also first developed by Keifer's group,[988,989] is significantly different. Rather than starting off with an NMR flow cell full of solvent and suffering the sensitivity losses caused by the ensuing dilution of the injected sample, the flow cell starts off empty in DI-NMR. A sample solution is injected directly into the NMR flow cell. The spectrometer then acquires the SCOUT scan, analyzes it, and acquires the signal-averaged data (with WET). When finished, the syringe pump pulls the sample back out of the NMR flow cell, in the reverse direction, and returns it to its original (or an alternative) sample container. Once the flow cell is emptied, clean solvent is injected and removed from the flow cell to rinse it. The solvents used for both the sample and the rinse must at least be miscible, and ideally should be identical, to avoid lineshape degradations caused by magnetic susceptibility discontinuities (due to liquid–liquid emulsions or precipitation of solutes in the flow cell) or any plugging of the transfer lines.

DI-NMR uses very different hardware than FIA-NMR. It uses the syringe pump in a standard Gilson 215 liquids handler to deliver samples to the NMR flow probe. Samples go directly into the NMR flow cell, through the bottom, via an unswitched Rheodyne injector port. The Gilson liquids handler is capable of accepting a wide variety of sample container formats, including both vials and microtiter plates.

One of the justifications for the development of DI-NMR was to improve the robustness and speed of automated NMR. The improvement in robustness has occurred, partly through the elimination of techniques such as sample spinning, automatic gain adjustment, automatic locking, and simplex shimming (PFG gradient shimming is used instead). This has allowed 96 samples at a time (stored in 96-well microtiter plates) to be run without error, and some installations have run tens of thousands of samples in a highly automated fashion. Sample turnaround times are from 1 min to 4 min per sample, but the speed at which it can be run depends quite heavily on the solvent being used. Although most solvents can be pumped in and out of the probe rapidly, dimethyl sulfoxide, which is used commonly, is viscous enough that sample flow rates are decreased and the resulting analyses may not be much faster than traditional robotic sample changers.

In comparison, FIA-NMR and DI-NMR have different advantages and disadvantages, and are designed for different purposes. The advantages of DI-NMR are that it generates the highest signal-to-noise ratio per sample and consumes less solvent. The disadvantages are that it has a minimum sample volume (if the sample volume is smaller than the NMR flow cell, the lineshape rapidly degrades), and no sample filtration is possible. The advantages of FIA-NMR are that is has no minimal sample volume (since the flow cell is always full) and can filter samples (with an in-line filter), and that it can rinse the NMR probe better in a shorter time (albeit by consuming more solvent). The disadvantages of FIA-NMR are that it generates a lower signal-to-noise ratio (because of sample dilution), and it consumes more solvent. Both techniques will always have some degree of carry-over (although it may be <0.1%) and be subject to blockages from solid particles. The only way to avoid these problems is to use a conventional robotic sample changer for glass tubes. In applications, FIA-NMR is more valuable for repetitive analyses or quality control functions, especially when there is plenty of sample and the samples are to be discarded. In contrast, DI-NMR is more valuable when there is a limited amount of sample and the samples must be recovered. DI-NMR, as implemented by Varian in the VAST instrument, is becoming well regarded as a tool for the analysis of single-compound combinatorial chemistry libraries,[956,1004,1041] biofluids, and for 'SAR-by-NMR' studies (described in Section 2.12.18.3.5.4).

2.12.18.3.5.3 Direct-injection nuclear magnetic resonance: data processing and analysis
Now that DI-NMR (VAST) is becoming routine, the next big challenge is the presentation and analysis of the data. As an example, we can examine how to display the spectral data acquired on the samples in a 96-well microtiter plate. Because these are combinatorial chemistry samples, there will be relationships between samples located in each row and each column of the plate, and we may choose to examine each of those relationships separately.

The most conventional way to present the data is as a stack of 96 spectra, plotted one spectrum per page. After acquiring data on several plates, this option tends to prove less popular to the chemists responsible for interpreting all of the information. Another option is to display all 96 spectra (and/or their integrals) on one piece of paper (**Figure 144**). This works best if only part of the chemical shift range is plotted, and serves better as a printed record of the data rather than as a means to analyze the data. An alternative, which is better for on-screen interactive analysis but is less useful as a printed record, is to display the data in a manner similar to that used to display LC–NMR data; namely, as a contour (or stacked) plot with the ^1H frequency axis along F_2 but in which the t_1 axis corresponds to individual wells (**Figure 145**). A third option is to display the spectral data as a stacked plot, either from one row, one column, or from a discrete list of well locations (**Figure 146**). The last option, which appeals to many, is to display integral information in one of three ways. First, one can extract integral intensities for different regions across the chemical shift range, for each of the 96 spectra, and list them serially in a text file. Second (and more usefully), one can take this text file and extract just one integral value per spectrum (all from the same region) and display it in a spreadsheet database. Third, one can display this same information as a color map, with color intensity representing integral intensity (or peak height or any other quantitative information). A more complete description of these various options appears elsewhere.[956]

Obviously, the most desirable way to interpret the data is automatically with a computer. Although the technology does not yet exist to determine the structure of an unknown compound de novo from its 1D ^1H spectrum, it is currently possible for a computer to determine if a given NMR spectrum is consistent with a suspected structure. This is relatively routine with ^{13}C data, but is harder for ^1H data because of the presence of homonuclear couplings (especially second-order couplings) and increased solvent effects, although software to accomplish this does exist.[1042] This capability is expected to be of great interest in combinatorial chemistry, since normally the expected structure already exists in an electronic database somewhere and can easily be submitted to the analysis software along with the experimental NMR spectral data.

2.12.18.3.5.4 Other techniques, applications, and hardware: SAR-by-NMR
Rational drug design programs have also been using DI-NMR. Starting in 1996, a group at Abbott Laboratories published a series of articles about 'SAR-by-NMR.'[1043] Structure–activity relationship (SAR) mapping, has been around

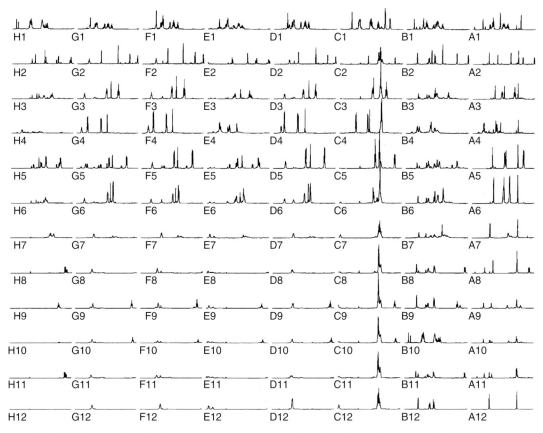

Figure 144 DI-NMR data automatically acquired on samples stored in a 96-well microtiter plate using VAST. This is an 8 × 12 matrix plot of the 96 1D ^{1}H NMR spectra acquired on a plate of samples dissolved in DMSO-h_6. This represents one way to plot the data from an entire titer plate on one page. Each spectrum was acquired, processed, and plotted with identical parameters, and is labeled with its alphanumeric position coordinates. For clarity, only an expansion of the proton spectrum (6–8 ppm) is displayed here. Data were acquired using WET solvent suppression as set up by the SCOUT scan experiment. Although the data presentation differs, these data are identical to those in **Figure 141**.

for many years, but the premise of SAR-by-NMR is to use changes in NMR chemical shifts to do the mapping. This is accomplished by first acquiring and assigning a ^{1}H{^{15}N} 2D correlation map (typically a PFG HSQC) of an ^{15}N-labeled receptor (a protein). Then, ^{1}H{^{15}N} correlation maps are acquired on a series of samples made by adding different ligands (small molecules) to different aliquots of this receptor solution. If the ligand binds to the receptor, individual ^{1}H–^{15}N correlations within the ^{1}H{^{15}N} HSQC spectrum will change positions, with the largest changes usually occurring for those nuclei that are closest to the active site of the receptor. (If the ligand does not bind to the receptor, the ^{1}H{^{15}N} HSQC spectrum will remain unchanged. The ligand itself is not ^{15}N labeled and will not appear in the HSQC spectrum.) By correlating the resonances that change position with the structures of the ligands, one can literally map the active site of the receptor. This technique has stimulated a flurry of interest in the primary literature.[1044,1045] One of the drawbacks to this technique is that, because a number of ligand–receptor complexes must be studied, a significant amount of purified ^{15}N-labeled protein is needed, which is not always easy to obtain.

Another issue is that, because a significant number of 2D NMR spectra must be acquired, automated techniques for handling the samples and acquiring the data are desirable. Although conventional automated NMR sample changers can be used, more recently DI-NMR systems have also proven useful. A DI-NMR system uses less sample than is required by a 5 mm tube, and the robotic liquids handler can be programmed to prepare the samples as well.

2.12.18.3.6 Mixture analysis: diffusion-ordered and relaxation-ordered experiments

One of the challenges of NMR is working with mixtures. In the past, NMR has always focused on the bulk properties of the solution, and – with the exception of LC–NMR – few attempts have been made to resolve the individual components of mixtures. The following three methods show that this is changing.

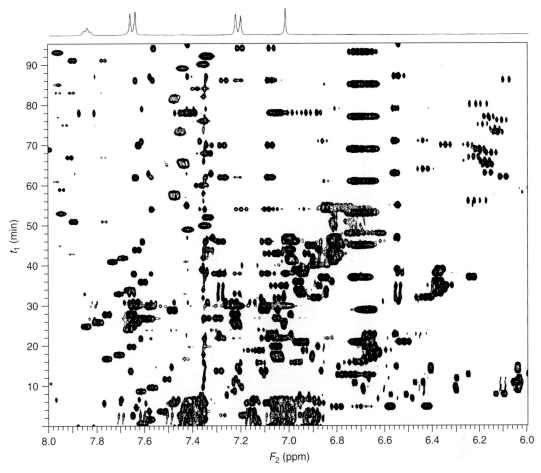

Figure 145 A contour plot of the 6–8 ppm region of the proton spectra of 96 1D ^{1}H NMR spectra acquired on the DMSO-h_6 microtiter plate using VAST. Although the data presentation differs, these data are identical to those in **Figure 140**. This represents an alternative way to plot the data from an entire titer plate on one page. This style of data presentation is especially useful when it can be displayed on the computer screen in an interactive mode, and a user can select the 1D spectra for individual wells at the top of the screen. Each spectrum was acquired, processed, and plotted with identical parameters.

The most well-known method for analyzing mixtures by NMR is diffusion-ordered spectroscopy (DOSY). It uses PFG to separate compounds based on their diffusion rates. PFG has been used since 1965 to measure diffusion rates;[957] but starting in 1992 it was shown how this technique could be used as a separation tool as well.[1046–1048] DOSY data are presented as a 2D contour (or stacked) plot where F_2 is the ^{1}H chemical shift axis and F_1 is the diffusion rate axis. Compounds having different diffusion rates are separated along F_1. Multiple components within a 5 mm tube have been separated.[1049] The technique can also be combined with other experiments, and extended to multiple dimensions, to create COSY-DOSY, DOSY-TOCSY, DOSY-NOESY, ^{13}C-detected INEPT-DOSY and DEPT-DOSY, and DOSY-HMQC, among others.[1047]

In a related technique, Shapiro and co-workers combined PFG-based diffusion analysis with TOCSY to analyze mixtures in which the diffusion coefficients for each component are similar or in which there is extensive chemical shift overlap. Calling their version 'diffusion-encoded spectroscopy' (DECODES),[958,959] they combined the standard longitudinal encode and decode (LED) sequence[1050] with a 2D TOCSY sequence. Multiple 2D TOCSY data sets are acquired, each with a different diffusion delay. The TOCSY component resolves the chemical shift overlaps, and the peak intensities of resolved resonances are plotted against the diffusion delay, to identify different chemical species in solution. (Each different chemical species in a mixture is likely to exhibit at least slightly different diffusion rates.) A variation of this technique, called 'affinity NMR,' was developed in which an unlabeled receptor was added to a mixture of ligands and a 1D or 2D DECODES spectrum was acquired.[1051] If the diffusion delay is set long enough, only the resonances of the receptor and the bound ligands are visible. The TOCSY component then helps to resolve the

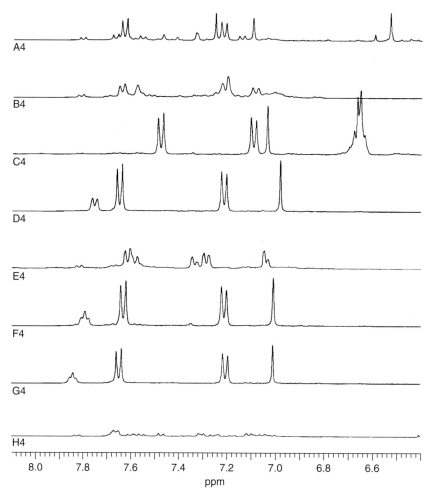

Figure 146 A stacked plot of the proton spectra (6.4–8.1 ppm) acquired on the '4' wells (wells A4 through H4 – top to bottom) of the DMSO-h_6 microtiter plate using VAST. These data are a subset of those in **Figures 140** and **141**. This style of data presentation allows comparisons to be made between spectra acquired on different wells. Each spectrum was acquired, processed, and plotted with identical parameters.

chemical shifts of the detected (actively bound) individual components. A diffusion-filtered NOESY version of this technique has also been developed.[1052] Unlike SAR-by-NMR, binding is detected by observing the ligand resonances and not by changes in the receptor resonances.

The third method for analyzing mixtures is relaxation-edited (or relaxation-resolved) spectroscopy. Large molecules have longer rotational correlation times than small molecules, and this results in shorter T_2 relaxation times (and larger NMR linewidths). These differences in T_2 relaxation can be used to edit the NMR spectrum of a mixture, based on the different relaxation rates of the individual components, in a manner analogous to diffusion ordering or diffusion editing.[1053–1055] A combination of diffusion and relaxation edited experiments (DIRE) was also reported, and applied to the analysis of biofluids.[1054]

A big advantage of LC–NMR is that the chromatography column can take a mixture and separate it into individual chromatographic peaks. However, one disadvantage is that no single combination of mobile phase and column can completely resolve all compounds, and a single chromatographic peak may still contain multiple components. Often, this can still be resolved by NMR because the individual components usually will have different chemical shifts, but this is only valid if the compounds and their chemical shifts assignments are already known in advance.[1037] The combination technique LC-DOSY-NMR (or DI-DOSY-NMR) would be a powerful way to analyze a mixture using as many as three different physical characteristics. Although this experiment has not yet been performed and the limited sample quantities used in LC–NMR may preclude its success, it remains an intriguing idea.

2.12.18.3.7 Other probes: superconducting probes, supercooled probes, and microcoils

Two other probe developments have also sprung from efforts to extract better signal-to-noise ratios from a given sample. The first is the use of probes that employ low-temperature RF coils to reduce the noise level. The RF coils are typically maintained at about 20 K while the samples are left at room temperature. Probes with coils of either superconducting metal[1056,1057] or normal metal[1058] have been made, although probes that have full PFG and indirect-detection capability have only become available in recent years. (The term 'cryoprobe' is often loosely applied to both kinds of probes as a group.) The advantage of these probes is that they exhibit higher sensitivities, and at a cost lower than that of the purchase of a higher field magnet (although without the increased dispersion). The current disadvantages are their significantly higher cost and limited availability. (They have also been associated with somewhat increased levels of t_1 noise because of their use of mechanical pumps.) The cost and delivery issues have precluded them from being widely used, yet. Although that is changing rapidly, there are still a very limited number of published reports of their use.[1059,1060] One very good illustration of their usefulness was made by using an ^{19}F version of a superconducting probe to study the kinetics of protein folding.[1061] Here, the higher sensitivity provided was critical because the reaction rate was too fast to allow signal averaging, so spectra acquired in only one scan each had to be used to follow the reaction.

The other probe development is the use of small-scale microcoils for the data acquisition of very small samples. This work, done by Sweedler and Webb at the University of Illinois starting in 1994,[1062] is allowing NMR data to be acquired on samples ranging in size from roughly 1 µL down to 1 nL.[1063,1064] The primary justification is to enable applications in flow NMR, that is, to allow NMR data to be acquired on capillary HPLC and capillary electrophoresis effluents.[1065] The biggest problem in using microcoils is dealing with the magnetic susceptibility interfaces;[1063] however, probes have developed to the point that 2D indirect-detection data have been generated.[1066]

2.12.18.3.8 Combination experiments

It is now routine in modern NMR spectroscopy to find that many of the above-mentioned techniques are combined in an ever-increasing number of ways. Many such examples can be observed. The resolution and sensitivity of the HMBC experiment for poorly resolved multiplets was improved significantly by using semiselective pulses.[1067] The 3D version of HMBC has been advocated for the elucidation of natural product structures.[1068] Selective excitation has been combined with a PFG version of a 3D triple-resonance experiment, to make a 2D version (SELTRIP) for the study of small biomolecules.[1069]

PFGs and selective excitation have been combined into a powerful tool called 'double-pulse field gradient spin echo' (DPFGSE), also known as 'excitation sculpting.'[902] This tool offers a lot of flexibility in selecting desirable (or discriminating against undesirable) NMR signals, to produce both cleaner and more sensitive NMR data. As such, it has been used in a number of applications. One example is the DPFGSE-NOE experiment,[902] which produces significantly cleaner data than the conventional 1D-difference NOE experiment. Another example is the HETGOESY indirect-detection experiment used for detecting heteronuclear NOEs.[1070] DPFGSE sequences have been used to perform isotopic filtering to select only ^{13}C-bound protons[1071] and for removing t_1 noise in 2D experiments.[1072] It has also been used in band-selective homonuclear-decoupled (BASHD) TOCSY[1073] and ROESY[1074] experiments.

2.12.18.3.9 Software

Software tools to massage and analyze data are constantly becoming more powerful and more important. Digital signal processing (DSP) techniques that use low-pass filters are being used to enhance the spectral signal-to-noise ratio (through oversampling) and to reduce data size.[1075] High-pass DSP filters, or notch filters, are being used to remove solvent resonances.[908,1076] Linear prediction techniques are being used to flatten baselines and remove broad resonances.[1077] Bayesian data analysis is providing more powerful and statistically meaningful spectral deconvolution, which can facilitate quantitative analysis.[1078] FRED software is providing automated interpretation of INADEQUATE data,[998,1079] and is allowing complete carbon skeletons of unknown molecules to be determined both automatically and with higher effective signal-to-noise ratios.[997] ACD software[1080] is not only generating organic structures from 1D ^{1}H and ^{13}C data, but is also helping analyze NMR data acquired on combinatorial chemistry libraries.[1042] The filter diagonalization method shows promise for very rapid 2D data acquisition, and has greatly reduced data storage requirements.[1081,1082] This too should prove useful for combinatorial chemistry.

2.12.18.3.10 Quantification

One of the big applications for NMR that seems to be on the verge of a growth spurt is in the area of quantitative analysis, especially in high-throughput techniques such as combinatorial chemistry. Part of this seems to arise from the difficulty of performing accurate quantification with mass spectrometric techniques, especially for samples still bound to SPS resin beads.[1083] NMR data can be highly quantitative, but certain precautions must be observed.[1084,1085] These include the need to use broadbanded excitation, generate flat baselines, minimize spectral overlap, minimize NOE interferences, and, in particular, ensure complete relaxation.[1086]

The other issue is the use of internal standards. Most internal standards have some drawbacks, such as too much volatility to be reliable (e.g., tetramethylsilane (TMS)) or too little volatility for easy removal from the solute after measurement (e.g., (trimethylsilyl)propionate (TSP)). Some standards have also been shown to interact with glass sample containers.[1087] To minimize spectral overlap, the NMR resonance(s) of the standard should be sharp singlets (e.g., TMS, TSP, hexamethyldisiloxane, or $CHCl_3$), yet such resonances often exhibit very long T_1 relaxation rates (which complicates quantification). Recently, 2,5-dimethylfuran was proposed as an internal standard because the integrals of the two resonances can be compared to verify complete relaxation.[1088]

2.12.18.3.11 Automation

Of course, tools are needed to automate the acquisition and data analysis of all these experiments in a repetitive fashion. There are different levels and kinds of automation for NMR. The most primitive level lets a user change the sample manually but use software macros to automate the setup, acquisition, or processing of one or more data sets. The next level of automation entails the use of robotic devices to change samples automatically, and this is usually done in combination with software automation to acquire and plot (but not necessarily analyze) the data automatically. Some open-access NMR spectrometers run in this latter mode, although more open-access systems probably run in the former mode. Many service facilities operate in a highly manual mode.

Conventional tube-based NMR spectrometers equipped with robotic sample changers never run at 100% reliability. Estimates of reliability probably range from 75% to 99%, with autospin, autolock, and autoshim software failures being the most common hazards. The strengths of tube-based robotics is that the probe and spectrometer are always intact, and any failed sample is usually just ignored, so that the spectrometer can move on to the next sample. Automation of sample preparation is often neglected.

The increased sample-throughput requirements of modern drug discovery programs are driving the development of more robust methods of NMR automation. Flow NMR techniques (the VAST DI-NMR system in particular) are being used to avoid failure-prone steps. This has pushed reliability up to 99.9% or higher, and this is important when thousands of samples are being analyzed. One advantage of DI-NMR is that sample preparation is minimal. One often forgotten limitation of DI-NMR is that dirty samples (those that contain precipitates, solids, emulsions, or immiscible mixtures) can clog up flow cells and degrade NMR lineshapes and solvent suppression performance. This suggests that DI-NMR is not ready for use in open-access environments and that tube-based NMR spectrometers are more appropriate for seeing nonfiltered samples.

2.12.18.3.12 Research nuclear magnetic resonance versus analytical nuclear magnetic resonance

This brings up a final point about the dual nature of NMR spectroscopy. NMR started off as a tool for research. For several decades it continued to serve in that role, and to a large extent it still does (and will for the foreseeable future). However, eventually it will also become an analytical tool, serving to provide answers to repetitive questions. The current drive to develop automation and high-throughput techniques is a reflection of the beginning of this trend, which is expected to grow.

2.12.18.4 Overview

The field of NMR spectroscopy has matured significantly, although this does not mean that the pace of development has slowed; rather, it means that this is a time of great opportunity. Many hardware and software tools have been developed to solve specific problems, and the key now is to apply these tools to new problems. One might even go so far as to say that the hardware and pulse sequence tools we have available are ahead of the potential applications, meaning that what remains are for users to find new ways to use the tools that have already been developed. Some of the most exciting possibilities undoubtedly lie at the boundaries of interdisciplinary problems; that is, scientists who are unfamiliar with NMR often have problems that could be solved with current technology if only these problems could be matched up with both people of vision and people with the correct technical expertise. As a result, it behooves all scientists in the field of drug discovery to – as much as possible – stay abreast of these many developments in NMR spectroscopy.

2.12.19 Materials Management

2.12.19.1 Introduction

During the rational drug design phase of the 1980s, effective materials management was an important, but not crucial, aspect of the drug discovery process. However, during the 1990s, as drug discovery methods became increasingly automated and probabilistic, efficient management of materials became a critical component of the drug discovery process.

This section will define materials management, describe its current role in drug discovery, and discuss future trends in the field. With a focus on the management of chemical reagents, natural products, and proprietary compounds, the section will explore both automation technology, which is used for management of the physical materials, and materials management information technology, which is central to an overall drug discovery informatics environment.

2.12.19.2 Materials Management: A Definition

Materials management refers to the systems, work processes, facilities, and technology involved in selecting, acquiring, receiving, storing, tracking, controlling, safely handling, distributing, and disposing of the materials defined below:

- Reagents – materials, usually available commercially from a chemical supplier, that are used throughout the performance of an experimental method. Within this section, the term 'reagents' refers to the specialty organic chemicals used by chemists as starting materials (including catalysts and solvents) in the synthesis of novel compounds.
- Compounds – proprietary chemicals synthesized or acquired for the purpose of testing for biological activity. A compound may be synthesized by an in-house medicinal chemist through manual or combinatorial methods or may be acquired from an external organization, such as a university or a government agency, or from an alliance partner. The term 'compound' may refer to one known chemical structure or to a mixture of multiple chemical structures.
- Natural products – naturally occurring materials that a drug discovery organization acquires and screens for biological activity.

Materials management processes are designed to:

- maximize the throughput of chemical synthesis by streamlining the process of selecting, acquiring, and managing reagents;
- minimize the costs involved in managing reagents by identifying duplicate inventory and excessive carrying and disposal costs;
- maximize screening throughput by streamlining the process of preparing proprietary samples for screening;
- reduce the risk of materials handling errors by creating an integrated informatics environment that obviates the collection and entry of incorrect data; and
- improve safety and regulatory compliance by providing tools for environmental, health, and safety personnel to monitor hazardous material usage and support preparation of regulatory reports.

In summary, materials management processes are designed to reduce cycle times, maximize productivity and efficiency, control costs, and enhance environmental health and safety.

2.12.19.3 Role of Materials Management in Drug Discovery

Figure 147 shows how materials management fits into the drug discovery process. Reagents are selected and acquired by medicinal chemists to support synthesis processes during both lead generation and lead optimization. To support screening processes, compounds are received, stored, and distributed for screening. Following results analysis, leads are requested for follow-up. Central to the overall materials management process is a materials management system that provides the foundation of a drug discovery informatics environment.

Figure 148 shows how SciQuest's Enterprise Substance Manager solution integrates the overall discovery informatics environment. The informatics environment shown in the figure breaks down as follows:

- Supply-chain data, which includes such information from reagent suppliers as product description, package quantity, and pricing information.

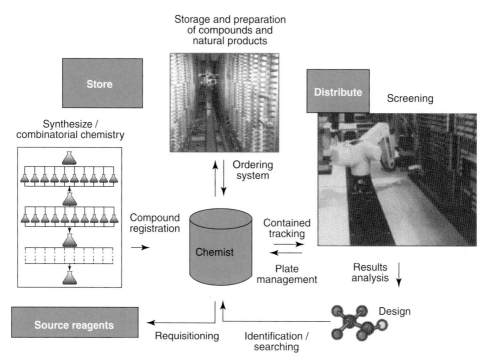

Figure 147 Materials management in drug discovery.

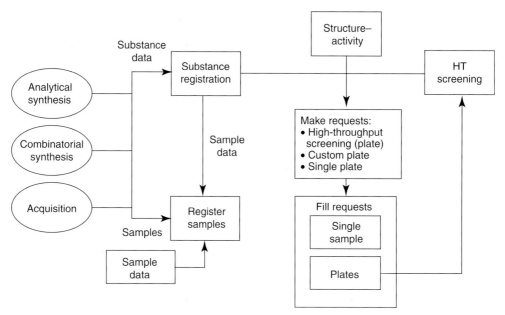

Figure 148 Materials management system.

- Enterprise resource planning systems, which provide purchasing and payment of reagents acquired from reagent suppliers.
- Cheminformatics and bioinformatics systems, which provide chemical, physical property, and biological activity data associated with materials. Compound registration systems and biological activity management systems are all integrated through a centralized inventory or materials management system.
- Process automation, such as robotic liquid-handling equipment and automated sample storage and retrieval systems.

After establishing the physical sample as the primary object with which all components of an informatics framework are integrated, a materials management system matches the attributes of materials with the attributes of containers, to provide an underlying data structure that integrates all of the components. As described by Frank Brown[1113]:

> "Using the sample object allows for simple change and additions to the description of the sample without a complete rewrite of the data structures. The hub of the chemical information system is the inventory system. The registration number should be considered a name and not a primary key for building relationship given the complexity of parsing needed to break the registration number into its pieces."

2.12.19.4 Job Functions Involved in Materials Management

Now that we have discussed how materials management fits into the drug discovery process and the informatics environment, it is important to discuss the way various job functions are involved in the materials management processes. Following is a description of these job functions:

- Medicinal chemists – select, order, track, and dispose of reagents as well as synthesize, register, and distribute proprietary compounds.
- Purchasing personnel – place orders for reagents, interact with suppliers on pricing and availability, and provide status updates to requesters of materials.
- Receiving/stockroom personnel – receive reagents at the facility, label containers of reagents, maintain and replenish stock, and provide delivery services to scientists for requested reagents.
- Waste management personnel – collect waste, prepare waste manifests for waste haulers, and provide reports of what material has been removed from a facility by what mechanism.
- Compound management personnel – receive, store, and track proprietary compounds and prepare samples for screening.
- Registrar – ensures integrity of chemical and physical property data entered into the corporate compound registration database. May or may not be part of the compound management group.
- HTS scientists – request samples for screening and perform plate management functions (including plate replication and combination) to support HTS activities.
- Therapeutic area biologists – select compounds of interest and place requests for samples for follow-up screening.
- Information technology personnel – assist with the design, development, integration, maintenance, and support of materials management systems.
- Environmental health and safety personnel – maintain lists of materials that require monitoring; track training requirements of personnel handling hazardous materials; and monitor and provide reports for regulatory purpose on acquisition, usage, and disposal of hazardous materials. They are responsible for the chemical assessment and clearance procedures.

2.12.19.5 Materials Management Work Processes

Figure 149 provides a summary of the key work processes – by job function – involved in materials management. The five primary work processes are (1) reagent sourcing; (2) reagent receiving and tracking; (3) new compound registration and sample submission; (4) master plate preparation and distribution for HTS and (5) follow-up screening. Each is discussed in detail below.

2.12.19.5.1 Reagent sourcing

Reagent sourcing refers to the selection and sourcing of reagents to support synthesis processes. The steps in the process are as follows:

(1) A medicinal chemist will establish a synthesis plan and design a virtual library of chemical structures.

(2) The chemist conducts a search for commercially available reagents that can be used as starting materials for the synthesis of the library. The 'hit list' may be further refined using chemical structure analysis tools to eliminate undesirable functional groups and maximize structural diversity; it ensures that the molecular weights of the desired products are within acceptable ranges.

(3) The chemist conducts an inventory search for reagents on his on her list that are already available in-house, and researches paper and/or electronic catalogs for products that can be purchased. For desired reagents that are not available, the chemist may consider custom synthesis. In some organizations, environmental health and safety personnel are part of the approval process for orders of hazardous materials but not part of the order approval process; they passively monitor orders, prepare reports of hazardous material usage, and identify any regulatory issues.

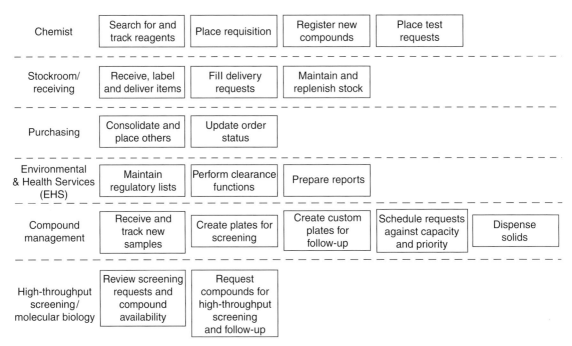

Figure 149 Materials management work processes.

(4) For items available in-house, the chemist will retrieve the item from another laboratory or place an internal requisition to a stockroom. For items not available in-house, the chemist will submit a purchase requisition. Depending on the organization, the purchase order is placed by routing a paper requisition, through integration with a purchasing system or directly to suppliers via the internet or electronic data interchange mechanisms. Some companies have preferred relationships with suppliers, and allow chemists to place orders directly, using credit cards. In other cases, the purchasing department or a third-party purchasing agent may be responsible for placing orders with suppliers and communicating the order status from the suppliers back to the original requester; this process can be automated as part of a business-to-business e-commerce application.

2.12.19.5.2 Reagent receiving and tracking

Once reagents are received by a facility, they are delivered to the requester, tracked, and disposed of. The steps of the process are as follows:

(1) The supplier delivers the ordered items to the loading dock.

(2) Receiving personnel unpack the box and match the packing slip to the original order and note any discrepancies. In some organizations, all orders for chemicals are processed by a chemical stockroom and are delivered directly by receiving personnel to the stockroom. When the item is unpacked, the purchasing system is updated to note receipt and process payment. Most organizations then update an internal tracking system to generate bar code labels and start tracking individual containers.

(3) Receiving personnel deliver the item to the original requester.

(4) The original requester uses the reagent and then places the remainder in a disposal bin, stores it in the laboratory, or returns the item to a centralized storeroom for reuse. The storeroom may update the quantity-available information.

(5) When items are ready to be disposed of, the individual responsible for waste management removes the material and prepares a manifest in conjunction with the waste hauler.

2.12.19.5.3 New compound registration and sample submission

Once the chemist completes the synthesis of a new compound or library of compounds, the compound has to be registered and the physical sample submitted for storage and distribution. The steps in the process are as follows:

(1) Once the chemist synthesizes new proprietary compounds (or compounds are received from outside), he or she enters the chemical and physical attributes of the compounds into the corporate database or registration system.

In some organizations, the processes of registering the compound and submitting it for analytical testing are integrated. In some cases, the chemist does the registering; in others, a centralized registration group handles the registration process. At a minimum, the centralized registration group will have final quality control, to ensure that the chemical structure of the compound is consistently entered into the corporate database for consistent data retrieval. Following registration of the compound(s), a company code number is assigned, to provide an internal naming convention. The registration database is then used as the chemical structure/information source in support of structure and biological activity database searches. In some companies, registration of the physical sample precedes compound registration. This happens more and more often in high-throughput processes where the structure characterization of the compound is done following completion of biological screening activities. Furthermore, as part of collaboration agreements, some companies might do 'blind' screening of plates without knowing the structures of the screened compounds.

(2) In conjunction with the registration of both the chemical and physical attributes of the new compound, the physical sample is usually distributed to a centralized compound management group for storage, distribution, and inventory management.

(3) Along with logging the new sample into the materials management system and placing the sample in storage, the compound management group may be responsible for the initial registration of the compounds and natural products received from external sources.

2.12.19.5.4 Master plate preparation and distribution for high-throughput screening
Once compounds are synthesized, they are plated and distributed for screening. The steps of the process are as follows:

(1) On receiving the compounds from the chemist, the compound management group will create master plates from the source compounds at standardized concentrations and volumes. This may be an automated process, using liquid-handling automation equipment and robotic weighing equipment, or it may be a semiautomated process, with technicians manually flicking small amounts of material to support plate preparation.

(2) Once the master plates are prepared, they are replicated and stored, usually in cold storage.

(3) Replicates are distributed for screening, either on demand or as part of a standard work process. Usually, when plates are distributed for screening, the contents of the plates are also entered into the biological activity database for future association with screening results. There may be manual population of plate data from the biological database or automated population via integration between the materials management system and the biological screening database.

2.12.19.5.5 Follow-up screening
Once hits are identified, samples must be prepared for follow-up screening. The steps of the process are as follows:

(1) When screening is completed, the results are entered into the biological activity database.

(2) After the hits are identified via searches against the biological activity database, a request for samples is submitted to the materials management system based on the hit list.

(3) The materials management system logs the request and identifies source containers that can be used to fill the request. Frequently, custom plates are prepared by 'cherry picking,' or removing materials from individual wells across multiple source plates. Often, serial dilutions are performed to generate dose–response results.

(4) When follow-up screening yields positive results, additional follow-up screens and solid samples are requested from the materials management system and obtained from solid storage.

2.12.19.6 Materials Management Technologies

A variety of technologies are in use to support the above work processes. Following is a description of available technologies.

2.12.19.6.1 Chemical structure-based reagent selection tools and products
Chemists select reagents via chemical structure criteria. A variety of chemical structure-based software tools are available on the market to search chemical databases by chemical structure, substructure, structural similarity, and other mechanisms. Numerous tools are also available to provide chemical structure searching capabilities. Among these are MDL's ISIS products, Accelrys, Inc.'s RS3 product, Daylight, Tripos tools, MSI's tools, CAS, Scifinder, and CambridgeSoft's tools.

In the realm of materials management, these tools are at their most powerful when incorporated into an overall materials management system. SciQuest has incorporated chemical structure searching tools into their reagent manager materials management product line, just as MDL has incorporated chemical structure searching tools into their product. Another example of how chemical structure searching information technology is applied to materials management is MDL's Reagent Selector. Both SciQuest and MDL products evaluate result sets of reagent–product chemical structures, based on criteria such as molecular weight and structure diversity, and easily allows for the elimination of structures with undesirable functional groups or elements.

2.12.19.6.2 ERP systems, supplier catalogs, and electronic commerce

Chemical structure searching tools are only as good as the content data they are searching against – which in turn are only as good as the level of integration with purchasing applications. The most widely used electronic catalog of reagents is MDL's Available Chemicals Directory. This product is a compilation of several catalogs providing chemical structures, as well as packaging and pricing information. Periodically, MDL provides these data to its customers in the form of a database that can be searched by chemical structure tools. Other vendors, such as CAS, provide similar products. Suppliers such as Sigma-Aldrich provide product information either through CDs or directly through their website. Some suppliers (e.g., Sigma-Aldrich) also provide detailed chemical information and safety information through their website.

Whereas chemists will search through electronic catalogs to select reagents, enterprise resource planning systems, such as SAP, are used to place orders to suppliers. Technology is also available to place orders directly with suppliers by electronic data interchange mechanisms or via suppliers' websites. Companies such as SciQuest provide third-party mechanisms to aggregate catalogs of available materials and support sourcing and procurement functions.

2.12.19.6.3 Robotic weighing and liquid-handling workstations

Plate preparation processes involve a number of repetitive steps in weighing samples and transferring liquids to solubilize samples, as well as preparing plates at desired concentrations and volumes. A number of robotics weighing systems, such as those provided by Zymark, are on the market to tare-weigh empty vials and gross-weigh vials full of solid material. The robotic workstations are integrated with laboratory balances to collect weight information and transfer data to formatted files.

Similarly, liquid-handling systems automate repetitive plate preparation tasks, including solubilizing samples, transferring liquid samples from a source container to a plate, adding top-off volumes to achieve desired concentrations, creating replicate plates for distribution and screening, and combining multiple plates into single plates. Examples of liquid-handling systems are the workstations provided by TECAN and Hamilton. Both robotic weighing and liquid transfer workstations exchange data to and from materials management systems, usually via formatted file transfers.

2.12.19.6.4 Materials management systems

As mentioned earlier, materials management systems provide the foundation for an integrated discovery research informatics architecture. The physical sample is the key data element that integrates supplier data, process automation systems, and chemical and bioinformatics systems. A robust materials management system supports such functions as:

- selecting reagents by chemical structure;
- placing orders, either directly with suppliers or via enterprise resource-planning systems;
- receiving items from suppliers;
- tracking container usage;
- submitting sample proprietary materials;
- submitting and fulfilling requests for samples in support of screening; and
- preparing, replicating, and combining plates.

Such a system requires integration with supplier catalogs, purchasing systems, compound registration systems, biological results databases, robotic plate preparation equipment, laboratory balances, and automated sample storage and retrieval systems.

SciQuest's product line is one such materials management system currently on the market. SciQuest refers to its offerings as research supply chain solutions, given its foundation level within the informatics environment and its focus on logistics functions. MDL's SMART product also provides some materials management capabilities across tracking containers of reagents and supporting capabilities to track proprietary samples. Some organizations have also built their own materials management systems, with Affymax being an example of one that is reported in the literature.

2.12.19.6.5 Automated sample storage and retrieval systems

As the numbers of samples begin to stretch into millions, the larger drug discovery organizations are procuring sample storage and retrieval systems that will automate their materials management functions. Such systems integrate a wide variety of functions, including submitting samples for storage, solubilizing and replicating plates, retrieving samples from storage, and weighing out and dispensing solid samples – all within an environmentally controlled secure, usually cold storage area. Usually provided by systems integration firms, these systems encompass both the equipment and technology to handle robotic weighing and liquid-handling equipment, laboratory balances, storage units (such as Kardex's Lektriever), and software to integrate all of these components. Usually, such systems integrate tightly with an external materials management system or have scaled-down versions of these systems as part of the overall solution. Examples of these solutions include TAP's Haystack system as well as systems provided by RTS Thurnall, REMP, and Aurora.

2.12.19.7 Future Directions

The future in drug discovery materials management technology is all about miniaturization, factory automation, leveraging the internet, and object-oriented information technology. As screening technology gets smaller, sample sizes and plate formats also continue to get smaller – meaning the use of 1536-well plates and microfluidic technology for the management of samples in the chip format.

Automation technology will improve throughput, further automating the discovery process and resulting in the application of factory automation principles to the discovery research process. Discreet automated components will be integrated, leading to tighter integration of automated storage and retrieval systems, and ultrahigh-throughput screening technology. Materials management systems will extend to provide production management and work scheduling functions to support automation processes. This high degree of automation and the unimaginable volumes of data that will be generated will introduce even more downstream bottlenecks.

The internet will be leveraged to integrate drug discovery organizations more tightly with their suppliers and partners. Chemists will be able to obtain real-time product, pricing, and availability information from their suppliers, minimizing out-of-date information and reducing bottlenecks. Furthermore, as drug discovery organizations continue to partner with other companies, the internet will support the establishment of virtual-project-based organizations to seamlessly share, track, and handle materials management information.

With the increase in throughput, the need for seamless informatics integration will continue. The migration to object-oriented systems using standards such as CORBA will allow discrete systems to integrate tightly, provided that an overall data architecture is established by information management professionals.

References

1. Drews, J. *Nat. Biotechnol.* **1999**, *17*, 406.
2. Drews, J. *Drug Disc. Today* **2000**, *5*, 2–4.
3. Wilson, S.; Bergsma, D. J.; Chambers, J. K.; Muir, A. I.; Fantom, K. G.; Ellis, C.; Murdock, P. R.; Herrity, N. C.; Stadel, J. M. *Br. J. Pharmacol.* **1998**, *125*, 1387–1392.
4. Bikker, J. A.; Trumpp-Kallmeyer, S.; Humblet, C. *J. Med. Chem.* **1998**, *41*, 2911–2927.
5. Valdenaire, O.; Giller, T.; Breu, V.; Ardati, A.; Schweizer, A.; Richards, J. G. *FEBS Lett.* **1998**, *424*, 193–196.
6. Hinuma, S.; Onda, H.; Fujino, M. *J. Mol. Med.* **1999**, *77*, 495–504.
7. McKusick, V. A. *Genomics* **1997**, *45*, 244–249.
8. Durick, K.; Mendlein, J.; Xanthopoulos, K. G. *Genome Res.* **1999**, *9*, 1019–1025.
9. Collins, F. S. *N. Engl. J. Med.* **1999**, *341*, 28–37.
10. Kruglyak, L. *Nat. Genet.* **1999**, *22*, 139–144.
11. Winzeler, E. A.; Richards, D. R.; Conway, A. R.; Goldstein, A. L.; Kalman, S.; McCullough, M. J.; McCusker, J. H.; Stevens, D. A.; Wodicka, L.; Lockhart, D. J. et al. *Science* **1998**, *281*, 1194–1197.
12. Cargill, M.; Altshuler, D.; Ireland, J.; Sklar, P.; Ardlie, K.; Patil, N.; Lane, C. R.; Lim, E. P.; Kalayanaraman, N.; Nemesh, J. et al. *Nat. Genet.* **1999**, *22*, 231–238.
13. Lifton, R. P. *Science* **1996**, *272*, 676–680.
14. Halushka, M. K.; Fan, J. B.; Bentley, K.; Hsie, L.; Shen, N.; Weder, A.; Cooper, R.; Lipshutz, R.; Chakravarti, A. *Nat. Genet.* **1999**, *22*, 239–247.
15. Rust, S.; Rosier, M.; Funke, H.; Real, J.; Amoura, Z.; Piette, J. C.; Deleuze, J. F.; Brewer, H. B.; Duverger, N.; Denefle, P. et al. *Nat. Genet.* **1999**, *22*, 352–355.
16. Curran, M. E. *Curr. Opin. Biotechnol.* **1998**, *9*, 565–572.
17. Burmeister, M. *Bas. Biol. Psychiatr.* **1999**, *45*, 522–532.
18. Wright, A. F.; Carothers, A. D.; Pirastu, M. *Nat. Genet.* **1999**, *23*, 397–404.
19. Chen, Q.; Kirsch, G. E.; Zhang, D.; Brugada, R.; Brugada, J.; Brugada, P.; Potenza, D.; Moya, A.; Borggrefe, M.; Breithardt, G. et al. *Nature* **1998**, *392*, 293–296.
20. Harris, T. *Med. Res. Rev.* **2000**, *20*, 203–211.
21. Chagnon, Y. C.; Perusse, L.; Bouchard, C. *Curr. Opin. Lipidol.* **1997**, *8*, 205–211.

22. Plomin, R. *Nature* **1999**, *402*, C25–C29.
23. Puranam, R. S.; McNamara, J. O. *Nat. Genet.* **1998**, *9*, 313–314.
24. Friedman, J. M.; Halaas, J. L. *Nature* **1998**, *395*, 763–770.
25. West, D. B.; Iakougova, O.; Olsson, C.; Ross, D.; Ohmen, J.; Chatterjee, A. *Med. Res. Rev.* **2000**, *20*, 216–230.
26. Carver, E. A.; Stubbs, L. *Genome Res.* **1997**, *7*, 1123–1137.
27. Castellani, L. W.; Weinreb, A.; Bodnar, J.; Goto, A. M.; Doolittle, M.; Mehrabian, M.; Demant, P.; Lusis, A. *J. Nat. Genet.* **1998**, *18*, 374–377.
28. Passarge, E.; Horsthemke, B.; Farber, R. A. *Nat. Genet.* **1999**, *23*, 387.
29. Lembertas, A. V.; Perusse, L.; Chagnon, Y. C.; Fisler, J. S.; Warden, C. H.; Purcell-Huynh, D. A.; Dionne, F. T.; Gagnon, J.; Nadeau, A.; Lusis, A. J. et al. *J. Clin. Invest.* **1997**, *100*, 1240–1247.
30. Roses, A. D. *Am. J. Med. Genet.* **1998**, *81*, 49–57.
31. Duff, K. *Curr. Opin. Biotechnol.* **1998**, *9*, 561–564.
32. Brent, R. *Cell* **2000**, *100*, 169–183.
33. Martin, G. R.; Eglen, R. M.; Hamblin, M. W.; Hoyer, D.; Yocca, F. *Trends Pharmacol. Sci.* **1998**, *19*, 2–4.
34. Strauss, E. J.; Falkow, S. *Science* **1997**, *276*, 707–712.
35. Blattner, F. R.; Plunkett, G., III; Bloch, C. A.; Perna, N. T.; Burland, V.; Riley, M.; Collado-Vides, J.; Glasner, J. D.; Rode, C. K.; Mayhew, G. F. et al. *Science* **1997**, *277*, 1453–1474.
36. Snel, B.; Bork, P.; Huynen, M. A. *Nat. Genet.* **1999**, *21*, 108–110.
37. Allsop, A. E. *Curr. Opin. Biotechnol.* **1998**, *9*, 637–642.
38. Saunders, N. J.; Moxon, E. R. *Curr. Opin. Biotechnol.* **1998**, *9*, 618–623.
39. Garber, K. *Bioinformatics* **2000**, *4*, 4–6.
40. Battey, J.; Jordan, E.; Cox, D.; Dove, W. *Nat. Genet.* **1999**, *21*, 73–75.
41. Bentley, D. R. *Med. Res. Rev.* **2000**, *20*, 189–196.
42. Dunham, I.; Shimizu, N.; Roe, B. A.; Chissoe, S.; Hunt, A. R.; Collins, J. E.; Bruskiewich, R.; Beare, D. M.; Clamp, M.; Smink, L. J. et al. *Nature* **1999**, *402*, 489–495.
43. O'Brien, S. J.; Menotti-Raymond, M.; Murphy, W. J.; Nash, W. G.; Wienberg, J.; Stanyon, R.; Copeland, N. G.; Jenkins, N. A.; Womack, J. E.; Marshall Graves, J. A. *Science* **1999**, *286*, 458–462, 479–481.
44. Campbell, P. *Nature* **1999**, *402*, C7–C9.
45. Freeman, T. *Med. Res. Rev.* **2000**, *20*, 197–202.
46. Watson, J. B.; Margulies, J. E. *Dev. Neurosci.* **1993**, *15*, 77–86.
47. Diatchenko, L.; Lau, Y. F.; Campbell, A. P.; Chenchik, A.; Moqadam, F.; Huang, B.; Lukyanov, S.; Lukyanov, K.; Gurskaya, N.; Sverdlov, E. D. et al. *Proc. Natl. Acad. Sci. USA* **1996**, *93*, 6025–6030.
48. Liang, P.; Pardee, A. B. *Science* **1992**, *257*, 967–971.
49. Nakai, M.; Kawamata, T.; Taniguchi, T.; Maeda, K.; Tanaka, C. *Neurosci. Lett.* **1996**, *211*, 41–44.
50. Ohtaka-Maruyama, C.; Hanaoka, F.; Chepelinsky, A. B. *Dev. Biol.* **1998**, *202*, 125–135.
51. Eberwine, J.; Yeh, H.; Miyashiro, K.; Cao, Y.; Nair, S.; Finnell, R.; Zettel, M.; Coleman, P. *Proc. Natl. Acad. Sci. USA* **1992**, *89*, 3010–3014.
52. Dixon, A. K.; Richardson, P. J.; Lee, K.; Carter, N. P.; Freeman, T. C. *Nucleic Acids Res.* **1998**, *26*, 4426–4431.
53. Velculescu, V. E.; Zhang, L.; Vogelstein, B.; Kinzler, K. W. *Science* **1995**, *270*, 484–487.
54. Fodor, S. P. A.; Read, J. L.; Pirrung, M. C.; Stryer, L.; Tsai Lu, A.; Solas, D. *Science* **1991**, *251*, 767–773.
55. Chu, S.; DeRisi, J.; Eisen, M.; Mulholland, J.; Botstein, D.; Brown, P. O.; Herskowitz, I. *Science* **1998**, *282*, 699–705.
56. Editorial. *Nat. Genet.* **1998**, *18*, 195–197.
57. Eisen, M. B.; Spellman, P. T.; Brown, P. O.; Botstein, D. *Proc. Natl. Acad. Sci. USA* **1998**, *95*, 14863–14868.
58. Ermolaeva, O.; Rastogi, M.; Pruitt, K. D.; Schuler, G. D.; Bittner, M. L.; Chen, Y.; Simon, R.; Meltzer, P.; Trent, J. M.; Boguski, M. S. *Nat. Genet.* **1998**, *20*, 19–23.
59. Editorial review. *Science* **1998**, *282*, 2012–2018.
60. White, K. P.; Rifkin, S. A.; Hurban, P.; Hogness, D. S. *Science* **1999**, *286*, 2179–2184.
61. Butler, D. *Nature* **1999**, *402*, 705–706.
62. Singh-Gasson, S.; Green, R. D.; Yue, Y.; Nelson, C.; Blattner, F.; Sussman, M. R.; Cerrina, F. *Nat. Biotechnol.* **1999**, *17*, 974–978.
63. Golub, T. R.; Slonim, D. K.; Tamayo, P.; Huard, C.; Gaasenbeek, M.; Mesirov, J. P.; Coller, H.; Loh, M. L.; Downing, J. R.; Caligiuri, M. A. et al. *Science* **1999**, *286*, 531–537.
64. Lee, C. K.; Klopp, R. G.; Weindruch, R.; Prolla, T. A. *Science* **1999**, *285*, 1390–1393.
65. Geiss, G. K.; Bumgarner, R. E.; An, M. C.; Agy, M. B.; van't Wout, A. B.; Hammersmark, E.; Carter, V. S.; Upchurch, D.; Mullins, J. I.; Katze, M. G. *Virology* **2000**, *266*, 8–16.
66. Cole, K. A.; Krizman, D. B.; Emmert-Buck, M. R. *Nat. Genet.* **1999**, *21*, 38–41.
67. Velculescu, V. E.; Madden, S. L.; Zhang, L.; Lash, A. E.; Yu, J.; Rago, C.; Lal, A.; Wang, C. J.; Beaudry, G. A.; Ciriello, K. M. et al. *Nat. Genet.* **1999**, *23*, 387–388.
68. Luo, L.; Salunga, R. C.; Guo, H.; Bittner, A.; Joy, K. C.; Galindo, J. E.; Xiao, H.; Rogers, K. E.; Wan, J. S.; Jackson, M. R. et al. *Nat. Med.* **1999**, *5*, 117–122.
69. Naggert, J. K.; Fricker, L. D.; Varlamov, O.; Nishina, P. M.; Rouille, Y.; Steiner, D. F.; Carroll, R. J.; Paigen, B. J.; Leiter, E. H. *Nat. Genet.* **1995**, *10*, 135–142.
70. Marton, M. J.; DeRisi, J. L.; Bennett, H. A.; Iyer, V. R.; Meyer, M. R.; Roberts, C. J.; Stoughton, R.; Burchard, J.; Slade, D.; Dai, H. et al. *Nat. Med.* **1998**, *4*, 1293–1301.
71. Berns, A. *Nat. Med.* **1999**, *5*, 989–990.
72. Longman, R. *In Vivo* **1999**, 1–8.
73. Dvorin, J. *Technol. Strat.* **1999**, 34–41.
74. Pavletich, N. P.; Pabo, C. O. *Science* **1991**, *252*, 809–817.
75. Yoon, K.; Cole-Strauss, A.; Kmiec, E. B. *Proc. Natl. Acad. Sci. USA* **1996**, *93*, 2071–2076.
76. Cole-Strauss, A.; Yoon, K.; Xiang, Y.; Byrne, B. C.; Rice, M. C.; Gryn, J.; Holloman, W. K.; Kmiec, E. B. *Science* **1996**, *273*, 1386–1389.
77. Xiang, Y.; Cole-Strauss, A.; Yoon, K.; Gryn, J.; Kmiec, E. B. *J. Mol. Med.* **1997**, *75*, 829–835.

78. Kren, B. T.; Cole-Strauss, A.; Kmiec, E. B.; Steer, C. J. *Hepatology* **1997**, *25*, 1462–1668.
79. Kren, B. T.; Bandyopadhyay, P.; Steer, C. J. *Nat. Med.* **1998**, *4*, 285–290.
80. Alexeev, V.; Yoon, K. *Nat. Biotechnol.* **1998**, *16*, 1343–1346.
81. Santana, E.; Peritz, A. E.; Iyer, S.; Uitto, J.; Yoon, K. *J. Invest. Dermatol.* **1998**, *111*, 1172–1177.
82. Yoon, K. *Biog. Amines* **1999**, *15*, 137–167.
83. Vitali, T.; Sossi, V.; Tiziano, F.; Zappata, S.; Giuli, A.; Paravatou-Petsotas, M.; Neri, G.; Brahe, C. *Hum. Mol. Genet.* **1999**, *8*, 2525–2532.
84. Pear, W.; Scott, M.; Nolan, G. P. Generation of High Titre, Helper-Free Retroviruses by Transient Transfection. In *Methods in Molecular Medicine: Gene Therapy Protocols*; Humana Press: Totowa, NJ, 1997, pp 1–57.
85. Holzmayer, T. A.; Pestov, D. G.; Roninson, I. B. *Nucleic Acids Res.* **1992**, *20*, 711–717.
86. Gudkov, A. V.; Kazarov, A. R.; Thimmapaya, R.; Axenovich, S. A.; Mazo, I. A.; Roninson, I. B. *Proc. Natl. Acad. Sci. USA* **1994**, *91*, 3744–3748.
87. Onishi, M.; Kinoshita, S.; Morikawa, Y.; Shibuya, A.; Phillips, J.; Lanier, L. L.; Gorman, D. M.; Nolan, G. P.; Miyajima, A.; Kitamura, T. *Exp. Hematol.* **1996**, *24*, 324–329.
88. Roninson, I. B.; Levenson, V. V.; Lausch, E.; Schott, B.; Kirschling, D. J.; Zuhn, D. L.; Tarasewicz, D.; Kandel, E. S.; Iraj, E. S.; Fedosova, V. et al. *Anti-Cancer Drugs* **1996**, *7*, 83–91.
89. Kinoshita, S.; Chen, B. K.; Kaneshima, H.; Nolan, G. P. *Cell* **1998**, *95*, 595–604.
90. Casadaban, M. J.; Cohen, S. N. *Proc. Natl. Acad. Sci. USA* **1979**, *76*, 4530–4533.
91. Bellofatto, V.; Shapiro, L.; Hodgson, D. *Proc. Natl. Acad. Sci. USA* **1984**, *81*, 1035–1039.
92. Schell, J. *Science* **1987**, *237*, 1176–1183.
93. Hope, I. A. *Development* **1991**, *113*, 399–408.
94. O'Kane, C. J.; Gehring, W. J. *Proc. Natl. Acad. Sci. USA* **1987**, *84*, 9123–9127.
95. Zambrowicz, B. P.; Friedrich, G. A.; Buxton, E. C.; Lilleberg, S. L.; Person, C.; Sands, A. T. *Nature* **1998**, *392*, 608–611.
96. Amsterdam, A.; Burgess, S.; Golling, G.; Chen, W.; Sun, Z.; Townsend, K.; Farrington, S.; Haldi, M.; Hopkins, N. *Genes Dev.* **1999**, *13*, 2713–2724.
97. Ross-Macdonald, P.; Coelho, P. S.; Roemer, T.; Agarwal, S.; Kumar, A.; Jansen, R.; Cheung, K. H.; Sheehan, A.; Symoniatis, D.; Umansky, L. et al. *Nature* **1999**, *402*, 413–418.
98. Whitney, M.; Rockenstein, E.; Cantin, G.; Knapp, T.; Zlokarnik, G.; Sanders, P.; Durick, K.; Craig, F. F.; Negulescu, P. A. *Nat. Biotechnol.* **1998**, *16*, 1329–1333.
99. Morris, S. C. *Cell* **2000**, *100*, 1–11.
100. Hodgkin, J.; Herman, R. K. *Trends Genet.* **1998**, *14*, 352–357.
101. Plasterk, R. H. *Nat. Genet.* **1999**, *21*, 63–64.
102. Bell, D. W.; Varley, J. M.; Szydlo, T. E.; Kang, D. H.; Wahrer, D. C.; Shannon, K. E.; Lubratovich, M.; Verselis, S. J.; Isselbacher, K. J.; Fraumeni, J. F. et al. *Science* **1999**, *286*, 2528–2531.
103. Cismowski, M. J.; Takesono, A.; Ma, C.; Lizano, J. S.; Xie, X.; Fuernkranz, H.; Lanier, S. M.; Duzic, E. *Nat. Biotechnol.* **1999**, *17*, 878–883.
104. Martzen, M. R.; McCraith, S. M.; Spinelli, S. L.; Torres, F. M.; Fields, S.; Grayhack, E. J.; Phizicky, E. M. *Science* **1999**, *286*, 1153–1155.
105. Simonsen, H.; Lodish, H. F. *Trends Pharmacol. Sci.* **1994**, *15*, 437–441.
106. Goodwin, S. F. *Curr. Opin. Neurobiol.* **1999**, *9*, 759–765.
107. Bargmann, C. I. *Science* **1998**, *282*, 2028–2033.
108. Fire, A.; Xu, S.; Montgomery, M. K.; Kostas, S. A.; Driver, S. E.; Mello, C. C. *Nature* **1998**, *391*, 806–811.
109. Zheng, Y.; Brockie, P. J.; Mellem, J. E.; Madsen, D. M.; Maricq, A. V. *Neuron* **1999**, *24*, 347–361.
110. Zhen, M.; Jin, Y. *Nature* **1999**, *401*, 371–375.
111. Tabara, H.; Sarkissian, M.; Kelly, W. G.; Fleenor, J.; Grishok, A.; Timmons, L.; Fire, A.; Mello, C. C. *Cell* **1999**, *99*, 123–132.
112. Metzstein, M. M.; Stanfield, G. M.; Horvitz, H. R. *Trends Genet.* **1998**, *14*, 410–416.
113. Martin, G. M.; Austad, S. N.; Johnson, T. E. *Nat. Genet.* **1996**, *13*, 25–34.
114. Migliaccio, E.; Giorgio, M.; Mele, S.; Pelicci, G.; Reboldi, P.; Pandolfi, P. P.; Lanfrancone, L.; Pelicci, P. G. *Nature* **1999**, *402*, 309–313.
115. Walhout, A. J.; Sordella, R.; Lu, X.; Hartley, J. L.; Temple, G. F.; Brasch, M. A.; Thierry-Mieg, N. *Science* **2000**, *287*, 116–122.
116. Mathey-Prevot, B.; Perrimon, N. *Cell* **1998**, *92*, 697–700.
117. Editorial review. *Science* **1999**, *285*, 1829.
118. Schafer, W. R. *Cell* **1999**, *98*, 551–554.
119. Andretic, R.; Chaney, S.; Hirsh, J. *Science* **1999**, *285*, 1066–1068.
120. Editorial. *Science* **1999**, *285*, 981.
121. Bedell, M. A.; Jenkins, N. A.; Copeland, N. G. *Genes Dev.* **1997**, *11*, 1–10.
122. Bedell, M. A.; Largaespada, D. A.; Jenkins, N. A.; Copeland, N. G. *Genes Dev.* **1997**, *11*, 11–43.
123. Copeland, N. G.; Jenkins, N. A.; Gilbert, D. J.; Eppig, J. T.; Maltais, L. J.; Miller, J. C.; Dietrich, W. F.; Weaver, A.; Lincoln, S. E.; Steen, R. G. et al. *Science* **1993**, *262*, 57–66.
124. Crabbe, J. C.; Belknap, J. K.; Buck, K. J. *Science* **1994**, *264*, 1715–1723.
125. Frankel, W. N. *Trends Genet.* **1995**, *11*, 471–477.
126. Purcell-Huynh, D. A.; Weinreb, A.; Castellani, L. W.; Mehrabian, M.; Doolittle, M. H.; Lusis, A. J. *J. Clin. Invest.* **1995**, *96*, 1845–1858.
127. Mogil, J. S.; Wilson, S. G.; Bon, K.; Lee, S. E.; Chung, K.; Raber, P.; Pieper, J. O.; Hain, H. S.; Belknap, J. K.; Hubert, L. et al. *Pain* **1999**, *80*, 83–93.
128. Spearow, J. L.; Doemeny, P.; Sera, R.; Leffler, R.; Barkley, M. *Science* **1999**, *285*, 1259–1261.
129. Ahmad, W.; Faiyaz ul Haque, M.; Brancolini, V.; Tsou, H. C.; ul Haque, S.; Lam, H.; Aita, V. M.; Owen, J.; deBlaquiere, M.; Frank, J. et al. *Science* **1998**, *279*, 720–724.
130. Chemelli, R. M.; Willie, J. T.; Sinton, C. M.; Elmquist, J. K.; Scammell, T.; Lee, C.; Richardson, J. A.; Williams, S. C.; Xiong, Y.; Kisanuki, Y. et al. *Cell* **1999**, *98*, 437–451.
131. Mohn, A. R.; Gainetdinov, R. R.; Caron, M. G.; Koller, B. H. *Cell* **1999**, *98*, 427–436.
132. Sadelain, M.; Blasberg, R. G. *J. Clin. Pharmacol.* **1999**, 34S–39S.
133. Stokstad, E. *Science* **1999**, *286*, 1068–1071.
134. Tsodyks, M.; Kenet, T.; Grinvald, A.; Arieli, A. *Science* **1999**, *286*, 1943–1946.
135. Bultman, S. J.; Michaud, E. J.; Woychik, R. P. *Cell* **1992**, *71*, 1195–1204.

136. Zhang, Y.; Proenca, R.; Maffei, M.; Barone, M.; Leopold, L.; Friedman, J. M. *Nature* **1994**, *372*, 425–432.
137. Tartaglia, L. A.; Dembski, M.; Weng, X.; Deng, N.; Culpepper, J.; Devos, R.; Richards, G. J.; Campfield, L. A.; Clark, F. T.; Deeds, J. et al. *Cell* **1995**, *83*, 1263–1271.
138. Noben-Trauth, K.; Naggert, J. K.; North, M. A.; Nishina, P. M. *Nature* **1996**, *380*, 534–538.
139. Crabbe, J. C.; Wahlsten, D.; Dudek, B. C. *Science* **1999**, *284*, 1670–1672.
140. Francis, D.; Diorio, J.; Liu, D.; Meaney, M. J. *Science* **1999**, *286*, 1155–1158.
141. Abbott, A. *Nature* **1999**, *402*, 715–720.
142. Anderson, L.; Seilhamer, J. *Electrophoresis* **1997**, *18*, 533–537.
143. Editorial. *Nature* **1999**, *402*, 703.
144. Garvin, A. M.; Parker, K. C.; Haff, L. *Nat. Biotechnol.* **2000**, *18*, 95–97.
145. Kreider, B. L. *Med. Res. Rev.* **2006**, in press.
146. Roberts, R. W.; Szostak, J. W. *Proc. Natl. Acad. Sci. USA* **1997**, *94*, 12297–12302.
147. Kay, B. K.; Kurakin, A. V.; Hyde-DeRuyscher, R. *Drug Disc. Today* **1998**, *3*, 370–378.
148. Gygi, S. P.; Rist, B.; Gerber, S. A.; Turecek, F.; Gelb, M. H.; Aebersold, R. *Nat. Biotechnol.* **1999**, *17*, 994–999.
149. Bieri, C.; Ernst, O. P.; Heyse, S.; Hofmann, K. P.; Vogel, H. *Nat. Biotechnol.* **1999**, *17*, 1105–1108.
150. Chakravarti, A. *Nat. Genet.* **1999**, *21*, 56–60.
151. Fikes, B. J. *Bioventure View.* **1999**, *14*, 1–4.
152. Taillon-Miller, P.; Gu, Z.; Li, Q.; Hillier, L.; Kwok, P. Y. *Genome Res.* **1998**, *8*, 748–754.
153. Wang, D. G.; Fan, J. B.; Siao, C. J.; Berno, A.; Young, P.; Sapolsky, R.; Ghandour, G.; Perkins, N.; Winchester, E.; Spencer, J. et al. *Science* **1998**, *280*, 1077–1082.
154. Nickerson, D. A.; Taylor, S. L.; Weiss, K. M.; Clark, A. G.; Hutchinson, R. G.; Stengard, J.; Salomaa, V.; Vartiainen, E.; Boerwinkle, E.; Sing, C. F. *Nat. Genet.* **1998**, *19*, 233–240.
155. Chakravarti, A. *Nat. Genet.* **1998**, *19*, 216–2177.
156. Housman, D.; Ledley, F. D. *Nat. Biotechnol.* **1998**, *16*, 492–493.
157. Sherry, S. T.; Ward, M.; Sirotkin, K. *Genome Res.* **1999**, *9*, 677–679.
158. Lipshutz, R. J.; Fodor, S. P.; Gingeras, T. R.; Lockhart, D. J. *Nat. Genet.* **1999**, *21*, 20–24.
159. Steemers, F. J.; Ferguson, J. A.; Walt, D. R. *Nat. Biotechnol.* **2000**, *18*, 91–94.
160. Bowtell, D. D. *Nat. Genet.* **1999**, *21*, 25–32.
161. Chervitz, S. A.; Aravind, L.; Sherlock, G.; Ball, C. A.; Koonin, E. V.; Dwight, S. S.; Harris, M. A.; Dolinski, K.; Mohr, S.; Smith, T. et al. *Science* **1998**, *282*, 2022–2028.
162. Burley, S. K.; Almo, S. C.; Bonanno, J. B.; Capel, M.; Chance, M. R.; Gaasterland, T.; Lin, D.; Sali, A.; Studier, F. W.; Swaminathan, S. *Nat. Genet.* **1999**, *23*, 151–157.
163. Lenski, R. E.; Ofria, C.; Collier, T. C.; Adami, C. *Nature* **1999**, *400*, 661–664.
164. Editorial. *Nat. Genet.* **1999**, 249–252.
165. Pennisi, E. *Science* **1998**, *280*, 1692–1693.
166. Shaw, I. *Pharm. Sci. Technol. Today* **1999**, *2*, 345–347.
167. Gelbart, W. M. *Science* **1998**, *282*, 659–661.
168. Hieter, P.; Boguski, M. *Science* **1997**, *278*, 601–602.
169. Drews, J. *Science* **2000**, *287*, 1960–1964.
170. Evans, W. E.; Relling, M. V. *Science* **1999**, *286*, 487–491.
171. Desai, A.; Mitchison, T. J. *Bioessays* **1998**, *20*, 523–527.
172. Lockless, S. W.; Ranganathan, R. *Science* **1999**, *286*, 295–299.
173. Hamburger, Z. A.; Brown, M. S.; Isberg, R. R.; Bjorkman, P. J. *Science* **1999**, *286*, 291–295.
174. Tatusov, R. L.; Koonin, E. V.; Lipman, D. J. *Science* **1997**, *278*, 631–637.
175. Green, P.; Lipman, D.; Hillier, L.; Waterston, R.; States, D.; Claverie, J. M. *Science* **1993**, *259*, 1711–1716.
176. Courseaux, A.; Nahon, J. L. *Science* **2001**, *291*, 1293–1297.
177. Tupler, R.; Perini, G.; Green, M. R. *Nature* **2001**, *409*, 832–833.
178. Weinstein, J. N.; Myers, T. G.; O'Connor, P. M.; Friend, S. H.; Fornace, A. J.; Kohn, K. W.; Fojo, T.; Bates, S. E.; Rubinstein, L. V.; Anderson, N. L. et al. *Science* **1997**, *275*, 343–349.
179. Caron, H.; Schaik Bv, B.; Mee Mv, M.; Baas, F.; Riggins, G.; Sluis Pv, P.; Hermus, M. C.; Asperen Rv, R.; Boon, K.; Voûte, P. A. et al. *Science* **2001**, *291*, 1289–1292.
180. Lander, E. S.; Linton, L. M.; Birren, B.; Nusbaum, C.; Zody, M. C.; Baldwin, J.; Devon, K.; Dewar, K.; Doyle, M.; FitzHugh, W. et al. *Nature* **2001**, *409*, 860–921.
181. Davies, D. E.; Djukanovic, R.; Holgate, S. T. *Thorax* **1999**, *54*, 79–81.
182. Patthy, L. *Cell* **1985**, *41*, 657–663.
183. Aravind, L.; Dixit, V. M.; Koonin, E. V. *Science* **2001**, *291*, 1279–1284.
184. Clayton, R. A.; White, O.; Ketchum, K. A.; Venter, J. C. *Nature* **1997**, *387*, 459–462.
185. Nicholas, H. B.; Deerfield, D. W.; Ropelewski, A. J. *Biotechniques* **2000** *28* 1174–1178, *1180*, 1182.
186. Friend, S. H.; Oliff, A. *N. Engl. J. Med.* **1998**, *338*, 125–126.
187. Henikoff, S.; Greene, E. A.; Pietrokovski, S.; Bork, P.; Attwood, T. K.; Hood, L. *Science* **1997**, *278*, 5338.
188. Arnone, M. I.; Davidson, E. H. *Development* **1997**, *124*, 1851–1864.
189. Heller, R. A.; Schena, M.; Chai, A.; Shalon, D.; Bedilion, T.; Gilmore, J.; Woolley, D. E.; Davis, R. W. *Proc. Natl. Acad. Sci. USA* **1997**, *94*, 2150–2155.
190. Marx, J. *Science* **2000**, *287*, 2390.
191. Schena, M.; Shalon, D.; Davis, R. W.; Brown, P. O. *Science* **1995**, *270*, 467–470.
192. Lipshutz, R. J.; Morris, D.; Chee, M.; Hubbell, E.; Kozal, M. J.; Shah, N.; Shen, N.; Yang, R.; Fodor, S. P. *Biotechniques* **1995**, *19*, 442–447.
193. Arnott, D.; O'Connell, K. L.; King, K. L.; Stults, J. T. *Anal. Biochem.* **1998**, *258*, 1–18.
194. Shevchenko, A.; Jensen, O. N.; Podtelejnikov, A. V.; Sagliocco, F.; Wilm, M.; Vorm, O.; Mortensen, P.; Shevchenko, A.; Boucherie, H.; Mann, M. *Proc. Natl. Acad. Sci. USA* **1996**, *93*, 14440–14445.

195. Roepstorff, P. *Curr. Opin. Biotechnol.* **1997**, *8*, 6–13.
196. Wilkins, M. R.; Williams, K. L.; Appel, R. D.; Hochstrasser, D. F. *Proteome Research: New Frontiers in Functional Genomics*; Springer-Verlag: Berlin, Germany, 1997.
197. Celis, J. E.; Oestergaard, M.; Rasmussen, H. H.; Gromov, P.; Gromova, I.; Varmark, H.; Palsdottir, H.; Magnusson, N.; Andersen, I.; Basse, B. et al. *Electrophoresis* **1999**, *20*, 300–309.
198. Dunn, M. J. *Biochem. Soc. Trans.* **1997**, *25*, 248–254.
199. Humphery-Smith, I.; Cordwell, S. J.; Blackstock, W. P. *Electrophoresis* **1997**, *18*, 1217–1242.
200. Wasinger, V. C.; Cordwell, S. J.; Cerpa-Poljak, A.; Yan, J. X.; Gooley, A. A.; Wilkins, M. R.; Duncan, M. W.; Harris, R.; Williams, K. L.; Humphery-Smith, I. *Electrophoresis* **1995**, *16*, 1090–1099.
201. O'Farrell, P. H. *J. Biol. Chem.* **1975**, *250*, 4007–4021.
202. Ingram, L.; Tombs, M. P.; Hurst, A. *Anal. Biochem.* **1967**, *20*, 24–29.
203. Weber, K.; Osborn, M. *J. Biol. Chem.* **1969**, *244*, 4406–4412.
204. Klose, J.; Kobalz, U. *Electrophoresis* **1995**, *16*, 1034–1059.
205. Lopez, M. F.; Patton, W. F. *Electrophoresis* **1997**, *18*, 338–343.
206. Yamada, Y. *J. Biochem. Biophys. Methods* **1983**, *8*, 175–181.
207. Bjellqvist, B.; Ek, K.; Righetti, P. G.; Gianazza, E.; Goerg, A.; Westermeier, R.; Postel, W. *J. Biochem. Biophys. Methods* **1982**, *6*, 317–339.
208. Bjellqvist, B.; Sanchez, J. C.; Pasquali, C.; Ravier, F.; Paquet, N.; Frutiger, S.; Hughes, G. J.; Hochstrasser, D. *Electrophoresis* **1993**, *14*, 1375–1378.
209. Gorg, A. IPG-D_{alt} of Very Alkaline Proteins. In *Methods in Molecular Biology: 2D Proteome Analysis Protocols*; Link, A. J., Ed.; Humana Press: Totowa, NJ, 1999; Vol. 3, pp 197–209.
210. Rabilloud, T. *Electrophoresis* **1996**, *17*, 813–829.
211. Link, A. J., Eds. *Methods in Molecular Biology: 2D Proteome Analysis Protocols*. Humana Press: Totowa, NJ, 1998; Vol. 112.
212. Laemmli, U. K. *Nature* **1970**, *227*, 680–685.
213. Wirth, P. J.; Romano, A. *J. Chromatogr. A* **1995**, *698*, 123–143.
214. Meyer, T. S.; Lamberts, B. L. *Biochim. Biophys. Acta* **1965**, *107*, 144–145.
215. Goerg, A.; Postel, W.; Guenther, S. *Electrophoresis* **1988**, *9*, 531–546.
216. Switzer, R. C., III; Merril, C. R.; Shifrin, S. *Anal. Biochem.* **1979**, *98*, 231–237.
217. Rabilloud, T. Silver staining of 2D electrophoresis gels. In ; Humana Press: Totowa, NJ, 1999; Vol. 112, pp 297–305.
218. Rabilloud, T. *Electrophoresis* **1992**, *13*, 429–439.
219. Patras, G.; Qiao, G. G.; Solomon, D. H. *Electrophoresis* **1999**, *20*, 2039–2045.
220. Hochstrasser, D. F.; Patchornik, A.; Merril, C. R. *Anal. Biochem.* **1988**, *173*, 412–423.
221. Giometti, C. S.; Gemmell, M. A.; Tollaksen, S. L.; Taylor, J. *Electrophoresis* **1991**, *12*, 536–543.
222. Gharahdaghi, F.; Weinberg, C. R.; Meagher, D. A.; Imai, B. S.; Mische, S. M. *Electrophoresis* **1999**, *20*, 601–605.
223. Steinberg, T. H.; Haugland, R. P.; Singer, V. L. *Anal. Biochem.* **1996**, *239*, 238–245.
224. Unlu, M.; Morgan, M. E.; Minden, J. S. *Electrophoresis* **1997**, *18*, 2071–2077.
225. Johnston, R. F.; Pickett, S. C.; Barker, D. L. *Electrophoresis* **1990**, *11*, 355–360.
226. Towbin, H.; Staehelin, T.; Gordon, J. *Proc. Natl. Acad. Sci. USA* **1979**, *76*, 4350–4354.
227. Durrant, I.; Fowler, S. Chemiluminescent Detection Systems for Protein Blotting. In *Protein Blotting: A Practical Approach*; Dunbar, B. S., Ed.; IRL Press: Oxford, UK, 1994, pp 141–152.
228. Watkins, C.; Sadun, A.; Marenka, S. *Modern Image Processing: Warping, Morphing and Classical Techniques*; Academic Press: San Diego, CA, 1993.
229. Sutherland, J. C. *Adv. Electrophor.* **1993**, *6*, 1–42.
230. Garrels, J. I.; Farrar, J. T.; Burwell, C. B., IV. The QUEST System for Computer-Analyzed Two-Dimensional Electrophoresis of Proteins. In *Two-Dimensional Gel Electrophoresis of Proteins: Methods and Applications*; Celis, J. E., Bravo, R., Eds.; Academic Press: San Diego, CA, 1984, pp 37–91.
231. Kraemer, E. T.; Ferrin, T. E. *Bioinformatics* **1998**, *14*, 764–771.
232. Tufte, E. R. *The Visual Display of Quantitative Information*; Graphic Press: Cheshire, CN, 1983.
233. Gooley, A. A.; Ou, K.; Russell, J.; Wilkins, M. R.; Sanchez, J. C.; Hochstrasser, D. F.; Williams, K. L. *Electrophoresis* **1997**, *18*, 1068–1072.
234. Wilkins, M. R.; Ou, K.; Appel, R. D.; Sanchez, J.-C.; Yan, J. X.; Golaz, O.; Farnsworth, V.; Cartier, P.; Hochstrasser, D. F.; Williams, K. L. et al. *Biochem. Biophys. Res. Commun.* **1996**, *221*, 609–613.
235. Shevchenko, A.; Wilm, M.; Vorm, O.; Mann, M. *Anal. Chem.* **1996**, *68*, 850–858.
236. Houthaeve, T.; Gausepohl, H.; Ashman, K.; Nillson, T.; Mann, M. *J. Protein Chem.* **1997**, *16*, 343–348.
237. Ashman, K.; Houthaeve, T.; Clayton, J.; Wilm, M.; Podtelejnikov, A.; Jensen, O. N.; Mann, M. *Lett. Pept. Sci.* **1997**, *4*, 57–65.
238. Ashman, K. *Am. Biotechnol. Lab.* **1999**, *17*, 92–93.
239. Karas, M.; Bahr, U.; Ingendoh, A.; Nordhoff, E.; Stahl, B.; Strupat, K.; Hillenkamp, F. *Anal. Chim. Acta* **1990**, *241*, 175–185.
240. Fenn, J. B.; Mann, M.; Meng, C. K.; Wong, S. F.; Whitehouse, C. M. *Science* **1989**, *246*, 64–71.
241. Cotter, R. J. *Anal. Chem.* **1999**, *71*, 445A–451A.
242. James, P. *Biochem. Biophys. Res. Commun.* **1997**, *231*, 1–6.
243. Yates, J. R., III. *J. Mass Spectrom.* **1998**, *33*, 1–19.
244. Jonscher, K. R.; Yates, J. R., III. *Anal. Biochem.* **1997**, *244*, 1–15.
245. Dongre, A. R.; Eng, J. K.; Yates, J. R., III. *Trends Biotechnol.* **1997**, *15*, 418–425.
246. McLuckey, S. A.; Van Berkel, G. J.; Goeringer, D. E.; Glish, G. L. *Anal. Chem.* **1994**, *66*, 737A–743A.
247. Clauser, K. R.; Baker, P.; Burlingame, A. L. *Anal. Chem.* **1999**, *71*, 2871–2882.
248. Opiteck, G. J.; Ramirez, S. M.; Jorgenson, J. W.; Moseley, M. A., III. *Anal. Biochem.* **1998**, *258*, 349–361.
249. Camilleri, P. *Capillary Electrophoresis: Theory and Practice*, 2nd ed.; CRC Press: Boca Raton, FL, 1998.
250. Ding, J.; Vouros, P. *Anal. Chem.* **1999**, *71*, 378A–385A.
251. Yang, Q.; Tomlinson, A. J.; Naylor, S. *Anal. Chem.* **1999**, *71*, 183A–189A.
252. Yang, L.; Lee, C. S.; Hofstadler, S. A.; Pasa-Tolic, L.; Smith, R. D. *Anal. Chem.* **1998**, *70*, 3235–3241.
253. Link, A. J.; Carmack, E.; Yates, J. R., III. *Int. J. Mass Spectrom. Ion. Proc.* **1997**, *160*, 303–316.
254. Oda, Y.; Huang, K.; Cross, F. R.; Cowburn, D.; Chait, B. T. *Proc. Natl. Acad. Sci. USA* **1999**, *96*, 6591–6596.
255. Pasa-Tolic, L.; Jensen, P. K.; Anderson, G. A.; Lipton, M. S.; Peden, K. K.; Martinovic, S.; Tolic, N.; Bruce, J. E.; Smith, R. D. *J. Am. Chem. Soc.* **1999**, *121*, 7949–7950.

256. Gygi, S. P.; Rist, B.; Gerber, S. A.; Turecek, F.; Gelb, M. H.; Aebersold, R. *Nat. Biotechnol.* **1999**, *17*, 994–999.
257. Petersen, K. E. *Proc. IEEE* **1982**, *70*, 420–457.
258. Licklider, L.; Wang, X.-Q.; Desai, A.; Tai, Y.-C.; Lee, T. D. *Anal. Chem.* **2000**, *72*, 367–375.
259. Madou, M. *Fundamentals of Microfabrication*; CRC Press: Boca Raton, FL, 1997.
260. Kovacs, G. T. A. *Micromachined Transducers Sourcebook*; WCB/McGraw-Hill: New York, 1998.
261. Campbell, S. A. *The Science and Engineering of Microelectronic Fabrication*; Oxford University Press: Oxford, UK, 1996.
262. Williams, K. R.; Muller, R. S. *J. Microelectromech. Syst.* **1996**, *5*, 256–269.
263. Shoji, S.; van der Schoot, B.; de Rooij, N.; Esahi, M. *Smallest Dead Volume Microvalves for Integrated Chemical Analyzing Systems*; Proceedings of the 1991 International Conference on Solid-State Sensors and Actuators, San Francisco, CA, June 24–27, 1991, pp 1052–1055.
264. Desai, A.; Tai, Y. C.; Davis, M. T.; Lee, T. D. *A MEMS Electrospray Nozzle for Mass Spectrometry*; Proceedings of the 1997 International Conference on Solid-State Sensors and Actuators, Chicago, IL, June 16–19, 1997, pp 927–930.
265. Becker, E. W.; Ehrfeld, W.; Muchmeyer, D.; Betz, H.; Heuberger, A.; Pongratz, S.; Glashauser, W.; Michel, H. J.; Siemens, V. R. *Nature* **1982**, *69*, 520–523.
266. Schomburg, W. K.; Fahrenberg, J.; Maas, D.; Rapp, R. *J. Micromech. Microeng.* **1993**, *3*, 216–218.
267. Rapp, R.; Schomburg, W. K.; Maas, D.; Schulz, J. W. *Sensors Actuators A* **1994**, *40*, 57061.
268. Vollmer, J.; Hein, H.; Menz, W.; Walter, F. *Sensors Actuators A* **1994**, *43*, 330–334.
269. Ehrfeld, W.; Einhaus, R.; Munchmeyer, D.; Strathmann, H. *J. Membr. Sci.* **1988**, *36*, 67–77.
270. Uhlir, A. *Bell. Syst. Tech. J.* **1956**, *35*, 333–347.
271. Service, R. *Science* **1997**, *278*, 806.
272. Mourlas, N. J.; Jaeggi, D.; Maluf, N. I.; Kovacs, G. T. A. *Reusable Microfluidic Coupler with PDMS Gasket*; Proceedings of 10th International Conference on Solid-State Sensors and Actuators, June 7–10, 1999, pp 1888–1889.
273. Bings, N. H.; Wang, C.; Skinner, C. D.; Colyer, C. L.; Thibault, P.; Harrison, D. J. *Anal. Chem.* **1999**, *71*, 3292–3296.
274. Walt, D. R. *Acc. Chem. Res.* **1998**, *31*, 267–278.
275. Taylor, L. C.; Walt, D. R. *Anal. Biochem.* **2000**, *278*, 132–142.
276. Walt, D. R. *Science* **2000**, *287*, 152–154.
277. Michael, K. L.; Taylor, L. C.; Schultz, S. L.; Walt, D. R. *Anal. Chem.* **1998**, *70*, 1242.
278. Illumina. http://www.illumina.com (accessed Aug 2006).
279. Fodor, S.; Read, J. L.; Pirrung, M. C.; Stryer, L.; Tsai, L. A.; Solas, D. *Science* **1991**, *251*, 767–773.
280. Sampsell, J. B. *J. Vac. Sci. Technol. B* **1994**, *12*, 3242–3246.
281. Singh-Gasson, S.; Green, R. D.; Yue, Y.; Nelson, C.; Blattner, F.; Sussman, M. R.; Cerrina, F. *Nat. Biotechnol.* **1999**, *17*, 974–978.
282. Jackman, R. J.; Duffy, D. C.; Ostuni, E.; Willmore, N. D.; Whitesides, G. M. *Anal. Chem.* **1998**, *70*, 2280–2287.
283. Drmanac, R.; Drmanac, S.; Strezoska, Z.; Paunesku, T.; Labat, I.; Zeremski, M.; Snoddy, J.; Funkhouser, W. K.; Koop, B.; Hood, L. et al. *Science* **1993**, *260*, 1649–1652.
284. Drmanac, R.; Labat, I.; Brukner, I.; Crkvenjakov, R. *Genomics* **1989**, *4*, 114–128.
285. Cheng, J.; Sheldon, E. L.; Wu, L.; Uribe, A.; Gerrue, L. O.; Carrino, J.; Heller, M. J.; O'Connell, J. P. *Nat. Biotechnol.* **1998**, *16*, 541–546.
286. Cheng, J.; Sheldon, E. L.; Wu, L.; Heller, M. J.; O'Connell, J. P. *Anal. Chem.* **1998**, *70*, 2321–2326.
287. Jacobson, S. C.; Ramsey, J. M. *Anal. Chem.* **1997**, *69*, 3212–3217.
288. Jacobson, S. C.; McKnight, T. E.; Ramsey, J. M. *Anal. Chem.* **1999**, *71*, 4455–4459.
289. Schrum, D. P.; Culbertson, C. T.; Jacobson, S. C.; Ramsey, J. M. *Anal. Chem.* **1999**, *71*, 4173–4177.
290. Belmont-Hebert, C.; Tercier, M. L.; Buffle, J.; Fiaccabrino, G. C.; deRooij, N. F.; Koudelka-Hep, M. *Anal. Chem.* **1998**, *70*, 2949–2956.
291. von Heeren, F.; Verpoorte, E.; Manz, A.; Thormann, W. *Anal. Chem.* **1996**, *68*, 2044–2053.
292. Oleschuk, R. D.; Schultz-Lockyear, L. L.; Ning, Y.; Harrison, J. D. *Anal. Chem.* **2000**, *72*, 585–590.
293. Li, J.; Kelly, J. F.; Chernushevich, I.; Harrison, D. J.; Thibault, P. *Anal. Chem.* **2000**, *72*, 599–609.
294. Manz, A.; Harrison, D. J.; Verpoorte, E. M. J.; Fettinger, J. C.; Paulus, A.; Lüdi, H.; Widmer, H. M. *J. Chromatogr.* **1992**, *593*, 253–258.
295. McCormick, R. M.; Nelson, R. J.; Alonso-Amigo, M. G.; Benvegnu, D. J.; Hooper, H. H. *Anal. Chem.* **1997**, *69*, 2626–2630.
296. Effenhauser, C. S.; Bruin, G. J. M.; Paulus, A.; Ehrat, M. *Anal. Chem.* **1997**, *69*, 3451–3457.
297. Orchid Cellmark. http://www.orchidbio.com (accessed Aug 2006).
298. Micronics.com. http://www.micronics.com (accessed Aug 2006).
299. Nyrén, P.; Lundin, A. *Anal. Biochem.* **1985**, *151*, 504–509.
300. Nordström, T.; Nourizad, K.; Ronaghi, M.; Nyren, P. *Anal. Biochem.* **2000**, *282*, 186–193.
301. World Technology Evaluation Center. http://itri.Loyola.edu (accessed Aug 2006).
302. Schaffer, A. J.; Hawkins, J. R. *Nat. Biotechnol.* **1998**, *16*, 33–39.
303. Brookes, A. J. *Gene* **1999**, *234*, 177–186.
304. Ghosh, S.; Collins, F. S. *Annu. Rev. Med.* **1996**, *47*, 333–353.
305. Terwilliger, J. D.; Weiss, K. M. *Curr. Opin. Biotechnol.* **1998**, *9*, 578–594.
306. Weinshilboum, R. M.; Otterness, D. M.; Szumlanski, C. L. *Annu. Rev. Pharmacol. Toxicol.* **1999**, *39*, 19–52.
307. Evans, W. E.; Relling, M. V. *Science* **1999**, *286*, 487–491.
308. Picoult-Newberg, L.; Ideker, T. E.; Pohl, M. G.; Taylor, S. L.; Donaldson, M. A.; Nickerson, D. A.; Boyce-Jacino, M. *Genome Res.* **1999**, *9*, 167–174.
309. Gu, Z. J.; Hillier, L.; Kwok, P. Y. *Hum. Mutat.* **1998**, *12*, 221–225.
310. Rieder, M. J.; Taylor, S. L.; Tobe, V. O.; Nickerson, D. A. *Nucleic Acids Res.* **1998**, *26*, 967–973.
311. Kwok, P. Y.; Deng, Q.; Zakeri, H.; Taylor, S. L.; Nickerson, D. A. *Genomics* **1996**, *31*, 123–126.
312. Collins, F. S.; Patrinos, A.; Jordan, E.; Chakravarti, A.; Gesteland, R.; Walters, L. *Science* **1998**, *282*, 682–689.
313. Collins, F. S.; Brooks, L. D.; Chakravarti, A. *Genome Res.* **1998**, *8*, 1229–1231.
314. Cotton, R. *Trends Genet.* **1997**, *13*, 43–46.
315. Landegren, U.; Nilsson, M.; Kwok, P. Y. *Genome Res.* **1998**, *8*, 769–776.
316. Wang, D. G.; Fan, J. B.; Siao, C. J.; Berno, A.; Young, P.; Sapolsky, R.; Ghandour, G.; Perkins, N.; Winchester, E.; Spencer, J. et al. *Science* **1998**, *280*, 1077–1082.
317. Nielsen, P. E. *Curr. Opin. Biotechnol.* **1999**, *10*, 71–75.
318. Uhlmann, E.; Peyman, A.; Breipohl, G.; Will, D. W. *Angew. Chem. Int. Ed.* **1998**, *37*, 2797–2823.

319. Newton, C. R.; Graham, A.; Heptinstall, L. E.; Powell, S. J.; Summers, C.; Kalsheker, N.; Smith, J. C.; Markham, A. F. *Nucleic Acids Res.* **1989**, *17*, 2503–2516.
320. Sarkar, G.; Cassady, J.; Bottema, C. D. K.; Sommer, S. S. *Anal. Biochem.* **1990**, *186*, 64–68.
321. Patrushev, L. I.; Zykova, E. S.; Kayushin, A. L.; Korosteleva, M. D.; Miroshnikov, A. I.; Bokarew, I. N.; Leont'ev, S. G.; Koshkin, V. M.; Severin, E. S. *Thrombosis Res.* **1998**, *92*, 251–259.
322. Bunce, M.; Fanning, G. C.; Welsh, K. I. *Tissue Antigens* **1995**, *45*, 81–90.
323. Fullerton, S. M.; Buchanan, A. V.; Sonpar, V. A.; Taylor, S. L.; Smith, J. D.; Carlson, C. S.; Salomaa, V.; Stengard, J. H.; Boerwinkle, E.; Clark, A. G. et al. *Am. J. Hum. Genet.* **63**, 595–612.
324. Pearson, S. L.; Hessner, M. J. A. *Br. J. Haematol.* **1998**, *100*, 229–234.
325. Pastinen, T.; Raitio, M.; Lindroos, K.; Tainola, P.; Peltonen, L.; Syvanen, A. C. *Genome Res.* **2000**, *10*, 1031–1042.
326. Tonisson, N.; Kurg, A.; Kaasik, K.; Lohmussaar, E.; Metspalu, A. *Clin. Chem. Lab. Med.* **2000**, *38*, 165–170.
327. Syvanen, A. C.; Aaltosetala, K.; Harju, L.; Kontula, K.; Soderlund, H. *Genomics* **1990**, *8*, 684–692.
328. Syvanen, A. C. *Hum. Mutat.* **1999**, *13*, 1–10.
329. Tobe, V. O.; Taylor, S. L.; Nickerson, D. A. *Nucleic Acids Res.* **1996**, *24*, 3728–3732.
330. Barany, F. *Proc. Nat. Acad. Sci. USA* **1991**, *88*, 189–193.
331. Lizardi, P. M.; Huang, X.; Zhu, Z.; Bray-Ward, P.; Thomas, D. C.; Ward, D. C. *Nat. Genet.* **1998** *19*, 225–232.
332. Baner, J.; Nilsson, M.; Mendelhartvig, M.; Landegren, U. *Nucleic Acids Res.* **1998**, *26*, 5073–5078.
333. Livak, K.; Marmaro, J.; Todd, J. A. *Nat. Genet.* **1995**, *9*, 341–342.
334. Lyamichev, V.; Mast, A. L.; Hall, J. G.; Prudent, J. R.; Kaiser, M. W.; Takova, T.; Kwiatkowski, R. W.; Sander, T. J.; de Arruda, M.; Arco, D. A. et al. *Nat. Biotechnol.* **1999**, *17*, 292–296.
335. Pastinen, T.; Partanen, J.; Syvanen, A. C. *Clin. Chem.* **1996**, *42*, 1391–1397.
336. Tully, G.; Sullivan, K. M.; Nixon, P.; Stones, R. E.; Gill, P. *Genomics* **1996**, *34*, 107–113.
337. Livak, K. J.; Hainer, J. W. *Hum. Mutat.* **1994**, *3*, 379–385.
338. Tyagi, S.; Bratu, D. P.; Kramer, F. R. *Nat. Biotechnol.* **1998**, *16*, 49–53.
339. Chen, X. N.; Livak, K. J.; Kwok, P. Y. *Genome Res.* **1998**, *8*, 549–556.
340. Chen, X. N.; Kwok, P. Y. *Gen. Anal.-Biomol. Eng.* **1999**, *14*, 157–163.
341. Laken, S. J.; Jackson, P. E.; Kinzler, K. W.; Vogelstein, B.; Strickland, P. T.; Groopman, J. D.; Friesen, M. D. *Nat. Biotechnol.* **1998**, *16*, 1352–1356.
342. Haff, L. A.; Smirnov, I. P. *Nucleic Acids Res.* **1997**, *25*, 3749–3750.
343. Fei, Z. D.; Ono, T.; Smith, L. M. *Nucleic Acids Res.* **1998**, *26*, 2827–2828.
344. Chen, X.; Fei, Z. D.; Smith, L. M.; Bradbury, E. M.; Majidi, V. *Anal. Chem.* **1999**, *71*, 3118–3125.
345. McKenzie, S. E.; Mansfield, E.; Rappaport, E.; Surrey, S.; Fortina, P. *Eur. J. Hum. Genet.* **1998**, *6*, 417–429.
346. Hacia, J. G.; Brody, L. C.; Collins, F. S. *Mol. Psychiatr.* **1998**, *3*, 483–492.
347. Kettman, J. R.; Davies, T.; Chandler, D.; Oliver, K. G.; Fulton, R. J. *Cytometry* **1998**, *33*, 234–243.
348. Gerry, N. P.; Witkowski, N. E.; Day, J.; Hammer, R. P.; Barany, G.; Barany, F. *J. Mol. Biol.* **1999**, *292*, 251–262.
349. Cai, H.; White, P. S.; Torney, D.; Deshpande, A.; Wang, Z.; Keller, R. A.; Marrone, B.; Nolan, J. P. *Genomics* **2000**, *68*, 135–143.
350. Edwards, B. S.; Kuckuck, F.; Sklar, L. A. *Cytometry* **1999**, *37*, 156–159.
351. Yang, G.; Olson, J. C.; Pu, R.; Vyas, G. N. *Cytometry* **1995**, *21*, 197–202.
352. Fulton, R. J.; McDade, R. L.; Smith, P. L.; Kienker, L. J.; Kettman, J. R. *Clin. Chem.* **1997**, *43*, 1749–1756.
353. Iannone, M. A.; Taylor, J. D.; Chen, J.; Li, M. S.; Rivers, P.; Slentz-Kesler, K. A.; Weiner, M. P. *Cytometry* **2000**, *39*, 131–140.
354. Chen, J.; Iannone, M. A.; Li, M. S.; Taylor, J. D.; Rivers, P.; Nelsen, A. J.; Slentz-Kesler, K. A.; Roses, A.; Weiner, M. P. *Genome Res.* **2000**, *10*, 549–557.
355. Cullen, D. C.; Brown, R. G. W.; Rowe, C. R. *Biosensors* **198788**, *3*, 211–225.
356. Wood, R. W. *Phil. Mag.* **1902**, *4*, 396–402.
357. Ritchie, R. H. *Phys. Rev.* **1957**, *106*, 874–881.
358. Neviere, M. The Homogeneous Problem. In *Electromagnetic Theory of Gratings*; Petit, R., Ed.; Springer-Verlag: Berlin, 1980, pp 123–157.
359. Maystre, D. General Study of Grating Anomalies from Electromagnetic Surface Modes. In *Electromagnetic Surface Modes*; Boardman, A. D., Ed.; John Wiley: Chichester, UK, 1982, pp 661–724.
360. Leatherbarrow, R. J.; Edwards, P. R. *Curr. Opin. Biotechnol.* **1999**, *3*, 544–547.
361. Malmquist, M. *Biochem. Soc. Trans.* **1999**, *27*, 335–340.
362. Lowe, P.; Clark, T.; Davies, R.; Edwards, P.; Kinning, T.; Yang, D. *J. Mol. Recognit.* **1998**, *11*, 194–199.
363. Kurrat, R.; Prenosil, J.; Ramsden, J. *J. Colloid. Interf. Sci.* **1997**, *185*, 1–8.
364. Johnsson, B.; Löfås, S.; Lindquist, G. *Anal. Biochem.* **1991**, *198*, 268–277.
365. Martin, J.; Langer, T.; Boteva, R.; Schramel, A.; Horwich, A. L.; Hartle, F. U. *Nature* **1991**, *352*, 36–42.
366. Cuatrecasas, P.; Praikh, I. *Biochemistry* **1972**, *12*, 2291–2299.
367. Raghavan, M.; Chen, M. Y.; Gastinel, L. N.; Bjorkman, P. J. *Immunity* **1994**, *1*, 303–315.
368. Corr, M.; Slanetz, A. E.; Boyd, L. F.; Jelonek, M. T.; Khilko, S.; Al-Ramadi, B. K.; Kim, Y. S.; Maher, S. E.; Bothwell, A. L. M.; Margulies, D. H. *Science* **1994**, *265*, 946–949.
369. Laminet, A. A.; Ziegelhoffer, T.; Georgopoulos, C.; Plückthun, A. *EMBO J.* **1990**, *9*, 2315–2319.
370. Zahn, R.; Axmann, S. E.; Rycknagel, K.-P.; Jaeger, E.; Laminet, A. A.; Plückthun, A. *J. Mol. Biol.* **1994**, *242*, 150–164.
371. Lin, Z.; Eisenstein, E. *Proc. Natl. Acad. Sci. USA* **1996**, *93*, 1977–1981.
372. Kragten, E.; Lalande, I.; Zimmermann, K.; Roggo, S.; Schindler, P.; Müller, P.; van Ostrum, J.; Waldmeier, P.; Fürst, P. *J. Biol. Chem.* **1998**, *273*, 5821–5828.
373. MaKenzie, C. R.; Hirama, T.; Deng, S.-J.; Bundle, D. R.; Narang, S. A.; Young, N. M. *J. Biol. Chem.* **1996**, *271*, 1527–1533.
374. Fisher, R. J.; Fivash, M.; Casas-Finet, J.; Erickson, J. W.; Kondoh, A.; Bladen, S. V.; Fisher, C.; Watson, D. K.; Papas, T. *Protein Sci.* **1994**, *3*, 257–266.
375. Spanopoulou, E.; Zaitseva, F.; Wang, F.-H.; Santagata, S.; Baltimore, D.; Panayotou, G. *Cell* **1996**, *87*, 263–276.
376. Haruki, M.; Noguchi, E.; Kanaya, S.; Crouch, R. J. *J. Biol. Chem.* **1997**, *272*, 22015–22022.
377. Lea, S. M.; Poewll, R. M.; McKee, T.; Evans, D. J.; Brown, D.; Stuart, D. I.; van der Merve, P. A. *J. Biol. Chem.* **1998**, *273*, 30443–30447.
378. Gill, A.; Bracewell, D. G.; Maule, C. H.; Lowe, P. A.; Hoare, M. *J. Biotechnol.* **1998**, *65*, 69–80.

379. Borrebaeck, C. A. K.; Malmborg, A.-C.; Furebring, C.; Michaelsson, A.; Ward, S.; Danielsson, L.; Ohlin, M. *Biotechnology* **1992**, *10*, 697–698.
380. Altschuh, D.; Dubs, M.-C.; Weiss, E.; Zeder-Lutz, G.; Van Regenmortel, M. H. V. *Biochemistry* **1992**, *31*, 6298–6304.
381. O'Shannessy, D. J.; Brigham, B. M.; Soneson, K. K.; Hensley, P.; Brooks, I. *Anal. Biochem.* **1993**, *212*, 457–468.
382. Cardone, M. H.; Roy, N.; Stennicke, H. R.; Salveson, G. S.; Franke, T. F.; Stanbridge, E.; Frisch, S.; Reed, J. C. *Science* **1998**, *282*, 1318–1321.
383. Sadir, R. S.; Forest, E.; Lortat-Jacob, H. *J. Biol. Chem.* **1998**, *273*, 10919–10925.
384. Ladbury, J. E.; Lemmon, M. A.; Zhou, M.; Green, J.; Botfield, M. C.; Schlessinger, J. *Proc. Natl. Acad. Sci. USA* **1995**, 3199–3203.
385. Thomas, C.; Surolia, A. *FEBS Lett.* **1999**, *445*, 420–424.
386. Stuart, J.; Myszka, D.; Joss, L.; Mitchell, R.; McDonald, S.; Xie, A.; Takayama, S.; Reed, J.; Ely, K. *J. Biol. Chem.* **1998**, *273*, 22506–22514.
387. Asensio, J.; Dosanjh, H.; Jenkins, T.; Lane, A. *Biochemistry* **1998**, *37*, 15188–15198.
388. Glaser, R. W. *Anal. Biochem.* **1993**, *213*, 152–161.
389. Schuck, P.; Minton, A. P. *Anal. Biochem.* **1996**, *240*, 262–272.
390. Schuck, P. *Biophys. J.* **1996**, *70*, 1230–1249.
391. Edwards, P.; Lowe, P.; Leatherbarrow, R. *J. Mol. Recognit.* **1997**, *10*, 128–134.
392. Müller, K. M.; Arndt, K. M.; Plückthun, A. *Anal. Biochem.* **1998**, *261*, 149–158.
393. Yu, Y.-Y.; Van Wie, B. J.; Koch, A. R.; Moffett, D. F.; Davies, W. C. *Anal. Biochem.* **1998**, *263*, 158–168.
394. Myszka, D. G.; Jonsen, M. D.; Graves, B. J. *Anal. Biochem.* **1998**, *265*, 326–330.
395. Schuck, P.; Millar, D. B.; Kortt, A. A. *Anal. Biochem.* **1998**, *265*, 79–91.
396. O'Shanessey, D. J. *Curr. Opin. Biotechnol.* **1994**, *5*, 65–71.
397. Cotter, R. J. *Time-of-flight Mass Spectrometry: Instrumentation and Applications in Biological Research*; American Chemical Society: Washington, DC, 1997.
398. Tanaka, K.; Waki, H.; Ido, Y.; Akita, S.; Yoshida, Y.; Yoshida, T. *Rapid Commun. Mass Spectrom.* **1988**, *2*, 151–153.
399. Karas, M.; Hillenkamp, F. *Anal. Chem.* **1988**, *60*, 2299–2301.
400. Papac, D. I.; Hoyes, J.; Tomer, K. B. *Anal. Chem.* **1994**, *66*, 2609–2613.
401. Hutchens, T. W.; Yip, T.-T. *Rapid Commun. Mass Spectrom.* **1993**, 576–580.
402. Ching, J.; Viovodov, K. I.; Hutchens, T. W. *J. Org. Chem.* **1996**, *67*, 3582–3583.
403. Ching, J.; Viovodov, K. I.; Hutchens, T. W. *Bioconj. Chem.* **1996**, *7*, 525–528.
404. Viovodov, K. I.; Ching, J.; Hutchens, T. W. *Tetrahedron Lett.* **1996**, *37*, 5669–5672.
405. Stennicke, H. R.; Salveson, G. S.; Franke, T. F.; Stanbridge, E.; Frisch, S.; Reed, J. C. *Science* **1998**, *282*, 1318–1321.
406. Austen, B.; Davies, M.; Stephens, D. J.; Frears, E. R.; Walters, C. E. *NeuroReport* **1999**, *10*, 1699–1705.
407. Thiede, B.; Wittmann-Liebold, B.; Bienert, M.; Krause, E. *FEBS Lett.* **1995**, *357*, 65–69.
408. Nelson, R. W.; Jarvik, J. W.; Taillon, B. E.; Kemmons, A. T. *Anal. Chem.* **1999**, *71*, 2858–2865.
409. Sonksen, C. P.; Nordhoff, E.; Jasson, O.; Malmquist, M.; Roepstorff, P. *Anal. Chem.* **1998**, *70*, 2731–2736.
410. Nelson, R. W.; Krone, J. R. *J. Mol. Recognit.* **1999**, *12*, 77–93.
411. Gygi, S. P.; Han, D. K. M.; Gingras, A. C.; Sonnenberg, N.; Aebersold, R. *Electrophoresis* **1999**, *20*, 310–319.
412. Wasinger, V. C.; Cordwell, S. J.; Cerpa-Poljak, A.; Yan, J. X.; Gooley, A. A.; Wilkins, M. R.; Duncan, M. W.; Harris, R.; Williams, K. L.; Humphery-Smith, I. *Electrophoresis* **1995**, *16*, 1090–1099.
413. Goodfellow, P. *Nat. Genet.* **1997**, *16*, 209–210.
414. Dawson, E. *Mol. Med.* **1999**, *5*, 280.
415. Sidransky, D. *Science* **1997**, *278*, 1054–1058.
416. Gygi, S. P.; Rochon, Y.; Franza, B. R.; Aebersold, R. *Mol. Cell. Biol.* **1999**, *3*, 1720–1730.
417. Parekh, R. P.; Lyall, A. *J. Commercial Biotechnol.* **2006**, in press.
418. Zong, Q.; Schummer, M.; Hood, L.; Morris, D. R. *Proc. Natl. Acad. Sci. USA* **1999**, *96*, 10632–10636.
419. Hunter, T. *Cell* **1997**, *88*, 333–346.
420. Hakomori, S. *Cancer Res.* **1996**, *56*, 5309–5318.
421. Sayed-Ahmed, N.; Besbas, N.; Mundy, J.; Muchaneta-Kubara, E.; Cope, G.; Pearson, C.; el Nahas, M. *Exp. Nephrol.* **1996**, *4*, 330–339.
422. Hasselgreen, P. O.; Fischer, J. E. *Ann. Surg.* **1997**, *225*, 307–316.
423. Beaulieu, M.; Brakier-Gingras, L.; Bouvier, M. *J. Mol. Cell. Cardiol.* **1997**, *29*, 111–119.
424. Makino, I.; Shibata, K.; Ohgami, Y.; Fujiwara, M.; Furukawa, T. *Neuropeptides* **1996**, *30*, 596–601.
425. Yao, G. L.; Kato, H.; Khalil, M.; Kiryu, S.; Kiyama, H. *Eur. J. Neurosci.* **1997**, *9*, 1047–1054.
426. Mansour, S. J.; Matten, W. T.; Hermann, A. S.; Candia, J. M.; Rong, S.; Fukasawa, K.; Vande Woude, G. F.; Ahn, N. G. *Science* **1994**, *265*, 966–970.
427. Woodburn, J. R.; Barker, A. J.; Gibson, K. H.; Ashton, S. E.; Wakeling, A. E.; Curry, B. J.; Scarlett, L.; Henthorn, L. R. *ZD1839, An Epidermal Growth Factor Tyrosine Kinase Inhibitor Selected For Clinical Development*; Proceedings of the American Association for Cancer Research, 1997; Vol. 38, p 633.
428. Shak, S. *Semin. Oncol.* **1999**, *4*, 71–77.
429. Gibbs, J. B.; Oliff, A. *Annu. Rev. Pharmacol. Toxicol.* **1997**, *37*, 143–166.
430. Monia, B. P.; Johnston, J. F.; Geiger, T.; Muller, M.; Fabbro, D. *Nat. Med.* **1996**, *2*, 668–675.
431. Glazer, R. I. *Antisense Nucleic Acid Drug Dev.* **1997**, *7*, 235–238.
432. Sebolt-Leopold, J. S.; Dudley, D. T.; Herrera, R.; Van Becelaere, K.; Wiland, A.; Gowan, R. C.; Tecle, H.; Barret, S. D.; Bridges, A.; Przybranowski, S. et al. *Nat. Med.* **1999**, *7*, 810–816.
433. Duesberry, N. S.; Webb, C. P.; Vande Woude, G. F. *Nat. Med.* **1999**, *7*, 736–737.
434. Alessandrini, A.; Namura, S.; Moskowitz, M. A.; Bonventre, J. V. *Proc. Natl. Acad. Sci. USA* **1999**, *96*, 12866–12869.
435. Milczarek, G. J.; Martinez, J.; Bowden, G. T. *Life Sci.* **1997**, *60*, 1–11.
436. Shaw, P.; Freeman, J.; Bovey, R.; Iggo, R. *Oncogene* **1996**, *12*, 921–930.
437. Maki, C. G. *J. Biol. Chem.* **1999**, *275*, 16531–16535.
438. Lukas, J.; Sørensen, C. S.; Lukas, C.; Santoni-Rugiu, E.; Bartek, J. *Oncogene* **1999**, *18*, 3930–3935.
439. Balint, E.; Bates, S.; Vousden, K. H. *Oncogene* **1999**, *18*, 3923–3930.
440. Robertson, K. D.; Jones, P. A. *Oncogene* **1999**, *18*, 3810–3820.
441. Taniguchi, K.; Kohsaka, H.; Inoue, N.; Terada, Y.; Ito, H.; Hirokawa, K.; Miyasaka, N. *Nat. Med.* **1999**, *7*, 760–767.

442. Kilpatrick, G. J.; Dautzenberg, F. M.; Martin, G. R.; Eglen, R. M. *Trends Pharmacol.* **1999**, *20*, 294–301.
443. Parekh, R.; Rohlff, C. *Curr. Opin. Biotechnol.* **1997**, *8*, 718–723.
444. Link, A. *2D Proteome Analysis Protocols: Methods in Molecular Biology*; Humana Press: Totowa, NJ, 1999; Vol. 112.
445. Page, M. P.; Amess, B.; Townsend, R. R.; Parekh, R.; Herath, A.; Brusten, L.; Zvelebil, M. J.; Stein, R. C.; Waterfield, M. D.; Davies, S. C. et al. *Proc. Natl. Acad. Sci. USA* **1999**, *96*, 12589–12594.
446. Rosenfeld, J.; Capdevielle, J.; Guillemot, J. C.; Ferrara, P. *Anal. Biochem.* **1992**, *203*, 173–179.
447. Shevchenko, A.; Wilm, M.; Vorm, O.; Mann, M. *Anal. Chem.* **1996**, *68*, 850–858.
448. Aebersold, R. H.; Leavitt, J.; Saavedra, R. A.; Hood, L. E.; Kent, S. B. H. *Proc. Natl. Acad. Sci. USA* **1987**, *84*, 6970–6974.
449. Hess, D.; Covey, T. C.; Winz, R.; Brownsey, R. W.; Aebersold, R. *Protein Sci.* **1993**, *2*, 1342–1351.
450. Van Oostveen, L.; Ducret, A.; Aebersold, R. *Anal. Biochem.* **1997**, *247*, 310–318.
451. Lui, M.; Tempst, P.; Erdjument-Bromage, H. *Anal. Biochem.* **1996**, *241*, 156–166.
452. Mann, M.; Wilm, M. *Anal. Chem.* **1994**, *66*, 4390–4399.
453. Courchesne, P. L.; Patterson, S. D. *Methods Mol. Biol.* **1999**, *112*, 487–511.
454. Haynes, P.; Miller, I.; Aebersold, R.; Gemeiner, M.; Eberini, I.; Lovati, R. M.; Manzoni, C.; Vignati, M.; Gianaza, E. *Electrophoresis* **1998**, *19*, 1484–1492.
455. Ducret, A.; Gu, M.; Haynes, P. A.; Yates, J. R., III; Aebersold, R. *Simple Design for a Capillary Liquid Chromatography-Microelectrospray-Tandem Mass Spectrometric System for Peptide Mapping at the Low Femtomole Sensitivity Range*; Proceedings of the Association of Biomolecular Resource Facilities (ABRF) International Symposium., Baltimore, MD, February 9–12, 1997, p 69.
456. Haynes, P. A.; Gyigi, S. P.; Figeys, D.; Aebersold, R. *Electrophoresis* **1998**, *19*, 1862–1871.
457. Gyigi, S. P.; Rist, B.; Gerber, S. A.; Turecek, F.; Gelb, M. H.; Aebersold, R. *Nat. Biotechnol.* **1999**, *10*, 994–999.
458. Emmert-Buck, M. R.; Bonner, R. F.; Smith, P. D.; Chuaqui, R. F.; Zhuang, Z.; Goldstein, S. R.; Weiss, R. A.; Liotta, L. *Science* **1996**, *274*, 998–1001.
459. Luo, L.; Salunga, R. C.; Guo, H.; Bittner, A.; Joy, K. C.; Galindo, J. E.; Xiao, H.; Rogers, K. E.; Wan, J. S.; Jackson, M. R. et al. *Nat. Med.* **1999**, *5*, 117–122.
460. Banks, R. E.; Dunn, M. J.; Forbes, M. A.; Stanley, A.; Pappin, D.; Naven, T.; Gough, M.; Harnden, P.; Selby, P. J. *Electrophoresis* **1999**, *20*, 689–700.
461. Bibby, M. C. *Br. J. Cancer* **1999**, *79*, 1633–1640.
462. Elkeles, A.; Oren, M. *Mol. Med.* **1999**, *5*, 334–335.
463. Gura, T. *Science* **1997**, *278*, 1041–1042.
464. Sanchez, J. C.; Hochstrasser, D. *Methods Mol. Biol.* **1999**, *112*, 87–93.
465. Argiles, A. *Nephrologie* **1987**, *8*, 51–54.
466. Janssen, P. T.; Van Bijsterveld, Q. P. *Clin. Chim. Acta* **1981**, *114*, 207.
467. Scheele, G. A. *J. Biol. Chem.* **1975**, *250*, 5375–5385.
468. Cassara, G.; Gianazza, E.; Righetti, P. G.; Poma, S.; Vicentini, L.; Scortecci, V. *J. Chromatogr.* **1980**, *221*, 279.
469. Jones, M. I.; Spragg, S. P.; Webb, T. *Biol. Neonate* **1981**, *39*, 171–177.
470. Toussi, A.; Paquet, N.; Huber, O.; Frutiger, S.; Tissot, J. D.; Hughes, G. J.; Hochstrasser, D. F. *J. Chromatogr.* **1992**, *582*, 87–92.
471. Dermer, G. B. *Clin. Chem.* **1982**, *28*, 881–887.
472. Brunet, J. F.; Berger, F.; Gustin, T.; Laine, M.; Benabid, H. L. *J. Neuroimmunol.* **1993**, *47*, 63–72.
473. Marshall, T.; Williams, K. M. *Anal. Biochem.* **1984**, *139*, 506–509.
474. Anderson, N. G.; Powers, M. T.; Tollaksen, S. L. *Clin. Chem.* **1982**, *28*, 1045–1050.
475. Edwards, J. J.; Tollaksen, S. L.; Anderson, N. G. *Clin. Chem.* **1981**, *27*, 1335–1340.
476. Muzio, M.; Chinnaiyan, A. M.; Kischkel, F. C.; O'Rourke, K.; Shevchenko, A.; Ni, J.; Scaffidi, C.; Bretz, J. D.; Zhang, M.; Gentz, R. et al. *Cell* **1996**, *85*, 817–827.
477. Mercurio, F.; Zhu, H.; Murray, B. W.; Shevchenko, A.; Bennett, B. L.; Li, J.; Young, D. B.; Barbosa, M.; Mann, M.; Manning, A.; Rao, A. *Science* **1997**, *278*, 860–866.
478. Neubauer, G.; Gottschalk, A.; Fabrizio, P.; Séraphin, B.; Lührmann, R.; Mann, M. *Proc. Natl. Acad. Sci. USA* **1997**, *94*, 385–390.
479. Chubet, R. G.; Brizzard, B. L. *Biotechniques* **1996**, *1*, 136–141.
480. Hunter, T.; Hunter, G. J. *Biotechniques* **1998**, *2*, 194–196.
481. Blagosklonny, M.; Giannakakou, P.; El-Deiry, W. S.; Kingston, D. G. I.; Higgs, P. I.; Neckers, L.; Fojo, T. *Cancer Res.* **1997**, *57*, 130–155.
482. Imazu, T.; Shimizu, S.; Tagami, C.; Matsushima, M.; Nakamura, Y.; Miki, T.; Okuyama, A.; Tsujimoto, Y. *Oncogene* **1999**, *18*, 4523–4529.
483. Harper, J. D.; Wong, S. S.; Lieber, C. M.; Lansbury, P. T., Jr. *Chem. Biol.* **1997**, *4*, 119–125.
484. Walsh, D. M.; Lomakin, A.; Benedek, G. B.; Condron, M. M.; Teplow, D. B. *J. Biol. Chem.* **1997**, *272*, 22364–22374.
485. Selkoe, D. J. *Nature* **1999**, *399*, A23–A31.
486. Lemere, C. A.; Lopera, F.; Kosik, K. S.; Lendon, C. L.; Ossa, J.; Saido, T. C.; Yamaguchi, H.; Ruiz, A.; Martinez, A.; Madrigal, L. et al. *Nat. Med.* **1996**, *2*, 1146–1148.
487. Mann, D. M.; Iwatsubo, T.; Cairns, N. J.; Lantos, P. L.; Nochlin, D.; Sumi, S. M.; Bird, T. D.; Poorkaj, P.; Hardy, J.; Hutton, M. et al. *Ann. Neurol.* **1996**, *40*, 149–156.
488. Higgins, L. S.; Cordell, B. *Methods Enzymol.* **1996**, *10*, 384–391.
489. Duff, K. *Curr. Opin. Biotechnol.* **1998**, *9*, 561–564.
490. Higgins, L. S. *Mol. Med.* **1999**, *5*, 274–276.
491. Hsiao, K.; Chapman, P.; Nilsen, S.; Eckman, C.; Harigay, Y.; Younkin, S.; Yang, F.; Cole, G. *Nature* **1996**, *274*, 99–102.
492. Moran, P. M.; Higgins, L. S.; Cordell, B.; Moser, P. C. *Proc. Natl. Acad. Sci. USA* **1995**, *92*, 5341–5345.
493. Olumi, A. F.; Dazin, P.; Tlsty, T. D. *Cancer Res.* **1998**, *58*, 4525–4530.
494. Wouters, B. G.; Giaccia, A. J.; Denko, N. C.; Brown, J. M. *Cancer Res.* **1997**, *57*, 4703–4706.
495. Waldman, T.; Zhang, Y.; Dillehay, L.; Yu, J.; Kinzler, K.; Vogelstein, B.; Williams, J. *Nat. Med.* **1997**, *9*, 1034–1036.
496. DePinho, R.; Jacks, T. *Oncogene* **1999**, *18*.
497. Yamamura, K. *Prog. Exp. Tumor Res.* **1999**, *35*, 13–24.
498. Aguzzi, A.; Raeber, A. J. *Brain Pathol.* **1998**, *8*, 695–697.
499. Sturchler-Pierrat, C.; Sommer, B. *Rev. Neurosci.* **1999**, *10*, 15–24.

500. Macleod, K. F.; Jacks, T. *J. Pathol.* **1999**, *187*, 43–60.
501. Plück, A. *Int. J. Exp. Pathol.* **1996**, 77, 269–278.
502. Stricklett, P. K.; Nelson, R. D.; Kohan, D. E. *Am. J. Physiol.* **1999**, *276*, F651–F657.
503. Xu, X.; Wagner, K. U.; Larson, D.; Weaver, Z.; Li, C.; Ried, T.; Hennighausen, L.; Wynshaw-Boris, A.; Deng, C. X. *Nat. Genet.* **1999**, *22*, 37–43.
504. Chen, J.; Kelz, M. B.; Zeng, G.; Sakai, N.; Steffen, C.; Shockett, P. E.; Picciotto, M. R.; Duman, R. S.; Nestler, E. J. *Mol. Pharmacol.* **1998**, *54*, 495–503.
505. Federspiel, M. J.; Bates, P.; Young, J. A. T.; Varmus, H. E.; Hughes, S. H. *Proc. Natl. Acad. Sci. USA* **1994**, *91*, 11241–11245.
506. Klose, J. *Electrophoresis* **1999**, *20*, 643–652.
507. Janke, C.; Holzer, M.; Goedert, M.; Arendt, T. *FEBS Lett.* **1996**, *379*, 222–226.
508. Selkoe, D. J. *Science* **1996**, *275*, 630–631.
509. Lane, D. *Lancet* **1998**, *351*, 17–20.
510. Senior, K. *Mol. Med.* **1999**, *5*, 326–327.
511. Devlin, J. P. (Chairman). *The First Forum on Data Management Technologies in Biological Screening*; SRI International, Menlo Park, CA, April 22–24, 1992.
512. Forum on Data Management Technologies in Biological Screening archives. http://www.msdiscovery.com (accessed Aug 2006).
513. Becton, Dickinson and Company. http://www.bd.com (accessed Aug 2006).
514. Dynex Technologies. http://www.dynextechnologies.com (accessed Aug 2006).
515. Greineramerica.com. http://www.greineramerica.com (accessed Aug 2006).
515a. Water filter search engine. http://www.polyfiltronics.com (accessed Aug 2006).
515b. Nalge Nunc International. http://www.nalgenunc.com (accessed Aug 2006).
517. Society for Biomolecular Sciences. http://www.sbsonline.org (accessed Aug 2006).
518. Tripos. http://www.tripos.com (accessed Aug 2006).
518a. Molecular Simulations. http://www.msi.com (accessed Aug 2006).
518b. Daylight Chemical Information Systems. http://www.daylight.com (accessed Aug 2006).
518c. Elsevier MDL. http://www.mdli.com (accessed Aug 2006).
518d. Accelrys. http://www.accelrys.com (accessed Aug 2006).
518e. DMR News Review Service. http://www.synopsis.co.uk (accessed Aug 2006).
519. ArQule. http://www.arqule.com (accessed Aug 2006).
519a. Pharmacopeia Drug Discovery. http://www.pcop.com (accessed Aug 2006).
520. Devlin, J. P. Chemical Diversity and Genetic Equity: Synthetic and Naturally Derived Compounds. In *The Discovery of Bioactive Substances: High Throughput Screening*; Devlin, J. P., Ed.; Marcel Dekker: New York, 1997, pp 3–48.
521. Gaia Chemical Corporation. http://www.gaiachem.com (accessed Aug 2006).
522. Aramburu, J.; Yaffe, M. B.; Lopez-Rodriguez, C.; Cantley, L. C.; Hogan, P. G.; Rao, A. *Science* **1999**, *285*, 2129.
523. Amersham. http://www.amersham.com (accessed Aug 2006).
524. Axyspharm.com. http://www.axyspharm.com (accessed Aug 2006).
524a. Merck. http://www.merck.com (accessed Aug 2006).
525. Invitrogen. http://www.invitrogen.com (accessed Aug 2006).
528. BD Biosciences. http://www.gentest.com (accessed Aug 2006).
530. Babiak, J. Management and Service Issues in a Centralized Robotics HTS Core. In *The Discovery of Bioactive Substances: High Throughput Screening*; Devlin, J. P., Ed.; Marcel Dekker: New York, 1997, pp 461–470.
531. Beggs, M.; Major, J. S. Flexible Use of People and Machines. In *The Discovery of Bioactive Substances: High Throughput Screening*; Devlin, J. P., Ed.; Marcel Dekker: New York, 1997, pp 3–48.
532. Monogram Biosciences. http://www.aclara.com (accessed Aug 2006).
533. Caliper Life Sciences. http://www.caliperls.com (accessed Aug 2006).
534. *Proceedings of Cyprus 1998*: New Technologies and Frontiers in Drug Research, Limassol, Cyprus, May 4–8, 1998.
536. Houston, J. G.; Banks, M. A. *Curr. Opin. Biotechnol.* **1997**, *8*, 734–740.
537. Turner, R.; Sterrer, S.; Wiesmüller, K. H. High Throughput Screening. In *Ullmann's Encyclopedia of Industrial Chemistry*; VCH: Weinheim, Germany, 2006, in press.
538. Turner, R.; Ullmann, D.; Sterrer, S. Screening in the Nanoworld: Single Molecule Spectroscopy and Miniaturized HTS. In *Handbook of Screening: Dr. James Swarbrick's the Drugs and the Pharmaceutical Sciences Series*, 2006, in press.
539. Auer, M.; Moore, K. J.; Meyer-Almes, F. J.; Guenther, R.; Pope, A. J.; Stoeckli, K. A. *Drug Disc. Today* **1998**, *3*, 457–465.
540. Eigen, M.; Rigler, R. *Proc. Natl. Acad. Sci. USA* **1994**, *91*, 5740–5747.
541. Kask, P.; Palo, K.; Ullmann, D.; Gall, K. *Proc. Natl. Acad. Sci. USA* **1999**, *96*, 13756–13761.
542. Horn, F.; Vriend, G. *J. Mol. Med.* **1998**, *76*, 464–468.
543. Fox, S. *Genet. Eng. News.* **1996**, 16.
544. Upham, L. *Biomed. Prod.* **1999**, 10–11.
545. Park, Y. W.; Garyantes, T.; Cummings, R. T.; Carter-Allen, K. *Optimization of 33P Scintillation Proximity Assays Using Cesium Chloride Bead Suspension: TopCount Topics TCA-030*; Packard Instrument Company: Meriden, CT, 1997.
546. Lakowicz, J. R. *Principles of Fluorescence Spectroscopy*, 2nd ed.; Plenum Publishing: New York, 1999.
547. Mathis, G. *Clin. Chem.* **1993**, *39*, 1953.
548. Packard Instrument Company. *The Principles of Time-Resolved Fluorescence: Application Note HTRF-001*; Packard Instrument Company, Meriden, CT, 1997.
549. Mabile, M.; Mathis, G.; Jolu, E. J.-P.; Pouyat, D.; Dumont, C. Method of Measuring the Luminescence Emitted in a Luminescent Assay. US Patent 5,527,684, June 18, 1996.
550. Hemmilä, I., Hill, S., Elcock, C. *Drug Disc. Technol.* **1998**.
551. Alpha, B.; Lehn, J.; Mathis, G. *Angew. Chem. Int. Ed. Engl.* **1987**, *26*, 266.
552. Mathis, G. *J. Biomol. Screen.* **1999**, *4*, 309–313.
553. Kolb, A.; Burke, J.; Mathis, G. Homogeneous, Time-Resolved Fluorescence Method for Drug Discovery. In *Assay Technologies and Detection Methods*; Marcel Dekker: New York, 1997, pp 345–360.
554. Forster, T. *Ann. Physik* **1948**, *2*, 55.

555. Mathis, G. *Amplification by Energy Transfer and Homogeneous Time-Resolved Immunoassay*; Proceedings of the Sixth International Symposium on Quantitative Luminescence Spectrometry in BioMedical Sciences, University of Ghent, Belgium, May 25–27, 1993.
556. Upham, L. *HTRF Epitope Tag Antibody Reagents for Assay Development and High Throughput Screening: Application note AN4003-DSC*; Packard Instrument Company: Wellesley, MA, 1999.
557. Wild, D. *The Immunoassay Handbook*; Stockton Press: New York, 1994.
558. Upham, L. V. *Lab. Robot. Automat.* **1999**, *11*, 324–329.
559. Hanks, S. T.; Quinn, A. M.; Hunter, T. *Science* **1988**, *241*, 42–51.
560. Hanks, S. T.; Hunter, T. *FASEB J.* **1995**, *9*, 576–596.
561. Park, Y. W.; Cummings, R. T.; Wu, L.; Zheng, S.; Cameron, P. M.; Woods, A.; Zaller, D. M.; Marcy, A. I.; Hermes, J. D. *Anal. Biochem.* **1999**, *269*, 94–104.
562. Rogers, M. V. *Drug Disc. Today* **1997**, *2*, 156–160.
563. Kolb, A. J.; Kaplita, P. V.; Hayes, D. J.; Park, Y. W.; Pernell, C.; Major, J. S.; Mathis, G. *Drug Disc. Today* **1998**, *3*, 333–342.
564. Cummings, R. T.; McGovern, H. M.; Zheng, S.; Park, Y. W.; Hermes, J. D. *Anal. Biochem.* **1999**, *269*, 79–93.
565. Yabuki, N.; Watanabe, S. I.; Kudoh, T.; Nihira, S. I.; Miyamoto, C. *Comb. Chem. High-Throughput Screening* **1999**, *2*, 279–287.
566. Lehmann, J. M.; Moore, L. B.; Smith-Oliver, T. A.; Wilkison, W. O.; Wilson, T. M.; Kliewer, S. A. *J. Biol. Chem.* **270**, 12953–12956.
567. Zhou, G.; Cummings, R. T.; Li, Y.; Mitra, S.; Wilkinson, H. A.; Elbrecht, A.; Hermes, J. D.; Schaeffer, J. M.; Smith, R. G.; Moller, D. E. *Mol. Endocrinol.* **1998**, *12*, 1594–1604.
568. Mathis, G.; Preaudat, M.; Trinquet, E.; Pernelle, C.; Trouillas, M. *A New Homogeneous Method Using Rare Earth Cryptate and Amplification by Energy Transfer for the Characterization of the FOS and JUN Leucine Zipper Peptides Dimerization*; Proceedings of the Fourth Annual Conference of the Society for Biomolecular Screening, Philadelphia, Sept. 1994; 2001.
569. Ellis, J. H.; Burden, M. N.; Vinogradov, D. V.; Linge, C.; Crowe, J. S. *J. Immunol.* **1996**, *56*, 2700.
570. June, C. H.; Bluestone, J. A.; Nadler, L. M.; Thompson, C. B. *Immunol. Today* **1994**, *15*, 321.
571. Linsley, P. S. *J. Exp. Med.* **1995**, *182*, 289.
572. Lenschow, D. J.; Walunas, T. L.; Bluestone, J. A. *Annu. Rev. Immunol.* **1996**, *14*, 233.
573. Mellor, G. W.; Burden, M. N.; Preaudat, M.; Joseph, Y.; Cooksley, S. B.; Ellis, J. H.; Banks, M. N. *J. Biomol. Screen.* **1998**, *3*, 91–99.
574. Farrar, S. J.; Whiting, P. J.; Bonnert, T. P.; McKernan, R. *J. Biol. Chem.* **1999**, *274*, 15, 10100–10104.
575. Alpha-Bazin, B.; Mathis, G. *New Homogeneous Assay Formats for the Measurement of Nuclease and Reverse Transcriptase Activity*; Proceedings of the Fourth Annual Conference of the Society for Biomolecular Screening, Baltimore, Sept. 21–24, 1998.
576. Alpha-Bazin, B.; Bazin, H.; Boissy, L.; Mathis, G. *Europium Cryptate Tethered Nucleoside Triphosphate for Nonradioactive Labeling and HTRF Detection of DNA and RNA*; Proceedings of the Fifth Annual Conference of the Society for Biomolecular Screening, Edinburgh, Scotland, Sept. 13–16, 1999.
577. Wilson, E. B. Introduction. In *General Cytology: A Textbook of Cellular Structure and Function for Students of Biology and Medicine*; Cowdry, E., Ed.; University of Chicago Press: Chicago, IL, 1924, pp 3–11.
578. Edström, J. E. *Nature* **1953**, *172*, 809.
579. Haynes, P. A.; Gygi, S. P.; Figeys, D.; Aebersold, R. *Electrophoresis* **1998**, *19*, 1862–1871.
580. Celis, J. E.; Gromov, P. *Curr. Opin. Biotechnol.* **1999**, *10*, 16–21.
581. Giuliano, K. A.; DeBiasio, R. L.; Dunlay, R. T.; Gough, A.; Volosky, J. M.; Zock, J.; Pavlakis, G. N.; Taylor, D. L. *J. Biomol. Screen.* **1997**, *2*, 249–259.
582. Ding, G. J. F.; Fischer, P. A.; Boltz, R. C.; Schmidt, J. A.; Colaianne, J. J.; Gough, A.; Rubin, R. A.; Miller, D. K. *J. Biol. Chem.* **1998**, *273*, 28897–28905.
583. Kain, S. R. *Drug Disc. Today* **1999**, *4*, 304–312.
584. Conway, B. R.; Minor, L. K.; Xu, J. Z.; Gunnet, J. W.; DeBiasio, R.; D'Andrea, M. R.; Rubin, R.; DeBiasio, R.; Giuliano, K.; Zhou, L. et al. *J. Biomol. Screen.* **1999**, *4*, 75–86.
585. Matz, M. V.; Fradkov, A. F.; Labas, Y. A.; Savitsky, A. P.; Zaraisky, A. G.; Markelov, M. L.; Lukyanov, S. A. *Nat. Biotechnol.* **1999**, *17*, 969–973.
586. Giuliano, K. A.; Post, P. L.; Hahn, K. M.; Taylor, D. L. *Annu. Rev. Biophys. Biomol. Struct.* **1995**, *24*, 405–434.
587. Giuliano, K. A.; Taylor, D. L. *Trends Biotechnol.* **1998**, *16*, 135–140.
588. Reers, M.; Smith, T. W.; Chen, L. B. *Biochemistry* **1991**, *30*, 4480–4486.
589. Smiley, S. T.; Reers, M.; Mottola-Hartshorn, C.; Lin, M.; Chen, A.; Smith, T. W.; Steele, G. D. J.; Chen, L. B. *Proc. Natl. Acad. Sci. USA* **1991**, *88*, 3671–3675.
590. Kapur, R.; Giuliano, K. A.; Campana, M.; Adams, T.; Olson, K.; Jung, D.; Mrksich, M.; Vasudevan, C.; Taylor, D. L. *Biomed. Microdev.* **1999**, *2*, 99–109.
591. Service, R. F. *Science* **1999**, *282*, 399–401.
592. Service, R. F. *Science* **1998**, *282*, 400.
593. Fodor, S. P. A.; Read, J. L.; Pirrung, M. C.; Stryer, L.; Lu, A. T.; Solas, D. *Science* **1991**, *251*, 767–773.
594. Carlson, R.; Brent, R. *Nat. Biotechnol.* **1999**, *17*, 536–537.
595. Marshall, S. *R&D Mag.* **1999**, 18–22.
596. Lehninger, A. L.; Nelson, D. L.; Cox, M. M. *Principles of Biochemistry*; Worth Publishers: New York, 1993.
596a. NC-IUB Panel on Enzyme Kinetics. *Eur. J. Biochem.* **1982**, *128*, 281–291 (erratum **1993**, *213*, 1).
597. Seneci, P. *Solid-Phase Synthesis and Combinatorial Technologies*; John Wiley: New York, 2001.
598. Merrifield, B. *Methods Enzymol.* **1997**, *289*, 3–13.
599. Merrifield, R. B. *J. Am. Chem. Soc.* **1963**, *85*, 2149–2154.
600. Merrifield, R. B. *Angew. Chem. Int. Ed.* **1985**, *24*, 799–810.
601. Merrifield, R. B. *Life During a Golden Age of Peptide Chemistry: The Concept and Development of Solid-Phase Peptide Synthesis*; American Chemical Society: Washington, DC, 1993.
602. Letsinger, R. L.; Mahadevan, V. *J. Am. Chem. Soc.* **1966**, *88*, 5319–5324.
603. Leznoff, C. C.; Wong, J. Y. *Can. J. Chem.* **1972**, *50*, 2892–2893.
604. Leznoff, C. C. *Can. J. Chem.* **2000**, *78*, 167–183.
605. Frank, R.; Heikens, W.; Heisterberg-Moutsis, G.; Blocker, H. *Nucleic Acid Res.* **1983**, *11*, 4365–4377.
606. Houghten, R. A. *Proc. Natl. Acad. Sci. USA* **1985**, *82*, 5131–5135.
607. Geysen, H. M.; Meloen, R. H.; Barteling, S. J. *Proc. Natl. Acad. Sci. USA* **1984**, *81*, 3998–4002.

608. Tuerk, C.; Gold, L. *Science* **1990**, *249*, 505–510.
609. Ellington, A. D.; Szostak, J. W. *Nature* **1990**, *346*, 818–822.
610. Ellington, A. D.; Szostak, J. W. *Nature* **1992**, *355*, 850–852.
611. Smith, G. P. *Science* **1985**, *228*, 1315–1317.
612. Smith, G. P.; Petrenko, V. A. *Chem. Rev.* **1997**, *97*, 391–410.
613. Geysen, H. M.; Rodda, S. J.; Mason, T. J. *Mol. Immunol.* **1986**, *23*, 709–715.
614. Furka, A.; Sebestyen, F.; Asgedom, M.; Dibo, G. *Highlights of Modern Biochemistry: Cornucopia of Peptides by Synthesis*. Proceedings of the 14th International Congress of Biochemistry, Prague; July, 1988; VSP: Ultrecht; p 47.
615. Furka, A., Sebestyen, F., Asgedom, M., Dibo, G. *More Peptides by Less Labour*. Xth International Symposium on Medicinal Chemistry, Budapest, Aug. 15–19, 1988; Poster.
616. Lam, K. S.; Salmon, S. E.; Hersh, E. M.; Hruby, V. J.; Kazmierski, W. M.; Knapp, R. J. *Nature* **1991**, *354*, 82–84.
617. Houghten, R. A.; Pinilla, C.; Blondelle, S. E.; Appel, J. R.; Dooley, C. T.; Cuervo, J. H. *Nature* **1991**, *354*, 84–86.
618. Dooley, C. T.; Houghten, R. A. *Life Sci.* **1993**, *52*, 1509–1517.
619. Houghten, R. A.; Pinilla, C.; Appel, J. R.; Blondelle, S. E.; Dooley, C. T.; Eichler, J.; Nefzi, A.; Ostresh, J. M. *J. Med. Chem.* **1999**, *42*, 3743–3778.
620. Lam, K. S.; Lebl, M.; Krchnak, V. *Chem. Rev.* **1997**, *97*, 411–448.
621. Lebl, M.; Krchnak, V.; Sepetov, N. F.; Seligmann, B.; Strop, P.; Felder, S.; Lam, K. S. *Biopolymers* **1995**, *37*, 177–198.
622. Salmon, S. E.; Lam, K. S.; Lebl, M.; Kandola, A.; Khattri, P. S.; Wade, S.; Patek, M.; Kocis, P.; Krchnak, V.; Thorpe, D. et al. *Proc. Natl. Acad. Sci. USA* **1993**, *90*, 11708–11712.
623. Salmon, S. E.; Liu-Stevens, R. H.; Zhao, Y.; Lebl, M.; Krchnak, V.; Wertman, K.; Sepetov, N.; Lam, K. S. *Mol. Divers.* **1996**, *2*, 57–63.
624. Bunin, B. A.; Ellman, J. A. *J. Am. Chem. Soc.* **1992**, *114*, 10997–10998.
625. DeWitt, S. H.; Kiely, J. S.; Stankovic, C. J.; Schroeder, M. C.; Cody, D. M. R.; Pavia, M. R. *Proc. Natl. Acad. Sci. USA* **1993**, *90*, 6909–6913.
626. Simon, R. J.; Kaina, R. S.; Zuckermann, R. N.; Huebner, V. D.; Jewell, D. A.; Banville, S.; Ng, S.; Wang, L.; Rosenberg, S.; Marlowe, C. K. et al. *Proc. Natl. Acad. Sci. USA* **1992**, *89*, 9367–9371.
627. Zuckermann, R. N.; Martin, E. J.; Spellmeyer, D. C.; Stauber, G. B.; Shoemaker, K. R.; Kerr, J. M.; Figliozzi, G. M.; Goff, D. A.; Siani, M. A.; Simon, R. J. et al. *J. Med. Chem.* **1994**, *37*, 2678–2685.
628. Dolle, R. E. *J. Comb. Chem.* **2000**, *2*, 383–433.
629. Franzén, R. G. *J. Comb. Chem.* **2000**, *2*, 195–214.
630. Guillier, F.; Orain, D.; Bradley, M. *Chem. Rev.* **2000**, *100*, 2057–2091.
631. Sammelson, R. E.; Kurth, M. J. *Chem. Rev.* **2001**, *101*, 137–202.
632. Kerr, J. M.; Banville, S. C.; Zuckermann, R. N. *J. Am. Chem. Soc.* **1993**, *115*, 2529–2531.
633. Nestler, H. P.; Bartlett, P. A.; Still, W. C. *J. Org. Chem.* **1994**, *59*, 4723–4724.
634. Nielsen, J.; Brenner, S.; Janda, K. D. Peptides 94, Proc.23. EPS. In *Implementation of Encoded Combinatorial Chemistry*; Maia, H. L. S., Ed.; ESCOM: Leiden, Germany, 1995, pp 92–93.
635. Nikolaiev, V.; Stierandova, A.; Krchnak, V.; Seligmann, B.; Lam, K. S.; Salmon, S. E.; Lebl, M. *Pept. Res.* **1993**, *6*, 161–170.
636. Briceno, G.; Chang, H.; Sun, X.; Schulz, P. G.; Xiang, X. D. *Science* **1995**, *270*, 273–275.
637. Sun, X. D.; Gao, C.; Wang, J. S.; Xiang, X. D. *Appl. Phys. Lett.* **1997**, *70*, 3353–3355.
638. Takeuchi, I.; Chang, H.; Gao, C.; Schultz, P. G.; Xiang, X. D.; Sharma, R. P.; Downes, M. J.; Venkatesan, T. *Appl. Phys. Lett.* **1998**, *73*, 894–896.
639. Wang, J.; Yoo, Y.; Gao, C.; Takeuchi, I.; Sun, X.; Chang, H.; Xiang, X. D.; Schultz, P. G. *Science* **1998**, *279*, 1712–1714.
640. Xiang, X. D.; Sun, X.; Briceno, G.; Lou, Y.; Wang, K. A.; Chang, H.; Wallace-Freedman, W. G.; Chen, S. W.; Schultz, P. G. *Science* **1995**, *268*, 1738–1740.
641. Nicolaou, K. C.; Xiao, X. Y.; Parandoosh, Z.; Senyei, A.; Nova, M. P. *Angew. Chem. Int. Ed.* **1995**, *34*, 2289–2291.
642. Moran, E. J.; Sarshar, S.; Cargill, J. F.; Shahbaz, M. M.; Lio, A.; Mjalli, A. M. M.; Armstrong, R. W. *J. Am. Chem. Soc.* **1995**, *117*, 10787–10788.
643. Xiao, X. Y.; Nicolaou, K. C. Combinatorial Chemistry: A Practical Approach. In *High-Throughput Combinatorial Synthesis of Discrete Compounds in Multimilligram Quantities: Nonchemical Encoding and Directed Sorting*; Fenniri, H., Ed.; Oxford University Press: Oxford, UK, 2004, pp 75–94.
644. Xiao, X.; Zhao, C.; Potash, H.; Nova, M. P. *Angew. Chem. Int.* **1997**, *36*, 780–782.
645. Lebl, M., Krchnak, V., Ibrahim, G., Pires, J., Burger, C., Ni, Y., Chen, Y., Podue, D., Mudra, P., Pokorny, V., et al. *Synthesis-Stuttgart* **1999**, 1971–1978.
646. Lebl, M. *Bioorg. Med. Chem. Lett.* **1999**, *9*, 1305–1310.
647. Krchnak, V.; Vagner, J. *Pept. Res.* **1990**, *3*, 182–193.
648. Krchnak, V.; Padera, V. *Bioorg. Med. Chem. Lett.* **1998**, *8*, 3261–3264.
649. Lebl, M. *J. Comb. Chem.* **1999**, *1*, 3–24.
650. Hudson, D. *J. Comb. Chem.* **1999**, *1*, 333–360.
651. Hudson, D. *J. Comb. Chem.* **1999**, *1*, 403–457.
652. Bannwarth, W.; Felder, E.; Mannhold, R.; Kubinyi, H.; Timmerman, H., Eds. Combinatorial Chemistry: A Practical Approach. In *Methods and Principles in Medicinal Chemistry*, Wiley-VCH: Weinheim, Germany, 2000.
653. Burgess, K., Ed. *Solid-Phase Organic Synthesis*; John Wiley: New York, 2000.
654. Dorwald, F. Z., Ed. *Organic Synthesis on Solid Phase: Supports, Linkers, Reactions*; Wiley-VCH: New York, 2000.
655. Fenniri, H., Ed. *Combinatorial Chemistry: A Practical Approach*; Oxford University Press: Oxford, UK, 2000.
656. Jung, G., Ed. *Combinatorial Chemistry: Synthesis, Analysis, Screening*; Wiley-VCH: Weinheim, Germany, 1999.
657. Miertus, S.; Fassina, G., *Combinatorial Chemistry and Technology: Principles, Methods, and Applications*; Marcel Dekker: New York, 1999.
658. Moos, W. H.; Pavia, M. R., Eds., *Annual Reports in Combinatorial Chemistry and Molecular Diversity*; Kluwer: Dordrecht, Germany, 1999; Vol. 2.
659. Bunin, B. A., Ed. *The Combinatorial Index*; Academic Press: San Diego, CA, 1998.
660. Gordon, E. M.; Kerwin, J. F. J.; *Combinatorial Chemistry and Molecular Diversity in Drug Discovery*; John Wiley: New York, 1998.
661. Terrett, N. K., Ed. *Combinatorial Chemistry*; Oxford University Press: New York, 1998.
662. Cabilly, S., Ed. *Combinatorial Peptide Library Protocols*; Humana Press: Totowa, NJ, 1997.
663. Czarnik, A. W.; DeWitt, S. H.; *A Practical Guide to Combinatorial Chemistry*; American Chemical Society: Washington, DC, 1997.
664. Devlin, J. P., Ed. *High Throughput Screening: The Discovery of Bioactive Substances*; Marcel Dekker: New York, 1997.
665. Fields, G. B.; Colowick, S. P.; *Solid-Phase Peptide Synthesis*; Academic Press: San Diego, CA, 1997.

666. Wilson, S. R.; Czarnik, A. W.; *Combinatorial Chemistry: Synthesis and Applications*; John Wiley: New York, 1997.
667. Abelson, J. N., Ed. *Combinatorial Chemistry*; Academic Press: San Diego, CA, 1996.
668. Chaiken, I. M.; Janda, K. D., Eds. Molecular Diversity and Combinatorial Chemistry: Libraries and Drug Discovery. ACS Conference Proceedings; American Chemical Society: Washington, DC, 1996.
669. Cortese, R., Ed. *Combinatorial Libraries: Synthesis, Screening and Application Potential*; Walter de Gruyter: New York, 1996.
670. Jung, G., Ed. *Combinatorial Peptide and Nonpeptide Libraries: A Handbook*; VCH: New York, 1996.
671. Epton, R., Ed. *Innovation and Perspectives in Solid Phase Synthesis and Combinatorial Libraries*; Mayflower Scientific Ltd: Birmingham, UK, 1999.
672. Kassel, D. B. *Chem. Rev.* **2001**, *101*, 255–267.
673. Reetz, M. T. *Angew. Chem. Int. Ed.* **2001**, *40*, 284–310.
674. An, H.; Cook, P. D. *Chem. Rev.* **2000**, *100*, 3311–3340.
675. Barnes, C.; Balasubramanian, S. *Curr. Opin. Chem. Biol.* **2000**, *4*, 346–350.
676. Brase, S. *Chim. Oggi* **2000**, *18*, 14–19.
677. Brase, S.; Dahmen, S. *Chem. Eur. J.* **2000**, *5*, 1899–1905.
678. Domling, A.; Ugi, I. *Angew. Chem. Int. Ed.* **2000**, *39*, 3169–3210.
679. Enjalbal, C.; Martinez, J.; Aubagnac, J. L. *Mass Spectrom. Rev.* **2000**, *19*, 139–161.
680. Gauglitz, G. *Curr. Opin. Chem. Biol.* **2000**, *4*, 351–355.
681. Hewes, J. D. *Chim. Oggi* **2000**, *18*, 20–24.
682. Kopylov, A. M.; Spiridonova, V. A. *Mol. Biol.* **2000**, *34*, 940–954.
683. Nestler, H. P. *Curr. Org. Chem.* **2000**, *4*, 397–410.
684. Porco, J. A., Jr. *Comb. Chem. High Throughput Screening* **2000**, *3*, 93–102.
685. Camps Diez, F.; Castells Guardiola, J.; Pi Sallent, J. *Solid Phase Synthesis of 1,4-Benzodiazepine Derivatives*; Proceedings of Patronato de Investigacion Cientifica y Tecnica, 'Juan de la Cierva', Spain, 1977.
686. Leznoff, C. C. *Chem. Soc. Rev.* **1974**, *3*, 65–85.
687. Leznoff, C. C. *Acc. Chem. Res.* **1978**, *11*, 327–333.
688. Frechet, J. M. J. *Tetrahedron* **1981**, *37*, 663–683.
689. Bunin, B. A.; Ellman, J. A. *J. Am. Chem. Soc.* **1992**, *114*, 10997–10998.
690. DeWitt, S. H.; Kiely, J. S.; Stankovic, C. J.; Schroeder, M. C.; Cody, D. M. R.; Pavia, M. R. *Proc. Natl. Acad. Sci. USA* **1993**, *90*, 6909–6913.
691. Geysen, H. M.; Meloen, R. H.; Barteling, S. J. *Proc. Natl. Acad. Sci. USA* **1984**, *81*, 3998–4002.
692. Houghten, R. A. *Proc. Natl. Acad. Sci. USA* **1985**, *82*, 5131–5135.
693. Furka, A.; Sebestyen, F.; Asgedom, M.; Dibo, G. *Int. J. Pept. Protein Res.* **1991**, *37*, 487–493.
694. Sebestyen, F.; Dibo, G.; Kovacs, A.; Furka, A. *Bioorg. Med. Chem. Lett.* **1993**, *3*, 413–418.
695. Balkenhohl, F.; von dem Bussche-Hünnefeld, C.; Lansky, A.; Zechel, C. *Angew. Chem. Int. Ed. Engl.* **1996**, *35*, 2288–2337.
696. Gordon, E. M.; Gallop, M. A.; Patel, D. V. *Acc. Chem. Res.* **1996**, *29*, 144–154.
697. Thompson, L. A.; Ellman, J. A. *Chem. Rev.* **1996**, *96*, 555–600.
698. Hermkens, P. H. H.; Ottenheijm, H. C. J.; Rees, D. *Tetrahedron* **1996**, *52*, 4527–4554.
699. Früchtel, J. S.; Jung, G. *Angew. Chem. Int. Ed. Engl.* **1996**, *35*, 17–42, 11–39.
700. Czarnik, A. W.; DeWitt, S. H. *A Practical Guide to Combinatorial Chemistry*; American Chemical Society: Washington, DC, 1997.
701. Hermkens, P. H. H.; Ottenheijm, H. C. J.; Rees, D. C. *Tetrahedron* **1997**, *53*, 5643–5678.
702. Kaldor, S. W.; Siegel, M. G. *Curr. Opin. Chem. Biol.* **1997**, *1*, 101–106.
703. Wilson, S.; Czarnik, A. W. *Combinatorial Chemistry: Synthesis and Application*; John Wiley: New York, 1997.
704. Gravert, D. J.; Janda, K. D. *Chem. Rev.* **1997**, *97*, 489–509.
705. Ugi, I.; Dömling, A.; Gruber, B.; Almstetter, M. *Croat. Chem. Acta* **1997**, *70*, 631–647.
706. Flynn, D. L.; Devraj, R. V.; Naing, W.; Parlow, J. J.; Weidner, J. J.; Yang, S. *Med. Chem. Res.* **1998**, *8*, 219–243.
707. Czarnik, A. W. *Anal. Chem.* **1998**, *70*, 378A–386A.
708. Dolle, R. E. *Mol. Divers.* **1998**, *3*, 199–233.
709. Flynn, D. L.; Devraj, R. V.; Parlow, J. J. *Curr. Opin. Drug Disc. Dev.* **1998**, *1*, 41–50.
710. Brown, R. C. D. *J. Chem. Soc. Perkin Trans.* **1998**, 3293–3320.
711. Andres, C. J.; Whitehouse, D. L.; Deshpande, M. S. *Curr. Opin. Chem. Biol.* **1998**, *2*, 353–362.
712. Nefzi, A.; Dooley, C.; Ostresh, J. M.; Houghten, R. A. *Bioorg. Med. Chem. Lett.* **1998**, *8*, 2273–2278.
713. Booth, S.; Hermkens, P. H. H.; Ottenheijm, H. C. J.; Rees, D. C. *Tetrahedron* **1998**, *54*, 15385–15443.
714. Merritt, A. T. *Comb. Chem. High Throughput Screening* **1998**, *1*, 57–72.
715. Houghten, R. A.; Pinilla, C.; Appel, J. R.; Blondelle, S. E.; Dooley, C. T.; Eichler, J.; Nefzi, A.; Ostresh, J. M. *J. Med. Chem.* **1999**, *42*, 3743–3778.
716. Dolle, R. E.; Nelson, K. H., Jr. *J. Comb. Chem.* **1999**, *1*, 235–282.
717. Drewry, D. H.; Coe, D. M.; Poon, S. *Med. Res. Rev.* **1999**, *19*, 97–148.
718. Hall, S. E. *Mol. Divers.* **1999**, *4*, 131–142.
719. Kobayashi, S. *Chem. Soc. Rev.* **1999**, *28*, 1–15.
720. Nuss, J. M.; Renhowe, P. A. *Curr. Opin. Drug Disc. Dev.* **1999**, *2*, 631–650.
721. Coe, D. M.; Storer, R. *Mol. Divers.* **1999**, *4*, 31–38.
722. Wentworth, P., Jr.; Janda, K. D. *Chem. Commun.* **1999**, 1917–1924.
723. Baldino, C. M. *J. Comb. Chem.* **2000**, *2*, 89–103.
724. Merrifield, R. B. *J. Am. Chem. Soc.* **1963**, *85*, 2149–2154.
725. Nestler, H. P.; Bartlett, P. A.; Still, W. C. *J. Org. Chem.* **1994**, *59*, 4723–4724.
726. Kaiser, E.; Colescott, R. L.; Bossinger, C. D.; Cook, P. I. *Anal. Biochem.* **1970**, *34*, 595–598.
727. Vojkovsky, T. *Pept. Res.* **1995**, *8*, 236–237.
728. Gallop, M. A.; Fitch, W. L. *Curr. Opin. Chem. Biol.* **1997**, *1*, 94–100.
729. Yan, B. *Acc. Chem. Res.* **1998**, *31*, 621–630.
730. Fitch, W. L. *Mol. Divers.* **1999**, *4*, 39–45.
731. Combs, A. P.; Saubern, S.; Rafalski, M.; Lam, P. Y. S. *Tetrahedron Lett.* **1999**, *40*, 1623–1626.

732. Rafalski, M.; Saubern, S.; Lam, P. Y. S.; Combs, A. P. *Cupric Acetate-Mediated N-Arylation by Arylboronic Acids: Solid-Supported C-N Cross-Coupling Reaction*; Abstract of 218th ACS National Meeting, New Orleans, Aug. 22–26, 1999; ORGN-342.
733. Tadesse, S.; Rafalski, M.; Lam, P. Y. S.; Combs, A. P. *Copper Acetate-Mediated N-Arylation of Cyclic and Acyclic Secondary Amines on Solid Support*; Abstract in 218th ACS National Meeting, New Orleans, Aug. 22–26, 1999; ORGN-343.
734. Combs, A. P.; Rafalski, M. *J. Comb. Chem.* **2000**, *2*, 29–32.
735. Kick, E. K.; Ellman, J. A. *J. Med. Chem.* **1995**, *38*, 1427–1430.
736. Haque, T. S.; Skillman, A. G.; Lee, C. E.; Habashita, H.; Gluzman, I. Y.; Ewing, T. J. A.; Goldberg, D. E.; Kuntz, I. D.; Ellman, J. A. *J. Med. Chem.* **1999**, *42*, 1428–1440.
737. Thompson, L. A.; Moore, F. L.; Moon, Y.-C.; Ellman, J. A. *J. Org. Chem.* **1998**, *63*, 2066–2067.
738. Lorsbach, B. A.; Kurth, M. J. *Chem. Rev.* **1999**, *99*, 1549–1581.
739. Dax, S. L.; McNally, J. J.; Youngman, M. A. *Curr. Med. Chem.* **1999**, *6*, 255–270.
740. Corbett, J. W. *Org. Prep. Proc. Int.* **1998**, *30*, 491–550.
741. Franzén, R. G. *Tetrahedron* **2000**, *56*, 685–691.
742. Kulkarni, B. A.; Ganesan, A. *J. Comb. Chem.* **1999**, *1*, 373–378.
743. Nicolaou, K. C.; Winssinger, N.; Pastor, J.; Ninkovic, S.; Sarabia, F.; He, Y.; Vourloumis, D.; Yang, Z.; Li, T.; Giannakakou, P.; Hamel, E. *Nature* **1997**, *387*, 268–272.
744. Peng, G.; Sohn, A.; Gallop, M. A. *J. Org. Chem.* **1999**, *64*, 8342–8349.
745. Lee, D.; Sello, J. K.; Schreiber, S. L. *J. Am. Chem. Soc.* **1999**, *121*, 10648–10649.
746. Boeijen, A.; Kruijtzer, J. A. W.; Liskamp, R. M. J. *Bioorg. Med. Chem. Lett.* **1998**, *8*, 2375–2380.
747. Shao, H.; Colucci, M.; Tong, S. J.; Zhang, H. S.; Castelhano, A. L. *Tetrahedron Lett.* **1998**, *39*, 7235–7238.
748. Bhalay, G.; Blaney, P.; Palmer, V. H.; Baxter, A. D. *Tetrahedron Lett.* **1997**, *38*, 8375–8378.
749. Cuny, G. D.; Cao, J. R.; Hauske, J. R. *Tetrahedron Lett.* **1997**, *38*, 5237–5240.
750. Franzén, R. G. *J. Comb. Chem.* **2000**, *2*, 195–214.
751. Combs, A. P. *Synthesis and Structure: Activity Relationship Analysis of Dehydroamino Acid Derivatives Related to the Azinomycins*; Department of Chemistry, University of California at Los Angeles: Los Angeles, CA, 1995.
752. Ugi, I. *Proc. Estonian Acad. Sci. Chem.* **1995**, *44*, 237–273.
753. Armstrong, R. W.; Combs, A. P.; Tempest, P. A.; Brown, S. D.; Keating, T. *Acc. Chem. Res.* **1996**, *29*, 123–131.
754. Kiselyov, A. S.; Armstrong, R. W. *Tetrahedron Lett.* **1997**, *38*, 6163–6166.
755. Harriman, G. C. B. *Tetrahedron Lett.* **1997**, *38*, 5591–5594.
756. Mjalli, A. M. M.; Sarshar, S.; Baiga, T. J. *Tetrahedron Lett.* **1996**, *37*, 2943–2946.
757. Sarshar, S.; Siev, D.; Mjalli, A. M. M. *Tetrahedron Lett.* **1996**, *37*, 835–838.
758. Zhang, C.; Moran, E. J.; Woiwode, T. F.; Short, K. M.; Mjalli, A. M. M. *Tetrahedron Lett.* **1996**, *37*, 751–754.
759. Hulme, C.; Ma, L.; Romano, J.; Morrissette, M. *Tetrahedron Lett.* **1999**, *40*, 7925–7928.
760. Hulme, C.; Peng, J.; Louridas, B.; Menard, P.; Krolikowski, P.; Kumar, N. V. *Tetrahedron Lett.* **1998**, *39*, 8047–8050.
761. Hamuro, Y.; Scialdone, M. A.; DeGrado, W. F. *J. Am. Chem. Soc.* **1999**, *121*, 1636–1644.
762. Gravel, M.; Bérubé, C. D.; Hall, D. G. *J. Comb. Chem.* **2000**, *2*, 228–231.
763. Merrifield, R. B. *Angew. Chem. Int. Ed.* **1985**, *24*, 799–810.
764. Baneyx, F. *Curr. Opin. Biotechnol.* **1999**, *10*, 411–421.
765. Liu, D. R.; Magliery, T. J.; Pasternak, M.; Schultz, P. G. *Proc. Natl. Acad. Sci. USA* **1997**, *94*, 10092–10097.
766. Bodansky, M. *Peptide Chemistry: A Practical Textbook*; Springer-Verlag: New York, 1988.
767. Pennington, M. W.; Dunn, B. M. Peptide Synthesis Protocols. In *Methods in Molecular Biology*; Walker, J. M., Ed.; Humana Press: Totowa, NJ, 1994; Vol. 35.
768. Dooley, C. T.; Ny, P.; Bidlack, J. M.; Houghten, R. A. *J. Biol. Chem.* **1998**, *273*, 18848–18856.
769. Stewart, J. M.; Young, J. D. *Solid Phase Peptide Synthesis*; Pierce Chemical Company: Rockford, IL, 1984.
770. Marshall, W. S.; Boymel, J. L. *Drug Disc. Today* **1998**, *3*, 34–42.
771. Ito, Y.; Manabe, S. *Curr. Opin. Chem. Biol.* **1998**, *2*, 701–708.
772. Osborn, H. M. I.; Khan, T. H. *Tetrahedron* **1999**, *55*, 1807–1850.
773. Seeberger, P. H.; Haase, W.-C. *Chem. Rev.* **2000**, *100*, 4349–4393.
774. Arsequell, G.; Valencia, G. *Tetrahedron Asymmetr.* **1999**, *10*, 3045–3094.
775. Hojo, H.; Nakahara, Y. *Curr. Protein Pept. Sci.* **2000**, *1*, 23–48.
776. Zheng, C.; Seeberger, P. H.; Danishefsky, S. J. *J. Org. Chem.* **1998**, *63*, 1126–1130.
777. Nielsen, P. E.; Egholm, M.; Berg, R. H.; Buchardt, O. *Science* **1991**, *254*, 1497–1500.
778. Egholm, M.; Buchardt, O.; Nielsen, P. E.; Berg, R. H. *J. Am. Chem. Soc.* **1992**, *114*, 1895–1897.
779. Nielsen, P. E. *Acc. Chem. Res.* **1999**, *32*, 624–630.
780. Dean, D. A. *Adv. Drug Deliv. Rev.* **2000**, *44*, 81–95.
781. Smith, A. B., III; Keenan, T. P.; Holcomb, R. C.; Sprengeler, P. A.; Guzman, M. C.; Wood, J. L.; Carroll, P. J.; Hirschmann, R. *J. Am. Chem. Soc.* **1992**, *114*, 10672–10674.
782. Hagihara, M.; Anthony, N. J.; Stout, T. J.; Clardy, J.; Schreiber, S. L. *J. Am. Chem. Soc.* **1992**, *114*, 6568–6570.
783. Gennari, C.; Salom, B.; Potenza, D.; Williams, A. *Angew. Chem. Int. Ed. Engl.* **1994**, *33*, 2067–2069.
784. Hamuro, Y.; Geib, S. J.; Hamilton, A. D. *Angew. Chem. Int. Ed. Engl.* **1994**, *33*, 446–448.
785. Moree, W. J.; van der Marel, G. A.; Liskamp, R. J. *J. Org. Chem.* **1995**, *60*, 5157–5169.
786. Paikoff, S. J.; Wilson, T. E.; Cho, C. Y.; Schultz, P. G. *Tetrahedron Lett.* **1996**, *37*, 5653–5656.
787. Soth, M. J.; Nowick, J. S. *Curr. Opin. Chem. Biol.* **1997**, *1*, 120–129.
788. Hamper, B. C.; Kolodziej, S. A.; Scates, A. M.; Smith, R. G.; Cortez, E. *J. Org. Chem.* **1998**, *63*, 708–718.
789. Appella, D. H.; Christianson, L. A.; Karle, I. L.; Powell, D. R.; Gellman, S. H. *J. Am. Chem. Soc.* **1999**, *121*, 6206–6212.
790. Bunin, B. A.; Plunkett, M. J.; Ellman, J. A. *Proc. Natl. Acad. Sci. USA* **1994**, *91*, 4708–4712.
791. DeWitt, S. H.; Czarnik, A. W. *Acc. Chem. Res.* **1996**, *29*, 114–122.
792. MacDonald, A. A.; DeWitt, S. H.; Hogan, E. M.; Ramage, R. *Tetrahedron Lett.* **1996**, *37*, 4815–4818.
793. Ostresh, J. M.; Schoner, C. C.; Hamashin, V. T.; Nefzi, A.; Meyer, J.-P.; Houghten, R. A. *J. Org. Chem.* **1998**, *63*, 8622–8623.
794. Morales, G. A.; Corbett, J. W.; DeGrado, W. F. *J. Org. Chem.* **1998**, *63*, 1172–1177.

795. Vojkovsky', T.; Weichsel, A.; Pátek, M. *J. Org. Chem.* **1998**, *63*, 3162–3163.
796. Dragoli, D. R.; Thompson, L. A.; O'Brien, J.; Ellman, J. A. *J. Comb. Chem.* **1999**, *1*, 534–539.
797. Wessjohann, L. A. *Curr. Opin. Chem. Biol.* **2000**, *4*, 303–309.
798. Wang, H.; Ganesan, A. *J. Comb. Chem.* **2000**, *2*, 186–194.
799. Meseguer, B.; Alonso-Diáz, D.; Griebenow, N.; Herget, T.; Waldmann, H. *Angew. Chem. Int. Ed.* **1999**, *38*, 2902–2906.
800. Tan, D. S.; Foley, M. A.; Shair, M. D.; Schreiber, S. L. *J. Am. Chem. Soc.* **1998**, *120*, 8565–8566.
801. Lindsley, C. W.; Chan, L. K.; Goess, B. C.; Joseph, R.; Shair, M. D. *J. Am. Chem. Soc.* **2000**, *122*, 422–423.
802. Curran, D. P. *Med. Res. Rev.* **1999**, *19*, 432–438.
803. Kim, R. M.; Manna, M.; Hutchins, S. M.; Griffin, P. R.; Yates, N. A.; Bernick, A. M.; Chapman, K. T. *Proc. Natl. Acad. Sci. USA* **1996**, *93*, 10012–10017.
804. Kaldor, S. W.; Siegel, M. G.; Fritz, J. E.; Dressman, B. A.; Hahn, P. J. *Tetrahedron Lett.* **1996**, *37*, 7193–7196.
805. Flynn, D. L.; Crich, J. Z.; Devraj, R. V.; Hockerman, S. L.; Parlow, J. J.; South, M. S.; Woodard, S. *J. Am. Chem. Soc.* **1997**, *119*, 4874–4881.
806. Parlow, J. J.; Devraj, R. V.; South, M. S. *Curr. Opin. Chem. Biol.* **1999**, *3*, 320–336.
807. Thompson, L. A. *Curr. Opin. Chem. Biol.* **2000**, *4*, 324–337.
808. Booth, R. J.; Hodges, J. C. *Acc. Chem. Res.* **1999**, *32*, 18–26.
809. Sucholeiki, I.; Perez, J. M. *Tetrahedron Lett.* **1999**, *40*, 3531–3534.
810. Parlow, J. J.; Case, B. L.; Dice, T. A.; South, M. S. *High-Throughput Purification of Solutionphase Periodinane-Mediated Oxidation Reactions Utilizing a Novel Thiosulfate Resin*; Abstract of 218th ACS National Meeting, New Orleans, Aug. 22–26, 1999; ORGN-424.
811. Parlow, J. J.; Naing, W.; South, M. S.; Flynn, D. L. *Tetrahedron Lett.* **1997**, *38*, 7959–7962.
812. Thompson, L. A.; Combs, A. P.; Trainor, G. L.; Wang, Q.; Langlois, T. J.; Kirkland, J. J. *Comb. Chem. High Throughput Screening* **2000**, *3*, 107–115.
813. Adamczyk, M.; Fishpaugh, J. R.; Mattingly, P. G. *Tetrahedron Lett.* **1999**, *40*, 463–466.
814. Salvino, J. M.; Groneberg, R. D.; Airey, J. E.; Poli, G. B.; McGeehan, G. M.; Labaudiniere, R. F.; Clerc, F.-f.; Bezard, D. N. A. Fluorophenyl Resin Compounds. WO, 1999, 113; Vol. 99/67288 A1.
815. Drew, M.; Orton, E.; Krolikowski, P.; Salvino, J. M.; Kumar, N. V. *J. Comb. Chem.* **2000**, *2*, 8–9.
816. Hahn, H.-G.; Kee, H. C.; Kee, D. N.; Bae, S. Y.; Mah, H. *Heterocycles* **1998**, *48*, 2253–2261.
817. Smith, R. A.; Bobko, M. A.; Lee, W. *Bioorg. Med. Chem. Lett.* **1998**, *8*, 2369–2374.
818. Chang, Y.-T.; Schultz, P. G. *Bioorg. Med. Chem. Lett.* **1999**, *9*, 2479–2482.
819. Kenner, G. W.; McDermott, J. R.; Sheppard, R. C. *Chem. Commun.* **1971**, 636–637.
820. Backes, B. J.; Virgilio, A. A.; Ellman, J. A. *J. Am. Chem. Soc.* **1996**, *118*, 3055–3056.
821. Backes, B. J.; Ellman, J. A. *J. Org. Chem.* **1999**, *64*, 2322–2330.
822. Scialdone, M. A.; Shuey, S. W.; Soper, P.; Hamuro, Y.; Burns, D. M. *J. Org. Chem.* **1998**, *63*, 4802–4807.
823. Katritzky, A. R.; Belyakov, S. A.; Tymoshenko, D. O. *J. Comb. Chem.* **1999**, *1*, 173–176.
824. Smith, E. M. *Tetrahedron Lett.* **1999**, *40*, 3285–3288.
825. Hu, Y.; Porco, J. A., Jr. *Tetrahedron Lett.* **1999**, *40*, 3289–3292.
826. Caldarelli, M.; Habermann, J.; Ley, S. V. *Bioorg. Med. Chem. Lett.* **1999**, *9*, 2049–2052.
827. Ley, S. V.; Schucht, O.; Thomas, A. W.; Murray, P. J. *J. Chem. Soc. Perkin Trans.* **1999**, 1251–1252.
828. Habermann, J.; Ley, S. V.; Scott, J. S. *J. Chem. Soc. Perkin Trans.* **1999**, 1253–1256.
829. Merrifield, R. B. *J. Am. Chem. Soc.* **1963**, *85*, 2149.
830. Gallop, M. A.; Barrett, R. W.; Dower, W. J.; Fodor, S. P. A.; Gordon, E. M. *J. Med. Chem.* **1994**, *37*, 1233.
831. Thompson, L. A.; Ellman, J. A. *Chem. Rev.* **1996**, *96*, 555.
831a. Balkenhohl, F.; von dem Bussche-Hunnefeld, C.; Lansky, A.; Zechel, C. *Angew. Chem. Int. Ed. Engl.* **1996**, *35*, 2288.
832. Leznoff, C. C. *Acc. Chem. Res.* **1978**, *11*, 327.
833. Hodge, P. *Chem. Soc. Rev.* **1997**, *26*, 417.
834. Sherrington, D. C. *Chem. Commun.* **1998**, 2275.
835. MacDonald, A. A.; Dewitt, S. H.; Ghosh, S.; Hogan, F. M.; Kieras, L.; Czarnik, A. W.; Ramage, R. *Mol. Divers.* **1996**, *1*, 183.
836. Hori, M.; Gravert, D. J.; Wentworth, P.; Janda, K. D. *Bioorg. Med. Chem. Lett.* **1998**, *8*, 2363.
837. Atherton, E.; Sheppard, R. C. *J. Chem. Soc. Perkin Trans.* **1981**, 529.
838. Zalipsky, S.; Chang, J. L.; Albericio, F.; Barany, G. *React. Polymers* **1994**, *22*, 243.
839. Bayer, E.; Angew, E. *Chem. Int.* **1991**, *30*, 113–129.
840. Bayer, E.; Albert, K.; Willish, H.; Rapp, W.; Hemmasi, B. *Macromolecules* **1990**, *23*, 1937.
841. Rapp, W. In *Combinatorial Libaries*; Jung, G. Ed.; VCH: Weinheim, Germany; 1996; Chapter 16.
842. Gooding, O. W.; Baudart, S.; Deegan, T. L.; Heisler, K.; Labadie, J. W.; Newcomb, W. S.; Porco, J. A.; Van Eikeren, P. *J. Comb. Chem.* **1999**, *1*, 113.
843. Kempe, M.; Barany, G. *J. Am. Chem. Soc.* **1996**, *118*, 7083.
844. Renil, M.; Nagaraj, R.; Pillai, V. N. R. *Tetrahedron* **1994**, *50*, 6681.
844a. Renil, M.; Pillai, V. N. R. *J. Appl. Polym. Sci.* **1996**, *61*, 1585.
845. Willson, M. E.; Paech, K.; Zhou, W.-J.; Kurth, M. J. *J. Org. Chem.* **1998**, *63*, 5094.
846. Meldal, M. *Tetrahedron Lett.* **1992**, *33*, 3077–3080.
846a. Meldal, M.; Auzanneau, F. I.; Hindsgaul, O.; Palcic, M. M. *J. Chem. Soc. Chem. Commun.* **1994**, 1849.
847. Renil, M.; Meldal, M. *Tetrahedron Lett.* **1995**, *36*, 4647–4650.
847a. Renil, M.; Ferreras, M.; Delaisse, J. M.; Foged, N. T.; Meldal, M. *J. Pept. Sci.* **1998**, *4*, 195.
848. Tam, J. P. *Proc. Natl. Acad. Sci. USA* **1988**, *85*, 5409.
849. Swali, V.; Wells, N. J.; Langley, G. J.; Bradley, M. *J. Org. Chem.* **1997**, *62*, 4902.
850. Wells, N. J.; Davies, M.; Bradley, M. *J. Org. Chem.* **1998**, *63*, 6430.
851. Fromont, C.; Bradley, M. *Chem. Commun.* **2000**, 283.
852. Houghten, R. A. *Proc. Natl. Acad. Sci. USA* **1985**, *82*, 5131.
853. Geysen, H. M.; Meloen, R. B.; Barteling, S. J. *Proc. Natl. Acad. Sci. USA* **1984**, *81*, 3998.
853a. Geysen, H. M.; Barteling, S. J.; Meloen, R. B. *Proc. Natl. Acad. Sci. USA* **1985**, *82*, 178.

854. Bray, A. M.; Maeji, N. J.; Geysen, H. M. *Tetrahedron Lett.* **1990**, *31*, 5811.
854a. Bray, A. M.; Maeji, N. J.; Jhingran, A. G.; Valerio, R. M. *Tetrahedron Lett.* **1991**, *32*, 6163.
855. Maeji, N. J.; Valerio, R. M.; Bray, A. M.; Campbell, R. A.; Geysen, H. M. *Int. React. Polym.* **1994**, *22*, 203.
856. Valerio, R. M.; Bray, A. M.; Campbell, R. A.; Dipasquale, A.; Margellis, C.; Rodda, S. J.; Geysen, H. M.; Maeji, N. J. *Int. J. Pept. Protein Res.* **1993**, *42*, 1.
857. Bray, A. M.; Maeji, N. J.; Valerio, R. M.; Campbell, R. A.; Geysen, H. M. *J. Org. Chem.* **1991**, *56*, 6659.
858. Bray, A. M.; Jhingran, A. J.; Valerio, R. M.; Maeji, N. J. *J. Org. Chem.* **1994**, *59*, 2197.
859. Valerio, R. M.; Bray, A. M.; Maeji, N. J. *Int. J. Pept. Protein Res.* **1994**, *44*, 158.
860. Bray, A. M.; Chiefari, D. S.; Valerio, R. M.; Maeji, N. J. *Tetrahedron Lett.* **1995**, *36*, 5081–5084.
860a. Valerio, R. M.; Bray, A. M.; Patsiouras, H. *Tetrahedron Lett.* **1996**, *37*, 3019.
861. Virgilio, A. A.; Ellman, J. A. *J. Am. Chem. Soc.* **1994**, *116*, 11580.
861a. Bunin, B. A.; Plunkett, M. J.; Ellman, J. A. *New J. Chem.* **1997**, *21*, 125.
861b. Virgilio, A. A.; Bray, A. A.; Zhang, W.; Trinh, L.; Snyder, M.; Morrissey, M. M.; Ellman, J. A. *Tetrahedron Lett.* **1997**, *91*, 4708.
861c. Bunin, B. A.; Ellman, J. A. *J. Am. Chem. Soc.* **1992**, *114*, 10997–10998.
861d. Bunin, B. A.; Plunkett, M. J.; Ellman, J. A. *Proc. Natl. Acad. Sci. USA* **1994**, *91*, 4708.
862. Zhao, C.; Shi, S.; Mir, D.; Hurst, D.; Li, R.; Xiao, X.-y.; Lillig, J.; Czarnik, A. W. *J. Comb. Chem.* **1999**, *1*, 91.
863. Berg, R. H.; Almdal, K.; Pederson, W. B.; Holm, A.; Tam, J. P.; Merrifield, R. B. *J. Am. Chem. Soc.* **1989**, *111*, 8024.
864. Daniels, S. C.; Bernatowicz, M. S.; Coull, J. M.; Köster, H. *Tetrahedron Lett.* **1989**, *30*, 4345.
864a. Gao, B., Esnouf, M. P. *J. Immunol.* **1996**, 183.
865. Frank, R. *Tetrahedron* **1992**, *48*, 9217–9232.
865a. Gao, B.; Esnouf, M. P. *J. Biol. Chem.* **1996**, *271*, 24634.
866. Terrett, N. K.; Gardner, M.; Gordon, D. W.; Kobylecki, R. J.; Steele, J. *Chem. Eur. J.* **1997**, *3*, 1917.
867. Devivar, R. V.; Koontz, S. L.; Peltier, W. J.; Pearson, J. E.; Guillory, T. A.; Fabricant, J. D. *Bioorg. Med. Chem. Lett.* **1999**, *9*, 1239.
868. Luo, K. X.; Zhou, P.; Lodish, H. F. *Proc. Natl. Acad. Sci. USA* **1995**, *92*, 11761.
868a. Zhao, C.; Shi, S.; Mir, D.; Hurst, D.; Li, R.; Xiao, X.-y.; Lillig, J.; Czarnik, A. W. *J. Comb. Chem.* **1999**, *1*, 91.
869. Fodor, S. P. A.; Read, J. L.; Pirrung, M. C.; Stryer, L.; Lu, A. T.; Solas, D. *Science* **1991**, 767.
869a. Pirrung, M. C. *Chem. Rev.* **1997**, *97*, 473.
870. Pease, A. C.; Solas, D.; Sullivan, E. J.; Cronin, M. T.; Holmes, C. P.; Fodor, S. P. A. *Proc. Natl. Acad. Sci. USA* **1994**, *91*, 5022.
871. McGall, G.; Labadie, J.; Brock, P.; Wallraff, G.; Nguyen, T.; Hinsberg, W. *Proc. Natl. Acad. Sci. USA* **1996**, *93*, 13555.
871a. Pirrung, M. C.; Fallon, L.; McGall, G. *J. Org. Chem.* **1998**, *63*, 241.
872. Wallraff, G. M.; Hinsberg, W. D. *Chem. Rev.* **1999**, *99*, 1801.
873. Maskos, U.; Southern, E. M. *Nucleic Acids Res.* **1993**, *21*, 2267–2269.
874. Southern, E. M.; Case-Green, S. C.; Elder, J. K.; Johnson, M.; Mir, U.; Wang, L.; Williams, J. C. *Nucleic Acids Res.* **1994**, *22*, 1368.
875. Grant, D. M.; Harris, R. K. Eds. *Encyclopedia of NMR*, Vol. 1. *Historical Perspectives*; John Wiley: Chichester, UK, 1996.
876. Croasmun, W. R.; Carlson, R. M. K. *Two-Dimensional NMR Spectroscopy: Applications for Chemists and Biochemists*; 2nd ed.; VCH: New York, 1994.
877. Martin, G. E.; Zektzer, A. S. *Two-Dimensional NMR Methods for Establishing Molecular Connectivity*; VCH: New York, 1988.
878. Kessler, H.; Gehrke, M.; Griesinger, C. *Angew. Chem.* **1988**, *27*, 490–536.
879. Clore, G. M.; Gronenborn, A. M. *Spectroscopy* **1991**, *23*, 43–92.
880. Oschkinat, H.; Mueller, T.; Dieckmann, T. *Angew. Chem. Int. Ed. Engl.* **1994**, *33*, 277–293.
881. Clore, G. M.; Gronenborn, A. M. *Science* **1991**, *252*, 1390–1399.
882. Shaka, A. J. Decoupling Methods. In *Encyclopedia of Nuclear Magnetic Resonance*; Grant, D. M., Harris, R. K., Eds.; John Wiley: Chichester, UK, 1996, pp 1558–1564.
883. Shaka, A. J.; Keeler, J. *Prog. Nucl. Magn. Reson. Spectrosc.* **1987**, *19*, 47–129.
884. Fujiwara, T.; Nagayama, K. *J. Magn. Reson.* **1988**, 77, 53–63.
885. Kupce, E.; Freeman, R. *J. Magn. Reson.* **1995**, *115*, 273–276.
886. Bendall, M. R. *J. Magn. Reson.* **1995**, *112*, 126–129.
887. Silver, M. S.; Joseph, R. I.; Hoult, D. I. *J. Magn. Reson.* **1984**, *59*, 347–351.
888. Bothner-By, A. A.; Stephens, R. L.; Lee, J. M.; Warren, C. D.; Jeanloz, R. W. *J. Am. Chem. Soc.* **1984**, *106*, 811–813.
889. Bax, A.; Davis, D. G. *J. Magn. Reson.* **1985**, *63*, 207–213.
890. Kessler, H.; Griesinger, C.; Kerssebaum, R.; Wagner, K.; Ernst, R. R. *J. Am. Chem. Soc.* **1987**, *109*, 607–609.
891. Braunschweiler, L.; Ernst, R. R. *J. Magn. Reson.* **1983**, *53*, 521–528.
892. Bax, A.; Davis, D. G. *J. Magn. Reson.* **1985**, *65*, 355–360.
893. Kupce, E.; Schmidt, P.; Rance, M.; Wagner, G. *J. Magn. Reson.* **1998**, *135*, 361–367.
894. Messerle, B. A.; Wider, G.; Otting, G.; Weber, C.; Wuethrich, K. *J. Magn. Reson.* **1989**, *85*, 608–613.
895. Otting, G.; Wuethrich, K. *J. Magn. Reson.* **1988**, *76*, 569–574.
896. Vold, R. L.; Waugh, J. S.; Klein, M. P.; Phelps, D. E. *J. Chem. Phys.* **1968**, *48*, 3831–3832.
897. McDonald, S.; Warren, W. S. *Concepts Magn. Reson.* **1991**, *3*, 55–81.
898. Bauer, C.; Freeman, R.; Frenkiel, T.; Keeler, J.; Shaka, A. J. *J. Magn. Reson.* **1984**, *58*, 442–457.
899. Davis, D. G.; Bax, A. *J. Am. Chem. Soc.* **1985**, *107*, 7197–7198.
900. Anet, F. A. L.; Bourn, A. J. R. *J. Am. Chem. Soc.* **1965**, *87*, 5250–5251.
901. Stott, K.; Keeler, J.; Van, Q. N.; Shaka, A. J. *J. Magn. Reson.* **1997**, *125*, 302–324.
902. Stott, K.; Stonehouse, J.; Keeler, J.; Hwang, T.-L.; Shaka, A. J. *J. Am. Chem. Soc.* **1995**, *117*, 4199–4200.
903. Berger, S. *J. Magn. Reson.* **1989**, *81*, 561–564.
904. Crouch, R. C.; Martin, G. E. *J. Magn. Reson.* **1991**, *92*, 189–194.
905. Berger, S. *Angew. Chem.* **1988**, *100*, 1198–1199.
906. Smallcombe, S. H. *J. Am. Chem. Soc.* **1993**, *115*, 4776–4785.
907. Liu, H.; Weisz, K.; James, T. L. *J. Magn. Reson.* **1993**, *105*, 184–192.
908. Smallcombe, S. H.; Patt, S. L.; Keifer, P. A. *J. Magn. Reson.* **1995**, *117*, 295–303.
909. Hore, P. J. *J. Magn. Reson.* **1983**, *54*, 539–542.

910. Piotto, M.; Saudek, V.; Sklenar, V. *J. Biomol. Nucl. Magn. Reson.* **1992**, *2*, 661–665.
911. Morris, G. A.; Freeman, R. *J. Magn. Reson.* **1978**, *29*, 433–462.
912. Live, D. H.; Greene, K. *J. Magn. Reson.* **1989**, *85*, 604–607.
913. Wang, A. C.; Grzesiek, S.; Tschudin, R.; Lodi, P. J.; Bax, A. *J. Biomol. Nucl. Magn. Reson.* **1995**, *5*, 376–382.
914. McCoy, M. A.; Mueller, L. *J. Magn. Reson.* **1993**, *101*, 122–130.
915. McCoy, M. A.; Mueller, L. *J. Am. Chem. Soc.* **1992**, *114*, 2108–2112.
916. Bendall, M. R. *J. Magn. Reson. A* **1995**, *116*, 46–58.
917. Hwang, T.-L.; Van Zijl, P. C. M. *J. Magn. Reson.* **1999**, *138*, 173–177.
918. Hwang, T.-L.; Van Zijl, P. C. M.; Garwood, M. *J. Magn. Reson.* **1998**, *133*, 200–203.
919. Patt, S. L. *J. Magn. Reson.* **1992**, *96*, 94–102.
920. Mueller, L. *J. Am. Chem. Soc.* **1979**, *101*, 4481–4484.
921. Bax, A.; Ikura, M.; Kay, L. E.; Torchia, D. A.; Tschudin, R. *J. Magn. Reson.* **1990**, *86*, 304–318.
922. Reynolds, W. F.; McLean, S.; Tay, L.-L.; Yu, M.; Enriquez, R. G.; Estwick, D. M.; Pascoe, K. O. *Magn. Reson. Chem.* **1997**, *35*, 455–462.
923. Rinehart, K. L.; Holt, T. G.; Fregeau, N. L.; Stroh, J. G.; Keifer, P. A.; Sun, F.; Li, L. H.; Martin, D. G. *J. Org. Chem.* **1990**, *55*, 4512–4515.
924. Mukherjee, R.; Da Silva, B. A.; Das, B. C.; Keifer, P. A.; Shoolery, J. N. *Heterocycles* **1991**, *32*, 985–990.
925. Mukherjee, R.; Das, B. C.; Keifer, P. A.; Shoolery, J. N. *Heterocycles* **1994**, *38*, 1965–1970.
926. Crews, P.; Farias, J. J.; Emrich, R.; Keifer, P. A. *J. Org. Chem.* **1994**, *59*, 2932–2934.
927. Vervoort, H. C.; Fenical, W.; Keifer, P. A. *J. Nat. Prod.* **1999**, *62*, 389–391.
928. Lauterbur, P. C. *Nature* **1973**, *242*, 190–191.
929. Barker, P.; Freeman, R. *J. Magn. Reson.* **1985**, *64*, 334–338.
930. Vuister, G. W.; Boelens, R.; Kaptein, R.; Hurd, R. E.; John, B.; Van Zijl, P. C. M. *J. Am. Chem. Soc.* **1991**, *113*, 9688–9690.
931. Hurd, R. E.; John, B. K. *J. Magn. Reson.* **1991**, *91*, 648–653.
932. Hurd, R. E. Field Gradients and Their Applications. In *Encyclopedia of Nuclear Magnetic Resonance*; Grant, D. M., Harris, R. K., Eds.; John Wiley: Chichester, UK, 1996, pp 1990–2005.
933. Price, W. S. Gradient NMR. In *Annual Reports on NMR Spectroscopy*; Webb, G. A., Ed.; Academic Press: San Diego, CA, 1996; Vol. 32, pp 51–142.
934. Sarkar, S. K.; Kapadia, R. D. Magnetic Resonance Imaging in Drug Discovery Research. In *Magnetic Resonance Microscopy*; Bluemich, B., Kuhn, W., Eds.; VCH: Weinheim, Germany, 1992, pp 513–531.
935. Sukumar, S.; Johnson, M. O. N.; Hurd, R. E.; Van Zijl, P. C. M. *J. Magn. Reson.* **1997**, *125*, 159–162.
936. Barjat, H.; Chilvers, P. B.; Fetler, B. K.; Horne, T. J.; Morris, G. A. *J. Magn. Reson.* **1997**, *125*, 197–201.
937. Rinaldi, P. L.; Keifer, P. *J. Magn. Reson.* **1994**, *108*, 259–262.
938. Rinaldi, P. L.; Ray, D. G.; Litman, V. E.; Keifer, P. A. *Polym. Int.* **1995**, *36*, 177–185.
939. Altieri, A. S.; Miller, K. E.; Byrd, R. A. *Magn. Reson. Rev.* **1996**, *17*, 27–81.
940. Ruiz-Cabello, J.; Vuister, G. W.; Moonen, C. T. W.; Van Zijl, P. C. M. *J. Magn. Reson.* **1992**, *100*, 282–302.
941. Tolman, J. R.; Chung, J.; Prestegard, J. H. *J. Magn. Reson.* **1992**, *98*, 462–467.
942. Davis, A. L.; Keeler, J.; Laue, E. D.; Moskau, D. *J. Magn. Reson.* **1992**, *98*, 207–216.
943. Shaw, A. A.; Salaun, C.; Dauphin, J.-F.; Ancian, B. *J. Magn. Reson. A* **1996**, *120*, 110–115.
944. Hurd, R. E. *J. Magn. Reson.* **1990**, *87*, 422–428.
945. Sheng, S.; Van Halbeek, H. *J. Magn. Reson.* **1998**, *130*, 296–299.
946. Marek, R.; Kralik, L.; Sklenar, V. *Tetrahedron Lett.* **1997**, *38*, 665–668.
947. Reif, B.; Koeck, M.; Kerssebaum, R.; Schleucher, J.; Griesinger, C. *J. Magn. Reson.* **1996**, *112*, 295–301.
948. Reif, B.; Kock, M.; Kerssebaum, R.; Kang, H.; Fenical, W.; Griesinger, C. *J. Magn. Reson.* **1996**, *118*, 282–285.
949. Krishnamurthy, V. V. *J. Magn. Reson.* **1996**, *121*, 33–41.
950. Wagner, R.; Berger, S. *Magn. Reson. Chem.* **1998**, *36*, S44–S46.
951. Von Kienlin, M.; Moonen, C. T. W.; Van der Toorn, A.; Van Zijl, P. C. M. *J. Magn. Reson.* **1991**, *93*, 423–429.
952. Kay, L.; Keifer, P.; Saarinen, T. *J. Am. Chem. Soc.* **1992**, *114*, 10663–10665.
953. Uhrin, D.; Barlow, P. N. *J. Magn. Reson.* **1997**, *126*, 248–255.
954. Holt, R. M.; Newman, M. J.; Pullen, F. S.; Richards, D. S.; Swanson, A. G. *J. Mass Spectrom.* **1997**, *32*, 64–70.
955. Ehlhardt, W. J.; Woodland, J. M.; Baughman, T. M.; Vandenbranden, M.; Wrighton, S. A.; Kroin, J. S.; Norman, B. H.; Maple, S. R. *Drug Metab. Dispos.* **1998**, *26*, 42–51.
956. Keifer, P. A.; Smallcombe, S. H.; Williams, E. H.; Salomon, K. E.; Mendez, G.; Belletire, J. L.; Moore, C. D. *J. Comb. Chem.* **2000**, *2*, 151–171.
957. Stejskal, E. O.; Tanner, J. E. *J. Chem. Phys.* **1965**, *42*, 288–292.
958. Lin, M.; Shapiro, M. J.; Wareing, J. R. *J. Am. Chem. Soc.* **1997**, *119*, 5249–5250.
959. Lin, M.; Shapiro, M. J.; Wareing, J. R. *J. Org. Chem.* **1997**, *62*, 8930–8931.
960. Van Zijl, P. C. M.; Moonen, C. T. W. *J. Magn. Reson.* **1990**, *87*, 18–25.
961. Cavanagh, J.; Fairbrother, W. J.; Palmer, A. G., III.; Skelton, N. J. *Protein NMR Spectroscopy – Principles and Practice*; Academic Press: San Diego, CA, 1996.
962. Kay, L. E.; Ikura, M.; Tschudin, R.; Bax, A. *J. Magn. Reson.* **1990**, *89*, 496–514.
963. Ikura, M.; Kay, L. E.; Bax, A. *Biochemistry* **1990**, *29*, 4659–4667.
964. Marion, D.; Kay, L. E.; Sparks, S. W.; Torchia, D. A.; Bax, A. *J. Am. Chem. Soc.* **1989**, *111*, 1515–1517.
965. Wider, G. *Prog. Nucl. Magn. Reson. Spectrosc.* **1998**, *32*, 193–275.
966. Biamonti, C.; Rios, C. B.; Lyons, B. A.; Montelione, G. T. *Adv. Biophys. Chem.* **1994**, *4*, 51–120.
967. Montelione, G. T.; Emerson, S. D.; Lyons, B. A. *Biopolymers* **1992**, *32*, 327–334.
968. Zwahlen, C.; Gardner, K. H.; Sarma, S. P.; Horita, D. A.; Byrd, R. A.; Kay, L. E. *J. Am. Chem. Soc.* **1998**, *120*, 7617–7625.
969. Venters, R. A.; Huang, C.-C.; Farmer, B. T. I.; Trolard, R.; Spicer, L. D.; Fierke, C. A. *J. Biomol. Nucl. Magn. Reson.* **1995**, *5*, 339–344.
970. Pervushin, K.; Riek, R.; Wider, G.; Wuthrich, K. *Proc. Natl. Acad. Sci. USA* **1997**, *94*, 12366–12371.
971. Wuthrich, K. *Nat. Struct. Biol.* **1998**, *5*, 492–495.
972. Pervushin, K.; Ono, A.; Fernandez, C.; Szyperski, T.; Lainosho, M.; Wüthrich, K. *Proc. Natl. Acad. Sci. USA* **1998**, *95*, 14147–14151.

973. Dingley, A. J.; Grzesiek, S. *J. Am. Chem. Soc.* **1998**, *120*, 8293–8297.
974. Cordier, F.; Grzesiek, S. *J. Am. Chem. Soc.* **1999**, *121*, 1601–1602.
975. Cornilescu, G.; Hu, J.-S.; Bax, A. *J. Am. Chem. Soc.* **1999**, *121*, 2949–2950.
976. Dingley, A. J.; Masse, J. E.; Peterson, R. D.; Barfield, M.; Feigon, J.; Grzesiek, S. *J. Am. Chem. Soc.* **1999**, *121*, 6019–6027.
977. Ikura, M.; Bax, A. *J. Am. Chem. Soc.* **1992**, *114*, 2433–2440.
978. Zwahlen, C.; Legault, P.; Vincent, S. J. F.; Greenblatt, J.; Konrat, R.; Kay, L. E. *J. Am. Chem. Soc.* **1997**, *119*, 6711–6721.
979. Tjandra, N.; Omichinski, J. G.; Gronenborn, A. M.; Clore, G. M.; Bax, A. *Nat. Struct. Biol.* **1997**, *4*, 732–738.
980. Tjandra, N.; Bax, A. *Science* **1997**, *278*, 1111–1114.
981. Tjandra, N.; Grzesiek, S.; Bax, A. *J. Am. Chem. Soc.* **1996**, *118*, 6264–6272.
982. Gueron, M.; Plateau, P.; Decorps, M. *Prog. Nucl. Magn. Reson. Spectrosc.* **1991**, *23*, 135–209.
983. Gueron, M.; Plateau, P. Water Signal Suppression in NMR of Biomolecules. In *Encyclopedia of Nuclear Magnetic Resonance*; Grant, D. M., Harris, R. K., Eds.; John Wiley: Chichester, UK, 1996, pp 4931–4942.
984. Crouch, R. C.; Martin, G. E. *J. Nat. Prod.* **1992**, *55*, 1343–1347.
985. Crouch, R. C.; Martin, G. E. *Magn. Reson. Chem.* **1992**, *30*, S66–S70.
986. Shoolery, J. N. *Spectroscopy* **1979**, *3*, 28–38.
987. Martin, G. E.; Guido, J. E.; Robins, R. H.; Sharaf, M. H. M.; Schiff, P. L., Jr.; Tackie, A. N. *J. Nat. Prod.* **1998**, *61*, 555–559.
988. Keifer, P. A. *Drug Disc. Today* **1997**, *2*, 468–478.
989. Keifer, P. A. *Drugs Future* **1998**, *23*, 301–317.
990. Keifer, P. A.; Baltusis, L.; Rice, D. M.; Tymiak, A. A.; Shoolery, J. N. *J. Magn. Reson. A* **1996**, *119*, 65–75.
991. Springer, C. S., Jr. Physiochemical Principles Influencing Magnetopharmaceuticals. In *Physiology and Biomedicine*; Gillies, N. M. R., Ed.; Academic Press: San Diego, CA, 1994, pp 75–99.
992. Barbara, T. M. *J. Magn. Reson. A* **1994**, *109*, 265–269.
993. Manzi, A.; Salimath, P. V.; Spiro, R. C.; Keifer, P. A.; Freeze, H. H. *J. Biol. Chem.* **1995**, *270*, 9154–9163.
994. Manzi, A. E.; Keifer, P. A. New Frontiers in Nuclear Magnetic Resonance Spectroscopy: Use of a nanoNMR Probe for the Analysis of Microgram Quantities of Complex Carbohydrates. In *Techniques in Glycobiology*; Townsend, R. R., Hotchkiss, A. T., Jr., Eds.; Marcel Dekker: New York, 1997, pp 1–16.
995. Klein, D.; Braekman, J. C.; Daloze, D.; Hoffman, L.; Demoulin, V. *Tetrahedron Lett.* **1996**, *37*, 7519–7520.
996. Klein, D.; Braekman, J. C.; Daloze, D.; Hoffman, L.; Castillo, G.; Demoulin, V. *Tetrahedron Lett.* **1999**, *40*, 695–696.
997. Chauret, D. C.; Durst, T.; Arnason, J. T.; Sanchez-Vindas, P.; Roman, L. S.; Poveda, L.; Keifer, P. A. *Tetrahedron Lett.* **1996**, *37*, 7875–7878.
998. Harper, J. K.; Dunkel, R.; Wood, S. G.; Owen, N. L.; Li, D.; Cates, R. G.; Grant, D. M. *J. Chem. Soc. Perkin Trans. 2* **1996**, *1*, 91–100.
999. MacKinnon, S. L.; Keifer, P.; Ayer, W. A. *Phytochemistry* **1999**, *51*, 215–221.
1000. Delepierre, M.; Prochnicka-Chalufour, A.; Possani, L. D. *Biochemistry* **1997**, *36*, 2649–2658.
1001. Roux, P.; Delepierre, M.; Goldberg, M. E.; Chaffotte, A. F. *J. Biol. Chem.* **1997**, *272*, 24843–24849.
1002. Delepierre, M.; Roux, P.; Chaffotte, A. F.; Goldberg, M. E. *Magn. Reson. Chem.* **1998**, *36*, 645–650.
1003. Delepierre, M.; Porchnicka-Chalufour, A.; Boisbouvier, J.; Possani, L. D. *Biochemistry* **1999**, *38*, 16756–16765.
1004. Keifer, P. A. *Curr. Opin. Biotechnol.* **1999**, *10*, 34–41.
1005. Doskocilová, D.; Schneider, B. *Chem. Phys. Lett.* **1970**, *6*, 381–384.
1006. Doskocilová, D.; Dang, D. T.; Schneider, B. *Czech. J. Phys. B* **1975**, *25*, 202–209.
1007. Fuks, L. F.; Huang, F. S. C.; Carter, C. M.; Edelstein, W. A.; Roemer, P. B. *J. Magn. Reson.* **1992**, *100*, 229–242.
1008. Doty, F. D.; Entzminger, G.; Yang, Y. A. *Concepts Magn. Reson.* **1998**, *10*, 239–260.
1009. Keifer, P. A. *J. Org. Chem.* **1996**, *61*, 1558–1559.
1010. Keifer, P. A.; Sehrt, B. *A Catalog of ^1H NMR Spectra of Different SPS Resins with Varying Solvents and Experimental Techniques – An Exploration of Nano-Nmr Probe Technology*; Varian NMR Instruments: Palo Alto, CA, 1996.
1011. Giralt, E.; Rizo, J.; Pedroso, E. *Tetrahedron* **1984**, *40*, 4141–4152.
1012. Epton, R.; Goddard, P.; Ivin, K. J. *Polymer* **1980**, *21*, 1367–1371.
1013. Hochlowski, J. E.; Whittern, D. N.; Sowin, T. J. *J. Comb. Chem.* **1999**, *1*, 291–293.
1014. Wehler, T.; Westman, J. *Tetrahedron Lett.* **1996**, *37*, 4771–4774.
1015. Fitch, W. L.; Detre, G.; Holmes, C. P.; Shoolery, J. N.; Keifer, P. A. *J. Org. Chem.* **1994**, *59*, 7955–7956.
1016. Sarkar, S. K.; Garigipati, R. S.; Adams, J. L.; Keifer, P. A. *J. Am. Chem. Soc.* **1996**, *118*, 2305–2306.
1017. Cheng, L. L.; Chang, I. W.; Louis, D. A.; Gonzalez, R. G. *Cancer Res.* **1998**, *58*, 1825–1832.
1018. Moka, D.; Vorreuther, R.; Schicha, H.; Spraul, M.; Humpfer, E.; Lipinski, M.; Foxall, P. J. D.; Nicholson, J. K.; Lindon, J. C. *J. Pharm. Biomed. Anal.* **1998**, *17*, 125–132.
1019. Weybright, P.; Millis, K.; Campbell, N.; Cory, D. G.; Singer, S. *Magn. Reson. Med.* **1998**, *39*, 337–344.
1020. Kupce, E.; Keifer, P. A.; Delepierre, M. *J. Magn. Reson.* **2001**, *148*, 115–120.
1021. Dorn, H. C. *Anal. Chem.* **1984**, *56*, 747A–758A.
1022. Albert, K.; Bayer, E. *Trends Anal. Chem.* **1988**, *7*, 288–293.
1023. Albert, K. *J. Chromatogr.* **1995**, *703*, 123–147.
1024. Lindon, J. C.; Nicholson, J. K.; Wilson, I. D. *Prog. Nucl. Magn. Reson. Spectrosc.* **1996**, *29*, 1–49.
1025. Lindon, J. C.; Nicholson, J. K.; Sidelmann, U. G.; Wilson, I. D. *Drug Metab. Rev.* **1997**, *29*, 705–746.
1026. Mistry, N.; Ismail, I. M.; Smith, M. S.; Nicholson, J. K.; Lindon, J. C. *J. Pharm. Biomed. Anal.* **1997**, *16*, 697–705.
1027. Roberts, J. K.; Smith, R. J. *J. Chromatogr.* **1994**, *677*, 385–389.
1028. Chin, J.; Fell, J. B.; Jarosinski, M.; Shapiro, M. J.; Wareing, J. R. *J. Org. Chem.* **1998**, *63*, 386–390.
1029. Wolfender, J.-L.; Rodriguez, S.; Hostettmann, K. *J. Chromatogr.* **1998**, *794*, 299–316.
1030. Spring, O.; Heil, N.; Vogler, B. *Phytochemistry* **1997**, *46*, 1369–1373.
1031. Wolfender, J.-L.; Rodriguez, S.; Hostettmann, K.; Hiller, W. *Phytochem. Anal.* **1997**, *8*, 97–104.
1032. Strohschein, S.; Rentel, C.; Lacker, T.; Bayer, E.; Albert, K. *Anal. Chem.* **1999**, *71*, 1780–1785.
1033. Strohschein, S.; Schlotterbeck, G.; Richter, J.; Pursch, M.; Tseng, L.-H.; Haendel, H.; Albert, K. *J. Chromatogr.* **1997**, *765*, 207–214.
1034. Strohschein, S.; Pursch, M.; Handel, H.; Albert, K. *Fresenius J. Anal. Chem.* **1997**, *357*, 498–502.
1035. Schlotterbeck, G.; Tseng, L.-H.; Haendel, H.; Braumann, U.; Albert, K. *Anal. Chem.* **1997**, *69*, 1421–1425.

1036. Sidelmann, U. G.; Lenz, E. M.; Spraul, M.; Hoffmann, M.; Troke, J.; Sanderson, P. N.; Lindon, J. C.; Wilson, I. D.; Nicholson, J. K. *Anal. Chem.* **1996**, *68*, 106–110.
1037. Albert, K.; Schlotterbeck, G.; Braumann, U.; Haendel, H.; Spraul, M.; Krack, G. *Angew. Chem.* **1995**, *34*, 1014–1016.
1038. Griffiths, L.; Horton, R. *Magn. Reson. Chem.* **1998**, *36*, 104–109.
1039. Pullen, F. S.; Swanson, A. G.; Newman, M. J.; Richards, D. S. *Rapid Commun. Mass Spectrom.* **1995**, *9*, 1003–1006.
1040. Burton, K. I.; Everett, J. R.; Newman, M. J.; Pullen, F. S.; Richards, D. S.; Swanson, A. G. *J. Pharm. Biomed. Anal.* **1997**, *15*, 1903–1912.
1041. Hamper, B. C.; Synderman, D. M.; Owen, T. J.; Scates, A. M.; Owsley, D. C.; Kesselring, A. S.; Chott, R. C. *J. Comb. Chem.* **1999**, *1*, 140–150.
1042. Williams, A.; Bakulin, S.; Golotvin, S. NMR Prediction Software and Tubeless NMR – An Analytical Tool for Screening of Combinatorial Libraries. First Annual SMASH Small Molecule NMR Conference, Argonne, IL, Aug. 15–18, 1999; Poster 2.
1043. Shuker, S. B.; Hajduk, P. J.; Meadows, R. P.; Fesik, S. W. *Science* **1996**, *274*, 1531–1534.
1044. Hajduk, P. J.; Dinges, J.; Miknis, G. F.; Merlock, M.; Middleton, T.; Kempf, D. J.; Egan, D. A.; Walter, K. A.; Robins, T. S.; Shuker, S. B. et al. *J. Med. Chem.* **1997**, *40*, 3144–3150.
1045. Hajduk, P. J.; Sheppard, G.; Nettesheim, D. G.; Olejniczak, E. T.; Shuker, S. B.; Meadows, R. P.; Steinman, D. H.; Carrera, G. M., Jr.; Marcotte, P. A.; Severin, J. et al. *J. Am. Chem. Soc.* **1997**, *119*, 5818–5827.
1046. Morris, K. F.; Johnson, C. S., Jr. *J. Am. Chem. Soc.* **1992**, *114*, 776–777.
1047. Johnson, C. S., Jr. *Prog. Nucl. Magn. Reson. Spectrosc.* **1999**, *34*, 203–256.
1048. Morris, G. A.; Barjat, H. High Resolution Diffusion Ordered Spectroscopy. In *Methods for Structure Elucidation by High Resolution NMR*; Koever, K.; Batta, G.; Szantay, C., Jr., Eds.; Elsevier: Amsterdam, The Netherlands, 1997, pp 209–226.
1049. Barjat, H.; Morris, G. A.; Smart, S.; Swanson, A. G.; Williams, S. C. R. *J. Magn. Reson. B* **1995**, *108*, 170–172.
1050. Gibbs, S. J.; Johnson, C. S., Jr. *J. Magn. Reson.* **1991**, *93*, 395–402.
1051. Anderson, R. C.; Lin, M.; Shapiro, M. J. *J. Comb. Chem.* **1999**, *1*, 69–72.
1052. Ponstingl, H.; Otting, G. *J. Biomol. NMR* **1997**, *9*, 441–444.
1053. Rabenstein, D. L.; Millis, K. K.; Strauss, E. J. *Anal. Chem.* **1988**, *60*, 1380A–1392A.
1054. Liu, M.; Nicholson, J. K.; Lindon, J. C. *Anal. Chem.* **1996**, *68*, 3370–3376.
1055. Hajduk, P. J.; Olejniczak, E. T.; Fesik, S. W. *J. Am. Chem. Soc.* **1997**, *119*, 12257–12261.
1056. Hill, H. D. W. *Trans. Appl. Supercond.* **1997**, *7*, 3750–3756.
1057. Anderson, W. A.; Brey, W. W.; Brooke, A. L.; Cole, B.; Delin, K. A.; Fuks, L. F.; Hill, H. D. W.; Johanson, M. E.; Kotsubo, V. Y.; Nast, R. et al. *Bull. Magn. Reson.* **1995**, *17*, 98–102.
1058. Styles, P.; Soffe, N. F.; Scott, C. A.; Cragg, D. A.; Row, F.; White, D. J.; White, P. C. J. *J. Magn. Reson.* **1984**, *60*, 397–404.
1059. Logan, T. A.; Murali, N.; Wang, G.; Jolivet, C. *Magn. Reson. Chem.* **1999**, *37*, 512–515.
1060. Flynn, P. F.; Mattiello, D. L.; Hill, H. D. W.; Wang, A. J. *J. Am. Chem. Soc.* **2000**, *122*, 4823–4824.
1061. Hoeltzli, S. D.; Frieden, C. *Biochemistry* **1998**, *37*, 387–398.
1062. Wu, N.; Peck, T. L.; Webb, A. G.; Magin, R. L.; Sweedler, J. V. *Anal. Chem.* **1994**, *66*, 3849–3857.
1063. Webb, A. G. *Prog. Nucl. Magn. Reson. Spectrosc.* **1997**, *31*, 1–42.
1064. Olson, D. L.; Lacey, M. E.; Sweedler, J. V. *Anal. Chem.* **1998**, *70*, 257A–264A.
1065. Wu, N.; Peck, T. L.; Webb, A. G.; Magin, R. L.; Sweedler, J. V. *J. Am. Chem. Soc.* **1994**, *116*, 7929–7930.
1066. Subramanian, R.; Sweedler, J. V.; Webb, A. G. *J. Am. Chem. Soc.* **1999**, *121*, 2333–2334.
1067. Bax, A.; Farley, K. A.; Walker, G. S. *J. Magn. Reson.* **1996**, *119*, 134–138.
1068. Furihata, K.; Seto, H. *Tetrahedron Lett.* **1996**, *37*, 8901–8902.
1069. Wagner, R.; Berger, S. *J. Magn. Reson.* **1996**, *120*, 258–260.
1070. Stott, K.; Keeler, J. *Magn. Reson. Chem.* **1996**, *34*, 554–558.
1071. Emetarom, C.; Hwang, T.-L.; Mackin, G.; Shaka, A. J. *J. Magn. Reson.* **1995**, *115*, 137–140.
1072. Van, Q. N.; Shaka, A. J. *J. Magn. Reson.* **1998**, *132*, 154–158.
1073. Krishnamurthy, V. V. *Magn. Reson. Chem.* **1997**, *35*, 9–12.
1074. Kaerner, A.; Rabenstein, D. L. *Magn. Reson. Chem.* **1998**, *36*, 601–607.
1075. Rosen, M. E. *J. Magn. Reson. A* **1994**, *107*, 119–125.
1076. Marion, D.; Ikura, M.; Bax, A. *J. Magn. Reson.* **1989**, *84*, 425–430.
1077. Reynolds, W. F.; Yu, M.; Enriquez, R. G.; Leon, I. *Magn. Reson. Chem.* **1997**, *35*, 505–519.
1078. Kotyk, J. J.; Hoffman, N. G.; Hutton, W. C.; Bretthorst, G. L.; Ackerman, J. J. H. *J. Magn. Reson.* **1995**, *116*, 1–9.
1079. Foster, M. P.; Mayne, C. L.; Dunkel, R.; Pugmire, R. J.; Grant, D. M.; Kornprobst, J. M.; Verbist, J. F.; Biard, J. F.; Ireland, C. M. *J. Am. Chem. Soc.* **1992**, *114*, 1110–1111.
1080. Advanced Chemistry Development. http://www.acdlabs.com (accessed Aug 2006).
1081. Mandelshtam, V. A.; Van, Q. N.; Shaka, A. J. *J. Am. Chem. Soc.* **1998**, *120*, 12161–12162.
1082. Hu, H.; Van, Q. N.; Mandelshtam, V. A.; Shaka, A. J. *J. Magn. Reson.* **1998**, *134*, 76–87.
1083. Czarnik, A. W.; Dewitt, S. H. *A Practical Guide to Combinatorial Chemistry*; American Chemical Society: Washington, DC, 1997.
1084. Rabenstein, D. L.; Keire, D. A. *Pract. Spectrosc.* **1991**, *11*, 323–369.
1085. Holzgrabe, U.; Diehl, B. W.; Wawer, I. *J. Pharm. Biomed. Anal.* **1998**, *17*, 557–616.
1086. Traficante, D. D.; Steward, L. R. *Concepts Magn. Reson.* **1994**, *6*, 131–135.
1087. Larive, C. K.; Jayawickrama, D.; Orfi, L. *Appl. Spectrosc.* **1997**, *51*, 1531–1536.
1088. Gerritz, S. W.; Sefler, A. M. *J. Comb. Chem.* **2000**, *2*, 39–41.
1089. National Human Genome Research Institute. http://www.nhgri.nih.gov/HGP (accessed Aug 2006).
1090. Computational Biology at ORNL (Protein Informatics Group). http://compbio.ornl.gov/structure (accessed Aug 2006).
1091. Computational Biology at ORNL (Genome Channel). http://compbio.ornl.gov/channel (accessed Aug 2006).
1092. Clontech Laboratories. http://www.clontech.com (accessed Aug 2006).
1093. Genzyme. http://www.genzyme.com (accessed Aug 2006).
1094. Affymetrix. http://www.affymetrix.com (accessed Aug 2006).
1095. Synteni.com (DNA testing). http://www.synteni.com (accessed Aug 2006).
1096. Third Wave Technologies. http://www.twt.com (accessed Aug 2006).
1097. Sangamo. http://www.sangamo.com (accessed Aug 2006).

1098. The Jackson Laboratory. http://www.jax.org (accessed Aug 2006).
1099. Genomic Solutions. http://www.genomicsolutions.com (accessed Aug 2006).
1100. Structural Research Consortium (Brookhaven National Laboratory). http://proteome.bnl.gov (accessed Aug 2006).
1101. DeRisi Lab – ArrayMaker. http://derisilab.ucsf.edu/arraymaker.shtml (accessed Aug 2006).
1102. BioRad Laboratories. http://www.bio-rad.com (accessed Aug 2006).
1103. Nonlinear Dynamics. http://www.phoretix.com (accessed Aug 2006).
1104. Cellomics. http://www.cellomics.com (accessed Aug 2006).
1105. Ruffolo, R. R. et al. *Drug News Persp.* 1991, May, 217–222.
1106. Ogawa, H. et al. *J. Med. Chem.* 1993, *36*, 2011–2017.
1107. Evans, B. E.; Lundell, G. F.; Gilbert, K. F.; Bock, M. G.; Rittle, K. E.; Carroll, L. A.; Williams, P. D.; Pawluczyk, J. M.; Leighton, J. L.; Young, M. B. et al. *J. Med. Chem.* 1993, *36*, 3993–4055.
1108. Salituro, G. M. et al. *Bioorg. Med. Lett.* 1993, *3*, 337–340.
1109. Hughes, J.; Boden, P.; Costall, B.; Domeney, A.; Kelly, E.; Horwell, D. C.; Hunter, J. C.; Pinnock, R. D.; Woodruff, G. N. *Proc. Natl. Acad. Sci. USA* 1990, *87*, 6728–6732.
1110. Horwell, D. C.; Birchmore, B.; Boden, P. R.; Higginbottom, M.; Ping Ho, Y.; Hughes, J.; Hunter, J. C.; Richardson, R. S. *Eur. J. Med. Chem.* 1990, *25*, 53–60.
1111. Vara Prasad, J. V. N.; Boyer, F. E.; Domagala, J. M.; Ellsworth, E. L.; Gajda, C.; Hagen, S. E.; Markoski, L. J.; Tait, B. D.; Lunney, E. A.; Tummino, P. J. *Bioorg. Med. Chem. Lett.* 1999, *9*, 1481–1486.
1112. Venter, J. C.; Adams, M. D.; Myers, E. W.; Li, P. W.; Mural, R. J.; Sutton, G. G.; Smith, H. O.; Yandell, M.; Evans, C. A.; Holt, R. A. et al. *Science* 2001, *291*, 1304–1351.
1113. Brown, F. K. *Annu. Rep. Med. Chem.* 1998, *33*, 375–384.
1114. The Whitaker Biomedical Engineering Institute at Johns Hopkins. http://www.bme.jhu.edu (accessed Sep 2006).

Biographies

Anthony W Czarnik received his BS degree (cum laude) from the University of Wisconsin-Madison in 1977, majoring in biochemistry. He received his graduate training at the University of Illinois at Urbana/Champaign, obtaining both an MS degree (1980) in biochemistry and a PhD (1981) in organic chemistry. From 1981 to 1983, he was an NIH Postdoctoral Fellow at Columbia University, working with Prof Ronald Breslow on the design of artificial enzymes. He began his academic career in 1983 at Ohio State University, in the Faculty of the Department of Chemistry.

Czarnik has received both DuPont and Merck awards for new faculty, and in 1986 was presented with an American Cyanamid award in recognition of excellence in the advancement of science and the art of chemical synthesis. He was named an Eli Lilly awardee in 1988, a Fellow of the Alfred P Sloan Foundation in 1989, and a Teacher-Scholar Fellow of the Camille and Henry Dreyfus Foundation in 1990. He is the author of over 100 scientific publications and an editor of seven books, and serves on the editorial board of a number of journals.

From 1993 to 1996, Czarnik served as the Director of BioOrganic Chemistry at Parke-Davis Pharmaceutical Research. In 1996, he accepted a position at IRORI in San Diego, CA, serving as the Senior Director and Vice President, Chemistry. From 1998 to 2000, he was the Co-Founder and Chief Scientific Officer of Illumina, Inc., which is involved in the discovery and development of ways of using bead-based combinatorial libraries of fluorescent receptors to invent and fabricate sensors. From 2001 to 2003, he served as the Chief Scientific Officer of Sensors for Medicine and Science, Inc. in Germantown, MD.

Czarnik is currently the Founder and Manager of a Nevada-based startup company, Protia, LLC, and is a Visiting Professor of Chemistry at the University of Nevada, Reno. Additionally, he is a member of the Scientific Advisory Board of Biopraxis, Inc.

Czarnik's current research interests include combinatorial chemistry as a tool for drug discovery, nucleic acids as targets for small molecule intervention, and fluorescent chemosensors of ion and molecule recognition.

Houng-Yau Mei is the President and Chief Executive Officer of Neural Intervention Technologies, a leading medical device company developing liquid embolic systems for treating endovascular diseases. Mei has extensive management and operations experience in both biotechnology and pharmaceutical industries. In his 10 year tenure with Parke-Davis and Pfizer Pharmaceuticals, he took on increasing responsibilities in drug discovery research and technology management. He was a principal investigator in anti-infective therapeutics and novel technology development for combinatorial chemistry and therapeutics against RNA targets. As the Director of Discovery Technologies at Parke-Davis/Pfizer, Mei played a critical role in global strategic planning, and was the program manager for implementing the US $25 million ultrahigh-throughput screening system and the automated compound management system at the Ann Arbor facility. Prior to joining Neural Intervention Technologies, Mei was the Vice President of Operations and Officer at Rubicon Genomics, responsible for operations, technology validation, commercialization of new products, and contract services for genome and transcriptome amplification applicable to pharmacogenomic, cancer diagnostic and biodefense research.

Mei received his PhD from Columbia University, his postdoctoral training at the University of California in Santa Barbara, and his MBA from the Executive Program at Michigan State University.

2.13 How and When to Apply Absorption, Distribution, Metabolism, Excretion, and Toxicity

R J Zimmerman, Zimmerman Consulting, Orinda, CA, USA

© 2007 Elsevier Ltd. All Rights Reserved.

2.13.1	**Introduction**	**560**
2.13.1.1	Attrition Rates: Shifting Clinical Failures to Nonclinical Failures	561
2.13.1.2	Intervention Points for Improvements: Process and Management	562
2.13.2	**Walking Backward to Leap Forward**	**563**
2.13.2.1	Write the Package Insert First	563
2.13.2.2	Identify Critical Pharmaceutical Qualities	563
2.13.3	**Quality Absorption, Distribution, Metabolism, Excretion, and Toxicology Data and Decisions**	**565**
2.13.3.1	High-Content Absorption, Distribution, Metabolism, Excretion, and Toxicology Data: The Challenges	565
2.13.4	**Gathering High-Content Absorption, Distribution, Metabolism, Excretion, and Toxicology Data: Developing Technologies**	**566**
2.13.4.1	Chemogenomics Applied to Pharmacokinetics and Absorption, Distribution, Metabolism, and Excretion Modeling	567
2.13.4.2	Metabonomics: Predicting Absorption, Distribution, Metabolism, and Excretion Behavior In Vivo	567
2.13.4.3	Predictive Toxicology Models: Toxicogenomics	568
2.13.5	**Integration: Absorption, Distribution, Metabolism, Excretion, and Toxicology, and Chemogenomics**	**569**
2.13.6	**Strategic Application of the New Absorption, Distribution, Metabolism, Excretion, and Toxicology Technologies**	**569**
2.13.7	**Conclusion**	**571**
	References	**571**

For the past 5 years or so, a major trend in the industry has been to move away from the straight-line, serial R&D strategy that simply moved compounds in a preordained, generally linear fashion from internal department to internal department as more information was gained and internal criteria for promotion were met at each stage. This notion of a fixed R&D path for every new screening hit from the chemistry bench to clinical testing is all but dead in today's highly cost-conscious and competitive biopharmaceutical world. The key strategic dictums for R&D today, as well as into the foreseeable future, are flexible, creative, individually designed project plans, with clinically oriented go/no-go decision points applied as early in the R&D process as possible, and an uncompromising focus on the critical pharmaceutical qualities that a new lead candidate must have in order to be commercially successful.

Many of these market- and clinically oriented pharmaceutical qualities can be effectively evaluated during the nonclinical, ADMET (absorption, distribution, metabolism, excretion, and toxicity) phase of R&D. Of equal importance, the evaluation of these critical pharmaceutical qualities is not only tractable during nonclinical development; the increasing application of in silico and in vitro methodologies offers the potential to reduce both the costs and time required to assess these criteria compared to more traditional in vivo approaches. Some of these in silico and in vitro technologies are relatively mature today and readily applied. On the other hand, substantial amounts of additional data remain to be acquired in order to compare and validate these techniques against the in vivo databases of not only the industry, but also the various worldwide regulatory bodies. Clearly, in this regulated environment, the industry and the regulators must partner and reach consensus on the place and role of these new development technologies in the drug application and review process, as well as at all phases of clinical and nonclinical development.

The focus of this chapter will not be on these emerging technologies and methodologies, which are presented and reviewed elsewhere in this work (*see* Volumes 2, 3, and 5), but rather on the strategic incorporation of these new approaches, as well as the more typical ADMET procedures and technologies, into the most efficient nonclinical evaluation plan possible. From the bench scientist to the senior management team, there must be an organizational willingness to think beyond their company's traditional, internally established and politically acceptable R&D pathways and project milestones. It should be obvious to any participant or observer that these low-risk, often culturally embedded compound-vetting paradigms have proven themselves to be highly inefficient, very expensive, and nonsustainable.

While many members of senior management, project managers, and even team members typically regard the formal, Investigational New Drug (IND)-enabling Good Laboratory Practice (GLP) toxicology studies as the slowest and most expensive part of the project plan, in reality these studies are only the last phase of nonclinical development. In a creative, individually designed project plan focused on the critical pharmaceutical qualities of a given lead compound, many other potential go/no-go milestones should have been passed by the time these studies are undertaken.

Making the right decision as to how and when to evaluate ADMET parameters during R&D can lead to key strategic advantages for an organization, with the potential to decrease both resource utilization and the time required to eventually reach the market with a high-quality compound. The economic realities of pharmaceutical R&D today demand nothing less.

2.13.1 Introduction

Generally discussed numbers for the achievement of a successful marketing application from the initiation of clinical testing are in the range of 10–20%. Coupled with the substantial time (7–15 years) and resources devoted to bringing a new molecular entity to market (estimates range from $0.8 billion to twice that amount), the strategic advantage to be gained by improving these numbers is obvious. Of course, the cost of a single success is a direct function of the large number of failed candidates in big pharma's pipeline. Certainly smaller companies have brought drugs to market with far fewer resources available to them. In fact, biotechnology-derived products have typically had higher success rates and shorter development times compared to new chemical entities. Nonetheless, while there may be debate around the exact numbers, there is no debate concerning the desirability to increase the success rate while reducing the time and costs required to bring a new drug on to the commercial market.[1–3] Not only will just the commercial imperatives of the industry be addressed, but so will the medical needs of patients as well. There are numerous examples of clinical indications for which the standard of care is inadequate. An industry able to bring higher-quality drugs to patients more quickly serves everyone's interests better.

There are several conceptual strategies that might be considered reasonable options to achieve this goal, independent of the regulatory review and approval process: (1) more efficient clinical testing in order to shorten the time to establish efficacy and/or toxicity profiles; (2) increase the throughput rate of new leads into clinical testing with minimal or no increases in the upfront costs; or (3) increase the quality of the leads going into clinical testing in order to reduce the drop-out rate and thereby improve the overall clinical success rate.

The subject of this chapter is option 3: strategies to increase the quality and efficiency of the preclinical candidate-vetting process in order to select higher-quality clinical candidates with a greater chance of clinical success. Option 1, improving the efficiency of clinical development, is beyond the scope of this chapter and reference work. Nonetheless, this approach certainly appears to be an achievable goal, particularly given the expressed cooperation of the US Food and Drug Administration (FDA) and other regulatory bodies to use the latest in medical sciences regarding study design, patient enrollment, and approvable endpoints.

Option 2 cannot be seriously considered in and of itself in today's R&D world. The evaluation of larger numbers of weak candidates, obviously, only increases overall R&D costs. There was a phase, perhaps 7–10 years ago, during which the industry experimented with moving more clinical candidates into early development in an effort to gather human data quickly, with limited success. This move was initiated by double-digit increases in R&D spending during this period (a trend which has only slightly slowed in recent years). Further, as industry-wide mergers and acquisitions created larger and larger organizations, business analysts called on these companies to launch multiple newly approved drugs annually in order to be economically viable.

At present, as a whole, the industry has now opted for the strategy of finding higher-quality clinical candidates based on state-of-the-art biomedical science, over the strategy of simply increasing the quantity of leads. Therefore, this chapter will explore strategies to achieve this goal of taking better leads into clinical development through the strategic application of ADMET during the nonclinical R&D phase.

2.13.1.1 Attrition Rates: Shifting Clinical Failures to Nonclinical Failures

Despite significant annual increases in R&D spending over the past 10 years or more in the pharmaceutical industry, the US FDA recently reported that the number of both drug and biological product submissions has been steadily following a downward trend since the peak observed in 1996 and 1997 (**Figure 1**). They report similar trends at their peers' offices around the world.[4] However, the newest data from 2004 at the FDA show an increase in the number of New Drug Applications (NDAs) and biologics license application (BLAs) received, from 109 in 2003, to 137 during 2004, a 5-year high.[5] Whether or not this increase in 2004 represents a sustainable increase will be an interesting part of the human genome-sequencing story and the variety of emerging '-omics' technologies associated with this period and their potential ultimately to improve the efficiency of the R&D process.

Approximately 20–30% of new drugs entering clinical trials fail because of unexpected toxicities in humans. Fortunately, these adverse events are generally recognized relatively early in clinical development. About twice that percentage of drug failures are attributed to a lack of efficacy, in either phase II or III. These late terminations are obviously much more expensive and time-consuming, and may be not seen until the end of phase III. A key strategic goal of any state-of-the-art ADMET evaluation program must be to address failures associated with human toxicity. Further, a lack of human efficacy due to an unfavorable pharmacokinetic-pharmacodynamic (PK/PD) relationship or other human exposure issues is again something that should be aggressively addressed whenever possible with nonclinical studies. The role of pharmacology and pharmacogenomics in effectively modeling human disease is discussed in Volumes 1, 2, and 3 of this reference work.

The time from discovery to market has been increasing for the past decade, despite the remarkable achievements in basic biomedical sciences over this period and increasingly larger investments in pharmaceutical R&D. Notwithstanding these advancements in technology, knowledge, and commitment, there have not yet been observable decreases in either the discovery or the development phases of pharmaceutical R&D. Most available data would in fact suggest that just the opposite has occurred – all phases of R&D have lengthened in the past decade compared to the previous one.[4,6,7] As noted by several commentators, it is quite clear that new, innovative technologies must be created and applied, particularly to the development phase. The need for advances in translational research has never been greater, as the gap widens between basic biomedical knowledge and expertise and innovation in all phases of development. Many of the tools used today for ADMET evaluation are decades old. The goal of these efforts in translational research is to identify potential issues prior to a drug entering the clinic. Obviously, it is far cheaper to end an unpromising project during the nonclinical phase than after it enters clinical testing.

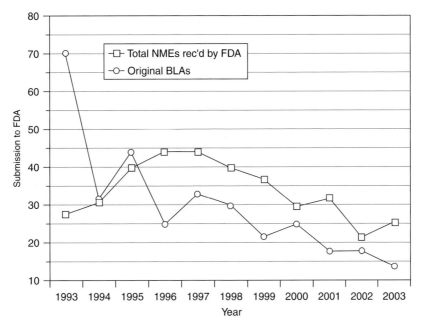

Figure 1 10-year trends in major drug and biological product submissions to the US FDA. Number of submissions of new molecular entities (NMEs) – drugs with a novel chemical structure – and the number of BLA submissions to the FDA over a 10-year period.

2.13.1.2 Intervention Points for Improvements: Process and Management

Figure 2 illustrates some of the fundamental activities that typically occur, and approximates their relative timing, between the time of target-hit identification in ultra-high-throughput screening (UHTS) and entry into clinical development. Another critical topic related to this discussion surrounds the selection of targets for screening, the application of human (and animal) genomic and proteomic information, and the presumptive relationship of these screening targets to causative features of human disease. This subject is presented in detail in Volumes 3 and 5 of this work. However, the need to understand the characteristics and behavior of potential lead compounds on human and appropriate animal drug targets is self-evident, as related to the selection of pharmacology and toxicology models, and should be included in any strategic planning discussion.

Perhaps the most important strategic goal of reengineering the R&D project plan is to effectively stop unpromising projects as soon as possible. R&D budgets are a zero-sum game: the sooner projects that have unacceptable pharmaceutical qualities are recognized and stopped, the more rapidly those resources can be shifted to the next most promising program. This willingness to stop projects with fatal flaws must become part of the entire corporate culture, not just among the nonclinical departments. It is widely acknowledged that termination of a project is one of the most difficult things to do in the biopharmaceutical industry. Nonetheless, it is essential to reduce resource utilization, including both scientific talent as well as research dollars.

The difficulty surrounds the issue of why a project is stopped, and effectively communicating those reasons to the team and the organization. Stories abound, that are retold until they have legend status, of the dead projects that were kept alive by impassioned scientists working underground at night until the problems were solved, and now the compound is an approved drug with a large market. No doubt, projects have been terminated for poor reasons that perhaps should have been allowed to continue. Sometimes unexpected budget shortfalls terminate good programs. Whatever the reasons, the decision-makers must have buy-in from the team, and effective communication. Senior management must support the concept of completely stopping a project on scientific or commercial grounds, and strongly encourage the team to move on to the next best one as soon as possible.

Another key element of this decision paradigm is to focus on learning from the biology as soon as possible. Biomedical science is a series of exercises to uncover information that is present in biological systems, yet currently unknown to the investigator. One must be fully focused on discovering the reality of the biological information, not

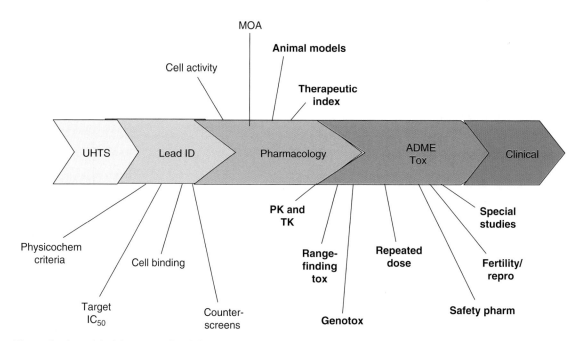

Figure 2 A model of the conventional, linear progress of compounds from hit identification through the initiation of clinical testing with selected R&D activities shown. Activities typically associated with ADMET are set in **bold**. ID, identification; MOA, mechanism of action; TK, toxicokinetics; IC_{50}, inhibitory concentration$_{50}$ (the concentration of drug that inhibits 50% of an activity).

wasting time being hopeful that nature might cooperate and support a hypothesis favored by the project team. It is absolutely essential to identify the one or two key biological questions as soon as possible, and to design and perform the best possible experiments to test those hypotheses or questions. Project team members should be rewarded for creating the fastest, most efficient experiment that most clearly tests the critical biological assumptions underpinning a project. Such an experiment will allow the biology to provide the answer.

Too often, team members in industry are only rewarded when projects move forward or achieve milestones. This mistake is easy to make, as everyone in the organization knows that when projects move further down the development path these achievements are positive for the organization. In reality, however, unquestionably the most valuable work is that which addresses the go/no-go biological issues as quickly as possible, irrespective of the outcome. After all, one cannot change nature's rules, but only hope to discover them more rapidly than anyone else. This philosophy must become part of an organization's strategic culture in order for it to become ruthlessly efficient in nonclinical R&D.

For the remainder of this chapter, a variety of potential scientific strategies to improve the nonclinical development process from the ADMET perspective will be presented and evaluated. Aspects of the conventional, linear R&D process shown in **Figure 2** will ultimately be rearranged in order to illustrate how newly emerging technologies allow the application of ADMET criteria at increasingly earlier stages of the cycle.

2.13.2 Walking Backward to Leap Forward

The most efficient nonclinical R&D strategy is the one created and driven by the critical pharmaceutical qualities needed in the targeted drug in order to make it a commercial success. 'Write the package insert first' is the mantra behind this concept. This strategic methodology must include input from marketing and business development, clinical and clinical laboratory functions, regulatory affairs, and research. This strategy takes into account the targeted indication's standard of care (present and anticipated) and evaluates competitive clinical candidates known to be in clinical development. Whether one is looking to be a fast-follower with essentially comparable benchmarked activities, or to develop a novel product with a clear strategic advantage, experience has shown that this must be the starting place of the project plan.

2.13.2.1 Write the Package Insert First

Even with what are believed to be completely novel mechanisms of action or products, it is vitally important to define what one ultimately wants to have available for patients and physicians as soon as possible in a program. Of course, it cannot be fully fleshed out in some cases, but the exercise is illuminating no matter how incomplete the end product may be. By attempting to write the package insert at the beginning of a project, it is often readily apparent how to evaluate the candidate most efficiently in ADMET testing. Out of this process comes the identification of critical pharmaceutical qualities that the lead candidate must possess in order to be acceptable in the clinic. Many of these pharmaceutical qualities are related to a drug's nonclinical behavior, and therefore highly tractable by the nonclinical scientist.

2.13.2.2 Identify Critical Pharmaceutical Qualities

Many larger companies have created tables or lists of various criteria that present generally desirable pharmaceutical qualities for new projects in a given indication. Often, these criteria have been obtained by benchmarking against competitors' drugs in the same indication, and by marketing assessments of what healthcare consumers, providers, and insurers will accept and pay for. These criteria are the underlying assumptions used in marketing and commercial assessments of how large the market might be for a given product possessing a given set of these criteria. Such benchmarking is an important exercise, and should be done by all new project teams if such information is not already available. However, these general criteria should be only a start.

Further detailed work (beginning with a draft of the package insert) must be done in order to accomplish the overall goal of improving the efficiency of the biopharmaceutical R&D process through picking the best programs prior to clinical entry, and conversely, and more importantly, by quickly stopping unpromising projects as soon as possible. From this exercise, one must identify the critical pharmaceutical qualities: those whose presence is absolutely essential for project success, and whose absence would be unacceptable to its commercial application.

From the ADMET perspective, there are often a number of such critical pharmaceutical qualities. Examples of these are presented in **Table 1**.

Table 1 Examples of critical pharmaceutical qualities that impact ADMET drug development strategy: conventional approaches compared to state-of-the-art strategies

Critical pharmaceutical quality	Conventional ADMET strategy/experiment	SOTA ADME strategy/experiment	Experimental criteria or goals
Clinical regimen: patient compliance is key	Evaluate viability of proposed clinical regimen: establish PK/PD and/or TK/TD relationships	In silico PBPK modeling linked to pharmacogenomics and toxicogenomics analyses. Evaluate potential for SNPs in humans, with either favorable or unfavorable effect	Establish PK predictions of anticipated exposure in humans and relate to PD or TD markers and/or effects. Evaluate efficacy at dose levels below those expected to induce adverse patient compliance. Screen patients and enroll only those for which the SNP data suggest higher probabilities of compliance
Clinical regimen: must be more convenient than competitors' drugs	Establish PK parameters; evaluate new formulations or delivery technologies. Link PK/PD to ensure efficacy maintained	In silico PBPK modeling	Establish appropriate predictive scaling to model anticipated blood levels and duration of exposure in humans. Predict human regimen needed to achieve these activities
Must be an oral drug	Evaluate oral bioavailability; consider role of nonplasma compartments as related to target tissue distribution. Evaluate impact of formulations or drug modifications if needed	In silico PBPK computer modeling, microsampling of tissue compartments, AMS-based distribution studies, chemogenomics analysis	Oral availability must be linked to activities, both PD and TD. Distribution to relevant target tissues should be evaluated
Cannot impair fertility or reproductive function	Evaluate appropriate species for fertility/repro endpoints	In silico structural chemistry scan, chemogenomics analysis, in vitro cell assays, perform fertility/repro tox very early	Incorporate into early compound selection criteria/lower chances of late failure in vivo
Efficacy is x-fold greater than competitor's drug	Evaluate in pharmacology models of human disease, establish therapeutic index	Establish PK/PD and TK/TD relationships, predict therapeutic index	Determine if exposure required for efficacy can be safely achieved
Toxicities typically associated with target X were not observed	Evaluate in appropriate in vivo toxicology model(s)	Toxicogenomics analysis, pathways analysis	Establish QSAR/SAR around avoidable activities
No drug–drug interactions were observed	Evaluate in vitro and in animal models	Evaluate in in vitro models and/or in silico models	Establish QSAR/SAR around drug–drug interactions
No metabolic liabilities	Labeled drug in ADME animal studies, in vitro assays	Metabonomics in silico analysis, including human polymorphism analysis	Predict likely metabolite profiles. Establish QSAR/SAR. Evaluate potential for influence of human SNPs

SOTA, State-of-the-art; TK, Toxicokinetics; TD, Toxicodynamics; PBPK, Physiological-based pharmacokinetics; SNP, Single nucleotide polymorphism; QSAR, Quantitative structure–activity relationship; SAR, Structure–activity relationship; AMS, Accelerator mass spectrometry.

2.13.3 Quality Absorption, Distribution, Metabolism, Excretion, and Toxicology Data and Decisions

As illustrated in **Table 1**, there are clearly situations in which the draft package insert identifies critical pharmaceutical qualities whose absence or presence would result in a nonviable program. Application of the conventional, linear drug development model shown in **Figure 2** would be a costly mistake in such an instance, as the typical linear order of hand-offs will likely burn both time and money until the most important ADMET experiment(s) was conducted much later than necessary in the project's life.

A major strategic challenge is how best to approach these ADMET issues as early as possible in the R&D cycle. It could be argued that so long as a decision to cut an unpromising program was made prior to clinical trial initiation, then that decision certainly was a better one than if it had been made on the basis of clinical data. Certainly that logic is hard to argue with; however it could also be applied in the case of the data safety-monitoring board closing down enrollment in a Phase III trial with only half the patients treated. This situation is perhaps somewhat better than finding the problem at the end of the trial, but it is still a very expensive decision at that point in a program's life. The strategic imperative is to make critical go/no-go decisions as early as possible in the R&D cycle, based on rich, high-content, high-quality data.

2.13.3.1 High-Content Absorption, Distribution, Metabolism, Excretion, and Toxicology Data: The Challenges

Broadly speaking, the tremendous strides that have been made in target-based screening and other discovery research tools in the past 5–7 years have today far outstripped the throughput capacity of the development scientists. Computer-driven database applications and the automation of many heretofore manual laboratory tasks and assays have vastly improved both the speed of acquisition and the density of the data available during discovery, hit identification, and lead optimization. Rich, high-content data are created at such rates that laboratory information management systems (LIMS) must often be in place simply to organize, search, and analyze these data. At the other extreme of the R&D cycle are the in vivo experimentalists, conducting a handful of studies with decades-old technology over the same time it takes the screening labs to analyze a million-compound chemical library against the next generation of potential drug targets.

Conventional in vivo ADMET models are of course very data-rich and can be used to make excellent decisions, but they are much too low in throughput to be of value during lead optimization and selection. 'Pathologists know everything, but too late,' as the saying goes. As shown in **Figure 3**, the funnel-shaped diagram used to illustrate the R&D screening cascade also clearly portrays the problem.

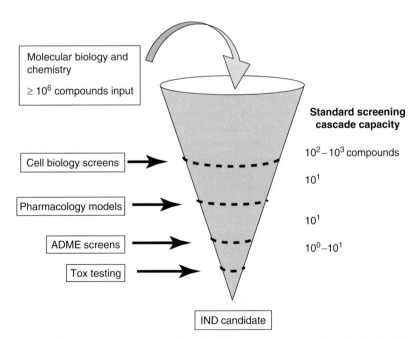

Figure 3 The capacity of the screening cascade at the ADMET level is currently orders of magnitude less than the input capacity at the discovery level.

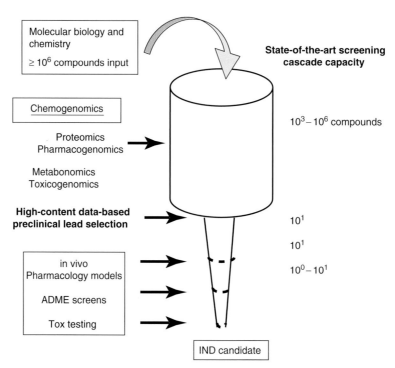

Figure 4 Application of chemogenomics technologies to improve preclinical lead selection vastly increases the throughput capacity and ADMET data content of lead prior to in vivo testing.

What is the value of generating millions of data points on over 10^6 compounds at one end of the screening cascade while, at the other end, compounds are being administered to animals one at a time? The most desirable screening cascade would allow the chemists and ADMET scientists to work side by side in real time with more comparable screening capacities. Strategically, these evaluations should be focused on the most relevant critical pharmaceutical quality required for success. It would be a significant improvement to introduce higher-throughput, development-oriented, in silico and in vitro procedures much earlier in the screening cascade.

Naturally these automated, computer-driven processes cannot be intended to replace in vivo testing in the foreseeable future. While this may certainly be the ultimate goal, and substantial strides have been made in recent years in computer models of various physiological processes, it is presently difficult to imagine that a validated, completely virtual physiological model of any mammalian species will be available in less than 10 years. In the meantime, 'in vivo veritas' must remain the guiding principle of the drug development scientist, as this principle is certainly likely to continue to guide the regulatory authorities as well for some time.

Rather than using a funnel to illustrate the overall screening cascade, the challenge for ADMET development is to make it much more cylindrical before the in vivo experiments are undertaken. As discussed later in this chapter, there are computer programs available today that have successfully modeled many ADME properties at the chemical structure level that could, in principle, be applied as far back as at the computational chemistry compound selection and/or design stages. Using these technologies, collectively referred to as chemogenomics, and those still under development, it should not be difficult in the very near future to show this much more desirable shape of the state-of-the-art screening cascade, as shown in **Figure 4**.

This screening cascade introduces in silico modeling of ADMET properties (chemogenomics) at the compound screening, hit identification, and lead optimization stages of the R&D cycle. As discussed below, currently available computer models have already become reasonably good at correlating chemical structure with ADMET features of molecules, including at the level of quantitative structure–activity relationship (QSAR).

2.13.4 Gathering High-Content Absorption, Distribution, Metabolism, Excretion, and Toxicology Data: Developing Technologies

There are a number of computer algorithms currently available that claim to be able to derive various ADMET parameters on the basis of chemical structure and previous in vivo experience. The technologies are presented and reviewed in detail in Volume 5 of this reference work. They will only be very briefly mentioned here in order to illustrate the direction these technologies are headed, and their current impact, as well as probable future impact, on the drug-screening cascade from the ADMET perspective.

The quality of the decisions that can be taken during drug development directly correlates with the density and richness of the available data. It should be appreciated that any new technologies that simply create reams of computer-generated data without linking these data to practical utility will not have a place in the new molecule ADMET development strategy. The essence of quality drug development work is the generation of applied data that have practical applications in the real world of preclinical testing and clinical development. In order to address this issue, many of the in silico chemogenomics approaches link structural features of molecules, and classes of molecules, with previously obtained animal and human in vivo data.

While the ultimate goal is to have fully validated expression data able to predict biological changes, examples of these outcomes are presently rather rare. Therefore, at present, there is an ongoing intermediate validation stage of chemogenomics that combines molecular expression profiling (mRNA and protein), chemical structural information, and in vivo data acquisition. For example, it may be possible to observe changes in the expression pattern in the liver of a treated rat weeks before the later appearance of a histopathologically confirmed lesion. With sufficient validation, these expression-profiling experiments may pick up those changes earlier, accurately predict the formation of the lesion, and thereby save time in the toxicology screen decision-making process. Therefore, while in vivo veritas remains a critical guidepost today, increasingly better decisions can be made based on in silico modeling of many ADMET elements of the R&D cycle using these existing validated tools, with newly validated data constantly being generated. Short summaries of some of these approaches and their strategic roles in ADMET screening are presented below.

2.13.4.1 Chemogenomics Applied to Pharmacokinetics and Absorption, Distribution, Metabolism, and Excretion Modeling

For many years, empirically based predictions of a molecule's pharmacokinetic behavior in humans have been made based on allometric scaling from animal data. A number of drugs scale on the basis of body weight rather well. Protein drugs, in particular, scale well on the basis of body surface area. Knowledge of physiology, such as the fact that the proportion of blood distributed to a given organ system scales with body surface area because mammals have solved the surface-area-to-volume ratio problem in effectively the same way, was the underpinning of these allometric projections. However, there are now many examples of molecules for which these empirical methods were not sufficient to predict pharmacokinetic behavior across species, and in particular in humans. As a result, more complex predictive models have been built that incorporate much more physiology data into the analysis parameters. These models are grouped under the term PBPK, or physiological-based pharmacokinetic, models. An example of the architecture of one such ADME model is presented in **Figure 5**.[8]

As suggested by **Figure 5**, there are various computer-based models available in each of the key knowledge areas. Some of these models are now commercially available, off-the-shelf programs that are relatively user-friendly, whereas others are custom modifications of existing programs to fit specific needs and applications.[9–16]

The overall experience with these predictive PBPK approaches can be judged to have been encouraging to date. As more training sets are introduced into these models, their predictive values will improve. In general, the availability of training sets is a current issue with many chemogenomics techniques. Large pharmaceutical companies have access to confidential information on thousands of compounds, and thus can build proprietary databases to incorporate this experience into their predictive models. While this ability may be viewed as one of their competitive advantages, at another level regulatory agencies such as the FDA and European Agency for the Evaluation of Medicinal Products (EMEA) are expected to protect the public health. Obviously, access to these corporate databases would broadly extend their ability to extract chemogenomic data and evaluate potential health risks of new and existing drugs. Public/private consortia have been assembled to address these issues, and their impact on predictive toxicology technologies and drug safety will be discussed later in this chapter. Inevitably, the publication of information from these databases will have a major impact on future ADMET development strategies.

2.13.4.2 Metabonomics: Predicting Absorption, Distribution, Metabolism, and Excretion Behavior In Vivo

Metabonomics is the name given to computer-based integrated analysis tools that attempt to predict the in vivo metabolite profile of a compound.[17] These models, of which there are several, produce information to predict potential

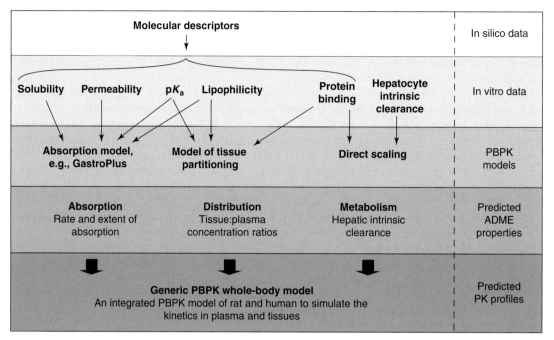

Figure 5 An example of an integrated PBPK model for prediction of ADME in a whole-body model. (Reproduced with permission from Parrott, N.; Jones, H.; Paquereau, N.; Lave, T. *Basic Clin. Pharm. Tox.* **2005**, *96*, 193–199.)

metabolic risks,[18] drug–protein-binding,[19,20] drug–drug interactions,[21] or similar metabolic features that might impact the disposition of the parent molecule and perhaps lead to systemic toxicities.[22] The recognition that human gene polymorphisms in certain metabolic enzymes markedly impact metabolism of certain classes of drugs has resulted in a focus on human pharmacogenomics and biomarkers in these models.

This concept is an extension of knowledge that began to appear in the mid-1990s regarding the presence of genetic polymorphisms associated with certain drug-metabolizing enzymes, in particular CYP2D6 and CYP2C19. In situations where drugs are metabolized by these enzymes, which are relatively common, genetic variations of these enzymes were found to result in widely varying drug exposure in humans.[18] A logical extension of this observation is to employ the tools of pharmacogenomics to these, as well as other, human drug-metabolizing enzymes that are less well characterized, such as P-glycoprotein, in order to assess the frequency of polymorphisms in humans and the effects these protein modifications may have on drug metabolism. As is now done almost routinely with CYP2D6, such knowledge could be easily inserted into the screening criteria rather early in the screening and optimization process.

2.13.4.3 Predictive Toxicology Models: Toxicogenomics

mRNA and protein expression analysis techniques have been increasingly applied to the characterization of toxic signs since their introduction over a decade ago. These tools provide the promise of molecularly characterizing the biology behind toxicologic events and outcomes. Combined with QSAR, SAR, and chemogenomics, the integration of molecular toxicology data with structural information holds the potential of someday performing initial toxicology screens in silico.[23–27] The strategic advantages of screening out undesirable compound characteristics before an animal is ever dosed are clear, and a long-sought-after goal in strategic drug development. There are well-recognized challenges, however, to the attainment of this long-term goal which should not be overlooked in the enthusiasm to embrace these technologies.[28–30]

As described by Waters *et al.*,[23] one of the primary goals of toxicogenomics is to establish 'phenotypic anchors' that link particular expression profiles or signatures to structural classes and biological changes and endpoints, with the goal of being able to predict such activities in untested compounds. A simple description of which genes or proteins are up- or down-regulated is only the very beginning of the work, because initially it will be unknown which of those altered expression patterns are directly related to the action of the compound, or due to the reaction of the animal to that damage, for example. In theory, the principles behind toxicogenomics are not much different from those highly validated principles that support the collection of clinical pathology data from peripheral blood. Patterns of changes in certain clinical chemistry or hematology data are known to be highly predictive of particular types of pathologic changes, based on decades of experience. The same type of validation may ultimately come to predictive toxicogenomics tools using molecular surrogates rather than blood markers.

2.13.5 Integration: Absorption, Distribution, Metabolism, Excretion, and Toxicology, and Chemogenomics

As occurred during the race to sequence the human genome, the developers of chemogenomics technologies in both the private and public sector have a variety of interests in accumulating large databases of structures and biological information. Such databases are essential in order to properly address the challenges posed by drug development efficiency, environmental health and safety, and the development of the appropriate regulatory environment to assess and utilize these new data. The competitive advantage of having large internal chemical libraries and biological data may drive some decisions in the private sector that are not completely aligned with the interests of the public sector.

However, as was seen with the single nucleotide polymorphism (SNP) Consortium and the Human Genome Project, public/private consortiums can work together to address key issues without compromising their competitive needs. Likewise, it appears that initiatives are currently underway with the promise to address database access issues in order to provide much larger sets of training data than might be possible through any one organization, whether public or private. These data may eventually level the playing field in the drug development world as higher-quality, validated models of various biological processes associated with ADMET characteristics will become more broadly available.

For example, the FDA launched the Critical Path Initiative in the spring of 2004 in order to help facilitate the growth of novel drug development tools. Having watched the growth on the discovery side of the modern techniques of molecular biology, the Agency has expressed concern that the lack of similar tools on the development side is hurting not only drug development, but patient access to effective new medicines, as the number of marketing applications has been steadily declining since the early 1990s.[4] The FDA, similar to the EMEA, is uniquely positioned to contribute to these consortium databases by having access to decades of information concerning chemical structures and biological effects. Hopefully, confidentiality issues and concerns will be allayed by the proposals made by FDA to industry, which will greatly facilitate moving forward on these initiatives.

In addition, the National Center for Toxicogenomics was created in September 2000 by the US National Institute of Environmental Health Services (NIEHS) in order to apply gene expression technologies to understand the relationships between gene expression and drug exposure, and the biological consequences of that exposure, among other related goals. As part of this initiative, the NIEHS Toxicogenomics Research Consortium was formed, which is structured to include corporate research members (CRMs). Through this consortium, microarray and other technical issues will be addressed and standardized, and data will be deposited into the Chemical Effects in Biological Systems (CEBS) database. This CEBS development process, which has many of the same elements as other chemogenomics and toxicogenomics projects, is illustrated in **Figure 6**.

With these types of public/private consortiums working together, there will hopefully be few barriers presented to any pharmaceutical organization, from start-up to multinational, to the incorporation of such state-of-the-art information into their ADMET R&D cycles.

2.13.6 Strategic Application of the New Absorption, Distribution, Metabolism, Excretion, and Toxicology Technologies

It follows quite clearly from the above discussion that various crucial elements of the R&D cycle can include ADMET measures far earlier in that process than is typically done. There are two ways to accomplish this: (1) through the incorporation of state-of-the-art chemogenomics-based studies that are focused on ADMET issues known to be critical to the success of the project (for examples, see **Table 1**); or (2) simply through the identification of, and testing for, critical pharmaceutical qualities using standard models earlier in the R&D cycle through the use of nonlinear development strategies.

The incorporation of modern computer-based technologies holds the greatest potential both to shorten the R&D cycle, and to drive early termination decisions on programs with potentially fatal flaws, thereby improving the overall efficiency of the process. The strategic goal of using these technologies can be illustrated in **Figure 7**.

With the barely adequate amount of validation currently available, many of these technologies already have the potential to improve the yield of the R&D process by introducing ADMET-based criteria much earlier in the pathway.

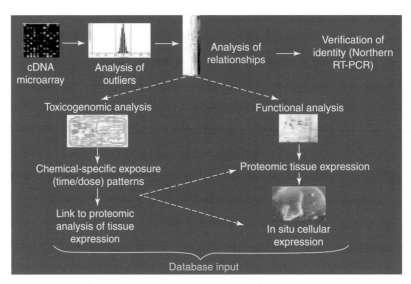

Figure 6 Microarray data are first verified, then coupled with biological data, in the process of building validated relationships between expression profiles and biological changes. RT-PCR, reverse transcriptase polymerase chain reaction. (Reproduced with permission from Waters, M. D.; Olden, K.; Tennant, R. W. *Mut. Res.* **2003**, *544*, 415–424.)

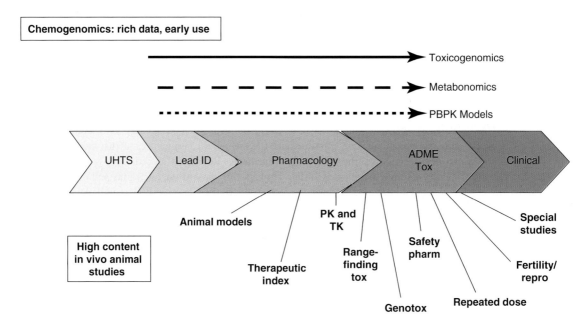

Figure 7 Incorporation of chemogenomics technologies early in the lead selection stage has the potential to vastly improve decision-making during the R&D cycle by gathering critical ADMET data much earlier in the process. ID, identification; TK, toxicokinetics.

As increasingly more validation, or 'phenotypic anchoring,' becomes available, the value of the data output from these systems will be correspondingly higher. There should be no doubt concerning the limitations of these technologies today. On the other hand, there can likewise be no doubt as to the direction they are headed over the next 5–7 years, and as to their ultimate utility.

That being the case, it must be acknowledged that it is still a long road from microarray data to biological endpoints. Starting with the microarray systems themselves, it is known that different technologies can give quite different output from the same input mRNA. Standardization of these systems and cross-technology comparisons are underway to understand these differences. Similarly, the input material, the mRNA preparations, can also be sources of significant variation. It is absolutely essential to have high-quality mRNA input that allows for the interrogation of both high- and low-frequency messages and that is fully representative of the cells or organs from which it was prepared.

An additional piece of information must also be sought when organ sources are used for these mRNA or protein preparations: which cell type in the organ is responsible for the observed changes? Hence the need to incorporate relatively sophisticated histopathology techniques such as in situ expression analyses on tissue preparations.

Caution must also be used in the interpretation of microarray expression profiling data that are not linked to protein expression (proteomics) and tissue distribution data. It is of course well known that there is not a 1:1 correspondence between the up- or down-regulation of the expression of mRNA and protein translation. It is not even enough, in some cases, to know that particular proteins are overexpressed without also knowing the status of their cognate receptor. If a cell is producing a lot of a particular protein but there is no receptor to acknowledge that production, this overproduction is irrelevant to the biology of the system.

A final problem that is likely to take considerable effort to resolve is that each of the mRNA or protein preparations taken from a particular biological system is only a snapshot of the movie of biology. The kinetics problem, the interpretation of changes in mRNA or protein expression levels over time, and how these expression changes ultimately impact biological endpoints, will be complicated at best. The solution may be one of the more distant goals of chemogenomics. A series of proof-of-principle experiments using known compounds is one of the standard ways to begin to address this challenge. The ability to link the kinetics of changes in molecular signatures to biological endpoints and more conventional toxicology parameters is obviously an important factor in selecting when and how to use these technologies during development.

2.13.7 Conclusion

There can be no doubt of the need to reduce the overall failure rate of new compounds, to increase the quality of pipeline candidates, and to reduce the time and costs of pharmaceutical R&D. Several strategic shifts from conventional drug development thinking have the potential to improve the efficiency of the R&D cycle from the ADMET perspective: (1) increased use of the new state-of-the-art tools associated with the term chemogenomics much earlier in compound vetting; (2) a stronger focus in nonclinical R&D on the marketing and clinical goals of the project as defined by the draft package insert and the critical pharmaceutical qualities that the successful drug candidate must possess; and (3) a willingness to perform the most critical experiment that addresses the key biological assumption as soon as feasible. While the application of these strategies may actually increase the failure rate of compounds early in the R&D cycle, dropping an unpromising candidate at these stages is much less expensive, resource-intensive, and time-consuming than any stage of clinical development failures. More importantly, higher-quality candidates are likely to be taken into clinical testing, thereby increasing the success rate to market.

These strategic shifts must be accompanied by a management team with the will to rapidly terminate unpromising projects based on sound scientific and commercial criteria, and to effectively reassign those resources on to the next best program. Similarly, the organization's staff members at all levels must become a part of this strategic paradigm for it to be successful. Small groups committed to keeping 'dead' projects alive can sometimes undermine a strongly committed management team. Therefore, the organization must strive to develop a strong culture that is committed to these principles.

The technology revolution that swept through the discovery end of the R&D process during the past decade must now be applied to the development piece of the puzzle. There is no doubt that such applications will be developed; indeed, the first generation of these tools is already available. The competitive advantage will come to those organizations that are smart and nimble. They must first understand these technologies, and how and when to apply them judiciously, as they must be sufficiently validated such that the potential risks are balanced by the rewards. Further, with the increasing knowledge of human pharmacogenomics and molecular disease mechanisms, those nimble organizations that are willing to shift their commercial strategy to consider launching higher-quality new products designed for smaller, more defined patient populations, may develop a competitive advantage through the use of these tools.

References

1. DiMasi, J. A. *Pharmacoeconomics* **2002**, *20*, 1–10.
2. DiMasi, J. A.; Hansen, R. W.; Grabowski, H. G. *J. Health Econ.* **2003**, *22*, 151–185.
3. *Outlook 2005*, Tufts Center for the Study of Drug Development, Tufts University, Boston, MA, USA, 2005.
4. *Innovation or Stagnation: Challenge and Opportunity on the Critical Path to New Medicinal Products*, US FDA white paper; FDA: Rockville, MD, USA, 2004.
5. *FY2004 Performance Report to Congress*, US FDA; FDA: Rockville, MD, USA, 2005.

6. Baumann, M.; Bentzen, S. M.; Doerr, W.; Joiner, W. C.; Saunders, M.; Tannock, I. F.; Thames, H. D. *Int. J. Radiat. Oncol. Biol. Phys.* **2001**, *49*, 345–351.
7. Gilbert, J.; Henske, P.; Singh, A. *Bus. Med. Rep.* **2003**, *17*, 73.
8. Parrott, N.; Jones, H.; Paquereau, N.; Lave, T. *Basic Clin. Pharm. Tox.* **2005**, *96*, 193–199.
9. Poulin, P.; Theil, F. P. *J. Pharm. Sci.* **2002**, *91*, 129–156.
10. Poulin, P.; Theil, F. P. *J. Pharm. Sci.* **2002**, *91*, 1358–1370.
11. Nestorov, I. *Clin. Pharmacokinet.* **2003**, *42*, 883–908.
12. Paranjpe, P. V.; Grass, G. M.; Sinko, P. J. *Am. J. Drug Deliv.* **2003**, *1*, 133–148.
13. Theil, F. P.; Guentert, T. W.; Haddad, S.; Poulin, P. *Toxicol. Lett.* **2003**, *138*, 29–49.
14. Wolohan, P. R.; Clark, R. D. *J. Comput.-Aided Mol. Descr.* **2003**, *17*, 65–76.
15. Klopman, G.; Chakravarti, S. K.; Zhu, H.; Inavov, J. M.; Saiakhov, R. D. *J. Chem. Inf. Comput. Sci.* **2004**, *44*, 704–715.
16. Stoner, C. L.; Gifford, E.; Stankovic, C.; Lepsy, C. S.; Brodfuehrer, J.; Vara Prasad, J. V. N.; Surendran, N. *J. Pharm. Sci.* **2004**, *93*, 1131–1141.
17. Nicholson, J. K.; Connelly, J.; Lindon, J. C.; Holmes, E. *Nat. Rev. Drug Disc.* **2002**, *1*, 153–161.
18. Walker, D. K. *Br. J. Clin. Pharm.* **2004**, *58*, 601–608.
19. Evans, D. C.; Watt, A. P.; Nicoll-Griffith, D. A.; Baillie, T. A. *Chem. Res. Toxicol.* **2004**, *17*, 316.
20. Yamazaki, K.; Kanaoka, M. *J. Pharm. Sci.* **2004**, *93*, 1480–1494.
21. Bachmann, K. A.; Lewis, J. D. *Ann. Pharmacother.* **2005**, *39*, 1064–1072.
22. Evans, D. C.; Baillie, T. A. *Curr. Opin. Drug Disc. Dev.* **2005**, *8*, 44–50.
23. Waters, M. D.; Olden, K.; Tennant, R. W. *Mut. Res.* **2003**, *544*, 415–424.
24. Steiner, G.; Suter, L.; Boess, F.; Gasser, R.; de Vera, M. C.; Albertini, S.; Ruepp, S. *Environ. Health Perspect.* **2004**, *112*, 1236–1248.
25. Fielden, M. R.; Pearson, C.; Brennan, R.; Kolaja, K. L. *Am. J. Pharmacogenet.* **2005**, *5*, 161–171.
26. Natsoulis, G.; El Ghaoui, L.; Lanckriet, G. R.; Tolley, A. M.; Leroy, F.; Dunlea, S.; Eynon, B. P.; Pearson, C. I.; Tugendreich, S.; Jarnagin, K. *Genome Res.* **2005**, *15*, 724–736.
27. Ganter, B.; Tugendreich, S.; Pearson, C. I.; Ayanoglu, E.; Baumhueter, S.; Bostian, K. A.; Brady, L.; Browne, L. J.; Calvin, J. T.; Day, G. J. et al. *J. Biotechnol.* **2005**, *119*, 219–244.
28. Farr, S.; Dunn, R. T. *Toxicol. Sci.* **1999**, *50*, 1–9.
29. Hughes, T. R.; Marton, M. J.; Jones, A. R.; Roberts, C. J.; Stoughton, R.; Armour, C. D.; Bennett, H. A.; Coffey, E.; Dai, H.; He, Y. D. et al. *Cell* **2000**, *102*, 109–126.
30. Fielden, M. R.; Zacharewski, T. R. *Toxicol. Sci.* **2001**, *60*, 6–10.

Biography

Robert J Zimmerman has worked in the biotechnology and pharmaceutical industry for over 20 years. Since 2003 he has been a consultant to the industry on drug development and R&D strategies. Prior to that, he held a number of senior-level R&D positions in the industry, including at Signature BioSciences, Bayer's Biotechnology Center in Berkeley, Chiron, Cetus, and Charles River Biotechnology Services. He has a Doctor of Science degree in Physiology/Radiobiology from the Harvard School of Public Health.

2.14 Peptide and Protein Drugs: Issues and Solutions

J J Nestor, TheraPei Pharmaceuticals, Inc., San Diego, CA, USA

© 2007 Elsevier Ltd. All Rights Reserved.

2.14.1	**Introduction**	**573**
2.14.1.1	Major Therapeutic Target Classes	574
2.14.1.1.1	G protein-coupled receptors	574
2.14.1.1.2	Receptor protein tyrosine kinases	576
2.14.1.1.3	Cytokine receptors	576
2.14.1.1.4	Protein–protein interactions	576
2.14.1.1.5	Proteases	577
2.14.2	**Current Peptide Pharmaceuticals**	**578**
2.14.3	**Initial Approaches**	**580**
2.14.3.1	Efficacy	580
2.14.3.2	Agonism	581
2.14.3.3	Antagonism	583
2.14.4	**Biological Half-Life Solutions**	**583**
2.14.4.1	Proteolysis	584
2.14.4.2	Glomerular Filtration	585
2.14.4.2.1	Hydrophobic depoting	585
2.14.4.2.2	Hydrophilic depoting	587
2.14.4.2.3	PEGylation	588
2.14.4.2.4	Protein conjugation	589
2.14.4.3	Structural Motif Replacement	590
2.14.4.3.1	Expression-modified proteins	591
2.14.5	**Drug Delivery to Overcome Bioavailability Issues**	**591**
2.14.5.1	Injection	592
2.14.5.1.1	Controlled release	592
2.14.5.1.2	Self-forming depots	593
2.14.5.2	Intranasal	593
2.14.5.3	Inhalation	595
2.14.5.4	Oral Administration	595
2.14.6	**Peptidomimetics**	**595**
2.14.7	**Conclusions and Outlook**	**596**
References		**596**

2.14.1 Introduction

Polypeptides and proteins play critical roles throughout the human body, indeed throughout all living systems. Many of these roles developed as the primordial system of controls evolved for the coordination of multicellular organism function and thus control the most basic physiological activities. The role of peptide signaling molecules has expanded to fill a vast array of control and modulatory systems in vertebrates. For example, the peptide hormones of just the pituitary gland control the fundamental aspects of gamete production, birth, suckling, fluid and salt balance, growth, response to stress, etc. Peptide hormones from the gut control insulin response, gut motility, and appetite, at a minimum (see incretins, below). Thus peptides have powerful signaling and control functions (e.g., autocrine, endocrine, paracrine, neuromodulatory, host defense roles, etc.) in all tissues, typically displaying high potency and selectivity for their target receptor classes. However despite these highly favorable characteristics as ligands, they are considered to be underrepresented as commercially viable targets and pharmaceuticals. Despite this view, roughly 20% of the top 200 commercial drugs in 2004 are proteins or peptides. This chapter will focus on the advantages and principal shortcomings of polypeptides as drug candidates, and on approaches that have been used to overcome these deficits successfully. With the completion of the sequencing of the human genome, it is certain that many more opportunities to evaluate new polypeptides and proteins will be recognized. Additionally, advances in peptide synthesis[1–4] and protein production by recombinant DNA methods or chemospecific chemical ligation[5,6] of protein fragments (meant broadly[7]) mean that it is becoming far easier to produce (to metric ton scale) and test such agents.

For the purposes of this chapter, peptide or polypeptide will be used interchangeably and will mean a polypeptide structure of <40 residues or so while 'protein' means larger than that. A better distinction might be that a protein has a unique tertiary structure in solution and therefore is subject to denaturation. The hundreds of monoclonal antibodies in clinical development will not be addressed as we focus on the novel medicinal chemistry lessons that we can learn from manipulation of the structures of polypeptides and proteins (especially the former). The exceptional diversity of structure and function that can be derived from assembling the 20 proteinogenic amino acids into protein and peptide structures, and the creativity of scientists working in this field, has generated a tremendous diversity in approaches. This complexity has resulted in excellent book-length reviews on synthesis[1,8] and design approaches.[2] There also are reviews of peptide drug design that focus on aspects such as constrained analogs,[9–12] an important route toward antagonism and receptor specificity. Although a great deal of progress has been made on the design of peptidomimetics (mimicking specific peptide interactions) or ligands that bind to peptide receptors (an important distinction), to date these are mostly antagonists. Agonistic action at a receptor requires many interactions to happen with great specificity and peptide analogs remain the most straightforward route to obtaining potent agonists. This appears to be especially true for the mid-sized peptide hormones of >30 residues.

The various approaches to 'peptidomimetics' that are remote from the starting structures or nonpeptide ligands for peptide receptors will not be addressed in detail (however see 2.15 Peptidomimetic and Nonpeptide Drug Discovery: Receptor, Protease, and Signal Transduction Therapeutic Targets; and reviews[13,14]). Due to space limitations, this chapter will focus on broad themes of practical use in specifically addressing the shortcomings of peptides and proteins as pharmaceutical agents (see reviews[15,16]).

Broadly, the shortcomings of peptides and proteins as drug candidates relate to their typically short biological $t_{1/2}$ due to rapid clearance, to low oral bioavailability, and to a lack of receptor subtype specificity. There are now very useful approaches to each of these challenges and they will be reviewed below. Several very exciting drug candidates in current development are polypeptides and this chapter will analyze how these challenges have been overcome. The summary message will be that, despite the issues outlined above, polypeptide research remains an exciting area for the identification of important therapeutics because of the approaches outlined below.

2.14.1.1 Major Therapeutic Target Classes

Each of the polypeptide and protein therapeutic families has developed in somewhat different fashions, depending on particular requirements of ligand or receptor structure, need for agonists or antagonists, and creativity of the practitioners. For ease of synthesis reasons, smaller peptide ligands have offered more opportunity for incorporation of unnatural amino acids to accomplish changes in physical properties (see Sections 2.14.4.1 and 2.14.4.2). Large proteins have been modified by PEGylation (see Section 2.14.4.2.3) and protein conjugation (see Section 2.14.4.2.4). Interestingly, the lessons from each class are now being brought to the other. How does the character of the therapeutic target's receptor structure affect drug design?

2.14.1.1.1 G protein-coupled receptors

G protein-coupled receptors (GPCRs) are unique integral membrane proteins comprising seven transmembrane helical regions with an N-terminal region (extracellular), a C-terminal region (cytosolic), and three loops on each of the interior and exterior faces of the cell membrane. The GPCR typically binds a ligand on the external face and transduces a signal by conformational change of the receptor protein assemblage to affect interactions with the heterotrimeric G protein (GTP-binding protein; Gαβγ) assemblage bound to its cytosolic face (**Figure 1**). Thus an agonistic ligand must accomplish a binding and an activation step. Activation of the GPCR causes exchange of the G-protein-bound guanosine diphosphate (GDP) molecule for a guanosine triphosphate molecule, dissociation of the heterotrimeric Gαβγ protein from the receptor and into the Gα (18 forms) and Gβγ (5 β forms; 11 γ forms) proteins. These proteins transduce the signal to multiple activation signaling pathways to cause changes in cellular function. The activated GPCR then typically binds an arrestin molecule and is internalized in a desensitization process that culminates with

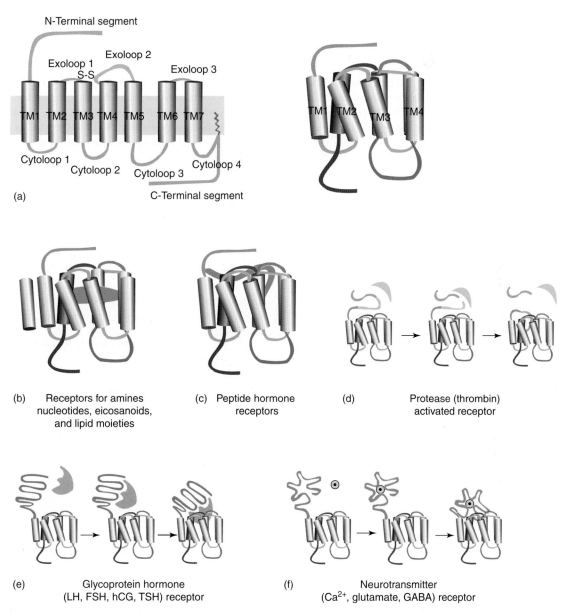

Figure 1 Schematic presentation of the general structure of GPCRs and receptor–ligand interactions. (a) General structure and terminology. Several distinct modes have been observed for ligand binding and signal generation. (b)–(f) indicate the manner in which different ligand classes make use of interactions in the core, using core plus extracellular loops, using proteolysis, or using the N-terminal extension plus the loops and core. (Reproduced with permission from Ji, T. H.; Grossmann, M.; Ji, I. *J. Biol. Chem.* **1998**, *273*, 17299–17302 © The American Society for Biochemistry & Molecular Biology.)

eventual return to the surface. A caveat for this discussion is that the specifics of these processes are constantly being shown to have variants which specialist reviews will discuss.

The members (>1000) of this family have been grouped into various classes.[17] Many receptors appear to be olfactory receptors and, while not of primary interest to this discussion, represent a very active area of research of which much will be heard in the future. Of most interest to this chapter are those members of the rhodopsin family (Type A receptor; e.g., angiotensin, CCK, endothelin, galanin, neurokinin, neurotensin, vasopressin, etc.) and secretin/glucagon family (Type B receptor; e.g., calcitonin, GLP-1, PTH, etc.) that bind to endogenous peptide hormones. Within each family there are multiple subfamilies and within subfamilies there typically are subtypes.[18] For example, within the Type B family, the somatostatin family has five known subtype receptors (sst1, sst2, sst3, sst4, sst5). Different subtypes may be expressed on different tissues, exhibit differential affinity for ligands, and activate different signaling pathways.[18] Thus subtype selectivity is an important direction for future research on peptide ligands. In addition, there are different capabilities for association with other receptors, different desensitization pathways, etc. The number of receptors and of the G proteins may change with further research and there appears to be even greater variability possible due to the multiple splice and editing variations.

Of most significance to this chapter is the mechanism of binding of polypeptide ligands to these GPCRs. Peptides typically bind to GPCRs through the use of the N-terminal extension and extracellular loops of this transmembrane protein assemblage (**Figure 1**). The type A peptide receptors have short N-terminal extensions and their ligands typically are small polypeptides, thought to use the extracellular loops and transmembrane core for critical binding and activation interactions. The type B polypeptide receptors have a large N-terminal region containing multiple disulfide bonds to generate stable structure. The typically larger polypeptide ligands for these latter receptors are believed to have important interactions in their C-terminal regions with this large N-terminal extension, while the N-terminus of the ligand interacts with the transmembrane region to activate the receptor. The significance of these aspects will be addressed below under 'Efficacy.' Much of the discussion of drug design that follows will focus on GPCR ligands.

2.14.1.1.2 Receptor protein tyrosine kinases

The transmembrane receptor protein tyrosine kinases (RPTKs; 58 members in 20 subfamilies), targeted by many growth factors (epidermal growth factor, insulin, fibroblast growth factor, etc.), are again a highly diverse family and differ fundamentally from GPCRs in the manner and mechanisms of their interaction and activation in response to ligands. A prototypical growth factor receptor will have a single transmembrane domain with a ligand-binding region on the extracellular face and various domains for dimerization, protein binding, activation, and kinase activity on the cytosolic side. Typically ligand binding will cause homo- or heterodimerization of the receptors and activation of the signaling mechanism by trans- or autophosphorylation of domains in the receptor dimer partners (see diagrams[19]). Receptor binding modes have been worked out in detail for some receptors.

These receptors typically bind large protein hormones and the mimicking of such large protein surfaces by polypeptides or small molecules has been expected to be difficult. Aside from insulin, the emphasis in this class has been on blocking function, since abnormal activation of this class is the most common source of oncogene-driven carcinogenesis.[19] Thus pharmaceuticals targeting ErbB2 (HER2/Neu in breast cancer; antibodies or kinase inhibitors) are important new cancer pharmaceuticals (trastuzumab, cetuximab, erlotinib, gefitinib, etc.).[20] Long-acting peptide/protein antagonists could be effective, but monoclonal antibodies appear to be a quicker approach, currently. The issues addressed for these proteins (duration of action, drug delivery) are relevant to the polypeptide drugs and will be addressed in the section on PEGylation.

2.14.1.1.3 Cytokine receptors

Cytokine ligands and their receptors have similarities to the growth factors discussed above.[21,22] Typically these ligands cause heterodimerization (or dimers of heterodimers) of a type 1 and a type 2 receptor (single transmembrane proteins), which initiates activation of the Janus kinase (Jak/Stat) signal transduction system by transphosphorylation. In many cases the heterodimers comprise a 'private,' type 1, ligand-specific receptor and a 'public,' class-specific signal transducer that functions with multiple type 1 receptors. Again there is substantial diversity of receptors and pathways for activation and signaling. These bind large protein ligands, some of the best known of which are growth hormone (GH), tumor necrosis factor alpha (TNF-α), the interferons (IFN), and the interleukins. In this large family of receptors, both agonistic (e.g., IFNs, GH) and antagonistic functions[23] (antagonism of TNF-α; soluble TNF receptor–IgG1 (immunoglobulin G1) conjugate, below) have current pharmaceutical roles. Lessons from these protein ligands in improving pharmaceutical characteristics (PEGylation, protein conjugation) are applicable to smaller peptide ligands, as we shall see. The research on ligand receptor binding interactions for this class of proteins is best thought of in general terms as protein–protein interactions.

2.14.1.1.4 Protein–protein interactions

Protein–protein interactions are an attractive area for the application of polypeptide drug discovery. An important aspect of this general area is the understanding of the interaction of the protein ligands for the RPTK and cytokine receptors above. This can be approached by minimizing protein structures through truncation (with the aid of phage display to increase affinity or to select new binding elements) with retention of binding and function[24] or from the perspective of Ala scanning of receptors or ligands to develop a detailed understanding of protein surface interactions.[25] The former would appear to be more straightforward to contemplate while the latter, when accomplished and generalized, appears to offer unlimited potential. An example of the former is the use of a recombinant display vector to allow selection of a small peptide (20 residues) that is capable of binding to and dimerizing the erythropoietin receptor.[26,27] A related peptide was reported to have agonistic activity in phase I and to be moving into phase II clinical trials. In a related study, a dimeric, 28-residue peptide with agonistic activity on the thrombopoietin receptor equivalent to the native protein ligand (332 residues) was reported.[28]

Re the second approach above, an intriguing observation was that a single GH molecule bound to and formed a homodimer of two of its receptors using completely different binding regions on the GH molecule. This was a completely unexpected result at the time[29] but led to the observation that the site for binding the second receptor could be modified to yield a monovalent ligand for the GH receptor incapable of binding the second receptor, thus yielding a potent antagonist. Analysis of this receptor ligand interaction led to the concept that the critical binding feature was a hydrophobic patch that generates most of the binding energy. Extensive study of this interaction has led to the development of an intriguing 'hot spot' approach[30–33] to understanding the critical binding interactions of large protein surfaces.[34] Although this further study has again demonstrated that nature is characterized by tremendous diversity and that general rules are hard to find,[34] much of the developing knowledge on rational design of ligands for binding to large protein surfaces is being generated in this area and it appears to have tremendous potential for the discovery of new peptide pharmaceutical leads. These will then need to be modified using the medicinal chemical approaches described below.

2.14.1.1.5 Proteases

Proteases would seem to be a rich target class and, although there are a number of blockbuster drugs inhibiting members of this class (see enzyme inhibitor product review [35]), most of the action has centered around a few enzymes with multiple products in each class: HIV protease (8 products), peptidyl-dipeptidase A (angiotensin-converting enzyme; 12 products), and thrombin (10 products). While these inhibitor classes began as peptidelike structures they have developed into areas that focus on peptidomimetics. However, initial inhibitors typically make use of a peptide-like framework and initial products have had substantial peptide character with high potency. For example, the initial ACE and HIV protease inhibitors had clear peptidic character (**Figure 2**). The modifications that led to high potency can be analyzed in terms both of binding efficacy and balanced physical properties that affect clearance by glomerular filtration (Section 2.14.4.2).

An additional major success with a peptide-based structure is the 28S proteosome inhibitor, bortezomib (**Figure 3**). This is a relatively nonspecific, dipeptide boronic acid serine protease inhibitor with chymotrypsin-like selectivity that appears to have not had a lot of medicinal chemistry optimization. It is proving to have widespread use in cancer treatment, and it is interesting that duration of protease inhibition is an important factor here. Apparently tumor cells are more sensitive to generalized protease inhibition than normal cells.[36] It is toxic to tumors but not to normal cells when the duration of protease inhibition is 24 h or less. However, longer periods of inhibition cause toxicity in normal cells as well, so episodic administration is required and very prolonged duration of action would be detrimental.

A substantial amount of work has been done on caspase inhibitors (see reviews [37,38]) because of the involvement of caspases in programmed cell death (apoptosis). Both computational and structure-based approaches to design of peptide[39] and peptidomimetic[40,41] inhibitors has been reported. Positional scanning to select optimal peptide frameworks as a starting point for optimized drug design is a powerful approach[41] that can be used broadly to accelerate programs. Since the caspase apoptosis machinery is present in all cells, however, selectivity or side effects are a concern. Modified and constrained peptide-based inhibitors have been investigated here and compounds are moving forward in clinical development.[37]

Certainly one of the most active research areas at present is the design of inhibitors of the dipeptidyl peptidase IV (DPPIV) enzyme, because of its role in terminating the effect of GLP-1 ($t_{1/2}$ on the order of 1 min in vivo) and GIP, both incretin hormones that potentiate the pancreatic islets' response to glucose.[42,43] Potent nonpeptide DPPIV inhibitors are the major focus and are in advanced clinical studies. It seems certain that pharmaceutical products will result, despite concerns that the DPPIV also terminates the function of multiple other polypeptide hormones[44] and side effects are to be expected.

In the area of protease inhibitor research, the HIV protease design area may offer the most medicinal chemistry lessons re peptide–protein interactions, but this is best left to specialist reviews.[45,46] The general lessons that have come out relate to the common binding conformation found for peptide substrates in proteases (bound as a beta sheet conformation) and the power of structure-based design of peptidomimetics.[45] In this research, actual peptide–protein interactions have actually been possible to mimic because of the availability of x-ray structures. Although much work continues on peptide protease inhibitors,[47,48] because of the applicability of structure-based drug design, in a broad sense, the protease inhibitor area appears to be ideal for nonpeptide rather than peptide inhibitor design and will not be treated in detail in this chapter.

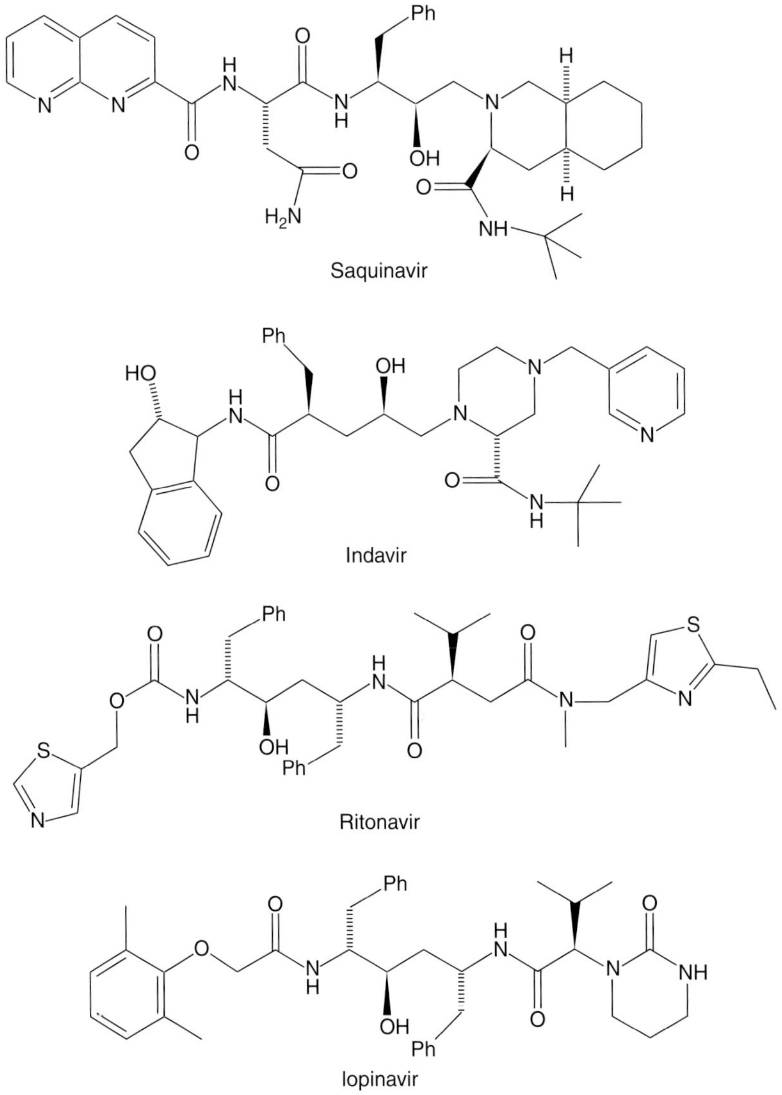

Figure 2 Peptidelike HIV protease inhibitors. (Reprinted with permission from Jiang, G.; Stalewski, J.; Galyean, R.; Dykert, J.; Schteingart, C.; Broqua, P.; Aebi, A.; Aubert, M. L.; Semple, G.; Robson, P. et al. *J. Med. Chem.* **2001**, *44*, 453–467 © American Chemical Society.)

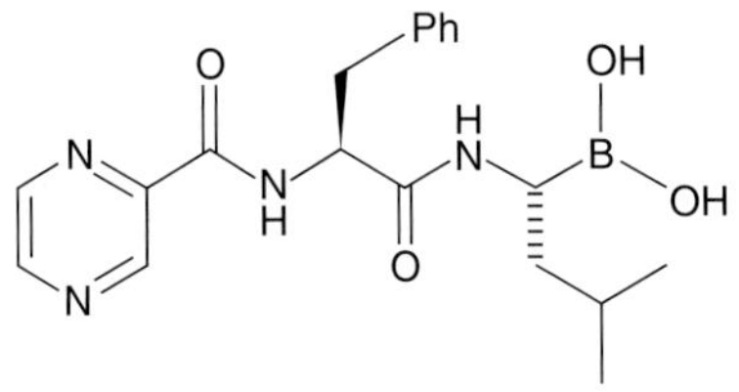

Figure 3 Proteosome inhibitor bortezomib.

2.14.2 Current Peptide Pharmaceuticals

Table 1 lists the Food and Drug Administration (FDA) approved 'polypeptide pharmaceuticals' that are available in the USA in 2005, using the definition given earlier. This limitation is imposed since this list is to be used for illustrative

Table 1 USFDA-approved Peptide Pharmaceuticals (<40 residues) in 2005[a]

Target receptor or binding protein (native ligand size)	Chemical name	Indication	Comment
Amylin	Pramlintide	Diabetes	Substitution with 3 Pro residues breaks up beta-sheet aggregation
Corticotropin-releasing hormone (41 residues)	Human CRH* Cortirelin	Diagnostic – ACTH release	Native ovine CRH
Calcitonin (32 residues)	Human calcitonin*	Paget's disease, osteoporosis	Native human
	Salmon calcitonin		Extended duration
Cyclophilin (11 residues)	Cyclosporin	Immunosuppressive	Fungal product, many D-residues
Gonadotropin hormone-releasing hormone (10 residues) (All have extensive modifications; see text)	Leuprolide	Prostatic cancer	Controlled release
	Nafarelin	Endometriosis	Intranasal
	Triptorelin		Pamoate salt
	Goserelin	Prostatic cancer	Controlled release
	Histrelin	Precocious puberty	Controlled release
	Ganirelix (antagonist)	Controlled ovulation	
	Cetrorelix (antagonist)	Controlled ovulation	
	Abarelix (antagonist)	Prostatic cancer	Controlled release
Growth hormone-releasing hormone (44 residues)	Sermorelin*		Native human
Glucagon (29 residues)	Glucagon*	Hypoglycemia, radiography	Native human
Glucagon-like peptide-1 (39 residues)	Exenatide	Type 2 diabetes	Incretin mimetic; Gila monster saliva hormone; native sequence
HIV protease	Saquinavir	HIV-1 infection	All are modified peptide/peptidomimetics, typically containing a statin-like residue for interaction with the catalytic Asp residues
	Indinavir		
	Ritonavir		
	Nelfinavir		
	Amprenavir		
	Lopinavir		
	Atazanavir		
Natriuretic peptide (32 residues)	Nesiritide*	Congestive heart failure	Native sequence
Oxytocin receptor (9 residues)	Oxytocin*	Induction of labor	Native sequence
Parathyroid hormone (34 residues)	Teriparatide*	Postmenopausal osteoporosis	Native sequence
Platelet receptor glycoprotein IIb/IIIa	Eptifibatide	Acute coronary syndrome	Cyclic binding fragment, all L
28S proteosome	Bortezomib	Multiple myeloma	Dipeptidyl boronic acid; chymotrypsin-like inhibitor

continued

Table 1 Continued

Target receptor or binding protein (native ligand size)	Chemical name	Indication	Comment
Secretin (27 residues)	Secretin*	Diagnostic	Native human
Somatostatin	Octreotide	Acromegaly, carcinoid tumors	Cyclic binding fragment, multi-D substituted
Thrombin	Bivalirudin	Anticoagulant for coronary angioplasty	D-Phe^1
Vasopressin (9 residues)	[Arg^8]Vasopressin*	Diabetes insipidus, roentgenography	Native human
	Desmopressin	Diabetes insipidus, enuresis	1-deamino,D-Arg^8
Viral fusion structure	Enfuvirtide	Blockade of HIV-1 fusion with CD4$^+$ cells	All L residues

^{a} Depsipeptide antibiotics like daptomycin etc. are considered outside the scope of this table and review, since they are natural products without medicinal chemistry input.
*Native sequence.

purposes, rather than as a comprehensive listing of products available worldwide. The unnatural sequences offer examples of a number of approaches to the achievement of increased potency and improved pharmaceutical properties, which will be treated in detail below.

A number of these products are unchanged from the native peptide and therefore exhibit the advantages/disadvantages of the parent structure. Thus molecules like oxytocin (induction of labor) and glucagon (transient stasis of the gut for clearer x-ray photos; transient elevation of blood glucose to reverse acute hypoglycemia) are applied for acute applications and, with their high potency with low toxicity, are well accepted. Teriparatide (native huPTH 1–34) for postmenopausal osteoporosis is used for chronic administration; however, the application requires episodic administration (once daily), since prolonged administration results in bone loss rather than the desired anabolic effect. Thus a short-acting molecule is acceptable. Teriparatide's $t_{1/2}$ is 5 min following intravenous and 1 h following subcutaneous administration.

Use of a natural analog for increased potency is indicated by the use of salmon calcitonin rather than human calcitonin for some applications. The salmon sequence is altered at a number of residues (e.g., adding positively charged residues in its amphiphilic alpha-helical region), resulting in a 10-fold increase in potency[49] and slowed metabolism. However it also causes generation of antibodies (typically nonneutralizing) in 50% of patients.

A step toward a sequence modification to improve pharmaceutical characteristics is represented by pramlintide, an analog of the pancreatic hormone, amylin. Amylin has low stability in solution because of its tendency toward beta-sheet formation which causes aggregation.[50] Replacement of one Ala and two Ser residues in its C-terminal region by Pro residues improved the stability by reducing this amyloid-forming tendency, resulting in the recently marketed pramlintide (diabetes; slows gastric emptying, inhibits glucagon secretion, inhibits appetite). Early research on vasopressin analogs, where agonistic activity is desired, used a kind of prodrug approach (terlipressin), wherein Gly-Gly-Gly was attached to the N-terminus of the native sequence in order to provide slowed release. Desmopressin ([D-Arg8]1-deaminovasopressin) illustrates two modifications designed to block proteolysis (trypsin, aminopeptidase). We will address far more radical modifications and their medicinal chemistry lessons in the sections below.

Polypeptide hormones and neuromodulators address many types of targets.[51] Among the important receptor classes are the GPCRs and cytokine classes. Peptides also play a neuromodulatory role where the target frequently is the GPCR class, although ion channels, proteases, etc. may also be affected. Since polypeptide signaling represents the primordial control system, the 'creativity' of natural selection means that all niches of signaling and control are filled with polypeptides. In this review we will focus primarily on GPCR and cytokine receptor ligands to examine medicinal chemistry approaches that are presumed to have wide application.

2.14.3 Initial Approaches

2.14.3.1 Efficacy

An important type of peptide or protein drug is that which has agonistic or antagonistic action at peptide hormone GPCRs. Knowledge developed from the modeling of small-molecule ligands interacting with GPCRs, where penetration to the lumen of the GPCR is the critical element required for activation of the receptor, suggests that a similar requirement may exist for peptide ligands (**Figure 1c**) to have an agonistic effect.[52] This implies that certain portions of the ligand must be presented in a very specific conformation in order for the receptor to transduce the agonistic signal through the G-protein effector system. A pharmacologist's perspective would be that the ligand actually stabilizes a receptor structure that is either an agonistic, antagonistic, or 'inverse agonistic' conformation, thereby allowing the receptor to activate the G-protein complex that transmits signal, or maintain the inactive conformation.

An early concept was that peptides comprise an address and a message region.[53–55] In modern terms this may be thought of as a region required for binding with high affinity to the receptor and a region required for receptor activation, respectively. A remarkable insight in that analysis was the suggestion that small peptides, which tend not to have an organized structure in water/plasma, might adopt the bioactive conformation on the cell membrane, prior to loading the integral membrane receptor.[54] A peptide with both regions acting in a functional manner would then be an agonist, while one lacking a functional message region but with a high-affinity address region would be an antagonist. Thus one would expect that deletion, rigidification, or altering of the conformation of particular regions of a peptide ligand might generate receptor subtype specific agonists and/or potent antagonists. Nature, in its diversity, has examples for these message regions being confined to various localized positions in the peptide (N-terminal, C-terminal, and in the middle) and some examples are examined below.

2.14.3.2 Agonism

A central issue in peptide design is a preliminary determination of (1) the required size of the peptide structure, and (2) an understanding of the critical residues required for function. The first is usually accomplished by N- and C-terminal truncation to find the active length, followed by various 'scans' of the structure to find the critical residues. Many peptides can to be shortened with retention of activity while others can yield antagonists when shortened (*see* Section 2.14.3.3). In general the longer peptide ligands are more likely to survive shortening, but much depends on the site of truncation.

An example of truncation to the active core of a peptide hormone and cyclization is the discovery of octreotide (**Table 1**), a somatostatin agonist. Somatostatin is found in 14- or 28-residue residue forms with a cystine disulfide-imposed cyclic structure (12 residue ring) and both forms are enzymatically processed from the same precursor. Somatostain-14 is found throughout the body and its functions in the pancreas (suppression of glucagon secretion), gut (suppression of endocrine tumor secretion), and pituitary (suppression of growth hormone release) have attracted the most interest. A series of peptide analog syntheses involving truncation, Ala-scan, and D-amino acid scan (see below) generated potent shortened forms, and structural studies using nuclear magnetic resonance (NMR) indicated interacting Phe residues across the disulfide ring. This led to smaller cyclic structures[56] with substantially greater potency and duration of action compared to the parent hormone (MK-678) (**Figure 4**). These data[57] illustrate the

Structure	Designation	K_i (nM) cloned hu sst				
		h1	h2	h3	h4	h5
Ala–Gly–Cys–Lys–Asn–Phe–Phe–Trp | | Cys–Ser←Thr←Phe←Thr←Lys	Somatostatin-14	0.38	0.04	0.66	1.76	2.32
Pro–Tyr–D-Trp | | Phe←Thr←Lys	L-383,377	5654	0.49	3072	>10 000	2000
N(Me)Ala–Tyr–D-Trp | | Phe← Val←Lys	MK-678	>10 000	0.05	230	4949	232
D–Phe–Cys–Phe–D-Trp | | Thr(ol)–Cys ← Thr ← Lys	Octreotide	230	0.27	45	2191	137

Figure 4 Receptor subtype potencies of somatostatin and several important analogs. (Reproduced with permission from Yang, L.; Berk, S. C.; Rohrer, S. P.; Mosley, R. T.; Guo, L.; Underwood, D. J.; Arison, B. H.; Birzin, E. T.; Hayes, E. C.; Mitra, S. W. et al. *Proc. Natl. Acad. Sci. USA* **1998**, *95*, 10836–10841. Copyright (1998) National Academy of Sciences, USA.)

selectivity issue discussed above: the high potency but lack of receptor subtype specificity of the native hormone as well as the effect of constraint of the peptide structure on receptor subtype specificity. The wide range of in vivo effects of the native hormone is related to this lack of specificity and interest in subtype specific analogs is driven by such observations. For example, evidence suggests that receptor subtype sst2 is responsible for suppression of growth hormone release from the pituitary and glucagon secretion from the pancreas, but insulin release is not suppressed by action at this receptor. Thus a highly sst2-specific agonist might be a useful diabetes drug.[57]

This drug discovery program illustrates several features that are relevant to a medicinal chemistry analysis. In the case of somatostatin, the agonistic trigger appears to be in the center of the molecule, focused on a beta turn region containing Trp and Lys. Stabilization of the beta turn by D-amino acid substitution (D-amino acid scan[58]) resulted in substantially higher potency and has remained the focus of most of the subsequent work, which resulted in the design and launch of octreotide (**Table 1**; **Figure 4**). Examination of octreotide shows several interesting features: (1) D-Phe at the N-terminus to block aminopeptidase cleavage, (2) D-Trp to stabilize the beta turn, and also to block chymotryptic action, and (3) a reduced C-terminal residue, threoninol, to block carboxypeptidase action. Explicit attention in the issues raised in the section below on biological $t_{1/2}$ could have resulted in selection of an analog of greater potency (see an example of hydrophobic depoting, lanreotide: Section 2.14.5.1.2). Much recent work has been devoted to searching for ligands exhibiting selectivity among the growing number of somatostatin receptor subtypes.

As illustrated above for somatostatin, scans of a peptide structure can identify critical information for analog design. For proteins or larger peptides this is sometimes carried out by a homolog scan,[59] using sequences from related hormones or receptor binding sequences spliced into the structure being analyzed. A more detailed analysis is usually carried out through use of sequential replacement of each residue in turn by an Ala residue (Ala scan,[60] see overview [61] and cautions[62,63]) or by divergent side chain types (e.g., a Lys scan). Ala scans are now used very frequently in the dissection of binding elements in the protein–protein interaction space.[25] A D-Ala scan[9] or D-residue scan (illustrated for somatostatin [58]) can provide useful information since D-residues are known to stabilize beta turn structures,[64] block proteolysis, and generate antagonists (see Section 2.14.3.3). Scanning with N-methyl amino acids is another approach that adds protection from proteolysis and altered flexibility in the peptide side chain.[65] The use of azaamino acids to scan structures[66] has been used less frequently, but this residue offers several potential benefits, including resistance to proteolysis. Incorporation of scan residues can be done synthetically[67–71] or by recombinant DNA procedures[60] for native amino acids in longer peptides. A more comprehensive approach has been advocated, substituting all positions by all proteinogenic amino acids using high-throughput analoging, illustrated by synthesis of 532 fluorescently labeled VIP analogs on spots on cellulose membranes.[68]

Use of β-diamino acids[72–74] to scan peptide structures was advocated ('betide' scan; substituted β-aminoglycine). These amino acids typically were used as β-acylaminoglycine or β-acyl,β-methylaminoglycine residues that were designed to be roughly isosteric with the proteinogenic amino acids while offering increased rigidity.[72] This offers an additional route to identification of receptor subtype selective ligands.[74]

Cyclization of peptides through amino acid side chains is the subject of many studies. Cyclization rigidifies the structure, can block proteolysis, may lead to receptor subtype selectivity, and may result in higher receptor affinity by reducing the entropy cost of binding. For example, rigidification of calcitonin (Ct) with an $i-i+4$ side chain cyclization (cyclo 17–21[Asp17,Orn21]huCt), gave a 400-fold increase in potency.[75] Similar studies have been carried out in additional members of the mid-sized peptide hormone families,[76] where an helical conformation is predicted for much of the mid to C-terminal region of the molecule. Thus sequential cyclization scans[11] can provide much information about active conformation and receptor selectivity as well as advancing the identification of a development candidate. Cyclization scan $(i-i+3)$ in corticotrophin-releasing factor (CRF) resulted in a stabilization of the alpha-helical character and dramatically improved the potency of a lead antagonist series (Astressin).[77] As a further example of cyclization scanning, rolling-loop scanning has been proposed, making use of olefin metathesis to join alkylation points on a peptide structure for global rigidification.[78] Other examples that illustrate several of the above approaches are the use of cyclization scans and D-amino acid scans to examine requirements for receptor subtype selectivity in the CRF[77] and MSH family of peptides.[79] Side chain to side chain cyclization modifies structures that may be critical for receptor interaction, so backbone to backbone cyclization was proposed[80] and tools were developed.[81,82] Examples of these approaches are given below.

An additional and quite interesting approach is to use 'positional scanning' to identify binding requirements at each site in a peptide sequence.[83–85] For example, if one were analyzing a tripeptide sequence, one would make three groups of libraries (O–X–X; X–O–X; X–X–O) wherein each group represents 20 libraries where the position labeled O is fixed as a single amino acid, but the positions labeled X contain a mixture of all 20 natural amino acids. Thus each grouping contains 20 libraries containing 400 tripeptides in each. Readout by the activity measure of interest has been demonstrated to indicate the favored amino acid(s) in each of the three positions. There have been many applications[41,86] of this approach to finding the preferred amino acid substitutions.

2.14.3.3 Antagonism

As discussed above, mid-sized peptide ligands (>30 residues) need to bind to the receptor's extracellular elements and activate the receptor. If penetration of the lumen of the receptor is required for activation, one might expect that either the N-terminus or C-terminus might be needed to carry out this function. Thus one might expect that truncation or modification of the terminus might generate antagonism or increased receptor selectivity. In fact there are a number of clear examples of such N-terminal truncation leading to antagonists, discussed below (RANTES, VIP, PACAP, GLP-1, PTH, CRF, etc.).

Chemokines are important hormonal signals that control lymphocyte trafficking into tissues through their GPCRs. In many cases they are important proinflammatory signals and potential drug discovery targets. Modification of the N-terminus of several chemokines has resulted in highly potent antagonists with unusual receptor selectivities.[87] For example, modification of the N-terminus of the chemokine RANTES (truncation of seven residues, extension of one residue, or extension of N-terminus with aminoxypentane) results in sub-nM antagonists. An interpretation of this result is that the more C-terminal region gives high-affinity binding but the N-terminus is now unable to trigger agonism.

Similarly, in vasoactive intestinal polypeptide (VIP),[88] truncation of the N-terminus results in antagonism[89] while acylation (i.e., extension) of the N-terminus can lead to increased receptor selectivity or antagonism, depending on the length of the chain. Thus the N-terminally truncated VIP(10-28) and N-terminally acylated myristoyl[Lys^{12}]VIP-(1-26)-Lys-Lys-Gly-Gly-Thr or N-myristoyl[Nle^{17}]VIP are high-affinity receptor binders[90] (VPAC1 and VPAC2) with partial agonistic action (low cAMP stimulation). In contrast, N-acetylVIP is a high potency agonist. Substitution of D-amino acids in the N-terminus also generates potent partial agonists (e.g., [D-Phe^2]VIP).[91] An Ala scan of VIP was used to elucidate the residues that had critical receptor binding interactions.[92] Thus many of the themes described above re generation of antagonism or partial agonism are demonstrated for this hormone.

Pituitary adenylate cyclase activating peptide (PACAP) is a 38 residue, C-terminally amidated, neuromodulatory peptide with homology to VIP. It has a number of roles in the body, one of which is potentiation of glucose-dependent insulin secretion, similar to the incretins (see below). Like VIP, it binds nonspecifically to multiple receptors (PAC1, VPAC1, VPAC2).[93] While activation of VPAC2 has desirable effects in causing potentiated insulin secretion, PAC1 and VPAC1 activation causes side effects.[94] Thus strong selectivity and strong agonistic efficacy are desirable for agents to treat type 2 diabetes. Acylation of the N-terminus of PACAP by long acyl chains results in antagonism, while acylation with a shorter chain (hexanoic acid) results in an analog with increased potency and selectivity for the VPAC2 receptor.[95] Thus long-chain acylation yields partial agonism or antagonism while short-chain acylation yields an agonist with increased potency and selectivity. Again, N-terminal truncation of PACAP is found to yield a potent antagonist, PACAP-(6-38).[96]

Analogs of the incretins (peptide hormones secreted, e.g., by L and K cells lining the gut) are an important new class of therapeutic agents for treatment of type 2 diabetes.[97,98] These hormones prime the pancreas to make more insulin mRNA and to give a rapid and potentiated response to glucose coming from a meal. First to be pursued was glucagon-like peptide 1 (GLP-1) and this was followed by elucidation of a similar, and perhaps predominant, role for glucose-dependent insulinotropic polypeptide (GIP). These molecules have extremely short duration of action in vivo (see Section 2.14.3) but a peptide hormone in Gila monster saliva was found to bind to the GLP-1 receptor with high affinity and longer duration of action in vivo. This molecule, exendin-4, was recently approved for treatment of type 2 diabetics (exenatide, Byetta). Truncation by removal of the first two residues from the N-terminus of exendin-4 yields a potent antagonist.[99]

As mentioned above, incorporation of D-amino acids has resulted in antagonistic ligands on a number of occasions. A few examples are the antagonists [D-Phe^2]LH-RH; [D-Phe^7]bradykinin; [D-Phe^6]bombesin(6–13)OMe. From these initial leads, very extensive analog programs have used medicinal chemistry modifications to improve potency and duration of action (see below).

2.14.4 Biological Half-Life Solutions

Rapid clearance from the circulation is one of the principal issues cited against peptide drug candidates and proteolysis is routinely cited as the central issue. Much attention has been paid to this issue and advances have led to important new drug candidates. However, while proteolysis is an important route for peptide degradation, it is not the only challenge to prolonged duration of action. The glomeruli, capillary nets in in the kidney, allow water and plasma components below a molecular-weight of about 60 kDa to filter out and form the initial dilute urine. Serum albumin and other high-molecular-weight proteins are not lost, but peptides and other components free in the plasma are lost. Glomerular filtration in the kidney can result in very rapid clearance and strategies to overcome this serious problem have been devised (see below). To reach its full potential, a peptide pharmaceutical thus needs both a degree of protection from rapid proteolysis and a mechanism for protection from rapid clearance by other routes, such as glomerular filtration.

2.14.4.1 Proteolysis

As an example of current interest, the incretin GLP-1 (see discussion above) has a $t_{1/2}$ in humans of only around 0.9 min, due to rapid clearance by DPPIV, a widely dispersed, membrane-bound, serine aminodipeptidase that cleaves many protein hormones in the plasma.[44] DPPIV favors an Ala or Pro residue as the amino acid in the second position from the N-terminus, so incorporation of other residues here can provide substantial protection from proteolysis. Similarly, D-residues and α-methyl amino acids (e.g., Aib) can completely block DPPIV degradation.[100] Pyroglutamic acid at the N-terminus is known to block aminopeptidase action, and this structure also blocks DPPIV degradation.[101] Similarly blocking the N-terminus of GLP-1 with an Ac function effectively blocks DDPIV degradation while retaining agonistic function.[101] The GLP-1 series thus illustrates a number of the most popular and effective routes to block N-terminal proteolysis. Some additional routes for proteolysis are summarized in **Table 2**.

There are a number of general routes to protection from proteolysis that have been used: (1) N- and C-blockade; (2) D-amino acid substitution; (3) unnatural amino acid side chain substitution; (4) unnatural peptide backbone substitution (e.g., azaamino acids,[102,103] N-methyl amino acids,[65] pseudopeptide linkages[104,105]); (5) constrained peptides[10,11] (cyclized, side chain constrained,[106] Cα-alkyl); (6) peptoids (N-alkylglycines[107]); (7) beta and gamma amino acid substitution,[108] etc. Again, there are book-length reviews of these individual areas, so just a few of the most successful examples will be sampled.

N-Terminal blockade by pGlu or N-acylation is typically well accepted, with the former being found on many natural peptide sequences. The latter may result in changes to receptor selectivity or function (*see* Section 2.14.3.3). D-Amino acid substitution has been the most successful substitution re commercially available peptides (LHRH agonists and antagonists, somatostatin agonists, vasopressin agonists) (**Table 1**). A further type of rapid proteolysis is carboxypeptidase cleavage and C-terminal amidation is routinely used to block this action, both in nature and in the lab. Many peptide hormones are cleaved from their prohormone forms at Arg/Lys-Arg-X sites[109] by the enzyme furin to yield the C-terminal amide form. Elaboration of this concept to alkyl amides has yielded significant potency increases both due to proteolytic protection and also likely due to hydrophobic interactions (see leuprolide, **Table 1**). Additional examples are given in **Table 2**.

Pseudopeptide linkages have not been a first line type of modification, but are favored in peptidomimetics. Nonetheless, examples of active peptide analogs exist and the reduced amide bond is particularly interesting in this class. This results in the introduction of a positive charge, which may result in concentration on negatively charged membranes (increased Gouy–Chapman potential).[15] This may have relevance re Schwyzer's concept for receptor selectivity and loading by membrane interaction.[110,111] The use of azaamino acids has several potential benefits.

Table 2 Principal proteolysis sites and commonly used protection

Enzyme class	*Selectivity*	*Protection examples*
Aminopeptidase	H-X-	pGlu¹, D-AA¹ or D-AA², N-terminal acylation
DPPIV	H-X-Ala	Other residue, pGlu¹, D-AA², Aib², Acyl-X-
Neutral endopeptidase	Multiple	Case by case
Trypsin	-Arg/Lys-X	Arg(R₂), Lys(iPr)
Chymotrypsin	-Hydrophobic-X	D-residue, α-azaaminoacid
Post-proline cleaving enzyme	Pro-X	Pro-NH-alkyl, aza-Gly
Carboxypeptidase	-X-OH	-X-NH₂

Azaamino acid residues have a rigidified conformational preference and are not attacked by proteases.[102] In addition, they are thought to stabilize a type VI beta turn structure.[112] Perhaps most importantly, they have resulted in potent, biologically active peptide analogs (see review[103]) and at least one peptide product (goserelin, **Table 1**).[113]

Scanning with *N*-methyl amino acids is another approach that adds protection from proteolysis and altered flexibility in the peptide chain.[65] This approach has provided potent analogs[114] and shifts in expected structure–activity relationships.[115] For example, replacement of the customary D-Phe residue found at position 7 in potent bradykinin antagonists by D-*N*-methylPhe converted these structures to potent agonists.[115]

Modifications to constrain peptides are an exceptionally rich area of research going forward. Such constraint may block proteolysis, but, importantly, may also limit the receptor subtypes to which a ligand can bind. Constraint has the theoretical benefit of reducing the entropic price paid as a ligand binds to a receptor as well. Constraint is also invoked as a mechanism to cause conversion of ligands from agonists to antagonists.[9] As peptide research moves forward to subtype selective drug candidates, constrained analogs will certainly contribute in a major fashion.

Beta amino acids have high stability to proteolysis and instances of high biological activity for beta peptide analogs have been seen.[116] Although these novel amino acids do form stable structures in solution,[117] experience in design of active peptides is at an early stage and it is expected that such analogs will be a fruitful future direction. Their novelty means that proprietary analogs of existing active peptide motifs may be readily obtained. Peptoids (*N*-alkylglycines) are another type of unnatural amino acid that can develop structure and has been explored extensively, generating some molecules with biological activity.[118]

2.14.4.2 Glomerular Filtration

Although proteolysis is an important factor, as outlined above, any polar molecule free in the plasma with a molecular weight lower than about 60 kDa is rapidly cleared by filtration from the plasma into the urine in the kidney. About 25% of such low-molecular-weight, soluble material is cleared on each pass through the kidney. As an example of the significance of this clearance, inulin, a carbohydrate that is not metabolized, nonetheless has a rapid clearance with a $t_{1/2}$ of around 1 h. This would not be acceptable in a drug candidate. The parallel to a metabolically stabilized peptide suggests that even a peptide with no susceptibility to proteolysis that was exclusively free in plasma would also be cleared rapidly, based on its size and physical properties. This is a very important consideration that has become a fertile route to important new peptide and protein drugs. Several methods are now are used to avert this rapid clearance with clearcut benefits. The medicinal chemistry approaches using unnatural amino acids to modify physical properties are addressed below as 'hydrophobic depoting' and 'hydrophilic depoting.'

Another approach, conjugation of active species to prolong duration of action, has become one of the most important mechanisms to prolonging the duration of effectiveness for protein and peptide drugs. Since glomerular filtration is so important,[119] increasing the effective molecular weight of a drug active species can be a very effective approach. Two methods stand out as having been successful in the clinic and one method is clearly successful on the market. PEGylation of protein products has resulted in a rapidly growing list of currently marketed pharmaceuticals and this has led to increasingly creative and complex applications.[120] Conjugation to a higher-molecular-weight protein has been promoted as well, with albumin being the principal carrier protein used. These approaches are discussed below (Sections 2.14.4.2.3 and 2.14.4.2.4).

2.14.4.2.1 Hydrophobic depoting

Perhaps the earliest explicit approach to addressing glomerular filtration as a critical factor in peptide clearance, and therefore a subject for drug design, was based on the concept of 'hydrophobic depoting' of the peptide.[16,121] Thus a substantial increase in the hydrophobicity of the peptide might cause it to form a depot in the body, dissolving slowly and reversibly binding to cell membranes and to hydrophobic carrier proteins throughout the body (whole body depot effect).[16] While bound to hydrophobic carrier proteins (e.g., serum albumin, molecular weight 66 kDa) the peptide would not be cleared by glomerular filtration (molecular weight cutoff for filtration into the urine is below the molecular weight of serum albumin) or by membrane bound proteases lining the circulatory system. In this manner it would be continuously released back to the plasma compartment over a period of hours, thereby prolonging its duration of action. 'Hydrophobic depoting' is, in fact, a powerful method to obtain increased duration of action for peptides.[15] Following this idea, a group of 15 highly hydrophobic, unnatural amino acids of the D-configuration were incorporated into position 6 of GnRH.[121] The potencies varied from 80 to 190 times the potency of the parent hormone, when tested in vivo in a stringent 2-week antiovulatory assay in rats. Nafarelin ([D-Nal(2)^6]GnRH, **Table 1**) was the most potent member of this series at 190 times the native hormone.[121] This analog also had the very substantially increased $t_{1/2}$ of 5–6 h in humans, compared to several minutes for GnRH.

This series bears further examination for medicinal chemical insight, because additional analogs were used to fine-tune the hydrophobicity. Following completion of this work, a quantitative structure–activity relationship (QSAR) was reported, suggesting a relationship between potency and hydrophobicity of the side chains in position 6 of published GnRH analogs.[122] Such relationships are frequently cited for small-molecule drugs and generally exhibit a parabolic or bilinear relationship, reaching a maximum and then falling off in potency as the molecules become too hydrophobic for easy partitioning in the tissues. Analysis of the much more hydrophobic series above indicated a bilinear relationship, with D-Nal(2) in position 6 being close to the optimum hydrophobicity for the monosubstituted GnRH analogs series. Combination of D-Nal(2) with the more polar aza-Gly modification[113] in position 10 increased the potency[123] to 230 times that of GnRH while combination with the more hydrophobic Pro-NHEt in this position decreased the potency to 80 times that of GnRH.[121] Both C-terminal modifications had previously been shown to yield highly potent GnRH analogs, but here the relative additive hydrophobicity, when near the optimal, was shown to have had a substantial negative effect on potency, allowing fine-tuning to yield the highest potency analog at a slightly less hydrophobic position.

Examination of the mechanism by which this duration of action was achieved for nafarelin indicated that there was strong binding to plasma protein. An equilibrium dialysis experiment was carried out with one chamber containing buffer and the other containing plasma.[124] The peptide was added to the buffer side and subsequent analysis indicated an excess of peptide on the plasma side for the parent hormone of 23% but for nafarelin the preference for the plasma side resulted in an excess of 80%. The identity of the binding protein was determined by removing serum albumin by a chromatographic step and demonstrating that the preference for the plasma chamber was abolished. Thus there was a substantial increase in binding of nafarelin for serum albumin and one can visualize that albumin can thus function as both a carrier and delivery protein, as projected. This type of protection can be a very important factor for extending peptide duration of action and another example is given below (see acylated GLP-1 analogs). Thus hydrophobic amino acid substitution appears to be an important and general approach to increasing the potency and duration of action of peptide analogs, when properly positioned in the structure.[15]

A similar example in the GnRH field was the design of [D-His(im-Bzl)^6,Pro^9-NHEt]GnRH, wherein the protected His residue could be protonated at lower pH for higher solubility for formulation purposes, but would become unprotonated and very hydrophobic at physiological pH.[125] This analog is a potent GnRH agonist histrelin; precocious puberty (**Table 1**) with 200 times the potency of the parent hormone in vitro. This is an interesting modification that was used also in the GnRH antagonist area, but has not seen much use since. Its behavior is dependent on the basic properties of the imidazole side chain, which is deprotonated near physiological pH (pK_b about 6.5). This type of reversible protonation appears to be highly attractive and worthy of wider exploitation. Although an analysis of the reasons for its potency has not been put forward, it seems likely that hydrophobic depoting in the injection site and on serum albumin also is at work for this hydrophobic molecule.

Another powerful approach to the prevention of clearance by glomerular filtration through hydrophobic depoting is exemplified by approaches using hydrophobic acylation of larger peptides. An example of this comes from research on insulin where acylation with fatty acids resulted in protracted biological efficacy due to the binding of the fatty acid chains to serum albumin.[126] As we have seen, serum albumin is a general carrier protein for a number of molecules including fatty acids. Much as described above for nafarelin, where strong affinity for serum albumin was found, this type of 'hydrophobic depoting' for insulin has several advantages. The authors cite slow uptake into the plasma phase due to the hydrophobicity at the injection site and depoting on albumin in the injection site. When circulating the serum albumin bound form will be protected from glomerular filtration and proteolysis. The bound insulin will be slowly released to the plasma phase to have its biological effect. [LysB^{29}-tetradecanoyl des-(B30)]insulin (insulin detemir, Levemir) had a clearance $t_{1/2}$ of 14 h[127] and a relative affinity for the insulin receptor of 50%. Importantly, the degree of albumin binding was found to vary 50-fold between species (human albumin affinity was lowest of those tested).[128] Importantly, the interinjection variability also was found to be reduced and this drug was approved by the FDA in late 2005 as a long-acting, basal insulin for once or twice daily administration. The authors point out that this approach may be a general one[126] for increasing the duration of action of peptide hormones and, shortly, investigated it on another peptide hormone, GLP-1.

Albumin binding by acylation with fatty acids also provides highly potent and very long-acting analogs of GLP-1. As described above, the incretins (peptide hormones from the gut lining cells that cause potentiated glucose-dependent insulin secretion)[97] represent an exciting new class of treatment for type-2 diabetes, but duration of action due to cleavage by DPPIV was a critical impediment. Building on the studies that showed that acylation of insulin with aliphatic chains could prolong duration of action, the effect of acylation of the C-terminus of GLP-1 with aliphatic chains and spacers was examined.[129] These workers found that a relatively wide range of chain lengths on a variety of spacers would give potent analogs with a substantially increased duration of action. Acylation near the C-terminus was acceptable broadly, but acylation near the N-terminus caused losses in binding affinity. This spatial sensitivity would fit with the notion that the N-terminal region is required for the specific receptor interactions required for agonistic action, while the C-terminus has more of an 'address' function. In fact, examination of the structure of peptides of this class shows a rigid helical region near the C-terminus and a flexible N-terminal region, suitable for adopting a receptor bound conformation.[130]

The prolonged duration of action was specifically related to the ability of such analogs to avoid rapid glomerular filtration due to serum albumin binding. A peptide analog from the series described (liraglutide, [(γ-L-glutamyl(N-α-hexadecanoyl))-Lys26, Arg34]GLP-1(7-37)) has a $t_{1/2}$ in humans of 8 h following subcutaneous administration and is progressing through clinical trials using once daily injection.[97] This is a dramatic improvement over the native hormone ($t_{1/2}$ < 1 min).

In a similar study, the other well-studied incretin (GIP) also was modified by acylation near the C-terminus with fatty acids. GIP is released by K cells in the gut, has a similar structure to that of GLP-1, and is considered an excellent candidate for investigation as a new diabetic therapy.[131] Investigation of the solution structure by NMR indicates a helical structure from residue 6 at least through 28.[132] It binds to a different receptor than GLP-1, but similarly, is very rapidly cleaved by DPPIV ($t_{1/2}$ c. 5 min). The framework used in these studies relies on pGlu at the N-terminus to block DPPIV cleavage and the structure was modified with palmitic acid at different positions (i.e., [pGlu^1, Lys(palmitoyl)^{16}]GIP and [pGlu^1, Lys(palmitoyl)^{37}]GIP).[133] Both of these analogs appear to be more potent in vitro than GIP and show prolonged duration of action. For example, the second peptide still shows strong glucose lowering activity 4 h post administration in mouse models.[133] A similar analog[134] also is undergoing testing (N-Ac[Lys(Palmitoyl)^{16}]GIP) and it appears likely that a member of these series will be carried forward into clinical studies.

Thus we have seen that 'hydrophobic depoting' of peptides can be a very powerful route to prolonged duration of action; however, it has not received much discussion and analysis in the literature as such. This seems likely to change. These approaches of hydrophobic amino acid substitution[15,16,125] and fatty acid acylation[126] (for larger peptides) appear to be quite general for peptide hormones and many more such analogs certainly will be seen. A similar approach involves acylation with 2-(sulfo)-9-fluorenylmethyloxycarbonyl residues to provide slow delivery of active substance by hydrolysis.[135]

2.14.4.2.2 Hydrophilic depoting

Amino acid substitution in small peptides was recognized as a route to modified physical properties and, thereby, to altered pharmacokinetic behavior.[16] Thus substitution by Nal(2) was used as a way to increase hydrophobicity and duration of action by hydrophobic depoting (see above) for GnRH agonists. The use of GnRH agonists relied on down regulation of receptors and signaling systems due to continuous hyperstimulation by 'superagonists.' Thus treatment resulted in biochemical and physiological sequelae that maintained a suppressed state for an extended period of time (perhaps days). However the field of peptide antagonists is more challenging since one must compete with the native hormone that is being presented, typically in large pulses, throughout the day. Since as little as 10% receptor occupancy of the GnRH receptor can result in a stimulatory signal,[136] very tight binding, slow receptor off-rate, long-acting peptide antagonists are needed. The Nal(2) modification above was therefore incorporated into GnRH antagonists and lead to some of the most potent antagonists[16,137] at that time. Unfortunately, these analogs became so hydrophobic that they were administered to animals by subcutaneous injection in corn oil.

In an attempt to increase water solubility, polar amino acids (Glu, Arg, Lys) were incorporated into position 6 of the receptor series [N-Ac-D-pCl-Phe^1,D-pCl-Phe^2,D-Trp^3,D-X^6,D-Ala^{10}] LRHR and, in the case of the D-Arg^6 substitution, unexpectedly high potency and duration of action were observed.[138] We developed the hypothesis that a form of 'hydrophilic depoting' was responsible, rather than specific receptor interactions, and that this followed from interaction of the positively charged Arg side chain interacting with the negatively charged phosphate head groups on the phospholipids making up cell membranes throughout the body. Further stabilization of this interaction by the addition of alkyl chains led to a novel class of amino acids (**Figure 5**), the Ng,Ng'-dialkylarginines and -homoarginines.[139] These amino acids are readily made by guanylation of the ε-amino group on the side chain of Lys or Orn to make the corresponding hArg or Arg analogs.[140] A perhaps related class of modification is the use of Nε-alkyl-lysine residues to generate potent GnRH analogs. Although originally introduced[141] to be a basic amino acid that might improve oral absorption, it would be expected to show the type of phospholipids interaction suggested for the hArg(R$_2$) residue. It has been used in many analogs because it also has a shielded basic function that reduces potential for histamine release.

Incorporation of these amino acids in place of the Arg residue in the analog cited above produced a peptide with dramatically prolonged duration of action (>24 h in the castrated male rat model) compared to the parent structure. Detirelix ([N-Ac-2Nal^1,D-pCl-Phe^2,D-Trp^3,D-hArg(Et$_2$)^6,D-Ala^{10}]GnRH) had very prolonged duration of action in man[142] with blood levels detectable at >72 h after a 1 mg subcutoneous dose ($t_{1/2}$ = 29 h). This long duration of action is believed to be due to both the 'whole body depot effect' alluded to earlier and depoting in the site of injection.

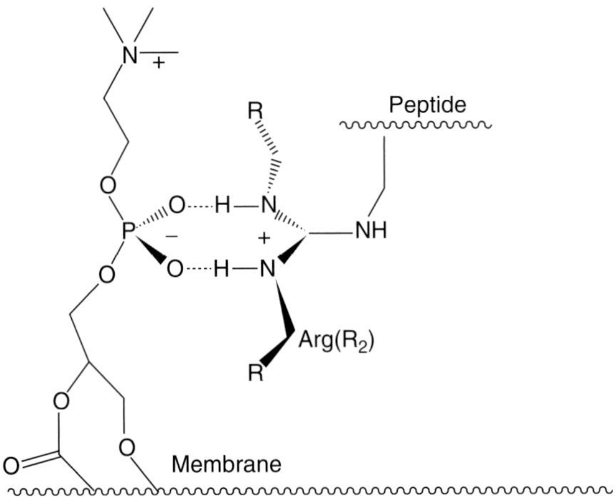

Figure 5 Proposed interaction between Arg(R₂) and phospholipid cell membrane.

Examples of much longer duration of action caused by depot formation at the injection site have been seen more recently and are discussed in Section 2.14.5.1.2.

Further modification of the structure (3 Pal for Trp replacement) to give a sharper profile of compound influx from the injection site and higher C_{max} led to the more soluble ganirelix ([N-Ac-D-2Nal^1, D-pCl-Phe^2, D-3 Pal^3, D-hArg(Et$_2$)^6,hArg(Et$_2$)^8,D-Ala^{10}]GnRH) (**Table 1**), the first GnRH antagonist registered in the USA (Antagon). This analog has a $t_{1/2}$ in humans[143] of 15 h and a threefold higher C_{max} than detirelix.[15] Ganirelix is suitable for once daily administration and is used for controlled ovulation. This same aminoacid modification was carried into additional analog studies. For example when incorporated into an enalapril analog (ACE inhibitor) it controls hypertension in the spontaneously hypertensive rat model for roughly fivefold longer (>30 h) than enalapril.[15]

These studies illustrate the concept of using unnatural amino acids to modify the physical properties of a small peptide in order to alter its pharmacokinetic profile. This is a very rich area for research and offers the opportunity to generate analogs with designed pharmacokinetic behavior and strong patent protection. This type of design work has further evolved to generate peptides with exceptionally prolonged duration of action and unusual properties, as can be seen in Section 2.14.5.1.2 on self-forming depots.

2.14.4.2.3 PEGylation

Prevention of rapid clearance by glomerular filtration in the kidney is now being recognized as a critical, in some cases predominant, factor in the development of long-acting peptides and proteins.[119] For example, the finding that modification of the larger peptide hormones with polyethyleneglycol moieties (PEGylation) is a route to increased duration of action and increased potency is widely recognized.[144] Although a detailed discussion is outside of the scope of this chapter, there are several mechanisms that have been invoked to explain this observation: (1) protection from immune surveillance,[145,146] (2) protection from proteolysis, and (3) prevention from glomerular filtration.[119,144] Recent thinking has given greater importance to the prolongation by reduction in glomerular filtration as a major factor. This is illustrated by detailed studies of the relationship between PEG molecular weight and $t_{1/2}$ in the blood which show a discontinuous relationship with a sharp increase in $t_{1/2}$ at about 34 kDa (**Figure 6**). The $t_{1/2}$ of PEG chains in the bloodstream increased from 18 min to 1 day as the molecular weight increased from 3 to 190 kDa.[119] The cut-off for allowing filtration into the urine is below the molecular weight of serum albumin (66 kDa) and the hydrated PEG chain behaves like a corresponding protein structure severalfold larger than its actual molecular weight. Thus all PEGs with a molecular weight above 35 kDa show similarly prolonged duration in the bloodstream, with little relationship to actual molecular weight. Similarly, their protein conjugates show substantially prolonged increases in $t_{1/2}$ once this range is reached.

The effects on duration of action and potency can be quite substantial.[147] For Interferon α2a, PEGylation increased the $t_{1/2}$ from 3–8 h to 65 h, with a 12–135-fold increase in antiviral potency. Similarly for IL6, PEGylation increased the $t_{1/2}$ from 2 min to >200 min with a 500-fold increase in thrombopoetic potency.

There is also evidence that the immunogenicity of proteins is reduced and that specific aspects of the PEGylation can be important in masking this propensity. For example, in early studies serum albumin's immunogenicity was shown to be suppressed by PEGylation[145] and this was extended to other proteins.[148] However, the first-generation

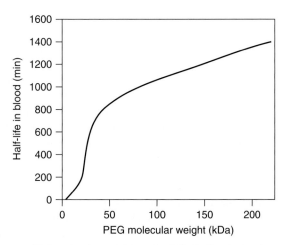

Figure 6 Correlation plot between PEG molecular weight and half-life in blood after intravenous administration. Note the discontinuity near 35 kDa. (Reprinted with permission from Caliceti, P.; Veronese, F. M. *Adv. Drug Deliv. Rev.* **2003**, *55*, 1261–1277 © Elsevier.)

PEGylated dimer of the soluble tumor necrosis factor receptor I (sTNF-RI, denoted TNFbp) was highly antigenic. A subsequent construction, modified with a monomeric 30 kDa PEG chain at the N-terminal amino acid, was found to be highly efficacious and nonimmunogenic.[149]

Thus while the early studies focused on multiple short PEG chains (e.g., molecular weight 5 kDa), examples in later work focus on single chains of larger molecular weight, in a number of cases resulting in the retention of higher activity. For example, interferon-α (19 kDa) coupled to a single PEG chain of 12 kDa has been commercialized for treatment of hepatitis C and has a $t_{1/2}$ that is eight times that of native Ifn-α. Thus PEGylation to yield 31 kDa will effectively block clearance by the glomerular filtration route, as described above. More recently, much larger PEG chains (e.g., to 40 kDa) are being used preferentially since single site modification results in better characterization and biological activity.

PEGylation has been extended to smaller peptides. For example modification of salmon calcitonin with one or two 12 kDa PEG chains results in a prolongation of $t_{1/2}$; 10-fold for the bis-PEGylated form.[150] Extension to releasable conjugates and small-molecule PEGylated drugs is an area of current/future development,[151,152] but it is already clear that this type of modification can be useful in the design of peptide drugs with prolonged duration of action. One area that needs to be considered relates to the finding that large proteins with large PEG groups may have restricted areas of permeation and certain applications may be either favored (e.g., permeation through the 'leaky vasculature' of tumors[153]) or disfavored (lack of permeation into sites for antiviral[154] or other action).

2.14.4.2.4 Protein conjugation

Another approach to prevent rapid clearance of a ligand by glomerular filtration is to attach it to a high-molecular-weight protein carrier. Attachment of ligands to serum albumin now has been used successfully to increase duration of action substantially for a number of peptide ligands, while retaining biological activity.[155,156] Initial studies examined the attachment of a dynorphin A (a 17-residue μ/κ opioid peptide) analog to the Cys residue at position 34 of human serum albumin by means of an ε-(3-maleimidopropionyl) function on a Lys residue at the C-terminus.[156] Use of the mouse acetic acid writhing assay showed that morphine, dynorphin A, and this conjugate (CCI-1035) showed activity at 5 min, but only CCI-1035 showed antinociceptive activity at 4 and 8 h, and does so at doses comparable to that of dynorphin A. This study demonstrated the retention of high agonistic potency and a very long duration of action.

Similarly, GLP-1 analogs suitable for conjugation were constructed through the substitution of an Nε-(2-{2-(3-maleimidopropylamido)ethoxy}ethoxyacetyl)lysine residue at various positions on the chain.[157] As expected from earlier studies with GLP-1 analogs, the favored position for substitution was at the C-terminus and a D-Ala residue was substituted at the second residue to block DPPIV activity. This analog, CJC-1131, reacts with human serum albumin (HSA) in vivo to form the circulating, long-acting form. The sequence had somewhat reduced potency when coupled to HSA, but a long duration of action in vivo is to be expected during ongoing in vivo trials. An analog that accomplishes a similar goal uses a GLP-1 sequence spliced directly onto that of HSA and is produced recombinantly (albugon). Albugon exhibits reduced acute potency, but much prolonged duration of action.[158] An extension of this approach proposes the use of transferrin as a carrier protein.

Thus this approach appears to have substantial potential for broad application to peptide and small protein ligands. This type of conjugation also was extended successfully to C-terminally conjugated analogs of human growth hormone-releasing factor[159] and atrial natriuretic factor.[160] Prior to contemplating its use, one may want to compare it to the use of PEGylation for a similar application. Conjugation to either gives prolonged effect and, in the case of PEGylation, immunogenicity is suppressed. Conjugation of ligands to proteins previously has been used as a way to raise antibodies against haptens, and this remains a concern for the protein conjugation approach, although researchers have explicitly stated that it has not been a problem with specific and single sites of conjugation. It seems clear that, since use of protein carriers and amino acid polymers in the past has been immunogenic, there is potential to observe that here. However, given this caveat, protein conjugation remains an intriguing and apparently general approach to increasing $t_{1/2}$ of peptides and proteins.

Another type of protein conjugation is represented by the use of an antibody Fc domain and taking advantage of its long $t_{1/2}$. For example, a number of cytokine receptors are found in a cleaved, soluble form in the plasma with the hypothesized role of modulating the function of that inflammatory system. Etanercept is a genetically engineered, dimerized, fusion protein composed of such a 'soluble TNF receptor 2' (endogenous antagonist of TNF function) linked to an IgG1 antibody Fc domain. This conjugate binds tightly to TNF-α and is highly effective in treating rheumatoid arthritis. This agent is long-acting and is administered twice weekly.

2.14.4.3 Structural Motif Replacement

As we have seen above, the class of peptide ligands in the range between 20 residues and 40 residues holds many attractive drug candidates and approaches to increased efficacy and potency will result in additional, important drugs. Some examples are the VIP/secretin family and PTH. Many of these hormones appear to have structural similarities, where this information is known. A typical finding is that the N-terminus is flexible and required for efficacy, while the more C-terminal region adopts a helical conformation and has regions that contribute to binding potency or just to helical structure.[161,162] Thus modifications of the correct portion of the C-terminus might yield agonistic activity with modified binding and/or pharmacokinetic behavior. We have seen that PEGylation and acylation with fatty chains can result in very prolonged duration of action and high potency for GLP-1. Cyclization of more C-terminal regions of PTH can result in improved biological activity.[163]

PTH offers an interesting challenge, because it must be present for a period of time, but not too long, in order to have a beneficial effect. Once daily injection of PTH or PTHrp has an agonistic effect on the PTH receptor on the osteoblast. This episodic treatment stimulates osteoblastic function[164] (laying down of matrix for new bone) and stimulates bone-lining cells to differentiate into active osteoblasts. However continuous treatment with PTH causes the osteoblasts to secrete a signal that activates osteoclastic action, with a resulting net bone loss. Interestingly, it is the time during which PTH levels remain elevated, rather than the total dose, that determines whether an anabolic or catabolic effect is seen on bone in the animal.[165] Therefore a superior PTH based drug for osteoporosis would have increased potency and longer duration of action, but not too long.

Human PTH is present in the body as an 84-residue polypeptide. SAR studies determined that the N-terminal 2 amino acids are required for agonistic action and the N-terminal 34 residue peptide is fully active. PTHrp is a related structure, frequently found as a secretion product from tumors (but in a developmental role in many tissues), which binds to the same receptor with similar potency and efficacy. A project in the author's laboratory directed at single amino acid substitutions in the C-terminus to achieve hydrophobic or hydrophilic depoting, as described above, did not yield substantial effects on potency for analogs of PTHrp.[166] PTHrp had been chosen for modification because a triple basic sequence, hypothesized to be critical for function, was located closer to the N-terminal region and it was felt that more extensive C-terminal truncation could be accepted with retention of activity. However the incorporation in place of residues 22–31 by a 10-residue model amphiphilic polypeptide helical structure,[167,168] originally designed as a lung surfactant peptide, led to remarkable increases in efficacy in vivo.[169,170] Although the increased efficacy may be due to the use of a stabilized helical motif, the compound (semparatide) may also exhibit altered behavior due to the amphiphilicity of the modification. Further study confirmed its enhanced helical character[171] and bone anabolic efficacy, but suggested that it had sharply reduced affinity for the PTH receptor[172] (see however studies[173] indicating high affinity for the G-protein-bound, receptor high-affinity state). This compound was carried into phase III clinical studies with excellent efficacy. One would surmise that a molecule that combined the features of somewhat prolonged duration of action, but with increased receptor affinity, might be a very desirable clinical development candidate.

The comparison of the lack of strong effect on protein function in this series by single amino acid substitutions, but strong effects by motif incorporation, suggests to the author that this may be a paradigm for future research on protein ligands of this size. While stabilization of the helix of a ligand may yield favorable entropic effects re binding thermodynamics (not seen here), the ability to modify protein behavior and PK thereby has obvious benefits.

Study of analogs of cortocotrophin-releasing factor (CRF) showed that cyclization[77] and incorporation of residues to optimize alpha-helical character (C-alpha amino acids)[174] could result in increased potency. This is another example of a mid-sized polypeptide hormone where the biologically active conformation appears to have a predominantly alpha-helical motif.

2.14.4.3.1 Expression-modified proteins

An approach that shows great promise for incorporation of unnatural amino acids into proteins and has the potential for extension to the incorporation of larger motifs is that developed by Schultz and co-workers.[175] It makes use of in vitro selected, unique tRNA/aminoacyl tRNA-synthase pairs that load specific unnatural amino acids and that make use of the amber nonsense codon, TAG. Some examples of this approach demonstrated the incorporation of amino acids with side chains that have sugar functionalities,[176] redox potential,[177] p-iodo-phenylalanine for x-ray phasing (single wave-length anomalous dispersion),[178] PEG for site-specific PEGylation,[179] and alkyne functionality (for 3+2 electrocyclic cyclizations, CLICK chemistry),[180] etc. Expansion to a four-base codon system[181] offers the opportunity to increase greatly the number of types of amino acids that can be incorporated successfully into proteins. Thus the use of multiple, or critically placed, unnatural amino acid substitutions offers the opportunity to incorporate substantially modifed motifs, and thereby proteins with significantly altered physical properties, with attendant beneficial effects on protein behavior in vivo as described in the previous section. As usual, the question of potential immunogenicity for modified proteins of >50 residues must be considered and even native proteins can be immunogenic.[182] However the frequent experience is that such antibodies are not neutralizing,[183] although serious instances of immunogenicity are known. Nonetheless, this approach appears to have great promise and flexibility for future development.

2.14.5 Drug Delivery to Overcome Bioavailability Issues

The most important challenge that faces a peptide or protein drug candidate is that of achieving the bioavailability required for therapeutic effect in the body and drug delivery technologies offer very powerful ways to add value to peptide and protein products. The challenges of duration of action and receptor selectivity have many solutions as we have seen above, but the physical barriers that the body erects to keep unwanted agents (viruses, toxins, antigens, etc.) outside, also prevents peptides or proteins from finding easy access. The routes across the mucosal barriers are either transmembrane (dissolving in the membrane, crossing the cell, and dissolving in the basolateral side membrane to exit), paracellular (passing through the 12 Å radius pores[184] between the intercellular tight junctions), or by active transport (e.g., the peptide[185] or amino acid transporters). The small-molecule drugs with a balanced hydrophobic/hydrophilic character are able to equilibrate across membrane barriers driven by a concentration gradient. Small molecules and salts can pass through the paracellular route. Some polar molecules like amino acids, and up to tripeptides,[185] are actively transported, as are some drugs (e.g., penicillins, cephalosporins) and prodrugs (e.g., valganciclovir, valcyte). It is apparent then that the bulk of peptide and protein drugs normally are not able to make use of any one of these three transport routes. Thus the challenge has been approached by injection, transient breaching of the barrier (mucosal, dermal, enterocyte), and inhalation. There are many books and specialist reviews of these areas, and what follows is an overview of selected areas that appear to have high promise or clear efficacy. Briefly, these routes are intranasal, inhalation, and controlled-release injectable.

For agonistic applications, episodic administration may be favored since many hormonal systems become desensitized (receptor downregulation, signaling system uncoupling) with continuous stimulation. In the case of GnRH, a well-studied case, continual administration of highly potent agonists actually leads to a 'medical castration' that is the basis of the therapeutic benefit. Thus for treatments where an antagonistic effect is desired, controlled-release formulations with continual delivery of the active agent have clear potential advantages. Some other hormonal systems appear to be less sensitive to this type of desensitization. For example, administration of GLP-1 or exendin-4 (Byetta) to cells in culture induces a clear state of receptor desensitization, but treatment in vivo appears not to.[186] It is possible that the variable levels seen during in vivo treatment (blood levels rapidly drop following the initial pulse) offer the opportunity for the system to resensitize. Early reports say that exenatide, a potent, longer-acting GLP-1 receptor agonist, does not exhibit diminished response following treatment periods of moderate length. However, longer, higher-dose treatment with more potent analogs might well cause diminished response. An extreme case is seen with PTH. As we have seen above, for treatment with PTH, episodic treatment is required. Thus the form of administration, episodic or continual release, may be critical to the pharmacological effect seen and desired. Careful evaluation of these considerations is needed on a case-by-case basis.

2.14.5.1 Injection

The oldest, most straightforward, and most widely used form of administration of peptide and protein drugs is by injection. This route has its own issues but typically one can achieve high bioavailability and deliver episodic pulses of the agent of interest. The issues for simple injection are well known (solubility, stability, denaturation, immunogenicity, etc.), but relatively straightforward for experienced pharmaceutical groups. We will focus on alternative formulations.

2.14.5.1.1 Controlled release

Controlled release of peptide drugs using polymeric matrices has become a widespread and successful approach, with effective delivery times of 3 months and more. Although controlled release from polymer matrices initially began with hydrophobic small molecules diffusing out of silicone rubber,[187] many different types of polymers[188] have been evaluated. Polyamino acids, polyesters, polyorthoesters, polyacetals, polyketals, polyurethanes, etc. are some of the most frequently cited. This type of research offers a route to improved behavior, ease of processing, or just proprietary formulations. Lactic/glycolic copolymers have been the most successful and most widely- used for peptide and protein delivery, and a brief illustration of development of this area is given below.

The observation that copolymers of lactic acid and glycolic acid (poly-D,L-lactic-co-glycolic acid; PLGA) could be used to form sutures that slowly eroded and resorbed with time to yield components that are naturally present in the body led to their application in the development of early formulations for agents such as naltrexone,[189] sulfadiazine,[190] and contraceptive agents.[191–193] Extension to the more polar polypeptides required some modifications and was carried out in studies with nafarelin.[194,195]

Characteristics of such a formulation include hydration and homogenous erosion of the polyester matrix. The hydration rate, and therefore the release rate, depends on the hydrophobicity of the polymer matrix, so variation in the percentage of lactic acid can vary the duration of release. In early studies, 50/50 lactic glycolic copolymer microcapsules (i.e., injectable by standard syringe) gave 21 days of essentially linear release of nafarelin in female rats with prompt return to estrus cycling at day 24.[196] This type of microcapsule formulation for GnRH analogs was subsequently applied to [D-Trp⁶]GnRH[197] and others. Eventually it was out-licensed to become the formulation (with modifications in manufacture) used in Leupron Depot, a peptide 'blockbuster.' The importance of formulation for peptide delivery is illustrated in this latter product because leuprorelin is relatively short-acting and required up to six times daily injection for treatment in some early rat models. In contrast, in an early microcapsule formulation, only once-monthly delivery was required.

The duration of release is dependent not just on the percentage lactic acid in these formulations, but also on the molecular weight of the polymer or mixture of polymers used and on the use of homochiral or racemic polylactic acid. In addition, the size of the structure determines the release rate. Formulation of zoladex into rods (1 mm × 3–6 mm) and implantation subcutaneously released the analog over a period of >28 days.[198] Thus longer-term release patterns were also possible with the same underlying chemical components and later modifications (dosage increased from 3.6 to 10.8 mg) have resulted in microcapsule formulations requiring administration once every 3 months.[199]

A very interesting approach, that may simplify the development of controlled release formulations, is the development of soluble formulations[200] for injection that contain the active ingredient and a mixture of polymers in a biocompatible solvent. This solution is then injected to the subcutaneous space where the solvent is leached away to precipitate the polymer and provide a solid polymer matrix containing the drug. Formulations for leuprolide (Eligard) which last for 1 month (7.5 mg), 3 months (22.5 mg), and 4 months (30 mg) have been approved by the FDA. The potential advantage is that some of the variables in the manufacture of a controlled-release formulation, which give rise to variations in the release profile, presumably can be avoided, perhaps saving many months in the development process.

The logical extension of these formulation design efforts is Vantas (histrelin hydrogel) for once-yearly administration. This formulation is a 3-cm tube containing 50 mg of the hydrophobic GnRH analog histrelin and stearic acid. On the negative side, a small incision is required and the tube must be removed at the end of its use. Another very long-term formulation is Viadur (4 × 45 mm implant), which is a once-yearly implantable device for osmotically driven release of leuprolide over a period of 1 year.[201] The use of polylactic/glycolic microspheres for the delivery of therapeutic proteins (serum albumin,[202] IL2[203]) and immunogens (e.g., CTL vaccine against Her-2/neu[204]) also has been demonstrated.

2.14.5.1.2 Self-forming depots

A relatively new approach to the design of long-acting forms of delivery for peptides is the use of self-forming depots. This approach can be broken into the use of solublilized polymeric depots that solidify at the injection site when the solubilizing agent leaches away (see above, Eligard) or active agents that have had this property designed into their structures. These latter agents typically exhibit characteristics that lead to the formation of intermolecular aggregates, combining hydrophobic and hydrophilic elements. Evolution of the latter agents can best be demonstrated in the GnRH and SRIF areas.

Early work on compounds designed with prolonged duration of action resulted in nafarelin[121] and detirelix,[139] very hydrophobic GnRH analogs that exhibited a propensity to form liquid crystals in solution.[205] It was felt that the prolonged duration of action of such molecules derived in part from depoting on serum albumin and cell membranes, but also from retention in the injection site.[206,207] Clearly such molecules showed the potential for the self-assembly of complex structures in solution. Thus detirelix showed a $t_{1/2}$ of 7 h following intravenous injection in monkeys (0.08 mg kg^{-1}) but a much-prolonged $t_{1/2}$ of 32 h from a subcutaneous injection of 1 mg kg^{-1}. For Ganirelix (**Table 1**) in monkeys, the $t_{1/2}$ was 19 h following a 1 mg kg^{-1} subcutaneous dose.[207] In vitro plasma binding was >80%, similar to nafarelin. Depot formation was indicated by the retention of 12% and 4% of the dose of radiolabeled analog in the injection site after 3 and 10 days, respectively.[207]

A similar set observations have developed around the hydrophobic somatostatin analog, lanreotide (cyclic [D-2Nal-Cys-Tyr-D-Trp-Lys-Val-Cys-Thr-NH2]).[208] This analog, which is essentially a D-2Nal analog of octreotide, springs from the observation that N-terminal Nal substitution in GnRH analogs and internal substitution in agonists provided increased potency. Like octreotide, this analog is selective for the sst2 receptor, which is present on >80% of digestive tract tumors.[209] This substitution led to a decrease in potency acutely, but to a substantial increase in duration of action. This was hypothesized to be due to slowed uptake from the site of injection and subsequent work detailed a self-association process that proceeded from the formation of beta-sheets to development of large nanotubes.[210] This self-assembly into nanotubes occurs through the formation of beta-sheets driven by amphiphilicity and a systematic aromatic/aliphatic side chain segregation.[211] The physical state is concentration dependent and drives toward nanotube formation at higher concentrations. Thus this product is not a polymer type, controlled-release formulation; the depoting character is built into the molecule through this ability to form a slow release, self-associated form that further assembles as stable nanotubes, with a terminal $t_{1/2}$ in humans of 4 weeks. Remarkably, lanreotide acetate hydrogel (30% w/w) thus has a very simple formulation in water and is administered every 4 weeks by deep subcutaneous injection.[209]

While the analogs above form a type of aggregation driven by hydrophobic character, another class of analogs with unusual properties spring from observations related to side chain urea functional groups such as that on citrulline. Thus cetrorelix ([N-Ac-D-2Nal^1,D-pClPhe^2,D-3 Pal^3,D-Cit^6,D-Ala^{10}]GnRH) is an approved pharmaceutical that appeared to have somewhat prolonged action. Other GnRH analogs containing urea functionality have been demonstrated to exhibit 'gelling' and delayed release from the injection site (2.6% of radioactive dose present at injection site at 264 h post dosing[212]), and in some cases local tissue reaction has been documented.[213] Further studies directed at long-acting GnRH analogs containing ureido functionality resulted in molecules such as azeline B, which showed very substantial potential for gelling in the injection site. Such analogs exhibited prolonged duration of action (14 days for azeline B following 2 mg kg^{-1} subcutaneous in rats).

A very detailed study of structure–activity relationships related to structures with side chain urea functional groups[214] resulted in the selection of degarelix (FE200486; [N-Ac-D-2-Nal^1,D-pClPhe^2,D-3-Pal^3,Aph(Hor)^5, D-Aph(Cbm)^6,Lys(iPr)^9,D-Ala^{10}]GnRH) (**Figure 7**) for preclinical development. Very dramatically prolonged action was demonstrated for this material when injected subcutaneously, in contrast to that obtained by intravenous injection implicating a depoting effect at the injection site. It is hypothesized that the urea functional group has unique properties that favor a 'slow diffusion from the subcutaneous sites of administration.'[214] While a detailed type of biophysical study[210] like that done with lanreotide autogel does not appear to have been published, the prolonged duration of action seen here may have a similar source in the self association to elaborate structures that 'do not mount to detectability as gels.'

2.14.5.2 Intranasal

Transmucosal administration of peptides and proteins has been successful and there are multiple products on the market. The advantage of using the nasal passages are that they are readily available, have a thin squamous cell lining, have a relatively large surface area (150 cm^2), allow the drug to avoid first-pass clearance, and minimize overdosage since drainage is to the gut where the peptide is destroyed. Several smaller peptides are commercially administered in this fashion: [D-Arg^8]vasopressin, nafarelin, and desmopressin. In addition, and, somewhat surprisingly considering its

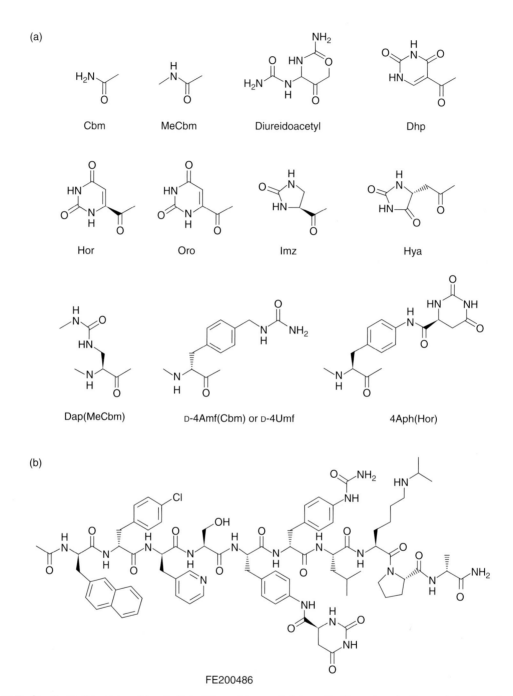

Figure 7 Structures of some p-ureido substituted Phe residues and of degarelix. (Reproduced with permission from Jiang, G.; Stalewski, J.; Galyean, R.; Dykert, J.; Schteingart, C.; Broqua, P.; Aebi, A.; Aubert, M. L.; Semple, G.; Robson, P. et al. J. Med. Chem. **2001**, 44, 453–467 © American Chemical Society.)

size, salmon calcitonin is reported to have relatively good bioavailability by this route, without added enhancers (however, see below).

In early studies with GnRH analogs, they were typically found to have bioavailablities in the 2–5% range. In animal studies with nafarelin, the bioavailability was found to be nonlinear, with higher absorption being seen at higher concentrations of peptide. Thus a 2.2-fold increase in concentration resulted in a 6.5-fold increase in bioavailability and a further 2-fold increase in concentration resulted in a fourfold increase in bioavailability (solubility limited to 2.5 mg mL^{-1}). The addition of protease inhibitors like Trasylol did not increase absorption.[215] Studies in humans using various enhancers found that instillation of a solution of 1.75 mg mL^{-1} in saline containing 2% sodium glycocholate showed an average bioavailability of 21%.[216] Unfortunately, detailed study of such enhancers in the past have turned up evidence of irritation and unsuitability for continuous use.[217]

Looking forward, there are new approaches to enhancers for nasal delivery. Studies with the peptide PYY and peptide enhancers designed to interfere with the protein structures that comprise the tight junctions between cells[218] have resulted in a clinical program that is moving forward with the intent to cause appetite suppressing effects in order to treat obesity. In a similar vein, very substantial improvements in nasal bioavailability of peptides and proteins have been reported with alkyl sugar enhancers.[219,220] For example, these researchers report that although the bioavailability of intranasal salmon calcitonin is highly variable (0.3% to 30%) in the commercially available formulation or in saline, it is 53% when administered in the presence of 0.125% tetradecylmaltoside.[219] Similar bioavailability values were demonstrated for insulin and leptin with this system with minimal effects on the structure of the nasal cilia.[220] This is a remarkable result and, if reproducible in humans without irritation, opens the way for additional mid-sized peptide drugs to be developed by intranasal administration.

2.14.5.3 Inhalation

The evaluation of the lungs for the administration of peptide and protein therapeutics[221] has been considered attractive because of the very large surface area (140 m^2), thin epithelium, good blood flow, and immunotolerant nature.[222] Many peptides and proteins are in evaluation and insulin has been studied in the most detail. Insulin inhaled as a dry powder has a bioavailability of 6–10% and a rapid onset of action (T_{max} 24 min versus 106 min for subcutaneous regular insulin). The duration of action is somewhat longer and it appears to be well accepted by patients.[223] This formulation is now approved by the FDA(Exubera) and validates the inhalation route for administration of peptides and even small proteins.

2.14.5.4 Oral Administration

Although oral administration of peptides and proteins has been studied for many years there still appear to be no reproducible data confirming substantial bioavailability.[222,224] Desmopressin (nonapeptide) is available for oral administration, but the bioavailability is very low. A similar statement could be made for the rectal and vaginal routes of administration. As noted above, the three normal routes of permeation are closed to peptides and proteins. Transient disruption of the barrier function of the tight junction between the enterocytes would seem to have more serious consequences than doing so in the nose or lungs, due to the potential for uptake of endotoxin due to the bacteria in the lower gut. It would appear that intranasal, inhalation, and controlled release injectable remain the best options for delivery of peptides.

2.14.6 Peptidomimetics

The detailed analysis of peptidomimetic approaches is outside of the scope of this chapter (see 2.15 Peptidomimetic and Nonpeptide Drug Discovery: Receptor, Protease, and Signal Transduction Therapeutic Targets; and reviews[1,13,14,106,225–232]), but is highly relevant, given that drug discovery relating to peptides is moving strongly in that direction. There are some important points that need clarification: (1) not all ligands that compete with a peptide in binding with its receptor represent true 'peptidomimetics' and (2) the results to date have largely yielded antagonists when applied to GPCRs. The latter is particularly true when addressing mid-sized peptide hormones (>20 residues).

Although initial interpretations were that the small-molecule ligands binding to peptide GPCR receptors were true peptidomimetics and specific features were hypothesized as representing specific amino acid side chains, site-specific mutagenesis studies soon demonstrated that there was no commonality in the receptor features required for binding the native ligand and the small molecule antagonist. For example, a comprehensive study of the critical interaction residues of the substance P/neurokinin (NK1) receptor for the native ligand and a small-molecule antagonist showed that different regions of the receptor, with little overlap, were involved in binding to the two different molecules.[233] Similarly, the critical interaction residues of the bradykinin 2 receptor for the native peptide ligand and for the structurally quite similar undecapeptide antagonist HOE-140 (three residues differ) show very little overlap,[234] suggesting very different critical binding interactions. It seems clear that in most cases of GPCR ligands, the small molecule antagonists should be assumed to not have the same binding mode or faithfully represent the critical amino acid residues characteristic of the native ligand. Nonetheless, a structurally related molecule to that mentioned above has now reached the market as an NK1 antagonist (aprepitant) for emesis.

'Peptidomimetics' have been classified, based on which were characterized first,[225] as Type I (backbone modifications to block proteolysis), Type II (functional mimetic, without structural overlap), and Type III (topographical mimics with the relevant amino acid side chains faithfully represented on a novel scaffold). By this definition, Type III is the type that has long been sought, but for antagonistic function, any type may be efficacious. An alternative to this type of classification was the term suggested by Veber and co-workers,[235] 'limetic,' for that class of small-molecule ligands for a peptide GPCR that is not truly mimicking a peptide's critical binding interactions, which would be peptidomimetic.

In general it has been felt that peptides interact strongly with both the N-terminal region of the GPCR receptor and the extracellular loops, but that interactions with the transmembrane region are critical for agonistic action. Thus a relatively large and complex assembly of interactions is required for agonistic action. Presumably this is the reason that there have been many more antagonists than agonists found in peptidomimetic research.

Research generating enzyme inhibitors and cytokine class receptors appear to be areas where the design of true, or Type III, peptidomimetics has been fruitful more directly. In these cases, x-ray structures allow the direct comparison of the peptide bound structure with that of the peptidomimetics. Particularly impressive have been the results of structure-based drug design programs in the HIV and matrix metalloproteinases (MMP) areas where many examples of direct design of potent inhibitors have been published.

2.14.7 Conclusions and Outlook

As we have seen above, the issues of duration of peptide/protein action in vivo, receptor subtype selectivity, and bioavailability are being answered effectively with medicinal chemical and drug delivery approaches. The diversity of chemical structural topologies available from the 20 natural and hundreds of unnatural amino acids is virtually unlimited. Approaches to increasing the speed of synthesis and testing of such structures have been developed and instituted broadly. Thus, despite advances in peptidomimetics and small-molecule drug design, peptide and protein chemistry would seem to be poised to multiply their impact on medicine.

An aspect of peptide and protein chemistry that is a powerful driver for future development is the tremendous improvement in production methods. One of the ongoing concerns re development of peptide and protein drugs has been cost of goods. However the rapid synthesis of even quite large polypeptides is now routine and the scale-up to production for clinical trials[3] or the remarkable achievement of ton-scale production[4] of enfuvirtide for commercialization has enabled us to think very creatively, with few remaining synthetic boundaries. In the case of enfuvirtide, the cost of synthesis has come down to the range of <$1 per gram per amino acid residue.[4]

An area of growth for protein-based therapeutics is chronic inflammation. Although this has been a major area of research for many years, it is now becoming generally recognized that inflammation lies at the core of the 'metabolic syndrome,' sepsis, and cachexia. This grouping is perhaps the major healthcare crisis facing the developed countries worldwide. For example, the metabolic syndrome encompasses obesity, type 2 diabetes, atherosclerosis, hypertension, and the prothrombotic state. Congestive heart failure may also have a related causation. One-half of cancer patients experience cachexia, an adipose and muscle-wasting syndrome that leads to death. The most recent thinking in each area is that a type of chronic inflammation lies at the heart of the disease. Inflammation is driven by protein signals (cytokines, chemokines, proteases[236]) that are still being elucidated and antagonists beyond the current anti-TNF approaches will likely be important new pharmaceuticals.

Interactions at large protein surfaces are being analyzed in detail[34] and this area offers almost unlimited potential to uncover novel pharmaceutical candidates. The use of protein ligand motifs, rational design, or phage display[24,237] to identify peptide ligands that can be optimized with the medicinal chemical approaches discussed above appears to be extremely attractive.

References

1. Houben-Weyl, *Synthesis of Peptides and Peptidomimetics*, 4th ed.; Thieme: New York, 2004.
2. Sewald, N.; Jakubke, H.-D., Eds. *Peptides: Chemistry and Biology*; Wiley-VCH Verlag GmbH: Weinheim, Germany, 2002.
3. Andersson, L.; Blomberg, L.; Flegel, M.; Lepsa, L.; Nilsson, B.; Verlander, M. *Biopolymers* **2000**, *55*, 227–250.
4. Bray, B. L. *Nat. Rev. Drug Disc.* **2003**, *2*, 587–593.
5. Dawson, P. E.; Muir, T. W.; Clark-Lewis, I.; Kent, S. B. *Science* **1994**, *266*, 776–779.
6. Nilsson, B. L.; Soellner, M. B.; Raines, R. T. *Annu. Rev. Biophys. Biomol. Struct.* **2005**, *34*, 91–118.

7. David, R.; Richter, M. P.; Beck-Sickinger, A. G. *Eur. J. Biochem.* **2004**, *271*, 663–677.
8. Lloyd-Williams, P.; Albericio, F.; Giralt, E., Eds. *Chemical Approaches to the Synthesis of Peptides and Proteins*; CRC Press: Boca Raton, FL, 1997.
9. Hruby, V. J. *Nat. Rev. Drug Disc.* **2002**, *1*, 847–858.
10. Cowell, S. M.; Lee, Y. S.; Cain, J. P.; Hruby, V. J. *Curr. Med. Chem.* **2004**, *11*, 2785–2798.
11. Reissmann, S.; Imhof, D. *Curr. Med. Chem.* **2004**, *11*, 2823–2844.
12. Li, P.; Roller, P. P. *Curr. Top. Med. Chem.* **2002**, *2*, 325–341.
13. Freidinger, R. M. *J. Med. Chem.* **2003**, *46*, 5553–5566.
14. Freidinger, R. M. *Curr. Opin. Chem. Biol.* **1999**, *3*, 395–406.
15. Nestor, J. J., Jr. Improved Duration of Action of Peptide Drugs. In *Peptide-Based Drug Design*; Taylor, M. D., Amidon, G. L., Eds.; American Chemical Society: Washington, DC, 1995, pp 449–471.
16. Nestor, J. J., Jr.; Ho, T. L.; Tahilramani, R.; Horner, B. L.; Simpson, R. A.; Jones, G. H.; McRae, G. I.; Vickery, B. H. LHRH Agonists and Antagonists Containing Very Hydrophobic Amino Acids. In *LHRH and Its Analogs: Contraceptive and Therapeutic Applications*; Vickery, B. H., Nestor, J. J., Jr., Hafez, E. S. E., Eds.; MTP Press: Lancaster, PA, 1984, pp 23–33.
17. Kroeze, W. K.; Sheffler, D. J.; Roth, B. L. *J. Cell Sci.* **2003**, *116*, 4867–4869.
18. Rashid, A. J.; O'Dowd, B. F.; George, S. R. *Endocrinology* **2004**, *145*, 2645–2652.
19. Blume-Jensen, P.; Hunter, T. *Nature* **2001**, *411*, 355–365.
20. Britten, C. D. *Mol. Cancer Ther.* **2004**, *3*, 1335–1342.
21. Schein, C. H. *Curr. Pharm. Des.* **2002**, *8*, 2113–2129.
22. Kotenko, S. V.; Langer, J. A. *Int. Immunopharmacol.* **2004**, *4*, 593–608.
23. Stevceva, L. *Curr. Med. Chem.* **2002**, *9*, 2201–2207.
24. Cunningham, B. C.; Wells, J. A. *Curr. Opin. Struct. Biol.* **1997**, *7*, 457–462.
25. Thorn, K. S.; Bogan, A. A. *Bioinformatics* **2001**, *17*, 284–285.
26. Livnah, O.; Stura, E. A.; Johnson, D. L.; Middleton, S. A.; Mulcahy, L. S.; Wrighton, N. C.; Dower, W. J.; Jolliffe, L. K.; Wilson, I. A. *Science* **1996**, *273*, 464–471.
27. Dower, W. J. *Curr. Opin. Chem. Biol.* **1998**, *2*, 328–334.
28. Cwirla, S. E.; Balasubramanian, P.; Duffin, D. J.; Wagstrom, C. R.; Gates, C. M.; Singer, S. C.; Davis, A. M.; Tansik, R. L.; Mattheakis, L. C.; Boytos, C. M. et al. *Science* **1997**, *276*, 1696–1699.
29. Cunningham, B. C.; Ultsch, M.; De Vos, A. M.; Mulkerrin, M. G.; Clauser, K. R.; Wells, J. A. *Science* **1991**, *254*, 821–825.
30. Cunningham, B. C.; Wells, J. A. *J. Mol. Biol.* **1993**, *234*, 554–563.
31. Bogan, A. A.; Thorn, K. S. *J. Mol. Biol.* **1998**, *280*, 1–9.
32. Clackson, T.; Wells, J. A. *Science* **1995**, *267*, 383–386.
33. Clackson, T.; Ultsch, M. H.; Wells, J. A.; de Vos, A. M. *J. Mol. Biol.* **1998**, *277*, 1111–1128.
34. DeLano, W. L. *Curr. Opin. Struct. Biol.* **2002**, *12*, 14–20.
35. Robertson, J. G. *Biochemistry* **2005**, *44*, 5561–5571.
36. Adams, J. *Oncologist* **2002**, *7*, 9–16.
37. Reed, J. C. *Nat. Rev. Drug Disc.* **2002**, *1*, 111–121.
38. O'Brien, T.; Lee, D. *Mini Rev. Med. Chem.* **2004**, *4*, 153–165.
39. Yoshimori, A.; Takasawa, R.; Tanuma, S. *BMC Pharmacol.* **2004**, *4*, 7.
40. Becker, J. W.; Rotonda, J.; Soisson, S. M.; Aspiotis, R.; Bayly, C.; Francoeur, S.; Gallant, M.; Garcia-Calvo, M.; Giroux, A.; Grimm, E. et al. *J. Med. Chem.* **2004**, *47*, 2466–2474.
41. Rano, T. A.; Timkey, T.; Peterson, E. P.; Rotonda, J.; Nicholson, D. W.; Becker, J. W.; Chapman, K. T.; Thornberry, N. A. *Chem. Biol.* **1997**, *4*, 149–155.
42. McIntosh, C. H.; Demuth, H. U.; Pospisilik, J. A.; Pederson, R. *Regul. Pept.* **2005**, *128*, 159–165.
43. Weber, A. E. *J. Med. Chem.* **2004**, *47*, 4135–4141.
44. Zhu, L.; Tamvakopoulos, C.; Xie, D.; Dragovic, J.; Shen, X.; Fenyk-Melody, J. E.; Schmidt, K.; Bagchi, A.; Griffin, P. R.; Thornberry, N. A.; Sinha Roy, R. *J. Biol. Chem.* **2003**, *278*, 22418–22423.
45. Wlodawer, A.; Vondrasek, J. *Annu. Rev. Biophys. Biomol. Struct.* **1998**, *27*, 249–284.
46. Ohtaka, H.; Muzammil, S.; Schon, A.; Velazquez-Campoy, A.; Vega, S.; Freire, E. *Int. J. Biochem. Cell Biol.* **2004**, *36*, 1787–1799.
47. Johansson, P. O.; Lindberg, J.; Blackman, M. J.; Kvarnstrom, I.; Vrang, L.; Hamelink, E.; Hallberg, A.; Rosenquist, A.; Samuelsson, B. *J. Med. Chem.* **2005**, *48*, 4400–4409.
48. Shie, J. J.; Fang, J. M.; Kuo, C. J.; Kuo, T. H.; Liang, P. H.; Huang, H. J.; Yang, W. B.; Lin, C. H.; Chen, J. L.; Wu, Y. T. et al. *J. Med. Chem.* **2005**, *48*, 4469–4473.
49. Findlay, D. M.; Michelangeli, V. P.; Martin, T. J.; Orlowski, R. C.; Seyler, J. K. *Endocrinology* **1985**, *117*, 801–805.
50. Weyer, C.; Maggs, D. G.; Young, A. A.; Kolterman, O. G. *Curr. Pharm. Des.* **2001**, *7*, 1353–1373.
51. Uings, I. J.; Farrow, S. N. *Mol. Pathol.* **2000**, *53*, 295–299.
52. Ji, T. H.; Grossmann, M.; Ji, I. *J. Biol. Chem.* **1998**, *273*, 17299–17302.
53. Schwyzer, R. *Proc. R. Soc. Lond. B* **1980**, *210*, 5–20.
54. Schwyzer, R. *Biochemistry* **1986**, *25*, 6335–6342.
55. Schwyzer, R. *Biopolymers* **1995**, *37*, 5–16.
56. Veber, D. F.; Saperstein, R.; Nutt, R. F.; Freidinger, R. M.; Brady, S. F.; Curley, P.; Perlow, D. S.; Paleveda, W. J.; Colton, C. D.; Zacchei, A. G. *Life Sci.* **1984**, *34*, 1371-1378.
57. Yang, L.; Berk, S. C.; Rohrer, S. P.; Mosley, R. T.; Guo, L.; Underwood, D. J.; Arison, B. H.; Birzin, E. T.; Hayes, E. C.; Mitra, S. W. et al. *Proc. Natl. Acad. Sci. USA* **1998**, *95*, 10836–10841.
58. Rivier, J.; Brown, M.; Vale, W. *Biochem. Biophys. Res. Commun.* **1975**, *65*, 746–751.
59. Cunningham, B. C.; Jhurani, P.; Ng, P.; Wells, J. A. *Science* **1989**, *243*, 1330–1336.
60. Cunningham, B. C.; Wells, J. A. *Science* **1989**, *244*, 1081–1085.
61. Raffa, R. B. *J. Theor. Biol.* **2002**, *218*, 207–214.
62. Di Cera, E. *Chem. Rev.* **1998**, *98*, 1563–1592.
63. Holst, B.; Zoffmann, S.; Elling, C. E.; Hjorth, S. A.; Schwartz, T. W. *Mol. Pharmacol.* **1998**, *53*, 166–175.
64. Monahan, M. W.; Amoss, M. S.; Anderson, H. A.; Vale, W. *Biochemistry* **1973**, *12*, 4616–4620.

65. Sagan, S.; Karoyan, P.; Lequin, O.; Chassaing, G.; Lavielle, S. *Curr. Med. Chem.* **2004**, *11*, 2799–2822.
66. Melendez, R. E.; Lubell, W. D. *J. Am. Chem. Soc.* **2004**, *126*, 6759–6764.
67. Beck-Sickinger, A. G.; Wieland, H. A.; Wittneben, H.; Willim, K. D.; Rudolf, K.; Jung, G. *Eur. J. Biochem.* **1994**, *225*, 947–958.
68. Bhargava, S.; Licha, K.; Knaute, T.; Ebert, B.; Becker, A.; Grotzinger, C.; Hessenius, C.; Wiedenmann, B.; Schneider-Mergener, J.; Volkmer-Engert, R. *J. Mol. Recognit.* **2002**, *15*, 145–153.
69. Tam, J. P.; Liu, W.; Zhang, J. W.; Galantino, M.; Bertolero, F.; Cristiani, C.; Vaghi, F.; de Castiglione, R. *Peptides* **1994**, *15*, 703–708.
70. Boyle, S.; Guard, S.; Hodgson, J.; Horwell, D. C.; Howson, W.; Hughes, J.; McKnight, A.; Martin, K.; Pritchard, M. C.; Watling, K. J.; *Bioorg. Med. Chem.* **1994**, *2*, 101–113.
71. Kramer, A.; Keitel, T.; Winkler, K.; Stocklein, W.; Hohne, W.; Schneider-Mergener, J. *Cell* **1997**, *91*, 799–809.
72. Rivier, J. E.; Jiang, G.; Koerber, S. C.; Porter, J.; Simon, L.; Craig, A. G.; Hoeger, C. A. *Proc. Natl. Acad. Sci. USA* **1996**, *93*, 2031–2036.
73. Rivier, J.; Erchegyi, J.; Hoeger, C.; Miller, C.; Low, W.; Wenger, S.; Waser, B.; Schaer, J. C.; Reubi, J. C. *J. Med. Chem.* **2003**, *46*, 5579–5586.
74. Erchegyi, J.; Penke, B.; Simon, L.; Michaelson, S.; Wenger, S.; Waser, B.; Cescato, R.; Schaer, J. C.; Reubi, J. C.; Rivier, J. *J. Med. Chem.* **2003**, *46*, 5587–5596.
75. Kapurniotu, A.; Kayed, R.; Taylor, J. W.; Voelter, W. *Eur. J. Biochem.* **1999**, *265*, 606–618.
76. Ahn, J. M.; Gitu, P. M.; Medeiros, M.; Swift, J. R.; Trivedi, D.; Hruby, V. J. *J. Med. Chem.* **2001**, *44*, 3109–3116.
77. Gulyas, J.; Rivier, C.; Perrin, M.; Koerber, S. C.; Sutton, S.; Corrigan, A.; Lahrichi, S. L.; Craig, A. G.; Vale, W.; Rivier, J. *Proc. Natl. Acad. Sci. USA* **1995**, *92*, 10575–10579.
78. Reichwein, J. F.; Wels, B.; Kruijtzer, J. A.; Versluis, C.; Liskamp, R. M. *Angew. Chem. Int. Ed. Engl.* **1999**, *38*, 3684–3687.
79. Balse-Srinivasan, P.; Grieco, P.; Cai, M.; Trivedi, D.; Hruby, V. J. *J. Med. Chem.* **2003**, *46*, 4965–4973.
80. Gilon, C.; Halle, D.; Chorev, M.; Selinger, Z.; Byk, G. *Biopolymers* **1991**, *31*, 745–750.
81. Gazal, S.; Gellerman, G.; Glukhov, E.; Gilon, C. *J. Pept. Res.* **2001**, *58*, 527–539.
82. Gazal, S.; Gellerman, G.; Gilon, C. *Peptides* **2003**, *24*, 1847–1852.
83. Houghten, R. A.; Pinilla, C.; Blondelle, S. E.; Appel, J. R.; Dooley, C. T.; Cuervo, J. H. *Nature* **1991**, *354*, 84–86.
84. Pinilla, C.; Appel, J. R.; Blanc, P.; Houghten, R. A. *Biotechniques* **1992**, *13*, 901–905.
85. Pinilla, C.; Appel, J. R.; Houghten, R. A. *Meth. Mol. Biol.* **1996**, *66*, 171–179.
86. Houghten, R. A.; Wilson, D. B.; Pinilla, C. *Drug Disc. Today* **2000**, *5*, 276–285.
87. Proudfoot, A. E. I.; Buser, R.; Borlat, F.; Alouani, S.; Soler, D.; Offord, R. E.; Schroder, J. M.; Power, C. A.; Wells, T. N. *J. Biol. Chem.* **1999**, *274*, 32478–32485.
88. Gozes, I.; Furman, S. *Curr. Pharm. Des.* **2003**, *9*, 483–494.
89. Turner, J. T.; Jones, S. B.; Bylund, D. B. *Peptides* **1986**, *7*, 849–854.
90. Gourlet, P.; Rathe, J.; De Neef, P.; Cnudde, J.; Vandermeers-Piret, M. C.; Waelbroeck, M.; Robberecht, P. *Eur. J. Pharmacol.* **1998**, *354*, 105–111.
91. Robberecht, P.; Coy, D. H.; De Neef, P.; Camus, J. C.; Cauvin, A.; Waelbroeck, M.; Christophe, J. *Eur. J. Biochem.* **1986**, *159*, 45–49.
92. Nicole, P.; Lins, L.; Rouyer-Fessard, C.; Drouot, C.; Fulcrand, P.; Thomas, A.; Couvineau, A.; Martinez, J.; Brasseur, R.; Laburthe, M. *J. Biol. Chem.* **2000**, *275*, 24003–24012.
93. Harmar, A. J.; Arimura, A.; Gozes, I.; Journot, L.; Laburthe, M.; Pisegna, J. R.; Rawlings, S. R.; Robberecht, P.; Said, S. I.; Sreedharan, S. P. et al. *Pharmacol. Rev.* **1998**, *50*, 265–270.
94. Tsutsumi, M.; Claus, T. H.; Liang, Y.; Li, Y.; Yang, L.; Zhu, J.; Dela Cruz, F.; Peng, X.; Chen, H.; Yung, S. L. et al. *Diabetes* **2002**, *51*, 1453–1460.
95. Langer, I.; Gregoire, F.; Nachtergael, I.; De Neef, P.; Vertongen, P.; Robberecht, P. *Peptides* **2004**, *25*, 275–278.
96. Robberecht, P.; Gourlet, P.; De Neef, P.; Woussen-Colle, M. C.; Vandermeers-Piret, M. C.; Vandermeers, A.; Christophe, J. *Eur. J. Biochem.* **1992**, *207*, 239–246.
97. Knudsen, L. B. *J. Med. Chem.* **2004**, *47*, 4128–4134.
98. Nielsen, L. L. *Drug Disc. Today* **2005**, *10*, 703–710.
99. Goke, R.; Fehmann, H. C.; Linn, T.; Schmidt, H.; Krause, M.; Eng, J.; Goke, B. *J. Biol. Chem.* **1993**, *268*, 19650–19655.
100. Deacon, C. F.; Knudsen, L. B.; Madsen, K.; Wiberg, F. C.; Jacobsen, O.; Holst, J. J. *Diabetologia* **1998**, *41*, 271–278.
101. Green, B. D.; Mooney, M. H.; Gault, V. A.; Irwin, N.; Bailey, C. J.; Harriott, P.; Greer, B.; O'Harte, F. P.; Flatt, P. R. *J. Endocrinol.* **2004**, *180*, 379–388.
102. Dutta, A. S.; Giles, M. B. *J. Chem. Soc. [Perkin 1]* **1976**, 244–248.
103. Zega, A. *Curr. Med. Chem.* **2005**, *12*, 589–597.
104. Spatola, A. F.; Agarwal, N. S.; Bettag, A. L.; Yankeelov, J. A., Jr.; Bowers, C. Y.; Vale, W. W. *Biochem. Biophys. Res. Commun.* **1980**, *97*, 1014–1023.
105. Venkatesan, N.; Kim, B. H. *Curr. Med. Chem.* **2002**, *9*, 2243–2270.
106. Hruby, V. J.; Li, G.; Haskell-Luevano, C.; Shenderovich, M. *Biopolymers* **1997**, *43*, 219–266.
107. Simon, R.; Kania, R. S.; Zuckermann, R. N.; Huebner, V. D.; Jewell, D. A.; Banville, S.; Ng, S.; Wang, L.; Rosenberg, S.; Marlowe, C. K.; *Proc. Natl. Acad. Sci. USA* **1992**, *89*, 9367–9371.
108. Frackenpohl, J.; Arvidsson, P. I.; Schreiber, J. V.; Seebach, D. *Chembiochemistry* **2001**, *2*, 445–455.
109. Rozan, L.; Krysan, D. J.; Rockwell, N. C.; Fuller, R. S. *J. Biol. Chem.* **2004**, *279*, 35656–35663.
110. Sargent, D. F.; Bean, J. W.; Kosterlitz, H. W.; Schwyzer, R. *Biochemistry* **1988**, *27*, 4974–4977.
111. Sargent, D. F.; Schwyzer, R. *Proc. Natl. Acad. Sci. USA* **1986**, *83*, 5774–5778.
112. Zhang, W. J.; Berglund, A.; Kao, J. L.; Couty, J. P.; Gershengorn, M. C.; Marshall, G. R. *J. Am. Chem. Soc.* **2003**, *125*, 1221–1235.
113. Dutta, A. S.; Furr, B. J.; Giles, M. B.; Valcaccia, B. *J. Med. Chem.* **1978**, *21*, 1018–1024.
114. Rajeswaran, W. G.; Hocart, S. J.; Murphy, W. A.; Taylor, J. E.; Coy, D. H. *J. Med. Chem.* **2001**, *44*, 1305–1311.
115. Reissmann, S.; Schwuchow, C.; Seyfarth, L.; Pineda De Castro, L. F.; Liebmann, C.; Paegelow, I.; Werner, H.; Stewart, J. M. *J. Med. Chem.* **1996**, *39*, 929–936.
116. Gademann, K.; Kimmerlin, T.; Hoyer, D.; Seebach, D. *J. Med. Chem.* **2001**, *44*, 2460–2468.
117. DeGrado, W. F.; Schneider, J. P.; Hamuro, Y. *J. Pept. Res.* **1999**, *54*, 206–217.
118. Goodson, B.; Ehrhardt, A.; Ng, S.; Nuss, J.; Johnson, K.; Giedlin, M.; Yamamoto, R.; Moos, W. H.; Krebber, A.; Ladner, M. et al. *Antimicrob. Agents Chemother.* **1999**, *43*, 1429–1434.
119. Caliceti, P.; Veronese, F. M. *Adv. Drug Deliv. Rev.* **2003**, *55*, 1261–1277.

120. Greenwald, R. B.; Choe, Y. H.; McGuire, J.; Conover, C. D. *Adv. Drug Deliv. Rev.* **2003**, *55*, 217–250.
121. Nestor, J. J., Jr.; Ho, T. L.; Simpson, R. A.; Horner, B. L.; Jones, G. H.; McRae, G. I.; Vickery, B. H. *J. Med. Chem.* **1982**, *25*, 795–801.
122. Nadasdi, L.; Medzihradszky, K. *Biochem. Biophys. Res. Commun.* **1981**, *99*, 451–457.
123. Ho, T. L.; Nestor, J. J., Jr.; McCrae, G. I.; Vickery, B. H. *Int. J. Pept. Prot. Res.* **1984**, *24*, 79–84.
124. Chan, R. L.; Chaplin, M. D. *Biochem. Biophys. Res. Commun.* **1985**, *127*, 673–679.
125. Vale, W. W.; Rivier, C.; Perrin, M.; Rivier, J. LRF Agonists and Fertility Regulation in the Male. In *Peptides: Structure and Biological Function*; Gross, E. A. M., Ed.; Pierce: Rockford, IL, 1979, pp 781–793.
126. Kurtzhals, P.; Havelund, S.; Jonassen, I.; Kiehr, B.; Larsen, U. D.; Ribel, U.; Markussen, J. *Biochem. J.* **1995**, *312*, 725–731.
127. Markussen, J.; Havelund, S.; Kurtzhals, P.; Andersen, A. S.; Halstrom, J.; Hasselager, E.; Larsen, U. D.; Ribel, U.; Schaffer, L.; Vad, K.; Jonassen, I. *Diabetologia* **1996**, *39*, 281–288.
128. Kurtzhals, P.; Havelund, S.; Jonassen, I.; Kiehr, B.; Ribel, U.; Markussen, J. *J. Pharm. Sci.* **1996**, *85*, 304–308.
129. Knudsen, L. B.; Nielsen, P. F.; Huusfeldt, P. O.; Johansen, N. L.; Madsen, K.; Pedersen, F. Z.; Thogersen, H.; Wilken, M.; Agerso, H. *J. Med. Chem.* **2000**, *43*, 1664–1669.
130. Neidigh, J. W.; Fesinmeyer, R. M.; Prickett, K. S.; Andersen, N. H. *Biochemistry* **2001**, *40*, 13188–13200.
131. Green, B. D.; Gault, V. A.; O'Harte, F. P.; Flatt, P. R. *Curr. Pharm. Des.* **2004**, *10*, 3651–3662.
132. Alana, I.; Hewage, C. M.; Malthouse, J. P.; Parker, J. C.; Gault, V. A.; O'Harte, F. P. *Biochem. Biophys. Res. Commun.* **2004**, *325*, 281–286.
133. Irwin, N.; Green, B. D.; Gault, V. A.; Greer, B.; Harriott, P.; Bailey, C. J.; Flatt, P. R.; O'Harte, F. P. *J. Med. Chem.* **2005**, *48*, 1244–1250.
134. Irwin, N.; Green, B. D.; Mooney, M. H.; Greer, B.; Harriott, P.; Bailey, C. J.; Gault, V. A.; O'Harte, F. P.; Flatt, P. R. *J. Pharmacol. Exp. Ther.* **2005**, *314*, 1187–1194.
135. Gershonov, E.; Goldwaser, I.; Fridkin, M.; Sechter, Y. *J. Med. Chem.* **2000**, *43*, 2530–2537.
136. Clayton, R. N. *Endocrinology* **1982**, *111*, 152–161.
137. Rivier, J.; Rivier, C.; Perrin, M.; Porter, J.; Vale, W. LHRH Analogs as Antiovulatory Agents. In *LHRH and its Analogs: Contraceptive and Therapeutic Applications*; Vickery, B. H., Nestor, J. J., Jr., Hafez, E. S. E., Eds.; MTP Press: Lancaster, PA, 1984, pp 11–22.
138. Coy, D. H.; Horvath, A.; Nekola, M. V.; Coy, E. J.; Erchegyi, J.; Schally, A. V. *Endocrinology* **1982**, *110*, 1445–1447.
139. Nestor, J. J., Jr.; Tahilramani, R.; Ho, T. L.; McRae, G. I.; Vickery, B. H. *J. Med. Chem.* **1988**, *31*, 65–72.
140. Nestor, J. J., Jr.; Tahilramani, R.; Ho, T. L.; Goodpasture, J. C.; Vickery, B. H.; Ferrandon, P. *J. Med. Chem.* **1992**, *35*, 3942–3948.
141. Prasad, K. U.; Roeske, R. W.; Weitl, F. L.; Vilchez-Martinez, J. A.; Schally, A. V. *J. Med. Chem.* **1976**, *19*, 492–495.
142. Andreyko, J. L.; Monroe, S. E.; Marshall, L. A.; Fluker, M. R.; Nerenberg, C. A.; Jaffe, R. B. *J. Clin. Endocrinol. Metab.* **1992**, *74*, 399–405.
143. Rabinovici, J.; Rothman, P.; Monroe, S. E.; Nerenberg, C.; Jaffe, R. B. *J. Clin. Endocrinol. Metab.* **1992**, *75*, 1220–1225.
144. Greenwald, R. B. *J. Control. Rel.* **2001**, *74*, 159–171.
145. Abuchowski, A.; van Es, T.; Palczuk, N. C.; Davis, F. F. *J. Biol. Chem.* **1977**, *252*, 3578–3581.
146. Abuchowski, A.; Kazo, G. M.; Verhoest, C. R., Jr.; Van Es, T.; Kafkewitz, D.; Nucci, M. L.; Viau, A. T.; Davis, F. F. *Cancer Biochem. Biophys.* **1984**, *7*, 175–186.
147. Harris, J. M.; Chess, R. B. *Nat. Rev. Drug. Disc.* **2003**, *2*, 214–221.
148. Abuchowski, A.; McCoy, J. R.; Palczuk, N. C.; van Es, T.; Davis, F. F. *J. Biol. Chem.* **1977**, *252*, 3582–3586.
149. Edwards, C. K., III; Martin, S. W.; Seely, J.; Kinstler, O.; Buckel, S.; Bendele, A. M.; Ellen Cosenza, M.; Feige, U.; Kohno, T. *Adv. Drug Deliv. Rev.* **2003**, *55*, 1315–1336.
150. Lee, K. C.; Tak, K. K.; Park, M. O.; Lee, J. T.; Woo, B. H.; Yoo, S. D.; Lee, H. S.; DeLuca, P. P. *Pharm. Dev. Technol.* **1999**, *4*, 269–275.
151. Greenwald, R. B.; Conover, C. D.; Choe, Y. H. *Crit. Rev. Ther. Drug Carrier Syst.* **2000**, *17*, 101–161.
152. Peleg-Shulman, T.; Tsubery, H.; Mironchik, M.; Fridkin, M.; Schreiber, G.; Shechter, Y. *J. Med. Chem.* **2004**, *47*, 4897–4904.
153. Seymour, L. W.; Miyamoto, Y.; Maeda, H.; Brereton, M.; Strohalm, J.; Ulbrich, K.; Duncan, R. *Eur. J. Cancer* **1995**, *31A*, 766–770.
154. Caliceti, P. *Dig. Liver Dis.* **2004**, *36*, S334–S339.
155. Krantz, A.; Ezrin, A. M.; Song, Y. Methods for Producing Novel Conjugates of Thrombin Inhibitors. U.S. Patent 5,942,620, 1999.
156. Holmes, D. L.; Thibaudeau, K.; L'Archeveque, B.; Milner, P. G.; Ezrin, A. M.; Bridon, D. P. *Bioconjug. Chem.* **2000**, *11*, 439–444.
157. Leger, R.; Thibaudeau, K.; Robitaille, M.; Quraishi, O.; van Wyk, P.; Bousquet-Gagnon, N.; Carette, J.; Castaigne, J. P.; Bridon, D. P. *Bioorg. Med. Chem. Lett.* **2004**, *14*, 4395–4398.
158. Baggio, L. L.; Huang, Q.; Brown, T. J.; Drucker, D. J. *Diabetes* **2004**, *53*, 2492–2500.
159. Jette, L.; Leger, R.; Thibaudeau, K.; Benquet, C.; Robitaille, M.; Pellerin, I.; Paradis, V.; van Wyk, P.; Pham, K.; Bridon, D. P. *Endocrinology* **2005**, *146*, 3052–3058.
160. Leger, R.; Robitaille, M.; Quraishi, O.; Denholm, E.; Benquet, C.; Carette, J.; van Wyk, P.; Pellerin, I.; Bousquet-Gagnon, N.; Castaigne, J. P. et al. *Bioorg. Med. Chem. Lett.* **2003**, *13*, 3571–3575.
161. Barden, J. A.; Kemp, B. E. *Biochemistry* **1993**, *32*, 7126–7132.
162. Jin, L.; Briggs, S. L.; Chandrasekhar, S.; Chirgadze, N. Y.; Clawson, D. K.; Schevitz, R. W.; Smiley, D. L.; Tashjian, A. H.; Zhang, F. *J. Biol. Chem.* **2000**, *275*, 27238–27244.
163. Willick, G. E.; Morley, P.; Whitfield, J. F. *Curr. Med. Chem.* **2004**, *11*, 2867–2881.
164. Reeve, J.; Meunier, P. J.; Parsons, J. A.; Bernat, M.; Bijvoet, O. L.; Courpron, P.; Edouard, C.; Klenerman, L.; Neer, R. M.; Renier, J. C. et al. *Br. Med. J.* **1980**, *280*, 1340–1344.
165. Frolik, C. A.; Black, E. C.; Cain, R. L.; Satterwhite, J. H.; Brown-Augsburger, P. L.; Sato, M.; Hock, J. M. *Bone* **2003**, *33*, 372–379.
166. Nestor, J. J., Jr.; Ho, T. L.; McRae, G.; Vickery, B., unpublished results.
167. Krstenansky, J. L.; Owen, T. J.; Hagaman, K. A.; McLean, L. R. *FEBS Lett.* **1989**, *242*, 409–413.
168. McLean, L. R.; Lewis, J. E.; Krstenansky, J. L.; Hagaman, K. A.; Cope, A. S.; Olsen, K. F.; Matthews, E. R.; Uhrhammer, D. C.; Owen, T. J.; Payne, M. H. *Am. Rev. Respir. Dis.* **1993**, *147*, 462–465.
169. Krstenansky, J. L.; Nestor, J. J., Jr.; Ho, T. L.; Vickery, B. H.; Bach, C. T. Analogs of Parathyroid Hormone and Parathyroid Hormone Related Peptide: Synthesis and Use for the Treatment of Osteoporosis. U.S. Patent 5,589,452, 1996.
170. Leaffer, D.; Sweeney, M.; Kellerman, L. A.; Avnur, Z.; Krstenansky, J. L.; Vickery, B. H.; Caulfield, J. P. *Endocrinology* **1995**, *136*, 3624–3631.
171. Pellegrini, M.; Bisello, A.; Rosenblatt, M.; Chorev, M.; Mierke, D. F. *J. Med. Chem.* **1997**, *40*, 3025–3031.
172. Frolik, C. A.; Cain, R. L.; Sato, M.; Harvey, A. K.; Chandrasekhar, S.; Black, E. C.; Tashjian, A. H., Jr.; Hock, J. M. *J. Bone Miner. Res.* **1999**, *14*, 163–172.
173. Hoare, S. R.; Gardella, T. J.; Usdin, T. B. *J. Biol. Chem.* **2001**, *276*, 7741–7753.
174. Rivier, J.; Gulyas, J.; Corrigan, A.; Martinez, V.; Craig, A. G.; Tache, Y.; Vale, W.; Rivier, C. *J. Med. Chem.* **1998**, *41*, 5012–5019.

175. Wang, L.; Schultz, P. G. *Angew. Chem. Int. Ed. Engl.* **2004**, *44*, 34–66.
176. Liu, H.; Wang, L.; Brock, A.; Wong, C. H.; Schultz, P. G. *J. Am. Chem. Soc.* **2003**, *125*, 1702–1703.
177. Alfonta, L.; Zhang, Z.; Uryu, S.; Loo, J. A.; Schultz, P. G. *J. Am. Chem. Soc.* **2003**, *125*, 14662–14663.
178. Xie, J.; Wang, L.; Wu, N.; Brock, A.; Spraggon, G.; Schultz, P. G. *Nat. Biotechnol.* **2004**, *22*, 1297–1301.
179. Deiters, A.; Cropp, T. A.; Summerer, D.; Mukherji, M.; Schultz, P. G. *Bioorg. Med. Chem. Lett.* **2004**, *14*, 5743–5745.
180. Deiters, A.; Schultz, P. G. *Bioorg. Med. Chem. Lett.* **2005**, *15*, 1521–1524.
181. Anderson, J. C.; Wu, N.; Santoro, S. W.; Lakshman, V.; King, D. S.; Schultz, P. G. *Proc. Natl. Acad. Sci. USA* **2004**, *101*, 7566–7571.
182. Amin, T.; Carter, G. *Curr. Drug Disc.* **2004**, *Nov*, 20-24.
183. Chamberlain, P.; Mire-Sluis, A. R. *Dev. Biol.* **2003**, *112*, 3–11.
184. Adson, A.; Burton, P. S.; Raub, T. J.; Barsuhn, C. L.; Audus, K. L.; Ho, N. F. *J. Pharm. Sci.* **1995**, *84*, 1197–1204.
185. Leibach, F. H.; Ganapathy, V. *Annu. Rev. Nutr.* **1996**, *16*, 99–119.
186. Baggio, L. L.; Kim, J. G.; Drucker, D. J. *Diabetes* **2004**, *53*, S205–S214.
187. Chien, Y. W.; Lambert, H. J.; Grant, D. E. *J. Pharm. Sci.* **1974**, *63*, 365–369.
188. Uhrich, K. E.; Cannizzaro, S. M.; Langer, R. S.; Shakesheff, K. M. *Chem. Rev.* **1999**, *99*, 3181–3198.
189. Schwope, A. D.; Wise, D. L.; Howes, J. F. *Life Sci.* **1975**, *17*, 1877–1885.
190. Wise, D. L.; McCormick, G. J.; Willet, G. P.; Anderson, L. C.; Howes, J. F. *J. Pharm. Pharmacol.* **1978**, *30*, 686–689.
191. Beck, L. R.; Pope, V. Z.; Flowers, C. E., Jr.; Cowsar, D. R.; Tice, T. R.; Lewis, D. H.; Dunn, R. L.; Moore, A. B.; Gilley, R. M. *Biol. Reprod.* **1983**, *28*, 186–195.
192. Pitt, C. G.; Marks, T. A.; Schindler, A. Biodegradable Drug Delivery Systems Based on Aliphatic Polyesters: Application to Contraceptives and Narcotic Antagonists. In *Controlled Release of Bioactive Materials*; Lewis, D., Ed.; Academic Press: London, 1980, pp 19–43.
193. Pitt, C. G.; Marks, T. A.; Schindler, A. *NIDA Res. Monogr.* **1981**, *28*, 232–253.
194. Sanders, L. M.; Kent, J. S.; McRae, G. I.; Vickery, B. H.; Tice, T. R.; Lewis, D. H. *J. Pharm. Sci.* **1984**, *73*, 1294–1297.
195. Anik, S.; Sanders, L. M.; Chaplin, M. D.; Kushinsky, S.; Nerenberg, C. A. Delivery Systems for LHRH and Analogs. In *LHRH and its Analogs: Contraceptive and Therapeutic Applications*; Vickery, B. H., Nestor, J. J., Jr., Hafez, E. S. E., Eds.; MTP Press: Lancaster, PA, 1984, pp 421–435.
196. Kent, J. S.; Sanders, L. M.; McRae, G. I.; Vickery, B. H.; Tice, T. R.; Lewis, D. H. *Contracept. Deliv. Syst.* **1982**, *3*, 58.
197. Redding, T. W.; Schally, A. V.; Tice, T. R.; Meyers, W. E. *Proc. Natl. Acad. Sci. USA* **1984**, *81*, 5845–5848.
198. Furr, B. J.; Hutchinson, F. G. *Prog. Clin. Biol. Res.* **1985**, *185A*, 143–153.
199. Dijkman, G. A.; del Moral, P. F.; Plasman, J. W.; Kums, J. J.; Delaere, K. P.; Debruyne, F. M.; Hutchinson, F. J.; Furr, B. J. *J. Steroid Biochem. Mol. Biol.* **1990**, *37*, 933–936.
200. Sartor, O. *Urology* **2003**, *61*, 25–31.
201. Fowler, J. E.; Flanagan, M.; Gleason, D. M.; Klimberg, I. W.; Gottesman, J. E.; Sharifi, R. *Urology* **2000**, *55*, 639–642.
202. Hora, M. S.; Rana, R. K.; Nunberg, J. H.; Tice, T. R.; Gilley, R. M.; Hudson, M. E. *Pharm. Res.* **1990**, *7*, 1190–1194.
203. Hora, M. S.; Rana, R. K.; Nunberg, J. H.; Tice, T. R.; Gilley, R. M.; Hudson, M. E. *Biotechnology* **1990**, *8*, 755–758.
204. Mossman, S. P.; Evans, L. S.; Fang, H.; Staas, J.; Tice, T.; Raychaudhuri, S.; Grabstein, K. H.; Cheever, M. A.; Johnson, M. E. *Vaccine* **2005**, *23*, 3545–3554.
205. Bennett, D. B.; Tyson, E.; Nerenberg, C. A.; Mah, S.; de Groot, J. S.; Teitelbaum, Z. *Pharm. Res.* **1994**, *11*, 1048–1055.
206. Chan, R. L.; Ho, W.; Webb, A. S.; LaFargue, J. A.; Nerenberg, C. A. *Pharm. Res.* **1988**, *5*, 335–340.
207. Chan, R. L.; Hsieh, S. C.; Haroldsen, P. E.; Ho, W.; Nestor, J. J., Jr. *Drug Metab. Dispos.* **1991**, *19*, 858–864.
208. Murphy, W. A.; Coy, D. H. *Pept. Res.* **1988**, *1*, 36–41.
209. Oberg, K.; Kvols, L.; Caplin, M.; Delle Fave, G.; de Herder, W.; Rindi, G.; Ruszniewski, P.; Woltering, E. A.; Wiedenmann, B. *Ann. Oncol.* **2004**, *15*, 966–973.
210. Valery, C.; Artzner, F.; Robert, B.; Gulick, T.; Keller, G.; Grabielle-Madelmont, C.; Torres, M. L.; Cherif-Cheikh, R.; Paternostre, M. *Biophys. J.* **2004**, *86*, 2484–2501.
211. Valery, C.; Paternostre, M.; Robert, B.; Gulik-Krzywicki, T.; Narayanan, T.; Dedieu, J. C.; Keller, G.; Torres, M. L.; Cherif-Cheikh, R.; Calvo, P. et al. *Proc. Natl. Acad. Sci. USA* **2003**, *100*, 10258–10262.
212. Schwahn, M.; Schupke, H.; Gasparic, A.; Krone, D.; Peter, G.; Hempel, R.; Kronbach, T.; Locher, M.; Jahn, W.; Engel, J. *Drug Metab. Dispos.* **2000**, *28*, 10–20.
213. Jiang, G.; Gavini, E.; Dani, B. A.; Murty, S. B.; Schrier, B.; Thanoo, B. C.; DeLuca, P. P. *Int. J. Pharm.* **2002**, *233*, 19–27.
214. Jiang, G.; Stalewski, J.; Galyean, R.; Dykert, J.; Schteingart, C.; Broqua, P.; Aebi, A.; Aubert, M. L.; Semple, G.; Robson, P. et al. *J. Med. Chem.* **2001**, *44*, 453–467.
215. Vickery, B. H. Intranasal Administration of LHRH and its Analogs. In *LHRH and Its Analogs: Contraceptive and Therapeutic Applications, Part 2*; Vickery, B. H., Nestor, J. J., Jr., Eds.; MTP Press: Lancaster, PA, 1987, pp 547–556.
216. Chan, R. L.; Henzl, M. R.; LePage, M. E.; LaFargue, J.; Nerenberg, C. A.; Anik, S.; Chaplin, M. D. *Clin. Pharmacol. Ther.* **1988**, *44*, 275–282.
217. Davis, S. S.; Illum, L. *Clin. Pharmacokinet.* **2003**, *42*, 1107–1128.
218. Johnson, P. H.; Quay, S. C. *Exp. Opin. Drug Del.* **2005**, *2*, 281–298.
219. Ahsan, F.; Arnold, J.; Meezan, E.; Pillion, D. J. *Pharm. Res.* **2001**, *18*, 1742–1746.
220. Arnold, J. J.; Ahsan, F.; Meezan, E.; Pillion, D. J. *J. Pharm. Sci.* **2004**, *93*, 2205–2213.
221. O'Callaghan, C. *Pediatr. Pulmonol. Suppl.* **2004**, *26*, 91.
222. Owens, D. R.; Zinman, B.; Bolli, G. *Diabet. Med.* **2003**, *20*, 886–898.
223. Odegard, P. S.; Capoccia, K. L. *Ann. Pharmacother.* **2005**, *39*, 843–853.
224. Hamman, J. H.; Enslin, G. M.; Kotze, A. F. *Bio Drugs* **2005**, *19*, 165–177.
225. Ripka, A. S.; Rich, D. H. *Curr. Opin. Chem. Biol.* **1998**, *2*, 441–452.
226. Bursavich, M. G.; Rich, D. H. *J. Med. Chem.* **2002**, *45*, 541–558.
227. Freidinger, R. M. *Prog. Drug Res.* **1993**, *40*, 33–98.
228. Jones, R. M.; Boatman, P. D.; Semple, G.; Shin, Y. J.; Tamura, S. Y. *Curr. Opin. Pharmacol.* **2003**, *3*, 530–543.
229. Hoesl, C. E.; Nefzi, A.; Ostresh, J. M.; Yu, Y.; Houghten, R. A. *Meth. Enzymol.* **2003**, *369*, 496–517.
230. Steer, D. L.; Lew, R. A.; Perlmutter, P.; Smith, A. I.; Aguilar, M. I. *Curr. Med. Chem.* **2002**, *9*, 811–822.
231. Hruby, V. J.; Ahn, J. M.; Liao, S. *Curr. Opin. Chem. Biol.* **1997**, *1*, 114–119.

232. Ahn, J. M.; Boyle, N. A.; MacDonald, M. T.; Janda, K. D. *Mini Rev. Med. Chem.* **2002**, *2*, 463–473.
233. Strader, C. D.; Fong, T. M.; Graziano, M. P.; Tota, M. R. *FASEB J.* **1995**, *9*, 745–754.
234. Jarnagin, K.; Bhakta, S.; Zuppan, P.; Yee, C.; Ho, T.; Phan, T.; Tahilramani, R.; Pease, J. H.; Miller, A.; Freedman, R. *J. Biol. Chem.* **1996**, *271*, 28277–28286.
235. Veber, D. F. *Merrifield Award Address*; Escom: Leiden, 1991, pp 1–14.
236. Moss, M. L.; Bartsch, J. W. *Biochemistry* **2004**, *43*, 7227–7235.
237. Fairbrother, W. J.; Christinger, H. W.; Cochran, A. G.; Fuh, G.; Keenan, C. J.; Quan, C.; Shriver, S. K.; Tom, J. Y.; Wells, J. A.; Cunningham, B. C. *Biochemistry* **1998**, *37*, 17754–17764.

Biography

John J Nestor, Jr, PhD – CSO, Founding CEO; TheraPei Pharmaceuticals, Inc., has more than 25 years of pharmaceutical industry experience in drug discovery and scientific management, much of that time at the VP level in major pharma. He is co-inventor of 10 compounds selected for clinical development, including three that are now marketed drugs (Synarel, Antagon, Valcyte).

Prior to founding TheraPei Pharmaceuticals, John was Executive Vice President, Drug Discovery at Sequenom, leading the target validation and drug discovery groups. *Prior to* joining Sequenom, he served as President and Chief Scientific Officer at Consensus Pharmaceuticals. John was Vice President, Syntex Discovery Research and Director of the Institute of Bio-Organic Chemistry. Following the acquisition of Syntex by Roche, he served as Distinguished Scientist in the Roche Bioscience organization, until co-founding a chemistry-focused, drug discovery company, Helios Pharmaceuticals.

John obtained his BS in chemistry from the Polytechnic Institute of Brooklyn and a PhD in organic chemistry from the University of Arizona. After post-doctoral research at Cornell University with the Nobelist, Prof Vincent du Vigneaud, Dr Nestor joined Syntex Research to initiate peptide drug discovery. Dr Nestor is co-inventor of more than 40 issued US patents, author of more than 60 scientific articles, co-editor of 2 books. He also is an Associate Editor of the journal *Letters in Drug Design and Discovery* and on the Editorial Advisory Boards of Current Medicinal Chemistry and Medicinal Chemistry (Bentham Publishers).

2.15 Peptidomimetic and Nonpeptide Drug Discovery: Receptor, Protease, and Signal Transduction Therapeutic Targets

T K Sawyer, ARIAD Pharmaceuticals, Cambridge, MA, USA

© 2007 Published by Elsevier Ltd.

2.15.1	**Introduction**	603
2.15.2	**Peptide Chemical Diversity and ϕ–ψ–χ Space**	604
2.15.3	**Therapeutic Targets for Peptides, Peptidomimetics, and Nonpeptides**	605
2.15.4	**Peptide Structure-Based Drug Design and Peptidomimetic Lead Compounds**	606
2.15.4.1	Peptide Structure-Based Drug Design of Peptidomimetics	606
2.15.4.2	Peptidomimetic Lead Compounds: Receptor Agonists and Antagonists	608
2.15.4.3	Peptidomimetic Lead Compounds: Protease Substrates and Inhibitors	613
2.15.4.4	Peptidomimetic Lead Compounds: Signal Transduction Modulators	615
2.15.5	**Therapeutic Target Screening, Chemical Diversity, and Nonpeptide Lead Compounds**	618
2.15.5.1	Therapeutic Target Screening and Chemical Diversity of Nonpeptides	618
2.15.5.2	Nonpeptide Lead Compounds: Three-Dimensional Pharmacophore Relationships to Peptides	619
2.15.6	**Therapeutic Target Structure-Based Drug Design: Peptidomimetic and Nonpeptide Lead Compounds**	622
2.15.6.1	G Protein-Coupled Receptors: Peptidomimetic and Nonpeptide Lead Compounds	623
2.15.6.2	Aspartyl Proteases: Peptidomimetic and Nonpeptide Lead Compounds	625
2.15.6.3	Serinyl Proteases: Peptidomimetic and Nonpeptide Lead Compounds	628
2.15.6.4	Cysteinyl Proteases: Peptidomimetic and Nonpeptide Lead Compounds	629
2.15.6.5	Metalloproteases: Peptidomimetic and Nonpeptide Lead Compounds	631
2.15.6.6	Src Homology-2 Domains: Peptidomimetic and Nonpeptide Lead Compounds	633
2.15.6.7	Tyrosine Phosphatases: Peptidomimetic and Nonpeptide Lead Compounds	638
2.15.7	**Recent Progress and Future Perspectives**	638
References		640

2.15.1 Introduction

Peptidomimetic and nonpeptide drug discovery has progressed tremendously over the last 25 years to become an extraordinary field of multidisciplinary research. It has engaged synthetic, computational, and biophysical chemists, biochemists, pharmacologists, and drug development scientists worldwide to advance novel small-molecule lead compounds, clinical candidates, and breakthrough medicines. The foundation for peptidomimetic drug discovery has evolved by way of countless academic, industrial, and government campaigns to achieve the rational transformation of first-generation peptide lead compounds to highly modified analogs possessing minimal peptide-like chemical structure.[1–18] Such work is typically illustrated by systematic backbone and/or side chain modifications, transformation to macrocyclic compounds, structure–conformation analysis (e.g., x-ray crystallography and nuclear magnetic resonance (NMR) spectroscopy), and a medley of sophisticated in silico drug design methodologies. Thus, a plethora of prototype peptidomimetic second-generation lead compounds has arisen from initial peptide lead compounds and extensive structure–conformation–activity studies.

Independently of peptide structure-based drug design strategies, noteworthy success using high-throughput screening of corporate chemical collections, natural product, or synthetic chemical libraries has dramatically impacted nonpeptide drug discovery. Furthermore, examples of nonpeptides having intriguing three-dimensional pharmacophore relationships with known peptide ligands have been proposed for a few receptor therapeutic targets, and a number of case studies providing detailed comparative molecular mapping of nonpeptide and peptide (or peptidomimetic) ligands at either protease or signal transduction therapeutic targets have been determined experimentally. Indubitably, such increased understanding of the three-dimensional structures and the identification of intermolecular interactions between peptide, peptidomimetic, or nonpeptide ligands and their therapeutic targets has greatly transformed the paradigm of such drug discovery. Overall, the emergence of three-dimensional structural information on therapeutic targets (e.g., receptors, proteases, and signal transduction proteins) has contributed dramatically to peptidomimetic and nonpeptide drug discovery strategies. Most noteworthy is the accomplishment of the de novo design of small-molecule, proof-of-concept lead compounds that possess very limited peptide-like substructure, but include structure-based functional group modifications exploiting unique intermolecular interactions with the therapeutic target (versus a first-generation peptide lead compound).

In this chapter a synopsis of peptidomimetic and nonpeptide drug discovery is given, providing a conceptual framework for drug design and describing the chemical diversity of such second- and third-generation small molecules. Particular emphasis is given to a few superclasses of therapeutic targets (e.g., receptors, proteases, and signal transduction proteins).

2.15.2 Peptide Chemical Diversity and ϕ–ψ–χ Space

Relative to all known classes of biologically active ligands, peptides exhibit extraordinary chemical diversity by virtue of their variety of primary structures (**Figure 1**). For many peptides (e.g., hormones, neurotransmitters, and secretagogs) the substructures of the amino acids that contribute to their molecular recognition (binding) and biological activity (signal transduction) at their therapeutic targets have been determined. Indeed, investigations focused on receptors have led to the generation of three-dimensional pharmacophore models of either agonist or antagonist peptide ligands and, in some cases, peptidomimetics. Yet, for most peptide growth factors, cytokines, and very large (>50 amino acids) peptide hormones, the identification of the amino acid substructure that accounts for binding and signal transduction has been difficult, and proposals for three-dimensional pharmacophore models remain challenging. In this regard the term 'three-dimensional pharmacophore' is defined as the collection of the relevant groups (substructure) of a ligand that are arranged in three-dimensional space in a manner complementary to the therapeutic target (e.g., receptor, protease, or signal transduction protein) and are responsible for the functional properties of the ligand as a result of its binding to the therapeutic target.[8]

The three-dimensional structural properties of peptides (**Figure 2**) are defined in terms of the torsion angles (ψ, ϕ, ω, χ) between the backbone amine nitrogen atom (N^α), the backbone carbonyl carbon atom (C'), the backbone methine carbon atom (C^α), and the side chain hydrocarbon functionalization (e.g., C^β, C^γ, C^δ, C^ϵ of Lys) derived from the amino acid sequence. A Ramachandran plot (ψ versus ϕ) may define the preferred combinations of torsion angles for ordered secondary structures (conformations), such as the α-helix, β-turn, γ-turn, or β-sheet. With respect to the amide-bond torsion angle (ω) the *trans* geometry is more energetically favored for most typical dipeptide substructures. However, when the carboxyl terminal (C-terminal) partner is Pro or some other N-alkylated (including cyclic) amino acid, the *cis* geometry is possible and may further stabilize the β-turn or γ-turn conformations. Molecular flexibility is directly related to covalent and/or noncovalent bonding interactions within a particular peptide. Even modest chemical modifications at N^α-methyl, C^α-methyl, or C^β-methyl (see the Phe analogs in **Figure 2**) can have significant consequences on the resulting conformation.

The peptide N^α–C^α–C' scaffold may be further transformed by introduction of alkene substitution (e.g., C^α–$C^\beta \rightarrow C=C$, or dehydroamino acid[19]) or insertion (e.g., C^α–$C' \rightarrow C^\alpha$α–$C=C$–C', or vinylogous amino acid[20]). Furthermore, the C^β atom may be substituted to create 'chimeric' amino acids. Finally, with respect to N-substituted amides it is also noteworthy to point out the intriguing approach of replacing the traditional peptide scaffold by achiral N-substituted glycine building blocks.[21] Overall, such N^α–C^α–C or C^α–C^β scaffold modifications provide significant opportunities to enhance peptide-like chemical diversity as well as to expand our knowledge of the three-dimensional structure of peptide (protein) $\phi - \psi - \chi$ space.

Peptide	Primary structure
Thyrotropin-releasing hormone	<Glu^1-His-Pro^3-NH$_2$
Enkephalin(Met)	Tyr^1-Gly-Gly-Phe-Met^5
Cholecystokinin-8	Asp^1-Tyr[SO$_3$H]-Met-Gly-Trp-Met-Asp-Phe^8-NH$_2$
Angiotensin II	Asp^1-Arg-Val-Tyr-Ile-His-Pro-Phe^8
Oxytocin	Cys^1-Tyr-Ile-Gln-Asn-Cys-Pro-Leu-Gly^9-NH$_2$ (Cys^1–Cys^6 disulfide)
Vasopressin	Cys^1-Tyr-Phe-Gln-Asn-Cys-Pro-Arg-Gly^9-NH$_2$ (Cys^1–Cys^6 disulfide)
Bradykinin	Lys^1-Arg-Pro-Gly-Phe-Ser-Pro-Phe-Arg^9
Gonadotropin-releasing hormone	<Glu^1-His-Trp-Ser-Tyr-Gly-Leu-Arg-Pro-Gly^{10}-NH$_2$
Substance P	Arg^1-Pro-Lys-Pro-Gln-Gln-Phe-Phe-Gly-Leu-Met^{11}-NH$_2$
α-Melanotropin	Ac-Ser^1-Tyr-Ser-Met-Glu-His-Phe-Arg-Trp-Gly-Lys-Pro-Val^{13}-NH$_2$
Neurotensin	<Glu^1-Leu-Tyr-Glu-Asn-Lys-Pro-Arg-Arg-Pro-Tyr-Ile-Leu^{13}
Somatostatin	Ala^1-Gly-Cys-Lys-Asn-Phe-Phe-Trp-Lys-Thr-Phe-Thr-Ser-Cys^{14} (Cys^3–Cys^{14} disulfide)
Endothelin	Cys^1-Ser-Cys-Ser-Ser-Leu-Met-Asp-Lys-Glu-Cys-Val-Tyr-Phe-Cys-His-Leu-Asp-Ile-Ile-Trp^{21} (Cys^1–Cys^{15}, Cys^3–Cys^{11} disulfides)
Vasoactive intestinal peptide	His^1-Ser-Asp-Ala-Val-Phe-Thr-Asp-Asn-Tyr-Thr-Arg-Leu-Arg-Lys-Gln-Met-Ala-Val-Lys^{21}-Tyr-Leu-Asn-Ser-Ile-Leu-Asn^{28}-NH$_2$
Glucagon	His^1-Ser-Gln-Gly-Thr-Phe-Thr-Ser-Asp-Tyr-Ser-Lys-Tyr-Leu-Asp-Ser-Arg-Arg-Ala-Gln-Asp^{21}-Phe-Val-Gln-Trp-Leu-Met-Asp-Thr^{29}
Galanin	Gly^1-Trp-Thr-Leu-Asn-Ser-Ala-Gly-Tyr-Leu-Leu-Gly-Pro-His-Ala-Ile-Asp-Asn-His-Arg-Ser^{21}-Phe-His-Asp-Lys-Tyr-Gly-Leu-Ala^{29}-NH$_2$
Corticotropin	Ser^1-Tyr-Ser-Met-Glu-His-Phe-Arg-Trp-Gly-Lys-Pro-Val-Gly-Lys-Lys-Arg-Arg-Pro-Val-Lys^{21}-Val-Tyr-Pro-Asn-Gly-Ala-Glu-Asp-Glu-Ser-Ala-Glu-Ala-Phe-Pro-Leu-Glu-Phe^{39}
Neuropeptide-Y	Tyr^1-Pro-Ser-Lys-Pro-Asp-Asn-Pro-Gly-Glu-Asp-Ala-Pro-Ala-Glu-Asp-Leu-Ala-Arg-Tyr-Tyr^{21}-Ser-Ala-Leu-Arg-His-Tyr-Ile-Asn-Leu-Met-Thr-Arg-Gln-Arg-Tyr^{36}-NH$_2$
Corticotropin-releasing factor	Ser^1-Gln-Glu-Pro-Pro-Ile-Ser-Leu-Asp-Leu-Thr-Phe-His-Leu-Leu-Arg-Glu-Val-Leu-Glu-Met^{21}-Thr-Lys-Ala-Asp-Gln-Leu-Ala-Gln-Gln-Ala-His-Ser-Asn-Arg-Lys-Leu-Leu-Asp-Ile^{40}-Ala^{41}-NH$_2$

Figure 1 Some naturally occurring peptides (including cyclic and bicyclic examples).

2.15.3 Therapeutic Targets for Peptides, Peptidomimetics, and Nonpeptides

Peptide, peptidomimetic, and nonpeptide drugs comprise a major area of pharmaceutical research and development, including work on receptor agonists and antagonists, protease inhibitors, and intracellular signal transduction protein modulators (**Table 1**). Of such therapeutic targets, major strides in drug discovery have been achieved for G protein-coupled receptors (GPCRs) with respect to agonists and antagonists (see below). Nevertheless, it is important to mention that the impact of high-throughput screening in the discovery of nonpeptide ligands (typically antagonists) at GPCRs has also been highly successful. Although screening-based nonpeptide drug discovery is not reviewed extensively in this chapter, it is possible that common pharmacophores exist, with respect to receptor binding, between peptide and nonpeptide ligands. Interestingly, however, many receptor mutagenesis studies suggest the existence of different binding pockets for peptide and nonpeptide ligands, regardless of whether they are functionally similar in terms of agonism or antagonism.[22] With respect to protease and signal transduction therapeutic targets, the emergence of three-dimensional structural information from high-resolution molecular maps of the catalytic (or noncatalytic) domains has given rise incredible opportunities for structure-based drug design to advance peptidomimetic and nonpeptide drug discovery.

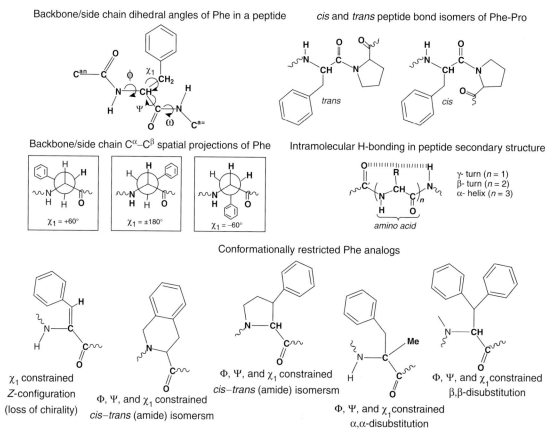

Figure 2 Three-dimensional structural properties of peptides (backbone and side chain).

2.15.4 Peptide Structure-Based Drug Design and Peptidomimetic Lead Compounds

A major focus of peptidomimetic design first evolved from receptor-focused drug discovery, and has not been directly impacted by an experimentally determined three-dimensional structure of the target protein. From such pioneering studies, a hierarchical approach of peptide → peptidomimetic drug design concepts and synthetic chemical modifications were advanced to focus on a systematic transformation of a peptide ligand using iterative analysis of the structure–activity relationships and three-dimensional structural determination (**Figure 3**). Such work has typically integrated biophysical techniques (x-ray crystallography and/or NMR spectroscopy), in silico drug design methodologies, and biological testing to advance the discovery of peptidomimetic drugs.

2.15.4.1 Peptide Structure-Based Drug Design of Peptidomimetics

Many synthetic chemistry methodologies have been utilized to advance the elucidation of novel peptidomimetics relative to peptide structure-based drug design. Such approaches include unnatural amino acids and dipeptide surrogates that incorporate backbone or side chain modifications to impart increased stability against peptidases, enhance biological potency, and afford improved bioavailability and/or other properties important to the genesis of key proof-of-concept lead compounds.[1–18] Examples of amide bond replacements (**Figure 4**) include: aminomethylene or $\Psi[CH_2NH]$ (**1**), ketomethylene or $\Psi[COCH_2]$ (**2**), ethylene or $\Psi[CH_2CH_2]$ (**3**), alkene or $\Psi[CH=CH]$ (**4**), ether or $\Psi[CH_2O]$ (**5**), thioether or $\Psi[CH_2S]$ (**6**), tetrazole or $\Psi[CN_4]$ (**7**), thiazole or $\Psi[thz]$ (**8**), retroamide or $\Psi[NHCO]$ (**9**), thioamide or $\Psi[CSNH]$ (**10**), and ester or $\Psi[CO_2]$ (**11**). Overall, such amide bond surrogates provide insight into the conformational and hydrogen-bonding properties that may be necessary for peptide molecular recognition (binding) and/or biological activity at receptor targets. Other nonhydrolyzable amide bond isosteres that have had a particular impact on the structure-based drug design of protease inhibitors include: hydroxymethylene or

Table 1 Receptor, protease, and signal transduction therapeutic target superfamilies

Receptor superfamily	Protease superfamily	Signal transduction superfamily
G protein-coupled receptor	**Aspartic protease**	**Tyrosine kinase**
AT_1 (angiotensin II)	Pepsin	Abl (SH_3–SH_2–kinase)
AT_2 (angiotensin II)	Renin	Src (SH_3–SH_2–kinase)
B_1 (bradykinin)	Cathepsin-D	Lck (SH_3–SH_2–kinase)
B_2 (bradykinin)	Cathepsin-E	Fyn (SH_3–SH_2–kinase)
CCKA (cholecystokinin)	HIV-1 protease	Syk (SH_3–SH_2–kinase)
CCKB (gastrin; cholecystokinin)	HIV-2 protease	Zap-70 (SH_3–SH_2–kinase)
ET_A (endothelin)		
ET_B (endothelin)	**Serinyl protease**	**Serine/threonine kinase**
MC1R (α-melanotropin)	Trypsin	cAMP-dependent protein kinase
MC2R (adrenocorticotropin)	Thrombin	Raf
NK1 (substance P)	Chymotrypsin-A	CDK-2
NK2 (neurokinin-A)	Kallikrein	
NK3 (neurokinin-B)	Elastase	**Dual specificity kinases**
δ-Opioid (enkephalin)	Tissue plasminogen activator	Mitogen-activated protein kinase
μ-Opioid (endorphin)	Cathepsin-A	
κ-Opioid (dynorphin)	Cathepsin-G	**Tyrosine phosphatases**
OT (oxytocin)	Cathepsin-R	PTP1BVH1
V_{1A} (vasopressin)		Syp (SH_2–SH_2–phosphatase)
V_{1B} (vasopressin)	**Cysteinyl protease**	
V_2 (vasopressin)	Papain	**Serine/threonine phosphatase**
Y_1 (neuropeptide-Y)	Cathepsin-B	PP-1
Y_2 (neuropeptide-Y)	Cathepsin-H	Calcineurin
	Cathepsin-L	
Growth factor receptor	Cathepsin-M	**Dual specificity phosphatases**
EGF (epidermal growth factor)	Cathepsin-N	VH1
FGF (fibroblast growth factor)	Cathepsin-S	
Insulin	Cathepsin-T	**Adapter proteins**
Insulin-like growth factor	Proline endopeptidase	Grb2
NGF (nerve growth factor)	Interleukin-converting enzyme	Crk
PDGF (platelet-derived growth factor)	Apopain (CPP-32)	IRS-1 (PTB)
	Picornavirus C3 protease	Shc (SH_2–PTB)
Cytokine receptor	Calpains	
IL1 (interleukin-1)		**Transferase**
IL2 (interleukin-2)	**Metalloprotease**	Farnesyl transferase
IL3 (interleukin-3)	Exopeptidase group	Geranyl–geranyl transferase
IL4 (interleukin-4)	Peptidyl dipeptidase-A (ACE)	
IL5 (interleukin-5)	Aminopeptidase-M	***Ras* exchange factors**
IL6 (interleukin-6)	Carboxypeptidase-A	GAP
IL7 (interleukin-7)	Endopeptidase group	SOS
IL8 (interleukin-8)	Thermolysin	
	Endopeptidase 24.11	**Proline *cis–trans* isomerases**
Cell adhesion integrin receptor	Endopeptidase 24.15	Cyclophilin
αVβ3 (Fibrinogen)	Stromelysin	FKBP-12
αIIbβ3 or gpIIbIIIa (fibrinogen)	Gelatinase-A	
α5β1 (fibronectin)	Gelatinase-B	**Lipase**
α4β1 (VCAM-1)	Collagenase	Phospholipase-C (SH_2–SH_2–lipase)

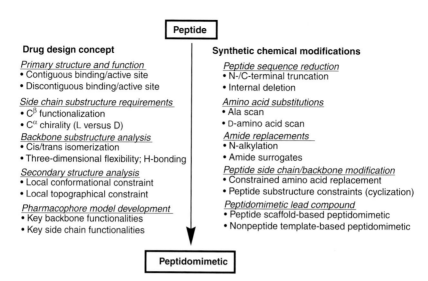

Figure 3 Hierarchical approach in peptidomimetic drug design from peptide ligands.

Ψ[CH(OH)] (**12**), hydroxyethylenes or Ψ[CH(OH)CH$_2$] (**13**) and Ψ[CH$_2$CH(OH)] (**14**), dihydroxyethylene or Ψ[CH(OH)CH(OH)] (**15**), hydroxyethylamine or Ψ[CH(OH)CH$_2$N] (**16**), C$_2$-symmetric hydroxymethylene (**17**), and dihydroxyethylene (**18**). Of particular importance to aspartyl protease inhibitor drug design (see below), these backbone modifications may be viewed as transition-state bio-isosteres of the hypothetical tetrahedral intermediate (e.g., Ψ[C(OH)$_2$NH]) for substrate peptide cleavage.

Peptide backbone and side chain modifications may also align with drug-design strategies focused on secondary structure mimetics.[23–33] More specifically, substitution of D-amino acids, N$^\alpha$-Me-amino acids, C$^\alpha$-Me-amino acids, and/or dehydroamino acids within a peptide lead compound may induce or stabilize regiospecific β-turn, γ-turn, β-sheet or α-helix conformations. Historically, a variety of secondary structure mimetics have been designed and incorporated in peptides or peptidomimetics (**Figure 5**), and the β-turn has perhaps been of the greatest interest with respect to peptidomimetic drug discovery for receptor therapeutic targets. The β-turn secondary structural motif exists within a tetrapeptide sequence, whereas the first and fourth C$^\alpha$ atoms are separated by <7 Å and there is a ten-membered intramolecular hydrogen bond between the ith and $i+4$ amino acid residues. One of the initial approaches of significance to the design of β-turn mimetics was the monocyclic dipeptide-based template (**19**),[23] which employs a side chain to backbone constraint at the $i+1$ and $i+2$ sites. Over the past decade a variety of monocyclic or bicyclic templates have been developed as β-turn mimetics (e.g., **20**,[24] **21**,[25] **22**,[26] **23**,[27] and **24**[28]). The monocyclic β-turn mimetic **25**[29] illustrates the potential opportunity to design scaffolds that may incorporate each of the side chains ($i, i+1, i+2,$ and $i+3$ positions) as well as five of the eight NH or C=O functionalities within the parent tetrapeptide sequence. Similarly, the benzodiazepine template **26**[30,31] provides a β-turn mimetic scaffold that may be multi-substituted to replicate side-chain functionalization (particularly at the i and $i+3$ positions) of the corresponding tetrapeptide sequence modeled in type I–VI β-turn conformations. The γ-turn mimetic **27**[32] illustrates an approach to incorporate a retroamide surrogate between the i and $i+1$ amino acid residues, with an ethylene bridge between the N′ (i.e., a nitrogen atom replacing the carbonyl C′ atom) and N atoms at the i and $i+2$ positions to allow for the possibility of all three side chains of the parent tripeptide sequence. Finally, the β-sheet mimetic **28**[33] provides a template that constrains the backbone of a peptide to one simulating an extended secondary structure conformation; the β-sheet mimicry is of particular interest to peptidomimetic drug discovery focused on protease therapeutic targets.

The discovery of peptidomimetic drugs follows convergent pathways (**Figure 6**), which may involve both native and foreign (of biological or chemical origin) peptides as well as nonpeptides (see below), and may provide additional insight into three-dimensional pharmacophore models and further development of novel small-molecule lead compounds.

2.15.4.2 Peptidomimetic Lead Compounds: Receptor Agonists and Antagonists

Both peptide-scaffold-based and nonpeptide-template-based drug design strategies have been successfully exploited to advance novel peptidomimetic agonist or antagonist lead compounds of receptor therapeutic targets (**Figure 7**). Such

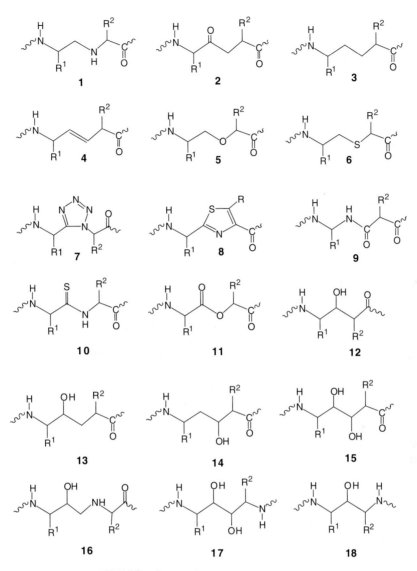

Figure 4 Peptide bond surrogates: ψ[CONH] replacements.

compounds include the μ-opioid endorphin (END) agonist **29**,[34] thyrotropin-releasing hormone (TRH) agonist **30**,[13] fibrinogen (GPIIa/IIIb) antagonists **31**[35] and **32**,[36] CCKA antagonist **33**,[37] CCKB/gastrin antagonist **34**,[38] endothelin antagonist **35**,[39] growth hormone secretagog (GHRP) **36**,[40] somatostatin agonist (partial) **37**,[41] substance-P (NK$_1$) antagonist **38**,[42] neurokinin-A (NK$_2$) antagonist **39**,[43] and neurokinin-B (NK$_3$) antagonist **40**.[44] These peptidomimetics have typically resulted from extensive structure–activity studies and focused three-dimensional structural studies using NMR on a conformationally constrained (linear or cyclic) peptide ligand. Most importantly, such structure–conformation–activity studies have inspired the development of three-dimensional pharmacophore models to guide iterative peptide structure-based drug design strategies.

Integrin receptor gpIIb/IIIa antagonists[45] are structurally derived from the tripeptide sequence Arg-Gly-Asp, a sequence that is common to several gpIIb/IIIa protein ligands (e.g., fibrinogen, vitronectin, fibronectin, von Willebrand factor, osteopontin, thrombospondin, and the collagens). Specifically, transformations of the linear peptide ligand Arg-Gly-Asp-Phe by both peptide scaffold (at the Arg-Gly backbone) modification and substitution of the Arg side chain by a benzamidine moiety provided the peptidomimetic lead compound **31** (**Figure 8**), which has been shown to be active in vivo as an antiplatelet agent.[35] In contrast, other known peptidomimetics, such as **32**, illustrate nonpeptide-template-based design strategies that have resulted from iterative transformations of a cyclic peptide Ac-Cys-N-Me-Arg-Gly-Asp-Pen-NH$_2$, in which a γ-turn about the Asp residue was conceptualized in a three-dimensional

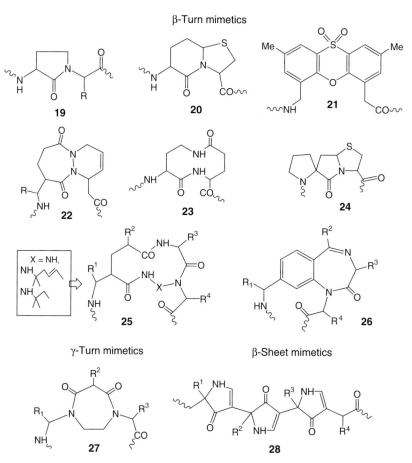

Figure 5 Peptide secondary structure modifications: β- or γ-turn and β-sheet replacements.

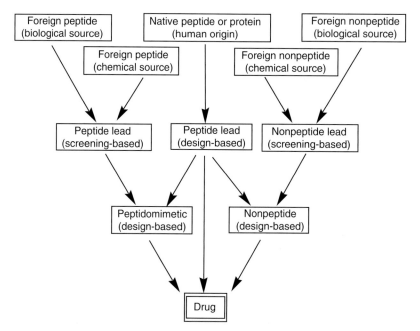

Figure 6 Convergent pathways in peptide, peptidomimetic, and nonpeptide drug discovery.

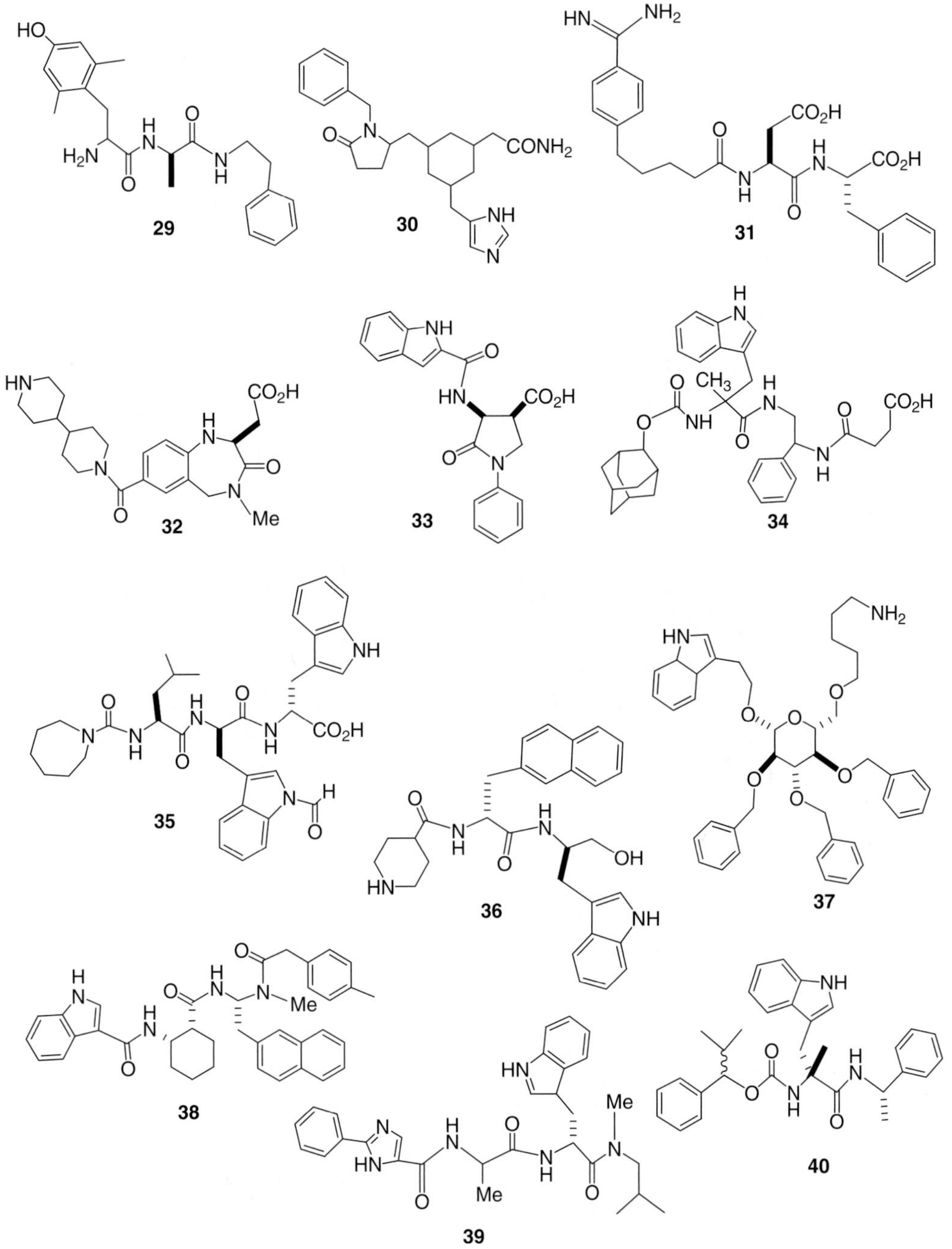

Figure 7 Some known peptidomimetics exemplifying receptor agonists or antagonists.

pharmacophore model for the bioactive conformation of the peptide ligand.[36] Specifically, the benzodiazepinone template of **32** may effectively replace this predicted γ-turn conformation about the Asp residue, and replacement of the N-Me-Arg sequence with piperidine has been shown to correlate with high-affinity receptor binding.

A glucopyranoside nonpeptide template has been shown[41,46] to yield novel peptidomimetics for the somatostatin (SRIF) and substance-P (NK$_1$) receptors. Specifically, a hexapeptide SRIF agonist provided a macrocyclic lead compound that was correlated structurally with a glucopyranoside template designed to substitute for a postulated β-turn about the Tyr-D-Trp-Lys-Thr substructure of the parent peptide ligand (**Figure 9**). The prototype small-molecule **37** was found to be a moderately potent SRIF-like agonist (partial) in cell assays.[41] Such studies extended previous work on TRH,[13] utilizing a cyclohexane ring system as a nonpeptide template to functionalize with the pGlu

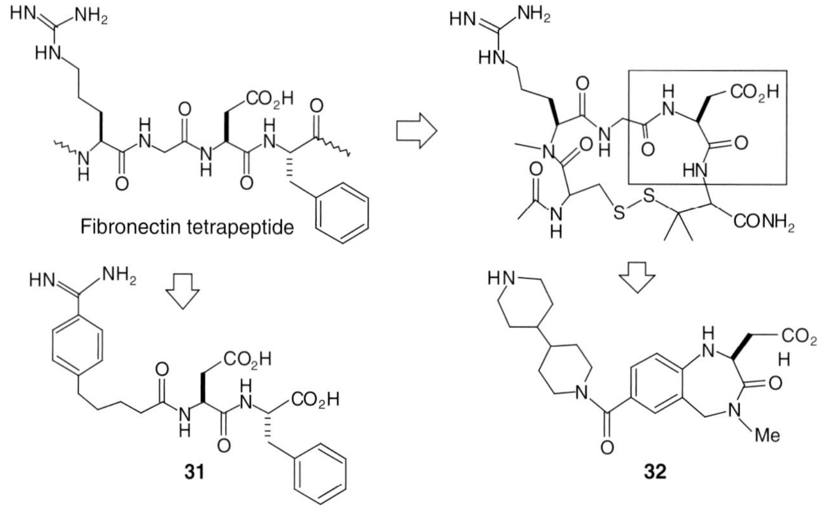

Figure 8 Integrin gpIIb/IIIa receptor: peptidomimetic and nonpeptide antagonists.

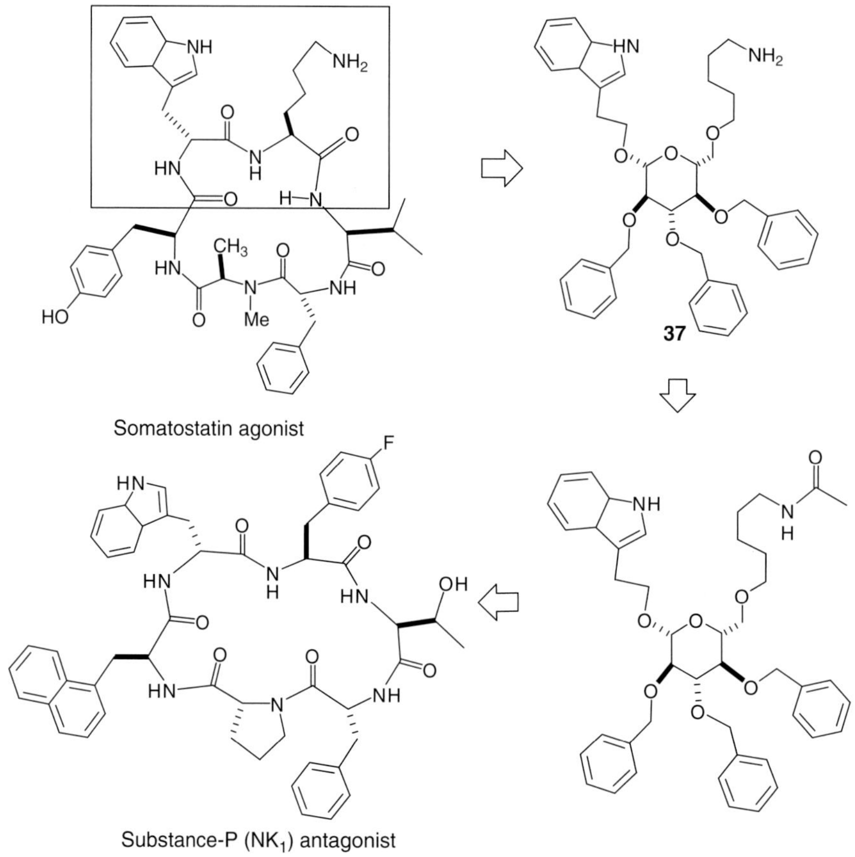

Figure 9 Somatostatin and neurokinin-1 receptors: nonpeptide agonists and antagonists.

and His side chains as well as the C-terminal carboxamide group of the parent peptide ligand **30** (**Figure 7**). A comparative analysis of analogs of the SRIF-mimetic **37** showed that N-acetylation of the alkylamine side chain moiety yields a potent NK_1 receptor (substance-P) antagonist. Interestingly, a 'reverse-design' strategy to correlate the latter glucopyranoside-based NK_1 ligand to a cyclic peptide (**Figure 9**) successfully led to the discovery of a novel cyclic peptide ligand that also exhibits potent NK_1 receptor binding and antagonism.[46]

The examples of peptide-scaffold-based or nonpeptide-template-based peptidomimetic agonists or antagonists illustrate various strategies for elucidating bioactive conformations and/or three-dimensional pharmacophore models of peptide ligands at their receptors. Such studies also include the biological screening of synthetic-chemical libraries of highly modified peptide molecules (e.g., N-substituted Gly oligomers[21]) and expanded three-dimensional pharmacophore modeling of novel small-molecule agonists or antagonists at GPCRs and other receptor types. Although the three-dimensional structures of GPCRs remain elusive, it is noteworthy that intriguing three-dimensional homology models have been constructed from experimental three-dimensional structures of bacteriorhodopsin or rhodopsin (see below). These models are providing insight into peptide, peptidomimetic, and nonpeptide structure–activity relationships and are assisting in drug discovery endeavors.

2.15.4.3 Peptidomimetic Lead Compounds: Protease Substrates and Inhibitors

Both peptide-scaffold-based and nonpeptide-template-based drug design strategies have been successfully exploited to advance novel peptidomimetic inhibitors of protease therapeutic targets (**Figure 10**), including: the renin inhibitors **41**[47] and **42**,[48] HIV protease inhibitors **43**[49] and **44**,[50] angiotensin-converting enzyme inhibitors **45**[51] and **46**,[52] collagenase inhibitor **47**,[53] gelatinase inhibitor **48**,[54] stromelysin inhibitor **49**,[55] elastase inhibitor **50**,[56] thrombin inhibitors **51**[57] and **52**,[58] and interleukin-converting enzyme inhibitor **53**.[59] For many of these peptidomimetics, drug-design strategies have focused on the natural substrate (primary structure), the mechanism of substrate cleavage, and the molecular recognition between peptide ligands, in terms of both hydrophobic and hydrogen-bonding interactions with the protease active site. Mechanistically, the traditional approach has been to incorporate nonhydrolyzable dipeptide surrogates (see above and **Figure 4**) as P_1–$P_{1'}$ modifications. Furthermore, substitution of the scissile amide (substrate) by 'transition state' bioisosteres or electrophilic ketomethylene moieties has provided tight-binding peptide or first-generation peptidomimetic inhibitor lead compounds.

Peptide-scaffold-based drug design strategies have been successfully used in the development of potent HIV protease inhibitors. For example, the natural product peptide pepstatin (**Figure 11**), an archetypical inhibitor of the aspartyl protease family of enzymes, was determined to be the first relatively potent HIV-1 protease inhibitor.[60] The central P_1–$P_{1'}$ statine (i.e., Sta or LeuΨ[CH(OH)]Gly) moiety of pepstatin was further evaluated within the context of an 'optimized' N- and C-terminal amino acid sequence using a chemical library strategy,[61] to provide a potent tetrapeptide inhibitor, Ac-Trp-Val-Sta-D-Leu-NH$_2$ (**Figure 11**). In other approaches, a renin inhibitor **55** (**Figure 11**) was determined to be a highly potent HIV protease inhibitor.[62] Further optimization studies led to the first reported[49] peptidomimetic inhibitor of HIV protease (Tba-ChaΨ[CH(OH)CH$_2$]Val-Ile-Amp; **43**). This compound exhibited cellular anti-HIV activity and provided a key proof-of-concept for HIV protease as a key therapeutic target for AIDS. Replacement of the peptide scaffold by the pyrrolidinone-type β-sheet mimetic **28** in a chemically related P_1–$P_{1'}$ PheΨ[CH(OH)CH$_2$]Phe-modified lead compound was shown[63] to yield effective peptidomimetic inhibitors of the HIV-1 protease **56** (**Figure 12**). The pyrrolidinone-type lead has been shown to have enhanced cellular permeability relative to its peptide backbone-type counterparts. In yet another approach guided by HIV substrate-based drug design, the cleavage site dipeptide Phe-Pro was substituted by 'transition state' bioisosteres to provide the highly potent and selective HIV protease inhibitor **57**, a P_1–$P_{1'}$ PheΨ[CH(OH)CH$_2$N]Pro modified heptapeptide.[64] Such work led to a series of highly potent, selective, and cellularly active HIV protease inhibitors,[50] as represented by the first US Food and Drug Administration (FDA) approved anti-HIV drug **44** (saquinavir). Finally, innovative drug-design strategies for HIV protease inhibitors incorporating a C_2-symmetric scaffold and/or cyclized P_1–$P_{1'}$ 'transition state' bioisosteres have been successfully advanced (for a review see Kempf and Sham[65]), and such work has been validated by three-dimensional structure determinations by x-ray crystallography (see below).

A second case of protease inhibitor design exemplifying a peptide-scaffold-based approach is that for peptidomimetic inhibitors of thrombin, a serine protease that cleaves a number of substrates (e.g., fibrinogen) and activates its platelet receptor (a GPCR) by proteolysis of the extracellular N-terminal domain to effect self-activation (for a review see Das and Kimball[66]). Initial thrombin inhibitor lead compounds were substrate-based, including the fibrinogen P_3–P_1 Phe-Pro-Arg sequence[67] and simple Arg derivatives such as Tos-Arg-OMe.[68] In addition, the natural product cyclothreonide-A, a macrocyclic peptide containing a Pro-Arg ketoamide sequence, was determined to be an effective lead compound.[69] Compounds **52** and **58–60** (**Figure 12**) illustrate different strategies to advancing the work on thrombin inhibitors. Particularly noteworthy within these studies is the thrombin inhibitor **52** (agatroban), a sulfonamide-modified Arg derivative,[70] which incorporates an unusual cyclic amino acid substituent C-terminal to the P_1 moiety, rather than reactive electrophilic groups (e.g., ketone, aldehyde, or boronic acid). Interestingly, replacement of the Arg side chain moiety within a structurally similar analog **60** by an amidinobenzyl group was determined[71] to be optimal when the stereochemistry at the P_1 α-carbon has a D-configuration, suggesting different binding modes for **52**

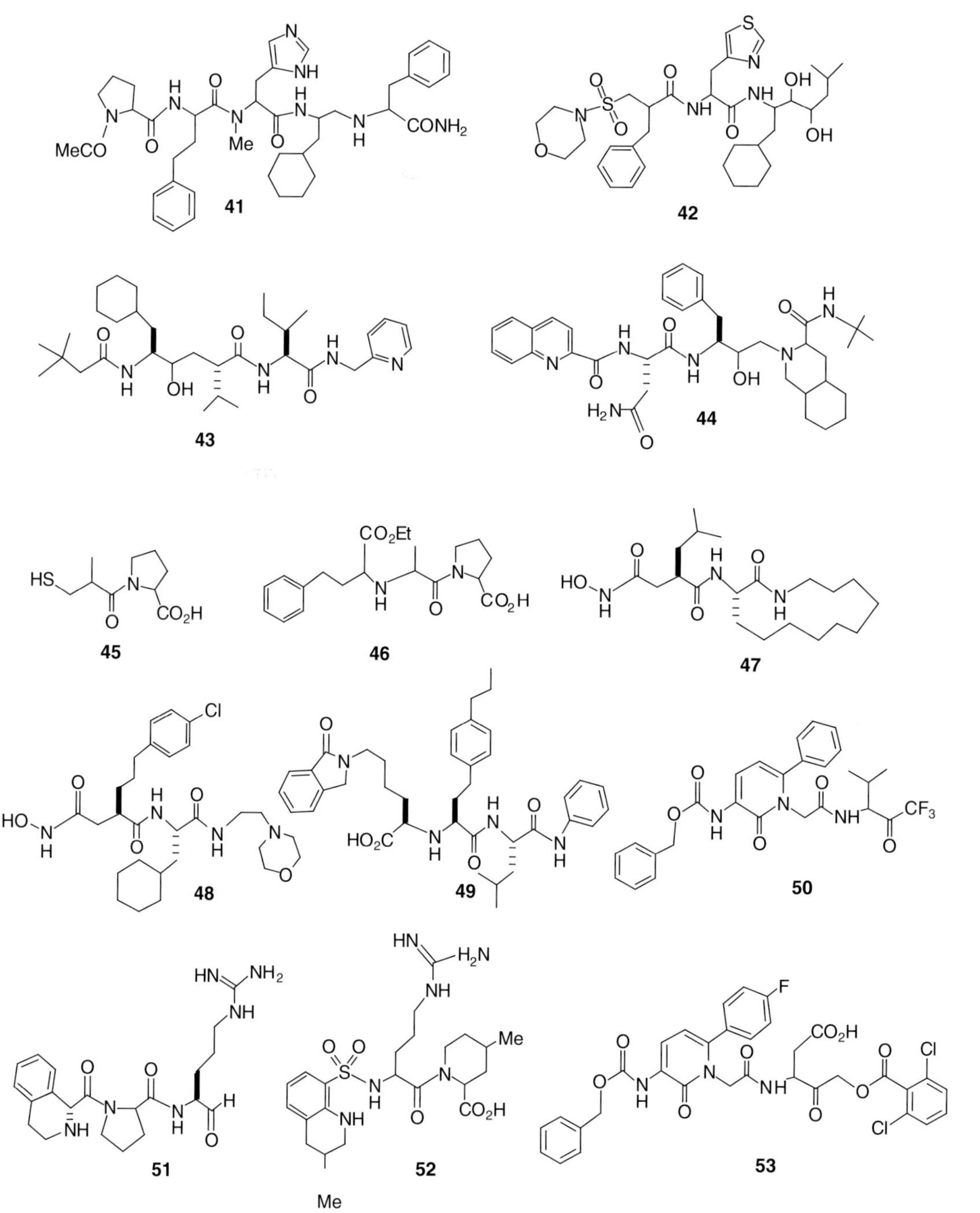

Figure 10 Some known peptidomimetics exemplifying protease inhibitors.

and **60**. In this regard, x-ray crystallographic analyses of thrombin–inhibitor complexes (see below) have provided insight into the structure–activity relationships of such peptidomimetic lead compounds.

Beyond thrombin, peptidomimetic inhibitors of the serine protease TTP-II (tripeptidyl peptidase-II) further illustrate peptide-scaffold-based design strategies.[72] Specifically, relative to a known TTP-II cleavage site on the endogenous neuropeptide CCK-8 (i.e., Asp-Tyr[SO$_3$H]-Met-Gly-Trp-Met-Asp-Phe-NH$_2$), the design of a highly potent inhibitor **61** (**Figure 12**) was successfully achieved by iterative structure-based optimization of the P$_3$–P$_1$ sequence. It is noteworthy that the TTP-II inhibitor **61** contains no functional group C-terminal to the P$_1$ α-carbon atom. This exemplifies a unique case of a substrate-based inhibitor of a protease not incorporating an electrophilic moiety, 'transition state' bioisostere, or other type of nonhydrolyzable amide replacement.

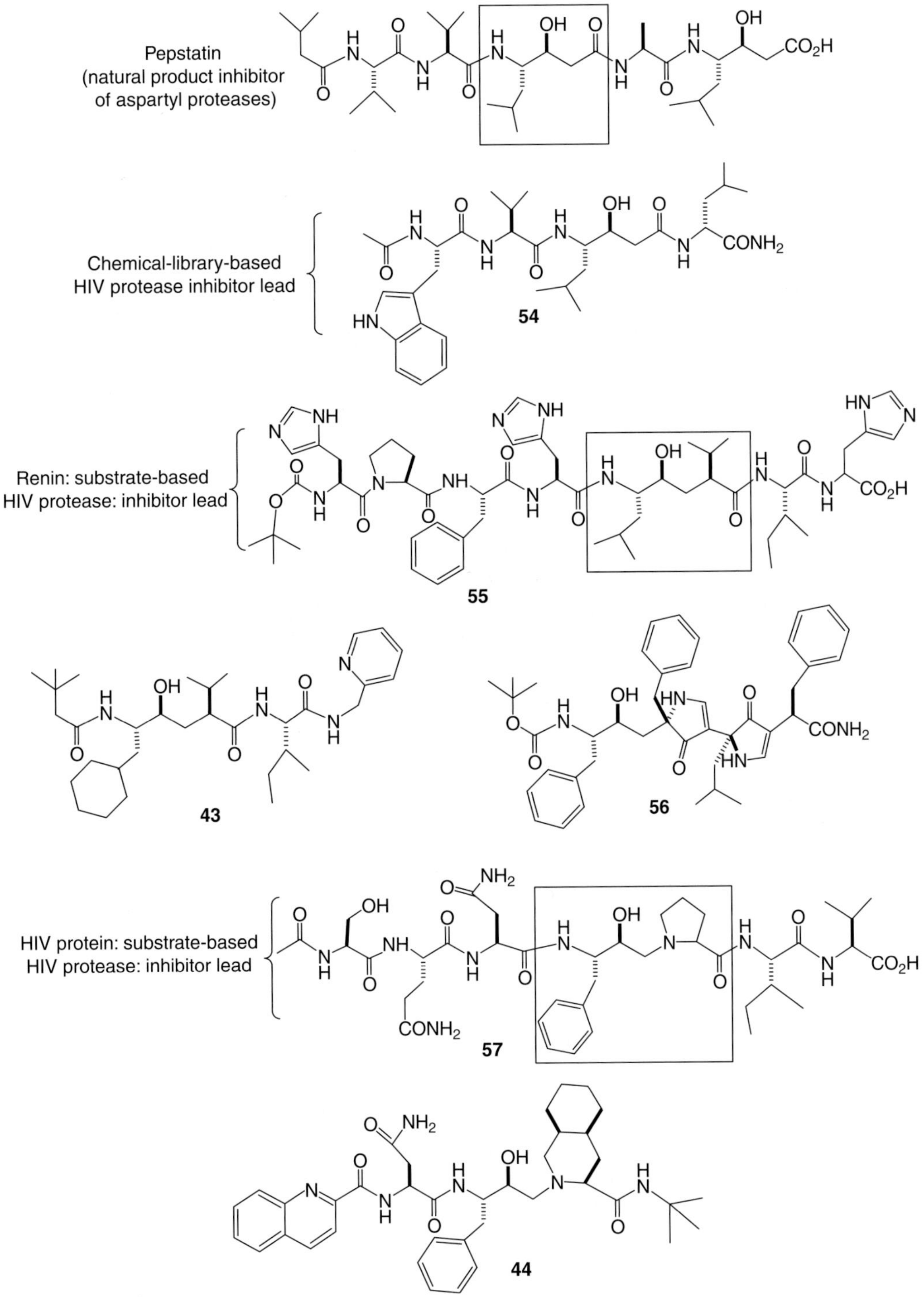

Figure 11 HIV protease: peptide and peptidomimetic inhibitors.

2.15.4.4 Peptidomimetic Lead Compounds: Signal Transduction Modulators

A multitude of catalytic and noncatalytic proteins have been identified that are critical components of intracellular signal transduction pathways. These signal transduction proteins provide the molecular basis for communication from extracellular 'effectors' (e.g., hormones, neurotransmitters, growth factors, and cytokines) to modulate cellular activity in a specific and regulated manner. Signal transduction pathways often involve protein–protein interactions, including examples of enzyme–substrate (e.g., kinases, phosphatases, transferases, and isomerases) as well as nonenzymatic

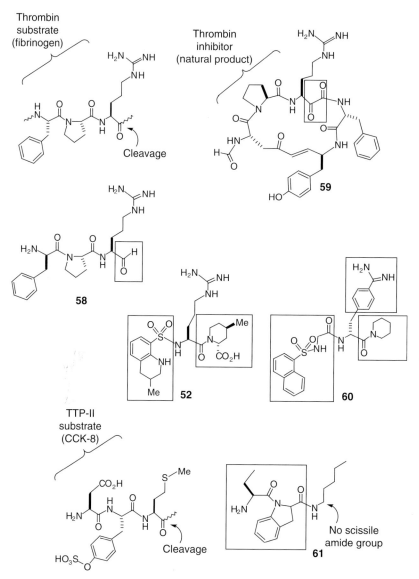

Figure 12 Thrombin: peptide and peptidomimetic inhibitors.

(e.g., 'adapter' proteins, exchange factors, and transcription factors) complex formation. Both peptide-scaffold-based and nonpeptide-template-based drug design strategies have been successfully exploited to advance novel peptidomimetic inhibitors of signal transduction therapeutic targets (**Figure 13**), including: the *Ras* farnesyl transferase inhibitors **62**,[73] **63**,[74] **64**,[75] and **65**;[76] the Src SH$_2$ domain inhibitors **66**,[77] **67**,[78] and **68**;[79] and the protein tyrosine phosphatase PTP1B inhibitor **69**.[80]

The discovery of *Ras* farnesyl transferase inhibitor drugs was originally perceived to be a promising new approach for treating *Ras*-related carcinogenesis.[81] Known substrate sequences for farnesyl transferase have the ~Cys-AA$_1$-AA$_2$-Met motif (AA refers to Val or Ile), and substrate-based inhibitors have led to potent lead compounds.[73–75,82,83] Of note, a series of non-Cys-containing peptide inhibitors have been identified.[76,84] A 'collected' substrate-based inhibitor **70** (**Figure 14**) illustrates drug-design strategies in which a farnesyl moiety is covalently linked to a peptide via a phosphinic acid moiety.[82] Relative to peptide substrate, structure-based design efforts, peptidomimetics incorporating Ψ[CH$_2$NH] substitutions (e.g., **62**[73]) or replacement of the central dipeptide moiety by a benzodiazepinone (e.g., **63**[74]) have yielded high-affinity inhibitors of Ras farnesyl protein transferase. Another series of potent peptidomimetic inhibitors have been designed in which the central dipeptide is substituted by various isomeric and/or homologous derivatives of aminobenzoic acid (e.g., **64**[75]), including a particularly effective analog biphenyl derivative **71**.[83] Interestingly, although compounds

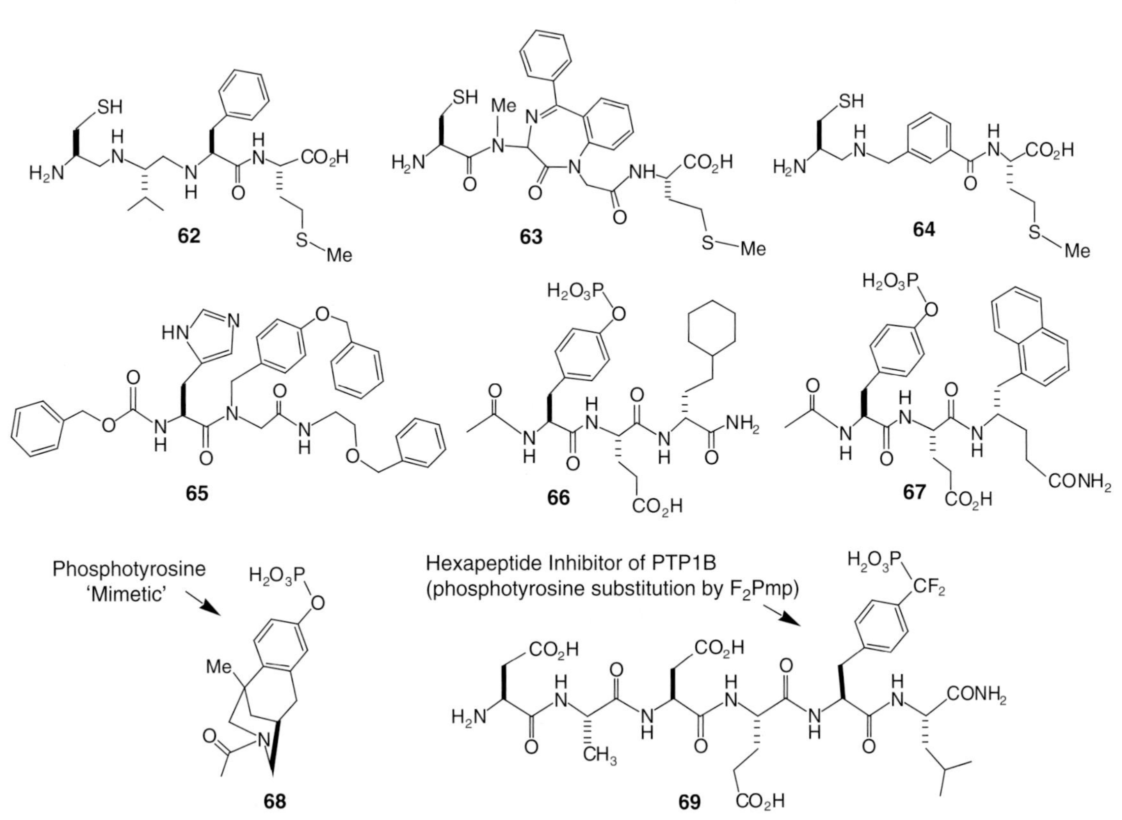

Figure 13 Some known peptidomimetics exemplifying signal transduction modulators.

such as **62–64** have 'free' sulfydryl groups (i.e., Cys), there is no evidence that these become farnesylated, and therefore the binding mode and the effect on the catalytic function of the target enzyme must be unique compared with their peptide substrate counterparts. Finally, a novel Ras farnesyl protein transferase inhibitor exemplified by Cbz-His-Tyr(O-benzyl)-Ser(O-benzyl)-Trp-D-Ala-NH$_2$ is a particularly surprising peptide lead compound.[76] These inhibitors do not contain a Cys residue and iterative drug-design efforts successfully led to a series of peptidomimetics (e.g., **65**) that have only a single chiral center. The surprising aspect of this lead compound series is the competitive inhibition of farnesyl pyrophosphate binding, rather than that of the peptide substrate. Nevertheless, peptide-substrate-based inhibitors that incorporate substitution of Cys by His (cf., **72** and **73** in **Figure 14**) have also been reported.[84] In the above cases, a divalent metal ion (e.g., Zn^{2+}) likely coordinates with the inhibitor sulfhydryl or imidazole groups in terms of their binding interactions at the farnesyl protein transferase active site.

The elucidation of Src homology-2 (SH2) domain inhibitor drugs was first conceptualized as a promising new strategy for blocking signal transduction with respect to several diseases.[85] The Src SH2 domain is a prototype of a large family of structurally homologous SH2 domains that specifically recognize cognate phosphoproteins in a sequence-dependent manner relative to a critical phosphotyrosine (pTyr) residue (i.e, ~pTyr-AA$_1$-AA$_2$-AA$_3$~). Furthermore, x-ray crystallographic studies of a several SH2 protein targets (and phosphopeptide complexes thereof) have led to opportunities for iterative structure-based drug design (see below[86]). With respect to Src SH2 domain inhibitor drug discovery, peptide library studies[87] have shown that ~pTyr-Glu-Glu-Ile~ is a preferred consensus sequence. Peptide-scaffold-based approaches to replace the internal dipeptide, Glu-Glu, by both flexible and rigid linkers were initially investigated,[88] but such studies failed to yield potent lead compounds. However, the prototype peptidomimetics **66**[77] and **67**[78] exemplified a successful drug design strategy relative to stereoinversion at the second residue (P$_{+2}$ relative to the pTyr) to the D-configuration, and side chain substitution for hydrophobic functionalities (e.g., cyclohexyl and naphthyl) conferred accessibility to the P$_{+3}$ hydrophobic binding pocket (**Figure 15**). Substitution of the pTyr residue of **66** by a difluoromethylphosphonate-modified analog (F$_2$Pmp) provided a metabolically stable derivative **74** (**Figure 15**) that could be used to advance cellularly active second-generation compounds. Finally, x-ray crystallographic structures of Src SH2 complexed with such peptidomimetic inhibitors[89] have provided insight into iterative drug-design strategies (see below).

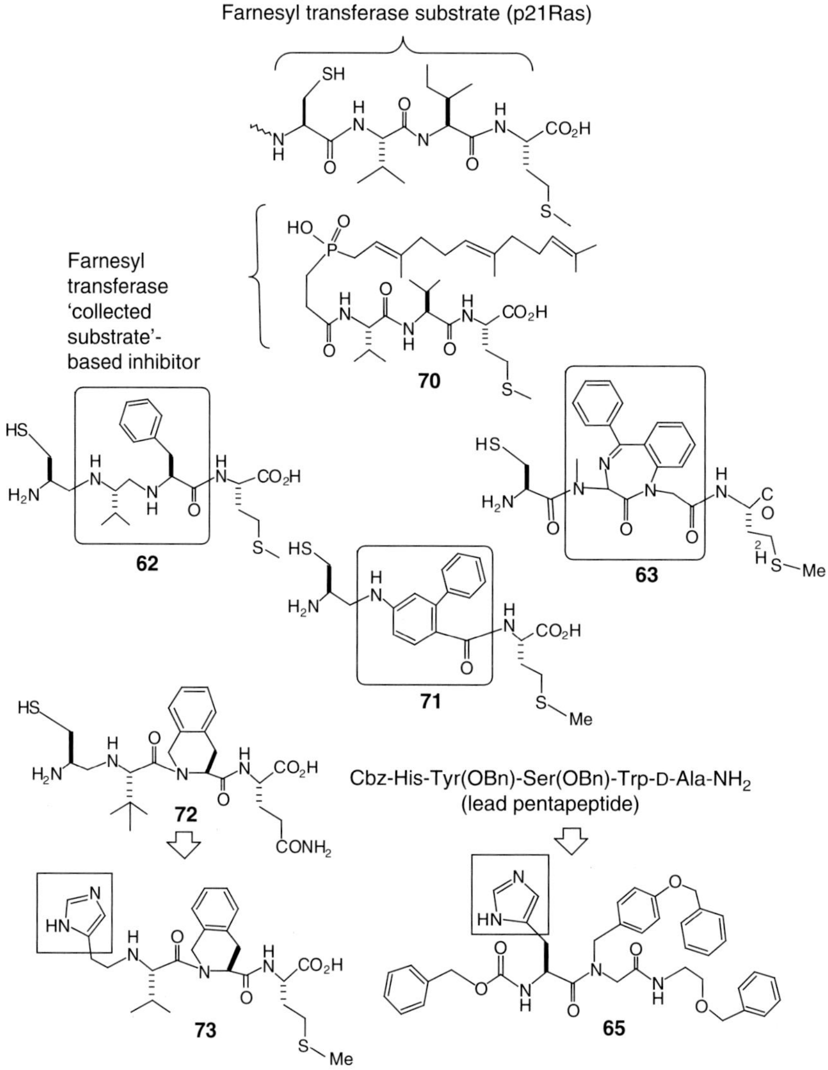

Figure 14 Ras farnesyl protein transferase: peptide and peptidomimetic inhibitors.

2.15.5 Therapeutic Target Screening, Chemical Diversity, and Nonpeptide Lead Compounds

As mentioned above, the convergent pathways in drug discovery have enabled the generation and optimization of peptides, peptidomimetics, and nonpeptides. In the case of nonpeptides, it is important to point out that both biological (e.g., natural products) and chemical (e.g., corporate collections or combinatorial libraries) compounds have provided lead compounds for drug discovery (for reviews see [12,15]). Furthermore, the three-dimensional pharmacophore relationship between peptides and peptidomimetic and nonpeptide ligands has provided some intriguing opportunities (see below).

2.15.5.1 Therapeutic Target Screening and Chemical Diversity of Nonpeptides

The initial success of nonpeptide drug discovery can be traced to the identification of morphine (**75**; **Figure 16**) as a nonpeptide natural product agonist at μ-opioid peptide receptors.[90] Since this initial discovery, the momentum of nonpeptide drug has continued. The range of drugs elucidated illustrates the significant chemical diversity and range of

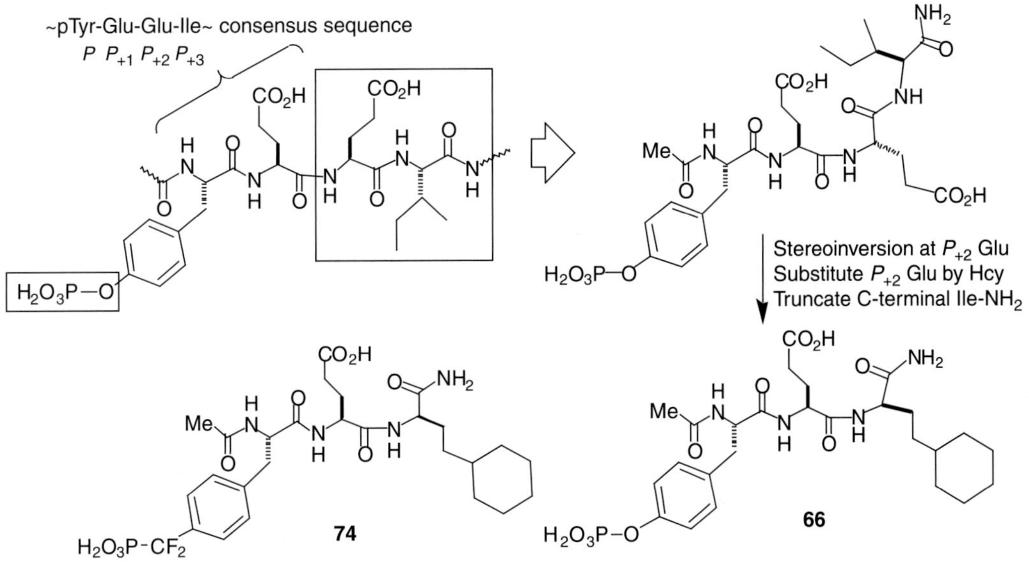

Figure 15 Src SH2 domain: peptide and peptidomimetic inhibitors.

therapeutic targets of screening-based nonpeptide lead compounds (**Figure 17**). Such compounds include the substance-P (NK$_1$) antagonist **76**,[91] angiotensin AT$_1$ antagonist **77**,[92] growth-hormone-releasing peptide (GHRP) agonist **78**,[93] cholecystokinin CCKA antagonist **79**,[94] CCKB/gastrin antagonist **80**,[95] CCKA agonist **81**,[96] endothelin antagonist **82**,[97] gonadotropin-releasing hormone (GnRH) antagonist **83**,[98] vasopressin V$_1$ antagonist **84**,[99] gastrin-releasing peptide antagonist **85**,[100] glucagon antagonist **86**,[101] neurotensin antagonist **87**,[102] angiotensin AT$_1$ agonist **88**,[103] oxytocin antagonist **89**,[104] and HIV protease inhibitor **90**.[105] Interestingly, a significant number of screening-derived nonpeptide lead compounds have been identified for GPCRs and, with few exceptions, they are competitive antagonists. The use of three-dimensional homology models and site-directed mutagenesis studies on GPCRs provides an opportunity to understand further the structure–activity relationships of nonpeptide ligands, including comparative analysis with peptide or peptidomimetic ligands (see below).

2.15.5.2 Nonpeptide Lead Compounds: Three-Dimensional Pharmacophore Relationships to Peptides

A vast number of screening-derived nonpeptide leads are multifunctionalized five- to seven-membered ring heterocycles (e.g., alkaloids and benzodiazepines) and contain conformationally rigid substructural elements (e.g., biphenyls, spirobicyclic rings, and N-substituted amide or amine linkages). This suggests that such compounds are likely to bind with highly favorable entropic driving forces than would more conformationally flexible peptide ligands. In this regard, efforts to 'rigidify' peptide-based scaffold or replace it with nonpeptide templates has been the underlying aim of many strategies of peptidomimetic drug design, and such concepts date back to the proposed use of multifunctionalized cycloaliphatic ring systems to create 'topographically designed' peptidomimetics.[106] Screening-based, nonpeptide-drug discovery has provided a treasure trove of structure–function information with which to investigate structure-based drug design and molecular recognition,[12,15,107] and in some cases there is the possibility of similar pharmacophores or substructural elements between such nonpeptides and their candidate peptide ligand counterparts (**Figure 17**).

Historically, opioid GPCR receptor (e.g., μ, δ, and κ) drug discovery has provided insight to explore the pharmacophores of both agonists and antagonists, including endogenous peptides (e.g., enkephalin, endorphin, and dynorphin) and nonpeptides (e.g., the μ-receptor selective agonist morphine and its N-allyl substituted antagonist derivative naloxone). It is likely that the N-terminal Tyr side chain and α-amino functionalities of Met-enkephalin show the same pharmacophoric features as the N-methyltyramine substructure of morphine (**Figure 17**) in terms of μ-receptor binding.[108] In the case of angiotensin II receptor antagonists, both peptide and nonpeptide (screening derived) studies have taken into consideration the likelihood of common three-dimensional pharmacophores. The

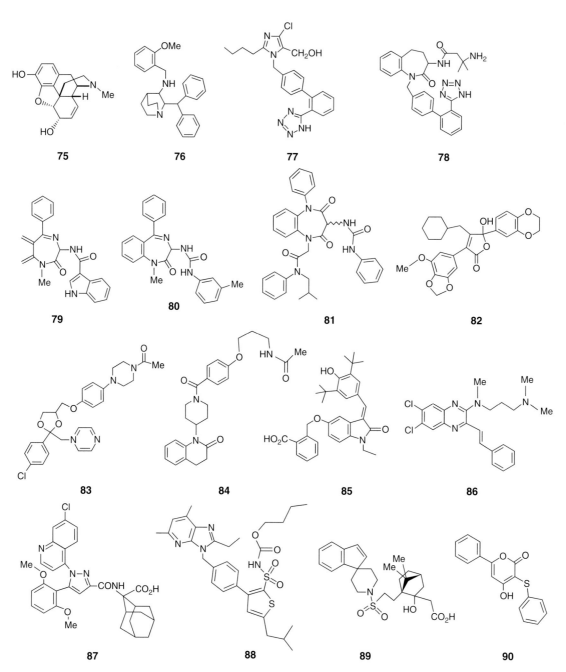

Figure 16 Some known nonpeptides exemplifying screening-based lead identification.

proposed possible substructure relationships between the C-terminal His-Pro-Phe-OH sequence of angiotensin II and the nonpeptide **91** (**Figure 17**)[109] inspired drug-design strategies that culminated in the clinical candidate **77** (lorsartan). Another case for correlating peptide and nonpeptide three-dimensional pharmacophore models is that for neuropeptide-Y (NPY) versus the benextramine-based derivative **92**[110] or the arpromidine-based derivative **93**.[111] In both cases the C-terminal Arg-Gln-Arg-Tyr-NH$_2$ sequence of NPY may be correlated with the nonpeptide ligands (**Figure 18**) such that the guanido functionalities of NPY are superimposed on the corresponding basic (i.e., guanido or imidazole) moieties of either **85** or **91**. Finally, three-dimensional pharmacophore models of a cyclic hexapeptide oxytocin antagonist and a conformationally constrained, tolylpiperazine camphorsulfonamide nonpeptide antagonist (**89**) exemplify an interesting study lead compound optimization.[112] Specifically, three-dimensional pharmacophore

Peptidomimetic and Nonpeptide Drug Discovery: Receptor, Protease, and Signal Transduction Therapeutic Targets

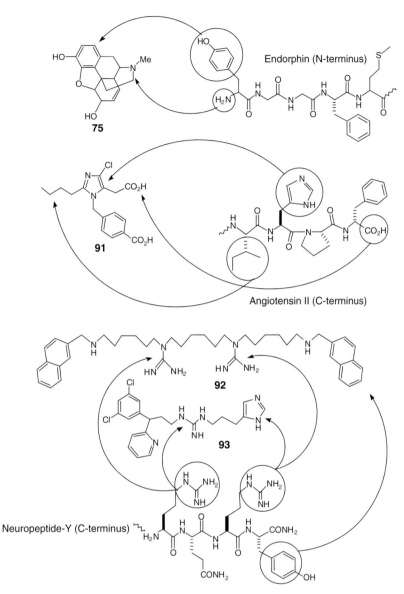

Figure 17 Comparative pharmacophores of peptide and nonpeptide ligands: μ-opioid receptor agonists, angiotensin receptor antagonists, and neuropeptide-Y receptor antagonists.

models (**Figure 18**) of the peptide and nonpeptide ligands suggest a common substructural element for molecular recognition at the oxytocin receptor, and drug-design strategies were undertaken that led to a highly potent second-generation nonpeptide lead compound **94**.

Nevertheless, such comparative three-dimensional pharmacophore 'mapping' studies are relatively limited because the three-dimensional structures of the receptor therapeutic targets (and, more importantly, the ligand complexes thereof) are not known. The creation of chemical diversity to augment nonpeptide agonist drug discovery would provide critical information to understand the molecular recognition (binding) and activation of receptors. In this regard, the nonpeptides **78** (growth-hormone-releasing peptide receptor agonist) and **81** (cholecystokinin receptor CCKA agonist) are noteworthy lead compounds in terms of exemplifying screening-based agonists for their respective GPCR therapeutic targets. Also noteworthy is the screening-based nonpeptide HIV protease inhibitor **90**,[105] which incorporates a pyrone template and functional-group elaboration that matches a known peptidomimetic inhibitor of HIV protease (**Figure 18**). Iterative structure-based drug design and lead-compound optimization has advanced this novel series of nonpeptide inhibitors of HIV protease into clinical candidates (see below).

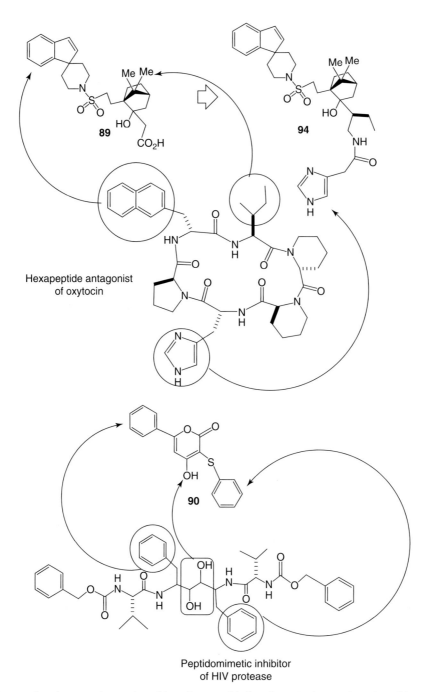

Figure 18 Comparative pharmacophores of peptide and nonpeptide ligands: oxytocin receptor antagonists and HIV protease inhibitors.

2.15.6 Therapeutic Target Structure-Based Drug Design: Peptidomimetic and Nonpeptide Lead Compounds

Peptidomimetic and nonpeptide drug discovery has evolved significantly as a result of the determination of the three-dimensional structures of an increasing number of therapeutic targets by x-ray crystallography or NMR spectroscopy[113–119]) and the exploitation of sophisticated in silico drug-design methodologies (e.g., quantitative structure–activity relationship (QSAR), three-dimensional QSAR/comparative molecular field analysis (CoMFA), ligand docking, virtual screening, two-/three-dimensional database searching, and de novo structure-based drug design).[115,120–122] Overall, the iterative cycle of structure-based drug design (**Figure 19**) has led to a successful 'engine of invention' for several examples of peptidomimetic and nonpeptide drug discovery, as exemplified by the studies described below.

2.15.6.1 G Protein-Coupled Receptors: Peptidomimetic and Nonpeptide Lead Compounds

Among the known types of cell membrane anchored receptors is the GPCR superfamily (**Table 1**), which consists of seven transmembrane-spanning (TM) helices. The structural homology between TM helices of GPCR is approximately 25–30%, and between some subtypes 50–70% homology is known.[22,123–132] The initial development of three-dimensional structural models of GPCR targets emerged from homology-building methodologies using a relatively low-resolution structure of bacteriorhodopsin.[133] A few studies of GPCR agonist and antagonist drug discovery that provide insight into three-dimensional pharmacophore modeling and structure-based drug design of peptide, peptidomimetic, and nonpeptide ligands are discussed below.

Early investigations to explore therapeutic target receptor binding interactions (e.g., site-directed mutated or chimeric receptors) and/or to probe the three-dimensional pharmacophore models of peptide, peptidomimetic, and/or nonpeptide ligands relative to the three-dimensional structural models of GPCRs include those of: angiotensin II AT_1 and AT_2 subtypes,[134] neurokinin NK_1 and NK_2 subtypes,[135] cholecystokinin/gastrin CCKA and CCKB subtypes,[136] opioid μ-, δ-, and κ-subtypes,[137] vasopressin V_{1A} subtype,[138] bradykinin B_2 subtype,[139] neurotensin,[140] and α-melanotropin MC_1 subtype.[141] Collectively, this work has implicated that different binding-site interactions exist for peptide and nonpeptide antagonists in terms of the different sensitivities of the ligands to site-directed mutants of the native GPCR. Furthermore, this work has provided a correlation between the structure–activity relationships and the experimentally determined (NMR) structures of ligands and their predicted therapeutic target molecular contacts.

In one study, the neurotensin C-terminal octapeptide was subjected to conformational searching (~Arg-Pro-Tyr~ sequences from the Brookhaven Protein Databank), manual docking to the homology-built neurotensin GPCR receptor model, and constrained molecular dynamics simulation to provide a three-dimensional structure of the ligand–receptor complex.[140] A compact structure of the peptide in its complexed conformation was consistent with a type-1 β-turn, as had been previously determined in structural and structure–activity studies. Key molecular contacts predicted from this neurotensin GPCR model include hydrophobic interactions with the C-terminal Ile and Leu side chains, π-cation

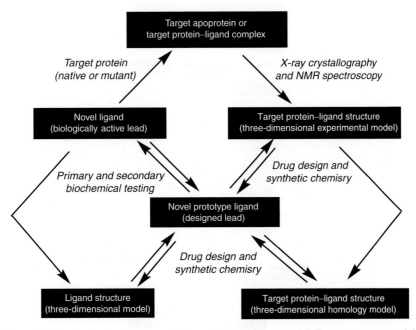

Figure 19 Iterative cycle for structure-based drug design: x-ray crystallography, NMR spectroscopy, and three-dimensional models (ligand or protein–ligand complexes).

interactions with each Arg residue side chain, and a 'cluster' of aromatic–aromatic interactions with the Tyr side chain (**Figure 20**). Finally, both mutagenesis analysis and structure–activity studies provide opportunities to further challenge such three-dimensional drug design studies.

In another study, the α-melanotropin (MC_1) GPCR model was constructed[141] by homology-building methods relative to both bacteriorhodopsin and rhodopsin fingerprint maps, and the MSH superagonist peptides [Nle^4, D-Phe^7]-MSH and Ac-cyclo[Nle^4, Asp^5, D-Phe^7, Lys^{10}]-MSH_{4-10}-NH_2 were modeled in conformations derived in previous experimental studies (i.e., a type-II′ β-turn at the common tetrapeptide sequence ∼His-D-Phe-Arg-Trp∼). Of the alternative binding modes that were described for the above the two MSH peptide ligands, one predicts the possibility of a network of aromatic–aromatic and hydrophobic interactions between the MC_1 receptor and the D-Phe and Trp side chains of the MSH ligand (**Figure 20**). In addition, this MC_1 receptor model predicts multiple electrostatic and π-cation interactions between the MC_1 receptor and the Arg side chain of the peptide ligand. The primary contact residues of this particular MC_1 receptor model were all derived from the transmembrane domain, and lie within 4–7.5 Å (centroid to centroid). Such three-dimensional models are being further challenged by mutagenesis studies and structure–activity relationships of an increasing range of agonist and antagonist ligands (e.g., peptide, peptidomimetic, and nonpeptide).

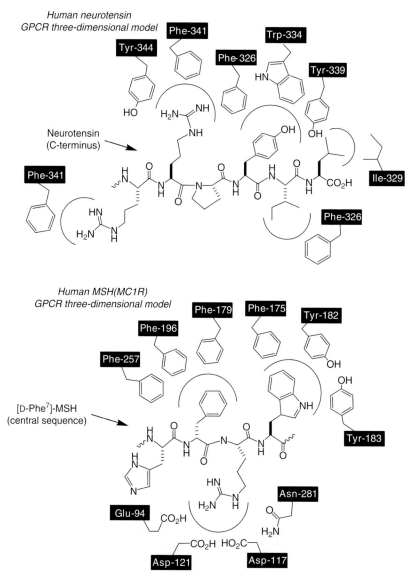

Figure 20 Three-dimensional models of neurotensin and the melanocortin receptor binding sites: peptide agonists.

2.15.6.2 Aspartyl Proteases: Peptidomimetic and Nonpeptide Lead Compounds

Both mechanistic and structure-based research into protease inhibitor drugs has been significantly successful for each of the representative compound classes (i.e., aspartyl, serinyl, metallo, and cysteinyl) (see **Table 2**; for a review see Leung et al.[142]). The pioneering advances in the design of peptidomimetic inhibitors of angiotensin-converting enzyme, including compounds **45** (captopril[51]) and **46** (enalapril[52]), provided the impetus to for protease inhibitor drug discovery. Over the past two decades there has been a pervasive effort in research in this area, integrating substrate-based inhibitor design, x-ray crystallography of protease–inhibitor complexes, high-throughput mass screening, and combinatorial chemistry technologies.

The aspartyl proteases include pepsin, renin, cathepsin-D, chymosin, and gastricsin, as well as microbial enzymes (e.g., penicillopepsin, rhizopuspepsin, and endothiapepsin) and retroviral proteases (e.g., HIV protease). The first high-resolution x-ray crystallographic structures of the aspartyl protease family were determined for penicillopepsin,[143] rhizopuspepsin,[144] endothiapepsin,[145] pepsinogen,[146] and pepsin.[147] Based on homology-building strategies, three-dimensional structural models of renin were subsequently constructed[148] to guide the initial structure-based drug design of peptidomimetic inhibitors (for a review see Abdel-Meguid[149]). Furthermore, in several cases the x-ray crystallographic structures of renin inhibitors were determined[150] as enzyme–ligand complexes with rhizopuspepsin, endothiapepsin, or pepsin. Eventually, the x-ray crystallographic structures of renin (apo/complexes) were achieved to provide high-resolution molecular maps of the target enzyme.[151] Substrate-based inhibitors, such as the highly potent peptidomimetic **95**,[151] showed well-defined hydrophobic pockets for the P_3–$P_{1'}$ side chains and hydrogen-bonding to the backbone of the inhibitor, which exists in a β-sheet type extended conformation (**Figure 21**). Although structure-based, drug-design strategies for renin inhibitors have mostly focused on systematic transformation of the substrate (angiotensinogen) relative to its tetrapeptide substructure ~ Phe^8-His-Leu-Val^{11} ~, there have been some noteworthy exceptions, such as the topographically designed $P_{1'} \rightarrow P_3$ inhibitors **96**,[152] **97**,[153] **98**,[154] and **99**.[155] The latter macrocyclic inhibitor is novel in terms of having two D-aromatic amino acids and lacking a 'transition state' bioisostere replacement at the P_1–$P_{1'}$ site.

In contrast to renin, the x-ray crystallography studies of HIV protease inhibitors show a high degree of synchronization with the iterative structure-based, drug-design efforts, and the binding mode of nonpeptide ligands has been identified by mass screening (e.g., coumarins and pyrones) or three-dimensional computational searching (e.g., haloperidol) to advance what has become a milestone achievement in structure-based drug design[113] (for a review on HIV protease see Kempf and Sham[156]). Representative of the range of the drugs discovered with respect to both peptidomimetic and nonpeptide inhibitors of HIV protease (**Figure 22**) are the C_2-symmetric inhibitors **100** (157) and **101** (158), the nonsymmetric peptidomimetcs **44**[50] and **102–106**,[159–163] and a series of nonpeptides derived originally from either three-dimensional computational searching (**107**[164]) or high-throughput screening (**108–110**[165–168]). Of these compounds, the US FDA has approved **102** (indinivar), **103** (ritonavir), and **44** (saquinavir).

Currently, there exist >100 x-ray crystallographic structures of HIV protease–inhibitor complexes, not including mutated forms of the target enzyme, which have also been identified as being important to the development of inhibitors effective against HIV resistant strains. The initial series of x-ray crystallographic structures of HIV protease included the apo[169] and enzyme–inhibitor complexes derived from substrate-based analogs being substituted at P_1–$P_{1'}$ by NleΨ[CH_2NH]Nle,[170] LeuΨ[$CH(OH)CH_2$]Val,[171] PheΨ[$CH(OH)CH_2N$]Pro,[172] and LeuΨ[$CH(OH)$]Gly or statine.[173] The first reported HIV protease–inhibitor complex[170] with the pseudopeptide **111** also provided a high-resolution map of the active site of the enzyme (**Figure 23**) as formed in a C_2-symmetric fashion by the homodimer, and the 'flaps' of each monomeric subunit (i.e., residues 35–57), which undergo intermolecular interactions with the backbone of the inhibitor by both direct hydrogen-bonding and through a structural water molecule (i.e., W301). Relative to the C_2-symmetry of HIV protease, the discovery of C_2-symmetric inhibitors was achieved by the design of PheΨ[$CH(OH)$]gPhe- and PheΨ[$CH(OH)CH(OH)$]gPhe-modified peptidomimetics (gPhe refers to gem-diamino-Phe, wherein the C^α–CO_2H moiety was replaced by C^α–NH_2), as exemplified by **101**.[157] Among the plethora of other structure-based, drug-design strategies focused on HIV protease inhibitor discovery, it is noteworthy to highlight the nonpeptide leads **100** and **108–111**, as each of these displaced a key structural water molecule (i.e., W301) and, in contrast to all previous substrate-based inhibitors, were each capable of direct hydrogen-bonding interactions with the HIV protease 'flap' regions (**Figure 23**). As a matter of precedence, previous x-ray crystallographic structural studies on a strepavidin–biotin complex[174] showed that structural water displacement (relative to the apoprotein) by the ligand (biotin) was possible. It is further noted that HIV protease is unique among the members of the aspartyl protease family in terms of the role of structural water W301 in the substrate/inhibitor binding. In contrast, the catalytic water, which is critical to the mechanism of substrate cleavage for all aspartyl proteases, has been a key feature in the design of a plethora of transition state, biostere-modified inhibitors and various tetrahedral hydroxymethyl-containing,

Table 2 Some known three-dimensional structures of proteases (Apo and/or complexes thereof with peptide, peptidomimetic, and nonpeptide inhibitors)

Proteases	Apo or inhibitor complex	Method (resolution)	Reference
Aspartic protease			
Penicillopepsin	Apo	X-ray (1.8 Å)	James et al.[143]
	Pepstatin fragment	X-ray (1.8 Å)	James et al.[287]
Endothiapepsin	Apo	X-ray (2.1 Å)	Blundell et al.[145]
	Ψ[CH(OH)CH$_2$]-modified renin inhibitor	X-ray (1.6 Å)	Veerapandian et al.[288]
	Ψ[CH$_2$NH]-modified renin inhibitor	X-ray (2.1 Å)	Cooper et al.[289]
Rhizopuspepsin	Apo	X-ray (2.5 Å)	Suguna et al.[144]
	Ψ[CH$_2$NH]-modified renin inhibitor	X-ray (1.8 Å)	Suguna et al.[290]
Pepsin	Apo	X-ray (2.0 Å)	Abad-Zapatero et al.[291]
	Ψ[CH(OH)CH(OH)]-modified renin inhibitors	X-ray (1.8 Å)	Abad-Zapatero et al.[292]
Renin	Apo	X-ray (2.5 Å)	Sielecki et al.[151]
	Ψ[CH(OH)CH$_2$]-modified substrate analog	X-ray (2.4 Å)	Rahuel et al.[293]
HIV protease	Apo	X-ray (2.8 Å)	Wlodawer et al.[169]
	Ψ[CH$_2$NH]-modified substrate analog	X-ray (2.3 Å)	Miller et al.[170]
	Ψ[CH(OH)CH$_2$]-modified substrate analog	X-ray (2.5 Å)	Jaskolski et al.[171]
	Ψ[CH(OH)CH$_2$N]-modified substrate analog	X-ray (2.4 Å)	Swain et al.[172]
	Acetyl-pepstatin	X-ray (2.0 Å)	Fitgerald et al.[173]
	C$_2$-symmetric Ψ[CH(OH)]-modified inhibitors	X-ray (2.8 Å)	Erickson et al.[157]
	C$_2$-symmetric Ψ[CH(OH)CH(OH)]-modified inhibitor	X-ray (2.8 Å)	Kempf et al.[294]
	C$_2$-symmetric, cyclic Ψ[CH(OH)CH(OH)]-modified inhibitor	X-ray (1.8 Å)	Lam et al.[158]
	Haloperidol	X-ray (NR)	Rutenbar et al.[164]
	Pyrone-based inhibitor	X-ray (NR)	Vara Prasad et al.[165]
Serinyl protease			
Thrombin	Apo	X-ray (NR)	Rydel et al.[295]
	d-Phe-Pro-Arg-chloromethylketone inhibitor	X-ray (1.9 Å)	Bode et al.[296]
	Ac-d-Phe-Pro-boroArg-OH inhibitor	X-ray (2.0 Å)	Weber et al.[176]
	d-Phe-Pro-agmatine inhibitor	X-ray (2.0 Å)	Wiley et al.[180]
	Arg-based inhibitor (agatroban)	X-ray (NR)	Banner et al.[297]
	Amidinophenylalanine-based inhibitor (NAPAP)	X-ray (NR)	Banner et al.[297]
	Amidinopiperadine-based inhibitors	X-ray (2.6 Å)	Hilpert et al.[182]
	Cyclothreonamide-A inhibitor	X-ray (2.3 Å)	Maryanoff et al.[179]
	Hirudin antagonist	X-ray (NR)	Rydel et al.[295]
Factor Xa	Apo	X-ray (2.2 Å)	Padmanabhan et al.[183]
Trypsin	Apo	X-ray (NR)	Walter et al.[298]
	Amidinophenylalanine-based inhibitor	X-ray (NR)	Bode et al.[299]
	bis-Amidinoaryl type factor Xa inhibitor	X-ray (1.9 Å)	Stubbs et al.[184]

Table 2 Continued

Proteases	Apo or inhibitor complex	Method (resolution)	Reference
Kallikrein-A	Trypsin inhibitor protein	X-ray (2.5 Å)	Chen et al.[185]
Elastase	Apo	X-ray (NR)	Meyer et al.[300]
	Peptide–benzoxazole inhibitor	X-ray (NR)	Edwards et al.[188]
	Peptide–trifluoromethylketone inhibitor	X-ray (NR)	Navia et al.[189]
	Peptide–α,α-difluoro-β-ketoamide inhibitor	X-ray (1.8 Å)	Takahashi et al.[190]
	Pyridone–trifluoromethylketone inhibitor	X-ray (NR)	Edwards et al.[301]
	N-Trifluoromethylacetyl–peptidomimetic inhibitor	X-ray (NR)	Peisach et al.[191]
Cysteinyl protease			
Papain	Apo	X-ray (2.8 Å)	Drenth et al.[193]
	Peptide–chloromethylketone inhibitor	X-ray (2.8 Å)	Drenth et al.[194]
	Leupeptin (peptide-aldehyde) inhibitor	X-ray (2.1 Å)	Schroeder et al.[195]
Cathepsin-B	Apo	X-ray (2.2 Å)	Musil et al.[197]
Picornaviral 3C protease	Apo	X-ray (NR)	Allair et al.[198]
Interleukin-converting enyzme (ICE)	Peptide–aldehyde inhibitor	X-ray (NR)	Wilson et al.[200]
	Peptide–alkyl-SH inhibitor	X-ray (2.5 Å)	Walker et al.[201]
CPP32 (an ICE homolog)	Peptide–aldehyde inhibitor	X-ray (2.5 Å)	Rotonda et al.[205]
Metalloprotease			
Thermolysin	Apo	X-ray (1.6 Å)	Holmes et al.[207]
	Phosphoramidon (hydroxyphosphinyl–peptide) inhibitor	X-ray (2.3 Å)	Weaver et al.[208]
	Ψ[P=O(OH)NH]-modified peptide inhibitor	X-ray (1.7 Å)	Holden et al.[209a]
	N-Carboxyalkyl-modified peptide inhibitor	X-ray (1.9 Å)	Monzingo et al.[210]
	N-Hydroxamate-modified peptide inhibitor	X-ray (2.3 Å)	Holmes et al.[211]
	N-Mercaptoalkyl-modified peptide inhibitor	X-ray (1.9 Å)	Monzingo et al.[212]
Carboxypeptidase-A	Apo	X-ray (NR)	Kester et al.[302]
	Ψ[P=O(OH)NH]-modified peptide inhibitor	X-ray (NR)	Christianson et al.[213]
Collagenase	Apo	X-ray (1.6 Å)	Spurlino et al.[303]
	N-Hydroxamate-modified peptide inhibitor	X-ray (2.1 Å)	Stams et al.[218]
	N-Hydroxamate-modified peptide inhibitor	X-ray (2.2 Å)	Borkakoti et al.[219]
	N-Carboxyalkyl-modified peptide inhibitor	X-ray (2.4 Å)	Lovejoy et al.[220]
Stromelysin	Apo	NMR (NR)	Van Doren et al.[221]
	N-Carboxyalkyl-modified peptide inhibitor	NMR (NR)	Gooley et al.[222]
	N-Carboxyalkyl-modified peptide inhibitor	X-ray (1.9 Å)	Becker et al.[223]
	N-Hydroxamate-modified peptide inhibitor	X-ray (2.0 Å)	Becker et al.[223]
Astacin	Apo	X-ray (1.8 Å)	Gromis-Ruth et al.[225]
	Ψ[P=O(OH)CH$_2$]-modified peptide inhibitor	X-ray (2.1 Å)	Grams et al.[226]

NR, not reported.

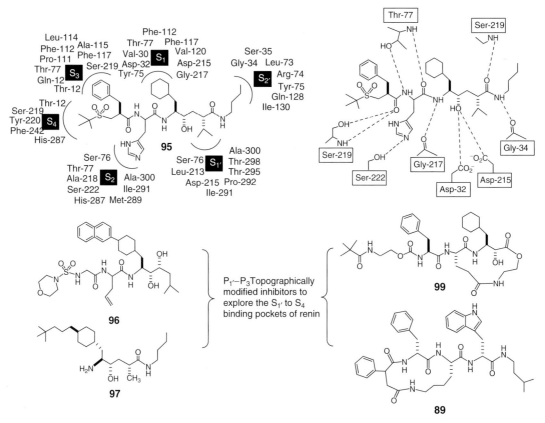

Figure 21 Models of the active site of renin: peptidomimetic inhibitors.

nonhydrolyzable surrogates of the scissile amide bond (e.g., Ψ[CH(OH)], Ψ[CH(OH)CH$_2$], and Ψ[CH(OH)CH$_2$N]; see above) have been created. In addition, there exist examples of potent inhibitors of renin (e.g., the Ψ[CH$_2$NH]-modified **41**[47] and the macrocyclic peptide **99**[155]) and HIV protease (e.g., the pyrone-based series **108–111**[165–168]) that do not possess a tetrahedral CH(OH) moiety per se.

2.15.6.3 Serinyl Proteases: Peptidomimetic and Nonpeptide Lead Compounds

The serinyl proteases include trypsin, chymotrypsin-A, elastase, thrombin, kallikrein, cathepsins-A, G, and R, factor VII, factors IXa–XIIa; and tissue plasminogen factor. High-resolution x-ray crystallographic structures of this protease family have been determined, including those of thrombin (for a review see Stubbs and Bode[175]; also see **Table 2**[176–182]), factor Xa,[183] trypsin,[184] kallikrein-A,[185] and elastase.[186–190] For example, a substrate-based inhibitor of thrombin having a boronic acid, B(OH)$_2$ (**Figure 25**), substitution for the scissile amide was determined by x-ray crystallography to form a covalent bond with the active site Ser-195 residue.[176] The N-terminal Ac-D-Phe-Pro moiety of this inhibitor binds in a β-sheet-type extended conformation that involves hydrogen-bonds with the enzyme and well-defined hydrophobic and aromatic–aromatic (edge-to-face) stacking interactions. The inhibitor Arg side chain binds in an extended conformation, and the guanidino moiety forms bidentate hydrogen-bonding interactions with an Asp-189 residue at the base of the S$_1$ 'specificity' pocket and additional hydrogen-bonds to the enzyme, one of which is mediated through a structural water. Related to the substrate-based peptidomimetic inhibitors of thrombin having C-terminal electrophilic groups (e.g., aldehyde, ketone, and boronic acid), a series of nonpeptide inhibitors devoid of P$_1$ electrophilic functionalization have been successfully advanced, as represented by **52**,[70] **60**,[71] and **113**.[182] The design of a highly potent and selective amidinopiperidine-based thrombin inhibitor **113** (**Figure 24**) was derived from analysis of the x-ray crystallographic structures of thrombin complexed with inhibitors **52** and **60**. The latter two compounds showed different trajectories of their P$_1$ side chains (i.e., guanidinoalkyl and amidinophenyl moieties, respectively) into the S$_1$ pocket, which accounts for the observed opposing chiral preferences at the C$^\alpha$-position of the P$_1$ amino acid residues. Furthermore, the C-terminal cycloalkyl moieties of both **52** and **60** were observed to bind to the so-called inhibitor 'P-pocket' (i.e., the P$_2$ substrate pocket), and these compounds therefore provide a unique model of binding versus the substrate-like

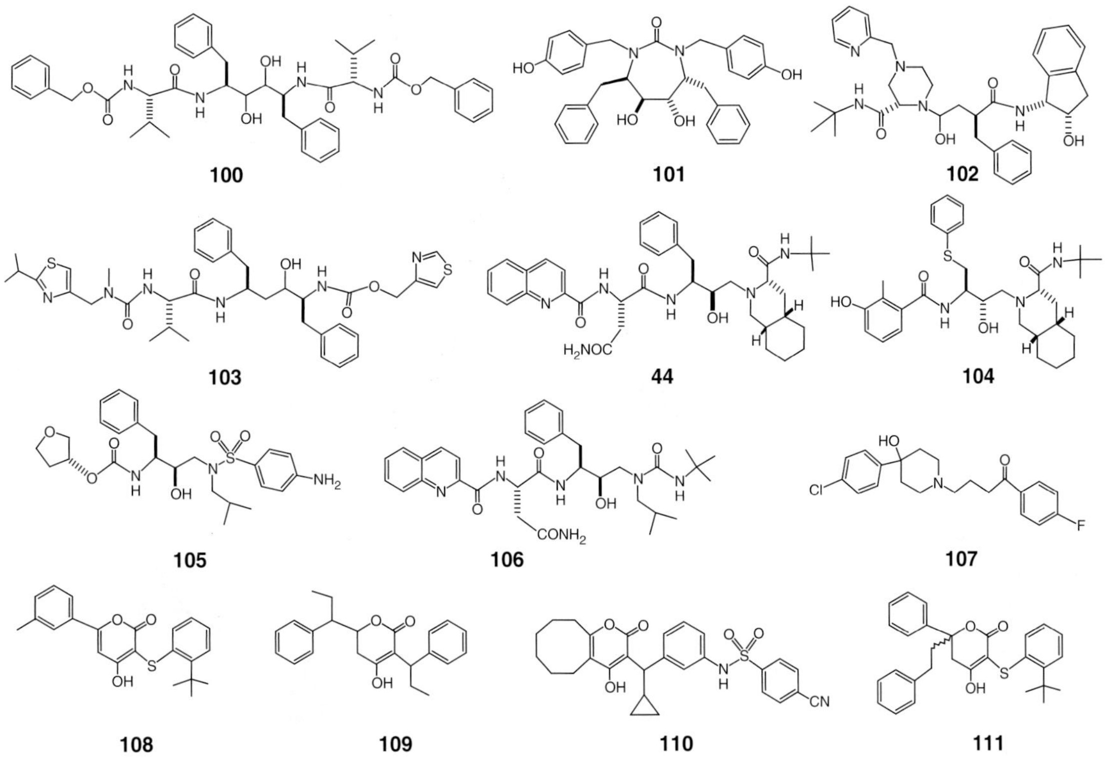

Figure 22 Models of the active site of HIV protease: peptidomimetic and nonpeptide inhibitors.

conformation adopted by the peptidomimetic inhibitors Ac-D-Phe-Pro-boroArg-OH described above. The design of the novel amidinopiperidine-based inhibitor **113** illustrates a 'transposition' of the P-pocket binding group to an N-substituted Gly-β-Asp scaffold.

Other serinyl proteases, such as factor Xa, kallikrein, and elastase, exemplify the availability of x-ray crystallographic structures (apo/complexes) for structure-based drug design (see **Table 2**). In particular, research into elastase inhibitor drugs is highlighted in terms of substrate-based peptidomimetic inhibitor drug-design strategies, which have focused on key P_2–P_3 side chain and backbone hydrogen-bonding interactions with the enzyme (for a review see Peisach et al.[191]). Several x-ray crystallographic structures have been determined for elastase[186–190], including complexes with peptide-substrate-based inhibitors having P_1 electrophilic functionalities such as benzoxazole,[188] trifluoromethyl ketone,[56,187,189] and α,α-difluoro-β-ketoamide.[190] Peptidomimetic inhibitors of elastase incorporating nonpeptidyl P_2–P_3 replacements led to highly potent compounds.[56,187] A lead compound series of highly potent trifluoromethylketone-based inhibitors of human leukocyte elastase incorporating a N-carboxymethyl-3-amino-6-arylpyridone template (**50** and **116**; **Figure 25**) was shown by x-ray crystallography to exhibit backbone hydrogen-bonding and a novel trajectory of a P_2 group from the pyridone ring to the enzyme. Further modification of the pyridone ring to give the bicyclic pyridopyrimidine derivative **117** was predicted from molecular modeling studies to provide additional hydrogen-bonding to the enzyme, as well as another to site on the bicyclic heteroaromatic ring system, for tethering various hydrophobic or hydrophilic groups. Finally, the design of a series of novel dipeptide-based inhibitors (e.g., trifluoromethyacetyl-Leu-Phe-isopropylanilide and a peptidomimetic derivative) is particularly intriguing because analysis of the x-ray crystallography structures of their complexes with pancreatic elastase were determined to bind 'backwards'.[191,192]

2.15.6.4 Cysteinyl Proteases: Peptidomimetic and Nonpeptide Lead Compounds

The cysteinyl proteases include papain, calpains I and II, cathepsins B, H, and L, proline endopeptidase, and interleukin-converting enzyme (ICE), and homologs thereof. The first highly investigated cysteinyl protease was papain, and the first x-ray crystallographic structures of papain, including apo[193] and a peptide chloromethylketone inhibitor complex,[194] provided early high-resolution molecular maps of its active site. Pioneering studies in the discovery of papain substrate peptide-based inhibitors having P_1 electrophilic moieties, such as aldehydes,[195] ketones

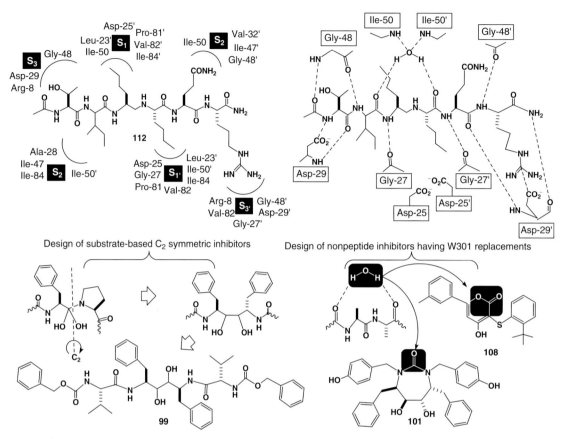

Figure 23 Models of the active site of HIV protease: peptidomimetic and nonpeptide inhibitors.

(e.g., fluoromethylketone, which has been determined[196] to exhibit selectivity for cysteinyl proteases versus serinyl proteases), semicarbazones, and nitriles, are noteworthy in that ^{13}C-NMR spectroscopic studies have shown that the active site Cys-25 of papain forms a reversible covalent bond with these electrophiles (for a review see Leung *et al.*[142]). The x-ray crystallographic structures of apo cathepsin-B[197] and apo picornaviral 3C protease[198] have also been determined. With respect to ICE (for a review see Ator and Dolle[199]), a number of high-resolution x-ray crystallographic structures of ICE–inhibitor complexes have been determined.[200,201] This enzyme is unique among the cysteinyl proteases in that it has a heterodimeric architecture in which two subunits form the catalytically active enzyme site. More specifically, two p10/p20 heterodimers apparently create a tetrameric form of the competent protease. An x-ray crystallographic structure of ICE complexed with a substrate peptide-based chloromethylketone inhibitor (**118**[201]) shows that is irreversibly bonded at the active site Cys-285 and that the P_1 Asp specificity pocket comprises two Arg residues at the base of the S_1 binding pocket (**Figure 26**). Relative to other side chain binding pockets, a hydrophobic 'channel'-type S_4 site exists for the P_4 Tyr of the inhibitor, whereas the P_3–P_2 Val-Ala side chains are well exposed to solvent. With respect to the peptide backbone of the inhibitor, hydrogen-bonding interactions are predicted between the P_3 Val (both NH and CO) and the P_1 Asp (NH) and the p10 monomer.

An x-ray crystallographic structure of ICE complexed with a reversible inhibitor Ac-Tyr-Val-Ala-Asp-aldehyde (**119**[202]) provided correlative structure–activity relationships with respect to *N*-methyl-amino acid substitutions and the three-dimensional mapping of hydrogen-bonding interactions between the inhibitor and the enzyme. These studies predicted that only N-Me-Ala replacement at the P_2 site would be tolerated. Subsequent ICE inhibitor design strategies, including C-terminal modification of P_1 Asp by alkylating groups such as the aryloxymethyl ketone analog **120**,[203] led to the first reported peptidomimetic inhibitor of ICE (**121**[204]), which incorporated a pyridone template as a P_2–P_3 replacement. Finally, the x-ray crystallographic structure of the ICE homolog referred to as apopain, or CPP32, as a complex with a peptide–aldehyde inhibitor[205] provided additional insight into the specificity of substrate recognition as related to both the S_1 (P_1 Asp) and S_4 (P_4 Asp) subsites. Overall, these studies provided the framework for iterative structure-based drug design to advance novel peptidomimetic and nonpeptide inhibitors of ICE and/or its homologs.

Peptidomimetic and Nonpeptide Drug Discovery: Receptor, Protease, and Signal Transduction Therapeutic Targets

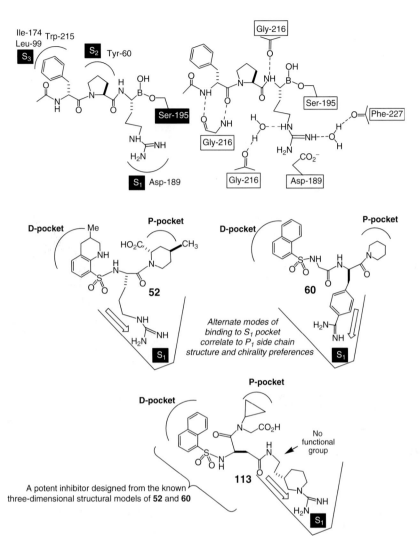

Figure 24 Model of the active site of thrombin: peptide and peptidomimetic inhibitors.

2.15.6.5 Metalloproteases: Peptidomimetic and Nonpeptide Lead Compounds

The metalloproteases include both exopeptidases (e.g., angiotensin-converting enzyme, aminopeptidase-M, and carboxypeptidase-A) and endopeptidases (e.g., thermolysin, endopeptidase 24.11 or NEP, collagenase, gelatinase, and stromelysin). The first x-ray crystallographic structures[207] were provided for thermolysin (for a review see Matthews[206] with several structurally distinct peptide inhibitors (see **Table 2**[208–212]), providing insight into the mechanistic roles of the metal ion for substrate hydrolysis. These studies exemplify inhibitor binding interactions of the P_1–$P_{1'}$ 'transition state' amide bond isosteres (e.g., $\Psi[P=O(OH)NH]$) as well as the metal chelating functionalities (e.g., hydroxamate, carboxylate, and sulfhydryl groups) introduced as N-substitutions on the P_1–P_2 peptide scaffolds.[208–212] Furthermore, at the time such structural and mechanistic information provided insight into drug-design strategies focused on other metalloprotease inhibitors, including the angiotensin-converting enzyme (ACE) (for a review see Leung et al.[142]). An x-ray crystallographic structure of carboxypeptidase, a related metalloprotease of the exopeptidase group, for both the apo and a $Gly\Psi[P=O(OH)NH]Phe$-modified inhibitor complex have been reported.[213] An x-ray crystallographic structure of the $Phe\Psi[P=O(OH)NH]Leu$-modified peptidomimetic inhibitor **122** complexed with thermolysin (**Figure 27**) shows the Zn^{2+} coordination and $P_{1'}$–$P_{2'}$ (or P_1–$P_{2'}$ collected product) mode of binding. Such three-dimensional structural studies enable the molecular interactions between ACE and its inhibitors (e.g. captopril,[51] enalaprilat,[52] and fosfinoprilat[214]) and between 'dual specific' inhibitors of ACE and NEP (e.g., **123**[215]) to be predicted.

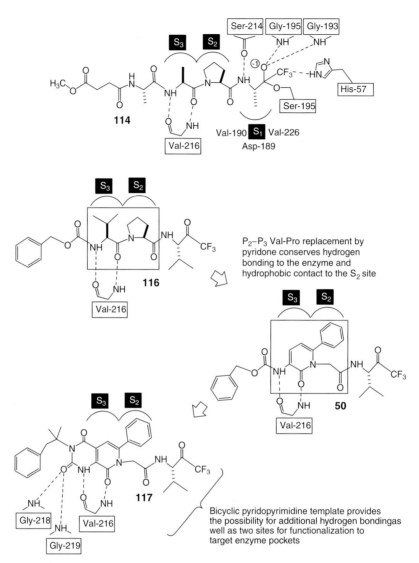

Figure 25 Model of the active site of elastase: peptide and peptidomimetic inhibitors.

The discovery of specific inhibitors of matrix metalloproteases (MMPs) is rapidly developing (for a review see Beckett et al.[216]), and is focused on several key MMP targets, including fibroblast collagenase (MMP-1), gelatinase-A (MMP-2), and stromelysin (MMP-3). Both x-ray crystallography and NMR spectroscopy have provided three-dimensional structural information about several MMPs as well as MMP–inhibitor complexes (see **Table 2**[217–226]). For collagenase, early substrate-based inhibitor design strategies focused on modifying the P_1–$P_{1'}$ cleavage site (e.g., Gly-Phe, Ala-Tyr, and Ala-Phe) by N-terminal functionalities capable of Zn^{2+} coordination (e.g., sulfhydryl, carboxyl, phosphonoalkyl, and hydroxamate[227–229]). An x-ray crystallographic structure of the N-hydroxamate-modified peptide **124** complexed with fibroblast collagenase (**Figure 28**) provided a molecular map of both its hydrogen-bonding interactions with the enzyme active site and its binding to bound Zn^{2+}.[219] Potent MMP-1 inhibitors have been designed which tether the $P_{2'}$ side chain to the inhibitor's C-terminus as macrocyclic rings.[53,230] In the example of gelatinase-A, potent inhibitors have been designed (e.g., **48**, **Figure 10**[54]) by modifying the $P_{1'}$–$P_{3'}$ peptide scaffold at the N-terminal hydroxamate and the $P_{1'}$ extended aromatic side chain.[54] Finally, in the example of stromelysin-1, potent inhibitors have been designed (e.g., **49**, **Figure 10**[55]) by N-terminal carboxyalkylamino functionalization relative to a P_1 substituent. An x-ray crystallographic structure of a related MMP-3 inhibitor **125** shows (**Figure 28**) the hydrogen-bonding interactions at the active site and coordination of the carboxylate with the Zn^{2+}.[223] Finally, the x-ray crystallographic structure of an PheΨ[P=O(OH)CH$_2$]Ala-modified peptide inhibitor **126** complexed with astacin (**Figure 28**) reveals the extensive hydrogen-bonding network between the inhibitor, enzyme, Zn^{2+}, and a structural water group.[226]

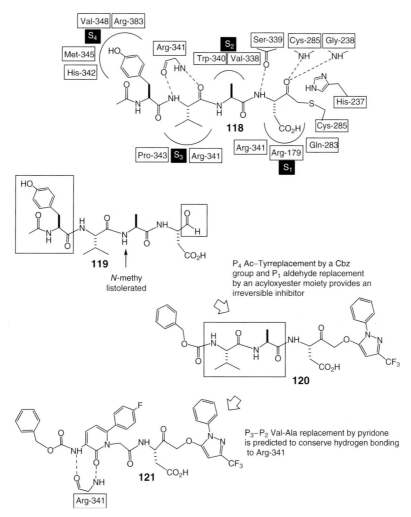

Figure 26 The active site of interleukin-converting enzyme: peptide and peptidomimetic inhibitors.

2.15.6.6 Src Homology-2 Domains: Peptidomimetic and Nonpeptide Lead Compounds

Beyond proteases, the opportunity for structure-based drug design is being realized in the rapidly developing area of signal transduction research (e.g., intracellular protein and nucleic acid targets). Both x-ray crystallography and NMR spectroscopy have significantly contributed to a widescope database of three-dimensional structural information for various catalytic and noncatalytic signal transduction protein targets (**Table 3**). These include tyrosine kinases (e.g., growth factor receptor kinases and Src family kinases[231]), serine/threonine and dual-specificity kinases (e.g., mitogen-activated protein kinases, CDK2- and cAMP-dependent protein kinases[232]), phosphotyrosine phosphatases (e.g., PTP1B and CD45 Syp[233]), phosphoserine/phosphothreonine and dual-specificity phosphatases (e.g., VH1 and CDC25[234]), noncatalytic 'adapter' proteins (e.g., Crk, Grb2, Shc, and IRS-1[85]), transferases (e.g., Ras farnesyl transferase[81]), proline *cis–trans* isomerases (e.g., FKBP-12 and cyclophilin A[235]), and GTP-binding proteins (e.g., p21 Ras[236] and the α-β/γ-heterotrimeric G-protein for the GPCR superfamily[237]). Indeed, it was recognized more than decade ago that the opportunity for signal transduction drug discovery was remarkable, in terms of both the mechanistic diversity of the therapeutic targets (e.g., enzyme–substrate or regulatory protein–protein interaction) and their emerging structural biology.[238–267]

The identification of noncatalytic regulatory domains referred to as Src homology (SH) domains has advanced rapidly over recent years as a critical link in deconvoluting both enzyme–substrate and regulatory protein–protein interactions for a number of signal transduction pathways (for a review see Cohen et al.[268]). This emerging superfamily of proteins includes SH2 and SH3 domains, the so-called 'choreographers of multiple signaling pathways'.[85] The SH2 domains have been determined to bind cognate phosphotyrosine (pTyr) containing proteins in a sequence-dependent

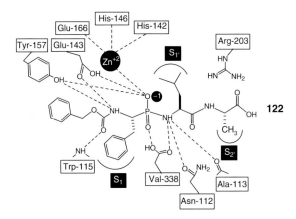
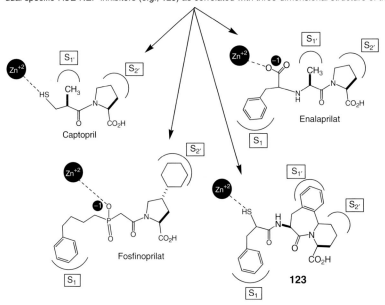

Figure 27 Models of the active sites of thermolysin and ACE: peptide and peptidomimetic inhibitors.

manner relative to the amino acids contiguous to the C-terminal side of the pTyr residue (e.g., for Src SH2 a preferred sequence is ∼pTyr-Glu-Glu-Ile∼ versus ∼pTyr-Tyr-Asn-Tyr for Grb2[87,269]).

With respect to SH2 domain structure-based drug design, the first x-ray crystallographic structures of pTyr-containing peptide ligands complexed with the Src SH2 domain[239] have been utilized to design the first peptidomimetic antagonists.[89] As illustrated in **Figure 29**, a molecular map of the tetrapeptide sequence ∼pTyr-Glu-Glu-Ile∼ complexed with Src SH2[239] shows the pTyr binding pocket and a second binding site for the P_{+3} Ile residue. As described previously, a prototypic peptidomimetic Ac-pTyr-Glu-D-Hcy-NH$_2$ (**66**) was first discovered using a peptide-scaffold-based drug-design strategy (see **Figure 15**[77]) that took into account the x-ray crystallographic structure of the Src SH2-phosphopeptide (Glu-Pro-Gln-pTyr-Glu-Glu-Ile-Pro-Ile-Tyr-Leu, **127**) complex. Further structure-based, drug-design modifications led to the discovery of a series of potent peptidomimetics having novel C-terminal functionalization (e.g., the 'transposed' side chain of the P_{+1} Glu, or conformational constraint using a pyrrolidine ring; see **Figure 29**), as represented by **128–130**.[89,270] Studies focused on Src SH2 have shown that the phosphate ester of pTyr is particularly critical for molecular recognition, and that significant loss in binding occurs by replacement with sulfate, carboxylate, nitrosyl, hydroxy, and amino groups.[88] However, backbone modifications of pTyr, which replace its acylated amino functionality with aromatic rings, designed to form π-cation type interactions with the Arg-αA2 were effective substitutions.[270]

Peptidomimetic and Nonpeptide Drug Discovery: Receptor, Protease, and Signal Transduction Therapeutic Targets 635

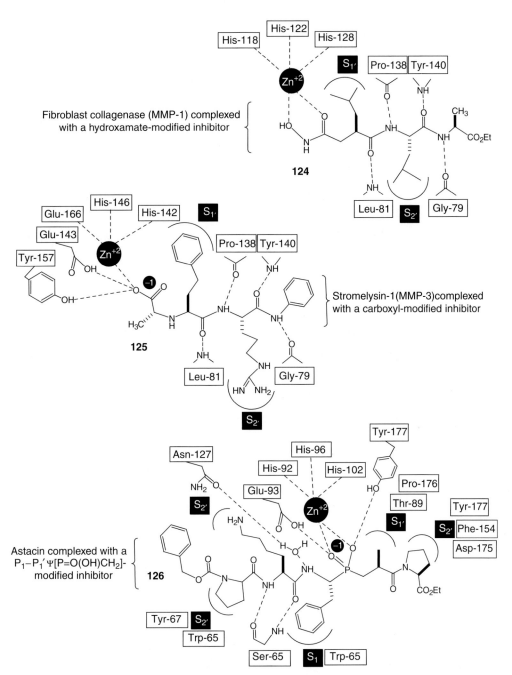

Figure 28 Models of the active sites of stromelysin and collagenase: peptide and peptidomimetic inhibitors.

Src SH2 structure-based drug design has also led to a series of nonpeptides utilizing a *m*-aminomethylbenzamide template, as first exemplified by **131** (**Figure 30**), and validated by x-ray structural studies[271] that revealed binding of both its pY + 3 group and the benzamide $CONH_2$ moiety, as predicted. Modifications to incorporate pTyr (**132**) and further elaboration of the template led to a second-generation series[272–274] of bicyclic benzamide analogs (**133–136**) that have increased potency and, in the case of **134** (AP22408), the first in vivo active lead compound.[272] Of note is the bicyclic benzamide template, which provides increased hydrophobic binding interactions with Src SH2 as well as the entropic advantage of locking the conformation of the benzamide via the fused cycloaliphatic ring system.[272–274] Furthermore, the pTyr mimetics incorporating phosphonate moieties (e.g. **133** and **134**) were designed to confer tissue selectivity by virtue of bone-targeting.[272,274] Another series of cyclic lactam-based, nonpeptide inhibitors of Src SH2,

Table 3 Some known three-dimensional structures of signal transduction proteins (Apo and/or complexes thereof with peptide, peptidomimetic and nonpeptide inhibitors)

Signal transduction protein	Apo or inhibitor complex	Method (resolution)	Reference
Src homology domains			
Abl SH2	Apo	NMR (NR)	Overduin et al.[238]
Src SH2	Apo	X-ray (2.5 Å)	Waksman et al.[239]
	Phosphopeptide	X-ray (2.7 Å)	Waksman et al.[239]
	Phosphopeptide	NMR (NR)	Xu et al.[240]
	Peptidomimetic	X-ray (2.4 Å)	Plummer et al.[304]
	Nonpeptide	X-ray (2.5 Å)	Lunney et al.[271]
Lck SH2	Phosphopeptide	X-ray (2.2 Å)	Eck et al.[305]
Lck (S164C) SH2	Nonpeptide	X-ray (1.95 Å)	Shakespeare et al.[273]
Lck (S164C) SH2	Nonpeptide	X-ray (2.4 Å)	Shakespeare et al.[272]
PLCγ(C) SH2	Phosphopeptide	NMR (NR)	Pascal et al.[241]
Shc SH2	Phosphopeptide	NMR (NR)	Zhou et al.[242]
p85 (N) SH2	Phosphopeptide	X-ray (2.0 Å)	Nolte et al.[244]
Syp SH2	Phosphopeptide	X-ray (2.0 Å)	Lee et al.[245]
Grb2 SH2	Phosphopeptide	X-ray (2.1 Å)	Rahuel et al.[246]
Syk SH2	Phosphopeptide	NMR (NR)	Narula et al.[247]
Zap70 SH2-SH2	Phosphopeptide	X-ray (1.9 Å)	Hatada et al.[248]
Src SH3	Apo	NMR (NR)	Yu et al.[249]
	Peptide	NMR (NR)	Yu et al.[249]
Abl SH3	Peptide	X-ray (2.0 Å)	Musacchio et al.[250]
Crk SH3	Peptide	X-ray (1.5 Å)	Wu et al.[251]
Grb2 SH3	Peptide	NMR (NR)	Goudreau et al.[252]
Grb2 SH3-SH2-SH3	Apo	X-ray (3.1 Å)	Maignan et al.[253]
Tyr and Ser/Thr kinases			
Insulin receptor kinase	Apo	X-ray (2.1 Å)	Hubbard et al.[254]
cAMP-dependent protein kinase	Apo	X-ray (3.9 Å)	Zheng et al.[255]
	Peptide	X-ray (2.9 Å)	Karlsson et al.[306]
Cyclin-dependent protein kinase-2	Apo	X-ray (2.4 Å)	De Bondt et al.[307]
	p27^{Kip1} inhibitory protein	X-ray (2.3 Å)	Russo et al.[256]
Mitogen activated protein kinase	Apo	X-ray (2.3 Å)	Zhang et al.[257]
Tyr and Ser/Thr phosphatases			
PTP1B (Tyr)	Apo	X-ray (2.8 Å)	Barford et al.[308]
	Peptide (C215S mutant enzyme)	X-ray (2.6 Å)	Jia et al.[258]
	Nonpeptide	X-ray (2.6 Å)	Iversen et al.[309]
VH1 (Tyr and Ser/Thr")	Apo	X-ray (2.1 Å)	Yuvaniyama et al.[259]
PP-1 (Ser/Thr)	Apo	X-ray (2.1 Å)	Goldberg et al.[260]
Pro cis–trans isomerases			
Cyclophilin	Peptide (cyclosporin-A)	X-ray (2.8 Å)	Pflugl et al.[263]
FKBP-12	FK-506 (macrolide antibiotic)	X-ray (1.7 Å)	Van Duyne et al.[264]

NR, not reported.

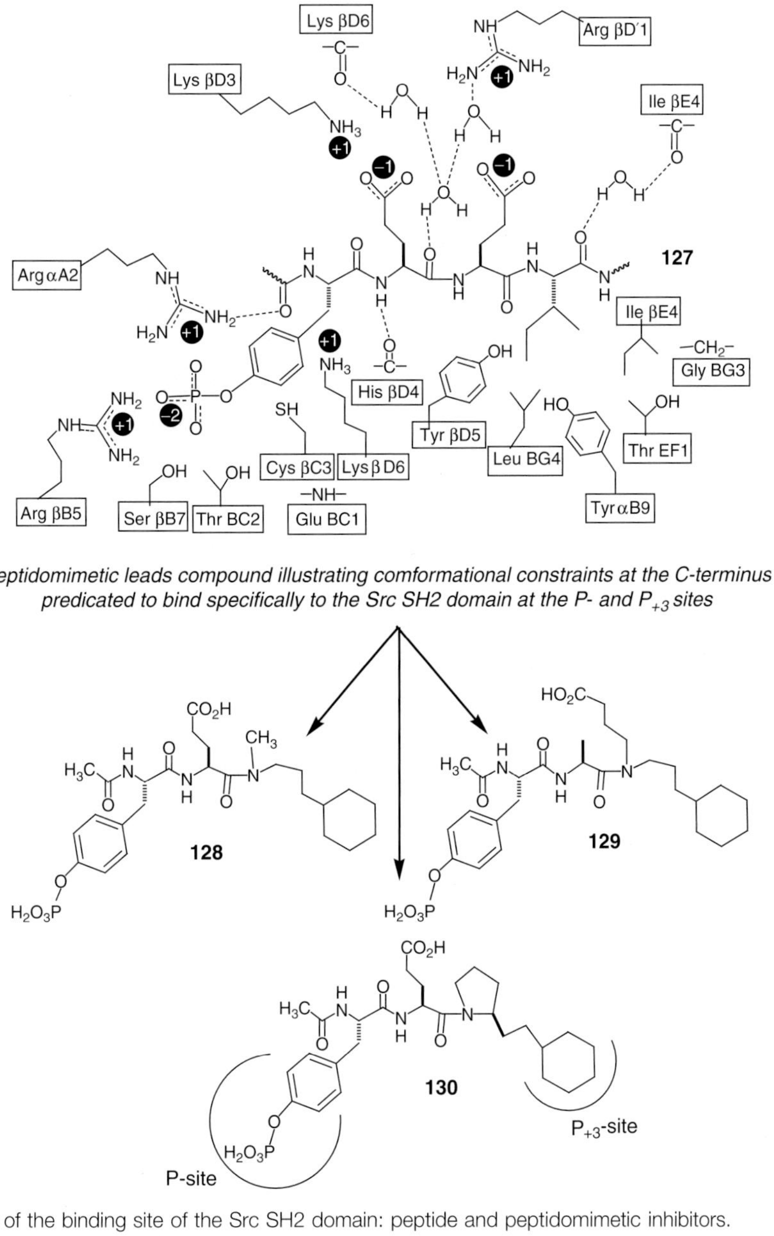

Figure 29 Models of the binding site of the Src SH2 domain: peptide and peptidomimetic inhibitors.

as exemplified by **137–139**, has also been described.[275–277] This series of nonpeptide inhibitors incorporated a caprolactam template in which the carbonyl moiety was designed, and confirmed by x-ray structure, to displace structural water (i.e., the same as displaced by the aforementioned benzamide template inhibitor **131**). Furthermore, the tricarboxy-modified pTyr mimic incorporated in the nonpeptide **139** yielded a high-affinity Src SH2 inhibitor.

High-resolution three-dimensional structures have been described for the noncatalytic 'adapter' protein Grb2 with respect to the apoprotein (SH3–SH2–SH3[153]) as well as the individual SH2 and SH3 domains.[246,252] In the case of the SH2 domain of Grb2, an x-ray crystallographic structure of a phosphopeptide complex provided insight to the molecular basis of the specificity of the Grb2 SH2 binding of $\sim$pTyr-Xxx-Asn-Yyy$\sim$ sequences. The binding interactions of Lys-Pro-Phe-pTyr-Val-Asn-Val (**140**) showed that the phosphopeptide adopts a β-turn conformation about the P–P$_{+3}$ residues, and that the P$_{+2}$ Asn side chain carboxamide moiety is extensively hydrogen-bonded to the protein (**Figure 31**). In contrast to the well-defined binding pocket for the P$_{+3}$ Ile of **127** to bind Src SH2, the P$_{+3}$ Val of **140** engages in limited surface hydrophobic interactions because the Trp-121 residue of Grb2 SH2 sterically blocks the phosphopeptide from attaining a similar binding mode. First-generation peptidomimetic and nonpeptide Grb2 SH2

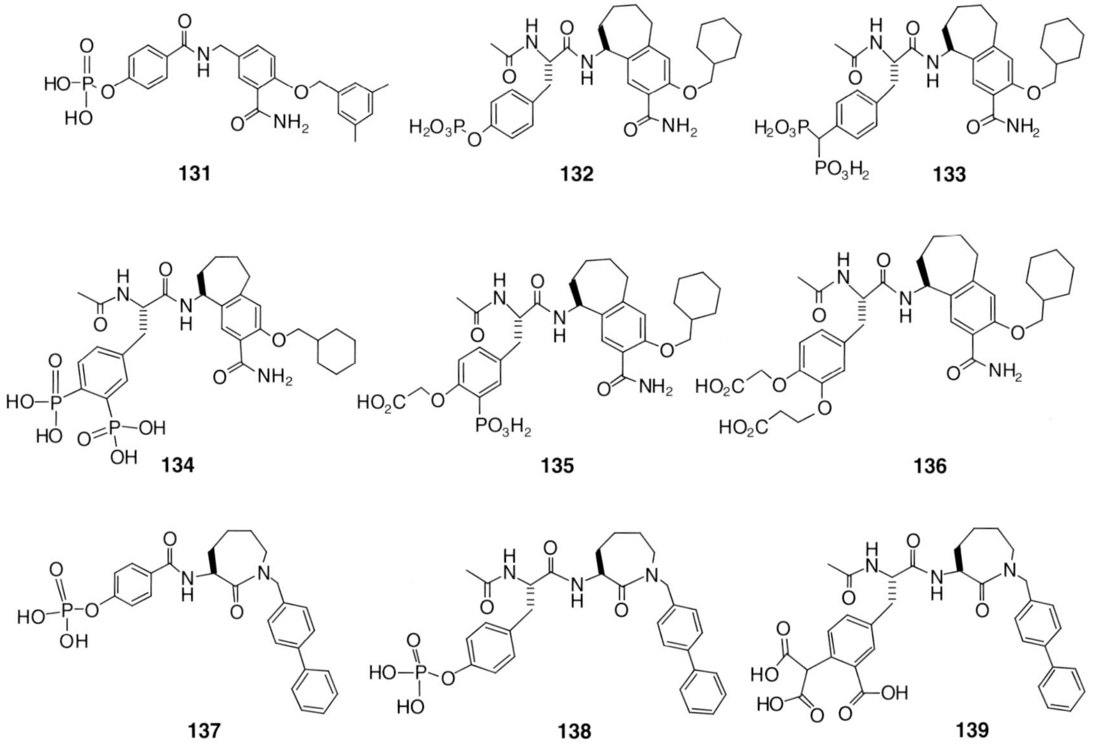

Figure 30 Some known peptidomimetic and nonpeptide inhibitors of Src SH2 domain.

inhibitors **141–143**[278–281] illustrate drug-design strategies exploiting the known β-turn conformation of cognate phosphopeptide ligands complexed with Grb2 SH2, as determined from x-ray structural studies. The peptidomimetic **142** was cellularly effective and provided the first proof-of-concept lead compound relative to disrupting the Ras-Grb2 pathway by the inhibition of Grb2 SH2. The nonpeptide **143** was a moderately potent Grb SH2 inhibitor and provided a template for further lead compound optimization.[281]

2.15.6.7 Tyrosine Phosphatases: Peptidomimetic and Nonpeptide Lead Compounds

PTP1B was the first member of the tyrosine phosphatase superfamily to be discovered and have its structure determined by x-ray crystallography, both as the apo catalytic domain and as a complex using a catalytically inactive Cys215Ser PTP1B mutant and the phosphopeptide Asp-Ala-Asp-Glu-pTyr-Leu-NH2 (**144**).[258] The molecular interactions between the tyrosine phosphatase and the phosphopeptide are dominated by electrostatic (i.e., the pTyr, P-1 Glu and P-2 Asp residues) and hydrogen-bonding contacts to key amide functionalities of the backbone of **144** (**Figure 32**). The P_{+1} Leu side chain forms hydrophobic contacts with several PTP1B residues at the surface proximate to the well-defined pTyr binding pocket. This three-dimensional structural information provided a useful molecular map of PTP1B active site for drug-design strategies focused on peptidomimetic and nonpeptide inhibitors such as compounds **145**,[282] **146**,[283,284] and **147**[285,286] (**Figure 32**). The peptidomimetic inhibitor **145** was generated using a combinatorial library strategy, and was found to be a potent and selective inhibitor of PTP1B.[282] The peptidomimetic inhibitor **146** was designed from the tripeptide Ac-Asp-Tyr[SO$_3$H]-Nle–NH$_2$ and was also found to be a potent and selective inhibitor of PTP1B.[283,284] The nonpeptide **147** is a tetrahydrothienopyridine template-based inhibitor of PTP1B, and provided a prototype lead compound for further drug development.[285,286]

2.15.7 Recent Progress and Future Perspectives

Beyond the many key pioneering studies described in this chapter, there has been significant further progress in peptidomimetic and nonpeptide drug discovery in the last few years. Such work includes the integration of increasingly sophisticated in silico drug-design, structural biology (x-ray crystallography and NMR spectroscopy), chemical diversity,

Peptidomimetic and Nonpeptide Drug Discovery: Receptor, Protease, and Signal Transduction Therapeutic Targets

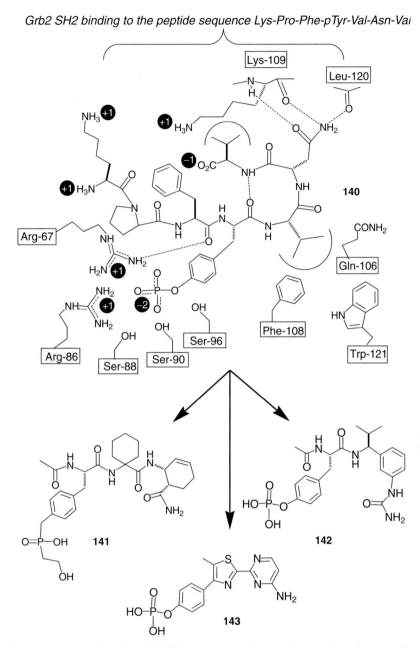

Figure 31 Models of the binding site of the Grb2 SH2 domain: peptide, peptidomimetic, and nonpeptide inhibitors.

and high-throughput screening technologies, which continue to facilitate innovative peptidomimetic and nonpeptide drug discovery. In terms of therapeutic targets for peptidomimetic and nonpeptide drug discovery, genome sequencing and unraveling of the proteome will also build upon the knowledge gained from past scientific endeavors. With respect to synthetic chemistry, the use of combinatorial chemistry is expected to expand dramatically the chemical space of known peptidomimetics and nonpeptides, including exploiting key functional groups and building blocks (e.g., D-amino acids, N_α-alkyl or C_α-alkyl amino acids, N-substituted Gly, and dipeptide surrogates (XxxΨ[Z]Yyy), and providing a collection of multifunctionalized templates (monocyclic or bicyclic) to mimic the three-dimensional pharmacophores of peptide ligands or complement three-dimensional binding sites of therapeutic targets. Finally, chemoinformatics and bioinformatics will significantly impact peptidomimetic and nonpeptide drug discovery in the future, in terms of harnessing the extraordinary amount of structure–property relationships that have emerged to date and those which are yet to come.

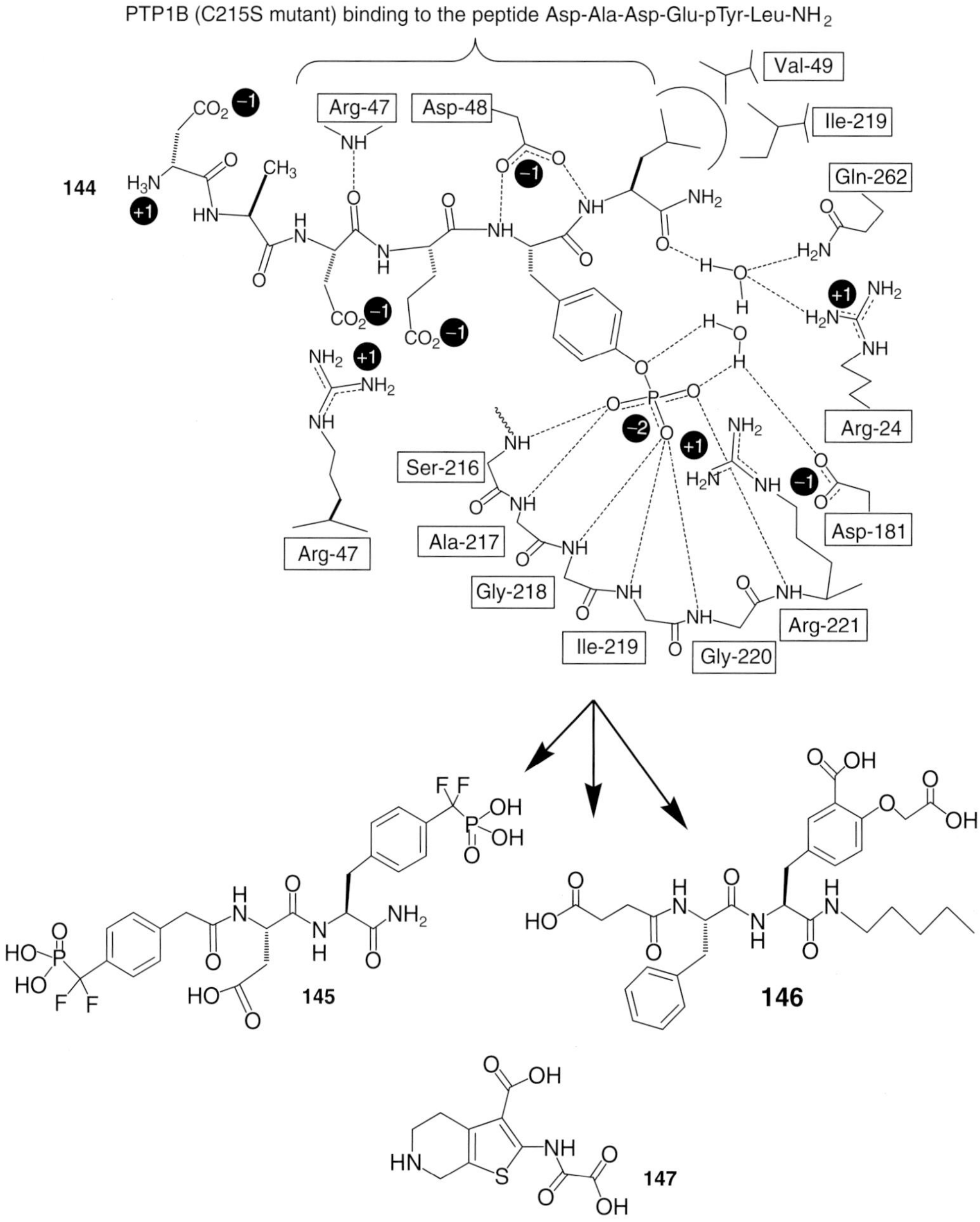

Figure 32 Model of the active site of PTP1b tyrosine phosphatase: peptide, peptidomimetic, and nonpeptide inhibitors.

References

1. Hruby, V. J. *Nat. Rev. Drug Disc.* **2002**, *1*, 847–858.
2. Cowell, S. M.; Gu, X.; Vagner, J.; Hruby, V. J. *Methods Enzymol. Part B* **2003**, *369*, 288–297.
3. Hruby, V. J. *J. Med. Chem.* **2003**, *46*, 4215–4231.
4. Hruby, V. J.; Al-Obeidi, F.; Kazmierski, W. *Biochem. J.* **1990**, *268*, 249–262.
5. Toniolo, C. *Int. J. Pept. Prot. Res.* **1990**, *35*, 287–300.
6. Goodman, M.; Zhang, J. *Chemtracts* **1997**, *10*, 629–645.
7. Diercks, T.; Coles, M.; Kessler, H. *Curr. Opin. Chem. Biol.* **2001**, *5*, 285–291.
8. Marshall, G. R. *Tetrahedron* **1993**, *49*, 3547–3558.
9. Fletcher, S.; Hamilton, A. D. *Curr. Opin. Chem. Biol.* **2005**, *9*, 632–638.
10. McDowell, R. S.; Artis, D. R. *Annu. Rep. Med. Chem.* **1995**, *30*, 265–274.
11. Ripka, A. S.; Rich, D. H. *Curr. Opin. Chem. Biol.* **1998**, *2*, 441–452.

12. Fairlie, D. P.; Abbenante, G.; March, D. R. *Curr. Med. Chem.* **1995**, *2*, 654–686.
13. Olson, G. L.; Bolin, D. R.; Bonner, M. P.; Bos, M.; Cook, C. M.; Fry, D. C.; Graves, B. J.; Hatada, M.; Hill, D. E.; Kahn, M. et al. *J. Med. Chem.* **1993**, *36*, 3039–3049.
14. Wiley, R. A.; Rich, D. H. *Med. Res. Rev.* **1993**, *13*, 284–327.
15. Giannis, A.; Kolter, T. *Angew. Chem. Int. Ed. Engl.* **1993**, *32*, 1244–1267.
16. Gante, J. *Angew. Chem. Int. Ed. Engl.* **1994**, *33*, 1699–1720.
17. Al-Obeidi, F.; Hruby, V. J.; Sawyer, T. K. *Mol. Biotechnol.* **1998**, *9*, 205–223.
18. Sawyer, T. K. *Drugs Pharm. Sci.* **2000**, *101*, 81–114.
19. Gupta, A.; Chauhan, V. S. *Int. J. Pept. Prot. Res.* **1993**, *41*, 421–426.
20. Haghihara, M.; Anthony, N. J.; Stout, T. J.; Clardy, J.; Schreiber, S. L. *J. Am. Chem. Soc.* **1992**, *114*, 6568–6570.
21. Simon, R. J.; Kania, R. S.; Zuckermann, R. N.; Huebner, V. D.; Jewell, D. A.; Banville, S.; Ng, S.; Wang, L.; Rosenberg, S.; Marlowe, C. K. et al. *Proc. Natl. Acad. Sci. USA* **1992**, *89*, 9367–9371.
22. Fong, T. M.; Strader, C. D. *Med. Res. Rev.* **1994**, *14*, 389–399.
23. Freidinger, R. M.; Veber, D. F.; Perlow, D. S.; Brooks, J. R.; Saperstein, R. *Science* **1980**, *210*, 656–658.
24. Nagai, U.; Sato, K. *Tetrahedron Lett.* **1985**, *26*, 647–650.
25. Feigl, M. *J. Am. Chem. Soc.* **1986**, *108*, 181–182.
26. Kahn, M.; Wilke, S.; Chen, B.; Fujita, K. *J. Am. Chem. Soc.* **1988**, *110*, 1638–1639.
27. Kemp, D. S.; Stites, W. E. *Tetrahedron Lett.* **1988**, *29*, 5057–5060.
28. Genin, M. J.; Johnson, R. L. *J. Am. Chem. Soc.* **1992**, *114*, 8316–8318.
29. Kahn, M.; Wilke, S.; Chen, B.; Fujita, K.; Lee, Y. H.; Johnson, M. E. *J. Mol. Recognition* **1988**, *1*, 75–79.
30. Ripka, W. C.; DeLucca, G. V.; Bach, A. C., II; Pottorf, R. S.; Blaney, J. M. *Tetrahedron* **1993**, *49*, 3593–3608.
31. Ripka, W. C.; DeLucca, G. V.; Bach, A. C., II; Pottorf, R. S.; Blaney, J. M. *Tetrahedron* **1993**, *49*, 3609–3628.
32. Callahan, J. F.; Newlander, K. A.; Burgess, J. L.; Eggleston, D. S.; Nichols, A.; Wong, A.; Huffman, W. F. *Tetrahedron* **1993**, *49*, 3479–3488.
33. Smith, A. B.; Knight, S. D.; Sprengeler, P. A.; Hirschmann, R. *Bioorg. Med. Chem.* **1996**, *4*, 1021–1034.
34. Pitzele, B. S.; Hamilton, R. W.; Kudla, K. D.; Taymbalov, S.; Stapefield, A.; Savage, M. A.; Clare, M.; Hammond, D. L.; Hansen, D. W., Jr. *J. Med. Chem.* **1994**, *37*, 888–896.
35. Zablocki, J. A.; Miyano, M.; Garland, B.; Pireh, D.; Schretzman, L.; Rao, S. N.; Lindmark, R. J.; Panzer-Knodle, S.; Nicholson, N.; Taite, B. et al. *J. Med. Chem.* **1993**, *36*, 1811–1819.
36. Samanen, J. M.; Ali, F. E.; Barton, L. S.; Bondinell, W. E.; Burgess, J. L.; Callahan, J. F.; Calvo, R. R.; Chen, W.; Chen, L.; Erhard, K. et al. *J. Med. Chem.* **1996**, *39*, 4867–4870.
37. Flynn, D. L.; Villamil, C. I.; Becker, D. P.; Gullikson, G. W.; Moummi, C.; Yang, D.-C. *Bioorg. Med. Chem. Lett.* **1992**, *2*, 1251–1256.
38. Horwell, D. C. *Neuropeptides* **1991**, *19*, 57–64.
39. Ishikawa, K.; Ihara, M.; Noguchi, K.; Mase, T.; Mino, N.; Saki, T.; Fukuroda, T.; Fukami, T.; Ozaki, S.; Nagase, T. et al. *Proc. Natl. Acad. Sci. USA* **1994**, *91*, 4892–4896.
40. McDowell, R. S.; Elias, K. A.; Stanley, M. S.; Burdick, D. J.; Burnier, J. P.; Chan, K. S.; Fairbrother, W. J.; Hammonds, R. G.; Ingle, G. S.; Jacobsen, N. E. et al. *Proc. Natl. Acad. Sci. USA* **1995**, *92*, 1165–11169.
41. Hirschmann, R.; Nicolaou, K. C.; Pietranico, S.; Salvino, J.; Leahy, E. M.; Sprengeler, P. A.; Furst, G.; Smith, A. B., III; Strader, C. D.; Cascieri, M. A. et al. *J. Am. Chem. Soc.* **1992**, *114*, 9217–9218.
42. Cirillo, R.; Astolfi, M.; Conte, B.; Lopez, G.; Parlani, M.; Terraccian, R.; Fincham, C. I.; Manzini, S. *Eur. J. Pharmacol.* **1998**, *341*, 201–209.
43. Smith, P. W.; McElroy, A. B.; Pritchard, J. M.; Deal, M. J.; Ewan, G. B.; Hagen, R. M.; Ireland, S. J.; Ball, D.; Beresford, I.; Sheldrick, R. et al. *Bioorg. Med. Chem. Lett.* **1993**, *3*, 931–935.
44. Boden, P.; Eden, J. M.; Hodgson, J.; Horwell, D. C.; Hughes, J.; McKnight, A. T.; Lewthwaite, R. A.; Prichard, M. C.; Raphy, J.; Meecham, K. et al. *J. Med. Chem.* **1996**, *39*, 1664–1665.
45. Zablocki, J. A.; Rao, S. N.; Baron, D. A.; Flynn, D. L.; Nicholson, N. S.; Feigen, L. P. *Curr. Pharm. Design* **1996**, *1*, 533–558.
46. Hirschmann, R.; Yao, W.; Cascieri, M. A.; Strader, C. D.; Maechler, L.; Cichy-Knight, M. A.; Hynes, J., Jr.; van Rijn, R. D.; Sprengeler, P. A.; Smith, A. B. *J. Med. Chem.* **1996**, *39*, 2441–2448.
47. Sawyer, T. K.; Maggiora, L. L.; Liu, L.; Staples, D. J.; Bradford, V. S.; Mao, B.; Pals, D. T.; Dunn, B. M.; Poorman, R.; Hinzmann, J. et al. In *Peptides, Chemistry, Biology*; Marshall, G. R., Rivier, J., Eds.; Escom Science: Leiden, 1990, pp 855–857.
48. Rosenberg, S. H.; Spina, K. P.; Condon, S. L.; Polakowski, J.; Yao, Z.; Kovar, P.; Stein, H. H.; Cohen, J.; Barlow, J. L.; Klinghofer, V. et al. *J. Med. Chem.* **1993**, *36*, 460–467.
49. McQuade, T. J.; Tomasselli, A. G.; Liu, L.; Karacostas, V.; Moss, B.; Sawyer, T. K.; Heinrikson, R. L.; Tarpley, W. G. *Science* **1990**, *247*, 454–456.
50. Roberts, N. A.; Martin, J. A.; Kinchington, D.; Broadhurst, A. V.; Craig, J. C.; Duncan, I. B.; Galpin, S. A.; Handa, B. K.; Kay, J.; Krohn, A. et al. *Science* **1990**, *248*, 358–361.
51. Ondetti, M. A.; Rubin, B.; Cushman, D. W. *Science* **1977**, *196*, 441–444.
52. Patchett, A. A.; Harris, E.; Tristam, E. W.; Wyvratt, M. J.; Wu, M. T.; Taub, D.; Peterson, E. R.; Ikeler, T. J.; TenBroeke, J.; Payne, L. G. et al. *Nature* **1980**, *288*, 280–283.
53. Bird, J.; Harper, G. P.; Hughes, I.; Hunter, D. J.; Karran, E. H.; Markwell, R. E.; Miles-Williams, A. J.; Rahman, S. S.; Ward, R. W. *Bioorg. Med. Chem. Lett.* **1995**, *5*, 2593–2598.
54. Porter, J. R.; Beeley, N. R. A.; Boyce, B. A.; Mason, B.; Millican, A.; Miller, K.; Leonard, J.; Morphy, J. R.; O'Connell, J. P. *Bioorg. Med. Chem. Lett.* **1994**, *4*, 2741–2746.
55. Chapman, K. T.; Durette, P. L.; Caldwell, C. G.; Sperow, K. M.; Niedzwiecki, I. M.; Harrison, R. K.; Saphos, C.; Christen, A. J.; Olszewski, J. M.; Moore, V. L. et al. *Bioorg. Med. Chem. Lett.* **1996**, *6*, 803–806.
56. Edwards, P. D.; Andiskik, D. W.; Strimpler, A. M.; Gomes, B.; Tuthil, P. A. *J. Med. Chem.* **1996**, *39*, 1112–1124.
57. Shuman, R. T.; Rothenberger, R. B.; Campbell, C. S.; Smith, G. F.; Gifford-Moore, D. S.; Gesellechen, P. D. *J. Med. Chem.* **1993**, *36*, 314–319.
58. Kikumoto, R.; Tamao, Y.; Ohkubo, K.; Tezuka, T.; Tonomura, S.; Okamoto, S.; Hijikata, A. *J. Med. Chem.* **1980**, *23*, 1293–1299.
59. Dolle, R. E.; Prouty, C. P.; Prasad, C. V. C.; Cook, E.; Saha, A.; Ross, T. M.; Salvino, J. M.; Helaszek, C. T.; Ator, M. A. *J. Med. Chem.* **1996**, *39*, 2438–2440.
60. Krausslich, H.-G.; Schneider, H.; Zybarth, G.; Carter, C. A.; Wimmer, E. *J. Virol.* **1988**, *2*, 4394–4397.
61. Owens, R. A.; Gesellchen, P. D.; Houchins, B. J.; DiMarchi, R. D. *Biochem. Biophys. Res. Commun.* **1991**, *181*, 401–408.
62. Richards, A. D.; Roberts, R.; Dunn, B. M.; Graves, M. C.; Kay, J. *FEBS Lett.* **1989**, *247*, 113–117.

63. Smith, A. B.; Hirschmann, R.; Pasternak, A.; Guzman, M. C.; Yokoyama, A.; Sprengeler, P. A.; Darke, P. L.; Emini, E. A.; Schleif, W. A. *J. Am. Chem. Soc.* **1995**, *117*, 11113–11123.
64. Rich, D. H.; Green, J.; Toth, M. V.; Marshall, G. R.; Kent, S. B. H. *J. Med. Chem.* **1990**, *33*, 1288–1295.
65. Kempf, J.; Sham, H. L. *Curr. Pharm. Des.* **1996**, *2*, 225–246.
66. Das, J.; Kimball, S. D. *Bioorg. Med. Chem.* **1995**, *3*, 999–1007.
67. Bajusz, S.; Szell, E.; Bagdy, D.; Barbas, E.; Horvath, G.; Dioszegi, M.; Fittler, Z.; Szabo, G.; Juhasz, A.; Tomori, E. et al. *J. Med. Chem.* **1990**, *33*, 1729–1735.
68. Sherry, S.; Alkjaersig, N.; Fletcher, A. P. *Am. J. Physiol.* **1995**, *209*, 577–583.
69. Fusetani, N.; Matsunaga, S.; Matsumoto, H.; Takebayashi, Y. *J. Am. Chem. Soc.* **1990**, *112*, 7053–7054.
70. Okamoto, S.; Hijikata, A. *J. Med. Chem.* **1980**, *23*, 1293–1299.
71. Sturzebecher, J.; Markwardt, F.; Voigt, B.; Wagner, G.; Walsmann, P. *Thromb. Res.* **1983**, *29*, 635–642.
72. Rose, C.; Vargas, F.; Facchinetti, P.; Bourgeat, P.; Babal, R. B.; Bishop, P. B.; Chan, S. M. T.; Moore, A. N. J.; Ganellin, C. R.; Schwartz, J.-C. *Nature* **1996**, *380*, 403–409.
73. Kohl, N. E.; Mosser, S. D.; deSolms, S. J.; Giuliani, E. A.; Pompliano, D. L.; Graham, S. L.; Smith, R. L.; Scolnick, E. M.; Oliff, A.; Gibbs, J. B. *Science* **1993**, *260*, 1934–1936.
74. James, G. L.; Goldstein, J. L.; Brown, M. S.; Rawson, T. E.; Somers, T. C.; McDowell, R. S.; Crowley, C. W.; Lucas, B. K.; Levinson, A. D.; Marsters, J. C., Jr. *Science* **1993**, *260*, 1937–1942.
75. Qian, Y.; Blaskovich, M. A.; Seong, C.-M.; Vogt, A.; Hamilton, A. D.; Sebti, S. M. *Bioorg. Med. Chem. Lett.* **1994**, *21*, 2579–2584.
76. Scholten, J. D.; Zimmerman, K.; Oxender, G. M.; Sebolt-Leopold, J.; Gowan, R.; Leonard, D.; Hupe, D. J. *Bioorg. Med. Chem.* **1996**, *4*, 1537–1543.
77. Plummer, M. S.; Lunney, E. A.; Para, K. S.; Vara Prasad, J. V. N.; Shahripour, A.; Singh, J.; Stankovic, C. J.; Humblet, C.; Fergus, J. H.; Marks, J. S. et al. *Drug Des. Disc.* **1996**, *13*, 75–81.
78. Rodriguez, M.; Crosby, R.; Alligood, K.; Gilmer, T.; Berman, J. *Lett. Pept. Sci.* **1995**, *2*, 1–6.
79. Burke, T. R., Jr.; Barchi, J. J., Jr.; George, C.; Wolf, G.; Shoelson, S. E.; Yan, X. *J. Med. Chem.* **1995**, *38*, 1386–1396.
80. Burke, T. R., Jr.; Kole, H. K.; Roller, P. P. *Biochem. Biophys. Res. Commun.* **1994**, *204*, 129–134.
81. Buss, J. E.; Marsters, J. C., Jr. *Chem. Biol.* **1995**, *2*, 787–791.
82. Manne, V.; Yan, N.; Carboni, J. M.; Tuomari, A. V.; Ricca, C. S.; Brown, J. G.; Andahazy, M. L.; Schmidt, R. J.; Patel, D.; Zahler, R. et al. *Oncogene* **1995**, *10*, 1763–1779.
83. Lerner, E. C.; Qian, Y.; Blaskovich, M. A.; Fossum, R. D.; Vogt, A.; Sun, J.; Cox, A. D.; Der, J. D.; Hamilton, A. D.; Sebti, S. M. *J. Biol. Chem.* **1995**, *270*, 26802–26806.
84. Hunt, J. T.; Lee, V. G.; Leftheria, K.; Seizinger, B.; Carboni, J.; Mabus, J.; Ricaa, C.; Yan, N.; Manne, V. *J. Med. Chem.* **1996**, *39*, 353–358.
85. Botfield, M. C.; Green, J. *Annu. Rep. Med. Chem.* **1995**, *30*, 227–237.
86. Kuriyan, J.; Cowburn, D. *Curr. Biol.* **1993**, *3*, 828–837.
87. Songyang, Z.; Shoelson, S. E.; Chaudhuri, M.; Gish, G.; Pawson, T.; Haser, W. G.; King, F.; Roberts, T.; Ratnofsky, S.; Lechleider, R. J. et al. *Cell* **1993**, *72*, 767–778.
88. Gilmer, T.; Rodriguez, M.; Jordan, S.; Crosby, R.; Alligood, K.; Green, M.; Kimery, M.; Wagner, C.; Kinder, D.; Charifson, P. et al. *J. Biol. Chem.* **1994**, *269*, 31711–31719.
89. Plummer, M. S.; Lunney, E. A.; Shahripour, A.; Para, K. S.; Stankovic, C. J.; Humblet, C.; Fergus, J.; Marks, J. S.; Herrera, R.; Hubbell, S. E. et al. *Bioorg. Med. Chem.* **1996**, *5*, 41–47.
90. Lord, J. A. H.; Waterfield, A. A.; Hughes, J.; Kosterlitz, H. W. *Nature* **1977**, *267*, 495–499.
91. Snider, M. R.; Constantine, J. W.; Lowe, J. A.; Longo, K. P.; Lebel, W. S.; Woody, H. A.; Drozda, S. E.; Desaia, M. C.; Vinick, F. J.; Spencer, R. W. et al. *Science* **1991**, *251*, 435–437.
92. Chiu, A. T.; McCall, D. E.; Aldrich, P. E.; Timmermans, P. B. M. W. M. *Biochem. Biophys. Res. Commun.* **1990**, *172*, 1195–1202.
93. Smith, R. G.; Cheng, K.; Schoen, W. R.; Pong, S.-S.; Hickey, G.; Jacks, T.; Butler, B.; Chan, W. W.-S.; Chaung, L.-Y. P.; Judith, F. et al. *Science* **1993**, *260*, 1640–1643.
94. Evans, B. E.; Rittle, K. E.; Bock, M. G.; DiPardo, R. M.; Freidinger, R. M.; Whitter, W. L.; Lundell, G. F.; Veber, D. F.; Anderson, P. S.; Chang, R. S. L. et al. *J. Med. Chem.* **1988**, *31*, 2235–2246.
95. Bock, M. G.; DiPardo, R. M.; Evans, B. E.; Rittle, K. E.; Whitter, W. L.; Veber, D. F.; Andeson, P. S.; Freidinger, R. M. *J. Med. Chem.* **1989**, *32*, 13–16.
96. Aquino, C. J.; Armour, D. R.; Berman, J. M.; Birkemo, L. S.; Carr, R. A. E.; Croom, D. K.; Dezube, M.; Dougherty, R. W.; Ervin, G. N.; Grizzle, M. K. et al. *J. Med. Chem.* **1996**, *39*, 562–569.
97. Doherty, A. M.; Patt, W. C.; Edmunds, J. J.; Berryman, K. A.; Reisdorph, B. R.; Plummer, M. S.; Shahripour, A.; Lee, C.; Cheng, X.-M.; Walker, D. M. et al. *J. Med. Chem.* **1995**, *38*, 1259–1263.
98. De, B.; Plattner, J. J.; Bush, E. N.; Jae, H.-S.; Diaz, G.; Johnson, E. S.; Perun, T. J. *J. Med. Chem.* **1989**, *32*, 2038–2041.
99. Yamamura, Y.; Ogawa, H.; Chihara, T.; Kondo, K.; Onogawa, T.; Nakamura, S.; Mori, T.; Tominaga, M.; Yabuuchi, Y. *Science* **1991**, *252*, 572–574.
100. Valentine, J. J.; Nakanishi, S.; Hageman, D. L.; Snider, R. M.; Spencer, R. W.; Vinick, F. J. *Bioorg. Med. Chem. Lett.* **1992**, *2*, 333–338.
101. Collins, J. L.; Dambek, P. J.; Goldstein, S. W.; Faraci, W. S. *Bioorg. Med. Chem. Lett.* **1992**, *2*, 915–918.
102. Gully, D.; Canton, M. M.; Boigegrain, R.; Jeanjean, F.; Molimard, J. C.; Poncelete, M.; Gueudet, C.; Heaulem, M.; Leris, R.; Brouard, A. et al. *Proc. Natl. Acad. Sci. USA* **1993**, *229*, 23–28.
103. Perlman, S.; Schambye, H. T.; Rivero, R. A.; Greenlee, W. J.; Hjorth, S. A.; Schwartz, T. W. *J. Biol. Chem.* **1995**, *270*, 1493–1496.
104. Evans, B. E.; Leighton, J. J.; Rittle, K. E.; Gilbert, K. F.; Lundell, G. F.; Gould, N. P.; Hobbs, D. W.; DiPardo, R. M.; Veber, D. F.; Pettitbone, D. J. et al. *J. Med. Chem.* **1992**, *35*, 3919–3927.
105. Vara Prasad, J. V. N.; Para, K. S.; Lunney, E. A.; Ortwine, D. F.; Dunbar, J. B.; Ferguson, D.; Tummino, P. J.; Hupe, D.; Tait, B. D.; Domagala, J. M. et al. *J. Am. Chem. Soc.* **1995**, *116*, 6989–6990.
106. Farmer, P. S.; Ariens, E. J. *Trends Pharmacol. Sci.* **1982**, *3*, 362–365.
107. Horwell, D. C. *Bioorg. Med. Chem. Lett.*, **1993**, *3*(5).
108. Rees, D. C. *Curr. Med. Chem.* **1994**, *1*, 145–158.
109. Wexler, R. R.; Greenlee, W. J.; Irvin, J. D.; Goldberg, M. R.; Prendergast, K.; Smith, R. D.; Timmermans, P. B. *J. Med. Chem.* **1996**, *39*, 625–656.
110. Chaurasia, C.; Misse, G.; Tessel, R.; Doughty, M. B. *J. Med. Chem.* **1994**, *37*, 2242–2248.
111. Nieps, S.; Michel, M. C.; Dove, S.; Buschauer, A. *Bioorg. Med. Chem. Lett.* **1995**, *5*, 2065–2070.

112. Williams, P. D.; Anderson, P. S.; Ball, R. G.; Bock, M. G.; Carroll, L. A.; Lee Chiu, S.-H.; Clineschmidt, B. V.; Culberson, J. C.; Erb, J. M.; Evans, B. E. et al. *J. Med. Chem.* **1994**, *37*, 565–571.
113. Greer, J.; Erickson, J. W.; Baldwin, J. J.; Varney, M. D. *J. Med. Chem.* **1994**, *37*, 1035–1054.
114. Colman, P. M. *Curr. Opin. Struct. Biol.* **1994**, *4*, 868–874.
115. Verlinde, L. M. J.; Hol, W. G. J. *Structure* **1994**, *2*, 577–587.
116. Navia, M. A.; Murcko, M. A. *Curr. Opin. Struct. Biol.* **1992**, *2*, 202–210.
117. Appelt, K.; Bacquet, R. J.; Barlett, C. A.; Booth, C. L. J.; Freer, S. T.; Fuhry, M. A.; Gehring, M. R.; Herrmann, S. M.; Howland, E. F.; Janson, C. A. et al. *J. Med. Chem.* **1991**, *34*, 1925–1934.
118. Hruby, V. J. *Acc. Chem. Res.* **2001**, *34*, 389–397.
119. Roe, D.; Kuntz, I. D. *Pharm. News* **1995**, *2*, 13–15.
120. Ajay, A.; Murko, M. A. *J. Med. Chem.* **1995**, *38*, 4967–4973.
121. Klebe, G. *Curr. Med. Chem.* **2000**, *7*, 861–888.
122. Gane, P. J.; Dean, P. M. *Curr. Opin. Struct. Biol.* **2000**, *10*, 401–404.
123. Schwartz, T. W.; Gether, U.; Schamby, H. T.; Hjorth, S. A. *Curr. Pharm. Design* **1995**, *1*, 355–372.
124. Humblet, C.; Mirzadegan, T. *Annu. Rep. Med. Chem.* **1992**, *27*, 291–300.
125. Findley, J.; Eliopoulos, E. *Trends Pharmacol. Sci.* **1990**, *11*, 492–499.
126. Hilbert, M. F.; Trumpp-Kallmeyer, S.; Hoflack, J.; Bruinvels, A. *Trends Pharmacol. Sci.* **1993**, *14*, 7–12.
127. Kontoyianni, M.; Lybrand, T. P. *Med. Chem. Res.* **1993**, *3*, 407–418.
128. Probst, W. C.; Snyder, L. A.; Schuster, D. I.; Brosius, J.; Sealfon, S. C. *DNA Cell Biol.* **1992**, *11*, 1–20.
129. Moereels, H.; Jannsen, P. A. J. *Med. Chem. Res.* **1993**, *3*, 335–343.
130. Attwood, T. K.; Findlay, J. B. C. *Protein Eng.* **1994**, *7*, 195–203.
131. Savares, T. M.; Fraser, C. M. *Biochem. J.* **1992**, *283*, 1–19.
132. Marshall, G. R. *Biopolymers* **2001**, *60*, 246–277.
133. Henderson, R.; Baldwin, J. M.; Ceska, T. A.; Semlin, F.; Beckman, E.; Downing, K. H. *J. Mol. Biol.* **1990**, *213*, 899–929.
134. Underwood, D. J.; Strader, C. D.; Rivero, R.; Patchett, A.; Greenlee, W.; Pendergrast, K. *Chem. Biol.* **1994**, *1*, 211–221.
135. Fong, T. M.; Yu, H.; Cascieri, M. A.; Underwood, D.; Swain, C. J.; Strader, C. D. *J. Biol. Chem.* **1994**, *269*, 14957–14961.
136. Kopin, A. S.; McBride, E. W.; Quinn, S. M.; Kolakowski, L. F.; Beinborn, M. *J. Biol. Chem.* **1995**, *270*, 5019–5023.
137. Kong, H.; Raynor, K.; Yano, H.; Takeda, J.; Bell, G. I.; Reisine, T. *Proc. Natl. Acad. Sci. USA* **1994**, *91*, 8042–8046.
138. Mouillac, B.; Chin, B.; Balestre, M.-N.; Elands, J.; Trumpp-Kallmeyer, S.; Hoflack, J.; Hibert, M.; Jard, S.; Barberis, C. *J. Biol. Chem.* **1995**, *270*, 25771–25777.
139. Kyle, D. J.; Chakravarty, S.; Sinako, J. A.; Stormann, T. M. *J. Med. Chem.* **1994**, *37*, 1347–1354.
140. Pang, Y.-P.; Cusack, B.; Groshan, K.; Richelson, E. *J. Biol. Chem.* **1996**, *271*, 15060–15068.
141. Yang, Y.; Dickinson, C.; Haskell-Luevano, C.; Gantz, I. *J. Biol. Chem.* **1997**, *272*, 23000–23010.
142. Leung, D.; Abbenante, G.; Fairlie, D. P. *J. Med. Chem.* **2000**, *43*, 305–341.
143. James, M. N.; Sielecki, A. R. *J. Mol. Biol.* **1983**, *163*, 299–361.
144. Suguna, K.; Bott, R. R.; Padlan, E. A.; Subramanian, E.; Sheriff, S.; Cohen, G. H.; Davis, D. R. *J. Mol. Biol.* **1987**, *196*, 877–900.
145. Blundell, T. L.; Jenkins, J. A.; Sewell, B. T.; Pearl, L. H.; Cooper, J. B.; Tickle, I. J.; Veerpandian, B.; Wood, S. P. *J. Mol. Biol.* **1990**, *211*, 919–941.
146. Sielecki, A. R.; Fujinaga, M.; Read, R. J.; James, M. N. *J. Mol. Biol.* **1991**, *219*, 671–692.
147. Sielecki, A. R.; Fedorov, A. A.; Boodhoo, A.; Andreeva, N. S.; James, M. N. *J. Mol. Biol.* **1990**, *214*, 143–170.
148. Sham, H. L.; Bolis, G.; Stein, H. H.; Fesik, S. W.; Marcotte, P. A.; Plattner, J. J.; Remple, C. A.; Greer, J. *Med. Chem.* **1988**, *31*, 284–295.
149. Abdel-Meguid, S. S. *Med. Res. Rev.* **1993**, *13*, 731–778.
150. Cooper, J. B.; Foundling, S. I.; Watson, F. I.; Sibanda, B. L.; Blundell, T. L. *Biochem. Soc. Trans.* **1987**, *15*, 751–754.
151. Sielecki, A. R.; Hayakawa, K.; Fujinaga, M.; Murphy, M. E.; Fraser, M.; Muir, A. K.; Carilli, C. T.; Lewicki, J. A.; Baxter, J. D.; James, M. N. *Science* **1989**, *243*, 1346–1351.
152. Plummer, M. S.; Shahripour, A.; Kaltenbronn, J. S.; Lunney, E. A.; Steinbaugh, B. A.; Hamby, J. M.; Hamilton, H. W.; Sawyer, T. K.; Humblet, C.; Doherty, A. E. et al. *J. Med. Chem.* **1995**, *38*, 2893–2905.
153. Rasetti, V.; Cohen, N. C.; Rueger, H.; Boschke, R.; Maibaum, J.; Cumin, F.; Fuhrer, W.; Wood, J. M. *Bioorg. Med. Chem. Lett.* **1996**, *6*, 1589–1594.
154. Weber, A. E.; Halgren, T. A.; Doyle, J. J.; Lynch, R. J.; Siegel, P. K. S.; Parsons, W. H.; Greenlee, W. J.; Patchett, A. A. *J. Med. Chem.* **1991**, *34*, 2692–2701.
155. Dutta, A. S.; Gormley, J. J.; McLachlan, P. F.; Major, J. S. *J. Med. Chem.* **1990**, *33*, 2560–2568.
156. Kempf, J.; Sham, H. L. *Curr. Pharm. Des.* **1996**, *2*, 225–246.
157. Erickson, J.; Neidhart, D. J.; VanDrie, J.; Kempf, D. J.; Wang, X. C.; Norbeck, D. W.; Plattner, J. J.; Rittenhouse, J. W.; Turon, M.; Wideburg, N. et al. *Science* **1990**, *249*, 529–533.
158. Lam, P. Y. S.; Jadhav, P. K.; Eyermann, C. J.; Hodge, C. N.; Ru, Y.; Bacheler, L. T.; Meek, J. L.; Otto, M. J.; Rayner, M. M.; Wong, Y. N. et al. *Science* **1994**, *263*, 380–383.
159. Kempf, D. J.; March, K. C.; Denissen, J. F.; McDonald, E.; Vasavanonda, S.; Flentge, C. A.; Green, B. E.; Fino, L.; Park, C. H.; Kong, X.-P. et al. *Proc. Natl. Acad. Sci. USA* **1995**, *92*, 2484–2488.
160. Vacca, J. P.; Dorsey, B. D.; Schlief, W. A.; Levin, R. B.; McDaniel, S. L.; Darke, P. L.; Zugay, J.; Quintero, J. C.; Blahy, O. M.; Roth, E. et al. *Proc. Natl. Acad. Sci. USA* **1994**, *91*, 4096–4100.
161. Kalish, V. J.; Tatlock, J. H.; Davies, J. F.; Kaldor, S. W., II; Dressman, B. A.; Reich, S.; Pino, M.; Nyugen, D.; Appelt, K.; Musick, L. et al. *Bioorg. Med. Chem. Lett.* **1995**, *5*, 727–732.
162. Kim, E. E.; Baker, C. T.; Dwyer, M. D.; Murko, M. A.; Rao, B. G.; Tung, R. D.; Navia, M. A. *J. Am. Chem. Soc.* **1995**, *117*, 1181–1182.
163. Vazquez, M. L.; Bryant, M. L.; Clare, M.; Decrescenzo, G. A.; Doherty, E. M.; Freskos, J. N.; Getman, D. P.; Houseman, K. A.; Julien, J. A.; Kocan, G. P. et al. *J. Med. Chem.* **1995**, *38*, 581–584.
164. Rutenbar, E.; Fauman, E. B.; Keenan, R. J.; Fong, S.; Furth, P. S.; Ortiz de Montellano, P. R.; Meng, E.; Kuntz, I. D.; DeCamp, D. L.; Salto, R. et al. *J. Biol. Chem.* **1993**, *268*, 15343–15346.
165. Vara Prasad, J. V. N.; Para, K. S.; Lunney, E. A.; Ortwine, D. F.; Dunbar, J. B.; Ferguson, D., Jr.; Tummino, P. J.; Hupe, D.; Tait, B. D.; Domagala, J. M. et al. *J. Am. Chem. Soc.* **1994**, *116*, 6989–6990.

166. Thaisrivongs, S.; Watenpaugh, K. D.; Howe, W. J.; Tomich, P. K.; Dolak, L. A.; Chong, K. T.; Tomich, C. S. C.; Tomaselli, A. G.; Turner, S. R.; Strohbach, J. W. et al. *J. Med. Chem.* **1995**, *38*, 3624–3637.
167. Skulnick, H. I.; Johnson, P. D.; Howe, W. J.; Tomich, P. K.; Chong, K. T.; Watenpaugh, K. D.; Janakiraman, M. N.; Dolak, L. A.; McGrath, J. P.; Lynn, J. C. et al. *J. Med. Chem.* **1995**, *38*, 4968–4971.
168. Tummino, P. J.; Vara Prasad, J. V. N.; Ferguson, D.; Nouhan, C.; Graham, N.; Domagala, J. M.; Ellsworth, E.; Gadja, C.; Hagen, S. E.; Lunney, E. A. et al. *Bioorg. Med. Chem.* **1996**, *4*, 1401–1410.
169. Wlodawer, A.; Miller, M.; Jaskolski, M.; Sathyanarayana, B. K.; Baldwin, E.; Weber, I. T.; Selk, L. M.; Clawson, L.; Schneider, J.; Kent, S. B. *Science* **1989**, *245*, 616–621.
170. Miller, M.; Schneider, J.; Sathyanarayana, B. K.; Toth, M. V.; Marshall, G. R.; Clawson, L.; Selk, L.; Kent, S. B. H.; Wlodawer, A. *Science* **1989**, *246*, 1149–1152.
171. Jaskolski, M.; Tomasselli, A. G.; Sawyer, T. K.; Staples, D. J.; Heinrikson, R. L.; Schneider, J.; Kent, S. B. H.; Wlodawer, A. *Biochemistry* **1991**, *30*, 1600–1609.
172. Swain, A. L.; Miller, M. M.; Green, J.; Rich, D. H.; Schneider, J.; Kent, S. B. H.; Wlodawer, A. *Proc. Natl. Acad. Sci. USA* **1990**, *87*, 8805–8809.
173. Fitzgerald, P. M. D.; McKeever, B. M.; Van Middlesworth, J. F.; Springer, J. P.; Heimbach, J. C.; Leu, C.-T.; Herber, W. K.; Dixon, A. F.; Darke, P. L. *J. Biol. Chem.* **1990**, *265*, 14209–14219.
174. Weber, P. C.; Ohlendorf, D. H.; Wendoloski, J. J.; Salemme, F. R. *Science* **1989**, *243*, 85–88.
175. Stubbs, M. T.; Bode, W. *Trends Cardiovasc. Med.* **1995**, *5*, 157–166.
176. Weber, P. C.; Lee, S.-L.; Lewandowski, F. A.; Schadt, M. C.; Chang, C.-H.; Kettner, C. A. *Biochemistry* **1995**, *34*, 3750–3757.
177. Bode, W.; Turk, D.; Karshikov, A. *Protein Sci.* **1992**, *1*, 426–471.
178. Arni, R. K.; Padmanabhan, K.; Padmanabhan, K. P.; Wu, T. P.; Tulinsky, A. *Biochemistry* **1993**, *32*, 4727–4737.
179. Maryanoff, B. E.; Qiu, X.; Pamanhabhan, K. P.; Tulinsky, A.; Almond, H. R., Jr.; Andrade-Gordon, P.; Greco, M. N.; Kauffman, J. A.; Nicolaou, K. C.; Liu, A. et al. *Proc. Natl. Acad. Sci. USA* **1993**, *90*, 8052–8084.
180. Wiley, M. R.; Chirgadze, N. Y.; Clawson, D. K.; Craft, T. J.; Gifford-Moore, D. S.; Jones, N. D.; Olkowski, J. L.; Schacht, A. L.; Weir, L. C.; Smith, G. F. *Bioorg. Med. Chem. Lett.* **1995**, *5*, 2835–2840.
181. Brandstetter, H.; Turk, D.; Hoeffken, H. W.; Grosse, D.; Sturzebecher, J.; Martin, P. D.; Edwards, B. F. P.; Bode, W. *J. Mol. Biol.* **1992**, *266*, 1085–1099.
182. Hilpert, K.; Ackermann, J.; Banner, D. W.; Gast, A.; Gubernator, K.; Hadvary, P.; Labler, L.; Muller, K.; Schmid, G.; Tschopp, T. B. et al. *J. Med. Chem.* **1994**, *37*, 3889–3901.
183. Padmanabhan, K.; Padmanabhan, K. P.; Tulinsky, A. *J. Mol. Biol.* **1993**, *232*, 947–966.
184. Stubbs, M. T.; Huber, R.; Bode, W. *FEBS Lett.* **1995**, *375*, 103–107.
185. Chen, Z. G.; Bode, W. *J. Mol. Biol.* **1983**, *164*, 283–311.
186. Bode, W.; Meyer, E., Jr.; Powers, J. C. *Biochemistry* **1989**, *28*, 1951–1963.
187. Bernstein, P. R.; Andiski, D.; Bradley, P. K.; Bryant, C. B.; Ceccarelli, C.; Damewood, J. R., Jr.; Earley, R.; Edwards, P. D.; Feeney, S.; Gomes, B. C. et al. *J. Med. Chem.* **1994**, *37*, 3313–3326.
188. Edwards, P. D.; Meyer, E. F., Jr.; Vijayalakshmi, J.; Tuthill, P. A.; Andisik, D. A.; Gomes, B.; Strimpler, A. *J. Am. Chem. Soc.* **1992**, *114*, 1854–1863.
189. Navia, M. A.; McKeever, B. M.; Springer, J. P.; Lin, T.-Y.; Williams, H. R.; Fluder, E. M.; Dorn, C. P.; Hoogsteen, K. *Proc. Natl. Acad. Sci. USA* **1989**, *86*, 7–11.
190. Takahashi, L. H.; Radhakrishnan, R.; Rosenfield, R. E., Jr.; Meyer, E. F.; Trainor, D. A. *J. Am. Chem. Soc.* **1989**, *111*, 3368–3374.
191. Peisach, E.; Casebier, D.; Gallion, S. L.; Furth, P.; Petsko, G. A.; Hogan, J. C., Jr.; Ringe, D. *Science* **1995**, *269*, 66–69.
192. Edwards, P. D.; Bernstein, P. R. *Med. Res. Rev.* **1994**, *14*, 127–194.
193. Drenth, J.; Jansonius, J. N.; Koekoek, R.; Swen, H. M.; Wolthers, B. G. *Nature* **1968**, *218*, 929–932.
194. Drenth, J.; Kalk, K. H.; Swen, H. M. *Biochemistry* **1976**, *15*, 3731–3738.
195. Schroeder, E.; Phillips, C.; Gaman, E.; Harlos, K.; Crawford, C. *FEBS Lett.* **1993**, *315*, 38–42.
196. Rauber, P.; Angliker, H.; Walker, B.; Shaw, E. *Biochem. J.* **1986**, *239*, 633–640.
197. Musil, D.; Zucic, D.; Turk, D.; Engh, R. A.; Mayr, I.; Huber, R.; Popovic, T.; Turk, V.; Towatari, T.; Katunuma, N. et al. *EMBO J.* **1991**, *10*, 2321–2330.
198. Allair, M.; Chernaia, M.; Malcolm, B. A.; James, M. N. G. *Nature* **1994**, *369*, 72–77.
199. Ator, M. A.; Dolle, R. E. *Curr. Pharm. Des.* **1995**, *1*, 191–210.
200. Wilson, K. P.; Black, J. F.; Thomson, J. A.; Kim, E. E.; Griffith, J. P.; Navia, M. A.; Murcko, M. A.; Chambers, S. P.; Aldape, R. A.; Raybuck, S. S. et al. *Nature* **1994**, *370*, 270–275.
201. Walker, N. P. C.; Talanian, R. V.; Brady, K. D.; Dang, L. C.; Bump, N. J.; Ferenz, C. R.; Franklin, S.; Ghayur, T.; Hackett, M. C.; Hammill, L. D. et al. *Cell* **1994**, *78*, 343–352.
202. Mullican, M. D.; Lauffer, D. J.; Gillespie, R. J.; Matharu, S. S.; Kay, D.; Porritt, G. M.; Evans, P. L.; Golec, J. M. C.; Murcko, M. A.; Luong, Y.-P. et al. *Bioorg. Med. Chem. Lett.* **1994**, *4*, 2359–2364.
203. Dole, R. E.; Singh, J.; Rinker, J.; Hoyer, D.; Prasad, C. V. C.; Graybill, T. L.; Salvino, J. M.; Helaszek, C. T.; Miller, R. E.; Ator, M. A. *J. Med. Chem.* **1994**, *37*, 3863–3866.
204. Dole, R. E.; Prouty, C. P.; Prasad, C. V. C.; Cook, E.; Saha, A.; Morgan Ross, T.; Salvino, J. M.; Helaszek, C. T.; Ator, M. A. *J. Med. Chem.* **1996**, *39*, 2438–2440.
205. Rotonda, J.; Nicholson, D. W.; Fazil, K. M.; Gallant, M.; Gareau, Y.; Labelle, M.; Peterson, E. P.; Rasper, D. M.; Ruel, R.; Vaillancourt, J. P. et al. *Nat. Struct. Biol.* **1996**, *3*, 619–625.
206. Matthews, B. W. *Acc. Chem. Res.* **1988**, *21*, 333–340.
207. Holmes, M. A.; Matthews, B. W. *J. Mol. Biol.* **1982**, *160*, 623–639.
208. Weaver, L. H.; Kester, W. R.; Matthews, B. W. *J. Mol. Biol.* **1977**, *114*, 119–132.
209. (a) Holden, H. M.; Tronrud, D. E.; Monzingo, A. F.; Weaver, L. H.; Matthews, B. W. *Biochemistry* **1987**, *26*, 8542; (b) Bartlett, P. A.; Marlowe, C. K. *Biochemistry* **1987**, *26*, 8553–8561.
210. Monzingo, A. F.; Matthews, B. W. *Biochemistry* **1984**, *23*, 5724–5729.
211. Holmes, M. A.; Matthews, B. W. *Biochemistry* **1981**, *20*, 6912–6920.
212. Monzingo, A. F.; Matthews, B. W. *Biochemistry* **1982**, *21*, 3390–3394.
213. Christianson, D. W.; Lipscomb, W. N. *J. Am. Chem. Soc.* **1988**, *110*, 5560–5565.

214. Krapcho, J.; Turk, C.; Cushman, D. W.; Rubin, B.; Powell, J. R.; DeForrest, J. M.; Spitzmiller, E. R.; Karanewsky, D. S.; Duggan, M.; Rovnyak, G. et al. *J. Med. Chem.* **1988**, *31*, 1148–1160.
215. Flynn, G. A.; Beight, D. W.; Mehdi, S.; Koehl, J. R.; Giroux, E. L.; French, J. F.; Hake, P. W.; Dage, R. C. *J. Med. Chem.* **1993**, *36*, 2420–2423.
216. Beckett, R. P.; Davidson, A. H.; Drummond, A. H.; Huxley, P.; Whittaker, M. *Drug Disc. Today* **1996**, *1*, 16–26.
217. Lovejoy, B.; Hassell, A. M.; Luther, M. A.; Weigl, D.; Jordon, S. R. *Biochemistry* **1994**, *33*, 8207–8217.
218. Stams, T.; Spurlino, J. C.; Smith, D. L.; Wahl, R. C.; Ho, T. F.; Qoronfleh, M. W.; Banks, T. M.; Rubin, B. *Nat. Struct. Biol.* **1994**, *1*, 119–123.
219. Borkakoti, N.; Winkler, F. K.; Williams, D. H.; D'Arcy, A.; Broadhurst, M. J.; Brown, P. A.; Johnson, W. H.; Murray, E. J. *Nat. Struct. Biol.* **1994**, *1*, 106–110.
220. Lovejoy, B.; Cleasby, A.; Hassell, A. M.; Longley, K.; Luther, M. A.; Weigl, D.; McGeehan, G.; McElroy, A. B.; Drewry, D.; Lambert, M. H. et al. *Science* **1994**, *263*, 375–377.
221. Van Doren, S. R.; Kurochkin, A. V.; Qi-Zhuang, Ye; Johnson, L. L.; Hupe, D. J.; Zuiderweg, E. R. P. *Biochemistry* **1993**, *32*, 13109–13122.
222. Gooley, P. R.; O'Connell, J. F.; Marcy, A. I.; Cuca, G. C.; Salowe, S. P.; Bush, B. L.; Hermes, J. D.; Esser, C. K.; Hagmann, W. K.; Springer, J. P. et al. *Nat. Struct. Biol.* **1994**, *1*, 111–119.
223. Becker, J. W.; Marcy, A. I.; Rokosz, L. L.; Axel, M. G.; Burbaum, J. J.; Fitzgerald, P. M. D.; Cameron, P. J.; Esser, C. K.; Hagmann, W. K.; Hermes, J. D. et al. *Protein Sci.* **1995**, *4*, 1966–1976.
224. Dhanaraj, V.; Ye, Q.-Z.; Johnson, L. L.; Hupe, D. J.; Ortwine, D. F.; Dunbar, J. B., Jr.; Rubin, J. R.; Pavlosky, A.; Humblet, C.; Blundell, T. L. *Structure* **1996**, *4*, 375–386.
225. Gromis-Ruth, F. X.; Stocker, W.; Huber, R.; Zwilling, R.; Bode, W. *J. Mol. Biol.* **1993**, *229*, 945–968.
226. Grams, R.; Diver, V.; Yiotakis, A.; Yiallouros, I.; Vassiliou, S.; Zwilling, R.; Bode, W.; Stocker, W. *Nat. Struct. Biol.* **1996**, *3*, 671–675.
227. Beszant, B.; Bird, J.; Gaster, L. M.; Harper, G. P.; Hughes, I.; Karran, E. H.; Markwell, R. E.; Miles-Williams, A. J.; Smith, S. A. *J. Med. Chem.* **1993**, *36*, 4030–4039.
228. Bird, J.; De Mello, R. C.; Harper, G. P.; Hunter, D. J.; Karran, E. H.; Markwell, R. E.; Miles-Williams, A. J.; Rahman, S. S.; Ward, R. W. *J. Med. Chem.* **1994**, *37*, 158–169.
229. Brown, F. K.; Brown, P. J.; Bickett, D. M.; Chambers, C. L.; Davies, H. G.; Deaton, D. N.; Drewry, D.; Foley, F.; McElroy, A. B.; Gregson, M. et al. *J. Med. Chem.* **1994**, *37*, 674–688.
230. Castelhano, A. L.; Billedeau, R.; Dewdney, N.; Donnelly, S.; Horne, S.; Kurz, L. J.; Liak, T. J.; Martin, R.; Uppington, R.; Yuan, Z. et al. *Bioorg. Med. Chem. Lett.* **1995**, *5*, 1415–1420.
231. Sawyer, T. K. *Curr. Med. Chem. Anticancer Agents* **2004**, *4*, 449–455.
232. Taylor, S. S.; Radzio-Andzelm, E. *Structure* **1994**, *2*, 345–355.
233. Mauro, L. J.; Dixon, J. E. *Trends Biochem. Sci.* **1994**, *19*, 152–155.
234. Shenolikar, S. *Annu. Rev. Cell Biol.* **1994**, *10*, 55–86.
235. O'Keefe, S. J.; O'Neill, E. A. *Perspect. Drug Disc. Des.* **1994**, *2*, 57–84.
236. Marshall, M. S. *Trends Biochem. Sci.* **1993**, *18*, 250–255.
237. Neer, E. J. *Protein Sci.* **1994**, *3*, 3–14.
238. Overduin, M.; Rios, C. B.; Mayer, B. J.; Baltimore, D.; Cowburn, D. *Cell* **1992**, *70*, 697–704.
239. Waksman, G.; Shoelson, S. E.; Pant, N.; Cowburn, D.; Kuriyan, J. *Cell* **1993**, *72*, 779–790.
240. Xu, R. X.; Word, J. M.; Davis, D. G.; Rink, M. J.; Willard, M. J., Jr.; Gampe, R. T., Jr. *Biochemistry* **1995**, *34*, 2107–2121.
241. Pascal, S. M.; Singer, A. U.; Gish, G.; Yamazaki, T.; Shoelson, S. E.; Pawson, T.; Kay, L. E.; Forman-Kay, J. D. *Cell* **1994**, *77*, 461–472.
242. Zhou, M.-M.; Meadows, R. P.; Logan, T. M.; Yoon, H. S.; Wade, W. S.; Ravichandran, K. S.; Burakoff, S. J.; Fesik, S. W. *Proc. Natl. Acad. Sci. USA* **1995**, *92*, 7784–7788.
243. Tong, L.; Warren, T. C.; King, J.; Betageri, R.; Rose, J.; Jakes, S. *J. Mol. Biol.* **1996**, *256*, 601–610.
244. Nolte, R. T.; Eck, M. J.; Schlessinger, J.; Shoelson, S. E.; Harrison, S. C. *Nat. Struct. Biol.* **1996**, *3*, 364–374.
245. Lee, C.-H.; Kominos, D.; Jacques, S.; Margolis, B.; Schlessinger, J.; Shoelson, S. E.; Kuriyan, J. *Structure* **1994**, *2*, 423–438.
246. Rahuel, J.; Gay, B.; Erdmann, D.; Strauss, A.; Garcia-Echeverria, C.; Furet, P.; Caravatti, G.; Fretz, H.; Schoepfer, J.; Grutter, M. G. *Nat. Struct. Biol.* **1996**, *3*, 586–589.
247. Narula, S. S.; Yuan, R. W.; Adams, S. E.; Green, O. M.; Green, J.; Philips, T. B.; Zydowsky, L. D.; Botfield, M. C.; Hatada, M.; Laird, E. R. et al. *Structure* **1995**, *3*, 1061–1073.
248. Hatada, M. H.; Lu, X.; Laird, E. R.; Green, J.; Morgenstern, J. P.; Lou, M.; Marr, C. S.; Phillips, T. B.; Ram, M. K.; Theriault, K. et al. *Nature* **1995**, *377*, 32–38.
249. Yu, H.; Rosen, M. K.; Shin, T. B.; Seidel-Dugan, C.; Brugge, J. S.; Schreiber, S. L. *Science* **1992**, *258*, 1655–1668.
250. Musacchio, A.; Saraste, M.; Wilmann, M. *Nat. Struct. Biol.* **1994**, *1*, 546–551.
251. Wu, X.; Knudsen, B.; Feller, S. M.; Zheng, J.; Sali, A.; Cowburn, D.; Hanafusa, H.; Kuriyan, J. *Structure* **1995**, *3*, 215–226.
252. Goudreau, N.; Cornille, F.; Duchesne, M.; Parker, F.; Tocque, B.; Garbay, C.; Roques, B. P. *Nat. Struct. Biol.* **1994**, *1*, 898–907.
253. Maignan, S.; Guilloteau, J.-P.; Fromage, N.; Arnoux, B.; Becquart, J.; Ducruix, A. *Science* **1995**, *268*, 291–293.
254. Hubbard, S. R.; Wei, L.; Ellis, L.; Hendrickson, W. A. *Nature* **1994**, *372*, 746–754.
255. Zheng, J.; Knighton, D. R.; Xuong, H.-H.; Taylor, S. S.; Sowadski, J. M.; Ten Eyck, L. F. *Protein Sci.* **1993**, *2*, 1559–1573.
256. Russo, A. A.; Jeffrey, P. D.; Patten, A. K.; Massague, J.; Pavletich, N. P. *Nature* **1996**, *38*, 325–331.
257. Zhang, F.; Strand, A.; Robbins, D.; Cobb, M. H.; Goldsmith, E. J. *Nature* **1994**, *367*, 704–710.
258. Jia, Z.; Barford, D.; Flint, A. J.; Tonks, N. K. *Science* **1995**, *268*, 1754–1758.
259. Yuvaniyama, J.; Denu, J. M.; Dixon, J. E.; Saper, M. A. *Science* **1996**, *272*, 1328–1331.
260. Goldberg, J.; Huang, H.-B.; Kwon, Y.-G.; Greengard, P.; Nairn, A. C.; Kuriyan, J. *Nature* **1995**, *376*, 745–753.
261. Eck, M. J.; Dhe-Paganon, S.; Trub, T.; Nolte, R. T.; Shoelson, S. E. *Cell* **1996**, *65*, 695–705.
262. Zhou, M.-M.; Huang, B.; Olejniczak, E. T.; Meadows, R. P.; Shuker, S. B.; Miyazaki, M.; Trub, T.; Shoelson, S. E.; Fesik, S. W. *Nat. Struct. Biol.* **1996**, *3*, 388–393.
263. Pflugl, G.; Kallen, J.; Scuirmer, T.; Jansonius, J. N.; Zurini, M. G. M.; Walkinshaw, M. D. *Nature* **1993**, *361*, 91–94.
264. Van Duyne, G. D.; Standaert, R. G.; Karplus, P. M.; Schreiber, S. L.; Clardy, J. *Science* **1991**, *252*, 839–842.
265. Saltiel, A. R. *Sci. Am. (Sci. Med.)* **1995**, *2*, 58–67.
266. Bridges, A. J. *Chemtracts (Org. Chem.)* **1995**, *8*, 73–107.
267. Huber, H. E.; Koblan, K. S.; Heimbrook, D. C. *Curr. Med. Chem.* **1994**, *1*, 13–34.
268. Cohen, G. B.; Ren, R.; Baltimore, D. *Cell* **1995**, *80*, 237–248.

269. Cantley, L. C.; Songyang, Z. *J. Cell Sci.* **1994**, *18*, 121–126.
270. Shahripour, A.; Plummer, M. S.; Lunney, E. A.; Para, K. S.; Stankovic, C. J.; Rubin, J. R.; Humblet, C.; Fergus, J. H.; Marks, J. S.; Herrera, R. et al. *Bioorg. Med. Chem. Lett.* **1996**, *6*, 1209–1214.
271. Lunney, E. A.; Para, K. S.; Rubin, J. R.; Humblet, C.; Fergus, J. H.; Marks, J. S.; Sawyer, T. K. *J. Am. Chem. Soc.* **1997**, *119*, 12471–12476.
272. Shakespeare, W. C.; Yang, M.; Bohacek, R.; Cerasoli, F.; Stebbins, K.; Sundaramoorthi, R.; Azimioara, M.; Vu, C.; Pradeepan, S.; Metcalf, C. et al. *Proc. Natl. Acad. Sci. USA* **2000**, *97*, 9373–9378.
273. Shakespeare, W. C.; Bohacek, R. S.; Azimioara, M. D.; Macek, K. J.; Luke, G. P.; Dalgarno, D. C.; Hatada, M. H.; Lu, X.; Violette, S. M.; Bartlett, C. et al. *J. Med. Chem.* **2000**, *43*, 3815–3819.
274. Sundaramoorthi, R.; Kawahata, N.; Yang, M.; Shakespeare, W. C.; Metcalf, C. A., III; Wang, Y.; Merry, T.; Eyermann, C. J.; Bohacek, R.; Narula, S. et al. *Peptide Sci.* **2004**, *71*, 717–729.
275. Lesuisse, D.; Deprez. E. Albert, P.; Duc, T. T.; Sortais, B.; Goffe, D.; Jean-Baptiste, V.; Marquette, J.-P.; Schoot, B.; Sarubbi, E.; Lange, G. et al. *Bioorg. Med. Chem. Lett.* **2001**, *11*, 2127–2131.
276. Lesuisse, D.; Deprez, P.; Bernard, D.; Broto, P.; Delettre, G.; Jean-Baptiste, V.; Marquette, J. P.; Sarubti, E.; Schoot, B.; Mandine, E. et al. *Chim. Nouv.* **2001**, *19*, 3240–3241.
277. Lesuisse, D.; Lange, G.; Deprez, P.; Benard, D.; Schoot, B.; Delettre, G.; Marquette, J. P.; Broto, P.; Jean-Baptiste, V.; Bichet, P. et al. *J. Med. Chem.* **2002**, *45*, 2379–2387.
278. Furet, P.; Gay, B.; Caravatti, G.; Garcia-Echeverria, C.; Gay, B.; Rahuel, J.; Schoepfer, J.; Fretz, H. *J. Med. Chem.* **1998**, *41*, 3442–3449.
279. Gay, B.; Suarez, S.; Weber, C.; Rahuel, J.; Fabbro, D.; Furet, P.; Caravatti, G.; Schoepfer, J. *J. Biol. Chem.* **1999**, *274*, 23311–23315.
280. Furet, P.; Caravatti, G.; Denholm, A. A.; Faessler, A.; Fretz, H.; Garcia-Echeverria, C.; Gay, B.; Irving, E.; Press, N. J.; Rahuel, J. et al. *Bioorg. Med. Chem. Lett.* **2000**, *10*, 2337–2341.
281. Schoepfer, J.; Gay, B.; Caravatti, G.; Garcia-Echeverria, C.; Fretz, H.; Rahuel, J.; Furet, P. *Bioorg. Med. Chem. Lett.* **1998**, *8*, 2865–2870.
282. Shen, K.; Keng, Y. F.; Wu, L.; Guo, X. L.; Lawrence, D. S.; Zhang, Z. Y. *J. Biol. Chem.* **2001**, *276*, 47311–47319.
283. Larsen, S. D.; Barf, T.; Liljebris, C.; May, P. D.; Ogg, D.; O'Sullivan, T. J.; Palazuk, B. J.; Schostarez, H. J.; Stevens, F. C.; Bleasdale, J. E. *J. Med. Chem.* **2002**, *45*, 598–622.
284. Bleasdale, J. E.; Ogg, D.; Palazuk, B. J.; Jacob, C. S.; Swanson, M. L.; Wang, X. Y.; Thompson, D. P.; Conradi, R. A.; Mathews, W. R.; Laborde, A. L. et al. *Biochemistry* **2001**, *40*, 5642–5654.
285. Andersen, H. S.; Olsen, O. H.; Iversen, L. F.; Sorensen, A. L.; Mortensen, S. B.; Christensen, M. S.; Branne, S.; Hansen, T. K.; Lau, J. F.; Jeppesen, L. et al. *J. Med. Chem.* **2002**, *45*, 4443–4459.
286. Lund, I. K.; Andersen, H. S.; Iversen, L. F.; Olsen, O. H.; Moller, K. B.; Pedersen, A. K.; Ge, Y.; Holsworth, D. D.; Newman, M. J.; Axe, F. U. et al. *J. Biol. Chem.* **2004**, *279*, 24226–24235.
287. James, M. N. G.; Sielecki, A. R.; Hayakawa Kotoko, M. H.; Gelb, H. *Biochemistry* **1992**, *31*, 3872–3886.
288. Veerapandian, B.; Cooper, J. B.; Sali, A.; Blundell, T. L. *J. Mol. Biol.* **1990**, *216*, 1017–1029.
289. Cooper, J. B.; Foundling, S. I.; Hemmings, A.; Blundell, T.; Jones, D. M.; Hallett, A.; Szelke, M. *Eur. J. Biochem.* **1987**, *169*, 215–221.
290. Suguna, K.; Padlan, E. A.; Smith, C. W.; Carlson, W. D.; Davies, D. R. *Proc. Natl. Acad. Sci. USA* **1987**, *84*, 7009–7013.
291. Abad-Zapatero, C.; Rydel, T. J.; Erickson, J. *Proteins* **1990**, *8*, 62–81.
292. Abad-Zapatero, C.; Rydel, T. J.; Neidhart, D. J.; Luly, J.; Erickson, J. W. *Adv. Exp. Med. Biol.* **1991**, *306*, 9–21.
293. Rahuel, J.; Priestle, J. P.; Grutter, M. G. *J. Struct. Biol.* **1991**, *107*, 227–236.
294. Kempf, D. J.; Norbeck, D. W.; Codacovi, L.; Wang, X. C.; Kohlbrenner, W. E.; Wideburg, N. E.; Paul, D. A.; Knigge, M. F.; Vasavanonda, S.; Craig-Kennard, A. et al. *J. Med. Chem.* **1990**, *33*, 2687–2689.
295. Rydel, T. J.; Tulinsky, A.; Bode, W.; Huber, R. *J. Mol. Biol.* **1991**, *221*, 583–601.
296. Bode, W.; Mayr, I.; Baumann, U.; Huber, R.; Stone, S. R.; Hofsteenge, J. *EMBO J.* **1989**, *8*, 3467–3477.
297. Banner, D. W.; Hadvary, P. *J. Biol. Chem.* **1991**, *266*, 20085–20093.
298. Walter, J.; Steigemann, W.; Singh, T. P.; Bartunik, H.; Bode, W.; Huber, R. *Acta Crystallogr. Part B* **1982**, *38*, 1462–1472.
299. Bode, W.; Turk, D.; Sturzebecher, J. *Eur. J. Biochem.* **1990**, *193*, 175–182.
300. Meyer, E.; Cole, G.; Radhakrishnan, R.; Epp, O. *Acta Crystallogr. Part B* **1988**, *44*, 26–38.
301. Edwards, P. D.; Andisik, D. W.; Strimpler, A. M.; Gomes, B.; Tuthill, P. A. *J. Med. Chem.* **1996**, *39*, 1112–1124.
302. Kester, W. R.; Matthews, B. W. *J. Biol. Chem.* **1977**, *252*, 7704–7710.
303. Spurlino, J. C.; Smallwood, A. M.; Carlton, D. D.; Banks, T. M.; Vavra, K. J.; Johnson, J. S.; Cook, E. R.; Falvo, J.; Wahl, R. C.; Pulvino, T. A. et al. *Proteins Struct. Funct. Genet.* **1995**, *19*, 98–109.
304. Plummer, M. S.; Holland, D. R.; Shahripour, A.; Lunney, E. A.; Fergus, J. H.; Marks, J. S.; McConnell, P.; Mueller, W. T.; Sawyer, T. K. *J. Med. Chem.* **1997**, *40*, 3719–3725.
305. Eck, M. J.; Shoelson, S. E.; Harrison, S. C. *Nature* **1993**, *362*, 87–91.
306. Karlsson, R.; Zheng, J.; Xuong, N.; Taylor, S. S.; Sowadski, J. M. *Acta Crystallogr. Part D* **1993**, *49*, 381–388.
307. De Bondt, H. L.; Rosenblatt, J.; Jancarik, J.; Jones, H. D.; Morgan, D. O.; Kim, S. H. *Nature* **1993**, *363*, 595–602.
308. Barford, D.; Flint, A. J.; Tonks, N. K. *Science* **1995**, *263*, 1397–1404.
309. Iversen, L. F.; Andersen, H. S.; Branner, S.; Mortensen, S. B.; Peters, G. H.; Norris, K.; Olsen, O. H.; Jeppesen, C. B.; Lundt, B. F.; Ripka, W. et al. *J. Biol. Chem.* **2000**, *275*, 10300–10307.

Biography

Tomi K Sawyer received a BSc degree in Chemistry at Moorhead State University (now Minnesota State University – Moorhead) and PhD in Organic Chemistry at the University of Arizona. His research has integrated synthetic chemistry, drug design, structural biology, chemoinformatics, biochemistry, cell biology, and in vivo disease models with a focus on cancer. Tomi's drug discovery track record includes contributions to clinical candidates and/or noteworthy molecular tools for several therapeutic targets, including GPCRs (melanocortin), aspartyl proteases (renin and HIV protease), and protein kinases (Src and Abl). He has published more than 200 scientific articles, reviews, commentaries, monographs, and books. Tomi is an inventor of more than 50 issued patents and patent filings. He worked at Upjohn Company and Parke-Davis/Warner-Lambert (now both Pfizer Global Research & Development), and is currently Senior-Vice President, Drug Discovery, at ARIAD Pharmaceuticals. He is concurrently adjunct professor, Chemistry as well as Biochemistry & Molecular Biology, University of Massachusetts, and also adjunct professor, Cancer Biology, at University of Massachusetts School of Medicine. Tomi has served on the highlights advisory panel of Nature Reviews Drug Discovery and the editorial advisory boards of Trends in the Pharmacological Sciences, Expert Reviews in Molecular Medicine, Expert Opinion on Investigational Drugs, Journal of Medicinal Chemistry, Chemistry and Biology, Current Medicinal Chemistry (Anti-Cancer Agents), Current Organic Synthesis, Expert Reviews in Molecular Medicine, Expert Opinion on Therapeutic Patents (Oncology), Drug Design and Discovery, Pharmaceutical Research, Molecular Biotechnology, and Biopolymers (Peptide Science). Most recently, Tomi was appointed Editor-in-Chief, Chemical Biology & Drug Design.

2.16 Bioisosterism

C G Wermuth, P Ciapetti, B Giethlen, and P Bazzini, Prestwick Chemical, Illkirch, France

© 2007 Elsevier Ltd. All Rights Reserved.

2.16.1	**Introduction**	650
2.16.1.1	The Scope of Bioisosterism	650
2.16.1.2	The Importance of Analog Design	651
2.16.1.3	Analog Design and Bioisosterism	651
2.16.2	**Currently Encountered Bioisosteric Modifications**	652
2.16.2.1	Ring Equivalents	652
2.16.2.1.1	Introduction	652
2.16.2.1.2	Bioisosteres of pyridine	654
2.16.2.1.3	Imidazo[1,2-a]pyridine bioisosteres	656
2.16.2.1.4	Pyridazine bioisosteres	656
2.16.2.1.5	Bioisosteres of other heterocycles	657
2.16.2.2	Groups with Similar Polar Effects: Functional Equivalents	659
2.16.2.2.1	Carboxylic acid bioisosteres	659
2.16.2.2.2	Bioisosteres of carboxylic esters	663
2.16.2.2.3	Carboxamide bioisosteres	666
2.16.2.2.4	Urea and thiourea bioisosteres	668
2.16.2.2.5	Phenol and catechol bioisosteres	670
2.16.2.2.6	Sulfonamides	672
2.16.2.3	Reversal of Functional Groups	673
2.16.3	**Modification of the Core Structure**	675
2.16.3.1	Reorganization of Ring Systems	675
2.16.3.1.1	Spiro derivatives and bi- or tricyclic systems	675
2.16.3.1.2	Splitting benzo compounds (benzo cracking)	678
2.16.3.1.3	Restructuring ring systems	678
2.16.3.1.4	Ring dissociation	679
2.16.3.2	Scaffold Hopping	680
2.16.3.2.1	Successful examples of scaffold hopping	680
2.16.3.2.2	Computational approaches to scaffold hopping	681
2.16.4	**Analysis of the Modifications Resulting from Bioisosterism**	685
2.16.4.1	Structural Parameters	686
2.16.4.2	Electronic Parameters	687
2.16.4.3	Solubility Parameters	687
2.16.5	**Anomalies in Isosterism**	688
2.16.5.1	Fluorine–Hydrogen Isosterism	688
2.16.5.1.1	Steric aspects	688
2.16.5.1.2	Electronic aspects	688
2.16.5.1.3	Absence of d orbitals	689
2.16.5.1.4	Lipophilicity	690
2.16.5.1.5	Case studies	691
2.16.5.2	Exchange of Ether Oxygen and Methylene Group: The Friedman Paradox	692
2.16.6	**Minor Metalloids-Toxic Isosteres**	694
2.16.6.1	Carbon–Silicon Bioisosterism	694
2.16.6.1.1	Toxicological aspects	694
2.16.6.1.2	Silicon in medicinal chemistry	694
2.16.6.1.3	Reactivity toward S_N2 and metabolic attacks	695
2.16.6.1.4	Silicon diols	696

	2.16.6.1.5	Diphenylpiperidine analogs	696
	2.16.6.1.6	Clinical studies	697
	2.16.6.1.7	A few words about C/Si/Ge isostery	697
2.16.6.2		Carbon–Boron Isosterism	698
	2.16.6.2.1	Toxicity	698
	2.16.6.2.2	Interest and applications	698
	2.16.6.2.3	Boranophosphate salts	699
	2.16.6.2.4	Boron-containing molecule designed for boron neutron capture therapy	700
2.16.6.3		Bioisosteres Involving Selenium	700
	2.16.6.3.1	Toxicity	701
	2.16.6.3.2	Applications	703
References			**703**

2.16.1 Introduction

2.16.1.1 The Scope of Bioisosterism

The concept of isosterism is based on the replacement in an active molecule of an atom or a group of atoms by another atom or group of atoms, presenting a comparable electronic and steric arrangement. The physicist Langmuir,[1] who was mainly interested in the physicochemical relationships of isosteric molecules, introduced the term isosterism in 1919. When, in addition to their physicochemical analogy, compounds share some common biological properties, the term bioisosterism, introduced by Friedman in 1951,[2] is used, even if the physicochemical resemblance is only vague. The term 'nonclassical isosterism' is often used interchangeably with the term bioisosterism, for example, when one has to deal with compounds not possessing the same number of atoms, but having in common some key parameters of importance for the activity in a given series. Preparing isosteric and bioisosteric replacements represents one of the main activities of medicinal chemists. Indeed, in the long history of medicinal chemistry, as soon as new natural or synthetic active principles were discovered, numerous attempts to prepare their analogs were undertaken. The main objective of analog design is to end up with compounds with enough resemblance to the model in order to guarantee a similar kind of biological activity. However, some differences from the model compound are not only accepted but are also actually welcomed insofar that they often reflect an improvement. Iterative improvements obtained by bioisosteric changes can yield a gain in potency, selectivity, bioavailability, deletion of an unwanted side effect, ease of administration, chemical stability, or cost of production. Historical examples of analog design are: the progressive passage from the alkaloid cocaine to the local anesthetic procaine; the change from the natural antibiotic penicillin to the orally active large-spectrum antibiotic amoxicillin; or the synthesis of potent HMG-CoA reductase inhibitors ('statins') such as atorvastatin or rosuvastatin inspired from the mevinolin structure (**Figure 1**).

Similar evolutions are found in the development of neuroleptics, antidepressants, or drugs to treat hypertension. Instead of natural models, analog design can use marketed drug structures as leads.

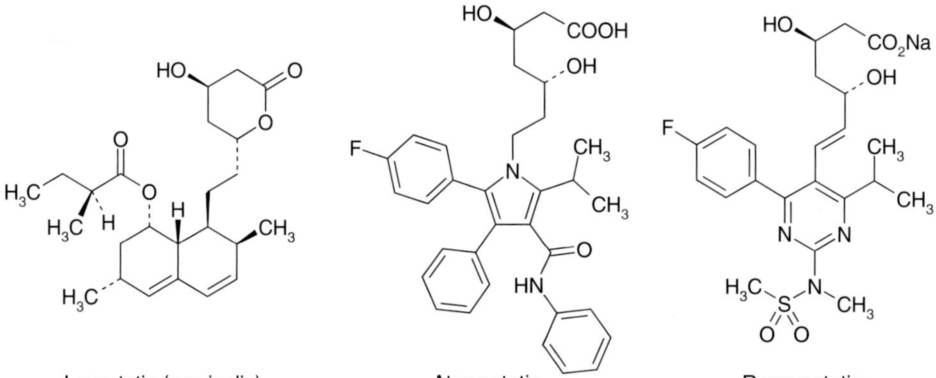

Figure 1 The passage from the natural lovastatin to the synthetic analogs atorvastatin and rosuvastatin.

2.16.1.2 The Importance of Analog Design

In the pharmaceutical industry, when a company launches a pioneer drug (i.e., resulting from an original invention and corresponding to a real therapeutic need), or considerably before then, competitors initiate immediately the search for similarly acting drugs in order to keep their market position. The analogs can be simple copies or display some improvements over the original drug molecule.

A survey of the novel therapeutic small molecules (molecular weight (MW) < 600) that were launched or approved for the first time during the period 2000–03[3,4] reveals that most of them resulted from the continued exploitation of already well-established structural classes (drug analogs) or the rapid expansion of newer structural classes (early phase analogs[5]). The market value of analog drugs can be estimated to be two-thirds of all small molecule sales. The continued preference for analog-based approaches has been criticized as being mainly restricted to the synthesis of 'me-too' compounds.[6] In reality, the situation is more subtle and needs to establish a distinction between three different types of analogs (see below).

2.16.1.3 Analog Design and Bioisosterism

The term 'analog' is derived from the term analogy, which is itself derived from the latin and greek *analogia*, and has been used in natural sciences since 1791 to describe structural and functional similarity.[7] Extended to drugs, this definition implies that the analog of an existing drug molecule shares chemical and therapeutic similarities with the original compound. Formally, this definition anticipates three categories of drug analogs: (1) those presenting chemical and pharmacological similarity; (2) those presenting only chemical similarity; and (3) those displaying similar pharmacological properties but presenting totally different chemical structures.

Analogs belonging to the first category showing both chemical and pharmacological similarities can be considered as direct analogs. These analogs correspond to the class of drugs often referred to as 'me-too' drugs. Usually they are improved versions of a pioneer drug over which they present a pharmacological, pharmacodynamic, or biopharmaceutical advantage. Examples are the angiotensin-converting enzyme (ACE) inhibitors derived from captopril, the histamine H_2 antagonists derived from cimetidine, and the hydroxymethyl-glutaryl coenzyme A (HMG-CoA) reductase inhibitors derived from mevinolin, etc. Such analogs are designed for the same industrial and marketing reasons as those that are valid for any other industrial products such as laptop computers or cars.

The second class of analogs made up of chemically similar molecules and for which we propose the term 'structural analogs,' contains compounds originally prepared as close and patentable analogs of a novel lead, but for which the biological assays revealed totally unexpected pharmacological properties. A historical example of the emergence of a new activity is provided by the discovery of the antidepressant properties of imipramine, which was originally designed as an analog of the potent neuroleptic drug chlorpromazine. Another example, illustrating that chemical similarity does not necessarily mean biological similarity, is that of the steroid hormones testosterone and progesterone, which although being chemically very similar have totally different biological functions, albeit both at nuclear receptors. Observation of an 'emergent' activity can be purely fortuitous or can result from a voluntary and systematic investigation.

For the third class of analogous compounds, chemical similarity is not observed; however, they share common biological properties. We propose the term 'functional analogs' for such compounds. Examples are the neuroleptics chlorpromazine and haloperidol or the tranquilizers diazepam and zopiclone. Despite their totally different chemical structures, they show similar affinities for the dopamine and benzodiazepine receptors, respectively. The design of such drugs is presently facilitated thanks to virtual screening of large libraries of diverse structures.

The search for analogs is helped very often by the concept of isosteric replacement and the derived concept of bioisosterism. Isosteric replacement requires substitution of one atom or group of atoms in the parent compound by another, with similar electronic and steric configuration. A typical example of isosteric replacement is given in **Figure 2** where four clozapine analogs derive one from the other by simple changes, the oxygen atom being successively replaced by the S and CH_2 isosteric groups. The claimed objective of isosterism is to produce an improved version of the pioneer drug; sometimes, however, compounds with modified pharmacological profiles and even with antagonistic properties are obtained.

The term bioisosterism is used to describe replacements of generally more important pieces of the original molecule. Thus, the replacement of the methylated carboxamide function of diazepam by its bioequivalent methylated triazole group in alprazolam (**Figure 3**) preserves the high affinity for the central benzodiazepine receptor.

Finally, taken in its broadest meaning, bioisosterism can consist of a complete replacement of the initial molecular structure by another one, isofunctional, but based on a different scaffold. This approach, called scaffold hopping, is illustrated by the passage of diazepam to zolpidem and to zopiclone (**Figure 4**).

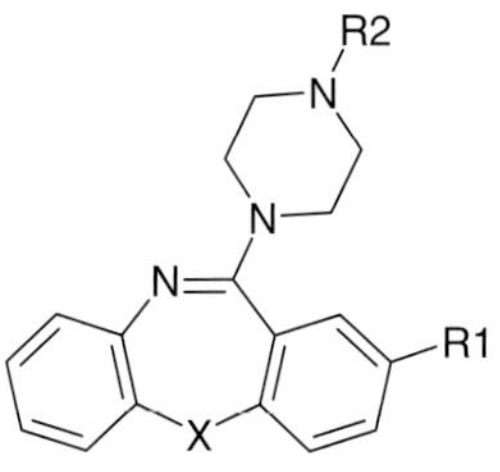

X	R^1	R^2	Compound
S	Cl	CH_3	Clotiapine
O	Cl	CH_3	Loxapine
O	Cl	H	Amoxapine
CH_2	H	CH_3	Perlapine

Figure 2 Isosteric replacements in clozapine analogs.

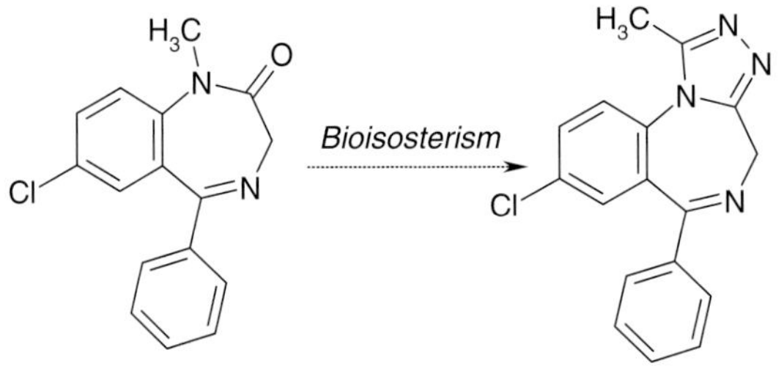

Diazepam Alprazolam

Figure 3 The passage from diazepam to alprazolam represents a typical bioisosteric change.

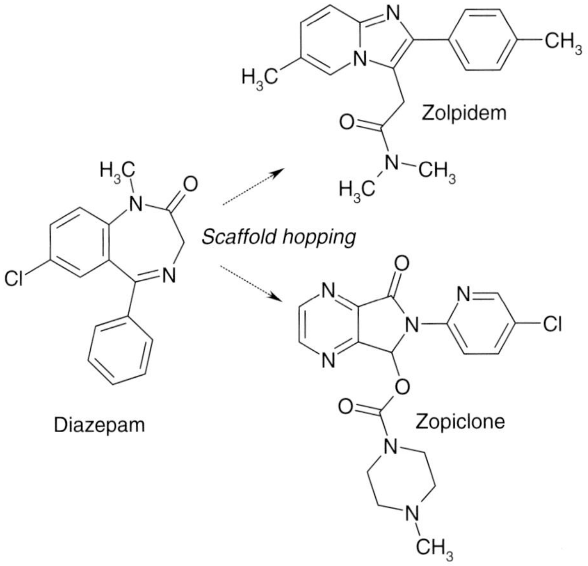

Figure 4 Scaffold hopping can be considered as an extreme application of the bioisosterism concept.

2.16.2 Currently Encountered Bioisosteric Modifications

2.16.2.1 Ring Equivalents

2.16.2.1.1 Introduction

Chemical rings are fundamental for drug molecules. A typical drug molecule consists of a combination of chemical rings, chains, and functional groups. Chemical rings represent an important portion of the drug structures. A brief look at the best-selling drugs in 2004[8] (**Table 1**) gives a better idea of the importance of this class of chemicals.

Table 1 Best selling drugs in 2004[8]

Structure	Drug	Purpose	Company	Global sales $billion
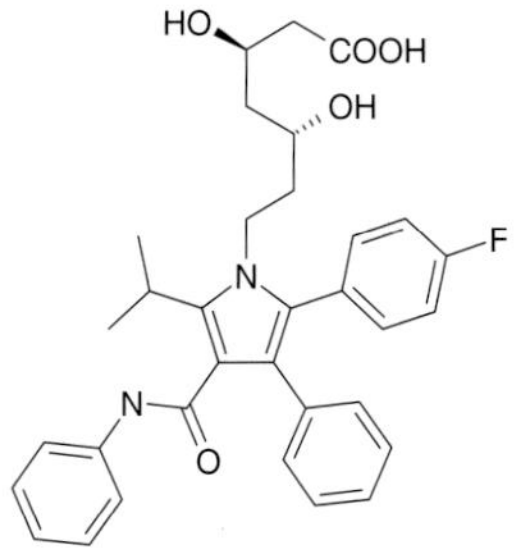	Atorvastatin	Lowers cholesterol	Pfizer	10.3
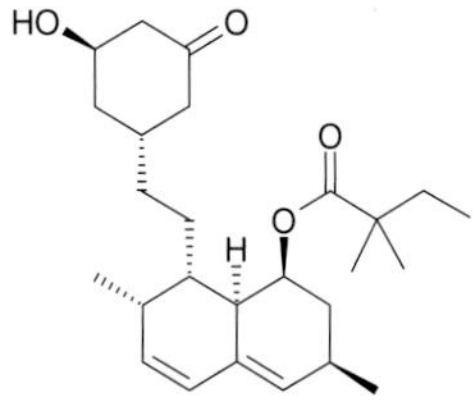	Simvastatin	Lowers cholesterol	Merck	6.1
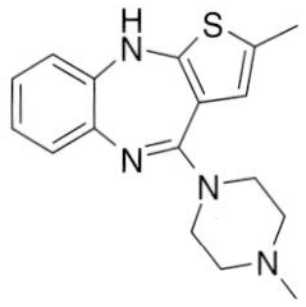	Olanzapine	Antipsychotic	Eli Lilly	4.8
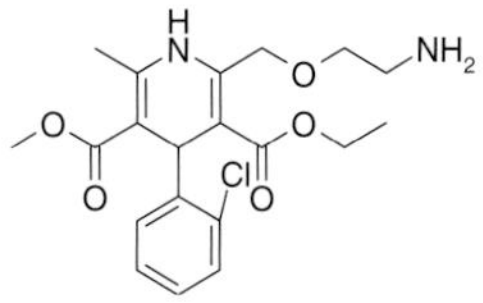	Amlodipine	Lowers blood pressure	Pfizer	4.5
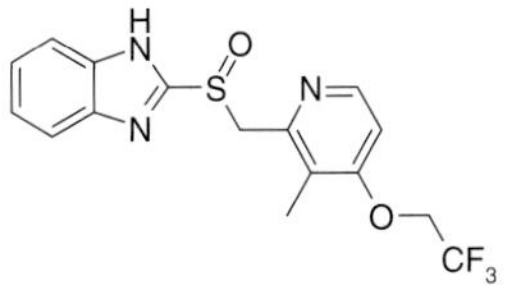	Lansoprazole	Treats ulcers	Takeda & Abbott Laboratories	4.0
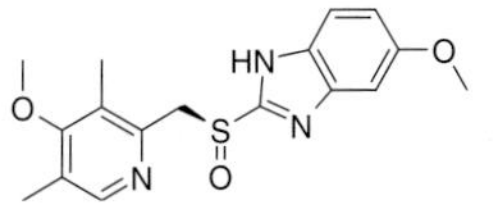	Esomeprazole	Treats ulcers	AstraZeneca	3.8
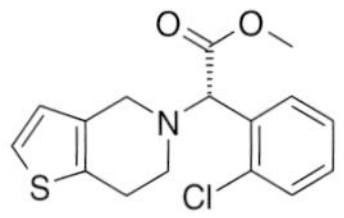	Clopidogrel	Bloodthinner	Sanofi-Aventis & Bristol-Myers Squibb	3.7
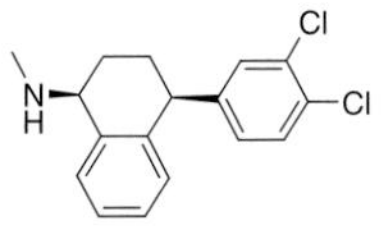	Sertraline	Antidepressant	Pfizer	3.4

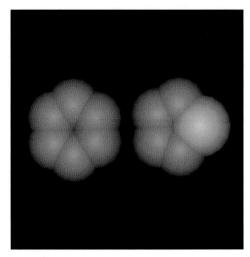

Figure 5 In thiophene the sulfur atom is approximately equivalent to an ethylenic group (size, mass, and capacity to provide an aromatic lone pair).

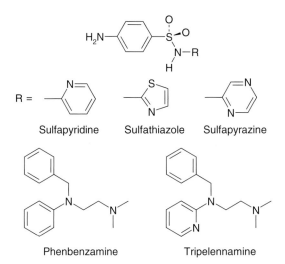

Figure 6 Classical ring equivalents.

The drugs reported in **Table 1** as well as almost all the other marketed drugs present at least one chemical ring. In some cases, the ring is used as a platform supporting the pharmacophoric elements to orient them in the right direction for optimal interaction with the receptor. In other examples, it is the ring itself that possesses intrinsic bioactive properties. There are also some examples where heterocyclic rings are used as bioisosteres of metabolically unstable functional groups.

The substitution of –CH= by –N= or –CH=CH– by –S– in aromatic rings has been one of the most successful applications of classical isosterism. Thus, the equivalence in size and lone pair participation to aromaticity between –CH=CH– (M = 26) and –S– (M = 32) explains the well-known analogy between benzene and thiophene (**Figure 5**). Early examples of ring equivalents are found in the sulfonamide antibacterials with the development of sulfapyridine, sulfapyrimidine, sulfathiazole, etc. (**Figure 6**).

The efforts to replace rings form a large part of medicinal chemistry practice. A typical lead optimization program often involves a trial and error process of replacing chemical functionalities including rings. In many cases this intellectual exercise is based on a chemist's knowledge and recollection of his or her past experience.

2.16.2.1.2 Bioisosteres of pyridine

One of the most used cyclic structures in medicinal chemistry is the pyridine heterocycle. The bioisosteres of this ring are well known as ligands for the central nicotinic cholinergic receptors. The pyridine ring of nicotine can be replaced by different rings like methyl-isoxazole or methyl-isothiazole[9–11] (**Figure 7a**). A novel series of nicotinic agonists was described by Olesen and co-workers.[11] In their paper, the bioisosteric replacement of the isoxazole ring in the (3-methyl-5-isoxazoly)methylene-azacyclic compound (**1**) with pyridine, pyrazine, oxadiazole, or an acyl group resulted in ligands with moderate to high affinity for the central nicotinic cholinergic receptors ($IC_{50} = 2.0$ nM to $IC_{50} > 1000$ nM) (**Figure 7b**).

Two other publications on the bioisosteric replacement of pyridine show similar results. The first paper reports the study of bioisosteric potential of diazines in the field of combined antithrombic thromboxane A_2 synthetase inhibitors and receptor antagonists.[12,13] On the basis of the structure–activity relationships observed in this study, it turned out that only the 2-pyrazinyl, 4-pyridazinyl, and 5-pyrimidinyl systems are appropriate bioisosteric moieties for the 3-pyridyl system in the dual active platelet antiaggregatory compound ridogrel (**Figure 8**). Gohlke and colleagues[12] also observed the bioisosteric potential of diazines in the structure–activity relationships (SARs) of the compound DUB-165. The replacement of the 3-pyridyl group of DUB-165 by a 4-pyridazinyl, 5-pyrimidinyl, or 2-pyrazinyl moiety resulted in ligands retaining affinity for nicotinic acetylcholine receptor (nAChR) subtypes, thus demonstrating that the three isomeric diazines are appropriate bioisosteres of the 3-pyridyl moiety (**Figure 8**).

N-(6-Chloronaphthalen-2-)sulfonylpiperazine derivatives **2** and **3** (**Figure 9**) are potent factor Xa inhibitors. Haginoya and co-workers[14] proposed to replace the pyridine-phenyl or the pyridine-piperidine residue by a fused bicyclic ring that contains an aliphatic amine and a pyridine to yield the compound **4**, which has an interesting factor Xa

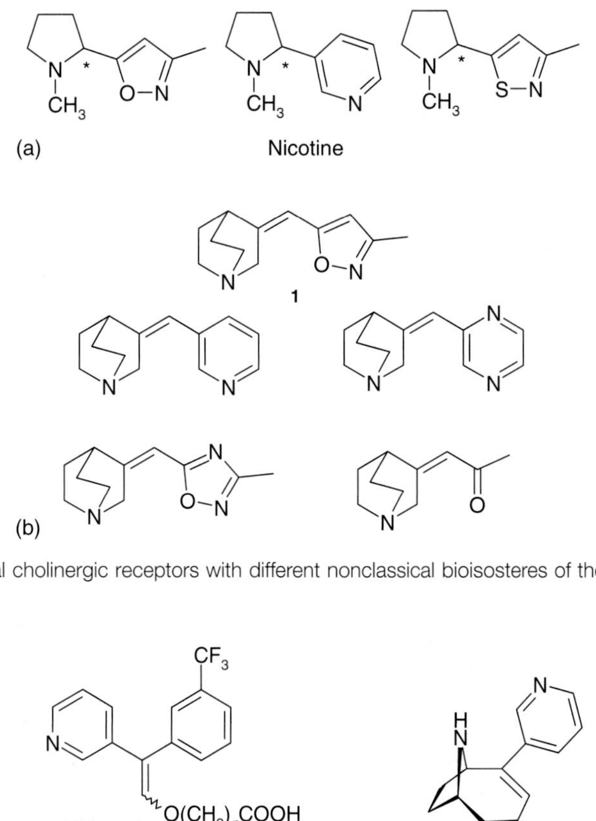

Figure 7 Ligands for central cholinergic receptors with different nonclassical bioisosteres of the pyridine ring.

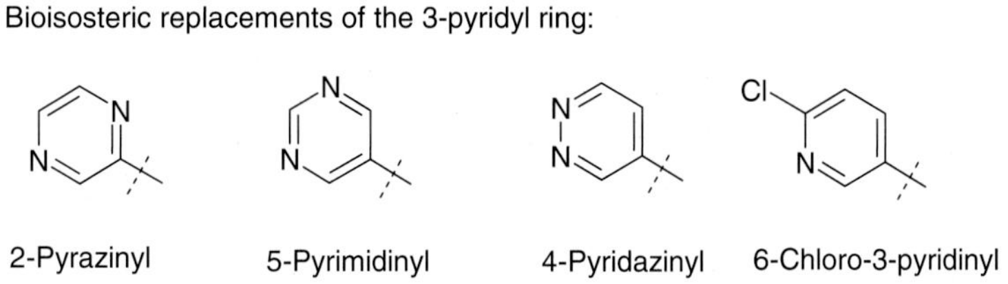

Figure 8 Bioisosteric moieties for the 3-pyridyl ring.

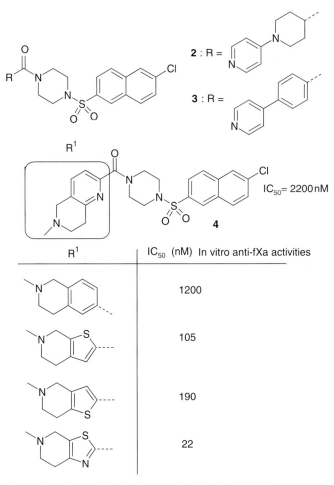

Figure 9 Bioisosteric replacements of the pyridine ring in a series of factor Xa inhibitors.

inhibitor activity. The bioisosteric replacement of the pyridine moiety of the 6-methyl-5,6,7,8-tetrahydro-[1,6]naphthyridine by phenyl, thiophene, or thiazole analogs yielded analogs with similar or better anti-factor Xa activity, but also conserved a moderate bioavailability.

2.16.2.1.3 Imidazo[1,2-a]pyridine bioisosteres

Abe and co-workers[15] reported that the imidazol[1,2-a]pyridine moiety of the basic framework of a class of the nonpeptide bradykinin B2 receptor antagonists (**5**) could be successfully replaced by several heterocyclic bioisosteres (**Figure 10**). Among those, the 1-methyl-2-methoxy-1H-benzimidazole (**8**), 2-methylquinoxaline (**6**), and 2-methylquinoline (**7**) derivatives showed potent B2 binding affinities against both human and guinea pig B2 receptors.

2.16.2.1.4 Pyridazine bioisosteres

Since the antidepressant minaprine was launched, several pyridazine analogs have been proposed and synthesized. Most of them showed different effects on the central nervous system. A series of these compounds (**Figure 11**) are acetylcholinesterase inhibitors with variable bioisosteres of the central pyridazine.[16,17] The replacement by pyridine, 1,2,4-thiadiazole, and triazines yields compounds with weaker but still acceptable activity.

As we have seen, there are several nonclassical bioisosteres that could be found for the pyridine system or the pyridazine system. It is important to have a good rationale to choose the most appropriate analogs to start with. An indication could be given by the comparison of the boiling points of these heterocycles (**Figure 12**) assuming that the more similar the boiling point is, the most appropriate the bioisostere is.[18] For example, while searching for a bioisostere of the pyridine ring, it appears that the best candidates are the pyrimidine, the pyrazine, the 1,3,4-oxadiazole, the

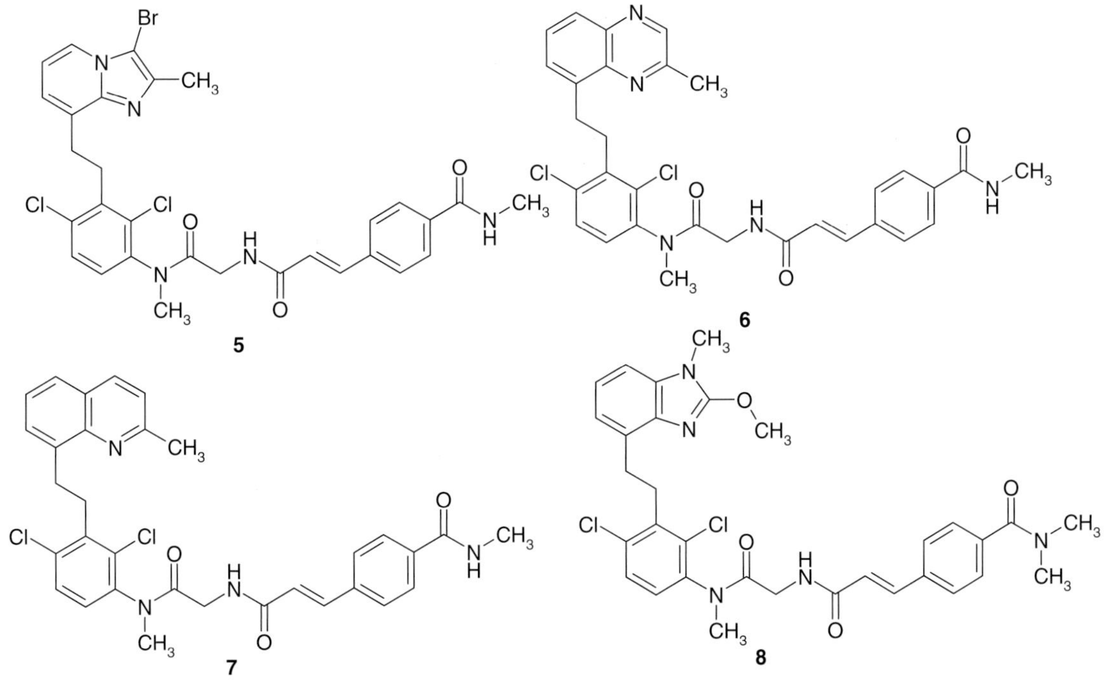

Figure 10 Novel class of orally active nonpeptide bradykinin B$_2$ receptor antagonists showing valuable examples of imidazo[1,2-a]pyridine analogs.

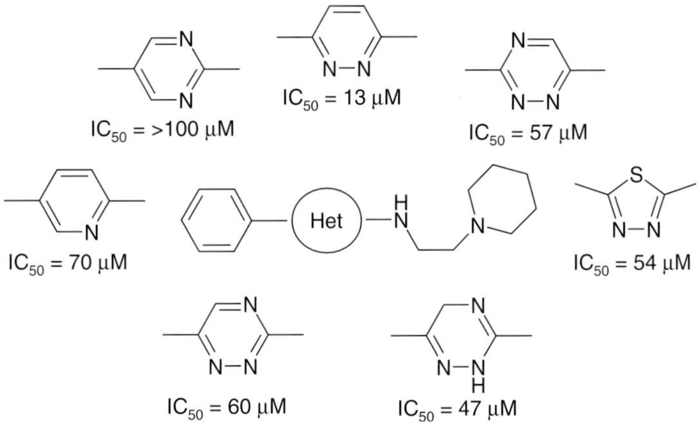

Figure 11 In vitro inhibition of acetylcholinesterase in rat striatum homogenates; pyridine, thiadiazole, and triazine replacement of the pyridazine ring.

1,2,4-oxadiazole, or the 1,2,4-thiadiazole systems. The same is true for the pyridazine ring where this comparison permits the selection of the 1,2,4-triazine or the 1,3,4-thiadiazole rings. In fact, those findings confirm the examples found in the literature and reported earlier in this chapter.

A possible interpretation of these results can be the fact that in the heterocyclic series, the boiling point is correlated to the dipolar moment of the molecule and that, for two heterocyclic rings having the same aromatic geometry, the similarity of the dipolar moments may represent the dominant feature.

2.16.2.1.5 Bioisosteres of other heterocycles

Chemical rings are fundamental to the structure of drug molecules, as a typical drug molecule consists of a combination of chemical rings, chains, and functional groups. The high proportion of chemical rings in drug molecules argues in favor

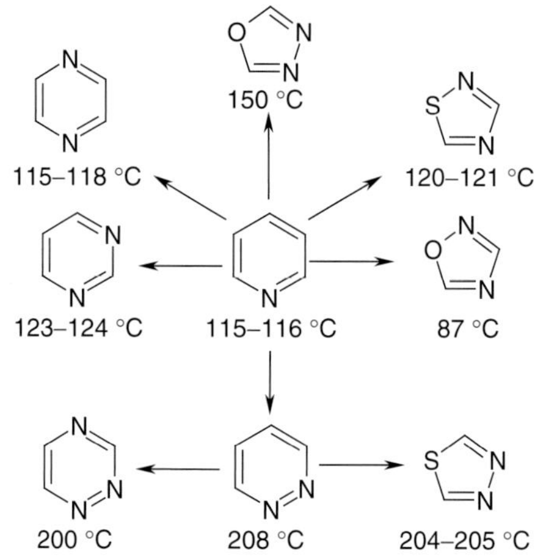

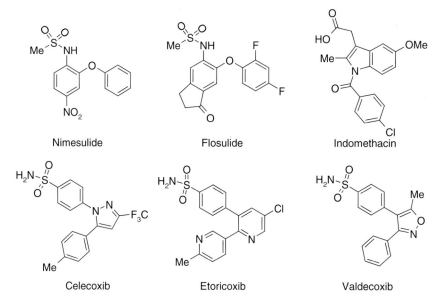

Figure 12 Structures and boiling points of pyridine and pyridazine isosteres.

Figure 13 Examples of selective cyclooxygenase-2 inhibitors showing different ring equivalents.

of the great need to synthesize and find new chemical ring equivalents, whether it is to design a new platform for a better patent position, replace a metabolically unstable moiety, look for a more favorable receptor interaction, or create a better pharmacodynamic profile.

There are many examples of the concept of ring equivalents and so far it seems impossible to rationalize a method that gives an indication of which heterocycle could be the best substitute for another one. The best approach is to look at the examples reported in the literature. One of the most fascinating heterocycle analogies is given by the class of selective cyclooxygenase-2 (COX-2) inhibitors. The comparison of the most potent selective COX-2 inhibitors (**Figure 13**) suggests that isoxazoles, pyridines, and pyrazoles are good bioisosteres of each other as well as nitrophenol, indole, and indanones.[19] Another nice example of heterocycle bioisosteres is given by the oxazolidinone antibacterials. Since scientists at Upjohn discovered linezolid as a potent antibacterial for Gram-positive organisms,[20] many efforts have been devoted to identifying new and potentially improved antibacterial agents. Some of these efforts have used the oxazolidinone silhouette as a starting point,[21] while others have taken advantage of the bioisosterism rationale to replace the central oxazolidinone heterocycle.[22]

The work of Snyder and co-workers, together with other examples found in the literature, shows that several levels of activity can be identified among the different isosteres. It should be kept in mind that the replacement of a ring by

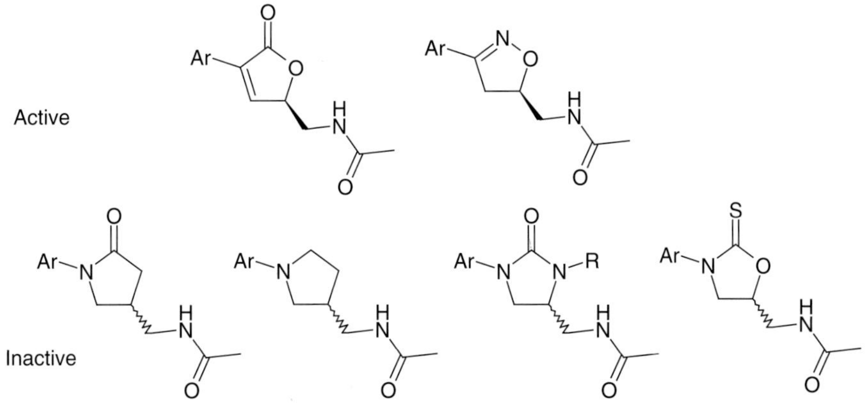

Figure 14 Oxazolidinone bioisosteres synthesized as antibacterial agents.

its isosteres does not always lead to an iso-active or more active compound (**Figure 14**), but, in general, it is possible to obtain at least one or more equivalent systems.[21] In **Table 2** we have listed some more 'exotic' examples of bioisosteric replacements of cyclic systems.[23–25]

2.16.2.2 Groups with Similar Polar Effects: Functional Equivalents

2.16.2.2.1 Carboxylic acid bioisosteres

The term nonclassical isosterism is often used interchangeably with the term bioisosterism, for example, when one has to deal with isosteres that do not possess the same number of atoms, but which have in common some key parameter of importance for the activity in a given series. Thus, the two GABAergic agonists isoguvacine and THIP (**Figure 15**) have similar pharmacological properties to GABA itself. The key parameters in these compounds are the acidic ($pK_a \approx 4$) and the basic (protonated nitrogen) functions with an intercharge distance of ≈ 5.1 Å.

For the carboxylic function of active compounds, three different classes of bioisosteres can be distinguished: the direct derivatives (**Table 3**) such as hydroxamic acids: R–CO–NH–OH, acyl-cyanamides: R–CO–NH–CN, and acyl-sulfonamides: R–CO–NH–SO2–R'; the planar acidic heterocycles (**Table 4**) such as tetrazoles, hydroxy-isoxazoles, etc.; and the nonplanar sulfur- or phosphorus-derived acidic functions (**Table 5**).

2.16.2.2.1.1 Direct derivatives
We consider as direct derivatives different functional groups that have similar specific interactions with the receptor as they maintain an acidic proton and a hydrogen bond acceptor. Among the most common direct derivatives are hydroxamic acids,[32,46] acyl-cyanamides,[40,41] and acyl-sulfonamides[42] in which an acidic NH group replaces the acidic OH group and the carbonyl is left unchanged.

2.16.2.2.1.1.1 Hydroxamic acids The hydroxamic acids have a high chelating power and they have found a wide application as inhibitors of enzymes having a Zn^{2+} in the active site, like matrix metalloproteinases (MMPs),[35–38] tumor necrosis factor alpha (TNF-α) converting enzyme,[39] and histone deacetylase (HDAC)[33,34] yielding potent and bioavailable compounds despite the hydroxamic acid group is particularly prone to hydrolysis, reduction, and glucuronidation.[47] Another interesting example of exploitation is given by the anti-inflammatory hydroxamates bufexamac,[46] ibuproxam,[48] and oxametacin[49] (**Figure 16**). While ibuproxam is classified as prodrug being metabolized to ibuprofen (CONHOH → COOH) in man,[50] oxametacin is a true bioisostere rather than a prodrug being metabolically stable in man.[51,52]

2.16.2.2.1.1.2 Acyl-sulfonamides Acyl-sulfonamides are also quite commonly used in drug discovery and have been employed in several therapeutic areas such as β3 adrenergic receptor agonists,[44] chemokine receptor 1 (CXCR1) inhibitors,[53] hepatitis C virus (HCV) inhibitors,[45] angiotensin receptor agonists,[54] antitumoral compounds with antiproliferative properties,[55,56] and prostaglandin EP4 receptor antagonists.[57]

Acyl-cyanamides are mainly of academic interest.

Table 2 Other ring bioisosteres

Original ring	Bioisostere	Activity	Reference
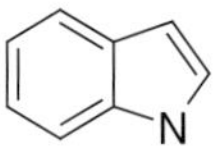 Indole	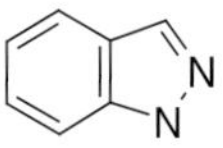 Indazole	5HT3 antagonists	Fludzinski et al.[26]
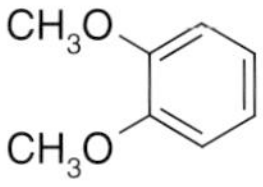 3,4-Dialkoxy-phenyl	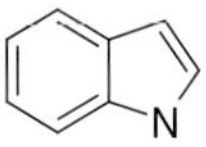 Indole	Phosphodiesterase inhibitors	Blaskó et al.[27]
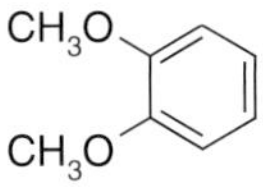 3,4-Dialkoxy-phenyl	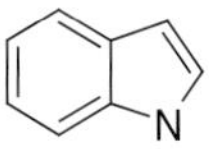 Indole	GABA uptake inhibitors	Kardos et al.[28]
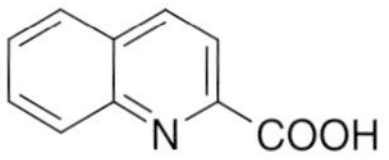 Quinoline-2-carboxylate	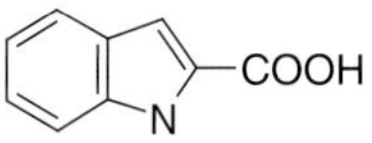 Indole-2-carboxylate	Glycine antagonists	Salituro et al.[29]
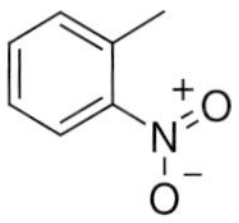 O-Nitro-phenyl	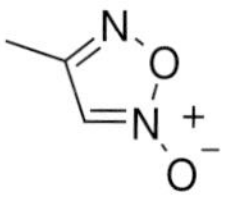 Furoxane	Calcium antagonist	Calvino et al.[30]
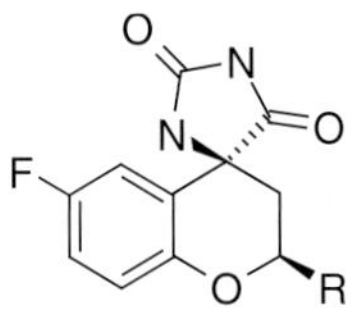 Spiro-hydantoin	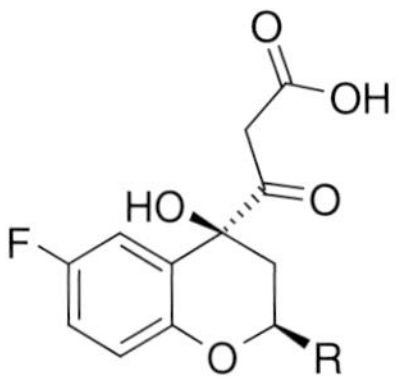 Spiro hydroxyacetic acid unit	Aldose reductase inhibitor	Lipinski et al.[31]
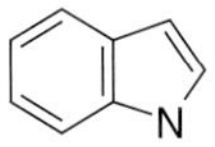 Indole	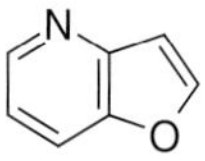 Furo[3,2-b] pyridine	5HT1F agonist	Mathes et al.[25]
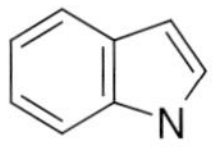 Indole	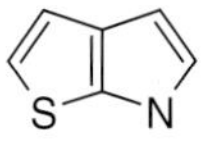 6H-thieno[2,3-b] pyrrole	Hallucinogen serotonin agonist	Blair et al.[23]
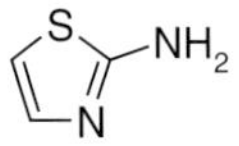 Thiazol-2-ylamine	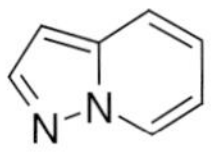 Pyrazolo[1,5-a] pyridine	Dopamine D3 receptor agonist	Löber et al.[24]

2.16.2.2.1.2 Planar acidic heterocycles and aryl derivatives

Table 4 gives the planar acidic heterocycles and aryl derivatives that are commonly used as carboxylic acid bioisosteres. The most commonly used is the tetrazole ring, which has found wide application in different therapeutic fields ranging from type 2 diabetes[69,70] to hepatitis C virus (HCV),[45] malaria,[71] Alzheimer's disease (AD),[72] anxiety treatment,[73] pain management,[74] erectile dysfunction,[74] and inflammation.[74] The tetrazole has been employed to fix different issues, for example, to improve bioavailability,[71] or to improve blood–brain barrier penetration,[73,74] to enhance potency,[70,75] to increase chemical stability,[72] to bring some selectivity (the GABA tetrazole analog inhibits GABA-transaminase, but not

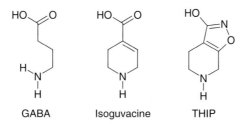

Figure 15 An example of bioisosterism, or nonclassical isosterism: GABA, isoguvacine, and THIP are all agonists for the GABA$_A$ receptor. The 3-hydroxy-isoxazole ring has a comparable acidity to that of a carboxylic acid function.

Table 3 Carboxylic acid isosteres: direct derivatives

Hydroxamic acids	High chelating power	Almquist et al.[32]	
	Histone deacetylase inhibitors	Massa et al.,[33] Lu et al.[34]	
	Matrix metalloproteinases inhibitors	Hanessian et al.,[35] Aranapakam et al.,[36,37] Noe et al.[38]	
	Tumor necrosis factor alpha converting enzyme inhibitors	Duan et al.[39]	
Acyl-cyanamides	Mainly academic interest	von Kohler et al.,[40] Kwon et al.[41]	
Acyl-sulfonamides	Glycine, GABA, and β-alanine analogs	Drummond & Johnson[42]	
	Antiatherosclerotics p$K_a \sim 4.5$	Albright et al.[43]	
	β$_3$-Adrenergic receptor agonist	Uehling et al.[44]	
	Hepatitis C virus inhibitors	Johansson et al.[45]	

succinic semialdehyde dehydrogenase)[76] or to be used as a prodrug.[77] However, in some instances, tetrazole analogs are poorly active.[78]

Hydroxy-isoxazoles and other cognate heterocyclic phenols encompassing an acidity range from 3.0 to 7.1 were incorporated in GABA agonists, antagonists, and uptake inhibitors.[79,80] The experience gained with 3-hydroxy-isoxazoles in the GABA field was also transferable to glutamate receptor ligands and led to selective antagonists for glutamic acid receptor subtypes.[80]

Thiazolidinediones are another class of heterocyclic carboxylic acid surrogates commonly used for PPAR agonists,[65,81] as potent antihyperglycemic and lipid activity modulators.

Other interesting, but less studied heterocyclic surrogates are: 3,5-dioxo-1,2,4-oxadiazolidine,[82] 3-hydroxy-1,2,5-thiadiazoles,[63] and 3-hydroxy-γ-pyrones.[64,83]

2.16.2.2.1.3 Nonplanar sulfur- or phosphorous-derived acidic functions

The most extensive use of phosphonates (**Table 5**) occurred in the design of amino acid neurotransmitter antagonists such as glutamate[86] and GABA$_B$ antagonists.[84] Among a set of cholecystokynin (CCK) antagonists derived from the nonpeptide CCK-B selective antagonist CI-988, Drysdale and colleagues[87] prepared a series of carboxylate surrogates spanning a pK_a range from <1 (sulfonic acid) to >9.5 (thio-1,2,4-triazole). The affinity and the selectivity of the compounds were rationalized by considering pK_a values, charge distribution, and geometry of the respective acid mimics (**Table 6**).

2.16.2.2.1.4 Comparison of pK_a and log P values

An interesting study comparing pK_a and log P values of some aryl phosphonic acids, aryl tetrazoles and aryl sulfonamides has been published by Franz.[88] The value of pK_a and log P are very important parameters in drug design. Several published examples show that the interchange of carboxylic acid for tetrazole and sulfonamide often results in useful drugs. This study indicates that the log P will be lowered by about 1 log unit substituting a phosphonic acid by a carboxylic acid group. There is an increase in acidity, which may be a limiting factor in the absence of active transport.

Table 4 Carboxylic acid isosteres: planar acidic heterocycles and aryl derivatives

Structure	Name	Notes	References
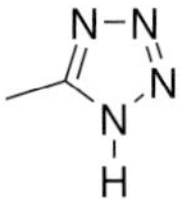	Tetrazoles	Very popular, great number of publications. $pK_a = 6.6–7.2$	Bovy et al.,[58] Marshall et al.[59]
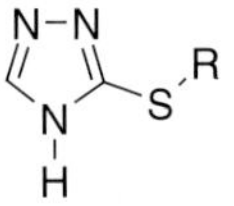	Mercaptoazoles, Sulfinylazoles, Sulfonylazoles	Phosphonate isosteres. pK_a mercapto: 8.2–11.5; pK_a sulfinyl: 5.2–9.8; pK_a sulfonyl: 4.8–8.7	Chen et al.[60]
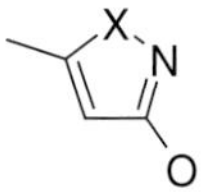 (X = O or S)	Isoxazoles, Isothiazoles	GABA and glutamic acid analogs	Krogsgaard-Larsen et al.,[61] Krogsgaard-Larsen[62]
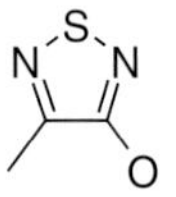	Hydroxy-thiadiazole	Isoxazole isostere. $pK_a \sim 5$	Lunn et al.[63]
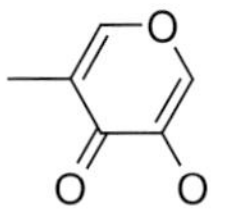	Hydroxy-chromones	Kojic acid derivatives (as GABA agonists)	Atkinson et al.[64]
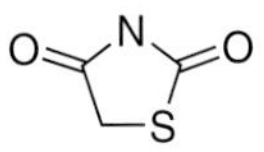	Thiazolidinediones	Dual PPAR α/γ agonists	Henke[65]
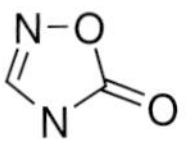	1,2,4-Oxadiazole-5(4H)-ones	Antimycobacterial	Gezginci et al.[66]
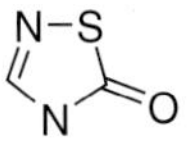	1,2,4-Oxadiazole-5(4H)-thiones	Antimycobacterial	Gezginci et al.[66]
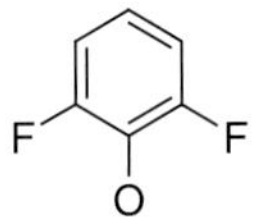	3,5-Difluoro-4-hydroxyphenyl	Aldose reductase inhibitors GABA analog	Nicolaou et al.,[67] Qiu et al.[68]

Table 5 Carboxylic acid isosteres: nonplanar sulfur- or phosphorus-derived acidic functions

Structure	Name	Notes	References
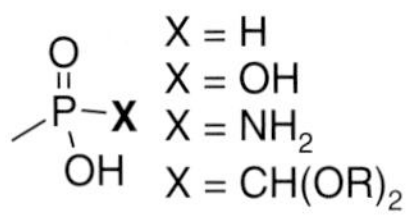 X = H, X = OH, X = NH₂, X = CH(OR)₂	Phosphinates, Phosphonates, Phosphonamides	Many examples in glutamate and in GABA$_B$ antagonist series	Froestl et al.[84]
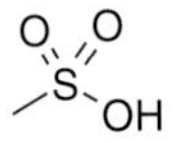	Sulfonates	Sulfonic analogs of GABA and glutamic acid	Rosowsky et al.[85]
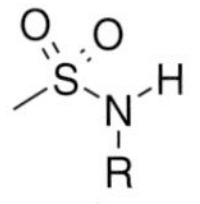	Sulfonamides	Weak acids, used rather as equivalents of phenolic hydroxyls in the design of catecholamine analogs	von Kohler et al.[40]

Replacing a carboxylic acid group by a phosphonic acid group gives a much more acidic compound; replacing it with a sulfonamide group results in a much less acidic compound; and with a tetrazole replacement, acidity is essentially unchanged (**Figure 17**). If one wants to lower the log P, and if increased acidity of the compound was not a limiting factor, substitution of a phosphonic acid group for a carboxylic acid would be a viable approach (**Figure 18**).

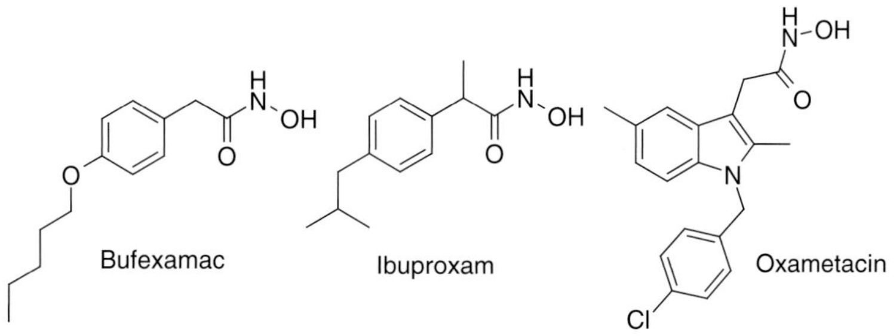

Figure 16 Hydroxamate isosteres of anti-inflammatory drugs.

2.16.2.2.1.5 Diamino-cyclobutene-diones as α-amino carboxylic acid surrogates
Diamino-cyclobutene-dione was proposed by Kinney and co-workers[89] as an original surrogate of the α-amino carboxylic acid function (**Figure 19**).

2.16.2.2.1.6 Carboxylic functions as phosphonate surrogates
Nonhydrolyzable phosphotyrosyl (pTyr) mimetics serve as important components of many competitive inhibitors of protein-tyrosine kinases. To date, the most potent of these inhibitors have relied on phosphonate-based structures to replace the 4-phosphoryl group of the parent pTyr residue (**Figure 20**). Interestingly, it was found that carboxy-based pTyr analogs can be utilized to introduce the anionic oxygen functionality of the parent phosphate. Particularly, when p-(2-malonyl)phenylalanine (Pmf) was incorporated as pTyr replacement in the high-affinity Grb2 SH2 domain binding sequence, potencies approaching those of phosphonate mimetics were obtained.[90]

The above example is an elegant illustration of the possibility to mimic the pyramidal structure of phosphates or phosphonates by means of two planar carboxylic groups, the three-dimensionality originating from the malonic methylenic carbon atom.

2.16.2.2.2 Bioisosteres of carboxylic esters

Ester-containing drug molecules are highly labile in vivo due to the widespread presence of esterases, which can be found in the blood, liver, kidney, and other organs. Bioisosteric substitution of this group has been extensively investigated in order to improve the metabolic stability of an ester-containing drug molecule, leading to improved pharmacokinetic and pharmacodynamic profiles.

The change from ester to amide (procaine → procainamide) is already illustrated above as an example of classical isosterism. Similarly, the lactone ring of the muscarinic agonist pilocarpine was changed into various, still active isosteres such as the corresponding thiolactone, lactam, lactol, and thiolactol.[91] A series of aspirin isosteres has been prepared by replacing the carboxylic ether oxygen successively by a nitrogen, sulfur, or carbon isosteric equivalent.[92] None of the isosteric compounds showed any activity. This result is readily understood since the particular role of aspirin as an acylating agent of the enzyme cyclooxygenase has been demonstrated.[93]

In addition to these classical changes, 1,2,4-oxadiazoles or 1,2,4,-thiadiazoles have been commonly used as carboxylic ester surrogates in series of benzodiazepine and muscarinic[94,95] receptor ligands (**Figure 21**). For muscarinic agonists, numerous successful attempts to replace the oxadiazole ring by other heterocyclic ring systems were published.[96–98] Pilocarpine is widely used as a topical miotic for controlling the elevated intraocular pressure associated with glaucoma. Besides its low lipophilicity, which stimulated the search for prodrugs,[99] pilocarpine has a short duration of action, its lactone ring being rapidly opened to yield pilocarpic acid. In pilocarpine, by substituting the lactone ester function by its carbamate equivalent, a much more stable analog, which is as effective as pilocarpine, was obtained.[100]

Several 1,2,4-oxadiazoles and the other five-membered heterocycles are employed also for other therapeutic indications such as monoamine transporter and opioid receptors,[101,102] 5HT agonists,[103] and bradykinin B$_1$ receptor antagonists.[104] Regarding this last study, it is noteworthy to observe the use of N-2 substituted methyl tetrazole as ester surrogates.

Another uncommon ester bioisostere is the N-methoxy imidoyl nitrile that was reported for a muscarinic M1 agonist (**Figure 22**). An interesting replacement of the ester group with a more metabolically stable one is given by the ethyl oxime ether. In the case of human rhinovirus, this change improved the oral bioavailability of the corresponding ester derivative[105] (**Figure 23**). The change in (−)-cocaine of the carbomethoxy substituent into carbethoxyisoxazole doubles the potency in [3H] mazindol binding and [3H] dopamine uptake (**Figure 24**). Astonishingly the replacement

Table 6 Exploration of the carboxyl isosterism possibilities in a series of CCK antagonists.

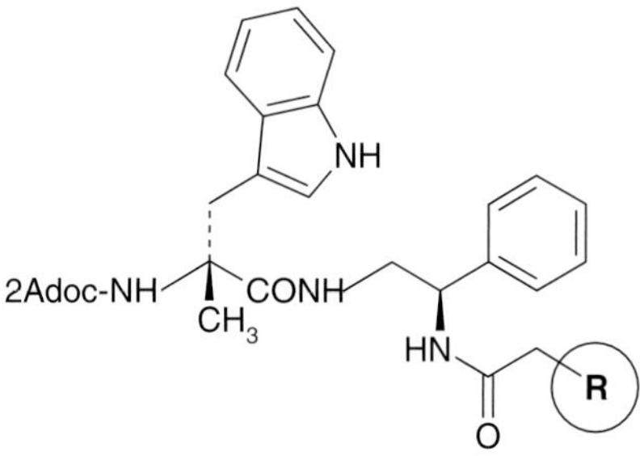

R	IC_{50} (nM) CCK-B	IC_{50} (nM) CCK-A	A/B Ratio	pK_a
–R–CH$_2$–COOH	1.7	4500	2500	5.6
Charge distributed monoanionic acid mimics				
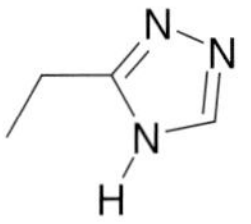 ethyl-triazole	6.0	970	160	5.4
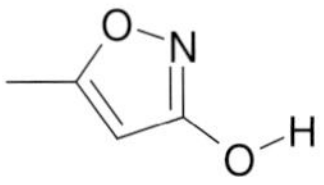 methyl-isoxazol-3-ol	2.6	1700	650	6.5
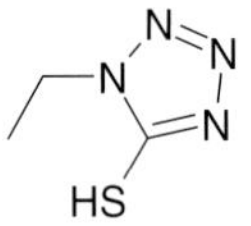 ethyl-mercaptotetrazole	2.4	620	260	4.3
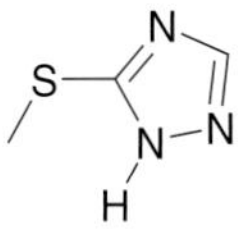 methylthio-triazole	2.5	680	270	>9.5
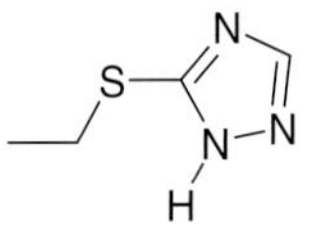 ethylthio-triazole	16	850	53	>9.5
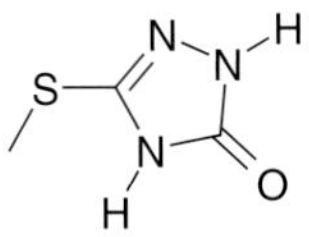 methylthio-triazolone	4.3	660	150	7.7
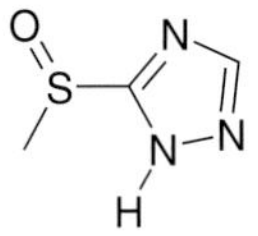 methylsulfinyl-triazole	1.7	940	550	7.0
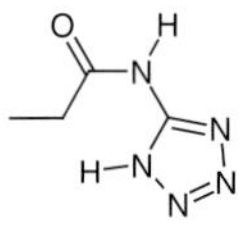 acylaminotetrazole	6.3	1300	200	5.2

Table 6 Continued

R	IC_{50} (nM) CCK-B	IC_{50} (nM) CCK-A	A/B Ratio	pK_a
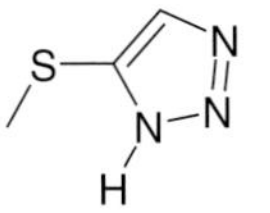 [S-linked 5-methylthio-1H-1,2,3-triazole]	18	600	33	>8.2
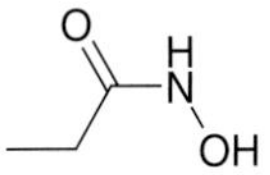 [CH₂-C(O)-NH-OH]	14	1300	93	>9.5
Point charge monoanionic acid mimics				
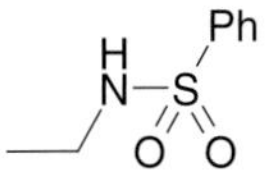 [CH₂-NH-S(O)₂-Ph]	70	300	4.3	>9.5
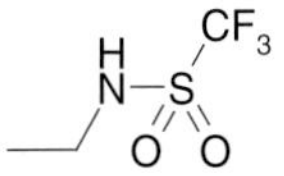 [CH₂-NH-S(O)₂-CF₃]	77	680	9	7.9
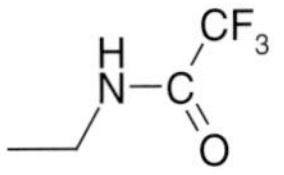 [CH₂-NH-C(O)-CF₃]	110	790	7	>9.5
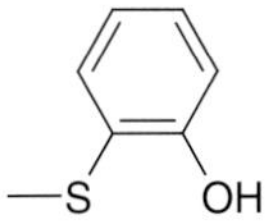 [CH₂-S-(2-hydroxyphenyl)]	80	510	6.4	>9.5
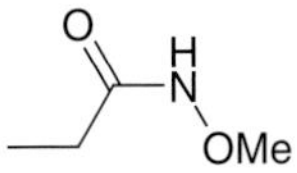 [CH₂-C(O)-NH-OMe]	21	1500	71	>9.5
Tetrahedral acid mimics				
P(O)(OH)₂	27	5200	190	3.4: 7.7
CH₂–P(O)(OH)₂	23	2700	120	3.4: 7.8
P(O)(OH)(OEt)	12	480	40	6.5
P(O)(OH)Me	12	1700	140	3.8
CH₂–P(O)(OH)Me	23	4400	190	3.7
CH₂–SO₃Na	1.3	1010	780	–

of the carbomethoxy group by a chlorovinyl moiety produces a comparable gain in potency thus arguing against the involvement of the carbomethoxy group in H-bonding.[106]

Another rather unusual example of ester isosterism is the replacement of the ether oxygen by a fluoro-nitrogen (**Figure 25a**) as mentioned by Lipinski.[107] Other uncommon examples are found in the replacement of the ester function of acetylcholine by exo-endo amidine functions of 3-aminopyridazines in muscarinic agonists (**Figure 25b**)[108] and of the carbomethoxy group of α-yohimbine (rauwolscine) by an *N*-methylsulfonamide function (**Figure 25c**).[109]

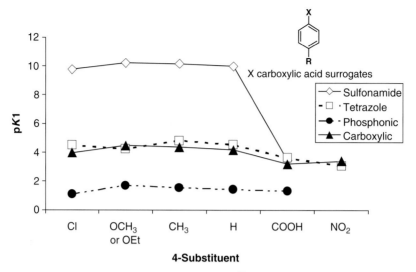

Figure 17 pK_1 values for aromatic compounds (redrawn after Franz[88]).

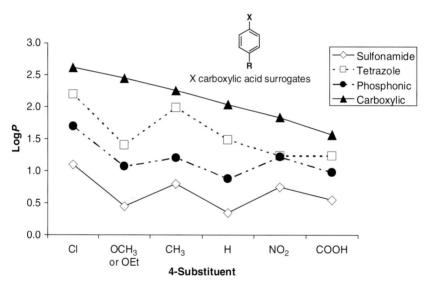

Figure 18 Log P values for aromatic compounds (redrawn after Franz[88]).

Figure 19 3,4-Diamino-3-cyclobutene-1,2-dione as surrogate of the α-amino carboxylic acid function.

2.16.2.2.3 Carboxamide bioisosteres

Carboxamides are usually converted to sulfonamides as illustrated by the synthesis of the hypoglycemic sulfonyl isostere of glybenclamide.[110] The isosteric replacements for peptidic bonds have been summarized by Spatola[111] and by Fauchère.[112] The most used and well-established modifications are: N-methylation, configuration change (D-configuration at Cα), formation of a retroamide or an α-azapeptide, use of aminoisobutyric or dehydroamino acids, replacement of the amide bond by an ester [depsipeptide], ketomethylene, hydroxyethylene or thioamide functional group, carba replacement of the amidic carbonyl, and use of an olefinic double bond (**Figure 26**).

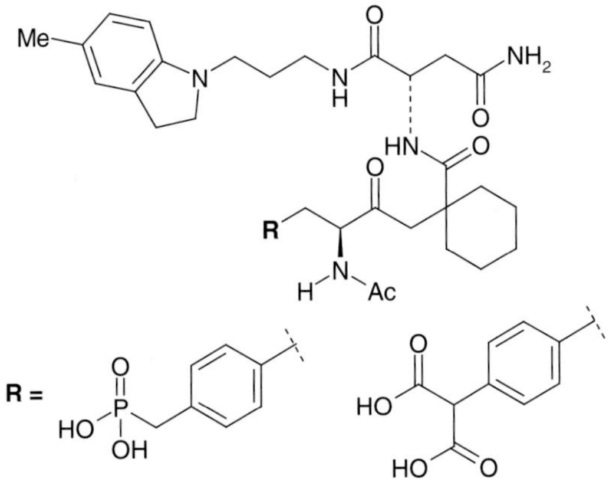

pTyr IC_{50} = 4 nM   Pmf IC_{50} = 12 nM

Figure 20 Malonates as surrogates of phosphonates.

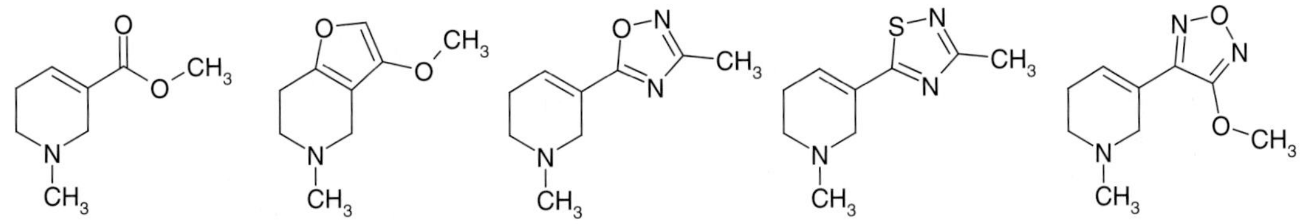

9 **10** **11** **12** **13**

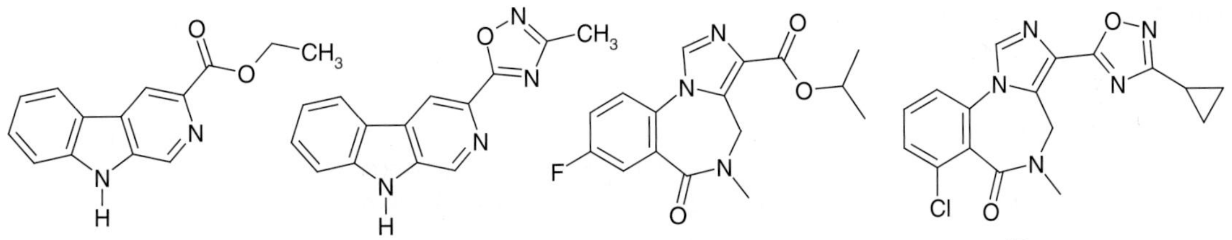

14 **15** **16** **17**

Figure 21 1,2,4-Oxadiazoles and related five-membered heterocycles as ester surrogates.

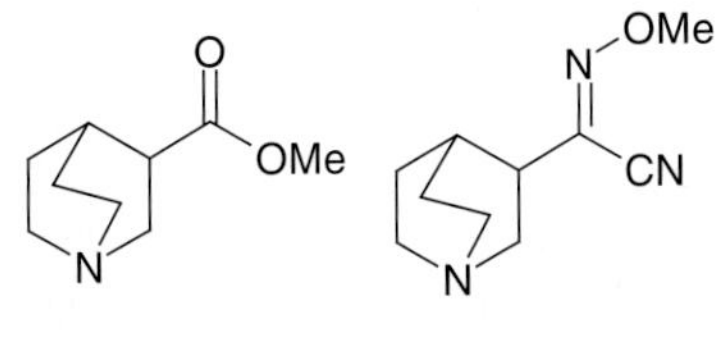

Figure 22 *N*-Methoxy imidoyl nitrile as ester surrogate.

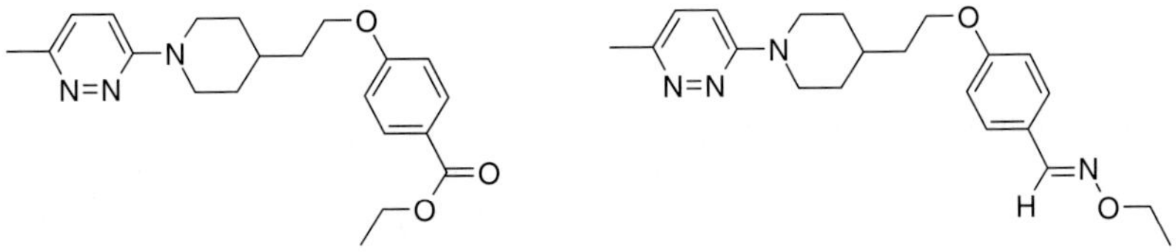

Figure 23 Ethyl oxime ether as ester replacement gave an improved bioavailability.

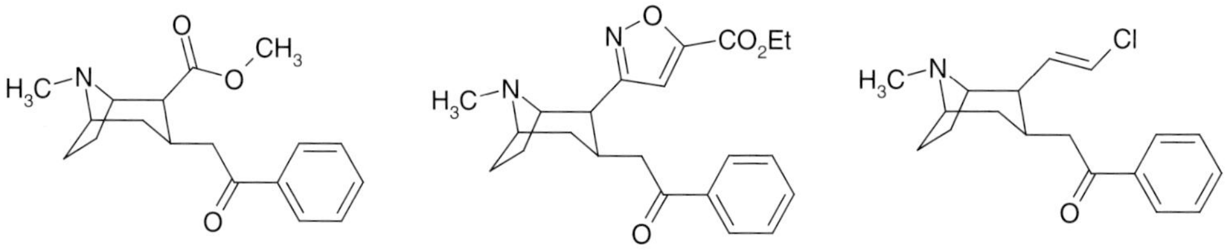

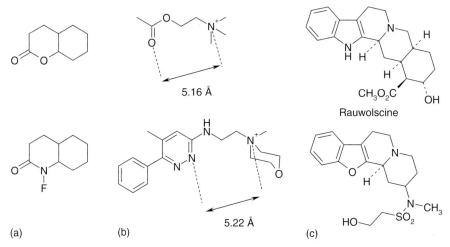

Figure 24 Replacement in (−)-cocaine of the carbomethoxy group by a carbethoxyisoxazole and a chlorovinyl moiety.

Figure 25 (a) Replacement of ester ether oxygen by a fluoro-nitrogen; (b) exo-endo amidine in place of a carboxylic ester functionality; and (c) N-methylsulfonamide analog of α-yohimbine (rauwolscine).

More unusual isosteric replacements for the peptidic bond were recently proposed (**Figure 27**). Among them, hydroxyethylureas were used in the design of a novel class of potent HIV-1 protease inhibitors, diacylcyclopropanes in the design of novel renin inhibitors, and pyrroline-3-ones for various proteolytic enzyme inhibitors.[113,114] Vinyl fluorides can probably be considered as representing the closest possible bioisosteres of the peptide bond. The synthetic methods available allow, by an appropriate selection of the precursors, the preparation of analogs of dipeptidic combinations of amino acids bearing no other functionalities in their side chains, e.g., Gly, Ala, Val, Phe, Pro.[115] Vinyl fluorides have been used in the design of bioisosteres of peptide bonds as in the case of the analgesic dipeptide 2,6,-dimethyl-L-tyrosyl-D-alanine-phenylpropionamide.[116]

Other structural modifications of the amide bond proposed to create more chemically stable and orally available molecules include heterocyclic rings such as 1,2,4-oxadiazoles and 1,3,4-oxadiazoles (**Table 7**). 1,2,4-Oxadiazoles have been used as bioisosteres of the carboxamide moiety in SH2 inhibitors of tyrosine kinase ZAP-70,[117] 5HT$_{1D}$ receptor agonists,[103] histamine H$_3$ receptor antagonists,[118] and 5HT$_3$ receptor antagonists.[119] An example worthy of note is given by the β$_3$ adrenergic receptor agonists where the replacement of the amide bond with 1,2,4-oxadiazoles led to compounds with improved oral bioavailability retaining the β$_3$ adrenergic receptor (AR) agonist activity.[120–124] 1,3,4-Oxadiazoles have also been used as amide bond surrogates in several therapeutic areas like benzodiazepine receptor agonists,[125] muscarinic receptor agonists,[126] neurokinin 1 (NK1) receptor antagonists,[127] and they have also been used as Phe-Gly peptidomimetics.[128]

Other heterocyclic replacements of the amide bond include oxazole rings,[129,130] cyclic amidines,[131,132] pyrrole rings,[133] and phenylimidazole[134] (**Table 7**).

2.16.2.2.4 Urea and thiourea bioisosteres

The very successful urea ↔ cyanoguanidine ↔ dimethylnitroethylene bioisosterism[135] was applied to the sulfonyl-urea function of torsemide,[136] a diuretic loop agent (**Figure 28**). This investigation led to the design and synthesis of the corresponding thiourea, cyanoguanidine, and dimethylnitroethylene group. The structure, the electronic features (charges, molecular electrostatic potential, molecular orbitals) and the lipophilic properties of these three analogs were

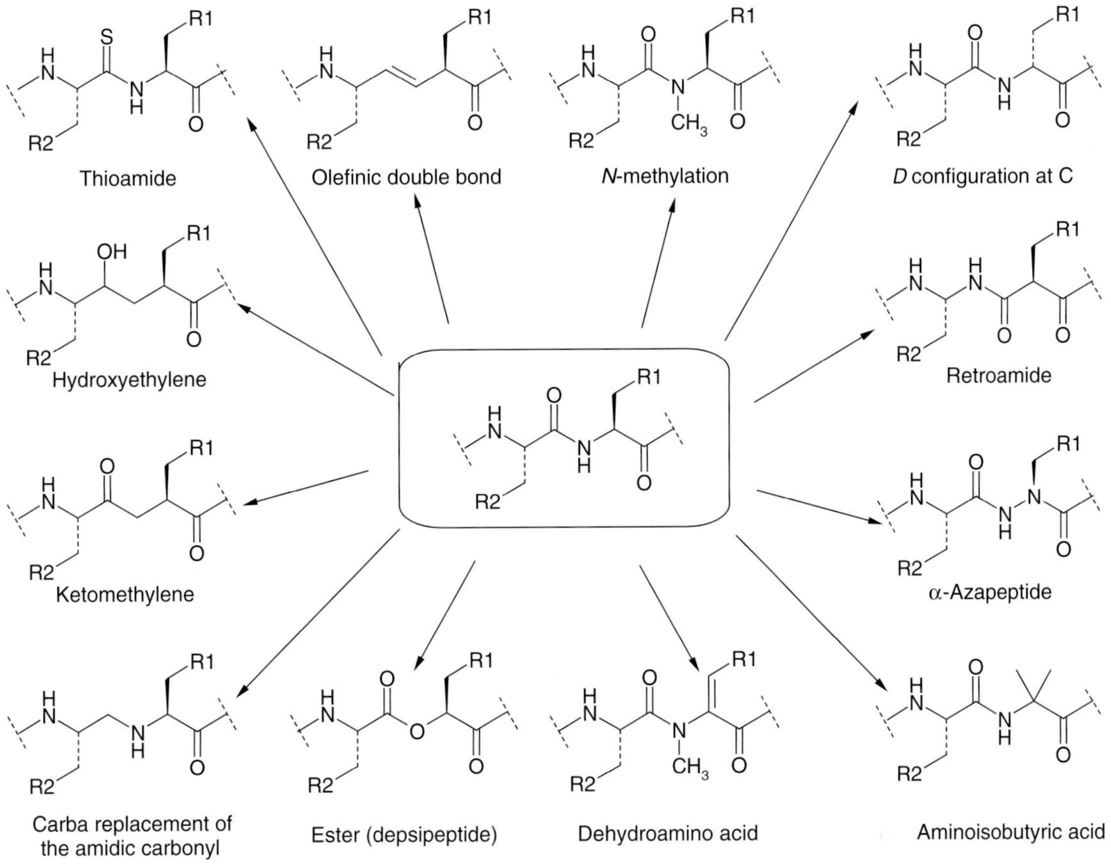

Figure 26 Well-established isosteric replacements for peptidic bonds.

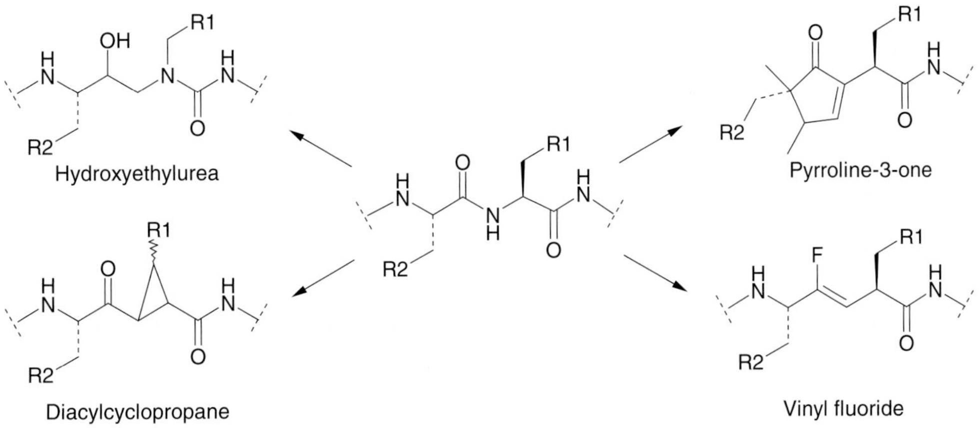

Figure 27 Unusual isosteric replacements for peptidic bonds.

determined. In conclusion, cyanoguanidine and diaminonitroethylene turned out to be valuable isosteres of the (thio)urea function as they share common geometric and electronic properties. In addition, they cover a range of lipophilicity that makes them suitable for compound pharmacomodulation as illustrated by the diuretic and antiepileptic properties of analogs of torsemide.

In a study undertaken on antagonists of the dihydropyridine neuropeptide Y_1 (NPY 1) receptor, common bioisosteres such as thiourea and cyanoguanidine and also heterocycles were examined as urea replacements (**Figure 29**).[137,138] Both cyanoguanidine and thiourea derivatives demonstrated potent binding affinity at the Y_1 receptor. The two known urea

Table 7 Heterocyclic surrogates of the amide bond

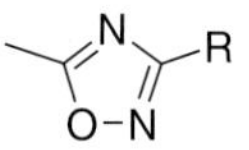

Structure	Name	Application	Reference
1,2,4-Oxadiazole	1,2,4-Oxadiazoles	β_3 Adrenergic receptor agonist	Naylor et al.[120–124]
		Fatty acid oxidation inhibitors	Elzein et al.,[129] Koltun et al.[130]
1,3,4-Oxadiazole	1,3,4-Oxadiazoles	Phe-Gly peptidomimetics	Borg et al.[128]
		NK1 receptor antagonists	Ladduwahetty et al.[127]
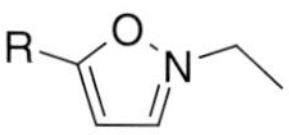 Oxazole	Oxazoles	Fatty acid oxidation inhibitors	Elzein et al.,[129] Koltun et al.[130]
Cyclic amidine	Cyclic amidines	Dopamine D_4 receptor agonists	Einsiedel et al.[131,132]
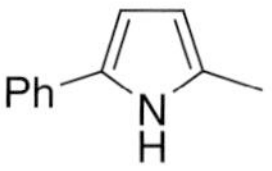 Pyrrole	Pyrroles	D_3 antagonists	Einsiedel et al.[133]
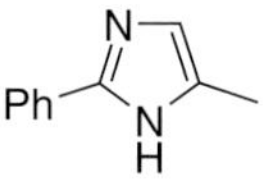 Phenylimidazole	Phenylimidazoles	D_4 receptor ligand	Thurkauf et al.[134]

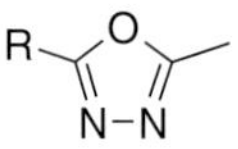

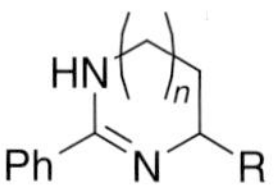

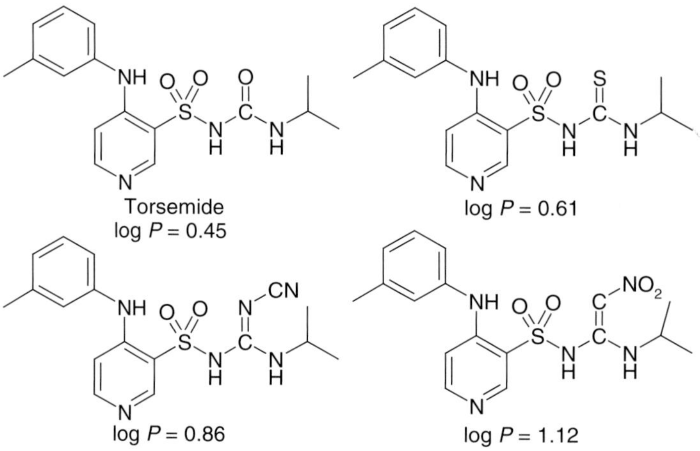

Figure 28 Structure of torsemide and its urea bioisosteres.

Torsemide log P = 0.45; log P = 0.61; log P = 0.86; log P = 1.12

heterocycle replacements, squaric acid and thiadiazole oxide, demonstrated good binding affinity although both were 10 times less active compared to parent urea.

Among more exotic surrogates, the 3,4-diamino thiadiazole dioxide moiety was proposed as a weakly acidic urea equivalent.[139] The similar thiatriazole dioxide is found in the H_2 antagonist tuvatidine (HUK 978). Other bioisosteres are exo-endo amidinic heterocyles bearing an electron-attracting function in the α position[140,141] (**Figure 30**).

2.16.2.2.5 Phenol and catechol bioisosteres
Phenol and catechol functions frequently use the same bioisosteric resources. However, due to the vicinity of the two hydroxylic functions in catechols, a few specific solutions are possible for them.

2.16.2.2.5.1 Bioisosteres of the phenol function
The bioisosteric groups of the phenolic function should not be much larger than the hydroxyl itself and should have approximately the same acidity range (weak acid) and be able to form hydrogen bonds. The most popular bioisoteric

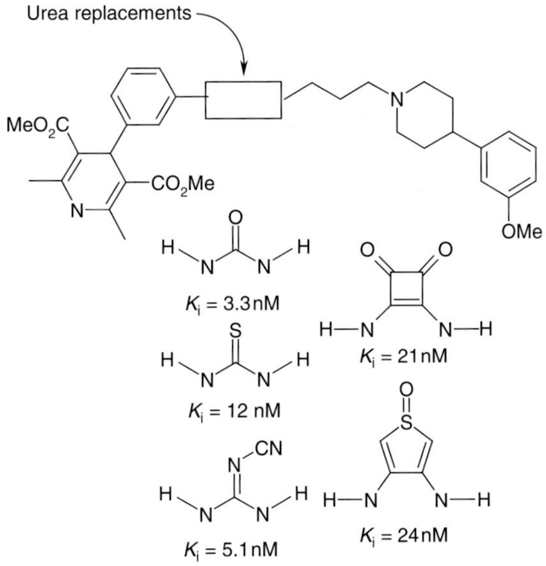

Figure 29 Binding affinities for various urea bioisosteres acting as antagonists at the NPY Y$_1$ receptor.

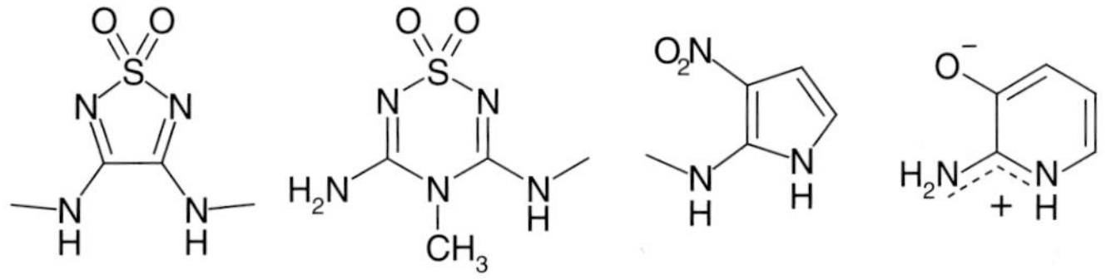

Figure 30 Less common urea equivalents.

substituents[142,143] are methanesulfamide (CH$_3$SO$_2$NH–), hydroxymethyl (HOCH$_2$–) or hydroxyisopropyl (HOC(CH$_3$)$_2$–), various amide groups (–NHCHO, –NHCOCH$_3$, –NHCOC$_6$H$_5$, methanesulfamidomethyl (CH$_3$SO$_2$NHCH$_2$–), dimethylaminosulfonamide ((CH$_3$)$_2$NSO$_2$NH–), and others with an ionizable proton next to or near an aromatic ring.

Using this analogy, Wu and co-workers[144] prepared a series of phenolic bioisosteres of benzazepine D$_1$/D$_5$ antagonists (**Figure 31**). Compared to the reference compound SCH 38393 they exhibit similar or more potent activities, high selectivity over D$_2$–D$_4$ receptors, and improved in vivo pharmacokinetics. The optimization of the hydrogen bond donating capacity of various heterocycles allowed the identification of several potent D$_1$/D$_5$ antagonists with high selectivity D$_2$–D$_4$, α$_{2a}$ adrenergic receptors, and the 5HT transporter, keeping at the same time excellent pharmacokinetic profiles.

The most popular surrogates for phenolic functions are NH groups rendered acidic through the presence of an electron-attracting group. **Table 8** shows an application of this bioisostery in the design of N-methyl-D-aspartate (NMDA) receptor antagonists.[145] In this case, the replacement of the phenol by heterocyclic NH-containing rings was performed in order to slow metabolism and hence to improve oral bioavailability. Indeed, the potent and NR1A/2B-receptor selective benzimidazolone analog was obtained demonstrating oral activity in a rodent model of Parkinson's disease at 10 and 30 mg kg^{-1}.

Hacksell and colleagues[146] described the compound (±)-3-(1-propyl-3-piperidinyl)phenol [(±)-3-PPP] as a highly selective agent for presynaptic brain dopamine (DA) receptors. However, the clinical potential of these molecules (**Figure 32**) as antischizophrenic or antiparkinsonian agents may be limited by their relatively low oral bioavailability and their short duration of action.[147] Thus, a heterocyclic analog of 3-PPP might retain some of its pharmacological activity while having improved oral bioavailability and duration of action.[148] In the case of talipexole (B-HT 920) (**Figure 32**), a DA agonist which has been reported to be selective for DA autoreceptor, 2-aminothiazolyl moiety was used as a replacement for the phenol ring. This successful example led to the discovery of the PD 118440, a dopamine autoreceptor agonist,[148] and pramipexole, a DA agonist with preference for the DA D$_3$ (over DA D$_2$) receptor.[149–151]

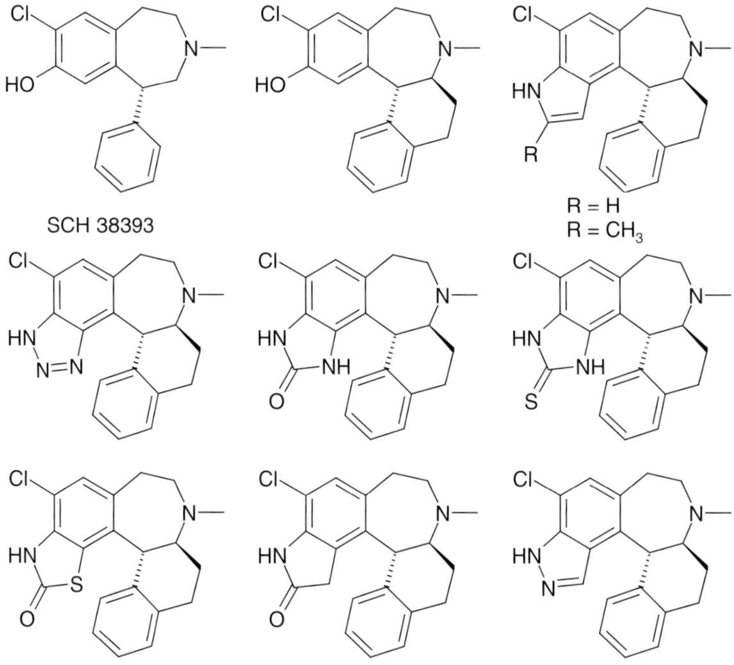

Figure 31 Phenol bioisosteres of benzazepine D_1/D_5 antagonists.

2.16.2.2.5.2 Catechol bioisosteres

All of the previous examples can be applied to the catechol family. Many of these bioisosteres share with the catechols the ability to chelate metal atoms and to form hydrogen-bonded second rings. In benzimidazole this H-bond ring is mimicked by way of a covalent ring structure (**Figure 33**). A recent case described the synthesis of more stable bioisosteres of inhibitors of the insulin-like growth factor-1 receptor kinase (IGF-1R; **Figure 34**). Based on the structure of AG 538,[152] which contains two catechol rings and is sensitive to oxidation in cellular environments, a new series of kinase inhibitors was developed. The catechol moiety was replaced by a benzoxazolone ring resistant to oxidation yielding two compounds GB19 and AGL2263, which maintain the same potency as AG 538.

2.16.2.2.5.3 Fancy catechol bioisosteres

Until now, 2-aminothiazole derivatives or other aromatic heterocyclic systems have been used as a bioisosteric surrogate for the catechol nucleus (see studies around the pramipexole). Hübner and colleagues[149,153–155] evaluated nonaromatic catechol bioisosteres (**Figure 35**). They were able to demonstrate that the π-electronic system of the nonaromatic enediyne FAUC 88 and 73 can be used as an efficient bioisostere for the catechol fragment of dopamine. Combined with conformational rigidization, this approach led to dopamine receptor agonists with high affinity for the subtypes of the D_2 family and especially for D_3.

2.16.2.2.6 Sulfonamides

The introduction of the serotonin $5HT_{1D}$ receptor agonist sumatriptan for the acute treatment of migraine has marked an intense research effort to discover more potent and selective $5HT_{1D}$ receptor agonists with improved pharmacokinetic profiles.[156] The H-bond acceptor ability of the 5-position is crucial for $5HT_{1D}$ receptor affinity and selectivity as it was revealed by reference compounds (5HT and sumatriptan). The first analog was an oxadiazole, L-695,894 (**Figure 36**), which had better oral bioavailability but had significant affinity for $5HT_{2A}$ and $5HT_{2C}$ receptors.[103] This work was then extended to explore alternative 5-membered heteroaromatic rings, which are also H-bond acceptors. Compounds were sought that had $\log Ds$ <0.5 in order to minimize central nervous system penetration.[156] Simple unsubstituted imidazoles, triazoles, and tetrazoles (**Table 9**) have high affinity and good selectivity for the $5HT_{1D}$ receptor. The 1,2,4-triazole 1, MK-462, was shown to have the optimal pharmacokinetic profile with rapid oral absorption and high bioavailability.

Table 8 Bioisosteric replacements of the phenol function

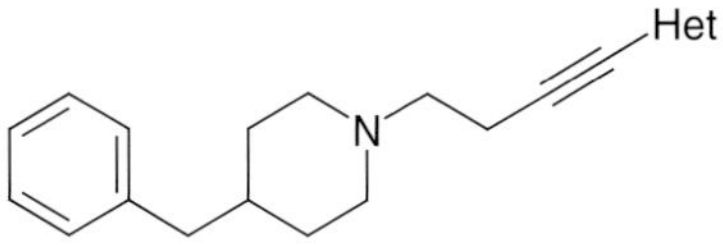

Name of «Het»	Structure of «Het»	IC_{50} values (μM) for the N-methyl-D-aspartate (NMDA) receptors NR1A/2B
4-Phenol	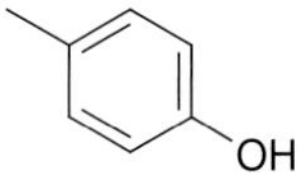	0.17
5-Indole	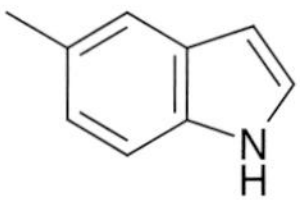	0.63
5-Indazole	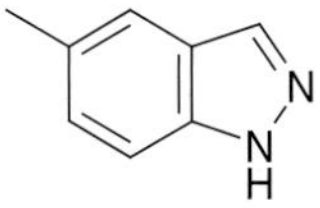	0.25
5-Benzotriazole	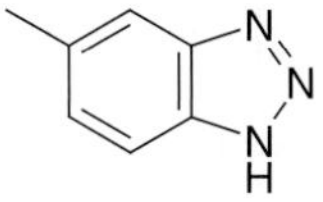	0.22
5-Indolone	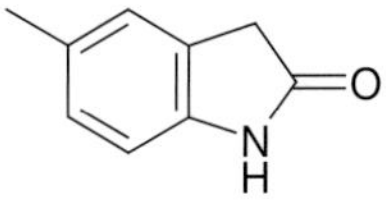	0.32
5-Imidazolone	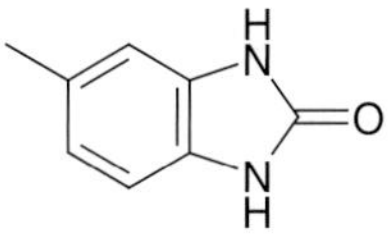	0.09
5-Imidazole-thione	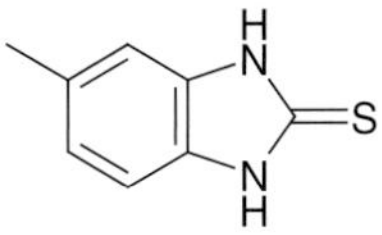	0.18
5-Benzoxazolone	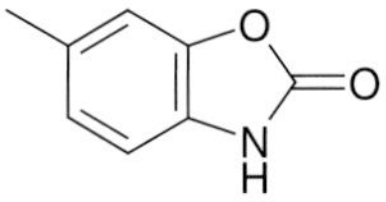	0.12

Using a different approach, Glen and co-workers[157] built a pharmacophore based on the pharmacological activities of a series of novel C- and N-linked hydantoin analogs. Their model was used as a framework for the design of a diverse series of analogs with good affinity and selectivity for $5HT_{1D}$. Suitable with the constraints required for good oral absorption, a potent selective $5HT_{1D}$ agonist, *S*-5-methyl-2-oxazolidinone analog, has been described. The dipolar azido group is a bioisostere[158] of the SO_2NH_2 and SO_2Me hydrogen bonding pharmacophores present in many selective COX-2 inhibitors and the azido analogs X and Y are useful biochemical agents for photoaffinity labeling of the COX-2 enzyme (**Figure 37**).

2.16.2.3 Reversal of Functional Groups

The reversal of the peptidic functional groups is often used in peptide chemistry. The obtained retropeptides are generally more resistant to enzymatic attacks (*see* Section 2.16.2.2.3). For thiorphan and *retro*-thiorphan an identical binding mode to the zinc protease thermolysin was demonstrated.[159] Similar inhibition values for thermolysin and

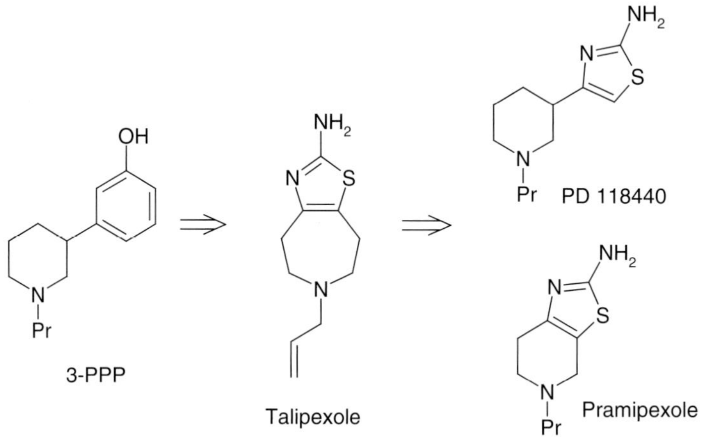

Figure 32 2-Aminothiazole bioisostery applied to the dopamine agonist 3-PPP.

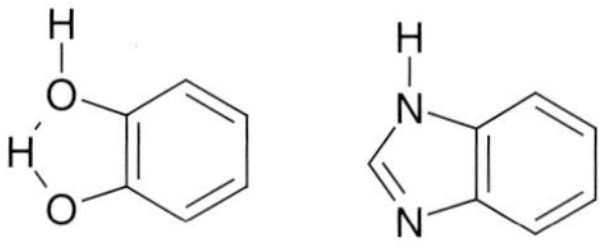

Figure 33 The bioisostery of catechol/benzimidazole.

Compounds	Name	IC$_{50}$ values (µM)			
		IGF-1R	IR	SRC	PKB
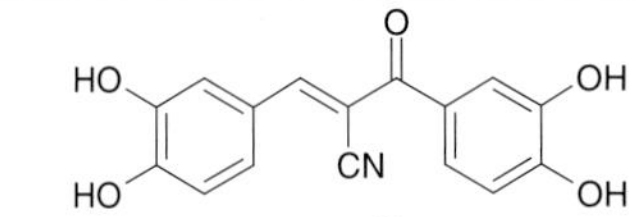 (HO,HO-phenyl)-CH=C(CN)-CO-(3,4-diOH-phenyl)	AG 538	0.06	0.12	2	76
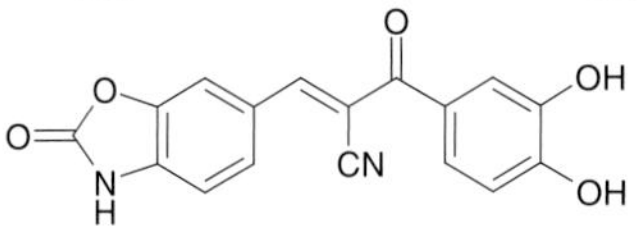 benzoxazolone-CH=C(CN)-CO-(3,4-diOH-phenyl)	GB 19	0.37	0.6	1.5	21
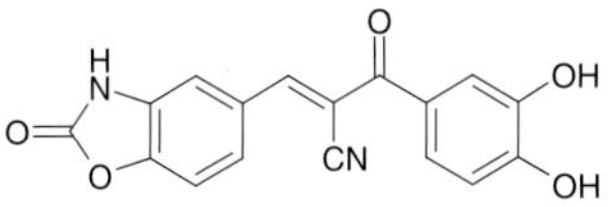 benzoxazolone isomer-CH=C(CN)-CO-(3,4-diOH-phenyl)	AGL 2263	0.43	0.4	2.2	55

IGF-1R: insulin-like growth factor-1 receptor kinase
IR: insulin receptor
SRC: cytoplasmic tyrosine kinase
PKB: protein-kinase B

Figure 34 Inhibition of IGF-1R and other kinases by AG 538 bioisosteres and analogs.

neutral endopeptidase were observed, whereas, for another zinc protease, ACE, noticeable differences for inhibition were found (**Figure 38**).

Nevertheless, the strategy of functional inversion can also be applied to nonpeptidic compounds. A historical example is the change from orthoform to neo-orthoform (orthocaine; **Figure 39**). The unwanted side effects, often encountered with aromatic para-amino substituted compounds ('para effects,' essentially of allergic origin) are abolished in the *meta* amino isomer whereas the local anesthetic activity is maintained. Similarly, the *meta* isomer of benoxinate has local anesthetic activity identical to that of benoxinate itself.[160]

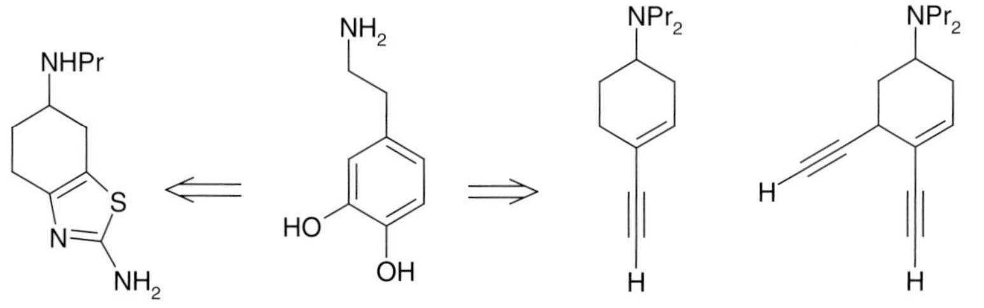

Figure 35 Aromatic and nonaromatic dopamine isosteres.

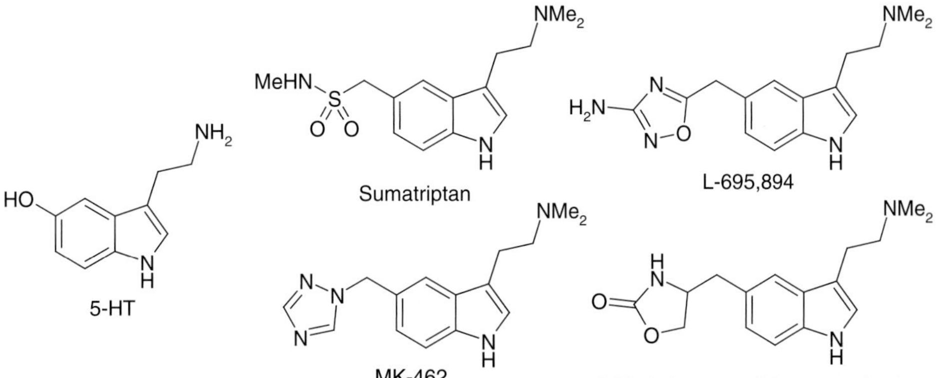

Figure 36 Replacement of the phenolic function of serotonin by various functions able to serve as H-bond acceptors.

The β-blocking agent practolol was one of the first cardioselective betablockers. It was rapidly replaced by its isomeric analog atenolol, which presents fewer side effects (**Figure 40**). The inversion of the ester function of meperidine leads to 1-methyl-4-phenyl-4-propionoxy piperidine (**Figure 41**), which is five times more potent as an analgesic drug than meperidine and represents the model compound of the inverted esters series.[161] Applied to serotonin, the reversal of a functional moiety yielded 5HT$_{2C}$ receptor selective agonists[162] (**Figure 42**).

2.16.3 Modification of the Core Structure

Most of the bioisosteric variations presented thus far in this chapter deal with functional and/or ring changes. The present paragraph deals with more profound modifications applied to the central architecture. This can notably be achieved in reorganizing the ring systems present in the active molecule or by computer-aided replacement of the initial molecular structure by another one that is isofunctional, but based on a different scaffold (scaffold hopping). In both cases, the functionalities necessary for the interaction with the receptor protein are maintained.

2.16.3.1 Reorganization of Ring Systems

The four bioisosteric variations described below deal with the reorganization of ring systems. These somewhat 'exotic' approaches may provide useful means to escape from overcrowded avenues of research.

2.16.3.1.1 Spiro derivatives and bi- or tricyclic systems

A first example is given by the guanethidine analogs.[163] As the original guanethidine patents covered ring sizes varying from five-membered to ten-membered rings, a possible way to get around them was the design of isolipophilic spiro systems (Dausse compounds a and b[164,165]; **Figure 43**). Another possibility, originating from Takeda scientists, involves the use of an azetidine surrogate for the ethylene-diamine chain.[166] Finally, polycyclic systems can replace the

Table 9 Displacement of [³H]-5HT to 5HT$_{1D}$ recognition sites in pig caudate membranes by N-linked imidazoles, triazoles and tetrazoles, and standard 5HT$_{1D}$ agonists

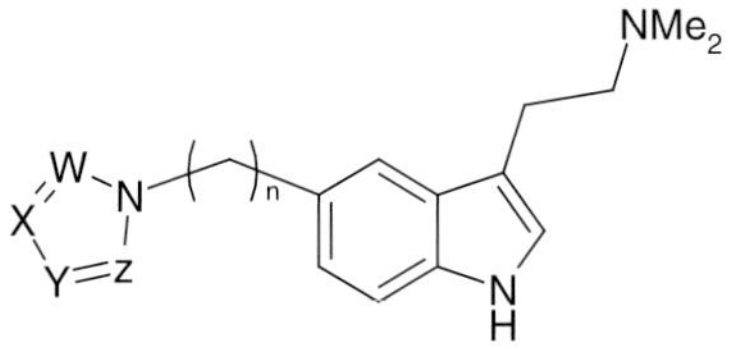

Compound	W	X	Y	Z	pIC_{50}	log D^a
5HT					8.0	
Sumatriptan					7.7	−1.17
L-695,894					7.6	−0.67
1 (MK-462)	N	C	N	C	7.3	−0.74
2	C	N	C	C	7.5	−0.53
3	C–Me	N	C	C	7.2	−0.74
4	N	N	C	C	7.3	−0.70
5	N	C	C	N	6.6	−0.20
6	N	N	N	C	7.4	−0.64
7	N	N	C	N	7.4	−0.34

^{a} log P measured at pH 7.4.

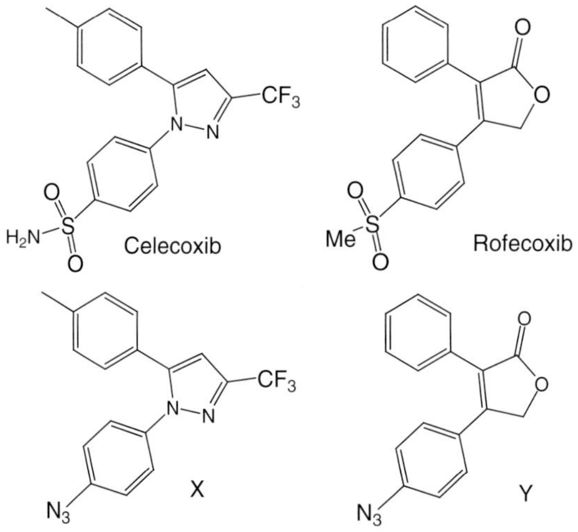

Figure 37 Selective COX-2 inhibitors and their azido bioisosteres.

octahydroazocine ring, as illustrated by the bicyclic compounds c and d from Dausse[167] or by the tricyclic compound from Lumière Laboratories.[168,169] Many other imaginative solutions have been proposed; these are well reviewed by Mull and Maxwell.[163]

More recently, similar variations were applied to the design of an impressive number of analogs of the anticonvulsant drug gabapentin[170] (**Figure 44**).

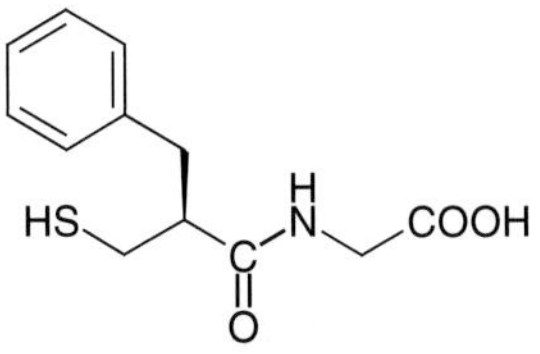

Figure 38 Inhibition values of thiorphan and *retro*-thiorphan for three zinc proteases.

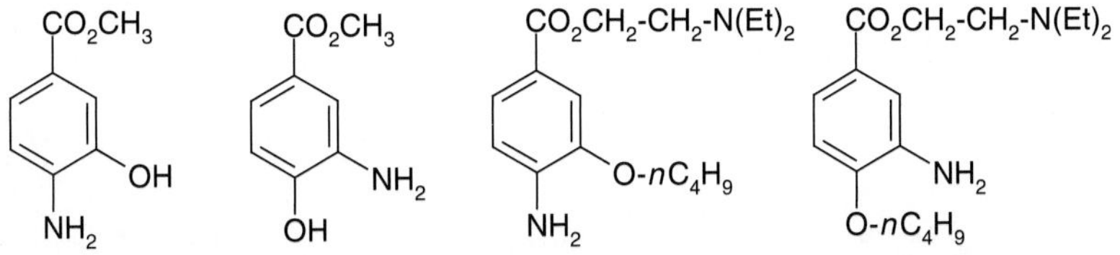

Enzyme	K_i value in μM
NEP 24.11	0.0019
Thermolysin	1.8
ACE	0.14

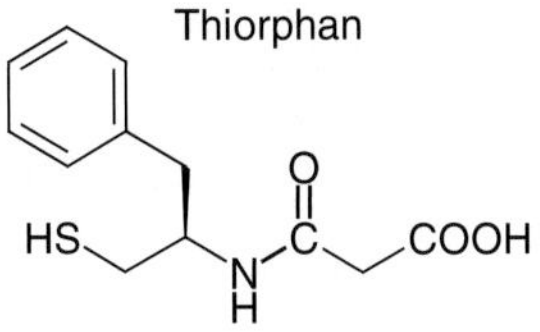

Thiorphan

Enzyme	K_i value in μM
NEP 24.11	0.0023
Thermolysin	2.3
ACE	>10

Retro-thiorphan

Figure 39 Positional isomery in local anesthetics.

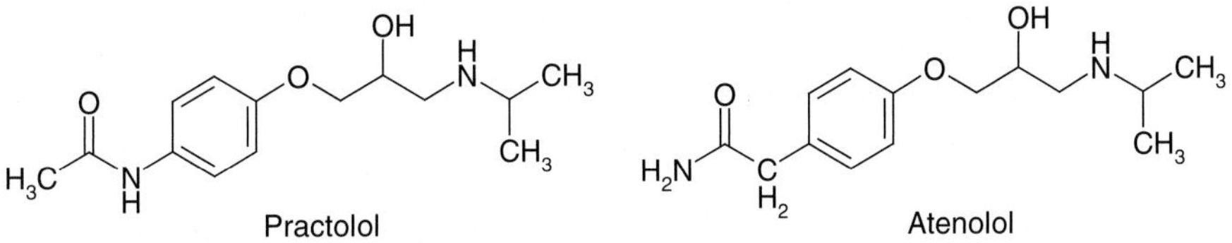

Orthoform old — Orthoform new — Benoxinate — Benoxinate isomer

Figure 40 Reversal of functional goups in practolol yields atenolol.

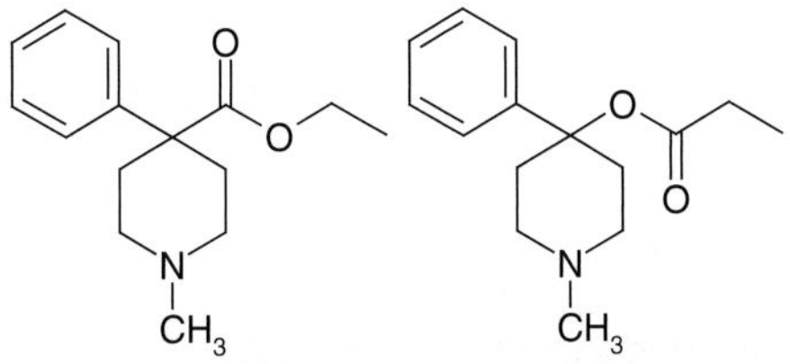

Practolol — Atenolol

Figure 41 Meperidine and the corresponding inverted ester.

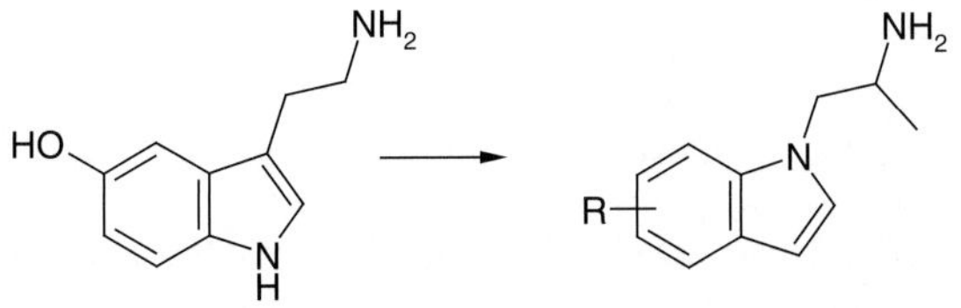

Meperidine — Inverted ester

Figure 42 Serotonin analogs resulting from a functional reversal.

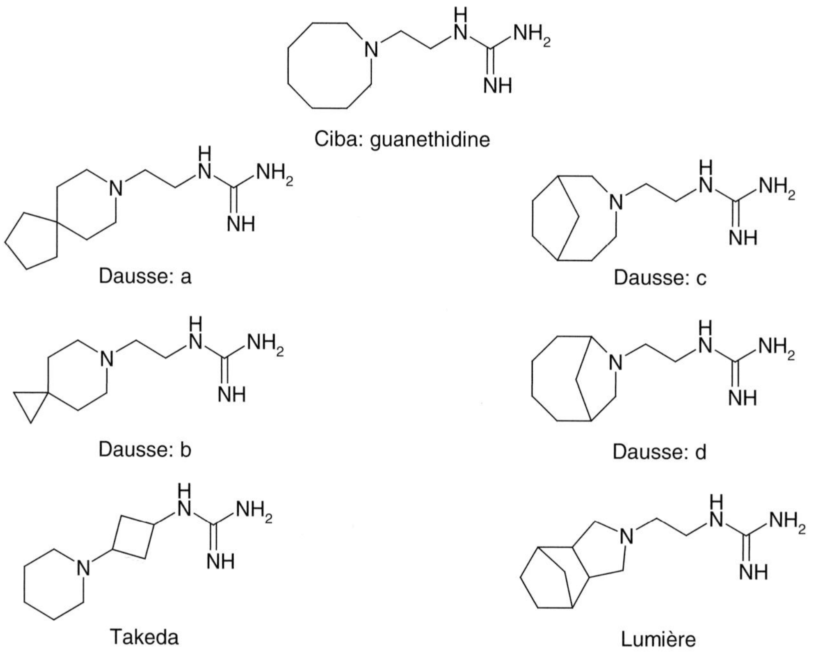

Figure 43 Bioisosteric possibilities in the design of guanethidine analogs.

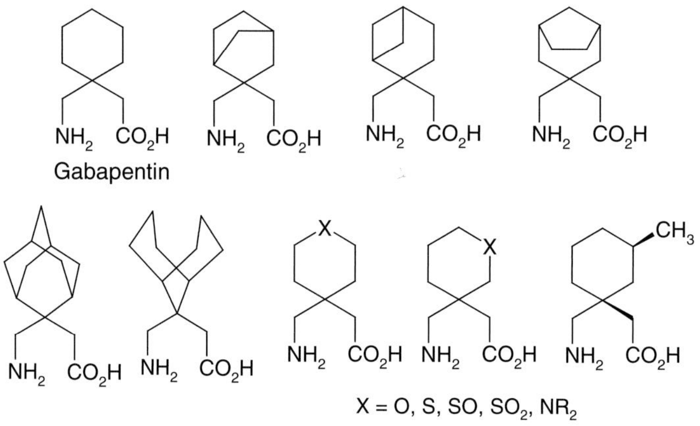

Figure 44 Bioisosteric possibilities in the design of gabapentin analogs.

2.16.3.1.2 Splitting benzo compounds (benzo cracking)
Dissociation of a fused ring system (**Figure 45**), particularly by splitting a benzo compound, can sometimes improve its solubility but only slightly alters its pharmacokinetic profile and its long-term toxicity.

2.16.3.1.3 Restructuring ring systems
Among the above-mentioned molecular variations on ring systems, some can be used simultaneously. Thus, the splitting of the benzimidazole heterocycle in the anthelmintic thiabendazole and the concomitant association of the two five-membered rings yields tetramizole (**Figure 46**). One of the two enantiomers of tetramizole, the L-(−)-form or levamizole, is also a potent anthelmintic.

The change from the D1 selective dopaminergic agonist DPTI 3-(3,4-dihydroxyphenyl)-1,2,3,4-tetrahydroisoquinoline) to the equally D1 selective compound SKF 38 393 is a combination of benzo cracking, a new benzo fusion, and ring enlargement (**Figure 46**). As a result, the compounds still resemble each other and are both recognized by the dopamine D1 receptor.[171] An interesting example of a restructured ring system was designed by Meyer and

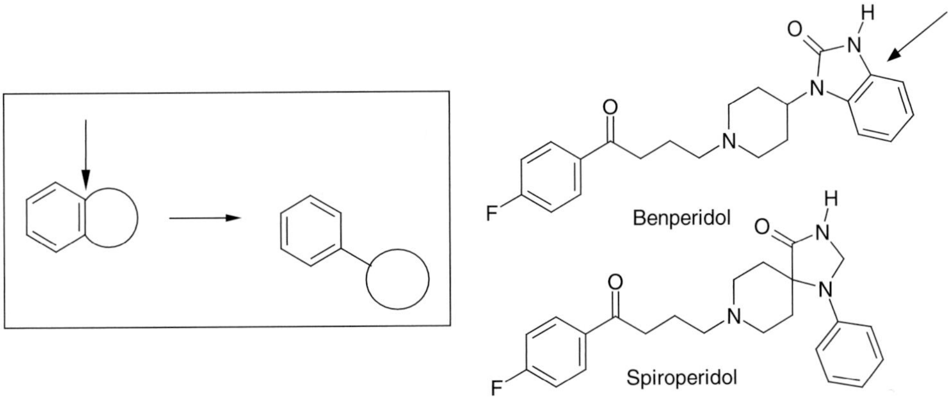

Figure 45 Splitting of fused rings often yields drugs with similar activity, sometimes with improved solubility and/or less toxicity.

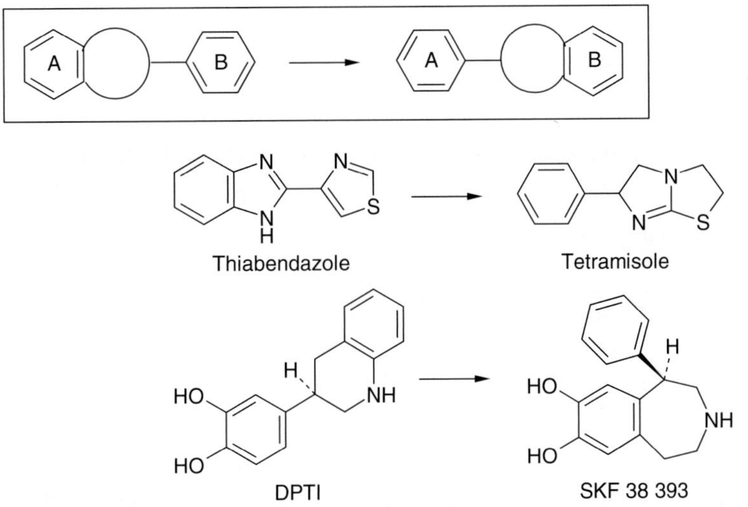

Figure 46 Restructured ring systems.

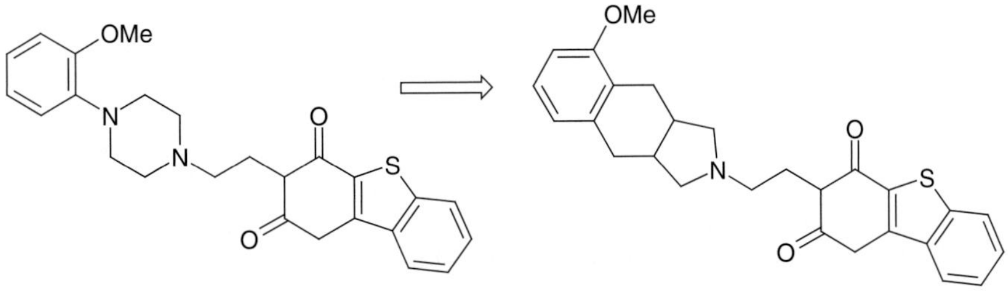

Figure 47 Hexahydro-isoindole as a surrogate of the very frequently used o-methoxy-N-arylpiperazine.

co-workers[172] for the design of 2-methoxybenz-[e]-hexahydro-isoindole as a surrogate of the very frequently used ortho-methoxy-N-arylpiperazine. It led to an enhancement of affinity and improved selectivity for α_{1A} receptor antagonism (**Figure 47**).

2.16.3.1.4 Ring dissociation

The natural compound khellin generated two families of cardioactive drugs: on one side the benzopyrones, illustrated by the 3-methyl-chromone[173] and chromonar (carbochromen), on the other side the benzofurans, illustrated by amiodarone (**Figure 48**). Both families possess antiarrhythmic and antianginal properties.

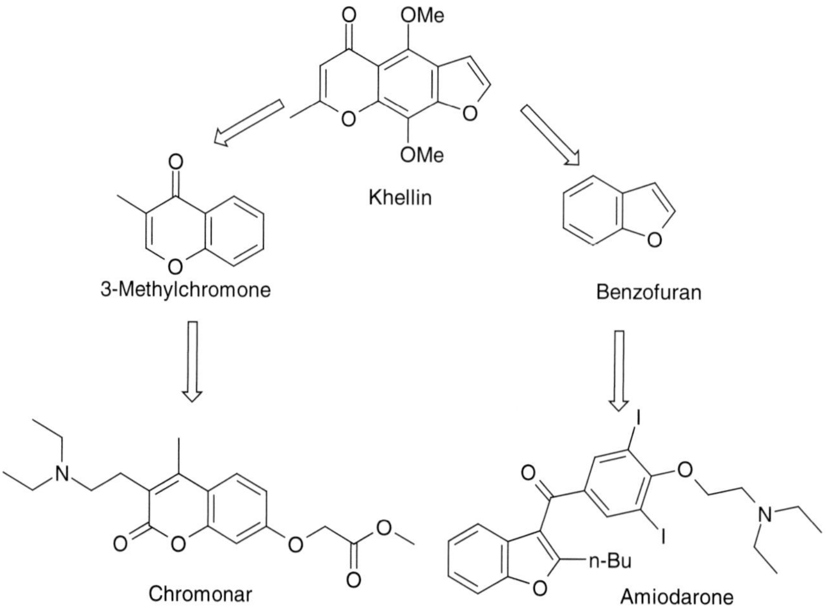

Figure 48 Cardioactive drugs obtained by dissociation of the khellin molecule into benzopyrones and benzofurans.

2.16.3.2 Scaffold Hopping

The goal of scaffold hopping is to design new structurally diverse compounds by a significant modification of the platform of a known active drug (the template). Other terms such as 'leapfrogging,' 'lead-hopping,' and 'structure morphing' have also been used to point out the discovery of novel molecules with biological activity comparable to a template and with a significantly different architecture. The concept is based on the assumption that structurally and chemically diverse structures interact with the same receptor eliciting similar biological activity. This can appear to be in contradiction to the usual way of performing the drug discovery process. In fact, for a long time, medicinal chemists have relied on the principle that structurally similar molecules have similar biological activity[174] to be inspired for modifying the structure of biologically active compounds in order to get new drug entities. However, there are also a large number of examples that contradict this principle[175] by showing that similar chemically related molecules have a different mechanism of action, interacting with different receptors. A notorious example of structurally similar molecules with different mode of action is given by the neuroleptic compound chlorpromazine, the antidepressant imipramine, and the antiallergic drug promethazine (**Figure 49**). It is remarkable that such structurally related compounds act on three different receptors. Speculation about chemical diversity and chemical similarity is beyond the scope of this chapter, and is considered here only in its broadest sense. Here, the scope and limitations of scaffold hopping in the drug discovery process is analyzed. Medicinal chemists usually adopt the scaffold hopping approach, which is in fact an enlargement of the bioisosterism concept, to fix some issues encountered during the lead optimization process; this has often successfully led to marketed drugs. The changes made in a scaffold hopping process may include the variation of the central core with a different heterocycle in order to get a better patent position, or the change of a lipophilic scaffold by a more polar one to improve the solubility and the pharmacological profile of the molecule. The improvement of the pharmacokinetic profile by changing a metabolically unstable scaffold for a more stable and less toxic one could also be considered as scaffold hopping or scaffold morphing.

2.16.3.2.1 Successful examples of scaffold hopping

The idea of scaffold hopping is not new. It is rather a new term to designate a very useful tool used quite often by medicinal chemists. In fact, an analysis of the marketed drugs reveals that many of them have different molecular structure though keeping the same substantial biological activity. Often, the different templates were suggested from the medicinal chemist's intuition and experience; others were derived from hits yielded from high-throughput screening (HTS) of different compound libraries.

Figure 50 shows some COX-2 preferential inhibitors. It is noteworthy that structurally different scaffolds such as nimesulide, indomethacin, and valdecoxib have been successfully used as nonsteroidal anti-inflammatory drugs (NSAIDs)

Imipramine Chlorpromazine Promethazine

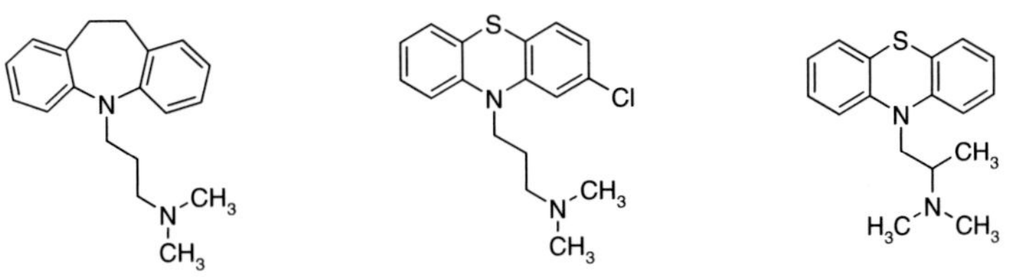

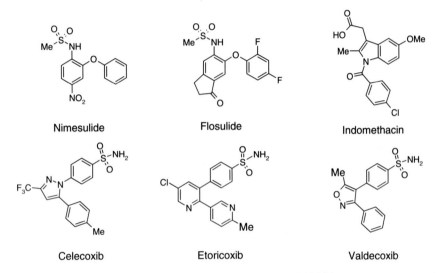

Figure 49 Highly structurally similar molecules with completely different mode of action: chlorpromazine is a neuroleptic drug acting as a dopamine antagonist, prometazine is a H_1 antagonist agent used as an antiallergic, and imipramine is an antidepressant acting as a neurotransmitter uptake inhibitor.

Nimesulide Flosulide Indomethacin

Celecoxib Etoricoxib Valdecoxib

Figure 50 A literature example of classical scaffold hopping: COX-2 preferential inhibitors.

with a preferential COX-2 inhibitor activity. On the other hand, all of the most recently approved ones have rather similar structures (see celecoxib, etoricoxib, and valdecoxib)[176,177] and represent a good example of bioisosterism among pyrazole, isoxazole, and pyridine.

Another well-known example of successful scaffold hopping is given by the $GABA_A$ ligands to the benzodiazepine site (**Figure 51**).[178] In this case two categories of scaffold hopping can be identified, one for the full agonist and the other one for the inverse agonist. It is interesting to notice that structurally diverse molecules like diazepam, zolpidem, zaleplon, and zopiclone all exert the same biological activity on the $GABA_A$ receptor as anxiolytics and sedatives acting as full agonists. The same comment can be extended to the inverse agonist's class of compounds such as DMCM, compounds **18** and **19**, FG 8094, and compound **20**, which, despite a different molecular architecture, maintain a similar anxiogenic and convulsant activity together with learning and memory enhancing properties. It is important to underline the dualism of those ligands keeping in mind that structurally related molecules such as DMCM and FG 7142 interact with the benzodiazepine site of the $GABA_A$ receptor but with different profiles, DMCM being an inverse agonist and FG 7142 a partial inverse agonist. A similar example is given by Ro 15–1788, FG 8094, and bretazenil where minor modifications of the structure led to an antagonist (Ro 15–1788), a selective inverse agonist (FG 8094), and a partial agonist (bretazenil).

2.16.3.2.2 Computational approaches to scaffold hopping

In the above paragraph some successful examples of classical scaffold hopping confirmed that structurally diverse molecules might be ligands of the same receptor inducing similar biological activity. Alternatively, one of the principles of modern medicinal chemistry is founded on the concept that structurally similar compounds will have similar modes of action. The corollary of this is that the more two molecules are structurally different, the fewer are the chances that these two molecules may have the same biological activity. The conclusion could be that a rational approach to scaffold

682 Bioisterism

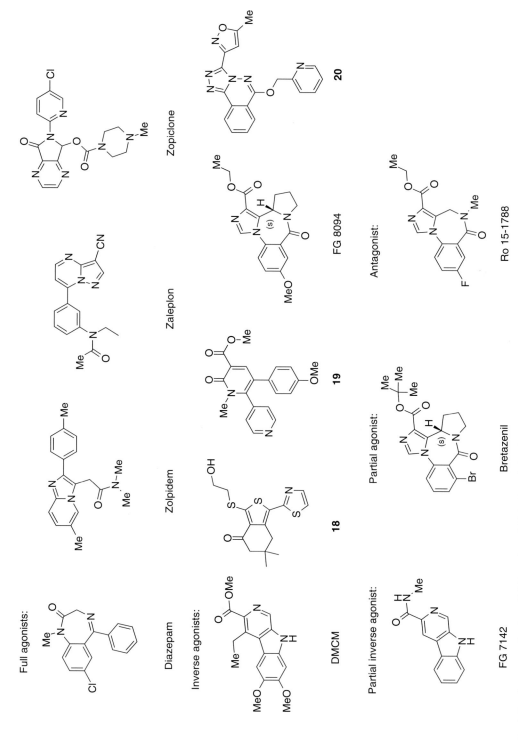

Figure 51 A literature example of scaffold hopping: GABA$_A$ ligands binding to the benzodiazepine site.

hopping is not achievable for a medicinal chemist and that the discovery of different templates eliciting similar biological activity relies upon a serendipitous discovery or upon HTS screening. Efforts from experts in molecular modeling and cheminformatics have supplied useful tools to address this paradox and many publications have proposed different computational approaches to solve the challenge of finding a rational way to perform scaffold hopping.[179] The four principal types of computational approaches for scaffold hopping are: (1) shape matching, (2) pharmacophore searching, (3) fragment replacement, and (4) similarity searching.

One of the crucial steps of the drug discovery process is to find an optimal chemical entity (a hit) that interacts with a specific target. The hit must then be validated to identify a more potent, better absorbed, and less toxic lead compound that undergoes the optimization process to yield a valuable drug candidate. Starting with a good lead can save a lot of money and can shorten significantly the entire drug discovery process. The most widely used method for hit discovery is HTS. However, this method turned out to have some drawbacks and not to be the panacea for the hit identification. In fact, a good hit search relies upon the chemical diversity and druglike features of the screened library, and the number of compounds contained in it. It is useless pointing out that HTS performed on a library of hundreds of thousands of similar molecules is going to yield similar hits. In addition, the HTS approach can be rather expensive and time consuming, particularly when it is applied to functional receptor assays. There are two possible alternatives to the HTS of large libraries with a poor drug likeness feature: one is to screen smaller but smarter libraries, as highlighted in the selective optimization of side activities (SOSA) approach[180]; the other is to perform a virtual screening based on pharmacophore models, shape matching, and similarity searching.

A successful example of topological pharmacophore search is given by Schneider[181] in a prospective test using the program 'CATS' (Chemically Advanced Template Search) to predict a novel cardiac T-type Ca^{2+} channel blocking agents. Starting from mibefradil (**21**), a known T-channel blocking agent from Roche ($IC_{50} = 1.7\,\mu M$) (**Figure 52**), as 'seed' for CATS, several molecules were identified. The 12 highest ranking molecules were tested for their ability to inhibit cellular Ca^{2+} influx. Nine compounds (75%) showed significant activity ($IC_{50} < 10\,\mu M$) of which one compound (**22**, clopimozid) had an $IC_{50} < 1\,\mu M$. This example shows that significantly diverse structural scaffolds maintain the same biological activity while preserving the essential function-determining points.

There are several computer-aided methods to build a pharmacophore model according to the information available on the ligand and/or on the receptor and several software programs exist on the market.[182] Once the pharmacophore has been determined, there are two ways to identify new molecules that share its features and can elicit the same biological response: one is the de novo design that is based on the fragment recombination of known ligands; the other is the 2D topology or the 3D database searching.

Recently, an application of the de novo generation of chemotypes has been reported in the absence of receptor information[183] with the use of the software Skelgen^T. In order to verify if the scaffold hopping for ligand generation could be achieved where the nature of the 'lock' is unknown, an attempt was made to generate ligands for the estrogen receptor in the absence of any consideration of the actual receptor protein structure.

Through use of ligand superposition or a single bound conformation of a known active, a pseudoreceptor can be generated as a design envelope, within which novel structures are readily assembled. A set of five known estrogen receptor active training ligands (**23a–23e**) were selected (**Figure 53**). A pharmacophoric representation for these ligands was obtained allowing full ligand flexibility in the superposition process, which was then coded as site-point design constraints for use in Skelgen structure generation. A total of 446 ligands were generated, which satisfied the volume and pharmacophoric constraints. **Figure 54** shows a few representative scaffolds that resulted from this study.

Virtual screening based exclusively on topological similarity (e.g., Daylight fingerprints) may produce sets of molecules that are structurally similar to the original query molecule. On the other hand, a pure 3D pharmacophore

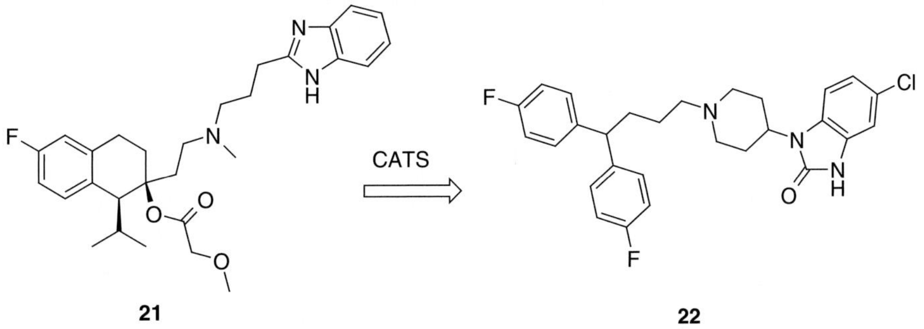

Figure 52 Query structure **1** (mibefradil) and a high-ranking isofunctional structure **2** (clopimozid) derived from **1** by CATS.

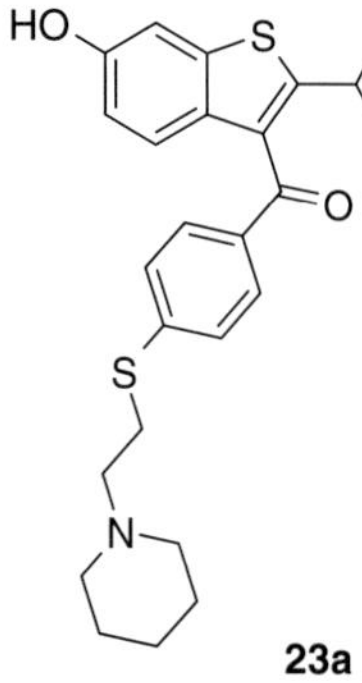

Figure 53 Known ER ligands used for pharmacophore generation.

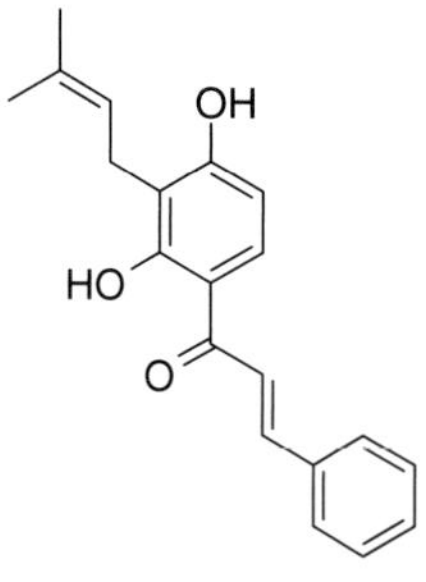

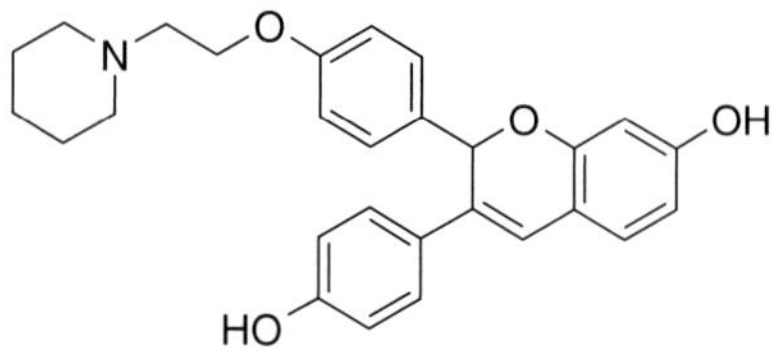

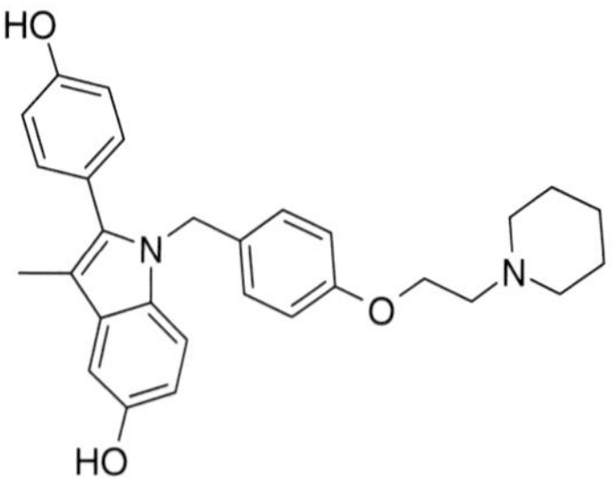

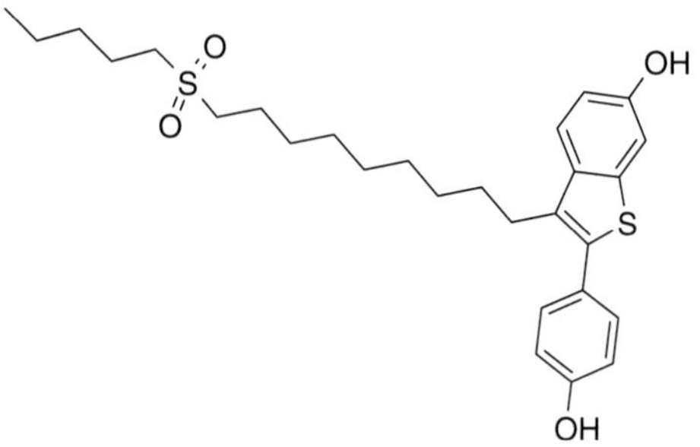

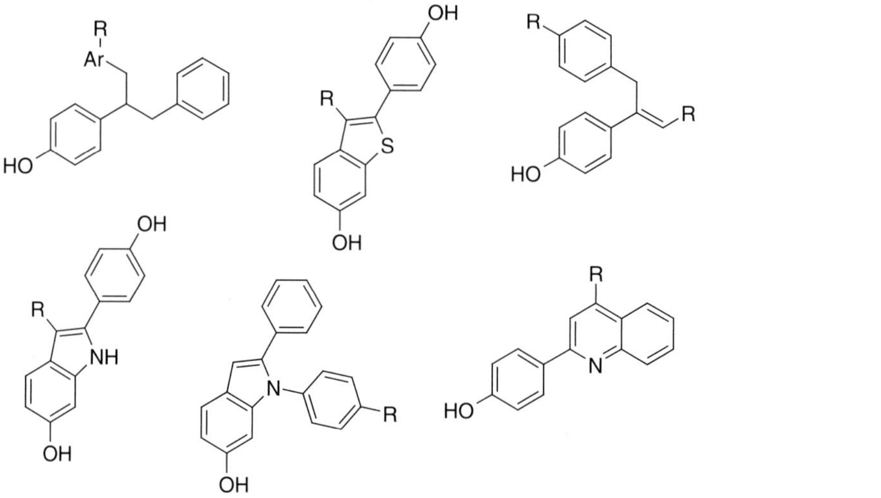

Figure 54 Representative scaffolds produced by Skelgen for the receptor for estrogens in de novo structure generation.

query has the potential to yield scaffolds with substantially different structures, which therefore risks losing binding affinity if compared to more topologically similar hits. A group from Hoffmann-La Roche has reported a successful example of virtual screening[184] based on a hybrid approach containing 3D pharmacophore information and topological elements using Catalyst software. In the quest for NPY5 receptor antagonists, three reference compounds from Banyu (**24**), Amgen (**25**), and Roche (**26**) (**Figure 55**) were chosen as seeds to build a 3D pharmacophore hypothesis containing topological elements also. Virtual screening of the Roche library yielded 632 molecules of which 31 compounds had an IC$_{50}$ <10 μM. The most interesting compound class is represented by **27** (**Figure 56**), which showed an IC$_{50}$ of 40nM at the mouse Y5 receptor. Moreover, the 4-aminothiazole hit class was free of patent claims. This hit class was submitted to a chemistry optimization process to establish a structure–activity relationship and in

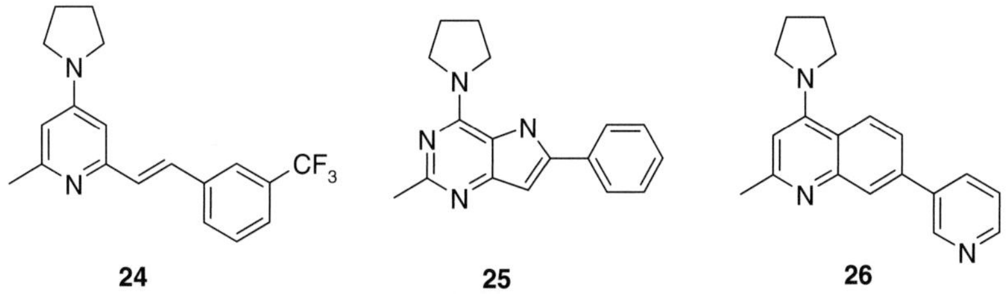

Figure 55 Seed compounds used to derive a 3D pharmacophore hypothesis including topological features.

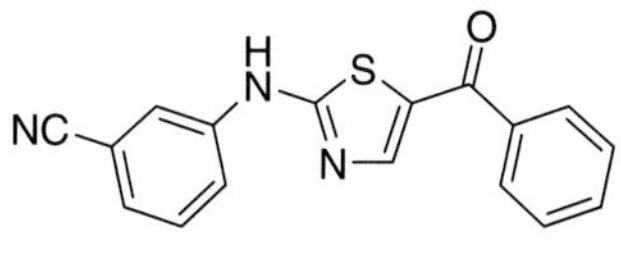

Figure 56 Aminothiazole (**27**) with IC$_{50}$ 40 nM.

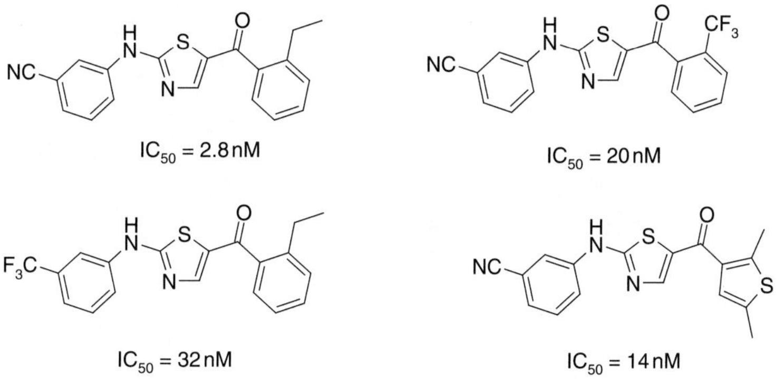

Figure 57 Some potent analogs issued from the first exploratory SAR around the aminothiazole hit **27**.

only two rounds of iterative optimization by parallel synthesis, a broader understanding of the factors influencing the potency was obtained. **Figure 57** shows some of the most potent analogs identified by this process.

Another application of virtual screening to the hit-to-lead optimization process for the design of optimized hit series is the rational morphing strategy based on the bioisosteric approach. Balakin and co-workers[185] used the program BIOISOSTER module of ChemoSoft to design a second-generation series against the *abl* tyrosine kinase in order to reduce some toxicological issues and to improve the intellectual property (IP) position of the initial hits. **Figure 58** illustrates a hit that showed low potential to generate new IP, and **Figure 59** shows a hit containing a potentially toxic *N*-aroylhydrazone group. The algorithm of BIOISOSTER found some bioisosteric analogs of these primary hits that were synthesized. The novel chemotypes, though structurally different from the primary hits, showed similar or slightly increased potency. In this case, a better patent position was developed (**Figure 58**) and molecules without toxic functionality were obtained (**Figure 59**).

2.16.4 Analysis of the Modifications Resulting from Bioisosterism

It is rare that the replacement of a part of a molecule by an isosteric or bioisosteric group leads to a strictly identical active principle. In practice, this is not actively sought, and it is preferable that the new compound produces a change as compared to the parent molecule. In general, the isosteric replacement, even though it represents a subtle structural change, results in a modified profile: the essential properties of the parent molecule will remain unaltered, while others will be changed. Bioisosterism will be productive if it increases the potency, selectivity, and bioavailability or decreases

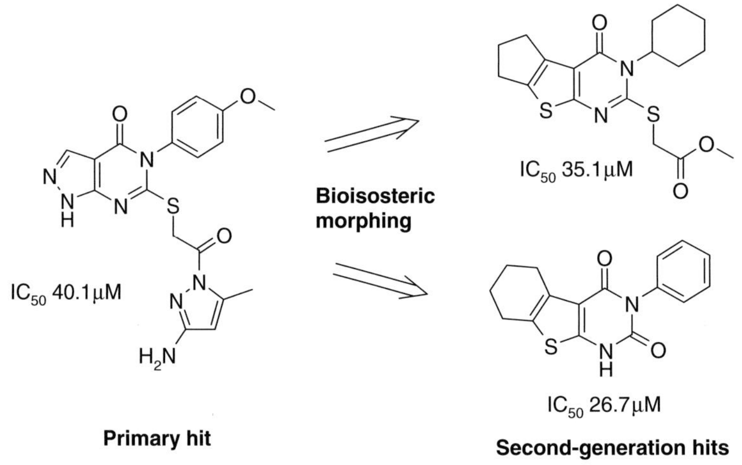

Figure 58 Bioisosteric transformations of the primary pyrazolopyrimidine scaffold and novel active chemotypes.

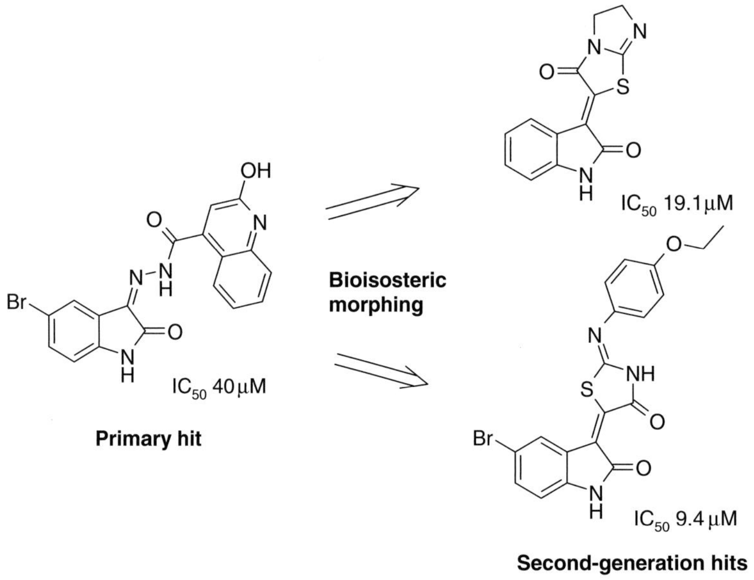

Figure 59 Bioisosteric transformations of an indol-2-one-hydrazone scaffold and novel active chemotypes.

the toxicity and undesirable effects of the compound. In proceeding to isosteric modifications, one will focus predominantly on a given parameter, structural, electronic, or hydrophilic, but this modification can lead simultaneously to the alteration of several other parameters.

2.16.4.1 Structural Parameters

These will be important when the portion of the molecule involved in the isosteric change serves to maintain other functions in a particular geometry. That is the case for tricyclic psychotropic drugs (**Figure 60**). In the two antidepressants (imipramine and maprotiline), the bioisosterism is geometrical insofar that the dihedral angle α formed by the two benzo rings is comparable: $\alpha = 65°$ for the dibenzazepine and $\alpha = 55°$ for the dibenzocycloheptadiene.[186] This same angle is only 25° for the neuroleptic phenothiazines and the thioxanthenes. In these examples, the part of the molecule modified by bioisosterism is not involved in the interaction with the receptor. It serves only to orient correctly the other elements of the molecule.

The structure of various bioisosteric retinoic acid receptor agonists highlights the dominantly geometric parameter of this bioisostery and thus can also be considered as scaffold hopping (**Figure 61**).[187]

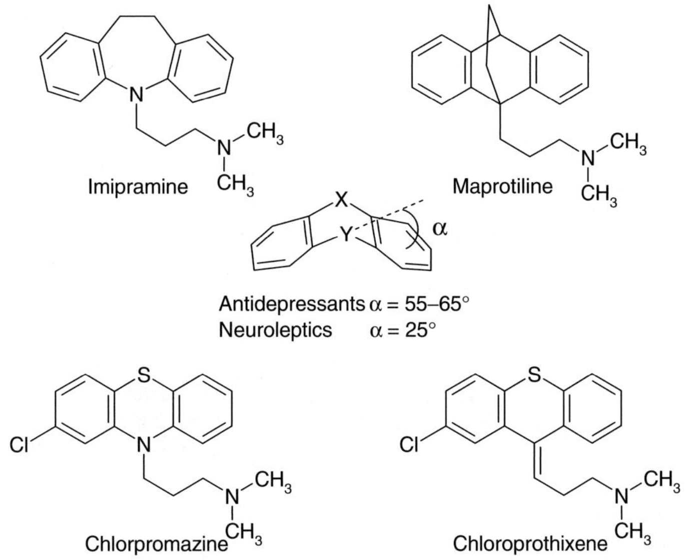

Figure 60 The tricyclic antidepressants (imipramine and maprotiline) are characterized by a dihedral angle of 55° to 65° between the two benzo rings, this angle is only 25° for the tricyclic neuroleptics (chlorpromazine, chlorprothixene).

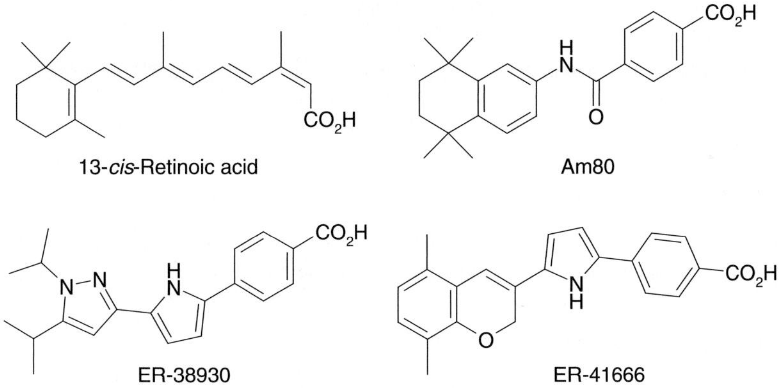

Figure 61 Bioisostery in its broadest sense can be considered as scaffold hopping.

2.16.4.2 Electronic Parameters

Electronic parameters govern the nature and the quality of ligand–receptor or ligand–enzyme interactions. The relevant parameters will be inductive or mesomeric effects, polarizability, pK_a, capacity to form hydrogen bonds, etc. Despite their very different substituents in the *meta* position, the two epinephrine analogs (**Figure 62**) exert comparable biological effects, both being alpha adrenergic agonists. In fact, the key parameter lies in the similar acidity as shown by their pK_a values.[188]

2.16.4.3 Solubility Parameters

When the functional group involved in the isosteric change plays a role in the absorption, the distribution, or the excretion of the active molecule, the hydrophilic–lipophilic parameters become important.

In order to better illustrate this concept, we can quote the example of the replacement of –CF3 ($\pi = +0.88$) by –CN ($\pi = -0.57$) in an active molecule (**Figure 63**). The electron-attracting effect of the two groups will be comparable, but the molecule with the cyano function will be clearly more hydrophilic. This loss in lipophilicity can then be corrected by attaching elsewhere on the molecule a propyl, isopropyl, or cyclopropyl group.

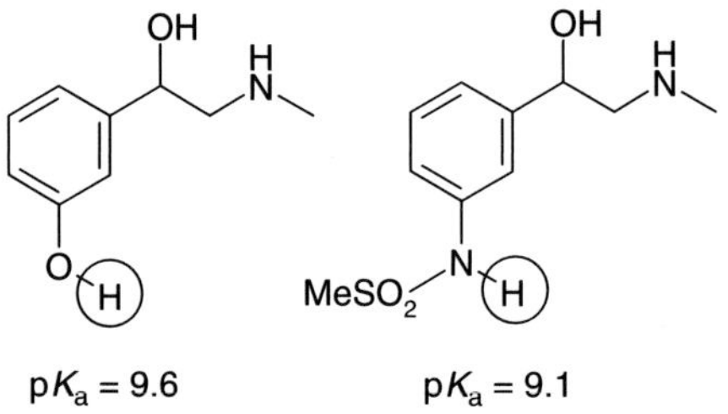

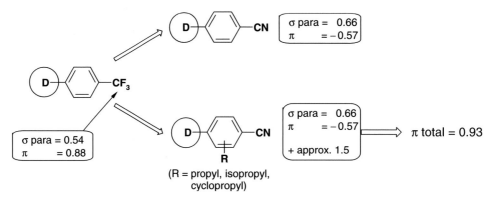

Figure 62 An example of bioisosterism, or nonclassical isosterism; the methylsulfonamide substituent has comparable acidity to the phenolic hydroxyl group.

Figure 63 The loss of lipophilicity resulting from the bioisosteric exchange of a CF_3 for a CN has to be compensated for by the equivalent of a three carbon residue.

2.16.5 Anomalies in Isosterism

In this section two unusual applications of the bioisosterism concept are developed. They concern the exchange between fluorine and hydrogen and between oxygen and methylene groups.

2.16.5.1 Fluorine–Hydrogen Isosterism

Fluorine is one of the most commonly used classical bioisosteric replacements. Although it neither resembles the other halogens nor hydrogen, it has been used as a reasonable hydrogen mimic.[189]

2.16.5.1.1 Steric aspects

It has been claimed for long time that fluorine is nearly the same size as hydrogen. This has turned out to be incorrect. If we take the commonly accepted van der Waals values of Bondi[190] or Williams and Houpt[191] instead of the obsolete Pauling values, fluorine is in actual fact close to the size of oxygen, and considerably smaller than the rest of the halogen atoms, and even than chlorine (**Table 10**).

The C–F bond length (1.39 Å) is also in the same range as the C–O bond length (1.40 Å), much bigger than C–H (1.09 Å). As a consequence, the effective van der Waals radius of trifluoromethyl (CF_3) is equal to isopropyl ($CH(CH_3)_2$) with 2.20 Å, and larger than methyl (CH_3, 1.80 Å).[192] However, there is a difference between the steric effect of fluorine, which is dependent on the transition state, and its steric size, which is an absolute value.[193] For example, if the internuclear distance is 2.5 Å, there is no H— (see similar examples)–>H or H—F steric effect (steric repulsion energy = 0 kcal mol^{-1}). But if the distance is 2.0 Å, the H—F steric repulsion energy is 1.3 kcal mol^{-1}, whereas the H—H steric repulsion energy is still negligible.

2.16.5.1.2 Electronic aspects

The electronic properties of fluorine make it a unique atom in its properties. It is the most electronegative atom in the periodic classification, and it forms particularly stable bonds with carbon atoms. The C–F bond energy is higher than

Table 10 Van der Waals radii (Å)

	Pauling	Bondi[190]	Williams and Houpt[191]
H	1.20	1.20	1.15
F	1.35	1.47	1.44
Cl	1.80	1.75	–
O	1.40	1.52	1.44

Reproduced with permission from Smart, B.E. *J. Fluor. Chem.* **2001**, *109*, 3–11.

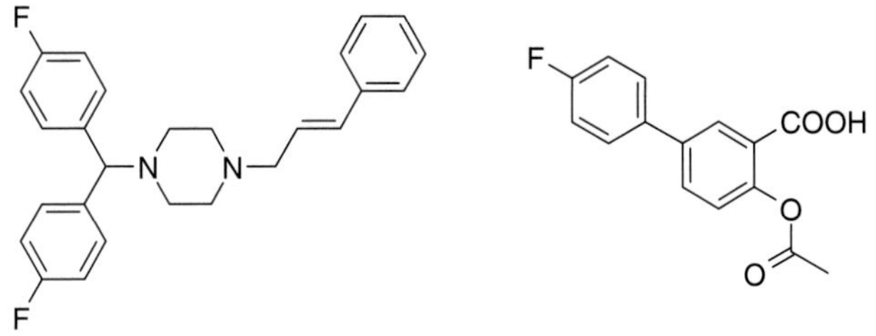

Figure 64 Fluorine atoms in the para position of the phenyl rings prevent metabolic hydroxylation.

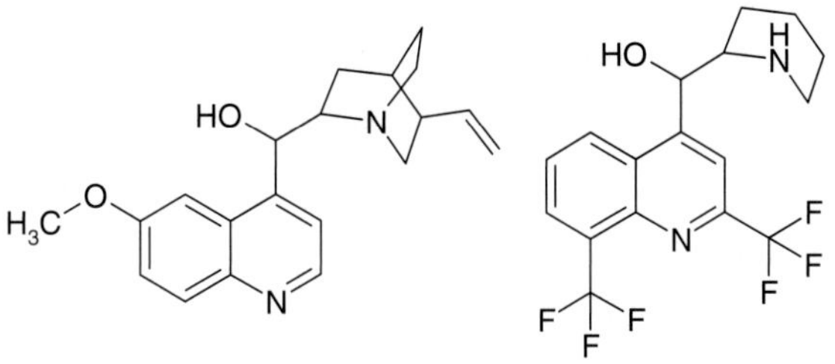

Figure 65 The half-life of quinine is approximately 11 h (oxidation in position 2 by CYP). For its analog, mefloquine, the half-life is 2–4 weeks.

the C–H or C–O bond (with 116, 99, and 85 kcal mol^{-1}, respectively), which makes the C–F bond less sensitive to metabolic degradation as found in flunarizide or in flufenisal (**Figure 64**). This particular stability explains the unaltered incorporation of fluoroacetic acid by living organisms in place of acetic acid[194] or the use of 5-fluoro-nicotinic acid and 5-fluoro-uracil as antimetabolites.

Fluorine also differs from the other halogens because it is rarely ionized or displaced. (An exception is made for aromatic systems if there is a para strong electron-withdrawing group.) Thus, for the β-haloalkylamines (nitrogen mustards), the alkylating activity is lost when chlorine or bromine are replaced by fluorine or by hydrogen.[195] This remarkable stability makes the C–F bond more resistant to direct chemical attack by cytochrome P450 (CYP) oxidases. It can also block oxidation at specific positions (**Figure 65**). A fluorine on a phenyl ring decreases the rate of interaction between the π-system and the CYP(FeO)$^{3+}$. Fluorine is also involved as a replacement for CH$_3$, which is prone to metabolism by CYP oxidases, whereas the CF$_3$ group is not susceptible to oxidation.[189]

Finally, hydrogen – fluorine bioisosterism is used to obtain analogs more resistant to metabolic degradation therefore increasing the biological half-life of the drug candidate and avoiding the formation of toxic metabolites.

2.16.5.1.3 Absence of d orbitals

The fact that fluorine has no d orbitals does not allow it to participate in resonance effects with a π electron donor (**Figure 66**). This explains why *p*-fluorophenol is slightly less acidic than phenol, while for other *p*-halogenated phenols the acidity changes in parallel with the atomic number[189] (**Table 11**).

Figure 66 The resonance between the OH lone pair and the X group is not possible if X = F.

Table 11 Dissociation constants of p-halogenated phenols[189]

Compound	Dissociation constant $K_a \times 10^{-10}$
Phenol	0.32
p-Fluorophenol	0.26
p-Chlorophenol	1.32
p-Bromophenol	1.55
p-Iodophenol	2.19

Table 12 Atomic and molecular physical properties

Parameter	H	F	Cl	CH_3	CF_3
Molecular refractivity	1.03	0.92	6.03	5.65	5.02
Electronic effect (para σ)[a]	0.00	0.06	0.23	−0.17	0.54
Resonance effect (R)[a]	0.00	−0.34	−0.15	−0.13	0.19
Electronic effect (σ^*)[b]	–	3.08	2.68	0.00	2.85
Ionization potential (kcal mol^{-1})	313.7	400.4	296.2	227.6	–
Atomic polarizability (Å^3)	0.80	0.50	2.20	–	–

[a] For aromatic systems.
[b] For aliphatic systems.

2.16.5.1.4 Lipophilicity

One interest of bioisosterism replacements is to change the physical properties of the analog. Fluorine has particular properties (**Table 12**) and its presence in a molecule could dramatically change its physical characteristics compared to the nonfluorinated analog. The effects of fluorination depend on the aromatic or the aliphatic nature of the fluorinated compound. In aromatic systems, fluorination always increases lipophilicity, and sometimes in a very significant manner (**Table 13**). When the fluorination takes place on aliphatic compounds, the prediction of the effect on lipophilicity is more difficult. Hereafter we formulate some general rules, and summarize the corresponding examples in **Table 14**[193]:

1. With alkanes, fluorination decreases the lipophilicity of the molecule.
2. With alcohols, the lipophilicity depends on the distance between the fluorine and the oxygen. If it is less than three C–C bonds in length, it increases, and if it is more, it decreases.
3. With acids and ketones, the presence of fluorine clearly modifies the electronic properties of the compounds. The hydration of aldehydes and ketones is increased, as is the acidity and hydrogen bonding capability of carboxylic acids. In these conditions, the mixture of solvents used to determine the lipophilicity plays a major role, and its transposition in biological systems is very difficult.

When the lipophilicity effect of fluorine is studied on ionizable compounds, it is important to consider other parameters such as pH. In these cases, the electronic effect of fluorine on the pK_a of these molecules is as important as its direct contribution to lipophilicity. **Table 15** summarizes the log D and pK_a values of fluorinated propranolol analogs.[196] For propranolol analogs the introduction of fluorinated amine substitutions (R = CHF_2, CF_3, CH_2CF_3) close

Table 13 Hydrophobic parameters of $C_6H_5\text{-}X$[192] $\pi_X = \log(P_X/P_H)$ (octanol–H_2O)

X	π_X
OH	−0.67
F	0.14
CH_3	0.56
CF_3	0.88
OCH_3	−0.02
OCF_3	1.04
SCH_3	0.61
SCF_3	1.44
$CH_3C(O)$	0.02
$CF_3C(O)$	0.55
CH_3SO_2	−1.63
CF_3SO_2	0.55

Reproduced with permission from Smart, B.E. *J. Fluor. Chem.* **2001**, *109*, 3–11.

Table 14 Lipophilicity of alkanes and alcohols

Molecule	*log P (octanol–H_2O)*
CH_3CH_3	1.81
CH_3CHF_2	1.75
CH_3CHCl_2	1.78
$CH_3(CH_2)_3CH_3$	3.11
$CH_3(CH_2)_3CH_2F$	2.33
CH_3CH_2OH	−0.32
CF_3CH_2OH	0.36
$CH_3(CH_2)_2OH$	0.34
$CF_3(CH_2)_2OH$	0.39
$CH_3(CH_2)_4OH$	1.19
$CF_3(CH_2)_4OH$	1.15

Reproduced with permission from Smart, B.E. *J. Fluor. Chem.* **2001**, *109*, 3–11.

to the basic nitrogen exerts a double effect on lipophilicity: (1) the fluorinated analogs are more lipophilic than their nonfluorinated analogs; and (2) fluoro substitution diminishes the basicity of the molecule and thus augments the proportion of the nonionized species and contributes to the elevation of the global lipophilicity expressed as $\log D$.

For an acid, the introduction of fluorine in the α-position enhances the acidic character of the molecule, because of the electron-withdrawing effect. So, the pK_a is decreased, as is the lipophilicity at physiological pH.[197]

2.16.5.1.5 Case studies
2.16.5.1.5.1 Tripelennamine analogs
Starting from tripelennamine (**Figure 67**, X = H), an antihistaminic drug, a series of 4-phenyl halogenated analogs have been synthesized. In vivo, *para*-hydroxylation of the phenyl ring seems to happen readily. To avoid it, the best candidate is the *para*-fluoro compound.

Table 15 Lipophilicity of propranolol analogs[196]

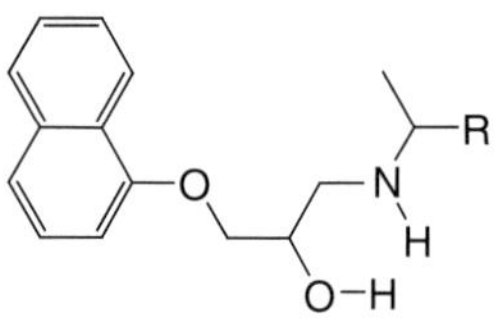

R	log P	log D (pH 7.4)	pK$_a$
CH$_3$	3.48	1.35	9.53
CH$_2$F	2.82	2.26	7.69
CHF$_2$	4.06	4.03	6.13
CF$_3$	3.66	3.66	4.37
CH(CH$_3$)$_2$	4.16	2.31	9.26
C(CH$_3$)$_3$	4.62	3.00	9.02
CH$_2$CF$_3$	3.14	3.14	4.38

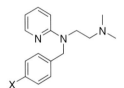

X	Activity
H	1
F	3–4
Cl	2–3
Br	1
I	0.3–0.5

Figure 67 Antihistaminic activity of *p*-halogenated compounds as a function of the halogen.[198]

2.16.5.1.5.2 Roflumilast analogs

Piclamilast and its difluorinated analog, roflumilast, are both selective phosphodiesterase inhibitors (PDE4) with the same potency in vitro. However, in vivo, roflumilast and its *N*-oxide, the primary metabolite, have superior potency, probably due to a weak oral bioavailability of piclamilast[199] (**Figure 68**). Roflumilast is in phase III clinical trials for asthma and in phase II for chronic obstructive pulmonary disease (COPD).

2.16.5.2 Exchange of Ether Oxygen and Methylene Group: The Friedman Paradox

Ether oxygen atoms and methylene groups possess a similar tetrahedral structure and should normally be isosteric. In fact O ↔ CH$_2$ isosterism very often yields anomalous results and brought Friedman[2] to the interesting observation "that the omission of the ether oxygen changes biological activity much less in some cases than the replacement by the isosteric methylene group" (**Figure 69**). In the meperidine series, for example, the change from the *N*-phenoxypropyl derivative to the isosteric phenoxybutyl decreases the analgesic potency by a factor of ten, whereas the omission of the ether oxygen yields a slightly more potent compound.[161] A possible explanation for this anomalous behavior may be that the omission of the ether oxygen yields a compound closer in terms of lipophilicity than its replacement by a methylene. Other examples are given by Schatz in the second edition of Burger's *Medicinal Chemistry*.[200]

Close examples, which do not correspond strictly to the Friedman paradox, can be found in the resemblance of the phenylethyl type β-blockers (e.g., sotalol, dichloroisoprenaline) to the phenoxypropanol type (e.g., practolol, acebutolol) and in the affinity for the same receptor of nicotine and the longer chain analogs of the ABT series[201,202] (**Figure 70**). According to docking experiments performed on a series of nicotinic agonists by Bisson *et al.*[203] on nicotine and the ABT compounds, the aromatic rings of the different molecules, overlap and interact with a same site in the receptor while the side chains of the different molecules do not overlap with each other and occupy a different part of the binding pocket.

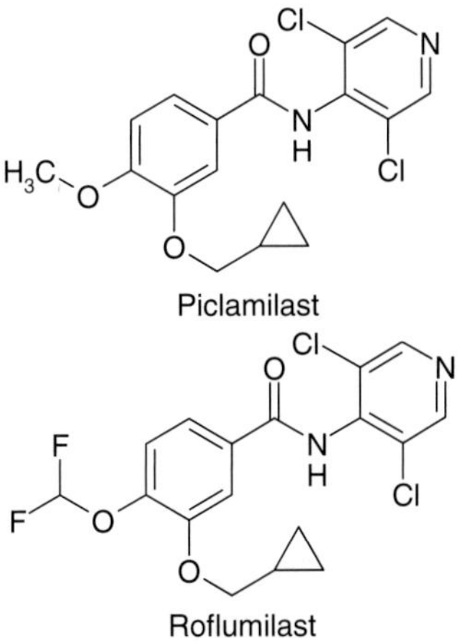

Compound	ED$_{50}$ p.o. (μmol kg^{-1})
Piclamilast	2.6
Roflumilast	0.3
N-oxide	0.3

Figure 68 Suppression by piclamilast, roflumilast, and its N-oxide metabolite of LPS-induced circulating TNFα in Sprague-Dawley rats.

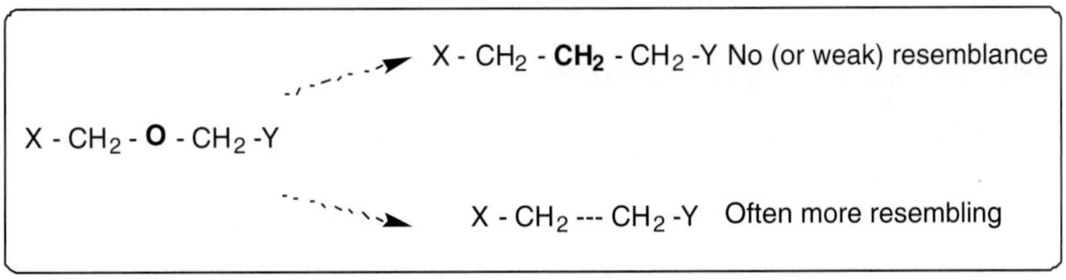

X - CH$_2$ - **O** - CH$_2$ -Y

→ X - CH$_2$ - **CH$_2$** - CH$_2$ -Y No (or weak) resemblance

→ X - CH$_2$ --- CH$_2$ -Y Often more resembling

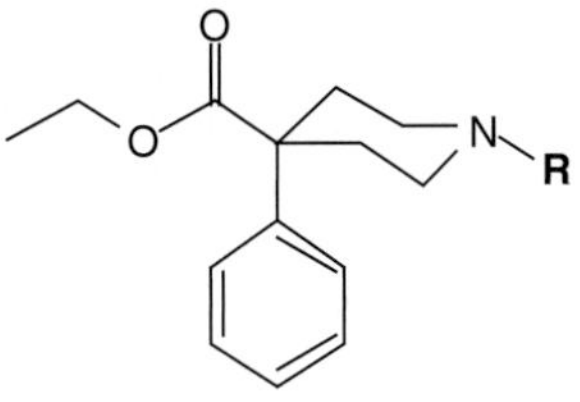

R =	Analgesic potency (meperidine = 1)
—(CH$_2$)$_3$—O—C$_6$H$_5$	15
—(CH$_2$)$_3$—CH$_2$—C$_6$H$_5$	1.5
—(CH$_2$)$_3$—C$_6$H$_5$	20

Figure 69 Friedman's ether oxygen–methylene group paradox.

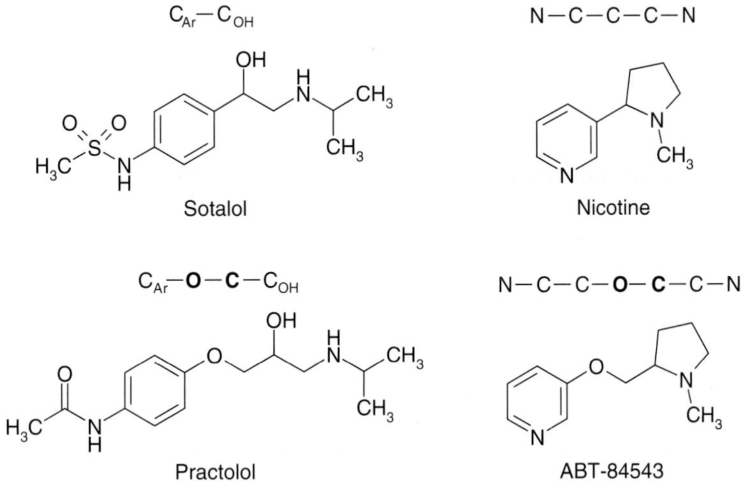

C$_{Ar}$—C$_{OH}$ Sotalol

N—C—C—C—N   Nicotine

C$_{Ar}$—O—C—C$_{OH}$ Practolol

N—C—C—O—C—C—N   ABT-84543

Figure 70 Two examples for which the omission (or, in the reverse way, the insertion) of a methylene-oxy group leads to similarly acting compounds.

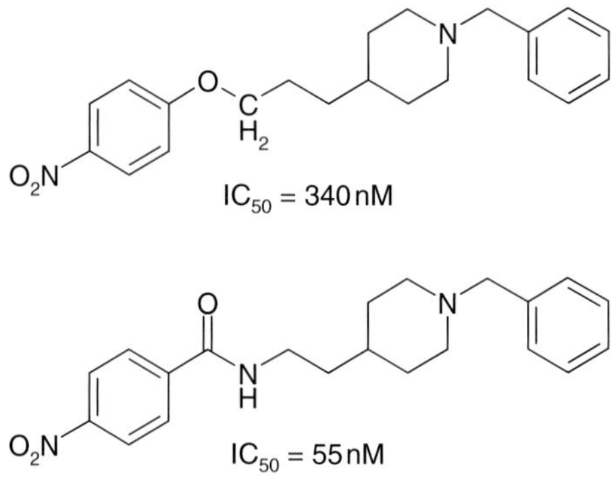

Figure 71 Unusual bioisosterism between a methylene-oxy group and a carboxamide.

A very unusual bioisosteric relationship between the methyleneoxy group and a carboxamide was observed by Sugimoto and colleagues[204] during early SAR studies, which led to the acetylcholinesterase inhibitor donepezil (Figure 71).

2.16.6 Minor Metalloids-Toxic Isosteres

This section is dedicated to more 'exotic' applications of the bioisosterism concept that involve the use of elements such as silicon, germanium, boron, and selenium. The use of these elements as carbon replacements in existing drugs is an approach for new drug-like candidates with new biological profiles and a strong IP position.

2.16.6.1 Carbon–Silicon Bioisosterism

In the periodic table of the elements, silicon is directly below carbon, which makes them, according to Erlenmeyer's expansion of the isosterism concept,[205] real isosteres. That is why the replacement of carbon by silicon in biologically active substances has been an attractive proposition for many organic chemists. However, this similarity remains limited. For reviews on the subject, see Fessenden and Fessenden,[206] Tacke,[207–209] Ricci et al.,[210] or more recently Liu et al.[211]

2.16.6.1.1 Toxicological aspects[212]
From a chemical point of view, the synthesis of silicon analogs can add substantial chemical diversity and also adds interesting biological functionalities for drug discovery. However, this has not been fully exploited, mainly because of the potential toxicity of silicon-containing molecules. However, most of the acute toxicological data (LD_{50} values) for organosilicon compounds are in the range of their carbon analogs. In 1971, Garson[213] said that "evidence is also now available to establish that organosilicon compounds *per se*, are not toxic." In a more recent perspective,[214] 74 compounds from different families were studied, and the conclusion was that, for chemically stable molecules, silicon conveys no systematic toxicological liability.

2.16.6.1.2 Silicon in medicinal chemistry
The silicon-containing bonds are always longer than their carbon analogs. For example, in the sp^3 hybridization state, the length of the C–Si bond is around 1.87 Å but the C–C bond is 1.54 Å.[215] The fact that silicon is more electropositive than carbon (and even more so compared to oxygen and nitrogen) allows the silanols to be stronger hydrogen bond donors than the corresponding carbinols.[216] This difference can produce stronger interactions with pharmacophores where carbinol acts as a hydrogen bond donor, and increases the potency of the compound.

Regarding lipophilicity, silicon-containing molecules are more lipophilic than their carbon bioisosteres. This modification of lipophilicity can be clearly observed in vivo.[217] A small increase in lipophilicity can dramatically increase the volume of distribution and therefore the tissue penetration. So, these silicon bioisosteres will be less prone to hepatic metabolism and will improve the plasma half-life when liver metabolism is significant. It will also help these drug candidates to cross the blood–brain barrier.[218]

From a chemical point of view, silicon-containing molecules are in general more sensitive to hydrolysis and nucleophilic attack. That is why, due to the chemical reactivity of silicon, carbon–silicon isosterism is generally

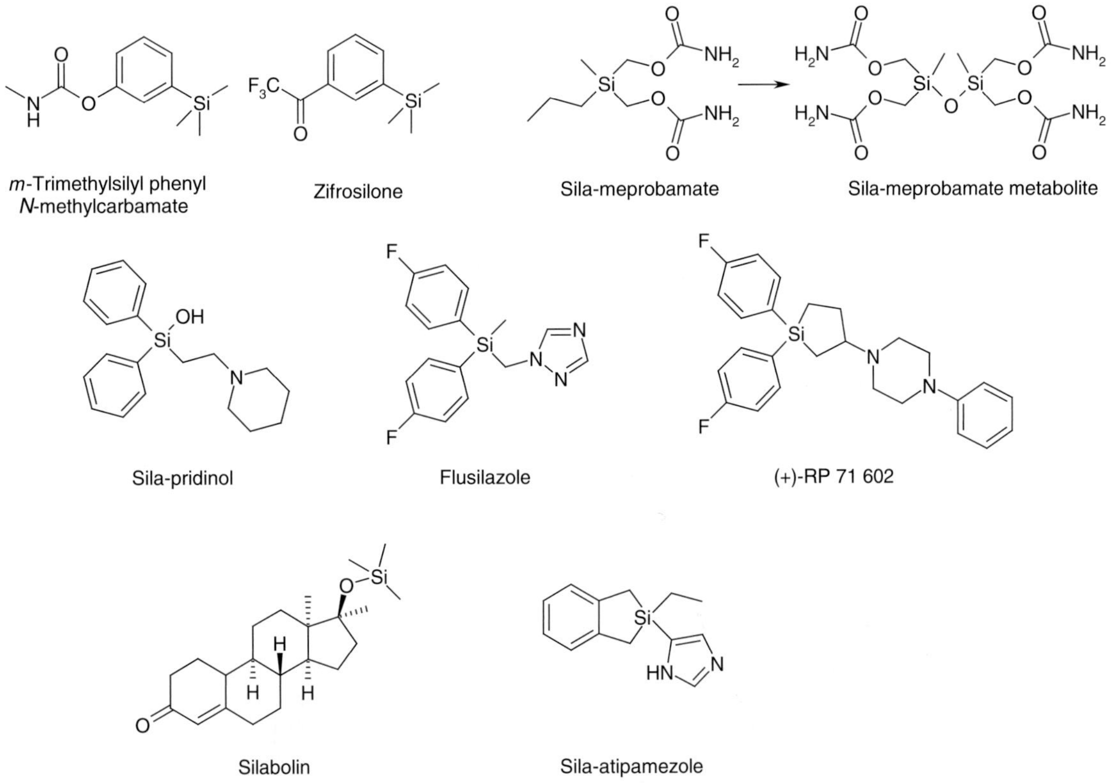

Figure 72 Organosilicon active substances.

practiced when the silicon can replace a carbon present in the center of a quaternary structure. These organosilicon compounds can exhibit various biological activities (**Figure 72**):

- acetylcholinesterase inhibitors (*m*-trimethylsilyl-phenyl *N*-methylcarbamate and *m*-trimethylsilyl-α-trifluoroacetophenone (zifrosilone))[219–221];
- CNS depressant (sila-meprobamate)[222];
- anticholinergic (sila-pridinol)[223];
- fungicide for agricultural use (flusilazole)[224];
- potent and selective 5HT2A antagonism ((+)-RP 71,602)[225];
- steroid analog (silabolin, tested for the treatment of muscle wasting disease, chronic 'cor pulmonale' (ischemic heart disease and chronic circulatory insufficiency))[226–228]; and
- α$_2$-adrenergic antagonism (sila-atipamezole),[229] etc.

2.16.6.1.3 Reactivity toward S$_N$2 and metabolic attacks

The reactivity of silicon is very different from that of carbon. Indeed, even when the silicon is tetrasubstituted, it can be easily attacked. This is due to the vacant d orbital of silicon that can still be hydroxylated by a lateral attack.[230] As an example, 1-chloro-1-sila-bicyclo-(2,2,1)-heptane can undergo an S$_N$2 reaction, whereas it is not possible with the corresponding carbon derivative (**Figure 73**). The fact that a lateral attack is possible explains why sila-meprobamate has a shorter duration of action than its carbon isostere on a model of tranquilizing activity in mice (rotarod test, potentiation of hexobarbital-induced sleep; intraperitoneal injection).[222] When sila-meprobamate is given orally, it is practically inactive. Indeed, it is quickly metabolized. The first metabolite formed has been characterized.[231] It is a di-siloxane (**Figure 72**), obtained by condensation in acidic conditions of unstable silanols.

Another metabolic oxidation known with organosilicon compounds is that of the trimethylsilyl group. In this case, the metabolism occurs not on the silicon but on one of methyl groups (Si-CH$_3$ → Si-CH$_2$-OH).[220] This is what happens for the two phenyl-trimethylsilyl-derived acetylcholinesterase (AChE) inhibitors, where the trimethyl-silyl group mimics the trimethyl-ammonium function present in acetylcholine.

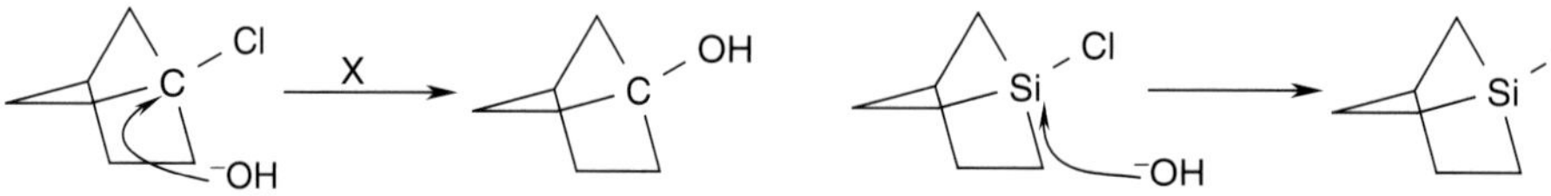

Figure 73 Owing to the presence of a vacant d orbital, a lateral attack can substitute for dorsal attacks in organo-silicon derivatives.

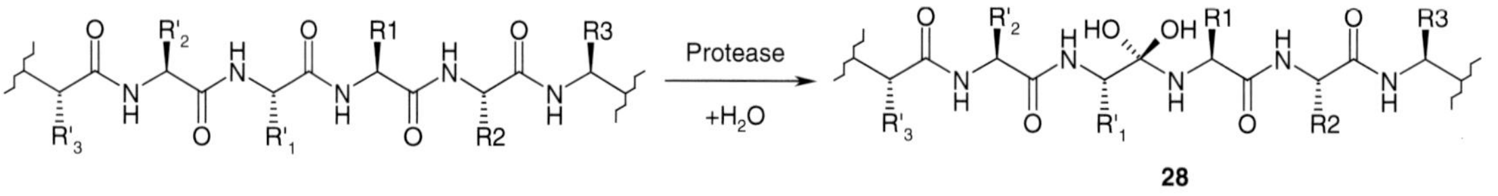

Figure 74 The tetrahedral intermediate **28** and the corresponding silicon-diol mimetic **29** are stabilized by the interaction with the protease.

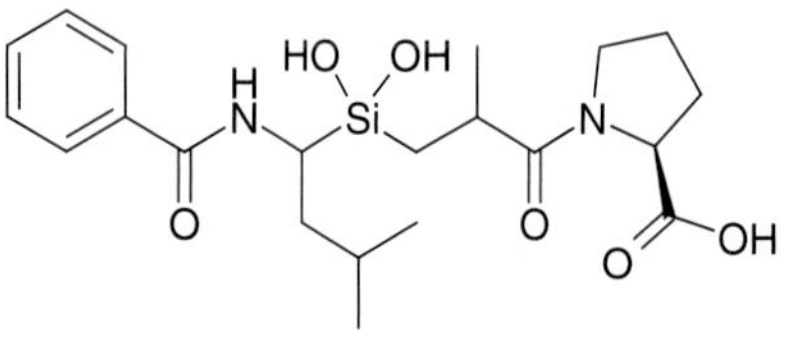

Angiotensin-converting enzyme inhibitor
IC_{50} = 14 nM

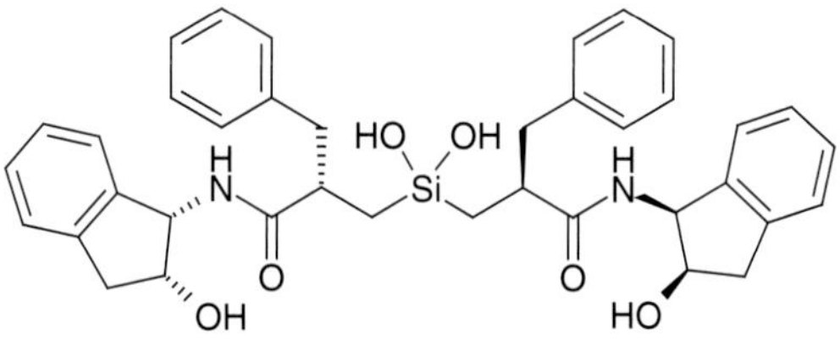

HIV protease inhibitor
IC_{50} = 2.7 nM

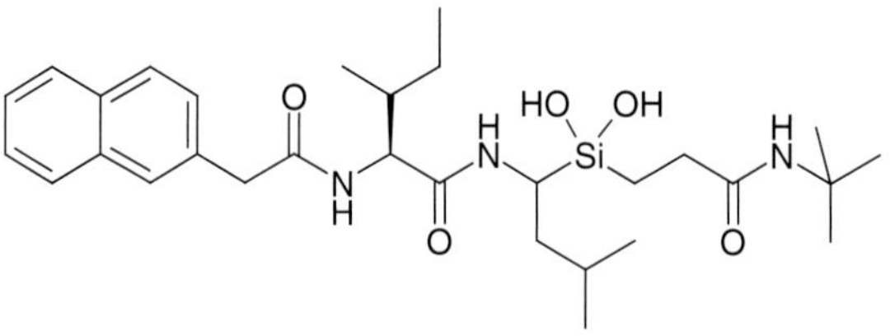

Human elastase inhibitor
IC_{50} = 300 nM

Figure 75 The structure of some silicon diol-based protease inhibitors.

2.16.6.1.4 Silicon diols[232]

Even if silicon-containing molecules are more sensitive toward nucleophiles, in some cases they are more stable than their carbon analogs, for example, regarding the interaction of a protease toward amino acids. The reaction mechanism of protease is to convert the carbonyl group of the target peptide bond into an enzyme-stabilized tetrahedral diol (**Figure 74**). Unfortunately, this diol collapses quickly to give two peptidic side products. Here, the replacement of the $C(OH)_2NH$ group of the transition state by a more stable $Si(OH)_2CH_2$ moiety represents an ideal, nonhydrolyzable analog of the tetrahedral transition state. Thanks to the stronger hydrogen bonding, it results in potent and selective inhibitors of metallo- (angiotensin-converting enzyme),[233] aspartic (HIV protease),[233] and serine (human elastase) protease[232] (**Figure 75**).

2.16.6.1.5 Diphenylpiperidine analogs[232]

The diphenylpiperidine moiety, with a quaternary carbon bearing two phenyl rings, can be considered to be a privileged structure, as it is found in many classes of biologically active molecules. The corresponding sila analog has been investigated as a possible bioisostere of quaternary carbon. In the case of budipine, studied as an agent to treat Parkinson's disease, the sila analog has shown increased lipophilicity that allows improved blood–brain barrier permeability.[234] Another example is given by niguldipine, where sila substitution of the quaternary carbon does not affect the in vitro pharmacological profile. These data suggest that sila-niguldipine derivatives may therefore be of potential benefit in the treatment of diseases such as benign prostatic hyperplasia[235] (**Figure 76**).

2.16.6.1.6 Clinical studies[212]

To date, no silicon-containing drug has been approved by the regulatory authorities in the US or Western Europe. However, some organosilicon compounds have been (or are still) studied in clinical phase II trials. Among them the following compounds can be cited (**Figure 77**):

- silperisone, a centrally acting muscle relaxant[236];
- Am-555S/TAC-101 from Taiho Pharmaceutical, an orally active retinoic acid receptor (RAR)-α-β-agonist for the treatment of lung cancer[237]; and
- karenitecin (BNP1350) from Bionumerik, a camptothecin analog.[238]

2.16.6.1.7 A few words about C/Si/Ge isostery

During the 1980s, germanium was listed as a possible anticancer drug, and even spirogermanyl and Ge-132 were sold (**Figure 78**). However, a decade of human clinical trials failed to conclusively show a benefit.[239] The major problem

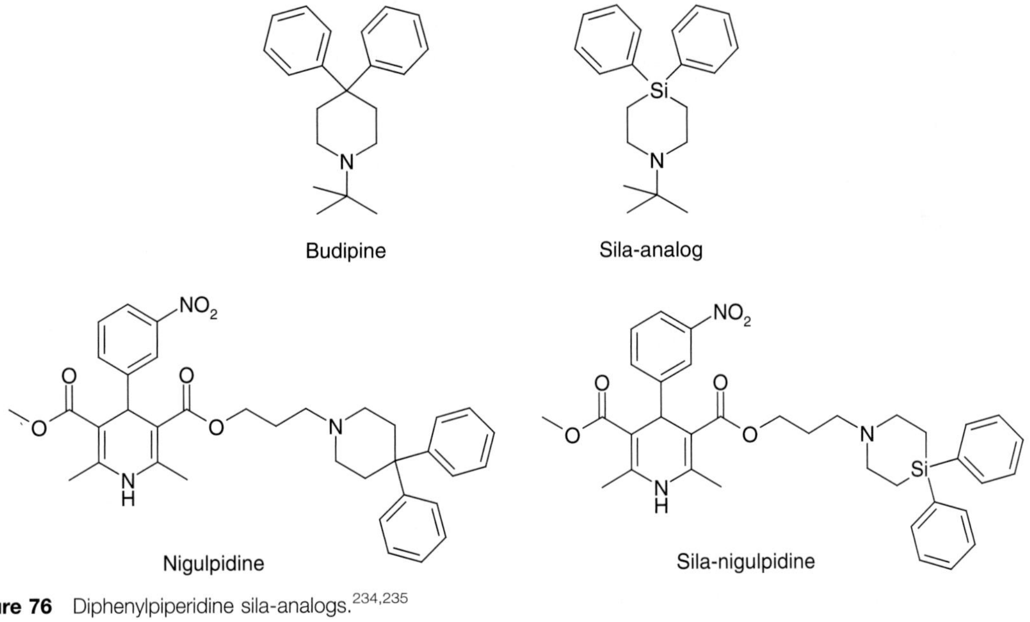

Figure 76 Diphenylpiperidine sila-analogs.[234,235]

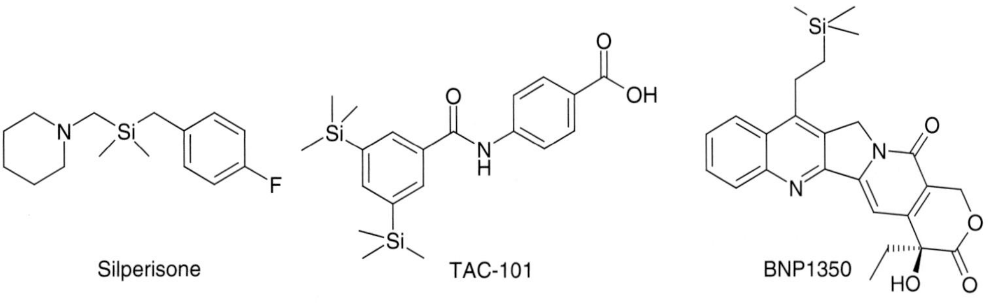

Figure 77 Silicon-containing drugs in phase II trials.

Figure 78 Organogermanium active substances.

with organogermanium compounds is their purity, as it is extremely difficult to separate the final compounds from the germanium salt used as starting material, and the germanium salts are known to be toxic.[240] Some examples of SARs with germanium-containing molecules are available in the literature,[241–243] and their biological results indicate strong bioisosterism among carbon, silicon, and germanium.

2.16.6.2 Carbon–Boron Isosterism

Boron is well known to be an essential element for higher plants. Indeed, its use in medicinal chemistry is generally limited to that of a coupling reagent. The potential of boron-containing compounds in medicine thus lags behind other elements. The most important medical use of boron derivatives today is the treatment of certain tumors by boron neutron capture therapy (BNCT),[244–246] which consists of delivering a ^{10}B nucleus directly inside the tumor cells. The major challenge of this therapy is to ensure effective concentrations of the product in the treated tumor (20–30 μg ^{10}B g^{-1})[247] – thereby keeping toxicity below the acceptable threshold.

2.16.6.2.1 Toxicity

In vivo, organoboron derivatives are sensitive to hydrolytic degradation, leading to the formation of boric acid. The long-term effects of boric acid in several species have been reported, with some toxicity on the male reproductive system and on developmental stages. The most sensitive toxicity endpoint for boric acid appears to be developmental toxicity in rats with a NOAEL (no observed adverse effect level) of 55 mg of boric acid kg^{-1} day^{-1},[248] which represents a daily consumption of 3.3 g of boric acid for humans.[249]

It has been shown that the use of boric acid at a high dose level has teratogenic effects in chickens, owing to a deficiency in riboflavin (vitamin B$_2$). The damage can be prevented by the administration of riboflavin.[250,251] In man, chronic utilization of boron-containing compounds can induce borism,[252] whose symptoms include dry skin, gastrointestinal disturbances, etc. Owing to the high quantities needed to observe a toxic effect, boric acid is classified by the World Health Organization (WHO) as very low in toxicity when ingested[253] (oral rat LD$_{50}$ = 2660 mg kg^{-1}; as an estimation, oral man LD$_{50}$ = 15–20 g kg^{-1} [254]). Investigations of the effects of boric acid on reproduction have not detected any effect on male fertility[255,256]; hence, boric acid is classified as a group E carcinogen,[257] which means that it is not considered to cause cancer (based on results from animal studies).

2.16.6.2.2 Interest and applications

Boron is peculiar in its ability to form trigonal as well as tetrahedral bonding patterns and to create complexes with organic functional groups, many of biological importance.[258]

Only a few medicines containing boron are known. The only two boron-containing natural products isolated so far are boromycin[259] (the first isolated) and aplasmomycin.[260] Both have a tetrahedral boron atom in the center of the structure, complexed with two vicinal diol groups,[261] and have antibiotic activity against Gram-positive bacteria. Most of the synthetic medicines in this class possess a boric acid or a boronic acid moiety used to esterify an α-diol or an o-diphenol. As an example, we can mention the antimony borotartrates used for injectable catecholamine solutions, like tolboxane,[262] a tranquilizer commercially available some decades ago. A more recent example is given by the phenylboronic esters of chloramphenicol.[263]

Boron-containing molecules, some examples of which are shown in **Figure 79**, have been tested on several targets and have shown different biological activities. Boronic analogs of amino acids have been identified as chymotrypsin and elastase inhibitors,[264] and more recently as an antineoplastic agent (Velcade, a proteosome inhibitor) (**Figure 79**).

Bortezomib (Velcade)
proteosome inhibitor

Diazaborinine
active against malaria

Oxazaborolidin
inhibitor of *Streptococcus mutans*

Boronic chalcone
antitumor agent

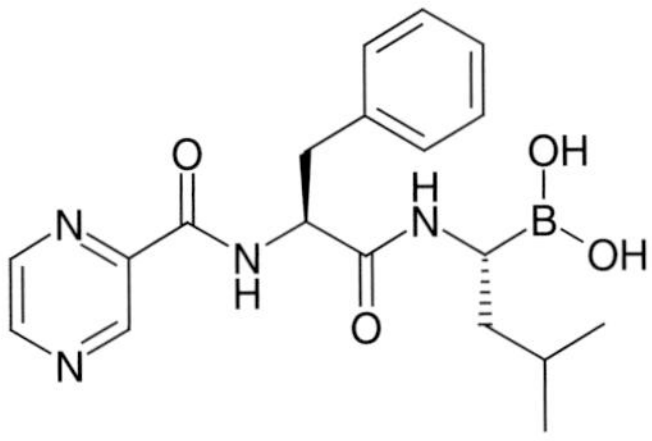

Figure 79 Organoboron active substances.

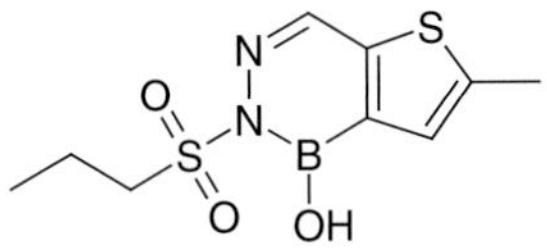

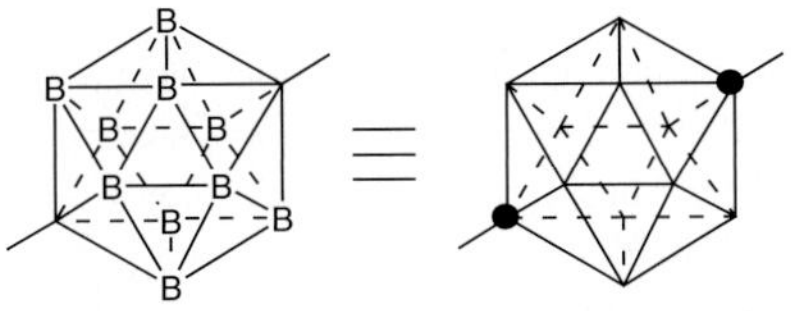

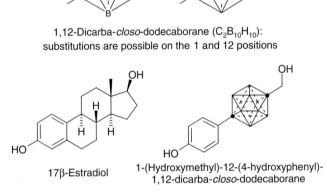

1,12-Dicarba-*closo*-dodecaborane ($C_2B_{10}H_{10}$):
substitutions are possible on the 1 and 12 positions

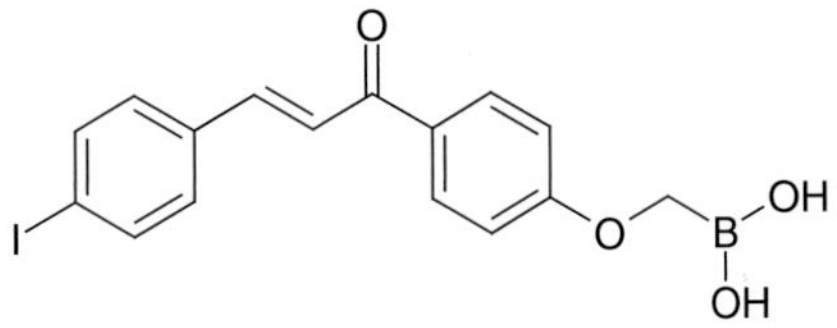

17β-Estradiol

1-(Hydroxymethyl)-12-(4-hydroxyphenyl)-1,12-dicarba-*closo*-dodecaborane

Figure 80 The dodecaborane analog has a potency 10-fold greater than that of the estradiol (luciferase reporter gene assay).[269]

Carboxyboranes complexed with a tertiary amine ($R_3N \cdot H_2B$–COOH) are considered as boron amino acid isosteres due to the isoelectronic features. Compounds containing carboxy boranes have shown anticancer, hypolipidemic, and antifungal activity.[265] Diazaborines are active against malaria,[266] and oxazaborolidines possess antibacterial activity.[267]

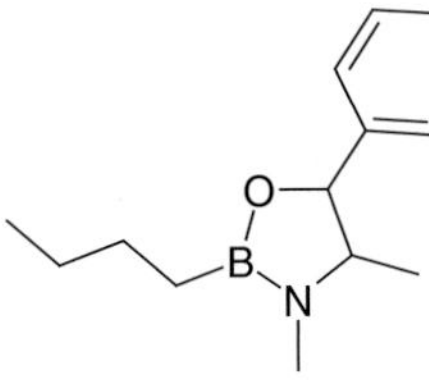

Boronic chalcones are reported to be antitumor agents. Some of them are able to inhibit the growth of human breast cancer cells and are 5- to 10-fold less toxic toward normal breast epithelial cells.[268] The remarkable hydrophobic character and geometry of carboranes (where both carbon and boron are hexacoordinated) allows their use for scaffold hopping in biologically active molecules that interact hydrophobically with receptors[269–271] (**Figure 80**).

2.16.6.2.3 Boranophosphate salts

Phosphate and phosphate-containing molecules play a major role in numerous biological systems.[272] However, the unwanted lability of the P–O bond has promoted the search for more stable bioisosteres.[273] The borane group (BH_3) is

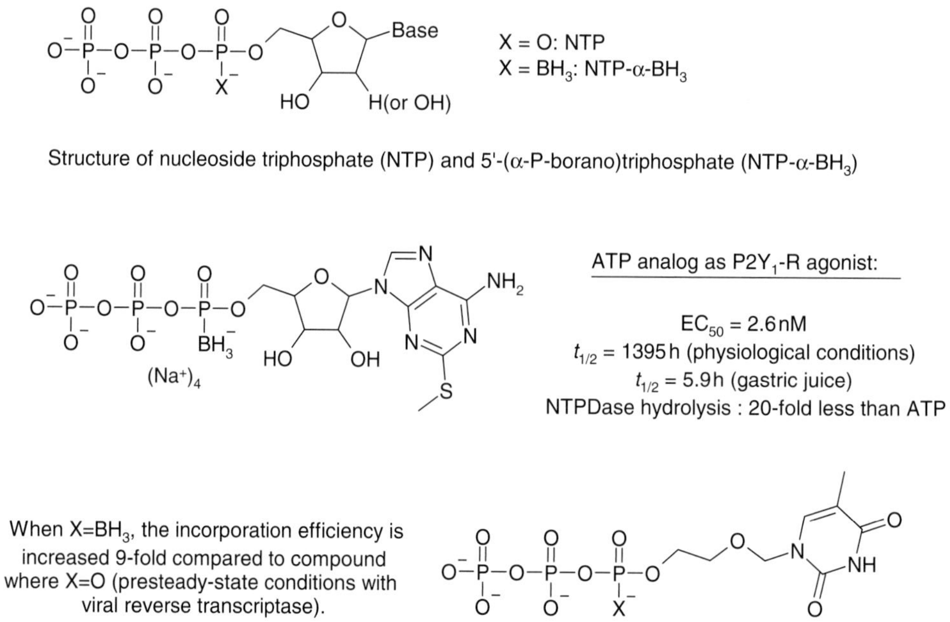

Figure 81 α-*P*-Boranotriphosphates.[274,276,277,281,282]

Structure of nucleoside triphosphate (NTP) and 5'-(α-P-borano)triphosphate (NTP-α-BH$_3$)

X = O: NTP
X = BH$_3$: NTP-α-BH$_3$

ATP analog as P2Y$_1$-R agonist:
EC$_{50}$ = 2.6 nM
$t_{1/2}$ = 1395 h (physiological conditions)
$t_{1/2}$ = 5.9 h (gastric juice)
NTPDase hydrolysis : 20-fold less than ATP

When X=BH$_3$, the incorporation efficiency is increased 9-fold compared to compound where X=O (presteady-state conditions with viral reverse transcriptase).

isoelectronic and isosteric with oxygen, but boranophosphate analogs have a different charge distribution and polarity than the corresponding natural nucleotides.[274] Nucleoside borano-phosphonate analogs are not expected to form H bonds, or to coordinate with metals,[275] but they are highly soluble in water because the borane group brings about a significant change in the nucleotide lipophilicity.[274]

The boranophosphate moiety is a new type of nucleotide modification expected to improve metabolic and chemical stability of the nucleotide derivative. A common example is given by the nucleoside 5'-(α-P-borano) triphosphate (NTP-α-BH$_3$),[274,276,277] in which a borane group (BH$_3$) substitutes one of the nonbridging α-phosphate oxygens in nucleoside 5'-triphosphate (NTP) (**Figure 81**). This group is nontoxic[274] though it is not a natural constituent of the organism. Moreover, the presence of the BH$_3$ group at the α-phosphate of clinically relevant dideoxy compounds, such as 3'-azido-3'-deoxythymidine,[278] 2',3'-didehydrodideoxythymidine (D4T),[278] and 2',3'-dideoxyadenosine (ddA),[279,280] improves both phosphorylation by nucleotide diphosphate kinase and incorporation by HIV-1 RTs.[278–280]

With the isosteric replacement of oxygen by borane, Nahum and co-workers have developed potent (in the range of ATP) and stable (under physiological conditions) purinergic P2Y$_1$-R agonists.[281] The same replacement on the phosphate chain of acyclothymidine triphosphate has led to more active compounds[282] (**Figure 81**).

2.16.6.2.4 Boron-containing molecule designed for boron neutron capture therapy

In this kind of biological approach, the goal is to deliver a ^{10}B nucleus directly inside the tumor cells. Accordingly, a variety of boron-containing molecules have been used: carboranes,[283,284] polyamines,[285] amino acids,[286] nucleoside,[287] antisense agents,[288] porphyrins,[289] and peptides[290] (**Figure 82**).

2.16.6.3 Bioisosteres Involving Selenium

In addition to the isostery between selenium and sulfur, as one appears below the other in the periodic table, these two atoms have very similar physical properties. The radius of selenium is only 12.5% bigger than that of sulfur, i.e., 1.17 Å for 1.04 Å (for comparison, the difference between oxygen and sulfur is about 43%!), and their electronegativity is rather similar. Selenium is an essential trace element because it has been found to be a component of mammalian glutathione peroxidase as selenocysteine (SeCys). SeCys is also found in the active site of two mammalian enzymes: phospholipid hydroxide glutathione-peroxidase (GSH-Px) and 5-deiodinase. Nowadays, selenium and its derivatives are generally studied as cancer chemopreventive agents in a variety of organs,[291–296] and not only as mineral supplements.

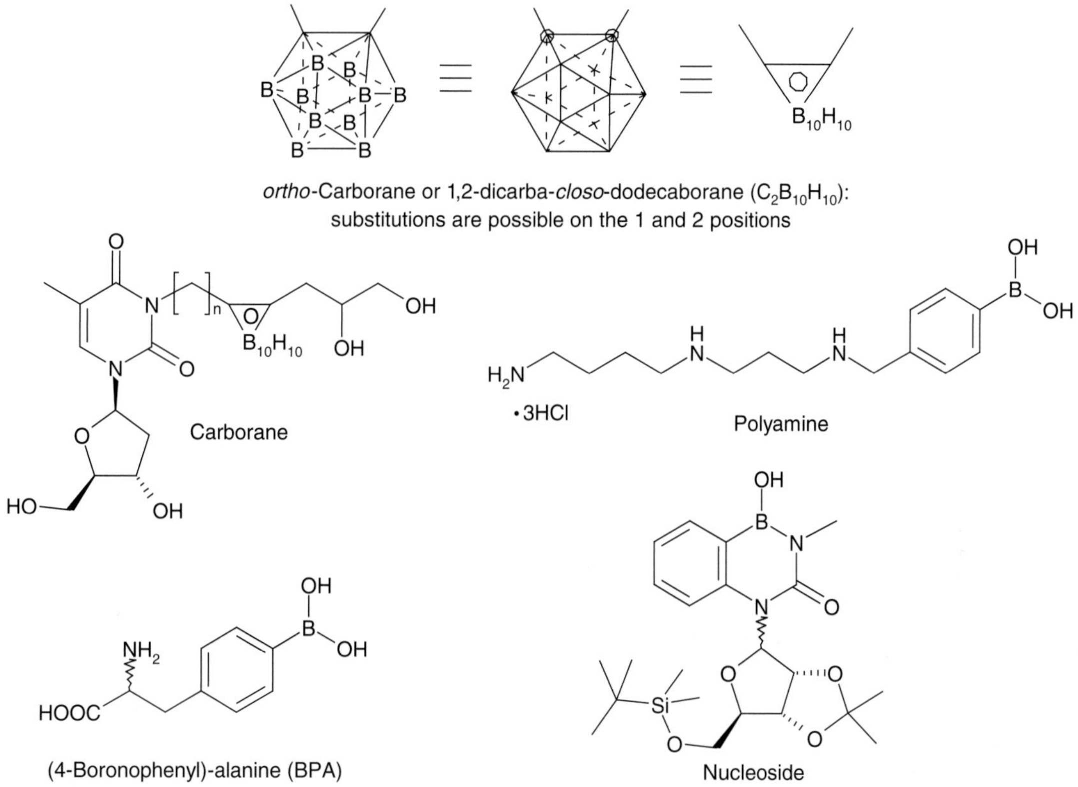

Figure 82 Examples of compounds designed for BNCT.

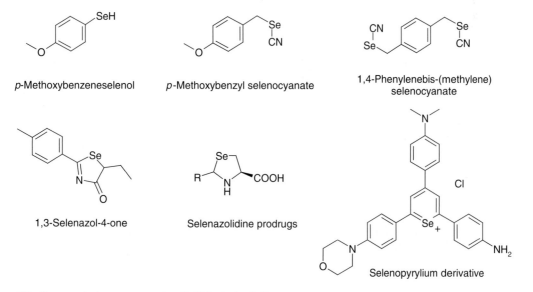

Figure 83 Organoselenium compounds with biological activities.

2.16.6.3.1 Toxicity

Before the 1950s, selenium was considered only as a poisonous substance. Then, its nutritional role was appreciated, and 30 years later, it was found to have a role in the antioxidant defense system.[297,298] Nevertheless, selenium and its inorganic derivatives are toxic for long-term or high-dose administration in humans[299,300] due to their metabolic conversion to H_2Se. The only exceptions are 75Se derivatives used for diagnostic purposes (e.g., 75Se-selenomethionine,

Table 16 Pharmacology of isonipecotic acid and its bioisosteres[314]

Compound		$GABA_C$ human p_1 receptors, K_i (μM)
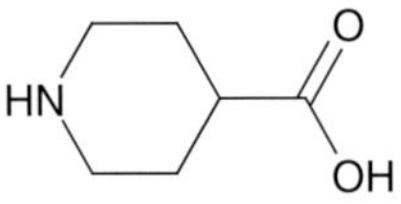 Isonipecotic acid		>300
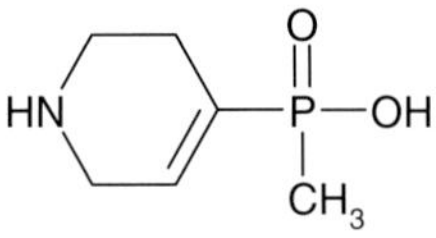 TPMPA		3.2
(piperidine-4-phosphinic acid)		2
(3-aminopropylseleninic acid)		5
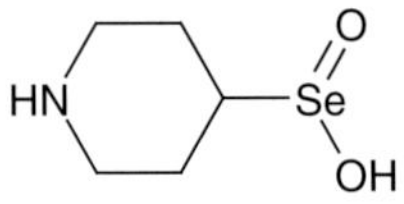 SEPI		0.95

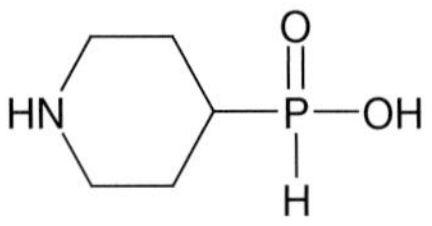

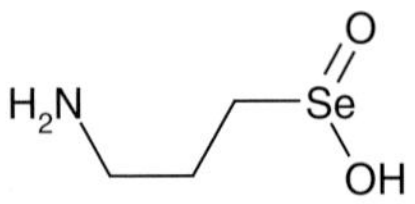

Reproduced with permission from Krehan, D.; Frolund, B.; Krogsgaard-Larsen, P.; Kehler, J.; Johnston, G. A.; Chebib, M. *Neurochem. Int.* **2003**, *42*, 561–565.

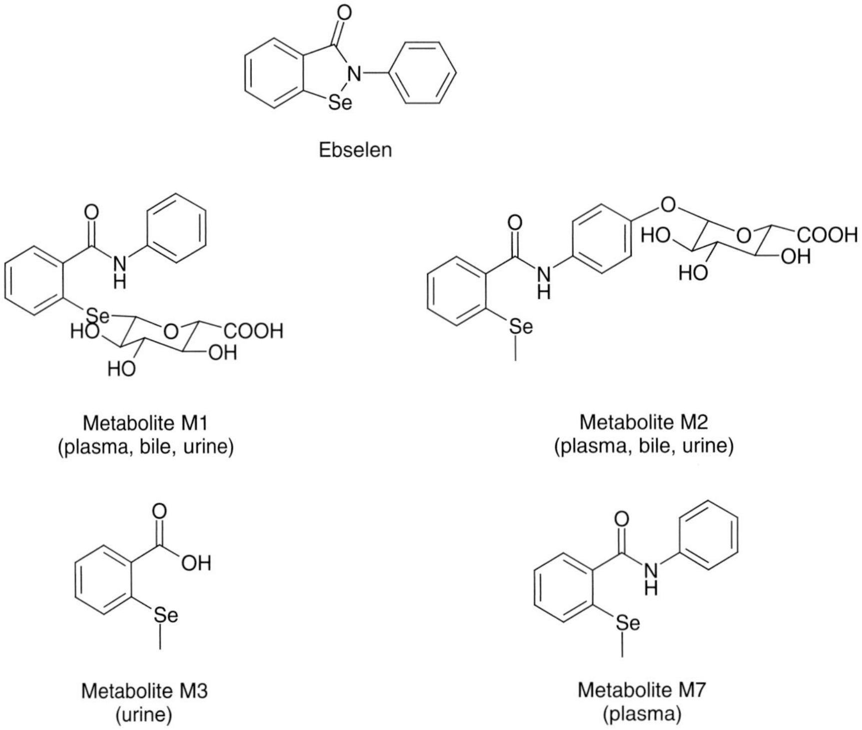

Ebselen

Metabolite M1
(plasma, bile, urine)

Metabolite M2
(plasma, bile, urine)

Metabolite M3
(urine)

Metabolite M7
(plasma)

Figure 84 Ebselen and its main metabolites.[316]

used as a radioactive imaging agent in pancreatic scanning). Organoselenium compounds are usually found to be less toxic than the inorganic forms of selenium, and they are comparatively more bioavailable.[301] Finally, on the basis of extensive studies, the upper limit of selenium per day was defined as 400 μg.[302–304] However, a consumption of 700 μg selenium per day has been reported in humans without adverse clinical symptoms.[304,305]

2.16.6.3.2 Applications

In 1973, a review containing a large number of selenium derivatives considered as chemotherapeutic agents was published.[306] Since then, organoselenium compounds have been mostly studied for their cancer preventive activities. We can cite the dietary *p*-methoxybenzeneselenol, found to inhibit azoxymethane-induced hepatocarcinogenesis in female rats without clinical toxicity,[307] or *p*-methoxybenzyl selenocyanate and 1,4-phenylenebis-(methylene)selenocyanate, which prevent both precancerous cell and tumor growth with no side effects.[308] Compared to sulfur compounds, the selenium isosteres appear to be much more effective in cancer prevention.[309]

Seleno-containing molecules have also been tested on other targets (**Figure 83**). 1,3-Selenazol-4-one derivatives are inhibitors of nitric oxide (NO) production, which plays a deleterious role in brain inflammation and neuronal death, and inducible NO synthase (iNOS) expression in LPS-induced BV-2 cells. Selenazolidine molecules can be considered prodrugs of SeCys used to deliver selenium.[310]

The relative toxicity of seleno-containing molecules is an interesting characteristic for photodynamic therapy (PDT) as a protocol for treating cancer, as well as other diseases.[311] The molecule should have a great ratio between toxicity in the dark (LD_{50}) and phototoxicity (EC_{50}). Selenopyrylium molecules are good candidates because of the heavy atom effect increasing singlet oxygen yield.[312,313] A recent study[314] on isonipecotic bioisosteres has shown that the seleno bioisosteric replacement can lead to compounds with interesting biological activity: selective $GABA_C$ receptor antagonists (**Table 16**).

The only selenium-containing drug candidate, currently in phase III trials, is ebselen (**Figure 84**). It possesses antioxidant and anti-inflammatory properties due to its ability to mimic the hydroperoxide reducing ability of the selenoenzyme glutathione-peroxidase.[315] Because of its strongly bound selenium moiety, only metabolites of low toxicity are formed.[316]

For a review on synthetic organoselenium compounds as mimics of GPx, see Mugesh and Singh.[317]

References

1. Langmuir, I. *J. Am. Chem. Soc.* **1919**, *41*, 1543–1559.
2. Friedman, H. L. In *Symposium on Chemical-Biological correlation*; Sciences, N. A. o., Ed.; National Research Council Publication: Washington DC, 1951; Vol. 206, p 295.
3. Proudfoot, J. R. In *Analog Based Drug Disc*; Fischer, J., Ganellin, C. R., Eds.; Wiley-VCH: Weinheim, 2006, pp 25–52.
4. Hegde, S.; Carter, J. *Annu. Rep. Med. Chem.* **2004**, *39*, 337–368.
5. Fischer, J.; Gere, A. *Pharmazie* **2001**, *56*, 675–682.
6. Angell, M. *The Truth About the Drug Companies: How They Deceive Us and What to do About It*; Random House: New York, 2004.
7. *Dictionnaire Historique de la Langue Française*; Rey, A., Ed.; Dictionnaires Le Robert: Paris, 1992; Vol. 1.
8. Herper, M. http://www.forbes (accessed Sep 2006).
9. Garvey, D. S.; Wasicak, J. T.; Elliott, R. L.; Lebold, S. A.; Hettinger, A. M.; Carrera, G. M.; Lin, N. H.; He, Y.; Holladay, M. W.; Anderson, D. J. et al. *J. Med. Chem.* **1994**, *37*, 4455–4463.
10. Garvey, D. S.; Wasicak, J. T.; Decker, M. W.; Brioni, J. D.; Buckley, M. J.; Sullivan, J. P.; Carrera, G. M.; Holladay, M. W.; Arneric, S. P.; Williams, M. *J. Med. Chem.* **1994**, *37*, 1055–1059.
11. Olesen, P. H.; Tønder, J. E.; Hansen, J. B.; Hansen, H. C.; Rimvall, K. *Bioorg. Med. Chem.* **2000**, *8*, 1443–1450.
12. Gohlke, H.; Gundisch, D.; Schwarz, S.; Seitz, G.; Tilotta, M. C.; Wegge, T. *J. Med. Chem.* **2002**, *45*, 1064–1072.
13. Heinisch, G.; Holzer, W.; Kunz, F.; Langer, T.; Lukavsky, P.; Pechlaner, C.; Weissenberger, H. *J. Med. Chem.* **1996**, *39*, 4058–4064.
14. Haginoya, N.; Kobayashi, S.; Komoriya, S.; Yoshino, T.; Suzuki, M.; Shimada, T.; Watanabe, K.; Hirokawa, Y.; Furugori, T.; Nagahara, T. *J. Med. Chem.* **2004**, *47*, 5167–5182.
15. Abe, Y.; Kayakiri, H.; Satoh, S.; Inoue, T.; Sawada, Y.; Inamura, N.; Asano, M.; Aramori, I.; Hatori, C.; Sawai, H.; Oku, T.; Tanaka, H. *J. Med. Chem.* **1998**, *41*, 4062–4079.
16. Contreras, J. M.; Parrot, I.; Sippl, W.; Rival, Y. M.; Wermuth, C. G. *J. Med. Chem.* **2001**, *44*, 2707–2718.
17. Contreras, J. M.; Rival, Y. M.; Chayer, S.; Bourguignon, J. J.; Wermuth, C. G. *J. Med. Chem.* **1999**, *42*, 730–741.
18. Wermuth, C. G. *The Practice of Medicinal Chemistry*, 2nd ed.; Academic Press: London, 2003.
19. Biava, M.; Porretta, G. C.; Cappelli, A.; Vomero, S.; Manetti, F.; Botta, M.; Sautebin, L.; Rossi, A.; Makovec, F.; Anzini, M. *J. Med. Chem.* **2005**, *48*, 3428–3432.
20. Barbachyn, M. R.; Ford, C. W. *Angew. Chem. Int. Ed. Engl.* **2003**, *42*, 2010–2023.
21. Nilius, A. M. *Curr. Opin. Investig. Drugs* **2003**, *4*, 149–155.
22. Snyder, L. B.; Meng, Z.; Mate, R.; D'Andrea, S. V.; Marinier, A.; Quesnelle, C. A.; Gill, P.; DenBleyker, K. L.; Fung-Tomc, J. C.; Frosco, M. et al. *J. Bioorg. Med. Chem. Lett.* **2004**, *14*, 4735–4739.
23. Blair, J. B.; Marona-Lewicka, D.; Kanthasamy, A.; Lucaites, V. L.; Nelson, D. L.; Nichols, D. E. *J. Med. Chem.* **1999**, *42*, 1106–1111.
24. Lober, S.; Hubner, H.; Gmeiner, P. *Bioorg. Med. Chem. Lett.* **2002**, *12*, 2377–2380.

25. Mathes, B. M.; Hudziak, K. J.; Schaus, J. M.; Xu, Y. C.; Nelson, D. L.; Wainscott, D. B.; Nutter, S. E.; Gough, W. H.; Branchek, T. A.; Zgombick, J. M.; Filla, S. A. *Bioorg. Med. Chem. Lett.* **2004**, *14*, 167–170.
26. Fludzinski, P.; Evrard, D. A.; Bloomquist, W. E.; Lacefield, W. B. *J. Med. Chem.* **1987**, *30*, 1535–1537.
27. Blaskó, G.; Major, E.; Blaskó, G.; Rózsa, I.; Szántay, C. *Eur. J. Med. Chem.* **1986**, *21*, 91–95.
28. Kardos, J.; Blaskó, G.; Simonyi, M.; Szántay, C. *Eur. J. Med. Chem.* **1985**, *21*, 151–154.
29. Salituro, F. G.; Harrison, B. L.; Baron, B. M.; Nyce, P. L.; Stewart, K. T.; Kehne, J. H.; White, H. S.; McDonald, I. *J. Med. Chem.* **1992**, *35*, 1791–1799.
30. Calvino, R.; Stilo, A. d.; Fruttero, R.; Gasco, A. M.; Sorba, G.; Gasco, A. *Il Farmaco.* **1993**, *48*, 321–334.
31. Lipinski, C. A.; Aldinger, C. E.; Beyer, T. A.; Bordner, J.; Burdi, D. F.; Bussolotti, D. L.; Inskeep, P. B.; Siegel, T. W. *J. Med. Chem.* **1992**, *35*, 2169–2177.
32. Almquist, R. G.; Chao, W. R.; Jennings-White, C. *J. Med. Chem.* **1985**, *28*, 1067–1071.
33. Massa, S.; Mai, A.; Sbardella, G.; Esposito, M.; Ragno, R.; Loidl, P.; Brosch, G. *J. Med. Chem.* **2001**, *44*, 2069–2072.
34. Lu, Q.; Yang, Y. T.; Chen, C. S.; Davis, M.; Byrd, J. C.; Etherton, M. R.; Umar, A.; Chen, C. S. *J. Med. Chem.* **2004**, *47*, 467–474.
35. Hanessian, S.; Moitessier, N.; Gauchet, C.; Viau, M. *J. Med. Chem.* **2001**, *44*, 3066–3073.
36. Aranapakam, V.; Grosu, G. T.; Davis, J. M.; Hu, B.; Ellingboe, J.; Baker, J. L.; Skotnicki, J. S.; Zask, A.; DiJoseph, J. F.; Sung, A. et al. *J. Med. Chem.* **2003**, *46*, 2361–2375.
37. Aranapakam, V.; Davis, J. M.; Grosu, G. T.; Baker, J.; Ellingboe, J.; Zask, A.; Levin, J. I.; Sandanayaka, V. P.; Du, M.; Skotnicki, J. S. et al. *J. Med. Chem.* **2003**, *46*, 2376–2396.
38. Noe, M. C.; Natarajan, V.; Snow, S. L.; Mitchell, P. G.; Lopresti-Morrow, L.; Reeves, L. M.; Yocum, S. A.; Carty, T. J.; Barberia, J. A.; Sweeney, F. J. *Bioorg. Med. Chem. Lett.* **2005**, *15*, 2808–2811.
39. Duan, J. J.; Chen, L.; Wasserman, Z. R.; Lu, Z.; Liu, R. Q.; Covington, M. B.; Qian, M.; Hardman, K. D.; Magolda, R. L.; Newton, R. C.; Christ, D. D.; Wexler, R. R.; Decicco, C. P. *J. Med. Chem.* **2002**, *45*, 4954–4957.
40. Kohler, H. v.; Eichler, B.; Salewski, R. *Z. Anorg. Chem.* **1970**, *379*, 183–192.
41. Kwon, C.-H.; Nagasawa, H. T.; DeMaster, E. G.; Shirota, F. N. *J. Med. Chem.* **1986**, *29*, 1922–1929.
42. Drummond, J. T.; Johnson, G. *Tetrahedron Lett.* **1988**, *29*, 1653–1656.
43. Albright, J. D.; DeVries, V. G.; Du, M. D.; Largis, E. E.; Miner, T. G.; Reich, M. F.; Shepherd, R. G. *J. Med. Chem.* **1983**, *26*, 1393–1411.
44. Uehling, D. E.; Donaldson, K. H.; Deaton, D. N.; Hyman, C. E.; Sugg, E. E.; Barrett, D. G.; Hughes, R. G.; Reitter, B.; Adkison, K. K.; Lancaster, M. E. *J. Med. Chem.* **2002**, *45*, 567–583.
45. Johansson, A.; Poliakov, A.; Akerblom, E.; Wiklund, K.; Lindeberg, G.; Winiwarter, S.; Danielson, U. H.; Samuelsson, B.; Hallberg, A. *Bioorg. Med. Chem.* **2003**, *11*, 2551–2568.
46. Buu-Hoï, N. P.; Lambelin, G.; Lepoivre, C.; Gillet, C.; Gautier, M.; Thiriaux, J. *C.R. Acad. Sci. (Paris)* **1965**, *261*, 2259–2262.
47. Noe, M. C.; Natarajan, V.; Snow, S. L.; Wolf-Gouveia, L. A.; Mitchell, P. G.; Lopresti-Morrow, L.; Reeves, L. M.; Yocum, S. A.; Otterness, I.; Bliven, M. A. *Bioorg. Med. Chem. Lett.* **2005**, *15*, 3385–3388.
48. Orzalesi, G.; Selleri, R. Ger. Offen. 2,400,531 (24 July 1974, to Societa Italo-Britannica L. Manetti & H. Roberts e C.). 1974, C.A. 1974; 81; 120272i.
49. De Martiis, F.; Corsico, N.; Franzone, J. S.; Tamietto, T. *Boll. Chim. Farm.* **1975**, *114*, 319–333.
50. Orzalesi, G.; Mari, F.; Bertol, E.; Selleri, R.; Pisaturo, G. *Arzneim.-Forsch.* **1980**, *30*, 1607–1609.
51. Demay, F.; De Sy, J. *Curr. Ther. Res.* **1982**, *31*, 113–118.
52. Vergin, H. v.; Ferber, H.; Brunner, F.; Kukovetz, W. R. *Arzneim.-Forsch.* **1981**, *31*, 513–518.
53. Allegretti, M.; Bertini, R.; Cesta, M. C.; Bizzarri, C.; Di Bitondo, R.; Di Cioccio, V.; Galliera, E.; Berdini, V.; Topai, A.; Zampella, G. et al. *J. Med. Chem.* **2005**, *48*, 4312–4331.
54. Wan, Y.; Wallinder, C.; Johansson, B.; Holm, M.; Mahalingam, A. K.; Wu, X.; Botros, M.; Karlen, A.; Pettersson, A.; Nyberg, F. et al. *J. Med. Chem.* **2004**, *47*, 1536–1546.
55. Lobb, K. L.; Hipskind, P. A.; Aikins, J. A.; Alvarez, E.; Cheung, Y. Y.; Considine, E. L.; De Dios, A.; Durst, G. L.; Ferritto, R.; Grossman, C. S. *J. Med. Chem.* **2004**, *47*, 5367–5380.
56. Mader, M. M.; Shih, C.; Considine, E.; Dios, A. D.; Grossman, C. S.; Hipskind, P. A.; Lin, H. S.; Lobb, K. L.; Lopez, B.; Lopez, J. E. et al. *Bioorg. Med. Chem. Lett.* **2005**, *15*, 617–620.
57. Hattori, K.; Tanaka, A.; Fujii, N.; Takasugi, H.; Tenda, Y.; Tomita, M.; Nakazato, S.; Nakano, K.; Kato, Y.; Kono, Y. et al. *J. Med. Chem.* **2005**, *48*, 3103–3106.
58. Bovy, P. R.; Reitz, D. B.; Collins, J. T.; Chamberlain, T. S.; Olins, G. M.; Corpus, V. M.; Mc Mahon, E. G.; Palomo, M. A.; Koepke, J. P.; McGraw, D. E.; Gaw, G. J. *J. Med. Chem.* **1993**, *36*, 101–110.
59. Marshall, W. S.; Goodson, T.; Cullinan, G. J.; Swanson-Bean, D.; Haisch, K. D.; Rinkema, L. E.; Fleisch, J. H. *J. Med. Chem.* **1987**, *30*, 682–689.
60. Chen, Y. L.; Nielsen, J.; Hedberg, K. D. A.; Jones, S.; Russo, L.; Johnson, J.; Ives, J.; Liston, D. *J. Med. Chem.* **1992**, *35*, 1429–1434.
61. Krogsgaard-Larsen, P.; Hjeds, H.; Falch, E.; Jørgensen, F. S.; Nielsen, L. *Adv. Drug Res.* **1988**, *17*, 381–456.
62. Krogsgaard-Larsen, P. In *Comprehensive Medicinal Chemistry*; Hansch, C., Sammes, P. G., Taylor, J. B., Emmet, J. C., Eds.; Pergamon Press: Oxford, UK, 1990; Vol. 3, pp 493–537.
63. Lunn, W. H.; Schoepp, D. D.; Lodge, D.; True, R. A.; Millar, J. D. *Book of Abstracts, 12th International Symposium on Medicinal Chemistry*, Basel, Switzerland, September 13–17, 1992; Swiss Chemical Society: Basel, pp 113–117.
64. Atkinson, J. G.; Girard, Y.; Rokach, J.; Rooney, C. S.; McFarlane, C. S.; Rackham, A.; Share, N. N. *J. Med. Chem.* **1979**, *22*, 90–106.
65. Henke, B. R. *J. Med. Chem.* **2004**, *47*, 4118–4127.
66. Gezginci, M. H.; Martin, A. R.; Franzblau, S. G. *J. Med. Chem.* **2001**, *44*, 1560–1563.
67. Nicolaou, I.; Zika, C.; Demopoulos, V. J. *J. Med. Chem.* **2004**, *47*, 2706–2709.
68. Qiu, J.; Stevenson, S. H.; O'Beirne, M. J.; Silverman, R. B. *J. Med. Chem.* **1999**, *42*, 329–332.
69. Liljebris, C.; Larsen, S. D.; Ogg, D.; Palazuk, B. J.; Bleasdale, J. E. *J. Med. Chem.* **2002**, *45*, 1785–1798.
70. Momose, Y.; Maekawa, T.; Odaka, H.; Ikeda, H.; Sohda, T. *Chem. Pharm. Bull. Tokyo* **2002**, *50*, 100–111.
71. Biot, C.; Bauer, H.; Schirmer, R. H.; Davioud-Charvet, E. *J. Med. Chem.* **2004**, *47*, 5972–5983.
72. Kimura, T.; Shuto, D.; Hamada, Y.; Igawa, N.; Kasai, S.; Liu, P.; Hidaka, K.; Hamada, T.; Hayashi, Y.; Kiso, Y. *Bioorg. Med. Chem. Lett.* **2005**, *15*, 211–215.
73. Roppe, J.; Smith, N. D.; Huang, D.; Tehrani, L.; Wang, B.; Anderson, J.; Brodkin, J.; Chung, J.; Jiang, X.; King, C. et al. *J. Med. Chem.* **2004**, *47*, 4645–4648.

74. Kozikowski, A. P.; Zhang, J.; Nan, F.; Petukhov, P. A.; Grajkowska, E.; Wroblewski, J. T.; Yamamoto, T.; Bzdega, T.; Wroblewska, B.; Neale, J. H. *J. Med. Chem.* **2004**, *47*, 1729–1738.
75. Valgeirsson, J.; Nielsen, E. O.; Peters, D.; Mathiesen, C.; Kristensen, A. S.; Madsen, U. *J. Med. Chem.* **2004**, *47*, 6948–6957.
76. Kraus, J. L. *Pharmacol. Res. Commun.* **1983**, *15*, 183–189.
77. Hallinan, E. A.; Tsymbalov, S.; Dorn, C. R.; Pitzele, B. S.; Hansen, D. W., Jr.; Moore, W. M.; Jerome, G. M.; Connor, J. R.; Branson, L. F.; Widomski, D. L. *J. Med. Chem.* **2002**, *45*, 1686–1689.
78. Schlewer, G.; Wermuth, C. G.; Chambon, J.-P. *Eur. J. Med. Chem.* **1984**, *19*, 181–186.
79. Krogsgaard-Larsen, P.; Rodolskov-Christiansen, T. *Eur. J. Med. Chem.* **1979**, *14*, 157–164.
80. Krogsgaard-Larsen, P.; Ferkany, J. W.; Nielsen, E. O.; Madsen, U.; Ebert, B.; Johansen, J. S.; Diemer, S. H.; Bruhn, T.; Beattie, D. T.; Curtis, D. R. *J. Med. Chem.* **1991**, *34*, 123–130.
81. Hulin, B.; McCarthy, P. A.; Gibbs, E. M. *Curr. Pharm. Des.* **1996**, *2*, 85–102.
82. Kraus, J. L. *Pharmacol. Res. Commun.* **1983**, *15*, 119–129.
83. Lichtenthaler, F. W.; Heidel, P. *Angew. Chem. Int. Ed. Engl.* **1969**, *8*, 978–979.
84. Froestl, W.; Furet, P.; Hall, R. G.; Mickel, S. J.; Strub, D.; Sprecher, G. v.; Baumann, P. A.; Bernasconi, R.; Brugger, F.; Felner, A. In *Perspectives in Medicinal Chemistry*; Testa, B., Kyburz, E., Fuhrer, W., Giger, R., Eds.; VHC: Weinheim, Germany, 1993, pp 259–272.
85. Rosowski, A.; Forsch, R. A.; Freisheim, J. H.; Moran, R. G.; Wick, M. *J. Med. Chem.* **1984**, *27*, 600–604.
86. Watkins, J. C.; Krogsgaard-Larsen, P.; Honoré, T. *Trends Pharmacol. Sci.* **1990**, *11*, 25–33.
87. Drysdale, M. J.; Pritchard, M. C.; Horwell, D. C. *J. Med. Chem.* **1992**, *35*, 2573–2581.
88. Franz, R. D. *AAPS PharmSci.* **2001**, *3*, E10.
89. Kinney, W. A.; Lee, N. E.; Garrison, D. T.; Podlesny Jr, E. J.; Simmonds, J. T.; Bramlet, D.; Notvest, R. R.; Kowal, D. M.; Tasse, R. P. *J. Med. Chem.* **1992**, *35*, 4720–4726.
90. Gao, Y.; Luo, Z.-J.; Yao, Z.-J.; Guo, R.; Zou, H.; Kelley, J.; Voigt, J. H.; Yang, D.; Burke, T. R., Jr. *J. Med. Chem.* **2000**, *43*, 911–920.
91. Shapiro, G.; Floersheim, P.; Boelsterli, J.; Amstutz, R.; Bolliger, G.; Gammenthaler, H.; Gmelin, G.; Supavilai, P.; Walkinshaw, M. *J. Med. Chem.* **1992**, *35*, 15–27.
92. Thompkins, L.; Lee, K. H. *J. Pharm. Sci.* **1975**, *64*, 760–763.
93. Roth, G. J.; Stanford, N.; Majerus, P. W. *Proc. Natl. Acad. Sci. USA* **1975**, *72*, 3073–3076.
94. Saunders, J.; Cassidy, M.; Freedman, S. B.; Harley, E. A.; Iversen, L. L.; Kneen, C.; MacLeod, A. M.; Merchant, K.; Snow, R. J.; Baker, R. *J. Med. Chem.* **1990**, *33*, 1128–1138.
95. Sauerberg, P.; Kindtler, J. W.; Nielsen, L.; Sheardown, M. J.; Honoré, T. *J. Med. Chem.* **1991**, *34*, 687–692.
96. Sauerberg, P.; Olesen, P. H.; Nielsen, S.; Treppendahl, S.; Sheardown, M. J.; Honoré, T.; Mitch, C. H.; Ward, J. S.; Pike, A. J.; Bymaster, F. P. et al. *J. Med. Chem.* **1992**, *35*, 2274–2283.
97. Wadsworth, H. J.; Jenkins, S. M.; Orlek, B. S.; Cassidy, F.; Clark, M. S. G.; Brown, F.; Riley, G. J.; Graves, D.; Hawkins, J.; Naylor, C. *J. Med. Chem.* **1992**, *35*, 1280–1290.
98. Street, L. J.; Baker, R.; Book, T.; Reeve, A. J.; Saunders, J.; Willson, T.; Marwood, R. S.; Patel, S.; Freedman, S. B. *J. Med. Chem.* **1992**, *35*, 295–305.
99. Bundgaard, H. *Design of Prodrugs*,; Elsevier: Amsterdam, the Netherlands, 1985.
100. Sauerberg, P.; Chen, J.; WoldeMussie, E.; Rapoport, H. *J. Med. Chem.* **1989**, *32*, 1322–1326.
101. Carroll, F. I.; Trisha, T. *J. Med. Chem.* **2003**, *46*, 1775–1794.
102. Petukhov, P. A.; Zhang, M.; Johnson, K. J.; Tella, S. R.; Kozikowski, A. P. *Bioorg. Med. Chem. Lett.* **2001**, *11*, 2079–2083.
103. Street, L. J.; Baker, R.; Castro, J. L.; Chambers, M. S.; Guiblin, A. R.; Hobbs, S. C.; Matassa, V. G.; Reeve, A. J.; Beer, M. S.; Middlemiss, D. N. et al. *J. Med. Chem.* **1993**, *36*, 1529–1538.
104. Kuduk, S. D.; Ng, C.; Feng, D. M.; Wai, J. M.; Chang, R. S.; Harrell, C. M.; Murphy, K. L.; Ransom, R. W.; Reiss, D.; Ivarsson, M. et al. *J. Med. Chem.* **2004**, *47*, 6439–6442.
105. Watson, K. G.; Brown, R. N.; Cameron, R.; Chalmers, D. K.; Hamilton, S.; Jin, B.; Krippner, G. Y.; Luttick, A.; McConnell, D. B.; Reece, P. A. et al. *J. Med. Chem.* **2003**, *46*, 3181–3184.
106. Kozikowski, A. P.; Roberti, M.; Xiang, L.; Bergmann, J. S.; Callahan, P. M.; Cunningham, K. A.; Johnson, K. M. *J. Med. Chem.* **1992**, *35*, 4764–4766.
107. Lipinski, C. A. *Annu. Rep. Med. Chem.* **1986**, *21*, 283–291.
108. Wermuth, C. G. *Il Farmaco* **1993**, *48*, 253–274.
109. Huff, J. R.; Anderson, P. S.; Baldwin, J. J.; Clineschmidt, B. V.; Guare, J. P.; Lotti, V. J.; Pettibone, D. J.; Randall, W. C.; Vacca, J. P. *J. Med. Chem.* **1985**, *28*, 1756–1759.
110. Fournier, J.-P.; Moreau, R. C.; Narcisse, G.; Choay, P. *Eur. J. Med. Chem.* **1982**, *17*, 81–84.
111. Spatola, A. F. In *Chemistry and Biochemistry of Amino Acids, Peptides and Proteins*; Weinstein, B., Ed.; Marcel Dekker: New York, 1983; Vol. 7, pp 267–357.
112. Fauchère, J.-L. *Adv. Drug Res.* **1986**, *15*, 29–69.
113. Smith, A. B., III; Keenan, T. P.; Holcomb, R. C.; Sprengeler, P. A.; Guzman, M. C.; Wood, J. L.; Caroll, P. J.; Hirschmann, R. *J. Am. Chem. Soc.* **1992**, *114*, 10672–10674.
114. Smith, A. B., III; Holcomb, R. C.; Guzman, M. C.; Keenan, T. P.; Sprengeler, P. A.; Hirschmann, R. *Tetrahedron Lett.* **1993**, *34*, 63–66.
115. Allmendinger, T.; Felder, E.; Hungerbuehler, E. In *Selective Fluorination in Organic and Bioorganic Chemistry*; Weldi, J. T., Ed.; ACS Symposium Series 456; American Chemical Society: Washington, DC, 1991, pp 186–195.
116. Chandrakumar, N. S.; Yonan, P. K.; Stapelfeld, A.; Svage, M.; Rorbacher, E.; Contreras, P. C.; Hammond, D. *J. Med. Chem.* **1992**, *35*, 223–233.
117. Vu, C. B.; Corpuz, E. G.; Merry, T. J.; Pradeepan, S. G.; Bartlett, C.; Bohacek, R. S.; Botfield, M. C.; Eyermann, C. J.; Lynch, B. A.; MacNeil, I. A. et al. *J. Med. Chem.* **1999**, *42*, 4088–4098.
118. Clitherow, J. W.; Beswick, P.; Irving, W. J.; Scopes, D. I. C.; Barnes, J. C.; Clapham, J.; Brown, J. D.; Evans, D. J.; Hayes, A. G. *Bioorg. Med. Chem. Lett.* **1996**, *6*, 833–838.
119. Swain, C. J.; Baker, R.; Kneen, C.; Moseley, J.; Saunders, J.; Seward, E. M.; Stevenson, G.; Beer, M.; Stanton, J.; Watling, K. *J. Med. Chem.* **1991**, *34*, 140–151.
120. Naylor, E. M.; Colandrea, V. J.; Candelore, M. R.; Cascieri, M. A.; Colwell, L. F., Jr.; Deng, L.; Feeney, W. P.; Forrest, M. J.; Hom, G. J.; MacIntyre, D. E. *Bioorg. Med. Chem. Lett.* **1998**, *8*, 3087–3092.

121. Naylor, E. M.; Parmee, E. R.; Colandrea, V. J.; Perkins, L.; Brockunier, L.; Candelore, M. R.; Cascieri, M. A.; Colwell, L. F., Jr.; Deng, L.; Feeney, W. P. et al. *Bioorg. Med. Chem. Lett.* **1999**, *9*, 755–758.
122. Parmee, E. R.; Naylor, E. M.; Perkins, L.; Colandrea, V. J.; Ok, H. O.; Candelore, M. R.; Cascieri, M. A.; Deng, L.; Feeney, W. P.; Forrest, M. J. et al. *Bioorg. Med. Chem. Lett.* **1999**, *9*, 749–754.
123. Biftu, T.; Feng, D. D.; Liang, G. B.; Kuo, H.; Qian, X.; Naylor, E. M.; Colandrea, V. J.; Candelore, M. R.; Cascieri, M. A.; Colwell, L. F., Jr. et al. *Bioorg. Med. Chem. Lett.* **2000**, *10*, 1431–1434.
124. Feng, D. D.; Biftu, T.; Candelore, M. R.; Cascieri, M. A.; Colwell, L. F., Jr.; Deng, L.; Feeney, W. P.; Forrest, M. J.; Hom, G. J.; MacIntyre, D. E. et al. *Bioorg. Med. Chem. Lett.* **2000**, *10*, 1427–1429.
125. Tully, W. R.; Gardner, C. R.; Gillespie, R. J.; Westwood, R. *J. Med. Chem.* **1991**, *34*, 2060–2067.
126. Orlek, B. S.; Blaney, F. E.; Brown, F.; Clark, M. S.; Hadley, M. S.; Hatcher, J.; Riley, G. J.; Rosenberg, H. E.; Wadsworth, H. J.; Wyman, P. *J. Med. Chem.* **1991**, *34*, 2726–2735.
127. Ladduwahetty, T.; Baker, R.; Cascieri, M. A.; Chambers, M. S.; Haworth, K.; Keown, L. E.; MacIntyre, D. E.; Metzger, J. M.; Owen, S.; Rycroft, W. et al. *J. Med. Chem.* **1996**, *39*, 2907–2914.
128. Borg, S.; Vollinga, R. C.; Labarre, M.; Payza, K.; Terenius, L.; Luthman, K. *J. Med. Chem.* **1999**, *42*, 4331–4342.
129. Elzein, E.; Ibrahim, P.; Koltun, D. O.; Rehder, K.; Shenk, K. D.; Marquart, T. A.; Jiang, B.; Li, X.; Natero, R.; Li, Y. et al. *Bioorg. Med. Chem. Lett.* **2004**, *14*, 6017–6021.
130. Koltun, D. O.; Marquart, T. A.; Shenk, K. D.; Elzein, E.; Li, Y.; Nguyen, K.; Kerwar, S.; Zeng, D.; Chu, N.; Soohoo, D. et al. *Bioorg. Med. Chem. Lett.* **2004**, *14*, 549–552.
131. Einsiedel, J.; Hubner, H.; Gmeiner, P. *Bioorg. Med. Chem. Lett.* **2001**, *11*, 2533–2536.
132. Einsiedel, J.; Hubner, H.; Gmeiner, P. *Bioorg. Med. Chem. Lett.* **2003**, *13*, 851–854.
133. Einsiedel, J.; Thomas, C.; Hubner, H.; Gmeiner, P. *Bioorg. Med. Chem. Lett.* **2000**, *10*, 2041–2044.
134. Thurkauf, A.; Yuan, J.; Chen, X.; He, X. S.; Wasley, J. W.; Hutchison, A.; Woodruff, K. H.; Meade, R.; Hoffman, D. C.; Donovan, H.; Jones-Hertzog, D. K. *J. Med. Chem.* **1997**, *40*, 1–3.
135. Ganellin, C. R. In *Medicinal Chemistry – The Role of Organic Chemistry in Drug Research*; Ganellin, C. R., Roberts, S. M., Eds.; Academic Press: London, 1993, pp 227–255.
136. Wouters, J.; Michaux, C.; Durant, F.; Dogne, J. M.; Delarge, J.; Masereel, B. *Eur. J. Med. Chem.* **2000**, *35*, 923–929.
137. Luo, G.; Mattson, G. K.; Bruce, M. A.; Wong, H.; Murphy, B. J.; Longhi, D.; Antal-Zimanyi, I.; Poindexter, G. S. *Bioorg. Med. Chem. Lett.* **2004**, *14*, 5975–5978.
138. Poindexter, G. S.; Bruce, M. A.; Breitenbucher, J. G.; Higgins, M. A.; Sit, S. Y.; Romine, J. L.; Martin, S. W.; Ward, S. A.; McGovern, R. T.; Clarke, W.; Russell, J.; Antal-Zimanyi, I. *Bioorg. Med. Chem.* **2004**, *12*, 507–521.
139. Lumma, W. C., Jr.; Anderson, P. S.; Baldwin, J. J.; Bolhofer, W. A.; Habecker, C. N.; Hirshfield, J. M.; Pietruszkewicz, A. M.; Randall, W. C.; Torchiana, M. L.; Britcher, S. F. et al. *J. Med. Chem.* **1982**, *25*, 207–210.
140. Young, R. C.; Ganellin, C. R.; Graham, M. J.; Grant, E. H. *Tetrahedron* **1982**, *38*, 1493–1497.
141. Young, R. C.; Ganellin, C. R.; Graham, M. J.; Roantree, M. J.; Grant, E. H. *Tetrahedron Lett.* **1985**, *26*, 1897–1900.
142. Uloth, R. H.; Kirk, J. R.; Gould, W. A.; Larsen, A. A. *J. Med. Chem.* **1966**, *9*, 88–97.
143. Burger, A. *Prog. Drug Res.* **1991**, *37*, 287–371.
144. Wu, W. L.; Burnett, D. A.; Spring, R.; Greenlee, W. J.; Smith, M.; Favreau, L.; Fawzi, A.; Zhang, H.; Lachowicz, J. E. *J. Med. Chem.* **2005**, *48*, 680–693.
145. Wright, J. L.; Gregory, T. F.; Kesten, S. R.; Boxer, P. A.; Serpa, K. A.; Meltzer, L. T.; Wise, L. D. *J. Med. Chem.* **2000**, *43*, 3408–3419.
146. Hacksell, U.; Arvidsson, L. E.; Svensson, U.; Nilsson, J. L.; Sanchez, D.; Wikstrom, H.; Lindberg, P.; Hjorth, S.; Carlsson, A. *J. Med. Chem.* **1981**, *24*, 1475–1482.
147. Bhaird, N. N.; Fowler, C. J.; Thorberg, O.; Tipton, K. F. *Biochem. Pharmacol.* **1985**, *34*, 3599–3601.
148. Jaen, J. C.; Wise, L. D.; Caprathe, B. W.; Tecle, H.; Bergmeier, S.; Humblet, C. C.; Heffner, T. G.; Meltzer, L. T.; Pugsley, T. A. *J. Med. Chem.* **1990**, *33*, 311–317.
149. Mierau, J.; Schneider, F. J.; Ensinger, H. A.; Chio, C. L.; Lajiness, M. E.; Huff, R. M. *Eur. J. Pharmacol.* **1995**, *290*, 29–36.
150. Schneider, C. S.; Mierau, J. *J. Med. Chem.* **1987**, *30*, 494–498.
151. van Vliet, L. A.; Rodenhuis, N.; Wikstrom, H.; Pugsley, T. A.; Serpa, K. A.; Meltzer, L. T.; Heffner, T. G.; Wise, L. D.; Lajiness, M. E.; Huff, R. M. *J. Med. Chem.* **2000**, *43*, 3549–3557.
152. Blum, G.; Gazit, A.; Levitzki, A. *J. Biol. Chem.* **2003**, *278*, 40442–40454.
153. Hubner, H.; Haubmann, C.; Utz, W.; Gmeiner, P. *J. Med. Chem.* **2000**, *43*, 756–762.
154. Lenz, C.; Boeckler, F.; Hubner, H.; Gmeiner, P. *Bioorg. Med. Chem.* **2004**, *12*, 113–117.
155. Lenz, C.; Haubmann, C.; Hubner, H.; Boeckler, F.; Gmeiner, P. *Bioorg. Med. Chem.* **2005**, *13*, 185–191.
156. Street, L. J.; Baker, R.; Davey, W. B.; Guiblin, A. R.; Jelley, R. A.; Reeve, A. J.; Routledge, H.; Sternfeld, F.; Watt, A. P.; Beer, M. S. et al. *J. Med. Chem.* **1995**, *38*, 1799–1810.
157. Glen, R. C.; Martin, G. R.; Hill, A. P.; Hyde, R. M.; Woollard, P. M.; Salmon, J. A.; Buckingham, J.; Robertson, A. D. *J. Med. Chem.* **1995**, *38*, 3566–3580.
158. Habeeb, A. G.; Praveen Rao, P. N.; Knaus, E. E. *J. Med. Chem.* **2001**, *44*, 3039–3042.
159. Roderick, S. L.; Fournié-Zaluski, M. C.; Roques, B. P.; Matthews, B. W. *Biochemistry* **1989**, *28*, 1493–1497.
160. Büchi, J.; Stünzi, E.; Flury, M.; Hirt, R.; Labhart, P.; Ragaz, L. *Helv. Chim. Acta* **1951**, *34*, 1002–1013.
161. Janssen, P. A. J.; Van der Eycken, C. A. M. In *Drugs Affecting the Central Nervous System*; Burger, A., Ed.; Marcel Dekker: New York, 1968, pp 25–60.
162. Bös, M.; Jenck, F.; Martin, J. R.; Moreau, J.-L.; Sleight, A. J.; Wichmann, J.; Widmer, U. *J. Med. Chem.* **1997**, *40*, 2762–2769.
163. Mull, R. P.; Maxwell, R. A. In *Antihypertensive Agents*; Schlittler, E., Ed.; Academic Press: New York, 1967, pp 115–149.
164. Giudicelli, R.; Najer, H.; Lefèvre, F. *Compt. Rend. Acad. Sci.* **1965**, *260*, 726–729.
165. Najer, H.; Giudicelli, R.; Sette, J. *Bull. Soc. Chim. France* **1964**, 2572–2581.
166. Toda, N.; Usui, H.; Shimamoto, K. *Jpn J. Pharmacol.* **1972**, *22*, 125–135.
167. Najer, H.; Giudicelli, R.; Sette, J. *Bull. Soc. Chim. France* **1962**, 1593–1597.
168. Anonymous. British Patent 972,088 (Feb 19, 1962, Laboratoire Lumière, S.A.), 1964.
169. Anonymous. British Patent 973,533 (Feb 21, 1962, Laboratoire Lumière, S.A.), 1964.

170. Bryans, J. S.; Davies, N.; Gee, N. S.; Dissanayake, V. U. K.; Ratcliffe, G. S.; Horwell, D. C.; Kneen, C. O.; Morrell, A. I.; Oles, R. J.; O'Toole, J. C. et al. *J. Med. Chem.* **1998**, *41*, 1838–1845.
171. Seiler, M. P.; Bölsterli, J. J.; Floersheim, P.; Hagenbach, A.; Markstein, R.; Pfäffli, P.; Widmer, A.; Wüthrich, H. In *Perspectives in Medicinal Chemistry*; Testa, B., Kyburz, E., Fuhrer, W., Giger, R., Eds.; Verlag Helvetica Chemica Acta and VCH: Basel and Weinheim, 1993, pp 221–237.
172. Meyer, M. D.; Altenbach, R. J.; Basha, F. Z.; Carroll, W. A.; Condon, S.; Steven W. Elmore, S. W.; Kerwin, J., J. F.; Sippy, K. B.; Tietje, K. *J. Med. Chem.* **2000**, *43*, 1586–1603.
173. Jongebreur, G. *Arch. Intern. Pharmacodyn.* **1952**, *90*, 384–411.
174. Martin, Y. C.; Kofron, J. L.; Traphagen, L. M. *J. Med. Chem.* **2002**, *45*, 4350–4358.
175. Kubinyi, H. *Perspect. Drug Disc. Des.* **1998**, *11*, 225–252.
176. de Leval, X.; Julemont, F.; Benoit, V.; Frederich, M.; Pirotte, B.; Dogne, J. M. *Mini Rev. Med. Chem.* **2004**, *4*, 597–601.
177. Julemont, F.; de Leval, X.; Michaux, C.; Renard, J. F.; Winum, J. Y.; Montero, J. L.; Damas, J.; Dogne, J. M.; Pirotte, B. *J. Med. Chem.* **2004**, *47*, 6749–6759.
178. Street, L. J.; Sternfeld, F.; Jelley, R. A.; Reeve, A. J.; Carling, R. W.; Moore, K. W.; McKernan, R. M.; Sohal, B.; Cook, S.; Pike, A. et al. *J. Med. Chem.* **2004**, *47*, 3642–3657.
179. Böhm, H.-J.; Flohr, A.; Stahl, M. *Drug Disc. Today: Technol.* **2004**, *1*, 217–224.
180. Wermuth, C. G. *J. Med. Chem.* **2004**, *47*, 1303–1314.
181. Schneider, G.; Neidhart, W.; Giller, T.; Schmid, G. *Angew. Chem. Int. Ed. Engl.* **1999**, *38*, 2894–2896.
182. Langer, T.; Wolber, G. *Drug Disc. Today: Technol.* **2004**, *1*, 203–207.
183. Lloyd, D. G.; Buenemann, C. L.; Todorov, N. P.; Manallack, D. T.; Dean, P. M. *J. Med. Chem.* **2004**, *47*, 493–496.
184. Guba, W.; Neidhart, W.; Nettekoven, M. *Bioorg. Med. Chem. Lett.* **2005**, *15*, 1599–1603.
185. Balakin, K. V.; Tkachenko, S. E.; Okun, I.; Skorenko, A. V.; Ivanenkov, Y. A.; Savchuk, N. P.; Ivashchenko, A. A.; Nikolosky, Y. *Chimica Oggi-Chemistry Today* **2004**, *22*, 15–18.
186. Wilhelm, M. *Pharm. J.* **1975**, *214*, 414–416.
187. Yoshimura, H.; Kikuchi, K.; Hibi, S.; Tagami, K.; Satoh, T.; Yamauchi, T.; Ishibahi, A.; Tai, K.; Hida, T.; Tokuhara, N.; Nagai, M. *J. Med. Chem.* **2000**, *43*, 2929–2937.
188. Larson, A. A.; Lish, P. M. *Nature (London)* **1964**, *203*, 1283–1285.
189. Chenoweth, M. B.; McCarthy, L. P. *Pharmacol. Rev.* **1963**, *15*, 673–707.
190. Bondi, A. *J. Phys. Chem.* **1964**, *64*, 441.
191. Williams, D. E.; Houpt, D. J. *Acta. Crystallogr.* **1986**, *B42*, 286.
192. Hansch, C.; Leo, A.; Hoekman, D. *Am. Chem. Soc.* **1995**.
193. Smart, B. E. *J. Fluor. Chem.* **2001**, *109*, 3–11.
194. Goldman, P. *Science* **1969**, *164*, 1123–1130.
195. Chapman, N. B.; James, J. W.; Graham, J. D. P.; Lewis, G. P. *Chem. Ind. (London)* **1952**, 805–807.
196. Upthagrove, A. L.; Nelson, W. L. *Drug Metab. Dispos.* **2001**, *29*, 1389–1395.
197. Tang, W.; Palaty, J.; Abbott, F. S. *J. Pharmacol. Exp. Ther.* **1997**, *282*, 1163–1172.
198. Vaughan, J. R. J.; Anderson, G. W.; Clapp, R. C.; Clark, J. H.; English, J. P.; Howard, K. L.; Marson, H. W.; Sutherland, L. H.; Denton, J. J. *J. Org. Chem.* **1949**, *14*, 228–234.
199. Bundschuh, D. S.; Eltze, M.; Barsig, J.; Wollin, L.; Hatzelmann, A.; Beume, R. *J. Pharmacol. Exp. Ther.* **2001**, *297*, 280–290.
200. Schatz, V. B. In *Medicinal Chemistry*; Burger, A., Ed.; Interscience Publishers, Inc.: New York, 1963, pp 72–88.
201. Holladay, M. W.; Bai, H.; Li, Y.; Lin, N. H.; Daanen, J. F.; Ryther, K. B.; Wasicak, J. T.; Kincaid, J. F.; He, Y.; Hettinger, A. M. et al. *Bioorg. Med. Chem. Lett.* **1998**, *8*, 2797–2802.
202. Holladay, M. W.; Wasicak, J. T.; Lin, N. H.; He, Y.; Ryther, K. B.; Bannon, A. W.; Buckley, M. J.; Kim, D. J.; Decker, M. W.; Anderson, D. J. et al. *J. Med. Chem.* **1998**, *41*, 407–412.
203. Bisson, W. H.; Scapozza, L.; Westera, G.; Mu, L.; Schubiger, P. A. *J. Med. Chem.* **2005**, *48*, 5123–5130.
204. Sugimoto, H.; Yamanishi, Y.; Iimura, Y.; Kawakami, Y. *Curr. Med. Chem.* **2000**, *7*, 303–339.
205. Erlenmeyer, H. *Bull. Soc. Chem. Biol.* **1948**, *30*, 792–805.
206. Fessenden, R. J.; Fessenden, J. S. *Adv. Drug Res.* **1967**, *4*, 95–132.
207. Tacke, R.; Handmann, V. I.; Bertermann, R.; Bruschka, C.; Penka, M.; Seyfried, C. *Organometallics* **2003**, *22*, 916–924.
208. Tacke, R.; Zilch, H. *Endeavour* **1986**, *10*, 191–197.
209. Tacke, R.; Zilch, H. *L'actualité Chimique* **1986**, 75–82.
210. Ricci, A.; Seconi, G.; Taddei, M. *Chimica Oggi-Chemistry Today* **1989**, *7*, 15–21.
211. Liu, X.-M.; Liao, R.-A.; Xie, Q.-L. *Chem. Abstr.* **1999**, *130*, 25102a.
212. Showell, G. A.; Mills, J. S. *Drug Disc. Today* **2003**, *8*, 551–556.
213. Garson, L. R.; Kirchner, L. K. *J. Pharm. Sci.* **1971**, *60*, 1113–1127.
214. Bains, W.; Tacke, R. *Curr. Opin. Drug Disc. Dev.* **2003**, *6*, 526–543.
215. Tacke, R.; Wannagat, U. *Top. Curr. Chem.* **1979**, *84*, 1–75.
216. Reichstat, M. M.; Mioč, U. B.; Bogunovic, L. J.; Ribnikar, S. V. *J. Mol. Struct.* **1991**, *244*, 283–290.
217. Woo, D. V.; Christian, J. E. et al. *Can. J. Pharm. Sci* **1979**, *14*, 12–14.
218. Clark, D. E. *Drug Disc. Today* **2003**, *8*, 927–933.
219. Metcalf, R. L.; Fukuto, T. R. *J. Econ. Entomol.* **1965**, *58*, 1151.
220. Anonymous. *Drugs Future* **1994**, *19*, 854–855.
221. Hornsperger, J. M.; Collard, J. N.; Heydt, J. G.; Giacobini, E.; Funes, S.; Dow, J.; Schirlin, D. *Biochem. Soc. Trans.* **1994**, *22*, 758–763.
222. Fessenden, R. J.; Coon, M. D. *J. Med. Chem.* **1965**, *8*, 604–608.
223. Tacke, R. *Chem. Ber.* **1980**, *113*, 1962–1980.
224. Moberg, W. K. *US Patent 4,510,136 to DuPont* 1985.
225. Damour, D.; Dutruc-Rosset, G.; Doble, A.; Piot, O.; Mignani, S. *Bioorg. Med. Chem. Lett.* **1994**, *4*, 415–420.
226. Bondarenko, I. P. *Vrach Delo* **1986**, *6*, 20–23.
227. Bondarenko, I. P. *Vrach Delo* **1988**, *10*, 33–35.
228. Bondarenko, I. P. *Vrach Delo* **1988**, *3*, 31–34.

229. Heinonen, P.; Sipilä, H.; Neuvonen, K.; Lönnberg, H.; Cockcroft, V. B.; Wurster, S.; Virtanen, R.; Savola, M. K. T.; Salonen, J. S.; Savola, J. M. *Eur. J. Med. Chem.* **1996**, *31*, 725–729.
230. Sommer, L. H.; Bennet, O. F.; Campbell, P. G.; Weyenberg, D. R. *J. Am. Chem. Soc.* **1957**, *79*, 3295–3296.
231. Fessenden, R. J.; Ahlfors, C. *J. Med. Chem.* **1967**, *10*, 810–812.
232. Mills, J. S.; Showell, G. A. *Expert Opin. Investig. Drugs* **2004**, *13*, 1149–1157.
233. Sieburth, S. M.; Nittoli, T.; Mutahi, A. M.; Guo, L. *Angew. Chem. Int. Ed.* **1998**, *37*, 812–814.
234. Stasch, J. P.; Russ, H.; Schacht, U.; Witteler, M.; Neuser, D.; Gerlach, M.; Leven, M.; Kuhn, W.; Jutzi, P.; Przuntek, H. *Arzneimittelforschung* **1988**, *38*, 1075–1078.
235. Heinrich, T.; Burschka, C.; Warneck, J.; Tacke, R. *Organometallics* **2004**, *23*, 361–366.
236. Farkas, S.; Kocsis, P.; Bielik, N.; Gemesi, L.; Trafikant, G. *Fundam. Clin. Phamacol.* **1999**, *13*, 144.
237. Rizvi, N. A.; Marshall, J. L.; Ness, E.; Hawkins, M. J.; Kessler, C.; Jacobs, H.; Brenckman, W. D., Jr.,; Lee, J. S.; Petros, W.; Hong, W. K.; Kurie, J. M. *J. Clin. Oncol.* **2002**, *20*, 3522–3532.
238. Hausheer, F. H.; Berghorn, E.; Liu, Z.; Sullivan, D.; Centeno, B.; Derderian, J.; Deconti, R.; Daud, A.; Sullivan, P.; Andrews, S. *J. Clin. Oncol. (Meeting Abstracts)* **2004**, *22*, 7554–7555.
239. Tao, S. H.; Bolger, P. M. *Regul. Toxicol. Pharmacol.* **1997**, *25*, 211–219.
240. Asaka, T.; Nitta, E.; Makifuchi, T.; Shibazaki, Y.; Kitamura, Y.; Ohara, H.; Matsushita, K.; Takamori, M.; Takahashi, Y.; Genda, A. *J. Neurol. Sci.* **1995**, *130*, 220–223.
241. Lukevics, E.; Ignatovich, L. *Appl. Organomet. Chem.* **1992**, *6*, 113–126.
242. Tacke, R.; Reichel, D.; Jones, P. G.; Hou, X.; Waelbroeck, M.; Mutschler, E.; Lambrecht, G. *J. Organomet. Chem.* **1996**, *640*, 140–165.
243. Tacke, R.; Kornek, T.; Heinrich, T.; Bruschka, C.; Penka, M.; Pülm, M.; Keim, C.; Mutschler, E.; Lambrecht, G. *J. Organomet. Chem.* **2001**, *640*, 140–165.
244. Alam, F.; Soloway, A. H.; Bapat, B. V.; Barth, R. F.; Adams, D. M. *Basic Life Sci.* **1989**, *50*, 107–111.
245. Gabel, D. *Basic Life Sci.* **1989**, *50*, 233–241.
246. Kahl, S. B.; Joel, D. D.; Finkel, G. C.; Micca, P. L.; Nawrocky, M. M.; Coderre, J. A.; Slatkin, D. N. *Basic Life Sci.* **1989**, *50*, 193–203.
247. El-Zaria, M. E.; Dorfler, U.; Gabel, D. *J. Med. Chem.* **2002**, *45*, 5817–5819.
248. Murray, F. J. *Biol. Trace Elem. Res.* **1998**, *66*, 331–341.
249. Hubbard, S. A. *Biol. Trace Elem. Res.* **1998**, *66*, 343–357.
250. Landauer, W. *J. Cell Physiol.* **1954**, *43*, 261–305.
251. Landauer, W.; Clark, E. M. *J. Exp. Zool.* **1964**, *156*, 307–312.
252. Browning, E. *Toxicity of Industrial Metals*, 2nd ed.; Butterworths and Co: London, 1969.
253. World Health Organization. *Boron, Environmental Health Criteria*; Geneva, Switzerland, 1998; Vol. 204.
254. Dixon, R. L.; Lee, I. P.; Sherins, R. J. *Environ. Health Perspect.* **1976**, *13*, 59–67.
255. Whorton, D.; Haas, J.; Trent, L. *Environ. Health Perspect.* **1994**, *102*, 129–132.
256. Whorton, M. D.; Haas, J. L.; Trent, L.; Wong, O. *Occup. Environ. Med.* **1994**, *51*, 761–767.
257. US Enviromental Protection Agency, O. o. P. P. 1997.
258. Woods, W. G. *Environ. Health Perspect.* **1994**, *102*, 5–11.
259. Hutter, R.; Keller-Schierlein, W.; Knusel, F.; Prelog, V.; Rodgers, G. C., Jr.; Suter, P.; Vogel, G.; Voser, W.; Zahner, H. *Helv. Chim. Acta* **1967**, *50*, 1533–1539.
260. Okami, Y.; Okazaki, T.; Kitahara, T.; Umezawa, H. *J. Antibiot. (Tokyo)* **1976**, *29*, 1019–1025.
261. Dunitz, J. D.; Hawley, D. M.; Miklos, D.; White, D. N.; Berlin, Y.; Marusic, R.; Prelog, V. *Helv. Chim. Acta* **1971**, *54*, 1709–1713.
262. Caujolle, F.; Pham Huu, C. *Arch. Int. Pharmacodyn. Ther.* **1968**, *172*, 467–474.
263. Mubarak, S. I.; Stanford, J. B.; Sugden, J. K. *Drug. Dev. Ind. Pharm.* **1984**, *10*, 1131–1160.
264. Kinder, D. H.; Katzenellenbogen, J. A. *J. Med. Chem.* **1985**, *28*, 1917–1925.
265. Dembitsky, V. M.; Srebnik, M. *Tetrahedron* **2003**, *59*, 579–593.
266. Surolia, N.; RamachandraRao, S. P.; Surolia, A. *Bioessays* **2002**, *24*, 192–196.
267. Jabbour, A.; Steinberg, D.; Dembitsky, V. M.; Moussaieff, A.; Zaks, B.; Srebnik, M. *J. Med. Chem.* **2004**, *47*, 2409–2410.
268. Kumar, S. K.; Hager, E.; Pettit, C.; Gurulingappa, H.; Davidson, N. E.; Khan, S. R. *J. Med. Chem.* **2003**, *46*, 2813–2815.
269. Endo, Y.; Iijima, T.; Yamakoshi, Y.; Yamaguchi, M.; Fukasawa, H.; Shudo, K. *J. Med. Chem.* **1999**, *42*, 1501–1504.
270. Ohta, K.; Iijima, T.; Kawachi, E.; Kagechika, H.; Endo, Y. *Bioorg. Med. Chem. Lett.* **2004**, *14*, 5913–5918.
271. Endo, Y.; Yoshimi, T.; Iijima, T.; Yamakoshi, Y. *Bioorg. Med. Chem. Lett.* **1999**, *9*, 3387–3392.
272. Westheimer, F. H. *Science* **1987**, 1173–1178.
273. Nahum, V.; Fischer, B. *Eur. J. Inorg. Chem.* **2004**, *20*, 4124–4131.
274. Shaw, B. R.; Madison, J.; Sood, A.; Spielvogel, B. F. *Meth. Mol. Biol.* **1993**, *20*, 225–243.
275. Summers, J. S.; Roe, D.; Boyle, P. D.; Colvin, M.; Shaw, B. R. *Inorg. Chem.* **1998**, *37*, 4158–4159.
276. Shaw, B. R.; Sergueev, D.; He, K.; Porter, K.; Summers, J.; Sergueeva, Z.; Rait, V. *Meth. Enzymol.* **2000**, *313*, 226–257.
277. Summers, J. S.; Shaw, B. R. *Curr. Med. Chem.* **2001**, *8*, 1147–1155.
278. Meyer, P.; Schneider, B.; Sarfati, S.; Deville-Bonne, D.; Guerreiro, C.; Boretto, J.; Janin, J.; Veron, M.; Canard, B. *EMBO J.* **2000**, *19*, 3520–3529.
279. Deval, J.; Selmi, B.; Boretto, J.; Egloff, M. P.; Guerreiro, C.; Sarfati, S.; Canard, B. *J. Biol. Chem.* **2002**, *277*, 42097–42104.
280. Selmi, B.; Boretto, J.; Sarfati, S. R.; Guerreiro, C.; Canard, B. *J. Biol. Chem.* **2001**, *276*, 48466–48472.
281. Nahum, V.; Zundorf, G.; Levesque, S. A.; Beaudoin, A. R.; Reiser, G.; Fischer, B. *J. Med. Chem.* **2002**, *45*, 5384–5396.
282. Li, P.; Dobrikov, M.; Liu, H.; Shaw, B. R. *Org. Lett.* **2003**, *5*, 2401–2403.
283. Al-Madhoun, A. S.; Johnsamuel, J.; Yan, J.; Ji, W.; Wang, J.; Zhuo, J. C.; Lunato, A. J.; Woollard, J. E.; Hawk, A. E.; Cosquer, G. Y.; Blue, T. E. et al. *J. Med. Chem.* **2002**, *45*, 4018–4028.
284. Cappelli, A.; Pericot Mohr, G.; Gallelli, A.; Giuliani, G.; Anzini, M.; Vomero, S.; Fresta, M.; Porcu, P.; Maciocco, E.; Concas, A. et al. *J. Med. Chem.* **2003**, *46*, 3568–3571.
285. Martin, B.; Posseme, F.; Le Barbier, C.; Carreaux, F.; Carboni, B.; Seiler, N.; Moulinoux, J. P.; Delcros, J. G. *J. Med. Chem.* **2001**, *44*, 3653–3664.
286. Nakamura, H.; Fujiwara, M.; Yamamoto, Y. *J. Org. Chem.* **1998**, *63*, 7529–7530.
287. Zhuo, J. C.; Soloway, A. H.; Beeson, J. C.; Ji, W.; Barnum, B. A.; Rong, F. G.; Tjarks, W.; Jordan, G. T., IV; Liu, J.; Shore, S. G. *J. Org. Chem.* **1999**, *64*, 9566–9574.

288. Olejiniczak, A. B.; Koziolkiewcz, M.; Lesnikowski, Z. J. *Antisense Nucleic Acid Drug Dev.* **2002**, *12*, 79.
289. Isaac, M. F.; Kahl, S. B. *J. Organomet. Chem.* **2003**, *680*, 232.
290. Ivanov, D.; Bachovchin, W. W.; Redfield, A. G. *Biochemistry* **2002**, *41*, 1587–1590.
291. Clark, L. C.; Combs, G. F., Jr.; Turnbull, B. W.; Slate, E. H.; Chalker, D. K.; Chow, J.; Davis, L. S.; Glover, R. A.; Graham, G. F.; Gross, E. G. et al. *JAMA* **1996**, *276*, 1957–1963.
292. Nelson, M. A.; Porterfield, B. W.; Jacobs, E. T.; Clark, L. C. *Semin. Urol. Oncol.* **1999**, *17*, 91–96.
293. Kelloff, G. J.; Crowell, J. A.; Steele, V. E.; Lubet, R. A.; Malone, W. A.; Boone, C. W.; Kopelovich, L.; Hawk, E. T.; Lieberman, R.; Lawrence, J. A. et al. *J. Nutr.* **2000**, *130*, 467S–471S.
294. Kelloff, G. J.; Lieberman, R.; Steele, V. E.; Boone, C. W.; Lubet, R. A.; Kopelovich, L.; Malone, W. A.; Crowell, J. A.; Higley, H. R.; Sigman, C. C. *Urology* **2001**, *57*, 46–51.
295. Witschi, H. *Exp. Lung Res.* **2000**, *26*, 743–755.
296. Duffield-Lillico, A. J.; Reid, M. E.; Turnbull, B. W.; Combs, G. F., Jr.,; Slate, E. H.; Fischbach, L. A.; Marshall, J. R.; Clark, L. C. *Cancer Epidemiol. Biomarkers Prev.* **2002**, *11*, 630–639.
297. Oldfield, J. E. *J. Nutr.* **1987**, *117*, 2002–2008.
298. Foster, L. H.; Sumar, S. *Crit. Rev. Food Sci. Nutr.* **1997**, *37*, 211–228.
299. Mahan, D. C.; Moxon, A. L. *J. Anim. Sci.* **1984**, *58*, 1216–1221.
300. O'Toole, D.; Raisbeck, M. F. *J. Vet. Diagn. Invest.* **1995**, *7*, 364–373.
301. Cantor, A.; Scot, M.; Noguch, T. *J. Nutr.* **1976**, *105*, 96–105.
302. Rayman, M. P. *Lancet* **2000**, *356*, 233–241.
303. Levander, O. A.; Whanger, P. D. *J. Nutr.* **1996**, *126*, 2427S–2434S.
304. Yang, G.; Yin, S.; Zhou, R.; Gu, L.; Yan, B.; Liu, Y.; Liu, Y. *J. Trace Elem. Electrolytes Health Dis.* **1989**, *3*, 123–130.
305. Longnecker, M. P.; Taylor, P. R.; Levander, O. A.; Howe, M.; Veillon, C.; McAdam, P. A.; Patterson, K. Y.; Holden, J. M.; Stampfer, M. J.; Morris, J. S. et al. *Am. J. Clin. Nutr.* **1991**, *53*, 1288–1294.
306. Klayman, D. L.; Günther, W. H. H. *Organic Selenium Compounds: Their Chemistry and Biology*; Wiley: New York, 1973.
307. Tanaka, T.; Reddy, B. S.; el-Bayoumy, K. *Jpn. J. Cancer Res.* **1985**, *76*, 462–467.
308. El-Bayoumy, K. *Drugs Future* **1997**, *22*, 539–545.
309. Thompson, H. J.; Wilson, A.; Lu, J.; Singh, M.; Jiang, C.; Upadhyaya, P.; el-Bayoumy, K.; Ip, C. *Carcinogenesis* **1994**, *15*, 183–186.
310. Short, M. D.; Xie, Y.; Li, L.; Cassidy, P. B.; Roberts, J. C. *J. Med. Chem.* **2003**, *46*, 3308–3313.
311. Sharman, W. M.; Allen, C. M.; van Lier, J. E. *Drug Disc. Today* **1999**, *4*, 507–517.
312. Brennan, N. K.; Hall, J. P.; Davies, S. R.; Gollnick, S. O.; Oseroff, A. R.; Gibson, S. L.; Hilf, R.; Detty, M. R. *J. Med. Chem.* **2002**, *45*, 5123–5135.
313. Leonard, K. A.; Hall, J. P.; Nelen, M. I.; Davies, S. R.; Gollnick, S. O.; Camacho, S.; Oseroff, A. R.; Gibson, S. L.; Hilf, R.; Detty, M. R. *J. Med. Chem.* **2000**, *43*, 4488–4498.
314. Krehan, D.; Frolund, B.; Krogsgaard-Larsen, P.; Kehler, J.; Johnston, G. A.; Chebib, M. *Neurochem. Int.* **2003**, *42*, 561–565.
315. Parnham, M. J.; Graf, E. *Biochem. Pharmacol.* **1987**, *36*, 3095–3102.
316. Fischer, H.; Terlinden, R.; Lohr, J. P.; Romer, A. *Xenobiotica* **1988**, *18*, 1347–1359.
317. Mugesh, G.; Singh, B. H. *Chem. Soc. Rev.* **2000**, *29*, 347–357.

Biographies

Camille Georges Wermuth, PhD, Professor and Founder of Prestwick Chemical, was Professor of Organic Chemistry and Medicinal Chemistry at the Faculty of Pharmacy, Louis Pasteur University, Strasbourg, France from 1969 to 2002. He founded the Prestwick Chemical Inc. in 1999 in which he acts as President and CSO. Professor Wermuth's main research themes focus on the chemistry and the pharmacology of pyridazine derivatives. Particularly, the 3-aminopyridazine pharmacophore allowed him to accede to an impressive variety of biological activities. Among them one can cite: antidepressant and anticonvulsant molecules; inhibitors of enzymes such as monoamine oxidases, phosphodiesterases, and acetylcholinesterase; ligands for neuro-receptors: GABA-A receptor antagonists, serotonin 5-HT$_3$ receptor antagonists, dopaminergic and muscarinic agonists.

Besides about 300 scientific papers and about 60 patents, Prof Wermuth is co-author or editor of several books. Particularly, the second edition of his book *The Practice of Medicinal Chemistry* was published by Academic Press in May 2003. The first and the second editions were translated into Japanese and the first edition into Italian. Together with Dr P H Stahl, he also published the *Handbook of Pharmaceutical Salts, Properties Selection and Use* by Wiley-VCH in 2002.

Professor Wermuth has been President of the Medicinal Chemistry Section of the International Union of Pure and Applied Chemistry (IUPAC) from 1988 to 1992 and from January 1998 to January 2000 was President of the IUPAC Division on Chemistry and Human Health.

Paola Ciapetti, PhD, Head of Medicinal Chemistry, joined Prestwick Chemical, Inc. in October 2001 as Senior Chemist in charge of the Medicinal Chemistry Department with Bruno Giethlen. Paola came to Prestwick Chemical, Inc. from the Medicinal Chemistry Department of Albany Molecular Research, Inc. (Albany, NY) where she worked for more than 1 year as Senior Research Chemist. While at AMRI she brought a relevant contribution to an oncology project developed in collaboration with Eli Lilly & Co. She is one of the co-inventors of the patent issued from that work. Prior to this experience Paola spent more than 2 years as Visiting Scientist at Eli Lilly & Co. (Indianapolis, IN). During this tenure she worked on CNS-related projects. Paola has acquired a broad and deep experience in both organic and medicinal chemistry. She is fluent in three languages (Italian, English, and French).

Paola graduated Magna Cum Laude with an MSc in chemistry from the University of Florence (Italy) in 1994. In 1998, she received a PhD in organic chemistry from the same university under the supervision of Prof Maurizio Taddei. During her thesis she completed a tenure at the University of Strasbourg under the supervision of Dr André Mann. Her studies have focused on the development of new methodologies for the synthesis of nonnatural alpha amino acids, inhibitors of proteases, and nucleic acids analogs. She is author of eight peer-reviewed scientific papers and 10 patents.

Bruno Giethlen, PharmD, PhD, is Head of Medicinal Chemistry in charge of the Medicinal Chemistry Department with Paola Ciapetti, PhD at Prestwick Chemical, Inc. in Strasbourg. Prior to joining Prestwick Chemical, he was Senior Research Chemist at Albany Molecular Research, Inc. (Albany, NY) for 1 year where he worked on oncology programs in collaboration with Eli Lilly & Co. Before this period, he spent 3 years at Eli Lilly & Co. (Indianapolis, IN) as a Visiting Scientist working on CNS projects in medicinal chemistry. His contribution to several serotonin projects has been relevant. Bruno is co-inventor of one patent and co-author of one peer-reviewed scientific paper. He has attained a broad knowledge of medicinal chemistry and a wide experience in organic chemistry.

Bruno graduated from the Faculty of Pharmacy of the Louis Pasteur University (ULP) in Strasbourg, France in 1990. He obtained an MSc in pharmacology and medicinal chemistry from the same university in 1991, before his PhD in medicinal chemistry under the supervision of Prof C G Wermuth. His thesis, developed in collaboration with Sanofi-Synthelabo, has set up an exhaustive SAR of the first nonpeptidic antagonists of Corticotropin Releasing Factor. This work has yielded three patents. After having spent 1 year as assistant professor at the ULP, he engaged himself for a 2-year postdoctoral fellowship at the same University in collaboration with Eli Lilly & Co. He spent part of this tenure at Eli Lilly in Indianapolis. Bruno is co-inventor of 14 patents and co-author of one peer-reviewed scientific paper.

Patrick Bazzini, PhD, is Research Chemist at Prestwick Chemical, Inc. in Strasbourg. Prior to joining Prestwick, Dr Bazzini prepared a PhD at the University of Aix-Marseille III, under the supervision of Prof M Santelli and in collaboration with Pierre Fabre MÈdicament. During this period, his interests focused on the development of a total synthesis of new steroids with interesting biological activities. After his PhD, he joined Prof C G Wermuth's research group at the Louis Pasteur University in Strasbourg for a postdoctoral fellowship. There, with a grant from Eli Lilly, he developed an SAR study around muscarinic receptor ligands potentially useful as symptomatic treatment of Alzheimer's disease. After his postdoctoral fellowship, he worked on anticancer project in collaboration with Sanofi-Synthelabo.

2.17 Chiral Drug Discovery and Development – From Concept Stage to Market Launch

H-J Federsel, AstraZeneca, Södertälje, Sweden

© 2007 Elsevier Ltd. All Rights Reserved.

2.17.1	**Scientific Rationale for Chiral Drugs**	713
2.17.2	**How to Obtain Chiral Compounds in Stereoisomeric Form**	715
2.17.3	**Supplying Stereoisomers: What the Market Can Offer**	718
2.17.4	**Stereospecific Analyses: An Overview**	720
2.17.4.1	Determination of Absolute Configuration	720
2.17.4.1.1	X-ray diffraction (XRD)	720
2.17.4.1.2	Vibrational circular dichroism (VCD)	720
2.17.4.1.3	Other methods	720
2.17.4.2	Determination of Enantiomeric Purity	721
2.17.4.2.1	Polarimetry	721
2.17.4.2.2	Nuclear magnetic resonance (NMR)	721
2.17.4.2.3	Direct enantioselective chromatography/electrophoresis	721
2.17.4.2.4	Indirect enantioselective chromatography/electrophoresis	721
2.17.5	**From Drawing Board to Consumer: Progression of Drug Compounds in Process R&D**	721
2.17.6	**The Regulatory Environment**	725
2.17.7	**Drugs at Any Cost?**	726
2.17.8	**Chiral Switches: From Racemate to Enantiomer**	728
2.17.9	**Successes and Failures: What Are the Discriminating Factors?**	730
2.17.10	**Major Conclusions and Projections for the Future**	731
References		733

2.17.1 Scientific Rationale for Chiral Drugs

One generation ago, that is roughly 25–30 years back in time which takes us to the mid to late 1970s, hardly anyone could have imagined the enormous changes that the pharmaceutical industry would undergo. On the macroscopic level we have seen a virtually continuously ongoing concentration of activity creating companies of ever-bigger size by series of acquisitions and mergers.[1] We have also witnessed the evolution of a versatile and blooming biotechnology sector capable of producing antibodies and proteins to be utilized for treatment of diseases with high-unmet medical need, many of which are not accessible via small-molecule therapy.[2,3] Furthermore a considerably less pleasant development has been the upward spiraling costs involved in taking a drug from the drawing-board through all stages and on to the market,[4–7] the declining rate of achieving approval from regulatory authorities to launch a new product,[8] and the inability to come to grips with the business-wide attrition, which keeps lingering around an average value of about 90% (9 out of 10 projects will be discontinued somewhere along the timeline from nomination of candidate drug to launch).[9–12] Also as the timeline itself has become more and more the focus of attention with particular emphasis on how long it takes to make new drugs available to patients – the current industrial norm is about 7.2 years[13] from first toxicology dosing to first file submitted for marketing approval, which means that easily another 3–5 years will have to be added on to cover the entire range from start of project to commercial launch – there have been many initiatives toward speeding up the work process and productivity.[14–16] Thus, as one of the rate-limiting activities is concerned with the identification of developable molecules, methods (often highly automated and robotized) for generating comprehensive compound libraries using parallel synthesis and combinatorial techniques are seen as major advancements.[17,18] Coupled with tailor-made screening cascades, notably in high-throughput architecture, considerable streamlining has been achieved in the low probability search for hits, leads, and later on candidate drugs (CDs). From a

Table 1 In 2004 7.5 (1/2 stems from the mixture of two actives in Advair) of the best-selling pharmaceuticals on the world market are represented by single enantiomers according to IMS (www.imshealth.com)

Rank	Name (generic/trade)	Company	Molecular characteristics of API
1	Atorvastatin/Lipitor	Pfizer	Single enantiomer
2	Simvastatin/Zocor	Merck	Single enantiomer
3	Clopidogrel/Plavix	Sanofi-Aventis	Single enantiomer
4	Salmeterol + fluticasone/Advair	GSK	Racemate + single enantiomer
5	Amlodipine/Norvasc	Pfizer	Racemate
6	Olanzapine/Zyprexa	Lilly	Achiral
7	Paroxetine/Paxil	GSK	Single enantiomer
8	Esomeprazole/Nexium	AstraZeneca	Single enantiomer
9	Sertraline/Zoloft	Pfizer	Single enantiomer
10	Celecoxib/Celebrex	Pfizer	Single enantiomer

microscopic perspective there has also been a revolutionary change in our understanding of many of the complex phenomena occurring at the medicine–biology–biochemistry–pharmacology–chemistry interface. At the beginning of the twenty-first century the mapping of the entire human genome, the linking of diseases to particular genes, and the identification of a multitude of biological targets and receptors prone to artificial interactions[19] would probably stand out as being among the most prominent areas. A lot of our previous thinking has been revised or even turned upside down based on current knowledge. On the molecular level this period has seen the definite breakthrough and acceptance of the three-dimensional nature of drug–target interactions. While the number of stereochemically pure products in therapeutic use was 20–25% 30 years ago – the vast majority of which represented by various antibiotics produced in fermentation processes – the bulk of the marketed portfolio was built up by racemates and achirals. The situation has changed dramatically and in 2003 the global top 10 list of drug sales contained six small-molecule active pharmaceutical ingredients (APIs) as single enantiomers and just two as racemic products (besides one achiral and one protein).[20] In the future this strong dominance of developing pure stereoisomers whenever the compound is chiral will remain or be even further enhanced[21] and it is only in cases where a clear benefit can be demonstrated from the antipode(s) that a racemic mixture will be approved. As a matter of fact the use of enantiomerically pure compounds for medical purposes is now so established that patenting authorities have signaled their reluctance to accept new intellectual property (IP) filings on stereoisomers of known racemic molecules. In other words, the link from racemate to antipodes falls under the obviousness headline and can, therefore, not be (easily) patented any more. **Table 1** shows the 2004 market statistics for the top 10 selling drugs worldwide.[21]

While on the surface of things it appears rather logical that molecular stereochemistry is deeply involved in the creation of a biologic response, there has been an ongoing discussion over decades on the exact mechanistic causes for this and, maybe more importantly, how to interpret the findings. The debate has by no means seen an end and as recently as in 2003 a paper on the subject[22] was entitled "Three-dimensional pharmacology, a subject ranging from ignorance to overstatements," which can hardly be misinterpreted as it points toward diametrically opposing views being expressed about this topic. Actually, this controversy was initiated some decades ago and here Ariëns' famous paper from 1984[23] probably stands out as one of the most important discussions on the subject catching a lot of attention. The perceived reluctance in disciplines like clinical pharmacology and pharmacokinetics to realize the stereoselective nature of drug–target interactions are addressed in a very critical and outspoken manner (by using phrases such as 'sophisticated nonsense'). Instead the then newly coined expression 'isomeric ballast' was introduced as a means to describe the therapeutically useless 50% (or potentially even more in cases with several stereogenic centers) of impurity present in a racemic API. The situation is extremely complex, however, which is underpinned by many notable exceptions to the stereospecific nature of molecular interactions with bioreceptors,[24] for example at mammalian taste buds where there is no discrimination between L- and D-sugars.[25] From a regulatory perspective, the governing authorities in the USA[26] and Europe dealing with approval of new medicines have provided clear guidance on how to deal with molecules displaying features of chirality. Thus, it is requested that the stereoisomeric composition of such compounds is quantified and, if possible, safety and pharmacological data have to be recorded for both racemate and pure enantiomers. It is important to point out, by no means is the development of racemic mixtures or of those represented by any other ratio between antipodes forbidden, but the documentary demands have discouraged most from making efforts in this area. The trend over the past 20 years has, therefore, been strongly in favor of only taking single stereoisomers into clinical studies in humans while aiming to reduce residual amounts of nondesired enantiomer (including diastereomers whenever more than one stereogenic center is present) to levels as low as technically feasible (often <1%). It is not surprising then that during this period we have witnessed the transformation of known marketed racemic products to their enantiopure, or at least enantiomerically enriched, counterparts. An entire industrial branch pursuing these so-called chiral switches has been formed building on the often highly praised benefits – both in a medical sense and from a marketing point of view – but in spite of some successes far from all practically possible cases have, for various reasons, survived this transition.[27,28]

2.17.2 How to Obtain Chiral Compounds in Stereoisomeric Form

In order to satisfy the ever-increasing demand of compounds in stereochemically pure or at least highly enriched state a whole raft of methods have been put at our disposal. These range from the simplest possible technique where the antipode can be separated off mechanically by the sheer difference in macroscopic appearance as pioneered by Pasteur in his renowned experiment in 1848 when picking one of the tartaric acid enantiomers with a pair of pincers. A highly sophisticated transition metal catalyzed process using a tailor-made chiral ligand, which is capable of operating at high productivity and efficiency, will represent another extreme. These options as well as other methodologies are schematically displayed in **Figure 1**.

For the end user it is of course very tempting and attractive to simply resort to purchase of starting materials in enantiopure form. And with the tremendous development that the market has seen over the past 10 to 20 years, to a large extent driven by the commercialization of various asymmetric technologies, many useful and versatile compounds are now available, increasingly so also in bulk quantities.[33–37] In cases where the stereochemical information – more precisely expressed as the absolute configuration of the stereogenic center(s) in the starting material – is retained in the target molecule there is, in theory, no need to be worried about enantiomeric purity as long as the reaction conditions applied along the synthetic pathway are not causing any (appreciable) racemization. **Figure 2** shows an industrially significant example describing this procedure.[38]

From a practical point of view this is rarely the case, simply because the stereochemical purity of the acquired starting material does not match the demand but also because of the risk of having to face a sacrifice in purity to a higher or lesser degree during the synthesis, in particular as the sequence gets longer and longer. For example when preparing a polypeptide it is of utmost importance to carefully control each coupling reaction between the growing chain and every newly added amino acid to avoid ending up with a messy mixture of stereoisomers that might become very troublesome and yield-consuming to purify. If, however, the original stereocenter(s) can be manipulated, meaning that transformations will be carried out directly at the stereogenic position, then it is more a matter of ensuring that the conditions utilized for this purpose guarantee that the reactions proceed in a selective fashion in order to always exert full stereocontrol. This control is also essential when operating in a mode where the stereochemistry is retained, but the risks of causing detrimental racemization are likely to be much higher in the former case. An illustration of this is when during a nucleophilic substitution on a particular substrate believed to follow a clean S_N2 pathway which would result in the expected full inversion of configuration instead some retention product is formed due to competition from a reaction following the S_N1 mechanism. In this case the overall outcome would be partial racemization, which demands a purification procedure to regain the stereochemical purity.

The workhorse when it comes to generating products with desired absolute stereochemistry still remains the classical diastereomer resolution.[39,40] Exact figures describing the current situation in industry are hard to access, but indications point in the range of as much as roughly half of what is produced having been made by this method.[20] And this is in spite of the normally perceived low technical grade of the technology, which, superficially, requires one antipode to be precipitated by an auxiliary of high enantiomeric purity. The waste of 50% of the material can, if environmentally or economically justified, be compensated for by an in-built process loop allowing racemization of the undesired isomer. Repetition of this cycle can in principle drive the yield close to quantitative and this concept has seen some real advances recently with the introduction and refinement of dynamic kinetic resolution (DKR).[41–44] Thus, in an in situ arrangement the wanted stereoisomer is accumulated while its antipode is continually being transformed either bio- or chemocatalytically so that at full conversion, in theory, only the desired product remains in the system. An elegant application[45] of this technique is shown in **Figure 3**.

With the third approach – stereoselective syntheses – we enter a field that has experienced a tremendous development since the mid-1960s when it was demonstrated that reductions could be effected in homogeneous phase

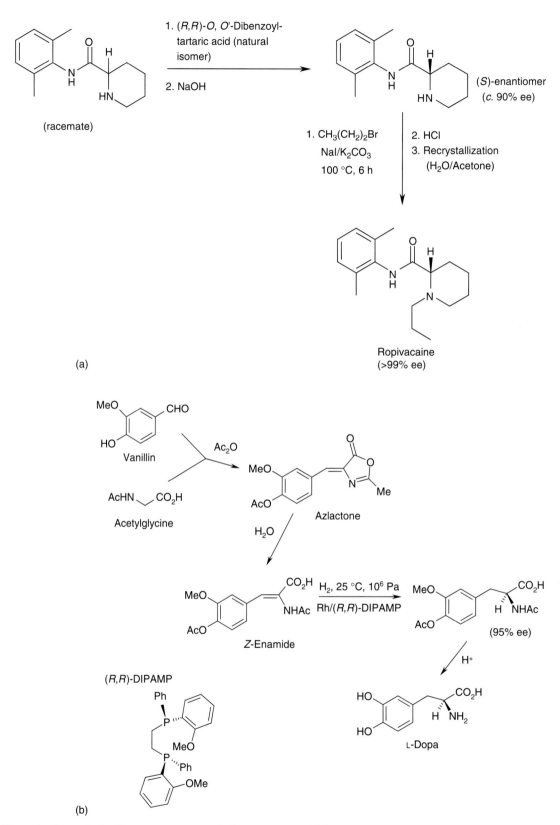

Figure 1 Conceptually different ways to create single stereoisomers. (a) A racemic material is subjected to resolution and then taken forward to the desired enantiomer represented by the local anesthetic ropivacain[29]; (b) a prochiral substrate is selectively transformed to one enantiomer in excellent purity as exemplified with L-Dopa synthesis.[30–32]

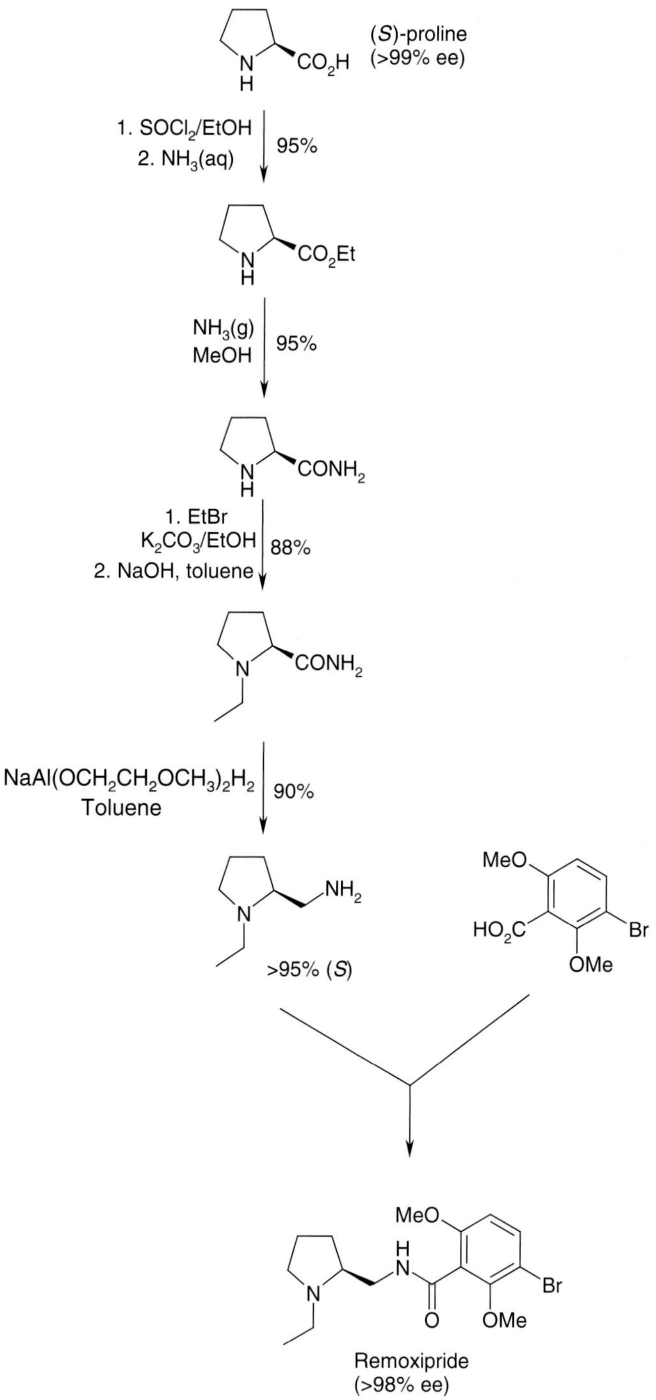

Figure 2 Starting from (S)-proline, an abundantly available natural chiral building block, a key intermediate is manufactured with preserved stereochemistry en route to the class of benzamide neuroleptics.

using transition metals in combination with chiral phosphorous compounds as ligands.[46–48] The versatility of what asymmetric reactions can achieve is just breathtaking and the whole area has been extremely innovative and prosperous.[49–67] Therefore it might seem somewhat surprising that not very many of these processes have in fact been scaled up to commercial production. Latest surveys indicate that less than 20 chemocatalytic processes are operative at full industrial scale, the majority of which are represented by various reductive transformations (hydrogenations).[68,69]

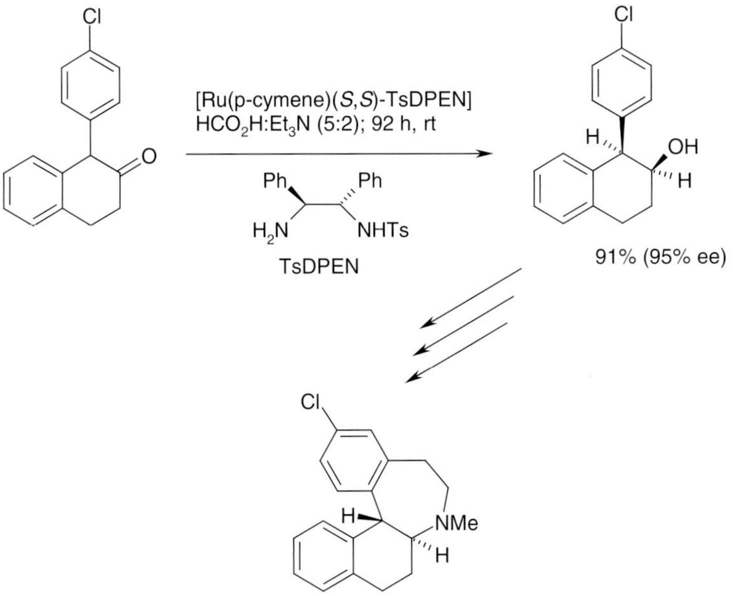

Figure 3 Dynamic kinetic resolution coupled with an asymmetric transfer hydrogenation using formic acid (HCO$_2$H) as hydrogen equivalent to form intermediates useful in the synthesis of benzazepine class of dopamine D1 antagonists.

The volumes produced per annum vary considerably depending on the intended usage of the product and so the span is from less than 1 metric ton (t) for certain intermediates used to prepare fine chemicals or pharmaceuticals up to the stunning level of more than 10 000 t of (*S*)-metolachlor, the active ingredient in a widely used herbicide; **Figure 4** shows details of this process.[70,71]

Given this situation of infrequent application of asymmetric technology at large scale there is an interesting disjunction observable between the high attention received by this area in academia and its practical use in industry. Thus, while figures based on journal citations concerning chiral technologies show that 72% fall into the arena of asymmetric methods, the corresponding figure is less than half (35%) when looking at what is actually utilized.[20] In fairness, however, it has to be said that the dominant role played by processes building on resolutions and starting materials from the chiral pool is underpinned by the fact that, in the latter case, Mother Nature has provided us with an enormously rich and abundant reservoir of useful compounds, exemplified by amino acids, carbohydrates (sugars), terpenes, and alkaloids.

2.17.3 Supplying Stereoisomers: What the Market Can Offer

Never have there been more opportunities for a buyer to find the chiral material of interest in enantiomerically pure or enriched form from a commercial source. This is not to say that there will always be a chiral building block available to serve your needs at any time, but the selection of compound classes and their breadth of structural variation is simply enormous. A word of caution, however, is justified when examining what the term 'availability' means. Namely, does it mean virtually any quantity being accessible in bulk from the shelf for immediate purchase or do we have to take long delivery times into account because the particular product has to be manufactured first? It could even be the case that the market only supplies small quantities for laboratory use and these amounts are prepared using a method unsuitable for upscaling. The following case story nicely describes this situation. For a certain project the nonproteinogenic amino acid (*S*)-azetidine-2-carboxylic acid (**Figure 5**) was required as a key raw material for the preparation of an API. This compound, naturally available from hydrolysates of plant proteins[72] albeit in fractions of 1%, was on offer in gram quantities for laboratory use[73] and given its structural simplicity it was, not surprisingly, believed to be also available in bulk amounts. The reality turned out to be rather different. After huge efforts trying to identify a source with capacity to deliver kilogram quantities it was serendipitously found that a mere 75 kg was located in a warehouse and this was all that could be rounded up on the entire globe. And this was not the only discouraging news. As it turned out the manufacture of this lot had been achieved by a tremendously effort-consuming extraction process starting off from hundreds of tons of molasses and as this was seen as a nonviable route for making this material there was in fact not any commercial route in operation that could support the large-scale needs of the project. Consequently, it became an extra

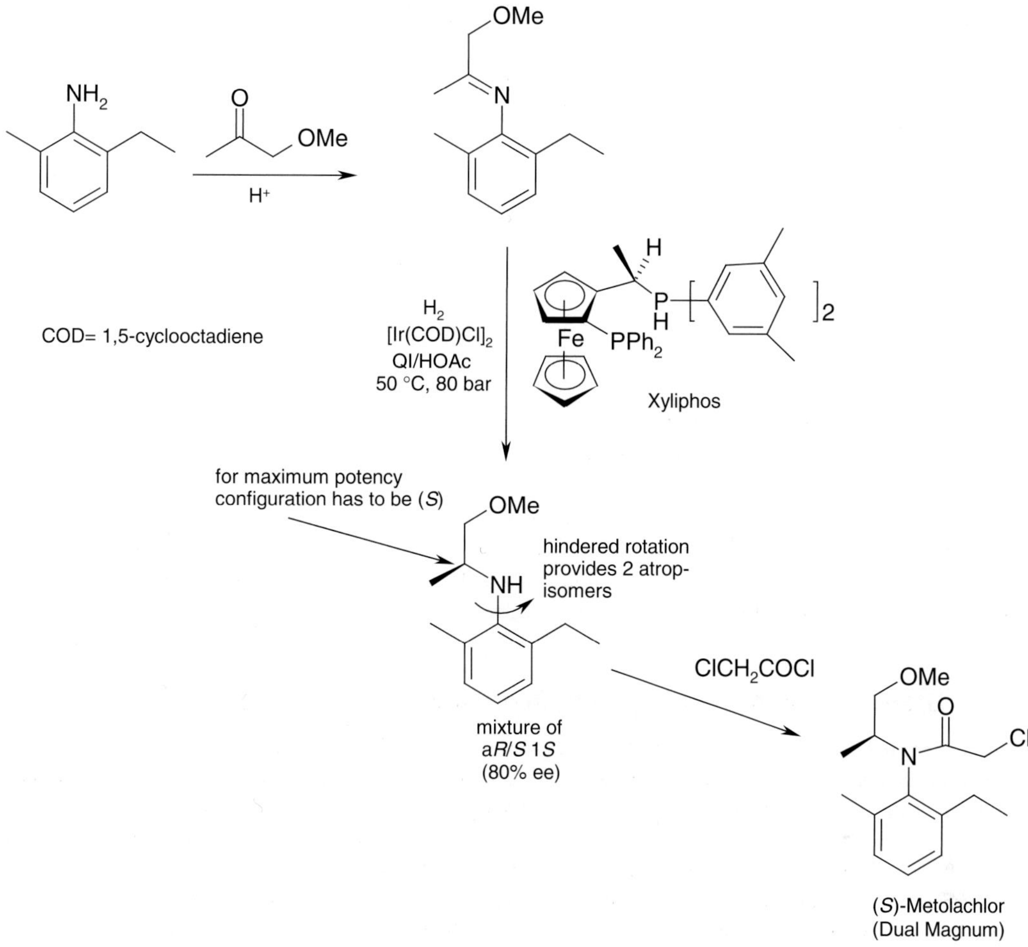

Figure 4 The largest volume enantioselective process to date produces (S)-metolachlor at an impressive catalytic efficiency of >10^6 expressed as mol product over mol catalyst (also called turnover number, TON).

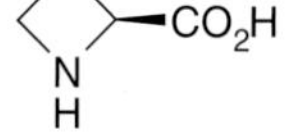

Figure 5 (S)-Azetidine-2-carboxylic acid.

task to devise, as quickly as possible, such a route with the potential to be scaled up to production size, a challenge that was far from trivial but which, fortunately, was concluded in a successful manner.[58,59]

Looking at the market situation it is striking to see how many useful compounds are available.[74,75] Besides the already mentioned classical chiral building blocks, notably amino acids and carbohydrates, there now exist a whole raft of new materials that have been put at our disposal many of which produced via asymmetric methodologies or kinetic resolution protocols. Thus, the C-3 synthons and other short chain or ring-containing compounds (**Figure 6**) have become particularly versatile and useful by virtue of a stereogenic center (or occasionally several) and a functionality that allows further chemical manipulations. Using these molecules displaying their linear carbon core or cyclic architecture offers easy access to a defined stereochemistry as well as possibilities for an array of substitutions as part of a wider synthetic scheme.

It is important to emphasize that many of the chiral compounds reported in the literature, in databases, and in product catalog have been made in only small quantities awaiting a market to be established which will eventually drive demand upward. Therefore, no guarantee can be given that a particular molecule will be accessible in the relatively

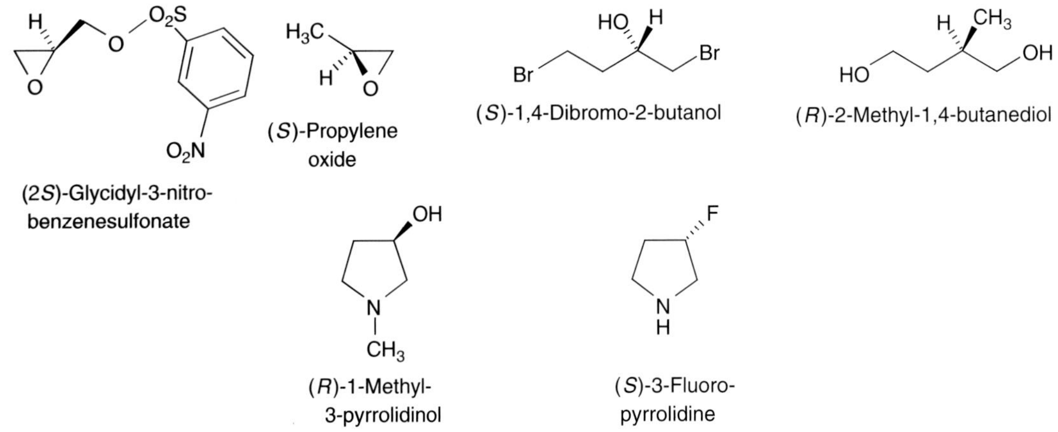

Figure 6 Structures of a few C-3 and other synthons.

short term (within weeks or a few months at the most) and this may have a profound influence on the design of the synthetic route as the time element is crucial when it comes to delivering your own product. This will not allow for extended lead times before receiving your starting material and, furthermore, the effort to quickly define the final sequence will also be influenced by what is available at scale and what is not.

2.17.4 Stereospecific Analyses: An Overview

A prerequisite for the breathtaking advances we have seen in the field of chirality – both from a chemistry point of view as well as in pharmacology, biology, and related areas – has certainly been the tremendous development in analytical methodologies achieved since the late 1970s and early 1980s.[76] Starting from a point many years back when the measurement of optical rotation was the sole source of determining enantiomeric purity, the situation today allows very precise identification and quantification of single stereoisomers using the plethora of techniques at hand. These advances offer the possibility of determining not only the intrinsic configuration of a compound but also of measuring its presence in sample matrices down to trace amounts, which often equals levels below 0.1% and into the ppm and ppb ranges. Partly as a consequence of this trend there have been continuously increasing demands from regulatory authorities to document and identify chiral impurities (and achiral as well) in active drug substances to at least the $\geq 0.1\%$ limit.

A list of major techniques in current practical use to determine stereochemical integrity and purity would include the following.

2.17.4.1 Determination of Absolute Configuration

2.17.4.1.1 X-ray diffraction (XRD)
This is the recognized standard for a priori determination of the absolute configuration of chiral molecules. Bijvoet[77] technique relies upon the interpretation of anomalous scattering of x-rays caused by atoms present in a crystal lattice. This is a highly refined technique today with many technical improvements since its introduction.[78,79] The major drawback with this technique is the requirement for a pure single crystal.

2.17.4.1.2 Vibrational circular dichroism (VCD)
This is a recently introduced technique offering a novel alternative to XRD. There is no requirement for derivatization or preparation of a single crystal. It is defined as the differential absorption of a molecule for left versus right circularly polarized infrared (IR) during a vibrational transition. Absolute stereochemistry is established by comparing solution-phase VCD spectra to results of ab initio quantum chemistry calculations.[80]

2.17.4.1.3 Other methods
Other methods under development and exploration include the use of optical rotation, electronic circular dichroism, and Raman optical activity.

2.17.4.2 Determination of Enantiomeric Purity

2.17.4.2.1 Polarimetry
Polarimetry is the oldest and most widely used method constituting a nonseparation technique. It relies upon measurement of a net effect. Data from pure enantiomers are required for comparison. The technique is based on the fact that chiral molecules rotate the plane of polarization of plane polarized light.[81]

2.17.4.2.2 Nuclear magnetic resonance (NMR)
This represents a nonseparation technique that relies upon the use of chiral derivatizing agents or a chiral solvent to form diastereomers, which may have different chemical shifts in the NMR spectrum.[82,83]

2.17.4.2.3 Direct enantioselective chromatography/electrophoresis
This is the most widely used to determine enantiomeric purity. Enantioselective separations have been realized in all possible separation techniques: high-performance liquid chromatography (HPLC), gas chromatography (GC), thin layer chromatography (TLC), supercritical fluid chromatography (SFC), capillary electrophoresis (CE), capillary electrochromatography (CEC).[84–92] This technique depends on the formation of transient diastereomeric complexes with either a chiral bonded phase or chiral additive to afford separation. Many phases/additives are now available.[93–95]

2.17.4.2.4 Indirect enantioselective chromatography/electrophoresis
This technique involves the use of chiral derivatizing agents to form diastereomers, which can be resolved without the need for chiral stationary phases or chiral additives.[96] The approach is still used for analytes such as amino acids.

Perhaps needless to say, but it is worth mentioning that in spite of the tremendous advances seen over the past couple of decades the development and improvement of techniques such as the ones mentioned above have not come to a standstill. In all likelihood we will be witnessing continuing efforts in these areas increasing our possibilities to analyze complex chiral chemical moieties with even higher accuracy.

2.17.5 From Drawing Board to Consumer: Progression of Drug Compounds in Process R&D

The complex and multifunctional task represented by discovering and developing a drug from initial idea through all stages to market approval and launch requires the input of enormous resources, in terms of both people and money. As many of the diverse activities throughout the various disciplines in R&D are strongly time-driven it is of the utmost importance that a free flow of information across functional boundaries is insured in order not to slow down any actions or transfer of knowledge. In other words, in order to keep up the speed in a project streamlined internal work processes need to be adopted as well as securing minimum slowdown at the interfaces with other departments. For Process R&D offering a skill base that covers the whole range from small-scale glassware experimentation up to pilot plant operation in several hundred or thousand liter vessels and bridging the entire gap between Medicinal Chemistry and commercial production, there is an early entry in the lifetime of a project and a late departure unless there is a premature discontinuation. For this purpose a model of a global work process has been constructed spanning the entire timeline from start to finish with the different Process R&D specific activities aligned in sequential order; see **Figure 7** for a graphical representation of the model.[58,59,97,98]

From the viewpoint of the wider pharmaceutical business it is entirely relevant to challenge a technically oriented function like Process R&D with questions like: How can you provide better support to our drug projects, and what would allow you to progress an increased number of CDs through your pipeline at a constant resource level? To sensibly address these queries, though, it is essential that the general comprehension and awareness of the main contributions from a Process R&D department is much higher. Here, the tighter integration of the various components in drug R&D that we have witnessed over recent years has supplied a remedy.[99] Today a good and robust process that can deliver the desired pharmaceutical compound with the required quality and at an acceptable cost of goods level is appreciated, and, as a matter of fact, is expected as never before. Historical precedents are available and a classical yet still intriguing case of successful methods development demonstrating the power of Process R&D is constituted by the synthesis of cortisone. Thus starting at a mind-boggling cost of $200 g^{-1} the level was drastically reduced to a mere $3.50 g^{-1} over a period of only a few years in the beginning of the 1950s, which equals a price drop of more than 98%.[98,100]

Contributions in the form of quantities of active molecule, a manufacturing method, and regulatory documentation to support clinical studies and marketing applications are widely acknowledged and valued. To be able to quickly respond to new compounds coming through the pipeline, a back-integration into the earlier stages of Discovery is

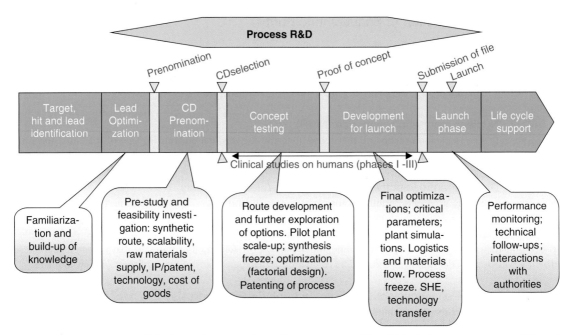

Figure 7 The value chain of R&D in the pharmaceutical industry is stretched from biological target to commercial launch. Stages of Process R&D involvement are specifically indicated.

essential, not stopping at the prenomination stage but pushing further back into lead optimization (LO) and maybe even lead selection. If this mode of operation is not present there will, inevitably, be a loss of valuable time once the new chemical entity (NCE) is appointed as CD, which, as a knockon effect, could delay market launch later on in cases where the project is successful. The strong need for speedy and timely manufacture of a first scaled-up batch is further underpinned by the desire to quickly document the compounds, mainly in the form of extended safety studies in various animal species, in order to allow entry into the first-time-in-human phase. From a purely statistical viewpoint, on the average roughly three out of four projects will face discontinuation up to the proof of principle point, which is where signs of desired pharmacological effect have been demonstrated in patients after single dosing. Understandably, it is of prime interest to reach this stage where the majority of attrition will have affected the pipeline and to successfully achieve this as fast as possible, API delivery cannot be tolerated as a bottleneck. Therefore, an early start to route design and development of methods is crucial, which still maintaining a balanced view on the actual resource input depending on the nature of the project – first in class, best in class, or backup – as the scientific and economic risks may vary widely including your own risk preparedness as well as the speed with which the process will be conducted.

As an aspirational target, the work should preferably be initiated to allow route freeze – the point where the synthetic sequence is decided, that is the timely order from starting material via intermediates to product is fully defined – to be reached prior to the first time-critical upscaling of the synthesis alluded to previously. The benefits from this are twofold: a quick concentration on a single synthetic path enabling all resources to be devoted only to this task and an advantageous regulatory situation as safety and human studies, from this point onward, will be conducted using active ingredient prepared by a fixed sequence, which, however, does not preclude changes in the manufacturing process such as switching of reagents or solvents and altering conditions. At the same time it must be made absolutely clear that this task is very stretched and to succeed a sophisticated set of tools needs to be at the disposal of highly skilled experts in route design of organic molecules. The data mining capabilities and information retrieval systems have to be of top-notch quality to aid in the expedient and mostly theoretical exercise of shortlisting plausible synthetic sequences. As the literature in the field of organic chemistry, which has its roots as far back as maybe 180 years, is too vast for any individual to grasp, huge efforts have gone into the construction of artificial systems to provide support. This in silico approach has seen a tremendous development over recent years and today several commercial data packages are available that to varying degrees can perform retrosynthetic analysis of given target compounds and propose ways of assembling the desired product, with cross-referencing to published sources. Specifically for application in the Process R&D environment the route design feature should, preferably, be complemented with safety, health, and environment (SHE) directed information as well as to give data on commercial availability of starting materials and reagents, in particular if prices are quoted together with delivery lead times. The former will present a vital part when making judgements on potential scale-up challenges and the latter could steer the selection of route of synthesis by discarding alternatives where there are no bulk quantities of the chemicals needed available on the market or where the cost of goods target would not be met. A model representing the desired state to aim for when it comes to computer directed use of databases is shown in **Figure 8**.[101]

At the Medicinal Chemistry–Process R&D interface the diversity expressed in the form of a multitude of chemical structures being prepared and investigated by the former is matched by an exploration of a variety of syntheses leading to the target molecule by the latter. From both ends it is essential to cover enough space so that important aspects are not left out, but also to narrow down the number of options to enable focusing of efforts. **Figure 9** shows an illustration of the context thus described and highlights the need for functional overlap.[101–104]

Thus far there has been no particular mention per se of structural features displayed by the molecules up for scrutiny. However, they do have an influence, and a fairly strong one as well for that matter. Focusing on the sheer number of new CDs to be brought forward is one thing, albeit of tremendous importance to the success criteria of the drug industry, but from a strict Process R&D perspective the number of steps that it takes to make each one of them is of considerably higher relevance. Superficially speaking, the efforts needed to pursue a single CD requiring 20 synthetic steps would be equal to running five CDs at four steps each. Hence, a major task in Process R&D during the phase of optimizing the structural architecture is to analyze and communicate the consequences that various changes will have on the synthesis. We could be looking at an overly complex substitution pattern on a cyclic core structure, the incorporation of substituents that are not easily compatible with each others (from a synthetic point), making use of esoteric or unusual heterocyclic units difficult to prepare, or incorporation of too large a number of stereochemical features such as several stereogenic centers or even torsional isomerism. The aim should not be to disqualify promising compounds per se because they would pose some rather awkward synthetic hurdles. Quite the contrary, but in cases where options in fact do exist pros and cons have to be pointed out for each alternative when examining the routes envisaged for their assembly. In the selection procedure, this can then be taken into appropriate account amongst all the other parameters that will have to be evaluated. And there, biology, pharmacology, and toxicology should be given preference, though without ignoring realities, capabilities, and limitations in chemistry and pharmaceutical formulation.

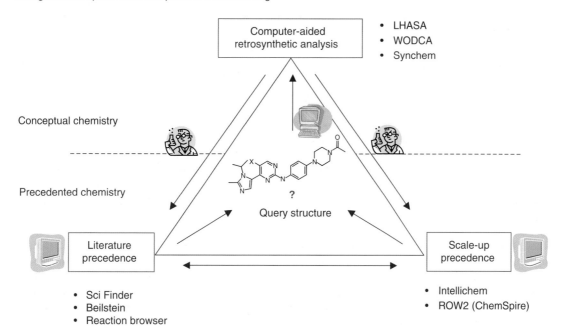

Figure 8 Database architecture tailored for Process R&D purposes. Note that the listed software provides examples of largely commercial systems, which have had varying degrees of industrial application.

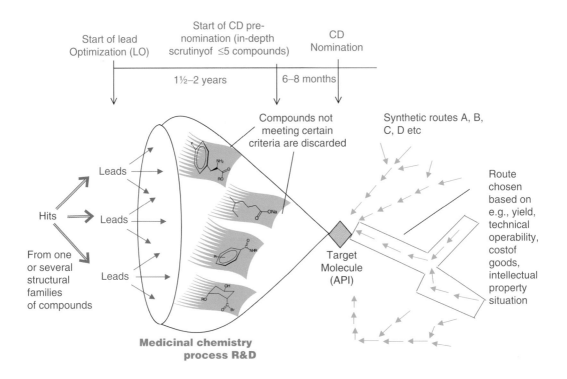

Figure 9 The crucial integration of Process R&D with the pre-CD work during the Discovery phase, notably at the Medicinal Chemistry interface, extends back into lead optimization.

Admittedly, in the life science industry, communicating matters of the technological nature of process science and concerning bulk production is not the easiest thing, which is altogether understandable given the focus of the business (to provide products capable of improving quality of life and addressing diseases where there are unmet medical needs). Perhaps one way forward to better reach out with the messages would be to create a mimic of the renowned Lipinski's[105] rule of five, but now describing different parameters of the process.[101]

So far the discussion has been strictly generic in the sense that specific structural features of the API have not been explicitly highlighted. The question is thus obvious whether other characteristics besides the length of the synthesis, for example presence of one or more stereogenic centers will exert a major impact on Process R&D? The answer is to a surprisingly small extent today compared with the situation one or two decades ago. The tremendous advances seen in the methodologies available to conduct a great variety of stereoselective transformations have no doubt triggered this change in attitude toward increasingly complex molecules. A striking example of how far the capabilities in this regard have reached in less than three decades is given by comparison of L-Dopa (**Figure 1b**) and (+)-discodermolide (**Figure 10**). The former, a famous anti-Parkinson compound consisting of one stereogenic center, became industrially available during the 1970s via the seminal asymmetric hydrogenation technology masterminded by Knowles[30–32] and his team at Monsanto (**Figure 1**). With the latter, a highly potent inhibitor of tumor cell growth (via microtubule stabilization), we are facing no fewer than 13 such centers, which leads to the theoretical existence of 2^{13} (= 8192!) unique stereoisomers (or 4096 pairs of enantiomers). And yet, in spite of its scarce availability from the natural source (the Caribbean marine sponge *Discodermia dissoluta*) pilot production on several tenths of grams scale has been successfully achieved and was to be succeeded by an intended batch size of several hundred grams.[106–108] Unfortunately, this wholehearted endeavor has now been discontinued (during 2004) allegedly due to toxic side effects and, therefore, we will never know whether a commercially viable process on multi-kilogram scale would have been possible for this intriguing molecule. Regardless of this setback, relentless synthetic efforts are being deployed with the aim of devising better and better procedures.[109] This level of complexity is, however, not the end, as literature sources indicate commercial development of the macrocyclic lactone apoptolidin is now under way – a molecule with 25 stereocenters displaying potent cytotoxic and tumor-inhibiting effects.[56,110–113] Not disregarding these stunning achievements, the fact still remains that probably about 50% of enantiomerically pure products on industrial scale are

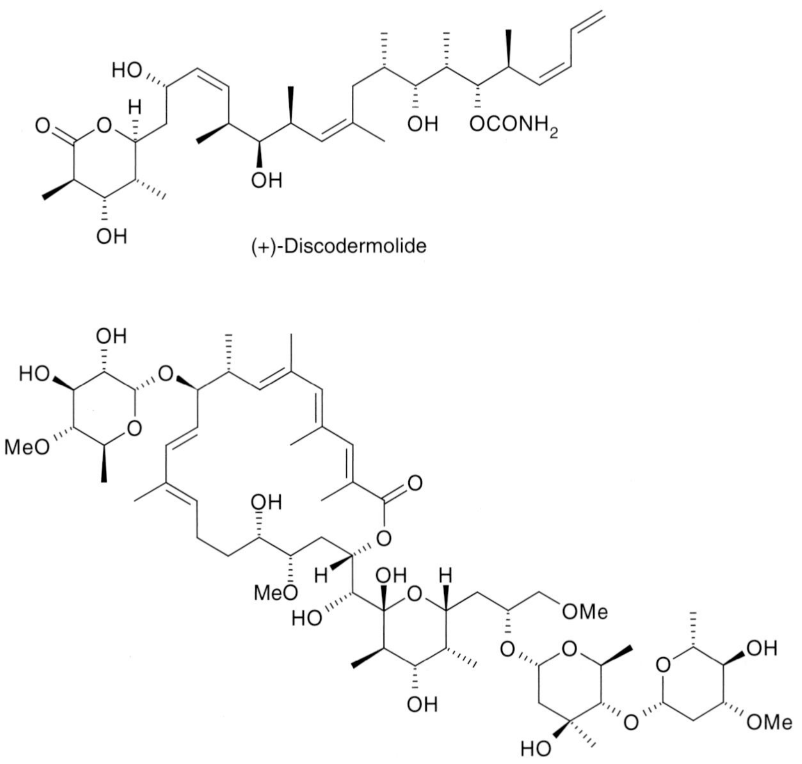

Figure 10 Structures of (+)-discodermolide and apoptolidin.

prepared via classical diastereomer resolution at least if one takes their value into account. In the future, as more and more robust and reliable 'competing' methodologies are developed and validated this figure will inevitably be reduced step by step. It should be noted that the discussion here is restricted to small molecule APIs, that is, the increasingly abundant monoclonal antibodies (MAbs) and protein-based drugs have been excluded. Commercial manufacturing of these products is making use of entirely different technologies such as fermentation or cell line expression.

2.17.6 The Regulatory Environment

Representing the industrial sector that has by far the strictest rules and legislations to adhere to, the pharmaceutical companies must develop a high degree of expertise in the regulatory field. Moreover, the fact that any drug to be marketed has to gain approval from the regulating body in charge before it can be made publicly available demands that forward-looking regulatory strategies are implemented in the very early stages of an R&D project. Also as has been seen in recent years, there is a pronounced mutual trend to engage in an open dialogue between both parties along the timeline – probably pioneered by the US Food and Drug Administration (FDA) who in an attempt to improve on performance measures such as approval rate and efficiency of throughput of new drug applications encourages data to be debated and discussed – to allow dissemination of different types of problems and challenges. As technological breakthroughs such as new therapeutic targets, biomarkers, and pharmacogenomics have not seen a corresponding translation into increased numbers of NCEs (on the contrary there has been a dramatic decline over the past 8 years), something needs to be done.[8] Furthermore, the burning issue of the reasons for more failures in late stage development (notably in phase III) as well as for market withdrawal has to be taken very seriously.[16] Thus, based on the insight in the FDA that the current drug development process is not operated in an adequately efficient way a strategy needs to be devised, which addresses the shortcomings. In a recently published 'white paper'[114] on the critical path to new medical products three dimensions stand out as being of key importance:

- Assessing safety
- Demonstrating medical utility
- Industrialization.

Each one of these brings its own set of activities, so for example in safety the product has to demonstrate clinical reliability; the utility has to be proven by achieving clinical effectiveness; and in industry there is a demand for the design of a high-quality product and the capacity for mass production under strict quality control. All in all it is strongly advocated in this paper that "a better product development toolkit is urgently needed" and a discussion is conducted on what these tools might be in the specific areas mentioned above as well as devising suitable standards and paths forward.

Focusing on chirality in this context it is interesting to notice the outcome of a recent survey on what kind of compounds have received worldwide approval as new drugs during the period 1983–2002 coupled with similar statistics from the FDA covering the years 1991–2002.[115] This clearly shows that drugs using single enantiomers have now surpassed those that have an achiral compound as API, putting racemates as a definite minority. The picture thus presented is further emphasized by the 15 NCEs approved by the FDA in 2003 where 64% are enantiomerically pure, 22% are achiral, and 14% are racemates. In actual fact, the gradually evolving awareness of biological and pharmacological issues connected with chiral drug molecules and their significance that started during the 1970s and accelerated in the 1980s brought about a surprisingly quick response from various stakeholders. Thus, the FDA launched a policy statement addressing the development of new stereoisomeric drugs in May 1992, which was later on complemented by a corrections note early in 1997.[116]

Furthermore, pharmaceutical industry bodies – both nationwide and on international committee levels – have picked up this urgent and complex question to try and formulate well-thought-through and adequate guidelines.[117,118] A common theme is the desire to demystify the concept of stereoisomerism[119] and instead apply a scientifically sound approach to chirality and the fact that mirror images of chiral compounds often but not always display different properties.[27] In general terms the recommendations speak of the necessity that in cases where both antipodes are present, for example in the form of a racemate, it is unambiguously demonstrated that such a drug is safe for the patient. Moreover, when this is not the case and instead one enantiomer has unwanted side effects then this component will be regarded in the same way as any impurity and has to be treated accordingly. This means that it rests upon the sponsor (i.e., the responsible pharmaceutical company) to ensure that the remaining amounts are brought down to residual levels dictated by the findings in toxicological assessments and other studies.[120] For new drugs branded as enantiopure this will today most likely imply around or below 1% of undesired antipode (indicating a purity of 98% ee or better) and with current manufacturing technology this is entirely feasible on a routine basis in many instances. As quality carries a price it is essential in this context to point out that, when methods and techniques are used to achieve extremely stereomerically pure materials ($\geq 99.9\%$ ee), this will inevitably result in a drastic increase in cost. Therefore, such ultra-high qualities should only be demanded when unambiguous and validated scientific data makes this indisputably necessary.

From a regulatory Chemistry, Manufacturing, and Control perspective the argument is made that the three-dimensional integrity (i.e., the molecular configuration) of the API has to be proven in a stereochemically selective identity test. And, furthermore, a selective assay method is required to accurately determine its stereochemical composition, which means that a reliable low-level quantitation of stereoisomers present should also be possible. Even if this set of criteria is applicable to all stereoismers – irrespective if in enantiomeric or diastereomeric relationship – the problems that one faces with the latter are considerably smaller as an upgrade of a mixture of diastereomers can, in most cases, be achieved relatively easily due to their differences in physical properties (e.g., solubility, melting and boiling points). In summary, the previous default position for developing and marketing racemic compounds has now been replaced by a situation where the clinically most efficacious and safe enantiomer is taken forward, unless the racemate has some intrinsically beneficial properties.

2.17.7 Drugs at Any Cost?

The pharmaceutical industry has, in general, a reputation of being a very lucrative business offering its owners good if not excellent profit margins. While this might be true in many cases it is important to be mindful of a few facts,[4,6,7,10,15] some of which are provided below:

- The vast majority of marketed drugs – maybe as many as 90% – earn less than $180 million per year
- As many as 70% fail to recover their overall R&D expenditures during the entire life cycle
- Only a small proportion ($\approx 12\%$) provide what might be called a satisfactory return on investment
- The total out-of-pocket R&D investment to take an NCE through to registration has seen a truly explosive development in recent years and was reported to be $403 million in 2003, which translates to the often-cited figure of capitalized pre-approval cost of $802 million. This number is now probably outdated and we are most likely

rapidly approaching the stunning level of $1 billion, especially in the light of recent setbacks and failures of some high-profile products
- Cost of drug marketing has accelerated tremendously over a period of years and has reached a level comparable to the R&D component if not higher
- Running clinical trials is becoming increasingly complex as more and more complicated diseases are addressed
- Safety concerns from health authorities demand more and more patients to be enrolled in the trials and when aiming at demonstrating superiority over current therapy (competitor drug) the number of patients needed in outcomes studies could mount to 20 000 or even 30 000
- The 'classical' pharmaceutical industry is under severe attack from generic competition and its patent portfolio is aggressively challenged from all angles
- The capability to deliver sustainable annual double-digit growth rates with regard to turnover and profit might be hampered as a consequence of fewer potential mega brand products (sales > $1 billion per year) coming through the pipeline
- Attrition is maintained at a high level (90% on average) partly because of poorly understood diseases being investigated (e.g., Alzheimer's, obesity, stroke, severe cancer forms, rheumatoid arthritis, human immunodeficiency virus (HIV)), and the desire to move from symptom relief to disease modification.

In particular the numerical data presented here (originating from the late 1990s and the early years of the new millennium) do not pretend to show a picture that is entirely correct in an absolute sense. This is mostly due to the inaccessibility of industry-wide intelligence that manages to provide real-time 'as-is' information, confidentiality and business secrets being main obstacle in this regard. Instead the information intends to highlight some intriguing relationships and globally observed trends. Taken in a broader context, the scenario alluded to above has confronted the drug industry with some unforeseen challenges. And on top of this has been added the sharply increasing resistance from authorities and the market to accept the prices that historically have been regarded as feasible to finance necessary reinvestments in new and continuously risky ventures. Furthermore, several newly introduced products have been aggressively attacked as policy-makers and various pressure groups have questioned their usefulness and safety. Also there is a clear reluctance from the (political) health officials to recommend that doctors should prescribe wider use of new, more expensive and efficacious drugs rather than the old less costly ones with lower efficacy. This has all of a sudden changed the once so prosperous, stable, and highly esteemed pharmaceutical sector into, if not a crisis, certainly one that is grappling with its self-image and severely curtailed public reputation. It is probably not too far-fetched to project drastic changes in our industry that in the end can affect the fundamental principles for operating this kind of business: **Figure 11** clearly indicates the sort of challenges that one faces with regard to R&D expenditures, new molecular entity (NME) output, and overall development time lines.

How does the process that is developed in order to produce API at commercial scale come into all of this? Is this component of the multifaceted activities required to successfully take a product to market launch really capable of playing any noticeable role, especially from a cost point of view? The outright answer to this is: previously not to a very

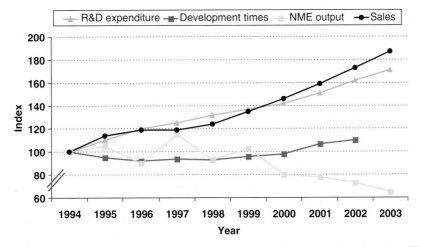

Figure 11 This index-based graph shows accumulated trends from 1994 through 2003 for sales, R&D expenditure, development times, and output of new molecular entities (NMEs). (*Source*: CMR international and IMS Health.)

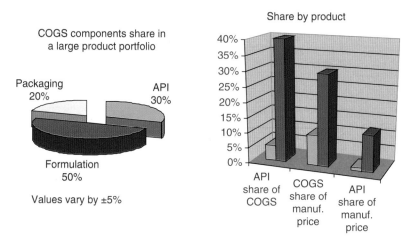

Figure 12 Pie chart: relative share of the cost of goods (COGS) components as represented in a large-sized product portfolio. Bar chart: product-wise minimum to maximum distribution of interrelated cost contributing parameters.

great extent, but in today's climate increasingly so. Looking at some recent industry estimates reveals that in a large product portfolio the API represents about one-third of the entire cost of goods (COGS), the other contributors being formulation and packaging. Extending the view to an inspection of product-by-product distribution the spread in API share goes from below 1% up to 40%. Similarly COGS and API, respectively, as share of manufacturing price display a wide variation as is evident from the graphics shown in **Figure 12**.

In the old regime where COGS rose to ≤30% of the manufacturing price the drive for process improvements were not very pronounced:

- Reducing the API cost by 50% (e.g., via improved yields, cheaper starting materials) resulted in a 15% COGS reduction at an API share of 30%
- This outturn brings a payback of a 1.5–4.5% lower manufacturing price.

In future with growing COGS share as dictated by severe downward-directed price pressure the situation is dramatically changed:

- At 50%, API share alone represents 15% of manufacturing price
- Cutting the API cost into half improves profit margin by 7.5%
- This increases the impact of API on manufacturing price by 67%.

These simple and rather generic calculations should suffice to convince even a skeptic that the industry is moving away from a view where the API is considered to be of inferior importance to one where much emphasis is placed on driving its cost down as much as possible. The knockon effect that this has is of course the need to develop more efficient and cost-conscious processes. By the way, in order to keep profits up in an era of severe price pressure all links in the chain – R&D, Operations, Sales & Marketing – contributing to the cost of the labeled drug will be forced to improve on their way of working and the methodologies applied. Only then can the pharmaceutical companies stay in business.

2.17.8 Chiral Switches: From Racemate to Enantiomer

As pointed out previously, the main drivers when contemplating the transformation of an existing drug in racemic form to a single enantiomer are improvement in clinical efficacy, elimination of undesired side effects (better toxicity profile), and eventually the possibility of reducing the chemical load on the body's metabolic system (in theory by 50% given that nothing else is changed).[27] In this context the interpretation of what constitutes a significant enough improvement to warrant a switch to be taken forward is up for debate. There are no clear guidelines for how this should be handled, but a general feeling is that in order to demonstrate a high enough innovation over the racemate to ensure IP protection, a prerequisite for wanting to take the inherent financial risk involved in the development of a new drug, an approximately 2:1 activity ratio between antipodes would probably be regarded as necessary. One further aspect of this is perhaps more irrational, namely that a marketing strategy could lie behind a company's decision to replace a racemic drug with one of its enantiomers as a means of counteracting generic competition (on the racemate). To repeat the words of warning made earlier in this chapter: this is a very complicated area and, therefore, each case has to be treated on its own merits. There is no law of nature that says an enantiopure compound is always the best, nor that we should focus only on the extremes with either 100% stereoisomeric purity or 50% as is the case in racemic mixtures. Instead the optimum from a pharmacological and medicinal point of view could be somewhere in between.

How successful has this transition from racemate to enantiomer been? When this business opportunity emerged around 1990 a rather large number of compounds were of course in use as racemates on the market. Unsurprisingly this raised expectations, and rosy pictures were painted envisioning more or less a switch of the entirely accessible portfolio of products. This has not been achieved. On the contrary, the outturn has been a number that is far from the optimistic target. Thus, available statistics reveals that in the period from 1994 to 2002 only about 10 instances occurred which resulted in the actual launch of a pure enantiomer drug.[28] And this is the case in spite of the continuous approval by regulatory authorities of racemic NCEs throughout the 1990s with on the average about 10% of the annually approved drugs worldwide, a trend that has actually proceeded into the new millennium according to statistics from both the FDA and the the UK Medicines Control Agency (MCA). A few pertinent examples from this exclusive list are shown in **Figure 13**.

Irrespective of the medical success or nonsuccess the chiral switch concept has, no doubt, spurred a lot of attention to scalable methodologies enabling the preparation of the desired antipode of previously manufactured racemates. Thus, one is likely to find all the conceptually different approaches outlined previously (*see* Section 2.17.2) represented amongst the cases known so far. In that regard this has been helpful in pushing the technology frontier forward and by doing so setting an example for how to deal with challenges of such a kind. A few years into the twenty-first century there is still a fairly well filled pipeline of potential chiral switch candidates around – some in late stage clinical phase, some in earlier, and some even in a state of suspension – but the projection that we are approaching the final act of this play in its original version is not very far-fetched. The reason being, of course, that the likelihood of developing chiral NMEs as racemates will continue to stay at a very low level, if not to virtually disappear, and, therefore, business opportunities are bound to be scarce. However, a new spin on this might be the growing interest to thoroughly investigate not only the natural product as it is found in nature per se,[121] but also the unnatural/low abundance enantiomer of natural or nearly natural products.[28] With the increasing knowledge about disease mechanisms and molecular biology it is entirely feasible to discover new and clinically significant pharmacological activities in the antipodes (or quasiantipodes in cases where several stereogenic centers are present and only some of them inverted) of

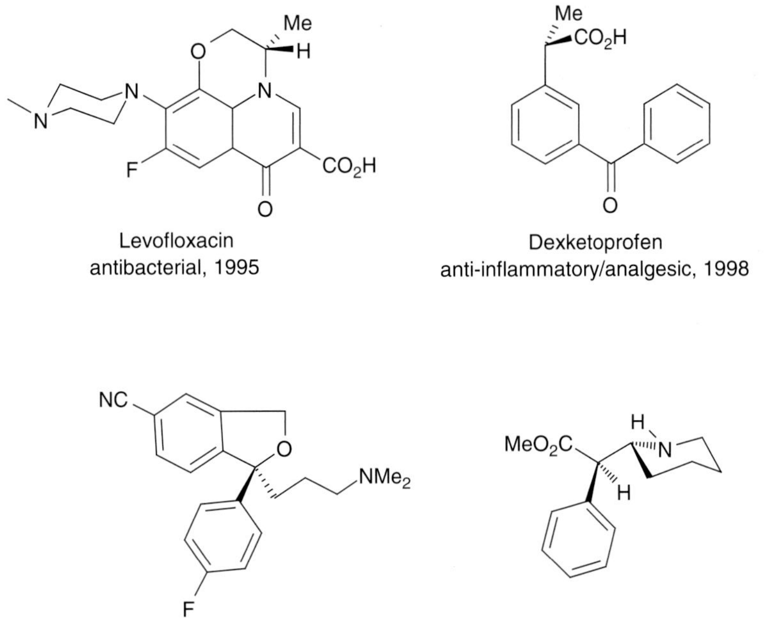

Levofloxacin
antibacterial, 1995

Dexketoprofen
anti-inflammatory/analgesic, 1998

Escitalopram
antidepressant, 2001

Dexmethylphenidate
attention deficit/hyperactivity disorder, 2001

Figure 13 Molecules that have gone through the transition from racemate to single enantiomer, listed alongside their medical indication and year of launch.

known single enantiomers. One early example illustrating this would be the reputable gossypol case from the 1970s. The lead role here was played by the yellow pigment gossypol present in cotton plants and mainly composed of the (+)-isomer,[122] which was found to exhibit male contraceptive activity. Later on it was proven that this spectacular property resided almost exclusively in the (−)-form, present only in minute quantities in the natural source. Due to systemic toxicity and incomplete reversibility of action this promising program had to be disbanded,[123] but hopefully without entirely disqualifying or compromising the general pros of the concept as such.

2.17.9 Successes and Failures: What Are the Discriminating Factors?

The heading expresses a classical and often debated question: is there a way for us to identify early on those cases that will become successful? Looking at the drug approval statistics of recent years there is, sad to say, little pointing in that direction. As alluded to previously the level of attrition is stubbornly staying at the annoying figure of roughly 90%, in spite of all the knowledge gathered about causes for diseases, mechanistic understanding of biological events,[124] and, not forgetting the wealth of data generated by unraveling the human genome. Two main bottlenecks often mentioned in this regard where the scientific frontline expertise will rarely give the correct guidance are validation of the biological target and biomarkers, respectively. In fact these 'components' stand at the very center of drug R&D with the former accountable for identifying the specific target (enzyme, protein, cell membrane, organ, etc.) responsible for an illness and the latter for what to measure and how to ensure that certain treatments work and others do not. This situation and all the frustrations caused by it have probably forced many pharmaceutical companies to re-evaluate their business strategies in the direction of a more humble and hopefully realistic approach with less emphasis on speed and more on quality. It takes some courage to say that although we know a lot about how to develop drugs, there is still much more that we don't know!

What we can be certain about, however, is that there will be progress – slowly but steadily. And at every new failure there will be lessons to take in which will improve the likelihood that, ideally speaking, a repeat of the same mistake should not occur. One of the intrinsic properties of computers besides speed of operation is their capacity to handle enormous amounts of data that human beings cannot really cope with any more. This invaluable feature will allow us to analyze the information matrix from various angles and pick out significant areas of similarities or discrepancies. Most importantly it will be a tool for identifying trends, which in turn will provide powerful support to make predictions on expected outcome. Here we have started to move into new territory where the available in silico technologies by virtually drawing from all accumulated experience can direct our move into the future. Science fiction perhaps but in areas where these models have been used and at least partly validated, for example predicting absorption, distribution, metabolism, and excretion (ADME) properties, cardiac toxicity, or human oral drug absorption, it is absolutely clear that the concept is standing on firm ground. Also in the more chemistry-oriented parts of the business these methods are being utilized more and more frequently. The by far best example is Lipinski's[105] rule of five focusing on physicochemical properties, which is now rather rigorously applied in the early discovery phase to select potential candidate compounds. It will most likely not tell you the absolute truth but instead will provide guidance when you have a wide variety of possible selections. Coupling this sort of predictability tool with for example a statistical analysis of the impact of API molecular weight on survival probability is a step toward extracting even more information out of the data. A comprehensive survey[125] has shown that the mean molecular weight for marketed oral drugs is somewhere in the interval 330 to 350 and that the population rapidly decreases when the value approaches 500 or above. Thus heavier molecules stand a much greater risk of failure than lighter ones, as projected by this study. Another striking result is that compounds featuring a more lipophilic character (measured as $\log P$ or $\log D_{7.4}$) are more frequently being discontinued than the ones with less. All in all, wise use of predictive tools as the ones mentioned here together with other knowledge might offer a good chance that attrition could start to creep down. It is, however, important to stress that, simply speaking, all tools are built up around models describing how the connection between variables can be expressed and translated into mathematical algorithms. And it is the quality of the latter that will decide how successful a prediction in a particular case will turn out. Furthermore, as most of these tools operate in isolation focusing on their respective field and fail to take other 'external' parameters into account, it is highly unlikely that they will all point in the same direction – for example to one particular API as being the optimum – when forecasts are compared.

Pharmacological activity alone is not sufficient for the success of a new drug but, moreover, purely molecular features which for convenience can be grouped together as the aforementioned physicochemical properties have to be given a high degree priority. This has brought the question about 'druglikeness' of compounds under evaluation to the forefront and their chance of being druggable.[126] In individual cases this empirical assessment is of course hampered by the far from 100% precision offered by the predictive tools. Nonetheless it has forced a change of the decision process from having been more or less entirely activity-driven to also encompass parameters like solubility, lipophilicity, molecular weight, number of hydrogen bonds, and oral bioavailability.[127] That on top of this expanded evaluatory procedure there is a demand for input from a Process R&D[58,59,97–99,101–104] and biopharmaceutics/preformulation perspective comes as quite a natural extension. The former to make statements about plausible process design, potential risks during scale-up, availability of raw materials in bulk, and estimated cost of API, while the latter focuses on perceived problems when trying to identify a suitable salt, the foreseeable presence of polymorphs, and eventual technical constraints to achieve the desired formulated drug product. Combining all these sets of data with results from in vitro and animal studies is bound to have a beneficial impact on the quality of the compounds selected and their suitability to be developed into novel drugs.

An interesting observation is that in this dissemination of best practice when it comes to weighing in relevant influencing factors to ensure that the most promising molecule is picked as candidate drug the words chirality or stereochemistry have not been mentioned at all. The reason is not ignorance; instead it shows how deeply this property has become embedded and integrated in the discovery process. Stereoisomers of chiral compounds will inevitably be scrutinized and only if they can be judged equivalent, notably from a safety and efficacy standpoint, might it be considered to take them forward as a racemic mixture or even in a nonequimolar composition. So if stereochemistry is seen as a normal parameter to deal with when hunting for new drug molecules it still has the potential to present some really challenging synthetic problems. Therefore, mastering adequate technologies and methods capable of achieving the desired single enantiomer not only in laboratory amounts but also on scale remains a key asset. And with the declared ambition to drive cost of goods down as a consequence of strong price pressure on medicines (see Section 2.17.7) it is even more important that best available technology has to be applied when designing synthetic routes and processes. Furthermore, there is and will always be a desire to minimize waste generation in chemical production, that is to move more and more toward a 'green' manufacturing regimen.[128,129] One industrially validated way to respond to these kinds of demands is to shift from stoichiometric or superstoichiometric (run with excess reactant and/or reagent) chemistry to catalytic procedures, irrespective of whether the task is to conduct asymmetric or nonstereoselective reactions.[130,131] In particular the latter problem focusing on environmental impact has been successfully addressed and amply demonstrated, but, equally, the cost of APIs and fine chemicals can undergo substantial improvement. The main contributing factors here are the often better yields attainable, a shorter synthetic sequence, higher productivity, and the move away from expensive reagents. Thus the current trend toward catalytic reactions becomes quite obvious and this approach is now evaluated on virtually a routine basis when considering the preparation of pharmaceutical intermediates and actives. The applicability of catalysis is very broad, allowing a wide range of substrates to be used, but the most spectacular achievements have perhaps been documented in the area of stereoselective synthesis. A renown example has already been mentioned ((S)-metolachlor in Section 2.17.2) and another one would be esomeprazole (active ingredient in antiulcer drug Nexium),[132,133] which is produced commercially by a highly efficient asymmetric sulfide oxidation (Figure 14).[58,59,98,134–136]

The tough challenge when aiming at designing and developing a catalytic process – irrespective of whether in asymmetric or achiral mode (exemplified by the increasingly popular and versatile C–C, C–N, and C–O coupling reactions following the Suzuki and Buchwald–Hartwig protocols) – is to identify which catalyst will be best for a given substrate. So far this has resulted in tedious trial-and-error work but much hope is placed in finding a prediction model that could dramatically narrow down the list of options in analogy to what has been alluded to earlier. Instead of randomly screening catalyst libraries or probing different experimental conditions such a tool would allow fine-tuning and optimization of a system (a metal in conjunction with the appropriate ligand) to quickly establish its performance on scale-up. Superficially the task seems possible to solve but with the multitude of parameters that can be used to describe such systems – electronic, geometric, and even energetic – it will most likely present a formidable challenge to condense multivariate data sets to their most significant linear combinations allowing a rank order to be calculated with the help of principal component analysis (PCA) technique or other statistical tools.[137] If successful, however, this could revolutionize the way in whch chemical processes are designed in the future with a default position to operate as much as possible under catalytic conditions and where traditional stoichiometry-based transformations appear as rare exceptions.

2.17.10 Major Conclusions and Projections for the Future

It is hoped that this account has provided a comprehensive overview of the current situation and challenges perceived in pharmaceutical R&D, both in a general context and more specifically with an emphasis on the issue of chirality and chiral drugs as seen through the eyes of Process R&D. On top of the wide variety of problems faced by the industry there is also a geographical dimension. This can be illustrated by the fact that in the late 1980s 41% of the top 50 list of innovative drugs originated from the USA, a figure which had risen to 62% a decade later. Furthermore, 1992 saw six out of the 10 best-selling medicines worldwide having been developed in Europe and this measure has fallen sharply to

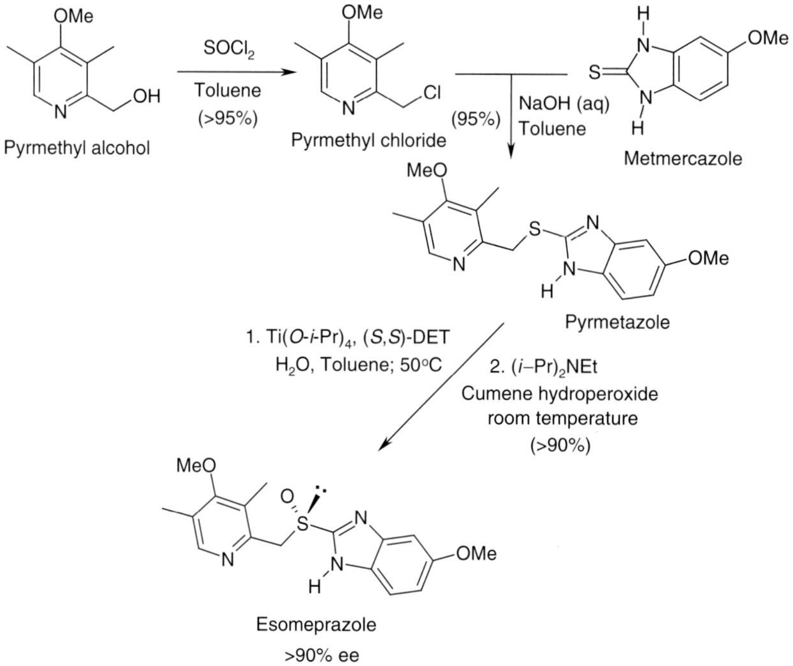

Figure 14 Efficacious production of the (S)-enantiomer sulfoxide esomeprazole on ton scale using titanium catalyzed oxidation.

only two over the succeeding 10 years. Without knowing exactly where the balance should lie, these figures reveal that some rather dramatic changes have taken place over only a short period of time. The root cause for this is open for debate, but will most certainly include socioeconomic factors, political agendas, variation in the organization of healthcare systems, and differences in the innovative climate. In particular the questions about innovation in the drug industry in general and about whether major pharma is still innovating have been the topics for in-depth analyses.[15,138]

Having devoted a fair number of pages to discussing the current landscape in the pharmaceutical industry[139] – its problems, challenges, and opportunities – and trying to explain some of the factors responsible for this as well as pointing out a roadmap for the future,[140] it seems best to summarize the key observations and messages in a condensed format with an emphasis on R&D matters:

- Portfolio composition to maintain an optimal balance between early–late and easy–difficult R&D projects
- Application of right level of risk taking along the Discovery–Development timeline
- Drive down the level of attrition or alternatively ensure a higher success rate
- Reach proofs of principle (PoP) or concept (PoC) as rapidly as possible
- Aim at using validated targets and establish biomarkers that enable unequivocal readout of results
- Explore how maximum use of prediction models and other in silico tools can support selection of more successful drug candidates or best technical methods
- Aspire to use cutting-edge technology wherever needed, but especially to overcome hitherto unsurpassed challenges
- Ensure top-class performance in all aspects along the entire value chain
- Keep cost of goods (API and formulated product) in focus with a declared policy of minimization to ensure profit margins are kept up
- Go for quality in all aspects of drug R&D and especially when appointing a new candidate drug
- Maintain time as a strong driving force when bringing the project forward to decision points and milestones
- Identify scientific and technical must-win areas and ensure the necessary competence level is available
- Debottleneck the value chain, particularly the R&D portion, to avoid unnecessary delays or misjudgements
- Apply strict cost control and decision-making
- Foster the skills and abilities to work across disciplinary and cultural boundaries in a global environment.

It would not be impossible to come up with another list of at least as many topics as are included in the one above, but depending on personal priorities and circumstances in individual companies there will always be variations, not to speak of the wide spread of opinions that one would face when asking for a prioritization from most to least important. In all likelihood company A would approach the challenges differently from company B, which however does not mean that their respective solutions are so far apart. It is important to stress that in as complex an environment as represented by the healthcare sector and the pharmaceutical industry it is not sufficient if one or a few areas operate effectively and efficiently. This modus operandi has to be adopted throughout the whole arena.

To finish off with a question as food for thought: where will the concept of personalized medicine[141] take us in the future? Will this, as it seems, revolutionary therapeutic approach be the dominating medicinal practice 10, 15, or 20 years from now? Impossible to say, but certainly something to speculate about and have your own visions for!

References

1. Frantz, S. *Nat. Rev. Drug Disc.* **2004**, *3*, 193–194. See also an editorial on the topic in *Nat. Rev. Drug Disc.* **2004**, *3*, 633.
2. Glassman, R. H.; Sun, A. Y. *Nat. Rev. Drug Disc.* **2004**, *3*, 177–183.
3. Reichert, J. M.; Pavlou, A. *Nat. Rev. Drug Disc.* **2004**, *3*, 383–384.
4. DiMasi, J. A.; Hansen, R. W.; Grabowski, H. G. *J. Health Econ.* **2003**, *22*, 151–185.
5. Goozner, M. *The $800 Million Pill: The Truth Behind the Cost of New Drugs*; University of California Press: Berkeley, CA, 2004.
6. Rawlins, M. D. *Nat. Rev. Drug Disc.* **2004**, *3*, 360–363.
7. Dickson, M.; Gagnon, J. P. *Nat. Rev. Drug Disc.* **2004**, *3*, 417–429.
8. Reichert, J. M. *Nat. Rev. Drug Disc.* **2003**, *2*, 695–702.
9. Kuhlmann, J. *Int. J. Clin. Pharm. Ther.* **1997**, *35*, 541–552.
10. Shillingford, C. A.; Vose, C. W. *Drug Disc. Today* **2001**, *6*, 941–946.
11. Federsel, H.-J. *PharmaChem* **2003**, *2*, 28–32.
12. Kola, I.; Landis, J. *Nat. Rev. Drug Disc.* **2004**, *3*, 711–715.
13. Centre for Medicines Research. http://www.cmr.org (accessed July 2006).
14. Cohen, C. M. *Nat. Rev. Drug Disc.* **2003**, *2*, 751–753.
15. Booth, B.; Zemmel, R. *Nat. Rev. Drug Disc.* **2004**, *3*, 451–456.
16. Preziosi, P. *Nat. Rev. Drug Disc.* **2004**, *3*, 521–526.
17. Kubinyi, H. *Nat. Rev. Drug Disc.* **2003**, *2*, 665–667.
18. Chapman, T. *Nature* **2004**, *430*, 109–115.
19. Arkin, M. R.; Wells, J. A. *Nat. Rev. Drug Disc.* **2004**, *3*, 301–317.
20. Rouhi, A. M. *Chem. Eng. News* **2004**, *82*, 47–62.
21. Maggon, K. *Drug Disc. Today* **2005**, *10*, 739–742.
22. Waldeck, B. *Pharmacol. Toxicol.* **2003**, *93*, 203–210.
23. Ariëns, E. J. *Eur. J. Clin. Pharmacol.* **1984**, *26*, 663–668.
24. Crossley, R. *Tetrahedron* **1992**, *48*, 8155–8178.
25. Shallenberger, R. S.; Acree, T. E.; Lee, C. Y. *Nature* **1969**, *221*, 555–556.
26. Strong, M. *Food Drug Law J.* **1999**, *54*, 463–487.
27. Szelenyi, I.; Geisslinger, G.; Polymeropoulos, E.; Paul, W.; Herbst, M.; Brune, K. *Drug News Perspect.* **1998**, *11*, 139–160.
28. Agranat, I.; Caner, H.; Caldwell, J. *Nat. Rev. Drug Disc.* **2002**, *1*, 753–768.
29. Federsel, H.-J.; Jaksch, P.; Sandberg, R. *Acta Chem. Scand. B* **1987**, *41*, 757–761.
30. Knowles, W. S. *Acc. Chem. Res.* **1983**, *16*, 106–112.
31. Knowles, W. S. *Angew. Chem. Int. Ed. Engl.* **2002**, *41*, 1998–2007.
32. Knowles, W. S. Asymmetric Hydrogenations – The Monsanto L-Dopa Process. In *Asymmetric Catalysis on Industrial Scale*; Blaser, H.-U., Schmidt, E., Eds.; Wiley-VCH: Weinheim, Germany, 2004, pp 23–38.
33. Crosby, J. Chirality in Industry – An Overview. In *Chirality in Industry; The Commercial Manufacture and Applications of Optically Active Compounds*; Collins, A. N., Sheldrake, G. N., Crosby, J., Eds.; Wiley: Chichester, UK, 1992, pp 1–66.
34. Blaser, H.-U. *Chem. Rev.* **1992**, *92*, 935–952.
35. Sheldon, R. A. The Chirality Pool. In *Chirotechnology: Industrial Synthesis of Optically Active Compounds*; Marcel Dekker: New York, 1993, pp 143–171.
36. Caldwell, J. *Chem. Ind.* **1995**, 176–179.
37. Bols, M. *Carbohydrate Building Blocks*; Wiley: New York, 1996.
38. Federsel, H.-J. Development of a Full-scale Process for a Chiral Pyrrolidine. In *Chirality in Industry, Vol. II, Developments in the Manufacture and Applications of Optically Active Compounds*; Collins, A. N., Sheldrake, G. N., Crosby, J., Eds.; Wiley: Chichester, UK, 1997, pp 225–244.
39. Sheldon, R. A. Racemate Resolution via Crystallization. In *Chirotechnology: Industrial Synthesis of Optically Active Compounds*; Marcel Dekker: New York, 1993, pp 173–204.
40. Bruggink, A. Rational Design in Resolutions. In *Chirality in Industry, Vol. II, Developments in the Manufacture and Applications of Optically Active Compounds*; Collins, A. N., Sheldrake, G. N., Crosby, J., Eds.; Wiley: Chichester, UK, 1997, pp 81–98.
41. Ward, R. S. *Tetrahedron: Asymmetry* **1995**, *6*, 1475–1490.
42. Stecher, H.; Faber, K. *Synthesis* **1997**, 1–16.
43. Keith, J. M.; Larrow, J. F.; Jacobsen, E. N. *Adv. Synth. Catal.* **2001**, *343*, 5–26.
44. Huerta, F. F.; Minidis, A. B. E.; Bäckvall, J. E. *Chem. Soc. Rev.* **2001**, *30*, 321–331.
45. Alcock, N. J.; Mann, I.; Peach, P.; Wills, M. *Tetrahedron: Asymmetry* **2002**, *13*, 2485–2490.
46. Nozaki, H.; Moriuti, S.; Takaya, H.; Noyori, R. *Tetrahedron Lett.* **1966**, 5239–5244.
47. Horner, L.; Siegel, H.; Büthe, H. *Angew. Chem., Int. Ed. Engl.* **1968**, *7*, 942.

48. Knowles, W.S.; Sabacky, M.J. *J. Chem. Soc., Chem. Commun.* **1968**, 1445–1446.
49. Corey, E. J.; Helal, C. J. *Angew. Chem., Int. Ed. Engl.* **1998**, *37*, 1986–2012.
50. Jacobsen, E. N.; Pfaltz, A.; Yamamoto, H.; *Comprehensive Asymmetric Catalysis, Vols. I–III*; Springer-Verlag: Heidelberg, Germany, 1999.
51. Ojima, I., Ed. *Catalytic Asymmetric Synthesis*, 2nd ed.; Wiley-VCH: Chichester, UK, 2000.
52. O'Brien, M. K.; Vanasse, B. *Curr. Opin. Drug Disc. Dev.* **2000**, *3*, 793–806.
53. Pfaltz, A. *Chimia* **2001**, *55*, 708–714.
54. Sharpless, K. B. *Angew. Chem., Int. Ed. Engl.* **2002**, *41*, 2024–2032.
55. Noyori, R. *Angew. Chem., Int. Ed. Engl.* **2002**, *41*, 2008–2022.
56. Hillier, M. C.; Reider, P. J. *Drug Disc. Today* **2002**, *7*, 303–314.
57. Rouhi, A. M. *Chem. Eng. News* **2003**, *81*, 45–55.
58. Federsel, H.-J. *Chirality* **2003**, *15*, S128–S142.
59. Federsel, H.-J. *Curr. Opin. Drug Disc. Dev.* **2003**, *6*, 838–847.
60. Enantioselective Catalysis; Bolm, C., Ed.; *Chem. Rev.* **2003**, *103*, 2761–3400.
61. Breuer, M.; Ditrich, K.; Habicher, T.; Hauer, B.; Kesseler, M.; Stürmer, R.; Zelinski, T. *Angew. Chem., Int. Ed. Engl.* **2004**, *43*, 788–824.
62. Hawkins, J. M.; Watson, T. J. N. *Angew. Chem., Int. Ed. Engl.* **2004**, *43*, 3224–3228.
63. Halpern, J., Trost, B.M., Eds. *Asymmetric Catalysis: Special Feature; Proc. Natl. Acad. Sci. USA* **2004**, *101*, Part I, 5347–5487; Part II, 5716–5850.
64. Houk, K.N., List, B., Eds. *Asymmetric Organocatalysis; Acc. Chem. Res.* **2004**, *37*, 487–631.
65. Blaser, H.-U.; Schmidt, E.; *Asymmetric Catalysis on Industrial Scale: Challenges, Approaches and Solutions*; Wiley-VCH: Weinheim, Germany, 2004.
66. Malhotra, S. V., Ed. *Methodologies in Asymmeric Catalysis*; American Chemical Society: Washington, DC, 2004.
67. Federsel, H.-J. *Nat. Rev. Drug Disc.* **2005**, *4*, 685–697.
68. Blaser, H.-U.; Spindler, F.; Studer, M. *Appl. Catal. A: General* **2001**, *221*, 119–143.
69. Blaser, H.-U.; Pugin, B.; Spindler, F. *J. Mol. Catal. A: Chemical* **2005**, *231*, 1–20.
70. Blaser, H.-U. *Adv. Synth. Catal.* **2002**, *344*, 17–31.
71. Blaser, H.-U.; Hanreich, R.; Schneider, H.-D.; Spindler, F.; Steinacher, B. The Chiral Switch of Metolachlor: The Development of a Large-Scale Enantioselective Catalytic Process. In *Asymmetric Catalysis on Industrial Scale. Challenges, Approaches and Solutions*; Blaser, H.-U., Schmidt, E., Eds.; Wiley-VCH: Weinheim, 2004, pp 55–70.
72. Fowden, L. *Nature* **1955**, *176*, 347–348.
73. Seebach, D.; Vettiger, T.; Müller, H.-M.; Plattner, D. A.; Petter, W. *Liebigs Ann. Chem.* **1990**, 687–695.
74. Ager, D. J., Ed. *Handbook of Chiral Chemicals*; Marcel Dekker: New York, 1999, 382pp.
75. Ager, D. J., Ed. *Handbook of Chiral Chemicals*, 2nd ed.; CRC Press: Boca Raton, FL, 2005, 646pp.
76. Wainer, I. W., Ed. *Drug Stereochemistry: Analytical Methods and Pharmacology*, 2nd ed., Clinical Pharmacology Series/18; Marcel Dekker: New York, 1993, 424pp.
77. Bijvoet, J. M.; Peerdeman, A. F.; van Bommel, A. J. *Nature* **1951**, *168*, 271–272.
78. Ladd, M. F. C.; Palmer, R. A. *Structure Determination by X-Ray Crystallography*, 4th ed.; Kluwer Academic/Plenum Publishers: New York, 2003.
79. Massa, W. *Crystal Structure Determination*, 2nd ed.; Springer: New York, 2004.
80. Freedman, T. B.; Cao, X.; Dukor, R. K.; Nafie, L. A. *Chirality* **2003**, *15*, 743–758.
81. Mason, S. F. *Molecular Optical Activity and Chiral Discrimination*; Cambridge University Press: Cambridge, 1982.
82. Parker, D. *Chem. Rev.* **1991**, *91*, 1441–1457.
83. Wenzel, T. J.; Wilcox, J. D. *Chirality* **2003**, *15*, 256–270.
84. Allenmark, S. G. *Chromatographic Enantioseparation: Methods and Applications. 2*; Elis Horwood: New York, 1991.
85. Subramanian, G., Ed. *A Practical Approach to Chiral Separations by Liquid Chromatography*, VCH: Weinheim, 1994; 405pp.
86. Chankvetadze, B. *Capillary Electrophoresis in Chiral Analysis*; Wiley: Chichester, 1997.
87. Beesley, T. E.; Scott, R. P. W. *Chiral Chromatography*; Wiley: Chichester, 1998.
88. Wren, S. A. C. *Chromatographia* **2001**, *54*, S1–S95.
89. Schurig, V. *J. Chromatogr. A* **2001**, *906*, 275–299.
90. Gubitz, G.; Schmid, M. G.; *Chiral Separations: Methods and Protocols*; Humana Press: Totowa, NJ, 2004.
91. Patel, B. K.; Hanna-Brown, M.; Hadley, M. R.; Hutt, A. J. *Electrophoresis* **2004**, *25*, 2625–2656.
92. Zhang, Y.; Wu, D.-R.; Wang-Iverson, D. B.; Tymiak, A. A. *Drug Disc. Today* **2005**, *10*, 571–577.
93. Pirkle, W. H.; Pochapsky, T. C. *Chem. Rev.* **1989**, *89*, 347–362.
94. Li, S.; Purdy, W. C. *Chem. Rev.* **1992**, *92*, 1457–1470.
95. Ward, T. J. *Anal. Chem.* **2002**, *74*, 2863–2872.
96. Toyooka, T. *Modern Derivatization Methods for Separation Science*; Wiley: Chichester, 1998.
97. Federsel, H.-J. *Nat. Rev. Drug Disc.* **2002**, *1*, 1013.
98. Federsel, H.-J. *Nat. Rev. Drug Disc.* **2003**, *2*, 654–664.
99. Federsel, H.-J. *Pharm. Sci. Technol. Today* **2000**, *3*, 265–272.
100. Pines, S. H. *Org. Process Res. Dev.* **2004**, *8*, 708–724.
101. Federsel, H.-J. *Comb. Chem. High Throughput Screen.* **2006**, *9*, 79–86.
102. Federsel, H.-J. *Drug Disc. Today* **2001**, *6*, 397–398.
103. Federsel, H.-J. *Drug Disc. Today* **2001**, *6*, 1047.
104. Federsel, H.-J. *Chim. Oggi/Chem. Today* **2004**, *22*, 9–12.
105. Lipinski, C. A.; Lombardo, F.; Dominy, B. W.; Feeney, P. J. *Adv. Drug Deliv. Rev.* **1997**, *23*, 3–25.
106. Mickel, S.J. et al. *Org. Process Res. Dev.* **2004**, *8*, 92–100; 101–106; 107–112; 113–121; 122–130.
107. Loiseleur, O.; Koch, G.; Wagner, T. *Org. Process Res. Dev.* **2004**, *8*, 597–602.
108. Loiseleur, O.; Koch, G.; Cercus, J.; Schürch, F. *Org. Process Res. Dev.* **2005**, *9*, 259–271.
109. Smith, A. B., III; Freeze, B. S.; Xian, M.; Hirose, T. *T. Org. Lett.* **2005**, *7*, 1825–1828.
110. Salomon, A. R.; Voehringer, D. W.; Herzenberg, L. A.; Khosla, C. *Chem. Biol.* **2001**, *8*, 71–80.
111. Wender, P. A.; Jankowski, O. D.; Tabet, E. A.; Seto, H. *Org. Lett.* **2003**, *5*, 2299–2302.

112. Nicolaou, K.C. et al. *J. Am. Chem. Soc.* **2003**, *125*, 15433–15442; 15443–15454.
113. Wehlan, H.; Dauber, M.; Mujica Fernaud, M.-T.; Schuppan, J.; Mahrwald, R.; Ziemer, B.; Juarez Garcia, M.-E.; Koert, U. *Angew. Chem. Int. Ed. Engl.* **2004**, *43*, 4597–4601.
114. U.S. Food and Drug Administration. http://www.fda.gov/oc/initiatives/criticalpath/whitepaper.html (accessed June 2006).
115. Caner, H.; Groner, E.; Liron, L.; Agranat, I. *Drug Disc. Today* **2004**, *9*, 105–110.
116. U.S. Food and Drug Administration. http://www.fda.gov/cder/guidance/stereo.html (accessed June 2006).
117. Allen, M. E. *Reg. Affairs J.* **1991**, *1*, 93–95.
118. Gross, M.; Cartwright, A.; Campbell, B.; Bolton, R.; Holmes, K.; Kirkland, K.; Salmonson, T.; Robert, J.-L. *Drug Inform. J.* **1993**, *27*, 453–457.
119. Mislow, K. *Chirality* **2002**, *14*, 126–134.
120. Blumenstein, J. J. Chiral Drugs: Regulatory Aspects. In *Chirality in Industry II. Developments in the Manufacture and Applications of Optically Active Compounds*; Collins, A. N., Sheldrake, G. N., Crosby, J., Eds.; Wiley: Chichester, 1997, pp 11–18.
121. Koehn, F. E.; Carter, G. T. *Nat. Rev. Drug Disc.* **2005**, *4*, 206–220.
122. Jaroszewski, J. W.; Stromhansen, T.; Honorehansen, S.; Thastrup, O.; Kofod, H. *Planta Med.* **1992**, *58*, 454–458.
123. Matlin, S. A. *Drugs* **1994**, *48*, 851–863.
124. Prescher, J. A.; Bertozzi, C. R. *Nat. Chem. Biol.* **2005**, *1*, 13–21.
125. Wenlock, M. C.; Austin, R. P.; Barton, P.; Davis, A. M.; Leeson, P. D. *J. Med. Chem.* **2003**, *46*, 1250–1256.
126. Kerns, E. H.; Di, L. *Drug Disc. Today* **2003**, *8*, 316–323.
127. Curatolo, W. *Pharm. Sci. Technol. Today* **1998**, *1*, 387–393.
128. Constable, D. J. C.; Curzons, A. D.; Cunningham, V. L. *Green Chem.* **2002**, *4*, 521–527.
129. Andraos, J. *Org. Process Res. Dev.* **2005**, *9*, 149–163.
130. Sheldon, R. A. *J. Chem. Tech. Biotechnol.* **1997**, *68*, 381–388.
131. Sheldon, R. A. *C.R. Acad. Sci. Paris, Série IIc, Chimie/Chemistry* **2000**, *3*, 541–551.
132. Carlsson, E.; Lindberg, P.; von Unge, S. *Chem. Brit.* **2002**, *38*, 42–45.
133. Olbe, L.; Carlsson, E.; Lindberg, P. *Nat. Rev. Drug Disc.* **2003**, *2*, 132–139.
134. Cotton, H.; Elebring, T.; Larsson, M.; Li, L.; Sörensen, H.; von Unge, S. *Tetrahedron: Asymmetry* **2000**, *11*, 3819–3825.
135. Federsel, H.-J. *Chim. Oggi/Chem. Today* **2002**, *20*, 57–62.
136. Federsel, H.-J.; Larsson, M. An Innovative Asymmetric Sulfide Oxidation: The Process Development History Behind the New Antiulcer Agent Esomeprazole. In *Asymmetric Catalysis on Industrial Scale. Challenges, Approaches and Solutions*; Blaser, H.-U., Schmidt, E., Eds.; Wiley-VCH: Weinheim, 2004, pp 413–436.
137. Burello, E.; Rothenberg, G. *Adv. Synth. Catal.* **2003**, *345*, 1334–1340.
138. Cohen, F. J. *Nat. Rev. Drug Disc.* **2005**, *4*, 78–84.
139. Frantz, S. *Nat. Rev. Drug Disc.* **2006**, *5*, 92–93.
140. Federsel, H.-J. *Drug Disc. Today*, in press.
141. Gurwitz, D.; Lunshof, J. E.; Altman, R. B. *Nat. Rev. Drug Disc.* **2006**, *5*, 23–26.

Biography

Hans-Jürgen Federsel, PhD, is a renowned specialist in the field of process R&D where he has spent his entire professional career spanning a period of 30 years. Starting off as bench chemist in former Astra at the major Swedish site in Södertälje he has climbed the ranks occupying positions both as line and project manager as well as being Director of a global project and portfolio management function. After the merger that formed AstraZeneca in 1999 he became Head of Projects Management at Södertälje and was then appointed to a newly created role as Director of Science in Global Process R&D from the beginning of 2004. In connection with this he was also given the prestigious title Senior Principal Scientist. Strong academic links have been further developed throughout the years after obtaining his PhD in organic chemistry in 1980 at the Royal Institute of Technology in Stockholm; his award of an Associate Professorship there in 1990 recognized this. Furthermore, his long-lasting links to this Institute have recently brought him a seat on the Board of the School of Chemical Science and Engineering. Publishing in peer-reviewed journals and books and frequent lecturing has made Dr Federsel's name well known far beyond the limits of his own company, and he enjoys invitations from all over the world to share learning and experience from his broad knowledge base on process R&D.

Comprehensive Medicinal Chemistry II
ISBN (set): 0-08-044513-6

ISBN (Volume 2) 0-08-044515-2; pp. 713–736

2.18 Promiscuous Ligands

S L McGovern, M.D. Anderson Cancer Center, Houston, TX, USA

© 2007 Elsevier Ltd. All Rights Reserved.

2.18.1	**Introduction**	737
2.18.2	**Promiscuous Inhibition by Aggregate-Forming Compounds**	738
2.18.3	**Experimental Identification of Promiscuous Inhibitors**	739
2.18.3.1	Biochemical Assays for Promiscuity	740
2.18.3.1.1	Detergent sensitivity	740
2.18.3.1.2	Sensitivity to enzyme concentration	741
2.18.3.1.3	Inhibition of diverse model enzymes	742
2.18.3.1.4	Time-dependent inhibition	742
2.18.3.1.5	Steep inhibition curves	743
2.18.3.2	Light Scattering	743
2.18.3.3	Microscopy	745
2.18.3.4	Additional Confirmation Assays	745
2.18.4	**Computational Prediction of Promiscuous Compounds**	746
2.18.4.1	Frequent Hitters	746
2.18.4.2	Aggregators	746
2.18.4.3	Privileged Substructures	748
2.18.4.4	Reactive Species	748
2.18.4.5	Experimental Artifact	749
2.18.5	**Future Directions**	749
References		750

2.18.1 Introduction

Virtual and experimental high-throughput screening (HTS) are widely used to identify novel ligands for drug development.[1–4] Despite their successes, these methods are haunted by false positives, compounds that initially appear to have a desired biological activity but on closer evaluation are found to act by spurious mechanisms. Often, these counterfeit screening hits have peculiar behaviors that cannot be reconciled with canonical modes of inhibition. These compounds are ultimately found to be developmental dead-ends and are discarded, typically after a great deal of time and resources has been devoted to them.

The frustration caused by such molecules is compounded by their prevalence; it has been estimated that in any given experimental HTS, the ratio of false positives to true positives is at best 1:1 and more likely 10:1.[5] This enrichment of phony inhibitors has led to increased scrutiny of hit lists and an increased interest in developing filters and counterscreens to rapidly recognize and eliminate 'non-lead-like' compounds from further consideration.[6–11]

Toward this end, work over the last decade has been devoted to understanding the origins of false-positive screening hits. Experimental artifact, such as compound impurity[12] or assay interference,[7,13] is a well-known cause. Inhibitor promiscuity due to reactive groups[14,15] or privileged scaffolds[16–22] has also been described. More recently, aggregate formation has emerged as a potential mechanism of promiscuous inhibition.[23,24] This model proposes that some nonspecific inhibitors form aggregates in solution and that the aggregate species is responsible for enzyme inhibition (**Figure 1**). Preliminary studies suggest that aggregate-forming promiscuous compounds may be common among screening libraries and hit lists. This chapter will initially focus on nonspecific enzyme inhibition as caused by aggregate-forming molecules and will review experimental methods for the identification of such compounds. Computational algorithms for the prediction of all types of promiscuous inhibitors will then be discussed and outstanding questions regarding promiscuous inhibitors in drug discovery will be considered.

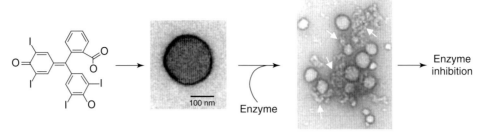

Figure 1 Model of promiscuous enzyme inhibition. The promiscuous compounds form aggregates in solution and enzyme molecules (white arrows) adsorb onto the surface of the aggregate particle, resulting in enzyme inhibition. Although initial microscopy images suggest surface adsorption of enzyme molecules, absorption of enzyme into the aggregate interior cannot be excluded. (Reprinted with permission from McGovern, S. L.; Caselli, E.; Grigorieff, N.; Shoichet, B. K. *J. Med. Chem.* **2002**, *45*, 1712–1722. Copyright (2002) American Chemical Society; from McGovern, S. L.; Helfand, B. T.; Feng, B.; Shoichet, B. K. *J. Med. Chem.* **2003**, *46*, 4265–4272. Copyright (2003) American Chemical Society.)

2.18.2 Promiscuous Inhibition by Aggregate-Forming Compounds

By the late 1990s, experimental and virtual HTS were commonly used to identify new leads for drug design. Although various screening techniques and algorithms had been developed, it became apparent that the output from these large-scale methods was suboptimal. Hit lists were populated, if not dominated, by nonspecific compounds with peculiar properties, such as steep inhibition curves, flat structure–activity relationships (SARs), and complex time-dependent behavior. Despite the use of filters for reactivity,[14] chemical swill,[11] and druglikeness,[25] problematic molecules continued to appear on screening hit lists. These compounds caused much frustration and were often abandoned after significant effort had been invested in them.

To explore the underlying mechanism responsible for this perplexing behavior, 115 compounds were initially investigated. This included 45 screening hits,[23,24] 15 leads used as experimental tools,[26] and 55 clinically prescribed drugs.[27] Of the 115 compounds studied, 53 were found to inhibit diverse model enzymes, including β-lactamase, chymotrypsin, dihydrofolate reductase, and β-galactosidase. These promiscuous compounds showed time-dependent inhibition that was sensitive to enzyme concentration, the presence of bovine serum albumin (BSA), and ionic strength. Based on these observations, it was hypothesized that the nonspecific compounds formed aggregates in solution, and the aggregate particles caused enzyme inhibition (**Figure 1**).[23] Consistent with this hypothesis, dynamic light scattering (DLS) and transmission electron microscopy revealed that the promiscuous compounds formed particles on the order of 30–1000 nm in diameter; these particles were absent from solutions of nonpromiscuous compounds.[23]

How did the aggregate particles interact with enzyme molecules to cause inhibition?[24] Centrifugation experiments suggested a direct interaction between aggregate-forming inhibitors and enzyme molecules; this interaction was disrupted by the addition of detergent such as Triton X-100.[24] Aggregate formation and enzyme inhibition by the promiscuous compounds were also prevented or rapidly reversed by the addition of Triton X-100.[24] Additional microscopy studies showed that protein molecules were adsorbed onto the surface of the aggregate particles, although absorption into the aggregate interior could not be excluded. This interaction was also prevented by the addition of detergent such as Triton X-100.[24] To account for these observations, it was proposed that some promiscuous compounds form aggregates in solution and enzyme molecules adsorb onto the surface of the aggregate particle, resulting in reversible enzyme inhibition (**Figure 1**).[24]

Since the first proposal of this model, a growing number of research groups have identified potential aggregate-based inhibitors among their screening hits (**Table 1**). Their observations, in addition to the initial, small-scale evaluation of the prevalence of aggregators suggest that they occur among screening hits,[23,28–31] small-molecule reagents used in the lab,[26] and, surprisingly, even clinically used drugs.[27,32] A recent study evaluating 1030 random druglike molecules suggests that up to 19% show experimental signatures of aggregate-based promiscuity.[33] Because of the potentially widespread occurrence of aggregate formers, there is much interest in the development of rapid experimental methods for the identification of these promiscuous compounds. There has also been a growing interest in the development of computational models to predict compounds likely to act as aggregate formers because such filters could be used to remove suspect molecules from screening libraries or from hit lists. These experimental and computational efforts are considered in turn in the following sections.

Table 1 Promiscuous compounds from virtual and high-throughput screening

Structure	Type of screen	IC_{50} (μM) versus target	Experimental evidence for promiscuity	Reference
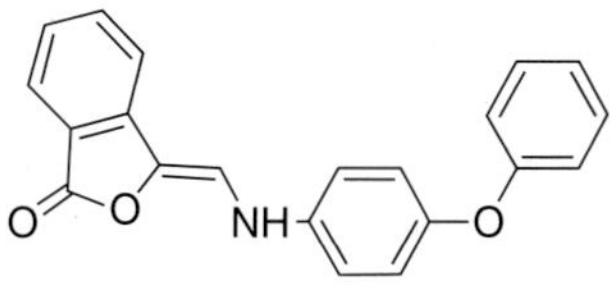	Virtual screen	5 β-Lactamase	• Decreased inhibition with detergent • Decreased inhibition with increased enzyme concentration • Inhibition of diverse enzymes • Time-dependent inhibition • Particles observed by DLS	23,24
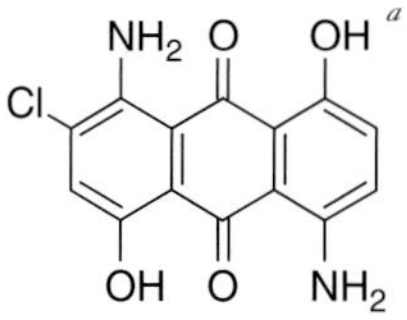 [a]	Virtual screen	5 PMM/PGM[b]	• Decreased inhibition with detergent • Decreased inhibition with increased enzyme concentration • Time-dependent inhibition • Particles observed by DLS	28
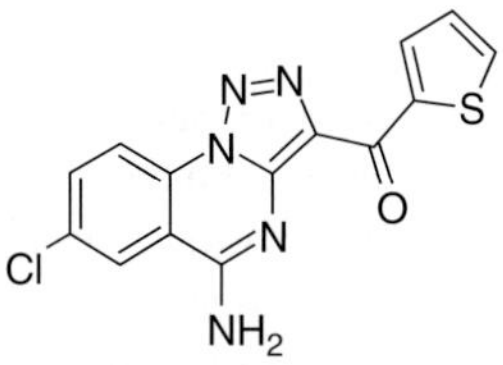	Virtual screen	25 Edema factor	• Decreased inhibition with increased enzyme concentration • Time-dependent inhibition	29
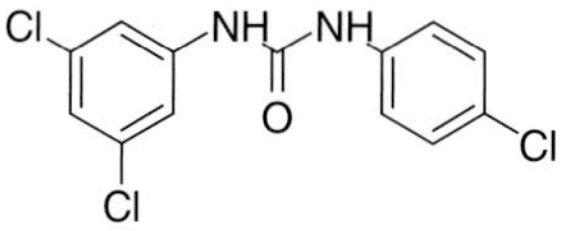	Cell-based HTS	ND[c] EF-CaM	• Inhibition of diverse enzymes	31
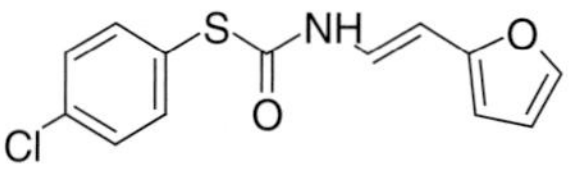	FRET-based HTS	7.4 Coronavirus proteinase	• Decreased inhibition with BSA	30
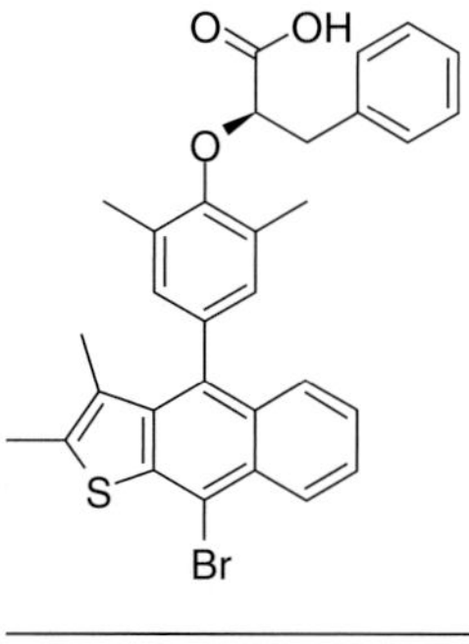	Phase II clinical trial[d]	1.6 PTP1b	• Decreased inhibition with detergent • Decreased inhibition with increased enzyme concentration • Time-dependent inhibition	32

[a] Disperse Blue.
[b] PMM/PGM, phosphomannomutase/phosphoglucomutase.
[c] Not determined, compound observed to inhibit edema toxin-induced change in cellular morphology.
[d] The compound, ertiprotafib, had been studied in phase II trials prior to evaluation for promiscuity.[32]

2.18.3 Experimental Identification of Promiscuous Inhibitors

The peculiar experimental signature of aggregate-forming compounds can be used to distinguish them from specific, well-behaved inhibitors,[23,24,28] and several research groups have used these characteristics to triage their screening hits for potential promiscuity.[12,28–31,34–42] The appropriate experiments range in complexity and intensity from a rapid biochemical assay to a much more laborious microscopy study. Depending on the purposes of the research effort, these experiments can be tailored appropriately. For most early-stage discovery programs, the following experiments are performed to answer the question: Is this hit likely inhibiting the screening target as an aggregate? If the answer is yes, it is unlikely that the hit will be a favorable starting point for lead design and it should be dropped from further consideration immediately, as the pain of throwing out compounds early in the process pales in comparison to the agony of abandoning leads that have been through several nonproductive development cycles. If the answer to the above question is no, then one element of the hit's potential has been reassuringly established. The necessary experiments and examples of their recent application in drug discovery projects are described below; detailed recipes for performing these assays have been described elsewhere.[43]

2.18.3.1 Biochemical Assays for Promiscuity

Fortunately for screening programs faced with evaluation of tens to thousands of hits, the most informative experiments are among the easiest to perform and are the most amenable to high-throughput methods. Often, the experimental assay that initially identified the screening hit can be easily modified to accomplish many of the initial tests of promiscuity. Additionally, these protocols can be used to assay for other mechanisms of promiscuity, such as chemical reactivity. If the original assay lacks sufficient dynamic range or is otherwise difficult to alter, it may be simpler to use easily purchased model enzyme systems such as chymotrypsin or β-lactamase. Colorimetric, kinetic assays have proven to be especially robust for these purposes.

The particular components of the assay can affect the outcome. For instance, the ionic strength[23] and pH[44] of the buffer can alter the aggregation and inhibition properties of the compounds; 50 mM potassium phosphate, pH 7.0, has previously been found useful. Increased enzyme concentration[23] or the presence of excess protein such as albumin[23,30] can decrease the apparent inhibition by aggregators; a starting enzyme concentration of 1–10 nM is typical. Ideally, solutions of inhibitors and substrate are prepared from fresh buffer or dimethyl sulfoxide (DMSO) stocks and diluted into buffer so that the final DMSO concentration in the assay is as low as possible, preferably less than 5%. Substrate concentrations are typically greater than the K_m for the reaction to allow for maximal velocities.

2.18.3.1.1 Detergent sensitivity

In the presence of detergents such as Triton X-100, inhibition by aggregate-forming inhibitors is markedly attenuated. For instance, the inhibition of β-lactamase by a 3 μM solution of tetraiodophenolphthalein (I4PTH), a prototypical aggregator, decreases from 78% in the absence of detergent to 3% in the presence of 0.01% Triton X-100.[24] Indeed, Triton X-100 rapidly reverses inhibition of β-lactamase by I4PTH without affecting a specific, well-behaved inhibitor of the enzyme (**Figure 2**).

To date, detergent sensitivity is the easiest way to discriminate between aggregators and nonaggregators.[24,45] Therefore, the most rapid method of identifying aggregate-based inhibitors is to repeat the screening assay with the addition of 0.01% Triton X-100. A typical protocol consists of incubating fresh detergent, inhibitor at the IC$_{50}$ (concentration of inhibitor that reduces enzyme activity by 50%), and enzyme for 5 min and initiating the reaction with substrate.[24,43,45] If inhibition is due to aggregation, inhibition should decrease from 50% to less than 40% in the presence of detergent. It may also be useful to consider adding 0.01% Triton X-100 to the baseline assay buffer as a prophylactic measure to prevent aggregate-forming compounds from appearing as inhibitors at all.

The first control for this experiment requires establishing that the addition of Triton X-100 does not significantly affect the reaction rate or any of the other assay components. A negative control experiment will show that a well-behaved inhibitor of the system is not affected by the detergent, and a positive control will show that a known aggregator, such as I4PTH,[23] is attenuated by the detergent.

It is worth noting the discussion in this chapter emphasizes the use of Triton X-100, as it has been shown to reverse enzyme inhibition by aggregators (**Figure 2**).[24] Other detergents have been employed as agents to study promiscuity, such as saponin,[27,38] Tween-20,[45] and CHAPS.[34,45]

Because of its relative simplicity, detergent sensitivity has been used in a growing number of drug discovery projects to triage compounds for promiscuity (**Table 1**).[28,34–36,38,39] For instance, using a library of 147 compounds derived from the Cdc25A phosphatase inhibitor dysidiolide, Koch and co-workers found selective inhibitors of three enzymes that were structurally similar to Cdc25A. All of the library compounds were assayed against each target enzyme in the presence and absence of 0.001 or 0.01% Triton X-100 to ultimately identify the specific inhibitors.[35] Detergents have also been used to study the inhibition of PTP1b by ertiprotafib, a compound that had reached phase II clinical trials (**Table 1**).[32] The IC$_{50}$ of ertiprotafib against 1 nM PTP1b increased 10-fold with the addition of 0.01% Triton X-100. This observation, in combination with other properties of the compound,[32] was consistent with an aggregate-based mechanism of inhibition.

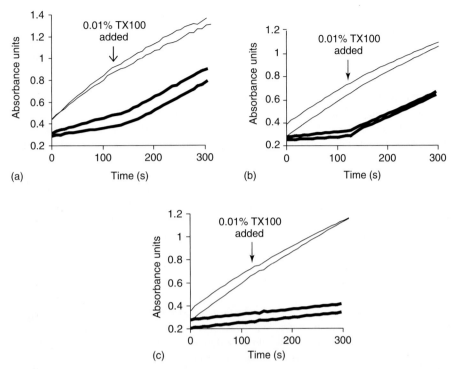

Figure 2 The effect of 0.01% Triton X-100 added during a β-lactamase inhibition assay. (a) Inhibitor is 10 μM I4PTH. (b) Inhibitor is 5 μM rottlerin. (c) Inhibitor is 0.6 μM BZBTH2B, a specific β-lactamase inhibitor.[59] In all panels, thick lines (—) denote reactions containing inhibitor and thin lines (—) denote reactions containing DMSO control. TX100, Triton X-100. (Reprinted with permission from McGovern, S. L.; Helfand, B. T.; Feng, B.; Shoichet, B. K. *J. Med. Chem.* **2003**, *46*, 4265–4272. Copyright (2003) American Chemical Society.)

2.18.3.1.2 Sensitivity to enzyme concentration

The key observation leading to the development of the aggregate hypothesis was that the promiscuous inhibitors were sensitive to the concentration of enzyme. For instance, a 10-fold increase in the concentration of β-lactamase from 1 to 10 nM was sufficient to increase the IC_{50} of 35 aggregate-forming compounds by 3- to over 50-fold, even when the inhibitor was present at micromolar concentrations.[23] The ability of nanomolar enzyme to titrate micromolar inhibitor suggested that the active form of the inhibitor might be an aggregate of many individual inhibitor molecules; although the ratio of individual inhibitor molecules to enzyme molecules might be 1000:1 or greater, the ratio of aggregate particles to enzyme molecules would be much lower.

To evaluate screening hits for this property, the concentration of the target enzyme is increased 10-fold and the assay is repeated with the inhibitor at the IC_{50} obtained at the baseline enzyme concentration. If aggregate formation is involved, inhibition ought to decrease from 50% to less than 40% against the 10-fold greater concentration of enzyme. This assay requires that the ratio of inhibitor to enzyme remains elevated, on the order of 1000:1 or greater, even after the enzyme concentration has been increased. Many assays may not be able to tolerate increases in enzyme concentration due to cost, limited reagent availability, or an increased reaction rate that makes it technically difficult to obtain accurate velocity measurements. In such cases, it may be helpful to decrease the initial assay concentration of the enzyme or to use alternative substrates with slower reaction rates.[23]

As in the detergent assay described above, a well-behaved inhibitor should not be affected by the increase in enzyme concentration and can serve as a negative control.[23] A known aggregator that has been shown to be sensitive to enzyme concentration should be used as a positive control.[23,26,27]

This protocol has also been used by several research groups to evaluate screening hits.[13,28,36,37] For example, Disperse Blue against 3.3 μg mL^{-1} phosphomannomutase/phosphoglucomutase (PMM/PGM) had an IC_{50} of 5 μM; when the enzyme concentration was increased to 33 μg mL^{-1}, the compound did not show any detectable inhibition up to 10 μM.[28] Similarly, the IC_{50} of ertiprotafib against PTP1b increased from 1.6 to 9 μM as the concentration of the enzyme increased from 10 to 100 nM.[32]

The inclusion of BSA also decreases the potency of aggregate-based inhibitors through a similar mechanism to increasing the enzyme concentration, except enzyme molecules are displaced from inhibitor particles by inert protein instead of active enzyme.[23] Consequently, addition of BSA to screening assays has been used to accomplish the equivalent purpose; for examples, see [30,36] and **Table 1**.

2.18.3.1.3 Inhibition of diverse model enzymes

One of the earliest warning signs leading to the discovery of aggregators was their ability to inhibit dissimilar targets.[23] For example, rottlerin, a widely used kinase inhibitor,[46] has an IC_{50} of 3 μM against PKCδ, 1.2 μM against β-lactamase, 2.5 μM against chymotrypsin, and 0.7 μM against malate dehydrogenase.[26] It has since become clear that aggregators can inhibit a variety of enzymes, ranging from β-lactamase (40 kDa monomer) to β-galactosidase (540 kDa tetramer).[23]

Ideally, for each screening hit under consideration, at least one counterscreen is performed against an enzyme that is unrelated to the target enzyme in terms of structure, function, and ligand recognition. Depending on the level of certainty required, testing against one dissimilar enzyme may be sufficient; testing against two or more provides additional reassurance. Screening across receptors also allows for investigation of other mechanisms of promiscuity, such as chemical reactivity or privileged scaffolds; nonspecific inhibition from any mechanism may be sufficient to drop the compound from further optimization studies.

For secondary screening to be practical, an experimentally robust system with readily available components and a straightforward assay readout should be used. Beta-lactamase, chymoptrypsin, and malate dehydrogenase have been used successfully in the past.[26,27,29,31,45] It may be more convenient to use enzymes already known to the research group; for instance, Fattorusso and colleagues assayed a novel caspase inhibitor discovered in a combination NMR and computational screening project against a metalloprotease that was also under investigation in their laboratory.[37]

Control experiments for these assays are straightforward. Specific inhibitors of the target enzyme and the enzymes used for counterscreening should not cross-react. For the positive control experiment, a known aggregator should be shown to inhibit each enzyme under study.

In practice, hits are often tested against related enzymes to evaluate for specificity within a protein class or across species.[46] For instance, Blanchard and co-workers used HTS of 50 000 compounds to identify novel coronavirus proteinase inhibitors. After a series of filters, the 572 hits yielded five compounds with IC_{50} values of 0.5–7 μM for the target, and two of those compounds inhibited coronavirus protease but not four other proteases.[30] Baldwin and colleagues screened a 220 000 compound library to find novel inhibitors of dihydroorotate dehydrogenase from *Plasmodium falciparum*; follow-up screens identified a competitive inhibitor with an IC_{50} of 16 nM against the parasite enzyme and 200 μM against the human isozyme.[12] These approaches have been extended by recent technological developments that allow for testing compounds against multiple enzymes in parallel. For instance, proteomics-based methods[47,48] and enzyme microarrays[49] have been described. These techniques provide an evaluation of inhibitor specificity against several enzymes simultaneously; future developments along these lines can be expected.

2.18.3.1.4 Time-dependent inhibition

Preincubation of aggregate-based inhibitors with enzyme increases the apparent IC_{50} of the compounds 2- to over 50-fold.[23,26,27] Several mechanisms could account for this observation, such as slow on-rates or covalent binding. To explore the origin of time-dependent behavior by aggregators, Tipton's group conducted a thorough analysis of the kinetic behavior of Disperse Blue.[28] Their results suggested that slow-binding did not occur when this compound interacted with its target enzyme, PMM/PGM. Instead, they found that the compound behaved as a parabolic, noncompetitive inhibitor, and they proposed that the kinetic signature of time-dependent inhibition with a nonlinear dependence on inhibitor concentration may be a marker for aggregation.[28] Future studies to evaluate the generalization of this observation are certainly anticipated. In the meantime, it should be noted that time-dependent inhibition is consistent with several modes of enzyme inhibition and is not singularly sufficient to denote a compound as an aggregator.

In a typical test for time dependence, an IC_{50} is determined by incubating enzyme with inhibitor for 5 min and initiating the reaction with substrate. The assay is then repeated, except substrate is first mixed with the inhibitor at the IC_{50} concentration and the reaction is initiated by the addition of enzyme. If inhibition without incubation decreases from 50% to less than 40%, the result is consistent with the time-dependent behavior observed of aggregators. Control experiments include repeating the assay with a specific, nontime-dependent inhibitor of the target enzyme as well as a known aggregator. These protocols are usually performed in conjunction with other assays to evaluate inhibitor behavior (e.g., see [32]).

2.18.3.1.5 Steep inhibition curves

It has been observed that steep inhibition curves are often associated with undesirable screening hits.[50] This experimental property has been used as a filter to triage compounds; for example, in their search for coronavirus inhibitors, Blanchard and co-workers eliminated 54 of 126 compounds because they did not have a sigmoidal semilogarithmic dose–response curve.[30] Aggregate-forming inhibitors have also been shown to have a steep inhibition curve relative to classically behaved inhibitors,[26] although mechanistic understanding of this experimental observation awaits further study.

2.18.3.2 Light Scattering

The biochemical experiments described above provide phenomenological support for the identification of an aggregate-based inhibitor, but they do not yield direct evidence for the presence of aggregate particles. To obtain such data, a biophysical method such as DLS is necessary. DLS has a growing number of applications in the biological and material sciences, but this discussion will focus on the use of DLS to determine if particles are present in a solution of a promiscuous inhibitor and, if so, what is their size? In conjunction with the assays previously described, DLS completes the series of experiments typically needed to characterize a promiscuous, aggregate-based inhibitor.

DLS is most commonly used to analyze particles with a diameter of 1–1000 nm. In a standard setup, a sample of the solution under study is placed in a chamber and exposed to a laser. Particles in the sample cause the laser light to scatter. A detector at a fixed angle relative to the chamber records the scattered photons over a period of time, usually tens of seconds to minutes. Because particles in the sample undergo Brownian motion, the intensity of the scattered light measured by the detector will fluctuate, typically on the microsecond timescale. These fluctuations reflect the rate of diffusion by the particle in solution, which in turn depends on the hydrodynamic radius of the particle according to the Stokes–Einstein equation. The necessary calculations are performed by software provided with the DLS instrument, and given certain assumptions about particle shape and distribution, the hydrodynamic radius can be used as a measure of particle size. This is only a brief explanation of the method; for more details on the theory and practice of DLS, see descriptions by Santos[51] and Wyatt.[52]

For analysis of a promiscuous inhibitor, the first task is to determine if particles are present. Graphically, this can be determined from the shape of the autocorrelation function obtained during the DLS experiment. For instance, rottlerin is a known promiscuous, aggregate-forming inhibitor,[26] and it yields an autocorrelation function with well-defined decay on the microsecond timescale as shown in **Figure 3a**. The same is true for K-252c, another promiscuous kinase inhibitor (**Figure 3b**). Surprisingly, suramin, a compound that is known to inhibit multiple targets, produces an autocorrelation function with a poorly defined decay, suggesting that it does not form aggregates in solution (**Figure 3c**).[26]

Additionally, it is useful to note the average intensity of the scattered light at the detector during the experiment. This is usually reported as counts per second (cps) or kilocounts per second (kps). The absolute value of the intensity will depend on the particular instrument used, but solutions that are known to be particle free, such as filtered water, should yield average intensities at least an order of magnitude less than that of solutions known to contain particles, such as latex beads or albumin. Therefore, it is worthwhile to conduct both positive and negative control experiments to evaluate the signal-to-noise ratio of the instrument.

If particles are present, the next step is to determine their size. Typically, this is calculated by software that accompanies the instrument. Results between different algorithms, even in the same software package, can vary markedly based on assumptions about particle shape. Again, it is worthwhile to test the system using particles of known size, such as latex bead standards or albumin.

More recent developments in DLS technology allow for coupling of DLS with a high-performance liquid chromatography (HPLC) column to provide rapid eluent analysis. Of particular interest for screening purposes, at least two companies (Wyatt Technology Corporation and Precision Detectors, Inc.) have developed high-throughput DLS, increasing the number of compounds that can be studied by this method.

A growing number of research groups have incorporated DLS into their evaluation of hit or lead compounds. For instance, Klebe's group used DLS to rule out aggregation as a contributor to the inhibition of tRNA-guanine transglycosylase by several hits identified in a FlexX-based screen.[40] Similarly, Wang and co-workers used light scattering to rule out aggregation as contributor to the activity of a novel non-nucleoside reverse transcriptase inhibitor (NNRTI).[39] The technique has also been used to confirm the presence of particles in solutions of suspected promiscuous inhibitors. Soelaiman and colleagues used DOCK to identify new inhibitors of edema factor, a toxin secreted by *Bacillus anthracis*. Three active compounds were evaluated by DLS, and one was found to form particles greater than 1000 nm in diameter.[29] Tipton's group also used DLS in their analysis of Disperse Blue; at 10 μM,

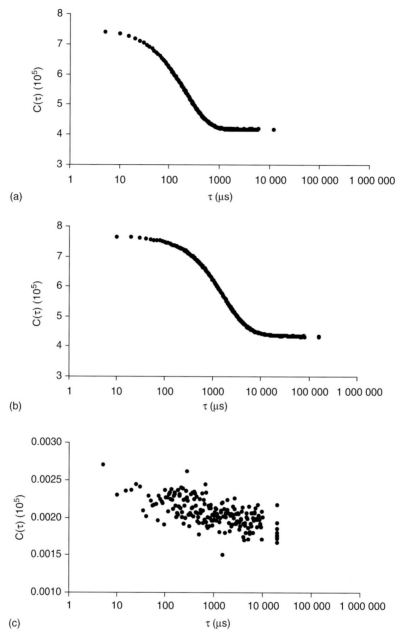

Figure 3 Representative autocorrelation functions from DLS experiments. (a) 15 μM rottlerin, (b) 10 μM K-252c, and (c) 400 μM suramin. (Reprinted with permission from McGovern, S. L.; Shoichet, B. K. *J. Med. Chem.* **2003**, *46*, 1478–1483. Copyright (2003) American Chemical Society.)

the compound was observed to form particles with a mean radius of 43.4 nm and average scattering intensity of 92.4 kcps.[28]

DLS has also been used to explore the factors influencing aggregation by small molecules. In a report from Arnold's group, 15 diaryltriazine and diarylpyrimidine NNRTIs were studied by DLS to evaluate the effect of pH and compound concentration on aggregate size.[44] All of the compounds were found to form particles in solution. For three compounds studied in detail, aggregate size increased as the pH increased from 1.5 to 6.5. Aggregate size was also found to increase as the inhibitor concentration increased from 0.001 to 10 mM. Intriguingly, compounds with particle radii of 30–110 nm had more favorable absorption than compounds with radii greater than 250 nm, and it was proposed that the systemic adsorption of the drugs depended on the formation of appropriately sized aggregates.[44]

2.18.3.3 Microscopy

Confocal fluorescence and transmission electron microscopy have been used to visualize small molecule aggregates and to explore the interaction of these particles with protein.[23,24,44] Low-resolution imaging has been obtained by confocal study of green fluorescent protein (GFP) in the presence of I4PTH aggregates.[24] Initial images obtained by this method suggest that the inhibitor aggregates cause GFP to accumulate in clusters; these clusters dissolve upon addition of Triton X-100 (**Figure 4**). Notably, GFP retains its fluorescence signal in the presence of I4PTH, suggesting that the protein is not denatured by the aggregates (**Figure 4b**). Higher resolution imaging of similar solutions of β-galactosidase and I4PTH obtained by transmission electron microscopy show that the protein adsorbs onto the surface of the inhibitor aggregate (**Figure 4e**), and with the addition of detergent, this interaction is prevented (**Figure 4f**). These observations, in combination with biochemical and biophysical studies, suggest that enzyme is adsorbed onto the surface of the aggregate particles and thereby inhibited, although absorption into the aggregate interior cannot be excluded.[24] Furthermore, the interaction between aggregate and enzyme can be prevented or reversed by the addition of Triton X-100.[24]

Electron microscopy is an intensive process not typically amenable to screening, although there has been at least one report of a research program using transmission electron microscopy to evaluate a novel drug lead for aggregation.[39] The technique was also used to visualize aggregates formed by NNRTIs studied by Arnold's group; although these compounds were not considered promiscuous, aggregate formation was suggested to mediate bioavailability of the drugs.[44]

2.18.3.4 Additional Confirmation Assays

Beyond the experiments described above, there are several tests one could consider performing to evaluate screening hits for peculiar behavior. Most simply, the original assay could be repeated under exactly the same conditions to evaluate for trivial experimental errors.[13] More rigorously, it has been proposed that when the same compound library is screened against the same target in three different assay protocols, the agreement between assays is only 35%.[1]

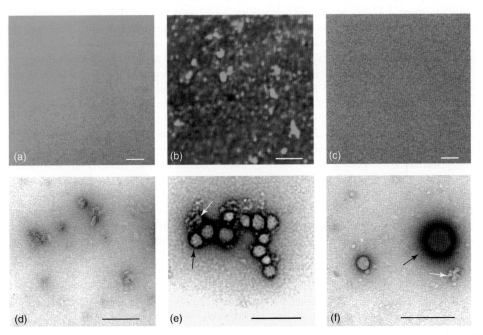

Figure 4 Visualization of I4PTH aggregates and GFP molecules by confocal fluorescence microscopy (a–c) and TEM (d–f). (a) 0.33 mg mL^{-1} GFP alone. (b) 0.25 mg mL^{-1} GFP with 500 μM I4PTH. (c) 0.25 mg mL^{-1} GFP with 500 μM I4PTH and 0.0075% Triton X-100. (d) 0.5 mg mL^{-1} GFP alone. (e) 0.1 mg mL^{-1} GFP with 100 μM I4PTH. (f) 0.1 mg mL^{-1} GFP with 100 μM I4PTH and 0.001% Triton X-100. Representative I4PTH aggregates are marked with black arrows, GFP molecules with white arrows. Bar = 5 μm in panels a–c; bar = 200 nm in panels d–f. (Reprinted with permission from McGovern, S. L.; Helfand, B. T.; Feng, B.; Shoichet, B. K. *J. Med. Chem.* **2003**, *46*, 4265–4272. Copyright (2003) American Chemical Society.)

Consequently, it may be useful to design an alternative assay protocol for the target and then rescreen the entire library or, less laboriously, only the initial hits with the alternative format.[13]

Depending on the particular screening protocol used, it may be informative to test the compound set for interference with the assay readout. For instance, Jenkins and colleagues used a fluorescence-based HTS to identify angiogenin inhibitors.[13] The initial set of hits was re-plated with activated fluorescent substrate to evaluate the hits for interference with the fluorescent readout; strikingly, 73% of the hits were found to interfere.[13]

Screening for chemical reactivity may also be productive. The ability of the same screening hit to inhibit diverse receptors, as described above, may be the first sign of reactivity. Time-dependent behavior may also suggest an irreversible reaction. More directly, some authors have found it useful to repeat the screening assay in the presence of dithiothreitol (DTT) to evaluate hits for thiol reactivity.[15,30] Alternatively, mass spectrometry (MS) has been used to evaluate compounds for adduct formation.[15]

Less glamorously but no less importantly, it is often useful to evaluate the purity of screening hits because the observed activity may be due to a contaminant. A variety of approaches are possible; most simply, a new solution of the compound can be prepared from fresh powder or with powder from a different lot.[15] Alternatively, the compound can be repurified and then reassayed.[15] For a more rigorous analysis, HPLC or MS can be employed to detect impurities.[12]

2.18.4 Computational Prediction of Promiscuous Compounds

Given the expense of experimental work and the desire to minimize efforts spent on false positives, there is a considerable interest in the development of computational methods to predict compounds likely to behave promiscuously. Since Lipinski's seminal work showed that 90% of drugs share a common set of easily identified properties,[25] there has been much effort devoted to predicting druglike behavior based on chemical structure. The following section will review recent computational approaches for identifying various classes of promiscuous compounds.

2.18.4.1 Frequent Hitters

As defined by the Roche group, 'frequent hitters' are compounds that appear as actives in multiple screening projects, across a range of targets.[53] These include reactive species, compounds that interfere with the assay, and privileged scaffolds. To predict compounds likely to act as frequent hitters, Roche and colleagues first developed a database of known frequent hitters. These compounds were culled from compounds that had hit in at least eight different HTS assays, compounds from an in-house depository that had been requested by at least six different discovery projects, and additional compounds from various sites within Roche.[53] Eleven teams of medicinal chemists voted on the structures to identify molecules that according to their expert opinion and experience were likely to be frequent hitters. To obtain a set of nonfrequent hitters, a diverse set of compounds from the Roche human drug database were selected. The final data set contained 479 frequent hitters and 423 nonfrequent hitters.[53]

Structural analysis with LeadScope revealed that no single substructure was sufficient to identify a frequent hitter. To extend the analysis, Ghose and Crippen descriptors were calculated for all compounds in the data set and used to define a partial least squares (PLS) multivariate linear regression model. The PLS model correctly predicted 92% of frequent hitters and 88% of nonfrequent hitters. The group then developed a nonlinear model using a neural network, again based on the Ghose and Crippen descriptors. The final model correctly classified 90% of frequent hitters and 91% of nonfrequent hitters.[53] As the authors explained, 'frequent hitter' is not synonymous with 'undesired structure;' for instance, ligands that bind a similar set of receptors will cross-react because of common substructures. Therefore, this model was not intended for elimination of compounds but for prioritization of compounds for purchase, library design, or testing.

A subsequent effort to identify frequent hitters was described by Merkwirth and co-workers.[54] Various single and ensemble methods, including k-nearest neighbors classifiers, support vector machines (SVM), and single ridge regression models, were derived for binary classification of compounds. These models were then trained against the same frequent hitters data set described above. The best models had cross-validated correlation coefficients of up to 0.92 and misclassification rates of 4–5%, an improvement from the 10% misclassification rate produced by the neural network approach.[54]

2.18.4.2 Aggregators

Recent efforts have been directed toward the problem of identifying aggregating promiscuous inhibitors. Seidler and colleagues screened 50 clinically prescribed drugs and found that seven showed aggregate-based inhibition of β-lactamase, chymotrypsin, and malate dehydrogenase.[27] These experimental results were used to develop a data set of 111 compounds containing 48 aggregators and 63 nonaggregators. Simple cutoffs based on solubility or ClogP classified 87% or 81% of the 111 compounds correctly, respectively. To obtain a more precise model, a recursive partitioning algorithm based on 260 physicochemical descriptors was developed (**Figure 5**). As shown in **Table 2**, the model correctly classified 43 of 48 aggregators (90%) and 61 of 63 nonaggregators (97%).[27]

The same neural network[53] and SVM method[54] described in the preceding section were then tested against the 111 compounds in the aggregator test set; the results are described fully in [54] and briefly here (**Table 2**). The neural net identified 30 of 48 aggregators (63%) and 53 of 63 nonaggregators (84%) with a Matthew's correlation coefficient of 0.48. The SVM method correctly predicted 32 of 48 aggregators (67%) and 58 of 63 nonaggregators (92%) with a Matthew's correlation coefficient of 0.63. In summary, these methods did not discriminate between aggregators and nonaggregators as well as the recursive partitioning model. One obvious cause is that these models were trained against a data set containing frequent hitters acting by a variety of mechanisms, not simply aggregation. Nonetheless, it

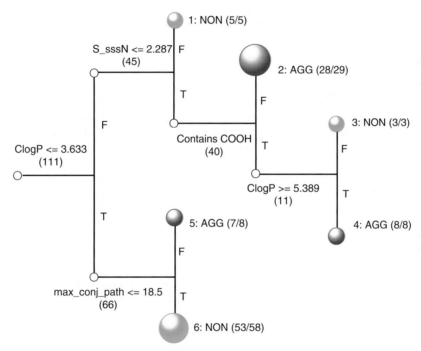

Figure 5 Recursive partitioning analysis of 111 aggregators and nonaggregators. Each branch contains the physicochemical criterion used to split a group of compounds; T indicates compounds that satisfy this criterion, and F indicates compounds that do not. Terminal nodes are green and coded as 'NON' if they consist predominantly or completely of nonaggregators; terminal nodes are red and coded as 'AGG' if they are predominantly or completely aggregators. Nodes with more compounds are identified by larger circles. (Reprinted with permission from Seidler, J.; McGovern, S. L.; Doman, T. N.; Shoichet, B. K. *J. Med. Chem.* **2003**, *46*, 4477–4486. Copyright (2003) American Chemical Society.)

Table 2 Classification rates for three different algorithms tested on the same data set of 48 aggregators and 63 nonaggregators

Algorithm	Aggregators correctly classified (total = 48)		Nonaggregators correctly classified (total = 63)		
	No.	%	No.	%	Reference
Recursive partitioning	43	90	61	97	27
Neural network	30	63	53	84	53
Ensemble method (SVM)[a]	32	67	58	92	53

[a] SVM, support vector machines.

is remarkable that these algorithms performed as well as they did, given that the molecules on which they were trained were not specifically designated aggregators or nonaggregators. Indeed, it is intriguing to consider that the frequent hitters data set may include (as yet unidentified) aggregators, thereby contributing to the success of these models.

Recently, an additional 1030 compounds have been tested for aggregation[33]; the results increase the number of publicly reported compounds studied for this property by an order of magnitude. As the database of compounds explicitly found to be aggregators or nonaggregators grows, it is hoped that these data will provide greater insight into the nature of aggregation and enable the development of more precise models.

2.18.4.3 Privileged Substructures

Certain molecular scaffolds encode for recognition of diverse receptors. These privileged substructures are often starting points for the design of ligands with improved specificity for a particular receptor. This section will review two pharmacophore-based studies describing the use of privileged substructures for designing G protein-coupled receptor (GPCR) targeted compound libraries[17] and for identifying promiscuous kinase inhibitors.[16]

Mason and co-workers introduced a method for generating 4-point pharmacophores that could be modified to include a pharmacophore for a particular substructure such as a privileged scaffold.[17] Using Chem-X, atom types were automatically generated and assigned to a feature type used to generate the pharmacophore. Atom types or dummy centroid atoms unique to a privileged substructure were designated as a special 'privileged' feature type. This feature type was designated as one of the four points used to generate the pharmacophore map for the compound. Because all of the subsequent pharmacophore descriptor sets contained the privileged substructure, molecular similarity and diversity could be compared relative to their shared privileged scaffold. The authors proposed that this could be used as an aid for library design, either to imitate features of known active compounds or to explore 'missing diversity' among existing compounds containing the desired scaffold. The latter goal could be accomplished by using molecules containing the privileged substructure as reagents to obtain desired chemically accessible products.[17]

In addition, the method was shown to identify pharmacophores enriched among a class of ligands sharing similar biological activity. For instance, structural features shared by GPCR ligands were distinguished from those in a set of small molecule enzyme inhibitors and in a set of random compounds.[17] The method could also be used to address ligand selectivity by comparing pharmacophore keys across sets of ligands for related enzymes, such as trypsin, thrombin, and factor Xa. More recent work has explored the use of pharmacophores for describing related binding sites, for instance see [55].

In a subsequent report, Aronov and Murcko derived a five-point pharmacophore to describe kinase frequent hitters.[16] In this work, frequent hitters were defined as compounds with K_i values less than 50 nM against two or more of the following diverse kinases: protein kinase A, Src, Cdk2, Erk2, and Gsk3. Selective inhibitors were defined as compounds that had a K_i value less than 50 nM against only one of the five kinases and greater than 50 nM against the other four. Both the promiscuous and selective compounds were known to exhibit ATP-competitive kinase inhibition at nanomolar concentrations. In total, the authors compiled a set of 43 frequent hitters and 209 selective inhibitors.[16]

Ligand-bound structures of four of the promiscuous compounds had been solved in-house; these structures were aligned and used to generate a five-point pharmacophore composed of two hydrogen bond donors, two hydrogen bond acceptors, and one aromatic feature.[16] These results were compared against the frequent hitter and selective inhibitor data sets described above; 38 of 43 frequent hitters (88%) but only 10 of 209 selective inhibitors (5%) matched the pharmacophore. Considering only selective inhibitors with a K_i value less than 2 nM for one of the kinases (25-fold greater specificity), 0 of 61 compounds (0%) matched the pharmacophore. Therefore, the pharmacophore captured some of the features that distinguished frequent hitters from specific inhibitors.[16] By placing the pharmacophore features in the ATP binding site, it was possible to identify interactions between conserved kinase residues and the frequent hitters that contributed to the inhibitors' promiscuity. The authors proposed that either by addition of features beyond those five associated with promiscuity or subtraction of at least one of the substructures represented by the pharmacophore, it might be possible to engineer ligands with greater specificity for a given kinase.[16]

2.18.4.4 Reactive Species

Chemically reactive groups are a well-known source of false-positive screening hits.[14] Several commercial and in-house filters for identifying and removing reactive species are currently in use.[50,56] Despite the widespread use of filters, reactive compounds remain a significant challenge for discovery efforts. For example, Blanchard and co-workers reported that of 69 HTS hits against coronavirus protease, 64 (93%) were discarded after repeat assays with DTT suggested thiol reactivity.[30]

To address this persistent problem, Hajduk's group at Abbott has proposed the thiol reactivity index (TRI) for estimating the probability that a given compound will be reactive.[15] Their recently developed ALARM NMR method for detecting thiol reactivity found that of 476 lead compounds experimentally shown to be reactive, only 156 (33%) were predicted as such by computational filters. To better define the relevant moieties, structural descriptors were calculated for a series of reactive compounds. The probability of reactivity (P_R) for any given structure was then defined as the sum over all structural descriptors of the number of occurrences of each descriptor in that compound weighted by the TRI for that descriptor. If the P_R for a compound was greater than 0.3, the compound was predicted to be reactive. The TRI for each descriptor was determined by using nonlinear regression to maximize the agreement between observed and predicted reactivity as described by a scoring function.[15]

The final set of 75 nonzero TRIs was applied to a data set of 3504 compounds studied by ALARM NMR.[15] Using a P_R of 0.3, 486 of 509 experimentally reactive compounds (95%) were correctly predicted and 2005 of 2995 experimentally nonreactive compounds (67%) were correctly predicted. As described, the method has a high sensitivity and a low number of false negatives, making it a useful flag for identifying compounds for follow-up experimental reactivity testing.[15]

2.18.4.5 Experimental Artifact

Interference with the assay readout is another familiar source of screening false positives.[7,14] Few computational approaches for removing actives due to experimental artifact have been developed, but Jenkins and colleagues proposed that virtual screening can be used as a filter for such compounds.[13] Their underlying hypothesis was that true positives should be enriched by virtual screening and false positives due to experimental artifact should be de-enriched. To explore this idea, the National Cancer Institute Diversity Set and the ChemBridge DIVERSet were combined to produce a library of 18 111 compounds that was initially docked against angiogenin in a fluorescence-based HTS. To distinguish true hits from false positives, HTS hits were rescreened in a luciferase assay and an HPLC assay. Of 178 initial HTS hits, 12 (6.7%) were confirmed as true hits by the follow-up assays.

The same database of compounds was also tested against angiogenin in two separate virtual screens, one using DockVision for docking and Ludi for scoring, and a second using GOLD for both docking and scoring. The computational methods were then evaluated for their ability to enrich for true hits among the total docked library and also among the HTS hit list. The DockVision/Ludi combination ranked 33% of the true hits in the top 10% of the total database for a 3.3-fold enrichment over random and the GOLD algorithm placed 42% of the true hits in the top 2% to yield a 21-fold enrichment. When only the HTS hits were docked, the highest enrichment was observed by a consensus scoring method that counted compounds ranked in the top 25% by both virtual screening methods. This approach placed four of the true hits among the top 9 (5%) of the 178 HTS hits for an enrichment of 6.6-fold over HTS alone. The authors concluded that virtual screening was best used as a post-HTS filter to prioritize compounds for further testing.

2.18.5 Future Directions

Despite the growing interest in promiscuous compounds, several outstanding questions remain. For instance, what is the prevalence of aggregate-based inhibitors, particularly among small molecules that are most interesting for drug design? Early observations suggest that these compounds may be widespread among druglike compounds[23,33]; further work to explore the distribution of aggregators is clearly needed.

Even if aggregators are found to occupy substantial regions of relevant chemical space, what is the significance of this behavior? Is aggregation a feature or a bug? Since clinically used drugs have been found capable of inhibiting enzymes by forming aggregates,[27] one could make the argument that aggregation is not a significant problem for drug design. Indeed, recent studies have proposed that aggregation may be a necessary step in the systemic uptake of NNRTIs,[44] suggesting that aggregation may even be a desirable property.

Clearly, it is possible for drugs to show aggregation-based inhibition, as a compound that is specific at nanomolar concentrations for a pharmacologic target may also aggregate at micromolar concentrations. Indeed, it is interesting to speculate on possible consequences of aggregation by drugs. But aggregation at micromolar concentrations leading to inhibition of nontarget receptors does not change the fact that the drug also inhibits its target at nanomolar concentrations. This is the important difference between a drug that inhibits off-pathway targets via aggregation and a hit that inhibits its screening target via aggregation. The purpose of screening is to find new leads for inhibitor design. A screening hit that is only active because it inhibits its target receptor via aggregation is likely not a good starting point for further design work and should be abandoned. Furthermore, even though the screening hit shares the same pathology (aggregation) as a real drug, it does not necessarily follow that it may also share the same desired feature (classical inhibition of a target). Conversely, aggregation at micromolar concentrations does not exclude the possibility that the compound may also act as a specific inhibitor of a given target at nanomolar concentrations.

Efforts have been made to explore the mechanism of aggregate-mediated inhibition,[24,28,45] but these experiments have only raised more questions. What is the nature of the physicochemical interaction between enzyme molecules and aggregate particles? Recent studies of nanoparticle-mediated inhibition of chymotrypsin[57,58] may provide insight. Functionalized amphiphilic nanoparticles have been found to inhibit chymoptrypsin via an electrostatic interaction followed by denaturation; up to 50% of enzyme activity was restored by the addition of surfactant.[57] Depending on the particular functional groups exposed on the surface of the nanoparticle, various modes and degrees of inhibition were observed.[58] Various biophysical experiments including fluorescence, fluorescence anisotropy, DLS, and circular dichroism were performed to study the conformational changes induced in the enzyme due to nanoparticle binding.[57,58] Similar experiments to explore the nature of aggregate-mediated enzyme inhibition may be worthwhile.

It is also interesting to consider the scope of receptors that are targets for small molecule aggregates. For instance, some enzymes are more susceptible to inhibition by aggregators than other enzymes. Is this due to variations in protein size, shape, surface charge distribution, or other physicochemical property? Additionally, most work to date has been done with soluble enzymes, but aggregators have also been identified in screens using membrane-bound receptors.[23] Do aggregating compounds generally inhibit surface receptors? If so, how does inhibition compare to that of soluble enzymes? Can aggregators act as agonists as well as antagonists? These questions are active areas of research, and results from these studies are eagerly anticipated.

A final area of interest concerns the SARs among aggregators. What distinguishes aggregators from nonaggregators? What factors determine the size of the particle formed by each aggregator? Preliminary observations have been made by a few groups. For instance, Frenkel and colleagues analyzed SARs among 15 aggregating NNRTIs and found that compounds containing oxygen atoms tended to form larger particles.[44] The authors proposed that oxygen's ability to serve as a hydrogen bond donor could result in different interactions from a similarly placed carbon or nitrogen atom and consequently lead to different aggregation patterns. In a separate study, Seidler and co-workers studied a series of azole antifungals, four of which were aggregate formers and two of which were not.[27] Interestingly, the two nonaggregators are prescribed as orally available medications, but the four aggregators are only administered as topical agents. These compounds could be distinguished from each other based on their ClogP values, with the aggregators having a ClogP greater than 5 and the nonaggregators having a value less than 5.[27] In addition to physical properties, potential structural descriptors have been suggested by the various computational models described above. Additionally, pharmacophore-based methods may be a useful approach for studying SARs among these compounds. It is hoped that as more small molecules are explicitly studied for aggregation, the structural influences leading to aggregation will be better defined and the mechanism itself will be better understood.

Acknowledgment

This work was supported by the Medical Scientist Training Program at Northwestern University (D. Engman, PI).

References

1. Lipinski, C.; Hopkins, A. *Nature* **2004**, *432*, 855–861.
2. Shoichet, B. K. *Nature* **2004**, *432*, 862–865.
3. Alvarez, J. C. *Curr. Opin. Chem. Biol.* **2004**, *8*, 365–370.
4. Erlanson, D. A.; Braisted, A. C.; Raphael, D. R.; Randal, M.; Stroud, R. M.; Gordon, E. M.; Wells, J. A. *Proc. Natl. Acad. Sci. USA* **2000**, *97*, 9367–9372.
5. Diller, D. J.; Hobbs, D. W. *J. Med. Chem.* **2004**, *47*, 6373–6383.
6. Hann, M. M.; Oprea, T. I. *Curr. Opin. Chem. Biol.* **2004**, *8*, 255–263.
7. Rishton, G. M. *Drug Disc. Today* **2003**, *8*, 86–96.
8. Muegge, I.; Heald, S. L.; Brittelli, D. *J. Med. Chem.* **2001**, *44*, 1841–1846.
9. Frimurer, T. M.; Bywater, R.; Naerum, L.; Lauritsen, L. N.; Brunak, S. *J. Chem. Inf. Comput. Sci.* **2000**, *40*, 1315–1324.

10. Ajay, A.; Walters, W. P.; Murcko, M. A. *J. Med. Chem.* **1998**, *41*, 3314–3324.
11. Walters, W. P.; Stahl, M. T.; Murcko, M. A. *Drug Disc. Today* **1998**, *3*, 160–178.
12. Baldwin, J.; Michnoff, C. H.; Malmquist, N. A.; White, J.; Roth, M. G.; Rathod, P. K.; Phillips, M. A. *J. Biol. Chem.* **2005**, *280*, 21847–21853.
13. Jenkins, J. L.; Kao, R. Y.; Shapiro, R. *Proteins* **2003**, *50*, 81–93.
14. Rishton, G. M. *Drug Disc. Today* **1997**, *2*, 382–384.
15. Huth, J. R.; Mendoza, R.; Olejniczak, E. T.; Johnson, R. W.; Cothron, D. A.; Liu, Y.; Lerner, C. G.; Chen, J.; Hajduk, P. J. *J. Am. Chem. Soc.* **2005**, *127*, 217–224.
16. Aronov, A. M.; Murcko, M. A. *J. Med. Chem.* **2004**, *47*, 5616–5619.
17. Mason, J. S.; Morize, I.; Menard, P. R.; Cheney, D. L.; Hulme, C.; Labaudiniere, R. F. *J. Med. Chem.* **1999**, *42*, 3251–3264.
18. Hajduk, P. J.; Bures, M.; Praestgaard, J.; Fesik, S. W. *J. Med. Chem.* **2000**, *43*, 3443–3447.
19. Bemis, G. W.; Murcko, M. A. *J. Med. Chem.* **1996**, *39*, 2887–2893.
20. Klabunde, T.; Hessler, G. *ChemBioChem* **2002**, *3*, 928–944.
21. Evans, B. E.; Rittle, K. E.; Bock, M. G.; DiPardo, R. M.; Freidinger, R. M.; Whitter, W. L.; Lundell, G. F.; Veber, D. F.; Anderson, P. S.; Chang, R. S. et al. *J. Med. Chem.* **1988**, *31*, 2235–2246.
22. Bondensgaard, K.; Ankersen, M.; Thogersen, H.; Hansen, B. S.; Wulff, B. S.; Bywater, R. P. *J. Med. Chem.* **2004**, *47*, 888–899.
23. McGovern, S. L.; Caselli, E.; Grigorieff, N.; Shoichet, B. K. *J. Med. Chem.* **2002**, *45*, 1712–1722.
24. McGovern, S. L.; Helfand, B. T.; Feng, B.; Shoichet, B. K. *J. Med. Chem.* **2003**, *46*, 4265–4272.
25. Lipinski, C. A.; Lombardo, F.; Dominy, B. W.; Feeney, P. J. *Adv. Drug Deliv. Rev.* **1997**, *23*, 3–25.
26. McGovern, S. L.; Shoichet, B. K. *J. Med. Chem.* **2003**, *46*, 1478–1483.
27. Seidler, J.; McGovern, S. L.; Doman, T. N.; Shoichet, B. K. *J. Med. Chem.* **2003**, *46*, 4477–4486.
28. Liu, H. Y.; Wang, Z.; Regni, C.; Zou, X.; Tipton, P. A. *Biochemistry* **2004**, *43*, 8662–8669.
29. Soelaiman, S.; Wei, B. Q.; Bergson, P.; Lee, Y. S.; Shen, Y.; Mrksich, M.; Shoichet, B. K.; Tang, W. J. *J. Biol. Chem.* **2003**, *278*, 25990–25997.
30. Blanchard, J. E.; Elowe, N. H.; Huitema, C.; Fortin, P. D.; Cechetto, J. D.; Eltis, L. D.; Brown, E. D. *Chem. Biol.* **2004**, *11*, 1445–1453.
31. Lee, Y. S.; Bergson, P.; He, W. S.; Mrksich, M.; Tang, W. J. *Chem. Biol.* **2004**, *11*, 1139–1146.
32. Erbe, D. V.; Wang, S.; Zhang, Y. L.; Harding, K.; Kung, L.; Tam, M.; Stolz, L.; Xing, Y.; Furey, S.; Qadri, A. et al. *Mol. Pharmacol.* **2005**, *67*, 69–77.
33. Feng, B. Y.; Shelat, A.; Doman, T. N.; Guy, R. K.; Shoichet, B. K. *Nat. Chem. Biol.* **2005**, *1*, 146–148.
34. Teklu, S.; Gundersen, L. L.; Larsen, T.; Malterud, K. E.; Rise, F. *Bioorg. Med. Chem.* **2005**, *13*, 3127–3139.
35. Koch, M. A.; Wittenberg, L. O.; Basu, S.; Jeyaraj, D. A.; Gourzoulidou, E.; Reinecke, K.; Odermatt, A.; Waldmann, H. *Proc. Natl. Acad. Sci. USA* **2004**, *101*, 16721–16726.
36. Brogan, A. P.; Verghese, J.; Widger, W. R.; Kohn, H. *J. Inorg. Biochem.* **2005**, *99*, 841–851.
37. Fattorusso, R.; Jung, D.; Crowell, K. J.; Forino, M.; Pellecchia, M. *J. Med. Chem.* **2005**, *48*, 1649–1656.
38. Li, C.; Xu, L.; Wolan, D. W.; Wilson, I. A.; Olson, A. J. *J. Med. Chem.* **2004**, *47*, 6681–6690.
39. Wang, L. Z.; Kenyon, G. L.; Johnson, K. A. *J. Biol. Chem.* **2004**, *279*, 38424–38432.
40. Brenk, R.; Meyer, E. A.; Reuter, K.; Stubbs, M. T.; Garcia, G. A.; Diederich, F.; Klebe, G. *J. Mol. Biol.* **2004**, *338*, 55–75.
41. Lee, K. H.; Shin, B. H.; Shin, K. J.; Kim, D. J.; Yu, J. *Biochem. Biophys. Res. Commun.* **2005**, *328*, 816–823.
42. Forino, M.; Jung, D.; Easton, J. B.; Houghton, P. J.; Pellecchia, M. *J. Med. Chem.* **2005**, *48*, 2278–2281.
43. McGovern, S. L. Experimental Identification of Promiscuous, Aggregate-Forming Screening Hits. In *Virtual Screening in Drug Discovery*; CRC Press: Boca Raton, FL, 2005, pp 107–124.
44. Frenkel, Y. V.; Clark, A. D., Jr.; Das, K.; Wang, Y. H.; Lewi, P. J.; Janssen, P. A.; Arnold, E. *J. Med. Chem.* **2005**, *48*, 1974–1983.
45. Ryan, A. J.; Gray, N. M.; Lowe, P. N.; Chung, C. W. *J. Med. Chem.* **2003**, *46*, 3448–3451.
46. Davies, S. P.; Reddy, H.; Caivano, M.; Cohen, P. *Biochem. J.* **2000**, *351*, 95–105.
47. Leung, D.; Hardouin, C.; Boger, D. L.; Cravatt, B. F. *Nat. Biotechnol.* **2003**, *21*, 687–691.
48. Godl, K.; Wissing, J.; Kurtenbach, A.; Habenberger, P.; Blencke, S.; Gutbrod, H.; Salassidis, K.; Stein-Gerlach, M.; Missio, A.; Cotten, M. et al. *Proc. Natl. Acad. Sci. USA* **2003**, *100*, 15434–15439.
49. Funeriu, D. P.; Eppinger, J.; Denizot, L.; Miyake, M.; Miyake, J. *Nat. Biotechnol.* **2005**, *23*, 622–627.
50. Walters, W. P.; Namchuk, M. *Nat. Rev. Drug Disc.* **2003**, *2*, 259–266.
51. Santos, N. C.; Castanho, M. A. R. B. *Biophys. J.* **1996**, *71*, 1641–1650.
52. Wyatt, P. J. *Anal. Chim. Acta* **1993**, *272*, 1–40.
53. Roche, O.; Schneider, P.; Zuegge, J.; Guba, W.; Kansy, M.; Alanine, A.; Bleicher, K.; Danel, F.; Gutknecht, E. M.; Rogers-Evans, M. et al. *J. Med. Chem.* **2002**, *45*, 137–142.
54. Merkwirth, C.; Mauser, H.; Schulz-Gasch, T.; Roche, O.; Stahl, M.; Lengauer, T. *J. Chem. Inf. Comput. Sci.* **2004**, *44*, 1971–1978.
55. Arnold, J. R.; Burdick, K. W.; Pegg, S. C.; Toba, S.; Lamb, M. L.; Kuntz, I. D. *J. Chem. Inf. Comput. Sci.* **2004**, *44*, 2190–2198.
56. Hann, M.; Hudson, B.; Lewell, X.; Lifely, R.; Miller, L.; Ramsden, N. *J. Chem. Inf. Comput. Sci.* **1999**, *39*, 897–902.
57. Fischer, N. O.; Verma, A.; Goodman, C. M.; Simard, J. M.; Rotello, V. M. *J. Am. Chem. Soc.* **2003**, *125*, 13387–13391.
58. Hong, R.; Fischer, N. O.; Verma, A.; Goodman, C. M.; Emrick, T.; Rotello, V. M. *J. Am. Chem. Soc.* **2004**, *126*, 739–743.
59. Weston, G. S.; Blazquez, J.; Baquero, F.; Shoichet, B. K. *J. Med. Chem.* **1998**, *41*, 4577–4586.

Biography

Susan L McGovern received her BSc in biochemistry and mathematics from the University of Notre Dame in 1997. Later that year, she joined the NIH-sponsored Medical Scientist Training Program of Northwestern University in Chicago, IL. In 2003, she earned her PhD in Brian Shoichet's laboratory for her work on molecular docking and aggregate-forming promiscuous inhibitors. In 2005, she finished her MD at Northwestern and moved to the University of Texas/MD Anderson Cancer Center to begin residency training in radiation oncology. Her long-term research goals are to use small molecules as tools for exploring the molecular basis of cancer and to develop pharmacologic agents for use in radiation oncology. Her research has been supported by two NIH training grants, a fellowship from the PhRMA Foundation, and a Presidential Fellowship from Northwestern University.

2.19 Diversity versus Focus in Choosing Targets and Therapeutic Areas

D A Giegel, Celgene Corporation, San Diego, CA, USA
A J Lewis, Novocell, Inc., Irvine, CA, USA
P Worland, Celgene Corporation, San Diego, CA, USA

© 2007 Elsevier Ltd. All Rights Reserved.

2.19.1	**Introduction**	753
2.19.2	**Definitions**	754
2.19.2.1	Therapeutic Area Focus versus Target Class Focus	754
2.19.2.2	Therapeutic Target versus Molecular Target	754
2.19.2.3	Target 'Druggability'	754
2.19.3	**The Impact of the 'Genomics Bubble' on Drug Discovery**	754
2.19.3.1	Payoff – Is it Here Yet or On the Horizon?	756
2.19.4	**Focusing on a Target Class**	756
2.19.4.1	G Protein-Coupled Receptors	758
2.19.4.2	Protein Kinases	758
2.19.4.3	Ion Channels	758
2.19.4.4	Intellectual Property (IP) Landscape	758
2.19.5	**Companies Built around Compound Classes or Particular Chemistries**	759
2.19.5.1	The Example of IMiDs Compounds	759
2.19.5.2	Strategies for Chemical Libraries	760
2.19.5.3	Structure-Based Drug Design	760
2.19.5.4	Ribonucleic Acid (RNA)-Based Strategies (ISIS[60] and Sirna[61])	761
2.19.6	**Companies Built with a Therapeutic Focus**	762
2.19.7	**Challenges to Moving Compounds into the Clinic**	763
2.19.8	**Models for Start-Up Companies**	764
2.19.8.1	Classical Biotechnology with a Twist	764
2.19.8.2	The Spin Cycle	765
2.19.8.3	The Virtual Company	766
2.19.8.4	The Bottom Line	767
References		767

2.19.1 Introduction

The pharmaceutical industry is confronted with several important challenges today, including the need to sustain their pipelines and offset patent expirations and inevitable generic competition. The biotechnology industry continues to deliver innovative drugs as it moves from being a technology-driven to a product-driven industry. Not unexpectedly, the pharmaceutical industry has aggressively partnered with biotechnology companies to fill their pipeline gaps.

The biotechnology industry began the revolution in drug discovery by creating protein biologicals such as human insulin, growth hormone, and erythropoeitin. In addition to the biologicals, the biotechnology companies have focused on using many cloned proteins as targets for small-molecule drugs. Throughout the last decade, there has been a dramatic increase in the number of potential molecular targets for potential disease states. Small companies, in general, do not have access to market research data to help guide their selection of disease indications. Therefore, these companies tend to pursue disease indications based on a perceived market or based on their own experiences. This chapter outlines some of the progress and technologies that are being used by the biotechnology industry that make a vital contribution to the healthcare industry. The industry has grown remarkably since its inception in the early 1980s, and it is a continual driver of both innovation and execution.

2.19.2 Definitions

2.19.2.1 Therapeutic Area Focus versus Target Class Focus

In the past decade, there have been two predominant themes in how to pursue drug discovery. One is to build an organization to pursue a particular disease area (therapeutic area) such as inflammatory or cardiovascular diseases. This permits the company to deeply explore one area of disease biology at the expense of having a cursory knowledge of many different target classes.

Another common theme for building a biotechnology company has been to focus on a particular class of molecular target (such as kinases or proteases). This is a particularly important way to focus an organization that is dealing with protein targets that have been recently discovered. For any new target there is extensive work that needs to be done to decipher the pathway around that target and how it links to a disease state. As with the first theme, there is a significant downside to this approach as well. While the company would have a deep understanding of a particular type of target, they would only have a very cursory understanding of several different therapeutic areas.

2.19.2.2 Therapeutic Target versus Molecular Target

The goal of a molecularly targeted approach to drug discovery is that when a modulator of the protein is developed, it will have an overall effect on the course of the disease in question. This assumes that the chosen molecular target has a pathophysiological link to the disease of interest. Molecular target-based approaches may also have advantages in terms of drug discovery, high-throughput screening (HTS), and preclinical–clinical safety predictions based on a logical extension of the effect of the drug candidate on the selected target. The overall goal of any drug discovery operation is to choose a disease indication (therapeutic target) that currently has no (or suboptimal) drugs in the market to effectively treat that condition, or to have a substantial therapeutic advantage over existing drugs via a new mechanism, even if the established drugs work well. Such advantages may include reduced side effects or greater safety. For example, the kinase bcr-abl is the molecular target for the compound imatinib that is used as a therapeutic to treat chronic myelogenous leukemia (CML).

2.19.2.3 Target 'Druggability'

There are several phrases that are sometimes used interchangeably when the issues of compounds and molecular targets are discussed. In this chapter, the term 'drug target' will be used to mean a biological macromolecule that has the potential to bind a small-molecule organic compound. It needs to be stressed that by this definition it is meant that a compound should be able to bind to the target of interest, but there is no guarantee that this binding will modulate the underlying biology, or that the compound will be selective for that target. Druggability, on the other hand, typically refers to a family of proteins where there are representative therapeutics on the market that modulate a member of this family and are effective therapeutics.

Finally, in the strictest definition, a validated protein target is one for which a drug that modulates the target is already on the market. With this in mind, there are degrees with which new targets have been validated as potential druggable targets. The least validated is one in which only a gene sequence is known and nothing about the role of that particular gene in a biological process. Nearly all of the projects being pursued in the pharmaceutical/biotechnology industry use targets that fall somewhere within these two boundaries.

2.19.3 The Impact of the 'Genomics Bubble' on Drug Discovery

The 'genomics revolution' has had an enormous impact on the practice of drug discovery, but the impact in terms of new therapeutic agents has yet to be realized to the same degree. The advances in DNA sequencing technology have paved the way for the scientific community to realize that, at last, rapid sequencing was available and that it would be possible to imagine obtaining the complete sequence of a particular organism. With this realization there were several companies created aimed at identifying, at the level of the DNA sequence, those genes that would be important for the initiation or perpetuation of disease states. The rapidity with which this developed created a naïve expectation that the drug discovery industry would change tomorrow (almost literally). However, the publicity surrounding the Human Genome Project and the genomics revolution did have the benefit of providing an impetus for these various projects and the funding that they required for success. There were variations in the approach, from a direct 'sequence everything' in competition with the US government-sponsored effort coordinated through the National Institutes of Health (e.g., Human Genome Sciences and Celera), to more specific disease-oriented sequencing efforts of genes that appeared to be upregulated in specific disease tissues compared with normal tissues (e.g., Incyte and Millennium Pharmaceuticals).

Commercializing the information coming from these massive sequencing projects was aided by the entities involved applying for patents on their sequences. These patents often were filed on the sequence itself (or fragments of the sequence) without any mention being made of the utility of that particular gene sequence. During this period there was significant concern among many companies that a great advantage would be lost by not being part of this 'genomics revolution,' either as a direct participant or as a paying customer to obtain rights for the screening of particular gene product. There were efforts by large pharmaceutical companies to participate in well-funded collaborations (such as Bayer with Millennium and Curagen) from which they received a quota of genomically defined 'targets.' In some circumstances this interest led to the acquisition of a biotechnology company that appeared to have significant numbers of potential drug targets (e.g., Upjohn (now Pfizer) and Sugen). Through the 1990s, this was a model that sustained many of the genomics companies and simultaneously spurred improvements in the capacity of drug discovery in other areas.

The large number of apparent drug targets being discovered on a daily basis provoked greater interest and emphasis on those efforts focused at improving the rate at which new compounds were discovered. Prior to the advent of the genomics revolution, the drug discovery industry had begun to increase their HTS capacity. Other companies began to pioneer the rapid synthesis of compound libraries utilizing solid phase synthesis approaches (e.g., Alanex, Trega, and Pharmacopeia, *see* Section 2.19.5.2). However, increased screening capabilities and the development of larger specialized libraries were not the only developments to create a means to capitalize on the genomics revolution. With the enormous amount of sequencing data being produced, there was a tremendous need for robust technology to handle and interpret the data. Improvements in the existing software and development of new programs either within companies or by specific specialty software companies, were necessary. This would allow the raw genome sequencing data to be correlated with functional data mined from the literature. Correlations could then be made between the raw sequence data and the molecular nature of disease and disease processes. These developments allowed the generation of biological databases that induced hypothesis generation in ways that were not possible previously. While the above are enabling technologies and processes, the number of new gene sequences initially identified had very little information that would help the understanding of their biological context either within normal or pathological states. The tools to understand the role of a particular gene product were lagging behind the requirement of having sufficient validation of a protein for it to be considered as a drug target. Whether bioinformatics, systems biology, or bio-information technology approaches are able to accelerate such validation in silico remains to be seen.

The translation of genomically defined drug targets into new approved therapeutics developed against those targets has been slow to date. Much of this relates to the difficulty in obtaining validation of many potential targets for a particular disease state and being able to use this information for selecting the best molecular targets to pursue. Recent estimates indicate that novel targets result in a 50% lower success rate than clinically validated targets.[1,2] Coupled with rising costs and an inherent reticence within the industry for working on unproven targets due to lower success rates, the low number of therapeutics directed against novel targets is perhaps not surprising. Thus, the industry continues to work on targets for which there is substantial preclinical and clinical validation in the literature. Companies are also creating small research groups on an 'experimental' basis to work on the validation aspect of 'novel' targets that have been identified through genomics (and, more recently, proteomics) efforts. As the understanding of these targets increases through efforts within companies and from academic work, the prevalence of these targets within discovery pipelines will increase.

The use of genetic models has accelerated the target validation process, and companies have been founded on the applications of these models to explore the role of gene products in both developmental and disease states. These companies are using model systems from simple organisms such as yeast and nematodes to gene knockout and transgenic strains in mice. These efforts have been complemented by academic research, which may be smaller on an individual scale, but larger in the aggregate. Also, the academic research groups have an imperative to publish their findings, which yields a larger public database. A significant advance in validating molecular targets has come from small interfering RNA (siRNA or RNAi) technology (*see* Section 2.19.5.4) that has recently become a widespread technique for selectively manipulating gene expression. This technology has been used in cellular systems as well as with in vivo models, and some companies have started clinical trials based on this technology. Gene expression analysis continues to improve both in speed and in the number of different organisms on which analyses can be performed (due to the availability of the complete genome sequences of other organisms in addition to humans). Coupled with increasingly sophisticated bioinformatics tools, the role of particular molecular targets allows the validation of these targets to be accomplished in far shorter time frames than even a decade past.

2.19.3.1 Payoff – Is it Here Yet or On the Horizon?

There was a serious misconception regarding the immediate benefits of genome sequencing relating to drug discovery. Stratospheric claims were made regarding the benefits and the enormity of riches to be immediately delivered into drug discovery pipelines consequent to the genomics efforts of the 1990s. The naïveté for the genomics pioneers was centered in the real, but not realized, difficulty of determining the biological functionality for these hundreds of gene sequences. Without solid biology relating to a new target, companies are reluctant to make the decision to spend tens of millions of dollars in a drug discovery effort relating to that target. There was an apparent lack of understanding from the leaders of these genomics companies on what it takes to make a drug, beyond identification of the gene product and a rudimentary biochemical assay.

However, as the 'land rush' for potential targets has largely ended with the completion of the human genome sequence, many of the companies involved in more specific 'genomics' activities have been attempting to transition into biopharmaceutical companies. This means that in order to survive they will have to produce compounds with a demonstrable therapeutic benefit. This has proven to be a significant challenge for these companies, particularly as expectations from the business community have become more exacting. Companies of the genomics era have had to rapidly transform and implement more rigorous drug discovery models as well as regulatory and clinical development functions. The processes associated with these activities are quite different from their previous genomic biotechnology era. Many companies have had to struggle with changes in company skill sets and, internally, with the way their company business operates. While attempting to manage personnel and process changes, these companies have also become aware that drug discovery and development include significant challenges quite apart from what they had previously encountered. Not the least of these challenges is the length of time it takes for a clinical trial – be it for therapeutic benefit or to characterize biomarkers of a response in the clinical population. The toll this has taken in terms of successful transition (and it is an ongoing experiment) can be determined from the listings of those biotechnology companies with genomics as a significant part of their platform technology between 10 years ago and today. There may be one or two companies that survive and continue to be successful following a transition from genomics to biotechnology–pharmaceutical (e.g., HGS and Millennium). The larger impact of genomics on drug discovery and development has not yet been realized in new therapeutics, but will continue to increase its impact for at least the next couple of decades.

2.19.4 Focusing on a Target Class

There are four main types of macromolecule that can be modulated by therapeutic agents: proteins, polysaccharides, lipids, and nucleic acids. However, the vast majority of successful drugs achieve their activity by binding to, and then modulating, the activity of proteins. Thus, proteins have provided the best source for drug targets, and nearly half of these targets fall into seven gene families: G protein-coupled receptors (GPCRs), ion channels, protein kinases, zinc metalloproteases, serine proteases, nuclear hormone receptors, and phosphodiesterases.[3,4] There has been much speculation as to the number of molecular targets that are 'druggable.' The druggable portion of the human genome ($\sim 30\,000$ genes) has been calculated to be between 5000 and 10 000 targets, although this may be even greater.[5,6] Current drugs bind to approximately 500 targets. While there may be a large number of drug targets, it is unclear how many of these are linked to disease, and the clinical evaluation of modulators of truly novel targets is the only definitive means to establish disease relevance.

In the 1980s, molecular biology technologies allowed the cloning of protein targets for use in biological screens for drug discovery. These new technologies spawned an increase in the number of biotechnology companies targeting either proteins themselves as therapeutics or discovering small-molecule modulators of these protein targets. In order to create a differentiating advantage, many biotechnology companies chose to focus on target classes or even individual targets. This focus has several advantages, including allowing the company to develop a deep understanding of the biological and disease relevance of the targets, to focus medicinal chemistry and structural knowledge of the target(s), and to select from multiple therapeutic areas associated with the target class. **Table 1** reflects a partial list of such companies and their target class focus.

Table 1 Biotechnology companies focusing on a specific therapeutic target

Target class	Company	Therapeutic focus	Stock symbol
GPCRs	Arena[7]	Metabolic	ARNA
		CNS	
		Cardiovascular	
		Inflammation	
	Acadia[8]	CNS	ACAD
	ChemoCentryx[9]	Inflammation	Private
	Synaptic Pharmaceuticals[10]	CNS	Acquired by Lundbeck
	Affectis[11]	CNS	Private
Kinases	Vertex[12]	Cancer	VRTX
		Inflammation	
		Antivirals	
	Sugen[13]	Cancer	Now part of Pfizer (PFE-N)
	Signal[14]	Cancer	Acquired by Celgene (CELG)
		Inflammation	Acquired by Celgene (CELG)
	Ariad[15]	Cancer	ARIA
		Inflammation	
	OSI[16]	Cancer	OSIP
	Onyx[17]	Cancer	ONXX
	Ambit Biosciences[18]	Cancer	Private
		Stroke	
Nuclear receptors	Ligand[19]	Cancer	LGND
		Metabolic diseases	
	Xceptor[20]	Metabolic	Now part of Exelixis (EXEL)
	Tularik[21]	Cardiovascular	Now part of Amgen (AMGN)
		Metabolic diseases	
		Cancer	
	KaroBio[22]	Metabolic diseases	KARBF
Ion channels	Icagen[23]	Sickle cell disease	Private
		CNS	
	Hydra Biosciences[24]	Cardiovascular	Private
	Biofocus[25]	Pain	Private
	Scion[26]	CNS	Private
	Targacept[27]	CNS	Private
	Lectus[28]	Infectious disease	Private
		CNS inflammation	
	Vertex[29]	Cancer	VRTX
		Inflammation	
		Antivirals	
	Celltech[30]	Inflammation	Now part of UCB Group
		Airway disease	
Phosphodiesterases	ICOS[31]	Inflammation	ICOS
		Airway diseases	
		Erectile dysfunction	
	Inflazyme[32]	Inflammation	IZP-T
		Airway diseases	
	Memory[33]	CNS	MEMY-Q
Proteases	Khepri[34]	Inflammation	Acquired by Axys (now Celera) CRA-N
	Collagenex[35]	Bone diseases	CPGI-Q
	Chiroscience/Celltech[36]	Inflammation	Acquired by Celltech/UCB
	Vertex[37]	Cancer	VRTX
		Inflammation	
		Antivirals	

CNS, central nervous system.

2.19.4.1 G Protein-Coupled Receptors

It is estimated that 50% of current drugs and approximately 25% of the top 200 best-selling drugs modulate GPCRs.[38] Drugs targeting GPCRs treat numerous diseases, including cardiovascular, CNS, and gastrointestinal diseases, along with asthma/allergies and cancer. There are between 800 and 1000 GPCR genes, but endogenous ligands are only known for approximately 210 of them (see the GPCR database[39]).[40,41] Consequently, there is tremendous interest in exploring these orphan receptors and considering their use for future drug discovery efforts. Concomitant with the increase in the number of GPCR targets is recognition that GPCR responses to ligands can be complex, leading to the identification of classic agonists and antagonists, as well as inverse agonists and allosteric modulators.

One strategy that has aided the drug discovery process for GPCRs is the availability of compound libraries tailored to interacting with this class of proteins. These libraries were constructed based on a pharmacophore model for GPCRs, and many companies have their own proprietary focused libraries. There is also direct structural information for rhodopsin and a model for the neurotensin type 1 receptor bound to a peptide agonist that have been used to design libraries.[42] The latter may represent viable templates for pharmacophore-based searches of chemical libraries for nonpeptide ligands that might be of therapeutic value.

2.19.4.2 Protein Kinases

Kinase-induced phosphorylation of cellular proteins is a universal mechanism to control virtually all cellular processes, including cell growth and differentiation, metabolism, membrane transport and apoptosis. The protein kinase family is comprised of two major subfamilies: the protein tyrosine kinase (PTK) subfamily and the serine–threonine kinase subfamily. Targeting PTKs has already generated important drugs to treat cancer, including imatinib and erlotinib. More than 530 kinase-related sequences have been identified in the human genome, and three-dimensional information is available for many of them.[43] This has greatly facilitated the discovery of selective kinase inhibitors, since all protein kinases contain a structurally conserved catalytic domain. ATP binds in a cleft between the two major portions (N- and C-terminal) of the catalytic domain, and the majority of kinase inhibitors are ATP-competitive. The commonality of the ATP binding sites does create selectivity issues for the development of kinase inhibitors. However, there are several ATP site-directed and highly selective inhibitors that are being developed, and a number of kinase inhibitors are also under development in various cancers that modulate multiple kinases.

2.19.4.3 Ion Channels

Ion channels are ubiquitous pore-forming proteins that allow the passive diffusion of ions (such as Na^+, K^+, Ca^{2+}, and Cl^-) across cell membranes. Information on some of these proteins is collected in a target-specific database, the Ligand-Gated Ion Channel Database.[44] Numerous therapies modify ion channel function, and 15 of the top 100 best-selling drugs target ion channels and include antihypertensive Ca^{2+} channel blockers. The list includes amlodipine, nifedipine, and the antiepileptics such as the Na^+ channel blocker carbamazepine. These ion channel-modulating drugs were discovered serendipitously before the ion channel superfamily was appreciated. In humans, approximately 650 ion channels have been identified to date.

These targets have proved challenging, and have only recently benefited from higher-throughput electrophysiology techniques, as well as advances in structural knowledge of the proteins. Greater numbers of ion channel-targeted drugs are anticipated to enter clinical development as these technologies improve screening campaigns. Also, natural products (especially toxins from the spider, bee, scorpion, snake, puffer fish, and coral) have helped in classifying channel subtypes and probing their functions. A proprietary ion channel ligand design tool called Helical Domain Recognition Analysis has been developed by BioFocus, to facilitate small-molecule library design for this target class.

Additional target focused companies continue to emerge; example, ProScript (targeting the proteasome), in Cambridge, MA (now part of Millennium), Conforma (targeting heat shock protein 90) in San Diego, CA, MethylGene (targeting histone deacetylases) in Montreal, Idun (targeting the caspase cysteine protease family) in San Diego (now part of Pfizer), and the Coley Pharmaceutical Group (targeting toll-like receptors) in Wellesley, MA. These and other biotechnology companies are hoping they have chosen targets that will soon be clinically validated by small-molecule modulators.

2.19.4.4 Intellectual Property (IP) Landscape

In order to work on targets or target classes it is essential to clarify the IP landscape. Gene sequence patents can provide incentives crucial to downstream investment, and have been central to the development of the biotechnology industry. According to a recent review, 20% of human genes are claimed as US IP.[45]

This represents 4382 of the 23 688 genes in the National Center for Biotechnology Information gene database. Major patent assignees include the University of California, ISIS Pharma, Human Genome Sciences, and Incyte Pharmaceuticals/Incyte Genomics. Much of the genome is unpatented, but some genes have numerous patents asserting rights to gene uses, cell lines, and constructs containing the gene.

2.19.5   Companies Built around Compound Classes or Particular Chemistries

2.19.5.1   The Example of IMiDs Compounds

Thalidomide (α-(N-phthalimide)glutarimide) was marketed in Germany in the mid-1950s as a safe, nonbarbiturate sedative – a so-called pure hypnotic – that was also an effective antiemetic in pregnancy. Unfortunately, the drug caused birth defects, and it was withdrawn in 1961. The compound was later found to be highly efficacious in patients with erythema nodosum leprosum (ENL), a potentially life-threatening inflammatory complication of leprosy. Thalidomide was only given US Food and Drug Administration (FDA) approval for the treatment of acute ENL in 1998 after additional immunological properties were discovered.

Apart from its approved use in ENL, thalidomide has been evaluated in over 150 clinical trials in the USA for various oncological, dermatological, and inflammatory conditions. In addition to the teratogenic liability, thalidomide treatment is associated with somnolence, constipation, rash, peripheral neuropathy, and deep vein thrombosis.[46] As a consequence, Celgene has synthesized thalidomide analogs that are structurally similar but functionally different from the parent compound. Lenalidomide has been approved by the FDA for the treatment of transfusion-dependent anemia due to low- and intermediate-1–risk myelodysplastic syndromes associated with a deletion 5q cytogenic abnormality with or without additional abnormalities. Two other IMiDs compounds (proprietary compounds having immunomodulatory properties) are CC-4047 and CC-11006, which are in clinical trials for various diseases (**Figure 1**).

While these compounds have several biological activities (notably lowering tumor necrosis factor α, antiangiogenic and T cell co-stimulatory properties, and direct antitumor activity including pro-apoptotic activity), the precise mechanism of action (MOA) of the IMiDs compounds is not known. Consequently, the development of these compounds is rather unusual in the modern day, since the determination of MOA may be made after the drugs have proven to be safe and effective. However, in much of the history of pharmaceuticals, the MOA was proven after marketing, or even not fully understood to this day (e.g., penicillin, methotrexate, and aspirin).

Clinical activity for the IMiDs compounds has been observed in various hematological and solid tumor cancers, including multiple myeloma, myelodysplastic syndrome, renal cell carcinoma, pancreatic cancer, and androgen-independent prostate cancer. The FDA is currently reviewing an application for the approval of thalidomide for newly diagnosed multiple myeloma. Lenalidomide is also being reviewed by the FDA for the treatment of relapsed or refractory multiple myeloma. Additional clinical studies are being conducted on a variety of other hematological diseases and solid tumor cancers. The resurrection and rehabilitation of thalidomide and the development of other IMiDs compounds has provided Celgene with a drug chemistry and IP platform for major unmet medical conditions.

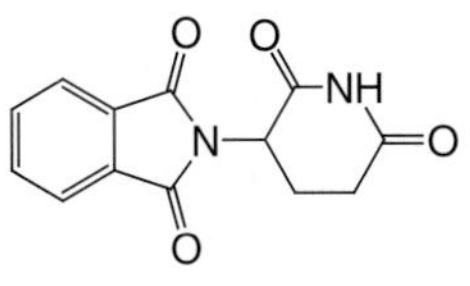

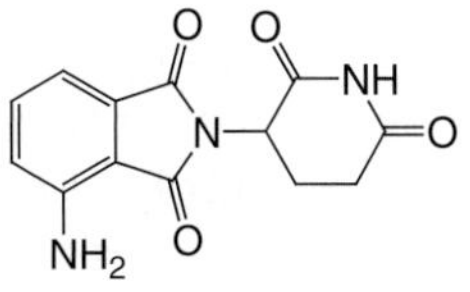

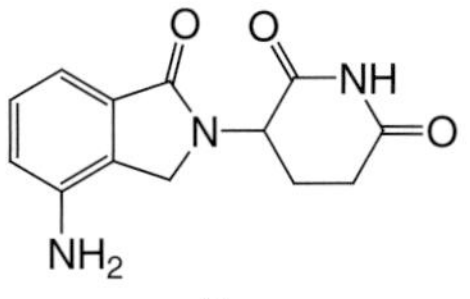

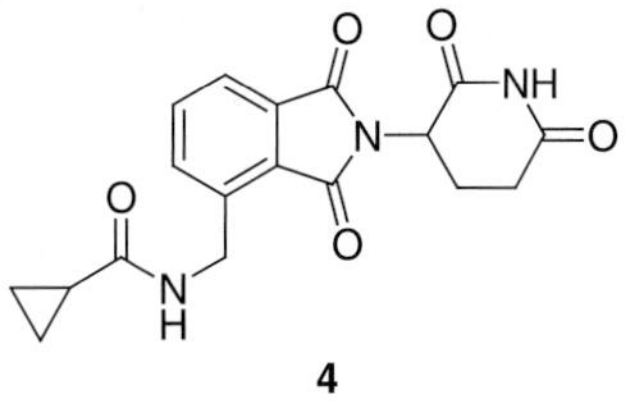

Figure 1   IMiDs structures. **1**, thalidomide; **2**, CC-4047; **3**, lenalidomide; **4**, CC-11006.

Table 2 Select combinational chemistry companies

Company	Location	Stock symbol
Trega[48]	San Diego, CA	Acquired by Lion (LIOG)
Pharmacopeia[49]	Princeton, NJ	PCOP
Alanex[50]	San Diego, CA	Acquired by Agouron now Pfizer (PFE)
Arqule[51]	Cambridge, MA	ARQL
Array[52]	Boulder, CO	ARRY
CombiChem[53]	San Diego, CA	Was part of Dupont then BMS (BMY)
Discovery Partners International[54]	San Diego, CA	DPII
Albany Molecular Research[55]	Albany, NY	AMRI
3-Dimensional Pharma[56]	Exton, PA	Acquired by J&J (JNJ)
Chembridge Corp[57]	San Diego, CA	Private

2.19.5.2 Strategies for Chemical Libraries

Combinational chemistry, born in the early 1990s, promised to revolutionize drug discovery, allowing even small biotechnology companies access to chemical libraries from thousands to millions of compounds. Many of the companies listed in **Table 2** that originally generated libraries that were licensed to pharmaceutical and biotechnology companies have changed their business models, and now perform their own drug discovery.

There was considerable interest in accessing chemical libraries with large numbers of members in order to increase the likelihood of obtaining a hit against any target screened. These combinational screening libraries, while very large (often in the millions), did not appear to increase successful identification of drug candidates, unfortunately. This may be attributed to their lack of true diversity (or at least pharmaceutically relevant diversity) and the manner in which they were synthesized. It is estimated that the number of theoretically possible small molecules is approximately 10^{60}, which may be lowered to 10^{28} by applying constraints of druglikeness and synthetic capability.[47] This is vastly greater than the largest compound collections (in the millions) and the largest virtual library (10^{12}). The diversity of many of the early combinatorial libraries was further hampered by having been synthesized on a solid support, or by being too peptide-like (and thus unlikely to be druglike, at least insofar as oral bioavailability and other drug characteristics are concerned). This limited the types of chemistries that could be used to create members of the library, thereby decreasing the diversity of the library. Another problem with the design of these early libraries was that the compounds made were never purified. Once a particular well of a library was identified as a hit, an attempt was made to purify the active component. Typically, the compound that was supposed to have been in the well comprised less than 40% of the material present. Often, the desired compound was not present at all. This model did not turn out to be very effective in most laboratories (with a few notable exceptions, e.g., Affymax, Chiron, and Pharmacopeia), which is why the industry (and the library suppliers) have moved away from libraries synthesized in this manner.

In an effort to increase the hit rate of HTS, and to find better-quality leads, library synthesis efforts have moved to a new model. In this model, libraries are generated based on known pharmacophores to the different target classes. These focused (or targeted libraries) are built around knowledge of target classes or individual hits. For example, as mentioned above, GPCR libraries and kinase libraries have been constructed.[58]

2.19.5.3 Structure-Based Drug Design

Devising new ways to identify chemical leads for specific protein targets has been a constant challenge for the biotechnology and pharmaceutical industries. Since many of the targets are primarily enzymes, x-ray crystallography has been increasingly used to accomplish this goal. Originally a slow and tedious process, several biotechnology companies have embraced the challenge and created HTS-based drug design platforms (**Table 3**). These have been particularly useful to identify kinase inhibitors.

Fragment-based lead discovery has also been developed whereby low molecular weight (MW) compounds (typically MW < 200) are screened against a target at high concentrations (~ 1 mM). The initial hits have very low affinities, but

Table 3 Structure-based drug design companies

Company	Location	Status
Agouron[67]	San Diego, CA	Acquired by WL/PD (now Pfizer)
Syrrx[68]	San Diego, CA	Acquired by Takeda
SGX Pharmaceuticals[69]	San Diego, CA	Private
Vertex[70]	Cambridge, MA	VRTX
Plexxikon[71]	S. San Francisco, CA	Private
Astex[72]	Cambridge, UK	Private

with the aid of x-ray crystallography or macromolecular structure determination by nuclear magnetic resonance spectroscopy, these low-affinity hits can be rapidly improved in potency. A typical nonfragment-based screening lead could have a MW > 350. Often, in order to improve the potency of this type of lead, functional groups will have to be removed (causing a decrease in potency) then others added to give the final compound. The fragment approach starts with the low-affinity hit, and a bound structure of the hit (or hits) in the target protein. This approach removes one portion of the synthetic cycle (i.e., having to remove unnecessary functional groups on the screening lead). It is becoming apparent that the speed to a final compound can be quicker with the fragment-based approach due to removal of excess synthetic design cycles.[59]

2.19.5.4 Ribonucleic Acid (RNA)-Based Strategies (ISIS[60] and Sirna[61])

Since the control of protein expression by antisense RNA was first described,[62,63] a number of different RNA-based therapeutics have been used in clinical trials. The first and only antisense RNA therapeutic to receive FDA approval and enter the market is fomivirsen (ISIS Pharmaceuticals), which is approved for cytomegalovirus retinitis.[64,65] There are three different mechanisms by which RNA-based inhibitors work, involving anti-sense RNA, ribozymes, and siRNA. New biotechnology companies have been started around each of these technologies, and a number of modified oligonucleotides have entered clinical trials (see the TriLink Biotechnologies website for an article listing many of the ongoing clinical trials with oligonucleotide-based therapeutics).[66] Two companies are good examples of the history of the different RNA interference strategies, ISIS Pharmaceuticals and Sirna Therapeutics (formerly Ribozyme Pharmaceuticals).

ISIS was founded to exploit the new field of RNA antisense technology in 1989. Over the years since its founding, ISIS has engaged in collaborations with over two dozen different entities, including academic institutions, biotechnology companies, and large pharmaceutical companies. The first generation of antisense therapeutics encountered physiological problems that included an innate immune response caused by certain dinucleotide sequences in the oligonucleotide.[65] The first generation of antisense RNA oligonucleotides incorporated phosphorothioate linkages between the bases (**Figure 2**) to prevent nuclease digestion in vivo.

Fomivirsen falls into the phosphorothioate class of oligonucleotides. ISIS had also been developing a number of other 'first-generation' antisense compounds, and the latest data from the Investigational Drugs database[73] indicates that nearly all of them have been discontinued.

The second generation of antisense oligonucleotides focused on increasing the stability of the compound and decreasing the immune response. The new compounds had a mixture of phosphorothioate linkages and ribonucleosides containing $2'$-OCH_3 modifications at the $5'$ and $3'$ ends. Several of these mixed-backbone oligonucleotides have entered clinical trials, with two of them having advanced as far as Phase II (GEM 231 from Hybridon and MG98 from MethylGene and Hybridon).

Sirna based their early technology on the ability of ribozymes (stretches of oligonucleotide that can catalytically cleave their target messenger (mRNA)) to inactivate mRNAs resulting from gene transcription. Three ribozymes were taken into Phase I clinical trials by Sirna in either cancer indications (RPI-4610, Angiozyme, and Herzyme) or hepatitis B infections (LY 466700). Based on the data from these Phase I trials, Sirna discontinued development of all three compounds and switched their focus to RNAi technologies (see below).

The most recent technology for regulating translation of mRNA to protein is called RNAi. This technology differs from the two previous technologies to silence RNA in that it is a natural process in many eukaryotic species and uses double-stranded oligonucleotides instead of single-stranded oligonucleotides.[74] The sequence of the RNAi is still

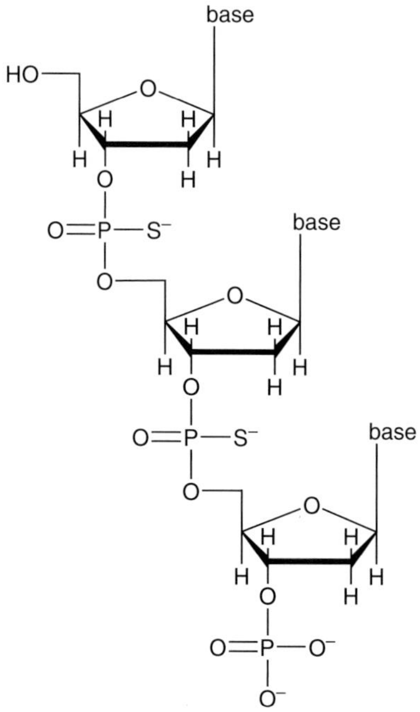

Figure 2 Phosphorothioate oligodeoxynucleotide.

complementary to the target sequence, but since it is double-stranded material it can either be delivered to the cell directly or by the use of viruses encoding the desired sequence.[75] The RNAi becomes part of a silencing complex that includes several proteins. Once the RNAi finds and binds to its targeted sequence, the mRNA is rapidly degraded.

A number of companies are using RNAi technology to move therapeutics into the clinic. As mentioned previously, Sirna discontinued their work on commercializing ribozyme therapeutics and began working on drugs based on RNAi. They currently have one compound in Phase I clinical trials for macular degeneration (Sirna-027).[75] Even though the technology is significantly newer than antisense oligonucleotides or ribozymes, there are approximately 10 products in late preclinical/early clinical development. These products come from about eight different biotechnology companies.

To date, there has been very little success with developing therapeutics based on disrupting the translation of mRNA to protein via nucleic acid technologies. Even the latest technology (RNAi) will have to overcome similar hurdles that have been faced by the older technologies (e.g., short in vivo half-lives, dosing directly to affected tissue, and potential immune response). For RNAi that will be delivered virally, there is the added burden of the lack of success with other types of gene therapy trials. What was hoped to be a 'magic bullet' for disease modulation has not fulfilled its promise to date, but has encountered similar levels of success as disease modulation with small, organic compounds.

2.19.6 Companies Built with a Therapeutic Focus

Two of the largest biotechnology companies in the world (Genentech[76] and Amgen[77]) had very similar beginnings. Both companies started out by trying to harness new genetic engineering technologies for making therapeutically important human proteins in other organisms or cells on industrial scales. Genentech was the first biotechnology company to get a product to market when it showed that human insulin could be made via recombinant DNA technology in bacteria. The project was licensed to Eli Lilly, which made the protein on an industrial scale and got it approved for marketing in 1982.[76]

Amgen was started in 1980, also with the goal of making human therapeutic proteins. During the 1980s, Genentech and Amgen pioneered efforts to switch from producing proteins in bacteria to producing them in mammalian cells. This was an important improvement in producing therapeutic proteins. Mammalian cells chemically modify many of the proteins that they make, which can alter their overall three-dimensional structure and/or the activity of the protein. Many of these modification pathways are not available in bacteria.

Both of these early biotechnology companies matured from their focus on applications of new technology to production of human therapeutics. Amgen and Genentech came by their therapeutic foci as a result of their early efforts to produce therapeutic proteins. Amgen's initial success was with epoetin alfa and filgrastim, two products that have an effect on blood function. Both of these products each had 2004 revenues in excess of US $2.5 billion. One of the indications for epoetin alfa is anemia induced by chemotherapy. For filgrastim, the primary indication is to boost white blood cell counts, to prevent infection in patients undergoing chemotherapy treatment. Building on their experience in the area of modulating blood cell function (including white blood cells), Amgen expanded their efforts in targeting immunological diseases with the purchase of Immunex.[78]

Genentech's current focus is on cancer, immunology, and vascular biology.[79] The company's expertise in producing and marketing proteins such as human growth factor and insulin has allowed it to expand into producing therapeutic antibodies for various cancer indications. Genentech currently either solely or with other companies markets a number of therapeutic proteins, including bevacizumab, trastuzumab, and rituximab in oncology and efalizumab in immunology.

ILEX, on the other hand, started with a focus in oncology. The company grew out of the nonprofit Cancer Therapy and Research Center in San Antonio, TX, and was formerly spun off into a for-profit enterprise in 1994. The founders felt that 'there was a big, unmet medical need' in cancer.[80] Prior to its acquisition by Genzyme, the company was split into two businesses. One half of the business was devoted to drug development services in oncology while the other subsidiary was devoted to discovering novel drugs for the treatment and prevention of cancer.

ILEX placed into trials and marketed alemtuzumab (a monoclonal antibody for the treatment of chronic lymphocytic leukemia), and was completing trials with clorfarabine for acute leukemia in children[81] when Genzyme bought the company in 2004 in order to strengthen its oncology portfolio.

The examples sited above show two different models by which companies have brought focus to their drug discovery efforts. Two of the oldest biotechnology companies (Amgen and Genentech) matured into a narrow focus in two or three narrow therapeutic areas. ILEX, on the other hand, began with a narrow focus, and continued to guide the company with that focus. There are other companies that have used this model successfully as well, one of which is Esperion, focusing on atherosclerotic diseases (see below).

2.19.7 Challenges to Moving Compounds into the Clinic

There have been a number of reviews written in the past several years indicating that there has been a decline in the number of new chemical entities being approved by the FDA for marketing.[82–84] There have been almost as many suggestions on how the drug discovery process can be fixed.[82,83,85–89] Drugs for oncology have one of the lowest success rates from first-in-human to registration of any of the therapeutic areas.[83,90] Many companies launched efforts into the development of oncology drugs, attempting to respond to the great unmet medical need. Additionally, the severity of the disease would require shorter clinical trials, and the tolerance for side effects in this disease is greater than for other therapeutic areas. However, many of these efforts have foundered on lack of sufficient efficacy and unacceptable side effects.

A number of concepts are beginning to mature, facilitated by improving technologies that are predicted to increase the success rate of compounds moving into clinical testing. The concept of biomarkers has become almost universal, and returns to center stage the study of pharmacodynamics. During the drug discovery phase, increased effort and emphasis has been placed on developing assays that will allow the determination of target inhibition in a functional assay. A simple example is to monitor the phosphorylation of a kinase substrate within the cell environment after exposure to the respective kinase inhibitor. Elaboration of this assay to in vivo and then the clinical setting would allow monitoring of target inhibition following dosing. These types of assays will allow the assessment of the magnitude, duration, and consequence of inhibition, leading to a more rapid conclusion of whether inhibition of the target will have therapeutic benefit. However, often with such assays, translation from discovery to development can be difficult, and alternatives need to be considered. The increasing use of gene chip arrays has prompted some to consider global gene expression of some readily sampled tissue in response to treatment as an alternative. Moving to this type of assay should more generally be considered a surrogate of response, rather than a marker of inhibition. Additionally, as diseases are becoming more molecularly defined, both in etiology and progression, the opportunity for more specific therapies is presented. A recent example of this would be imatinib, a compound approved for CML.[83] Ninety-five percent of CML patients have a particular chromosomal translocation that creates a mutant tyrosine kinase called bcr-abl. Imatinib was developed to target this particular mutation. Looking to the future, the potential of defining a disease by the molecular pathology and employing specific antagonists to create a 'tailor-made' therapy can be imagined.

The major difficulty with any new technologies introduced into drug discovery is the long time frames before the results are realized by new drugs on the market.[91] There are two facets of drug discovery where improvements in the individual processes would speed the approval process and decrease the amount of money required to bring drugs to market. The first area where improvements need to be made is in the overall attrition rates of compounds from first-in-human to approval. Recent estimates put the overall success rate for new chemical entities at 11%.[1,83,90] By reducing attrition rates, the expense for the whole process would decrease due to failures in clinical trials, which are the most expensive part of the discovery process.[91]

The second area that would improve the overall process is by reducing the time that drugs spend in the R&D process. This would potentially reduce the overall amount spent on any single drug, giving it a longer patent protection period once it entered the market. Some estimates indicate that about 40% of all drugs entering clinical trials fail due to poor pharmacokinetic/pharmacodynamic properties.[92] Being able to get an early indication if a new chemical entity is being extensively metabolized will decrease the amount spent on further clinical development. Micro-dosing (Phase 0) pharmacokinetic studies could be done in healthy volunteers, at subtherapeutic doses of compound labeled with an appropriate tracer. With the increased sensitivity of mass spectroscopic detection, levels of compound in the blood and even metabolites of the parent compound can be detected.[93] This process will also allow comparisons of two pharmacologically similar compounds in humans in order to determine which compound would have the better metabolic profile,[92] and should lower attrition rates due to poor pharmacokinetic/pharmacodynamic parameters discovered in Phase I safety testing.

An early determination of the efficacy of a drug in clinical testing would be a major benefit in the design of subsequent trials. The ideal test for early efficacy would be minimally invasive and have a high correlation with disease remediation. This desire has spawned the field of biomarker discovery and utilization in the discovery and clinical testing of new therapies. A recent review by Frank and Hargreaves[94] defines the different types of biomarkers that are used, and covers examples of direct and indirect markers of disease/compound efficacy.

In the most straightforward case, the measurement that identifies patients requiring therapy is the same measurement made to monitor the results of the therapy. An example of this would be plasma cholesterol levels, which indicate which patients should take cholesterol-lowering therapies. This is the same measurement that is made to indicate the success of the drug.[94] Not all cases are this simplistic, however. Many diseases do not have a biomarker that can be measured directly. This is especially true of diseases such as Alzheimer's disease, where affected tissue in a living patient is not accessible. In these cases, surrogate endpoints are required.[93] A surrogate endpoint is a biomarker, but not one that is directly linked to the therapeutic outcome. An example of this would be the effect that a kinase inhibitor has on the phosphorylation status of the protein substrate for that kinase. The hypothesis being tested in the clinic, then, is that inhibition of the particular kinase correlates with inhibition or remission of the disease. An example of this would be imatinib and its efficacy in CML. The surrogate biomarker would then be inhibition of bcr-abl activity in patient blasts.[89,95]

Whether direct or indirect, the biomarker field could have a positive impact on the speed and course of clinical trials. The power of clinical trials will be enhanced for any particular therapeutic if patients are preselected for the trial based on the presence or absence of a relevant biomarker. Most likely, these biomarkers will be mechanism based and be responsive to the compound being administered. The correlation between the biomarker level and disease response will need to be addressed prior to the beginning of the clinical trial for a particular compound. The regulatory agencies (such as the FDA) will have to accept that the mechanistic biomarker is a surrogate for an effect on the clinical response of the patient. A clinical trial could be ended early if the compound, when dosed in humans, does not modify the biomarker. This will permit less money to be spent on compounds that do not have a chance of succeeding. However, clinical trials may need to enroll fewer patients, since those only likely to respond to the therapy will be selected for inclusion in the trial. This has the potential to reduce the overall cost of bringing a compound to market.

2.19.8 Models for Start-Up Companies

If one were to start a biotechnology/drug discovery company today, there are several different models that could be used. The classical biotechnology model starts with target discovery, and progresses through drug discovery to compounds in the clinic. A second model for a start-up company is one where a project (along with a compound or compounds and the scientists involved in the project) is spun out of a large pharmaceutical company. A third model is a virtual company that is built with the intention of being sold as soon as there is sufficient clinical data to indicate a reasonable chance of success. In the current funding environment, each model has its plusses and minuses.

2.19.8.1 Classical Biotechnology with a Twist

Over the years, many companies have been started based on a molecular target class or a technology. Examples of the former include Exelixis, Plexxicon, and Arena. Examples of the latter include the structural biology companies such as Structural Genomics (now SGX Pharmaceuticals) and Syrrx, and the genome companies such as Millennium, Human Genome Sciences, and Curagen. The company of today would have to have equal emphasis on chemistry as well as on biology. It would be built from the ground up, but instead of spreading itself broadly across a target class, it would go deep into a particular region of biology. An example of this would be a particular cell lineage that is known to be involved in many different disease states (such as a monocyte/macrophage). Another example is a particular signaling pathway that has been shown clinically to be involved in a particular disease state. This intense focus will allow the company to drive deep into the biology of the target cell/pathway, and create a significant body of knowledge in a relatively confined area of biology.

The second requirement would be to have medicinal chemistry on board at the inception of the company. Many biotechnology companies have been started with a strong biology focus and have added chemistry much later in the game. With the tight funding for start-up companies, the process of drug discovery needs to move more quickly, and bringing trained medicinal chemists on at company inception will promote this speed. The remainder of the organization needs to be built with an eye to products that will be entering the clinic 3–4 years from now. An innovative preclinical pharmacology group that can bring on-line novel animal disease models that utilize specific biomarkers to track compound effectiveness is helpful. A translation medicine group would become involved with the project early on, in order to examine how preclinical biomarkers could be used for first-in-human studies (even Phase 0 work). The speed with which a pharmacokinetic/pharmacodynamic relationship can be established during the development and clinical testing phase of an experimental drug will have a direct impact on stopping work on compounds early in the process and saving money in the long run. Venture capitalists these days are looking for major clinical milestones in order to fund another round for a new company. They also argue for focus, though this can work against a company if the focal area is not successful. Indeed, companies need to diversify their portfolio of projects, much as portfolio managers need to diversify their holdings.

Once proof-of-principle has been determined for a particular compound, collaborating with a company that has the capabilities to move the compound all the way to market is paramount. This will require business development staff who can determine the worth of a particular compound in light of its fit in the marketplace. The business development staff ideally should have experience from working in a company that has taken projects all the way through the clinic and onto the market. They will have a good appreciation of what is entailed in going to the next steps in the process and have sufficient contacts in the industry to open doors. Crafting the deal must be on realistic terms due to the difficulty in getting funding from a larger partner. Realistic expectations on both sides of the negotiating table will go a long way to speed up the process of getting the compound tested and into the market.

2.19.8.2 The Spin Cycle

Over the past decade, large drug discovery operations have been spinning off noncore areas of research. This has led to significant opportunities for starting up new companies that may have projects close to the clinic. One of the examples of this would be Esperion Pharmaceuticals, which was started by a group of scientists from the Atherosclerosis Department (including the former director of the department) at Warner-Lambert/Parke-Davis (WL/PD, now part of Pfizer) in the 1990s. The team that started Esperion had all taken part in the successful discovery program that resulted in the new statin atorvastatin.[96] As a class, the statins work to lower low-density lipoprotein levels, and hence lower 'bad cholesterol' in the body. The focus of Esperion was on raising the levels of high-density lipoprotein (HDL, or 'good cholesterol'). The team licensed a project from Pharmacia-Upjohn, with Pharmacia-Upjohn retaining an exclusive option for co-development and marketing outside the USA and Canada. The project focused on the development of a protein that had been shown to promote rapid cholesterol removal from atherosclerotic plaques.[97]

The licensed technology (ApoA1-Milano or ETC-216) was not a high priority for Pharmacia-Upjohn. Additionally, the effect of artificially raising HDL levels on atherosclerotic plaques was still speculative. The founders of Esperion, however, had extensive experience in atherosclerosis, and were willing to take a chance on developing a protein therapeutic. ETC-216 was in Esperion's main area of focus (i.e., modulating cholesterol levels and effecting atherosclerotic lesions). Esperion was founded in 1998, and by 2003 was publishing results of a Phase II clinical trial with ETC-216.[98] Esperion also was developing a pipeline of other therapeutics for atherosclerosis and cardiovascular disease.

The published Phase II results for ETC-216 were very impressive, showing a statistically significant reduction in atherosclerosis after 6 weeks of treatment. The results were impressive enough that it generated interest in Esperion from larger companies. Finally, at the end of 2003, Pfizer agreed to buy Esperion for US $1.3 billion.[99] This brought several of the founders full circle, having left WL/PD years before it was acquired by Pfizer, then to have Pfizer bring them back into the company after showing potential success with a new therapeutic protein. The circle was fully closed on Esperion, since Pfizer also bought Pharmacia-Upjohn, from whom they had licensed the technology from in the first place.

The above example illustrates a number of advantages of building a company along this model. There are other examples of companies that have been started with a similar model, such as Barrier Therapeutics (whose founders worked together at Johnson & Johnson) and Actelion (whose founders worked together at Hoffmann-La Roche). The first (and one of the most important) is that these companies were started by groups of scientists that had been working together already. They did not have to spend as much time integrating the initial personalities, as happens with many start-up companies. Another advantage of this model is that the company was started with a fairly advanced project that had already been of interest to a major pharmaceutical company. The scientists themselves were already familiar with the steps necessary for putting a therapeutic into development, since many of them had been on development teams during their careers.

The model here, then, is one that relies on the current efforts at focusing the research areas at major pharmaceutical companies. This results in significant talent being let go by these large companies, many of whom are well schooled in the drug discovery process. Instead of research teams going their separate ways, however, there could be a big advantage to keeping the team together and starting a new company. The team would have to choose the project to work on with care (the more advanced, the better). The project could be one that the large company does not feel would be significantly profitable for them to pursue but could be profitable for a smaller organization or may not be in the large company's core area of marketing or development expertise.

Since the project that has been licensed will be more advanced than starting with a therapeutic target from the beginning, the infrastructure that needs to be built will be different as well. More effort should be devoted to building up a development organization (late-stage preclinical to first-in-human) than to building up the early discovery organization. Concentration on an advanced project will make raising venture capital easier, at least in today's funding environment, since the investors will be potentially able to realize a quicker exit strategy.

Once a compound or therapeutic is advanced enough to give a strong indication of success (through the end of Phase II), the rest of the organization can be backfilled. This would be done in the opposite way to a company that was starting with a target or technology but no compound. In vivo pharmacology and pharmacokinetic/drug metabolism would be the next departments to be staffed. This would allow the company to begin looking for other indications where the clinical compound may be of use. The focus would still be on the program that is in the clinic, but the thought process of building an early discovery organization would begin. Questions that would need to be addressed would include the therapeutic area for the company to concentrate on (if there is a compelling reason to change from the direction it started with), small-molecule discovery versus biological therapeutics, the target class to pursue, and how many programs should be run at one time. The goal down the road would be to become a fully integrated drug discovery operation, and potentially have a revenue stream with which to build the company.

2.19.8.3 The Virtual Company

In this model, the long-term up-front costs are kept low. Most of the work is outsourced to contract research organizations (CROs), with a few specialists on staff to interpret the data. There is some question as to whether this model is sustainable in the long run, however. The virtual company can be geared to take a candidate compound that has failed in one indication and push it forward in another indication that may have a lower return but could be quicker into the clinic. The second goal would be to give investors a relatively quick return on their investment. The exit strategy would be aimed toward getting sufficient proof of concept for a compound in order to make it interesting to a potential acquirer. By outsourcing most of the research, fixed costs could be kept low. The exchange for this reduction in fixed costs, however, is loss of control of the project. The company is dependent on time-lines from the CROs, and a significant amount of time is spent looking for CROs that can reliably deliver high-quality data.

The advantages of operating a biotechnology company with a small internal staff of experts with most of the data being generated by CROs was beginning to be touted in the early 1990s.[100] In this scenario, projects would be in-licensed, then the company would act as a group of study directors to obtain the necessary data for presentation to the FDA. Taken to the extreme, this type of operation would generate no data internally, but collect and collate the data generated by other organizations. While this may increase an organization's flexibility, there is no a priori reason why it should increase the success rate of bringing drugs to market.

Perhaps a more workable term that encompasses different degrees of outsourcing is the extended enterprise.[101] Nearly all current biotechnology companies, as well as large pharmaceutical companies, engage in some form of an extended enterprise arrangement. Specific portions of the process are outsourced, due to either time constraints of building the effort internally or access to technologies/intellectual property that are proprietary to another company. An article by Cavalla[101] lists some of the advantages and disadvantages of the extended enterprise model. Major disadvantages of this model would have to be the loss of control of the time-lines of the project, and difficulties in managing external resources. For access to new projects or promising compounds, there is significant competition in the marketplace currently. Large pharmaceutical companies have long had very active licensing operations. With the current downturn in the rate of new chemical entity approvals, the search for quality external programs has increased. This makes it difficult for a small company to compete at in-licensing discovery stage projects/compounds for eventual clinical trials.

An example of this type of company would be Salmedix, located in San Diego, CA (now part of Cephalon). The company has identified several oncology compounds that had been launched overseas but not in the USA or had been in clinical trials previously and failed. They are taking the compounds into trials designed to achieve FDA registration, and are partnering with other organizations to perform the clinical trials.

2.19.8.4 The Bottom Line

If a biotechnology company was being started now, in order for it to be successful it would have to combine portions of all three of the above models. The company should focus its efforts in one particular therapeutic area.[102] This focus will allow the staff of the new company to drill deep into the biology and chemistry of the therapeutic area. Even further focus in one part of the biology of the therapeutic area would be even better (such as focusing only on preventing atherosclerotic plaque build-up instead of all cardiovascular diseases). Instead of looking to license a project or compound from another small biotechnology company, it should look to license in a project from large pharmaceutical company that is outside of its expertise or does not have the potential return on investment for a large company.

Probably the quickest way for the new company to move forward would be to hire a research and development group that has already worked extensively together. Several of the members should have experience bringing compounds through development and into clinical trials. With the recent down-sizing that has happened in large pharmaceutical companies, there are not only compounds/projects available, but also even whole research teams. This process has two advantages. The first is that the project is one that a large company had initially thought worth pursuing, and the second is that the company starts with a seasoned research team. The further along that the licensed project/compound is in the discovery/development process, the potentially shorter time will be required to reach a decision on its feasibility.

Finally, by starting with a later-stage project, the new company will focus on hiring individuals with late-stage preclinical, early clinical experience. Any studies that need to be performed that require expertise in earlier parts of the drug discovery process can initially be farmed out to a CRO. As the company matures (i.e., a potential payoff from a compound making it to market is looming), the discovery stage staff can be brought on-board to begin to build an early pipeline.

With the current mood in the venture capital community of seeking more immediate returns on their investment, the above model has the potential to succeed. The new company must continue to seek more efficient methods to get early reads on clinical compounds. This can be done either through Phase 0 microdosing trials to obtain an early read on pharmacokinetic/drug metabolism issues or early proof-of-principal studies using relevant disease biomarkers to give an early indication if the therapy has a chance to succeed. Work in this area could potentially limit the cash drain of a compound/therapeutic that does not have a reasonable chance of success. This process could also help to raise additional capital for the company if these small-scale trials add data that speak to the chance of the therapeutic succeeding.

References

1. Booth, B.; Zemmel, R. *Nat. Rev. Drug Disc.* **2004**, *3*, 451–456.
2. Ma, P.; Zemmel, R. *Nat. Rev. Drug Disc.* **2002**, *1*, 571–572.
3. Hopkins, A. L.; Groom, C. R. *Nat. Rev. Drug Disc.* **2002**, *1*, 727–730.
4. Knowles, J.; Gromo, G. *Nat. Rev. Drug Disc.* **2003**, *2*, 63–69.
5. Bailey, D.; Zanders, E.; Dean, P. *Nat. Biotechnol.* **2001**, *19*, 207–209.
6. Drews, J. *Science* **2000**, *287*, 1960–1964.
7. Arena. http://www.arena.pharm.com (accessed April 2006).
8. Acadia. http://www.acadia-pharm.com (accessed April 2006).
9. ChemoCentryx. http://www.chemocentryx.com (accessed April 2006).
10. Synaptic Pharmaceuticals. http://www.lundbeck.com (accessed April 2006).
11. Affectis. http://www.affectis.com (accessed April 2006).
12. Vertex. http://www.vrtx.com (accessed April 2006).
13. Sugen. http://www.pfizer.com (accessed April 2006).
14. Signal. http://www.celgene.com (accessed April 2006).
15. Ariad. http://www.ariad.com (accessed April 2006).
16. OSI. http://www.osip.com (accessed April 2006).
17. Onyx. http://www.onyx-pharm.com (accessed April 2006).
18. Ambit Biosciences. http://www.ambitbio.com (accessed April 2006).
19. Ligand. http://www.ligand.com (accessed April 2006).
20. Xceptor. http://www.exelixis.com (accessed April 2006).
21. Tularik. http://www.amgen.com (accessed April 2006).
22. KaroBio. http://www.karobio.com (accessed April 2006).
23. Icagen. http://www.icagen.com (accessed April 2006).
24. Hydra Biosciences. http://www.hydrabiosciences.com (accessed April 2006).
25. Biofocus. http://www.biofocus.com (accessed April 2006).

26. Scion. http://www.scionresearch.com (accessed April 2006).
27. Targacept. http://www.targacept.com (accessed April 2006).
28. Lectus. http://www.lectustherapeutics.com (accessed April 2006).
29. Vertex. http://www.vrtx.com (accessed April 2006).
30. Celltech. http://www.ucb-group.com (accessed April 2006).
31. ICOS. http://www.icos.com (accessed April 2006).
32. Inflazyme. http://www.inflazyme.com (accessed April 2006).
33. Memory. http://www.memorypharma.com (accessed April 2006).
34. Khepri. http://www.celera.com (accessed April 2006).
35. Collagenex. http://www.collagenex.com (accessed April 2006).
36. Chiroscience/Celltech. http://www.celltechgroup.com, http://www.ucb-group.com (accessed April 2006).
37. Vertex. http://www.vrtx.com (accessed April 2006).
38. Behan, D. P.; Chalmers, D. T. *Curr. Opin. Drug Disc. Dev.* **2001**, *4*, 548–560.
39. Molecular Class-Specific Information System (MCSIS) Project. http://www.gpcr.org (accessed April 2006).
40. Lander, E. S.; Linton, L. M.; Birren, B.; Nusbaum, C.; Zody, M. C.; Baldwin, J.; Devon, K.; Dewar, K.; Doyle, M.; Fitzhugh, W. et al. *Nature* **2001**, *409*, 860–921.
41. Venter, J. C.; Adams, M. D.; Myers, E. W.; Li, P. W.; Mural, R. J.; Sutton, G. G.; Smith, H. O.; Yandell, M.; Evans, C. A.; Holt, R. A. et al. *Science* **2001**, *291*, 1304–1351.
42. Luca, S.; White, J. F.; Sohal, A. K.; Filippov, D. V.; van Boom, J. H.; Grisshammer, R.; Baldus, M. *Proc. Natl. Acad. Sci. USA* **2003**, *100*, 10706–10711.
43. Manning, G.; Whyte, D. B.; Martinez, R.; Hunter, T.; Sudarsanam, S. *Science* **2002**, *298*, 1912–1934.
44. Ligand-Gated Ion Channel Database. http://www.ebi.ac.uk/compneur-srv/LGICdb/LGICdb.php (accessed April 2006).
45. Jensen, K.; Murray, F. *Science* **2005**, *310*, 239–240.
46. Ghobrial, I. M.; Rajkumar, S. V. *J. Support Oncol.* **2003**, *1*, 194–205.
47. Crossley, R. *Chim. Oggi* **2003**, Mar/April, 59–61.
48. Trega. http://www.lionbioscience.com (accessed April 2006).
49. Pharmacopeia. http://www.pharmacopeia.com (accessed April 2006).
50. Alanex. http://www.pfizer.com (accessed April 2006).
51. Arqule. http://www.arqule.com (accessed April 2006).
52. Array. http://www.arraybiopharma.com (accessed April 2006).
53. CombiChem. http://www.bms.com (accessed April 2006).
54. Discovery Partners International. http://www.discoverypartners.com (accessed April 2006).
55. Albany Molecular Research. http://www.albmolecular.com (accessed April 2006).
56. 3-Dimensional Pharma. http://www.jnj.com (accessed April 2006).
57. Chembridge Corp. http://www.chembridge.com (accessed April 2006).
58. Klabunde, T.; Hessler, G. *ChemBioChem* **2002**, *3*, 928–944.
59. Rees, D. C.; Congreve, M.; Murray, C. W.; Carr, R. *Nat. Rev. Drug Disc.* **2004**, *3*, 660–672.
60. ISIS Pharmaceuticals. http://www.isispharm.com (accessed April 2006).
61. Sirna Therapeutics. http://www.sirna.com (accessed April 2006).
62. Stephenson, M. L.; Zamecnik, P. C. *Proc. Natl. Acad. Sci. USA* **1978**, *75*, 285–288.
63. Zamecnik, P. C.; Stephenson, M. L. *Proc. Natl. Acad. Sci. USA* **1978**, *75*, 280–284.
64. Wallace, R. W. *Drug Disc. Today* **1999**, *4–5*, 4.
65. Winkler, K. E. *Nat. Rev. Drug Disc.* **2004**, *3*, 823.
66. Trilink Biotechnologies. http://www.trilinkbiotech.com/tech/antisense.asp (accessed April 2006).
67. Agouron. http://www.pfizer.com (accessed April 2006).
68. Syrrx. http://www.syrrx.com (accessed April 2006).
69. SGX Pharmaceuticals. http://www.stromix.com (accessed April 2006).
70. Vertex. http://www.vrtx.com (accessed April 2006).
71. Plexxikon. http://www.plexxikon.com (accessed April 2006).
72. Astex. http://www.astex-therapeutics.com (accessed April 2006).
73. Investigational Drugs Database. http://www.iddb3.com (accessed April 2006).
74. Fire, A.; Xu, S.; Montgomery, M. K.; Kostas, S. A.; Driver, S. E.; Mello, C. C. *Nature* **1998**, *391*, 806–811.
75. Beal, J. *Drug Disc. Today* **2005**, *10*, 169–172.
76. Genentech. http://www.gene.com/gene/index.jsp (accessed April 2006).
77. Amgen. http://www.amgen.com (accessed April 2006).
78. Fletcher, L. *Nat. Biotechnol.* **2002**, *20*, 105–106.
79. Weintraub, A. *Business Week* **2003**, Oct. 6, 72–80.
80. Moseley, M. *Pharm. Exec.* **1998**, *18*, 80–83.
81. Datamonitor. *PharmaWatch: Biotechnology*; Datamonitor: New York, 2004.
82. Butcher, E. C. *Nat. Rev. Drug Disc.* **2005**, *4*, 461–467.
83. Kola, I.; Landis, J. *Nat. Rev. Drug Disc.* **2004**, *3*, 711–715.
84. Schmid, E. F.; Smith, D. A. *Drug Disc. Today* **2005**, *10*, 1031–1039.
85. Chanda, S. K.; Caldwell, J. S. *Drug Disc. Today* **2003**, *8*, 168.
86. Drews, J. *Drug Disc. Today* **2000**, *5*, 547–553.
87. Gershell, L. J.; Atkins, J. H. *Nat. Rev. Drug Disc.* **2003**, *2*, 321–327.
88. Koehn, F. E.; Carter, G. T. *Nat. Rev. Drug Disc.* **2005**, *4*, 206–220.
89. Kola, I.; Hazuda, D. *Curr. Opin. Biotechnol.* **2005**, *16*, 644–646.
90. DiMasi, J. A.; Grabowski, H. G.; Vernon, J. *Drug Inform. J.* **2004**, *38*, 211–223.
91. Schmid, E. F.; Smith, D. A. *Drug Disc. Today* **2004**, *9*, 18–26.
92. Lappin, G.; Garner, R. C. *Nat. Rev. Drug Disc.* **2003**, *2*, 233.

93. Preziosi, P. *Nat. Rev. Drug Disc.* **2004**, *3*, 521.
94. Frank, R.; Hargreaves, R. *Nat. Rev. Drug Disc.* **2003**, *2*, 566.
95. Capdeville, R.; Buchdunger, E.; Zimmermann, J.; Matter, A. *Nat. Rev. Drug Disc.* **2002**, *1*, 493–502.
96. Winslow, R. *Wall Street Journal.* **2000**, Nov. 9, A.4.
97. Anonymous. *Lab Business Week* **2004**, March 1, 61.
98. Nissen, S. E.; Tsunoda, T.; Tuzcu, E. M.; Schoenhagen, P.; Cooper, C. J.; Yasin, M.; Eaton, G. M.; Lauer, M. A.; Sheldon, W. S.; Grines, C. L. et al. *JAMA* **2003**, *290*, 2292.
99. Gallagher, J. *Detroit Free Press* **2003**, Dec. 22.
100. Abrams, P. *Biotechnology* **1993**, *11*, 775.
101. Cavalla, D. *Drug Disc. Today* **2003**, *8*, 267–274.
102. Levinson, A. D. *Nat. Biotechnol.* **1998**, *16*, 45–46.

Biographies

David A Giegel currently heads the biochemistry department at Celgene, Signal Research Division. Prior to joining Celgene in October of 2003, he was the director of lead discovery at MitoKor, Inc. There, he worked on mitochondrial targets involved in central nervous system diseases and diabetes. From 1991 to 2000, Giegel was in the department of biochemistry at the Parke-Davis Research Division of Warner-Lambert (now Pfizer) in Ann Arbor, MI. He rose to the level of director of high-throughput screening technologies, where he was responsible for introducing several new screening platforms and assay technologies. He has spent considerable time investigating serine and cysteine proteases as drug discovery targets in inflammatory and cardiovascular diseases. He was a project leader of a collaboration team between Parke-Davis and BASF Bioresearch (now Abbott Bioresearch) that examined the role of IL-1b-converting enzyme (caspase I) in inflammatory disease. Prior to joining Parke-Davis, Giegel worked at the NutraSweet Division of Monsanto in Chicago, IL. Giegel is trained as an enzymologist, and received his PhD in biological chemistry under Dr Vincent Massey and Dr Charles Williams at the University of Michigan. He did his postdoctoral training, also at the University of Michigan, for the same advisors that he worked with for his PhD.

Alan J Lewis, PhD, joined Novocell, Inc. in February 2006 as President and Chief Executive Officer. Novocell is a biotechnology company focused on commercializing cellular therapies to cure intractable diseases. Alan was previously President of Celgene (San Diego), a biopharmaceutical company focused on the discovery, development, and commercialization of small-molecule drugs for the treatment of cancer and immunological diseases. He served as Chief Executive Officer and Director of Signal Pharmaceuticals since 1996 and as President of the company since 1994. Prior to joining the company, Dr Lewis worked for 15 years at the Wyeth-Ayerst Research division of American Home Products Corporation, where he served as Vice President of Research from 1990 to 1994. At Wyeth-Ayerst, Dr Lewis was responsible for research efforts in CNS, cardiovascular, inflammatory, allergy and bone metabolism diseases. Dr Lewis currently serves as a Director of Discovery Partners International, BioMarin Pharmaceutical, Inc., and Cytochroma, Inc. Dr Lewis holds a PhD in pharmacology from the University of Wales in Cardiff and completed his postdoctoral training at Yale University.

Peter Worland is the executive director of experimental therapeutics at Celgene, responsible for both inflammation and oncology. Worland was previously at Millennium Pharmaceuticals, Inc., where he was the senior director and head of discovery oncology, and the senior director of the cancer pharmacology department. Worland has worked in both biotechnology and large pharmaceutical companies (Celgene, Millennium, Mitotix, and Pharmacia). He has worked on cell cycle and mitotic kinases in the past, leading the CDK inhibitor program at Mitotix and the Aurora program at Pharmacia, and overseeing several discovery oncology programs at Millennium Pharmaceuticals as the head of the discovery oncology effort.

# 2.20 G Protein-Coupled Receptors

W J Thomsen and D P Behan, Arena Pharmaceuticals, Inc., San Diego, CA, USA

© 2007 Elsevier Ltd. All Rights Reserved.

2.20.1	**G Protein-Coupled Receptor (GPCR) Therapeutic Relevance, Structure, and Families**	**772**
2.20.1.1	Therapeutic Potential of G Protein-Coupled Receptors	772
2.20.1.2	Prototypic Drugs Targeting G Protein-Coupled Receptors	772
2.20.1.3	Genetic Studies of G Protein-Coupled Receptors and Disease Relevance	772
2.20.1.4	Conserved Structural Features of G Protein-Coupled Receptor	774
2.20.1.5	Families of Human G Protein-Coupled Receptors	777
2.20.2	**G Protein-Coupled Receptor-Mediated Activation of G Proteins**	**779**
2.20.2.1	The G Protein Cycle	779
2.20.2.2	General Structure of G Proteins	780
2.20.2.3	Families of G Proteins and Effectors Regulated by α-Subunits	780
2.20.2.4	$\beta\gamma$-Regulation of Effectors	784
2.20.2.5	G Protein-Coupled Receptor-Mediated Activation of Mitogen-Activated Protein Kinases	784
2.20.2.6	Regulators of G Protein Signaling	788
2.20.3	**Updated Models of G Protein-Coupled Receptor Activation**	**789**
2.20.3.1	Constitutively Active G Protein-Coupled Receptors	789
2.20.3.2	Models of G Protein-Coupled Receptor Activation and G Protein Coupling	789
2.20.3.3	The Multistate Model of G Protein-Coupled Receptor Activation	794
2.20.3.4	Organization of G Protein-Coupled Receptors and Signaling Molecules	796
2.20.3.5	Models of Class A G Protein-Coupled Receptor Activation and G Protein Coupling Based on Rhodopsin	796
2.20.3.6	G Protein-Coupled Receptor Oligomerization and Heterodimerization	798
2.20.3.7	G Protein-Coupled Receptor Desensitization	800
2.20.3.8	Allosteric Modulation of G Protein-Coupled Receptors	802
2.20.4	**Screening G Protein-Coupled Receptors in Functional Assays**	**805**
2.20.4.1	Impacts of Multiplicity of G Protein-Coupled Receptor Signaling, Agonist Trafficking, and G Protein-Coupled Receptor Homodimerization and Heterodimerization on G Protein-Coupled Receptor Drug Discovery and Screen Development	805
2.20.4.2	Functional Assays for G Protein-Coupled Receptors	806
2.20.4.3	GTPγS Binding Assays	807
2.20.4.4	cAMP Assays	807
2.20.4.4.1	Radiometric cAMP assays	808
2.20.4.4.2	Fluorescence-based cAMP assays	809
2.20.4.4.3	Chemiluminescent cAMP assays	809
2.20.4.5	Inositol Phosphate Accumulation-IP$_3$ Assays	809
2.20.4.6	Intracellular Calcium Assays	811
2.20.4.7	Reporter Gene Assays	812
2.20.4.8	Promiscuous and Chimeric G Proteins and Design of Universal Screens for G Protein-Coupled Receptors	812

2.20.4.9	G Protein-Coupled Receptor Biosensors	813
2.20.4.10	G Protein-Coupled Receptors and High-Content Imaging	813
2.20.5	**Target Validation of G Protein-Coupled Receptors**	814
References		**814**

2.20.1 G Protein-Coupled Receptor (GPCR) Therapeutic Relevance, Structure, and Families

2.20.1.1 Therapeutic Potential of G Protein-Coupled Receptors

G protein-coupled receptors (GPCRs) are a large protein superfamily comprising greater than 1% of the total human genome. Recent market analyses indicate that 40–50% of all modern drugs and almost 25% of the top 200 best-selling drugs target GPCRs.[1–3] GPCR-targeted drugs are used to treat a wide variety of diseases including cardiovascular and gastrointestinal diseases, CNS and immune disorders, diabetes, and cancer. Endogenous ligands have been identified for approximately 210 GPCRs, many of which have been important drug discovery targets. However, evaluation of the human genome suggests that there are between 800 and 1000 genes coding for GPCRs. With the exclusion of olfactory GPCRs, over 100 'orphan' GPCRs remain for which the endogenous ligands have not been identified. Orphan GPCRs undoubtedly represent the future provisional drug discovery targets representing lucrative economic opportunities for the pharmaceutical industry.[4]

2.20.1.2 Prototypic Drugs Targeting G Protein-Coupled Receptors

There are a wide variety of drugs that act directly to activate (agonists) or block (antagonists) GPCRs (**Table 1**) and most of these drugs target class A GPCRs. The largest number of drugs targeting GPCRs are receptor antagonists and examples include β_1-adrenergic antagonists used to treat hypertension, angiotensin AT_1 antagonists used to treat hypertension, and histamine H1 antagonists used to treat allergies. Some of the drugs targeting GPCRs are receptor agonists and include mu-opiate agonists for treatment of pain, β_2-adrenergic agonists used to treat asthma, and α_2-adrenergic agonists to treat glaucoma. A common issue with the therapeutic use of GPCR agonists is that they can lose their efficacy with repeated dosing, commonly referred to as tachyphylaxis. There are also a number of drugs not listed in **Table 1** that indirectly target GPCRs. For example, serotonin reuptake inhibitors (e.g., Prozac, Paxil, and Zoloft) used to treat depression act indirectly at serotonin GPCRs by increasing synaptic levels of serotonin and facilitating activation of both presynaptic ($5HT_{1A}$) and postsynaptic ($5HT_2$) serotonin receptors. The efficacy of most drugs targeting GPCRs is due to a selective interaction with one GPCR whereas the efficacy of some drugs, for example, risperidone or olanzapine, which are used for treatment of schizophrenia, is due to blockade of several different classes of GPCRs at therapeutic doses including select dopamine, serotonin, muscarinic cholinergic, histamine, and adrenergic receptors. Most drugs targeting GPCRs are small molecules that have oral bioavailability. However, there are also recent examples of the therapeutic use of peptides to target GPCRs. An example is the recent approval of the use of the glucagons-like peptide-1 (GLP-1) receptor analog, exenitide, for treatment of type 2 diabetes.

2.20.1.3 Genetic Studies of G Protein-Coupled Receptors and Disease Relevance

There are a number of examples in which naturally occurring mutations of known GPCRs are directly related to human diseases thus providing direct evidence for their therapeutic relevance. It appears that loss-of-function mutations are typically autosomal recessive whereas gain-of-function mutations are autosomal dominant. Furthermore, gain-of-function mutations typically involve mutations that result in ligand-independent activation. Many functionally relevant GPCR polymorphisms result in changes in ligand binding, receptor G protein coupling, or in receptor phosphorylation and desensitization. Evaluation of the literature concerning GPCR polymorphisms indicates that for many, the linkage of the mutation with disease is often not statistically supported. However, considerable evidence now indicates that several adrenergic receptor polymorphisms may confer susceptibility to congestive heart failure. **Table 2** lists some examples of human GPCRs for which mutations have been found and may be implicated in human disease. For many of these examples, the evidence is not compelling and further detailed studies are required to firmly establish their connection to disease. Furthermore, some GPCR mutations may affect receptor sequence without promoting a change

Table 1 Top selling drugs targeting GPCRs

Brand name	Generic name	No. prescriptions* (millions)	Target/pharmacology	Primary use
Hydrocodone	Hydrocodone	92.7	μ-agonist	Pain
Atenolol	Atenolol	44.2	β_1-antagonist	Hypertension
Toprol	Metoprolol	32.8	β_1-antagonist	Hypertension
Proventil	Albuterol	31.2	β_2-agonist	Asthma
Zyrtec	Cetirizine	22.4	H_1-antagonist	Allergy
Singular	Montelukast	22.0	CysLT1-antagonist	Asthma
Lopressor	Metoprolol tartrate	20.8	β_1-antagonist	Hypertension
Allegra	Fexofenadine	18.8	H_1-antagonist	Allergy
Plavix	Clopidogrel	18.7	$P2Y_{12}$-antagonist	Coronary artery disease, thrombosis
Darvon	Propoxyphene	17.9	μ-agonist	Pain
Zantac	Ranitidine	14.9	H_2-antagonist	Ulcers
Desyrel	Trazodone	14.5	$5HT_{2/1A}$ and α-antagonist	Depression
Flexeril	Cyclobenzaprine	13.3	$5HT_2$-antagonist	Muscle relaxant
Diovan	Valsartan	13.1	AT_1-antagonist	Hypertension
Ultram	Tramadol	11.9	μ-agonist	Pain
Phenergan	Promethazine	9.7	D_2-antagonist	Motion sickness, nausea, vomiting
Cozaar	Losartan	9.5	AT_1-antagonist	Hypertension
Clonidine	Clonidine	9.5	α_2-agonist	Hypertension, CNS disorders
Risperdal	Risperidone	8.5	$D_2/5HT_2$-antagonist	Schizophrenia
Seroquel	Quetiapine	7.6	$D_2/5HT_2$-antagonist	Schizophrenia
Clarinex	Desloratadine	7.6	H_1-antagonist	Allergy
Xalatan	Latanoprost	7.1	$PGF_2\alpha$-agonist	Glaucoma
Coreg	Cavedilol	7.0	$\beta_{1/2}$-agonist	Heart failure
Combivent	Ipratropium/albuterol	6.9	Muscarinic antagonist and β_2-agonist	Chronic obstructive pulmonary disease
Zyprexa	Olanzapine	6.5	$D_2/5HT_2$-antagonist	Schizophrenia
Cardura	Doxazosin	6.5	α_1-antagonist	Urinary disorders
Oxycontin	Oxycodone	6.4	μ-agonist	Pain
Regulan	Metoclopramide	6.0	Dopamine $D_{2/3}/5HT_3$ antagonist	Heartburn
Avapro	Irbesartan	5.9	AT_1-antagonist	Hypertension
Imitrex	Sumatriptan	5.8	$5HT_{1B/D}$-agonist	Migrane
Detrol	Tolterodine	5.7	Muscarinic antagonist	Overactive bladder

continued

Table 1 Continued

Brand name	Generic name	No. prescriptions* (millions)	Target/pharmacology	Primary use
Antivert	Meclizine	5.6	H_1-antagonist	Nausea, vomiting, motion sickness
Ziac	Bisoprodol	5.0	β_1-antagonist	Hypertension
Remeron	Mirtazapine	4.7	$\alpha_2/5HT_{1A/2}/H_1$-antagonist	Depression
Inderal	Propanolol	4.6	$\beta_{1/2}$-antagonist	Hypertension
Hytrin	Terazosin	4.6	α_1-antagonist	Benign prostatic hyperplasia
Duragesic	Fentanyl	4.4	μ-agonist	Pain
Pepsid	Famotidine	4.3	H_2-antagonist	Ulcers
Claritin	Loratidine	4.0	H_1-antagonist	Allergy
Patanol	Olopatadine	3.6	H_1-antagonist	Allergy
Benicar	Olmesartan	3.9	AT_1-antagonist	Hypertension
Avalide	Irbesartan	3.5	AT_1-antagonist	Hypertension
Ditropan	Oxybutynin	3.5	Muscarinic antagonist	Over reactive bladder
Zanaflex	Tizanidine	3.4	α_2-agonist	Antispasmotic
Timolol	Timolol	3.4	$\beta_{1/2}$-antagonist	Glaucoma
Sinemat	Carbidopa, Levodopa	3.3	Dopamine precursor	Parkinson's disease
Levsin	Hyoscyamine	3.0	Muscarinic antagonist	Antispasmodic
Dicyclomine	Dicyclomine	2.8	Muscarinic antagonist	Antispasmodic
Alphagan	Brimonidine	2.8	α_2-agonist	Glaucoma

*Based on number of prescriptions dispensed in 2004 in the US. NDC Health Pharmaceutical Audit Suite (PHAST).

in phenotype or in disease incidence but may alter the efficacy of drugs used to treat the disease. It is likely that further genotyping of humans with various diseases will reveal additional orphan GPCRs that may directly play a role in human diseases. In addition, screening for various GPCR polymorphisms may enable development of more personalized drug treatments in the future.

2.20.1.4 Conserved Structural Features of G Protein-Coupled Receptor

GPCRs share several conserved structural features even though they are activated by chemically diverse ligands including biogenic amines, purines, nucleotides, amino acids, lipids, phospholipids, peptides, proteins, glycoproteins, ions, and sensory stimuli. First, the N-terminus of all GPCRs is located extracellularly while the C-terminus is located intracellularly (**Figure 1**). Second, all members possess seven highly conserved antiparallel transmembrane (TM) helices consisting of 20–28 amino acids in length (TM1–7) surrounding a central cleft and connected by three intracellular (IC1, IC2, IC3) and three extracellular (EC1, EC2, and EC3) loops. Other highly conserved regions include a disulfide bond between EC2 and EC3 loops, the presence of a (D/E)RY motif at the junction of TM3 and the IC2 loop, and a NpxxY motif in TM7.[5,6] GPCRs interact with G proteins via IC3 and IC2 loops and their C-terminal tails.[7–11] The N-terminus of many GPCRs contains several glycosylation sites and highly conserved C-terminal cysteine residues that can undergo palmitoylation, and these posttranslational modifications can influence both receptor expression and function.[12–16] It has also been proposed that palmitoylated C-terminal cysteine residues form a hydrophobic attachment to the cell membrane and, thereby, create a fourth IC loop or eighth helix.[17–19] The IC3 loop and C-terminal tail of GPCRs both undergo phosphorylation by specific kinases called GRKs (G protein-coupled

Table 2 Examples of GPCR polymorphisms that may contribute to human diseases

GPCR	Potential disease relationships
α_{1A}-adrenergic	Alzheimer's disease
α_{2A}-adrenergic	ADHD, obesity, motion sickness
α_{2B}-adrenergic	Hypertension, coronary artery disease, obesity
α_{2C}-adrenergic	Heart failure
Adensoine A_{2A}	Panic disorder
ACTH	Familial ACTH resistance
Angiotensin AT_1	Diabetic nephropathy
Angiotensin AT_2	Hypertension, renal disease
β_1-adrenergic	Obesity, anxiety, heart failure, hypertension, Alzheimer's disease
β_2-adrenergic	Increased heart failure with β-blocker treatment, asthma, obesity, metabolic syndrome, glaucoma, hypertension, myocardial infarction, COPD, rheumatoid arthritis, type 2 diabetes, cystic fibrosis, arrhythmias
β_3-adrenergic	Obesity, type 2 diabetes, hypertension
Bradykinin B1	Coronary artery disease
Bradykinin B2	Coronary artery disease, asthma
$GABA_B$	Epilepsy
Ca^{2+} sensing GPCR	Familial hypocalciuric hypercalcemia, neonatal hyperparathyroidism
Calcitonin	Hypertension, osteoporosis
Cannabinoid CB_1	Substance abuse, schizophrenia
Cannabinoid CB_2	Autoimmune disease susceptibility
CCK_A	Irritable bowel syndrome, panic disorder, Parkinson's disease, schizophrenia, substance abuse, gallstone formation, type 2 diabetes
CCK_B	Panic disorder, Parkinson's disease, type 2 diabetes
Chemokine CCR2	Type 1 diabetes, endometriosis, arteriosclerosis, hypertension, osteoporosis
Chemokine CCR3	Asthma
Chemokine CCR5	Type 1 diabetes, HIV progression, hypertension
Chemokine CCR5	HIV progression, type 1 diabetes, diabetic nephropathy
CRTH2	Asthma
CRF1	Obesity
Chemokine CX3CR1	Macular degeneration, acute coronary syndrome, arteriosclerosis
Chemokine CXCR6	HIV progression
Leukotriene CysLTR	Asthma
Dopamine D_1	Bipolar disease
Dopamine D_2	Extrapyramidal adverse effects to typical antipsychotics, schizophrenia, substance abuse, Parkinson's disease, obesity, hypertension
Dopamine D_3	Obsessive compulsive disorder, tardive dykinesia, schizophrenia, personality disorder, tardive dyskinesia
Dopamine D_4	Obsessive compulsive disorder, ADHD, schizophrenia, personality disorder, tic disorders

continued

Table 2 Continued

GPCR	Potential disease relationships
Dopamine D_5	ADHD, blepharospasm
Endothelin ETA	Migraine
Endothelin ETB	Hirschsprung disease
FPR1	Periodontitis
FSH	Infertility
GHRH	Growth hormone deficiency
Glucagon	Obesity, hypertension
GnRH	Central hypogonadism
*GPR10	Hypertension
*GPR50	Bipolar disease
*GPR54	Hypogonadotropic hypogonadism
*GPR56	Bilateral frontoparetal polymicrogyria
*GPR154	Asthma
Ghrelin	Type 2 diabetes
Histamine H_1	Autoimmune disease
$5HT_{1A}$	Anxiety, depression
$5HT_{1B}$	ADHD
$5HT_{1D}$	ADHD, anorexia nervosa
$5HT_{2A}$	Alzheimer's disease, irritable bowel syndrome, addiction, obsessive compulsive disorder, panic disorder, myocardial infarction, schizophrenia, Alzheimer's disease, hypertension, tardive dyskinesia, Tourette syndrome
$5HT_{2C}$	Tardive dyskinesia, antipsychotic-induced weight gain, schizophrenia, anorexia nervosa, obesity, migraine, panic disorder
$5HT_{5A}$	Schizophrenia
$5HT_6$	Schizophrenia, Parkinson's disease, Alzheimer's disease
IL-8	COPD and asthma
LH	Male pseudohermaphrodism
Melanocortin MC_1	Skin cancer
Melanocortin MC_3	Obesity
Melanocortin MC_4	Obesity
Melatonin MT1a	Rheumatoid arthritis
Melatonin concentrating hormone MCHR1	Obesity
Muscarinic M_1	Alzheimer's disease
Muscarinic M_2	Depression, substance abuse, depression
Muscarinic M_3	Asthma
δ-opiate	Anorexia nervosa
μ-opiate	Obsessive compulsive disorder, substance abuse, epilepsy

Table 2 Continued

GPCR	Potential disease relationships
Oxytocin	Autism
PTH/PTHrP	Blomstrand chondrodysplasia
Purinergic P2Y1	Thrombosis
Purinergic P2Y12	Arteriosclerosis
Rhodopsin	Retinitis pigmentosa
Somatostatin sst5	Bipolar disease
Thromboxane A_2	Asthma, atopic dermatitis
Urotensin II	Type 2 diabetes
Vasopressin 1B	Depression
Vasopressin V2	Nephrogenic diabetes insipidus

ACTH, adrenocorticotropic hormone; ADHD, attention deficit and hyperactivity disease; COPD, chronic obstructive pulmonary disease; HIV, human immunodeficiency virus; PTH, parathyroid hormone.
*Orphan GPCRs.

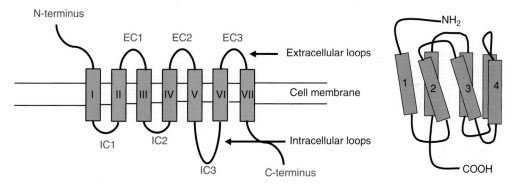

Figure 1 GPCR membrane topology and TM bundles.

receptor kinases), which is important for initiating the process of functional desensitization (*see* Section 2.20.3.8) and also influences G protein coupling.[20–22] Furthermore, the C-terminal tail of GPCRs serves as an important docking site for GPCR interacting proteins such as scaffold proteins, which play important roles in organizing signaling molecules, and beta-arrestins, which mediate receptor internalization and desensitization.[23–26]

2.20.1.5 Families of Human G Protein-Coupled Receptors

The GPCR superfamily can be divided into six subfamilies based on endogenous ligand structure, sequence similarity, and/or G protein-coupling characteristics.[27–32] A regularly updated classification list of all known and orphan GPCRs is currently available for downloading from the International Union of Pharmacology (IUPHAR) database.[33,666] Members of each GPCR subfamily generally share greater than 20% sequence identity in their TM regions and are considered to have evolved from a common ancestor. The focus of this chapter will be confined to class A, B, and C GPCRs as these are the predominant subfamilies targeted for drug discovery.

Class A GPCRs, represented by rhodopsin and the β_2-adrenergic receptors, are the largest subfamily of GPCRs and are activated by a wide variety of different ligands including biogenic amines (e.g., norepinephrine, histamine, dopamine, serotonin), purines (e.g., adenosine), nucleotides (e.g., ADP, ATP), eicosanoids (e.g., prostaglandins, leukotrienes), lipids (e.g., lysophosphatidic acid, shingosine-1-phosphate, endocannabinoids), small peptides (e.g., vasopressin, angiotensin), glycoproteins (e.g., TSH, LH, FSH), and sensory stimuli (e.g., photons). The ligand-binding pocket for class A GPCRs activated by smaller ligands such as biogenic amines is located deep in the TM core formed primarily by TM3, TM4, TM5, TM6, and TM7 (**Figure 2**).[34] In contrast, the ligand-binding pocket for class A GPCRs activated by small peptides is composed of intracellular loops and the N-terminus of the receptor with some contribution by the upper region of the TM core.[35,36] The ligand-binding pocket of class A GPCRs activated by glycoproteins is located on the large N-terminus of the receptor and is also comprised of regions in extracellular loops ECL1 and ECL3.[37,38] The class A subfamily of protease-activated GPCRs, represented by the thrombin receptors, is uniquely activated by proteolysis of their N-terminus, which allows the remaining N-terminus to act as a covalently bound ligand. Mutational studies of human thrombin receptors also suggest that the tethered N-terminal peptide interacts with amino acids contained within the ECL2 loop.[39] Phylogenetic analyses of the human genome sequence indicate that 79% of the total GPCRs in the genome are class A members and that half of these are orphan GPCRs.[40]

Class B GPCRs, also referred to as the secretin family, typically contain a N-terminal hydrophobic signal peptide, a large N-terminus (100–160 amino acids), and are activated by large peptides (e.g., 30–40 amino acids). Examples of class B GPCRs include the PTH (PTH1R and PTH2R), glucagon-like peptide (GLP-1R and GLP-2R), glucagon, calcitonin, corticotropin-releasing factor (CRFR1 and CRFR2), and vasoactive peptide (VPAC1R, VPAC2R, PAC1R) receptors. Class B GPCRs have overall morphology similar to class A GPCRs activated by glycoproteins but do not share any sequence homology. Furthermore, the overall sequence homology between class B members is relatively low.[41,42] Mutational studies suggest that the binding pocket of these GPCRs is discontinuous and is comprised of contact points from the extended N-terminus and the three extracellular loops. Class B GPCRs also contain six conserved cysteine residues involved in intradomain disulfide bridges and also contain a conserved region within the N-terminal domain close to TM1, which is involved in ligand binding.[43,44] A simple low-resolution 'two domain' model of the class B GPCR ligand binding pocket suggests that the N-terminal region of the hormone binds to the N-terminal domain and that this initial interaction acts as an affinity trap supporting an additional interaction of the N-terminal region of the ligand with the TM bundle region.[45]

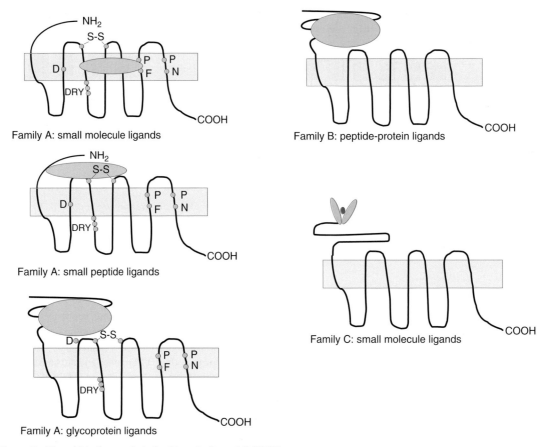

Figure 2 Ligand binding pockets for Class A, B, and C GPCRs.

A novel subclass of class B GPCRs, referred to as large N-terminal GPCRs or EGF-TM7 receptors, shares homology in the TM regions with secretin receptors, but also contains several structural features that are unique for this subclass. These features include a long N-terminus with variable numbers of EGF-like motifs, a mucin-rich stalk region enriched in serine and threonine residues, and a GPCR proteolytic site (called the GPS domain), which, after cleavage, becomes noncovalently associated with the TM regions of the receptor.[46] These GPCRs can also contain additional N-terminal domains including arginine-glycine-aspartic acid (RGD) motifs, cadherin domains, and immunoglobulin domains. The physiological functions of members of this subfamily are still largely unknown but some evidence suggests that they play roles in mammalian development and cellular adhesion. Examples of this subclass include the latrotoxin and CD97 receptors. Within this subfamily, only CD97 has an identified binding partner, membrane protein CD55, and activation of this receptor appears to play a role in leukocyte migration.[47] Recently, a total of 30 different orphan GPCRs belonging to this subclass have been identified but little is known concerning the function of these new GPCRs.[31] Since many of these GPCRs most likely bind large proteins, their small molecule 'druggability' remains to be established.

Class C GPCRs, represented by metabotropic glutamate receptors, contain a large N-terminal extracellular domain, followed by a cysteine-rich domain (except the $GABA_B$ receptor), and a C-terminus of variable length; and they have low but significant homology with bacterial periplasmic proteins. Endogenous ligands for class C GPCRs include amino acids (glutamate), γ-amino-butyric acid (GABA), magnesium and calcium ions, putative pheromones, and taste molecules. The metabotropic glutamate and calcium-sensing receptors share approximately 20 conserved cysteine residues in the extracellular region of the TM bundle. Class C GPCRs generally form homodimers (see Section 2.20.3.6) and each subunit has a large extracellular domain called a venus flytrap module (VFTM). The ligand binding pocket of class C GPCRs is situated in a cleft between two lobes of the VFTM and x-ray crystallography studies indicate that the VFTM lobes close upon ligand binding leading to activation of the TM core domain.[48–51] Class C GPCRs make up only a small proportion of mammalian GPCRs (1–2%).

2.20.2 G Protein-Coupled Receptor-Mediated Activation of G Proteins

2.20.2.1 The G Protein Cycle

Ligand binding to the extracellular surface of GPCRs initiates a set of conformational changes culminating in the exposure of receptor IC2 and IC3 loops to participate in G protein coupling and subsequent activation of G proteins. Activated G proteins in turn regulate a variety of specific effector proteins and signal transduction pathways ultimately leading to a variety of cellular responses. G proteins are heterotrimeric proteins containing α-, β-, and γ-subunits and function by a cycle of α- and βγ-subunit dissociation superimposed by GPCR-mediated guanine nucleotide exchange on the α-subunit (**Figure 3**). The sequential events occurring in receptor-mediated activation of G-proteins are as follows: (1) agonist binds to the receptor promoting a series of conformational changes favoring receptor coupling to G protein(s); (2) formation of the agonist–receptor–G protein ternary complex induces a G protein conformational change that facilitates the exchange of α-subunit bound guanosine diphosphate (GDP) for guanosine triphosphate (GTP); (3) GTP binding destabilizes the G protein heterotrimer and promotes dissociation into

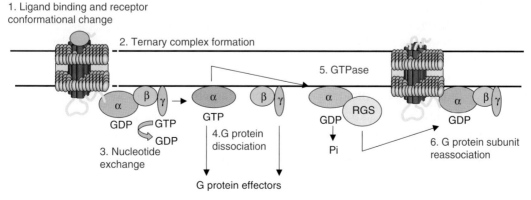

Figure 3 G protein activation cycle.

GTP bound α-subunit and the βγ-complex; (4) GTP bound α-subunit and the βγ-complex regulate the activity of select intracellular effector proteins (enzymes or ion channels) leading to changes in levels of second messengers (e.g., cAMP and calcium) and regulation of select signal transduction pathways; (5) activity of GTP bound α-subunit is terminated by hydrolysis of GTP to GDP, referred to as the GTPase reaction; (6) inactivated GDP bound α-subunit reassociates with the βγ-complex completing the G protein cycle. The G protein activation/deactivation cycle is generally rapid and ranges from milliseconds to seconds in duration. Some reports in the literature suggest that G proteins can activate effectors without subunit dissociation but this claim currently remains controversial.[52–55] In addition, GPCRs can also modulate signaling pathways by G protein-independent mechanisms (see Section 2.20.3.8).

2.20.2.2 General Structure of G Proteins

Mammalian G proteins are divided into four subfamilies (G_s, $G_{i/o}$, G_q, and $G_{12/13}$) based on the structural and functional similarities of their α-subunits. There are several isoforms of G protein subunits with over 16 different α-subunits, 5 β-subunits, and 14 γ-subunits expressed in mouse and human genomes.[56–60] The α-subunits are typically palmitoylated on conserved N-terminal cysteine residues (with the exception of $G\alpha_t$) or myristoylated on N-terminal glycine residues (only $G\alpha_{i/o/z}$)[61–64] both of which play an important role in membrane localization as well as regulation of interactions with other proteins including βγ-complexes.[64–67] Unlike myristoylation, palmitoylation is a reversible process and it appears to be highly regulated. Structural studies including x-ray crystallography of $G\alpha_{i1}$ indicate that G protein α-subunits are composed of two important structural domains, a Ras-like nucleotide-binding domain (called the GTPase domain) that mediates GTP binding and hydrolysis, and an alpha-helical domain that buries GTP within the core of the G protein α-subunit.[68–71] The α-subunit helical portion is most structurally divergent between G protein families and plays an important role in determining the specificity of receptor– and effector–G protein coupling. The N-terminal region of α-subunits is ordered by their interaction with the β-subunit.[69–71] Domains of α-subunits involved in GPCR recognition include the α-subunit C-terminus, which contacts receptor IC2 and IC3 loops, the extreme N-terminus, a region between helices α4 and α5, and a region within the loop linking the N-terminal α-helix to the β1-strand of the Ras-like domain.[71–73] Unfortunately, several attempts to determine G protein selectivity for various GPCRs empirically based on GPCR sequences have not been successful to date.

The βγ-complex is made up of two polypeptides, but is functionally a monomer because this complex does not dissociate in cells. All β-subunits contain seven WD40 repeat motifs and a tryptophan-aspartic acid sequence that repeats about every 40 amino acids and forms small antiparallel beta-strands.[57,70,74] X-ray crystal studies indicate that the seven WD40 repeats fold to form a seven-bladed beta-propeller structure and the N-terminus forms an α-helix.[74] With the exception of $β_5$, which is exclusively expressed in the central nervous system, all β-subunit isoforms display 79–90% sequence identity. In contrast, the γ-subunits are much more heterogeneous suggesting that they have unique functional roles. Structural studies indicate that the γ-subunit folds into two helices: the N-terminal helix forms a coiled-coil with the α-helix of the β-subunit, while the C-terminal helix makes extensive contacts with the base of the β-subunit.[69,70,74–76] All γ-subunits are alpha-helical and isoprenylated with either farnesyl- or geranylgeranyl-groups that are important for function and facilitate membrane localization of the βγ-complex.[77–82] βγ-complex interacts with a hydrophobic pocket residing in the GDP-bound α-subunit and GTP binding eliminates this pocket promoting a reduction in affinity between the GTP-bound α-subunit and the βγ-complex resulting in G protein subunit dissociation.[69] Biochemical approaches have suggested that different βγ-complex combinations are functionally similar but more recent studies using gene targeting approaches have demonstrated that GPCRs have a preference for certain subunit combinations to link to perspective effectors. For example, it has been reported that dopamine D_1 receptors require βγ-complexes with the $γ_7$ to stimulate adenylyl cyclase in discrete brain areas.[83]

2.20.2.3 Families of G Proteins and Effectors Regulated by α-Subunits

The $G\alpha_s$ G protein subfamily includes the members $G\alpha_s$ and $G\alpha_{olf}$, which primarily act to stimulate adenylyl cyclase, an enzyme responsible for converting ATP to the second messenger, cAMP (**Figure 4**; **Table 3**).[84] Increases in cAMP levels, in turn, promote activation of protein kinase A (PKA) and phosphorylation of specific cellular substrates leading to cellular responses. The activation of PKA is terminated by phosphodiesterase-mediated conversion of cAMP to AMP. Cholera toxin promotes activation of adenylyl cyclase by catalyzing ADP ribosylation of a conserved C-terminal arginine residue of $G\alpha_s$ resulting in an inhibition of its GTPase activity. Although $G\alpha_s$ and $G\alpha_{olf}$ both stimulate adenylyl cyclase, $G\alpha_{olf}$ has been shown to differ from $G\alpha_s$ by having a lower affinity for GDP and a more rapid deactivation

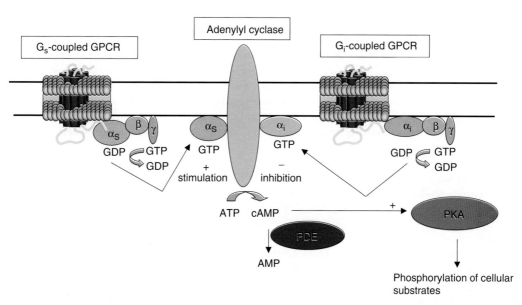

Figure 4 G protein-mediated regulation of adenylyl cyclase. PDE, phosphodiesterase; PKA, protein kinase A.

Table 3 Effectors directly regulated by different G protein α-subunits

Subfamily	Members	Effectors
G_s	$G\alpha_s$	Stimulation of adenylyl cyclase, src activation
	$G\alpha_{olf}$	Stimulation of adenylyl cyclase
$G_{i/o}$	$G\alpha_{i1}$	Inhibition of adenylyl cyclase
	$G\alpha_{i2}$	Inhibition of adenylyl cyclase
	$G\alpha_{i3}$	Inhibition of adenylyl cyclase
	$G\alpha_{oA}$	Inhibition of adenylyl cyclase
	$G\alpha_{oB}$	Inhibition of adenylyl cyclase
	$G\alpha_z$	Inhibition of adenylyl cyclase; GRINI, Eya2
	$G\alpha_{t1}$	Stimulation of cGMP-phosphodiesterase
	$G\alpha_{t2}$	Stimulation of cGMP-phosphodiesterase
G_q	$G\alpha_q$	Stimulation of PLCβ, PLD; activation of RhoGEF
	$G\alpha_{11}$	Stimulation of PLCβ; stimulation of Brunton's tyrosine kinase; activation of RhoGEF
	$G\alpha_{14}$	Stimulation of PLCβ
	$G\alpha_{15/16}$	Stimulation of PLCβ
$G_{12/13}$	$G\alpha_{12}$ $G\alpha_{13}$	Activation of p115RhoGEF, PDZ-RhoGEF, lbc-RhoGEF, LARG, interaction with glutamate transporter associated protein GTRAP4, cell adhesion molecules, PLCε activation, A-kinase anchoring protein 110, heat shock protein 90, and protein phosphatase 5

cycle.[85] There are four known splice variants of $G\alpha_s$, two shorter variants and two larger variants, with the latter being predominantly expressed in neuroendocrine cells.[86] Constitutively activating mutations of $G\alpha_s$ in which GTPase activity is diminished are present in endocrine tumors, fibrous bone dysplasia, and McCune–Albright syndrome, whereas a loss-of-function mutation leads to Albright hereditary osteodystrophy.[87] Biochemical and genetic evaluations suggest that $G\alpha_s$ is involved in diverse physiological processes including regulation of bone growth,[88,89] ion channel function[90–92] meiotic arrest,[93] and insulin sensitivity[94] as well as recognized roles in regulation of the immune response, adipose metabolism, glycogen synthesis, and cardiac function. $G\alpha_{olf}$ is expressed in olfactory neurons, striatum, and pancreatic beta-cells.[95–97] Besides playing a fundamental role in olfaction, activation of $G\beta_{olf}$ in beta-cells has been reported to be involved in regulation of insulin status and beta-cell survival and regeneration during development and diabetes.[98] Striatal dopamine D_1 and adenosine A_{2A}-mediated increases in cAMP are reported to be mediated by $G\alpha_{olf}$[99] and prolonged elevation of $G\alpha_{olf}$ levels in human Parkinson's disease patients may lead to a persistent D_1 receptor hypersensitivity and contribute to the genesis of long-term complications of L-dopa treatment.[100]

The larger $G\alpha_{i/o}$ G protein subfamily includes $G\alpha_{i1}$, $G\alpha_{i2}$, $G\alpha_{i3}$, $G\alpha_{oA}$, $G\alpha_{oB}$, $G\alpha_z$, $G\alpha_{t1}$, $G\alpha_{t2}$, and $G\alpha_g$. All family members (except $G\alpha_z$) are substrates for pertussis toxin (PTX) ADP ribosyl transferase and ADP ribosylation of C-terminal arginine residues inhibits receptor-G protein coupling. The cellular focus of activated $G\alpha_i$, $G\alpha_o$, and $G\alpha_z$ subunits is primarily inhibition of adenylyl cyclase[101] (**Figure 4**). Of the nine different isoforms of mammalian adenylyl cyclase identified, the V, VI, and I isoforms are directly inhibited by $G\alpha_{i/o}$ subunits.[102,103]

$G\alpha_i$ proteins are the most widely expressed G protein subfamily and are hypothesized to serve as an important source of G protein βγ-complexes. The high structural similarity of $G\alpha_i$ proteins suggests that they may have at least partially redundant physiological functions. $G\alpha_{i1}$ subunits have been reported to play an important role in regulation of calcium currents,[104] platelet activation,[105] and activation of K^+ channels.[106] $G\alpha_{i2}$ subunits have been demonstrated to play physiological roles in regulation of calcium currents,[107] control of hormone release from pancreatic alpha-cells,[107–109] activation of K^+ channels in pancreatic alpha-cells,[108] control of proliferation of neural progenitor cells,[110] facilitation of GLUT4 translocation[111] and insulin signaling in skeletal muscle and adipose tissue, regulation of platelet activation and adipocyte differentiation,[112,113] regulation of immune cell development[114] and function,[115–118] and regulation of cardiac cell apoptosis and cardiac hypertrophy.[119–121] The $G\alpha_{i3}$ subunit plays key roles in regulation of calcium currents[122] and inflammatory responses of immune cells.[123] Interestingly, $G\alpha_{i2}$ is reported to be required for direct inhibition of adenylyl cyclase whereas $G\alpha_{i3}$ is required to inhibit $G\alpha_s$-activated adenylyl cyclase.[124] The rate of guanine nucleotide exchange for $G\alpha_i$ is one of the most rapid of all G proteins and the turnover rate for $G\alpha_z$ is one of the slowest.[125]

$G\alpha_o$ is predominantly expressed throughout the nervous system with lower levels expressed in heart, pituitary, and pancreas.[126] The physiological functions of $G\alpha_o$ isoforms ($G\alpha_{oA}$ and $G\alpha_{oB}$) other than inhibition of adenylyl cyclase, regulation of ion channel activity, and regulation of retinal function remain to be further elucidated. Recent evaluations of this G protein subfamily also suggest that they play roles in modulation of voltage-gated calcium channels and in exocytosis in chromaffin cells, pituitary cells, and melanotrophs. Most of the actions of $G\alpha_o$ are thought to be mediated primarily by βγ-complexes.

$G\alpha_z$ is predominantly expressed in cells exhibiting regulated exocytosis and include hypothalamic neurons, adrenal medulla cells, platelets, and pancreatic beta-cells. $G\alpha_z$ participates in several cellular processes including cellular and neuronal development, survival, proliferation, differentiation, pituitary hormone secretion,[127–129] apoptosis,[130] and pancreatic beta-cell signaling.[131] Association of $G\alpha_z$ with a number of different signaling molecules, some of which are neuronal specific proteins, suggests that it may play a role in cellular development, survival, proliferation, differentiation, and apoptosis. These proteins include G protein-regulated inducers of neurite outgrowth (GRINI1, 2, and 3)[132] and Eya2 transcription factor, both of which are involved in embryonic neuronal development.[133] $G\alpha_{t1}$ and $G\alpha_{t2}$, known as transducins, are well-characterized retinal G proteins that interact with specific phosphodiesterases that hydrolyze cGMP.[134] $G\alpha_g$, also known as gustducin, is highly related to transducin and is primarily expressed in taste tissue and plays a role in taste signal transduction.[135–137]

The $G\alpha_q$ G protein subfamily includes the members $G\alpha_q$, $G\alpha_{11}$, $G\alpha_{14}$, $G\alpha_{15}$, or $G\alpha_{16}$, all of which stimulate phospholipase Cβ, an enzyme that catalyzes hydrolysis of membrane-associated phosphatidylinositol species (PI, PIP, and PIP2) to form inositol phosphates (IP, IP_2, and IP_3) and diacylglycerol (DAG)[138,139] (**Figure 5**). Elevation of intracellular IP_3 levels in turn leads to activation of an IP_3 receptor located on the endoplasmic reticulum membrane resulting in a transient elevation of intracellular Ca^{2+} and activation of a number of calcium-dependent proteins and enzymes including protein kinase C (PKC). $G\alpha_q$ and $G\alpha_{11}$ are ubiquitously expressed in human tissues while $G\alpha_{14}$ is predominantly expressed in spleen, lung, testes, and kidney and $G\alpha_{16}$ expression is restricted to hematopoietic lineages.[140] Although $G\alpha_{14}$ and $G\alpha_{15}$ are expressed in mice, only the $G\alpha_{16}$ subunit is expressed in humans. Receptors activating the G_q family in mammalian systems do not generally distinguish between $G\alpha_q$ and $G\alpha_{11}$.[141] Targeted knockout of $G\alpha_q$ and $G\alpha_{11}$ in mice results in impairment of normal hypothalamic functions resulting in somatotroph hypoplasia, dwarfism, and anorexia.[142] Several cardiac G_q-coupled receptors (e.g., angiotensin and endothelin receptors) mediate cardiac inotropy and chronotropy as well as development of hypertrophy.[119] Recently, GPCR-mediated

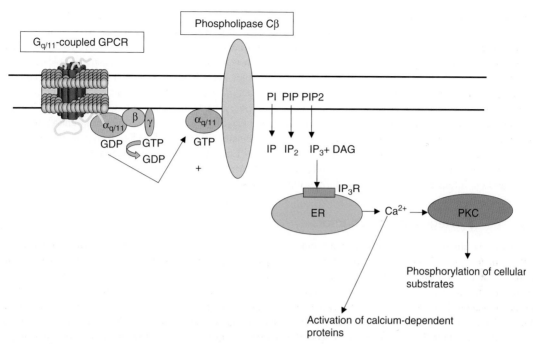

Figure 5 G protein-mediated activation of phospholipase C. PKC, protein kinase C; ER, endoplasmic reticulum.

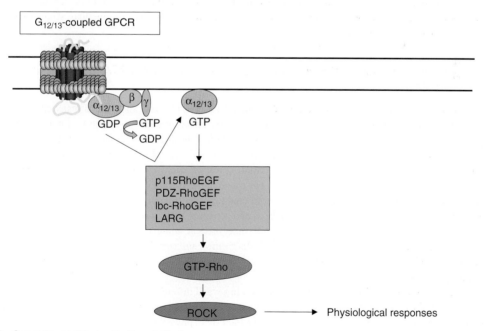

Figure 6 G protein-mediated activation of RhoA.

activation of a K$^+$ channel by Gα_q subunits has been reported.[143] Stimulation of some G$_q$-coupled GPCRs has been observed to induce phosphorylation of G$\alpha_{q/11}$ but the significance of this is unclear.[144]

The Gα_{12} G protein subfamily has two members, Gα_{12} and Gα_{13}.[145–147] It is currently unclear which signaling pathways are shared between these two subunits and which are controlled more exclusively by one or the other. The effector pathways regulated by Gα_{12} and Gα_{13} have recently been elucidated and activation of both G protein α-subunits leads to activation of the small G protein RhoA through the guanine nucleotide exchange factor RhoGEF or related proteins (PDZ-RhoEGF, ibc-RhoGEF, LARG) (**Figure 6**). Indeed, in cell-free studies, Gα_{13} directly stimulates p115RhoGEF to enhance guanine nucleotide exchange for RhoA and that p115RhoGEF binding with RhoA accelerates the rate of GTP hydrolysis by both $G\alpha_{12}$ and $G\alpha_{13}$.[148,149] Activated RhoA in turn activates RhoA kinase (ROCK) leading to regulation of a variety of important cellular activities including cell cycle progression, cellular proliferation, cellular adhesion, chemotaxis, cytoskeletal rearrangements (such as stress fiber formation), smooth muscle contraction, and gene expression.[147–149] Furthermore, $G\alpha_{12}$ and $G\alpha_{13}$ knockout mice deficient in neural crest–cell-derived cardiac cells display significant cardiac malformations.[150] In addition, $G\alpha_{12/13}$ has been shown to possess oncogenic activity.[151] The majority of GPCRs coupling to $G\alpha_{12/13}$ also couple to $G\alpha_q$, which has complicated determination of cellular processes selectively regulated by $G\alpha_{12/13}$ (**Table 4**). $G\alpha_{12}$ and $G\alpha_{13}$ have also been reported to interact directly with other effector proteins besides p115RhoGEF including glutamate transporter EAAT4-associated protein, radixin, cortactin-interacting protein Hax-1, cell adhesion molecules, PLCε, A-kinase anchoring protein 110 (AKAP), heat shock protein 90, Ser/Thr phosphatase type 5, and protein phosphatase 2A.[152–162] The physiological importance of most of these protein interactions remains to be further elucidated.

2.20.2.4 βγ-Regulation of Effectors

The original functional roles that G protein βγ-complexes were thought to play in GPCR signal transduction were to facilitate the attachment of G protein heterotrimers to the plasma membrane as well as to complex with inactive GDP-bound α-subunits. However, it is now well substantiated that βγ-complexes regulate the activity of a wide variety of effectors. In fact, the list of different effector proteins regulated by βγ-complexes is now larger than those regulated by α-subunits (**Tables 3** and **5**).[58–60,163–166] Many effector proteins regulated by βγ-complexes contain conserved protein–protein interaction domains called PH domains. However, not all proteins containing PH domains are βγ-complex effectors. An important factor concerning βγ-complex-mediated signaling is that the βγ composition of any given cell will determine the quality and efficiency of effector regulation and will also influence receptor – G protein coupling specificity similar to α-subunits. For any GPCR, activation of one G protein generates not only activated α-subunits but βγ-complexes as well, which contributes to the potential complexity of GPCR signaling. For example, some $G_{i/o}$-coupled receptors may not only inhibit adenylyl cyclase activity, but may also mediate stimulation of PLCβ-pathway by a βγ-mediated mechanism.

2.20.2.5 G Protein-Coupled Receptor-Mediated Activation of Mitogen-Activated Protein Kinases

GPCR-mediated control of cellular growth, proliferation, differentiation, and survival in many instances involves activation of a family of cytoplasmic serine/threonine kinases called mitogen activated protein (MAP) kinases (MAPKs). Mammalian cells express three major classes of MAPKs: the extracellular signal-regulated kinases (ERK1 and ERK2); c-Jun terminal kinase/stress-activated protein kinases (JNKs/SAPKs); and p38/HOG1.[217–222] All MAPKs are activated by a series of parallel kinase phosphorylation cascades, each pathway comprising a collection of three distinct kinases that are successively phosphorylated. For example, activation of the ERK1/2 cascade involves c-Raf-1 kinase and B-Raf kinase-mediated phosphorylation of the dual function threonine/tyrosine kinases, MAP kinase kinase (MEK1 and MEK2), which subsequently phosphorylates and activates ERK1 and ERK2. Ultimately, activated MAPKs then in turn phosphorylate a variety of membrane, cytoplasmic, nuclear, and cytoskeletal substrates leading to a diverse set of cellular responses. One common endpoint of MAPK activation and signaling involves their translocation from the cytosol into the nucleus where they phosphorylate and activate specific nuclear transcription factors that control DNA synthesis and cell division.

GPCR-mediated activation of MAPKs is clearly involved in several different human diseases such as certain forms of cancer, cardiac hypertrophy, neointimal hyperplasia of vascular smooth muscle, and prostate hypertrophy.[223–226] Activation of MAPK appears to play a role in a number of important physiological processes including smooth muscle relaxation, contraction, and proliferation,[227,228] control of learning and memory processes,[229] protection against ischemia/reperfusion injury,[230–232] and control of immune cell function.[233] GPCRs that couple to G_s, $G_{i/o}$, and $G_{q/11}$ can, depending on the GPCR and the cell type, activate different MAPKs. Although GPCRs are known to activate JNKs and p38 families of MAPKs, the following discussion of GPCR regulation of MAPKs is restricted to activation of ERK1/2 as these pathways are more established than pathways leading to activation of the later two families of MAPKs. However, GPCR-mediated activation of JNK and p38 MAPK families is undoubtedly important both physiologically and pathophysiologically.[220–222]

GPCR-mediated activation of ERK1/2 generally occurs by three different mechanisms; signals initiated by classical G protein signaling molecules such as PKA and PKC; signals initiated by GPCR-mediated activation

Table 4 Examples of different GPCRs activating $G\alpha_{12/13}$

GPCR	Cells	Assay cellular change
Serotonin 5HT$_{2C}$	NIH3T3	RhoA activation
Serotonin 5HT$_{4A}$	NIE-115	Neurite retraction
Serotonin 5HT$_{4A}$	NIH 3T3	Serum response element activation
Adenosine A$_3$	Cadiomyocytes	Phospholipase D activation
α_{1A}-adrenergic	Cardiomyocytes	Hypertrophy
Angiotensin AT$_2$	Smooth muscle	Contraction
Bradykinin B$_2$	132N1	Phospholipase D activation
Bombesin	Swiss 3T3	Actin polymerization
Complement C5a	Neutrophils	RhoA activation
Cholecystokinin (CCK)	Intestinal smooth muscle	Phospholipase D activation
Chemokine CXCR3	BML	RhoA activation and chemotaxis
Leukotriene CysLT(1)	Epithelial	Stress fiber formation
EDG3	CHO	RhoA activation
EDG5	CHO	RhoA activation
Prostaglandin EP$_3$	PC12	RhoA activation and neurite retraction
Endothelin ET$_{A/B}$	Smooth muscle	Actin polymerization
FMLP	Neutrophils	F-actin polymerization
G2A	Swiss 3T3	Rho A activation, stress fiber formation, SRE activation
Galanin GAL2	Swiss 3T3	Stress fiber formation
Gastrin-releasing peptide	Isrecol	Cellular invasion
Histamine H$_1$	1321N1	PLD activation
LPA	Swiss 3T3	Stress fiber formation
LPA	Endothelial	Stimulate calcium current
LPA	Fibroblasts	Membrane depolarization-chloride channel activation
LPA	NEI-115	RhoA activation and neurite retraction
Muscarinic M$_1$	Jurkat	Serum response element activation
Muscarinic M$_1$	HEK293	Rho activation
Muscarinic M$_3$	CHO	Stress fiber formation
Metabotropic mGluR1a	CHO	Phospholipase D activation
Tachykinin NK1	U-373-M6	IL6 and IL8 production
Purinergic P2Y	C6	MAP kinase activation
Purinergic P2Y	Swiss 3T3	RhoA activation, stress fiber formation
Purinergic P2Y2	Swiss 3T3	RhoA activation, stress fiber formation
Purinergic P2Y4	Swiss 3T3	RhoA activation, stress fiber formation
Purinergic P2Y6	Swiss 3T3	RhoA activation, stress fiber formation

continued

Table 4 Continued

GPCR	Cells	Assay cellular change
SDF-1alpha	Lymphocytes	Chemotaxis
Thrombin	132N1	Phospholipase D activation
Thrombin	Endothelial	Cell contraction and rounding
Thrombin	Fibroblasts	Membrane depolarization, Cl^- channel activation
Thrombin	Platelets	RhoA activation and shape change
Thrombin	NIH3T3	Focus formation and transformation
Thyroid-releasing hormone	GH4	Inhibition of ERG channels
Thyroid-stimulating hormone	FTRL-5	Phospholipase D activation
Urotensin II	Arterial smooth muscle	Contraction

Table 5 Effectors directly interacting with βγ-subunits

Effector	Regulation	References
Adenylyl cyclase I	Inhibition	167, 101, 168–171
Adenylyl cyclase II, IV, and VII	Stimulation	167, 101, 169–171
Akt/protein kinase B	Stimulation	172, 173
Brunton's tyrosine kinase	Stimulation	174
Calcium channels (L, N, P, Q, R type)	Inhibition	175–180
Calmodulin	Inhibition of CAM kinase	181
Dynamin I	Increased GTPase activity	182
GIRK 1, 2, 4 ion channels	Stimulation	183–188
G protein receptor kinases	Membrane recruitment	189–195
KSR-1 (stress responsive kinase)	Sequestration of βγ	196
PI3-kinase	Stimulation	197–200
PLCβ	Stimulation	201–205
PLCε	Stimulation	206
P-Rex1	Stimulation	207
Protein kinase D	Stimulation	208
RACK1 (receptor for activated C kinase)	Inhibition of βγ-mediated stimulation of effectors	209
Raf-1 protein kinase	Sequestration of βγ	210, 211
Sodium channels	Stimulation	212
Src kinase	Stimulation	213, 214
Tsk tyrosine kinase	Stimulation	215
Tubulin	Increased GTPase activity	216

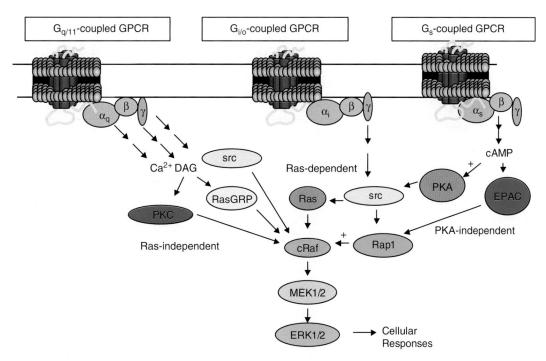

Figure 7 GPCR-mediated regulation of ERKs by classical pathways.

of receptor tyrosine kinases (RTKs) (a process called transactivation); and signals generated by the direct interaction of beta-arrestin and components of the MAPK phosphorylation cascade.[234–239] The overall mechanism(s) and the different signaling molecules involved in GPCR-mediated activation of ERK1/2 are still poorly defined. Furthermore, the mechanism involved in a given cell will be dictated by the complement and cellular orientation of signaling proteins present, which is regulated by interaction with a number of cellular scaffolding proteins.[236,238,240]

Classical activation of ERKs by GPCRs involves G protein-mediated activation of PKA or PKC (**Figure 7**). For example, in neuronal cells, G_s-coupled receptors stimulate ERKs via PKA mediated phosphorylation of the small G proteins, Rap-1 and RapB.[241–244] In addition, cAMP can act directly to stimulate the guanine nucleotide exchange factor EPAC in some cells, which in turn stimulates the Ras-like GTPase Rap-1 (**Figure 7**).[245–248] Activation of PKC by G_q-coupled receptors via PLCβ-mediated increases in intracellular diacylglycerol and Ca^{2+} can also lead to MAPK activation through both Ras-dependent and independent mechanisms.[249–251] In addition, PKC can directly phosphorylate Raf-1 resulting in MAPK activation.[252] G_i-coupled receptors can activate MAPKs via increases in PKC activation or through a mechanism involving βγ-mediated activation of the cytoplasmic tyrosine kinase, c-src. As previously stated, the mechanism of GPCR-mediated activation of ERK1/2 is highly dependent upon the receptor and the cell type that is investigated.

The second mechanism of GPCR-mediated MAPK activation involves activation of the small G protein Ras, via a pathway involving transactivation of specific RTKs such as the epidermal growth factor receptor (EGFR) (**Figure 8**).[253–255] The first step in the process of transactivation involves GPCR-mediated G protein heterotrimer dissociation and βγ-complex-mediated activation of c-src, which along with the βγ-complex, activates specific membrane-associated metalloprotease(s) (primarily members of the ADAM (a disintegrin and metalloprotease) family). These metalloproteases in turn act to cleave membrane-associated growth factor precursors to their active form, a process termed ectodomain shedding. The processed growth factor then binds to its receptor leading to receptor autophosphorylation, activation of a series of cytoplasmic adapter proteins culminating in activation of Ras, which in turn triggers activation of ERK1/2.[256] The differential involvement of RTKs and downstream signaling molecules activated in response to GPCR-mediated stimulation allows fine-tuning of GPCR signaling.[223] GPCR-mediated transactivation of other receptor tyrosine kinases, such as PDGF (platelet-derived growth factor), FGF (fibroblast growth factor), and Trk (a neurotrophic growth factor), has been reported, but it is unclear if ADAM proteases are involved as has been established for EGF receptor transactivation. It appears that GPCR-associated cancers involve activation of EGF and Ras signaling pathways.

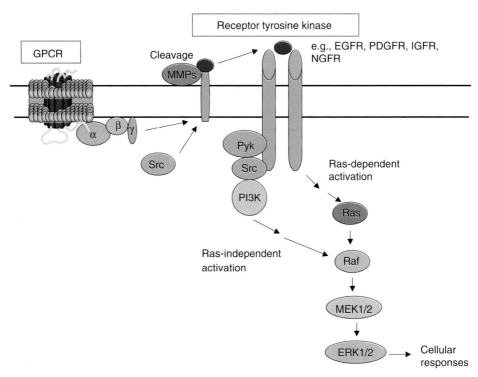

Figure 8 GPCR-mediated transactivation of RTKs and MAPK activation.

The third mechanism of MAPK activation involves the scaffolding protein, beta-arrestin. Initially, beta-arrestin was thought to primarily play a role in GPCR desensitization and internalization. However, beta-arrestin has clearly been demonstrated to play an important role in GPCR-mediated activation of ERK1/2. Specifically, it is now appreciated that beta-arrestin acts to recruit signaling proteins such as src and MAPK cascade components to the membrane as well as modulate GPCR endocytosis, trigger ERK1/2 activation, and target MAPKs and associated signaling molecules to specific locations in the cell.[222] In some cases, receptor-mediated activation of ERK1/2 via beta-arrestin is G protein independent. The precise mechanism(s) of beta-arrestin-mediated activation of ERKs is currently unclear but appears to require recruitment and interaction with c-src kinase.[257–258] For certain GPCRs, receptor internalization is a prerequisite for MAPK activation and for others it is not.[243] It appears that for many GPCRs, beta-arrestin-mediated activation of MAPKs results in retention of the kinase in the cytoplasm and results in phosphorylation of predominately cytoplasmic substrates. In contrast, activation of MAPKs through G protein-dependent pathways primarily promotes translocation of MAPKs into the nucleus resulting in activation of nuclear transcription factors.

2.20.2.6 Regulators of G Protein Signaling

Regulators of G protein signaling (RGS proteins) are cytoplasmic proteins that modulate signaling of heterotrimeric G proteins by facilitating the GTPase activity of G protein α-subunits.[259–262] Currently, there are more than 30 different members of this family divided into five separate subfamilies. All RGS proteins share a conserved RGS domain of 120–130 amino acids flanked by N- and C-terminal domains of variable length. RGS proteins also contain a γ-subunit-like domain important for interaction with the G protein β-subunits.[263] In addition to increasing G protein GTPase activity, RGS proteins can accelerate the kinetics of G protein α-subunit-mediated responses without compromising steady-state signaling strength.[264–267] One explanation for this latter effect may be that RGS proteins act as important scaffolds facilitating the assembly of signaling complexes between GPCRs and G proteins. Alternatively, RGS proteins may promote entry into the G protein cycle by facilitating dissociation of trimeric G proteins into prospective GTP-α and βγ subunits, thus serving as 'kinetic' scaffolds.[268,269] Recently, a novel model for the action of RGS proteins on the G protein cycle has been described in which RGS proteins enhance the kinetic efficiency of the ternary complex between agonist, GPCR, and G protein by a direct association with the G protein α-subunits.[270]

Evidence now suggests that RGS proteins can directly interact with GPCRs to modulate their activity. For example, RGS2 has been reported to directly interact with the IC3 loop of the α_{1A}-adrenergic receptor and is recruited by the unstimulated receptor to the plasma membrane to inhibit $G_{q/11}$ signaling.[271]

2.20.3 Updated Models of G Protein-Coupled Receptor Activation

2.20.3.1 Constitutively Active G Protein-Coupled Receptors

Many GPCRs are capable of undergoing ligand-independent activation and G protein coupling, commonly referred as constitutive activation. One of the first reports regarding constitutive activity of a GPCR came from a landmark study by Costa and Herz (1998) of delta-opiate receptors naturally expressed in NG108–15 cells.[272] They observed that certain delta-opiate antagonists displayed physiological actions that were opposite to those observed with delta-opiate agonists (termed 'negative intrinsic activity'), while other antagonists displayed no intrinsic activity. Furthermore, these unexpected actions could also be blocked with delta-opiate antagonists lacking negative intrinsic activity. Compounds displaying this novel activity were subsequently called 'inverse agonists' and compounds not displaying negative intrinsic activity are now called 'neutral antagonists.' There have been a large number of papers describing constitutively activate GPCRs in which ligand-independent signaling was observed in guanine nucleotide exchange assays or assays measuring second messengers such as cAMP or inositol phosphates (**Table 6**). There are three primary experimental approaches to generating constitutively active GPCRs including overexpression of native receptors in recombinant expression systems, overexpression of select G proteins,[273–276] or by mutation of amino acids primarily contained within TM regions and IC loops (see **Table 6**). Pharmacological characterization of a wide variety of previously characterized antagonists for several different GPCRs using constitutively active receptors has revealed that a large proportion of these GPCR ligands (greater than 85%) have now been reclassified as inverse agonists.[277] Furthermore, studies of different constitutively activating mutations of GPCRs have helped to elucidate the regions of GPCRs that are important in receptor activation and G protein coupling (see Section 2.20.3.2). An impressive list of different mutations that promote GPCR constitutive activation are available on-line.[667]

The pathophysiological significance of constitutively active GPCRs has been clearly confirmed by the existence of several naturally occurring activating mutations of different GPCRs that are clearly associated with several different human diseases (**Table 7**). Furthermore, a variety of GPCRs display constitutive activity in cell lines and tissues suggesting that this property is not simply an artifact of GPCR overexpression and that some GPCRs display constitutive activity when expressed at physiological densities (**Table 8**). Recently, a protein isolated from the CNS called agouti-related protein, has been shown to be an endogenous melanocortin MC3 and MC4 receptor inverse agonist. It is likely that other natural inverse agonists may exist for other constitutively active human GPCRs. One caution in studies of endogenous constitutively active GPCRs is that trace amounts of agonist must be removed from cell suspensions or plasma membrane preparations before constitutive activity can be confirmed. Another requirement for proving the existence of naturally occurring constitutively active GPCRs is the availability of a neutral antagonist to prevent inverse agonist as well as agonist actions. Several excellent reviews addressing agonist-independent activation of GPCRs have appeared in recent years.[278–289]

Methodologies to promote constitutive activation of GPCRs have enabled development of screens that allow identification of inverse agonists. If constitutively active GPCRs are associated with human disease, inverse agonists should be superior therapeutics to neutral antagonists.[288] However, if constitutively active GPCRs are not associated with a particular disease, neutral antagonists would be predicted to be as therapeutically effective as inverse agonists. Thus, the therapeutic importance of inverse agonists will be dependent upon instances in which the basal signaling by the receptor is controlled mainly by its intrinsic activity rather than by the endogenous ligand. A large number of inverse agonists, particularly for biogenic amine receptors, are widely used drugs. However, the clinical importance of their inverse agonist properties remains to be further established. In addition, the question remains as to whether inverse agonists can be designed rather than found in primary screening of small-molecule compound libraries.[290]

2.20.3.2 Models of G Protein-Coupled Receptor Activation and G Protein Coupling

The original model of GPCR activation and G protein coupling, called the 'ternary complex model' (**Figure 9a**), suggests that GPCR activation and receptor–G protein coupling involves an agonist-induced receptor conformational change that facilitates receptor and G protein coupling leading to formation of an active ternary complex between agonist, receptor, and G protein.[291,292] In the presence of GDP, agonist binding promotes the formation of a stable ternary complex displaying high affinity for agonists. However, in the presence of GTP, the ternary complex is only transiently formed, resulting in receptor-mediated G protein activation followed by ternary complex dissociation and

Table 6 Examples of constitutively active recombinant expressed human GPCRs

GPCR	Constitutive activation	Assay	GPCR	Constitutive activation	Assay
α_{1A}-adrenergic	Wild-type	IP	CCK_2	Mutation (TM3; ICIII)	IP
α_{1B}-adrenergic	Mutation (TM3 and TM5)	IP	Chemokine CCR2	Mutation (TM2)	GTPγS
α_{1B}-adrenergic	Wild-type	IP	Chemokine CCR3	Wild-type	GTPγS
α_{1B}-adrenergic	Mutation (IC3 loop)	IP	Chemokine CCR5	Mutation (TM2)	GTPγS
α_{1B}-adrenergic	Mutation (DRY)	IP			
α_{1D}-adrenergic	Wild-type	IP	Chemokine CXCR4	Mutation (TM3)	GTPγS
α_{2A}-adrenergic	Wild-type	cAMP	Leukotriene CysLT1	Mutation (TM2 and TM3)	IP
α_{2A}-adrenergic	Mutation (IC3 loop)	cAMP; GTPγS			
α_{2B}-adrenergic	Wild-type	cAMP	Dopamine D_1	Mutation (TM6)	cAMP
α_{2C}-adrenergic	Mutation		Dopamine D_{1A}	Wild-type	cAMP
Adenosine A_1	Complete	GTPγS	Dopamine D_{1B}	Mutation (IC3)	cAMP
Adenosine A_{2B}	Mutation (IC1 loop)	cAMP	Dopamine D_2	Mutation (IC3 loop)	cAMP
Adenosine A_3	Mutation (TM3 and TM6)	IP, cAMP	Dopamine D_2	Mutation (TM6)	cAMP
Angiotensin AT_{1A}	Mutation (TM3)	IP	Dopamine D_{2S}	Wild-type	cAMP; GTPγS
Angiotensin AT_{1A}	Mutation (TM7)	IP	Dopamine D_3	Wild-type	GTPγS;
β_1-adrenergic	Wild-type	cAMP	Dopamine $D_{4.2}$	Wild-type	GTPγS
β_2-adrenergic	Mutation (IC3 loop)	cAMP	Dopamine $D_{4.4}$	Overexpression	GTPγS
β_2-adrenergic	Wild-type	cAMP; ERK1/2	Prostaglandin EP_3	C-terminal truncation	cAMP; Rho
β_2-adrenergic	Mutation (TM3)	cAMP	Prostaglandin EP_3	C-terminus	Rho
β_2-adrenergic	Mutation (DRY)	cAMP	Prostaglandin EP_3	Wild-type	cAMP
β_2-adrenergic	Mutation (TM6)	cAMP			
Bradykinin B_1	Wild-type	IP	FPR1	Wild-type	GTPγS
Bradykinin B_2	Mutation (TM3 and TM6)	IP	FSH	Mutation (TM3)	cAMP
Bradykinin B_2	Mutation (IC2 loop)	IP, arachidonic acid release	$GABA_B$	Wild-type	cAMP
			Ghrelin (GHSR1)	Wild-type	IP
Bradykinin B_2	Wild-type	IP	Glucagon	Mutation (TM6)	cAMP
Calcitonin	Wild-type	cAMP	GPR3	Wild-type	cAMP
Calcium sensing	Mutation (N-terminus)	IP	GPR6	Wild-type	cAMP
CB_1 cannabinoid	Wild-type	GTPγS; cAMP	GPR12	Wild-type	cAMP
CB_1 cannabinoid	Mutation (IC3 loop)	cAMP	Gastric releasing peptide	Wild-type	IP
CB_2 cannabinoid	Wild-type	GTPγS, cAMP			
CB_2 cannabinoid	Wild-type	Reporter (CRE)	Histamine H_1	Wild-type	IP; NFκB
CCK	Mutation (DRY)	IP	Histamine H_2	Wild-type	cAMP

Table 6 Continued

GPCR	Constitutive activation	Assay	GPCR	Constitutive activation	Assay
Histamine H$_3$	Mutation (IC3 loop)	cAMP; GTPγS	Melatonin MT$_{1a}$	Wild-type	cAMP
Histamine H$_3$	Wild-type	GTPγS	Melatonin MT$_1$	Wild-type	GTPγS
Serotonin 5HT$_{1A}$	Mutation (IC3 loop)	GTPγS	Melatonin MT$_2$	Wild-type	GTPγS
Serotonin 5HT$_{1A}$	Wild-type	GTPγS	Neurotensin NT$_1$	Mutation (TM7)	IP
Serotonin 5HT$_{1B}$	Mutation (IC3 loop)	GTPγS; cAMP	Neurotensin NT$_2$	Wild-type	IP
Serotonin 5HT$_{1B}$	Wild-type	GTPγS	δ-opiate	Mutation (TM3 and TMV7)	GTPγS
Serotonin 5HT$_{1D}$	Wild-type	GTPγS	δ-opiate	Wild-type	cAMP; GTPγS
Serotonin 5HT$_{2A}$	Mutation (IC3 loop)	IP	κ-opiate	Wild-type	GTPγS
Serotonin 5HT$_{2A}$	Wild-type	RSAT	μ-opiate	Wild-type	GTPγS; cAMP
Serotonin 5HT$_{2C}$	Wild-type	IP; thymidine incorporation	μ-opiate	Mutation (TM3)	GTPγS
Serotonin 5HT$_{2C}$	Mutation (IC3 loop)	IP	μ-opiate	Mutation (DRY)	GTPγS
Serotonin 5HT$_{2C}$	Mutation (TM7)	IP	OLR-1	Mutation (TM3)	cAMP
Serotonin 5HT$_4$	Mutation (IC3 loop and C-terminus)	cAMP	Opsin	Mutation (TM6)	?
Serotonin 5HT$_4$	Mutation	cAMP	PTH/PTHrP	N-terminal truncation	cAMP
Serotonin 5HT$_6$	Mutation (IC3 loop)	cAMP	PTH/PTHrP	Mutation (IC1 and IC3 loops, TM2)	cAMP
Serotonin 5HT$_7$	Mutation (IC3 loop)	cAMP	PTH/PTHrP	Mutation (TM2, TM6, TM7)	cAMP
Serotonin 5HT$_7$	Wild-type	cAMP	PAF	Mutation (IC3 loop)	IP
Serotonin 5HT$_{7a}$	Wild-type	cAMP	Secretin	Mutation of conserved cysteine residues in EC loops	cAMP
Serotonin 5HT$_{7b}$	Wild-type	cAMP			
Serotonin 5HT$_{7d}$	Wild-type	cAMP			
LHR	Wild-type	cAMP	Secretin	Mutation (IC2 and IC3 loops, TM2)	cAMP
LHR	Mutation (TM4 and TM6)	cAMP	SIP5	Wild-type	cAMP, Erk
Melanocortin MC$_1$	Wild-type	cAMP	THR	C-terminus truncation and mutation	IP
Melanocortin MC$_4$	Wild-type	cAMP	TSHR	Wild-type	cAMP
Muscarinic M$_2$	Mutation (TM6)	cAMP	TSHR	Deletion (TM3)	cAMP
Muscarinic M$_3$	Mutation (TM5, TM6, TM2)	Yeast cell growth	TSHR	Mutation (TM6)	cAMP; IP
Muscarinic M$_5$	Mutation (TM6)	Proliferation	TSHR	C-terminal truncation	cAMP
mGluR1a	Wild-type	IP	TXA2	Mutation (ERY)	IP
mGluR5	Mutation (N-terminally truncated)	IP	Vasopressin V2	Mutation (DRY)	cAMP
			VIP$_1$	Mutation (IC1 loop)	cAMP
mGluR8	Mutation (IC2 loop)	cAMP	VIP$_1$	Mutation (IC2 loop)	cAMP

Information taken from [278–289].

Table 7 Human diseases caused by naturally mutated constitutively active GPCRs

GPCR	Disease
Ca^{2+} sensing	Familial hypocalcemia
LH	Familial male precocious puberty
LH	Sporadic Leydig cell tumors
PTH/PTHrP	Jensen's chondrodysplasia
Rhodopsin	Congenital night blindness
TSH	Familial nonautoimmune hyperthyroidism
TSH	Sporadic hyperfunctional thyroid adenomas

Information taken from [278–289].

conversion of the receptor to a low-affinity conformation. Although this model has been widely accepted for several years, it does not account for constitutively active GPCRs and inverse agonism.

To accommodate these new phenomena, the 'extended ternary complex model' has been recently adopted in place of the ternary complex model[293] (**Figure 9b**). This new model has incorporated the ability of GPCRs to spontaneously and conformationally isomerize (called the J constant) to an active state. A simplification of the extended ternary complex model, called the 'two-state model,' is shown in **Figure 9c**. This model suggests that GPCRs exist in an equilibrium between inactive (R) and active (R*) conformations and only the R* state is capable of interacting with G protein to promote a physiological response. Full agonists act to shift this equilibrium from the inactive R state to the active R* state and stabilize the active conformation. Partial agonists promote a similar directional equilibrium shift but are less effective than full agonists. Conversely, inverse agonists are capable of shifting the equilibrium from the active R* state to the inactive R state and stabilize the receptor in its inactive conformation. Similarly, partial inverse agonists promote a similar directional shift but are less effective than full inverse agonists. Inverse agonists also block agonist actions and this explains why many inverse agonists have been initially classified as antagonists. Neutral antagonists do not promote a shift in equilibrium between R and R* but are capable of blocking the actions of either agonists or inverse agonists. Ultimately, this model predicts that the concentration of active receptor G protein complex [R*G] will determine the level of constitutive activation. Several constitutively active mutant GPCRs display enhanced agonist-independent signaling compared to the native wild-type receptor expressed at similar levels. These observations suggest that the mutated receptor is a better representation of the natural active conformation of GPCRs, although this is likely an oversimplification. For example, in some instances, GPCR mutations have been reported that signal at considerably higher levels than agonist-occupied receptor suggesting that some constitutive activating mutations may not represent the agonist-occupied active R* state. Consistent with the two-state model, constitutively active GPCRs will display increased agonist affinity, increased potency for agonists and partial agonists, and, in some cases, increased receptor instability and desensitization.[294–299] Indeed, several reports regarding constitutively active mutated GPCRs suggest that they also display constitutive phosphorylation and undergo constitutive internalization[300–308] and, in some cases, the majority of receptor is internalized at steady state. However, there are also reports that inverse agonists can prevent constitutive receptor internalization.[307,308] Na^+ ions, similar to inverse agonists, stabilize the inactive R form of GPCRs and the precise molecular locus of the action of Na^+ ions is presumably a highly conserved aspartate residue in TM2.[309–313]

The extended ternary complex model adequately addresses most of the molecular pharmacology of GPCRs but it does not account for unproductive receptor and G protein coupling. For example, antagonists for opioid, histamine H_2, pheromone, and CB_1 receptors have been reported to promote formation of a nonsignaling ternary complex.[314–316] To accommodate these observations, the 'cubic ternary complex model' (**Figure 9d**) has been formulated. This model incorporates an allowance for the interaction of G protein not only with the active conformation of the receptor (R*) but also with the inactive conformation (R).[317–319] There are a number of excellent reviews that discuss the detailed theoretical considerations of the different models presented in **Figure 9**.[319–325]

Protean agonism is another theoretical prediction of the extended ternary complex model.[321] For constitutively active GPCRs, some compounds can display either agonist or inverse agonist activity depending upon the level of receptor constitutive activity. The reversal from agonism to inverse agonism will occur if the agonist-promoted active

Table 8 Examples of constitutively active GPCRs in cell lines and tissues and inverse agonist actions

Receptor	Tissue or cell	Inverse agonist effect
α_1-adrenergic	Rat aorta	Decreased contraction
α_{2A}-adrenergic	H.E.L.92.1.17	Increased cAMP levels
α_{2D}-adrenergic	Rat RIN5AH cells	Decreased basal GTPγS binding
β_1-adrenergic	Human ventricle	Decrease in contraction
$\beta_{1/2}$-adrenergic	Guinea pig and human cardiomyocytes	Decreased calcium current
$\beta_{1/2}$-adrenergic	Turkey erythrocyte membranes	Decreased adenylyl cyclase activity
$\beta_{1/2}$-adrenergic	Rat atria	Reduction in contraction
β_2-adrenergic	Human mast cells	Opening of iKCa1 ion channel
β_2-adrenergic	Guinea pig trachea	Increased contraction
β_2-adrenergic	Bovine tracheal smooth muscle	Increased sensitivity to muscarinic agonists
β_2-adrenergic	BC3H1 cells	Decrease in basal cAMP levels
Bradykinin B	Human H69 cells	Decrease in basal thymidine incorporation
Bradykinin B_1	Rat and bovine myometrial membranes	Decreased IP accumulation
CB_1 cannabinoid	Rat cerebellar granule cells	Increased aspartate release
CB_1 cannabinoid	Mouse vas deferens	Increased norepinephrine release
CB_1 cannabinoid	Human neocortex	Increased acetylcholine release
CB_1 cannabinoid	Rat superior cervical ganglion	Increased calcium current
CB_1 cannabinoid	Mammalian brain	Increase in cAMP levels
CB_1 cannabinoid	Rat brain	Increased basal cAMP; decreased basal GTPγS binding
CB_1 cannabinoid	Rat hippocampal synaptosomes	Increase in acetylcholine release
Dopamine D_2	GH4C1 cells	Increases in cAMP levels
GLP-1	Murine betaTC-Tet cells	Decreased basal insulin release
Histamine H_2	U-937 cells	Decrease in basal cAMP levels
Histamine H_3	Rat synaptosomes	Increased histamine release
Serotonin $5HT_{2C}$	Rat striatum and nucleus accumbens	Increased dopamine release
Serotonin $5HT_{2C}$	Rat choroid plexus	Decrease in basal IP accumulation
Melatonin MT_1	Rat caudal arteries	Decreased GTPγS binding
Muscarinic	Rat cardiomyocytes	Increase in cAMP levels
Muscarinic	Porcine atrial membranes	Reduction in GTPγS binding
Muscarinic	Frog/rat cardiac cells	Decreased calcium current
Muscarinic	Rat thoracic aorta strips	Decreased calcium current
Muscarinic M_2	Pig atrial sarcolemma	Decreased GTPase activity
Muscarinic M_2	Rat cardiomyocytes	Increase in cAMP levels
δ-opiate	NG-108-15	Decreased GTPase activity; decreased basal GTPγS binding
δ-opiate	GH3 cells	Reduction in basal GTPγS binding
κ-opiate	Guinea pig ileum	Reduced contraction
μ-opiate	GH3	Reduction in GTPγS binding

Information taken from [278–289].

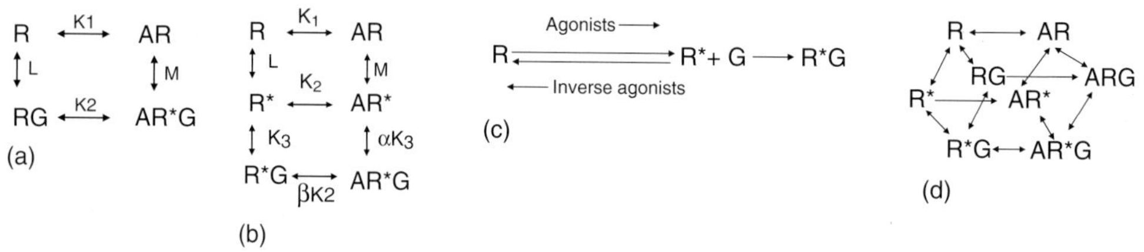

Figure 9 Models of GPCR and G protein activation. (a) Ternary complex model; (b) extended ternary complex model; (c) simplified two-state model; and (d) the cubic extended ternary complex model (without equilibrium dissociation constants). R, receptor; A, agonist or endogenous ligand; G, G protein; AR, inactive agonist receptor complex; AR*G, active ternary complex between agonist, receptor, and G protein; ARG, inactive ternary complex between agonist, receptor, and G protein; R*G, active receptor-G protein complex; RG, inactive receptor-G protein complex.

conformation has lower efficacy for G protein coupling than the constitutively active receptor. A clear example of protean agonism concerns the histamine H_3 ligand proxyfan, which acts as a protean agonist at recombinant H_3 receptors expressed in Chinese hamster ovary (CHO) cells.[326] The therapeutic value of protean agonists has not been established.

2.20.3.3 The Multistate Model of G Protein-Coupled Receptor Activation

The various forms of the extended ternary complex described above adequately explain the properties of agonism, antagonism, and inverse agonism for GPCRs, but all of these models assume that there are only two functional or active states of GPCRs. However, a considerable amount of experimental evidence now suggests that most GPCRs are not just on/off switches for activating G proteins and that there are at least three or more active conformations of GPCRs. An alternative model of GPCR activation accommodating multiple active GPCR conformations is referred to as the 'three state' or 'multistate' model.[327]

One important observation supporting this model is the simple fact that several GPCRs clearly couple to multiple G proteins and activate multiple signaling pathways.[328] Although, in many cases, this multiplicity of G protein coupling can be an artifact of overexpression of GPCRs in recombinant expression systems, several examples of GPCRs coupling to multiple G proteins and signaling pathways in cell lines and tissues expressing endogenous levels of receptor exist.[328–335] Another important observation supporting this new model is that mutations in the TM regions of certain GPCRs have been shown to result in a conversion of activation of multiple G proteins and signaling pathways to a single G protein and associated signaling pathway.[336–339] Since TM regions are not involved in influencing selectivity of receptor and G protein coupling, these specific mutations should not alter receptor G protein interactions. Furthermore, functional G protein reconstitution studies with GPCRs expressed and purified from insect cells and different G protein α-subunits indicate that different agonists can produce different patterns of G protein activation.[340]

Multiple agonist-induced active GPCR conformations can also be inferred from studies in which different agonists have differential efficacy in activation of different G proteins and signaling pathways,[341–344] and that certain agonists appear to prefer activation of one G protein signaling pathway over another, a phenomenon referred to as 'agonist trafficking.'[320–325] The multistate model thus also provides the theoretical basis for the concept of signaling-selective agonism. There are numerous examples of agonist trafficking for several different GPCRs in the literature. For example, agonist activation of the $5HT_{2C}$ receptor results in the activation of two signaling pathways, G_q-mediated activation of inositol phosphate accumulation and G_q-independent activation of phospholipase A_2 resulting in cellular arachidonic acid accumulation. Some agonists (e.g., trifluoromethylphenyl-piperazine) were found to be more efficacious in stimulating inositol phosphate accumulation whereas others (e.g., LSD) favored activation of the phospholipase A_2.[341] Agonist trafficking has been demonstrated for several surrogate ligands, but only a few examples exist with endogenous GPCR ligands. In some cases, agonist trafficking may really occur through a strong signal-type mechanism in which a high-efficacy agonist could activate multiple signaling pathways whereas a weaker efficacy agonist may stimulate only one.[342] Other examples of ligand-selective GPCR regulation include ligands that promote coupling to one G protein pool while inhibiting coupling to another. For example, the GnRH receptor antagonist Ant135-25 acts as an agonist in the G_i signaling pathway and as an antagonist in the G_q pathway.[343] Signal-selective antagonism has also been reported for GPCRs including the cholecystokinin CCK_B[344] and tachykinin NK1 receptors.[345]

Dual efficacy ligands have also recently been reported to exist for certain GPCRs and their existence also supports the concept of multiple active conformations of GPCRs. For example, propanolol and ICI118551, two well-established β_2-adrenergic inverse agonists in the G_s pathway, have been reported to be partial agonists in activating the MAPK pathway in the same cells.[346,347] Furthermore, the activation of the MAPK pathway was demonstrated to be G protein-independent and required recruitment of the scaffold protein, beta-arrestin. Dual efficacy ligands have also been discovered for delta-opiate receptors that act as inverse agonists in the cAMP pathway (reverse agonist-mediated decreases in cAMP) and agonists in the MAPK pathway. The discovery of dual efficacy ligands adds even more complexity to the molecular pharmacology of GPCRs that already exists with the advent of agonist trafficking.

The most convincing data regarding the existence of multiple agonist-induced conformations of GPCRs comes from several important fluorescence labeling studies performed in Brian Kobilka's laboratory. In an early study, a fluorescent reporter was chemically incorporated into the interface of the IC3 loop and TM3 of the β_2-adrenergic receptor and fluorescence changes around this reporter were evaluated with exposure to chemically different receptor agonists.[348] These studies indicated that agonists promote formation of two different conformations of the β_2-adrenergic receptor IC3 loop, a receptor domain critical in mediating G protein coupling. In addition, it was observed that receptor conformations promoted by partial agonists were distinguishable from those promoted by full agonists. In another later study, catechol derivatives were designed to mimic the endogenous ligand (epinephrine, norepinephrine) and were observed to induce two kinetically distinguishable active conformations of the β_2-adrenergic receptor.[349] A more recent study from this laboratory has also demonstrated differences in receptor conformations promoted by structurally related agonists and partial agonists for the β_2-adrenergic receptor.[350] Recently, studies of both opiate[351,352] and β_2-adrenergic receptors[353] using plasmon resonance spectroscopy, a method used to evaluate ligand–receptor interactions that avoids the requirement of receptor labeling, have also indicated that agonists, partial agonists, and inverse agonists may promote formation of different receptor conformations. Collectively, these studies support the multistate model of GPCR activation.

Several recent studies of GPCR phosphorylation, internalization, and desensitization also indicate that different agonists for a given GPCR can promote different levels of receptor phosphorylation and subsequent internalization supporting the notion that GPCRs can also adopt different deactivated conformations.[354–361] For example, both enkephalins and morphine stimulate delta- and mu-opiate receptors but only the former promotes receptor desensitization. Similarly, endogenous ligands for the CCR7 receptor including CCL19 and CCL21 display equivalent potency and efficacy in stimulating increases in intracellular calcium but dramatically differ in their ability to promote desensitization of the CCR7 receptor.[362] Recent mass spectroscopy studies of the β_2-adrenergic receptor have shown that selective ligands can determine the degree of phosphorylation of different intracellular regions of the receptor. Phosphorylation of the IC3 loop is increased to a similar degree with the agonists isoproterenol, dopamine, and epinephrine. However, differences in agonist-mediated phosphorylation of the C-terminal tail were detected with dopamine being less capable of promoting phosphorylation than isoproterenol or epinephrine.[363] A recent study of three constitutively active mutants of the C5a receptor also provides evidence for multiple conformations of GPCRs. One receptor mutant exhibited both constitutive activation of G protein and endocytosis, another mutant exhibited constitutive activation of G protein but was found to undergo endocytosis only in the presence of agonist, and a third mutant only activated G protein with agonist present and was found to be constitutively endocytosed.[364] Vauquelin and Van Liefde,[327] Perez and Karnik,[365] and Maudsley and co-workers[342] have recently reviewed the accumulating evidence supporting multiple active conformations of GPCRs and their potential impact on GPCR drug discovery. These reviews suggest the theoretical possibility that each chemically distinct agonist will impart upon the receptor its own unique activated conformation.

The potential existence of multiple GPCR conformations of endogenously expressed GPCRs (possibly several) in disease relevant cells and tissues could have an enormous impact on GPCR drug discovery programs. Their existence suggests that it may be possible to selectively target different receptor conformations with new drugs to increase their efficacy and reduce on-target associated side effects. The case for the multiple active states for several different GPCRs is now quite strong, but the most important question concerns what receptor conformation(s) and associated signaling pathway(s) are related to GPCR-associated diseases. If the signaling pathway associated with a therapeutic target can be linked to the disease of interest or to unwanted side effects, GPCR signaling-biased screens should be developed and implemented for drug discovery efforts. For GPCR drug discovery programs, it is possible that important compounds may be missed in a screening campaign because the appropriate signaling pathway was not captured in the functional screen used. Furthermore, if signaling pathways associated with the therapeutic endpoint are unknown, it may be necessary to perform multiple functional screens relevant to all known GPCR-activated signal transduction pathways activated by the GPCR target. Some GPCR experts have recently suggested that the potential multiplicity of G protein coupling and signaling may require that GPCR drug discovery should start with a primary emphasis on receptor binding screens followed by a battery of relevant functional assays.[366] This is a departure from recent years when the industry embraced predominantly functional cell-based assays over ligand binding assays. The development of fluorescence-based technologies to evaluate different agonist or surrogate ligand-induced conformational states upon ligand binding will become a new component to successful GPCR drug discovery programs in the future.

2.20.3.4 Organization of G Protein-Coupled Receptors and Signaling Molecules

Over the last 4 years, it has become clearly established that many GPCRs, G proteins, and effectors are not randomly distributed in the plasma membrane but form discrete signaling complexes involving a variety of accessory proteins referred to as GPCR interacting proteins (GIPs) (Table 9). Specifically, GIPs act to assemble GPCRs, G proteins, and effectors into prearranged signaling complexes and involve receptor C-terminal tail PDZ domain motifs and SH3 motifs.[23–26,342,371–375] In general, organization of GPCRs in microdomains and their interaction with GIPs is thought mainly to control the speed and specificity of GPCR signaling.[240,367] GIPs also play roles in GPCR regulation such as desensitization and internalization. Potential GPCR–protein interactions are typically evaluated using yeast two-hybrid screening and immunoprecipitation methodologies in recombinant cells[368] or by using the C-terminal tail GST-fusions as bait to isolate the interacting proteins.[369] Once identified, it is important to establish that a putative GPCR–protein interaction be demonstrated in vivo. Nowhere has the importance of the organization of GPCRs and associated signaling molecules been more clearly demonstrated than in the spatial and temporal activation of the MAPK pathway.[370] Activation of a nuclear pool of ERK1/2 as a consequence of GPCRs transactivation of receptor tyrosine kinases may provide a mitogenic stimulus whereas activation of ERK1/2 localized at the plasma membrane or in endosomal vesicles via scaffolding proteins such as beta-arrestin may spatially restrict ERK1/2 activity favoring phosphorylation of non-nuclear ERK substrates.[370] Proteomics approaches should enable identification and quantification of additional GIPs associated with various GPCRs so that a better understanding of macromolecular signaling complexes can be delineated.

Several GIPs reported to interact with GPCRs and their potential effects on GPCR function, signaling, and regulation are listed in Table 9. GIPs may represent a new locus to pharmacologically modulate GPCR function and may not only provide alternative ways to modify receptor activity but also to exploit new chemical space for drug-like molecules.[376] The relevance of GIPs to GPCR drug discovery is currently unclear but we anticipate that cellular content and GPCR–GIP interactions will become increasingly more important variables to be considered in developing new GPCR functional screens. Targeting GPCR and protein interactions may be a reasonable therapeutic endpoint, but our ability to selectively block protein–protein interactions with small molecules has been limited up to this point.

2.20.3.5 Models of Class A G Protein-Coupled Receptor Activation and G Protein Coupling Based on Rhodopsin

Elucidation of the structural basis for ligand-mediated activation of GPCRs by x-ray crystallography has been limited because it is difficult to produce and purify large quantities of protein and it is difficult to form suitable crystals of integral membrane proteins. However, the x-ray crystal structure of the light-activated GPCR rhodopsin has been solved primarily as a result of the availability of large quantities of purified receptor from bovine retina and the stabilization of solubilized receptor with the inverse agonist 11-cis-retinal resulting in suitable formation of crystals for x-ray.[417,418] The x-ray crystal structure of rhodopsin confirms the seven TM bundle arrangement of GPCRs proposed many years earlier. Collectively, analyses of x-ray crystallography data, biochemical approaches including cysteine cross-linking, spin labeling and scanning accessibility, and analyses of retinal movement by photoaffinity labeling have been valuable methodologies to determine the mechanism of light activation of the rhodopsin receptor.[419–421] Current models suggest that photons induce retinal isomerization, which triggers an outward movement of TM helices 3, 4, and 6, and which, in turn, exposes rhodopsin intracellular loops to couple to the G protein transducin.[422–425] Recent reviews suggest that light activation of rhodopsin involves reorientation of the cytoplasmic end of TMVI and changes in the relative disposition of TMVI and TMIII.[426,427] Light-induced conformational changes have also been observed in the cytoplasmic domain spanning TM1 and TM2 and the cytoplasmic end of TM7 and helix 8. Since the crystal structure of rhodopsin has been generated for only inverse agonist bound receptor, it has provided only limited information concerning activated conformations of other GPCRs.

Table 9 GPCR interacting proteins

GPCR interacting protein	GPCRs	Functional significance	References
AKAP79/250 (Gravin)	β_2-adrenergic	Organizes GPCRs, kinases, phosphatases, ion channels, and cytoskeletal elements	377, 378
β-Arrestin	Most GPCRs	Organizes numerous signaling molecules (e.g., src) into receptosomes	379
ATRAP	Angiotensin AT_1	Promotes receptor downregulation	380
Calmodulin	$mGluR_7$	Regulation of receptor phosphorylation and signaling	381, 382
Calnexin	LHR, FSHR, vasopressin V_2	Regulation of endoplasmic reticulum export	383, 384
Caveolin-1	Serotonin $5HT_{2A}$	Regulates G_q signaling	385
CIPP	Serotonin $5HT_{2A}$	Regulation of ion channel activity?	386
Clathrin adaptor complex 2	α_{1B}-adrenergic	Role in receptor endocytosis	387
CortBP1	Somatostatin $SSTR_2$	Unknown	388
Endophilin	β_1-adrenergic	Promotes receptor internalization	389
Filamin-A	CaR, dopamine $D_{2/3}$	Scaffold protein for MAPK components	390–392
GASP-1/2	δ-opiate	May regulate receptor sorting in the golgi	393
GIPC	Dopamine $D_{2/3}$, LHR	Regulates receptor trafficking, recruits RGS proteins	394, 395
Homer	$mGluR_{1, 5}$	Regulates receptor interactions with signaling proteins	396
JM4	Chemokine CCR_5	Regulation of receptor trafficking	397
MAGI-2	β_1-adrenergic	Regulation of receptor internalization	398
MMP-3	Serotonin $5HT_{2C}$	Unknown	386
MRAP	Melanocortin MCR2	May play a role in receptor trafficking to the plasma membrane	399
MUPP1	Serotonin $5HT_{2C}$	Assembly of signaling complexes,	400
Muskelin	Prostaglandin EP_3	Controls cytoskeletal organization	401
NHERF-1/2	β_2-adrenergic	Role in receptor recycling versus lysosomal degradation	402
NHERF	κ-opiate	Facilitates receptor homodimerization	403
Nm23-H2	Prostanoid TP receptor	Modulates receptor internalization	404
Periplakin	Opiate, MCHR	Limits receptor and G protein interaction	405, 406
PICK	Prolactin-releasing peptide, $mGluR_7$	Regulates PKC phosphorylation of receptors, receptor clustering	407, 408
Protein phosphatase 1C, 2C	$MGluR_{1a, 5a, 5b, 7b}$	Regulates receptor internalization	409
PSD-95	β_1-adrenergic, serotonin $5HT_{2C}$, $5HT_{2A}$	Augments signal transduction and inhibits receptor internalization	410
RCP	CGRP	Regulates receptor signal transduction	411
Siah1A	$mGluR_{1/5}$	Mediates degradation of internalized receptors	412
Spinophilin	$\alpha_{2A/B}$-adrenergic	Stabilizes receptor expression; regulates calcium signaling via interaction with RGS proteins	413, 414
Tamalin	$mGluR_{1,2,3,5}$	Regulates receptor signaling	415
Tubulin	$mGluR_7$	Regulation of receptor trafficking	416

The mechanism by which the binding of 'diffusible ligands' leads to GPCR activation and G protein coupling has only recently begun to be elucidated. The x-ray crystal structure of rhodopsin has been successfully used as a template to create homology models for several GPCRs including the dopamine D_2, muscarinic M_1, and chemokine CCR5 receptors. However, crystal structures of other GPCRs have been difficult to obtain due to limitations in producing milligram quantities of pure receptor, poor methods of crystallization, and inadequate methods to reconstitute GPCRs into lipid bilayers so that active conformations can be studied. Fluorescence spectroscopic studies of the β_2-adrenergic receptor using fluorescent probes indicate that activation of the β_2-adrenergic receptor involves rotation and tilting of TM6 upon agonist binding.[428–433] The change in conformation of this core domain generally affects the conformation of the IC2 and IC3 intracellular loops that are directly linked to TM3 and TM6 and form one of the principle sites involved in G protein recognition and activation. Other regions of class A GPRCs playing a role in receptor activation include an aspartate residue in TM2 and the DRY region at the interface of TM3 and the IC2 loop. Recently, light-driven activation of β_2-adrenergic receptor signaling has been demonstrated using a chimeric rhodopsin receptor containing β_2-adrenergic cytoplasmic loops.[434] Site-directed mutational and biochemical studies of dopamine D_2[435,436] and muscarinic M_1[437,438] receptors have also supported this model of GPCR activation. However, cysteine cross-linking studies of the muscarinic M_3 receptor also suggest that receptor activation involves both rotation of TM7 and an increase in the proximity of the C-terminal ends of TM1 and TM7.[439] Recently, Baneres and colleagues (2005) have purified the serotonin 5-HT$_{4a}$ receptor and evaluated conformational changes associated with receptor activation by both circular dichroism (CD) and steady-state fluorescence measurements.[440] They reported a specific change in the near UV CD band associated with the disulfide bond connecting TMII to the IC2 loop. New developments in NMR technology should allow further elucidation of ligand-mediated activation of GPCRs and provide information complementary to x-ray crystallography studies. In general, activation of class A GPCRs by biogenic amines is considered to involve a conformational change resulting in release of a constrained inactive conformation resulting in G protein coupling.[441] In addition, numerous GPCR mutational studies resulting in constitutive activation also support this model of activation (see Section 2.20.3.1).

Activation of class A GPCRs by peptide-hormone receptors does not appear to involve salt bridge disruption, but displacement of residues in TM3 appears to be involved in activation.[441–444] In addition to TM3 movements, activation of the angiotensin II receptor has also been reported to involve movement of TM2 and has led to the hypothesis that receptor activation involves an interaction of TM2 and TM7.[444] The EC3 loop connects TM6 and TM7 and at the cytoplasmic terminus of these two helices are the IC3 loop and helix 8, both of which have been implicated in G protein coupling. The EC3 loop is generally short and, as such, may constrain their motion preserving the inactive state. Substitutions of the EC3 loop relieving this constraint lead to constitutive activation of the delta-opiate receptor suggesting that the EC3 loop is important for receptor activation.[445] Activation of glycoprotein binding GPCRs such as the TSH and LH receptors appears to involve initial binding to the long N-terminus of the receptor and activation also involves TM helices.[446,447] For example, mutation of TM6 and TM7 of the LH receptor results in constitutive activation. Thus, conformational changes associated with extracellular loops are also associated with ligand-dependent activation of some GPCRs. Our understanding of GPCR activation and associated conformational changes is still in its infancy and is mostly limited to class A receptors. It will be interesting to see if the x-ray crystal structures of additional GPCRs will be solved, especially class B and class C members.

2.20.3.6 G Protein-Coupled Receptor Oligomerization and Heterodimerization

The view that GPCRs function as monomeric proteins has recently been challenged by numerous recent studies and it is now well established that many, if not most GPCRs, exist as dimers or even higher structure oligomers.[448–452] A commonly used methodology employed in early studies of GPCR dimerization has been coimmunoprecipitation of epitope-tagged GPCRs after solubilization with appropriate detergents. However, potential issues with this approach include concern that solubilization of hydrophobic proteins with detergents can promote formation of artifactual protein aggregates and the use of antibodies can promote artifactual receptor clustering. A newer methodology used to study GPCR dimerization in living cells is bioluminescence resonance energy transfer (BRET), which is based on the transfer of energy between recombinant expressed GPCR fusion proteins in which, for example, Renilla luciferase and green fluorescent protein (GFP) are recombinantly expressed on the C-terminal tail. Energy transfer can only take place between luciferase-tagged GPCR and GFP-tagged GPCR if both molecules are in close proximity after addition of the luciferase substrate coelenterazine.[453–455] Another related methodology used to study GPCR dimerization in living cells is fluorescence resonance energy transfer (FRET) and is based on transfer of fluorescence energy between recombinant expressed GPCR fused to cyan fluorescent protein (CFP) and the same or a distinct GPCR fused to yellow fluorescent protein (YFP) when they are in close spatial proximity.[456]

In addition, time resolved FRET[457,458] using europium chelate technology has also been used to study GPCR dimerization. Furthermore, fluorescent-labeled ligands have been used to study GPCR dimerization as well.[459,460] One limitation of early BRET and FRET studies was that they were not able to discern whether receptor dimerization occurred at the cell surface or in intracellular compartments. However, the marriage of BRET/FRET technologies with confocal microscopy has helped to resolve the existence of GPCR dimers in the plasma membrane.

Considerable evidence now suggests that many GPCRs undergo dynamic dimerization.[460–462] A major concern with dimerization studies using recombinant expressed GPCRs is the possibility that artifactual dimerization may take place primarily due to GPCR overexpression and may not be physiologically relevant. In addition, it is often difficult to determine the proportion of receptors existing as monomers and dimers. A study of BRET saturation in which different ratios of Renilla luciferase and GFP-tagged forms of the β_2-adrenergic receptor were recombinantly coexpressed has indicated that 80% of the total receptor population is present as dimers.[463] The percentage of monomers and dimers of any given GPCR may be quite variable and will most likely be different for different cells. Important GPCR domains mediating dimer formation include TM regions and the C-terminal tail.[464–470] Current evidence suggests that at least for the metabotropic glutamate receptors, GPCR dimers can interact with only one G protein and that, in some cases, only one dimer binds the endogenous ligand.[471]

One of the primary cellular effects of GPCR dimerization appears to be regulation of GPCR expression and trafficking. For example, dimerization of the muscarinic M_3 receptor has recently been reported to be required for beta-arrestin recruitment to the plasma membrane during the process of endocytosis.[472] In addition, GPCR dimerization is important for proper protein folding and for transport of GPCRs from the endoplasmic reticulum and the golgi apparatus to the plasma membrane during biosynthesis.[473] A recent report by Cao and co-workers (2005) suggests that dimerization of the β_2-adrenergic receptor when expressed in human embryonic kidney cell line (HEK293) cells may be important for determining the fate of the receptor upon internalization (recycling to the plasma membrane versus lysosomal degradation).[474] Currently, it is unclear whether oligomerization is ligand regulated or a general prerequisite for GPCR signaling.[450]

GPCRs also undergo heterodimerization, a process in which different but usually related GPCRs form physical complexes that, in some cases, can result in changes in GPCR ligand binding specificity, trafficking, and signaling.[448–450,475] Heterodimerization is absolutely required for functional activity for some GPCRs such as GABA$_B$ and taste receptors. As with studies of dimerization in recombinant cells, it is often unclear as to whether heterodimerization represents a physiologically relevant interaction or whether it is an artifact of GPCR overexpression. Unfortunately, there is limited information regarding GPCR heterodimerization in endogenous cells and tissues. That said, there are many reports of GPCR heterodimerization in a number of different cell lines. Putative GPCR domains important for heterodimerization, similar to those mediating homodimerization, include TM regions and the C-terminal tail.[476–478] One of the most striking examples of heterodimerization involves delta- and mu-opiate receptors. Coexpression of these two receptors results in pharmacology that is distinct from either of the monomers and the delta/mu heterodimer also displays strong positive cooperativity in ligand binding.[479] Several other reports also suggest that other GPCR heterodimers display a unique pharmacology compared to their constitutive monomers and this property could significantly contribute to the functional diversity of GPCRs.[480] For example, a study of melatonin MT1/MT2 heterodimers suggests that heterodimers have distinct ligand interaction properties compared to monomers.[481] Similarly, heterodimerization of β_1-adrenergic and β_2-adrenergic receptors in cardiac myocytes promotes a higher potency of isoproterenol to stimulate cAMP accumulation than with either receptor alone.[482]

One of the most common functional consequences of GPCR heterodimerization is change in receptor internalization and desensitization. In a study of the functional consequences of heterodimerization of $G_{i/o}$-coupled opiate receptor and G_s-coupled β_2-adrenergic receptor, heterodimerization did not affect ligand binding or G protein coupling but did affect receptor internalization as indicated by the ability of the β_2-agonist isoproterenol to promote delta-opiate receptor internalization and, conversely, the opiate agonist etorphine promotes β_2-adrenergic receptor internalization.[483] There are several other reports also indicating that GPCR heterodimerization can prevent internalization and endocytosis of associated receptor monomer.[484–450] For example, coexpression of β_1-adrenergic and β_2-adrenergic receptors in HEK293 cells results in heterodimer formation and agonist-mediated internalization of β_2-receptor was inhibited by heterodimerization with the β_1-receptor. Recently, Uberti and co-workers (2005) have reported that heterodimerization of α_{1D}-adrenergic receptors with β_2-adrenergic receptors promotes membrane surface expression of α_{1D}-adrenergic receptors as well as their ability to mobilize intracellular calcium and that heterodimerization with β_2-adrenergic receptors also confers the ability of the α_{1D}-adrenergic receptor to become internalized with the β_2-adrenergic receptor.[484]

Heterodimerization may also affect G protein coupling and GPCR signaling. For example, heterodimerization of the ORL-1 receptor and mu-opiate receptors when expressed in HEK293 cells results in formation of a heterodimer that selectively cross-desensitizes the mu-opiate receptor leading to impairment of its signaling and may account for ORL-1 receptor-mediated antiopioid effects in the brain.[485] In addition, formation of cannabinoid CB_1 and dopamine D_2 receptor heterodimers has been reported to result in a shift from CB_1-mediated inhibition of adenylyl cyclase to stimulation of adenylyl cyclase and MAPK cascades.[486] The question as to whether receptor ligands induce or inhibit GPCR heterodimerization is currently still debated and there are examples of both effects.[448–450] Overall, the functional and pharmacological consequences of GPCR heterodimerization appear to be diverse and can be highly influenced by what cells are used for expression. Additional studies of GPCR heterodimers in endogenous cells and tissues should help to further elucidate the pharmacological and functional consequences of GPCR heterodimerization.

The relevance of GPCR heterodimerization to GPCR drug discovery remains to be further established. Despite the accumulating in vitro data supporting physiological heterodimerization, supporting in vivo information from animals is lacking. Furthermore, the role of GPCR heterodimers in human diseases needs to be further established. Limited clinical evidence for a role of GPCR heterodimerization in human disease comes from a study in which hypertension in pre-eclampsia has been found to be associated with an increase in heterodimerization of angiotensin AT_1 and bradykinin B_2 receptors.[487] The greatest impact of GPCR heterodimerization for drug discovery will be targeting therapeutically relevant constitutive GPCR heterodimers that exhibit novel pharmacology compared to receptor monomers. A recent report indicates that the opiate analgesic agonist 6′GNTI has the unique property of selectively activating only opioid receptor heterodimers (κ/δ) and not momomers.[488] The therapeutic relevance of heterodimerization of a large number of GPCR pairs in native tissues still needs to be fully elucidated. What complicates this evaluation is that many therapeutically interesting cells or tissues express up to 40–50 different GPCRs making the potential number GPCR monomer combinations enormous.[489] It will also be interesting to determine which existing orphan GPCRs do not have their own personal endogenous ligand; these might work by endowing novel pharmacological and functional properties on other known GPCRs through heterodimerization.[490] A combinatorial expression approach, although laborious, may help to establish this possibility. GPCR heterodimerization has also led to the evaluation of novel bivalent ligands that interact with binding pockets of both monomers but their therapeutic utility remains to be established.

2.20.3.7 G Protein-Coupled Receptor Desensitization

Determination of the molecular events and players associated with GPCR desensitization, internalization, recycling, and degradation has been the focus of several laboratories. A major facilitator of these studies has been the marriage of the use of GFP-tagged GPCRs or beta-arrestin proteins, BRET or FRET technologies, and confocal microscopy.[491,492] GPCR function is generally terminated by three mechanisms: desensitization (defined as a loss of agonist response), endocytosis or internalization, and downregulation or a loss of receptor number due to degradation. A widely accepted canonical model of GPCR desensitization involves the following steps: (1) agonist binding promotes a conformational change in the receptor leading to phosphorylation of serine and threonine residues located within IC2 and IC3 loops and the C-terminal tail by PKA, PKC, or one of a collection of seven G protein-coupled receptor kinases called GRKs; (2) GPCR phosphorylation leads to recruitment of beta-arrestin from the cytosol to the plasma membrane leading to inhibition of receptor–G protein coupling and facilitating targeting of the phosphorylated GPCR to clathrin-coated pits; (3) GPCR is removed from the plasma membrane surface via a dynamin-mediated endocytosis; (4) agonist dissociates from the GPCR within endosomes; and (5) internalized GPCR is either chaperoned back to the plasma membrane subsequent to beta-arrestin dissociation or is degraded by lysosomes[493] (**Figure 10**). Important GPCR domains associated with desensitization and endocytosis include IC2 and IC3 loops and the C-terminal tail, all of which are also important for G protein coupling.[494–499] Specific C-terminal amino acid motifs present in some GPCRs are important for determining whether the receptor is recycled to the plasma membrane or is degraded[500] and are also critical for beta-arrestin-independent internalization.[501]

Ligand-dependent GPCR internalization and endocytosis occurs primarily by two different mechanisms. The most widely employed mechanism involves GPCR internalization into clathrin-coated pits.[502] Clathrin-coated pit formation requires recruitment of adapter protein complexes and accessory proteins to the membrane, assembly of clathrin coats, induction of membrane curvature, and fission of the mature endosomal bud.[503–505] Beta-arrestin-associated delivery of GPCRs to clathrin-coated pits appears to be mediated by its direct interaction with the adapter protein.[506] An alternative but controversial method of internalization utilized by chemokine, sphingosine-1-phosphate, endothelin, and cannabinoid receptors involves interaction of receptors with lipid rafts enriched in cholesterol, sphingolipids, and

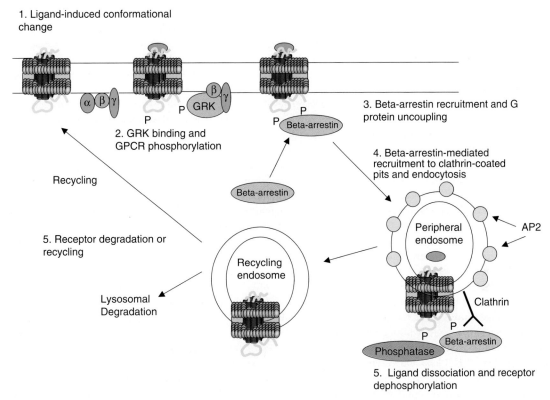

Figure 10 GPCR desensitization, internalization, degradation, and recycling.

caveolin, a protein associated with caveolae.[507–510] For some GPCRs, a C-terminal NpxxY motif appears to be essential for internalization.[511–513]

In most cases, the efficacy of an agonist to promote G protein activation and signaling correlates with its ability to desensitize and internalize the receptor. However, there are some exceptions to this general rule that also lend support to the multistate model of agonist activation and multiple active GPCR conformations. First, several GPCRs including metabotropic glutamate receptor mGluR5, galanin GAL2, and α_{1A}-adrenergic receptors have been reported to undergo beta-arrestin-dependent constitutive internalization in the absence of agonist.[514–517] Furthermore, internalization and desensitization of some GPCRs has been shown to be independent of receptor phosphorylation[518–520] or beta-arrestin interaction.[521–526] Beta-arrestin-independent internalization of certain GPCRs appears to rely on serine residues on the C-terminal tail.[527] Angiotensin AT_2 receptors with mutations of their DRY sequence leading to reduced coupling to G protein have been reported to internalize normally in the presence of AngII indicating that G protein-independent internalization can occur.[528] In addition, certain β_2-adrenergic ligands that do not activate G proteins are able to recruit beta-arrestin to the plasma membrane.[347] Similarly, truncated forms of apelin activate the apelin receptor and subsequent signaling but only the full-length version of the peptide is capable of efficiently inducing receptor internalization.[529] Furthermore, the hypotensive effects of apelin have been reported to correlate with their propensity to promote receptor internalization. Comparison of various peptide and nonpeptide somatostatin receptor agonists ability to promote inhibition of adenylyl cyclase and receptor endocytosis has indicated that peptides promote similar potencies for both events, but nonpeptide agonists are more potent in inhibiting adenylyl cyclase than promoting receptor internalization.[530] Thus, for some GPCRs, the process of desensitization and internalization deviates from the canonical model that generally applies to many GPCRs.

Beta-arrestins are ubiquitously expressed in cells and the recruitment of this protein to the plasma membrane can be visualized using confocal microscopy and GFP-tagged beta-arrestins. There are two forms of nonvisual beta-arrestins (beta-arrestin2 and beta-arrestin3) that mediate GPCR responses and endocytosis. The x-ray crystal structure of bovine arrestin2 in its resting state has been published and indicates that the protein has N- and C-terminal domains flanked by a polar core.[531,532] As previously stated, beta-arrestins are major players in the process of receptor desensitization and internalization as well as activation of MAPK cascades. The interaction of beta-arrestin with phosphorylated GPCRs results in inhibition of receptor and G protein coupling as well as the initiation of recruitment of the receptor into clathrin-coated pits.[533,534] There are a number of adapter proteins that facilitate the endocytosis of the receptor–beta-arrestin complex with clathrin and adapter protein-2 plays a paramount role.[535] In some cases, c-src interaction with beta-arrestin is also important for clathrin-coated pit-mediated receptor endocytosis.[536] Important GPCR contact sites involved in beta-arrestin recruitment include IC2 and IC3 loops as well as the C-terminal tail.[537] Based on differences in the stability and trafficking of receptor–beta-arrestin complexes, there are two primary classes of receptors: one in which this interaction is transient resulting in beta-arrestin dissociation and rapid recycling of GPCR back to the membrane (class A, e.g., β_2-adrenergic receptor); and another (class B, e.g., vasopressin V_2 receptor) in which the receptor and beta-arrestin complex are stable and receptors either recycle more slowly or are degraded in lysosomes. Introduction of serine and threonine residues in the C-terminal tail from class B receptors into class A receptors results in prolonged retention of the receptor and reduces the rate of recycling.[538] In some cases, disruption of receptor and beta-arrestin interactions does not alter receptor desensitization, endocytosis, and G protein-dependent signaling.[539] For certain GPCRs, beta-arrestin-mediated receptor endocytosis requires homodimerization,[540] and several reports suggest that GPCR heterodimerization can influence receptor internalization.[541,542] Dissociation of beta-arrestin from GPCRs such as the bradykinin B_2 receptor during endocytosis has been reported to be necessary for receptor cycling back to the plasma membrane.[543] Whether this is a general requirement for GPCR recycling remains to be determined.

GRKs have been widely studied and there are a total of seven different isoforms of this kinase. The main GRKs associated with phosphorylation and desensitization of GPCRs are GRK2, GRK3, GRK4, GRK5, and GRK6. Some GRKs appear to shuttle between the cytoplasm and the plasma membrane while others such as GRK5 appear to associate exclusively with the plasma membrane. GRKs contain three important functional domains, an RGS homology domain, a protein kinase domain, and a PH domain that mediates interaction with $\beta\gamma$-complex.[544,545] GRK-mediated phosphorylation is initiated when, upon receptor activation, a conformational change in the receptor causes exposure of IC3 loop and C-terminal tail phosphorylation sites become exposed, which facilitates GPCR and GRK interaction.[546–550] For example, agonist occupation of the α_{2A}-adrenergic receptor leads to phosphorylation of four sites on the IC3 loop, and peptide-blocking studies indicate that GRK2 interacts with basic residues in both IC2 and IC3 loops.[550] There is considerable evidence that PKA and PKC also initiate and mediate GPCR internalization. GRK2 and GRK3 have been shown to interact with $G\alpha_q$ via its RGS homology domain and it appears that GRKs in addition to beta-arrestin act to prevent receptor and G protein coupling. This may explain how GPCRs may uncouple from G protein activation via phosphorylation-independent mechanisms.[551] The interaction of the human oxytocin receptor with GRK2 and beta-arrestin has been measured in real time in living cells using BRET, indicating that receptor and GRK interaction takes place temporally prior to beta-arrestin-mediated desensitization.[552] Furthermore, FRET studies have indicated that the kinetics of β_2-adrenergic receptor and beta-arrestin interaction are limited by the kinetics of GRK-mediated phosphorylation of the receptor and that repeated agonist stimulation results in the accumulation of phosphorylated receptor that can interact rapidly with beta-arrestin upon reactivation.[553]

Subsequent to receptor endocytosis, the events associated with receptor degradation or recycling to the plasma membrane are not well understood. Endocytic sorting of several different internalized GPCRs has been shown to be dependent upon a signal residing on the GPCR C-terminal tail.[554] Different C-terminal GPCR sorting motifs include PDZ ligands, tyrosine-based motifs, and lysine residues that can become ubiquinated.[555] A recent report also suggests that palmitoylation of the LH receptor may regulate recycling of the receptor back to the plasma membrane.[556] Furthermore, Mialet-Perez and co-workers (2004) report that N-terminal glycosylation of the β_2-adrenergic receptor is important for promoting lysosomal degradation of the receptor.[557] Conjugation of GPCRs such as the β_2-adrenergic receptor with ubiquitin appears also to play a role in sorting internalized receptors to lysosomes for degradation,[558] but not all GPCRs require this modification for lysosomal degradation.[559]

2.20.3.8 Allosteric Modulation of G Protein-Coupled Receptors

Historically, the majority of GPCR drug discovery has been focused on drugs that bind to the same site as the endogenous ligand, commonly referred to as the 'orthosteric site' As such, most of the drugs on the market are orthosteric agonists, neutral antagonists, or inverse agonists. With the recent shift in GPCR screens from competitive radioligand binding assays to functional screens, an increasing number of new compounds has been discovered from screening small molecule libraries that do not interact with orthosteric sites but rather interact with distinct allosteric sites. Allosteric modulators have now been discovered for a variety of different GPCRs including adenosine, adrenergic, chemokine, dopamine, endothelin, glutamate, muscarinic, neurokinin, purinergic, and serotonin receptors[560–563] (Table 10).

Table 10 Examples of allosteric modulators of GPCRs

Receptor	Modulator	Actions
α_{2A}-adrenergic	SCH-202676	Negative modulator, reduces the density of orthosteric agonist and antagonist binding sites
Adenosine A_1	LUF5484	Positive modulator, enhances orthosteric agonist binding
Adenosine A_1	SCH-202676	Negative modulator, reduces orthosteric ligand binding and accelerates dissociation
Adenosine A_1	PD81723	Positive modulator, increases binding and function of orthosteric agonists and inhibits binding of antagonists and inverse agonists
Adenosine A_{2A}	Amiloride	Negative modulator, accelerates orthosteric agonist dissociation kinetics
Adenosine A_{2A}	SCH-202676	Negative modulator, reduces orthosteric ligand binding and accelerates dissociation
Adenosine A_3	DU124183	Positive modulator, increases orthosteric agonist efficacy
Adenosine A_3	SCH-202676	Negative modulator, inhibits orthosteric ligand binding and increases dissociation kinetics
Adenosine A_3	VUF5455	Positive modulator, decreases orthosteric agonist dissociation kinetics
β_2-adrenergic	Zn(II)	Positive modulator, increases orthosteric ligand affinity
Ca^{2+} sensing GPCR	Calcindol	Positive modulator, increases calcium signaling (IP accumulation)
Ca^{2+} sensing GPCR	Calhex 231	Negative modulator, decreases calcium-induced signaling (IP accumulation)
Ca^{2+} sensing GPCR	Cinacalcet (AMG073)	Positive modulator, increases orthosteric agonist signaling (IP accumulation)
Ca^{2+} sensing GPCR	NPS R-568	Positive modulator
Ca^{2+} sensing GPCR	NPS 2143	Negative modulator, decreases calcium-mediated signaling (calcium)
CXCR1/2	Repertaxin	Negative modulator, prevents orthosteric ligand signaling
CRF_1	NBI35965	Negative modulator, inhibits orthosteric agonist binding
CRF_1	DMP696; DMP904	Negative modulator, blocks CRF signaling and decreases CRF binding
Endothelin ET_A	Aspirin, sodium salicylate	Negative modulator, decreases endothelin signaling and accelerates ET-1 dissociation kinetics
$GABA_B$	CGP7930	Positive modulator, enhances orthosteric agonist functional potency
$GABA_B$	GS39783	Positive modulator, increases functional potency and efficacy of GABA (GTP-S binding), increases agonist affinity, promotes G protein coupling
mGluR1	EM-TBPC	Negative modulator, decreases orthosteric ligand efficacy (calcium)
mGluR1	BAY36-7620	Negative modulator, decreases efficacy of glutamate
mGluR1	R214127	Negative modulator, decreases glutamate signaling
mGluR2	LY487379	Positive modulator, enhances potency and efficacy of glutamate
mGluR4	PHCC	Positive modulator, enhances glutamate potency and efficacy
mGluR5	CDPPB	Positive modulator, increases efficacy and potency of glutamate
mGluR5	CPPHA	Positive modulator, increases efficacy and potency of glutamate
mGluR5	DFB	Positive modulator, increases glutamate potency (calcium assay)

continued

Table 10 Continued

Receptor	Modulator	Actions
mGluR5	DMeOB	Negative modulator, decreases orthosteric ligand potency (calcium assay)
mGluR5	MPEP	Negative modulator, inhibits orthosteric ligand binding
Serotonin 5HT$_{1B/D}$	Moduline (Leu-Ser-Ala-Leu)	Negative modulator, decreases orthosteric agonist binding sites, reduces function (GTP-S binding)
Serotonin 5HT$_{2C}$	PNU-69176E	Positive modulator, increases serotonin binding sites (no effect on antagonist sites) and increases G protein coupling (GTP-S binding assay)
Serotonin 5HT$_7$	Oleamide	Positive modulator, increases 5HT binding signaling (cAMP)
Muscarinic M$_1$	AF-DX 384	Positive modulator, increases orthosteric ligand binding
Muscarinic M$_1$	Anandamide	Negative modulator, reduces orthosteric ligand antagonist binding sites
Muscarinic M$_1$	Brucine	Negative modulator, decreases ACh binding, enhanced binding of other agonists
Muscarinic M$_1$	N-desmethylclozapine	Allosteric partial agonist, no requirement for orthosteric ligand
Muscarinic M$_1$	Mamba toxin MT-7	Negative modulator, slows orthosteric antagonist and accelerates orthosteric agonist dissociation kinetics
Muscarinic M$_1$	SCH-202676	Negative, inhibits orthosteric antagonist binding
Muscarinic M$_1$	Tacrine	Negative modulator, decreases orthosteric agonist binding
Muscarinic M$_1$	W84	Negative modulator, decreases orthosteric agonist binding
Muscarinic M$_1$	Xanomeline	Negative modulator, inhibits orthosteric antagonist binding
Muscarinic M$_2$	Allcuronim	
Muscarinic M$_2$	Heptane-1,7-bis (dimethyl-3′-phthalimidopropyl)-ammonium bromide	Negative modulator, decreased orthosteric antagonist binding
Muscarinic M$_2$	(-)Eburamonine	Negative modulator, decreases acetylcholine binding but increases binding of other agonists
Muscarinic M$_2$	Gallamine	Negative modulator, decreases orthosteric antagonist affinity
Muscarinic M$_2$	W84	Negative modulator, decreases orthosteric antagonist binding
Muscarinic M$_3$	Brucine	Negative modulator, decreases ACh binding, enhanced binding of other agonists
Muscarinic M$_3$	Gallamine	Negative modulator, decreases orthosteric antagonist affinity
Muscarinic M$_3$	PG987	Negative modulator, accelerates orthosteric antagonist dissociation kinetics
Muscarinic M$_3$	WIN 62577	Positive modulator, increases acetylcholine and antagonist affinity
Muscarinic M$_4$	(-)Eburamonine	Negative modulator, decreases orthosteric agonist
Muscarinic M$_4$	Thiochrome	Positive modulator, increases orthosteric ligand affinity

The majority of these examples are taken from [560–568].

The simplest and most common type of allosteric modulation occurs when the binding of an allosteric modulator results in changes in the affinity (positively or negatively) of the orthosteric ligand for its binding site. For example, synthetic small molecule allosteric modulators of the cannabinoid CB$_1$ have been discovered that increase affinity of the orthosteric ligand by reducing its dissociation kinetics.[564] Another type of allosteric modulator is one that modifies GPCR signaling properties in addition to, or instead of, affecting the affinity of the orthosteric ligand. This type of allosteric modulator can act to alter the efficacy of the orthosteric ligand or activate the receptor in the absence of the orthosteric ligand.[560–563] A third type of GPCR allosteric modulator is one that acts to either increase or decrease the number of orthosteric sites. For example, SCH-202676 is a compound that acts to decrease the number of orthosteric ligand binding sites of α_{2A}-receptors as well as other GPCRs.[565] In contrast, PNU-69176E, a positive allosteric modulator of the $5HT_{2C}$ receptor, acts to increase the number of $5HT_{2C}$ agonist binding sites and also promotes constitutive activation of the receptor.[566] Allosteric modulators have been discovered for most families of GPCRs. Interestingly, a relatively large number have been discovered for class C GPCRs, presumably because they possess a very long N-terminal tail that is linked by a spacer region to TM domains. For example, a family of small molecule allosteric modulators acting at mGluR5 receptors has been discovered, which act as either positive or negative allosteric modulators.[567,568] Both radioligand binding studies (kinetic and equilibrium) as well as functional assays have been used to characterize the mechanisms of GPCR allosteric modulation. These studies can be complicated and several good reviews describe the theoretical basis of GPCR allosteric modulation well and how it is experimentally evaluated.[560–563]

Allosteric modulators may possess certain properties that may make them superior drugs compared to those interacting with orthosteric sites. One of the largest benefits would be that allosteric modulators may confer better receptor selectivity within a related family of GPCRs (e.g., muscarinic M_1, M_2, M_3, M_4, M_5) than orthosteric ligands resulting in better efficacy and fewer side effects.[560–563] Orthosteric binding pockets are highly conserved for receptors activated by the same endogenous ligand and allosteric modulators would interact with potentially less conserved regions of GPCRs. Second, allosteric modulators displaying limited cooperativity with the orthosteric ligand may be safer than orthosteric ligands because they have the potential for a 'ceiling effect' resulting in less potential for patient overdosing.[560–563] This property can be capitalized upon to provide an increased duration of action by increasing drug doses without producing unwanted side effects. Third, a majority of allosteric modulators discovered to date act directly to influence orthosteric binding and/or signaling properties.[560–563] Thus, they will only possess pharmacological activity in the presence of the endogenous ligand and thus are likely to mimic more closely the pharmacological effects of the orthosteric agonist. Another potential benefit of screening for allosteric modulators is that there is potentially a greater probability of discovering allosteric modulators acting at GPCRs activated by large peptides, for many of which it has been difficult to discover small molecules acting at orthosteric sites. We anticipate that the pharmaceutical industry will focus more on discovery of selective allosteric modulators in known and orphan GPCR drug discovery programs.

2.20.4 Screening G Protein-Coupled Receptors in Functional Assays

To have an appropriate appreciation of GPCRs as drug discovery targets, the following sections describe commonly used functional assays used to screen small molecule libraries to identify novel pharmacophores as well as to direct GPCR structure–activity relationship (SAR) programs. There are a number of choices for GPCR functional assays and several companies sell assay kits that can be utilized for rapid screen development and validation. However, each of these different assays can differ in price, assay throughput, and reliability, and different assay platforms can sometimes give significantly different experimental results. In the first part of this section, the impact of several new paradigms concerning GPCR function and regulation on assay development and screening outcomes are discussed. Several functional assay platforms are then described that are widely applied to GPCR drug discovery. An understanding of these assays will benefit medicinal chemists and facilitate a better understanding of functional screening data generated for a GPCR drug discovery program.

2.20.4.1 Impacts of Multiplicity of G Protein-Coupled Receptor Signaling, Agonist Trafficking, and G Protein-Coupled Receptor Homodimerization and Heterodimerization on G Protein-Coupled Receptor Drug Discovery and Screen Development

GPCR functional assays have greatly facilitated GPCR drug discovery. Traditionally, drug discovery was initiated by screening small molecule libraries for competition for binding of a trace radioligand to the GPCR of interest. Although binding assays are still employed in drug discovery, they have largely been replaced by functional assays designed to detect agonists, antagonists, inverse agonists, and allosteric modulators up front. GPCR functional assays are generally membrane- or cell-based assays. Membrane-based assays are operationally useful as large preparations of cells can be scaled up intermittently and large membrane preparations can be made and stored at $-80\,°C$ before use. Cell-based assays are currently more popular but require continual cell culture activity to supply adequate quantities of cells. Recombinant expressed GPCRs are commonly used for development of functional screens and generally allow the expression of high levels of GPCRs to increase assay signal and provide a homogeneous population of cells expressing the GPCR of interest.

However, there are several important considerations regarding the use of recombinant expressed GPCRs for functional screen development. First, the levels of receptor expression in recombinant expression systems should be measured using N-terminally epitope-tagged GPCRs and commercially available antibodies. This is important for several reasons. First, a threshold of receptor expression is required for generation of a suitable assay signal that is dependent upon expression of sufficient levels of receptor. Second, G protein coupling can vary as the density of receptor is increased and nonphysiological coupling to multiple G proteins can occur. Third, the level of constitutive activity may be important for optimizing the assay signal. For example, constitutive activity may need to be minimized for development of agonist-biased screens and may need to be maximized for optimizing inverse agonist screens. It addition, as the levels of constitutive activity increase, the ability to detect partial agonists is reduced and these will actually appear as full agonists.

Another important consideration for developing GPCR functional assays is that the GPCR-mediated signal event chosen for measurement should be as proximal to receptor activation as possible. This is because ligand efficacies and potencies measured at the level of G proteins or effector proteins can often be quite different. Although GPCR functional assays involving downstream signaling events have the potential for favorable signal amplification, they can also become mechanistically very complicated and assay 'hits' can be difficult to interpret. Furthermore, the more signal transduction steps captured subsequent to receptor activation, the higher the probability of obtaining false-positive hits in which the activity of the test compound is not due to an effect at the level of the receptor but at a locus downstream from the receptor. However, this type of assay hit can simply be re-evaluated in the same assay using cells that do not express the receptor of interest to determine if the hit is receptor mediated.

The choice of cells used for developing functional GPCR screens is also an important consideration. Commonly used mammalian cells for recombinant expression of GPCRs for assay development (e.g., CHO, HEK293, COS7, HeLa) may not have the relevant signaling molecules required or their concentrations may be different from those in endogenous cells or tissues, both of which can influence receptor pharmacology. In addition, functionally and pharmacologically important GIPs may not be present in recombinant expression systems. The possibility of therapeutically relevant heterodimerization partners being absent in these systems is generally not considered. Furthermore, the posttranslational machinery required for functionally significant posttranslational modification of the receptor (e.g., glycosylation, palmitoylation) may not be present. If therapeutically relevant cells endogenously expressing the GPCR target of interest exist, their use for screen development should be considered. Many investigators suggest that expression of physiological levels of GPCRs in a therapeutically relevant cell line or use of cells that express the GPCR of interest endogenously is most desirable for a relevant cell-based screen.

Development of functional screens for GPCRs has demonstrated that coupling to multiple G proteins and activating multiple signaling pathways can be a challenge for a drug discovery program. If the disease-relevant GPCR signaling pathway is known, then it should be captured in the relevant signaling-biased screen. Furthermore, activation of certain signaling pathways associated with a particular GPCR target may also be associated with undesired side effects and these may need to be captured in secondary assay evaluations of primary assay hits. If therapeutically relevant signaling pathways for a given GPCR target are unknown and multiple signaling pathways are activated in therapeutically relevant cells or tissues, then it may be important to develop and utilize multiple screens in GPCR drug discovery programs. This can be an essential consideration for agonist screens for receptors in which agonist trafficking is a possibility. G protein activation by a given GPCR target can be established experimentally by immunoprecipitation of G proteins in a relevant cell line after labeling activated G protein with [^{35}S]GTPγS or using photoactive guanine nucleotide analogs followed by Western blotting.[569–577] Definition of therapeutically important signaling pathways for GPCR targets and the use of appropriate signaling-biased screens will hopefully lead to the future development of drugs that possess desired downstream effects while minimizing unwanted side effects. Furthermore, if different agonists promote different rates and extent of receptor internalization and desensitization, one may be therapeutically more valuable because it will be less apt to promote tachyphylaxis. Signal-biased screens can in theory be designed by coexpression of a desired G protein with constitutively active GPCRs to selectively enrich the system for the desired pathway for assay development and screening.

2.20.4.2 Functional Assays for G Protein-Coupled Receptors

Before describing different screening assays, there are several general considerations that should be made in designing functional screens. Cost-effective high-throughput functional screening platforms capable of efficiently evaluating large compound libraries (and SAR compounds) are critical for development of a successful GPCR drug discovery program.[578] An ideal GPCR screen should be simple and robust (high signal to noise), should not involve separations, should contain a minimum of assay additions or steps, and should be amenable to an automation-compatible microtiter plate format (96-, 384-, or 1536-well). Radiometric assays are less desirable currently because of safety issues and significant expenses associated with radioactive waste disposal.

2.20.4.3 GTPγS Binding Assays

GPCR-mediated guanine nucleotide exchange is a proximal step in GPCR signaling (see **Figure 3**). This event can be captured in an assay in which binding of the nonhydrolyzable GTP analog, radiolabeled GTPγS ([^{35}S]GTPγS), is measured by using crude plasma membranes prepared from cells expressing the GPCR of interest.[579–581] Typically, membranes are incubated for 30–60 min with test compound, [^{35}S]GTPγS, and an assay buffer containing NaCl, MgCl$_2$, and GDP. Then free and bound [^{35}S]GTPγS are separated by a filtration step followed by scintillation counting. Unfortunately, this assay has been generally limited to use for G$_{i/o}$-coupled GPCRs because G$_{i/o}$ is the most abundant G protein in most cells and has faster kinetics of GDP-GTP exchange rate compared to other G proteins. In addition, filtration significantly limits assay throughput and generally these assays are performed in a 96-well microtiter format. Owing to the slow turnover of [^{35}S]GTPγS for G$_s$- and G$_q$-coupled receptors, assay backgrounds are generally high but can be reduced by immunoprecipitation of select G proteins using G-protein α-subunit selective antibodies. The SPA (scintillation proximity assay, GE Healthcare) technology has been used to develop homogeneous [^{35}S]GTPγS binding assays to increase assay throughput. With a combination of these approaches, [^{35}S]GTPγS binding assays have been developed for G$_q$- and G$_s$-coupled GPCRs.[582–585] In addition, [^{35}S]GTPγS binding assays have been widely used to demonstrate constitutive activity of several different G$_{i/o}$-coupled GPCRs (see **Table 6**). Commonly, a comparison of K_d values for agonists derived from radioligand binding evaluations are lower than EC$_{50}$ values obtained in [^{35}S]GTPγS binding assays because of the requirement for NaCl and GDP in the latter assay, both of which will reduce agonist potency. Another consideration for these assays is that high NaCl concentrations are required for optimization of agonist assay windows and low concentrations of NaCl are required for maximizing windows for inverse agonist screens.

Perkin Elmer has recently released a nonradioactive time-resolved fluorescence (TRF) GTP binding assay that measures GPCR-mediated binding of europium-labeled GTP rather than [^{35}S]GTPγS to G proteins. TRF is a sensitive platform because it allows excitation and a delay in emission reading to allow deletion of short-lived background fluorescence contributed by test compounds and membranes because the europium chelate has a slow rate of fluorescence emission decay. Optimization and validation of this assay involves evaluation of similar variables as required for [^{32}S]GTPγS binding assays (**Table 11**). The throughput of this assay is still limited because it requires both filtration and washing steps, which at a cost, can be automated. This new assay has been validated using membranes expressing recombinant α$_{2A}$-adrenergic,[586] motilin, serotonin 5HT5a, neurotensin, muscarinic M1 receptors,[668] and NPFF2[587] receptors. Agonist EC$_{50}$ values obtained with the TRF and [^{32}S]GTPγS binding assays are reported to be comparable. **Figure 11** lists the assay steps associated with the different guanine nucleotide binding assays used in GPCR drug discovery.

2.20.4.4 cAMP Assays

Several commercially available cAMP kits can be used to conveniently develop whole cell-based cAMP or membrane-based adenylyl cyclase assays for screening either G$_s$- or G$_i$-coupled GPCRs.[588,589] Cell-based assays are easy to develop

Table 11 Important variables for optimizing nonhomogeneous membrane [^{35}S]GTPγS or Eu-GTP binding assays

Variable	Test range	Reason
Membrane	1–20 μg/well	Optimize S/N, reduce background, limited capacity due to filtration unless antibody capture is used
[^{35}S]GTPs or Eu-GTP	0.1–1 nM for [^{35}S]GTPγS; 0.1–10 nM for Eu-GTP	Optimize S/N, reduce background
MgCl$_2$	0.1 μM–10 mM	Important variable because different G proteins have different MgCl$_2$ requirements
NaCl	0–300 mM	Has an influence on assay background, agonist screens optimal at high NaCl, inverse agonist screens optimal with low NaCl
GDP	0.1 M–100 μM	Has an influence on assay background by enhancing inactive G protein state

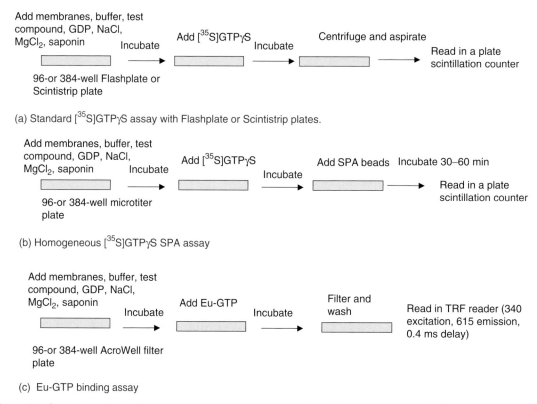

Figure 11 Steps associated with guanine nucleotide binding assays. Eu-GTP, europium-labeled GTP.

and optimize because the only assay variables are number of cells, the incubation conditions, and the concentration of phosphodiesterase inhibitor (usually IBMX or rolipram) to include. Membrane-based adenylyl cyclase assays are also widely used in screening GPCRs, but require addition and optimization of additional assay reagents such as GTP, $MgCl_2$, ATP, a regenerating system for ATP, and phosphodiesterase inhibitors. Screening small molecule libraries against G_s-coupled receptors in cAMP assays is generally straightforward. However, screening $G_{i/o}$-coupled receptors in cAMP assays is more complicated because activation of adenylyl cyclase with the direct adenylyl cyclase activator forskolin is required to optimize the window of GPCR-mediated inhibition. In addition, use of cells in which the GPCR is transiently transfected rather than stably expressed can result in a low signal to noise. This is because adenylyl cyclase in all cells added to the assay will be stimulated by forskolin, but not all cells will express receptor unless high transfection efficiency is obtained. For this reason, cells stably expressing the GPCR of interest are more commonly used to develop cAMP assays for $G_{i/o}$-coupled receptors. Another important consideration for cAMP assays is that levels generated in the assay need to be in the linear portion of standard curves used to extrapolate cAMP levels.

2.20.4.4.1 Radiometric cAMP assays

The first commercially available homogeneous assay kits used for measurement of cAMP consist of radiometric ELISA-based assays and include the GE Healthcare SPA[590] and the Perkin Elmer FlashPlate[591] assays. The SPA assay is based on the competition of endogenously produced cAMP and trace radiolabeled [^{125}I]cAMP for binding to a scintillant-containing bead conjugated to a cAMP antibody. Only [^{125}I]cAMP in close proximity with the SPA bead can activate scintillant contained within the beads leading to an assay signal. Conversely, the FlashPlate assay utilizes microtiter plates coated with scintillant to which a cAMP antibody is attached. Endogenously generated cAMP competes for binding of trace labeled [^{125}I]cAMP to the immobilized cAMP antibody. Both of these assay platforms are sensitive, can be configured into 96- and 384-well formats, are compatible for use in both membrane- and cell-based assays, and cAMP increases result in a decreased assay signal. However, safety considerations and the high expense associated with radioactive waste disposal have led to their replacement with nonradioactive assay alternatives.

2.20.4.4.2 Fluorescence-based cAMP assays
2.20.4.4.2.1 Time-resolved fluorescence resonance energy transfer cAMP assay
A high-throughput cAMP assay kit based on time-resolved fluorescence energy transfer technology is available from Cisbio (HTRF). The basis of this assay consists of a reduction in assay signal as endogenous cAMP acts to inhibit the interaction of cAMP labeled with the acceptor allophycanin (XL665) and an anti-cAMP antibody labeled with europium cryptate resulting in a decrease in energy transfer.[574,575] Laser light is used to excite at Eu-cryptate 337 nm and energy transfer occurs to the allophycocyanin acceptor and emission is read at 665 nm. This assay platform is amenable to assay miniaturization (96-, 384, and 1536-well formats available), has low test-compound endpoint detection interference, and is sensitive. Furthermore, both whole cells as well as membranes can be used in this assay. This assay has been validated for both G_s-coupled (β_2-adrenergic, histamine H_2, melanocortin MC_4, CGRP, dopamine D_1) and G_i-coupled histamine H_3 receptors.[669] With high assay sensitivity, this assay is suitable for use in cells expressing lower densities of GPCR.

2.20.4.4.2.2 Fluorescence polarization cAMP assay
Fluorescence polarization (FP)-based cAMP measurement kits are now commercially available from Perkin Elmer,[592,593] Molecular Devices,[594] and Amersham (GE Healthcare). In this assay, as endogenous cAMP increases, it competes for trace fluorescent-labeled cAMP binding to a cAMP antibody. As endogenous levels of cAMP increase, the polarization of light becomes less because it competes for binding of the larger fluorescent-labeled cAMP to the cAMP antibody.[595] This assay platform has been successfully miniaturized to a 1536-well format and it has been used for both membrane and whole cell assays. The FP cAMP assay has been reported to have a lower signal-to-noise ratio and sensitivity than other platforms, and a potential for endpoint detection interference. The release of new red-shifted fluorescent dye-labeled cAMP analogs has been reported to reduce compound interference and increase assay sensitivity.[589] This assay platform is more compatible for assays in which high densities of recombinant GPCRs are expressed.

2.20.4.4.3 Chemiluminescent cAMP assays
2.20.4.4.3.1 AlphaScreen cAMP assay
AlphaScreen (amplified luminescent proximity homogeneous assay) is another homogeneous high-throughput cAMP measurement assay platform developed by Packard Bioscience/Perkin Elmer that has been widely used in the pharmaceutical industry for functional screening of GPCRs.[578,588,589] The assay is based on competition of endogenous cAMP with biotin-conjugated cAMP for binding to high-affinity streptavidin-coated donor beads and a decrease in assay signal due to reduced donor bead and acceptor bead proximity. Following excitation of donor beads by laser at 680 nm, singlet oxygen is generated, which in turn reacts with a chemiluminescent compound contained in the acceptor bead if it is in close proximity to the donor bead. Activation of the acceptor beads then leads to generation of a chemiluminescent signal at 320 nm. The AlphaScreen assay has been successfully configured into 96-, 384-, and 1536-well formats, is sensitive, and is cost effective. It is most widely used for cell-based cAMP measurement. Potential issues with this assay format include acceptor bead light and temperature sensitivity and the potential for color quenching by test compounds.

2.20.4.4.3.2 HitHunter cAMP assay
DiscoveRX offers a homogeneous high-throughput cAMP assay kit called HitHunter that is based on a patented enzyme complementation technology.[596–598] This technology is based upon an engineered beta-galactosidase enzyme consisting of two subunits which, when joined, display enzymatic activity. Specifically, the assay utilizes a small complementation peptide conjugated with cAMP that is recognized by a cAMP antibody and a complementation peptide recognized by an acceptor fragment of the engineered beta-galactosidase enzyme. As endogenously produced cAMP increases, it binds to the cAMP antibody and displaces the small complementation peptide cAMP conjugate resulting in an increase in assay signal with increased endogenous cAMP levels. This assay can be performed using either a chemiluminescent or a fluorescent beta-galactosidase substrate. This assay kit can be purchased and performed in a 96-, 384-, or 1536-well format[599] and kits can be purchased for either adherent or nonadherent cells.

2.20.4.5 Inositol Phosphate Accumulation-IP$_3$ Assays
Cell-based radiometric inositol phosphate accumulation assays have been widely used to develop reliable functional screens for G_q-coupled GPCRs (see **Figures 5** and **12**). The basis of this assay is to initially incorporate [^{3}H]inositol

into membrane-associated inositol phospholipids and measure GPCR-mediated PLCβ activation by the measurement of generated [^{3}H]inositol phosphates (IP$_1$, IP$_2$, and IP$_3$). Typical assays first consist of overnight loading of cells with [^{3}H]inositol, which becomes incorporated into membrane-associated phospholipids to form phosphatidyl inositol species (PI, PIP, PIP$_2$). The next day, cells are washed and then test compounds are added followed by cell lysis and an ion exchange step to resolve [^{3}H]inositol and [^{3}H]inositol phosphates. This step can be accomplished in a 96-well microtiter plate containing ion exchange resin and elutions are performed under vacuum pressure.[600,601] However, the ion exchange step still significantly limits assay throughput and disposal costs associated with radioactivity add considerable expense to the assay. The SPA technology has also been applied to develop higher throughput homogeneous assays for inositol phosphate that avoid the separation step.[602] One approach utilizes metal ions immobilized to the SPA bead to bind [^{3}H]inositol phosphates via their phosphate groups and has been validated using the neurokinin NK$_1$ receptor.[603] Another approach utilizes yttrium silicate immobilized to SPA beads that bind inositol phosphates but not inositol.[604] Although IP accumulation assays are not widely used for primary screening of G$_q$-coupled GPCRs and have generated excellent pharmacological data, they are commonly used as secondary assays to confirm results obtained using higher throughput assays such as intracellular calcium assays.

Assay kits designed to measure IP$_3$ instead of total inositol phosphate species have been commercially available for some time and are based on the use of a crude IP$_3$ receptor preparation and competition of endogenous IP$_3$ with trace-labeled IP$_3$ (typically radiolabeled). These assays are difficult to automate because IP$_3$ generated becomes rapidly dephosphorylated to form IP$_2$ and IP$_1$ and assay incubations need to be very short (15–30s). A nonradioactive assay method using the AlphaScreen platform has been developed in which donor and acceptor beads are brought into close proximity with biotinylated IP$_3$ and signal declines as IP$_3$ is formed.[596] Echelon has also developed an AlphaScreen-compatible assay using a different IP binding protein that binds both IP$_2$ and IP$_4$.[596] In addition, DiscoverX has also developed and released a nonradioactive IP$_3$ assay that is based on fluorescence polarization.[596] This assay is based on the competition of in assay endogenous IP$_3$ and trace fluorescence labeled IP$_3$ molecule binding to an IP$_3$ antibody. Compound interference or quenching is minimized by use of a series of novel 'red' tracers in the assay. Validation of this assay for high-throughput screening of G$_q$-coupled GPCRs has not appeared in the literature to date.

Cisbio has recently released a new homogeneous time-resolved fluorescence assay for measuring inositol phosphate accumulation, called the IP-One assay.[669] The basis of this assay platform is competition of endogenous inositol phosphate for trace acceptor-labeled inositol phosphate binding to europium-conjugated inositol phosphate antibody (**Figure 12**). This assay has been configured into a 384-well microtiter plate format and has been validated with muscarinic M1, vasopressin V1A, oxytocin, histamine H$_2$, purogenic P2Y$_1$, chemokine CCR5, metabotropic mGluR1, and metabotropic MGluR5 receptors.[669] Since intracellular calcium assays are not suitable for screening for inverse

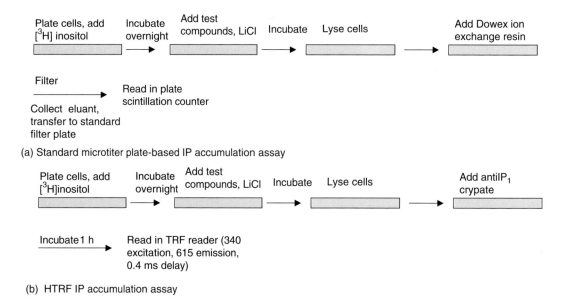

Figure 12 Steps associated with radiometric and TRF IP accumulation assays.

agonists, the IP-One assay will be a very useful alternative to traditional assays that require separation of total [^{3}H] inositol phosphates from unincorporated [^{3}H]inositol. There have been no reports in the literature concerning the use of the IP-One assay for screening G_q-coupled GPCR targets and industrial validation of this platform remains to be published.

2.20.4.6 Intracellular Calcium Assays

A popular homogeneous high-throughput functional assay applicable for screening G_q-coupled GPCRs is the measurement of receptor-mediated increases in intracellular calcium ([Ca^{2+}]$_i$) (see **Figure 5**) using calcium-sensitive fluorescent dyes such as Fura-2, Fluo-3, Fluo-4, and Calcium Green-1. GPCR-mediated increases in [Ca^{2+}]$_i$ are typically transient and return to basal levels in less than 1 min. Thus, typical fluorescence plate readers that read one row of wells at a time could not be used to capture this signal for a whole assay plate. Fortunately, the availability of automated real-time CCD-based fluorescence plate readers such as fluorometric imaging plate reader (FLIPR) (Molecular Devices), Cell Lux (Perkin Elmer), or the FDSS6000 (Hamamastu), allowing real-time fluorescence reading of all wells simultaneously on a subsecond time scale, has dramatically increased the throughput of this assay platform. These instruments also contain automated liquid handling capacity allowing simultaneous addition of reagents into all microtiter plate wells. For typical assays, cells expressing the GPCR of interest (adherent or nonadherent) are initially loaded with a fluorescent dye that becomes trapped intracellularly because of cellular esterase activity and unincorporated dye is removed by washing the cells. Cells are then loaded into special microtiter plates, test compounds are added by automated liquid handling, and changes in [Ca^{2+}]$_i$ are measured over a period of a few minutes. Measurement of [Ca^{2+}]$_i$ is now the method of choice for the functional screening of G_q-coupled receptors and is a sensitive, cost-effective format.[605–607] One consideration when using this assay platform is that increases in [Ca^{2+}]$_i$ are not only a function of IP$_3$-mediated release from the endoplasmic reticulum but are a reflection of several homeostatic mechanisms that act to control [Ca^{2+}]$_i$, including calcium influx through calcium-permeable ion channels such as the L-type calcium channels, and by cellular processes involved in restoring calcium levels to resting levels such as cellular Ca^{2+} ATPases and Na^+/Ca^{2+} antiporter activities. It is not uncommon to observe right-shifts in dose–response curves obtained in [Ca^{2+}]$_i$ assays compared to IP accumulation assays, which is likely to be due to these other factors regulating [Ca^{2+}]$_i$. One potential disadvantage of fluorescence-based assays is that test compounds can either autofluoresce or quench the assay signal resulting in false-negative or positive hits.

Two recent advances have further increased the capacity of [Ca^{2+}]$_i$ assays for high-throughput screening. First, a new fluorescent dye kit called Calcium 3 has been developed by Molecular Devices that allows fluorescent dye loading of cells without the requirement for removal of unincorporated dye by repeated cell washing.[608] Elimination of the cell washing step not only saves time but also reduces several potential problems associated with cell washing such as cell lifting and lowered cell competency, and incomplete dye washing leading to assay variability. A comparison of agonist potency data for serotonin 5HT$_{2C}$ and mGluR5 glutamate receptors obtained using a conventional fluorescent calcium dye and the Calcium 3 assay kit suggests that both methods give comparable results.[608] However, the contents of this kit are proprietary making it difficult to determine any potential pharmacological effects of the assay ingredients.

The second advance is that the FLIPR instrument can now be configured using 96-well, 384-well, and 1536-well pipetting heads, thus enabling considerable assay miniaturization.[609,610] Issues with the 1536-well format include concerns regarding assay mixing and evaporation of reagents. Although the measurement of changes in intracellular calcium has been used for high-throughput agonist and antagonist screening, this method cannot be used to screen for inverse agonists because increases in basal intracellular calcium are not observed for constitutively active G_q-coupled GPCRs. One possible explanation for this observation is that the endoplasmic reticulum IP$_3$ receptor becomes desensitized due to increased basal IP$_3$ levels.

A related high-throughput platform for measuring GPCR-mediated changes in intracellular calcium utilizes recombinant expressed jellyfish photoprotein aequorin instead of calcium-sensitive dyes.[611–617] This assay platform is based on the use of aequorin, a calcium-sensitive reporter protein that generates a luminescent signal when a coelenterazine derivative is added. Euroscreen (Brussels) offers cell lines that express receptors, G proteins, and apoaequorin (AequoScreen) that can be utilized for efficient development of high-throughput [Ca^{2+}]$_i$ assays for GPCRs.[618] The Hamamatsu Photonics FDSS600 plate reader is commonly used to measure the bioluminescent signal.

A variant of the intracellular calcium assay has been developed recently for screening G_s- and G_i-coupled GPCRs. This assay is based on increases in live cell intracellular cAMP, which act to open a recombinant expressed (HEK293 cells) modified (lacks the calmodulin regulatory site involved in channel closing) olfactory cyclic nucleotide-gated (CNG) calcium channel leading to increases in [Ca^{2+}]$_i$, which can be measured using FLIPR.[619] This assay platform has been validated in a 96-well format for G_s-coupled GPCRs including β_2-adrenergic, adenosine A_{2B}, dopamine D_1, and trace amine TA1 receptors and G_i-coupled ORL-1 and kappa-opiate receptors and yields results similar to those obtained in cAMP platforms. BD Biosciences is offering this technology to the screening community.

2.20.4.7 Reporter Gene Assays

Cell-based reporter gene assays are a widely used high-throughput functional homogeneous assay platform for screening GPCRs and are based on the ability of GPCR-mediated second messenger molecules such as cAMP or $[Ca^{2+}]_i$ to activate or inhibit a responsive element placed upstream of a minimal promoter, which, in turn, regulates the expression of a selected reporter protein.[589] For G_s- and G_i-coupled receptors, the cAMP responsive element (CRE) is commonly used to develop reporter assays. For G_q-coupled receptors, activating protein-1 (AP-1) or nuclear factor of activated cells (NFAT) promoters are employed. Performing these assays is very simple and consists of adding assay reagents, incubating for 8–24 h, lysing cells, and reading plates on an endpoint detector. Specific protocols for these assays may differ in the number of response elements and the reporter gene chosen.[589] GFP and beta-galactosidase reporters have been widely used but have the disadvantage of being less sensitive than other reporters due to high background noise.[589] The luciferase reporter, which has a much shorter half-life, allows greater assay sensitivity. However, automation of luciferase assays is difficult unless newer commercially available long-lived luciferase substrates are employed (glow versus flash substrates).[589] Primary advantages of this assay platform include successful miniaturization to a 1536-well format, which is primarily due to signal amplification and full robotic automation, and it is relatively inexpensive to perform. Potential disadvantages of reporter gene assays include long incubation periods (8–24 h) and the requirement for a cell lysis step. The lengthy incubation times required can promote compound degradation and cytotoxicity can lead to false-positive or negative hits.

The beta-lactamase reporter gene assay has become very popular in recent years and has been successfully configured into a 3456-well microtiter plate format. Use of cell-permeable fluorescent substrates has also allowed omission of cell lysis prior to endpoint reading. Furthermore, the beta-lactamase reporter provides high sensitivity due to the absence of endogenous beta-lactamase activity in mammalian cells.[620–625] Another live-cell reporter assay has been described in which GFP was used as a reporter gene under transcriptional control of the CREs, which also does not require cell lysis.[626]

2.20.4.8 Promiscuous and Chimeric G Proteins and Design of Universal Screens for G Protein-Coupled Receptors

Several attempts have been made to design a 'universal' GPCR screen that utilizes a common assay endpoint. This would allow better comparison of the activity of screening compounds across a set of GPCRs. In addition, this would reduce costs of running different assays utilizing different assay reagents and endpoint readers. One approach to accomplish this goal has been to utilize promiscuous G proteins such as $G\alpha_{14}$, $G\alpha_{15}$, or $G\alpha_{16}$ to force signaling of G_s- and G_i-coupled GPCRs to changes in IP accumulation and, more importantly, to changes in $[Ca^{2+}]_i$.[627–631] This has been successfully accomplished for several GPCRs, but these promiscuous G proteins do not couple to all GPCRs. Additional engineering of promiscuous G proteins has improved the coupling efficiency of these G proteins. For example, coupling of both G_s- and $G_{i/o}$-coupled receptors to $G\alpha_{15/16}$ can be further improved by production of a G protein α-subunit chimera ($G\alpha_{16z25}$) of $G\alpha_{16}$ is combined with the C-terminal tail of $G\alpha_z$.[632–634] Stable cell lines expressing these modified promiscuous G proteins can be generated and coexpression of GPCR and $G\alpha_{16z25}$ does not interfere with the receptor's ability to activate endogenous signaling pathways. The increases in GPCR coupling observed with these new chimeric G proteins is presumably due to greater association of the $G\alpha_{16}$ backbone with the plasma membrane. Similarly, chimeric G proteins in which the last five amino acids of $G\alpha_{16}$ are replaced with various lengths of $G\alpha_s$ have also been reported to increase the promiscuity of $G\alpha_{16}$ toward G_s-coupled receptors.[635]

Several studies have indicated that the C-terminus of G proteins is an important molecular determinant conferring selectivity for GPCR coupling to G proteins. Indeed, Coward et al. reported that replacement of the last five amino acids of $G\alpha_q$ with the three C-terminal amino acid residues of $G\alpha_i$ could switch coupling of G_i-coupled receptors to stimulation of phospholipase C and increases in intracellular calcium.[636] The nomenclature of this chimera is $G\alpha_{q/i3}$, which denotes that the backbone of $G\alpha_q$ is combined with the three C-terminal amino acids of $G\alpha_i$. Based on these results, it was suggested that a C-terminal turn, centered around a glycine residue, plays an important part in specifying receptor interactions with G proteins. Several G protein chimeras have been produced by a number of groups and have been extensively used to evaluate GPCR–G protein coupling specificity. Typically, maximally effective chimeras are ones in which between four and five C-terminal amino acids are replaced to alter G protein specificity. Molecular

Devices offers expression vectors or cells stably expressing $G_{q/s}$, $G_{q/i}$, $G_{q/o}$, and $G_{q/z}$ chimeras for easy assay development. Although a significant number of G_i- and G_s-coupled receptors have been configured to couple to these chimeric G proteins, this approach does not work for all GPCRs. Chimeric G proteins have been widely used for development of drug discovery screens for GPCRs[637–641] as well as endogenous ligand identification for orphan GPCRs.[607,617,631,639] One caution concerning the use of chimeric G proteins in GPCR functional assays is that the efficiency of coupling of GPCRs for chimeric G proteins may be less than that for the G protein(s) to which it couples physiologically.

At Arena Pharmaceuticals, we utilize the Melanophore platform for screening known and orphan GPCR targets. This assay platform is suitable for G_s-, G_i-, and G_q-coupled GPCRs, and is based on the ability of transiently expressed GPCRs to alter the distribution of melanin-containing melanosomes in *Xenopus* melanophores. Specifically, transfected melanophores will either aggregate (absorb less light) or disperse (absorb more light) the melanosomes depending on whether the GPCR is $G_{i/o}$- or $G_{s/q/11}$-coupled, respectively. This assay allows a simple readout based on light absorption, is robust and cheap (5 cents/well), does not require cell lysis before absorbance reading, has been fully automated, and is adapted to 384- and 1536-microtiter plate formats as well as open lawns.[642–647] One advantage of this platform is that expression of native as well as mutated GPCRs in melanophores consistently results in constitutive activity that can often be titrated by altering the level of GPCR expression.

2.20.4.9 G Protein-Coupled Receptor Biosensors

There are several recent reports concerning the construction and use of GPCR 'biosensors' to study GPCR function and regulation in real time and in live cells. These new tools have helped to further progress our understanding of GPCR activation, protein interactions, and signal transduction in a 'natural' environment. GPCR biosensors have initially been developed by academic groups and may find an application for development of new functional screening platforms for GPCR drug discovery.

BRET (bioluminescence resonance energy transfer) has been used for several years to study GPCR protein interactions such as homodimerization, heterodimerization, and beta-arrestin interactions. BRET studies of beta-arrestin-mediated GPCR internalization typically utilize recombinant expressed GPCRs in which green fluorescent protein (GFP) is fused to the C-terminus and recombinant expressed beta-arrestin2 is fused with Renilla luciferase in living cells.[648] For example, this approach was used to demonstrate the important functional role of vasopressin V_2 receptor C-terminal palmitoylation on beta-arrestin2 membrane recruitment, receptor-mediated ERK1/2 activation, and receptor endocytosis. Recently, this technology has been used to develop high-throughput homogeneous GPCR screens but early applications were limited by low signal-to-noise ratios. However, Vrecl and co-workers (2004) have reported using BRET and GFP-tagged beta-arrestin2 mutants that are either phosphorylation-independent in their interaction with GPCRs or lack sites important for interaction with clathrin-coated pits.[649,650] By introduction of these mutations, a reasonable signal-to-noise ratio has been obtained allowing the assay to be configured into a microtiter-plate format. Similarly, Hamdan and colleageues (2005) have reported development of a BRET-based screening assay using a stable cell line expressing both beta-arrestin-Renilla luciferase and GPCR with a C-terminal fused acceptor EYFP (VENUS).[651] BRET has also been used to evaluate the kinetics of the interaction of the oxytocin receptor with the G protein kinase GRK2 and this assay could possibly be configured as a potential screen depending upon the level of the assay signal compared to background.[552] BRET has also been utilized recently to study the interaction of GPCRs and G proteins in living cells and this assay could be potentially used as a functional screen for GPCRs.[652] Validation of this assay consisted of screening a small molecule compound library for novel antagonists of the chemokine receptor CCR5.

FRET has also been used to measure GPCR and beta-arrestin interactions[653–655] or GPCR and G protein interaction in real time with live cells.[656] These assays typically utilize two variants of GFP, cyan (CFP) and yellow (YFP) fluorescent proteins. For example, live cell studies of G protein subunit dissociation typically utilize CFP-tagged Gα-subunit and YFP-tagged β-subunits. Using a different approach, Hoffman and co-workers have introduced both CFP and a small membrane-permeable fluorescein derivative called FlAsH, tagged into a short tetracysteine sequence of the human adenosine A_{2A} receptor, and report that they can monitor receptor activation in live cells by FRET.[657] Although FRET biosensors have not been used for screening GPCRs, they may become more popular if the assay signal-to-noise ratio is sufficient and the assay can be automated.

2.20.4.10 G Protein-Coupled Receptors and High-Content Imaging

High-content imaging is a new assay technology that has continued to mature in recent years and is almost ripe for application in high-throughput screening for GPCR targets. High-content GPCR assays are defined as assays in which pieces of information about the behavior of GPCRs can be gathered in parallel because they offer a method to multiplex several assay endpoints into one assay (usually up to five colors).[658–660] High-content screening platforms generally combine confocal microscopy and charge coupled device (CCD) camera imaging systems to analyze multiple cells in a microtiter plate. High-content assays designed specifically for GPCRs generally rely on fluorescence measurement of protein translocation within cells (e.g., cytosol to the plasma membrane or cytosol to the nucleus) or receptor internalization and trafficking.[661–664] One of the first GPCR functional assays based on high content imaging was developed by Norak and is called the Transfluor technology (now owned by Molecular Devices). This assay platform is based on the redistribution of fluorescent-labeled beta-arrestin (GFP-tagged) from the cytosol to the plasma membrane and subsequent internalization of the GPCR–beta-arrestin–GFP complex into clathrin-coated pits to measure receptor agonist activation or inactivation.[664,665] This platform has been used to screen small molecule libraries to discover vasopressin V2 agonists.[662] To date, the Transfluor technology has been validated for over 85 different GPCRs (G_s-, $G_{i/o}$-, and $G_{q/11}$-coupled) and is currently available by license from Molecular Devices (Sunnyvale, CA). Several high-content imaging instruments including the IN Cell Analyzer (GE Healthcare), ArrayScan (Cellomics), Explorer (Accumen), Opera (Evotec), and MetaXpress (Molecular Devices) have been shown to be compatible for use with the Transfluor technology. Alternatively, BioImage has developed a GPCR translocation assay utilizing GPCRs in which EGFP (enhanced green fluorescent protein) is fused to the C-terminus of the receptor of interest, such as the chemokine CXCR4 receptor, and a granularity algorithm is used to quantify receptor internalization using the IN Cell Analyzer.[661] This platform may have some advantage over the Transfluor platform in that beta-arrestin-independent internalization of GPCRs can be measured using the latter methodology. Important assay variables for development of high-content GPCR internalization assays include plate type used, cell plating conditions, number of cells used, kinetics of GPCR internalization, and dimethyl sulfoxide (DMSO) sensitivity.

Assay throughput continues to be an issue that limits the use of high-content imaging for larger primary screening campaigns. However, throughputs of around 100 plates per day are now claimed by some instrument companies, allowing consideration as a primary screening tool for carefully selected targets such as orphan GPCRs with poorly defined G protein signaling properties and 7-TM proteins more distantly related to traditional GPCRs. For example, Borchert and co-workers recently reported the development of a high-content assay for activators of the Wnt-Frizzled pathway based on translocation of beta-Catenin.[665] This assay has been used to screen a library of 51 000 compounds and produced a reasonable 0.6% hit rate. High-content screening will undoubtedly be used to a greater extent for screening GPCRs in the future.

2.20.5 Target Validation of G Protein-Coupled Receptors

The pharmaceutical industry is now focusing on mining the human genome sequence to identify new GPCR drug discovery targets. Analyses of the human genome suggest that there are currently over 100 human orphan GPCRs (oGPCRs) for which the endogenous ligand remains to be identified. Thus, oGPCRs represent a vast opportunity for development of new therapeutics. The current challenge facing the pharmaceutical industry concerns rapid implementation of new strategies to enable efficient target validation of oGPCRs. An integrated approach in which well-established strategies (such as reverse pharmacology as well as gene knockdown) are combined in parallel with newer strategies (such as chemical genomic and microarray analysis) should help to achieve this goal. We have recently published a review on target validation of oGPCRs that describes such an integrated approach.[4]

References

1. Drews, J. *Science* **2000**, *287*, 1960–1964.
2. Klabunde, T.; Hessler, G. *ChemBioChem* **2002**, *3*, 928–944.
3. Brink, C. B.; Harvey, B. H.; Bodenstein, J.; Venter, D. P.; Oliver, D. W. *Br. J. Clin. Pharmacol.* **2004**, *57*, 373–387.
4. Thomsen, W.; Leonard, J.; Behan, D. P. *Curr. Opin. Mol. Ther.* **2004**, *6*, 640–656.
5. Karnik, S. S.; Gogonea, C.; Patil, S.; Saad, Y.; Takezako, T. *Trends Endocrinol. Metab.* **2003**, *14*, 431–437.
6. Kristiansen, K. *Pharmacol. Ther.* **2004**, *103*, 21–80.
7. Gudermann, T.; Schoneberg, T.; Schultz, G. *Annu. Rev. Neurosci.* **1997**, *20*, 399–427.
8. Wess, J.; Brann, M. R.; Bonner, T. I. *FEBS Lett.* **1989**, *258*, 133–136.
9. Wade, S. M.; Scribner, M. K.; Dalman, H. M.; Taylor, J. M.; Neubig, R. R. *Mol. Pharmacol.* **1996**, *50*, 351–358.
10. Gether, U.; Kobilka, B. K. *J. Biol. Chem.* **1998**, *273*, 17979–17982.
11. Chakir, K.; Xiang, Y.; Zhang, S. J.; Cheng, H.; Kobilka, B. K.; Xiao, R. P. *Mol. Pharmacol.* **2003**, *64*, 1048–1058.
12. O'Dowd, B. F.; Hnatowich, M.; Caron, M. G.; Lekowitz, R. J.; Bouvier, M. *J. Biol. Chem.* **1989**, *264*, 7564–7569.
13. Blanpain, C.; Wittamer, V.; Vanderwinden, J. M.; Boom, A.; Renneboog, B.; Lee, B.; Le Poul, E.; El Asmar, L.; Govaerts, C.; Vassart, G. M. et al. *J. Biol. Chem.* **2001**, *276*, 23795–23804.
14. Kraft, K.; Olbrich, H.; Majoul, I.; Mack, M.; Proudfoot, A.; Oppermann, M. *J. Biol. Chem.* **2001**, *276*, 34408–34418.

15. Ponimaskin, E.; Dumuis, A.; Gaven, F.; Barthet, G.; Heine, M.; Glebov, K.; Richter, D. W.; Oppermann, M. *Mol. Pharmacol.* **2005**, *67*, 1434–1443.
16. Qanbar, R.; Bouvier, M. *Pharmacol. Ther.* **2003**, *97*, 1–33.
17. Zhong, M.; Navratil, A. M.; Clay, C.; Sanborn, B. M. *Biochemistry* **2004**, *43*, 3490–3498.
18. Smotrys, J. E.; Linder, M. E. *Annu. Rev. Biochem.* **2004**, *73*, 559–587.
19. Choi, G.; Guo, J.; Makriyannis, A. *Biochim. Biophys. Acta* **2005**, *1668*, 1–9.
20. Freedman, N. J.; Lefkowitz, R. *J. Rec. Prog. Horm. Res. 51*, 319–351.
21. Fergusson, S. S. *Pharmacol. Rev.* **2001**, *53*, 1–24.
22. Gurpreet, K. D.; Dale, L. B.; Anborgh, P. H.; O'Connor-Halligan, K. E.; Sterne-Marr, R.; Ferguson, S. S. G. *J. Biol. Chem.* **2004**, *279*, 16614–16620.
23. Brady, A. E.; Limbird, L. E. *Cell Signal.* **2002**, *14*, 297–309.
24. Bockaert, J.; Dumuis, A.; Fagni, L.; Marin, P. *Curr. Opin. Drug Disc. Dev.* **2004**, *7*, 649–657.
25. Bockaert, J.; Fagni, A.; Dumuis, A.; Marin, P. *Pharmacol. Ther.* **2004**, *103*, 203–221.
26. Tan, C. M.; Brady, A. E.; Nickols, H. H.; Wang, Q.; Limbird, L. E. *Annu. Rev. Pharmacol. Toxicol.* **2004**, *44*, 559–609.
27. Bockaert, J.; Pin, J. P. *EMBO J.* **1999**, *18*, 1723–1729.
28. Josefsson, L. G. *Gene* **1999**, *239*, 333–340.
29. Graul, R. C.; Sadee, W. *AAPS Pharm. Sci.* **2001**, *3*, E12.
30. Joost, P.; Methner, A. *Genome Biol.* **2002**, *3*, Research0063.
31. Fredriksson, R.; Lagerstrom, M. C.; Lundin, L. G.; Schioth, H. B. *Mol. Pharmacol.* **2003**, *63*, 1256–1272.
32. Schioth, H. B.; Fredriksson, R. *Gen. Comp. Endocrinol.* **2005**, *142*, 94–101.
33. Foord, S. M.; Bonner, T. L.; Neubig, R. R.; Rosser, E. M.; Pin, J. P.; Davenport, A. P.; Spedding, M.; Harmar, A. *J. Pharmacol. Rev.* **2005**, *57*, 279–288.
34. Schwartz, T. W.; Frimurer, T. M.; Holst, B.; Rosenkilde, M. M.; Elling, C. E. *Annu. Rev. Pharmacol. Toxicol.* **2006**, *46*, 481–519.
35. Trumpp-Kallmeyer, S.; Chini, B.; Mouillac, B.; Barberis, C.; Hoflack, J.; Hibert, M. *Pharm. Acta Helv.* **1995**, *70*, 255–262.
36. Ji, T. H.; Grossman, M.; Ji, I. *J. Biol. Chem.* **1998**, *273*, 17299–17302.
37. Ji, I.; Ji, T. H. *J. Biol. Chem.* **1995**, *270*, 15970–15973.
38. Fernandez, L. M.; Puett, D. *Mol. Cell Endocrinol.* **1997**, *128*, 161–169.
39. Mcfarlane, E.; Seatter, M. J.; Kanke, T.; Hunter, G. D.; Plevin, R. *Pharmacol. Rev.* **2001**, *53*, 245–282.
40. Chalmers, D. T.; Behan, D. P. *Nat. Rev. Drug Disc.* **2002**, *1*, 599–608.
41. Vilardaga, J. P.; Di Paolo, E.; Bialek, C.; De Neef, P.; Waelbroeck, M.; Bollen, A.; Robberecht, P. *Eur. J. Biochem.* **1997**, *246*, 173–180.
42. Asmann, Y. W.; Dong, M.; Ganguli, S.; Hadac, E. M.; Miller, L. *J. Mol. Pharmacol.* **2000**, *58*, 911–919.
43. Lisenbee, C. S.; Dong, M.; Miller, L. J. *J. Biol. Chem.* **2005**, *280*, 12330–12338.
44. Harmar, A. J. *Genome Biol.* **2001**, *2*, Reviews3013.
45. Hoare, S. R. *Drug Disc. Today* **2005**, *10*, 417–427.
46. Lin, H. H.; Chang, G. W.; Davies, J. Q.; Stacy, M.; Harris, J.; Gordon, S. *J. Biol. Chem.* **2004**, *279*, 31823–31832.
47. Hamann, J.; Vogel, B.; van Schijndel, G. M.; van Lier, R. A. *J. Exp. Med.* **1996**, *184*, 1185–1189.
48. Kunishima, N.; Shimada, Y.; Tsuji, Y.; Sato, T.; Yamamoto, M.; Kumasaka, T.; Nakanishi, S.; Jingami, H.; Morikawa, K. *Nature* **2000**, *407*, 971–977.
49. Bessis, A. S.; Rondard, P.; Gaven, F.; Brabet, I.; Triballeau, N.; Prezeau, L.; Acher, F.; Pin, J. P. *Proc. Natl. Acad. Sci. USA* **2002**, *99*, 11097–11102.
50. Pin, J. P.; Galvez, T.; Prezeau, L. *Pharmacol. Ther.* **2003**, *98*, 325–354.
51. Kubo, Y.; Tateyama, M. *Curr. Opin. Neurobiol.* **2005**, *15*, 289–295.
52. Levitski, A.; Klein, S.; Reuveni, H. *ChemBioChem* **2002**, *3*, 815–818.
53. Peleg, S.; Varon, D.; Ivanina, T.; Dessauer, C. W.; Dascal, N. *Neuron* **2002**, *33*, 87–99.
54. Bunemann, M.; Frank, M.; Lohse, M. *J. Proc. Natl. Acad. Sci. USA* **2003**, *100*, 16077–16082.
55. Frank, M.; Thumer, L.; Lohse, M. J.; Bunemann, M. *J. Biol. Chem.* **2005**, *280*, 24584–24590.
56. Simon, M. I.; Strathmann, M. P.; Gautam, N. *Science* **1991**, *252*, 802–808.
57. Hepler, J. R.; Gilman, A. G. *Trends Biochem. Sci.* **1992**, *17*, 383–387.
58. Clapham, D. E.; Neer, E. J. *Annu. Rev. Pharmacol. Toxicol.* **1997**, *37*, 167–203.
59. Downes, G. B.; Gautam, N. *Genomics* **1999**, *62*, 544–552.
60. Cabrera-Vera, T. M.; Vanhauwe, J.; Thomas, T. O.; Medkova, M.; Preininger, A.; Mazzoni, M. R.; Hamm, H. E. *Endocrinol. Rev.* **2003**, *24*, 765–781.
61. Mumby, S. M.; Kleuss, C.; Gilman, A. G. *Proc. Natl. Acad. Sci. USA* **1994**, *91*, 2800–2804.
62. Ross, E. M. *Curr. Biol.* **1995**, *5*, 107–109.
63. Iiri, T.; Backlund, P. S.; Jones, T. L.; Wedgaertner, P. B.; Bourne, H. R. *Proc. Natl. Acad. Sci. USA* **1996**, *93*, 14592–14597.
64. Morales, J.; Fishburn, C. S.; Wilson, P. T.; Bourne, H. R. *Mol. Biol. Cell* **1998**, *9*, 1–14.
65. Moffett, S.; Brown, D. A.; Linder, M. E. *J. Biol. Chem.* **2000**, *275*, 2191–2198.
66. Peitzsch, R. M.; McLaughlin, S. *Biochemistry* **1993**, *32*, 10436–10443.
67. Chen, C. A.; Manning, D. R. *Oncogene* **2001**, *20*, 1643–1652.
68. Sprang, S. R. *Curr. Opin. Struct. Biol.* **1997**, *7*, 849–856.
69. Lambright, D. G.; Noel, J. P.; Hamm, H. E.; Sigler, P. B. *Nature* **1994**, *369*, 621–628.
70. Rens-Domianao, S.; Hamm, H. E. *FASEB J.* **1995**, *9*, 1059–1066.
71. Wall, M. A.; Coleman, D. E.; Lee, F.; Inguez-Lluhi, J. A.; Posner, B. A.; Gilman, A. G. *Cell* **1995**, *83*, 1047–1058.
72. Slessareva, J. E.; Ma, H.; Depree, K. M.; Flood, L. A.; Bae, H.; Cabrera-Vera, T. M.; Hamm, H. E.; Graber, S. G. *J. Biol. Chem.* **2003**, *278*, 50530–50536.
73. Heydron, A.; Ward, R. J.; Jorgensen, R.; Rosenkilde, M. M.; Frimurer, T. M.; Milligan, G.; Kostenis, E. *Mol. Pharmacol.* **2004**, *66*, 250–259.
74. Sondek, J.; Bohm, A.; Lambright, D. G.; Hamm, H. E.; Sigler, P. B. *Nature* **1996**, *379*, 369–374.
75. Bohm, A.; Gaudet, R.; Sigler, P. B. *Curr. Opin. Biotechnol.* **1997**, *8*, 480–487.
76. Loew, A.; Ho, Y. K.; Blundell, T.; Bax, B. *Structure* **1998**, *15*, 1007–1019.
77. Myung, C. S.; Yasuda, H.; Liu, W. W.; Harden, T. K.; Garrison, J. C. *J. Biol. Chem.* **1999**, *274*, 16595–16603.
78. Lukov, G. L.; Myung, C. S.; McIntire, W. E.; Shao, J.; Zimmerman, S. S.; Garrison, J. C.; Willardson, B. M. *Biochemistry* **2004**, *43*, 5651–5660.

79. Wedegaertner, P. B.; Wilson, P. T.; Bourne, H. R. *J. Biol. Chem.* **1995**, *270*, 503–506.
80. Casey, P. J.; Seabra, M. C. *J. Biol. Chem.* **1996**, *271*, 5289–5292.
81. Sondek, J.; Siderovski, D. P. *Biochem. Pharmacol.* **2001**, *61*, 1329–1337.
82. Takida, S.; Wedegaertner, P. B. *J. Biol. Chem.* **2003**, *278*, 17284–17290.
83. Schwindinger, W. F.; Betz, K. S.; Giger, K. E.; Sabol, A.; Bronson, S. K.; Robishaw, J. D. *J. Biol. Chem.* **2003**, *278*, 6575–6579.
84. Taussig, R.; Gilman, A. G. *J. Biol. Chem.* **1995**, *270*, 1–4.
85. Liu, H. Y.; Wenzel-Seifert, K.; Seifert, R. *J. Neurochem.* **2001**, *78*, 325–338.
86. Frey, U. H.; Nuckel, H.; Dobrev, D.; Manthey, I.; Sandalcioglu, I. E.; Eisenhardt, A.; Worm, K.; Hauner, H.; Siffert, W. *Gene Expr.* **2005**, *12*, 69–81.
87. Weinstein, L. S.; Chen, M.; Liu, J. *Ann. NY Acad. Sci.* **2002**, *968*, 173–197.
88. Bastepe, M.; Weinstein, L. S.; Ogata, N.; Kawaguchi, H.; Juppner, H.; Kronenberg, H. M.; Chung, U. I. *Proc. Natl. Acad. Sci. USA* **2004**, *101*, 14794–14799.
89. Samamoto, A.; Chen, M.; Nakamura, T.; Xie, T.; Karsenty, G.; Weinstein, L. S. *J. Biol. Chem.* **2005**, *280*, 21369–21375.
90. Lu, T.; Lee, H. C.; Kabat, J. A.; Shibata, E. F. *J. Physiol.* **1999**, *518*, 371–384.
91. Duffy, S. M.; Cruse, G.; Lawley, W. J.; Bradding, P. *FASEB J.* **2005**, *19*, 1006–1008.
92. Van der Heyden, M. A.; Wijnhoven, T. J.; Opthof, T. *Cardiovasc. Res.* **2005**, *65*, 28–39.
93. Kalinowski, R. R.; Berlot, C. H.; Jones, T. L.; Ross, L. F.; Jaffe, L. A.; Mehlmann, L. M. *Dev. Biol.* **2004**, *267*, 1–13.
94. Yu, S.; Castle, A.; Chen, M.; Lee, R.; Takeda, K.; Weinstein, L. S. *J. Biol. Chem.* **2001**, *276*, 19994–19998.
95. Zhuang, X.; Belluscio, L.; Hen, R. *J. Neurosci.* **2000**, *20*, RC91.
96. Corvol, J. C.; Studler, J. M.; Schonn, J. S.; Girault, J. A.; Herve, D. *J. Neurochem.* **2001**, *76*, 1585–1588.
97. Regnauld, K. L.; Leterurtre, E.; Gutkind, S. J.; Gespach, C. P.; Emami, S. *Am. J. Regul. Integr. Comp. Physiol.* **2002**, *282*, R870–R880.
98. Astesano, A.; Regnauld, K.; Ferrand, N.; Gingras, D.; Bendayan, M.; Rosselin, G.; Emmai, S. *J. Histochem. Cytochem.* **1999**, *47*, 289–302.
99. Corvol, J. C.; Studler, J. M.; Schonn, J. S.; Girault, J. A.; Herve, D. *J. Neurochem.* **2001**, *76*, 1585–1588.
100. Corvol, J. C.; Muriel, M. P.; Valjent, E.; Feger, J.; Hanoun, N.; Girault, J. A.; Hirsch, E. C.; Herve, D. *J. Neurosci.* **2004**, *24*, 7007–7014.
101. Gilman, A. G. *Annu. Rev. Biochem.* **1987**, *56*, 615–649.
102. Taussig, R.; Iniguez-Lluhi, J. A.; Gilman, A. G. *Science* **1993**, *261*, 218–221.
103. Taussig, R.; Tang, W. J.; Helper, J. R.; Gilman, A. G. *J. Biol. Chem.* **1994**, *269*, 6093–7000.
104. Jeong, S. W.; Ikeda, S. R. *J. Neurosci.* **1999**, *19*, 4755–4761.
105. Patel, Y. M.; Patel, K.; Rahman, S.; Smith, M. P.; Spooner, G.; Sumathipala, R.; Mitchell, M.; Flynn, G.; Aitken, A.; Savidge, G. *Blood* **2003**, *101*, 4828–4835.
106. Ivanina, T.; Rishal, I.; Varon, D.; Mullner, C.; Frohnwiesser-Steinecke, B.; Screibmayer, W.; Dessauer, C. W.; Dascal, N. *J. Biol. Chem.* **2003**, *278*, 29174–29183.
107. Hoy, M.; Bokvist, K.; Xiao-Gang, W.; Hansen, J.; Juhl, K.; Berggren, P. O.; Buschard, K.; Gromada, J. *J. Biol. Chem.* **2001**, *276*, 924–930.
108. Gromada, J.; Hoy, M.; Olsen, H. L.; Gotfredsen, C. F.; Buschard, K.; Rorsman, P.; Bokvist, K. *Pflügers Arch.* **2001**, *442*, 19–26.
109. Guo, J. H.; Wang, H. Y.; Malbon, C. C. *J. Biol. Chem.* **1998**, *273*, 16487–16493.
110. Shinohara, H.; Udagawa, J.; Morishita, R.; Ueda, H.; Otani, H.; Semba, R.; Kato, K.; Asano, T. *J. Biol. Chem.* **2004**, *279*, 41141–41148.
111. Song, X.; Zheng, X.; Malbon, C. C.; Wang, H. *J. Biol. Chem.* **2001**, *276*, 34651–34658.
112. Gordeladze, J. O.; Hovik, K. E.; Merendino, J. J.; Hermouet, S.; Gutkind, S. *J. Cell. Biochem.* **1997**, *64*, 242–257.
113. Hovik, K. E.; Wu, P.; Gordeladze, J. O. *Lipids* **1999**, *34*, 355–362.
114. Dalwadi, H.; Wei, B.; Schrage, M.; Spicher, K.; Su, T. T.; Birnbaumer, L.; Rawlings, D. J.; Braun, J. *J. Immunol.* **2003**, *170*, 1707–1715.
115. Kim, M. H.; Agarwal, D. K. *J. Asthma* **2002**, *39*, 441–448.
116. Huang, T. T.; Zong, Y.; Dalwadi, H.; Chung, C.; Miceli, M. C.; Spicher, K.; Birnbaumer, L.; Braun, J.; Aranda, R. *Int. Immunol.* **2003**, *15*, 1359–1367.
117. Skokowa, J.; Ali, S. R.; Felda, O.; Kumar, V.; Konrad, V.; Shushakova, N.; Schmidt, R. E.; Piekorz, R. P.; Nurnberg, B.; Spicher, K. et al. *J. Immunol.* **2005**, *174*, 3041–3050.
118. Han, S. B.; Moratz, C.; Huang, N. N.; Kelsall, B.; Cho, H.; Shi, C. S.; Schwartz, O.; Kehrl, J. H. *Immunity* **2005**, *22*, 343–354.
119. Kiltz, J. D.; Akazawa, T.; Richardson, M. D.; Kwatra, M. M. *J. Biol. Chem.* **2002**, *277*, 31257–31262.
120. Foerster, K.; Groner, F.; Matthes, J.; Koch, W. J.; Birnbaumer, L.; Herzig, S. *Proc. Natl. Acad. Sci USA* **2003**, *100*, 14475–14480.
121. Rau, T.; Nose, M.; Remmers, U.; Weil, J.; Weissmuller, A.; Davia, K.; Harding, S.; Peppel, K.; Koch, W. J.; Eschenhagen, T. *FASEB J.* **2003**, *17*, 523–525.
122. Jeong, S. W.; Ikeda, S. R. *J. Neurosci.* **1999**, *19*, 4755–4761.
123. Fan, H.; Zingarelli, B.; Peck, O. M.; Teti, G.; Tempel, G. H.; Halushka, P. V.; Cook, J. A. *Am. J. Cell Phyisol.* **2005**, *289*, C293–C301.
124. Ghahremani, M. H.; Cheng, P.; Lembo, P. M.; Albert, P. R. *J. Biol. Chem.* **1999**, *274*, 9238–9245.
125. Casey, P. J.; Fong, H. K.; Simon, M. I.; Gilman, A. G. *J. Biol. Chem.* **1990**, *265*, 2383–2390.
126. Strittmatter, S. M.; Valenzuela, D.; Kennedy, T. E.; Neer, E. J.; Fishman, M. C. *Nature* **1990**, *344*, 836–841.
127. Avigan, J.; Murtagh, J. J., Jr.; Stevens, L. A.; Angus, C. W.; Moss, J.; Vaughn, M. *Biochemistry* **1992**, *31*, 7736–7740.
128. Meng, J.; Casey, P. J. *J. Biol. Chem.* **2002**, *277*, 43417–43424.
129. Andric, S. A.; Zivadinovic, D.; Gonzalez-Iglesias, A. E.; Lachowicz, A.; Tomic, M.; Stojilkovic, S. S. *J. Biol. Chem.* **2005**, *280*, 26896–26903.
130. Ho, M. K.; Wong, Y. H. *Oncogene* **2001**, *20*, 1615–1625.
131. Kimple, M. E.; Nixon, A. B.; Kelly, P.; Bailey, C. L.; Youg, K. H.; Fields, T. A.; Casey, P. J. *J. Biol. Chem.* **2005**, *280*, 31708–31713.
132. Iida, N.; Kozasa, T. *Methods Enzymol.* **2004**, *390*, 475–483.
133. Embry, A. C.; Glick, J. L.; Linder, M. E.; Casey, P. J. *Mol. Pharmacol.* **2004**, *66*, 1325–1331.
134. Stryer, L. *J. Biol. Chem.* **1991**, *266*, 10711–10714.
135. McLaughlin, S. K.; McKinnon, P. J.; Margolskee, R. F. *Nature* **1992**, *357*, 563–569.
136. Hoon, M. A.; Northrup, J. K.; Margolskee, R. F.; Ryba, N. *Biochem. J.* **1995**, *309*, 626–636.
137. Hoon, M. A.; Adler, E.; Lindemeier, J.; Battey, J. F.; Ryba, N. J.; Zuker, C. S. *Cell* **1999**, *96*, 541–551.
138. Berridge, M. J. *Nature* **1993**, *365*, 388–389.
139. Exton, J. *Annu. Rev. Pharmacol. Toxicol.* **1996**, *36*, 481–509.
140. Wilke, T. M.; Scherle, P. A.; Strathmann, M. P.; Slepak, V. Z.; Simon, M. I. *Proc. Natl. Acad. Sci. USA* **1991**, *88*, 10049–10053.
141. Offermanns, S. *Prog. Biophys. Mol. Biol.* **2003**, *83*, 101–130.

142. Wettschureck, N.; Moers, A.; Wallenwein, B.; Parlow, A. F.; Maser-Gluth, C.; Offermanns, S. *Mol. Cell Biol.* **2005**, *25*, 1942–1948.
143. Shi, H.; Wang, H.; Yang, B.; Xu, D.; Wang, Z. *J. Biol. Chem.* **2004**, *279*, 21774–21778.
144. Baldwin, M. R.; Pullinger, G. D.; Lax, A. J. *J. Biol. Chem.* **2003**, *278*, 32719–32725.
145. Hall, A. *Science* **1998**, *280*, 2074–2075.
146. Offermanns, S.; Shultz, G. *Mol. Cell Endocrinol.* **1994**, *100*, 71–74.
147. Gohla, A.; Schultz, G.; Offermanns, S. *Circ. Res.* **2000**, *87*, 221–227.
148. Riobo, N. A.; Manning, D. R. *Trends Pharmacol. Sci.* **2005**, *26*, 146–154.
149. Hart, M. J.; Jiang, X.; Kozasa, T.; Roscoe, W.; Singer, W. D.; Gilman, A. G.; Sternweis, P. C.; Bollag, G. *Science* **1998**, *280*, 2112–2114.
150. Kozasa, T.; Jiang, X.; Hart, M. J.; Sternweis, P. M.; Singer, W. D.; Gilman, A. G.; Bollag, G.; Sternweis, P. C. *Science* **1998**, *280*, 2109–2111.
151. Chan, A. M.; Flemming, T. P.; McGovern, E. S.; Chedid, M.; Miki, T.; Aronson, S. A. *Mol. Cell Biol.* **1993**, *13*, 762–768.
152. Jackson, M.; Song, W.; Liu, M. Y.; Jin, L.; Dykes-Hoberg, M.; Lin, C. I.; Bowers, W. J.; Federoff, H. J.; Sternweis, P. C.; Rothstein, J. D. *Nature* **2001**, *410*, 89–93.
153. Vaiskunaite, R.; Adarichev, V.; Furthmayr, H.; Kozasa, T.; Gudkov, A.; Yoyno-Yasenetskaya, T. A. *J. Biol. Chem.* **2000**, *275*, 26206–26212.
154. Radhika, V.; Onesime, D.; Ha, J. H.; Dhanasekaran, N. *J. Biol. Chem.* **2004**, *279*, 49406–49413.
155. Meigs, T. E.; Fields, T. A.; McKee, D. D.; Casey, P. J. *Proc. Natl. Acad. Sci. USA* **2001**, *98*, 519–524.
156. Krakstad, B. F.; Ardawatia, V. V.; Aragay, A. M. *Proc. Natl. Acad. Sci. USA* **2004**, *101*, 10314–10319.
157. Lopez, I.; Mak, E. C.; Ding, J.; Hamm, H. E.; Lomasney, J. W. *J. Biol. Chem.* **2001**, *276*, 2758–2765.
158. Niu, J.; Vaiskunaite, R.; Suzuki, N.; Kozasa, T.; Carr, D. W.; Dulin, N.; Voyno-Yasenetskaya, T. A. *Curr. Biol.* **2001**, *11*, 1686–1690.
159. Vaiskunaite, R.; Kozawa, T.; Voyno-Yasenetskaya, T. A. *J. Biol. Chem.* **2001**, *276*, 46088–46093.
160. Yamaguchi, Y.; Katoh, H.; Mori, K.; Negishi, M. *Curr. Biol.* **2002**, *12*, 1353–1358.
161. Yamazaki, M.; Zhang, Y.; Watanabe, H.; Yokozeki, T.; Ohno, S.; Kaibuchi, K.; Shibata, H.; Mukai, H.; Ono, Y.; Frohman, M. A. et al. *J. Biol. Chem.* **1999**, *274*, 6035–6038.
162. Zhu, D.; Kosik, K. S.; Meigs, T. E.; Yanamadala, V.; Denker, B. M. *J. Biol. Chem.* **2004**, *279*, 54983–54986.
163. Gautam, N.; Downes, G. B.; Yan, K.; Kisselev, O. *Cell Signal.* **1998**, *10*, 447–455.
164. Vanderbeld, B.; Kelly, G. M. *Biochem. Cell Biol.* **2000**, *78*, 537–550.
165. Jones, M. B.; Siderovski, D. P.; Hooks, S. B. *Mol. Interv.* **2004**, *4*, 200–214.
166. McCudden, C. R.; Hains, M. D.; Kimple, R. J.; Siderovski, D. P.; Willard, F. S. *Cell Mol. Life Sci.* **2005**, *62*, 551–577.
167. Iyengar, R. *FASEB J.* **1993**, *7*, 768–775.
168. Sunahara, R. K.; Dessauer, C. W.; Gilman, A. G. *Annu. Rev. Pharmacol. Toxicol.* **1996**, *36*, 461–480.
169. Bayewitch, M. L.; Avidor-Reiss, T.; Levy, R.; Pfeuffer, T.; Nevo, I.; Simonds, W. F.; Vogel, Z. *FASEB J.* **1998**, *12*, 1019–1025.
170. Bayewitch, M. L.; Avidor-Reiss, T.; Levy, R.; Pfeuffer, T.; Nevo, I.; Simonds, W. F.; Vogel, Z. *J. Biol. Chem.* **1998**, *273*, 2273–2276.
171. Chen, Y.; Weng, G.; Li, J.; Harry, A.; Pieroni, J.; Dingus, J.; Hildebrandt, J. D.; Guarnieri, F.; Weinstein, H.; Iyengar, R. *Proc. Natl. Acad. Sci. USA* **1997**, *94*, 2711–2714.
172. Bommakanti, R. K.; Vinayak, S.; Simonds, W. F. *J. Biol. Chem.* **2000**, *275*, 38870–38876.
173. Liu, S.; Premont, R. T.; Kontos, C. D.; Huang, J.; Rockney, D. C. *J. Biol. Chem.* **2003**, *278*, 49929–49935.
174. Tsukada, S.; Simon, M. I.; Witte, O. N.; Katz, A. *Proc. Natl. Acad. Sci. USA* **1994**, *91*, 11256–11260.
175. Herlitze, S.; Garcia, D. E.; Mackie, K.; Hille, B.; Scheuer, T.; Catterall, W. A. *Nature* **1996**, *380*, 258–262.
176. Ikeda, S. R. *Nature* **1996**, *380*, 255–258.
177. Catterall, W. A. *Annu. Rev. Cell Dev. Biol.* **2000**, *16*, 521–555.
178. Zamponi, G. W. *Cell Biochem. Biophys.* **2001**, *34*, 79–94.
179. Dolphin, A. C. *Pharmacol. Rev.* **2003**, *55*, 607–627.
180. Wolfe, J. T.; Wang, H.; Howard, J.; Garrison, J.; Barrett, P. Q. *Nature* **2003**, *424*, 209–213.
181. Liu, M.; Yu, B.; Nakanishi, O.; Wieland, T.; Simon, M. *J. Biol. Chem.* **1997**, *272*, 18801–18807.
182. Lin, H. C.; Gilman, A. G. *J. Biol. Chem.* **1996**, *271*, 27979–27982.
183. Kofuji, P.; Davidson, N.; Lester, H. A. *Proc. Natl. Acad. Sci. USA* **1995**, *92*, 6542–6546.
184. Kunkel, M. T.; Perallta, E. G. *Cell* **1995**, *83*, 443–449.
185. Albsoul-Younes, A. M.; Sternweis, P. M.; Zhao, P.; Nakata, H.; Nakajima, S.; Nakajima, Y.; Kozasa, T. *J. Biol. Chem.* **2001**, *276*, 12712–12717.
186. Huang, C. L.; Feng, S.; Hilgemann, D. W. *Nature* **1998**, *391*, 803–806.
187. Kawano, T.; Chen, T.; Watanabe, S. Y.; Yamauchi, J.; Kaziro, Y.; Nakajima, Y.; Nakajima, S.; Itoh, H. *FEBS Lett.* **1999**, *463*, 355–359.
188. Lei, Q.; Jones, M. B.; Talley, E. M.; Schrier, A. D.; McIntire, W. E.; Garrison, J. C.; Bayliss, D. A. *Proc. Natl. Acad. Sci. USA* **2000**, *97*, 9771–9776.
189. Pitcher, J. A.; Inglse, J.; Higgins, J. B.; Arriza, J. L.; Casey, P. J.; Kim, C.; Benovic, J. L.; Kwatra, M. M.; Caron, M. G.; Lefkowitz, R. J. *Science* **1992**, *257*, 1264–1267.
190. Eichmann, T.; Lorenz, K.; Hoffman, M.; Brockman, J.; Krasel, C.; Lohse, M. J.; Quitterer, U. *J. Biol. Chem.* **2003**, *278*, 8052–8057.
191. Lefkowitz, R. J. *Cell* **1993**, *13*, 409–412.
192. Li, Z.; Laugwitz, K. L.; Pinkernell, K.; Pragst, I.; Baumgartner, C.; Hoffman, E.; Rosport, K.; Munch, G.; Moretti, A.; Humrich, J. et al. *Gene Ther.* **2003**, *10*, 1354–1361.
193. Lodowski, D. T.; Barnhill, J. F.; Pitcher, J. A.; Capel, W. D.; Lefkowitz, R. J.; Tesmer, J. *J. Acta Crystallogr. Sect. D Biol. Crystallogr.* **2003**, *59*, 936–939.
194. Lodowski, D. T.; Barnhill, J. F.; Pyskadlo, R. M.; Ghirlando, R.; Sterne-Marr, R.; Tesmer, J. *J. Biochem.* **2005**, *44*, 6958–6970.
195. Kozasa, T. *Trends Pharmacol. Sci.* **2004**, *25*, 61–63.
196. Bell, B.; Xing, H.; Yan, K.; Gautam, N.; Muslin, A. J. *J. Biol. Chem.* **1999**, *274*, 7982–7986.
197. Tang, X.; Downes, C. P. *J. Biol. Chem.* **1997**, *272*, 14193–14199.
198. Maier, U.; Babich, A.; Nurnberg, B. *J. Biol. Chem.* **1999**, *274*, 29311–29317.
199. Kerchner, K. R.; Clay, R. L.; McCleery, G.; Watson, N.; McIntire, W. E.; Myung, C. S.; Garrison, J. C. *J. Biol. Chem.* **2004**, *279*, 44554–44562.
200. Goel, R.; Phillips-Mason, P. J.; Gardner, A.; Raben, D. M.; Baldassare, J. J. *J. Biol. Chem.* **2004**, *279*, 6701–6710.
201. Boyer, J. L.; Waldo, G. L.; Harden, T. K. *J. Biol. Chem.* **1992**, *267*, 25451–25456.
202. Panchenko, M. P.; Saxena, K.; Li, Y.; Charnecki, S.; Sternweis, P. M.; Smith, T. F.; Giman, A. G.; Kozasa, T.; Neer, E. J. *J. Biol. Chem.* **1998**, *273*, 28298–28304.
203. Akgoz, M.; Azpiazu, I.; Kalyanaraman, V.; Gautam, N. *J. Biol. Chem.* **2002**, *277*, 19573–19578.
204. Bonacci, T. M.; Ghosh, M.; Malik, S.; Smrcka, A. V. *J. Biol. Chem.* **2005**, *280*, 10174–10181.

205. Feng, J.; Roberts, M. F.; Drin, G.; Scarlata, S. *Biochemistry* **2005**, *44*, 2577–2584.
206. Wing, M. R.; Houston, D.; Kelly, G. G.; Der, C. J.; Siderovski, D. P.; Harden, T. K. *J. Biol. Chem.* **2001**, *276*, 448257–448261.
207. Welsh, H. C.; Coadwell, W. J.; Ellson, C. D.; Ferguson, G. J.; Andrews, S. R.; Erdjument-Bromage, H.; Tempst, P.; Hawkins, P. T.; Stephans, L. R. *Cell* **2002**, *108*, 809–821.
208. Jamora, C.; Yamanouye, N.; Van Lint, J.; Laudenslager, J.; Vandenheede, J. R.; Faulkner, D. J.; Malhotra, V. *Cell* **1999**, *98*, 59–68.
209. Dell, E. J.; Connor, J.; Chen, S.; Stebbins, E. G.; Skiba, N. P.; Mochly-Rosen, D.; Hamm, H. E. *J. Biol. Chem.* **2002**, *277*, 49888–49895.
210. Slupsky, J. R.; Quitterer, U.; Weber, C. K.; Gierschlik, P.; Lohse, M. J.; Rapp, U. R. *Curr. Biol.* **1999**, *9*, 971–974.
211. Ehrenreiter, K.; Piazzolla, D.; Velamoor, V.; Sobczak, I.; Small, J. V.; Takeda, J.; Leung, T.; Baccarini, M. J. *Cell Biol.* **2005**, *168*, 955–964.
212. Mantegazza, M.; Yu, F. H.; Powell, A. J.; Clare, J. J.; Catterall, W. A.; Scheuer, T. *J. Neurosci.* **2005**, *25*, 3341–3349.
213. Luttrell, L. M.; Hawes, B. E.; van Biesen, T.; Luttrell, D. K.; Lansing, T. J.; Lefkowitz, R. J. *J. Biol. Chem.* **1996**, *271*, 19443–19450.
214. Shajahan, A. N.; Tiruppathi, C.; Smrcka, A. V.; Malik, A. B.; Minshall, R. D. *J. Biol. Chem.* **2004**, *279*, 48055–48062.
215. Langhans-Rajasekaran, S. A.; Wan, Y.; Huang, X. Y. *Proc. Natl. Acad. Sci. USA* **1995**, *92*, 8601–8605.
216. Roychowdhury, S.; Rasenick, M. M. *J. Biol. Chem.* **1997**, *272*, 31576–31581.
217. Kolch, W.; Calder, M.; Gilbert, D. *FEBS Lett.* **2005**, *579*, 1891–1895.
218. Werry, T. D.; Sexton, P. M.; Christopoulos, A. *Trends Endocrinol. Metab.* **2005**, *16*, 26–33.
219. Bogoyevitch, M. A.; Court, N. W. *Cell Signal.* **2004**, *16*, 1345–1354.
220. Nishina, H.; Wada, T.; Katada, T. *J. Biochem. (Tokyo)* **2004**, *136*, 123–126.
221. A Bogoyevitch, M.; Boehm, I.; Oakly, A.; Ketterman, A. J.; Barr, R. K. *Biochim. Biophys. Acta* **2004**, *1697*, 89–101.
222. Zarubin, T.; Han, J. *Cell Res.* **2005**, *15*, 11–18.
223. Shah, B. H.; Catt, K. J. *Trends Pharmacol. Sci.* **2003**, *24*, 239–244.
224. Benoit, M. J.; Rindt, H.; Allen, B. G. *Biochem. Cell Biol.* **2004**, *82*, 719–727.
225. Schafer, B.; Gschwind, A.; Ullrich, A. *Oncogene* **2004**, *23*, 991–999.
226. Luttrell, D. K.; Luttrell, L. M. *Oncogene* **2004**, *23*, 7969–7978.
227. Uchiba, M.; Okajima, K.; Oike, Y.; Ito, Y.; Fukudome, K.; Isobe, H.; Suda, T. *Circ. Res.* **2004**, *95*, 34–41.
228. Yang, C. M.; Lin, M. I.; Hsieh, H. L.; Sun, C. C.; Ma, Y. H.; Hsiao, L. D. *J. Cell Physiol.* **2005**, *203*, 538–546.
229. Gallagher, S. M.; Daly, C. A.; Bear, M. F.; Huber, K. M. *J. Neurosci.* **2004**, *24*, 4859–4864.
230. Kis, A.; Baxter, G. F.; Yellon, D. M. *Cardiovasc. Drugs Ther.* **2003**, *17*, 415–425.
231. Schulte, G.; Sommerschild, H.; Yang, J.; Tokuno, S.; Goiny, M.; Lovdahl, C.; Johannsson, B.; Fredholm, B. B.; Valen, G. *Acta Physiol. Scand.* **2004**, *182*, 133–143.
232. Narkar, V.; Kunduzova, O.; Hussain, T.; Cambon, C.; Parini, A.; Lokhandwala, M. *Kidney Int.* **2004**, *66*, 633–640.
233. Chi, D. S.; Fitzgerald, S. M.; Pitts, S.; Cantor, K.; King, E.; Lee, S. A.; Huang, S. K.; Krishnaswamy, G. *BMC Immunol.* **2004**, *22*.
234. Daub, H.; Weiss, F. U.; Wallasch, C.; Ullrich, A. *Nature* **1996**, *379*, 557–560.
235. Lefkowitz, R. J.; Pierce, K. L.; Luttrell, L. M. *Mol. Pharmacol.* **2002**, *62*, 971–974.
236. Luttrell, D. K.; Luttrell, L. M. *Assay Drug Develop. Technol.* **2003**, *1*, 327–338.
237. Gutkind, J. S. *J. Biol. Chem.* **1998**, *273*, 1839–1842.
238. Luttrell, L. M. *Can. J. Physiol. Pharmacol.* **2002**, *80*, 375–382.
239. Shah, B. H.; Catt, K. J. *Trends Neurosci* **2004**, *27*, 48–53.
240. Neubig, R. R. *FASEB J.* **1994**, *8*, 939–946.
241. Vossler, M. R.; Yao, H.; York, R. D.; Pan, M. G.; Rim, C. S.; Stork, P. J. *Cell* **1997**, *89*, 73–82.
242. Grewal, S. S.; Fass, D. M.; Yao, H.; Ellig, C. L.; Goodman, R. H.; Stork, P. J. *J. Biol. Chem.* **2000**, *275*, 34433–34441.
243. Fujita, T.; Meguro, T.; Fukuyama, R.; Nakamuta, H.; Koida, M. *J. Biol. Chem.* **2002**, *277*, 22191–22200.
244. Schmitt, J. M.; Stork, P. J. *J. Biol. Chem.* **2002**, *277*, 43024–43032.
245. Keiper, M.; Stope, M. B.; Szatkowski, D.; Bohm, A.; Tysack, K.; Vom Dorp, F.; Saur, O.; Oude Weernink, P. A.; Evellin, S.; Jakobs, K. H. et al. *J. Biol. Chem.* **2004**, *279*, 46497–46508.
246. De Rooji, J.; Zwartkruis, F. J.; Verheijen, M. H.; Cool, R. H.; Nijman, S. M.; Wittinghofer, A.; Bos, J. L. *Nature* **1998**, *396*, 416–417.
247. Lin, S. J.; Johnson-Farley, N. N.; Lubinsky, D. R.; Cowen, D. S. *J. Neurochem.* **2003**, *87*, 1076–1085.
248. Chen, C.; Koh, A. J.; Datta, N. S.; Zhang, J.; Keller, E. T.; Xiao, G.; Franceschi, R. T.; D'Silva, N. J.; McCauley, L. K. *J. Biol. Chem.* **2004**, *279*, 29121–29129.
249. Grobben, B.; Claes, P.; Van Kolen, K.; Roymans, D.; Frasen, P.; Sys, S. U.; Slegers, H. *J. Neurochem.* **2001**, *78*, 1325–1338.
250. Masri, B.; Lahlou, H.; Mazarguil, H.; Knibiehler, B.; Audigier, Y. *Biochem. Biophys. Res. Commun.* **2002**, *290*, 539–545.
251. Duca, L.; Lambert, E.; Debret, R.; Rothhut, B.; Blanchevoye, C.; Delacoux, F.; Hornbeck, W.; Martiny, L.; Debelle, L. *Mol. Pharmacol.* **2005**, *67*, 1315–1324.
252. Kolch, W.; Heidecker, G.; Kochs, G.; Hummel, R.; Vahidi, H.; Mischak, H.; Finkenzeller, G.; Marme, D.; Rapp, U. R. *Nature* **1993**, *364*, 249–252.
253. Oak, J.; Lavine, N.; Van Tol, H. H. *Mol. Pharmacol.* **2001**, *60*, 92–103.
254. Wetzker, R.; Bohmer, F. D. *Nat. Rev. Mol. Cell Biol.* **2003**, *4*, 651–657.
255. Luttrell, D. K.; Luttrell, L. M. *Oncogene* **2004**, *23*, 7969–7978.
256. Luttrell, L. M.; Ferguson, S. S. G.; Daaka, Y.; Miller, W. E.; Maudsley, S.; Della Rocca, G. J.; Lin, F. T.; Kawakatsu, H.; Owada, K.; Luttrell, D. K. et al. *Science* **1999**, *283*, 655–661.
257. DeFea, K. A.; Vaughn, Z. D.; O'Bryan, E. M.; Nishijima, D.; Dery, O.; Bunnett, N. W. *Proc. Natl. Acad. Sci. USA* **2000**, *97*, 11086–11091.
258. Miller, W. E.; Maudsley, S.; Ahn, S.; Kahn, K. D.; Luttrell, L. M.; Lefkowitz, R. J. *J. Biol. Chem.* **2000**, *275*, 11312–11319.
259. Huess, C.; Gerber, U. *Trends Neurosci.* **2000**, *10*, 469–475.
260. Ross, E. M.; Wilkie, T. M. *Annu. Rev. Biochem.* **2000**, *69*, 795–827.
261. Neubig, R. R.; Siderovski, D. P. *Nat. Rev. Drug Disc.* **2002**, *1*, 187–197.
262. Siderovski, D. P.; Willard, F. S. *Int. J. Biol. Sci.* **2005**, *1*, 51–66.
263. Snow, B. E.; Betts, L.; Mangion, J.; Sondek, J.; Siderovski, D. P. *Proc. Natl. Acad. Sci. USA* **1999**, *96*, 6489–6494.
264. Saitoh, O.; Kubo, Y.; Miyatani, Y.; Asano, T.; Nakata, H. *Nature* **1997**, *390*, 525–529.
265. Doupnik, C. A.; Davidson, N.; Lester, H. A.; Kojuji, P. *Proc. Natl. Acad. Sci. USA* **1997**, *94*, 10461–10466.
266. Zerangue, N.; Jan, L. Y. *Curr. Biol.* **1998**, *8*, 313–316.
267. Hollinger, S.; Hepler, J. R. *Pharmacol. Rev.* **2002**, *54*, 527–559.

268. Biddlecome, G. H.; Bernstein, G.; Ross, E. M. *J. Biol. Chem.* **1996**, *271*, 7999–8007.
269. Zhong, H.; Wade, S. M.; Woolf, P. J.; Linderman, J. J.; Traynor, J. R.; Neubig, R. R. *J. Biol. Chem.* **2003**, *278*, 7278–7284.
270. Benians, A.; Nobles, M.; Hosny, S.; Tinker, A. *J. Biol. Chem.* **2005**, *280*, 13383–13394.
271. Hague, C.; Bernstein, L. S.; Ramineni, S.; Chen, Z.; Minneman, K. P.; Hepler, J. R. *J. Biol. Chem.* **2005**, *280*, 27289–27295.
272. Costa, T.; Herz, A. *Proc. Natl. Acad. Sci. USA* **1989**, *86*, 7321–7325.
273. Burstein, E. S.; Spalding, T. A.; Brauner-Osborne, H.; Brann, M. R. *FEBS Lett.* **1995**, *363*, 261–263.
274. Francken, B. J.; Josson, K.; Lijnen, P.; Jurzak, M.; Luyten, W. H.; Leysen, J. E. *Mol. Pharmacol.* **2000**, *57*, 1034–1044.
275. Dupre, D. J.; Le Gouill, C.; Rola-Pleszczynski, M.; Stankova, J. *J. Pharmacol. Exp. Ther.* **2001**, *299*, 358–365.
276. Scragg, J. L.; Warburton, P.; Ball, S. G.; Balmforth, A. *J. Biochem. Biophys. Res. Commun.* **2005**, *334*, 134–139.
277. Kenakin, T. *Mol. Pharmacol.* **2004**, *65*, 2–11.
278. Lefkowitz, R. J.; Cotecchia, S.; Samama, P.; Costa, T. *Trends Pharmacol. Sci.* **1993**, *14*, 303–307.
279. Parma, J.; Duprez, L.; Van Sande, J.; Paschke, R.; Tonacchera, M.; Dumont, J.; Vassart, G. *Mol. Cell. Endocrinol.* **1994**, *100*, 159–162.
280. Scheer, A.; Cotecchia, S. *J. Recept. Signal Transduct. Res.* **1997**, *17*, 57–73.
281. Milligan, G.; Bond, R. A. *Trends Pharmacol. Sci.* **1997**, *18*, 468–474.
282. Leurs, R.; Smit, M. J.; Alewijnse, A. E.; Timmerman, H. *Trends Biochem. Sci.* **1998**, *23*, 418–422.
283. Pauwels, P. J.; Wurch, T. *Mol. Neurobiol.* **1998**, *17*, 109–135.
284. de Ligt, R. A.; Kourounakis, A. P.; Ijzerman, A. P. *Br. J. Pharmacol.* **2000**, *130*, 1–12.
285. Behan, D. P.; Chalmers, D. T. *Curr. Opin. Drug Disc. Dev.* **2001**, *4*, 548–560.
286. Parnot, C.; Miserey-Lenkei, S.; Bardin, S.; Corvol, P.; Clauser, E. *Trends Endocrinol. Metab.* **2002**, *13*, 336–343.
287. Seifert, R.; Wenzel-Seifert, K. *Naunyn-Schmiedeberg's Arch. Pharmacol.* **2002**, *366*, 381–416.
288. Cotecchia, S.; Fanelli, F.; Costa, T. *Assay Drug Dev. Technol.* **2003**, *1*, 311–316.
289. Costa, T.; Cotecchia, S. *Trends Pharmacol. Sci.* **2005**, *26*, 618–624.
290. Soudijn, W.; van Wijngaarden, I.; Ijzerman, A. P. *Med. Res. Rev.* **2005**, *25*, 398–426.
291. De Lean, A.; Stadel, J. M.; Lefkowitz, R. J. *J. Biol. Chem.* **1980**, *255*, 7108–7117.
292. Wreggett, K. A.; De Lean, A. *Mol. Pharmacol.* **1984**, *26*, 214–227.
293. Samama, P.; Cottechia, S.; Costa, T.; Lefkowitz, R. J. *J. Biol. Chem.* **1993**, *268*, 4625–4636.
294. Milligan, G.; Bond, R. A. *Trends Pharmacol. Sci.* **1997**, *18*, 468–474.
295. Gether, U.; Ballesteros, J. A.; Seifert, R.; Sanders-Bush, E.; Weinstein, H.; Kobilka, B. K. *J. Biol. Chem.* **1997**, *272*, 2587–2590.
296. Alewijnse, A. E.; Smit, M. J.; Hoffmann, M.; Verzijl, D.; Timmerman, H.; Leurs, R. *J. Neurochem.* **1998**, *71*, 799–807.
297. Alewijnse, A. E.; Timmerman, H.; Jakobs, E. H.; Smit, M. J.; Roovers, E.; Cottechia, S.; Leurs, R. *Mol. Pharmacol.* **2000**, *57*, 890–898.
298. Pei, G.; Samama, P.; Lohse, M.; Wang, M.; Codina, J.; Lefkowtiz, R. J. *Proc. Natl. Acad. Sci. USA* **1994**, *91*, 2699–2702.
299. Shapiro, M. J.; Trejo, J.; Zeng, D.; Coughlin, S. R. *J. Biol. Chem.* **1996**, *271*, 32874–32880.
300. Tarasova, N. I.; Stauber, R. H.; Hudson, E. A.; Czerwinski, G.; Miller, J. L.; Pavlakis, G. N.; Michejda, C. J.; Wank, S. A. *J. Biol. Chem.* **1997**, *272*, 14817–14824.
301. Mhaouty-Kodja, S.; Barak, L. S.; Scheer, A.; Abuin, L.; Diviani, D.; Caron, M. G.; Cottechia, S. *Mol. Pharmacol.* **1999**, *55*, 339–347.
302. Min, L.; Ascoli, M. *Mol. Endocrinol.* **2000**, *14*, 1797–1810.
303. Miserey-Lenkei, S.; Parnot, C.; Bardin, S.; Corvol, P.; Clauser, E. *J. Biol. Chem.* **2002**, *277*, 5891–5901.
304. Morris, D. P.; Price, R. R.; Smith, M. P.; Lei, B.; Schwinn, D. A. *Mol. Pharmacol.* **2004**, *66*, 843–854.
305. Leterrier, C.; Bonnard, D.; Carrel, D.; Rossier, J.; Lenkei, Z. *J. Biol. Chem.* **2004**, *279*, 36013–36021.
306. Smit, M. J.; Leurs, R.; Alewijnse, A. E.; Blauw, J.; Van Nieuw Amerongen, G. P.; Van De Vrede, Y.; Roovers, S.; Timmerman, H. *Proc. Natl. Acad. Sci. USA* **1996**, *93*, 6802–6807.
307. Lee, T. W.; Cotecchia, S.; Milligan, G. *Biochem. J.* **1997**, *325*, 733–739.
308. Stevens, P. A.; Bevan, N.; Rees, S.; Milligan, G. *Mol. Pharmacol.* **2000**, *58*, 438–448.
309. Cetani, F.; Tonacchera, M.; Vassart, G. *FEBS Lett.* **1996**, *378*, 27–31.
310. Cosi, C.; Koek, W. *Eur. J. Pharmacol.* **2000**, *401*, 9–15.
311. Pauwels, P. J.; Tardif, S. *Naunyn-Schmiedeberg's Arch. Pharmacol.* **2002**, *366*, 134–141.
312. Munshi, U. M.; Pogozheva, I. D.; Menon, K. M. *Biochemistry* **2003**, *42*, 3708–3715.
313. Ceresa, B. P.; Limbird, L. E. *J. Biol. Chem.* **1994**, *269*, 29557–29564.
314. Brown, G. P.; Pasternak, G. W. *J. Pharmacol. Exp. Ther.* **1998**, *286*, 376–381.
315. Monczor, F.; Fernandez, N.; Legnazzi, B. L.; Riveiro, M. E.; Baldi, A.; Shayo, C.; Davio, C. *Mol. Pharmacol.* **2003**, *64*, 512–520.
316. Vasquez, C.; Lewis, D. L. *J. Neurosci.* **1999**, *19*, 927–980.
317. Weiss, J. M.; Morgan, P. H.; Lutz, M. W.; Kenakin, T. P. *J. Theor. Biol.* **1996**, *181*, 381–397.
318. Kenakin, T. *Trends Pharmacol. Sci.* **1997**, *18*, 416–417.
319. Kenakin, T. *FASEB J.* **2001**, *15*, 598–611.
320. Kenakin, T. *Annu. Rev. Pharmacol. Toxicol.* **2002**, *42*, 349–379.
321. Kenakin, T. *Nat. Rev. Drug Disc.* **2002**, *1*, 103–110.
322. Kenakin, T. *Trends Pharmacol. Sci.* **2003**, *24*, 346–354.
323. Kenakin, T. *Mol. Pharmacol.* **2004**, *65*, 2–11.
324. Kenakin, T. *Nat. Rev. Drug Disc.* **2005**, *4*, 919–927.
325. Kenakin, T. *Nat. Methods* **2005**, *2*, 163–164.
326. Gbahou, F.; Rouleau, A.; Morisset, S.; Parmentier, R.; Crochet, S.; Lin, J.-S.; Ligneau, X.; Tardivel-Lacombe, J.; Stark, H.; Schnack, W. et al. *Proc. Natl. Acad. Sci. USA* **1993**, *100*, 11086–11091.
327. Vauquelin, G.; Van Liefde, I. *Fund. Clin. Pharmacol.* **2005**, *19*, 45–56.
328. Hermans, E. *Pharmacol. Ther.* **2003**, *99*, 25–44.
329. Jin, L. Q.; Wang, H. Y.; Friedman, E. *J. Neurochem.* **2001**, *78*, 7178–7188.
330. Kilts, J. D.; Gerhardt, M. A.; Richardson, M. D.; Sreeram, G.; Mackensen, G. B.; Grocott, H. P.; White, W. D.; Davis, R. D.; Newman, M. F.; Reves, J. G. et al. *Circ. Res.* **2000**, *87*, 705–709.
331. Klein, J.; Reymann, K. G.; Riedel, G. *Neuropharmacology* **1997**, *36*, 261–263.
332. Laugwitz, K. L.; Allgeier, A.; Offermanns, S.; Spicher, K.; Van-Sande, J.; Dumont, J. E.; Schultz, G. *Proc. Natl. Acad. Sci. USA* **1996**, *93*, 116–120.

333. Santos-Alverez, J.; Sanchez-Margalet, V. *J. Cell. Biochem.* **1999**, *73*, 469–477.
334. Shi, L. C.; Wang, H. Y.; Horwitz, J.; Friedman, E. *J. Neurochem.* **1996**, *67*, 1478–1484.
335. Stanislaus, D.; Ponder, S.; Ji, T. H.; Connn, P. M. *Biol. Reprod.* **1998**, *59*, 579–586.
336. Perez, D. M.; Hwa, J.; Gaivan, R.; Mathar, M.; Brown, F.; Graham, R. M. *Mol. Pharmacol.* **1996**, *49*, 112–122.
337. Ishii, T.; Izumi, T.; Tsukamoto, H.; Umeyama, H.; Ui, M.; Shimizu, T. *J. Biol. Chem.* **1997**, *272*, 7846–7854.
338. Noda, K.; Feng, Y.-H.; Lui, X.-P.; Saad, Y.; Husain, A.; Karnik, S. S. *Biochemistry* **1996**, *35*, 16435–16442.
339. Zuscik, M. J.; Porter, J. E.; Gavain, R.; Perez, D. M. *J. Biol. Chem.* **1998**, *273*, 3401–3407.
340. Gazi, L.; Wurch, T.; Lopez-Gimenez, F.; Pauwels, P. J.; Strange, P. G. *FEBS Lett.* **2003**, *545*, 155–160.
341. Berg, K. A.; Maayani, S.; Goldfarb, J.; Scaramellini, C.; Leff, O.; Clarke, W. P. *Mol. Pharmacol.* **1998**, *54*, 94–104.
342. Maudsley, S.; Martin, B.; Luttrell, L. M. *J. Pharmacol. Exp. Ther.* **2005**, *314*, 485–495.
343. Maudsley, S.; Davidson, L.; Pawson, A. J.; Chan, R.; de Maturana, R. L.; Millar, R. P. *Cancer Res.* **2004**, *15*, 7533–7544.
344. Pommier, B.; Da Nascimento, S.; Dumont, S.; Bellier, B.; Million, E.; Garbay, C.; Roques, B. P.; Noble, F. *J. Neurochem.* **1999**, *73*, 281–288.
345. Sagan, S.; Chassaing, G.; Pradier, L.; Lavielle, S. *J. Pharmacol. Exp. Ther.* **1996**, *276*, 1039–1048.
346. Baker, J. G.; Hall, I. P.; Hill, S. J. *Mol. Pharmacol.* **2003**, *64*, 1357–1369.
347. Azzi, M.; Charest, P. G.; Angers, S.; Rousseau, G.; Kohout, T.; Bouvier, M.; Pineyro, G. *Proc. Natl. Acad. Sci. USA* **2003**, *100*, 11406–11411.
348. Ghanouni, P.; Gryczynski, Z.; Steenhuis, J. J.; Lee, T. W.; Farrens, D. L.; Lefkowitz, R. J.; Kobilka, B. K. *J. Biol. Chem.* **2001**, *276*, 24433–24436.
349. Swaminath, G.; Xiang, Y.; Lee, T. W.; Steenhuis, J.; Parnot, C.; Kobilka, B. K. *J. Biol. Chem.* **2004**, *279*, 686–691.
350. Swaminath, G.; Deupi, X.; Lee, T. W.; Zhu, W.; Thian, F. S.; Kobilka, T. S.; Kobilka, B. *J. Biol. Chem.* **2005**, *280*, 22165–22171.
351. Alves, I. D.; Cowell, S. M.; Salamon, Z.; Devanathan, S.; Tollin, G.; Hruby, V. *J. Mol. Pharmacol.* **2004**, *65*, 1248–1257.
352. Alves, I. D.; Ciano, K. A.; Boguslavski, V.; Varga, E.; Salamon, Z.; Yamamura, H. I.; Hruby, V. J.; Tollin, G. *J. Biol. Chem.* **2004**, *279*, 44673–44682.
353. Devanathan, S.; Yao, Z.; Salamon, Z.; Kobilka, B.; Tollin, G. *Biochemistry* **2004**, *43*, 3280–3288.
354. Arden, J. R.; Segredo, V.; Wang, Z.; Lameh, J.; Sadee, W. *J. Neurochem.* **1995**, *65*, 1636–1645.
355. Zhang, L.; Yu, Y.; Mackin, S.; Weight, F. F.; Uhl, G. R.; Wang, J. B. *J. Biol. Chem.* **1996**, *271*, 11449–11454.
356. Keith, D. E.; Murray, S. R.; Zaki, P. A.; Chu, P. C.; Lissin, D. V.; Kang, L.; Evans, C. J.; von Zastrow, M. *J. Biol. Chem.* **1996**, *271*, 19021–19024.
357. Rittano, D.; Werge, T. M.; Costa, T. *J. Biol. Chem.* **1997**, *272*, 7646–7655.
358. Chakrabarti, S.; Law, P. Y.; Loh, H. H. *J. Neurochem.* **1998**, *71*, 231–239.
359. Wiens, B. L.; Nelson, C. S.; Neve, K. A. *Mol. Pharmacol.* **1998**, *54*, 435–444.
360. Roettger, B. F.; Ghanekar, D.; Rao, R.; Toledo, C.; Yingling, J.; Pinon, D.; Miller, L. *J. Mol. Pharmacol.* **1997**, *51*, 357–362.
361. Thomas, W. G.; Qian, H.; Chang, C. S.; Karnik, S. *J. Biol. Chem.* **2000**, *275*, 2893–2900.
362. Kohout, T. A.; Nicholas, S. L.; Perry, S. J.; Reinhart, G.; Junger, S.; Struthers, R. S. *J. Biol. Chem.* **2004**, *279*, 23214–23222.
363. Trester-Zedlitz, M.; Burlingame, A.; Kobilka, B.; von Zastrow, M. *Biochemistry* **2005**, *44*, 6133–6143.
364. Whistler, J. L.; Gerber, B. O.; Meng, E. C.; Baranski, T. J.; von Zastrow, M. *Traffic* **2002**, *3*, 866–877.
365. Perez, D. M.; Karnik, S. S. *Pharmacol. Rev.* **2005**, *57*, 147–161.
366. Simmons, M. A. *Mol. Interv.* **2005**, *5*, 154–157.
367. Ostrom, R. S.; Post, S. R.; Insel, P. A. *J. Pharmacol. Exp. Ther.* **2000**, *294*, 407–412.
368. Tanowitz, M.; von Zastrow, M. *Methods Mol. Biol.* **2004**, *259*, 353–369.
369. Bockaert, J.; Roussignol, G.; Becamel, C.; Gavarini, S.; Joubert, L.; Dumuis, A.; Fagni, L.; Marin, P. *Biochem. Soc. Trans.* **2004**, *32*, 851–855.
370. Luttrell, M. L. *J. Mol. Neurosci.* **2005**, *26*, 253–264.
371. Hall, R. A.; Lefkowitz, R. J. *Circ. Res.* **2002**, *91*, 672–680.
372. He, J.; Bellini, M.; Inuzuka, H.; Xu, J.; Xiong, Y.; Yang, X.; Castleberry, A. M.; Hall, R. A. *J. Biol. Chem.* **2006**, *281*, 2820–2827.
373. Gavarini, S.; Becamel, C.; Chanrion, B.; Bockaret, J.; Marin, O. *Biol. Cell* **2004**, *96*, 373–381.
374. Tilakaratne, N.; Sexton, P. M. *Clin. Exp. Pharmacol. Physiol.* **2005**, *32*, 979–987.
375. Ciruela, F.; Canela, L.; Burgueno, J.; Soriguera, A.; Cabello, N.; Canela, E. I.; Casado, V.; Cortes, A.; Mallol, J.; Woods, A. S. et al. *J. Mol. Neurosci.* **2005**, *26*, 277–292.
376. Presland, J. *Biochem. Soc. Trans.* **2004**, *32*, 888–891.
377. Fan, G.; Shumay, E.; Wang, H.; Malbon, C. C. *J. Biol. Chem.* **2001**, *276*, 24005–24014.
378. Malbon, C. C.; Tao, J.; Wang, H. Y. *Biochem. J.* **2004**, *379*, 1–9.
379. Miller, W. E.; Maudsley, S.; Ahn, S.; Khan, K. D.; Luttrell, L. M.; Lefkowitz, R. J. *J. Biol. Chem.* **2000**, *275*, 11312–11319.
380. Tanaka, Y.; Tanura, K.; Koide, Y.; Sakai, M.; Tsurumi, Y.; Noda, Y.; Umemura, M.; Ishagami, T.; Uchino, K.; Kimura, K. et al. *FEBS Lett.* **2005**, *579*, 1579–1586.
381. El Far, O.; Betz, H. *Biochem. J.* **2002**, *365*, 329–336.
382. Turner, J. H.; Gelasco, A. K.; Raymond, J. R. *J. Biol. Chem.* **2004**, *279*, 17027–17037.
383. Rozell, T. G.; Davis, D. P.; Chai, Y.; Segaloff, D. L. *Endocrinology* **1998**, *139*, 1588–1593.
384. Morello, J. P.; Salahpour, A.; Petaja-Repo, U. E.; Laperriere, A.; Lonergan, M.; Arthus, M. F.; Nabi, I. R.; Bichet, D. G.; Bouvier, M. *Biochemistry* **2001**, *12*, 6766–6775.
385. Bhatnagar, A.; Sheffler, D. J.; Kroeze, W. K.; Compton-Toth, B.; Roth, B. L. *J. Biol. Chem.* **2004**, *279*, 34614–34623.
386. Becamel, C.; Gavarini, S.; Chanrion, B.; Alonso, G.; Galeotti, N.; Dumuis, A.; Bockaert, J.; Marin, P. *J. Biol. Chem.* **2004**, *279*, 20257–20266.
387. Diviani, D.; Lattion, A. L.; Abuin, L.; Staub, O.; Cottechia, S. *J. Biol. Chem.* **2003**, *278*, 19331–19340.
388. Zitzer, H.; Richter, D.; Kreienkamp, H. J. *J. Biol. Chem.* **1999**, *274*, 18153–18156.
389. Tang, Y.; Hu, L. A.; Miller, W. E.; Ringstad, N.; Hall, R. A.; Pitcher, J. A.; DeCamilli, P.; Lefokowitz, R. J. *Proc. Natl. Acad. Sci. USA* **1999**, *96*, 12559–12564.
390. Li, M.; Bermak, J. C.; Wang, Z. W.; Zhou, Q. Y. *Mol. Pharmacol.* **2000**, *57*, 446–452.
391. Hjalm, G.; MacLeod, R. J.; Kifor, O.; Chattopadhyay, N.; Brown, E. M. *J. Biol. Chem.* **2001**, *276*, 34880–34887.
392. Zhang, M.; Breitwieser, G. E. *J. Biol. Chem.* **2005**, *280*, 11140–11146.
393. Simonin, F.; Karcher, P.; Boeuf, J. J.; Matifas, A.; Kieffer, B. L. *J. Neurochem.* **2004**, *89*, 766–775.
394. Jeanneteau, F.; Guillin, O.; Diaz, J.; Griffon, N.; Sokoloff, P. *Mol. Biol. Chem.* **2004**, *15*, 4926–4937.
395. Hirakawa, T.; Galet, C.; Kishi, M.; Ascoli, M. *J. Biol. Chem.* **2003**, *278*, 49348–49357.
396. Ango, F.; Robbe, D.; Tu, J. C.; Xiao, B.; Worley, P. F.; Pin, J. P.; Bockaert, J.; Fagni, L. *Mol. Cell Neurosci.* **2002**, *20*, 323–329.
397. Schweneker, M.; Bachman, A. S.; Moelling, K. *FEBS Lett.* **2005**, *579*, 1751–1758.
398. Xu, J.; Paquet, M.; Lau, A. G.; Wood, J. D.; Ross, C. A.; Hall, R. A. *J. Biol. Chem.* **2001**, *276*, 41310–41317.

399. Metherell, L. A.; Chapple, J. P.; Cooray, S.; David, A.; Becker, C.; Ruschendorf, F.; Naville, D.; Begeot, M.; Khoo, B.; Nurnberg, P. et al. *Nat. Genet.* **2005**, *37*, 166–170.
400. Becamel, C.; Figge, A.; Poliak, S.; Dumuis, A.; Peles, E.; Bockaert, J.; Lubbert, H.; Ullmer, C. *J. Biol. Chem.* **2001**, *276*, 12974–12982.
401. Hasegawa, H.; Katoh, H.; Fujita, H.; Mori, K.; Negishi, M. *Biochem. Biophys. Res. Commun.* **2000**, *276*, 350–354.
402. Hall, R. A.; Premont, R. T.; Chow, C. W.; Blitzer, J. T.; Picher, J. A.; Claing, A.; Stoffel, R. H.; Barak, L. S.; Shenolikar, S.; Weinman, E. J. et al. *Nature* **1998**, *392*, 626–630.
403. Huang, P.; Steplock, D.; Weinman, E. J.; Hall, R. A.; Ding, Z.; Li, J.; Wang, Y.; Liu-Chen, L. Y. *J. Biol. Chem.* **2004**, *279*, 25002–25009.
404. Rochdi, M. D.; Laroche, G.; Dupre, E.; Giguere, P.; Lebel, A.; Waiter, V.; Hamelin, E.; Lepine, M. C.; Dupuis, G.; Parent, J. L. *J. Biol. Chem.* **2004**, *279*, 18981–18989.
405. Feng, G. J.; Kellet, E.; Scorer, C. A.; Wilde, J.; White, J. H.; Milligan, G. *J. Biol. Chem.* **2003**, *278*, 33400–33407.
406. Milligan, G.; Murdoch, H.; Kellet, E.; White, J. H.; Feng, G. J. *Biochem. Soc. Trans.* **2004**, *32*, 878–880.
407. Boudin, H.; Craig, A. M. *J. Biol. Chem.* **2001**, *276*, 30270–30276.
408. Enz, R.; Croci, C. *Biochem. J.* **2003**, *372*, 183–191.
409. Flajolet, M.; Rakhilin, S.; Wang, H.; Starkova, N.; Nuangchamnong, N.; Narin, A. C.; Greengard, P. *Proc. Natl. Acad. Sci. USA* **2003**, *100*, 16006–16011.
410. Hu, L. A.; Tang, Y.; Miller, W. E.; Cong, M.; Mau, A. G.; Lefkowitz, R. J.; Hall, R. A. *J. Biol. Chem.* **2000**, *275*, 38659–38666.
411. Evans, B. N.; Rosenblatt, M. I.; Mnayer, L. O.; Oliver, K. R.; Dickerson, I. M. *J. Biol. Chem.* **2000**, *275*, 31438–31443.
412. Moriyoshi, K.; Iijima, K.; Fujii, H.; Ito, H.; Cho, Y.; Nakanishi, S. *Proc. Natl. Acad. Sci. USA* **2004**, *101*, 8614–8619.
413. Brady, A. E.; Wang, Q.; Colbran, R. J.; Allen, P. B.; Greengard, P.; Limbird, L. E. *J. Biol. Chem.* **2003**, *278*, 32405–32412.
414. Wang, Q.; Limbird, L. E. *J. Biol. Chem.* **2002**, *277*, 50589–50596.
415. Kitano, J.; Kimura, K.; Yamazaki, Y.; Soda, T.; Shigemoto, R.; Nakajima, Y.; Nakanishi, S. *J. Neurosci.* **2002**, *22*, 1280–1289.
416. Saugstad, J. A.; Yang, S.; Pohl, J.; Hall, R. A.; Conn, P. J. *J. Neurochem.* **2002**, *80*, 980–988.
417. Palczewski, K.; Kumasaka, T.; Hori, T.; Behnke, C. A.; Motoshima, H.; Fox, B. A.; Le Trong, D. C.; Okada, T.; Stenkamp, R. E.; Yamamoto, M. et al. *Science* **2000**, *289*, 739–745.
418. Teller, D. C.; Okada, T.; Behnke, C. A.; Palczewski, K.; Stenkamp, R. E. *Biochemistry* **2001**, *40*, 7761–7772.
419. Okada, T.; Fujiyoshi, Y.; Silow, M.; Navarro, J.; Landau, E. M.; Shichida, Y. *Proc. Natl. Acad. Sci USA* **2002**, *99*, 5982–5987.
420. Sakmar, T. P. *Curr. Opin. Cell Biol.* **2002**, *14*, 189–195.
421. Meng, E. C.; Bourne, H. R. *Trends Pharmacol. Sci.* **2001**, *22*, 587–593.
422. Borhan, B.; Souto, M. L.; Imai, H.; Shichida, Y.; Nakanishi, K. *Science* **2000**, *288*, 2209–2212.
423. Farahbakhsh, Z. T.; Ridge, K. D.; Khorana, H. G.; Hubbell, W. L. *Biochemistry* **1995**, *34*, 8812–8819.
424. Altenbach, C.; Yang, K.; Farrens, D. L.; Farabakhsh, Z. T.; Khorana, H. G.; Hubbell, W. L. *Biochemistry* **1996**, *35*, 12470–12478.
425. Farrens, D. L.; Altenbach, C.; Yang, K.; Hubbell, W. L.; Khorana, H. G. *Science* **1996**, *274*, 768–770.
426. Sakmar, T. P.; Menon, S. T.; Marin, E. P.; Awad, E. S. *Annu. Rev. Biophys. Biomol. Struct.* **2002**, *31*, 443–484.
427. Hubbell, W. L.; Altenbach, C.; Hubbell, C. M.; Khorana, H. G. *Adv. Protein Chem.* **2003**, *63*, 243–290.
428. Ballesteros, J.; Palczewski, K. *Curr. Opin. Drug Disc. Dev.* **2001**, *4*, 561–574.
429. Gether, U.; Lin, S.; Ghanouni, P.; Ballesteros, J. A.; Weinstein, H.; Kobilka, B. K. *EMBO J.* **1997**, *16*, 6737–6747.
430. Rasmussen, S. G.; Jensen, A. D.; Liapakis, G.; Ghanouni, P.; Javitch, J. A.; Gether, U. *Mol. Pharmacol.* **1999**, *56*, 175–184.
431. Jensen, A. D.; Guarnieri, F.; Rasmussen, S. G.; Asmar, F.; Ballesteros, J. A.; Gether, U. *J. Biol. Chem.* **2001**, *276*, 9279–9290.
432. Gether, U. *Endocrinol. Rev.* **2000**, *21*, 90–113.
433. Gether, U.; Asmar, F.; Meinild, A. K.; Rasumssen, S. G. *Pharmacol. Toxicol.* **2002**, *91*, 304–312.
434. Kim, J. M.; Hwa, J.; Garriga, P.; Reeves, P. J.; RajBhandary, U. L.; Khorana, H. G. *Biochemistry* **2005**, *44*, 2284–2292.
435. Javitch, J. A.; Shi, L.; Simpson, M. M.; Chen, J.; Chiappa, V.; Visiers, I.; Weinstein, H.; Ballesteros, J. A. *Biochemistry* **2000**, *39*, 12190–12199.
436. Shi, L.; Simpson, M. M.; Ballesteros, J. A.; Javitch, J. A. *Biochemistry* **2001**, *40*, 12339–12348.
437. Splading, T. A.; Burstein, E. S.; Henderson, S. C.; Brann, M. R. *J. Biol. Chem.* **1998**, *273*, 21563–21568.
438. Lu, Z. L.; Saldanha, J. W.; Hulme, E. C. *J. Biol. Chem.* **2001**, *276*, 34098–34104.
439. Han, S. J.; Hamdan, F. F.; Kim, S. K.; Jacobson, K. A.; Brichta, L.; Bloodworth, L. M.; Li, J. H.; Wess, J. *J. Biol. Chem.* **2005**, *280*, 24870–24879.
440. Baneres, J. L.; Mesnier, D.; Martin, A.; Joubert, L.; Dumis, A.; Bockaert, J. *J. Biol. Chem.* **2005**, *280*, 20253–20260.
441. Miura, S.; Feng, Y. H.; Husain, A.; Karnik, S. S. *J. Biol. Chem.* **1999**, *274*, 7103–7110.
442. Miura, S.; Karnik, S. S. *J. Biol. Chem.* **2002**, *277*, 24299–24305.
443. Langerstrom, M. C.; Klovins, J.; Fredriksson, R.; Fridmanis, D.; Haitina, T.; Ling, M. K.; Berglund, M. M.; Schioth, H. B. *J. Biol. Chem.* **2003**, *278*, 51521–51526.
444. Miura, S.; Zhang, J.; Boros, J.; Karnik, S. S. *J. Biol. Chem.* **2003**, *278*, 3720–3725.
445. Decaillot, F. M.; Befort, K.; Filliol, D.; Yue, S.; Walker, P.; Kieffer, B. L. *Nat. Struct. Biol.* **2003**, *10*, 629–636.
446. Govaerts, C.; Lefort, A.; Costagliola, S.; Wodak, S. J.; Ballesteros, J. A.; Van Sande, J.; Pardo, L.; Vassart, G. *J. Biol. Chem.* **2001**, *276*, 22991–22999.
447. Zhang, M.; Mizzachi, D.; Fanelli, F.; Segaloff, D. L. *J. Biol. Chem.* **2005**, *280*, 26169–26176.
448. Angers, S.; Salahpour, A.; Boouvier, M. *Annu. Rev. Pharmacol. Toxicol.* **2002**, *42*, 409–435.
449. Terrillon, S.; Bouvier, M. *EMBO J.* **2004**, *5*, 30–34.
450. Milligan, G. *Mol. Pharmacol.* **2004**, *66*, 1–7.
451. Bulenger, S.; Marullo, S.; Bouvier, M. *Trends Pharmacol. Sci.* **2005**, *26*, 131–137.
452. Maggio, R.; Novi, F.; Scarselli, M.; Corsini, G. U. *FEBS J.* **2005**, *272*, 2939–2946.
453. Angers, S.; Salahpour, A.; Joly, E.; Hilairet, S.; Chelsky, D.; Dennis, M.; Bouvier, M. *Proc. Natl. Acad. Sci. USA* **2000**, *97*, 3684–3689.
454. McVey, M.; Ramsay, D.; Kellett, E.; Rees, S.; Wilson, S.; Pope, A. J.; Milligan, G. *J. Biol. Chem.* **2001**, *276*, 14092–14099.
455. Kroeger, K. M.; Hanyaloglu, A. C.; Seeber, R. M.; Miles, L. E.; Eidne, K. A. *J. Biol. Chem.* **2001**, *276*, 12736–12743.
456. Overton, M. C.; Blummer, K. *J. Curr. Biol.* **2000**, *10*, 341–344.
457. Rocheville, M.; Lange, D. C.; Kumar, U.; Sasi, R.; Patel, R. C.; Patel, Y. C. *J. Biol. Chem.* **2000**, *275*, 7862–7869.
458. Rocheville, M.; Lange, D. C.; Kumar, U.; Patel, S. C.; Patel, Y. C. *Science* **2000**, *288*, 65–67.
459. Roess, D. A.; Brady, C. J.; Barisas, B. G. *Biochim. Biophys. Acta* **2000**, *1464*, 242–250.
460. Roess, D. A.; Horvat, R. D.; Munnelly, H.; Barisas, B. G. *Endocrinology* **2000**, *141*, 4518–4523.
461. Gazi, L.; Lopez-Gimenez, J. F.; Strange, P. G. *Curr. Opin. Drug Disc. Dev.* **2002**, *5*, 756–763.

462. Park, P. S.; Filipek, S.; Wells, J. W.; Palczewski, K. *Biochemistry* **2004**, *43*, 15643–15656.
463. Pfleger, K. D.; Eidne, K. A. *Biochem. J.* **2005**, *385*, 625–637.
464. Mercier, J. F.; Salahpour, A.; Angers, S.; Breit, A.; Bouvier, M. *J. Biol. Chem.* **2002**, *277*, 44925–44931.
465. Klco, J. M.; Lassere, T. B.; Baranski, T. J. *J. Biol. Chem.* **2003**, *278*, 35345–35353.
466. Stanasila, L.; Perez, J. B.; Vogel, H.; Cottechia, S. *J. Biol. Chem.* **2003**, *278*, 40239–40251.
467. Lee, S. P.; O'Dowd, B. F.; Rajaram, R. D.; Nguyen, T.; George, S. R. *Biochemistry* **2003**, *42*, 1023–1031.
468. Hernaz-Falcon, P.; Rodriguez-Frade, J. M.; Serrano, A.; Juan, D.; del Sol, A.; Soriano, S. F.; Rocal, F.; Gomez, L.; Valencia, A.; Martinez, A. C. et al. *Nat. Immunol.* **2004**, *5*, 216–223.
469. Grant, M.; Patel, R. C.; Kumar, U. *J. Biol. Chem.* **2004**, *279*, 38636–38643.
470. Thevenin, D.; Lazarova, T.; Roberts, M. F.; Robinson, C. R. *Protein Sci.* **2005**, *14*, 2177–2186.
471. Hlavackova, V.; Goudet, C.; Kniazeff, J.; Zikova, A.; Damien, M.; Vol, C.; Trojanova, J.; Prezeau, L.; Pin, J.; Blahos, J. *EMBO J.* **2005**, *24*, 499–509.
472. Novi, F.; Stanasila, L.; Giorgi, F.; Corsini, G. U.; Cotecchia, S.; Maggio, R. *J. Biol. Chem.* **2005**, *280*, 19768–19776.
473. Salapour, A.; Angers, S.; Mercier, J. F.; Lagace, M.; Marullo, S.; Bouvier, M. *J. Biol. Chem.* **2004**, *279*, 3390–3397.
474. Cao, T. T.; Brelot, A.; von Zastrow, M. *Mol. Pharmacol.* **2005**, *67*, 288–297.
475. Dean, M. K.; Higgs, C.; Smith, R. E.; Bywater, R. P.; Snell, C. R.; Scott, P. D.; Upton, G. J.; Howe, T. J.; Reynolds, C. A. *J. Med. Chem.* **2001**, *44*, 4595–4614.
476. Lamey, M.; Thompson, M.; Varghese, G.; Chi, H.; Sawzdargo, M.; George, S. R.; O'Dowd, B. F. *J. Biol. Chem.* **2002**, *277*, 9415–9421.
477. Grant, M.; Patel, R. C.; Kumar, U. *J. Biol. Chem.* **2004**, *279*, 38636–38643.
478. Milligan, G.; Wilson, S.; Lopez-Gimenez, J. F. *J. Mol. Neurosci.* **2005**, *26*, 161–168.
479. Jordan, B. A.; Devi, L. A. *Nature* **1999**, *7*, 697–700.
480. George, S. R.; O'Dowd, B. F.; Lee, S. P. *Nat. Rev. Drug Disc.* **2002**, *1*, 808–820.
481. Ayoub, M. A.; Levoye, A.; Delagrange, P.; Jockers, R. *Mol. Pharmacol.* **2004**, *66*, 312–321.
482. Zhu, W. Z.; Chakir, K.; Zhang, S.; Yang, D.; Lavoie, C.; Bouvier, M.; Hebert, T. E.; Lakatta, E. G.; Cheng, H.; Xiao, R. P. *Circ. Res.* **2005**, *97*, 244–251.
483. Lavoie, C.; Mercier, J. F.; Salahpour, A.; Umapathy, D.; Breit, A.; Villeneuve, L. R.; Zhu, W. Z.; Xiao, R. P.; Lakatta, E. G.; Bouvier, M. et al. *J. Biol. Chem.* **2002**, *277*, 35402–35410.
484. Uberti, M. A.; Hague, C.; Oller, H.; Minneman, K. P.; Hall, R. A. *J. Pharmacol. Exp. Ther.* **2005**, *313*, 16–23.
485. Wang, H. L.; Hsu, C. Y.; Huang, P. C.; Kuo, Y. L.; Li, A. H.; Yeh, T. H.; Tso, A. S.; Chen, Y. L. *J. Neurochem.* **2005**, *92*, 1285–1294.
486. Kearn, C. S.; Blake-Palmer, K.; Daniel, E.; Mackie, K.; Glass, M. *Mol. Pharmacol.* **2005**, *67*, 1697–1704.
487. AbdAlla, S.; Lother, H.; el Massiery, A.; Quitterer, U. *Nat. Med.* **2001**, *7*, 1003–1009.
488. Waldhoer, M.; Fong, J.; Jones, R. M.; Lunzer, M. M.; Sharma, S. K.; Kostenis, E.; Portoghese, P. S.; Whistler, J. L. *Proc. Natl. Acad. Sci. USA* **2005**, *102*, 9050–9055.
489. Hakak, Y.; Shrestha, D.; Goegel, M. C.; Behan, D. P.; Chalmers, D. T. *FEBS Lett.* **2003**, *550*, 11–17.
490. Lefkowitz, R. J. *J. Mol. Neurosci.* **2005**, *26*, 293–294.
491. Barak, L. S.; Zhang, J.; Ferguson, S. S.; Laporte, S. A.; Caron, M. G. *Meth. Enzymol.* **1999**, *302*, 153–171.
492. Charest, P. G.; Terrillon, S.; Bouvier, M. *EMBO J.* **2005**, *6*, 334–340.
493. Ferguson, S. S. *Pharmacol. Rev.* **2001**, *53*, 1–24.
494. Raman, D.; Osawa, S.; Gurevich, V. V.; Weiss, E. R. *J. Neurochem.* **2003**, *84*, 1040–1050.
495. Bhaskaran, R. S.; Min, L.; Krishnamurthy, H.; Ascoli, M. *Biochemistry* **2003**, *42*, 13950–13959.
496. Schmidlin, F.; Roostermann, D.; Bunnet, N. W. *Am. J. Physiol. Cell Physiol.* **2003**, *285*, C945–C958.
497. Milasta, S.; Evans, N. A.; Ormiston, L.; Wilson, S.; Lefkowitz, R. J.; Milligan, G. *Biochem. J.* **2005**, *387*, 573–584.
498. DeGraff, J. L.; Gurevich, V. V.; Benovic, J. L. *J. Biol. Chem.* **2002**, *277*, 43247–43252.
499. Kishi, H.; Krishnamurthy, H.; Galet, C.; Bhaskaran, R. S.; Ascoli, M. *J. Biol. Chem.* **2002**, *277*, 21939–21946.
500. Vargas, G. A.; von Zastrow, M. *J. Biol. Chem.* **2004**, *279*, 37461–37469.
501. Ronacher, K.; Matsiliza, N.; Nkwanyana, N.; Pawson, A. J.; Adam, T.; Flanagan, C. A.; Millar, R. P.; Katz, A. A. *Endocrinology* **2004**, *145*, 4480–4488.
502. Smythe, E. *Biochem. Soc. Trans.* **2003**, *31*, 736–739.
503. Brodsky, F. M.; Chen, C. Y.; Knuehl, C.; Towler, M. C.; Wakeham, D. E. *Annu. Rev. Cell Dev. Biol.* **2001**, *17*, 517–568.
504. Takei, K.; Haucke, V. *Trends Cell Biol.* **2001**, *11*, 385–391.
505. Mousavi, S. A.; Malerod, L.; Berg, T.; Kjeken, R. *Biochem. J.* **2004**, *377*, 1–16.
506. Laporte, S. A.; Oakly, R. H.; Holt, J. A.; Barak, L. S.; Caron, M. G. *J. Biol. Chem.* **2000**, *275*, 23120–23126.
507. Carman, C. V.; Lisanti, M. P.; Benovic, J. L. *J. Biol. Chem.* **1999**, *274*, 8858–8864.
508. Ostrom, R. S.; Insel, P. A. *Br. J. Pharmacol.* **2004**, *143*, 235–245.
509. Chini, B.; Parenti, M. *J. Mol. Endocrinol.* **2004**, *32*, 325–338.
509a. Insel, P. A.; Head, B. P.; Ostrom, R. S.; Patel, H. H.; Swaney, J. S.; Tang, C. M.; Roth, D. M. *Ann. NY Acad. Sci.* **2005**, *1047*, 166–172.
510. Barnett-Norris, J.; Lynch, D.; Reggio, P. H. *Life Sci.* **2005**, *77*, 1625–1639.
511. Bohm, S. K.; Khitin, L. M.; Smeekens, S. P.; Grady, E. F.; Payan, D. G.; Bunnett, N. W. *J. Biol. Chem.* **1997**, *272*, 2363–2372.
512. Bouley, R.; Sun, T. X.; Chenard, M.; McLaughlin, M.; McKee, M.; Lin, H. Y.; Brown, D.; Ausiello, D. A. *Am. J. Cell Physiol.* **2003**, *285*, C750–C762.
513. Paing, M. M.; Temple, B. R.; Trejo, J. *J. Biol. Chem.* **2004**, *279*, 21938–21947.
514. Fourgeaud, L.; Bessis, A. S.; Rossignol, F.; Pin, J. P.; Olivo-Marin, J. C.; Hemar, A. *J. Biol. Chem.* **2003**, *278*, 12222–122230.
515. Xia, S.; Kjaer, S.; Zheng, K.; Hu, P. S.; Bai, L.; Rigler, R.; Pramanik, A.; Xu, T.; Hokfelt, T.; Xu, Z. Q. *Proc. Natl. Acad. Sci. USA* **2004**, *101*, 15207–15212.
516. Morris, D. P.; Price, R. R.; Smith, M. P.; Lei, B.; Schwinn, D. A. *Mol. Pharmacol.* **2004**, *66*, 843–854.
517. Pediani, J. D.; Colston, J. F.; Calwell, D.; Milligan, G.; Daly, C. J.; McGrath, J. C. *Mol. Pharmacol.* **2004**, *67*, 992–1004.
518. Chen, C. H.; Paing, M. M.; Trejo, J. *J. Biol. Chem.* **2004**, *279*, 10020–10031.
519. Rasmussen, T. N.; Novak, I.; Nielsen, S. M. *Eur. J. Biochem.* **2004**, *271*, 4366–4374.
520. Jala, V. R.; Shao, W. H.; Haribabu, B. *J. Biol. Chem.* **2005**, *280*, 4880–4887.
521. Lee, K. B.; Pals-Rylaarsdam, R.; Benovic, J. L.; Hosey, M. M. *J. Biol. Chem.* **1998**, *273*, 12967–12972.

522. Bhatnagar, A.; Willins, D. L.; Gray, J. A.; Woods, J.; Benovic, J. L.; Roth, B. L. *J. Biol. Chem.* **2001**, *276*, 8269–8277.
523. Gilbert, T. L.; Bennett, T. A.; Maestas, D. C.; Cimino, D. F.; Prossnitz, E. R. *Biochemistry* **2001**, *40*, 3467–3475.
524. Fraile-Ramos, A.; Kohout, T. A.; Waldhoer, M.; Marsh, M. *Traffic* **2003**, *4*, 243–253.
525. van Koppen, C. J.; Jakobs, K. H. *Mol. Pharmacol.* **2004**, *66*, 365–367.
526. Chen, Z.; Gaudreau, R.; Le Gouill, C.; Rola-Pleszczynski, M.; Stankova, J. *Mol. Pharmacol.* **2004**, *66*, 377–386.
527. Ronacher, K.; Matisiliza, N.; Nkwanyana, N.; Pawson, A. J.; Adam, T.; Flanagan, C. A.; Millar, R. P.; Katz, A. A. *Endocrinology* **2004**, *145*, 4480–4488.
528. Feng, Y. H.; Ding, Y.; Ren, S.; Zhou, L.; Xu, C.; Karnik, S. S. *Hypertension* **2005**, *46*, 419–425.
529. El Messari, S.; Iturrioz, X.; Fasot, C.; De Mota, N.; Roesch, D.; Llorens-Cortes, C. *J. Neurochem.* **2004**, *90*, 1290–1301.
530. Liu, Q.; Cesato, R.; Dewi, D. A.; Rivier, J.; Rebi, J. C.; Schonbrunn, A. *Mol. Pharmacol.* **2005**, *68*, 90–101.
531. Han, M.; Gurevich, V. V.; Vishnivetsky, S. A.; Sigler, P. B.; Schubert, C. *Structure* **2001**, *9*, 869–880.
532. Milano, S. K.; Pace, H. C.; Kim, Y. M.; Brenner, C.; Benovic, J. L. *Biochemistry* **2002**, *41*, 3321–3328.
533. Ferguson, S. S.; Zhang, J.; Barak, L. S.; Caron, M. G. *Life Sci.* **1998**, *62*, 1561–1565.
534. Barak, L. S.; Ferguson, S. S.; Zhang, J.; Caron, M. G. *J. Biol. Chem.* **1997**, *272*, 27497–27500.
535. Goodman, O. B.; Krupnick, J. G.; Santini, F.; Gurevich, V. V.; Pen, R. B.; Gagnon, A. W.; Keen, J. H.; Benovic, J. L. *Nature* **1996**, *383*, 447–450.
536. Fessart, D.; Simaan, M.; Laporte, S. A. *Mol. Endocrinol.* **2005**, *19*, 491–503.
537. Hirsch, J. A.; Schubert, C.; Gurevich, V. V.; Sigler, P. B. *Cell* **1999**, *97*, 257–269.
538. Oakley, R. H.; Laporte, S. A.; Holt, J. A.; Barak, L. S.; Caron, M. G. *J. Biol. Chem.* **2001**, *276*, 19452–19460.
539. Estall, J. L.; Koechler, J. A.; Yusta, B.; Drucker, D. J. *J. Biol. Chem.* **2005**, *280*, 22124–22134.
540. Novi, F.; Stanasila, L.; Giorgi, F.; Corsini, G. U.; Cotecchia, S.; Maggio, R. *J. Biol. Chem.* **2005**, *280*, 19768–19776.
541. Pfeiffer, M.; Koch, T.; Schroder, H.; Laugsch, M.; Hollt, V.; Schulz, S. *J. Biol. Chem.* **2002**, *277*, 19762–19772.
542. Pfeiffer, M.; Kirscht, S.; Stumm, R.; Koch, T.; Wu, D.; Laugsch, M.; Schroder, H.; Hollt, V.; Schulz, S. *J. Biol. Chem.* **2003**, *78*, 51630–51637.
543. Simaan, M.; Bedard-Goulet, S.; Fessart, D.; Gratton, J. P.; Laporte, S. A. *Cell Signal.* **2005**, *17*, 1074–1083.
544. Lodowski, D. T.; Pitcher, J. A.; Capel, W. D.; Lefkowitz, R. J.; Tesmer, J. J. *Science* **2003**, *300*, 1256–1262.
545. Lodowski, D. T.; Barnhill, J. F.; Pyskadlo, R. M.; Ghirlando, R.; Sterne-Marr, R.; Tesmer, J. J. *Biochemistry* **2005**, *44*, 6958–6970.
546. Benovic, J. L.; Onorato, J.; Lohse, M. L.; Dohlman, H. G.; Staniszewski, C.; Caron, M. G.; Lefowitz, R. J. *Br. J. Clin. Pharmacol.* **1990**, *30*, 3S–12S.
547. Palczewski, K.; Buczylko, J.; Kaplan, M. W.; Polans, A. S.; Crabb, J. W. *J. Biol. Chem.* **1991**, *266*, 12949–12955.
548. Haga, K.; Kameyama, K.; Haga, T. *J. Biol. Chem.* **1994**, *269*, 12594–12599.
549. Teli, T.; Markovic, D.; Levine, M. A.; Hillhouse, E. W.; Grammatopoulos, D. K. *Mol. Endocrinol.* **2005**, *19*, 474–490.
550. Pao, C. S.; Benovic, J. L. *J. Biol. Chem.* **2005**, *280*, 11052–11058.
551. Sterne-Marr, R.; Dhami, G. K.; Tesmer, J. J.; Ferguson, S. S. *Meth. Enzymol.* **2004**, *390*, 436.
552. Hasbi, A.; Devost, D.; Laporte, S. A.; Zingg, H. H. *Mol. Endocrinol.* **2004**, *18*, 1277–1286.
553. Krasel, C.; Bunemann, M.; Lorenz, K.; Lohse, M. J. *J. Biol. Chem.* **2005**, *280*, 9528–9535.
554. Paasche, J. D.; Attramadal, T.; Kristiansen, K.; Oksvold, M. P.; Johansen, H. K.; Huitfeldt, H. S.; Dahl, S. G.; Attramadal, H. *Mol. Pharmacol.* **2005**, *67*, 1581–1590.
555. Trejo, J. *Mol. Pharmacol.* **2005**, *67*, 1388–1390.
556. Munshi, U. M.; Clouser, C. L.; Peegel, H.; Menon, K. M. *Mol. Endocrinol.* **2005**, *19*, 749–758.
557. Mialet-Perez, J.; Green, S. A.; Miller, W. E.; Liggett, S. B. *J. Biol. Chem.* **2004**, *279*, 38603–38607.
558. Shenoy, S. K.; McDonald, P. H.; Kohout, T. A.; Lefkowitz, R. J. *Science* **2001**, *294*, 1307–1313.
559. Hislop, J. N.; Marley, A.; von Zastrow, M. *J. Biol. Chem.* **2004**, *279*, 22522–22531.
560. Christopoulos, A. *Nat. Rev. Drug Disc.* **2002**, *1*, 198–210.
561. Kenakin, T. *Receptors Channels* **2004**, *10*, 51–60.
562. May, L. T.; Avlani, V. A.; Sexton, P. M.; Christopoulos, A. *Curr. Pharm. Des.* **2004**, *10*, 2003–2013.
563. Kew, J. N. *Pharmacol. Ther.* **2004**, *104*, 233–244.
564. Price, M. R.; Baillie, G.; Thomas, A.; Stevenson, L. A.; Easson, M.; Goodwin, R.; Walker, G.; Westwood, P.; Marrs, J.; McLean, A. et al. *Mol. Pharmacol.* **2005**, *68*, 1484–1495.
565. Fawzi, A. B.; Macdonald, D.; Benbow, L. L.; Smith-Torhan, A.; Zhang, H.; Weig, B. C.; Ho, G.; Tulshian, D.; Linder, M. E.; Graziano, M. P. *Mol. Pharmacol.* **2001**, *59*, 30–37.
566. Im, W. B.; Chio, C. L.; Alberts, G. L.; Dinh, D. M. *Mol. Pharmacol.* **2003**, *64*, 78–84.
567. O'Brien, J. A.; Lemarie, W.; Chen, T. B.; Chang, R. S.; Jacobson, M. A.; Ha, S. N.; Lindsley, C. W.; Schaffhauser, H. J.; Sur, C.; Pettibone, D. J. et al. *Mol. Pharmacol.* **2003**, *64*, 731–740.
568. O'Brien, J. A.; Lemarie, W.; Wittmann, M.; Jacobson, M. A.; Ha, S. N.; Wisnoski, D. D.; Lindsley, C. W.; Schaffhauser, H. J.; Rowe, B.; Sur, C. et al. *Mol. Pharmacol.* **2003**, *64*, 731–740.
569. Alberts, G. L.; Pregenzer, J. F.; Im, W. B.; Zaworski, P. G.; Gill, G. S. *Eur. J. Pharmacol.* **1999**, *131*, 514–520.
570. Allgeier, A.; Offermanns, S.; Van-Sande, J.; Schultz, G.; Dumont, J. E. *J. Biol. Chem.* **1994**, *269*, 13733–13735.
571. Brydon, L.; Roka, F.; Petit, L.; de Coppet, P.; Tissot, M.; Barrett, P.; Morgan, P. J.; Nanoff, C.; Strosberg, A. D.; Jockers, R. *Mol. Endocrinol.* **1999**, *13*, 202–208.
572. Chakrabarti, S.; Prather, P. L.; Yu, L.; Law, P. Y.; Loh, H. H. *J. Neurochem.* **1999**, *64*, 2534–2543.
573. Eason, M. G.; Ligget, S. B. *J. Biol. Chem.* **1995**, *270*, 24753–24760.
574. Hermans, E.; Saunders, R.; Selkirk, J. V.; Mistry, R.; Nahorski, S. R.; Challiss, R. R. *Mol. Pharmacol.* **2000**, *58*, 352–360.
575. Kuhn, B.; Gundermann, T. *Biochemistry* **1999**, *38*, 12490–12498.
576. Lawler, O. A.; Miggin, S. M.; Kinsella, B. T. *J. Biol. Chem.* **2001**, *276*, 33596–33607.
577. Offermanns, S.; Weiland, T.; Homann, D.; Sandmann, J.; Bombien, E.; Spicer, K.; Shultz, G.; Jakobs, K. H. *Mol. Pharmacol.* **1994**, *45*, 890–898.
578. Thomsen, W. J.; Gatlin, J.; Unett, D. J.; Behan, D. P. *Curr. Drug Discov.* **2004**, 13–18.
579. Harrison, C.; Traynor, J. R. *Life Sci.* **2003**, *74*, 489–508.
580. Milligan, G. *Trends Pharmacol. Sci.* **2003**, *24*, 87–90.
581. Birdlack, J. M.; Parkhill, A. L. *Meth. Mol. Biol.* **2004**, *237*, 135–143.

582. Ferrer G. D. Kolodin, M.; Zuck, P.; Peltier, R.; Berry, K.; Mandala, S. M.; Rosen, H.; Ota, H.; Ozaki, S.; Inglese, J.; Strulovici, B. *Assay Drug Dev. Technol.* **2003**, *1*, 261–273.
583. DeLapp, N. W.; McKinzie, J. H.; Sawyer, B. D.; Vandergriff, A.; Falcone, J.; McClure, D.; Felder, C. C. *J. Pharmacol. Exp. Ther.* **1999**, *289*, 946–955.
584. Cussac, D.; Newman-Tancredi, A.; Duqueyroix, D.; Pasteau, V.; Millan, M. *J. Mol. Pharmacol.* **2002**, *62*, 578–589.
585. DeLapp, N. *Trends Pharmacol. Sci.* **2004**, *25*, 400–401.
586. Frang, H.; Mukkala, V. M.; Syysto, R.; Ollikka, P.; Hurskainen, P.; Scheinin, M.; Hemmila, I. *Assay Drug Dev. Technol.* **2003**, *1*, 275–280.
587. Engstrom, M.; Narvanen, A.; Savola, J. M.; Wuster, S. *Peptides* **2004**, *25*, 2099–2104.
588. Gabriel, D.; Vernier, M.; Pfeifer, M. J.; Dasen, B.; Tenaillon, L.; Bouhelal, R. *Assay Drug Dev. Technol.* **2003**, *1*, 291–303.
589. Williams, C. *Nat. Rev. Drug Disc.* **2004**, *3*, 125–135.
590. Hancock, A. A.; Vodenlich, A. D.; Maldonado, C.; Janis, R. *J. Recept. Signal Transduct. Res.* **1995**, *15*, 557–579.
591. Kariv, I. I.; Stevens, M. E.; Behrens, D. L.; Oldenburg, K. R. *J. Biomol. Screen.* **1999**, *4*, 27–32.
592. Prystay, L.; Gagne, A.; Kasila, P.; Yeh, L. A.; Banks, P. *J. Biomol. Screen.* **2001**, *6*, 75–82.
593. Hesley, J.; Daijo, J.; Ferguson, A. T. *Biotechniques* **2002**, *33*, 691–694.
594. Sportsman, J. R.; Daijo, J.; Gaudet, E. A. *Comb. Chem. High Throughput Screen.* **2003**, *6*, 195–200.
595. Banks, P.; Gosselin, M. *L. Prystay. J. Biomol. Screen.* **2000**, *5*, 159–168.
596. Eglen, R. *Comb. Chem. High Throughput Screen.* **2005**, *8*, 311–318.
597. Eglen, R. M. *Assay Drug Dev. Technol.* **2002**, *1*, 97–104.
598. Golla, R.; Seethala, R. *J. Biomol. Screen.* **2002**, *7*, 515–525.
599. Weber, M.; Ferrer, M.; Zheng, W.; Inglese, J.; Strulovici, B.; Kunapuli, P. *Assay Drug Dev. Technol.* **2004**, *2*, 39–49.
600. Chengalvala, M.; Kostek, B.; Frail, D. E. *J. Biochem. Biophys. Meth.* **1999**, *38*, 163–170.
601. Benjamin, E. R.; Haftl, S. L.; Xanthos, D. N.; Crumley, G.; Hachicha, M.; Valenzano, K. L. *J. Biomol. Screen.* **2004**, *9*, 343–353.
602. Rodgers, G.; Hubert, C.; McKinzie, J.; Suter, T.; Stanick, M.; Emmerson, P.; Stancato, L. *Assay Drug Dev. Technol.* **2003**, *1*, 627–636.
603. Liu, J. J.; Hartman, D. S.; Bostwick, J. R. *Anal. Biochem.* **2003**, *318*, 91–99.
604. Brandish, P. E.; Hill, L. A.; Zheng, W.; Skolnick, E. M. *Anal. Biochem.* **2003**, *313*, 311–318.
605. Sullivan, E.; Tucker, E. M.; Dale, I. L. *Meth. Mol. Biol.* **1999**, *114*, 125–133.
606. Simpson, P. B.; Villullas, I. R.; Schurov, I.; Millard, R.; Haldon, C.; Beer, M. S.; McAllister, G. *Assay Drug Dev. Technol.* **2002**, *1*, 31–40.
607. Chambers, C.; Smith, F.; Williams, C.; Marcos, S.; Liu, Z. H.; Hayter, P.; Ciaramella, G.; Keighley, W.; Gribbon, P.; Sewing, A. *Comb. Chem. High Throughput Screen.* **2003**, *6*, 355–362.
608. Zhang, Y.; Kowal, D.; Kramer, A.; Dunlop, J. *J. Biomol. Screen.* **2003**, *8*, 571–577.
609. Hodder, P.; Mull, R.; Cassaday, J.; Berry, K.; Strulovici, B. *J. Biomol. Screen.* **2004**, *9*, 417–426.
610. Gopalakrishnan, S. M.; Mammen, B.; Schmidt, M.; Otterstaetter, B.; Amberg, W.; Wernet, W.; Kofron, J. L.; Burns, D. J.; Warrior, U. *J. Biomol. Screen.* **2005**, *10*, 46–55.
611. Stables, J.; Green, A.; Marshall, F.; Fraser, N.; Knight, E.; Sautel, M.; Milligan, G.; Lee, M.; Rees, S. *Anal. Biochem.* **1997**, *252*, 115–126.
612. George, S. E.; Schaeffer, M. T.; Cully, D.; Beer, M. S.; McAllister, G. *Anal. Biochem.* **2000**, *286*, 231–237.
613. Stables, J.; Mattheakis, L. C.; Chang, R.; Rees, S. *Meth. Enzymol.* **2000**, *327*, 456–471.
614. Ungrin, M. D.; Carriere, C.; Denis, D.; Lamontagne, S.; Sawyer, N.; Stocco, R.; Tremblay, N.; Metters, K. M.; Abramovitz, M. *Mol. Pharmacol.* **2001**, *59*, 1446–1456.
615. Dupriez, V. J.; Maes, K.; Le Poul, E.; Burgeon, E.; Detheux, M. *Receptors Channels* **2002**, *8*, 319–330.
616. Knight, P. J.; Pfeifer, T. A.; Grigliatti, T. A. *Anal. Biochem.* **2003**, *320*, 88–103.
617. Nidernberg, A.; Tunaru, S.; Blaukat, A.; Harris, B.; Kostenis, E. *J. Biomol. Screen.* **2003**, *8*, 500–510.
618. Le Poul, E.; Hisada, S.; Mizuguchi, Y.; Dupriez. E. Burgeon, V. J.; Detheux, M. *J. Biomol. Screen.* **2002**, *7*, 57–65.
619. Reinscheid, R. K.; Kim, J.; Zeng, J.; Civelli, O. *Eur. J. Pharmacol.* **2003**, *478*, 27–34.
620. Zlokarkik, G.; Negulescu, P. A.; Knapp, T. E.; Mere, L.; Burres, N.; Feng, L.; Whitney, M.; Roemer, K.; Tsien, R. Y. *Science* **1998**, *279*, 84–88.
621. Kunapuli, P.; Ransom, R.; Murphy, K. L.; Pettibone, D.; Kerby, J.; Grimwood, S.; Zuck, P.; Hodder, P.; Lacson, R.; Hoffman, I.; Inglese, J.; Strulovici, B. *Anal. Biochem.* **2003**, *314*, 16–29.
622. Bresnick, J. N.; Skynner, H. A.; Chapman, K. L.; Jack, A. D.; Zamiara, E.; Negulescu, P.; Beaumont, K.; Patel, S.; McAllister, G. *Assay Drug Dev. Technol.* **2003**, *1*, 239–249.
623. Oosterom, J.; van Doornmalen, E. J.; Lobregt, S.; Blomenrohr, M.; Zaman, G. J. *Assay Drug Dev. Technol.* **2005**, *3*, 143–154.
624. Kornienko, O.; Lacson, R.; Kunapuli, P.; Schneeweis, J.; Hoffman, I.; Smith, T.; Alberts, M.; Inglese, J.; Strulovici, B. *J. Biomol. Screen.* **2004**, *9*, 186–195.
625. Cacase, A.; Banks, M.; Spicer, T.; Civoli, F.; Watson, J. *Drug Disc. Today* **2003**, *17*, 785–792.
626. Dinger, M. C.; Beck-Sickinger, A. G. *Mol. Biotechnol.* **2002**, *21*, 9–18.
627. Offermanns, S.; Simon, M. I. *J. Biol. Chem.* **1995**, *270*, 15175–15180.
628. Joshi, S.; Lee, J. W.; Wong, Y. H. *Eur. J. Neurosci.* **1999**, *11*, 383–388.
629. Lee, J. W.; Joshi, S.; Chan, J. S.; Wong, Y. H. *J. Neurochem.* **1998**, *70*, 2203–2211.
630. Ho, M. K.; Yung, L. Y.; Chan, J. H.; Chen, J. H.; Wong, C. S.; Wong, Y. H. *Br. J. Pharmacol.* **2001**, *132*, 1431–1440.
631. Mody, S. M.; Ho, M. K.; Joshi, S. A.; Wong, Y. H. *Mol. Pharmacol.* **2000**, *57*, 13–23.
632. Kostenis, E. *Trends Pharmacol. Sci.* **2002**, *22*, 560–564.
633. Liu, A. M.; Ho, M. K.; Wong, C. S.; Chan, J. H.; Pau, A. H.; Wong, Y. H. *J. Biomol. Screen.* **2003**, *8*, 39–49.
634. New, D. C.; Wong, Y. H. *Assay Drug Dev. Technol.* **2004**, *2*, 269–280.
635. Hazari, A.; Lowes, V.; Chan, J. H.; Wong, C. S.; Ho, M. K.; Wong, Y. H. *Cell Signal.* **2004**, *16*, 51–62.
636. Coward, P.; Chan, S. D.; Wada, H. G.; Humphries, G. M.; Conklin, B. R. *Anal. Biochem.* **1999**, *270*, 242–248.
637. Yokoyama, T.; Kato, N.; Yamada, N. *Neurosci. Lett.* **2003**, *344*, 45–48.
638. Zhang, J. K.; Nawoschik, S.; Kowal, D.; Smith, D.; Spangler, R.; Ochalski, R.; Schechter, L.; Dunlop, J. *Eur. J. Pharmacol.* **2003**, *472*, 33–38.
639. Milligan, G.; Rees, S. *Trends Pharmacol. Sci.* **1999**, *20*, 118–124.
640. Niederberg, A.; Tunaru, S.; Blaukat, A.; Ardati, A.; Kostenis, E. *Cell Signal.* **2003**, *15*, 435–446.
641. Robas, N. M.; Fidock, M. D. *Meth. Mol. Biol.* **2005**, *306*, 17–26.
642. Potenza, M. N.; Lerner, M. R. *Pigment Cell Res.* **1992**, *5*, 372–378.
643. Graminski, G. F.; Jayawickreme, C. K.; Potenza, M. N.; Lerner, M. R. *J. Biol. Chem.* **1993**, *268*, 5957–5964.

644. Potenza, M. N.; Graminski, G. F.; Schmauss, C.; Lerner, M. R. *J. Neurosci.* **1994**, *14*, 1463–1476.
645. Lerner, M. R. *Trends Neurosci.* **1994**, *17*, 142–146.
646. Jayawickreme, C. K.; Sauls, H.; Bolio, N.; Ruan, J.; Moyer, M.; Burkhart, W.; Marron, B.; Rimele, T.; Shaffer, J. *J. Pharmacol. Toxicol. Meth.* **1999**, *42*, 189–197.
647. Chen, G.; Jayawickreme, C.; Way, J.; Armour, S.; Queen, K.; Watson, C.; Ignar, D.; Chen, W. J.; Kenakin, T. *J. Pharmacol. Toxicol. Meth.* **1999**, *42*, 199–206.
648. Charest, P. G.; Bouvier, M. *J. Biol. Chem.* **2003**, *278*, 41541–41551.
649. Vrecl, M.; Jorgensen, R.; Pogacnik, A.; Heding, A. *J. Biomol. Screen.* **2004**, *9*, 322–333.
650. Heding, A. *Expert Rev. Mol. Diagn.* **2004**, *4*, 403–411.
651. Hamdan, F. F.; Audet, M.; Garneau, P.; Pelletier, J.; Bouvier, M. *J. Biomol. Screen.* **2005**, *10*, 463–475.
652. Gales, C.; Rebois, R. V.; Hogue, M.; Trieu, P.; Breit, A.; Hebert, T. E.; Bouvier, M. *Nat. Meth.* **2005**, *2*, 177–184.
653. Kraft, K.; Olbrich, H.; Majoul, I.; Mack, M.; Proudfoot, A.; Oppermann, M. *J. Biol. Chem.* **2001**, *276*, 34408–34418.
654. Eidne, K. A.; Kroeger, K. M.; Hanyaloglu, A. C. *Trends Endocrinol. Metab.* **2002**, *13*, 415–421.
655. Krasel, C.; Bunemann, M.; Lorenz, K.; Lohse, M. J. *J. Biol. Chem.* **2005**, *280*, 9528–9535.
656. Azpiazu, I.; Gautam, N. *J. Biol. Chem.* **2004**, *279*, 27709–27718.
657. Hoffmann, C.; Gaietta, G.; Bunemann, M.; Adams, S. R.; Oberdorff-Maass, S.; Behr, B.; Vilardaga, J. P.; Tsien, R. Y.; Ellisman, M. H.; Lohse, M. J. *Nat. Meth.* **2005**, *2*, 171–176.
658. Milligan, G. *Drug Disc. Today* **2003**, *8*, 579–585.
659. Giuliano, K. A.; Haskins, J. R.; Taylor, D. L. *Assay Drug Dev. Technol.* **2003**, *1*, 565–577.
660. Gurwitz, D.; Haring, R. *Drug Disc. Today* **2003**, *8*, 1108–1109.
661. Granas, C.; Lundholt, B. K.; Heydorn, A.; Linde, V.; Pedersen, H.-C.; Krog-Jensen, C.; Rosenkilde, M. M.; Pagliaro, L. *Comb. Chem. High Throughput Screen.* **2005**, *8*, 301–309.
662. Ghosh, R. N.; DeBiasio, R.; Hudson, C. C.; Ramer, E. R.; Cowan, C. L.; Oakley, R. H. *J. Biomol. Screen.* **2005**, *10*, 476–484.
663. Barak, L. S.; Ferguson, S. S.; Zhang, J.; Caron, M. G. *J. Biol. Chem.* **1997**, *272*, 27497–27550.
664. Oakley, R. H.; Hudson, C. C.; Cruickshank, R. D.; Meyers, D. M.; Payne, R. E.; Rhem, S. M.; Loomis, C. R. *Assay Drug Dev. Technol.* **2002**, *1*, 21–30.
665. Borchert, K. M.; Galvin, R. J.; Frolik, C. A.; Hale, L. V.; Halladay, D. L.; Gonyier, R. J.; Trask, O. J.; Nickischer, D. R.; Houck, K. A. *Assay Drug Dev. Technol.* **2005**, *3*, 133–141.
666. The IUPHAR Database. http://www.IUPHAR.org (accessed July 2006).
667. MCSIS Project. Mutations promoting GPCR constitutive activation http://www.gpcr.org (accessed April 2006).
668. http://www.Perkin Elmer.com (accessed July 2006).
669. Cisbio International HTRF Site. http://www.htrf-assays.com (accessed April 2006).

Biographies

William J Thomsen is currently Director of Molecular Pharmacology at Arena Pharmaceuticals, Inc., and has also served as Research Fellow (2003–04), Associate Director of Assay Development and Screening (2001–03), and Manager of Assay Development and Screening (1998–2001). Prior to joining Arena Pharmaceuticals, Dr Thomsen was Director of Research at CellPath, Inc. (1997–98) and Director of Research at Lasure and Crawford, Inc. (1994–97), both in Seattle. From 1992 to 1994, Dr Thomsen served as a Senior Project Manager in Drug Discovery at Panlabs, Inc. in Seattle. Prior to entering industry, Dr Thomsen was a Postdoctoral Fellow in the Department of Pharmacology at the University of Washington between 1988 and 1991 (William Catterall's laboratory) and he received his PhD in Pharmacology from the University of Michigan (Rick Neubig's laboratory) in 1988. Dr Thomsen also received an MS degree in Biochemistry and a BSc in Chemistry from the University of South Dakota.

Dominic P Behan is a cofounder of Arena Pharmaceuticals, Inc. and has served as a director since April 2000, and as Senior Vice President and Chief Scientific Officer since June 2004. Dr Behan served as Vice President of Research from April 1997 to June 2004. From 1993 to January 1997, Dr Behan directed various research programs at Neurocrine Biosciences, Inc., a public biopharmaceutical company. From 1990 to 1993, Dr Behan was engaged in research at the Salk Institute. Dr Behan holds a BSc in Biochemistry from Leeds University, England and a PhD in Biochemistry from Reading University, England.

2.21 Ion Channels – Voltage Gated

J G McGivern, Amgen Inc., Thousand Oaks, CA, USA
J F Worley III, Vertex Pharmaceuticals Inc., San Diego, CA, USA

© 2007 Elsevier Ltd. All Rights Reserved.

2.21.1	**Introduction**	**827**
2.21.2	**Sodium Channels**	**831**
2.21.2.1	Current Nomenclature	831
2.21.2.2	Molecular Biology	832
2.21.2.3	Protein Structure	834
2.21.2.4	Physiological Functions	835
2.21.2.5	Regulation	836
2.21.2.6	Prototypical Pharmacology	837
2.21.2.7	Prototypical Therapeutics	839
2.21.2.8	Genetic Diseases	840
2.21.3	**Calcium Channels**	**841**
2.21.3.1	Current Nomenclature	841
2.21.3.2	Molecular Biology	844
2.21.3.3	Protein Structure	845
2.21.3.4	Physiological Functions	846
2.21.3.5	Regulation	848
2.21.3.6	Prototypical Pharmacology	849
2.21.3.7	Prototypical Therapeutics	850
2.21.3.8	Genetic Diseases	852
2.21.4	**Potassium Channels**	**853**
2.21.4.1	Current Nomenclature	853
2.21.4.2	Molecular Biology Classification	857
2.21.4.3	Protein Structure	858
2.21.4.4	Physiological Functions	859
2.21.4.5	Regulation	861
2.21.4.6	Prototypical Pharmacology	861
2.21.4.7	Prototypical Therapeutics	863
2.21.4.8	Genetic Diseases	864
2.21.5	**Challenges and New Directions for Ion Channel Drug Discovery**	**865**
References		**866**

2.21.1 Introduction

Ion channels form a large superfamily of integral membrane proteins that are expressed almost ubiquitously in excitable and nonexcitable cells.[1] They take advantage of transmembrane potential and ionic concentration differences to regulate the passive and rapid exchange of ions across cell membranes. Consequently, ion channels control electrical and biochemical signaling, and regulate a multitude of diverse physiological processes, including modulation of neuronal excitability, muscle contraction, neurotransmitter and hormone release, cell proliferation, and cell volume. An ion channel is minimally defined as an integral membrane protein that provides a regulated ion-permeable route through the cell membrane. Most ion channels are in fact multisubunit protein complexes, and several key features are critical for their proper operation. To form a functional ion channel, at least one of the subunits must be anchored in the

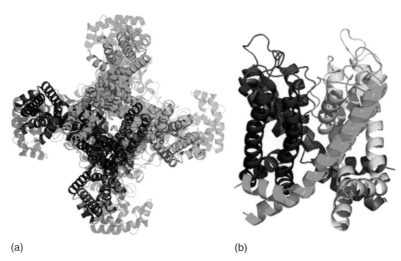

Figure 1 (a) The structure of the $K_V1.2$ homotetramer, as viewed from the extracellular side of the membrane. The crystallographic coordinates of $K_V1.2$ were taken from the Protein Data Bank entry 2A79[49] and visualized with PyMol. Each of the four monomers, which are colored differently, contributes to the formation of the central pore. There is a striking degree of interdigitation between the monomers. (b) This view differs from (a) by a 90° rotation of the axis that goes through the pore. In this view, only the S5–S6 segments from each monomer are displayed for clarity. It shows again the high degree of interdigitation between the monomers, and illustrates in more detail the interactions that create the minimum functional unit. (The assistance of Carlos Faerman of Vertex Pharmaceuticals Inc. is gratefully acknowledged in generating Figure 1.)

cell membrane by two or more membrane-spanning segments (S1, S2, through Sn), and must contain the necessary pore-forming elements. Typically, the subunits containing the pore-forming elements will be positioned in a circular arrangement with the pore located in the center (**Figure 1**). The pore must be capable of opening and closing in response to physiological stimuli, such as a change in the transmembrane potential or the binding and unbinding of extracellular (e.g., neurotransmitters) or intracellular (e.g., calcium) ligands (*see* 2.22 Ion Channels – Ligand Gated). Ion selectivity is a key attribute, and is provided by a theoretical filter that may form the narrowest part of the pore. The direction of ion flux across the cell membrane is governed by the electrochemical driving force on the permeant ion. Ion channels are often categorized by their physiological regulators of gating and by their permeant ionic species. Thus, there are the large families of voltage-gated sodium (Na^+), calcium (Ca^{2+}), and potassium (K^+) ion channels as well as the diverse families of ligand-gated anion and cation channels. Isoforms representing each of the voltage-gated ion channel families are often co-expressed in excitable cells, where they work in concert as initiators and effectors of electrical and biochemical signaling.

Most voltage-gated ion channel subunits are characterized by the presence of multiple positively charged amino acids in one of the transmembrane segments. This part of the ion channel is capable of moving in response to changes in the transmembrane potential of the cell, and effectively functions as its voltage sensor. Voltage-gated ion channels can exist in a variety of different states, which may either be conducting (open) or nonconducting (closed), and may switch between these different states by theoretical voltage-dependent gating processes. The properties of gating underlie the complex behavior of individual ion channel subtypes. Importantly, there are two major closed states, namely resting and inactivated, a crucial difference being that resting channels can open, whereas most inactivated channels do not have the ability to open until they have returned to a resting state. In quiescent excitable cells, the membrane potential normally sits at a relatively hyperpolarized level of $-60\,mV$ to $-90\,mV$ (cell type-dependent), and the majority of the ion channels will be in a resting state. Membrane depolarization increases the probability that resting channels will open or activate. The membrane potential at which a channel first opens is its threshold potential, and often this is a defining parameter. With maintained depolarization, open channels will eventually close or inactivate. Other key biophysical parameters that can be used to differentiate the members of an ion channel family are the duration of the open state, the rate and type of inactivation, and the rate at which inactivated channels recover to the resting state.

Voltage-gated ion channels are key regulators of cellular function, and are widely considered to be viable and attractive targets for therapeutic modulation in the treatment of human diseases, particularly those diseases that affect excitable cells. Several drugs that target voltage-gated ion channels are currently marketed for the treatment of a variety of cardiovascular and neurological diseases (**Table 1**). Together, these ion channel drugs have been estimated to

Table 1 Ion channel drugs used to treat cardiovascular and neurological diseases

Primary mechanism of action	Drug name	Molecular structure	Primary indications
Cardiovascular diseases			
L-type Ca^{2+} channel blockade	Nitrendipine	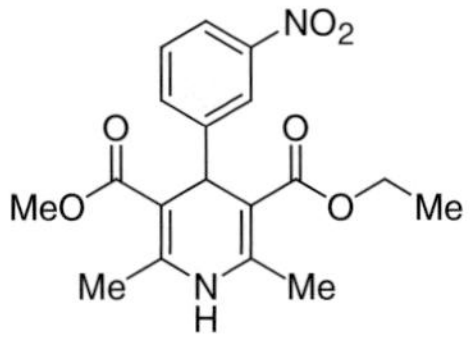	Hypertension
	Nimodipine	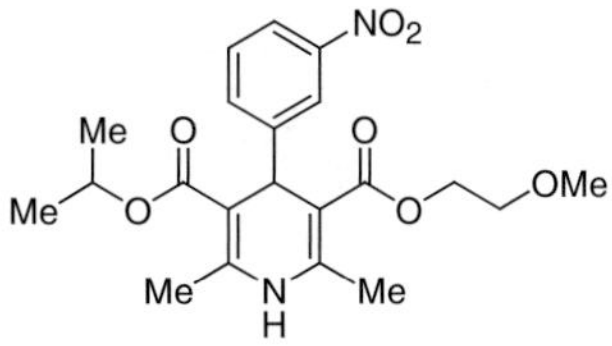	Subarachnoid hemorrhage
	Nicardipine	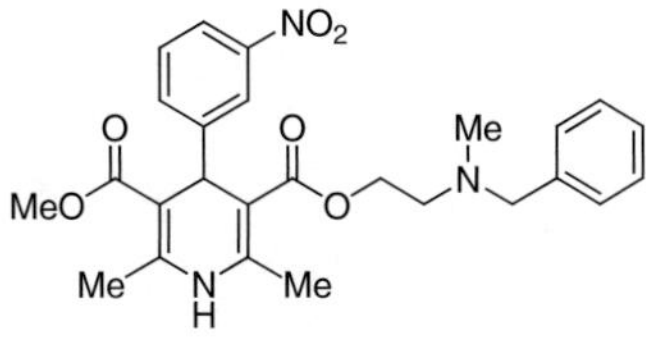	Hypertension and angina pectoris
	Diltiazem	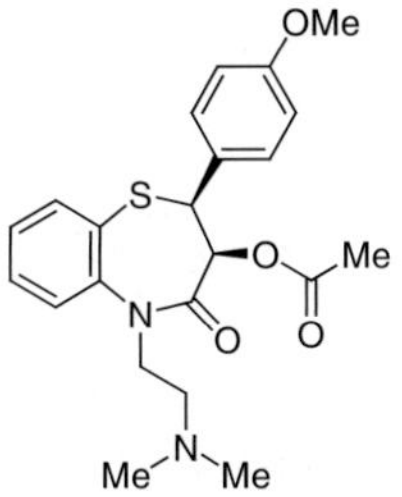	Hypertension, angina pectoris and cardiac arrhythmia
	Verapamil	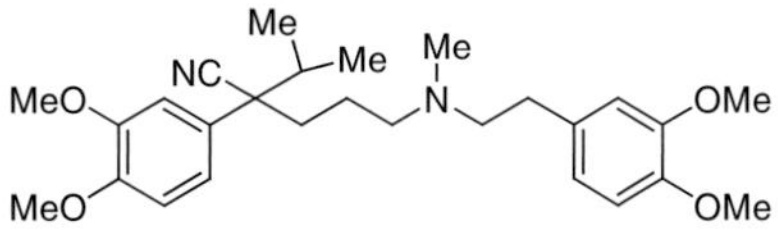	Hypertension, angina pectoris and cardiac arrhythmia
Mixed Na^+ and L-type Ca^{2+} channel blockade	Amiodarone	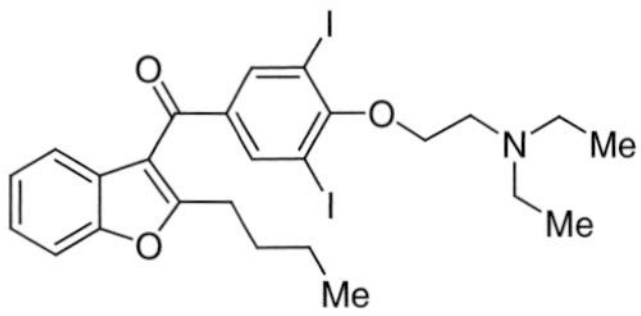	Cardiac arrhythmia
Na^+ channel blockade	Disopyramide	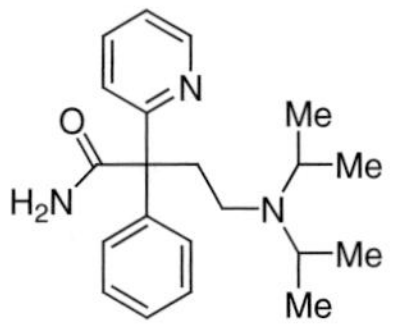	Cardiac arrhythmia
	Flecainide	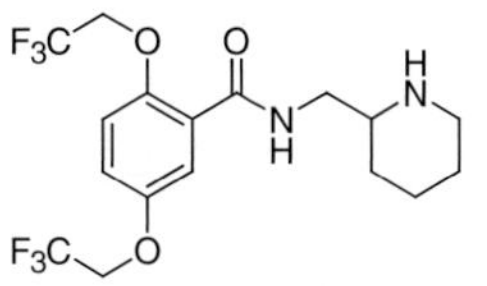	Cardiac arrhythmia

continued

Table 1 Continued

Primary mechanism of action	Drug name	Molecular structure	Primary indications
	Lidocaine	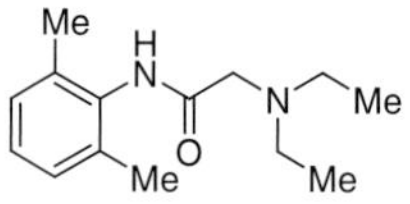	Cardiac arrhythmia
	Procainamide	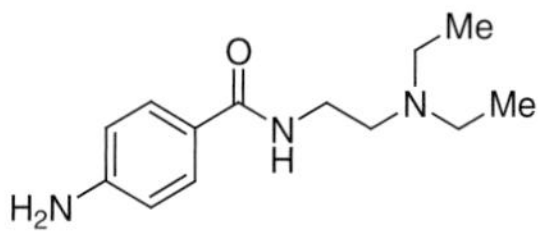	Cardiac arrhythmia
	Propafenone	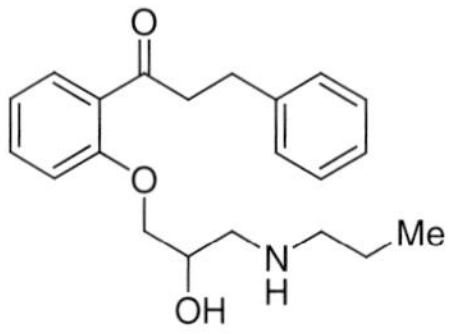	Cardiac arrhythmia
	Quinidine	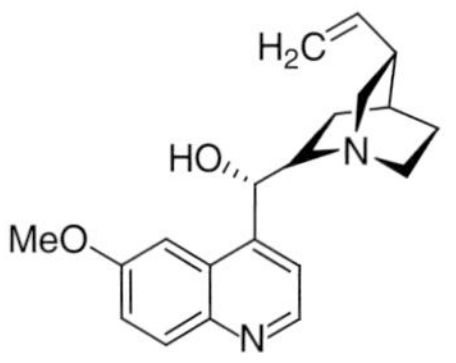	Cardiac arrhythmia
Neurological diseases			
Na$^+$ channel blockade	Phenytoin	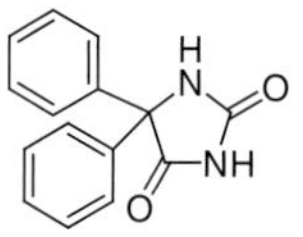	Epilepsy
	Fosphenytoin	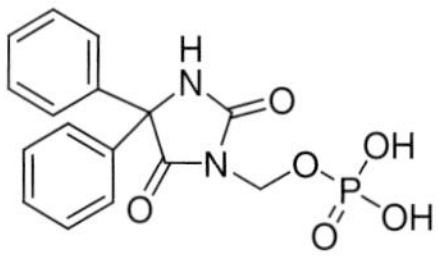	Epilepsy
	Oxcarbazepine	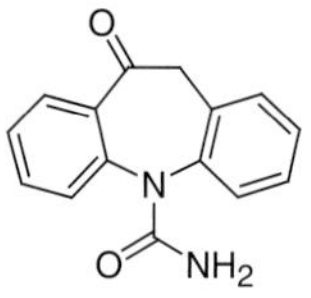	Epilepsy
	Carbamazepine	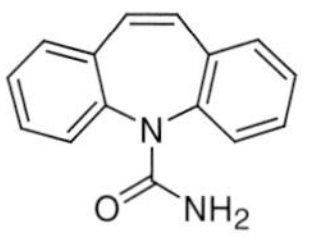	Epilepsy and pain
	Topiramate	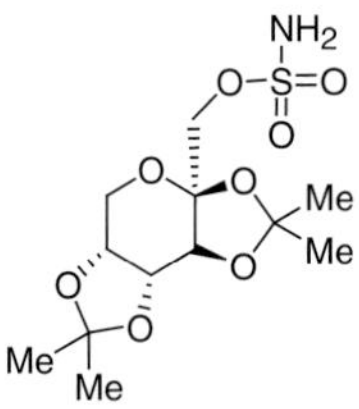	Epilepsy and pain

Table 1 Continued

Primary mechanism of action	Drug name	Molecular structure	Primary indications
	Lamotrigine	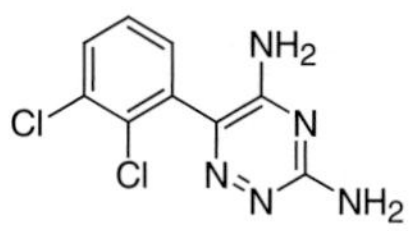	Epilepsy and bipolar disorder
	Amitriptyline	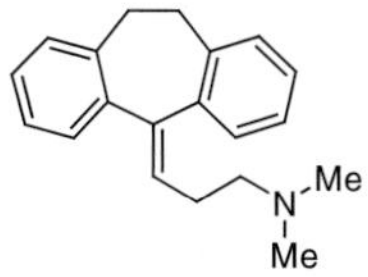	Pain and depression
Mixed Na^+ and T-type Ca^{2+} channel blockade	Zonisamide	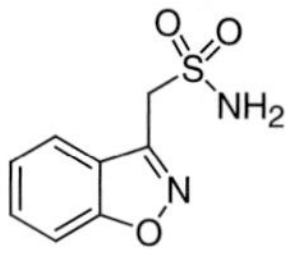	Epilepsy
T-type Ca^{2+} channel blockade	Ethosuximide	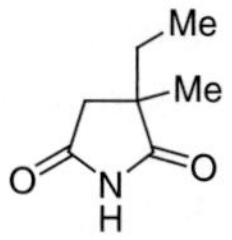	Epilepsy and pain
N-type Ca^{2+} channel blockade	Levetiracetam	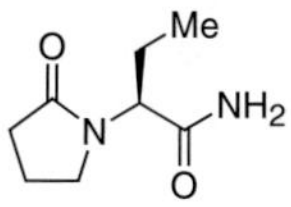	Epilepsy
N- and/or P-type Ca^{2+} channel modulation(?)	Gabapentin	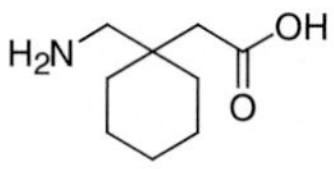	Epilepsy and pain

generate US $6–15 billion in total annual sales. The majority of these marketed drugs were discovered and developed many years ago using traditional pharmacological approaches. However, it is widely believed that the ion channel family remains underexploited because all of these drugs target only a handful of distinct ion channel subtypes. Thus, the ion channel drug class is likely to grow in the future as a direct consequence of recent advances in our understanding of ion channel structure–function relationships, the role of ion channels in disease, and the development of new technologies, such as automated electrophysiology platforms. Together, these biological and technological advances are expected to have a major impact on the discovery of new chemical entities that modulate ion channels.

The large families of ion channels that selectively pass Na^+, Ca^{2+}, or K^+ and which share the common property of being gated by changes in the transmembrane potential are the subjects of this chapter. This chapter will review our current understanding of the many genes that encode voltage-gated ion channel subunits, addressing important topics such as the physiological and pathological roles that are associated with these subunits in normal and diseased states, their prototypical pharmacology, and therapeutic opportunities of subtype-selective drugs.

2.21.2 Sodium Channels

2.21.2.1 Current Nomenclature

Voltage-gated Na^+-permeable ion channels are members of the diverse ion channel superfamily that also comprises Ca^{2+}- and K^+-permeable channels.[2] Native voltage-gated Na^+ channels are composed of a large pore-forming

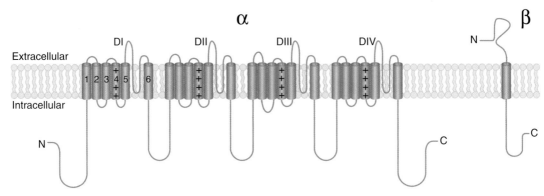

Figure 2 Pore-forming (α) and auxiliary subunits of a voltage-gated Na$^+$ channel. The α subunit consists of four highly homologous domains (DI through DIV), each containing six transmembrane segments (S1 through S6). The pore loop between S5 and S6 determines ion selectivity, S5 and S6 line the pore, and positively charged amino acids in S4 confer voltage dependence on the protein. The β subunits have a single transmembrane segment, with an extracellular N-terminus and an intracellular C-terminus. The $β_1$ and $β_3$ subunits are noncovalently linked to the α subunit, whereas the $β_2$ and $β_4$ subunits are capable of forming a covalent disulfide linkage (not shown).

α subunit that is associated with one or two smaller auxiliary β subunits (**Figure 2**). It is the α subunit that provides the ion permeation pathway, although the β subunits may modulate its function, pharmacology, and membrane expression. Voltage-gated Na$^+$ channels in different cell types are functionally quite similar, yet, with good reason, various groups of researchers have used their own classification systems for categorizing them. As with all ion channels, the nomenclature of Na$^+$ channels has evolved in parallel with the development of new technologies that could be applied to their study. Typically, physiological, biophysical, pharmacological, and/or tissue distribution properties were measured, and distinct putative subtypes were often characterized according to differences in these properties. More recently, genomic approaches have led to the identification of the pore-forming α and auxiliary subunits of Na$^+$ channels as well as orphan Na$^+$ channel-like proteins, which have not been functionally expressed but that have been classified solely on the basis of sequence similarity to known subunits. By the end of the 1990s, the varied nomenclature coupled with the fact that most excitable cells express multiple subtypes of Na$^+$ channels had produced an apparently confusing state of affairs and so a new, International Union of Pharmacology (IUPHAR)-approved naming system was proposed.[3,518] The new naming system takes advantage of knowledge gained from recent advances in the identification and sequencing of several genes that encode Na$^+$ channel subunits. Consequently, the current Na$^+$ channel nomenclature reflects an integrated numerical-based system that parallels those used for Ca^{2+} and K$^+$ channels.

For the family of Na$^+$ channel α subunits, the current terminology (**Table 2**) indicates the permeant ionic species (Na) along with the physiological regulator, voltage (subscripted V), to yield Na$_V$. Currently, there is only one subfamily of Na$^+$ channel α subunits, and this is denoted as Na$_V$1. The Na$_V$1 subfamily contains nine subtypes (i.e., Na$_V$1.1 through Na$_V$1.9), each displaying varying levels of amino acid similarity. Na$^+$ channel α subunits are subject to extensive splice variation, and, where appropriate, these are indicated by a letter after the subtype number (e.g., Na$_V$1.1a). In the future, it is possible that additional Na$^+$ channel subtypes and/or subfamilies will be identified among those orphan Na$^+$ channel-like genes that make up the Na$_X$ group. There is a parallel naming system for the genes that encode the various Na$^+$ channel α subunits, although the gene number is not always the same as the isoform number (see **Table 2**). Consistent with their common evolutionary origins, several Na$^+$ channel α subunit genes are proximally located on the same chromosomes. Most of the genes encoding tetrodotoxin-sensitive (TTX$_S$) Na$^+$ channel α subunits are found on human chromosome 2. The exceptions are the Na$_V$1.6 gene (*SCN8A*), which is found on chromosome 12 and the skeletal muscle Na$_V$1.4 gene (*SCN4A*), which is found on chromosome 17 in close proximity to the skeletal muscle Ca^{2+} channel $γ_1$ subunit (see **Table 4**). The TTX-insensitive (TTX$_I$) and TTX-resistant (TTX$_R$) Na$^+$ channel genes are found on chromosome 3.

For the Na$^+$ channel auxiliary β subunits, the current nomenclature distinguishes different subtypes by using a subscripted number (e.g., $β_1$ – see **Table 2**). Currently, four distinct Na$^+$ channel β subunit isoforms have been identified,[4–9] with at least one isoform subject to splice variation.[10] The Na$^+$ channel β subunits are not related to the auxiliary subunits (β or other) of Ca^{2+} and K$^+$ channels, but they are related to certain cell adhesion molecules. Similar to the Na$^+$ channel α subunit genes, there is a parallel naming system for the β subunit genes (see **Table 2**). The gene encoding the $β_1$ subunit is located on chromosome 19, whereas the three other β subunit genes are located on chromosome 11.

Table 2 Voltage-gated Na$^+$ channel subunit classification

Subunit type and name	Human gene symbol name	Human chromosome	General description
Pore-forming α subunit			
Na$_V$1.1	SCN1A	2q24	TTX$_S$ Na$^+$ channels found in central and peripheral neurons
Na$_V$1.2	SCN2A	2q23	
Na$_V$1.3	SCN3A	2q24	
Na$_V$1.4	SCN4A	17q23	TTX$_S$ Na$^+$ channel found in skeletal muscle
Na$_V$1.5	SCN5A	3p21	TTX$_I$ Na$^+$ channel found in heart and denervated skeletal muscle
Na$_V$1.6	SCN8A	12q13	TTX$_S$ Na$^+$ channels found in central and peripheral neurons
Na$_V$1.7	SCN9A	2q24	
Na$_V$1.8	SCN10A	3p21	TTX$_R$ Na$^+$ channels found in peripheral neurons
Na$_V$1.9	SCN11A	3p21	
Auxiliary β subunits			
β$_1$	SCN1B	19q13	This subunit is noncovalently linked to the α subunit and is found mainly in central neurons, heart, and skeletal muscle
β$_2$	SCN2B	11q23	This subunit forms a covalent linkage with the α subunit and is found mainly in neurons
β$_3$	SCN3B	11q23	This subunit is noncovalently linked to the α subunit and is found mainly in central neurons and skeletal muscle
β$_4$	SCN4B	11q23	This subunit forms a covalent linkage with the α subunit and is found mainly in neurons

2.21.2.2 Molecular Biology

Each of the nine Na$^+$ channel α subunit isoforms has a distinct pattern of tissue and subcellular distribution. In general, they are expressed highly in excitable cells that generate fast action potentials. Na$^+$ channel α and β subunit expression patterns are developmentally regulated, and may also become altered in complex ways in pathological conditions. The changes in expression can be causative in diseases that are associated with genetic mutations in Na$^+$ channel genes (see Section 2.21.2.8) or they can represent a component of the compensatory response of the cell to an ongoing pathological state.

The TTX$_S$ subtypes, Na$_V$1.1, Na$_V$1.2, Na$_V$1.3, and Na$_V$1.6, are expressed predominantly in neurons of the central nervous system (CNS),[11–13] whereas the Na$_V$1.7, Na$_V$1.8, and Na$_V$1.9 subtypes are expressed primarily in neurons of the peripheral nervous system (PNS).[14–17] At the subcellular level, neuronal Na$^+$ channel subtypes may be distributed differentially on dendrites, soma, and/or axons, suggesting that each isoform serves distinct yet complementary functions in integrating input signals and controlling the output responses of the neuron.[18,19] In addition, Na$^+$ channel density varies significantly across the neuronal cell membrane, and, generally speaking, this reflects the localized membrane excitability. Na$^+$ channel density is highest at specialized regions, such as the axon initial segment,[20] where action potential initiation occurs, and also at nodes of Ranvier in myelinated axons to support saltatory (jumping) conduction. Saltatory conduction is a rapid process that requires both Na$^+$ and K$^+$ channels to be concentrated at axonal nodes of Ranvier, which are breaks in the insulating myelin sheath provided by oligodendrocytes in the CNS or Schwann cells in the PNS. Neurological diseases that induce axonal demyelination, such as multiple sclerosis, will permit Na$^+$ and K$^+$ channel redistribution away from the nodes of Ranvier, leading ultimately to impairment or even complete blockade of action potential conduction.[21] In contrast, ectopic action potential generation and overexcitability of peripheral and central neurons in response to nerve injury involves up- and downregulation of certain Na$^+$ channel α and β subunits as well as their redistribution at the subcellular level.[22–25] The preferential targeting of individual isoforms to specific subdomains under normal and diseased conditions suggests that isoform-specific signaling may guide these processes. The protein trafficking and compartmentalization processes that govern the precise distribution of Na$^+$ channels in neuronal cell membrane subdomains are far from being completely understood. Protein–protein interactions between Na$^+$ channel α and β subunits and other molecules, either within the neuron itself or in supporting cells such as astrocytes and oligodendrocytes, are sure to be involved, and several mechanisms have been proposed recently.[26–31]

The TTX$_S$ subtype Na$_V$1.4 is found primarily in innervated skeletal muscle cells,[32] with its highest density being at the motor end-plate.[33] Interestingly, Na$^+$ channel expression in mammalian skeletal muscle undergoes several well-documented phenotypic changes upon tissue denervation. The first change of note is a reorganization of Na$^+$ channel expression away from the end-plate to a more uniform distribution pattern.[34] The other major changes are a transient downregulation of the Na$_V$1.4 transcript coupled with a parallel upregulation of the Na$_V$1.5 transcript (TTX$_I$ subtype).[35] In addition to their differences in TTX sensitivity, these two Na$^+$ channel subtypes have distinct biophysical properties.[36] Consequently, the changes in transcript levels can explain earlier electrophysiological studies on the contrasting biophysical properties and TTX sensitivity of Na$^+$ currents in innervated and denervated skeletal muscle fibers.[37]

Studies on Na$^+$ channel expression in cardiac myocytes have revealed that Na$_V$1.5[38] is the predominant subtype and that several TTX$_S$ subtypes such as Na$_V$1.1, Na$_V$1.2, Na$_V$1.3, and Na$_V$1.6 are also present at lower levels.[13,39] Subcellularly, Na$^+$ channels are found in high density in the plasma and transverse tubule membranes of atria and ventricles. The Na$_V$1.5 isoform is preferentially localized, along with the β$_2$ auxiliary subunit, in the plasma membrane and intercalated disks at the ends of adjacent ventricular myocytes.[40,41] In contrast, the brain-type Na$^+$ channel isoforms appear to be preferentially localized, along with the β$_1$ and β$_3$ auxiliary subunits, in the transverse tubule membrane of ventricular myocytes[41] and in the sinoatrial (SA) node and atrial myocytes.[42] The mechanisms underlying the expression and subcellular distribution patterns of Na$^+$ channels in the heart have not been fully elucidated. Ubiquitin–protein ligases, such as Nedd4-2, which is expressed in heart, can downregulate the expression of Na$_V$1.5.[43] Many additional factors that control Na$^+$ channel trafficking in the heart have been reviewed recently.[44] Following myocardial infarction, tissue remodeling takes place in the heart. This is associated with ventricular hypertrophy and altered Na$^+$ channel expression. In particular, upregulation of Na$_V$1.1 appears to alter electrical activity in remodeled myocytes, perhaps leading to arrhythmogenicity.[45] In contrast, there appears to be a downregulation of Na$^+$ channel expression in end-stage heart failure.[46]

Regarding the general expression of the auxiliary subunits, the β$_1$ subunit exhibits a widespread distribution, and has been identified in the neurons of brain, spinal cord, and dorsal root ganglia (DRG) as well as skeletal muscle and cardiac myocytes.[47] In contrast, expression of the β$_2$, β$_3$, and β$_4$ subunits is restricted mainly to neurons of the CNS and PNS,[6,8,9] although some evidence suggests that β$_3$ may also be found in skeletal muscle.[48]

2.21.2.3 Protein Structure

Since the Na$^+$ channel α subunit has not yet been crystallized, very little is known about its tertiary structure. However, given the similarity in primary and predicted secondary structures of voltage-gated Na$^+$, Ca^{2+}, and K$^+$ channels, it is likely that the overall tertiary structure of the Na$^+$ channel α subunit resembles that of the tetrameric K$^+$ channel K$_V$1.2, which was crystallized recently in its open state (see **Figure 1**).[49,50] Therefore, the mechanisms that are involved in activation and inactivation of the various voltage-gated ion channels are likely to be variations on a common functional theme, and the recent ground-breaking work toward understanding the structure of K$^+$ channels can be developed to provide models for other subtypes of voltage-gated ion channels.

The Na$^+$ channel α subunit is an integral membrane protein (approximately 2000 amino acid residues and 260 kDa) with intracellular N and C termini (see **Figure 2**). The α subunit is comprised of four highly homologous domains (DI through DIV) that are linked sequentially by intracellular loops, some of which contain multiple sites for phosphorylation by protein kinase A (PKA) and protein kinase C (PKC).[51] Several portions of the α subunit contain key structural elements that are involved in determining its localization as well as its functional and pharmacological properties. The first and second intracellular loops are relatively long, whereas the third intracellular loop is short. The second intracellular loop contains a motif that directs axonal clustering of Na$^+$ channels.[52] The third intracellular loop is of particular importance because it contains a hydrophobic IFM motif. This motif is believed to represent the Na$^+$ channel fast inactivation gate that binds within the inner mouth of the pore to block ion permeation.[53] Within each homologous domain, there are six transmembrane segments (S1 through S6) that are linked by short intracellular and extracellular loops. In addition, there are four short segments, or pore loops (P-loops), which are positioned between S5 and S6 of each domain. These short segments re-enter (but do not span) the membrane from its extracellular side, and line the outer entrance to the ion permeation pathway.[54] The S5 and S6 segments appear to line the inner entrance to the Na^+ channel pore. The S4 segments are believed to represent the voltage sensor of the Na^+ channel. According to some ion channel gating models, coordinated outward movement of the S4 segments in a helical screw-like motion within the plane of the membrane changes the position of the activation gate in S5 and S6, leading to Na^+ channel opening. The S4 segments contain regularly spaced positively charged amino acid residues such as arginine or lysine at every third position. Mutagenesis experiments have confirmed that these amino acid residues determine the voltage-dependence of Na^+ channel activation, and are essential for the channel to sense and then open in response to small changes in the local electric field within the cell membrane.[55,56] The ion selectivity and permeation characteristics of the Na^+ channel are determined by EEDD and DEKA motifs, with one amino acid residue in each motif being contributed by each of the four P-loops.[57–59] The negatively charged amino acid residues that make up these motifs are believed to create two rings that are located adjacent to the outer and inner entrances of the pore, respectively.

The Na^+ channel β subunits are smaller but heavily glycosylated proteins, containing around 200 amino acid residues (approximately 33–36 kDa). Each β subunit contains a large extracellular N-terminal domain, a single transmembrane α helical segment, and a short cytoplasmic C terminus (see **Figure 2**). The extracellular domains of the four known β subunits have multiple cysteine residues, some of which pair to form an immunoglobulin (Ig)-like fold, similar to that found in cell adhesion molecules. The $β_1$ and $β_3$ subunits are noncovalently linked to the α subunit. In contrast, the extracellular domains of the $β_2$ and $β_4$ subunits have an unpaired cysteine residue that is capable of forming a covalent disulfide linkage with the Na^+ channel α subunit.

2.21.2.4 Physiological Functions

Voltage-gated Na^+ channels are critical regulators of electrical activity in excitable cells such as neurons as well as skeletal muscle, cardiac muscle, and neuroendocrine cells. They are found in the plasma membrane of these cells, where their major function is to regulate the flux of positively charged Na^+ ions along their electrochemical gradient. Under physiological conditions, the direction of Na^+ flux is from the extracellular milieu into the cell. The physiological roles of Na^+ channels are to integrate postsynaptic potentials, influence subthreshold electrical activity, and amplify small depolarizing changes in membrane potential into larger events called action potentials. Regardless of the particular combination of α and β subunits in an excitable cell, all voltage-gated Na^+ channels serve basically this same physiological role. Nevertheless, subtle differences in the voltage-dependent behavior and kinetic properties of each subtype, coupled to specific subcellular distribution patterns, can conspire to impact cell responsiveness and action potential firing rates in profound ways. This is particularly relevant in disease states, such as epilepsy, cardiac arrhythmias, and chronic neuropathic pain, which may involve phenotypic switches in expression, altered membrane insertion, and subcellular redistribution of certain Na^+ channel subunits.

At negative membrane potentials, most Na^+ channels exist in resting states, and can open in response to depolarization of the cell membrane. Under physiological conditions, the excitatory input that drives the membrane depolarization normally originates from the opening of other cation-permeable ion channels. In neurons, these cation channels may include neurotransmitter-gated ionotropic receptors such as the α-amino-3-hydroxyl-5-methyl-4-isoxazolepropionate (AMPA) and N-methyl-D-aspartate (NMDA) subtypes of the glutamatergic receptor, which are localized postsynaptically on dendritic spines. In addition, low-voltage-activated (LVA) Ca^{2+} channels may cause complex Na^+ channel activation patterns under conditions that promote neuronal burst firing (*see* Section 2.21.3.4). In cardiac myocytes, the excitatory drive may be provided by hyperpolarization-activated, cyclic nucleotide-gated (HCN) cation channels, which serve a pacemaker function in the SA node. In skeletal muscle cells, it is acetylcholine-gated nicotinic receptors that excite the membrane at the neuromuscular junction, leading to Na^+ channel activation. Once open, Na^+ channels cause further depolarization of the cell membrane by permitting an influx of positively charged Na^+ ions. In this way, voltage-gated Na^+ channels amplify the initial excitatory input in a voltage-dependent, nonlinear fashion. When the local depolarizing currents spread to activate adjacent Na^+ channels, the membrane depolarizes even further, and may reach its threshold level, at which point an all-or-nothing Na^+-dependent action potential will be generated and then propagated throughout all electrically excitable regions of the cell membrane. In muscle cells and unmyelinated axons, action potentials are conducted relatively slowly in a smooth wave-like manner, whereas in myelinated axons, they travel more rapidly by saltatory conduction. Normally, after a short period of time (typically a few milliseconds), Na^+ channels inactivate. This terminates the Na^+ influx and allows K^+ channel-dependent membrane repolarization to proceed, signifying the end of the action potential. More complex Na^+ channel gating modes are apparent in cerebellar Purkinje neurons, where resurgent Na^+ currents may be generated upon membrane repolarization, reflecting the reopening of inactivated $Na_V1.6$ channels.[60]

Na$^+$ channels play a central role in the generation of cardiac action potentials and in the propagation of electrical impulses throughout the heart. Consistent with the isoform-specific distribution patterns, the TTX$_I$ subtype Na$_V$1.5 appears to be the critical determinant of action potential initiation and propagation throughout the heart. Indeed, Na$_V$1.5 heterozygous knockout mice display reduced Na$^+$ channel expression in SA node and atrial tissue along with a correspondingly depressed heart rate, slowed conduction, and frequent conduction block.[61] Given their relatively low abundance in heart, some TTX$_S$ Na$^+$ channels appear to play only a supporting role in the control of nodal pacemaker rate[42] and excitation–contraction coupling in the ventricles.[62] Certain subpopulations of TTX$_S$ Na$^+$ channels may fail to inactivate after activation in nodal and ventricular tissue. This gives rise to persistent Na$^+$ currents that may contribute to the pacemaker current in nodal tissue, thereby influencing the heart rate.[63] In addition, persistent Na$^+$ currents may contribute to the plateau phase of the action potential in ventricular myocytes,[64] thereby influencing the effective refractory period.

The β subunits differentially influence several characteristics of the Na$^+$ channel.[65] In general, when they associate with the α subunit, they modify its functional properties, probably through an allosteric effect on the gating machinery. As a result, the voltage dependence of channel inactivation can be shifted to more negative membrane potentials, and the Na$^+$ current kinetics accelerated. Courtesy of their extracellular Ig-like fold, the β subunits also function as cell adhesion molecules, and they have been shown to interact with other cell adhesion molecules such as contactin[29] and neurofascin[66] as well as the extracellular matrix molecules tenascin-C and tenascin-R.[27] In this way, the β subunits modulate insertion of the α subunit into the cell membrane and contribute to subcellular localization of the Na$^+$ channel (e.g., at nodes of Ranvier in myelinated axons). Interestingly, the α subunit is also modulated indirectly by contactin[29,30,67] and the glycoprotein tenascin-R,[68] possibly through their respective interactions with the β subunits. The protein–protein interactions between the α subunit and the β subunits and cell adhesion molecules may represent a novel point for therapeutic intervention in order to control Na$^+$ channel function, membrane excitability, and cellular response characteristics in diseases that affect excitable cells.

In addition to their direct effects on membrane potential, Na$^+$ ions are an effective physiological second messenger, and are involved in the opening of large conductance, Na$^+$-activated K$^+$ channels in cardiac myocytes and neurons.[69–71] These channels require high concentrations of intracellular Na$^+$ for activation, and may be functionally important under conditions of rapid, perhaps pathological, firing as well as ischemia.[72,73] Two candidate genes have been identified (*KCNT1* (*Slack* or *Slo2.2*) and *KCNT2* (*Slick* or *Slo2.1*)) that, when expressed in mammalian cells, produce Na$^+$-activated K$^+$ currents.[74,75] Both genes display differential tissue expression, with Slick being expressed in brain, heart, and skeletal muscle, and Slack being found predominantly in neurons. These channels are related to the large conductance Ca^{2+}-activated K$^+$ channel commonly referred to as Slo1, BK$_{Ca}$, or maxi-K (KCNMA1).

2.21.2.5 Regulation

Regulation of Na$^+$ channel function by intracellular second messenger-dependent signaling cascades is a fundamental process that is implicated in the control of input-output relationships in excitable cells. Modulation of Na$^+$ channel function produces a flexible system for the fine tuning of electrical excitability in neurons and cardiac muscle cells. Several Na$^+$ channel α subunits have multiple intracellularly accessible serine, threonine, and tyrosine residues that are subject to phosphorylation by PKC and PKA. Electrophysiological studies have been critical in determining how phosphorylation affects the properties of the channel. PKA and PKC typically exert inhibitory effects on TTX$_S$ Na$^+$ currents in central neurons. Neurotransmitters that activate the PKC pathway induce phosphorylation of residues in the DI–DII (S554, S573, and S576) and the DII–DIII (S1506) intracellular linkers of the Na$^+$ channel α subunit. Phosphorylation of S1506 appears to be necessary for PKC to induce all of its effects, including reduced peak Na$^+$ current amplitude and slowed channel inactivation.[76] Both of these effects have been observed following activation of muscarinic acetylcholine receptors in hippocampal neurons.[77] In addition, neurotransmitters that activate the cAMP-dependent protein kinase pathway also reduce Na$^+$ current amplitudes by inducing phosphorylation of amino acids (probably S687) in the DI–DII linker.[78,79] This effect has been observed following activation of D$_1$-like dopaminergic receptors in hippocampal neurons.[80] Interestingly, it seems that phosphorylation of the S576 residue by PKC is required before phosphorylation of S687 by PKA can inhibit the Na$^+$ currents.[79] In rat striatal neurons, phosphatase-1 inhibition leads to reduced Na$^+$ current amplitudes, probably as a result of enhanced α subunit phosphorylation.[81] Thus, the function of neuronal Na$^+$ channels appears to be regulated through dynamic control of their phosphorylation state. Indeed, the appearance of the resurgent Na$^+$ current in repolarizing cerebellar Purkinje neurons seems to depend on the phosphorylation state of the Na$_V$1.6 TTX$_S$ Na$^+$ channel.[82]

A change in the phosphorylation state of Na$^+$ channels is an important mechanism of neuromodulation in both peripheral and central neurons. Inflammatory mediators such as prostaglandin E$_2$ (PGE$_2$) can induce hyperalgesia in humans and animals, and this is associated with increased primary sensory neuron excitability. Mechanistically, the effect of PGE_2 appears to involve activation of E-prostanoid subtypes of G protein-coupled receptors (GPCRs) with downstream activation of cAMP-dependent PKA. Subsequently, PKA-mediated phosphorylation of the TTX_R Na^+ channel induces an enhancement of the TTX_R Na^+ current amplitude in sensory neurons, and this is associated with more rapid current inactivation and a negative shift in the voltage-dependence of activation (reduces the threshold for action potential firing).[83] Studies with heterologously expressed $Na_V1.8$ have suggested that PKA-dependent phosphorylation occurs at sites that are located in the DI–DII linker.[84] Interestingly, the PKA-dependent phosphorylation of the TTX_R Na^+ channel seems to require prior phosphorylation by PKC at a different site.[85]

The synchronized propagation of Na^+ channel-dependent cardiac action potentials is critical for coordinated atrial and ventricular contraction. Cardiac Na^+ channels are regulated differentially by both PKC- and PKA-mediated phosphorylation in response to GPCR activation. In the heart, β-adrenergic receptor activation speeds up action potential conduction and induces a faster heart rate. Consistent with this, voltage clamp experiments on cardiac myocytes have shown that β-adrenergic receptor activation by agonists, such as isoproterenol, enhances Na^+ current amplitudes and induces faster Na^+ channel inactivation.[86] Studies with heterologously expressed $Na_V1.5$ have suggested that the effects of PKA on the cardiac Na^+ channel involve amino acid residues that are located in the large cytoplasmic DI–DII (S526 and S529) linker.[87,88] In contrast, PKC-mediated phosphorylation of $Na_V1.5$ (S1505) inhibits the Na^+ current amplitude, stabilizes inactivation, and induces a negative shift in the steady state inactivation relationship.[89] In summary, both neuronal and cardiac Na^+ channels are subject to protein kinase-mediated phosphorylation, although the pattern of modulation is isoform-specific in each tissue.

2.21.2.6 Prototypical Pharmacology

Ion permeation through the Na^+ channel can be modulated by a variety of pharmacologically active substances, including animal and plant toxins (peptides and alkaloids) as well as synthetic therapeutically useful small molecules (**Table 3**). All of these substances exert their effects on Na^+ channel function by binding directly to the α subunit. Some of them bind within or close to the pore and block ion permeation whereas others bind to regions that are coupled to the voltage-gating machinery, and can allosterically activate or inactivate the channel in the absence of membrane potential changes. So far, seven pharmacologically relevant receptor sites have been described on the Na^+ channel, although not all α subunits contain all sites.

Within the Na^+ channel pore loops, the amino acids forming the EEDD and DEKA selectivity motifs are important for defining neurotoxin receptor site 1.[57,59] This is where TTX (from the puffer fish *Fugu*), saxitoxin (STX, from the algae *Alexandrium catenella*), and μ-conotoxins (e.g., μ-PIIIA from the marine cone snail *Conus geographus*) all bind. Site 1 toxins have served as valuable research tools to identify and describe native and recombinant Na^+ channels. These toxins occlude the pore and directly interfere with Na^+ permeation. They do not exhibit any preference for open or inactivated states of the channel, and so their mechanism of action displays little or no use- or voltage-dependence. Na^+ currents are often characterized and described in terms of their sensitivities to inhibition by TTX and STX, which are extremely potent and selective blockers of many mammalian Na^+ channels. Most native Na^+ currents in the CNS, PNS, and skeletal muscle are blocked by very low concentrations of TTX (single-to-double-digit nanomolar IC_{50} values), and, consistent with these data, the Na^+ channel subtypes that predominate in these tissues ($Na_V1.1–1.4$, $Na_V1.6$, and $Na_V1.7$) also have high sensitivity to TTX. In contrast, the predominant native cardiac Na^+ channel, $Na_V1.5$, has been classified as TTX_I (single-digit micromolar IC_{50} value). A number of kinetically distinct Na^+ currents with low sensitivity to TTX (single-to-double-digit micromolar IC_{50} values) have been described in isolated primary sensory neurons. Consistent with these findings, the Na^+ channel subtypes $Na_V1.8$ and $Na_V1.9$ are expressed in DRG neurons, and display correspondingly low sensitivities to TTX.[14,17,90] While the EEDD and DEKA motifs are important for the binding of site 1 toxins, the distinct TTX sensitivities of the different Na^+ channel subtypes are determined primarily by the identity of a specific amino acid in the P-loop of domain I.[14,91] In TTX_S Na^+ channels, the amino acid residue that is located two positions before the first E residue of the EEDD motif is a tyrosine, which contains an aromatic ring. In contrast, the equivalent residue in $Na_V1.5$ is a cysteine, and in $Na_V1.8$ it is a serine. Mutagenesis studies have confirmed that an aromatic ring at this location is essential for the high-affinity binding of TTX.[92]

As already mentioned, there are many natural peptide and alkaloid toxins that interfere with the activation and inactivation processes of the Na^+ channel by binding to neurotoxin receptor sites 2–6, which are located in close proximity or are allosterically coupled to the gating machinery. Toxins such as veratridine (from the seed of the lily *Schoenocaulon officinale*), aconitine (from the root of monkshood, *Aconitum napellus*), batrachotoxin (from the poison dart frog *Phyllobates terribilis*) (all site 2), α scorpion toxins (e.g., Lqh II from *Leiurus quinquestriatus hebraeus*) (site 3),

Table 3 Voltage-gated ion channel pharmacology

Ion channel family	Ion channel grouping	Blockers	Activators
Na_V1	TTX_S Na^+ channels	Tetrodotoxin	α and β scorpion toxins
		μ-Conotoxins (e.g., μGIIIA)	Type II pyrethroids
		μO-Conotoxins (e.g., MrVIA)	Veratridine
		Local anesthetics/antiarrhythmics (e.g., lidocaine)	Aconitine
		Antiepileptics (e.g., lamotrigine)	
	TTX_R Na^+ channels	μO-Conotoxins (e.g., MrVIA)	Type II pyrethroids
		Local anesthetics/antiarrhythmics (e.g., lidocaine)	
Ca_V1	L-type Ca^{2+} channels	Divalent cations (e.g., Cd^{2+})	Bay K-8644
		Classical Ca^{2+} channel blockers:	
		(1) dihydropyridines (e.g., nifedipine)	
		(2) phenylalkylamines (e.g., verapamil)	
		(3) benzothiazepines (e.g., diltiazem)	
Ca_V2	P/Q-type Ca^{2+} channel	ω-Agatoxin IVA	
		ω-Conotoxin MVIIC	
		Gabapentin and pregabalin?	
	N-type Ca^{2+} channel	ω-Conotoxin GVIA	
		ω-Conotoxin MVIIA	
		Gabapentin and pregabalin?	
	R-type Ca^{2+} channel	SNX-482	
Ca_V3	T-type Ca^{2+} channels	Divalent cations (e.g., Ni^{2+})	
		Tetralols (e.g., mibefradil)	
		Antiepileptics (e.g., ethosuximide)	
K_V		Class III antiarrhythmics	Retigabine
		Tetraethylammonium	BMS-204352 (MaxiPost)
		4-Aminopyridine	
		Charybdotoxin	
		Dendrotoxin	
		Psora-4	
		PAP-1	
K_{Ca}	SK_{Ca}[a]	Apamin	
		Tamapin	
		Fluoxetine	
	IK_{Ca}[a]	Iberiotoxin	EBIO
		Clotrimazole	
		TRAM-34	
	BK_{Ca}	Charybdotoxin	NS1619
			BMS-204352 (MaxiPost)
			Pimaric acid
			Dehydrosoyasaponin-I

[a] SK_{Ca} and IK_{Ca} channels are not covered in this chapter.

β scorpion toxins (e.g., toxin IV from *Centruroides suffusus suffusus*) (site 4), brevetoxins (from the algae *Ptychodiscus brevis*), ciguatoxin (from the dinoflagellate *Gambierdiscus toxicus*) (both site 5), and δ-conotoxins (e.g., δ-EVIA from the marine cone snail *Conus ermineus*) (site 6) may directly activate Na^+ channels and may slow or prevent their inactivation. Depending on the extent and time-course of their effects on gating processes, these toxins may enhance Na^+ influx, depolarize cell membranes, prolong action potential durations, induce repetitive firing, reduce membrane excitability, and modulate neurotransmitter release. Paralysis of skeletal muscle, tachycardia in the heart, and convulsions in the brain are commonly observed toxicities. Since neurotoxin sites 2–6 are not found in all Na^+ channel subtypes,[93,94] several of the site 2–6 toxins can be used as research tools to discriminate between the various native neuronal, skeletal muscle, and cardiac channels.

Voltage-gated Na^+ channels have been established as the molecular target for several classes of clinically useful drugs, including local anesthetics as well as certain antiepileptic, antiarrhythmic, and analgesic agents. This includes older drugs such as lidocaine, phenytoin, and mexiletine as well as newer ones such as ropivacaine, lamotrigine, and ralfinimide (NW-1029). Many of these small-molecule drugs block ion permeation through the Na^+ channel, and inhibit electrical activity in excitable cells by binding to the local anesthetic (LA) receptor site (site 7), which is thought to be located on the cytoplasmic side of the Na^+ channel α subunit. Several amino acid residues in the S6 segments of the four homologous domains have been implicated in forming the LA receptor site,[95,96] although only two residues appear to be critical (i.e., F1764 and Y1771). Since the amino acid residues that define site 7 are highly conserved between the nine different TTX_S, TTX_I, and TTX_R Na^+ channel α subunit isoforms, the majority of these drugs display little or no subtype selectivity, and so the challenge for the discovery novel subtype-selective drugs is huge. Most clinically useful local anesthetics and antiarrhythmics inhibit action potential conduction in a frequency-dependent manner (i.e., the extent of block varies in direct proportion to the firing rate). This is because local anesthetic-like drugs block Na^+ channels by binding preferentially to open and/or inactivated states and then stabilizing the channels in drug-bound, nonconducting states. The mechanism by which this occurs can be explained in terms of the modulated receptor hypothesis (MRH)[97] or the guarded receptor hypothesis (GRH).[98] Simply stated, the MRH proposes that the affinity of drugs for the LA receptor site varies with the conformational state of the Na^+ channel, whereas the GRH proposes that the affinity of drugs for the LA receptor site is constant, but that access of the drug to the site is controlled or guarded by a Na^+ channel gate. According to these hypotheses, circumstances that increase the probability of the channel being in open and/or inactivated states will enhance Na^+ channel block and will therefore cause greater inhibition of action potential conduction. Frequency-dependent block of Na^+ channels is particularly relevant in diseases that involve abnormal action potential firing, such as may be observed in the heart during arrhythmias and in neurons during epilepsy and neuropathic pain. Interestingly, the pharmacological sensitivity of the Na^+ channel may be impacted by the particular β subunit that is present, possibly due to the effect of the latter on inactivation behavior of the channel.[99]

2.21.2.7 Prototypical Therapeutics

Voltage-gated Na^+ channels represent well-validated targets for local anesthesia as well as a number of disease states that are associated with overactivity of excitable cells. Changes in the expression levels, distribution patterns, or functional properties of Na^+ channels are observed in many diseases that affect excitable cells. Thus, Na^+ channel blockers will have therapeutic utility in situations where there is a need to interfere with action potential generation or conduction in normal or pathological cells. Na^+ channel inhibition represents an attractive therapeutic approach to the treatment of cardiac arrhythmias and certain forms of epilepsy, as well as neuropathic pain and chronic inflammatory pain.

Conduction anesthesia comprises a variety of local and regional anesthetic techniques that are often applied to peripheral or spinal nerves during surgery. Local anesthesia is restricted to a small part of the body such as a tooth or an area of skin, whereas regional anesthesia affects a larger part of the body such as a leg or an arm. Application of local anesthetic drugs to specific nerve pathways may lead to loss of all sensation, including pain, as well as paralysis. Conduction anesthesia involves the inhibition of spontaneous and evoked electrical activity in normal nerves, and is a clinically important and frequently used procedure in minor surgical settings. In addition, due to superior pain relief and fewer side effects, conduction anesthesia is sometimes used to relieve nonsurgical pain.[100] Most amide- and ester-based local anesthetics are weak bases, and both protonated and unprotonated forms exist in equilibrium at normal pH. While it is the unprotonated form that easily crosses the cell membrane, only the protonated form is able to block Na^+ channels. Thus, local anesthetic potency is impacted by pH changes, and local acidosis caused by infection may reduce the effectiveness of these drugs in the dental surgery setting.

A variety of evidence has linked voltage-gated Na^+ channels to the development and maintenance of various epilepsies. First, electrophysiological studies in brain slices have demonstrated that experimental manipulations that increase Na$^+$ channel activity can induce epileptiform-like discharging.[101,102] Second, the levels of Na$^+$ channel α and β subunit mRNA and protein as well as Na$^+$ current amplitudes may be differentially altered in an isoform-specific manner in the brains of epileptic patients[103,104] and epilepsy-prone animals.[105–108] Finally, many CNS-penetrant antiepileptic drugs such as carbamazepine, phenytoin, and lamotrigine appear to act by modifying Na$^+$ channels in neurons. The mechanism of action of these drugs is use-dependent, as a result of inactivated state-dependent interactions with the Na$^+$ channel α subunit. Therefore, they are capable of reducing late Na$^+$ channel openings[109] and limiting sustained high-frequency neuronal discharges.

Physiological, pharmacological, and gene knock-down approaches have also implicated voltage-gated Na$^+$ channels in the development and maintenance of chronic pain states associated with tissue inflammation and nerve injury. In particular, subcellular changes in Na$^+$ channel expression can induce repetitive spontaneous and evoked action potential firing patterns, which may underlie the abnormal sensory phenomena in affected patients. The symptoms of chronic pain include abnormal sensory disturbances such as spontaneous pain, allodynia, and hyperalgesia, all of which reflect the increased excitability of peripheral and spinal neurons.[110,111] Neuronal hyperexcitability may be explained by increased Na$^+$ current levels, either as a result of altered expression per se,[112,113] or as a result of G protein modulation of function.[114] Electrophysiological and molecular biology approaches have revealed that both TTX$_S$ and TTX$_R$ Na$^+$ channels are affected, and, at a molecular level, the neuronal hyperexcitability may also involve phenotypic switches whereby Na$^+$ channel α and β subunits that are normally expressed at lower levels become upregulated either in direct response to the nerve injury or in response to ongoing neuronal overactivity.[25,115–122] Moreover, gene deletion of Na$_V$1.7, Na$_V$1.8, or Na$_V$1.9 in mice has predictable effects on Na$^+$ currents in DRG neurons, and can alleviate the behavioral symptoms of neuropathic pain in nerve-injured animals.[123–128] In addition, there may be changes in the subcellular distribution pattern of neuronal Na$^+$ channels in primary sensory neurons, including redistribution and accumulation at ectopic sites in the membrane, leading to aberrant action potential firing.[22,129] Importantly, nonselective use-dependent Na$^+$ channel blockers, including some local anesthetics, antiepileptics, and tricyclic antidepressants, can reduce neuronal hyperexcitability, and this may explain their proven efficacy in the alleviation of neuropathic pain symptoms in humans.[130]

With the recent identification of the molecular components of voltage-gated Na$^+$ channels, there has been a surge of research to better understand and describe the tissue-specific distribution of Na$^+$ channel isoforms as well as the subtype-specific changes in expression that occur in disease. For instance, individual Na$^+$ channel subtypes that are selectively expressed in sensory neurons suggest opportunities for the development of potent and safer alternatives to currently available analgesics, although targeting of the LA receptor site will probably not be a fruitful approach. Rather, success may come in the form of novel Na$^+$ channel gating modifiers (e.g., inactivation gate mimetics that bind to the gate receptor, or molecules that bind to the voltage sensor and cause positive shifts in the voltage dependence of activation). An understanding of these areas will underpin efforts to discover novel molecules displaying improved Na$^+$ channel subtype selectivity in addition to well-understood frequency-dependent mechanisms of action. It is anticipated that these novel molecules will be developed as more efficacious and safer drugs for the treatment of diseases associated with membrane hyperexcitability.

2.21.2.8 Genetic Diseases

The cell surface density and functional properties of voltage-gated Na$^+$ channels are key determinants regulating neuronal and myocyte activity. It has long been known that certain diseases involving electrically excitable cells have strong genetic predispositions. Perhaps not surprisingly, analysis of DNA samples taken from affected individuals has revealed that mutations in certain ion channel genes are associated with several of these inherited diseases. Relevant mutations may cause amino acid exchanges, amino acid deletions, or protein truncations that induce abnormal operation of certain key components of the ion channel, including the voltage sensor and inactivation gate. Mutations in multiple ion channel genes have been associated with various monogenic epilepsy disorders. However, many of these diseases no doubt are heterogeneous in origin (i.e., multiple genes and possibly environmental factors are likely involved). Collectively, ion channel-associated diseases are referred to as channelopathies.[131]

Cell membrane overexcitability is a feature of epilepsy and cardiac arrhythmias, and is consistent with genetic mutations that might produce a Na$^+$ channel gain of function phenotype. In particular, mutations in Na$^+$ channel α and β subunit genes have been reported, and often these are associated with either subtle or obvious abnormalities of Na$^+$ channel function, including altered biophysical properties or overexpression in the cell membrane.[106] Mutations in the Na$^+$ channel genes *SCN1A* and *SCN1B* have been implicated in at least two forms of familial childhood epilepsy (i.e., generalized epilepsy with febrile seizures plus (GEFS+) and severe myoclonic epilepsy of infancy (SMEI)). More than 100 single-point mutations have been documented in *SCN1A*, making it the most clinically relevant epilepsy gene identified to date. Yet, electrophysiological studies show no consistent relationship between alterations in Na$^+$ channel properties and clinical phenotype, suggesting heterogeneity of underlying causes. Nevertheless, functional analysis of Na$_V$1.1 subunits that have been engineered to include specific epilepsy relevant mutations has revealed several gain of function phenotypes that may contribute to epileptiform activity in affected patients. Typically, these mutations affect the Na$^+$ channel gating processes and promote neuronal hyperexcitability. The mutation D188V endows the Na$^+$ channel with an increased ability to sustain high-frequency activity.[132] The mutation R1648C produces a persistent or noninactivating (window) current, which would tend to increase Na$^+$ influx at negative membrane potentials and exert a continuous depolarizing influence on the cell.[133] Furthermore, the D1866Y amino acid exchange in Na$_V$1.1 weakens the ability of the α subunit to interact with β subunits, perhaps leading to increased Na$^+$ channel availability at hyperpolarized membrane potentials.[134] Conversely, several of these effects on Na$^+$ channel gating are replicated by the GEFS+-associated mutation C121W in *SCN1B*.[135] This mutation affects the extracellular Ig-like fold of the Na$^+$ channel auxiliary β$_1$ subunit, interfering with its ability to bond noncovalently with the α subunit. Therefore, mutations either in α or β subunit genes can lead to the gain of function phenotype that may underlie neuronal overexcitability and seizure activity. Of significant clinical importance, even though these mutations may not be located at the LA site, is the fact that they may indirectly alter the pharmacological sensitivities of the Na$^+$ channel by altering the gating kinetics or the voltage-dependent equilibrium that exists between the different states of the channel.[136]

Primary erythermalgia is an autosomal-dominant painful neuropathy that is characterized by intermittent burning pain with redness and heat in the extremities.[137] This condition has been associated with mutations in the α subunit gene *SCN9A*,[138,139] which encodes a TTX$_S$ Na$^+$ channel that is expressed primarily in primary sensory and sympathetic neurons.[15] Several mutations have been detected, including I848T, L858H, and F1449V (close to the IFM motif in the loop connecting domains III and IV).[140,141] These mutations alter the activation characteristics of Na$_V$1.7, lowering the threshold for action potential generation and promoting neuronal burst firing in the DRG.

Several cardiac arrhythmias, including the LQT3 form of congenital long-QT syndrome, Brugada syndrome, and conduction system disease, have been associated with more than 100 mutations in the cardiac Na$^+$ channel α subunit gene *SCN5A*. Interestingly, mutations in the TTX$_S$ Na$^+$ channel genes do not appear to be associated with cardiac conduction abnormalities, underscoring the dominant role that Na$_V$1.5 plays in controlling heart function. The majority of the *SCN5A* mutations are missense, but deletions, insertions, frameshifts, nonsense, and splice donor errors have also been observed with distinct cellular and clinical electrophysiological phenotypes. In congenital long-QT syndrome, prolongation of the cardiac action potential predisposes individuals to syncope and sudden death as a result of ventricular arrhythmias. The LQT3 syndrome is associated with a gain of function mutations in *SCN5A*, and typically, the mutant channels flicker abnormally between open and closed states, producing sustained inward Na$^+$ current during ventricular myocyte depolarization. This serves to prolong the cardiac action potential, and provides a molecular mechanism for LQT3 syndrome. Brugada syndrome is an inherited cardiac disorder that involves ventricular fibrillation that may lead to sudden death. Brugada syndrome appears to be associated with mutations in the *SCN5A* gene, including R282H[142] and the double-mutant R1232W/T1620M.[143] Electrophysiological investigation of mutant channels reveals a large reduction in Na$^+$ current density and a shift of Na$^+$ channel activation to more positive membrane potentials. The phenotypes are closely related to the clinical observations in Brugada syndrome, which includes conduction block at the atrioventricular (AV) node.

Mutations in the skeletal muscle voltage-gated Na$^+$ channel α subunit gene (*SCN4A*) have been associated with a variety of inherited nondystrophic myotonias and periodic paralyses, including paramyotonia congenita (PC)[144,145] and hyperkalemic periodic paralysis (HYPP).[146] Both conditions are autosomal-dominant disorders that are associated with muscle weakness. In PC the weakness may be induced by low muscle temperature, whereas in HYPP the weakness is associated with high levels of serum K$^+$. Many of the disease-associated alleles include missense mutations in *SCN4A* that result in amino acid substitutions in transmembrane regions of the Na$_V$1.4 channel, including S4. Electrophysiological methods have demonstrated that many of these mutations delay Na$^+$ channel inactivation and induce a persistent Na$^+$ current. Open- and inactivated-state Na$^+$ channel blockers (e.g., phenytoin and tocainide) may have therapeutic benefit in patients suffering from PC and/or HYPP.[147,148]

2.21.3 Calcium Channels

2.21.3.1 Current Nomenclature

In common with other members of the voltage-gated ion channel superfamily, Ca^{2+} channels are integral membrane protein multisubunit complexes.[2] Native Ca^{2+} channels are composed of a large α$_1$ subunit that is commonly

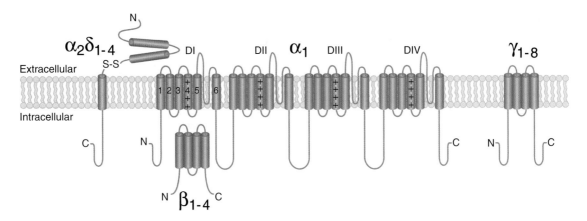

Figure 3 The large pore-forming α_1 subunit has four homologous domains (DI through DIV), each with six transmembrane segments. The $\alpha_2\delta$ subunit is the product of a single gene, but the α_2 and δ portions are post-translationally cleaved and then linked by disulfide bonds. The α_2 portion is located extracellularly, and is essential for modifying the function of the α_1 subunit. The δ portion contains a transmembrane segment that serves to anchor the subunit in the cell membrane. The γ subunit has four transmembrane segments. The β subunit is located in the cytosol, where it interacts with the intracellular loop connecting DI to DII in the α_1 subunit.

associated with auxiliary $\alpha_2\delta$, β, and γ subunits (**Figure 3**). The α_1 subunit provides the ion permeation pathway, and contains most of the regulatory and pharmacologically relevant binding sites. The roles of the auxiliary subunits include modulation of Ca^{2+} channel kinetics, pharmacology, and membrane expression.[149]

Voltage-gated Ca^{2+} channels are highly diverse, and the terminologies that have been used to categorize them over the years are correspondingly complex. Ca^{2+} channels in native tissues were originally classified alphabetically on the basis of qualitative descriptors of their biophysical and pharmacological properties. Thus, the term L-type has been assigned to high-threshold Ca^{2+} channels that underlie high-voltage-activated (HVA), dihydropyridine-sensitive Ca^{2+} currents in neurons and muscle cells. Similarly, the name T-type has been applied to those low-threshold Ca^{2+} channels that produce LVA Ca^{2+} currents in neurons as well as cardiac and smooth muscle cells. For both L-type and T-type Ca^{2+} channels, evidence was apparent for the existence of distinct subtypes of each channel, often within the same cell. Neuronal N-type channels were defined as those that generate ω-conotoxin GVIA-sensitive HVA Ca^{2+} currents. Additional Ca^{2+} channels underlying neuronal HVA Ca^{2+} currents include the ω-agatoxin IVA-sensitive P-type channel, which was first identified in cerebellar Purkinje neurons, and the ω-conotoxin MVIIC-sensitive Q-type channel, which is in fact an ω-agatoxin IVA-insensitive variant of the P-type channel. Finally, the R-type channel was named because of its pharmacological resistance to all known Ca^{2+} channel blockers, although it was recently shown to be sensitive to the peptide SNX-482.[150] Individual cells in many native tissues often display multiple Ca^{2+} currents (e.g., some neurons have five distinct Ca^{2+} current components).[151] This alphabetical terminology remains in widespread use today because it helps to classify Ca^{2+} currents in native tissue where the exact genetic composition of expressed channels is unknown.

With the advent of the gene-cloning era, a letter-based naming system was proposed for the α_1 subunits.[152] This sequential naming system began at α_{1A}, has been extended through to α_{1I} and also includes α_{1S}, for a total of 10 separate gene products (**Table 4**). Ca^{2+} currents that result from the expression of cloned α_1 subunits in heterologous systems can be defined using the same functional and pharmacological descriptors that are used for native Ca^{2+} currents. The α_{1A} subunit gives rise to the P/Q-type Ca^{2+} current, the α_{1B} subunit to the N-type current, and the α_{1E} subunit to the R-type current. Also, it is universally accepted that the α_{1C}, α_{1D}, α_{1F} and α_{1S} subunits correspond to the predominant L-type Ca^{2+} currents in neurons, cardiac muscle, retinal cells, and skeletal muscle, respectively. Finally, there is little dispute that the α_{1G}, α_{1H}, and α_{1I} subunits underlie T-type Ca^{2+} currents in neurons, muscle, and endocrine cells.

More recently, a number-based IUPHAR-approved naming convention was proposed to better coordinate the α_1 subunit classification and also pre-empt the confusing scenario that a future α_{1L} subunit may not encode an L-type Ca^{2+} channel.[153,154,518] The new numerical nomenclature for Ca^{2+} channel α_1 subunits mirrors that of the voltage-gated Na^+ and K^+ channel families, by systematically grouping Ca^{2+} channel α_1 subunits into subfamilies according to their structural relationships (see **Table 4**). The permeant ionic species is denoted (Ca) along with the physiological activator (subscripted V for voltage), to yield Ca_V. Currently, three subfamilies have been described. Subfamily 1

Table 4 Voltage-gated Ca^{2+} channel subunit classification

Subunit type and name	Human gene symbol name	Human chromosome	General description
Pore-forming α_1 subunits			
$Ca_V1.1$	CACNA1S	1q32	Skeletal muscle L-type Ca^{2+} channel/voltage sensor
$Ca_V1.2$	CACNA1C	12p13	Cardiac L-type Ca^{2+} channel
$Ca_V1.3$	CACNA1D	3p14	Neuronal L-type Ca^{2+} channel
$Ca_V1.4$	CACNA1F	Xp11	Retinal L-type Ca^{2+} channel
$Ca_V2.1$	CACNA1A	19p13	Neuronal P/Q-type Ca^{2+} channel
$Ca_V2.2$	CACNA1B	9q34	Neuronal N-type Ca^{2+} channel
$Ca_V2.3$	CACNA1E	1q25	Neuronal R-type Ca^{2+} channel
$Ca_V3.1$	CACNA1G	17q22	Neuronal and cardiac T-type Ca^{2+} channel
$Ca_V3.2$	CACNA1H	16p13	Neuronal and cardiac T-type Ca^{2+} channel
$Ca_V3.3$	CACNA1I	22q12	Neuronal T-type Ca^{2+} channel
High-threshold Ca^{2+} channel auxiliary subunits			
$\alpha_2\delta_1$	CACNA2D1	7q21	The auxiliary subunits regulate the membrane insertion of the α_1 subunits and modulate the functional properties, including biophysical and pharmacological properties, of the assembled Ca^{2+} channels
$\alpha_2\delta_2$	CACNA2D2	3p21	
$\alpha_2\delta_3$	CACNA2D3	3p21	
$\alpha_2\delta_4$	CACNA2D4	12p13	
β_1	CACNB1	17q11	
β_2	CACNB2	10p12	
β_3	CACNB3	12q13	
β_4	CACNB4	2q22	
γ_1	CACNG1	17q24	
γ_2	CACNG2	22q13	
γ_3	CACNG3	16p12	
γ_4	CACNG4	17q24	
γ_5	CACNG5	17q24	
γ_6	CACNG6	19q13	
γ_7	CACNG7	19q13	
γ_8	CACNG8	19q13	

comprises $Ca_V1.1$–1.4 (corresponding to α_{1C}, α_{1D}, α_{1F}, and α_{1S}), and represents the L-type Ca^{2+} channels. Subfamily 2 contains the toxin-sensitive subunits $Ca_V2.1$ (α_{1A}; P/Q-type), $Ca_V2.2$ (α_{1B}; N-type), and $Ca_V2.3$ (α_{1E}; R-type). Subfamily 3 contains three members (i.e., $Ca_V3.1$–3.3 – corresponding to α_{1G}, α_{1H}, and α_{1I}, respectively), and represents the T-type channels. Ca^{2+} channel α_1 subunits are subject to extensive splice variation, and, where necessary, lowercase letters are used to distinguish unique variants. A parallel gene symbol nomenclature exists where *CACNA1* forms the root of the name and the letters A through I plus S denote the 10 members of the three subfamilies. (Caution: this gene naming system still reflects the old letter-based α_1 subunit naming convention, and so the possibility of a future *CACNA1L* being non-L-type still exists.)

The naming of the Ca^{2+} channel auxiliary subunits utilizes numbers to distinguish different isoforms and letters to distinguish splice variants (see **Table 4**). The $\alpha_2\delta$ subunit family (denoted by the gene symbol *CACNA2D*) has four members ($\alpha_2\delta_{1-4}$) that display 30–60% amino acid similarity with each other. The β subunit family (denoted by the gene symbol *CACNB*) has four members (β_{1-4}), whereas the γ subunit family (denoted by the gene symbol *CACNG*) has eight members (γ_{1-8}). As with the α_1 subunits, these auxiliary subunits are subject to extensive splice variation.

2.21.3.2 Molecular Biology

Voltage-gated Ca^{2+} channels exhibit significant molecular and functional diversity, and often an individual cell will express multiple subtypes. The expression of Ca^{2+} channels in neurons as well as in endocrine, cardiac muscle, smooth muscle, skeletal muscle, fibroblasts, photoreceptor, and kidney cells has been supported by a combination of electrophysiological and pharmacological studies. However, molecular biology approaches, including immunohistochemical and in situ hybridization methods, have proven most useful in the localization of individual Ca^{2+} channel subunits to different tissues. At the molecular level, the primary source of Ca^{2+} channel diversity is that 26 genes are known to encode the pore-forming and auxiliary subunits. Further molecular diversity exists because most of these Ca^{2+} channel subunits are subject to splice variation. Theoretically, the array of potential subunit combinations could therefore extend to several hundred distinct possibilities. The subunit composition of the Ca^{2+} channels underlying neuronal HVA Ca^{2+} currents likely includes two or more auxiliary subunits (i.e., $\alpha_2\delta$, β, and, possibly, γ). Some Ca^{2+} channel auxiliary subunits have been shown to affect the function of recombinant Ca_V3 subunits,[155,156] although anti-β subunit antisense strategies failed to reveal any consequence of β subunit depletion on T-type Ca^{2+} channel function in neuroblastoma cells.[157] Therefore, the exact subunit composition of neuronal T-type Ca^{2+} channels remains a matter of speculation. At the functional level, each α_1 subunit displays unique biophysical and pharmacological properties that have enabled them to be correlated with native Ca^{2+} currents in various tissues. However, this is not a conclusive exercise because the functional properties of pore-forming subunits can be modulated in an isoform-specific manner by association with auxiliary subunits, indicating that functional diversity may also reflect heterogeneity of subunit combinations.[158,159] Additional complexity of function arises from the fact that many Ca^{2+} channels are subject to regulation by intracellular second messengers and enzymes.

In various types of muscle cells, excitation–contraction coupling is dependent on the L-type (Ca_V1) subfamily of Ca^{2+} channels, although T-type (Ca_V3) Ca^{2+} channels also play a role. The molecular composition of L-type Ca^{2+} channels has been widely studied in skeletal muscle because of their high density in this tissue. The $Ca_V1.1$ subtype is expressed almost exclusively in skeletal muscle, where it is associated with the Ca^{2+} channel $\alpha_2\delta_1$, β_{1a}, and γ_1 auxiliary subunits[160] as well as the large (>5000 amino acids) ryanodine receptor type 1 (RyR_1).[161] The $Ca_V1.2$ subtype is found in contractile cells of the cardiovascular system,[162] and several $\alpha_2\delta$ and β subunits are also expressed, raising the possibility that multiple molecularly distinct L-type Ca^{2+} channels may be expressed in atrial and ventricular cells.[163,164] The cardiac $Ca_V1.2$-containing L-type Ca^{2+} channels are localized in both the plasma and transverse tubule membranes, where they are functionally coupled to the ryanodine receptor type 2 (RyR_2).[161] Interestingly, the $Ca_V1.2$ splice variant in cardiac muscle cells appears to be different from that in vascular smooth muscle cells.[165] The $Ca_V1.3$ subtype is also present in heart, being expressed in SA node cells, where, due to its relatively low activation threshold, it may play a pacemaking role.[166] The $Ca_V1.4$ subtype was originally thought to be retinal-specific, but is now known to have a broader distribution, including certain cells of the immune system.[167] This L-type Ca^{2+} channel controls neurotransmitter release in retinal bipolar cells, and appears to be the only physiologically active Ca^{2+} channel present.[168] Gene deletion studies in mice have shown that the β_2 subunit forms an essential component of the retinal L-type Ca^{2+} channel.[169] At a molecular level, the $Ca_V3.1$ and $Ca_V3.2$ subunits are expressed widely but at relatively low levels in the cardiovascular system, and their relative expression changes with development.[170] The molecular composition of cardiac T-type Ca^{2+} channels has not been determined biochemically, so it remains unclear if they resemble the L-type channels in being multisubunit complexes or whether they consist only of the α_1 subunit. In vascular smooth muscle, the high sensitivity of the T-type Ca^{2+} current to nickel is consistent with expression of $Ca_V3.2$,[171] although its functional role is unclear.[172]

Almost all of the known Ca^{2+} channel α_1 subunits appear to be expressed in neurons. Neuronal L-type Ca^{2+} channels likely contain either $Ca_V1.2$ or $Ca_V1.3$, and are localized mainly on neuronal cell bodies and proximal dendrites.[173,174] Ca^{2+} influx through neuronal L-type Ca^{2+} channels appears to be involved in the regulation of gene expression[175] as well as the activation of Ca^{2+}-dependent K^+ channels[176,177] and enzymes.[178] Immunohistochemical studies with P/Q-type ($Ca_V2.1$-containing)[179] and N-type ($Ca_V2.2$-containing)[180] selective antibodies have demonstrated that these channels are present at high levels in the presynaptic terminals of central and peripheral neurons, where they may be complexed with a variety of proteins that are involved in neurotransmitter secretion, including syntaxin, synaptotagmin, and SNAP-25.[181] The P-type Ca^{2+} channel is also found on cell bodies and dendrites of cerebellar Purkinje neurons as well as cell bodies of pyramidal neurons in CA1–CA4 of the hippocampus.[182] The $Ca_V2.3$ subunit is a likely candidate to explain the neuronal SNX-482-sensitive R-type Ca^{2+} current,[183] which has a somatic, proximal dendritic, and presynaptic distribution in certain neurons of the brain,[174,184] and may play a role in controlling neurotransmitter release.[185] Consistent with the widespread distribution of LVA Ca^{2+} currents throughout the nervous system, the protein and mRNA for the Ca_V3 α_1 subunits have been found throughout the brain, spinal cord, and peripheral sensory ganglia.[186–188]

2.21.3.3 Protein Structure

As with voltage-gated Na^+ channels, the Ca^{2+} channel α_1 subunit has not yet been crystallized, and therefore its overall tertiary structure remains a matter of speculation. Nevertheless, it is very probable that the tertiary structure of the Ca^{2+} channel α_1 subunit is grossly similar to that of the tetrameric K^+ channel $K_V1.2$, which has been crystallized in its open state (see **Figure 1**).[49,50] Regardless of whether they underlie HVA or LVA Ca^{2+} currents, all of the Ca_V α_1 subunits exhibit highly similar primary and predicted secondary structures. Each α_1 subunit contains approximately 2000 amino acid residues (190–250 kDa). Within each Ca^{2+} channel subfamily, the primary amino acid sequences are approximately 70% identical, whereas between subfamilies the similarity is typically 40% or less. Each α_1 subunit has intracellular N and C termini, and four sets of transmembrane α-helical segments (S1 through S6), which are grouped in four homologous domains (DI through DIV). These domains are linked sequentially by intracellular loops, the first of which (DI–DII) contains a so-called α-interaction domain (AID) that is required for the cytosolic β subunit to bind and exert its modulatory effects on activation and inactivation rates, G protein modulation, pharmacological sensitivity, and cell surface expression.[189,190] Several amino acid residues that are involved in or are critical for this interaction have been defined in both the α_1 and β subunit.[191,192] The fourth transmembrane segment (S4) in each domain of the α_1 subunit contains several positively charged amino acid residues, and is believed to represent the voltage sensor of the channel. Surprisingly, the amino acid sequences of the S4 segments do not appear to explain the different voltage sensitivities of high- and low-threshold channels.[193]

The cation selectivity of voltage-gated Ca^{2+} channels is controlled by a putative ring structure that is formed by four negatively charged amino acids contributed by the pore loops of each domain. In high-threshold Ca^{2+} channels, each pore loop contributes a glutamate residue, to form an EEEE motif,[194] whereas in low-threshold Ca^{2+} channels the third and fourth pore loops each contribute an aspartate residue, to form an EEDD motif.[195–197] These motifs are analogous to the DEKA motif of the voltage-gated Na^+ channels. The exact spatial arrangement of these four amino acid residues in the pore is responsible for determining the divalent cation selectivity and permeation mechanism of the channel. That the motifs differ between the high- and low-threshold Ca^{2+} channels may explain their distinct selectivity profiles: high-threshold channels typically display a higher permeability to Ba^{2+} while low-threshold channels are equally permeable to both cations.

Voltage-dependent inactivation of Ca^{2+} channels appears to proceed via a hinged lid mechanism. In this model of Ca^{2+} channel gating, the DI–DII connecting loop acts as an inactivation gate, and appears to occlude the pore by physically docking at the intracellular ends of the S6 segments in DII and DIII.[198] The hinged lid model provides a mechanism for auxiliary β subunits to influence the inactivation process by altering the mobility of the DI–DII inactivation gate. A Ca^{2+}-dependent inactivation process has also been described for Ca^{2+} channels in cardiac cells, and this represents a negative-feedback mechanism for regulating L-type Ca^{2+} channel activity.[199] It appears that a Ca^{2+}-binding EF-hand motif in the C terminus of the α_1 subunit contributes to the Ca^{2+}-dependent inactivation process.

The $\alpha_2\delta$ subunit is a medium-sized (140–170 kDa) and heavily glycosylated protein that regulates the trafficking, function, and pharmacology of voltage-gated Ca^{2+} channels.[200] This subunit is composed of two domains, α_2 and δ, which are in fact products of the same gene. After translation, the $\alpha_2\delta$ protein is cleaved to produce the larger α_2 and the smaller δ domains that then become linked by disulfide bonds. The $\alpha_2\delta$ subunit is anchored in the membrane by the δ domain, which has one transmembrane segment and contains the intracellular C terminus of the protein.[201] In contrast, the α_2 domain is located extracellularly, and it is this part of the protein that interacts with the α_1 subunit to modulate membrane insertion and functional properties of the resultant Ca^{2+} channel. Interestingly, both α_2 and δ domains are required for gabapentin binding.[202]

Unlike the Na$^+$ channel β subunits, which are anchored in the cell membrane, the Ca^{2+} channel β subunits are soluble proteins (52–78 kDa) that are found in the cytosol.[203] Crystallization studies with recombinant protein revealed that Ca^{2+} channel β subunits exhibit a high-affinity interaction with the AID in the DI–DII linker of the α$_1$ subunit by virtue of multiple protein-interacting modules.[204] Homology modeling of the β subunits has identified Src homology 3 (SH3) and guanylate kinase (GK) motifs, both of which are essential for producing normal Ca^{2+} currents.[205] An additional role of the β subunit is to mask an endoplasmic retention signal in the DI–DII loop of the α$_1$ subunit, thereby facilitating cell surface expression of the channel.[206] The AID region of the α$_1$ subunit does not appear to mediate the functional effects of the β subunit directly. Rather, its purpose appears to be to ensure correct orientation of the β subunit. For instance, when the β subunit interacts with the AID, it is positioned close to the intracellular end of S6 in DI, and may influence its ability to move, resulting in altered channel inactivation kinetics.[207] Furthermore, isoform-specific effects of β subunits on the function of the channel occur through contacts in the α$_1$ subunit that reside outside of the AID,[208] including the N terminus.[209] The distinct regulatory properties of the β$_{2a}$ subunit appear to be controlled by its state of palmitoylation, and so this mechanism contributes to the functional diversity within the Ca^{2+} channel family.[210]

The γ subunit is an integral component of some voltage-gated Ca^{2+} channels.[211] Ca^{2+} channel γ subunits are glycosylated proteins (25 kDa) that contain four transmembrane segments with intracellular N and C termini, as well as a conserved N-linked glycosylation site and a signature GLWXXC motif in the first extracellular loop. The γ subunits typically have an inhibitory effect on Ca^{2+} channel function, but do not modulate channel insertion into the membrane.

2.21.3.4 Physiological Functions

Ca^{2+} is a ubiquitous intracellular second messenger that activates a multitude of effector molecules and downstream processes in excitable cells, including neurons, myocytes, and neuroendocrine cells, as well as nonexcitable cells such as osteoblasts, lymphocytes, mast cells, and endothelial cells. Normally, the cytoplasmic concentration of free Ca^{2+} is kept low (0.1 μM) by Ca^{2+}-binding proteins and Ca^{2+} pumps that transport Ca^{2+} ions across the plasma membrane or the membranes of intracellular organelles.[212] However, the cytoplasmic Ca^{2+} level may become elevated 10-fold or more during cellular activity. While voltage-gated Na$^+$ and K$^+$ channels govern action potential initiation and propagation, it is the voltage-gated Ca^{2+} channels that represent the most significant and obvious signal transduction mechanism that links membrane electrical activity to intracellular events. Voltage-gated Ca^{2+} channels are important dynamic regulators of intracellular Ca^{2+} levels because when they are open, positively charged Ca^{2+} ions can move passively into the cell according to the driving force provided by their electrochemical gradient. The elevation of intracellular Ca^{2+} that accompanies Ca^{2+} channel opening contributes to many physiologically significant events such as excitation–secretion coupling in neurons and endocrine cells, excitation–contraction coupling in muscle cells, as well as regulation of cell growth, proliferation, and differentiation.

Ca^{2+} channels operate by undergoing voltage-dependent gating transitions between conducting and nonconducting states in response to changes in the transmembrane voltage of the cell. The voltage dependencies and rates of these transitions are defining characteristics of different Ca^{2+} channel subtypes, and, ultimately, voltage-dependent Ca^{2+} influx will be shaped by the kinetics of channel activation, inactivation, and deactivation processes. The T-type Ca^{2+} channels are low-threshold in nature (i.e., they begin to open when the membrane potential depolarizes from -90 mV to beyond -70 mV) and their currents are transient in nature as a result of relatively fast activation and inactivation processes. However, their deactivation process is slow, and, consequently, in excitable cells they permit most Ca^{2+} entry during the action potential repolarization phase.[213] In contrast, most L-, N-, and P/Q-type channels are high-threshold in nature, and do not begin to open until the membrane depolarizes beyond -30 mV. Compared with T-type Ca^{2+} channels, their activation and inactivation processes are relatively slow, whereas their deactivation process tends to be more rapid. Therefore, most of their Ca^{2+} influx occurs during depolarization, and longer action potentials will evoke more Ca^{2+} influx.[213] Many Ca^{2+} channels display pronounced overlap in the voltage dependence of activation and inactivation, and, consequently, a small percentage of channels may fail to inactivate at certain membrane potentials, leading to the appearance of a so-called window current.[167,214] The window current is important because it permits physiologically relevant Ca^{2+} entry even under depolarized conditions in vascular smooth muscle cells[215] and neurons.[216] Ca^{2+} influx generally exerts a depolarizing influence on the cell membrane, and also acts as a second messenger for the initiation of downstream processes. In electrically excitable cells, the direct depolarizing influence of Ca^{2+} entering the cell can result in the activation of other voltage-gated ion channels, such as Na$^+$ channels, and these may serve to amplify and accelerate the response of the membrane to depolarization. The second messenger effects of Ca^{2+} include activation of Ca^{2+}-activated K$^+$ channels, which may exert counteracting inhibitory effects on membrane excitation through hyperpolarizing and/or current-shunting actions. Therefore, a consequence of Ca^{2+} channel activation in excitable cells may be an increased complexity of membrane electrical responsiveness. The extensive diversity in the molecular composition, tissue distribution, and functional properties of Ca^{2+} channels is consistent with the diversity in the physiological roles that they play in different cells.

Skeletal muscle contracts in response to acetylcholine release from the terminals of motor neurons at the neuromuscular junction. Acetylcholine activates nicotinic cholinergic receptors on the membrane of the skeletal muscle, leading to the generation of an endplate potential that may be large enough to induce a Na^+ channel-dependent action potential. This action potential pervades the muscle cell membrane, including the transverse tubule system that projects deeply into the cell. Although the skeletal muscle L-type Ca^{2+} channel complex is localized to both the plasma and transverse tubule membranes, Ca^{2+} influx through this channel is not required for skeletal muscle contraction. Rather, the L-type Ca^{2+} channel is tightly associated with RyR_1, which is localized on the membrane of internal Ca^{2+} stores, where it functions as a Ca^{2+} release channel. The L-type Ca^{2+} channel acts as a voltage-sensor, and effectively couples membrane depolarization to activation of RyR_1. Functional coupling of L-type Ca^{2+} channel activation to RyR_1 activation in skeletal muscle provides a mechanism by which plasma membrane depolarization can evoke rapid release of Ca^{2+} from the sarcoplasmic reticulum. This causes the large increase in intracellular Ca^{2+} concentration that is required for the contraction of skeletal muscle fibers.

In contrast to skeletal muscle, the contraction of cardiac muscle is not initiated by the release of a neurotransmitter from nerve terminals, and does not rely solely on the L-type Ca^{2+} channel acting as a voltage sensor. Instead, cardiac muscle contraction is initiated by the spread of spontaneous, regularly spaced single action potentials that originate in the pacemaker tissue of the SA node. Cardiac action potentials spread from the SA node to the AV node, and then to the working cells of the heart via specialized cells that constitute the conduction system. While these action potentials are Na^+-dependent, gene knockout and pharmacological studies have suggested that $Ca_V1.3$-containing L-type[166] and $Ca_V3.2$-containing T-type[217] Ca^{2+} channels contribute to the control of heart rate. Although no T-type Ca^{2+} current has ever been recorded from human atrial cells, the biophysical properties of T-type channels and the clinical effects of mibefradil, a moderately selective T-type Ca^{2+} channel blocker, are also consistent with a role in pacemaking. When action potentials invade the plasma membrane and transverse tubule membrane of cardiac myocytes, the ensuing activation of $Ca_V1.2$-containing L-type channels is critical for cardiac contractility. L-type channels are functionally coupled to RyR_2, which unlike RyR_1 is activated by Ca^{2+}. Therefore, Ca^{2+} influx through L-type Ca^{2+} channels represents a crucial step in the excitation–contraction process in cardiac myocytes. Intracellular Ca^{2+} serves as a second messenger to activate RyR_2, which in turn causes the release of Ca^{2+} from the sarcoplasmic reticulum. The influx of positively charged Ca^{2+} ions, coordinated with sustained Na^+ influx through persistent Na^+ channels and delayed K^+ efflux through voltage-gated K^+ channels, supports the generation of a relatively long cardiac action potential. Prolonged membrane depolarization is significant in the heart because it drives further Ca^{2+} influx to maintain the ventricles in a contracted state for long enough to eject most of the blood during systole. Furthermore, the prolonged membrane depolarization sets the effective refractory period by maintaining most of the voltage-gated Na^+ channels in inactivated states, thereby reducing membrane excitability. Unlike skeletal muscle, the ventricle cannot usually respond to premature action potentials, and can therefore avoid entering a dangerous state of fibrillation. As a consequence of the multiple roles that the various L-type channels play in cardiac function, certain Ca^{2+} channel blockers may exert both negative chronotropic and negative inotropic effects on the heart. Furthermore, neurons of the autonomic nervous system release neurotransmitters that act through adrenergic and cholinergic receptors to modulate the function of cardiac ion channels, thereby regulating the rate of depolarization in the SA node, action potential conduction velocity, as well as the force of cardiac muscle contraction.

Vascular smooth muscle cells in resistance arteries and arterioles express L- and T-type Ca^{2+} channels.[218] Both subtypes of channel permit physiologically significant Ca^{2+} influx that leads to activation of calmodulin and myosin light chain kinase. In turn, this leads to enhanced myosin light chain phosphorylation and smooth muscle contraction.[219] Both types of Ca^{2+} channel are involved not only in maintaining basal myogenic tone but also in regulating the reactivity of vascular smooth muscle cells to increased intraluminal tension and the presence of vasoactive substances. The expression of voltage-gated Ca^{2+} channels in vascular smooth muscle also correlates with cell proliferation rates. In particular, neointima formation, which occurs in response to vascular injury and hypertension, involves excessive smooth muscle cell proliferation and Ca^{2+} influx through T-type channels has been implicated in this pathological process.[220]

All three subfamilies of voltage-gated Ca^{2+} channel are well-represented in central and peripheral neurons.[221] At the subcellular level, both high- and low-threshold Ca^{2+} channels are found at postsynaptic and somatic sites, where they control the integration of synaptic inputs and the responsiveness of the neuron.[222] High-threshold Ca^{2+} channels are also found at presynaptic sites, where they regulate neurotransmitter release in response to action potential invasion of the nerve terminal. The N-type Ca^{2+} channels are of special interest because of their exclusive presynaptic localization at many central synapses.[180] This includes the central terminals of primary afferent neurons in the superficial laminae of the dorsal horn in the spinal cord,[174] where they control the release of pronociceptive neurotransmitters.[223] In contrast, P-type Ca^{2+} channels provide the Ca^{2+} influx that is necessary to trigger acetylcholine release at the neuromuscular junction.[224,225] P-type channels also contribute to the control of neurotransmitter release at multiple synapses throughout the CNS and PNS.[226–228] Interestingly, N- and P-type Ca^{2+} channels may coexist at certain synapses, where the distinct functional properties of both subtypes enable them to modulate differentially the pattern of action potential-evoked neurotransmitter release.[229] Furthermore, circumstances that alter the expression level of one subtype often cause compensatory changes in the expression level of other Ca^{2+} channels.[230,231] While $Ca_V2.2$ knockout mice are viable, disruption of $Ca_V2.1$ expression leads to a severe neurological deficit and, ultimately, death. Therefore, P-type Ca^{2+} channels appear to be more important than N-type Ca^{2+} channels in the control of critical neurological and life functions. Interestingly, the so-called stargazer mouse is an ataxic and epileptic mutant mouse that is associated with an abnormal neuronal Ca^{2+} channel auxiliary subunit. The affected protein in stargazer is a homolog of the γ_1 auxiliary subunit of the skeletal muscle L-type Ca^{2+} channel, and is called stargazin or γ_2.[232] Stargazin is enriched in synaptic plasma membranes, and has been shown to interact not only with voltage-gated Ca^{2+} channel α_1 subunits, including $Ca_V2.1$,[233,234] but also with AMPA glutamatergic receptors and the synaptic scaffolding protein, PSD-95.[235] Stargazin clearly represents a multifunctional protein that, despite its relatively small size, plays an exceptionally important role in the normal functioning of the brain.

The biophysical properties and distribution patterns of T-type Ca^{2+} channels allow them to exert profound influences on action potential firing patterns in both peripheral and central neurons. Depending on the membrane potential, some neurons may respond to excitation by firing action potentials either in a tonic or a phasic manner.[236–238] This switch is dependent on the presence and availability of T-type Ca^{2+} channels. At relatively depolarized resting membrane potentials (e.g., −50 mV), T-type channels are inactivated, and membrane excitation typically leads to the generation of a single Na^+-dependent action potential (tonic firing). In contrast, at more negative resting membrane potentials (e.g., −90 mV), T-type Ca^{2+} channels are resting, and, in this case, membrane excitation can activate the T-type channels. This leads to the generation of a low-threshold spike (LTS) that briefly maintains the membrane potential in a depolarized state and allows the repetitive activation of voltage-gated Na^+ and K^+ channels. Thus, the T-type channel-dependent LTS promotes the generation of high-frequency action potential bursts (phasic firing). The action potential burst terminates partly as a consequence of time-dependent inactivation of the T-type channels and also because of the opening of Ca^{2+}-activated K^+ channels. Burst firing is an important electrophysiological phenomenon because it is intimately involved in the promotion of rhythmic/oscillatory behavior in neuronal circuits and also in the generation of sleep patterns, epileptiform activity, and nociception.[239]

2.21.3.5 Regulation

The function of voltage-gated Ca^{2+} channels is subject to complex modulation by several mechanisms, including GPCR-activated pathways.[240] Ca^{2+} channel modulation may occur via direct influences on the biophysical properties of the α_1 subunit or via influences on its ability to associate with Ca^{2+} channel auxiliary subunits and other regulatory proteins.[241] Furthermore, through their modulatory actions on G_α, regulators of G protein signaling (RGS) molecules add an additional level of complexity to Ca^{2+} channel modulation.[242] Activation of phospholipase C-coupled $G_{\alpha q}$ leads to the production of inositol triphosphate (IP_3) and diacylglycerol (DAG). In turn, DAG activates PKC, which can enhance L- and N-type Ca^{2+} channel function.[243,244] In addition, activation of adenylate cyclase-coupled $G_{\alpha s}$ and $G_{\alpha i}$ can regulate (in opposite directions) the intracellular level of cAMP. Manipulations that increase intracellular cAMP cause activation of PKA, which can enhance L-type Ca^{2+} channel function in the heart and vascular smooth muscle.[245,246] In contrast, and apparently in a G protein isoform-specific manner,[247] $G_{\beta\gamma}$ exerts inhibitory effects on both HVA and LVA Ca^{2+} currents. The $Ca_V3.2$ subunit is subject to $G_{\beta\gamma}$ modulation via a site that is located on the DII–DIII linker,[248] whereas multiple loci have been identified where $G_{\beta\gamma}$ may interact with $Ca_V2.1$ and $Ca_V2.2$, including the DI-DII intracellular linker and portions of the N- and C-terminal domains.[249,250] Biochemical studies suggest that $G_{\beta\gamma}$ competes with the Ca^{2+} channel auxiliary β subunit for binding to the DI–DII linker. Therefore, it is possible that some of the inhibitory effects of $G_{\beta\gamma}$ on N-type Ca^{2+} channel function may be mediated by displacement of the auxiliary β subunit.[251] In the cases of N- and P/Q-type Ca^{2+} channels, $G_{\beta\gamma}$ appears to stabilize the channels in a reluctant gating mode, thereby inhibiting the Ca^{2+} influx required for neurotransmitter release.[252] This is a key mechanism in the spinal analgesic actions of opioid receptor agonists. The voltage-gated Ca^{2+} channel α_1 subunit may serve as an integrator of multiple signals, since both strong membrane depolarization and PKC activation stabilize the channels in a willing gating mode, and therefore can prevent inhibition of the channel by $G_{\beta\gamma}$.[253] In conclusion, voltage-gated Ca^{2+} channels are subject to complex regulation by intracellular second-messenger signaling pathways.

2.21.3.6 Prototypical Pharmacology

The pharmacology of Ca^{2+} channels has been studied extensively in native cells and heterologous expression systems. While potency and selectivity determinations at recombinant channels are important, it is the pharmacology of native channels that is of prime importance in elucidating the mechanism of action of a drug. Ion permeation through Ca^{2+} channels can be modulated by a wide variety of substances, suggesting that these channels contain multiple ligand-binding sites. The three main types of blockers that will be discussed are inorganic metal cations, invertebrate-derived peptide toxins, and organic small molecules (see **Table 3**). With the exception of the small-molecule drugs gabapentin and pregabalin, both of which bind with high affinity to the Ca^{2+} channel $\alpha_2\delta$ subunit, most Ca^{2+} channel modulators bind to the α_1 subunit to modulate channel function. While selective blockers of N-, P/Q-, and L-type Ca^{2+} channels exist, no ligand that is truly specific for T-type channels has been discovered. Inorganic divalent (e.g., cadmium, nickel, and zinc) and trivalent (e.g., yttrium and holmium) metal cations are potent blockers of voltage-gated Ca^{2+} channels, and are used routinely to probe the functional roles of Ca^{2+} channels in various isolated tissues. For example, they are capable of interfering with Ca^{2+} channel-dependent physiological processes such as neuronal synaptic transmission,[254] hormone secretion,[255] and cardiac pacemaking.[256] Due to their generally poor selectivity profiles, metal cations have limited application in defining the subunit composition of native Ca^{2+} channels, although nickel is interesting because it is relatively selective for T-type Ca^{2+} channels. Furthermore, within the Ca_V3 family, it displays moderate selectivity for the $Ca_V3.2$ subtype, and can therefore be used as a tool to dissect certain components of complex native LVA Ca^{2+} currents.[257] Mechanistically, the metal cations appear to compete with Ca^{2+} for its binding site in the pore of the channel, and so increasing the concentration of Ca^{2+} can overcome metal cation-induced block. A multitude of ion channel-modulating peptide toxins have been purified from the venoms of several invertebrate species. Many of these toxins display great selectivity for certain voltage-gated Ca^{2+} channels, particularly the N and P/Q types, and, as such, they have proven to be extremely useful in investigating the expression and functional roles of Ca^{2+} channels in a variety of neuronal tissues. Of particular interest are the N-type channel-selective ω-conotoxins GVIA,[258] MVIIA,[259] and CVID,[260] which were isolated from the venoms of the marine snails *Conus geographus*, *Conus magus*, and *Conus catus*, respectively. These ω-conotoxins selectively block ion permeation through N-type Ca^{2+} channels by binding with high affinity to the S5–S6 region of DIII, and thereby occluding the pore. They inhibit the presynaptic Ca^{2+} influx that is necessary to trigger exocytotic vesicle fusion, and have been instrumental in substantiating the role of N-type Ca^{2+} channels in controlling neurotransmitter release at many central and peripheral synapses. The toxin MVIIA is of immense therapeutic interest because a synthetic version (SNX-111 or ziconotide (Prialt)) has been approved by regulatory organizations in several countries for the symptomatic treatment of intractable severe pain. This drug likely alleviates pain by inhibiting nociceptive signal transmission in the spinal cord.[261] Invertebrate-derived peptide toxins have also proven useful in the investigation of neuronal P/Q-type Ca^{2+} channels. Even though these channels are products of the same gene (i.e., $Ca_V2.1$),[262] they actually exhibit defining differences in their pharmacological profiles. While both subtypes are blocked by ω-conotoxin *MVIIC*, the Q-type channel is distinguished from the P-type channel by its lower sensitivity to block by the funnel-web spider (*Agelenopsis aperta*) peptide ω-agatoxin IVA.[151] Together, all of these marine snail and spider toxins have been used to determine the relative contributions of N-, P-, and Q-type Ca^{2+} channels to the control of neurotransmission at mammalian neuronal and neuromuscular synapses. In contrast to the plethora of peptides that selectively target high-threshold Ca^{2+} channels, less progress has been made toward the discovery of T-type channel-selective peptides. Kurtoxin, which was isolated from scorpion venom (*Parabuthus transvaalicus*),[263] reduces T-type Ca^{2+} currents in thalamic neurons[264] and spermatogenic cells.[265] Electrophysiological studies with multiple recombinant Ca^{2+} channels have demonstrated that it is somewhat selective for Ca_V3-containing channels. Within the T-type Ca^{2+} channel family, the $Ca_V3.1$ and $Ca_V3.2$ subtypes are more sensitive to block than $Ca_V3.3$, and so kurtoxin may help define the molecular identity of complex T-type Ca^{2+} currents in native tissue. In contrast to the high-threshold Ca^{2+} channel pore-blocking toxins, kurtoxin appears to reduce Ca^{2+} currents by interfering with the channel activation mechanism. Other toxins that inhibit LVA Ca^{2+} currents by interfering with gating include the tarantula venom (*Thrixopelma pruriens*) peptides ProTx-I and ProTx-II.[266]

A diverse collection of small molecules, belonging to several distinct chemical classes, displays Ca^{2+} channel-modulating abilities. The classical Ca^{2+} channel blockers are represented by the dihydropyridines (e.g., nicardipine), phenylalkylamines (e.g., verapamil), and benzothiazepines (e.g., diltiazem) (see **Table 3**). These drugs are normally considered to be selective for L-type Ca^{2+} channels, although there are examples of dihydropyridines that also block T-type channels. The binding sites for these Ca^{2+} channel modulators have been localized to various regions of the α_1 subunit, and at least some of the amino acid residues that are involved in these interactions have been described. Most of their binding sites are in close proximity to or are allosterically coupled to the channel pore and gating machinery. In particular, the S5, S6, and pore regions of DIII and DIV are critically involved in the interaction between the L-type Ca^{2+} channel and the dihydropyridines,[267,268] phenylalkylamines,[269,270] and benzothiazepines,[271,272] although distinct groups of amino acids appear to be involved in the binding of each class. Even though they share a common molecular target, there are important mechanistic differences in these drugs. Many of the newer dihydropyridines display higher affinity for the inactivated states of the Ca^{2+} channel,[273] and, so, in accordance with the MRH and GRH, their blocking potency is predicted to be greater under depolarized conditions. In contrast, the phenylalkylamines appear to target the open state preferentially.[274]

The tetralol derivative, mibefradil is differentiated from these classical Ca^{2+} channel blockers on the basis of its novel antihypertensive mechanism of action. Mibefradil is a potent inhibitor of LVA Ca^{2+} currents in a variety of cell types, including vascular smooth muscle cells[275] and cardiac pacemaking cells,[276] as well as peripheral[277] and central[278,279] neurons. Mibefradil also potently inhibits Ca^{2+} currents in mammalian cells expressing recombinant Ca_V3 subunits, although it displays no subtype selectivity.[214,280,281] Mibefradil has high affinity for inactivated states of the T-type Ca^{2+} channel, and therefore its apparent potency increases under depolarized conditions, possibly explaining the favorable vascular to cardiac tissue selectivity ratio of this drug. Despite its widespread acceptance as a selective blocker of T-type Ca^{2+} channels, mibefradil also interacts with L-type Ca^{2+} channels at a site that is distinct from the classical Ca^{2+} channel blocker-binding sites. Mibefradil displays approximately 10-fold selectivity for Ca_V3 over $Ca_V1.2$, although analogs with improved potency and selectivity profiles have also been described.[282]

Modulation of neuronal Ca^{2+} channels appears to be at least a component of the mechanism of action of a large number of diverse drugs that are used to treat nervous system disorders. This includes antiepileptic drugs such as phenytoin, gabapentin, pregabalin, ethosuximide, and zonisamide as well as diphenylbutylpiperidine derivatives such as pimozide and penfluridol that, owing to their high-affinity antagonism at dopamine D_2 receptors, are commonly used as neuroleptic drugs. Also included are the diphenylmethylpiperazines such as flunarizine and cinnarizine that are neuroprotective and may be used to treat a variety of conditions, including migraine, vertigo, and epilepsy. Most of the antiepileptic drugs are relatively nonselective Ca^{2+} channel blockers, although ethosuximide and zonisamide appear to work by a distinct molecular mechanism involving a reduction in membrane responsiveness through preferential block of neuronal T-type Ca^{2+} channels.[283,284] Neither drug discriminates between the three different T-type Ca^{2+} channel α_1 subunits,[285] and, like the other antiepileptic drugs mentioned, the interaction of ethosuximide and zonisamide with T-type Ca^{2+} channels is state-dependent, although differences exist. Whereas ethosuximide behaves as an open-state blocker,[285] zonisamide interacts preferentially with inactivated states.[286] Both classes of diphenyl-containing drugs are noteworthy because several representatives are potent Ca^{2+} channel blockers that display moderate selectivity for T-type and N-type channels.[287,288] In general, these drugs preferentially interact with and stabilize inactivated states of the channel. Substitution at the diphenyl group, to produce 4-arylpiperidine or 4-arylpiperidinol compounds, is an effective method for reducing the affinity of binding at the dopamine D_2 receptor while at the same time maintaining potency at neuronal T-type Ca^{2+} channels.[289,290] Compounds with this profile remain active in models of neuroprotection and epilepsy, and, if they were to be developed clinically, they could potentially be devoid of adverse extrapyramidal side effects that are associated with D_2 receptor antagonism. As with mibefradil, it has been proposed that the diphenyl-containing drugs bind to a site that is distinct from those of the classical Ca^{2+} channel blockers. The precise amino acid residues that comprise this unique site have not yet been fully defined, but the site seems to be located intracellularly and may include the DI–DII linker.[291] This happens to be where both the auxiliary β subunit and G protein $\beta\gamma$ subunits bind to regulate Ca^{2+} channel function. Consequently, the diphenyl pharmacology of the Ca^{2+} channel may be modulated by the β subunit isoform contained in the channel complex and by the extent of G protein activation in the cell.

2.21.3.7 Prototypical Therapeutics

Dysregulation of Ca^{2+} handling underlies a variety of pathological conditions that may be associated with abnormal gene transcription, uncontrolled cell proliferation, and aberrant electrical activity. Therefore, ion channels that contribute to the regulation of intracellular Ca^{2+} levels represent an attractive group of targets for potentially treating disorders that are characterized by abnormal regulation of Ca^{2+}-dependent signaling. In particular, voltage-gated Ca^{2+} channels represent well-validated targets for a number of diseases that involve overactivity of excitable cells. Indeed, a large number of commonly prescribed drugs that block Ca^{2+} channels have significant beneficial impact in the treatment of cardiovascular and nervous system disorders, although clearly there is room for improvement. An increased understanding of the specific voltage-gated Ca^{2+} channels that are functionally important in relevant cell types is expected to aid not only in the discovery and development of new drugs for treating cardiovascular and nervous system diseases but also in the identification of novel disease opportunities.

Ca^{2+} channel blockers have been used successfully for many years in the clinical management of cardiovascular diseases such as hypertension, angina pectoris, and cardiac arrhythmias.[292] This therapeutic profile is consistent with the physiological roles that voltage-gated Ca^{2+} channels play in the blood vessels and heart, including initiation of the muscle contraction process in both tissues and generation of the prolonged action potential in the heart. The dihydropyridine drugs exert their therapeutic effects primarily through blockade of L-type Ca^{2+} channels in vascular smooth muscle cells. This promotes arterial dilation and leads to a reduction in the resistance to circulation of the blood. Arterial relaxation in the peripheral vasculature is therapeutically beneficial in conditions of hypertension, whereas relaxation of coronary arteries is a successful approach to the treatment of angina pectoris. Even though L-type Ca^{2+} channels are also expressed in the heart, some dihydropyridine drugs display a favorable vascular to cardiac tissue selectivity ratio,[293] and this may help them to avoid negative inotropism. The molecular basis of the vascular selectivity is twofold. First, the membrane potential tends to be more depolarized in vascular smooth muscle cells than in cardiac muscle cells. Therefore, drugs that interact preferentially with inactivated states of the channel will tend to be more potent at relaxing vascular smooth muscle. Second, different splice variants of $Ca_V1.2$ are expressed in the cardiovascular system, and the smooth muscle specific variant appears to have a higher sensitivity than the cardiac variant to inhibition by certain dihydropyridines.[294] Thus, differences in membrane potential coupled with the tissue-specific expression of splice variants displaying different drug sensitivities represent a plausible explanation for the favorable vascular selectivity displayed by novel dihydropyridines. In addition, it is fortuitous that the $Ca_V1.3$[295] and $Ca_V1.4$[296,297] subtypes also appear to be less sensitive to inhibition by many commonly used Ca^{2+} channel blockers, and so the potential side effects of bradycardia and vision impairment are not expected with these drugs.

Some L-type Ca^{2+} channel blockers have proven to be useful in the treatment of cardiac arrhythmias of supraventricular origin, including those associated with AV node re-entry.[298] Open-state blockers of L-type Ca^{2+} channels, such as the phenylalkylamine verapamil, are particularly effective. The antiarrhythmic effect of verapamil occurs largely as a result of its actions on the SA and AV nodes, where impulse conduction is Ca^{2+} channel-dependent. Verapamil shortens the duration of nodal action potentials, depresses AV node conduction, and prolongs the effective refractory period, without altering normal atrial or ventricular conduction time.

Selective T-type Ca^{2+} channel blockers represent a novel and potentially improved approach to the treatment of cardiovascular disease.[299] Mibefradil was marketed briefly as an antihypertensive drug (Posicor), and, as exemplified by its clinical utility, selective T-type Ca^{2+} channel blockers are expected to be useful in the treatment of hypertension and angina pectoris, as well as in reducing cardiac ischemia-induced infarct damage. If mibefradil is indicative, then selective T-type Ca^{2+} channel blockers may also protect against the tachycardia-induced atrial remodeling that plays a role in maintenance and recurrence of atrial fibrillation. Unlike dihydropyridine drugs, selective T-type channel blockers have the desirable potential of lowering blood pressure without unwanted negative inotropic effects on the heart.

Ca^{2+} channels contribute to the appearance of diseases associated with neuronal hyperexcitability, such as epilepsy and certain sensory abnormalities. Neuronal hyperexcitability may occur due to extrinsic mechanisms such as enhanced excitatory synaptic input or reduced inhibitory input, as well as intrinsic mechanisms such as changes in the electrical properties of the neuronal membrane. Perhaps indicative of similarities in the underlying molecular mechanisms of such diseases, many Ca^{2+} channel modulators (e.g., phenytoin, gabapentin, pregabalin, ethosuximide, and zonisamide) are used successfully to treat certain forms of epilepsy and chronic pain in humans. These drugs have in common an ability to decrease neuronal excitability by inhibiting the function of Ca^{2+} channels that are involved either in increasing neuronal membrane responsiveness (e.g., T-type channels) or in promoting neurotransmitter release (e.g., N- and P/Q-type channels). Open- and inactivated-state interactions are common among the Ca^{2+} channel-blocking antiepileptic drugs, and this state dependence permits the preferential targeting of pathological neurons. The mechanism of action of the T-type channel blockers likely involves inhibition of Ca^{2+}-dependent LTS, interruption of neuronal burst firing, and abolition of rhythmic electrical behavior in affected neuronal circuits. The mechanism of action of the N-type Ca^{2+} channel blockers likely involves the reduction of excitatory and, possibly, inhibitory neurotransmitter release in the CNS. Interestingly, α_{1B} and $\alpha_2\delta$ subunits are upregulated in peripheral sensory neurons following nerve injury or tissue inflammation, suggesting that N-type Ca^{2+} channels continue to play a dominant role in controlling nociceptive signal transmission under pathological conditions.[300–305] Indeed, blockade of N-type Ca^{2+} channels that control the release of excitatory neurotransmitters involved in the processing of nociceptive information at the level of the spinal cord is a clinically proven mechanism for the management of pain in humans. Ziconotide, which contains the active ingredient ω-conotoxin MVIIA (SNX-111), is effective at treating a wide range of intractable pain conditions in humans,[306–308] and has been approved in several countries worldwide for the treatment of intractable severe chronic pain.

As indicated already, most Ca^{2+} channel-modulating drugs bind directly to the pore-forming α_1 subunits to inhibit ion permeation. Gabapentin (Neurontin) and pregabalin (Lyrica) are two structurally related drugs that display similar but not identical antiepileptic and analgesic properties in a wide range of clinical settings. While their mechanism of action remains to be fully clarified, these drugs are unique in that they bind with high affinity to the Ca^{2+} channel $\alpha_2\delta$ subunit.[309,310] In particular, the $\alpha_2\delta_1$ subtype appears to be essential for the analgesic actions of pregabalin.[311] Gabapentin and pregabalin have been reported to inhibit HVA Ca^{2+} currents in central and peripheral neurons,[312–316] although there are some reports to the contrary.[317,318] A possible explanation for the discrepancies could be the precise subunit composition of native Ca^{2+} channels in different neuronal populations.[319] It has been postulated that the mechanism of action of these drugs actually involves binding to the Ca^{2+} channel $\alpha_2\delta$ subunit, blockade of presynaptic Ca^{2+} influx, with inhibition of neurotransmitter release in brain[320–322] and in the spinal cord.[323–326] Given the non-pore-forming role of the $\alpha_2\delta$ subunit and the small physical size of gabapentin and pregabalin, it may be the case that these drugs modulate ion permeation through a subtle mechanism, perhaps by modifying Ca^{2+} channel gating processes and/or the kinetics of resting–open–inactivated state transitions.[327,328] Interestingly, mutations in the gene encoding $\alpha_2\delta_2$ have been found to underlie the mouse *ducky* phenotype, which is characterized by spike wave seizures and cerebellar ataxia, and is considered to be a model for absence epilepsy.

2.21.3.8 Genetic Diseases

Many diseases have strong genetic predispositions. Not surprisingly, given their integral role in the control of cellular function, mutations in Ca^{2+} channel pore forming and auxiliary subunits have been implicated in certain diseases involving electrically excitable cells. Most Ca^{2+} channelopathies are associated with mutations in the *CACNA1A* gene, leading to abnormalities of the P/Q-type channel, although mutations in L-type Ca^{2+} channel genes are also known.

CACNA1A-associated disorders include familial hemiplegic migraine (FHM), spinocerebellar ataxia type 6 (SCA6), episodic ataxia type 2 (EA2), and absence epilepsy.[329] FHM is an autosomal-dominant disorder that is sometimes associated with a mild permanent cerebellar ataxia. FHM is associated with more than 10 different missense mutations in the *CACNA1A* gene, and these induce a variety of functional changes in $Ca_V2.1$-containing channels. Most of these mutations affect the biophysical properties of P/Q-type currents, and a common mechanism appears to be the alteration of channel gating. The genetic mutations that are associated with FHM sometimes affect voltage-sensing regions of the channel,[330] leading to gain of function phenotypes where Ca^{2+} influx is enhanced at negative membrane potentials.[331] In addition, the inactivation process may be affected,[332] and some mutations appear to relieve G protein $\beta\gamma$ subunit-mediated inhibition of channel activity.[333] The P/Q-type Ca^{2+} channel phenotypes appear to favor a persistent state of neuronal membrane hyperexcitability, potentially leading to an increase in action potential-triggered vasoactive neurotransmitter release from neurons in the trigeminal ganglia. Consequently, Ca^{2+} channel blockers such as verapamil may represent effective therapies for the treatment of this inherited form of migraine.[334] SCA6 is a dominantly inherited degenerative disorder of the cerebellum, characterized by progressive death of Purkinje cells. SCA6 is associated with an expansion of a trinucleotide CAG repeat in the 3′ region of the *CACNA1A* gene, leading to an expanded tract of glutamine residues in $Ca_V2.1$ subunits that are expressed in cerebellar Purkinje cell bodies and dendrites. This results in a gain of function phenotype, leading to excessive Ca^{2+} entry and Purkinje cell degeneration.[335] EA2 is an autosomal-dominant channelopathy. A few EA2-associated genetic mutations have been described, and these tend to affect regions of the $Ca_V2.1$ α_1 subunit that are important for the normal permeation and gating of the channel.[336] Often, these mutations lead to $Ca_V2.1$ protein truncation, and they have similar functional consequences in that they cause a complete loss of P/Q-type Ca^{2+} current.[337–339]

A couple of inherited diseases are associated with dysfunction of the L-type Ca^{2+} channel-RyR_1 complex in skeletal muscle. Hypokalemic periodic paralysis is an autosomal-dominant skeletal muscle disorder manifested by episodes of weakness under conditions of reduced serum K^+ concentrations. Hypokalemic periodic paralysis seems to be caused by mutations in *CACNA1S* that result in loss of function of the skeletal muscle L-type Ca^{2+} channel. This channel acts as a voltage sensor in skeletal muscle, coupling membrane depolarization to the release of intracellular Ca^{2+}. Mutations have been identified that affect positively charged amino acids in the voltage-sensing S4 of DII and DIV.[340,341] Although not very well understood, the loss of function in the voltage sensor may be associated with muscle weakness, because the skeletal muscle would be unable to respond normally to membrane depolarization. Malignant hyperthermia is an autosomal-dominant disorder of skeletal muscle. It is manifested as a potentially fatal hypermetabolic condition that is triggered by commonly used volatile anesthetics, such as halothane. During exposure to such agents, an increase in intracellular Ca^{2+} concentration leads to sustained muscle contractions with increased muscle metabolism and heat production, triggering hyperthermia. Malignant hyperthermia appears to be caused by mutations in the gene encoding the RyR_1 Ca^{2+} release channel of the sarcoplasmic reticulum.[342] The dysfunctional RyR_1 has mutations in the cytosolic part of the protein (i.e., the foot), and is sensitive to activation by volatile anesthetics, resulting in unwanted muscle contractures. Dantrolene, an allosteric inhibitor of Ca^{2+} release from the sarcoplasmic reticulum, has been used successfully to abort fulminant crises and reduce the mortality rate.

Electroretinogram abnormalities and visual impairments such as congenital X-linked stationary night blindness, abnormal color vision, and hyperopia are all symptoms that may be associated with mutations in *CACNA1F*, which is the gene that encodes the slowly inactivating α_1 subunit $Ca_V1.4$.[343,344] Synaptic transmission in the retina depends on graded changes in the membrane potential of retinal photoreceptor cells. These cells become depolarized in the absence of light stimuli, and tonically secrete the neurotransmitter glutamate in response to Ca^{2+} entry, mainly through L-type Ca^{2+} channels. Normally, light stimuli hyperpolarize the photoreceptor cell membrane, which closes the retinal L-type Ca^{2+} channels, inhibiting further Ca^{2+} influx and reducing glutamate release from the photoreceptors. Single point mutations leading to an amino acid change in DII–S6 leads to a gain of function phenotype that is characterized by profound effects on the voltage-dependence of L-type Ca^{2+} channel activation and the kinetics of inactivation.[345] Consequently, mutated $Ca_V1.4$ subunits may fail to close in response to the membrane hyperpolarization that normally accompanies light stimuli, and so neurotransmitter release continues unabated, leading to vision impairment.

2.21.4 Potassium Channels

2.21.4.1 Current Nomenclature

K^+ channels are members of a large family of transmembrane proteins that allow K^+ to cross biological membranes selectively. Like Ca^{2+} and Na^+ channels, voltage-gated K^+ channels undergo conformational changes to open and close a gate in response to membrane depolarization.[1] The K^+ channel family is formed by a complex and diverse group of proteins that is known to exist in all three domains of organisms, eubacteria, archaebacteria, and eukaryotes, and this continues to stimulate discussions on ion channel evolution. K^+ channels have a wide range of functions that includes setting the resting membrane potential, modulating electrical excitability, and regulating cell volume. It has become increasingly apparent that a greater understanding of the impact of K^+ channels on cellular function will come from insightful integration of molecular biology, cell biology, physiology, and pharmacology data (**Figure 4**). The following sections will attempt to summarize and provide highlights of a very broad and complex body of literature. Several detailed reviews have addressed specific attributes of voltage-gated K^+ channels, and will be referenced to guide further clarification and understanding of this complex ion channel family.

K^+ channels are a large and diverse family of membrane proteins that provide a selective pore for K^+ to cross the lipid bilayer. This attribute of selective K^+ ion permeation is the central feature classifying an ion channel as belonging to the K^+ channel family, and is easily recognized by a highly conserved amino acid motif called the K^+ channel pore signature sequence.[346,347] This signature sequence forms the structural element commonly referred to as the selectivity filter, and contains the amino acid residues TTVGYG (TMxTVYG), with some variations. In 2003, a collaboration between IUPHAR and the American Society for Pharmacology and Experimental Therapeutics (ASPET) provided the current standard classification for K^+ channels.[348,349,517,518]

The nomenclature of K^+ channel families primarily reflects two broad groups: the 6TM voltage-gated and the 2TM inward rectifier.[1,350] For the most part, these are referred to as the α subunits because, in common with the Na^+ and

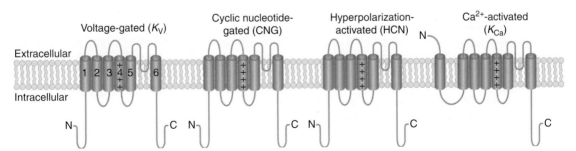

Figure 4 Voltage-gated K^+ channels are the most diverse group of voltage-gated ion channels. The K_V subunits represent the archetypal K^+ channel, and the other families appear to have evolved from them through acquisition of C-terminal sensor domains for various intracellular molecules (e.g., Ca^{2+}, Na^+, Cl^-, pH, and cyclic nucleotides). In terms of membrane topology and structural features, K^+ channel α subunits resemble each of the four homologous domains of the Na^+ and Ca^{2+} channels: the K_V, CNG, HCN, and most K_{Ca} subunits have six transmembrane segments (S1 through S6), whereas $K_{Ca}1.1$ has seven transmembrane segments (S0–S6). A functional and properly localized K^+ channel is formed by the co-assembly of four α subunits along with appropriate auxiliary subunits and accessory proteins (not shown). Furthermore, within each subfamily, heteromultimeric K^+ channels can be formed by mixing and matching appropriate α subunits.

Ca^{2+} channels, they contain the pore-forming elements of the channel. As more information has been forthcoming in relation to the structural and functional diversity of K^+ channels, the underpinnings of these classifications have been challenged, and this has contributed to the seemingly confusing historical nomenclature for these channels. For voltage-gated K^+ channels, there are three families that we will address in this chapter: the delayed rectifier K^+ channel family (K_V); the Ca^{2+}-activated K^+ channel family (K_{Ca}); and the hyperpolarization-activated, cyclic nucleotide-gated cation channel family (HCN), covered here because of their relative selectivity for K^+.[351–356] Other members of the K^+ channel family, such as K_{IR} and certain subtypes of K_{Ca}, will not be discussed in this chapter because these K^+ channels are not regulated by transmembrane voltage changes.

The K_V family of K^+ channels represents the delayed rectifier channels, as first described by Hodgkin and Huxley.[357] These channels allow excitable cells to repolarize efficiently after an action potential, returning Na^+ and Ca^{2+} channels to their resting state. Hodgkin and Huxley first used the term 'delayed rectifier' to describe the voltage-dependent K^+ current that developed after the Na^+ current in response to depolarization of squid giant axon. Since then, the term has remained in use to describe K^+ channels that exhibit a delayed onset of activation followed by little or slow inactivation. The original nomenclature of the K_V family was based on pioneering studies in *Drosophila melanogaster*, where four related K^+ channel genes were identified (i.e., *Shaker, Shab, Shaw,* and *Shal*).[1] Since then, mammalian homologs have been identified for each one of these genes, and, furthermore, the K_V family has expanded greatly as novel genes have been discovered.[348] All K_V channels comprise 6TM subunits that contain the pore signature sequence between S5 and S6 (see **Figure 4**). The homo- or heteromultimeric assembly of four 6TM K_V subunits forms a functional K^+ channel. As masterfully summarized by Gutman and colleagues,[348] the subfamilies are numbered 1–12, and most subfamilies contain several members. This is a very diverse group of K^+ channels with a variety of common names, such as *Shaker, Shab, Shaw, Shal*, KCNQ, human ether-a-go-go related gene (hERG), erg, elk, etc. (**Table 5**).

Numerous cytosolic and membrane-spanning auxiliary and accessory subunits for K_V channels exist. In common with the Na^+ and Ca^{2+} channel auxiliary subunits, association of the K_V α subunit with auxiliary or accessory proteins often leads to changes in protein trafficking and/or function. For the most part, these proteins differentially co-assemble in tetramers, producing a channel complex that consists of four α subunits and four auxiliary subunits.[358,359] Some K_V family members such as K_V5, K_V6, K_V8, and K_V9 appear to function only as auxiliary subunits that modulate the activity of other K_V family members.[358,359] While these proteins display the 6TM topology, they are silent and do not generate functional channels when expressed alone, possibly due to their retention in the endoplasmic reticulum. A number of K_V accessory or auxiliary subunits that are not 6TM in nature may play a role in the generation of functional delayed rectifier K^+ channels (**Table 6**). These subunits include IsK,[360] $K_V\beta$,[361–363] K^+ channel-associated protein (KChAP),[364,365] K^+ channel-interacting proteins (KChIP),[366–368] and dipeptidyl aminopeptidase-like (DPPL) proteins, such as dipeptidyl aminopeptidase-X (DPPX)[369] and DPP10.[370] The IsK-related (*KCNE*) gene family encodes single transmembrane proteins, such as minK and MiRP1–4, which are essential for the formation of delayed rectifier K^+ channels in heart. The $K_V\beta$ subunits are cytoplasmic proteins that share sequence similarity with aldo-keto reductases. There are three known members of this gene family (*KCNAB1–3*), and the proteins they encode appear to interact only with K_V1 channels. KChAP is a cytoplasmic protein that belongs to a class of transcription factor interacting proteins, although its effects on specific K^+ channels (i.e., $K_V1.3$, K_V2, and $K_V4.3$) are transcription-independent. Interestingly, KChAP is not required for proper functioning of K_V-containing channels. Rather, it interacts with the N terminus, and enhances K^+ channel density, suggesting a pure chaperone function. KChIP auxiliary subunits are members of a class of multiple EF-hand-containing neuronal Ca^{2+}-sensor proteins that appear to bind to an intracellular domain of K_V4 subunits. DPPL proteins are transmembrane proteins that also interact with K_V4 subunits. Both KChIP and DPPL proteins appear to be critical components in controlling the trafficking and operation of K_V4 subunits that underlie transient outward currents in neurons (I_A)[368,371,372] and heart (I_{to}).[367,373] This is a burgeoning area, and regrettably we have provided only a brief summary, but it is necessary to highlight that these auxiliary and accessory proteins appear to endow cloned K_V subunits with many of the properties ascribed to native K^+ channels.

The family of so-called K_{Ca} channels is divided into three groups that may be differentiated on the basis of their single-channel conductance (small (SK_{Ca}), intermediate (IK_{Ca}), or large (BK_{Ca})). There are five subfamilies of K_{Ca} channels that can be distinguished on the basis of sequence similarity, but, surprisingly, not all of them are Ca^{2+}-sensitive. It would appear that the K_{Ca} family evolved from the 6TM K_V channels through acquisition of sensors for a variety of intracellular factors, including Ca^{2+},[374] Na^+,[75] chloride (Cl^-),[74] or pH (see **Figure 4**).[375] Even among the Ca^{2+}-sensitive members, different mechanisms appear to have evolved for the sensing of intracellular Ca^{2+}.[376,377] Despite the putative common ancestral origins of these channels, only two members of this family appear to be voltage-gated (i.e., $K_{Ca}1.1$[374] and $K_{Ca}5.1$).[375] $K_{Ca}1.1$ underlies the large-conductance Ca^{2+}-activated K^+ channel, which is commonly referred to as Slo1, BK_{Ca}, or maxi-K.[378] This channel is interesting because it is sensitive to allosteric

Table 5 K$^+$ channel α subunit classification

α subunit name	Human gene symbol name	Human chromosome	General description
K$_v$1.1	KCNA1	12p13.3	*Shaker*-related, delayed-rectifier K$^+$ channel family
K$_v$1.2	KCNA3	1p13	
K$_v$1.3	KCNA3	1p13.3	
K$_v$1.4	KCNA4	11p14.3–15.2	
K$_v$1.5	KCNA5	12p13.3	
K$_v$1.6	KCNA6	12p13.3	
K$_v$1.7	KCNA7	19p13.3	
K$_v$1.8	KCNA10	1p 13.3–22.1	
K$_v$2.1	KCNB1	20q13.11–13.3	*Shab*-related, delayed-rectifier K$^+$ channel family
K$_v$2.2	KCNB2	8q12–13	
K$_v$3.1	KCNC1	11p14.3–15.2	*Shaw*-related, delayed-rectifier K$^+$ channel family
K$_v$3.2	KCNC2	19q13.3–13.4	
K$_v$3.3	KCNC3	19q13.3	
K$_v$3.4	KCNC4	1p21	
K$_v$4.1	KCND1	Xp11.23	*Shal*-related, delayed-rectifier K$^+$ channel family
K$_v$4.2	KCND2	7p31	
K$_v$4.3	KCND3	1p13	
K$_v$5.1	KCNF1	2p25	Auxiliary subunits
K$_v$6.1	KCNG1	20q13	Auxiliary subunits
K$_v$6.2	KCNG2	13q22–23	
K$_v$6.3	KCNG3	2p21	
K$_v$6.4	KCNG4	16q24.1	
K$_v$7.1	KCNQ1	11p15.5	KQT-related, delayed-rectifier K$^+$ channel family
K$_v$7.2	KCNQ2	20p13.3	
K$_v$7.3	KCNQ3	8q24.22–24.3	
K$_v$7.4	KCNQ4	1p34	
K$_v$7.5	KCNQ5	6p14	
K$_v$8.1	KCNV1/KCNB3	8q22.3–22.4	Auxiliary subunits
K$_v$8.2	KCNV2	9p24.2	
K$_v$9.1	KCNS1	20q112–13.2	Auxiliary subunits
K$_v$9.2	KCNS2	8q22	
K$_v$9.3	KCNS3	2p24	

continued

Table 5 Continued

α subunit name	Human gene symbol name	Human chromosome	General description
$K_v10.1$	KCNH1	1q32.1–32.3	*Eag*-related, delayed-rectifier K^+ channel family
$K_v10.2$	KCNH5	14q24.3	
$K_v11.1$	KCNH2	7q35–36	*Erg*-related, delayed-rectifier K^+ channel family
$K_v11.2$	KCNH6	17q23.3	
$K_v11.3$	KCNH7	2q24.3	
$K_v12.1$	KCNH8	3p24.3	*Elk*-related, delayed-rectifier K^+ channel family
$K_v12.2$	KCNH3	12q13	
$K_v12.3$	KCNH4	17q21.31	
$K_{Ca}1.1$	KCNMA1/Slo1	10q22	*Slo*-related, large-conductance, Ca^{2+}-activated K^+ channel, BK_{Ca}, maxi-K
$K_{Ca}4.1$	KCNT1/KCNMB2/Slack/Slo2.2	9q34.3	*Slo*-related, large-conductance, Na^+-activated K^+ channel family
$K_{Ca}4.2$	KCNT2/Slick/Slo2.1	1q31.3	
$K_{Ca}5.1$	KCNU1/KCNMC1/Slo3	8p11.22	*Slo*-related, large-conductance, pH-activated K^+ channel
HCN1	HAC2, BCNG1	5p12	Hyperpolarization-activated, cyclic nucleotide-gated cation channel family
HCN2	HAC1, BCNG2	19p13.3	
HCN3	HAC3, BCNG4	1q22	
HCN4	HAC4, BCNG3	15q24–25	

Table 6 K^+ channel β subunit classification

β subunit name	Human gene symbol name	Human chromosome	General description
$K_V\beta1$	KCNAB	3q26.1	β subunits of the K_V1, *Shaker*-related, voltage-gated K^+ channel family
$K_V\beta2$	KCNAB	1p36.3	
$K_V\beta3$	KCNAB	17p13.1	
$K_{Ca}\beta1$	KCNMB1	5q34	β subunits of $K_{Ca}1.1$, the *Slo*-related, large-conductance, Ca^{2+}-activated K^+ channel, BK_{Ca}, maxi-K
$K_{Ca}\beta2$	KCNMB	3q26.2–27.1	
$K_{Ca}\beta3$	KCNMB	3q26.3–27	
$K_{Ca}\beta4$	KCNMB	12q14.1–15	

activation by both voltage and physiological Ca^{2+}.[379] BK_{Ca} is a multisubunit K^+ channel that contains four $K_{Ca}1.1$ α subunits (KCNMA1) in combination with four β subunits, of which there are four subtypes (KCNMB1–4).[380,381]

The family of cyclic nucleotide-modulated K^+ channels comprise two groups: cyclic nucleotide-gated (CNG) and HCN channels.[352,382,383] These channels are characterized by K_V-like 6TM topology, and are believed to have originated through evolution from K_V channels through acquisition of a cyclic nucleotide-binding domain (CNBD) near the C terminus. There are four known members of this class (HCN1–4), and there are no known auxiliary or accessory subunits.[355] Like K_V and K_{Ca} subunits, CNG and HCN subunits form tetramers, and their activity is influenced by transmembrane potential, with HCN subunits demonstrating a greater dependence on voltage for activation. However, HCN channels open in response to membrane hyperpolarization and close upon depolarization.[355,384] Like K_{Ca} channels, CNG and HCN opening can also be influenced by intracellular signaling events. In the case of CNG and HCN, cyclic nucleotides may interact directly with the α subunit. In comparison with K_V and K_{Ca} channels, HCN channels are less selective for K^+, and hence they can be referred to as cation-selective channels. Due to their structural and functional similarities to K_V and K_{Ca} channels, we have included the HCN channels in this chapter.

2.21.4.2 Molecular Biology Classification

Voltage-gated K^+ channels display significant molecular diversity as a result of the expression of a multitude of pore-forming and auxiliary subunit genes. The first *Drosophila melanogaster* K^+ channel genes were characterized in 1987[385,386] and 1988,[387] and, since then, almost 100 different mammalian K^+ channel subunit genes have been identified.[348,349,518,519] However, there still exists only a basic understanding of the molecular biology aspects of these genes.[348,350,358,388] K^+ channel genes are similar to Na^+ and Ca^{2+} channel genes in that they encode for proteins with presumed 6–7TM configurations. In contrast to Na^+ and Ca^{2+} channel genes, K^+ channel genes encode only one homologous domain, supporting the widely held view that functional K^+ channels are tetrameric in nature.[49,50] K_V, K_{Ca}, and HCN K^+ channels are protein complexes, often composed of several cytosolic and polytopic membrane subunits. Many mammalian voltage-gated K^+ channel α subunits are encoded in a single exon, whereas others are segmented into multiple exons. An additional factor determining the molecular diversity of K^+ channels is the alternative splicing of gene transcripts.[389] In the case of K_{Ca}-containing channels, alternatively spliced variants can differ in their sensitivity to Ca^{2+} and/or voltage. Auxiliary subunits also arise from separate genes that display alternative splicing.

Voltage-gated K^+ channels are especially diverse proteins that determine patterns of cellular excitability.[1] In general, the voltage-gated K_V, K_{Ca}, and HCN channels are abundantly expressed in mammalian cells. Functional channels can either be homo- or heteromultimeric, and this increases the apparent diversity of K^+ channels in native tissues. Due to the ubiquitous distribution patterns of the multitude of K_V and K_{Ca} subunits and the complexity of subunit associations, it is difficult to summarize adequately their distribution and localization patterns in this chapter, and, therefore, only a few salient features will be highlighted. As one can imagine, the use of genetic tools, such as transgenic and gene knockout animals, as well as delivery of small interfering RNA (siRNA) and gene regulatory agents has led to a very mixed understanding of K^+ channel distribution and contribution to cell function, but this will not be reviewed here.

K^+ channels are found ubiquitously in cell membranes, but evidence is accumulating to suggest that they are not randomly distributed. The precise subcellular localization of K^+ channels is certainly a necessary feature of cells to ensure rapid and efficient integration of intracellular and extracellular signaling events. Indeed, specific subcellular distribution patterns of voltage-gated K^+ channel subunits have been documented, but little is known about the mechanisms responsible, although significant gene regulation clearly occurs. Most of the proposed mechanisms have fallen under the umbrella of protein trafficking; however, other factors have recently been examined.[390,391] Knowledge of the processes that direct the specific placement of ion channel pore-forming and auxiliary subunits in subcellular compartments is necessary for a complete understanding of the role of ion channels in normal physiology and pathophysiology.

Distinct cellular distribution patterns can occur through protein–protein interactions such as between K^+ channel α and auxiliary subunits.[391–393] An excellent example of the complex patterns of voltage-gated K^+ channel distribution and the influence of this on cell function is illustrated in a recent review by Misonou and Trimmer.[390] Neurons are morphologically complex cells, and the precise control of electrical activity underpins neurological function. K_V, K_{Ca}, and HCN channels are very important contributors to the control of membrane excitability. For the K_V subunits, K_V1 are mostly localized on axons, whereas K_V2 and K_V4 are mostly somatodendritic. The K_V3 family is expressed more ubiquitously in subcellular compartments of neurons. Other excitable cell types, such as cardiac and smooth muscle, demonstrate equally complex expression patterns of K_V,[394] K_{Ca},[395] and HCN subunits.[383,396] The differential cell surface expression of K^+ channels in polarized and nonpolarized cells as well as morphologically complex cells suggests isoform-specific mechanisms exist to direct subcellular localization.[358] For example, K_V1 channel trafficking can be regulated by chaperone proteins such as calnexin.[397] The oxidoreductase activity of K^+ channel β subunits may also play a role in K_V subunit trafficking, potentially linking cellular metabolic states to channel function.[398]

In addition to the role of accessory proteins in controlling the distribution of K^+ channels, there is also increasing information that supports the involvement of membrane lipids.[399] The existence of plasma membrane microdomains has extended interest in lipid interactions of a variety of proteins, including ion channels. Lipid rafts represent specialized microdomains within the plasma membrane. They are rich in cholesterol and sphingolipids, and can be associated with cellular enzymes and cytoskeletal proteins. Lipid rafts have been shown to assist in the localization of various subtypes of K^+ channels[400] while excluding others.[392,399] In addition, their lipid components, such as cholesterol, have been shown to modulate K_V channel function.[401] Moreover, these and other specialized lipid domains have been shown to organize other cellular proteins, suggesting that lipid rafts in excitable and nonexcitable cell membranes form a scaffold where selected proteins can be brought in close proximity with others in order to achieve specific functions.

2.21.4.3 Protein Structure

Structural perspectives on ion channels have been of great interest since the first descriptions of their function more than 60 years ago. However, information about the molecular architecture of K^+ channels has advanced significantly only over the last several years as higher resolution structural information has accumulated. As discussed above, the proteins produced by the K_V, K_{Ca}, and HCN genes have provided an early topological map of the 6–7TM ion channel subunits. More recently, a variety of elegant studies have focused on increasing our understanding of the relationship of protein structure to the three main functions of voltage-gated channels, namely subunit assembly, ion permeation/selectivity, and gating. These efforts have generated much excitement and debate as structural information has been applied to gain an understanding of functional measurements of voltage-gated ion channels.[350,402,403]

In voltage-gated K^+ channels, the intracellular region from the N terminus to S1 is an important determinant of K_V subunit assembly. This region is called the T1 domain, and it contains signals that promote tetramerization of α subunits within the same subfamily.[404] One interesting aspect of the T1 domain is that it appears to be positioned directly over the intracellular mouth of the pore, suggesting that ions moving through the pore must pass close to it.[50] During the process of N-type inactivation, the T1 domain has been shown to interact with another polypeptide region of the channel to occlude the pore.[405,406] There is also evidence that the T1 domain may interact with the $K_V\beta$ subunits.[50] The T1 domain appears to be multifunctional, since it can also serve to dock auxiliary subunits, modulate gating of the channel, and guide the channel to preferred subcellular locations such as axons. Undoubtedly, the T1 domain will be the subject of more studies to elucidate its roles in controlling K^+ channel structure–function relationships.

For the permeation pathway, K^+ channels exhibit the properties of high conduction rates and high selectivity. The first high-resolution crystal structure of the permeation pathway came from the prokaryotic K^+ channel, KcsA, through the ground-breaking work of MacKinnon and associates.[347] This bacterial channel differs from the mammalian K_V channels in that its topology is 2TM, with the pore loop residing between S1 and S2. In this study, the permeation pathway was best represented by an inverted tepee structure, with the pore lined by the elements of signature sequence described earlier. Crystal structures have now been determined for four different prokaryotic channels as well as the mammalian $K_V1.2$ channel (see **Figure 1**).[49,50] Collectively, these studies demonstrate a growing consensus view of the common architecture of the pore domain as well as the basic principles underlying permeation and selectivity for K^+ ions. The channel uses the pore signature sequence to present electronegative oxygen moieties that are arranged to interact strongly with a dehydrated permeating K^+ ion. This structure varies with conformational changes as the pore opens and the T1 domain communicates with the pore. Selectivity for K^+ ions has been described as a matter of fit. Compared with K^+, Na^+ and other ions are presented in energetically unfavorable states with respect to their fit in the pore, and this underpins the preference of the channels for K^+ ions.[347] High permeation rates while maintaining high selectivity is believed to arise from three ion-binding sites that are arranged sequentially and in close proximity along the axis of the pore so that ions will electrostatically repel each other.[50,407]

The structural view of ion channel gating has been more challenging, and has stimulated much debate.[403,408–410] Since the studies of gating charge movements in *Shaker* K_V channels, it has been known that some 13 electron equivalent charges are moved through the lipid membrane environment.[1] These observations have stimulated many other crystallographic, computational, and amino acid charge-mapping studies aimed at understanding how the subunits are positioned in the membrane and how they move to expose the pore to permeating ions. In addition, they have addressed the role of lipids and the membrane environment in controlling these ion channel functions. Much attention has been focused on S4, which is a common structural feature among all voltage-gated ion channels. One common principle for understanding the voltage-sensing role of S4 domain is its helical structure with regularly spaced positively charged amino acid residues. The positive charges of S4 lie within the transmembrane electric field, and are capable of sensing changes in the transmembrane voltage. These charges respond by moving within the lipid environment, and this movement causes the ion channel pore to open, although it is clear that other parts of the channel are also necessary. The first high-resolution picture that provided a working model came from the prokaryotic channels MthK[411] and K$_V$AP.[412,413] Comparison of the closed-pore structure of KcsA and the open-pore structure of MthK provided evidence that S5 and S6 are involved in channel opening. The structure of K$_V$AP provided evidence that S4 may be highly mobile, although this was in disagreement with other studies suggesting smaller S4 movements.[414–418] Recent studies with rat K$_V$1.2[49,50] have confirmed the importance of S3, S4, the S4–S5 linker, and the T1 domain in modulating ion channel gating. As more studies are now focused on examining structural movements within ion channels, the expectation of the ion channel community is that our understanding of how channels gate will progress rapidly, adding an important structural dimension to ion channel knowledge.

The Ca^{2+}-activated K^+ channel subunit K$_{Ca}$1.1 has an extracellular N terminus, and comprises a core, which contains the voltage-sensing and pore-forming elements, a linker, and an intracellular C-terminal tail, which contains the Ca^{2+}-sensing elements.[419–422] The linker joins the core to the tail, although covalent linkage is not essential for the tail to regulate the function of the core.[423] The core has nine hydrophobic segments, seven of which are believed to be membrane-spanning (S0–S6) and the other two to be intracellular (see **Figure 4**). The segments S1 through S6 are analogous to the corresponding regions of the K$_V$ subunits, and, in common with most voltage-gated ion channels, S4 serves as the primary voltage sensor. Ion selectivity is determined by signature residues in the pore loop, and the ion conduction pathway is lined by S5 and S6. The tail appears to contain two hydrophobic segments and a so-called Ca^{2+} bowl. The Ca^{2+} bowl contains multiple aspartate residues, and appears to be critically involved in Ca^{2+} sensing.[424,425] Nevertheless, there remains some controversy over how many domains actually participate in Ca^{2+} sensing.[426,427] Interestingly, both the core and the tail are required to produce functioning K^+ channels,[423] suggesting that the tail not only senses Ca^{2+} but that somehow it also regulates the voltage sensitivity of the core. Therefore, a gating model has been proposed in which the tail acts as a negative regulator of K$_{Ca}$1.1, and, in the absence of Ca^{2+}-binding, exerts a tonic inhibitory effect on voltage-dependent channel gating. In this model, Ca^{2+} binding induces conformational changes in the protein that reduce the association of the tail with the core. In this way, Ca^{2+} binding relieves the tonic inhibition, resets the voltage sensitivity of the core, and permits the channel to open in response to membrane potential changes.[428]

HCN channels have many structural features in common with K$_V$ and K$_{Ca}$ channels, yet they function differently. Like K$_V$ channels, HCN channels contain positively charged amino acid residues in the voltage-sensing S4. However unlike K$_V$ channels, HCN channels are activated by membrane hyperpolarization, and are not as sensitive to changes in membrane potential.[355,429] In order to open, HCN channels require only 4–6 equivalent charges to move across the electric field, compared with 13 charges for K$_V$ channels. Other differences are that HCN channels open quite slowly compared with K$_V$ channels, and the gate that controls channel activation and inactivation is located on the intracellular side of the channel.[430] Together, these data strongly suggest that voltage-dependent gating of HCN channels involves novel mechanisms of transmembrane segment movements.[355,429] Perhaps hyperpolarization may remove HCN channel inactivation, leading to channel opening, or the linkage mechanism by which the voltage sensor is coupled to the activation gate is reversed. Ion selectivity of HCN channels also remains a puzzle. They are selective for K^+ over Na^+ (ratio of about 4:1), possibly due to a CYG motif in the pore structure,[355,429] yet they can also conduct inward Na^+ and Ca^{2+} currents under physiological conditions.[356] Another interesting feature of HCN channels is the intracellular CNBD, which is located in the C terminus. The CNBD is functionally analogous to the Ca^{2+}-binding regions of K$_{Ca}$1.1. Deletion of the CNBD mimics the effect of cAMP (i.e., it enhances channel activity). This suggests that the CNBD tonically inhibits HCN channels and that cyclic nucleotide binding enhances gating by relieving the inhibitory action of the CNBD. Recently, a model for ligand-dependent channel modulation was proposed based on the crystal structure of a C-terminal fragment of HCN2 that was obtained in the presence of cyclic nucleotides.[431] Comparisons of the HCN channel structure–function relationship to K$_V$ and K$_{Ca}$ channels and the CNBD of other proteins will undoubtedly be of great interest in the coming years.

2.21.4.4 Physiological Functions

K^+ channels are ubiquitously expressed. They are most notable for their role in buffering cell membrane excitability and in setting the resting membrane potential in all cells.[432,433] Voltage-gated K^+ channels have an intracellular gate that serves to control the flow of ions through the pore. Upon membrane depolarization, these K^+ channels transition through multiple closed states before finally opening. Thus, K^+ channels undergo voltage-dependent gating that involves resting, activated-not-open, open, and inactivated states. As a general principle, K^+ channel opening promotes K^+ efflux and causes cell membrane hyperpolarization. Conversely, K^+ channel closing reduces K^+ efflux and promotes cell membrane depolarization.

To accomplish their physiological roles, K$^+$ channels must be expressed in relevant regions of the cell at a density that is appropriate to supply the required K$^+$ conductance. The proper functioning of K$^+$ channels is determined by the biochemical and biophysical properties of the component subunits. K$^+$ channel operation must be tightly coordinated in time with the activity of other ion channels, most notably the faster activating Na$^+$ and Ca^{2+} channels. Despite the plethora of work concerning the cellular expression of K$^+$ channel genes, considerable uncertainty remains in correlating K$^+$ channel genes to native K$^+$ currents. As described earlier, several auxiliary and accessory subunits can associate with K$^+$ channel α subunits, thereby increasing the molecular and functional diversity within the K$^+$ channel family. Indeed, some of these auxiliary and accessory subunits are required in order to reconstitute certain properties of native channels. In terms of the impact of voltage-gated K$^+$ channels on cellular function, it must be appreciated that various physiological stimuli and pathophysiological conditions can regulate K$^+$ channel expression and function. To illustrate this, several studies have demonstrated that metabolic conditions[434,435] and changes in lipid composition[436] can regulate K$^+$ channel activity. Moreover, diseases such as hypertension can impact K$_V$ and K$_{Ca}$ channel subunits.[437,438] The translation from gene function to native currents will clearly continue to be approached using a variety of research tools and the blending of several scientific disciplines.

Native voltage-gated K$^+$ channels in excitable cells are normally activated by membrane depolarization. Voltage-gated K$^+$ channels open typically in response to Na$^+$- or Ca^{2+}-dependent action potentials that are generated in neurons, skeletal muscle, and secretory cells.[1] This reactive behavior represents a common feature of K$^+$ channels. When activated, K$^+$ channels control not only the generation and propagation of action potentials but also their duration and frequency. Consequently, K$^+$ channels participate in the control of excitation–secretion coupling in neurons and endocrine cells and excitation-contraction coupling in muscle cells.[439] Neuronal M channels are characterized as low-threshold, noninactivating, voltage-gated K$^+$ channels, and are defined by their physiological regulation by muscarinic receptors.[440] M channels have slow opening and closing kinetics, and are thought to limit repetitive firing of action potentials. Notable K$^+$ channels that control cardiac function are I$_{to}$, I$_{Kur}$, I$_{Ks}$, and I$_{Kr}$, which are distributed in specific regions of the heart where they contribute to complex and carefully orchestrated cardiac excitablility.[441] To accomplish their functions in excitable cells such as neurons, different K$^+$ channels tend to be localized in specific subcellular compartments (e.g., axonal or somatodendritic regions).[390] Due to their delayed activation kinetics, the K$_V$ subunits are believed to underlie the delayed rectifier K$^+$ currents. Like Na$^+$ and Ca^{2+} channels, some K$^+$ channels inactivate during maintained depolarization. The kinetics of inactivation can either be rapid or slow, and often they are voltage-dependent, with inactivation being speeded up by stronger membrane depolarization. Still other K$_V$ channels inactivate little during maintained depolarization. Thus, as a class, K$_V$ channels demonstrate a wide range of functional properties, and defy a universal and general characterization as illustrated for Na$^+$ and Ca^{2+} channels.

Activity of BK$_{Ca}$ channels is dependent not only on membrane depolarization but also on increased intracellular Ca^{2+}. Overall, these channels have been implicated in the control of neuronal excitability, muscle contractility, and secretory events.[442–444] BK$_{Ca}$ channels require >1 μM intracellular Ca^{2+} to be activated,[442,443] and studies have suggested that this can in fact occur under physiological conditions as a result of localized concentration changes.[445] In neurons, BK$_{Ca}$ channels have been implicated in action potential repolarization and generation of after-hyperpolarizations. BK$_{Ca}$ activity can alter the duration and frequency of action potential firing in secretory cells, and thus they participate in excitation–secretion coupling. In smooth muscle cells, increases in intracellular Ca^{2+} occur at specific intracellular sites independent of membrane depolarization. This may cause BK$_{Ca}$ activation, thereby dampening membrane electrical activity and limiting myocyte contraction.[395,446,447]

The K$^+$ current produced by HCN channels has been termed I$_h$, I$_f$, or I$_q$, and is found in a variety of neurons, pacemaker cells, and photoreceptors.[353,354,448] The best understood function of HCN is as the pacemaker current (I$_h$) in cardiac SA node cells.[355,356] When the cardiac action potential terminates, the membrane potential hyperpolarizes, and HCN channels open. By virtue of their cation selectivity, HCN channels exert a depolarizing influence on the cell. As the membrane depolarizes, Na$^+$ and Ca^{2+} channels activate, and this leads to the initiation and propagation of another action potential. Subsequently, the Na$^+$ and Ca^{2+} channels begin to inactivate, and then delayed rectifier K$^+$ channels open, allowing the repolarization phase to proceed. As the action potential terminates and additional K$^+$ channels become activated, the membrane potential returns to a hyperpolarized level again, and the process repeats. The pacemaker function of HCN channels can also be observed in neurons,[449,450] as well as in signal transduction cells, such as taste receptors, photoreceptors, and auditory receptors.

K$^+$ channels are functional in nonexcitable cells, such as lymphocytes and epithelial cells, where they provide an electrochemical gradient to support many cell functions. An essential signaling step following T cell receptor activation in lymphocytes is Ca^{2+} entry through the Ca^{2+} release-activated Ca^{2+} channel, which underlies the native current, I$_{crac}$, the molecular identity of which is incompletely understood. This Ca^{2+} influx depends on a hyperpolarized membrane potential, and is essential for driving downstream signaling cascades that ultimately lead to increased T cell proliferation and cytokine production. The voltage-gated K^+ channel $K_V1.3$ (along with the voltage-insensitive channel IK_{Ca}) regulates the membrane potential and indirectly controls Ca^{2+} entry. In renal and other epithelial cell types, various K^+ channels have specialized functions in the apical and/or basolateral membranes.[451] In the apical membrane of renal epithelial cells, K_V channels are responsible for stabilizing the membrane potential, whereas in the basolateral membrane, Ca^{2+}-activated K^+ channels play a role in flow-dependent K^+ secretion. $K_V1.1$ and $K_V1.3$ channels have been identified in epithelial cells of the choroid plexus, where they contribute to secretion of cerebral spinal fluid.[452] As has been discussed, K^+ channels exhibit diverse functions, and these are determined not only by their biophysical properties but also by their subcellular placement.

2.21.4.5 Regulation

Central to their physiological roles, voltage-gated K^+ channels are both initiators and effectors of cell signaling. In common with Na^+ and Ca^{2+} channels, K_V, K_{Ca}, and HCN channels can interact with a variety of other cellular molecules, and are subject to complex modulation by multiple mechanisms (e.g., enzymes, membrane lipids, scaffolding proteins, and intracellular second messengers can regulate the expression and function of these channels). There are numerous articles in the literature that provide a wealth of information about the regulation of K_V, K_{Ca}, and HCN channels as well as auxiliary subunits. Many reports address the modulation of both native and cloned channels, and provide a clear demonstration of direct interactions between ion channel subunits and other cellular signaling molecules within a multicomponent protein complex.

Modulation of K^+ channels by second messengers is a dynamic way to fine-tune the excitability of neurons and muscle cells.[1] In particular, K^+ channel modulation influences the ability of neurons to encode information, and is exemplified by the following examples. $K_V1.3$ is a substrate for multiple tyrosine kinases, which can regulate its activity and impact cell function. This channel is highly expressed in rodent olfactory bulb, where it is localized in mitral and granule neurons.[453] It has been shown that $K_V1.3$ phosphorylation modulates not only the K^+ current amplitude but also the voltage-dependence and kinetics of inactivation.[453] Two adaptor proteins, Grb10 and n-Shc, differentially modulate the degree of $K_V1.3$ phosphorylation and the biophysical properties of the channel.[454] Regulation of $K_V1.3$ phosphorylation by adaptor proteins is most likely via protein–protein interactions, suggesting a protein complex comprising K^+ channel, kinase, and adaptor molecules. Regulation of $K_V1.3$ by molecules involved in intracellular signaling has also shown to be an important regulator of its role in T cell function. $K_V2.1$ localization has been shown to depend on the phosphorylation state of the channel. $K_V2.1$ is often found in large clusters on the cell bodies and dendrites of hippocampal pyramidal neurons. However, neuronal activity enhances dephosphorylation of $K_V2.1$, which induces its translocation from clusters to a more uniform distribution pattern.[455] This effect is regulated by the Ca^{2+}-dependent phosphatase calcineurin, and appears to involve serine and threonine residues in the cytoplasmic tail of $K_V2.1$. Interestingly, $K_V2.1$ translocation can be mimicked by glutamatergic receptor activation and induced in vivo by motor seizures, suggesting a link between neurotransmission and intrinsic neuronal excitability. $K_V2.1$ phosphorylation can also induce a hyperpolarizing shift in the voltage dependence of activation, suggesting that this mechanism could dynamically regulate dendritic excitability.[456] In contrast, $K_V2.1$ can be downregulated by protein-tyrosine phosphatase ϵ and upregulated by Src, with both effects occurring through the T1 domain.[457] Similarly, the functional modulation of $K_V1.5$ currents is complex and may involve multiple mechanisms, including SH3 domain binding-dependent and binding-independent phosphorylation, both of which lead to suppression of $K_V1.5$.[458] Many reports have suggested modulation of native K^+ channels by physiological agents. For instance, activation of pancreatic β cell glucagon-like peptide-1 receptors by exendin-4 causes a hyperpolarizing shift in the voltage dependence of steady state inactivation, effectively suppressing the K^+ current. This effect is dependent on phosphorylation by PKA and phosphatidylinositol 3-kinase.[459] Enzymatic modulation K_{Ca} channels have been studied extensively. PKA-dependent phosphorylation of residue S899 in all four α subunits of the BK_{Ca} channel is required to increase channel activity.[460] In addition, specific PKC isozymes can increase BK_{Ca} currents in vascular smooth muscle cells through a cyclic guanosine monophosphate (cGMP)-dependent protein kinase.[461]

2.21.4.6 Prototypical Pharmacology

The pharmacological characterization of voltage-gated K^+ channels has benefited from the availability of a broad collection of relatively selective modulators, and these have fueled a large body of literature, some of which was reviewed recently.[432] Thus, native and cloned K^+ channels have been characterized according to the effects of inorganic ions (e.g., cesium), toxins, and antiarrhythmic drugs (see **Table 3**). However, lack of specificity of many of these agents continues to spark debate in the literature, as attempts are made to deduce the subunit composition of native K^+ channel by comparison with the pharmacological characteristics of cloned channels. Nevertheless, pharmacological modulators of K^+ channels have proven to be valuable tools in elucidating the physiological roles of K^+ channels in many cells and tissues. Here, we will present only a brief overview of K^+ channel pharmacology.

As with Na^+ and Ca^{2+} channel modulators, most K^+ channel modulators either act through a pore-blocking mechanism or via an allosteric effect on channel gating processes. Nonpeptide K^+ channel pore blockers including 4-aminopyridine (4-AP) and quaternary ammonium ion derivatives such as tetraethylammonium (TEA) appear to bind competitively to open K^+ channels. Despite the conserved sequences in the pores of most K^+ channel subtypes, blockers such as TEA often display a wide range of potencies ($IC_{50} < 1$ mM to > 20 mM). 4-AP enters the K^+ channel pore after the channel has opened, and appears to promote closure of the intracellular gate, thereby stabilizing the activated-not-open state. In addition, various peptide toxins, including charybdotoxin, dendrotoxin, margatoxin, noxiustoxin, agatoxin, and kaliotoxin,[462–464] have been isolated from the venoms of species such as scorpions, snakes, spiders, and snails. Like 4-AP, many of these toxins block the channel by direct occlusion of the pore, yet at the same time they often exhibit significant selectivity for certain K^+ channel subtypes. In general, the selectivity of the K^+ channel-blocking toxins derives from tight interactions not just with the pore but also with adjacent domains. Charybdotoxin (from the scorpion *Leiurus quinquestriatus hebraeus*) is a potent but relatively nonselective blocker of some K_V and K_{Ca} channels, whereas the dendrotoxins, which are a group of proteins that are found in the venom of various green mamba snakes (e.g., *Dendroaspis angusticeps* and *Dendroaspis viridis*), appear to be very selective for some K_V1 channels. Still other toxins from tarantula venom appear to display remarkable specificity for individual K^+ channel subtypes. For instance, hanatoxin (from *Grammostola spatulata*) is a potent blocker of $K_V2.1$,[449] while phrixotoxin-1, phrixotoxin-2 (from *Phrixotrichus auratus*),[465,466] and *Theraphosa leblondi* toxin 1[467] appear to inhibit K_V4 channels specifically.

Small-molecule modulators of voltage-gated K^+ channels appear to act by allosterically interfering with the gating processes, but, unfortunately, most of these agents are not selective among the different K_V channels. Thus, while small molecules have been good starting points to characterize K_V currents, proper caution must be exercised when dissecting complex native voltage-activated K^+ currents. Due to the therapeutic interests of pharmaceutical and biotechnology companies, the most novel compounds tend to be those that target K_V and K_{Ca} channels in cardiac myocytes, smooth muscle cells, and neurons, although $K_V1.3$ blockers for the treatment of immune disorders are also attractive. Antiarrhythmic drugs are a heterogeneous group that includes molecules with actions on voltage-gated Na^+ (class I), K^+ (class III), and Ca^{2+} (class IV) channels. These classifications are not mutually exclusive (i.e., some Na^+ channel blockers also block Ca^{2+} and/or K^+ channels). Consistent with the wide repertoire of K^+ channels expressed in the heart, class III antiarrhythmic drugs are a structurally and mechanistically diverse group that includes tedisamil, dofetilide, ibutilide, clofilium, flecainide, quinidine, and sotalol. Few of these drugs display selectivity for individual K_V channels, or indeed other ion channel subtypes. Courtesy of its key role in the repolarization phase of the cardiac action potential and its propensity to block by many drugs, the hERG ($K_V11.1$) K^+ channel has gained notoriety, and has been classified by the US Food and Drug Administration and other regulatory bodies worldwide as a safety hurdle in drug development. Many class III antiarrhythmics as well as other (noncardiac) drugs, such as cisapride, astemizole, terfenadine, and sertindole, have been shown to block hERG, and are associated with significant adverse consequences for cardiac function.[441] In addition to blocking hERG, many class III antiarrhythmic drugs also inhibit K_V4 channels.[468] Diverse voltage-gated K^+ channels are also sensitive to block by many Na^+ and Ca^{2+} channel modulators, including local anesthetics (e.g., lidocaine and tetracaine), phenylalkylamines (e.g., verapamil), and diarylaminopropylamines (e.g., bepridil). The only K_V channels that appear to enjoy a distinguishing pharmacology are the K_V7 (KCNQ) subtypes. These pore-forming α subunits associate with minK-related auxiliary subunits, and are believed to underlie the K^+ current, I_{Ks} ($K_V7.1$) in cardiac cells and the so-called M current ($K_V7.2$, $K_V7.3$, and $K_V7.5$) in neurons. Channels containing these subunits have distinct sensitivities to TEA (e.g., $K_V7.2$ is more sensitive than $K_V7.3$), and can also be blocked by the cognition enhancers XE991 and linopirdine.[469,470] Interestingly, the pharmacology of K_V7 may be altered significantly in the presence of auxiliary subunits. Finally, neuronal K_V7-containing K^+ channels are activated by the antiepileptic drug retigabine[471] and the antiischemic drug BMS-204352.[472]

A number of pharmacological agents that either open or block BK_{Ca} channels have been described. The prototypical pharmacology for $K_{Ca}1.1$ channels is block by toxins (iberiotoxin, charybdotoxin, and noxiustoxin) and indole-diterpenes (penitrem-A and paxilline).[432,444] Unlike most K_V channels, $K_{Ca}1.1$ channels are quite sensitive to low concentrations of TEA. Like the K_V7 channels, openers of $K_{Ca}1.1$ have been identified, and these are of great therapeutic importance. The first opener of this channel that was identified was the glycosylated triterpene, dehydrosoyasaponin-1.[473] Subsequently, several small molecule activators (e.g., NS1608[474] and NS1619[475]) have been described, although these agents are not selective. It has been difficult to separate the effects of these agents on the voltage and Ca^{2+} dependence of the channel, underscoring the close working relationship of these mechanisms in channel activation.

Although HCN channels play a key role in pacemaking activity of neurons and heart cells, there are few reports of specific modulation by pharmacological agents. The most extensively studied blocker of HCN channels is ZD7288,[449,476] which specifically blocks native I_h and cloned HCN channels in a voltage-dependent manner. Other agents that block HCN channels are the bradycardic agents ivabradine, zatebradine, and alinidine, although these are less selective than ZD7288.[355,356,477]

2.21.4.7 Prototypical Therapeutics

Many pharmacological agents that target voltage-gated K^+ channels are of proven therapeutic utility or are in various stages of clinical development. The rationale for targeting specific voltage-gated K^+ channel subtypes for disease management arises from several forms of target validation. Data from genetic and molecular biology studies, including knockout or overexpression of various pore-forming and auxiliary subunits and accessory proteins, have revealed novel roles for K^+ channels in animal physiology. Moreover, pharmacological characterization of K^+ channels in normal and pathological cells has been essential in supporting the validation of K^+ channels in disease. Consequently, agents that modulate various voltage-gated or Ca^{2+}-activated K^+ channels are displaying potential in many phases of drug discovery and early clinical development.

As a class, most voltage-gated K^+ channel modulators are primarily focused on the management of cardiac and airway disorders,[478] although additional therapeutic indications are becoming apparent. Disruptions of normal cardiac rhythm are the leading cause of serious illness and premature death. The key to developing treatments for these conditions is an understanding of the molecular basis of electrical activity in the heart. The heart is a highly specialized organ that maintains a closely choreographed sequence of depolarization, repolarization, and muscle contraction.[441] There are multiple K^+ channels that contribute to membrane repolarization and hyperpolarization during the cardiac action potential, and, consequently, several potential drug targets have been suggested for the treatment of cardiac electrical or contractile abnormalities. In particular, $K_V1.5$ subunits are believed to underlie the ultra-rapid delayed rectifier K^+ current (I_{Kur}) in atrial myocytes.[479] Moreover, many class III antiarrhythmic drugs inhibit this channel, and so it has emerged as an attractive drug target for the management of atrial arrhythmias.[480] Consequently, several novel inhibitors of $K_V1.5$ are currently being investigated in the clinic as potential antiarrhythmics. Adding to the pharmacological complexity, it has been shown that the association of $K_V\beta1.3$ subunits with $K_V1.5$ can modulate drug block.[481] Other studies have established the importance of the transient outward K^+ current (I_{to}) on cardiac function.[482] This current is relatively large in atrial cells, and its blockade by antiarrhythmic drugs causes a significant increase in the duration of atrial depolarization. The proteins that underlie I_{to} have been the focus of many studies, and the channel identity is highly species-dependent. In most mammals, I_{to} appears to be a protein complex containing $K_V4.2$ and/or $K_V4.3$, KChIP,[483] and, perhaps, additional proteins.[373,484] However in the rabbit, I_{to} appears to result primarily from $K_V1.4$ expression.[485] Adding to the difficulty of validating I_{to} is the observation that phrixotoxin, which inhibits $K_V4.2$ and $K_V4.3$ channels, affects mainly AV node and ventricular function but does not appear to affect atrial repolarization significantly.[465] Other important delayed rectifier currents in the heart are I_{Kr} and I_{Ks}. I_{Kr} results from hERG association with MiRP1, whereas I_{Ks} results from $K_V7.1$ association with minK. However, again there are considerable species differences, and so there is uncertainty surrounding the identity of the actual subunits underlying these native channels. In addition to the complexities of composition and function of native voltage-gated K^+ channel in heart cells,[486,487] the additional problem of species differences underscores the challenges that are faced in the preclinical development K^+ channel-modulating antiarrhythmic drugs. Perhaps as a result of these difficulties, many drugs that target voltage-gated K^+ channels have had limited success in the clinical setting, and this has enhanced the desire to invoke novel strategies and technologies in the search for new class III antiarrhythmic drugs.[394,478]

T cells of the immune system express $K_V1.3$, and blockade of this channel may be a possible treatment in autoimmune diseases such as multiple sclerosis. Due to the critical role of $K_V1.3$-containing channels in controlling the membrane potential of T cells, blockade of $K_V1.3$ has been shown to be effective in suppressing Ca^{2+} influx and T cell activation. Moreover, expression of $K_V1.3$ is upregulated in T cells derived from rodent models of autoimmune diseases as well as from multiple sclerosis patients.[488] Several pharmacological agents have been identified that can block $K_V1.3$ and impact T cell activation.[489–491] Currently, agents that block $K_V1.3$ are in various stages of clinical development, and may represent a new treatment paradigm for immunosuppression. $K_V1.3$ has also gained interest as an interesting therapeutic target for the management of insulin resistance and type II diabetes.[492] Recent data indicate that $K_V1.3$ inhibition increases insulin sensitivity, suggesting that this channel may be an important regulator of glucose metabolism.[493] It has been proposed that $K_V1.3$ may increase translocation of the glucose transporter GLUT4 to the plasma membrane by decreasing inflammatory cytokines.

Voltage-gated K^+ channel modulators are also relevant in other areas, including oncology, neurology, the circulatory system, and the genitourinary system. K^+ channel genes are amplified and protein subunits are upregulated in many cancerous conditions. Consistent with a role for this upregulation, K^+ channel blockers have been shown to suppress the proliferation of uterine cancer cells[494] and hepatocarcinoma cells,[495] presumably through inhibition of Ca^{2+} influx and effects on Ca^{2+}-dependent gene expression. Voltage-gated K^+ channels are well established as critical for determining the resting membrane potential, action potential firing frequency, and neurotransmitter release in neurons. Both activators and inhibitors are in various preclinical and clinical phases of development as antiepileptics[496,497] and analgesics.[498] In airway,[499,500] vascular,[501–503] and bladder[439] smooth muscle cells, voltage-gated K^+ channels (K_V and K_{Ca}) are important for setting basal myogenic tone, suggesting opportunities to develop novel pharmacological therapies for treating the symptoms of asthma, hypertension, and incontinence.[395,432]

Due to the role of HCN channels in controlling pacemaker potentials in heart and neurons, there has been a growing interest to identify pharmacological agents. Although the focus of HCN channels has been in the heart,[355,356] two recent studies have reported that the HCN blocker ZD7288 can modulate pain responses in several animal models of pain.[450,504] As a result of perineural administration of ZD7288, mechanical allodynia was reduced, presumably through nerve block. Similarly, intraperitoneal administration of ZD7288 reduced paw withdrawal thresholds in a test of mechanical allodynia. While ZD7288 inhibited neuronal action potential firing, it also caused a marked bradycardia.[476] In order to overcome the inherent safety concerns, identification of HCN subtype-selective agents will be essential.

2.21.4.8 Genetic Diseases

Complex genetic factors contribute to the pathogenesis of many types of diseases. For those genetic mutations related to voltage-gated K^+ channelopathies, many appear to be due to gain or loss of function, and are primarily determined by surface expression abnormalities.[131,505–510] K^+ channelopathies are primarily implicated in diseases of cardiac, skeletal muscle, and neuronal systems. For cardiac diseases, over 200 mutations have been characterized in the seven so-called long QT (LQT) genes, and these account for 50–60% of the clinically manifested LQT syndromes.[507] In nervous system disorders, K^+ channel mutations have been associated with various forms of epilepsy as well as neuromuscular ataxia and deafness. However, in each disease, the relationship between genetic mutations and the clinical presentation in patients is extremely complex and only partially understood.

The best known yet least understood cardiac K^+ channel genetic mutations inhibit the current I_{Kr},[441] which is believed to provide the main repolarizing influence on the action potential. It is believed that the *KCNH2* gene encodes the α subunit (hERG; $K_V11.1$), and the *KCNE2* gene encodes the auxiliary subunit (MiRP1), of the channel that underlies this current. Loss of function mutations in both genes may cause a variety of inherited LQT syndromes. These mutations may delay action potential repolarization and induce the electrical abnormality torsade de pointes, which in turn may lead to cardiac fibrillation. In addition, a variety of drugs can prolong the cardiac action potential and induce LQT, apparently by inhibiting the function of I_{Kr} through blockade of hERG-containing K^+ channels. This has forced most drug development initiatives to include hERG pharmacology measurements in their integrated safety evaluations.[511] The misunderstood component of I_{Kr}-related LQT channelopathies arises from the fact that the cardiac action potential is the result of a delicate balance and close coordination between depolarizing and repolarizing ionic currents. Therefore, it is not surprising that not all mutations in KCNH2 and KCNE2 lead to LQT and not all drugs that inhibit I_{Kr} (or block hERG) cause QT prolongation.[512]

There are other forms of cardiac disease associated with K^+ channel genetic mutations. Some forms of LQT have been associated with loss of function of other subtypes of voltage-gated K^+ channels.[478,512] The implicated proteins include the pore-forming α subunit of $K_V7.1$ (KCNQ1) and the auxiliary subunit KCNE1 (minK/IsK). $K_V7.1$ associates with minK, to form the channel that underlies the slowly activating I_{Ks} in heart. I_{Ks} plays a role in action potential repolarization as well as rate-dependent shortening of the action potential. There are also genetic mutations in K^+ channels that lead to short QT syndromes.[513] These include gain of function mutations in hERG and $K_V7.1$. These syndromes are associated with sudden cardiac death, syncope, palpitations, and paroxysmal atrial fibrillation. In addition, loss of function mutations in HCN2 and HCN4 has been related to sick sinus syndrome, familial atrial fibrillation, and sudden cardiac death.[514,515]

Several nervous system disorders have been associated with loss of function of voltage-gated K^+ channels. Epilepsy has been defined as an imbalance in the electrical harmony of neuronal circuits, and, as such, it has much in common with cardiac arrhythmias. This imbalance has been associated with the abnormal functioning of several subtypes of voltage-gated K^+ channels. Of these, the loss of function mutation in the pore-forming $K_V7.2$ (KCNQ2) and $K_V7.3$ (KCNQ3) subunits have been linked to various forms of epilepsy.[508] Episodic ataxia type 1 is a rare disabling condition that begins in early childhood and is associated with uncoordinated movement of the head, arms, and legs. Attacks are provoked by sudden postural change or emotional stimulus, and have been associated with $K_V1.1$ (KCNA1) mutations. $K_V7.1$ and $K_V7.4$ subunits are expressed in the ear, and mutations in both have been associated with deafness.[516,517] Loss of function mutations in $K_V7.1$ is associated with aberrant endolymph secretion, whereas mutations in $K_V7.4$ are associated with abnormal electrical signaling.

2.21.5 Challenges and New Directions for Ion Channel Drug Discovery

In recent years, ion channel biology has developed into one of the most challenging, exciting, and rewarding fields in life science and drug discovery. As discussed above, Na^+, K^+, and Ca^{2+} channels represent key components of the cellular machinery that controls diverse physiological functions. However, despite years of intensive research there are many levels on which our understanding of ion channel biology remains rudimentary. Several hundred genes encoding ion channel subunits have been identified, yet there remains a large gap in our understanding not only of the molecular composition of individual ion channel subtypes but also of how the properties of individual subunits relate to the physiological functions of native ion channels in different highly specialized systems such as the brain, kidney, and heart. The challenge has been to correlate sometimes incomplete information about ion channel subunit expression and function in certain populations of cells with ion channel properties that have been studied in most detail in heterologous

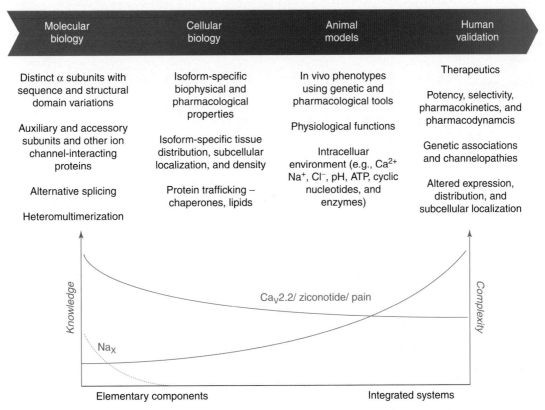

Figure 5 Schematic representation of the multidisciplinary challenges that are faced in ion channel basic research and in drug discovery and development. The inherent level of complexity tends to increase (the blue line) as one moves from describing the elementary molecular basis of ion channels toward elucidating their functional roles in complicated systems such as the human body. However, sufficient knowledge and understanding (the red lines) may permit the complexities to be addressed. For instance, the N-type ($Ca_V2.2$-containing) Ca^{2+} channel represents a target where a wealth of molecular and cellular data exist and significant knowledge has been obtained about its role in human physiology and disease. Furthermore, validation of this target has been achieved in clinical trials of the N-type Ca^{2+} channel-specific drug ziconotide, which has been approved for the treatment of chronic pain in humans. In contrast, the orphan Na^+ channel-like genes (Na_X) have been categorized solely on the basis of sequence similarity, since their functional properties have not been determined in native cells or in heterologous expression systems. Perhaps through development and implementation of new technologies, the frontiers of knowledge about Na_X genes might advance, and eventually a more complete understanding of their role in human physiology and disease may emerge.

expression systems. Furthermore, the ways in which ion channel expression and function are modified in response to changing physiological environments present many barriers to translating ion channel gene functions to whole-animal physiology and pathology. Pharmacological and genetic approaches have proven to be invaluable in the study of ion channel function in native cells and in promoting our understanding of the role of ion channels as potential therapeutic targets. Nevertheless, it is becoming increasingly evident that an integrated approach is required in order to correlate information and knowledge obtained from in vitro cell biology and in vivo physiology approaches (**Figure 5**).

Looking to the future of drug discovery for voltage-gated ion channel targets, the development of automated planar chip-based electrophysiology systems will have an impact on success. Despite their current limitations, these systems have enabled a paradigm shift in ion channel research, and expectations are high for them to promote a better understanding of structure–activity relationships within multiple series of novel compounds. Over the next few years, it will be exciting to observe these new technologies improve and advance to enable true high-throughput, parallel, electrophysiology-based screening of ion channels. Of significance, automated electrophysiology-based screening systems offer a high-content advantage over other screening systems because multiple measurements can be made on each cell. At the macroscopic current level, the impact of an ion channel modulator on important parameters, such as peak current amplitude, as well as the kinetics of activation and inactivation, may be examined. These electrophysiology-based systems are also able to provide an additional level of information aimed at the early elucidation of the mechanism of action of an increased number of compounds than previously possible. Careful design of voltage clamp protocols enables several characteristics of the interaction of a compound with voltage-gated ion channels to be described (e.g., tonic versus phasic inhibition of ionic currents reflecting use dependence, as well as variable channel block at different holding potentials, reflecting state dependence). These valuable mechanistic data will enhance the ion channel drug discovery process and will be useful for prioritizing series of compounds that may appear similarly behaved when considered solely from a potency and selectivity perspective.

References

1. Hille, B. *Ion Channels of Excitable Membranes*, 3rd ed; Sinauer Associates: Sunderland, MA, 2001.
2. Yu, F. H.; Catterall, W. A. *Sci. STKE* **2004**, *253*, re15. http://stke.sciencemag.org (accessed July 2006).
3. Catterall, W. A.; Goldin, A. L.; Waxman, S. G. *Pharmacol. Rev.* **2003**, *55*, 575–578.
4. Isom, L. L.; De Jongh, K. S.; Patton, D. E.; Reber, B. F.; Offord, J.; Charbonneau, H.; Walsh, K.; Goldin, A. L.; Catterall, W. A. *Science* **1992**, *256*, 839–842.
5. McClatchey, A. I.; Cannon, S. C.; Slaugenhaupt, S. A.; Gusella, J. F. *Hum. Mol. Genet.* **1993**, *2*, 745–749.
6. Isom, L. L.; Ragsdale, D. S.; De Jongh, K. S.; Westenbroek, R. E.; Reber, B. F.; Scheuer, T.; Catterall, W. A. *Cell* **1995**, *83*, 433–442.
7. Eubanks, J.; Srinivasan, J.; Dinulos, M. B.; Disteche, C. M.; Catterall, W. A. *Neuroreport* **1997**, *8*, 2775–2779.
8. Morgan, K.; Stevens, E. B.; Shah, B.; Cox, P. J.; Dixon, A. K.; Lee, K.; Pinnock, R. D.; Hughes, J.; Richardson, P. J.; Mizuguchi, K. et al. *Proc. Natl. Acad. Sci. USA* **2000**, *97*, 2308–2313.
9. Yu, F. H.; Westenbroek, R. E.; Silos-Santiago, I.; McCormick, K. A.; Lawson, D.; Ge, P.; Ferriera, H.; Lilly, J.; DiStefano, P. S.; Catterall, W. A. et al. *J. Neurosci.* **2003**, *23*, 7577–7585.
10. Qin, N.; D'Andrea, M. R.; Lubin, M. L.; Shafaee, N.; Codd, E. E.; Correa, A. M. *Eur. J. Biochem.* **2003**, *270*, 4762–4770.
11. Westenbroek, R. E.; Merrick, D. K.; Catterall, W. A. *Neuron* **1989**, *3*, 695–704.
12. Tzoumaka, E.; Tischler, A. C.; Sangameswaran, L.; Eglen, R. M.; Hunter, J. C.; Novakovic, S. D. *J. Neurosci. Res.* **2000**, *60*, 37–44.
13. Thimmapaya, R.; Neelands, T.; Niforatos, W.; Davis-Taber, R. A.; Choi, W.; Putman, C. B.; Kroeger, P. E.; Packer, J.; Gopalakrishnan, M.; Faltynek, C. R. et al. *Eur. J. Neurosci.* **2005**, *22*, 1–9.
14. Sangameswaran, L.; Delgado, S. G.; Fish, L. M.; Koch, B. D.; Jakeman, L. B.; Stewart, G. R.; Sze, P.; Hunter, J. C.; Eglen, R. M.; Herman, R. C. *J. Biol. Chem.* **1996**, *271*, 5953–5956.
15. Sangameswaran, L.; Fish, L. M.; Koch, B. D.; Rabert, D. K.; Delgado, S. G.; Ilnicka, M.; Jakeman, L. B.; Novakovic, S.; Wong, K.; Sze, P. et al. *J. Biol. Chem.* **1997**, *272*, 14805–14809.
16. Toledo-Aral, J. J.; Moss, B. L.; He, Z. J.; Koszowski, A. G.; Whisenand, T.; Levinson, S. R.; Wolf, J. J.; Silos-Santiago, I.; Halegoua, S.; Mandel, G. *Proc. Natl. Acad. Sci. USA* **1997**, *94*, 1527–1532.
17. Dib-Hajj, S. D.; Tyrrell, L.; Black, J. A.; Waxman, S. G. *Proc. Natl. Acad. Sci. USA* **1998**, *95*, 8963–8968.
18. Whitaker, W. R.; Faull, R. L.; Waldvogel, H. J.; Plumpton, C. J.; Emson, P. C.; Clare, J. J. *Brain Res. Mol. Brain Res.* **2001**, *88*, 37–53.
19. Boiko, T.; Van Wart, A.; Caldwell, J. H.; Levinson, S. R.; Trimmer, J. S.; Matthews, G. *J. Neurosci.* **2003**, *23*, 2306–2313.
20. Catterall, W. A. *J. Neurosci.* **1981**, *1*, 777–783.
21. Rasband, M. N.; Kagawa, T.; Park, E. W.; Ikenaka, K.; Trimmer, J. S. *J. Neurosci. Res.* **2003**, *73*, 465–470.
22. Novakovic, S. D.; Tzoumaka, E.; McGivern, J. G.; Haraguchi, M.; Sangameswaran, L.; Gogas, K. R.; Eglen, R. M.; Hunter, J. C. *J. Neurosci.* **1998**, *18*, 2174–2187.
23. Blackburn-Munro, G.; Fleetwood-Walker, S. M. *Neuroscience* **1999**, *90*, 153–164.
24. Shah, B. S.; Stevens, E. B.; Gonzalez, M. I.; Bramwell, S.; Pinnock, R. D.; Lee, K.; Dixon, A. K. *Eur. J. Neurosci.* **2000**, *12*, 3985–3990.
25. Hains, B. C.; Klein, J. P.; Saab, C. Y.; Craner, M. J.; Black, J. A.; Waxman, S. G. *J. Neurosci.* **2003**, *23*, 8881–8892.
26. Noebels, J. L.; Marcom, P. K.; Jalilian-Tehrani, M. H. *Nature* **1991**, *352*, 431–434.
27. Srinivasan, J.; Schachner, M.; Catterall, W. A. *Proc. Natl. Acad. Sci. USA* **1998**, *95*, 15753–15757.
28. Jenkins, S. M.; Bennett, V. *J. Cell Biol.* **2001**, *155*, 739–746.
29. Kazarinova-Noyes, K.; Malhotra, J. D.; McEwen, D. P.; Mattei, L. N.; Berglund, E. O.; Ranscht, B.; Levinson, S. R.; Schachner, M.; Shrager, P.; Isom, L. L. et al. *J. Neurosci.* **2001**, *21*, 7517–7525.

30. Shah, B. S.; Rush, A. M.; Liu, S.; Tyrrell, L.; Black, J. A.; Dib-Hajj, S. D.; Waxman, S. G. *J. Neurosci.* **2004**, *24*, 7387–7399.
31. Rougier, J. S.; Van Bemmelen, M. X.; Bruce, M. C.; Jespersen, T.; Gavillet, B.; Apotheloz, F.; Cordonier, S.; Staub, O.; Rotin, D.; Abriel, H. *Am. J. Physiol. Cell Physiol.* **2005**, *288*, C692–C701.
32. George, A. L., Jr.; Komisarof, J.; Kallen, R. G.; Barchi, R. L. *Ann. Neurol.* **1992**, *31*, 131–137.
33. Beam, K. G.; Caldwell, J. H.; Campbell, D. T. *Nature* **1985**, *313*, 588–590.
34. Lupa, M. T.; Krzemien, D. M.; Schaller, K. L.; Caldwell, J. H. *J. Physiol.* **1995**, *483* (Part 1), 109–118.
35. Yang, J. S.; Sladky, J. T.; Kallen, R. G.; Barchi, R. L. *Neuron* **1991**, *7*, 421–427.
36. Chahine, M.; Deschene, I.; Chen, L. Q.; Kallen, R. G. *Am. J. Physiol.* **1996**, *271*, H498–H506.
37. Pappone, P. A. *J. Physiol.* **1980**, *306*, 377–410.
38. Gellens, M. E.; George, A. L., Jr.; Chen, L. Q.; Chahine, M.; Horn, R.; Barchi, R. L.; Kallen, R. G. *Proc. Natl. Acad. Sci. USA* **1992**, *89*, 554–558.
39. Haufe, V.; Camacho, J. A.; Dumaine, R.; Gunther, B.; Bollensdorff, C.; von Banchet, G. S.; Benndorf, K.; Zimmer, T. *J. Physiol.* **2005**, *564*, 683–696.
40. Cohen, S. A. *Circulation* **1996**, *94*, 3083–3086.
41. Maier, S. K.; Westenbroek, R. E.; McCormick, K. A.; Curtis, R.; Scheuer, T.; Catterall, W. A. *Circulation* **2004**, *109*, 1421–1427.
42. Maier, S. K.; Westenbroek, R. E.; Yamanushi, T. T.; Dobrzynski, H.; Boyett, M. R.; Catterall, W. A.; Scheuer, T. *Proc. Natl. Acad. Sci. USA* **2003**, *100*, 3507–3512.
43. van Bemmelen, M. X.; Rougier, J. S.; Gavillet, B.; Apotheloz, F.; Daidie, D.; Tateyama, M.; Rivolta, I.; Thomas, M. A.; Kass, R. S.; Staub, O. et al. *Circ. Res.* **2004**, *95*, 284–291.
44. Herfst, L. J.; Rook, M. B.; Jongsma, H. J. *J. Mol. Cell. Cardiol.* **2004**, *36*, 185–193.
45. Huang, B.; El-Sherif, T.; Gidh-Jain, M.; Qin, D.; El-Sherif, N. *J. Cardiovasc. Electrophysiol.* **2001**, *12*, 218–225.
46. Borlak, J.; Thum, T. *FASEB J.* **2003**, *17*, 1592–1608.
47. Sutkowski, E. M.; Catterall, W. A. *J. Biol. Chem.* **1990**, *265*, 12393–12399.
48. Stevens, E. B.; Cox, P. J.; Shah, B. S.; Dixon, A. K.; Richardson, P. J.; Pinnock, R. D.; Lee, K. *Pflügers Arch.* **2001**, *441*, 481–488.
49. Long, S. B.; Campbell, E. B.; Mackinnon, R. *Science* **2005**, *309*, 897–903.
50. Long, S. B.; Campbell, E. B.; Mackinnon, R. *Science* **2005**, *309*, 903–908.
51. Vijayaragavan, K.; Boutjdir, M.; Chahine, M. *J. Neurophysiol.* **2004**, *91*, 1556–1569.
52. Garrido, J. J.; Giraud, P.; Carlier, E.; Fernandes, F.; Moussif, A.; Fache, M. P.; Debanne, D.; Dargent, B. *Science* **2003**, *300*, 2091–2094.
53. West, J. W.; Patton, D. E.; Scheuer, T.; Wang, Y.; Goldin, A. L.; Catterall, W. A. *Proc. Natl. Acad. Sci. USA* **1992**, *89*, 10910–10914.
54. Yamagishi, T.; Li, R. A.; Hsu, K.; Marban, E.; Tomaselli, G. F. *J. Gen. Physiol.* **2001**, *118*, 171–182.
55. Fleig, A.; Fitch, J. M.; Goldin, A. L.; Rayner, M. D.; Starkus, J. G.; Ruben, P. C. *Pflügers Arch.* **1994**, *427*, 406–413.
56. Kontis, K. J.; Rounaghi, A.; Goldin, A. L. *J. Gen. Physiol.* **1997**, *110*, 391–401.
57. Terlau, H.; Heinemann, S. H.; Stuhmer, W.; Pusch, M.; Conti, F.; Imoto, K.; Numa, S. *FEBS Lett.* **1991**, *293*, 93–96.
58. Heinemann, S. H.; Terlau, H.; Stuhmer, W.; Imoto, K.; Numa, S. *Nature* **1992**, *356*, 441–443.
59. Schlief, T.; Schonherr, R.; Imoto, K.; Heinemann, S. H. *Eur. Biophys. J.* **1996**, *25*, 75–91.
60. Raman, I. M.; Bean, B. P. *J. Neurosci.* **1997**, *17*, 4517–4526.
61. Lei, M.; Goddard, C.; Liu, J.; Leoni, A. L.; Royer, A.; Fung, S. S.; Xiao, G.; Ma, A.; Zhang, H.; Charpentier, F. et al. *J. Physiol.* **2005**, *567*, 387–400.
62. Maier, S. K.; Westenbroek, R. E.; Schenkman, K. A.; Feigl, E. O.; Scheuer, T.; Catterall, W. A. *Proc. Natl. Acad. Sci. USA* **2002**, *99*, 4073–4078.
63. Ju, Y.; Gage, P. W.; Saint, D. A. *Pflügers Arch.* **1996**, *431*, 868–875.
64. Sakmann, B. F.; Spindler, A. J.; Bryant, S. M.; Linz, K. W.; Noble, D. *Circ. Res.* **2000**, *87*, 910–914.
65. Qu, Y.; Curtis, R.; Lawson, D.; Gilbride, K.; Ge, P.; DiStefano, P. S.; Silos-Santiago, I.; Catterall, W. A.; Scheuer, T. *Mol. Cell. Neurosci.* **2001**, *18*, 570–580.
66. Ratcliffe, C. F.; Westenbroek, R. E.; Curtis, R.; Catterall, W. A. *J. Cell Biol.* **2001**, *154*, 427–434.
67. Liu, C. J.; Dib-Hajj, S. D.; Black, J. A.; Greenwood, J.; Lian, Z.; Waxman, S. G. *J. Biol. Chem.* **2001**, *276*, 46553–46561.
68. Xiao, Z. C.; Ragsdale, D. S.; Malhotra, J. D.; Mattei, L. N.; Braun, P. E.; Schachner, M.; Isom, L. L. *J. Biol. Chem.* **1999**, *274*, 26511–26517.
69. Kameyama, M.; Kakei, M.; Sato, R.; Shibasaki, T.; Matsuda, H.; Irisawa, H. *Nature* **1984**, *309*, 354–356.
70. Dryer, S. E.; Fujii, J. T.; Martin, A. R. *J. Physiol.* **1989**, *410*, 283–296.
71. Bischoff, U.; Vogel, W.; Safronov, B. V. *J. Physiol.* **1998**, *510* (Part 3), 743–754.
72. Dryer, S. E. *Trends Neurosci.* **1994**, *17*, 155–160.
73. Veldkamp, M. W.; Vereecke, J.; Carmeliet, E. *Cardiovasc. Res.* **1994**, *28*, 1036–1041.
74. Bhattacharjee, A.; Joiner, W. J.; Wu, M.; Yang, Y.; Sigworth, F. J.; Kaczmarek, L. K. *J. Neurosci.* **2003**, *23*, 11681–11691.
75. Yuan, A.; Santi, C. M.; Wei, A.; Wang, Z. W.; Pollak, K.; Nonet, M.; Kaczmarek, L.; Crowder, C. M.; Salkoff, L. *Neuron* **2003**, *37*, 765–773.
76. Numann, R.; Catterall, W. A.; Scheuer, T. *Science* **1991**, *254*, 115–118.
77. Cantrell, A. R.; Ma, J. Y.; Scheuer, T.; Catterall, W. A. *Neuron* **1996**, *16*, 1019–1026.
78. Smith, R. D.; Goldin, A. L. *J. Neurosci.* **1996**, *16*, 1965–1974.
79. Cantrell, A. R.; Tibbs, V. C.; Yu, F. H.; Murphy, B. J.; Sharp, E. M.; Qu, Y.; Catterall, W. A.; Scheuer, T. *Mol. Cell. Neurosci.* **2002**, *21*, 63–80.
80. Cantrell, A. R.; Smith, R. D.; Goldin, A. L.; Scheuer, T.; Catterall, W. A. *J. Neurosci.* **1997**, *17*, 7330–7338.
81. Schiffmann, S. N.; Desdouits, F.; Menu, R.; Greengard, P.; Vincent, J. D.; Vanderhaeghen, J. J.; Girault, J. A. *Eur. J. Neurosci.* **1998**, *10*, 1312–1320.
82. Grieco, T. M.; Afshari, F. S.; Raman, I. M. *J. Neurosci.* **2002**, *22*, 3100–3107.
83. Gold, M. S.; Zhang, L.; Wrigley, D. L.; Traub, R. J. *J. Neurophysiol.* **2002**, *88*, 1512–1522.
84. Fitzgerald, E. M.; Okuse, K.; Wood, J. N.; Dolphin, A. C.; Moss, S. J. *J. Physiol.* **1999**, *516* (Part 2), 433–446.
85. Gold, M. S.; Levine, J. D.; Correa, A. M. *J. Neurosci.* **1998**, *18*, 10345–10355.
86. Matsuda, J. J.; Lee, H.; Shibata, E. F. *Circ. Res.* **1992**, *70*, 199–207.
87. Murphy, B. J.; Rogers, J.; Perdichizzi, A. P.; Colvin, A. A.; Catterall, W. A. *J. Biol. Chem.* **1996**, *271*, 28837–28843.
88. Frohnwieser, B.; Chen, L. Q.; Schreibmayer, W.; Kallen, R. G. *J. Physiol.* **1997**, *498* (Part 2), 309–318.
89. Qu, Y.; Rogers, J. C.; Tanada, T. N.; Catterall, W. A.; Scheuer, T. *J. Gen. Physiol.* **1996**, *108*, 375–379.
90. Renganathan, M.; Dib-Hajj, S.; Waxman, S. G. *Brain Res. Mol. Brain Res.* **2002**, *106*, 70–82.
91. Chen, L.-Q.; Chahine, M.; Kallen, R. G.; Barchi, R. L.; Horn, R. *FEBS Lett.* **1992**, *309*, 253–257.

92. Leffler, A.; Herzog, R. I.; Dib-Hajj, S. D.; Waxman, S. G.; Cummins, T. R. *Pflügers Arch.* **2005**.
93. Barbier, J.; Lamtanh, H.; Le Gall, F.; Favreau, P.; Benoit, E.; Chen, H.; Gilles, N.; Ilan, N.; Heinemann, S. H.; Gordon, D. et al. *J. Biol. Chem.* **2004**, *279*, 4680–4685.
94. Saab, C. Y.; Cummins, T. R.; Dib-Hajj, S. D.; Waxman, S. G. *Neurosci. Lett.* **2002**, *331*, 79–82.
95. Kondratiev, A.; Tomaselli, G. F. *Mol. Pharmacol.* **2003**, *64*, 741–752.
96. Nau, C.; Wang, S. Y.; Wang, G. K. *Mol. Pharmacol.* **2003**, *63*, 1398–1406.
97. Hille, B. *J. Gen. Physiol.* **1977**, *69*, 497–515.
98. Starmer, C. F.; Grant, A. O.; Strauss, H. C. *Biophys. J.* **1984**, *46*, 15–27.
99. Lenkowski, P. W.; Shah, B. S.; Dinn, A. E.; Lee, K.; Patel, M. K. *Eur. J. Pharmacol.* **2003**, *467*, 23–30.
100. Abram, S. E. *Clin. J. Pain* **2000**, *16*, S56–S61.
101. Tian, L. M.; Otoom, S.; Alkadhi, K. A. *Brain Res.* **1995**, *680*, 164–172.
102. Otoom, S.; Tian, L. M.; Alkadhi, K. A. *Brain Res.* **1998**, *789*, 150–156.
103. Whitaker, W. R.; Faull, R. L.; Dragunow, M.; Mee, E. W.; Emson, P. C.; Clare, J. J. *Neuroscience* **2001**, *106*, 275–285.
104. Vreugdenhil, M.; Hoogland, G.; van Veelen, C. W.; Wadman, W. J. *Eur. J. Neurosci.* **2004**, *19*, 2769–2778.
105. Sashihara, S.; Yanagihara, N.; Kobayashi, H.; Izumi, F.; Tsuji, S.; Murai, Y.; Mita, T. *Neuroscience* **1992**, *48*, 285–291.
106. Klein, J. P.; Khera, D. S.; Nersesyan, H.; Kimchi, E. Y.; Waxman, S. G.; Blumenfeld, H. *Brain Res.* **2004**, *1000*, 102–109.
107. Ellerkmann, R. K.; Remy, S.; Chen, J.; Sochivko, D.; Elger, C. E.; Urban, B. W.; Becker, A.; Beck, H. *Neuroscience* **2003**, *119*, 323–333.
108. Chen, C.; Bharucha, V.; Chen, Y.; Westenbroek, R. E.; Brown, A.; Malhotra, J. D.; Jones, D.; Avery, C.; Gillespie, P. J., III; Kazen-Gillespie, K. A. et al. *Proc. Natl. Acad. Sci. USA* **2002**, *99*, 17072–17077.
109. Segal, M. M.; Douglas, A. F. *J. Neurophysiol.* **1997**, *77*, 3021–3034.
110. Coderre, T. J.; Katz, J. *Behav. Brain Sci.* **1997**, *20*, 404–419 (discussion 435–513).
111. Hains, B. C.; Saab, C. Y.; Klein, J. P.; Craner, M. J.; Waxman, S. G. *J. Neurosci.* **2004**, *24*, 4832–4839.
112. Beyak, M. J.; Ramji, N.; Krol, K. M.; Kawaja, M. D.; Vanner, S. J. *Am. J. Physiol. Gastrointest. Liver Physiol.* **2004**, *287*, G845–G855.
113. Coggeshall, R. E.; Tate, S.; Carlton, S. M. *Neurosci. Lett.* **2004**, *355*, 45–48.
114. Rush, A. M.; Waxman, S. G. *Brain Res.* **2004**, *1023*, 264–271.
115. Coward, K.; Plumpton, C.; Facer, P.; Birch, R.; Carlstedt, T.; Tate, S.; Bountra, C.; Anand, P. *Pain* **2000**, *85*, 41–50.
116. Shembalkar, P. K.; Till, S.; Boettger, M. K.; Terenghi, G.; Tate, S.; Bountra, C.; Anand, P. *Eur. J. Pain* **2001**, *5*, 319–323.
117. Craner, M. J.; Lo, A. C.; Black, J. A.; Waxman, S. G. *Brain* **2003**, *126*, 1552–1561.
118. Takahashi, N.; Kikuchi, S.; Dai, Y.; Kobayashi, K.; Fukuoka, T.; Noguchi, K. *Neuroscience* **2003**, *121*, 441–450.
119. Black, J. A.; Liu, S.; Tanaka, M.; Cummins, T. R.; Waxman, S. G. *Pain* **2004**, *108*, 237–247.
120. Casula, M. A.; Facer, P.; Powell, A. J.; Kinghorn, I. J.; Plumpton, C.; Tate, S. N.; Bountra, C.; Birch, R.; Anand, P. *Neuroreport* **2004**, *15*, 1629–1632.
121. Hong, S.; Morrow, T. J.; Paulson, P. E.; Isom, L. L.; Wiley, J. W. *J. Biol. Chem.* **2004**, *279*, 29341–29350.
122. Renton, T.; Yiangou, Y.; Plumpton, C.; Tate, S.; Bountra, C.; Anand, P. *BMC Oral Health* **2005**, *5*, 5.
123. Lai, J.; Gold, M. S.; Kim, C. S.; Bian, D.; Ossipov, M. H.; Hunter, J. C.; Porreca, F. *Pain* **2002**, *95*, 143–152.
124. Laird, J. M.; Souslova, V.; Wood, J. N.; Cervero, F. *J. Neurosci.* **2002**, *22*, 8352–8356.
125. Roza, C.; Laird, J. M.; Souslova, V.; Wood, J. N.; Cervero, F. *J. Physiol.* **2003**, *550*, 921–926.
126. Nassar, M. A.; Stirling, L. C.; Forlani, G.; Baker, M. D.; Matthews, E. A.; Dickenson, A. H.; Wood, J. N. *Proc. Natl. Acad. Sci. USA* **2004**, *101*, 12706–12711.
127. Priest, B. T.; Murphy, B. A.; Lindia, J. A.; Diaz, C.; Abbadie, C.; Ritter, A. M.; Liberator, P.; Iyer, L. M.; Kash, S. F.; Kohler, M. G. et al. *Proc. Natl. Acad. Sci. USA* **2005**, *102*, 9382–9387.
128. Yeomans, D. C.; Levinson, S. R.; Peters, M. C.; Koszowski, A. G.; Tzabazis, A. Z.; Gilly, W. F.; Wilson, S. P. *Hum. Gene Ther.* **2005**, *16*, 271–277.
129. Yiangou, Y.; Birch, R.; Sangameswaran, L.; Eglen, R.; Anand, P. *FEBS Lett.* **2000**, *467*, 249–252.
130. McCleane, G. *CNS Drugs* **2003**, *17*, 1031–1043.
131. Celesia, G. G. *Clin. Neurophysiol.* **2001**, *112*, 2–18.
132. Cossette, P.; Loukas, A.; Lafreniere, R. G.; Rochefort, D.; Harvey-Girard, E.; Ragsdale, D. S.; Dunn, R. J.; Rouleau, G. A. *Epilepsy Res.* **2003**, *53*, 107–117.
133. Rhodes, T. H.; Lossin, C.; Vanoye, C. G.; Wang, D. W.; George, A. L., Jr. *Proc. Natl. Acad. Sci. USA* **2004**, *101*, 11147–11152.
134. Spampanato, J.; Kearney, J. A.; de Haan, G.; McEwen, D. P.; Escayg, A.; Aradi, I.; MacDonald, B. T.; Levin, S. I.; Soltesz, I.; Benna, P. et al. *J. Neurosci.* **2004**, *24*, 10022–10034.
135. Meadows, L. S.; Malhotra, J.; Loukas, A.; Thyagarajan, V.; Kazen-Gillespie, K. A.; Koopman, M. C.; Kriegler, S.; Isom, L. L.; Ragsdale, D. S. *J. Neurosci.* **2002**, *22*, 10699–10709.
136. Lucas, P. T.; Meadows, L. S.; Nicholls, J.; Ragsdale, D. S. *Epilepsy Res.* **2005**, *64*, 77–84.
137. van Genderen, P. J.; Michiels, J. J.; Drenth, J. P. *Am. J. Med. Genet.* **1993**, *45*, 530–532.
138. Waxman, S. G.; Dib-Hajj, S. D. *Ann. Neurol.* **2005**, *57*, 785–788.
139. Michiels, J. J.; te Morsche, R. H. M.; Jansen, J. B. M. J.; Drenth, J. P. H. *Arch. Neurol.* **2005**, *62*, 1587–1590.
140. Cummins, T. R.; Dib-Hajj, S. D.; Waxman, S. G. *J. Neurosci.* **2004**, *24*, 8232–8236.
141. Dib-Hajj, S. D.; Rush, A. M.; Cummins, T. R.; Hisama, F. M.; Novella, S.; Tyrrell, L.; Marshall, L.; Waxman, S. G. *Brain* **2005**, *128*, 1847–1854.
142. Itoh, H.; Shimizu, M.; Mabuchi, H.; Imoto, K. *J. Cardiovasc. Electrophysiol.* **2005**, *16*, 378–383.
143. Baroudi, G.; Acharfi, S.; Larouche, C.; Chahine, M. *Circ. Res.* **2002**, *90*, E11–E16.
144. Bendahhou, S.; Cummins, T. R.; Kwiecinski, H.; Waxman, S. G.; Ptacek, L. J. *J. Physiol.* **1999**, *518* (Part 2), 337–344.
145. Sasaki, R.; Takano, H.; Kamakura, K.; Kaida, K.; Hirata, A.; Saito, M.; Tanaka, H.; Kuzuhara, S.; Tsuji, S. *Arch. Neurol.* **1999**, *56*, 692–696.
146. Sillen, A.; Wadelius, C.; Sundvall, M.; Ahlsten, G.; Gustavson, K. H. *Genet. Couns.* **1996**, *7*, 267–275.
147. Streib, E. W. *Muscle Nerve* **1987**, *10*, 155–162.
148. Fredericson, M.; Kim, B. J.; Date, E. S. *Foot Ankle Int.* **2004**, *25*, 510–512.
149. Arikkath, J.; Campbell, K. P. *Curr. Opin. Neurobiol.* **2003**, *13*, 298–307.
150. Newcomb, R.; Szoke, B.; Palma, A.; Wang, G.; Chen, X.; Hopkins, W.; Cong, R.; Miller, J.; Urge, L.; Tarczy-Hornoch, K. et al. *Biochemistry* **1998**, *37*, 15353–15362.
151. Randall, A.; Tsien, R. W. *J. Neurosci.* **1995**, *15*, 2995–3012.

152. Birnbaumer, L.; Campbell, K. P.; Catterall, W. A.; Harpold, M. M.; Hofmann, F.; Horne, W. A.; Mori, Y.; Schwartz, A.; Snutch, T. P. et al. *Neuron* **1994**, *13*, 505–506.
153. Ertel, E. A.; Campbell, K. P.; Harpold, M. M.; Hofmann, F.; Mori, Y.; Perez-Reyes, E.; Schwartz, A.; Snutch, T. P.; Tanabe, T.; Birnbaumer, L. et al. *Neuron* **2000**, *25*, 533–535.
154. Catterall, W. A.; Striessnig, J.; Snutch, T. P.; Perez-Reyes, E. *Pharmacol. Rev.* **2003**, *55*, 579–581.
155. Hansen, J. P.; Chen, R. S.; Larsen, J. K.; Chu, P. J.; Janes, D. M.; Weis, K. E.; Best, P. M. *J. Mol. Cell. Cardiol.* **2004**, *37*, 1147–1158.
156. Lacinova, L.; Klugbauer, N. *Arch. Biochem. Biophys.* **2004**, *425*, 207–213.
157. Leuranguer, V.; Bourinet, E.; Lory, P.; Nargeot, J. *Neuropharmacology* **1998**, *37*, 701–708.
158. Jones, L. P.; Wei, S. K.; Yue, D. T. *J. Gen. Physiol.* **1998**, *112*, 125–143.
159. Takahashi, S. X.; Mittman, S.; Colecraft, H. M. *Biophys. J.* **2003**, *84*, 3007–3021.
160. Flucher, B. E.; Obermair, G. J.; Tuluc, P.; Schredelseker, J.; Kern, G.; Grabner, M. *J. Muscle Res. Cell Motil.* **2005**, 1–6.
161. Protasi, F. *Front. Biosci.* **2002**, *7*, d650–d658.
162. Nargeot, J.; Lory, P.; Richard, S. *Eur. Heart J.* **1997**, *18*, 15–26.
163. Chu, P. J.; Best, P. M. *J. Mol. Cell. Cardiol.* **2003**, *35*, 207–215.
164. Hullin, R.; Khan, I. F.; Wirtz, S.; Mohacsi, P.; Varadi, M.; Schwartz, A.; Herzig, S. *J. Biol. Chem.* **2003**, *278*, 21623–21630.
165. Saada, N.; Dai, B.; Echetebu, C.; Sarna, S. K.; Palade, P. *Biochem. Biophys. Res. Commun.* **2003**, *302*, 23–28.
166. Platzer, J.; Engel, J.; Schrott-Fischer, A.; Stephan, K.; Bova, S.; Chen, H.; Zheng, H.; Striessnig, J. *Cell* **2000**, *102*, 89–97.
167. McRory, J. E.; Hamid, J.; Doering, C. J.; Garcia, E.; Parker, R.; Hamming, K.; Chen, L.; Hildebrand, M.; Beedle, A. M.; Feldcamp, L. et al. *J. Neurosci.* **2004**, *24*, 1707–1718.
168. Berntson, A.; Taylor, W. R.; Morgans, C. W. *J. Neurosci. Res.* **2003**, *71*, 146–151.
169. Ball, S. L.; Powers, P. A.; Shin, H. S.; Morgans, C. W.; Peachey, N. S.; Gregg, R. G. *Invest. Ophthalmol. Vis. Sci.* **2002**, *43*, 1595–1603.
170. Niwa, N.; Yasui, K.; Opthof, T.; Takemura, H.; Shimizu, A.; Horiba, M.; Lee, J. K.; Honjo, H.; Kamiya, K.; Kodama, I. *Am. J. Physiol. Heart Circ. Physiol.* **2004**, *286*, H2257–H2263.
171. Petkov, G. V.; Fusi, F.; Saponara, S.; Gagov, H. S.; Sgaragli, G. P.; Boev, K. K. *Acta Physiol. Scand.* **2001**, *173*, 257–265.
172. Chen, C. C.; Lamping, K. G.; Nuno, D. W.; Barresi, R.; Prouty, S. J.; Lavoie, J. L.; Cribbs, L. L.; England, S. K.; Sigmund, C. D.; Weiss, R. M. et al. *Science* **2003**, *302*, 1416–1418.
173. Hell, J. W.; Westenbroek, R. E.; Warner, C.; Ahlijanian, M. K.; Prystay, W.; Gilbert, M. M.; Snutch, T. P.; Catterall, W. A. *J. Cell Biol.* **1993**, *123*, 949–962.
174. Westenbroek, R. E.; Hoskins, L.; Catterall, W. A. *J. Neurosci.* **1998**, *18*, 6319–6330.
175. Hardingham, G. E.; Chawla, S.; Johnson, C. M.; Bading, H. *Nature* **1997**, *385*, 260–265.
176. Bowden, S. E.; Fletcher, S.; Loane, D. J.; Marrion, N. V. *J. Neurosci.* **2001**, *21*, RC175.
177. Hallworth, N. E.; Wilson, C. J.; Bevan, M. D. *J. Neurosci.* **2003**, *23*, 7525–7542.
178. Rittenhouse, A. R.; Zigmond, R. E. *J. Neurobiol.* **1999**, *40*, 137–148.
179. Westenbroek, R. E.; Sakurai, T.; Elliott, E. M.; Hell, J. W.; Starr, T. V.; Snutch, T. P.; Catterall, W. A. *J. Neurosci.* **1995**, *15*, 6403–6418.
180. Westenbroek, R. E.; Hell, J. W.; Warner, C.; Dubel, S. J.; Snutch, T. P.; Catterall, W. A. *Neuron* **1992**, *9*, 1099–1115.
181. Zamponi, G. W. *J. Pharmacol. Sci.* **2003**, *92*, 79–83.
182. Nakanishi, S.; Fujii, A.; Kimura, T.; Sakakibara, S.; Mikoshiba, K. *J. Neurosci. Res.* **1995**, *41*, 532–539.
183. Smith, S. M.; Piedras-Rentera, E. S.; Namkung, Y.; Shin, H. S.; Tsien, R. W. *Ann. NY Acad. Sci.* **1999**, *868*, 175–198.
184. Hanson, J. E.; Smith, Y. *J. Comp. Neurol.* **2002**, *442*, 89–98.
185. Gasparini, S.; Kasyanov, A. M.; Pietrobon, D.; Voronin, L. L.; Cherubini, E. *J. Neurosci.* **2001**, *21*, 8715–8721.
186. Lambert, R. C.; McKenna, F.; Maulet, Y.; Talley, E. M.; Bayliss, D. A.; Cribbs, L. L.; Lee, J. H.; Perez-Reyes, E.; Feltz, A. *J. Neurosci.* **1998**, *18*, 8605–8613.
187. Talley, E. M.; Cribbs, L. L.; Lee, J. H.; Daud, A.; Perez-Reyes, E.; Bayliss, D. A. *J. Neurosci.* **1999**, *19*, 1895–1911.
188. Yunker, A. M.; Sharp, A. H.; Sundarraj, S.; Ranganathan, V.; Copeland, T. D.; McEnery, M. W. *Neuroscience* **2003**, *117*, 321–335.
189. Pragnell, M.; De Waard, M.; Mori, Y.; Tanabe, T.; Snutch, T. P.; Campbell, K. P. *Nature* **1994**, *368*, 67–70.
190. De Waard, M.; Witcher, D. R.; Pragnell, M.; Liu, H.; Campbell, K. P. *J. Biol. Chem.* **1995**, *270*, 12056–12064.
191. De Waard, M.; Scott, V. E.; Pragnell, M.; Campbell, K. P. *FEBS Lett.* **1996**, *380*, 272–276.
192. Leroy, J.; Richards, M. S.; Butcher, A. J.; Nieto-Rostro, M.; Pratt, W. S.; Davies, A.; Dolphin, A. C. *J. Neurosci.* **2005**, *25*, 6984–6996.
193. Li, J.; Stevens, L.; Klugbauer, N.; Wray, D. *J. Biol. Chem.* **2004**, *279*, 26858–26867.
194. Sather, W. A.; McCleskey, E. W. *Annu. Rev. Physiol.* **2003**, *65*, 133–159.
195. Perez-Reyes, E.; Cribbs, L. L.; Daud, A.; Lacerda, A. E.; Barclay, J.; Williamson, M. P.; Fox, M.; Rees, M.; Lee, J. H. *Nature* **1998**, *391*, 896–900.
196. Cribbs, L. L.; Lee, J. H.; Yang, J.; Satin, J.; Zhang, Y.; Daud, A.; Barclay, J.; Williamson, M. P.; Fox, M.; Rees, M. et al. *Circ. Res.* **1998**, *83*, 103–109.
197. Lee, J. H.; Daud, A. N.; Cribbs, L. L.; Lacerda, A. E.; Pereverzev, A.; Klockner, U.; Schneider, T.; Perez-Reyes, E. *J. Neurosci.* **1999**, *19*, 1912–1921.
198. Stotz, S. C.; Hamid, J.; Spaetgens, R. L.; Jarvis, S. E.; Zamponi, G. W. *J. Biol. Chem.* **2000**, *275*, 24575–24582.
199. de Leon, M.; Wang, Y.; Jones, L.; Perez-Reyes, E.; Wei, X.; Soong, T. W.; Snutch, T. P.; Yue, D. T. *Science* **1995**, *270*, 1502–1506.
200. Felix, R. *Receptors Channels* **1999**, *6*, 351–362.
201. Wiser, O.; Trus, M.; Tobi, D.; Halevi, S.; Giladi, E.; Atlas, D. *FEBS Lett.* **1996**, *379*, 15–20.
202. Wang, M.; Offord, J.; Oxender, D. L.; Su, T. Z. *Biochem. J.* **1999**, *342* (Part 2), 313–320.
203. Perez-Reyes, E.; Castellano, A.; Kim, H. S.; Bertrand, P.; Baggstrom, E.; Lacerda, A. E.; Wei, X. Y.; Birnbaumer, L. *J. Biol. Chem.* **1992**, *267*, 1792–1797.
204. Chen, Y. H.; Li, M. H.; Zhang, Y.; He, L. L.; Yamada, Y.; Fitzmaurice, A.; Shen, Y.; Zhang, H.; Tong, L.; Yang, J. *Nature* **2004**, *429*, 675–680.
205. Takahashi, Y.; Hashimoto, K.; Tsuji, S. *J. Pain* **2004**, *5*, 192–194.
206. Bichet, D.; Cornet, V.; Geib, S.; Carlier, E.; Volsen, S.; Hoshi, T.; Mori, Y.; De Waard, M. *Neuron* **2000**, *25*, 177–190.
207. Van Petegem, F.; Clark, K. A.; Chatelain, F. C.; Minor, D. L., Jr. *Nature* **2004**, *429*, 671–675.
208. Maltez, J. M.; Nunziato, D. A.; Kim, J.; Pitt, G. S. *Nat. Struct. Mol. Biol.* **2005**, *12*, 372–377.
209. Stephens, G. J.; Page, K. M.; Bogdanov, Y.; Dolphin, A. C. *J. Physiol.* **2000**, *525* (Part 2), 377–390.
210. Qin, N.; Platano, D.; Olcese, R.; Costantin, J. L.; Stefani, E.; Birnbaumer, L. *Proc. Natl. Acad. Sci. USA* **1998**, *95*, 4690–4695.
211. Kang, M. G.; Campbell, K. P. *J. Biol. Chem.* **2003**, *278*, 21315–21318.

212. Berridge, M. *J. Neuron* **1998**, *21*, 13–26.
213. McCobb, D. P.; Beam, K. G. *Neuron* **1991**, *7*, 119–127.
214. Monteil, A.; Chemin, J.; Bourinet, E.; Mennessier, G.; Lory, P.; Nargeot, J. *J. Biol. Chem.* **2000**, *275*, 6090–6100.
215. Imaizumi, Y.; Muraki, K.; Takeda, M.; Watanabe, M. *Am. J. Physiol.* **1989**, *256*, C880–C885.
216. Williams, S. R.; Toth, T. I.; Turner, J. P.; Hughes, S. W.; Crunelli, V. *J. Physiol.* **1997**, *505*, 689–705.
217. Satoh, H. *Gen. Pharmacol.* **1995**, *26*, 581–587.
218. VanBavel, E.; Sorop, O.; Andreasen, D.; Pfaffendorf, M.; Jensen, B. L. *Am. J. Physiol. Heart Circ. Physiol.* **2002**, *283*, H2239–H2243.
219. Zou, H.; Ratz, P. H.; Hill, M. A. *Am. J. Physiol.* **1995**, *269*, H1590–H1596.
220. Schmitt, R.; Clozel, J. P.; Iberg, N.; Buhler, F. R. *Arterioscler. Thromb. Vasc. Biol.* **1995**, *15*, 1161–1165.
221. Catterall, W. A. *Cell Calcium* **1998**, *24*, 307–323.
222. D'Angelo, E.; De Filippi, G.; Rossi, P.; Taglietti, V. *J. Neurophysiol.* **1997**, *78*, 1631–1642.
223. Maggi, C. A.; Tramontana, M.; Cecconi, R.; Santicioli, P. *Neurosci. Lett.* **1990**, *114*, 203–206.
224. Uchitel, O. D.; Protti, D. A.; Sanchez, V.; Cherksey, B. D.; Sugimori, M.; Llinas, R. *Proc. Natl. Acad. Sci. USA* **1992**, *89*, 3330–3333.
225. Protti, D. A.; Uchitel, O. D. *Neuroreport* **1993**, *5*, 333–336.
226. Luebke, J. I.; Dunlap, K.; Turner, T. *J. Neuron* **1993**, *11*, 895–902.
227. Takahashi, T.; Momiyama, A. *Nature* **1993**, *366*, 156–158.
228. Wright, C. E.; Angus, J. A. *Br. J. Pharmacol.* **1996**, *119*, 49–56.
229. Currie, K. P.; Fox, A. P. *J. Physiol.* **2002**, *539*, 419–431.
230. Pagani, R.; Song, M.; McEnery, M.; Qin, N.; Tsien, R. W.; Toro, L.; Stefani, E.; Uchitel, O. D. *Neuroscience* **2004**, *123*, 75–85.
231. Takahashi, E.; Nagasu, T. *Exp. Anim.* **2005**, *54*, 29–36.
232. Letts, V. A.; Felix, R.; Biddlecome, G. H.; Arikkath, J.; Mahaffey, C. L.; Valenzuela, A.; Bartlett, F. S., II; Mori, Y.; Campbell, K. P.; Frankel, W. N. *Nat. Genet.* **1998**, *19*, 340–347.
233. Rousset, M.; Cens, T.; Restituito, S.; Barrere, C.; Black, J. L., III; McEnery, M. W.; Charnet, P. *J. Physiol.* **2001**, *532*, 583–593.
234. Moss, F. J.; Dolphin, A. C.; Clare, J. *J. BMC Neurosci.* **2003**, *4*, 23.
235. Chen, L.; Chetkovich, D. M.; Petralia, R. S.; Sweeney, N. T.; Kawasaki, Y.; Wenthold, R. J.; Bredt, D. S.; Nicoll, R. A. *Nature* **2000**, *408*, 936–943.
236. Suzuki, S.; Rogawski, M. A. *Proc. Natl. Acad. Sci. USA* **1989**, *86*, 7228–7232.
237. White, G.; Lovinger, D. M.; Weight, F. F. *Proc. Natl. Acad. Sci. USA* **1989**, *86*, 6802–6806.
238. Huguenard, J. R.; Prince, D. A. *J. Neurosci.* **1992**, *12*, 3804–3817.
239. Llinas, R. R.; Ribary, U.; Jeanmonod, D.; Kronberg, E.; Mitra, P. P. *Proc. Natl. Acad. Sci. USA* **1999**, *96*, 15222–15227.
240. Wickman, K.; Clapham, D. E. *Physiol. Rev.* **1995**, *75*, 865–885.
241. Zamponi, G. W.; Bourinet, E.; Nelson, D.; Nargeot, J.; Snutch, T. P. *Nature* **1997**, *385*, 442–446.
242. Mark, M. D.; Wittemann, S.; Herlitze, S. *J. Physiol.* **2000**, *528* (Part 1), 65–77.
243. Yang, J.; Tsien, R. W. *Neuron* **1993**, *10*, 127–136.
244. Stea, A.; Soong, T. W.; Snutch, T. P. *Neuron* **1995**, *15*, 929–940.
245. Zhong, J.; Dessauer, C. W.; Keef, K. D.; Hume, J. R. *J. Physiol.* **1999**, *517* (Part 1), 109–120.
246. Kamp, T. J.; Hell, J. W. *Circ. Res.* **2000**, *87*, 1095–1102.
247. Zhou, J. Y.; Siderovski, D. P.; Miller, R. J. *J. Neurosci.* **2000**, *20*, 7143–7148.
248. Wolfe, J. T.; Wang, H.; Howard, J.; Garrison, J. C.; Barrett, P. Q. *Nature* **2003**, *424*, 209–213.
249. Wyatt, C. N.; Page, K. M.; Berrow, N. S.; Brice, N. L.; Dolphin, A. C. *J. Physiol.* **1998**, *510*, 347–360.
250. Li, B.; Zhong, H.; Scheuer, T.; Catterall, W. A. *Mol. Pharmacol.* **2004**, *66*, 761–769.
251. Canti, C.; Bogdanov, Y.; Dolphin, A. C. *J. Physiol.* **2000**, *527* (Part 3), 419–432.
252. Herlitze, S.; Zhong, H.; Scheuer, T.; Catterall, W. A. *Proc. Natl. Acad. Sci. USA* **2001**, *98*, 4699–4704.
253. Lee, J. J.; Hahm, E. T.; Min, B. I.; Cho, Y. W. *Brain Res.* **2004**, *1017*, 108–119.
254. Seabrook, G. R.; Adams, D. J. *Br. J. Pharmacol.* **1989**, *97*, 1125–1136.
255. Lorenson, M. Y.; Robson, D. L.; Jacobs, L. S. *J. Biol. Chem.* **1983**, *258*, 8618–8622.
256. Srivastava, R. D. *Indian J. Physiol. Pharmacol.* **1985**, *29*, 239–244.
257. Lee, J. H.; Gomora, J. C.; Cribbs, L. L.; Perez-Reyes, E. *Biophys. J.* **1999**, *77*, 3034–3042.
258. Olivera, B. M.; McIntosh, J. M.; Cruz, L. J.; Luque, F. A.; Gray, W. R. *Biochemistry* **1984**, *23*, 5087–5090.
259. Olivera, B. M.; Cruz, L. J.; de Santos, V.; LeCheminant, G. W.; Griffin, D.; Zeikus, R.; McIntosh, J. M.; Galyean, R.; Varga, J. et al. *Biochemistry* **1987**, *26*, 2086–2090.
260. Lewis, R. J.; Nielsen, K. J.; Craik, D. J.; Loughnan, M. L.; Adams, D. A.; Sharpe, I. A.; Luchian, T.; Adams, D. J.; Bond, T.; Thomas, L. et al. *J. Biol. Chem.* **2000**, *275*, 35335–35344.
261. Bowersox, S. S.; Gadbois, T.; Singh, T.; Pettus, M.; Wang, Y. X.; Luther, R. R. *J. Pharmacol. Exp. Ther.* **1996**, *279*, 1243–1249.
262. Bourinet, E.; Soong, T. W.; Sutton, K.; Slaymaker, S.; Mathews, E.; Monteil, A.; Zamponi, G. W.; Nargeot, J.; Snutch, T. P. *Nat. Neurosci.* **1999**, *2*, 407–415.
263. Chuang, R. S.; Jaffe, H.; Cribbs, L.; Perez-Reyes, E.; Swartz, K. *J. Nat. Neurosci.* **1998**, *1*, 668–674.
264. Sidach, S. S.; Mintz, I. M. *J. Neurosci.* **2002**, *22*, 2023–2034.
265. Olamendi-Portugal, T.; Garcia, B. I.; Lopez-Gonzalez, I.; Van Der Walt, J.; Dyason, K.; Ulens, C.; Tytgat, J.; Felix, R.; Darszon, A.; Possani, L. D. *Biochem. Biophys. Res. Commun.* **2002**, *299*, 562–568.
266. Middleton, R. E.; Warren, V. A.; Kraus, R. L.; Hwang, J. C.; Liu, C. J.; Dai, G.; Brochu, R. M.; Kohler, M. G.; Gao, Y. D.; Garsky, V. M. et al. *Biochemistry* **2002**, *41*, 14734–14747.
267. Mitterdorfer, J.; Wang, Z.; Sinnegger, M. J.; Hering, S.; Striessnig, J.; Grabner, M.; Glossmann, H. *J. Biol. Chem.* **1996**, *271*, 30330–30335.
268. Sinnegger, M. J.; Wang, Z.; Grabner, M.; Hering, S.; Striessnig, J.; Glossmann, H.; Mitterdorfer, J. *J. Biol. Chem.* **1997**, *272*, 27686–27693.
269. Hockerman, G. H.; Johnson, B. D.; Abbott, M. R.; Scheuer, T.; Catterall, W. A. *J. Biol. Chem.* **1997**, *272*, 18759–18765.
270. Motoike, H. K.; Bodi, I.; Nakayama, H.; Schwartz, A.; Varadi, G. *J. Biol. Chem.* **1999**, *274*, 9409–9420.
271. Kraus, R.; Reichl, B.; Kimball, S. D.; Grabner, M.; Murphy, B. J.; Catterall, W. A.; Striessnig, J. *J. Biol. Chem.* **1996**, *271*, 20113–20118.
272. Bodi, I.; Koch, S. E.; Yamaguchi, H.; Szigeti, G. P.; Schwartz, A.; Varadi, G. *J. Biol. Chem.* **2002**, *277*, 20651–20659.
273. Bean, B. P. *Proc. Natl. Acad. Sci. USA* **1984**, *81*, 6388–6392.
274. McDonald, T. F.; Pelzer, D.; Trautwein, W. *J. Physiol.* **1984**, *352*, 217–241.

275. Mishra, S. K.; Hermsmeyer, K. *Circ. Res.* **1994**, *75*, 144–148.
276. Masumiya, H.; Kase, J.; Tanaka, Y.; Tanaka, H.; Shigenobu, K. *Res. Commun. Mol. Pathol. Pharmacol.* **1999**, *104*, 321–329.
277. Todorovic, S. M.; Lingle, C. J. *J. Neurophysiol.* **1998**, *79*, 240–252.
278. Viana, F.; Van den Bosch, L.; Missiaen, L.; Vandenberghe, W.; Droogmans, G.; Nilius, B.; Robberecht, W. *Cell Calcium* **1997**, *22*, 299–311.
279. McDonough, S. I.; Bean, B. P. *Mol. Pharmacol.* **1998**, *54*, 1080–1087.
280. Martin, R. L.; Lee, J. H.; Cribbs, L. L.; Perez-Reyes, E.; Hanck, D. A. *J. Pharmacol. Exp. Ther.* **2000**, *295*, 302–308.
281. Monteil, A.; Chemin, J.; Leuranguer, V.; Altier, C.; Mennessier, G.; Bourinet, E.; Lory, P.; Nargeot, J. *J. Biol. Chem.* **2000**, *275*, 16530–16535.
282. Huang, L.; Keyser, B. M.; Tagmose, T. M.; Hansen, J. B.; Taylor, J. T.; Zhuang, H.; Zhang, M.; Ragsdale, D. S.; Li, M. *J. Pharmacol. Exp. Ther.* **2004**, *309*, 193–199.
283. Coulter, D. A.; Huguenard, J. R.; Prince, D. A. *Ann. Neurol.* **1989**, *25*, 582–593.
284. Suzuki, S.; Kawakami, K.; Nishimura, S.; Watanabe, Y.; Yagi, K.; Seino, M.; Miyamoto, K. *Epilepsy Res* **1992**, *12*, 21–27.
285. Gomora, J. C.; Daud, A. N.; Weiergraber, M.; Perez-Reyes, E. *Mol. Pharmacol.* **2001**, *60*, 1121–1132.
286. Kito, M.; Maehara, M.; Watanabe, K. *Seizure* **1996**, *5*, 115–119.
287. Sah, D. W.; Bean, B. P. *Mol. Pharmacol.* **1994**, *45*, 84–92.
288. Santi, C. M.; Cayabyab, F. S.; Sutton, K. G.; McRory, J. E.; Mezeyova, J.; Hamming, K. S.; Parker, D.; Stea, A.; Snutch, T. P. *J. Neurosci.* **2002**, *22*, 396–403.
289. Annoura, H.; Nakanishi, K.; Uesugi, M.; Fukunaga, A.; Miyajima, A.; Tamura-Horikawa, Y.; Tamura, S. *Bioorg. Med. Chem. Lett.* **1999**, *9*, 2999–3002.
290. Annoura, H.; Nakanishi, K.; Uesugi, M.; Fukunaga, A.; Imajo, S.; Miyajima, A.; Tamura-Horikawa, Y.; Tamura, S. *Bioorg. Med. Chem.* **2002**, *10*, 371–383.
291. Zamponi, G. W. *J. Membr. Biol.* **1999**, *167*, 183–192.
292. Romero, M.; Sanchez, I.; Pujol, M. D. *Curr. Med. Chem. Cardiovasc. Hematol. Agents* **2003**, *1*, 113–141.
293. Sarsero, D.; Fujiwara, T.; Molenaar, P.; Angus, J. A. *Br. J. Pharmacol.* **1998**, *125*, 109–119.
294. Welling, A.; Ludwig, A.; Zimmer, S.; Klugbauer, N.; Flockerzi, V.; Hofmann, F. *Circ. Res.* **1997**, *81*, 526–532.
295. Koschak, A.; Reimer, D.; Huber, I.; Grabner, M.; Glossmann, H.; Engel, J.; Striessnig, J. *J. Biol. Chem.* **2001**, *276*, 22100–22106.
296. Baumann, L.; Gerstner, A.; Zong, X.; Biel, M.; Wahl-Schott, C. *Invest. Ophthalmol. Vis. Sci.* **2004**, *45*, 708–713.
297. Cia, D.; Bordais, A.; Varela, C.; Forster, V.; Sahel, J. A.; Rendon, A.; Picaud, S. *J. Neurophysiol.* **2005**, *93*, 1468–1475.
298. Akhtar, M.; Tchou, P.; Jazayeri, M. *Circulation* **1989**, *80*, IV31–IV39.
299. Ertel, S. I.; Ertel, E. A.; Clozel, J. P. *Cardiovasc. Drugs Ther.* **1997**, *11*, 723–739.
300. Luo, Z. D.; Chaplan, S. R.; Higuera, E. S.; Sorkin, L. S.; Stauderman, K. A.; Williams, M. E.; Yaksh, T. L. *J. Neurosci.* **2001**, *21*, 1868–1875.
301. Newton, R. A.; Bingham, S.; Case, P. C.; Sanger, G. J.; Lawson, S. N. *Brain Res. Mol. Brain Res.* **2001**, *95*, 1–8.
302. Abe, M.; Kurihara, T.; Han, W.; Shinomiya, K.; Tanabe, T. *Spine* **2002**, *27*, 1517–1524.
303. Cizkova, D.; Marsala, J.; Lukacova, N.; Marsala, M.; Jergova, S.; Orendacova, J.; Yaksh, T. L. *Exp. Brain Res.* **2002**, *147*, 456–463.
304. Luo, Z. D.; Calcutt, N. A.; Higuera, E. S.; Valder, C. R.; Song, Y. H.; Svensson, C. I.; Myers, R. R. *J. Pharmacol. Exp. Ther.* **2002**, *303*, 1199–1205.
305. Yokoyama, K.; Kurihara, T.; Makita, K.; Tanabe, T. *Anesthesiology* **2003**, *99*, 1364–1370.
306. tanassoff, A. P. G.; Hartmannsgruber, M. W.; Thrasher, J.; Wermeling, D.; Longton, W.; Gaeta, R.; Singh, T.; Mayo, M.; McGuire, D.; Luther, R. R. *Reg. Anesth. Pain Med.* **2000**, *25*, 274–278.
307. Taqi, D.; Gunyea, I.; Bhakta, B.; Movva, V.; Ward, S.; Jenson, M.; Royal, M. *Pain Med.* **2002**, *3*, 180–181.
308. Staats, P. S.; Yearwood, T.; Charapata, S. G.; Presley, R. W.; Wallace, M. S.; Byas-Smith, M.; Fisher, R.; Bryce, D. A.; Mangieri, E. A.; Luther, R. R. et al. *JAMA* **2004**, *291*, 63–70.
309. Gee, N. S.; Brown, J. P.; Dissanayake, V. U.; Offord, J.; Thurlow, R.; Woodruff, G. N. *J. Biol. Chem.* **1996**, *271*, 5768–5776.
310. Schwarz, J. B.; Gibbons, S. E.; Graham, S. R.; Colbry, N. L.; Guzzo, P. R.; Le, V. D.; Vartanian, M. G.; Kinsora, J. J.; Lotarski, S. M.; Li, Z. et al. *J. Med. Chem.* **2005**, *48*, 3026–3035.
311. Field, M. J. Pregabalin's analgesic actions are mediated by its binding to the α2δ type 1 subunit of voltage-gated calcium channels. Spring Pain Research Conference 24 April–1 May 2004, Grand Cayman, BWI, 2004.
312. Stefani, A.; Spadoni, F.; Bernardi, G. *Neuropharmacology* **1998**, *37*, 83–91.
313. Stefani, A.; Spadoni, F.; Giacomini, P.; Lavaroni, F.; Bernardi, G. *Epilepsy Res.* **2001**, *43*, 239–248.
314. Alden, K. J.; Garcia, J. *J. Pharmacol. Exp. Ther.* **2001**, *297*, 727–735.
315. Sutton, K. G.; Martin, D. J.; Pinnock, R. D.; Lee, K.; Scott, R. H. *Br. J. Pharmacol.* **2002**, *135*, 257–265.
316. McClelland, D.; Evans, R. M.; Barkworth, L.; Martin, D. J.; Scott, R. H. *BMC Pharmacol.* **2004**, *4*, 14.
317. Rock, D. M.; Kelly, K. M.; Macdonald, R. L. *Epilepsy Res.* **1993**, *16*, 89–98.
318. Schumacher, T. B.; Beck, H.; Steinhauser, C.; Schramm, J.; Elger, C. E. *Epilepsia* **1998**, *39*, 355–363.
319. Martin, D. J.; McClelland, D.; Herd, M. B.; Sutton, K. G.; Hall, M. D.; Lee, K.; Pinnock, R. D.; Scott, R. H. *Neuropharmacology* **2002**, *42*, 353–366.
320. Fink, K.; Meder, W.; Dooley, D. J.; Gothert, M. *Br. J. Pharmacol.* **2000**, *130*, 900–906.
321. Meder, W. P.; Dooley, D. J. *Brain Res.* **2000**, *875*, 157–159.
322. Fink, K.; Dooley, D. J.; Meder, W. P.; Suman-Chauhan, N.; Duffy, S.; Clusmann, H.; Gothert, M. *Neuropharmacology* **2002**, *42*, 229–236.
323. Fehrenbacher, J. C.; Taylor, C. P.; Vasko, M. R. *Pain* **2003**, *105*, 133–141.
324. Maneuf, Y. P.; Blake, R.; Andrews, N. A.; McKnight, A. T. *Br. J. Pharmacol.* **2004**, *141*, 574–579.
325. Bayer, K.; Ahmadi, S.; Zeilhofer, H. U. *Neuropharmacology* **2004**, *46*, 743–749.
326. Coderre, T. J.; Kumar, N.; Lefebvre, C. D.; Yu, J. S. *J. Neurochem.* **2005**, *94*, 1131–1139.
327. Kang, M. G.; Felix, R.; Campbell, K. P. *FEBS Lett.* **2002**, *528*, 177–182.
328. Alden, K. J.; Garcia, J. *Am. J. Physiol. Cell Physiol.* **2002**, *283*, C941–C949.
329. Jouvenceau, A.; Eunson, L. H.; Spauschus, A.; Ramesh, V.; Zuberi, S. M.; Kullmann, D. M.; Hanna, M. G. *Lancet* **2001**, *358*, 801–807.
330. Battistini, S.; Stenirri, S.; Piatti, M.; Gelfi, C.; Righetti, P. G.; Rocchi, R.; Giannini, F.; Battistini, N.; Guazzi, G. C.; Ferrari, M. et al. *Neurology* **1999**, *53*, 38–43.
331. Tottene, A.; Fellin, T.; Pagnutti, S.; Luvisetto, S.; Striessnig, J.; Fletcher, C.; Pietrobon, D. *Proc. Natl. Acad. Sci. USA* **2002**, *99*, 13284–13289.
332. Kraus, R. L.; Sinnegger, M. J.; Koschak, A.; Glossmann, H.; Stenirri, S.; Carrera, P.; Striessnig, J. *J. Biol. Chem.* **2000**, *275*, 9239–9243.
333. Melliti, K.; Grabner, M.; Seabrook, G. R. *J. Physiol.* **2003**, *546*, 337–347.

334. Yu, W.; Horowitz, S. H. *Neurology* **2001**, *57*, 1732–1733.
335. Restituito, S.; Thompson, R. M.; Eliet, J.; Raike, R. S.; Riedl, M.; Charnet, P.; Gomez, C. M. *J. Neurosci.* **2000**, *20*, 6394–6403.
336. Denier, C.; Ducros, A.; Durr, A.; Eymard, B.; Chassande, B.; Tournier-Lasserve, E. *Arch. Neurol.* **2001**, *58*, 292–295.
337. Denier, C.; Ducros, A.; Vahedi, K.; Joutel, A.; Thierry, P.; Ritz, A.; Castelnovo, G.; Deonna, T.; Gerard, P.; Devoize, J. L. et al. *Neurology* **1999**, *52*, 1816–1821.
338. Guida, S.; Trettel, F.; Pagnutti, S.; Mantuano, E.; Tottene, A.; Veneziano, L.; Fellin, T.; Spadaro, M.; Stauderman, K.; Williams, M. et al. *Am. J. Hum. Genet.* **2001**, *68*, 759–764.
339. Wan, J.; Khanna, R.; Sandusky, M.; Papazian, D. M.; Jen, J. C.; Baloh, R. W. *Neurology* **2005**, *64*, 2090–2097.
340. Lapie, P.; Goudet, C.; Nargeot, J.; Fontaine, B.; Lory, P. *FEBS Lett.* **1996**, *382*, 244–248.
341. Lerche, H.; Klugbauer, N.; Lehmann-Horn, F.; Hofmann, F.; Melzer, W. *Pflügers Arch* **1996**, *431*, 461–463.
342. Wallace, A. J.; Wooldridge, W.; Kingston, H. M.; Harrison, M. J.; Ellis, F. R.; Ford, P. M. *Anaesthesia* **1996**, *51*, 16–23.
343. Jacobi, F. K.; Hamel, C. P.; Arnaud, B.; Blin, N.; Broghammer, M.; Jacobi, P. C.; Apfelstedt-Sylla, E.; Pusch, C. M. *Am. J. Ophthalmol.* **2003**, *135*, 733–736.
344. Hope, C. I.; Sharp, D. M.; Hemara-Wahanui, A.; Sissingh, J. I.; Lundon, P.; Mitchell, E. A.; Maw, M. A.; Clover, G. M. *Clin. Exp. Ophthalmol.* **2005**, *33*, 129–136.
345. Hemara-Wahanui, A.; Berjukow, S.; Hope, C. I.; Dearden, P. K.; Wu, S. B.; Wilson-Wheeler, J.; Sharp, D. M.; Lundon-Treweek, P.; Clover, G. M.; Hoda, J. C. et al. *Proc. Natl. Acad. Sci. USA* **2005**, *102*, 7553–7558.
346. Heginbotham, L.; Lu, Z.; Abramson, T.; MacKinnon, R. *Biophys. J.* **1994**, *66*, 1061–1067.
347. Doyle, D. A.; Morais Cabral, J.; Pfuetzner, R. A.; Kuo, A.; Gulbis, J. M.; Cohen, S. L.; Chait, B. T.; MacKinnon, R. *Science* **1998**, *280*, 69–77.
348. Gutman, G. A.; Chandy, K. G.; Adelman, J. P.; Aiyar, J.; Bayliss, D. A.; Clapham, D. E.; Covarrubias, M.; Desir, G. V.; Furuichi, K.; Ganetzky, B. et al. *Pharmacol. Rev.* **2003**, *55*, 583–586.
349. Li, B.; Gallin, W. J. *BMC Bioinformatics* **2004**, *5*, 3.
350. Miller, C. *Genome Biol.* **2000**, *1*, 1–5.
351. Biel, M.; Ludwig, A.; Zong, X.; Hofmann, F. *Rev. Physiol. Biochem. Pharmacol.* **1999**, *136*, 165–181.
352. Biel, M.; Zong, X.; Ludwig, A.; Sautter, A.; Hofmann, F. *Rev. Physiol. Biochem. Pharmacol.* **1999**, *135*, 151–171.
353. Santoro, B.; Tibbs, G. R. *Ann. NY Acad. Sci.* **1999**, *868*, 741–764.
354. Santoro, B.; Chen, S.; Luthi, A.; Pavlidis, P.; Shumyatsky, G. P.; Tibbs, G. R.; Siegelbaum, S. A. *J. Neurosci.* **2000**, *20*, 5264–5275.
355. Baruscotti, M.; Difrancesco, D. *Ann. NY Acad. Sci.* **2004**, *1015*, 111–121.
356. DiFrancesco, D. *Curr. Med. Res. Opin.* **2005**, *21*, 1115–1122.
357. Hodgkin, A. L.; Huxley, A. F. *J. Physiol.* **1952**, *117*, 500–544.
358. Trimmer, J. S.; Rhodes, K. *J. Annu. Rev. Physiol.* **2004**, *66*, 477–519.
359. Kerschensteiner, D.; Soto, F.; Stocker, M. *Proc. Natl. Acad. Sci. USA* **2005**, *102*, 6160–6165.
360. McCrossan, Z. A.; Abbott, G. W. *Neuropharmacology* **2004**, *47*, 787–821.
361. Chouinard, S. W.; Wilson, G. F.; Schlimgen, A. K.; Ganetzky, B. *Proc. Natl. Acad. Sci. USA* **1995**, *92*, 6763–6767.
362. Pongs, O.; Leicher, T.; Berger, M.; Roeper, J.; Bahring, R.; Wray, D.; Giese, K. P.; Silva, A. J.; Storm, J. F. *Ann. NY Acad. Sci.* **1999**, *868*, 344–355.
363. Aimond, F.; Kwak, S. P.; Rhodes, K. J.; Nerbonne, J. M. *Circ. Res.* **2005**, *96*, 451–458.
364. Ible, W. B. A.; Yang, Q.; Kuryshev, Y. A.; Accili, E. A.; Brown, A. M. *J. Biol. Chem.* **1998**, *273*, 11745–11751.
365. Kuryshev, Y. A.; Gudz, T. I.; Brown, A. M.; Wible, B. A. *Am. J. Physiol. Cell Physiol.* **2000**, *278*, C931–C941.
366. An, W. F.; Bowlby, M. R.; Betty, M.; Cao, J.; Ling, H. P.; Mendoza, G.; Hinson, J. W.; Mattsson, K. I.; Strassle, B. W.; Trimmer, J. S. et al. *Nature* **2000**, *403*, 553–556.
367. Rosati, B.; Pan, Z.; Lypen, S.; Wang, H. S.; Cohen, I.; Dixon, J. E.; McKinnon, D. *J. Physiol.* **2001**, *533*, 119–125.
368. Rhodes, K. J.; Carroll, K. I.; Sung, M. A.; Doliveira, L. C.; Monaghan, M. M.; Burke, S. L.; Strassle, B. W.; Buchwalder, L.; Menegola, M.; Cao, J. et al. *J. Neurosci.* **2004**, *24*, 7903–7915.
369. Strop, P.; Bankovich, A. J.; Hansen, K. C.; Garcia, K. C.; Brunger, A. T. *J. Mol. Biol.* **2004**, *343*, 1055–1065.
370. Jerng, H. H.; Qian, Y.; Pfaffinger, P. J. *Biophys. J.* **2004**, *87*, 2380–2396.
371. Nadal, M. S.; Ozaita, A.; Amarillo, Y.; Vega-Saenz de Miera, E.; Ma, Y.; Mo, W.; Goldberg, E. M.; Misumi, Y.; Ikehara, Y.; Neubert, T. A. et al. *Neuron* **2003**, *37*, 449–461.
372. Zagha, E.; Ozaita, A.; Chang, S. Y.; Nadal, M. S.; Lin, U.; Saganich, M. J.; McCormack, T.; Akinsanya, K. O.; Qi, S. Y.; Rudy, B. *J. Biol. Chem.* **2005**, *280*, 18853–18861.
373. Radicke, S.; Cotella, D.; Graf, E. M.; Ravens, U.; Wettwer, E. *J. Physiol.* **2005**, *565*, 751–756.
374. Dworetzky, S. I.; Trojnacki, J. T.; Gribkoff, V. K. *Brain. Res. Mol. Brain. Res.* **1994**, *27*, 189–193.
375. Schreiber, M.; Wei, A.; Yuan, A.; Gaut, J.; Saito, M.; Salkoff, L. *J. Biol. Chem.* **1998**, *273*, 3509–3516.
376. Magleby, K. L. *J. Gen. Physiol.* **2003**, *121*, 81–96.
377. Stocker, M. *Nat. Rev. Neurosci.* **2004**, *5*, 758–770.
378. Cui, J.; Cox, D. H.; Aldrich, R. W. *J. Gen. Physiol.* **1997**, *109*, 647–673.
379. Rothberg, B. S. *Sci. STKE* **2004**, pe16. http://stke.sciencemag.org (accessed July 2006).
380. Toro, L.; Wallner, M.; Meera, P.; Tanaka, Y. *News Physiol. Sci.* **1998**, *13*, 112–117.
381. Jiang, Z.; Wallner, M.; Meera, P.; Toro, L. *Genomics* **1999**, *55*, 57–67.
382. Hofmann, F.; Biel, M.; Kaupp, U. B. *Pharmacol. Rev.* **2003**, *55*, 587–589.
383. Robinson, R. B.; Siegelbaum, S. A. *Annu. Rev. Physiol.* **2003**, *65*, 453–480.
384. Rothberg, B. S.; Shin, K. S.; Yellen, G. *J. Gen. Physiol.* **2003**, *122*, 501–510.
385. Kamb, A.; Iverson, L. E.; Tanouye, M. A. *Cell* **1987**, *50*, 405–413.
386. Tempel, B. L.; Papazian, D. M.; Schwarz, T. L.; Jan, Y. N.; Jan, L. Y. *Science* **1987**, *237*, 770–775.
387. Pongs, O.; Kecskemethy, N.; Muller, R.; Krah-Jentgens, I.; Baumann, A.; Kiltz, H. H.; Canal, I.; Llamazares, S.; Ferrus, A. *EMBO J.* **1988**, *7*, 1087–1096.
388. Armstrong, C. M. *Sci. STKE* **2003**, re10. http://stke.sciencemag.org (accessed July 2006).
389. Tseng-Crank, J.; Foster, C. D.; Krause, J. D.; Mertz, R.; Godinot, N.; DiChiara, T. J.; Reinhart, P. H. *Neuron* **1994**, *13*, 1315–1330.
390. Misonou, H.; Trimmer, J. S. *Crit. Rev. Biochem. Mol. Biol.* **2004**, *39*, 125–145.
391. Heusser, K.; Schwappach, B. *Curr. Opin. Neurobiol.* **2005**, *15*, 364–369.
392. Martens, J. R.; Sakamoto, N.; Sullivan, S. A.; Grobaski, T. D.; Tamkun, M. M. *J. Biol. Chem.* **2001**, *276*, 8409–8414.

393. Deutsch, C. *Annu. Rev. Physiol.* **2002**, *64*, 19–46.
394. Sanguinetti, M. C.; Bennett, P. B. *Circ. Res.* **2003**, *93*, 491–499.
395. Gollasch, M.; Lohn, M.; Furstenau, M.; Nelson, M. T.; Luft, F. C.; Haller, H. *Z. Kardiol.* **2000**, *89*, 15–19.
396. Kaupp, U. B.; Seifert, R. *Annu. Rev. Physiol.* **2001**, *63*, 235–257.
397. Manganas, L. N.; Trimmer, J. S. *Biochem. Biophys. Res. Commun.* **2004**, *322*, 577–584.
398. Campomanes, C. R.; Carroll, K. I.; Manganas, L. N.; Hershberger, M. E.; Gong, B.; Antonucci, D. E.; Rhodes, K. J.; Trimmer, J. S. *J. Biol. Chem.* **2002**, *277*, 8298–8305.
399. Martens, J. R.; O'Connell, K.; Tamkun, M. *Trends Pharmacol. Sci.* **2004**, *25*, 16–21.
400. Bravo-Zehnder, M.; Orio, P.; Norambuena, A.; Wallner, M.; Meera, P.; Toro, L.; Latorre, R.; Gonzalez, A. *Proc. Natl. Acad. Sci. USA* **2000**, *97*, 13114–13119.
401. Hajdu, P.; Varga, Z.; Pieri, C.; Panyi, G.; Gaspar, R., Jr. *Pflügers Arch.* **2003**, *445*, 674–682.
402. MacKinnon, R. *FEBS Lett.* **2003**, *555*, 62–65.
403. Swartz, K. *J. Nat. Rev. Neurosci.* **2004**, *5*, 905–916.
404. Shen, N. V.; Pfaffinger, P. J. *Neuron* **1995**, *14*, 625–633.
405. Hoshi, T.; Zagotta, W. N.; Aldrich, R. W. *Science* **1990**, *250*, 533–538.
406. Zagotta, W. N.; Hoshi, T.; Aldrich, R. W. *Science* **1990**, *250*, 568–571.
407. Noskov, S. Y.; Berneche, S.; Roux, B. *Nature* **2004**, *431*, 830–834.
408. Cohen, B. E.; Grabe, M.; Jan, L. Y. *Neuron* **2003**, *39*, 395–400.
409. Ahern, C. A.; Horn, R. *Trends Neurosci.* **2004**, *27*, 303–307.
410. Bezanilla, F. *Trends Biochem. Sci.* **2005**, *30*, 166–168.
411. Jiang, Y.; Lee, A.; Chen, J.; Cadene, M.; Chait, B. T.; MacKinnon, R. *Nature* **2002**, *417*, 523–526.
412. Jiang, Y.; Lee, A.; Chen, J.; Ruta, V.; Cadene, M.; Chait, B. T.; MacKinnon, R. *Nature* **2003**, *423*, 33–41.
413. Jiang, Y.; Ruta, V.; Chen, J.; Lee, A.; MacKinnon, R. *Nature* **2003**, *423*, 42–48.
414. Horn, R. *J. Gen. Physiol.* **2002**, *120*, 449–453.
415. Gandhi, C. S.; Isacoff, E. Y. *J. Gen. Physiol.* **2002**, *120*, 455–463.
416. Ahern, C. A.; Horn, R. *Neuron* **2005**, *48*, 25–29.
417. Chanda, B.; Asamoah, O. K.; Blunck, R.; Roux, B.; Bezanilla, F. *Nature* **2005**, *436*, 852–856.
418. Posson, D. J.; Ge, P.; Miller, C.; Bezanilla, F.; Selvin, P. R. *Nature* **2005**, *436*, 848–851.
419. Meera, P.; Wallner, M.; Song, M.; Toro, L. *Proc. Natl. Acad. Sci. USA* **1997**, *94*, 14066–14071.
420. Bian, S.; Favre, I.; Moczydlowski, E. *Proc. Natl. Acad. Sci. USA* **2001**, *98*, 4776–4781.
421. Schmalhofer, W. A.; Sanchez, M.; Dai, G.; Dewan, A.; Secades, L.; Hanner, M.; Knaus, H. G.; McManus, O. B.; Kohler, M.; Kaczorowski, G. J. et al. *Biochemistry* **2005**, *44*, 10135–10144.
422. Zeng, X. H.; Xia, X. M.; Lingle, C. J. *J. Gen. Physiol.* **2005**, *125*, 273–286.
423. Wei, A.; Solaro, C.; Lingle, C.; Salkoff, L. *Neuron* **1994**, *13*, 671–681.
424. Bao, L.; Kaldany, C.; Holmstrand, E. C.; Cox, D. H. *J. Gen. Physiol.* **2004**, *123*, 475–489.
425. Moczydlowski, E. G. *J. Gen. Physiol.* **2004**, *123*, 471–473.
426. Schreiber, M.; Salkoff, L. *Biophys. J.* **1997**, *73*, 1355–1363.
427. Piskorowski, R.; Aldrich, R. W. *Nature* **2002**, *420*, 499–502.
428. Schreiber, M.; Yuan, A.; Salkoff, L. *Nat. Neurosci.* **1999**, *2*, 416–421.
429. Craven, K. B.; Zagotta, W. N. *Annu. Rev. Physiol.* **2006**, *68*, 375–401.
430. Shin, K. S.; Maertens, C.; Proenza, C.; Rothberg, B. S.; Yellen, G. *Neuron* **2004**, *41*, 737–744.
431. Zagotta, W. N.; Olivier, N. B.; Black, K. D.; Young, E. C.; Olson, R.; Gouaux, E. *Nature* **2003**, *425*, 200–205.
432. Shieh, C. C.; Coghlan, M.; Sullivan, J. P.; Gopalakrishnan, M. *Pharmacol. Rev.* **2000**, *52*, 557–594.
433. Korn, S. J.; Trapani, J. G. *IEEE Trans. Nanobiosci.* **2005**, *4*, 21–33.
434. Michelakis, E. D.; Thebaud, B.; Weir, E. K.; Archer, S. L. *J. Mol. Cell. Cardiol.* **2004**, *37*, 1119–1136.
435. Tang, X. D.; Santarelli, L. C.; Heinemann, S. H.; Hoshi, T. *Annu. Rev. Physiol.* **2004**, *66*, 131–159.
436. Oliver, D.; Lien, C. C.; Soom, M.; Baukrowitz, T.; Jonas, P.; Fakler, B. *Science* **2004**, *304*, 265–270.
437. Cox, R. H. *Vascul. Pharmacol.* **2002**, *38*, 13–23.
438. Navarro-Antolin, J.; Levitsky, K. L.; Calderon, E.; Ordonez, A.; Lopez-Barneo, J. *Circulation* **2005**, *112*, 1309–1315.
439. Gopalakrishnan, M.; Shieh, C. C. *Expert Opin. Ther. Targets* **2004**, *8*, 437–458.
440. Surti, T. S.; Jan, L. Y. *Curr. Opin. Investig. Drugs* **2005**, *6*, 704–711.
441. Brown, A. M. *Cell Calcium* **2004**, *35*, 543–547.
442. Kaczorowski, G. J.; Knaus, H. G.; Leonard, R. J.; McManus, O. B.; Garcia, M. L. *J. Bioenerg. Biomembr.* **1996**, *28*, 255–267.
443. Weiger, T. M.; Hermann, A.; Levitan, I. B. *J. Comp. Physiol. A Neuroethol. Sens. Neural Behav. Physiol.* **2002**, *188*, 79–87.
444. Wu, S. N. *Curr. Med. Chem.* **2003**, *10*, 649–661.
445. Orio, P.; Rojas, P.; Ferreira, G.; Latorre, R. *News Physiol. Sci.* **2002**, *17*, 156–161.
446. McCarron, J. G.; Bradley, K. N.; Muir, T. C. *Novartis Found. Symp.* **2002**, *246*, 52–64 (discussion 64–70 221–227).
447. Tanaka, Y.; Koike, K.; Toro, L. *J. Smooth Muscle Res.* **2004**, *40*, 125–153.
448. Delisle, B. P.; Slind, J. K.; Kilby, J. A.; Anderson, C. L.; Anson, B. D.; Balijepalli, R. C.; Tester, D. J.; Ackerman, M. J.; Kamp, T. J.; January, C. T. *Mol. Pharmacol.* **2005**, *68*, 233–240.
449. Lee, H. C.; Wang, J. M.; Swartz, K. J. *Neuron* **2003**, *40*, 527–536.
450. Dalle, C.; Eisenach, J. C. *Reg. Anesth. Pain Med.* **2005**, *30*, 243–248.
451. Wang, W. *Curr. Opin. Nephrol. Hypertens.* **2004**, *13*, 549–555.
452. Speake, T.; Kibble, J. D.; Brown, P. D. *Am. J. Physiol. Cell Physiol.* **2004**, *286*, C611–C620.
453. Fadool, D. A.; Levitan, I. B. *J. Neurosci.* **1998**, *18*, 6126–6137.
454. Cook, K. K.; Fadool, D. A. *J. Biol. Chem.* **2002**, *277*, 13268–13280.
455. Misonou, H.; Mohapatra, D. P.; Park, E. W.; Leung, V.; Zhen, D.; Misonou, K.; Anderson, A. E.; Trimmer, J. S. *Nat. Neurosci.* **2004**, *7*, 711–718.
456. Murakoshi, H.; Shi, G.; Scannevin, R. H.; Trimmer, J. S. *Mol. Pharmacol.* **1997**, *52*, 821–828.
457. Tiran, Z.; Peretz, A.; Attali, B.; Elson, A. *J. Biol. Chem.* **2003**, *278*, 17509–17514.
458. Nitabach, M. N.; Llamas, D. A.; Thompson, I. J.; Collins, K. A.; Holmes, T. C. *J. Neurosci.* **2002**, *22*, 7913–7922.

459. MacDonald, P. E.; Wang, X.; Xia, F.; El-kholy, W.; Targonsky, E. D.; Tsushima, R. G.; Wheeler, M. B. *J. Biol. Chem.* **2003**, *278*, 52446–52453.
460. Tian, L.; Coghill, L. S.; McClafferty, H.; MacDonald, S. H.; Antoni, F. A.; Ruth, P.; Knaus, H. G.; Shipston, M. *J. Proc. Natl. Acad. Sci. USA* **2004**, *101*, 11897–11902.
461. Barman, S. A.; Zhu, S.; White, R. E. *Am. J. Physiol. Lung Cell. Mol. Physiol.* **2004**, *286*, L1275–L1281.
462. Harvey, A. L.; Robertson, B. *Curr. Med. Chem.* **2004**, *11*, 3065–3072.
463. Jouirou, B.; Mouhat, S.; Andreotti, N.; De Waard, M.; Sabatier, J. M. *Toxicon* **2004**, *43*, 909–914.
464. Rodriguez de la Vega, R. C.; Possani, L. D. *Toxicon* **2004**, *43*, 865–875.
465. Diochot, S.; Drici, M. D.; Moinier, D.; Fink, M.; Lazdunski, M. *Br. J. Pharmacol.* **1999**, *126*, 251–263.
466. Chagot, B.; Escoubas, P.; Villegas, E.; Bernard, C.; Ferrat, G.; Corzo, G.; Lazdunski, M.; Darbon, H. *Protein Sci.* **2004**, *13*, 1197–1208.
467. Ebbinghaus, J.; Legros, C.; Nolting, A.; Guette, C.; Celerier, M. L.; Pongs, O.; Bahring, R. *Toxicon* **2004**, *43*, 923–932.
468. Friederich, P.; Solth, A. *Anesthesiology* **2004**, *101*, 1347–1356.
469. Lamas, J. A.; Selyanko, A. A.; Brown, D. A. *Eur. J. Neurosci.* **1997**, *9*, 605–616.
470. Wang, H. S.; Brown, B. S.; McKinnon, D.; Cohen, I. S. *Mol. Pharmacol.* **2000**, *57*, 1218–1223.
471. Wua, Y. J.; Dworetzky, S. I. *Curr. Med. Chem.* **2005**, *12*, 453–460.
472. Dupuis, D. S.; Schroder, R. L.; Jespersen, T.; Christensen, J. K.; Christophersen, P.; Jensen, B. S.; Olesen, S. P. *Eur. J. Pharmacol.* **2002**, *437*, 129–137.
473. McManus, O. B.; Harris, G. H.; Giangiacomo, K. M.; Feigenbaum, P.; Reuben, J. P.; Addy, M. E.; Burka, J. F.; Kaczorowski, G. J.; Garcia, M. L. *Biochemistry* **1993**, *32*, 6128–6133.
474. Strobaek, D.; Christophersen, P.; Holm, N. R.; Moldt, P.; Ahring, P. K.; Johansen, T. E.; Olesen, S. P. *Neuropharmacology* **1996**, *35*, 903–914.
475. Edwards, G.; Niederste-Hollenberg, A.; Schneider, J.; Noack, T.; Weston, A. H. *Br. J. Pharmacol.* **1994**, *113*, 1538–1547.
476. BoSmith, R. E.; Briggs, I.; Sturgess, N. C. *Br. J. Pharmacol.* **1993**, *110*, 343–349.
477. Bois, P.; Bescond, J.; Renaudon, B.; Lenfant, J. *Br. J. Pharmacol.* **1996**, *118*, 1051–1057.
478. Varro, A.; Biliczki, P.; Iost, N.; Virag, L.; Hala, O.; Kovacs, P.; Matyus, P.; Papp, J. G. *Curr. Med. Chem.* **2004**, *11*, 1–11.
479. Feng, J.; Wible, B.; Li, G. R.; Wang, Z.; Nattel, S. *Circ. Res.* **1997**, *80*, 572–579.
480. Brendel, J.; Peukert, S. *Curr. Med. Chem. Cardiovasc. Hematol. Agents* **2003**, *1*, 273–287.
481. Gonzalez, T.; Navarro-Polanco, R.; Arias, C.; Caballero, R.; Moreno, I.; Delpon, E.; Tamargo, J.; Tamkun, M. M.; Valenzuela, C. *Mol. Pharmacol.* **2002**, *62*, 1456–1463.
482. Sah, R.; Ramirez, R. J.; Oudit, G. Y.; Gidrewicz, D.; Trivieri, M. G.; Zobel, C.; Backx, P. H. *J. Physiol.* **2003**, *546*, 5–18.
483. Kim, L. A.; Furst, J.; Butler, M. H.; Xu, S.; Grigorieff, N.; Goldstein, S. A. *J. Biol. Chem.* **2004**, *279*, 5549–5554.
484. Patel, S. P.; Campbell, D. L. *J. Physiol.* **2005**, *569*, 7–39.
485. McKinnon, D. *Circ. Res.* **1999**, *84*, 620–622.
486. Xu, X.; Salata, J. J.; Wang, J.; Wu, Y.; Yan, G. X.; Liu, T.; Marinchak, R. A.; Kowey, P. R. *Am. J. Physiol. Heart Circ. Physiol.* **2002**, *283*, H664–H670.
487. Ehrlich, J. R.; Pourrier, M.; Weerapura, M.; Ethier, N.; Marmabachi, A. M.; Hebert, T. E.; Nattel, S. *J. Biol. Chem.* **2004**, *279*, 1233–1241.
488. Rus, H.; Pardo, C. A.; Hu, L.; Darrah, E.; Cudrici, C.; Niculescu, T.; Niculescu, F.; Mullen, K. M.; Allie, R.; Guo, L. et al. *A. Proc. Natl. Acad. Sci. USA* **2005**, *102*, 11094–11099.
489. Vennekamp, J.; Wulff, H.; Beeton, C.; Calabresi, P. A.; Grissmer, S.; Hansel, W.; Chandy, K. G. *Mol. Pharmacol.* **2004**, *65*, 1364–1374.
490. Bao, J.; Miao, S.; Kayser, F.; Kotliar, A. J.; Baker, R. K.; Doss, G. A.; Felix, J. P.; Bugianesi, R. M.; Slaughter, R. S.; Kaczorowski, G. J. et al. *Bioorg. Med. Chem. Lett.* **2005**, *15*, 447–451.
491. Beeton, C.; Pennington, M. W.; Wulff, H.; Singh, S.; Nugent, D.; Crossley, G.; Khaytin, I.; Calabresi, P. A.; Chen, C. Y.; Gutman, G. A. et al. *Mol. Pharmacol.* **2005**, *67*, 1369–1381.
492. Desir, G. V. *Expert Opin. Ther. Targets* **2005**, *9*, 571–579.
493. Xu, J.; Wang, P.; Li, Y.; Li, G.; Kaczmarek, L. K.; Wu, Y.; Koni, P. A.; Flavell, R. A.; Desir, G. V. *Proc. Natl. Acad. Sci. USA* **2004**, *101*, 3112–3117.
494. Suzuki, T.; Takimoto, K. *Int. J. Oncol.* **2004**, *25*, 153–159.
495. Zhou, Q.; Kwan, H. Y.; Chan, H. C.; Jiang, J. L.; Tam, S. C.; Yao, X. *Int. J. Mol. Med.* **2003**, *11*, 261–266.
496. Yogeeswari, P.; Ragavendran, J. V.; Thirumurugan, R.; Saxena, A.; Sriram, D. *Curr. Drug Targets* **2004**, *5*, 589–602.
497. Errington, A. C.; Stohr, T.; Lees, G. *Curr. Top. Med. Chem.* **2005**, *5*, 15–30.
498. Ocana, M.; Cendan, C. M.; Cobos, E. J.; Entrena, J. M.; Baeyens, J. M. *Eur. J. Pharmacol.* **2004**, *500*, 203–219.
499. Coghlan, M. J.; Carroll, W. A.; Gopalakrishnan, M. *J. Med. Chem.* **2001**, *44*, 1627–1653.
500. Pelaia, G.; Gallelli, L.; Vatrella, A.; Grembiale, R. D.; Maselli, R.; De Sarro, G. B.; Marsico, S. A. *Life Sci.* **2002**, *70*, 977–990.
501. Mayhan, W. G.; Mayhan, J. F.; Sun, H.; Patel, K. P. *Microcirculation* **2004**, *11*, 605–613.
502. Chai, Q.; Liu, Z.; Chen, L. *Chin. J. Physiol.* **2005**, *48*, 57–63.
503. Zhang, Y.; Gao, Y. J.; Zuo, J.; Lee, R. M.; Janssen, L. J. *Eur. J. Pharmacol.* **2005**, *514*, 111–119.
504. Lee, D. H.; Chang, L.; Sorkin, L. S.; Chaplan, S. R. *J. Pain* **2005**, *6*, 417–424.
505. Felix, R. *J. Med. Genet.* **2000**, *37*, 729–740.
506. Avanzini, G.; Franceschetti, S.; Avoni, P.; Liguori, R. *Expert Rev. Neurother.* **2004**, *4*, 519–539.
507. Delisle, B. P.; Anson, B. D.; Rajamani, S.; January, C. T. *Circ. Res.* **2004**, *94*, 1418–1428.
508. George, A. L., Jr. *Epilepsy Curr.* **2004**, *4*, 65–70.
509. Jurkat-Rott, K.; Lehmann-Horn, F. *J. Clin. Invest.* **2005**, *115*, 2000–2009.
510. Kass, R. S. *J. Clin. Invest.* **2005**, *115*, 1986–1989.
511. Sanguinetti, M. C.; Chen, J.; Fernandez, D.; Kamiya, K.; Mitcheson, J.; Sanchez-Chapula, J. A. *Novartis Found. Symp.* **2005**, *266*, 159–166, (discussion 166–170).
512. Sarkozy, A.; Brugada, P. *J. Cardiovasc. Electrophysiol.* **2005**, *16*, S8–S20.
513. Schimpf, R.; Wolpert, C.; Gaita, F.; Giustetto, C.; Borggrefe, M. *Cardiovasc. Res.* **2005**, *67*, 357–366.
514. Schulze-Bahr, E.; Neu, A.; Friederich, P.; Kaupp, U. B.; Breithardt, G.; Pongs, O.; Isbrandt, D. *J. Clin. Invest.* **2003**, *111*, 1537–1545.
515. Ueda, K.; Nakamura, K.; Hayashi, T.; Inagaki, N.; Takahashi, M.; Arimura, T.; Morita, H.; Higashiuesato, Y.; Hirano, Y.; Yasunami, M. et al. *J. Biol. Chem.* **2004**, *279*, 27194–27198.
516. Kubisch, C.; Schroeder, B. C.; Friedrich, T.; Lutjohann, B.; El-Amraoui, A.; Marlin, S.; Petit, C.; Jentsch, T. *J. Cell* **1999**, *96*, 437–446.
517. Mohammad-Panah, R.; Demolombe, S.; Neyroud, N.; Guicheney, P.; Kyndt, F.; van den Hoff, M.; Baro, I.; Escande, D. *Am. J. Hum. Genet.* **1999**, *64*, 1015–1023.
518. IUPHAR Compendium of Voltage-gated Ion Channels 2005. http://www.iuphar-db.org/iuphar-ic (accessed Aug 2006).
519. Voltage-Gated Potassium Channel Database. http://vkcdb.biology.ualberta.ca (accessed Aug 2006).

Biographies

Joseph G McGivern is the associate director of research in the HTS-Molecular Pharmacology Department at Amgen Inc., Thousand Oaks, CA. He obtained his PhD in physiology from the School of Medicine at the Queen's University of Belfast, UK, in 1992. Following the award of his PhD, he held research positions at Syntex Pharmaceuticals in Edinburgh (1991–95) and Roche Bioscience in Palo Alto, CA (1995–2000), before moving to Amgen in 2000. During his career in industry, McGivern has been involved in many aspects of ion channel drug discovery, including target validation, assay development, high-throughput screening, and technology development.

Jennings F Worley is the associate director of discovery operations at Vertex Pharmaceuticals Inc., San Diego, CA. He has been involved in ion channel research and drug discovery throughout his career. He received his PhD from the University of Maryland, School of Medicine in 1985 from the Department of Physiology. He then held postdoctoral positions at the University of Miami, Department of Pharmacology, and the University of Vermont, Department of Pharmacology. As a faculty member at the West Virginia University Department of Pharmacology and Toxicology, he continued his work on the relationship between ion channel structure and function and the role of ion channels in controlling cell function. Since 1990, Worley has been contributing to drug discovery, first at GlaxoSmithKline for over 12 years, then at Amphora Discovery Corporation and, more recently, at Vertex. During this time he has been involved in many phases of drug discovery, including target identification and validation, high-throughput screening methodology and technology development, as well as advancing compounds through preclinical studies for candidate selection.

2.22 Ion Channels – Ligand Gated

C A Briggs and M Gopalakrishnan, Abbott Laboratories, Abbott Park, IL, USA

© 2007 Elsevier Ltd. All Rights Reserved.

2.22.1	**Introduction**	**878**
2.22.2	**Cys-Loop Superfamily**	**879**
2.22.2.1	Nicotinic Acetylcholine Receptors	879
2.22.2.1.1	Current nomenclature	879
2.22.2.1.2	Molecular biology	884
2.22.2.1.3	Protein structure	885
2.22.2.1.4	Physiological functions	886
2.22.2.1.5	Regulation	886
2.22.2.1.6	Prototypical pharmacology	887
2.22.2.1.7	Prototypical therapeutics	887
2.22.2.1.8	Diseases	888
2.22.2.2	$5HT_3$ Receptors	888
2.22.2.2.1	Current nomenclature	888
2.22.2.2.2	Molecular biology	888
2.22.2.2.3	Protein structure	889
2.22.2.2.4	Physiological functions	889
2.22.2.2.5	Regulation	889
2.22.2.2.6	Prototypical pharmacology	889
2.22.2.2.7	Prototypical therapeutics	890
2.22.2.2.8	Diseases	891
2.22.2.3	$GABA_A$ and $GABA_C$ Receptors	891
2.22.2.3.1	Current nomenclature	891
2.22.2.3.2	Molecular biology	891
2.22.2.3.3	Protein structure	891
2.22.2.3.4	Physiological functions	892
2.22.2.3.5	Regulation	892
2.22.2.3.6	Prototypical pharmacology	892
2.22.2.3.7	Prototypical therapeutics	893
2.22.2.3.8	Diseases	894
2.22.2.4	Glycine Receptors	894
2.22.2.4.1	Current nomenclature	894
2.22.2.4.2	Molecular biology	894
2.22.2.4.3	Protein structure	894
2.22.2.4.4	Physiological functions	894
2.22.2.4.5	Regulation	895
2.22.2.4.6	Prototypical pharmacology	895
2.22.2.4.7	Prototypical therapeutics	896
2.22.2.4.8	Diseases	896
2.22.2.5	Zinc-Gated Ion Channel	896
2.22.3	**Glutamate-Gated Family**	**897**
2.22.3.1	Current Nomenclature	897
2.22.3.2	Molecular Biology	897
2.22.3.3	Protein Structure	897
2.22.3.3.1	α-Amino-3-hydroxy-5-methylisoxazole-4-propionic acid-sensitive receptors	898
2.22.3.3.2	Kainate-sensitive receptors	898
2.22.3.3.3	N-methyl-D-aspartate-sensitive receptors	898
2.22.3.4	Regulation	899
2.22.3.5	Physiological Functions	899

2.22.3.6	Prototypical Pharmacology	900
2.22.3.7	Prototypical Therapeutics	902
2.22.3.8	Diseases	903
2.22.4	**Adenosine Triphosphate-Gated Family**	**903**
2.22.4.1	Current Nomenclature	903
2.22.4.2	Molecular Biology	904
2.22.4.3	Protein Structure	904
2.22.4.4	Physiological Functions	904
2.22.4.5	Regulation	905
2.22.4.6	Prototypical Pharmacology	905
2.22.4.7	Prototypical Therapeutics	905
2.22.4.8	Diseases	907
2.22.5	**Transient Receptor Potential Family**	**907**
2.22.5.1	Current Nomenclature	907
2.22.5.2	Molecular Biology	907
2.22.5.3	Protein Structure	907
2.22.5.4	Physiological Functions	908
2.22.5.5	Regulation	908
2.22.5.6	Prototypical Pharmacology	908
2.22.5.7	Prototypical Therapeutics	908
2.22.5.8	Diseases	909
2.22.6	**Acid-Sensitive Ion Channel Family**	**910**
2.22.6.1	Current Nomenclature	910
2.22.6.2	Molecular Biology	910
2.22.6.3	Protein Structure	910
2.22.6.4	Physiological Functions	911
2.22.6.5	Regulation	911
2.22.6.6	Prototypical Pharmacology	911
2.22.6.7	Prototypical Therapeutics	911
2.22.6.8	Diseases	912
2.22.7	**Future Directions**	**912**
References		**913**

2.22.1 Introduction

Ion channels expressed in the plasma membrane have been classified broadly as voltage-gated ion channels (VGICs; *see* 2.21 Ion Channels – Voltage Gated) and ligand-gated ion channels (LGICs). In case of the latter, the ligand is generally considered to be an extracellular messenger such as a neurotransmitter. Often, the same transmitter acts upon an array of G protein-coupled receptors (GPCRs) as well as LGIC, for example, acetylcholine acting via the muscarinic GPCR as well as nicotinic acetylcholine receptor (nAChR) LGIC. A few notable exceptions include glycine which, to date, is known to act on LGIC only and neuropeptides and histamine which appear to act exclusively via GPCRs. The boundaries between LGIC and VGIC become less marked when one considers intracellular ligands, for example Ca^{2+}/calmodulin or G protein-regulated channels, or the fact that some channels may be regulated by both ligand and voltage. In this chapter, however, we will consider the term 'ligand' to refer to extracellular messengers that are meant, physiologically, to exert predominant control over the opening or closing of the ion channel. Typically, this would comprise neurotransmitter receptor-associated LGIC hosting an ion channel domain that is integral to the transmembrane region and an extracellular ligand binding domain. The chemical signal to which an LGIC responds depends on the specificity of ligand binding whereas the nature of the electrical signal generated depends on both selectivity of the ion channel and the electrochemical gradient across the membrane. Generally, cation-selective channels generate a net depolarizing or excitatory signal and anion-selective channels generate a net hyperpolarizing or inhibitory signal.

LGIC activated by extracellular ligands may be divided into five families. Three LGIC families that have initially been recognized based upon protein sequence and predicted receptor structure are (1) the Cys-loop superfamily (nAChRs, 5-hydroxytryptamine $5HT_3$ receptors, γ-amino-butyric acid ($GABA_{A/C}$) receptors, glycine receptor (GlyR), and a newly discovered Zn^{2+}-gated channel, (2) the glutamate (Glu)-gated family, and (3) the ATP (purinergic)-gated family. Two additional LGIC that are increasingly being recognized and whose better-known roles lie outside the synapse are (1) the transient receptor potential (TRP) channel family where members have been shown to respond to specific extracellular messengers, and (2) the acid-sensitive ion channel (ASIC) family whose members are gated by proton interaction. TRP and ASIC generally appear to function as sensory channels, but some family members are also expressed in central neural pathways, including brain, where the roles remain to be further elucidated. Many LGIC are expressed not only in neurons and muscle, but also in nonneuronal, 'nonexcitable' cells such as glia, lymphocytes, keratinocytes, endothelial cells, pancreatic cells, sperm, and others. Thus, LGIC undoubtedly serve many roles apart from neurotransmission, but these roles are only beginning to be elucidated. This chapter will present an overview of the properties of various LGIC. For more detailed information and discussion, the reader is referred to recent review articles cited in the following topical sections. Topography models of LGIC are depicted in **Figure 1**, and a general summary of functional and pharmacological properties are presented in **Table 1**.

2.22.2 Cys-Loop Superfamily

All members of the Cys-loop superfamily are thought to possess a similar architecture. These LGICs are pentameric, comprising five membrane-spanning subunits arrayed psuedosymmetrically around a central ion-conducting channel that is opened and closed depending upon agonist binding and allosteric transitions associated with activation and desensitization. Each subunit has four putative transmembrane (TM) regions. Functional Cys-loop LGIC require subunits with a large N-terminal domain, wherein agonist is bound, and at least one (usually two) subunits must contain the signature vicinal cysteine pair that defines one of the peptide loops important for ligand binding and/or channel gating. Typically, a subunit containing the Cys-loop signature sequence is given the α appellation, whether or not it has been demonstrated to participate directly in channel gating.

The Cys-loop superfamily contains both cationic and anionic LGIC. Cation-conducting LGIC are excitatory – depolarizing the cell membrane to activate voltage-dependent channels and, in many but not all instances, gating Ca^{2+} to activate intracellular Ca^{2+}-dependent signaling pathways. Other LGIC are anion-conducting and generally inhibit neuronal firing in adult mammals (depending upon the Cl^- ion gradient that differs among cells and with development). Within the Cys-loop superfamily, nAChR and $5HT_3$ receptor channels are cationic and excitatory whereas $GABA_A$ and Gly receptor channels are anionic and generally inhibitory. Although structure–function studies of LGIC have revealed a great deal about the molecular determinants critical to agonist binding and ion selectivity, a comprehensive understanding of how LGIC function is limited; high-resolution positioning and dynamics to designate these determinants in a three-dimensional context is an increasingly active area of research.[1–5] The more recent discovery and structural resolution of a soluble acetylcholine binding protein (AChBP) by Sixma and colleagues has provided a framework to model the extracellular domain of all Cys-loop LGIC in three-dimensional terms, and chimeric constructs of AChBP with Cys-loop LGIC (e.g., $5HT_3$) have begun to determine how agonist binding is converted to channel opening in this family.[1,6]

2.22.2.1 Nicotinic Acetylcholine Receptors

nAChRs are widely characterized transmembrane allosteric proteins involved in the physiological responses to acetylcholine. These receptors are composed of homologous subunits encoded by a large multigene family, and are expressed in muscle, nerve, and sensory cells where they play critical roles in neuronal transmission and modulation. Several excellent reviews describing biophysical, physiological, and pharmacological properties of nAChR subtypes have appeared recently.[7–11]

2.22.2.1.1 Current nomenclature

nAChRs comprise the ionotropic side of acetylcholine receptors, with muscarinic acetylcholine receptors (mAChRs) comprising the GPCR side. Classically, GPCR is the metabotropic receptor. However, nAChR and other LGIC also can

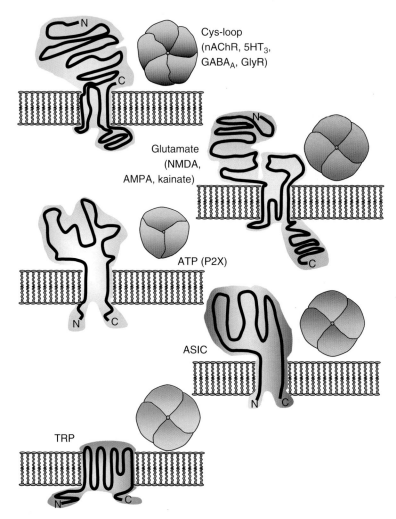

Figure 1 Topography models of ligand-gated ion channels (LGICs). Schematic representation of transmembrane topology of LGIC – Cys-loop (nAChR, 5HT$_3$R, GABA$_A$, and GlyR), glutamate (NMDA, AMPA, and kainate), ATP (P2X), ASIC, and TRP members. N and C indicate the N- and C-terminus, respectively. The heterooligomeric (trimeric, tetrameric, or pentameric) organization is also depicted.

influence metabotropic processes through Ca^{2+} influx mediated directly through the LGIC pore and indirectly through depolarization and activation of voltage-gated Ca^{2+} channels. Classically, nAChRs were categorized as 'muscle-type' (skeletal muscle and the related electroplaque of *Torpedo* and other electric fish) and ganglionic (sympathetic, parasympathetic, and enteric ganglia) based upon pharmacologic distinctions such as selective blockade of ganglionic receptors with hexamethonium and selective neuromuscular desensitization by decamethonium.[12] In the past 15 years, however, molecular biology studies have made it clear that nAChRs are widespread in brain as well as ganglia and skeletal muscle. The term neuronal nAChRs is meant to include those found in the central nervous system (CNS – brain and spinal cord) and in the peripheral nervous system (PNS – sensory as well as autonomic and enteric), but not skeletal muscle/electroplaque. Indeed, no muscle nAChR subunit has yet been found to be expressed in brain or ganglion. However, neuronal nAChRs are not exclusively neuronal; several of these subunits are found in various nonneuronal cells such as endothelial cells, keratinocytes, macrophage, and lymphocytes where their roles remain to be determined.

nAChR subunits containing the Cys-loop signature vicinal cysteines in the appropriate position are designated as α. In skeletal muscle, the other subunits were named β, δ, ε (embryonic), and γ (adult substitute for ε). Since the cloning of neuronal nAChR, muscle α has become α1 and neuronal α are numbered in sequence of discovery. Neuronal nAChR subunits that did not contain the signature vicinal cysteines were designated as non-α in the earlier literature. Today, all neuronal non-α are designated as β and numbered in order of discovery with the muscle subunit being β1. These

Table 1 Ligand-gated ion channels: functional and pharmacological distinctions

Transmitter class	Receptor subtype	Ion selectivity[a]	Agonists	Antagonists	Modulators
ACh nicotinic	$\alpha1\beta1\gamma\delta$ / muscle, Torpedo	Cation; low PCa^{2+}	Nicotine, decamethonium, succinylcholine	Pancuronium, vecuronium, α-bungarotoxin, α-conotoxin MI	Physostigmine, galantamine, proadifen, steroids, e.g., promgestone, cholesterol
	$\alpha3\beta4$ / ganglionic nAChR		Nicotine, lobeline, DMPP	Hexamethonium, trimetaphan	
	$\alpha4\beta2$	Cation; moderate PCa^{2+}	Nicotine, cytisine, ispronicline, ABT-418, SIB-1508Y	Dihydro-β-erythroidine	
	$\alpha6$		Epibatidine, A-85380, cytisine	α-Conotoxin MII	
	$\alpha7$	Cation; high PCa^{2+}	Choline, PNU-282987	Methyllycaconitine	PNU-120596, 5-hydroxy indole
	$\alpha9$, $\alpha9\alpha10$		ACh	α-Conotoxin PeIA, nicotine, strychnine	
Serotonin	$5HT_{3A}$	Cation; moderate PCa^{2+}	2-methyl-5-hydroxytryptamine, phenylbiguanide, m-chlorophenylbiguanide, SR 57227A	Ondansetron; Granisetron; MDL72222	Ethanol, anesthetics (volatile)
GABA $GABA_A$	$GABA_A$ $\alpha1\beta2\gamma2$	Anion (Cl^-)	Muscimol > GABA, THIP (selective for $GABA_A$ versus $GABA_C$)	Bicuculline	Benzodiazepines, barbiturates, neuroactive steroids, ethanol
	$GABA_A$ $\alpha4$, $\alpha6$				Relatively insensitive to benzodiazepines
GABA $GABA_C$	$\rho1$, $\rho2$, $\rho3$		GABA > muscimol, (+)-CAMP	TPMPA, CGP36742, 3-ACPBuPA	Glycine potentiates, Zn^{2+} inhibits, loreclezole and (+)-ROD188 potentiate $GABA_A$, inhibit $GABA_C$; μM neurosteroids potentiate $GABA_C$ versus nM for $GABA_A$
Glycine		Anion	Glycine, β-alanine, taurine	Strychnine, nipecotic acid, ginkgolide B, tropisetron	Positive ivermectin, ethanol and volatile anesthetics Negative progesterone, pregnenolone (inhibition)
Zn^{2+}	ZAC	Cation	Zn^{2+}	d-Tubocurarine	
Glutamate[b]	GluR1	Cation, moderate PCa^{2+}	(S)-AMPA, (S)-ACPA, (S)-5-fluorowillardiine	ATPO, GYKI 53655, GYKI 52466, SYM-2206, Spider toxins (iGluR subunit and	Cyclothiazide, PEPA, AMPAkines

continued

Ion Channels – Ligand Gated

Table 1 Continued

Transmitter class	Receptor subtype	Ion selectivity[a]	Agonists	Antagonists	Modulators
AMPA-sensitive	GluR2	Cation and low Cl$^-$ (0.1 PCl$^-$/PNa$^+$) Q/R RNA editing reduces conductance and PCa^{2+}/PNa^{+c}		Q/R editing dependent), CNQX, DNQX	
Glutamate kainate-sensitive	Generic GluR5-GluR6			Spider toxins (Q/R dependent), CNQX/DNQX (also block AMPA-sensitive iGluR)	Concanavalin A
	GluR5	Cation	(S)-5-iodowillardiine, (S)-ATPA, LY339434	LY382884, LY294486, NS-102, SYM-2081 (desensitizer)	
	GluR6	Cation, Q/R RNA editing reduces PCa^{2+}/PNa$^+$ and increases PCl$^-$/PNa$^+$		NS-102, SYM-2081 (desensitizer)	
	GluR6/KA		5-Iodowillardiine (weak)		
Glutamate NMDA-sensitive	NR1 (glycine coagonist site)			Kynurenic acid derivatives, quinoxalinediones, phthalazinediones, benzazepinediones	
	NR1/NR2 generic	Cation; high PCa^{2+}	NMDA, aspartate	Mg^{2+}, AP5, AP7; MK801	Positive: arachidonic acid, PACAP Negative: dynorphin, H$^+$, Zn^{2+}
			Coagonist: glycine	10-fold less sensitive to glycine site antagonists	
	NR1/NR2A	Cation; moderate-high PCa^{2+}, low PCl$^-$			High Zn^{2+} sensitivity
	NR1/NR2B	Cation; high PCa^{2+}		CGP 61594 glycine site antagonist, Ifenprodil, Ro 25-6981, CP 101,606	
	NR1/NR2C	Cation; moderate PCa^{2+}			Low Zn^{2+} sensitivity
	NR1/NR2D				
Unknown	NR3A	Cation	Glycine alone is agonist		
	NR3B				
ATP (purines)	P2X1	Cation, high PCa^{2+}, (PCa^{2+}/Na$^+$ ~3.9)	BzATP, α,β-MeATP	Ip$_5$I, TNP-ATP, suramin analogs, PPADS analogs	
	P2X1/5	Cation, moderate PCa^{2+}	α-MeSATP, α,β-MeATP	Weak sensitivity to TNP-ATP	

Ion Channels – Ligand Gated

	Receptor	Properties	Ligands/Activators	Antagonists/Blockers	Modulators
	P2X2	Cation, moderate PCa^{2+}, large pore formation	2-MeSATP; ATPγS (insensitive to α,β-MeATP)	PPADS	Zn^{2+} potentiates, H$^+$ potentiates
	P2X2/3	Cation, moderate PCa^{2+}	BzATP, α,β-MeATP little response to Ap5A	TNP-ATP, A-317491	H$^+$ potentiates
	P2X3	Cation, moderate PCa^{2+}	BzATP, α,β-MeATP, Ap5A	TNP-ATP, A-317491, suramin, PPADS	Zn^{2+} potentiates, H$^+$ inhibits
	P2X4	Cation, large pore formation	BzATP	(Insensitive to suramin, PPADS)	Zn^{2+}; ivermectin
	P2X5	Cation	ATP (insensitive to α,β-MeATP)	(Insensitive to suramin, PPADS)	
	P2X6		α,β-MeATP		
	P2X7	Cation, large pore formation		Calmidazolium; KN-62	
Sensory stimuli	TRPA1		Mustard oil, cinnamon oil		
	TRPM8	Cation, moderate PCa^{2+} (PCa^{2+}/Na$^+$ 1–3)	Menthol, Icilin		
	TRPV1	Cation, high PCa^{2+} (PCa^{2+}/Na$^+$ ~10), PMg^{2+} (PMg^{2+}/Na$^+$ ~5)	Capsaicin, resiniferatoxin, anandaminde, olvanil, SDZ-249665 Endogenous: anandaminde, 12-HPETE, leukotriene B4, N-arachidonoyl dopamine	Capsazepine, Iodo-resiniferatoxin, NDT-9525276, ruthenium red	'Agonists', protons, temperature, may act in concert as modulators
	TRPV3		Camphor		
Proton	ASIC1a	Cation; Ca^{2+}	Proton	PcTx1	Zn^{2+} (nM inhibition), FMRFamide (potentiation)
	ASIC1b	Cation	Proton		FMRFamide (potentiation)
	ASIC2a	Cation, low PCa^{2+}	Proton		Zn^{2+} (μM potentiation)
	ASIC2b	Cation	Proton		
	ASIC3	Cation	Proton	APETx2; Gd^{3+}	FMRFamide, neuropeptide SF, neuropeptide FF[d]
	ASIC4		Proton		

[a] Calcium permeability (PCa^{2+}): low PCa^{2+}: PCa^{2+}/Na$^+$ ≤1; moderate PCa^{2+}: PCa^{2+}/Na$^+$ 1–3; high PCa^{2+}: PCa^{2+}/Na$^+$ 3–20.
[b] Glutamate LGIC are subject to RNA editing and alternative splicing at sites in or near the channel region.[145]
[c] RNA-edited forms tend to be predominant in the adult (for further discussion, see text).
[d] Lower concentrations of neuropeptide FF appear selective for ASIC2a/3 heteromeric receptors.[228]

numeric designations are not sub- or superscripted. Subscripted numbers are meant to refer specifically to the number of like subunits present in the pentameric nAChR. Thus, for example, the adult muscle nAChR is $(\alpha 1)_2 \beta 1 \delta \gamma$, meaning two $\alpha 1$, one $\beta 1$, one δ, and one γ subunit. However, one should be cautious in that the same muscle nAChR might also be called $\alpha 2 \beta \delta \gamma$ or $\alpha_2 \beta \delta \gamma$, especially before additional nAChR α became known. Muscle nAChRs do not contain $\alpha 2$ or any α other than $\alpha 1$, nor any β other than $\beta 1$. Likewise, muscle nAChR subunits are not known to be expressed in neurons nor any other cell type. The lone exception is that 'neuronal' $\alpha 7$ nAChR, which in fact is found in many nonneuronal cells, can be transiently expressed in embryonic or denervated skeletal muscle, in addition to the classical $\alpha 1$-containing muscle nAChR.[13–15]

Lastly, use of the asterisk (*) wild card should be noted. In most instances outside the neuromuscular junction, the exact composition and stoichiometry of nAChR is not known with certainty. Characteristics of the physiologic response may be similar to a recombinant nAChR expressed, for example, in *Xenopus* oocytes or transfected HEK293 cells, but often differences also can be found and it is not clear whether the differences are due to assay, expression context, or receptor structure. The asterisk wild card is used to indicate the possibility that other subunits may also contribute to the nAChR, or may be read to indicate that the response is similar but not necessarily identical to the nAChR designated. Thus, for example, $\alpha 4 \beta 2^*$ would mean $\alpha 4 \beta 2$-like nAChR that may contain other subunits.

2.22.2.1.2 Molecular biology

To date, 17 nAChR subunits have been cloned. Of these, five ($\alpha 1$, $\beta 1$, δ, ε, and γ) are expressed in skeletal muscle or the related fish electroplaque, and apparently only in those tissues. The other 12 subunits are found in neurons, in endocrine cells (adrenal chromaffin cells), and to some extent, also in classically nonexcitable cells such as endothelial, epithelial, and immune system cells. These are the $\alpha 2$ through $\alpha 10$ and $\beta 2$ through $\beta 4$ subunits. Among these, $\alpha 7$ nAChR subunits appear to be one of the most widespread throughout the body, including brain, autonomic ganglia, adrenal chromaffin cells, nonneuronal cells, and even skeletal muscle during development or following denervation. The roles of $\alpha 7$ nAChR are not well understood, but are of considerable interest because of implications in diseases with large unmet medical need.

The nAChRs principally responsible for autonomic synaptic transmission, including sympathetic/adrenal transmission, are the $\alpha 3^*$ nAChRs. These include $\alpha 3 \beta 2^*$ as well as $\alpha 3 \beta 4^*$ nAChR. In recombinant expression systems the $\alpha 3 \beta 4$ nAChR correlates more closely with ganglionic receptors expressed by neuroblastoma IMR-32 or pheochromocytoma PC-12 cells assayed in the same system. Other nAChR subunits expressed in autonomic neurons and endocrine cells are $\alpha 5$ and $\alpha 7$. There are many possible combinations for these five subunits in a pentameric complex, but exactly what combinations are formed and whether they differ in any disease or pathological state is not clear. The $\alpha 5$ subunit does not function alone or with any known β subunit, but $\alpha 5$ may combine with other subunits to modify their function biophysically or pharmacologically, for example, in $\alpha 3 \alpha 5 \beta 4$. The $\alpha 7$ subunit, in contrast to $\alpha 5$, can form functional homomeric nAChR in recombinant systems, and it is believed that native $\alpha 7$ nAChRs also are homomeric, at least, in part. There is some evidence that $\alpha 7$ may combine with $\beta 2$ or other subunits, but further research is needed. Ganglionic $\alpha 3^*$ nAChRs also are expressed in CNS and other areas of pharmaceutic interest. However, potential adverse effect liability resulting from interaction with autonomic and endocrine nAChR may be an obvious problem.

CNS receptors of particular interest are $\alpha 4 \beta 2^*$, $\alpha 7^*$, and $\alpha 6^*$. The $\alpha 4 \beta 2$ nAChR comprises 90% of the high-affinity nicotine-binding sites in rodent brain and may be involved in attentional and cognitive functions as well as mediating the nicotine cue (reward). In recombinant systems, $\alpha 4 \beta 4$ also can be expressed and are particularly sensitive to many agonists, but there is little evidence for physiological expression of $\alpha 4 \beta 4$ except perhaps in retina.[16,17] The $\alpha 7^*$ nAChRs have among the lowest affinity to nicotine, but also may contribute to cognition and $\alpha 7$ deficits have been implicated in schizophrenia and Alzheimer's disease (AD). While $\alpha 7$ expression is widespread, its roles are largely unknown and knockout mice have shown surprisingly subtle deficits.[18,19] The $\alpha 6^*$ nAChRs have been difficult to express functionally in recombinant systems and therefore development of $\alpha 6^*$ pharmacology has been slow. However, recent studies in mice have made it clear that $\alpha 6^*$ nAChR is expressed in brain and can stimulate dopamine release.[20–25] Unlike many other nAChR subunits, $\alpha 6$ and $\alpha 4$ appear to be selectively expressed in CNS and are thereby an enticing target if highly selective ligands can be identified.

The $\alpha 9$ subunit has a restricted distribution, being found primarily in the inner ear,[26–31] keratinocytes,[32] and lymphocytes.[33] Additionally there is some evidence for $\alpha 9$ expression in sensory dorsal root ganglia.[34] In recombinant systems, $\alpha 10$ increases the functional expression of $\alpha 9$ without marked influence on the unique $\alpha 9$ pharmcology.[32,35] Whether $\alpha 10$ can combine with other subunits is not clear.

The $\alpha 2$ subunit forms functional nAChR when combined with $\beta 2$ or $\beta 4$, and $\alpha 2$ is expressed in brain, particularly monkey.[36–38] However, at present, there are few tools available to address $\alpha 2^*$ nAChR and its roles are unclear.

Two subunits, α5 and β3, have been enigmatic in expression systems. The α5 subunit clearly is expressed throughout the nervous system and likely plays a role in nAChR expression and function. In recombinant systems, however, it does not form functional nAChR either by itself or when combined with any known β subunit. There are reports that α5 may modify the function of other nAChR, particularly as a third party in α4β2* and α3β4*. However, functional effects are subtle and further work is needed.

Likewise, the β3 subunit does not form a functional nAChR when combined with any known α subunit in a mixture limited to two subunit types. However, β3 again may modify expression or function of other nAChR. In brain, elegant recent studies using gene deletion and selective ligands including *Conus* toxins indicate that β3 may be particularly important in α6* nAChR.[21,23,39,40]

The α8 subunit is known to be expressed only in avian brain and retina, not in mammals. Pharmacologically and biophysically it is similar to α7.

2.22.2.1.3 Protein structure

All nAChRs, like other members of the Cys-loop superfamily, are believed to be pentameric (assembled from exactly five homologous subunits). The α7, α8, and α9 nAChR subunits each are able to form functional homomeric LGIC (comprising one and only one subunit type) in recombinant expression systems. Among these, α9 functional expression is boosted markedly by α10 making it seem likely that α9α10 is a prominent heteromer, α8 has been found in chick only, and thus α7 appears to be the predominant homomeric nAChR in mammals. However, conceivably α7 could also participate in heteromeric assemblies and some evidence suggests expression of α7β2 by some neurons.[41]

Other nAChRs are heteromeric and assembled from at least two different subunits including at least one from the α group and at least one from the non-α group. The binding site for ACh lies at the interface(s) between α and non-α subunit, and peptide loops from both subunits contribute to binding and allosteric channel activation. While the α subunit may have the more prominent influence, the non-α subunit also contributes to nAChR pharmacology. The existence of transcripts for a range of α and β subunits in neurons raises the potential for a wide array of receptor combinations, although one would expect that a smaller subset of combinations would be preferred in vivo. Accordingly, the analysis of native nAChR composition is an area of intense research guided by both pharmacologic studies with subtype selective ligands and evaluation of gene knockout animals. For example, it has been determined that a variety of α/β subunit combinations underlie α-conotoxin MII-sensitive and -insensitive nAChRs modulating dopamine release in the striatum[22] whereas a unique combination has been elucidated in retinal afferents.[21]

Initial evidence that nAChRs are pentamers is derived from electron microscopy of tubular crystals prepared from *Torpedo* electric organ postsynaptic membranes.[4] More recently, additional insight into nAChR structure has emerged from x-ray diffraction studies with an invertebrate soluble ACh binding protein (AChBP).[1,2,42–45] AChBP also is a pentamer and shows considerable similarity to the extracellular N-terminal ligand binding domain of nAChR, with conservation of amino acid residues involved in the formation of agonist binding site.

The muscle-type nAChR, derived from assembly of two α1 subunits and one each β1, δ, and γ (ε in fetus), is organized in a clockwise α1γα1β1δ arrangement. ACh binds to two orthosteric sites located at the interfaces of α1–γ and α1–δ. The segments hosting the ligand binding domain are organized around two sets of beta-sheets and joined via disulfide bridges forming the Cys-loop, as demonstrated by the structure of AChBP.[46] The subunits in the membrane-spanning domains are made from four alpha-helical segments TM1–TM4 that are arranged symmetrically, forming an inner ring of helices (TM2) which shapes the pore and an outer shell of helices (TM1, TM3, and TM4), which coil around to shield the inner ring from lipids.[4] The ion-conducting channel makes a narrow water-filled path across the membrane and contains the ion gate, which opens when ACh occupies both binding sites. Utilizing the structural model of the *Torpedo* ACh receptor, a series of molecular links that couple the neurotransmitter binding to channel gating has been identified.[47] A pair of arginine and glutamate residues in each α subunit electrostatically links peripheral and inner beta-sheets from the binding domain and positions them to engage with the channel. The glutamate, along with the flanking valine residues, energetically couples to conserved proline and serine residues in the pore-forming alpha-helix. These series of interresidue couplings serve as a primary pathway for linking ligand binding to channel gating. Mutual interactions between residues in the extracellular agonist binding domain and the TM2–TM3 loop has also been shown to be critical for fast gating properties.[48]

The TM3–TM4 intracellular loop in nAChR and other Cys-loop members is relatively long, exhibits relatively low homology among subunits, and contains a number of potential phosphorylation sites. It is thought that this loop may be important for receptor assembly, trafficking, and/or modulation by kinase signaling cascades. However, this is an active research area with much remaining to be learned. The extracellular C-terminus beyond TM4 is short, for example, only eight amino acids in rat and human α4. Nevertheless, it remains a region that may participate in LGIC pharmacology, and for example has been implicated in the potentiation of α4β2 nAChR by estradiol.[49]

Neuronal nAChRs are believed to have a similar pentameric structure. Most are thought to comprise at least two functional α subunits (five in the case of homomeric nAChR) and three β or combination of β plus α5 (α7α8 in chick and α9α10 in mammal would be exceptions lacking β but not homomeric). Thus, α4β2*, thought to comprise 90% of the high-affinity nicotine binding sites in rodent brain, would be expected to be $(α4)_2(β2)_3$ or possibly $(α4)_2α5(β2)_3$. However, more complex combinations can be envisioned and even with just α4 and β2, at least two pharmacologically distinct nAChR can be expressed in recombinant systems.[50] The composition of muscle nAChR is fairly well known, those of neuronal nAChRs are beginning to come to light, and those of other nAChR are not known but generally thought to follow the neuronal pattern. Pharmacologically, one may think of nAChR in six categories: (1) skeletal muscle (α1*); (2) α3β4* (correlating with the more prominent ganglionic nAChR pharmacology although other nAChR also are expressed in autonomic ganglia); (3) α3β2* and α6β2* (these likely will be subdivided in the near future, but there are few pharmacologic tools at present); (4) α4β2*; (5) α7*; and (6) α9*. Other receptors likely are expressed, for example α4β4* in retina[16] but are yet to attract as much therapeutic interest as others.

In addition to primary subunits, a number of accessory proteins that coassociate or interact with nAChRs continue to emerge. A family of genes encoded by the neuregulin-1 gene can modulate the expression of both muscle and neuronal-type nAChRs. In addition to their well-described roles at the neuromuscular junction, neuregulins can also increase expression of nAChRs in neurons such as those from interpeduncular nucleus and hippocampal interneurons.[51] In interaction of rapsyn with muscle, but not neuronal, nAChRs appear to be important for clustering nAChRs at the neuromuscular junction. RIC-3, originally identified in *Caenorhabditis elegans* as a protein encoded by the gene resistance to inhibitors of cholinesterase, can enhance the expression of multiple nAChRs including both homomeric and heteromeric receptors in transfected mammalian cells, possibly by enhancing subunit folding and assembly.[52] Other novel proteins have been identified by tandem affinity purification of the levamisole-sensitive nAChRs from *C. elegans*.[53]

2.22.2.1.4 Physiological functions

nAChRs are involved in a range of synaptic and extra synaptic functions.[9,54,55] The muscle-type nAChR is a key mediator of electrical transmission at the neuromuscular junction. In the PNS, nAChRs mediate ganglionic neurotransmission whereas in the CNS, nicotinic cholinergic innervation mediates synaptic transmission and regulates processes such as transmitter release, synaptic plasticity, and neuronal network integration by providing modulatory input to a range of other neurotransmitter systems. Thus, nAChR subtypes are implicated in a range of physiological and pathophysiological functions related to cognitive functions, learning and memory, reward, motor control, arousal, and analgesia. For example, the central α4β2 containing nAChRs are strongly implicated in the analgesic effects as revealed by knockout mice where antinociceptive effects to nicotine are reduced in both α4 and β2 null mutant mice.[56] Results from the gain-of-function α4-nAChR knockin mice suggest that this nAChR subunit is also important for nicotine-induced reward, tolerance, and sensitization.[57] On the other hand, α7 nAChRs are more implicated in cognitive and psychotic disorders. There is also growing evidence that nAChRs can modulate cellular functions beyond synaptic transmission in CNS and PNS. Acetylcholine can inhibit the release of macrophage tumor necrosis factor (TNF-α) and high mobility group box 1 (HMGB1) principally via the α7 nAChRs.[58]

2.22.2.1.5 Regulation

Upregulation of nAChRs, particularly the α4β2 subtype in the CNS, is a notable feature of chronic nicotine exposure, both in vitro and in vivo including in smokers.[59] This phenomenon has been studied in a number of heterologous expression systems. At least with α4β2 nAChRs, the extent, magnitude, and functional consequences have been attributed to factors such as ratio of heteromeric subunits, ligand-induced chaperon activity that accelerates pentamer assembly, and stabilization of receptors in a predominantly high-affinity state.[60,61] Upregulation of nAChRs derived from other subunits including α7, α3, and α6 has been reported, albeit to a lesser extent compared to α4β2 nAChRs.[62,63] nAChR regulation by kinases and phosphatases also is known, with a major role for tyrosine phosphorylation. The density and functional properties of α7 nAChRs are influenced by tyrosine phosphorylation status, where the net effects may be the result of balance between SRC family kinases and tyrosine phosphatases.[64,65] α7 nAChRs on chick ganglion neurons undergo calcium-dependent trafficking that requires functional soluble

N-ethylmaleimide-sensitive factor attachment protein receptors (SNARE) without which these receptors loose their ability to activate the transcription factor cAMP response element binding protein (CREB) when re-exposed to agonist.[66]

2.22.2.1.6 Prototypical pharmacology

In addition to the endogenous transmitter acetylcholine and the prototypical agonist nicotine, a range of selective agonists, competitive antagonists, and allosteric modulators of nAChRs are known to exist (**Figure 2**). Many systematic reviews of published nAChR agonists and antagonists of nAChRs are available.[7,8,11] In the area of agonists, medicinal chemistry efforts have focused mostly on analogs of nicotine, epibatidine, anabaseine, and cytisine. Analogs of nicotine and metanicotine analogs, cytisine (e.g., varenicline), epibatidine, and various classes of diamines and pyridyl ethers (e.g., A-85380, ABT-594) have been described with varying degrees of agonist efficacy and selectivity. A major objective in the optimization of these analogs is to enhance selectivity to specific heteromeric nAChR combinations (α4*, α6*, etc.) versus ganglionic nAChRs. At the α7 nAChR, a variety of quinuclidine analogs have been described as selective agonists with little activity at α4β2 and other heteromeric receptors whereas hydroxyindole, urea, and tetrahydroquinoline derivatives have been claimed as positive allosteric modulators. Like agonists, most of the reference antagonists are also obtained from natural sources. These include α-bungarotoxin, methyllycaconitine, dihydro-β-erythroidine, and D-tubocurarine with varying levels of discrimination among different nAChR subtypes. *Conus* peptides targeted to three different families of ligand-gated ion channels including nAChRs also are described, of which the most widely examined are the alpha-conotoxins exemplified by alpha-conotoxin ImI and alpha-conotoxin MII.[67]

2.22.2.1.7 Prototypical therapeutics

Neuromuscular nAChR antagonists have been marketed as muscle relaxants (antispasmodics) including surgical adjuncts, and in some instances, these compounds may reduce headache or muscle pain by reducing muscle spasm. Examples include afloqualone, atracurium, cisatracurium, doxacurium, eperisone, mivacurium, oxantel, pipecuronium, rocuronium, and vecuronium. Mecamylamine, a CNS-penetrating nAChR antagonist, was once marketed for

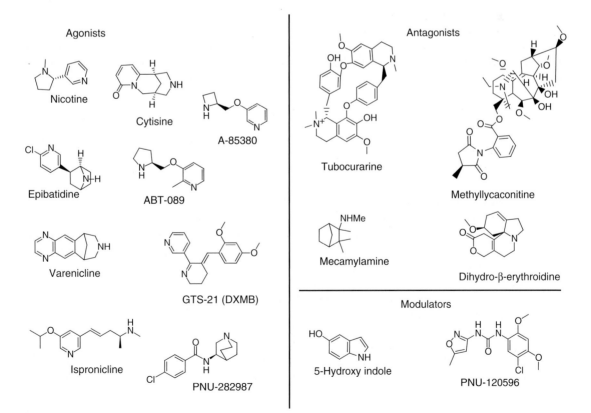

Figure 2 Agonists, antagonists, and modulators of nAChR.

hypertension, and currently may be of interest for neuropsychiatric disorders such as Tourette's syndrome, attention deficit hyperactivity disorder (ADHD), and depression. Other nAChR-related compounds in the market include ondansetron for emesis and galantamine for AD/cognitive impairment. However, neither of these compounds was targeted toward nAChR initially. Ondansetron is a 5HT$_3$ antagonist, which undoubtedly could explain its antiemetic action, but it also acts as an antagonist at muscle and neuronal nAChRs.[68–70] The alkaloid galantamine is an acetylcholinesterase inhibitor, but some laboratories have argued that it may also act as an allosteric modulator of certain nAChRs.[71–77] However, the latter remains somewhat controversial since the modulatory effects are rather modest compared to more recently discovered compounds.

Compounds with varying degrees of nAChR subtype selectivity have more recently entered clinical trials. These include varenicline, an α4β2 partial agonist for smoking cessation; ABT-089 and ispronicline (TC-1734), α4β2 partial agonists for ADHD, AD, and other cognitive deficits; GTS-21, an α7/α4β2 ligand for cognitive deficits in schizophrenia; MEM-3454 and TC-1698, α7 selective ligands for cognitive impairment and schizophrenia. Other agents that have been or continue to be investigated clinically include gantacurium, a neuromuscular nicotinic antagonist as anesthesia adjunct for muscle relaxation; rivanicline, an nAChR agonist for ulcerative colitis; ACV-1, an nAChR antagonist isolated from *Conus victoriae*, for neuropathic pain; CGX-1007, a *Conus* peptide, as an nAChR antagonist for epilepsy; and the alkaloid lobeline, an inhibitor of nicotine-evoked dopamine release for ADHD and methamphetamine abuse. Finally, another approach under study is the use of nicotine immunization to aid smoking cessation. Substances in clinical trials include NicQb (CYT-002) in phase II, Nic Vax, and TA-NIC (IPC-1020).

2.22.2.1.8 Diseases

The importance of nAChRs in pathology is illustrated by the findings that muscle nAChR is a target for autoimmunity in myasthenia gravis. Myasthenia gravis is caused by antibodies against the muscle nAChR, which compromise the end-plate potential reducing effective synaptic transmission.[78] Neuronal nAChR mutations are associated with a rare form of genetically transmissible epilepsy, the autosomal dominant nocturnal frontal lobe epilepsy (ADNFLE).[79] Mutations in the α4 and β2 genes in individuals suffering from ADNFLE give rise to single nucleotide polymorphisms, which cause single amino acid mutations or insertions. Heteromeric expression of the mutant subunit with its wild-type α or β counterpart demonstrated increased sensitivity to ACh and reduced Ca^{2+} potentiation response.[80] Another β2 mutation in the TM3 segment, outside the known ADNFLE cluster, also resulted in increased sensitivity to ACh.[81] This mutation is associated not only with epilepsy, but also with defects in cognitive tasks such as nAChR verbal memory tasks. The α7 nAChR has been implicated in schizophrenia, in particular for the auditory sensory processing deficit (P50 deficit) through genetic linkage findings in chromosome 15q14 locus, which corresponds to the locus of the gene encoding α7 subunit.[82]

2.22.2.2 5HT$_3$ Receptors

Serotonergic signaling plays key roles in the generation and modulation of various cognitive and behavioral functions such as pain, mood, sleep, addiction, depression, anxiety, and learning, and disruption of serotonergic systems has been implicated in the etiology of various psychiatric disorders.

2.22.2.2.1 Current nomenclature

Serotonin (5-hydroxytryptamine, 5HT) exerts its diverse actions by binding to receptors that are classed based on sequence similarity, second messenger signaling pathways, and pharmacological properties. The wide range of biological functions of 5HT is reflected in the existence of a large family of receptor subtypes for this neurotransmitter.[83] The majority of 5HT receptors couple to effector molecules through GPCRs.[84,85] However, the 5HT$_3$ receptor functions as a rapidly activating LGIC with a nonselective cation channel.[86–88]

2.22.2.2.2 Molecular biology

The human genome contains five genes (*5HT3A–E*) of which two subunits, *5HT3A* and *5HT3B*, have been well described and shown to be involved in the formation of functional ion channels. The first to be cloned, *5HT3A*, expresses as a functional homomer, whereas the second, *5HT3B*, does not express alone, but expresses in the presence of 5HT$_{3A}$ subunit. The genes for these two subunits are located in close proximity on chromosome 11. Although the exact subunit composition of receptors in vivo is not fully known, there does appear to be region-specific expression of the different subunits.[89] Long and short splice variants of 5HT$_{3A}$ receptors are known. In the case of the short splice variant, five or six amino acid residues located within the putative large intracellular loop between the TM3 and TM4 domains are deleted. The shorter form is the most abundant in the CNS. Cloned human forms correspond to the shorter form. In neuroblastoma cells and rodents, the proportions of short and long forms are regulated by development and/or differentiation. Studies in rat hippocampal CA1 interneurons demonstrate that the primary 5HT$_3$ receptor is the short subunit with 5HT$_{3B}$ and 5HT$_{3A}$ long variants undetected.[90]

2.22.2.2.3 Protein structure

Like other members of this superfamily, the 5HT$_3$ receptor is pentameric in nature, with the five subunits surrounding a central ion channel. Purified 5HT$_3$ receptor from NIE-115 neuroblastoma reveals a protein band of ~55 kDa. The receptor shares structural features common to Cys-loop LGIC, such as a large extracellular N-terminus region, four transmembrane domains (TM1–TM4), and a relatively long intracellular loop between TM3 and TM4. Studies of epitope-tagged 5HT$_3$ subunits using atomic force microscopy show that the subunit stoichiometry is 2A:3B and that the arrangement around the receptor rosette is B-B-A-B-A.[91] Like other LGIC, the agonist-binding site is located in the N-terminal region. A conserved proline residue at the apex of the TM2–TM3 loop of 5HT$_3$ has been shown to be critical for gating and mutation of this residue yielded receptors that could bind radiolabeled ligands, but were nonfunctional.[92] Recent studies of 5HT$_3$ in conjunction with other members of the Cys-loop family have helped identify key determinants in TM2 and the intracellular TM2–TM3 loop that affect channel conductance and ion selectivity.[93]

2.22.2.2.4 Physiological functions

5HT$_3$ receptors are expressed in native central and peripheral neurons where they are thought to play important roles in sensory processing and control of autonomic reflexes. Activation of the presynaptic 5HT$_3$ receptors is associated with modulation in the release of neurotransmitters and neuropeptides such as acetylcholine, GABA, dopamine, glutamate, and vasoactive intestinal peptide (VIP). 5HT$_3$ receptors play a role in nociceptive processing consistent with their expression on primary sensory afferents in the dorsal root ganglion and in the dorsal horn of the spinal cord. Overexpression of 5HT$_3$ receptors in mouse resulted in enhanced hippocampal-dependent learning and attention.[94] 5HT$_3$ receptors on vagal sensory afferents modulate visceral afferent and efferent pathways in the gastrointestinal and cardiovascular systems. In the enteric nervous system, 5HT$_3$ receptors also regulate gut motility and peristalsis whereas within the lower urinary tract, presynaptic 5HT$_3$ receptors have been implicated in parasympathetic transmission to the bladder.[95]

Native and cloned receptors evoke current responses to 5HT that reverse close to 0 mV, and desensitize in the continued presence of the agonist. Under physiological conditions, the response to 5HT is carried by inward movement predominantly of Na$^+$ ions, and divalent cations such as Ca^{2+} and Mg^{2+} suppress 5HT$_3$-mediated inward currents. The biophysical properties of the cloned 5HT$_{3A}$ and 5HT$_{3A-B}$ heteromers exhibit significant differences. Coexpression with the 5HT$_{3B}$ subunit results in about 10-fold higher single-channel conductance receptors derived from the 5HT$_{3A}$ homomer alone. By constructing chimeric 5HT$_{3A}$ and 5HT$_{3B}$ subunits, the 'HA-stretch' region within the large cytoplasmic loop of the receptor was identified as markedly influencing the single-channel conductance.[96] Although evidence for 5HT$_{3A/3B}$ heteromeric combinations cannot be ruled out in vivo, single-cell reverse-transcriptase polymerase chain reaction (RT-PCR) and whole-cell patch-clamp recording studies of rat hippocampal neurons suggest that the 5HT$_{3A}$ short subunit is most likely the primary 5HT$_3$ subunit contributing to the formation of functional 5HT$_3$ receptors in vivo.[90]

2.22.2.2.5 Regulation

In addition to the observation that coexpression with the 5HT$_{3B}$ subunit contributes to tissue-specific functional and pharmacological differences in 5HT$_3$ mediated signaling,[97] there is also evidence that the nicotinic α$_4$ subunit can coassemble with the 5HT$_{3A}$ subunit to form a native heteromeric ion channel in CA1 hippocampal neurons in vivo.[90] Biochemical studies and site-directed mutagenesis to selectively eliminate N-linked glycosylation showed that all three sites on the N-terminal of the murine 5HT$_{3A}$ receptor are critical for maximal plasma membrane targeting, ligand binding, and calcium influx.[98] The human homolog of the *ric-3* gene, *hRIC-3*, which enhances expression of α7 and possibly other nAChRs in heterologous expression systems, reportedly abolishes the expression of 5HT$_3$ receptors.[99]

2.22.2.2.6 Prototypical pharmacology

5HT$_3$ receptor antagonists are clinically utilized for preventing nausea and vomiting that commonly occur during cytotoxic cancer chemotherapy and radiation therapy.[100] These agents, also known as 'setrons', are regarded as one of the most efficacious antiemetics available and serve as agents of first choice to control nausea and vomiting. These

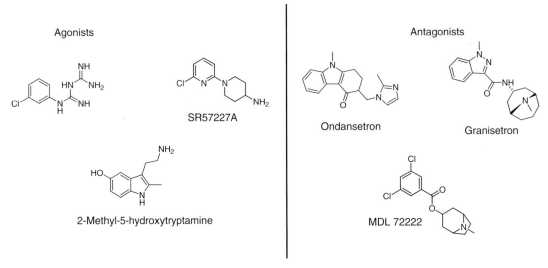

Figure 3 Agonists and antagonists of 5HT$_3$.

agents inhibit 5HT activity in the small intestine, vagus nerve, and the chemoreceptor trigger zone of the medulla-vomiting center. Examples of 5HT$_3$ antagonists include ondansetron, dolaesetron, tropisetron, palonosetron, and granisetron (**Figure 3**). Currently approved 'setrons' are all effective antiemetics, but differ in terms of pharmacodynamic and pharmacokinetic profiles influencing dosing regimens, cardiovascular safety, and potential drug–drug interactions. Notable differences in the pharmacological profiles for some 5HT$_3$ antagonists also exist. For example, granisetron is a potent and highly selective 5HT$_3$ receptor antagonist and has little or no affinity for other 5HT receptors whereas ondansetron, palonosetron, and tropisetron display affinity for other neurotransmitter receptors and LGIC, although the relative contributions of these interactions to the efficacy and/or adverse event profile are unclear.

The plant alkaloid picrotoxin, known as an antagonist of GABA$_A$ and glycine receptors, also inhibits 5HT$_3$ receptors. Picrotoxin displays notable selectivity between homomeric and heteromeric 5HT$_3$ receptors, with reduced sensitivity following coexpression with 5HT$_{3B}$ subunit. The TM2 segment is a major determinant in picrotoxin sensitivity.[101] Halogenated volatile anesthetics such as chloroform, isoflurane, and sevoflurane and n-alcohols such as octanol can modulate heteromeric human 5HT$_{3A/B}$ and 5HT$_{3A}$ receptors, albeit differentially. The effects (increase or decrease) on 5HT$_{3A/3B}$ currents are dependent on the molecular volume of the anesthetic agent.[102] An allosteric interaction of certain cannabinoids at 5HT$_{3A}$ receptor was suggested on the basis that compounds such as tetrahydrocannabinol (Δ^9-THC) and anandamide inhibited currents through recombinant 5HT$_{3A}$ receptors independently of cannabinoid receptors.[103]

Besides 5HT, other known agonists include 2-methyl-5-hydroxytryptamine, phenylbiguanide, m-chlorophenylbiguanide, and SR-57227A. More recently, 5HT$_3$ receptor partial agonists also have been interrogated as agents with potential utility in the control of gastroenteric motility without completely blocking 5HT$_3$ sensitized nerve function. No potent positive allosteric agents have yet been developed for 5HT$_3$ receptors.[104] Anesthetic alcohols such as trichloroethanol potentiate 5HT$_3$ function at millimolar concentrations whereas ifenprodil, a ligand of the N-methyl-D-aspartate (NMDA) sensitive NR2B N-terminal domain, and certain L-type calcium channel blockers such as verapamil, inhibit 5HT$_3$ binding in a noncompetitive manner.

2.22.2.2.7 Prototypical therapeutics

5HT$_3$ antagonists are routinely used as antiemetics postoperatively and during chemotherapy. These include numerous launched products – azasetron, granisetron, indisetron, ondansetron, palonosetron, dolasetron, ramosetron, and tropisetron. Some of the 5HT$_3$ antagonists reportedly are in clinical trials for other disorders such as arrhythmias (tropisetron), irritable bowel syndrome (ramosetron), and schizophrenia (dolasetron). Alosetron has been launched for irritable bowel syndrome but its use is restricted because of side effects. Mosapride is a 5HT$_4$ agonist and a 5HT$_3$ antagonist launched as a prokinetic for gastritis. Additional compounds reportedly in development for irritable bowel syndrome include AGI-003, DDP-22, pumosetrag, and cilansetron. Clearly 5HT$_3$ receptors are present in the CNS, but besides dolasetron there appear to be no 5HT$_3$ targeted therapeutics in clinical trials for neuropsychiatric disorders. The potential and limitations for CNS indications were recently reviewed.[88]

2.22.2.2.8 Diseases

Serotonergic dysfunction has been implicated in a variety of neuropsychiatric disorders, largely in relation to metabotropic (GPCR) 5HT systems. $5HT_3$ LGIC have been examined to a more limited extent. In patients with bipolar disorders, polymorphism and missense mutations were observed in the upstream open reading frame of the *HTR3A* gene. However, no significant association was found between the C195T polymorphism and bipolar affective disorder. In contrast, a C178T (Pro16Ser) variant showed functional differences versus wild-type alleles and it was suggested that this might affect susceptibility to bipolar disorders.[105,106] Although polymorphisms were detected in $5HT_{3A}$ and $5HT_{3B}$ receptors in patients with Tourette's syndrome, these do not appear to be directly disease-related.[107] A valine to serine mutation in the TM2 domain of the $5HT_{3A}$ receptor subunit was previously shown to produce a homomeric receptor ~70-fold more sensitive to 5HT than the wild-type receptor when expressed in oocytes.[108] Characterization of gene knockin mice expressing this mutation showed morphologic, pharmacologic, and cystometric changes in the urinary bladder and urethral outlet tissues, characteristic of urinary bladder outlet obstruction as seen in patients with benign prostatic hyperplasia (BPH) or neuropathic lesions.[95]

2.22.2.3 GABA$_A$ and GABA$_C$ Receptors

GABA$_A$ receptors are widely distributed in the mammalian CNS where they are typically thought to mediate fast synaptic inhibition. This inhibition is generally via enhancement of GABA$_A$-mediated Cl^- conductance that leads to postsynaptic hyperpolarization and additional dampening of excitatory inputs due to the electrical shunt. The net overall effect depends upon a number of variable factors, including resting membrane potential, subcellular localization of the receptors, and synaptic drive. In addition to distribution at inhibitory synapses on dendrites and cell bodies of neurons, evidence for GABA$_A$ receptor mediated tonic transmission at extrasynaptic locations has also been documented.[109] Of the diverse LGIC, GABA$_A$ receptors have received heightened attention as they serve as targets for clinically important drugs such as sedatives, hypnotics, anticonvulsants, anxiolytics, muscle relaxants, and anesthetics. For recent reviews, see [110–113].

2.22.2.3.1 Current nomenclature

Being the major inhibitory neurotransmitter in the CNS, GABA plays a key role in modulating neuronal activity via distinct receptor systems – the ionotropic GABA$_A$ and GABA$_C$ and metabotropic GABA$_B$ receptors. Unlike GABA$_B$ receptors that act via the binding and activation of guanine nucleotide-binding proteins,[114] GABA$_A$ and GABA$_C$ receptors directly gate chloride conductance and mediate majority of the rapid inhibitory synaptic transmission.[109,115]

2.22.2.3.2 Molecular biology

Multiple subunit associations underlie the pharmacological and biochemical diversity of GABA$_A$ receptors. At least 19 different mammalian GABA$_{A/C}$ receptor subunit genes have been identified and divided into subfamilies. These include GABA$_A$ receptor subunits – alpha ($\alpha 1$–$\alpha 6$), beta ($\beta 1$–$\beta 3$), gamma ($\gamma 1$–$\gamma 3$), delta (δ), epsilon (ϵ), theta (θ), and pi (π) and GABA$_C$ receptor subunits – rho ($\rho 1$–$\rho 3$). Although a multitude of combinations in theory is possible, it appears that an α, β, and one other subunit type such as γ, δ, or ϵ are required for fully functional GABA$_A$ receptor. Coexpression of α and β subunits produces GABA-gated channels, but inclusion of a γ subunit provides sensitivity to benzodiazepine modulation. Although the exact subunit composition of GABA$_A$ receptor subtype in the CNS remains to be determined and complex patterns of subunit distribution in different brain regions have been observed, studies with recombinant receptors indicate a likely subunit ratio of $2\alpha:2\beta:1\gamma$. In brain, the predominant GABA$_A$ is $\alpha 1\beta 2\gamma 2$, comprising about one-third of the receptors.[116] Another third is $\alpha 2\beta 3\gamma 2$ and $\alpha 3\beta 3\gamma 2$ together, and other less well-defined combinations make up the remainder.

GABA$_C$ receptors, highly enriched in the vertebrate retina, are biochemically and pharmacologically distinct from GABA$_A$ receptors.[117–119] GABA$_C$ receptors have a simpler organization derived from three different ρ subunits, and are insensitive to bicuculline and baclofen, unlike GABA$_A$ receptors.

2.22.2.3.3 Protein structure

GABA$_A$ receptors are heteropentameric proteins with a total molecular mass of 230–270 kDa. All of the GABA$_A$ subunits are predicted to have a similar overall domain structure. This includes a signal sequence, a large N-terminal region, four transmembrane domains (TM1–TM4), and a relatively large and variable intracellular loop between TM3 and TM4. Models of the extracellular and transmembrane domains of GABA$_A$ receptor based on AChBP and *Torpedo* nAChR along with putative drug binding pockets within and between subunits have been described. By overlaying the α and γ subunits onto the model of AChBP, the pockets involved in benzodiazepine binding sites have been modeled with previously identified amino acids.[120,121]

2.22.2.3.4 Physiological functions

GABA$_A$ receptors are critical for diverse physiological processes including the regulation of vigilance, anxiety, epileptogenic activity, and memory functions. Activation of GABA$_A$ receptors increases inward chloride currents, which hyperpolarizes the postsynaptic cell and inhibits synaptic activity. Activation of GABA$_A$ receptors underlies fast synaptic inhibitory postsynaptic potentials (IPSPs) and thus influences impulse initiation, duration and amplitude of the excitatory postsynaptic potentials (EPSPs). However, the frequency of synaptic transmission is a critically important determinant of neural communication, as one may appreciate in considering the difference between a thump and a symphony. GABAergic transmission does not simply turn the system on or off, but is critical for controlling the frequency, as in electroencephalogram (EEG) rhythms associated with learning and memory, and for preventing excessive activity as observed in seizures.[122,123] Enhancement of GABA$_A$ receptor mediated fast synaptic inhibition is the basis for pharmacotherapy of various neurological and psychiatric disorders. Both GABA$_A$ and GABA$_C$ receptors have high Cl$^-$ sensitivity, with permeability ratios of K$^+$ to Cl$^-$ of <0.05. However, GABA$_C$ receptors are pharmacologically and biophysically distinct from GABA$_A$ receptors. Major differences include about 10-fold higher sensitivity to GABA, steeper Hill slopes for the agonist concentration response, weaker desensitization even in presence of high agonist concentration, and low single-channel conductance, relative to GABA$_A$ receptor. Studies on GABA$_A$ receptor subunit knockout and knockin mice have provided valuable insights into the physiological, pharmacological, and behavioral role for the various subunits.[124] Knockin mice where specific GABA$_A$ receptor mutations are insensitive to diazepam and general anesthetics have related specific roles of individual subunits to pharmacology.

2.22.2.3.5 Regulation

GABA$_A$ receptors are regulated by neurosteroids and by ethanol, and under pathological conditions such as neurodegenerative diseases and epilepsy. The regulated expression of the GABA$_A$ receptor subunit can be influenced by conditions that alter subunit mRNA levels, promoters that control such levels, and multiple transcription response elements.[125] Other proteins such as gephyrin can also modulate expression of GABA$_A$ receptors. Gephyrin, a 93 kDa membrane protein, is a major component of the postsynaptic apparatus that is involved in the stabilization of GABA$_A$ and glycine receptor clusters at the synapse. In several brain regions, gephyrin is selectively associated with GABAergic synapses, whereas in brainstem and spinal cord, it appears to be localized at glycinergic synapses.[126] Intracellular domains of GABA$_A$ receptor subunits can be phosphorylated by a variety of kinases at serine, threonine, and tyrosine residues. GABA$_A$ receptors are associated with both protein kinase C βII and RACK1 (receptor for activated C kinase) that together mediates phosphorylation of β receptor subunits, thereby facilitating trafficking and functional modulation of GABA$_A$ receptors. GABA$_A$ receptors also undergo dynamin-dependent, clathrin-mediated endocytosis.[127]

2.22.2.3.6 Prototypical pharmacology

A variety of agonists, competitive and noncompetitive antagonists, and a wide array of allosteric modulators are known. GABA site ligands include muscimol, 3-aminopropane sulfonate, imidazole acetic acid, and piperidine-4-sulfonate, whereas antagonists include bicuculline and gabazine (**Figure 4**). Subunit-selective fine-tuning of pharmacological activity of GABA$_A$ receptors can be achieved via allosteric rather than orthosteric ligands. This has been realized with several classes of clinically important drugs acting in a positive allosteric manner at the GABA$_A$ receptors such as benzodiazepines, barbiturates, neurosteroids, and anesthetic compounds. GABA is about 10 times less potent at GABA$_A$ than GABA$_C$ receptors. Whereas GABA$_A$ receptors are selectively blocked by the alkaloid bicuculline and modulated by benzodiazepines, steroids, and barbiturates, these compounds are much weaker at GABA$_C$ receptors. The conformationally restricted GABA analog *cis*-amino crotonic acid (CACA) is an early, moderately selective GABA$_C$ partial agonist, but the compound does inhibit GABA transporters, potentially producing indirect effects through alteration of GABA levels, and CACA can activate α6-containing GABA$_A$ LGIC. Another conformationally restricted analog, (+)-CAMP [(*1S,2R*)-2-aminomethylcyclopropanecarboxylic acid] appears to be more selective and is more efficacious (full agonist) at ρ1 and ρ2 GABA$_C$. TPMPA is a selective antagonist. Muscimol is commonly used as a GABA$_A$ agonist in physiological studies, but it is important to note that muscimol also activates GABA$_C$ LGIC. Thus, THIP (gaboxadol; 4,5,6,7-tetrahydroisoxazolo[5,4-c]pyridin-3-ol), which also activates GABA$_A$ but inhibits GABA$_C$, and TPMPA, the GABA$_C$-selective antagonist, are important additional tools. Among modulators, GABA$_C$ receptors are insensitive to barbiturates and benzodiazepines, compounds well known to potentiate common GABA$_A$ LGIC. Neurosteroids also differentiate, potentiating GABA$_C$ at micromolar concentrations and GABA$_A$ at nanomolar concentrations. Two other modulators – loreclezole and (+)-ROD188 – inhibit GABA$_C$ while potentiating GABA$_A$, or at least those GABA$_A$ containing β2 or β3 subunits. The differential pharmacology of GABA$_C$ and GABA$_A$ receptors has been reviewed.[117]

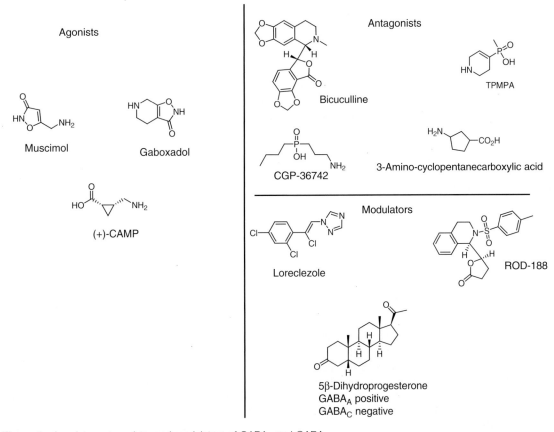

Figure 4 Agonists, antagonists, and modulators of GABA$_A$ and GABA$_C$.

2.22.2.3.7 Prototypical therapeutics

Benzodiazepines are well-known CNS therapeutics whose sedative/hypnotic, antiepileptic, and anxiolytic activities are primarily due to positive allosteric modulation of GABA$_A$ LGIC. Examples of clinically available compounds include clonazepam, prazepam, tofisopam, cloxazolam, pinazepam, estazolam, flunitrazepam, camazepam, temazepam, nimetazepam, delorazepam, clobazam, ketazolam, lormetazepam, alprazolam, halazepam, oxazolam, doxefazepam, flutazolam, mexazolam, quazepam, flutoprazepam, metaclazepam, and cinolazepam. Many were launched in the mid 1970s to mid 1980s. Flumazenil is a benzodiazepine antagonist marketed for benzodiazepine or alcohol overdose. Nonbenzodiazepines that appear to act through the benzodiazepine GABA$_A$ modulatory site include thiazoles such as clomethiazole, thienodiazepines like clotiazepam, etizolam, and brotizolam, benzoxazines like etifoxine, the benzothienodiazepine bentazepam, the piperazinecarboxylic acid zopiclone, the pyrazolopyrimidine zaleplon, and the cyclopyrrolone pagoclone.

Another approach to potentiation of GABA$_A$ LGIC takes advantage of the neurosteroid example, as exemplified with alphadolone and ganaxolone. Other compounds thought to exert a component of their actions at GABA$_A$ LGIC include: pregabalin (diabetic neuropathy, epilepsy), a GABA analog that acts as a GABA$_A$ LGIC agonist (also α2δ Ca^{2+} channel ligand); substituted thiophenes (e.g., Merck's 6,6-dimethyl-3-(2-hydroxyethyl)thio-1-(thiazol-2-yl)-6,7-dihydro-2-benzothiophen-4(5H)-one) and naphthyridenes (AC-3933) that are GABA$_A$ inverse agonists for cognitive impairment; TP-003 that is a GABA$_A$ α3 selective modulator for cognitive impairment, anxiety, and epilepsy; ELB-139, a GABA$_A$ α3 selective modulator/inverse agonist for anxiety, psychoses, and epilepsy; THIP, a muscimol analog and GABA$_A$ agonist sedative–hypnotic. Other GABA$_A$ agonist anxiolytic molecules include ocinaplon, SL 651498, and PNU 101017. Agents launched primarily for the treatment of epilepsy include tiagabine, a GABA reuptake inhibitor that acts indirectly by increasing extracellular GABA levels, and vigabatrin, a compound that acts as both a GABA aminotransferase inhibitor, which would increase GABA levels, and a dopamine receptor antagonist.

2.22.2.3.8 Diseases

Mutations in GABA$_A$ receptor α1, γ2, and δ subunits have been associated with different idiopathic generalized epilepsy syndromes.[128] These mutations could result in altered receptor gating, expression, or trafficking and contribute to neuronal disinhibition, thereby predisposing affected individuals to epilepsy. For example, in a family with juvenile myoclonic epilepsy, a mutation in the *GABRA1* gene predicting a single amino acid substitution A322D results in loss of receptor function attributed to reduced surface expression, reduced GABA sensitivity, and accelerated deactivation.[129] Association studies implicate an involvement of the *GABRA4* gene in the etiology of autism and a potential increase in autism risk through interaction with another clustering *GABRB1* gene.[130] Another linkage analysis in families with migraine identified a susceptibility locus in the 15q11–q13 genomic region, which harbors genes encoding three GABA$_A$ receptor subunits.[131]

2.22.2.4 Glycine Receptors

2.22.2.4.1 Current nomenclature

Glycine receptors are responsible for fast inhibitory neurotransmission in the CNS, predominantly in the spinal cord and brainstem. Similar to GABA, glycine also activates anion channels (Cl$^-$ and HCO$_3^-$ conducting) that lead to hyperpolarization, thereby suppressing neuronal firing. These receptors share structural homology with the nAChRs, GABA$_{A/C}$, and 5HT$_3$ receptors. Several reviews[132–135] have dealt in detail with the structure–function, physiology, and pharmacology of glycine receptors.

2.22.2.4.2 Molecular biology

Four ligand binding glycine receptor α subunits (α1–α4) and one β subunit have been identified. The α subunits can form functional homomeric LGIC in recombinant expression systems, but the β subunit can not.[136] Studies on native and recombinant receptors demonstrate that glycine receptors in the adult CNS are probably composed of α and β subunits, and a subunit stoichiometry of 2α1:3β, as per nAChR, has been proposed.[137] Coexpression of β subunit with α1 subunits does not appear to influence macroscopic properties such as whole-cell current amplitudes or kinetics, but results in lower single-channel conductance properties. Alternative splicing within the intracellular loop between TM3 and TM4 has been observed with α1 and α3 subunits, the latter resulting in differences in channel desensitization.

2.22.2.4.3 Protein structure

The glycine receptor was one of the first neurotransmitter receptors to be isolated from mammalian CNS, exploiting its high affinity to the convulsant alkaloid, strychnine. Like other Cys-loop receptors, the glycine subunits are regarded as pentameric oligomers, composed of five subunits, each consisting of a large extracellular N-terminus, four TM domains, and a large intracellular domain between TM3 and TM4. The extracellular N-terminus comprises approximately 50% of the total protein and contains regions that are critical for ligand binding and associated steps leading to channel gating, as well as for subunit assembly. Because the sequence of AChBP is about 20–24% identical to the extracellular domain of nAChR and some 15–18% identical to the aligned extracellular domain sequences of glycine, GABA$_{A/C}$, and 5HT$_3$ subunits, the AChBP structure provides a structural template for the extracellular domain of Cys-loop receptors. By a combination of electrophysiological and molecular modeling approaches, two conserved, oppositely charged residues located on adjacent subunit interfaces were identified as crucial for agonist binding. A ring of threonine residues from each of the five α subunits is critical for picrotoxin sensitivity; the presence of β subunits reduces sensitivity of the glycine receptor to picrotoxin. With regard to transmembrane architecture, a high-resolution structure of the 61-residue TM2–TM3 segment of GlyR α1 subunit was reported.[138] Well-defined domain structures were identified for TM2, TM3, and interconnecting extracellular loop regions. Contrary to the popular model of alpha-helical structure for the pore-lining TM2 domain for the Cys-loop receptor family, the last three residues of the TM2 domain and the first eight residues of the loop appear to be intrinsically nonhelical and highly flexible. The six remaining residues of the TM2–TM3 loop and most of the TM3 domain exhibit helical structures.

2.22.2.4.4 Physiological functions

Functional glycinergic synapses have been characterized in various brainstem nuclei, spinal cord motor reflex pathways and pain sensory pathways, and in the retina. The glycine receptor plays a fundamental role in mediating inhibitory neurotransmission in the spinal cord and brainstem, although recent evidence suggests it may also have other physiological roles, including excitatory neurotransmission in embryonic neurons.[134] Additionally, activation of glycine receptors together with GABA$_A$ receptors in the sperm plasma membrane appears to be essential for acrosome reaction, an important step in the fertilization process. Glycine binding to the receptors opens anion channels, mostly leading to Cl$^-$ influx, which hyperpolarizes the neuron and thereby inhibits neuronal activity. In general, glycine receptors have permeability and conductance properties similar to GABA$_A$ receptors – both being anion selective with a PK$^+$/PCl$^-$ <0.05. The α1 subunit exhibits five conductance states ranging from 20 to 90 pS, with the 90 pS state occurring with the greatest frequency. Coexpression with the β subunit eliminates the highest conducting levels leaving a 45 pS state as the most frequently occurring. Substitution of two to three residues of the glycine receptor can switch the α7 nAChR pore from cation selective to anion selective suggesting that these residues in TM2 sequences are important determinants of the selectivity filter.[48]

2.22.2.4.5 Regulation

As with GABA$_A$ receptors, gephyrin also plays a key role in regulating the postsynaptic clustering of the glycine receptor. Gephyrin mediates synaptic clustering by binding to glycine receptor β subunits, and receptor activation is required for initiation of the anchoring process. In rats, a developmental switch from α2 to predominantly α1β heteromers is observed around postnatal day 20, an event that may be important for, or reflective of, the β subunit dependent receptor anchoring via gephyrin. Glycine receptor subunits also carry consensus motifs for phosphorylation, glycosylation, and ubiquination.

2.22.2.4.6 Prototypical pharmacology

In addition to glycine, taurine and beta-alanine have been proposed as endogenous glycine receptor agonists in early development (**Figure 5**). Native and recombinant glycine receptors are activated by amino acid agonists with a rank order of potency: glycine > beta-alanine > taurine > GABA. When α1 receptors are expressed in HEK-293 cells, glycine, beta-alanine, and taurine all behave as full agonists, with glycine exhibiting an EC$_{50}$ value of 20–50 μM.

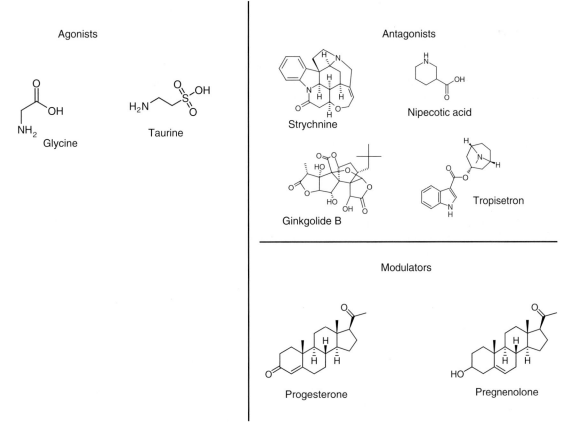

Figure 5 Agonists, antagonists, and modulators of glycine receptors.

Glycine receptors derived from α2, α3, and α4 subunits are thought to have similar agonist sensitivities to that derived from the α1 subunit. The plant alkaloid strychnine is a selective and potent competitive antagonist versus glycine, beta-alanine, and taurine with a K_D value of 5–10 nM. Strychnine also inhibits α9 nAChR, but GlyR currents (anionic) are readily distinguished from α9 nAChR currents (cationic) by other means. Thus, strychnine sensitivity is one of the most definitive means of discriminating glycinergic from GABAergic anionic synaptic currents. Radioligand binding and electrophysiological studies suggest that glycine and strychnine bind to partially overlapping, but not identical, sites on the receptor. The proconvulsant alkaloid picrotoxin also inhibits glycine receptors albeit with lower potency than at $GABA_A$ receptors. The GlyR α/β heteromers are much less sensitive to inhibition by picrotoxin than α homomers, independently of the type of α subunit (α1, α2, or α3). The picrotoxin components picrotin and picrotoxinin show similar potencies in inhibiting α1 glycine receptors. Tropisetron, known as a $5HT_3$ antagonist, also interacts with GlyR. At low concentrations (0.01–1 μM), tropisetron potentiates GlyR α1 and α1β whereas it suppresses currents at high (>10 μM) concentrations. The antiparasitic agent ivermectin is a potentiator of GlyR currents, by interaction with a site distinct from the glycine and strychnine binding site. Several volatile anesthetics such as isoflurane, halothane, and enflurane can potentiate α1 GlyR, inducing a leftward shift in the EC_{50} value of glycine.

2.22.2.4.7 Prototypical therapeutics

A number of agents that have other primary targets have been reported to interact with the glycine receptor. Besides $5HT_3$ antagonists, calcium channel antagonists such as nitrendipine, the neuroprotective agent riluzole, and the microtubule-depolymerizing agent colchicine all antagonize glycine-induced currents in the micromolar range. Clinically relevant concentrations of riluzole have been found to accelerate the desensitization of the α1/β glycine receptor while having no effect on maximal current amplitude. The estrogen receptor modulator tamoxifen potentiates effects of submaximal glycine responses. Development of therapeutic agents that potentiate glycine receptor responses could have significant utility as muscle relaxants and analgesic agents, as these receptors are involved in motor reflex circuits and nociceptive sensory pathways.[135]

2.22.2.4.8 Diseases

Inherited disorders that have been ascribed to mutations in GlyR subunits include the familiar startle disease (hyperekplexia) in humans, spastic, spasmodic, and oscillator phenotypes in mouse, and myoclonus in bovine TM1. Aberrant splicing leading to >90% loss of functional β subunits underlies the murine phenotype *spastic*. In human startle disease, many of the mutations reduce glycine sensitivity and/or GlyR channel gating and most occur in GlyR α1 TM1, TM2 the intracellular loop between TM1 and TM2, and the α helical region of the extracellular TM2–TM3 loop near TM2.[134] Additionally, a startle mutation inserts a stop codon prior to TM1, and one other substitutes an arginine for a serine in TM1 to alter receptor insertion into the plasma membrane.

2.22.2.5 Zinc-Gated Ion Channel

A distinct member of the Cys-loop LGIC family, designated as a Zn^{2+}-activated ion channel, was uncovered using searches of expressed sequence tag (EST) databases using a consensus peptide sequence of $5HT_3$ and nAChRs.[139] The subunit (ZAC, also called L2 protein) encodes a polypeptide of 411 amino acids, has a signal sequence, cys–cys motif, four predicted transmembrane domains, and several invariant residues that underpin the conserved secondary structure of the Cys-loop LGIC superfamily. Sequence analysis reveals that a common ancestral gene gave rise to the ZAC, nAChR, and $5HT_3$ receptor subunit genes. Expression studies indicate that ZAC subunit genes are absent in mouse and rat, but present in dog, chimpanzee, cow, opossum, and human.[140] Messenger RNA (mRNA) levels have been detected in various tissues including pancreas, brain, liver, lung, heart, and kidney. In the brain, immunoreactivity was dominantly detected in the hippocampal CA3 pyramidal cells and in the polymorphic layer of the dentate gyrus, an area of the brain where neuronal Zn^{2+} is particularly high. The ZAC subunit, expressed in HEK-293 cells, formed homomeric channels that were activated by application of Zn^{2+} (1 mM) but not by agonists of other LGIC. Zn^{2+}-activated currents are inhibited by relatively high concentrations of tubocurarine, but not by any of the other LGIC antagonists. These effects are of interest because at ZAC Zn^{2+} behaves as an agonist, unlike its modulatory effects at other LGIC, and because it adds to the evidence that endogenous Zn^{2+} may act like a neurotransmitter in an area where Zn^{2+} is present at high concentrations in synaptic vesicles and may be released synaptically. These effects remain to be further investigated using tissues from nonrodent species.

2.22.3 Glutamate-Gated Family

Glutamatergic transmission is the predominant form of excitatory, rapid synaptic transmission in the CNS. Other excitatory amino acids that may act as neurotransmitters include aspartate and N-acetylaspartylglutamate (NAAG), but it is not clear that these substances act upon receptors distinct from those known as glutamate receptors. Aspartate can activate NMDA-sensitive glutamate receptors selectively, and NAAG can activate metabotropic glutamate receptors.[141] For recent reviews, see [142–151].

2.22.3.1 Current Nomenclature

Glutamate receptors (GluR) refer to both metabotropic GPCR and ionotropic LGIC and are often termed mGluR and iGluR, respectively, to make the distinction. However, this terminology has not been applied universally, and in many reports the LGIC may be referred to as simply GluR. Among the iGluR, there are three broad families that share a common structure but are differentiated according to their subunit composition, functional properties, and pharmacology. The three main families are the AMPA (α-amino-3-hydroxy-5-methylisoxazole-4-propionic acid)-sensitive receptors, the kainate-sensitive receptors, and the NMDA-sensitive receptors. These often are referred to simply as AMPA receptors, kainate receptors, and NMDA receptors, although this is not strictly accurate because AMPA, kainate, and NMDA are not the physiological ligands. Additionally, a misconception that arises from this terminology is that AMPA and kainate themselves define distinct iGluR. In fact, kainate can also activate AMPA-sensitive receptors and AMPA can activate kainate-sensitive receptors. The term 'AMPA/kainate receptor' often is used in physiological research addressing native receptors, in recognition of the difficulty in distinguishing kainate-sensitive from AMPA-sensitive receptors. In addition, the $\delta1/\delta2$ and NR3A/NR3B subunits may be considered to identify subfamilies within the NMDA-sensitive iGluR, but a clear nomenclature on this aspect has not arisen.

2.22.3.2 Molecular Biology

Eighteen iGluR subunits or homologous proteins have been cloned, not including splice variants and other edited products. The iGluR share a common structure and subunit composition, but there is a degree of specificity to the subunit assembly. AMPA-sensitive iGluR comprise iGluR1, iGluR2, iGluR3, and iGluR4 subunits. Kainate-sensitive iGluR comprise iGluR5, iGluR6, iGluR7, KA-1, and KA-2 subunits. NMDA-sensitive iGluR comprise NR1, NR2A, NR2B, NR2C, and NR2D (all separate gene products despite the nomenclature).

Additionally, NR3A and NR3B can combine with other subunits to modify the properties of NMDA-sensitive iGluR in heterologous expression systems.[152–154] Indeed, it has been reported that NR3A and NR3B combined with the glycine-binding subunit NR1 can form a novel cationic LGIC gated by glycine, but expression of such a construct is yet to be confirmed in situ. Nevertheless, NR3A and NR3B are expressed in CNS where they may alter the properties of native NMDA-sensitive iGluR,[155–158] and there is some evidence that altered NR3A expression may be involved in schizophrenia and bipolar disorders.[159]

Two 'orphan' gene products have been identified by homology cloning: $\delta1$ and $\delta2$.[160–162] These proteins are $\leqslant 25\%$ homologous to other iGluR subunits and it appears that $\delta1$ and $\delta2$ do not combine with other iGluR in situ.[163] With or without other known iGluR subunits, however, it does appear that the δ subunits play important roles in CNS physiology, as $\delta2$ is selectively expressed in cerebellar Purkinje neurons where it appears to play key roles in synaptic plasticity and motor control.[164–168]

2.22.3.3 Protein Structure

While the iGluR families share a similar structure overall, they may display significantly different functional properties owing to differences in subunit composition. Each iGluR comprises four subunits arranged around the central ion channel pore. The AMPA- and kainate-sensitive iGluR subunits are approximately 900 amino acids or 100 kDa in size. The NMDA-sensitive iGluR subunits tend to be somewhat larger with approximately 1000–1200 amino acids, although NR3B is about 900 amino acids. The N-terminal region is extracellular, the C-terminus is intracellular, and there are three transmembrane segments. Additionally, between TM1 and TM2 in the intracellular membrane face (as opposed to the extracellular face in VGIC) there is a re-entrant 'P-loop' that contributes to the channel and is at least partly intramembraneous, but does not completely traverse the membrane. It was initially thought that the P-loop was another transmembrane segment, making a total of four transmembranes per subunit. This would place both N- and C-terminal regions on the same (extracellular) side of the membrane with the N-terminal and TM3–TM4 loop on opposite sides of the membrane. However, a number of structure–function studies and the application of membrane-impermeant antibodies selective for various regions of the iGluR subunit have clarified that the C-terminal region is intracellular, the N-terminal region and the intertransmembrane loop are extracellular, and the region initially defined as TM2 is, in fact, a 'P-loop' that lines the ion channel, but does not traverse the membrane.

The ligand binding domain is formed as an interface between a portion of the N-terminal region closer to the membrane and the extracellular loop between TM2 and TM3. To some extent, this may be considered similar to ligand binding at the interface between subunits in the Cys-loop LGIC family, but the iGluR ligand binding domain is formed within a subunit, not between subunits. Further toward the N-terminus is a large N-terminal domain or N-terminal domain that is fairly well conserved among subunits. The function of the N-terminal domain is not entirely clear for many of the iGluR, but it may be involved in modulatory interactions and/or receptor assembly.

2.22.3.3.1 α-Amino-3-hydroxy-5-methylisoxazole-4-propionic acid-sensitive receptors

Four of the iGluR subunits (iGluR1, iGluR2, iGluR3, and iGluR4) combine to form AMPA-sensitive iGluR. Receptors are formed as pairs of dimers with like subunits facing one another across the central channel. The AMPA-sensitive receptor agonist binding site is relatively well characterized since it has been possible to separately express and crystallize individual receptor subunits and the ligand binding core with and without ligands bound. The binding site is made from two domains of the subunit, one being an ~ 150 amino acid segment between the N-terminal domain and TM1, and the other being the extracellular domain between TM2 and TM3. The alpha-carboxyl of glutamate binds to the first domain (D1 or S1, that preceding TM1) and the gamma-carboxyl binds to the second domain (D2 or S2). However, the amino acid residues involved are not necessarily the same for every ligand. Among the ligands glutamate, AMPA, and kainate (which is also an AMPA agonist), the alpha-carboxyl of each binds to the same residue in D1 but at the other end of the ligand, the isoxazole of AMPA does not bind to the same residue in D2 as do the gamma-carboxyls of glutamate and kainate. In the NMDA-sensitive NR1 subunit, which binds glycine but not glutamate, the binding pocket is particularly small, about one-fourth the size of iGluR pockets, and binding of the glutamate gamma-carboxyl is blocked by a tryptophan residue.

How ligand binding is coupled to channel gating is less clear, although important clues have been derived from crystallization studies. The interactions of partial agonists remain somewhat enigmatic. Some of the data seem to fit the induced-fit model, which predicts a variable degree of conformational transition according to agonist efficacy, i.e., a partial agonist produces a smaller transition than does a full agonist. Other data fit the Monod–Wyman–Changeux model, which assumes discrete conformation states (i.e., either closed, open, or desensitized) and variable population of these states according to relative ligand affinity.

2.22.3.3.2 Kainate-sensitive receptors

Kainate-sensitive LGIC are structurally similar to AMPA-sensitive receptors, but are formed from different combinations among the five known subunits. In expression systems, recombinant iGluR5, iGluR6, and iGluR7 can form functional homomeric or heteromeric assemblies. Two other subunits – KA1 and KA2 – bind ligands, but do not themselves form functional LGIC. However, KA1 and KA2 can be coexpressed with iGluR5, iGluR6, or iGluR7 to form functional LGIC with properties that can be differentiated from the respective iGluR without KA. Often, one is not clear whether the complex created in recombinant expression systems is physiological, but in brain there is evidence for iGluR6 coassembly with KA1 and KA2.

2.22.3.3.3 N-methyl-D-aspartate-sensitive receptors

NMDA-sensitive LGICs are believed to be tetrameric, like the AMPA- and kainate-sensitive counterparts, but formed from a different set of subunits: NR1 (which has eight splice variants), NR2A, NR2B, NR2C, NR2D (these NR2s are four different gene products, not splice variants of one another), and NR3. Unlike the iGluR subunits, which can form functional homomeric assemblies in expression systems, NMDA-sensitive receptors require at least two different types of subunit – NR1 together with either NR2 or NR3. Recently the ligand binding core of the NMDA-sensitive receptor, like that of the AMPA-sensitive receptor, was crystallized and the structures determined with glutamate and glycine bound.[169]

NR1 is not a glutamate receptor per se. NR1 binds the coagonist glycine, but does not itself form a functional LGIC in heterologous expression systems. NR2 subunits bind glutamate and coassemble with NR1 to form NMDA-sensitive receptors that require both glycine and glutamate for activation. However, in physiological saline and at normal neuronal resting potentials (example at $\sim -70\,\text{mV}$), the NMDA-sensitive receptor channel is blocked by extracellular

Mg^{2+} and little current flows through the channel despite the binding of glutamate and glycine. The Mg^{2+} blockade is voltage dependent and if the cell is depolarized by glutamatergic or other types of inputs, then the NMDA-sensitive channel block by Mg^{2+} is relieved, allowing ion flux consequent to receptor activation by agonist. Thus, NMDA-sensitive LGICs are considered 'coincidence detectors' because full activity requires appropriately timed coincident signals – depolarization plus agonist and coagonist. Although prevailing levels of glycine in cerebrospinal fluid may be sufficient for coagonist activity when glutamate is released as transmitter, nevertheless the chemical signal also requires an electrical signal in order to gate ion flux through the channel.

NR3A and NR3B subunits[170,171] also require NR1 coexpression and do not combine functionally with NR2, but also do not generate the typical NMDA-sensitive glutamate receptor. LGIC composed of NR1 and NR3 are gated by glycine alone, not glutamate. The NR1/NR2 glycine site coagonists D-alanine, D-cycloserine, and 1-aminocyclopropanecarboxylic acid (ACPC) are weak inhibitors of NR1/NR3 activation by glycine, while the NR1/NR2 coagonist D-serine is an NR1/NR3 weak partial agonist that can block the response to glycine. These glycine-sensitive LGIC are cation permeable and excitatory, unlike GlyR in the Cys-loop family, but the NR1/NR3 channel is not blocked by Mg^{2+} as is the NR1/NR2 NMDA-sensitive LGIC.

2.22.3.4 Regulation

Glutamate LGICs are subject to a remarkable degree of physiological regulation, and various subunit isoforms could create a wide array of receptors.[145,150] Some of this includes RNA editing, a function not known to occur with other classes of LGIC. In AMPA- and kainate-sensitive LGIC, iGluR2, iGluR5, and iGluR6 are subject to glutamine/arginine (Q/R) editing at a specific site in the P-loop that affects channel gating and ion selectivity. Q is specified genomically, but after the rat 'second trimester' (E14), almost all iGluR2 is found to have R in this position, thereby reducing single-channel conductance, inward rectification, and relative Mg^{2+} permeability. This is accomplished by conversion of adenosine to inosine (deamination) at the mRNA level, thus changing the relevant codon to one specifying arginine instead of glutamine. Similar effects are seen in the kainate-sensitive iGluR5 and iGluR6 subunits, although the overall degree of editing may be lower. Another site, near the flip/flop region (below) in AMPA-sensitive iGluR2, iGluR3, and iGluR4 can be edited from arginine to glycine (R/G), reducing desensitization and increasing resensitization. Additionally, in TM1 of kainate-sensitive subunits iGluR5 and iGluR6, isoleucine can be edited to valine (I/V) and tyrosine to cysteine (Y/C). The effect depends upon the subunit assembled, but has been found to affect ion permeability when iGluR6 is assembled with an iGluR containing Q in the edited P-loop site. In adult, 50–100% of the Q/R and R/G sites are edited in iGluR, depending upon site, subunit, and in the case of iGluR4 whether it is flip (50% edited) or flop (80–90% edited). However, the degree of editing can be altered under pathological conditions such as ischemia.

Flip and flop are alternative splicing variants found in AMPA-sensitive subunits iGluR1, iGluR2, iGluR3, and iGluR4. The site is in the TM3–TM4 extracellular loop just prior to TM4. This tends to affect desensitization, with intrinsic rates and/or extent of desensitization being moderately slower/lesser in the flip forms. Additionally, effects of desensitization modulators may be flip/flop dependent, e.g., cyclothiazide has greater effect on flip and 4-[2(phenylsulfonylamino)ethylthio]-2,6-difluorophenoxyacetamide (PEPA) has greater effect on flop. Additional alternative splicing variants are found in the C-terminal region of iGluR2-7 and NR1 and in the N-terminal region of iGluR5 and NR1.

Expression of AMPA-sensitive receptors in the synaptic region can be upregulated through increased surface expression, synaptic trafficking, and stabilization of turnover. Ca^{2+}/calmodulin regulated processes and PDZ [postsynaptic density-95 (PSD-95)/Discs large (Dlg)/zona occludens-1 (ZO-1)] protein domain interactions are involved. Kainate-sensitive receptor expression does not appear to be as immediately upregulated as are AMPA-sensitive receptors but the PDZ domain in the C-terminal region of kainate-sensitive subunits may be involved in receptor localization and does appear to affect synaptic efficacy.

2.22.3.5 Physiological Functions

AMPA-, kainate-, and NMDA-sensitive LGIC, like the metabotropic glutamate receptors, respond to glutamate, the most broadly expressed excitatory neurotransmitter in the vertebrate CNS. AMPA-sensitive receptors mediate rapid high-amplitude synaptic transmission. Kainate-sensitive receptors also mediate fast synaptic transmission in hippocampus, but these synaptic events tend to be smaller and relatively slower than those mediated by the AMPA-sensitive receptors. NMDA-sensitive receptors tend to play important modulatory roles as opposed to immediate signal transfer. NMDA-sensitive LGIC are about two orders of magnitude more sensitive to glutamate than are AMPA- or kainate-sensitive receptors, are about two orders of magnitude slower to desensitize, and are about one to three orders of magnitude slower to deactivate upon agonist removal. Thus, NMDA-sensitive LGIC can respond to glutamate over a longer period of time and/or distance (e.g., extrasynaptic) than can AMPA- and kainate-sensitive LGIC. These kinetics together with the voltage-dependent Mg^{2+} block allows NMDA-sensitive LGIC to integrate glutamate, glycine, and electrical signals over specific time domains. With multiple signals occurring in the appropriate frequency, or timing coincidence, the channel is opened and can partially depolarize the cell and, more importantly, allow sufficient Ca^{2+} inflow to trigger a host of Ca^{2+}-dependent signaling processes. Such processes may include positive modulation of other LGIC through kinase cascades, increased synaptic expression of AMPA-sensitive receptors, activation of nuclear transcription factors such as CREB ultimately to augment synaptic function or morphology, or adverse consequences including cytotoxicity with excess stimulation.

NMDA-sensitive LGIC containing NR2B have been considered as a target of interest for pain, in part due to their dominant expression in spinal dorsal horn relative to other NR2 subunits and upregulation of NR2B in pain states. NMDA-sensitive receptors are involved in nociceptive sensitization, a process like memory, thought to be dependent upon long-term potentiation (LTP) in central synapses. However, NR2B subunits also are expressed in brain. During maturation of the rodent CNS, NR2B is downregulated while NR2A is upregulated. Nevertheless, NR2B is not eliminated and synaptic receptors are thought to comprise NR1/NR2A/NR2B while extrasynaptic receptors, which still could be activated by transmitters, are derived from NR1/NR2B. Thus, NR2B-selective antagonists may inhibit brain as well as spinal NMDA-sensitive receptors with potential adverse effects on memory formation, cognitive performance, and psychoses.

2.22.3.6 Prototypical Pharmacology

The initial and still commonly used prototypic agonists for AMPA-sensitive, kainate-sensitive, and NMDA-sensitive LGIC are AMPA, kainate, and NMDA themselves. However, historically it has been difficult to distinguish AMPA from kainate responses in tissues. Kainate, for example, has only moderate selectivity for kainate- compared to AMPA-sensitive receptors.

Structure–activity relationship (SAR) studies based upon the isoxazole moiety of AMPA led to a series of compounds useful in cocrystallization studies of the AMPA-sensitive iGluR subunits.[172,173] (S)-ACPA was identified as one of the most potent agonists in the series. For both ACPA and AMPA, stereoselectivity favors the (S) configuration by >100-fold. (S)-5-Fluorowillardiine is a nonisoxazole selective agonist for AMPA-sensitive iGluR.[173] Within AMPA-sensitive subtypes, (S)-5-fluorowillardiine exhibits an order of magnitude selectivity for iGluR1 and iGluR2 relative to iGluR4. Ibotenic acid, a naturally occurring isoxazole, is an agonist at mGluR and NMDA-sensitive iGluR. However, derivatives of homoibotenic acid (HIBO) display interesting selectivity for iGluR1 and iGluR2 relative to iGluR3, iGluR4, and iGluR6 while selectivity against iGluR5 varies.[173,174] Representative structures are depicted in **Figure 6**.

At the kainate-sensitive receptors, SYM-2081 ((2S,4R)-4-methylglutamate) shows two to three orders of magnitude selectivity relative to AMPA-sensitive receptors.[145] The isoxazole derivatives (S)-ATPA and (S)-5-iodowillardiine also are selective agonists, particularly for iGluR5.[173] The nonisoxazole LY 339434, in contrast, shows three orders of magnitude selectivity for iGluR6 (**Figure 7**).

The competitive antagonists CNQX and DNQX often are used as AMPA/kainate-sensitive blockers, but they have little selectivity between the two subtypes. AMPA-selective antagonists include NBQX (competitive), GYKI 52466 and GYKI 53655 (noncompetitive), and joro spider toxin (channel blocker). At kainate-sensitive receptors, LY 294486 is selective for kainate-sensitive iGluR5 relative to iGluR6 or iGluR7. LY 294486 also inhibits AMPA-sensitive iGluR, although with less potency than at iGluR5. NS-102, on the other hand, is an iGluR6 selective antagonist.

Modulation of AMPA-sensitive receptors has been another direction of interest with respect to positive allosteric modulators ('AMPAkines') such as CX-516 and LY 395153 for treatment of cognitive disorders such as in AD, and with respect to negative allosteric modulators such as talampanel for treatment of disorders such as epilepsy and Parkinson's disease.[173]

For NMDA-sensitive receptors, prototypic selective agonists include NMDA itself, aspartate (which may function as a signaling molecule in some areas of the CNS), and quinolinic acid (**Figure 8**). Additionally, glycine or D-serine is required for the glycine coagonist NR1 site, but in in vivo or in tissue studies, sufficient glycine is generally present. Commonly used competitive selective antagonists are AP5 (also known as APV), AP7, CPP, and CGS19755. 7-Chlorokynurenate, 5,7-dichlorokynurenate, L-701,324, and MNQX are selective glycine-site inhibitors. NMDA-sensitive iGluR channel blockers include (+)-MK-801 (dizocilpine), ketamine, PCP (phencyclidine), dextromethorphan, and memantine. MK-801 is a very selective, potent, and long-lasting inhibitor owing in part to entrapment in the NMDA-sensitive iGluR channel. It is a widely used research tool to identify NMDA-sensitive iGluR in vitro. In vivo, the ability of MK-801 and other NMDA-sensitive iGluR channel blockers to induce psychosis or psychosis-like behavior is both a limitation and a research tool. Memantine also inhibits NMDA-sensitive iGluR, but it is less potent

Ion Channels – Ligand Gated

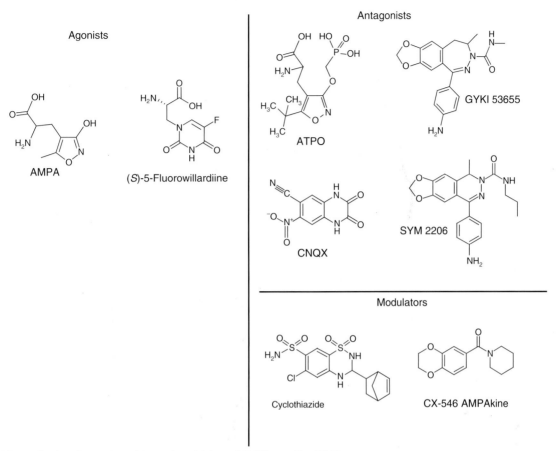

Figure 6 Agonists, antagonists, and modulators of AMPA-sensitive iGluR.

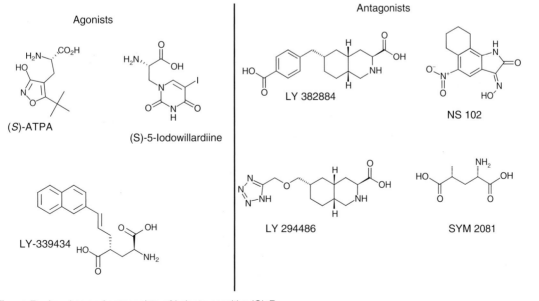

Figure 7 Agonists and antagonists of kainate-sensitive iGluR.

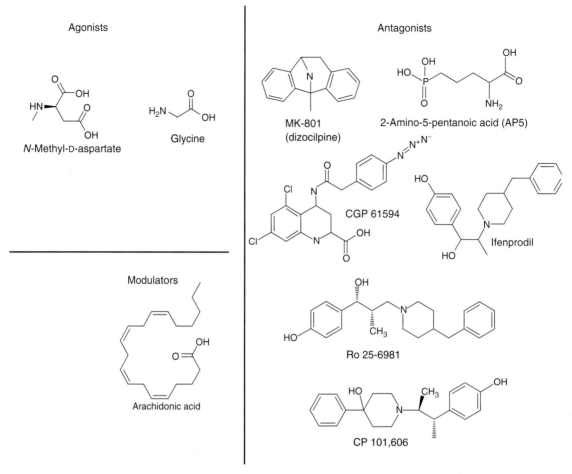

Figure 8 Agonists, antagonists, and modulators of NMDA-sensitive iGluR.

and reversible, and this is thought to be advantageous in allowing the compound to inhibit a low level of NMDA-sensitive iGluR activation while not blocking phasic synaptic transmission. Other NMDA-sensitive iGluR inhibitors of interest, particularly for neuropathic pain, are the NR2B-selective compounds such as ifenprodil, Ro 25-6981, Ro 8-4304, and CP 101,606.[151]

2.22.3.7 Prototypical Therapeutics

NMDA-sensitive LGIC has been of considerable therapeutic interest in light of its role in cognition (memory/LTP), ischemia/stroke (excitotoxicity), and pain/hyperalgesia. MK-801 is potent and selective NMDA-sensitive LGIC antagonist, but it induces psychosis-like behavior indicating a potential antipsychotic-role of NMDA-sensitive LGIC. On the other hand, excess NMDA-sensitive LGIC activation may be neurotoxic. Modulation of the glycine coagonist site has been of interest as a means of regulating NMDA-sensitive LGIC activity without excessive activation or blockade. Recently, memantine has emerged as an NMDA-sensitive LGIC antagonist, paradoxically, for the treatment of AD.[175–177] The paradox is that NMDA-sensitive LGIC blockade should, in principle, inhibit LTP induction and thereby inhibit memory formation and aggravate the disease. The rationale with memantine, however, is that in the aged population there is excess glutamatergic transmission and/or increased neuronal depolarization and diminished intracellular Ca^{2+} buffering such that the excitotoxic capacity of NMDA-sensitive LGIC is enhanced. Memantine, as a weak competitive inhibitor, could reduce excess Ca^{2+} influx through NMDA-sensitive LGIC while having less impact on synaptic transmission where the agonist (glutamate) is present at higher concentration. However, memantine is certainly a weak antagonist (micromolar IC_{50}) and is not entirely selective for NMDA-sensitive receptors. For example, it has been reported to inhibit α7 nAChR with potency better than at NMDA-sensitive LGIC.[178] Memantine appears to offer some benefit in a CNS disorder still starved for effective therapy, but the true target of the drug may be not yet known. In addition to AD, memantine also is of clinical interest for diabetic neuropathy and glaucoma.

AMPA/kainate-sensitive receptors may be involved in epilepsy and neurodegeneration, and antagonists have been found to be anticonvulsant in animal models.[179–182] However, these receptors also are involved in physiological transmission throughout the brain and an antiepileptic therapy based upon AMPA/kainate-sensitive blockade has not emerged. As LTP is thought to be a principal neural mechanism underlying memory, and enhancement of AMPA-sensitive receptor transmission is thought to underlie LTP, then a positive allosteric modulator of AMPA-sensitive receptors could be an effective approach for cognitive enhancement (e.g., mild cognitive impairment, AD). The nootropic aniracetam potentiates AMPA-sensitive iGluR and is effective in tissue models.[183–185] A metabolite of aniracetam has been found to potentiate α7 nAChR.[186] Aniracetam was launched for cognitive impairment in Japan, Italy, and Greece (1993), but subsequently withdrawn in Japan due to insufficient efficacy. Nefiracetam, a structurally similar nootropic, potentiates NMDA-sensitive iGluR and nAChR.[187,188] Ampakines represent another class of AMPA-sensitive iGluR modulator.[189–191] Examples include CX516 and CX717 for AD and ORG-24448 for schizophrenia. In addition to memantine, launched products include budipine, an NMDA-sensitive receptor antagonist for Parkinson's dyskinesia, also of interest for cluster headache and migraine, premarin for menopausal syndrome and postmenopausal osteoporosis, and ifenprodil, an NMDA-sensitive receptor antagonist/α-adrenergic receptor antagonist vasodilator.

Numerous other iGluR-based compounds have been evaluated clinically. These include: NMDA-sensitive receptor antagonists BIO-176 (a ketamine formulation) for restless leg syndrome; CHF-3381 for neuropathic pain, epilepsy, and cognitive impairment; CNS 5161 for neuropathic pain; dimebon, which also may have cholinergic agonist and histaminergic antagonist properties, for cognitive impairment; perzinfotel for neuropathic pain and stroke; PMI-150 (another ketamine formulation) for pain; traxoprodil (NR2B selective) for pain and stroke; delucemine for depression, epilepsy, neuroprotection, and pain; neramexane for cognitive impairment, ischemia, alcohol dependence, pain, and osteoarthritis; dexanabinol for stroke and other forms of CNS injury; and neurodex, which also has sigma-1 receptor agonist and CYP2D6 inhibitor properties, for emotional liability in amyotrophic lateral sclerosis and multiple sclerosis, neuropathic pain, and cough. D-Serine and neboglamine are glycine site coagonists of interest for schizophrenia, while GW 468816 is a glycine site antagonist of interest for smoking cessation. SYM-1010, a conantokin-G analog, is an NMDA-sensitive receptor partial agonist that was of interest for epilepsy, although intrathecal administration is required. AMPA/kainate-sensitive receptor inhibitors include NGX-424 for pain and migraine; LY 293558 for epilepsy/seizures, neuroprotection, pain, and migraine; AMP-397 for epilepsy; E-2007 for Parkinson's disease, multiple sclerosis, and epilepsy; NS 1209 for epilepsy and stroke; and talampanel for epilepsy, Parkinson's disease, and neuroprotection.

2.22.3.8 Diseases

In mouse, the *lurcher* and *hotfoot* mutations in iGluR-δ2 cause disruption in cerebellar Purkinje synaptic transmission.[192,193] The *lurcher* mutation can cause iGluR-δ2 to be constitutively open when expressed in *Xenopus* oocytes, but this probably is influenced by association with an endogenous *Xenopus* iGluR subunit in oocytes and does not explain the *lurcher* mechanism in the mammalian Purkinje neuron where δ2 appears to function independently of LGIC.

2.22.4 Adenosine Triphosphate-Gated Family

There is now abundant evidence that ATP functions as a neurotransmitter and/or cotransmitter in addition to being a critical player in kinase signaling pathways and energy metabolism. Purinergic transmission is that which utilizes a purine nucleotide (typically ATP) or its metabolite as the extracellular messenger. ATP is rapidly degraded to ADP, AMP, and adenosine by ectonucleotidases, and therefore these substances may serve as messengers consequent to ATP release and metabolism as well as being released directly. Indeed, adenosine acts as an extracellular messenger through P1 GPCR. The purinergic LGIC are termed P2X and are selective for ATP. For reviews see [194–202].

2.22.4.1 Current Nomenclature

Extracellular nucleotides such as ATP and UTP modulate cellular function by activation of membrane-bound P2-receptors, which are divided into two families of P2 receptors: ionotropic P2X LGIC and metabotropic P2Y GPCR. The idea that ATP is a neurotransmitter was controversial for many years, but was supported by pioneering research principally in Burnstock's laboratory and ultimately widely accepted following the cloning of ATP receptors. Initially, the P2X family was subclassified and members named according to the cell type in which the receptor was first found and according to the pharmacology of the receptor. This nomenclature became cumbersome and confusing, and present nomenclature is based primarily upon amino acid sequence. Up to now seven mammalian P2X receptor subtypes ($P2X_{1-7}$) and eight mammalian P2Y receptor subtypes ($P2Y_{1,2,4,6,11,12,13,14}$) have been cloned and functionally defined as P2 receptors.[198,203]

2.22.4.2 Molecular Biology

The seven P2X receptor subunits cloned to date, $P2X_{1-7}$, form a new structural family of ligand-gated cation channels.[196,201] The $P2X_1$ to $P2X_6$ receptors have 379–472 amino acids, with a predicted tertiary structure of two TM segments, a large extracellular loop with 10 conserved cysteine residues, and intracellular N and C termini. The $P2X_7$ receptor (595 amino acids) is similar, but with a much larger intracellular C terminus. This is different from that of the other known ligand-gated ionotropic receptors, such as the nicotinic, glutamate, glycine, $GABA_A$, or $5HT_3$ receptors.

2.22.4.3 Protein Structure

Currently known P2X subunits can combine to form homomeric or heteromeric LGIC channels in a trimeric arrangement. Like other LGIC, the P2X subunits are thought to span the membrane and to assemble in such a way as to form a centrally located gated ion channel. However, the trimeric structure, relatively short N- and C-termini in many of the subunits, and presence of only two putative transmembrane regions are departures from the Cys-loop and iGluR LGIC noted above. Each subunit has a large extracellular domain (~280 amino acids) between the two TM segments. In the extracellular domain are 10 conserved cysteine residues thought to be important for proper folding or loop formation via disulfide bonds. Within two of these loops are conserved histidine, proline, and glycine residues associated with modulation by metal binding and with receptor flexibility. The ATP-binding motifs found in other ATP-sensitive proteins are not found in P2X. Rather, the phosphate portion of ATP appears to interact with basic residues near the channel mouth – a lysine near TM1, and a lysine and slightly upstream asparagine-phenylalanine-arginine near TM2 – while the adenine portion of ATP interacts with asparagine-phenylalanine-threonine in a separate loop approximately in the middle of the extracellular sequence. Additionally, near the phosphate-binding loop are two conserved histidine residues that appear to be involved in modulation by protons. Intracellularly, the N-terminal region is short but includes a protein kinase C consensus site, and the C-terminal region varies considerably in length among subtypes (27–239 residues) but includes a trafficking motif near the intracellular portion of TM2.

Known P2X receptors are trimeric assemblies comprising one or two of the seven subunits. Combined pharmacological and electrophysiological evaluation has provided evidence that heteromeric assemblies exist natively, as well as interesting clues related to channel function. For example, $P2X_3$ homomeric receptors are sensitive to α,β-methylene-ATP (α,β-meATP) and desensitize rapidly upon continued agonist application, while $P2X_2$ homomers are insensitive to α,β-meATP and desensitize slowly to ATP or other agonists. Heteromeric $P2X_{2/3}$ receptors can be clearly observed as LGIC that are activated by α,β-meATP but desensitize slowly. Similar currents are observed in sensory ganglia, suggesting that the native LGIC are $P2X_{2/3}$ heteromers. This also suggests that ligand binding occurs within the subunit, as in glutamate-gated channels rather than at the subunit interfaces, as in nAChR. $P2X_7$ is the subunit with the longest intracellular C-terminal region, and is the only subunit that appears not to form functional heteromeric assemblies. $P2X_6$, in contrast, appears to function only in heteromeric assemblies.

P2X members are nonselective cation channels, permeable to Ca^{2+} as well as monovalent cations. Some homomeric LGIC, notably $P2X_1$ and $P2X_3$, desensitize rapidly while others display slower desensitization kinetics. Yet others – $P2X_2$, $P2X_4$, and $P2X_7$ – display a remarkable, and unusual, macromolecular pore formation in which agonist initially gates cation channel opening, but after minutes of continued agonist application, the channel is opened sufficiently wide to be permeable to dyes such as YO-PRO-1. This has been of particular interest in $P2X_7$ (P2Z in older literature) function where it is thought that pore formation may mediate apoptotic or anti-infection processes. The answer is not resolved, but this remains an active area of research.

2.22.4.4 Physiological Functions

Purinergic receptors and responses are found throughout nervous system and in smooth muscle of various types. Characterization of the subtypes involved has been hampered by a lack of highly selective agonists and antagonists, but has, to some extent, benefited from gene knockout studies and from unique properties of certain heteromeric combinations. $P2X_1$ is found in smooth muscle and platelets as well as sensory ganglia, sympathetic nervous system, and CNS. It functions in purinergic transmission in arterial, bladder, gut, and vas deferens smooth muscle. One of its roles is exemplified by the finding that male $P2X_1$ knockout mice are infertile owing to the lack of $P2X_1$ mediated contraction in vas deferens. $P2X_1$ also is found in platelets and may play a role in platelet aggregation. On the other hand, $P2X_2$ and $P2X_3$ are expressed in sensory neurons (e.g., dorsal root ganglia) and appear to function in a heteromeric ($P2X_{2/3}$) complex. Given the predominant localization of $P2X_3$ in sensory neurons and upregulation of $P2X_2$ and $P2X_3$ expression with inflammation, $P2X_3$ has been a target of interest for pain and urological disorders. The effects of $P2X_3$ agonists, antagonists, and intrathecal antisense treatment are consistent with a role for $P2X_3$ in pain, although results with $P2X_3$ knockout mice have been modest. Interestingly, capsaicin can induce a Ca^{2+}-dependent cross-desensitization of the $P2X_{2/3}$ response in dorsal root ganglia, suggesting the possibility that efficacy of capsaicin or other TRPV1 agonists could, in part, be due to desensitization of P2X as well as TRPV1 receptors. $P2X_{1/5}$ and $P2X_{4/6}$ also may be expressed in spinal cord, along with $P2X_7$. Disruption of $P2X_7$ receptors abolished chronic inflammatory and neuropathic pain and it has been hypothesized that the $P2X_7$ receptor, via regulation of mature interleukin-1β (IL-1β) production, has a common upstream transductional role in the development of neuropathic and inflammatory pain.[204]

P2X subunits of all types also are expressed in the sympathetic nervous system, where in fact evidence for purinergic transmission was initially developed.[205] ATP and ACh are copackaged in cholinergic synaptic vesicles (as are ATP and catecholamines in secretory granules) and might function as cotransmitters, although the physiological/pathological role for this remains unclear. Interestingly, there is evidence that P2X receptors and Cys-loop receptors (e.g., nAChR) may inhibit one another apparently through protein–protein interaction. $P2X_7$ has attracted attention for its potential role in mycobacterial killing and other potential functions in lymphocytes (mitogenesis) and microglia where $P2X_7$ is highly expressed. However, expression or $P2X_7$ is widespread, including airway epithelium, vascular and other smooth muscle, pancreas, pituitary, spinal cord, retina, mast cells, skin, bone, and salivary glands.

2.22.4.5 Regulation

P2X receptors have the potential to activate Ca^{2+}-dependent intracellular signaling by admitting Ca^{2+} through the P2X channel and by activating voltage-dependent Ca^{2+} channels through depolarization of the cell. In turn, P2X receptors may be modulated through the conserved protein kinase C consensus site in the N-terminal region. The role of this process is not clearly understood, but regulation of desensitization kinetics is suspected.

2.22.4.6 Prototypical Pharmacology

Prototypic P2X agonists are ATP derivatives,[195,201,206] such as α,β-MeATP ($P2X_1$ and $P2X_3$ selective), BzATP ($P2X_1$, $P2X_2$, and $P2X_7$), and 2-Me-thioATP (active at all but most potent at $P2X_3$ and $P2X_1$). The dyes suramin and PPADS are commonly used antagonists, which are not selective among $P2X_1$, $P2X_2$, $P2X_3$, and $P2X_5$ but are selective for these P2X (micromolar IC_{50}) relative to $P2X_4$, $P2X_6$, and $P2X_7$ where they have little activity. $P2X_7$, on the other hand, is selectively blocked by calmidazolium. TNP-ATP is a potent (nanomolar) antagonist selective for $P2X_1$ and $P2X_3$, but does display micromolar IC_{50} for $P2X_2$ and $P2X_4$. Most compounds do little to distinguish $P2X_1$ and $P2X_3$. However, the antagonist IP_5I is three orders of magnitude selective for $P2X_1$ versus $P2X_3$ (~ 3 nM versus $\sim 3\,\mu M$ IC_{50}). A-317491, on the other hand, selectively inhibits $P2X_2$, $P2X_3$, and $P2X_{2/3}$ receptors and exhibits analgesic-like effects in animal models (see **Figure 9** for representative structures).[199,200,207]

2.22.4.7 Prototypical Therapeutics

P2X receptors appear to be involved in numerous physiological systems, and recently $P2X_3$ and $P2X_7$ have been of particular interest for pain and infection, respectively. However, development of a successful therapeutic, as well as verification of the suspected role in vivo, has been hampered by issues such as selectivity and suitable pharmacokinetic/physicochemical properties. Pyridoxal phosphate (MC-1) is a naturally occurring substance evaluated as a cytoprotectant in myocardial infarction, ischemia, and stroke.[208,209] MC-1 combined with other substances such as an angiotensin receptor antagonist (MC-4262) or lisinopril as an angiotensin converting enzyme inhibitor (MC-4232) are of interest in hypertension. However, pyridoxal phosphate may have a variety of actions and its specificity for any particular P2X receptor is not clear. YM-529 is a novel bisphosphonate investigated as a potential therapeutic in

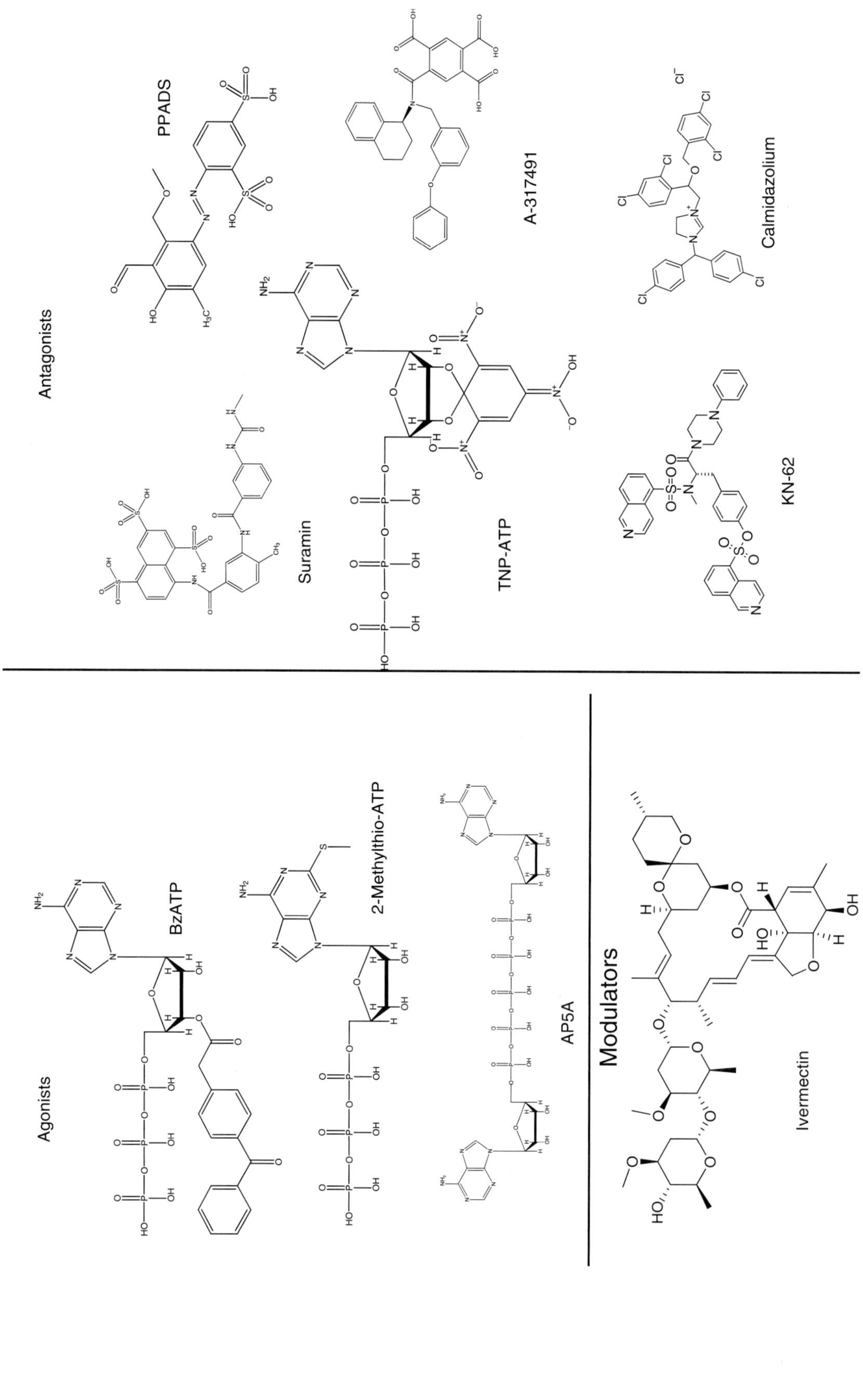

Ion Channels – Ligand Gated

Figure 9 Agonists, antagonists, and modulators for P2X receptors.

osteoporosis, various types of cancer, and pain. It has been reported to inhibit $P2X_{2/3}$ responses in an expression system, but in that regard it is not particularly potent nor is its selectivity for $P2X_{2/3}$ clear. AZD 9056 is a $P2X_7$ antagonist with potential for the treatment of rheumatoid arthritis and chronic obstructive pulmonary disease.

2.22.4.8 Diseases

Although a P2X1 dominant negative mutation has been reported, the role of that mutation in the patient's bleeding disorder remains uncertain. Nevertheless, P2X1 as well as certain P2Y receptors may be of interest in platelet disorders such as thrombosis.

2.22.5 Transient Receptor Potential Family

There are seven transient receptor potential (TRP) channel families playing a variety of roles from *Drosophila* to human: TRPC (canonical), TRPV (vanilloid), TRPM (melastatin), TRPP (polycystin), TRPML (mucolipin), and TRPA (ankyrin). Many roles appear to be in the sensory system, and various channels are activated by diverse stimuli, including temperature and possibly other physical parameters such as mechanical stimuli. However, some of the TRP also appear to be affected by exogenous and endogenous ligands, and therefore may be considered LGIC. For reviews, see [210–220].

2.22.5.1 Current Nomenclature

TRP ion channels are so named because *Drosophila* photoreceptors lacking TRP exhibit a transient voltage response to light stimulation. Unlike other ion channels, TRP channels are not named according to ligand, function, and/or selectivity. TRP channel mechanistic functions are disparate and sequence homology among family members may be as low as 20%. Nevertheless, TRP channels share overall structural similarity, and are named and grouped into subfamilies based upon structure. The nomenclature also reflects a mix of structure, mechanism, and pharmacology. TRPC are the classical or canonical subfamily with structure similar to *Drosophila* photoreceptor proteins where TRP channels were first identified.[210] TRPV are named for sensitivity of TRPV1 (formerly known as VR1 or vanilloid receptor 1) to vanilloid compounds such as capsaicin found in chili peppers. TRPM1 is named for melastatin, found in metastatic melanoma cells. However, this should not be taken to indicate that TRPM channels are involved in cancer induction or metastasis. TRPM8, for example, is distributed in sensory neurons and responds to cold and to menthol. TRPP are related to the polycystic kidney disease proteins PKD2 (TRPP2), PKD2L1 (TRPP3), and PKD2L2 (TRPP5). TRPML are related to the founding mucolipins, intracellular vesicular membrane proteins identified through mutation of TRPML1 (MCOLN1) in mucolipidosis. TRPA1 (formerly ANKTM1) is named for its large number of ankyrin repeats ($\geqslant 14$) and thus far is the only TRPA subfamily member. TRPA1 responds to temperature (cold in mammals, warmth in *Drosophila*). TRPN1 (formerly NompC) also has a large number of ankyrin repeats but nevertheless constitutes a subfamily different from TRPA1.[213,215] TRPN1 and TRPA1 appear to be required for mechanotransduction in hair cells, perhaps explaining the high content of ankyrin repeats.

2.22.5.2 Molecular Biology

The primary structure of TRP channels predicts six TM-spanning domains with a channel-forming pore domain (P-loop) between the TM5 and TM6 segments. Both the C- and the N-termini are located intracellularly. This architecture is a common theme for ion channels such as voltage-gated K^+ channels present in life forms ranging from bacteria to mammals. The mammalian TRP channel family is united primarily by structural homology within the TM-spanning domains. Other features include a 25-amino acid motif (TRP domain) containing TRP box (EWKFAR) just C-terminal to the sixth TM segment. The TRP domain and box are present in all other TRPC channel genes but not in all other TRP channel genes. The N-terminal cytoplasmic domain of TRPA1, TRPC, TRPN1, and TRPV contain ankyrin repeats, whereas the TRPC and TRPM contain proline-rich regions just C-terminal to the predicted sixth TM segment. At present, no one feature other than overall six-TM architecture and homology defines the TRP family. Thus, we expect that the definition of TRP channels will evolve as functions and structures are clarified. Updated summary tables of the molecular and biophysical properties of various TRP channels including tissue distribution can be found elsewhere.[221]

2.22.5.3 Protein Structure

Functional TRP assemblies are derived from four subunits. While the subunit and quaternary structure bears an overall similarity to voltage-gated K^+ channels, TRP channels are not K^+-selective but rather are cation channels permeable to Ca^{2+} as well as Na^+ and K^+. The Ca^{2+} permeability is significant with PCa^{2+}/PNa^+ typically up to 10 with a few exceptions: TRPM4 and TRPM5 are not permeable to Ca^{2+} (monovalent cation selective) while TRPV6 is Ca^{2+} selective. Studies have focused upon homomeric assemblies, but heteromers may be formed within each family. The N- and C-termini of TRP channels demonstrate a variety of regulatory sites including potential sites of protein–protein interaction PDZ domain in C-terminal region (TRPV), and ankyrin repeats in the N-terminal region (3–4 in TRPV, 14 in TRPA1).

2.22.5.4 Physiological Functions

The gating of TRP channels is not well understood and is an active area of research. Although TRPV1 is activated by ligands such as capsaicin applied extracellularly, the binding site is thought to be intracellular. Mutagenesis studies have suggested that TM3 and the N- and C-termini are important to vanilloid binding, with the hydrophobic area of the ligand interacting with TM3 and the hydrophilic area interacting with N- and C-termini. TRPV1–4 are polymodal sensory receptors responsive to elevated temperature, and acid and endogenous substances such as arachidonate lipoxygenase products as well as exogenous substances such as capsaicin or camphor. Protons appear to act as modulators, increasing thermal and ligand sensitivity. TRPV1–4 appear to function as temperature sensors and cover specific temperature ranges according to receptor subtype. Additionally, extracellular messengers and low pH associated with inflammation can activate TRPV1 and produce burning pain. TRPV1 knockout mice show reduced thermal sensitivity in inflammation models, although there appears to be less effect on mechanical hyperalgesia.

Most studies of recombinant TRP subunits have used homomeric receptors, but some may function as heteromers in vivo and this could alter the pharmacology of the receptors with important impact on therapeutic development. TRPV3 and TRPV1, for example, share some elements of function and of neuronal distribution, but it is not clear whether they coassemble and how that might influence their response to ligands. TRPV5 and TRPV6, on the other hand, are found in renal and intestinal epithelia and have an exceptionally high Ca^{2+} permeability ($PCa^{2+}/Na^+ > 100$). It has been proposed that these TRPV may function in Ca^{2+} transport.

TRPM5 seems to be a taste transducer although it is not clear whether these responses reflect direct or modulatory interactions. TRPM8 responds to cold, menthol, and icilin and is thought to function as cold sensor in C-fiber neurons, but is not localized specifically to sensory ganglia and may have additional functions.

TRPA1, characterized by 14 ankyrin repeats in the N-terminal region, is coexpressed with TRPV1 but not with TRPM8 in mammalian sensory neurons. The role of TRPA1 in mammals is not clear. Activation of TRPA1 by cold is controversial, but it apparently is effectively activated by mustard oil and cinnamaldehyde. THC also has been reported to activate the channel. It has been suggested that TRPA1 may be involved in mammalian hearing, in analogy with related TRP in zebrafish and *Drosophila*.

2.22.5.5 Regulation

TRPV and other TRP channels are clearly regulated by protein kinase A and C, and some responses to exogenous substances and extracellular messengers may in fact be indirect. However, TRPV1 responses to vanilloids and arachidonate derivatives do appear to be mediated by ligand interaction. Principal intracellular activators of TRP channels include calcium and diacylglycerol, products from the phospholipase C pathway.[216]

2.22.5.6 Prototypical Pharmacology

The majority of the ligand pharmacology currently available pertains to TRPV1. Many of the ligands are based upon vanilloid- or anandamide-like structures. The potent antagonist NDT-9525276 and other similar compounds are a departure. Other recently disclosed selective antagonists that block diverse modes of TRPV1 activation (by protons, heat, capsaicin, anandamide, etc.) with nanomolar affinities include isoquinoline urea (A-425619), pyridinylpiperazine ureas, biarylcarboxybenzamides, and phenylacrylamide (AMG9810) analogs (**Figure 10**).

2.22.5.7 Prototypical Therapeutics

Therapeutic interest has focused largely upon TRPV1 in the sensory system, particularly in pain and inflammation (including airway inflammation) and bladder control. Capsaicin activates and desensitizes TRPV1. In neonates, capsaicin can kill sensory neurons, but in adults it is limited to receptor desensitization. Capsaicin is used in some

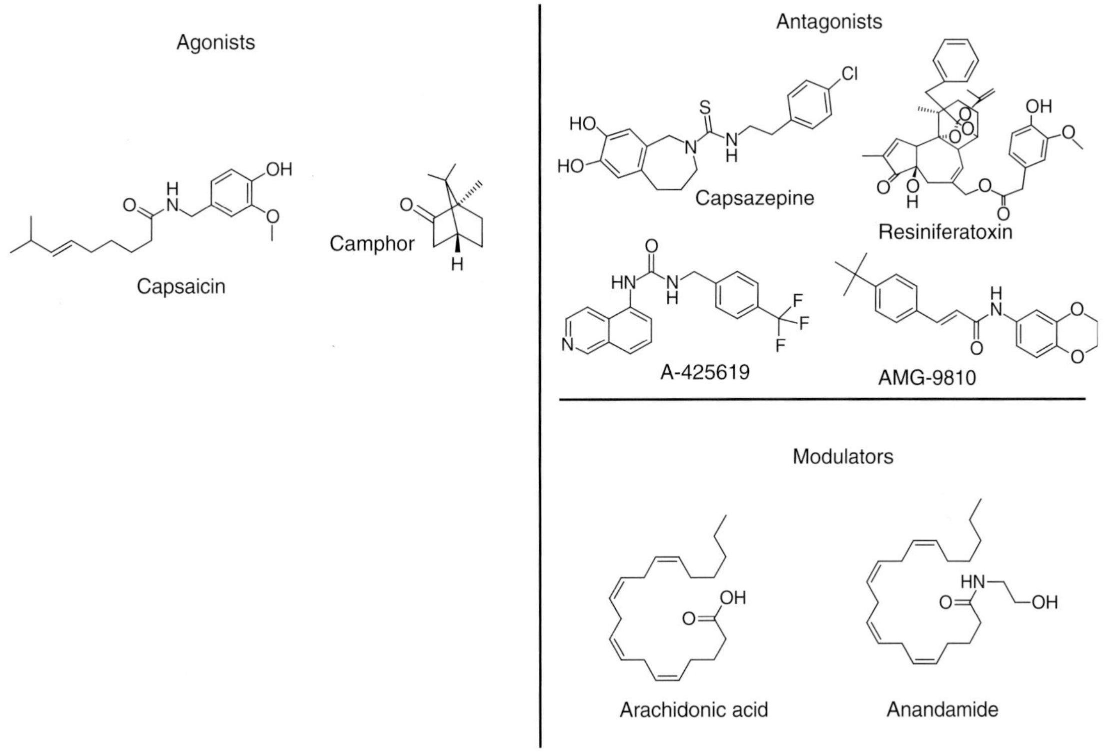

Figure 10 Agonists, antagonists, and modulators of TRP channels.

topical treatments to alleviate burning pain and itch, but its efficacy has not been impressive. However, capsaicin and another TRPV1 agonist, resiniferatoxin, have been reported to be more effective in bladder hyperreflexia when administered by bladder installation.[222,223]

Launched capsaicin-derived products include Capsidol and GenDerm as analgesic topicals. Other modifications evaluated in clinical trials include NasoCap, as an intranasal capsaicin for cluster headache and migraine; Transdolor, a dermal patch for neuralgia and neuropathy; Civamide, a capsaicin isomer for neuralgia/osteoarthritis or cluster headache/migraine; ALGRX-4975, an injectable capsaicin for pain and osteoarthritis; and resiniferatoxin for overactive bladder. Clearly, an antagonist rather than a desensitizing agonist would be preferred for indications such as pain. The antagonist capsazepine has been a useful tool in vitro, but has not been effective as an analgesic, presumably because of low potency and limited utility against activation by protons. However, I-resiniferatoxin is potent but not effective in vivo.[224] AMG-517 and GlaxoSmithKline 705498 are TRPV1 antagonists reportedly in clinic trials for inflammatory pain. With the emerging roles of TRP channels as intrinsic cellular sensors in a variety of processes including vision, pain, taste, temperature sensation, and hearing along with links to cancer and apoptosis, selective molecules targeting other TRP channels are expected to advance into clinical proof of concept studies in the not too distant future.

2.22.5.8 Diseases

TRP channels play key roles as receptors for irritants, inflammation products, and xenobiotic toxins, and their involvement in various diseases is an area of active investigation. At least four channelopathies are known at present and the list continues to grow. TRPP containing ion channels play important roles in epithelial cells and tubule formation, and mutations in TRPPs, particularly TRPP1 and TRPP2, are responsible for polycystic kidney disease. Loss-of-function mutations in TRPM6 cause the rare autosomal recessive disorder hypomagnesemia with secondary hypocalcemia manifested as a multitude of neurological symptoms including seizures and muscle spasms during infancy. Mutations in the TRPML1 gene cause mucolipidosis type IV, a neurodegenerative lysosomal storage disorder characterized by severe psychomotor retardation and other abnormalities. Defective TRPML1 alters endocytotic transport of membrane components by attenuating the fusion between late endosome–lysosome hybrid vesicles. As reviewed recently,[225] other indications of the involvement of TRPs in diseases have emerged from correlations between the levels of TRP channel expression and disease symptoms or from the mapping of TRP-encoding genes to susceptible chromosome regions. Changes in TRPV1 levels and regulation are implicated in dysfunctional pain sensations such as allodynia and hyperalgesia, gastrointestinal diseases such as inflammatory bowel disease and Crohn's disease, and urinary bladder diseases such as overactive bladder. TRPV4 is widely expressed in airways, and may be involved in bronchial hyperresponsiveness, as in chronic asthma. It has been suggested that TRPV4 provides the calcium influx pathway for regulatory volume decrease in response to hypotonic stress, a process that is defective in airway epithelia of individuals with cystic fibrosis.[226] A role in hearing impairment also is implicated, as one of the gene loci associated with autosomal dominant nonsyndromic hearing loss has been mapped to a region in chromosome 12q21–24 where the TRPV4 gene is located. Although precise involvement remains to be determined, TRPM genes such as the TRPM2 and TRPM3 gene locus has been linked to bipolar disorders, amyotrophic lateral sclerosis, and retinal diseases.

2.22.6 Acid-Sensitive Ion Channel Family

To the extent that protons may be considered ligands, then acid-sensitive ion channels (ASICs) are LGIC. Protons can act as extracellular messengers. Tissue acidifies to pH level < 6 under conditions of ischemia and inflammation, and these conditions are detected by the sensory nervous system. Additionally, transmitters are packaged in a low-pH environment in synaptic vesicles and secretory granules and, therefore, protons may be co-released during exocytosis. Furthermore, catabolic hydrolysis of some neurotransmitters increases acid content (ACh to choline and acetic acid, ATP to ADP, AMP, and phosphoric acid). Indeed, there is direct evidence for localized acidification during synaptic activity.[227] Thus, protons may signal pathologic conditions (ischemia, inflammation) and may be released as cotransmitters during physiologic activity. ASICs are highly expressed in sensory neurons and in brain and apparently serve as receptors for these proton messengers. The present challenge, however, is to define the binding site SAR to develop competitive ligands, or to identify selective channel blockers. For recent reviews see [228,229].

2.22.6.1 Current Nomenclature

ASICs are one of the seven subfamilies in the ENaC (epithelial sodium channel)/DEG (degenerin) family of cation channels. Earlier nomenclature labeled these channels in analogy to ENaC or DEG and in recognition of the source tissue, for example BNaC for brain (acid-sensitive) sodium channel, DRASIC for dorsal root acid-sensitive channel, etc. These nomenclatures have now been consolidated as ASIC1, which exists as two isoforms, ASIC1a (formerly ASICα or BNaC2α) and ASIC1b (formerly ASICβ or BNaC2β), ASIC2 along with its two isoforms, ASIC2a (MDEG1, BnaC1α, or BNC1) and ASIC2b (MDEG2 or BNaC1β), ASIC3 (DRASIC), and ASIC4 (SPASIC).

2.22.6.2 Molecular Biology

The three principal genes, *ASIC1*, *ASIC2*, and *ASIC3*, and their alternatively spliced transcripts (ASIC1a, ASIC1b, ASIC2a, and ASIC2b) produce subunits that combine to form multimeric channels activated by protons. Neurons express ASIC subunits in variable and overlapping distributions. ASICs are localized principally to neurons in central and peripheral systems, or their target organs. ASIC1a, ASIC2a, and ASIC2b are found in CNS with widespread distribution and particularly high levels in hippocampus, cortex, cerebellum, amygdala, olfactory bulb, and habenula. ASIC1b and particularly ASIC3 are highly expressed in sensory ganglia. ASIC4 also is found in brain and in pituitary, but its function remains to be clarified. As discussed below, much of the current understanding of the functional roles of ASIC subunits is derived from gene knockout experiments.

2.22.6.3 Protein Structure

Each ASIC subunit has two transmembrane domains with relatively short intracellular N- and C-termini, and a large extracellular region between TM1 and TM2. While that is vaguely reminiscent of P2X subunits, functional ASICs are thought to be comprised of four subunits rather than three in the case of P2X. ASIC1–3 can form homomeric or heteromeric LGIC in the typical expression systems, but ASIC4 has not yet been shown to participate in either homomeric or heteromeric assemblies. The site and mechanism of proton gating remain areas of investigation, and may differ among ASIC subtypes. In ASIC3, it appears that gating is due to proton displacement of Ca^{2+}, apparently bound to the receptor so as to block the channel[230] – a 'dechelation'. However, the amino acid residues contributing to this interaction have yet to be identified.[231] The intracellular N-terminal region is short but contributes to channel function by regulating Mg^{2+} ion selectivity.[232]

2.22.6.4 Physiological Functions

Elucidation of the physiological roles of ASICs is hampered by the limited availability of pharmacological tools. However, gene knockout studies are consistent with the ideas that ASIC3 is involved in pain,[233,234] and that ASIC2 functions in retinal activity.[235] An early study suggested that ASIC2a participates in touch mechanosensory transmission, but there is some uncertainty.[236–238] In any case, it does not appear to be involved in pain. The Welsh laboratory has used gene knockout approach to study ASIC subunit function in behavior as well as in in vitro CNS tissue physiology. ASIC1 was found to mediate amygdala acid responses and amygdalar behavior such as fear conditioning,[239,240] and ASIC1a/2a heteromer was implicated in hippocampal function including induction of LTP as well as the adverse effects of ischemia.[241–243] The hippocampal studies in particular suggest that ASIC may be involved in synaptic transmission or in modulating synaptic transmission.

ASIC1 enjoys a neuronal distribution, including Zn^{2+}-rich areas of hippocampus.[244] While ASIC1a and ASIC2a may modulate synaptic function in hippocampus, these LGIC themselves (above) are modulated in a complex manner by Zn^{2+},[245–247] which is concentrated in specific hippocampal and spinal dorsal horn neurons. At micromolar concentrations, Zn^{2+} selectively potentiates ASIC2a-containing LGIC.[247] However, at nanomolar concentrations, Zn^{2+} inhibits ASIC1a-containing LGIC.[245] Low, nanomolar concentrations of Zn^{2+} are maintained in vivo, but in artificial 'physiological' saline Zn^{2+} contamination may be sufficiently high to block ASIC1a unless selective chelators are added, raising questions as to the extent to which ASIC1a and other Zn^{2+}-blocked channels have been underappreciated in in vitro studies. In ASIC1a/ASIC2a heteromers, as expressed in the hippocampus, it has been suggested that while nanomolar concentrations of Zn^{2+} may block by interacting with the LGIC complex at one site, micromolar Zn^{2+} may interact at another site to relieve the block.[245] Thus, the roles of ASIC in hippocampus may be a complex interplay between Zn^{2+}, potentially released from dentate gyrus/CA3 mossy fibers upon specific neuronal input, and protons.

2.22.6.5 Regulation

There is considerable heterogeneity in the nature of pH-evoked currents within and between individual neurons attributable to differences in ASIC subunit make-up, and it is to be expected that channels derived from diverse ASIC subunits could be regulated under normal physiologic and pathologic conditions. For example, transient global ischemia upregulates ASIC2a expression in hippocampal and cortical neurons that survive ischemia.[248] Complete Freund's adjuvant-induced inflammation resulted in increased expression of both ASIC1a and ASIC2a in the dorsal horn, and this may play important roles in inflammation-related persistent pain.[249]

2.22.6.6 Prototypical Pharmacology

As has often been the case, nature leads the way. By far the most potent and specific ASIC blockers are toxins – the ASIC3 blocker APETx2 from sea anemone[250,251] and the ASIC1a inhibitor psalmotoxin-1 from spider.[252–254] Structures of these toxins are known, and perhaps this will lead to better understanding of the LGIC structure and pathways to synthetic ASIC blockers. Amiloride has been typically used to demonstrate that acid-induced effects are receptor-mediated and not via nonspecific interactions. At submicromolar concentrations, amiloride does inhibit ENaC. However, ASIC sensitivity to amiloride is much lower (mid micromolar), and the inhibition is not selective. At such high concentrations, amiloride may inhibit a number of ion channels apart from ENaC and ASIC. More recently, a nonamiloride blocker A-317567 was reported to inhibit ASIC responses in dorsal root ganglion neurons,[255] with modest potency and selectivity among the dorsal root ganglion ASIC – perhaps an order of magnitude better than amiloride. As noted above, Zn^{2+} inhibits or potentiates specific ASIC. Gadolinium in micromolar concentrations inhibits ASIC3 homomeric and ASIC2a/ASIC3 heteromeric LGIC.[256]

ASIC3 is modulated by three related neuropeptides: FMRFamide, which is expressed in invertebrates and neuropeptides FF and SF, which are expressed in mammals including human.[257–260] This may represent another avenue for pain therapeutics, either by inhibiting the LGIC potentiation or by controlling expression and/or release of the endogenous modulator. Potentiation of ASIC3 appears to involve reduced desensitization. The LGIC site of action is not known.[261]

2.22.6.7 Prototypical Therapeutics

No ASIC-based therapeutics are known. However, interest in ASIC blockers for the treatment of ischemia (cardiac and CNS) and pain is growing. A collaboration between PainCeptor (Canada) and NeuroSearch (Denmark) claims to have identified potent ASIC3 and ASIC1a antagonists from a screening approach and is interested in optimizing these leads for treatment of pain associated with cardiac ischemia, muscle, and bone pain.

2.22.6.8 Diseases

Detailed functions of ASICs in both peripheral and central nervous systems remain to be fully elucidated. In peripheral sensory neurons, ASICs have been implicated in mechanosensation and perception of pain particularly during tissue acidosis as in ischemic myocardium where ASICs transduce anginal pain. The presence of ASICs in the brain suggests that these channels have other functions as well and studies have indicated that ASIC1a is involved in synaptic plasticity, learning/memory, and fear conditioning. Gain-of-function mutations of ASIC1 and ASIC2 have been proposed to participate in neurodegenerative diseases. Additionally, increased ASIC3 expression in inflamed human intestine has been reported.[262]

2.22.7 Future Directions

The fundamental 'specialties' of the nervous system are spatial and temporal resolution. Neurons use a variety of extracellular messengers, as do the endocrine and immune systems, but further deliver those messengers to a spatially defined volume (synapse) with diffusion-limited catabolic and reuptake processes. Additionally, the messenger is delivered rapidly and in a discrete time window measured in milliseconds to seconds. Thus, the nervous system achieves a high degree of specificity with relatively few transmitters compared to the variety of signals processed. In principle, a systemic therapeutic agent cannot approach this degree of resolution with present technology, and thus the term 'neurotransmitter mimic', which has sometimes been applied to LGIC agonists, may be a misnomer. Nevertheless, some LGIC agonists, such as those for nAChR and $GABA_A$ receptors, have appeared in clinical trials with demonstrated proof of concept. In theory, however, one might expect an LGIC antagonist or a modulator to have a better chance of success. For an antagonist, the issue of temporal resolution is reduced since the antagonist would have its effect principally when transmitter is released, but spatial resolution remains a potential problem and a source of adverse effects. The utility is limited to disorders where transmission is excessive, and where blockade of transmission at targeted synapses and pharmacologically similar synapses can be tolerated. Such restrictions can be problematic and may limit therapeutic utility. Furthermore, a number of important CNS disorders, for example, neurodegenerative disorders such as AD and Parkinson's disease, are due to deficits in transmission, at least in part.

Thus, modulation of transmission would seem a favorable path for neuronal therapeutics, and perhaps LGIC therapeutics in general. Positive modulators could increase the gain on what remains of the physiological signal without causing an inappropriate tonic signal. Negative modulators may be able to reduce the gain without completely blocking transmission. Therapeutic examples of transmission modulators may include biogenic amine reuptake inhibitors (antidepressants), acetylcholinesterase inhibitors (AD), and L-dopa (Parkinson's disease); all of which, by reduced catabolism or increased synthesis, boost synaptic levels of transmitter contingent upon physiological release of the transmitter. One might also achieve success by modulating the receptor, and perhaps improve selectivity by specifically targeting the receptor compromised in the disorder. Benzodiazepines and related 'benzodiazepine receptor agonists' are examples of LGIC ($GABA_A$) modulator therapeutics. Presumably modulators for other LGIC, particularly homologous LGIC from the Cys-loop family, could be found and developed into successful therapeutics. Indeed, some LGIC modulators may already be in clinical or advanced preclinical phases, for example AMPA-sensitive iGluR and $\alpha 7$ nAChR positive allosteric modulators. However, the binding sites and molecular mechanisms behind these modulators are not clear. Some immediate challenges will be to clarify modulator sites so as to aid rational drug design, to optimize modulator efficacy and selectivity, to identify lead modulators for other LGIC involved in disease processes or disorders, and to identify those LGIC that are appropriate targets.

A pitfall is that our high-throughput, best-identified approaches utilize predefined hetero- or homomeric subunit combinations that may not be strictly reflective of native receptors under physiological and/or pathological conditions. Molecular biology has provided a means of isolating specific LGIC through cloning and functional expression. But, largely unknown are the exact subunit composition of the native LGIC, whether that composition may vary among cells or between health and disease, or whether the expression system provides posttranslational processing and cytoskeletal coupling similar to that of the native LGIC. Clearly, molecular cloning and heterologous expression techniques have provided keys to significant advances in pharmacology and physiology and accelerated initial stages of drug discovery efforts. However, novel compounds discovered with these tools need to be further verified and understood in native systems. LGIC need to be understood in their contextual roles. For example in neuroscience, there is a particularly wide gulf between recombinant LGIC and behavioral readout, and a considerable challenge is to bridge that divide with physiologically relevant systems. Likewise, as new compounds advance through clinical trials, it becomes important to relate efficacy and adverse effects for further target validation by establishing predictability and correlations with preclinical parameters such as receptor occupancy, behavioral measures, and therapeutic window.

Finally, we must better understand the roles of LGIC outside classical synaptic transmission. In the past two decades, it has become increasingly apparent that neuronal LGIC are involved not only in synaptic depolarization/hyperpolarization, but also in mediating signaling processes, particularly Ca^{2+}-mediated processes, important to neuronal regulation. Further, we are beginning to better appreciate that many LGIC are present and functional not only in neuron, muscle, and endocrine cells but also in so-called nonexcitable cells such as endothelial cells, epithelial cells, and immunocompetent cells. A challenge will be to better elucidate the expression and roles of LGIC in such tissues as well as in nervous system in order to better anticipate potential adverse effects of CNS-targeted therapeutics as well as to expand the opportunity to identify novel therapeutics through nonneuronal LGIC.

References

1. Hansen, S. B.; Sulzenbacher, G.; Huxford, T.; Marchot, P.; Taylor, P.; Bourne, Y. *EMBO J.* **2005**, *24*, 3635–3646.
2. Bouzat, C.; Gumilar, F.; Spitzmaul, G.; Wang, H. L.; Rayes, D.; Hansen, S. B.; Taylor, P.; Sine, S. M. *Nature* **2004**, *430*, 896–900.
3. Miyazawa, A.; Fujiyoshi, Y.; Unwin, N. *Nature* **2003**, *423*, 949–955.
4. Unwin, N. *J. Mol. Biol.* **2005**, *346*, 967–989.
5. Sine, S. M.; Wang, H. L.; Gao, F. *Curr. Med. Chem.* **2004**, *11*, 559–567.
6. Barry, P. H.; Lynch, J. W. *IEEE Trans. Nanobiosci.* **2005**, *4*, 70–80.
7. Bunnelle, W. H.; Decker, M. W. *Exp. Opin. Ther. Pat.* **2003**, *13*, 1003–1021.
8. Gundisch, D. *Exp. Opin. Ther. Pat.* **2005**, *15*, 1221–1239.
9. Gotti, C.; Clementi, F. *Prog. Neurobiol.* **2004**, *74*, 363–396.
10. Hogg, R. C.; Raggenbass, M.; Bertrand, D. *Rev. Physiol. Biochem. Pharmacol.* **2003**, *147*, 1–46.
11. Jensen, A. A.; Frolund, B.; Liljefors, T.; Krogsgaard-Larsen, P. *J. Med. Chem.* **2005**, *48*, 4705–4745.
12. Paton, W. D. M.; Perry, W. L. M. *J. Physiol.* **1953**, *119*, 43–57.
13. Tsuneki, H.; Salas, R.; Dani, J. A. *J. Physiol.* **2003**, *547*, 169–179.
14. Fischer, U.; Reinhardt, S.; Albuquerque, E. X.; Maelicke, A. *Eur. J. Neurosci.* **1999**, *11*, 2856–2864.
15. Romano, S. J.; Pugh, P. C.; Mcintosh, J. M.; Berg, D. K. *J. Neurobiol.* **1997**, *32*, 69–80.
16. Barabino, B.; Vailati, S.; Moretti, M.; Mcintosh, J. M.; Longhi, R.; Clementi, F.; Gotti, C. *Mol. Pharmacol.* **2001**, *59*, 1410–1417.
17. Vailati, S.; Moretti, M.; Longhi, R.; Rovati, G. E.; Clementi, F.; Gotti, C. *Mol. Pharmacol.* **2003**, *63*, 1329–1337.
18. Wang, N. S.; Orr-Urtreger, A.; Korczyn, A. D. *Prog. Neurobiol.* **2002**, *68*, 341–360.
19. Cordero-Erausquin, M.; Marubio, L. M.; Klink, R.; Changeux, J. P. *Trends Pharmacol. Sci.* **2000**, *21*, 211–217.
20. Bohr, I. J.; Ray, M. A.; McIntosh, J. M.; Chalon, S.; Guilloteau, D.; McKeith, I. G.; Perry, R. H.; Clementi, F.; Perry, E. K.; Court, J. A. et al. *Exp. Neurol.* **2005**, *191*, 292–300.
21. Gotti, C.; Moretti, M.; Zanardi, A.; Gaimarri, A.; Champtiaux, N.; Changeux, J. P.; Whiteaker, P.; Marks, M. J.; Clementi, F.; Zoli, M. *Mol. Pharmacol.* **2005**, *68*, 1162–1171.
22. Salminen, O.; Murphy, K. L.; McIntosh, J. M.; Drago, J.; Marks, M. J.; Collins, A. C.; Grady, S. R. *Mol. Pharmacol.* **2004**, *65*, 1526–1535.
23. Quik, M.; Vailati, S.; Bordia, T.; Kulak, J. M.; Fan, H.; Mcintosh, J. M.; Clementi, F.; Gotti, C. *Mol. Pharmacol.* **2005**, *67*, 32–41.
24. Azam, L.; Winzer-Serhan, U. H.; Chen, Y. L.; Leslie, F. M. *J. Comp. Neurol.* **2002**, *444*, 260–274.
25. Kulak, J. M.; Mcintosh, J. M.; Quik, M. *Mol. Pharmacol.* **2002**, *61*, 230–238.
26. Luo, L.; Bennett, T.; Jung, H. H.; Ryan, A. F. *J. Comp. Neurol.* **1998**, *393*, 320–331.
27. Morley, B. J.; Li, H. S.; Hiel, H.; Drescher, D. G.; Elgoyhen, A. B. *Mol. Brain Res.* **1998**, *53*, 78–87.
28. Anderson, A. D.; Troyanovskaya, M.; Wackym, P. A. *Brain Res.* **1997**, *778*, 409–413.
29. Park, H. J.; Niedzielski, A. S.; Wenthold, R. J. *Hearing Res.* **1997**, *112*, 95–105.
30. Hiel, H.; Elgoyhen, A. B.; Drescher, D. G.; Morley, B. J. *Brain Res.* **1996**, *738*, 347–352.
31. Glowatzki, E.; Wild, K.; Brandle, U.; Fakler, G.; Fakler, B.; Zenner, H. P.; Ruppersberg, J. P. *Proc. R. Soc. London, B* **1995**, *262*, 141–147.
32. Arredondo, J.; Nguyen, V. T.; Chernyavsky, A. I.; Bercovich, D.; Orr-Urtreger, A.; Kummer, W.; Lips, K.; Vetter, D. E.; Grando, S. A. *J. Cell Biol.* **2002**, *159*, 325–336.
33. Peng, H.; Ferris, R. L.; Matthews, T.; Hiel, H.; Lopez-Albaitero, A.; Lustig, L. R. *Life Sci.* **2004**, *76*, 263–280.
34. Lips, K. S.; Pfeil, U.; Kummer, W. *Neuroscience* **2002**, *115*, 1–5.
35. Baker, E. R.; Zwart, R.; Sher, E.; Millar, N. S. *Mol. Pharmacol.* **2004**, *65*, 453–460.
36. Moretti, M.; Vailati, S.; Zoli, M.; Lippi, G.; Riganti, L.; Longhi, R.; Viegi, A.; Clementi, F.; Gotti, C. *Mol. Pharmacol.* **2004**, *66*, 85–96.
37. Han, Z. Y.; Zoli, M.; Cardona, A.; Bourgeois, J. P.; Changeux, J. P.; Le Novere, N. *J. Comp. Neurol.* **2003**, *461*, 49–60.
38. Han, Z. Y.; Le Novere, N.; Zoli, M.; Hill, J. A.; Champtiaux, N.; Changeux, J. P. *Eur. J. Neurosci.* **2000**, *12*, 3664–3674.
39. Gotti, C.; Moretti, M.; Clementi, F.; Riganti, L.; Mcintosh, J. M.; Collins, A. C.; Marks, M. J.; Whiteaker, P. *Mol. Pharmacol.* **2005**, *67*, 2007–2015.
40. Grinevich, V. P.; Letchworth, S. R.; Lindenberger, K. A.; Menager, J.; Mary, V.; Sadieva, K. A.; Buhlman, L. M.; Bohme, G. A.; Pradier, L.; Benavides, J. et al. *J. Pharmacol. Exp. Ther.* **2005**, *312*, 619–626.
41. Azam, L.; Winzer-Serhan, U.; Leslie, F. M. *Neuroscience* **2003**, *119*, 965–977.
42. Criado, M.; Mulet, J.; Bernal, J. A.; Gerber, S.; Sala, S.; Sala, F. *Mol. Pharmacol.* **2005**, *68*, 1669–1677.
43. Grutter, T.; Changeux, J. P. *Trends Biochem. Sci.* **2001**, *26*, 459–463.
44. Harel, M.; Kasher, R.; Nicolas, A.; Guss, J. M.; Balass, M.; Fridkin, M.; Smit, A. B.; Brejc, K.; Sixma, T. K.; Katchalski-Katzir, E. et al. *Neuron* **2001**, *32*, 265–275.
45. Smit, A. B.; Syed, N. I.; Schaap, D.; van Minnen, J.; Klumperman, J.; Kits, K. S.; Lodder, H.; van der Schors, R. C.; van Elk, R.; Sorgedrager, B. et al. *Nature* **2001**, *411*, 261–268.
46. Smit, A. B.; Brejc, K.; Syed, N.; Sixma, T. K. *Ann. N. Y. Acad. Sci.* **2003**, *998*, 81–92.
47. Lee, W. Y.; Sine, S. M. *Nature* **2005**, *438*, 243–247.
48. Grutter, T.; de Carvalho, L. P.; Dufresne, V.; Taly, A.; Edelstein, S. J.; Changeux, J. P. *Proc. Natl. Acad. Sci. USA* **2005**, *102*, 18207–18212.
49. Curtis, L.; Buisson, B.; Bertrand, S.; Bertrand, D. *Mol. Pharmacol.* **2002**, *61*, 127–135.

50. Nelson, M. E.; Kuryatov, A.; Choi, C. H.; Zhou, Y.; Lindstrom, J. *Mol. Pharmacol.* **2003**, *63*, 332–341.
51. Liu, Y.; Ford, B.; Mann, M. A.; Fischbach, G. D. *J. Neurosci.* **2001**, *21*, 5660–5669.
52. Lansdell, S. J.; Gee, V. J.; Harkness, P. C.; Doward, A. I.; Baker, E. R.; Gibb, A. J.; Millar, N. S. *Mol. Pharmacol.* **2005**, *68*, 1431–1438.
53. Gottschalk, A.; Almedom, R. B.; Schedletzky, T.; Anderson, S. D.; Yates, J. R., III.; Schafer, W. R. *EMBO J.* **2005**, *24*, 2566–2578.
54. Hogg, R. C.; Bertrand, D. *Curr. Drug Targets CNS Neurol. Disord.* **2004**, *3*, 123–130.
55. Le Novere, N.; Corringer, P. J.; Changeux, J. P. *J. Neurobiol.* **2002**, *53*, 447–456.
56. Champtiaux, N.; Changeux, J. P. *Prog. Brain Res.* **2004**, *145*, 235–251.
57. Tapper, A. R.; McKinney, S. L.; Nashmi, R.; Schwarz, J.; Deshpande, P.; Labarca, C.; Whiteaker, P.; Marks, M. J.; Collins, A. C.; Lester, H. A. *Science* **2004**, *306*, 1029–1032.
58. Ulloa, L. *Nat. Rev. Drug Disc.* **2005**, *4*, 673–684.
59. Buisson, B.; Bertrand, D. *Trends Pharmacol. Sci.* **2002**, *23*, 130–136.
60. Kuryatov, A.; Luo, J.; Cooper, J.; Lindstrom, J. *Mol. Pharmacol.* **2005**, *68*, 1839–1851.
61. Vallejo, Y. F.; Buisson, B.; Bertrand, D.; Green, W. N. *J. Neurosci.* **2005**, *25*, 5563–5572.
62. Visanji, N. P.; Mitchell, S. N.; O'Neill, M.; Duty, S. *Neuropharmacology* **2005**, *7*, 7.
63. Molinari, E. J.; Delbono, O.; Messi, M. L.; Renganathan, M.; Arneric, S. P.; Sullivan, J. P.; Gopalakrishnan, M. *Eur. J. Pharmacol.* **1998**, *347*, 131–139.
64. Charpantier, E.; Wiesner, A.; Huh, K. H.; Ogier, R.; Hoda, J. C.; Allaman, G.; Raggenbass, M.; Feuerbach, D.; Bertrand, D.; Fuhrer, C. *J. Neurosci.* **2005**, *25*, 9836–9849.
65. Cho, C. H.; Song, W.; Leitzell, K.; Teo, E.; Meleth, A. D.; Quick, M. W.; Lester, R. A. J. *J. Neurosci.* **2005**, *25*, 3712–3723.
66. Liu, Z.; Tearle, A. W.; Nai, Q.; Berg, D. K. *J. Neurosci.* **2005**, *25*, 1159–1168.
67. Terlau, H.; Olivera, B. M. *Physiol. Rev.* **2004**, *84*, 41–68.
68. Paul, M.; Callahan, R.; Au, J.; Kindler, C. H.; Yost, C. S. *Anesth. Analg.* **2005**, *101*, 715–721.
69. Papke, R. L.; Porter Papke, J. K.; Rose, G. M. *Bioorg. Med. Chem. Lett.* **2004**, *14*, 1849–1853.
70. Macor, J. E.; Gurley, D.; Lanthorn, T.; Loch, J.; Mack, R. A.; Mullen, G.; Tran, O.; Wright, N.; Gordon, J. C. *Bioorg. Med. Chem. Lett.* **2001**, *11*, 319–321.
71. Akk, G.; Steinbach, J. H. *J. Neurosci.* **2005**, *25*, 1992–2001.
72. Smulders, C. J. G. M.; Zwart, R.; Bermudez, I.; van Kleef, R. G. D. M.; Groot-Kormelink, P. J.; Vijverberg, H. P. M. *Eur. J. Pharmacol.* **2005**, *509*, 97–108.
73. Texido, L.; Ros, E.; Martin-Satue, M.; Lopez, S.; Aleu, J.; Marsal, J.; Solsona, C. *Br. J. Pharmacol.* **2005**, *145*, 672–678.
74. Fayuk, D.; Yakel, J. L. *Mol. Pharmacol.* **2004**, *66*, 658–666.
75. Kihara, T.; Sawada, H.; Nakamizo, T.; Kanki, R.; Yamashita, H.; Maelicke, A.; Shimohama, S. *Biochem. Biophys. Res. Commun.* **2004**, *325*, 976–982.
76. Maelicke, A.; Schrattenholz, A.; Samochocki, M.; Radina, M.; Albuquerque, E. X. *Behav. Brain Res.* **2000**, *113*, 199–206.
77. Pereira, E. F. R.; Alkondon, M.; Reinhardt, S.; Maelicke, A.; Peng, X.; Lindstrom, J.; Whiting, P.; Albuquerque, E. X. *J. Pharmacol. Exp. Ther.* **1994**, *270*, 768–778.
78. Hughes, B. W.; Moro De Casillas, M. L.; Kaminski, H. J. *Semin. Neurol.* **2004**, *24*, 21–30.
79. Steinlein, O. K. *Prog. Brain Res.* **2004**, *145*, 275–285.
80. Rodrigues-Pinguet, N. O.; Pinguet, T. J.; Figl, A.; Lester, H. A.; Cohen, B. N. *Mol. Pharmacol.* **2005**, *68*, 487–501.
81. Bertrand, D.; Elmslie, F.; Hughes, E.; Trounce, J.; Sander, T.; Bertrand, S.; Steinlein, O. K. *Neurobiol. Dis.* **2005**, *20*, 799–804.
82. Martin, L. F.; Kem, W. R.; Freedman, R. *Psychopharmacology* **2004**, *174*, 54–64.
83. Hoyer, D.; Clarke, D. E.; Fozard, J. R.; Hartig, P. R.; Martin, G. R.; Mylecharane, E. J.; Saxena, P. R.; Humphrey, P. P. *Pharmacol. Rev.* **1994**, *46*, 157–203.
84. Pucadyil, T. J.; Kalipatnapu, S.; Chattopadhyay, A. *Cell. Mol. Neurobiol.* **2005**, *25*, 553–580.
85. Sari, Y. *Neurosci. Biobehav. Rev.* **2004**, *28*, 565–582.
86. Wolf, H. *Scand. J. Rheumatol.* **2000**, *29*, 37–45.
87. Reeves, D. C.; Lummis, S. C. R. *Mol. Mem. Biol.* **2002**, *19*, 11–26.
88. Costall, B.; Naylor, R. J. *Curr. Drug Targets CNS Neurol. Disord.* **2004**, *3*, 27–37.
89. Morales, M.; Wang, S. D. *J. Neurosci.* **2002**, *22*, 6732–6741.
90. Sudweeks, S. N.; van Hooft, J. A.; Yakel, J. L. *J. Physiol.* **2002**, *544*, 715–726.
91. Barrera, N. P.; Herbert, P.; Henderson, R. M.; Martin, I. L.; Edwardson, J. M. *Proc. Natl. Acad. Sci. USA* **2005**, *102*, 12595–12600.
92. Lummis, S. C. R.; Beene, D. L.; Lee, L. W.; Lester, H. A.; Broadhurst, R. W.; Dougherty, D. A. *Nature* **2005**, *438*, 248–252.
93. Peters, J. A.; Hales, T. G.; Lambert, J. *Trends Pharmacol. Sci.* **2005**, *26*, 587–594.
94. Harrell, A. V.; Allan, A. M. *Learn.* **2003**, *10*, 410–419.
95. Bhattacharya, A.; Dang, H.; Zhu, Q. M.; Schnegelsberg, B.; Rozengurt, N.; Cain, G.; Prantil, R.; Vorp, D. A.; Guy, N.; Julius, D. et al. *J. Neurosci.* **2004**, *24*, 5537–5548.
96. Kelley, S. P.; Dunlop, J. I.; Kirkness, E. F.; Lambert, J. J.; Peters, J. A. *Nature* **2003**, *424*, 321–324.
97. Dubin, A. E.; Huvar, R.; D'Andrea, M. R.; Pyati, J.; Zhu, J. Y.; Joy, K. C.; Wilson, S. J.; Galindo, J. E.; Glass, C. A.; Luo, L. et al. *J. Biol. Chem.* **1999**, *274*, 30799–30810.
98. Quirk, P. L.; Rao, S.; Roth, B. L.; Siegel, R. E. *J. Neurosci. Res.* **2004**, *77*, 498–506.
99. Castillo, M.; Mulet, J.; Gutierrez, L. M.; Ortiz, J. A.; Castelan, F.; Gerber, S.; Sala, S.; Sala, F.; Criado, M. *J. Biol. Chem.* **2005**, *280*, 27062–27068.
100. Constenla, M. *Ann. Pharmacother.* **2004**, *38*, 1683–1691.
101. Das, P.; Dillon, G. H. *J. Pharmacol. Exp. Ther.* **2005**, *314*, 320–328.
102. Stevens, R.; Rusch, D.; Solt, K.; Raines, D. E.; Davies, P. A. *J. Pharmacol. Exp. Ther.* **2005**, *314*, 338–345.
103. Barann, M.; Molderings, G.; Bruss, M.; Bonisch, H.; Urban, B. W.; Gothert, M. *Br. J. Pharmacol.* **2002**, *137*, 589–596.
104. Maksay, G.; Biro, T.; Bugovics, G. *Eur. J. Pharmacol.* **2005**, *514*, 17–24.
105. Niesler, B.; Flohr, T.; Nothen, M. M.; Fischer, C.; Rietschel, M.; Franzek, E.; Albus, M.; Propping, P.; Rappold, G. A. *Pharmacogenetics* **2001**, *11*, 471–475.
106. Niesler, B.; Weiss, B.; Fischer, C.; Nothen, M. M.; Propping, P.; Bondy, B.; Rietschel, M.; Maier, W.; Albus, M.; Franzek, E. et al. *Pharmacogenetics* **2001**, *11*, 21–27.
107. Niesler, B.; Frank, B.; Hebebrand, J.; Rappold, G. *Psychiatr. Genet.* **2005**, *15*, 303–304.

108. Dang, H.; England, P. M.; Farivar, S. S.; Dougherty, D. A.; Lester, H. A. *Mol. Pharmacol.* **2000**, *57*, 1114–1122.
109. Kullmann, D. M.; Ruiz, A.; Rusakov, D. M.; Scott, R.; Semyanov, A.; Walker, M. C. *Prog. Biophys. Mol. Biol.* **2005**, *87*, 33–46.
110. Foster, A. C.; Kemp, J. A. *Curr. Opin. Pharmacol.* **2005**, *21*, 21.
111. Rudolph, U.; Mohler, H. *Curr. Opin. Pharmacol.* **2005**, *20*, 20.
112. Whiting, P. J. *Curr. Opin. Drug Disc. Dev.* **2003**, *6*, 648–657.
113. Whiting, P. J. *Drug Disc. Today* **2003**, *8*, 445–450.
114. Bettler, B.; Kaupmann, K.; Mosbacher, J.; Gassmann, M. *Physiol. Rev.* **2004**, *84*, 835–867.
115. Farrant, M.; Nusser, Z. *Nat. Rev. Neurosci.* **2005**, *6*, 215–229.
116. Wafford, K. A. *Curr. Opin. Pharmacol.* **2005**, *5*, 47–52.
117. Johnston, G. A.; Chebib, M.; Hanrahan, J. R.; Mewett, K. N. *Curr. Drug Targets CNS Neurol. Disord.* **2003**, *2*, 260–268.
118. Johnston, G. A. *Curr. Topics Med. Chem.* **2002**, *2*, 903–913.
119. Bormann, J. *Trends Pharmacol. Sci.* **2000**, *21*, 16–19.
120. Ernst, M.; Bruckner, S.; Boresch, S.; Sieghart, W. *Mol. Pharmacol.* **2005**, *68*, 1291–1300.
121. Chou, K. C. *Biochem. Biophys. Res. Commun.* **2004**, *316*, 636–642.
122. Hajos, M.; Hoffmann, W. E.; Orban, G.; Kiss, T.; Erdi, P. *Neuroscience* **2004**, *126*, 599–610.
123. Hajos, N.; Palhalmi, J.; Mann, E. O.; Nemeth, B.; Paulsen, O.; Freund, T. F. *J. Neurosci.* **2004**, *24*, 9127–9137.
124. Rudolph, U.; Mohler, H. *Annu. Rev. Pharmacol. Toxicol.* **2004**, *44*, 475–498.
125. Steiger, J. L.; Russek, S. J. *Pharmacol. Ther.* **2004**, *101*, 259–281.
126. Sassoe-Pognetto, M.; Fritschy, J. M. *Eur. J. Neurosci.* **2000**, *12*, 2205–2210.
127. Thomas, P.; Mortensen, M.; Hosie, A. M.; Smart, T. G. *Nat. Neurosci.* **2005**, *8*, 889–897.
128. Macdonald, R. L.; Gallagher, M. J.; Feng, H. J.; Kang, J. *Biochem. Pharmacol.* **2004**, *68*, 1497–1506.
129. Krampfl, K.; Maljevic, S.; Cossette, P.; Ziegler, E.; Rouleau, G. A.; Lerche, H.; Bufler, J. *Eur. J. Neurosci.* **2005**, *22*, 10–20.
130. Ma, D. Q.; Whitehead, P. L.; Menold, M. M.; Martin, E. R.; Ashley-Koch, A. E.; Mei, H.; Ritchie, M. D.; Delong, G. R.; Abramson, R. K.; Wright, H. H. et al. *Am. J. Hum. Genet.* **2005**, *77*, 377–388.
131. Russo, L.; Mariotti, P.; Sangiorgi, E.; Giordano, T.; Ricci, I.; Lupi, F.; Chiera, R.; Guzzetta, F.; Neri, G.; Gurrieri, F. *Am. J. Hum. Genet.* **2005**, *76*, 327–333.
132. Cascio, M. *J. Biol. Chem.* **2004**, *279*, 19383–19386.
133. Colquhoun, D.; Sivilotti, L. G. *Trends Neurosci.* **2004**, *27*, 337–344.
134. Lynch, J. W. *Physiol. Rev.* **2004**, *84*, 1051–1095.
135. Laube, B.; Maksay, G.; Schemm, R.; Betz, H. *Trends Pharmacol. Sci.* **2002**, *23*, 519–527.
136. Breitinger, H. G.; Becker, C. M. *ChemBioChem* **2002**, *3*, 1042–1052.
137. Grudzinska, J.; Schemm, R.; Haeger, S.; Nicke, A.; Schmalzing, G.; Betz, H.; Laube, B. *Neuron* **2005**, *45*, 727–739.
138. Ma, D.; Liu, Z.; Li, L.; Tang, P.; Xu, Y. *Biochemistry* **2005**, *44*, 8790–8800.
139. Davies, P. A.; Wang, W.; Hales, T. G.; Kirkness, E. F. *J. Biol. Chem.* **2003**, *278*, 712–717.
140. Houtani, T.; Munemoto, Y.; Kase, M.; Sakuma, S.; Tsutsumi, T.; Sugimoto, T. *Biochem. Biophys. Res. Commun.* **2005**, *335*, 277–285.
141. Neale, J. H.; Olszewski, R. T.; Gehl, L. M.; Wroblewska, B.; Bzdega, T. *Trends Pharmacol. Sci.* **2005**, *26*, 477–484.
142. Waxman, E. A.; Lynch, D. R. *Neuroscientist* **2005**, *11*, 37–49.
143. Lerma, J. *Nat. Rev. Neurosci.* **2003**, *4*, 481–495.
144. Isaac, J. T.; Mellor, J.; Hurtado, D.; Roche, K. W. *Pharmacol. Ther.* **2004**, *104*, 163–172.
145. Dingledine, R.; Borges, K.; Bowie, D.; Traynelis, S. F. *Pharmacol. Rev.* **1999**, *51*, 7–62.
146. Palmer, C. L.; Cotton, L.; Henley, J. M. *Pharmacol. Rev.* **2005**, *57*, 253–277.
147. Mayer, M. L. *Curr. Opin. Neurobiol.* **2005**, *15*, 282–288.
148. Jaskolski, F.; Coussen, F.; Mulle, C. *Trends Pharmacol. Sci.* **2005**, *26*, 20–26.
149. Huettner, J. E. *Prog. Neurobiol.* **2003**, *70*, 387–407.
150. Erreger, K.; Chen, P. E.; Wyllie, D. J.; Traynelis, S. F. *Crit. Rev. Neurobiol.* **2004**, *16*, 187–224.
151. Chizh, B. A.; Headley, P. M. *Curr. Pharm. Des.* **2005**, *11*, 2977–2994.
152. Eriksson, M.; Nilsson, A.; Froelich-Fabre, S.; Akesson, E.; Dunker, J.; Seiger, A.; Folkesson, R.; Benedikz, E.; Sundstrom, E. *Neurosci. Lett.* **2002**, *321*, 177–181.
153. Matsuda, K.; Kamiya, Y.; Matsuda, S.; Yuzaki, M. *Brain Res. Mol. Brain Res.* **2002**, *100*, 43–52.
154. Bendel, O.; Meijer, B.; Hurd, Y.; von Euler, G. *Neurosci. Lett.* **2005**, *377*, 31–36.
155. Brody, S. A.; Nakanishi, N.; Tu, S.; Lipton, S. A.; Geyer, M. A. *Biol. Psychiatr.* **2005**, *57*, 1147–1152.
156. Fukaya, M.; Hayashi, Y.; Watanabe, M. *Eur. J. Neurosci.* **2005**, *21*, 1432–1436.
157. Mueller, H. T.; Meador-Woodruff, J. H. *J. Chem. Neuroanat.* **2005**, *29*, 157–172.
158. Paarmann, I.; Frermann, D.; Keller, B. U.; Villmann, C.; Breitinger, H. G.; Hollmann, M. *J. Neurochem.* **2005**, *93*, 812–824.
159. Mueller, H. T.; Meador-Woodruff, J. H. *Schizophr. Res.* **2004**, *71*, 361–370.
160. Araki, K.; Meguro, H.; Kushiya, E.; Takayama, C.; Inoue, Y.; Mishina, M. *Biochem. Biophys. Res. Commun.* **1993**, *197*, 1267–1276.
161. Tolle, T. R.; Berthele, A.; Zieglgansberger, W.; Seeburg, P. H.; Wisden, W. *J. Neurosci.* **1993**, *13*, 5009–5028.
162. Yamazaki, M.; Araki, K.; Shibata, A.; Mishina, M. *Biochem. Biophys. Res. Commun.* **1992**, *183*, 886–892.
163. Mayat, E.; Petralia, R. S.; Wang, Y. X.; Wenthold, R. J. *J. Neurosci.* **1995**, *15*, 2533–2546.
164. Katoh, A.; Yoshida, T.; Himeshima, Y.; Mishina, M.; Hirano, T. *Eur. J. Neurosci.* **2005**, *21*, 1315–1326.
165. Takeuchi, T.; Miyazaki, T.; Watanabe, M.; Mori, H.; Sakimura, K.; Mishina, M. *J. Neurosci.* **2005**, *25*, 2146–2156.
166. Murai, N.; Tsuji, J.; Ito, J.; Mishina, M.; Hirano, T. *Eur. Arch. Otorhinolaryngol.* **2004**, *261*, 82–86.
167. Cesa, R.; Morando, L.; Strata, P. *J. Neurosci.* **2003**, *23*, 2363–2370.
168. Ikeno, K.; Yamakura, T.; Yamazaki, M.; Sakimura, K. *Neurosci. Res.* **2001**, *41*, 193–200.
169. Furukawa, H.; Singh, S. K.; Mancusso, R.; Gouaux, E. *Nature* **2005**, *438*, 185–192.
170. Chatterton, J. E.; Awobuluyi, M.; Premkumar, L. S.; Takahashi, H.; Talantova, M.; Shin, Y.; Cui, J.; Tu, S.; Sevarino, K. A.; Nakanishi, N. et al. *Nature* **2002**, *415*, 793–798.
171. Andersson, O.; Stenqvist, A.; Attersand, A.; von Euler, G. *Genomics* **2001**, *78*, 178–184.
172. Burkhart, D. J.; Natale, N. R. *Curr. Med. Chem.* **2005**, *12*, 617–627.
173. Johansen, T. N.; Greenwood, J. R.; Frydenvang, K.; Madsen, U.; Krogsgaard-Larsen, P. *Chirality* **2003**, *15*, 167–179.

174. Madsen, U.; Pickering, D. S.; Nielsen, B.; Brauner-Osborne, H. *Neuropharmacology* **2005**, *49*, 114–119.
175. Danysz, W.; Parsons, C. G. *Int. J. Geriat. Psychiat.* **2003**, *18*, S23–S32.
176. Doggrell, S. A.; Evans, S. *Exp. Opin. Invest. Drugs* **2003**, *12*, 1633–1654.
177. Doraiswamy, P. M. *CNS Drugs* **2002**, *16*, 811–824.
178. Aracava, Y.; Pereira, E. F. R.; Maelicke, A.; Albuquerque, E. X. *J. Pharmacol. Exp. Ther.* **2005**, *312*, 1195–1205.
179. Fisahn, A. *J. Physiol.* **2005**, *562*, 65–72.
180. Stutzmann, J. M.; Vuilhorgne, M.; Mignani, S. *Mini Rev. Med. Chem.* **2004**, *4*, 123–140.
181. Conti, P.; De Amici, M.; De Micheli, C. *Mini Rev. Med. Chem.* **2002**, *2*, 177–184.
182. Madsen, U.; Stensbol, T. B.; Krogsgaard-Larsen, P. *Curr. Med. Chem.* **2001**, *8*, 1291–1301.
183. Jin, R.; Clark, S.; Weeks, A. M.; Dudman, J. T.; Gouaux, E.; Partin, K. M. *J. Neurosci.* **2005**, *25*, 9027–9036.
184. Gouliaev, A. H.; Senning, A. *Brain Res. Rev.* **1994**, *19*, 180–222.
185. Isaacson, J. S.; Nicoll, R. A. *Proc. Natl. Acad. Sci. USA* **1991**, *88*, 10936–10940.
186. Miyamoto, H.; Yaguchi, T.; Ohta, K.; Nagai, K.; Nagata, T.; Yamamoto, S.; Nishizaki, T. *Mol. Brain Res.* **2003**, *117*, 91–96.
187. Narahashi, T.; Moriguchi, S.; Zhao, X.; Marszalec, W.; Yeh, J. Z. *Biol. Pharm. Bull.* **2004**, *27*, 1701–1706.
188. Moriguchi, S.; Marszalec, W.; Zhao, X.; Yeh, J. Z.; Narahashi, T. *J. Pharmacol. Exp. Ther.* **2003**, *307*, 160–167.
189. Porrino, L. J.; Daunais, J. B.; Rogers, G. A.; Hampson, R. E.; Deadwyler, S. A. *PLoS Biol.* **2005**, *3*, e299.
190. Rex, C. S.; Kramar, E. A.; Colgin, L. L.; Lin, B.; Gall, C. M.; Lynch, G. *J. Neurosci.* **2005**, *25*, 5956–5966.
191. Xia, Y. F.; Kessler, M.; Arai, A. C. *J. Pharmacol. Exp. Ther.* **2005**, *313*, 277–285.
192. Sacchetti, B.; Scelfo, B.; Strata, P. *Neuroscientist* **2005**, *11*, 217–227.
193. Yuzaki, M. *Neurosci. Res.* **2003**, *46*, 11–22.
194. Burnstock, G.; Knight, G. E. *Int. Rev. Cytol.* **2004**, *240*, 31–304.
195. Khakh, B. S.; Burnstock, G.; Kennedy, C.; King, B. F.; North, R. A.; Seguela, P.; Voigt, M.; Humphrey, P. P. *Pharmacol. Rev.* **2001**, *53*, 107–118.
196. Egan, T. M.; Cox, J. A.; Voigt, M. M. *Curr. Topics Med. Chem.* **2004**, *4*, 821–829.
197. Illes, P.; Ribeiro, J. A. *Curr. Topics Med. Chem.* **2004**, *4*, 831–838.
198. Vial, C.; Roberts, J. A.; Evans, R. J. *Trends Pharmacol. Sci.* **2004**, *25*, 487–493.
199. Jarvis, M. F. *Exp. Opin. Ther. Targets* **2003**, *7*, 513–522.
200. Jacobson, K. A.; Jarvis, M. F.; Williams, M. *J. Med. Chem.* **2002**, *45*, 4057–4093.
201. North, R. A. *Physiol. Rev.* **2002**, *82*, 1013–1067.
202. Gachet, C. *Annu. Rev. Pharmacol. Toxicol.* **2006**, *46*, 277–300.
203. von Kugelgen, I. *Pharmacol. Ther.* **2006**, *110*, 415–432.
204. Chessell, I. P.; Hatcher, J. P.; Bountra, C.; Michel, A. D.; Hughes, J. P.; Green, P.; Egerton, J.; Murfin, M.; Richardson, J.; Peck, W. L. et al. *Pain* **2005**, *114*, 386–396.
205. Dunn, P. M.; Zhong, Y.; Burnstock, G. *Prog. Neurobiol.* **2001**, *65*, 107–134.
206. Jacobson, K. A.; Costanzi, S.; Ohno, M.; Joshi, B. V.; Besada, P.; Xu, B.; Tchilibon, S. *Curr. Topics Med. Chem.* **2004**, *4*, 805–819.
207. Jarvis, M. F.; Burgard, E. C.; McGaraughty, S.; Honore, P.; Lynch, K.; Brennan, T. J.; Subieta, A.; Van Biesen, T.; Cartmell, J.; Bianchi, B. et al. *Proc. Natl. Acad. Sci. USA* **2002**, *99*, 17179–17184.
208. Kandzari, D. E.; Dery, J. P.; Armstrong, P. W.; Douglas, D. A.; Zettler, M. E.; Hidinger, G. K.; Friesen, A. D.; Harrington, R. A. *Exp. Opin. Invest. Drugs* **2005**, *14*, 1435–1442.
209. Wang, C. X.; Yang, T.; Noor, R.; Shuaib, A. *J. Neurosurg.* **2005**, *103*, 165–169.
210. Clapham, D. E. *Nature* **2003**, *426*, 517–524.
211. Tominaga, M.; Caterina, M. J. *J. Neurobiol.* **2004**, *61*, 3–12.
212. Szallasi, A.; Blumberg, P. M. *Pharmacol. Rev.* **1999**, *51*, 159–212.
213. Fleig, A.; Penner, R. *Trends Pharmacol. Sci.* **2004**, *25*, 633–639.
214. Krause, J. E.; Chenard, B. L.; Cortright, D. N. *Curr. Opin. Invest. Drugs* **2005**, *6*, 48–57.
215. Lin, S. Y.; Corey, D. P. *Curr. Opin. Neurobiol.* **2005**, *15*, 350–357.
216. Moran, M. M.; Xu, H.; Clapham, D. E. *Curr. Opin. Neurobiol.* **2004**, *14*, 362–369.
217. van der Stelt, M.; Di Marzo, V. *Eur. J. Biochem.* **2004**, *271*, 1827–1834.
218. Wang, H.; Woolf, C. J. *Neuron* **2005**, *46*, 9–12.
219. Lopez-Rodriguez, M. L.; Viso, A.; Ortega-Gutierrez, S.; Diaz-Laviada, I. *Mini Rev. Med. Chem.* **2005**, *5*, 97–106.
220. Szallasi, A.; Appendino, G. *J. Med. Chem.* **2004**, *47*, 2717–2723.
221. Clapham, D. E.; Julius, D.; Montell, C.; Schultz, G. *Pharmacol. Rev.* **2005**, *57*, 427–450.
222. Evans, R. J. *Curr. Urol. Rep.* **2005**, *6*, 429–433.
223. de Groat, W. C.; Yoshimura, N. *Prog. Brain Res.* **2005**, *152*, 59–84.
224. Seabrook, G. R.; Sutton, K. G.; Jarolimek, W.; Hollingworth, G. J.; Teague, S.; Webb, J.; Clark, N.; Boyce, S.; Kerby, J.; Ali, Z. et al. *J. Pharmacol. Exp. Ther.* **2002**, *303*, 1052–1060.
225. Nilius, B.; Voets, T.; Peters, J. *Sci. STKE* **2005**, *2005*, re8.
226. Arniges, M.; Vazquez, E.; Fernandez-Fernandez, J. M.; Valverde, M. A. *J. Biol. Chem.* **2004**, *279*, 54062–54068.
227. Miesenbock, G.; De Angelis, D. A.; Rothman, J. E. *Nature* **1998**, *394*, 192–195.
228. Krishtal, O. *Trends Neurosci.* **2003**, *26*, 477–483.
229. Kellenberger, S.; Schild, L. *Physiol. Rev.* **2002**, *82*, 735–767.
230. Immke, D. C.; McCleskey, E. W. *Neuron* **2003**, *37*, 75–84.
231. Coric, T.; Zheng, D.; Gerstein, M.; Canessa, C. M. *J. Physiol.* **2005**.
232. Coscoy, S.; De Weille, J. R.; Lingueglia, E.; Lazdunski, M. *J. Biol. Chem.* **1999**, *274*, 10129–10132.
233. Sluka, K. A.; Price, M. P.; Breese, N. M.; Stucky, C. L.; Wemmie, J. A.; Welsh, M. J. *Pain* **2003**, *106*, 229–239.
234. Chen, C. C.; Zimmer, A.; Sun, W. H.; Hall, J.; Brownstein, M. J.; Zimmer, A. *Proc. Natl. Acad. Sci. USA* **2002**, *99*, 8992–8997.
235. Ettaiche, M.; Guy, N.; Hofman, P.; Lazdunski, M.; Waldmann, R. *J. Neurosci.* **2004**, *24*, 1005–1012.
236. Drew, L. J.; Rohrer, D. K.; Price, M. P.; Blaver, K. E.; Cockayne, D. A.; Cesare, P.; Wood, J. N. *J. Physiol.* **2004**, *556*, 691–710.
237. Roza, C.; Puel, J. L.; Kress, M.; Baron, A.; Diochot, S.; Lazdunski, M.; Waldmann, R. *J. Physiol.* **2004**, *558*, 659–669.
238. Price, M. P.; Lewin, G. R.; McIlwrath, S. L.; Cheng, C.; Xie, J.; Heppenstall, P. A.; Stucky, C. L.; Mannsfeldt, A. G.; Brennan, T. J.; Drummond, H. A. et al. *Nature* **2000**, *407*, 1007–1011.

239. Wemmie, J. A.; Coryell, M. W.; Askwith, C. C.; Lamani, E.; Leonard, A. S.; Sigmund, C. D.; Welsh, M. J. *Proc. Natl. Acad. Sci. USA* **2004**, *101*, 3621–3626.
240. Wemmie, J. A.; Askwith, C. C.; Lamani, E.; Cassell, M. D.; Freeman, J. H., Jr.; Welsh, M. J. *J. Neurosci.* **2003**, *23*, 5496–5502.
241. Xiong, Z. G.; Zhu, X. M.; Chu, X. P.; Minami, M.; Hey, J.; Wei, W. L.; MacDonald, J. F.; Wemmie, J. A.; Price, M. P.; Welsh, M. J. et al. *Cell* **2004**, *118*, 687–698.
242. Askwith, C. C.; Wemmie, J. A.; Price, M. P.; Rokhlina, T.; Welsh, M. J. *J. Biol. Chem.* **2004**, *279*, 18296–18305.
243. Wemmie, J. A.; Chen, J.; Askwith, C. C.; Hruska-Hageman, A. M.; Price, M. P.; Nolan, B. C.; Yoder, P. G.; Lamani, E.; Hoshi, T.; Freeman, J. H., Jr. et al. *Neuron.* **2002**, *34*, 463–477.
244. Alvarez de la Rosa, D.; Krueger, S. R.; Kolar, A.; Shao, D.; Fitzsimonds, R. M.; Canessa, C. M. *J. Physiol.* **2003**, *546*, 77–87.
245. Chu, X. P.; Wemmie, J. A.; Wang, W. Z.; Zhu, X. M.; Saugstad, J. A.; Price, M. P.; Simon, R. P.; Xiong, Z. G. *J. Neurosci.* **2004**, *24*, 8678–8689.
246. Gao, J.; Wu, L. J.; Xu, L.; Xu, T. L. *Brain Res.* **2004**, *1017*, 197–207.
247. Baron, A.; Schaefer, L.; Lingueglia, E.; Champigny, G.; Lazdunski, M. *J. Biol. Chem.* **2001**, *276*, 35361–35367.
248. Johnson, M. B.; Jin, K.; Minami, M.; Chen, D.; Simon, R. P. *J. Cereb. Blood Flow Metab.* **2001**, *21*, 734–740.
249. Wu, L. J.; Duan, B.; Mei, Y. D.; Gao, J.; Chen, J. G.; Zhuo, M.; Xu, L.; Wu, M.; Xu, T. L. *J. Biol. Chem.* **2004**, *279*, 43716–43724.
250. Chagot, B.; Escoubas, P.; Diochot, S.; Bernard, C.; Lazdunski, M.; Darbon, H. *Protein Sci.* **2005**, *14*, 2003–2010.
251. Diochot, S.; Baron, A.; Rash, L. D.; Deval, E.; Escoubas, P.; Scarzello, S.; Salinas, M.; Lazdunski, M. *EMBO J.* **2004**, *23*, 1516–1525.
252. Chen, X.; Kalbacher, H.; Grunder, S. *J. Gen. Physiol.* **2005**, *126*, 71–79.
253. Escoubas, P.; Bernard, C.; Lambeau, G.; Lazdunski, M.; Darbon, H. *Protein Sci.* **2003**, *12*, 1332–1343.
254. Escoubas, P.; De Weille, J. R.; Lecoq, A.; Diochot, S.; Waldmann, R.; Champigny, G.; Moinier, D.; Menez, A.; Lazdunski, M. *J. Biol. Chem.* **2000**, *275*, 25116–25121.
255. Dube, G. R.; Lehto, S. G.; Breese, N. M.; Baker, S. J.; Wang, X.; Matulenko, M. A.; Honore, P.; Stewart, A. O.; Moreland, R. B.; Brioni, J. D. *Pain* **2005**, *117*, 88–96.
256. Babinski, K.; Catarsi, S.; Biagini, G.; Seguela, P. *J. Biol. Chem.* **2000**, *275*, 28519–28525.
257. Lilley, S.; LeTissier, P.; Robbins, J. *J. Neurosci.* **2004**, *24*, 1013–1022.
258. Deval, E.; Baron, A.; Lingueglia, E.; Mazarguil, H.; Zajac, J. M.; Lazdunski, M. *Neuropharmacology* **2003**, *44*, 662–671.
259. Catarsi, S.; Babinski, K.; Seguela, P. *Neuropharmacology* **2001**, *41*, 592–600.
260. Askwith, C. C.; Cheng, C.; Ikuma, M.; Benson, C.; Price, M. P.; Welsh, M. J. *Neuron* **2000**, *26*, 133–141.
261. Ostrovskaya, O.; Moroz, L.; Krishtal, O. *J. Neurochem.* **2004**, *91*, 252–255.
262. Yiangou, Y.; Facer, P.; Smith, J. A.; Sangameswaran, L.; Eglen, R.; Birch, R.; Knowles, C.; Williams, N.; Anand, P. *Eur. J. Gastroenterol. Hepatol.* **2001**, *13*, 891–896.

Biographies

Clark Briggs is a neuropharmacologist/electrophysiologist in the Neuroscience Research Area within the Global Pharmaceutical Research and Development Division of Abbott. Dr Briggs received his ScB (Chemistry) from Brown University and PhD (Pharmacology) from Yale University. Following a postdoctoral fellowship and Research Scientist position at the Beckman Research Institute at the City of Hope Medical Center (Duarte, CA), he joined Abbott in 1986 where he has been engaged in drug discovery programs pertaining to G protein-coupled receptors and ligand- and voltage-gated ion channel targets for neuropsychiatric, analgesic, and urologic indications. Currently, Dr Briggs is engaged in nicotinic acetylcholine receptor research as an Associate Volwiler Research Fellow at Abbott.

Murali Gopalakrishnan is currently working in the Neuroscience Research Area within the Global Pharmaceutical Research and Development Division of Abbott. Dr Gopalakrishnan received his BPharm degree from Banaras Hindu University and PhD degree in Biochemical Pharmacology at the State University of New York at Buffalo. Following a postdoctoral fellowship at the Department of Physiology at Baylor College of Medicine, he joined the Neuroscience Research Area at Abbott. Over the past 13 years, he has been engaged in drug discovery programs in the areas of central nervous system, urology, and pain, focusing on voltage- and ligand-gated ion channels. Currently, Dr Gopalakrishnan is Project Leader and Associate Volwiler Research Fellow at Abbott.

2.23 Phosphodiesterases

D P Rotella, Wyeth Research, Princeton, NJ, USA

© 2007 Elsevier Ltd. All Rights Reserved.

2.23.1	**Current Nomenclature**	**920**
2.23.2	**Gene/Molecular Biology Classification**	**920**
2.23.3	**Protein Structure**	**920**
2.23.4	**Physiological Function**	**923**
2.23.4.1	PDE1	923
2.23.4.2	PDE2	923
2.23.4.3	PDE3	924
2.23.4.4	PDE4	924
2.23.4.5	PDE5	927
2.23.4.6	PDE6	928
2.23.4.7	PDE7	929
2.23.4.8	PDE8	929
2.23.4.9	PDE9	930
2.23.4.10	PDE10	930
2.23.4.11	PDE11	931
2.23.5	**Second Messengers**	**931**
2.23.6	**Prototypical Pharmacology**	**931**
2.23.6.1	PDE1 Inhibitors	931
2.23.6.2	PDE2 Inhibitors	933
2.23.6.3	PDE3 Inhibitors	933
2.23.6.4	PDE4 Inhibitors	935
2.23.6.5	PDE5 Inhibitors	938
2.23.6.6	PDE6 Inhibitors	942
2.23.6.7	PDE7 Inhibitors	943
2.23.6.8	PDE8 Inhibitors	945
2.23.6.9	PDE9 Inhibitors	945
2.23.6.10	PDE10 Inhibitors	946
2.23.6.11	PDE11 Inhibitors	946
2.23.7	**Prototypical Therapeutics**	**947**
2.23.7.1	PDE1	947
2.23.7.2	PDE2	947
2.23.7.3	PDE3	947
2.23.7.4	PDE4	948
2.23.7.5	PDE5	948
2.23.7.6	PDE6	949
2.23.7.7	PDE7	949
2.23.7.8	PDE8	949
2.23.7.9	PDE9	950
2.23.7.10	PDE10	950
2.23.7.11	PDE11	950

2.23.8	Genetic Diseases	950
2.23.9	New Directions in Phosphodiesterase Enzyme Biology and Inhibition	950
References		951

2.23.1 Current Nomenclature

Phosphodiesterase enzymes (PDEs) are so named because of the nature of the reaction they catalyze, as shown in eqn [1]. The enzyme activates a molecule of water to hydrolyze the phosphorus–oxygen bond in the substrate, a cyclic phosphate ester nucleotide to produce a nucleoside monophosphate. The first PDE enzymes, and their substrates, were identified in 1972.[1]

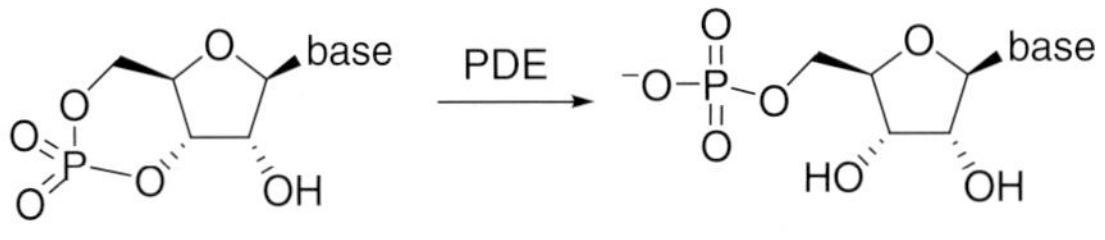

Cyclic AMP or GMP → Nucleoside monophosphate [1]

This family of intracellular enzymes modulates levels of the essential second messengers, cyclic adenosine monophosphate (cAMP) and cyclic guanosine monophosphate (cGMP), which are produced in response to various stimuli. Inhibition of enzymatic activity raises the level of the cyclic nucleotide, and results in maintained or enhanced physiological response to the stimulus. For example, it has been reported that a two- to fourfold increase in cGMP concentration results in a maximal biological response.[2] The acyclic nucleotide product of eqn [1] is generally inactive as a second messenger, and, as a result, the physiological response associated with stimulus of the signaling pathway is terminated.

A total of 11 different types of mammalian PDEs have been identified to date (PDE1–PDE11) in class I of the phosphodiesterase superfamily. Class II phosphodiesterases have no recognizable sequence similarity to enzymes in class I. Each of the subfamilies in class I have between one and four distinct gene product(s) (designated by a capital letter (A–D)), and in some cases there are also splice variants of these gene products (designated by a roman numeral). Enzymes in the family have individual and distinct degrees of substrate specificity (based on K_m (Michaelis–Menten affinity) values), as shown in **Table 1**. As noted in Section 2.23.4, V_{max} (maximal velocity) values for the two cyclic nucleotides can differ substantially, rendering differences in K_m values insignificant. The cAMP-specific PDEs include PDE4, PDE7, and PDE8, and the cGMP-specific enzymes are PDE5, PDE6, and PDE9. The remaining family members can process both cyclic nucleotides, although they prefer one or the other to varying degrees.

2.23.2 Gene/Molecular Biology Classification

Table 2 lists the human chromosome location for each of the known PDE enzymes.

2.23.3 Protein Structure

The structural organization of the mammalian PDE superfamily shows a common theme. Each enzyme has a C-terminal catalytic domain and an N-terminal regulatory domain. Most commonly, splice variants occur in the N-terminal region, and regulate the subcellular localization of the enzyme. In the catalytic domain, there is a consensus sequence, $HD(X_2)H(X_4)N$, which includes binding domains for divalent cations (e.g., Zn^{2+} and Mg^{2+}), needed for structural or catalytic functions. The mechanism by which the cyclic phosphate ester is cleaved involves nucleophilic activation of a water molecule and S_N2 displacement at the phosphorus atom, with inversion of configuration.[3,4] A representation of the structural organization for the 11 PDEs is shown in **Figure 1**.

The C-terminal catalytic domain of all 11 mammalian family members displays the closest similarities in terms of amino acid sequence (up to 45%). An early prediction by Beavo and co-workers that the catalytic domain resided in this region of the enzyme[5] was subsequently shown to be correct when PDE1, PDE2, and PDE3 were subjected to partial proteolytic cleavage (to remove regions of the protein outside the C-terminal domain), and these protein

Table 1 Substrate specificity of PDEs

PDE	Substrate preference (K_m)
PDE1A–C	1A: cGMP > cAMP
	1B, 1C: no preference
PDE2A	cGMP > cAMP
PDE3A–B	No preference
PDE4A–D	cAMP
PDE5A	cGMP
PDE6A–C	cGMP
PDE7A–B	cAMP
PDE8A–B	cAMP
PDE9A	cGMP
PDE10A	cAMP > cGMP
PDE11A	No preference

Table 2 Chromosome number and location for class I human PDE enzymes

PDE	Entrez gene no.	Chromosome	Location
PDE1A	5136	2	2q32.1
PDE1B	5153	12	12q13
PDE1C	5137	7	7p14.3
PDE2A	5138	11	11q13.4
PDE3A	5139	12	12p12
PDE3B	5140	11	11p15.1
PDE4A	5141	19	19p13.2
PDE4B	5142	19	1p31
PDE4C	5143	19	19p13.11
PDE4D	5144	5	5q12
PDE5A	8654	4	4q25–q27
PDE6A	5145	5	5q31.2–q34
PDE6B	5158	4	4p16.3
PDE6C	5146	10	10q24
PDE7A	5150	8	8q13
PDE7B	27115	6	6q23–q24
PDE8A	5151	15	15q25.3
PDE8B	8622	5	5q13.3
PDE9A	5152	21	21q22.3
PDE10A	10846	6	6q26
PDE11A	50940	2	2q31.2

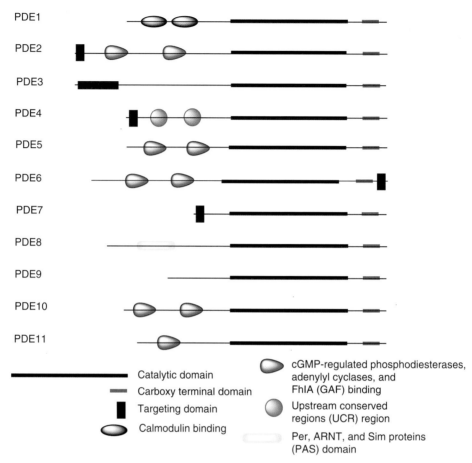

Figure 1 Structural organization of the 11 PDEs.

fragments demonstrated PDE activity. In addition, expression of mutant PDE3, PDE4, and PDE5, engineered to delete N-terminal sequences, also displayed PDE activity in vitro.[6–13] These structural modifications did not significantly influence the binding affinity of cGMP or inhibition by compounds in the case of PDE5, while effects using PDE4 depended on the specific changes in the protein outside of the catalytic domain.

Mutagenesis and hydropathic analysis of PDE5, in an attempt to understand the substrate selectivity of cGMP-specific enzymes, was carried out by Turko et al.[14] Substitution of Tyr602, or Glu775 in PDE5 substantially reduced substrate binding, and other amino acid mutations in the sequence around Glu775, including Ala769, Trp762, Gln765, and Leu771, increased the K_m value of cGMP and decreased the K_m value of cAMP, without influencing the K_D (dissociation constant) value of cGMP, or the respective IC_{50} (50% inhibitory concentration) values for the standard inhibitors zaprinast and rolipram. The alanine and leucine mutations employed in this experiment substituted more polar residues, while tryptophan and glutamine were replaced by leucine and tyrosine, respectively. Hydropathic analysis of a 30-amino-acid sequence surrounding the putative catalytic site of PDE1–PDE7 revealed that the cAMP-selective enzymes are uniformly more hydrophilic in a short sequence of residues on the N-terminal side of a key glutamic acid residue that is conserved in all known mammalian PDEs. In the cGMP-selective enzymes, this same sequence is always hydrophobic. Enzymes that can process both cyclic nucleotides show intermediate hydrophobicity in this region of the catalytic site. These differences in amino acid characteristics may be associated with the distinct hydrogen bond donor/acceptor characteristics of purine and pyrimidine bases. This study did not attempt to separate the structural effects of amino acid mutations (on the shape or size of the substrate binding site) from electronic effects that may be associated with substrate binding.

There are much more substantial variations in the N-terminal domains of PDEs, in terms of size, amino acid composition, and organization. Differences in this region of the molecule may be associated with specific subcellular localization requirements for individual PDEs that are associated with distinct cell types and physiological functions for the enzyme in a particular type of cell. As an example, the N-terminal domain of PDE1 has evolved with a Ca^{2+}/calmodulin-binding domain, while PDE3 has evolved with a distinct transmembrane region, and PDE7 and PDE9 have simple, low-molecular-weight N-terminal domains.

2.23.4 Physiological Function

Phosphodiesterases are highly regulated enzymes as a result of their critical role in second-messenger metabolism. At least three different means have been identified by which enzymatic activity can be controlled or influenced in cells: (1) regulation of substrate availability that can occur by alteration of the rate of cyclic nucleotide hydrolysis in response to competition by other phosphodiesterases, or by elevation of substrate levels; (2) extracellular stimuli that influence intracellular signaling, such as nitric oxide, insulin, or calcium ions; (3) feedback regulation by allosteric binding of a cyclic nucleotide or phosphorylation in response to elevation of the cyclic nucleotide substrate of the PDE. Phosphorylation can occur by a variety of protein kinases, including protein kinase A (PKA), protein kinase B (PKB), mitogen-activated protein kinase, and calmodulin kinase.[15,16] It is important to keep in mind that all of these processes can also be influenced by the specific subcellular localization of the PDE of interest, because this can limit substrate availability, or influence protein–protein interactions.[2]

2.23.4.1 PDE1

There are three known genes that encode for PDE1, five different splice variants of PDE1C, and two splice variants of PDE1A. The molecular weight range for enzymes in this family is 60–75 kDa. The PDE1 family hydrolyzes both cAMP and cGMP, and is stimulated by the calcium/calmodulin system. The degree of stimulation depends on the specific isoform. The PDE1A family prefers cGMP (twofold, based on K_m), while the B and C gene products have comparable K_m values for both cyclic nucleotides. Two different types of activity are known: one is a so-called 'low-affinity' form, which has been extensively studied.[16] A second, less well studied, 'high-affinity' form appears to be localized in the brain and peripheral nervous system.[17] PDE1A is highly expressed in the cortex and hippocampus in the brain. However, a specific physiological role has not yet been conclusively established. PDE1B is found predominantly in the striatal region of the brain and in the granule layer of the dentate gyrus.[18] PDE1C can be found in olfactory epithelial cells, and plays a role in signaling.[19] PDE1C is also found in proliferating cells in smooth muscle vasculature, where it may contribute to vascular tone.[20–21] Antisense inhibition of the enzyme in cell culture results in decreased cell proliferation.

In addition to these roles, PDE1 may be involved in the integration and cross-talk of signaling pathways associated with insulin secretion. In the pancreas, β cells respond to an increase in Ca^{2+} concentration associated with glycolysis by secreting insulin. This results in an increase in the ATP/ADP ratio that activates PDE1. The resulting increase in cAMP levels leads to a decrease in insulin secretion.[22]

In intact cells, Ca^{2+} levels serve as the primary means of activation of PDE1. There are a variety of means by which calcium ion levels are modulated in cells, including ion channels (voltage gated, as well as channels activated by a variety of intra- and extracellular second messengers), and second messengers, such as inositol phosphate, that are themselves activated by other systems.[23] Given the complexity of this system, it is not completely understood if the enzyme can be regulated differently by calcium ions derived from these distinct sources. This could lead to a dynamic and variable degree of cAMP levels in different subcellular compartments.

2.23.4.2 PDE2

This phosphodiesterase subfamily consists of a single known gene product, with three known splice variants. The enzymes are homodimers, with molecular weights of approximately 240 kDa. PDE2 is known to be both a cytosolic as well as membrane-bound enzyme. The subcellular localization is thought to be determined by the hydrophobicity of the N-terminal domain. Both cyclic nucleotides are cleaved, with an approximately twofold difference in K_m values. The V_{max} values for cGMP and cAMP are similar. Each substrate can substantially stimulate (up to 50-fold) the hydrolysis of the other. Hill values range from 1.2 to 1.6 for cGMP, and 1.6–1.9 for cAMP.[24–27] Stimulation of catalysis is due to an increase in affinity, rather than an increase in V_{max}. Occupation of the allosteric nucleotide binding site induces a conformational change in the protein that in addition to enhancing activity also increases susceptibility to proteolytic degradation. From a practical perspective, in cells, the activation constant for cGMP is much lower than that for cAMP (as low as 0.2 μM), meaning that the preferred substrate is most likely cAMP. PDE2 is most highly expressed in the central nervous system (CNS) and in the adrenal cortex, and is also found in the liver and in platelets. In the adrenal glands, ACTH acts via its cell surface receptor to elevate levels of cAMP, which leads to the increased synthesis of aldosterone. In addition, the peptide hormone atrial naturietic peptide (ANP), which is involved in maintenance of plasma volume, also acts on cells in the adrenal glands, where activation of its receptors lead to an increase in cGMP synthesis. The binding of cGMP to allosteric sites on PDE2 activates the enzyme toward cAMP hydrolysis.[28] In addition, PDE2 also plays a role in platelet aggregation[29] and in olfactory signaling pathways.[30] More recently, it has been reported that PDE2 can play a role in learning and memory, and that inhibition of PDE2 can lead to increased levels of cGMP in cell culture and in test animals. In addition, improved performances in T maze and social and object recognition assays was observed following administration of a PDE2 inhibitor to rats and mice.[31]

2.23.4.3 PDE3

This PDE exists as two known gene products, each of which is expressed in different tissues. The molecular weights of the enzymes are estimated to be approximately 122–125 kDa.[16] The K_m value for cyclic nucleotides is comparable (ranging from 100 to 800 nM); however, the V_{max} for cAMP is up to 10-fold higher than that observed for cGMP. This substantial rate difference results in selective hydrolysis of cAMP.[32,33] Some have hypothesized that because cGMP binds tightly to PDE3, but is not cleaved efficiently by the enzyme, it may function as an endogenous inhibitor, and thereby act as a key regulator of cAMP signaling mediated by PDE3.[34] Interestingly, the competition between these two substrates can play an important role in mediating the biological actions by enhancing signaling through one pathway at the expense of the other.

PDE3A is present in cardiac tissues, platelets, and smooth muscle, and one of the first physiological actions associated with PDE3 was a link with platelet function. It is known that both cyclic nucleotides inhibit platelet aggregation, and because elevated levels of cGMP will result in inhibition of cAMP hydrolysis by PDE3A, elevation of these cyclic nucleotides can result in an 'antiplatelet' effect on blood coagulation.[35–36] A recent study suggests a role for PDE3A in cardiomyocyte apoptosis.[37] Using primary cultured human myocytes derived from patients with congestive heart failure, decreased expression and activity of PDE3A was measured. Similar data were obtained from cardiac tissue of mice in which chronic pressure overload was experimentally induced. Chronic inhibition of PDE3A by adenovirus-delivered antisense and the PDE3 inhibitor milrinone led to the same result in cell culture.

PDE3B is highly expressed in adipocytes and the liver. The protein appears to be membrane associated.[38] This localization has stimulated research into the role the enzyme plays in metabolism and glucose homeostasis. Corbin and co-workers have demonstrated that PDE3-selective inhibitors can effectively eliminate the lipolytic response of adipocytes in cell culture when stimulated by insulin.[39,40] There is recent evidence for species-specific regulation of lipolysis. Murine 3T3-L1 and rat adipocytes in culture, when exposed to cilostamide, showed moderate levels of stimulation of lipolysis, while human adipocytes gave a much more substantial increase in lipolysis under similar circumstances. Interestingly, this same study demonstrated that a selective PDE1 inhibitor suppressed lipolysis in murine cells, and had no effect on human cells, while a selective PDE2 inhibitor had no effect on lipolysis in either human or murine cell culture. Rolipram, a PDE4 inhibitor, had no effect on lipolysis by human adipocytes, but did stimulate the process in the murine cell lines. A synergistic effect between cilostamide and rolipram was demonstrated only in the murine cell lines.[41] This study illustrates the need to make certain that cellular effects of PDEs and their inhibitors are similar in both animal models and human target tissues.

The regulation of insulin secretion by PDE3B involves both insulin-like growth factor 1 (IGF-1) and leptin. The role of the latter hormone was established with cell-permeable cAMP analogs, one of which was a substrate for PDE3B, and one that was not.[42,43] Leptin signaling requires activation of phosphoinositide-3-kinase, which activates PDE3B. In pancreatic β cells, leptin causes activation of PDE3B, leading to substantial inhibition of glucagon-like peptide 1 (GLP-1)-mediated insulin secretion.[44] The role of IGF-1 was established by Beavo and co-workers, who elucidated the signaling mechanism involved.[45] Using both pancreatic islet cells and insulin-secreting HIT-T15 cells, it was shown that the addition of IGF-1 administration decreased cAMP levels in parallel with a reduction in insulin secretion. Milrinone, a PDE3-subtype selective inhibitor, reversed the effects of IGF-1 in cell culture.

Both PDE3 isoforms are activated by phosphorylation. A variety of different protein kinases are known to have PDE3 as a substrate including PKA.[16,34]

2.23.4.4 PDE4

This family of cAMP-specific PDE enzymes is the largest in the class I PDE superfamily. There are four distinct gene products, A–D, and a variety of splice variants in each gene product. The K_m value for cAMP ranges from 0.6 μmol to 4 μM. These values were obtained using isoforms that were transiently expressed in COS7 cells that had been transfected with the appropriate genes.[46] The molecular weight of enzymes in the PDE4 family ranges from 68 to 75 kDa for the so-called 'short' form, to 85–110 kDa for the 'long' form. The distinction is based on the presence of one or two upstream conserved regions (UCRs, see below). Molecular weight variability in each of these groups is associated with the particular isoform and splice variant. It is interesting to note that there is a great deal of similarity between human, rat, and mouse forms of each of these gene products. This homology extends beyond amino acid sequence and function to include the genomic organization.[46,47] The catalytic domain of each PDE4 gene product is very similar, with greater than 80% sequence homology. However, it is possible to achieve selective inhibition of some of these isoforms with small molecules.

PDE4 enzymes exist as both soluble and membrane-bound forms. This feature is associated primarily with the nature of the regulatory N-terminal domain of the protein. The specific N-terminal domain of each PDE4 is also involved in determining specific subcellular localization, catalytic activity, response to other ligands, and temperature stability.[48–51] As one example, Houslay and co-workers investigated a variety of rat PDE4A isoforms.[52–54] They have shown that a 25-amino-acid peptide in the N-terminal domain is responsible for membrane localization, and that the structure of this domain contains an amphipathic α helix located adjacent to a tryptophan-rich region.[55] Distinct functional properties of the cytosolic and membrane-bound forms of PDE4B and PDE4D have also been elucidated.[56,57] The long form of rat PDE4A5 has been shown to interact with the v-Src SH3 domain in the N-terminal region, while the short-form splice variant of the same enzyme does not.[58,59]

PDE4 has been hypothesized to exist in different conformations that have different affinity for rolipram (see below).[60] As a part of this hypothesis, distinct functional activity has been associated with the so-called 'high-affinity'- and 'low-affinity'-rolipram-binding conformers. Emesis and nausea have been proposed to be due to interaction with the former, and anti-inflammatory activity has been proposed by binding to the latter. Detailed characterization of a specific PDE4 isoform, PDE4B2B, was carried out in an attempt to better characterize these potential sites.[61] A nonequilibrium filter binding assay showed a binding site for (R)-rolipram with a K_d value of 1.5 nM. Equilibrium binding dialysis gave a binding constant of 140 nM. This enantiomer of rolipram had a K_i value of 600 nM for this enzyme. Proteolysis experiments allowed characterization of a 43 kDa fragment of the enzyme that retained catalytic activity and was inhibited by rolipram. This work suggests that these two binding sites are located in distinct regions of the enzyme. A second report described the preparation of recombinant PDE4B (a truncated form similar to that described by Rocque et al.[61]) that could be efficiently assayed using a microtiter plate scintillation proximity protocol. This assay was claimed to be useful to identify compounds that preferentially inhibit the low-affinity conformer.[62] Others have developed a recombinant cell-based system that allows for evaluation of a variety of PDE4 isoforms. Chinese hamster ovary cells (CHO) endogenously express low levels of PDE4 activity, and do not have β$_2$-adrenergic receptors, and were transfected with this receptor and various PDE4 isoforms. It was shown that PDE4A4, PDE4B2, PDE4C2, PDE4D2, and PDE4D3 all could exist in a form that contained the high-affinity rolipram-binding site, while PDE4$_{330}$ demonstrated a low-affinity binding conformation in this cell system.[63] Using human monocytes and the PDE4 inhibitor RP73401 (piclamilast, 1, see below), it was shown that tumor necrosis factor α (TNF-α) generation could be suppressed by binding to the low-affinity binding conformation of PDE4.[64] In recent years, it has become more commonly accepted that the description 'low-affinity binding conformation' refers more specifically to inhibition of the catalytic activity of the enzyme, and the term 'high-affinity binding conformation' refers to a distinct rolipram-binding site on PDE4.

There is another unique structural feature of PDE4 enzymes in their respective regulatory domain that also influences enzyme function. All PDE4 enzymes (long and short) contain one or two conserved regions, UCR1 and UCR2. These regions differ in their amino acid composition and length. UCR1 is shorter (55 amino acids), and is located further away from the catalytic domain. A short linker connects this region to UCR2, which is made up of 75 amino acids, and is connected to the catalytic domain by a hydrophobic linker that varies in length.[65,66] While the direct functional relevance of these regions has not yet been conclusively established, there is some evidence to suggest that the UCR regions can influence the nature of regulation by small molecules such as rolipram.[67] In addition, the UCR1 region of PDE4D contains a serine residue (Ser54) that can be phosphorylated by PKA, resulting in a rapid activation of the enzyme. Mutagenesis experiments suggest that hydrogen bonding in the sequence surrounding this serine residue is an important structural feature associated with enzyme regulation. Replacement of Glu53 with an alanine did not affect phosphorylation of Ser54, but did influence enzyme activation.[68]

Phosphorylation provides a short term means by which PDE4 activity can be regulated. PDE4 enzyme activity can also be regulated by transcription. It has been shown that prolonged exposure of rat Sertoli cells to cAMP can result in a substantial increase in PDE4D2, and a smaller increase in PDE4B activity.[69,70]

Because PDE4 has been the subject of intense interest as a potential target for the treatment of a variety of inflammatory and CNS disorders, structural features in the protein necessary for catalytic activity and inhibition by small molecules have been heavily investigated. Point mutation of selected histidine residues in PDE4D2 (His278 and His311) to either alanine or proline, or substitution of alanine or proline for Thr349 completely eliminates catalytic activity.[71] Similarly, individual replacement of histidines at position 433, 437, 473, or 477 in a truncated version of human PDE4A4B resulted in either complete or substantial loss of enzymatic activity.[72] Binding of the prototypical PDE4 inhibitor rolipram (**2**) was also reduced by these mutations. There was a parallel change in rolipram affinity and catalytic activity of the enzyme. Mutant versions of PDE4B1 have been isolated with single amino acid replacements at residues 457, 625, and 667 that are no longer sensitive to rolipram inhibition.[73] When serine replaced histidines at residue 505 or 506 (PDE4B numbering), catalytic activity was reduced by 90%, without affecting rolipram binding or IC_{50}. The IC_{50} of another prototypical PDE inhibitor, IBMX (**3**) was reduced. This information suggests that these histidines are important for cAMP binding. Furthermore, it indicates that IBMX binds at the same site, or in a closely related one, while rolipram does not. Circumstantial support for this hypothesis follows from inspection of the structures of these two compounds.[46]

PDE4 enzymes have also been shown to be associated with β-arrestins, and to play a role in cytoplasmic signaling via protein complexes associated with the arrestins.[74,75] In this system, the arrestins modulate signaling of β$_2$-adrenergic receptors by recruiting PDE4D family members to the plasma membrane where the receptor is located. The phosphodiesterase is available to cleave cAMP in this specific subcellular location, and thereby attenuate the signal that is produced by activation of the β$_2$ receptor in response to an extracellular stimulus. It was shown that in HEK 293 cells that overexpressed the β$_2$-adrenergic receptors, administration of isoproterenol led to a time-dependent recruitment of PDE4D3 and PDED5 to the cell membrane. This corresponded to an increase in PDE4 enzyme activity in this region. Recent evidence, gained by use of siRNA technology, provided additional information to indicate that PDE4D5 is the functionally most relevant isoform associated with this process.[76]

The intracellular and tissue-specific distribution of the various family members of PDE4 has been studied in some detail.[46] Most of this work has been carried out using rodent sources of enzyme. For example, different splice variants of PDE4A have been found to localize to particulate fractions. The specific N-termini in these isoforms determines the precise subcellular location of the protein. Selected PDE4D forms are found in the cytosolic compartment, while others can be found in membrane fractions. The distribution of PDE4B and PDE4C has been less extensively studied. PDE4B, like PDE4D, exists in both cytosolic and membrane fractions. It has been shown that PDE4B2 plays an important role in T cell activation, and this role involves distribution of the enzyme to the immunological synapse during the early stages of T cell signaling.[77] This complex distribution of PDE4 within cells must have evolved to permit precise control of cAMP in specific regions of the cell in order to control or result in specific signaling pathways and biological response to distinct stimuli.

PDE4 enzymes are expressed in the CNS and other nervous system tissues, smooth muscle, inflammatory and endothelial cells, and in the heart. The expression of PDE4A in the cerebellum suggested a role in neurotransmitter release.[78,79] More recently, it has been shown that PDE4D knockout mice showed better long-term memory in spatial memory tests, compared with wild-type counterparts, and mice lacking PDE4B did not show similar improvements.[80] All four PDE4 isoforms have been localized and identified in human brains,[81] and rolipram has been shown to demonstrate efficacy in a variety of cognition and antidepressant animal models.[82,83] The relative potency in these screens of racemic, (S)- and (R)-rolipram correspond to their respective potency in vitro as PDE4 inhibitors, and to their affinity for the high-affinity rolipram-binding site.[80] PDE4 is widely distributed in tissues associated with immunologic and inflammatory responses, including T cells as noted above, macrophages, eosinophils, monocytes, and lymphocytes.[84] A variety of isoforms have been identified, with the specific enzymes dependent on the source of the cells or tissue used in the experiment. cAMP is well known to play a role in maintaining vascular tone in smooth muscle tissue, and, in humans and rats, PDE4D is the primary isoform present.[85,86] It has been recently shown that specific short-form versions of PDE4D, PDE4D1, and PDE4D2 play a key role in regulating the cAMP-mediated desensitization of activated vascular smooth muscle cells, leading to the suggestion that isoform-selective inhibitors of this PDE4 variant might be useful in the treatment of atherosclerosis, and/or vascular inflammation associated with cardiac procedures such as stent installation.[87] PDE4 activity has also been measured in osteoblasts,[88,89] in mouse skeletal muscle,[90] human articular chondrocytes,[91] adipocytes,[41] and olfactory tissue.[92]

2.23.4.5 PDE5

The emergence of PDE5 as a validated drug target based on the commercial success of sildenafil (4), vardenafil (5), and tadalafil (6) has stimulated substantial interest in this particular PDE in the last 10 years. More details on these and many other PDE5 inhibitors can be found in Section 2.23.6.

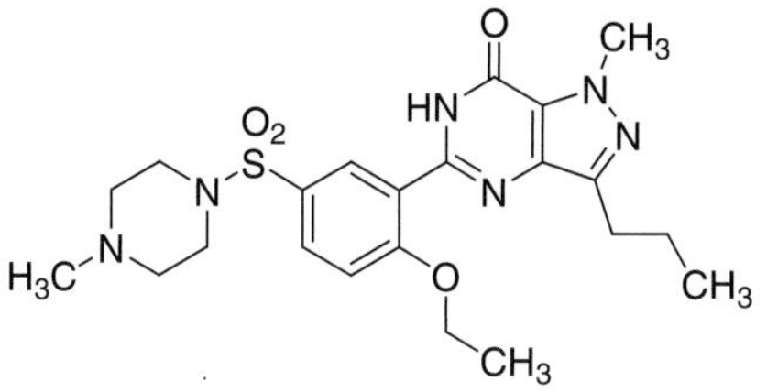

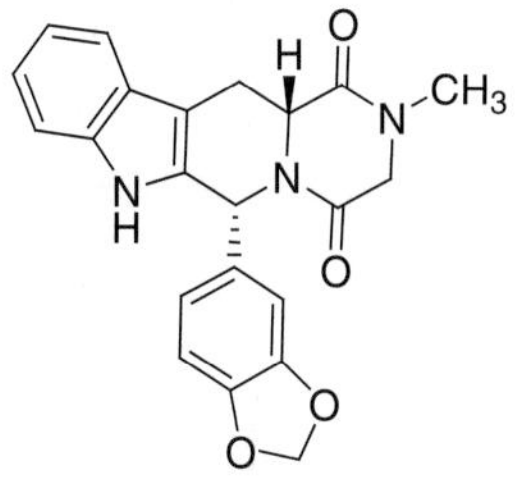

PDE5 is cGMP-specific, and only a single gene product is known. Three splice variants have been identified that differ in their tissue distribution.[93,94] The catalytic properties ($K_m \approx 0.6$–$5\,\mu M$, $V_{max} < 2.6\,\mu mol\,min^{-1}$ per milligram of protein) and inhibitor sensitivity of these different forms are equivalent.[95–97] The molecular weight of these enzymes ranges from 95 to 100 kDa. PDE5 is found in vascular and tracheal smooth muscle, platelets, gastrointestinal epithelial cells, spleen, kidney, prostate gland, and in the Purkinje cells of the cerebellum,[98–100] and it exists as a homodimer, where each monomeric unit is composed of a catalytic domain and a regulatory domain.

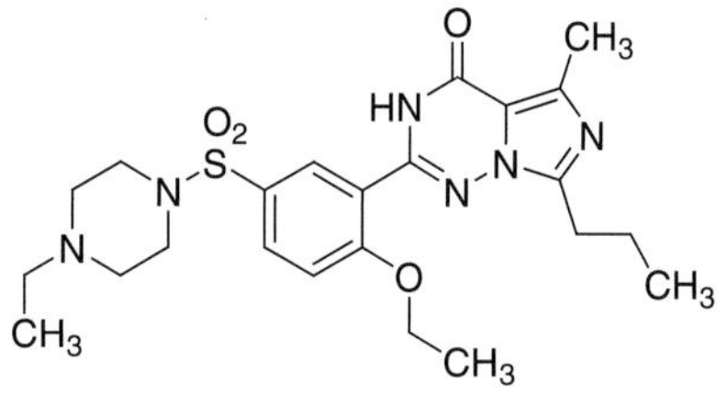

The regulatory domain contains allosteric binding sites for cGMP, and cGMP binding at these sites increases the rate of cGMP hydrolysis.[101,102] These sites are kinetically distinct, and occupation of both is required for activation of the enzyme by phosphorylation at Ser92 by either PKA or protein kinase G (PKG).[103–105] It has been demonstrated that there appears to be a positive correlation between the level of phosphorylation and the catalytic activity of PDE5 in cultured vascular smooth muscle cells.[106] The affinity of these allosteric sites for cGMP is comparatively high (K_m as low as 200 nM), and the high selectivity of the site for cGMP has been attributed to the ionization state of Asp289.[107]

Phosphorylation of the isolated PDE5 regulatory domain causes a conformational change in this portion of the protein, and this conformational change contributes to the increased binding affinity of the cyclic nucleotide (about 10-fold).[108] Addition of cGMP has been shown to enhance the affinity of tritiated tadalafil and vardenafil. This interesting observation suggests the possibility that PDE5 inhibitors may potentiate their own binding to the enzyme as cyclic nucleotide levels increase as a result of inhibition.[109] The physiological relevance of this information was supported by experiments using rabbit corpus cavernosum tissue, where binding of cGMP to PDE5 and PKG was found to greatly exceed the free concentration of the cyclic nucleotide in the tissue.[110]

The regulatory subunit of PDE5 also contains two GAF domains. These regions are contributors to cGMP binding, monomer dimerization, and enzyme regulation.[111] Structural analysis of the portions of the regulatory domain that contribute to each of these features was carried out by Corbin and Francis.[112] It was shown that the GAF *a* modules can homodimerize with high affinity ($K_D < 30\,nM$), while the GAF *b* modules have even higher affinity ($\sim 20\,pM$). Dimer stability is controlled by the amino acid sequence linking the two regions. GAF *a* has higher affinity for cGMP ($K_D < 40\,nM$), and a 176-amino-acid portion of the sequence in this region is sufficient for cyclic nucleotide binding. The sequence containing GAF *b* and the amino acids immediately flanking it are critical for cGMP-mediated stimulation of Ser102 phosphorylation by PKA and PKG. As noted above, PDE5 phosphorylation is an important activating step for enzymatic activity.

Section 2.23.3 describes some of the work that has been carried out to identify key residues for cGMP binding and catalysis. As noted, a divalent cation, particularly zinc, is important for both of these functions. Point mutations in the enzyme indicate that two key histidine residues, His607 and His643, are critical for metal cation binding. Substitution of these amino acids with alanine resulted in a substantial drop in enzyme activity, and a decrease in binding by Mg^{2+} or Mn^{2+}. It was interesting to note that mutation of several other highly conserved residues within the putative metal binding motif present in PDE5 (e.g., Asn604, His675, or aspartic acids at positions 644, 714, or 754) did not have a substantial influence on catalysis.[113]

PDE5 is widely known as a key regulator of vascular tone. In this role, it controls blood flow in the corpus cavernosum tissue of the penis, where it is the primary cGMP-hydrolyzing activity present. Inhibition of enzymatic activity reduces the outflow of blood, leading to engorgement and, ultimately, an erection following sexual stimulation. Recent work using knockout mice lacking in endothelial nitric oxide synthase illustrated that PDE5 expression and activity were downregulated, resulting in priapism in the knockout mice.[114] Increased PDE5 activity in the inner medullary collecting duct in the kidney has been demonstrated in pregnant rats. This leads to excessive sodium retention, and resistance to the effects of ANP. ANP signals via cGMP, and resistance to the diuretic effects of the peptide hormone are directly related to enzymatic activity.[115] In pulmonary endothelial cells, PDE5 has been recently shown to regulate growth and apoptosis. This process is regulated in part by ANP, as administration of the peptide to pulmonary microvasculature endothelial cells in culture leads to an increase in cGMP concentration, and results in growth inhibition and apoptosis.[116]

2.23.4.6 PDE6

This PDE is a highly cGMP-specific enzyme that, unlike most other members of the mammalian PDE family, is found essentially in a single tissue, the eye. Some expression can be detected in the pineal gland as well. There are three distinct gene products that encode the catalytic domain of the enzyme (A–C), and three corresponding regulatory domain gene products (D, G, and H). These regulatory domains are composed largely of two distinct GAF binding domains for cGMP binding. PDE6 is localized in the rod and cone cells in the retina, where the enzyme plays a key role in the response to light. Rod PDE6 is organized as a tetramer consisting of two catalytic subunits and two corresponding regulatory subunits (which incorporate the GAF domains mentioned above). Cone PDE6 differs in that the catalytic subunits are identical, rather than homologous. The regulatory subunits are cone-specific and distinct from the rod regulatory subunits. The molecular weight of these various complexes is approximately 200 kDa.[117]

Specific posttranslational modifications (farnesylation or geranylgeranylation at a CAAX sequence) of the regulatory subunit direct the enzyme to the membrane fraction of the cell. PDE6 activity also exists in the cytosolic fraction of cells. The biological basis for the evolution of these different mixtures of subunits, and their (potential) for distinct function and regulation is not well understood.[118] PDE6 has moderate affinity for cGMP ($K_m = 20$–$25\,\mu M$).[119,120] The enzyme has a substantial degree of sequence homology to PDE5 in the catalytic subunit[121]; however, PDE6 operates at a near-diffusion-controlled rate to hydrolyze cGMP, unlike PDE5, which processes substrate at a rate approximately 1000-fold less rapidly.[122–124]

The extremely high catalytic rate of the enzyme is thought to be evolutionarily driven, in response to its role in light transduction associated with vision. PDE6 is activated by displacement of the inhibitory/regulatory subunit by activated transducin as a part of the visual photoreceptor system in the eye. In the dark, the cellular concentration of cGMP is very high (estimated to be up to $60\,\mu M$), most of which is bound to allosteric cGMP-binding sites on PDE6. Unbound cyclic nucleotide concentrations are sufficient to maintain the cGMP-gated ion channel in an open conformation. Absorption of a single photon of light initiates a highly amplified process whereby transducin is activated by photoexcitation of rhodopsin. Transducin interacts with the inhibitory subunits of the PDE6 tetramer, resulting in their dissociation. This results in a very rapid increase in enzyme activity. As cGMP is cleaved locally by PDE6, this reduces the concentration of the cyclic nucleotide, and ultimately leads to closure of cGMP-gated ion channels, resulting in membrane hyperpolarization and termination of the light signal.[125]

PDE6 also contains allosteric binding sites for cGMP. These sites contribute to substrate binding at the active site of the enzyme. In the heterotetramer there are two cGMP-binding sites per complex, one on each regulatory subunit. Cyclic nucleotide binding at each allosteric site appears to be coupled. The binding affinity at these allosteric sites is influenced by a variety of factors, including the relative activity state of the tetramer, protein–protein interactions that are in play locally, as well as the presence or absence of the inhibitory Pγ subunit, which may be associated with the tetramer.

There are a number of visual defects, including retinitis pigmentosum, associated with mutations in PDE6.[126] Because there is high sequence homology between PDE5 and PDE6 in the active site, enzyme selectivity is an important feature that must be assessed when evaluating novel PDE5 inhibitors. Isolation of native human PDE6 is known to be difficult, and as a potential substitute, recombinant human enzyme has been used in some cases. A recent study suggests that native and recombinant versions of catalytically active PDE6 exhibit differential sensitivity to PDE5 inhibitors, and as a result may provide potentially incorrect information on PDE selectivity. For example, sildenafil demonstrated comparable K_i values for both rod and cone forms ($\sim 95\,nM$) of the recombinant enzyme, and a K_i of 25 nM when the native human preparation was used. In contrast, zaprinast and E4021 demonstrated 30–80-fold lower affinity for the recombinant enzyme.[127] As an alternative, bovine PDE6 has been used to assess selectivity. However, a similar detailed comparison of inhibitor sensitivity between the native human and bovine forms has not yet been reported.

2.23.4.7 PDE7

Two different gene products are known for this cAMP-specific PDE. The enzyme has high affinity for cAMP ($K_m \approx 200$ nM), and is most closely related to PDE4, although the sequence similarity is not high. Enzymatic activity is not affected by calcium ions, calmodulin, or cGMP. Three distinct splice variants of the PDE7A isoform have been identified with molecular weights in the range of 50–57 kDa.[128] These isoforms differ only in their N-terminal sequences, which may influence their subcellular localization.[129] PDE7A1 mRNA has been reported to be associated with both particulate and cytosolic fractions of skeletal muscle and cardiac tissue, and enzymatic activity has been detected in adult skeletal muscle. More abundant protein levels have been measured in T lymphocyte cell lines. PDE7A2 mRNA has been localized to only the particulate portions derived from these tissues, and, as of this writing, no active PDE7A2 enzymatic activity in adult tissues has been detected. The physiological basis for these observations is not yet clear. In lymphocytes that have been activated by both CD3 and CD28, PDE7 activity is markedly stimulated, and inhibition of PDE7 by antisense reduces T cell proliferation.[130,131]

However, later work suggested that T cell proliferation, using a series of small-molecule PDE7 inhibitors, was equally inhibited in T cells derived from wild type and PDE7A knockout mice.[132] Further work regarding the influence of PDE7A on T cells was reported using a different and less potent compound, BRL 50481. In three different cell types implicated in the pathogenesis of chronic obstructive lung disease (CD8+ T lymphocytes, monocytes, and lung macrophages), PDE7A1 and PDE7A2 could be readily detected by Western blotting. BRL 50481, at concentrations of up to 30 μM, had little effect on the biosynthesis of TNF-α, and had no effect on the proliferation of CD8+ lymphocytes. However, when the compound was tested, along with either rolipram or ORG 9935 (a PDE3 inhibitor using 'aged' monocytes in vitro), the addition of BRL 50481 had an additive and concentration-dependent effect on TNF-α levels.[133] The potential role of PDE7 in human T cell function was assessed in a third paper that employed two compounds that were very potent PDE4 inhibitors, with IC$_{50}$ values less than 1 nM, but with very different activities against PDE7 (~ 2 μM and greater than 10 μM). The former, T-2585, inhibited cytokine synthesis, proliferation, and CD25 expression over a concentration range where PDE7A activity was suppressed (1–10 μM). The latter, RP73401, had little or no effect at 10 μM, leading the authors to conclude that PDE7 may play a role in the biosynthesis of these inflammatory cytokines.[134] However, in view of the other papers cited, the very potent PDE4 inhibition associated with these compounds may be the primary feature responsible for the observed activity.

PDE7A expression has been measured by Western blot analysis in normal human B lymphocytes, as well as in primary chronic lymphocytic leukemia (CLL) cells and in a CLL-derived cell line in culture. Protein expression was upregulated in these lines by elevation of cAMP, and also by a PDE inhibitor, IC242. The PKA inhibitor H-89 blocked the IC242-mediated upregulation of PDE7 expression in these cell lines in culture.[135] In addition to the aforementioned tissues, PDE7A1 and PDE7A2 mRNA has also been measured in rat brain extracts by in situ hybridization histochemistry. The highest levels of expression were observed in the olfactory bulb and tubercle, hippocampus, cerebellum, medial habenula nucleus, pineal gland, area postrema, and choroid plexus.[136]

PDE7B has approximately 70% identity to PDE7A at the amino acid level in the catalytic domain of the protein.[137–139] The enzyme has a mass of approximately 50 kDa, and a K_m value of 30 nM for cAMP, nearly an order of magnitude lower than lowest reported value for PDE7A, and, like PDE7A, cGMP did not affect the activity of the 7B isoform. Northern blot analysis of mRNA expression of the enzyme revealed the highest expression in the pancreas, followed by brain, heart, skeletal muscle, eye, thyroid, ovary, testis, submaxillary gland, epididymus, and liver. When compared with the 7A isoform, PDE7B shows a similar pattern of inhibitor sensitivity. IBMX, papaverine, dipyridamole, and SCH51866 show broadly similar inhibition. It may be possible that isoform-selective inhibitors could be identified, based on this preliminary data, because IBMX and dipyridamole give lower IC$_{50}$ values for PDE7B compared with PDE7A (2.1 versus 4.5 μM and 9 versus 42 μM, respectively). Papaverine shows somewhat greater inhibition of PDE7A compared with PDE7B (IC$_{50}$: 12.5 μM for PDE7A, 22 μM for PDE7B). However, more potent compounds will be required to establish the validity and potential utility of this possibility.[140,141]

2.23.4.8 PDE8

This cAMP-selective ($K_m \approx 70$ nM) PDE exists as two gene products. Five different splice variants of the 8A isoform have been identified to date.[142] At the amino acid level, the proteins vary in length from 449 amino acids (PDE8A3) to 829 amino acids (PDE8A1). The enzyme requires comparatively high concentrations of Mn^{2+} or Mg^{2+} (1 mM) for maximal in vitro activity, while calcium ions have little effect on enzymatic activity. When compared with PDE4, the PDE8A enzyme expressed in a baculovirus system processes cAMP with a V_{max} approximately an order of magnitude lower than PDE4.[143] Expression of PDE8A, measured by Northern blot analysis, in mice appears to be highest in the testis, followed by the eye, liver, skeletal muscle, and heart. Lower levels can be found in the kidney, ovary and brain.[144] Tissue distribution is similar in humans.[145] This tissue distribution has led to the hypothesis that the enzyme plays a role (as yet undefined) in germ cell development.[146] Among the commonly used PDE inhibitors, only dipyridamole shows significant activity as a PDE8A inhibitor, with an IC_{50} value of 4.5 μM. The presence of a PAS domain in the protein suggests that this domain may control specific protein–protein interactions that are associated with subcellular localization, as well as provide a binding site for small-molecule or peptide/protein ligand binding.

The PDE8B isoform is very highly expressed in the thyroid gland,[147] with lower levels in the brain. Dipyridamole inhibits this isoform weakly (50% at 40 μM).

When the enzymatic properties of the 8A and 8B isoforms were compared, PDE8B had a slightly higher K_m value for cAMP (100 nM, versus 40 nM for PDE8A), but a higher V_{max}.[148]

2.23.4.9 PDE9

PDE9 exists as a single, known gene product, is highly cGMP-specific, and has been cloned and expressed from both human and murine sources.[149,150] Four different splice variants have been identified, based on Northern blot analysis, encoding proteins with between 465 and 593 amino acids.[151] The K_m value for cGMP has been reported to be as low as 70 nM, and the V_{max} value of 4.9 nmol min^{-1} per microgram of protein is twofold faster than the rate at which PDE4 processes cAMP. When the recombinant protein is assayed in vitro, the activity is strongest in the presence of Mn^{2+} (1–10 mM), compared to Mg^{2+} or Ca^{2+}. Unlike other cGMP-specific PDEs, PDE9 lacks an allosteric binding site for cGMP. When the amino acid sequence is compared with other PDEs, greatest homology (34%) is observed with the cAMP-selective enzyme PDE8A, while the cGMP-specific PDE5 has the lowest (28%) amino acid identity. It has been proposed that because this PDE has very high affinity for cGMP, it functions to maintain basal levels of the cyclic nucleotide in cells.[152] PDE9 is more broadly expressed than other cGMP-specific PDEs, and this observation has been cited as supporting circumstantial evidence of this hypothesis. The enzyme is particularly abundantly expressed in the brain, heart, spleen, kidney, and intestine. Because of its presence in the kidney, a potential role in ANP-mediated diuresis has been proposed.[153] More recently, Hendrix has shown evidence that PDE9 plays a role in memory acquisition, consolidation, and retrieval in rats, and restoring memory deficits induced by MK801 in mice, based on the activity of a PDE9 inhibitor.[154] The inhibitor sensitivity of PDE9 is distinct from other cGMP-selective enzymes. For example, sildenafil and zaprinast are comparatively weak inhibitors, with IC_{50} values of 7 and 29 μM, respectively, while rolipram, dipyridamole, IBMX, and *erythro*-9-(2-hydroxy-3-nonyl)adenine (EHNA) do not inhibit enzymatic activity at concentrations as high as 200 μM. SCH-51866 has better affinity for PDE9, with an IC_{50} value of approximately 1.5 μM.

2.23.4.10 PDE10

This PDE acts on both cGMP and cAMP, and exists as a single gene product with three known splice variants. The A1 and A2 variants are derived from human sources, while the A3 variety was identified in rats. The enzyme has been cloned and expressed from mouse, rat, and human cDNA libraries.[155–158] The protein has between 779 and 794 amino acids, and although it has a higher affinity for cAMP ($K_m = 50$–280 nM) than cGMP ($K_m = 3$–14 μM), PDE10A more efficiently hydrolyzes cGMP (two- to fourfold). Each cyclic nucleotide serves as an inhibitor of the binding and hydrolysis of the other. The N-terminus of the enzyme has a GAF binding domain, like PDE2, PDE5, and PDE6, where cGMP may bind and regulate enzymatic activity. The catalytic domain has low to moderate sequence homology (∼16–47%) with other known PDEs.

The enzyme is highly expressed in specific regions of the CNS in both humans and rats. In particular, it is found in the putamen and caudate nucleus. More detailed studies to investigate the subcellular localization revealed that PDE10A2 is associated with membrane fractions derived from rat striatum, and immunocytochemical analysis further localized the enzyme to the Golgi apparatus. PDE10A1 and PDE10A3 were found primarily in the cytosolic fractions in the same preparation.[159] It was also shown that PDE10A2 was phosphorylated by PKA at a particular threonine residue. This posttranslational modification was hypothesized to guide the localization of the enzyme to the Golgi membrane. Lower levels of expression were detected in the thyroid gland and testes.[160] The inhibitor sensitivity of PDE10A is unique, compared with other PDEs. Dipyridamole and SCH51866 are comparatively potent ($IC_{50} \approx 1$ μM) inhibitors of the enzyme, while sildenafil, rolipram, zaprinast, enoximone, and vinpocetine are much less potent ($IC_{50} > 10$ μM, or inactive). IBMX has moderate activity as an inhibitor of PDE10A ($IC_{50} \approx 3$–10 μM).

The primary physiological relevance of PDE10A to date appears to be in CNS diseases, specifically in Huntington's disease and in schizophrenia. The potential for a role in the former was suggested by monitoring the rate of reduction of PDE10A mRNA in transgenic mice. It was shown that levels decrease with time, but before the onset of symptoms, suggesting a cause-and-effect relationship in Huntington's disease.[161] Papaverine has demonstrated dose-dependent inhibition of hyperactivity induced by phencyclidine and amphetamine in rats, without affecting dopamine release.[162,163] PDE10A knockout mice treated with the compound do not show the same activity. Papaverine has been reported to be a 40 nM inhibitor of PDE10A in vitro, and elevate cAMP levels in both brain slices and in cell culture.[164]

2.23.4.11 PDE11

The most recently identified member of the superfamily processes both cyclic nucleotides, and only a single gene product has been identified to date. PDE11A has been cloned and expressed from both human and murine cDNA, and a total of four different splice variants have been identified.[165–168] The molecular weight range for these enzymes was reported to be 56–105 kDa. The N-terminus of the protein includes up to two complete GAF domains, depending on the specific variant. Their affinity for cAMP and cGMP is similar; however, one publication cites very different V_{max} values for each isoform.[168] Tissue-selective expression of the A3 (testes) and A4 isozymes (prostate) has been shown. Other tissues in which PDE11A mRNA has been detected include skeletal muscle, kidney, liver, and pituitary glands. A recent publication attempted to establish PDE11A protein expression in human tissues. The enzyme (the A4 isoform) was found only in the glandular epithelium of the prostate and, to a much lesser extent, in neuronal cells in the parasympathetic ganglia in the heart. No active protein was found by immunohistochemistry in skeletal muscle, testis, or penis.[169] Dipyridamole and zaprinast are comparatively potent inhibitors of PDE11A, with IC_{50} values of 379 nM and 5 μM, respectively. In an attempt to establish a physiological role for PDE11, mice lacking the enzyme were produced and evaluated.[170] It was found that ejaculated sperm from knockout animals had a reduced rate of forward progression, and a lower percentage and concentration of live sperm. Pre-ejaculated sperm displayed increased spontaneous and premature capacitation. This study also states that these observations are consistent with human data. However, another recent paper suggests that PDE11 inhibition by tadalafil does not alter semen parameters.[171] At the time of this writing, the issue remains unresolved. Another recent paper examined the expression of PDE11A in a range of normal and malignant human tissues using a polyclonal antibody that recognizes all four splice variants.[172] The protein was detected to varying degrees in a wide range of epithelial, endothelial, and smooth muscles cells. Interestingly, this paper reports expression in spermatogenic cells of human testes. Expression of the protein in different human carcinoma cell lines was also measured.

2.23.5 Second Messengers

As noted, the phosphodiesterase enzymes act to regulate levels of the key intracellular second messengers, cAMP and cGMP. These cyclic nucleotides are produced in response to a variety of cellular stimuli, such as release of nitric oxide, activation of G-protein-coupled receptors (GCPRs), and other receptors on the cell surface. Intracellular levels of these second messengers initiate and control a wide range of processes, such as cell growth and differentiation, tissue function, and ion channels. Some receptors (e.g., the vasopressin receptor) signal by elevating levels of the cyclic nucleotide,[115] while others signal by negatively affecting the activity of the nucleotide cyclase that generates the cyclic nucleotide (e.g., the CB1 receptor for cannabinoids).[173]

2.23.6 Prototypical Pharmacology

2.23.6.1 PDE1 Inhibitors

Comparatively little work has been carried out in an attempt to prepare PDE1 inhibitors. The alkaloid vinpocetine (**7**) is a very weak inhibitor in vitro, with an IC_{50} value of approximately 20 μM.[174] Most of the published work has described preparation of combination PDE1/PDE5 inhibitors. Structurally, many compounds in this group are tetracyclic guanine derivatives, and researchers at Schering Plough have exemplified it with a variety of derivatives, such as compounds **8–11** that have in vitro IC_{50} values less than 20 nM.[174,175] A less potent derivative in this family, SCH51866 (**12**), had an IC_{50} value of 70 nM, and demonstrated antihypertensive activity and platelet antiaggregatory effects in vivo.[176]

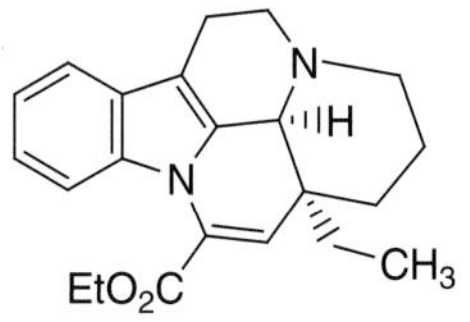

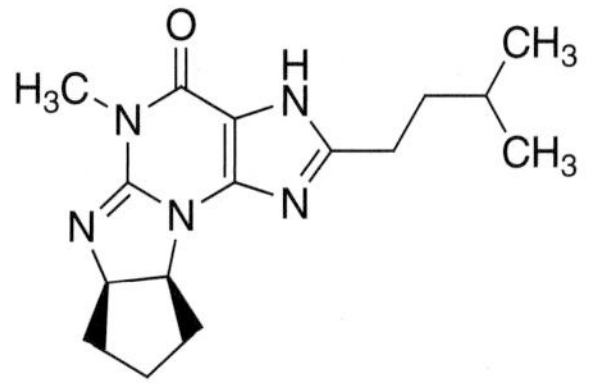

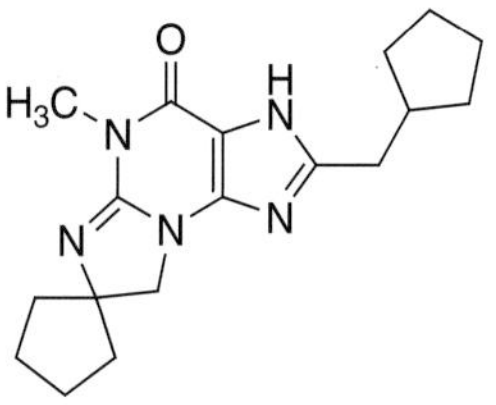

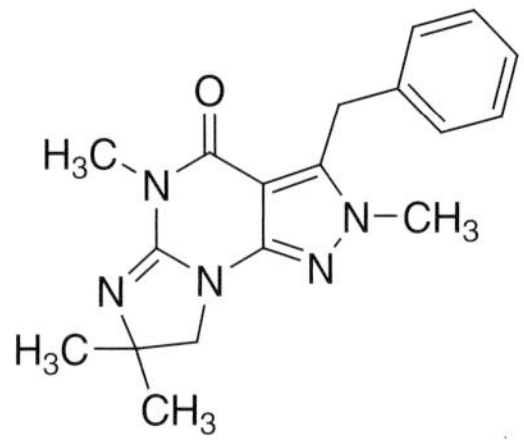

10 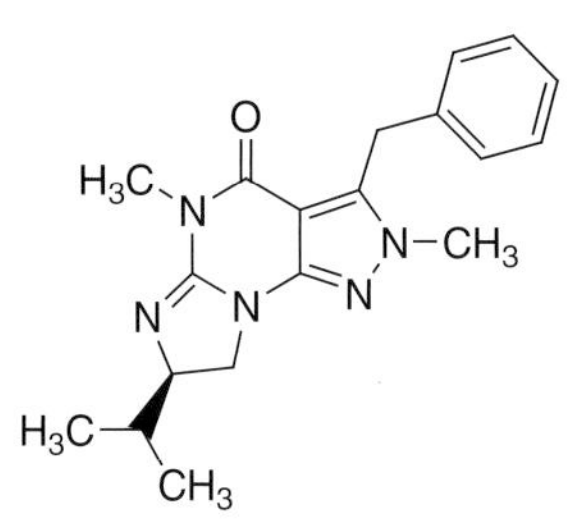 **11** 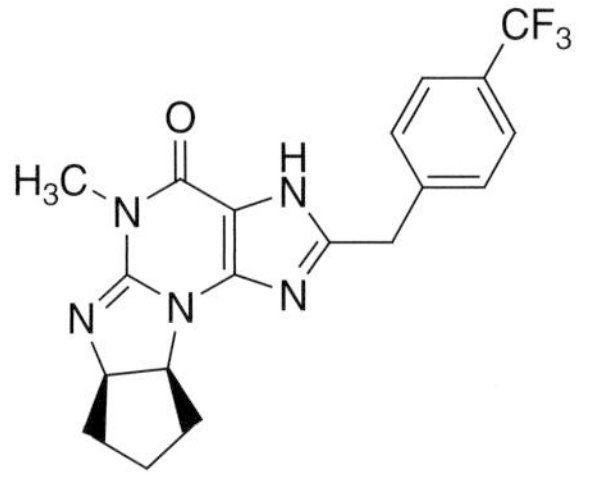 **12**

Another example of guanine-based inhibitors with good potency (PDE1 IC$_{50}$ = 38 nM) was disclosed in a patent application (**13**).[177]

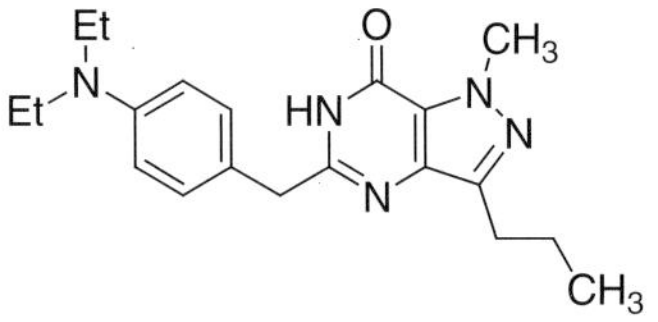

13

Older work has summarized the activity of early PDE1 inhibitors, including HA-558 (**14**), ICI 74917 (**15**), and MY 5445 (**16**). These compounds have low micromolar activity against PDE1.[178]

Another series of dual PDE1/PDE5 inhibitors was reported by Dan and co-workers (e.g., **17** and **18**). This series is structurally distinct from other PDE1 inhibitors, and is based on the presence of a hydroxamic acid moiety. The most potent compounds in this group had in vitro IC$_{50}$ values between 40 and 90 nM. Selected compounds in this group demonstrated ex vivo vasodilatory activity, with an EC$_{50}$ of 900 nM.[179]

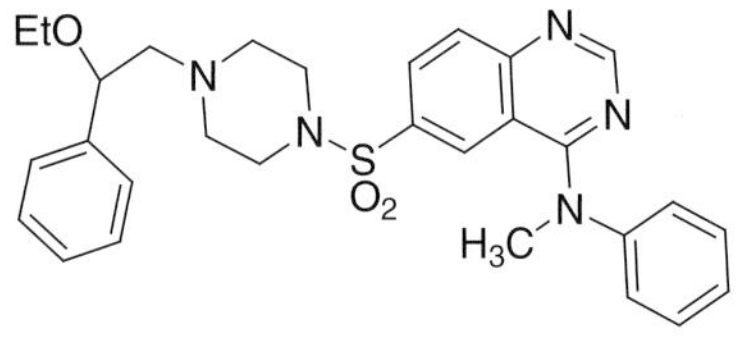

14 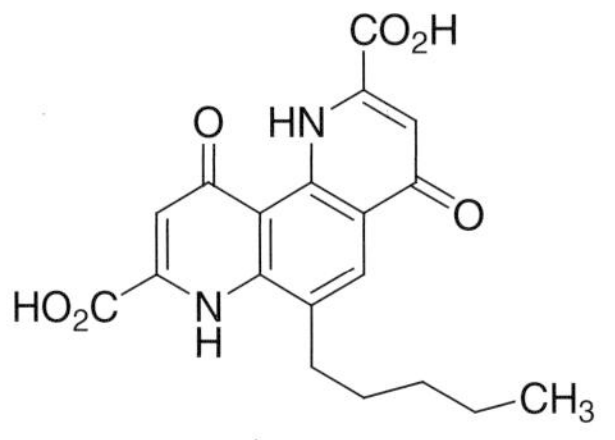 **15** 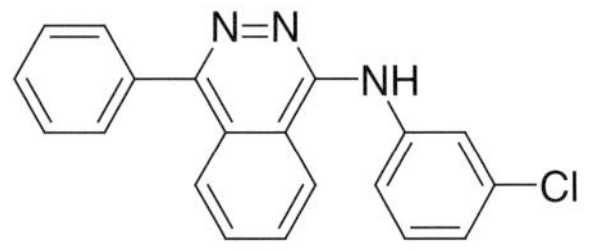 **16**

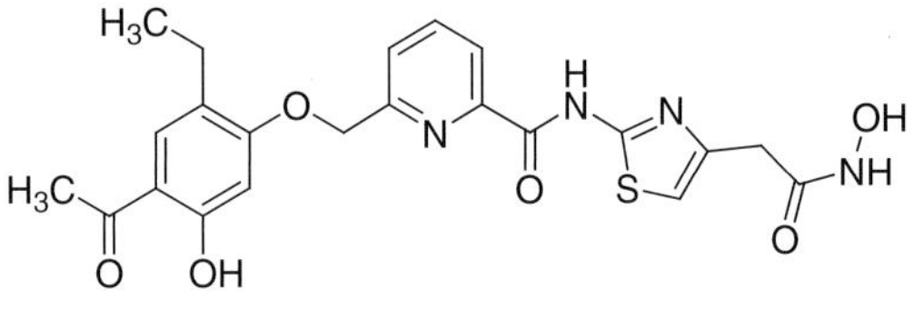

17 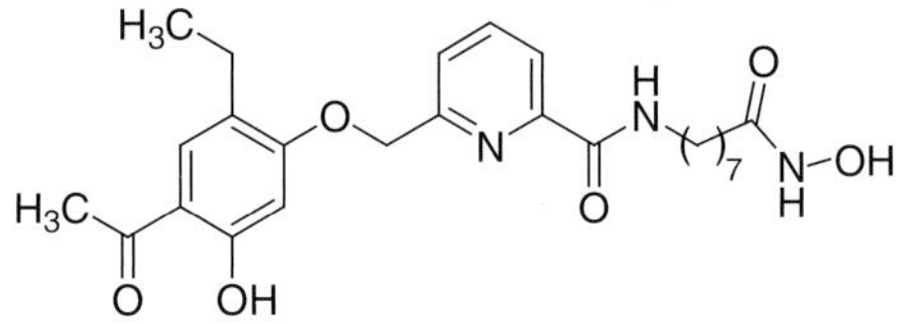 **18**

Another distinct series that demonstrated good selectivity for PDE1 was derived from known PDE inhibitors such as milrinone and zaprinast. This group of imidazotriazinones, exemplified by compound **19**, had in vitro IC$_{50}$ values as low as 85 nM.[180]

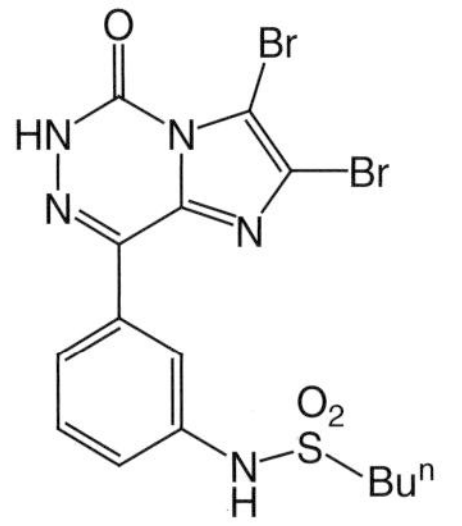

19

2.23.6.2 PDE2 Inhibitors

Interest in developing inhibitors of this phosphodiesterase is growing, based in part on indications that this may lead to novel treatments for CNS disorders such as cognitive deficits.[31,181] There are a variety of different chemotypes with demonstrated PDE2 activity. Among the least potent are a series of morpholinochromones (**20**) related to a PI3 kinase inhibitor, LY294002 (**21**). Activity in this group of compounds was measured as the percentage inhibition at 50 μM, and ranged from 0% to 93%.[182]

As noted in Section 2.23.4.2, a PDE2 inhibitor has demonstrated good selectivity and potency, as well as in vivo activity in animal models of cognition. This compound, BAY 60-7550 (**22**), is an imidazotriazinone derivative, and inhibits human PDE2A with an IC_{50} value of 4.7 nM. Selectivity against other PDEs ranges from 50-fold (PDE1) to 800-fold (PDE3B, PDE7B, PDE8A, PDE9A, and PDE11A).[31]

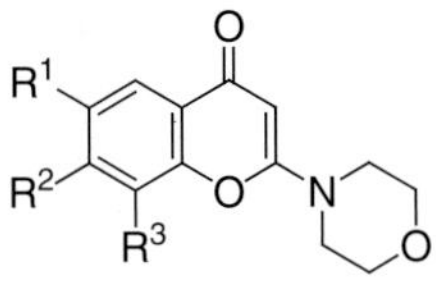

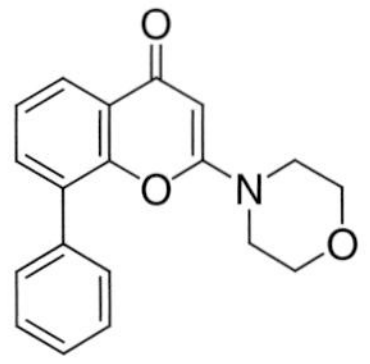

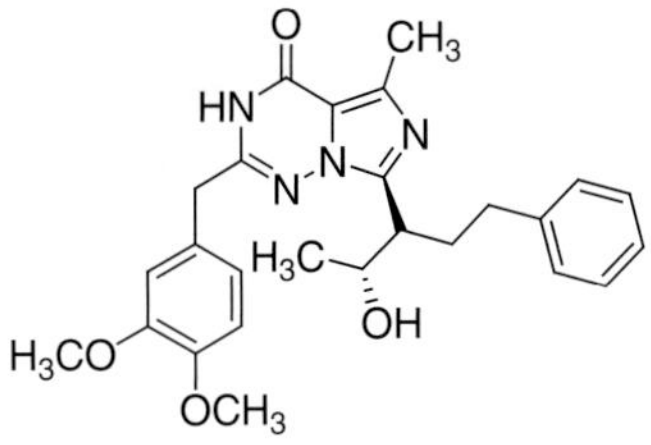

Pyrido[2,3-*d*]pyrimidine derivatives (e.g., **23**) were claimed in a patent application where the IC_{50} value was reported to be less than 50 nM for this compound.[183] Tricyclic oxindoles such as **24** were claimed in another patent application, and the activity of one example was cited as less than 200 nM (IC_{50}).[184] Benzodiazepinone compounds such as **25** also show PDE2 inhibition; however, no detailed description of activity is given.[185]

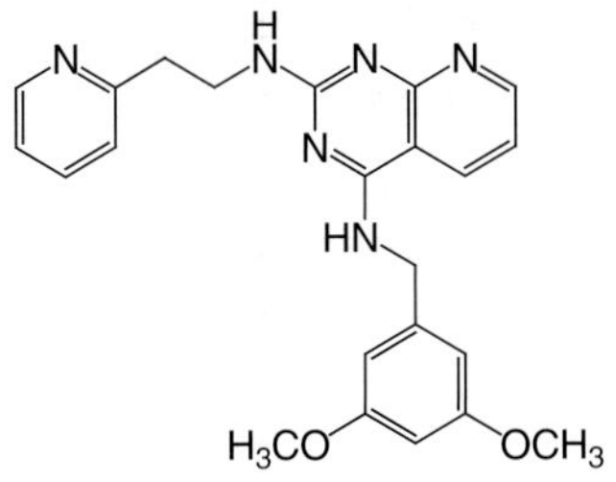

An x-ray crystal structure of the catalytic domain of PDE2 has recently been reported. In this same work, a unique cell-free system was employed to rapidly produce and evaluate active site mutant versions of human PDE2 in order to more rapidly identify key residues in the active site associated with substrate and inhibitor binding.[186] Four key amino acids were identified (Asp811, Gln812, Ile826, and Tyr827); each of these participates in key interactions with the substrate, and potentially with inhibitors.

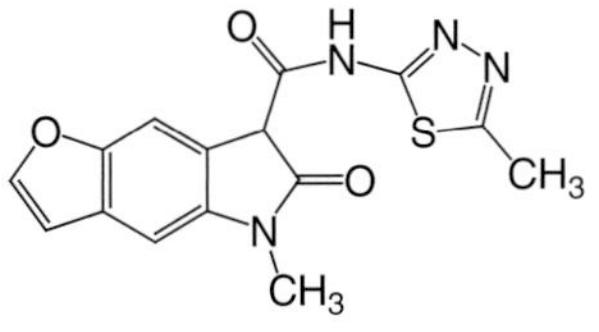

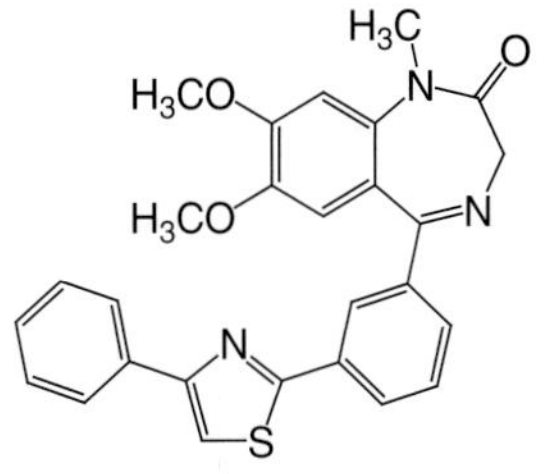

The adenosine deaminase inhibitor EHNA (**26**) has been used as a selective PDE2 inhibitor in experiments to determine the role of phosphodiesterases in tissues, or to attempt to understand the sensitivity of a newly discovered PDE to inhibitors.[187,187a]

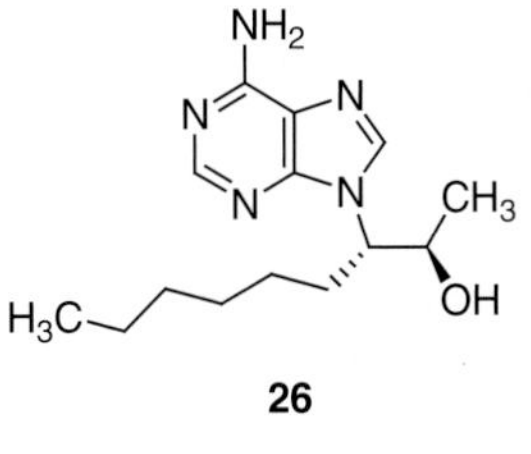

2.23.6.3 PDE3 Inhibitors

Unlike other phosphodiesterases, the structural variety of compounds that are reported to be 'potent' and 'selective' inhibitors of PDE3 is comparatively limited. Many PDE3 inhibitors contain a substituted pyridizanone or valerolactam ring to which a substituted aromatic ring is attached. Examples of prototypical compounds are **27–37**.

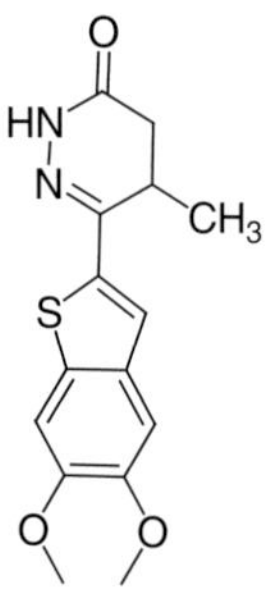

Org 9935
27

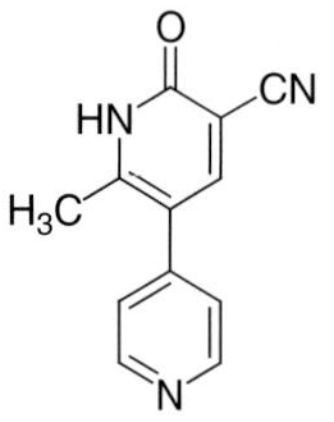

Milrinone
28

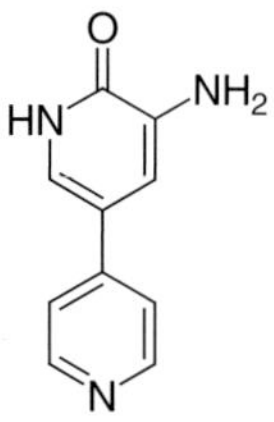

Amrinone
29

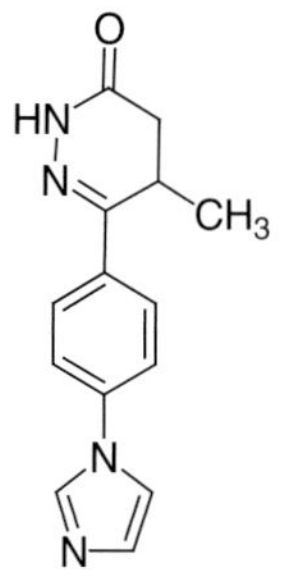

CI-930
30

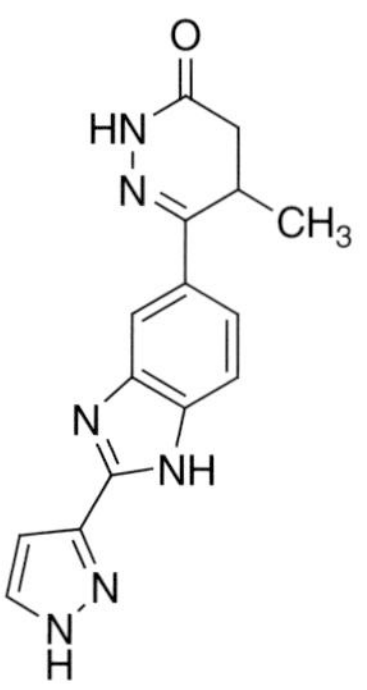

Meribendan
31

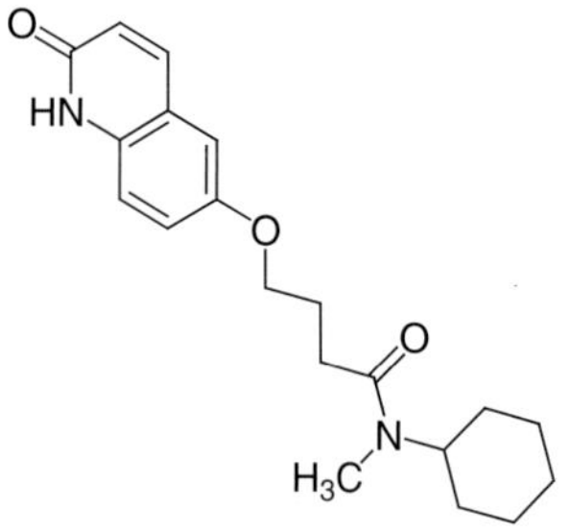

Cilostamide
32

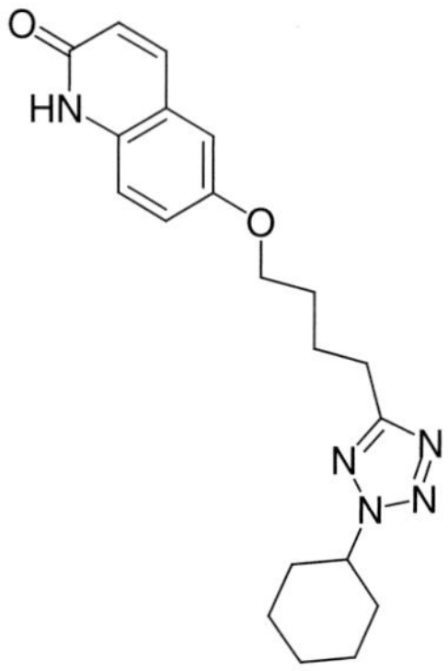

Cilostazol
33

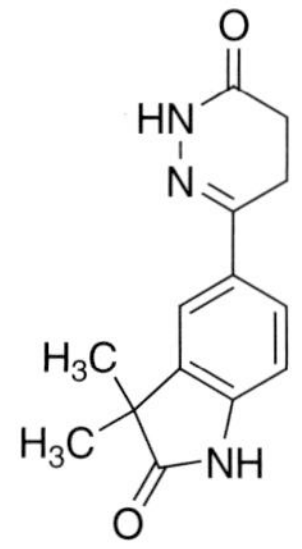

Indolidan
34

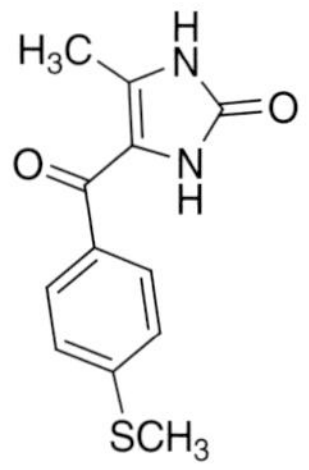

Enoximone
35

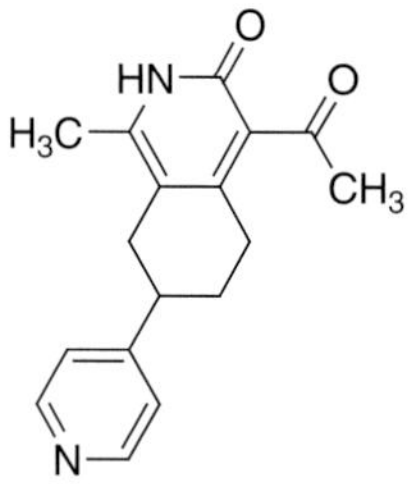

MS-857
36

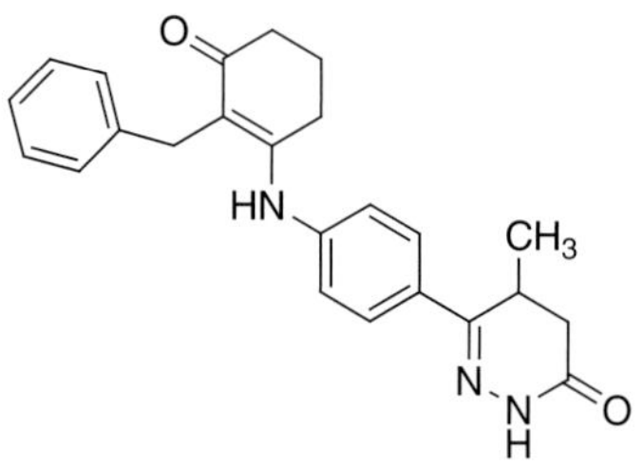

37

The potency of these compounds as PDE3 inhibitors varies widely. In this group, the most potent compound is cilostamide (**32**), with an IC_{50} of 5 nM, and the least potent is amrinone (**29**), with an IC_{50} of 17 μM.[188–191] Edmondson and co-workers have reported the preparation of a very potent PDE3B inhibitor that was investigated for potential use in vivo.[192] This compound (**37**) proved to be moderately selective for PDE3B (10-fold) versus PDE3A. Other compounds in this series lacking the phenyl moiety on the cyclohexenone ring were approximately 10-fold less potent, without improved selectivity for PDE3B.

Fossa and co-workers carried out a modeling study in an attempt to understand structural requirements for potent PDE3 inhibition using available PDE4 crystal structure data, as well as PDE3 site-directed mutagenesis and computational methods. This analysis revealed four residues that were essential for inhibitor binding: Thr908, Lys947, Gln1001, and Phe1004. A variety of interactions occur between inhibitors and these residues, including π stacking and hydrogen bond formation. In addition to these amino acids, Phe989 occupies a potentially important position in the inhibitor-binding site in a hydrophobic region of pocket A on the enzyme. PDE3 has a serine residue at position 1003, and at a similar position in PDE4, a glycine residue is most commonly found. This creates the possibility of a hydrogen bond between the inhibitor and a serine hydroxyl group that is not present in PDE4.[193]

In addition to this work, Scapin *et al.* have reported the crystal structure of the PDE3B catalytic domain complexed with a derivative of **37**.[194] This showed that the modest level of subtype specificity obtained with **37** cannot be readily explained by an examination of the structure. This issue may be resolved if a crystal structure of the full-length protein can be obtained. The structure did reveal features that were important for the high affinity of the inhibitor, including hydrophobic interactions of the aryl substituent on the cyclohexenone ring. The excellent degree of PDE3 selectivity can be attributed to a hydrogen bond interaction between the carbonyl group of the pyridazinone ring and the side chain of His948.

2.23.6.4 PDE4 Inhibitors

The prototypical PDE4 inhibitor is rolipram (**2**).[46] It is widely used to investigate the potential role of the enzyme in a given cell or tissue preparation because it inhibits all known PDE4 isozymes (although to varying degrees). The structure of rolipram has been used as a template for the discovery of a number of other PDE4 inhibitors. Examples of these structures are **38–45**.

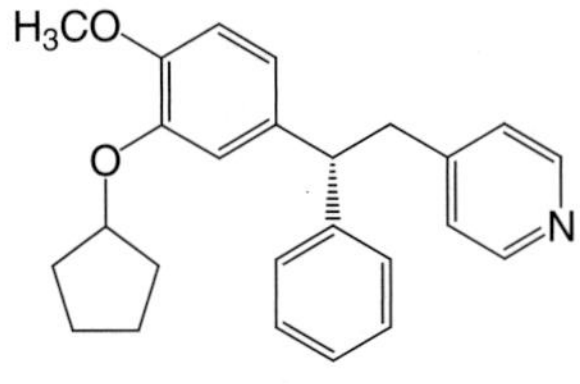

CDP840
38

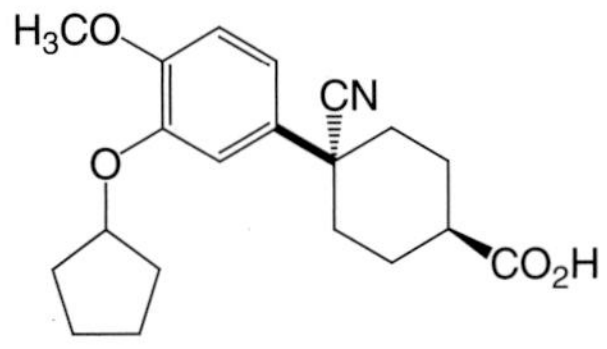

Cilomilast
39

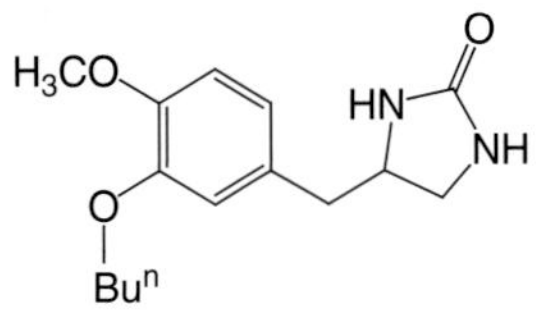

RO 20-174
40

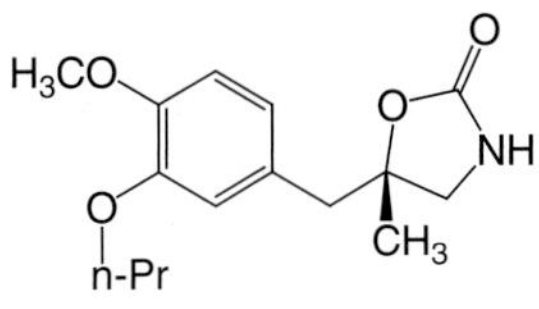

Mesopram
41

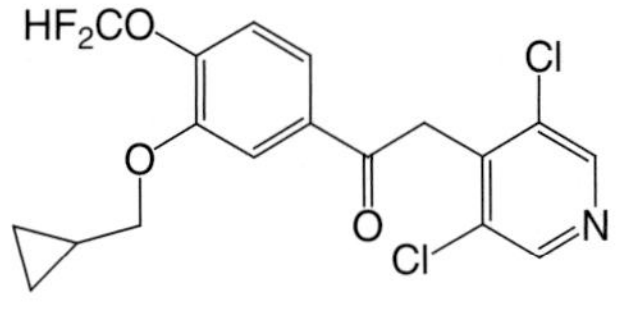

Roflumilast
42

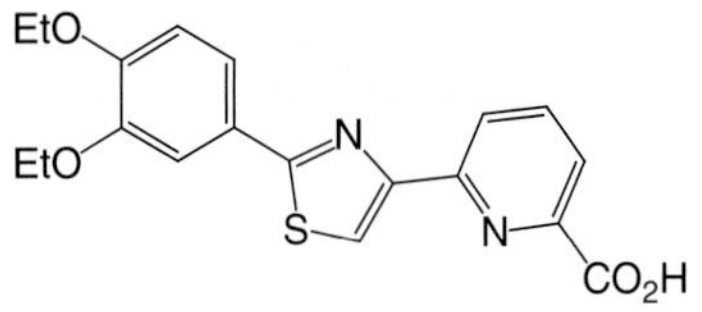

Tetomilast
43

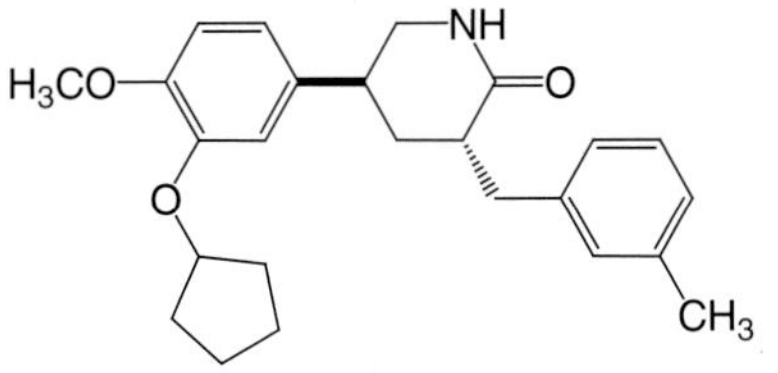

HT-0712
44

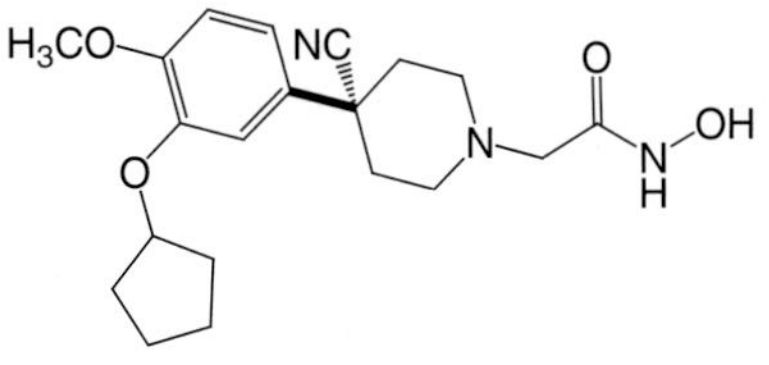

ONO 6126
45

Rolipram and several related compounds have undergone extensive clinical evaluation for a variety of applications, including asthma, chronic obstructive pulmonary disease (COPD), psoriasis, memory impairment, ulcerative colitis, and multiple sclerosis. None of these compounds have been fully cleared for use, although cilomilast has received an approvable letter in the USA.[195] None of these compounds have demonstrated substantial PDE4 subtype selectivity. Other chemotypes have been discovered with good PDE4 potency. Examples of some of these structures are **46–50**. In this group, both AWD 12-281 and YM976 have undergone clinical evaluation. The former is a potent ($IC_{50} \approx 10$ nM) inhibitor of PDE4 with low affinity for the so-called high-affinity rolipram-binding site. AWD 12-281 was specifically designed for topical/inhalation use in the treatment of asthma, topical rhinitis, and COPD.[196] YM976 is a potent PDE4 inhibitor, with an IC_{50} of 2.2 nM against PDE4 purified from human peripheral leukocytes. When studied in animal models of inflammation, it was possible to establish a window between the beneficial effect and emesis in rats and ferrets. In rats, the oral ED_{30} in a carrageenan-induced pleurisy model was 9.1 mg kg^{-1}, and no emesis was noted at doses up to 10 mg kg^{-1}. In ferrets, there was a dose-dependent effect at 1, 3, and 10 mg kg^{-1} in a similar model without emesis. In contrast, both rolipram and cilomilast induced vomiting at doses comparable to, or less than, their respective efficacious doses.[197]

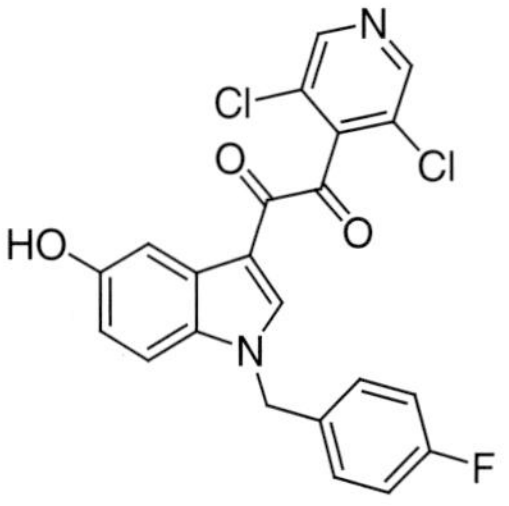

AWD 12-281
46

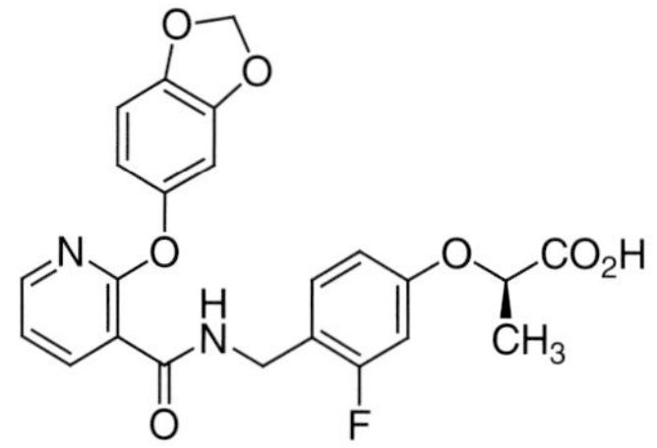

CP 671305
47

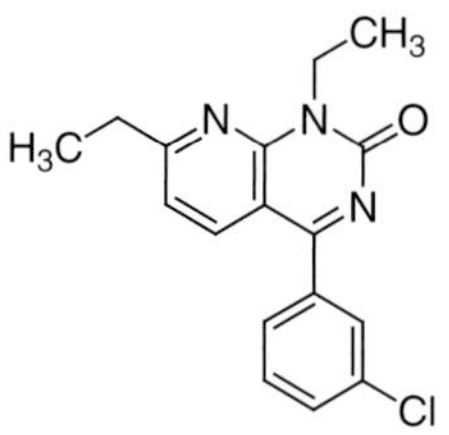

YM 976
48

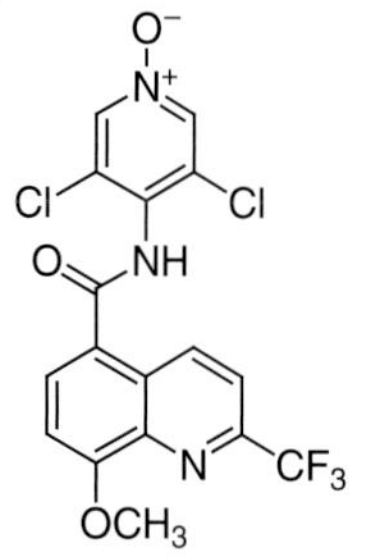

SCH-351591
49

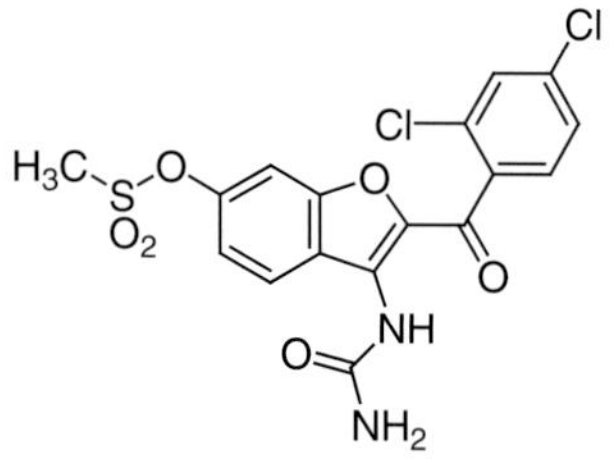

BAY 19-8004
50

Other noncatechol-based PDE4 inhibitors that have been evaluated in clinical trials include SCH-351591 (**49**) and lirimilast (BAY 19-8004, **50**). SCH-351591 was shown in a ferret emesis model to not lead to vomiting at doses up to 5 mg kg^{-1}, in spite of very high plasma concentration (up to 3.5 μM). This compound is highly selective for PDE4 compared with PDE1–PDE3 and PDE5–PDE7, and inhibits PDE4 with an IC_{50} of approximately 60 nM.[198]

A catechol derivative discovered by workers at Merck, L791943 (**51**), was shown in a similar ferret model to be nonemetic at doses up to 30 mg kg^{-1}, where the compound achieved plasma concentrations of 14 μM. L791943 was derived from CDP840 (**38**), and was designed to be metabolically more stable than the parent compound.[199] This required substantial changes in chemical structure in order to achieve and maintain good potency against PDE4 (IC_{50} = 4 nM), and acceptable activity in whole-cell screens and pharmacokinetic parameters. Additional optimization was required to eliminate the affinity of this chemotype for the hERG channel. This led to the identification of L869298 (**52**), a compound with potent PDE4 activity (IC_{50} = 0.5 nM), excellent whole-cell activity (EC_{50} = 90 nM), and low emetic potential in squirrel monkeys.[200]

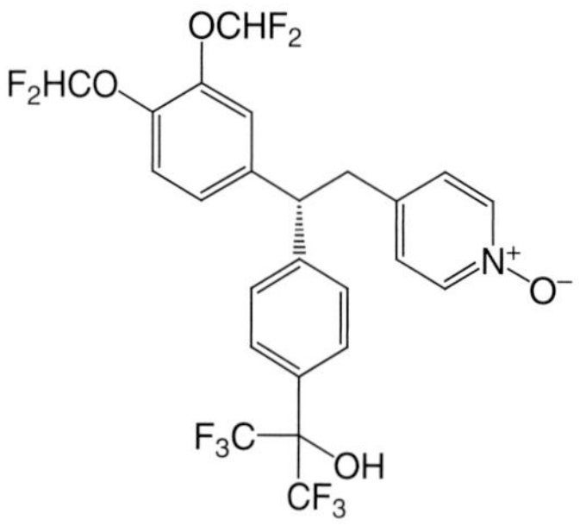

51

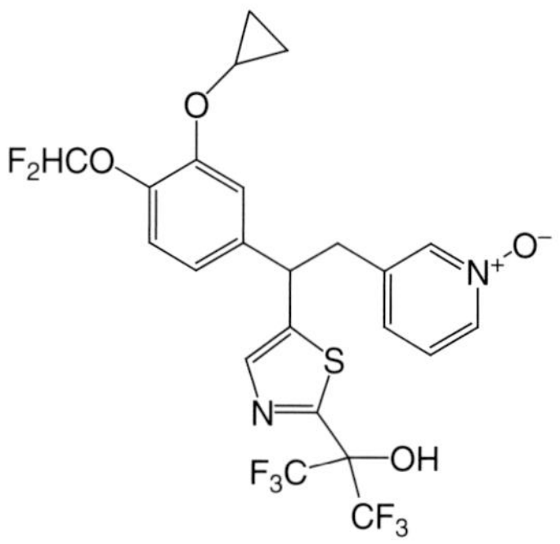

52

The development of PDE4 inhibitors is continuing, and recent publications have shown that a wide variety of chemical structures are tolerated by the enzyme. For example, Whitehead and co-workers have identified a series of substituted catechol derivatives that incorporate a substituted adenine ring system with good activity against PDE4 isozymes.[201] One of the most potent compounds in this series (**53**) has a reported IC_{50} of 6 nM against human PDE4D5, and shows some subtype specificity (e.g., $IC_{50} = 2.6\,\mu M$ for hPDE4D2).[201] The structure–activity relationships associated with the discovery of ONO-6126 (**45**) were reported recently.[202] The objective was to identify a more hydrophilic compound compared with rolipram to explore the hypothesis that such a compound would penetrate the CNS to a lesser extent and thereby reduce the emetic potential. This work illustrated that the hydroxamic acid moiety found in **45** resulted in a substantial increase in PDE4 affinity ($IC_{50} = 0.05\,nM$), compared with the corresponding carboxylic acid ($IC_{50} = 65\,nM$). ONO-6126 was tested in a ferret emesis model at doses of 3 and 10 mg kg^{-1}, and did not cause any animals in these dose groups to vomit. Pyridylisoquinolines, such as **54**, were reported by Ukita and co-workers to be potent PDE4 inhibitors with similarly high affinity for the HARBS site.[203] A naphthyridine derivative, NVP-ABE171 (**55**), is a PDE4 inhibitor that demonstrates some subtype specificity. The IC_{50} values against PDE4A, PDE4B, PDE4C, and PDE4D are, respectively, 600, 34, 1230, and 1.5 nM. The compound is active in a variety of preclinical models of inflammation, but no information on the emetic potential of **55** is available.[204] The pteridine derivative **56** (DC-TA-46) was shown to inhibit PDE4, to elevate cAMP levels in cells, and inhibit the growth of tumor cells in culture.[205] Phthalazinone derivatives such as **57** have been reported by Van der May and co-workers to be yet another series of potent PDE4 inhibitors with in vivo activity in animal models of inflammation.[206] In this group of compounds, only the *cis* isomers were active, and it was found that a lipophilic substituent on the phthalazinone ring, such as the adamantyl derivative shown, was more potent than an analog with either a benzoic acid or benzoic ester at the same position.[206] The amino diazepinoindole CI-1044 (**58**) was discovered by Burnouf and co-workers.[207] The compound had good selectivity for PDE1–PDE3 and PDE5, and moderate potency for PDE4 ($IC_{50} = 270\,nM$) obtained from human U937 cells. In spite of this modest potency, the compound demonstrated good activity in vivo ($ED_{50} < 3\,mg\,kg^{-1}$) in animal models of inflammation, and no emesis in ferrets following intravenous administration, and low affinity for the HARBS PDE4 site. No subtype specificity was observed in this series of compounds. More potent derivatives of the compound were identified, but these analogs led to emesis in ferrets.

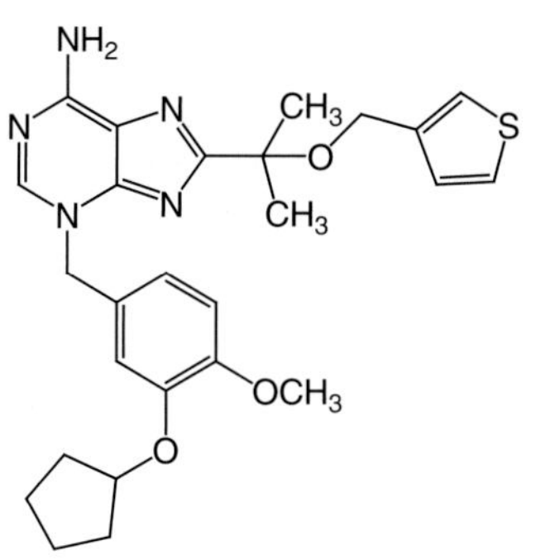

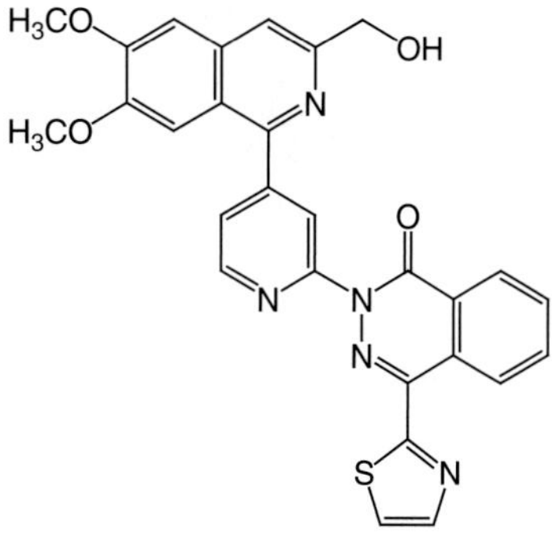

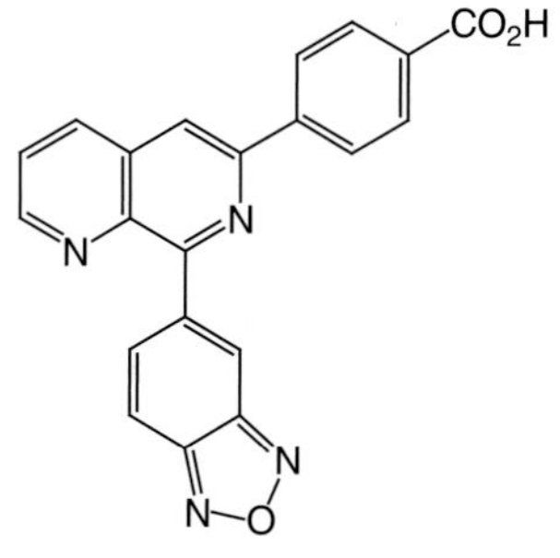

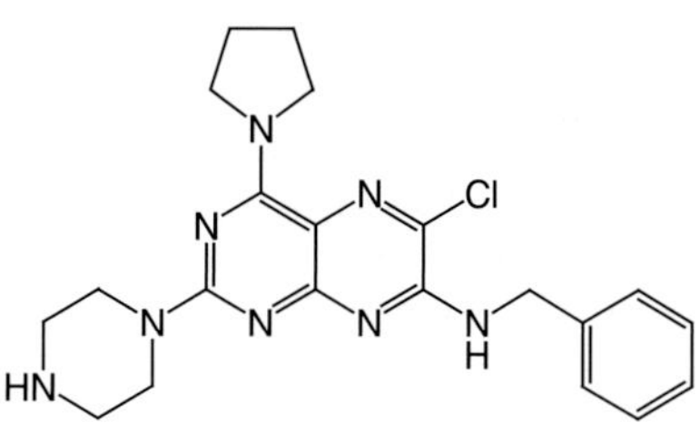

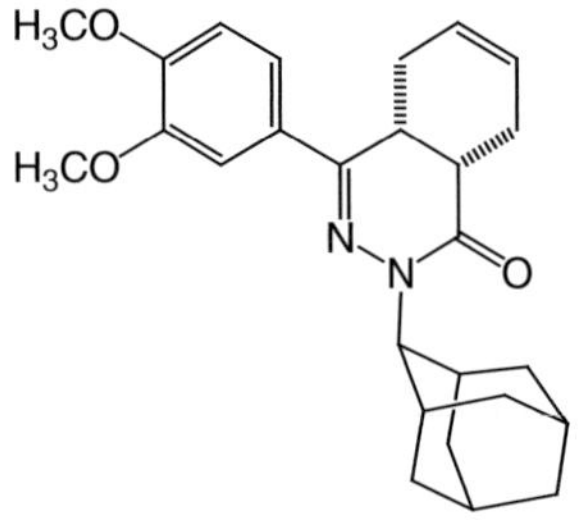

57 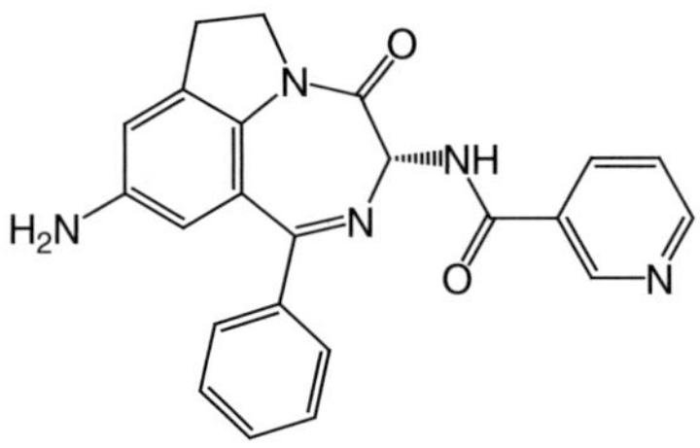 **58**

Very recently, two papers reported distinct approaches to the discovery of novel PDE4 chemotypes. One method, reported by Card and co-workers, utilizes high-throughput x-ray crystallography in combination with a scaffold-based approach to convert fragments with low affinity into molecules with high affinity for a given target.[208] The N-1 substituents on the pyrazole ring that were evaluated in this work included isoxazoles, quinoline, phenyl, and tetrahydrobenzothienopyrimidine. Of these, the phenyl-based derivatives proved to be the most potent. Analysis of the structural features that are thought to influence association of the small molecule with the enzyme allowed for the rapid optimization of a series of pyrazole-based compounds, the most potent of which (**59**) had IC_{50} values of 19 and 56 nM for PDE4B and PDE4D, respectively. Card and co-workers have also published a more comprehensive review of the structural basis underlying their scaffold-based approach to PDE inhibitor discovery.[209] In the second approach, a particular scaffold with known, but modest, PDE4 activity was chosen as a starting point.[210] This example used zardavarine, a PDE3/PDE4 inhibitor with a PDE4 IC_{50} of 800 nM. Using a carefully designed set of linkers and functional groups (selected based on a set of rules derived from physicochemical properties), docking of candidates into PDE4 was scored using three different protocols. Three different points of attachment to zardaverine were considered, along with five simple alkyl linkers and 15 different functional groups chosen with a variety of physical properties. This small library produced a compound (**60**) with a PDE4 IC_{50} value of 0.9 nM, a 4-log increase from the starting point.

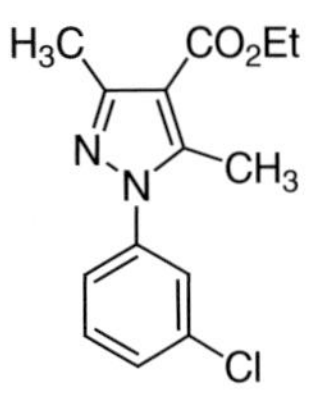

59 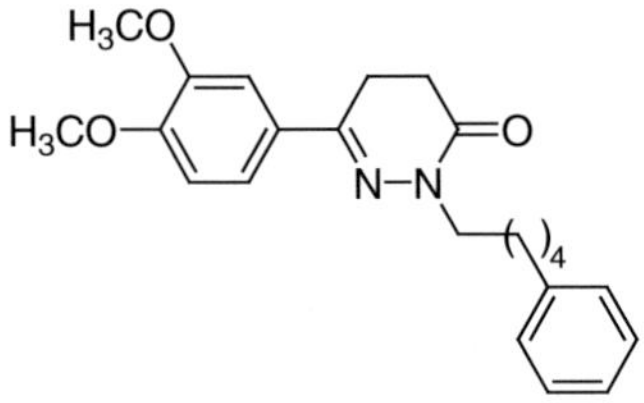 **60**

There have been several crystal structures of PDE4 enzymes reported complexed with inhibitors.[211] Analysis of these structures shows that compounds bind to the enzyme in a variety of ways. A combination of hydrophobic and hydrogen bond interactions are important. None of the crystal structures to date reveal significant interactions between the inhibitor and the essential cations bound within the active site. However, as noted above with **45**, the presence of a hydroxamic acid moiety, which is known to coordinate strongly with certain divalent cations such as zinc, significantly improves potency, possibly by interacting with the bound cation.

Two recent reviews give additional examples of PDE4 inhibitors.[212,213]

2.23.6.5 PDE5 Inhibitors

The commercial success of sildenafil (Viagra) stimulated a great deal of interest in the pharmaceutical industry in an attempt to develop additional compounds that might be efficacious for the treatment of PDE5-related conditions. The discovery of sildenafil[214] (**4**) was based on the optimization of zaprinast (**61**). The second PDE5 inhibitor to be approved for use, vardenafil (Levitra **5**), is structurally related to sildenafil.[215] Interestingly, the rearrangement of nitrogen atoms in the bicyclic core of vardenafil leads to a substantial increase in PDE5 potency, as exemplified by the activity of **62** ($IC_{50} = 40$ nM) and **63** ($IC_{50} = 5$ nM). The other structural differences between the two compounds do not contribute to the difference in potency between sildenafil and vardenafil.[216] Haning and co-workers carried out a more detailed investigation of heterocyclic scaffolds as PDE5 inhibitors, where their research showed that the enzyme does bind certain scaffolds more tightly than others, and that a number of other core structures give single-digit nanomolar PDE5 inhibitors.[217]

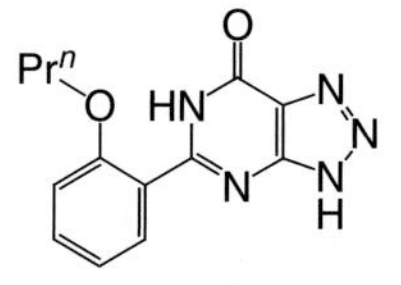

61 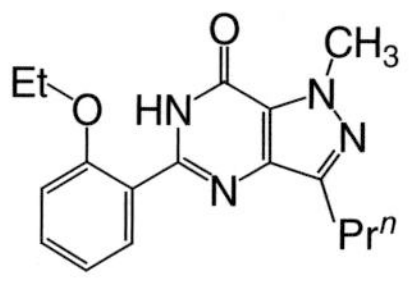 **62** 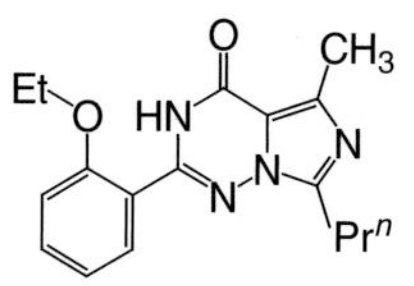 **63**

Tadalafil (Cialis, **6**) is a carbazole analog that clearly is structurally distinct from the other approved PDE5 inhibitors.[218] It was developed from a distinct carbazole starting point. The structural difference contributes to a distinct binding mode (see the discussion of the x-ray crystal structures below), which contributes to the substantial difference in PDE6 selectivity between tadalafil and the other approved PDE5 inhibitors.[219]

Much of the initial research that lead to the discovery of PDE5 inhibitors focused on chemotypes that resembled the guanine template upon which sildenafil and vardenafil were based. Previous reviews have summarized this work very well, and readers are referred to these citations for a thorough discussion and description of this portion of the effort invested in PDE5 inhibitor discovery.[220–222] This section will focus on developments in the field since 2000.

The quinazoline derivative E4021 (**64**) is a well-known, potent PDE5 inhibitor with an IC_{50} value of 3.5 nM. Just as was done with vardenafil and sildenafil, changes to the bicyclic core of this compound were investigated. This showed that the quinazoline nucleus could be replaced by a phthalazine structure, such as **65**, which, like vardenafil, provided even more potent PDE5 inhibition ($IC_{50} = 0.56$ nM) compared with E4021. Interestingly, the structure–activity relationships of the quinazoline and phthalazine cores are distinct, and substituents on the respective nuclei lead to compounds with distinct rank order potency as PDE5 inhibitors. The PDE6 potency of this compound was not described in this paper.[223]

A series of pyrazolopyridines was described by Yu et al.,[224] who used parallel synthesis techniques to rapidly optimize this series of compounds. Like Watanabe and co-workers, Yu found that the 3-chloro-4-methoxybenzylamino substituent was optimal for PDE5 activity. The preferred compound in this series (**66**) was a subnanomolar PDE5 inhibitor ($IC_{50} = 0.8$ nM, sildenafil $IC_{50} = 1.6$ nM) with superior PDE6 selectivity compared with sildenafil. This compound demonstrated equivalent activity to sildenafil in a rabbit corpus cavernosum strip assay, which measures the functional activity of PDE5 inhibitors, and also showed good oral bioavailability in rats and dogs.[224] A series of naphthyridines related to T-1032[225] (**67**, $IC_{50} = 1$ nM) was investigated in an effort to improve upon the oral pharmacokinetic properties of the original compound. It was hypothesized that by adding a nitrogen atom in the benzene ring of **67**, the resulting significantly lower the log P value would improve these properties. In this case, the optimal compound in the series (**68**) was more potent than both sildenafil and T-1032 in vitro ($IC_{50} = 0.23$ nM), and equally efficacious in a secondary functional assay. Oral bioavailability was not discussed in this paper. In this series, it was noted that 1,7-naphthyridine isomers were equipotent PDE5 inhibitors compared with the 2,7-naphthyridines such as **68**, but did not have an improved PDE6 selectivity profile, and were less efficacious in functional assays.[226] A third compound in this series, T-0156 (**69**) was extensively characterized. Like **68**, T-0156 was a potent ($IC_{50} = 0.23$ nM) PDE5 inhibitor with good selectivity for PDE6. It increased cGMP levels in corpus cavernosum tissue in a dose–response manner, and relaxed the tissue in a functional assay. In a dog model of erectile dysfunction, **69** potentiated the pelvic nerve stimulation-induced tumescence in the penis.[227]

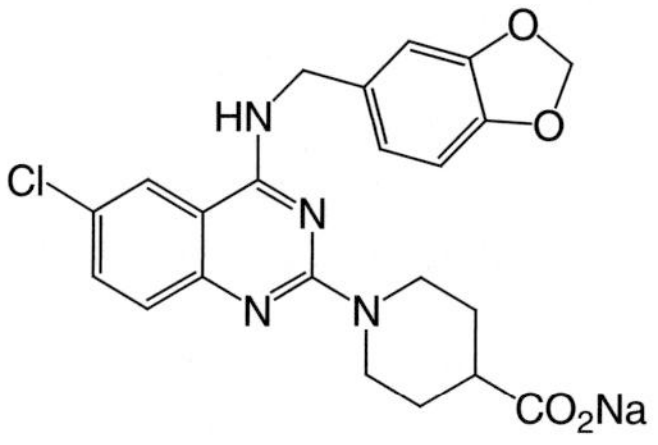

64 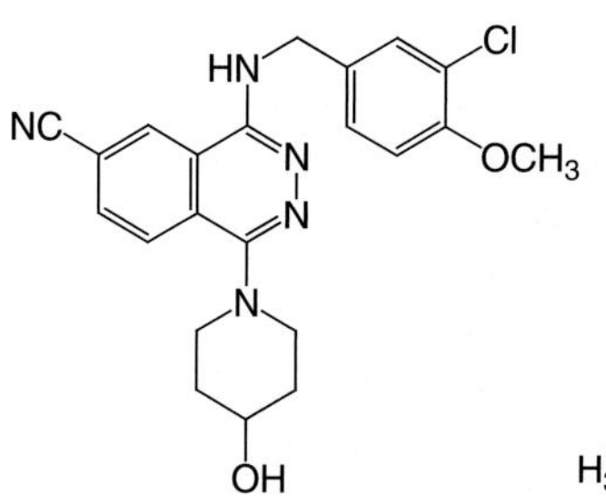 **65** 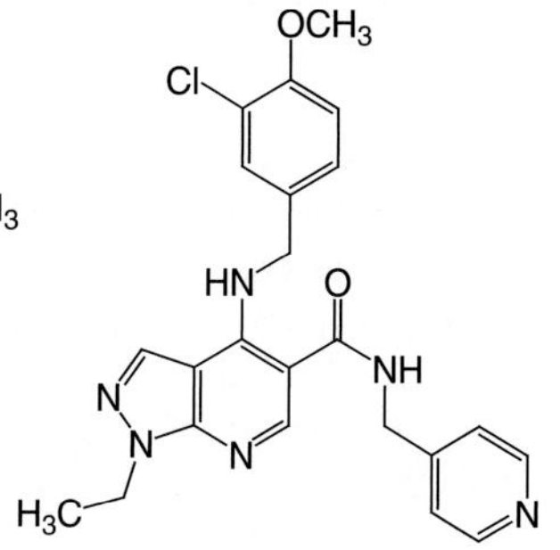 **66**

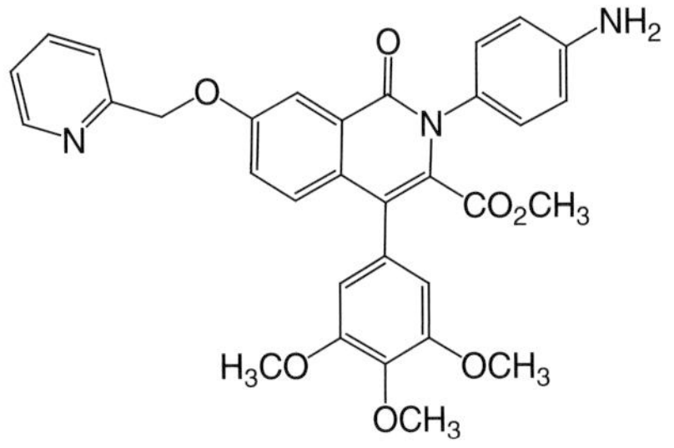

67, 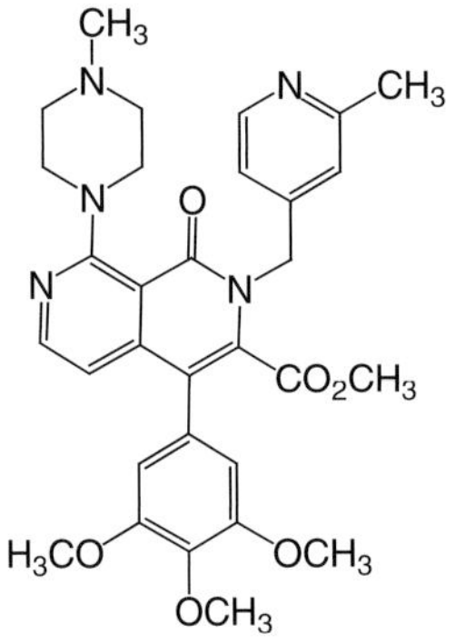 **68**, 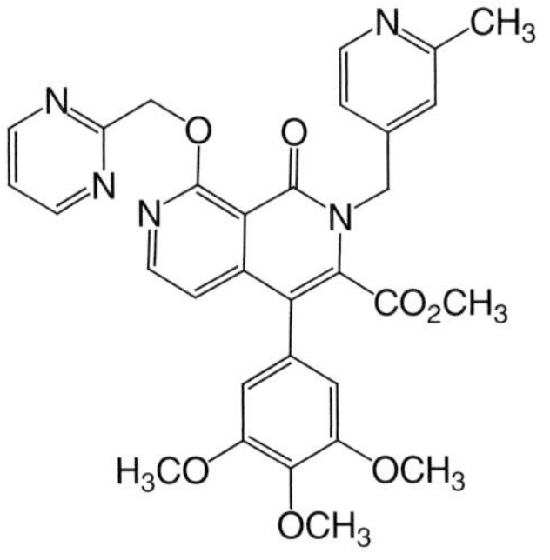 **69**

Using the basic bicyclic guanine ring system of cGMP as a template, addition of a third ring to the system has been investigated by a number of groups. One approach inserted a benzene ring between the two heterocycles, leading to two distinct series of PDE5 inhibitors, exemplified by carboxamide (**70**, $IC_{50} = 0.48$ nM) and **71** ($IC_{50} = 0.62$ nM). Both series of compounds demonstrated excellent selectivity for PDE5 over PDE1–PDE4, and both showed superior PDE6 selectivity compared with sildenafil. Activity in a functional assay proved difficult to achieve in the sulfonamide series based on **71**, as this was the sole analog in the series that demonstrated efficacy in this screen. It was hypothesized that molecular weight was a contributing factor, as this series had molecular weights greater than 550. Carboxamide (**70**) proved to be as efficacious as sildenafil in the corpus cavernosum functional assay, but had poor oral bioavailability. PDE6 selectivity in both of these series was dictated by the N-3 benzyl moiety. The respective carboxamide and sulfonamides also showed distinct structure activity relationships with regard to their respective amine substituents.[228,229]

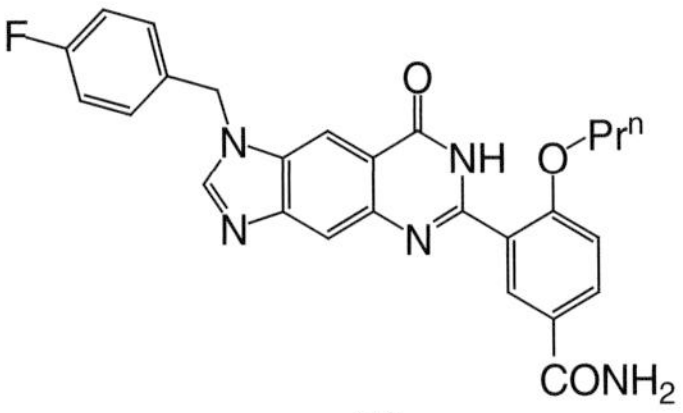

70, 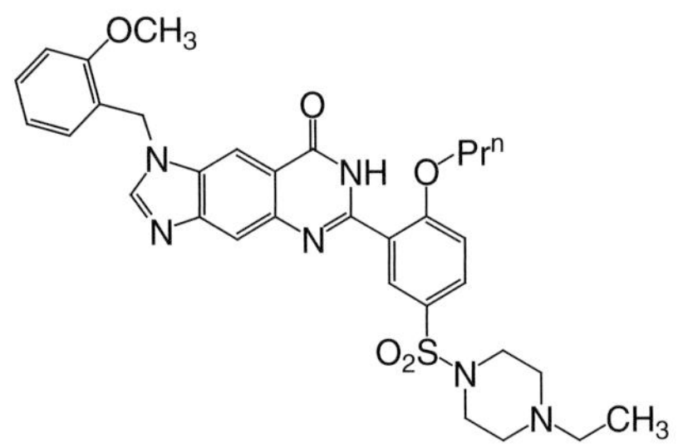 **71**

In place of a linear tricyclic core, the imidazoquinazoline could be converted to an angular one. This furnished a group of pyrazolopyridopyrimidines, such as **72**. These compounds, like the linear analogs **70** and **71**, proved to be potent ($IC_{50} < 1$ nM) and highly selective (>150-fold) for PDE5 over PDE6, with a benzylamino moiety added as indicated on the pyridine ring. The activity of this series in functional assays was not described.[230] Another version of this approach was exemplified by Feixas and co-workers, who prepared a series of pyrazolopyrimidopyridazines, such as **73**.[231] In this case, two changes to the angular nucleus of **72** were made: the pyrimidine ring was converted to a pyridazine, and the pyrazolopyridine was isomerized to a pyrimidinylpyrazole. The result was a compound that was approximately twofold less potent than sildenafil, with improved PDE6 specificity. A variety of substituents on the pyrimidine ring were investigated, and these had significant effects on both potency and selectivity, with n-alkyl substituents furnishing optimal effects.

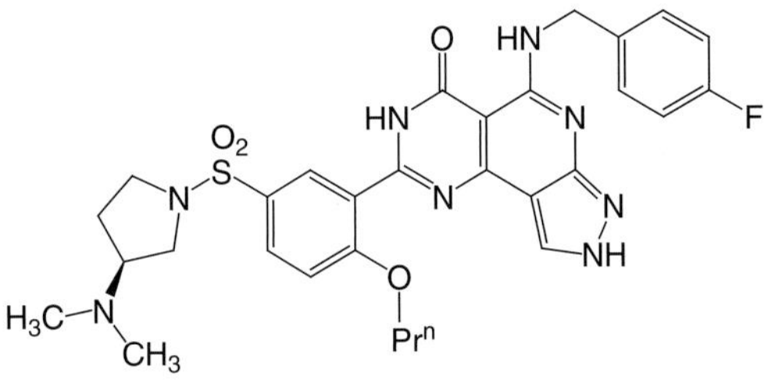

72, 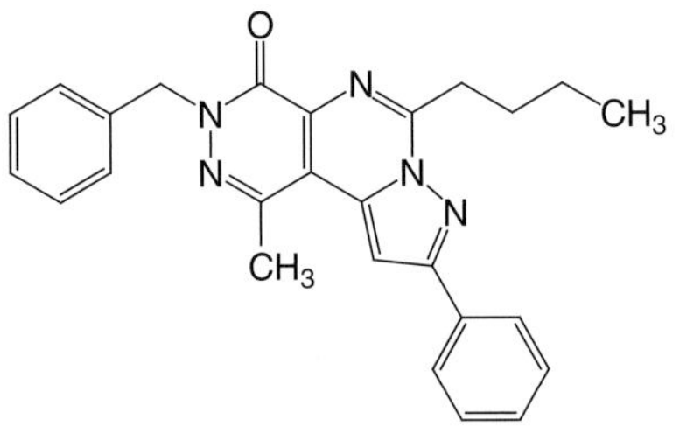 **73**

An alternative to these strategies was described by Xia and co-workers.[232] In this instance, the angular system fused a six membered ring to the pyrimidine core. Structure–activity studies showed that the optimal substituent on the added fused ring was a methyl group. Other moieties that were investigated included halogens, the cyano group, hydrogen, and aminoalkyl groups. This optimal derivative, **74**, had an IC_{50} of 9 nM and an activity comparable to sildenafil in a corpus cavernosum functional assay. Selectivity for PDE6 was approximately 20-fold.[232]

In yet another example of this line of PDE5 inhibitor design, Yu and colleagues described the pyrazolopyridopyridazine **75**. This compound was designed as a second-generation version of pyrazolopyridine **66**. In the x-ray crystal structure of **66**, a hydrogen bond between the benzylamino NH and amide carbonyl substituents formed a pseudo-six-membered ring. In **75**, the pyridazine formalized this conformation. In many ways, structure–activity studies of the amino substituents on the pyridazine ring mirrored those seen with the phthalazine **65**, and with **66**. The imidazocarboxamide moiety proved to be superior to other substituents, giving a compound with an IC_{50} of 0.3 nM against PDE5, excellent selectivity for PDE5 over PDE1–PDE4, and 150-fold selectivity for PDE5 over PDE6, good functional activity, and good oral bioavailability. In healthy human volunteers, **75** showed no evidence of visual disturbances.[233]

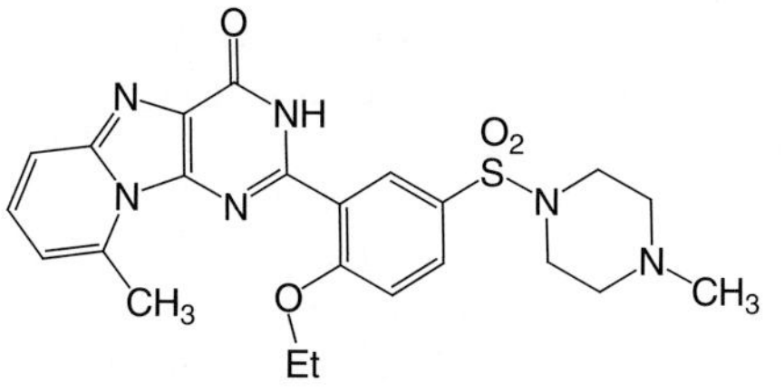

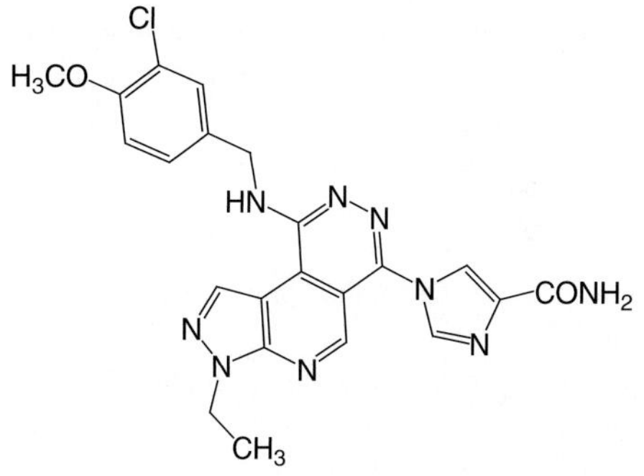

At least one group has reported the synthesis of analogs of tadalafil, where the methyl substituent of the diketopiperazine ring was replaced by a wide variety of other substituents such as cyclic amines, alkyl, alkylcarboxamides, aryl/heteroaryl, and benzyl/heterobenzyl moieties. Some of these, such as benzyl, imidazolylmethyl, and carboxamido groups, furnished potent (IC_{50} as low as 0.7 nM (e.g., **76**), compared with tadalafil with an IC_{50} of 6.7 nM), but somewhat less PDE6-selective, derivatives. One objective of this study was to also investigate PDE11 selectivity in the series. None of the compounds reported in this work demonstrated a significant difference from tadalafil.[234]

Researchers at Johnson and Johnson have extensively investigated a group of pyrroloquinolines as PDE5 inhibitors. In the first paper in the series, structure–activity relationships in the rings pendant on the pyrroloquinoline nucleus were investigated. It was found that the number and nature of oxygenated positions played an important role in providing both PDE5 potency, and in one case (**77**) excellent (>300-fold) selectivity for PDE6. This series demonstrated the ability to increase cGMP levels in cells more potently than sildenafil, and efficacy in dogs, as measured by an increase in mean arterial pressure, when administered intravenously.[235] This was followed by an investigation of amide derivatives (**78**) that demonstrated similar PDE5 potency, variable levels of PDE6 selectivity, with less efficacy in the dog model of erectile dysfunction.[236] Substitution of a carbamate for the amide in **78** produced another potent series (**79**) with lower selectivity for PDE5 than for PDE6, and similar results in functional and animal models of efficacy compared with **78**.[237] In a series of pyrroloquinoline analogs, such as **80**, that resembled tadalafil and those reported by Maw and co-workers, markedly lower potency in vitro and in vivo was observed.[238] The final paper in this series resulted in the identification of a compound with properties suitable for clinical evaluation. This was accomplished by converting the pyrimidine ring of **77** to a pyridine, and separation of the enantiomers to furnish the preferred (*R*)-isomer (**81**), leading to a PDE5 inhibitor with good potency ($K_i = 0.12$ nM), sufficient aqueous solubility, efficacy in erectile dysfunction models, and oral bioavailability in animals.[239]

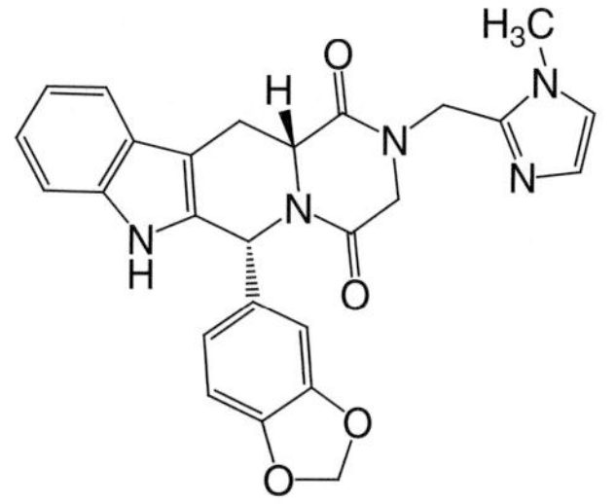

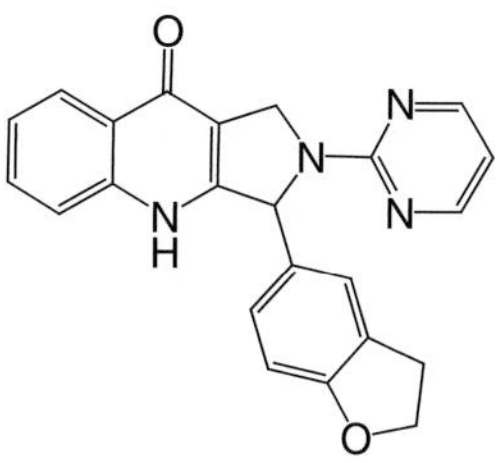

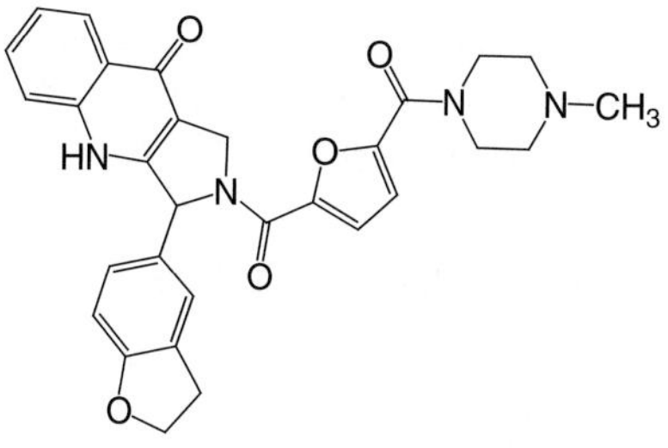

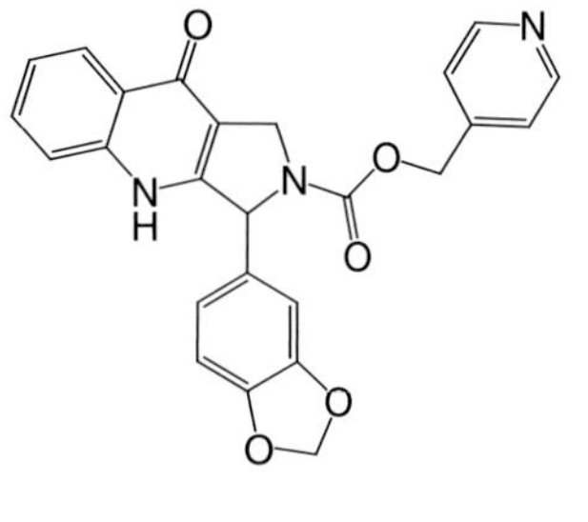

79 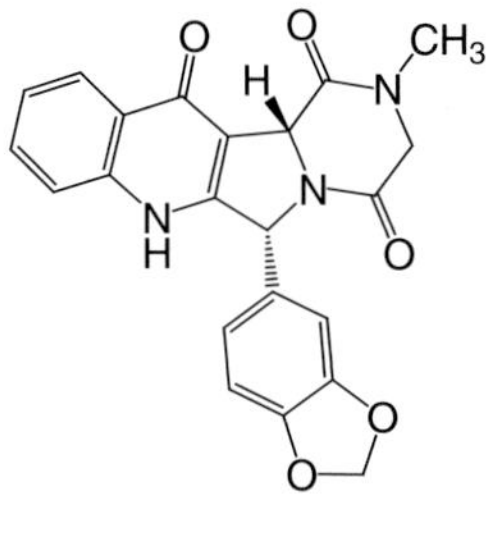 **80** 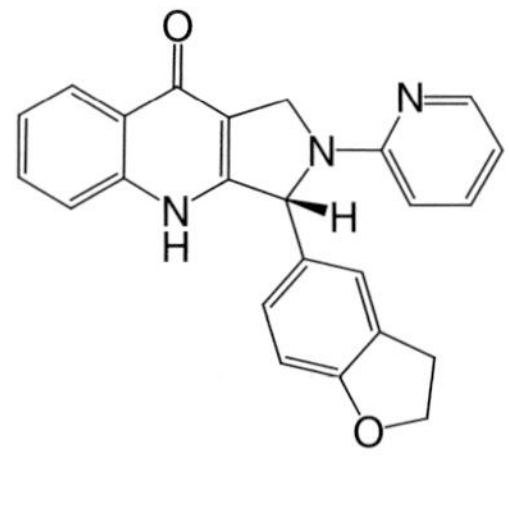 **81**

An alternative tricyclic template was reported by Boyle and co-workers. In this case, a five-membered ring was fused to the six-membered heterocycle of the guanine nucleus. This allowed for more expedient exploration of structure–activity relationships in the series previously reported by Ahn.[174] It was found that an ethyl moiety at N-5 of the guanine was preferred with respect to both potency and PDE6 selectivity. Further, a benzyl group at C-7 was superior to other groups in conferring potency and selectivity for PDE5 over PDE1. As observed by others,[228] N-3 benzyl substitution gave more potent and selective PDE5 inhibitors. The optimal compound in the series (**82**) showed an IC_{50} of 2.5 nM against PDE5, 90-fold selectivity for PDE6, and 200-fold selectivity for PDE11. This compound also demonstrated oral bioavailability comparable with sildenafil (**4**) in rats and dogs, and efficacy in a dog model of erectile dysfunction.[240]

The pyrazole ring in **66** could be replaced by a substituted benzene ring, as shown by Bi and co-workers.[241] Interestingly, it was necessary to also replace the carboxamide in **66** with a simple hydroxymethyl group to retain potent PDE5 inhibition. The ethyl moiety at C-8 of the quinoline core was an important contributor to both potency and PDE6 selectivity, and the cyano group at C-6 was preferred. The optimal compound in this series (**83**) was shown to be highly potent ($IC_{50} = 0.05$ nM) as a PDE5 inhibitor with excellent (~8000-fold) PDE6 selectivity.[241]

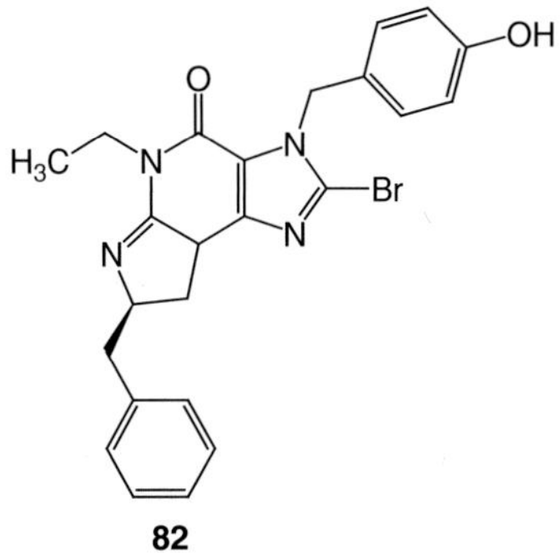

82 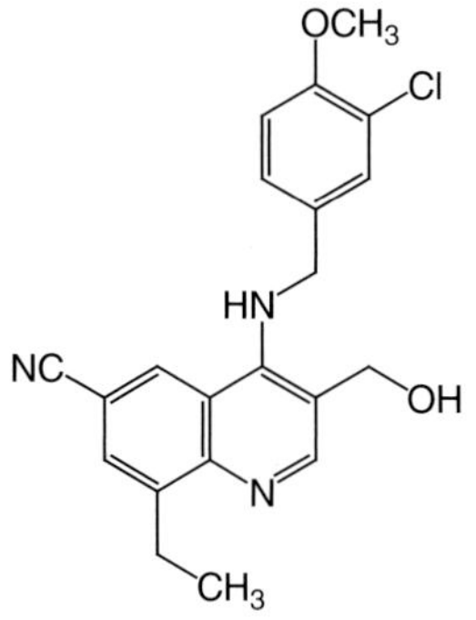 **83**

At least two crystal structures of the catalytic domain of PDE5 with inhibitors bound have been reported. One employed IBMX, and it was shown that it bound in a pocket composed of hydrophobic residues including Val782, Phe786, Phe820, and the previously mentioned key glutamine at position 817. The so-called H and M loops of the enzyme are important structural features for selective inhibitor binding, and differences in this region in PDE5 help to account for subtype specificity compared with PDE4.[242] The second structure showed how sildenafil, vardenafil, and tadalafil bound within the active site of PDE5.[243] It showed that sildenafil and vardenafil bound in the same pocket (Q pocket) as IBMX. The sulfonamide moiety in both compounds does not participate in key attractive interactions with the enzyme. Other groups, such as the piperazine ring, alkoxy substituents, and the heterocyclic core, do make important contacts and undergo hydrophobic, as well as hydrogen bonding and π-stacking, interactions with neighboring amino acids on the protein. Tadalafil fills other pockets on the enzyme, in particular the hydrophobic H pocket, and does not interact, like the other two compounds, with the L region of the enzyme.

2.23.6.6 PDE6 Inhibitors

No selective inhibitors of PDE6 have been reported. However, as a part of the development of selective PDE5 inhibitors, this enzyme is often part of the panel of PDEs to be screened, given the high degree of sequence homology and observation that sildenafil, as a prototypical PDE5 inhibitor, does inhibit PDE6 with an IC_{50} of 37 nM.[219] This level of in vitro activity is sufficient to result in inhibition of the enzyme at therapeutically relevant doses in humans and produce transient visual disturbances. Vardenafil is also a potent PDE6 inhibitor, with an IC_{50} value of 3.5 nM, and tadalafil has a reported IC_{50} value of 1.2 μM.[219] Clinical evidence gained to date indicates a lower incidence of visual changes with vardenafil and tadalafil.

2.23.6.7 PDE7 Inhibitors

The first inhibitors of PDE7 were a group of benzothieno[3,2-a]thiadiazine 2,2-dioxides. These compounds (e.g., **84**) proved to be weak inhibitors of the enzyme ($IC_{50} \approx 20$ μM, little selectivity for PDE7 over PDE3 or PDE4).[244] A comparative molecular field analysis of this series of compounds was carried out in an attempt to improve the potency. This analysis suggested that hydrogen-bonding interactions between the inhibitor and enzyme were important determinants of activity. Based on this suggestion, an analog was prepared that had improved activity as a PDE7 inhibitor (**85**). This change also improved the selectivity of the series for PDE7 over PDE3 and PDE4.[245] More potent guanine-based compounds were described by Barnes and co-workers, where the sugar moiety of the cyclic nucleotide was replaced with hydrophobic bicyclic substituents.[246] This furnished compounds (e.g., **86**) with low single-digit micromolar potency, and good selectivity for PDE7 over PDE3 and PDE4.[246] Another series of purine-based compounds was described where instead of a lipophilic substituent at N-3 of the purine ring system, an amino thiazole was added at C-2. Along with a substituted benzylamine at C-6, this furnished compounds with IC_{50} values as low as 10 nM. The most potent analogs in this series (**87**) also proved to be potent PDE5 inhibitors.[247] Further modification of the C-6 substituent in this series, coupled with alteration of the purine heterocycle, significantly improved the PDE selectivity profile and maintained good PDE7 potency (e.g., **88**).[248]

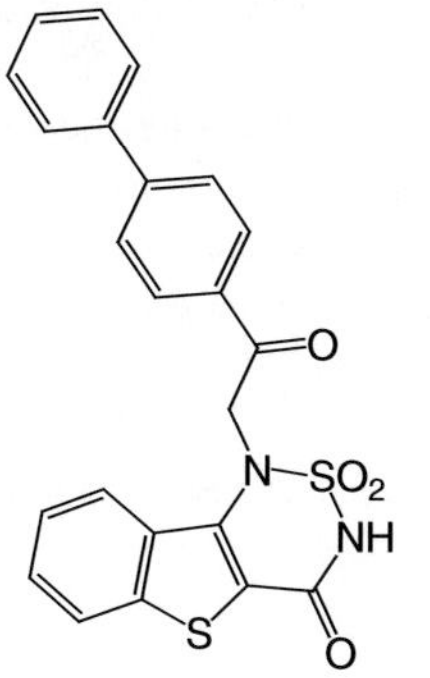

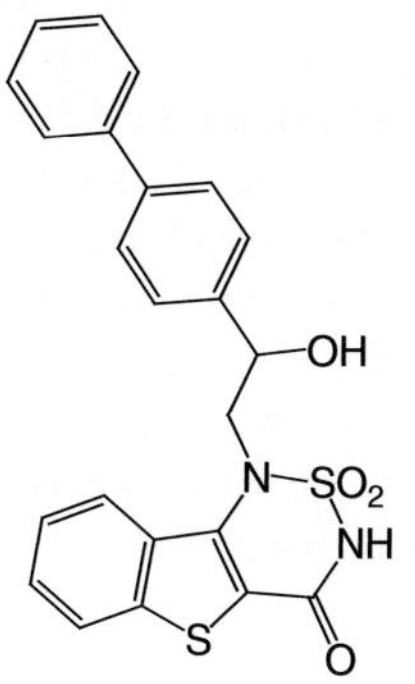

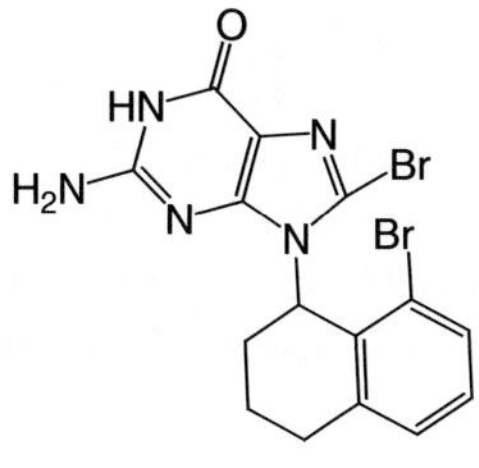

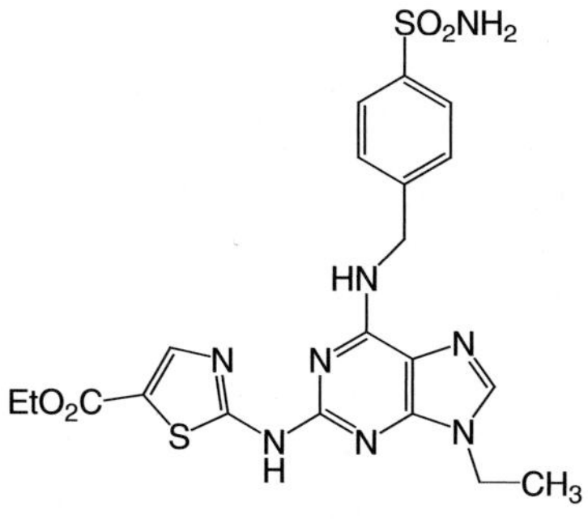

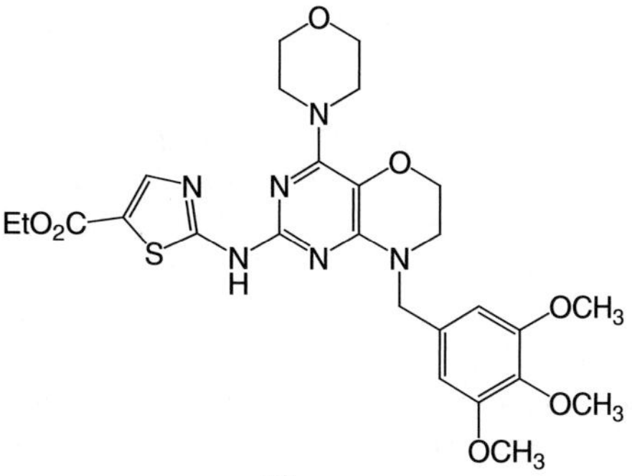

A series of spiroquinazolinones was found to give potent PDE7 inhibitors. A high-throughput screen identified a simple spiroquinazoline lead with good potency that was optimized to provide a compound (**89**) with an order of magnitude increase in activity.[249] This was further improved to yield the oxadiazole analog **90**, which had an IC_{50} of 3.5 nM, and good to excellent (at least 100-fold) selectivity against PDE1–PDE6. Selected derivatives in this group had measurable oral bioavailability.[250]

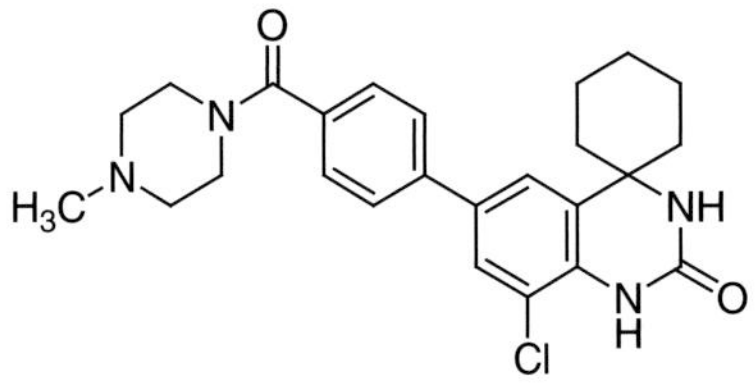

89 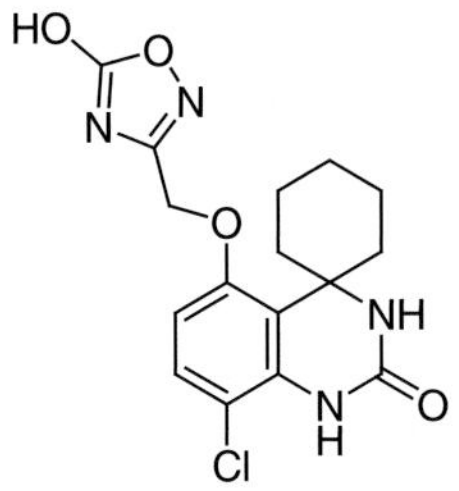 **90**

The same group identified and optimized a series of thiadiazoles that were discovered by high-throughput screening. Once again, the potency of the initial lead was significantly improved, to furnish a compound (**91**) with an IC_{50} of 30 nM for PDE7. Selectivity against PDE4D3 was excellent, as this compound had an IC_{50} of over 8 μM.[251] Further improvements in this series focused on oral pharmacokinetic characteristics, and this yielded a compound (**92**) with slightly lower potency (IC_{50} = 60 nM), but with excellent oral bioavailability.[252]

A simple sulfonamide derivative, BRL 50481 (**93**) was reported to be a moderately potent PDE7 inhibitor (K_i = 180 nM), with good selectivity against PDE3, and moderate selectivity against PDE4.[133]

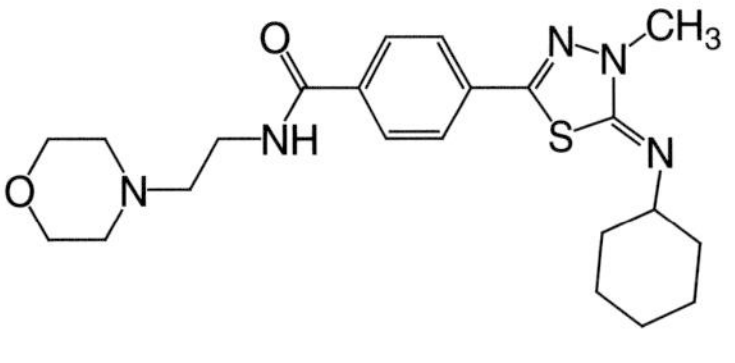

91 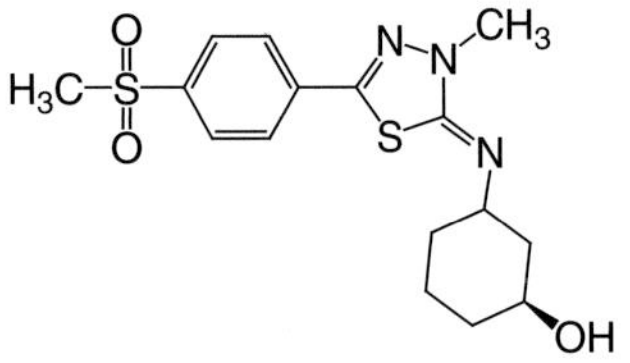 **92** 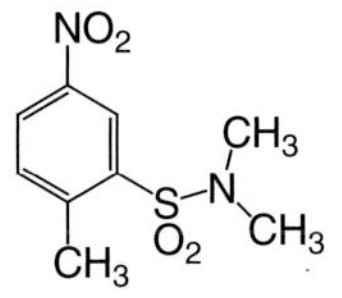 **93**

Several patents have described other chemotypes that have claimed activity as PDE7 inhibitors. In some cases, IC_{50} values were reported, and in other cases ranges of IC_{50} values were reported. Representative examples from these patents are shown below. Pyrazolopyrimidinones (**94**) were claimed by Daiichi. These compounds had excellent PDE7 potency, with IC_{50} values as low as 3.5 nM.[253] A series of isoquinolines was claimed by workers at Ono. PDE7 activity was specified for a dihydroisoindole derivative (**95**), with an IC_{50} of 23 nM.[254] Phthalazinones (**96**) were claimed by Altana as a series of dual PDE4/PDE7 inhibitors with respective –log IC_{50} values of 8.64 and 7.64,[255] and dihydroisoquinolines (**97**) were claimed by Byk Gulden with a –log IC_{50} value of 7.49.[256] Sulfonamides **98** were claimed by Celltech, with an IC_{50} range reported as less than 1 μM.[257]

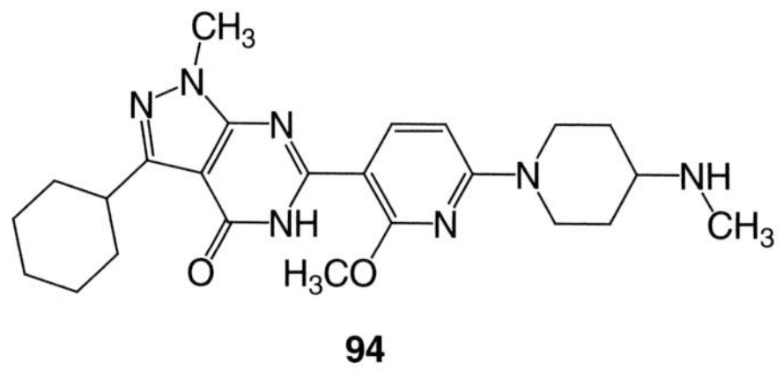

94 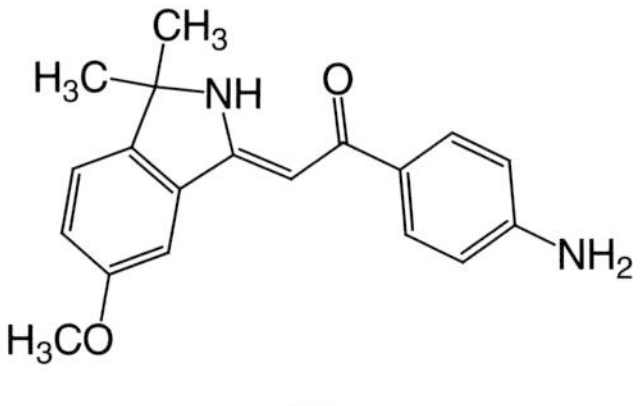 **95**

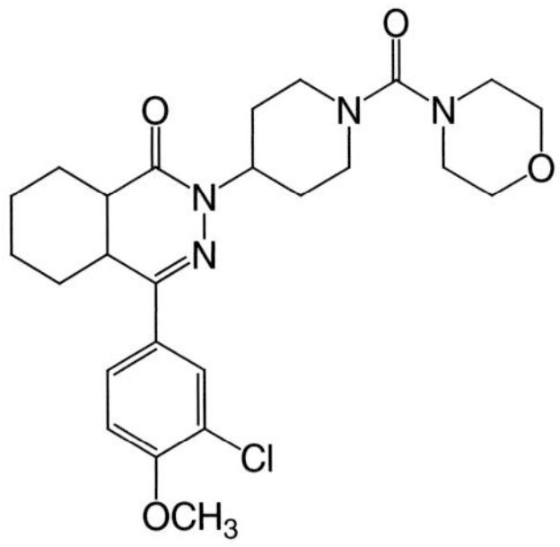

96 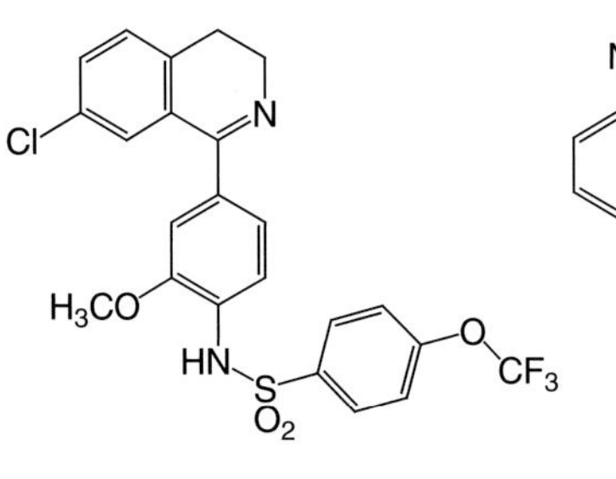 **97** 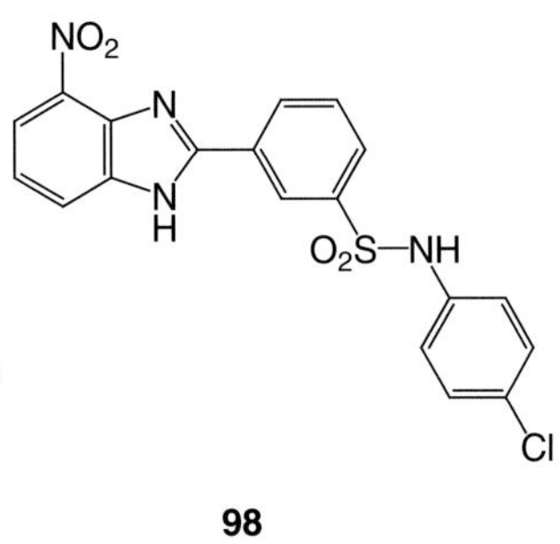 **98**

In a recently reported x-ray crystal structure of the catalytic domain of PDE7, it was found that multiple elements of the structure contribute to selectivity between PDE7 and PDE4.[258] The nonselective PDE inhibitor IBMX was employed as the ligand in these studies. Analysis of the binding revealed that the conformation and position of an invariant glutamine residue (Gln413 in PDE7) formed a critical hydrogen bond with IBMX, but clashed sterically with rolipram. A serine residue in PDE7 (Ser377) interacts with this key glutamine to control its orientation and conformation in the active site, and Ile412 occupies a key site that influences the size and shape of the IBMX-binding pocket in PDE7.

2.23.6.8 PDE8 Inhibitors

No selective inhibitors of this enzyme have been reported. Dipyridamole (**99**) is reported to be a weak inhibitor of the enzyme in vitro, with an IC_{50} as low as 4.5 µM.[259] Other standard PDE inhibitors such as rolipram, zaprinast, IBMX, and vinpocetine do not inhibit PDE8.

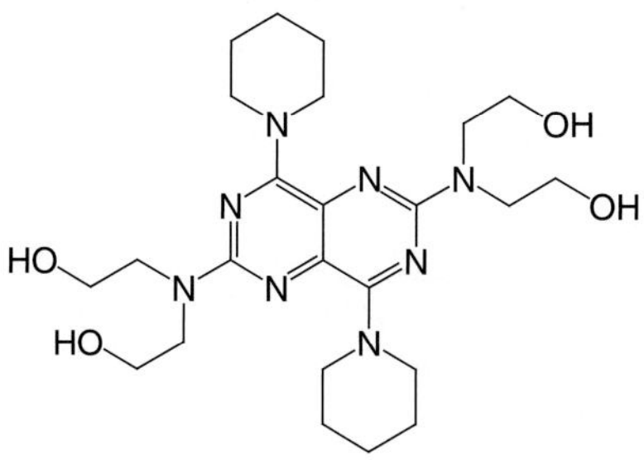

99

2.23.6.9 PDE9 Inhibitors

All activity reported to date in the discovery of PDE9 inhibitors has been in the patent literature. Pfizer has been active in the field, with three applications that have published in the last few years. Examples of compounds and chemotypes claimed as PDE9 inhibitors are **100–102**.[260–262] Compound **102** was reported to lower blood glucose in a mouse model.[262]

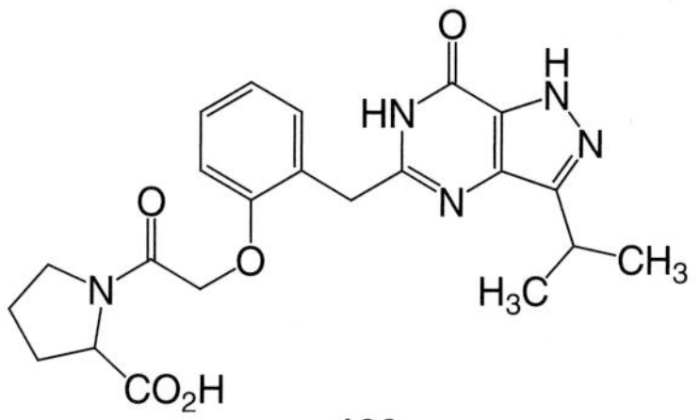

100 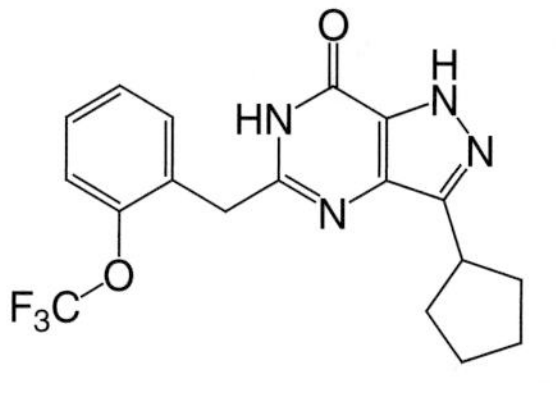   **101** 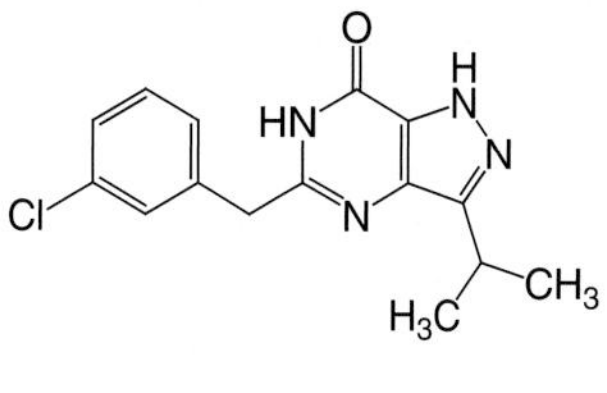   **102**

Another related active series of PDE9 inhibitors was reported by Hendrix. This compound (**103**) had moderate activity ($IC_{50} = 50$ nM) against PDE9, and excellent selectivity against all other PDEs except PDE11, against which it demonstrated 50-fold selectivity.[263]

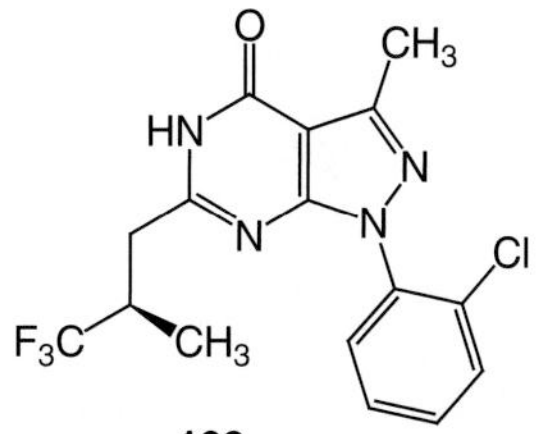

103

The x-ray crystal structure of the catalytic domain of PDE9A2 has been solved using the inhibitor IBMX. Analysis of the structure showed that IBMX binds in a pocket similar to that found in the active sites of PDE4 and PDE5, but with a substantially different orientation. Common binding features include stacking of the xanthine ring of IBMX with conserved phenylalanine residues, and a hydrogen bond between the N-7 NH of IBMX and a conserved glutamine residue. However, the xanthine ring of IBMX is rotated considerably in the PDE9A2 active site, compared with the other enzymes. This may be due to the larger volume of the PDE9A2 active site compared with the other enzymes, or to more subtle conformational differences in the structure of the H and M loops of the PDE9A2.[264]

2.23.6.10 PDE10 Inhibitors

There is increasing interest in the discovery of PDE10 inhibitors, as reflected by some recent patent applications claiming a variety of chemotypes for CNS and metabolic disease applications. Examples of dihydroisoquinolines are given in a patent application from Bayer Pharmaceuticals (**104–107**). These compounds were reported to have IC_{50} values of approximately 30 nM in a scintillation proximity assay.[265] Structurally related compounds have also been disclosed by others.[266] Imidazotriazines such as **107** were claimed for the treatment of neurodegenerative disorders. The compound was stated to have an IC_{50} value of 410 nM.[267]

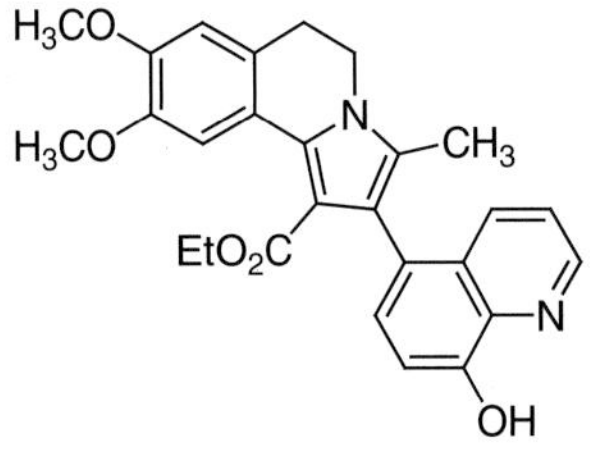

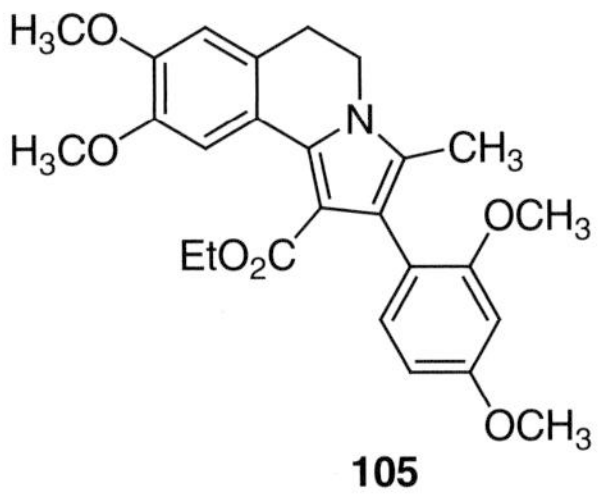

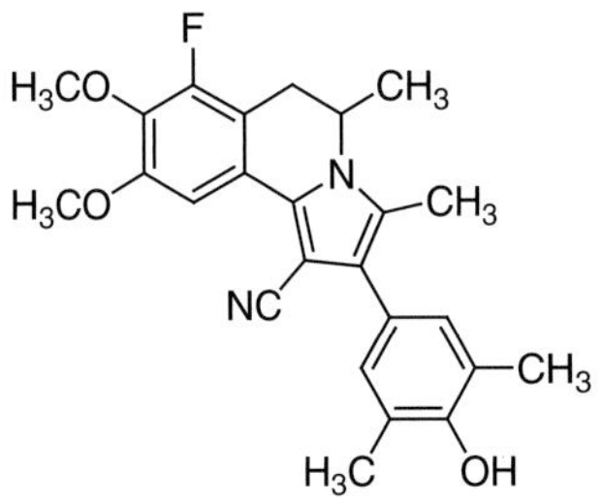

The isoquinoline alkaloid papaverine **108** has also been reported to be a PDE10A inhibitor, with an IC_{50} of 40 nM. The compound was used in a mouse model of schizophrenia, where, at oral doses of 3 and 10 mg kg^{-1} daily, it demonstrated activity comparable with haloperidol. When added to brain tissue slices and striatal spiny neurons in cell culture, papaverine was shown to increase the level of cAMP in the cells. When papaverine was administered to PDE10A knockout mice and examined in the same model, no evidence for efficacy was seen. This suggests that the enzyme may play a role in the disease.[268]

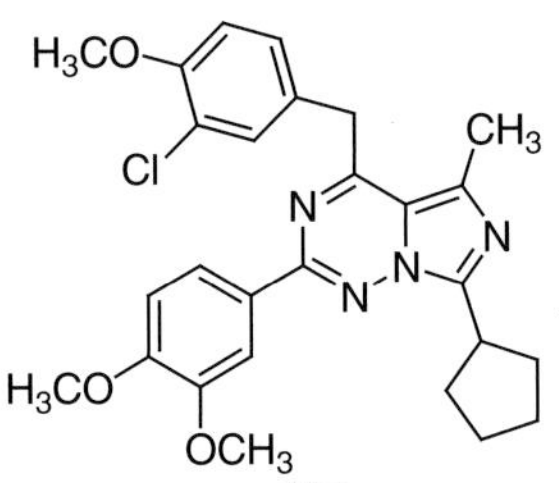

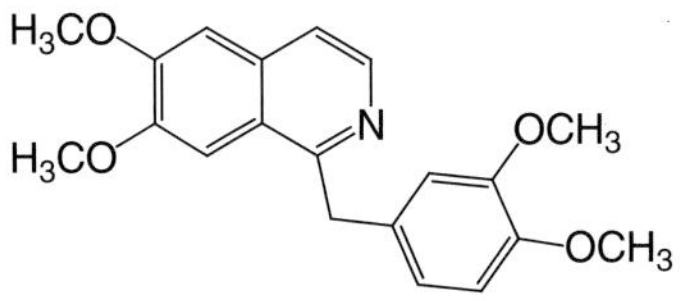

2.23.6.11 PDE11 Inhibitors

No selective inhibitors of PDE11 have been disclosed at this time. Tadalafil, sildenafil, and vardenafil have been reported to inhibit the enzyme in vitro. In this group, tadalafil is the most potent inhibitor, with an IC_{50} value of 37 nM.[219] Dipyridamole (**99**) and E4021 (**64**) have also been reported to be a moderately potent PDE11 inhibitors using both cAMP and cGMP as substrates, with IC_{50} values of 0.35–1.8 μM, when studied with PDE11A3 and PDE11A4.[166]

2.23.7 Prototypical Therapeutics

Table 3 lists the generic names of PDE inhibitors approved for use, along with the PDE(s) targeted and the primary indication(s) for use of the compound.

2.23.7.1 PDE1

At the present time, there are no specific therapeutic applications for which a PDE1 inhibitor is indicated. There are, however, different lines of evidence to suggest a role for PDE1 in a number of physiological processes. The issue in these cases is whether or not the enzyme plays a critical part in controlling the signaling pathway. One example can be found in the investigation of combined PDE1/PDE5 inhibitors for vascular injury. SCH51866 (**12**) at a dose of 17 mg kg^{-1} day^{-1} for 1 month has been shown to reduce by approximately 50% the degree of vascular injury in a porcine model.[269] Compounds related to the combined PDE1/PDE5 inhibitor **11** showed good antihypertensive activity in a spontaneously hypertensive rat model.[175] As noted in Section 2.23.4.1, PDE1 is a part of a multicomponent signaling pathway that can influence insulin secretion from.

The PDE1 inhibitor vinpocetine **7** was reported to be in clinical trials for the management of urinary urgency and incontinence.[270,271] This was based on the demonstration that PDE1 plays a key role in maintaining tone in the detrusor muscle.

2.23.7.2 PDE2

Like PDE1, there are no approved therapeutic applications for a PDE2 inhibitor. Unlike PDE1, the availability of highly potent and PDE-subtype selective inhibitors has made it easier to identify and understand the benefits associated with inhibition of this enzyme. As noted in Section 2.23.4.2, a PDE2 inhibitor (**22**) has shown activity in animal models of cognition.[31] Furthermore, because PDE2 is closely associated with signaling by the ANP receptor, it is conceivable that PDE2 may play a role in maintaining plasma volume. The enzyme has also been shown to have a role in platelet aggregation.[29]

2.23.7.3 PDE3

PDE3 inhibitors such as milrinone have been used for acute treatment of heart failure. However, the safety of their long-term use in treatment of this disease has been questioned.[272] Cardiac arrhythmias have occurred, possibly as a result of cAMP accumulation in the heart. Cilostazol is approved for use as a platelet antiaggregatory agent. Like milrinone, it is not widely used because of safety concerns. The association in vitro between PDE3 inhibition and platelet antiaggregatory action has been studied in detail, and it was shown that more potent cilostazol analogs, such as OPC-33540, were more potent as inhibitors of platelet aggregation.[273] When combined with β-adrenoreceptor antagonists, there is some recent clinical evidence to suggest the use of PDE3 inhibitors as an adjunct in the treatment

Table 3 PDE inhibitors approved for use

Generic name	Target PDE(s)	Primary indication(s)
Theophylline	PDE3, PDE4	Asthma, emphysema, bronchitis
Aminophylline	PDE3, PDE4	Asthma, emphysema, bronchitis
Olprinone	PDE3	Acute cardiac insufficiency
Milrinone	PDE3	Acute cardiac failure
Amrinone	PDE3	Hypertension, acute cardiac failure
Cilostazol	PDE3	Intermittent claudication
Sildenafil	PDE5	Erectile dysfunction, pulmonary hypertension
Vardenafil	PDE5	Erectile dysfunction
Tadalafil	PDE5	Erectile dysfunction
Dipyridamole	PDE1, PDE5	Platelet aggregation

of dilated cardiomyopathy.[274] Cilostamide (**32**) has been shown to inhibit cell proliferation in a KB squamous cell carcinoma cell line.[275] Other investigations into the potential role of PDE3 for use in the treatment of other types of cancer (e.g., malignant melanoma) showed little role for the enzyme.[276] PDE3B has been shown in multiple laboratories to play a key role in insulin release, and, because it is expressed primarily in the liver, it has been hypothesized to be a target for control of metabolism.[194,277,278] However, the lack of subtype-selective (3A versus 3B) agents has limited the investigation to largely model studies. Cilostamide has also shown activity in a rat balloon injury model, suggesting that PDE3 inhibitors could find use in preventing intimal hyperplasia following angioplasty.[279] Olprinone has shown in vivo efficacy in humans with mild, stable asthma when infused intravenously. It does not act additively or synergistically with aminophylline, a standard drug for the treatment of asthma.[280]

2.23.7.4 PDE4

PDE4 inhibitors have been widely investigated for the treatment of asthma and COPD because of their well-established anti-inflammatory activity in a variety of cellular and animal models. No PDE4 inhibitors have been approved for clinical use; however, cilomilast (Ariflo, **39**) has received an approvable letter from the US Food and Drugs Administration. The potential value of PDE4 inhibitors for these conditions is associated with the expression of the enzyme in the cell types involved, including neutrophils, eosinophils, mast cells, and macrophages. In both cellular and animal models, it is possible to correlate PDE4 inhibition with decreased production of inflammatory cytokines such as TNF-α.[46] Several PDE4 inhibitors are in clinical development[195] for a variety of conditions, as outlined in Section 2.23.6.4. More recently, PDE4 inhibitors have begun to be investigated for CNS diseases, including anxiety, depression, and cognition disorders.[80,195] At this point in time, one of the issues that continues to plague PDE4 inhibitor clinical development is the separation between the dose required for the desired therapeutic effect and the dose at which nausea and emesis become intolerable.

In addition to these applications, various PDE4 inhibitors have shown activity in animals for other conditions. For example, rolipram has shown the capacity to reduce aspirin-induced gastric mucosal injury in rats when administered before aspirin.[281] Rolipram in combination with prostacyclin analogs works synergistically to attenuate acute hypoxic vasoconstriction in the lungs.[282] It has been shown that PDE4B2 is overexpressed in human myometrium at the end of pregnancy, and inhibition of the enzyme in cell culture reduces the production of inflammatory cytokines. Thus, PDE4 may be a potential mechanism by which preterm labor can be avoided.[283] Because of the well-established utility of PDE4 inhibitors in inflammatory diseases, they have also been proposed for use in the treatment of inflammatory bowel disease.[284] It has been known for several years that high levels of cAMP are cytotoxic. Investigation of the potential for PDE4 inhibitors to act as potential anticancer agents has been reported. It has been shown, for example, that the general PDE inhibitors aminophylline and theophylline can reduce the cytotoxic does of cisplatin and gemcitabine in cell culture, when tested against lines derived from human lung, ovarian, and prostate tumors. The PDE inhibitors themselves are mildly cytotoxic when used alone. It appears that the PDE inhibitors exert their actions by inducing apoptosis in the tumor cells.[285] Rolipram has shown the ability to synergize with glucocorticoids, to cause apoptosis in human B cell chronic lymphocytic leukemia cells. There is no synergistic effect on T cell apoptosis. Based on this activity, clinical investigation of the potential for use in chronic lymphocytic leukemia has been proposed.[286]

2.23.7.5 PDE5

PDE5 inhibitors have been approved for use for the treatment of erectile dysfunction since 1998. More recently, sildenafil has also been approved for use in the treatment of pulmonary hypertension. Clinical studies are underway for the potential use of PDE5 inhibitors in benign prostatic hypertrophy, based on the high level of expression of the enzyme in the prostate, and the demonstration of the functional relevance of the enzyme in the tissue.[287] As noted in Section 2.23.4.5, PDE5 is widely expressed in a number of different smooth muscle tissue types. Because of this, PDE5 inhibitors have been investigated for functional applications in these tissues. For example, in rats, it has been shown that sildenafil dilates fetal and neonatal ductus. It was effective at lower doses in preterm rat pups, and at much higher doses in newborn rats.[288] In spontaneously hypertensive rats, a PDE5 inhibitor in combination with the angiotensin-converting enzyme inhibitor enalapril produces a synergistic effect on renal, mesenteric, and hindquarter vascular beds. When given alone, the PDE5 inhibitor produced a smaller effect on the renal vasculature.[289] In guinea pig models of airway disease, sildenafil has shown the ability to inhibit inflammation in response to both lipopolysaccharide- and ovalbumin-induced stimuli.[290]

PDE5 is also expressed in endothelial tissues. For this reason, PDE5 inhibitors have been studied in male smokers for their ability to reverse smoking-induced pulmonary endothelial dysfunction. At a dose of 50 mg, sildenafil completely eliminated the decrease in flow-mediated dilatation of the brachial artery that was acutely induced by smoking.[291]

In patients with congestive heart failure, sildenafil at a dose of 50 mg, in combination with 10 mg of ramipril, an angiotensin-converting enzyme inhibitor, increased flow-mediated dilation. Each drug alone was also able to improve this measure of endothelial function.[292] Additional clinical studies have shown a beneficial effect of sildenafil on pulmonary hemodynamics, diffusion capacity, oxygen uptake kinetics, and exercise ventilatory efficiency in patients with chronic heart failure. It has been proposed that these effects are due to the effect of the compound on endothelial function.[293]

As noted above, PDE4 inhibitors have been shown to reduce aspirin-induced gastric damage. Sildenafil has shown that it too can prevent indomethacin-induced gastropathy in rats. Acting via nitric oxide-dependent mechanisms, sildenafil decreased leukocyte adhesion, prevented the decrease in gastric blood flow induced by the nonsteroidal anti-inflammatory agent, and also reduced the direct gastric damage caused by indomethacin.[294] Delayed gastric emptying, associated with diabetes-induced gastroparesis, can be reversed by sildenafil in mice. Two different murine models of the disease, using nonobese genetically diabetic mice, and in mice where diabetes was caused by streptozotocin administration, were studied, and the compound was efficacious in both cases.[295] Animal studies suggested that sildenafil could limit damage to the heart, induced by prolonged ischemia. This was investigated in humans, and in a double-blind, placebo-controlled cross-over study, 50 mg of sildenafil was administered to healthy male volunteers. Flow-mediated dilatation of the radial artery was measured following 15 min of ischemia/reperfusion. It was found that patients given sildenafil were effectively protected against damage. More detailed study of the mechanism using a blocker of K_{ATP} channels, glibenclamide, showed that this ion channel was involved in the process, as glibenclamide reversed the beneficial effects of sildenafil.[296] This beneficial effect on cardiac tissue was shown in mice to contribute to attenuation of doxorubicin-induced left ventricular function, and to reduce cardiomyocyte apoptosis.[297] Very recently, clinical studies with tadalafil have been reported to show that it too improves endothelial function in men with increased cardiovascular risk.[298] In a murine model of cardiac hypertrophy, sildenafil deactivated multiple hypertrophy signaling pathways, and reversed pre-established hypertrophy induced by chronic pressure overload, and also restored chamber function to normal.[299]

In cases of nephrogenic diabetes insipidus, sildenafil has shown the ability to stimulate insertion of the AQP2 membrane into renal epithelial cells in rats. This suggests that PDE5 inhibitors may be useful in the treatment of this condition in humans.[300]

In addition to these potential peripheral applications, there is increasing evidence to suggest that PDE5 inhibitors may be useful in cognitive disorders as well. It is well known that nitric oxide–cGMP pathways are involved in memory and learning, and that elevated levels of cGMP in the hippocampus contribute to improved performance of rats in object recognition and passive avoidance learning. Addition of sildenafil to rat hippocampal brain slices elevated cGMP levels in this tissue, and in vivo the compound improved the memory performance of rats.[301,302] Sildenafil has also been shown to attenuate the learning impairment induced by cholinergic blockade in rats.[303]

Sildenafil and vardenafil induce caspase-dependent apoptosis in B-cell-derived chronic lymphocytic leukemia cells in vitro. In this study, the less potent PDE5 inhibitor zaprinast had no effect on apoptosis inducement.[304] Thus, similar to PDE4 inhibitors, PDE5 inhibitors may find a use in certain types of cancer chemotherapy. This effect has also been demonstrated in a human colon tumor cell line.[305]

2.23.7.6 PDE6

No beneficial therapeutic indications have been associated with inhibition of PDE6. However, as noted above, inhibition of this enzyme by selected PDE5 inhibitors leads to reversible changes in color perception.

2.23.7.7 PDE7

Inhibition of PDE7 was originally hypothesized to be useful for the treatment of immunological/inflammatory diseases because the enzyme is substantially upregulated in activated T cells. However, the report that PDE7A knockout mice have functionally competent T cells calls this potential application into question. At the time of this writing, the issue remains unresolved. There is no substantive evidence published on small-molecule inhibitors of the enzyme that beneficial actions are associated with this activity.

2.23.7.8 PDE8

The lack of potent PDE8 inhibitors has limited investigations of the role of enzyme inhibitors for therapeutic benefit. Perez-Torres and co-workers have shown that in the brains of patients with Alzheimer's disease, PDE8B mRNA levels in the hippocampus were significantly higher than in the brains of age-matched controls, and was higher still in patients with advanced disease.[306]

2.23.7.9 PDE9

As noted in Section 2.23.6.9, PDE9 may play a role in the regulation of glucose homeostasis in type 2 diabetes, based on the ability of a compound (**102**) to lower plasma glucose in an *ob/ob* mouse model of the disease at a dose of 10 mg kg^{-1}.

In addition, there is evidence to indicate that PDE9 plays a role in memory in mice and rats.[154]

2.23.7.10 PDE10

In the patents cited in Section 2.23.6.10, PDE10 inhibitors are claimed to be useful for the management of glucose homeostasis by stimulating insulin release from pancreatic β cells. It was possible, using compounds **104–106**, to demonstrate increases in insulin secretion from β cells derived from lean Sprague–Dawley rats in cell culture. Furthermore, PDE10 inhibitors may be useful in the treatment of CNS disorders such as schizophrenia, anxiety, and/or depression, based on the activity of papaverine in animal models, as noted in Section 2.23.6.10.

2.23.7.11 PDE11

No specific applications have been demonstrated for PDE11 inhibitors. However, as outlined in Section 2.23.6.11, it is possible the enzyme plays a role in sperm function and motility, and a recent report identified a distinct pattern of enzyme expression in a variety of normal and malignant human tissues.[307]

2.23.8 Genetic Diseases

Mutations in the PDE6 gene are known to be associated with retinitis pigmentosum and other retinal dystrophies.[126,308–310] These mutations are rare.

2.23.9 New Directions in Phosphodiesterase Enzyme Biology and Inhibition

From a therapeutic (enzyme inhibitor) perspective, advances in the field of PDEs will be dependent in part on the discovery of novel role(s) for the enzymes. With the sequence of the human genome in hand, the discovery of new enzymes in the class I family may occur, although no new PDEs have been reported since PDE11 was cloned in the late 1990s. From a purely enzymologic perspective, the enzymes about which the most information is available are those that have attracted the most interest as therapeutic targets: PDE4 and PDE5. Researchers in the field can expect the same to hold true as roles for the more recently discovered PDEs become better defined. The availability of high-resolution x-ray structures for many of the PDEs should prove useful for the discovery of novel inhibitors. Comparison of the structures with inhibitors bound may provide insights for the discovery of highly selective agents, as well as for molecules that can potently inhibit two or more PDEs. The existence of a range of distinct binding modes and subtle differences in the primary sequence in the active sites of the PDEs, along with the wide structural variety of the inhibitors shown in Section 2.23.6, should help to ensure that chemical diversity will allow for the discovery of a range of inhibitors with distinct chemical and physical properties.

Whenever there is more than one family member in a group of enzymes, the question of selectivity arises and must be addressed. This is necessary to first understand the role the enzyme plays in the potential target tissue and disease model, and second to allow researchers to clarify safety issues that may emerge during preclinical and clinical evaluation of a given compound. In the case of PDEs, the fact that within a family, different members may be responsible for distinct physiological responses (as noted, it is possible that this is true for the PDE4 family) creates yet another level of complexity that can complicate the interpretation of in vivo results. A PDE3B-selective compound has not yet been reported, making detailed evaluation of the potential of this enzyme as a target for obesity and type 2 diabetes problematic because of the potential safety issues associated with chronic inhibition of PDE3A.

One area that has not been extensively explored is the development of compounds that potently inhibit two or more PDEs. Given that these enzymes are widely distributed, but also often subcellularly localized and therefore probably tightly regulated, the availability of dual/multiple PDE inhibitors could allow better understanding of the role the enzyme(s) play in cell biology and the signaling pathways they influence. These efforts can go hand in hand with gene knockout technology to rapidly advance our understanding of the role PDEs play in signal transduction and regulation.

Among the currently known PDEs, interest in PDE2, PDE9, and PDE10 is slowly increasing because of reports that these enzymes may play a role in CNS disorders such as anxiety, depression, and cognition. PDE7 has been a target of interest for immunologic disorders. However, this interest has waned in recent years, perhaps because (as noted in

Section 2.23.4.7) there is some controversy surrounding the role the enzyme plays in activated T cells. The use of gene knockout and siRNA technology should allow researchers to better understand the role(s) an individual PDE can play in whole animals, and in disease models.

The possibility that two or more PDEs can play a role in modulating cell biology is particularly intriguing. In endocrine biology, for example, glucose homeostasis in response to certain stimuli appears to involve both PDE3 and PDE1. It is possible that other PDEs have similar 'cross-talk' roles. The observation that PDE4 plays a key role in arrestin-mediated signaling raises the possibility that other PDEs might be components of similar or other distinct multicomponent pathways.

The study of PDEs is a complex and very active field of research. Even for those enzymes that have been comparatively heavily studied, new information on structure, function, and inhibition is being generated on a regular basis. The field can look ahead to sustained interest and continued significant developments in the future.

References

1. Sutherland, E. W. *Science* **1972**, *177*, 401–408.
2. Corbin, J. D.; Francis, S. H. *J. Biol. Chem.* **1999**, *274*, 13729–13732.
3. Burgers, P. M.; Eckstein, F.; Hunneman, D. H.; Baraniak, J.; Kinas, R. W.; Lesiak, K.; Stec, W. J. *J. Biol. Chem.* **1979**, *254*, 9959–9961.
4. Jarvest, R. L.; Lowe, G.; Baraniak, J.; Stec, W. J. *Biochem. J.* **1982**, *203*, 461–470.
5. Charbonneau, H.; Beier, N.; Walsh, K. A.; Beavo, J. A. *Proc. Natl. Acad. Sci. USA* **1986**, *83*, 9308–9312.
6. Kincaid, R. L.; Stith-Coleman, I. E.; Vaughan, M. *J. Biol. Chem.* **1985**, *260*, 9009–9015.
7. Stroop, S. D.; Charbonneau, H.; Beavo, J. A. *J. Biol. Chem.* **1989**, *264*, 13718–13725.
8. Charbonneau, H.; Kumar, S.; Novack, J. P.; Blumenthal, D. K.; Griffin, P. R.; Shabanowitz, J.; Hunt, D. F.; Beavo, J. A.; Walsh, K. A. *Biochemistry* **1991**, *30*, 7931–7940.
9. Novack, J. P.; Charbonneau, H.; Bentley, J. K.; Blumenthal, D. K.; Walsh, K. A.; Beavo, J. A. *Biochemistry* **1991**, *30*, 7940–7947.
10. Lin, S. L.; Swinnen, J. V.; Conti, M. *J. Biol. Chem.* **1992**, *267*, 18929–18939.
11. Meacci, E.; Taira, M.; Moos, M.; Smith, C. J.; Mousesian, M. A.; Degerman, E.; Belfrage, P.; Manganiello, V. *Proc. Natl. Acad. Sci. USA* **1992**, *89*, 3721–3725.
12. Cheung, P. P.; Xu, H.; McLaughlin, M. M.; Ghazaleh, F. A.; Livi, G. P.; Colman, R. W. *Blood* **1996**, *88*, 1321–1329.
13. Jacobitz, S.; McLaughlin, M. M.; Livi, G. P.; Burman, M.; Torphy, T. J. *Mol. Pharmacol.* **1996**, *50*, 891–896.
14. Turko, I. V.; Francis, S. H.; Corbin, J. D. *Biochemistry* **1998**, *37*, 4200–4205.
15. Conti, M. *Mol. Endocrinol.* **2000**, *14*, 1317–1327.
16. Wang, J. H.; Sharma, R. K.; Mooibroek, M. J. Calmodulin Stimulated Cyclic Nucleotide Phosphodiesterases. In *Cyclic Nucleotide Phosphodiesterases: Structure, Function, and Drug Action*; Beavo, J., Houslay, M., Eds.; John Wiley: New York, 1990; pp 19–60.
17. Kincaid, R. L.; Balaban, C. D.; Billingsley, M. L. *Adv. Second Messenger Phosphoprotein Res.* **1992**, *25*, 111–122.
18. Gupta, R.; Kumar, G.; Sunil Kumar, R. *Methods Find. Exp. Clin. Pharmacol.* **2005**, *27*, 101–118.
19. Boekhoff, I.; Kroner, C.; Breer, H. *Cell. Signal.* **1996**, *8*, 167–171.
20. Rybalkin, S. D.; Bornfeldt, K. E.; Sonneburg, W. K.; Rybalkina, I. G.; Kwak, K. S.; Hanson, K.; Krebs, E. G.; Beavo, J. A. *J. Clin. Invest.* **1997**, *100*, 2611–2621.
21. Rybalkin, S. D.; Rybalkina, I.; Beavo, J. A.; Bornfeldt, K. E. *Circ. Res.* **2002**, *90*, 151–157.
22. Han, P.; Werber, J.; Surana, M.; Fleischer, N.; Michaeli, T. *J. Biol. Chem.* **1997**, *274*, 22337–22344.
23. Goraya, T. A.; Cooper, D. M. F. *Cell. Signal.* **2005**, *17*, 789–797.
24. Erneux, C.; Couchie, D.; Dumont, J. E.; Baraniak, J.; Stec, W. J.; Abbad, E. G.; Petridis, G.; Jasturoff, B. *Eur. J. Biochem.* **1981**, *115*, 503–510.
25. Martins, T. J.; Mumby, M. C.; Beavo, J. A. *J. Biol. Chem.* **1982**, *257*, 1973–1979.
26. Yamamoto, T.; Manganiello, V. C.; Vaughan, M. *J. Biol. Chem.* **1983**, *258*, 12526–12533.
27. Yamamoto, T.; Yamamoto, S.; Osborne, J. C., Jr.; Manganiello, V. C.; Vaughan, M.; Hidaka, H. *J. Biol. Chem.*, *258*, *258*, 14173–14177.
28. MacFarland, R. T.; Zelus, B. D.; Beavo, J. A. *J. Biol Chem.* **1991**, *266*, 136–142.
29. Dickinson, N. T.; Jang, E. K.; Haslam, R. J. *Biochem. J.* **1997**, *323*, 371–377.
30. Juilfs, D. M.; Fulle, H. J.; Zhao, A. Z.; Houslay, M. D.; Garbers, D. L.; Beavo, J. A. *Proc. Natl. Acad. Sci. USA* **1997**, *97*, 3388–3395.
31. Boess, F. G.; Hendrix, M.; van der Staay, F.-J.; Erb, C.; Schreiber, R.; van Staveren, W.; de Vente, J.; Prickaerts, J.; Blokland, A.; Koenig, G. *Neuropharmacology* **2004**, *47*, 1081–1092.
32. Degerman, E.; Belfrage, P.; Newman, A. H.; Rice, K. C.; Manganiello, V. C. *J. Biol. Chem.* **1987**, *262*, 5797–5807.
33. Rascon, A.; Lindgren, S.; Stavenow, L.; Belfrage, P.; Andersson, K. E.; Manganiello, V. C.; Degerman, E. *Biochim. Biophys. Acta* **1992**, *1134*, 149–156.
34. Beavo, J. A. *Physiol. Rev.* **1995**, *75*, 725–748.
35. Grant, P. G.; Colman, R. W. *Biochemistry* **1984**, *23*, 1801–1807.
36. MacPhee, C. H.; Harrison, S. H.; Beavo, J. A. *Proc. Natl. Acad. Sci. USA* **1986**, *83*, 6660–6663.
37. Ding, B.; Abe, J.; Wei, H.; Huang, Q.; Walsh, R. A.; Molina, C. A.; Zhao, A.; Sadoshima, J.; Blaxall, B. C.; Berk, B. C. et al. *Circulation* **2005**, *111*, 2469–2475.
38. Scapin, G.; Patel, S. B.; Chung, C.; Varnerin, J. P.; Edmondson, S. D.; Mastracchio, A.; Parmee, E. A.; Singh, S. B.; Becker, J. W.; Van der Ploeg, L. H. T. et al. *Biochemistry* **2004**, *43*, 6091–6100.
39. Beebe, S. J.; Beasley, L. A.; Corbin, J. D. *Methods Enzymol.* **1988**, *159*, 531–540.
40. Gettys, T. W.; Blackmore, P. F.; Corbin, J. D. *Am. J. Physiol.* **1988**, *254*, E449–E453.
41. Synder, P. B.; Esselstyn, J. M.; Loughney, K.; Wolda, S. L.; Florio, V. A. *J. Lipid. Res.* **2005**, *46*, 494–503.
42. Lester, L. B.; Langeberg, L. K.; Scott, J. D. *Proc. Natl. Acad. Sci. USA* **1997**, *94*, 14942–14947.
43. Fraser, I. D.; Tavalin, S. J.; Lester, L. B.; Langeberg, L. K.; Westphal, A. M.; Dean, R. A.; Marrion, N. V.; Scott, J. D. *EMBO J.* **1998**, *17*, 2261–2272.

44. Zhao, A. Z.; Bornfeldt, K. E.; Beavo, J. A. *J. Clin. Invest.* **1998**, *102*, 869–873.
45. Zhao, A. Z.; Zhao, H.; Teague, J.; Fujimoto, W.; Beavo, J. A. *Proc. Natl. Acad. Sci. USA* **1997**, *94*, 3223–3228.
46. Houslay, M. D.; Sullivan, M.; Bolger, G. B. *Adv. Pharmacol.* **1998**, Aug., 225–342.
47. Conti, M.; Jin, S.-L. C. *Nucl. Acid. Res. Mol. Biol.* **1999**, *63*, 1–38.
48. Houslay, M. D.; Scotland, G.; Pooley, L.; Spence, S.; Wilkinson, I.; McCallum, F.; Julien, P.; Rena, N. G.; Michie, A. M.; Erdogan, S. *Biochem. Soc. Trans.* **1995**, *23*, 393–398.
49. Bolger, G. B.; Erdogan, S.; Jones, R. E.; Loughney, K.; Scotland, G.; Hoffman, R.; Wilkinson, I.; Farrell, C.; Houslay, M. D. *Biochem. J.* **1997**, *328*, 539–548.
50. Jin, S. L.; Bushnik, T.; Lan, L.; Conti, M.; Swinnen, J. V. *J. Biol. Chem.* **1998**, *273*, 19672–19678.
51. Grange, M.; Picq, M.; Prigent, A. F.; Lagarde, M.; Nemoz, G. *Cell Biochem. Biophys.* **1998**, *29*, 1–17.
52. Shakur, Y.; Pryde, J. G.; Houslay, M. D. *Biochem. J.* **1993**, *292*, 677–686.
53. Scotland, G.; Houslay, M. D. *Biochem. J.* **1995**, *308*, 673–681.
54. McPhee, I.; Pooley, L.; Lobban, M.; Bolger, G.; Houslay, M. D. *Biochem. J.* **1995**, *310*, 964–974.
55. Smith, K. J.; Scotland, G.; Beattie, J.; Trayer, I. P.; Houslay, M. D. *J. Biol. Chem.* **1996**, *271*, 16703–16711.
56. Bolger, G. B.; Erdogan, S.; Jones, R. E.; Loughney, K.; Scotland, G.; Hoffman, R.; Wilkinson, I.; Farrell, C.; Houslay, M. D. *Biochem. J.* **1997**, *328*, 539–548.
57. Huston, E.; Lumb, S.; Russell, A.; Catterall, C.; Ross, A. H.; Steele, M. R.; Bolger, G. B.; Perry, M. J.; Owens, R. J.; Houslay, M. D. *Biochem. J.* **1997**, *328*, 549–558.
58. O'Connell, J. C.; McCallum, J. F.; McPhee, I.; Wakefield, J.; Houslay, E. S.; Wishart, W.; Bolger, G.; Frame, M.; Houslay, M. D. *Biochem. J.* **1996**, *318*, 255–261.
59. Houslay, M. D.; Scotland, G.; Erdogan, S.; Huston, E.; Mackenzie, S.; McCallum, J. F.; McPhee, I.; Pooley, L.; Rena, G.; Ross, A. et al. *Biochem. Soc. Trans.* **1997**, *25*, 374–381.
60. Souness, J. E.; Rao, S. *Cell. Signal.* **1997**, *9*, 227–236.
61. Rocque, W. J.; Holmes, W. D.; Patel, I. R.; Dougherty, R. W.; Ittoop, O.; Overton, L.; Hoffman, C. R.; Wisely, G. B.; Willard, D. H.; Luther, M. A. *Protein Expr. Purif.* **1997**, *9*, 191–202.
62. Bardelle, C.; Smales, C.; Ito, M.; Nomoto, K.; Wong, E. Y. M.; Kato, H.; Saeki, T.; Staddon, J. M. *Anal. Biochem.* **1999**, *275*, 148–155.
63. Allen, R. W.; Merriman, M. W.; Perry, M. J.; Owens, R. J. *Biochem. Pharmacol.* **1999**, *57*, 1375–1382.
64. Souness, J. E.; Griffin, M.; Maslen, C.; Ebsworth, K.; Scott, L. C.; Pollock, K.; Palfreyman, M. N.; Karlsson, J.-A. *Br. J. Pharmacol.* **1996**, *118*, 649–658.
65. Bolger, G.; Michaeli, T.; Martins, T.; St. John, T.; Steiner, B.; Rodgers, L.; Riggs, M.; Wigler, J.; Ferguson, K. *Mol. Cell. Biol.* **1993**, *13*, 6558–6571.
66. Bolger, G. *Cell. Signal.* **1994**, *6*, 851–859.
67. Owens, R. J.; Catterall, C.; Batty, D.; Jappy, J.; Russell, A.; Smith, B.; O'Connell, J.; Perry, M. J. *Biochem. J.* **1997**, *326*, 53–60.
68. Hoffmann, R.; Wilkinson, I. R.; McCallum, J. F.; Engels, P.; Houslay, M. D. *Biochem. J.* **1998**, *333*, 139–149.
69. Swinnen, J. V.; Joseph, D. R.; Conti, M. *Proc. Natl. Acad. Sci. USA* **1989**, *86*, 8197–8201.
70. Swinnen, J. V.; Tsikalas, K. E.; Conti, M. *J. Biol. Chem.* **1996**, *266*, 18370–18377.
71. Jin, S. L. C.; Swinnen, J. V.; Conti, M. *J. Biol. Chem.* **1992**, *267*, 18929–18939.
72. Jacobitz, S.; Ryan, M. D.; McLaughlin, M. M.; Livi, G. P.; DeWolf, W. E.; Torphy, T. J. *Mol. Pharmacol.* **1997**, *50*, 891–899.
73. Pillai, R.; Staub, S. F.; Colicelli, J. *J. Biol. Chem.* **1994**, *269*, 30676–30681.
74. Lefkowitz, R. J.; Shenoy, S. K. *Science* **2005**, *308*, 512–517.
75. Perry, S. J.; Baillie, G. S.; Kohout, T. A.; McPhee, I.; Magiera, M. M.; Ang, K. L.; Miller, W. E.; McLean, A. J.; Conti, M.; Houslay, M. D. et al. *Science* **2002**, *298*, 834–836.
76. Lynch, M. J.; Baillie, G. S.; Mohamed, A.; Li, X.; Maisonneuve, C.; Klussmann, E.; van Heeke, G.; Houslay, M. D. *J. Biol. Chem.* **2005**, *280*, 33178–33189.
77. Arp, J.; Kirchof, M. G.; Baroja, M. L.; Nazarian, S. H.; Chau, T. A.; Strathdee, C. A.; Ball, E. H.; Madrenas, J. *Mol. Cell Biol.* **2003**, *23*, 8042–8057.
78. Eker, P.; Holm, P. K.; Van Deurs, B.; Sandvig, K. *J. Biol. Chem.* **1994**, *269*, 18607–18615.
79. Engels, P.; AbdelAl, S.; Hulley, P.; Lubbert, H. *J. Neurosci. Res.* **1995**, *41*, 169–178.
80. Rose, G. M.; Hopper, A.; De Vivo, M.; Tehim, A. *Curr. Pharm. Des.* **2005**, *11*, 3329–3334.
81. Perez-Torres, S.; Miro, X.; Palacios, J. M.; Cortes, R.; Puigdomenech, P.; Mengod, G. *J. Chem. Neuroanat.* **2000**, *20*, 349–374.
82. O'Donnell, J. M.; Zhang, H.-T. *Trends Pharmcol. Sci.* **2003**, *25*, 158–163.
83. Sarter, M.; Hagan, J.; Dudchenko, P. *Psychopharmacology (Berlin)* **1992**, *107*, 461–473.
84. Soto, F. J.; Hanania, N. A. *Curr. Opin. Pulm. Med.* **2005**, *11*, 129–134.
85. Liu, H.; Maurice, D. H. *J. Biol. Chem.* **1999**, *274*, 10557–10565.
86. Palmer, D.; Maurice, D. H. *Mol. Pharmacol.* **2000**, *58*, 247–252.
87. Tilley, D. G.; Maurice, D. H. *Mol. Pharmacol.* **2005**, *68*, 596–605.
88. Takami, M.; Cho, E. S.; Lee, S. Y.; Kamijo, R.; Yim, M. *FEBS Lett.* **2005**, *579*, 832–838.
89. Wakabayshi, S.; Tsutsumimoto, T.; Kawasaki, S.; Kinoshita, T.; Horiuchi, H.; Takaoka, K. *J. Bone Miner. Res.* **2002**, *17*, 249–256.
90. Bloom, T. J. *Can. J. Physiol. Pharmacol.* **2002**, *80*, 1132–1135.
91. Tenor, H.; Hedbom, E.; Häuselmann, H.-J.; Schudt, C.; Hatzelmann, A. *Br. J. Pharmacol.* **2002**, *135*, 609–618.
92. Cherry, J. A.; Pho, V. *Chem. Senses* **2002**, *27*, 643–652.
93. Kotera, J.; Fuijshige, K.; Akatsuka, H.; Imai, Y.; Yanaka, H.; Omori, K. *J. Biol. Chem.* **1998**, *273*, 26982–26990.
94. Lin, C.-S.; Lau, A.; Tu, R.; Lue, T. F. *Biochem. Biophys. Res. Commun.* **2000**, *268*, 628–635.
95. Francis, S. H.; Corbin, J. D. *Methods Enzymol.* **1988**, *159*, 722–729.
96. Lin, C.-S.; Lau, A.; Tu, R.; Lue, T. F. *Biochem. Biophys. Res. Commun.* **2000**, *268*, 628–635.
97. Wang, P.; Wu, P.; Myers, J. G.; Stamford, A.; Egan, R. W.; Billah, M. M. *Life Sci.* **2001**, *68*, 1977–1987.
98. Francis, S. H.; Turko, I. V.; Corbin, J. D. *Prog. Nucl. Acid Res. Mol. Biol.* **2001**, *65*, 1–52.
99. Shimizu-Albergine, M.; Rybalkin, S. D.; Rybalkina, I. G.; Feil, R.; Wolfsgruber, W.; Hoffmann, F.; Beavo, J. A. *J. Neurosci.* **2003**, *23*, 6452–6459.
100. Ückert, S.; Küthe, A.; Jonas, U.; Stief, C. G. *J. Urol.* **2001**, *166*, 2484–2490.
101. Corbin, J. D.; Francis, S. H. *J. Biol. Chem.* **1999**, *274*, 13729–13732.

102. Okada, D.; Asakawa, S. *Biochemistry* **2002**, *41*, 9672–9679.
103. Corbin, J. D.; Francis, S. H. *J. Biol. Chem.* **1999**, *274*, 13729–13732.
104. Thomas, M. K.; Francis, S. H.; Corbin, J. D. *J. Biol. Chem.* **1990**, *265*, 14971–14978.
105. Turko, I. V.; Francis, S. H.; Corbin, J. D. *Biochem. J.* **1998**, *329*, 505–510.
106. Wyatt, T. A.; Naftilan, A. J.; Francis, S. H.; Corbin, J. D. *Am. J. Physiol. Heart Circ. Physiol.* **1998**, *274*, H448–H455.
107. Turko, I. V.; Francis, S. H.; Corbin, J. D. *J. Biol. Chem.* **1999**, *274*, 29038–29041.
108. Francis, S. H.; Bessay, E. P.; Kotera, J.; Grimes, K. A.; Liu, L.; Thompson, W. J.; Corbin, J. D. *J. Biol. Chem.* **2002**, *277*, 47581–47587.
109. Blount, M. A.; Beasley, A.; Zoraghi, R.; Sekhar, K. R.; Bessay, E. P.; Francis, S. H.; Corbin, J. D. *Mol. Pharmacol.* **2004**, *66*, 144–152.
110. Gopal, V. K.; Francis, S. H.; Corbin, J. D. *Eur. J. Biochem.* **2001**, *268*, 3304–3312.
111. Zoraghi, R.; Corbin, J. D.; Francis, S. H. *Mol. Pharmacol.* **2004**, *65*, 267–287.
112. Zoraghi, R.; Bessay, E. P.; Francis, S. H.; Corbin, J. D. *J. Biol. Chem.* **2005**, *280*, 12051–12063.
113. Francis, S. H.; Turko, I. V.; Grimes, K. A.; Corbin, J. D. *Biochemistry* **2000**, *39*, 9591–9596.
114. Champion, H. C.; Bivalacqua, T. J.; Takimoto, E.; Kass, D. A.; Burnett, A. L. *Proc. Natl. Acad. Sci. USA* **2005**, *102*, 1661–1666.
115. Ni, X.-P; Safai, M.; Rishi, R.; Baylis, C.; Humphreys, M. H. *J. Am. Soc. Nephrol.* **2004**, *15*, 1254–1260.
116. Bing, Z.; Strada, S.; Stevens, T. *Am. J. Physiol.* **2005**, *289*, L196–L206.
117. Kameni Tcheudji, J. F.; Lebeau, L.; Virmaux, N.; Maftei, C. G.; Cote, R. H.; Lugnier, C.; Schultz, P. *J. Mol. Biol.* **2001**, *310*, 781–791.
118. Cote, R. H. Structure, Function and Regulation of Photoreceptor Phosphodiesterase (PDE6). In *Handbook of Cell Signaling*; Bradshaw, R., Dennis, E. A., Eds.; Elsevier: Amsterdam, The Netherlands, 2003; Vol. 2, pp 453–457.
119. D'Amours, M. R.; Granovsky, A. E.; Artemyev, N. O.; Cote, R. H. *Mol. Pharmacol.* **1999**, *55*, 508–514.
120. Kameni Tcheudji, J. F.; Lebeau, L.; Virmaux, N.; Maftei, C. G.; Cote, R. H.; Lugnier, C.; Schultz, P. *J. Mol. Biol.* **2001**, *310*, 781–791.
121. McAllister-Lucas, L.; Sonnenburg, W. K.; Kadlecek, A.; Seger, D.; Le Trong, H.; Colbran, J. L.; Thomas, M. K.; Walsh, K. A.; Francis, S. H.; Corbin, J. D. et al. *J. Biol. Chem.* **1993**, *268*, 22863–22873.
122. D'Amours, M. R.; Cote, R. H. *Biochem. J.* **1999**, *340*, 863–869.
123. Mou, H.; Grazio, H. J., III.; Cook, T. A.; Beavo, J. A.; Cote, R. H. *J. Biol. Chem.* **1999**, *274*, 18813–18820.
124. Thomas, M. K.; Francis, S. H.; Corbin, J. D. *J. Biol. Chem.* **1990**, *265*, 14964–14970.
125. Arshavsky, V. Y.; Lamb, T. D.; Pugh, E. N., Jr. *Annu. Rev. Physiol.* **2002**, *64*, 153–187.
126. Muradov, K. G.; Granovsky, A. E.; Artemyev, N. O. *Biochemistry* **2003**, *42*, 3305–3310.
127. Zhang, J.; Kuvelkar, R.; Wu, P.; Egan, R. W.; Billah, M. M.; Wang, P. *Biochem. Pharmacol.* **2004**, *68*, 867–873.
128. Bloom, T. J.; Beavo, J. A. *Proc. Natl. Acad. Sci. USA* **1996**, *93*, 14188–14192.
129. Han, P.; Zhu, X.; Michaeli, T. *J. Biol. Chem.* **1997**, *272*, 16152–16157.
130. Li, L.; Yee, C.; Beavo, J. A. *Science* **1999**, *283*, 848–851.
131. Glavas, N. A.; Ostenson, C.; Schaefer, J. B.; Vasta, V.; Beavo, J. A. *Proc. Natl. Acad. Sci. USA* **2001**, *98*, 6319–6324.
132. Yang, G.; McIntyre, K. W.; Townsend, R. M.; Shen, H. H.; Pitts, W. J.; Dodd, J. H.; Nadler, S. G.; McKinnon, M.; Watson, A. J. *J. Immunol.* **2003**, *171*, 6414–6420.
133. Smith, S. J.; Cieslinksi, L. B.; Newton, R.; Donnelly, L. E.; Fenwick, P. S.; Nicholson, A. G.; Barnes, P. J.; Barnette, M. S.; Giembycz, M. A. *Mol. Pharmacol.* **2004**, *66*, 1679–1689.
134. Nakata, A.; Ogawa, K.; Sasaki, T.; Koyama, N.; Wada, K.; Kotera, J.; Kikkawa, H.; Omori, K.; Kaminuma, O. *Clin. Exp. Immunol.* **2002**, *128*, 460–466.
135. Lee, R.; Wolda, S.; Moon, E.; Esselstyn, J.; Hertel, C.; Lerner, A. *Cell. Signal.* **2002**, *14*, 277–284.
136. Miro, X.; Perez-Torres, S.; Palacios, J. M.; Puigdomenech, P.; Mengod, G. *Synapse* **2001**, *40*, 201–214.
137. Hetman, J. M.; Soderling, S. H.; Glavas, N. A.; Beavo, J. A. *Proc. Natl. Acad. Sci. USA* **2000**, *97*, 472–473.
138. Sasaki, T.; Kotera, J.; Yuasa, K.; Omori, K. *Biochem. Biophys. Res. Commun.* **2000**, *271*, 575–583.
139. Gardner, C.; Robas, N.; Cawkill, D.; Fidock, M. *Biochem. Biophys. Res. Commun.* **2000**, *272*, 186–192.
140. Hetman, J. M.; Soderling, S. H.; Glavas, N. A.; Beavo, J. A. *Proc. Natl. Acad. Sci. USA* **2000**, *97*, 472–473.
141. Sasaki, T.; Kotera, J.; Yuasa, K.; Omori, K. *Biochem. Biophys. Res. Commun.* **2000**, *271*, 575–583.
142. Wang, P.; Wu, P.; Egan, R. W.; Billah, M. M. *Gene* **2001**, *280*, 183–194.
143. Fisher, D. A.; Smith, J. F.; Pillar, J. S.; St Denis, S. H.; Cheng, J. B. *Biochem. Biophys. Res. Commun.* **1998**, *246*, 570–577.
144. Soderling, S. H.; Bayuga, S. J.; Beavo, J. A. *Proc. Natl. Acad. Sci. USA* **1998**, *95*, 8991–8996.
145. Fisher, D. A.; Smith, J. F.; Pillar, J. S.; St Denis, S. H.; Cheng, J. B. *Biochem. Biophys. Res. Commun.* **1998**, *246*, 570–577.
146. Soderling, S. H.; Beavo, J. A. *Curr. Opin. Cell Biol.* **2000**, *12*, 174–179.
147. Hayashi, M.; Matsushima, K.; Ohashi, H.; Tsunoda, H.; Murase, M.; Kawarada, Y.; Tanaka, T. *Biochem. Biophys. Res. Commun.* **1998**, *250*, 751–756.
148. Gamanuma, M.; Yuasa, K.; Sasaki, T.; Sakurai, N.; Kotera, J.; Omori, K. *Cell. Signal.* **2003**, *15*, 565–574.
149. Fisher, D. A.; Smith, J. F.; Pillar, J. S.; St Denis, S. H.; Cheng, J. B. *J. Biol. Chem.* **1998**, *273*, 15559–15564.
150. Soderling, S. H.; Bayuga, S. J.; Beavo, J. A. *J. Biol. Chem.* **1998**, *273*, 15553–15558.
151. Guipponi, M.; Scott, H. S.; Kudoh, J.; Kawasaki, K.; Shibuya, K.; Shintani, A.; Asakawa, S.; Chen, H.; Lalioti, M. D.; Rossier, C. et al. *Hum. Genet.* **1998**, *103*, 386–392.
152. Fisher, D. A.; Smith, J. F.; Pillar, J. S.; St Denis, S. H.; Cheng, J. B. *J. Biol. Chem.* **1998**, *273*, 15559–15564.
153. Soderling, S. H.; Bayuga, S. J.; Beavo, J. A. *J. Biol. Chem.* **1998**, *273*, 15553–15558.
154. Hendrix, M. *PDE Inhibitors for AD: Influencing Learning and Memory via the cGMP-PDE Pathway.* 2nd Annual Phosphodiesterases in Drug Discovery and Development, Philadelphia, PA, November 8–9, 2004.
155. Fujishige, K.; Kotera, J.; Michibata, H.; Yuasa, K.; Takebayashi, S.; Okumura, K.; Omori, K. *J. Biol. Chem.* **1999**, *274*, 18438–18445.
156. Kotera, J.; Fujishige, K.; Yuasa, K.; Omori, K. *Biochem. Biophys. Res. Commun.* **1999**, *261*, 551–557.
157. Loughney, K.; Synder, P. B.; Uher, L.; Rosman, G. J.; Ferguson, V. A. *Gene* **1999**, *234*, 109–117.
158. Soderling, S. H.; Bayuga, S. J.; Beavo, J. A. *Proc. Natl. Acad. Sci. USA* **1999**, *96*, 7071–7076.
159. Kotera, J.; Sasaki, T.; Kobayashi, T.; Fujishige, K.; Yamashita, Y.; Omori, K. *J. Biol. Chem.* **2004**, *279*, 4366–4375.
160. Fujishige, K.; Kotera, J.; Omori, K. *Eur. J. Biochem.* **1999**, *266*, 1118–1127.
161. Hebb, A. L.; Robertson, H. A.; Denovan-Wright, E. M. *Neuroscience* **2004**, *123*, 967–981.
162. Miyamoto, S.; Duncan, G. E.; Marx, C. E.; Lieberman, J. A. *Mol. Psychiatry* **2005**, *10*, 79–104.
163. Rodefer, J. S.; Murphy, E. R.; Baxter, M. G. *Eur. J. Neurosci.* **2005**, *21*, 1070–1076.
164. Kleinman, R. J. *The Role of PDE10A in Modulation of Basal Ganglia Function.* 2nd Annual Phosphodiesterases in Drug Discovery and Development, Philadelphia, PA, November 8–9, 2004.

165. Fawcett, L.; Baxendale, R.; Stacey, P.; McGrouther, C.; Harrow, I.; Soderling, S.; Hetman, J.; Beavo, J. A.; Phillips, S. C. *Proc. Natl. Acad. Sci. USA* **2000**, *97*, 3702–3707.
166. Yuasa, K.; Kotera, J.; Fujishige, K.; Michibata, H.; Sasaki, T.; Omori, K. *J. Biol. Chem.* **2000**, *275*, 31469–31479.
167. Hetman, J. M.; Robas, N.; Baxendale, R.; Fidock, M.; Phillips, S. C.; Soderling, S. H.; Beavo, J. A. *Proc. Natl. Acad. Sci. USA* **2000**, *97*, 12891–12895.
168. Yuasa, K.; Ohgaru, T.; Asahina, M.; Omori, K. *Eur. J. Biochem.* **2001**, *268*, 4440–4448.
169. Loughney, K.; Taylor, J.; Florio, V. A. *Int. J. Impot. Res.* **2005**, *17*, 320–325.
170. Wayman, C.; Phillips, S.; Lunny, C.; Webb, T.; Fawcett, L.; Baxendale, R.; Burgess, G. *Int. J. Impot. Res.* **2005**, *17*, 216–223.
171. Francis, S. H. *Int. J. Impot. Res.* **2005**, *17*, 5–9.
172. D'Andrea, M. R.; Qiu, Y.; Haynes-Johnson, D.; Bhattacharjee, S.; Kraft, P.; Lundeen, S. *J. Histochem. Cytochem.* **2005**, *53*, 895–903.
173. Xiang, J.-N.; Lee, J. C. *Annu. Rep. Med. Chem.* **1999**, *34*, 199–208.
174. Ahn, H.-S.; Bercovici, A.; Boykow, G.; Bronnenkant, A.; Chackalamannil, S.; Chow, J.; Cleven, R.; Cook, J.; Czarniecki, M.; Domalski, C. et al. *J. Med. Chem.* **1997**, *40*, 2110–2196.
175. Xia, Y.; Chackalamannil, S.; Czarniecki, M.; Tsai, H.; Vaccaro, H.; Cleven, R.; Cook, J.; Fawzi, A.; Watkins, R.; Zhang, H. *J. Med. Chem.* **1997**, *40*, 4372–4377.
176. Vemulapalli, S.; Watkins, R. W.; Chintala, M.; Davis, H.; Ahn, H. S.; Fawzi, A.; Tulshian, D.; Chiu, P.; Chatterjee, M.; Lin, C. C. et al. *J. Cardiovasc. Pharmacol.* **1996**, *28*, 862–869.
177. Bell, A. S.; Terrett, N. K. Preparation of Pyrazolo[4,3-*d*]pyrimidine Derivatives as Inhibitors of Phosphodiesterase 1 and Pharmaceutical Compositions Containing Them. European Patent 911333, April 10, 2002.
178. Weishaar, R. E.; Cain, M. H.; Bristol, J. A. *J. Med. Chem.* **1985**, *28*, 537–545.
179. Dan, A.; Shiyama, T.; Yamazaki, K.; Kusunose, N.; Fujita, K.; Sato, H.; Matsui, K.; Kitano, M. *Bioorg. Med. Chem. Lett.* **2005**, *15*, 4085–4090.
180. Hlasta, D. J.; Bode, D. C.; Court, J.; Desai, R. C.; Pagani, E. D.; Silver, P. J. *Bioorg. Med. Chem. Lett.* **1997**, *7*, 89–94.
181. Domek-Łopacińska, K.; Strosznajder, J. B. *J. Physiol. Pharmacol.* **2005**, *56*, 15–34.
182. Abbott, B. M.; Thompson, P. E. *Bioorg. Med. Chem. Lett.* **2004**, *14*, 2847–2851.
183. Bayer, T. A.; Chambers, R. J.; Lam, K.; Li, M.; Morrell, I.; Thompson, D. D. Preparation of Pyrido[2,3-*d*]pyrimidine-2,4-diamines as PDE2 Inhibitors. International Patent WO 2005061497, July 7, 2005.
184. Chambers, R. J.; Lam, K. T. Preparation of Oxindole Derivatives and their use as Phosphodiesterase Type 2 Inhibitors. International Patent WO 2005041957, May 12, 2005.
185. Bourguignon, J.-J.; Lugnier, C.; Abaraghaz, M.; Lagouge, Y.; Wagner, P.; Mondadori, C.; Macher, J.-P.; Schultz, D.; Raboisson, P. Preparation of Benzodiazepinones as Cyclic Nucleotide Phosphodiesterase, in Particular, PDE2 Inhibitors, for Treating Central and Peripheral Nervous System Disorders. International Patent WO 2004041258, September 23, 2004.
186. Iffland, A.; Kohls, D.; Low, S.; Luan, J.; Zhang, Y.; Kothe, M.; Cao, Q.; Kamath, A. V.; Ding, Y.-H.; Ellenberger, T. *Biochemistry* **2005**, *44*, 8312–8325.
187. van Staveren, W. C. G.; Markerink-van Ittersum, M.; Steinbusch, H. W. M.; de Vente, J. *Brain Res.* **2001**, *888*, 275–286.
187a. Pauvert, O.; Salvail, D.; Rousseau, E.; Lugnier, C.; Marthan, R.; Savineau, J. P. *Biochem. Pharmacol.* **2002**, *63*, 1763–1772.
188. Fossa, P.; Boggia, R.; Mosti, L. *J. Comput. Aided Mol. Des.* **1998**, *12*, 361–372.
189. Hidaka, H.; Endo, T.; Ito, H. *Trends Pharmacol. Sci.* **1984**, *5*, 237–239.
190. Kaufman, R. F.; Crowe, V. G.; Utterback, B. G.; Robertson, D. W. *Mol. Pharmacol.* **1986**, *30*, 609–616.
191. Bethke, T.; Eschenhagen, T.; Klimkiewicz, A.; Kohl, C.; von der Leyen, H.; Mehl, H.; Mende, U.; Meyer, W.; Neumann, J.; Rosswag, S. et al. *Arzn. Forschung* **1992**, *42*, 437–445.
192. Edmondson, S. D.; Mastracchio, A.; He, J.; Chung, C. C.; Forrest, M. J.; Hofsess, S.; MacIntyre, E.; Metzger, J.; O'Connor, N.; Patel, K. et al. *Bioorg. Med. Chem. Lett.* **2003**, *13*, 3983–3987.
193. Fossa, P.; Giordanetto, F.; Menozzi, G.; Mosti, L. *Quant. Struct.–Act. Relat.* **2002**, *21*, 267–275.
194. Scapin, G.; Patel, S. B.; Chung, C.; Varnerin, J. P.; Edmondson, S. D.; Mastracchio, A.; Parmee, E. R.; Singh, S. B.; Becker, J. W.; Van der Ploeg, L. H. T. et al. *Biochemistry* **2004**, *43*, 6091–6100.
195. Odingo, J. O. *Exp. Opin. Ther. Patents* **2005**, *15*, 773–787.
196. Draheim, R.; Egerland, U.; Rundfeldt, C. *J. Pharmacol. Exp. Ther.* **2004**, *308*, 555–563.
197. Aoki, M.; Kobayashi, M.; Ishikawa, J.; Saita, Y.; Terai, Y.; Takayama, K.; Miyata, K.; Yamada, T. *J. Pharmacol. Exp. Ther.* **2000**, *295*, 255–260.
198. Billah, M.; Cooper, N.; Cuss, F.; Davenport, R. J.; Dyke, H. J.; Egan, R.; Ganguly, A.; Gowers, L.; Hannah, D. R.; Haughan, A. F. et al. *Bioorg. Med. Chem. Lett.* **2002**, *12*, 1621–1623.
199. Guay, D.; Hamel, P.; Blouin, M.; Brideau, C.; Chan, C. C.; Chauret, N.; Ducharme, Y.; Huang, Z.; Girard, M.; Jones, T. R. et al. *Bioorg. Med. Chem. Lett.* **2002**, *12*, 1457–1461.
200. Friesen, R. W.; Ducharme, Y.; Ball, R. G.; Blouin, M.; Boulet, L.; Côté, B.; Frenette, R.; Girard, M.; Guay, D.; Huang, Z. et al. *J. Med. Chem.* **2003**, *46*, 2413–2426.
201. Whitehead, J. W. F.; Lee, G. P.; Gharagozloo, P.; Hofer, P.; Gehrig, A.; Wintergerst, P.; Smyth, D.; McCoull, W.; Hachicha, M.; Patel, A. et al. *J. Med. Chem.* **2005**, *48*, 1237–1243.
202. Ochiai, H.; Ohtani, T.; Ishida, A.; Kusumi, K.; Kato, M.; Kohno, H.; Odagaki, Y.; Kishikawa, K.; Yamamoto, S.; Takeda, H. et al. *Bioorg. Med. Chem.* **2004**, *12*, 4645–4665.
203. Ukita, T.; Sugahara, M.; Terakawa, Y.; Kuroda, T.; Wada, K.; Nakata, A.; Kikkawa, H.; Ikezawa, K.; Naito, K. *Bioorg. Med. Chem. Lett.* **2003**, *13*, 2347–2350.
204. Trifilieff, A.; Wyss, D.; Walker, C.; Mazzoni, L.; Hersperger, R. *J. Pharmacol. Exp. Ther.* **2002**, *301*, 241–248.
205. Wagner, B.; Jakobs, S.; Habermeyer, M.; Hippe, F.; Cho-Chung, Y. S.; Eisenbrandt, G.; Marko, D. *Biochem. Pharmacol.* **2002**, *63*, 659–668.
206. Van der May, M.; Boss, H.; Couwenberg, D.; Hatzelmann, A.; Sterk, G. J.; Goubitz, K.; Schenk, H.; Timmerman, H. *J. Med. Chem.* **2002**, *45*, 2526–2533.
207. Burnouf, C.; Auclair, E.; Avenel, N.; Bertin, B.; Bigot, C.; Calvet, A.; Chan, K.; Durand, C.; Fasquelle, V.; Feru, F. et al. *J. Med. Chem.* **2000**, *43*, 4850–4857.
208. Card, G. I.; Blasdel, L.; England, B. P.; Zhang, C.; Suzuki, Y.; Gillette, S.; Fong, D.; Ibrahim, P. N.; Artis, D. R.; Bollag, G. et al. *Nat. Biotechnol.* **2005**, *23*, 201–207.
209. Card, G. L.; England, B. P.; Suzuki, Y.; Fong, D.; Powell, B.; Lee, B.; Luu, C.; Tabrizizad, M.; Gillette, S.; Ibrahim, P. N. et al. *Structure* **2004**, *12*, 2233–2247.

210. Krier, M.; de Araújo-Júnior, J. X.; Schmitt, M.; Duranton, J.; Justiano-Basaran, H.; Lugnier, C.; Bourguignon, J.-J.; Rognan, D. *J. Med. Chem.* **2005**, *48*, 3816–3822.
211. Manallack, D. T.; Hughes, R. A.; Thompson, P. E. *J. Med. Chem.* **2005**, *48*, 3449–3462.
212. Gupta, R.; Kumar, G.; Kumar, R. S. *Methods Find. Exp. Clin. Pharmacol.* **2005**, *27*, 101–118.
213. Castro, A.; Jerez, M. J.; Gil, C.; Martinez, A. *Med. Res. Rev.* **2004**, *25*, 229–244.
214. Terrett, N. K.; Bell, A. S.; Brown, D.; Ellis, P. *Bioorg. Med. Chem. Lett.* **1996**, *6*, 1819–1824.
215. Haning, H.; Niewöhner, U.; Schenke, T.; Es-Sayed, M.; Schmidt, G.; Lampe, T.; Bischoff, E. *Bioorg. Med. Chem. Lett.* **2002**, *12*, 865–868.
216. Corbin, J. D.; Beasley, A.; Blount, M. A.; Francis, S. H. *Neurochem. Int.* **2004**, *45*, 859–863.
217. Haning, H.; Niewöhner, U.; Schenke, T.; Lampe, T.; Hillisch, A.; Bischoff, E. *Bioorg. Med. Chem. Lett.* **2005**, *15*, 3900–3907.
218. Daugan, A.; Grondin, P.; Le Monier de Gouville, A.-C.; Coste, H.; Linget, J. M.; Kirilovsky, J.; Hyafil, F.; Labaudinière, R. *J. Med. Chem.* **2003**, *46*, 4533–4542.
219. Corbin, J. D.; Francis, S. H. *Int. J. Clin. Pract.* **2002**, *56*, 453–459.
220. Haning, H.; Niewöhoner, U.; Bischoff, E. *Progr. Med. Chem.* **2000**, *41*, 249–306.
221. Rotella, D. P. *Nat. Rev. Drug Disc.* **2002**, *1*, 674–682.
222. Sui, Z. *Exp. Opin. Ther. Patents* **2003**, *13*, 1373–1388.
223. Watanabe, N.; Adachi, H.; Takase, Y.; Ozaki, H.; Matsukura, M.; Miyazaki, K.; Ishibashi, K.; Ishihara, H.; Kodama, K.; Nishino, M. et al. *J. Med. Chem.* **2000**, *43*, 2523–2529.
224. Yu, G.; Mason, H. J.; Wu, X.; Wang, J.; Chong, S.; Dorough, G.; Henwood, A.; Pongrac, R.; Seliger, L.; He, B. et al. *J. Med. Chem.* **2001**, *44*, 1025–1027.
225. Ukita, T.; Nakamura, Y.; Kubo, A.; Yamamoto, Y.; Moritani, Y.; Saruta, K.; Higashijima, T.; Kotera, J.; Takagi, M.; Kikkawa, K. et al. *J. Med. Chem.* **2001**, *44*, 2204–2218.
226. Ukita, T.; Nakamura, Y.; Kubo, A.; Yamamato, Y.; Moritani, Y.; Saruta, K.; Higashijima, T.; Kotera, J.; Fujishige, K.; Takagi, M. et al. *Bioorg. Med. Chem. Lett.* **2003**, *13*, 2341–2345.
227. Mochida, H.; Takagi, M.; Inoue, H.; Noto, T.; Yano, K.; Fujishige, K.; Sasaki, T.; Yuasa, K.; Kotera, J.; Omori, K. et al. *Eur. J. Pharmacol.* **2002**, *456*, 91–98.
228. Rotella, D. P.; Sun, Z.; Zhu, X.; Krupinski, J.; Pongrac, R.; Seliger, L.; Normandin, D.; Macor, J. E. *J. Med. Chem.* **2000**, *43*, 1257–1263.
229. Rotella, D. P.; Sun, Z.; Zhu, X.; Krupinski, J.; Pongrac, R.; Seliger, L.; Normandin, D.; Macor, J. E. *J. Med. Chem.* **2000**, *43*, 5037–5043.
230. Bi, Y.; Stoy, P.; Adam, L.; He, B.; Krupinski, J.; Normandin, D.; Pongrac, R.; Seliger, L.; Watson, A.; Macor, J. E. *Bioorg. Med. Chem. Lett.* **2001**, *11*, 2461–2464.
231. Feixas, J.; Giovannoni, M. P.; Vergelli, C.; Gavaldà, A.; Cesari, N.; Graziano, A.; Dal Paz, V. *Bioorg. Med. Chem. Lett.* **2005**, *15*, 2381–2384.
232. Xia, G.; Li, J.; Peng, A.; Lai, S.; Zhang, S.; Shen, J.; Liu, Z.; Chen, X.; Ji, R. *Bioorg. Med. Chem. Lett.* **2005**, *15*, 2790–2794.
233. Yu, G.; Mason, H. J.; Wu, X.; Wang, J.; Chong, S.; Beyer, B.; Henwood, A.; Pongrac, R.; Seliger, L.; He, B. et al. *J. Med. Chem.* **2003**, *46*, 457–460.
234. Maw, G. N.; Allerton, C. M. N.; Gbekor, E.; Million, W. A. *Bioorg. Med. Chem. Lett.* **2003**, *13*, 1425–1428.
235. Sui, Z.; Guan, J.; Macielag, M. J.; Jiang, W.; Zhang, S.; Qui, Y.; Kraft, P.; Bhattacharjee, S.; John, T. M.; Haynes-Johnson, D. et al. *J. Med. Chem.* **2002**, *45*, 4094–4096.
236. Jiang, W; Sui, Z.; Macielag, M. J.; Walsh, S. P.; Fiordeliso, J. J.; Lanter, J. C.; Guan, J.; Qiu, Y.; Kraft, P.; Bhattacharjee, S. et al. *J. Med. Chem.* **2003**, *46*, 441–444.
237. Lanter, J. C.; Sui, Z.; Macielag, M. J.; Fiordeliso, J. J.; Jiang, W.; Qiu, Y.; Bhattacharjee, S.; Kraft, P.; John, T. M.; Haynes-Johnson, D. et al. *J. Med. Chem.* **2004**, *47*, 656–662.
238. Jiang, W.; Alford, V. C.; Qiu, Y.; Bhattacharjee, S.; John, T. M.; Haynes-Johnson, D.; Kraft, P. J.; Lundeen, S. G.; Sui, Z. *Bioorg. Med. Chem.* **2004**, *12*, 1505–1515.
239. Jiang, W.; Guan, J.; Macielag, M. J.; Zhang, S.; Qiu, Y.; Kraft, P.; Bhattacharjee, S.; John, T. M.; Haynes-Johnson, D.; Lundeen, S. et al. *J. Med. Chem.* **2005**, *48*, 2126–2133.
240. Boyle, C. D.; Xu, R.; Asberom, T.; Chackalamannil, S.; Clader, J. W.; Greenlee, W. J.; Guzik, H.; Hu, Y.; Hu, Z.; Lankin, C. M. et al. *Bioorg. Med. Chem. Lett.* **2005**, *15*, 2365–2369.
241. Bi, Y.; Stoy, P.; Adam, L.; He, B.; Krupinski, J.; Normandin, D.; Pongrac, R.; Seliger, L.; Watson, A.; Macor, J. E. *Bioorg. Med. Chem. Lett.* **2004**, *14*, 1577–1580.
242. Huai, Q.; Liu, Y.; Francis, S. H.; Corbin, J. D.; Ke, H. *J. Biol. Chem.* **2004**, *279*, 13095–13101.
243. Sung, B.-J.; Hwang, K. Y.; Jeon, Y. H.; Lee, J. I.; Heo, Y.-S.; Kim, J. H.; Moon, J.; Yoon, J. M.; Hyun, Y.-L.; Kim, E. et al. *Nature* **2003**, *425*, 98–102.
244. Martinez, A.; Castro, A.; Gil, C.; Miralpeix, M.; Segarra, V.; Domènech, T.; Beleta, J.; Palacios, J. M.; Ryder, H.; Miró, X. et al. *J. Med. Chem.* **2000**, *43*, 683–689.
245. Castro, A.; Abasolo, M. I.; Gil, C.; Segarra, V.; Martinez, A. *Eur. J. Med. Chem.* **2001**, *36*, 333–338.
246. Barnes, M. J.; Cooper, N.; Davenport, R. J.; Dyke, H. J.; Galleway, F. P.; Galvin, F. C. A.; Gowers, L.; Haughan, A. F.; Lowe, C.; Meissner, J. W. G. et al. *Bioorg. Med. Chem. Lett.* **2001**, *11*, 1081–1083.
247. Pitts, W. J.; Vaccaro, W.; Huynh, T.; Leftheris, K.; Roberge, J. Y.; Barbosa, J.; Guo, J.; Brown, B.; Watson, A.; Donaldson, K. et al. *Bioorg. Med. Chem. Lett.* **2004**, *14*, 2955–2958.
248. Kempson, J.; Pitts, W. J.; Barbosa, J.; Guo, J.; Omotoso, O.; Watson, A.; Stebbins, K.; Starling, G. C.; Dodd, J. H.; Barrish, J. C. et al. *Bioorg. Med. Chem. Lett.* **2005**, *15*, 1829–1833.
249. Lorthiois, E.; Bernardelli, P.; Vergne, F.; Oliveira, C.; Mafroud, A.-K.; Proust, E.; Heuze, L.; Moreau, F.; Idrissi, M.; Tertre, A. et al. *Bioorg. Med. Chem. Lett.* **2004**, *14*, 4623–4626.
250. Bernardelli, P.; Lorthiois, E.; Vergne, F.; Oliveira, C.; Mafroud, A.-K.; Proust, E.; Pham, N.; Ducrot, P.; Moreau, F.; Idrissi, M. et al. *Bioorg. Med. Chem. Lett.* **2004**, *14*, 4627–4631.
251. Vergne, F.; Bernardelli, P.; Lorthiois, E.; Pham, N.; Proust, E.; Oliveira, C.; Mafroud, A.-K.; Royer, F.; Wrigglesworth, R.; Schellhaas, J. K. et al. *Bioorg. Med. Chem. Lett.* **2004**, *14*, 4607–4613.
252. Vergne, F.; Bernardelli, P.; Lorthiois, E.; Pham, N.; Proust, E.; Oliveira, C.; Mafroud, A.-K.; Ducrot, P.; Wrigglesworth, R.; Berlioz-Seux, F. et al. *Bioorg. Med. Chem. Lett.* **2004**, *14*, 4615–4621.
253. Inoue, H.; Murafuji, H.; Hayashi, Y. Preparation of (Pyridinyl)pyrazolopyrimidinone Derivatives as PDE7 Inhibitors. International Patent WO 2004111054, December 23, 2004.
254. Ohhata, A.; Takaoka, Y.; Ogawa, M.; Nakai, H.; Yamamoto, S.; Ochiai, H. Preparation of Isoquinoline Derivatives as Phosphodiesterase (PDE) 7 Inhibitors. International Patent WO 2003064389, August 7, 2003.

255. Hatzelman, A.; Marx, D.; Steinhilber, W.; Sterk, G. J. Preparation of Phthalazinones as Phosphodiesterase 4/7 Inhibitors. International Patent WO 2002085906, October 31, 2002.
256. Bundschuh, D.; Kley, H.-P.; Steinhilber, W.; Grundler, G.; Gutterer, B.; Natzelmann, A.; Stadwiesser, J.; Sterk, G. J. Preparation of (Dihydro)isoquinolines as Phosphodiesterase Inhibitors. International Patent WO 2002040450, May 23, 2002.
257. Haughan, A. F.; Lowe, C.; Buckley, G. M.; Dyke, H. J.; Glavin F. C. A.; Mack, S. R.; Meissner, J. W. G.; Morgan, T.; Watson, R. J.; Picken, C. L. et al. A. Preparation of Sulfonamides as Potent Inhibitors of PDE7. International Patent WO 2001098274, December 27, 2001.
258. Wang, H.; Liu, Y.; Chen, Y.; Robinson, H.; Ke, H. *J. Biol. Chem.* **2005**, *280*, 30949–30955.
259. Soderling, S. H.; Bayuga, S. J.; Beavo, J. A. *Proc. Natl. Acad. Sci. USA* **1998**, *95*, 8991–8996.
260. Bell, A. S.; Deninno, M. P.; Palmer, M. J.; Visser, M. S. Preparation of Pyrazolopyrimidinones as Phosphodiesterase 9 (PDE9) Inhibitors for Treating Type 2 Diabetes. US Patent 2004220186, November 4, 2004.
261. Deninno, M. P.; Hughes, B.; Kemp, M. I.; Palmer, M. J.; Wood, A. Preparation of Pyrazolo[4,3-*d*]pyrimidin-7-ones as PDE9 Inhibitors for Treating Cardiovascular Disorders. International Patent WO 2003037899, October 16, 2003.
262. Fryburg, D. A.; Gibbs, E. M. Treatment of Insulin Resistance Syndrome and Type 2 Diabetes with PDE9 Inhibitors. International Patent WO 03037432, May 8, 2003.
263. Hendrix, M. PDE Inhibitors for AD: Influencing Learning and Memory via the cGMP-PDE Pathway. 2nd Annual Phosphodiesterases in Drug Discovery and Development, Philadelphia, PA, November 8–9, 2004.
264. Huai, Q.; Wang, H.; Zhang, W.; Colman, R. W.; Robinson, H.; Ke, H. *Proc. Natl. Acad. Sci. USA* **2004**, *101*, 9624–9629.
265. Sweet, L. Methods for Treating Diabetes and Related Disorders using PDE10A Inhibitors. International Patent WO 2005012485, February 10, 2005.
266. Maier, T.; Graedler, U.; Gimmnich, P.; Quintini, G.; Ciapetti, P.; Contreras, J.-M.; Wermuth, C. G.; Vennemann, M.; Baier, T.; Braunger, J. Preparation of Pyrrolodihydroisoquinolines as Cyclic 3′,5′-Nucleotide Phosphodiesterase (PDE10) Inhibitors International Patent WO 2005003129, January 13, 2005.
267. Ergueden, J. K.; Bauser, M.; Burkhardt, N.; Flubacher, D.; Friedl, A.; Gerlach, I.; Hinz, V.; Jork, R.; Naab, P.; Niewoehner, U. et al. Preparation of Imidazotriazines as Phosphodiesterase 10A Inhibitors for the Treatment of Neurodegenerative Diseases. International Patent WO 2003000693, October 3, 2003.
268. Kleinman, R. PDE10: A Potential Target for CNS Diseases. 2nd Annual Phosphodiesterases in Drug Discovery and Development, Philadelphia, PA, November 8–9, 2004.
269. Czarniecki, M.; Ahn, H.-S.; Sybertz, E. J. *Annu. Rep. Med. Chem.* **1996**, *31*, 61–70.
270. Truss, M. C.; Stief, C. G.; Uckert, S.; Becker, A. J.; Wefer, J.; Schultheiss, D.; Jonas, U. *World J. Urol.* **2001**, *19*, 344–350.
271. Truss, M. C.; Stief, C. G.; Uckert, S.; Becker, A. J.; Schultheiss, D.; Machten, S.; Udo, J. *World J. Urol.* **2000**, *18*, 439–443.
272. Stoclet, J.-C.; Keravis, T.; Komas, N.; Lugnier, C. *Exp. Opin. Ther. Patents* **1995**, *4*, 1081–1100.
273. Sudo, T.; Tachibana, K.; Toga, K.; Tochizawa, S.; Inoue, Y.; Kimura, Y.; Hidaka, H. *Biochem. Pharmacol.* **2000**, *59*, 347–356.
274. Alharethi, R.; Movesesian, M. A. *Exp. Opin. Invest. Drugs* **2002**, *11*, 1529–1536.
275. Shimizu, K.; Murata, T.; Okumura, K.; Manganiello, V. C.; Tagawa, T. *Anticancer Drugs* **2002**, *13*, 875–880.
276. Murata, T.; Shimizu, K.; Narita, M.; Manganiello, V. C.; Tagawa, T. *Anticancer Res.* **2002**, *22*, 3171–3174.
277. Harndahl, L.; Jing, X.-J.; Ivarsson, R.; Degerman, E.; Ahren, B.; Manganiello, V. C.; Renstrom, E.; Holst, L. S. *J. Biol. Chem.* **2002**, *277*, 37446–37455.
278. Ahmad, M.; Abdel-Wahab, Y. H. A.; Tate, R.; Flatt, P. R.; Pyne, N. J.; Furman, B. L. *Br. J. Pharmacol.* **2000**, *129*, 1228–1234.
279. Inoue, Y.; Toga, K.; Sudo, T.; Tachibana, K.; Tochizawa, S.; Kimura, Y.; Yoshida, Y.; Hidaka, H. *Br. J. Pharmacol.* **2000**, *130*, 231–241.
280. Myou, S.; Fujimura, M.; Kamio, Y.; Hirose, T.; Kita, T.; Tachibana, H.; Ishiura, Y.; Watanabe, K.; Hashimoto, T.; Nakao, S. *Br. J. Clin. Pharmacol.* **2003**, *55*, 341–346.
281. Odashima, M.; Otaka, M.; Jin, M.; Komatsu, K.; Konishi, N.; Wada, I.; Horikawa, Y.; Matsuhashi, T.; Ohba, R.; Oyake, J. et al. *Digest. Dis. Sci.* **2005**, *50*, 1097–1102.
282. Phillips, P. G.; Long, L.; Wilkins, M. R.; Morrell, N. W. *Am. J. Physiol.* **2005**, *288*, L103–L115.
283. Mehats, C.; Oger, S.; Leroy, M.-J. *Eur. J. Obstet. Gynecol.* **2004**, *117*, S15–S17.
284. Banner, K. H.; Trevethick, M. A. *Trends Pharmacol. Sci.* **2004**, *25*, 430–436.
285. Hirsh, L.; Dantes, A.; Suh, B.-S.; Yoshida, Y.; Hosokawa, K.; Tajima, K.; Kotsuji, F.; Merimsky, O.; Amsterdam, A. *Biochem. Pharmacol.* **2004**, *68*, 981–988.
286. Tiwari, S.; Dong, H.; Kim, E. J.; Weintraub, L.; Epstein, P. M.; Lerner, A. *Biochem. Pharmacol.* **2005**, *69*, 473–483.
287. Ückert, S.; Küthe, A.; Jonas, U.; Stief, C. G. *J. Urol.* **2001**, *166*, 2484–2490.
288. Momma, K.; Toyoshima, K.; Imamura, S.; Nakanishi, T. *Pediatr. Res.* **2005**, *58*, 42–45.
289. Gardiner, S. M.; March, J. E.; Kemp, P. A.; Ballard, S. A.; Hawkeswood, E.; Hughes, B.; Bennett, T. *J. Pharmacol. Exp. Ther.* **2005**, *312*, 265–271.
290. Toward, T. J.; Smith, N.; Broadley, K. J. *Am. J. Respir. Crit. Care Med.* **2004**, *169*, 227–234.
291. Vlachopoulos, C.; Tsekoura, D.; Alexopoulos, N.; Panagiotakos, D.; Aznaouridis, K.; Stefanadis, C. *Am. J. Hypertens.* **2004**, *17*, 1040–1044.
292. Hryniewicz, K.; Dimayuga, C.; Hudaihed, A.; Androne, A. S.; Zheng, H.; Jankowski, K.; Katz, S. D. *Clin. Sci.* **2005**, *108*, 331–338.
293. Guazzi, M.; Tumminello, G.; Di Marco, F.; Fiorentini, C.; Guazzi, M. D. *J. Am. Coll. Cardiol.* **2004**, 2339–2348.
294. Santos, C. L.; Souza, M. H. L. P.; Gomes, A. S.; Lemos, H. P.; Santos, A. A.; Cunha, F. Q.; Wallace, J. L. *Br. J. Pharmacol.* **2005**, *146*, 481–486.
295. Watkins, C. C.; Sawa, A.; Jaffrey, S.; Blackshaw, S.; Barrow, R. K.; Snyder, S. H.; Ferris, C. D. *J. Clin. Invest.* **2000**, *106*, 373–384.
296. Gori, T.; Sicuro, S.; Dragoni, S.; Donati, G.; Forconi, S.; Parker, J. C. *Circulation* **2005**, *111*, 742–746.
297. Fisher, P. W.; Salloum, F.; Das, A.; Hyder, H.; Kukreja, R. C. *Circulation* **2005**, *111*, 1601–1610.
298. Rosano, G. M.; Aversa, A.; Vitale, C.; Fabbri, A.; Fini, M.; Spera, G. *Eur. Urol.* **2005**, *47*, 214–220.
299. Takimoto, E.; Champion, H. C.; Li, M.; Belardi, D.; Ren, S.; Rodriguez, E. R.; Bedja, D.; Gabrielson, K. L.; Wang, Y.; Kass, D. A. *Nat. Med.* **2004**, *11*, 214–222.
300. Bouley, R.; Pastor-Soler, N.; Cohen, O.; McLaughlin, M.; Breton, S.; Brown, D. *Am. J. Physiol. Renal Physiol.* **2005**, *288*, F1103–F1112.
301. Prickaerts, J.; Şik, A.; van Staveren, W. C. G.; Koopmans, G.; Steinbusch, H. W. M.; van der Staay, F. J.; de Vente, J.; Blokland, A. *Neurochem. Int.* **2004**, *45*, 915–928.
302. Prickaerts, J.; Şik, A.; van der Staay, F. J.; de Vente, J.; Blokland, A. *Psychopharmacology* **2004**, *177*, 381–390.
303. Devan, B. D.; Sierra-Mercado, D.; Jimenez, M.; Bowker, J. L.; Duffy, K. B.; Spangler, E. L.; Ingram, D. K. *Pharmacol. Biochem. Behav.* **2004**, *79*, 691–698.
304. Sarfati, M.; Mateo, V.; Baudet, S.; Rubio, M.; Fernandez, C.; Davi, F.; Binet, J.-L.; Delic, J.; Merle-Béral, H. *Blood* **2003**, *101*, 265–269.

305. Zhu, B.; Vemavarapu, L.; Thompson, W. J.; Strada, S. J. *J. Cell. Biochem.* **2005**, *94*, 336–350.
306. Perez-Torres, S.; Cortes, R.; Tolnay, M.; Probst, A.; Palacios, J. M.; Mengod, C. *Exp. Neurol.* **2003**, *182*, 322–334.
307. D'Andrea, M. R.; Qiu, Y.; Haynes-Johnson, D.; Bhattacharjee, S.; Kraft, P.; Lundeen, S. *J. Histochem. Cytochem.* **2005**, *53*, 895–903.
308. Bell, C.; Converse, C. A.; Hammer, H. M.; Osborne, A.; Haites, N. E. *Br. J. Ophthalmol.* **1994**, *78*, 933–938.
309. Meins, M.; Janecke, A.; Marschke, C.; Denton, M. J.; Kumaramanickavel, G.; Pittler, S.; Gal, A. *Mutations in the PDE6A Gene Encoding the α-Subunit of Rod Photoreceptor cGMP-Specific Phosphodiesterase, are rare in Autosomal Recessive Retinitis Pigmentosum*, Proceedings of the 7th International Symposium on Retinal Degeneration, Sendai, Japan, October 5–9, 1996; Plenum Press: New York, 1997.
310. Veske, A.; Orth, U.; Ruether, K.; Zrenner, E.; Rosenberg, T.; Baehr, W.; Gal, A. *Mutations in the Gene for the B-Subunit of Rod Photoreceptor cGMP-Specific Phosphodiesterase (PDEB) in Patients with Retinal Dystrophies and Dysfunctions*, Proceedings of the 6th International Symposium on Retinal Degeneration, Jerusalem, Israel, November 4–9, 1994; Plenum Press: New York, 1994.

Biography

David P Rotella earned his BS degree (*magna cum laude*) in pharmacy at the University of Pittsburgh, PA, in 1981, and a PhD in 1985 in medicinal chemistry at The Ohio State University College of Pharmacy under the direction of Donald T Witiak. The topic for his dissertation research was the synthesis and evaluation of novel platinum antitumor complexes and potential antimetastatic agents. He was a postdoctoral scholar from 1985 to 1987 in the Department of Chemistry at the Pennsylvania State University under the tutelage of Ken S Feldman, where he was engaged in the development of a novel [6 + 2] photocycloaddition reaction for the synthesis of cyclooctane rings.

His first independent position (1987–91) was as an assistant professor of pharmacognosy at the University of Mississippi School of Pharmacy, where his research interests included synthetic methods and the development of novel enzyme inhibitors. In 1991, he moved to Cephalon, Inc., where he led drug discovery programs for the discovery of potential neuroprotective agents. After accepting a position at Bristol-Myers Squibb in 1997, he became involved in the discovery of phosphodiesterase 5 inhibitors, and while at Lexicon Pharmaceuticals (2003–05), he directed drug discovery programs with metabolic disease applications. At Wyeth Research (2005 to the present) he is a team leader for the discovery of agents that may be useful for the treatment of schizophrenia and depression.

Rotella has authored approximately 25 peer-reviewed articles, is a co-inventor on seven patents or patent applications, and has been invited to speak at a variety of meetings on the topic of phosphodiesterase enzyme medicinal chemistry.

2.24 Protein Kinases and Protein Phosphatases in Signal Transduction Pathways

Tomi K Sawyer, ARIAD Pharmaceuticals, Cambridge, MA, USA

© 2007 Published by Elsevier Ltd.

2.24.1	**Introduction**	**959**
2.24.2	**Molecular, Cellular, and Structural Biology of Protein Phosphorylation**	**960**
2.24.2.1	Functional Genomics and Signal Transduction Pathways	960
2.24.2.2	Structural Genomics and Signal Transduction Pathways	964
2.24.3	**Protein Kinase Therapeutic Targets and Drug Discovery**	**964**
2.24.3.1	Receptor and Nonreceptor Tyrosine Kinases	967
2.24.3.1.1	Epidermal growth factor receptor family kinases	967
2.24.3.1.2	Vascular endothelial growth factor receptor, Flt-3, and KIT family kinases	969
2.24.3.1.3	Bcr-Abl kinases	972
2.24.3.1.4	Src family kinases	975
2.24.3.2	Receptor and Nonreceptor Serine/Threonine Kinases	977
2.24.3.2.1	TGFβR kinases	977
2.24.3.2.2	Cyclin-dependent kinase family kinases	979
2.24.3.2.3	Raf-MEK-MAPK pathway kinases	980
2.24.3.2.4	PI3K-Akt/PKB-mTOR pathway kinases	980
2.24.4	**Protein Phosphatase Therapeutic Targets and Drug Discovery**	**981**
2.24.4.1	Receptor and Nonreceptor Tyrosine Phosphatases	981
2.24.4.2	Receptor and Nonreceptor Serine/Threonine Phosphatases	982
2.24.5	**Phosphoprotein-Interacting Domain Containing Cellular Protein Therapeutic Targets and Drug Discovery**	**982**
2.24.5.1	Phosphotyrosine-Interacting Domains	983
2.24.5.2	Phosphoserine/Phosphothreonine-Interacting Domains	985
2.24.6	**Concluding Remarks**	**985**
References		**985**

2.24.1 Introduction

Protein phosphorylation has become a central focus of drug discovery as the result of the identification and validation of promising therapeutic targets such as protein kinases, protein phosphatases, and phosphoprotein binding domains.[1–24] With respect to such protein phosphorylation therapeutic targets, significant progress has been made in mapping the signal transduction pathways involved in disease, including in particular the molecular genesis and/or sustainment of various cancers, and several inflammatory, immune, metabolic, and bone diseases. Protein phosphorylation is an extraordinarily important component of life processes, including various signal transduction pathways underlying cellular proliferation, differentiation, metabolism, survival, motility, and gene transcription. Protein phosphorylation is a complex phenomenon, with molecular triggering, compensatory mechanisms, and both spatial and temporal factors contributing to the biological specificity and functional endpoints within the cellular milieu. The global phosphorylation state of proteins in cells is fundamentally impossible to decipher at high resolution, even though sophisticated methods (e.g., mass spectrometry, functional interaction traps, affinity chromatography) are emerging that can be used to analyze the so-called phosphoproteome. In this chapter, protein phosphorylation is described relative to our current understanding of the biology of protein kinase, protein phosphatase, and the phosphoprotein-interacting domain containing intracellular proteins. In addition a few noteworthy examples of drug discovery focused on developing small-molecule inhibitors of such therapeutic targets are described.

2.24.2 Molecular, Cellular, and Structural Biology of Protein Phosphorylation

Protein phosphorylation is controlled by superfamilies of protein kinases (e.g., tyrosine and serine/threonine protein kinases), protein phosphatases (e.g., tyrosine and serine/threonine protein phosphatases), and phosphoprotein-interacting domains (e.g., SH2, PTB, WW, FHA, and 14–3–3) of various catalytic/noncatalytic intracellular proteins to provide a multidimensional, dynamic and reversible orchestration of the phosphoprotein formation, hydrolysis, and molecular recognition at specific sites involving Tyr, Ser, and/or Thr residues (**Figure 1**).

A significant part (approximately 30%) of proteins encoded by the human genome are probably phosphorylated by ATP as the phosphoryl donor. Relative to Tyr phosphorylation, it is noted that pTyr represents only approximately 0.05% of total phosphorylated protein in normal cells (the remaining 99.95% being pSer and pThr), whereas oncogenic constitutively activated tyrosine kinases (see below) may increase this percentage to 0.5% within transformed cells.[25] Since the first observations of protein kinase activity about 50 years ago, there have been significant advances in basic research and, in recent years, drug development.[26] Key milestones in basic research into protein phosphorylation include the discovery of Src[27,28] and epidermal growth factor receptor (EGFR),[28] as the first determined nonreceptor and growth factor receptors and tyrosine kinases, respectively, the discovery of PTP1B[29] as the first determined protein tyrosine phosphatase, and the discovery of the Src SH2 domain,[30] as the first determined phosphoprotein-interacting motif conferring molecular recognition to pTyr-containing cellular proteins. Such studies have led to a plethora of knowledge about a large and diverse superfamily of protein kinases, protein phosphatases, and phosphoprotein-interacting domains containing cellular proteins (**Table 1**).

2.24.2.1 Functional Genomics and Signal Transduction Pathways

Functional genomics and signal transduction pathway studies[1,7,10,12–17,31–33] have correlated the possible relationships between gene deletion, fusion, mutation and/or overexpression of several known protein kinases, protein phosphatases, and phosphoprotein-interacting domains containing cellular proteins with aberrant cellular activities and in vivo phenotypes correlating to known cancers or other diseases (**Table 2**).

Cancer is one of the most complex diseases, and it is becoming increasingly well understood due to progress in deciphering several of its causative factors, including imbalances in cell-cycle progression, cell growth, and programmed cell death (apoptosis). Of the nearly 300 cancer genes that have been reported to date, protein kinases represent the largest group (nearly 10%) having structural homology. Furthermore, many cancers have been correlated with somatic

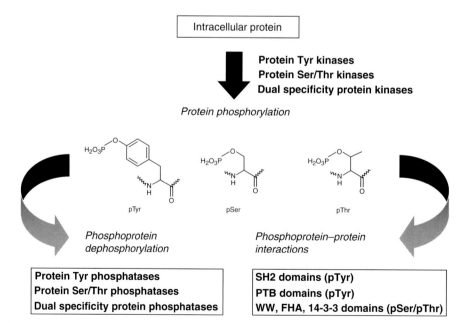

Figure 1 Protein phosphorylation at specific Tyr, Ser, and Thr sites. The relationship of protein kinases, protein phosphatases, and phosphoprotein-interacting domains containing cellular proteins.

Table 1 Some examples of protein kinases, protein phosphatases, and phosphoprotein-interaction domain containing cellular proteins

Therapeutic target

Receptor tyrosine kinases

Epidermal growth factor receptor (EGFR)

Fibroblast growth factor receptor (FGFR)

Vascular endothelial growth factor receptor (VEGFR)

Platelet-derived growth factor receptor (PDGFR)

Stem cell receptor (KIT)

Hematopoietic class III receptor (Flt)

Insulin receptor (IRK)

Insulin-like growth factor receptor (IGFR)

Colony stimulating factor receptor (CSFR)

Nerve growth factor receptor (NGFR)

Hepatocyte growth factor receptor (Met)

Glial-derived neurotrophic factor receptor (RET)

Nonreceptor tyrosine kinases

Src and Src-family kinase (SFK)

Abl

FAK, Pyk2

Janus kinase (JAK) family

Receptor serine/threonine kinases

Transforming growth factor receptor-β (TGFR-β)

Nonreceptor serine/threonine kinases and dual specificity kinases

cAMP-dependent protein kinase (PKA)

Phosphoinositide-3-kinase (PI3K)

Cyclin-dependent kinase (CDK)

Mitogen activated protein kinase (MAPK)

MAPKK (ERK)

MAPKKK (MEK)

Raf kinase

Aurora kinases

Protein kinase C (PKC)

Protein kinase B (PKB/Akt)

mTor (FRAP)

Polo-like kinases (Plk)

Integrin-linked kinase (ILK)

Glycogen synthase kinase-4

continued

Table 1 Continued

Therapeutic target

Protein tyrosine phosphatases and dual specificity phosphatases

Protein tyrosine phosphatase-1B (PTP1B)

SH2-containing tyrosine phosphatase (SHP2)

Protein tyrosine phosphatase-α (PTPα)

Cyclin-dependent kinase/cyclin-25 (CDC25)

Protein tyrosine phosphatase receptor-C (PTPRC/CD45)

Phosphatase and tensin homolog on chromosome 10 (PTEN)

Protein serine/threonine phosphatases

Protein phosphatase-1 (PP1)

Protein phosphatase-2B (PP2A)

Protein phosphatase-2B (PP2B/calcineurin)

Protein phosphatase-2C (PP2C)

Protein phosphatase-3 (PP3)

SH2 and PTB domain containing cellular proteins

Src and Src family (SH2 domain)

Abl (SH2-domain)

Grb2 (SH2 domain)

Shc (PTB and SH2 domains)

IRS-1 (PTB domain)

WW, FHA, and 14-3-3 domain cellular proteins

Pin1 (WW domain

Rad53 (FHA domain)

14-3-3 family (14-3-3 domain)

---

mutations of protein kinases, of which both receptor and nonreceptor tyrosine kinases have emerged as particularly significant therapeutic targets for cancer drug discovery. In humans, oncogenic transformation of protein kinases may arise from the fusion products of genomic rearrangements (e.g., chromosomal translocations), mutations (e.g., gain-of-function), deletions, and overexpression resulting from gene amplification. Typically, such transformations result in enhanced or constitutive kinase activity, which then effects subsequent altered downstream signal transduction. Some examples of oncogenic protein tyrosine kinases (see below) are EGFR, HER-2, HER-3, VEGFR-1 (Flt-1), VEGFR-2 (Flk-1/KDR), Flt-3, Flt-4, PDGFR-α, PDGFR-ββ, KIT, RET, MET, IGF-1R, Abl, Src and the Src family, FAK, Pyk2, and the JAK family. Some examples of protein serine/threonine kinases that have been identified as key therapeutic targets with respect to oncogenic signaling (see below) are TGFββR, the CDK family, Raf, MEK, Erk, MAPK, PKC, PI3K, Akt/PKB, mTOR, and aurora kinases.

Functional genomics and signal transduction pathway studies have also correlated possible relationships between key protein phosphatases and the phosphoprotein-interacting domain containing cellular proteins and certain diseases (**Table 2**).[12–17] Some examples of protein phosphatase therapeutic targets (see below) are PTP1B, CD45, and

Table 2 Some functional genomic relationships of protein kinases, protein phosphatases, and phosphoprotein-interacting domain containing cellular proteins with disease phenotypes

Therapeutic target	Gene modification	Cancer (or other disease)
Protein kinase		
EGFR/ErbB1	Overexpression, point mutations	Breast, NSCL, ovarian, glioblastoma
ErbB2/HER2/Neu	Overexpression, point mutations	Breast, ovarian, gastric, NSCL, colon
ErbB3/HER3	Overexpression	Breast
IGF-1R	Overexpression	Cervical, sarcomas
PDGFR-α	Overexpression	Glioma, glioblastoma, ovarian, GIST
PDGFR-β	Fusions (Tel-PDGFR-β)	Leukemias
	Overexpression	Glioma
Flt-3, Flt-4	Point mutation	Leukemias, angiosarcoma
KIT	Point mutations, overexpression	GIST, AML, myelodysplastic syndromes
MET	Point mutations, overexpression	Renal, hepatocellular
RET	Point mutations, fusions	Thyroid, parathyroid, adrenal
VEGFR1/Flt1	Expression	Tumor angiogenesis
VEGFR2/Flk1	Expression	Tumor angiogenesis
Src	C-terminal truncation, point mutations, overexpression	Colon, breast, pancreatic; metastasis
	Deletion (KO)	Osteopetrosis
Yes	Overexpression	Colon, melanoma
FAK	Overexpression	Metastases, adhesion, invasion
Pyk2	Overexpression	Metastases, adhesion, invasion
Abl	Fusions (Bcr-Abl), point mutations	CML, ALL
JAK1, JAK3	Overexpression	Leukemias
JAK2	Translocation	Leukemias
Protein phosphatase		
PTP1B	Deletion (KO)	Increased sensitivity to insulin, resistance to weight gain
Phosphoprotein-interacting domain (cellular protein)		
SH2 (SAP)	Point mutations	X-linked lymphoproliferative disease
SH2 (Btk)	Point mutation	X-linked agammaglobulinemia
SH2 (SHP-2)	Point mutations	Noonan's syndrome
PTB (IRS-1/2)	Point mutations	Type-2 diabetes mellitus
PTB (ARH)	Point mutations	Autosomal recessive hypercholesterolemia

CDC25A for diabetes (and obesity), immune and inflammatory disease, and cancer, respectively. Some examples of phosphoprotein-interacting domains containing cellular protein therapeutic targets (see below) are SH2 (e.g., Grb2, Shc, Src, and ZAP-70) and PTB (e.g., Shc and IRS-1), which are related to cancer, autoimmune disease, osteoporosis, type-2 diabetes mellitus, and autosomal recessive hypercholesterolemia.

2.24.2.2 Structural Genomics and Signal Transduction Pathways

Structural genomics has also provided insight into the molecular architecture and mechanistic properties of protein kinases, protein phosphatases, and phosphoprotein-interacting domains containing cellular proteins intimately involved in signal transduction pathways. Several hundred x-ray crystallographic structures have been determined for this collective group of signal transduction therapeutic targets.[4–6,13–17,19,34–39]

In the case of protein kinases, both structural biology and drug design studies of small-molecule inhibitors has led to the determination of various modes of enzyme inhibition (i.e., both active and inactive conformations of the catalytic domain), and opportunities to exploit binding interactions beyond those existing for ATP and peptide substrates.[5,6,9,34–50] Such findings have had an extraordinary impact both on understanding and on overcoming the challenges of resistance to small-molecule inhibitors resulting from mutations of the protein kinase (e.g., imatinib-resistant Bcr-Abl mutants).[51] The number of x-ray crystallographic structures deposited in the Protein Data Bank for protein kinases and inhibitor complexes thereof has continued to increase at an extraordinary rate over the past decade (**Table 3**). Such three-dimensional structural information provides the incentive for the development of high-powered in silico drug design technologies to analyze computationally, visualize, and compare the binding sites of such protein kinases relative to their complexes with chemically diverse ligands, including ATP derivatives (e.g., nonhydrolyzable ATP analogs, ADP, AMP, and adenosine), small-molecule ATP mimetics (e.g., substituted purines, pyrrolopyrimidines, pyrazolopyrimidines, indolinones, pyridopyrimidinones, quinazolines, and quinolines), natural products (e.g., staurosporine), and an emerging class of kinase inactive conformation binding ligands (e.g., phenylaminopyrimidines and diarylureas).

Structural genomics provides the opportunity to exploit high-resolution x-ray structures as well as three-dimensional molecular models of protein kinases, protein phosphatases, and phosphoprotein-binding domains to guide and support drug discovery efforts focused on inhibitors. In silico drug design technologies may further be used to identify candidate inhibitors through virtual screening and de novo construction with respect to various templates and functional group elaborations thereof. As shown in a simplified perspective of the prototypic protein kinase Src complexed with ATP and substrate (**Figure 2**), key hydrogen-bonding and hydrophobic and ionic (e.g., Mg^{2+} coordination) interactions confer molecular recognition between the protein and ligands. Furthermore, an auxiliary binding site, which is often referred to as the 'specificity pocket', exists proximate to the purine ring of ATP, and this site has become a key focus of novel small-molecule inhibitor drug discovery for oncogenic protein kinases. The size, shape, and functional group composition of the specificity pocket varies between the different protein kinase gene families. Similar drug design approaches have been used to advance potent inhibitors of other protein kinases as well as inhibitors of protein phosphatases and phoshoprotein-binding domains.

2.24.3 Protein Kinase Therapeutic Targets and Drug Discovery

More than 500 protein kinase genes (excluding additional protein kinase pseudogenes) have been identified from the human genome and constitute nearly 2% of all human genes. Furthermore, chromosomal mapping studies have revealed that 244 protein kinases map to disease loci or cancer amplicons.[2] It is interesting to note that there may exist several times more proteins (in human) that utilize ATP than the estimated number of protein kinases. However, protein kinases are the largest family of structurally homologous proteins in the human genome that are critical components of the protein phosphorylation in cells. The catalytic domains of most protein kinases contain approximately 250–300 amino acids and have defined binding pockets for ATP (and typically complexed Mg^{2+} or Mn^{2+}) and substrate proteins. The classification of protein kinases by virtue of the alignment of catalytic domain sequences includes several major groups, as exemplified by AGC (e.g., PKA, PKG, and PKC family protein kinases), CaMK (e.g., calcium-/calmodulin-regulated protein kinases), CMGC (e.g., CDK family kinases, MAPK, GSK, and CKII protein kinase), TK (e.g., EGFR and Src tyrosine kinases), and OPK (e.g., other protein kinases). There is a website devoted to protein kinase genomic analysis.[2,2a]

Since the identification of the nonreceptor tyrosine kinase, Src, about 25 years ago,[26a] a plethora of biological studies have steadily unraveled the complex orchestration of signal transduction pathways involving protein kinases of varying structural architecture and functional properties.[1,7–10] The superfamily of protein kinases is exemplified by growth factor receptor tyrosine kinases (e.g., the EGFR family, VEGFR family, PDGFR family, and FGFR family), nonreceptor tyrosine kinases (e.g., Abl, the Src family, and FAK), growth factor receptor serine/threonine kinases (e.g., TGFβR), and nonreceptor serine/threonine as well as dual specificity kinases (e.g., CDK family, Raf, MEK, MAPK, PI3K, PKC, PKB/Akt, and mTOR). A number of such protein kinases have proven to be particularly significant as therapeutic targets for cancer drug discovery (see below).

Protein Kinases and Protein Phosphatases in Signal Transduction Pathways

Table 3 Summary of >100 human kinase x-ray structures in the Protein Data Bank (PDB). Entries are listed relative to the number of coordinates deposited[a]

Gene name[b]	Family name[c]	Subfamily name[c]	No. of PDB entries[d]	No. of ligands (non-ATP)[e]	Year first deposited[f]	PDB access codes for related human kinase structures[g]
CDK2	CMGC	CDK, CDC2	41	22	1996	1b38, 1b39, 1e1v, 1e1x, 1fin, 1hck, 1hcl, 1aq1, 1ckp, 1buh, 1dm2, 1di8, 1f5q, 1fvt, 1fvv, 1jsv, 1g5s, 1jvp, 1gih, 1ke5, 1ke6, 1ke7, 1ke8, 1ke9, 1e9h, 1jst, 1jsu, 1fq1, 1qmz, 1gy3, 1h1p, 1h1q, 1h1r, 1h1s, 1h24, 1h25, 1h26, 1h27, 1h28, 1gii, 1gij
PRKACA	AGC	PKA	17	5	1992	1ctp, 1stc, 1cmk, 1cdk, 2cpk, 1atp, 1apm, 1fmo, 1jlu, 1bkx, 1bx6, 1jbp, 1l3r, 1ydt, 1ydr, 1yds, 1j3h
MAPK14	CMGC	MAPK, p38	13	8	1996	1ian, 1wfc, 1m7q, 1a9u, 1bl6, 1bl7, 1di9, 1kv1, 1kv2, 1lew, 1lez, 1p38, 1bmk
PIK3CG	Lipid	PI3	8	5	2000	1e8y, 1e8z, 1he8, 1e7v, 1e8w, 1e8x, 1e90, 1e7u
MAPK1	CMGC	MAPK, ERK	6	4	1996	1erk, 3erk, 4erk, 1gol, 2erk, 1pme
ABL1	TK	Abl	6	6	2000	1fpu, 1iep, 1m52, 1opj, 1opl, 1opk
LCK	TK	Src	5	3	1997	1qpe, 1qpj, 1qpc, 3lck, 1qpd,
CDK6	CMGC	CDK, CDK4	5	0	1998	1bi8, 1bi7, 1blx, 1g3n, 1jow
AKT2	AGC	AKT	5	0	2002	1gzk, 1gzo, 1gzn, 1o6k, 1o6l
DAPK1	CAMK	DAPK	5	0	2001	1ig1, 1jkl, 1jks, 1jkt, 1jkk
INSR	TK	InsR	4	0	1995	1irk, 1i44, 1ir3, 1gag
CHEK1	CAMK	CAMKL, CHK1	4	3	2001	1ia8, 1nvq, 1nvr, 1nvs
FGFR1	TK	FGFR	4	3	1997	1fgk, 1fgi, 1agw, 2fgi
SRC	TK	Src	4	0	1997	1fmk, 2src, 1ksw, 2ptk
IGF1R	TK	InsR	4	0	2001	1m7n, 1jqh, 1p4o, 1k3a
GSK3B	CMGC	GSK	3	0	2001	1i09, 1gng, 1h8f
HCK	TK	Src	3	2	1997	1qcf, 1ad5, 2hck
PHKG1	CAMK	PHK	3	0	1996	1phk, 1ql6, 2phk
CSNK1D	CK1	CK1	2	0	1995	1ckj, 1cki
CSK	TK	Csk	2	1	1998	1byg, 1k9a
EGFR	TK	EGFR	2	1	2002	1m14, 1m17
TGFBR1	TKL	STKR, type1	2	0	1999	1b6c, 1ias
MUSK	TK	Musk	1	0	2002	1luf

continued

Table 3 Continued

Gene name[b]	Family name[c]	Subfamily name[c]	No. of PDB entries[d]	No. of ligands (non-ATP)[e]	Year first deposited[f]	PDB access codes for related human kinase structures[g]
MAPKAPK2	CAMK	MAPKAPK	1	0	2002	1kwp
STK6	Other	AUR	1	0	2002	1muo
ADRBK1	AGC	GRK, BARK	1	0	2003	1omw
EPHB2	TK	Eph	1	0	2001	1jpa
PAK1	STE	STE20, PAKA	1	0	2000	1f3m
CAMK1	CAMK	CAMK1	1	0	1997	1a06
MAPK12	CMGC	MAPK, p38	1	0	1999	1cm8
CSNK2A1	Other	CK2	1	0	2001	1jwh
TTN	CAMK	MLCK	1	0	1998	1tki
MAPK10	CMGC	MAPK, JNK	1	0	1998	1jnk
KDR	TK	VEGFR	1	0	1998	1vr2
FGFR2	TK	FGFR	1	0	2001	1gjo
CDK5	CMGC	CDK, CDK5	1	0	2001	1h4l
BTK	TK	Tec	1	0	2001	1k2p
TEK	TK	Tie	1	0	2000	1fvr

[a] Adapted from Vieth, M.; Higgs, R. E.; Robertson, D. H.; Shapiro, M.; Gragg, E. A.; Hemmerle, H. *Biochim. Biophys. Acta.* **2004**, *1697*, 243–257.
[b] Human gene name corresponding to x-ray structures deposited in PDB (criterion is 95% sequence identity).
[c] Family and subfamily for the human gene as defined by Sugen.[34a]
[d] Number of PDB entries corresponding to the human gene.
[e] Number of non-ATP ligands corresponding to the human gene. ATP ligands include ATP, ADP, AMP, and adenosine.
[f] Year that an x-ray structure corresponding to the human gene was first deposited in the Protein Data Bank.
[g] Noncomprehensive listing of PDB codes of x-ray structures corresponding to the human kinase.

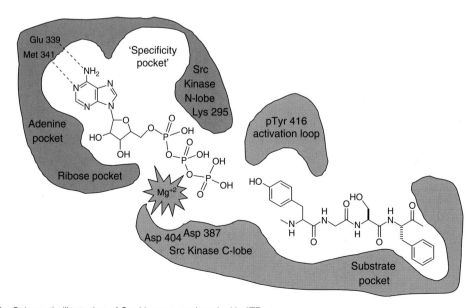

Figure 2 Schematic illustration of Src kinase complexed with ATP.

2.24.3.1 Receptor and Nonreceptor Tyrosine Kinases

The seminal identifications of Src and EGFR as oncogenic protein tyrosine kinases launched a campaign in both basic research and drug discovery that has resulted in the creation of first-generation therapies, including small-molecule inhibitors and monoclonal antibodies. Of the approximately 60 receptor tyrosine kinases presently known, each has cell membrane-spanning sequences with extracellular ligand-binding domains and intracellular catalytic (kinase) domains (**Figure 2**). Ligand (e.g., growth factor) binding induces dimerization or an increased association of already dimerized subunits and subsequent autophosphorylation or cross-phosphorylation of proximate receptors at specific tyrosine residues on the intracellular carboxyl terminal sequences. With respect to signal transduction regulation, the resultant phosphotyrosine (pTyr) sequences of such activated receptor tyrosine kinases may then interact with various intracellular proteins by virtue of pTyr molecular recognition by SH2 domains (e.g., Src, phospholipase-Cγ, and phosphatidykinositol-3-kinase, Grb2) or PTB domains (e.g., IRS-1) to modulate various downstream biological pathways. As representatives of nonreceptor tyrosine kinases, Src and Abl each illustrate multiple domain architectures, which include noncatalytic SH2 and SH3 domains (**Figure 3**).

Some noteworthy examples of receptor and nonreceptor protein tyrosine kinases described further below.

2.24.3.1.1 Epidermal growth factor receptor family kinases

The EGFR family of growth factor receptor tyrosine kinases has been the major focus of drug discovery efforts over the past decade to advance new therapies for the treatment of several cancers (e.g., lung, breast, colon, prostate, ovarian, head and neck, and gliomas). Extracellular-mediated triggering of EGFR by EGF (or other cognate ligands) results in the activation of a network of signal transduction pathways that modulate cell division, apoptosis, motility, and adhesion.[52–55] EGFR (HER-1/erbB), HER-2 (Neu/erbB-2), HER-3 (erbB-3), and HER-4 (erbB-4) constitute four members of the EGFR family. Dysregulation of EGFR family signaling can result from gene amplification and/or mutations, effecting overexpression and, in some cases, constitutive activity correlating with certain cancers (e.g., lung, gliomas, breast, and ovarian). Recently, two small-molecule inhibitors have advanced clinically to become the first approved EGFR kinase inhibitors for the treatment of specific cancers. These first-generation EGFR tyrosine kinase inhibitors are ZD1839 (Iressa, gefitinib) and OSI-774 (Tarceva, erlotinib, CP-358774); they represent noteworthy ATP-competitive, small molecules based on a quinazoline template (**Figure 4**).

ZD1839[56–61] is a potent EGFR tyrosine kinase inhibitor ($IC_{50} = 23$ nM) with high selectivity relative to other protein tyrosine kinases (e.g., HER-2, and VEGFR-1 and VEGFR-2) and serine/threonine kinases. ZD1839 inhibits the proliferation of several cancer cell lines (e.g., NSCLC, breast, ovarian, and colon), shows synergistic enhancement of the inhibitory action of single-agent cytotoxic drugs, and effects dose-dependent antitumor activity in vivo in nude

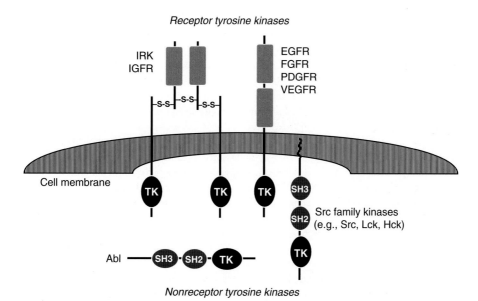

Figure 3 Schematic representation of the molecular architecture of receptor tyrosine kinases (e.g., the IRK, IGFR, EGFR, FGFR, PDGFR, and VEGFR families) and nonreceptor tyrosine kinases (e.g., the Src family and Abl) showing the catalytic tyrosine kinase (TK) as well as noncatalytic (SH2 and SH3) domains.

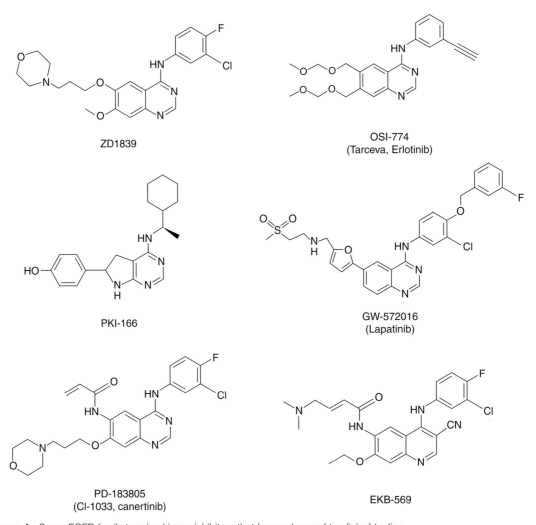

Figure 4 Some EGFR family tyrosine kinase inhibitors that have advanced to clinical testing.

mice bearing several human cancer cell xenografts. Very recently, a proteomics study identified cellular targets (beyond EGFR) for ZD1839, and showed that several other protein tyrosine kinases (e.g., Brk, Met, Yes, Hck, and EphB4) as well as protein serine/threonine kinases (e.g., RICK, GAK, Aurora-B, and p38 MAPK) were inhibited.[62] ZD1839 has been clinically tested and was first approved (in Japan) for the treatment of nonsmall cell lung cancer (NSCLC) as related to inoperable or recurrent disease; it has gained further approvals for previously treated NSCLC in the USA and other countries.

OSI-774[63–66] is also a potent inhibitor of EGFR tyrosine kinase ($IC_{50} = 20$ nM). It has high selectivity for other protein tyrosine kinases and high efficacy in vivo in nude mice bearing human cancer cell xenografts. Recently determined x-ray crystal structures of the EGFR tyrosine kinase domain, including both apoprotein and a complex with OSI-774, have provided the first experimental analysis of the EGFR family. OSI-774 effects a long-lasting inhibition of EGFR autophosphorylation in tumor xenografts (ex vivo analysis) following high-dose single administration. In combination with cisplatin, OSI-774 showed additive antitumor activity and enhanced apoptosis via suppression of the PKB/Akt survival pathway. Tarceva has been clinical tested and it has just been approved in the USA for the treatment of patients with locally advanced or metastatic NSCLC after failure of at least one prior chemotherapy.

Significant clinical attention has recently been focused on identifying the activating mutations in EGFR tyrosine kinase that underlie the responsiveness of NSCLC (primary tumors from patients) to Iressa.[67,68] In brief, this study indicated a correlation with NSCLC patients having marked response to Iressa, with a preponderance of somatic mutations in the ATP-binding domain of the EGFR gene. In this particular scenario, such EGFR mutations may stabilize the binding of some ATP-competitive inhibitors, such as ZD-1839, to afford greater sensitivity relative to wild-type EGFR. Recent clinical studies have also shown that acquired resistance to either Iressa or Tarceva may be correlated with a specific mutation of EGFR at the ATP binding site, namely Thr-790 substitution by Met (T790 M).[69] Overall, the above clinical results have a significant impact on drug discovery related to EGFR kinase inhibitors within the scope of profiling both wild-type and mutant EGFR kinases.

Relative to ZD1839 and OSI-774, some other noteworthy second-generation EGFR kinase inhibitors that have advanced to clinical testing include GW-572016, PKI-166, PD183805 (CI-1033, canertinib), and EKB-569 (**Figure 4**). These compounds have different biological profiles and mechanisms of action versus ZD-1839 and OSI-774. GW-572016[70–73] is a potent quinazoline-based inhibitor of EGFR ($IC_{50} = 11$ nM) and HER-2 ($IC_{50} = 9.2$ nM); it is able to block signaling from either receptor, which may be overexpressed in certain cancers (e.g., breast). GW-572016 inhibits proliferation in EGFR/HER-2 overexpressing cancer cells and was effective in vivo at high dose to suppress tumor growth in breast as well as head and neck cancer xenograft mice. PKI-166[74–77] illustrates a potent pyrrolopyrimidine-based dual inhibition of EGFR ($IC_{50} = 1$ nM) and HER-2 ($IC_{50} = 11$ nM), with additional inhibitor selectivity against Abl, Src, and VEGFR-2 tyrosine kinases ($IC_{50} = 26$, 103, and 327 nM, respectively). CI-1033[78–80] is a potent quinazoline-based inhibitor of EGFR ($IC_{50} = 1.5$ nM) and other EGFR family members. In contrast to the aforementioned quinazoline-based compounds, CI-1033 is chemically novel due to its incorporation of an acrylamide electrophilic moiety that was designed to interact covalently (as a Michael acceptor) with a proximate cysteine residue to the ATP binding pocket. This hypothesis was confirmed experimentally, and the uniqueness of this cysteine relative to other receptor or nonreceptor tyrosine kinases may account for the extraordinary selectivity properties of CI-1033 to inhibit EGFR. CI-1033 was a potent inhibitor in vitro and in vivo, including in several EGFR/HER-2-dependent cancer xenograft models, and it has been found to synergize with various cytotoxic agents and radiation. Like CI-1033, EKB-569[81–84] is also an acrylamide-containing molecule; however, EKB-569 is a cyanoquinoline template-based compound that has been shown to be a potent inhibitor of both EGFR ($IC_{50} = 1.3$ nM) and HER-2 ($IC_{50} = 15$ nM). EKB-569 inhibits EGFR and HER-2 expressing cell proliferation in vitro, and it is also found to be effective in vivo in human squamous cancer xenografts in mice. Furthermore, EKB-569, alone or in combination with a nonsteroidal, anti-inflammatory agent was determined to be effective in vivo in a mouse model of human colon cancer.

2.24.3.1.2 Vascular endothelial growth factor receptor, Flt-3, and KIT family kinases

In addition to the EGFR family kinases there are many other promising growth factor receptor kinase therapeutic targets for drug discovery efforts focused on cancer.[34–50] These include VEGFR family kinases (e.g., VEGFR1 to VEGFR3), PDGFR family kinases (e.g., PDGFR-α, PDGFR-β, CSF-1R, KIT, and Flt-3), insulin family kinases (e.g., IGF-1R and IRK), FGFR family kinases (e.g., FGFR-1 to FGFR-4), NGFR family kinases (e.g., TrkA, TrkB, and TrkC), HGFR family kinases (e.g., Met and Ron), EPHR family kinases (e.g., EPHA1 to EPHA8 and EPHB1 to EPHB6) and TIE family kinases (e.g., TIE, TEK, RYK, DDR, RET, and ROS).

Furthermore, there has emerged an increasing effort to understand the biology of tumors with respect to angiogenesis and lymphangiogenesis in order to identify therapeutic targets critical for the recruitment and formation of blood vessels and lymphatics.[85–87] One of the major pathways involved in both processes is the VEGF family of ligands and receptors (VEGFR-1/Flt-1 and VEGFR-2/Flk-1/KDR for angiogenesis, and VEGFR-3/Flt-4 for lymphangiogenesis). Overexpression of VEGF ligands has been associated with various cancers (e.g., colorectal, gastric, pancreatic, breast, prostate, lung, and melanoma). Upregulation of VEGFR expression occurs in activated endothelium and is high in blood vessels surrounding tumor tissue that is growing and invading. Some examples of first-generation VEGFR tyrosine kinase inhibitors that are noteworthy ATP-competitive, small-molecules based on a variety of different templates include SU-5416 (semaxanib), SU-6668, SU-11248, PTK-787 (CGP79787, vatalanib), ZD-6474, CEP-7055, and CP-547632 (**Figure 5**).

SU-5416,[88–91] a prototype lead compound of a series of indolinone template-based small-molecule inhibitors of VEGFR, was the first to advance to clinical testing with a focus on metastatic colorectal cancer in combination with chemotherapy. Unfortunately, the Phase II/III clinical studies showed no survival advantage for SU-5416, and thus further drug development was discontinued. A second indolinone analog, SU-6668,[91–93] was determined to be a very potent inhibitor of PDGFR ($IC_{50} = 8$ nM) relative to VEGFR-2 ($IC_{50} = 2.1$ μM) and it was quite effective in inhibiting VEGF-driven mitogenesis in cells. SU-6668 effected significant inhibition of a panel of human tumor xenografts by oral administration in nude mice, and it was further shown to exhibit in vivo antiangiogenic and antimetastatic activities in colon-cancer-cell injected mice. SU-6668 is in Phase II clinical trials for solid tumors. A third indolinone analog, SU-11248,[94–97] was determined to be a highly potent inhibitor of VEGFR-2 ($IC_{50} = 9$ nM) and PDGFR-β ($IC_{50} = 8$ nM) as well as relatively potent inhibitor of other receptor tyrosine kinases (FGFR-1, Flt-3, and KIT). SU-11248 effects potent inhibition of cell proliferation as driven by VEGF, PDGF-β, and FGF. SU-11248 was determined to be orally effective and showed potent in vivo antitumor activity in mice relative to several human xenografts. SU-11248 has also shown

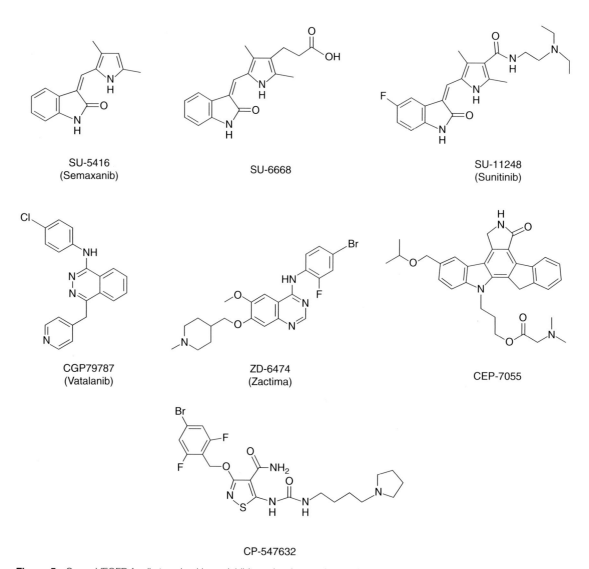

Figure 5 Some VEGFR family tyrosine kinase inhibitors that have advanced to clinical testing.

in vivo antiangiogenic activity in a lung cancer metastasis model. SU-11248 is undergoing Phase III clinical trials for various cancers. PTK-787,[98–101] a novel phthalazine-based small-molecule inhibitor, is a potent inhibitor of VEGFR-1 ($IC_{50} = 77$ nM) and VEGFR-2 ($IC_{50} = 37$ nM). PTK-787 effects antiangiogenic activity in vitro, including inhibition of VEGF-driven cell proliferation, survival, and cell migration. PTK-787 demonstrates effective in vivo antiangiogenic and antitumor activity in several animal models, and it has further shown in vivo antimetastatic activity, in both orthotopic renal cell carcinoma and human pancreatic xenograft models. ZD-6474[102–106] is an inhibitor of VEGFR-2 ($IC_{50} = 40$ nM), VEGFR-3 ($IC_{50} = 110$ nM), EGFR ($IC_{50} = 500$ nM), and RET ($IC_{50} = 100$ nM). ZD-6474 effectively inhibits endothelial cell proliferation driven by VEGF, EGF, and FGF. ZD-6474 has been shown to inhibit the growth of various human tumor xenografts in a dose-dependent manner, and it was found also to inhibit primary tumor growth in a spontaneous metastasis model of breast cancer at high dose. ZD-6474 is in Phase II clinical trials for solid tumors. CEP-7055,[107,108] a water-soluble prodrug of a indenopyrrolocarbazole template-based parent molecule (CEP-5214), is a potent inhibitor of VEGFR-1, VEGFR-2, and VEGFR-3 (IC_{50} in the range 12–18 nM). CEP-7055 is a potent inhibitor of tumor growth in several subcutaneous tumor xenografts, including decreasing the number of metastases. CEP-7055 is in Phase I clinical trials, with indications for solid tumors. CP-547632,[109] an isothiazole template-based molecule, is a potent inhibitor of VEGFR-2 ($IC_{50} = 11$ nM) and FGFR-2 ($IC_{50} = 9$ nM). CP-547632 was found to be an orally active inhibitor of VEGF-induced corneal angiogenesis in mice, and it was also effective in significantly inhibiting human xenograft in nude mice. CP-547632 is in Phase I clinical trials, with indications for solid tumors.

The functional relationship between Flt-3 receptor tyrosine kinase and acute myeloid leukemia (AML) has been implied by the fact that about 30% of AML patients have constitutively activating internal tandem duplications of the juxtamembrane domain or mutations in the activation loop of the Flt-3 receptor kinase.[110–112] First-generation Flt-3 kinase inhibitors, PKC-412 (CGP41251, midostaurin), CEP-701 (KT-5555), SU-11248 (see above), and MLN-518 (CT-53518), are particularly noteworthy in terms of their chemical diversity (**Figure 6**) as well as the fact that they have each advanced to clinical testing in patients with AML harboring Flt-3 activating mutations.

PKC-412,[113,114] a derivative of the natural product staurosporine, is an inhibitor of Flt-3 and several other protein kinases, including VEGFR-2, PDFGR, KIT, and Syk. CEP-701,[115] an indolocarbazole derivative, is a potent inhibitor of

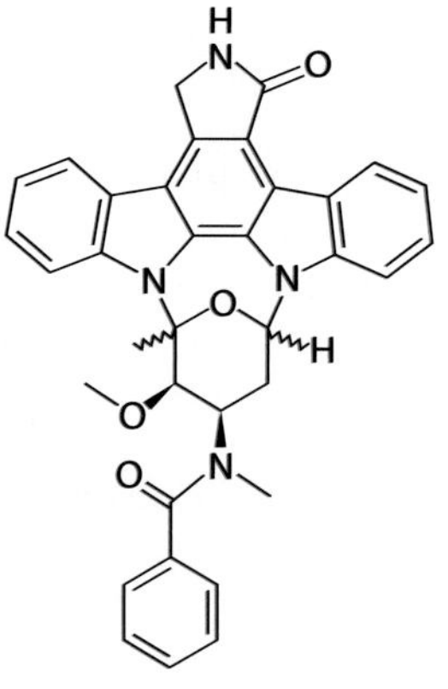

PKC-412
(CGP41251, midostaurin)

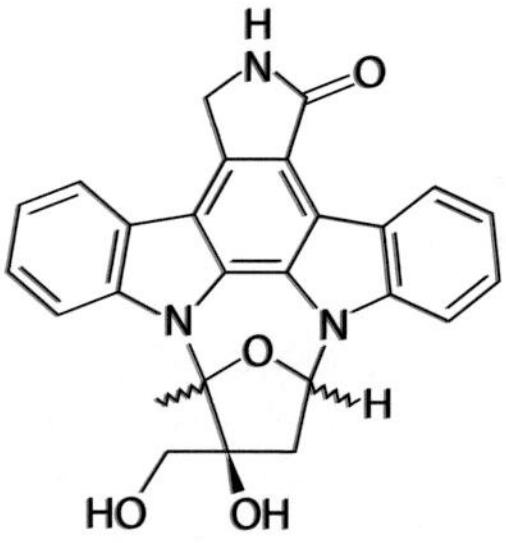

CEP-701
(KT-5555)

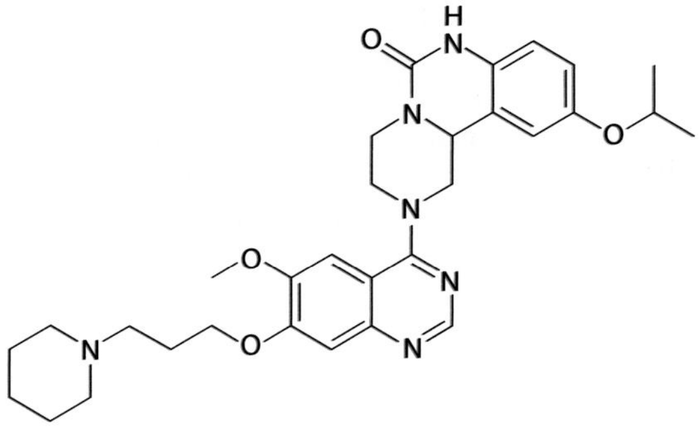

MLN-518
(CT-53518)

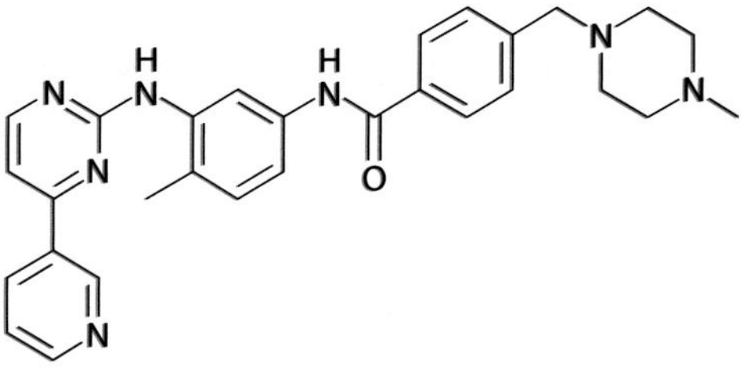

STI-571
(Gleevec/Glivec, imatinib)

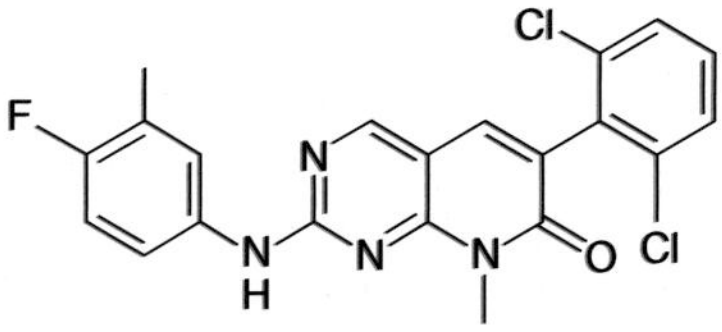

PD180970

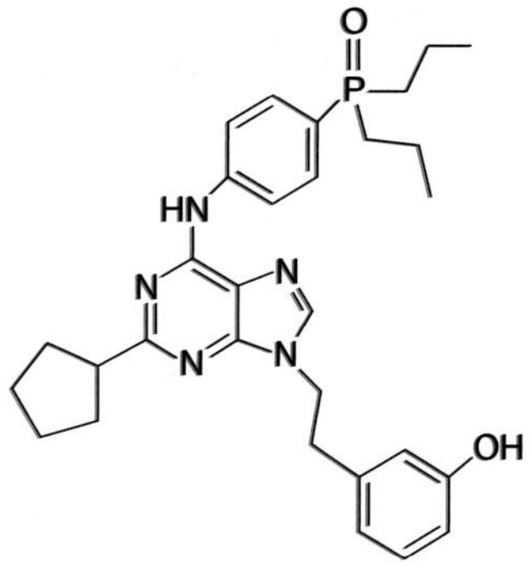

AP23848

Figure 6 Some Flt-3 and KIT tyrosine kinase inhibitors that have advanced to clinical testing.

Flt-3 and Trk tyrosine kinases. SU-11248 (see above) is also a potent inhibitor of Flt-3 tyrosine kinase. MLN-518,[116] a quinazoline template-based molecule, is a potent inhibitor of Flt-3, PDGFR, and KIT tyrosine kinases. Several studies have been described[113–119] that show the effectiveness of these compounds in vitro and in vivo, including the induction of apoptosis in cell lines with Flt-3 activating mutations and prolonging the survival of mice expressing mutant Flt-3 in their bone marrow cells. Prediction of resistance to some of the above small-molecule Flt-3 tyrosine kinase inhibitors using an in vitro screen designed to identify mutations of the ATP-binding pocket has shown several key amino acid mutations, including G697R, that confer resistance and potential clinical problems with a single-agent Flt-3 drug therapy.[120] In this regard, mutations and translocations involving both receptor (e.g., Flt-3 and KIT) and nonreceptor tyrosine kinases (e.g., Abl, see below) are now well recognized within the scope of transformed hematopoietic cells and underlying myeloid and acute leukemias.[121] Like the aforementioned clinical outcomes from studies on EGFR kinase inhibitors, such mutations enhance the complexity of small-molecule drug discovery relative to lead optimization as well as both in vitro and in vivo testing in disease models, which must reflect the inclusion of known mutant tyrosine kinase therapeutic targets.

KIT is a member of a family of growth factor receptor tyrosine kinases that includes RET and PDGFR. KIT is known to contribute to the growth and survival of cancers, especially those such as gastrointestinal stromal tumors (GISTs) in which it is frequently mutated and activated.[122] Co-expression of KIT and its ligand, stem cell factor (SCF), are known to exist in breast cancer, small-cell lung carcinoma, gynecological tumors, and malignant glioma cells. More than 30 gain-of-function mutations of KIT have been identified.[10] Juxtamembrane domain mutations of KIT are common in GISTs, while kinase activation loop mutations (e.g., D816V) are common in both systemic mastocytosis and acute myelogenous leukemia (AML). First-generation KIT tyrosine kinase inhibitors STI-571 (CGP57148, Gleevec/Glivec, imatinib), SU-6668 (see above), SU-11248 (see above), PTK-787 (see above), ZD-6474[11] (see above), PKC-412 (see above), MLN-518 (see above), PD180970, and AP23848, are particularly noteworthy small molecules based on a variety of different templates (**Figure 6**) as well as the fact that several compounds, including MLN518 and PKC-412 (each also being Flt-3 tyrosine kinase inhibitors, see above), are in clinical testing for AML.

Initial studies to investigate a KIT tyrosine kinase inhibitor with respect to KIT-driven cell lines, including both wild-type and mutant KIT, which provided proof-of-concept were achieved with STI-571.[123–125] Consistent with preclinical data showing that KIT wild-type and juxtadomain mutants were highly sensitive to STI-571, clinical testing of imatinib revealed that GIST patients having these mutations had a 30–70% response rate to drug therapy.[126] However, cells expressing KIT activation loop mutant D816V are resistant to imatinib, and patients with systemic mastocytosis involving this mutant are completely resistant to imatinib therapy.[123,127–129] With regard to identifying small-molecule inhibitors of such KIT tyrosine kinase mutants (i.e., imatinib-resistant), some progress has been achieved with several compounds, including SU-11652, MLN-518, PD180979, and AP23848.[130–132] For example, AP23848 has very recently been shown to be a highly potent inhibitor of KIT D816V mutant-expressing cell proliferation as well as to induce cell cycle arrest and apoptosis in vitro, and it was further effective in vivo in inhibiting KIT D816V mutant expressing tumor growth at high dose.

2.24.3.1.3 Bcr-Abl kinases

Of the nonreceptor tyrosine kinases that have emerged as highly promising therapeutic targets for cancer therapy, Bcr-Abl provides a fascinating story of mechanistic and structural biology involving mutations and the discovery of small-molecule inhibitors binding to either the so-called 'active' or 'inactive' conformations of the catalytic domain. Bcr-Abl is encoded by the Bcr-Abl oncogene, a translocation gene product found on the Philadelphia (Ph) chromosome in hematopoietic stem cells, and is a causative factor in chronic myelogenous leukemia (CML) and Ph+ acute lymphoblastic leukemia (ALL).[133] The chimeric Bcr-Abl protein is expressed with a constitutively active tyrosine kinase, and it has been found to be essential for Bcr-Abl driven transformation and excessive proliferation of Ph+ cells of myeloid and lymphoid lineage.

STI-571 (**Figure 6**, see above)[134,135] illustrates the first successful development of small-molecule inhibitor of Bcr-Abl kinase to advance into clinical testing and the first-in-class compound approved in the USA for the treatment of CML (and an increasing number of other cancer indications, including GIST, see above). STI-571 is a relatively potent inhibitor of Bcr-Abl ($IC_{50} = 250$ nM) and several other protein kinases, including KIT ($IC_{50} = 100$ nM) and PDGFR ($IC_{50} = 100$ nM). STI-571 has a novel chemical structure relative to previously known ATP-mimetic templates and it has a novel mechanism of inhibition (binding to an inactive conformation of the enzyme), as revealed by recent x-ray structures of STI-571 and analogs complexed with the Abl tyrosine kinase.[136,137] STI-571 was determined to bind to an inactive conformation of Abl kinase in which the active loop mimics bound peptide substrate. The potency of STI-571 against activated forms of Bcr-Abl probably results from the dynamic nature of protein kinases relative to switching between active and inactive conformations, thus allowing STI-571 to gain entry and effect inhibition. STI-571 was first determined to be effective in vivo with respect to the inhibition of tumor growth in a model developed using syngeneic mice injected with Bcr-Abl transformed cells,[138] and a plethora of both in vitro and in vivo studies have further examined in cellular mechanisms of action and pharmacological properties relative to Bcr-Abl as well as other therapeutic targets.[134] However, despite the successful drug development of imatinib, the issue of Bcr-Abl mutations effecting resistance in CML and ALL patients has confounded the overall effectiveness of STI-571, and this situation has provided an opportunity for second-generation, small-molecule inhibitors having superior potency to Bcr-Abl wild-type and many of the known clinically relevant mutants of Bcr-Abl tyrosine kinase.

Some second-generation Bcr-Abl kinase inhibitors that exemplify advances in potency and/or selectivity toward Bcr-Abl mutants relative to STI-571 include AMN-107, BMS-354825, ON012380, SKI-606, PD180970 (see above), PP1, CGP-76030, and AP23464 (**Figure 7**). Several compounds, including SKI-606, AZM-475271, PD-180970, PP1, CGP-76030, and AP23464, have been initially described as Src or Src family kinase inhibitors (see below). Both AMN-107 and BMS-354825 are in clinical trials for Bcr-Abl-dependent leukemias.

AMN-107,[139] a second-generation analog of STI-571, is a more potent inhibitor of Bcr-Abl kinase ($IC_{50} < 30$ nM) as well as being active against several STI-571-resistant mutants (e.g., E255 K, E255 V, F317L, and M351 T) except T315I. AMN-107 inhibits the proliferation of STI-571-resistant Bcr-Abl-expressing cells (except the T315I mutant). An x-ray structure of AMN-107 complexed with Abl kinase mutant M351 T showed that it binds in a similar mode as STI-571; however, the protein α-helix containing M351 T shifts slightly to accommodate binding of AMN-107 relative to its unique disubstituted aniline moiety, reflecting the chemical modification relative to STI-571. In contrast, the Bcr-Abl T315I mutant presumably blocks interaction of AMN-107 by sterically preventing the ability of the inhibitor to penetrate into the so-called 'hydrophobic specificity pocket' that exists proximate to the ATP binding site. AMN-107 is in Phase I clinical trials with indications for CML and ALL. BMS-354825,[140–142] a novel pyrimidinylaminothiazole template-based molecule, is a highly potent inhibitor of Bcr-Abl ($IC_{50} < 1$ nM), Src ($IC_{50} = 0.5$ nM), and KIT ($IC_{50} = 5$ nM), and it was determined to be further effective in inhibiting both PDGFR-β ($IC_{50} = 28$ nM) and EGFR ($IC_{50} = 180$ nM). An x-ray structure of BMS-354825 revealed its binding to the ATP pocket relative to a number of hydrogen-bonding interactions between the inhibitor and protein, with the disubstituted benzamide moiety fitting well into the hydrophobic specificity pocket (and further implicating the lack of potency against the T315I mutant). BMS-354825 has been described to be an effective inhibitor of several Bcr-Abl mutants (except for T315I) in vitro, and in vivo studies have shown that it prolonged survival of mice infused with Bcr-Abl wild-type or M351 T mutants. Further comparative analysis of BMS-354825 and STI-571 using saturation mutagenesis experiments showed that each molecule gave rise to an overlapping but different population of mutants, and that combination of the two inhibitors significantly reduced recovery of drug-resistant clones. BMS-354825 is in Phase II clinical trials for CML.

ON012380,[143] a novel small molecule that is not ATP competitive, is a highly potent inhibitor of Bcr-Abl ($IC_{50} = 1.5$ nM). Preliminary biochemical analysis indicates that ON012380 is substrate-competitive, and if further supported by structural studies, ON012380 would constitute an important breakthrough in small-molecule inhibitor drug discovery for protein kinases. ON012380 induces apoptosis of cells expressing Bcr-Abl as well as a number of mutants (including T315I) at low nanomolar concentrations. ON012380, at high dose, was effective in vivo in causing regression of leukemias induced by Bcr-Abl T315I mutant expressing cells injected into mice. SKI-606,[144,145] a novel quinoline–carbonitrile template-based molecule, is a highly potent inhibitor of Abl kinase ($IC_{50} = 1$ nM) as well as Src ($IC_{50} = 1.1$ nM). SKI-606 effects potent antiproliferative activity against CML cells in vitro, and in vivo studies further showed SKI-606, at high dose, to cause complete regression of CML xenografts in nude mice. PD180970[146–148] is a highly potent inhibitor of Abl ($IC_{50} = 2.2$ nM). PD180970 induces apoptosis in CML cells in vitro, and it was further shown to block Stat5 signaling and induce apoptosis in a Bcr-Abl high-expressing cell line that is resistant to STI-571. PD180970 was also determined to inhibit several STI-571-resistant Bcr-Abl mutants in vitro, with the exception of T315I. A related pyridopyrimidine, PD166326,[149,150] has been shown to be highly potent against Bcr-Abl kinase and several Bcr-Abl mutants in vitro. PP1[151,152] and a related pyrazolopyrimide template-based analog, PP2,[153] were found to be effective inhibitors of Bcr-Abl kinase in vitro. CGP76030,[153] a pyrrolopyrimidine template-based molecule, has been shown to be an effective inhibitor of Bcr-Abl kinase and STI-571-resistant Bcr-Abl mutants (except for T315I) in vitro. AP23464,[154] a purine template-based compound, has been determined to be a highly potent inhibitor of Abl kinase ($IC_{50} = 1$ nM) and to be effective in inhibiting Bcr-Abl and STI-571-resistant Bcr-Abl cell lines (except for T315I) in vitro. AP23464 effectively ablated Bcr-Abl tyrosine phosphorylation, blocked cell cycle progression, and promoted apoptosis in Bcr-Abl-expressing cells.

Beyond the development of Bcr-Abl kinase inhibitors capable of inhibiting essentially all clinically relevant Bcr-Abl mutants, including the challenging T315I mutant, a combination of Bcr-Abl tyrosine kinase inhibitors may provide a strategy to provide a potentially fail-safe therapy for CML or Ph + ALL patients. Furthermore, combination strategies

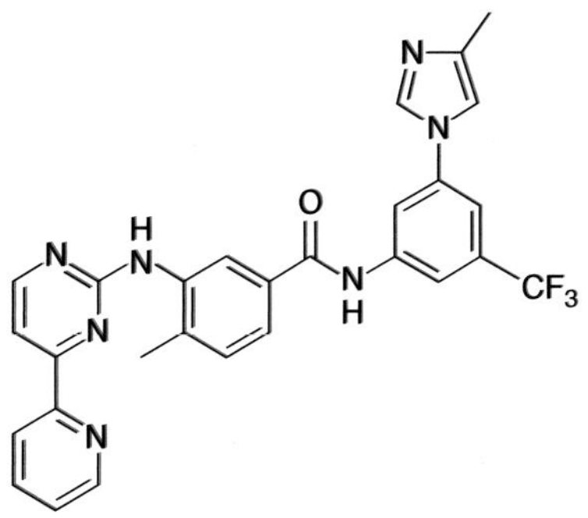

AMN-107

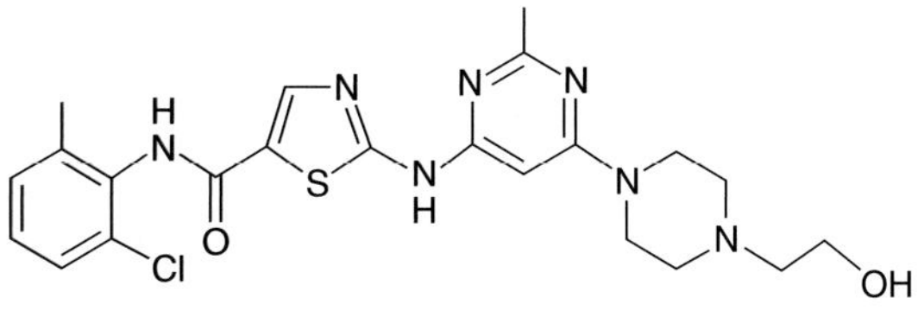

BMS-354825
(Dasatinib)

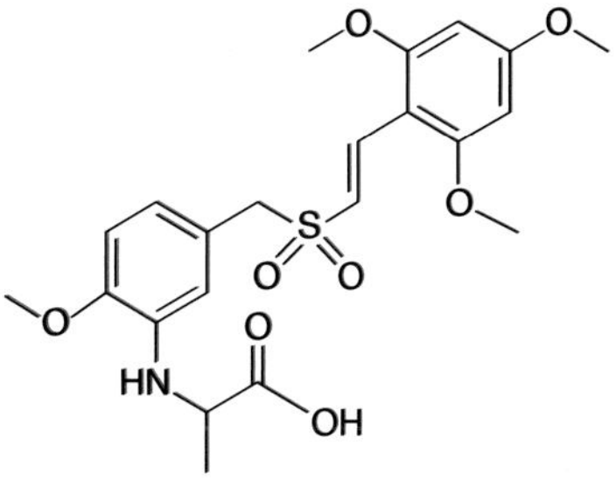

ON012380

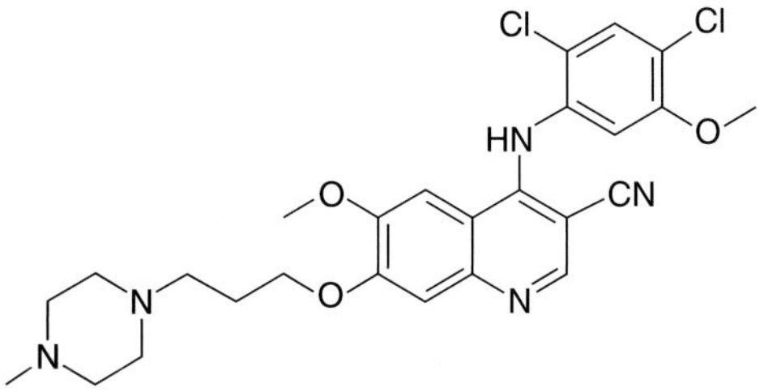

SKI-606

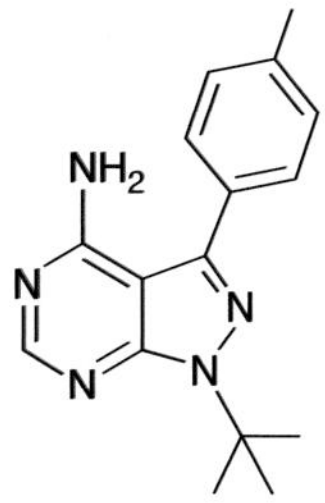

PP1

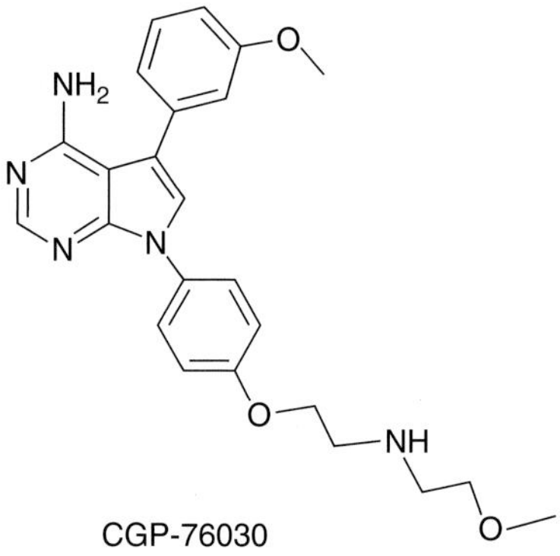

CGP-76030

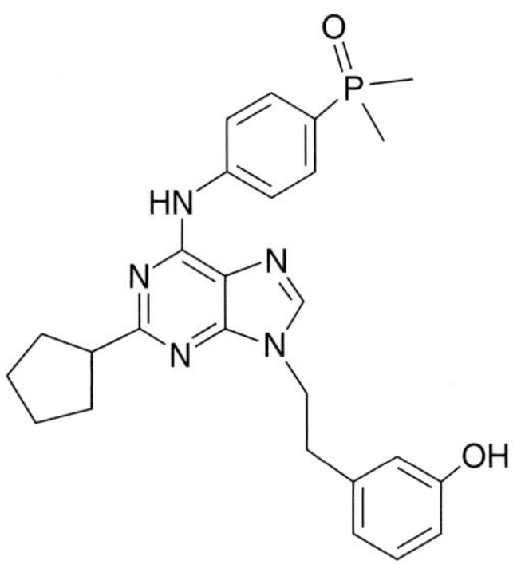

AP23464

Figure 7 Some Bcr-Abl and dual Src and Bcr-Abl tyrosine kinase inhibitors, including several that have advanced to clinical testing.

may include imatinib (or other Bcr-Abl kinase inhibitor) together with another drug that acts on a different, but mechanistically related, signal transduction pathway therapeutic target to overcome STI-571 resistance.[155] In this regard, the farnesyl protein transferase inhibitor SCH66336[156,157] and PI3K inhibitor OSU-013012[158] have recently been shown to enhance the antiproliferative properties of imatinib in both Bcr-Abl- and STI-571-resistant Bcr-Abl-expressing cells (including T315I) in vitro.

2.24.3.1.4 Src family kinases

As a prototype of the superfamily of protein tyrosine kinases to be identified and characterized, Src tyrosine kinase has been the focus of a great many molecular, cellular, and structural biology studies designed to understand its role in signal transduction pathways as well as in disease processes, including cancer, osteoporosis, and both tumor- and inflammation-mediated bone loss.[10,36,158–167] Such pioneering studies on Src have provided some of the first evidence correlating protein kinase activity and substrate protein phosphorylation in the regulation of various signal transduction pathways critical to cell growth, cell cycle, cell migration, cell survival, and malignant transformation of mammalian cells. Src family kinases include Fyn, Yes, Yrk, Blk, Fgr, Hck, Lyn, and the Frk subfamily members Frk/Rak and Iyk/Bsk. These Src family kinases exhibit a broad spectrum of functional properties, including cell growth, differentiation, survival, cytoskeletal alterations, adhesion, and migration. As the result of many landmark studies,[168–190] the molecular mechanisms of Src-dependent cancer cell biology (e.g., cell–cell and cell–matrix adhesion, cell motility, and invasion), including cancer metastasis and angiogenesis, have been revealed (**Figure 8**), as described below. Studies using Src-specific antisense DNA have shown that elevated Src expression and/or activity is correlated with tumor growth in specific cancers having HER2 or c-Met receptors. Elevated Src expression and/or activity has been found in breast cancer cell lines and malignant breast tumors. Src has been implicated in metastatic colon cancer, head and neck cancers, and pancreatic cancer. Activating Src mutations in advanced human colon cancer have also been identified. Src has been implicated in malignant transformations for certain cancers, such as breast cancer and multiple myeloma, via EGF-R or interleukin-6 receptor (IL6-R) signaling pathways, respectively, which commonly activate the transcription factor known as 'signal transducing activators of transcription-3' (STAT3). Aberrant activation of STAT signaling pathways have been linked to oncogenesis with respect to the prevention of apoptosis. Relative to integrin receptors and the dynamic relationship between cell–cell and cell–matrix interactions, Src-induced deregulation of E-cadherin in colon cancer cells has been determined to require integrin signaling, and Src kinase activity is required for adhesion

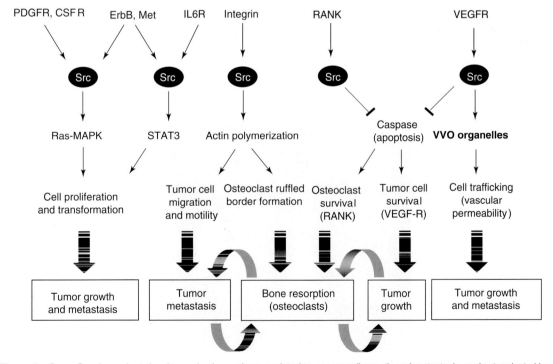

Figure 8 Some Src-dependent signal transduction pathways related to cancer cell growth and metastasis, and osteoclast-driven bone resorption. (Adapted from Sawyer, T.; Boyce, B.; Dalgarno, D.; Iuliucci, J. *Expert Opin. Investig. Drugs* **2001**, *10*, 1327–1344.)

turnover associated with cancer cell migration. Collectively, such findings are consistent with the correlation of Src kinase activity (via overexpression of CSK and/or dominant-negative mutants of Src) with cellular and in vivo metastasis. With respect to VEGFR, Src is intimately involved in VEGF-mediated angiogenesis and vascular permeability. In particular, the ability of VEGF to disrupt endothelial barrier function, which has been correlated with tumor cell extravasation and metastasis, is mediated through Src kinase. In recent years, gene expression fingerprinting and RNA interference-based analysis of Src-transformed in cancer cells have provided insight to explore further the Src-induced signal transduction and gene activation pathways.

With respect to its structural architecture, Src possesses both catalytic and noncatalytic regulatory motifs (i.e., SH3 and SH2 domains; see below), and these are functionally important in signal transduction processes. The molecular basis of Src activation has been revealed further by structural biology studies, including x-ray structures of near full-length Src (i.e., SH3-SH2–tyrosine kinase).[191–193] From such investigations, it has been shown that Src can exist in an assembled, inactive conformation (**Figure 9**) involving intramolecular binding of the SH2 domain with the carboxyl terminal tail (phosphorylated at Tyr-527), as well as intramolecular binding of the SH3 domain with a linker sequence between the SH2 domain and the amino terminal lobe of the tyrosine kinase. Src activation is believed to involve displacement of the imperfect intramolecular SH3 and SH2 interactions within the inactive conformation by intermolecular binding with SH3 and/or SH2 cognate proteins, and subsequent phosphorylation at Tyr-416 (kinase domain) and dephosphorylation at Tyr-527.

Of relevance to Src family kinases, an x-ray structures of Lck have been determined[194] as complexes of the tyrosine kinase domain with AMP-PNP, staurosporine, and PP2. The Lck tyrosine kinase–PP2 complex revealed the detailed molecular interactions of the ATP-mimetic pyrazolopyrimidine template with respect to conserved hydrogen bonding (experimentally comparable with the AMP–PNP complex). The inhibitor PP2 binds into the large, hydrophobic, specificity pocket that is proximate, but not accessed by, AMP–PNP (or ATP). It is noteworthy with respect to the entire superfamily of protein kinases, that this hydrophobic specificity pocket varies in size, shape, and amino acid composition with further relationship to the conformation (active or inactive) of the enzyme and, therefore, provides insight for future drug discovery. Src kinase has been exploited as a prototype model for protein engineering approaches to mutate the ATP-binding pockets of protein kinase, with the objective of enhancing selectivity for synthetic ATP analogs or inhibitors.[195–197] In brief, mutation of a conserved amino acid in the ATP binding pocket was made to create a unique new site that would accommodate a synthetic ATP substrate analogue, namely, [γ-^{32}P]-N^6-(benzyl)-ATP, which then provides a matched set of enzyme–substrate to explore signal transduction pathways with respect to the identification of cellular substrates under varying experimental conditions. The design of Src kinase inhibitors has focused on a number of strategies,[5,35,36,81,162,163,198,199] including ATP template-related mimetics, novel heterocyclic leads (from screening corporate chemical collections and/or combinatorial libraries), natural products, and peptide substrate-based compounds.

Some first- and second-generation Src (and Src family) kinase inhibitors that exemplify a variety of different templates include BMS-354825 (see above), SKI-606 (see above), PD180970 (see above), PP1 (see above),

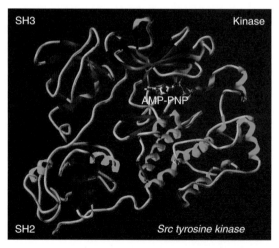

Figure 9 X-ray structure of Src SH3-SH2-tyrosine kinase complexed with AMP-PNP, showing a downregulated, inactive conformation of the protein. (Adapted from Xu, W.; Harrison, S. C.; Eck, M. J. *Nature* **1997**, *285*, 595–602.)

CGP-76030 (see above), AP23464 (see above), AZM-475271, SU-6656, and AP23451 (**Figure 10**). BMS-354825 and AZD-530 (an analog of AZM-475271 with unpublished chemical structure) are in clinical trials for Src-dependent cancers. BMS-354825[140] is a highly potent inhibitor of Src kinase as well as Bcr-Abl kinase (see above). There are as yet no reports that describe the biological properties of BMS-35825 or AZD-530 with respect to Src-dependent disease models in vitro or in vivo.

SKI-606,[144,145,200–202] is a highly potent inhibitor of Src kinase ($IC_{50} = 1.1$ nM). SKI-606 exhibits antiproliferative efficacy against Src-transformed fibroblasts in vitro and in vivo in a xenograft model. PD180970 and related pyridopyrimidine analogs[146–148,203–205] have been advanced as potent inhibitors of Src kinase with varying selectivities to PDGFR, FGFR, and EGFR tyrosine kinases. PP1 and the related pyrazolopyrimidine analog PP2 were first described[206] as potent inhibitors of Src-family kinases with marked selectivity versus ZAP-70, JAK2, EGF-R, and PKA kinases. PP1 has provided a key Src kinase inhibitor to determine the role of Src in VEGF-mediated angiogenesis and vascular permeability.[186,187] PP1 is an effective inhibitor of Src-driven human breast cancer cells with respect to heregulin-dependent or independent growth.[207] pp1 has also been reported[208] to inhibit collagen type-I-induced E-cadherin downregulation and the consequent effects on cell proliferation and metastatic properties. CGP-76030 and related pyrrolopyrimidine analogs have been described as potent and selective inhibitors of Src kinase in vitro and in vivo in animal models of osteoporosis.[209,210] CGP-76030 has been determined to reduce growth, adhesion, motility, and invasion of prostate cancer cells.[211] AP23464,[35] a purine template-based compound, is a highly potent inhibitor of Src tyrosine kinase ($IC_{50} < 1$ nM), and it has recently been utilized[212] to examine the functional relationship of Src and FAK in adhesion turnover associated with migration of colon cancer cells. AZM-475271,[213] a quinazoline template-based compound, is a potent inhibitor of Src kinase and has been determined to be effective in inhibiting tumor growth in Src-transformed cell xenograft mice as well as in orthotopically implanted pancreatic cancer cells in nude mice. Such studies on AZM-475271 have supported the concept of the therapeutic potential of a Src kinase inhibitor for cancer invasion and metastasis. The combination of AZM-475271 and gemcitabin has also been shown to be particularly effective in the pancreatic cancer model. SU-6656[214] has been described as a potent inhibitor of Src tyrosine kinase (as well as Lck, Fyn, and Yes tyrosine kinases) and to inhibit PDGF-stimulated DNA synthesis and Myc induction in a fibroblast cell line. AP23451,[36,185,215] a purine template-based compound, is a potent inhibitor of Src kinase ($K_i = 8$ nM) and incorporates a bone-targeting moiety that confers to it a unique property to bind to bone surface (hydroxyapatite) and inhibit Src-dependent osteoclast activity. AP23451 analogs have been described[216] that detail the structure–activity properties of various bone-targeting substituents, and other templates (e.g., pyridopyrimidine, pyrrolopyrimidine, and pyrazolopyrimidine) have also been elaborated with bone-targeting groups to advance novel Src inhibitors for use in osteolytic bone metastasis and bone diseases.[217,218] A recent determination[219] of the x-ray structures of Src kinase complexed with AP23464 and AP23451 provides the first detailed molecular mapping of this therapeutic target with small-molecule inhibitors. Such three-dimensional information also provides insight to understand the various modes of binding and the comparative similarities and differences in molecular recognition amongst the aforementioned Src kinase inhibitors and other chemically novel molecules, including natural products, combinatorial library-based compounds, and substrate-based analogs.[220–223]

2.24.3.2 Receptor and Nonreceptor Serine/Threonine Kinases

Protein serine/threonine and dual specificity protein kinases are represented by both receptor types (e.g., TGFβR) and nonreceptor types (e.g., the CDK family, Raf, MEK, MAPK, PI3K, PKB/Akt, and mTOR). Due to the fact that, to an extraordinary degree, the primary type of protein phosphorylation in cells is at serine/threonine, there have been identified numerous promising cancer therapeutic targets from this major subfamily of protein kinases. A few examples of receptor and nonreceptor serine/threonine kinase inhibitors are highlighted below to illustrate recent advances in basic research and drug discovery.

2.24.3.2.1 TGFβR kinases

The transforming growth factor ligand TGFβ is critically involved in the progression of fibrosis as well as tumor invasiveness and metastasis[224,225] through binding to the receptor serine kinase TGFβR, of which two subtypes (TGFβR1 and TGFβR2) exist. TGFβR1 activation subsequently leads to the phosphorylation of transcription factors (e.g., SMADs). Uncontrolled cell growth due to decreased growth-inhibiting activity is a major feature of a defect in TGFβ function, and evidence is emerging that common variants of the TGFβ pathway (ligand and receptors) that alter TGFβ signaling may modify cancer risk. For example, a variant of the TGFβR1 gene known as *TGFβR1*6A* has been correlated with decreased TGFβ-mediated growth inhibition and increased cancer risk. Interestingly, while decreased TGFβ signaling increases cancer risk, TGFβ secretion and activated TGFβ signaling enhances the aggressiveness of

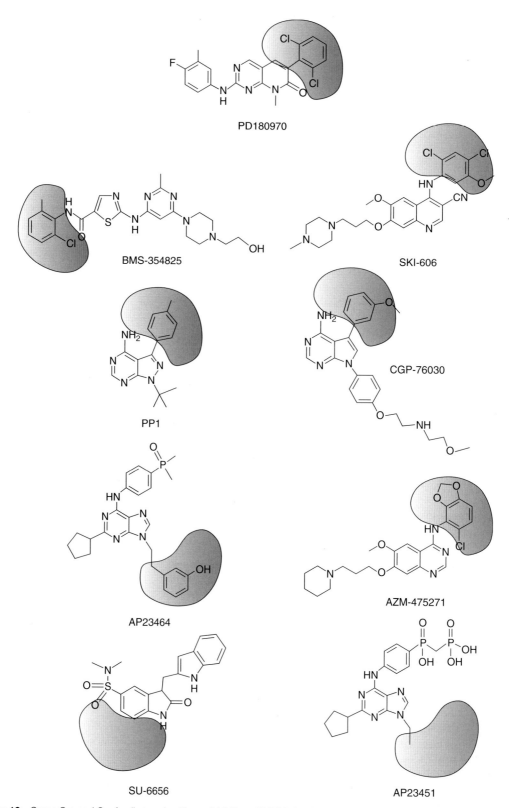

Figure 10 Some Src and Src family tyrosine kinase inhibitors, highlighting the structural groups that are predicted to bind to the hydrophobic specificity pocket.

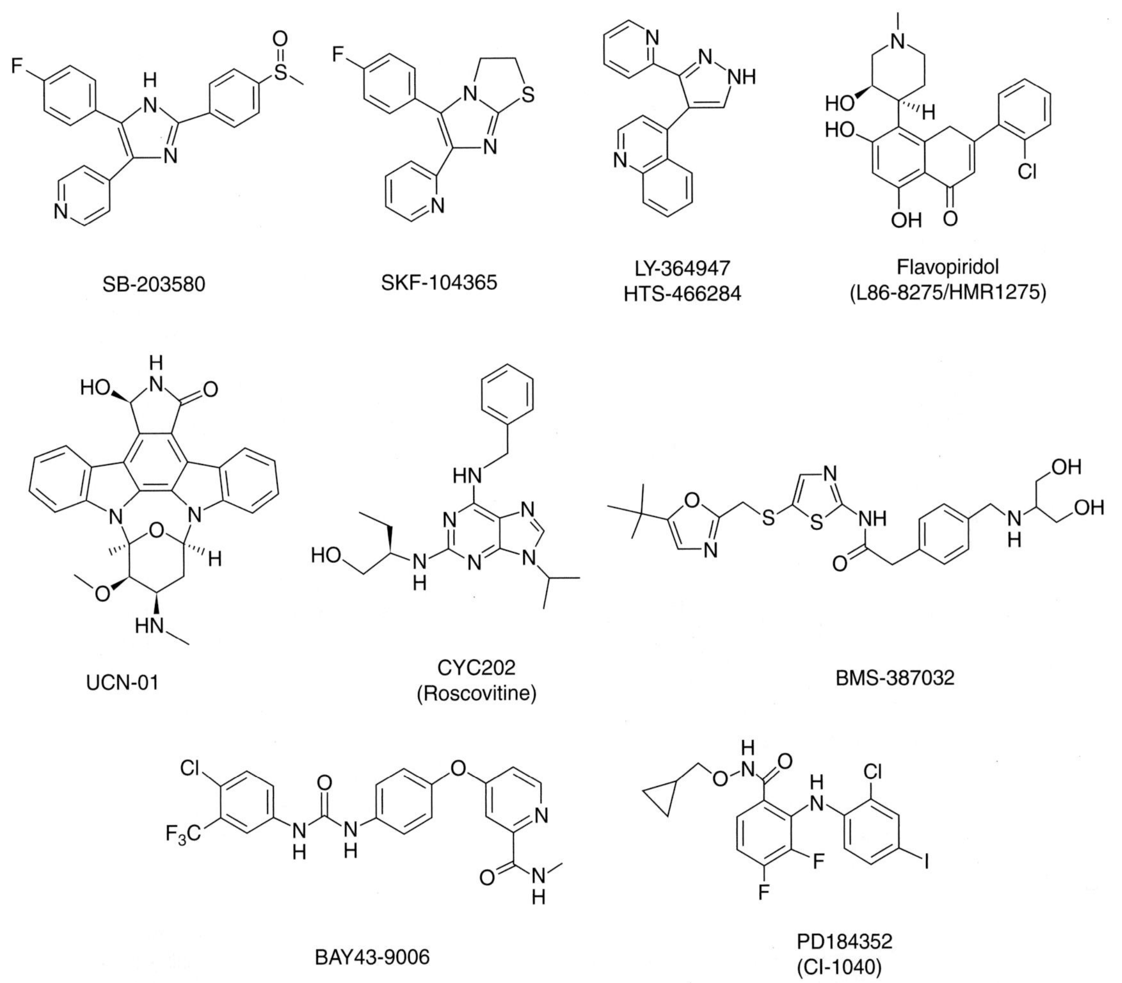

Figure 11 Some TGFβR1, CDK, PKC, Raf, and MEK kinase inhibitors, including several that have advanced to clinical testing.

several types of tumors. Noteworthy in this respect are the first-generation TGFβR1 serine kinase inhibitors SB-203580, SKF-104365, and LY-364947/HTS-466284, which exemplify small molecules based on substituted imidazole or pyrazole templates (**Figure 11**). SB-203580,[226] a trisubstituted imidazole template-based compound, moderately inhibits TGFβR1 serine kinase autophosphorylation ($IC_{50} = 20\,\mu M$) and was discovered by cross-screening small-molecule inhibitors of p38 MAPK. SKF-104365[227] was found to be a relatively potent inhibitor of TGFβR1 phosphorylation of SMAD3 ($IC_{50} = 1.6\,\mu M$) and was selective relative to p38 MAPK. LY-364947/HTS-466284[228,229] was independently discovered by screening- and structure-based strategies and is a highly potent inhibitor of TGFβR1 autophosphorylation ($IC_{50} = 51\,nM$). Finally, an x-ray structure of LY-36494/HTS-466284 complexed with TGFβR1 serine kinase was also independently determined in these studies.

2.24.3.2.2 Cyclin-dependent kinase family kinases

In a majority of cancers there are aberrations in cell cycle progression, and it has been determined that the tumor suppressor retinoblastoma (Rb) family of proteins act as a master switch to regulate cell cycling. Cell proliferation is driven by Rb proteins via their phosphorylation by cyclin-dependent kinases (CDKs).[230–233] The CDK family of serine kinases include those activated by D-type cyclins (D1, D2, and D3) and cyclin-E, and are inhibited by two families of CDK inhibitors (CKIs), which include Ink and the Cip/Kip subfamilies. Rb is an important regulator of G1/S transition, and its function is abnormal in most cancers. Loss in Rb function occurs by CDK hyperactivation, hence implicating CDK inhibition as a promising strategy for drug discovery. CKIs have further been shown to have additional functions such as the regulation of Rho signaling and the control of cytoskeletal organization and cell migration. In addition, upregulated cytoplasmic CKIs appear to be involved in tumor invasion and metastasis.[234] With respect to the CDK family, the prominent therapeutic targets for small-molecule inhibitor drug discovery have been CDK1, CDK2, CDK4, and CDK5. The first- and second-generation CDK serine kinase inhibitors flavopiridol (L86-8275/HMR1275), UCN-01, CYC202 (roscovitine), and BMS-387032 are particularly noteworthy small molecules that have advanced to clinical testing (**Figure 11**, see above).

Flavopiridol[231,235–238] is a semisynthetic flavonoid natural product derivative that is a potent inhibitor of CDKs (IC_{50} values ~ 100 nM). Flavopiridol effects G1/S and G2/M arrest by inhibition of CDK1 and CDK2. Furthermore, it has been shown to induce apoptosis and to effect antiangiogenesis properties. An x-ray structure of a flavopiridol analog complexed with CDK2 has revealed its binding to the ATP pocket. UCN-01,[239–241] a staurosporine analogue initially developed as a protein kinase C (PKC) inhibitor, is a potent inhibitor of CDK1 ($IC_{50} = 31$ nM) and CDK2 ($IC_{50} = 30$ nM), as well as PKC ($IC_{50} = 6.8$ nM). UCN-01 inhibits the production of phosphorylated Rb, and induces apoptosis in cancer cells. Recently, UCN-01 has been shown to be a potent inhibitor of Pdk1 serine/threonine kinase, which then results in the inactivation of Akt/PKB, thus modulating the PI3K survival pathway. UCN-01 effects antitumor activities in xenograft mice models of breast cancer, renal cancer, and leukemias. CYC202,[242,243] a purine template-based compound, is a potent inhibitor of CDK2/cyclin E kinase ($IC_{50} = 100$ nM), and it effects cytotoxicity against many human cancer cell lines (IC_{50} ~ 10 µM range). CYC202 induces cell death from all compartments of the cell cycle, as was shown in colorectal cancer cells in vitro; in vivo studies in nude mice bearing this cancer showed that CYC202 effected significant antitumor activity at high dose. BMS-387032,[244] a substituted aminothiazole-template-based molecule, is a potent and selective inhibitor of CDK2/cyclin-E ($IC_{50} = 48$ nM). BMS-387032 effects potent antiproliferative activity in cancer cells in vitro, and in vivo it produced significant antitumor activity in a human ovarian carcinoma xenograft model. Collectively, such CDK serine kinase inhibitors have been shown to be quite effective, and combination studies with either conventional cytotoxic drugs or other novel signal transduction modulatory drugs are expected to be useful in the clinical treatment of certain cancers.

2.24.3.2.3 Raf-MEK-MAPK pathway kinases

The well-established functional relationship between Ras signal transduction and cancer[245,246] was hallmarked by the first identification of oncogenic and mutationally activated forms of *Ras* genes more than two decades ago. Mutated Ras alleles are known to exist in about 30% of human cancers, especially those with poor survival indications (e.g., lung and pancreatic carcinomas). Ras can also be abnormally activated by other mechanisms, such as overexpressed or mutationally activated EGFR. Among the therapeutic targets intimately involved in the Ras signal transduction are those intimately related to the Raf-MEK-MAPK pathway, including Ras farnesyl protein transferase, Raf,[247] and MEK kinase inhibitors.[248] EGFR inhibitors (both small-molecule tyrosine kinase inhibitors and monoclonal antibodies to the extracellular receptor) and inhibitors of gene expression (i.e., H-*Ras* and *Raf*) provide additional upstream and downstream drug discovery strategies for the Raf-MEK-MAPK pathway. First-generation Raf serine kinase and MEK dual specificity kinase inhibitors BAY43-9006 and PD184352 (CI-1040) are particularly noteworthy small molecules that have advanced to clinical testing (**Figure 11**; see above). BAY43-9006,[249,250] a novel biaryl-urea template-based molecule, is a potent inhibitor of Raf-1 serine kinase as well as several receptor tyrosine kinases, including VEGFR-2, PDGFR-β, Flt-3, and KIT. BAY 43-9006 effectively inhibits the MAPK pathway in cancer cell lines (e.g., colon, pancreatic, and breast) expressing mutant K-RAS or wild-type or mutant BRAF. BAY 43-9006 was effective in vivo against a broad-spectrum antitumor activity in colon, breast, and NSCLC xenograft models, and additional evidence showing its inhibition of neovascularization in these xenograft models implicates BAY43-9006 as an in vivo dual inhibitor of both Raf and VEGFR. PD184352,[251,252] a novel substituted anthranilic acid template-based molecule, is a potent and highly specific inhibitor of MEK ($IC_{50} = 1$ nM), blocking the phosphorylation of ERK and downstream signal transduction thereof. PD184352 is not ATP-competitive, but is an allosteric inhibitor of MEK. PD184352 effects significant antitumor activity in vitro and in vivo (e.g., pancreas, colon, and breast cancers), and a correlation with the inhibition of ERK phosphorylation has been determined.

2.24.3.2.4 PI3K-Akt/PKB-mTOR pathway kinases

Like the Ras signal transduction and its biological relevance to cancer, the relationship between the PI3K-Akt/PKB-mTOR pathway and cancer has been investigated intensively,[253–260] and significant evidence exists that dysregulation of the PI3K-Akt/PKB-mTOR pathway may include loss of the PI3K suppressor protein PTEN, PI3K mutations (constitutive activating), and/or other downstream complications involving Akt and mTOR, of which the latter integrates signal transduction from both growth factors and nutrients. Activation of the PI3K-Akt/PKB-mTOR pathway potentiates cell survival and proliferation as well as cytoskeletal function and motility to impact on tumor invasion. Furthermore, the PI3K-Akt/PKB-mTOR pathway is implicated in angiogenesis, including upregulation of angiogenic

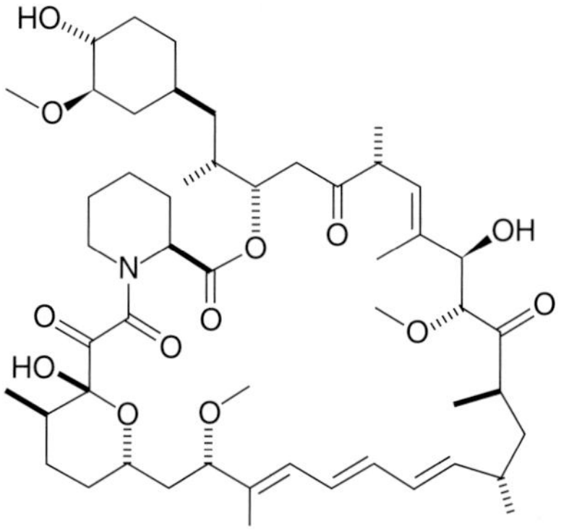

Figure 12 mTOR inhibitors that have advanced to clinical testing.

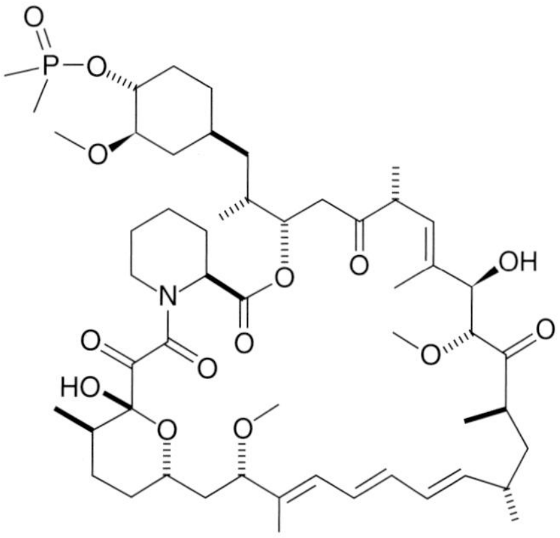

cytokines. The first-generation mTOR inhibitor rapamycin (**39**, sirolimus) and the second-generation rapamcyin analog AP23573 (**40**) are noteworthy natural product-based molecules that have advanced to clinical testing (**Figure 12**). The antiproliferative properties of rapamycin were identified more than two decades ago, and rapamycin has been developed as an immunosuppressive and antifungal agent. Rapamycin has provided a powerful proof-of-concept lead compound in terms of exploiting its potent and specific inhibitory properties against mTOR[253–255,258,259] in both in vitro and in vivo cancer models.

AP23573[253–255] exemplifies a highly potent analog of rapamycin, which incorporates a novel modification of the macrolide template with a dimethylphosphoryl moiety. AP23573 and other rapamycin analogs (e.g., CCI-779 and RAD-001)[253–255] have advanced to clinical testing for the treatment of many cancers.

2.24.4 Protein Phosphatase Therapeutic Targets and Drug Discovery

More than 100 protein phosphatases have been identified[11–14] from the human genome, and these constitute a superfamily including protein tyrosine phosphatases (e.g., PTP1B, SHP2, PEST, PTPH1, PTPα and CD45), dual-specificity protein phosphatases (e.g., VHR, CDC25, PTEN, and MKP-4), and protein serine/threonine phosphatases (e.g., PP1, PP2A, and PP2B/calcineurin). Both receptor and nonreceptor tyrosine phosphatases have vital roles in signal-transduction pathways to regulate cell growth and proliferation, cell-cycle progression, cytoskeletal integrity, differentiation, and metabolism. In particular, there is intense interest in the tyrosine phosphatase PTP1B, which dephosphorylates activated insulin receptor kinase, as a result of PTP1B gene knockout. Other studies have shown its functional role as a negative regulator of insulin receptor signaling.[261–263] Such work suggests that novel small-molecule inhibitors might improve insulin sensitivity in type II diabetes and/or be effective in the treatment of obesity.[264–266] Dual-specificity phosphatases are generally categorized as a subclass of protein tyrosine phosphatases, and they are uniquely able to hydrolyze a phosphate ester bond on either a tyrosine, serine, or threonine residue of a substrate phosphoprotein. Dual-specificity phosphatases have crucial roles in intracellular signal-transduction pathways, and are most prominently known for regulating the mitogen activated protein kinase (MAPK) signal transduction pathways and cell-cycle progression. Key dual-specificity phosphatases include the MAPK phosphatase VHR and the CDK phosphatase cell-division cycle 25 (CDC25). Protein tyrosine phosphatases differ both in terms of substrate structural recognition and the catalytic mechanism of substrate hydrolysis from protein serine/threonine phosphatases. Protein serine/threonine phosphatases are metalloenzymes, whereas protein tyrosine phosphatases hydrolyze phosphotyrosine substrates in a two-step mechanism involving a conserved cysteine residue to form a thiol–phosphate intermediate, followed by water-mediated hydrolysis assisted by key active site residues in the enzyme.

2.24.4.1 Receptor and Nonreceptor Tyrosine Phosphatases

Some noteworthy first-generation PTP1B tyrosine phosphatase inhibitors include compounds **1–3** (**Figure 13**). Compound **1**,[267] which was generated using a combinatorial library strategy, is a potent inhibitor of PTP1B ($K_i = 2.4$ nM)

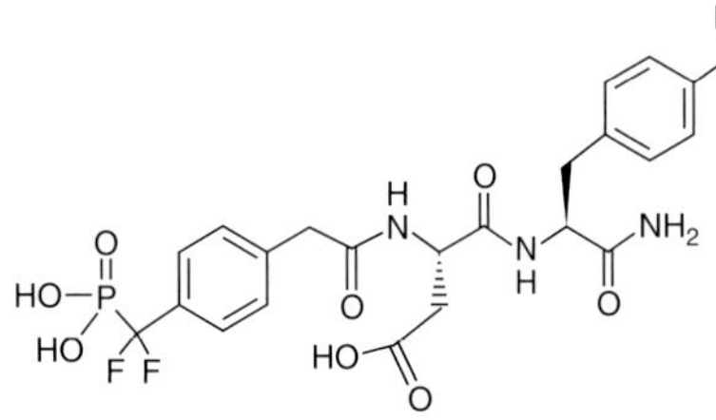

Figure 13 Some PTP1B, PP1, PP2A, and PP2B phosphatase inhibitors.

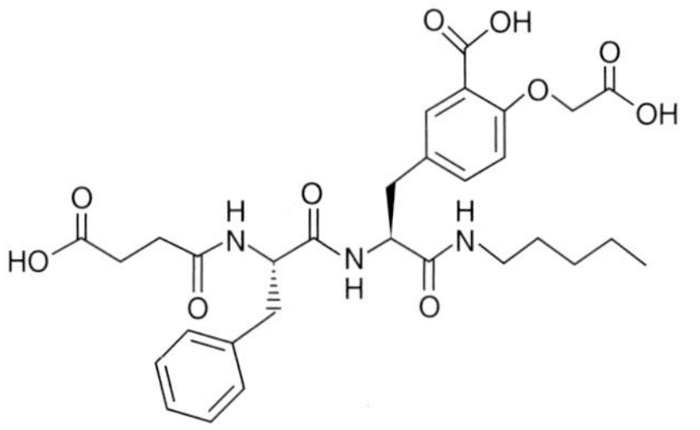

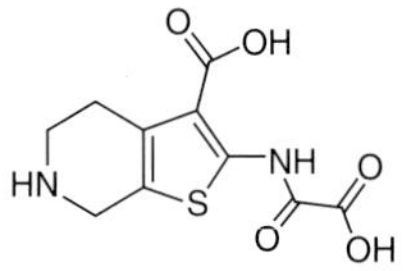

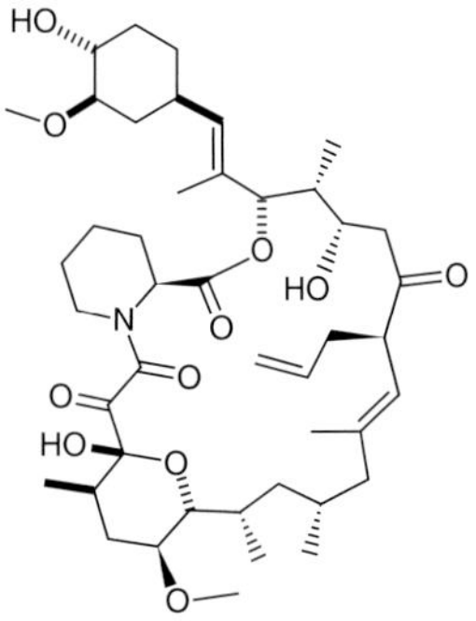

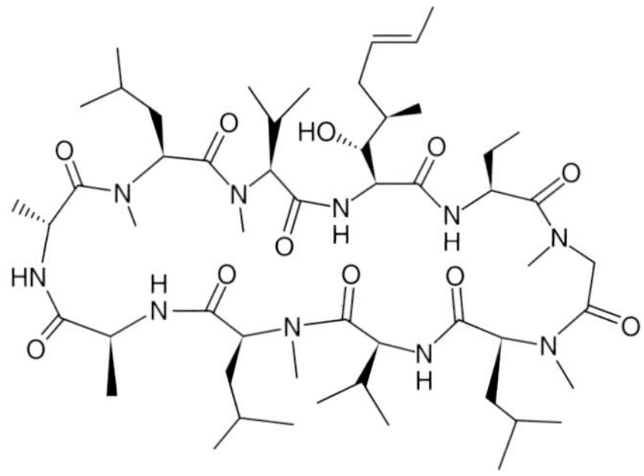

and is selective relative to other protein phosphatases (e.g., Yersinia PTPase, SHP1, SHP2, LAR, HePTP, PTPα, CD45, VHR, MKP3, Cdc25A, Stp1, and PP2C). Compound **2**,[268,269] designed from a lead tripeptide (Ac-Asp-Tyr[SO$_3$H]-Nle-NH$_2$), is a moderately potent inhibitor of PTP1B tyrosine phosphatase (K_i = 220 nM) and is selective relative to other protein phosphatases (e.g., SHP-2, LAR, CDC25b, and calcineurin). Prodrug modifications provided cellular activity as measured in terms enhanced 2-deoxyglucose uptake in cells (relative to the parent molecule) with concomitant augmentation of the tyrosine phosphorylation levels of insulin-signaling molecules. Compound **3**,[270,271] a tetrahydro-thienopyridine template-based molecule, is a less potent inhibitor of PTP1B (K_i = 5.1 μM), but provided a key lead compound for further optimization to achieve oral bioavailability. It is noted that significant efforts to develop PTP1B tyrosine phosphatase inhibitors are ongoing, and the aforementioned small-molecule lead compounds are only representative of such progress in the field of protein phosphatase inhibitor drug discovery.

2.24.4.2 Receptor and Nonreceptor Serine/Threonine Phosphatases

Some noteworthy first-generation PP1, PP2A, and PP2B phosphatase inhibitors include FK506, cyclosporin, and cantharidin (**Figure 13**, see above). These exemplify both natural-product and small-molecule lead compounds for therapeutic treatment of several diseases, including neurological and metabolic disorders, respiratory disease, immunosuppression, and cancer (e.g., cell-cycle modulation).[272] FK506, a macrolide isolated from bacteria, is a highly potent inhibitor of PP2B (IC$_{50}$ = 0.5 nM). Cyclosporin-A, a cyclic peptide isolated from fungus, is also a highly potent inhibitor of PP2B (IC$_{50}$ = 5 nM). Cantharidin, a natural product isolated from insects, is a highly potent inhibitor of PP1 and PP2A (IC$_{50}$ = 500 and 200 nM, respectively).

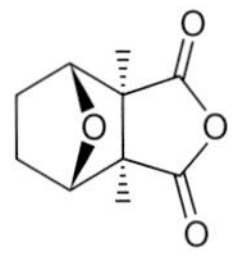

2.24.5 Phosphoprotein-Interacting Domain Containing Cellular Protein Therapeutic Targets and Drug Discovery

The complex orchestration of signal transduction and related cellular activities that have been unraveled with respect to both protein kinases and protein phosphatases also includes the intimate participation of a relatively large number of cellular proteins that contain phosphoprotein-interacting domains,[3,4,15–20] including those recognizing phosphotyrosine, such as SH2 (e.g., Grb2, Shc, Src, and ZAP-70) and PTB (e.g., IRS-1 and ARH), as well as others that recognize phosphoserine/phosphothreonine, such as WW, FHA, and 14-3-3. Collectively, such phosphoprotein-interacting domains underlie cellular mechanisms that have been related to cancer, autoimmune disease, osteoporosis, autoimmune disease, type-2 diabetes mellitus, and autosomal recessive hypercholesterolemia.

2.24.5.1 Phosphotyrosine-Interacting Domains

With respect to small-molecule-drug discovery, the overwhelming focus on such phosphoprotein-interacting motifs has been on the SH2 domain.[5,15,36,273–277] SH2 domains are noncatalytic motifs of approximately 100 amino acids, and have been determined to be one of the top 25 most frequently occurring protein structural types that have been identified from the human genome. Numerous x-ray and/or nuclear magnetic resonance (NMR) structures have been determined for SH2 domains (e.g., Src, Grb2, and Zap70) and complexes thereof with phosphopeptide, peptidomimetic, or nonpeptide inhibitors. As the prototype SH2 domain, the Src SH2 domain and phosphopeptide (e.g., pTyr-Glu-Glu-Ile sequences) complexes were first determined by x-ray structural studies to provide detailed molecular maps, especially of the pTyr moiety.

A major challenge of SH2 inhibitor drug design was the pTyr moiety, in terms of developing metabolically stable pTyr mimics that exhibit high affinity to a SH2 domain. Another major challenge was the transformation of the peptide ligand scaffold to peptidomimetic or nonpeptide templates that would afford more drug-like properties with respect to cellular and in vivo activity. Two successful drug design campaigns that have advanced key proof-of-concept lead compounds against Src SH2 and Grb2 SH2 domains illustrate endeavors to overcome such pTyr and peptide scaffold challenges. First-generation nonpeptide Src SH2 inhibitors PD-157934, AP22161 and AP22408 are particularly noteworthy small-molecules that have advanced the first proof-of-concept in vivo effective lead compound (Figure 14).

PD-157934,[278] a benzamide template-based molecule, was created using de novo drug design strategies and was determined to be a moderately potent inhibitor of the Src SH2 domain ($IC_{50} = 6.6\,\mu M$). AP22161,[279] a bicyclic benzamide template-based molecule incorporating a novel pTyr mimic and having an electrophilic aldehyde designed to interact with a cysteine residue in the pTyr binding pocket of Src SH2, was determined to be a potent inhibitor of Src SH2 ($IC_{50} = 240\,nM$) and to be effective in cell-based assays to block Src-dependent activities in vitro. AP22408,[280] a bicyclic benzamide template-based molecule incorporating a novel pTyr mimetic designed to increase binding affinity and confer bone-targeting properties, is a potent inhibitor of the Src SH2 domain ($IC_{50} = 300\,nM$) and is effective in inhibiting osteoclast-dependent bone resorption in vitro and in vivo. It is noteworthy that the overall structure-based design of AP22408, with respect to both the pTyr and peptide scaffold replacement strategies, and further studies[281] have extended the scope of other pTyr mimetics by a strategy of multiple functional group replacement. Other drug

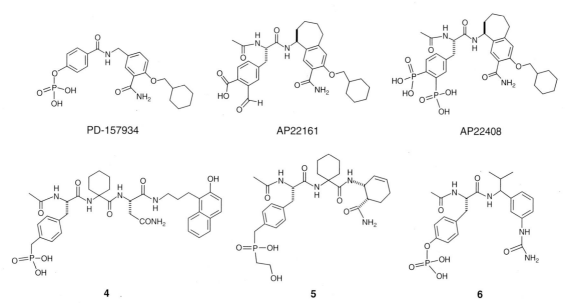

Figure 14 Some Src SH2 and Grb2 SH2 inhibitors.

Table 4 Summary of clinically investigated small-molecule protein kinase inhibitors[a]

Compound	Protein kinase(s)	Clinical status (cancer indications)	Company
STI-571 (Imatinib)	Bcr-Abl, PDGFR, Kit	Approved (chronic myelogenous leukemia/CML)	Novartis
		Approved (gastrointestinal stromal cell tumors/GIST)	
AMN-107	Bcr-Abl, PDGFR, Kit	Phase I (CML, acute lymphocytic leukemia/ALL)	Novartis
BMS-354825	Bcr-Abl, SFK	Phase II (CML)	Bristol-Myers Squibb
SKI-166	Bcr-Abl, SFK	Phase II (CML, solid tumors)	Wyeth
AZD-530	SFK, Bcr-Abl	Phase I (solid tumors)	AstraZeneca
ZD-1839 (Iressa)	EGFR	Approved (NSCLC)	AstraZeneca
OSI-774 (Tarceva)	EGFR	Approved (NSCLC)	OSI Pharmaceuticals
CI-1033	EGFR, HER2	Phase II (various solid tumors)	Pfizer
PKI-166	EGFR, HER2	Phase III (solid tumors)	Novartis
EKB-569	EGFR, HER2	Phase II (NSCLC, colorectal)	Wyeth
GW-572016	EGFR, HER2	Phase III (NSCLC, solid tumors)	GlaxoSmithKline
PTK787/ZK222584	VEGFR, PDGFR, Kit	Phase III (colorectal, solid tumors)	Novartis/Schering AG
AMG-706	VEGFR, PDGFR, Kit, RET	Phase II (NSCLC, colorectal, GIST)	Amgen
SU-6668	VEGFR, PDGFR, Kit	Phase II (solid tumors)	Pfizer (Sugen)
SU-11248	VEGFR, PDGFR, Kit, Flt-3	Phase III (GIST, solid tumors, acute myeloid leukemia/AML)	Pfizer (Sugen)
CGP53716	PDGFR	Phase III (brain tumors)	Novartis
ZD-6474	VEFGR, Kit	Phase II (solid tumors)	AstraZeneca
CEP-7055	VEGFR	Phase I (solid tumors)	Cephalon/Sanofi
CP-547632	VEGFR, FGFR	Phase I/II (ovarian, NSCLC)	OSI Pharmaceuticals/Pfizer
MLN-518	Flt-3, PDGFR, Kit	Phase II (AML)	Millennium Pharmaceuticals
PKC-412	Flt-3, VEGFR, PDGFR, Kit	Phase II (AML, myelodysplastic syndrome)	Novartis
CEP-701	VEGFR, Flt-3, Trk	Phase II (AML)	Cephalon/Kyowa Hakko Kogyo
BAY-43-9006	Raf	Phase II/III (kidney, breast, lung)	Onyx Pharmaceuticals/Bayer
PD-184352	MEK	Phase II (pancreatic, breast)	Pfizer
Flavopiridol	CDK	Phase II (head and neck cancer, solid tumors)	Aventis
CYC202	CDK	Phase II (NSCLC, lymphoma)	Cyclacel
BMS-387032	CDK	Phase I (metastatic refractory solid tumors)	Bristol-Myers Squibb
UCN-01	PKC, CDK	Phase I/II (refractory solid tumors and lymphomas)	Kyowa Hakko Kogyo
AP23573	mTOR	Phase II (hematologic malignancies, various solid tumors)	ARIAD Pharmaceuticals
CCI-779	mTOR	Phase III (renal cell carcinoma, breast, prostate)	Wyeth
RAD001	mTOR	Phase II (GIST, relapsed/refractory CML or AML)	Novartis

[a] Adapted from Garcia-Echevaria and Fabbro,[54] Laird and Cherrington,[41] Sawyer,[35] and other sources of information available as of July 2005.

discovery efforts have advanced Src SH2 inhibitors having nonpeptide templates.[282–284] First-generation nonpeptide Grb2 SH2 inhibitors 4–6 are particularly noteworthy small molecules that have advanced the first proof-of-concept cellularly effective lead compound (**Figure 14**). Compounds 4 and 5[285–287] are peptidomimetics that exploit a β-turn conformation as determined from x-ray structural studies of cognate ligand phosphopeptides complexed with Grb2 SH2, and each is a potent inhibitor of Grb2 SH2(IC_{50} = 43 nM and 1.6 μM, respectively). Compound 5 was effective in cells, and provided the first proof-of-concept lead compound relative to disrupting the Ras-Grb2 pathway by the inhibition of Grb2 SH2. Compound 6[288] is a novel nonpeptide molecule that is a moderately potent Grb SH2 inhibitor (IC_{50} = 6.2 μM) and provides a template for further lead compound optimization.

2.24.5.2 Phosphoserine/Phosphothreonine-Interacting Domains

More recently, the identification and characterization of phosphoserine/phosphothreonine (pSer/pThr)-interacting domains has provided new insights into the modulation of signal transduction pathways and cellular activities in terms of the integration of protein serine/threonine kinases and protein serine/threonine phosphatases.19, 20]. The 14-3-3 proteins were the first shown to possess pSer/pThr molecular recognition properties, and there are more than 100 known phosphoprotein ligands identified to bind to 14-3-3 domains. Such phosphoprotein ligands include a vast number of protein kinases, protein phosphatases, cell surface receptors, ion channels, and a number of proteins involved in cell-cycle control as well as transcriptional control of gene expression, metabolism, and apoptosis. The WW domain containing cellular proteins includes examples (e.g., proline isomerase Pin1) that specifically recognize pSer-Pro and Thr-Pro motifs in their phosphoprotein ligands. Forkhead-associated (FHA) domains have been found in a number of transcriptional control proteins, DNA-damage-activated protein kinases, cell-cycle checkpoint proteins, and phosphatases. FHA domains bind to pThr sites on phosphoprotein ligands, and there are no known examples of FHA domains that bind to pSer motifs. Small-molecule drug discovery for such pSer/pThr-interacting domains has not risen to the level that has existed for SH2 proteins. However, some noteworthy progress has been achieved relative to probing 14-3-3, WW, and FHA domains with phosphopeptide libraries to provide optimal pSer/pThr-containing sequences recognized as ligands as well as enabling structural biology studies.

2.24.6 Concluding Remarks

Protein kinases and protein phosphatases are key superfamilies of enzymes that modulate protein phosphorylation and signal transduction pathways critical to various cellular functions. The functional relationship of protein kinases to cancer has translated to significant advances in basic research and drug discovery over the past several decades, especially within the last few years. Highlighting such progress in protein kinase drug discovery is an emerging armamentarium of powerful molecular probes, proof-of-concept lead compounds, and clinical candidates (**Table 4**). Significant progress in drug discovery has also been made for both protein phosphatase and phosphoprotein-interacting domain containing proteins, albeit relatively less than that for protein kinases.

From a medicinal chemistry perspective, the chemical and mechanistic diversity of small-molecule inhibitors for such therapeutic targets is extraordinary. Key concepts and technologies related to structure-based drug design, chemoinformatics, combinatorial libraries, and high-throughput screening have impacted drug discovery efforts with respect to the generation, optimization, and development of lead compounds. Nevertheless, a major challenge for this multidisciplinary field of signal transduction research is biological selectivity (molecular and cellular). As exemplified by some protein kinase inhibitors, multiple inhibition of key therapeutic targets that may synergize with respect to signal transduction pathways converging in a particular disease, may have advantages relative to the modulation of a single therapeutic target. Biological selectivity has also transcended to include an ever-increasing number of mutations of signal transduction therapeutic targets, especially protein kinases (e.g., EGFR, Bcr-Abl, KIT, and Flt-3). Combination drug therapy is an option to overcome resistance due to mutations that select against a particular small-molecule inhibitor. Finally, biological selectivity at the cellular level provides new opportunities to advance novel cancer therapies focused on metastasis and angiogenesis (see above). Future advances in the discovery of small-molecule drugs involving protein phosphorylation in signal transduction pathways will probably address a wider range of diseases and take advantage of progress that has been highlighted in this review.

References

1. Hahn, W. C.; Weinberg, R. A. *Nat. Rev. Cancer* **2002**, *2*, 331–341.
2. Manning, G.; Whyte, D. B.; Martinez, R.; Hunter, T.; Sudarsanam, S. *Science* **2002**, *298*, 1912–1934.

2a. http://www.kinase.com/human/kinome (accessed Aug 2006).
3. Scott, J. D.; Pawson, T. *Sci. Am.* **2000**, *282*, 72–79.
4. Pawson, T. *Cell* **2004**, *116*, 191–203.
5. Dalgarno, D. C.; Metcalf, C. A., III; Shakespeare, W. C.; Sawyer, T. K. *Curr. Opin. Drug Disc. Dev.* **2000**, *3*, 549–564.
6. Nam, N.-H.; Paranga, K. *Curr. Drug Targets* **2003**, *4*, 159–179.
7. Hanahan, D.; Weinberg, R. A. *Cell* **2000**, *100*, 57–70.
8. Hunter, T. *Cell* **2000**, *100*, 113–127.
9. Cohen, P. *Nat. Rev. Drug Disc.* **2002**, *1*, 309–315.
10. Blume-Jensen, P.; Hunter, T. *Nature* **2001**, *411*, 355–365.
11. Alonso, A.; Sasin, J.; Bottini, N.; Friedberg, I.; Friedberg, I.; Osterman, A.; Godzik, A.; Hunter, T.; Dixon, J.; Mustelin, T. *Cell* **2004**, *117*, 699–711.
12. Anderson, J. N.; Jansen, P. G.; Echwald, S. M.; Mortensen, O. H.; Fukada, T.; Vecchio, R. D.; Tonks, N. K.; Moller, N. P. H. *FASEB J.* **2004**, *18*, 8–30.
13. Wang, W.-Q.; Sun, J.-P.; Zhang, Z.-Y. *Curr. Top. Med. Chem.* **2003**, *3*, 739–748.
14. Zhang, Z.-Y.; Zhou, B.; Xie, L. *Pharma. Ther.* **2002**, *93*, 307–317.
15. Machida, K.; Mayer, B. J. *Biochim. Biophys. Acta* **2005**, *1747*, 1–25.
16. Waksman, G.; Kumaran, S.; Lubman, O. *Expert Rev. Mol. Med.* **2004**, *6*, 1–18.
17. Uhlik, M. T.; Temple, B.; Bencharit, S.; Kimple, A. J.; Siderovski, D. P.; Johnson, G. L. *J. Mol. Biol.* **2005**, *345*, 1–20.
18. Vetter, S. W.; Zhang, Z. Y. *Curr. Protein Pept. Sci.* **2002**, *3*, 365–397.
19. Yaffe, M. B.; Elia, A. E. H. *Curr. Opin. Cell Biol.* **2001**, *13*, 131–138.
20. Yaffe, M. B.; Smerdon, S. J. *Annu. Rev. Biophys. Biomol. Struct.* **2004**, *33*, 225–244.
21. Peters, E. C.; Brock, A.; Ficarro, S. B. *Mini-Rev. Med. Chem.* **2004**, *4*, 313–324.
22. Machida, K.; Mayer, B. J.; Nollau, P. *Mol. Cell. Proteom.* **2003**, *2*, 215–233.
23. Sharma, A.; Antoku, S.; Fujiwara, K.; Mayer, B. J. *Mol. Cell. Proteomics* **2003**, *2*, 1217–1224.
24. Lim, Y.-P.; Diong, L.-S.; Qi, R.; Druker, B. J.; Epstein, R. J. *Mol. Cancer Ther.* **2003**, *2*, 1369–1377.
25. Hunter, T. *Curr. Opin. Cell Biol.* **1989**, *1*, 1168–1181.
26. Cohen, P. *Nat. Cell Biol.* **2002**, *4*, E127–E130.
26a. Collett, M. S.; Erickson, R. L. *Proc. Natl. Acad. Sci. USA* **1978**, *75*, 2021–2024.
27. Hunter, T.; Sefton, B. M. *Proc. Natl. Acad. Sci. USA* **1980**, *77*, 1311–1315.
28. Ushiro, H.; Cohen, S. *J. Biol. Chem.* **1980**, *255*, 8363–8365.
29. Tonks, N. K.; Diltz, C. D.; Fischer, E. H. *J. Biol. Chem.* **1988**, *263*, 6722–6730.
30. Sadowski, I.; Stone, J. C.; Pawson, T. *Mol. Cell. Biol.* **1986**, *6*, 4398–4408.
31. Futreal, P. A.; Coin, L.; Marshall, M.; Down, T.; Hubbard, T.; Wooster, R.; Rahman, N.; Stratton, M. R. *Nat. Rev. Cancer* **2004**, *4*, 177–183.
32. Vogelstein, B.; Kinzler, K. W. *Nat. Med.* **2004**, *10*, 789–799.
33. Evan, G. I.; Vousden, K. H. *Nature* **2001**, *411*, 342–348.
34. Vieth, M.; Higgs, R. E.; Robertson, D. H.; Shapiro, M.; Gragg, E. A.; Hemmerle, H. *Biochim. Biophys. Acta.* **2004**, *1697*, 243–257.
34a. See http://198.202.68.14/human/kinome/phylogeny.html (accessed Aug 2006).
35. Sawyer, T. K. *Curr. Med. Chem. (Anti-Cancer Agents)* **2004**, *4*, 449–455.
36. Sawyer, T. K.; Wang, Y.; Shakespeare, W. C.; Metcalf, C. A., III; Sundaramoorthi, R.; Narula, S.; Dalgarno, D. C. In *Proteomics and Protein–Protein Interactions: Biology, Chemistry, Bioinformatics and Drug Design*, Waksman, G., Ed.; Series on Protein Reviews, Vol. 3, Springer Publishers, 2006, pp 219–253.
37. Cherry, M.; Williams, D. H. *Curr. Med. Chem.* **2004**, *11*, 663–673.
38. Sawyer, T. K.; Bohacek, R. S.; Metcalf, C. A., III; Shakespeare, W. C.; Wang, Y.; Sundaramoorthi, R.; Keenan, T.; Narula, S.; Weigele, M.; Dalgarno, D. C. *BioTechniques* **2003**, *June* (Suppl.), 2–15.
39. Scapin, G. *Drug Disc. Today* **2002**, *11*, 601–611.
40. Fischer, P. M. *Curr. Med. Chem.* **2004**, *11*, 1563–1583.
41. Laird, A. D.; Cherrington, J. M. *Expert Opin. Investig. Drugs* **2003**, *12*, 51–64.
42. Madhusudan, S.; Granesan, T. S. *Clin. Biochem.* **2004**, *37*, 618–635.
43. Kumar, C. C.; Madison, V. *Expert Opin. Emerging Drugs* **2001**, *6*, 303–315.
44. Dumas, J. *Expert Opin. Ther. Patents* **2001**, *11*, 405–429.
45. Huse, M.; Kuriyan, J. *Cell* **2002**, *109*, 275–282.
46. Muegge, I.; Enyedy, I. J. *Curr. Med. Chem.* **2004**, *11*, 693–707.
47. Chuaqui, C.; Deng, Z.; Singh, J. *J. Med. Chem.* **2005**, *48*, 121–133.
48. Aronov, A. M.; Murko, M. A. *J. Med. Chem.* **2004**, *47*, 5616–5619.
49. Diller, D. J.; Li, R. *J. Med. Chem.* **2003**, *46*, 4638–4647.
50. Hartshorn, M. J.; Murray, C. W.; Cleasby, A.; Frederickson, M.; Tickle, I. J.; Jhoti, H. *J. Med. Chem.* **2005**, *48*, 403–413.
51. Daub, H.; Specht, K.; Ullrich, A. *Nat. Rev. Drug Disc.* **2004**, *3*, 1001–1010.
52. Yarden, Y.; Sliwkowski, M. X. *Nat. Rev. Mol. Cell Biol.* **2001**, *2*, 127–137.
53. de Bono, J. S.; Rowinski, E. K. *Trends Mol. Med.* **2002**, *8*, S19–S26.
54. Garcia-Echevaria, C.; Fabbro, D. *Mini-Rev. Med. Chem.* **2004**, *4*, 273–283.
55. Grunwald, V.; Hidalgo, M. *J. Natl. Cancer Inst.* **2003**, *95*, 851–867.
56. Barker, A. J.; Gibson, K. H.; Grundyk, W.; Godfrey, A. A.; Barlow, J. J.; Healy, M. P.; Woodburn, J. R.; Ashton, S. E.; Curry, B. J.; Scarlett, L. et al. *Bioorg. Med. Chem. Lett.* **2001**, *11*, 1911–1914.
57. Ward, W. H.; Cook, P. N.; Slater, A. M.; Davies, D. H.; Holdgate, G. A.; Green, L. R. *Biochem. Pharmacol.* **1994**, *48*, 659–666.
58. Wakeling, A. E. *Cancer Res.* **2002**, *62*, 5749–5754.
59. Ciardiello, F.; Caputo, R.; Bianco, R.; Damiano, V.; Pomatico, G.; De Placido, S.; Bianco, A. R.; Tortora, G. *Clin. Cancer Res.* **2000**, *6*, 2053–2063.
60. Sirotnak, F. J.; Zakowski, M. F.; Miller, V. A.; Scher, H. I.; Kris, M. G. *Clin. Cancer Res.* **2000**, *6*, 4885–4892.
61. Herbst, R. S.; Fukuoka, M.; Baselga, J. *Nat. Rev. Cancer* **2004**, *4*, 956–965.
62. Brehmer, D.; Greff, Z.; Godl, K.; Blencke, S.; Kurtenbach, A.; Weber, M.; Muller, S.; Klebl, B.; Cotton, M.; Keri, G. et al. *Cancer Res.* **2005**, *65*, 379–382.

63. Moyer, J. D.; Barbacci, E. G.; Iwata, K. K.; Arnold, L.; Boman, B.; Cunningham, A.; DiOrio, C.; Doty, Y.; Morin, M. J.; Moyer, M. et al. *Cancer Res.* **1997**, *57*, 4838–4848.
64. Pollack, V. A.; Savage, D. M.; Baker, D. A.; Tsaparikos, K. E.; Sloan, D. E.; Moyer, J. D.; Barbacci, E. G.; Pustilnik, L. R.; Smolarek, T. A.; Davis, J. A. et al. *J. Pharmacol. Exp. Ther.* **1999**, *291*, 739–748.
65. Gieseg, M. A.; DeBlock, C.; Ferguson, L. R.; Denny, W. A. *Anti-Cancer Drugs* **2001**, *12*, 681–690.
66. Stamos, J.; Sliwkowski, M. X.; Eigenbrot, C. *J. Biol. Chem.* **2002**, *277*, 46265–46272.
67. Lynch, T. J.; Bell, D. W.; Sordella, R.; Gurubhagavatula, S.; Okimoto, R. A.; Brannigan, B. W.; Harris, P. L.; Haserlat, S. M.; Supko, J. G.; Haluska, F. G. et al. *N. Engl. J. Med.* **2004**, *350*, 2129–2139.
68. Paez, J. G.; Jainne, P. A.; Lee, J. C.; Tracey, S.; Greulich, H.; Gabriel, S.; Herman, P.; Kaye, F. J.; Lindeman, N.; Boggon, T. J. et al. *Science* **2004**, *304*, 1497–1500.
69. Pao, W.; Miller, V. A.; Politi, K. A.; Riely, G. J.; Somwar, R.; Zakowski, M. F.; Kris, M. G.; Varmus, H. *PLoS Med.* **2005**, *2*, e73.
70. Rusnak, D. W.; Lackey, K.; Affleck, K.; Wood, E. R.; Alligood, K. J.; Rhodes, N.; Keith, B.; Murray, D. M.; Mullin, R. J.; Knight, W. B. et al. *Mol. Cancer Ther.* **2001**, *1*, 85–94.
71. Rusnak, D. W.; Affleck, K.; Cockerill, S. G.; Stubberfield, C.; Harris, R.; Page, M.; Smith, K. J.; Guntrip, S. B.; Carter, M. C.; Shaw, R. J. et al. *Cancer Res.* **2001**, *61*, 7196–7203.
72. Xia, W.; Mullin, R. J.; Keith, B. R.; Liu, L. H.; Ma, H.; Rusnak, D. W.; Owens, G.; Alligood, K. J.; Spector, N. L. *Oncogene* **2002**, *21*, 6255–6263.
73. Wood, E. R.; Truesdale, A. T.; McDonald, O. B.; Yuan, D.; Hassell, A.; Dickerson, S. H.; Ellis, B.; Pennisi, C.; Horne, E.; Lackey, K. et al. *Cancer Res.* **2004**, *64*, 6652–6659.
74. Bruns, C. J.; Solorzano, C. C.; Harbison, M. T.; Ozawa, S.; Tsan, R.; Fan, D.; Abbruzzese, J.; Traxler, P.; Buchdunger, E.; Radinsky, R. et al. *Cancer Res.* **2000**, *60*, 2926–2935.
75. Brandt, R.; Wong, A. M. L.; Hynes, N. E. *Oncogene* **2001**, *20*, 5459–5465.
76. Baker, C. H.; Solorzano, C. C.; Fidler, I. J. *Cancer Res.* **2002**, *62*, 1996–2003.
77. Cohen, M. S.; Hussain, H. B.; Moley, J. F. *Surgery* **2002**, *132*, 960–966.
78. Smaill, J. B.; Showalter, H. D.; Zhou, H.; Bridges, A. J.; McNamara, D. J.; Fry, D. W.; Nelson, J. M.; Sherwood, V.; Vincent, P. W.; Roberts, B. J. et al. *J. Med. Chem.* **2000**, *43*, 1380–1397.
79. Nelson, J. M.; Fry, D. W. *J. Biol. Chem.* **2001**, *276*, 14842–14847.
80. Allen, L. F.; Eiseman, I. A.; Fry, D. W.; Lenehan, P. F. *Semin. Oncol.* **2003**, *30*, 65–78.
81. Boschelli, D. H. *Curr. Top. Med. Chem.* **2002**, *2*, 1051–1063.
82. Wissner, A.; Overbeek, E.; Reich, M. F.; Floyd, M. B.; Johnson, B. D.; Mamuya, N.; Rosfjord, E. C.; Discafani, C.; Davis, R.; Shi, X. et al. *J. Med. Chem.* **2003**, *46*, 49–63.
83. Nunes, M.; Shi, C.; Greenberger, L. M. *Mol. Cancer Ther.* **2004**, *3*, 21–27.
84. Torrance, C. J.; Jackson, P. E.; Montgomery, E.; Kinzler, K. W.; Vogelstein, B.; Wissner, A.; Nunes, M.; Frost, P.; Discafani, C. M. *Nat. Med.* **2000**, *6*, 1024–1028.
85. Carmeliet, P.; Jain, R. K. *Nature* **2000**, *407*, 249–257.
86. McColl, B. K.; Stacker, S. A.; Achen, M. G. *AMPIS* **2004**, *112*, 463–480.
87. Hicklin, D. J.; Ellis, L. M. *J. Clin. Oncol.* **2005**, *23*, 1–17.
88. Shawver, L. K.; Siamon, D.; Ullrich, A. *Cancer Cell* **2002**, *1*, 117–123.
89. Fong, T. A. T.; Shawver, L. K.; Sun, L.; Tang, C.; App, H.; Powell, T. J.; Kim, Y. H.; Schreck, R.; Wang, X.; Risau, W. et al. *Cancer Res.* **1999**, *59*, 99–106.
90. Stopeck, A.; Sheldon, M.; Vahedian, M.; Cropp, G.; Gosalia, R.; Hannah, A. *Clin. Cancer Res.* **2002**, *8*, 2798–2805.
91. Mendel, D. B.; Laird, A. D.; Smolich, B. D.; Blake, R. A.; Liand, C.; Hannah, A. L.; Shaheen, R. M.; Ellis, L. M.; Weitman, S.; Shawver, L. K. et al. *Anti-Cancer Drug Des.* **2000**, *15*, 29–41.
92. Laird, A. D.; Vajkoczy, P.; Shawver, L. K.; Thurnher, A.; Liang, C.; Mohammadi, M.; Schlessinger, J.; Ullrich, A.; Hubbard, S. R.; Blake, R. A. et al. *Cancer Res.* **2000**, *60*, 4152–4160.
93. Laird, A. D.; Christensen, J. G.; Carver, J.; Smith, K.; Xin, X.; Moss, K. G.; Louie, S. G.; Mendel, D. B.; Cherrington, J. M. *FASEB J.* **2002**, *16*, 681–690.
94. Mendel, D. B.; Laird, A. D.; Xin, X.; Louie, S. G.; Christensen, J.; Li, G.; Schreck, R. E.; Abrams, T. J.; Ngal, T. J.; Lee, L. B. et al. *Clin. Cancer Res.* **2003**, *9*, 327–337.
95. Osusky, K. L.; Hallahan, D. E.; Fu, A.; Ye, F.; Shyr, Y.; Geng, L. *Angiogenesis* **2004**, *7*, 225–233.
96. Morimoto, A. M.; Tan, N.; West, K.; McArthur, G.; Tone, G. C.; Manning, W. C.; Smolich, B. D.; Cherrington, J. M. *Oncogene* **2004**, *23*, 1618–1626.
97. Murray, L. J.; Abrams, T. J.; Long, K. R.; Ngai, T. J.; Olson, L. M.; Hong, W.; Keast, P. K.; Brassard, J. A.; O'Farrell, A. M.; Cherrington, J. M. et al. *Clin. Exp. Metastasis* **2003**, *20*, 757–766.
98. Bold, G.; Altmann, K.-H.; Frei, J.; Lang, M.; Manley, P. W.; Traxler, P.; Wietfeld, B.; Bruggen, J.; Buchdunger, E.; Cozens, R. et al. *J. Med. Chem.* **2000**, *43*, 2310–2323.
99. Wood, J. M.; Bold, G.; Buchdunger, E.; Cozens, R.; Ferrari, S.; Frei, J.; Hofmann, F.; Mestan, J.; Mett, H.; O'Reilly, T. et al. *Cancer Res.* **2000**, *60*, 2178–2189.
100. Lin, B.; Podar, K.; Gupta, D.; Tai, Y.-T.; Li, S.; Weller, E.; Hideshima, T.; Lentzsch, S.; Davies, F.; Li, C. et al. *Cancer Res.* **2002**, *62*, 5019–5026.
101. Drevs, J.; Hofmann, I.; Hugenschmidt, H.; Wittig, C.; Madjar, H.; Muller, M.; Wood, J.; Martiny-Baron, G.; Unger, C.; Marme, D. *Cancer Res.* **2000**, *60*, 4819–4824.
102. Wedge, S. R.; Ogilvie, D. J.; Dukes, M.; Kendrew, J.; Chester, R.; Jackson, J. A.; Boffey, S. J.; Valentine, P. J.; Curwen, J. O.; Musgrove, H. L. et al. *Cancer Res.* **2002**, *62*, 4645–4655.
103. Hennequin, L. F.; Stokes, E. S.; Thomas, A. P.; Johnstone, C.; Ple, P. A.; Ogilvie, D. J.; Dukes, M.; Wedge, S. R.; Kendrew, J.; Curwen, J. O. *J. Med. Chem.* **2002**, *43*, 1300–1312.
104. Carlomagno, F.; Vitagliano, D.; Guida, T.; Ciardiello, F.; Tortora, G.; Vecchio, G.; Ryan, A. J.; Fontanini, G.; Fusco, A.; Santoro, M. *Cancer Res.* **2002**, *63*, 7284–7290.
105. McCarty, M. F.; Wey, J.; Stoeltzing, O.; Liu, W.; Fan, F.; Bucan, C.; Mansfield, P. F.; Ryan, A. J.; Ellis, L. M. *Mol. Cancer Ther.* **2004**, *3*, 1041–1048.
106. Ciardiello, F.; Bianco, R.; Caputo, R.; Caputo, R.; Damiano, V.; Troiani, T.; Melisi, D.; De Vita, F.; De Placido, S.; Bianco, R. et al. *Clin. Cancer Res.* **2004**, *10*, 784–793.

107. Gingrich, D. E.; Reddy, D. R.; Iqbal, M. A.; Singh, J.; Aimone, L. D.; Angeles, T. S.; Albom, M.; Yang, S.; Ator, M. A.; Meyer, S. L. et al. *J. Med. Chem.* **2003**, *46*, 5375–5488.
108. Ruggeri, B.; Singh, J.; Gingrich, D.; Angeles, T.; Albom, M.; Chang, H.; Robinson, C.; Hunter, K.; Dobrzanski, P.; Jones-Bolin, S. et al. *Cancer Res.* **2003**, *63*, 5978–5991.
109. Beebe, J. S.; Jani, J. P.; Knauth, E.; Goodwin, P.; Higdon, C.; Rossi, A. M.; Emerson, E.; Finkelstein, M.; Floyd, E.; Harriman, S. et al. *Cancer Res.* **2003**, *63*, 7301–7309.
110. Stirewalt, D. L.; Radich, J. P. *Nat. Rev. Cancer* **2003**, *3*, 650–665.
111. Sawyers, C. L. *Cancer Cell* **2002**, *1*, 413–415.
112. Gilliland, D. G.; Griffin, J. D. *Blood* **2002**, *100*, 1532–1542.
113. Weisberg, E.; Boulton, C.; Kelly, L. M.; Manley, P.; Fabbro, D.; Meyer, T.; Gilliland, D. G.; Griffin, J. D. *Cancer Cell* **2002**, *1*, 433–443.
114. Fabbro, D.; Ruetz, S.; Bodis, S.; Pruschy, M.; Csermak, K.; Man, A.; Campochiaro, P.; Wood, J.; O'Reilly, T.; Meyer, T. *Anti-Cancer Drug Des.* **2000**, *15*, 17–28.
115. Pinski, J.; Weeraratna, A.; Uzgare, A. R.; Arnold, J. T.; Denmeade, S. R.; Issacs, J. T. *Cancer Res.* **2002**, *62*, 986–989.
116. Kelly, L. M.; Yu, J.-C.; Boulton, C. L.; Apatira, M.; Li, J.; Sullivan, C. M.; Williams, I.; Amaral, S. M.; Curley, D. P.; Duclos, N. et al. *Cancer Cell* **2002**, *1*, 421–438.
117. Levis, M.; Allechach, J.; Tse, K. F.; Zheng, R.; Balwin, B. R.; Smith, B. D.; Jones-Bolin, S.; Ruggeri, B.; Dionne, C.; Small, D. *Blood* **2002**, *99*, 3885–3891.
118. Yee, K. W.; O'Farrell, A. M.; Smolich, B. D.; Cherrington, J. M.; McMahon, G.; Wait, C. L.; McGreevey, L. S.; Griffith, D. J.; Henirich, M. C. *Blood* **2002**, *100*, 2941–2949.
119. O'Farrel, A. M.; Abrams, T.; Yuen, H. A.; Ngai, T. J.; Louie, S. G.; Yee, K. W. H.; Wong, L. M.; Hong, W.; Lee, L. B.; Town, A. et al. *Blood* **2003**, *101*, 3605–3957.
120. Cools, J.; Mentens, N.; Furet, P.; Fabbro, D.; Clark, J. J.; Griffin, J. D.; Marynen, P.; Gilliland, D. G. *Cancer Res.* **2004**, *64*, 6385–6389.
121. Banerji, L.; Sattler, M. *Expert Opin. Ther. Targets* **2004**, *8*, 1–19.
122. Sattler, M.; Salgia, R. *Leukemia Res.* **2004**, *28*, S11–S20.
123. Frost, M. J.; Ferrao, P. T.; Hughes, T. P.; Ashman, L. K. *Mol. Cancer Ther.* **2002**, *1*, 1115–1124.
124. Buchdunger, E.; Cioffi, C. L.; Lawe, N.; Stover, D.; Ohno-Jones, S.; Druker, B. J.; Lydon, N. B. *J. Pharmacol. Exp. Ther.* **2002**, *295*, 139–145.
125. Heinrich, M. C.; Griffith, D. J.; Druker, B. J.; Wait, C. L.; Ott, K.; Zigler, A. J. *Blood* **2000**, *96*, 925–932.
126. Demetri, G. D.; von Mehren, M.; Blanke, C. D.; Van Den Abbeele, A. D.; Eisenberg, B.; Roberts, P. J.; Heinrich, M. C.; Tuvenson, D. A.; Singer, S.; Janicek, M. et al. *N. Engl. J. Med.* **2002**, *347*, 472–480.
127. Ma, Y.; Zeng, S.; Metcalfe, D. D.; Akin, C.; Dimitrijevic, S.; Butterfield, J. H.; McMahon, G.; Longley, B. J. *Blood* **2002**, *99*, 1741–1744.
128. Zermati, Y.; De Sepulveda, P.; Feger, F.; Letard, S.; Kersual, J.; Casteran, N.; Gorochov, G.; Dy, M.; Ribadeau Dumas, A.; Dorgham, K. et al. *Oncogene* **2003**, *22*, 660–664.
129. Pardanani, A.; Elliott, M.; Reeder, T.; Li, C. Y.; Baxter, E. J.; Cross, N. C.; Tefferi, A. *Lancet* **2003**, *362*, 535–536.
130. Liao, A. T.; Chien, M. B.; Shenoy, N.; Mendel, D. B.; McMahon, G.; Cherrington, J. M.; London, C. A. *Blood* **2002**, *100*, 585–593.
131. Corbin, A. S.; Griswold, I. J.; La Rosee, P.; Yee, K. W. H.; Heinrich, M. C.; Reimer, C. L.; Druker, B. J.; Deininger, M. W. N. *Blood* **2004**, *104*, 3754–3757.
132. Corbin, A. S.; Demehri, S.; Griswold, I. J.; Wang, Y.; Metcalf, C. A., III; Sundaramoorthi, R.; Shakespeare, W. C.; Snodgrass, J.; Wardwell, S.; Dalgarno, D. et al. *Blood* **2005**, *106*, 227–234.
133. Ren, R. *Nat. Rev. Cancer* **2005**, *5*, 172–183.
134. Capdeville, R.; Buchdunger, E.; Zimmermann, J.; Matter, A. *Nat. Rev. Drug Disc.* **2002**, *1*, 493–502.
135. Druker, B. J.; Talpaz, M.; Resta, D. J.; Peng, B.; Buchdunger, E.; Ford, J. M.; Lydon, N. B.; Kantarjian, H.; Capdeville, R.; Ohno-Jones, S. et al. *N. Engl. J. Med.* **2001**, *344*, 1031–1037.
136. Schindler, T.; Bornmann, W.; Pellicena, P.; Miller, W. T.; Clarkson, B.; Kuriyan, J. *Science* **2000**, *289*, 1938–1942.
137. Nagar, B.; Bornmann, W. G.; Pellicena, P.; Schindler, T.; Veach, D. R.; Miller, W. T.; Clarkson, B.; Kuriyan, J. *Cancer Res.* **2002**, *62*, 4236–4243.
138. Druker, B. J.; Tamura, S.; Buchdunger, E.; Ohno, S.; Segal, G. M.; Fanning, S.; Zimmerman, J.; Lydon, N. B. *Nat. Med.* **1996**, *2*, 561–566.
139. Weisberg, E.; Manley, P. W.; Breitenstein, W.; Bruggen, J.; Cowan-Jacob, S. W.; Ray, A.; Huntly, B.; Fabbro, D.; Fendich, G.; Hall-Meyers, E. et al. *Cancer Cell* **2005**, *7*, 129–141.
140. Shah, N. P.; Tran, C.; Lee, F. Y.; Chen, P.; Norris, D.; Sawyers, C. L. *Science* **2004**, *305*, 399–401.
141. Lombardo, L. J.; Lee, F. Y.; Chen, P.; Norris, D.; Barrish, J. C.; Behnia, K.; Castaneda, S.; Cornelius, L. A. M.; Das, J.; Doweyko, A. M. et al. *J. Med. Chem.* **2004**, *47*, 6658–6661.
142. Burgess, M. R.; Skaggs, B. J.; Shah, N. P.; Lee, F. Y.; Sawyers, C. L. *Proc. Natl. Acad. Sci. USA* **2005**, *102*, 3395–3400.
143. Gumireddy, K.; Baker, S. J.; Cosenza, S. C.; John, P.; Kang, A. D.; Robell, K. A.; Reddy, M. V. R.; Reddy, E. P. *Proc. Natl. Acad. Sci. USA* **2005**, *102*, 1992–1997.
144. Boschelli, D. H.; Ye, F.; Wang, Y. D.; Dutia, M.; Johnson, S. L.; Wu, B.; Miller, K.; Powell, D. W.; Yaczko, D.; Young, M. et al. *J. Med. Chem.* **2001**, *44*, 3965–3977.
145. Golas, J. M.; Arndt, K.; Etienne, C.; Lucas, J.; Nardin, D.; Gibbons, J.; Frost, P.; Ye, F.; Boschelli, D.; Boschelli, F. *Cancer Res.* **2003**, *63*, 375–381.
146. Dorsey, J. F.; Jove, R.; Kraker, A. J.; Wu, J. *Cancer Res.* **2000**, *60*, 3127–3131.
147. Huang, M.; Dorsey, J. F.; Epling-Burnette, P. K.; Nimmanapalli, R.; Landowski, T. H.; Mora, L. B.; Niu, G.; Sinibaldi, D.; Bai, F.; Kraker, A. et al. *Oncogene* **2002**, *21*, 8804–8816.
148. La Rosee, P.; Corbin, A. S.; Stoffregen, E. P.; Deininger, M. W.; Druker, B. J. *Cancer Res.* **2002**, *62*, 7149–7173.
149. Huron, D. R.; Gorre, M. E.; Kraker, A. J.; Sawyers, C. L.; Rosen, N.; Moasser, M. M. *Clin. Cancer Res.* **2003**, *9*, 1267–1273.
150. Von Bubnoff, N.; Veach, D. R.; Miller, W. T.; Li, W.; Sanger, J.; Peschel, C.; Bornmann, W. G.; Clarkson, B.; Duyster, J. *Cancer Res.* **2003**, *63*, 6395–6404.
151. Tatton, L.; Morley, G. M.; Chopras, R.; Khwaja, A. *J. Biol. Chem.* **2003**, *278*, 4847–4853.
152. Warmuth, M.; Simon, N.; Mitina, O.; Mathes, R.; Fabbro, D.; Manley, P. W.; Buchdunger, E.; Forster, K.; Moarefi, I.; Hallek, M. *Blood* **2003**, *101*, 664–672.
153. Wilson, M. B.; Schreiner, S. J.; Choi, H.-J.; Kamens, J.; Smithgall, T. S. *Oncogene* **2002**, *21*, 8075–8088.
154. O'Hare, T.; Pollack, R.; Stoffregen, E. P.; Keats, J. A.; Abdullah, O. M.; Moseson, E. M.; Rivera, V. M.; Tang, H.; Metcalf, C. A., III; Bohacek, R. S. et al. *Blood* **2004**, *104*, 2532–2539.
155. Druker, B. J. *Mol. Cancer Ther.* **2003**, *2*, 225–226.

156. Hoover, R. R.; Mahon, F. X.; Melo, J. V.; Daley, G. Q. *Blood* **2002**, *100*, 1068–1071.
157. Tseng, P.-H.; Lin, H.-P.; Zhu, J.; Chen, K.-F.; Hade, E. M.; Young, D. C.; Byrd, J. C.; Grever, M.; Johnson, K.; Druker, B. J.; Chen, C.-S. *Blood* **2005**, Jan 21 (E-pub ahead of print).
158. Yeatman, T. J. *Nat. Rev. Cancer* **2004**, *4*, 470–480.
159. Roskoski, R. *Biochem. Biophys. Res. Commun.* **2004**, *324*, 1155–1164.
160. Summy, J. M.; Gallick, G. E. *Cancer Metastasis Rev.* **2003**, *22*, 337–358.
161. Warmuth, M.; Damoiseaux, R.; Liu, Y.; Fabbro, D.; Gray, N. *Curr. Pharm. Des.* **2004**, *9*, 2043–2059.
162. Sawyer, T.; Boyce, B.; Dalgarno, D.; Iuliucci, J. *Expert Opin. Investig. Drugs* **2001**, *10*, 1327–1344.
163. Susa, M.; Missbach, M.; Green, J. *Trends Pharmacol. Sci.* **2000**, *21*, 489–495.
164. Parsons, S. J.; Parsons, J. T. *Oncogene* **2004**, *18*, 7906–7909.
165. Schlessinger, J. *Cell* **2000**, *100*, 293–296.
166. Biscardi, J. S.; Tice, D. A.; Parsons, S. J. *Adv. Cancer Res.* **1999**, *76*, 61–119.
167. Thomas, S. M.; Brugge, J. B. *Annu. Rev. Cell Dev. Biol.* **1997**, *13*, 513–609.
168. Maa, M.-C.; Leu, T. H.; McCarley, D. J.; Schatzman, R. C.; Parsons, S. J. *Proc. Natl. Acad. Sci. USA* **1995**, *92*, 6981–6985.
169. Mao, W.; Irby, R.; Coppola, D.; Fu, L.; Wloch, M.; Turner, J.; Yu, H.; Garcia, R.; Jove, R.; Yeatman, T. J. *Oncogene* **1997**, *15*, 3083–3090.
170. Staley, C. A.; Parikh, N. U.; Gallick, G. E. *Cell Growth Differ.* **1997**, *8*, 269–274.
171. Ellis, L. M.; Staley, C. A.; Liu, W.; Fleming, R. Y.; Parikh, N. U.; Bucana, C. D.; Gallick, G. E. *J. Biol. Chem.* **1998**, *273*, 1052–1057.
172. Irby, R. B.; Mao, W.; Coppola, D.; Kang, J.; Loubeau, J. M.; Trudeau, W.; Karl, R.; Fujita, D. J.; Jove, R.; Yeatman, T. J. *Nature Genet.* **1998**, *21*, 187–190.
173. Egan, C.; Pang, A.; Durda, D.; Cheng, H. C.; Wang, J. H.; Fujita, D. J. *Oncogene* **1999**, *18*, 1227–1237.
174. Verbeek, B. S.; Vroom, T. M.; Adriaansen-Slot, S. S.; Ottenhoff-Kalff, A. E.; Geertzema, J. G.; Hennipman, A.; Rijksen, G. *J. Pathol.* **1996**, *180*, 383–388.
175. Talamonti, M. S.; Roh, M. S.; Curley, S. A.; Gallick, G. E. *J. Clin. Invest.* **1993**, *91*, 3–60.
176. Van Oijen, M. G.; Rijkseng, G.; Ten Broek, F. W.; Slootweg, P. J. *J. Oral Pathol. Med.* **1998**, *27*, 147–152.
177. Lutz, M. P.; Esser, I. B.; Flossmann-Kast, B. B.; Vogelmann, R.; Luhrs, H.; Friess, H.; Buchler, M. W.; Adler, G. *Biochem. Biophys. Res. Commun.* **1998**, *243*, 503–508.
178. Karni, R.; Jove, R.; Levitzki, A. *Oncogene* **1999**, *18*, 4462–4654.
179. Tsai, Y. T.; Su, Y. H.; Fang, S. S.; Huang, T. N.; Qiu, Y.; Jou, Y. S.; Shih, H. M.; Kung, H. J.; Chen, R. H. *Mol. Cell. Biol.* **2000**, *20*, 2043–2054.
180. Turkson, J.; Bowman, T.; Garcia, R.; Caldenhoven, E.; De Groot, R. P.; Jove, R. *Mol. Cell. Biol.* **1998**, *18*, 2545–2552.
181. Avizienyte, E.; Wyke, A. W.; Jones, R. J.; Mclean, G. W.; Westhoff, M. A.; Brunton, V. G.; Frame, M. C. *Nat. Cell Biol.* **2002**, *8*, 632–638.
182. Nakagawa, T.; Tanaka, S.; Suzuki, H.; Takayanagi, H.; Miyazaki, T.; Nakamura, K.; Tsuruo, T. *Int. J. Cancer* **2000**, *88*, 384–391.
183. Sakamoto, M.; Takamuri, M.; Ino, Y.; Miuru, A.; Genda, T.; Hirohasi, S. *Jpn. J. Cancer Res.* **2001**, *92*, 941–946.
184. Boyer, B.; Bourgeois, Y.; Poupon, M.-R. *Oncogene* **2002**, *21*, 2347–2356.
185. Weis, S.; Cui, J.; Barnes, L.; Cheresh, D. *J. Cell Biol.* **2004**, *167*, 223–229.
186. Eliceiri, B. P.; Paul, R.; Schwartzbert, P. I.; Hood, J. D.; Leng, J.; Cheresch, D. A. *Mol. Cell* **1999**, *4*, 915–924.
187. Paul, R.; Zhang, Z. G.; Eliceiri, B. P.; Jiang, Q.; Boccia, A. D.; Zhang, R. L.; Chopp, M.; Cheresh, D. A. *Nat. Med.* **2001**, *7*, 222–227.
188. Malek, R. L.; Irby, R. B.; Guo, Q. M.; Lee, K.; Wong, S.; He, M.; Tsai, J.; Frank, B.; Liu, E. T.; Quackenbush, J. et al. *Oncogene* **2002**, *21*, 7256–7265.
189. Dehm, S. M.; Bonham, K. *Biochem. Cell Biol.* **2004**, *82*, 263–274.
190. Irby, R. B.; Malek, R. L.; Bloom, G.; Tsai, J.; Letwin, N.; Fank, B. C.; Verratti, K.; Yeatman, T. J.; Lee, N. H. *J. Cancer Res.* **2005**, *65*, 1814–1821.
191. Xu, W.; Harrison, S. C.; Eck, M. J. *Nature* **1997**, *285*, 595–602.
192. Williams, J. C.; Weijland, A.; Gonfloni, S.; Thomson, A.; Courtneidge, S. A.; Superti-Furga, G.; Wierenga, R. K. *J. Mol. Biol.* **1997**, *274*, 757–775.
193. Xu, W.; Doshi, A.; Lei, M.; Eck, M.; Harrison, S. C. *Mol. Cell* **1999**, *3*, 629–636.
194. Zhu, X.; Kim, J. L.; Newcomb, J. R.; Rose, P. E.; Stover, D. R.; Toledo, L. M.; Zhao, H.; Morgenstern, K. A. *Structure* **1999**, *7*, 651–661.
195. Kraybill, B. C.; Elkin, L. L.; Blethrow, J. D.; Morgan, D. O.; Shokat, K. M. *J. Am. Chem. Soc.* **2002**, *124*, 12118–12128.
196. Bishop, A. C.; Kung, C.-Y.; Shah, K.; Witucki, L.; Shokat, K. M.; Liu, Y. *J. Am. Chem. Soc.* **1999**, *121*, 627–631.
197. Liu, Y.; Shah, K.; Yang, F.; Witucki, L.; Shokat, K. M. *Chem. Biol.* **1998**, *5*, 91–101.
198. Shakespeare, W. C.; Metcalf, C. A., III; Wang, Y.; Sundaramoorthi, R.; Keenan, T.; Weigele, M.; Bohacek, R. S.; Dalgarno, D. C.; Sawyer, T. K. *Curr. Opin. Drug Disc. Dev.* **2003**, *6*, 729–741.
199. Susa, M.; Teti, A. *Drug News Perspect.* **2000**, *13*, 169–175.
200. Boschelli, D. H.; Wang, Y. D.; Johnson, S.; Wu, B.; Ye, F.; Barrios Sosa, A. C.; Golas, J. M.; Boschelli, F. *J. Med. Chem.* **2004**, *47*, 599–601.
201. Boschelli, D. H.; Wang, Y. D.; Ye, F.; Wu, B.; Zhang, N.; Dutia, M.; Powell, D. W.; Wissner, A.; Arndt, K.; Weber, J. M. et al. *J. Med. Chem.* **2001**, *44*, 822–833.
202. Wang, Y. D.; Miller, K.; Boschelli, D. H.; Ye, F.; Wu, B.; Floyd, M. B.; Powell, D. W.; Wissner, A.; Weber, J. M. et al. *Bioorg. Med. Chem. Lett.* **2000**, *10*, 2477–2480.
203. Hamby, J. M.; Connolly, C. J.; Schroeder, M. C.; Winters, R. T.; Showalter, H. D.; Panek, R. L.; Major, T. C.; Olsewski, B.; Ryan, M. J.; Dahring, T. et al. *J. Med. Chem.* **1997**, *40*, 2296–2303.
204. Klutchko, S. R.; Hamby, J. M.; Boschelli, D. H.; Wu, Z.; Kraker, A. J.; Amar, A. M.; Hartl, B. G.; Shen, C.; Klohs, W. D.; Steinkampf, R. W. et al. *J. Med. Chem.* **1998**, *41*, 3276–3292.
205. Kraker, A. J.; Hartl, B. G.; Amar, A. M.; Barvian, M. R.; Showalter, H. D.; Moore, C. W. *Biochem. Pharmacol.* **2000**, *60*, 885–898.
206. Hanke, J. H.; Gardner, J. P.; Dow, R. L.; Changelian, P. S.; Brissette, W. H.; Weringer, E. J.; Pollok, B. A.; Connelly, P. A. *J. Biol. Chem.* **1996**, *271*, 695–701.
207. Belsches-Jablonski, A. P.; Biscardi, J. S.; Peavy, D. R.; Tice, D. A.; Romney, D. A.; Parsons, S. J. *Oncogene* **2000**, *20*, 1464–1475.
208. Menke, A.; Philipp, C.; Vogelmann, R.; Seidel, B.; Lutz, M. P.; Adler, G.; Wedlich, D. *Cancer Res.* **2001**, *61*, 3508–3517.
209. Missbach, M.; Jeschke, M.; Feyen, J.; Muller, K.; Glatt, M.; Green, J.; Susa, M. *Bone* **1999**, *24*, 437–449.
210. Recchia, I.; Rucci, N.; Funari, A.; Migliaccio, S.; Taranta, A.; Longo, M.; Kneissel, M.; Susa, M.; Fabbro, D.; Teti, A. *Bone* **2004**, *34*, 65–79.
211. Recchia, I.; Rucci, N.; Festuccia, C.; Bologna, M.; MacKay, A. R.; Migliaccio, S.; Longo, M.; Susa, M.; Fabbro, D.; Teti, A. *Eur. J. Cancer* **2003**, *39*, 1927–1935.
212. Brunton, V. G.; Avizienyte, E.; Fincham, V. J.; Serrels, B.; Metcalf, C. A., III; Sawyer, T. K.; Frame, M. C. *Cancer Res.* **2005**, *65*, 1335–1342.

213. Ple, P. A.; Green, T. P.; Hennequin, I. F.; Curwen, J.; Fennel, M.; Allen, J.; Lambert-Van Der Brempt, C.; Costello, G. *J. Med. Chem.* **2004**, *47*, 871–878.
214. Blake, R. A.; Broome, M. A.; Liu, X.; Wu, J.; Gishizky, M.; Sun, L.; Courtneidge, S. A. *Mol. Cell. Biol.* **2000**, *20*, 9018–9027.
215. Boyce, B. F.; Xing, L.; Shakespeare, W.; Wang, Y.; Dalgarno, D.; Iuliucci, J.; Sawyer, T. *Kidney Intern.* **2003**, *85*, 52–55.
216. Wang, Y.; Metcalf, C. A., III; Shakespeare, W. C.; Sundaramoorthi, R.; Keenan, T. P.; Bohacek, R. S.; van Schravendijk, M. R.; Violette, S. M.; Narula, S. S.; Dalgarno, D. C. et al. *Bioorg. Med. Chem. Lett.* **2003**, *13*, 3067–3070.
217. Vu, C. B.; Luke, G. P.; Kawahata, N.; Shakespeare, W. C.; Wang, Y.; Sundaramoorthi, R.; Metcalf, C. A., III; Keenan, T. P.; Pradeepan, S.; Corpuz, E. et al. *Bioorg. Med. Chem. Lett.* **2003**, *13*, 3071–3074.
218. Sundaramoorthi, R.; Shakespeare, W. C.; Keenan, T. P.; Metcalf, C. A., III; Wang, Y.; Mani, U.; Taylor, M.; Liu, S.; Bohacek, R. S.; Narula, S. S. et al. *Bioorg. Med. Chem. Lett.* **2003**, *13*, 3063–3066.
219. Dalgarno, D.; Stehle, T.; Narula, S.; Schelling, P.; van Schravendijk, M.; Adams, S.; Keats, J.; Ram, M.; Jin, L.; Grossman, T. et al. *Chem. Biol. Drug Des.* **2006**, *67*, 46–57.
220. Yoneda, T.; Lowe, C.; Lee, C.-H.; Gutierrez, G.; Niewolna, M.; Williams, P. J.; Izbicka, E.; Uehara, Y.; Mundy, G. R. *J. Clin. Invest.* **1993**, *91*, 2791–2795.
221. Slate, D. L.; Lee, R. H.; Rodriguez, J.; Crews, P. *Biochem. Biophys. Res. Commun.* **1994**, *203*, 260–264.
222. Maly, D. J.; Choong, I. C.; Ellman, J. A. *Proc. Natl. Acad. Sci. USA* **2000**, *97*, 2419–2424.
223. Alfaro-Lopez, J.; Yuan, W.; Phan, B. C.; Kamath, J.; Lou, Q.; Lam, K. S.; Hruby, V. J. *J. Med. Chem.* **1998**, *41*, 2252–2260.
224. Singh, J.; Ling, L. E.; Sawyer, J. S.; Lee, W.-C.; Zhang, F.; Yingling, J. M. *Curr. Opin. Drug Disc. Dev.* **2004**, *7*, 437–445.
225. Kaklamani, V.; Pasche, B. *Expert Rev. Anticancer Ther.* **2004**, *4*, 649–661.
226. Eyers, P. A.; Craxton, M.; Morrice, N.; Cohen, P.; Goedert, M. *Chem. Biol.* **1998**, *5*, 321–326.
227. Callahan, J. F.; Burgess, J. L.; Fornwald, J. A.; Gaster, L. M.; Harling, J. D.; Harrington, F. P.; Heer, J.; Kwon, C.; Lehr, R.; Mathur, A. et al. *J. Med. Chem.* **2002**, *45*, 999–1001.
228. Sawyer, J. S.; Anderson, B. D.; Beight, D. W.; Campbell, R. M.; Jones, M. L.; Herron, D. K.; Lampe, J. W.; McCowan, J. R.; McMillen, W. T.; Mort, N. et al. *J. Med. Chem.* **2003**, *46*, 3953–3956.
229. Singh, J.; Chuaqui, C. E.; Boriack-Sjodin, P. A.; Lee, W. C.; Pontz, T.; Corbley, M. J.; Cheung, H. K.; Arduini, R. M.; Mead, J. N.; Newman, M. N. et al. *Bioorg. Med. Chem. Lett.* **2003**, *13*, 4355–4359.
230. Stewart, Z. A.; Westfall, M. D.; Pietenpol, J. A. *Trends Pharmacol. Sci.* **2003**, *24*, 139–145.
231. Sendererowicz, A. *Oncogene* **2003**, *22*, 6609–6620.
232. Sielecki, T. M.; Boylan, J. F.; Benfield, P. A.; Trainer, G. L. *J. Med. Chem.* **2000**, *43*, 2–18.
233. Dai, Y.; Grant, S. *Curr. Oncol. Rep.* **2004**, *6*, 123–130.
234. Besson, A.; Assoian, R. K.; Roberts, J. M. *Nature Cancer Rev.* **2004**, *4*, 948–955.
235. Senderowicz, A. M. *Invest. New Drugs* **1999**, *17*, 313–320.
236. Sedlacek, H. H.; Czech, J.; Naik, R.; Kaur, G.; Worland, P.; Losiewicz, M.; Parker, B.; Carlson, B.; Smith, A. *Int. J. Oncol.* **1996**, *9*, 1143–1168.
237. Newcomb, E. W. *Anticancer Drugs* **2004**, *15*, 411–419.
238. Shapiro, G. I. *Clin. Cancer Res.* **2004**, *10*, 4270S–4275S.
239. Seynaeve, C. M.; Stetler, S. M.; Sebers, S.; Kaur, G.; Sausville, E. A.; Worland, P. J. *Cancer Res.* **1993**, *53*, 2081.
240. Sato, S.; Fujita, N.; Tsuruo, T. *Oncogene* **2002**, *21*, 1727–1738.
241. Akinaga, S.; Sugiyama, K.; Akiyama, T. *Anticancer Drug Des.* **2000**, *15*, 43–53.
242. McClue, S. J.; Blake, D.; Clarke, R.; Cowan, A.; Cummings, L.; Fischer, P. M.; MacKenzie, M.; Melville, J.; Stewart, K.; Wang, S. et al. *Int. J. Cancer* **2002**, *102*, 463–468.
243. Whittaker, S. R.; Walton, M. I.; Garrett, M. D.; Workman, P. *Cancer Res.* **2004**, *64*, 262–272.
244. Misra, R. N.; Xiao, H. Y.; Kim, K. S.; Lu, S.; Han, W. C.; Barbosa, S. A.; Hunt, J. T.; Rawlins, D. B.; Shan, W.; Ahmed, S. Z. et al. *J. Med. Chem.* **2004**, *47*, 1719–1728.
245. Cox, A. D.; Der, C. J. *Cancer Biol. Ther.* **2002**, *1*, 599–606.
246. Chang, F.; Steelman, L. S.; Lee, J. T.; Shelton, J. G.; Navolanic, P. M.; Blalock, W. L.; Franklin, R. A.; McCubrey, J. A. *Leukemia* **2003**, *17*, 1263–1293.
247. Bollag, G.; Freeman, S.; Lyons, J. F.; Post, L. E. *Curr. Opin. Investig. Drugs* **2003**, *4*, 1436–1441.
248. Sebolt-Leopold, J. S. *Curr. Pharm. Des.* **2004**, *10*, 1907–1914.
249. Wilhelm, S. M.; Carter, C.; Tang, L.; Wilkie, D.; McNabola, A.; Rong, H.; Chen, C.; Zhang, X.; Vincent, P.; McHugh, M. et al. *Cancer Res.* **2004**, *64*, 7099–7109.
250. Wilhelm, S.; Chien, D.-S. *Curr. Pharm. Des.* **2002**, *8*, 2255–2257.
251. Allen, L. F.; Sebolt-Leopold, J.; Meyer, M. B. *Semin. Oncol.* **2003**, *30*, 105–116.
252. Sebolt-Leopold, J. S.; Dudley, D. T.; Herrera, R.; Van Becelaere, K.; Wiland, A.; Gowan, R. C.; Tecle, H.; Barrett, S. D.; Bridges, A.; Przybranowski, S. et al. *Nat. Med.* **1999**, *5*, 810–816.
253. Chan, S. *Br. J. Cancer* **2004**, *9*, 1420–1424.
254. Panwalkar, A.; Verstovsek, S.; Giles, F. J. *Cancer* **2004**, *100*, 657–666.
255. Mita, M. M.; Mita, A.; Rowinsky, E. K. *Clin. Breast Cancer* **2003**, *4*, 126–137.
256. Brader, S.; Eccles, S. A. *Tumori* **2004**, *90*, 2–8.
257. Parsons, R. *Semin. Cell Dev. Biol.* **2004**, *15*, 171–176.
258. Bjornsti, M.-A.; Houghten, P. *Nat. Rev. Cancer* **2004**, *4*, 335–348.
259. Sawyer, C. L. *Cancer Cell* **2003**, *4*, 343–348.
260. Kang, S.; Bader, A. G.; Vogt, P. K. *Proc. Natl. Acad. Sci. USA* **2005**, *102*, 802–807.
261. Elchebly, M.; Payette, P.; Michaliszyn, E.; Cromlish, W.; Collins, S.; Loy, A. L.; Normandin, D.; Cheng, A.; Himms-Hagen, J.; Chan, C.-C. et al. *Science* **1999**, *283*, 1544–1548.
262. Byon, J. C.; Kusari, H. A.; Kusari, J. *Mol. Cell. Biochem.* **1998**, *182*, 101–108.
263. Ahmad, F.; Azevedo, J. L.; Cortright, R.; Dohm, G. L.; Goldstein, B. J. *J. Clin. Invest.* **1997**, *100*, 449–458.
264. van Huijsduijnen, R. H.; Sauer, W. H. B.; Brobrun, A.; Swinnen, D. *J. Med. Chem.* **2004**, *47*, 4142–4146.
265. Pei, Z.; Liu, G.; Lubben, T. H.; Szczepankiewicz, B. G. *Curr. Pharm. Des.* **2004**, *10*, 3481–3504.
266. Johnson, T. O.; Ermolieff, J.; Jirouisek, M. R. *Nat. Rev. Drug Disc.* **2002**, *1*, 696–709.
267. Shen, K.; Keng, Y. F.; Wu, L.; Guo, X. L.; Lawrence, D. S.; Zhang, Z. Y. *J. Biol. Chem.* **2001**, *276*, 47311–47319.

268. Larsen, S. D.; Barf, T.; Liljebris, C.; May, P. D.; Ogg, D.; O'Sullivan, T. J.; Palazuk, B. J.; Schostarez, H. J.; Stevens, F. C.; Bleasdale, J. E. *J. Med. Chem.* **2002**, *45*, 598–622.
269. Bleasdale, J. E.; Ogg, D.; Palazuk, B. J.; Jacob, C. S.; Swanson, M. L.; Wang, X. Y.; Thompson, D. P.; Conradi, R. A.; Mathews, W. R.; Laborde, A. L. et al. *Biochemistry* **2001**, *40*, 5642–5654.
270. Andersen, H. S.; Olsen, O. H.; Iversen, L. F.; Sorensen, A. L.; Mortensen, S. B.; Christensen, M. S.; Branne, S.; Hansen, T. K.; Lau, J. F.; Jeppesen, L. et al. *J. Med. Chem.* **2002**, *45*, 4443–4459.
271. Lund, I. K.; Andersen, H. S.; Iversen, L. F.; Olsen, O. H.; Moller, K. B.; Pedersen, A. K.; Ge, Y.; Holsworth, D. D.; Newman, M. J.; Axe, F. U. et al. *J. Biol. Chem.* **2004**, *279*, 24226–24235.
272. McCluskey, A.; Sim, A. T. R.; Sakoff, J. A. *J. Med. Chem.* **2002**, *45*, 1151–1175.
273. Muller, G. *Top. Curr. Chem.* **2000**, *211*, 17–59.
274. Cody, W. L.; Lin, Z.; Panek, R. L.; Rose, D. W.; Rubin, J. R. *Curr. Pharm. Des.* **2000**, *6*, 59–98.
275. Fretz, H.; Furet, P.; Garcia-Echeverria, C.; Schoepfer, J.; Rahuel, J. *Curr. Pharm. Des.* **2000**, *6*, 1777–1796.
276. Burke, T. R., Jr.; Yao, Z.-J.; Liu, D.-G.; Voigt, J.; Gao, Y. *Biopolymers (Peptide Sci.)* **2001**, *60*, 32–44.
277. Sawyer, T. K.; Bohacek, R.; Dalgarno, D. C.; Metcalf, C. M., III; Shakespeare, W. C.; Sundarmoorthi, R.; Wang, Y. *Mini-Rev. Med. Chem.* **2002**, *2*, 475–488.
278. Lunney, E. A.; Para, K. S.; Rubin, J. R.; Humblet, C.; Fergus, J. H.; Marks, J. S.; Sawyer, T. K. *J. Am. Chem. Soc.* **1997**, *119*, 12471–12476.
279. Violette, S. M.; Shakespeare, W. C.; Bartlett, C.; Guan, W.; Smith, J. A.; Rickles, R. J.; Bohacek, R. S.; Holt, D. A.; Baron, R.; Sawyer, T. K. *Chem. Biol.* **2000**, *7*, 225–235.
280. Shakespeare, W.; Yang, M.; Bohacek, R.; Cerasoli, F.; Stebbins, K.; Sundaramoorthi, R.; Azimioara, M.; Vu, C.; Pradeepan, S.; Metcalf, C., III et al. *Proc. Natl. Acad. Sci. USA* **2000**, *97*, 9373–9378.
281. Sundaramoorthi, R.; Kawahata, N.; Yang, M. G.; Shakespeare, W. C.; Metcalf, C. A., III; Wang, Y.; Merry, T.; Eyermann, C. J.; Bohacek, R. S.; Narula, S. et al. *Biopolymers (Pept. Sci.)* **2003**, *71*, 717–729.
282. Lesuisse, D.; Deprez, P.; Albert, E.; Duc, T. T.; Sortais, B.; Goffe, D.; Jean-Baptiste, V.; Marquette, J.-P.; Schoot, B.; Sarubbi, E. et al. *Bioorg. Med. Chem. Lett.* **2001**, *11*, 2127–2131.
283. Lesuisse, D.; Deprez, P.; Bernard, D.; Broto, P.; Delettre, G.; Jean-Baptiste, V.; Marquette, J. P.; Sarubti, E.; Schoot, B.; Mandine, E. et al. *Chim. Nouv.* **2001**, *19*, 3240–3241.
284. Lesuisse, D.; Lange, G.; Deprez, P.; Benard, D.; Schoot, B.; Delettre, G.; Marquette, J. P.; Broto, P.; Jean-Baptiste, V.; Bichet, P. et al. *J. Med. Chem.* **2002**, *45*, 2379–2387.
285. Furet, P.; Gay, B.; Caravatti, G.; Garcia-Echeverria, C.; Gay, B.; Rahuel, J.; Schoepfer, J.; Fretz, H. *J. Med. Chem.* **1998**, *41*, 3442–3449.
286. Gay, B.; Suarez, S.; Weber, C.; Rahuel, J.; Fabbro, D.; Furet, P.; Caravatti, G.; Schoepfer, J. *J. Biol. Chem.* **1999**, *274*, 23311–23315.
287. Furet, P.; Caravatti, G.; Denholm, A. A.; Faessler, A.; Fretz, H.; Garcia-Echeverria, C.; Gay, B.; Irving, E.; Press, N. J.; Rahuel, J.; Schoepfer, J.; Waker, C. V. *Bioorg. Med. Chem. Lett.* **2000**, *10*, 2337–2341.
288. Schoepfer, J.; Gay, B.; Caravatti, G.; Garcia-Echeverria, C.; Fretz, H.; Rahuel, J.; Furet, P. *Bioorg. Med. Chem. Lett.* **1998**, *8*, 2865–2870.

Biography

Tomi K Sawyer received a BSc degree in Chemistry at Moorhead State University (now Minnesota State University-Moorhead) and PhD in Organic Chemistry at the University of Arizona. His research has integrated synthetic chemistry, drug design, structural biology, chemoinformatics, biochemistry, cell biology and in vivo disease models with a focus on cancer. Tomi's drug discovery track record includes contributions to clinical candidates and/or noteworthy molecular tools for several therapeutic targets, including GPCR receptors (melanocortin), aspartyl proteases (renin and HIV protease), and protein kinases (Src and Abl). He has published more than 200 scientific articles, reviews, commentaries, monographs and books. Tomi is an inventor of more than 50 issued patents and patent filings. He worked at Upjohn Company and Parke-Davis/Warner-Lambert (now both Pfizer Global Research & Development), and is currently Senior-Vice President, Drug Discovery, at ARIAD Pharmaceuticals. He is concurrently adjunct professor, Chemistry as well as Biochemistry & Molecular Biology, University of Massachusetts and also adjunct professor, Cancer

Biology, at University of Massachusetts School of Medicine. Tomi has served on the highlights advisory panel of Nature Reviews Drug Discovery and the editorial advisory boards of Trends in the Pharmacological Sciences, Expert Reviews in Molecular Medicine, Expert Opinion on Investigational Drugs, Journal of Medicinal Chemistry, Chemistry and Biology, Current Medicinal Chemistry (Anti-Cancer Agents), Current Organic Synthesis, Expert Reviews in Molecular Medicine, Expert Opinion on Therapeutic Patents (Oncology), Drug Design and Discovery, Pharmaceutical Research, Molecular Biotechnology, and Biopolymers (Peptide Science). Most recently, Tomi was appointed Editor-in-Chief, Chemical Biology & Drug Design.

Comprehensive Medicinal Chemistry II
ISBN (set): 0-08-044513-6

ISBN (Volume 2) 0-08-044515-2; pp. 959–992

2.25 Nuclear Hormone Receptors

N T Zaveri and B J Murphy, SRI International, Menlo Park, CA, USA

© 2007 Elsevier Ltd. All Rights Reserved.

2.25.1	**Introduction**	**994**
2.25.2	**Current Classification and Nomenclature**	**994**
2.25.2.1	Classification of Human Nuclear Receptors	994
2.25.2.2	Nomenclature of Nuclear Receptors	995
2.25.3	**Nuclear Receptor Protein Structure**	**995**
2.25.3.1	Three-Dimensional Nuclear Receptor Protein Structure	997
2.25.3.1.1	The ligand-binding domain	998
2.25.3.1.2	Ligand-independent activation function (AF-1) domain	1001
2.25.3.1.3	DNA-binding domain	1002
2.25.4	**Coregulators and Nuclear Receptor-Dependent Transcriptional Control**	**1002**
2.25.4.1	Coactivators	1002
2.25.4.2	Corepressors	1003
2.25.4.3	Modifications of Nuclear Receptors and Their Coregulators	1003
2.25.4.3.1	Ubiquitination	1003
2.25.4.3.2	Acetylation	1003
2.25.4.3.3	Methylation	1003
2.25.4.3.4	Phosphorylation	1004
2.25.5	**Physiological Functions of Nuclear Receptors**	**1004**
2.25.5.1	Class I Receptors	1004
2.25.5.1.1	Estrogen receptors	1004
2.25.5.1.2	Progesterone receptor	1005
2.25.5.1.3	Androgen receptor	1006
2.25.5.1.4	Glucocorticoid receptor	1006
2.25.5.1.5	Mineralocorticoid receptor	1007
2.25.5.2	Class II Receptors	1007
2.25.5.2.1	Vitamin D receptor	1007
2.25.5.2.2	Retinoic acid receptor	1007
2.25.5.2.3	Thyroid hormone receptor	1008
2.25.5.2.4	Peroxisome proliferator-activated receptors	1008
2.25.5.2.5	Liver X receptors	1010
2.25.5.2.6	Farnesoid X receptor	1010
2.25.5.2.7	Constitutive androstane receptor and pregnane X receptor	1011
2.25.6	**Pharmacology of Nuclear Receptor Ligands**	**1011**
2.25.6.1	Estrogen Receptor Ligands	1013
2.25.6.2	Progesterone Receptor Ligands	1014
2.25.6.3	Androgen Receptor Ligands	1015
2.25.6.4	Glucocorticoid Receptor Ligands	1016
2.25.6.5	Mineralocorticoid Receptor Ligands	1017
2.25.6.6	Vitamin D Receptor Ligands	1018
2.25.6.7	Thyroid Hormone Receptor Ligands	1019
2.25.6.8	Retinoid Receptor (RAR and RXR) Ligands	1020
2.25.6.9	Peroxisome Proliferator-Activated Receptor Ligands	1021
2.25.6.10	Liver X Receptor Ligands	1023

	2.25.6.11	Farnesoid X Receptor Ligands	1024
	2.25.6.12	Pregnane X Receptor and Constitutive Androstane Receptor Ligands	1024
	2.25.7	**Therapeutic Disease Applications for Nuclear Receptor Ligands**	1025
	2.25.8	**Genetic Diseases Associated with Nuclear Receptors**	1026
	2.25.9	**Perspectives and New Directions in Nuclear Receptor Research**	1028
	References		1028

2.25.1 Introduction

Nuclear receptors (NRs) are one of the largest known superfamilies of eukaryotic transcription factors that regulate the expression of genes involved in critical processes such as embryonic development, metabolism, and inflammation.[1–3] These receptors, upon binding a range of lipophilic extracellular ligands, control the transcription of target genes. Some well-known NR ligands include hormones such as steroids, retinoids, thyroid hormone, and vitamin D3. The NRs are now recognized as playing a pivotal role in critical physiologic and metabolic functions, and dysfunction of NR signaling has been implicated in major disease states such as cancer, infertility, diabetes, obesity, and hypercholesterolemia. As a result, NRs are major targets for drug discovery in several therapeutic areas. Indeed, 8 of the top 100 prescription drugs target an NR, including well-known drugs such as rosiglitazone (Avandia), a PPARγ ligand; pioglitazone (Actos), also a PPARγ ligand; raloxifene (Evista), an ERα ligand; and fluticasone (Flonase), a glucocorticoid receptor ligand.

2.25.2 Current Classification and Nomenclature

Genome sequence analysis has identified 49 NR sequences in the human genome, whereas 21 NRs have been found in *Drosophila melanogaster* and about 270 NRs in the *Caenorhabditis elegans* genome.[4–6] Through phylogenetic analysis and sequence matching of the identified NR sequences, six subfamilies have been defined among the NR superfamily.[7]

2.25.2.1 Classification of Human Nuclear Receptors

The human NR family is loosely grouped into three classes based on their ligand- and DNA-binding properties.[3] Class I NRs are the classic steroid hormone receptors and include the estrogen receptor (ER), androgen receptor (AR), progesterone receptor (PR), glucocorticoid receptor (GR), and the mineralocorticoid receptor (MR). In the absence of ligand, these NRs are sequestered in nonfunctional complexes with heat-shock proteins and are transcriptionally inactive. On ligand activation, Class I NRs bind, as homodimers, to palindromic (head-to-head arrangement) DNA sequences called NR response elements (NREs) (**Figure 1**). Class II receptors include thyroid hormone receptor (TR), vitamin D receptor (VDR), retinoid receptor (RAR), peroxisome proliferator-activated receptor (PPAR), liver X receptor (LXR), and the farnesoid X receptor (FXR). These receptors bind head-to-tail DNA repeat sequences as heterodimers with the retinoid X receptor (RXR), even in the absence of ligand, and exert a repressive silencing effect on basal promoter activity that is reversed on ligand binding. Class III NRs comprise the orphan nuclear receptors (oNRs) that have been identified using genomic methods. These oNRs are structurally related to the NRs, but they lacked known physiological ligands at the time the receptors were identified. As shown in **Figure 1** and **Table 1**, the oNRs are subclassified as those that are known to bind their DNA sequences as homodimers (e.g., the chicken ovalbumin upstream promoter transcription factor (COUP-TF) and the hepatocyte nuclear factor 4 (HNF-4)), as monomers (e.g., steroidogenic factor (SF-1)), and those that lack DNA-binding domains (DBDs) (e.g., small heterodimeric partner (SHP)).

oNRs present a huge opportunity to uncover new physiological regulatory systems that may be involved in human health and disease and, thus, have potential as future therapeutic targets. Several pharmaceutical companies have undertaken research programs to identify ligands for these oNRs. A combination of targeted screening of small, biased compound collections and high-throughput screening of large random libraries have successfully led to the identification of naturally occurring as well as synthetic ligands for a number of oNRs. Such oNRs have been termed 'adopted orphan receptors,' and include the pregnane X receptor (PXR), the constitutive androstane receptor (CAR), and the oxysterol-binding liver X receptor (LXR). Identification of naturally occurring ligands for these oNRs has proved to be one of the most powerful ways to identify the physiological function of the NR; this process is a classic example of reverse endocrinology.[8]

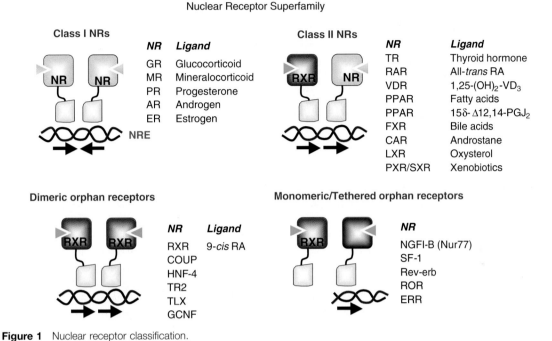

Figure 1 Nuclear receptor classification.

2.25.2.2 Nomenclature of Nuclear Receptors

The genomic revolution of the last decade resulted in the identification in various species of over 300 genes belonging to the NR superfamily. This necessitated the development of a logical and useful classification and nomenclature system that would not only integrate NR genes already identified, but also be flexible enough to allow inclusion of newly identified sequences for which no biological data is yet known. The new nomenclature, proposed in 1999, is based on the evolution of the two well-conserved domains of NRs (the DNA-binding C domain and the ligand-binding E domain, discussed below) and uses molecular phylogeny to connect and classify all known NR sequences.[9] This system divides the superfamily into six subfamilies and 26 groups of receptors. Each receptor/gene name is described by the letters 'NR' and a three-digit identifier containing an Arabic numeral for the subfamily, a capital letter for the group, and another Arabic numeral for the individual gene. Splice variants (isoforms) are designated by a final lowercase letter. **Table 1** shows the designated nomenclature for the known human NRs. For example, the estrogen receptor ERα which belongs to subfamily 3, group A, and is the first gene identified in that group, is named NR3A1. Detailed information about the evolution of this nomenclature can be found on the Nuclear Receptor Nomenclature Homepage.[377]

2.25.3 Nuclear Receptor Protein Structure

NRs are modular proteins containing three major functional domains,[10] composed of six regions labeled A through F (**Figure 2**). Each domain has a separate and crucial function in NR-mediated transactivation and signal transduction. The N-terminal domain (known as the A/B region) is highly variable in length and sequence among NRs, and contains an intrinsic transactivation domain, called activation function (AF-1) domain, that is involved in ligand-independent transcriptional activity. The central domain (C/D region) is the DBD. It is the most highly conserved region among the receptors. Although the DBD–DNA interactions vary from receptor to receptor, all DBDs contain two type II zinc-finger motifs that are common to the entire superfamily. These zinc fingers fold into a single structural domain with an α-helical reading head containing three amino acids (the P box) responsible for DNA sequence recognition in the hormone response elements (NREs).[11] The C-terminal domain is the ligand-binding domain (LBD) or E/F region, which contains regions for receptor dimerization, heat-shock protein interaction, transcriptional regulation, and, most importantly, ligand binding. LBDs have a conserved overall α-helical architecture, with sufficient sequence diversity to allow for selective ligand recognition among the various NRs. The LBD contains the second activation function (AF-2),

Table 1 Nuclear receptor gene and protein nomenclature

Nuclear receptor	Protein	Nomenclature	Gene
Estrogen receptor	ERα	NR3A1	*Esr1*
	ERβ	NR3A2	*Esr2*
Androgen receptor	AR	NR3C4	*Ar*
Glucocorticoid receptor	GR	NR3C1	*Gr*
Mineralocorticoid receptor	MR	NR3C2	*Nr3c2*
Progesterone receptor	PR	NR3C3	*Pgr*
Thyroid hormone receptor	TRα	NR1A1	*Thrα*
	TRβ	NR1A2	*Thrβ*
Retinoic acid receptor	RARα	NR1B1	*Rarα*
	RARβ	NR1B2	*Rarβ*
	RARγ	NR1B3	*Rarγ*
Vitamin D receptor	VDR	NR1I1	*Vdr*
Peroxisome proliferator-activated receptor	PPARα	NR1C1	*Pparα*
	PPARγ	NR1C2	*Pparγ*
	PPARβ	NR1C3	*Pparβ*
Liver X receptor	LXRα	NR1H3	*Nr1h3*
	LXRβ	NR1H2	*Nr1h2*
Farnesoid X receptor	FXRα	NR1H4	*Nr1h4*
	FXRβ	NR1H5	*Nr1h5*
Constitutive androstane receptor	CAR	NR1I3	*Nr1i3*
Pregnane X receptor	PXR	NR1I2	*Nr1i2*
Retinoid X receptor	RXRα	NR2B1	*Rxrα*
	RXRβ	NR2B2	*Rxrβ*
	RXRγ	NR2B3	*Rxrγ*
Estrogen receptor-related receptor	ERRα	NR3B1	*Esrrα*
	ERRβ	NR3B2	*Esrrβ*
	ERRγ	NR3B3	*Essrγ*
Steroidogenic factor 1	SF1	NR5A1	*Nr5a1*
Retnoid X receptor	RXRα	NR2B1	*Rxrα*
	RXRβ	NR2B2	*Rxrβ*
	RXRγ	NR2B3	*Rxrγ*
RAR-related orphan receptor	RORα	NR1F1	*Rorα*
	RORβ	NR1F2	*Rorβ*
	RORγ	NR1F3	*Rorγ*
Chicken ovalbumin upstream promoter-transcription factor	COUP-TFI	NR2F1	*Nr2f1*
	COUP-TFII	NR2F2	*Nr2f2*

Table 1 Continued

Nuclear receptor	Protein	Nomenclature	Gene
Human nuclear factor 4 receptor	HNF4α	NR2A1	*Hnf4α*
	HNF4γ	NR2A2	*Hnf4γ*
Testis receptor	TR2	NR2C1	*Nr2c1*
	TR4	NR2C2	*Nr2c2*
Reverse erbA receptor	Rev-erbα	NR1D1	*Nr1d1*
	Rev-erbβ	NR1D2	*Nr1d2*
Photoreceptor-specific nuclear receptor	PNR	NR2E3	*Nr2e3*
Germ cell nuclear factor	GCNF	NR6A1	*Nr6a1r*
Nur related factor 1	NURR1	NR4A2	*Nr4a2*
NGF-induced factor B (Nur77)	NGFIB	NR4A1	*Nr4a1*
Neuron-derived orphan receptor 1	NOR1	NR4A3	*Nr4a3*
Liver receptor homologous protein 1	LRH1	NR5A2	*Nr5a2*
DSS-AHC critical region on the chromosome, gene 1	DAX1	NR0B1	*Nr0b1*
Short heterodomeric partner	SHP	NR0B2	*Nr0b2*

Human FXRβ is a pseudogene; DSS-AHC, dosage-sensitive sex reversal-adrenal hypoplasia-congenita; NGF, nerve growth factor.

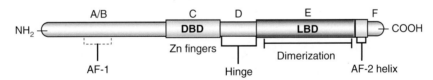

Figure 2 Domain structure of nuclear receptors.

which is critical for ligand-induced transcriptional regulation. The multistep process of gene transcription is orchestrated by the LBD through a complex interplay of other coregulatory proteins (coactivators and corepressors) that couple ligand-induced conformational changes to the regulation of gene transcription. The LBD contains the binding site for synthetic ligands as well as natural hormonal NR ligands, and has been the target for major drug discovery efforts in the pharmaceutical industry (see [12–15] for excellent in-depth reviews).

2.25.3.1 Three-Dimensional Nuclear Receptor Protein Structure

Three-dimensional (3D) structures of proteins are usually obtained through a combination of techniques, such as x-ray crystal structure and nuclear magnetic resonance (NMR) spectroscopy, and confirmed by site-directed mutagenesis of the proteins under study. Although structural studies have been carried out on isolated NR domains, no crystal structure of a full-length NR has yet been reported. The Gustafsson and Yamamoto laboratories[16] published the first 3D structure of an NR domain, the GR DBD, as determined by solution NMR spectroscopy. This landmark work set the stage for subsequent determinations of the 3D structures of NRs that have provided remarkable insights into the mechanism of NR action and the design of synthetic NR ligands to modulate these actions. The 3D structures of LBDs have been determined for a few unliganded (apo) receptors and many liganded (holo) NRs (**Table 2**),[17] and these structures provide a good understanding of the molecular dynamics of ligand binding, coregulator interaction, and transactivation (reviewed in [18]).

Table 2 Selected list of crystal structures of nuclear receptor ligand-binding domains deposited in the Protein Data Bank

NRLBD	Ligand	Ligand type	Cofactor	PDB file name	Reference
RXRα	–	–	–	1LBD	24
	9cRA	Agonist	–	1FBY	366
	BMS649	Agonist	GRIP1box2	–	367
RXRβ	LG100268	Agonist	–	1H9U	368
RARγ	TRA	Agonist	–	2LBD	25
TRα	T3	Agonist	–	–	369
TRβ	T3	Agonist	GRIP1box2	1BSX	42
VDR	1α,25(OH)$_2$D$_3$	Agonist	–	1DB1	370
	Calcipotriol	Agonist		1S19	371
PPARα	AZ242	Agonist	–	1I7G	372
	GW6471	Antagonist	SMRTbox2	1KKQ	21
PPARβ	GW2433	Agonist	–	1GWX	29
PPARγ	–	–	–	3PRG	373
	Rosiglitazone	Agonist	–	2PRG	26
RXRα/PPARγ	9cRA/rosiglitazone	Agonist/agonist	SRC1box2	1FM6	57
ERα	RAL (Raloxifene)	Antagonist	–	1ERR	22
	4-OHT	Antagonist	–	3ERT	23
ERβ	RAL	Antagonist	–	1QKN	53
	GEN (Genistein)	Partial agonist	–	1QKM	53
PR	Progesterone	Agonist	–	1A28	55
AR	R1881	Agonist	–	1E3G	374
LXRα/RXRα	GSK3987/9-cis RA	Agonist	–	ZACL	315
LXRβ	T0901317	Agonist	–	1UPV and 1UPW	375
RORα	Cholesterol	Agonist	–	–	376
PXR	SR12813	Agonist	–	1ILH	27

2.25.3.1.1 The ligand-binding domain

Structural studies show that the LBDs of all NRs share a common overall 3D structure, consisting of 12 α-helices (H1–H12), two or three short β-sheets and connecting loops, arranged as a three-layered sandwich (**Figure 3a**).[19] The top half of the LBD forms the structural core of the protein and is relatively invariant between receptors. In the lower half, the central helical layer is moved away to form a large cavity of the ligand-binding pocket (LBP). The size of the predominantly nonpolar ligand-binding site varies significantly between NRs, allowing for the recognition of unique ligands and hormones of varying sizes. The LBP is lined by residues from helices H3, H5, H6, and H11. Near the entrance of the LBP is the C-terminal helix, H12, that contains the residues of the ligand-dependent AF-2, which is crucial for coactivator binding. The position of helix 12 is highly flexible, and its structural transition and repositioning is essential for recruitment of transcriptional cofactors. On agonist binding, H12 moves over the entrance to the LBP, into a hydrophobic cleft formed by H3 and H4, and forms a charge clamp between a Lys at the C-terminal end of H3 and Glu on H12.[20] The hydrophobic cleft and charge clamp create a docking surface for coactivator interaction. The precise positioning of H12 depends on the size and shape of the binding ligand, thus allowing the ligand to regulate the

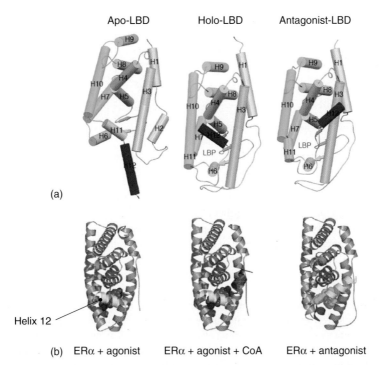

Figure 3 (a) Schematic representation of the different conformational states of ligand-binding domains (LBD) of retinoid NRs. The α-helices (H1–H12) are depicted as tubes and β-turns as broad arrows. The apo-LBD shows the three-layered sandwich structure of the 12 α-helices. In the agonist-bound holoLBD, H12 (red) moves over the LBP to form a binding site for coactivators. In the antagonist-bound LBD, H12 is moved away from the LBP and does not form the coactivator-binding site. The different colors of the tubes represent the different functions of the α-helices in the LBD: dimerization surface (green), coactivator-binding regions H3, H4, and H5 (orange), and the AF-2 function (H12) shown in red. (Adapted with permission from Bourguet, W.; Germain, P.; Gronemeyer, H. *Trends Pharmacol. Sci.* **2000**, *21*, 381–388.) (b) X-ray crystal structure of ERα complexed with agonist estradiol (PDB code 1ERE), with agonist + coactivator peptide (PDB code 1GWR) and with antagonist raloxifene (PDB code 1ERR). Note that in the antagonist structure, H12 (yellow) occupies the same position as the coactivator peptide (red) in the agonist + coactivator structure. (Adapted with permission from Nagy and Schwabe, *Trends Biochem Sci.* **2004**, *29*, 317–324.)

transcriptional activity of the NR. In one model, binding of an antagonist in the LBP may prevent H12 from docking in the hydrophobic groove. A corepressor binds in part of the groove, and H12 is forced out of its normal activating position, as is the case with antagonist binding to PPARα.[21] However, an antagonist may force H12 to dock in the coactivator-binding groove and block coactivator binding, as is the case when antagonist ligands tamoxifen and raloxifene bind the estrogen receptor ERα.[22,23]

2.25.3.1.1.1 Ligand-binding pocket in absence of ligand

The first x-ray crystal structure of an unliganded LBD was that of the apo-retinoid X receptor alpha (RXRα).[24] While it conformed to the common canonical 3D structure of NRs, it appeared to lack an obviously discernible LBP. Comparison with the structure of the liganded RAR receptor[25] suggested that NRs may undergo a conformational rearrangement of helices that allows for ligand binding. In contrast, however, NRs such as PPARs, PXR, and LXR have large LBPs that appear as cavities in the unliganded receptor.[26,27]

The size and 3D structure of the LBPs of various NRs offer several insights about the respective NR function.[15] In classical nuclear hormone receptors, such as the steroid, thyroid, and retinoid receptors that are activated by endocrine and paracrine signaling molecules, the LBP is relatively small (400–500 Å) and the ligands typically occupy most of it. This close binding interaction explains both the high affinity of such hormonal ligands and their high specificity for the NR.

In contrast to these classical nuclear hormone receptors, the more recently characterized classes of NRs, such as the PPARs, LXR, and FXR, have large LBPs suggesting that these NRs may have different modes of cellular function via low affinity and structurally diverse ligands. The Y-shaped PPAR ligand-binding site is the largest LBP among the NRs, at 1400 Å.[28] Ligand binding in such a large pocket is less discriminate and enables the PPARs to bind structurally diverse ligands such as fatty acids of various chain lengths[29] and synthetic ligands such as the thiazolidinediones that

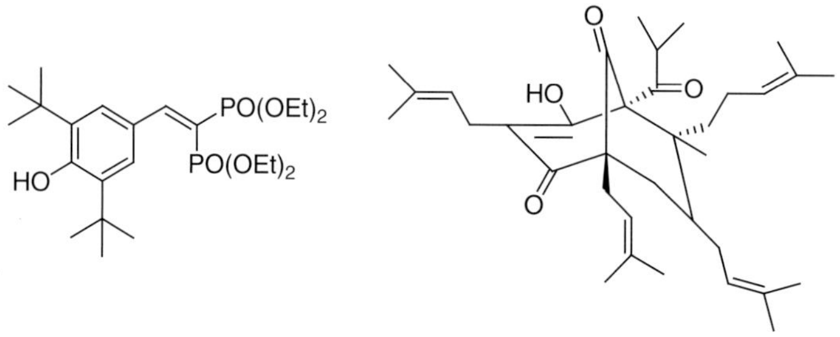

Figure 4 Structures of xenobiotics that bind PXR.

target the PPARγ NRs and are used as insulin-sensing pharmaceuticals.[30] The LBP of LXR is also large, at 700–800 Å, and it binds a variety of hydroxylated cholesterol metabolites.[31] Since the ligands for these receptors are naturally occurring metabolites of nutrients such as cholesterol and bile acids, it has been suggested that the PPARs, FXRs, and LXRs function primarily as nutritional sensors rather than as hormonal receptors for endocrine signaling ligands.[15,32]

Another receptor whose structure suggests that it may be a nutritional sensor is the PXR receptor. PXR is a xenobiotic sensor and it regulates genes, such as CYP3A and MDR1, that are involved in the metabolism and transport of xenobiotics and diet-derived natural compounds (discussed in detail in the following section). The crystal structure of its LBD reveals several features that indicate its ability to interact with diverse, structurally unrelated xenobiotics. Its large (1150 Å) LBP contains several distinguishing characteristics, such as more β-strands and a helical insert, that result in a spherical LBP. The spherical shape of the LBP allows ligands as small as the cholesterol-lowering agent SR 12813 (**Figure 4**) to bind in multiple orientations,[27] and ligands as large as rifampicin and hyperforin (**Figure 4**) to bind PXR. All these xenobiotics are notorious for inducing CYP3A4 and causing drug–drug interactions during pharmaceutical use.[15,33] Determining the crystal structures of the LBDs of oNRs, therefore, may offer important clues about the physiological function of oNRs and can be valuable in the search for the endogenous ligands for such NRs.

2.25.3.1.1.2 Ligands as structural cofactors

The 3D structures of NRs have also expanded our understanding of the role of ligands that bind NRs. The hepatocyte nuclear factors 4 (HNF4α and HNF4γ) have been shown to be constitutively active in cellular experiments.[34] The determination of the crystal structure of the native HNF4 revealed a relatively small LBP (625 Å) that was completely occupied by a fatty acid that cocrystallized with the NR, but did not interact with the AF-2 helix that capped the LBP, suggesting that the cocrystallized fatty acid may serve as a structural cofactor, rather than as a conventional ligand. Being constitutively active, HNF4 is not amenable to modulation by ligands through its LBP, but rather requires new approaches to regulate its activity via coactivator-binding sites or posttranslational modifications. Such approaches are being actively pursued in the pharmaceutical industry to target HNF4 as a potential treatment for Type 2 diabetes.[35] Another such example is the retinoid receptor-related orphan receptor RORβ, the crystal structure of which shows that it has a large LBP that cocrystallizes with a tightly bound stearic acid molecule, suggesting that RORβ may require fatty acid-related molecules as structural cofactors, rather than as regulatory ligands.[36]

A third aspect of the NR structure and function revealed by 3D analysis is the discovery of NRs that lack LBPs entirely. Nurr1, an oNR that has been studied extensively for its essential functions in dopamine neurons,[37] does not possess an LBP. Instead, bulky, hydrophobic amino acids occupy the space in the lower half of the LBD that normally contains the LBP cavity in other NRs.[38] Such NRs will therefore require new approaches, other than the traditional ligand-based approach, for regulating their function as targets for drug discovery.

2.25.3.1.1.3 Coregulator binding site in the ligand-binding domain and the structural determinants of transactivation (agonism) and repression (antagonism)

The C-terminal end of the LBD contains the AF-2 domain, a conserved region involved in ligand-induced coregulator interaction and transcriptional regulation by the LBD. Helix 12 of the LBD contains residues that play a pivotal role in the function of AF-2, and hence H12 is sometimes referred to as the AF-2 helix.[39,40] Regulation of NR-dependent transcriptional action requires recruitment of a complex series of proteins called coregulators (coactivators and corepressors) that are recruited by the AF-2 domain of the LBD. Most coactivators and corepressors contain multiple short NR interaction motifs, called NR boxes, of the general sequence LxxLL in coactivators (CoNR box)[41,42] and LxxxIxxx[L/I] in corepressors (CoRNR box).[43,44] These NR boxes directly interact with the LBD AF-2 domain.

Several crystal structures of LBD–coactivator peptide complexes (**Table 2**) reveal a general mode of coactivator binding to the LBD. The coactivators can possess multiple LxxLL motifs that adopt a two-turn helix, both ends of which are stabilized by a charge clamp formed with a conserved acidic residue from the AF-2 helix.[23,26,42] Given that there are a large number of coactivators (see Section 2.25.4), each containing multiple LxxLL motifs, questions arise about how specific coactivators are recruited to given NR LBDs.

A crystal structure of GR-LBD bound to the TIF-2 coactivator peptide sheds light on the selectivity of specific coactivators for some NRs.[45] This structure reveals that GR uses two charge clamps to define its sequence-specific binding to the LxxLL motifs of TIF-2. The first charge clamp is with the highly conserved residues of the AF-2 helix and H3. The second charge clamp is composed of Arg585 and Glu590 in the GR-LBD, which form hydrogen bonds with the third LxxLL motif of TIF-2, but not the first and second motifs. The residues of the GR-LBD involved in the second charge clamp are conserved across the steroid receptors of Class I that bind oxysteroids (GR, PR, AR, and MR) but are not present in the ER or in RXR obligate heterodimers such as PPARs. This second charge clamp accounts for the differential binding of coactivator motifs by many NRs.

Since the coactivators and corepressors bind to nearly overlapping surfaces of the LBD and AF-2 domain, the choice of which coregulator is bound to the NR is controlled by the NR ligand.[46] In Class II NRs such as VDR and TR, which are typically found in the nucleus bound to their NREs, the LBD is bound to corepressors in the absence of ligand. The CoRNR box of corepressors NcoR1 (nuclear receptor corepressor 1) and SMRT (silencing mediator of retinoid and thyroid hormone) dock in a hydrophobic groove formed by H3 and H4 of the LBD and recruit specific histone deacetylases (HDACs). These HDACs generate a condensed chromatin structure over the promoter at the NRE, resulting in transcriptional repression in the absence of ligand. Agonist binding to the LBP induces conformational allosteric effects, leading to the movement of H12 and disruption of the interaction of H3 and H4 with the corepressor. The corepressor complex dissociates from the ligand-bound NR, and a coactivator is now able to bind to the coactivator-binding surface generated by H12, H3, and H4.[47] The bound coactivator then recruits other proteins to form a complex that leads to initiation of transcription and gene activation.

Steroid NRs from Class I NRs are usually present in the cytoplasm, associated with chaperone proteins such as Hsp90 and the immunophilin p23[48] that bind to the LBD. Ligand binding releases the chaperone proteins by a conformational adjustment, allowing the ligand to access the LBP. The ligand-bound receptor then dimerizes and is translocated to the nucleus, binds to its cognate response elements, and initiates transcription.

Antagonist NR ligands can interfere with LBD-mediated transcriptional regulation via different modes, as revealed by crystal structures of LBDs complexed with antagonists (**Table 2**). In LBD/antagonist complexes such as ERα with raloxifene,[22] GR with RU486,[49] and RARα with BMS614,[50] the steric hindrance afforded by the bulky side chains of the antagonist ligands (dimethylaniline in RU486, phenoxyethylpiperidine in raloxifene and quinoline in BMD614, **Figure 3**) prevents H12 from adopting the active conformation. Instead, H12 is forced to occupy the coactivator groove, effectively blocking the coactivator-binding surface (**Figure 3b**). Indeed, the H12 of NRs also does contain an LxxLL-like motif that allows H12 to sit in the coactivator groove, to relieve steric contacts imposed by bulky antagonist ligands.[22] This mode of antagonism has been termed 'active antagonism.' It differs from the 'passive' mode of antagonism observed with antagonists such as MR antagonist spironolactone[51] or ERβ antagonist tetrahydrochrysenediol (THC).[52] In these NRs, the agonist ligands provide critical hydrogen bond interactions with residues of the LBP that stabilize the agonist H12 position.[51] When antagonists bind, they do not make these essential contacts, and therefore they destabilize the active H12 conformation, resulting in a transcriptionally inactive LBD. Partial agonists also fall into the passive mode of antagonism. In the ERβ–genistein complex[53] and the PPARγ–GW0072 complex,[54] the ligands are small, and they do not interact with or stabilize the AF-2 helix and result in a diminished transcriptional response. It is evident that the chemical structure of the NR ligand plays a significant role in the outcome of the NR interaction with natural and synthetic ligands.

2.25.3.1.1.4 Dimerization sites in the ligand-binding domain

Most NRs function either as homodimers (steroid receptors of Class I and RXR) or as heterodimers with RXR (Class II receptors). The residues that take part in the dimerization interaction are found in the LBD and are well characterized for the ER- and PR-LBD.[22,55] Crystal structures of two different RXR heterodimers, RXR-RAR and RXR-PPAR, also reveal a common mode of dimerization in which residues of H10 from one monomer pack against H10 residues of the heterodimeric partner.[56,57]

2.25.3.1.2 Ligand-independent activation function (AF-1) domain

The N-terminal AF-1 domain does not directly interact with the NR ligand, but it has been shown to be important in the recruitment of coactivators and formation of a transcriptionally active NR complex (see [58] for a detailed review).

In the AR, an interaction between the AF-1 domain and the ligand-bound AF-2 domain is important for gene activation.[59] Ligand-independent, AF-1 domain-mediated transcriptional activation is also implicated in the tissue-selective effects of the selective estrogen receptor modulators (SERMs), such as tamoxifen. The relative contribution of AF-1-mediated transcriptional activation is dependent on the cellular context and tissue type.

2.25.3.1.3 DNA-binding domain

The DBD in NRs is a highly conserved region comprising about 60 amino acids and containing eight conserved Cys residues that coordinate two Zn ions and form the two zinc-finger motifs characteristic of all NR DBDs. Although the first structures of NR DBDs were solved by NMR (GR[16] and ER DBD[60]), crystal structures of several DBDs complexed with their consensus DNA response element sequences have also been solved and are reviewed by Renault and Moras.[18]

2.25.4 Coregulators and Nuclear Receptor-Dependent Transcriptional Control

As discussed above, the functional activity of an NR is not only regulated by hormones and other endogenous lipophilic ligands, but also by a complex array of regulatory proteins, including corepressors, coactivators, and chromatin modifiers. Recent studies indicate that transcription of NR-dependent genes is a dynamic process with continual exchange and turnover of NRs, coregulators, and other components of the transcriptional machinery.[61] Once recruited, coregulators and NRs cycle on and off the target promoter many times, interacting only briefly with the regulatory elements (see [61]). The NRs and their coregulators are also subjected to rapid modifications, such as ubiquitination, phosphorylation, acetylation, and methylation. The orderly and sequential recruitment of coactivators with different enzymatic activities has recently led to the concept of a 'transcriptional clock that directs and achieves the sequential and combinatorial assembly of a transcriptionally productive complex on a promoter.'[62]

Class I NRs (including the steroid hormone receptors) reside in the cellular cytoplasm, where they are held in an inactive complex with chaperone (heat shock) proteins. The binding of their ligands induces release of the chaperone molecules, dimerization (homodimers), transport into the nucleus, and binding to specific NREs within the promoter regions of target genes, resulting in the recruitment of coactivators and chromatin rearrangement (e.g., see [61]). However, the Class II NRs (e.g., VDR, TRs, RARs, PPARs, LXRs, and FXRs) are found primarily in the nucleus, bound to their respective NR response elements as heterodimers with RXR. These DNA–receptor complexes are normally associated with histone deacetylase (HDAC)-containing complexes tethered through the corepressors NCoR and SMRT. On ligand binding, the local chromatin structure is altered and the corepressors are dissociated from the DNA while coregulators are recruited.[63] Other recent studies suggest that both coactivators and corepressors can be found within the same complex, allowing for more efficient and more complex transcriptional control mechanisms.[64]

2.25.4.1 Coactivators

Ligand-activated NRs bind to NREs embedded within chromatin, and this event initiates an ordered recruitment of a series of coactivators. Four major classes of NR coactivators have been described:

1. The p160/steroid receptor coactivator (SRC) family (SRC-1, SRC-2, and SRC-3), CBP (cAMP response element binding protein (CREB)-binding protein), p300, and PCAF (CBP/p300-associated factor). These coactivators possess acetyltransferase activity, which involves posttranslational modification of lysine residues within the N-terminal tail of histone proteins (e.g., see [65]).
2. Protein arginine methyltransferase 1 (PRMT1) and coactivator-associated arginine methyltransferase (CARM). Coactivators in this class methylate several arginine residues within histone H3 and, along with acetylases, are believed to neutralize the electrostatic interaction between the negatively charged phosphate backbone of DNA, which relaxes the association between the DNA and histone 3 protein.
3. The multiprotein thyroid receptor-associated protein (TRAP), VDR-interacting protein (DRIP), and SRB-MED-containing complex (SMCC) complexes bridge NRs to the general transcription factors associated with NR activation of target genes.
4. The ATP-dependent chromatin remodelers, including the SWI/SNF family. These coactivators introduce superhelical torsion into chromatin, creating a change in the topology of nucleosomal DNA. Furthermore, they slide nucleosomes into the *cis* form, opening the compact structure of chromatin and making it more accessible to the general transcription factors (see [65,66] for reviews).

For a more detailed list of known coactivators, see [378].

2.25.4.2 Corepressors

The NR corepressors NCoR and SMRT associate with HDACs and DNA methyltransferases, and with histone methyltransferases on Class II NR target gene promoters. These enzymatic activities promote association of proteins important for maintaining heterochromatin and silenced regions of DNA.[67] Recent results have also made it clear that SMRT and NCoR are actually common repressors of all NRs, including the steroid hormone receptors ER, PR, and AR (see [68] for a review). With Class I receptors, corepressor binding occurs only when the steroid hormone receptors bind their corresponding antagonists. For example, binding of tamoxifen to ER enhances the affinity between ER and NCoR. Over the past few years both the SMRT and NCoR complexes have been shown to associate with many other proteins, including transducing beta-like protein 1 (TBL1), TBL1-related protein (TBLR1), and HDAC3. These three components appear to be critical for the transcriptional repression caused by NRs.[68]

Agonist binding to NRs can also recruit specific corepressors, as well as coactivators. These corepressors attenuate agonist-activated NR signaling by multiple mechanisms (see [69] for more details). Two of these agonist-binding-dependent corepressors have been identified as the ligand-dependent corepressor (LCoR) and RIP40.[69] LCoR is widely expressed in fetal and adult tissues and interacts with ER, GR, PR, and VDR. LCoR also interacts with HDAC3 and HDAC6, but not with HDAC1 or HDAC4. RIP40 interacts with ERs and some Class II NRs, including TRs, RARs, and PPARs.[68]

2.25.4.3 Modifications of Nuclear Receptors and Their Coregulators

2.25.4.3.1 Ubiquitination

Ubiquitin is a 76-amino acid polypeptide that is covalently linked to target proteins via a sequential enzymatic cascade involving an ubiquitin-activating enzyme, a ubiquitin-conjugation enzyme, and a ubiquitin ligase. Modification of proteins is central to the regulation of many biological pathways, including intracellular protein degradation, cell cycle regulation, signal transduction, and regulation of gene expression.[66] Current studies strongly suggest roles for ubiquitination in condensation of chromatin, regulation of histones, DNA repair, replication, and transcriptional control.[66] In keeping with the notion of an NR-dependent transcriptional clock, it is now believed that the cyclic interactions of NRs and their cofactors are most likely driven by the ubiquitin-proteasome pathway.[61] The most plausible mechanism for this process seems to involve the removal of corepressors and/or coactivators to enable other regulators to subsequently bind the promoter complex.[70] However, the ubiquitination system can also function in transcriptional activation independently of proteolysis (see [61] for a detailed review). For example, it has been suggested that histones may be ubiquitinated to create new platforms for binding additional chromatin modifiers and thereby influence other modifications such as acetylation and methylation.[71] Other studies have highlighted the involvement of the ubiquitination pathway in the elongation phase of transcription.[72]

2.25.4.3.2 Acetylation

Acetylation of both histone proteins and coregulators constitutes an important regulatory event for the NR transcriptional apparatus. For example, the hyperacetylation of histones 3 and 4 correlates with increased transcriptional activity. Many NR coregulators, including CBP, possess intrinsic histone acetylation activity.[66] CBP/p300 is also known to acetylate other coregulators, resulting in a disruption of the association between histone acetyltransferase (HAT) coactivator complexes and promoter-bound NR. This CBP acetylase process can therefore provide negative regulation of ligand-dependent transcription.[73] In addition, some nuclear hormone receptors can also be acetylated (e.g., ERα and AR). Acetylation of ERα and AR results in completely different outcomes: ERα transactivation is attenuated, but AR transcription activity is induced.[66] Acetylation is documented to exert other biological effects such as regulation of the translocation of nuclear transcription factors.

2.25.4.3.3 Methylation

Histone arginine methylation is commonly associated with activated gene expression[74] that strengthens coregulator functions through a mechanism that releases histone proteins from DNA.[65,66] This action complements the acetyltransferase activity of other coregulators. For example, the methyltransferase PRMT1 methylates histone H4, and this step facilitates subsequent acetylation of H4 by acetyltransferase p300. Crosstalk between another methyltransferase, CARM1, and CBP/p300 can also positively regulate the expression of estrogen-responsive genes.[75] Apparently acetylation enables tethering of CARM1 to the histone H3 tail, effectively producing a more efficient arginine methyltransferase. PRMT1 and CARM1 have also been shown to bind the p160 family of coactivators, resulting in increased receptor activation and NR-mediated gene transcription.[76]

2.25.4.3.4 Phosphorylation

Phosphorylation plays an integral role in regulating both NR coregulator activity and subsequent NR function,[61] adding yet another layer of complexity to the control of NR target gene regulation. Ongoing research has uncovered crosstalk between both Class I and II receptors and several kinase cascades.[61] For example, estrogen (E_2) and retinoic acid (RA) signaling causes rapid activation of mitogen-activated protein kinases (MAPKs), which can then phosphorylate the N-terminal AF-1 domains of both ERα and RARγ. This type of phosphorylation event controls chromatin rearrangement, in part by recruiting p160 coactivators. MAPK-induced phosphorylation also induces the ubiquitination (and degradation) of some NRs, including RARs and PR. The situation is even more complex because both corepressors and coactivators are also phosphorylated by MAPKs in response to steroidal and nonsteroidal hormones (e.g., see [77]). Phosphorylation of histones by MAPK highlights another important concept – the 'histone code hypothesis' – which states that a combinatorial and coordinated dynamic model exists whereby phosphorylation participates with acetylation, methylation, and ubiquintination to provide motifs for recruiting other chromatin-modifying or -remodeling complexes.[61]

Other new concepts relevant to the role of phosphorylation in regulating NR-mediated transcription have emerged over the past few years. For example, it now appears that the general transcription factor TFIIH is also capable of phosphorylating NRs.[78] In addition, phosphorylation in response to several extracellular signals, including growth factors, insulin, stress, and cytokines that end in kinases such as MAPK, can provide a mechanism to terminate the response to the NR ligand and ensure the escape of the NRs (e.g., PPARγ and RARs) from the transcription initiation complex.[79,80]

2.25.5 Physiological Functions of Nuclear Receptors

The NR superfamily regulates many aspects of development, metabolism, and inflammation. As described above, these transcription factors share a common molecular structure and therefore can be considered as a functional entity. Dr Ronald Evans (The Salk Institute, San Diego) recently speculated that the evolution of this superfamily is related to survival of the organism.[1] To survive, all organisms need to acquire and manage energy stores. Generally, the PPARs regulate dietary endogenous fat (and cholesterol), the LXRs and FXRs respond to cholesterol and its derivatives, GR to sugar mobilization, MR to salt, VDR to calcium, TR to basic metabolic rate, while new studies show that ERα increases mitochondrial efficiency.[81] The xenobiotic receptors (PXR and CAR) are specialized to respond to environmental toxins, many of which are present in the human diet. Reproduction is also a means of survival, and the gonadal steroid receptors (ER, AR, and PR) control this important biological function. The retinoid acid-dependent RARs are essential for embryonic development. Fertility depends on nutritional status, and thus these two regulatory branches of the family must communicate. Finally, the organism must protect itself from viral, bacterial, and fungal infections mainly through the inflammatory response, but also by depressing appetite, conserving fuel, and inducing sleep. On the other hand, GR signaling can also be anti-inflammatory, as can PPARs, RARs, and LXRs. This section outlines the physiological functions of each of the NRs for which endogenous ligands have been identified (Classes I and II). A discussion of the physiological functions of the less well-defined oNRs (Type III) is beyond the scope of this chapter (see [82,83] for reviews).

2.25.5.1 Class I Receptors

Class I receptors comprise the classical high affinity, steroid hormone receptors (ERs, PRs, AR, GR, and MR), that bind their DNA targets as homodimers and sometimes as monomers.

2.25.5.1.1 Estrogen receptors

Estrogen (E_2) is one of the most well-defined nuclear hormones from the point of view of biological effects and clinical implications. Estrogens mainly control three female biological processes: pubertal development, regulation of estrous cycles, and establishment and maintenance of mammary tissue (see [84]). The effects of E_2 are mediated predominantly by two distinct estrogen NRs, ERα and ERβ. The physiological consequences of ER signaling pathways extend beyond female reproductive tissues to include control over the skeletal system, cardiovascular system, and the central and peripheral nervous systems (see [84–86]). General cellular responses include cell cycle progression, cell growth and survival, energy, and cellular maintenance.

E_2 mediates its physiological and cellular effects through direct binding of liganded ERα to target genes, a process known as genomic signaling. ERβ is a weak transcription factor and its translocation to the nucleus is rarely reported. For a review of the known genomic targets of estrogen, see [87]. Briefly, ERα-regulated genes encode protein products that include nuclear hormone receptors (including ERα, RARα, and PR), non-NR transcription factors (c-fos), growth factors (e.g., vascular endothelial growth factor and transforming growth factor-α), hormones (e.g., insulin-like growth factor-1), enzymes (e.g., TERT), and various other important proteins, including the cell cycle-related cyclin D1 and the prosurvival Bcl-2. ERs can also indirectly associate with gene promoters through protein–protein interactions with other transcription factors such as activating protein-1 (AP-1) and Sp1.[87]

Recent studies have uncovered yet another level of genetic control that involves plasma membrane-associated ERs that, upon ligand binding, initiate signaling cascades of phosphorylation events, resulting in unique sets of ER-dependent genes (e.g., see [87,88]). In addition, a G protein-coupled receptor (GPR300), localized to the endoplasmic reticulum, specifically binds E_2 resulting in intracellular calcium mobilization and synthesis of phosphatidylinositol 3,4,5-triphosphate in the nucleus.[89] This nonclassical type of NR-dependent gene activation is termed nongenomic signaling. The relative levels of ERα and ERβ (ERβ often suppresses the effects to ERα) and available coactivators and corepressors further increase the complexity of E_2 response in a tissue- and gene-specific manner.[14,88] Relatively little information is available for the nongenomic targets of ERs. However, recent reports show that membrane-bound ERα activates the expression of the endothelial nitric oxide synthase (eNOS) enzyme.[90–92] Furthermore, nongenomic signaling of ERα (but not ERβ) is involved in the control of dopamine function and protection of astrocytes.[93] ERβ, however, has been localized to the mitochondria of a variety of cell types, suggesting estrogens can directly affect mitochondrial function through the modification of calcium influx, ATP production, apoptosis, and ROS generation.[94]

Transgenic mice harboring whole body-, tissue-, or cell-specific knockout of one or more specific genes have revolutionized the study of the physiological and pathological functions of NRs. For example, one study found that *Erα* knockout female mice are infertile because they are anovulatory, have disruption of luteinizing hormone (LH) regulation, and have estrogen-insensitive uteri.[95] The same study found that *Erβ* knockout female mice are subfertile and display defective ovulation similar to that found in human polycystic ovarian syndrome presented as lack ovarian function. Knockout mice lacking both genes are similar to *Erα* knockout mice, but exhibit unique ovarian pathology.[95] Mice with inactivated ERα and/or ERβ also display a number of other interesting phenotypes, including incomplete differentiated epithelium in tissues under steroid control (breast, ovary, prostate, and salivary glands), obesity, and insulin resistance.

ERα signaling is implicated in cardiovascular homeostasis, in part through stimulation of athero-protective prostaglandin 12 (PG12) formation via activation of cyclooxygenase-2 (COX-2).[96] The nongenomic induction of eNOS by ER signaling probably also contributes to vascular and myocardial homeostasis through the controlled production of nitric oxide (NO) (e.g., see [97–99]). The involvement of ERα in cardiovascular health is highlighted by the ability of estrogen to protect against coronary heart disease, including ischemia-reperfusion injury[100] and other atherosclerotic diseases (e.g., see [101]) and its well-characterized ability to decrease low-density lipoprotein cholesterol (LDL-C) and increase high-density lipoprotein cholesterol (HDL-C). Estrogens are also well known to have beneficial effects on bone, but the roles of ERs in mediating these effects are not fully understood. A recent study of *Erα* knockout mice confirmed earlier data indicating that these mice have shorter bones, as well as a total absence of growth plates in the tibia of aged (*Erα* knockout) females. However, trabecular bone volume and thickness were enhanced in both female and male knockout animals.[102]

Estrogens are also widely recognized for their neuroprotective effects in several models of neurotoxicity, both in vitro and in vivo.[103,104] Interestingly, however, in most of these studies, the naturally occurring estrogenic hormone 17β-estradiol (17βE2) is as neuroprotective as its inactive diastereomer 17α-estradiol, even though 17αE2 does not bind to ERα or ERβ to any significant extent. These observations led to the proposal that the neuroprotective effect of estrogens is achieved through a nongenomic mechanism, and not through the classical nuclear receptor-mediated gene transcription. Whether this effect is mediated through a plasma membrane-associated ER or other specific mechanisms such as mitochondrial stabilization [105,106] is still under active investigation. Nevertheless, this interesting dichotomy opens up the possibility of using 17αE2 and other estrogen derivatives that do not bind the classical ER, as selective neuroprotectants that are devoid of feminizing side effects of classical estrogens in the breast and uterus.[107] Such nonfeminizing estrogens have demonstrated efficacy in animal models of stroke and Alzheimer's disease[107] and are under development.[108]

ERβ is also believed to play a role in bone homeostasis in addition to contributing to prostate and colon homeostasis,[85,109] and controlling hematopoiesis.[110,111] Finally, ERs also influence the development, plasticity, survival, and regulation of neurons, including serotonergic, dopaminergic, and cholinergic neurons.[111]

2.25.5.1.2 Progesterone receptor

The expression of the progesterone receptor is activated by ERs, and progesterone cooperates with E_2 in the control of female pubertal development, regulation of estrous cycles, and establishment and maintenance of mammary tissue.[84] Whereas estrogens usually induce cellular proliferation, progesterone causes differentiation and inhibits cell growth.[112] Progesterone exerts its cellular effects through binding and activation of two receptors, PR-A and PR-B. Unlike ERs, the PRs are expressed from one gene, but alternative transcription from two promoters yields two distinct proteins.[113] The molecular activation of PR involves direct binding of the ligand-activated receptor to the promoter regions of each gene (genomic signaling). However, growing evidence indicates that plasma membrane-bound PRs exert some of progesterone's effects via nongenomic signaling pathways.[114]

Recent studies using knockouts of the *Pr* gene have provided further evidence not only for the essential role of PRs in a variety of female reproductive systems, but also for their importance in nonreproductive activities (e.g., see [115,116]). Female mice lacking both PRs exhibit impaired sexual behavior and neuroendocrine gonadotrophin regulation, anovulation, uterine dysfunction, and impaired ductal branching morphogenesis and lobuloalveolar differentiation of the mammary gland (for a review see [116]). The PRs also regulate thymic involution during pregnancy and the cardiovascular system through regulation of endothelial and vascular smooth muscle cell proliferation and response to injury.[116] PRs have also been detected in mouse ovary and are believed to participate in the regulation of ovulation, along with suppression of granulosa cell apoptosis and promotion of cell survival.[117] PRs have been identified in bone tissue and the central nervous system, where they may be involved in bone maintenance and cognitive function, respectively.[116] More recent studies have described isoform-specific knockouts for *Pr-a* and *Pr-b*[115] supporting previous in vitro observations. The knockout of PR-A function results in an infertile phenotype, whereas knockout of the PR-B protein did not affect fertility in mice. On the other hand, loss of PR-A has no effect on mammary morphogenesis, but pregnancy-associated mammary morphogenesis is severely impaired in *Pr-b* knockout mice. By virtue of its ability to suppress E_2-induced and PR-B-mediated uterine and mammary proliferation, the PR-A isoform is likely to be an attractive drug target for the next generation of selective PR modulators (SPRMs) in the treatment of uterine and mammary gland hyperplasia.

2.25.5.1.3 Androgen receptor

Androgens (e.g., testosterone) are steroid hormones necessary for normal male phenotype expression, including the outward development of secondary sex characteristics as well as the initiation and maintenance of spermatogenesis. AR directly regulates expression of an array of target genes that are important in male pubertal development and fertility. The androgen/AR system is also known to have a variety of other effects in many target organs, including brain, skeletal tissue, and heart (e.g., see [118–121]). Transgenic *Ar* knockout mouse studies have highlighted other physiological actions of AR, including promoting late-onset obesity of white adipose tissue, while mutant ARs (in a *Drosophila* model) show marked neurodegeneration.[122] In addition, AR may also modulate specific immune functions. Androgens can exert a cardiovascular protective effect, in part by protecting against angiostatin (Ang) II-stimulated cardiac fibrosis (impairment of both the concentric hypertrophic response and left ventricular function). The molecular underpinnings of this action involve activation of expression and signaling through transforming growth factor-β1, a well-known profibrotic cytokine, and subsequent upregulation of collagen I and III expression.[120] Interestingly, this same study suggests that transduction through an important signaling pathway, extracellular signaling-regulated kinases (ERK) 1/2 and ERK5, by Ang II was attenuated in the *Ar* knockout hearts compared with wild-type animals. These data imply that AR signaling, similar to other Class I NRs (ER and PR), may involve nongenomic pathways.

2.25.5.1.4 Glucocorticoid receptor

Glucocorticoids, such as cortisol, exert their effects by binding and activating the GR. New studies have identified a second GR, GRβ, which is a splice variant of the classical receptor GRα and functions as a dominant-negative transcriptional inhibitor of GRα.[123] The glucocorticoid/GR system is central to protecting the body against stress by regulating glucose metabolism and blood pressure (e.g., see [124] for a review). Glucocorticoids stimulate hepatic gluconeogenesis and reduce the ability of insulin to inhibit glucose production.[125] This NR is also involved in lipid metabolism and glycogen deposition in the liver. In addition, the glucocorticoid/GR system affects behavior and brain function, organ development, and calcium reabsorption.

GR is a potent anti-inflammatory and immunosuppressive modulator (e.g., see [124]). This control is achieved through an indirect negative regulation of gene expression (transrepression) whereby ligand-activated GR binds as a monomer to other transcription factors, such as the pro-inflammatory nuclear factor kappa B (NFκB) and AP-1 transcription factors.[124,126] The precise mechanism of action of GR transrepression remains unresolved, but it appears to involve the DBD of the GR protein that directly associates with the target transcription factor. The transcriptional induction of specific NFκB and AP-1 target genes (including a number of pro-inflammatory genes) is subsequently repressed.[124,126] GR also contributes to thymocyte and lymphocyte selection and survival.[124] Furthermore, activated GR is required for stress-induced red blood cell production,[127] and more recent data implicate the receptor in activation of human platelets.[128]

2.25.5.1.5 Mineralocorticoid receptor

Aldosterone, the most potent endogenous activator of the MR, has long been recognized as an important component of the secretions of the adrenal cortex and the prime regulator of sodium and potassium homeostasis. The main site of action of this nuclear hormone system is the sodium-transporting epithelia, such as the distal part of the nephron and the distal colon, as well as sweat and salivary glands, the cardiovascular system, the central nervous system (hippocampus), and the brown adipose tissue.[129] Even though the receptor was cloned more than 17 years ago, many basic features of MR function and regulation have yet to be fully characterized. Although glucocorticoids can bind MRs and are in much greater circulating abundance than aldosterone, a ligand-selective regulation exists.[130] In this model, aldosterone binding, but not cortisol binding, recruits the cellular transcriptional complex to the receptor and to native MR target gene promoters. MR is also capable of forming heterodimers with the GR, which can modulate gene expression in a manner that is distinct from the MR and GR monomers. Furthermore, aldosterone/GR effects can be antagonized by the PR system and by the TR, but it remains uncertain whether MR can form heterodimers with these receptors.[129] Taken together, the available literature indicates that crosstalk among these four nuclear hormone receptors influences the complex physiological effects of MR. This regulatory control is made more complex by the fact that these NR actions depend on a number of factors including the tissue site, target gene, the relative distribution of each NR, ligands, and coregulators.

2.25.5.2 Class II Receptors

Class II NRs typically form heterodimers with RXR, a lipid sensor receptor that can also form homodimers. Type II NRs can be divided into functionally distinct nonpermissive (VDR and TRs) and permissive (PPARs, LXRs, and FXR) groups.[61] The permissive partners generally function as low-affinity lipid sensors, while the nonpermissive partners are high-affinity hormone receptors.

2.25.5.2.1 Vitamin D receptor

The vitamin D–VDR system plays an essential role in calcium homeostasis, bone mineralization, and skeletal architecture. It also has antiproliferative, prodifferentiation, and immunomodulatory properties, and controls other hormonal systems both in cell culture and in animal models (e.g., see [131]). VDR is expressed in many tissues, including the intestine, bone, kidney, and parathyroid glands. VDR is also considered to be a bile acid sensor/receptor in both the liver and the intestine, that causes induction of cytochrome P450 (CYP)3A4 expression.[132,133] The control of bile acid homeostasis is complex and is known to involve other nuclear hormone receptors, including LXR, FXR, CAR, and PXR (see below).

Vitamin D, a steroid, is metabolically converted to the active metabolite, 1,25-dihydroxyvitamin D [$1,25(OH)_2D_3$] in the renal proximal tubules, after facilitated uptake and intracellular delivery of the precursor to the enzyme 1-α hydroxylase. The 1-α hydroxylase protein has also recently been detected in the parathyroid glands.[134] Transgenic *Vdr* (and/or *1-α hydroxylase*) knockout mice models have shown both $1,25(OH)_2D_3$-dependent and -independent actions of the VDR as well as VDR-dependent and -independent actions of $1,25(OH)_2D_3$.

Downstream physiological targets are influenced through VDR forming a heterodimeric transcriptional complex with RXR to exert its effects over target genes. Thus, not surprisingly, the vitamin D system most likely involves the participation of more than a single receptor and ligand. The presence of 1-α hydroxylase in many target cells indicates autocrine/paracrine functions for $1,25(OH)_2D_3$ in the control of cell proliferation and differentiation. Local production of $1,25(OH)_2D_3$ is dependent on circulating precursor levels, and fluctuations in these levels provide a potential explanation for the association of vitamin D deficiency with various cancers and autoimmune diseases (*see* Section 2.25.6 for a discussion of the pathological implications of VDR and other NRs).

2.25.5.2.2 Retinoic acid receptor

Retinoic acid (RA), the most biologically active metabolite of vitamin A (retinol), influences embryogenesis and cell proliferation, differentiation, and apoptosis in the adult.[135] All-*trans* and 9-*cis* RAs exert their biological effects through RARs and RXRs, respectively. The highly pleiotropic effects of signaling through RARs are caused, in part, by the combinatorial action of the three types of RAR receptors (RARα, RARβ, and RARγ) and their isoforms (α1, α2, β1-4, and γ1 and γ2) with the three RXRs (RXRα, -β, and -γ) and with coregulators. This subsection focuses on the physiological functions of RARs in general. For detailed reviews of RXRs, see [2,14,136,137].

The importance of retinoids and their receptors in development has been recognized for years.[138,139] Retinoid receptors are widely expressed in the vertebrate embryo, and there appears to be little redundancy among the RAR subtypes. Vitamin A or RAR deficiency has been reported to affect the developing embryo at the time of the first organ system initiation.[138] From that developmental time point onward, RARs control multiple stages of development for numerous tissues and organs. For example, RAR signaling is known to be essential for normal heart development, and loss of RAR expression leads to heart abnormalities in ventricular chambers and defects in the outflow tract. On the other hand, other studies now suggest that RARs are critical in embryo formation well before organ formation, and in a number of adult homeostatic functions.[137] For example, RA/RAR signaling was recently suggested to be essential in controlling the bilateral symmetry of vertebrate embryos.[140] RARs are also involved in heart asymmetry determinations – vitamin A and RARs provide a proper environment for the expression of adequate levels of heart asymmetry genes.[138] RARs play important roles in the development of the central nervous system and are required at the time of specification of the posterior hindbrain and for the subsequent specification of segments in that region and for neurite outgrowth and neural crest survival.[138,141] RA and RAR have long been linked to vertebrate limb morphogenesis. For example, an analysis of the expression of genes in embryo limb buds revealed a critical role for RA/RAR in anteroposterior and dorsoventral patterning.[142] Finally, RARs also contribute to the development of a normal respiratory tract.[143]

In the normal adult, RARs have long been implicated in the proliferation and differentiation of epithelial cells. A common model of RAR dysfunction that involves a deficiency in vitamin A (VAD) presented several symptoms associated with VAD: blindness, immune response (to infectious disease) dysfunction, and skin disorders. RA has also been shown to promote tissue regeneration in a RAR-dependent manner in adult, nonregenerating tissues such as lung, inner ear, hair, and spinal cord.[144] Retinoids are also believed to protect the human lung against injury from elastase.[145] Thus, RARs are vital for the development and maintenance of a number of physiological functions, including vision, hematopoiesis, reproduction, and contribute to the normal functioning of the nervous, pulmonary, and immune systems. Some of these physiological properties are directly related to the ability of this NR to control cellular proliferation and differentiation.[137]

New studies suggest that RAs/RARs are important inducers of genes involved in the reverse cholesterol transport (RCT) associated with increased macrophage lipid efflux, such as the cholesterol transporter adenosine triphosphate (ATP)-binding cassette A1 (ABCA1), ABCG1, and LXRα.[146] RCT is discussed in more detail in the PPAR and LXR discussions below. Finally, like GR and PPARs, RARs also display anti-inflammatory properties through transrepression (e.g., see [137]). Ligand-activated, monomeric RAR binds directly to pro-inflammatory transcription factor proteins (e.g., AP-1 and NFκB) causing downregulation of a number of important inflammatory genes.[126,137]

2.25.5.2.3 Thyroid hormone receptor

TRs are expressed throughout the body and are involved in normal growth, development, differentiation, and metabolism in vertebrates. The biologically active thyroid hormone (TH), L-3,5,3′-triiodo-L-thyronine [T_3], is derived from L-3,5,3′,5′-tetraiodo-L-thyronine (T_4) through the action of the D2 iodothyronine deiodinase, and this hormone exerts its effects mainly through two known TRs (TRα and TRβ). Recent studies indicate that the low-affinity TR-binding 3,5-diiodo-L-thyronine (T_2) is also active, involved in at least cellular-mitochrondrial respiration, through a possible TR-independent mechanism.[147,148] Alterations in the activities of the three selenocysteine-containing iodothyronine deiodinase enzymes have been reported during development and in specific cells and tissues in the adult. These deiodinases are envisaged as 'guardians to the gate of thyroid hormone action mediated by T_3 receptors'.[149]

The two subtypes of TRs are known to exist in a complex distribution pattern throughout the body. For example, TRα is highly abundant in the heart, and most effects of T_3 on the cardiovascular system, including mitochondrial maturation and substrate metabolism after birth, are mediated through this receptor.[150] TRβ preferentially regulates energy and cholesterol metabolism,[151] offering the possibility that isoform-specific TR synthetic ligands might safely increase energy expenditure. TRα is also essential for skeletal development, regulating bone turnover and mineralization in adults. These physiological effects are now believed to include TH/TR control over growth hormone receptor and insulin-like growth factor-1 receptor.[152] Finally, the TH/TR system is believed to play important roles in trophoblast function, early pregnancy maintenance, fetal neurodevelopment, and the adult central nervous system.[153]

2.25.5.2.4 Peroxisome proliferator-activated receptors

The three closely related PPAR isotypes, PPARα, PPARγ, and PPARδ, have recently been the focus of extensive research since the exciting discovery that PPARα and PPARγ were targets of major classes of drugs (the fibrates and glitazones, respectively) used to correct abnormalities of lipid and glucose homeostasis (see Section 2.25.6.9 for a discussion of the current status of the pharmacology of PPARs). The PPARs function as lipid sensors that coordinate the expression of large gene arrays and thereby modulate important metabolic events.[154] The endogenous, low-affinity ligands of PPARs include a wide range of fatty acids and oxidized fatty acid metabolites from the lipoxygenase and cyclo-oxygenase pathways. Each PPAR protein is encoded in a separate gene, and upon activation these ligand-dependent transcription factors form heterodimers with RXR. PPARs and RXRs can apparently function independently (in the absence of a hetero-partner) to modulate gene expression,[155] adding to the complexity of PPAR transcriptional control. PPARs are known to control multiple metabolic functions including fatty acid β-oxidation, cholesterol transport, and glucose and amino acid metabolism.[156] All three PPARs also participate in certain aspects of inflammation.[157] PPARs are able to regulate inflammatory responses as a result of their transactivation and transrepression capacities. Like GR and RARs,[126,137] PPARs achieve transrepression by antagonizing NFκB and AP-1.[158] In addition, PPARγ can transrepress the pro-inflammatory transcription factor, signal transducer and activator of transcription (STAT3). Conversely, PPARδ has been implicated in the activation of inflammatory macrophages through antagonism of BCL6, an inflammatory cytokine repressor.[159] The PPARs also control vascular wall biology, placental physiology, and immunity.[160,161] The impressive multifunctional effects of the PPAR family, as those of the RARs, result partly from the combinatorial action of each PPAR protein with the three RXR proteins.

PPARα is highly enriched in the liver, heart, muscle, brown adipose tissue, kidney, adrenal gland, and pancreas, where it primarily regulates fatty acid catabolism and glucose oxidation.[162,163] Mice with targeted disruption of PPARα exhibit low expression of a number of genes required for metabolism of fatty acids, including acyl-CoA oxidase,[164] and suffer from a variety of metabolic defects, including hypoketonemia, hypothermia, elevated plasma free fatty acid levels, and hypoglycemia.[155] Other studies have shown that activation of PPARα increases lipoprotein lipase activity in skeletal muscle, with a reciprocal decrease in adipose tissue lipase (reviewed in [160]). Therefore, PPARα agonists can reduce the uptake of triglycerides in adipose tissue while increasing uptake in muscle for fatty acid oxidation. PPARα also contributes to glucose-stimulated insulin secretion from pancreatic β-cells while increasing fatty acid uptake and β-oxidation.[165] Activation of PPARα (and PPARγ) also induces cholesterol removal from human macrophage foam (lipid-laden) cells through stimulation of ABCA1[166] through a mechanism involving PPARγ- and PPARα-dependent upregulation of LXRα-dependent transcription.[167,168] This process is a first step in RCT that involves net flux of cholesterol to the liver and disposal of it in the form of bile acids. PPARα ligands are also linked to improved myocardial homeostasis (e.g., reduction in incidence of myocardial infarction); the ligands probably act at the artery wall (in part through control of smooth muscle proliferation) and systemically to control lipid homeostasis.[160] Furthermore, PPARα is believed to be involved in wound healing through its effects on related inflammatory and keratinocyte responses.[169]

The expression of PPARγ is enriched in brown and white adipose tissues, and the receptor is also found in numerous other tissues, including macrophages, adipocytes, mammary epithelium, ovary, and lymphoid tissue.[170] This NR is a late marker of adipocyte differentiation, and its forced ectopic expression can push fibroblasts into the adipogenic program.[171] Although *Pparγ* mice are not viable (because of defects in placenta formation), genetic knockout models of tissue-specific deletion of *Pparγ* imply that this protein is essential for the survival of mature adipocytes in the adult.[171] PPARγ is also involved in fat storage not only in adipocytes, but also in liver,[155] smooth muscle cells, activated macrophages,[159,172] and pancreatic β cells.[165] For example, *Pparγ* heterozygous mice challenged by a high-fat diet are less prone to liver cell accumulation of fat (steatosis) than their wild-type counterparts.[173]

As discussed above, PPARγ and PPARα induce RCT.[167] In addition to its effect on ABCA1 via LXR synthesis, PPARγ also increases the production of LXR ligands.[160] Other recent studies demonstrate the involvement of PPARγ in the regulation of adipocyte cholesterol metabolism via oxidized low-density lipoprotein (LDL) receptor 1 (OLR1).[174] Although the physiological role of PPARγ in cholesterol and LDL remains to be established, the induction of OLR1 is a potential means by which PPARγ regulates lipid metabolism and insulin sensitivity in adipocytes.[174] Another seminal study found that muscle-specific *Pparγ* deletion causes insulin resistance.[175] Furthermore, PPARγ, in contrast with PPARα, attenuates glucose-stimulated insulin secretion in pancreatic β-cells.[165] The concept that PPARγ is an important regulator of glucose metabolism and improves insulin sensitivity in muscle offers an exciting link between adipocyte biology and peripheral insulin resistance.

PPARγ, although expressed in very low levels in the heart, appears to contribute to overall cardiac homeostasis by limiting fibrotic effects of angiotensin II.[160] PPARγ also controls cellular proliferation and possesses an antiproliferative effect in pre-adipocytes and mammary epithelial cells.[170] Interestingly, mice with knockout oocyte- and granulosa-specific *Pparγ* have low fertility rates.[170] Finally, PPARγ has been implicated in a number of processes associated with the central nervous system, including protection against oxidative stress and control of the amyloid cascade.[176]

PPARδ is ubiquitously expressed with high levels in colon, brain, skeletal muscle, heart, and skin.[177,178] Data on the physiological roles for PPARδ have just begun to emerge, but they underscore the multifunctional nature of this transcription factor. It plays a role in controlling not only lipid oxidation, but also many other physiological processes, including embryonic development, energy homeostasis, macrophage lipid and cholesterol levels, inflammation, cardiovascular homeostasis, wound healing, cell survival, differentiation, and proliferation.[162,177,179–181]

Activation of this transcription factor increases lipid catabolism in skeletal muscle, heart, and adipose tissue, and thus this NR is a target for weight reduction.[182] The major effects of PPARδ on liver appear to involve the promotion of glycolysis, leading to reduced hepatic glucose output and lowered glucose levels.[160] PPARδ also functions as a very low-density lipoprotein (VLDL) sensor in macrophages, regulating triglyceride accumulation. This important observation reveals a pathway through which dietary triglycerides and VLDL can directly regulate gene expression in atherosclerotic plaques.[183] In addition, PPARδ has been found to contribute to inflammation by controlling the inflammatory switch of foam cells (lipid-laden macrophages) through its association and dissociation with transcriptional repressors such as BCL6.[159]

Other studies suggest a role for PPARδ in macrophage accumulation of cholesterol. For example, activation of macrophage PPARδ with a high-affinity synthetic agonist (GW501516) resulted in increased expression of the cholesterol transporter ABCA1 and the induced apolipoprotein A1, and thereby led to cholesterol efflux (RCT).[184] This observation has been supported by in vivo studies showing PPARδ activation improved cholesterol homeostasis in obesity-prone mouse models.[181] Together, these data imply that PPARδ plays a role, not yet clearly specified, in vascular homeostasis and disease.

PPARδ has also been reported to play important roles in the control of cell proliferation, differentiation, and survival, especially in keratinocytes.[162] Other work has suggested that PPARδ, like PPARα, is involved in wound healing, a process involving inflammatory myeloid cells (e.g., macrophages) and keratinocytes.[155] The Evans' group found that forced overexpression of PPARδ in muscle led to a significant increase in the slow twitch fiber type and increased endurance in treadmill tests.[160] The same laboratory has also offered evidence that PPARδ actually controls the expression of the other two PPARs, and thus acts as "a gateway receptor whose relative levels of expression can modulate PPARα and PPARγ activity."[178] The many diverse physiological roles of this ubiquitous NR remain controversial, but the available data certainly highlight the potential importance of all three PPARs in a number of critical physiological (and pathological) processes.

2.25.5.2.5 Liver X receptors

The LXRs (α and β) bind oxidized cholesterol derivatives (oxysterols) such as 24(S),25-epoxycholesterol. LXRα is expressed primarily in liver, adipose tissue, intestine, macrophages, and kidney, whereas LXRβ is ubiquitous.[151] Activated LXRs are deeply involved in RCT. Specifically LXRs induce the expression of proteins that stimulate export of cholesterol from macrophages (ABCA1 and ABCG1); promote cholesterol transport in serum and uptake in liver (apolipoprotein E, phospholipids transfer protein, lipoprotein lipase, and cholesterol ester transfer protein); increase cholesterol catabolism into bile acids (CYP7A1); increase biliary secretion of cholesterol (ABCG5 and ABCG8); and inhibit absorption of cholesterol in the intestine (ABCG5, ABCG8, and ABCA1).[151,185,186] There are other routes for cholesterol efflux from macrophages, including loading onto mature high-density lipoprotein (HDL), involving the bidirectional scavenger receptor CLA-1/SR-B1, and oxidation of cholesterol to derivatives. LXRα and LXRβ have also been shown to control major genes involved in each of these pathways.[186] LXRs increase fatty acid and triglyceride synthesis by increasing the protein synthesis of sterol regulatory element-binding protein 1c (SREBP-1c), the rate-limiting enzyme in fatty acid synthesis.[187]

Like many of the other NRs, LXR also exhibits anti-inflammatory properties. For example, Joseph et al., in a DNA microarray analysis, found that activation of LXR (with the synthetic ligand GW3965) suppressed many of the inflammatory-related genes induced by treatment of mouse (or human) macrophages with lipopolysaccharide (LPS).[188] These studies demonstrated reciprocal interference between the LXR and the pro-inflammatory NFκB pathways,[186] a theme common among the NRs. Furthermore, LXRα is upregulated in macrophages in response to Listeria infection.[160] LXR induces genes that protect against programmed cell death (apoptosis) during bacterial infections in warm-bodied animals, contracted by consuming contaminated food. Not surprisingly, a new study (using high-throughput quantitative PCR analysis) documented the expression of 27 of the 48 known NRs in macrophages, and many of these proteins exhibited distinct temporal changes in expression following pro-inflammatory LPS stimulation (see [160]).

2.25.5.2.6 Farnesoid X receptor

FXR is a sensor of bile acids such as chenodeoxycholic acid and cholic acid.[151] Target genes of this NR regulate the secretion of bile acids and phospholipids into bile (bile salt efflux pump and multidrug resistance proteins 2 and 3), the intestinal re-absorption of bile acids (ileal bile acid-binding protein (IBABP)), and hepatic cholesterol uptake from serum HDL (high-density lipoprotein; phospholipids transfer protein). Activated FXR increases transcription of short heterodimeric partner (SHP) and of CYP7A1, a repressor of the rate-limiting enzyme in bile acid biosynthesis, and this mechanism effectively avoids negative feedback repression of bile acid synthesis. These properties were confirmed through the use of transgenic *Fxr* knockout mice. FXR-deficient mice display increased serum bile, total bile acid pool size, and fecal bile-acid excretion,[151] findings that are consistent with altered bile acid homeostasis in response to impaired feedback inhibition of hepatic synthesis.

2.25.5.2.7 Constitutive androstane receptor and pregnane X receptor

CAR and PXR are xenosensors that mediate xenobiotic drug-induced changes by increasing transcription of genes that are involved in drug clearance and disposition (e.g., drug-metabolizing cytochromes P450, CYPs). Both receptors are highly expressed in the liver and small intestine. More recently, CAR and PXR have been implicated in the homeostasis of cholesterol, bile acids, bilirubin, and other endogenous hydrophobic molecules in the liver. CAR and PXR thus form an intricate regulatory network with other members of the NR superfamily, foremost LXR, FXR, and PPARs.[189] Through coordinated regulation of transcriptional programs, these NRs regulate key aspects of cellular and whole-body sterol homeostasis, including cholesterol absorption, lipoprotein synthesis and remodeling, lipoprotein uptake by peripheral tissues, RCT, and bile acid synthesis and absorption.[190]

Table 3 summarizes some of the more pharmacologically relevant functions of Class I and II NRs and includes the specific receptors currently known to be involved in each of these mechanisms.

2.25.6 Pharmacology of Nuclear Receptor Ligands

The discussion in Section 2.25.5 clearly demonstrates that NRs are involved in several important physiological functions. The NR superfamily therefore offers a plethora of targets for pharmacological intervention and development of therapeutics. Several widely used clinical drugs were discovered to act on NR targets (e.g., fibrates and glitazones). On the other hand, the identification of natural ligands for oNRs and development of synthetic agonists and antagonists has enabled significant advances in our understanding of the roles of the NR superfamily in health and disease. Several synthetic NR ligands are used clinically in the treatment of cancer, metabolic syndromes, inflammation, and immunosuppression. Table 4 shows a list of approved drugs that target NRs.

An emerging concept in the area of NR pharmacology is the principle of selective NR modulators (SNRMs). SNRMs are NR ligands whose spectrum of activity can range from agonist to partial agonist to antagonist, in a tissue-selective manner. This phenomenon of tissue-selective action was first observed with the ER antagonist tamoxifen, widely used in the treatment and prevention of breast cancer. Tamoxifen has antagonist activity at the ER in breast tissue, but has protective effects in bone and cardiovascular tissue as an ER agonist. When the concept of SERMs was introduced, it sparked intense research to understand the molecular basis of the tissue- and cell-specific actions of

Table 3 Nuclear receptor involvement in human physiology

Physiological function	*Implicated NRs*
Reproduction	ER, PR, and AR
Fetal development	TR, RAR, and PPARδ
Cardiovascular homeostasis	ERα, PR, RAR, TR, PPARγ, and PPARδ
Skeletal homeostasis	ERα, VDR, TR
Central nervous system homeostasis	ER, PR, GR, RAR, TR, and PPARδ
Immunomodulation	GR, VDR, RAR, and PPAR
Hematopoiesis	ERβ and RAR
Inflammation	GR, RAR, PPAR, and LXR
Wound healing	PPARα and PPARδ
Glucose homeostasis	ER, GR, and PPAR
Lipid and cholesterol homeostasis	VDR, TR, PPAR, LXR, FXR, CAR, and PXR
Toxin detoxification	CAR and PXR

Table 4 Selected approved drugs targeting nuclear receptors

Target receptor	Drug	Compound class	Brand name	Indication
ERα	Tamoxifen	SERM	Nolvadex	Breast cancer treatment, prevention
	Raloxifene	SERM	Evista	Osteoporosis
	Conjugated equine estrogens	Natural hormone	Premarin	Menopausal symptoms
PR	Norgestimate	Progesterone-based PR agonist	OrthoCyclen	Oral contraceptive
	Norethisterone	Testosterone-based PR agonist	Micro-Novum	Oral contraceptive
AR	Bicalutamide	Nonsteroidal androgen antagonist	Casodex	Prostate cancer treatment
	Flutamide	Nonsteroidal androgen antagonist	Drogenil	Prostate cancer treatment
GR	Dexamethasone	Steroidal GR agonist	Decadron, Dexameth,	Anti-inflammatory
MR	Eplerenone	Nonsteroidal MR antagonist	Inspra	Hypertension
RXRα, RXRβ, and RXRγ	Bexarotene	Rexinoid	Targretin	Refractory cutaneous T-cell lymphoma
RARα, RARβ, and RARγ; RXRα, RXRβ, and RXRγ	Alitretinoin	Retinoid	Panretin	Kaposi's sarcoma (topical only)
RARα, RARβ, and RARγ	Tretinoin	Retinoid (e.g., all-*trans*-Retinoic acid)	Retin-A, Renova, Vesanoid	Acute promyelocytic leukemia, acne
	Isotretinoin	Retinoid (e.g., 13-*cis*-Retinoic acid)	Accutane	Severe nodular acne
Vitamin D receptor	Calcitriol	1,25-Dihydroxyvitamin D_3	Rocaltrol, Calcijex	Hypocalcemia due to chronic renal failure and hypoparathyroidism
	Ergocalciferol	Vitamin D_2	Calciferol	Vitamin D-resistant rickets, hypoparathyroidism, familial hypophosphatemia
TRα and TRβ	Levothyroxine	Thyroid hormone	Levo-T, Unithroid, Levothyroid, Levoxyl, Synthroid	Hypothyroidism, euthyroid goiters, Hashimoto's thyroiditis
PPARγ	Pioglitazone	Thiazolidinedione	Actos	Type 2 diabetes mellitus (monotherapy or combination therapy)
	Rosiglitazone	Thiazolidinedione	Avandia	Type 2 diabetes mellitus (monotherapy or combination therapy)
PPARα	Fenofibrate	Fibrate	Tricor	Types IIa, IIb, IV, and V hyperlipidemia
	Gemfibrozil	Fibrate	Gemcor, Lopid	Types IIb, IV, and V hyperlipidemia

SERMs in particular and NRs in general.[191,192] What we now know about the selective control of NR action by NR coactivators, corepressors, and other modifiers of NR action makes it possible to design NR ligands that dissociate desired NR effects from the undesired effects in a tissue-selective manner. Thus, the concept of SNRMs has been extended to other NRs such as the GR, PPAR, and AR. The development of SNRMs is an active area in pharmaceutical drug discovery and has yielded several potential clinical candidates.

2.25.6.1 Estrogen Receptor Ligands

The natural ligand for ERα and ERβ is the steroidal hormone 17β-estradiol (E2), which plays a critical role in the growth and function of reproductive tissues such as breast, uterus, and ovaries, as well as in maintaining bone mineral density, cardiovascular health, and cognitive function. E2 and synthetic estrogens such as ethinylestradiol (EE) and norethindrone acetate (**Figure 5**) have been widely used clinically for oral contraception and management of osteoporosis and postmenopausal symptoms in hormone replacement therapy (HRT). Although evidence from studies using animal models suggests that estradiol plays a critical neuroprotective role against multiple types of neurodegenerative diseases, recent clinical studies have reported either inconclusive or untoward effects of HRT on the brain.[193] HRT is a regimen generally including a variety of sex steroids such estrogens and progestins, given to

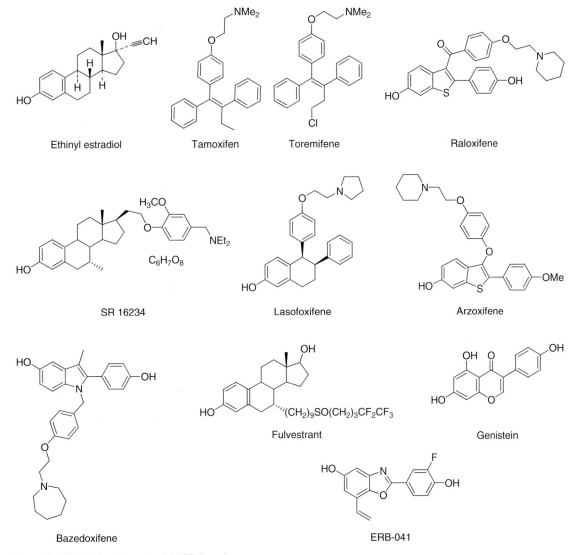

Figure 5 Steroidal and nonsteroidal ER ligands.

postmenopausal women for the relief of menopausal symptoms and prevention of osteoporosis, cardiovascular disease and dementia. However, the Women's Health Initiative (WHI), the largest randomized clinical trial of HRT to date, showed that the combination of conjugated equine estrogen and progesterone medroxyprogesterone acetate (PremPro) as well as estrogen given alone, resulted in an 'increased' risk of strokes, breast cancer and blood clots and provided no significant benefit.[194] Although skeletal benefits and fracture reduction were observed, the overall lack of benefit prompted the discontinuation of the long-term use of HRT for chronic disease prevention in postmenopausal women.[195] An ancillary arm (WHI Memory Study) of this randomized study also evaluated the effect of HRT on cognitive function and found no benefit but an increased risk of stroke and Alzheimer's disease.[196,197]

Synthetic estrogens function as ER agonists in all tissues that express ERα and ERβ. Since ER also has a proliferative effect on breast and uterine tissue, long-term estrogen use, particularly in HRT, has been implicated as a risk factor for breast and uterine cancer.[198] However, ER antagonists, also sometimes referred to as antiestrogens, have been used to treat and prevent breast cancer for decades. Tamoxifen, a nonsteroidal antiestrogen (Figure 5), has been the mainstay for therapy of ER-positive breast cancer and is now also used as adjuvant therapy in early breast cancer and as a chemopreventive to reduce breast cancer incidence in healthy women at high risk for the disease. Tamoxifen's activity profile of estrogen antagonist effects in breast tumor tissue but estrogen-like agonist effects on the endometrium and bone led to the concept of tissue-selective effects of antiestrogens and SERMs.[199] Since then, several nonsteroidal and steroidal SERMs have been developed and used clinically not only for breast cancer, but also for other indications such as osteoporosis and HRT.[200] Among the triphenylethylene class of SERMs similar to tamoxifen, toremifene (Fareston) is currently used in advanced breast cancer and has the same mixed agonist/antagonist profile as tamoxifen. Second- and third-generation SERMs have now been developed that have a much improved tissue profile than tamoxifen. Fixed-ring SERMs such as raloxifene are devoid of any estrogenic effects in the uterus, but are potent antiestrogens in the breast while retaining favorable estrogen-like agonist activity in the bone and cardiovascular system. Raloxifene is currently used for osteoporosis in postmenopausal women and is in clinical trials for breast cancer prevention. Newer SERMs such as lasofoxifene (Oporia), arzoxifene, and basedoxifene (Figure 5) are also under development for prevention of breast cancer and osteoporosis.[200–203] Among steroidal SERMs, SR16234 (TAS-108) (Figure 5), currently in clinical trials against advanced breast cancer, shows efficacy in tamoxifen-resistant breast cancer[204] and has a favorable SERM profile without estrogenic effects on uterine tissue.

Another class of ER antagonists distinguishable from the SERMs in terms of pharmacology and molecular mechanism of action comprises the selective estrogen receptor downregulators (SERDs). Of the three steroidal SERDs developed thus far, only fulvestrant (ICI 162384, Faslodex) is currently used clinically as second-line therapy in advanced breast cancer. Although SERDs also bind the ER, they inhibit receptor dimerization and disrupt nuclear localization, with a concomitant increase in ER turnover and subsequent reduction in the number of ER molecules in the cell.[205,206] Thus, SERDs inhibit the expression of estrogen-dependent genes by ER downregulation. This effect is independent of coregulator involvement and occurs in all ER-expressing tissues; thus these compounds are also referred to as 'pure antiestrogens,' in contrast to the tissue-selective SERMs.

SERMs also have the potential to be used for their cardiovascular benefits and favorable effects on cholesterol ratios. Indeed, lasofoxifene is in clinical trials for dyslipidemia. The beneficial effects of SERMs on cognitive function, however, have been underexplored, and an opportunity for future drug discovery in this area may exist.

The pharmacology of estrogen action mediated via the ERβ subtype is only now being unraveled. While both ERα and ERβ have significant homology in their DNA- and ligand-binding domains, the different expression patterns in tissues suggests that each receptor plays a distinct role in estrogen signaling.[207] Although phytoestrogens such as genistein have ERβ agonist activity and several other synthetic selective ligands have been identified, only a few have sufficiently high selectivity to be used to probe ERβ pharmacology.[208] The recent discovery of a highly selective ERβ agonist, ERB-041, with a 200-fold selectivity for ERβ led to the discovery of a new function for ERβ in modulation of the inflammatory response[209,210] in a model of chronic inflammation and in a model of inflammatory bowel syndrome. Thus, the discovery of ERβ-selective ligands will aid in understanding the role of ERβ in mammalian biology and discovering new therapeutic targets. Exploiting the role of ERβ in the prostate for the treatment of prostate cancer is another area that will benefit from the discovery of ERβ-selective ligands.[111]

2.25.6.2 Progesterone Receptor Ligands

Progesterone, the natural ligand for the PR, exerts its physiological effects through the PR, which is expressed as two isoforms, PR-A and PR-B. Progesterone plays an important role in female reproduction, controlling ovulation and maintaining pregnancy, and in the growth and differentiation of endometrial and myometrial cells in the uterus.

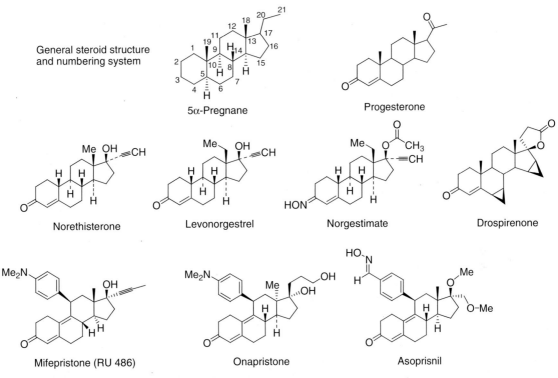

Figure 6 Progesterone receptor ligands.

Several PR agonist ligands (progestins) and antagonist ligands (antiprogestins) have been developed for use as oral contraceptives and HRT. The synthetic progestins are testosterone derivatives (norethisterone and levonorgestrel) or progesterone derivatives (desogestrel, etonogestrel, gestodene, and norgestimate), shown in **Figure 6**. Several new fourth-generation progestins, such as dienogest and drospirenone, have been synthesized in the past decade. These two new progestins have no androgenic effect but a partial antiandrogenic effect. Drospirenone also exerts antimineralocorticoid effects, leading to decreased salt and water retention and lowering of blood pressure in users of pills containing this progestin. The 19-norprogesterone derivatives such as trimegestone are specifically progestational and do not possess any androgenic, estrogenic, or glucocorticoid activity. They are referred to as 'pure' progestational molecules because they bind almost exclusively to the PR and do not interfere with the other steroid receptors.

The discovery of mifepristone (RU486), the first GR and PR antagonist, marked the beginning of drug discovery related to PR antagonists (antiprogestins). Several newer progesterone antagonists with reduced glucocorticoid activity (e.g., onapristone) (**Figure 6**) have also been synthesized. Most antiprogestins are used primarily for termination of pregnancy and postcoital contraception, and do possess undesirable side effects on the endometrium. A new class of SPRMs has now been developed that have antiproliferative effects on the endometrium and breast, but retain the protective effects of ovarian estrogen on bone and cardiovascular system. SPRMs are being developed for the treatment of uterine leiomyoma and endometriosis.[211] Asoprisnil (J867) is the first SPRM to reach an advanced clinical stage and demonstrates a high degree of endometrial selectivity and partial agonist/antagonist PR activity.[212,213] It is effective in reducing endometriosis-associated pain, shrinks uterine fibroids, and induces a reversible amenorrhea without estrogen deprivation, indicating that the endometrium is its preferred target tissue.

2.25.6.3 Androgen Receptor Ligands

The AR is expressed in several tissues, and androgens play a role in maintaining muscle mass and bone strength, decreasing fat tissue, and enhancing libido.[118] In addition, androgens, acting through ARs, play a critical role not only in the physiological development and function of the prostate, but also in the genesis of prostate cancer.

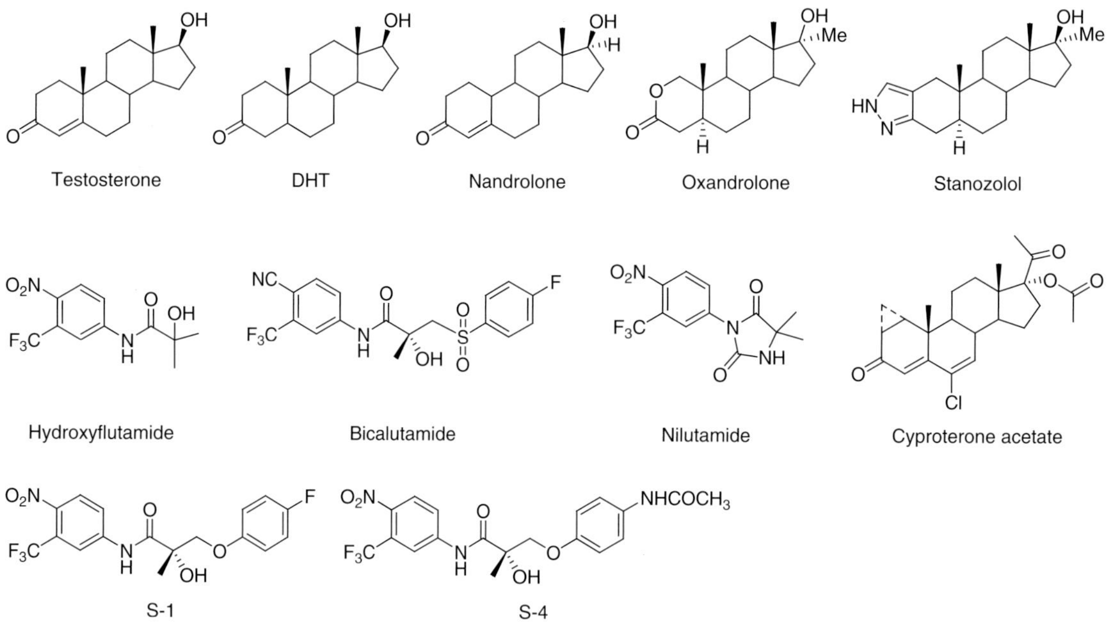

Figure 7 Androgen receptor ligands.

Natural AR agonists, principally testosterone and 5-α-dihydrotestosterone (DHT) and their synthetic derivatives nandrolone, oxandrolone, and stanozolol (**Figure 7**), have been used to treat a number of clinical conditions, such as hypogonadism, delayed puberty, and muscle wasting, and as anabolic steroids. However, use of androgen therapy has been limited by lack of efficacious compounds with easy delivery options and concerns about side effects such as stimulatory effects on the prostate, hirsutism, hepatic toxicity, and steroid abuse.

These problems sparked drug discovery efforts for selective androgen receptor modulators (SARMs).[214] Although SARMs are not yet in clinical use, several nonsteroidal SARMs are under development.[215] Ideally, a tissue-selective SARM would have the beneficial anabolic effect of androgens with reduced risk of prostate cancer in men or hirsutism and virilization in women. An SARM for the treatment of hypogonadism or osteoporosis would be an AR agonist in the muscle and bone, with minimal hypertrophic agonist effects in the prostate. The first generation of SARMs was designed by making structural modifications of the antiandrogen bicalutamide.[216] SARMs S-1 and S-4 (**Figure 7**) demonstrate tissue selectivity in animal models and function as AR antagonists in the prostate without affecting gonadotrophin release or the effects of androgens on bone and muscle mass.[217] These SARMs have shown efficacy in benign prostatic hyperplasia (BPH)[217] and as male contraceptives.[215] SARMs with greater AR antagonist activity but with agonist activity in the bone are also being explored as potential prostate cancer therapies.[218]

Approximately 80–90% of prostate cancers are androgen dependent at initial diagnosis, and prostate cancer therapy is directed toward reducing serum androgen (androgen ablation therapy) and inhibiting AR by antagonists (antiandrogens).[219] The four antiandrogens in clinical use are the nonsteroidal antiandrogens hydroxyflutamide, bicalutamide, and nilutamide and the steroidal antagonist cyproterone acetate. Antiandrogens bind to AR and downregulate the effects of endogenous circulating androgens and remain the first-line treatment for palliation of advanced prostate cancer.

2.25.6.4 Glucocorticoid Receptor Ligands

Glucocorticoids are the most effective and frequently used drugs to treat acute and chronic inflammatory and immunological diseases such as rheumatoid arthritis, lupus erythematosus, asthma, transplant rejection, and inflammatory bowel disease (IBD). Synthetic glucocorticoids such as hydrocortisone, prednisone, methylprednisone, and dexamethasone mediate their anti-inflammatory activity through the GR, whose endogenous ligand is the adrenal hormone cortisol. The use of glucocorticoids is associated with potentially severe side effects, including diabetes, osteoporosis, water retention, and psychosis.[220] The potent anti-inflammatory activity of glucocorticoids has been shown to be due to transrepression of pro-inflammatory cytokines and inflammation mediators such as TNF-α, IL-6,

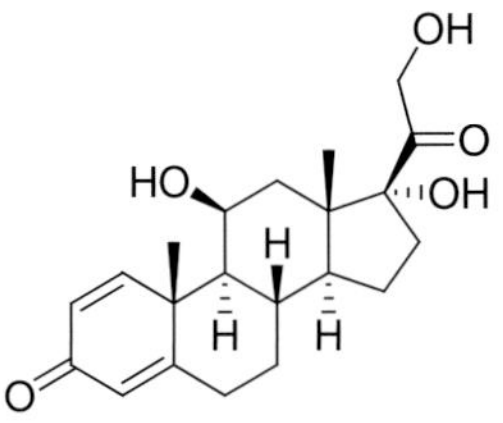

Prednisolone

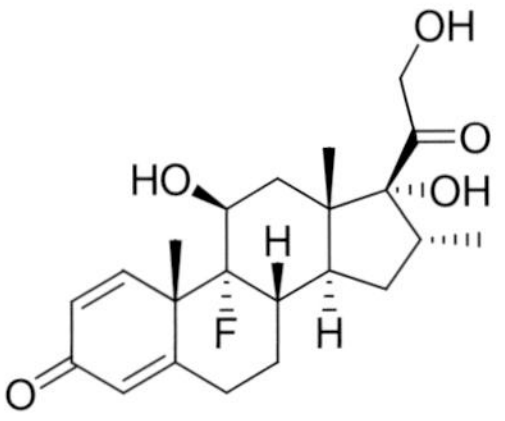

Dexamethasone

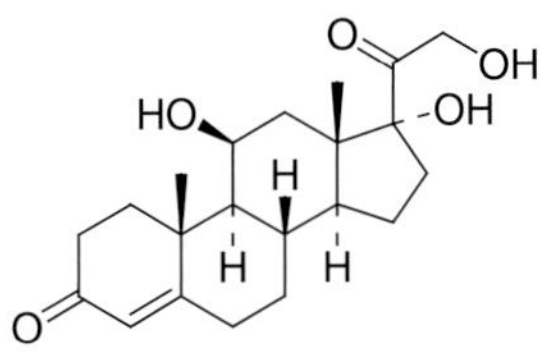

Hydrocortisone

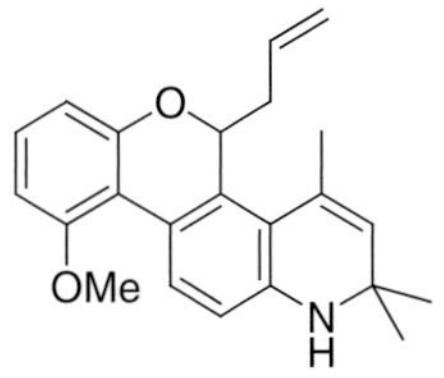

AL-438

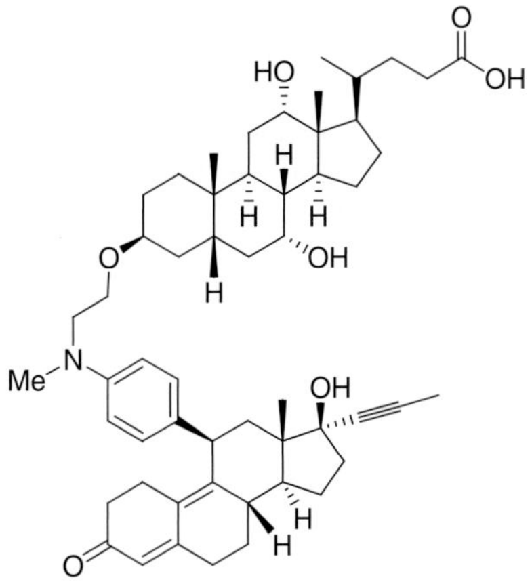

A-348441

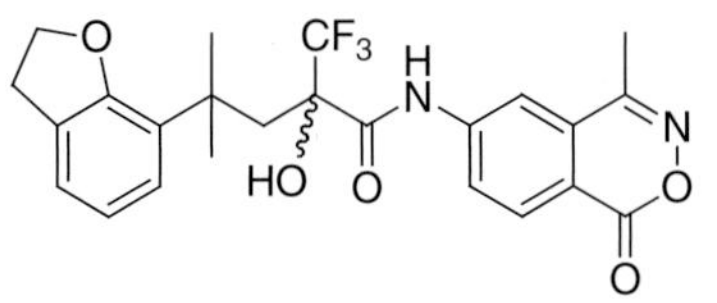

ZK-216348

Figure 8 Glucocorticoid receptor ligands.

IL-12, and PGE_2. The side effects of glucocorticoids are thought to arise from transactivation of enzymes involved in gluconeogenesis and muscle metabolism.[124] In addition to transrepression mediated via the GR response elements in pro-inflammatory gene promoters, a direct interaction of ligand-bound GR, as a monomer, with other transcription factors, such as NFκB and AP-1, has been shown to inhibit pro-inflammatory gene expression via these transcription factors.[221] These studies on the molecular mechanisms of the anti-inflammatory activities and side effects of glucocorticoids led to the concept of 'dissociated GR agonists' that preferentially induce transrepression, but not the transactivation functions of GR, and have an improved therapeutic profile with reduced side effects.[222] Active interest in developing such selective GR agonists has led to the identification of two nonsteroidal GR agonists, AL-438[223] (**Figure 8**) and ZK-216348,[126] that demonstrate anti-inflammatory activity comparable with that of prednisone, but with superior side-effect profiles in animal models. These are currently being used as tools to understand the complex pharmacology of the anti-inflammatory effects of glucocorticoids.

However, GR antagonists have been investigated for their potential to control peripheral glucose metabolism in Type 2 diabetes. As their name suggests, glucocorticoids raise blood glucose levels by antagonizing the action of insulin and increasing hepatic glucose output. Liver-selective GR antagonists could decrease blood glucose levels without adverse peripheral side effects. A-348441 (**Figure 8**) is a liver-targeting GR antagonist based on the prototype GR antagonist RU486 that shows improved glucose control in animal models of Type 2 diabetes.[224,225] Thus, GR ligands with selective profiles of agonist and antagonist activity against several therapeutic targets are the focus of many drug discovery programs.

2.25.6.5 Mineralocorticoid Receptor Ligands

Aldosterone is the endogenous ligand for the MR, which is sometimes also referred to as the aldosterone receptor. Aldosterone binding to MR results in widespread physiologic effects in the kidney, heart, and vasculature, such as sodium retention and potassium excretion, ventricular hypertrophy, fibrosis, vasoconstriction, inflammation, and endothelial dysfunction.[226] All these are risk factors for congestive heart failure (CHF). In addition, aldosterone production in the heart and plasma levels in blood are increased after myocardial infarction and CHF to an extent that correlates with the severity of the disease.[227] Aldosterone receptor (MR) antagonists have now emerged as a new paradigm in the treatment of CHF. Spironolactone was the first MR antagonist to be evaluated for hypertension, but its use is precluded by its poor specificity for the MR and the side effects from its antiandrogenic and progestational activities. Eplerenone (Inspra) (**Figure 9**), a spironolactone derivative, is the first selective aldosterone receptor antagonist to be developed, and it has recently been approved for the treatment of systemic hypertension. It has also been approved for the treatment of left-ventricular systolic dysfunction in acute myocardial infarction.[228] A combination of eplerenone with angiotensin-converting enzyme (ACE) inhibitors offers further advantages in this setting.[229]

2.25.6.6 Vitamin D Receptor Ligands

1α,25-Dihydroxyvitamin D_3 or calcitriol (**Figure 10**), the active metabolite of vitamin D, is the physiological VDR ligand, known for maintaining calcium and phosphate homeostasis. In addition to the gut and bone, VDR has now been shown to be present in virtually all tissues, and its natural and synthetic ligands are increasingly recognized for their noncalcemic actions, such as potent antiproliferative, prodifferentiative, and immunomodulatory activities. The cellular VDR protein level is upregulated in several disease states, providing a rationale for using VDR ligands to treat these

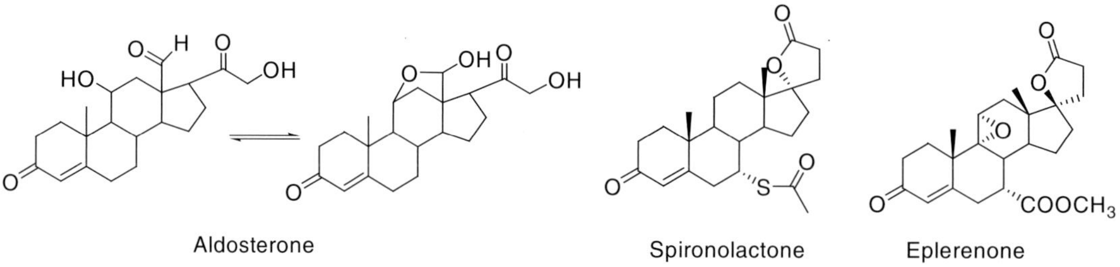

Figure 9 Mineralocorticoid receptor ligands.

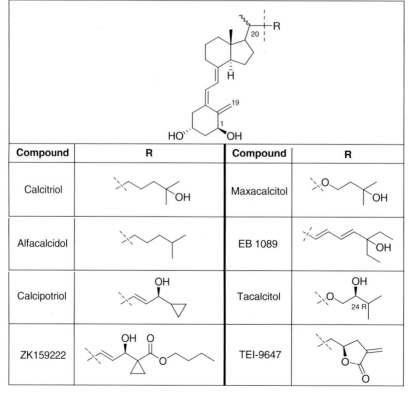

Figure 10 Vitamin D receptor ligands.

conditions. The utility of VDR ligands has been explored for various therapeutic indications, such as rheumatoid arthritis, psoriasis, actinic keratosis, osteoporosis, autoimmune diseases (type I diabetes, lupus erythematosus, and multiple sclerosis), and several cancers (prostate, colon, breast, leukemia, head, and neck). See [230,231] for excellent in-depth reviews of the clinical use of VDR ligands. However, the use of VDR agonists is hampered by their dose-related toxicity, manifested as hypercalcemia/hypercalciuria.

A large number of synthetic vitamin D_3 agonists have been synthesized, and several are used clinically. Calcitriol is used orally for osteoporosis and in topical form for psoriasis. Several other analogs with reduced calcemic effects are also used clinically. Alfacalcidol (**Figure 10**), a precursor prodrug of calcitriol, is used in osteoporosis and arthritis. Calcipotriol, a metabolically labile analog of vitamin D_3, was developed for topical use in the treatment of psoriasis and has significantly less calcemic activity. Other modified agonist analogs include EB 1089, ED-71, maxacalcidol (OCT), and tacalcitol (**Figure 10**), in clinical trials for hyperparathyroidism, osteoporosis, prostate cancer, and psoriasis.[231] In contrast, significantly fewer VDR antagonists have been developed, and these have mainly been used as tools to study the physiological function of VDR and its interactions with other NR pathways.[232] VDR antagonists ZK 159222 and TEI-9647 (**Figure 10**) have been useful in studying the molecular aspects of VDR activation and coregulator interactions.[233] TEI 9647 has also been tested in animal models of Paget's disease as a potential treatment for the excessive bone resorption and osteoclast formation seen in this disease.[234] Elucidation of the mechanism of action of the different vitamin D analogs will enhance our understanding of vitamin D pathways and improve uses of VDR-based therapies.

2.25.6.7 Thyroid Hormone Receptor Ligands

3,5,3′-Triiodo-L-thyronine (T_3) is the physiologically active thyroid hormone that exerts significant effects on growth and homeostasis through its actions on the TR. T_3 is synthesized in peripheral target tissues by metabolic deiodination of the prohormone T_4 (3,5,3′,5′-tetraiodo-L-thyronine), which is secreted from the thyroid gland. T_4 itself does not possess appreciable biological activity. The diverse actions of T_3 in regulating metabolic rate, cholesterol and triglyceride levels, cardiac output, and skeletal development are mediated through the two thyroid receptor subtypes, TRα and TRβ, products of two different genes, but widely expressed in most tissues.[235] Although T_3 binds with nearly equal affinity to both receptor subtypes, studies from knockout animals have shown that most effects of T_3 on the heart and cardiac rhythm are mediated through the TRα isoform, whereas most effects on the liver, skeleton, and other tissues are mediated through TRβ.[236,237] T_3 agonism, or high levels of T_3, can exert desirable effects, such as reduction in plasma LDL cholesterol, lipoprotein(a), and triglycerides and promotion of weight loss, all actions mediated by TRβ. However, these are accompanied by undesirable effects mediated by TRα, including tachycardia, atrial arrhythmias, and heart failure. Like with other NR ligands, selective thyroid receptor modulators (STRMs) that bind preferentially to one isoform or have tissue-selective uptake are being actively pursued,[238–240] particularly TRβ-selective agonists, as hepatic-selective, cardiac-sparing drugs for therapeutic use as cholesterol-lowering agents. The most studied T_3 agonist in this regard is GC-1 (**Figure 11**), an analog of T_3 that was shown to reduce plasma cholesterol in rats and monkeys with greater efficacy than atorvastatin, the HMG-CoA reductase inhibitor widely used clinically as a cholesterol-lowering agent.[241] Although GC-1 is selective for TRβ,[242] its decreased uptake into the heart compared with the liver actually accounts for its selective pharmacological activity on cholesterol metabolism.[243] GC-1 also reverses skeletal development and maturation defects of hypothyroidism in rats; this effect indicates that TRβ-mediated thyroid action plays a role in bone development.[244] Another promising STRM recently reported is KB-141 (**Figure 11**), which has a 14-fold higher affinity for TRβ over TRα.[245] Studies of KB-141 in monkeys, rats, and mice show reductions in plasma cholesterol, lipoprotein (a), and body weight, without significant cardiac side effects, demonstrating effective separation of potential therapeutic benefits of TRβ agonism while sparing TRα stimulation.[246] TRα-selective agonists may be developed for increasing cardiac performance and output in patients with CHF. For example, 3,5-diiodo-thyropropionic acid (DITPA) is a T_3 analog that has demonstrated clinical efficacy in increasing cardiac index and reducing serum cholesterol and triglyceride levels in patients with CHF and elevated cholesterol, suggesting a new therapeutic use for TRα-selective TR agonists.[247]

TR antagonists have a potential role in the treatment of hyperthyroidism, an area in which therapeutic choices are quite limited. The antiarrhythmic drugs amiodarone and the newer Dronedarone (**Figure 11**) have been reported to act by antagonizing T_3 binding to TR.[248] Dronedarone, which has a better clinical profile and lower side effects, is a TRα-selective competitive antagonist; its effectiveness suggests possible therapeutic applications for isoform-specific TR antagonists.[249] Several TR antagonists based on the GC-1 class of compounds have been designed.[250] Compound NH-3 (**Figure 11**) is the first potent TR antagonist with nanomolar affinity for TR that also inhibits TR action in an animal model.[251] The significant therapeutic potential of STRM ligands merits an active interest in their development.

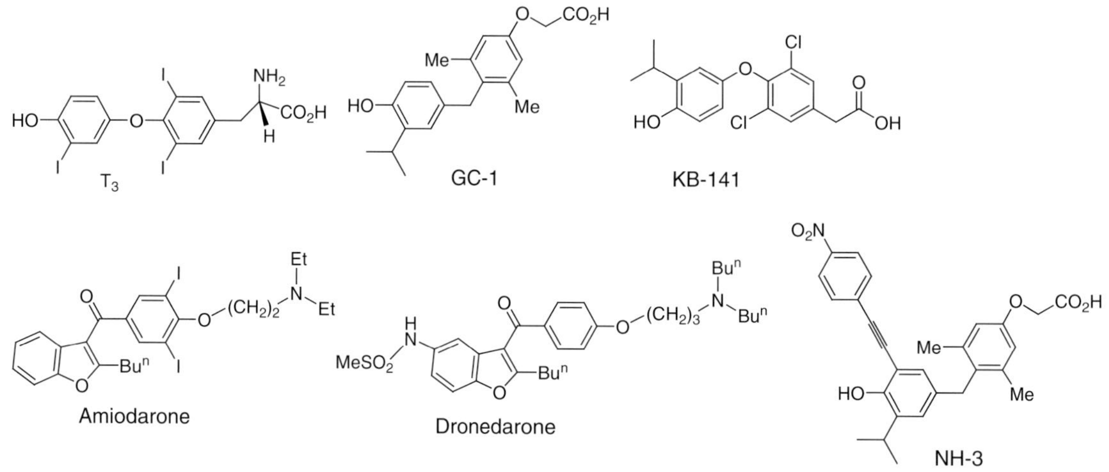

Figure 11 Thyroid receptor ligands.

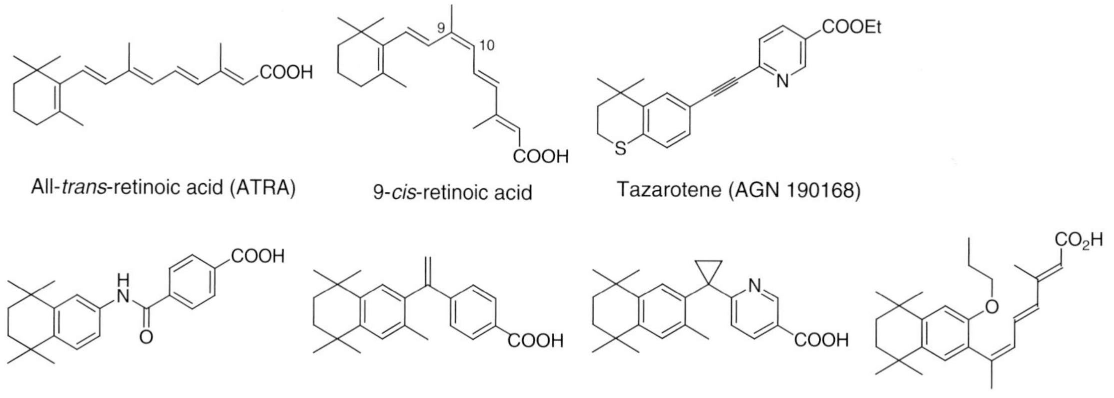

Figure 12 Retinoid acid receptor (RAR) and retinoid X receptor (RXR) ligands.

2.25.6.8 Retinoid Receptor (RAR and RXR) Ligands

Naturally occurring retinoids, all-*trans*-retinoic acid (ATRA) and 9-*cis* retinoic acid (9-*cis* RA), regulate a wide variety of physiological functions (*see* Section 2.25.5.2.2) through two NR subfamilies called the retinoic acid receptors (RARs) and the RXRs. Each of these subfamilies has three subtypes (α, β, and γ), which are encoded by a single gene each. Retinoids, acting through the RARs and RXRs, have also been used clinically for a broad range of therapeutic applications, that are poised to expand even further with increasing development of receptor-selective NR ligands and the concept of tissue-selective NR action.[151] The endogenous ligand for the RARs is ATRA (**Figure 12**), whereas 9-*cis* RA binds both RARs and RXRs. The RARs form heterodimers with RXR and bind to retinoic acid response elements (RAREs) to modulate target gene expression. RXRs, however, can either form homodimers or heterodimers with other NRs. This unique role of RXR as the preferred heterodimeric partner for several other NRs accounts for its ubiquitous actions in several physiological pathways and is the basis of the emerging expansion of the therapeutic applications of RXR-selective retinoids in several disease processes, as discussed below.

The RARs have been implicated in the regulation of cancer development and epithelial differentiation. Translocation of the RARα gene is responsible for the development of acute promyelocytic leukemia (APL).[252] RARα is also associated with elevations in triglyceride levels. RARγ is associated with skin, bone, and teratogenic toxicity. RARβ is considered to have a tumor suppressor role since its expression is lost during early carcinogenesis as a result of methylation-induced silencing of the promoter region of its gene.[253,254] In addition, RARβ contains a RARE in its promoter region that mediates retinoic acid (RA)-induced RARβ gene expression. This upregulation of the RARβ gene plays a critical role in RA-induced growth inhibition and apoptosis in breast and lung cancer cells.[255,256] ATRA, a RAR pan-agonist, has shown complete responses in APL patients,[257] but ATRA therapy is associated with significant side effects and development of resistance. RARα-selective ligands such as AM80 (**Figure 12**) have a high therapeutic index and have successfully induced remissions in ATRA-resistant patients, with a significantly milder side-effect profile.[258,259] 9-*cis* RA, another RAR pan-agonist, is currently used as a topical agent for cutaneous AIDS-related Kaposi's sarcoma.[260] Tazarotene (AGN 190168), a RARβ, γ-selective agonist, is currently used clinically for the treatment of psoriasis.[261] Several classes of subtype-selective RAR ligands have been developed; their discovery and selective pharmacology are extensively covered in two excellent reviews.[262,263]

RXR-selective synthetic retinoids have been termed 'rexinoids.'[264] One of the well-known rexinoids is LGD1069 (Bexarotene, Targretin) (**Figure 12**), clinically used in cutaneous T-cell lymphoma and in clinical trials against non-small cell lung cancer.[265] However, use of LGD1069 is associated with significant toxicity, such as hypothyroidism and hypertriglyceridemia, attributed to its weak RAR agonist activity.

A newer, more specific rexinoid, LG100268,[266] has been extensively evaluated to understand the pleiotropic pharmacology of RXR receptors in various metabolic pathways. NRs that form heterodimers with RXRs are subdivided into two functionally distinct groups – permissive and nonpermissive.[267] A permissive heterodimer can be activated by agonists for either RXR or the NR partner, or both, for synergistic activation. Examples of permissive heterodimers include RXR heterodimers with PPAR, LXR, and FXR. Heterodimers with nonpermissive partners can only be activated by the agonist of the binding partner, not by an RXR agonist. Examples of nonpermissive partners are TR and VDR, which function primarily as hormone receptors that keep gene expression under tight hormonal control. Thus, RXR agonists can regulate the gene expression of multiple permissive NR partners, and can have widespread pharmacological effects on several important metabolic pathways. Indeed, LG100268 has been shown to have insulin-sensitizing effects (related to the PPARγ-permissive partner) and to reduce atherosclerosis in animal models; the latter effect related to its LXR- and FXR-permissive activation.[268,269] Even more selective RXR modulators now under development are specific for certain permissive partners, but not all. For example, LGD100754 (**Figure 12**) is a selective antagonist of the RXR homodimers and an agonist of the RXR:PPAR heterodimer, but not of RXR:LXR or RXR:FXR heterodimers.[270] LGD100754 triggered PPAR-activated pathways such as adipocyte differentiation and also demonstrated insulin sensitization in db/db obese mice more effectively than the relatively nonselective rexinoid LGD100268.[269] Thus, it now appears possible to design specific RXR modulators that can provide specific regulation of metabolic pathways.[271,272]

2.25.6.9 Peroxisome Proliferator-Activated Receptor Ligands

The PPARs function primarily as lipid sensors and control several biological processes including energy and lipid homeostasis, cell differentiation and inflammation. Endogenous ligands for PPARα include polyunsaturated and saturated fatty acids, leukotriene B$_4$ and more recently, oleoylethanolamide, which was shown to regulate feeding and body weight through activation of PPARα.[273] PPARα binds to a wide range of physiological fatty acids albeit with low affinity, pointing to its role as a general lipid sensor. Its activation increases fatty acid catabolism and oxidation, modulates inflammatory response, and reduces triglyceride levels. These actions form the basis of the pharmacological lipid lowering and cardioprotective effects of the well-known hypolipidemic fibrate class of drugs, which were shown to be synthetic PPARα ligands. The fibrates clofibrate, fenofibrate, bezafibrate, and gemfibrozil were used clinically as hypolipidemics long before PPARα was identified as their pharmacological target.[274,275] However, the fibrates show only weak PPARα agonist activity and are 10-fold selective for PPARα over PPARγ.[275] They function by increasing the clearance and decreasing the synthesis of triglyceride-rich VLDL, and increasing HDL-cholesterol levels by modulating the genes that control these processes.[276] Recent data indicate that PPARα agonists have direct vasoprotective effects and reduce the progression of atherosclerosis, particularly in diabetic patients.[277] Cardiovascular disease is the major cause of mortality in Type 2 diabetes patients, a group in which dyslipidemia is also 2–3 times more prevalent than in the general population.[278] More selective PPARα ligands are now being actively sought and several new classes of PPARα ligands have now been reported. For example, ureidofibric acids GW9578 (EC$_{50}$ 50 nM)[279] and phenoxypropanoic acid LY518674 (EC$_{50}$ 42 nM)[280] have been reported as potent PPARα agonists with 20–200-fold selectivities. A group from Merck recently reported a novel class of highly potent (<10 nM) PPARα agonists, exemplified by compound I (**Figure 13**) that lowered cholesterol and triglyceride levels in animal models at doses significantly lower than fenofibrate.[281]

PPARγ, expressed in adipose cells, is activated by polyunsaturated fatty acids and by prostaglandin J2, and plays a crucial role in adipogenesis. PPARγ is perhaps best known as the target for the insulin-sensitizing drugs, the glitazones

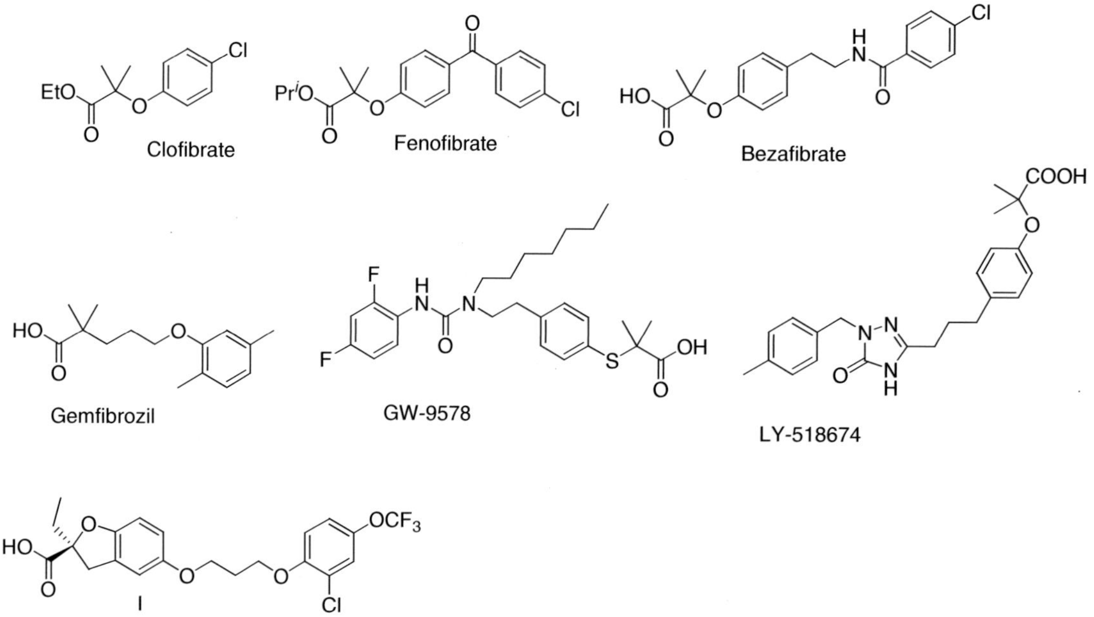

Figure 13 PPARα ligands.

(thiazolidinediones). Pioglitazone and rosiglitazone (**Figure 14**), used for Type 2 diabetes, are among the top 100 selling drugs in the United States. They exert their antidiabetic effects through the activation of PPARγ and improve insulin sensitivity in muscle and adipose tissue. However, their long-term use is associated with the development of side effects such as weight gain, edema, and anemia. The glitazones also have a deleterious effect on bone, since PPARγ activation causes bone osteoblasts to undergo terminal differentiation to adipocytes.[282,283] Netoglitazone (**Figure 14**) is a new selective glitazone that lowers blood glucose as effectively as rosiglitazone in diabetic animal models, at doses that do not affect the bone.[284,285] Epidemiological studies show that Type 2 diabetes is associated with an increased risk for coronary heart disease and ischemic stroke. A majority of Type 2 diabetic patients are obese and have elevated triglyceride levels and low HDL levels.[286,287] Since PPARα regulates lipid and triglyceride homeostasis, there has recently been considerable interest in combining the beneficial effects of PPARα and PPARγ activation. Several PPARα/γ dual agonists of different chemical classes other than the glitazones have been discovered, the most popular among which are the 'glitazars' farglitazar, ragaglitazar, and muraglitazar (**Figure 14**). Farglitazar,[288] a potent PPARγ agonist with moderate PPARα efficacy, was in clinical trials but was dropped due to emergence of edema. Ragaglitazar (originally DRF2725)[289,290] was also dropped from clinical development due to carcinogenicity in rodent toxicity models. The most recent addition to the glitazar family, muraglitazar,[291] is the only PPARα/γ dual agonist to successfully advance to an nondisclosure agreement (NDA) filing (December 2004). At the time of writing this chapter, this drug received a 'non-approvable' letter, pending further information on its cardiovascular safety profile. The muraglitazar-treated patient group showed an increased incidence of CHF, ischemic attack, and death compared with placebo.[292] Bristol-Myers Squibb and Merck, who are jointly developing Muraglitazar, are considering the option of dropping this compound from further development.[293] Regardless, the discovery of dual PPARα/γ agonists of diverse chemical classes distinct from the glitazars, still continues at an active pace.[294–296]

PPARδ is the least understood among the PPARs and is not yet a clinical target, unlike PPARα and γ. PPARδ is expressed ubiquitously in all tissues and also binds long-chain fatty acids with low affinity.[29] There are only a few selective PPARδ ligands known thus far. GW2433 (**Figure 15**) is a ureidofibrate that was potent and selective for PPARδ but also activated PPARα.[297] L-165041, a leukotriene antagonist, was also a PPARδ ligand with 10-fold selectivity over PPARα and γ.[298] The most potent and selective ligand reported thus far is GW501516 (**Figure 15**),[184] with a 1000-fold selectivity and subnanomolar potency at PPARδ. GW501516 promotes RCT and increase HDL-c levels in obese rhesus monkeys.[184] It also attenuates weight gain and insulin resistance, in mice on high fat diets, by promoting fatty acid oxidation in skeletal muscle[299] and adipose tissue.[182] Several studies with GW501516, now suggest that PPARδ agonists may have utility in metabolic syndrome.[300,301] However, the role of PPARδ in cell proliferation and differentiation is far from clear, and will have to be determined before PPARδ-selective ligands can be considered for therapeutic development.[162]

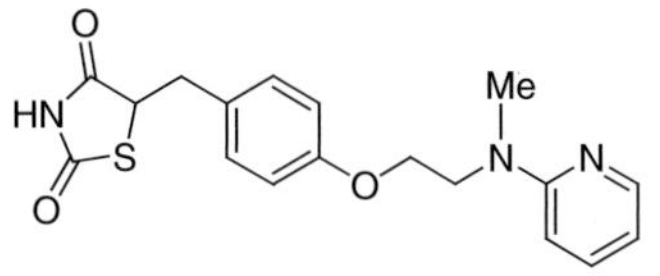

Rosiglitazone

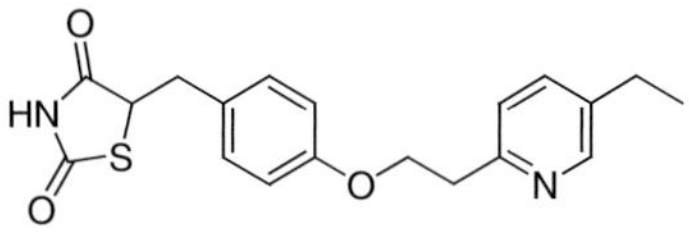

Pioglitazone

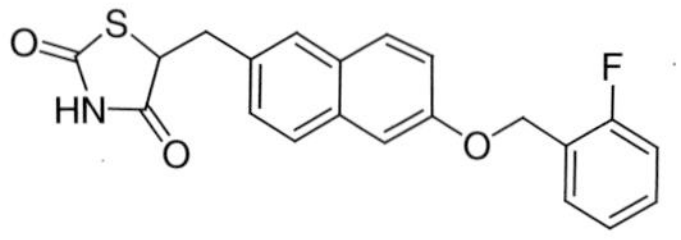

Netoglitazone

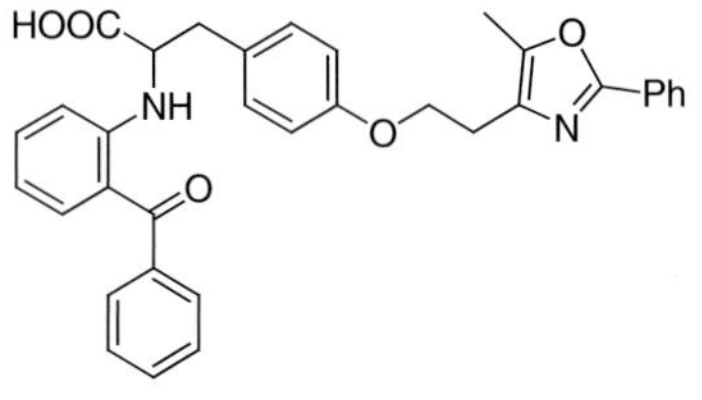

Farglitazar

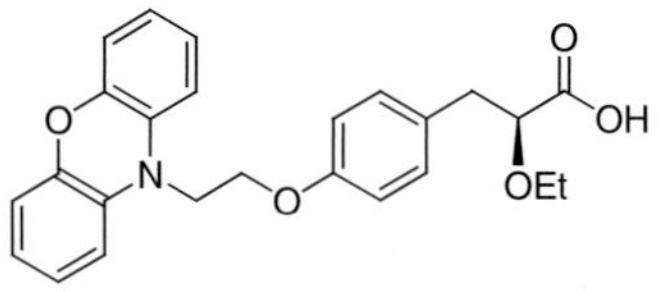

Ragaglitazar

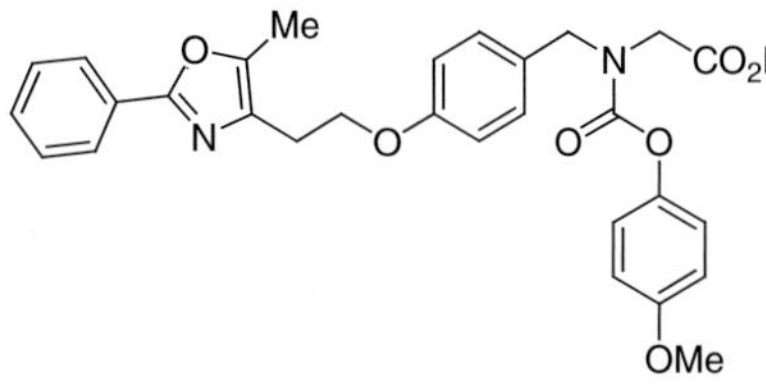

Muraglitazar

Figure 14 PPARγ ligands.

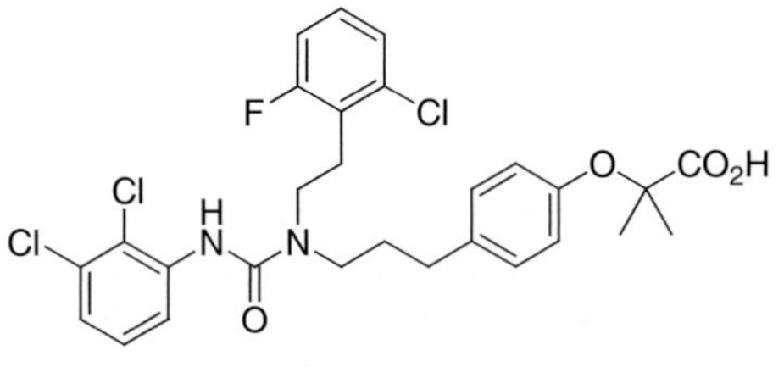

GW 2433

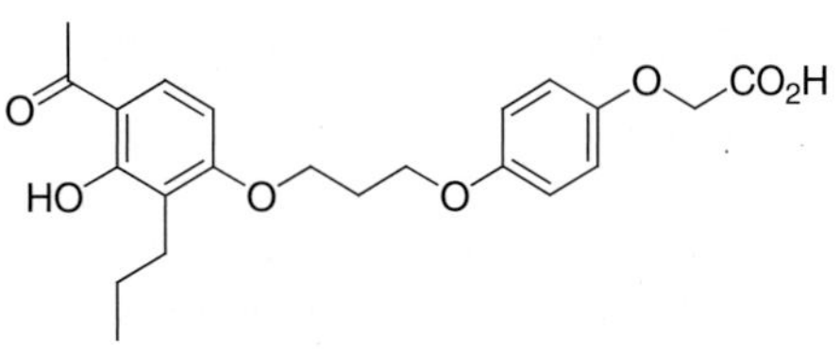

L-165041

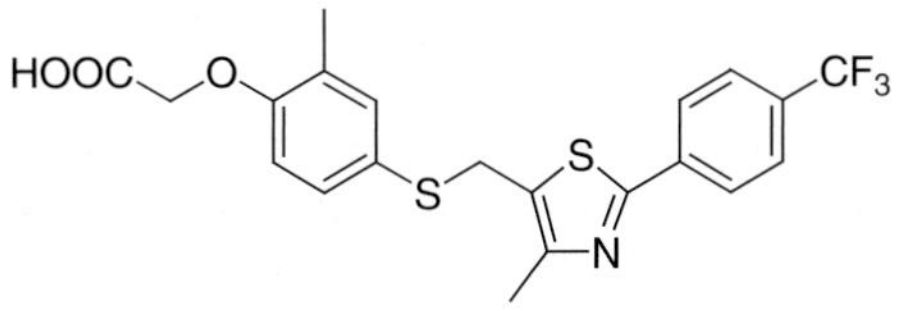

GW 501516

Figure 15 PPARδ ligands.

2.25.6.10 Liver X Receptor Ligands

The NRs LXRα and LXRβ regulate the expression of key genes in cholesterol, fatty acid, and glucose metabolism and function as cholesterol sensors.[302–304] Several oxidized forms of cholesterol (oxysterols), but not cholesterol itself, have been reported as endogenous ligands for LXRs.[31,305,306] 22(*R*)-hydroxycholesterol and 24(*S*), 25-epoxycholesterol (**Figure 16**) activate LXRs at their physiological concentrations. LXRs, as heterodimers with RXR, regulate the expression of various genes involved in cholesterol homeostasis, including *CYP7A1*, *CETP* (cholesterol ester transfer protein)[307] and most importantly, *ABCA1*.[187,308] The ATP-binding cassette transporter protein (ABCA1) is required for the first step in RCT of cholesterol from macrophage foam cells to nascent HDL particles, and plays a critical role in decreasing atherosclerotic plaques. Synthetic nonsteroidal LXR agonists T0901317[187] and GW3965[309] (**Figure 16**) show significant atheroprotective effects and increased plasma HDL-c levels in mouse models.[310,311] Synthetic LXR agonists also regulate the expression of genes involved in glucose metabolism and show antidiabetic activity in animal models of Type 2 diabetes.[312,313] However, LXR also activates several genes associated with lipogenesis, particularly SREBP-1c, which is the master regulator of de novo lipogenesis[314] and fatty acid synthase. In addition, hypertriglyceridemia is another significant liability. These effects compromise the use of LXR agonists as a treatment for atherosclerosis. There is considerable effort underway to discover LXR modulators that are devoid of lipogenic activity, and to generate subtype selective LXR ligands.[315]

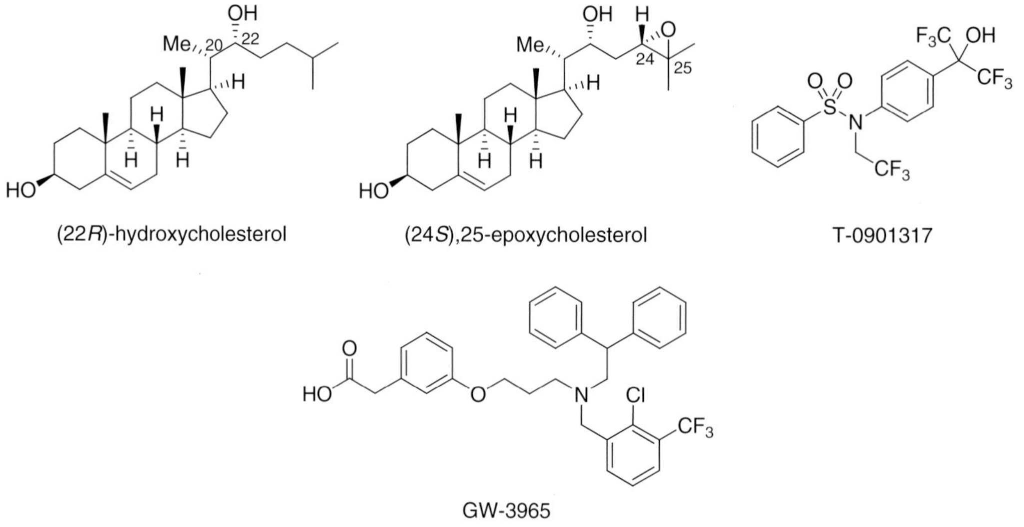

Figure 16 Liver X receptor (LXR) ligands.

2.25.6.11 Farnesoid X Receptor Ligands

FXR is also commonly known as the bile acid receptor (BAR) since bile acids were shown to be the physiologically relevant endogenous ligands for this receptor.[316–318] The bile acid chenodeoxycholic acid (CDCA) (**Figure 17**) has the highest affinity among bile acids at 10–30 μM, while lithocholic acid, deoxycholic acid, and glycine–taurine conjugates also bind at physiological concentrations. Ursodeoxycholic acid and cholic acid do not bind to FXR. Bile acids are end products of cholesterol catabolism and are important for the intestinal absorption of dietary fat and vitamins. FXR is important for maintaining bile acid homeostasis by modulation of genes involved in bile acid synthesis as well as transport (*see* Section 2.25.5.2.6). In response to bile acids, FXR represses the transcription of *CYP7A1*, the liver-specific cholesterol 7α-hydroxylase, which catalyzes the first rate-limiting step in cholesterol metabolism,[168] while inducing the transcription of bile acid transporter proteins.[319] The first synthetic FXR agonist, GW 4064 (**Figure 17**) was reported by GlaxoSmithKline to be an activator (nanomolar range) of FXR.[320] The same group later reported a series of potent FXR ligands based on CDCA, of which 6α-ethyl-CDCA is the most potent FXR agonist reported thus far[321] and has demonstrated reversal of cholestasis in an in vivo rat model.[322] The GlaxoSmithKline group have now reported a similar series of potent FXR ligands based on the CDCA platform, but with a range of functional activities, from full agonism to antagonism, which should be useful for evaluating the therapeutic benefit of FXR ligands in metabolic diseases.[323] Dussault *et al.*[324] also reported a series of FXR ligands based on an RXR ligand scaffold. They reported a series of agonists (AGN29 and AGN31) as well as an antagonist (AGN34) (**Figure 17**) that appeared to be a selective FXR gene modulator. AGN34 like an agonist repressed expression of CYP7A1, but potently antagonized the expression of bile acid transport protein IBABP. This selective regulation of genes related to cholesterol and bile acid homeostasis suggests that selective FXR ligands may be developed to selectively enhance cholesterol metabolism, without deleterious effects such as cholestasis or hypertriglyceridemia.[325]

2.25.6.12 Pregnane X Receptor and Constitutive Androstane Receptor Ligands

The orphan NRs CAR and PXR (and its human homolog SXR, steroid and xenobiotic receptor) function as xenobiotic sensors and regulate the expression of various Phase I and II drug-metabolizing enzymes as well as drug-transporter proteins, in response to xenobiotics.[326,327] As discussed in Section 2.25.3.1.1, the LBDs of these NRs are relatively large and can accommodate a diverse series of structurally unrelated xenobiotics. However, these NRs demonstrate species specificity in their induction of drug-metabolizing enzymes. Rifampicin (**Figure 18**) is an agonist for SXR and induces CYP3A4 in humans but not in rodents. Pregnenolone carbonitrile (PCN) (**Figure 18**), however, is an agonist for the rodent ortholog PXR and induces CYP3A in rodent assays but not in human hepatocytes.[328] CYP3A4 metabolizes a large number of pharmaceutical drugs and xenobiotic activation of PXR by one drug can lead to inadvertent increase in metabolism of another administered drug, causing a deleterious drug–drug interaction. PXR also transcriptionally regulates transporter proteins such as OATP2, MDR1 (multidrug resistance 1) and MRP2 (MDR-associated protein). Taxol was shown to induce MDR1 via PXR activation.

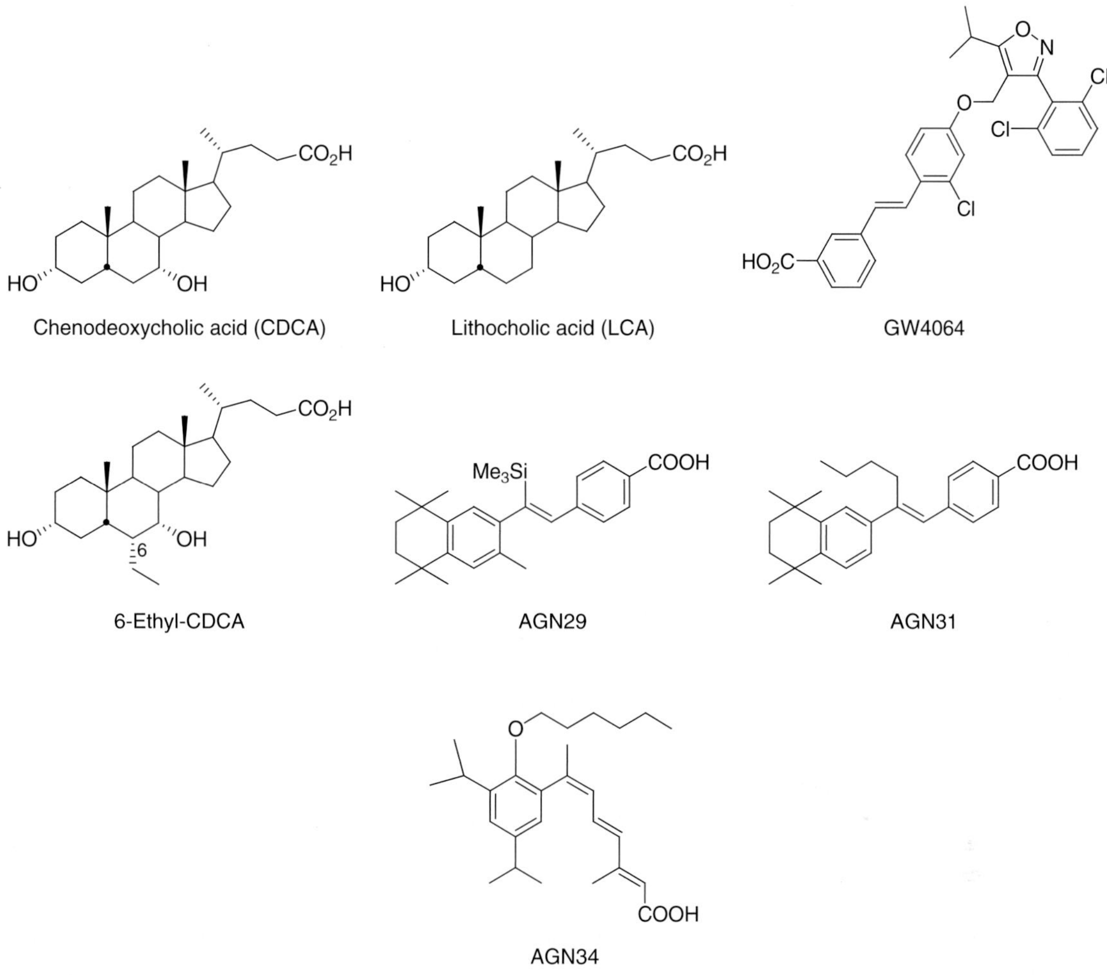

Figure 17 Farnesoid X receptor ligands.

Binding to CAR induces the CYP2B class of cytochrome P450 enzymes. Although CAR is constitutively active and is able to transactivate genes without a ligand, its activity is potentiated by CAR agonists such as phenobarbital and 1,4-bis[2-(3,5-dichloropyridoxyl)]benzene (TCPOBOP)[329] (**Figure 18**) and is inhibited by antagonists such as androstanols.[330] Recently, guggulsterone, a PXR and FXR ligand, and an active ingredient in the herbal Ayurvedic cholesterol-lowering remedy, guggulipid, was found to be an inverse agonist at CAR, and repressed the activation of CYP2B enzymes.[331] The recent discovery that CAR is also involved in thyroid metabolism suggests a potential role for CAR inverse agonists as antiobesity drugs.[332,333]

2.25.7 Therapeutic Disease Applications for Nuclear Receptor Ligands

The discussion in Section 2.25.6 clearly shows that drugs that target NRs offer therapies for a multitude of diseases and disorders, such as cancer, metabolic diseases (including diabetes, obesity, and cardiovascular pathologies), inflammation, and osteoporosis. Several NRs are established targets for clinical therapeutics, whereas other NRs and orphan receptors offer the possibility of new therapeutic targets. Among established targets, the steroid hormone receptors ERα and AR and the retinoid receptors RARβ and RXR are known targets for cancer therapeutics.[151,334,335] PPARs are known targets for metabolic diseases such as hyperlipidemia and Type 2 diabetes,[301] GR and MR are targets for anti-inflammatory therapies,[336] VDR is a target for osteoporosis and psoriasis, while ER and PR are targets for oral contraceptives and HRT.

One area that is of tremendous interest and has a high potential for generating the next blockbuster drug targeting an NR is the so-called metabolic syndrome. Metabolic syndrome is a collection of disorders typified by obesity, Type 2

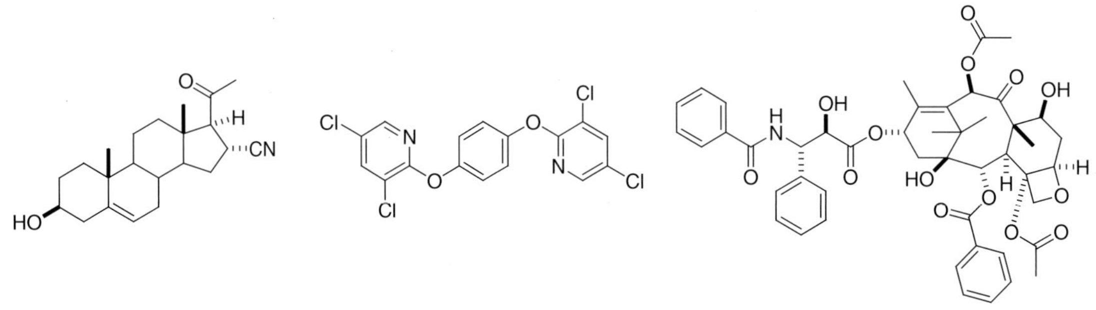

Figure 18 Pregnane X receptor and constitutive androstane receptor ligands.

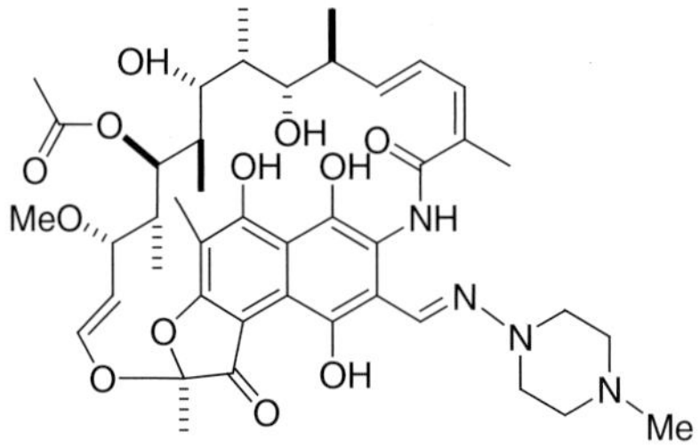

diabetes, hypertension, hyperlipidemia, and atherosclerosis, which tend to occur in the same predisposed population. It appears that dysregulation of lipid metabolism and homeostasis is the primary imbalance that leads to the metabolic syndrome.[337] Accordingly, NRs that are involved in maintaining lipid homeostasis are potential targets for diseases associated with the metabolic syndrome. A number of in-depth reviews discuss the involvement of several NRs, including LXR, FXR, PPARs, and LRH-1 in metabolic diseases. Some of these NRs are being actively pursued as targets for selective therapeutic modulation for the metabolic syndrome.[325,338–340]

2.25.8 Genetic Diseases Associated with Nuclear Receptors

NRs contribute to a broad spectrum of physiological processes, and it is therefore not surprising that genetic alteration of any individual NR could result in a disease state. This section outlines the pathological conditions ascribed to documented genetic alterations in a number of NRs.

The ERs are by far the most extensively studied NRs, and many research efforts have focused on identifying alterations within the coding sequences of both ERα and ERβ from clinical samples.[341] A number of groups have identified naturally occurring splice variants and mutations with constitutive activity, even in the absence of estrogen. ERα splice variants have been detected in various tumor types, including breast cancer, endometrial carcinomas, prolactinoma, and meningiomas (for a more extensive review see [341]). The ERα splice variant with a deleted exon 7 (ERα exon 7 deletion (Δ7)) is identified as the most frequently observed variant in breast cancer.[342] This variant encodes a protein lacking the AF-2 domain and a portion of the hormone-binding domain. It is not transcriptionally active, but it may act as dominant-negative isoform of both ERα and ERβ to inhibit expression.[343] ERαΔ7 has not only been found in breast tumors, but also meningiomas, endometrial hyperplasias, moderate- to well-differentiated endometrial adenocarcinomas, prostate cancer cell lines, and lupus erythematosus.[341] Although current studies suggest that the dominant effect of ERα Δ7 is not a significant cause of breast cancer, it nevertheless contributes to disease progression.

Compared with ER splice variants, only a handful of ER mutations have been identified or studied from human samples. However, some of the known mutants have been implicated in a variety of disease states, including breast and endometrial cancers, systemic lupus erythematosus, and psychiatric diseases.[341] There appear to be multiple hot spots for spontaneous ER mutations in the relatively short hinge region of the ER LBD. One of these mutations, ERα K303R (lysine to arginine substitution), is believed to lead to a receptor that is able to induce proliferation even when exposed to extremely low levels of hormone; the mutation is found with relatively high frequency in primary breast cancers.[344]

It has also been suggested that ER mutation frequency may be higher in metastatic breast tumors, and some ER mutations have been correlated with tamoxifen resistance and estrogen independence.[345,346] These mutations have become the focus of intense research that has yielded valuable insights into potential mechanisms of hormone-independent signaling. Other ERα and ERβ polymorphisms are also believed to contribute to osteoporosis, but these data await further confirmation.[347]

Not surprisingly, much less is known about ERβ splice variants and natural point mutations. ERβΔ5, like ERαΔ7, acts as a dominant-negative isoform of both ERs, and it is believed to contribute to tumor progression. Although very little is known about ERβ point mutations, a G250S mutation has been identified in obese male children and other silent mutations have been observed in both underweight and obese children.[348]

A number of gain of function mutations in the AR are known to contribute to prostate cancer.[349] For example, T877A (threonine to alanine switch) and H874Y (histidine to tyrosine) allow transcriptional activation of the AR by the adrenal androgen dehydroepiandrosterone (DHEA) which does not normally cause substantial activation of wild-type AR. Although both mutations occur within the helix 11 region of the LBP, amino acid residue 877 contacts the ligand directly while residue 874 is distal to the pocket and may affect binding of coactivator proteins. The T877A mutation alters the stereochemistry of the binding domain and broadens the specificity of ligand recognition.[350] The H874Y mutation also probably affects ligand binding, but indirectly. In support of these findings, it has been demonstrated that the AR-T877A mutant is activated by a broad range of other ligands including estrogen, progestins, pregnenolone (precursor of many steroids), and some antiandrogens (e.g., cyproterone and hydroxyflutamide).[351] The T877A mutation has been detected in a number of human prostate tumors.[349] Other mutations have been found in prostate cancer cell lines, and all point to a mechanism involving prostate tumor cells surviving in the absence (or at low levels) of androgens.

Amplification of the AR gene is another level of genetic control of AR that can result in prostate cancer. This possibility is highlighted by studies showing amplified AR in 30% of recurrent tumors.[349] Finally, genetic control over AR dysfunction appears to involve an N-terminal domain harboring from 8 to 39 glutamine residues. A decrease in glutamine-repeat length is associated with an increase in AR transactivation and a higher risk and a higher grade of prostate tumors.[349,352] In summary, these studies show that alterations in AR protein structure probably promote the development and progression of prostate tumors.

The glucocorticoid-resistant syndromes are another well-established class of NR-related genetic diseases.[353] Glucocorticoid resistance is a rare, familial or sporadic condition that is characterized by partial end-organ insensitivity to glucocorticoids. Affected patients exhibit compensatory elevations in circulating cortisol and ACTH concentrations. The clinical spectrum of the condition is broad, ranging from completely asymptomatic to severe hyperandrogenism or mineralocorticoid excess.[354] Fatigue, anxiety, and hypertension are also associated with this condition, with fatigue considered to be a potential manifestation of muscular or CNS glucocorticoid deficiency.[355] The molecular underpinnings for glucocorticoid-resistant syndromes involve mutations in the *GRα* gene that impair one or more of the molecular mechanisms of GR function (including reduced transactivation of GRα and therefore reduced target gene transcription and expression) and result in altered tissue sensitivity to glucocorticoids. A number of mutations within the LBD and DBD of the GR protein as well as 4-base pair deletion at the 3′-boundary of exon 6, have been described.[354-356] A polymorphism of GRα that is associated with obesity rather than generalized glucocorticoid resistance has also been reported.[356] Some of these polymorphisms have been shown to affect the interaction of GRα with the GR-interacting protein 1 coactivator.[354] Another recent study suggested that high levels of GRβ, which acts as a dominant-negative regulator of GRα, were found in some cases of Crohn's disease and may be caused by interleukin (IL)-18-dependent inappropriate mRNA splicing of the GR gene.[357] ILs are also believed to be involved in the development of glucocorticoid resistance in multiple sclerosis, and this disease appears to have a genetic component in some patients.[358]

Perhaps the best known example of a NR genetic event resulting in a significant disease is translocation of RARα in APL.[359] APL is a common form of acute myeloid leukemia (AML) characterized by attenuation of myeloid differentiation and accumulation of immature promyelocytes in patient bone marrow and peripheral blood. RARα is found fused to different partners, which often encode transcription factors or other chromatin modifying enzymes. These chimeric fusion oncogenes, generated by reciprocal chromosomal translocations (involving the RARα gene on chromosome 17), are responsible for chromatin alterations on target genes whose protein products are critical to stem cell development or lineage specification in hematopoiesis. Five different partners of RARα have been identified, including the promyelocyte gene (*PML*), the promyelocyte leukemia zinc finger gene (*PLZF*), the nucleophosmin gene (*NPM*), the nuclear mitotic apparatus gene (*NuMA*), and the signal transducer and activator of transcription 5b gene (*Stat5*). All translocations are reciprocal and generate X-RARα and RARα-X fusion proteins (X = RARα partner genes). APL presents as X-RARα chimeras. Unlike other types of AML, APL responds to RA treatment, which causes remissions highlighted by cellular differentiation of leukemic blasts.[360]

The final example of an NR-associated disease involves polymorphism of a PPARγ isoform that offers a protective advantage to the organism. PPARγ1 and PPARγ2 isoforms of PPARγ result from alternative splicing and are known to have both ligand-dependent and -independent activation domains. Insulin stimulates the ligand-independent activation of both proteins while obesity and nutritional factors only influence PPARγ2.[361] A relatively common amino acid substitution (Pro12Ala) of PPARγ2 was found to reduce body mass index and improve insulin sensitivity among middle-aged and elderly people.[361] Not surprisingly, subsequent studies found that this P12A PPARγ2 polymorphism protects against Type 2 diabetes.[362] Since the study of PPARs, and even the entire NR family, is still in its infancy, many more polymorphisms (and spice variants) of NR genes are likely to be identified and linked to other genetic diseases.

2.25.9 Perspectives and New Directions in Nuclear Receptor Research

It is clear that NRs are a rich source of therapeutic targets for the development of drugs against many important human diseases. With several oNRs waiting to be 'adopted' and understood, this class of receptors will continue to provide new therapeutic opportunities. However, monumental challenges need to be overcome to complete our understanding of NR pharmacology. With the advent of novel technologies such as selective genetic knockouts, gene microarrays and RNA interference (RNAi), we have begun to unravel the complexities of NR physiology and regulation of function in a tissue- and cell-specific context. Ideally, this knowledge can be utilized to discover new tissue-selective ligands to modulate NR function in disease processes. In this modern era of molecularly targeted and tailored medicine, the immense complexity of NR activation and function also makes the task of drug discovery more arduous. The trend toward developing SNRMs makes it important not only to characterize binding and functional activity of new ligands for a particular NR of interest, but to also examine gene expression profiles in response to ligands as functions of tissue and cellular contexts. Such data will be essential in understanding the tissue-selective action of the NR ligand, and allow educated judgments to be made regarding the pharmacological profile and side effects of each ligand in detail.

Another aspect of NR pharmacology that was not considered in great detail in this chapter is the nongenomic actions of NRs. In addition to directly modulating gene transcription (by binding to DNA response elements), many Class I and II NRs are now known to interact with other molecules in the signal transduction network either in the cytoplasm or by direct interaction with plasma membrane NR isoforms. This type of membrane-bound NR signaling therefore indirectly affects transcription of a number of mostly unidentified genes, adding yet another dimension to NRs as targets for drug discovery.

Finally, the many combinatorial coregulator complexes of ligand-bound NRs offer other sites that can be targeted by small-molecule or peptide ligands. Ligands that interfere with protein–protein interactions, while not a new concept per se, are being explored for disrupting NR–coactivator interactions.[363–365] While such ligands have mainly been small peptides, these new target sites provide the opportunity to develop nonpeptidic chemical scaffolds that can mimic the α-helical binding surfaces of NR coactivators to disrupt NR function. Given the impact of NRs on the pathophysiology of so many human diseases, the above approaches will likely lead to the development of selective pharmaceuticals that could revolutionize modern molecular medicine.

References

1. Evans, R. M. *Mol. Endocrinol.* **2005**, *19*, 1429–1438.
2. Chambon, P. *Mol. Endocrinol.* **2005**, *19*, 1418–1428.
3. Mangelsdorf, D. J.; Thummel, C.; Beato, M.; Herrlich, P.; Schutz, G.; Umesono, K.; Blumberg, B.; Kastner, P.; Mark, M.; Chambon, P.; Evans, R. M. *Cell* **1995**, *83*, 835–839.
4. Robinson-Rechavi, M.; Carpentier, A. S.; Duffraisse, M.; Laudet, V. *Trends Genet.* **2001**, *17*, 554–556.
5. Maglich, J. M.; Sluder, A.; Guan, X.; Shi, Y.; McKee, D. D.; Carrick, K.; Kamdar, K.; Willson, T. M.; Moore, J. T. *Genome Biol.* **2001**, *2*, research 0029.1–0029.7.
6. Enmark, E.; Gustafsson, J. A. *Trends Pharmacol. Sci.* **2001**, *22*, 611–615.
7. Laudet, V. *J. Mol. Endocrinol.* **1997**, *19*, 207–226.
8. Kliewer, S. A.; Lehmann, J. M.; Willson, T. M. *Science* **1999**, *284*, 757–760.
9. Nuclear Receptors Committee. *Cell* **1999**, *97*, 1–20.
10. Wrange, O.; Gustafsson, J. A. *J. Biol. Chem.* **1978**, *253*, 856–865.
11. Mader, S.; Kumar, V.; de Verneuil, H.; Chambon, P. *Nature* **1989**, *338*, 271–274.
12. Bourguet, W.; Germain, P.; Gronemeyer, H. *Trends Pharmacol. Sci.* **2000**, *21*, 381–388.
13. Greschik, H.; Moras, D. *Curr. Top. Med. Chem.* **2003**, *3*, 1573–1599.
14. Gronemeyer, H.; Gustafsson, J. A.; Laudet, V. *Nat. Rev. Drug Disc.* **2004**, *3*, 950–964.
15. Benoit, G.; Malewicz, M.; Perlmann, T. *Trends Cell Biol.* **2004**, *14*, 369–376.
16. Hard, T.; Kellenbach, E.; Boelens, R.; Maler, B. A.; Dahlman, K.; Freedman, L. P.; Carlstedt-Duke, J.; Yamamoto, K. R.; Gustafsson, J. A.; Kaptein, R. *Science* **1990**, *249*, 157–160.
17. Steinmetz, A. C.; Renaud, J. P.; Moras, D. *Annu. Rev. Biophys. Biomol. Struct.* **2001**, *30*, 329–359.

18. Renaud, J. P.; Moras, D. *Cell. Mol. Life Sci.* **2000**, *57*, 1748–1769.
19. Wurtz, J. M.; Bourguet, W.; Renaud, J. P.; Vivat, V.; Chambon, P.; Moras, D.; Gronemeyer, H. *Nat. Struct. Biol.* **1996**, *3*, 206.
20. Feng, W.; Ribeiro, R. C.; Wagner, R. L.; Nguyen, H.; Apriletti, J. W.; Fletterick, R. J.; Baxter, J. D.; Kushner, P. J.; West, B. L. *Science* **1998**, *280*, 1747–1749.
21. Xu, H. E.; Stanley, T. B.; Montana, V. G.; Lambert, M. H.; Shearer, B. G.; Cobb, J. E.; McKee, D. D.; Galardi, C. M.; Plunket, K. D.; Nolte, R. T. et al. *Nature* **2002**, *415*, 813–817.
22. Brzozowski, A. M.; Pike, A. C.; Dauter, Z.; Hubbard, R. E.; Bonn, T.; Engstrom, O.; Ohman, L.; Greene, G. L.; Gustafsson, J. A.; Carlquist, M. *Nature* **1997**, *389*, 753–758.
23. Shiau, A. K.; Barstad, D.; Loria, P. M.; Cheng, L.; Kushner, P. J.; Agard, D. A.; Greene, G. L. *Cell* **1998**, *95*, 927–937.
24. Bourguet, W.; Ruff, M.; Chambon, P.; Gronemeyer, H.; Moras, D. *Nature* **1995**, *375*, 377–382.
25. Renaud, J. P.; Rochel, N.; Ruff, M.; Vivat, V.; Chambon, P.; Gronemeyer, H.; Moras, D. *Nature* **1995**, *378*, 681–689.
26. Nolte, R. T.; Wisely, G. B.; Westin, S.; Cobb, J. E.; Lambert, M. H.; Kurokawa, R.; Rosenfeld, M. G.; Willson, T. M.; Glass, C. K.; Milburn, M. V. *Nature* **1998**, *395*, 137–143.
27. Watkins, R. E.; Wisely, G. B.; Moore, L. B.; Collins, J. L.; Lambert, M. H.; Williams, S. P.; Willson, T. M.; Kliewer, S. A.; Redinbo, M. R. *Science* **2001**, *292*, 2329–2333.
28. Xu, H. E.; Lambert, M. H.; Montana, V. G.; Plunket, K. D.; Moore, L. B.; Collins, J. L.; Oplinger, J. A.; Kliewer, S. A.; Gampe, R. T., Jr.; McKee, D. D. et al. *Proc. Natl. Acad. Sci. USA* **2001**, *98*, 13919–13924.
29. Xu, H. E.; Lambert, M. H.; Montana, V. G.; Parks, D. J.; Blanchard, S. G.; Brown, P. J.; Sternbach, D. D.; Lehmann, J. M.; Wisely, G. B.; Willson, T. M. et al. *Mol. Cell* **1999**, *3*, 397–403.
30. Lehmann, J. M.; Moore, L. B.; Smith-Oliver, T. A.; Wilkison, W. O.; Willson, T. M.; Kliewer, S. A. *J. Biol. Chem.* **1995**, *270*, 12953–12956.
31. Janowski, B. A.; Grogan, M. J.; Jones, S. A.; Wisely, G. B.; Kliewer, S. A.; Corey, E. J.; Mangelsdorf, D. J. *Proc. Natl. Acad. Sci. USA* **1999**, *96*, 266–271.
32. Francis, G. A.; Fayard, E.; Picard, F.; Auwerx, J. *Annu. Rev. Physiol.* **2003**, *65*, 261–311.
33. Lehmann, J. M.; McKee, D. D.; Watson, M. A.; Willson, T. M.; Moore, J. T.; Kliewer, S. A. *J. Clin. Invest.* **1998**, *102*, 1016–1023.
34. Wisely, G. B.; Miller, A. B.; Davis, R. G.; Thornquest, A. D., Jr.; Johnson, R.; Spitzer, T.; Sefler, A.; Shearer, B.; Moore, J. T.; Willson, T. M. et al. *Structure (Camb.)* **2002**, *10*, 1225–1234.
35. Viollet, B.; Kahn, A.; Raymondjean, M. *Mol. Cell. Biol.* **1997**, *17*, 4208–4219.
36. Stehlin, C.; Wurtz, J. M.; Steinmetz, A.; Greiner, E.; Schule, R.; Moras, D.; Renaud, J. P. *EMBO J.* **2001**, *20*, 5822–5831.
37. Wallen, A.; Perlmann, T. *Ann. NY Acad. Sci.* **2003**, *991*, 48–60.
38. Wang, Z.; Benoit, G.; Liu, J.; Prasad, S.; Aarnisalo, P.; Liu, X.; Xu, H.; Walker, N. P.; Perlmann, T. *Nature* **2003**, *423*, 555–560.
39. Barettino, D.; Vivanco Ruiz, M. M.; Stunnenberg, H. G. *EMBO J.* **1994**, *13*, 3039–3049.
40. Danielian, P. S.; White, R.; Lees, J. A.; Parker, M. G. *EMBO J.* **1992**, *11*, 1025–1033.
41. Heery, D. M.; Kalkhoven, E.; Hoare, S.; Parker, M. G. *Nature* **1997**, *387*, 733–736.
42. Darimont, B. D.; Wagner, R. L.; Apriletti, J. W.; Stallcup, M. R.; Kushner, P. J.; Baxter, J. D.; Fletterick, R. J.; Yamamoto, K. R. *Genes Dev.* **1998**, *12*, 3343–3356.
43. Hu, X.; Lazar, M. A. *Nature* **1999**, *402*, 93–96.
44. Webb, P.; Anderson, C. M.; Valentine, C.; Nguyen, P.; Marimuthu, A.; West, B. L.; Baxter, J. D.; Kushner, P. J. *Mol. Endocrinol.* **2000**, *14*, 1976–1985.
45. Bledsoe, R. K.; Montana, V. G.; Stanley, T. B.; Delves, C. J.; Apolito, C. J.; McKee, D. D.; Consler, T. G.; Parks, D. J.; Stewart, E. L.; Willson, T. M. et al. *Cell* **2002**, *110*, 93–105.
46. Benko, S.; Love, J. D.; Beladi, M.; Schwabe, J. W.; Nagy, L. *J. Biol. Chem.* **2003**, *278*, 43797–43806.
47. Perissi, V.; Aggarwal, A.; Glass, C. K.; Rose, D. W.; Rosenfeld, M. G. *Cell* **2004**, *116*, 511–526.
48. Pratt, W. B.; Toft, D. O. *Endocr. Rev.* **1997**, *18*, 306–360.
49. Kauppi, B.; Jakob, C.; Farnegardh, M.; Yang, J.; Ahola, H.; Alarcon, M.; Calles, K.; Engstrom, O.; Harlan, J.; Muchmore, S. et al. *J. Biol. Chem.* **2003**, *278*, 22748–22754.
50. Bourguet, W.; Vivat, V.; Wurtz, J. M.; Chambon, P.; Gronemeyer, H.; Moras, D. *Mol. Cell* **2000**, *5*, 289–298.
51. Bledsoe, R. K.; Madauss, K. P.; Holt, J. A.; Apolito, C. J.; Lambert, M. H.; Pearce, K. H.; Stanley, T. B.; Stewart, E. L.; Trump, R. P.; Willson, T. M. et al. *J. Biol. Chem.* **2005**, *280*, 31283–31293.
52. Shiau, A. K.; Barstad, D.; Radek, J. T.; Meyers, M. J.; Nettles, K. W.; Katzenellenbogen, B. S.; Katzenellenbogen, J. A.; Agard, D. A.; Greene, G. L. *Nat. Struct. Biol.* **2002**, *9*, 359–364.
53. Pike, A. C.; Brzozowski, A. M.; Hubbard, R. E.; Bonn, T.; Thorsell, A. G.; Engstrom, O.; Ljunggren, J.; Gustafsson, J. A.; Carlquist, M. *EMBO J.* **1999**, *18*, 4608–4618.
54. Oberfield, J. L.; Collins, J. L.; Holmes, C. P.; Goreham, D. M.; Cooper, J. P.; Cobb, J. E.; Lenhard, J. M.; Hull-Ryde, E. A.; Mohr, C. P.; Blanchard, S. G. et al. *Proc. Natl. Acad. Sci. USA* **1999**, *96*, 6102–6106.
55. Williams, S. P.; Sigler, P. B. *Nature* **1998**, *393*, 392–396.
56. Bourguet, W.; Andry, V.; Iltis, C.; Klaholz, B.; Potier, N.; Van Dorsselaer, A.; Chambon, P.; Gronemeyer, H.; Moras, D. *Protein Expres. Purif.* **2000**, *19*, 284–288.
57. Gampe, R. T., Jr.; Montana, V. G.; Lambert, M. H.; Miller, A. B.; Bledsoe, R. K.; Milburn, M. V.; Kliewer, S. A.; Willson, T. M.; Xu, H. E. *Mol. Cell* **2000**, *5*, 545–555.
58. Warnmark, A.; Treuter, E.; Wright, A. P.; Gustafsson, J. A. *Mol. Endocrinol.* **2003**, *17*, 1901–1909.
59. He, B.; Lee, L. W.; Minges, J. T.; Wilson, E. M. *J. Biol. Chem.* **2002**, *277*, 25631–25639.
60. Schwabe, J. W.; Neuhaus, D.; Rhodes, D. *Nature* **1990**, *348*, 458–461.
61. Rochette-Egly, C. *J. Biol. Chem.* **2005**, *280*, 32565–32568.
62. Metivier, R.; Penot, G.; Hubner, M. R.; Reid, G.; Brand, H.; Kos, M.; Gannon, F. *Cell* **2003**, *115*, 751–763.
63. Glass, C. K.; Rosenfeld, M. G. *Genes Dev.* **2000**, *14*, 121–141.
64. Kumar, R.; Gururaj, A. E.; Vadlamudi, R. K.; Rayala, S. K. *Clin. Cancer Res.* **2005**, *11*, 2822–2831.
65. Dennis, A. P.; O'Malley, B. W. *J. Steroid Biochem. Mol. Biol.* **2005**, *93*, 139–151.
66. Baek, S. H.; Rosenfeld, M. G. *Biochem. Biophys. Res. Commun.* **2004**, *319*, 707–714.
67. Yoon, H. G.; Chan, D. W.; Reynolds, A. B.; Qin, J.; Wong, J. *Mol. Cell* **2003**, *12*, 723–734.
68. Tsai, C. C.; Fondell, J. D. *Vitam. Horm.* **2004**, *68*, 93–122.

69. Fernandes, I.; Bastien, Y.; Wai, T.; Nygard, K.; Lin, R.; Cormier, O.; Lee, H. S.; Eng, F.; Bertos, N. R.; Pelletier, N. et al. *Mol. Cell* **2003**, *11*, 139–150.
70. Dennis, A. P.; Lonard, D. M.; Nawaz, Z.; O'Malley, B. W. *J. Steroid Biochem. Mol. Biol.* **2005**, *94*, 337–346.
71. Xiao, T.; Kao, C. F.; Krogan, N. J.; Sun, Z. W.; Greenblatt, J. F.; Osley, M. A.; Strahl, B. D. *Mol. Cell. Biol.* **2005**, *25*, 637–651.
72. Kinyamu, H. K.; Chen, J.; Archer, T. K. *J. Mol. Endocrinol.* **2005**, *34*, 281–297.
73. Chen, H.; Lin, R. J.; Xie, W.; Wilpitz, D.; Evans, R. M. *Cell* **1999**, *98*, 675–686.
74. Berger, S. L. *Curr. Opin. Genet. Dev.* **2002**, *12*, 142–148.
75. Daujat, S.; Bauer, U. M.; Shah, V.; Turner, B.; Berger, S.; Kouzarides, T. *Curr. Biol.* **2002**, *12*, 2090–2097.
76. Mahajan, M. A.; Samuels, H. H. *Endocr. Rev.* **2005**, *26*, 583–597.
77. Wu, R. C.; Smith, C. L.; O'Malley, B. W. *Endocr. Rev.* **2005**, *26*, 393–399.
78. Giglia-Mari, G.; Coin, F.; Ranish, J. A.; Hoogstraten, D.; Theil, A.; Wijgers, N.; Jaspers, N. G.; Raams, A.; Argentini, M.; van der Spek, P. J. et al. *Nat. Genet.* **2004**, *36*, 714–719.
79. Rochette-Egly, C. *Cell Signal.* **2003**, *15*, 355–366.
80. Bruck, N.; Bastien, J.; Bour, G.; Tarrade, A.; Plassat, J. L.; Bauer, A.; Adam-Stitah, S.; Rochette-Egly, C. *Cell Signal.* **2005**, *17*, 1229–1239.
81. Stirone, C.; Duckles, S. P.; Krause, D. N.; Procaccio, V. *Mol. Pharmacol.* **2005**, *68*, 959–965.
82. Willson, T. M.; Moore, J. T. *Mol. Endocrinol.* **2002**, *16*, 1135–1144.
83. Mohan, R.; Heyman, R. A. *Curr. Top. Med. Chem.* **2003**, *3*, 1637–1647.
84. Hewitt, S. C.; Harrell, J. C.; Korach, K. S. *Annu. Rev. Physiol.* **2005**, *67*, 285–308.
85. Imamov, O.; Lopatkin, N. A.; Gustafsson, J. A. *N. Engl. J. Med.* **2004**, *351*, 2773–2774.
86. Bai, C.; Schmidt, A.; Freedman, L. P. *Assay Drug Dev. Technol.* **2003**, *1*, 843–852.
87. O'Lone, R.; Frith, M. C.; Karlsson, E. K.; Hansen, U. *Mol. Endocrinol.* **2004**, *18*, 1859–1875.
88. Bjornstrom, L.; Sjoberg, M. *Mol. Endocrinol.* **2005**, *19*, 833–842.
89. Revankar, C. M.; Cimino, D. F.; Sklar, L. A.; Arterburn, J. B.; Prossnitz, E. R. *Science* **2005**, *307*, 1625–1630.
90. Chambliss, K. L.; Shaul, P. W. *Endocr. Rev.* **2002**, *23*, 665–686.
91. Chambliss, K. L.; Shaul, P. W. *Steroids* **2002**, *67*, 413–419.
92. Chambliss, K. L.; Yuhanna, I. S.; Anderson, R. G.; Mendelsohn, M. E.; Shaul, P. W. *Mol. Endocrinol.* **2002**, *16*, 938–946.
93. Pawlak, J.; Karolczak, M.; Krust, A.; Chambon, P.; Beyer, C. *Glia* **2005**, *50*, 270–275.
94. Yang, S. H.; Liu, R.; Perez, E. J.; Wen, Y.; Stevens, S. M., Jr.; Valencia, T.; Brun-Zinkernagel, A. M.; Prokai, L.; Will, Y.; Dykens, J. et al. *Proc. Natl. Acad. Sci. USA* **2004**, *101*, 4130–4135.
95. Hewitt, S. C.; Korach, K. S. *Reproduction* **2003**, *125*, 143–149.
96. Shah, B. H. *Trends Endocrinol. Metab.* **2005**, *16*, 199–201.
97. Massion, P. B.; Pelat, M.; Belge, C.; Balligand, J. L. *Comp. Biochem. Physiol. A Mol. Integr. Physiol.* **2005**, *142*, 144–150.
98. Schulz, R.; Rassaf, T.; Massion, P. B.; Kelm, M.; Balligand, J. L. *Pharmacol. Ther.* **2005**, *108*, 225–256.
99. Pelat, M.; Massion, P. B.; Balligand, J. L. *Arch. Mal. Coeur. Vaiss.* **2005**, *98*, 242–248.
100. Booth, E. A.; Obeid, N. R.; Lucchesi, B. R. *Am. J. Physiol. Heart Circ. Physiol.* **2005**, *289*, H2039–H2047.
101. Paranjape, S. G.; Turankar, A. V.; Wakode, S. L.; Dakhale, G. N. *Med. Hypotheses* **2005**, *65*, 725–727.
102. Parikka, V.; Peng, Z.; Hentunen, T.; Risteli, J.; Elo, T.; Vaananen, H. K.; Harkonen, P. *Eur. J. Endocrinol.* **2005**, *152*, 301–314.
103. Green, P. S.; Simpkins, J. W. *Int. J. Dev. Neurosci.* **2000**, *18*, 347–358.
104. Lee, S. J.; McEwen, B. S. *Annu. Rev. Pharmacol. Toxicol.* **2001**, *41*, 569–591.
105. Dykens, J. A.; Simpkins, J. W.; Wang, J.; Gordon, K. *Exp. Gerontol.* **2003**, *38*, 101–107.
106. Simpkins, J. W.; Wang, J.; Wang, X.; Perez, E.; Prokai, L.; Dykens, J. A. *Curr. Drug Targets CNS Neurol. Disord.* **2005**, *4*, 69–83.
107. Simpkins, J. W.; Wen, Y.; Perez, E.; Yang, S.; Wang, X. *Ann. NY Acad. Sci.* **2005**, *1052*, 233–242.
108. Dykens, J. A.; Moos, W. H.; Howell, N. *Ann. NY Acad. Sci.* **2005**, *1052*, 116–135.
109. Koehler, K. F.; Helguero, L. A.; Haldosen, L. A.; Warner, M.; Gustafsson, J. A. *Endocr. Rev.* **2005**, *26*, 465–478.
110. Jayachandran, M.; Karnicki, K.; Miller, R. S.; Owen, W. G.; Korach, K. S.; Miller, V. M. *J Gerontol. A Biol. Sci. Med. Sci.* **2005**, *60*, 815–819.
111. Imamov, O.; Shim, G. J.; Warner, M.; Gustafsson, J. A. *Biol. Reprod.* **2005**, *73*, 866–871.
112. Clarke, C. L.; Sutherland, R. L. *Endocr. Rev.* **1990**, *11*, 266–301.
113. Kastner, P.; Krust, A.; Turcotte, B.; Stropp, U.; Tora, L.; Gronemeyer, H.; Chambon, P. *EMBO J.* **1990**, *9*, 1603–1614.
114. Edwards, D. P. *Annu. Rev. Physiol.* **2005**, *67*, 335–376.
115. Fernandez-Valdivia, R.; Mukherjee, A.; Mulac-Jericevic, B.; Conneely, O. M.; DeMayo, F. J.; Amato, P.; Lydon, J. P. *Semin. Reprod. Med.* **2005**, *23*, 22–37.
116. Conneely, O. M.; Mulac-Jericevic, B.; Lydon, J. P. *Steroids* **2003**, *68*, 771–778.
117. Shao, R.; Markstrom, E.; Friberg, P. A.; Johansson, M.; Billig, H. *Biol. Reprod.* **2003**, *68*, 914–921.
118. Mooradian, A. D.; Morley, J. E.; Korenman, S. G. *Endocr. Rev.* **1987**, *8*, 1–28.
119. Wilson, J. D. *Endocr. Rev.* **1999**, *20*, 726–737.
120. Ikeda, Y.; Aihara, K.; Sato, T.; Akaike, M.; Yoshizumi, M.; Suzaki, Y.; Izawa, Y.; Fujimura, M.; Hashizume, S.; Kato, M. et al. *J. Biol. Chem.* **2005**, *280*, 29661–29666.
121. Isidori, A. M.; Giannetta, E.; Greco, E. A.; Gianfrilli, D.; Bonifacio, V.; Isidori, A.; Lenzi, A.; Fabbri, A. *Clin. Endocrinol. (Oxf)* **2005**, *63*, 280–293.
122. Matsumoto, T.; Takeyama, K.; Sato, T.; Kato, S. *J. Steroid Biochem. Mol. Biol.* **2003**, *85*, 95–99.
123. Charmandari, E.; Chrousos, G. P.; Ichijo, T.; Bhattacharyya, N.; Vottero, A.; Souvatzoglou, E.; Kino, T. *Mol. Endocrinol.* **2005**, *19*, 52–64.
124. De Bosscher, K.; Vanden Berghe, W.; Haegeman, G. *Endocr. Rev.* **2003**, *24*, 488–522.
125. Liu, Y.; Nakagawa, Y.; Wang, Y.; Sakurai, R.; Tripathi, P. V.; Lutfy, K.; Friedman, T. C. *Diabetes* **2005**, *54*, 32–40.
126. Schacke, H.; Schottelius, A.; Docke, W. D.; Strehlke, P.; Jaroch, S.; Schmees, N.; Rehwinkel, H.; Hennekes, H.; Asadullah, K. *Proc. Natl. Acad. Sci. USA* **2004**, *101*, 227–232.
127. Bauer, A.; Tronche, F.; Wessely, O.; Kellendonk, C.; Reichardt, H. M.; Steinlein, P.; Schutz, G.; Beug, H. *Genes Dev.* **1999**, *13*, 2996–3002.
128. Moraes, L.; Paul-Clark, M. J.; Rickman, A.; Flower, R. J.; Goulding, N. J.; Perretti, M. *Blood* **2005**, *106*, 4167–4175.
129. Galigniana, M. D.; Piwien Pilipuk, G.; Kanelakis, K. C.; Burton, G.; Lantos, C. P. *Mol. Cell. Endocrinol.* **2004**, *217*, 167–179.
130. Kitagawa, H.; Yanagisawa, J.; Fuse, H.; Ogawa, S.; Yogiashi, Y.; Okuno, A.; Nagasawa, H.; Nakajima, T.; Matsumoto, T.; Kato, S. *Mol. Cell. Biol.* **2002**, *22*, 3698–3706.
131. Dusso, A. S.; Brown, A. J.; Slatopolsky, E. *Am. J. Physiol. Renal Physiol.* **2005**, *289*, F8–F28.

132. Gascon-Barre, M.; Demers, C.; Mirshahi, A.; Neron, S.; Zalzal, S.; Nanci, A. *Hepatology* **2003**, *37*, 1034–1042.
133. Makishima, M.; Lu, T. T.; Xie, W.; Whitfield, G. K.; Domoto, H.; Evans, R. M.; Haussler, M. R.; Mangelsdorf, D. J. *Science* **2002**, *296*, 1313–1316.
134. Buchwald, P. C.; Westin, G.; Akerstrom, G. *Int. J. Mol. Med.* **2005**, *15*, 701–706.
135. Altucci, L.; Gronemeyer, H. *Trends Endocrinol. Metab.* **2001**, *12*, 460–468.
136. Mangelsdorf, D. J.; Evans, R. M. *Cell* **1995**, *83*, 841–850.
137. Lefebvre, P.; Martin, P. J.; Flajollet, S.; Dedieu, S.; Billaut, X.; Lefebvre, B. *Vitam. Horm.* **2005**, *70*, 199–264.
138. Zile, M. H. *J. Nutr.* **2001**, *131*, 705–708.
139. Niederreither, K.; Subbarayan, V.; Dolle, P.; Chambon, P. *Nat. Genet.* **1999**, *21*, 444–448.
140. Vermot, J.; Gallego Llamas, J.; Fraulob, V.; Niederreither, K.; Chambon, P.; Dolle, P. *Science* **2005**, *308*, 563–566.
141. Gale, E.; Zile, M.; Maden, M. *Mech. Dev.* **1999**, *89*, 43–54.
142. Stratford, T.; Logan, C.; Zile, M.; Maden, M. *Mech. Dev.* **1999**, *81*, 115–125.
143. Ross, S. A.; McCaffery, P. J.; Drager, U. C.; De Luca, L. M. *Physiol. Rev.* **2000**, *80*, 1021–1054.
144. Maden, M.; Hind, M. *Dev. Dyn.* **2003**, *226*, 237–244.
145. Nakajoh, M.; Fukushima, T.; Suzuki, T.; Yamaya, M.; Nakayama, K.; Sekizawa, K.; Sasaki, H. *Am. J. Respir. Cell. Mol. Biol.* **2003**, *28*, 296–304.
146. Schmitz, G.; Langmann, T. *Biochim. Biophys. Acta* **2005**, *1735*, 1–19.
147. Lanni, A.; Moreno, M.; Lombardi, A.; de Lange, P.; Silvestri, E.; Ragni, M.; Farina, P.; Baccari, G. C.; Fallahi, P.; Antonelli, A. et al. *FASEB J.* **2005**, *19*, 1552–1554.
148. Goglia, F. *Biochemistry (Mosc.)* **2005**, *70*, 164–172.
149. Kohrle, J. *Z. Arztl. Fortbild Qualitatssich* **2004**, *98*, 17–24.
150. Malm, J. *Curr. Pharm. Des.* **2004**, *10*, 3525–3532.
151. Shulman, A. I.; Mangelsdorf, D. J. *N. Engl. J. Med.* **2005**, *353*, 604–615.
152. O'Shea, P. J.; Bassett, J. H.; Sriskantharajah, S.; Ying, H.; Cheng, S. Y.; Williams, G. R. *Mol. Endocrinol.* **2005**, *19*, 3045–3059.
153. Calza, L.; Fernandez, M.; Giuliani, A.; D'Intino, G.; Pirondi, S.; Sivilia, S.; Paradisi, M.; Desordi, N.; Giardino, L. *Brain Res. Rev.* **2005**, *48*, 339–346.
154. Takada, I.; Kato, S. *Nippon Rinsho.* **2005**, *63*, 573–577.
155. Tan, N. S.; Michalik, L.; Desvergne, B.; Wahli, W. *J. Steroid. Biochem. Mol. Biol.* **2005**, *93*, 99–105.
156. Clarke, S. D.; Thuillier, P.; Baillie, R. A.; Sha, X. *Am. J. Clin. Nutr.* **1999**, *70*, 566–571.
157. Genolet, R.; Wahli, W.; Michalik, L. *Curr. Drug Targets Inflamm. Allergy* **2004**, *3*, 361–375.
158. Fujii, H. *Nippon Rinsho.* **2005**, *63*, 609–615.
159. Lee, C. H.; Chawla, A.; Urbizondo, N.; Liao, D.; Boisvert, W. A.; Evans, R. M.; Curtiss, L. K. *Science* **2003**, *302*, 453–457.
160. Lehrke, M.; Pascual, G.; Glass, C. K.; Lazar, M. A. *Genes Dev.* **2005**, *19*, 1737–1742.
161. Fujii, H. *Nippon Rinsho.* **2005**, *63*, 565–571.
162. Michalik, L.; Desvergne, B.; Wahli, W. *Nat. Rev. Cancer* **2004**, *4*, 61–70.
163. Lee, C. H.; Olson, P.; Evans, R. M. *Endocrinology* **2003**, *144*, 2201–2207.
164. Lee, S. S.; Pineau, T.; Drago, J.; Lee, E. J.; Owens, J. W.; Kroetz, D. L.; Fernandez-Salguero, P. M.; Westphal, H.; Gonzalez, F. J. *Mol. Cell. Biol.* **1995**, *15*, 3012–3022.
165. Ravnskjaer, K.; Boergesen, M.; Rubi, B.; Larsen, J. K.; Nielsen, T.; Fridriksson, J.; Maechler, P.; Mandrup, S. *Endocrinology* **2005**, *146*, 3266–3276.
166. Akiyama, T. E.; Sakai, S.; Lambert, G.; Nicol, C. J.; Matsusue, K.; Pimprale, S.; Lee, Y. H.; Ricote, M.; Glass, C. K.; Brewer, H. B., Jr. et al. *Mol. Cell. Biol.* **2002**, *22*, 2607–2619.
167. Chinetti, G.; Lestavel, S.; Bocher, V.; Remaley, A. T.; Neve, B.; Torra, I. P.; Teissier, E.; Minnich, A.; Jaye, M.; Duverger, N. et al. *Nat. Med.* **2001**, *7*, 53–58.
168. Chawla, A.; Boisvert, W. A.; Lee, C. H.; Laffitte, B. A.; Barak, Y.; Joseph, S. B.; Liao, D.; Nagy, L.; Edwards, P. A.; Curtiss, L. K. et al. *Mol. Cell* **2001**, *7*, 161–171.
169. Michalik, L.; Feige, J. N.; Gelman, L.; Pedrazzini, T.; Keller, H.; Desvergne, B.; Wahli, W. *Mol. Endocrinol.* **2005**, *19*, 2335–2348.
170. Cui, Y.; Miyoshi, K.; Claudio, E.; Siebenlist, U. K.; Gonzalez, F. J.; Flaws, J.; Wagner, K. U.; Hennighausen, L. *J. Biol. Chem.* **2002**, *277*, 17830–17835.
171. Desvergne, B.; Michalik, L.; Wahli, W. *Mol. Endocrinol.* **2004**, *18*, 1321–1332.
172. Lee, C. H.; Evans, R. M. *Trends Endocrinol. Metab.* **2002**, *13*, 331–335.
173. Yu, S.; Matsusue, K.; Kashireddy, P.; Cao, W. Q.; Yeldandi, V.; Yeldandi, A. V.; Rao, M. S.; Gonzalez, F. J.; Reddy, J. K. *J. Biol. Chem.* **2003**, *278*, 498–505.
174. Chui, P. C.; Guan, H. P.; Lehrke, M.; Lazar, M. A. *J. Clin. Invest.* **2005**, *115*, 2244–2256.
175. Hevener, A. L.; He, W.; Barak, Y.; Le, J.; Bandyopadhyay, G.; Olson, P.; Wilkes, J.; Evans, R. M.; Olefsky, J. *Nat. Med.* **2003**, *9*, 1491–1497.
176. d'Abramo, C.; Massone, S.; Zingg, J. M.; Pizzuti, A.; Marambaud, P.; Dalla Piccola, B.; Azzi, A.; Marinari, U. M.; Pronzato, M. A.; Ricciarelli, R. *Biochem. J.* **2005**, *391* (Pt 3), 693–698.
177. Bedu, E.; Wahli, W.; Desvergne, B. *Expert Opin. Ther. Targets* **2005**, *9*, 861–873.
178. Shi, Y.; Hon, M.; Evans, R. M. *Proc. Natl. Acad. Sci. USA* **2002**, *99*, 2613–2618.
179. Kersten, S.; Wahli, W. *EXS* **2000**, *89*, 141–151.
180. Basu-Modak, S.; Braissant, O.; Escher, P.; Desvergne, B.; Honegger, P.; Wahli, W. *J. Biol. Chem.* **1999**, *274*, 35881–35888.
181. Tan, N. S.; Michalik, L.; Desvergne, B.; Wahli, W. *Am. J. Clin. Dermatol.* **2003**, *4*, 523–530.
182. Wang, Y. X.; Lee, C. H.; Tiep, S.; Yu, R. T.; Ham, J.; Kang, H.; Evans, R. M. *Cell* **2003**, *113*, 159–170.
183. Chawla, A.; Lee, C. H.; Barak, Y.; He, W.; Rosenfeld, J.; Liao, D.; Han, J.; Kang, H.; Evans, R. M. *Proc. Natl. Acad. Sci. USA* **2003**, *100*, 1268–1273.
184. Oliver, W. R., Jr.; Shenk, J. L.; Snaith, M. R.; Russell, C. S.; Plunket, K. D.; Bodkin, N. L.; Lewis, M. C.; Winegar, D. A.; Sznaidman, M. L.; Lambert, M. H. et al. *Proc. Natl. Acad. Sci. USA* **2001**, *98*, 5306–5311.
185. Repa, J. J.; Mangelsdorf, D. J. *Nat. Med.* **2002**, *8*, 1243–1248.
186. Crestani, M.; De Fabiani, E.; Caruso, D.; Mitro, N.; Gilardi, F.; Vigil Chacon, A. B.; Patelli, R.; Godio, C.; Galli, G. *Biochem. Soc. Trans.* **2004**, *32*, 92–96.

187. Repa, J. J.; Liang, G.; Ou, J.; Bashmakov, Y.; Lobaccaro, J. M.; Shimomura, I.; Shan, B.; Brown, M. S.; Goldstein, J. L.; Mangelsdorf, D. J. *Genes Dev.* **2000**, *14*, 2819–2830.
188. Joseph, S. B.; Castrillo, A.; Laffitte, B. A.; Mangelsdorf, D. J.; Tontonoz, P. *Nat. Med.* **2003**, *9*, 213–219.
189. Handschin, C.; Meyer, U. A. *Arch. Biochem. Biophys.* **2005**, *433*, 387–396.
190. Ory, D. S. *Circ. Res.* **2004**, *95*, 660–670.
191. Jordan, V. C. *J. Med. Chem.* **2003**, *46*, 883–908.
192. Jordan, V. C. *J. Med. Chem.* **2003**, *46*, 1081–1111.
193. Wise, P. M.; Dubal, D. B.; Rau, S. W.; Brown, C. M.; Suzuki, S. *Endocr. Rev.* **2005**, *26*, 308–312.
194. Anderson, G. L.; Limacher, M.; Assaf, A. R.; Bassford, T.; Beresford, S. A.; Black, H.; Bonds, D.; Brunner, R.; Brzyski, R.; Caan, B. et al. *JAMA* **2004**, *291*, 1701–1712.
195. Brunner, R. L.; Gass, M.; Aragaki, A.; Hays, J.; Granek, I.; Woods, N.; Mason, E.; Brzyski, R.; Ockene, J.; Assaf, A. et al. *Arch. Intern. Med.* **2005**, *165*, 1976–1986.
196. Espeland, M. A.; Rapp, S. R.; Shumaker, S. A.; Brunner, R.; Manson, J. E.; Sherwin, B. B.; Hsia, J.; Margolis, K. L.; Hogan, P. E.; Wallace, R. et al. *JAMA* **2004**, *291*, 2959–2968.
197. Craig, M. C.; Maki, P. M.; Murphy, D. G. *Lancet Neurol.* **2005**, *4*, 190–194.
198. Rossouw, J. E.; Anderson, G. L.; Prentice, R. L.; LaCroix, A. Z.; Kooperberg, C.; Stefanick, M. L.; Jackson, R. D.; Beresford, S. A.; Howard, B. V.; Johnson, K. C. et al. *JAMA* **2002**, *288*, 321–333.
199. Jordan, V. C.; Phelps, E.; Lindgren, J. U. *Breast Cancer Res. Treat.* **1987**, *10*, 31–35.
200. Howell, S. J.; Johnston, S. R.; Howell, A. *Best Pract. Res. Clin. Endocrinol. Metab.* **2004**, *18*, 47–66.
201. Ke, H. Z.; Qi, H.; Crawford, D. T.; Chidsey-Frink, K. L.; Simmons, H. A.; Thompson, D. D. *Endocrinology* **2000**, *141*, 1338–1344.
202. Suh, N.; Glasebrook, A. L.; Palkowitz, A. D.; Bryant, H. U.; Burris, L. L.; Starling, J. J.; Pearce, H. L.; Williams, C.; Peer, C.; Wang, Y. et al. *Cancer Res.* **2001**, *61*, 8412–8415.
203. Gruber, C.; Gruber, D. *Curr. Opin. Investig. Drugs* **2004**, *5*, 1086–1093.
204. Buzdar, A. U. *Clin. Cancer Res.* **2005**, *11*, 906s–908s.
205. Parker, M. G. *Breast Cancer Res. Treat.* **1993**, *26*, 131–137.
206. Pink, J. J.; Jordan, V. C. *Cancer Res.* **1996**, *56*, 2321–2330.
207. Hall, J. M.; McDonnell, D. P. *Endocrinology* **1999**, *140*, 5566–5578.
208. Wallace, O. B.; Richardson, T. I.; Dodge, J. A. *Curr. Top. Med. Chem.* **2003**, *3*, 1663–1682.
209. Malamas, M. S.; Manas, E. S.; McDevitt, R. E.; Gunawan, I.; Xu, Z. B.; Collini, M. D.; Miller, C. P.; Dinh, T.; Henderson, R. A.; Keith, J. C., Jr. et al. *J. Med. Chem.* **2004**, *47*, 5021–5040.
210. Harris, H. A.; Albert, L. M.; Leathurby, Y.; Malamas, M. S.; Mewshaw, R. E.; Miller, C. P.; Kharode, Y. P.; Marzolf, J.; Komm, B. S.; Winneker, R. C. et al. *Endocrinology* **2003**, *144*, 4241–4249.
211. Chwalisz, K.; Perez, M. C.; Demanno, D.; Winkel, C.; Schubert, G.; Elger, W. *Endocr. Rev.* **2005**, *26*, 423–438.
212. Schubert, G.; Elger, W.; Kaufmann, G.; Schneider, B.; Reddersen, G.; Chwalisz, K. *Semin. Reprod. Med.* **2005**, *23*, 58–73.
213. Chwalisz, K.; Elger, W.; Stickler, T.; Mattia-Goldberg, C.; Larsen, L. *Hum. Reprod.* **2005**, *20*, 1090–1099.
214. Negro-Vilar, A. *J. Clin. Endocrinol. Metab.* **1999**, *84*, 3459–3462.
215. Chen, J.; Hwang, D. J.; Bohl, C. E.; Miller, D. D.; Dalton, J. T. *J. Pharmacol. Exp. Ther.* **2005**, *312*, 546–553.
216. Yin, D.; Gao, W.; Kearbey, J. D.; Xu, H.; Chung, K.; He, Y.; Marhefka, C. A.; Veverka, K. A.; Miller, D. D.; Dalton, J. T. *J. Pharmacol. Exp. Ther.* **2003**, *304*, 1334–1340.
217. Gao, W.; Kearbey, J. D.; Nair, V. A.; Chung, K.; Parlow, A. F.; Miller, D. D.; Dalton, J. T. *Endocrinology* **2004**, *145*, 5420–5428.
218. Gao, W.; Bohl, C. E.; Dalton, J. T. *Chem. Rev.* **2005**, *105*, 3352–3370.
219. Heinlein, C. A.; Chang, C. *Endocr. Rev.* **2004**, *25*, 276–308.
220. Stanbury, R. M.; Graham, E. M. *Br. J. Ophthalmol.* **1998**, *82*, 704–708.
221. Reichardt, H. M.; Tuckermann, J. P.; Gottlicher, M.; Vujic, M.; Weih, F.; Angel, P.; Herrlich, P.; Schutz, G. *EMBO J.* **2001**, *20*, 7168–7173.
222. Rosen, J.; Miner, J. N. *Endocr. Rev.* **2005**, *26*, 452–464.
223. Coghlan, M. J.; Jacobson, P. B.; Lane, B.; Nakane, M.; Lin, C. W.; Elmore, S. W.; Kym, P. R.; Luly, J. R.; Carter, G. W.; Turner, R. et al. *Mol. Endocrinol.* **2003**, *17*, 860–869.
224. von Geldern, T. W.; Tu, N.; Kym, P. R.; Link, J. T.; Jae, H. S.; Lai, C.; Apelqvist, T.; Rhonnstad, P.; Hagberg, L.; Koehler, K. et al. *J. Med. Chem.* **2004**, *47*, 4213–4230.
225. Jacobson, P. B.; von Geldern, T. W.; Ohman, L.; Osterland, M.; Wang, J.; Zinker, B.; Wilcox, D.; Nguyen, P. T.; Mika, A.; Fung, S. et al. *J. Pharmacol. Exp. Ther.* **2005**, *314*, 191–200.
226. Takeda, Y. *Hypertens. Res.* **2004**, *27*, 781–789.
227. Fraccarollo, D.; Galuppo, P.; Bauersachs, J. *Curr. Med. Chem. Cardiovasc. Hematol. Agents* **2004**, *2*, 287–294.
228. Barnes, B. J.; Howard, P. A. *Ann. Pharmacother.* **2005**, *39*, 68–76.
229. Thohan, V.; Torre-Amione, G.; Koerner, M. M. *Curr. Opin. Cardiol.* **2004**, *19*, 301–308.
230. Yee, Y. K.; Chintalacharuvu, S. R.; Lu, J.; Nagpal, S. *Mini-Rev. Med. Chem.* **2005**, *5*, 761–778.
231. Nagpal, S.; Na, S.; Rathnachalam, R. *Endocr. Rev.* **2005**, *26*, 662–687.
232. Toell, A.; Gonzalez, M. M.; Ruf, D.; Steinmeyer, A.; Ishizuka, S.; Carlberg, C. *Mol. Pharmacol.* **2001**, *59*, 1478–1485.
233. Kim, S.; Shevde, N. K.; Pike, J. W. *J. Bone Miner. Res.* **2005**, *20*, 305–317.
234. Ishizuka, S.; Kurihara, N.; Miura, D.; Takenouchi, K.; Cornish, J.; Cundy, T.; Reddy, S. V.; Roodman, G. D. *J. Steroid Biochem. Mol. Biol.* **2004**, *89–90*, 331–334.
235. Lazar, M. A. *Endocr. Rev.* **1993**, *14*, 184–193.
236. Forrest, D.; Vennstrom, B. *Thyroid* **2000**, *10*, 41–52.
237. O'Shea, P. J.; Williams, G. R. *J. Endocrinol.* **2002**, *175*, 553–570.
238. Yoshihara, H. A.; Scanlan, T. S. *Curr. Top. Med. Chem.* **2003**, *3*, 1601–1616.
239. Webb, P. *Expert Opin. Investig. Drugs* **2004**, *13*, 489–500.
240. Kraiem, Z. *Thyroid* **2005**, *15*, 336–339.
241. Grover, G. J.; Egan, D. M.; Sleph, P. G.; Beehler, B. C.; Chiellini, G.; Nguyen, N. H.; Baxter, J. D.; Scanlan, T. S. *Endocrinology* **2004**, *145*, 1656–1661.
242. Chiellini, G.; Apriletti, J. W.; Yoshihara, H. A.; Baxter, J. D.; Ribeiro, R. C.; Scanlan, T. S. *Chem. Biol.* **1998**, *5*, 299–306.

243. Trost, S. U.; Swanson, E.; Gloss, B.; Wang-Iverson, D. B.; Zhang, H.; Volodarsky, T.; Grover, G. J.; Baxter, J. D.; Chiellini, G.; Scanlan, T. S. et al. *Endocrinology* **2000**, *141*, 3057–3064.
244. Freitas, F. R.; Capelo, L. P.; O'Shea, P. J.; Jorgetti, V.; Moriscot, A. S.; Scanlan, T. S.; Williams, G. R.; Zorn, T. M.; Gouveia, C. H. *J. Bone Miner. Res.* **2005**, *20*, 294–304.
245. Ye, L.; Li, Y. L.; Mellstrom, K.; Mellin, C.; Bladh, L. G.; Koehler, K.; Garg, N.; Garcia Collazo, A. M.; Litten, C.; Husman, B. et al. *J. Med. Chem.* **2003**, *46*, 1580–1588.
246. Grover, G. J.; Mellstrom, K.; Ye, L.; Malm, J.; Li, Y. L.; Bladh, L. G.; Sleph, P. G.; Smith, M. A.; George, R.; Vennstrom, B. et al. *Proc. Natl. Acad. Sci. USA* **2003**, *100*, 10067–10072.
247. Morkin, E.; Ladenson, P.; Goldman, S.; Adamson, C. *J. Mol. Cell. Cardiol.* **2004**, *37*, 1137–1146.
248. Forini, F.; Nicolini, G.; Balzan, S.; Ratto, G. M.; Murzi, B.; Vanini, V.; Iervasi, G. *Thyroid* **2004**, *14*, 493–499.
249. Van Beeren, H. C.; Jong, W. M.; Kaptein, E.; Visser, T. J.; Bakker, O.; Wiersinga, W. M. *Endocrinology* **2003**, *144*, 552–558.
250. Webb, P.; Nguyen, N. H.; Chiellini, G.; Yoshihara, H. A.; Cunha Lima, S. T.; Apriletti, J. W.; Ribeiro, R. C.; Marimuthu, A.; West, B. L.; Goede, P. et al. *J. Steroid Biochem. Mol. Biol.* **2002**, *83*, 59–73.
251. Lim, W.; Nguyen, N. H.; Yang, H. Y.; Scanlan, T. S.; Furlow, J. D. *J. Biol. Chem.* **2002**, *277*, 35664–35670.
252. Warrell, R. P., Jr.; de The, H.; Wang, Z. Y.; Degos, L. *N. Engl. J. Med.* **1993**, *329*, 177–189.
253. Widschwendter, M.; Berger, J.; Hermann, M.; Muller, H. M.; Amberger, A.; Zeschnigk, M.; Widschwendter, A.; Abendstein, B.; Zeimet, A. G.; Daxenbichler, G. et al. *J. Natl. Cancer Inst.* **2000**, *92*, 826–832.
254. Sirchia, S. M.; Ferguson, A. T.; Sironi, E.; Subramanyan, S.; Orlandi, R.; Sukumar, S.; Sacchi, N. *Oncogene* **2000**, *19*, 1556–1563.
255. Liu, Y.; Lee, M. O.; Wang, H. G.; Li, Y.; Hashimoto, Y.; Klaus, M.; Reed, J. C.; Zhang, X. *Mol. Cell. Biol.* **1996**, *16*, 1138–1149.
256. Li, Y.; Dawson, M. I.; Agadir, A.; Lee, M. O.; Jong, L.; Hobbs, P. D.; Zhang, X. K. *Int. J. Cancer* **1998**, *75*, 88–95.
257. Ohno, R.; Asou, N.; Ohnishi, K. *Leukemia* **2003**, *17*, 1454–1463.
258. Takeshita, A.; Shibata, Y.; Shinjo, K.; Yanagi, M.; Tobita, T.; Ohnishi, K.; Miyawaki, S.; Shudo, K.; Ohno, R. *Ann. Intern. Med.* **1996**, *124*, 893–896.
259. Takeuchi, M.; Yano, T.; Omoto, E.; Takahashi, K.; Kibata, M.; Shudo, K.; Harada, M.; Ueda, R.; Ohno, R. *Leuk. Lymphoma* **1998**, *31*, 441–451.
260. Bodsworth, N. J.; Bloch, M.; Bower, M.; Donnell, D.; Yocum, R. *Am. J. Clin. Dermatol.* **2001**, *2*, 77–87.
261. Chandraratna, R. A. *Br. J. Dermatol.* **1996**, *135*, 18–25.
262. Thacher, S. M.; Vasudevan, J.; Chandraratna, R. A. *Curr. Pharm. Des.* **2000**, *6*, 25–58.
263. Dawson, M. I. *Curr. Med. Chem. Anti-Canc. Agents* **2004**, *4*, 199–230.
264. Hede, K. *J. Natl. Cancer Inst.* **2004**, *96*, 1807–1808.
265. Duvic, M.; Hymes, K.; Heald, P.; Breneman, D.; Martin, A. G.; Myskowski, P.; Crowley, C.; Yocum, R. C. *J. Clin. Oncol.* **2001**, *19*, 2456–2471.
266. Boehm, M. F.; Zhang, L.; Zhi, L.; McClurg, M. R.; Berger, E.; Wagoner, M.; Mais, D. E.; Suto, C. M.; Davies, J. A.; Heyman, R. A. et al. *J. Med. Chem.* **1995**, *38*, 3146–3155.
267. Forman, B. M.; Umesono, K.; Chen, J.; Evans, R. M. *Cell* **1995**, *81*, 541–550.
268. Claudel, T.; Leibowitz, M. D.; Fievet, C.; Tailleux, A.; Wagner, B.; Repa, J. J.; Torpier, G.; Lobaccaro, J. M.; Paterniti, J. R.; Mangelsdorf, D. J. et al. *Proc. Natl. Acad. Sci. USA* **2001**, *98*, 2610–2615.
269. Cesario, R. M.; Klausing, K.; Razzaghi, H.; Crombie, D.; Rungta, D.; Heyman, R. A.; Lala, D. S. *Mol. Endocrinol.* **2001**, *15*, 1360–1369.
270. Lala, D. S.; Mukherjee, R.; Schulman, I. G.; Koch, S. S.; Dardashti, L. J.; Nadzan, A. M.; Croston, G. E.; Evans, R. M.; Heyman, R. A. *Nature* **1996**, *383*, 450–453.
271. Michellys, P. Y.; Ardecky, R. J.; Chen, J. H.; Crombie, D. L.; Etgen, G. J.; Faul, M. M.; Faulkner, A. L.; Grese, T. A.; Heyman, R. A.; Karanewsky, D. S. et al. *J. Med. Chem.* **2003**, *46*, 2683–2696.
272. Michellys, P. Y.; Ardecky, R. J.; Chen, J. H.; D'Arrigo, J.; Grese, T. A.; Karanewsky, D. S.; Leibowitz, M. D.; Liu, S.; Mais, D. A.; Mapes, C. M. et al. *J. Med. Chem.* **2003**, *46*, 4087–4103.
273. Fu, J.; Gaetani, S.; Oveisi, F.; Lo Verme, J.; Serrano, A.; Rodriguez De Fonseca, F.; Rosengarth, A.; Luecke, H.; Di Giacomo, B.; Tarzia, G. et al. *Nature* **2003**, *425*, 90–93.
274. Issemann, I.; Green, S. *Nature* **1990**, *347*, 645–650.
275. Forman, B. M.; Chen, J.; Evans, R. M. *Proc. Natl. Acad. Sci. USA* **1997**, *94*, 4312–4317.
276. Berger, J.; Moller, D. E. *Annu. Rev. Med.* **2002**, *53*, 409–435.
277. Israelian-Konaraki, Z.; Reaven, P. D. *Cardiology* **2005**, *103*, 1–9.
278. Rubins, H. B.; Robins, S. J.; Collins, D.; Nelson, D. B.; Elam, M. B.; Schaefer, E. J.; Faas, F. H.; Anderson, J. W. *Arch. Intern. Med.* **2002**, *162*, 2597–2604.
279. Brown, P. J.; Winegar, D. A.; Plunket, K. D.; Moore, L. B.; Lewis, M. C.; Wilson, J. G.; Sundseth, S. S.; Koble, C. S.; Wu, Z.; Chapman, J. M. et al. *J. Med. Chem.* **1999**, *42*, 3785–3788.
280. Xu, Y.; Mayhugh, D.; Saeed, A.; Wang, X.; Thompson, R. C.; Dominianni, S. J.; Kauffman, R. F.; Singh, J.; Bean, J. S.; Bensch, W. R. et al. *J. Med. Chem.* **2003**, *46*, 5121–5124.
281. Shi, G. Q.; Dropinski, J. F.; Zhang, Y.; Santini, C.; Sahoo, S. P.; Berger, J. P.; Macnaul, K. L.; Zhou, G.; Agrawal, A.; Alvaro, R. et al. *J. Med. Chem.* **2005**, *48*, 5589–5599.
282. Rzonca, S. O.; Suva, L. J.; Gaddy, D.; Montague, D. C.; Lecka-Czernik, B. *Endocrinology* **2004**, *145*, 401–406.
283. Soroceanu, M. A.; Miao, D.; Bai, X. Y.; Su, H.; Goltzman, D.; Karaplis, A. C. *J. Endocrinol.* **2004**, *183*, 203–216.
284. Lecka-Czernik, B.; Moerman, E. J.; Grant, D. F.; Lehmann, J. M.; Manolagas, S. C.; Jilka, R. L. *Endocrinology* **2002**, *143*, 2376–2384.
285. Lazarenko, O. P.; Rzonca, S. O.; Suva, L. J.; Lecka-Czernik, B. *Bone* **2005**, *38*, 74–84.
286. Howard, B. V.; Howard, W. J. *Endocr. Rev.* **1994**, *15*, 263–274.
287. Hannah, J. S.; Howard, B. V. *J. Cardiovasc. Risk* **1994**, *1*, 31–37.
288. Henke, B. R.; Blanchard, S. G.; Brackeen, M. F.; Brown, K. K.; Cobb, J. E.; Collins, J. L.; Harrington, W. W., Jr.; Hashim, M. A.; Hull-Ryde, E. A.; Kaldor, I. et al. *J. Med. Chem.* **1998**, *41*, 5020–5036.
289. Lohray, B. B.; Lohray, V. B.; Bajji, A. C.; Kalchar, S.; Poondra, R. R.; Padakanti, S.; Chakrabarti, R.; Vikramadithyan, R. K.; Misra, P.; Juluri, S. et al. *J. Med. Chem.* **2001**, *44*, 2675–2678.
290. Ebdrup, S.; Pettersson, I.; Rasmussen, H. B.; Deussen, H. J.; Frost Jensen, A.; Mortensen, S. B.; Fleckner, J.; Pridal, L.; Nygaard, L.; Sauerberg, P. *J. Med. Chem.* **2003**, *46*, 1306–1317.

291. Devasthale, P. V.; Chen, S.; Jeon, Y.; Qu, F.; Shao, C.; Wang, W.; Zhang, H.; Cap, M.; Farrelly, D.; Golla, R. et al. *J. Med. Chem.* **2005**, *48*, 2248–2250.
292. Nissen, S. E.; Wolski, K.; Topol, E. J. *JAMA* **2005**, *294*, 2581–2586.
293. Bristol Myers Squibb Press Release. October 27, 2005. http://www.bms.com/news/press/data/fig_press_release_5965.html. (access date June 28, 2006).
294. Liu, K.; Xu, L.; Berger, J. P.; Macnaul, K. L.; Zhou, G.; Doebber, T. W.; Forrest, M. J.; Moller, D. E.; Jones, A. B. *J. Med. Chem.* **2005**, *48*, 2262–2265.
295. Xu, Y.; Rito, C. J.; Etgen, G. J.; Ardecky, R. J.; Bean, J. S.; Bensch, W. R.; Bosley, J. R.; Broderick, C. L.; Brooks, D. A.; Dominianni, S. J. et al. *J. Med. Chem.* **2004**, *47*, 2422–2425.
296. Shi, G. Q.; Dropinski, J. F.; McKeever, B. M.; Xu, S.; Becker, J. W.; Berger, J. P.; MacNaul, K. L.; Elbrecht, A.; Zhou, G.; Doebber, T. W. et al. *J. Med. Chem.* **2005**, *48*, 4457–4468.
297. Brown, P. J.; Smith-Oliver, T. A.; Charifson, P. S.; Tomkinson, N. C.; Fivush, A. M.; Sternbach, D. D.; Wade, L. E.; Orband-Miller, L.; Parks, D. J.; Blanchard, S. G. et al. *Chem. Biol.* **1997**, *4*, 909–918.
298. Brooks, C. D.; Summers, J. B. *J. Med. Chem.* **1996**, *39*, 2629–2654.
299. Tanaka, T.; Yamamoto, J.; Iwasaki, S.; Asaba, H.; Hamura, H.; Ikeda, Y.; Watanabe, M.; Magoori, K.; Ioka, R. X.; Tachibana, K. et al. *Proc. Natl. Acad. Sci. USA* **2003**, *100*, 15924–15929.
300. Luquet, S.; Gaudel, C.; Holst, D.; Lopez-Soriano, J.; Jehl-Pietri, C.; Fredenrich, A.; Grimaldi, P. A. *Biochim. Biophys. Acta* **2005**, *1740*, 313–317.
301. Berger, J. P.; Akiyama, T. E.; Meinke, P. T. *Trends Pharmacol. Sci.* **2005**, *26*, 244–251.
302. Peet, D. J.; Janowski, B. A.; Mangelsdorf, D. J. *Curr. Opin. Genet. Dev.* **1998**, *8*, 571–575.
303. Peet, D. J.; Turley, S. D.; Ma, W.; Janowski, B. A.; Lobaccaro, J. M.; Hammer, R. E.; Mangelsdorf, D. J. *Cell* **1998**, *93*, 693–704.
304. Tobin, K. A.; Ulven, S. M.; Schuster, G. U.; Steineger, H. H.; Andresen, S. M.; Gustafsson, J. A.; Nebb, H. I. *J. Biol. Chem.* **2002**, *277*, 10691–10697.
305. Lehmann, J. M.; Kliewer, S. A.; Moore, L. B.; Smith-Oliver, T. A.; Oliver, B. B.; Su, J. L.; Sundseth, S. S.; Winegar, D. A.; Blanchard, D. E.; Spencer, T. A. et al. *J. Biol. Chem.* **1997**, *272*, 3137–3140.
306. Forman, B. M.; Ruan, B.; Chen, J.; Schroepfer, G. J., Jr.; Evans, R. M. *Proc. Natl. Acad. Sci. USA* **1997**, *94*, 10588–10593.
307. Luo, Y.; Tall, A. R. *J. Clin. Invest.* **2000**, *105*, 513–520.
308. Venkateswaran, A.; Laffitte, B. A.; Joseph, S. B.; Mak, P. A.; Wilpitz, D. C.; Edwards, P. A.; Tontonoz, P. *Proc. Natl. Acad. Sci. USA* **2000**, *97*, 12097–12102.
309. Collins, J. L.; Fivush, A. M.; Watson, M. A.; Galardi, C. M.; Lewis, M. C.; Moore, L. B.; Parks, D. J.; Wilson, J. G.; Tippin, T. K.; Binz, J. G. et al. *J. Med. Chem.* **2002**, *45*, 1963–1966.
310. Joseph, S. B.; McKilligin, E.; Pei, L.; Watson, M. A.; Collins, A. R.; Laffitte, B. A.; Chen, M.; Noh, G.; Goodman, J.; Hagger, G. N. et al. *Proc. Natl. Acad. Sci. USA* **2002**, *99*, 7604–7609.
311. Terasaka, N.; Hiroshima, A.; Koieyama, T.; Ubukata, N.; Morikawa, Y.; Nakai, D.; Inaba, T. *FEBS Lett.* **2003**, *536*, 6–11.
312. Laffitte, B. A.; Chao, L. C.; Li, J.; Walczak, R.; Hummasti, S.; Joseph, S. B.; Castrillo, A.; Wilpitz, D. C.; Mangelsdorf, D. J.; Collins, J. L. et al. *Proc. Natl. Acad. Sci. USA* **2003**, *100*, 5419–5424.
313. Cao, G.; Liang, Y.; Broderick, C. L.; Oldham, B. A.; Beyer, T. P.; Schmidt, R. J.; Zhang, Y.; Stayrook, K. R.; Suen, C.; Otto, K. A. et al. *J. Biol. Chem.* **2003**, *278*, 1131–1136.
314. Schultz, J. R.; Tu, H.; Luk, A.; Repa, J. J.; Medina, J. C.; Li, L.; Schwendner, S.; Wang, S.; Thoolen, M.; Mangelsdorf, D. J. et al. *Genes Dev.* **2000**, *14*, 2831–2838.
315. Jaye, M. C.; Krawiec, J. A.; Campobasso, N.; Smallwood, A.; Qiu, C.; Lu, Q.; Kerrigan, J. J.; De Los Frailes Alvaro, M.; Laffitte, B.; Liu, W. S. et al. *J. Med. Chem.* **2005**, *48*, 5419–5422.
316. Wang, H.; Chen, J.; Hollister, K.; Sowers, L. C.; Forman, B. M. *Mol. Cell* **1999**, *3*, 543–553.
317. Parks, D. J.; Blanchard, S. G.; Bledsoe, R. K.; Chandra, G.; Consler, T. G.; Kliewer, S. A.; Stimmel, J. B.; Willson, T. M.; Zavacki, A. M.; Moore, D. D. et al. *Science* **1999**, *284*, 1365–1368.
318. Makishima, M.; Okamoto, A. Y.; Repa, J. J.; Tu, H.; Learned, R. M.; Luk, A.; Hull, M. V.; Lustig, K. D.; Mangelsdorf, D. J.; Shan, B. *Science* **1999**, *284*, 1362–1365.
319. Willson, T. M.; Jones, S. A.; Moore, J. T.; Kliewer, S. A. *Med. Res. Rev.* **2001**, *21*, 513–522.
320. Maloney, P. R.; Parks, D. J.; Haffner, C. D.; Fivush, A. M.; Chandra, G.; Plunket, K. D.; Creech, K. L.; Moore, L. B.; Wilson, J. G.; Lewis, M. C. et al. *J. Med. Chem.* **2000**, *43*, 2971–2974.
321. Pellicciari, R.; Fiorucci, S.; Camaioni, E.; Clerici, C.; Costantino, G.; Maloney, P. R.; Morelli, A.; Parks, D. J.; Willson, T. M. *J. Med. Chem.* **2002**, *45*, 3569–3572.
322. Mi, L. Z.; Devarakonda, S.; Harp, J. M.; Han, Q.; Pellicciari, R.; Willson, T. M.; Khorasanizadeh, S.; Rastinejad, F. *Mol. Cell* **2003**, *11*, 1093–1100.
323. Pellicciari, R.; Costantino, G.; Camaioni, E.; Sadeghpour, B. M.; Entrena, A.; Willson, T. M.; Fiorucci, S.; Clerici, C.; Gioiello, A. *J. Med. Chem.* **2004**, *47*, 4559–4569.
324. Dussault, I.; Beard, R.; Lin, M.; Hollister, K.; Chen, J.; Xiao, J. H.; Chandraratna, R.; Forman, B. M. *J. Biol. Chem.* **2003**, *278*, 7027–7033.
325. Claudel, T.; Sturm, E.; Kuipers, F.; Staels, B. *Expert Opin. Investig. Drugs* **2004**, *13*, 1135–1148.
326. Willson, T. M.; Kliewer, S. A. *Nat. Rev. Drug Disc.* **2002**, *1*, 259–266.
327. Xie, W.; Uppal, H.; Saini, S. P.; Mu, Y.; Little, J. M.; Radominska-Pandya, A.; Zemaitis, M. A. *Drug Disc. Today* **2004**, *9*, 442–449.
328. Jones, S. A.; Moore, L. B.; Shenk, J. L.; Wisely, G. B.; Hamilton, G. A.; McKee, D. D.; Tomkinson, N. C.; LeCluyse, E. L.; Lambert, M. H.; Willson, T. M. et al. *Mol. Endocrinol.* **2000**, *14*, 27–39.
329. Tzameli, I.; Pissios, P.; Schuetz, E. G.; Moore, D. D. *Mol. Cell. Biol.* **2000**, *20*, 2951–2958.
330. Forman, B. M.; Tzameli, I.; Choi, H. S.; Chen, J.; Simha, D.; Seol, W.; Evans, R. M.; Moore, D. D. *Nature* **1998**, *395*, 612–615.
331. Ding, X.; Staudinger, J. L. *J. Pharmacol. Exp. Ther.* **2005**, *314*, 120–127.
332. Maglich, J. M.; Watson, J.; McMillen, P. J.; Goodwin, B.; Willson, T. M.; Moore, J. T. *J. Biol. Chem.* **2004**, *279*, 19832–19838.
333. Qatanani, M.; Zhang, J.; Moore, D. D. *Endocrinology* **2005**, *146*, 995–1002.

334. Nilsson, M.; Dahlman-Wright, K.; Gustafsson, J. A. *Essays Biochem.* **2004**, *40*, 157–167.
335. Gottlieb, B.; Beitel, L. K.; Wu, J.; Elhaji, Y. A.; Trifiro, M. *Essays Biochem.* **2004**, *40*, 121–136.
336. Wilkstrom, A. C. *J. Endocrinol.* **2001**, *169*, 425–428.
337. Berkenstam, A.; Gustafsson, J. A. *Curr. Opin. Pharmacol.* **2005**, *5*, 171–176.
338. Pelton, P. D.; Patel, M.; Demarest, K. T. *Curr. Top. Med. Chem.* **2005**, *5*, 265–282.
339. Luquet, S.; Lopez-Soriano, J.; Holst, D.; Gaudel, C.; Jehl-Pietri, C.; Fredenrich, A.; Grimaldi, P. A. *Biochimie* **2004**, *86*, 833–837.
340. Lala, D. S. *Curr. Opin. Investig. Drugs* **2005**, *6*, 934–943.
341. Herynk, M. H.; Fuqua, S. A. *Endocr. Rev.* **2004**, *25*, 869–898.
342. Zhang, Q. X.; Hilsenbeck, S. G.; Fuqua, S. A.; Borg, A. *J. Steroid Biochem. Mol. Biol.* **1996**, *59*, 251–260.
343. Garcia Pedrero, J. M.; Zuazua, P.; Martinez-Campa, C.; Lazo, P. S.; Ramos, S. *Endocrinology* **2003**, *144*, 2967–2976.
344. Fuqua, S. A.; Wiltschke, C.; Zhang, Q. X.; Borg, A.; Castles, C. G.; Friedrichs, W. E.; Hopp, T.; Hilsenbeck, S.; Mohsin, S.; O'Connell, P. et al. *Cancer Res.* **2000**, *60*, 4026–4029.
345. Karnik, P. S.; Kulkarni, S.; Liu, X. P.; Budd, G. T.; Bukowski, R. M. *Cancer Res.* **1994**, *54*, 349–353.
346. Zhang, Q. X.; Borg, A.; Wolf, D. M.; Oesterreich, S.; Fuqua, S. A. *Cancer Res.* **1997**, *57*, 1244–1249.
347. Gennari, L.; Merlotti, D.; De Paola, V.; Calabro, A.; Becherini, L.; Martini, G.; Nuti, R. *Am. J. Epidemiol.* **2005**, *161*, 307–320.
348. Rosenkranz, K.; Hinney, A.; Ziegler, A.; Hermann, H.; Fichter, M.; Mayer, H.; Siegfried, W.; Young, J. K.; Remschmidt, H.; Hebebrand, J. *J. Clin. Endocrinol. Metab.* **1998**, *83*, 4524–4527.
349. Huang, H.; Tindall, D. J. *Crit. Rev. Eukaryot. Gene Expr.* **2002**, *12*, 193–207.
350. McDonald, S.; Brive, L.; Agus, D. B.; Scher, H. I.; Ely, K. R. *Cancer Res.* **2000**, *60*, 2317–2322.
351. Grigoryev, D. N.; Long, B. J.; Njar, V. C.; Brodie, A. H. *J. Steroid Biochem. Mol. Biol.* **2000**, *75*, 1–10.
352. Gottlieb, B.; Beitel, L. K.; Lumbroso, R.; Pinsky, L.; Trifiro, M. *Hum. Mutat.* **1999**, *14*, 103–114.
353. Werner, S.; Bronnegard, M. *Steroids* **1996**, *61*, 216–221.
354. Charmandari, E.; Raji, A.; Kino, T.; Ichijo, T.; Tiulpakov, A.; Zachman, K.; Chrousos, G. P. *J. Clin. Endocrinol. Metab.* **2005**, *90*, 3696–3705.
355. Chrousos, G. P.; Detera-Wadleigh, S. D.; Karl, M. *Ann. Intern. Med.* **1993**, *119*, 1113–1124.
356. Malchoff, C. D.; Malchoff, D. M. *Endocrinol. Metab. Clin.* **2005**, *34*, 315–326.
357. Towers, R.; Naftali, T.; Gabay, G.; Carlebach, M.; Klein, A.; Novis, B. *Clin. Exp. Immunol.* **2005**, *141*, 357–362.
358. Rzazewska-Makosa, B. *Postepy. Hig. Med. Dosw. (Online)* **2005**, *59*, 457–463.
359. Lin, R. J.; Sternsdorf, T.; Tini, M.; Evans, R. M. *Oncogene* **2001**, *20*, 7204–7215.
360. Huang, M. E.; Ye, Y. C.; Chen, S. R.; Chai, J. R.; Lu, J. X.; Zhoa, L.; Gu, L. J.; Wang, Z. Y. *Blood* **1988**, *72*, 567–572.
361. Deeb, S. S.; Fajas, L.; Nemoto, M.; Pihlajamaki, J.; Mykkanen, L.; Kuusisto, J.; Laakso, M.; Fujimoto, W.; Auwerx, J. *Nat. Genet.* **1998**, *20*, 284–287.
362. Hara, K.; Okada, T.; Tobe, K.; Yasuda, K.; Mori, Y.; Kadowaki, H.; Hagura, R.; Akanuma, Y.; Kimura, S.; Ito, C. et al. *Biochem. Biophys. Res. Commun.* **2000**, *271*, 212–216.
363. Hall, J. M.; Chang, C. Y.; McDonnell, D. P. *Mol. Endocrinol.* **2000**, *14*, 2010–2023.
364. McDonnell, D. P.; Chang, C. Y.; Norris, J. D. *J. Steroid Biochem. Mol. Biol.* **2000**, *74*, 327–335.
365. Rodriguez, A. L.; Tamrazi, A.; Collins, M. L.; Katzenellenbogen, J. A. *J. Med. Chem.* **2004**, *47*, 600–611.
366. Egea, P. F.; Mitschler, A.; Rochel, N.; Ruff, M.; Chambon, P.; Moras, D. *EMBO J.* **2000**, *19*, 2592–2601.
367. Egea, P. F.; Mitschler, A.; Moras, D. *Mol. Endocrinol.* **2002**, *16*, 987–997.
368. Love, J. D.; Gooch, J. T.; Benko, S.; Li, C.; Nagy, L.; Chatterjee, V. K.; Evans, R. M.; Schwabe, J. W. *J. Biol. Chem.* **2002**, *277*, 11385–11391.
369. Wagner, R. L.; Apriletti, J. W.; McGrath, M. E.; West, B. L.; Baxter, J. D.; Fletterick, R. J. *Nature* **1995**, *378*, 690–697.
370. Rochel, N.; Wurtz, J. M.; Mitschler, A.; Klaholz, B.; Moras, D. *Mol. Cell.* **2000**, *5*, 173–179.
371. Tocchini-Valentini, G.; Rochel, N.; Wurtz, J. M.; Moras, D. *J. Med. Chem.* **2004**, *47*, 1956–1961.
372. Cronet, P.; Petersen, J. F.; Folmer, R.; Blomberg, N.; Sjoblom, K.; Karlsson, U.; Lindstedt, E. L.; Bamberg, K. *Structure (Camb.)* **2001**, *9*, 699–706.
373. Uppenberg, J.; Svensson, C.; Jaki, M.; Bertilsson, G.; Jendeberg, L.; Berkenstam, A. *J. Biol. Chem.* **1998**, *273*, 31108–31112.
374. Matias, P. M.; Donner, P.; Coelho, R.; Thomaz, M.; Peixoto, C.; Macedo, S.; Otto, N.; Joschko, S.; Scholz, P.; Wegg, A. et al. *J. Biol. Chem.* **2000**, *275*, 26164–26171.
375. Hoerer, S.; Schmid, A.; Heckel, A.; Budzinski, R. M.; Nar, H. *J. Mol. Biol.* **2003**, *334*, 853–861.
376. Kallen, J. A.; Schlaeppi, J. M.; Bitsch, F.; Geisse, S.; Geiser, M.; Delhon, I.; Fournier, B. *Structure (Camb.)* **2002**, *10*, 1697–1707.
377. Nuclear Receptor Nomenclature Homepage. http://www.ens-lyon.fr/LBMC/laudet/nomenc.html (accessed May 2006).
378. Nuclear Receptor Signaling Atlas. http://www.nursa.org (accessed May 2006).

Biographies

Nurulain T Zaveri is the Program Director for the Drug Discovery Program at SRI International. Dr Zaveri received her PhD in Medicinal Chemistry from Duquesne University, Pittsburgh, PA (1994) and postdoctoral training at SRI International (1992–94), after which she continued her tenure at SRI International as a Staff Scientist and then Program Director. Her main research interests are in drug discovery and medicinal chemistry in the primary areas of oncology and CNS therapeutics. Some of her active research projects include the design of novel synthetic analogs of green tea catechins, small-molecule antimetastatic compounds, discovery of small-molecule ligands for the nicotinic and nociceptin receptors and in collaboration with Dr Murphy, the design of novel PPARδ ligands.

Brian J Murphy is Senior Molecular Biologist in the Biosciences Division at SRI International (Menlo Park, CA). He received his PhD in Comparative Biology from the University of British Columbia (Vancouver, Canada) (1980) and postdoctoral training in cellular and molecular biology at the Stanford University Medical Center (Palo Alto, CA) (1980–84) after which he joined SRI International. His main, NIH-funded, research interest relates to the involvement of the redox-sensitive metal transcription factor-1 (MTF-1) in tumor development, inflammation, and fibrosis. He is also involved in another research project, with Dr Nurulain Zaveri, focused upon the putative role of the peroxisome proliferator activated receptor-delta (PPARδ) nuclear receptor in breast and prostate cancer and in the development of novel PPARδ antagonists.

2.26 Nucleic Acids (Deoxyribonucleic Acid and Ribonucleic Acid)

E E Swayze, R H Griffey, and C F Bennett, ISIS Pharmaceuticals, Carlsbad, CA, USA

© 2007 Elsevier Ltd. All Rights Reserved.

2.26.1	**Introduction**	1037
2.26.2	**Deoxyribonucleic Acid**	1038
2.26.2.1	Structure and Nomenclature	1038
2.26.2.2	Prototypical Function and Pharmacology	1039
2.26.2.3	Prototypical Therapeutics	1039
2.26.2.3.1	Transition metal complexes	1040
2.26.2.3.2	Intercalation	1040
2.26.2.3.3	Minor groove binders	1040
2.26.2.3.4	Nonnatural nucleosides	1040
2.26.2.3.5	Other structures	1040
2.26.3	**Ribosomal Ribonucleic Acids**	1040
2.26.3.1	Structure and Nomenclature	1040
2.26.3.2	Physiological Function	1041
2.26.3.3	Prototypical Pharmacology	1041
2.26.3.4	Prototypical Therapeutics	1041
2.26.3.4.1	Aminoglycosides	1041
2.26.3.4.2	Macrolides	1042
2.26.3.4.3	Other ribosomal ribonucleic acid binding antimicrobials	1042
2.26.4	**Messenger Ribonucleic Acids**	1043
2.26.4.1	Structure and Nomenclature	1043
2.26.4.2	Physiological Function	1044
2.26.4.3	Prototypical Pharmacology	1045
2.26.4.4	Prototypical Therapeutics	1045
2.26.4.4.1	Antisense-based strategies	1045
2.26.4.4.2	Internal ribosomal entry site	1047
2.26.4.4.3	Human immunodeficiency virus transactivation response	1047
2.26.4.4.4	Berberine	1048
2.26.5	**New Directions**	1048
2.26.5.1	Riboswitches	1048
2.26.5.2	Signal Recognition Particle and tmRNA	1049
2.26.5.3	MicroRNA	1049
References		**1049**

2.26.1 Introduction

Nucleic acids present a broad palate of drug targets, based on their function and structure. Nucleic acids are composed of four nucleotide 'letters' arranged by sequence into strands. The nucleotides consist of three chemical components: the heterocyclic base that defines the 'letter'; the sugar; and the internucleotide phosphate diester linker (**Figure 1**). There are two fundamental types of nucleic acids: ribonucleic acids (RNAs) and deoxyribonucleic acids (DNAs), whose structures are determined by the removal of a hydroxyl group from the ribose sugar and methylation of uracil to give thymine. Cellular DNA exists in a double-stranded form in the chromosomes of the nucleus, serving as the storage media for a cell's information and programming instructions. The double-stranded DNA is opened transiently to allow

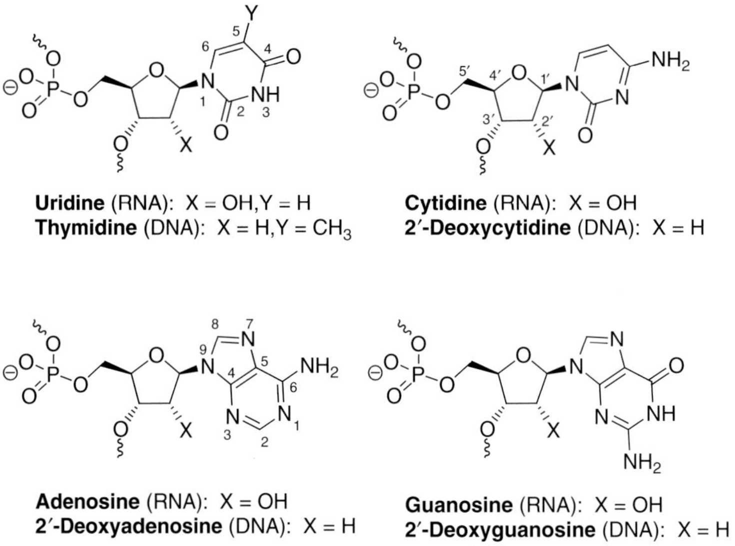

Figure 1 Common building-blocks of nucleic acids.

Uridine (RNA): X = OH, Y = H
Thymidine (DNA): X = H, Y = CH_3

Cytidine (RNA): X = OH
2′-Deoxycytidine (DNA): X = H

Adenosine (RNA): X = OH
2′-Deoxyadenosine (DNA): X = H

Guanosine (RNA): X = OH
2′-Deoxyguanosine (DNA): X = H

transcription into RNA molecules. RNAs have a large range of sizes (20 to >100 000 nucleotides) and shapes generated by combinations of single, double, and higher-order strands, allowing them to perform a variety of functions in the cell. For example, messenger RNA (mRNA) acts as the template for the linking of amino acids into peptide chains by the ribosome. The mRNAs are produced from longer pre-mRNA transcripts through a series of complex splicing reactions performed by the spliceosome. The ribosome in turn is composed of three ribosomal RNAs (rRNAs) that serve as the macromolecular machine and chemical catalyst for amino acid coupling. The amino acids are delivered to the ribosome by transfer RNAs (tRNAs), which contain a three-nucleotide anticodon sequence complementary to three nucleotides on the mRNA. Additional small RNAs and RNA domains that control and edit RNA and DNA through chemical modifications have been discovered over the last 10 years (e.g., small nucleolar RNAs and guide RNAs). Among the most exciting are the abundant microRNAs (miRNAs), which appear to function as an intermediate level of transcriptional and posttranscriptional control on gene expression. Mitochondrial RNA and DNA have both structural similarities and differences with bacterial and human nucleic acids.

The divergence of bacterial and eukaryotic life, the increase in gene complexity in higher organisms, and the central role that nucleic acids play in all aspects of cellular growth, proliferation, and differentiation make them attractive targets for drug discovery. A number of older drugs now are known to be directed against nucleic acids. However, only a limited number of strategies have been employed to target RNAs and DNA with any level of specificity. Gene therapy has been demonstrated as a method to add new DNA to a cell to produce a new protein product, replace a mutated form of a protein, or to alter the sequence of endogenous DNA. The DNA bases have modest reactivity, and can be altered chemically by alkylating agents and metal complexes such as cisplatin. The phosphodiester backbone can be cleaved chemically by compounds such as bleomycin or enzymatically by endogenous nucleases that recognize RNA:RNA or DNA:RNA duplexes produced by administered oligonucleotide agents. Binding of small molecules or oligonucleotides to a sequence of mRNA or DNA can produce steric blocking that prevents subsequent processing or translation. Finally, small molecules and oligonucleotides can induce rearrangements or stabilization of RNA or DNA structures that block specific biological functions. Examples of these mechanisms are described in subsequent sections.

2.26.2 Deoxyribonucleic Acid

2.26.2.1 Structure and Nomenclature

Double-stranded (duplex) DNA is organized into chromosomes in the cell nucleus of higher organisms. These chromosomes contain duplex DNA wrapped around the histone proteins. The DNA can be bound to histones and transcriptionally silent or bound to histones in conjunction with transcriptional factors and polymerases that mediate transcription of the DNA into RNA. The normal DNA double helix adopts a supercoiled form from binding to histones.

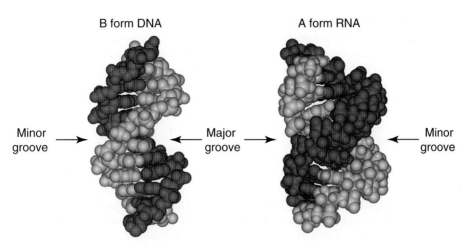

Figure 2 Models of helical B form DNA (left) and A form RNA (right). B DNA has a wide major groove and a narrow minor groove, while A RNA has a very narrow and deep major groove, and a wide and shallow minor groove.

Duplex DNA is a helical structure composed of two antiparallel strands linked through hydrogen bonds formed between the complementary donor/acceptor groups of deoxyadenosine (dA) and deoxythymidine (dT) residues and deoxyguanosine (dG) and deoxycytidine (dC) residues, respectively. These base pairs form a ladder that gives the DNA its natural right-handed twist (**Figure 2**). The DNA helix has two grooves, the deep and narrow minor groove and the shallow, wide major groove. As discussed below, these grooves present unique hydrophobic, electrostatic, and hydrogen bonding environments where small molecules are known to bind.

The nomenclature for nucleic acids has been specified in detail by the International Union of Pure and Applied Chemistry (IUPAC).[1] DNA bases, nucleosides, and nucleotides are referred to by their standard common names, and the numbering system for the bases and sugars are presented in **Figure 1**. The purine heterocycles (adenine and guanine) are attached to the 2′ deoxyribose sugars through N9, and the pyrimidine heterocycles (uracil and cytidine) are bonded through N1. In all natural DNA, the 3′ hydroxyl and the 5′ hydroxyl of the subsequent nucleotide are linked through a phosphate diester. Typically, DNA and RNA strand sequences are written from the 5′ end toward the 3′ terminus.

2.26.2.2 Prototypical Function and Pharmacology

As noted above, all RNAs and proteins in a cell are derived from information stored in the DNA sequence. In contrast to prokaryotes, the majority of DNA present in eukaryotic cells is not transcribed. Why this is the case is not well understood. Single base changes in DNA sequence are responsible for more than 100 human diseases.[2] Transcription of human DNA is regulated by both protein binding to DNA and methylation of specific sites in the DNA. Drugs that target DNA can alter cellular physiology by inhibiting transcription of RNA or by preventing DNA replication in dividing cells. In all cells, irreversible damage to the DNA (alkylation, strand cleavage, etc.) or inhibition of DNA replication leads to cell cycle arrest. In mammalian cells, cell cycle arrest may trigger apoptosis and cell death. A therapeutic index can be realized by exploiting the difference in rates of cell division between rapidly growing cells and normal human tissue. Hence, most agents targeting DNA are used in treatment of bacterial infection or cancer.

2.26.2.3 Prototypical Therapeutics

The DNA strands stored in a supercoiled form are cleaved to allow transcription or replication to initiate, and subsequently ligated back together by a topoisomerase protein. Many therapeutics target this process. As described above, the rate of RNA transcription is controlled by binding of activator and suppressor proteins with specificity for the DNA surfaces of certain sequences. Therapeutic agents such as intercalators can disrupt the shape of the DNA helix and alter transcription. DNA is copied by the enzyme DNA polymerase, and nucleoside analogs can prevent DNA chain extension. Maintaining the fidelity of the DNA sequence is critical, and the cell has multiple enzymes that inspect and repair the DNA duplex.

2.26.2.3.1 Transition metal complexes

The electron-rich environment at N7 of adjacent purine bases provides a good coordination site for transition metals such as platinum, whose affinity is sufficient to block proper DNA replication and transcription. Cisplatin and analogs are among the most effective anticancer drugs in treatment of solid tumors. New platinum complexes with different binding modes and ruthenium–imidazole complexes are in clinical and preclinical development.[3,4] Mithramycin has been demonstrated to form a 2:1 complex with Fe(II) that induces a conformational change in DNA which may facilitate strand cleavage.[5]

2.26.2.3.2 Intercalation

The DNA duplex can be altered through intercalation of small molecules between the base pairs or in the minor groove. The base pairs have modest separation and are known to open partially at 37 °C. Small, planar, aromatic groups can be inserted between the base pairs, resulting in a local distortion of the helical geometry. Many classes of drugs are known to work through intercalation, which inhibits the ability of DNA topoisomerase and polymerase enzymes to replicate the DNA. Doxorubicin traps a cleavage intermediate between topoisomerase II and torsionally strained DNA. Analogs of anthracycline drugs with attached saccharides have been shown to intercalate and position carbohydrates at specific sites in the groove.[6] DNA intercalating drugs also reduce the activity of DNA methyltransferases, and can induce apoptosis through this inhibition.[7] Many intercalating drugs reduce levels of mitochondrial DNA in mammalian cells through altered metabolism or inhibition of replication.[8]

2.26.2.3.3 Minor groove binders

The interactions between proteins and DNA are mediated by hydrogen bonding networks of side chains along the minor groove of the DNA. Distamycin-related molecules bind in the narrow minor groove of duplex DNA, often as dimers that widen and distort the shape of the groove. The ability of aromatic polyamides to bind in the hydrophobic minor groove has been exploited to develop compounds that form sequence-specific H-bond donor and acceptor interactions with the bases.[9,10] These compounds can have very high (<100 pM) affinity and selectivity for the target sequence, and have been shown to block the binding of transcription factors to the DNA. In addition, conjugation of regulatory peptides to these molecules produces molecules that upregulate gene expression or recruit additional transcription factors.[11]

2.26.2.3.4 Nonnatural nucleosides

Nucleoside analogs can be incorporated into DNA and block natural maturation processes. For example, enzymatic methylation (and demethylation) of dC residues at C5 is a natural tagging process used to silence transcription of genes that are no longer required for cell function. In some cancers, specific methylation of tumor suppressor genes promotes uncontrolled cell growth. Treatment of B cell leukemias with 5-azadeoxycytidine generates DNA containing dC residues that cannot be methylated, and the tumor suppressor genes are left in the 'on' state. Other nucleoside drugs can act as inhibitors of DNA replication.

2.26.2.3.5 Other structures

DNA can form higher-order structures, such as the G quartets present as telomeres at the ends of genes. These G quartet DNAs are very stable, and are recognized specifically by cellular proteins. Hence, artificial G quartet structures have been prepared and demonstrated to have anticancer activity.

DNA:RNA duplexes formed during transcription are potential targets for drug discovery. These structures are present during transcription, and may be long-lived. For example, the transcription of human immunodeficiency virus (HIV) RNA has stall sites where DNA:RNA structures are formed as sites for recruitment of additional protein factors. DNA:RNA duplexes also are formed during trailing strand DNA synthesis (Okazaki fragment) and subsequently cleaved.

2.26.3 Ribosomal Ribonucleic Acids

2.26.3.1 Structure and Nomenclature

Ribosomes are large ribonucleoprotein particles, with a bacterial ribosome from *Escherichia coli* having a molecular mass around 2700 kDa, a diameter of approximately 200 angstroms, and a sedimentation coefficient of 70S. The rRNA constitutes nearly two-thirds of the mass of the ribosome, and makes up nearly all of the key sites for ribosomal function. Because of this, ribosomes can be viewed as RNA enzymes, or ribozymes, and drugs which exert their effects via binding to ribosomes interact predominantly with rRNA.

Bacterial ribosomes are composed of a small (30S) and a large (50S) subunit, The small subunit (30S) consists of a large, highly structured RNA molecule of over 1500 nucleotides (the 16S rRNA), along with 21 proteins (referred to as S1 through S21). The large subunit consists of two RNAs, the 23S (about 2900 nucleotides) and the 5S (120 nucleotides), along with 34 different proteins (designated L1 through L34). The large and small subunits join to provide for three tRNA binding sites at the interface, the A (aminoacyl) site, the P (peptidyl) site, and the E (exit) site. The 'active site' of the ribosome, the peptidyl transferase center, is near the P site on the large subunit. A common structural feature to all ribosomes is a tunnel, from the interface at P site through the large subunit to the cytoplasmic side.

Eukaryotic ribosomes are slightly larger (80S) than their bacterial counterparts, with a mass of roughly 4200 kDa. The eukaryotic small subunit (40S) contains a rRNA (18S) which is homologous to the slightly smaller bacterial 16S rRNA. The large subunit (60S) contains 28S and 5S homologs of the bacterial rRNAs, as well as an additional RNA, the 5.8S rRNA. Mitochondria also contain ribosomes that contain similar structural features,[12] despite consisting of more protein than rRNA. These mitochondrial rRNAs must be considered as potential sites of side effects of antibacterial rRNA acting drugs, as they have rRNA components similar in structure to bacterial ribosomes.

2.26.3.2 Physiological Function

The ribosomal RNA is the cellular machine that synthesizes proteins by reading the genetic code of a particular mRNA. Recent cryoelectron microscopic and crystallographic studies have provided astounding insights into both the structure and function of ribosomes,[13] and illustrated how many antibiotics act to inhibit protein synthesis.[14] A recent review provides a detailed analysis of how the ribosome functions in light of these studies.[15]

The rRNA of the small subunit (16S in bacteria) functions primarily to decode the genetic message. This is accomplished by recognition of correct Watson–Crick base pairing between a particular three nucleotide sequence on an mRNA (the codon) with a three-nucleotide sequence on a tRNA which carries the appropriate amino acid. This recognition occurs at a portion of the small subunit termed the A site, which is the binding site of the aminoglycoside antibiotics.

The large subunit functions to append the selected amino acid on the A site tRNA onto the growing peptide chain, which releases the chain from the P site tRNA. This task involves movement of the A site tRNA to the peptidyl transferase center, which is the catalytic site for peptide bond formation. The appropriate amino acid is then transferred onto the growing peptide chain, which exits the ribosome through the tunnel. Many antibiotics bind at or near the peptidyl transferase center, including chloramphenicol and the macrolides such as erythromycin. The macrolides function by blocking the exit of the growing peptide chain through the tunnel.

A translocation of the P site tRNA to the E site, and the A site tRNA to the P site allows for the cycle to be repeated. The peptidyl transferase and decoding centers consist predominantly of RNA, and the key processes and reactions of protein synthesis are carried out by the rRNA. Prokaryotic 5S rRNA interacts with ribosomal proteins L5, L18, and L25 and enhances protein synthesis by stabilization of the ribosome structure but its exact role in protein synthesis is still not known, and 5S rRNA remains a largely unexplored target.

2.26.3.3 Prototypical Pharmacology

Because of the central importance of ribosomes to the synthesis of all proteins, rRNA is not a target well suited to host diseases. However, because of subtle differences in ribosome structure between prokaryotes and higher eukaryotes, rRNA is an excellent target for antibacterial chemotherapy. Most antibiotics which target the rRNA and inhibit bacterial protein synthesis prevent bacterial growth (bacteriostatic). These include the tetracyclines, chloramphenicol, and the macrolides. While each class of drug has a slightly different mechanism of action, they all interfere with the function of the ribosomal machinery and inhibit protein synthesis. This leads to a reduction or cessation of bacterial growth. In contrast, the aminoglycosides which act at the A site to induce miscoding are rapidly lethal to bacteria (bactericidal). This is likely due to a progressive build-up of aberrant proteins, which ultimately cause a disruption of multiple cellular processes leading to cell death.

2.26.3.4 Prototypical Therapeutics

2.26.3.4.1 Aminoglycosides

There are several related structural classes of aminoglycoside antibiotics, representative structures of which are shown in **Figure 3**.[16] Uniquely differentiated from the majority of aminoglycosides is streptomycin, which contains a

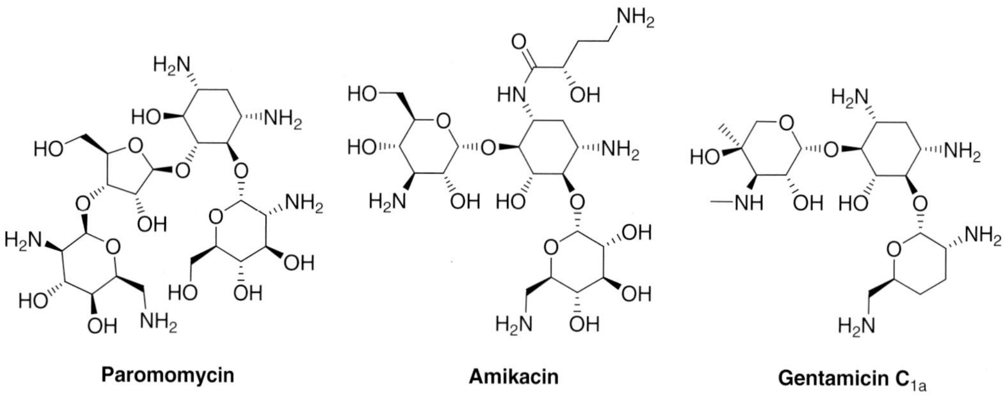

Figure 3 Aminoglycosides which bind at the bacterial 16S rRNA site.

guanylated streptamine core, as opposed to the deoxystreptamine core of the other aminoglycosides. Streptomycin binds near the A site at different location from that of the other aminoglycosides, but still interferes with decoding. The 2-deoxystreptamine based aminoglycosides can be classified into the 4,5-substituted analogs such as paromomycin and neomycin, and the 4,6-substituted analogs such as kanamycin, tobramycin, and gentamicin. These compounds all bind to the 16S rRNA A site, and force two conserved adenosine residues to swing into the minor groove formed by the mRNA/tRNA codon/anticodon base triplet formed during decoding. This stabilizes both cognate and noncognate tRNA pairings, and leads to miscoding and inhibition of protein synthesis.

Tobramycin long has been administered intravenously for treatment of bacterial infections, and has been approved recently as an inhaled formulation for the treatment of cystic fibrosis patients with *Pseudomonas aeruginosa* infections. Also in the kanamycin family is amikacin, which is a semisynthetic derivative that contains a 2-hydroxy-4-aminobutyryl chain at the 1-amino group of kanamycin. This substitution confers resistance to a large number of aminoglycoside deactivating enzymes, which modify various portions of the aminoglycoside skeleton and render it pharmacologically inactive. Because of this unique resistance to a broad array of resistance enzymes, amikacin is one of the broadest spectrum aminoglycosides known. It has broad-spectrum activity against a variety of Gram-positive and Gram-negative organisms. However, because of the availability of many other safer, orally acting classes of antibiotics, amikacin and other aminoglycosides such as gentamicin are primarily used against serious Gram-negative infections. In particular, amikacin is used either alone or in combination to treat multidrug resistant nosocomial infections such as *P. aeruginosa*.

All aminoglycosides suffer from potentially severe nephro- and ototoxicities, which limit the use of these excellent antibacterial agents. The nephrotoxicity presents as mild renal impairment in approximately 10–25% of patients receiving an aminoglycoside for more than several days. The nephrotoxicity is generally reversible, and can be monitored by common kidney function tests. In contrast, aminoglycoside-induced ototoxicity is not generally reversible. Both vestibular and auditory dysfunction are evident, and the degree and type of ototoxicity vary with the specific aminoglycoside. The ototoxicity has been correlated with mutations in the 12S mitochondrial A site rRNA, which supports aminoglycoside interaction with this rRNA as a source of aminoglycoside toxicity.[17]

2.26.3.4.2 Macrolides

The macrolide antibiotics are an important class of orally active antibiotics.[18] Major members of the class include erythromycin and azithromycin, as well as telithromycin, which was approved in 2004 (**Figure 4**). The macrolides are most commonly used against Gram-positive organisms, as they are weakly active against most Gram-negative bacilli. The macrolides are generally bacteriostatic agents that bind to the 23S rRNA on the large subunit and block the 'tunnel' through which the growing peptide chain exits. Resistance to macrolides arises from a methylase which modifies the ribosomal target and reduces binding, along with enzymes that chemically modify the drug structures and more general uptake/efflux resistance mechanisms. Unlike the aminoglycosides, the macrolides are generally well tolerated and give rise to few serious toxicity issues, and as such are used broadly.

2.26.3.4.3 Other ribosomal ribonucleic acid binding antimicrobials

Several other classes of antibacterial agents bind rRNA, including the tetracyclines, chloramphenicol, clindamycin, spectinomycin, and the oxazolidinones (**Figure 5**).[18] Tetracycline and its close relatives such as oxytetracycline,

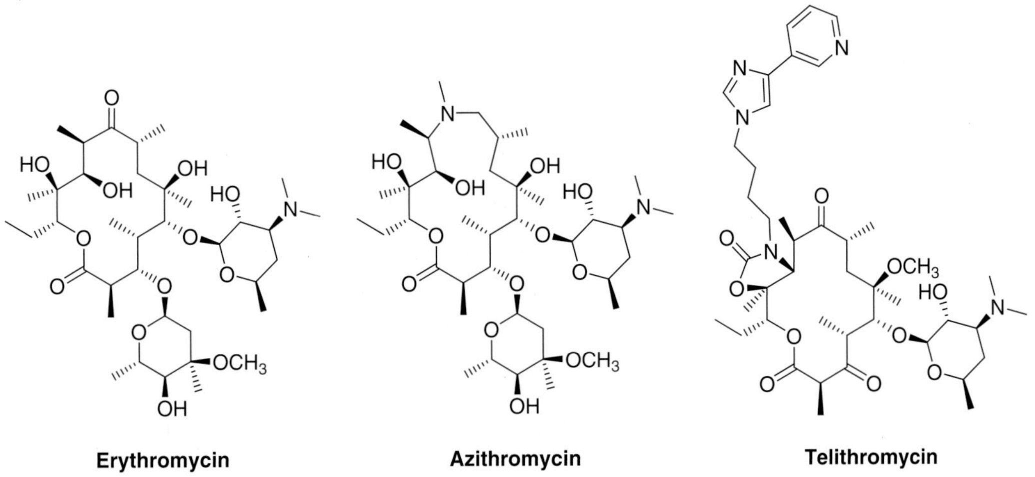

Figure 4 Macrolides that bind to the 50S subunit.

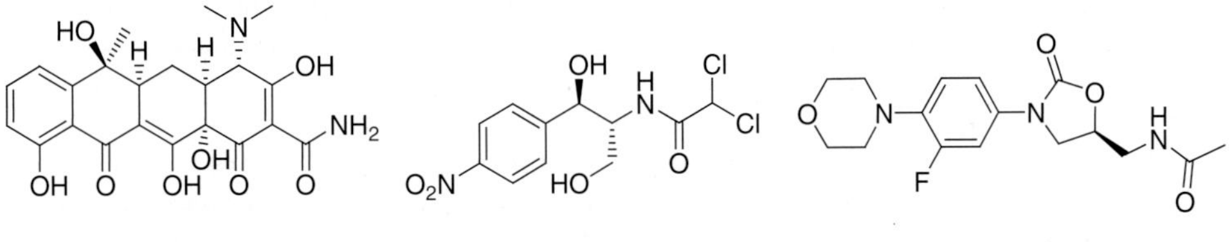

Figure 5 Other antibacterials that bind to the bacterial ribosome.

minocycline, and doxycycline are bacteriostatic agents which function by binding to the small subunit near the A site, hindering movement of the A site tRNA so that it cannot interact with the peptidyl transferase center. Spectinomycin is also bacteriostatic, and binds to the 16S rRNA of the small subunit in a location distinct from the aminoglycosides or tetracycline. It functions to inhibit translocation of the A site tRNA to the P site, presumably by inhibiting a conformational change required to accomplish this translocation. In contrast, clindamycin and chloramphenicol function by binding near the macrolide-binding sites on the rRNA of the large subunit near the peptidyl transferase center. Clindamycin has been useful for the treatment of anaerobes, as has chloramphenicol. However, the toxicity of chloramphenicol to bone marrow restricts its use to serious infections that cannot be otherwise treated.

Linezolid (the prototypical oxazolidinone) represents one of the newer chemical classes of antibacterial agents discovered. It binds to the 23S rRNA of the large subunit, and functions to prevent assembly of the functioning 70S ribosome. Linezolid is broadly active against Gram-positive organisms, and is used for the treatment of vancomycin and/or methicillin-resistant infections. Because of the unique structural class and mechanism of action, there is no cross-resistance with other antibacterials; however, resistance due to a mutation of the 23S rRNA has emerged in clinical practice in a surprisingly short period.[19,20]

2.26.4 Messenger Ribonucleic Acids

2.26.4.1 Structure and Nomenclature

All nucleic acid sequence information is collected and annotated by the National Center for Biotechnology Information (NCBI), and mRNAs are uniquely identified by their accession number from GenBank.[21] However, mRNAs are more generally referred to by their common gene name and species.

RNA is synthesized on a DNA template by RNA polymerases. Bacteria utilize a single RNA polymerase to make all RNA. Messenger RNA molecules are made as either monocistronic or polycistronic molecules in bacteria. Translation can occur on bacterial RNA coincident with transcription.

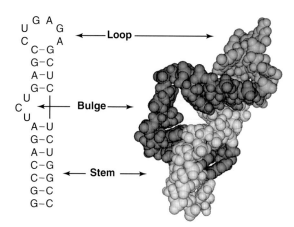

Figure 6 Two-dimensional representation and a model of the three-dimensional structure of the HIV-1 TAR stem loop RNA (PDB ID 1ANR). The 5′ strand of the stem is shown dark gray, and the 3′ strand is shown light gray. The loop is medium gray.

In contrast, eukaryotes utilize three distinct RNA polymerases to transcribe rRNA, mRNA, and tRNA plus other small RNAs: RNA polymerase I, II, and III, respectively. Eukaryotes also compartmentalize transcription and translation, with transcription occurring in the cell nucleus and translation occurring in the cytoplasm. Eukaryotic mRNA is transcribed as a monocistronic precursor molecule in the nucleus which must undergo several processing steps before it is translocated into the cytoplasm where it is translated.

The 5′ terminus of eukaryotic mRNAs is modified with a unique 5′ cap structure consisting of an inverted 5′-5′-triphoshate linkage between the first transcribed nucleotide and 7-methyguanine. Many eukaryotic mRNAs are further modified on their 5′ terminus by methylation of the 2′ oxygen of the ribose nucleotides. The presence of the methyl group adds stability, and also changes the sugar conformation such that it is predominantly (60%) in the C3′ endo conformation.[22] The 5′ cap on an mRNA contributes to export out of the nucleus to the cytoplasm, acts as a recognition element for translation initiation factors, and confers exonuclease stability to the 5′ termini of mRNAs.

The primary transcript for most eukaryotic mRNAs is significantly larger than the mature mRNA due to the presence of extraneous nucleotides called intervening sequences or introns. These introns are removed by a complex cleavage and reannealing process called RNA splicing.[23,24] The reason why eukaryotic RNA contain these intervening sequences is poorly understood. Most eukaryotic RNAs are also modified at their 3′ termini by the addition of 50 to 250 adenosines referred as the poly(A) tail. Finally a select few mRNA transcripts are modified by specific deaminases that either deaminate adenosines or cytosines.[25,26] Once each of these steps is complete the mRNA can then be exported out of the nucleus through nuclear pores to the cytoplasm. Thus in contrast to prokaryotes, eukaryotic mRNA must undergo a series of complex processing events before they can be translated.

Unlike DNA, mRNAs fold into complex three-dimensional structures. These structures tend to be dominated by Watson–Crick base pairing interactions, which drive the folding of mRNAs into self-complementary stemlike structures. Imperfections in the duplex structure lead to unique shapes which can interact with other portions of the mRNA to form more complex structures in a manner similar to protein folding. Expressed in two-dimensional space, RNA structures can be binned into several classes, including stem loops, internal loops, and bulges, junctions, and pseudoknots.[27] Stem loops consist of a self-complementary duplex portion, the stem, which is connected by a loop portion of unpaired nucleobases. Internal loops and bulges are formed as a result of imperfect pairing in a stem. This often allows the normally rigid duplex to bend, and the loop or bulge region can be an attractive binding pocket for a potential drug molecule. A representative structure of a stem loop with a bulge is the HIV transactivation response (TAR) element (**Figure 6**). Junctions typically join three or more stems together, and can form intricate structures.[28] Pseudoknots are complex elements in which two stems from nearby regions of sequence stack on top of each other. The stem regions are connected by necessary loops, and the resulting structure is usually tightly folded and stable. Pseudoknot structures are utilized by several viral genomes to promote frameshifting[29] and are also found in eukaryotic RNAs such as telomerase RNA.[30,31]

2.26.4.2 Physiological Function

Messenger RNA, as its name implies, has been well established as the intermediary in the expression of genetic information contained within DNA as proteins. In that every protein produced in a cell derives from a specific mRNA, targeting mRNA represents a viable alternative to targeting proteins as potential therapeutic agents. However, it is also clear that the role of mRNA extends beyond being simply an information carrier, and that many interactions with mRNAs are important determinants in the level of expression of the encoded protein. Many mRNAs contain complex structures outside of the coding sequence, e.g., the 5′ and 3′ untranslated regions (UTR). While the precise biochemical functions of these regions are only just being elucidated, it is becoming evident that many of these elements are involved in regulation of the ultimate expression of the encoded gene.

These mechanisms are illustrated by the example of iron regulation in eukaryotes. The mRNAs for ferritin, an iron storage protein, and transferrin receptor, a receptor for the iron carrier transferrin, both contain stem loop structures. A structure in the ferritin 5′ UTR termed the iron-responsive element (IRE) binds an IRE-binding protein which serves to prevent translation. In the presence of iron, the IRE-binding protein binds iron, which prevents it from binding the IRE, and translation can take place. This produces ferritin to store the excess iron. In contrast, transferrin receptor mRNA contains IREs in the 3′ UTR. In iron-free conditions, IRE-binding protein binds the IREs in the transferrin 3′ UTR, and stabilizes the mRNA, preventing degradation, increasing the levels of transferrin receptor. This increases the uptake of iron-rich transferrin from outside the cell. Iron levels are therefore tightly controlled by cells using this posttranscriptional regulatory process.

2.26.4.3 Prototypical Pharmacology

Modulating mRNA levels can in theory lead to dramatic pharmacological effects, as the expression of particular proteins can be reduced by inhibiting translation of or destroying the mRNA which codes for that protein. Conversely, the expression of a particular protein can be increased by stabilizing a particular message such that more is produced from each message. These effects can be observed for a great many mRNA targets by direct measurement of RNA levels by quantitative methods, or by quantitation of the encoded protein. The downstream pharmacological consequences are many, and depend on the particular mRNA target. Targeting mRNA, therefore, represents an orthogonal means to attack any potential protein target, and is especially attractive for 'undruggable' protein targets strongly implicated in a disease state.

Unfortunately the strategies for selectively modulating an mRNA inside a cell are still only nascent and have only started to be developed over the last 15 years. Messenger RNAs present unique challenges as drug target. Current RNA structure prediction programs are unlikely to predict accurately the structure of an mRNA as it exists in a cell. This is due in part to the length limitations of most computational programs, but more importantly mRNAs have dynamic three-dimensional structures as they are being processed and translated. Additionally, mRNAs are bound by cellular proteins that tend to exhibit low sequence selectivity. Thus it is not currently possible to predict structure of an mRNA in a cell nor to identify routinely which regions can be targeted with drug molecules.

2.26.4.4 Prototypical Therapeutics

2.26.4.4.1 Antisense-based strategies

One of the most direct methods for selectively targeting RNA is through antisense-based strategies, wherein oligonucleotides are designed to interact with RNA by Watson–Crick base pairing. Representative antisense drugs are shown in **Figure 7**. Antisense-based approaches are broadly used to modulate expression of genes in cell culture, in animal models, and in the clinic. There are two basic antisense mechanisms: oligonucleotide-induced cleavage of the targeted RNA and steric blocking of RNA. Examples of the former mechanism include ribonuclease (RNase) H mediated cleavage, small interfering RNAs, and ribozymes.[32–36] Examples of steric effects of antisense oligonucleotides include inhibition of translation, alternative RNA processing, and disruption of RNA regulatory structures.[37–39]

2.26.4.4.1.1 Oligodeoxynucleotides

The most widely utilized antisense mechanisms are RNase H based cleavage of RNA and the more recently described RNA interference (RNAi) mechanism. RNase H is a ubiquitously expressed enzyme, present in viruses, prokaryotes, and eukaryotes, which hydrolyzes the RNA strand of a DNA–RNA duplex. In general, oligonucleotides that support the RNase H based mechanism are 15–25 nucleotides in length and contain at least 7–10 consecutive deoxynucleotides. RNase H based oligonucleotides (**Figure 7**) are the most advanced with the approval of one drug that selectively targets a viral mRNA (fomivirsen[40]) and multiple drugs in clinical trials.[41]

Fomivirsen, as well the other early antisense agents to enter clinical trials, are phosphorothioate oligodeoxynucleotides, which are short (typically 20 bases) DNA oligomers that contain a phosphorothioate linkage in place of the phosphodiester linkage. These molecules activate RNase H cleavage of the target mRNA after binding, and reduce

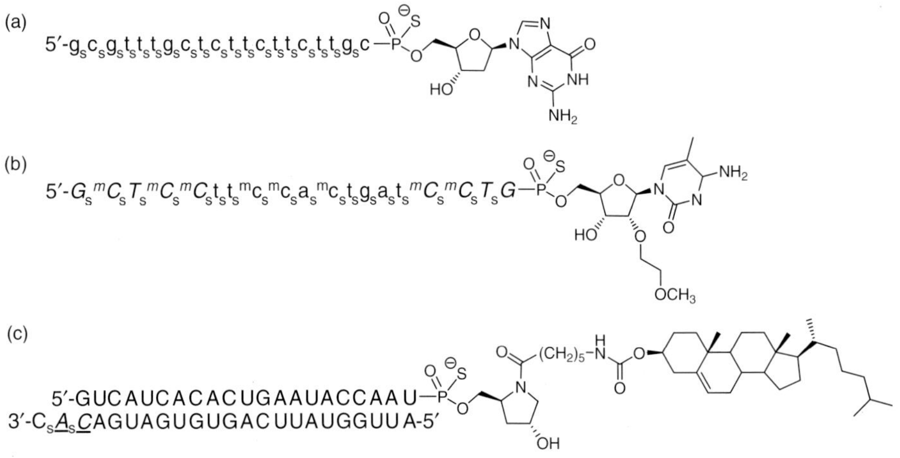

Figure 7 Representative antisense drugs. (a) Fomivirsen; (b) ISIS 113715; (c) cholesterol conjugated siRNA duplex. *Key*: lowercase, deoxyribose (DNA); UPPERCASE, ribose (RNA); *ITALIC*, 2′-O-methoxyethylribose; *ITALIC*, 2′-O-methylribose; m, 5-methyl substitution on cytosine; s, phosphorothioate linkage. All other linkages are phosphodiester, and the letters A, C, G, T, U refer to adenine, cytosine, guanine, thymine, and uracil, respectively.

levels of the target protein. The phosphorothioate serves to stabilize the molecule to nucleolytic degradation, and aid pharmacokinetics by improving binding to plasma proteins which prevents renal excretion, and leads to tissue uptake.[42]

The properties of the first-generation phosphorothioate oligodeoxynucleotides have been dramatically improved by incorporating nucleosides modified at the 2′ position on the sugar into the 5′ and 3′ ends of the molecule while keeping the 2′ deoxy core region, or gap, and the phosphorothioate linkage. These modifications serve to further improve metabolic stability, and also improve potency by increasing the binding affinity for the target mRNA.[43,44] They also activate RNase H cleavage of the target mRNA due to the presence of the 2′ deoxy gap. Uniformly 2′ modified oligonucleotides are not substrates for RNase H, and generally show reduced potency relative to the gapped versions. The most advanced of these chemistries, 2′-methoxyethyl (MOE), typically employs a design of 5 MOE residues at each end, flanking a 10-base deoxy gap. Several MOE oligonucleotides shown to produce the desired pharmacological endpoints in animal models have advanced to human clinical trials and have shown early evidence of efficacy. These include ISIS-113715, an inhibitor of protein-tyrosine phosphatase 1B (PTP1B) which has shown efficacy in models of diabetes,[45] and the human specific version (ISIS-301012) of a mouse-specific apolipoprotein B-100 (apoB-100) antisense drug, which has shown benefit in lowering low-density lipoprotein (LDL) cholesterol.[46]

2.26.4.4.1.2 Ribonucleic acid interference

Oligonucleotides based on RNA interference (RNAi), especially small duplexed RNAs of approximately 20 base pairs known as small interfering RNAs (siRNAs), are gaining widespread use as research reagents for cell culture-based experiments.[47,48] There are many similarities for RNase H and siRNA oligonucleotides with the key difference being that, in general, double-stranded RNA oligonucleotides are utilized for siRNA. Once delivered inside the cell, the double stranded oligonucleotide interacts with a protein complex called the RNA-induced silencing complex (RISC), which contains a helicase activity and the cleavage enzyme argonaute 2.[49] The helicase unwinds the two RNA strands retaining the antisense strand by an as-yet poorly characterized mechanism which may, in part, be dependent on the thermodynamic stability of the base pairs formed at the 5′ end of the RNA.[50] The RISC complex then facilitates interaction of the antisense strand with the targeted RNA where it binds by Watson–Crick base pairing. The enzyme argonaute 2, which shares structure and biochemical similarities to RNase H, cleaves the RNA approximately 10 nucleotides from the 5′ end of the antisense oligonucleotide.[51,52]

Although there has been tremendous progress in defining the biochemical mechanisms by which siRNA oligonucleotides promote specific degradation of RNA, progress in identifying strategies for systemic use of these molecules as drugs has been slower. Structure–activity relationship studies[53] and subsequent optimization have led to motifs with dramatically altered structures that have increased potency and metabolic stability relative to unmodified siRNA oligonucleotides[54]; however, rigorous in vivo pharmacokinetic studies have not been reported for any siRNA to date. There is one preliminary report demonstrating activity of systemically administered siRNA in vivo, in which conjugation of cholesterol to the sense strand enhanced delivery to mouse tissue.[55] Liposomal and particulate formulations are also being explored for systemic delivery of siRNA, and have shown in vivo effects.[56–58] However, more work is needed to further optimize siRNA constructs and/or formulations for broad systemic delivery, and most current therapeutic strategies are focused on local delivery of siRNA molecules.

2.26.4.4.1.3 5′ Cap of messenger ribonucleic acid

Several investigators have exploited the unique chemical reactivity of the 5′ cap structure, identifying molecules that selectively hydrolyze the triphosphate bond.[59,60] Indiscriminate interference with function of 5′ caps present on all mRNA would globally inhibit translation and would be predicted to be rather toxic. Such a strategy is utilized by several viruses as a strategy to compete for host cell translation factors.[61]

In contrast, selective interference with the function of the 5′ cap can be used to specifically modulate the production of a protein product. Antisense-based strategies represent the most direct way of doing so. Antisense oligonucleotides can be designed to the 5′ terminus of a transcript and sterically interfere with the recognition of the 5′ cap by translation factors.[38] Another strategy is to conjugate reactive groups to an oligonucleotide that induce cleavage of the 5′ cap structure, the oligonucleotide providing the specificity. Thus antisense-based strategies represent the most direct and efficient approach for modulating the function of the 5′ cap structure present on mammalian mRNAs.

2.26.4.4.1.4 Other antisense strategies

In addition to promoting the degradation of a target mRNA, oligonucleotides can be utilized to modulate RNA maturation selectively. The best-characterized example is use of oligonucleotides to selectively modulate RNA splicing.[62] Oligonucleotides can be used to correct aberrant splicing events such as occurs in thalassemia.[39] Alternatively oligonucleotides can be used to redirect splicing such that different alternative spliced variants are produced.[63,64] As the role of alternative spliced protein products in human disease becomes better characterized, the latter approach will be an increasingly important application for oligonucleotide-based drugs.

2.26.4.4.2 Internal ribosomal entry site

Translation of most eukaryotic mRNAs is initiated at the 5′ cap structure. However, some viral[65] and eukaryotic[66,67] mRNAs contain a sequence in the 5′ UTR which functions as an internal ribosome entry site (IRES), although the exact function of the IRES remains controversial.[68] IRES elements were first discovered in picornavirus and later in other viral mRNAs. More recently IRES-like sequences have been found in some eukaryotic mRNAs such as the mRNAs encoding vascular endothelial growth factor, fibroblast growth factor-2, c-myc, N-myc and the antiapoptotic protein Apaf-1.[69,70]

One of the best-characterized IRESs is that of extrahepatic hepatitis C virus (HCV). The HCV IRES occurs in the 5′ UTR of the positive strand polycistronic mRNA. It is approximately 360 nucleotides in length, extending approximately 30 nucleotides 3′ to the AUG translation initiation codon for the core protein. The IRES consists of three major structural domains which extend from a pseudoknot structure.[71] The pseudoknot and domain IV are centrally located with the other domains radiating out from the central region. Domain III is required for IRES function and contains a large four-way junction and two stem loop structures (IIId and IIIe). Domain IIId contains an E-loop motif presenting a unique narrowed major groove and a distorted phosphodiester backbone.[72,73] This structure is similar to the sarcin/ricin loop in 28 S rRNA. Domain IV includes the translation initiation codon.

As the HCV IRES is essential for translation of the HCV genome, it represents a unique structure for drug targeting. There are multiple approaches being pursued to interfere with the HCV IRES. Several investigators have designed synthetic oligonucleotides to bind to the IRES and disrupt its function.[74–76] These oligonucleotides were shown to block translation of HCV proteins in a cell culture model. In a related approach, oligonucleotide aptamers were selected to bind to the HCV IRES.[77] Aptamers selected to bind to domains II–IV had the highest affinity, and were shown to contain a consensus sequence ACCCA which base pairs to the apical loop of domain IIId.

Several efforts to identify small molecules have also been reported. Screening of chemical libraries has resulted in the identification of phenazine[78] and guanidine[79] small molecules capable of inhibiting HCV IRES dependent translation. In related work, a class of benzimidazoles (**Figure 8**) was found to bind to an internal loop region of the HCV 5′ UTR stem II, and subsequently to inhibit HCV replication in a cellular assay.[80]

2.26.4.4.3 Human immunodeficiency virus transactivation response

One of the best-characterized and explored structured RNA drug targets is the TAR within the HIV RNA genome (**Figure 6**). TAR is recognized and bound by a virally encoded protein, Tat, which promotes transcription of viral RNA. As a correlation between Tat binding to TAR and HIV mRNA transcription has been shown, the Tat–TAR interaction is

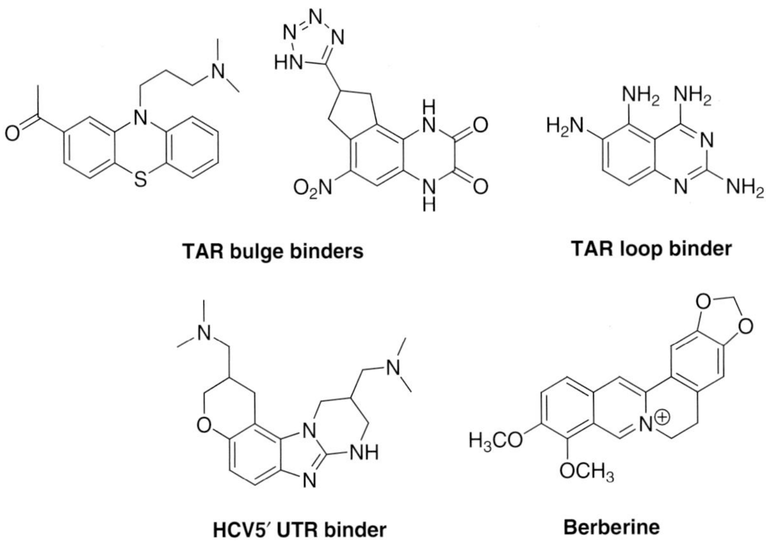

Figure 8 Small molecules that target RNA.

an interesting drug target for HIV chemotherapy. The TAR structure is a stem loop, containing an asymmetric internal loop, or bulge, of three bases. It therefore contains two unique elements for the recognition of drugs and the Tat protein: the terminal stem loop and the internal bulge.

To identify small molecule inhibitors of the Tat–TAR interaction, a corporate library of 150 000 compounds was screened in a high-throughput format. A 1–2% hit rate was found for molecules that showed significant inhibition at 20 μM in the screen. From these hits, approximately 20 compounds were identified that had dose–responsive activity in a cellular assay at nontoxic levels.[81] In continued investigations of the mode of action using mass spectrometry, two compounds, a quinoxaline-2,3-dione and a 2,4-diaminoquinazoline (**Figure 8**), were found to bind the TAR RNA in different locations, and one demonstrated antiviral activity in a cellular assay.[82] Although this work has not resulted in movement of an RNA-binding drug to clinical trials, it clearly demonstrates that high-throughput screening approaches can be successfully applied to structured RNA drug targets.

Computational screening has been used to identify compounds that should bind to the 5′ bulge of TAR RNA. An acetylpromazine (**Figure 8**) lead was identified that inhibited Tat protein binding to TAR RNA at a 100 nM concentration. The nuclear magnetic resonance (NMR) structure of TAR RNA with the ligand showed that binding of the ligand to the bulge induces a modest conformational change but prevents stacking of the upper and lower stems of the RNA via intercalation at the bulge.[83]

2.26.4.4.4 Berberine

Berberine (**Figure 8**) is an alkaloid present in a number of clinically important medicinal plants, extracts of which have been used in Chinese medicine for many years, predominantly as antimicrobial treatments. Recently, berberine has been shown to lower cholesterol in animal models of hyperlipidemia, as well as in humans.[84] This activity has been attributed to the stabilization of the mRNA for low-density lipoprotein receptor (LDLR), resulting in increased production of protein. The LDLR receptor regulates plasma LDL cholesterol levels, and increased expression of LDLR receptor results in increased clearance of LDL cholesterol. That berberine was having a direct effect on the mRNA was further shown by mutational studies in a reporter system where the stabilization effect was mapped to the 5′ proximal region of the LDLR 3′ UTR. The precise biochemical interactions that underlie berberine's effects are unknown. However, the observed clinical and mechanism of action data strongly support that increasing the expression of a protein via stabilization of an mRNA is a viable means to achieve a therapeutic result in patients.

2.26.5 New Directions

2.26.5.1 Riboswitches

Riboswitches are metabolite-binding domains present within certain mRNAs.[85,86] To date riboswitches have been identified as occuring in bacterial, archaeal, fungal and plant mRNAs. Although not identified to date, it is possible that they will also be found to occur in mammalian mRNAs as well. Riboswitches typically contain two domains, an aptameric domain responsible for binding the metabolite and another domain responsible for genetic regulation. Examples of riboswitch ligands include glycine, coenzyme B12, thiamine, flavin mononucleotides, *S*-adenosylmethionine, and guanine.[87,88] As such, riboswitches represent unique target opportunities for drugs. It should, in principle, be possible to identify agonists and antagonists which modulate riboswitch function in an analogous manner to protein receptors. However, studies validating such an approach have yet to be published.

2.26.5.2 Signal Recognition Particle and tmRNA

Additional large RNA targets that may be good therapeutic targets are the bacterial tmRNA and signal recognition particle (SRP) RNA. Both adopt complex tertiary folds stabilized through interactions with the ribosome or cognate proteins. The SRP RNA binds to a stalled ribosome through an unknown structure present in the mRNA for membrane proteins, and translocates the ribosomal machinery to the cell wall to facilitate insertion of hydrophobic proteins directly into the membrane. Knockout of SRP results in bacterial cell death. The tmRNA is a large (~340 nucleotide) RNA that recognizes bacterial ribosomes that have stalled on a damaged mRNA. The tmRNA provides a pseudo-mRNA that swaps into the 30S subunit to replace the damaged mRNA, then presents a tRNA-like domain to the ribosome to add a short peptide sequence that targets the mRNA for degradation.

2.26.5.3 MicroRNA

MicroRNAs (miRNAs) are a newly discovered class of 20–24 nucleotide RNAs critical to a variety of cellular processes in organisms from yeast to humans.[89] They appear to work as inhibitors of translation by binding imperfectly to the 3′ UTR of mRNAs as part of a complex with RISC. miRNAs have also been shown to result in reduction of certain target mRNAs, and it appears that they may also play roles in the transcriptional regulation of gene expression via DNA methylation[90,91] and heterochromatin formation.[92,93] miRNAs are generated from structured regions of transcribed RNAs called pri-miRNAs through the action of the double-stranded RNA nuclease Drosha. The resulting immature hairpin RNAs termed pre-miRNAs are then exported from the nucleus and cleaved in the cytoplasm by another double-stranded RNA nuclease (Dicer) to generate the mature miRNA.[49]

miRNAs are an attractive new class of therapeutic targets, as they are differentially expressed in tissues, and are associated with tumorigenesis, regulation of adipocyte differentiation, insulin secretion, and activation of immune cells.[94–96] Since miRNAs are similar in structure to siRNAs, it is possible to envision the use of siRNA analogs as miRNA mimetics, which would presumably inhibit the expression of the target genes. Perhaps more importantly, inhibition of miRNAs would be expected to increase the expression of its target genes. Inhibition of miRNAs can be achieved with antisense oligonucleotides,[97] which provides a straightforward way examine the potential of miRNAs as targets, as well as a means to develop therapeutics. This is exemplified by studies in which inhibition of a liver specific microRNA (miR-122) was achieved in vivo via the use of uniformly modified oligonucleotides, either with[98] or without[99] cholesterol conjugation. These studies showed upregulation of several target genes, and gave a phenotypic effect of lowered cholesterol and decreased steatohepatitis, suggesting a potential therapeutic indication for microRNA modulation. An additional therapeutic indication for inhibitors of miR-122 is suggested by a connection between miR-122 function and HCV RNA replication, and the observation that inhibition of miR-122 with an antisense oligonucleotide resulted in reduction of HCV RNA levels in an HCV replicon assay.[100] Furthermore, the pre-miRNA and pri-miRNA structures contain bulges and mismatched bases at key regions. These structures are potential binding pockets for small molecules which could interfere with their processing and alter miRNA levels.

References

1. Liebecq, C. *Compendium of Biochemical Nomenclature and Related Documents*, 2nd ed.; Portland Press: London, 1992.
2. Cartegni, L.; Krainer, A. R. *Nat. Struct. Biol.* **2003**, *10*, 120–125.
3. Zhang, C. X.; Lippard, S. J. *Curr. Opin. Chem. Biol.* **2003**, *7*, 481–489.
4. Zhao, G.; Lin, H. *Curr. Med. Chem. Anti-Canc. Agents* **2005**, *5*, 137–147.
5. Hou, M. H.; Wang, A. H. *Nucleic Acids Res.* **2005**, *33*, 1352–1361.
6. Temperini, C.; Cirilli, M.; Aschi, M.; Ughetto, G. *Bioorg. Med. Chem.* **2005**, *13*, 1673–1679.
7. Yokochi, T.; Robertson, K. D. *Mol. Pharmacol.* **2004**, *66*, 1415–1420.
8. Rowe, T. C.; Weissig, V.; Lawrence, J. W. *Adv. Drug Deliv. Rev.* **2001**, *49*, 175–187.
9. Dervan, P. B. *Bioorg. Med. Chem.* **2001**, *9*, 2215–2235.
10. Dervan, P. B.; Edelson, B. S. *Curr. Opin. Struct. Biol.* **2003**, *13*, 284–300.
11. Ansari, A. Z.; Mapp, A. K.; Nguyen, D. H.; Dervan, P. B.; Ptashne, M. *Chem. Biol.* **2001**, *8*, 583–592.
12. Sharma, M. R.; Koc, E. C.; Datta, P. P.; Booth, T. M.; Spremulli, L. L.; Agrawal, R. K. *Cell* **2003**, *115*, 97–108.

13. Yusupov, M. M.; Yusupova, G. Z.; Baucom, A.; Lieberman, K.; Earnest, T. N.; Cate, J. H.; Noller, H. F. *Science* **2001**, *292*, 883–896.
14. Harms, J. M.; Bartels, H.; Schlunzen, F.; Yonath, A. *J. Cell Sci.* **2003**, *116*, 1391–1393.
15. Wilson, D. N.; Nierhaus, K. H. *Angew. Chem. Int. Ed. Engl.* **2003**, *42*, 3464–3486.
16. Chambers, H. F. In *Goodman & Gilman's The Pharmacological Basis of Therapeutics*, 10th ed.; Hardman, J. G., Limbird, L. E., Eds.; McGraw-Hill: New York, 2001, pp 1219–1238.
17. Zhao, H.; Li, R.; Wang, Q.; Yan, Q.; Deng, J. H.; Han, D.; Bai, Y.; Young, W. Y.; Guan, M. X. *Am. J. Hum. Genet.* **2004**, *74*, 139–152.
18. Chambers, H. F. In *Goodman & Gilman's The Pharmacological Basis of Therapeutics*, 10th ed.; Hardman, J. G., Limbird, L. E., Eds.; McGraw-Hill: New York, 2001, pp 1239–1271.
19. Prystowsky, J.; Siddiqui, F.; Chosay, J.; Shinabarger, D. L.; Millichap, J.; Peterson, L. R.; Noskin, G. A. *Antimicrob. Agents Chemother.* **2001**, *45*, 2154–2156.
20. Bersos, Z.; Maniati, M.; Kontos, F.; Petinaki, E.; Maniatis, A. N. *J. Antimicrob. Chemother.* **2004**, *53*, 685–686.
21. Benson, D. A.; Karsch-Mizrachi, I.; Lipman, D. J.; Ostell, J.; Wheeler, D. L. *Nucleic Acids Res.* **2004**, *32*, D23–D26.
22. Kim, C. H.; Sarma, R. H. *Nature* **1977**, *270*, 223–235.
23. Collins, C. A.; Guthrie, C. *Nat. Struct. Biol.* **2000**, *7*, 850–854.
24. Kornblihtt, A. R.; de la Mata, M.; Fededa, J. P.; Munoz, M. J.; Nogues, G. *RNA* **2004**, *10*, 1489–1498.
25. Wedekind, J. E.; Dance, G. S.; Sowden, M. P.; Smith, H. C. *Trends Genet.* **2003**, *19*, 207–216.
26. Bass, B. L. *Annu. Rev. Biochem.* **2002**, *71*, 817–846.
27. Tamura, M.; Hendrix, D. K.; Klosterman, P. S.; Schimmelman, N. R. B.; Brenner, S. E.; Holbrook, S. R. *Nucleic Acids Res.* **2004**, *32*, D182–D184.
28. Kieft, J. S.; Zhou, K.; Grech, A.; Jubin, R.; Doudna, J. A. *Nat. Struct. Biol.* **2002**, *9*, 370–374.
29. Egli, M.; Sarkhela, S.; Minasov, G.; Rich, A. *Helv. Chim. Acta* **2003**, *86*, 1709.
30. Chen, J. L.; Greider, C. W. *Proc. Natl. Acad. Sci. USA* **2005**, *102*, 8080–8085.
31. Theimer, C. A.; Blois, C. A.; Feigon, J. *Mol. Cell* **2005**, *17*, 671–682.
32. Lima, W. F.; Wu, H.; Crooke, S. T. *Methods Enzymol.* **2001**, *341*, 430–440.
33. Elbashir, S. M.; Harborth, J.; Lendeckel, W.; Yalcin, A.; Weber, K.; Tuschl, T. *Nature* **2001**, *411*, 494–497.
34. Cech, T. R. *Curr. Opin. Struct. Biol.* **1992**, *2*, 605–609.
35. Wu, H.; Lima, W. F.; Zhang, H.; Fan, A.; Sun, H.; Crooke, S. T. *J. Biol. Chem.* **2004**, *279*, 17181–17189.
36. Usman, N.; Blatt, L. M. *J. Clin. Invest.* **2000**, *106*, 1197–1202.
37. Helene, C.; Toulme, J.-J. *Biochim. Biophys. Acta* **1990**, *1049*, 99–125.
38. Baker, B. F.; Lot, S. S.; Condon, T. P.; Cheng-Flournoy, S.; Lesnik, E. A.; Sasmor, H. M.; Bennett, C. F. *J. Biol. Chem.* **1997**, *272*, 11994–12000.
39. Sierakowska, H.; Sambade, M. J.; Agrawal, S.; Kole, R. *Proc. Natl. Acad. Sci. USA* **1996**, *93*, 12840–12844.
40. Vitravene Study Group. *Am. J. Ophthalmol.* **2002**, *133*, 475–483.
41. Bennett, C. F.; Swayze, E.; Geary, R.; Levin, A. A.; Mehta, R.; Teng, C. L.; Tillman, L.; Hardee, G. In *Gene and Cell Therapy: Therapeutic Mechanisms and Strategies*; Templeton, N. S., Ed.; Marcel Dekker: New York, 2004, pp 347–374.
42. Geary, R. S.; Yu, R. Z.; Levin, A. A. *Curr. Opin. Investig. Drugs* **2001**, *2*, 562–573.
43. Geary, R. S.; Watanabe, T. A.; Truong, L.; Freier, S.; Lesnik, E. A.; Sioufi, N. B.; Sasmor, H.; Manoharan, M.; Levin, A. A. *J. Pharmacol. Exp. Ther.* **2001**, *296*, 890–897.
44. Henry, S. P.; Geary, R. S.; Yu, R.; Levin, A. A. *Curr. Opin. Investig. Drugs* **2001**, *2*, 1444–1449.
45. Zinker, B. A.; Rondinone, C. M.; Trevillyan, J. M.; Gum, R. J.; Clampit, J. E.; Waring, J. F.; Xie, N.; Wilcox, D.; Jacobson, P.; Frost, L. et al. *Proc. Natl. Acad. Sci. USA* **2002**, *99*, 11357–11362.
46. Crooke, R. M.; Graham, M. J.; Lemonidis, K. M.; Whipple, C. P.; Koo, S.; Perera, R. J. *J. Lipid Res.* **2005**, *46*, 872–884.
47. Zamore, P. D. *Science* **2002**, *296*, 1265–1269.
48. Silva, J.; Chang, K.; Hannon, G. J.; Rivas, F. V. *Oncogene* **2004**, *23*, 8401–8409.
49. Tomari, Y.; Zamore, P. D. *Genes Dev.* **2005**, *19*, 517–529.
50. Tamari, Y.; Matranga, C.; Haley, B.; Martinez, N.; Zamore, P. D. *Science* **2004**, *306*, 1377–1380.
51. Song, J. J.; Smith, S. K.; Hannon, G. J.; Joshua-Tor, L. *Science* **2004**, *305*, 1434–1437.
52. Liu, J.; Carmell, M. A.; Rivas, F. V.; Marsden, C. G.; Thomson, J. M.; Song, J.-J.; Hammond, S. M.; Joshua-Tor, L.; Hannon, G. J. *Science* **2004**, *10*, 1–5.
53. Prakash, T. P.; Allerson, C. R.; Dande, P.; Vickers, T. A.; Sioufi, N.; Jarres, R.; Baker, B. F.; Swayze, E. E.; Griffey, R. H.; Bhat, B. *J. Med. Chem.* **2005**, *48*, 4247–4253.
54. Allerson, C. R.; Sioufi, N.; Jarres, R.; Prakash, T. P.; Naik, N.; Berdeja, A.; Wanders, L.; Griffey, R. H.; Swayze, E. E.; Bhat, B. *J. Med. Chem.* **2005**, *48*, 901–904.
55. Soutschek, J.; Akinc, A.; Bramlage, B.; Charisse, K.; Constien, R.; Donoghue, M.; Elbashir, S. M.; Geick, A.; Hadwiger, P.; Harborth, J. et al. *Nature* **2004**, *3121*, 1–9.
56. Schiffelers, R. M.; Ansari, A. M.; Xu, J.; Zhou, Q.; Storm, G.; Molema, G.; Lu, P. Y.; Scaria, P. V.; Woodle, M. C. *Nucleic Acids Res.* **2004**, *32*, e149.
57. Pal, A.; Ahmad, A.; Khan, S.; Sakabe, I.; Zhang, C.; Kasid, U. N.; Ahmad, I. *Int. J. Oncol.* **2005**, *26*, 1087–1091.
58. Morrissey, D. V.; Lockridge, J. A.; Shaw, L.; Blanchard, K.; Jensen, K.; Breen, W.; Hartsough, K.; Machemer, L.; Radka, S.; Jadhav, V. et al. *Nat. Biotechnol.* **2005**, *23*, 1002–1007.
59. Baker, B. F.; Ramasamy, K.; Kiely, J. *Bioorg. Med. Chem. Lett.* **1996**, *6*, 1647–1652.
60. Zhang, Z.; Lonnberg, H.; Mikkola, S. *Org. Biomol. Chem.* **2003**, *7*, 3404–3409.
61. LaGrandeur, T. E.; Parker, R. *Biochimie* **1996**, *78*, 1049–1055.
62. Kole, R.; Vacek, M.; Williams, T. *Oligonucleotides* **2004**, *14*, 65–74.
63. Mercatante, D. R.; Mohler, J. L.; Kole, R. *J. Biol. Chem.* **2002**, *277*, 49374–49832.
64. Taylor, J. K.; Dean, N. M. *Curr. Opin. Drug Disc. Dev.* **1999**, *2*, 147–151.
65. Dasgupta, A.; Das, S.; Izumi, R.; Venkatesan, A.; Barat, B. *FEMS Microbiol. Lett.* **2004**, *234*, 189–199.
66. Holcik, M. *Curr. Cancer Drug Targets* **2004**, *4*, 299–311.
67. Stoneley, M.; Willis, A. E. *Oncogene* **2004**, *23*, 3200–3207.
68. Kozak, M. *Gene* **2003**, *318*, 1–23.
69. Hellen, C. U. T.; Sarnow, P. *Genes Dev.* **2001**, *15*, 11593–11612.
70. Bonnal, S.; Boutonnet, C.; Prado-Lourenco, L.; Vagner, S. *Nucleic Acids Res.* **2003**, *31*, 427–428.
71. Spahn, C. M. T.; Kieft, J. S.; Grassucci, R. A.; Penczek, P.; Zhou, K.; Doudna, J. A.; Frank, J. *Science* **2001**, *291*, 1959–1962.

72. Klinck, R.; Westhof, E.; Walker, S.; Afshar, M.; Collier, A.; Aboul-Ela, F. *RNA* **2000**, *6*, 1423–1431.
73. Lukavsky, P. J.; Otto, G. A.; Lancaster, A. M.; Sarnow, P.; Puglisi, J. D. *Nat. Struct. Biol.* **2000**, 7, 1105–1110.
74. Hanecak, R.; Brown-Driver, V.; Fox, M. C.; Azad, R. F.; Furusako, S.; Nozaki, C.; Ford, C.; Sasmor, H.; Anderson, K. P. *J. Virol.* **1996**, *70*, 5203–5212.
75. Brown-Driver, V.; Fox, M. C.; Lesnik, E.; Hanecak, R.; Anderson, K. *Antisense Nucleic Acid Drug Dev.* **1999**, *9*, 145–154.
76. Tallet-Lopez, B.; Aldaz-Carroll, L.; Chabas, S.; Dausse, E.; Staedel, C.; Toulme, J. J. *Nucleic Acids Res.* **2003**, *31*, 734–742.
77. Kikuchi, K.; Umehara, T.; Fukuda, K.; Kuno, A.; Hasegawa, T.; Nicshikawa, S. *Nucleic Acids Res.* **2005**, *33*, 683–692.
78. Wang, W.; Preville, P.; Morin, N.; Mounir, S.; Cai, W.; Siddiqui, M. A. *Bioorg. Med. Chem. Lett.* **2000**, *10*, 1151–1154.
79. Jefferson, E. A.; Seth, P. P.; Robinson, D. E.; Winter, D. K.; Miyaji, A.; Osgood, S. A.; Swayze, E. E.; Risen, L. M. *Bioorg. Med. Chem. Lett.* **2004**, *14*, 5139–5143.
80. Seth, P. P.; Miyaji, A.; Jefferson, E. A.; Sannes-Lowery, K. A.; Osgood, S. A.; Propp, S. S.; Ranken, R.; Massire, C.; Sampath, R.; Ecker, D. J. et al. *J. Med. Chem.* **2005**, *48*, 7099–7102.
81. Mei, H.-Y.; Mack, D. P.; Galan, A. A.; Halim, N. S.; Heldsinger, A.; Loo, J. A.; Moreland, D. W.; Sannes-Lowery, K. A.; Sharmeen, L.; Truong, H. N. et al. *Bioorg. Med. Chem. Lett.* **1997**, *5*, 1173–1184.
82. Mei, H. Y.; Cui, M.; Heldsinger, A.; Lemrow, S. M.; Loo, J. A.; Sannes-Lowery, K. A.; Sharmeen, L.; Czarnik, A. W. *Biochemistry* **1998**, *37*, 14204–14212.
83. Du, Z.; Lind, K. E.; James, T. L. *Chem. Biol.* **2002**, *9*, 707–712.
84. Kong, W.; Wei, J.; Abidi, P.; Lin, M.; Inaba, S.; Li, C.; Wang, Y.; Wang, Z.; Si, S.; Pan, H. et al. *Nat. Med.* **2004**, *10*, 1344–1351.
85. Mandal, M.; Boese, B.; Barrick, J. E.; Winkler, W. C.; Breaker, R. R. *Cell* **2003**, *113*, 577–586.
86. Mandal, M.; Lee, M.; Barrick, J. E.; Weinberg, Z.; Emilsson, G. M.; Ruzzo, W. L.; Breaker, R. R. *Science* **2004**, *306*, 275–279.
87. Breaker, R. R. *Nature* **2004**, *432*, 838–845.
88. Winkler, W. C.; Nahvi, A.; Sudarsan, N.; Barrick, J. E.; Breaker, R. R. *Nat. Struct. Biol.* **2003**, *10*, 701–707.
89. Bartel, D. P.; Chen, C. Z. *Nat. Rev. Genet.* **2004**, *5*, 396–400.
90. Morris, K. V.; Chan, S. W.; Jacobsen, S. E.; Looney, D. J. *Science* **2004**, *305*, 1289–1292.
91. Kawasaki, H.; Taira, K. *Nature* **2004**, *431*, 211–217.
92. Cam, H. P.; Sugiyama, T.; Chen, E. S.; Chen, X.; FitzGerald, P. C.; Grewal, S. I. *Nat. Genet.* **2005**, *37*, 809–819.
93. Verdel, A.; Jia, S.; Gerber, S.; Sugiyama, T.; Gygi, S.; Grewal, S. I.; Moazed, D. *Science* **2004**, *303*, 672–676.
94. Chen, C.-Z.; Li, L.; Lodish, H. F.; Bartel, D. P. *Science* **2004**, *303*, 83–87.
95. Calin, G. A.; Sevignani, C.; Dumitru, C. D.; Hyslop, T.; Noch, E.; Yendamuri, S.; Shimizu, M.; Rattan, S.; Bullrich, F.; Negrini, M. et al. *Proc. Natl. Acad. Sci. USA* **2004**, *101*, 2999–3004.
96. Poy, M. N.; Eliasson, L.; Krutzfeldt, J.; Kuwajima, S.; Ma, X.; MacDonald, P. E.; Pfeffer, S.; Tuschl, T.; Rajewsky, N.; Rorsman, P. et al. *Nature* **2004**, *432*, 226–230.
97. Esau, C.; Kang, X.; Peralta, E.; Hanson, E.; Marcusson, E. G.; Ravichandran, L. V.; Sun, Y.; Koo, S.; Perera, R. J.; Jain, R. et al. *J. Biol. Chem.* **2004**, *279*, 52361–52365.
98. Krutzfeldt, J.; Rajewsky, N.; Braich, R.; Rajeev, K. G.; Tuschl, T.; Manoharan, M.; Stoffel, M. *Nature* **2005**, *43*, 685–689.
99. Esau, C.; Davis, S.; Murray, S. F.; Yu, X. X.; Pandey, S. K.; Pear, M.; Watts, L.; Booten, S. L.; Graham, M.; McKay, R. et al. *Cell Metab.* **2006**, *3*, 87–98.
100. Jopling, C. L.; Yi, M.; Lancaster, A. M.; Lemon, S. M.; Sarnow, P. *Science* **2005**, *309*, 1577–1581.

Biographies

Eric E Swayze received a BS in chemistry degree from the University of Michigan honors college in 1987, and a PhD in organic chemistry from the University of Michigan in 1994 under the guidance of Prof Leroy B Townsend. In 1994, Dr Swayze joined Isis Pharmaceuticals as a Sr Scientist, where he has focused on developing drugs which act on RNA targets, including both oligonucleotide and small molecule therapeutics. Dr Swayze is currently Vice President of Medicinal Chemistry at Isis Pharmaceuticals of Carlsbad, CA.

Rich H Griffey is currently a Senior Engineer at SAIC in San Diego. He received a BA in chemistry from Rice University in 1978. His PhD in organic chemistry was obtained in 1983 under the guidance of C Dale Pounter at the University of Utah. After a postdoctoral fellowship with Alfred Redfield, Dr Griffey has held jobs in both academia and industry. His interests include analytical and synthetic chemistry at the chemistry/biology interface, cycling, and astrophotography.

C Frank Bennett is Senior Vice President of Research at Isis Pharmaceuticals. He is responsible for preclinical antisense drug discovery research, manufacturing and pharmaceutics. Dr Bennett is one of the founding members of the company. He has been involved in the development of antisense oligonucleotides as therapeutic agents, including research on the application of oligonucleotides for inflammatory and cancer targets, oligonucleotide delivery, and pharmacokinetics. He also runs the company's antisense mechanism program which is focused on the development of RNase H, RNAi, micro-RNA, and splicing. Dr Bennett has published more than 120 papers in the field of antisense research and development and has more than 100 issued US patents.

Prior to joining Isis, Dr Bennett was Associate Senior Investigator in the Department of Molecular Pharmacology at Smith Kline and French Laboratories (currently GlaxoSmithKline).

Dr Bennett received his BS degree in pharmacy from the University of New Mexico, Albuquerque, New Mexico and his PhD in pharmacology from Baylor College of Medicine, Houston, Texas. Dr Bennett performed his postdoctoral research in the Department of Molecular Pharmacology at SmithKline and French Laboratories.

2.27 Redox Enzymes

J A Dykens, EyeCyte Therapeutics, Encinitas, CA, USA

© 2007 Elsevier Ltd. All Rights Reserved.

2.27.1	**Introduction**	**1054**
2.27.2	**Mitochondria**	**1055**
2.27.2.1	Biology	1055
2.27.2.2	Membranes and Electron Transport Complexes	1057
2.27.2.3	Electron Transfer Models	1057
2.27.2.4	Nernst Equation	1058
2.27.2.5	Inner Membrane	1059
2.27.2.6	Membrane Potential	1059
2.27.2.7	Mitochondrial Genome	1059
2.27.2.8	Mitochondrial Diseases	1060
2.27.2.9	Mitochondrial Failure and Radical Production	1060
2.27.3	**Mitoprotective Drug Development**	**1061**
2.27.3.1	Mitochondrial Permeability Transition as Target	1061
2.27.3.2	Caveats	1063
2.27.3.3	Novel Steroids with a Novel Mechanism	1063
2.27.3.4	Other molecules	1064
2.27.4	**Mitochondrial Radical Production**	**1064**
2.27.4.1	Biology and Sites of Production	1064
2.27.5	**Mitochondrial Diversity**	**1065**
2.27.5.1	Biology	1065
2.27.5.2	Electron Paramagnetic Resonance Studies	1065
2.27.6	**Mitochondrial Drug Targeting**	**1067**
2.27.6.1	Membrane Potential	1067
2.27.7	**Other Mitoactive Compounds**	**1068**
2.27.8	**Electron Transport Components**	**1068**
2.27.8.1	Complex I (EC 1.6.5.3)	1068
2.27.8.1.1	Biology	1068
2.27.8.1.2	Inhibitors	1068
2.27.8.1.3	Composition	1070
2.27.8.1.4	Catalysis	1070
2.27.8.1.5	Pathogenesis	1071
2.27.8.1.6	Drug development	1071
2.27.8.2	Complex II (EC 1.3.5.1)	1072
2.27.8.2.1	Biology	1072
2.27.8.2.2	Composition	1072
2.27.8.2.3	Inhibitors	1072
2.27.8.2.4	Pathology	1073
2.27.8.2.5	Huntington's Disease Model	1073
2.27.8.3	Ubiquinone (Coenzyme Q, Q_{10})	1073
2.27.8.3.1	Biology	1073
2.27.8.3.2	Pathology	1074
2.27.8.3.3	Drug development	1074
2.27.8.4	Complex III (EC 1.10.2.2)	1074
2.27.8.4.1	Biology and composition	1074

	2.27.8.4.2	Catalysis	1075
	2.27.8.4.3	Inhibitors	1075
	2.27.8.4.4	Pathology	1075
2.27.8.5		Cytochrome c	1076
	2.27.8.5.1	Biology	1076
	2.27.8.5.2	Apoptosis	1076
	2.27.8.5.3	Cardiolipin interactions	1077
2.27.8.6		Complex IV (EC1.9.3.1)	1077
	2.27.8.6.1	Composition and catalysis	1077
	2.27.8.6.2	Inhibitors	1077
	2.27.8.6.3	Oxidative regulation	1077
	2.27.8.6.4	Regulation versus pathology	1078
2.27.8.7		Complex V (EC 3.6.1.34)	1078
	2.27.8.7.1	Composition and catalysis	1078
	2.27.8.7.1	Regulation	1078
	2.27.8.7.2	Other inhibitors	1079
	2.27.8.7.3	Pathology	1079
2.27.9		**Other Mitochondrial Drug Development Strategies**	**1079**
2.27.9.1		Uncoupling OxPhos	1079
	2.27.9.1.1	Pharmacological approach	1079
	2.27.9.1.2	Uncoupling proteins	1080
	2.27.9.1.3	Caveat	1080
2.27.9.2		Manipulating Mitochondrial Biomass	1080
	2.27.9.2.1	Transcriptional regulation	1080
	2.27.9.2.2	Pharmacological approach	1081
2.27.9.3		Antiviral Effects on Mitochondria	1082
2.27.10		**Summary**	**1082**
References			**1083**

2.27.1 Introduction

It has been said that reduction–oxidation (redox) reactions are the foundation of life – an accurate reflection in that much of cellular anabolic and catabolic metabolism is based on enzyme-catalyzed redox reactions. It is impossible to characterize coherently the biochemical complexity, regulatory networks, and mechanistic insights of all redox enzymes in one chapter. However, comprehensive reviews and texts cover this ground, such as the recent text by Copeland.[1]

Readers interested in the redox biology of oxygen- and nitrogen-centered radicals, and their interactions, are encouraged to consult the comprehensive two-volume series edited by Sen and Packer in *Methods in Enzymology*,[2] the classic text by Halliwell and Gutteridge,[3] and the works by Cadenas and Packer,[4] Balaban *et al.*,[5] and Duncan and Heales.[6] Similarly, readers interested in mitochondrial biology and bioeneregetics would do well to consult the books by Nicholls and Ferguson,[7] Beal *et al.*,[8] Scheffler,[9] Lane,[10] and Berdanier.[11]

Although redox enzymes have received much attention from analytical and pharmaceutical perspectives, this is not the case for the medicinal chemistry of these systems in mitochondria. Indeed, mitochondrial dysfunction and failure are increasingly implicated in the etiology of both acute and chronic human diseases. For example, acute mitochondrial collapse and ensuing apoptosis cause cell death in the lesion penumbra resulting from transient ischemia in stroke, myocardial infarction, and other instances of ischemia–reperfusion.[12,13] On the other hand, accelerated mitochondrial senescence and dysfunction are central events in a host of neurodegenerative diseases, including Parkinson's, Huntington's (HD), amyotrophic lateral sclerosis (ALS), Friedreich's ataxia, and likely Alzheimer's (AD), many retinodegenerative diseases, including retinitis pigmentosa, glaucoma and age-related macular degeneration (AMD), and in metabolic diseases such as type II (noninsulin-dependent) diabetes (NIDDM).[14–24] This is in addition to the more than 75 syndromes and diseases caused by well-defined mutations in the mitochondrial DNA (mtDNA) and/or in nuclear genes of proteins destined for import into the mitochondria. The journal *Neuromuscular Disorders* publishes periodic updates on mitochondrial diseases, and the most recent review at the time of this writing of mitochondrial encephalomyopathies is from DiMauro and Hirano.[25]

It is clear that mitochondrial pathology is maturing into a major arena for therapeutic intervention, and hence a rapidly expanding field for medicinal chemists. As a result, this is a timely opportunity to summarize current understandings of mitochondrial function and dysfunction in disease, the chemistry of known mitochondrial inhibitors, and ongoing pharmaceutical development efforts targeting mitochondrial integrity as a therapeutic strategy. Given the complexity and diversity of mitochondrial redox systems, restricting perspective to mitochondrial biology does not correspondingly limit the breadth of topics anticipated in a review of redox or transporter biology; the redox systems in mitochondria provide a microcosm of the panoply of [Fe-S], heme, and other metalloprosthetic moieties found in nonmitochondrial biology.[26–29]

When mitochondria fail, the cell dies. In large measure, this dictum reflects why the study of mitochondrial physiology and pathology has been reinvigorated in recent years. Increased understanding of the separate redox reactions that constitute the electron transport system (ETS) has helped illuminate the sites and character of potential dysfunction, typically reflected by free radical production, collapse of membrane potential, and cellular energetic impairment.[15–17,30,31] Such electron flow does not occur without highly regulated spatial constraints, which entails an understanding not only of protein–protein interactions, but also of how the lipid domain of the inner membrane contributes to mitochondrial function and failure. Indeed, mitochondrial failure can be most basically defined as the loss of the inner membrane impermeability to protons, which imposes both oxidative and energetic crises that conspire to undermine cellular viability.

There has been a pervasive myth in the pharmaceutical realm that 'mitochondria are undruggable,' fostered in part because mitochondrial dysfunction characteristically yields multiorgan pathology that is often profoundly debilitating, and that seems to defy rational symptom-oriented drug development strategies. Many of the known mitochondrial diseases are rare by pharmaceutical industry standards, early-onset, and often lethal, which discourages many companies from even entering the arena. To some extent, the current paucity of mitochondrial medicines also reflects a lingering legacy of the fact that the first mitochondrial drug, dinitrophenol (DNP) was highly toxic, and in many cases, lethal. DNP was sold in the early part of the twentieth century for weight loss, and although it was effective at low doses, it was justifiably included in the US Food and Drug Administration's (FDA's) list of snake oils and injurious medicines that resulted in the 1938 Federal Food, Drug, and Cosmetic Act. This Act gave the FDA enforcement power to actively assess and pursue drug safety. Moreover, the 'undruggable' myth has been fueled by the inaccurate notion that 'all mitochondria are the same,' thereby precluding the opportunity to capitalize on tissue selectivity to limit potential drug side-effects and toxicity. It is our hope that this chapter will help dispel these myths that have previously impeded mitochondrial drug development.

Because no single chapter can do justice to the vast literature on mitochondrial physiology and pathology, citations have been skewed toward reviews that provide more comprehensive treatments of individual topics. In addition to directing the reader to the relevant literature, other goals here are to provide an overview of:

- mitochondrial biology and genetics, and how dysfunction contributes to human disease;
- the respiratory complexes, including electron flow, redox centers, and known inhibitors. It is hoped that consideration of inhibitors, many of which are natural products, will encourage exploration of structural variations;
- how oxidative phosphorylation (OxPhos) and oxidative pathology converge, and how these interactions may provide therapeutic opportunities;
- mitochondrial failure;
- strategies that have been used to develop mitochondrially directed therapeutics.

2.27.2 Mitochondria

2.27.2.1 Biology

Mitochondria are a double-membraned organelle of prokaryotic origin found in almost all eukaryotic organisms in direct proportion to the energy demand of the cell and tissue.[32–34] Highly active cells can contain hundreds to thousands of individual mitochondria, and in those rare instances where mitochondria are absent, they have been lost via secondary deletion with a corresponding increased reliance on anaerobic glycolysis. A number of independent lines of evidence indicate that mitochondria originated from a free-living bacterial endosymbiont some 2 billion years ago.[9,10] The transition from ancestral endosymbiont to modern mitochondrion has been accompanied by major changes in its proteome, extensive metabolic remodeling, and profound reduction of the endogenous genome via gene transfer to the nuclear genome.[35,36] Mitochondria reproduce independently of the cell cycle, contain their own genome with a genetic code that differs from that in the nuclear genome, and are dynamic structures within the cell, fusing into megamitochondrial networks in some instances, while budding into the classically shown bean shape, as shown in **Figure 1**, in others. The images in **Figure 1** use electron tomography to generate three-dimensional reconstructions from serial two-dimensional images, as described by Frey and Mannella.[37]

A more reticular structure of the mitochondrial population is shown in **Figure 2**, where staining with an anticytochrome c antibody reveals the mitochondrial network encircling the nucleus in the cell body, and extending into the dendrites and axon of a rat hippocampal neuron in primary culture.

The equilibrium between solo mitochondria and a reticulum is largely regulated by the mitofusin family of proteins which has become an area of active investigation, and may offer a viable avenue for pharmaceutical development.[38–41] The dynamic nature of this equilibrium is revealed via time-lapse imaging.

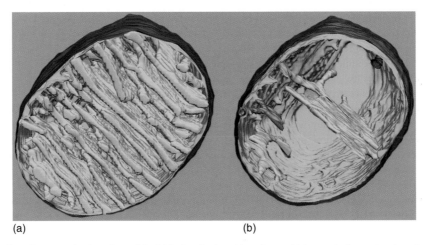

(a)   (b)

Figure 1 Electron tomography imaging of individual mitochondrion from chick cerebellum showing three-dimensional ultrastructure. (a) The outer membrane (dark blue) is reminiscent of the endoplasmic reticulum in composition. The inner membrane invaginates into the matrix space to form cristae (yellow): this increases surface area, and hence oxidative phosphorylation capacity. The cristae form 30 nm tubular junctions, with the inner boundary membrane (light blue), forming a continuous surface. (b) Four individual cristae are shown to highlight anatomical complexity ranging from simple tube (red), more elaborate examples with several cristal junctions variability (green and gray), to one with many junctions plus both lamellar and tubular components (yellow). Many of the proteins involved with transmembrane exchange of substrates and proteins, such as the adenine nucleotide translocator (ANT) and voltage-dependent anion transporter (VDAC), are localized at the cristal junctions. (Reproduced with permission from Frey, T. G.; Mannella, C. A. *Trends Biochem. Sci.* **2000**, *27*, 319–324.)

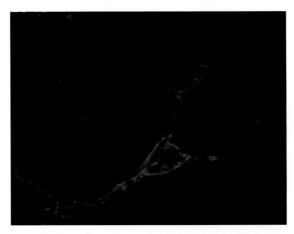

Figure 2 Characteristic reticular anatomy and perinuclear density of mitochondria. Budding into typically shown 'Bean' shape and aggregation of reticulum is dynamic. Mitochondria stained with an anticytochrome *c* antibody. (Reproduced with permission of Sandra Wiley, University of California at San Diego.)

Although mitochondria generate >90% of the energy required for aerobically poised cells to survive, in so doing they also generate >90% of the highly reactive oxygen-centered radicals that undermine cell viability.[2–6] Collapse of mitochondrial function is lethal for aerobically poised cells, leading to apoptosis or necrosis depending on the severity and duration of mitochondrial dysfunction. Mitochondria integrate a host of pro- and antiapoptotic and necrotic signaling pathways, and are the initial domino in several cell death cascades.[13,42] Indeed, mitochondrial dysfunction is rapidly becoming appreciated as an important etiological component in a wide variety of both acute and chronic human diseases,[43–45] as well as in toxicity of drugs, including antivirals and cisplatin.[46] This is increasingly reinforced by insights into free radical-mediated regulation of respiration, which in turn alters the oxidative poise of the organelle and the cell.[15,47,48]

2.27.2.2 Membranes and Electron Transport Complexes

The outer, relatively permeable, mitochondrial membrane is reminiscent of endoplasmic reticulum in both lipid composition and transport physiology. Not so the inner membrane surrounding the central matrix space which contains the Kreb's cycle enzymes. The inner membrane contains approximately twice the amount of phosphatidylethanolamine, and half the amount of phosphatidylcholine, of the outer membrane, plus about 20% by weight cardiolipin (diphosphatidylglycerin), an atypical membrane lipid with four highly unsaturated acyl chains found in the cell almost exclusively in the mitochondrial inner membrane.[49]

Over 80% (by weight) of the inner membrane is composed of proteins, the highest density within the cell. In large measure, these proteins are the four respiratory complexes: C-I, NADH-CoQ oxidoreductase; C-II, succinate dehydrogenase-CoQ oxidoreductase; C-III, CoQ-cytochrome c oxidoreductase; and C-IV, cytochrome c oxidase (COX), plus adenosine triphosphate (ATP) synthase, frequently referred to as Complex V, although unlike the other four complexes, it does not catalyze a redox reaction.

OxPhos includes two separate, but interdependent, processes: electron transfer system (ETS) and ATP synthase (**Figures 3** and **4**). The first entails the sequential redox transfer of electrons, removed from pyruvate in the Krebs cycle within the matrix, down a potential energy gradient via the four respiratory complexes, ultimately reducing molecular O_2 to H_2O. In so doing, protons are translocated out of the matrix to yield a transmembrane electrochemical potential. In the second phase of OxPhos, the chemiosmotic potential generated by ETS drives ADP phosphorylation via ATP synthase.[9]

2.27.2.3 Electron Transfer Models

The initial models of mitochondrial respiration described electron transfer as a 'chain' of respiratory centers, connoting a 1:1 linkage with a coordinated linear redox transfer. However, it was noted early on that the molar ratios of the various respiratory components were not uniform, and for every C-I, there were 3 mol C-III, 9 mol cytochrome c, 50 mol

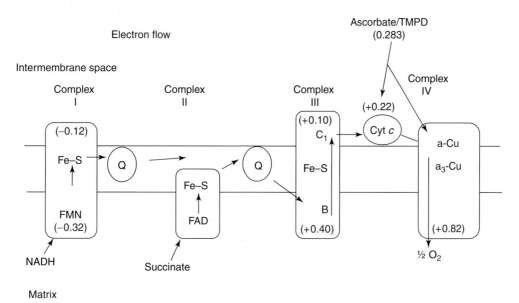

Figure 3 Schematic of the four respiratory complexes within the inner mitochondrial membrane, with electron flow and some of the mid-point redox potentials indicated.

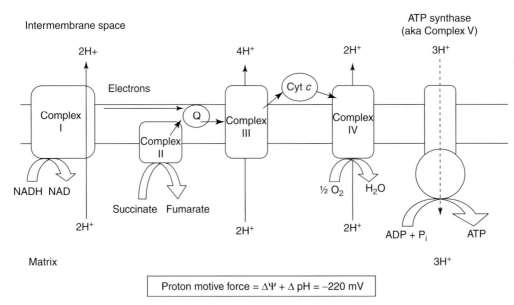

Figure 4 Three respiratory complexes couple redox catalysis with proton translocation, thereby generating a transmembrane potential of ~220 mV (matrix negative) due to both electrical and pH gradients. ATP synthase, also known as Complex V, couples the dissipation of the proton motive force to phosphorylation of ADP.

ubiquinone, and 7 mol C-IV.[50] Despite molar inequalities, factoring in the lateral diffusion coefficients of the respiratory complexes yielded a model of electron transfer much closer to unity, suggesting that catalysis of respiration was mediated by collisions between individual ETS components within the inner membrane.[50] More recent studies of *Escherichia coli*, *Saccharomyces cerevisiae* and mammals have revealed that individual respiratory complexes congregate into larger heteromeric assemblies, and that these in turn aggregate into higher-order supercomplexes called 'respiratosomes.'[51] These supercomplexes can be dissociated via detergent treatment, and the presence of cardiolipin is required to maintain quaternary structure, at least for the association between C-III/C-IV in yeast.[52] In mammals the respiratosomes appear to be constructed of two aggregates consisting of C-I and 2 C-II and 4 C-IV, and 2 C-III and 4 C-IV.[51] Note that the subunit composition of such supercomplexes reflects the molar concentrations noted above, and that it may be that the lateral diffusion coefficients of ubiquinone and cytochrome *c* serve as the proximate determinants of catalytic rates within these supercomplexes (although pool behavior may not always hold).[51]

2.27.2.4 Nernst Equation

Within the dictates of reactant concentration, and importantly pH, the equilibria of redox reactions follow the half-cell potential of the redox pair, as predicted by the Nernst equation:

$$E = E^0 \frac{RT}{nF} \ln \frac{a_{[red]}}{a_{[ox]}}$$

where E^0, the standard electrode potential, is the electrode potential at the standard conditions of temperature 298 K, 1 atm pressure, and at 1 mol of the activity of redox participants of the half-reaction. And where:

- R is the gas constant, equal to 8.31 J K^{-1} mol^{-1}
- T is the temperature in Kelvin
- a is the chemical activities on the reduced and oxidized side, respectively
- F is the Faraday constant, equal to 9.648×10^4 C mol^{-1}
- n is the number of electrons transferred in the half-reaction
- [*red*] is the concentration of the reduced species
- [*ox*] is the concentration of the oxidized species

Moreover, since:

$$\Delta G^0 = -nF\Delta E^0$$

it follows that an electron transfer is spontaneous (negative ΔG) if $E^{0\prime}$ (mid-point potential) of the e$^-$ donor is more negative than $E^{0\prime}$ of the e$^-$ acceptor, i.e., when there is a positive $\Delta E^{0\prime}$.

Thus, the mid-point potentials of the individual redox reactions within the ETS can, in large measure, predict the sequence of reactions, as shown in the schematic of ETS (**Figure 3**). However, from a more constrained biological perspective, redox chemistry need not dictate physiological outcome. For example, given the highly positive E^0 for the reduction of oxygen, it is apparent that any of the ETS components is capable of reducing O_2, but this is not the case biologically. Indeed, because of spatial and other considerations, only three sites on ETS are apparently capable of directly catalyzing O_2 reduction, namely C-I, ubiquinone, and C-IV. Similarly, a simplistic linear arrangement of redox potentials does not faithfully predict the Q-cycles entailed in proton translocation in C-III, and likely in C-I.[53]

2.27.2.5 Inner Membrane

The oxidation of glucose, amino acids, and fats to CO_2 and water coupled with generation of ATP via OxPhos yields 36 mol ATP per mole of glucose, 18-fold more than anaerobic glycolysis. This process is entirely dependent on both the impermeability of the inner mitochondrial membrane, and the catalytic function of the respiratory complexes, which is also dependent upon their orientation, quaternary assemblages, and lateral mobility within that membrane. As such, the inner mitochondrial membrane becomes a dominating determinant of mitochondrial function and failure, and thus an attractive target for therapeutic intervention.[54,55] Indeed, during pathogenic Ca^{2+} loading, Ca^{2+} binding directly to cardiolipin in the inner membrane leads to immobilized lipid clusters, particularly around the adenine nucleotide translocator (ANT), increases in free radical production, and accelerated lipid peroxidation.[56–58] As such, cardiolipin oxidation and Ca^{2+} binding are likely early, and crucial, events in mitochondrial failure under pathogenic conditions. This notion is reinforced by the association between repressed cardiolipin production and the mitochondrial disorder, Barth syndrome.[59]

2.27.2.6 Membrane Potential

The sequential redox reactions of the respiratory system are coupled stoichiometrically to translocation of protons from the matrix to the intermembrane space, resulting in a ~ 220 mV electrochemical potential (matrix-negative) (see **Figure 5**). The majority of this (~ 180 mV) is due to ion potential and is termed $\Delta\psi_m$, with the remainder due to a pH gradient (ΔpH). Peter Mitchell was awarded a 1978 Nobel Prize for his work showing that the potential energy of this gradient is coupled to ADP phosphorylation via activity of $F_0 - F_1$ ATP synthase, the molecular mechanisms of which have now been elucidated.[51] In any event, establishment and maintenance of $\Delta\psi_m$ is directly dependent not only on the ETS, but also on the impermeability of the inner membrane. Loss of inner membrane impermeability, with complete mitochondrial failure, can occur via several mechanisms discussed below, but the well-known ability of agents such as cyclosporine A and bongkrekic acid to moderate $\Delta\psi_m$ collapse, and conversely the ability of carboxyatractyloside to induce it, suggests that it is amenable to therapeutic intervention.[60]

2.27.2.7 Mitochondrial Genome

Mitochondrial assembly and function is the only cellular system regulated by two separate genomes, mtDNA and the nuclear DNA (nDNA). Human mtDNA is an intronless, circular, double-stranded molecule of DNA, containing 16 569 bp, and each organelle contains 5–15 copies.[61] Unlike the nDNA, which is characterized by vast sequences of unencoding 'junk' DNA, mtDNA is exceedingly concise; it encodes on both strands, depending on direction, 13 polypeptides that are key subunits of the OxPhos system, plus 22 tRNAs and 2 rRNAs required for translation. Of the respiratory complexes, only C-II is entirely encoded by nuclear genes, as are the remaining ~ 112 subunits of the OxPhos system, plus the several thousand other proteins and RNAs required to assemble and maintain the mitochondria, including Krebs cycle and beta oxidation pathways, as well as heme biosynthesis.[51,62]

Within a given cell or tissue, under normal conditions, all the mtDNA in the entire mitochondrial population is identical, a condition termed homoplasmy. When mutations occur, they are typically evident in only a subset of mtDNA molecules, resulting in two or more subpopulations of mtDNA within the cell, a condition termed heteroplasmy. Mitochondrial division and mtDNA replication are continuous, with a half-life of several weeks in human hepatocytes, and are independent of cell cycle. As such, in long-lived, terminally differentiated cells like neurons, the relative ratio of normal to mutated mtDNA can drift over time.[61] When the proportion of damaged copies of mtDNA exceeds a threshold, set as a function of cellular energetic demand, oxidative status and nature of the mutation, pathology emerges. Most cancers are homoplasmic with localized mutations,[63] and there are now over 100 diseases attributed to over 190 specific mutations in mtDNA, or in mutations in nuclear-encoded proteins that are secondarily transported into the mitochondria.[64] An example of the latter is frataxin, where a recessively inherited CAG repeat expansion disrupts mitochondrial iron homeostasis, exacerbating endogenous oxidative stress and assembly of [Fe-S] centers, resulting in the disease Friedreich's ataxia.[21,65]

2.27.2.8 Mitochondrial Diseases

Mitochondria are not only the key determinants of cellular energetic and oxidative status, thermogenesis, and Ca^{2+}-buffering capacity, but they also integrate the various pro- and antiapoptotic signals into actions that will ultimately either induce or repress apoptosis.[12,13,66] Given such crucial roles, it should not be surprising that mitochondrial dysfunction figures centrally in aging,[5,19,45,67–71] chronic neurodegenerative diseases,[72–74] and acute pathologies such as those associated with transient ischemia and reperfusion in a host of aerobically poised tissues, such as central nervous system (CNS) stroke and myocardial infarction.[16,17] Since the initial identification of a mitochondrial disease by Rolf Luft in 1959, dozens of diseases with a mitochondrial etiology have been described, including mitochondrial encephalopathy, lactic acidosis, and stroke-like episodes syndrome (MELAS), and myoclonus epilepsy with ragged-red fibers (MERRF). Many of these diseases are associated with known deletions or other defects in mtDNA.[25,75–77] For example, impairment of C-I in the electron transport system in systemic and neuronal tissues is found in Parkinson's disease patients,[78] whereas in HD, the extent of expanded CAG repeats in the mitochondrial protein huntingtin predicts the degree of mitochondrial impairment via membrane intercalation, and hence age of onset.[79]

Indeed, our insight into the molecular etiology of a host of neurodegenerative diseases, and acute conditions such as stroke or myocardial infarction, has been fundamentally altered by studies implicating mitochondrial dysfunction in Ca^{2+} dyshomeostasis, free radical production, and cell death.[15,80] Similarly, our understanding of the normal aging process has been substantially illuminated by genetic and bioenergetic studies implicating mitochondrial senescence and associated radical generation and energetic decline in gradual cellular deterioration.

Despite the strides made in characterizing the symptoms and etiology of these ailments, the continuing paucity of therapeutic agents remains a major hurdle, and thus an opportunity for drug development.

2.27.2.9 Mitochondrial Failure and Radical Production

The convergence of mitochondrial failure, free radical pathophysiology, and disease has also been fostered by evidence linking mitochondrial radical production and bioenergetic physiology as synergistic contributors to cell senescence and death.[15–18,81] For example, free radical exposure greatly potentiates the catastrophic loss of inner membrane impermeability and consequent mitochondrial permeability transition (MPT), that yields irreversible collapse of $\Delta\psi_m$.[82] MPT is readily induced by excessive Ca^{2+} exposure,[15–18,83] such as occurs when ionotropic dicarboxylate ion channels, including N-methyl-D-aspartate (NMDA), kainate and ibotenate receptors, are excessively stimulated. The major pathogenic receptor in ischemia–reperfusion and excitotoxicity is the NMDA channel. This excitotoxicity receptor has redox-susceptible disulfides that potentiate activation when exposed to reducing radicals and to various NO species which can, depending on redox status, exacerbate Ca^{2+} influx. The latter provides a facile explanation for the synergistic effects of oxidative stress and MPT in excitotoxicity.[84] Calcium ionophores such as A-23189, ionomycin, and others are used to mimic pathogenic NMDA receptor activation.[85] Despite its disastrous nature, MPT is moderated by cyclosporine A and bongkrekic acid, not only implicating the ANT in MPT formation, but also providing encouragement that small-molecule development strategies targeting MPT hold promise.[60]

A likely cause for MPT is an irreversible conformational shift in the ANT, a transmembrane carrier that uses $\Delta\psi_m$ to exchange ATP for ADP across the inner membrane.[86] ANT is usually associated with a number of other proteins, including cyclophilin D (the binding site of cyclosporine A), creatine kinase, and voltage-dependent anion channel (VDAC), which have all been implicated in the formation of the MPT 'megapore,'[87] although VDAC is also found in the plasma membrane.[88] Interestingly, for full catalytic capacity, ANT requires small amounts of cardiolipin, suggesting the possibility of allosteric modulation as a strategy for compound development with fewer potentially deleterious pitfalls than agents binding at the adenylate site.

Collapse of $\Delta\psi_m$ obviously abrogates ATP production, and also causes mitochondrial swelling as water enters down its osmotic gradient. In response to swelling, mitochondria release a number of proapoptotic proteins, such as cytochrome c, apoptosis-inducing factor (AIF), Smac/DIABLO, Omi/HtrA2, and endonuclease G, that activate various caspase and other apoptotic pathways.[13,89–91] Release of mitochondrial proapoptotic factors occurs upon osmotic swelling, but swelling is not necessarily required in all cases; for example, cytochrome *c* can be released prior to $\Delta\psi_m$ collapse.[92]

Under some circumstances, MPT also accelerates free radical production from mitochondria.[15] When provided with adequate oxidizable substrates and O_2, free radical production from coupled mitochondria isolated from rat cerebral cortex and cerebellum is below the levels of detection of electron paramagnetic resonance spin-trapping techniques. However, immediately upon exposure to $\mu mol\,L^{-1}$ Ca^{2+} concentrations akin to those occurring during pathogenesis, spin adducts with hyperfine splitting characteristics consistent with both carbon- and oxygen-centered radicals are readily detected.[15–18] Moreover, radical exposure (H_2O_2 plus Fe-EDTA) increases subsequent radical production almost 10-fold, suggesting that radicals arising from one mitochondrion can imperil neighboring organelles.[15–18,93] It bears reiteration that accelerated radical production after hydroxyl radical exposure was detected in the presence of excess catalase, implicating an alternative source for these radicals, such as mitochondrial peroxinitrite, and not H_2O_2.[93]

In contrast to the potentiation of MPT induced by radical exposure, high ATP-to-ADP ratios, and even the mere presence of adenylates, serve to protect against Ca^{2+}-mediated induction of MPT.[94,95] Indeed, mitochondria exposed to physiologically relevant concentrations ($mmol\,L^{-1}$) of ATP tolerate eightfold higher Ca^{2+} concentrations before undergoing MPT. As such, susceptibility to MPT results from an equilibrium between oxidative and energetic factors, suggesting that slightly shifting this equilibrium might be a more readily attainable therapeutic goal than complete MPT blockade, which may have undesirable consequences from the standpoint of adenylate exchange or $\Delta\psi_m$ regulation. Not unexpectedly, given the declines in mitochondrial function and corresponding increases in oxidative markers associated with normal aging, mitochondria in tissues from aged animals show increased susceptibility to MPT.[96,97]

2.27.3 Mitoprotective Drug Development

2.27.3.1 Mitochondrial Permeability Transition as Target

In acknowledgement of importance of mitochondrial integrity to cell viability, a number of drug development campaigns have been mounted to moderate MPT under pathogenic conditions such as acute Ca^{2+} loading. Using assays that monitor mitochondrial membrane potential, several groups have found molecules that inhibit MPT with varying potencies. For example, Fuks *et al*.[98] screened compounds against isolated liver mitochondria exposed to $200\,\mu mol\,L^{-1}$ Ca^{2+} and reported that 5-(benzylsulfonyl)-4-bromo-2-methyl(2H)-pyridazione (BBMP) moderates MPT with comparable sub-$\mu mol\,L^{-1}$ potency as minocycline,[99] and only somewhat less efficiently as cyclosporine A (**Figure 5**).

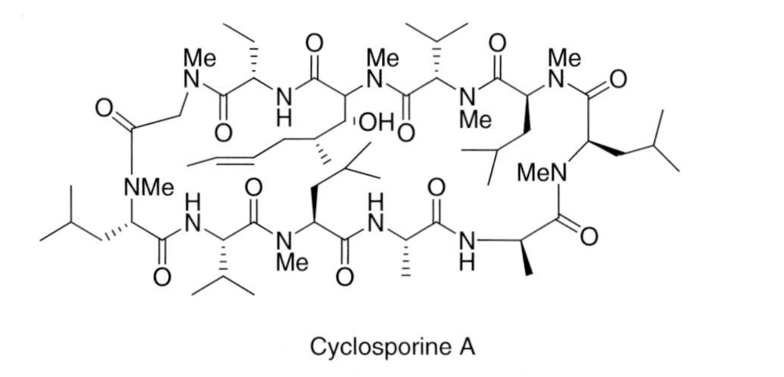

Figure 5 Several molecules that moderate mitochondrial permeability transition in isolated liver mitochondrial upon exposure to $200\,mol\,L^{-1}$ Ca^{2+}.

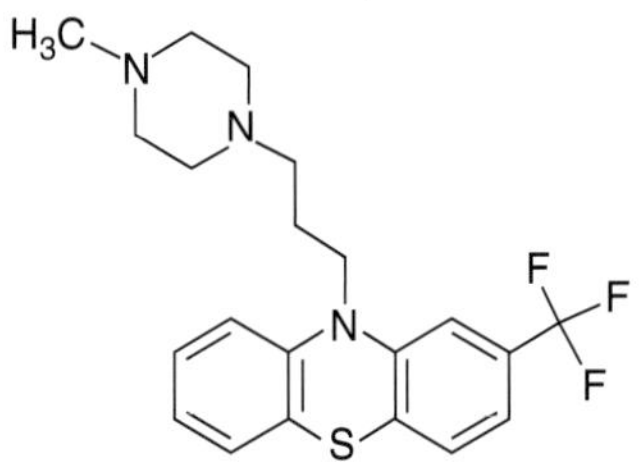

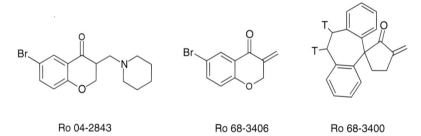

Figure 6 Screening of >1000 marketed drugs reveals that many N-substituted phenothiazine or dibenzazepine analogs moderate permeability transition in isolated mitochondrial exposed to 100 μmol L^{-1} Ca^{2+}.

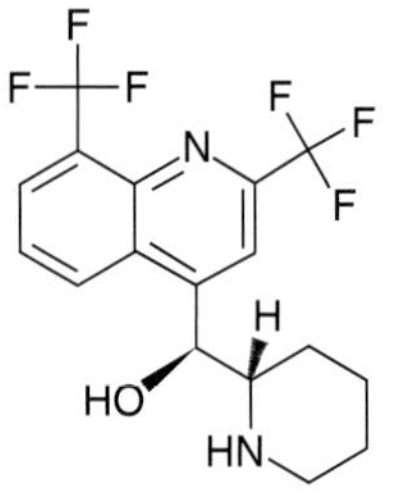

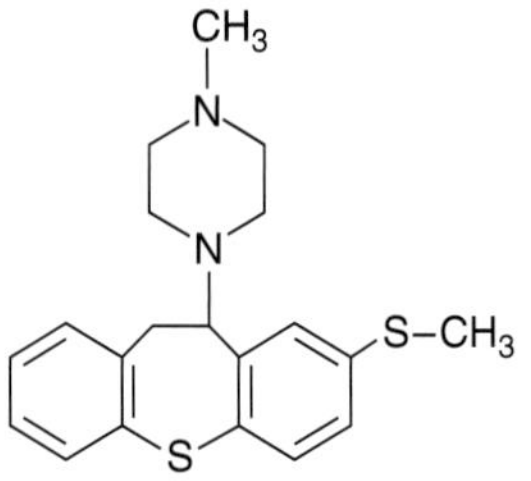

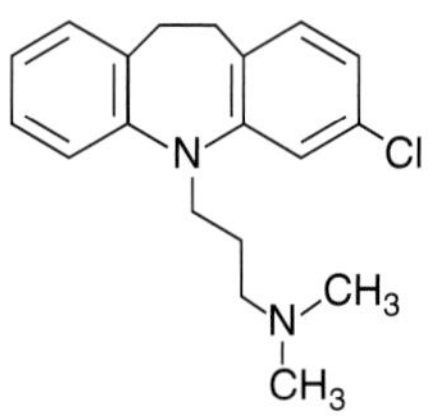

Figure 7 Heterocyclic and tricyclic motifs were also reported as potent mitochondrial permeability transition inhibitors by Hoffmann-La Roche. Ro-68-3400 is almost as potent an inhibitor of MPT as cyclosporine A (IC$_{50}$ ~0.5 μmol L^{-1}).

Similar screening strategies have yielded classes of molecules that are likely amenable to further chemical refinement. For example, Stavrovskaya et al.[100] screened >1000 FDA-approved pharmaceuticals against MPT in liver mitochondria exposed to elevated 100 μmol L^{-1} Ca^{2+}. A series of >25 heterocycles, marketed for a host of clinical indications, were shown to be effective MPT blockers (**Figure 6**). Interestingly, many are N-substituted phenothiazine or dibenz(b,f)azepine analogs. Most of these molecules were optimized and marketed for behavioral indications, not mitochondrial stabilization, which suggests that there is ample opportunity to refine these basic structures toward more potent MPT inhibition. Moreover, the structural similarities of these MPT inhibitors to inhibitors of the multidrug resistance transporter may be of some interest.[101] Among the most potent MPT inhibitors from this screen are those shown in **Figure 6**.[100]

This heterocyclic and tricyclic motif also emerged from Hoffmann-La Roche who screened against MPT (swelling) in liver mitochondria during a 40–80 μmol L^{-1} Ca^{2+} challenge.[102] Several MPT inhibitors were found (**Figure 7**), and one, Ro-68-3400, has activity comparable to cyclosporine A (IC$_{50}$ ~0.5 μmol L^{-1}). This group used Ro-68-3400 and protein cross-linking techniques to identify the binding partner as the 32 kDa isoform 1 of the VDAC, which had previously been associated 'megapore' formation.[87]

2.27.3.2 Caveats

The notion that MPT is a viable drug development target for a host of diseases is not new, but its acceptance as a viable strategy has been championed primarily by specialists.[15,16,103,104] There are important caveats. For example, bongkrekic acid, or exposure to uncouplers such as p-trifluoromethoxy carbonyl cyanide phenyl hydrazone (FCCP) to moderate Ca^{2+} uptake during excitotoxicity, correspondingly moderate mitochondrial dysfunction and cell death in the short term.[105,106] However, by preventing mitochondrial adenylate exchange, and by uncoupling OxPhos, respectively, both strategies are cytotoxic in the long run. As a result, screening programs based on improving binding affinities for potential ligands to ANT, VDAC, or any of the other components of MPT may yield potent MPT inhibitors, but in so doing may also undermine normal mitochondrial function such as ATP–ADP exchange with the cytosol.

2.27.3.3 Novel Steroids with a Novel Mechanism

In acknowledgement of the above, several successful drug development strategies have focused less on specific protein interactions to moderate MPT, and more on assays that report mitochondrial function. For example, by following mitochondrial membrane potential during acute Ca^{2+} load, a variety of steroid analogs known to be cytoprotectants were shown to be mitoprotectants (**Figure 8**).[54,55,107–109] Structure–activity relationship (SAR) studies with these molecules indicate that the A-ring phenol is the sine qua non for mito- and cytoprotection, and that planarity of the steroid ring structure, lipophilicity, and extended resonance all increase potency.[110] Conversely, substitutions at the 17′ position are relatively innocuous, separating drug development from hormonal activities.[109,111] The current mechanistic model, supported by cell-based studies of glutathione (GSH) turnover and cytoprotection, as well as by in vitro rotational-echo double-resonance nuclear magnetic resonance (REDOR NMR) and lipid peroxidation studies,[55,111] indicates that these molecules preserve mitochondrial integrity by preserving the impermeability of the inner membrane during pathogenic conditions, such as oxidative stress and excess Ca^{2+} loading.[55] They do not moderate Ca^{2+} uptake, but rather serve as catalysts to allow the reducing potential of mitochondrial GSH to terminate and repair lipid peroxidation that leads to membrane failure and MPT.[54,55] Because the mitochondrial GSH pool is metabolically maintained at extraordinarily high concentrations (8–10 mmol L^{-1}), the steroids can catalytically forestall lipid peroxidation at low nanomolar concentrations.

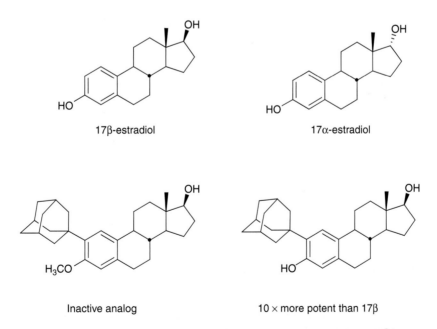

Figure 8 A-ring phenolic steroid analogs moderate mitochondrial collapse in response to Ca^{2+} and oxidative stress. Hormonal response of 17β-estradiol is reduced several hundred-fold by reversing chirality at the 17′ position to yield 17α-estradiol, while increasing lipophilicity via adamantyl addition increases potency. Planarity of the steroid is essential for cytoprotection.

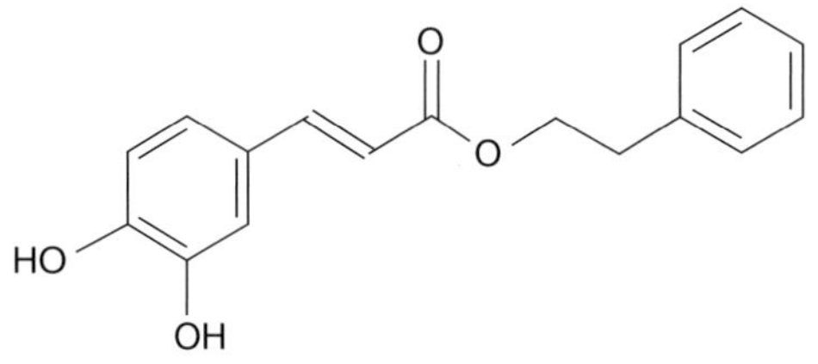

Figure 9 The antioxidant flavanoid caffeic acid phenethyl ester (CAPE) slows Ca^{2+}-induced cytochrome c release from isolated liver mitochondria, and provides neuroprotection in the 6-OH-dopamine model of Parkinson's disease.

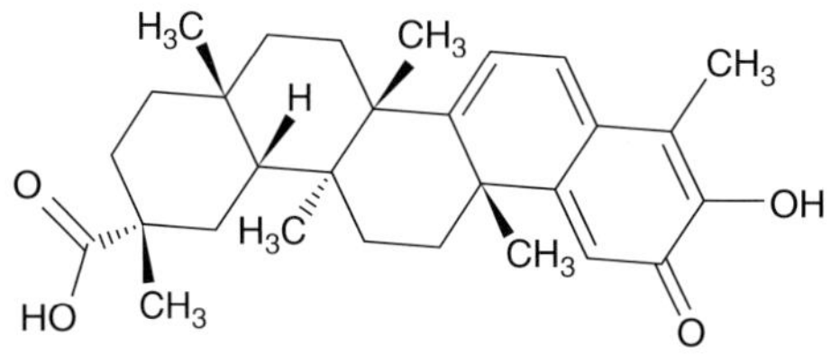

Figure 10 Celastrol, a planar quinone methide triterpene, inhibits mitochondrial membrane peroxidation with $IC_{50} \sim 7 \mu mol\,L^{-1}$. In vivo, this molecule significantly protects dopaminergic neurons from the C-I inhibitor MPTP, but also striatal neurons from the mitochondrial C-II inhibitor 3-nitropropionic acid, suggesting utility for both Parkinson's and Huntington's diseases.

2.27.3.4 Other molecules

Another recent report finds that the antioxidant flavanoid caffeic acid phenethyl ester (CAPE), derived from honey bee propolis, has antiviral, antiinflammatory, antioxidant, and immunomodulatory properties (**Figure 9**).[112] It has provided neuroprotection against ischemia and low potassium-induced apoptosis, and it has shown neuroprotection in the 6-OH-dopamine model of Parkinson's disease.[112] In isolated liver mitochondria, CAPE slowed Ca^{2+}-induced cytochrome c release, suggesting clinical utility for Parkinson's disease and perhaps other neurodegenerative diseases.[112] Although a mitochondrial mechanism has not been elucidated, the extended resonance and several redox-reactive moieties might encourage examination of its interactions with GSH similar to those that are responsible for mitochondrial stabilization by the phenolic steroids.[54,55]

Other polycycles similar to the A-ring phenolic steroids have shown similar cytoprotective effects in comparable mitochondrial assays. For example, celastrol (**Figure 10**), a quinone methide triterpene extracted from a member of the Celastraceae (e.g., staff tree and bittersweet), forestalls proinflammatory cytokine production, reduces inducible NO synthase (iNOS) expression, represses lipid peroxidation, and is a highly specific activator of heat shock transcription factor 1.[113,114] Importantly, celastrol has an $IC_{50} \sim 7\,\mu mol\,L^{-1}$ as an inhibitor of mitochondrial membrane peroxidation, which is some 15-fold more potent than α-tocopherol in this assay.[115] The mechanistic model proposes that the dienonephenol moiety moderates peroxidation by direct radical scavenging, while the anionic carboxyl repulses negatively charged oxygen radicals at the surface of the membrane.[116] In vivo, this molecule significantly protects not only dopaminergic neurons in the substantia nigra pars compacta from the C-I inhibitor 1-methyl-4-phenyl-1,2,3,6-tetra-hydropyridine (MPTP), indicating utility for Parkinson's disease, but also striatal neurons from the mitochondrial C-II inhibitor 3-nitropropionic acid, suggesting utility for HD.[113] Its induction of heat shock protein 70, coupled with decreased tumor necrosis factor-alpha and nuclear factor kappa B (NFκB) immunostaining, further recommend this class of molecules as potential mitotherapeutics.[114]

2.27.4 Mitochondrial Radical Production

2.27.4.1 Biology and Sites of Production

Free radicals are produced by the mitochondrial ETS in direct proportion to concentration of the reactants, i.e., reduced ETS components, and availability of O_2.[117] This mass action relationship was succinctly expressed by Turrens et al.[118] as:

$$\frac{d[O_2]}{dt} = k[O_2][R\bullet]$$

Mitochondrial radical production is induced not only by Ca^{2+} loading, and/or by transient ischemia in aerobically poised cells such as myocardial myocytes and neurons, but also by tumor necrosis factor-α.[15–18,83,119]

Although redox reactions within the normally functioning ETS occur via paired electron transfers, orbital spin restrictions dictate that O_2 reduction proceed univalently, the first product of which is superoxide anion ($O_2\cdot$) (reaction 1).[3] Superoxide either spontaneously or enzymatically (via superoxide dismutases) dismutes into peroxide, which is not a radical but is capable of diffusing long distances and through membranes.[3] In the presence of reduced transition metals, which abound in biological systems, peroxide can undergo Fenton chemistry to yield hydroxyl radical, the most reactive, and hence injurious, radical in biology.[3]

1. $O_2 + e \rightarrow O_2\cdot$ superoxide radical
2. $O_2\cdot + 2H+ \rightarrow H_2O_2$ hydrogen peroxide
3. $H_2O_2 + Fe^{2+} \rightarrow Fe^{3+} + OH^- + \cdot OH$ hydroxyl radical

Most studies of mitochondrial $O_2\cdot$ and H_2O_2 production assess the effect of various inhibitors of electron transfer based on the reasoning that 'downstream' blockade will increase the number of 'upstream' carriers in the reduced state, and correspondingly increase the probability of their undergoing autoxidation and redox cycling. For example, by using rotenone and antimycin A, inhibitors of electron transfer that block electron transfer between ubiquinone (coenzyme Q) and cytochrome b, and between cytochrome b and $c1$, respectively, it was shown that NADH-CoQ reductase (C-I) and ubiquinone were the two most likely sources for radical production.[69,81]

C-II has also been implicated as a radical generator, especially in the presence of mutation in one of the four constituent proteins.[120] However, it seems likely that such radical efflux is mediated by ubisemiquinone autoxidation rather than redox cycling of prosthetic groups within the protein. Moreover, production of oxyradicals from this region of the respiratory system is exacerbated by numerous pathologically relevant perturbations, such as transient hypoxia, exposure to $O_2\cdot$ or $\cdot OH$, and increases in mitochondrial Ca^{2+}. Of course, the toxicity of superoxide production is exacerbated by adventitious transition metals within the mitochondria, which certainly contributes to the etiology of Friedreich's ataxia.[21,121,122] As an aside, mitochondrial dysfunction is often assessed my monitoring H_2O_2 efflux from isolated organelles. Although this provides a reflection of superoxide production from ETS, it does not report oxidative damage occurring to ETS components or the membranes. For example, membrane lipid peroxidation is initiated by $\cdot OH$ generated via Fenton chemistry from H_2O_2.[3] As such, assessments of H_2O_2 efflux from mitochondria report only that fraction of H_2O_2 that did not initiate peroxidation, with no accounting for H_2O_2 that did. This may be an important caveat in the evaluation of compounds that moderate lipid peroxidation, such as the phenolic steroids.

2.27.5 Mitochondrial Diversity

2.27.5.1 Biology

Dogma holds that 'all mitochondria are the same' and hence do not provide the tissue- or disease-specificity that encourages modern drug development programs. However, mitochondria in neuronal subtypes collapse when confronted with Ca^{2+} transients normally occurring during cyclical myocardial depolarizations, which provides a priori evidence that 'not all mitochondria are the same.'

2.27.5.2 Electron Paramagnetic Resonance Studies

A growing literature is now documenting such functional differences between mitochondria from different tissues, and even between neuronal subtypes. For example, electron paramagnetic resonance spin-trap data show that conditions that elicit copious free radical production from isolated, respiring cerebral cortical mitochondria[15–18] fail to do so in mitochondria identically isolated from rat liver (**Figure 11**). This finding was recapitulated using mitochondria isolated simultaneously from brain and liver from the same animal (**Figure 12**). Upon addition of Ca^{2+} and Na^+ as before, a robust electron paramagnetic resonance signal is obtained from the cortical mitochondria (**Figure 11a**), but not liver mitochondria (**Figure 11b**).

By increasing the biomass of liver mitochondria within the electron paramagnetic resonance magnet (**Figure 11b**), the 10 G electron paramagnetic resonance signal from the membrane-bound semiquinone in ETS becomes apparent as a centerfield asymmetry. This signal is so readily visible not only because the mitochondrial density in the liver isolates

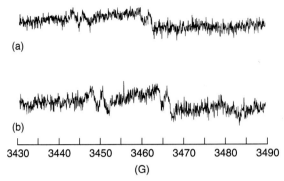

Figure 11 Electron paramagnetic resonance (EPR) spin-trap radical signals are barely visible from normoxic mitochondria isolated from rat liver exposed to 2.5 µmol L^{-1} Ca^{2+} and 14 mmol L^{-1} Na$^+$ (a). Increasing the biomass of mitochondria in the sample reveals the membrane bound semiquinone radical as a center field asymmetry, but no substantial increase in spin-trap adducts (b). Mitochondria are suspended in 100 mmol L^{-1} KCl containing 50 mmol L^{-1} KH$_2$PO$_4$, 10 mmol L^{-1} succinate, and 10 mmol L^{-1} ADP, and held in an oxygen-permeable Teflon tube with O$_2$ passing through the chamber thereby maintaining normoxia.[15–18]

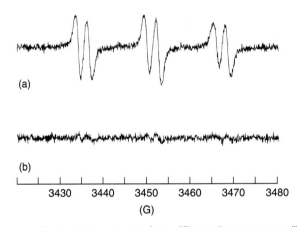

Figure 12 Differences between mitochondrial responses from different tissues are readily shown using mitochondria simultaneously isolated from the brain and liver of the same animal. Exposure to 2.5 µmol L^{-1} Ca^{2+} and 14 mmol L^{-1} Na$^+$ elicits robust POBN spin-trap signals from cerebral (a), but not liver (b), mitochondria. The POBN hyperfine splitting values of the signal from cerebral mitochondria range from $a_N = 15.2$ to 15.9 G and $a_H = 2.4$ to 3.0 G, indicative of a mixture of several radical species. Identical EPR experiments using DMPO instead of POBN reveal production of both OH and carbon-centered radicals, so that the most likely explanation is that the observed POBN signals arise from a mixture of hydroxyl radical adducts ($a_N = 14.96$ G, $a_H = 1.68$ G) plus a variety of short chain carbon-centered radicals with larger splitting constants, and not solely lipid peroxyl radicals ($a_N = 15.8$ G, $a_H = 2.6$ G for lipid peroxyl in aqueous media).[15–18]

it severalfold higher than in the brain samples, but also because the semiquinone is apparently not autooxidizing to yield spin-trappable radicals, as is the case from brain mitochondria. To some extent, the ability of liver mitochondria to tolerate Ca^{2+} as shown here artifactually results from a lower absolute Ca^{2+} concentration relative to the mitochondrial biomass; a full dose response of Ca^{2+} using identical mitochondrial density from brain versus liver would resolve this issue.

A parsimonious explanation of this difference is that mitochondrial antioxidant activities are higher in the liver compared to the brain, and this is indeed the case: Mn superoxide dismutase (MnSOD) activity in isolated liver mitochondria is threefold higher (149.9 U mg^{-1} protein + 20.1 SE, $n = 4$) than those in brain (40.8 + 7.1 SE, $n = 5$, $P < 0.001$) (Dykens, unpublished data). Clearly the antioxidant side of the oxidative equilibrium is only part of the equation. Nevertheless, these tissue differences in mitochondrial antioxidants and radical production, combined with the cytosolic paucity of both catalase and total SOD activity in brain (107.4 U mg^{-1} protein + 8.6 SE, $n = 4$) versus liver (967.5 + 75.5 U mg^{-1} SE, $n = 4$, $P < 0.001$) (Dykens, unpublished) may help explain why the CNS is the primary site of pathology in a host of neurodegenerative disorders, such as Parkinson's disease, where comparable mitochondrial defects are seen in both CNS and systemic tissues.[78]

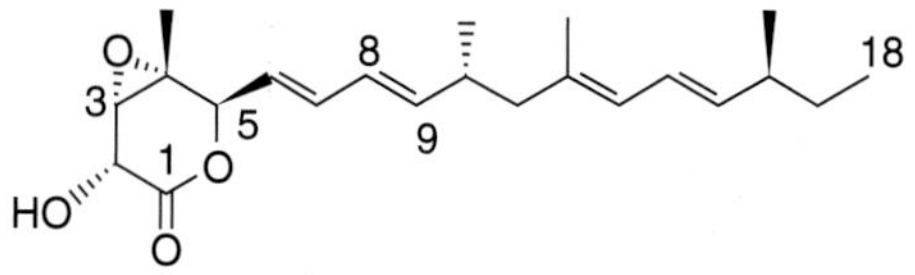

Figure 13 Parasite-specific respiratory systems offer potential targets for chemotherapy for trypanosome diseases such as African sleeping sickness, and Ascarial infection. Screening for inhibitors of parasite mitochondrial reductases has yielded nafuridin, a specific inhibitor of nematode complex I.

These electron paramagnetic resonance data also highlight diversity in the amount of Ca^{2+} tolerated by mitochondria from different tissues. Data directly addressing this point have been published by Kushnareva et al.,[83] who report that Ca^{2+} uptake capacity by mitochondria isolated from cerebral cortex is substantially lower compared to those isolated from cerebellar cortical neurons.[83] The justified inference is that the varying susceptibility among different neuronal subpopulations to a host of pathogenic stimuli is largely determined by mitochondrial constraints. The evidence indicates that, analogous to cellular differentiation, mitochondria contain a constitutive complement of proteins and other molecules required for the basic functions of replication, signaling, and OxPhos, but that there are subtle differences between tissues, and even between cellular subpopulations within a tissue, that could provide avenues for insightful drug development programs. Indeed, just this approach has been used to target selectively parasite mitochondrial pathways, while avoiding host mitochondria, in several diseases, and has led to compounds like nafuridin, a selective inhibitor of helminth C-I (**Figure 13**).[123]

2.27.6 Mitochondrial Drug Targeting

2.27.6.1 Membrane Potential

In a variation on the theme that mitochondria differ in different cell types, it was noted years ago that mitochondria within many tumor cells maintain $\Delta\psi_m$ 10–15 mV greater than surrounding, normal tissues,[124] and that this difference could be used to target tumor mitochondria selectively.[125] Trapp and Horobin report that the characteristics of molecules most likely to be taken up selectively by tumor mitochondria are polar acids with pK_a between 5 and 9, because of ion-trapping, lipophilic bases with $pK_a > 11$ or permanent cations, because of electrical attraction, and monovalent cations or strong bases with $\log K$ (octanol–water) between −2 and 2, with an optimum near 0.[126]

Several groups have followed this line of reasoning to design cationic drug delivery systems to target selectively cytoprotective or toxic molecules to mitochondria, a strategy that is likely also adaptable for gene therapy in mtDNA diseases.[127,128] Indeed, because of the Nernst equation, a monovalent cationic carrier is predicted to accumulate in the mitochondrial matrix a thousandfold over the cytoplasm, a factor that is raised another three orders of magnitude simply by using a divalent cation.[128] This strategy was tested by linking simple alkyltriphenylphosphonium cation with coenzyme Q and vitamin E derivatives.[129] It was found that mitochondrial accumulation of the antioxidants conferred significantly more cytoprotection against oxidative insult compared to the untargeted compounds. However, this same group also evaluated mitochondrially targeted free radical scavengers 4-hydroxy-2,2,6,6-tetramethylpiperidin-N-oxide (TEMPOL) and Salen-Mn(III) complex of o-vanillin and found that, although treatment delayed apoptosis after an exogenous oxidative insult, neither of the mito-targeted agents was any more effective than the untargeted compounds.[130]

A similar mito-targeting strategy was used to develop a selective antitumor compound, the rhodacyanine dye MKT-077 (**Figure 14**), which was evaluated in a phase I clinical trial.[131] They reported that, at toxicity, profile at the recommended dose was consistent with preferential accumulation of the agent within tumor cell mitochondria. Although MKT-077 was developed as a mitochondrial metabolic inhibitor, its exact mechanism was not well characterized at that time. Later, MKT-077 was found to have several nonmitochondrial effects, such as binding to mortalin (mot-2), an hsp70 family member, thereby preventing its interactions with p53.[132] This group reports MKT-077 released wild-type p53 from cytoplasmically sequestered p53-mot-2 complexes which rescued its transcriptional activity in cancer cells, but not normal cells. This activity suggests utility against cancers with wild-type p53.[132] Similar hsp binding was also reported by Tikoo et al., who also found that MKT-077 directly binds actin and by cross-linking, prevents membrane ruffling.[133] Nevertheless, despite technical hurdles and ancillary nonmitochondrial activities of MKT-077, the notion of selectively accumulating a therapeutic agent 10^3–10^6-fold higher within the mitochondria of tumor cells continues to be an attractive strategy for antitumor therapies, as well as possible mtDNA gene therapies.

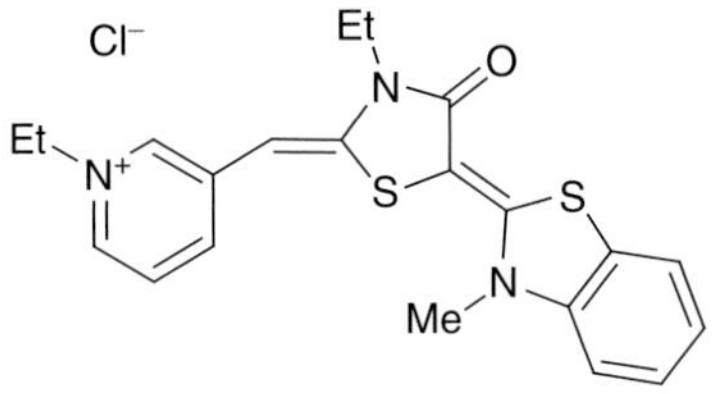

Figure 14 The rhodacyanine dye MKT-077 was developed with an eye toward preferential accumulation of the agent within tumor cell mitochondria because of elevated membrane potential there.

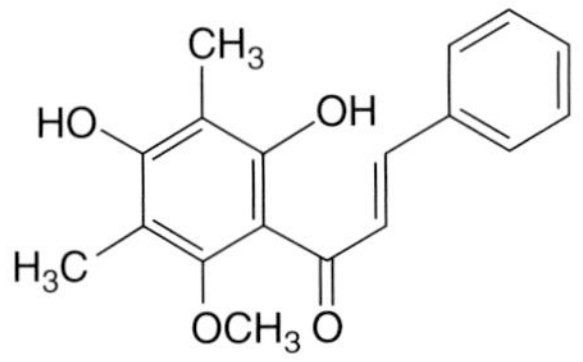

Figure 15 The flavanoids 2′,4′-dihydroxy-6′-methoxy-3′,5′-dimethylchalcone (DMC), isolated from the buds of *Cleistocalyx operculatus*, induces mitochondrial instability and ensuing apoptosis in human leukemia K562 cells via a downregulation of Bcl-2.

2.27.7 Other Mitoactive Compounds

Flavanoids such as the chalcones have also shown a variety of cytotoxic, and depending on substitutions, cytoprotective responses that involve mitochondrial pathways. For example, 2′,4′-dihydroxy-6′-methoxy-3′,5′-dimethylchalcone (DMC; **Figure 15**), isolated from the buds of *Cleistocalyx operculatus*, induces mitochondrial instability and ensuing apoptosis in human leukemia K562 cells via a downregulation of Bcl-2.[134] These structures raise questions about redox reactivity and other mitochondrial mechanisms similar to those for the A-ring phenolic steroids, a suggestion reinforced by SAR data showing that rigidity, lipophilicity, and conformation of rings A and B are crucial for the cytotoxicity shown by novel chalcone-derived pyrazoles such as 3,5-diphenyl,1*H*-pyrazole.[135,136]

In a similar vein, the bisphenol triclosan (2,4,4′-trichloro-2′–hydroxydiphenylether (TRN) **Figure 16**) is a synthetic, nonionic, broad-spectrum antibacterial, antifungal, and antiviral agent. It is widely used in hand soaps, skin care preparations, and mouthwashes:

TRN is a picomolar inhibitor of bacterial NADH-dependent enoyl-[acyl carrier protein] reductases and its homologs in several bacteria, and in *Plasmodium falciparum*, where it is a potent inhibitor of parasite development.[137] However, Newton *et al.* found that, at doses much lower than those generally used in commercial preparations, TRN uncouples OxPhos and induces mitochondrial depolarization.[137] They conclude that mitochondrial impairment is an unintended consequence of interfering with bacterial fatty acid metabolism that warrants further examination.

2.27.8 Electron Transport Components

2.27.8.1 Complex I (EC 1.6.5.3)

2.27.8.1.1 Biology

This is the entry point into ETS for most of the electrons from Kreb's cycle. It is first of three sites in ETS where redox transfer is coupled with translocation of protons, in this instance with a $4H+/2e^-$ stochiometry, to generate the membrane potential that drives ADP phosphorylation.[51] C-I is a huge, 1000 kDa holoenzyme composed of >40 proteins, fully justifying the name 'complex.' It is a major site of free radical production from ETS, especially under hypoxic conditions when the ETS components are more completely reduced and thus more susceptible to redox cycling to yield superoxide or carbon-centered radicals.

2.27.8.1.2 Inhibitors

From a medicinal chemist's perspective, over 60 inhibitors of C-I have been described, including the rotenoids, such as rotenone and deguelin, which can induce apoptosis via oxidative pathways (**Figure 17**).[138,139] There are a number of synthetic insecticides/acaricides with C-I as their target and they can be grouped into two main classes: (1) pyrazoles and substituted pyrimidines, and (2) pyridines and quinazolines.[123,138] Prominent examples of these two classes are

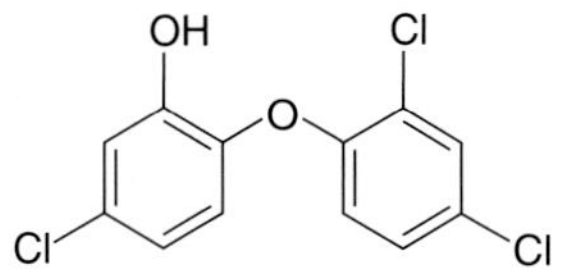

Figure 16 The bisphenol tricloscan 2,4,4′-trichloro-2′-hydroxydiphenylether is a synthetic, nonionic, broad-spectrum antibacterial, antifungal, and antiviral agent that induces mitochondrial depolarization. Such mitochondrial impairment may be an unintended consequence of interfering with bacterial fatty acid metabolism that warrants further examination.

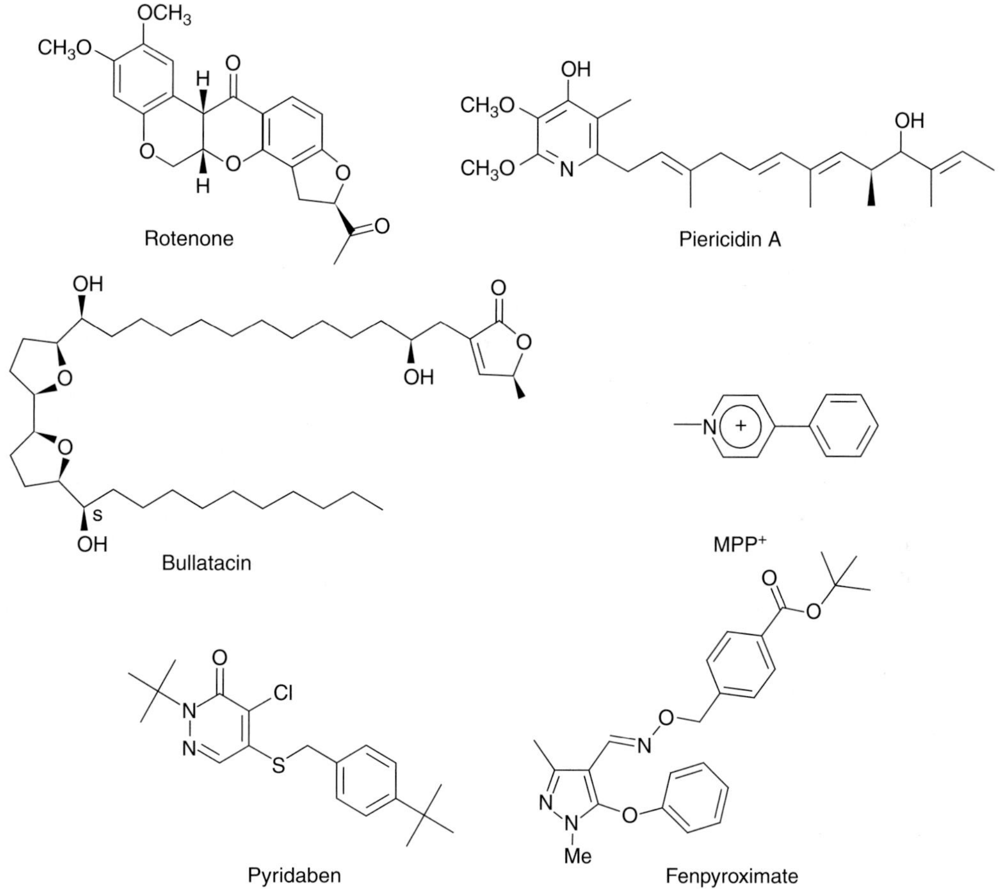

Figure 17 Over 60 inhibitors of respiratory Complex I have been described including the rotenoids, such as rotenone and deguelin. A number of synthetic insecticides/acaricides have C-I as their target and can be grouped into two main classes: (i) pyrazoles and substituted pyrimidines, and (ii) pyridines and quinazolines. Prominent examples of these two classes are fenpyroximate and 2-decyl-4-quinazolinyl amine, respectively. Other C-1 inhibitors include annonaceous acetogenins rolliniastatin-1, rolliniastatin-2 [bullatacin], vanilloids (capsaicin), amytal, and MPP⁺.

fenpyroximate and 2-decyl-4-quinazolinyl amine (DQA: formerly known as SAN 548A), respectively. Further study has suggested that all hydrophobic or amphipathic C-I inhibitors occupy the ubiquinone-binding domain, which is formed by apposition of several subunits, notably ND2.[140] Other inhibitors include annonaceous acetogenins (rolliniastatin-1, rolliniastatin-2 (bullatacin), vanilloids (capsaicin), amytal, and 1-methyl-4-phenyl pyridinium salt (MPP+), that is used to generate an animal model of Parkinson's disease.[138]

Although C-I has been considered resistant to inhibition via its NADH-binding domain, an NADH analog has shown potent ($K_i \sim 10$ nmol L^{-1}) inhibition of C-I and other NADH reductases.[141] This analog was generated via prolonged alkaline exposure under aerobic conditions. The mass and absorption spectroscopic studies suggest that the inhibitor is derived from attachment of two oxygen atoms to the nicotinamide moiety of NADH, with a tentative structure shown in **Figure 18**.[141]

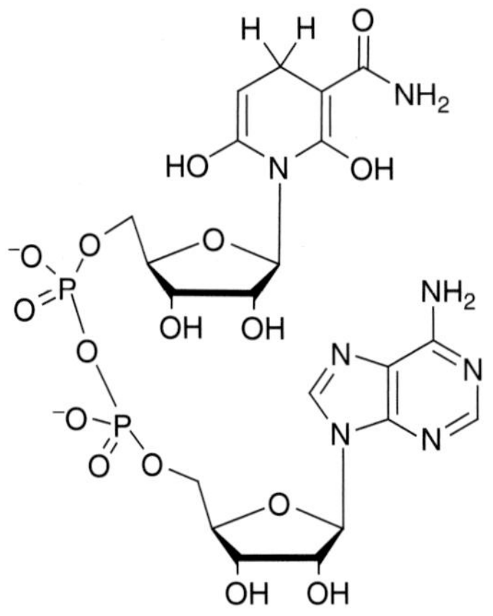

Figure 18 Inhibition of C-1 via its NADH-binding domain is shown by this NADH analog.

2.27.8.1.3 Composition

Bovine myocardial C-I consists of 45–46 subunits, 6 tetranuclear and 2 binuclear [Fe-S] clusters, only some of which are electron paramagnetic resonance-visible, probably more than one ubiquinone-binding site, and noncovalently bound flavin adenine nucleotide.[51] Of these 46 polypeptides, seven are encoded by the mtDNA. Fourteen subunits are essential for catalysis, with the remainder likely involved in regulation, assembly, and maintenance of its quaternary structure.[142–144] This is not ubiquitously the case in that *S. cerevisiae* lacks C-I, substituting a single polypeptide NADH-ubiquinone oxidoreductase lacking [Fe-S] centers but having an FAD cofactor.[144] Plants and fungi show much heterogeneity in that both the simpler oxidoreductase and the rotenone-sensitive form coexist in some species. Similarly, bacteria contain one of three different forms that catalyze electron transfer between NADH and a membrane-bound quinol, and these are variously inhibited by Ag^+, 2-*n*-heptyl-4-hydroxyquinoline-*N*-oxide (HQNO), and by the antibiotic korormicin.[51,145] This diversity of proteins and posttranslational modifications, and corresponding diversity of inhibitors, may offer an opportunity to develop selective fungal, plant, insect, or mammalian inhibitors of electron entry into ETS.[146]

Single-particle electron microscopy studies in several species, including the filamentous fungus *Neurospora crassa* and aerobic yeast *Yarrowia*, have revealed two major structural components that form an L-shape quaternary structure, with a hydrophobic membrane-bound leg containing all seven mitochondrially encoded proteins, and a more hydrophilic peripheral leg that faces the matrix containing all the nuclear-encoded proteins, all the redox active centers, and the ubiquinone-binding domain.[51,140]

2.27.8.1.4 Catalysis

C-I function has been difficult to parse, primarily because of a lack of crystal structure data and inaccessibility of some of the catalytic domains to electron paramagnetic resonance. Nevertheless, the electron transfer route has been generally established as starting from flavin adenine nucleotide and traversing all eight [Fe-S] clusters, ending at N2. Several models of the mechanism responsible for proton pumping have been proposed, but are essentially of two constructs: direct transfer of RedOx energy into a coupled vectoral proton translocation, and an indirect mechanism where such energy imposes a conformational change, resulting in proton translocation.[140] The latter seems more reasonable, and is supported by several lines of evidence.[140]

Given that the 'catalytic core' of 14 proteins in bacteria can competently catalyze the redox and proton translocation reactions, the function of the remaining 31 + 'accessory proteins' in mammalian and other species remains an area of active investigation.[51] Several of these proteins have enzymic activities quite separate from C-I function. For example, sequence homology places the 39 kDa subunit into the reductase isomerase family, while others are related to acyl carrier proteins involved with fatty acid anabolism.[51] Mutations in the latter quadrupled the phospholipid content of mitochondrial membranes, but also impaired C-I assembly in *N. crassa*.[147] Likewise, subunit B16.6 is highly homologous to GRIM19, a nuclear gene product involved in interferon/retinoic acid-induced apoptosis. Homologous deletion of GRIM19 is embryonic lethal, and it is required for proper assembly of C-I, thus providing yet another direct connection between ETS, mitochondrial function, and apoptotic pathways.[148] Of course, many of these proteins could also be serving a secondary service of simply insulating the reactive redox centers from interactions with bulk fluids in the matrix, thereby ensuring regulated and uninterrupted electron flow.

2.27.8.1.5 Pathogenesis

The association with C-I defects in human disease is striking, clearly defined, and presents numerous opportunities for pharmaceutical intervention. Indeed, as Nitjmans points out,[51] defects in C-I are the most frequently encountered of the human mitochondriopathies. Mutations in mtDNA-encoded ND4 are responsible for Leber's hereditary optic neuropathy (LHON), while mutations in the remaining mtDNA-encoded C-I subunits are associated with Leigh syndrome, MELAS, and a host of other syndromes. Although many of the defects that yield a phenotype are in subunits that contain a redox prosthetic group, some are also found in the assembly proteins, as is the case for C-IV.[51]

C-I activity is elevated in familial ALS patients,[149] yet it is diminished in Parkinson's disease patients, to some extent as a consequence of oxidative inactivation.[15,144,150] Rotenone exposure induces a Parkinson's syndrome, and the C-I inhibitor MPTP was found to be the cause of a drug-induced Parkinson's syndrome.[151] Indeed, depending on dose and route of administration, MPTP and rotenone selectively lesion the dopaminergic neurons of the substantia nigra pars compacta, the same cells that are at risk in Parkinson's disease. Animal models showing dyskinesia and dementia characteristic of Parkinson's disease are generated by various dosing regimes of MPTP and rotenone.[113,152] The gene *DJ-1* is linked epidemiologically to sporadic Parkinson's disease in humans, and in *Drosophila* deletion of the two orthologs of *DJ-1* renders them exceptionally susceptible to C-I inhibition by rotenone, and to oxidative stress induced by paraquat or H_2O_2.[153] As noted above, mitochondrial stabilizers will likely provide therapeutic benefit in Parkinson's disease.[113,154]

2.27.8.1.6 Drug development

Rasagiline (*N*-propargyl-1*R*-aminoindane) is a potent, irreversible inhibitor of monoamine oxidase (MAO) B used as treatment of Parkinson's disease. Unlike selegiline, it is not metabolized to a neurotoxic methamfetamine analog (**Figure 19**).[155,156] Importantly, the *S*-isomer of rasagiline is substantially inactive as an MAO inhibitor, yet shows comparable neuroprotection in both cell culture and in in vivo models of neurotoxicity such as anoxia and excitotoxicity.[156] SAR studies have shown that neuroprotection is afforded by mitochondrial protection, and is due to the propargyl group. For example, *N*-propargylamine prevented apoptosis of SH-SY5Y cells induced by the dopaminergic neurotoxin *N*-methyl(*R*)salsolinol, whereas *N*-methylpropargylamine and propiolaldehyde did not.[157] Such cytoprotection was via stabilization of mitochondrial membrane potential, as well as induction of antiapoptotic Bcl-2, and it occurred at concentrations substantially lower than those required for MAO inhibition.[155,158]

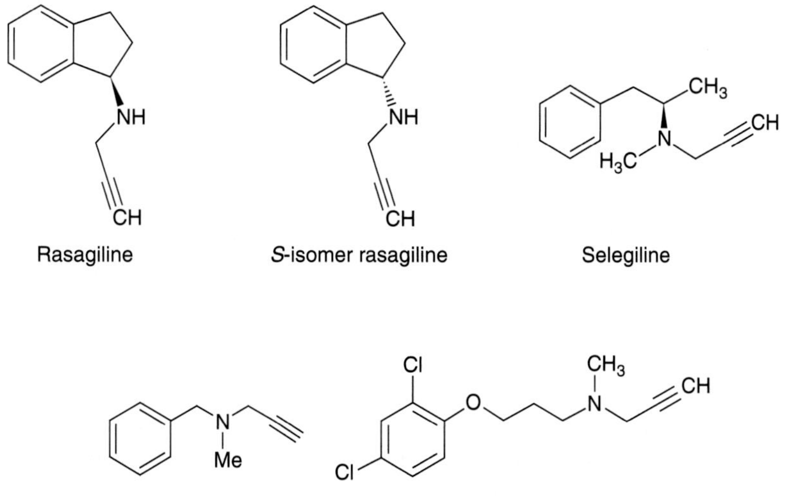

Figure 19 Rasagiline (*N*-propargyl-1*R*-aminoindane) is a potent, irreversible inhibitor of monoamine oxidase (MAO) B used as treatment of PD. Unlike selegeline, it is not metabolized to a neurotoxic methamfetamine analog. SAR studies have shown that neuroprotection is afforded by mitochondrial protection, and is due to the propargyl group.

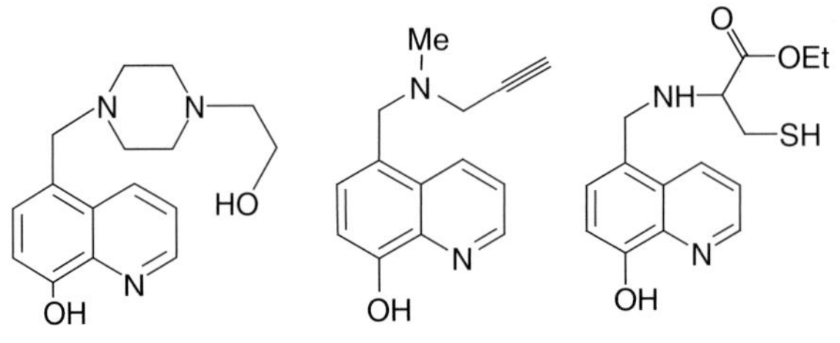

Figure 20 Addition of carbamate or morpholino moieties (VK-28) to repress Fenton chemistry to the methylpropargylamine core, increases cytoprotective potency. M-30 has both antioxidant and MAO inhibition activities.

Further modifications to the methylpropargylamine core, such as addition of carbamate or morpholino moieties to chelate iron and thus repress Fenton chemistry, increase potency in cytoprotection assays and in in vivo models of Parkinson's disease, such as 6-OH-dopamine (**Figure 20**).[159] Evaluation of additional compounds that ligand VK-28, a morpholino chelator that is able to cross the blood–brain barrier, to several MAO inhibitors show that some, like M-30, have both antioxidant activity and MAO inhibition, whereas others, like M-32, show potent antioxidant activity with no MAO effects (**Figure 20**).[156,158] It is interesting to note again the phenolic heterocycle, which has been identified with cytoprotective activities of nonhormonal estradiol-derived steroids.

2.27.8.2 Complex II (EC 1.3.5.1)

2.27.8.2.1 Biology

C-II (succinate-ubiquinone oxidoreductase) is the only membrane-bound component of the Kreb's cycle, and it consists of only four nuclear-encoded polypeptides containing, in mammals, a heme b_{558}, three linearly aligned [Fe-S] clusters, and a covalently bound FAD.[51] It is not a site of proton pumping, but mutations and dysfunction in C-II have become an active area of investigation.[160]

2.27.8.2.2 Composition

Subunits A and B are within the matrix and are anchored to the inner mitochondrial membrane by subunits C and D that each have three membrane-spanning domains.[51] Subunit A contains the FAD flavoprotein, while subunit B contains the three [Fe-S] clusters. When released from the membrane, A and B show succinate dehydrogenase (DH) activity when provided with an exogenous electron acceptor. The two membrane-binding proteins contain ubiquinone-binding sites, thereby effectively coupling matrix oxidations to membrane-bound ubiquinone reduction. C-II from various phyla can be parsed on the basis of the number of hydrophobic units and heme contents, with the mammalian subunits C and D both containing heme and ubiquinone-binding domains.[51]

The crystal structure of C-II at 0.24 nm resolution has been published,[120] which has fostered important insights. For example, the classical catalytic pathway proposed electrons sequentially traversing FAD, [2Fe-2S], [4Fe-4S], [3Fe-4S], heme b_{558} to ubiquinone. However, analysis of the edge-to-edge distances suggest that direct ubiquinone reduction by [3Fe-4S] is more likely than via heme b_{558} reduction, despite the predictions of midpoint-reducing potentials.[120] Further, binding of the inhibitor 3-nitropionate at the succinate site, and binding of 2-thenoyltrifluoroacetone (TTFA), has more clearly revealed the contralateral membrane positions of the two ubiquinone-binding sites.[120]

2.27.8.2.3 Inhibitors

While screening soil fungi extracts for ascaris NADH-fumarate reductase inhibitors, Miyadera *et al.* identified harzianopyridone (**Figure 21**), a highly selective C-II inhibitor with some 1600 fold greater potency than TTFA, with antifungal, antibacterial, and herbicidal activities.[161] Interestingly, this molecule, and even more potent aptenin analogs A4 and A5 (**Figure 21**), more effectively inhibit bovine than ascarid C-II. Given the structures, it is not surprising that inhibition appears to be by binding to a site that partially overlaps with the physiological ubiquinone-binding site(s). These molecules show nmol L^{-1} IC$_{50}$ values, yet are unable to inhibit completely C-II activity, even at 10 μmol L^{-1}, supporting the notion of incomplete occultation of the ubiquinone-binding site(s). Also of interest is that, despite their potent inhibition of C-II, neither A4 nor A5 inhibits C-III, suggesting that these molecules are not simply competitively occupying ubiquinone-binding domains, but rather are interacting with ancillary components peripheral to the ubiquinone-binding site.[161]

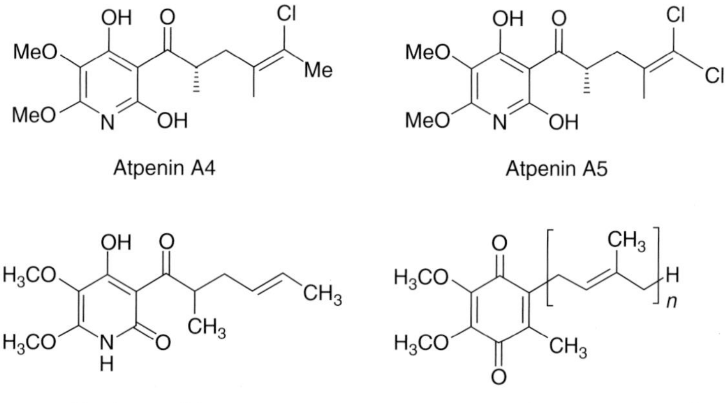

Figure 21 Harzianopyridone is a highly selective C-II inhibitor with antifungal, antibacterial, and herbicidal activities. Aptenin analogs A4 and A5 more effectively inhibit bovine than ascarid C-II. Given the structures, it is not surprising that inhibition appears to be via binding to a site that partially overlaps with the physiological ubiquinone-binding site(s).

2.27.8.2.4 Pathology

Over 20 nuclear mutations in C-II subunits have been linked to human disease.[120] Mutations in subunit A that impede succinate interactions with its binding site result in reduced C-II activity, and cause the neurological disease Leigh syndrome, an early-onset neurodegenerative disease characterized by ataxia, dystonia, blindness, and deafness, among other classical mitochondrial symptoms. Mutations in the B, C, and D subunits are all associated with familial paragangliomas and pheochromocytomas, typically in neural crest-derived tissues of the neck and head.[51,162] Some of these substitution mutations keep the redox centers intact, but greatly exacerbate free radical production, presumably by impeding ubiquinone binding and/or by exposing the flavin or [Fe-S] centers to ambient O_2. Physiologically, mutations leading to C-II impairment are the functional equivalent of chronic hypoxia (increased reliance on pyruvate/lactate), where ETS would preferentially rely on pyruvate as substrate rather than succinate, suggesting a role of heme b_{558} in O_2 sensing.[162] Regardless, this direct connection between cancer and mutations in mitochondrial proteins has reinvigorated the examination of likely links between mitochondrial dysfunction and tumorigenesis.

2.27.8.2.5 Huntington's Disease Model

In rodents and primates, the C-II inhibitor 3-nitropropionic acid (3-NP) produces the principal features of HD, including degeneration of striatal medium-sized spiny GABAergic neurons, abnormal movements, and cognitive deficits.[163,164] 3-NP induces mitochondrial dysfunction, accelerates both free radical production and release of proinflammatory cytokines, and induces gliosis.[113]

2.27.8.3 Ubiquinone (Coenzyme Q, Q_{10})

2.27.8.3.1 Biology

Ubiquinone is a parabenzoquinone with methoxy groups on carbons 2 and 3 and a methyl group on carbon 5, plus a polyisoprene chain the length of which varies among species (**Figure 22**). Almost 80% of the CoQ_{10} in the cell is within the inner mitochondrial membrane, with the remainder widely distributed in other cellular membranes.[51] Ubiquinone serves as the electron shuttle between C-I and C-II to C-III. It typically undergoes two reductions to form the ubiquinol via a single electron reduction to the radical semiquinone, autoxidation of which is believed to be a major source of superoxide production from ETS.[3,15–18] It is found in various forms in all living organisms, although the length of the acyl chain varies among groups, with rodents and other mammals having 6–10 isoprenoids that confine the molecule to the most lipophilic interior membrane domains.

In the reduced form, ubiquinol serves as an antioxidant by reducing lipid peroxyls to form the more stable semiquinone, which in turn is reduced by ETS.[3,51] The quinol can also reduce the vitamin E radical, tocopheroxyl, to allow continued redox cycling of the vitamin. Vitamin E is likely the more important of the two in terms of maintaining the impermeability of the inner mitochondrial membrane,[165] but, unlike the vitamin, ubiquinone is synthesized within cells and is obviously also serving as a key component in ETS.

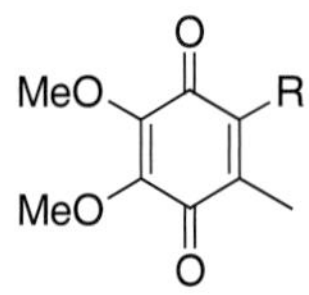

Idebenone R = $(CH_2)_{10}OH$
Ubiquinone R = $(CH_2CH=C(Me)CH_2)_{10}H$

Figure 22 Idebenone, a synthetic ubiquinone (CoQ_{10}) analog provides cardioprotective effects in Friedreich's ataxia patients, although its inability to cross the blood–brain barrier prevented beneficial effects in the CNS. It has shown therapeutic promise in Parkinson's disease.

2.27.8.3.2 Pathology

A number of pathologies are associated with CoQ_{10} deficiency, as outlined in the most recent review of mitochondrial encephalomyopathies in the ongoing series from the *Journal of Neuromuscular Disorders*.[25] For example, many of the clinical phenotypes entail myopathies with an associated CNS involvement, including seizures, ataxia, or mental retardation. Muscle biopsies often show ragged red fibers and lipid storage, and such patients often respond well to oral CoQ_{10} supplementation. On the other hand, a more common presentation is childhood-onset cerebellar ataxia, inconsistently accompanied by seizures, weakness, mental retardation, pyramidal signs, and peripheral neuropathy. Muscle biopsy in this group of patients can appear normal, and although ataxic patients also respond to CoQ_{10} supplementation, they improve less dramatically and less consistently than 'myopathic' patients.[25,75] CoQ_{10} deficiency is also associated with a severe, potentially fatal, encephalomyopathy of infancy, associated with renal disease, where CoQ_{10} supplementation can be lifesaving.[25]

2.27.8.3.3 Drug development

Tissue distribution of oral CoQ_{10} is skewed toward liver and away from brain or skeletal muscle,[166] but clinical trials of high-dose CoQ_{10} for Parkinson's disease have already shown some benefit,[167] and such treatment may well provide benefit in other neurodegenerative diseases with a mitochondrial etiology.[18,73,74] Indeed, CoQ_{20} was approved in Japan in 1974 as an adjunct therapy for congestive heart failure, and clinical trials have demonstrated improvement in cardiovascular performance.[168] Similarly, idebenone, a synthetic CoQ_{10} analog (**Figure 22**), affords cardioprotective effects in Friedreich's ataxia patients,[169] although its inability to cross the blood–brain barrier precluded effects on the CNS deterioration in this disease.[170]

Mevalonic acid is an obligatory intermediate in the synthesis of CoQ_{10}, and the cholesterol-lowering 3-hydroxy-3-methylglutaryl coenzyme A (HMG-CoA) reductase inhibitors (statins) target the synthesis of mevalonate, which raises the question of whether the iatrogenic condition of statin myopathy may be due to coincident CoQ_{10} depletion.[122,171,172] In one clinical trial to address this issue, 48 patients with hypercholesterolemia were randomly assigned to receive simvastatin, atorvastatin, or placebo for 8 weeks.[173] Although markers of endogenous cholesterol synthesis decreased in both statin groups, muscle ubiquinone concentration was significantly reduced by simvastatin, but not by atorvastatin or placebo. Respiratory enzyme and citrate synthase activities were assessed in a subset of patients with depressed ETS function associated with simvastatin treatment, leading to the conclusion that some high-dose statins may secondarily affect mitochondrial function.[173] Finally, although the possibility exists, the clinical data do not yet show that statin therapy, with an unintended consequence of lowering CoQ_{10} levels, worsens Parkinson's disease.[174]

2.27.8.4 Complex III (EC 1.10.2.2)

2.27.8.4.1 Biology and composition

C-III (ubiquinol:cytochrome *c* oxidoreductase; cytochrome bc_1 complex) is the second site of ETS where the free energy of ETS redox reactions is converted into the proton gradient that drives phosphorylation of ADP. The holoenzyme is composed of 11 proteins, described by their electrophoretic mobility, of which only subunit III is encoded by mtDNA.[51] Subunit III (aka cytochrome *b*) contains two of the three hemes in C-III, a low potential cytochrome b_{566}, and a high potential cytochrome b_{562}. Subunit IV contains the third heme, cytochrome c_1, and this assemblage of hemes has fostered the less formal name of cytochrome bc_1. In addition to the hemes, subunit V contains a [2Fe-2S] center, called the Rieske complex after its discoverer. Subunits I and II, the so-called 'core proteins,' are essential for proper assembly of the holocomplex, as are the remaining proteins, at least in yeast.[51] The primary sequences from several species have been collated. Crystal structure indicates a homodimer with a twofold axis perpendicular to the membrane plane.[175–178]

2.27.8.4.2 Catalysis

The coupling between the ubiquinone–cytochrome c_1 redox reaction and proton translocation, the so-called 'protonmotive Q cycle,' has been a focus of intense interest ever since Mitchell's initial proposal of the protonmotive OxPhos coupling models, in large measure because it personifies how ETS redox potential is transduced into the proton gradient.[179,180] The initial linear models of ETS based solely on redox potentials could not account for the observations, as outlined in the classic review by Trumpower.[53] The currently accepted model entails two separate ubiquinone reaction centers in subunit III that catalyze the oxidation of ubiquinol to reduce two molecules of cytochrome c, and hence to O_2 via C-IV, plus the translocation of four protons from the mitochondrial matrix (N, or negative side) into the intermembrane space (P, or positive side).

Although the details are beyond the scope of this chapter, an overview of the Q cycle is warranted, if only to put the diversity of C-III inhibitors into context. The first quinol is oxidized at the SQ_o site in subunit III, releasing two protons to the P side of the membrane, with the two electrons following divergent pathways. One follows a high potential path by reducing cytochrome c via the Rieske [Fe-S] and cytochrome c_1 centers. The other electron follows a low potential route, via transfer from cytochrome b_{556} to b_{562} to SQ_1, a second quinone reaction site in subunit II closer to the matrix, where it generates a semiquinone. A second oxidation of QH2 at SQ_o then reduces yet another cytochrome c_1 via the high potential pathway, and in so doing releases another two protons at the P side of the membrane, with the other electron following the low potential route to eventually reduce the semiquinone at the SQ_i site to the ubiquinone by recruitment of two protons from the nearby matrix. Thus one turn of the Q cycle entails oxidation of two molecules of ubiquinol to ubiquinone, with the subsequent regeneration of one of the ubiquinols. In so doing, the high potential reaction leading to O_2 reduction via cytochrome c occurs twice, while the reaction via the low potential route occurs only once. It is the redox flow from SQ_o to $SQ1$ that provides the electrogenic work to translocate two protons from the N to the P side of the membrane.

Much of the above has been summarized by Anthony Crofts in a review of Q-cycle research.[181]

2.27.8.4.3 Inhibitors

The known inhibitors function at either of the quinol reaction centers. The SQi inhibitors, such as antimycin, target electron transfer between heme b and ubiquinone/ubisemiquinol bound to the SQi site.[182] Ilicicolin H (**Figure 23**), an antibiotic isolated from the fungus *Cylindrocladium iliciola*, is another SQi site inhibitor with an IC_{50} of 200–250 nmol L^{-1} for bovine bc_1 complex, but its IC_{50} for *Saccharomyces cerevisiae* complex is 3–5 nmol L^{-1}.[183] The inhibitors of SQo can be divided into three classes that reflect both structural and binding diversity. The methoxyacrylate derivatives myxothiazol and MAO-stilbene bind close to heme$_{566}$ whereas the hydroxyquinine mimics 5-undecyl-6-hydroxy-4,7-dioxobenzothiazol (UHDBT) bind closer to the [Fe-S]. Binding of the third group of chromones like stigmatellin overlaps the other two.[183] This diversity might encourage investigation into selective C-III inhibitors for anticytoproliferative indications.

2.27.8.4.4 Pathology

Mutations in C-III were among the first to be identified with human disease, when Johns and Neufeld showed that, in addition to well-known mutations in various C-I genes, mtDNA mutations in subunit III also led to dysfunction and LHON.[184] Other mtDNA mutations in subunit III are associated with various phenotypes, including hypertrophic cardiomyopathy and exercise intolerance.[40] In light of the discussion above identifying autoxidation of ubisemiquinone as a primary source of mitochondrial radical production, any functional restriction of C-III activity would yield a more highly reduced ubiquinone pool, correspondingly accelerating radical production via autoxidation. In this light, observation of impaired C-I activity associated with C-III mutation may well result from semiquinone radical production and the corresponding oxidative susceptibilities of the [Fe-S] centers in C-I.[185] A deletion in the *QP-C* gene, one of the C-III nuclear-encoded proteins, yields C-III deficiency with hypoglycemia and lactic acidosis, while mutations in the assembly protein BCS1L are associated with tubulopathy, encephalopathy, and liver failure.[186]

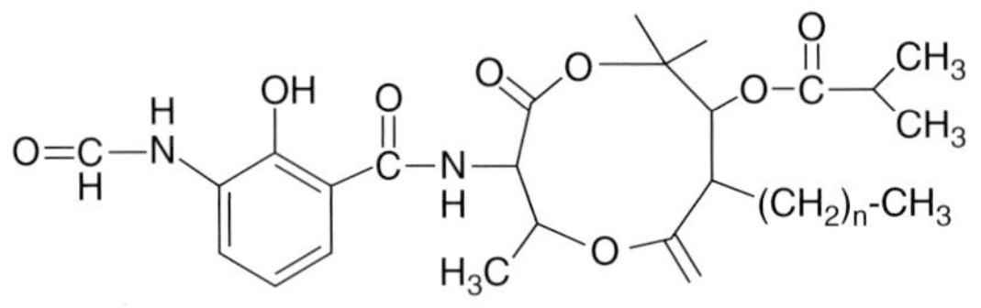

Figure 23 Complex III inhibitors function at either of the quinol reaction centers. Ilicicolin H has an IC_{50} of 200–250 nmol L^{-1} for bovine bc_1 complex, but some 75-fold lower for Saccharomyces cerevisiae. Binding of chromones like stigmatellin should encourage investigation into selective C-III inhibitors for anticytoproliferative indications.

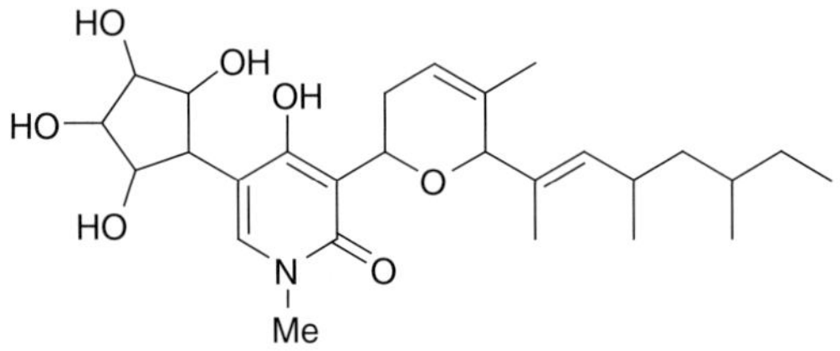

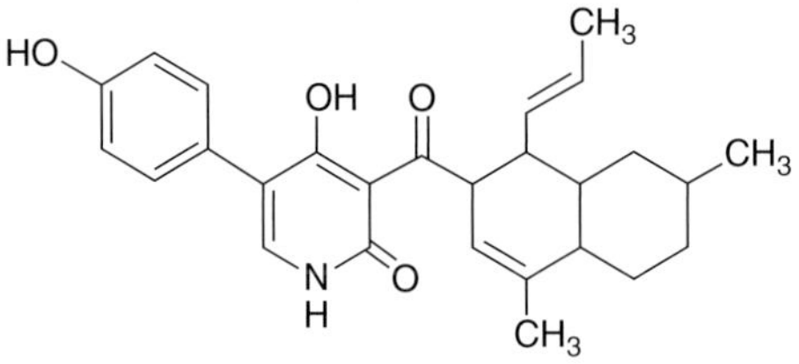

2.27.8.5 Cytochrome c

2.27.8.5.1 Biology

Cytochrome c is one of the two electron shuttles in ETS. It is a small, highly conserved protein of 104 amino acids, with a molecular weight of ~ 12 kDa. The heme in c-type cytochromes differs from b-type in that the two vinyl side chains are covalently bound to the protein in the former.[51,187] Cytochrome c does not bind O_2, it carries only a single electron, and its deletion is embryonic-lethal.[187] It is the only water-soluble component of ETS, and although it is loosely associated with the outer surface of the mitochondrial inner membrane facing the intermembrane space, it is readily extracted under high salt conditions. Indeed, this ionic interaction with the membrane has consequences for its release as a proapoptotic activator.[89]

It had been mainly studied from physiochemical and evolutionary perspectives. Because of its near-ubiquitous distribution in both obligate and facultative aerobic organisms, sequence analysis provides evidence for evolutionary divergence, and thus phylogenetic trees. Most mutations are quite conservative, such that despite numerous substitutions and many other mutations, human cytochrome c functions well in yeast ETS, even though there is only a 40% sequence homology.[51] The folding patterns and resulting tertiary structures have been well defined, and are also highly conserved.[188] In large measure this is because only about a third of the amino acids in the sequence are required for function.[189] Reassuringly, phylogenies based on cytochrome c sequences nicely recapitulate those derived from morphology and other criteria.[190]

2.27.8.5.2 Apoptosis

Study of cytochrome c has been reinvigorated by its identification as one of the main proapoptotic signals released from mitochondria that, along with others, activates the caspase and nuclease cascades that are the proximate mediators of apoptosis.[89,191,192] In cells overexpressing Bcl-2 (or Bcl-XL), the release of cytochrome c is blocked, thereby aborting the apoptotic response.[193] In most models, loss of $\Delta\psi_m$ and mitochondrial swelling precede cytochrome c release. However, this is not universally the case,[92] suggesting alternate pathways whereby mitochondrial failure can precipitate cell death. Indeed, alternate activities of cytosolic cytochrome c have been identified, such as increased release of Ca^{2+} from the endoplasmic reticulum after binding of cytochrome c to the inositol 1,4,5-triphosphate receptor.[194]

There are two separate populations of cytochrome c within the mitochondria, with ~15% within a compartment more readily available to participate in induction of apoptosis.[66] The evidence indicates that the initial release of cytochrome c during proapoptotoic stimulation does not directly undermine $\Delta\psi_m$ and OxPhos, but rather these mitochondrial factors are undermined via caspase activation and disruption of cytosolic Ca^{2+} homeostaisis.[66] Nevertheless, sustained depletion of cytochrome c inevitably leads to $\Delta\psi_m$ loss, C-I and C-III inactivation, increased mitochondrial free radical production, and cell death. It is of interest that the cytochrome c heme must be in the oxidized form in order to participate in the formation of the apoptosome, a heteromeric aggregation of apoptotic protease-activating factor 1 (Apaf-1), caspase 9, and other cofactors.[195]

2.27.8.5.3 Cardiolipin interactions

Cytochrome c has also been implicated in other pathways that highlight the interdependence of both the proteins and the lipid domains in establishing mitochondrial functional integrity, and in causing cell death. The presence of anionic phospholipids, especially cardiolipin and phosphatidylserine, weakens the heme–protein bonding in cytochrome c, thereby inducing a previously repressed peroxidase activity by the heme.[196] As a result, in the presence of H_2O_2 or $ONOO^-$, especially when accompanied by chloride anion,[197] tightly bound cytochrome c exacerbates cardiolipin oxidation and correspondingly undermines the integrity of the inner mitochondrial membrane, leading to cell death. In this way, cytochrome c redox cycling may well be initiating or facilitating apoptosis via destabilizing mitochondrial membrane integrity, independent of that portion that is released to the cytoplasm.[13]

2.27.8.6 Complex IV (EC1.9.3.1)

2.27.8.6.1 Composition and catalysis

Complex IV (COX) catalyses the oxidation of four electrons from cytochrome c to the tetravalent reduction of dioxygen to yield two molecules of water, coupled with the translocation of four protons.[51] Similar to the case for C-III, C-IV is composed of 13 proteins, with the three central proteins containing the catalytically essential prosthetic groups encoded by mtDNA, and the remaining 10 proteins encoded by the nuclear genome. C-IV exists as a homodimer, with each monomer containing two heme a groups, three Cu^{2+} atoms, one Zn^+, and one Mg^{2+}, the latter two likely serving structural rather than catalytic functions.[51] Subunit I contains both heme a groups, with heme a_3 forming a binuclear center with Cu_B to form the active site for O_2 reduction.[51]

Studies of the crystal structure of C-IV from several species have substantially clarified the biophysics of electron flow through C-IV and have raised the possibility of allosteric cooperativity akin to O_2 binding to polymeric hemoglobins.[198,199] Given the intense study and analytical techniques brought to bear, several models of redox energy coupled to proton translocation have evolved.[200,201] Briefly, cytochrome c shuttles single electrons to the bimetallic Cu_a that extends well above the membrane into the intermembrane space. Cu_a reduces heme$_a$ 1.9 nm away and embedded within the membrane, with no change in iron coordination, although conformational shifts have been detected that may be important for proton translocation.[201] Electrons are transferred laterally within the protein to heme a_3, some 1.4 nm distant, where they fully reduce the heme a_3-Cu_b center. Once the latter is fully reduced, O_2 is able to bind, and with an additional electron from heme a, is reduced to water.[198–201] Although complete reduction of the heme a_3-Cu_b center site is required for O_2 and CO binding, it is not the case for NO binding, which plays a regulatory role.

2.27.8.6.2 Inhibitors

The classical inhibitors of cyanide and H_2S act via irreversible interaction with Fe^{3+} in the heme moieties.[202] CO is a competitive inhibitor of O_2 by binding to the fully reduced heme a_3-Cu_b center, but, with a K_m some 100-fold higher, it is a low-affinity interaction. It has long been recognized that C-IV was a key regulatory center of respiration by responding to adenylate status, and that several mutations yield pathology.[203–205]

2.27.8.6.3 Oxidative regulation

More recently, interest in COX regulation has been rekindled by the observation that NO rapidly, selectively, and potently inhibits mitochondrial respiration by impeding electron flow through C-IV,[15,206] an observation first reported in the mid-1960s by Gibson and Greenwood.[207] This inhibition is readily reversible, a hallmark of physiological flux control. It appears that NO is a competitive ligand for the active O_2 site, with nanomolar IC_{50} values that increase in proportion to the square of oxygen concentration.[208] The K_m of O_2 for respiration is ~ 20–$30\,\mu mol\,L^{-1}$, such that respiration in the presence of physiological amounts of NO is exquisitely susceptible to decreasing O_2 availability.[206]

The initial observations of inhibition of respiration by NO were ascribed to reactions with O_2^- to yield peroxinitrite $ONOO^-$, and other reactive nitrogen species (RNS), with deleterious consequences for the [Fe-S] centers and other radical-susceptible centers in ETS.[15,209] These reactions do indeed occur, but the kinetics are in large measure dependent upon the coincident availability of O_2, which is the net result of production via redox cycling, competing reactions, and the efficiency of the antioxidant defenses.[144,210]

Rather, NO is enzymatically generated from L-arginine via three isozymes of NO synthase, the distributions of which are not limited to the tissues for which they are named: eNOS (endothelial), nNOS (neuronal), and an inducible (iNOS) form associated with the inflammatory response. Although several studies identified eNOS with mitochondrial outer membrane, more recent work has indicated a Ca^{2+}-dependent nNOS on the matrix side of the inner membrane.[4,211]

The interplay between NO as a normal physiological regulator of respiration that responds to P_{O_2}, and NO as a deleterious prooxidant involved in reactive oxygen species (ROS)/RNS-induced injury, is difficult to parse without considering the oxidative status within the mitochondrial compartment.[47,144,201,212,213] When GSH and other antioxidants are depleted, reactions favoring $ONOO^-$ formation are likely to dominate, whereas when O_2^- is low, regulatory activities likely dominate. This 'Janus-faced' complexity also raises questions about what were previously thought to be well-understood pharmaceuticals, such as nitroglycerin which, although clearly imparting beneficial vasodilator effects, may at the same time also be imposing unintended deleterious consequences at the level of mitochondrial respiration.[15,214,215]

2.27.8.6.4 Regulation versus pathology

Schon and DiMauro have proposed an intriguing model of the mitochondrial disease MELAS based on the duality of NO as a radical versus its regulatory roles at both C-IV and within the vasculature.[61] They note that MELAS, which is caused by a mutation in mtDNA gene for leucine tRNA, is atypical of most mitochondrial diseases because of its angiopathy and strokes. In many other mitoproliferative disorders like MERRF, mitochondrial accumulation results in histopathologically characteristic 'ragged red' muscle fibers where C-IV activity is diminished. However, in MELAS the hypertrophied mitochondrial populations contain elevated amounts of C-IV. This is the case for the vascular smooth muscle in MELAS, yielding a model wherein stroke-like episodes and cortical blindness arise from paradoxical vasoconstriction when NO, which should have induced vasodilatation, is preferentially taken up by the excess C-IV in the vascular muscle bed. As such, exogenous NO might provide therapeutic benefit to MELAS patients, a contention for which there are supportive data.[61]

Several groups are also pursuing strategies that combine NO donors with other pharmacological agents or medical devices.[216] For example, a postangioplasty stent coated with an NO donor is in development to improve restenosis outcome,[217] and COX-2 inhibitors have also been coupled with NO donors,[218] a strategy that may help address the vascular comorbidity that undermines some COX-2 inhibitors.

2.27.8.7 Complex V (EC 3.6.1.34)

2.27.8.7.1 Composition and catalysis

C-V (ATP synthase, F_1F_0-ATPase) is the transmembrane protein aggregation that couples the protonmotive force generated by ETS respiration to ADP phosphorylation. It is composed of two functional hydrophilic domains, F_0 and F_1, that regulate proton flow at the intermembrane space, and ADP phosphorylation within the matrix, respectively.[51] F_1 is composed of five proteins that catalytically function as a nanomotor akin to the bacterial flagellum.[51,86,199,219–221] The 'stalk' connecting the two catalytic subunits and the F_0 component are comprised of at least 10 proteins, of which two are encoded by mtDNA. It exists as a dimer, and the details of the catalytic mechanism are fascinating,[222] but attention here focuses on its physiological regulation and xenobiotic responses.

2.27.8.7.1 Regulation

The energy demand of tissues can vary widely depending on a host of variables. For example, ATP demand of mammalian heart can easily vary 5–10-fold depending on workload.[223] The classical models of respiratory regulation contended that C-V activity is directly and predominantly modulated by ADP availability within the mitochondrial matrix. Although this is indeed part of the regulatory mechanism(s), it has also been found that Ca^{2+} upregulates C-V activity, thereby increasing ATP production, via release of an endogenous inhibitor 'Ca^{2+}-binding inhibitor.'[223]

Similarly, under pathogenic conditions, such as neuronal excitotoxicity and/or prolonged hypoxia, $\Delta\psi_m$ is dissipated, thereby reversing the equilibrium across C-V. In response, C-V runs in reverse, with ATP hydrolysis being used to preserve $\Delta\psi_m$.[224] Clearly, such unregulated adenylate depletion will eventually become injurious to the cell, but aerotolerant organisms have a constitutive F_1-inhibitor protein, IF-1, that slows such reverse ATP hydrolysis upon binding to F_1.[224] The 84-amino-acid IF-1 protein has a well-defined binding domain, and crystallographic studies have also revealed how other inhibitors interact with the ATPase, including binding sites for aurovertin B, efrapeptin, polyphenolic phytochemicals such as resveratrol and piceatannol, nonpeptidyl lipophilic cations, and amphiphilic peptides like melittin inhibitors competing for the aurovertin B-binding site (or sites). The nonpeptidyl lipophilic cation rhodamine 6G acts at a separate unidentified site, indicating that there are at least five inhibitory sites in the F_1-ATPase.[225]

As noted previously, ATP is among the most potent inhibitors of MPT. From a medicinal chemistry perspective, development of small molecule mimetics of IF-1 might be useful for treatment of transient ischemia, such as stroke or myocardial infarction, to preserve mitochondrial function by slowing ATP depletion. Conversely, polycationic inhibitors of IF-1 might be a reasonable approach to exacerbate induction of apoptosis in antitumor therapies.

2.27.8.7.2 Other inhibitors

In addition to the inhibitors noted above that interact directly with subunits of F_1F_0-ATPase, indirect xenobiotic C-V inhibitors are also known, such as oligomycin, a macrolide antibiotic isolated from *Streptomyces*. Oligomycin, and analogs, interact with a separate protein, the 'oligomycin sensitivity-conferring protein' (OSCP), that binds to well-characterized subunits of F_1 and impedes proton flow.[226] Interestingly, OSCP also binds 17β-estradiol, 17α-estradiol, an isomer with substantially lower hormonal-transactivating potency, and diethylstilbestrol.[227] These data suggest that estradiol, and other related compounds, in addition to classical genomic mechanisms, may also interact with ATP synthase to modulate cellular energy metabolism.[227] Indeed, transient estrogen binding to OSCP could well inhibit reverse ATP hydrolysis during hypoxia or Ca^{2+} overload, thereby preserving ATP and correspondingly preventing MPT. This would be an additional mechanism whereby phenolic steroids like estrogen and its analogs provide cytoprotection during pathogenic ischemia, oxidative stress, excitotoxicity, and other stressors.[54,55,108]

Other groups have focused on OSCP as potential drug target, and Bz-423, a 1,4-benzodiazepine, has been shown to bind to OSCP, resulting in accelerated O_2^- production, which initiates apoptosis.[228] These data suggest the potential utility of Bz-423 as a cytotoxic agent against autoimmune lymphocytes, rendering OSCP a viable target for novel lupus therapeutics.[228]

2.27.8.7.3 Pathology

A point mutation in the mt-DNA-encoded subunit 6 of the F_0 is associated with the mitochondrial diseases neurogenic muscle weakness, ataxia, and retinitis pigmentosa (NARP) and Leigh syndrome, and other mutations have been identified.[229,230] Das has highlighted how inadequacies in physiological and genetic regulation of C-V may be contributing to a host of human pathologies, such as neuronal ceroidlipofuscinoses, as well as methylmalonic and glutaric acidurias.[223]

2.27.9 Other Mitochondrial Drug Development Strategies

2.27.9.1 Uncoupling OxPhos

2.27.9.1.1 Pharmacological approach

The notion of purposefully rendering the inner mitochondrial membrane permeable to protons, i.e., uncoupling ETS from OxPhos, is intriguing, but also disquieting. Such uncoupling increases heat production at the expense of caloric efficiency, a mechanism unknown when DNP was touted as a weight-loss drug in the early twentieth century. DNP did reduce patient weight, but the narrow therapeutic window was determined empirically by patients altering intake, not without a host of deleterious consequences, ranging from cataracts to death.

However, judicious and transient use of uncoupling agents has been shown to confer neuroprotection during acute Ca^{2+} overload associated with dicarboxylate excitotoxicity.[105,106] For example, Stout *et al.* used three different inhibitor treatments to dissipate the mitochondrial membrane potential, and correspondingly eliminate Ca^{2+} uptake, including the protonophore uncouplers carbonyl cyanide *p*-(trifluoromethoxy) phenyl-hydrazone (FCCP), and DNP, or the C-I inhibitor rotenone in combination with the ATP-synthase inhibitor oligomycin.[105] Glutamate exposure significantly increased intracellular free calcium concentrations, and this was exacerbated by mitochondrial inhibition. However, regardless of elevated cytoplasmic Ca^{2+}, transiently preventing mitochondrial Ca^{2+} uptake correspondingly prevented cell death. These beneficial effects of uncoupling OxPhos have also been demonstrated in vivo where DNP reduces the cerebral infarct volume approximately 40% in a rat model of focal ischemia–reperfusion.[106] Additional studies indicate that moderation of Ca^{2+} uptake preserves mitochondrial function, represses oxyradical generation, and moderates cytochrome c release.[106] These studies, among others, suggest that transiently uncoupling ETS may be a viable therapeutic strategy to reduce infarct volume during excitotoxicity associated with acute ischemia or trauma. Of course, although transient uncoupling serves to preserve mitochondrial function in the short term, the uncoupling agent must subsequently be removed in order to allow for resumption of ATP production.

2.27.9.1.2 Uncoupling proteins

Physiologically relevant uncoupling proteins have been identified and may serve as viable drug targets. For example, uncoupling protein UCP-1 (thermogenin) was isolated from rodent brown adipose tissue specialized for nonshivering thermogenesis, where it regulates proton leak.[231,232] Such UCPs have been found in bacteria, fungi, and plants, particularly in the arum family, where thermogenic activities are important for reproduction.[231,232] UCPs are members of the mitochondrial anion carrier superfamily, and five isoforms are now known from mammals. All have molecular weights 31–34 kDa, with UCPs 1–3 sharing much sequence homology, whereas UCPs 4 and 5 do not.[232] UCPs 1–3 function within the inner mitochondrial membrane as homodimers, and may be responsible for the well-known thermogenic responses to elevated thyroid hormones.[233] Interestingly, mitochondria UCP activity requires activation by superoxide exposure from the matrix side of the inner membrane, and such activation is impeded by mitochondrially targeted lipophilic antioxidants.[130,232] This could well reflect a tightly regulated feedback system, where oxyradicals generated by redox cycling of highly reduced ETS components serve to dissipate $\Delta\psi_m$, thereby accelerating electron flow and shifting ETS to a more oxidized state.[234] It would be instructive to monitor radical production and UCP activation under conditions unfavorable for acceleration of electron flow, such as anoxia or C-IV inhibition.

UCPs are expressed in brain, and UCP-2 mRNA is upregulated in response to excitotoxicity (kainate administration) and ischemia.[235] In accord with studies showing transient uncoupling and prevention of mitochondrial Ca^{2+} uptake is neuroprotective, overexpression of UCP-2 is also neuroprotective against oxidative stress in vivo and in vitro.[235] For example, UCP-2 is elevated twofold in the substantia nigra in a transgenic rodent model overexpressing UCP-2 in catecholaminergic neurons.[236] This was accompanied by partial mitochondrial uncoupling, reduced oxidative stress, and substantial neuroprotection against the C-I inhibitor MPTP, recommending UCP intervention as potential treatment for Parkinson's disease.[236] Although these data are convincing, pharmacological mitochondrial uncoupling may be too precipitous a strategy, at least until safety issues can be resolved. In the interim, it may be that small-molecule inducers or activators of UCPs can provide a more circuitous avenue to achieve the same result.

2.27.9.1.3 Caveat

Physiological caveats to any systemic uncoupling strategy are that prolonged myocardial uncoupling will carry catastrophic consequences, and that, even at lower doses, glucose-stimulated insulin secretion (GSIS) from pancreatic islet β cells is tightly coupled to adenylate charge. The ATP/ADP ratio is a key component in how β cells 'sense' glucose, and hence regulate insulin secretion.[232] Untoward systemic UCP activity might impede glucose-induced ATP increase, and correspondingly moderate insulin secretion. Again, tissue-specific differences between mitochondrial phenotypes may offer solutions to these conundrums.

2.27.9.2 Manipulating Mitochondrial Biomass

Endurance training increases the mitochondrial biomass, and recent studies have revealed a number of the signaling pathways involved.[237,238] Several such pathways are outlined here, if only to stimulate thought of pharmacological interventions designed to increase mitochondrial biogenesis in the contexts of human physical performance and the increasing prevalence of obesity and diabetes in the developed countries.

2.27.9.2.1 Transcriptional regulation

Two transcriptional regulators involved in the coordinated expression of nuclear-encoded respiratory genes and mitochondrial biogenesis in mammals are nuclear respiratory factors NRF-1 and NRF-2.[239] It is now clear that one or both regulate the majority of the nuclear genes encoding subunits of the respiratory chain complexes. They are also involved in the expression of mitochondrial transcription and replication factors, heme biosynthetic enzymes, and other proteins required for respiratory function.[239] For example, mtDNA transcription requires expression of mitochondrial

RNA polymerase and a mitochondrial transcription factor A (Tfam), and is greatly enhanced by the mitochondrial transcription specificity factors TFB1 M and TFB2 M.[239] NRF-1 and NRF-2 not only regulate human TFB1 M and TFB2 M promoters, but are also required for maximal *trans* activation by the PGC-1 family coactivators, PGC-1 (peroxisome proliferator-activated receptor gamma (PPAR-γ) coactivator-1) and PRC (PGC-1 related coactivator). The latter two can induce mitochondrial biogenesis when expressed ectopically in cultured cells or in transgenic mice.[240]

Taken in toto, the coordinated regulation and expression of nuclear- and mitochondriallyencoded proteins, and mitochondrial biosynthesis, are primarily under the control of PGC-1 and NRF family coactivators, via an ever-expanding series of interactions involving other transcription factors, such as Sp1, CREB, and YY1, among others. Nuclear genes for other mitochondrial functions such as beta-oxidation of fatty acids are controlled by transcription factors such as PPAR-α that are not involved with expression of the respiratory subunits.

2.27.9.2.2 Pharmacological approach

The notion that manipulation of mitochondrial biogenesis can be a therapeutic strategy amenable to pharmaceutical intervention is under evaluation. For example, coadministration of 17β-estradiol and flutamide improves cardiac outcome following trauma hemorrhage in rats by increasing PGC-1.[241] Estrogen and flutamide normalize cardiac PGC-1, NRF-2, Tfam, COX-4, beta-ATP synthase expression, mitochondrial ATP, cytochrome *c* oxidase activity, and mitochondrial DNA-encoded gene COX 1.[241] However, the estrogen receptor antagonist ICI 182,780 abolishes the flutamide-mediated PGC-1 upregulation, indicating that both estrogen and flutamide upregulate PGC-1 via classical estrogen receptors.[24,242] This is intriguing, given data localizing at least some of the classical estrogen receptor β to mitochondria.[243]

It has long been thought that the biguanide metformin and related thiazolidenediones (TZDs) such as troglitazone, pioglitazone, and rosiglitazone (**Figure 24**), improve insulin sensitivity in diabetic patients by several mechanisms, including increased uptake and metabolism of free fatty acids in adipose tissue, increased insulin-stimulated glucose uptake by skeletal muscle, and adipocyte differentiation via activation of PPAR-γ.[244] However, independent of PPAR-, TZDs also have direct deleterious effects on mitochondria such as reduced glutamate/malate oxidation and elevated lactate release.[244–246]

Direct inhibition of complex I by TZDs, and mitochondrial uncoupling, have been demonstrated in isolated mitochondria and skeletal muscle strip-preps.[247] Selective C-I inhibition by metformin was noted early on,[248] and the resulting elevated plasma lactic levels are a hallmark of mitochondrial inhibition and diseases. Metformin appears to interact directly with C-I, whereas C-I inhibition by TZDs appears to be indirect. For example, both PPAR-γ antagonists and agonists lacking a TZD structure inhibit C-I, whereas compounds with weak PPAR-γ affinity, such as the oxadiazolidinedione YM440, do not.[249] Burnmaier *et al.* interpret this to mean that complex I inhibition is not mediated by PPAR-γ activation per se, but rather that the properties responsible for PPAR-γ binding are also conducive to C-I inhibition.[247] Moreover, in addition to C-I inhibition, both rosiglitazone and pioglitazone decrease respiratory control and ADP-to-oxygen ratios due to increased oxygen consumption in state 4, a clear indication of uncoupled

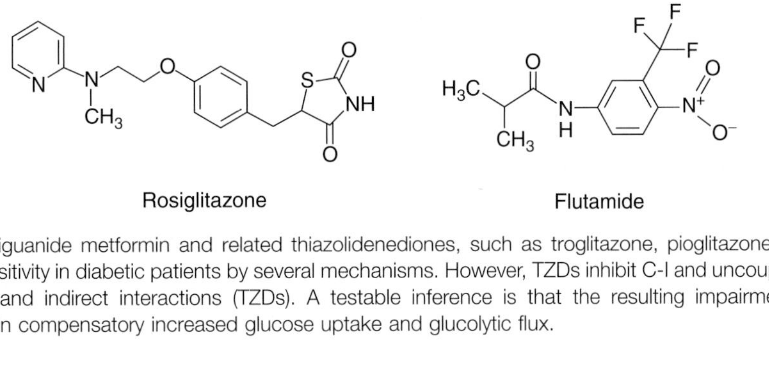

Figure 24 The biguanide metformin and related thiazolidenediones, such as troglitazone, pioglitazone, and rosiglitazone, improve insulin sensitivity in diabetic patients by several mechanisms. However, TZDs inhibit C-I and uncouple mitochondria via direct (metformin) and indirect interactions (TZDs). A testable inference is that the resulting impairment of aerobic ATP production results in compensatory increased glucose uptake and glucolytic flux.

OxPhos.[247] The reasonable supposition is that reduced aerobic ATP production due to C-I inhibition, and uncoupling in some cases, results in compensatory increased glucose uptake, glucolytic flux, and ensuing lactate production.[247]

Interestingly, at least one of these TZDs, pioglitazone, has been shown to increase mitochondrial biogenesis in adipose tissues of diabetic patients.[242] A number of studies have reported reduced expression of PPAR coactivator-1α (PGC-1α), a key regulator of mitochondrial biogenesis, and mitochondrial dysfunction in patients with insulin resistance, type 2 diabetes, and in morbidly obese prediabetic patients. Conversely, troglitazone treatment of Zucker diabetic rats restored PGC-1α and NRF-1 expression in muscle, while rosiglitazone treatment of 3T3-L1 adipocytes increased the number of mitochondria.[242] It would be illuminating to examine the possibility that increased mitochondrial biogenesis in these instances is actually a compensatory response to xenobiotically induced respiratory inhibition.

2.27.9.3 Antiviral Effects on Mitochondria

Secondarily discovered mitochondrial pathology also pertains to some antiviral therapeutics, where mitochondrial function is systemically undermined and causes several iatrogenic complications. For example, currently approved antiretroviral therapies focus on human immunodeficiency virus-1 (HIV-1) reverse transcriptase and HIV-1 protease as potential targets. Nucleoside reverse transcriptase inhibitors (NRTIs; nucleoside analogs) directly terminate DNA elongation, while nonnucleoside reverse transcriptase inhibitors (NNRTIs) inhibit enzyme activity.[250] The current standard of care, termed highly active antiretroviral therapy (HAART), is a combination of reverse transcriptase inhibitors and protease inhibitors. However, long-term toxicity and the lipodystrophy syndrome (LDS), consisting of dyslipidemia, metabolic abnormalities of insulin resistance, and redistribution of body fat, has emerged as a limiting factor for HAART and hence an important hurdle in HIV care.[250] This may also be the case for steatohepatitis and nonalcoholic steatohepatitis (NASH), which can also be induced by drugs such as amiodarone, tamoxifen, and some antiretroviral drugs. Increasingly, evidence is accumulating that respiratory chain deficiencies play a role in the pathophysiology of NASH.[251]

Many of the toxicities of various HAART regimes are due to tissue-selective mitochondrial effects.[251] For example, AZT (zidovudine) selectively reduces mitochondrial mass and function in subcutaneous adipose tissue, which surely contributes to lipodystrophy syndrome,[252] while high-dose AZT induces skeletal and cardiac myopathies where mitochondria are enlarged with disrupted cristae and paracrystalline inclusions, hallmarks of mitochondrial failure.[250] In retrospect, such pathologies could have been anticipated: NRTIs inhibit DNA polymerase γ, which is required for mtDNA replication. By undermining mitochondrial biogenesis, OxPhos is compromised, leading to clinical lactic acidosis, and apoptotic cell death is exacerbated.[250–252]

In contrast with the above, a combination of protease inhibitors nelfinavir and ritonavir reduced mitochondrially mediated apoptosis in vivo in models such as Fas-induced fatal hepatitis, enterotoxin B-induced shock, and middle cerebral artery occlusion model of stroke.[253] Computer modeling based on the crystal structure of ANT indicates that nelfinavir interacts directly with ANT to prevent permeability transition.[253] Although this seems reasonable, it remains to be seen whether such ANT inhibition also prevents adenylate exchange, which of course would have deleterious consequences for long-term cell viability, and may also contribute to the toxicity of current HAART regimes.

2.27.10 Summary

Mitochondrial dysfunction contributes to a host of well-characterized human diseases; mitochondrial failure will be increasingly implicated in widespread degenerative illnesses and aging, as well as in acute indications such as stroke and heart attack. Therapeutics that restore or maintain mitochondrial integrity will have profound and fundamental consequences in human health and well-being. In acute indications, transient and well-regulated mitochondrial uncoupling could provide novel treatments for acute ischemia–reperfusion pathology in brain, myocardium, and other aerobically poised tissues, not to mention organ transplantation. As a result, mitochondrial function and failure will increasingly fall within the purview of medicinal chemists. Contrary to received knowledge, mitochondrial physiology lends itself to targeted drug delivery strategies, and mitochondrial phenotypic diversity offers potential refinements to such strategies, including the possibility of tissue-selective targeting. Numerous inhibitors of the individual electron transfer components have been described, and crystal structures are providing mechanistic insights into both catalysis and inhibition. We know that mitochondria are proximate determinants of cell death and viability, and therefore offer therapeutic possibilities for both cytoprotection and cytotoxicity. The more concisely we characterize the underlying factors that determine and maintain mitochondrial integrity, and thus the equilibrium between cell death and life, the more effective, safe, and elegant will be the drugs of the future.

References

1. Copeland, R. A. *Evaluation of Enzyme Inhibitors in Drug Discovery: A Guide for Medicinal Chemists and Pharmacologists (Methods of Biochemical Analysis)*, Wiley-Interscience: Hoboken, NJ, 2005, pp 1–296.
2. Sen, C.; Packer, L. *Redox Cell Biology and Genetics, Methods in Enzymology*, Elsevier: London, 2002, Vols 352 and 353.
3. Halliwell, B.; Gutteridge, J. M. C. *Free Radicals in Biology and Medicine*. Oxford University Press: New York, 1999, pp 1–936.
4. Cadenas, E.; Packer, L. *Methods Enzymol.* **2002**, *359*, 3–514.
5. Balaban, R. S.; Nemoto, S.; Finkel, T. *Cell* **2005**, *120*, 483–495.
6. Duncan, A. J.; Heales, S. J. *Mol. Aspects Med.* **2005**, *26*, 67–96.
7. Nicholls, D. G.; Ferguson, S. J. *Bioenergetics 3*, Academic Press: London, 2002, pp 1–297
8. Beal, M. F.; Howell, N.; Bodis-Wollner, I. *Mitochondria and Free Radicals in Neurodegenerative Diseases*, John Wiley: New York, 1997, pp 1–630.
9. Scheffler, I. E. *Mitochondria*, Wiley-Liss: New York, 1999, pp 1–367.
10. Lane, N. *Power, Sex Suicide: Mitochondria and the Meaning of Life*, Oxford University Press: London, 2005, pp 1–354.
11. Berdanier, C. D. *Mitochondria in Health and Disease (Oxidative Stress and Disease)*, Taylor & Francis: Boca Raton, FL, 2005, pp 1–619.
12. Jiang, X.; Wang, X. *Annu. Rev. Biochem.* **2004**, *73*, 87–106.
13. Heck, D. E.; Kagan, V. E.; Shvedova, A. A.; Laskin, J. D. *Toxicology* **2005**, *208*, 259–271.
14. Briere, J. J.; Chretien, D.; Benit, P.; Rustin, P. *Biochim. Biophys. Acta* **2004**, *1659*, 172–177.
15. Dykens, J. A. *J. Neurochem.* **1994**, *63*, 584–591.
16. Dykens, J. A. Mitochondrial Radical Production and Mechanisms of Oxidative Excitotoxicity. In *Oxygen '93;* Davies, K. J. A., Ursini, F., Eds.; Cleup University Press: Padova, Italy, 1995, pp 453–468.
17. Dykens, J. A. Mitochondrial Free Radical Production and the Etiology of Neurodegenerative Disease. In *Neurodegenerative Diseases: Mitochondria and Free Radicals in Pathogenesis*; Beal, M. F., Bodis-Wollner, I., Howell, N., Eds.; John Wiley: New York, 1997, pp 29–55.
18. Dykens, J. A. Free Radicals and Mitochondrial Dysfunction in Excitotoxicity and Neurodegenerative Diseases. In *Cell Death and Diseases of the Nervous System*; Koliatos, V. E., Ratan, R. R., Eds.; Humana Press: New Jersey, 1999, pp 45–68.
19. Dykens, J. A.; Carroll, A. K.; Wright, A. C.; Fleck, B. Mitochondrial Dysfunction in Aging and Disease: Development of a Therapeutic Strategy. In *Oxidative Stress and Aging: Advances in Basic Science, Diagnostics, and Intervention;* Cutler, R. G., Rodriguez, H., Eds.; World Scientific Press: Hackensack, NJ, 2002, pp 1377–1392.
20. Valentine, J. S.; Doucette, P. A.; Potter, S. Z. *Annu. Rev. Biochem.* **2004**, *74*, 563–593.
21. Calabrese, V.; Lodi, R.; Tonon, C.; D'Agata, V.; Sapienza, M.; Scapagnini, G.; Mangiameli, A.; Pennisi, G.; Stella, A. M.; Butterfield, D. A. *J. Neurol. Sci.* **2005**, *233*, 145–162.
22. Cookson, M. R. *Annu. Rev. Biochem.* **2005**, *74*, 29–52.
23. Thiffault, C.; Bennett, J. P., Jr. *Mitochondrion* **2005**, *5*, 109–119.
24. Xu, Z.; Jung, C.; Higgins, C.; Levine, J.; Kong, J. *J. Bioenerg. Biomembr.* **2004**, *36*, 395–399.
25. DiMauro, S.; Hirano, M. *Neuromusc. Disord.* **2005**, *15*, 276–286.
26. Beinert, H.; Holm, R. H.; Munck, E. *Science* **1997**, *277*, 653–659.
27. Rees, D. C. *Annu. Rev. Biochem.* **2002**, *71*, 221–246.
28. Johnson, D. C.; Dean, D. R.; Smith, A. D.; Johnson, M. K. *Annu. Rev. Biochem.* **2005**, *74*, 247–281.
29. Decker, A.; Solomon, E. I. *Curr. Opin. Chem. Biol.* **2005**, *9*, 152–163.
30. Chernyak, B. V.; Pletjushkina, O. Y.; Izyumov, D. S.; Lyamzaev, K. G.; Avetisyan, A. V. *Biochemistry (Mosc)*. **2005**, *70*, 240–245.
31. Stefanis, L. *Neuroscientist* **2005**, *11*, 50–62.
32. Andersson, S. G.; Karlberg, O.; Canback, B.; Kurland, C. G. *Philos. Trans. R. Soc. Lond. B Biol. Sci.* **2003**, *358*, 165–177.
33. Emelyanov, V. V. *Eur. J. Biochem.* **2003**, *270*, 1599–1618.
34. Spring, J. *J. Struct. Funct. Genomics* **2003**, *3*, 19–25.
35. Gabaldon, T.; Huynen, M. A. *Science* **2003**, *301*, 609.
36. Rivera, M. C.; Lake, J. A. *Nature* **2004**, *431*, 152–155.
37. Frey, T. G.; Mannella, C. A. *Trends Biochem. Sci.* **2000**, *27*, 319–324.
38. O'Brien, T. W.; O'Brien, B. J.; Norman, R. A. *Gene* **2005**, *354*, 147–151.
39. Santel, A.; Frank, S.; Gaume, B.; Herrler, M.; Youle, R. J.; Fuller, M. T. *J. Cell Sci.* **2003**, *116*, 2763–2774.
40. Santel, A.; Fuller, M. T. *J. Cell Sci.* **2001**, *114*, 867–874.
41. Legros, F.; Lombes, A.; Frachon, P.; Rojo, M. *Mol. Biol. Cell.* **2002**, *13*, 4343–4354.
42. Chernomordik, L. V.; Kozlov, M. M. *Annu. Rev. Biochem.* **2003**, *72*, 175–207.
43. Newmeyer, D. D.; Ferguson-Miller, S. *Cell* **2003**, *112*, 481–490.
44. Ristow, M. *J. Mol. Med.* **2004**, *82*, 510–529.
45. Schriner, S. E.; Linford, N. J.; Martin, G. M.; Treuting, P.; Ogburn, C. E.; Emond, M.; Coskun, P. E.; Ladiges, W.; Wolf, N.; Van Remmen, H. et al. *Science* **2005**, *308*, 1909–1911.
46. Schwerdt, G.; Freudinger, R.; Schuster, C.; Weber, F.; Thews, O.; Gekle, M. *Toxicol. Sci.* **2005**, *85*, 735–742.
47. Brown, G. C.; Borutaite, V. *Free Radic. Biol. Med.* **2002**, *33*, 1440–1450.
48. Butler, A. R.; Flitney, F. W.; Williams, D. L. *Trends Pharmacol. Sci.* **1995**, *16*, 18–22.
49. Petit, J. M.; Huet, O.; Gallet, P. F.; Maftah, A.; Ratinaud, M. H.; Julien, R. *Eur. J. Biochem.* **1994**, *220*, 871–879.
50. Chazotte, B.; Hackenbrock, C. R. *J. Biol. Chem.* **1988**, *263*, 14359.
51. Nijtmans, L. G. J.; Ugalde, C.; Lambert, P.; Smeitink, A. M. *Topics Curr. Genetics* **2004**, *8*, 149–175.
52. Zhang, M.; Mileykovskaya, E.; Dowhan, W. *J. Biol. Chem.* **2005**, *280*, 29403–29438.
53. Trumpower, B. L. *J. Biol. Chem.* **1990**, *265*, 11409–11412.
54. Dykens, J. A.; Simpkins, J. W.; Wang, J.; Gordon, K. *Exp. Gerontol.* **2003**, *38*, 101–107.
55. Dykens, J. A.; Carroll, A. K.; Wiley, S. E.; Zhao, L.; Wen, R. *Biochem. Pharmacol.* **2004**, *68*, 1971–1984.
56. Ng, Y.; Barhoumi, R.; Tjalkens, R. B.; Fan, Y. Y.; Kolar, S.; Wang, N.; Lupton, J. R.; Chapkin, R. S. *Carcinogenesis* **2005**, *26*, 1914–1921.
57. Grijalba, M. T.; Vercesi, A. E.; Schreier, S. *Biochemistry* **1999**, *38*, 13279–13287.
58. Inoue, M.; Sato, E. F.; Nishikawa, M.; Park, A. M.; Kira, Y.; Imada, I.; Utsumi, K. *Curr. Med. Chem.* **2003**, *10*, 2495–2505.
59. Vreken, P.; Valianpour, F.; Nijtmans, L. G.; Grivell, L. A.; Plecko, B.; Wanders, R. J.; Barth, P. G. *Biochem. Biophys. Res. Commun.* **2000**, *279*, 378–382.

60. Li, P. A.; Kristian, T.; He, Q. P.; Siesjo, B. K. *Exp. Neurol.* **2000**, *165*, 153–163.
61. Schon, E. A.; DiMauro, S. *Curr. Med. Chem.* **2003**, *10*, 2523–2533.
62. Taylor, S. W.; Fahy, E.; Zhang, B.; Glenn, G. M.; Warnock, D. E.; Wiley, S.; Murphy, A. N.; Gaucher, S. P.; Capaldi, R. A.; Gibson, B. W. et al. *Nat. Biotechnol.* **2003**, *21*, 281–286.
63. Ohta, S. *Curr. Med. Chem.* **2003**, *10*, 2485–2494.
64. Pecina, P.; Houstkova, H.; Hansikova, H.; Zeman, J.; Houstek, J. *Physiol. Res.* **2004**, *53*, 213–223.
65. Gerber, J.; Muhlenhoff, U.; Lill, R. *EMBO Rep.* **2003**, *4*, 906–911.
66. Saelens, X.; Festjens, N.; Vande Walle, L.; van Gurp, M.; van Loo, G.; Vandenabeele, P. *Oncogene* **2004**, *23*, 2861–2874.
67. Melov, S.; Ravenscroft, J.; Malik, S.; Gill, M. S.; Walker, D. W.; Clayton, P. E.; Wallace, D. C.; Malfroy, B.; Doctrow, S. R.; Lithgow, G. J. *Science* **2000**, *289*, 1567–1569.
68. Melov, S. *Ann. NY Acad. Sci.* **2000**, *908*, 219–225.
69. Lenaz, G.; Bovina, C.; D'Aurelio, M.; Fato, R.; Formiggini, G.; Genova, M. L.; Giuliano, G.; Merlo Pich, M.; Paolucci, U.; Parenti Castelli, G. et al. *Ann. NY Acad. Sci.* **2002**, *959*, 199–213.
70. Sohal, R. S.; Sohal, B. H. *Mech. Aging Dev.* **1991**, *7*, 187–202.
71. Mecocci, P.; MacGarvey, U.; Kaufman, A. E.; Koontz, D.; Shoffner, J. M.; Wallace, D. C.; Beal, M. F. *Ann. Neurol.* **1993**, *34*, 609–616.
72. Schapira, A. H. *Biochem. Biophys. Acta* **1999**, *1410*, 159–170.
73. Beal, M. F. *Trends Neurosci.* **2000**, *23*, 298–304.
74. Beal, M. F. *J. Bioenerg. Biomembr.* **2004**, *36*, 381–386.
75. DiMauro, S.; Davidzon, G. *Ann. Med.* **2005**, *37*, 222–232.
76. Taylor, R. W.; Turnbull, D. M. *Nat. Rev. Genet.* **2005**, *6*, 389–402.
77. DiMauro, S. *Biochim. Biophys. Acta* **2004**, *1659*, 107–114.
78. Haas, R. H.; Nasirian, F.; Nakano, K.; Ward, D.; Pay, M.; Hill, R.; Shults, C. W. *Ann. Neurol.* **1995**, *37*, 714–722.
79. Jenkins, B. G.; Andreassen, O. A.; Dedeoglu, A.; Leavitt, B.; Hayden, M.; Borchelt, D.; Ross, C. A.; Ferrante, R. J.; Beal, M. F. *J. Neurochem.* **2005**, *95*, 553–562.
80. Dykens, J. A.; Stern, A.; Trenkner, E. *J. Neurochem.* **1987**, *49*, 1223–1228.
81. Chen, Y. R.; Chen, C. L.; Zhang, L.; Green-Church, K. B.; Zweier, J. L. *J. Biol. Chem.* **2005**, *280*, 37339–37348.
82. Verrier, F.; Mignotte, B.; Jan, G.; Brenner, C. *Ann. NY Acad. Sci.* **2003**, *1010*, 126–142.
83. Kushnareva, Y. E.; Wiley, S. E.; Ward, M. W.; Andreyev, A. Y.; Murphy, A. N. *J. Biol. Chem.* **2005**, *280*, 28894–28902.
84. Choi, Y. B.; Lipton, S. A. *Cell. Mol. Life Sci.* **2000**, *57*, 1535–1541.
85. Hamahata, K.; Adachi, S.; Matsubara, H.; Okada, M.; Imai, T.; Watanabe, K.; Toyokuni, S. Y.; Ueno, M.; Wakabayashi, S.; Katanosaka, Y. et al. *Eur. J. Pharmacol.* **2005**, *516*, 187–196.
86. Halestrap, A. P. *Nature* **2004**, *430*, 984.
87. Gincel, D.; Vardi, N.; Shoshan-Barmatz, V. *Invest. Oph. Vis. Sci.* **2002**, *43*, 2097–2104.
88. Lawen, A.; Ly, J. D.; Lane, D. J.; Zarschler, K.; Messina, A.; De Pinto, V. *Int. J. Biochem. Cell Biol.* **2005**, *37*, 277–282.
89. Uren, R. T.; Dewson, G.; Bonzon, C.; Lithgow, T.; Newmeyer, D. D.; Kluck, R. M. *J. Biol. Chem.* **2005**, *280*, 2266–2274.
90. Mendoza, F. J.; Henson, E. S.; Gibson, S. B. *Biochem. Biophys. Res. Commun.* **2005**, *331*, 1089–1098.
91. Li, L.; Thomas, R. M.; Suzuki, H.; De Brabander, J. K.; Wang, X.; Harran, P. G. *Science* **2004**, *305*, 1471–1474.
92. Andreyev, A. Y.; Fahy, B.; Fiskum, G. *FEBS Lett.* **1998**, *439*, 373–376.
93. Gizatullina, Z. Z.; Chen, Y.; Zierz, S.; Gellerich, F. N. *Biochim. Biophys. Acta* **2005**, *1706*, 98–104.
94. Crompton, M. *Biochem. J.* **1999**, *341*, 233–249.
95. Fontaine, E.; Bernardi, P. *J. Bioenerg. Biomembr.* **1999**, *31*, 335–345.
96. Miro, O.; Casademont, J.; Casals, E.; Perea, M.; Urbano-Marques, A.; Rustin, P.; Cardellach, F. *Cardiovasc. Res.* **2000**, *47*, 624–631.
97. Mather, M.; Rottenberg, H. *Biochem. Biophys. Res. Commun.* **2000**, *273*, 603–608.
98. Fuks, B.; Talaga, P.; Huart, C.; Henichart, J. P.; Bertrand, K.; Grimee, R.; Lorent, G. *Eur. J. Pharmacol.* **2005**, *519*, 24–30.
99. Wang, X.; Zhu, S.; Drozda, M.; Zhang, W.; Stavrovskaya, I. G.; Cattaneo, E.; Ferrante, R. J.; Kristal, B. S.; Friedlander, R. M. *Proc. Natl. Acad. Sci. USA* **2003**, *100*, 10483–10487.
100. Stavrovskaya, I. G.; Narayanan, M. V.; Zhang, W.; Krasnikov, B. F.; Heemskerk, J.; Young, S. S.; Blass, J. P.; Brown, A. M.; Beal, M. F.; Friedlander, R. M. et al. *J. Exp. Med.* **2004**, *200*, 211–222.
101. Wiese, M.; Pajeva, I. K. *Curr. Med. Chem.* **2001**, *8*, 685–713.
102. Cesura, A. M.; Pinard, E.; Schubenel, R.; Goetschy, V.; Friedlein, A.; Langen, H.; Polcic, P.; Forte, M. A.; Bernardi, P.; Kemp, J. A. *J. Biol. Chem.* **2003**, *278*, 49812–49818.
103. Kristal, B. S.; Stavrovskaya, I. G.; Narayanan, M. V.; Krasnikov, B. F.; Brown, A. M.; Beal, M. F.; Friedlander, R. M. *J. Bioenerg. Biomembr.* **2004**, *36*, 309–312.
104. Shanmuganathan, S.; Hausenloy, D. J.; Duchen, M. R.; Yellon, D. M. *Am. J. Physiol. Heart Circ. Physiol.* **2005**, *289*, 237–242.
105. Stout, A. K.; Raphael, H. M.; Kanterewicz, B. I.; Klann, E.; Reynolds, I. J. *Nat. Neurosci.* **1998**, *1*, 366–373.
106. Korde, A. S.; Pettigrew, L. C.; Craddock, S. D.; Maragos, W. F. *J. Neurochem.* **2005**, *94*, 1142–1149.
107. Wang, X.; Simpkins, J. W.; Dykens, J. A.; Cammarata, P. R. *Invest. Oph. Vis. Sci.* **2003**, *44*, 2067–2075.
108. Simpkins, J. W.; Wang, J.; Wang, X.; Perez, E.; Prokai, L.; Dykens, J. A. *Curr. Drug Targets: CNS Neurol. Disord.* **2005**, *4*, 69–83.
109. Kumar, D. M.; Perez, E.; Cai, Z. Y.; Aoun, P.; Brun-Zinkernagel, A. M.; Covey, D. F.; Simpkins, J. W.; Agarwal, N. *Free Radic. Biol. Med.* **2005**, *38*, 1152–1163.
110. Simpkins, J. W.; Yang, S. H.; Liu, R.; Perez, E.; Cai, Z. Y.; Covey, D. F.; Green, P. S. *Stroke*, **2004**, *35*, 2648–2651.
111. Cegelski, L.; Rice, C. V.; O'Connor, R. D.; Caruano, A. L.; Tochtrop, G. P.; Cai, Z. Y.; Covey, D. F.; Schaefer, J. *Drug Dev. Res.* **2006**, *66*, 93–102.
112. Noelker, C.; Bacher, M.; Gocke, P.; Wei, X.; Klockgether, T.; Du, Y.; Dodel, R. *Neurosci. Lett.* **2005**, *383*, 39–43.
113. Cleren, C.; Calingasan, N. Y.; Chen, J.; Beal, M. F. *J. Neurochem.* **2005**, *94*, 995–1004.
114. Westerheide, S. D.; Bosman, J. D.; Mbadugha, B. N.; Kawahara, T. L.; Matsumoto, G.; Kim, S.; Gu, W.; Devlin, J. P.; Silverman, R. B.; Morimoto, R. I. *J. Biol. Chem.* **2004**, *279*, 56053–56060.
115. Sassa, H.; Takaishi, Y.; Terada, H. *Biochem. Biophys. Res. Commun.* **1990**, *172*, 890–897.
116. Sassa, H.; Kogure, K.; Takaishi, Y.; Terada, H. *Free Rad. Biol. Med.* **1994**, *17*, 201–207.
117. Boveris, A.; Chance, B. *Biochem. J.* **1973**, *134*, 707–716.
118. Turrens, J. F.; Freeman, B. A.; Crapo, J. D. *Arch. Biochem. Biophys.* **1982**, *217*, 411–421.

119. Turrens, J. F. *J. Physiol.* **2003**, *552*, 335–344.
120. Sun, F.; Huo, X.; Zhai, Y.; Wang, A.; Xu, J.; Su, D.; Bartlam, M.; Rao, Z. *Cell* **2005**, *121*, 1043–1057.
121. Valko, M.; Morris, H.; Cronin, M. T. *Curr. Med. Chem.* **2005**, *12*, 1161–1208.
122. Hargreaves, I. P.; Duncan, A. J.; Heales, S. J.; Land, J. M. *Drug Saf.* **2005**, *28*, 659–676.
123. Kita, K.; Nihei, C.; Tomitsuka, E. *Curr. Med. Chem.* **2003**, *10*, 2535–2548.
124. Chen, L. B. *Annu. Rev. Cell Biol.* **1988**, *4*, 155–181.
125. Davis, S.; Weiss, M. J.; Wong, J. R.; Lampidis, T. J.; Chen, L. B. *J. Biol. Chem.* **1985**, *260*, 13844–13850.
126. Trapp, S.; Horobin, R. W. *Eur. Biophys. J.* **2005**, *34*, 959–966.
127. Murphy, M. P. *Trends Biotechnol.* **1997**, *15*, 326–330.
128. Dessolin, J.; Schuler, M.; Quinart, A.; De Giorgi, F.; Ghosez, L.; Ichas, F. *Eur. J. Pharmacol.* **2002**, *447*, 155–161.
129. Smith, R. A.; Porteous, C. M.; Gane, A. M.; Murphy, M. P. *Proc. Natl. Acad. Sci. USA* **2003**, *100*, 5407–5412.
130. Echtay, K. S.; Murphy, M. P.; Smith, R. A.; Talbot, D. A.; Brand, M. D. *J. Biol. Chem.* **2002**, *277*, 47129–47135.
131. Britten, C. D.; Rowinsky, E. K.; Baker, S. D.; Weiss, G. R.; Smith, L.; Stephenson, J.; Rothenberg, M.; Smetzer, L.; Cramer, J.; Collins, W. et al. *Clin. Cancer Res.* **2000**, *6*, 42–49.
132. Wadhwa, R.; Sugihara, T.; Yoshida, A.; Nomura, H.; Reddel, R. R.; Simpson, R.; Maruta, H.; Kaul, S. C. *Cancer Res.* **2000**, *60*, 6818–6821.
133. Tikoo, A.; Shakri, R.; Connolly, L.; Hirokawa, Y.; Shishido, T.; Bowers, B.; Ye, L. H.; Kohama, K.; Simpson, R. J.; Maruta, H. *Cancer J.* **2000**, *6*, 162–168.
134. Ye, C. L.; Qian, F.; Wei, D. Z.; Lu, Y. H.; Liu, J. W. *Leuk. Res.* **2005**, *29*, 887–892.
135. Bhat, B. A.; Dhar, K. L.; Puri, S. C.; Saxena, A. K.; Shanmugavel, M.; Qazi, G. N. *Bioorg. Med. Chem. Lett.* **2005**, *15*, 3177–3180.
136. Go, M. L.; Wu, X.; Liu, X. L. *Curr. Med. Chem.* **2005**, *12*, 481–499.
137. Newton, A. P.; Cadena, S. M.; Rocha, M. E.; Carnieri, E. G.; Martinelli de Oliveira, M. B. *Toxicol. Lett.* **2005**, *160*, 49–59.
138. Degli Esposti, M. *Biochim. Biophys. Acta* **1998**, *1364*, 222–235.
139. Li, N.; Ragheb, K.; Lawler, G.; Sturgis, J.; Rajwa, B.; Melendez, J. A.; Robinson, J. P. *J. Biol. Chem.* **2003**, *278*, 8516–8525.
140. Brandt, U.; Kerscher, S.; Drose, S.; Zwicker, K.; Zickermann, V. *FEBS Lett.* **2003**, *545*, 9–17.
141. Kotlyar, A. B.; Karliner, J. S.; Cecchini, G. *FEBS Lett.* **2005**, *579*, 4861–4866.
142. Carroll, J.; Fearnley, I. M.; Shannon, R. J.; Hirst, J.; Walker, J. E. *Mol. Cell. Proteomics* **2003**, *2*, 117–126.
143. Hirst, J.; Carroll, J.; Fearnley, I. M.; Shannon, R. J.; Walker, J. E. *Biochim. Biophys. Acta* **2003**, *1604*, 135–150.
144. Brown, G. C.; Borutaite, V. *Biochim. Biophys. Acta* **2004**, *1658*, 44–49.
145. Nakayama, Y.; Hayashi, M.; Yoshikawa, K.; Mochida, K.; Unemoto, T. *Biol. Pharm. Bull.* **1999**, *22*, 1064–1067.
146. Carroll, J.; Fearnley, I. M.; Skehel, J. M.; Runswick, M. J.; Shannon, R. J.; Hirst, J.; Walker, J. E. *Mol. Cell. Proteomics* **2005**, *4*, 693–699.
147. Schneider, R.; Brors, B.; Massow, M.; Weiss, H. *FEBS Lett.* **1997**, *407*, 249–252.
148. Huang, G.; Lu, H.; Hao, A.; Ng, D. C.; Ponniah, S.; Guo, K.; Lufei, C.; Zeng, Q.; Cao, X. *Mol. Cell. Biol.* **2004**, *24*, 8447–8456.
149. Bowling, A. C.; Schulz, J. B.; Brown, R. H., Jr.; Beal, M. F. *J. Neurochem.* **1993**, *61*, 2322–2325.
150. Schapira, A. H. *Biochem. Biophys. Acta* **1999**, *1410*, 159–170.
151. Langston, J. W. *Neurology* **1996**, *47*, S153–S160.
152. Kotake, Y.; Ohta, S. *Curr. Med. Chem.* **2003**, *10*, 2507–2516.
153. Meulener, M.; Whitworth, A. J.; Armstrong-Gold, C. E.; Rizzu, P.; Heutink, P.; Wes, P. D.; Pallanck, L. J.; Bonini, N. M. *Curr. Biol.* **2005**, *15*, 1572–1577.
154. Dykens, J. A.; Moos, W. H.; Howell, N. *Ann. NY Acad. Sci.* **2005**, *1052*, 1116–1135.
155. Mandel, S.; Weinreb, O.; Amit, T.; Youdim, M. B. *Brain Res. Rev.* **2005**, *48*, 379–387.
156. Zheng, H.; Weiner, L. M.; Bar-Am, O.; Epsztejn, S.; Cabantchik, Z. I.; Warshawsky, A.; Youdim, M. B.; Fridkin, M. *Bioorg. Med. Chem.* **2005**, *13*, 773–783.
157. Yi, H.; Maruyama, W.; Akao, Y.; Takahashi, T.; Iwasa, K.; Youdim, M. B.; Naoi, M. *J. Neural. Transm.* **2006**, *113*, 21–32, e-pub ahead of print Apr 22.
158. Zheng, H.; Gal, S.; Weiner, L. M.; Bar-Am, O.; Warshawsky, A.; Fridkin, M.; Youdim, M. B. *J. Neurochem.* **2006**, *13*, 773–783, e-pub ahead of print Jul 25.
159. Youdim, M. B.; Fridkin, M.; Zheng, H. *Mech. Aging Dev.* **2005**, *126*, 317–326.
160. Baysal, B. E.; Rubinstein, W. S.; Taschner, P. E. *J. Mol. Med.* **2001**, *79*, 495–503.
161. Miyadera, H.; Shiomi, K.; Ui, H.; Yamaguchi, Y.; Masuma, R.; Tomoda, H.; Miyoshi, H.; Osanai, A.; Kita, K.; Omura, S. *Proc. Natl. Acad. Sci. USA* **2003**, *100*, 473–477.
162. Cecchini, G. *Annu. Rev. Biochem.* **2003**, *72*, 77–109.
163. Brouillet, E.; Conde, F.; Beal, M. F.; Hantraye, P. *Prog. Neurobiol.* **1999**, *59*, 427–468.
164. Blum, D.; Galas, M. C.; Gall, D.; Cuvelier, L.; Schiffmann, S. N. *Neurobiol. Dis.* **2002**, *10*, 410–426.
165. Sohal, R. *Methods Enzymol.* **2004**, *378*, 146–151.
166. Turunen, M. *J. Nutr.* **1999**, *129*, 2113–2118.
167. Shults, C. W.; Oakes, D.; Kieburtz, K.; Beal, M. F.; Haas, R.; Plumb, S.; Juncos, J. L.; Nutt, J.; Shoulson, I.; Carter, J. et al. Parkinson Study Group. *Arch. Neurol.* **2002**, *59*, 1541–1550.
168. Hargreaves, I. P. *Ann. Clin. Biochem.* **2003**, *40*, 207–218.
169. Rustin, P.; Rotig, A.; Munnich, A.; Sidi, D. *Free Rad. Res.* **2002**, *36*, 467–469.
170. Artuch, R.; Aracil, A.; Mas, A.; Monros, E.; Vilaseca, M. A.; Pineda, M. *Neuropediatrics* **2004**, *35*, 95–98.
171. Owczarek, J.; Jasinska, M.; Orszulak-Michalak, D. *Pharmacol. Rep.* **2005**, *57*, 23–34.
172. Cenedella, R. J.; Neely, A. R.; Sexton, P. *Mol. Vis.* **2005**, *11*, 594–602.
173. Paiva, H.; Thelen, K. M.; Van Coster, R.; Smet, J.; De Paepe, B.; Mattila, K. M.; Laakso, J.; Lehtimaki, T.; von Bergmann, K.; Lutjohann, D. et al. *Clin. Pharmacol. Ther.* **2005**, *78*, 60–68.
174. Lieberman, A.; Lyons, K.; Levine, J.; Myerburg, R. *Parkinsonism Relat. Disord.* **2005**, *11*, 81–84.
175. Xia, D.; Yu, C. A.; Kim, H.; Xia, J-Z.; Kachurin, A. M.; Zhang, L.; Yu, L.; Deisenhofer, J. *Science* **1997**, *277*, 60–66.
176. Zhang, Z.; Huang, L.; Shulmeister, V. M.; Chi, Y-I.; Kim, K. K.; Hung, L. W.; Crofts, A. R.; Berry, E. A.; Kim, S. K. *Nature* **1998**, *392*, 677.
177. Iwata, S.; Lee, J. W.; Okada, K.; Lee, J. K; Iwata, M.; Rasmussen, B.; Link, T. A.; Ramaswamy, S.; Jap, B. K. *Science* **1998**, *281*, 64–71.
178. Hunte, C.; Koepke, J.; Lange, C.; Rossmanith, T.; Michel, H. *Struct. Fold. Des.* **2000**, *8*, 669–684.
179. Wilkström, M. K. F.; Berden, J. A. *Biochim. Biophys. Acta* **1972**, *283*, 403–420.

180. Mitchell, P. *FEBS Lett.* **1975**, *56*, 1–6.
181. Crofts, A. R.; Sohal, R. *Annu. Rev. Physiol.* **2004**, *66*, 689–733.
182. Kim, H.; Xia, D.; Yu, C. A.; Xia, J. Z.; Kachurin, A. M.; Zhang, L.; Yu, L.; Deisenhofer, J. *Proc. Natl. Acad. Sci. USA* **1998**, *95*, 8026–8033.
183. Gutierrez-Cirlos, E. B.; Merbitz-Zahradnik, T.; Trumpower, B. L. *J. Biol. Chem.* **2004**, *279*, 8708–8714.
184. Johns, D. R.; Neufeld, M. J. *Biochem. Biophys. Res. Commun.* **1991**, *181*, 1358–1364.
185. Lamantea, E.; Carrara, F.; Mariotti, C.; Morandi, L.; Tiranti, V.; Zeviani, M. *Neuromusc. Disord.* **2002**, *12*, 49–52.
186. de Lonlay, P.; Valnot, I.; Barrientos, A.; Gorbatyuk, M.; Tzagoloff, A.; Taanman, J. W.; Benayoun, E.; Chretien, D.; Kadhom, N.; Lombes, A. et al. *Nat. Genet.* **2001**, *29*, 57–60.
187. Stevens, J. M.; Daltrop, O.; Allen, J. W.; Ferguson, S. J. *Acc. Chem. Res.* **2004**, *37*, 999–1007.
188. Ptitsyn, O. B. *J. Mol. Biol.* **1998**, *278*, 655–666.
189. Dickerson, R. E.; Timkovich, R. Cytochrome *c*. In *The Enzymes*; Boyer, P. D., Ed.; Academic Press: New York; Vol. 11, pp 395–544.
190. McLaughlin, P. J.; Dayhoff, M. O. *J. Mol. Evol.* **1973**, *2*, 99–116.
191. Wallace, D. C.; Tanaka, Y. *J. Biochem. (Tokyo)* **1994**, *115*, 693–700.
192. Kluck, R. M.; Bossy-Wetzel, E.; Green, D. R.; Newmeyer, D. D. *Science* **1997**, *275*, 1132–1136.
193. Yang, J.; Liu, X.; Bhalla, K.; Kim, C. N.; Ibrado, A. M.; Cai, J.; Peng, T. I.; Jones, D. P.; Wang, X. *Science* **1997**, *275*, 1129–1132.
194. Boehning, D.; Patterson, R. L.; Sedaghat, L.; Glebova, N. O.; Kurosaki, T.; Snyder, S. H. *Nat. Cell Biol.* **2003**, *5*, 1051–1061.
195. Suto, D.; Sato, K.; Ohba, Y.; Yoshimura, T.; Fujii, J. *Biochem. J.* **2005**, *392*, 399–406.
196. Zucchi, M. R.; Nascimento, O. R.; Faljoni-Alario, A.; Prieto, T.; Nantes, I. L. *Biochem. J.* **2003**, *370*, 671–678.
197. Giuffre, A.; Stubauer, G.; Brunori, M.; Sarti, P.; Torres, J.; Wilson, M. T. *J. Biol. Chem.* **1998**, *273*, 32475–32478.
198. Michel, H. *Biochemistry* **1999**, *38*, 15129–15140.
199. Papa, S. *Biochemistry (Mosc.)* **2005**, *70*, 178–186.
200. Namslauer, A.; Brzezinski, P. *FEBS Lett.* **2004**, *567*, 103–110.
201. Brunori, M.; Giuffre, A.; Sarti, P. *J. Inorg. Biochem.* **2005**, *99*, 324–326.
202. Thompson, R. W.; Valentine, H. L.; Valentine, W. M. *Toxicology* **2003**, *188*, 149–159.
203. Villani, G.; Attardi, G. *Methods Cell Biol.* **2001**, *65*, 119–131.
204. Kadenbach, B.; Huttemann, M.; Arnold, S.; Lee, I.; Bender, E. *Free Rad. Biol. Med.* **2000**, *29*, 211–221.
205. Punter, F. A.; Glerum, D. M. *Top. Curr. Genet.* **2004**, *8*, 123–148.
206. Antunes, F.; Nunes, C.; Laranjinha, J.; Cadenas, E. *Toxicology* **2005**, *208*, 207–212.
207. Gibson, Q. H.; Greenwood, C. *J. Biol. Chem.* **1965**, *240*, 957–958.
208. Koivisto, A.; Matthias, A.; Bronnikov, G.; Nedergaard, J. *FEBS Lett.* **1997**, *417*, 75–80.
209. Vatassery, G. T.; Lai, J. C.; DeMaster, E. G.; Smith, W. E.; Quach, H. T. *J. Neurosci. Res.* **2004**, *75*, 845–853.
210. Mancardi, D.; Ridnour, L. A.; Thomas, D. D.; Katori, T.; Tocchetti, C. G.; Espey, M. G.; Miranda, K. M.; Paolocci, N.; Wink, D. A. *Curr. Mol. Med.* **2004**, *4*, 723–740.
211. Ghafourifar, P.; Cadenas, E. *Trends Pharmacol. Sci.* **2005**, *26*, 190–195.
212. Brunori, M.; Giuffre, A.; Forte, E.; Mastronicola, D.; Barone, M. C.; Sarti, P. *Biochim. Biophys. Acta* **2004**, *1655*, 365–371.
213. Shiva, S.; Oh, J. Y.; Landar, A. L.; Ulasova, E.; Venkatraman, A.; Bailey, S. M.; Darley-Usmar, V. M. *Free Radic. Biol. Med.* **2005**, *38*, 297–306.
214. Thatcher, G. R.; Nicolescu, A. C.; Bennett, B. M.; Toader, V. *Free Radic. Biol. Med.* **2004**, *37*, 1122–1143.
215. Csont, T.; Ferdinandy, P. *Pharmacol. Ther.* **2005**, *105*, 57–68.
216. Janero, D. R. *Free Radic. Biol. Med.* **2000**, *15*, 1495–1506.
217. Janero, D. R.; Ewing, J. F. *Free Radic. Biol. Med.* **2000**, *29*, 1199–1221.
218. Ranatunge, R. R.; Augustyniak, M.; Bandarage, U. K.; Earl, R. A.; Ellis, J. L.; Garvey, D. S.; Janero, D. R.; Letts, L. G.; Martino, A. M.; Murty, M. G. et al. *J. Med. Chem.* **2004**, *47*, 2180–2193.
219. Arechaga, I.; Jones, P. C. *FEBS Lett.* **2001**, *494*, 1–5.
220. Gaballo, A.; Zanotti, F.; Papa, S. *Curr. Prot. Pept. Sci.* **2002**, *3*, 451–460.
221. Papa, S.; Zanotti, F.; Gaballo, A. *J. Bioenerg. Biomembr.* **2000**, *32*, 401–411.
222. Minauro-Sanmiguel, F.; Wilkens, S.; Garcia, J. J. *Proc. Natl. Acad. Sci. USA* **2005**, *102*, 12356–12358.
223. Das, A. M. *Mol. Gen. Metab.* **2003**, *79*, 71–82.
224. Ichikawa, N.; Chisuwa, N.; Tanase, M.; Nakamura, M. *J. Biochem. (Tokyo)* **2005**, *138*, 201–207.
225. Gledhill, J. R.; Walker, J. E. *Biochem. J.* **2005**, *386*, 591–598.
226. Devenish, R. J.; Prescott, M.; Boyle, G. M.; Nagley, P. *J. Bioenerg. Biomembr.* **2000**, *32*, 507–515.
227. Zheng, J.; Ramirez, V. D. *J. Steroid Biochem. Mol. Biol.* **1999**, *68*, 65–75.
228. Johnson, K. M.; Chen, X.; Boitano, A.; Swenson, L.; Opipari, A. W., Jr.; Glick, G. D. *Chem. Biol.* **2005**, *12*, 485–496.
229. Nijtmans, L. G.; Henderson, N. S.; Attardi, G.; Holt, I. J. *J. Biol. Chem.* **2001**, *276*, 6755–6762.
230. Houstek, J.; Mracek, T.; Vojtiskova, A.; Zeman, J. *Biochim. Biophys. Acta* **2004**, *1658*, 115–121.
231. Bouillaud, F.; Couplan, E.; Pecqueur, C.; Ricquier, D. *Biochim. Biophys. Acta* **2001**, *1504*, 107–119.
232. Ito, K.; Ito, T.; Onda, Y.; Uemura, M. *Plant Cell Physiol.* **2004**, *45*, 257–264.
233. Krauss, S.; Zhang, C. Y.; Lowell, B. B. *Nat. Rev. Mol. Cell. Biol.* **2005**, *6*, 248–261.
234. Collin, A.; Cassy, S.; Buyse, J.; Decuypere, E.; Damon, M. *Domest. Anim. Endocrinol.* **2005**, *9*, 78–87.
235. Kim-Han, J. S.; Dugan, L. L. *Antioxid. Redox Signal.* **2005**, *7*, 1173–1181.
236. Conti, B.; Sugama, S.; Lucero, J.; Winsky-Sommerer, R.; Wirz, S. A.; Maher, P.; Andrews, Z.; Barr, A. M.; Morale, M. C.; Paneda, C. et al. *J. Neurochem.* **2005**, *93*, 493–501.
237. Terada, S.; Tabata, I. *Am. J. Physiol. Endocrinol. Metab.* **2004**, *286*, 208–216.
238. Tunstall, R. J.; Mehan, K. A.; Wadley, G. D.; Collier, G. R.; Bonen, A.; Hargreaves, M.; Cameron-Smith, D. *Am. J. Physiol. Endocrinol. Metab.* **2002**, *283*, 66–72.
239. Gleyzer, N.; Vercauteren, K.; Scarpulla, R. C. *Mol. Cell. Biol.* **2005**, *25*, 1354–1366.
240. Puigserver, P.; Spiegelman, B. M. *Endocrinol. Rev.* **2003**, *24*, 78–90.
241. Hsieh, Y. C.; Yang, S.; Choudhry, M. A.; Yu, H. P.; Rue, L. W.; Bland, K. I.; Chaudry, I. H. *Am. J. Physiol. Heart Circ. Physiol.* **2006**, *289*, H2665–H2672, e-pub Jul 29.
242. Bogacka, I.; Xie, H.; Bray, G. A.; Smith, S. R. *Diabetes* **2005**, *54*, 1392–1399.

243. Yang, S. H.; Liu, R.; Perez, E.; Wen, Y.; Stevens, S. M.; Valencia, T.; Brun-Zinkernagel, A. M.; Prokai, L.; Will, Y.; Dykens, J. A. et al. *Proc. Natl. Acad. Sci. USA* **2004**, *101*, 4130–4135.
244. Brunmair, B.; Gras, F.; Neschen, S.; Roden, M.; Wagner, L.; Waldhäusl, W.; Fürnsinn, C. *Diabetes* **2001**, *50*, 2309–2315.
245. Dello Russo, C.; Gavrilyuk, V.; Weinberg, G.; Almeida, A.; Bolanos, J. P.; Palmer, J.; Pelligrino, D.; Galea, E.; Feinstein, D. L. *J. Biol. Chem.* **2003**, *278*, 5828–5836.
246. Preininger, K.; Stingl, H.; Englisch, R.; Fürnsinn, C.; Graf, J.; Waldhäusl, W.; Roden, M. *Br. J. Pharmacol.* **1999**, *126*, 372–378.
247. Brunmair, B.; Staniek, K.; Gras, F.; Scharf, N.; Althaym, A.; Clara, R.; Roden, M; Gnaiger, E.; Nohl, H.; Waldhäusl, W. et al. *Diabetes* **2004**, *53*, 1052–1059.
248. Owen, M. R.; Doran, E.; Halestrap, A. P. *Biochem. J.* **2000**, *348*, 607–614.
249. Shimaya, A.; Kurosaki, E.; Nakano, R.; Hirayama, R.; Shibasaki, M.; Shikama, H. *Metabolism* **2000**, *49*, 411–417.
250. Petit, F.; Fromenty, B.; Owen, A.; Estaquier, J. *Trends Pharmacol. Sci.* **2005**, *26*, 254–258.
251. Fromenty, B.; Robin, M. A.; Igoudjil, A.; Mansouri, A.; Pessayre, D. *Diabetes Metab.* **2004**, *30*, 121–138.
252. Deveaud, C.; Beauvoit, B.; Hagry, S.; Galinier, A.; Carriere, A.; Salin, B.; Schaeffer, J.; Caspar-Bauguil, S.; Fernandez, Y.; Gordien, J. B. et al. *Biochem. Pharmacol.* **2005**, *70*, 90–101.
253. Weaver, J. G.; Tarze, A.; Moffat, T. C.; Lebras, M.; Deniaud, A.; Brenner, C.; Bren, G. D.; Morin, M. Y.; Phenix, B. N.; Dong, L. et al. *J. Clin. Invest.* **2005**, *115*, 1828–1838.

Biography

James A Dykens earned his PhD from the University of Maine studying photosensitized oxidative physiology of corals on Australia's Great Barrier Reef. He pursued postdoctoral research in the Department of Pharmacology at New York University Medical Center, where he studied tryptophan metabolism and hemoglobin biochemistry. His work at NYU was among the first to propose the oxidative pathology of excitotoxicity and its role in the etiology of neurodegenerative diseases. He was on faculty at Grinnell College prior to joining the Immunopathology Department at Warner-Lambert, Parke-Davis Pharmaceutical Research, in Ann Arbor (now Pfizer). At Parke-Davis, he continued his work on oxidative neuropathology, using electron paramagnetic resonance spectroscopy to characterize free radical production during mitochondrial failure. In 1996, he joined MitoKor in San Diego (now Migenix) in order to focus on the mitochondrial etiology of both chronic degenerative diseases and the cytotoxicity resulting from acute ischemia-reperfusion. Dr Dykens founded EyeCyte Therapeutics to develop novel, nonhormonal, cytoprotective steroids to preserve retinal function and vision in blinding diseases.

List of Abbreviations

11β-HSD	11β-Hydroxysteroid dehydrogenase
2/4/A1	Conditionally immortalized cell line derived from fetal rat intestine
2D-NMR	Two-dimensional nuclear magnetic resonance
3α-HSD	3α-Hydroxysteroid dehydrogenase
3D	Three-dimensional
3-HPPA	3-Hydroxyphenylpropionic acid
5-ASA	5-Aminosalicylic acid
5-FU	5-Fluorouracil
5HT	5-Hydroxytryptamine
5MC	5-Methylcytosine
AAE	Average absolute error
AAMU	5-Acetylamino-6-amino-3-methyl uracil
ABC	ATP-binding cassette
ABCB1	Gene symbol of human multidrug resistant P-glycoprotein
ABCC1	Gene symbol of human multidrug resistant associated protein MRP1
ABCC2	Gene symbol of human multidrug resistant associated protein MRP2
ABCG2	Gene symbol of human breast cancer resistance protein
ACAT	Coenzyme A-cholesterol-*O*-acyltransferase
ACE	Angiotensin-converting enzyme
ACN	Acetonitrile
ACS	American Chemical Society
ACTH	Adrenocorticotropic hormone
ADEPT	Antibody-directed enzyme prodrug therapy
ADH	Alcohol dehydrogenase
ADME	Absorption, distribution, metabolism, and excretion
ADMET	Absorption, distribution, metabolism, excretion, and toxicity
ADR	Adverse drug reaction
AE	Average error
AGP	α_1-Acid glycoprotein
AHLs	Acyl homoserine lactones
AHR	Arylhydrocarbon receptor
AIDS	Acquired immune deficiency syndrome
AKR	Aldo-keto reductase
ALDH	Aldehyde dehydrogenase
ALL	Acute lymphoblastic leukemia
AL-PEG	PEG aldehyde
ALS	Adaptive least squares
AMP	Adenosine monophosphate

AMP-PNP	Adenosine 5′-(β,γ-imino) triphosphate (nonhydrolyzable ATP analog)
AMT	Absorptive-mediated transcytosis
ANDA	Abbreviated new drug application
ANN	Artificial neural networks
ANNE	Artificial neural networks ensemble
ANOVA	Analysis of variance
AOT	Dioctyl sulfosuccinate
AOX	Aldehyde oxidase
AP	Absorption potential
AP-1	Activating protein-1
API	Active pharmaceutical ingredient
aq	Aqueous (environment)
ARC	Accurate radioisotope counting
ARDS	Adult respiratory distress syndrome
ARNT	Arylhydrocarbon receptor nuclear translocator
ASA	Apolar surface area
ASBT	Apical sodium-dependent bile acid transporter
ATP	Adenosine triphosphate
ATR/FTIR	Attenuated total reflectance/Fourier transform infrared spectroscopy
AZT	Zidovudine
β-FNA	Beta-funaltrexamine
BBM	Brush-border membrane
BCRP1 (ABCG2)	Breast cancer resistance protein (ABCG2)
BCS	Biopharmaceutics classification system
bFGF	Basic fibroblast growth factor
BMC	Biopartitioning micellar chromatography
BMP	Bone morphogenetic protein
BNPP	Bis-nitrophenyl phosphate
BPH	Benign prostatic hypertrophy or hyperplasia
BSA	Bovine serum albumin
BTC-PEG	PEG benzotriazolyl carbonate
BtuCD	ATP dependent vitamin B_{12} uptake transporter
BUI	Brain uptake index
C6	Rat glioma cell line
Caco-2	Adenocarcinoma cell line derived from human colon (used for estimation of human absorption)
CADD	Computer-assisted drug design
CAT	Compartment absorption transit
CBR1	Cytosolic carbonyl reductase
CDI-PEG	PEG carbonyldiimidazole
CDK	Cyclin-dependent kinase (CDK-7, CDK-9, etc.)
CDNB	1-Chloro-2,4-dinitrobenzene

CDR	Convective diffusion with simultaneous chemical reaction
CE	Capillary electrophoresis
CEDD	Centers of excellence in drug discovery
cEND	Immortalized mouse brain endothelial cell line
CFC	Chlorofluorocarbon
CGH	Comparative genomic hybridization
CHAPs	Cyclic hydroxamic acid-containing compounds
CHD	Coronary heart disease
CHO	Chinese hamster ovary cell line
CI	Clearance index
CLND	Chemical luminescence nitrogen detector
ClogP	Calculated log P according to the method by Leo and Hansch (sometimes generic for calculated partition coefficient)
CMC	Critical micelle concentration
CMR	Calculated molar refractivity
CMV	Cytomegalovirus
CNS	Central nervous system
CNS –	Drugs that are not taken up by the brain or have no central effect
CNS +	Drugs that are taken up into the brain and have a central effect
CNS +/–	Central nervous system availability
CoA	Coenzyme A
COMET	Consortium for metabonomic toxicology
CoMFA	Comparative molecular field analysis
CoMMA	Comparative molecular moment analysis
CoMSA	Comparative molecular surface analysis
CoMSIA	Comparative molecular similarity indices analysis
COMT	Catechol O-methyltransferase
COSY	Correlation spectroscopy
COX	Cyclooxygenase
COX-1	Cyclooxygenase-1
CP	Carboxypeptidase
CPA	N^6-Cyclopentyladenosine
cPCA	Consensus principal component analysis
CPK	Corey, Pauling and Koltun
CPR	Cytochrome P450 reductase
CPSA	Charged partial surface area
CR	Controlled release
CREB	cAMP response element binding protein
CRF	Corticotropin-releasing factor
CRO	Contract research organization
CSF	Cerebrospinal fluid
CT-2103	PGA-paclitaxel conjugate

CV	Cyclic voltammetry
CYP	Cytochrome P450
CYP2D6	Cytochrome P450 2D6
CYP3A4	Cytochrome P450 3A4
CZE	Capillary zone electrophoresis
DA	Discriminant analysis
DAD	Diode assay detector
Daevent	Dermally absorbed dose per event
DAF	Decay accelerating factor
DAT	Dopamine transporter
DDI	Drug–drug interaction
DF	Decision forest
DHFR	Dihydrofolate reductase
DIDR	Disk intrinsic dissolution rate
DIVEMA	Divinylethermaleic anhydride/acid copolymer
DMPK	Drug metabolism and pharmacokinetics
DMSO	Dimethyl sulfoxide
DNA	Deoxyribonucleic acid
DOX	Doxorubicin
DPI	Dry powder inhaler
DR	Dissolution rate
DSC	Differential scanning calorimetry
E. coli	*Escherichia coli*
ECF	Extracellular fluid
ECG	Surface electrocardiogram
ECV304	Cell line with endothelial-epithelial phenotype
ECVAM	European Centre for Validation of Alternative Methods
EEG	Electro encephalogram
EGF	Epidermal growth factor
EH	Epoxide hydrolase
EI	Electron impact
EIA	Enzyme immunometric assay
EMEA	European Medicines Agency (European Agency for the Evaluation of Medicinal Products)
EMS	Ethylmethane sulfonate
ENU	*N*-Ethyl-*N*-nitrosourea
EPHX1	Microsomal epoxide hydrolase
EPHX2	Soluble epoxide hydrolase
EPR	Enhanced permeability and retention
ESI	Electrospray ionization
EST	Expressed sequence tag
E-State	Electrotopological state
EVA	Eigen-values analysis

FAB	Fast atom bombardment
FAD	Flavin adenine dinucleotide
FDA	Food and Drug Administration (USA)
FFA	Free fatty acid
FIPCO	Fully integrated pharmaceutical company
FITC	Fluorescein isothiocyanate
FKBP12	FK506 binding protein 12
FL	Fuzzy logic
FMN	Flavin mononucleotide
FMO	Flavin-containing monooxygenase
FRET	Fluorescence resonance energy transfer
FSCPX	8-Cyclopentyl-N-[3-(4-fluorosulfonyl) (benzoyl)-oxy)-propyl]-1-N-propyl-xanthine
FTE	Full-time employee
GABA	γ-Amino-butyric acid
GC	Gas chromatography
G-CSF	Granulocyte colony stimulating factor
GDEPT	Gene-directed enzyme prodrug therapy
GDNF	Glial cell-derived neurotrophic factor
GFAcT	Genome functionalization through arrayed cDNA transduction
GFP	Green fluorescent protein
GFR	Glomerular filtration rate
GI	Gastrointestinal
GLP	Good laboratory practice
GM1	Ganglioside GM1
GMDs	Gel microdroplets
GMP	Good manufacturing practice
GP	Genetic programming
GPCR	G protein-coupled receptor
GRE	Glucocorticoid receptor element
GRNN	General regression neural network
GSH	Glutathione (reduced)
GSK	Glycogen synthase kinase (e.g., GSK-3)
GSSG	Glutathione (oxidized)
GST	Glutathione S-transferase
Hb	Hemoglobin
HBA	Number of hydrogen bond acceptors
hCP	Human carboxypeptidase
HCPSA	High-charged polar surface area
HCV	Hepatitis C virus
HDAC	Histone deacetylase (HDAC1, HDAC4, etc.)
HDACI	Histone deacetylase inhibitor
HDL	High-density lipoproteins

HDM	Hexadecane membrane (artificial membrane)
HEK293	Human embryonic kidney cell line
hERG	Human ether-a-go-go related gene
HFA	Hydrofluoroalkane
HGH	Human growth factor
HIA	Human intestinal absorption
HIF	Hypoxia inducible factor
HINT	Hydrophobic interaction field
HIP	Hydrophobic ion pairing
HIV	Human immunodeficiency virus
HLB	Hydrophilic-lipophilic balance
HLM	Human liver microsomes
HMBA	Hexamethylene bisacetamide
HMEC	Human mammary epithelial cell
HOMO	Highest occupied molecular orbital
HPAC	High-performance affinity chromatography
hPEPT1	Human oligo-peptide carrier for di- and tripeptides
HPGL	Number of hepatocytes per gram of liver
HPLC	High-performance liquid chromatography
HPMA	N-(2-Hydroxypropyl)methacrylamide copolymer
HQSAR	Hologram quantitative structure–activity relationship
HRMAS	High-resolution magic angle spinning
HSA	Human serum albumin
HSP	Heat shock protein (hsp90, hsp100, etc.)
HT	High throughput
HT29	Mucus-producing adenocarcinoma cell line (used for estimation of human absorption)
HTS	High-throughput screening
IC	Inhibitor concentration
IAM	Immobilized artificial membrane
IA-PEG	PEG-iodoacetamide
IBD	inflammatory bowel disease
IBS	Irritable bowel syndrome
ICH	International Conference on Harmonization
IDR	Intrinsic dissolution rate
IFN	Interferon
IGF	Insulin-like growth factor
I_{kr}	Rapidly activating delayed rectifier potassium current
I_{ks}	Slowly activating delayed rectifier potassium current
IL2	Interleukin-2
ILC	Immobilized liposome chromatography
in silico	Computational
IND	Investigative new drug application

IPR	Intellectual property rights
IQ	2-Amino-3-methylimidazo[4,5-*f*]quinoline
IR	Infrared
IRES	Internal ribosome entry site
ISEF	Intersystem extrapolation factor
ISF	Interstitial fluid (also called extracellular fluid)
IT	Information technology
IUBMB	International Union of Biochemistry and Molecular Biology
IUPAC	International Union of Pure and Applied Chemistry
i.v.	Intravenous(ly)
IVIVC	In vitro/in vivo correlation
JAK	Janus kinase
kNN	k-Nearest neighbor analysis
LC	Liquid chromatography
LC/MS	Combined liquid chromatography/mass spectrometry
LC/MS/MS	Combined liquid chromatography/tandem mass spectrometry
LCM	Laser capture microdissection
LDA	Linear discriminant analysis
LDL	Low-density lipoproteins
LEKC	Liposome electrokinetic chromatography
LFER	Linear free-energy relationship
LHRH	Luteinizing hormone-releasing hormone
LIMS	Laboratory information management system
LLC-PK1	Cell line derived from pig kidney epithelia
LmrA	*Lactococcus lactis* multidrug resistance ABC transporter
LOF	Lack-of-fit statistic
LQTS	Long QT syndrome
LSCM	Laser scanning confocal microscopy
LT	Low throughput
LUMO	Lowest unoccupied molecular orbital
M	Response modifier
mAb	Monoclonal antibody
MAD	Maximum absorbable dose
MAE	Mean absolute error
MAL-PEG	PEG-maleimide
MAO	Monoamine oxidase
MAP	Mitogen activated protein
MAS	Magic angle spinning
MC	3-Methylcholanthrene
MCA	Medicines Control Agency
MCG	Membrane-coating granule
MCT	Monocarboxylic acid cotransporter

MD	Molecular dynamics
MDCK	Madin–Darby canine kidney cell line (used to estimate human oral absorption)
mDDI	Metabolic drug–drug interaction
MDI	Metered dose inhaler
MDN	Metered dose nebulizer
MDR	Multidrug resistance
mEH	Microsomal epoxide hydrolase
MEP	Molecular electrostatic potential
MHBP	Molecular hydrogen-bonding potential
MHC	Major histocompatibility complex
MI	Membrane-interaction
MIF	Molecular interaction field
miRNAs	microRNAs
MLP	Molecular lipophilicity potential
MLR	Multiple linear regression
MMAD	Mass median aerodynamic diameter
MMPs	Matrix metalloproteinases (e.g., MMP-9)
MO	Molecular orbital(s)
MOA	Mechanism of action
mp	Melting point
MPDP$^+$	1-Methyl-4-phenyl-2,3-dihydropyridinium
mPEG	Monomethoxy-poly(ethylene glycol)
MPP$^+$	1-Methyl-4-phenyl pyridinium salt
MPTP	1-Methyl-4-phenyl-1,2,3,6-tetrahydropyridine
MQSM	Molecular quantum similarity measures
MR	Molar refractivity
MRI	Magnetic resonance imaging
MRM	Multiple reaction monitoring (mode in MS/MS)
MRP	Multidrug resistance associated protein (ABCC)
MRP1	Multidrug resistance related protein-1 (ABCC1)
MRT	Mean residence time
MS	Mass spectrometry
MS/MS	Tandem mass spectrometry
MsbA	Lipid flippase multidrug resistance ABC transporter
MT	Methyltransferase
MTD	Minimum topological difference
MTS	Methanethiosulfonate
MTSF	Molecular tree structured fingerprints
MTT	Cell vitality assay using 3-(4,5-dimethylthiazol-2-yl)-2,5-diphenyltetrazolium bromide
Mu	Amount of drug excreted in urine
MudPIT	Multidimensional protein identification technology
MVP	Major vault protein

MW	Molecular weight
Na$^+$,K$^+$-ATPase	Sodium-potassium ATPase
NADPH	Nicotinamide adenine dinucleotide phosphate (reduced)
NAS	National Academy of Sciences
NAT	*N*-Acetyltransferase
NBD	Nucleotide-binding domain
NBE	New biological entity
NC	Nomenclature Committee
NCA	Normal coronary artery
NCBI	National Center for Biotechnology Information
NCE	New chemical entity
NCI	National Cancer Institute
NCS	Neocarcinostatin
NDA	New drug application
NEFA	Non-esterified fatty acid
NFκB	Nuclear factor kappa B
NHS	*N*-Hydroxysuccinimide
NIH	National Institutes of Health (also not-invented-here)
NMDA	*N*-Methyl-D-aspartate
NMR	Nuclear magnetic resonance (spectroscopy)
NN	Neural network
NNK	4-(Methylnitrosamino)-1-(3-pyridyl)-butan-1-one
NO	Nitric oxide
NORD	National Organization of Rare Disorders
NPSA	Nonpolar surface area
NQO	NAD(P)H quinone oxidoreductase (aka DT-diaphorase)
NQOR	NAD(P)H quinone oxidoreductase (DT-diaphorase)
NRPs	Nonribosomal peptides
NSAID	Nonsteroidal anti-inflammatory drug
Oa-ToF	Orthogonal acceleration time-of-flight
Oatp/OATP	Organic anion transporting polypeptide (animal/human, resp.)
OCT	Organic cation transporter family
ODA	Orphan Drug Act
OECD	Organization of Economic Co-operation and Development
o-NPOE	*Ortho*-nitrophenyloctyl ether
OP	Organophosphates
OPSS-PEG	PEG-orthopyridyl-disulfide
ORMUCS	Ordered multicategorical classification method using the simplex technique
OSC	Orthogonal signal correction
OTC	Over-the-counter
p.o.	*Per os* = oral(ly)
P450	Cytochrome P450

PA	Phosphatidic acid
PacM	Poly(acroloylmorpholine)
PAH	Polycyclic aromatic hydrocarbon
PAMPA	Parallel artificial membrane permeability assay
PAPS	3′-Phosphoadenosine 5′-phosphosulfate
PATQSAR	Population analysis by topology-based QSAR
PB	Phenobarbital
PBL	Porcine brain lipid
PBPK	Physiologically-based pharmacokinetics
PCA	Principal component analysis
PCM	Polarized continuum method
PcV	Phenoxymethylpenicillin
PD	Pharmacodynamic(s)
PDA	Photo-diode array
PDB	Protein Data Bank
PDMS	Polydimethylsiloxane
PE	Phosphatidylethanolamine
PECAM-1	Platelet endothelial adhesion molecule-1
PEG	Polyethyleneglycol
PEI	Poly(ethyleneimine)
PEPT1	(Oligo)peptide transporter 1
PET	Positron emission tomography
PG	Phosphatidylglycerol
PGA	Polyglutamic acid
P-gp	P-glycoprotein (MDR1, ABCB1)
PGS	Prostaglandin synthase
PHAH	Polyhalogenated aromatic hydrocarbon
PHEG	Poly((N-hydroxyethyl)-L-glutamine)
PhIP	2-Amino-1-methyl-6-phenylimidazo[4,5-b]pyridine
PI	Phosphatidylinositol
PI3K	Phosphoinositide-3-kinase
PK	Pharmacokinetic(s)
PK/PD	Pharmacokinetic–pharmacodynamic
PK1	HPMA-(Gly-Phe-Leu-Gly-doxorubicin)$_n$ conjugate
PK2	(N-acylated galactosamine)$_m$-HPMA-(Gly-Phe-Leu-Gly-doxorubicin)$_n$ conjugate
PKC	Protein kinase C
PKS	Polyketide synthase enzyme
PMT	Photo multiplier tube
PNN	Probabilistic neural network
pNPC-PEG	PEG p-nitrophenyl carbonate
PON	Paraoxonase
POPC	Palmitoyloleilphosphatidylcholine

PPAR-γ	Peroxisome proliferator-activated receptor gamma
PSA_d	Dynamic polar surface area
PSI	Protein structure initiative
PTSA	Partitioned total polar surface area
PVA	Polyvinylalcohol
PVP	Poly(vinylpyrrolidone)
QC	Quality control
QM	Quantum mechanical
QSAR	Quantitative structure–activity relationship
QSPeR	Quantitative structure–permeability relationship
QSPkR	Quantitative structure–pharmacokinetic relationship
QSPR	Quantitative structure–property relationship
QTL	Quantitative trait loci
R&D	Research and development
RAF	Relative activity factor
RBF	Radial basis function
RFP	Request for proposal
RH	Relative humidity
rhCYP	Recombinant human cytochrome P450
rhGH	Recombinant human growth hormone
RIA	Radioimmunoassay
RISC	RNA-induced silencing complex
RLIP-76 (RALBP1)	Ral-interacting protein, non-ABC transporter
r-metHuG-CSF	Recombinant-methionine human G-CSF
RMSE	Root mean square error
RMT	Receptor-mediated transcytosis
RNA	Ribonucleic acid
RNAi	RNA interference
ROI	Return on investment
ROS	Reactive oxygen species
RP	Recursive partitioning (also called decision tree)
RSF	Relative substrate activity factor
RT-PCR	Reverse-transcriptase polymerase chain reaction
S/N	Signal to noise ratio
S9	Supernant from homogenation and 9000g sedimentation
SAM	S-Adenosyl-L-methionine
Saos-2	Osteosarcoma cell line
SAR	Structure–activity relationship
SARS	Severe acute respiratory syndrome
SAS	Solvent accessible surface
SASA	Solvent accessible surface area
SAT	Site acceptance test

List of Abbreviations

SATE	*S*-Acyl-2-thioethyl
SBDD	Structure-based drug design
SC	Stratum corneum
SC-PEG	PEG succinimidyl carbonate
SD	Standard deviation (see also s)
SDK	Substrate disappearance kinetics
SDR	Short-chain dehydrogenase/reductase
SERT	Serotonin transporter
SES	Solvent excluded surface
SF	Shake-flask
SGF	Simulated gastric fluid (USP)
SIF	Simulated intestinal fluid (USP)
siRNA	Small interfering RNA
SITT	Small intestinal transit time
SIWV	Small intestinal water volume
SLS	Sodium lauryl sulfate
SMA	Poly(styrene-co-maleic acid/anhydride)
SMILES	Simplified molecular input line entry system
SMM	Small molecule microarray
SNP	Single nucleotide polymorphism
SOM	Self-organizing map
SPA	Scintillation proximity assay
SPE	Solid phase extraction
SPEC	Solid phase extraction chromatography
SPECT	Single photon emission computed tomography
SRIF	Somatotropin-release inhibiting factor
SRP	Signal recognition particle
SRS	Substrate recognition site
SRS-A	Slow reacting substance of anaphylaxis
SSLM	Solid-supported lipid membrane
SS-PEG	PEG succinimidyl succinate
SSRI	Selective serotonin reuptake inhibitor
STAT	Signal-transducing activators of transcription
STI	Signal transduction inhibitor
SULT	Sulfotransferase
SVM	Support vector machine
TAP	Tumor-activated prodrug
TAR	Transactivation response
TCDD	2,3,7,8-Tetrachlorodibenzo-*p*-dioxin ('dioxin')
TC-NER	Transcription-coupled nucleotide excision repair
TCP-PEG	PEG trichlorophenyl carbonate
TDI	Time-dependent inhibition

TEER	Transendothelial electrical resistance
Tgase	Transglutaminase
TGFβ	Transforming growth factor beta
Thy-1	Thymus cell antigen-1 (CD90) expressed on pericytes/fibroblasts
TIAs	Tubulin interactive agents
TIC	Total ion current
TK	Tyrosine kinase
TMA	Trimethylamine
TMD	Transmembrane domain
TNF-α	Tumor necrosis factor alpha
TOCSY	Total correlation spectroscopy
ToF	Time-of-flight
TOPS-MODE	Topological substructural molecular design
TPA	Tissue plasminogen activator
TPGS	Tocopheryl polyethylene glycol succinate
TPSA	Topological polar surface area
TRD	Transcriptional repression domain
TRF	Time-resolved fluorescence
TVD	Triple vessel disease
UDP	Uridine diphosphate
UDPGA	UDP-glucuronic acid = uridine-5′-diphospho-α-D-glucuronic acid
UGT	UDP-glucuronosyltransferase = UDP-glucuronyltransferase
UPLC	Ultra performance liquid chromatography
URS	User requirement specification
US EPA or EPA (US)	United States Environmental Protection Agency
USP	United States Pharmacopeia
UTR	Untranslated region
UV	Ultraviolet
UWL	Unstirred water layer
V. cholerae	*Vibrio cholerae*
VEGFR	Vascular endothelial growth factor receptor
vHTS	Virtual high-throughput screening
VLDL	Very low-density lipoproteins
VS-PEG	PEG-vinylsulfone
WDI	World Drug Index
WHI	Women's Health Initiative
XED	Extended electron distribution
z	Charge number
ZO-1	Zonula occludens protein-1

Software and Vendors

Computer Programs *URL Vendor*

ACD	Advanced Chemistry Development	www.acdlabs.com
ALOGPS	Artificial neural network program to predict lipophilicicy (log P) and aqueous solubility (log S) of chemicals	
CART	Recursive classification regression tree	www.salford-systems.com
CHARMM	Chemistry at Harvard molecular mechanics	www.charmm.org
CLOGP	Calculation of octanol/water partition coefficients for neutral species	www.biobyte.com
DAYLIGHT	Toolkit for database applications	www.daylight.com
DERMWIN	Dermal permeability coefficient program	www.syrres.com
DISCO	Pharmacophore perception module in SYBYL software system	
DRAGON	Set of over 1800 molecular descriptors	www.talete.mi.it
GASP	Genetic algorithm similarity program	
GOLPE	Generating optimal linear PLS estimation	www.miasrl.com
GRIND	GRID-Independent (descriptors derived from the ALMOND program)	www.moldiscovery.com
HYBOT	Hydrogen bond thermodynamics	http://software.timtec.net/hybot-plus.htm
MMFF94	Merck Molecular Force Field 94	
SIMCA	Soft-independent modelling of class analogy	www.umetrics.com
SPARC	Computer program to calculate pK_a values	
WHIM	Weighted holistic invariant molecular descriptor	

List of Symbols

%ppb	Percentage of bound drug in plasma
%PSA	Percentage polar surface area
%T	Percentage transported through a membrane
%uptake	Per cent uptake efficiency of the trout gill
[...]	Square brackets used to indicate concentration of, e.g. [Na$^+$] is 'the concentration of Na$^+$'
[A]	Concentration of agonist
[AR]	Concentration of agonist–receptor complex
[I]$_{av}$	Average systemic plasma concentration of inhibitor after repeated oral administration
[I]$_{in}$	Maximum hepatic input concentration of inhibitor
[I]$_u$	Inhibitor concentration, unbound
[R_0]	Total receptor concentration
0D, 1D, 2D, 3D, 4D	Zero-, One-, two-, three-, four-dimensional
$^6\chi$	Sixth-order path molecular connectivity
A	Generic symbol for the total concentration of the substance A; the sum of the reactant concentration and the concentrations of all the associated species containing the A reactant; usually expressed in units of mol cm^{-3} in kinetic equations
a	Constant reflecting the slope of a parabolic stimulus–effect relationship
A%	Percentage oral absorption
A_D	Cross-sectional area of a molecule
A log P	Calculated log P according to an atomic contribution method
A_m	Membrane area
A_s	Molecular cross-sectional area
AUC	Area under the curve of a plasma concentration vs time profile
AUC$_i$	AUC for substrate in presence of inhibitor
B	Hydrogen-bonding basicity (Abraham equation)
b	Stimulus at maximal effect of a parabolic stimulus–effect relationship
BBB	blood–brain barrier
C_0	Initial total concentration of drug in plasma
C_a	Binding capacity
C_A	Concentration in the water phases of the acceptor
C_a and C_d	Free energy factor of proton acceptors and proton donors, respectively (HYBOT, Raevsky)
C_b	Protein-bound drug concentration
C_D	Concentration in the water phases of the donor
C_e	Drug concentration in hypothetical effect compartment
C_{fa}	Fetal artery concentration
C_{fv}	Fetal vein concentration
C_{in}	Concentration of inhibitor
C_j	jth associated species concentration in kinetic equations

List of Symbols

CL	Total in vivo plasma clearance
CL_b	In vivo blood clearance
CL_H	Hepatic clearance
$CL_{H,int}$	Hepatic intrinsic clearance
$CL_{H,int,u}$	Unbound hepatic intrinsic clearance
$CL_{H,u}$	Unbound hepatic clearance
CL_{int}	Intrinsic (metabolic, hepatic) clearance
$CL_{int,u}$	Unbound hepatic intrinsic clearance
$CL_{int,us}$	Intrinsic tubular secretion rate
CL_{max}	Maximum intrinsic clearance; when an enzyme is fully activated (used for enzymes which show positive cooperativity in vitro)
ClogP	Calculated log P according to the method by Leo and Hansch (sometimes generic for calculated partition coefficient)
CL_p	In vivo plasma clearance
CL_R	Renal clearance
C_{m0}	Concentration at position 0 in the membrane
C_{ma}	Maternal artery concentration
C_{max}	Maximum (peak) plasma concentration
$C_{max,ss}$	C_{max} at steady state
C_{mh}	Concentration at position h in the membrane
C_{min}	Minimum (peak) plasma concentration
$C_{min,ss}$	C_{min} at steady state
C_{on}	Concentration of surface activity onset
C_T	Concentration in tissue
C_u	Free (unbound) drug concentration
ΔE	Molecular energy difference
ΔG	(Change in) Gibbs free energy
$\Delta G(x)$	Free energy difference between the water phase and position x in the membrane
ΔH	(Change in) enthalpy
$\Delta \log P$	Difference log P (octanol/water) – log P (alkane/water)
$\Delta \log P_{oct\text{-}dce}$	Difference between the logarithms of the octanol/water and 1,2-dichloroethane/water partition coefficients
ΔM	Amount of diffusant transported through the membrane
δ	Binding affinity (CYP)
δ_m	Solubility parameter for the membrane
δ_v	Solubility parameter for the vehicle
$D(x)$	Local diffusion coefficient at x position
$D_{barrier}$	Diffusion coefficient in the barrier domain
D_j	Diffusivity coefficient of reactant or associated species
D_m	Diffusion coefficient in the membrane
D_{oct}	Octanol/buffer distribution coefficient
Dose	Dose
$D_{T/B}$	Tissue/blood distribution coefficient

$D_{T/P}$	Tissue/plasma distribution coefficient
$D_{vo/w}$	Vegetable oil/water distribution coefficient
e	Proportionality constant denoting the power of a drug to produce a response
ε	Intrinsic efficacy
ϵ	Dielectric constant
E_0	Baseline drug effect
EC_{25}	Effective concentration needed to observe 25% of the biological response
EC_{50}	Drug concentration at half maximal drug effect
$E_{inter(vdW)}$	Energy of the Van der Waals' penetrant–membrane interaction
E_m	Maximum system response
E_{max}	Maximum effect of hyperbolic concentration–effect relationship
$E_{SS}(tor)$	Torsional energy at the total intermolecular system minimum energy for the solute
E_{top}	Maximum effect of parabolic stimulus–effect relationship
E_V	Free energy of vaporization
f	Molar fraction
$F\%$	Percentage bioavailable
f_a	Fraction of drug absorbed from the gut into the hepatic portal vein after oral administration
F_G	Gut availability (fraction of drug which escapes elimination by the gut)
f_g	Fraction of drug metabolized by the gut wall
f_h	Fraction of drug metabolized by the liver
$f_{i(7.4)}$	Fraction of ionized drug at pH 7.4
F_{inh}	Pulmonary bioavailability
f_j	Fraction of drug at ionization state j
$f_{m(CYP)}$	Fraction metabolized by a CYP isozyme
f_P	Fractional content of extracellular fluid in plasma
f_r	Fraction of drug reabsorbed
F_{rel}	Relative bioavailability comparing two different pharmaceutical
f_{rotb}	Fraction of rotatable bonds
f_T	Fractional content of extracellular fluid in tissues
$f_{u,b}$	Fraction of drug unbound (free) in blood
$f_{u,mic}$	Fraction of drug unbound (free) in microsomes
$f_{u,p}$	Fraction of drug unbound (free) in plasma
$f_{u,t}$	Fraction of drug unbound (free) in tissue
f_z	Fraction of each charged species
g	Gaseous environment
γ	Catalytic efficacy (CYP)
γ^0	Surface tension of the bare interface
h_j	Thickness of the aqueous boundary layer of reactant or associated species j
IC_{50}	Concentration of inhibitor required to give 50% inhibition
J	Flux across a membrane or dissolution of a solid
J_{max}	Maximal flux

J_{PDMS}		Maximum steady-state flux through PDMS
J_{SS}		Steady-state flux
K		Tissue/medium partition coefficient
k		Boltzmann constant
k_{1e}		Rate constant for distribution into the hypothetical effect compartment
$\kappa 3$		Third-order kappa index (topological descriptor)
k_a		Absorption rate constant
K_A		In vivo dissociation equilibrium constant
K_{aw}		Air/water partition coefficient
$K_{barrier/water}$		Partition coefficient of a solute from water to the barrier domain
K_{BB}		The ratio of concentration in brain over blood at equilibrium
K_{blood}		The ratio of concentration in blood divided by concentration in air, at equilibrium
K_{brain}		The ratio of concentration in brain divided by concentration in air, at equilibrium
k_{deg}		Endogenous degradation rate constant
k_{degrad}		Degradation rate constant
k_{dep}		Substrate depletion rate constant
K_E		Value midpoint location of transducer function
$k_{eff,GI}$		Effective rate of absorption
k_{el}		Elimination rate constant
k_{eO}		Rate constant for distribution out of the hypothetical effect compartment
k_f		Constant of passive transfer
k_{gill}		Rate of absorption across the gills in the guppy
K_i		Inhibition equilibrium constant
K_I		Inhibition equilibrium constant for mechanism-based inhibition
k_{IAM}		Capacity factor of IAM column chromatography
K_{ic}		Competitive inhibition equilibrium constant
k_{ILC}		Capacity factor of immobilized liposome chromatography
K_{in}		Unidirectional influx coefficient
k_{in}		Zero-order rate constant for production of a physiological entity
k_{inact}		Maximal inactivation rate constant
K_{iu}		Uncompetitive inhibition equilibrium constant
k_{loss}		First-order rate constant for release of precursor in central compartment
K_m		Michaelis–Menten constant – substrate concentration at which rate of reaction is half V_{max}
k_m		Rate constant for formation and dissipation of a response modifier
K_{memb}		Membrane partition coefficient
K_{mp}		Milk to plasma ratio
K_n		Aggregation constant, where n is the degree of aggregation
k_{out}		First-order rate constant for loss of a physiological entity
k_p		First-order permeability rate constant
$K_{p(epi)}$		Human epidermis permeability coefficient
$K_{p(sil)}$		Silicone permeability coefficient

$K_{\text{p-corneal}}$	Corneal permeability coefficient
K_s	Binding constant
λ	Rate constant of enzyme inactivation
λ_1	Apparent elimination constant
LBF	Liver blood flow; also Q_H
log BB	log[brain:plasma ratio]
log D	Logarithm of the distribution coefficient at given pH, typically in *n*-octanol/water and at pH 7.4
log $D_{7.4}$	Logarithm of the distribution coefficient at given pH 7.4, typically in *n*-octanol/water
log D_{oct}	Logarithm of the distribution coefficient at given pH, in *n*-octanol/water
log k_s	Logarithm of the chromatography capacity factor
log k_{sdc}	Logarithm of the chromatography capacity factor (capillary electrophoresis)
log P	Logarithm of the partition coefficient for the neutral species, typically in *n*-octanol/water
log P_{hept}	Logarithm of the partition coefficient for the neutral species in *n*-heptane/water
log P_{ion}	Log P of the ionized species
log P_{neutral}	Logarithm of the partition coefficient for the neutral species, typically in *n*-octanol/water
log P_{oct}	Logarithm of the partition coefficient for the neutral species in *n*-octanol/water
μ	Dipole moment
M	Amount of drug
mp	Melting point
n	Number of datapoints (compounds or observations) used in the regression
N_A	Avogadro's number
O_{lumen}	Oversaturation number in the lumen
π^*	Dipolarity/polarizability
π_{memb}	Surface pressure of a lipid bilayer
ϕ	Membrane-to-water distribution coefficient
P	Partition coefficient between water and lipid or organic solvent (typically *n*-octanol) for a molecule in its unionized form (see also log P)
P_{chl}	Chloroform/buffer partition coefficient
$P(x)$	Local membrane permeability at position x
$p[\text{H}]$	$-\log_{10}[\text{H}^+]$, the degree of acidity of a solution, expressed on an concentration scale
P_0^{PAMPA}	Intrinsic PAMPA permeability of uncharged species
P_{app}	Apparent permeability coefficient
P_{cyc}	Water to cyclohexane partition coefficient
P_e	Permeability constant or coefficient
$P_{e,\text{PAMPA}}$	PAMPA permeability
P_{e0}	Intrinsic permeability of uncharged species
P_{e0}^{BBB}	Intrinsic BBB permeability of uncharged species
P_{eff}	Effective intestinal membrane permeability
$P_{\text{eg/hept}}$	Heptane to ethylene glycol partition coefficient
pH	Negative logarithm of the hydrogen ion concentration reflecting acidity/basicity
pK_a	Negative logarithm of the ionization/dissociation constant ($-\log_{10} K_a$)

pK_a''	Negative logarithm of the ionization constant determined by direct titration in KCl/water-saturated n-octanol
pK_a^0	In Hammett and Taft equations, the pK_a of the unsubstituted acid or base
PLS	Partial least squares (or projection to latent structures)
PLS-DA	Partial least squares–discriminant analysis
P_m	Membrane permeability
$P_{m,app}$	Apparent membrane permeability
P_{oct}	Partition coefficient between water and n-octanol for a molecule in its unionized form
P_{para}	Paracellular pathway permeability
PPB	Plasma protein binding (see also %ppb)
P_{UWL}	Unstirred water layer permeability
$P_{vo/w}$	Vegetable oil/water partition coefficient
q	Electric charge
$q\ (q^2)$	Cross-validated correlation coefficient (squared cross-validated correlation coefficient)
q^-	Lowest atomic charge in the molecule
q^2	Square of the cross-validated correlation coefficient for external validation data set
Q_f	Fetal flow rate
Q_H	Hepatic blood flow
Q_R	Renal blood flow
r	Molecular radius
R	Universal gas constant
RMSE	Root mean square error
$r\ (r^2)$	Correlation coefficient (squared correlation coefficient)
$R_{e/i}$	Ratio of extravascular to intravascular proteins
$R_{E/P}$	Erythrocyte/plasma ratio
r_{gyr}	Radius of gyration
r_{para}	Apparent pore radius of the paracellular pathway
σ	In Hammett and Taft equations, a constant that is characteristic of a particular substituent
s	Standard deviation/error of a regression equation
S_0	Intrinsic solubility, solubility of the uncharged species
S_{50}	Substrate concentration at half V_{max} (used for enzymes which show positive cooperativity in vitro)
S_{pH}	The concentration of a substance in solution at equilibrium with excess solid at a given pH
S_s	Solubility at the interface
$\Sigma MHBP_{do}$	Sum of the calculated molecular hydrogen bond donor potential
τ	Dosing interval
T	(Body) temperature
$t_{1/2}$	Half-life
$t_{1/2deg}$	Degradation half-life
t_l	Lag time
t_{max}	Time to reach maximal (peak) plasma concentration C_{max}

TR		Placental transfer ratio
T_{SP}		Set-point of body temperature
TUI		Testis uptake index
V		Volume in experimental setup such as plate well
v		Volume fraction of component in tissue or blood
$V(x)$		Axial velocity in the rotating disk apparatus, as a function of the distance (x) from the surface of the rotating disk and directed into the disk
V_{ac}		Volume of the acceptor wells
V_D		Volume of distribution
$V_{D,ss}$		Volume of distribution at steady state
$V_{D,u}$		Unbound volume of distribution
V_{do}		Volume of the donor wells
VdWS		Van der Waals' surface
V_e		Effective Van der Waals' volume
V_E		Extracellular volume
V_{ery}		Volume of erythrocytes
V_F		Sorption of solvents into an artificial membrane
VISEF		ISEF based on V_{max} values in HLM and rhCYP
V_P		Plasma volume
V_R		Volume in which the drug distributes minus the extracellular space
V_T		Volume of tissue compartment
wt%		Weight per cent
x		Thickness of a membrane

INDEX FOR VOLUME 2

Notes

Abbreviations

HTS – high-throughput screening
IP – intellectual property
PDB – Protein Data bank

Cross-reference terms in italics are general cross-references, or refer to subentry terms within the main entry (the main entry is not repeated to save space). Readers are also advised to refer to the end of each article for additional cross-references – not all of these cross-references have been included in the index cross-references.

The index is arranged in set-out style with a maximum of three levels of heading. Major discussion of a subject is indicated by bold page numbers. Page numbers suffixed by T and F refer to Tables and Figures respectively. *vs.* indicates a comparison.

This index is in letter-by-letter order, whereby hyphens and spaces within index headings are ignored in the alphabetization. Prefixes and terms in parentheses are excluded from the initial alphabetization.

Any method, model or other subject, associated with the name of the developer (e.g. name's model) does NOT imply that Elsevier, nor the indexers, have assumed the right to name models/methods after the authors of the papers in which they are described. This is merely a succinct phrase to refer to a model/method developed/described by the relevant author, so that the subentry could be alphabetized under the most pertinent name.

A

A-85380, structure 887F
A-348441 1017
  structure 1017F
A-425619, structure 909F
Abarelix 579T
ABC-1 transporter, positional cloning 306
ABC transporters 58, 59F
  polymorphisms 67
Abgenix, characteristics 226T
ABL1, PDB entries 965T
Abl tyrosine kinase
  in disease 963T
  ligands, scaffold hopping 685, 686F
  SH2, structure 636T
  SH3, structures 636T
Absence (petit-mal) seizures, calcium channels 852
Absolute configuration determination, chiral drug discovery 720
ABT 378, structure 302F
'Accessory proteins'
  mitochondria complex I 1070–1071
  nicotinic acetylcholine receptors 886

ACE inhibitors
  analog design 651
  peptidomimetics 613, 614F
Acetaminophen, structure 9F
N-Acetylation, solid support synthesis 466F, 467, 467F
O-Acetylation, solid support synthesis 466F, 467, 468F
Acetylcholinesterase, nicotinic acetylcholine receptor binding 885
Acetylcholinesterase inhibitors
  as 'old' target 274–275
  ring equivalent bioisosterism 656, 657F
Acetylsalicylic acid (aspirin)
  gastric damage
    phosphodiesterase-5 inhibitors 949
    rolipram 948
  structure 9F
Acid-sensitive ion channels (ASICs) 879, 910
  characteristics 881T
  diseases 912
    pain perception 911
  expression 879
  molecular biology 910

  nomenclature 910
  pharmacology 911
    neuropeptides 911
    toxins 911
  physiological functions 911
    neuronal distribution 911
    pain 911
    retinal activity 911
  regulation 911
  structure 880F, 910
  therapeutics 911
    ischemia 911
ACLARA Biosciences, microfluidic devices 350F, 359
Acquisitions, pharmaceutical companies 101
ACTH receptors, in human disease 775T
Activating protein 1 (AP-1), GPCR functional assay screening 812
Active transport, peptide/protein drugs 591
Activity cluster analysis, sourcing development 216, 216F, 217F
Activity determination *see* Sourcing development
Activity interdependency, sourcing development 214

1111

Index

ACT-UP, drug discovery 270–271
Acute lymphoblastic leukemia (ALL), Bcr-Abl kinases 972
Acute myelogenous leukemia (AML)
  PKC-412 971
  retinoic acid receptor-α (RAR-α) 1027
Acute vs. chronic treatment, drug development productivity 128
ACV-1, nicotinic acetylcholine receptor binding 888
Acylation, hydrophobic depoting 586
Acyl-cyanamides, carboxylic acids, bioisosterism 659
Acyl-sulfonamides, carboxylic acids, bioisosterism see Bioisosterism, carboxylic acids
Address/message region, peptide/protein drugs 581
Adenine nucleotide translocator (ANT) 60–61
  conformation shifts 1060
  mitochondria 1059
Adenosine, structure 1038F
Adenosine A_1 receptor
  allosteric modulators 790T
  constitutively active 790T
Adenosine A_2 receptor, in human disease 775T
Adenosine A_{2A} receptor, allosteric modulators 790T
Adenosine A_{2B} receptor, constitutively active 790T
Adenosine A_3 receptor
  activation 785T
  allosteric modulators 790T
  constitutively active 790T
Adenosine receptors, as 'old' target 275
Adenosine triphosphate synthetase 1057, 1058F
Adenylyl cyclases, GPCR interaction 786T
Adipocytes, phosphodiesterase-3 924
ADME
  attrition reduction 54
  fail fast vs. 'follow the science' 121
  lead optimization 139
  pharmaceutical R&D 105
ADMET
  databases 569
  data/decisions 565
  definition 559–560
  final drug characteristics 563
  high-content data 565
  see also Chemogenomics; Metabonomics; Toxicogenomics
  computer modeling 566
  laboratory information management systems 565
  links to practical utility 567
  parallel research 566, 566F
  QSAR 566

HTS see High-throughput screening (HTS)
  pharmaceutical quality identification 563, 564T
  strategic application **559–572**
  computer-based technologies 569, 570F
  microarray systems 570–571
  technique validation 559–560
Administrative responsibilities, leadership 243
ADRBK1, PDB entries 965T
Adrenal glands, phosphodiesterase-2 expression 923–924
α_1-Adrenergic receptors, constitutively active 793T
α_{1A}-Adrenergic receptors
  activation 785T
  constitutively active 790T
  in human disease 775T
α_{1B}-Adrenergic receptors, constitutively active 790T
α_{1D}-Adrenergic receptors, constitutively active 790T
α_{2A}-Adrenergic receptors
  allosteric modulators 790T
  constitutively active 790T, 793T
  in human disease 775T
α_{2B}-Adrenergic receptors
  constitutively active 790T
  in human disease 775T
α_{2C}-Adrenergic receptors
  constitutively active 790T
  in human disease 775T
α_{2D}-Adrenergic receptors, constitutively active 793T
β-Adrenergic receptors, constitutively active 793T
β-Adrenergic receptor antagonists see also specific drugs
  angina treatment 254
  carboxylic acid bioisosterism 659
  hypertension treatment 254
β_1-Adrenergic receptors
  constitutively active 790T, 793T
  in human disease 775T
β_2-Adrenergic receptors 777–778
  allosteric modulators 790T
  constitutively active 790T, 793T
  desensitization, recycling 802
  in human disease 775T
  multistate activation model
    deactivation conformations 795
    fluorescence labeling studies 795
  oligomerization/heterodimerization, internalization 799
  phosphodiesterase-4 926
  structure 798
β_3-Adrenergic receptors, in human disease 775T

Adverse drug reactions (ADRs)
  phase IV clinical trials see Clinical trials, phase IV (postmarketing)
  prediction, preclinical trials 179–180, 180F
AequoScreen, GPCR functional assay screening 811
AERES Biomedical, characteristics 226T
AF-1 domain, nuclear receptors 1001
Affinity, selection, combinatorial chemistry 39–41, 43F
Age-related macular degeneration (AMD), disease basis, mitochondria 1054–1055
Aggregate-forming compounds, promiscuous ligands 738
Aggregators see Promiscuous ligands, computational prediction
Aging, mitochondria dysfunction 1060
AGN29
  as farnesoid X receptor ligand 1024
  structure 1025F
AGN31
  as farnesoid X receptor ligand 1024
  structure 1025F
AGN34
  as farnesoid X receptor ligand 1024
  structure 1025F
Agonists, peptide/protein drugs see Peptide/protein drugs
Agouron 761T
Airway disorders, potassium channels 863
AKAP79/250, GPCR interaction 797T
AKT2, PDB entries 965T
Akt/PDK1, GPCR interaction 786T
AL-438 1016–1017
  structure 1017F
Alanex 760T
Ala scan, peptide/protein agonists 582
Albany Molecular Research 760T
Albright hereditary osteodystrophy, Gα_s subfamily mutations 780–782
Albumin binding, hydrophobic depoting 586
Albuterol, mechanism of action, GPCR targeting 773T
Aldosterone 1017–1018
  structure 1018F
Alfacalcidol
  structure 1018F
  as vitamin D receptor ligand 1019
Alitretinoin, characteristics 1012T
Alkene substitution, peptides 604
N-Alkylation, solid support synthesis 466F, 467, 467F
O-Alkylation, solid support synthesis 466F, 467, 468F
Allele-specific polymerase chain reaction, single nucleotide polymorphisms 365

Index 1113

Allodynia, transient receptor potential ion channels 909–910
Allosteric effects, potassium channels 862
Allosteric modulation, GPCRs see G protein-coupled receptors (GPCRs)
All-*trans*-retinoic acid (ATRA)
 as retinoic acid receptor ligand 1020
 structure 1020*F*
α-subunits, G protein structure 780
Alphadolone, GABA$_A$ receptors 893
AlphaScreen cAMP assay, GPCR functional assay screening 809
Alprazolam, structure 652*F*
Alzheimer's disease
 disease basis, mitochondria 1054–1055
 epidemiology 52–53
 positional cloning 307
 proteomics 388
 treatment, phosphodiesterase-8 inhibitors 949
 Women's Health Initiative Memory Study 53
AM80
 as retinoic acid receptor ligand 1020–1021
 structure 1020*F*
Ambient temperature stability, drug reference standard 167, 168*T*
American Association of Colleges of Pharmacy 85
American Inventors Protection Act (1999) 131
AMG-706, characteristics 984*T*
AMG-9810, structure 909*F*
Amgen
 foundation 762
 initial products 762–763
 licensing 236
 R&D spending 101*T*
Amide bond replacements, peptidomimetics 606–608, 609*F*
Amides, phenol bioisosterism 670–671
Amikacin
 applications 1042
 structure 1042*F*
Amino acids
 analysis, protein identification 336–337
 beta, proteolysis resistance 584, 585
 gamma, proteolysis resistance 584
D-Amino acid substitution, proteolysis resistance 584
Amino acid transporters 60*T*
3-Amino-cyclopentanecarboxylic acid, structure 893*F*
Aminoglycosides
 mechanism of action, ribosomal RNA (rRNA) interactions 1041, 1042*F*
 side effects 1042

Aminophylline, as phosphodiesterase inhibitor 947*T*
Amiodarone
 bioisosterism 680*F*
 characteristics 829*T*
 ring system reorganization (bioisosterism) 679, 680*F*
 structure 680*F*, 829*T*, 1020*F*
Amitriptyline
 characteristics 829*T*
 structure 63*F*, 829*T*
Amlodipine
 characteristics 653*T*, 829*T*
 structure 653*T*
AMN-107
 Bcr-Abl kinase inhibitor 973
 characteristics 984*T*
 structure 974*F*
AMP-397 903
AMPA, structure 901*F*
AMPA-sensitive receptors 898
 characteristics 881*T*
 diseases
  epilepsy 903
  neurodegeneration 903
 pharmacology 900, 901*F*
 physiological functions 899
 regulation 900
 splicing variants 899
 SAR 900
Ampholytes, isoelectric focusing 332
Amprenavir 579*T*
Amrinone
 as phosphodiesterase inhibitor 947*T*
 structure 933–934
AMT-089
 nicotinic acetylcholine receptor binding 888
 structure 887*F*
Amylin analogs, pramlintide 580
Amyloid precursor protein (APP) 388
Amyotrophic lateral sclerosis, mitochondria 1054–1055
Analogs
 definition/description 651
 design
  angiotensin-converting enzyme (ACE) inhibitors 651
  bioisosterism see Bioisosterism
 production, out-sourcing see Outsourcing
Analysis frequency, drug reference standard 168
Analytical methods, DCN 155
Anandamide, structure 909*F*
Androgen receptor (AR) 994, 1006
 genetic disease
  amplification of 1027
  gain of function mutants 1027
  prostate cancer 1027
 knockout animal models 1006

 nomenclature 996*T*
 pharmacology 1015
 see also specific drugs
Androgen receptor ligands 1015, 1016*F*
 prostate cancer treatment 1016
 selective androgen receptor modulators 1016
Angina pectoris, treatment, β-adrenergic receptor antagonists 254
Angiogenesis, Src kinases 975–976
Angiotensin II receptor(s), as targets 300
Angiotensin II receptor agonists, carboxylic acid bioisosterism 659
Angiotensin AT$_{1A}$ receptors, constitutively active 790*T*
Angiotensin AT$_1$ receptor agonists
 see also specific drugs
 nonpeptide drugs 618–619, 620*F*
Angiotensin AT$_1$ receptor antagonists
 see also specific drugs
 nonpeptide drugs 618–619, 620*F*
Angiotensin AT$_1$ receptors, in human disease 775*T*
Angiotensin AT$_2$ receptor antagonists
 see also specific drugs
 nonpeptide drugs 619–621, 621*F*, 622*F*
Angiotensin AT$_2$ receptors
 activation 785*T*
 desensitization 801
 in human disease 775*T*
Animal models
 drug discovery see Drug discovery
 lead optimization 139
 target identification 312
Animal toxicity studies, drug development 28–29, 29*T*
Aniracetam 903
Anisotropic etching see Lithography/photolithography
Anthracyclines
 see also specific drugs
 mechanism of action 1040
Antiarrhythmic drugs
 see also specific drugs
 mechanism of action, potassium channels 862
Antibacterial drugs
 see also specific drugs
 2′,4,4′-trichloro-2′-hydroxydiphenylether 1068
Antibody response, chemical *vs.* biological 117
Anticancer drugs
 see also specific drugs
 bortezomib 577
 phosphodiesterase-4 inhibitors 948

Antidepressant drugs 62
　see also specific drugs
　characteristics 64*T*
　side effects, transporter binding 62
Antiepileptic drugs
　see also specific drugs
　calcium-channel blockers 849–850
　sodium channels 839–840
Antifungal drugs
　see also specific drugs
　2′,4,4′-trichloro-2′-
　　hydroxydiphenylether 1068
Antigen-antibody reactions,
　nanotechnology 361
Anti-inflammatory actions
　glucocorticoid receptor 1006
　liver X receptor (LXR) 1010
Antimicrobial drugs
　see also specific drugs
　HTS 397
　mechanism of action, ribosomal RNA
　　(rRNA) interactions 1041
Antimycin
　mitochondria complex III 1075
　structure 1076*F*
Antisense-based therapeutics
　see also Messenger RNA (mRNA)
　drug discovery 35, 35*F*
　target selection 761
Antiviral drugs
　see also specific drugs
　lipodystrophy syndrome,
　　mitochondrial effects 1082
　mitochondria *see* Mitochondria
　2′,4,4′-trichloro-2′-
　　hydroxydiphenylether 1068
AP6 900
AP22161
　SH2 inhibitor 983
　structure 983*F*
AP22408
　SH2 inhibitor 983
　structure 983*F*
AP23451
　analogs 977
　Src kinase inhibitor 976–977
　structure 978*F*
AP23464
　Src kinase inhibitor 976–977
　structure 974*F*, 978*F*
AP23573
　characteristics 984*T*
　PI-3K–Akt/PKB–aTOR pathway
　　kinase inhibitors 980–981
　structure 981*F*
AP23644, Bcr-Abl kinase inhibitor 973
AP23848
　KIT inhibitor 972
　structure 971*F*
API stability, drug reference
　standard 167
Aplasmomycin, carbon–boron
　bioisosterism 698

Apoptolidin, structure 725*F*
Apoptosis
　Bcl-2 overexpression 1076–1077
　cytochrome c *see* Cytochrome c
　initiation, mitochondria complex
　　I 1068–1069
　mitochondria *see* Mitochondria
Appendix, DCN 155
Approval systems, R&D
　management 250
Arachidonic acid, structure 909*F*
Arglabin, mechanism of action 851–852
Aromatic para-amino substitutes,
　bioisosterism 674
Arqule 760*T*
Array Biosciences 760*T*
ArrayScan II system, high-content
　screening (HCS) 427–428
Arrestin *see* Beta-arrestin
β-Arrestin, phosphodiesterase-4
　association 926
N-Arylation, solid support
　synthesis 466*F*, 467, 467*F*
O-Arylation, solid support
　synthesis 466*F*, 467, 468*F*
Aryl derivatives, carboxylic acid
　bioisosterism 660, 662*T*
Arzoxifene
　as estrogen receptor ligand 1014
　structure 1013*F*
Ascaricides, mitochondria complex
　I 1068–1069
Ascorbic acid (vitamin C), discovery,
　Szent-Gyorgi, Albert 258
Asoprisnil
　progesterone receptor ligands 1015
　structure 1015*F*
Aspartic proteases
　see also specific enzymes
　members 625
　nonpeptide drugs 625
　peptidomimetics 625
　three-dimensional structures 626*T*
Asperlicin, structure 36*F*
Assays
　false positives 737
　HTS *see* High-throughput screening
　　(HTS)
　readout interference, promiscuous
　　ligands 746
　throughput, GPCR functional assay
　　screening 814
　variation, HTS 403
Assessment, project teams 146
Assessment responsibilities, best in
　class *vs.* first in class 119–120
Astex Molecules 761*T*
　characteristics 226*T*
Asthma treatment, phosphodiesterase-4
　inhibitors 935–936, 948
AstraZeneca, R&D spending 101*T*
Astressin, development 582

Atazanavir 579*T*
Atenolol
　bioisosterism 675, 677*F*
　mechanism of action, GPCR
　　targeting 773*T*
　structure 677*F*
Atorvastatin/atorvastatin calcium
　characteristics 653*T*
　structure 650*F*, 653*T*
ATP, generation
　mitochondria inner membrane 1059
　mitochondrial permeability transition
　　(MPT) 1061
(*S*-)ATPA, structure 901*F*
ATP-dependent chromatin remodelers,
　nuclear receptor
　coactivators 1002
Atpenin A4, structure 1073*F*
Atpenin A5, structure 1073*F*
ATP-gated ion channels 903
　characteristics 881*T*
　diseases 907
　molecular biology 904
　nomenclature 903
　pharmacology 905
　physiological functions 904
　　nervous system 904–905
　　smooth muscle 904–905
　　sympathetic nervous system 905
　regulation 905
　structure 904
　　heteromeric form 904
　　homomeric form 904
　　nonselectivity 904
　　trimeric form 904
　therapeutics 905
　　infection 905–907
　　pain 905–907
ATPO, structure 901*F*
ATRAP, GPCR interaction 797*T*
Atrial naturietic peptide (ANP)
　as peptide/protein drug, protein
　　conjugation 590
　phosphodiesterase-2
　　interaction 923–924
　phosphodiesterase-5 928
Atrial/ventricular contraction, sodium
　channels 837
Attrition, drug discovery/
　development 11, 270, 560, 561,
　　721–722
　biologics license applications 561,
　　561*F*
　causes of 11
　discovery to market time 561
　drug development 730
　costs 727
　lead optimization 205, 206*F*
　new drug applications 561, 561*F*
　numbers 561
　product submissions 561, 561*F*
　reduction, ADME 54
　target selection 763

Index

Authority, project teams 145
Autocorrelation, light scattering, promiscuous ligands 743
Autoimmune diseases, potassium channels 863
Automated proteolytic digestion *see* Protein identification
Automation
  biotechnology companies 103
  data management 449
  HTS data acquisition integration 449
  pharmaceutical productivity effects 132
  planar chip-based electrophysiology, voltage-gated ion channels 866
  sample storage/retrieval 537
Autonomic synaptic transmission, nicotinic acetylcholine receptors 884
Autoradiography, two-dimensional gel electrophoresis 335
Autoregulation, gene expression 326
Autosomal dominant nocturnal frontal lobe epilepsy (ADNFLE), nicotinic acetylcholine receptors 888
Auxiliary subunits
  calcium channels 844
  sodium channels 834
Average investment, drug development costs 726–727
Avoidance, conflict management 242, 248
Azaamino acids
  peptide/protein agonists 582
  proteolysis resistance 584–585
5-Azadeoxycytidine, B cell leukemia treatment 1040
Azasetron, 5-HT$_3$ receptor binding 890
AZD-530
  characteristics 984T
  Src kinase inhibitor 976–977
Azithromycin, structure 1043F
AZM-475271
  Src kinase inhibitor 976–977
  structure 978F

B

Background elimination, fluorescence correlation spectroscopy-related confocal fluorimetric techniques (FCS$^+$) 407
Backups/second generation candidates, drug discovery productivity 125
Bacteria
  genome sequencing 307
  mitochondria complex I 1070
Bacterial mimetics, SLC22 transporters 59–60, 61F

Barth syndrome, etiology 1059
Basedoxifene
  as estrogen receptor ligand 1014
  structure 1013F
Basic science, innovation 253, 257–258, 261
BAY 43-9006
  characteristics 984T
  Raf-MEK-MAPK pathway kinase inhibitors 980
  structure 979F
BAY 60-7550, phosphodiesterase-2 inhibitors 933
Bayh–Dole Act (1980) 277–278
  productivity effects 123–124
Baylis–Hilman reaction, solid support synthesis 469F
B-cell leukemias, 5-azadeoxtcytidine 1040
B cells, phosphodiesterase-7 929
Bcl-2 protein family, overexpression, apoptosis 1076–1077
Bcr-Abl kinases 972
  see also specific kinases
  acute lymphoblastic leukemia 972
  chronic myelogenous leukemia 972
  inhibitors 971F, 972–973, 974F
  Philadelphia chromosome 972
Bead-based fiberoptic arrays 356, 356F
  single nucleotide polymorphisms 357
Benign prostatic hyperplasia (BPH) treatment, phosphodiesterase-5 inhibitors 948
Benoxinate
  bioisosterism 677F
  structure 677F
Benzazepine antagonists, phenol bioisosterism 671, 672F
Benzimidazole compounds, hepatitis C virus inhibition 1047, 1048F
Benzo compound splitting (benzo cracking), ring system reorganization (bioisosterism) 678, 679F
Benzodiazepines
  see also specific drugs
  GABA receptor interactions
    GABA$_A$ receptors 893
    GABA$_C$ receptors 893
  phosphodiesterase-2 inhibitors 933
  receptor ligands, carboxylic ester bioisosterism 663, 667F
1,4-Benzodiazepines, solid support synthesis 479, 481F
Benzofurans
  bioisosterism 680F
  ring system reorganization (bioisosterism) 679
  structure 680F

Benzopiperazinones, solid support synthesis 479, 481F
Benzopyrones, ring system reorganization (bioisosterism) 679
Benzothiazepines, as calcium channel blocker 849–850
Benzothienol[3,2-a]thiadiazine 2,2-dioxides, phosphodiesterase-7 inhibitors 943
5-Benzotriazole, bioisosterism 673T
Benzoxazines, GABA receptors 893
5-Benzoxazolone, bioisosterism 673T
5-(Benzylsulfonyl)-4-bromo-2-methyl(2H)-pyridazione (BBMP) 1061, 1061F
Berberine 1048
  as antimicrobial 1048
  Chinese medicine 1048
  cholesterol-lowering properties 1048
  low-density lipoprotein receptor stabilization 1048
  structure 1048F
Bermexane 903
Best in class *vs.* first in class *see* Philosophy (of drug discovery/development)
Beta amino acids, proteolysis resistance 584, 585
Beta-arrestin
  GPCR desensitization 801
  GPCR interaction 788, 797T
  phosphodiesterase-4 association 926
βγ-regulation, G proteins 781T, 784, 786T
β-turn mimetics, peptidomimetics 608
β-lactamases, GPCR functional assay screening 812
Bexarotene, characteristics 1012T
Bezafibrate 1021
  structure 1022F
βγ-subunits, G protein structure 780
BIAcore *see* Surface plasmon resonance (SPR)
Bicalutamide
  characteristics 1012T
  structure 1016F
Bicuculline, structure 893F
Bicyclic guanidines, solid support synthesis 481F
Bicyclic systems, ring system reorganization (bioisosterism) 675
BIO-176 903
Biochemical assays, promiscuous ligands *see* Promiscuous ligands, identification
Biochemical pathways
  diagrams, Cellomics Knowledgebase 435
  proteomics 330
Biogen-IDEC, formation 102

Bioinformatics
  definition 317
  HTS see High-throughput screening (HTS)
  target identification see Target identification
Bioisosterism **649–711**
  see also specific drugs
  analog design 651
    market value 651
  anomalies 688
    ether–oxygen/methylene groups 692, 693F
    fluorine–hydrogen see Bioisosterism, fluorine–hydrogen
    Friedman's paradox 692
  carboxamides 666
    1,3,4-oxadiazoles 668, 670T
    cyclic amidines 668, 670T
    histamine H_3 receptor antagonists 668
    HIV-1 protease inhibitors 668
    hydroxyethylureas 668, 669F
    1,2,4-oxadiazoles 668, 670T
    oxazole rings 668, 670T
    phenylimidazole 668, 670T
    proteolytic enzyme inhibitors 668
    pyrrole rings 668, 670T
    renin inhibitors 668
    serotonin receptor antagonists 668
    sulfonamide conversion 666, 669F
    vinyl fluorides 668, 669F
  carboxylic esters 663, 668F
    benzodiazepine receptor ligands 663, 667F
    bioavailability 663–665, 667F
    bradykinin B_1 receptor antagonists 663
    ester to amide 663
    N-methoxy imidoyl nitrile 663–665, 667F
    monoamine transporter antagonists 663
    muscarinic receptor ligands 663, 667F
    opiate receptor antagonists 663
    1,2,4-oxadiazoles 663
    serotonin antagonists 663
    1,2,4-thiadiazoles 663
  core structure reorganization 675
    ring system reorganization see Bioisosterism, ring system reorganization
    scaffold hopping see Bioisosterism, scaffold hopping
  definition 650, 651
  'functional analogs' 659
    carboxamides see Bioisosterism, carboxamides
    carboxylic acids see Bioisosterism, carboxylic acids
    carboxylic esters see Bioisosterism, carboxylic esters
    catechols 672, 674F, 675F
    definition 651
    phenols see Bioisosterism, phenols
    sulfonamides see Bioisosterism, sulfonamides
    thiourea 668
    urea 668
  functional group reversal 673
    aromatic para-amino substitutes 674
    atenolol 675, 677F
    enzyme resistance 673–674, 677F
    meperidine 675
    1-methyl-4-phenyl-4-propionoxy piperidine 675, 677F
    neoorthoforms 674, 677F
    serotonin 675, 677F
  'me-to drugs' 650
  minor metalloids 694
    carbon–boron see Bioisosterism, carbon–boron
    carbon–silicon see Bioisosterism, carbon–silicon
    selenium see Bioisosterism, selenium
  modification analysis 685
    electronic parameters 687, 688F
    solubility 687, 688F
    structural parameters 686, 687F
  phenols 670
    amide groups 670–671
    benzazepine antagonists 671, 672F
    dimethylamino-sulfonamide 670–671
    dopamine receptor antagonists 671, 674F
    hydroxyisopropyl 670–671
    hydroxymethyl 670–671
    methanesulfamide 670–671
    N-methyl-D-aspartate (NMDA) receptor antagonists 671, 673T
  ring equivalents see Bioisosterism, ring equivalents
  'structural analogs,' definition 651
  sulfonamides 672
    sumatriptan 672–676
Bioisosterism, carbon–boron 698
  aplasmomycin 698
  applications 698
  boranophosphate salts 699, 700F
  bond lability 699–700
  boromycin 698
  boronic chalcones 698–699, 699F
  Boron Neutron Capture Therapy (BNCT) 698, 700, 701F
    compounds 701F
  carboxyboranes 698–699
  compounds 699F
  tolboxane 698
  toxicity 698
    riboflavin deficiency 698
  Velcade 698–699, 699F
Bioisosterism, carbon–silicon 694
  biological activity 694–695, 695F
  bond length 694
  clinical studies 697, 697F
    karenitecan 697
    silperisone 697
  diols 695, 696F
  diphenylpiperidine analogs 696, 697F
  germanium 697, 698F
  hydrolysis sensitivity 694
  lipophilicity 694
  metabolic attacks 695
  reactivity 695, 696F
  trimethylsilyl group 695
  silicon diols 696
  toxicology 694
Bioisosterism, carboxylic acids 659
  acyl-cyanamides 659
  acyl-sulfonamides 659
    angiotensin receptor agonists 659
    beta-adrenergic receptor antagonists 659
    chemokine receptor 1 (CXCR1) inhibitors 659
    hepatitis C inhibitors 659
  aryl derivatives 660, 662T
  diamino-cyclobutene-diones 663, 666F
  direct derivatives 659, 661T
  hydroxamic acids 659, 663F
    bufexamac 659
    histone deactylase inhibitors 659
    ibuproxam 659
    matrix metalloproteinase inhibitors 659
    oxametacin 659
    TNF-α converting enzyme inhibitors 659
  nonclassical isosterism 659
  nonplanar sulfur-derived acidic functions 661, 662T, 669F
  as phosphonate surrogates 663, 667F
  phosphorus-derived acidic functions 661, 662T, 664T, 669F
    CCK antagonists 661
    GABAB antagonists 661
    glutamate antagonists 661
    pka/logP values 661, 666F
  planar acidic heterocycles 659, 660, 662T
    acidity 661
    3,5-dioxo-1,2,4-oxadiazolidine 661
    3-hydroxy-γ-pyrones 661
    3-hydroxy-1,2,5-thiadiazoles 661
    therapeutic fields 660–661
    thiazolidinediones 661
Bioisosterism, fluorine–hydrogen 688
  d orbital absence 689, 690F, 690T
  electronic aspects 688
    bond energies 688–689, 689F
    stabilities 689, 689F

lipophilicity 690, 690T, 691T
  acids 668–669, 690
  alcohols 690, 691T
  alkanes 690, 691T
  ketones 690
  propranolol analogs 668, 692T
rofumilast analogs 692, 693F
steric aspects 688
  bond length 688, 689T
tripelennamine analogs 691, 692F
Bioisosterism, ring equivalents 652
  best-selling drugs 652, 653T, 660T
  heterocycles 657
    cyclooxygenase-2 (COX-2) inhibitors 658, 658F
    uncertainty of results 658–659, 659F
  imidazo[1,2-a]pyridines 656
    bradykinin receptor antagonists 656, 657F
  pyridazines 655F, 656
    acetylcholinesterase inhibitors 656, 657F
    boiling points 656–657, 658F
    N-(6-chloronaphthalen-2-)sulfonylpipazine derivatives 655–656, 656F
    minaprine 656, 657F
    ridogrel 655
    thromboxane A_2 synthetase inhibitors 655
  pyridine 654
  sulfonamide antibacterial drugs 654, 654F
Bioisosterism, ring system reorganization 675
  benzo compound splitting (benzo cracking) 678, 679F
  bicyclic systems 675
  dissociation 679
    amiodarone 679, 680F
    benzofurans 679
    benzopyrones 679
    chromonar 679
    khellin 679
    3-methyl-chromone 679
  restructuring 678
    2-methoxybenze-[e]-hexahydo-isoindole 678–679, 679F
    SKF 38 393 678–679, 679F
    tetramizole 678, 679F
  spiro derivatives 675
    gabapentin 676, 678F
    guanethidine analogs 675–676, 678F
  tricyclic systems 675
Bioisosterism, scaffold hopping 651, 652F, 680
  chlorpromazine 680, 681F
  computational approaches 681
    abl tyrosine kinase ligands 685, 686F
    BIOSTER software 685

estrogen receptor ligands 683, 684F, 685F
fragment replacement 681–683
pharmacophore searching 681–683, 683F
selective optimization offside activities (SOSA) 683
shape matching 681–683
similarity searching 681–683
Skelgen software 683
software 683
topological similarities 683–685, 685F
cyclooxygenase-2 inhibitors 680, 681F
$GABA_A$ ligands 681, 682F
imipramine 680, 681F
promethazine 680, 681F
Bioisosterism, selenium 700
  applications 701
    nitric oxide production inhibition 703
    photodynamic treatment 702T, 703
    1,3-selenazol-4-one derivatives 703
  toxicity 701
Biological activity, monitoring, career guidelines 19
Biological macromolecules, drug discovery 47
Biological membranes, depolarization, potassium channels 860
Biological production, chemical vs. biological 117
Biological products, small molecules vs., drug discovery productivity 126, 128F
Biologics license applications (BLAs) 561, 561F
Biology
  discipline changes 88
  pharmaceutical R&D 105
Bioluminescence energy transfer (BRET), GPCRs 798–799, 813
Bioluminescent assays, drug development 193
Biomarkers
  clinical trials 30–31, 185
  definition 185
  target selection 763
Biomolecule spectra, NMR 509
Biosensors, GPCR functional assay screening see G protein-coupled receptors (GPCRs)
BIOSTER software, scaffold hopping 685
Biotechnology, definition 225
Biotechnology companies 99, 102
  see also specific companies
  automation 103
  characteristics 225, 226T
  development 22, 23T

failure of 22, 24T
future problems 55
history 99
human genome sequence, effects of 103
manufacturing 102–103
partnering 102–103
products vs. technology platforms 231
  collaborations 231
  contracts 231
  form/formulation enhancement 232
  innovation decrease 231
  mechanization 231
protein therapeutics 102
R&D finances 225, 226F
size 110–111
small molecule drugs 102
target selection 753–754, 757T
venture capital funded 8
Bipolar affective disorder (BAPD), 5-HT_3 receptors 891
Bisoprolol, GPCR targeting 773T
Bivalirudin 579T
Black, James
  cimetidine, invention of 253, 254
  medicinal chemistry, definition of 92
  Nobel Prize 253
  propranolol, invention of 253, 254
  training 253
BLAST database, sequence searching 344–345
Blumberg, Baruch, management strategies 261
BMS-354825
  Bcr-Abl kinase inhibitor 973
  characteristics 984T
  Src kinase inhibitor 976–977
  structure 974F, 978F
BMS-387032
  CDK family kinase inhibitors 979–980
  characteristics 984T
  structure 979F
Body fluid analysis, proteomics see Proteomics
Boiling points, ring equivalent bioisosterism 656–657, 658F
Bok, Derek, universities in commerce 93
Bombesin receptors, activation 785T
Bond energies, fluorine–hydrogen bioisosterism 688–689, 689F
Bond lengths
  carbon–silicon bioisosterism 694
  fluorine–hydrogen bioisosterism 688, 689T
Bone mineralization, vitamin D receptor 1007
Boranophosphate salts, bioisosterism see Bioisosterism, carbon–boron

Boromycin, carbon-boron bioisosterism 698
Boronic chalcones
  carbon-boron bioisosterism 698–699, 699F
  structure 699F
Boron Neutron Capture Therapy (BNCT), bioisosterism *see* Bioisosterism, carbon–boron
(4-Boronophenyl)-alanine, structure 701F
Bortezomib 579T
  anti-tumor activity 577
  carbon-boron bioisosterism 698–699, 699F
  structure 578F, 699F
Bottleneck analysis, drug discovery productivity 104F, 124–125, 124T
Bradykinin B_1 receptor antagonists, carboxylic ester bioisosterism 663
Bradykinin B_1 receptors
  constitutively active 790T, 793T
  in human disease 775T
Bradykinin B_2 receptors
  activation 785T
  constitutively active 790T
  in human disease 775T
Bradykinin receptor antagonists
  see also specific drugs
  ring equivalent bioisosterism 656, 657F
Brain expression, uncoupling proteins 1080
Breast cancer, disease basis, estrogen receptors (ERs) 1026
Bretazenil
  scaffold hopping 682F
  structure 682F
Brimonidine, GPCR targeting 773T
Bristol-Myers Squibb, R&D spending 101T
Broadband decoupling, NMR 508
Broad corporate alliances, licensing 234
Brugada syndrome, sodium channels 841
Brunton's tyrosine kinase (BTK)
  GPCR interaction 786T
  PDB entries 965T
Budget constraints, fail fast *vs.* 'follow the science' 121
Bufexamac, carboxylic acid bioisosterism 659
Buffers, homogeneous time-resolved fluorescence 423, 424F
Build-to-own (BTO) models, definition 206
Bulk micromatching, lithography/ photolithography 349, 350F

Bullatacin
  mitochondria complex I 1068–1069
  structure 1069F
Bupidine
  sila analog 697F
  structure 697F
Burroughs Wellcome, history 100–101
Business model, pharmaceutical industry 99
Business unit, pharmaceutical industry organization 110
Bz-423, mitochondria complex V 1079
BzATP, structure 906F

C

Caenorhabditis elegans model system 312–313
Caffeic acid phenethyl ester (CAPE), mitoprotective drugs 1064, 1064F
Calcipotriol
  structure 1018F
  as vitamin D receptor ligand 1019
Calcitonin
  eel 579T
  human 579T
  salmon 579T
    intranasal administration 594–595
    PEGylation 589
  as peptide/protein drug 580
Calcitonin receptors
  constitutively active 790T
  in human disease 775T
Calcitriol
  characteristics 1012T
  structure 1018F
Calcium
  homeostasis, vitamin D receptor (VDR) 1007
  mitochondria inner membrane binding 1059
  mitochondrial permeability transition (MPT) 1060
  phosphodiesterase-1 regulation 923
  toleration, mitochondria diversity 1067
Calcium 3 assay, GPCR functional assay screening 811
Calcium-channel blockers 849–850
  characteristics 829T
  clinical use
    cardiac arrhythmias 851
    cardiovascular drugs 851
    epilepsy 851
  development, selection of 73
  mechanism of action, glycine receptors 896
  neuronal hyperexcitability 851

Calcium channels 841
  genetic disease 852
    absence epilepsy 852
    episodic ataxia 852
    familial hemiplegic migraine (FHM) 852
    hypokalemic periodic paralysis 852
    malignant hyperthermia 852
    spinocerebellar ataxia 852
    visual impairments 853
  GPCR interaction 786T
  molecular biology 844
    auxiliary subunits 844
    cardiac tissue 844
    muscle cells 844
    neurons 844–845
    tissue expression 844
  nomenclature 841, 843T
    IUPHAR-approved scheme 842, 843T
    N-type 842
    subfamily 1 842–844
    T-type 842
  pharmacology 838T, 849
    see also specific drugs
    inorganic metal ions 849
    small molecules 849–850
  physiological functions 846
    cardiac muscle contraction 847
    neurons 847–848
    N-type channels 847–848
    P-type channels 847–848
    second messenger effects 846–847
    skeletal muscle 847
    smooth muscle cells 847
    transmembrane voltage 846–847
  regulation 848
    GPCR 848
  structure 828F, 841–842, 845
    $\alpha_2\delta$ subunit 845
    α-interaction domain (AID) 845
    β subunit 846
    cation selectivity 845
    γ subunit 846
    hinged lid mechanism 845
    subunits 841–842
  target selection 73
  therapeutics 838T, 849–850
    see also specific drugs
Calcium sensing G protein-coupled receptors
  allosteric modulators 790T
  constitutively active 790T
  human disease 792T
  in human disease 775T
Calcium-sensitive dyes, GPCR functional assay screening 811
Caliper Technologies, microfluidic devices 359, 360T

Calmodulin, GPCR interaction 786T, 797T
Calnexin, GPCR interaction 797T
Cambridge Antibody Technology, characteristics 226T
CAMK1, PDB entries 965T
cAMP, phosphodiesterase(s) specific to 920
cAMP assays, GPCRs *see* G protein-coupled receptors (GPCRs)
cAMP dependent protein kinase, three-dimensional structures 636T
Camphor, structure 909F
cAMP response element-binding protein (CREB), nicotinic acetylcholine receptor regulation 886–887
cAMP responsive element (CRE), GPCR functional assay screening 812
Cancer
   disease basis
      EGFR family kinases 967
      nuclear receptors 1025
      proteomics 388
      signal transduction 960–962
      target selection 73
   treatment, cilostamide 947–948
Candidate parallel development, HTS 401, 402F
Candidate seeking, lead optimization 205
Cannabinoid CB_1 receptors
   constitutively active 790T, 793T
   in human disease 775T
Cannabinoid CB_2 receptors
   constitutively active 790T
   in human disease 775T
Cantharidin, structure 982F
Capability improvement, outsourcing 207
Capillary electrophoresis, proteomics 347, 347F
Capillary isoelectric focusing, proteomics 347
Capital investment, pharmaceutical industry 111
Capsaicin
   mitochondria complex I 1068–1069
   structure 909F
Capsaicin-derived products, transient receptor potential ion channels 909
Capsazepine, structure 909F
Captive offshoring, definition 206
Captopril
   analog design 651
   design/development 625
Carbamazepine
   characteristics 829T
   structure 829T
Carbidopa, GPCR targeting 773T

Carbon–carbon bond-forming reactions, solid support synthesis *see* Solid support synthesis
Carbon–heteroatom coupling reactions, solid support synthesis 466, 466F
Carborane, structure 701F
Carboxamides, phosphodiesterase-5 inhibitors 940
Carboxyboranes, carbon-boron bioisosterism 698–699
Carboxylic ester bioisosterism, benzodiazepine receptor ligands 667F
Carboxypeptidase-A, three-dimensional structures 626T
Cardiac action potentials, sodium channels 836
Cardiac arrhythmias
   calcium-channel blockers 851
   sodium channels 840–841
Cardiac disorders, potassium channels 863
Cardiac expression, peroxisome proliferator-activated receptor-γ (PPAR-γ) 1009
Cardiac muscle contraction, calcium channels 847
Cardiac myocytes, sodium channels 834
Cardiac tissue
   calcium channels 844
   phosphodiesterase-3 924
Cardiolipin
   cytochrome *c* interactions 1077
   mitochondria membranes 1057
   inner 1059
Cardiovascular disease treatment, calcium-channel blockers 851
Cardiovascular homeostasis, estrogen receptors (ERs) 1005
Career guidelines 17
   gender issues 244
   mission 17
   responsibilities 18
      biological activity monitoring 19
      compound submission 18
      compound synthesis 18
      experimental records 18
      goal awareness 19
      independence/initiative 18
      invention records 18
      laboratory equipment treatment 18
      literature work 18
      management 19
      oral/written reports 18
      personal interactions 19
      safe laboratory conduct 18
      timeliness 19
   salaries 19
   success keys 17
Carnitine, structure 67F

Carrier proteins, hydrophobic depoting 585
Carvedilol, GPCR targeting 773T
Case-control studies, phase IV clinical trials 191
Caspase inhibitors, mechanism of action 577
Catechol derivatives, phosphodiesterase-4 inhibitors 936
Catechols, 'functional analogs' 672, 674F, 675F
Cathepsin-B, three-dimensional structures 626T
Caveolin-1, GPCR interaction 797T
CC-4047 759
   structure 759F
CC-11006 759
   structure 759F
CCI-779, characteristics 984T
CCR2
   constitutively active 790T
   in human disease 775T
CCR3
   constitutively active 790T
   in human disease 775T
CCR5
   constitutively active 790T
   in human disease 775T
Ceftriaxone, repositioning 276
Celastrol, mitoprotective drugs 1064, 1064F
Celecoxib, structure 9F, 658F, 681F
Cell(s)
   potassium channel distribution 857, 860
   proliferation, peroxisome proliferator-activated receptor-δ (PPAR-δ) 1010
Cell-based assays
   drug discovery productivity 124–125
   fluorescence correlation spectroscopy-related confocal fluorimetric techniques (FCS^+) 411F
CellChip System *see* High-content screening (HCS)
Cell compartments, drug discovery 41–43
Cell membranes
   depolarization, sodium channels 836
   excitability, potassium channels 859
   hydrophobic depoting 585
Cellomics Knowledgebase *see* High-content screening (HCS)
Cell-substrate adhesion, Ultra-High-Throughput Screening System (Aurora Biosciences) 431, 441F
Cellular proteomics *see* Proteomics
Centers of excellence in drug discovery (CEDDS) 109–110
Centers of excellence proximity, pharmaceutical industry organization 110

Central domain, nuclear receptors 995–997
Central nervous system (CNS)
  nicotinic acetylcholine receptors 884
  phosphodiesterase-4 926–927
  sodium channels 833–834
Central nervous system diseases/disorders
  drugs 53T
  phosphodiesterase-4 925–926
  phosphodiesterase-10 930–931
  preclinical research 52
CEP-701
  acute myeloid leukemia 971
  characteristics 984T
  structure 971F
CEP-7055 969–970
  characteristics 984T
  structure 970F
  VEGFR family kinase inhibitors 969
Cerebral infarct, oxidative phosphorylation uncoupling 1079–1080
Cetirizine, GPCR targeting 773T
Cetrorelix 579T
  controlled release formulations 593
Cetus Corporation
  formation 102
  PCR invention 256
cGMP, phosphodiesterases 920
  allosteric binding sites 928
CGP-36742, structure 893F
CGP41251 see PKC-412
CGP-53716, characteristics 984T
CGP-76030
  Bcr-Abl kinase inhibitor 973
  Src kinase inhibitor 976–977
  structure 974F, 978F
CGP79787 see PTK-787
Chairpersons, project management 14
Chalcones, mitochondria, drug targeting 1068
CHAPS, promiscuous ligands 740
Charge-coupled devices (CCDs)
  GPCR functional assay screening 813–814
  two-dimensional gel electrophoresis 335
CHEK1, PDB entries 965T
Chembridge Corporation 760T
ChemBridge DIVERSet, promiscuous ligands, computational prediction 749
Chemical development, predevelopment 141
Chemical Effects in Biological Systems (CEBS) database 569, 570F
Chemical libraries, target selection see Target selection
Chemical reactivity, promiscuous ligands, identification 746
Chemical space, drug discovery 49, 51F

Chemiluminescent assays, GPCR functional assay screening 809
Chemistry
  DCN 154
  discipline changes 88
Chemogenomics 567
  advantages 567
  physiological-based pharmacokinetic models 567, 568F
  regulatory implications 567
Chemokine receptors, peptide/protein drugs, N-terminus modification 583
Chenodeoxycholic acid (CDCA)
  as farnesoid X receptor ligand 1024
  structure 1025F
CHF-3381 903
Chicken ovalbumin upstream promoter-transcription factor, nomenclature 996T
'Chimeric' amino acids, peptides 604
Chimeric G proteins see G protein-coupled receptors (GPCRs)
China, market dominance 94
Chinese medicine, berberine 1048
Chiral drug discovery 713–736
  see also specific drugs
  chiral switches 728
    scalable methodologies 729–730
    'significant' improvements 728–729
    success of 729, 729F
  clinical rationales 713
  future work 731
  historical perspective 713
  'isometric ballast' 714–715
  market delivery 718
    available compounds 719, 720F
    examples 718–719, 719F
  stereoisomer synthesis 715, 716F, 717F
    diastereomer resolution 715, 718F
    large scale synthesis 718
    purity 715
    stereoselective syntheses 715–718, 719F
  stereospecific analyses 720
    see also specific methods
    absolute configuration determination 720
Chiral switches see Chiral drug discovery
Chiron Corporation, formation 102
Chlomipramine, structure 1062F
Chloramphenicol
  mechanism of action, ribosome interactions 1042–1043
  structure 1043F
N-(6-Chloronaphthalen-2-)sulfonylpiperazine derivatives, ring equivalent bioisosterism 655–656, 656F

Chlorpromazine
  scaffold hopping 680, 681F
  structure 681F, 687F
Chlorprothixene, structure 687F
Cholecystokinin antagonists, carboxylic acid bioisosterism 661
Cholecystokinin-A receptor agonists, nonpeptide drugs 620F, 621
Cholecystokinin-A receptor antagonists
  see also specific drugs
  nonpeptide drugs 618–619, 620F
  peptidomimetics 608–609, 611F
Cholecystokinin-A receptors, in human disease 775T
Cholecystokinin-B receptor antagonists
  see also specific drugs
  nonpeptide drugs 618–619, 620F
  peptidomimetics 608–609, 611F
Cholecystokinin-B receptors, in human disease 775T
Cholecystokinin receptor-2 (CCKR$_2$), constitutively active 790T
Cholecystokinin receptors
  activation 785T
  constitutively active 790T
  as targets 300–301
Cholesterol, removal, peroxisome proliferator-activated receptor-α (PPAR-α) 1009
Cholesterol-lowering drugs, berberine 1048
Chromonar
  bioisosterism 680F
  ring system reorganization (bioisosterism) 679
  structure 680F
Chromosomes, definition 1038–1039
Chronic lymphocytic leukemia
  phosphodiesterase-5 inhibitors 949
  phosphodiesterase-7 929
Chronic myelogenous leukemia (CML), Bcr-Abl kinases 972
Chronic obstructive pulmonary disease (COPD)
  disease basis, phosphodiesterase-7 929
  treatment, phosphodiesterase-4 inhibitors 935–936, 948
Chronic pain, sodium channels 840
Chymotrypsin, promiscuous ligands 742
CI-1033, characteristics 984T
Ciclosporin A, structure 1061F
Cilomilast
  FDA approval 948
  structure 935
Cilostamide
  cancer treatment 947–948
  structure 934
Cilostazol
  as phosphodiesterase inhibitor 947T
  platelet aggregation 947–948
  structure 934

Index

Cimetidine
  analog design 651
  discovery/development 260–261
    Black, Sir James 253, 254
CIPP, GPCR interaction 797T
Cisplatin, mechanism of action 1040
Citalopram, structure 62F
Citrulline, controlled release
  formulations 593
Clarity, project teams 145
Class I nuclear receptors *see* Nuclear
  receptors
Class II nuclear receptors *see* Nuclear
  receptors
Class III nuclear receptors 994, 995F
Class A G protein-coupled receptors *see*
  G protein-coupled receptors
  (GPCRs)
Class B G protein-coupled receptors
  (GPCRs) *see* G protein-coupled
  receptors (GPCRs)
Clathrin adaptor complex-1, GPCR
  interaction 797T
Clathrin-coated pits, GPCR
  internalization 800–801
CLEAR resins, solid support
  synthesis 496, 497F
Clindamycin, ribosome
  interactions 1042–1043
Clinical development,
  predevelopment 141
Clinical evaluation attrition rates, drug
  discovery productivity 126
Clinically prescribed drugs, promiscuous
  ligands 746
Clinical trial(s) 9–10, 10F, 104–105,
  180
  biomarkers 30–31, 185
  control groups 181
  definitions 30T
  design 29–30, 30F
  diseases/disorders, clinical
    history 181
  drug development costs 727
  drug development productivity 130
  drugs, formivirsen 1045
  efficacy 764
  efficiency 560
  false negatives 30–31
  false positives 30–31
  Helsinki declaration 180
  models/techniques 181
    see also specific types
    crossover design 181
    historical control groups 182
    nonrandomized uncontrolled
      trials 182
    parallel multiple-arm studies
      181
    randomization 181
    sequential trials 182, 183F
    uncontrolled trials 182
  patient recruitment/enrollment 130

phases 187
  see also specific phases
placebo 181
protocol evaluation 186
  drug coadministration 187
  endpoint definitions 186
  ethical evaluation 187
  facilities 186–187
  inclusion/exclusion criteria
    186–187
  informed consent 187
  medical history 187
  recording/reporting process 187
risk factors 181
statistical analysis 184
  efficacy comparisons 184–185
  intention-to-treat 185
  interim analyses 185
  interval scales 184
  nominal result scale 184
  ordinal result scale 184
  result scale 184
  statistical significance 184
supply coordination 130
Clinical trials, phase 0 29, 188
Clinical trials, phase I 188
  definition 180–181
  development chemistry timeline 170
  dose 188
  drug development productivity 126–
    127
  exclusions 188
  genotyping 197
  information given 188
  regulatory requirements 155
  subject numbers 188
  written informed consent 188
Clinical trials, phase II 188
  definition 180–181
  development chemistry timeline 170
  drug comparisons 189
  drug development productivity 126–
    127
  genotyping 197
  information given 189
  participant characteristics 188
  two-stage designs 189
  written consent 189
Clinical trials, phase III 189
  definition 180–181
  development chemistry timeline 170
  drug development productivity 126–
    127
  drug efficacy 189
  drug safety 189
  genotyping 197
  optimal dose range 189
  pediatric patients 189
  subject numbers 189
Clinical trials, phase IV
  (postmarketing) 189
  adverse drug reactions 190–191
  causality 191

  adverse effects 189–190
    regulatory authorities 190
  case-control studies 191
  definition 180–181
  epidemiologic studies 191
    random/nonrandom sampling 191
  genotyping 197
  morbidity/mortality
    identification 189–190
  novel property identification 190
  patient numbers 190, 191
  prospective/retrospective
    assessments 191
Clofibrate 1021
  structure 1022F
Clonidine, mechanism of action, GPCR
  targeting 773T
Clopidogrel hydrosulfate
  characteristics 653T
  GPCR targeting 773T
  structure 653T
Clorgyline, structure 1071F
Clozapine
  analog design 651, 652F
  as multitarget drug 275–276
CNQX 900
  structure 901F
CNS 5161 903
Coadministered drugs, clinical
  trials 187
Cocaine
  dopamine transporter interaction 63
  structure 65F
Cognitive disorders/dysfunction
  animal models 926–927
  phosphodiesterase-5 inhibitors 949
Cohort trials, definition 30T
Cole, Jonathan, universities in
  commerce 93
Collaborations
  biotechnology companies 231
  functional discipline
    organization 109
  insular culture *vs*. 121–122
  pharma *vs*. biotech, drug
    discovery 230–231, 230F
  project management *see*
    Collaborations, project
    management
  start-up companies 765
Collaborations, project
  management 151
  initiation 151
    communication 151
    role/responsibility definitions 151
    logistical considerations 151
      communication 151
      security 151
  oversight 152
    joint management team 152
    responsibilities 152
Collagenase, three-dimensional
  structures 626T

Collagenase inhibitors, peptidomimetics 613, 614F
Colorimetric assays, promiscuous ligands, identification 740
Colpaert, Francis, risperidone, invention of 257–258
CombiCHEM system 760T
Combinatorial chemistry 460
  definition 460
  drug discovery see Drug discovery
  genomics 44, 46F
  history 461
    'divide–couple–recombine' 461–462
    Frank, Ronald 461
    Geysen, Mario 461
    Houghten, Richard 461
    Merrifield, Bruce 461
    'portioning–mixing' 461–462
    small organic molecules 462
    techniques 462
  invention of 256
    Hobbs DeWitt, Sheila 260
    Kiely, John 260
    Pavia, Mike 260
  pharmaceutical productivity effects 132
  strategies 464
  see also specific techniques
  success of 52
Combinatorial libraries, drug discovery 300, 300T
Commercial activities
  outsourcing 209
  universities 93
Commercial fits, drug development productivity 130
Commitment, project teams 145
Common technical document (CTD) 161
  see also Development chemistry
  drug nomenclature 161–162
  in-process controls (IPCs) 163
  manufacturer details 163
  packaging materials 163
  starting material definition 163
  synthesis steps 163
Communication
  collaborative project management 151
  diversity issues 244
  leadership strategies see Leadership
  offshore outsourcing 215
  outsourcing risks 220
  project team leaders 142–143
    assessment 146
Company size, pharmaceutical industry see Pharmaceutical industry
Comparative genomics, HTS 401
Competition
  DCN 154
  phosphodiesterase(s) regulation 923
  sourcing development 213

Competitive collaborations, culture 122
Complement C5a receptor, activation 785T
'Complexity-generating' reactions, solid support synthesis 479–480, 483F
Complex mixtures, peptide fingerprint searching 344
Complex molecules, drug development 724–725, 725F
Complex multistep synthesis, solid support synthesis 475
Compliance, ideal drug characteristics 564T
Composition of matter, intellectual property 130
Compounds
  in development, licensing 234, 234F
  impurity, false positives 737
  materials management 531
  resynthesis, outsourcing 211, 213
  submission 18
  synthesis 18
Compromise, conflict management 248
Computed tomography (CT), drug development 192
Computer-assisted drug design (CADD) 49, 50F
Computer-based technologies
  ADMET 566, 569, 570F
  design software, HTS 398, 398F
  scaffold hopping see Bioisosterism, scaffold hopping
  sequence searching, protein identification 341, 343F
'Concurrent validation,' process variable statistical evaluation 164–165
Conduction anesthesia, sodium channels 839
Conflict management, leadership strategies see Leadership
Confocal fluorescence, promiscuous ligands, identification 745, 745F
Confocal microscopy, GPCR functional assay screening 813–814
Conformational changes, GPCR desensitization 800
Confrontation, conflict management 249
Consistency, drug discovery 283
Constitutive androstane receptor (CAR) 1011
  nomenclature 996T
  pharmacology 1024
  see also specific drugs
Constrained peptides, proteolysis resistance 584, 585
Consumer advocacy groups, drug discovery see Drug discovery

Contract research organizations (CROs) 226
  changing roles 236
  by country 228T
  growth of 205, 227, 228T
  history 102
  importance of 104
  virtual companies 766
Contracts
  biotechnology companies 231
  pharma vs. biotech, drug discovery 230, 230F, 230T
Control groups, clinical trials 181
Controlled release formulations, peptide/protein drugs, injection
  hydration effects 592
  lactic acid/glycolic acid polymers 592
  polymer matrix 592
  side chain urea functional groups 593
  soluble formulations 592
Controls
  homogeneous time-resolved fluorescence 425
  HTS data management 453
Convenience, ideal drug characteristics 564T
Coomassie blue G-250 333
Coomassie blue R-250 333
Coomassie stains, two-dimensional gel electrophoresis 333
Copper-mediated N-arylation, solid support synthesis 470, 471F
CoQ-cytochrome c oxidoreductase complex see Mitochondria complex III
Coregulators, nuclear receptors 1002
Corepressors, nuclear receptors 1003
Core structure reorganization, bioisosterism 675
Core team definition, project teams 141
Corporate goals, innovation 254
Corporate mergers, innovation 254–255
Corporate strategies, innovation see Innovation
CortBPI, GPCR interaction 797T
Corticotropin-releasing factor (CRF) 579T
  cyclization scanning, drug design 582
  as peptide/protein drug, structural motif replacement 591
Cortirelin 579T
Cost differential, outsourcing risks 220
Cost effectiveness, drug discovery 276
Cost of goods (COGS), drug development costs 727–728, 728F

Costs
　DCN 155
　drug discovery *see* Drug discovery
　outsourcing scale-up 211
　pharmaceutical R&D 123
CP-547632 969–970
　characteristics 984*T*
　structure 970*F*
　VEGFR family kinase inhibitors 969
CRF1 receptors
　allosteric modulators 790*T*
　in human disease 775*T*
Critical Path Initiative 569
Critical process parameters, process
　　variable statistical
　　evaluation 164
Critical residues, peptide/protein
　　agonists 581
Criticism, communication 243
Crk SH3, three-dimensional
　　structures 636*T*
Crohn's disease
　glucocorticoid receptor 1027
　transient receptor potential ion
　　channels 909–910
Cross-assay reports, HTS data
　　analysis 455
Crossover trials 181
　definition 30*T*
Cross-sectional trials, definition 30*T*
Crowns *see* Solid supports
CRTH2/DP-1 receptor, in human
　　disease 775*T*
CSK, PDB entries 965*T*
CSNK2A1, PDB entries 965*T*
CSNKD1, PDB entries 965*T*
CT-53518 *see* MLN-518
C-terminal domain
　GPCR structure 774–777
　nuclear receptors 995–997
　phosphodiesterase(s) 920–922
　proteolysis resistance 584
'Cubic ternary complex model,' GPCR
　　activation models 792, 794*F*
Culture *see* Philosophy (of drug
　　discovery/development)
Curriculum recommendations 85, 86*T*,
　　87*T*
CX3CR1, in human disease 775*T*
CX3CR6, in human disease 775*T*
CX-516 900, 903
CX517 903
CX-546 AMPAkine, structure 901*F*
CXCR1, allosteric modulators 790*T*
CXCR1 inhibitors, carboxylic acid
　　bioisosterism 659
CXCR2, allosteric modulators
　　790*T*
CXCR4, constitutively active 790*T*
Cyanine dyes 334
CYC202 (roscovitine)
　CDK family kinase inhibitors 979–
　　980

characteristics 984*T*
　structure 979*F*
Cyclic amidines, carboxamide
　　bioisosterism 668, 670*T*
Cyclin-dependent kinase-5 (CDK5),
　　PDB entries 965*T*
Cyclin-dependent kinase-6 (CDK6),
　　PDB entries 965*T*
Cyclin-dependent kinases
　　(CDKs) 979
　see also specific kinases
　inhibitors 979–980
　PDB entries 965*T*
　three-dimensional structures 636*T*
Cyclization, peptide/protein
　　agonists 582
Cycloaddition reactions
　solid support synthesis 472, 473*F*
　solution phase synthesis 490, 491*F*
Cyclobenzaprine, GPCR
　　targeting 773*T*
Cyclooxygenase-2 (COX-2) inhibitors
　fate/markets 8
　ring equivalent bioisosterism 658,
　　658*F*
　scaffold hopping 680, 681*F*
Cyclophilin, three-dimensional
　　structures 636*T*
Cyclosporine A (CsA) 579*T*
　structure 46*F*, 982*F*
Cyclothiazide, structure 901*F*
Cyproterone acetate, structure 1016*F*
Cysteinyl proteases
　nonpeptide drugs 629
　peptidomimetics 629
　three-dimensional structures 626*T*
Cystic fibrosis, transient receptor
　　potential ion channels 909–910
Cytidine, structure 1038*F*
Cytisine, structure 887*F*
Cytochrome *c* 1076
　apoptosis 1076
　　Bcl-2 overexpression 1076–1077
　biology 1076
　　mutations 1076
　cardiolipin interactions 1077
Cytochrome *c* oxidase *see* Mitochondria
　　complex IV
Cytokine receptors
　heterodimerization 576
　Janus kinase signal transduction 576
　mechanism of action 576
　peptide/protein drugs 576
　peptidomimetics 596
Cytoprotective drugs
　see also specific drugs
　mitochondria complex I 1072

D

DAPK1, PDB entries 965*T*
Dapoxetine, development 236

Data
　HTS 402
　liquid chromatography-NMR 523,
　　523*F*
　management
　　HTS 454
　　two-dimensional gel
　　　electrophoresis 335–336
　processing
　　DI-NMR 525, 526*F*, 527*F*, 528*F*
　　homogeneous time-resolved
　　　fluorescence (HTRF) 416
　volume, HTS 449
Data acquisition, HTS *see* High-
　　throughput screening (HTS)
Data amounts, expression
　　profiling 311
Data analysis, surface plasmon
　　resonance 374
Databases
　ADMET 569
　proteomics 425–426
Data-dependent scanning, raw MS/MS
　　data searching 345
Data mining, Cellomics
　　Knowledgebase 436
Daunorubicin, structure 66*F*
Deafness, potassium channels 864
Decision making systems
　organizational alignment *vs. see*
　　Philosophy (of drug discovery/
　　development)
　R&D management 250
Decision points, project
　　management 16
Decision support, HTS data
　　management *see* High-
　　throughput screening (HTS),
　　data management
2-Decyl-4-quinazolinyl amine (DQA),
　　mitochondria complex I 1068–
　　1069
Deep reactive ion etching
　　(DRIE) 353
Degarelix
　controlled release formulations 593,
　　594*F*
Deguelin
　mitochondria complex I 1068–1069
Delegation, leadership 243
Deletion detection, functional
　　genomics 329
DELFIA, homogeneous time-resolved
　　fluorescence 412–413
Delucemine 903
Demographics, education 86
Dendrimers
　solid support synthesis 499
　solution phase synthesis 485, 485*F*
Department-driven projects, project
　　management 12–13, 12*F*
Desensitization, GPCRs *see* G protein-
　　coupled receptors (GPCRs)

Design of experiments (DoE), process variable statistical evaluation 164
Desloratadine, GPCR targeting 773T
Deslorelin 579T
Desmopressin 579T
  intranasal administration 593–594
  oral administration 595
  as vasopressin analog 580
Dess–Martin oxidation procedure, solution phase synthesis 487F, 488
Detergents, promiscuous ligands 738, 740
Detirelix
  hydrophilic depoting, duration of action 587–588
  self-forming depots 593
Development candidate nomination (DCN) 153
  appendix 155
  chemistry 154
  estimated human dose 154
  executive summation 152
  IND timeline and resources 155
  introduction 152
  major issues/assumptions/constraints 155
  metabolism/pharmacokinetics 154
    drug interaction profile 154
    in vitro metabolism 154
    in vitro profiling 154
    in vivo profiling 154
  pharmaceutical development 154
    analytical methods 155
    estimated cost 155
    formulation 155
    physiochemical properties 155
    process (scale-up) 155
    synthetic route 154
  pharmacology 154
    mechanism of action 154
    in vitro activity 154
    in vivo activity 154
  product profile 152
    competition 154
    IP position 154
    key distinguishing features 154
    rationale 153
  project team 155
  recommendation 155
  references 155
  toxicity 154
Development candidates (DCs), criteria 142, 143T
Development chemistry **159–172**
  see also Common technical document (CTD); Drug approval
  analytical methods 166
    validation 166
  cleaning 165
    'major cleaning' 165–166
    'minor cleaning' 165
  multi-use manufacturing equipment 165
  physical facility 165
  verification 166
drug reference standard 167
  ambient temperature stability 167, 168T
  analysis frequency 168
  API stability 167
  freezer storage stability 167–168, 168T
  intermediate/accelerated storage conditions 168
  refrigeration temperature stability 167–168, 168T
economics 160–161
environmental exposure 161
impurities 167
material supply requirements 160, 161
operator exposure 161
packaging 168
process variable statistical evaluation 164
  'concurrent validation' 164–165
  critical process parameters 164
  design of experiments (DoE) 164
  documentation 164
  Fisher, Ronald 164
  FusionPro 164
reprocessing 169
reworking 169
solvents 161, 162T
synthesis control 160, 161
timeline 169, 170F
  human evaluation step 170
  Occupational Safety and Health Administration 170
  phase I clinical trials 170
  phase II clinical trials 170
  phase III clinical trials 170
  process hazard analysis (PHA) 170
  process safety management (PSM) 170
Dexamethasone
  characteristics 1012T
  structure 1017F
Dexketoprofen, structure 729F
Dexmethylphenidate hydrochloride, structure 729F
DHT, structure 1016F
Diabetes insipidus, phosphodiesterase-5 inhibitors 949
Diabetes mellitus type 2
  fibrates 1021
  glucocorticoid receptor ligands 1017
  mitochondria 1054–1055
Diagnostics, pharmaceutical companies 100
3,4-Dialkoxy-phenyl, bioisosterism 660T
β-Diamino acids, peptide/protein agonists 582
Diamino-cyclobutene-diones, carboxylic acid bioisosterism 663, 666F
2,4-Diaminoquinazoline 1048
  structure 1048F
3,4-Diamino thiadiazole, bioisosterism 659, 670
Diasteromer resolution, stereoisomer synthesis 715, 718F
Diazaborinine, structure 699F
Diazepam
  analog design 652F
  scaffold hopping 652F, 682F
  structure 682F
Diazepinoindole, phosphodiesterase-4 inhibitors 936
Dibenz (b,f)azepine analogs, mitoprotective drugs 1062
(S)-1,4-Dibromo-2-butanol, structure 720F
Diclofenac, structure 9F
Dicyclomine, GPCR targeting 773T
Difference gel electrophoresis (DIGE) 334
Differentiation, best in class vs. first in class 119, 120T
Diffusion-encoded spectroscopy (DECODES), mixture analysis 527–528
Diffusion-ordered spectroscopy (DOSY), mixture analysis 527
3,5-Difluoro-4-hydroxyphenyl, bioisosterism 662T
Digital micromirror devices, surface micromachining 352
Digital signal processing, NMR 529
Dihydro-β-erythroidine, structure 887F
Dihydroisoquinolines, phosphodiesterase-10 inhibitors 946
Dihydropyridine neuropeptide Y_1 receptor antagonists, bioisosterism 669–670, 671F
Dihydropyridines, as calcium channel blocker 849–850
Dilated cardiomyopathy, phosphodiesterase-3 inhibitors 947–948
Diltiazem
  characteristics 829T
  structure 75F, 829T
Dimethylaminosulfonamide, phenol bioisosterism 670–671
2′,4′-Dihydroxy-6′methoxy-3′,5′-dimethylchalcone, mitochondria, drug targeting 1068, 1068F
DI-NMR see Nuclear magnetic resonance (NMR)
Diols, carbon-silicon bioisosterism 695, 696F
3,5-Dioxo-1,2,4-oxadiazolidine, carboxylic acid bioisosterism 661

Dipeptide surrogates, peptidomimetics 606–608
Dipeptidyl peptidase-IV, peptide/protein drugs 577
Diphenylpiperidine analogs, carbon-silicon bioisosterism 696, 697F
Dipyridamole, as phosphodiesterase inhibitor 931, 947T
Direct (real-time) dimension, two dimensional NMR 507
Direct enantioselective chromatography 721
Discarded research, innovation 259
Discodermolide, structure 28F
(+)-Discodermolide, structure 725F
Discovering, innovation vs. 255
Discovery Partners International 760T
Disease
  complexity 274
  drug development 175
  etiology, target vs. disease 112
  expression profiling 311
  history, clinical trials 181
  models, preclinical trials 177
Disease-specific proteins, proteomics 386
Disopyramide
  characteristics 829T
  structure 829T
Dispensing, Ultra-High-Throughput Screening System (Aurora Biosciences) 442–443, 443F
Distamycin, mechanism of action 1040
Divergent side chains, peptide/protein agonists 582
Diversified model companies 102
Diversity issues
  functional discipline organization 108
  leadership strategies see Leadership
  project teams 145
'Divide–couple–recombine,' combinatorial chemistry 461–462
Divinylbenzene, resin beads, solid support synthesis 493–494
DLS
  light scattering, promiscuous ligands 743
  promiscuous ligands, identification 743
DMDs, DNA microarrays 357
DMP323, structure 302F
DMP450, structure 302F
DNA 1038
  functions 1039
  hybridization, nanotechnology 361
  nomenclature 1038
    International Union of Pure and Applied Chemistry (IUPAC) 1039
  organization 425–426
  pharmacology 1039
  singe base change 1039
  RNA duplexes 1040
  sequencing
    nanotechnology 361
  pharmaceutical productivity effects 132
  structure 1037, 1038, 1038F, 1039F
    G-quartets 1040
    helix 1038–1039
    major groove 1038–1039, 1039F
    minor groove 1038–1039, 1039F
    residues 1038–1039
  therapeutic drugs 1039
    DNA polymerase interference 1040
    intercalation 1040
    minor groove binders 1040
    nonnatural nucleosides 1040
    replication interference 1039
    topoisomerase interference 1040
    transcription interference 1039
    transition metal complexes 1040
  transcription 1037
DNA-binding domain, nuclear receptors 1002
DNA microarrays 357
  DMDs 357
  electronically enhanced hybridization 357
  concentration/hybridization 359
  electronic addressing 358
  stringency control 359
  expression profiling 310
  Hi-Chip system 357
  lithographic techniques 357
  photolithographic masks 357
  limitations 357
  solid supports 504–505, 505F
DNA nuclease, homogeneous time-resolved fluorescence 422, 422F, 423F
DNA polymerases, interference 1040
DNQX 900
DockVision, promiscuous ligands, computational prediction 749
Documentation
  preclinical trials 178–179
  process variable statistical evaluation 164
Dolasetron mesylate, 5-HT$_3$ receptor binding 890
Domains, conservation, proteins 320, 321T
Domestic in-house sourcing, definition 205
Domestic outsourcing
  definition 205
  suitability 215
Domestic providers, definition 206
L-Dopa see Levodopa
Dopamine, structure 57F
Dopamine D_{1A} receptor, constitutively active 790T
Dopamine D_{1B} receptor, constitutively active 790T
Dopamine D_1 receptor
  constitutively active 790T
  in human disease 775T
Dopamine D_2 receptor
  constitutively active 790T, 793T
  in human disease 775T
  structure 798
Dopamine D_{2s} receptor, constitutively active 790T
Dopamine D_3 receptor
  constitutively active 790T
  in human disease 775T
Dopamine $D_{4.2}$ receptor, constitutively active 790T
Dopamine $D_{4.4}$ receptor, constitutively active 790T
Dopamine D_4 receptor, in human disease 775T
Dopamine D_5 receptor, in human disease 775T
Dopamine receptor antagonists, phenol bioisosterism 671, 674F
Dopamine transporters (DATs) 57, 59, 60T
  cocaine interaction 63
  mechanism of action 57
Dose/dosage
  phase I clinical trials 188
  preclinical trials 178, 179F
Dose–response curves, homogeneous time-resolved fluorescence 421, 422F
Dose response tests, HTS data analysis 452
Double-blind trials, definition 30T
'Double-pulse field gradient spin echo (DPFGSE), NMR 529
Doxazosin, GPCR targeting 773T
Doxepin, structure 63F
Doxycycline, ribosome interactions 1042–1043
Dronedarone
  structure 1020F
  as thyroid hormone receptor ligand 1019
Drosophila melanogaster, as model organism 312–313
Drospirenone, structure 1015F
Drug(s)
  characteristics 51
  costs 270
  efficacy
    expression profiling 311
    phase III clinical trials 189
  interactions
    DCN 154
    ideal drug characteristics 564T
    transporters 61

Drug(s) (continued)
  mechanism of action, target vs.
      disease   112
  molecular weight characteristics   52
  nomenclature, common technical
      document (CTD)   161–162
  responses, single nucleotide
      polymorphisms (SNPs)   362
  top-selling   714T
Drug approval   161
  see also Development chemistry
  FDA   159–160
  International Conference on
      Harmonization   159–160
  physical/chemical properties   161
    particle size   162–163
    polymorphic forms   161–162
  time reductions   8
Drug combinations
  clinical trials   187
  drug discovery   277
Drug delivery systems
  peptide/protein drugs see Peptide/
      protein drugs
  transporters   66, 66F
Drug development   26, **173–202**
  see also Clinical trials
  animal toxicity studies   28–29, 29T
  clinical trials see Clinical trials
  components   304
  costs   2, 199, 726
    attrition   727
    average investment   726–727
    clinical trials   727
    cost of goods (COGS)   727–728,
        728F
    failure rate   199
    generic competition   727
    marketing   727
    pricing   727, 727F
    process improvements   728
    safety   727
    time in development   199
  desired areas   173
  development candidate nomination
      (DCN) see Development
      candidate nomination (DCN)
  functional discipline
      organization   108–109
  general process   174F
  Good Laboratory Practice (GLP)   27
  Good Manufacturing Practice
      (GMP)   27
  high throughput organic
      synthesis   396, 396F
  hits/leads   175, 175F
    disease areas   175
    'druggability' assessment   176
    good manufacturing process   177
    HTS   177
    microarrays   177
    pharmacophore   176
    QSAR   177

    strategies   176, 176F
    target assessment   176
  imaging technology   192
    see also specific techniques
  important properties   75T, 76
  innovative improvement   174
  innovative research   174
  'me-too' projects   174
  mitochondria   1079
  mitochondria complex I see
      Mitochondria complex I
  philosophy see Philosophy (of drug
      discovery/development)
  preclinical trials see Preclinical trials
  process   721, 722F
    best in class vs. first in class   119
    complex molecules   724–725, 725F
    element integration   721
    manufacturing methods   721–722
    medicinal chemistry overlap   723,
        724F
    regulatory documentation   721–
        722
    safety, health and environment
        information   722–723, 723F
    structural features   723–724
    synthesis effects   723–724
  productivity   126
    acute vs. chronic treatment   128
    clinical trial organization   130
    commercial fits   130
    failure visibility   126
    pharmacokinetics   130
    scale-up   130
    success rates   126–127
    target types   128
    therapeutic area   128, 129T
    toxicology   130
  purity   27
  regulatory guidelines   27, 200,
      725
    approval by property   726
    guidelines   726
    industrialization   725
    medical utility   725
    molecular configuration   726
    safety assessment   725
  required decisions   174
  salt forms   27, 27T
  stages of   27, 29F
  success vs. failure   730
    attrition rate   730
    'druglikeness'   730–731
    molecular weight effects   730
    'rule-of-five'   730
  surface-enhanced laser desorption
      ionization (SELDI)   378, 379
  time lines   396, 396F
  transition from research   54
Drug discovery   31
  ADMET see ADMET
  animal models   277, 284
    human disease vs.   277

    productivity   125
    side effects   277
  attrition see Attrition, drug discovery/
      development
  bioisosterism see Bioisosterism
  chemical space   49, 51F
  chiral drugs see Chiral drug discovery
  combinatorial chemistry   31–32, 33,
      39, 40F, 40T
    affinity selection   39–41, 43F
    HTS development   39, 42F
    library preparation   39, 41F
    quantity vs. quality   39, 41F
  consumer advocacy groups   270
    ACT-UP   270–271
    stem cell research   271
  costs   270, 560
    attrition rate   270
    drug costs   270
    pre-approval costs   4, 6F
    research/development costs   270
  critical path   11, 11F
    efficacy   11
    safety   11
  cultural challenges   282
    individuality   282
  data focus   272
    independence   272–273
    job security   273
    qualitative vs. quantitative   273
    reductionism   272
  development candidates see
      Development candidate
      nomination (DCN)
  development chemistry see
      Development chemistry
  discipline changes   88
  disease complexity   274
  drug development see Drug
      development
  drug re-engineering   276
    drug combinations   277
  early cessation of poor leads   562,
      565
    biological basis   562–563
    communication to teams   562
    reward structure changes   563
  flower plots   31–32, 33F
  functional discipline
      organization   108–109
  future work   282
  genomics see Genomics
  government initiatives   271
  GPCRs   800
  historical perspective   3,
      266–267
    biochemistry (1948-1987)   267
    biotechnology (1987-2001)
        267
    empirical/physiological (1885-
        1948)   267
    genomics (2001-present)   269
  hit identification   31, 32F

hit-to-lead 139
  screening cascade 139
  target product profile (TPP) 139, 140T
Horrabin, David and 269
HTS 139
improvement intervention points 562
  clinical trials efficiency 560
  lead quality 560
    see also ADMET
  lead throughput rate 560
individuality 269
innovation see Innovation
intellectual property 277
  Bayh–Dole Act 277–278
  genes, patenting 278
  patent licensing 278
  patents, ease of 278
investment 4, 6F
knowledge requirements 89
lead optimization see Lead(s)
licensing see Licenses/licensing
linear process 562F
medicinal chemistry 267
methods 295, 296T
  combinatorial libraries 300, 300T
  known drug modification 295–296, 296T
  library screening 297, 297T
  natural products 296, 296T, 297
  proteins 297, 298T
  structure-based in-silico design 301, 301T
new model 32, 34F, 302–303, 303F, 304, 304F
  traditional methods vs. 87, 88F, 89, **265–287**
novel targets 274
old target opportunities 274
  target centric vs. multiple targets 275
operational challenges 283
  animal models 284
  consistency 283
  pharmacology re-introduction 283
organizational aspects see Organization, drug discovery
  leadership strategies see Leadership
  management see Leadership
  pharma. vs. biotech see Pharmaceutical industry vs. biotech, drug discovery
organizational effectiveness 271
  individualism 272
  leadership 271–272
pharmacology 267, 276
philosophy see Philosophy (of drug discovery/development)
pipeline filling 278
  in-house value 279
  in-licensing activities 278–279

pipeline phases 138, 138F
  see also specific phases
predevelopment 141
  chemical development 141
  clinical development 141
  formulation 141
  toxicology 141
process 266, 268F, 562F
productivity 124
  animal disease models 125
  backups/second generation candidates 125
  biologicals vs. small molecules 126, 128F
  bottleneck analysis 104F, 124–125, 124T
  cell-based screens 124–125
  clinical evaluation attrition rates 126
  compound libraries 124–125
  failure rates 124–125
  HTS 124–125
  improvement initiatives 125, 125T
  parallel development 125–126, 126F, 127F
  preclinical safety evaluation 125
  SAR 124–125
  target selection 126
project management see Project management
project teams see Project teams
promiscuous ligands see Promiscuous ligands
proteomics 266
research to development transition 54
safety 2
science simplification 273
  technocrats 273
  technology investments 274
scientific challenges 284
  marketing, replacement by 284
  'new,' focus on 284
scientific expertise 272
  employee characteristics 272
screening cascade 565, 565F
  mixture screening 39
in silico methods 48
  QSAR 48, 48F
single nucleotide polymorphisms 315–316
sources 31
specificity 298–299
  enhancement 299, 299T
  small molecules 299–300, 300T
  'specificity scale' 298, 299T
strategies 271
successes 7, 54
systems biology 276
  cost effectiveness 276
  SAR 276
target discovery see Target(s)
target validation 138

technologies 3, 31, 265, 279
  see also specific technologies
  antisense therapies 35, 35F
  cell compartments 41–43
  combinatorial chemistry 33
  data interpretation limitations 280
  molecular biology vs. native target 280
  nucleic acids 34
  revisionist approach 279–280
  traditional method exclusion 280
timelines 4
traditional methods 559
  new model vs. 87, 88F, 89, **265–287**
Drug discovery
  decline in 4, 6F, 266
  legislation 4
  reasons for 266
'Druggability'
  assessment, drug development 176
  target selection 754
'Druggable genome,' target vs. disease 113, 113F
'Druggable targets,' definition 112, 112F
Drug-induced microtubule reorganization, high-content screening (HCS) 428, 429F
Drug introductions, DCN 152
'Druglikeness,' drug development 730–731
Drug Price Competition and Patent Term Restoration Act (Hatch-Waxman Act: 1984), productivity effects 123–124
Drug reference standard see Development chemistry
Drug targeting, mitochondria see Mitochondria, drug targeting
Drug withdrawals 6–7
DTI-571, KIT inhibitor 972
Duncan, William, management strategies 260–261
Duration of action, hydrophilic depoting 587–588
Dynamic dimerization, GPCRs 799
Dynamin I, GPCR interaction 786T
Dynorphin A, protein conjugation 589

E

E-2007 903
EB 1089
  structure 1018F
  as vitamin D receptor ligand 1019
Ebselen, structure 702F
Economics
  development chemistry 160–161
  importance of 240
  peptide/protein drugs 596
  pharmaceutical R&D 100, 101T

ED-71, as vitamin D receptor
      ligand   1019
EDG-3 receptor, activation   785T
EDG-5 receptor, activation   785T
Edman sequencing   345
   protein identification   336–337
Education
   American Association of Colleges of
         Pharmacy   85
   curriculum recommendations   85,
         86T, 87T
   demography   86
   foreign national applicants   85–86
   income differences   85–86
   nature of   85
   range of fields   87
   science integration   88–89
   summer schools   89, 90T
Effective communication, project
      teams   145
Effective leaders see Leadership
Efficacy
   comparisons, clinical trials   184–185
   drug discovery   11
   ideal drug characteristics   564T
Efficiency
   discreditation, innovation   263
   outsourcing   228, 229F
Efflux transporters, drug activity   62
EKB-569
   characteristics   984T
   EGFR family kinase inhibitors   968
   structure   968F
Elastase inhibitors,
      peptidomimetics   613, 614F
Elastases, three-dimensional
      structures   626T
Electronically enhanced hybridization,
      DNA microarrays see DNA
      microarrays
Electronic circular dichroism, chiral
      drug analysis   720
Electronic commerce, materials
      management   536
Electronic parameters, bioisosterism
      modification analysis   687, 688F
Electron microscopy, promiscuous
      ligands, identification   745
Electron paramagnetic resonance
      studies, mitochondria
      diversity   1065
Electron shuttle, ubiquinone
      (coenzyme Q)   1073
Electron transfer models, mitochondria
      see Mitochondria
Electron transfer system (ETS),
      mitochondria see Mitochondria
Electrophoresis, single nucleotide
      polymorphisms (SNPs) see Single
      nucleotide polymorphisms
      (SNPs)
Element integration, drug
      development   721

Eli Lilly and Co.
   innovation   262
   R&D spending   101T
Embryogenesis, retinoic acid
      receptor   1007
EMEA, regulatory requirements   155
Emesis/nausea
   phosphodiesterase-4   925
   phosphodiesterase-4 inhibitors   936
Empowerment, project teams   145
Enalapril
   analogs, hydrophilic depoting   588
   design/development   625
Enantioselective chromatography
   chiral drug analysis   721
   direct   721
   indirect   721
Encephalomyopathies, ubiquinone
      (coenzyme Q)   1074
Endocrine tumors, Gα$_s$ subfamily
      mutations   780–782
Endometrial cancer, estrogen receptors
      (ERs)   1026–1027
Endophilin, GPCR interaction   797T
Endothelial nitric oxide synthase
      (eNOS)   1078
Endothelin ET$_{A/B}$ receptor,
      activation   785T
Endothelin ET$_A$ receptors
   allosteric modulators   790T
   in human disease   775T
Endothelin ET$_B$ receptors, in human
      disease   775T
Endothelin receptor antagonists
   nonpeptide drugs   618–619, 620F
   peptidomimetics   608–609, 611F
Endothiapepsin, three-dimensional
      structures   626T
Endpoints, clinical trials   186
Enfuvirtide   579T
Enjoyment, project teams   145
Enoximone, structure   934
Entrepreneurism, culture   123
Environmental effects
   development chemistry   161
   HTS data management   453
Enzyme(s)
   promiscuous ligands,
         identification   741
   resistance, bioisosterism   673–674,
         677F
Enzyme assays
   fluorescence correlation spectroscopy-
         related confocal fluorimetric
         techniques (FCS$^+$)   409, 409F
   homogeneous time-resolved
         fluorescence (HTRF) see
         Homogeneous time-resolved
         fluorescence (HTRF)
Enzyme-linked immunosorbent assays
      (ELISAs)
   homogeneous time-resolved
      fluorescence vs.   417

   surface-enhanced laser desorption
         ionization (SELDI) vs.   378–379
EPH2, PDB entries   965T
Epibatidine
   solution phase synthesis   492,
         493F
   structure   887F
Epidermal growth factor receptor
      (EGFR), posttranslational
      modifications, as drug
      targets   383
Epidermal growth factor receptor
      (EGFR) family kinases   967
   see also specific kinases
   in disease   963T
      cancer   967
   inhibitors   967, 968F
      see also specific drugs
   PDB entries   965T
   signal transduction   967
Epilepsy
   disease basis
      AMPA-sensitive receptors   903
      kainate-sensitive receptors   903
      potassium channels   864–865
      sodium channels   839–840, 841
   drug-resistant   67
   treatment, calcium-channel
         blockers   851
Epimaritidine, solution phase
      synthesis   491–492, 492F
Epinephrine, structure   57F
Episodic ataxia, calcium channels
      852
Epithelial cells
   differentiation, retinoic acid receptor
         (RAR)   1008
   potassium channels   860–861
Eplerenone   1017–1018
   characteristics   1012T
   structure   1018F
Epoetin alpha   762–763
Epothilone A, structure   28F
Epothilones, solid support
         synthesis   482, 484F
25-Epoxycholesterol   1023
   structure   1024F
Eptifibatide   579T
Equipment failure, HTS data
      management   453
Equity, motivation   246
ERB-041, structure   1013F
ErbB2/HER2/Neu, in disease   963T
ErbB3/HER3, in disease   963T
Erectile dysfunction,
      phosphodiesterase-5
      inhibitors   948
Ergocalciferol, characteristics   1012T
ERK1/2, GPCR MAP kinase
         activation   784–787, 787F
Erlotinib see OSI-774
ERP systems, materials
      management   536

Error sources, HTS data management *see* High-throughput screening (HTS), data management
Erythema nodosum, thalidomide 759
Erythromycin, structure 46F, 1043F
Erythropoietin receptor, phage display 576–577
Escitalopram oxalate, structure 729F
ESI, proteomics 347
Esomeprazole magnesium
 characteristics 653T
 structure 653T, 732F
 synthesis 732F
Established protocols, outsourcing scale-up 211
Estimated human dose, DCN 154
17α-Estradiol, structure 1063F
17β-Estradiol, structure 1063F
Estrogen
 neuroprotective effects 1005
 physiological effects 1004–1005
Estrogen receptors (ERs) 994, 1004
 cardiovascular homeostasis 1005
 genetic disease
  breast cancer 1026
  endometrial cancer 1026–1027
  psychiatric disease 1026–1027
  splice variants 1026
  systemic lupus erythematosus 1026–1027
 knockout animal models 1005
 nomenclature 996T
 plasma membrane-associated 1005
Estrogen receptor ligands 1013, 1013F
 see also specific drugs
 hormone replacement therapy (HRT) 1013–1014
 phytoestrogens 1014
 scaffold hopping 683, 684F, 685F
 selective estrogen receptor downregulators (SERDs) 1014
 selective estrogen receptor modulators (SERMs) 1014
Estrogen receptor-related receptor, nomenclature 996T
Etanercept, protein conjugation 590
Etch stopping, lithography/photolithography 352
Ether–oxygen/methylene groups, bioisosterism 692, 693F
Ethical evaluation, clinical trials 187
Ethinyl estradiol, structure 1013F
Ethosuximide 850
 characteristics 829T
 structure 829T
6-Ethyl-CDCA, structure 1025F
Etoricoxib, structure 658F, 681F
Eukaryotes, RNA polymerases 1044
European Agency for the Evaluation of Medicinal Products (EMEA), chemogenomics 567
Evaluation, innovation 260

Evanescent wave, surface plasmon resonance 371
Excitability, potassium channel patterns 857
Excitable cells, sodium channels 835
Excitatory amino acid transporters 60T
Exclusion, gender issues 245
Executive summation, DCN 152
Exenatide 579T
 GPCR targeting 772
Experimental artifacts, false positives 737
Experimental error detection, HTS data management 452
Experimental records, career guidelines 18
Expression levels, messenger RNA (mRNA) 1044–1045
Expression-modified proteins, peptide/protein drugs *see* Peptide/protein drugs
Expression monitoring, functional genomics 329
Expression profiling, target identification *see* Target identification, genomics
Expression/trafficking regulation, GPCRs 799
Extended enterprises 229
 definition 229
 virtual companies 766
Extended ternary complex model, GPCR activation models 792–794, 794F
External event monitoring, project team leaders 144
Extracellular stimuli, phosphodiesterase(s) regulation 923
Eye, phosphodiesterase-6 928

F

F_1-inhibitor protein (IF-1), mitochondria complex V 1078–1079
Facilities, clinical trials 186–187
Factor Xa, three-dimensional structures 626T
Fail fast *vs.* 'follow the science' *see* Philosophy (of drug discovery/development)
Failure rates
 drug development 199
 drug discovery 124–125
Failure visibility, drug development productivity 126
Fairness, motivation 246
False-negatives, clinical trials 30–31
False-positives, clinical trials 30–31
Familial amyotrophic lateral sclerosis 1071

Familial childhood epilepsy 840–841
Familial hemiplegic migraine (FHM) 852
Familial paragangliomas 1073
Famotidine, GPCR targeting 773T
Farglitazar
 as PPARγ ligand 1021–1022
 structure 1023F
Farnesoid X receptor (FXR) 1010
 nomenclature 996T
 pharmacology 1023, 1025F
 see also specific drugs
 three dimensional structure 999–1000
Fc domains, peptide/protein drug conjugation 590
Feedback, communication 243
Felodipine, characteristics 829T
Fenfluramine
 structure 65F
 transporter interactions 63
Fenofibrate 1021
 characteristics 1012T
 structure 1022F
Fenpyroximate
 mitochondria complex I 1068–1069
 structure 1069F
Fentanyl, GPCR targeting 773T
Fexofenadine
 GPCR targeting 773T
 structure 66F
Fibrates 1021
 see also specific drugs
 diabetes mellitus type 2 1021
Fibrinogen receptor (GPIIa/IIIb) antagonists, peptidomimetics 608–609, 611F
Fibroblast collagenase (MMP-1), peptidomimetics 632, 635F
Fibroblast growth factor receptor 1 (FGFR1), PDB entries 965T
Fibroblast growth factor receptor 2 (FGFR2), PDB entries 965T
Fibrous bone dysplasia, $G\alpha_s$ subfamily mutations 780–782
Filamin-A, GPCR interaction 797T
Filgrastim 762–763
Final drug characteristics, ADMET 563
Finances, outsourcing 207
Fisher, Ronald, process variable statistical evaluation 164
5′ cap structure, messenger RNA (mRNA) structure 1044
FKBP-12, three-dimensional structures 636T
Flat microarrays, single nucleotide polymorphisms (SNPs) *see* Single nucleotide polymorphisms (SNPs)
Flavonoids, mitochondria, drug targeting 1068

Flavopiridol
    CDK family kinase inhibitors 979–980
    characteristics 984T
    structure 979F
Flecainide
    characteristics 829T
    structure 829T
Flexibility, project teams 145
FLIPR instrumentation, GPCR functional assay screening 811
Flosulide, structure 658F, 681F
Flow cytometry
    microfluidic devices *see* Microfluidic devices
    single nucleotide polymorphisms 368
Flower plots, drug discovery 31–32, 33F
Flow nuclear magnetic resonance 521
Flt3 family kinases 969
    acute myeloid leukemia 971
    in disease 963T
Fluidics/electronics integration, microtechnologies 355
Fluorescence assays, GPCR functional assay screening 809
Fluorescence correlation spectroscopy (FCS) 404
Fluorescence correlation spectroscopy-related confocal fluorimetric techniques (FCS$^+$) 404
    advantages 406
    background elimination 407
    case studies 407
    cell assays 411F
    enzyme assays 409, 409F
    GPCR assays 410, 410F
    homogeneous assays 405T, 406
    miniaturization 406
        case studies 407, 408F
    multiple read-out modes 407
        flexibility 408
    multiple read-out parameters 407
        case studies 407, 408, 409F
    read-out 405, 427T, 456T
    reagent safety 407
    single component labeling 407
    throughput 407
Fluorescence detection systems
    HTS *see* High-throughput screening (HTS)
    microplate-based assays 367
    Ultra-High-Throughput Screening System (Aurora Biosciences) 445, 446F
Fluorescence imaging, drug development 193
Fluorescence-mediated tomography (FMT), drug development 193
Fluorescence polarization (FP), GPCR functional assay screening 809

Fluorescence reflectance imaging (FRI), drug development 193
Fluorescence resonance energy transfer (FRET)
    GPCRs 798–799
        functional assay screening 813
    homogeneous time-resolved fluorescence 413–414
    HTS 401
    single nucleotide polymorphisms 366
    Ultra-High-Throughput Screening System (Aurora Biosciences) 437
Fluorescent reagents, high-content screening (HCS) *see* High-content screening (HCS)
Fluorescent stains, two-dimensional gel electrophoresis 334
Fluorinated hydrocarbons, solution phase synthesis 485
(S)-3-Fluoropyrrolidine, structure 720F
(S)-5-Fluorowillardine 900
    structure 901F
Fluoxetine, structure 57F, 62F
Flusilazole, structure 695F
Flutamide
    characteristics 1012T
    structure 1081F
FMLP receptor, activation 785T
Fmoc peptide synthesis
    peptide synthesis 475–476
    sheets 503, 503F
Focal adhesion kinase (FAK), in disease 963T
Focused individuals, innovation 262
Follicle-stimulating hormone receptors
    constitutively active 790T
    in human disease 775T
Follow-up screening, materials management 535
Food and Drug Administration (FDA)
    approval 159–160
        cilomilast 948
    approval time 200
    chemogenomics 567
    clinical trials phase I, regulatory requirements 155
    fast-track approval 200
Food and Drug Administration Modernization Act (1997) (FDAMA), productivity effects 123–124
Forcing, conflict management 248
Foreign national applicants, education 85–86
Forkhead-associated domains, phosphoserine/phosphothreonine-interacting domains 985
Formal project reviews, project management teams 148

Form/formulation enhancement, biotechnology companies 232
Formivirsen sodium 761, 1045–1046
    clinical trials 1045
    structure 1046F
Formulations
    DCN 155
    predevelopment 141
Fosphenytoin sodium
    characteristics 829T
    structure 829T
FPR1 receptors, in human disease 775T
Fragment-based drug discovery 49, 51F
Fragment replacement, scaffold hopping 681–683
Frank, Ronald, combinatorial chemistry 461
Frataxin mutation, Friedreich's ataxia 1059–1060
Free radicals
    Friedreich's ataxia pathogenesis 1065
    mitochondria *see* Mitochondria
    mitochondria complex II 1065
    mitochondrial permeability transition (MPT) 1061
Freezer storage stability, drug reference standard 167–168, 168T
Frequency selective nuclear magnetic resonance 509
'Frequent hitters' *see* Promiscuous ligands, computational prediction
Friedman's paradox, bioisosterism 692
Friedreich's ataxia
    frataxin mutation 1059–1060
    free radical production 1065
    mitochondria 1054–1055
    ubiquinone (coenzyme Q) treatment 1074
FRR1 receptor, constitutively active 790T
Fully integrated pharmaceutical company (FIPCO) 100
Fulvestrant
    as estrogen receptor ligand 1014
    structure 1013F
'Functional analogs,' bioisosterism *see* Bioisosterism
Functional areas
    project management 149
    project team goals *see* Project teams
Functional departments, project teams *vs.* 145
Functional discipline organization *see* Pharmaceutical industry
Functional genomics 319
    definition 382T
    gene expression 326, 327F
        autoregulation 326
        techniques 326

Index

molecular pathways 324
  protein–protein interactions 325
  signal transduction 324
signal transduction 960, 963T
target identification 308, 329
techniques 326
  deletion detection 329
  expression monitoring 329
  HTS 328
  hybridization differential display 329
  microarrays 328
Functional group reversal, bioisosterism *see* Bioisosterism
Functionalized polystyrene resins, solution phase synthesis 485–486, 487F
Functional proteomics
  definition 381, 382T
  gene-related 389
Fungi, mitochondria complex I 1070
Funiculosin, structure 1076F
Furchgott, Robert, sildenafil citrate, invention of 257
Furo[3,2-b] pyridine, bioisosterism 660T
Furoxane, bioisosterism 660T
FusionPro, process variable statistical evaluation 164

G

G2A receptor, activation 785T
$G_{12/13}$ subfamily, G proteins 780
$G\alpha_2$ subfamily, G proteins 782
$G\alpha_3$ subfamily *see* G proteins
$G\alpha_{12}$ subfamily, G proteins 783–784, 783F, 785T
GABA, structure 661F
GABA$_A$ receptor 891
  characteristics 881T
  diseases 894
  heterodimerization 799
  molecular biology 891
  nomenclature 891
  pharmacology 892, 893F
  physiological functions 892
    synaptic transmission 892
  regulation 892
  structure 891
  therapeutics 893
    alphadolone 893
    benzodiazepines 893
    benzoxazines 893
    ganaxolone 893
    thiazoles 893
GABA$_A$ receptor ligands, scaffold hopping 681, 682F
GABA$_B$ receptor
  allosteric modulators 790T
  constitutively active 790T
  in human disease 775T

GABA$_B$ receptor antagonists
  see also specific drugs
  carboxylic acid bioisosterism 661
GABA$_C$ receptor 891
  characteristics 881T
  diseases 894
  molecular biology 891
  nomenclature 891
  pharmacology 892, 893F
  physiological functions 892
    synaptic transmission 892
  regulation 892
  structure 891
  therapeutics 893
    benzodiazepines 893
    benzoxazines 893
    thiazoles 893
GABA transporters 60T
Gabapentin
  mechanism of action 851–852
  ring system reorganization (bioisosterism) 676, 678F
Gaboxadol, structure 893F
GAF domains, phosphodiesterase-5 927
'Gain of function' mutants, androgen receptor (AR) 1027
$G\alpha_{i/o}$ subfamily, G proteins 781F, 782
$G\alpha_i$ subfamily, G proteins 782
Galanin GA2 receptor, activation 785T
Galantamine (galanthamine), nicotinic acetylcholine receptor binding 887–888
Gamma amino acids, proteolysis resistance 584
γ-scintigraphy, drug development 192
γ-turn mimetics, peptidomimetics 608
Ganaxolone, GABA$_A$ receptors 893
Ganirelix acetate 579T
  hydrophilic depoting 588
  self-forming depots 593
$G\alpha_o$ subfamily, G proteins 782
$G\alpha_q$ subfamily, G proteins 782–783, 783F
GASP-1/2, GPCR interaction 797T
Gastrin-releasing peptide receptor
  activation 785T
  antagonists, nonpeptide drugs 618–619, 620F
  constitutively active 790T
Gastrointestinal stromal tumors, KIT 972
GC-1
  structure 1020F
  as thyroid hormone receptor ligand 1019
Gefitinib (ZD-1839)
  characteristics 984T
  EGFR family kinase inhibitors 967–968, 969
  structure 71F, 968F

Gelatinase inhibitors, peptidomimetics 613, 614F
Gel electrophoresis 331
Gemfibrozil 1021
  characteristics 1012T
  structure 1022F
GenChem database, bioinformatics 317
Gender issues, leadership strategies *see* Leadership
Gene(s)
  activity, target identification 316–317
  expression levels, bioinformatics 318
  patenting 278
Gene expression
  expression profiling 308–309
  functional genomics *see* Functional genomics
  studies, drug development 197
Gene expression microarrays, drug development 194–195
Genentech 762
  characteristics 226T
  formation 102
  research focus 763
General Agreement on Tariffs and Trade (GATT), intellectual property protection 131
Generic drugs, drug development costs 727
Gene therapy 1038
  chemical *vs.* biological 117–118
  mitochondria, drug targeting 1067
Genetic diseases, transporters 66
Genetic mapping, target identification, genomics 306
Genetic models, target selection 755–756
Genetic polymorphisms, drug development 195, 195T
Gene trapping, expression profiling 312
Genex, formation 102
Genistein, structure 1013F
Genome-based medicine, drug discovery 266
Genomes
  definition 382T
  DNA organization 425–426
  druggable targets 4
  sequencing, target identification *see* Target identification, genomics
Genomics
  combinatorial biosynthesis 44, 46F
  definition 382T
  drug discovery 43, 44, 266, 303
  target opportunities 44

Index

Genomics (*continued*)
  druggable targets 68
  lead screening 425
  manipulations 44, 45F
  pharmaceutical productivity
    effects 132–133
  proteomics *vs.* 381, 382F
  stem cells 46
  synthetic biology 44–46
  target selection *see* Target selection
Genotyping
  drug development 197
  phase IV clinical trials 197
Gentamicin, structure 1041–1042
Gentamicin C$_{1a}$, structure 1042F
Germanium, carbon-silicon
    bioisosterism 697, 698F
Germ cell nuclear factor,
    nomenclature 996T
Geysen, Mario, combinatorial
    chemistry 461
Ghrelin receptors
  constitutively active 790T
  in human disease 775T
GHRH receptors, in human
    disease 775T
'Gift economy model,' universities
    91–92, 92F
Ginkgolide B, structure 895F
G$_{i/o}$ subfamily, G proteins 780
GIPC, GPCR interaction 797T
GIRK ion channels, GPCR
    interaction 786T
Glass, solid supports *see* Solid supports
Glaucoma, mitochondria 1054–1055
GlaxoSmithKline (GSK)
  centers of excellence in drug discovery
    (CEDDS) 109–110
  history 100–101
  multiple sites 19
  R&D spending 101T
  Theravance alliance 234
GlaxoWellcome, history
    100–101
Glitazones
  see specific drugs
Globalization of science 94
  country comparisons 94
  USA dominance 94
Global organization, hybrid
    organization 109–110
Glomerular filtration
  PEGylation 588
  peptide/protein drugs 585
GLP-1 receptor, constitutively
    active 793T
Glucagon, as peptide/protein
    drugs 579T, 580
Glucagon-like peptide 1 (GLP-1)
  analog development 583
    proteolysis resistance 583
  hydrophobic depoting 586–587
  as peptide/protein drug 589

Glucagon receptors
  antagonist, nonpeptide drugs
    618–619, 620F
  constitutively active 790T
  in human disease 775T
Glucocorticoid receptor (GR) 994,
    1006
  anti-inflammatory actions 1006
  genetic disease
    Crohn's disease 1027
    glucocorticoid-resistant
      syndrome 1027
  immunosuppressive actions 1006
  nomenclature 996T
  three-dimensional structure,
    coregulator binding site
    1001
Glucocorticoid receptor ligands 1016,
    1017F
  see also specific drugs
  diabetes mellitus type 2 1017
  immunological disease 1016–1017
  inflammation 1016–1017
  mechanism of action 1016–1017
Glucocorticoid-resistant syndrome,
  glucocorticoid receptor
  1027
Glucose/amino acid/fat oxidation,
    mitochondria inner
    membrane 1059
Glucose-dependent insulinotropic
    polypeptide (GIP)
  analog development 583
  hydrophobic depoting 587
Glucose homeostasis
  phosphodiesterase-9 inhibitors
    950
  phosphodiesterase-10 inhibitors
    950
Glucose-stimulated insulin secretion,
    uncoupling proteins 1080
Glutamate antagonists, carboxylic acid
    bioisosterism 661
Glutamate-gated ion channels 879,
    897
  see also specific types
  characteristics 881T
  diseases 903
  expression 879
  molecular biology 897
  nomenclature 897
  pharmacology 900
  physiological functions 899
    pain 900
  regulation 899
    RNA editing 899
  structure 880F, 897
    ligand-binding domain 898
  therapeutics 902
Glyantrypine, solid support
    synthesis 482
(2S)-Glycidyl-3-nitrobenzenesulfonate,
    structure 720F

Glycine, structure 895F
Glycine receptors 894
  characteristics 881T
  diseases 896
    hyperplexia 896
  molecular biology 894
  nomenclature 894
  pharmacology 895,
    895F
  physiological functions 894
  regulation 895
  structure 894
  therapeutics 896
    calcium channel antagonists
      896
Glycine transporters 60T
Glycine transporter type 2
    63
  inhibitors 65F
Glycopeptides, solid support
    synthesis 477, 478F
Goal awareness, career guidelines
    19
GOLD, promiscuous ligands,
    computational prediction
    749
Gonadotropin-releasing hormone
    (GnRH), analogs
    594–595
Gonadotropin-releasing hormone
    receptor
  agonists, hydrophilic depoting
    587
  antagonist, nonpeptide drugs
    618–619, 620F
  in human disease 775T
Good Laboratory Practice (GLP), drug
    development 27
Good Manufacturing Practice (GMP),
    drug development 27,
    177
Goserelin 579T
Governments, in drug discovery
    271
GPR3 receptor, constitutively
    active 790T
GPR6 receptor, constitutively
    active 790T
*GPR10 receptors, in human
    disease 775T
GPR12 receptor, constitutively
    active 790T
*GPR50 receptors, in human
    disease 775T
*GPR54 receptors, in human
    disease 775T
*GPR56 receptors, in human
    disease 775T
*GPR154 receptors, in human
    disease 775T
G protein-coupled receptor interacting
    proteins (GIPs) 796,
    797T

Index

G protein-coupled receptors
   (GPCRs) **771–826**
  see also G proteins
  allosteric modulation 802, 803*T*
     orthosteric site 802
     signaling properties 804–805
  amines, structure 575*F*
  biological functions, calcium channel
     regulation 848
  class A (rhodopsin family) 575–576,
     777–778, 778*F*
     activation 796
     see also Rhodopsin receptor
  class B (secretin/glucagon
     family) 575–576, 778
     large N-terminal GPCRs 779
  class C 779
  classification 575–576
  constitutively active 789, 790*T*
     in human disease 789, 792*T*, 793*T*
     inverse agonist identification 789
  definition 574–575
  desensitization 800, 801*F*
     agonist efficacy *vs.* 801
     beta-arrestin 801
     conformational changes 800
     GRKs 802
     internalization 800–801
     phosphorylation 800, 801
     recycling 802
  disease relevance 772
  drugs targeting 772, 773*T*
     see also specific drugs
  eicosanoids 575*F*
  families 777
     see also specific families
  genetic studies 772, 775*T*
     gain-of-function mutations 772
  glycoprotein hormones 575*F*
  ligands 576
  MAP kinase activation 784
     beta-arrestin 788
     ERK1/2 784–787, 787*F*
     human disease 784
     protein kinase A activation 787
     protein kinase C activation 787
     Ras activation 787, 788*F*
  mechanism of action 574–575
     heterotrimeric G protein (GTP-
        binding protein; Gαβγ) 574–
        575
  neurotransmitters 575*F*
  nucleosides 575*F*
  oligomerization/
     heterodimerization 798
     bioluminescence energy transfer
        (BRET) 798–799
     drug discovery relevance 800
     dynamic dimerization 799
     expression/trafficking
        regulation 799
     fluorescence energy transfer
        (FRET) 798–799
     internalization 799
     signaling 800
  peptide hormones 575*F*
  promiscuous ligands, computational
     prediction 748
  protease activated receptor 575*F*
  signaling molecules 796
     GPCR interacting proteins
        (GIPs) 796, 797*T*
  signal transduction 574–575
     regulation 788
  structure 574–575, 575*F*, 774, 777*F*
     C-terminus 774–777
     NpxxY motif 774–777
     N-terminus 774–777
     (D/E)RY motif 774–777
     three-dimensional structural
        models 623
     transmembrane helices 774–777
G protein-coupled receptors, drug
   design 70, 300, 757*T*, 758, 805
  see also specific drugs
  nonpeptide drugs 619–621, 623
     peptide/protein drugs *vs.* 623
  peptide/protein drugs 574, 580–581
     nonpeptide drugs *vs.* 623
     subtype selectivity 575–576
  peptidomimetics 595–596, 605, 623
  target validation 814
  therapeutic potential 772
G protein-coupled receptors, models
  activation models 789
     'cubic ternary complex
        model' 792, 794*F*
     extended ternary complex
        model 792–794, 794*F*
     ternary complex model 789–792,
        794*F*
     'two-state model' 792, 794*F*
  multistate activation model 794
     deactivation conformations 795
     definition 794
     drug design implications
        795–796
     dual efficacy ligands 795
     evidence 794
     fluorescence labeling studies
        795
G protein-coupled receptors (GPCRs),
   assays 805, 806
  biosensors 813
     bioluminescence energy transfer
        (BRET) 813
     fluorescence energy transfer
        (FRET) 813
  cAMP assays 807
     AlphaScreen cAMP assay
        809
     chemiluminescent 809
     fluorescence-based 809
     fluorescence polarization 809
     HitHunter cAMP assay 809
     radiometric 808
     receptor subtypes 807–808
     time-resolved fluorescence-
        based 809
  cell-based assays 806
  chimeric G proteins 812
     melanophore platform 813
  constitutive activity 806
  fluorescence correlation spectroscopy-
     related confocal fluorimetric
     techniques (FCS$^+$) 410, 410*F*
  GTPγS binding assays 778*F*, 807
     time-resolved fluorescence 807,
        807*T*
  high-content imaging 813
     assay throughput 814
     CCD cameras 813–814
     confocal microscopy 813–814
     Transfluor technology 813–814
  inositol phosphate accumulation-IP$_3$
     assays 809, 810*F*
  homogeneous time-resolved
     fluorescence assays 794*F*,
     810–811
  intracellular calcium assays 811
     AequoScreen 811
     Calcium 3 assay 811
     calcium-sensitive dyes 811
     FLIPR instrumentation 811
     HTS 811
     subtypes 811–812
  membrane-based 805
  multiple signaling 806
  receptor expression levels 806
  reporter gene assays 812
     activator protein 1 (AP-1) 812
     beta-lactamase 812
     cAMP responsive element 812
     luciferase 812
  signals measured 806
  universal screens 812
G protein cycle 779, 779*F*
G proteins
  activation 779
     α-subunit regulation 780, 781*T*
     adenylate cyclase 780–782, 781*F*
     βγ-regulation 781*T*, 784, 786*T*
  families 780
     G$_{12/13}$ subfamily 780
     Gα$_2$ subfamily 782
     Gα$_3$ subfamily 780–782
        associated diseases 780–782
     Gα$_{12}$ subfamily 783–784, 783*F*,
        785*T*
     Gα$_{i/o}$ subfamily 781*F*, 782
     Gα$_i$ subfamily 782
     Gα$_o$ subfamily 782
     Gα$_q$ subfamily 782–783,
        783*F*
     G$_{i/o}$ subfamily 780
     G$_q$ subfamily 780
     Gs subfamily 780
  structure 780
     βγ-subunits 780

G proteins (*continued*)
  palmitoylation 780
  α-subunits 780
G_q subfamily, G proteins 780
G-quartets, DNA structure 1040
Granisetron
  5-HT_3 receptor binding 890
  structure 890F
Graphs, HTS data management 454
Grb2 SH2, three-dimensional structures 636T
Green fluorescent protein (GFP)
  high-content screening (HCS) 427, 430F
  Ultra-High-Throughput Screening System (Aurora Biosciences) 437
Grignard reaction, solid support synthesis 467–468, 469F
GRKs, GPCR desensitization 802
Growth hormone releasing peptide (GHRP)
  nonpeptide drugs 618–619, 620F
  as peptide/protein drug 590
  peptidomimetics 608–609, 611F
Growth hormone releasing peptide (GHRP) receptor agonists 620F, 621
GSE, expression profiling 312
GSK3B, PDB entries 965T
Gs subfamily, G proteins 780
GTPγS binding assays *see* G protein-coupled receptors (GPCRs)
GTS-21
  nicotinic acetylcholine receptor binding 888
  structure 887F
Guanethidine analogs, ring system reorganization (bioisosterism) 675–676, 678F
Guanidine, hepatitis C virus inhibition 1047
Guanosine, structure 1038F
Guggulsterone, as nuclear receptor ligand 1010
Guidelines, regulatory problems 726
GW-404, as farnesoid X receptor ligand 1024
GW-2433
  as PPAR-δ ligand 1022
  structure 1023F
GW-3965 1023
  structure 1024F
GW-4064, structure 1025F
GW-9578, structure 1022F
GW-468816 903
GW-501516
  as PPAR-δ ligand 1022
  structure 1023F
GW-572016
  characteristics 984T

EGFR family kinase inhibitors 968
  structure 968F
GYKI-53655, structure 901F
Gyros Microlabs, microfluidic devices 350F, 361

H

Half-life ($t_{1/2}$)
  PEGylation effects 588, 589F
  peptide/protein drugs 583
Hardware developments, NMR 515
Harzianopyridone
  mitochondria complex II 1072, 1073F
  structure 1073F
HCK, PDB entries 965T
HCN ion channels *see* K^+ ion channels
Hearing impairments, transient receptor potential ion channels 909–910
Heart, phosphodiesterase-4 926–927
Heart failure, milrinone 947–948
Heck reaction, solid support synthesis 467–468, 470F
Helix, DNA structure 1038–1039
Helsinki declaration, clinical trials 180
Hepatitis C inhibitors
  benzimidazoles 1047, 1048F
  carboxylic acid bioisosterism 659
  guanidine 1047
  phenazine 1047
Hepatitis C virus, internal ribosomal entry site (IRES) 1047
Hepatocyte nuclear factors (HNFs), nuclear receptor interactions 1000
Heterocycles
  mitoprotective drugs 1062, 1062F
  ring equivalent bioisosterism *see* Bioisosterism, ring equivalents
  solid support synthesis *see* Solid support synthesis
Heterodimerization, cytokine receptors 576
Heterogenous approach, microplate-based assays 367
Heterotrimeric G protein (GTP-binding protein; Gαβγ), GPCRs 574–575
Hi-Chip system, DNA microarrays 357
High-content imaging *see* G protein-coupled receptors (GPCRs)
High-content screening (HCS) 425, 460
  advantages 426–427, 427T
  ArrayScan II system 427–428
  automation 428
  CellChip System 431, 432F, 433F, 434F
    advantages 432–433

Cellomics Knowledgebase 433, 435F
  biochemical pathway diagrams 435
  data mining 436
  definition 426–427
  drug-induced microtubule reorganization 428, 429F
  fluorescent reagents 428
  green fluorescent protein (GFP) 427, 430F
  JC-1 429, 430F, 431F
  future work 436
  living cells 426
Highly active antiretroviral therapy (HAART), mitochondria, effects on 1082
High-performance liquid chromatography (HPLC), combination, light scattering, promiscuous ligands 743
High-resolution magic angle spinning (HRMAS)
  drug development 194
  nanoprobes 519
High throughput organic synthesis (HTOS), HTS *see* High-throughput screening (HTS)
High-throughput screening (HTS) 391
  ADMET 393–394, 401
  candidate parallel development 401, 402F
  assay development 399
    cell-based 400
    lead validation 400
    relevance *vs.* cost 399, 399F
    target selection 399–400
  bioinformatics 402
    assay variation 403
    data generation 402
  comparative functional genomics 401
  data acquisition 449
    automation integration 449
    information system integration 450
    plate layouts 451
    well types 451
  development, combinatorial chemistry 39, 42F
  distribution, materials management 535
  drug development 177
  drug discovery 139
    productivity 124–125
  expression profiling 310, 311
  false positives 737
    see also Promiscuous ligands
    assay interference 737
    compound impurity 737
    experimental artifact 737

fluorescence detection systems 400
fluorescence resonance energy transfer 401
functional genomics 328
future work 403, 459
microfluidics 403
plate densities 459
ultraHTS 459
GPCR functional assay screening 811
high throughput organic synthesis 393, 395
drug development impact 396, 396F
history 395
peptide-based 390F, 395
solid supports 396
support industries 395
historical perspective 391
assay development 392
perception of 392
'random screening' 392
standardization 392
support industries 392–393
throughput changes 392
homogeneous time-resolved fluorescence 416
logistics 458
management/personnel 403
natural products 397, 397F
marine organisms 397
microbials 397
plants 397, 398
sensitivity issues 398
plate handling 458
consolidation 459
format changes 459
pooling/deconvolution 459
present day 393
ADMET 393–394
de novo HTS 393
technology integration 393
promiscuous ligands see Promiscuous ligands
structure-based drug design 397F, 398
computational design software 398, 398F
success of 52
target selection 399–400, 755
target vs. disease 114–115
test substance supply 394
see also Libraries
sources 394
see also specific sources
High-throughput screening (HTS), data analysis 451, 455
cross-assay reports 455
dose response tests 452
kinetics experiments 452
percent inhibition 451
variation 451

percent of control 451
variation 451
pivoted data 455
classical structure activity 455
ratios 452
visualization 455
High-throughput screening (HTS), data management 448
automation 449
data volume 449
decision support 453
controls 453
data integration 454
hit location 454
pattern searches 453
references 453
single run evaluation 454
systematic error detection 453
error sources 453
environmental issues 453
equipment failure 453
reagent stability 453
graphical display 454
inherent variability 449
reporting 455
result validation 452
experimental error detection 452
false positives vs. false negatives 453
tabular display 454, 456T, 458T
High-throughput x-ray crystallography, phosphodiesterase-4 inhibitors 938
Hirschmann, Ralph, medicinal chemistry, definition of 92
Histamine, structure 57F
Histamine H_1 receptor antagonists
see also specific drugs
as 'old' target 274
Histamine H_1 receptors
activation 785T
constitutively active 790T
in human disease 775T
Histamine H_2 receptor antagonists
see also specific drugs
analog design 651
as 'old' target 274
Histamine H_2 receptors, constitutively active 793T
Histamine H_3 receptor antagonists
see also specific drugs
carboxamide bioisosterism 668
Histamine H_3 receptors, constitutively active 793T
Histone deacetylase (HDAC) inhibitors, carboxylic acid bioisosterism 659
Historical control groups, clinical trials 182
Histrelin 579T
controlled release formulations 592
hydrophobic depoting 586

Hit(s)
follow up 205
identification, drug discovery 31, 32F
location, HTS data management 454
HitHunter cAMP assay, GPCR functional assay screening 809
Hit-to-lead see Drug discovery
HIV protease, structure, three-dimensional structures 626T
HIV protease inhibitors
see also specific drugs
carboxamide bioisosterism 668
discovery/development, compound library screening 301–302
nonpeptide drugs 618–619, 620F
peptide/protein drugs 577
structure 578F
peptidomimetics 613, 614F, 615F, 616F
structure, x-ray crystallography 625, 630F
HIV TAR 1044F, 1047
HMR1275 see Flavopiridol
Hobbs DeWitt, Sheila
combinatorial chemistry, invention of 260
management difficulties 260
Hoffman-La Roche drugs, mitoprotective drugs 1062, 1062F
Homogeneous approach, microplate-based assays 367
Homogeneous assays, fluorescence correlation spectroscopy-related confocal fluorimetric techniques (FCS^+) 405T, 406
Homogeneous time-resolved fluorescence (HTRF) 411
enzyme assays 418, 419F
sensitivity 419, 420F
GPCR functional assay screening 794F, 810–811
HTS applications 416
immunoassays 417, 418F
ELISA vs. 417
lanthanide cryptates 412, 413F
DELFIA 412–413
fluorescence resonance energy transfer 413–414
signal amplification 413
XL665 413, 414F
media, resistance to 412
nucleic acid hybridizations 422
DNA nuclease assays 422, 422F, 423F
reverse transcriptase assays 422, 423F, 424F
optimization 423
buffers 423, 424F
controls 425
label choice 424

Homogeneous time-resolved
fluorescence (HTRF) (continued)
protein–protein interactions 421,
421F
dose response curves 421,
422F
receptor-binding assays 419
nuclear receptors 420
peroxisome proliferator-activated
receptor-γ 420, 420F, 421F
signal measurement 414
data reduction 416
dual-wavelength 415, 416F, 417F
time-resolved 415, 415F
signal modulation 412
unique tracers 412
Homoplasty, mitochondria
genome 1059–1060
Honesty, project teams 145
Horizontal structure, R&D
management 249
Hormone replacement therapy (HRT),
estrogen receptor
ligands 1013–1014
Hormones, peptide/protein drugs 580
Horner–Emmons reaction, solid support
synthesis 469F
Horrabin, David, drug discovery 269
Houghten, Richard, combinatorial
chemistry 461
HPLC-MS see High-performance liquid
chromatography (HPLC)
5-HT (serotonin)
bioisosterism 675, 677F
structure 57F
5-HT receptor antagonists
see also specific drugs
carboxamide bioisosterism 668
carboxylic ester bioisosterism
663
repositioning 276
5-HT$_{1A}$ receptor, in human
disease 775T
5-HT$_{1B}$ receptor, in human
disease 775T
5-HT$_{1B/D}$ receptor, allosteric
modulators 790T
5-HT$_{1D}$ receptor, in human
disease 775T
5-HT$_{2A}$ receptor, in human
disease 775T
5-HT$_{2C}$ receptor
activation 785T
allosteric modulators 790T
constitutively active 793T
in human disease 775T
5-HT$_3$ receptor 888
diseases 891
bipolar disorders 891
neuropsychiatric disorders 891
Tourette's syndrome 891
molecular biology 888
nomenclature 888

pharmacology 889, 890
agonists 890
picrotoxin 890
'setrons' 889–890
physiological functions 889
neurotransmitter release 889
regulation 889
structure 889
therapeutics
see specific drugs
5-HT$_{4A}$ receptor, activation 785T
5-HT$_{5A}$ receptor, in human
disease 775T
5-HT$_6$ receptor, in human
disease 775T
5-HT$_7$ receptor, allosteric
modulators 790T
Human disease, GPCR MAP kinase
activation 784
Human evaluation step, development
chemistry timeline 170
Human Genome Project
genetic mapping objective 306
target selection 754–755
Human nuclear factor 4 receptor,
nomenclature 996T
Human Pharmaco DNP
Consortium 316
Huntington's disease
mitochondria 1054–1055
model, mitochondria complex II see
Mitochondria complex II
phosphodiesterase-10 930–931
Hybridization
differential display, functional
genomics 329
single nucleotide
polymorphisms 364, 365F
Hybrid oligomers, solid support
synthesis see Solid support
synthesis
Hybrid organization see Pharmaceutical
industry
Hybrid vendors, definition 206
Hydrochlorothiazide, structure 66F
Hydrocodone, GPCR targeting 773T
Hydrocortisone, structure 1017F
Hydrogen bond J coupling, NMR 514
Hydrolysis, carbon-silicon
bioisosterism 694
Hydrophilic depoting, peptide/protein
drugs see Peptide/protein drugs
Hydrophobic depoting, peptide/protein
drugs see Peptide/protein drugs
Hydroxamic acids, carboxylic acid
bioisosterism see Bioisosterism,
carboxylic acids
22(R)-Hydroxycholesterol 1023
structure 1024F
Hydroxychromones,
bioisosterism 662T
Hydroxyethylureas, carboxamide
bioisosterism 668, 669F

Hydroxyflutamide, structure
1016F
3-Hydroxy-γ-pyrones, carboxylic acid
bioisosterism 661
5-Hydroxy indoles, structure 887F
Hydroxyisopropyl, phenol
bioisosterism 670–671
Hydroxymethyl, phenol
bioisosterism 670–671
Hydroxyquinine, mitochondria complex
III 1075
Hydroxythiadiazoles,
bioisosterism 662T
3-Hydroxy-1,2,5-thiadiazoles, carboxylic
acid bioisosterism 661
5-Hydroxytryptamine see 5-HT
(serotonin)
Hyoscyamine, mechanism of action,
GPCR targeting 773T
Hyperalgesia, transient receptor
potential ion channels 909–910
Hyperkalemic periodic paralysis, sodium
channels 841
Hyperplexia, glycine receptors 896
Hypertension treatment
see also specific drugs
β-adrenergic receptor
antagonists 254
Hypokalemic periodic paralysis, calcium
channels 852

I

Ibuprofen, structure 9F
Ibuproxam, carboxylic acid
bioisosterism 659
Idebenone 1074
structure 1074F
Ignarro, Louis, sildenafil citrate,
invention of 257
ILEX 763
Ilicicolin H
mitochondria complex III
1075
structure 1076F
Image analysis
bioinformatics 318
two-dimensional gel electrophoresis
see 2D-PAGE
Imaging, drug development
192
Imatinib mesylate
as multitarget drug 275
structure 71F
Imidazo[1,2-a]pyridines, ring equivalent
bioisosterism see Bioisosterism,
ring equivalents
5-Imidazole-thione,
bioisosterism 673T
5-Imidazolone, bioisosterism 673T
Imidazotriazines, phosphodiesterase-10
inhibitors 946

Index

Imipramine
  scaffold hopping 680, 681F
  structure 63F, 681F, 687F
Immunoassays
  see specific assays
  homogeneous time-resolved
    fluorescence (HTRF) see
    Homogeneous time-resolved
    fluorescence (HTRF)
Immunogenicity, reduction,
  PEGylation 588–589
Immunosuppressant drugs,
  glucocorticoid receptor 1006
Import/export considerations, offshore
  outsourcing 215
Improvement initiatives, drug discovery
  productivity 125, 125T
Impurities, development
  chemistry 167
Inclusion/exclusion criteria, clinical
  trials 186–187
Income differences, education
  85–86
Incretin
  analogs 583
  hydrophobic depoting 587
Indavir, structure 578F
Indazole, bioisosterism 660T
5-Indazole, bioisosterism 673T
Independence/initiative
  career guidelines 18
  drug discovery 272–273
India, market dominance 94
Indinavir sulfate 579T
Indirect enantioselective
  chromatography 721
Indisetron, 5-HT$_3$ receptor
  binding 890
Individual evaluation, project
  teams 146
Individuality, drug discovery 269, 272, 282
Indolactam V, solid support
  synthesis 480–482, 483F
Indole-2-carboxylate,
  bioisosterism 660T
Indoles, bioisosterism 660T
5-Indoles, bioisosterism 673T
Indolidan, structure 934
5-Indolone, bioisosterism 673T
Indomethacin, structure 9F, 658F, 681F
IND timeline and resources,
  DCN 155
Inducible nitric oxide synthase
  (iNOS) 1078
Industrialization, regulatory
  problems 725
Ineffective leaders see Leadership
Inequality, motivation 246
Infections, ATP-gated ion
  channels 905–907
Infectious diseases, target selection 76

'Infinite-cylinder' approximation,
  nanoprobes 517
Inflammation
  glucocorticoid receptor
    ligands 1016–1017
  nuclear receptors 1025
  phosphodiesterase-4 925–926
  transient receptor potential ion
    channels 908–909, 910
Inflammatory bowel disease (IBD),
  transient receptor potential ion
  channels 909–910
Information systems
  integration, HTS data
    acquisition 450
  pharmaceutical R&D 131
  project management 156
  R&D budget amount 131
  R&D management 250
Informed consent, clinical trials 187
Infrastructure, pharmaceutical
  R&D 130
Inhalation, peptide/protein drugs 595
Inherent variability, HTS data
  management 449
Inhibition curves, promiscuous ligands,
  identification 743
In-house value, drug discovery 279
Injection, peptide/protein drugs see
  Peptide/protein drugs, injection
In-licensing activities
  drug discovery 278–279
  increase in 255
Inner membrane
  mitochondria see Mitochondria
  permeability, mitochondria 1059
Innovation 3, 20, **253–264**
  basic research 253
  characteristics of 253, 259, 259T
    management difficulties 259
  corporate strategies 261, 262T
    basic research 261
    efficiency discreditation 263
    Eli Lilly 262
    focused individuals 262
    multi-field individuals 262
    recruitment 262
    'skunk works' 262
  decline of 254–255
    biotechnology companies 231
    definition 255
    unplanned 255
  discovering vs. 255
  drug development 174
  fostering of 257
  importance of 240
  improvement, drug
    development 174
  individual members 22
  industry failure to 254
    corporate goals 254
    corporate mergers 254–255

ICI 254
  marketing control 254
  marketing efficiency 254–255
  managerial strategies 260, 261T
    evaluation 260
    isolation 260–261
  problem generation 255
    failure readdressing 256–257
    problem recognition 256
  research strategies 22, 257, 258T
    basic science 257–258
    discarded research 259
    new approaches 258
    questioning daily practice 258
    quick/cheap/simple 258
  technology 21, 22T, 23T
Inorganic metal ions, calcium channel
  pharmacology 849
Inositol phosphate accumulation-IP$_3$
  assays see G protein-coupled
  receptors (GPCRs)
In-process controls (IPCs), common
  technical document
  (CTD) 163
Insecticides, mitochondria complex
  I 1068–1069
in silico methods,
  peptidomimetics 622–623
Insoluble/soluble polymers, solid
  support synthesis 477
INSR, PDB entries 965T
Insulin
  genetically engineered, invention
    of 262
  hydrophobic depoting 586
  inhalation administration 595
  secretion
    phosphodiesterase-1 923
    phosphodiesterase-3 924
Insulin-like growth factor-1 receptor
  (IGF-1R)
  in disease 963T
  PDB entries 965T
Insulin receptor kinase, three-
  dimensional structures
  636T
Integrin receptor antagonists,
  peptidomimetics 609–611, 612F
Intellectuality, effective leaders 241
Intellectual property (IP) 130
  business development 26
  composition of matter 130
  DCN 154
  drug discovery see Drug discovery
  generation, sourcing
    development 214
  loss, outsourcing risks 219
  methods of use 130
    drug discovery tools 130–131
    new uses 131

Intellectual property (IP) (*continued*)
  outsourcing 208
    protection 212
    risk 215
    scale-up 211
  pharmaceutical R&D 130
  slippage, definition 219
  target selection 758
Intention-to-treat
  clinical trials 185
  definition 30*T*
Intercalation, DNA therapeutic drugs 1040
Interdependence, project teams 145
Interfaces, microtechnologies 355
Interferon α2a, PEGylation 588
Interim analyses, clinical trials 185
Interleukin-6 (IL-6), PEGylation 588
Interleukin-8 receptors, in human disease 775*T*
Interleukin-converting enzyme (ICE), three-dimensional structures 626*T*
Interleukin-converting enzyme inhibitors 625
  peptidomimetics 613, 614*F*
  SAR 630
  x-ray crystallography 625, 633*F*
Intermediate/accelerated storage conditions, drug reference standard 168
Internalization, GPCRs 799
  desensitization 800–801
Internal staff activity value, outsourcing 208
International Conference on Harmonization (ICH)
  clinical study guidelines 104–105
  drug approval 159–160
International Union of Pharmacology (IUPHR), sodium channel nomenclature 831–832
International Union of Pure and Applied Chemistry (IUPAC)
  DNA nomenclature 1039
  drug nomenclature 161–162
Interpersonal skills, leadership 243
Interval scales, clinical trials 184
Intracellular calcium assays, GPCR functional assay screening *see* G protein-coupled receptors (GPCRs)
Intracellular loop, sodium channels 834–835
Intracellular regions, potassium channels 858
Intramolecular cyclization-resin-cleavage, solid support synthesis 470, 471*F*, 472*F*

Intranasal administration, peptide/protein drugs *see* Peptide/protein drugs
Introns, messenger RNA (mRNA) structure 1044
Invader, single nucleotide polymorphisms 366
Invention records, career guidelines 18
Investigational drug applications (IND), pharmaceutical R&D 104
Investment
  chemical *vs.* biological 117
  drug discovery 4, 6*F*
  NME *vs.* life cycle management 115
in vitro activity, DCN 154
in vitro metabolism, DCN 154
in vitro profiling, DCN 154
in vivo activity, DCN 154
in vivo profiling, DCN 154
(*S*-)5-Iodowillardine, structure 901*F*
Ion beam milling, lithography/photolithography 354
Ion channels, as targets 757*T*, 758
  natural products 758
Ion channels, ligand-gated **877–918**
  see also specific ion channels
  ATP (purinergic) gated family 879
    expression 879
    structure 880*F*
    characteristics 881*T*
  Cys-loop superfamily 879
    anion-conducting 879
    cation-conducting 879
    expression 879
    structure 879, 880*F*
    definition 878
    future work 912
    HTS, changes in 912
    transmission modulation 912
Ion channels, voltage-gated **827–875**
  see also specific types
  cell function regulation 828–831
  definition 827
  drugs acting on 829*T*
  future work 865, 865*F*
    automated planar chip-based electrophysiology 866
    subunit expression 865–866
  gating properties 828
  structure 827, 828*F*
  transmembrane segments 828
Ionic strength, promiscuous ligands, identification 740
IPG gels, isoelectric focusing 332
Ipratropium bromide, GPCR targeting 773*T*
Iproniazid, structure 68*F*
Irbesartan, GPCR targeting 773*T*
Iron-responsive element (IRE) 1045

Ischemia
  acid-sensitive ion channels (ASICs) 911
  uncoupling proteins 1080
Ischemia–reperfusion injury, mitochondria 1054–1055
ISIS, foundation 761
Isoelectric focusing (IEF), two-dimensional gel electrophoresis *see* 2D-PAGE
Isoguvacine, structure 661*F*
Isolation, innovation 260–261
'Isometric ballast,' chiral drug discovery 714–715
Isoquinolines, phosphodiesterase-7 inhibitors 944
Isotope-coded affinity tags (ICAT), protein expression mapping 385
Isotopic labels, NMR 512
Isotretinoin, characteristics 1012*T*
Isotropic etching *see* Lithography/photolithography
Isoxazoles, bioisosterism 662*T*
Ispronicline
  nicotinic acetylcholine receptor binding 888
  structure 887*F*
Istradipine, characteristics 829*T*
IUPHAR-approved scheme, calcium channel nomenclature 842, 843*T*

J

Janus kinase (JAK), signal transduction 576
Janus kinase-1 (JAK1) 963*T*
Janus kinase-2 (JAK2) 963*T*
Janus kinase-3 (JAK3), in disease 963*T*
JC-1, high-content screening (HCS) 429, 430*F*, 431*F*
JM4, GPCR interaction 797*T*
Job functions, materials management 533
Job security 273
Joint management team 152
Junctions, messenger RNA (mRNA) structure 1044

K

Kainate-sensitive receptors 898
  characteristics 881*T*
  diseases
    epilepsy 903
    neurodegeneration 903
  pharmacology 900, 901*F*
  physiological functions 899

Kallikrein, three-dimensional structures 626T
Kanamycin, structure 1041–1042
Karenitecin, clinical studies 697
KB-141
    structure 1020F
    as thyroid hormone receptor ligand 1019
K_{Ca} ion channel 853–854, 853F
    pharmacology 862
    physiological functions 860
    structure 859
KDR, PDB entries 965T
Key distinguishing features, DCN 154
Khellin 679
Kiely, John
    combinatorial chemistry, invention of 260
    management difficulties 260
Kilby, Jack, innovation 255
Kinetics
    HTS data analysis 452
    promiscuous ligands, identification 740
    proteins 326
K^+ ion channels 853
    cyclic-nucleotide gated 856–857
    genetic diseases 864
        deafness 864
        epilepsy 864–865
        LQT syndromes 864
        nervous system disorders 864–865
        neuromuscular ataxia 864
    HCN 853–854, 856–857
        pacemaker function 864
        pharmacology 863
        physiological functions 860
        structure 859
    molecular biology classification 857
        cellular distribution patterns 857
        excitability patterns 857
        expression 857
    nomenclature 853, 853F
    pharmacology 838T, 861
        allosteric interference 862
        pore-blocking mechanism 862
    physiological functions 859
        cell membrane excitability 859
        cellular distribution patterns 860
        epithelial cells 860–861
        lymphocytes 860–861
        membrane depolarization 860
    regulation 861
        second messengers 861
    structure 853, 858
        intracellular region 858
        membrane lipid association 858
        permeation pathway 858
        S4 domain 858–859
        T1 domain 858
    therapeutics 863
    see also specific drugs

airway disorders 863
autoimmune disease 863
cardiac disorders 863
neurology 864
oncology 864
Kit family kinases 969, 972
see also specific kinases
    in disease 963T
    gastrointestinal stromal tumors 972
    inhibitors 972
KN-62, structure 906F
Knockout animal models
    androgen receptor (AR) 1006
    estrogen receptors (ERs) 1005
    proteomics 389
Known drug modification, drug discovery 295–296, 296T
Krebs' cycle enzymes, mitochondria membranes 1057
KSR-1, GPCR interaction 786T
KT-5555 see CEP-701
K_v ion channel 853–854, 855T
    auxillary/accessory subunits 854

L

L86-8275 see Flavopiridol
L-1065041
    as PPAR-δ ligand 1022
    structure 1023F
LabChip system, microfluidic devices 359
Laboratory equipment treatment, career guidelines 18
Laboratory information management systems, ADMET 565
Labor cost reduction, outsourcing 208
β-Lactamase
    promiscuous ligands 740, 741F
        enzyme concentration 741
        as model system 742
        tetraiodophenolphthalein, inhibition by 740, 741F
Lamotrigine
    characteristics 829T
    structure 829T
Langmuir, Irving, management strategies 261
Lanreotide acetate 579T
    self-forming depots 593
Lansoprazole
    characteristics 653T
    structure 653T
Lanthanide cryptates see Homogeneous time-resolved fluorescence (HTRF)
Lapatinib see GW-572016
Large scale synthesis, stereoisomer synthesis 718
Laser ablation, lithography/ photolithography 354

Lasofoxifene
    cardiovascular effects 1014
    as estrogen receptor ligand 1014
    structure 1013F
Latanoprost, GPCR targeting 773T
Lck kinase
    PDB entries 965T
    SH2, three-dimensional structures 636T
    structure 976
LCM technology, cellular proteomics 385–386
Lead(s)
    optimization 140, 205
        activity interdependency 214
        ADME 139
        animal models 139
        attrition 205, 206F
        candidate seeking 205
        hit follow up 205
        HTS 400
        lead development 205
        lead optimization 205
        pharmacokinetics 139
        pharmacology 139
        process outcome 213
        subactivities 213
        toxicology 139
    quality 560
    screening 425
    throughput rate 560
Leadership **239–252**, 271–272
    becoming a leader 242, 243
        administrative responsibilities 243
        delegation 243
        interpersonal skills 243
    behavioral aspects 240
    collaborative research 240
    communication 243, 247
        criticism 243
        definition 247
        feedback 243
        nonjudgemental 247
        open-mindedness 247
    conflict management 242, 243, 248
        avoidance 242, 248
        compromise 248
        confrontation 249
        forcing 248
        smoothing 248
        sources of conflict 248
        superordinate goals 249
    diversity issues 244
        communication 244
        nationalities 244
    effective leaders 241
        intellectuality 241
        management skills 241
        role models 241
        working climate 242
    gender issues 244
        being ignored 245
        career obstacles 244

Leadership (continued)
  exclusion 245
  judgment of 244–245
  male preconceptions 245
  male stereotyping 245
 importance of 239
  economic effects 240
  innovation 240
  issues 240
  social effects 240
 ineffective leaders 240, 241
  impact of 241
  long-term effects 240
  working climate 242
 laboratory climate 242
  positive feedback 242
 motivation 243, 245
  effort vs. outcome 246
  equity 246
  fairness 246
  inequality 246
  micromanagement dangers 247
  reasonable working situation 246
  recruitment 246
  target setting 246
  technical competencies 246
 narrow training 240
 project team leader assessment 146
 R&D management 249
  approval systems 250
  decision making systems 250
  horizontal structure 249
  information systems 250
  performance appraisal 250
  problem definitions 249
  recruitment 250
  reward systems 250
  size 250
  structure 249
  systems 250
  vertical structure 249
 scientific training 240
  behavioral aspects 240
  collaborative research 240
  narrow training 240
  'soft' science perception 240
 'soft' science perception 240
Learning/memory, phosphodiesterase-2 923–924
Leber's hereditary optic neuropathy (LHON), mitochondria complex I 1071
Legislation
 drug discovery, decline in 4
 pharmaceutical R&D 123–124
Leigh syndrome
 mitochondria complex I 1071
 mitochondria complex II 1073
 mitochondria complex V 1079
Lenalidomide, structure 759F
Leptin, phosphodiesterase-3 interaction 924

Leukocyte elastase inhibitors, peptidomimetics/nonpeptide drugs 629, 632F
Leukotriene CysLTR receptors
 activation 785T
 constitutively active 790T
 in human disease 775T
Leuprolide 579T
 controlled release formulations 592
Levetiracetam
 characteristics 829T
 structure 829T
Levodopa
 mechanism of action, GPCR targeting 773T
 synthesis 716F
Levofloxacin, structure 729F
Levonorgestrel, structure 1015F
Levothyroxine, characteristics 1012T
LG1069
 as retinoic acid receptor ligand 1020–1021
 structure 1020F
LG100268
 as retinoic acid receptor ligand 1021
 structure 1020F
LG100754
 as retinoic acid receptor ligand 1021
 structure 1020F
LHON, mitochondria complex III 1075
Libraries 394
 construction
  combinatorial chemistry 39, 41F
  peptide synthesis 476
  solid support synthesis 464
 diversity 394
 drug discovery, productivity 124–125
 guidelines 395
 management 394–395
  software 395
 screening, drug discovery 297, 297T
 size 394
Licenses/licensing 230F, 231
 broad corporate alliances 234
 competition 234
 growth of 234
  compounds in development 234, 234F
 product proportion 235, 235F
Lidocaine
 characteristics 829T
 structure 829T
Ligand(s)
 peroxisome proliferator-activated receptor-γ (PPAR-γ) 1021–1022
 receptor protein tyrosine kinases (RPTKs) 576
Ligand-binding domain (LBD), glutamate-gated ion channels 898

Ligand-Gated Ion Channel Database 758
Ligand-gated ion channels see Ion channels, ligand-gated
Light scattering, promiscuous ligands see Promiscuous ligands, identification
Light transduction, phosphodiesterase-6 928
Linezolid
 mechanism of action, ribosome interactions 1043
 structure 1043F
Lion Biosciences, structure 232
Lipodystrophy syndrome, antivirals 1082
Lipophilicity
 carbon-silicon bioisosterism 694
 fluorine-hydrogen bioisosterism see Bioisosterism, fluorine–hydrogen
Liquid chromatography-NMR see Nuclear magnetic resonance (NMR)
Liraglutide, hydrophobic depoting 587
Lisinopril, ATP-gated ion channels 905–907
Literature work, career guidelines 18
Lithocholic acid, structure 1025F
Lithography/photolithography 349
 anisotropic etching 352
  solutions 352, 352T
 bulk micromatching 349, 350F
 deep reactive ion etching (DRIE) 353
 DNA microarrays 357
 etch stopping 352
 ion beam milling 354
 isotropic etching 350, 351F
  solutions 351, 351T
 laser ablation 354
 metal electrodeposition 353
  electroless process 353
 plasma etching 353
 polymer replication 354
 porous silicon 355
 surface micromachining 352, 353F
  digital micromirror devices 352
 ultraprecision machining 354
 ultraviolet LIGA 354
 x-ray LIGA 353, 354F
  drawback 305
  polymethyl methacrylate (PMMA) 305
Liver, phosphodiesterase-3 924
Liver receptor homologous protein 1, nomenclature 996T
Liver X receptor (LXR) 1010
 anti-inflammatory properties 1010
 nomenclature 996T
 peroxisome proliferator-activated receptor-γ (PPAR-γ) effects 1009

Index

pharmacology 1023, 1024F
  see also specific drugs
three dimensional structure, ligand-binding pocket 999–1000
Local anesthesia, sodium channels 839
Logistics, HTS 458
Long QT syndrome (LQTS)
  potassium channels 864
  sodium channels 841
Long-term memory, phosphodiesterase-4 926–927
Lopinavir 579T
  structure 578F
Loratadine, GPCR targeting 773T
Loreclezole, structure 893F
Losartan potassium, mechanism of action, GPCR targeting 773T
Lovastatin
  analog design 651
  structure 650F
Low-density lipoprotein receptor, stabilization, berberine 1048
Loxapine, structure 63F
LPA receptor, activation 785T
Luciferase, GPCR functional assay screening 812
Luteinizing hormone (LH) receptors
  activation 798
  human disease 792T
  in human disease 775T
LY 293558 903
LY 294486, structure 901F
LY 339434, structure 901F
LY 364947
  structure 979F
  TGFβR kinase inhibitors 977–979
LY 382884, structure 901F
LY 395153 900
LY 518674, structure 1022F
Lymphocytes, potassium channels 860–861
Lypressin 579T

M

Macrolide antibiotics
  mechanism of action, ribosomal RNA (rRNA) interactions 1042
  structure 1043F
Macrophages, peroxisome proliferator-activated receptor-δ (PPAR-δ) effects 1010
Macroporous resins *see* Resin beads, solid support synthesis
MAGI-2, GPCR interaction 797T
Magic angle spinning (MAS)
  nanoprobes 518–519
  NMR 506
Magnetic resonance imaging (MRI), drug development 192
'Major cleaning,' development chemistry 165–166

Major groove, DNA structure 1038–1039, 1039F
Malate dehydrogenase, promiscuous ligands, as model system 742
Male preconceptions, gender issues 245
Male stereotyping, gender issues 245
Malignant hyperthermia, calcium channels 852
Management
  career guidelines 19
  communication, project team leaders 144
  effective leaders 241
  materials management 536
Management teams *see* Project management
Manganese superoxide dismutase, mitochondria diversity 1066
Manufacturer details, common technical document (CTD) 163
Manufacturing
  biotechnology companies 102–103
  drug development 721–722
MAO-stilbene, mitochondria complex III 1075
MAPKAP2, PDB entries 965T
MAP kinase (MAPK)
  activation, GPCR000s *see* G protein-coupled receptors (GPCRs)
  three-dimensional structures 636T
MAP kinase-1 (MAPK1), PDB entries 965T
MAP kinase-10 (MAPK10), PDB entries 965T
MAP kinase-12 (MAPK12), PDB entries 965T
MAP kinase-14 (MAPK14), PDB entries 965T
Mapping processes, primary chemistry activity definition 209
Maprotiline, structure 687F
Marine organisms, HTS 397
Market (drug/products)
  commoditization, outsourcing 212
  pharmaceutical companies 100
  prescription number measurement 7
  US 7
Marketing
  control, innovation 254
  drug development costs 727
  efficiency, innovation 254–255
  science, replacement of 284
Market research, culture 122–123
Mass spectrometry (MS)
  drug development 194
  single nucleotide polymorphisms (SNPs) *see* Single nucleotide polymorphisms (SNPs)
Mass transfer, surface plasmon resonance 373–374

Masterplate preparation, materials management 535
Materials management 531, 534F
  definitions 531
    compounds 531
    natural products 531
    reagents 531
  development chemistry 160, 161
  follow-up screening 535
  HTS distribution 535
  job functions 533
  masterplate preparation 535
  new compound registration 534
  processes 533
  reagent receiving/tracking 534
  reagent sourcing 533
  role 531, 532F
  sample submission 534
  technologies 535
    automated sample storage/retrieval 537
    chemical structure-based 535
    electronic commerce 536
    ERP systems 536
    management systems 536
    robotic liquid-handling 536
    robotic weighing 536
    supplier catalogues 536
Matrices, matrix-assisted laser desorption/ionization (MALDI)-MS 375
Matrix-assisted laser desorption ionization (MALDI) 374
  development 374–375
  matrices 375
  purity issues 375–376
  spectrum 375, 375F
  surface-enhanced laser desorption ionization (SELDI) *vs.* 377–378
  technology 375
Matrix-assisted laser desorption/ionization (MALDI)-MS
  protein identification *see* Protein identification
  with surface plasmon resonance *see* Surface plasmon resonance (SPR)
Matrix-driven organizations *see* Project management
Matrix-driven projects, project management 12–13, 13F
Matrix metalloproteinase(s) (MMPs)
  inhibitors, carboxylic acid bioisosterism 659
  peptidomimetics 596
  three-dimensional structures 626T
Matrix metalloproteinase-3 (MMP3), GPCR interaction 797T
Maxacalcitol
  structure 1018F
  as vitamin D receptor ligand 1019
McCune–Albright syndrome, Gα$_3$ subfamily mutations 780–782
MDL 72222, structure 890F

MDM2 inhibitors 383, 384F
MDMA
  structure 65F
  transporter interactions 63
Mecamylamine
  nicotinic acetylcholine receptor binding 887–888
  structure 887F
Mechanization, biotechnology companies 231
Meclizine, GPCR targeting 773T
Medical history, clinical trials 187
Medical utility, regulatory problems 725
Medicinal chemistry
  discipline changes 88
  drug discovery 267
Mefloquine, structure 1062F
MEK1 kinase, posttranslational modifications 383
Melanin-concentrating hormone receptors, in human disease 775T
Melanocortin MC_{1a} receptors, in human disease 775T
Melanocortin MC_1 receptors, in human disease 775T
Melanocortin MC_3 receptors, in human disease 775T
Melanocortin MC_4 receptors, in human disease 775T
Melanophore platform, GPCR functional assay screening 813
α_1-Melanotropin receptor, binding 624, 624F
Melatonin receptors
  constitutively active 793T
  heterodimerization 799
MEM-3454, nicotinic acetylcholine receptor binding 888
Memantine, NMDA-sensitive receptors 902–903
Membrane, inner, mitochondria see Mitochondria
Membrane-bound semiquinone, mitochondria diversity 1065–1066
Membrane potential, mitochondria see Mitochondria
Membranes, mitochondria see Mitochondria
Memory acquisition, phosphodiesterase-9 930
Memory impairment, phosphodiesterase-4 inhibitors 935–936
Meperidine
  bioisosterism 675
  structure 677F
Mercaptoazoles, bioisosterism 662T
Mergers, pharmaceutical companies 101
Meribendan, structure 933–934

Merits, best in class vs. first in class 118–119
MERRF syndrome, mitochondria complex IV 1078
Merrifield, Bruce
  combinatorial chemistry 461
  solid support synthesis 464–465
Merrifield resins, resin beads, solid support synthesis 493–494
Mesopram, structure 935
Messenger RNA (mRNA) 1043
  antisense-based therapeutics 1045, 1046F
  mRNA caps 1047
  oligodeoxynucleotides 1045
  RNA interference 1045, 1046
  RNA maturation 1047
  RNase H based cleavage 1045, 1046F
  splicing redirection 1047
caps 1047
function 1037, 1043
HIV TAR 1044F, 1047
internal ribosomal entry site (IRES) 1047
  hepatitis C virus 1047
microRNA 1049
National Center for Biotechnology Information (NCBI) 1043
pharmacology 1045
  mRNA degradation 1045
  translation inhibition 1045
physiological functions 1044
  expression levels 1044–1045
  protein production 1044–1045
  untranslated regions (UTRs) 1044–1045
riboswitches 1048–1049
signal recognition particle (SNP) 1049
structure 1043
  5' cap structure 1044
  internal loops and bulges 1044
  introns 1044
  junctions 1044
  pseudoknots 1044
  RNA splicing 1044
  stem loops 1044, 1044F
  three-dimensional structure 1044
synthesis 1043
tmRNA 1049
MET, in disease 963T
Meta-analysis, definition 30T
Metabolic attacks, carbon-silicon bioisosterism 695
Metabolic syndrome, nuclear receptors 1025–1026
Metabolism/pharmacokinetics, development candidate nomination (DCN) see Development candidate nomination (DCN)
Metabolomics, drug development 197

Metabonomics 567
  definition 567–568
  drug development 197, 198F
Metabotropic glutamate receptor-1 (mGluR1), allosteric modulators 790T
Metabotropic glutamate receptor-1a (mGluR1a), activation 785T
Metabotropic glutamate receptor-2 (mGluR2) receptor, allosteric modulators 790T
Metabotropic glutamate receptor-4 (mGluR4) receptor, allosteric modulators 790T
Metabotropic glutamate receptor-5 (mGluR5), allosteric modulators 790T
Metal electrodeposition see Lithography/photolithography
Metalloproteinases, nonpeptide drugs 631
Metastasis, Src kinases 975–976
Metformin
  mechanism of action 1081
  mitochondria complex I 1081–1082
  structure 1081F
Methanesulfamide, phenol bioisosterism 670–671
Methiothepin, structure 1062F
Methods of use, intellectual property see Intellectual property
Methotrexate, structure 66F
2-Methoxybenze-[e]-hexahydo-isoindole, ring system reorganization (bioisosterism) 678–679, 679F
p-Methoxybenzene selenol, structure 701F
p-Methoxybenzyl selenocyanate, structure 701F
2' Methoxyethyl (MOE), oligodeoxynucleotide modification 1046
N-Methoxy imidoyl nitrile, carboxylic ester bioisosterism 663–665, 667F
(R)-2-Methyl-1,4-butanediol, structure 720F
(R)-1-Methyl-3-pyrrolidinol, structure 720F
1-Methyl-4-phenylpyridinium (MPP+), mitochondria complex I 1068–1069
1-Methyl-4-phenyl-4-propionoxy piperidine, bioisosterism 675, 677F
2-Methyl-5-hydroxytryptamine, structure 890F
N-Methyl amino acid scans, peptide/protein agonists 582
Methylation, nuclear receptors 1003

3-Methylchromone
  bioisosterism 680F
  ring system reorganization
    (bioisosterism) 679
  structure 680F
Methyllycaconitine, structure
  887F
2-Methylthio-ATPBzATP,
  structure 906F
Metoclopramide, GPCR
  targeting 773T
'Me-to drugs'
  bioisosterism 650
  definition 651
  drug development 174
Metoprolol succinate, GPCR
  targeting 773T
Mevalonic acid 1074
Mevinolin see Lovastatin
Mexiletine, characteristics
  829T
Mibefradil 850, 851
Microarrays
  applications
    ADMET 570–571
    drug development 177
    functional genomics 328
  flat, single nucleotide polymorphisms
    (SNPs) see Single nucleotide
    polymorphisms (SNPs)
Microcells, NMR 516F, 517
Microdose studies, positron emission
  tomography (PET) 193
Microfluidic devices 359, 360T
  in CD 361
    Gyros Microlabs 350F, 361
  flow cytometry 360
    Micronics 350F, 360–361
    HTS 403
    Oak Ridge National Laboratories
      system 359, 360T
  separations 359
    ACLARA Biosciences 350F,
      359
    Caliper Technologies 359,
      360T
    LabChip system 359
  synthesis 360
    Orchid Biocomputer 350F,
      360
  Ultra-High-Throughput Screening
    System (Aurora Biosciences) see
    Ultra-High-Throughput
    Screening System (Aurora
    Biosciences)
Micromanagement dangers,
  motivation 247
Micronics, microfluidic devices 350F,
  360–361
Microplate-based assays, single
  nucleotide polymorphisms
  (SNPs) see Single nucleotide
  polymorphisms (SNPs)

Microplates
  densities 459
  layouts 451
Microprobes, NMR 516F, 517
MicroRNAs
  messenger RNA (mRNA) 1049
  structure 1037
Microscience, characteristics 226T
Microtechnologies 348
  see also specific techniques
  applications 356
  construction 348
  field use/portability 348–349
  system integration 355
    detection systems 355
    fluidics/electronics
      integration 355
    interfaces 355
    material choice 355, 355T
  traditional methods 349
Midostaurin see PKC-412
Mifepristone
  progesterone receptor ligands 1015
  structure 1015F
Milestones, pharmaceutical R&D 104
Millennium Pharmaceuticals Inc
  characteristics 226T
  structure 231
Milnacipran, structure 62F
Milrinone
  derivatives, phosphodiesterase-1
    inhibitors 932
  heart failure 947–948
  as phosphodiesterase inhibitor
    947T
  structure 933–934
Minaprine, ring equivalent
  bioisosterism 656, 657F
Mineralocorticoid receptor (MR) 994,
  1007
  ligands 1017, 1018F
    see also specific drugs
  nomenclature 996T
Miniaturization
  high-throughput biology 436
    see also Ultra-High-Throughput
    Screening System (Aurora
    Biosciences)
  pharmaceutical productivity
    effects 132
Minocycline
  mechanism of action, ribosome
    interactions 1042–1043
  structure 1061F
'Minor cleaning,' development
  chemistry 165
Minor groove
  binders 1040
  DNA structure 1038–1039, 1039F
Minor metalloids, bioisosterism 694
Mirtazapine, mechanism of action,
  GPCR targeting 773T
Mitchell, Peter, Nobel prize 1059

Mithramycin, mechanism of
  action 1040
Mitochondria 1055, 1056F
  see also Mitoprotective drugs
  adenine nucleotide translocator
    (ANT) 1059
  antiviral drug effects 1082
    AZT (zidovudine) 1082
    highly active antiretroviral therapy
      (HAART) 1082
    nelfinavir 1082
    nucleoside reverse transcriptase
      inhibitors (NRTIs) 1082
    ritonavir 1082
  apoptosis 1054–1055
    mechanism of action 1055
  biology 1055
  biomass manipulation 1080
    pharmacology 1081
    transcriptional regulation 1080
  cytochrome c see Cytochrome c
  diseases/disorders 1054–1055, 1060
    see also specific diseases
    age-related macular
      degeneration 1054–1055
    Alzheimer's disease 1054–1055
    amyotrophic lateral
      sclerosis 1054–1055
    diabetes mellitus type II 1054–
      1055
    Friedreich's ataxia 1054–1055
    glaucoma 1054–1055
    Huntington's disease 1054–1055
    ischemia–reperfusion injury 1054–
      1055
    myocardial infarction 1054–1055
    Parkinson's disease 1054–1055
    pro/anti-apoptotic signals 1060
    rarity 1055
    retinitis pigmentosa 1054–1055
    stroke 1054–1055
  diversity 1065
    brain vs. liver as source 1065,
      1066F
    calcium toleration 1067
    electron paramagnetic resonance
      studies 1065
    manganese superoxide dismutase
      activity 1066
    membrane-bound
      semiquinone 1065–1066
    parasites vs. host 1067, 1067F
  drug development 1079
  electron transfer models 1057
    complex agglomeration 1057–
      1058
    'respiratosomes' 1057–1058
  electron transfer system
    (ETS) 1057, 1057F, 1068, 1070
    reaction sequence 1057F, 1059
  electron transport complexes 1057,
    1068
    see also specific types

Mitochondria (continued)
  free radical production 1057, 1060, 1064
    induction 1065
    mitochondrial permeability transition (MPT) see Mitochondrial permeability transition (MPT)
    superoxide anions 1065
  genome 1059
    homoplasty 1059–1060
    mutations 1059–1060
    structure 1059
  inner membrane 1059
    ATP generation 1059
    calcium binding 1059
    cardiolipin oxidation 1059
    glucose/amino acid/fat oxidation 1059
  membrane potential 1059
    inner membrane permeability 1059
    proton translocation 1058F, 1059
  membranes 1057
    cardiolipin 1057
    Krebs' cycle enzymes 1057
    phosphatidylcholine 1057
    phosphatidylethanolamine 1057
  Nernst equation 1058
    redox reaction equilibria 1058
  origins 1055–1056
  oxidative phosphorylation uncoupling see Oxidative phosphorylation uncoupling
  reticular structure 1056, 1056F
  ubiquinone see Ubiquinone (coenzyme Q)
Mitochondria, drug targeting 1067
  see also specific drugs
  chalcones 1068
  2′,4′-dihydroxy-6′methoxy-3′,5′-dimethylchalcone 1068, 1068F
  flavonoids 1068
  membrane potential 1067
    gene therapy 1067
    rhodacyanine dye MKT-077 1067, 1068F
    4-hydroxy-2,2,6,6-tetramethylpiperidin-N-oxide (TEMPOL) 1067
    tumors 1067
  2′,4,4′-trichloro-2′-hydroxydiphenylether 1068, 1069F
    as antibacterial 1068
    as antifungal 1068
    as antiviral 1068
    mitochondrial depolarization 1068
Mitochondria complex I 1057, 1068
  biology 1068
  catalysis 1070
    'accessory proteins' 1070–1071
    electron transfer route 1070

composition 1070
  bacteria 1070
  fungi 1070
  plants 1070
  structure 1070
drug development 1071
  cytoprotection assays 1072
  modifications 1072, 1072F
  rasagiline 1071, 1071F
inhibitors 1068–1069
  apoptosis initiation 1068–1069
  metformin 1081–1082
  NAPH analogs 1069, 1070F
  pioglitazone 1081–1082
  rosiglitazone 1081–1082
pathogenesis 1071
  familial amyotrophic lateral sclerosis 1071
  Leber's hereditary optic neuropathy (LHON) 1071
  Leigh syndrome 1071
  MELAS 1071
  Parkinson's disease 1071
Mitochondria complex II 1057, 1072
  biology 1072
  composition 1072
    crystal structure 1072
  free radical production 1065
  Huntington's disease model 1073
    3-nitropropionic acid 1073
  inhibitors 1072
    harzianopyridone 1072, 1073F
  pathology 1073
    familial paragangliomas 1073
    Leigh syndrome 1073
    mutations 1073
    pheochromocytomas 1073
Mitochondria complex III 1057, 1074
  biology 1074
  catalysis 1075
    'proton motive Q cycle' 1075
  composition 1074
  inhibitors 1075
  pathology 1075
    LHON 1075
Mitochondria complex IV 1057, 1077
  catalysis 1077
    water production 1077
  composition 1077
    crystal structure 1077
  inhibitors 1077
  oxidative regulation 1077
    nitric oxide 1077–1078
  regulation vs. pathology 1078
    MELAS 1078
    MERRF 1078
Mitochondria complex V 1078
  catalysis 1078
  composition 1078
  inhibitors 1079
    Bz-423 1079
    oligomycin 1079

oligomycin sensitivity-conferring protein (OSCP) 1079
pathology 1079
  Leigh syndrome 1079
  neurogenic muscle weakness, ataxia and retinitis pigmentosa (NARP) 1079
regulation 1078
  ADP availability 1078–1079
  F_1-inhibitor protein (IF-1) 1078–1079
Mitochondrial encephalopathy, lactic acidosis and stroke-like episodes syndrome (MELAS) 1060
  mitochondria complex I 1071
  mitochondria complex IV 1078
Mitochondrial permeability transition (MPT)
  adenine nucleotide translocator conformation shifts 1060
  ATP production as protection 1061
  calcium exposure 1060
  free radical production 1061
  mitoprotective drugs see Mitoprotective drugs
Mitoprotective drugs 1061
  see also specific drugs
  caffeic acid phenethyl ester (CAPE) 1064, 1064F
  celastrol 1064, 1064F
  mitochondrial permeability transition as target 1061
    5-(benzylsulfonyl)-4-bromo-2-methyl(2H)-pyridazione (BBMP) 1061, 1061F
  cytotoxicity 1063
  dibenz (b,f)azepine analogs 1062
  heterocycles 1062, 1062F
  Hoffman-La Roche drugs 1062, 1062F
  N-substituted phenothiazines 1062
  novel steroids 1063, 1063F
  SAR 1063
Mixture analysis, nuclear magnetic resonance (NMR) see Nuclear magnetic resonance (NMR)
MK-801, NMDA-sensitive receptors 902–903
MLN-518
  acute myeloid leukemia 971
  characteristics 984T
  KIT inhibitor 972
  structure 971F
Moclobemide, structure 68F
Modification analysis, bioisosterism see Bioisosterism
Modulated receptor hypothesis, sodium channels 839
Molecular biology
  definition 87
  paradigms 87

Molecular configuration, regulatory problems 726
Molecular diversity, protein display chips 370
Molecular evolution, proteins *see* Protein(s)
Molecular pathways, functional genomics *see* Functional genomics
Molecular size/weight
 drug development 730
 NMR 512
Molecular target classes, target *vs.* disease 112, 112F
Monamine oxidase inhibitors 67
Monoamine transporter antagonists
 carboxylic ester bioisosterism 663
 structure 62F
Monoamine transporters 60T
Monoclonal antibodies (mAbs) 47
 proteomics 315
Monomers and templates (M&Ts), outsourcing scale-up 211
Montelukast sodium, GPCR targeting 773T
Morbidity, phase IV clinical trials 189–190
Morphine, structure 36F
Morpholinochromes, phosphodiesterase-2 inhibitors 933
Motivation *see* Leadership
Mouse genes, positional cloning 306
Mouse models 313
MPP+, structure 1069F
MRAP, GPCR interaction 797T
MS-based methods, protein identification 337, 338F
Mucolipidosis type IV, transient receptor potential ion channels 909–910
Mullis, Kary
 management difficulties 259–260
 PCR invention 256
Multi-center trials, definition 30T
Multidimensional chromatography, proteomics 346–347
Multidimensional profiling, chemical *vs.* biological 117
Multidisciplinarity, outsourcing 229
Multidisciplinary competencies, pharmaceutical research and development (R&D) *see* Pharmaceutical research and development (R&D)
Multi-field individuals, innovation 262
Multiple RF channel, NMR 512, 513F
Multiple sites, organization, drug discovery *see* Organization, drug discovery

Multiple well dispensing, Ultra-High-Throughput Screening System (Aurora Biosciences) 444, 444F
Multiplexing, single nucleotide polymorphisms 363
Multiprotein thyroid-associated protein (TRAP), nuclear receptor coactivators 1002
Multistate activation model, GPCRs *see* G protein-coupled receptors (GPCRs)
Multi-use manufacturing equipment, cleaning 165
MUPP1, GPCR interaction 797T
Murad, Ferid, sildenafil citrate, invention of 257
Muraglitazar
 as PPARγ ligand 1021–1022
 structure 1023F
Muscarinic M_1 receptors
 activation 785T
 allosteric modulators 790T
 drug discovery 302
 in human disease 775T
 structure 798
Muscarinic M_2 receptors
 allosteric modulators 790T
 constitutively active 793T
 in human disease 775T
Muscarinic M_3 receptors
 activation 785T
 allosteric modulators 790T
 in human disease 775T
Muscarinic M_4 receptors, allosteric modulators 790T
Muscarinic receptor(s)
 constitutively active 793T
 ligands, carboxylic ester bioisosterism 663, 667F
Muscimol, structure 893F
Muscle cells, calcium channels 844
'Muscle-type' nicotinic acetylcholine receptors 879–880
MUSK, PDB entries 965T
Muskelin, GPCR interaction 797T
Mutagenesis
 models, proteomics 389
 phosphodiesterase-5 922
Mutations
 cytochrome c 1076
 mitochondria complex II 1073
 mitochondria genome 1059–1060
Myasthenia gravis, pathogenesis, nicotinic acetylcholine receptors 888
Myocardial infarction (MI)
 mitochondria 1054–1055
 mitochondria dysfunction 1060
Myoclonus epilepsy with ragged-red fibers (MERRF) 1060
Myxothiazol, mitochondria complex III 1075

N

NADH-CoQ oxidoreductase complex *see* Mitochondria complex I
Nafarelin 579T
 hydrophobic depoting 585
 intranasal administration 593–594
 self-forming depots 593
Nafuridin, structure 1067F
Nandrolone decanoate, structure 1016F
Nanoprobes, nuclear magnetic resonance (NMR) *see* Nuclear magnetic resonance (NMR)
Nanotechnology 46–47, 361
 antigen-antibody reactions 361
 definition 349
 DNA hybridization 361
 DNA sequencing 361
NAPH analogs, mitochondria complex I 1069, 1070F
National Cancer Institute (NCI), Diversity Set 749
National Center for Biotechnology Information (NCBI), messenger RNA (mRNA) 1043
National Center for Toxicogenomics 569
Nationalities, diversity issues 244
Native peptides, peptide/protein drugs 580
Natural analogs, peptide/protein drugs 580
Natural products
 drug discovery 296, 296T, 297
 HTS *see* High-throughput screening (HTS)
 materials management 531
 outsourcing 205
Neboglamine 903
'Need to know' *vs.* 'nice to know,' culture 121–122
Nefiracetam 903
Nelfinavir mesylate 579T
 mitochondria, effects on 1082
Neomycin, structure 1041–1042
Neoorthoforms, bioisosterism 674, 677F
Nernst equation, mitochondria *see* Mitochondria
Nervous system
 ATP-gated ion channels 904–905
 phosphodiesterase-4 926–927
Nervous system diseases/disorders, potassium channels 864–865
Neseritide 579T
Netoglitazone
 as PPARγ ligand 1021–1022
 structure 1023F
Neurodegenerative diseases
 AMPA-sensitive receptors 903
 kainate-sensitive receptors 903
 mitochondria dysfunction 1060

Neurogenic muscle weakness, ataxia and retinitis pigmentosa (NARP), mitochondria complex V 1079
Neurokinin A receptor antagonist, peptidomimetics 608–609, 611F
Neurokinin B receptor antagonist, peptidomimetics 608–609, 611F
Neuroleptic drugs
see also specific drugs
calcium-channel blockers 849–850
Neurology, potassium channels 864
Neuromodulators, peptide/protein drugs 580
Neuromuscular ataxia, potassium channels 864
Neuronal distribution, acid-sensitive ion channels (ASICs) 911
Neuronal hyperexcitability, calcium-channel blockers 851
Neuronal nicotinic receptors 879–880
Neuronal nitric oxide synthase (nNOS) 1078
Neuron-derived orphan receptor 1, nomenclature 996T
Neurons, calcium channels 844–845, 847–848
Neuropeptides, acid-sensitive ion channels (ASICs) 911
Neuroprotective drugs
see also specific drugs
calcium-channel blockers 849–850
oxidative phosphorylation uncoupling 1079–1080
Neuropsychiatric disorders, 5-HT$_3$ receptors 891
Neurotensin receptor antagonists
binding 623–624, 624F
nonpeptide drugs 618–619, 620F
Neurotoxin receptor site 1, sodium channels 837
Neurotransmitters
release, 5-HT$_3$ receptors 889
sodium channels 836
Neurotransmitter transporters 59, 60T
New compound registration, materials management 534
New diseases, target *vs.* disease 113, 114T
New drug applications (NDAs) 561, 561F
pharmaceutical R&D 104–105
New technology, outsourcing 208
NGF-induced factor B, nomenclature 996T
NGX-424 903
NH-3
structure 1020F
as thyroid hormone receptor ligand 1019
NHERF, GPCR interaction 797T
Nicardipine
characteristics 829T
structure 829T

Nicotine, structure 655F, 887F
Nicotinic acetylcholine receptors (nAchRs) 879
characteristics 881T
diseases 888
autosomal dominant nocturnal frontal lobe epilepsy (ADNFLE) 888
myasthenia gravis 888
distribution 884
molecular biology 884
autonomic synaptic transmission 884
CNS receptors 884
nomenclature 879
'muscle-type' 879–880
neuronal 879–880
pharmacology 887, 887F
physiological functions 886
regulation 886
cAMP response element binding protein (CREB) 886–887
SNARE 886–887
structure 885
accessory proteins 886
ACh binding site 885
muscle-type 885
neuronal 886
pentameric 885
subunits 880–884
TM3-TM4 intracellular loop 885–886
therapeutics 887
see also specific drugs
Nifedipine
characteristics 829T
structure 75F
Niguldipine
sila analog 697F
structure 697F
Nilutamide, structure 1016F
Nimesulide, structure 658F, 681F
Nimodipine
characteristics 829T
structure 829T
Nipecotic acid, structure 895F
Nisoldipine, characteristics 829T
Nitrendipine
characteristics 829T
structure 829T
Nitric oxide
mitochondria complex IV 1077–1078
production inhibition, selenium bioisosterism 703
O-Nitro-phenyl, bioisosterism 660T
3-Nitropropionic acid, mitochondria complex II 1073
N-Methyl D-aspartate (NMDA) receptor *see* NMDA receptors
Nm23-H2, GPCR interaction 797T
NMDA receptors 898

antagonists, phenol bioisosterism 671, 673T
characteristics 881T
cognition 902–903
pharmacology 900–902, 902F
physiological functions 899
NME *vs.* life cycle management *see* Philosophy (of drug discovery/development)
Nobel Prize, Black, Sir James 253
Nominal result scale, clinical trials 184
Nonalcoholic steatohepatitis (NASH), antivirals 1082
Nonnatural nucleosides, DNA therapeutic drugs 1040
Nonpeptide drugs
aspartyl proteases as target 625
cysteinyl proteases as target 629
GPCRs as target 623
metalloproteinases as targets 631
serinyl proteases as target 628
Src homology domains as targets 633
target structure-based drug design 622
therapeutic targets 605, 607T
see also specific targets
screening 618, 621F
as three-dimensional pharmacophores 619, 621F, 622F
limitations 621
tyrosine phosphatases as targets 638, 640F
Nonplanar sulfur-derived acidic functions, carboxylic acid bioisosterism 661, 662T, 669F
Nonrandomized uncontrolled trials 182
Nonsteroidal anti-inflammatory drugs (NSAIDs), markets/sales 8
Nontechnical activities, research projects 149
Norepinephrine, structure 57F
Norepinephrine transporter (NET) 59, 60T
Norethisterone
characteristics 1012T
structure 1015F
Norgestimate
characteristics 1012T
structure 1015F
Novartis, sales/R&D expenditure 101T
Novel amino acids, hydrophilic depoting 587, 588F
Novel property identification, phase IV clinical trials 190
NpxxY motif, GPCR structure 774–777
NRDOs (no research, development only) companies 103
NS 102, structure 901F
N-terminal blockade, proteolysis resistance 584

Index

N-terminal domain
  GPCR structure 774–777
  modification/truncation, peptide/
    protein antagonists 583
  nuclear receptors 995–997
  phosphodiesterase(s) 920, 922–923
N-type calcium channels 842
Nuclear magnetic resonance
    (NMR) 506
  automation 530
    throughput 530
  broadband decoupling 508
  chiral drug analysis 721
  combination experiments 506, 529
    'double-pulse field gradient spin
      echo (DPFGSE) 529
  DI-NMR 523
    advantages/disadvantages 525
    data processing 525, 526F, 527F,
      528F
    development 525
    FIA-NMR vs. 524
    hardware 524
  drug development 192
  FIA-NMR 523
    advantages/disadvantages 525
    DI-NMR vs. 524
    hardware 524
  flow NMR 521
  four-dimensional 507
  high-field 508
  indirect detection 510
    advantage 510–511
  liquid chromatography-NMR 506, 522
    advantages 522, 523, 528
    data presentation 523, 523F
    hardware 522
    history 522
    solvent composition 522–523
    solvent resonance suppression 522
  magic angle spinning (MAS) 506
  mixture analysis 526
    diffusion-encoded spectroscopy
      (DECODES) 527–528
    diffusion-ordered spectroscopy
      (DOSY) 527
    relaxation-edited
      spectroscopy 528
  multi (n)-dimensional 507–508
  nanoprobes 516F, 517, 518F
    advantages 518
    high-resolution-magic angle
      spinning (MAS) 519
    'infinite-cylinder'
      approximation 517
    magic angle spinning (MAS) 518–
      519
  one-dimensional 506–507
  probes 529
    developments 515
    heating 516
    microcells 516F, 517
    microprobes 516F, 517
    pulse widths 515
    RF homogeneity 515
    salt tolerance 515–516
    sample volumes 516, 516F
    sensitivity 515
  pulsed-field gradients 511
    applications 511
    solvent resonance suppression 511
  quantification 529
    internal standards 530
    research vs. analytical 530
  SAR 525
  shaped pulses 509
    biomolecule spectra 509
    frequency selective 509
    shift laminar pulse (SLP) 510
    solvent suppression 509
  software
    digital signal processing 529
  spin locks 508
    TOCSY 508–509
  SPS resins 519
    observed nucleus 520
    probe choice 520, 521F
    resin/solvent choice 520
  three-dimensional 507
  two dimensional 507
    direct (real-time) dimension 507
    homonuclear vs. heteronuclear
      coupling 507
    indirect dimension 507
Nuclear magnetic resonance (NMR),
    structure determination 512
  dipolar coupling 514
  hardware developments 515
  hydrogen bond J coupling 514
  isotopic labels 512
  molecular weights 512
  multiple RF channel 512, 513F
  techniques 512
  transverse relaxation-optimized
    spectroscopy (TROSY) 514
  water (solvent) suppression 514
Nuclear magnetic resonance
    spectroscopy
  drug development 194
  nuclear receptors 997
  peptidomimetics 622–623
Nuclear receptors 757T, **993–1036**
  see also specific types
  acetylation 1003
  class I 994, 995F
    physiological functions 1004,
      1012T
    regulation 1002
  class II 994, 995F
    physiological functions 1007, 1012T
    regulation 1002
  class III 994, 995F
  classification 994
  coactivators 1002
    ATP-dependent chromatin
      remodelers 1002
    multiprotein thyroid-associated
      protein (TRAP) 1002
    p160/steroid receptor
      coactivator 1002
    protein arginine methyltransferase 1
      (PRMT1) 1002
  coregulators 1002
  corepressors 1003
  diseases 1025
    cancer 1025
    genetic disease 1026
    inflammation 1025
    metabolic syndrome 1025–1026
    osteoporosis 1025
  future work 1028
  homogeneous time-resolved
    fluorescence 420
  ligand-binding domain (LBD) 998,
    998T
    dimerization sites 1001
    ligand absence 999
    ligand as structural cofactor
      1000
  methylation 1003
  nomenclature 995, 996T
  orphan 994, 995F, 996T
    as drug targets 994
  pharmacology 1011, 1012T
    see also specific drugs
  phosphorylation 1004
  physiological functions 1004
  structure 995, 997F
    AF-1 domain 1001
    central domain 995–997
    C-terminal domain 995–997
    DNA-binding domain 1002
    N-terminal domain 995–997
  three-dimensional structure 997,
    998T
    coregulator binding site 998T,
      999F, 1000
  NMR spectroscopy 997
  ubiquitination 1003
Nuclease-based assays, single
    nucleotide polymorphisms
    (SNPs) see Single nucleotide
    polymorphisms (SNPs)
Nucleic acids **1037–1052**
  see also DNA; RNA
  drug discovery 34
  as drug targets 1037
  hybridizations, homogeneous time-
    resolved fluorescence (HTRF) see
    Homogeneous time-resolved
    fluorescence (HTRF)
  structure 1037, 1038F
Nucleoside reverse transcriptase
    inhibitors (NRTIs),
    mitochondria, effects on 1082
Nucleotide identification, single
    nucleotide polymorphisms 363
Nur related factor 1,
    nomenclature 996T

O

Oak Ridge National Laboratories system, microfluidic devices 359, 360T
OAT *(SLC22)* transporters 59
  renal clearance 61
  structure 59, 61F
    bacterial mimetics 59–60, 61F
Objectiveness, culture 122–123
Occupational Safety and Health Administration, development chemistry timeline 170
Octreotide 579T
  development 581–582
  structure 581–582
OCT *(SLC22)* transporters 59
  structure 59, 61F
    bacterial mimetics 59–60, 61F
Offshore outsourcing
  definition 205, 206
  determination of 215
    communication 215
    existence of 216
    import/export considerations 215
    IP risk 215
    suitability 215
Olanzapine
  characteristics 653T
  mechanism of action, GPCR targeting 772, 773T
  structure 653T
Olefin metathesis, solid support synthesis 469, 470F
Olfactory signaling, phosphodiesterase-2 923–924
Oligodeoxynucleotides, antisense-based therapeutics 1045
Oligomers, solid support synthesis 475
Oligomycin, mitochondria complex V 1079
Oligomycin sensitivity-conferring protein (OSCP), mitochondria complex V 1079
Oligonucleotide drugs, chemical *vs.* biological 117–118
Oligonucleotides
  ligation, single nucleotide polymorphisms (SNPs) 365F, 366
  solid support synthesis 476, 477F
Oligosaccharides, solid support synthesis *see* Solid support synthesis
Olmesartan medoxomil, GPCR targeting 773T
Olopatadine hydrochloride, mechanism of action, GPCR targeting 773T
Olprinone, as phosphodiesterase inhibitor 947T

Omeprazole
  mechanism of action 63
  structure 65F
ON012380
  Bcr-Abl kinase inhibitor 973
  structure 974F
Onapristone
  progesterone receptor ligands 1015
  structure 1015F
Oncogenesis, receptor protein tyrosine kinases (RPTKs) 576
Oncology, potassium channels 864
Ondansetron
  5-HT$_3$ receptor binding 890
  nicotinic acetylcholine receptor binding 887–888
  structure 890F
Open-label trials, definition 30T
Open-mindedness, communication 247
Operational activities, research projects 149
Operations, project team leader assessment 146
Operator exposure, development chemistry 161
Opiate receptor antagonists, carboxylic ester bioisosterism 663
δ-Opiate receptor
  constitutively active 793T
  in human disease 775T
μ-Opiate receptor
  constitutively active 793T
  in human disease 775T
μ-Opiate receptor agonists, peptidomimetics 608–609, 611F
κ Opiate receptors (KOR), constitutively active 793T
Optical rotation, chiral drug analysis 720
Oral administration, peptide/protein drugs 595
Oral bioavailability, carboxylic ester bioisosterism 663–665, 667F
Oral drugs, ideal drug characteristics 564T
Oral/written reports, career guidelines 18
Orchid Biocomputer, microfluidic devices 350F, 360
Ordinal result scale, clinical trials 184
Organic chemistry, pharmaceutical R&D 105
Organization, drug discovery 2, 269
  biotechnology companies *vs.* pharmaceutical companies 20
  business development 25
    intellectual property 26
    partnerships 25

    product development transition 26
    U-shaped value curve 25–26, 25F, 26F
  multiple sites 19, 21F
    disadvantages 20
    GlaxoSmithKline 19
  portfolio management 25
  project management 12
    chairpersons 14
    decision points 16
    department-driven projects 12–13, 12F
    matrix-driven projects 12–13, 13F
    project assessment scheme 14
    self-contained projects 12–13, 13F
    Strengths, Weaknesses, Opportunities and Threat analysis (SWOT) 14
    target validation 15
    team members 14, 14T
  recruitment/human resource management 2, 16
    interviews 17
    management transition 17
Orphan Drug Act (1983) 10
  productivity effects 123–124
Orthoform
  bioisosterism 677F
  structure 677F
Orthosteric site, GPCR allosteric modulation 802
OSI-774 (erlotinib)
  characteristics 984T
  EGFR family kinase inhibitors 968
  structure 71F, 968F
Osteoporosis, nuclear receptors 1025
Outsourcing 20, **203–223**
  see also Contract research organizations (CROs); Service companies; *specific types*
  advantages 207, 227, 228T
    capability improvement 207
    commercial considerations 209
    efficiency 228, 229F
    financial fluctuations 207
    intellectual property (IP) generation 208
    internal staff activity value 208
    labor cost reduction 208
    multidisciplinary 228, 229
    new technology acquisition 208
    performance improvement 208
    resource lack 227–228
    risk minimization 208
    staffing flexibility 207
    talent pool diversity 207
    time saving 228, 229F
  analog production (singletons) 211
    compounds prepared per unit time 218
    IP protection 212
    performance assessment 218

problem-solving 212
  reactions per unit time 218
basic research 20
chemistry vs. biotechnology 204–205
compound resynthesis 211, 213
context 204
  natural products 205
disadvantages 227, 228T
  project management 228
history 205
internal performance vs. 219
parallel chemistry-enabled
    synthesis 205, 212
  market commoditization 212
  performance assessment 218
  value chain migration 212
performance assessment 217
  metric drawbacks 218
risks 219, 220
  communication 220
  cost differential 220
  IP loss 219
  organizational complexity 220
  publicity 220
  quality 219
  staff morale 219
scale-up 211
  cost 211
  established protocols 211
  IP 211
  monomers and templates
      (M&Ts) 211
  performance assessment 218
  precandidates 211
  time 211
Ownership, culture 122
1,2,4-Oxadiazole-5(4-H)-ones,
    bioisosterism 662T
1,2,4-Oxadiazole-5(4-H)-thiones,
    bioisosterism 662T
1,2,4-Oxadiazoles
  carboxamide bioisosterism 668,
      670T
  carboxylic ester bioisosterism
      663
1,3,4-Oxadiazoles, carboxamide
    bioisosterism 668, 670T
Oxametacin, carboxylic acid
    bioisosterism 659
Oxandrolone, structure 1016F
Oxapiperazines, solid support
    synthesis 482F
Oxazaborolidin, structure 699F
Oxazole rings, carboxamide
    bioisosterism 668, 670T
Oxazolidinones, mechanism of action,
    ribosome interactions 1042–
    1043
Oxcarbazepine
  characteristics 829T
  structure 829T
Oxidative phosphorylation
    uncoupling 1079

pharmacology 1079
  cerebral infarct 1079–1080
  neuroprotection 1079–1080
  weight-loss drugs 1079
uncoupling proteins 1080
  brain expression 1080
  glucose-stimulated insulin
      secretion 1080
  ischemia 1080
  Parkinson's disease
      treatment 1080
  uncoupling protein-1
      (thermogenin) 1080
Oxomaritidine, synthesis, solution
    phase synthesis 491–492, 492F
Oxybutynin, mechanism of action,
    GPCR targeting 773T
Oxycodone, mechanism of action,
    GPCR targeting 773T
Oxytetracycline, mechanism of action,
    ribosome interactions 1042–
    1043
Oxytocin, as peptide/protein
    drugs 579T, 580
Oxytocin receptors
  antagonists, nonpeptide drugs 618–
      619, 620F
  desensitization, GRKs 802
  in human disease 775T
  as targets 300

P

p85 SH2, three-dimensional
    structures 636T
p160/steroid receptor coactivator,
    nuclear receptor
    coactivators 1002
Packaging
  development chemistry 168
  materials, common technical
      document (CTD) 163
Paclitaxel, structure 28F, 1026F
Paget, Edward, management
    strategies 260–261
Pain, ion channels/receptors
  acid-sensitive ion channels
      (ASICs) 911
  ATP-gated ion channels 905–907
  glutamate-gated ion channels 900
  transient receptor potential ion
      channels 908–909
PAK1, PDB entries 965T
Palladium-mediated N-arylation, solid
    support synthesis
    470, 471F
Palmitoylation, G protein
    structure 780
Palonosetron, 5-HT$_3$ receptor
    binding 890
Papain, three-dimensional
    structures 626T

Papaverine, phosphodiesterase-10
    inhibitors 946
Parallel-arm trials, definition
    30T
Parallel chemistry-enabled synthesis,
    outsourcing see Outsourcing
Parallel development
  chemical vs. biological 117, 117F
  drug discovery productivity 125–126,
      126F, 127F
Parallel multiple-arm studies, clinical
    trials 181
Parallel research, ADMET 566,
    566F
Paramyotonia congenita, sodium
    channels 841
Parathyroid hormone (PTH)
  receptor, in human disease 775T,
      792T
  structural motif replacement 590
Pargyline, structure 1071F
Parkinson's disease
  disease basis
    mitochondria 1054–1055
    mitochondria complex I 1071
  treatment
    ubiquinone (coenzyme Q)
        1074
    uncoupling proteins 1080
Paromomycin, structure 1041–1042,
    1042F
Paroxetine, structure 62F
Partial least squares (PLS),
    promiscuous ligands,
    computational prediction
    746
Particle size
  drug approval 162–163
  light scattering, promiscuous
      ligands 743
Partnerships
  biotechnology companies
      102–103
  business development 25
  philosophy (of drug discovery/
      development) 123
Passerini reaction, solid support
    synthesis 472–474, 474F
Passion, project teams 145
Patents
  breadth of 131
  ease of 278
  exclusivity periods 133
  filing, target selection 755
  licensing 278
Patients
  clinical trials
    numbers 190, 191
    recruitment/enrollment 130
    identification/diagnosis, target
        selection 764
Pattern searches, HTS data
    management 453

Pavia, Mike
  combinatorial chemistry, invention of 260
  management difficulties 260
PD-157934
  SH2 inhibitor 983
  structure 983F
PD-166326, Bcr-Abl kinase inhibitor 973
PD-180970
  Bcr-Abl kinase inhibitor 973
  KIT inhibitor 972
  Src kinase inhibitor 976–977
  structure 971F, 978F
PD-183805
  EGFR family kinase inhibitors 968
  structure 968F
PD-184352
  characteristics 984T
  Raf-MEK-MAPK pathway kinase inhibitors 980
  structure 979F
PDE3 inhibitors
  see specific inhibitors
PED1 inhibitors
  see specific inhibitors
PED2 inhibitors
  see specific inhibitors
PEGA-poly (N,N-)dimethylacrylamide (PEGA), solid support synthesis 497, 499F
Pegvisomant 579T
PEGylation, peptide/protein drugs see Peptide/protein drugs
Penicillin G, structure 46F
Penicillopepsin, three-dimensional structures 626T
Pepsin, three-dimensional structures 626T
Peptide fingerprint searching, protein identification see Protein identification
Peptide hormones 573
Peptide maps, matrix-assisted laser desorption/ionization (MALDI)-MS 340
Peptide nucleic acids (PNAs), solid support synthesis 478, 479F
Peptide/protein drugs 35, 35T, 573–601
  agonism 580–581
    Ala scan 582
    azaamino acids 582
    critical residues 581
    cyclization 582
    β-diamino acids 582
    divergent side chains 582
    N-methyl amino acid scans 582
    size 581
    structure determination 581–582
  antagonism 580–581, 583
    N-terminus modification/truncation 583

  current usage 578, 579T
    hormones 580
    modifications 580
    native peptides 580
    natural analogs 580
    neuromodulators 580
    stability 580
  definition 574
  drug delivery 591
    active transport 591
    agonists 591–592
    inhalation 595
    injection see Peptide/protein drugs, injection
    oral 595
    paracellular 591
    transmembrane 591
  efficacy 580
    address/message region 581
  expression-modified proteins 591
    amber nonsense codon 591
  future work 596
    economics 596
  glomerular filtration resistance 585
  half-life increase 583
    see also specific methods
  hydrophilic depoting 585, 587
    duration of action 587–588
    novel amino acids 587, 588F
    water solubility 587
  hydrophobic depoting 585
    acylation 586
    albumin binding 586
    carrier proteins 585
    cell membrane binding 585
    plasma protein binding 586
    QSAR 586
  intranasal 593
    tight junctions 595
  PEGylation 585, 588
    glomerular filtration protection 588
    half-life effects 588, 589F
    immunogenicity reduction 588–589
    proteolysis protection 588
  peptidomimetics see Peptidomimetics
  protein conjugation 589
    Fc domains 590
    serum albumin 589
  proteolysis resistance 584, 584T
    azaamino acids 584–585
    beta amino acids 584, 585
    constrained peptides 584, 585
    C-terminal blockade 584
    D-amino acid substitution 584
    gamma amino acids 584
    N-terminal blockade 584
    peptoids 584
    unnatural amino acids 584
    unnatural peptide backbone 584–585
  structural motif replacement 590

  technology advances 573
  therapeutic targets 574, 605, 607T
    see also specific targets
Peptide/protein drugs, injection 592
  controlled release 592
    duration 592
    hydration effects 592
    lactic acid/glycolic acid polymers 592
    polymer matrix 592
    side chain urea functional groups 593
    soluble formulations 592
  self-forming depots 593
Peptides
  alkene substitution 604
  chemical diversity 604, 605F
  'chimeric' amino acids 604
  molecular flexibility 604, 606F
  as pharmacophores 604
  solid support synthesis see Solid support synthesis
  three-dimensional structure 604
  torsion angles 604
Peptidomimetics 35–36, 595, 603–647
  see also specific drugs
  classification 596
  definition 574
  design 606, 608F
  development obstacles 36
  future work 638
  protease substrates/inhibitors 613, 614F
  receptors as targets 608
    conformation elucidation 613
    cytokine receptors 596
    GPCRs 595–596, 605, 623
    three-dimensional pharmacophores 613
  signal transduction modulators 615, 617F
    Src homology domains as targets 633
  structure-based design 606
    amide bond replacements 606–608, 609F
    β-turn mimetics 608
    convergent pathways 608, 610F
    dipeptide surrogates 606–608
    γ-turn mimetics 608
    secondary structure mimetics 608, 610F
    unnatural amino acids 606–608
  substance P 595–596
  target structure-based drug design 622
    aspartyl proteases 625
    cysteinyl proteases 629
    matrix metalloproteinases 596
    NMR spectroscopy 622–623
    serinyl proteases 628
    in silico methods 622–623

Index

tyrosine phosphatases 638, 640F
x-ray crystallography 622–623
therapeutic targets 605, 607T
see also specific targets
Peptidyl-dipeptidase A, peptide/protein drugs 577
Peptoid-based antibiotics
see also specific drugs
discovery of 259
Peptoid drugs
see also specific drugs
development 36, 36F, 37F
proteolysis resistance 584
Percent inhibition, HTS data analysis
see High-throughput screening (HTS), data analysis
Percent of control, HTS data analysis
see High-throughput screening (HTS), data analysis
Performance assessment
outsourcing 218
scale-up 218
R&D management 250
Performance improvement, outsourcing 208
Periplakin, GPCR interaction 797T
Peroxisome proliferator-activated receptor(s) (PPARs) 1008
crystal structure, ligand-binding pocket 999–1000
nomenclature 996T
pharmacology 1021
see also specific drugs
Peroxisome proliferator-activated receptor-α (PPAR-α) 1008–1009
cellular distribution 1009
cholesterol removal 1009
ligands 1021
see also Fibrates
Peroxisome proliferator-activated receptor-δ (PPAR-δ) 1008–1009
cell proliferation 1010
expression 1009
ligands 1022
liver, effects on 1010
macrophage accumulation 1010
Peroxisome proliferator-activated receptor-γ (PPAR-γ) 1008–1009
cardiac expression 1009
cell expression 1009
homogeneous time-resolved fluorescence 420, 420F, 421F
ligands 1021–1022
LXR expression 1009
polymorphisms 1028
Personal interactions, career guidelines 19
Personalized medicine 55

Pfizer
collaborations 123
sales/R&D expenditure 101T
P-glycoprotein (P-gp), inhibitors 63
Phage display, peptide/protein drugs 576–577
PharmaBio, drug development 236
Pharmaceutical industry 100, **99–135**
see also specific types of company
biotech *vs.* drug discovery
extended enterprises *see* Extended enterprises
licenses *see* Licenses/licensing
service companies *see* Service companies
business model 99
company size 110
capital investment 111
functional discipline organization 109
productivity *vs.* 111, 111F
definition 100
diagnostics 100
functional discipline organization 108, 108F
collaboration 109
company size 109
diversity 108
drug development 108–109
drug discovery 108–109
therapeutic area decisions 109
growth rate 4–6
history 100–101
acquisitions 101
mergers 101
hybrid organization 109–110
global organization 109–110
marketplace-driven 100
organization 106
see also specific types
business unit 110
centers of excellence proximity 110
centralized *vs.* decentralized 110
size 110
value drivers 106–108
organizational aspects *see* Organization, drug discovery
philosophy of drug discovery/development *see* Philosophy (of drug discovery/development)
public view of 10
societal expectations 269
drugs delivered 270
financial productivity 270
sourcing development 214
sustainability 6–7
therapeutic area organization 108, 108F
diseases 108
drug discovery targets 108
therapeutic drugs 100

vaccines 100
veterinary products 100
Pharmaceutical industry *vs.* biotech, drug discovery **225–238**
collaboration 230–231, 230F
contracts 230, 230F, 230T
definitions 225
types 232, 233F
deal value 232, 233F
development stage 232
virtual companies 229
project management 229
'Pharmaceutical paradigm' 105, 107T
Pharmaceutical research and development (R&D) 104
costs/expenditure 100, 101T
drug discovery 270
increase in spending 255
multidisciplinary competencies 105, 106T
ADME 105
biology 105
organic chemistry 105
pharmacodynamics 105
pharmaceutical paradigm 105, 107T
process 104F
clinical studies 104–105
investigational drug applications (IND) 104
milestones 104
new drug applications (NDAs) 104–105
phase duration 104–105, 104F
productivity 111F, 123
costs 123
information technology 131
infrastructure 130
intellectual property 130
see also Intellectual property
legislation 123–124
regulatory issues 131
value per drug 124
technology effects 132
Pharmacodynamics 105
Pharmacogenetics, target identification *see* Target(s)
Pharmacogenomics
definition 381, 382T
drug discovery/development 196, 568
pharmaceutical productivity effects 132–133
target identification *see* Target(s)
Pharmacokinetics
assessment, positron emission tomography (PET) 193
drug development productivity 130
lead optimization 139
preclinical trials 178
target selection 76

Pharmacology
  development candidate nomination (DCN) see Development candidate nomination (DCN)
  drug discovery 267, 276
  lead optimization 139
  studies, preclinical trials 177, 178T
Pharmacopeia 760T
Pharmacophores
  drug development 176
  nonpeptide drugs see Nonpeptide drugs
  peptides 604
  searching, scaffold hopping 681–683, 683F
  three-dimensional descriptors, receptors as targets 613
Pharmacoproteomics 390, 390F
  definition 381, 382T
Phase 0 clinical trials see Clinical trials, phase 0
Phase I clinical trials see Clinical trials, phase I
Phase II clinical trials see Clinical trials, phase II
Phase III clinical trials see Clinical trials, phase III
Phase IV clinical trials see Clinical trials, phase IV (postmarketing)
Phase duration, pharmaceutical R&D 104–105, 104F
Phenazine, hepatitis C virus inhibition 1047
Phenbanazamine, structure 654F
Phenelzine, structure 68F
1,4-Phenenebis-(methylene) selenocyanate, structure 701F
4-Phenol, bioisosterism 673T
N-Phenothiazines, mitoprotective drugs 1062
Phenotype, definition 382T
Phenylalkylamines, as calcium channel blocker 849–850
Phenylimidazole, carboxamide bioisosterism 668, 670T
Phenytoin
  characteristics 829T
  structure 829T
Pheochromocytomas, mitochondria complex II 1073
Philadelphia chromosome, Bcr-Abl kinases 972
Philosophy (of drug discovery/development) 111
  best in class vs. first in class 118, 118F, 119F
    assessment responsibilities 119–120
    development process effects 119
    differentiation 119, 120T
    merits 118–119
  chemical vs. biological 115
    antibody responses 117
    biological production 117

gene therapy 117–118
investment 117
multidimensional profiling 117
oligonucleotide drugs 117–118
parallel development 117, 117F
route of administration 117
sulfanilamides 115, 116F
culture 121
  collaborative vs. insular 121–122
  competitive collaborations 122
  entrepreneurism 123
  market research 122–123
  'need to know' vs. 'nice to know' 121–122
  objectiveness 122–123
  ownership 122
  transparency 122–123
decision making vs. organizational alignment 120
  small vs. large organizations 120
  speed definitions 120
  transparency 120
fail fast vs. 'follow the science' 121
  ADME 121
  budget constraints 121
  predictive screens 121
NME vs. life cycle management 115
  investments 115
partnering 123
target vs. disease 112
  disease etiology 112
  'druggable genome' 113, 113F
  'druggable target' definition 112, 112F
  drug mechanism of action 112
  HTS 114–115
  molecular target classes 112, 112F
  new diseases 113, 114T
  structure-based drug design (SBDD) 114–115
  targets by therapeutic area 114–115, 115T
PHKG1, PDB entries 965T
Phosphatidylcholine (PC), mitochondria membranes 1057
Phosphatidylethanolamine (PE), mitochondria membranes 1057
Phosphatidylinositol-3-kinase (PI3K), GPCR interaction 786T
Phosphinates, bioisosterism 662T
Phosphodiesterase(s) **919–957**
see also specific enzymes
biological activity 920, 923
cAMP levels 920
cGMP levels 920
classification 920, 921T
  class I 920
  class II 920
future work 950
genes/molecular biology 920, 921T
genetic diseases 950

inhibitors
  see also specific inhibitors
  multiple idiotype inhibition 950
  selectivity 950
nomenclature 920
pharmacology 931, 947, 947T
reaction 920
regulation 923
  competition 923
  extracellular stimuli 923
  phosphorylation 923
  second messengers 931
splice variants 920
structure 920, 922F
  C-terminal catalytic domain 920–922
  N-terminal regulatory domain 920, 922–923
  substrate specificity 922
subcellular localization 922–923
substrate specificity 920
as targets 757T
target selection 73
Phosphodiesterase-1 923
  genetics 921T, 923
  insulin secretion 923
  regulation, calcium by 923
  splice variants 923
  structure 922F
    C-terminal catalytic domain 920–922
    substrate specificity 921T, 923
  as therapeutic target 950–951
Phosphodiesterase-1 inhibitors 931
  see also specific drugs
  milrinone derivatives 932
  tetracyclic guanine derivatives 931–932
  therapeutic applications 947
  zaprinast derivatives 932
Phosphodiesterase-2 923
  atrial naturietic peptide interaction 923–924
  conformational changes 923–924
  genetics 921T, 923–924
  learning/memory 923–924
  olfactory signaling 923–924
  platelet aggregation 923–924
  splice variants 923–924
  structure 922F
    crystal structure 933
    C-terminal catalytic domain 920–922
  subcellular localization 923–924
  substrate specificity 921T
  as therapeutic target 950–951
  tissue expression 923–924
Phosphodiesterase-2 inhibitors 933
  see also specific drugs
  benzodiazepines 933
  morpholinochromes 933
  pyrido[2,3-d]pyrimidine derivatives 933

therapeutic applications 947
tricyclic oxindoles 933
Phosphodiesterase-3 924
  genetics 921T, 924
  insulin secretion 924
  leptin interaction 924
  phosphorylation 924
  platelet function 924
  structure 922F
    crystal structure 934–935
    C-terminal catalytic domain 920–922
  substrate specificity 921T, 924
  as therapeutic target 950–951
  tissue distribution 924
    adipocytes 924
    cardiac tissue 924
    liver 924
    platelets 924
    smooth muscle 924
Phosphodiesterase-3 inhibitors 933
  see also specific drugs
  structural requirements 934–935
  therapeutic applications 947
    dilated cardiomyopathy 947–948
Phosphodiesterase-4 924
  β-arrestin association 926
  β$_2$-adrenergic receptors 926
  CNS disorders 925–926
  conformations 925
  emesis/nausea 925
  genetics 921T, 924–925
  inflammation 925–926
  intracellular distribution 926
  long-term memory 926–927
  membrane-bound 925
  phosphorylation 925
  point mutation 925–926
  rolipram affinity 925–926
  soluble form 925
  splice variants 924–925
  structure 922F
  substrate specificity 921T, 924–925
  as therapeutic target 950
  tissue distribution 926
    CNS 926–927
    heart 926–927
    nervous system 926–927
    smooth muscle 926–927
  upstream conserved regions (UCRs) 924–925
Phosphodiesterase-4 inhibitors 935
  see also specific drugs
  amino diazepinoindole 937–938
  catechol derivatives 936
  high-throughput x-ray crystallography 938
  phthalazinone derivatives 936
  pyridylisoquinolines 936
  scaffold-based design 938
  therapeutic applications 948
    anticancer drugs 948
    asthma treatment 935–936, 948

chronic obstructive pulmonary disease 935–936, 948
    emesis 936
    memory impairment 935–936
    psoriasis 935–936
    ulcerative colitis 935–936
Phosphodiesterase-5 927
  atrial natriuretic peptide 928
  conformational chains 927
  genetics 921T
  mutagenesis 922
  phosphorylation 927
  point mutations 927
  regulatory domain 927
    GAF domains 927
  splice variants 927
  structure 922F
    crystal structure 942
  substrate specificity 921T, 922, 927
  as therapeutic target 950
  vascular tone regulator 928
Phosphodiesterase-5 inhibitors 938
  see also specific drugs
  carboxamide 940
  pyrazolopyridines 939–940
  pyrazolopyridopyridazine 941
  pyrazolopyrimidines 940
  pyrroloquinolines 941–942
  quinazoline derivatives 939
  sildenafil citrate, invention of 257
  therapeutic applications 948
    aspirin-induced gastric damage 949
    benign prostatic hypertrophy 948
    chronic lymphocytic leukemia 949
    cognitive disorders 949
    diabetes insipidus 949
    erectile dysfunction 948
    smoking-induced pulmonary endothelial dysfunction 948–949
Phosphodiesterase-6 928
  in eye 928
  genetics 921T, 928
  in light transduction 928
  posttranslational modification 928
  retinitis pigmentosum 928, 950
  structure 922F
    cGMP allosteric binding sites 928
    regulatory domain 928
  substrate specificity 921T
Phosphodiesterase-6 inhibitors 942
  see also specific drugs
  sildenafil 942–943
  therapeutic applications 949
Phosphodiesterase-7 929
  B cells 929
  chronic lymphocytic leukemia 929
  chronic obstructive lung disease 929
  genetics 921T, 929
  splice variants 929
  structure 922F
  substrate specificity 921T

T-cell proliferation 929
  as therapeutic target 950–951
  tissue distribution 929
Phosphodiesterase-7 inhibitors 943
  see also specific drugs
  benzothienol[3,2-a]thiadiazine 2,2-dioxides 943
  isoquinolines 944
  phthalazinones 944
  pyrazolopyrimidinones 944
  spiroquinazolinones 943–944
  therapeutic applications 949
  thiadiazoles 944
Phosphodiesterase-8 929
  genetics 921T, 929–930
  splice variants 929–930
  structure 922F
  substrate specificity 921T
  thyroid gland 930
  tissue expression 929–930
Phosphodiesterase-8 inhibitors 945
  see also specific drugs
  therapeutic applications 949
    Alzheimer's disease 949
Phosphodiesterase-9 930
  genetics 921T
  memory acquisition 930
  splice variants 930
  structure 922F
    crystal structure 945–946
  substrate specificity 921T
  as therapeutic target 950–951
  tissue expression 930
Phosphodiesterase-9 inhibitors 945
  see also specific drugs
  therapeutic applications 950
    glucose homeostasis 950
Phosphodiesterase-10 930
  CNS diseases 930–931
  genetics 921T, 930
  Huntington's disease 930–931
  inhibitor sensitivity 930
  schizophrenia 930–931
  splice variants 930
  structure 922F
  substrate specificity 921T
  as therapeutic target 950–951
  tissue expression 930
Phosphodiesterase-10 inhibitors 946
  see also specific drugs
  dihydroisoquinolines 946
  imidazotriazines 946
  papaverine 946
  therapeutic applications 950
    glucose homeostasis 950
Phosphodiesterase-11 931
  genetics 921T
  isoenzymes 931
  sperm characteristics 931
  structure 922F
  substrate specificity 921T
  tissue expression 931

Phosphodiesterase-11 inhibitors 946
see also specific drugs
  dipyridamole 931
  therapeutic applications 950
  zaprinast 931
Phosphodiesters, structure 35F
Phosphoprotein-interacting
    domains 961T
  genomic relationships 963T
Phosphorothioate, structure 35F
Phosphorus-derived acidic functions,
    carboxylic acid bioisosterism see
    Bioisosterism, carboxylic acids
Phosphorylation
  GPCR desensitization 800, 801
  nuclear receptors 1004
  phosphodiesterase-3 924
  phosphodiesterase-4 925
  phosphodiesterase-5 927
  phosphodiesterase regulation 923
  sodium channels 836
Phosphoserine/phosphothreonine-
    interacting domains 985
  forkhead-associated domains 985
Phosphotyrosine-interacting
    domains 983
  SH2 domain inhibitors 983, 983F
Photodynamic treatment (PDT),
    selenium bioisosterism 702T,
    703
Photolithographic masks
  DNA microarrays see DNA microarrays
  solid supports 503, 504F
Photoreceptor-specific nuclear receptor,
    nomenclature 996T
Phthalazinones
  phosphodiesterase-4 inhibitors 936
  phosphodiesterase-7 inhibitors 944
Physical facility, cleaning 165
Physicochemistry, DCN 155
Physiologically based pharmacokinetics
    (PBPK), chemogenomics 567, 568F
Phytoestrogens, estrogen receptor
    ligands 1014
PI-3K–Akt/PKB–aTOR pathway
    kinases 980
  see also specific kinases
  inhibitors 980–981, 981F
PICK, GPCR interaction 797T
Picornaviruses, 3C protease
    structure 626T
Picrotoxin, 5-HT$_3$ receptors 890
Piericidin A, structure 1069F
PIK3CG, PDB entries 965T
Pins see Solid supports
Pioglitazone hydrochloride
  characteristics 1012T
  mechanism of action 1081
    mitochondria complex I 1081–1082
    mitochondrial biogenesis 1082

  as PPARγ ligand 1021–1022
  structure 1023F, 1081F
Pipeline phases, drug discovery 138, 138F
Piroxicam, structure 9F
Pituitary adenylate cyclase activating
    peptide (PACAP), as peptide/
    protein drug 583
Pivoted data, HTS data analysis see
    High-throughput screening
    (HTS), data analysis
pka/logP values, carboxylic acid
    bioisosterism 661, 666F
PKC-412 971–972
  acute myeloid leukemia 971
  characteristics 984T
  KIT inhibitor 972
  structure 971F
PKI-166
  characteristics 984T
  EGFR family kinase inhibitors 968
  structure 968F
Placebo-controlled studies 181
Planar acidic heterocycles, carboxylic
    acid bioisosterism see
    Bioisosterism, carboxylic acids
Plants
  HTS 397, 398
  mitochondria complex I 1070
Plasma etching, lithography/
    photolithography 353
Plasma protein binding, hydrophobic
    depoting 586
Plate design/performance, Ultra-High-
    Throughput Screening System
    (Aurora Biosciences) see Ultra-
    High-Throughput Screening
    System (Aurora Biosciences)
Plate handling, HTS see High-
    throughput screening (HTS)
Platelet-derived growth factor receptor-
    α (PDGFR-α), in disease 963T
Platelet-derived growth factors
    receptor-β (PDGFR-β?, in
    disease 963T
Platelets
  aggregation
    cilostazol 947–948
    phosphodiesterase-2 923–924
  function 924
  phosphodiesterase-3 924
PLCβ, GPCR interaction 786T
PLCε, GPCR interaction 786T
PLCγ SH2, three-dimensional
    structures 636T
Plexxikon 761T
PMI-50 903
PNU-120596, structure 887F
PNU-282987, structure 887F
Point mutations
  phosphodiesterase-4 925–926
  phosphodiesterase-5 927
Polarimetry, chiral drug analysis 721

Polyamine transporters, clinical use 61
Poly(ethylene glycol)-containing resins
    see Resin beads, solid support
    synthesis
Polycystic kidney disease, transient
    receptor potential ion
    channels 909–910
Polymerase chain reaction (PCR)
  invention of 261
    Cetus Corporation 256
    Mullis, Kary 256
  single nucleotide
    polymorphisms 363, 369
Polymer-assisted solution phase
    synthesis (PASP), solution phase
    synthesis 477F, 489
Polymer-bound ^1H-benzotriazole,
    solution phase synthesis 490, 490F
Polymer replication, lithography/
    photolithography 354
Polymer-supported synthesis, solution
    phase synthesis see Solution
    phase synthesis
Polymethyl methacrylate (PMMA), x-
    ray LIGA 305
Polymorphic forms, drug
    approval 161–162
Polystyrene-PEG, solid support
    synthesis 496, 496F, 497T
Pore-blocking mechanism, potassium
    channels 862
Porous silicon, lithography/
    photolithography 355
Portfolio management, organization,
    drug discovery 25
'Portioning–mixing,' combinatorial
    chemistry 461–462
Positional cloning, target identification
    see Target identification,
    genomics
Positron emission tomography (PET)
  applications
    drug development 192
    transporters 65
  combination with other imaging
    technology 193
  mechanism of action 192–193
  microdose studies 193
  pharmacokinetic studies 193
Posttranslational modifications
  definition 381
  as drug targets 383
  pathogenesis associated 381
  phosphodiesterase-6 928
  protein display chips 370
PP-1
  Bcr-Abl kinase inhibitor 973
  Src kinase inhibitor 976–977
  structure 974F, 978F
  three-dimensional structures 636T
Practolol, structure 677F

Pragmatic innovation, project teams 145
Pramlintide 579*T*
  amylin analog 580
Pravastatin sodium, structure 66*F*
Pre-approval costs, drug discovery 4, 6*F*
Precandidates, outsourcing scale-up 211
Preclinical research 52
  bottlenecks 3
  central nervous system disease 52
  decision making 70
  safety evaluation, drug discovery productivity 125
  solid-phase chemistry 52
Preclinical trials 177
  adverse event prediction 179–180, 180*F*
  definition 177
  documentation 178–179
  dosage 178, 179*F*
  models 177
  pharmacokinetics 178
  pharmacological studies 177, 178*T*
  'rule-of-five' 178
  toxicology studies 178, 179*T*
Predevelopment, drug discovery *see* Drug discovery
Predictive screens, fail fast *vs.* 'follow the science' 121
Prednisolone, structure 1017*F*
5α-Pregnane, structure 1015*F*
Pregnane X receptor (PXR) 1011
  nomenclature 996*T*
  pharmacology 1024
  see also specific drugs
Pregnenolone, structure 895*F*
Pregnenolone carbonitrile (PCN)
  as nuclear receptor ligand 1010
  structure 1026*F*
Preincubation, promiscuous ligands, identification 742
Prescription Drug User Free Act (1992) (PDUFA), productivity effects 123–124
Prescription number, market measurement 7
Previous experience, project team leaders 142
P-Rex1, GPCR interaction 786*T*
Pricing, drug development costs 727, 727*F*
Primary chemistry activity definition, sourcing *see* Sourcing development
Primary erythermalgia, sodium channels 841
Pritchard, Brian, propranolol, invention of 254
PRKACA, PDB entries 965*T*
Pro/anti-apoptotic signals, mitochondrial diseases 1060

Probes, NMR *see* Nuclear magnetic resonance (NMR)
Problem definitions, R&D management 249
Problem recognition, innovation 256
Problem-solving, outsourcing 212
Procainamide
  characteristics 829*T*
  structure 829*T*
Process hazard analysis (PHA), development chemistry timeline 170
Process improvements, drug development costs 728
Process maturity, sourcing development 213
Process outcome control, sourcing development 213
Process safety management (PSM), development chemistry timeline 170
Product development transition, business development 26
Productivity
  drug development *see* Drug development
  drug discovery *see* Drug discovery
  pharmaceutical research and development (R&D) *see* Pharmaceutical research and development (R&D)
Product profile, development candidate nomination (DCN) *see* Development candidate nomination (DCN)
PROfusion technology, proteomics 315
Progesterone, structure 895*F*, 1015*F*
Progesterone receptor (PR) 994, 1005
  knockout animal models 1006
  nomenclature 996*T*
Progesterone receptor ligands 1014, 1015*F*
  see also specific drugs
  selective progesterone receptor modulators (SPRMs) 1015
Project assessment scheme, project management 14
Project commitment, project teams 145
Project decision points, project management teams 149
Project goals, project management 142
Project leaders, project manager *vs.* 149, 151*T*
Project management **137–158**, 149
  see also Organization, drug discovery; Project teams
  collaborative projects *see* Collaborations, project management

  into development 152
  see also Development candidate nomination (DCN)
  project team membership 155
  functional areas 149
  information technology 156
  management teams 147
    composition 147
    formal project reviews 148
    project decision points 149
    project team meetings 148
    responsibilities 147
    team member feedback 149
    written project reports 148
  matrix-driven organizations 146
    advantages 146
    disadvantages 146–147
  outsourcing 228
  pharma *vs.* biotech, drug discovery 229
  project goals 142
  research projects 149
    nontechnical activities 149
    operational activities 149
    project manager 149
    roles and responsibilities 149, 150*T*
  roles/responsibilities 150*T*
Project managers 149
  project leader *vs.* 149, 151*T*
Project team leaders 142
  assessment 146
  communication 146
  operations 146
  skills 146
  teamwork 146
  effective leadership 142
  interpersonal skills 144
    communication 142–143
  previous experience 142
  qualities 142
  roles/responsibilities 144, 150*T*
    external event monitoring 144
    management communication 144
    team development 144
    team motivation 144
  transitioning 144
Project team meetings, project management teams 148
Project teams 138, 145
  changes 138
  core team definition 141
  DCN 155
  definition 144
  development, project team leaders 144
  functional area goals 142
    development candidate nomination deadlines 142
    IND-enabling studies 142
  functional departments *vs.* 145
  high-performing 146

Project teams (*continued*)
  individual goals 142
    written 142
  leader *see* Project team leaders
  management, communication to 141
  meetings 141
  members
    feedback 149
    project management 14, 14T
  motivation, project team leaders 144
  productive 145
    authority 145
    clarity 145
    commitment 145
    diversity 145
    effective communication 145
    empowerment 145
    enjoyment 145
    flexibility 145
    honesty 145
    interdependence 145
    passion 145
    pragmatic innovation 145
    project commitment 145
    reliability 145
    responsibility 145
    strategic thinking 145
    supportiveness 145
    thoughtful responsiveness 145
  project goals 142
    timelines 141
  responsibilities 141
  rewards 146
    assessment 146
    individual evaluation 146
Prokaryotes, RNA polymerases 1043
Proline *cis-trans* isomerases, three-dimensional structures 636T
Promethazine
  GPCR targeting 773T
  scaffold hopping 680, 681F
  structure 681F
Promiscuous ligands **737–752**
  aggregate-forming compounds 738
  detergent response 738
  drug mechanism of action 749
  examples 739T
  as false positives in HTS 737
  future work 749
  history of 738
  mechanism of action 738, 750
  model of 738F
  receptor specificity 750
  SAR 750
Promiscuous ligands, computational prediction 738, 746
  aggregators 746, 747T
    clinically prescribed drugs 746
    recursive partitioning model 746, 747F
  experimental artifacts 749
    ChemBridge DIVERSet 749
    DockVision 749
    GOLD 749
    National Cancer Institute Diversity Set 749
  'frequent hitters' 746
    definition 746
    partial least squares model 746
    substructure 746
  privileged substructures 748
    GPCRs 748
    kinase inhibitors 748
    reactive species 748
    thiol reactivity index 749
Promiscuous ligands, identification 738, 739
  assay protocol changes 745–746
  assay readout interference 746
  biochemical assays 740
    see also specific assays
    buffer pH 740
    colorimetric assays 740
    ionic strength 740
    kinetic assays 740
    protein excess 740, 742
  chemical reactivity 746
  detergent sensitivity 740
    detergents used 740
  diverse model enzyme inhibition 742
    β-lactamase 742
    chymotrypsin 742
    malate dehydrogenase 742
    rottlerin 742
  enzyme concentration 741
    β-lactamase 741
  light scattering 743, 744F
    autocorrelation functions 743
    DLS 743
    HPLC combination 743
    particle size 743
    scattered light intensity 743
    small molecules 744
    Stock–Einstein equation 743
  microscopy 745
    confocal fluorescence 745, 745F
    electron microscopy 745
  purity issues 746
  time-dependent inhibition 742
    inhibition curves 743
    pre-incubation effects 742
Propafenone
  characteristics 829T
  structure 829T
Propoxyphene, GPCR targeting 773T
Propranolol
  analogs, fluorine-hydrogen bioisosterism 668, 692T
  discovery/development (by)
    Black, Sir James 253, 254
    Pritchard, Brian 254
    Stephensen, John 254
  GPCR targeting 773T
(S)-Propylene oxide, structure 720F
Prospective/retrospective assessments, phase IV clinical trials 191
Prospective trials, definition 30T
Prostaglandin E_2, sodium channels 836–837
Prostaglandin EP_3 receptor
  activation 785T
  constitutively active 790T
Prostaglandins, solid support synthesis 479, 482F
Prostate cancer
  androgen receptor (AR) 1027
  treatment, androgen receptor ligands 1016
Protease inhibitors, peptidomimetics 613, 614F
Proteases
  classification 577
  peptide/protein drugs 577
  as targets 757T
Protein(s)
  complexes, proteomic analysis 386, 387F
  display chips 370
    molecular diversity 370
    posttranslational modification 370
    recombinant protein tagging 370
    time-of-flight mass spectrometry 374
  drug discovery 297, 298T
    biotechnology companies 102
  excess, promiscuous ligands, identification 740, 742
  families 323, 323T
    sequence relationships 323, 324F
  function 319
    expression profiling 309
    single nucleotide polymorphisms 362
  kinetics 326
  molecular evolution 320, 320F
    domain conservation 320, 321T
  posttranslational modification *see* Posttranslational modifications
  structure 319
  synthesis, ribosomal RNA (rRNA) 1041
  tertiary structure/convergence 325, 325F
  translation 1037
Protein arginine methyltransferase 1 (PRMT1), nuclear receptor coactivators 1002
Protein conjugation, peptide/protein drugs *see* Peptide/protein drugs
Protein display chips *see* Surface plasmon resonance (SPR)
  time-of-flight mass spectrometry *see* Matrix-assisted laser desorption ionization (MALDI); Surface-enhanced laser desorption ionization (SELDI)

Protein expression
  mapping 384
    isotope-coded affinity tag methods (ICAT) 385
    sample preparation 384–385
    SEQUEST program 385
    two-dimensional gels 384–385
    zoom gels 385
  single nucleotide polymorphisms 362
Protein hormone binding, receptor protein tyrosine kinases (RPTKs) 576
Protein identification 336
  amino acid analysis 336–337
  automated proteolytic digestion 337
    robots 339
    throughput 339
  computer-based sequence searching 341, 343F
    see also specific methods
  Edman sequencing 336–337
  HPLC-MS 338F, 340, 342F
    quadrupole ion trap mass spectrometry 341, 343F
    matrix-assisted laser desorption/ionization (MALDI)-MS 339, 341F
    accuracy 340
    equipment 339
    peptide maps 340
    throughput 340
    time-of-flight technology 339
  MS-based methods 337, 338F
  peptide fingerprint searching 342, 344F
    complex mixtures 344
    ProteinProspector 343–344
    utility 344
  raw MS/MS data searching 342, 345
    data-dependent scanning 345
    SEQUEST 345
    strategy choices 346
    web-based programs 346
  sequence searching 342, 344
    BLAST 344–345
    Edman degradation 345
    software 345
  tandem MS (MS/MS) 340
Protein kinase(s) **959–992**, 961T
  see also Signal transduction; specific types
  actions 960F
  classification 964
  in drug discovery 959
  genomic relationships 963T
  historical perspective 960
  mechanism of action 71
  superfamilies 960
  as targets 71, 757T, 758, 964
  x-ray crystallography 964, 965T, 966F
Protein kinase A (PKA)
  GPCR MAP kinase activation 787

phosphodiesterase phosphorylation 923
sodium channels 836
Protein kinase B (PKB), phosphodiesterase phosphorylation 923
Protein kinase C (PKC)
  GPCR MAP kinase activation 787
  sodium channels 836
Protein kinase D (PKD), GPCR interaction 786T
Protein kinase inhibitors 71, 71F, 72F
  promiscuous ligands, computational prediction 748
Protein phosphatases **959–992**, 961T, 981
  see also Signal transduction; specific types
  dual-specificity 981
  genomic relationships 963T
  GPCR interaction 797T
Protein production, messenger RNA (mRNA) 1044–1045
ProteinProspector, peptide fingerprint searching 343–344
Protein–protein interactions
  functional genomics 325
  homogeneous time-resolved fluorescence (HTRF) see Homogeneous time-resolved fluorescence (HTRF)
  peptide/protein drugs 576
    phage display 576–577
    truncation 576–577
    second receptor binding 577
Protein tyrosine kinases (PTKs), as targets 758
Protein tyrosine phosphatase-1B inhibitors, peptidomimetics 615–616, 617F
Proteolysis resistance, peptide/protein drugs see Peptide/protein drugs
Proteolytic enzyme inhibitors
  carboxamide bioisosterism 668
  PEGylation 588
Proteome, definition 381, 382T
Proteomics 330, 380
  biochemical pathways 330
  body fluid analysis 386
    disease-specific proteins 386
  cellular proteomics 385
    LCM technology 385–386
  databases 425–426
  definition 314, 382T
  drug development 196
  drug discovery 43, 266, 303
    power of 330
  functional proteomics see Functional proteomics
  future work 390
  gene-related functional proteomics 389
  genomics vs. 381, 382F

in vivo target validation 388
  Alzheimer's disease 388
  cancer 388
  disease model relevance 388
  knockout models 389
  mutagenesis models 389
lead screening 425
monoclonal antibodies 315
pharmaceutical productivity effects 132–133
pharmacoproteomics see Pharmacoproteomics
posttranslational modification see Posttranslational modifications
PROfusion technology 315
protein arrays, complex analysis 386, 387F
protein expression mapping see Protein expression
publications 330, 331F
target identification 314
technologies
  see also specific technologies
  capillary electrophoresis 347, 347F
  capillary IEF 347
  ESI 347
  mRNA/protein expression analysis 387
  multidimensional chromatography 346–347
  quantification problems 348
terminology 381
Proton channels, characteristics 881T
'Proton motive Q cycle,' mitochondria complex III 1075
Proton translocation, mitochondria 1058F, 1059
PSD-95, GPCR interaction 797T
Pseudoknots, messenger RNA (mRNA) structure 1044
Psoriasis treatment, phosphodiesterase-4 inhibitors 935–936
Psychiatric diseases/disorders, estrogen receptors (ERs) 1026–1027
PTB, in disease 963T
PTK-787 969–970
  characteristics 984T
  KIT inhibitor 972
  structure 970F
  VEGFR family kinase inhibitors 969
PTP1B
  in disease 963T
  three-dimensional structures 636T
Publicity, outsourcing risks 220
Pulsed-field gradients, NMR see Nuclear magnetic resonance (NMR)
Pulse widths, NMR 515
Purinergic P2Y1 receptors
  activation 785T
  in human disease 775T

Purinergic P2Y2 receptors
  activation 785T
  in human disease 775T
Purinergic P2Y4 receptors,
  activation 785T
Purinergic P2Y6 receptors,
  activation 785T
Purity
  drug development 27
  stereoisomer synthesis 715
Pyk2, in disease 963T
Pyrazoles, mitochondria complex
  I 1068–1069
Pyrazolo[1,5-a]pyridine,
  bioisosterism 660T
Pyrazolopyridines, phosphodiesterase-5
  inhibitors 939–940
Pyrazolopyridopyridazine,
  phosphodiesterase-5
  inhibitors 941
Pyrazolopyrimidines, phosphodiesterase-
  5 inhibitors 940
Pyrazolopyrimidinones,
  phosphodiesterase-7
  inhibitors 944
Pyridaben, structure 1069F
Pyridazines, ring equivalent
  bioisosterism see Bioisosterism,
  ring equivalents
Pyridines
  mitochondria complex I 1068–1069
  ring equivalent bioisosterism 654
Pyrido[2,3-d]pyrimidine derivatives,
  phosphodiesterase-2
  inhibitors 933
Pyridylisoquinolines, phosphodiesterase-
  4 inhibitors 936
Pyrrole rings, carboxamide
  bioisosterism 668, 670T
Pyrroloquinolines, phosphodiesterase-5
  inhibitors 941–942

Q

Quadrupole ion trap mass
  spectrometry 341, 343F
Quality, outsourcing risks 219
Quality identification, ADMET 563,
  564T
Quantitative structure–activity
  relationship (QSAR)
  ADMET 566
  applications
    drug development 177
    drug discovery 48, 48F
  historical perspective 3
  hydrophobic depoting 586
  toxicogenomics combination 568
Questioning daily practice,
  innovation 258
QUEST system, two-dimensional gel
  electrophoresis 335

Quetiapine fumarate, GPCR
  targeting 773T
Quinazolines
  derivatives
    phosphodiesterase-5
      inhibitors 939
    sildenafil 939
  mitochondria complex I 1068–1069
Quinidine
  characteristics 829T
  structure 829T
Quinoline-2-carboxylate,
  bioisosterism 660T
Quinolone antibiotics, solid support
  synthesis 481F
Quinoxaline-2,3-dione 1048
  structure 1048F

R

RACk1, GPCR interaction 786T
RAD001, characteristics 984T
Radiation grafting, solid supports
  499–500, 501F
Radiometric assays, GPCR functional
  assay screening 808
Raf kinase, GPCR interaction
  786T
Raf-MEK-MAPK pathway kinases 980
  see also specific kinases
  inhibitors 980
Ragaglitazar
  as PPARγ ligand 1021–1022
  structure 1023F
Raloxifene hydrochloride
  characteristics 1012T
  as estrogen receptor ligand 1014
  structure 1013F
Raman spectroscopy, chiral drug
  analysis 720
Ramosetron, 5-HT$_3$ receptor
  binding 890
Randomization, clinical trials 181
Randomized clinical trials (RCTs)
  182
  definition 30T
  multiple arms 182
Random/nonrandom sampling, phase IV
  clinical trials 191
'Random screening,' HTS 392
Ranitidine bismuth citrate, mechanism
  of action, GPCR
  targeting 773T
Ras, GPCR MAP kinase
  activation 787, 788F
Rasagiline
  mitochondria complex I 1071, 1071F
  structure 1071F
Ras farnesyl transferase inhibitors,
  peptidomimetics 615–616, 617,
  617F
Rationale, DCN 153

Raw MS/MS data searching, protein
  identification see Protein
  identification
RCP, GPCR interaction 797T
R&D management see Leadership
Reagents
  receiving/tracking 534
  sourcing 533
  stability, HTS data management 453
Reasonable working situation,
  motivation 246
Reboxetine, structure 62F
Receptor-binding assays, homogeneous
  time-resolved fluorescence
  (HTRF) see Homogeneous time-
  resolved fluorescence (HTRF)
Receptor protein tyrosine kinases
  (RPTKs)
  ligand binding 576
  mechanism of action 576
  oncogene carcinogenesis 576
  peptide/protein drugs 576
  protein hormone binding 576
  structure 576
Receptors
  peptidomimetics see Peptidomimetics
  promiscuous ligands 750
Recombinant proteins, tagging, protein
  display chips 370
Recommendation, DCN 155
Recording/reporting process, clinical
  trials 187
Recruitment/human resources
  innovation 262
  management see Organization, drug
    discovery
  motivation 246
  R&D management 250
Recursive partitioning (RP),
  promiscuous ligands,
  computational prediction 746,
  747F
Recycling, GPCR desensitization
  802
Redox targets **1053–1087**
  see also Mitochondria; Mitoprotective
    drugs
Redrugs, materials management 531
Reductionism, drug discovery 272
References
  DCN 155
  HTS data management 453
Refrigeration temperature stability, drug
  reference standard 167–168,
  168T
Regulatory aspects 131
  chemogenomics 567
  clinical trials, phase I 155
  documentation 721–722
  drug development 200
  guidelines, drug development 27
Regulatory authorities, phase IV clinical
  trials 190

Regulatory elements,
	phosphodiesterase-6 928
Reimbursement policies,
	pharmaceutical productivity
	effects 133
Relaxation-edited spectroscopy, mixture
	analysis 528
Reliability, project teams 145
Remoxipride
	structure 717F
	synthesis 717F
Renin inhibitors
	active site binding 628F
	carboxamide bioisosterism 668
	nonpeptide drugs 625
	peptidomimetics 613, 614F, 615F,
		625
	three-dimensional structures 626T
Replication interference, DNA
	therapeutic drugs 1039
Reporter gene assays, GPCR functional
	assay screening see G protein-
	coupled receptors (GPCRs)
Reporting, HTS data
	management 455
Reprocessing, development
	chemistry 169
Research projects, project management
	see Project management
Research strategies, innovation 22
Residues, DNA structure 1038–1039
Resin beads, solid support
	synthesis 493
	macroporous resins 495
		definition 495
		Merrifield resins vs. 495, 495T
	poly(ethylene glycol)-containing
		resins 495, 496T
		CLEAR resins 496, 497F
		dendrimers 499
		polyethylene glycol-poly (N,N-
			)dimethylacrylamide
			(PEGA) 497, 499F
		polystyrene-PEG 496, 496F, 497T
		tetraethylene glycol diacrylate
			cross-linked polystyrene support
			(TTEGDA-PS) 497, 498F, 498T
	polystyrene-based resins 493
		divinylbenzene 493–494
		functional groups 493–494, 494F
		Merrifield resins 493–494
Resin-bound reagents, solution phase
	synthesis see Solution phase
	synthesis
Resin-bound scavengers/reagents,
	solution phase synthesis see
	Solution phase synthesis
Resiniferatoxin, structure 909F
Resin-to-resin transfer reactions, solid
	support synthesis 474, 475F
Resource lack, outsourcing 227–228
'Respiratosomes,' mitochondria 1057–
	1058

Responsibilities
	project management teams 147
	project teams 145
Resting state, sodium channels 835
Result scale, clinical trials 184
RET, in disease 963T
Retinal, acid-sensitive ion channels
	(ASICs) 911
Retinitis pigmentosum
	mitochondria 1054–1055
	phosphodiesterase-6 928, 950
9-cis-Retinoic acid, structure 1020F
Retinoic acid receptor-α (RAR-α)
	acute myeloid leukemia 1027
	three dimensional structure 999
Retinoic acid receptor-related orphan
	receptor, nomenclature 996T
Retinoic acid receptors (RARs) 1007
	in cancer 1020–1021
	embryogenesis 1007
	epithelial cell differentiation 1008
	ligands 1020, 1020F
	see also specific drugs
	nomenclature 996T
	reverse cholesterol transport 1008
Retinoid X receptors (RXRs)
	heterodimers 1001
	nomenclature 996T
	three dimensional structure, ligand-
		binding pocket 1000, 1000F
	vitamin D receptor
		heterodimerization 1007
Reuptake blockers, characteristics 64T
Reverse cholesterol transport, retinoic
	acid receptor 1008
Reverse erbA receptor,
	nomenclature 996T
Reverse transcriptase, homogeneous
	time-resolved fluorescence 422,
	423F, 424F
Reward systems, R&D
	management 250
Reworking, development
	chemistry 169
RF homogeneity, NMR 515
Rhizopuspepsin, three-dimensional
	structures 626T
Rhodacyanine dye MKT-077,
	mitochondria, drug
	targeting 1067, 1068F
Rhodopsin receptor 575–576, 777–778
	activation 796
		TM3 residue displacement 798
	in human disease 775T, 792T
	light-induced conformational
		changes 796
	structure 796
Riboflavin deficiency, carbon-boron
	bioisosterism toxicity 698
Ribosomal RNA (rRNA) 1040
	aminoglycosides 1041, 1042F
	antimicrobial drugs 1041
	bacterial 1041

eukaryotic 1041
macrolides 1042
pharmacology 1041
	antibacterial chemotherapy
	see specific antibacterials
	physiological function 1041
	protein synthesis 1041
	structure 1037, 1040
Ribosomes
	interactions
		chloramphenicol 1042–1043
		clindamycin 1042–1043
		linezolid 1043
		oxazolidinones interactions
			1042–1043
		oxytetracycline interactions
			1042–1043
		tetracycline interactions
			1042–1043
Riboswitches, messenger RNA
	(mRNA) 1048
Ridogrel
	ring equivalent bioisosterism 655
	structure 655F
Rifampicin
	as nuclear receptor ligand 1010
	structure 1026F
Rifamycin B, structure 46F
Risk factors, clinical trials 181
Risk minimization, outsourcing
	208
Risperidone
	invention of, Colpaert, Francis
		257–258
	mechanism of action, GPCR
		targeting 772, 773T
Ritonavir 579T
	mitochondria, effects on 1082
	structure 578F
Rivanicline, nicotinic acetylcholine
	receptor binding 888
RNA
	see also Messenger RNA (mRNA);
		Ribosomal RNA (rRNA)
	sizes 1037
	structure 1037, 1038F
RNA-based strategies, target selection
	see Target selection
RNA duplexes, DNA 1040
RNA editing, glutamate-gated ion
	channel regulation 899
RNAi (RNA interference) 47,
	761–762, 1045, 1046
	RNA-induced silencing complex
		(RISC) 1046
RNA-induced silencing complex
	(RISC) 1046
RNA maturation, antisense-based
	therapeutics 1047
RNA polymerases, mRNA
	synthesis 1043
	eukaryotes 1044
	prokaryotes 1043

RNase H based cleavage, antisense-based therapeutics 1045, 1046F
RNA splicing, messenger RNA (mRNA) structure 1044
Ro 04-2843, structure 1062F
Ro 68-3400
　mitoprotective activity 1062
　structure 1062F
Ro 68-3406, structure 1062F
Robotic liquid-handling, materials management 536
Robotic weighing, materials management 536
Roche, R&D spending 101T
ROD-188, structure 893F
Roflumilast, analogs
　fluorine-hydrogen bioisosterism 692, 693F
　structure 935
Role models, effective leaders 241
Role/responsibility definitions, collaborative project management 151
Rolipram 935
　animal models, cognition 926–927
　aspirin-induced mucosal injury 948
　clinical evaluation 935–936
　phosphodiesterase-4 affinity 925–926
Rolliniastatin-1, mitochondria complex I 1068–1069
Ropivacaine
　structure 716F
　synthesis 716F
Roscovitine see CYC202
Rosiglitazone maleate
　characteristics 1012T
　mechanism of action 1081
　　mitochondria complex I 1081–1082
　　as PPARγ ligand 1021–1022
　structure 1023F, 1081F
Rosuvastatin, structure 650F
Rotenone
　mitochondria complex I 1068–1069
　structure 1069F
Route design, process outcome 213
Routes of administration, chemical vs. biological 117
'Rule-of-five'
　drug development 730
　preclinical trials 178
(D/E)RY motif, GPCR structure 774–777

S

S4 domain, potassium channels 858–859
Safe laboratory conduct, career guidelines 18

Safety
　drug development costs 727
　drug discovery 2, 11
　monitoring 8–9
　regulatory problems 725
Safety, health and environment information, drug development 722–723, 723F
'Safety-catch' linkers, solution phase synthesis 489–490, 490F
Salmedix 767
Salmon calcitonin see Calcitonin
Salt forms, drug development 27, 27T
Salt tolerance, NMR probes 515–516
Sample analysis
　preparation, isoelectric focusing (IEF) (first dimension) 332
　single nucleotide polymorphisms 363
Sample volumes, NMR probes 516, 516F
Sanofi-Aventis, R&D spending 101T
Saponin, promiscuous ligands 740
Saquinavir mesylate 579T
　design/development, peptidomimetics 613
　structure 578F
SB-203580
　structure 979F
　TGFβR kinase inhibitors 977–979
Scaffold-based design, phosphodiesterase-4 inhibitors 938
Scaffold hopping see Bioisosterism, scaffold hopping
Scalable methodologies, chiral switches 729–730
Scale-up
　drug development productivity 130
　outsourcing see Outsourcing
Scattered light intensity, light scattering, promiscuous ligands 743
Scavenging resins, solution phase synthesis see Solution phase synthesis
Schizophrenia, disease basis, phosphodiesterase-10 930–931
Schultz, Theodore, universities in commerce 93
Science, marketing, replacement by 284
Science integration, education 88–89
Scientific training, leadership strategies see Leadership
Screening cascade
　drug discovery 565, 565F
　hit-to-lead 139
SDF-1 alpha receptor, activation 785T
SDS-PAGE, two-dimensional gel electrophoresis 332–333
Secondary structure mimetics, peptidomimetics 608, 610F

Second messengers
　potassium channels 861
　sodium channels 836
Secretin 575–576
　as peptide/protein drugs 579T
Secretin receptors 778
Security, collaborative project management 151
Selective androgen receptor modulators (SARMs) 1016
Selective estrogen receptor downregulators (SERDs) 1014
Selective estrogen receptor modulators (SERMs) 1014
Selective optimization offside activities (SOSA), scaffold hopping 683
Selective serotonin reuptake inhibitors (SSRIs) 62
　mechanism of action 67
Selegiline, structure 68F, 1071F
1,3-Selenazol-4-one
　selenium bioisosterism 703
　structure 701F
Selenazolidine prodrugs, structure 701F
Selenopyrylium derivatives, structure 701F
Self-contained projects, project management 12–13, 13F
Self-forming depots, peptide/protein drugs 593
Semaxanib see SU-5416
Sensitivity
　homogeneous time-resolved fluorescence 419, 420F
　NMR probes 515
Sensorgram, surface plasmon resonance 371–372, 372F
Sensory receptors, transient receptor potential ion channels 908
Sepracor, characteristics 226T
Sequence searching, protein identification see Protein identification
Sequencing, single nucleotide polymorphisms 364
Sequential clinical trials 182, 183F
SEQUEST
　protein expression mapping 385
　raw MS/MS data searching 345
Serine–threonine protein kinases 977
　see also specific kinases
　as targets 758, 960–962
　three-dimensional structures 636T
Serine–threonine protein phosphatases 981, 982
　see also Phosphoserine/phosphothreonine-interacting domains; specific types
　three-dimensional structures 636T
Serinyl proteases
　nonpeptide drugs 628

peptidomimetics 628
three-dimensional structures 626T
Serotonin see 5-HT (serotonin)
Serotonin-selective reuptake inhibitors (SSRIs) 56–57, 57F
  mechanism of action, GPCR targeting 772
Serotonin transporter (SERT) 59, 60T
Sertraline
  characteristics 653T
  structure 653T
Serum albumin, peptide/protein drug conjugation 589
Service companies 226
  see also Contract research organizations (CROs)
  history 226
  preclinical trials 226
Severe myoclonic epilepsy of infancy (SMEI), sodium channels 840–841
SGX Pharmaceuticals 761T
SH2 domain, in disease 963T
Shaped pulses, NMR see Nuclear magnetic resonance (NMR)
Shape matching, scaffold hopping 681–683
Sheets, solid supports see Solid supports
Shift laminar pulse (SLP), NMR 510
Shorter, Edward, innovation 258
Short interfering RNAs (siRNAs), definition 1046
Siah1A, GPCR interaction 797T
Siani, Mike, peptoid-based antibiotics, discovery of 259
Signal amplification, homogeneous time-resolved fluorescence 413
Signaling pathways, GPCRs 800
Signal recognition particle (SNP), messenger RNA (mRNA) 1049
Signal transduction
  in disease 960, 963T
    cancer 960–962
  EGFR family kinases 967
  functional genomics 324, 960, 963T
  GPCRs 574–575
  modulators, peptidomimetics 615, 617F
  structural genomics 964
  x-ray crystallography 964
Sila-atipamezole, structure 695F
Silabolin, structure 695F
Sila-meprobamate, structure 695F
Sila-pridinol, structure 695F
Sildenafil citrate 927
  derivatives 938–939
  discovery/development 257
  phosphodiesterase-5 blockers 257
  phosphodiesterase inhibition 942–943, 947T
  quinazoline derivatives 939
Silicon diols, carbon-silicon bioisosterism 696

Silperisone
  clinical studies 697
  structure 697F
Silver stain, two-dimensional gel electrophoresis 333
Similarity searching, scaffold hopping 681–683
Simvastatin
  characteristics 653T
  discovery/development 1074
  structure 653T
Single base extension (SBE), single nucleotide polymorphisms 365, 365F
Single Nucleotide Polymorphism (SNP) Consortium 569
Single nucleotide polymorphisms (SNPs) 362
  allele-specific chain extension 364, 365F
  allele-specific PCR 365
  applications 362
    drug responses 362
    protein expression 362
    protein function 362
  automation 363
  bead-based fiberoptic arrays 357
  detection 328
  discovery 362
  drug development 195, 195T
  drug discovery 315–316
  electrophoresis 364, 367, 369T
    limitations 367
  flat microarrays 364, 368, 369T
  hybridization assays, with 368
  Human Pharmaco DNP Consortium 316
  hybridization 364, 365F
  mass spectrometry 367, 369T
    limitations 367–368
  microplate-based assays 363, 367, 369T
    fluorescence detection 367
    heterogenous approach 367
    homogeneous approach 367
  multiplexing 363
  nuclease-based assays 366
    fluorescence-based energy transfer (FRET) 366
    Invader 366
    Taq-Man 366
  oligonucleotide ligation 365F, 366
  PCR 369
  platforms 367, 369T
    see also specific platforms
  sample analysis 363
  sample preparation 363
    nucleotide identification 363
    PCR 363
  sequencing 364
  single base extension (SBE) 365, 365F

soluble arrays 364, 368, 369T
  advantages 368
  flow cytometry 368
Single run evaluation, HTS data management 454
Sirna 761
Size
  pharmaceutical industry organization 110
  R&D management 250
Skeletal muscle
  calcium channels 847
  sodium channels 834
Skelgen software, scaffold hopping 683
SKF-38 393, ring system reorganization (bioisosterism) 678–679, 679F
SKF-104365
  structure 979F
  TGFβR kinase inhibitors 977–979
SKI-166, characteristics 984T
SKI-606
  Src kinase inhibitor 976–977
  structure 974F, 978F
'Skunk works,' innovation 262
SkyPharma, controlled release formulation technology 232
SLC transporters 58
Small molecule drugs
  biotechnology companies 102
  calcium channel pharmacology 849–850
  light scattering, promiscuous ligands 744
  specificity 299–300, 300T
Small organic molecules, combinatorial chemistry 462
SMEs (small/medium-sized enterprises), definitions 225
Smith Kline Beckman, history 100–101
Smith Kline Beecham, history 100–101
Smoking-induced pulmonary endothelial dysfunction, phosphodiesterase-5 inhibitors 948–949
Smoothing, conflict management 248
Smooth muscle
  ATP-gated ion channels 904–905
  calcium channels 847
  phosphodiesterase-3 924
  phosphodiesterase-4 926–927
SNARE, nicotinic acetylcholine receptor regulation 886–887
Social effects, importance of 240
Sodium channel blockers, characteristics 829T
Sodium channels 831
  genetic diseases 840
    Brugada syndrome 841
    cardiac arrhythmias 840–841
    epilepsy 840–841

Sodium channels (*continued*)
   familial childhood epilepsy 840–841
   hyperkalemic periodic paralysis 841
   long-QT syndrome 841
   paramyotonia congenita 841
   primary erythermalgia 841
   severe myoclonic epilepsy of infancy (SMEI) 840–841
  GPCR interaction 786T
  molecular biology 833
   TTX$_s$ subtypes 833–834
  nomenclature 831
   International Union of Pharmacology (IUPHR) 831–832
   subunits 832, 833T
  pharmacology 837, 838T
   clinical useful drugs 839
   modulated receptor hypothesis 839
   neurotoxin receptor site 1 837
   tetrodotoxin (TTX) 837
  physiological functions 835
   β subunits 836
   cardiac action potentials 836
   cell membrane depolarization 836
   excitable cells 835
   resting state 835
   second messenger functions 836
  regulation 836
   atrial/ventricular contraction 837
   neurotransmitters 836
   phosphorylation 836
   prostaglandin E_2 836–837
   protein kinase A 836
   protein kinase C 836
  structure 832F, 834
   β subunits 835
   intracellular loop 834–835
   ion selectivity/permeation 834–835
  subcellular distribution 833
  subunit classification 833T
  therapeutics 839
   antiepileptic drugs 839–840
   chronic pain 840
   conduction anesthesia 839
   epilepsy 839–840
   local anesthetics 839
   tissue distribution 840
  tissue distribution 833
   auxiliary subunits 834
   cardiac myocytes 834
   CNS 833–834
   skeletal muscle cells 834
Sodium codependent bile acid transporter inhibitors 48F, 63
Sodium estrone sulfate, structure 66F
Software
  see also specific programs
  scaffold hopping 683

Solenoid-controlled liquid dispensing, Ultra-High-Throughput Screening System (Aurora Biosciences) 444, 444F
Solid supports 493
  glass 503, 504F
   DNA arrays 504–505, 505F
   photolithographic masks 503, 504F
  high throughput organic synthesis 396
  pins and crowns 499
   improvements 500, 501F
   monomer alternatives 501–502, 502F
   radiation grafting 499–500, 501F
   removable crown 501–502, 501F
  resin beads *see* Resin beads, solid support synthesis
  sheets 502
   Fmoc peptide synthesis 503, 503F
Solid support synthesis 464
  see also Solid supports
  N-acetylation 466F, 467, 467F
  O-acetylation 466F, 467, 468F
  advantages/disadvantages 464
  N-alkylation 466F, 467, 467F
  O-alkylation 466F, 467, 468F
  N-arylation 466F, 467, 467F
  O-arylation 466F, 467, 468F
  carbon–carbon bond-forming reactions 467
   Baylis–Hilman reaction 469F
   Grignard reaction 467–468, 469F
   Heck reaction 467–468, 470F
   Horner–Emmons reaction 469F
   Stille reaction 467–468, 470F
   Suzuki reaction 467–468, 470F
   Wittig reaction 468–469, 469F
  carbon–heteroatom coupling reactions 466, 466F
  chemical libraries 760
  complex multistep synthesis 475
   see also specific reactions
  cycloadditions 472, 473F
  heterocycle/pharmacophores 479
   1,4-benzodiazepines 479, 481F
   benzopiperazinones 479, 481F
   bicyclic guanidines 481F
   'complexity-generating' reactions 479–480, 483F
   oxapiperazines 482F
   prostaglandins 479, 482F
   quinolones 481F
  history 464–465
   Merrifield, Bruce 464–465, 465F
  hybrid oligomers 478
   peptide nucleic acids 478, 479F
  intramolecular cyclization-resin-cleavage 470, 471F, 472F
  library construction 464

  multiple-component reactions 472, 473F
   Passerini reaction 472–474, 474F
  natural products 480, 484F
   epothilones 482, 484F
   glyantrypine 482
   indolactam V 480–482, 483F
  oligomers 475
  oligonucleotides 476, 477F
  oligosaccharides 477, 477F, 478F
   glycopeptides 477, 478F
   insoluble/soluble polymers 477
  peptides 475, 476F
   Fmoc-amino acid approach 475–476
   library generation 476
   tBoc-protected approach 476
  resin-to-resin transfer reactions 474, 475F
  solution phase synthesis *vs.* 465T
  transition metal-mediated reactions 469, 470F
   copper-mediated N-arylation 470, 471F
   olefin metathesis 469, 470F
   palladium-mediated N-arylation 470, 471F
  unnatural oligomers 478, 480F
Solubility, bioisosterism modification analysis 687, 688F
Soluble arrays, single nucleotide polymorphisms (SNPs) *see* Single nucleotide polymorphisms (SNPs)
Solution phase synthesis 482
  polymer-supported synthesis 482
   advantages 482–485
   dendrimers 485, 485F
   examples 482–485, 485F
   fluorinated hydrocarbons 485
  resin-bound reagents 488
   cycloaddition reactions 490, 491F
   polymer-assisted solution phase PASP chemistry 477F, 489
   polymer-bound ^1H-benzotriazole 490, 490F
   'safety-catch' linkers 489–490, 490F
  resin-bound scavengers/reagents 491, 491F
   epibatidine synthesis 492, 493F
   epimaritidine synthesis 491–492, 492F
   oxomaritidine synthesis 491–492, 492F
  scavenging resins 485
   Dess–Martin oxidization procedure 487F, 488
   disadvantages 486
   functionalized polystyrene resins 485–486, 487F
   product trapping 486, 486F
   sulfonic acid resins 488, 488F
  solid-phase synthesis *vs.* 465T

Solvent resonance suppression
   liquid chromatography-NMR 522
   NMR 511
Solvents
   categorization 162T
   development chemistry 161, 162T
   liquid chromatography-NMR 522–523
Solvent suppression, NMR 509
Somatoliberin 579T
Somatostatin, structure 581–582
Somatostatin receptor partial agonist, peptidomimetics 608–609, 611–612, 611F
Somatostatin receptors st5, in human disease 775T
Sourcing
   see also Outsourcing
   definitions 205
Sourcing development 209
   activity cluster analysis 216, 216F, 217F
   activity determination 212
      activity interdependency 214
      company/industry practice 214
      competitive advantage 213
      existence of 214
      IP generation 214
      offshore outsourcing see Offshore outsourcing
      process maturity 213
      process outcome control 213
   discrete activities 209
   primary chemistry activity
      definition 209
      mapping processes 209
      subactivities 209, 210F
      subcategories 210, 210F
Specialty pharma companies 102
Specificity, drug discovery see Drug discovery
'Specificity scale,' drug discovery 298, 299T
Spectinomycin, mechanism of action, ribosome interactions 1042–1043
Speed definitions, decision making vs. organizational alignment 120
Speedel, characteristics 226T
Spin cycle see Start-up companies
Spinocerebellar ataxia, calcium channels 852
Spinophilin, GPCR interaction 797T
Spiroderivatives, ring system reorganization (bioisosterism) see Bioisosterism, ring system reorganization
Spirogermanyl, structure 698F
Spiro-hydantoin, bioisosterism 660T
Spironolactone, structure 1018F
Spiroquinazolinones, phosphodiesterase-7 inhibitors 943–944

Splice variants
   AMPA-sensitive receptor regulation 899
   estrogen receptors (ERs) 1026
   phosphodiesterase-1 923
   phosphodiesterase-2 923–924
   phosphodiesterase-4 924–925
   phosphodiesterase-5 927
   phosphodiesterase-7 929
   phosphodiesterase-8 929–930
   phosphodiesterase-9 930
   phosphodiesterase-10 930
Splicing redirection, antisense-based therapeutics 1047
SPS resins, nuclear magnetic resonance (NMR) see Nuclear magnetic resonance (NMR)
SQi inhibitors, mitochondria complex III 1075
SR 5722A, structure 890F
SR 16234
   as estrogen receptor ligand 1014
   structure 1013F
Src homology domains
   identification 633–634
   members 633, 636T
   nonpeptide drugs 633
   peptidomimetics 633
   three-dimensional structures 635–637, 636T, 639F
   x-ray crystallography 634, 637F, 638F
Src kinases 975
   angiogenesis 975–976
   in disease 963T
   GPCR interaction 786T
   inhibitors 976
   members 975–976
   metastases 975–976
   PDB entries 965T
   SH2
      domain inhibitors 615–616, 617, 617F
      three-dimensional structures 636T
   SH3, three-dimensional structures 636T
   structure 976, 976F
   substrate complex, x-ray crystallography 964, 966F
   transduction pathways 975–976, 975F
Staffing flexibility, outsourcing 207
Staff morale, outsourcing risks 219
Stanozolol, structure 1016F
Starting material definition, common technical document (CTD) 163
Start-up companies 764
   chemistry vs. biology 764–765
   collaborations 765
   medicinal chemistry importance 765
   spin cycle 765
   technology licensing 765

   virtual companies 766
      advantages 766
      contract research organizations 766
      extended enterprises 766
Statins
   analog design 651
   bioisosterism 650, 650F
   discovery/development 1074
   polymorphism effects 67
Statistical analysis, clinical trials see Clinical trials
Statistical significance, clinical trials 184
Statistics, two-dimensional gel electrophoresis 335–336
Steatohepatitis, antivirals 1082
Stem cells
   consumer advocacy groups 271
   genomics 46
Stem loops, messenger RNA (mRNA) structure 1044, 1044F
Stephensen, John, propranolol, invention of 254
Stereoisomer synthesis see Chiral drug discovery
Stereoselectivity, stereoisomer synthesis 715–718, 719F
Sternbach, Leo
   management difficulties 259
   Valium, invention of 256–257
Steroidogenic factor 1, nomenclature 996T
STI-571 972
   Bcr-Abl kinase inhibitor 972–973
   characteristics 984T
   structure 971F
Stigmatellin, mitochondria complex III 1075
Stille reaction, solid support synthesis 467–468, 470F
STK6, PDB entries 965T
Stock–Einstein equation, light scattering, promiscuous ligands 743
Strategic thinking, project teams 145
Strengths, Weaknesses, Opportunities and Threat analysis (SWOT), project management 14
Streptomycin, mechanism of action 1041–1042
Stroke, mitochondria 1054–1055, 1060
Stromelysin (MMP-3) inhibitors, peptidomimetics 613, 614F, 632
Structural genomics
   definition 382T
   signal transduction 964
Structural motif replacement, peptide/protein drugs 590
Structural parameters, bioisosterism modification analysis 686, 687F
Structural proteomics, definition 382T

Structure, R&D management 249
Structure–activity relationships(SARs)
   drug discovery 276
      productivity 124–125
   NMR 525
   promiscuous ligands 750
   specific drugs, AMPA-sensitive
      receptors 900
   target development 139
   toxicogenomics combination 568
Structure-based drug design
      (SBDD) 760
   companies 761T
   fragment-based 760–761
   HTS see High-throughput screening
      (HTS)
   in-silico design, drug discovery 301,
      301T
   peptidomimetics see Peptidomimetics
   target vs. disease 114–115
Structure elucidation, nuclear magnetic
      resonance (NMR) see Nuclear
      magnetic resonance (NMR)
Strychnine, structure 895F
SU-5416 969–970
   structure 970F
   VEGFR family kinase inhibitors 969
SU-6656
   Src kinase inhibitor 976–977
   structure 978F
SU-6668 969–970
   characteristics 984T
   KIT inhibitor 972
   structure 970F
   VEGFR family kinase inhibitors 969
SU-11248 969–970
   acute myeloid leukemia 971
   characteristics 984T
   KIT inhibitor 972
   structure 970F
   VEGFR family kinase inhibitors 969
Subactivities
   lead optimization 213
   primary chemistry activity
      definition 209, 210F
Substance P receptor antagonist
   nonpeptide drugs 618–619, 620F
   peptidomimetics 595–596, 608–609,
      611–612, 611F, 612F
Substructure, promiscuous ligands,
      computational prediction 746
Subtypes, GPCR functional assay
      screening 811–812
Success rates, drug development
      productivity 126–127
Succinate dehydrogenase-CoQ
      oxidoreductase complex see
      Mitochondria complex II
Sulfanilamides, chemical vs.
      biological 115, 116F
Sulfapyrazine, structure 654F
Sulfapyridine, structure 654F
Sulfathiazole, structure 654F

Sulfonamides
   bioisosterism 662T
   conversion, carboxamide
      bioisosterism 666, 669F
   ring equivalent bioisosterism 654,
      654F
Sulfonates, bioisosterism 662T
Sulfonic acid resins, solution phase
      synthesis 488, 488F
Sumatriptan, GPCR targeting 773T
Summer schools, education 89, 90T
Sunitinib see SU-11248
Superordinate goals, conflict
      management 249
Superoxide anions, mitochondria 1065
Supplier catalogues, materials
      management 536
Supply coordination, clinical trials 130
Support industries
   high throughput organic
      synthesis 395
   HTS 392–393
Supportiveness, project teams 145
Suramin, structure 906F
Surface-enhanced laser desorption
      ionization (SELDI) 376
   applications 376
      preactivated surfaces 376–377
      profiling 379
   development 374–375, 376
   drug development 378, 379
   ELISA vs. 378–379
   matrix-assisted laser desorption/
      ionization (MALDI)-MS
      vs. 377–378
   proteomics 314
   readout 377F
   sample preparation 377, 378F
   sensitivity 378–379
Surface micromachining see
      Lithography/photolithography
Surface plasmon resonance (SPR) 371
   applications 372
   data analysis 374
   detection system 371, 371F
   evanescent wave 371
   kinetics 374
   limitations 371–372
      mass transfer 373–374
      purity of analyte 373–374
   with MALDI-TOF 380
      disadvantages 380
   sensorgram 371–372, 372F
Surrogate endpoints
   definition 185–186
   target selection 764
Suzuki couplings, solid support
      synthesis 467–468, 470F
Syk SH2, three-dimensional
      structures 636T
SYM-1010 903
SYM-2081 900
   structure 901F

SYM-2206, structure 901F
Sympathetic nervous system, ATP-gated
      ion channels 905
Synaptic transmission, GABA$_C$
      receptors 892
Synteny, positional cloning 306–307
Synthesis
   control, development chemistry 160,
      161
   steps, common technical document
      (CTD) 163
Synthetic biology, genomics 44–46
Synthetic route, DCN 154
Sypro fluorescent stains 334
Syp SH2, three-dimensional
      structures 636T
Syrrx 761T
Systematic error detection, HTS data
      management 453
Systemic lupus erythematosus (SLE),
      pathogenesis, estrogen receptors
      (ERs) 1026–1027
Systems, R&D management 250
Systems biology, pharmaceutical
      productivity effects 132–133
Szent-Gyorgi, Albert, vitamin C
      discovery 258

T

T1 domain, potassium channels 858
T0901317 1023
   structure 1024F
Tabular display, HTS data
      management 454, 456T, 458T
Tacalcitol
   structure 1018F
   as vitamin D receptor ligand 1019
Tachykinin NK1 receptor,
      activation 785T
Tacrolimus (rapamycin)
   PI-3K–Akt/PKB–aTOR pathway
      kinase inhibitors 980–981
   structure 46F, 981F, 982F
Tadalafil 927, 939
   analogs of 941
   as phosphodiesterase inhibitor
      947T
Talampanel 903
Talent pool diversity, outsourcing
      207
Tamalin, GPCR interaction 797T
Tamoxifen
   characteristics 1012T
   mechanism of action, as estrogen
      receptor ligand 1014
   structure 1013F
Tandem mass spectrometry (MSD/MS),
      protein identification 340
Taq-Man, single nucleotide
      polymorphisms 366
Tarceva see OSI-774

Target(s) 56, 73
  see also specific drug targets
  assessment
    drug development 176
    druggability 49
  classes 68
  definition 304
  development 139
    SAR 139
  resynthesis, process outcome 213
  synthesis, process outcome 213
  by therapeutic area 4
    target vs. disease 114–115, 115T
  types, drug development
    productivity 128
  validation 304
    drug discovery 138
    expression profiling 312
    genomics 266
    project management 15
    proteomics 266
Target identification 138, 304
  bioinformatics 317
    GenChem database 317
    gene expression levels 318
    image analysis 318
    variable data quality 317–318
  drug discovery 138
  druggability 3
  model organisms 312
  project selection 73
    disease vs. target area 73
Target identification, genomics 305
  'bottom-up' approach 305
  expression profiling
    data amounts 311
    disease biology 311
    DNA chips/microarrays 310
    drug effects 311
    gene expression 308–309
    gene trapping 312
    GSE 312
    HTS 310, 311
    issues 309
    methods 309, 310, 310F
    protein function 309
    tumors 311
    zinc finger proteins 312
  functional genomics 308, 329
  genetic mapping 306
  genome sequencing 307
    bacterial genetics 307
    purpose 308
  pharmacogenetics/
    pharmacogenomics 315
    gene activity variations 316–317
    polymorphisms vs. disease 316
  positional cloning 306
    Alzheimer's disease 307
    mouse genes 306
    synteny 306–307
  technologies 305
  'top-down' approach 305

Target product profile (TPP), hit-to-
    lead 139, 140T
Target selection 4, 70, 70T, 73, 74T,
    753–770
  see also specific drugs
  calcium channels 73
  cancer 73
  chemical libraries 760
    large 760
    solid support synthesis 760
  development challenges 763
    attrition rates 763
    biomarkers 763
    efficacy in clinical trials 764
    patient identification/
      diagnosis 764
    specific therapies 763
    surrogate endpoints 764
    time frames 763
  drug discovery productivity 126
  'druggability' 754
  genomics 754
    benefits 756
    genetic models 755–756
    HTS 755
    Human Genome Project 754–755
    patent filing 755
    progress 755
  infectious disease 76
  intellectual property 758
  molecular target, therapeutic target
    vs. 754
  pharmacokinetic considerations
    76
  phosphodiesterases 73
  RNA-based strategies 761
    antisense RNA 761
    success of 762
  target class 756
    see also specific targets
    cloning 756
    proteins 756
    therapeutic area vs. 754
  therapeutic focus 762
  therapeutic target
    molecular target vs. 754
    target class vs. 754
  transporters 56, 57
Target setting, motivation 246
Target structure-based drug design
  nonpeptide drugs 622
  peptidomimetics see Peptidomimetics
Taste receptors,
    heterodimerization 799
Taurine, structure 895F
Tazarotene
  as retinoic acid receptor
    ligand 1020–1021
  structure 1020F
tBoc-protected approach, peptide
    synthesis 476
TC-1687, nicotinic acetylcholine
    receptor binding 888

T cells, activation, phosphodiesterase-
    7 929
TCPOBOP
  as nuclear receptor ligand 1010
  structure 1026F
Teamwork, project team leader
    assessment 146
Technical competencies,
    motivation 246
Techniques, validation, ADMET 559–
    560
Technocrats, drug discovery 273
Technology
  innovation 21, 22T, 23T
  investments, drug discovery 274
  licensing, start-up companies 765
TEI-9647
  structure 1018F
  as vitamin D receptor ligand 1019
TEK, PDB entries 965T
Telithromycin, structure 1043F
Temperature sensors, transient receptor
    potential ion channels 908
Terazosin, GPCR targeting 773T
Teriparatide 579T
  as peptide/protein drug 580
Terlipressin 579T
  as vasopressin analog 580
Ternary complex model, GPCR
    activation models 789–792,
    794F
Testis receptor, nomenclature 996T
Testosterone, structure 1016F
Tetomilast, structure 935
Tetracyclic guanine derivatives,
    phosphodiesterase-1
    inhibitors 931–932
Tetracyclines
  mechanism of action, ribosome
    interactions 1042–1043
  structure 1043F
Tetraethylene glycol diacrylate cross-
    linked polystyrene support
    (TTEGDA-PS), solid support
    synthesis 497, 498F, 498T
Tetraiodophenolphthalein, β-lactamase,
    inhibition of 740, 741F
4-hydroxy-2,2,6,6-Tetramethylpiperidin-
    N-oxide (TEMPOL),
    mitochondria, drug
    targeting 1067
Tetramisole
  bioisosterism 678, 679F
  structure 679F
Tetrazoles, bioisosterism 662T
Tetrodotoxin (TTX), sodium
    channels 837
Thalidomide 759
  biological activity 759
  clinical activity 759
    erythema nodosum treatment 759
  clinical trials 759
  structure 759F

Theophylline, as phosphodiesterase inhibitor 947T
Therapeutic area
  decisions, functional discipline organization 109
  drug development productivity 128, 129T
  organization see Pharmaceutical industry
Theravance, GlaxoSmithKline alliance 234
Thermogenin (uncoupling protein-1) 1080
Thermolysin
  peptidomimetics 631, 634F
  three-dimensional structures 626T
Thiabendazole
  bioisosterism 679F
  structure 679F
Thiadiazoles, phosphodiesterase-7 inhibitors 944
1,2,4-Thiadiazoles, carboxylic ester bioisosterism 663
Thiazide diuretics, mechanism of action 63
Thiazol-2-ylamine, bioisosterism 660T
Thiazoles, GABA receptors 893
Thiazolidinediones 1021–1022
  bioisosterism 662T
  carboxylic acid bioisosterism 661
  mechanism of action 1081
6H-Thieno[2,3-b]pyrrole, bioisosterism 660T
Thiol reactivity index, promiscuous ligands, computational prediction 749
Thiorphan
  inhibition values 677F
  structure 677F
Thiourea, 'functional analogs' 668
Thoughtful responsiveness (attribute), project teams 145
3-Dimensional Pharma 760T
3D-structure, messenger RNA (mRNA) 1044
Thrombin
  peptide/protein drugs 577
  three-dimensional structures 626T
Thrombin inhibitors
  nonpeptide drugs 628–629, 631F, 632F
  peptidomimetics 613–614, 614F, 616F, 628–629, 631F, 632F
Thrombin receptor, activation 785T
Thromboxane A_2 receptors, in human disease 775T
Thromboxane A_2 synthetase inhibitors, ring equivalent bioisosterism 655
Throughput, NMR automation 530

Thyroid gland, phosphodiesterase-8 930
Thyroid hormone receptor (TR) 1008
  ligands 1019, 1020F
  nomenclature 996T
  subtypes 1008
  see also specific drugs
Thyroid-stimulating hormone (TSH) receptor
  activation 785T, 798
  human disease 792T
Thyroliberin 579T
Thyrotropin-releasing hormone (TRH) receptor, activation 785T
Thyrotropin-releasing hormone receptor agonists, peptidomimetics 608–609, 611F
Tianeptine, structure 62F
Tight junctions, intranasal drug administration 595
Time, outsourcing scale-up 211
Timeliness, career guidelines 19
Time-of-flight (ToF) detectors, matrix-assisted laser desorption/ionization (MALDI)-MS 339
Time-of-flight mass spectrometry, protein display chips 374
Time-resolved fluorescence-based assays, GPCR functional assay screening 807, 807T, 809
Time saving, outsourcing 228, 229F
Timolol, GPCR targeting 773T
Tipranavir, structure 302F
Tissue distribution, sodium channels 840
Tizanidine, GPCR targeting 773T
TM3-TM4 intracellular loop, nicotinic acetylcholine receptors 885–886
TNP-ATP, structure 906F
Tobramycin
  applications 1042
  structure 1041–1042
Tolboxane, carbon-boron bioisosterism 698
Tolterodine, GPCR targeting 773T
Topiramate
  characteristics 829T
  structure 829T
Topoisomerase inhibitors, DNA therapeutic drugs 1040
Topological similarities, scaffold hopping 683–685, 685F
Toremifene
  as estrogen receptor ligand 1014
  structure 1013F
Torsemide, bioisosterism 668–669, 670F
Torsion angles, peptides 604
Tourette's syndrome, 5-HT$_3$ receptors 891

Toxicity
  carbon-boron bioisosterism see Bioisosterism, carbon–boron
  DCN 154
  ideal drug characteristics 564T
  selenium bioisosterism 701
Toxicogenomics 568
  cost reduction 198
  drug development 197
  goals 568–569
  QSAR combination 568
  SAR combination 568
Toxicology
  carbon-silicon bioisosterism 694
  drug development productivity 130
  lead optimization 139
  preclinical trials 178, 179T
  predevelopment 141
Toxins, acid-sensitive ion channels (ASICs) 911
TPMA, structure 893F
Trade-Related Aspects of Intellectual Property (TRIPS) 94, 211
Tramadol, mechanism of action 773T
Transcription
  DNA 1037
  interference, DNA therapeutic drugs 1039
Transcription factors, modular design 301T, 323–324
Transcriptomics, drug development 196
Transfer RNA (tRNA), structure 1037
Transfluor technology, GPCR functional assay screening 813–814
Transforming growth factor-β (TGF-β), cancer inhibition, animal models 388–389
Transforming growth factor-β receptor kinases 977
  see also specific kinases
  inhibitors 977–979, 979F
  PDB entries 965T
TransForm Pharmaceuticals 232
Transient receptor potential (TRP) ion channels 879, 907
  diseases 909
    allodynia 909–910
    Crohn's disease 909–910
    cystic fibrosis 909–910
    hearing impairments 909–910
    hyperalgesia 909–910
    IBD 909–910
    inflammation 909–910
    mucolipidosis type IV 909–910
    polycystic kidney disease 909–910
  expression 879
  molecular biology 907
  nomenclature 907
  pharmacology 908, 909F
  physiological functions 908
    sensory receptors 908
    temperature sensors 908

regulation 908
structure 880F, 907
therapeutics 908
  capsaicin-derived products 909
  inflammation 908–909
  pain 908–909
Transition metal complexes, DNA therapeutic drugs 1040
Transition metal-mediated reactions, solid support synthesis *see* Solid support synthesis
Transmembrane domains
  GPCRs 774–777
  voltage-gated ion channels 828
Transmembrane voltage, calcium channels 846–847
Transparency
  culture 122–123
  decision making *vs.* organizational alignment 120
Transporter binding, antidepressant drugs 62
Transporters
  see also specific transporters
  classification 59
  drug delivery 66, 66F
  drug interactions 61
  expression 56
  genetic diseases 66
  imaging 63
    positron emission tomography 65
  information 58T
  mechanism of action 57, 58F
  nomenclature 58
  physiological function 60
  structure 59
  as targets 56, 57
  therapeutics 62
  see also specific drugs
Transverse relaxation-optimized spectroscopy (TROSY), NMR 514
Tranylcypromine, structure 68F
Traxoprodil 903
Trazodone
  GPCR targeting 773T
  structure 62F
Trega 760T
Tretinoin, characteristics 1012T
2′.4.4′-Trichloro-2′-hydroxydiphenylether (TRN), mitochondria, drug targeting *see* Mitochondria, drug targeting
Tricyclic antidepressants (TCAs)
  characteristics 64T
  structure 63F
Tricyclic oxindoles, phosphodiesterase-2 inhibitors 933
Tricyclic systems, ring system reorganization (bioisosterism) 675
Trifluoperazine, structure 1062F

Triiodothyronine (T_3), structure 1020F
Trimethylsilyl group, reactivity 695
m-Trimethylsilyl phenyl N-methylcarbamate, structure 695F
Tripelennamine
  analogs, fluorine-hydrogen bioisosterism 691, 692F
  structure 654F
Tripeptidyl peptidase-II (TTP-II) inhibitors, peptidomimetics 614, 616F
Triptorelin 579T
Triton X-100, promiscuous ligands 738, 740
Troglitazone, mechanism of action 1081
Tropisetron
  5-HT$_3$ receptor binding 890
  structure 895F
Truncation, peptide/protein drugs 576–577
Trypsin, three-dimensional structures 626T
Tsk tyrosine kinase, GPCR interaction 786T
TTN, PDB entries 965T
TTX$_s$ subtypes, sodium channels 833–834
T-type calcium channels 842
Tubocurarine, structure 887F
Tubulin, GPCR interaction 786T, 797T
Tumor necrosis factor-α converting enzyme inhibitors, carboxylic acid bioisosterism 659
Tumors
  expression profiling 311
  mitochondria, drug targeting 1067
Tween-20, promiscuous ligands 740
2D-PAGE 331
  imaging/image analysis 335
    autoradiography 335
    CCD devices 335
    data management 335–336
    QUEST system 335
    statistical analysis 335–336
  isoelectric focusing (IEF) (first dimension) 331–332
    ampholyte carriers *vs.* IPG gels 332
    sample preparation 332
  protein expression mapping 384–385
  protein spot excision 336
  proteomics 314
  SDS-PAGE (second dimension) 332–333
  staining 333
    Coomassie stains 333
    fluorescent stains 334
    silver staining 333

Two dimensional nuclear magnetic resonance *see* Nuclear magnetic resonance (NMR)
'Two-state model,' GPCR activation models 792, 794F
Tyrosine kinases 967
  see also specific kinases
  oncogenic 960–962
  structure 967F
  three-dimensional structures 636T
Tyrosine phosphatases 981
  see also Phosphotyrosine-interacting domains; *specific types*
  inhibitors 981–982, 982F
  nonpeptide drugs 638, 640F
  peptidomimetics 638, 640F
  three-dimensional structures 636T

U

Ubiquinone (coenzyme Q) 1073
  biology 1073
    as antioxidant 1073
    electron shuttle 1073
  drug development 1074
    Friedreich's ataxia treatment 1074
    idebenone 1074
    mevalonic acid 1074
    Parkinson's disease treatment 1074
    simvastatin 1074
    statins 1074
  pathology 1074
    encephalomyopathies 1074
  structure 1073F, 1074F
UCN-01
  CDK family kinase inhibitors 979–980
  characteristics 984T
  structure 979F
Ulcerative colitis (UC), phosphodiesterase-4 inhibitors 935–936
Ultrahigh-throughput screening, HTS data analysis 459
Ultra-High-Throughput Screening System (Aurora Biosciences) 436
  assay development 437
    fluorescence resonance energy transfer 437
    green fluorescent protein 437
  assay examples 447, 448F
  assay loading 438
  compound distribution 438
  detection 439
  fluorescence detection 445, 446F
  incubation 439
  key technologies 436
  microfluidics 441, 442F
    dispensing accuracy 442–443, 443F

Ultra-High-Throughput Screening System (Aurora Biosciences) (*continued*)
  multiple well dispensing 444, 444F
  plate design/performance 440, 441F
    cell-substrate adhesion 431, 441F
  results database 440
  solenoid-controlled liquid dispensing 444, 444F
Ultraprecision machining, lithography/photolithography 354
Ultraviolet (UV) radiation, LIGA, lithography/photolithography 354
Uncontrolled trials 182
Uncoupling protein-1 (thermogenin) 1080
Uncoupling proteins *see* Oxidative phosphorylation uncoupling
Universities
  academic environment 89–90
  challenges 91
  changes 91, 94
  commercial activities 93
  definition 89–90
  history 90
  science, role in 91–92
  'gift economy model' 91–92, 92F
  'use-inspired research' 91
University-based providers, definition 206
Unnatural amino acids
  peptidomimetics 606–608
  proteolysis resistance 584
Unnatural oligomers, solid support synthesis 478, 480F
Unnatural peptide backbone, proteolysis resistance 584–585
Untranslated regions (UTRs), messenger RNA (mRNA) 1044–1045
Upstream conserved regions (UCRs), phosphodiesterase-4 924–925
Urea, 'functional analogs' 668
Uric acid, structure 67F
Uridine, structure 1038F
Urotensin II receptors
  activation 785T
  in human disease 775T
USA
  dominance 94
  markets 7
US Adopted Name (USAN) 161–162
'Use-inspired research,' universities 91
U-shaped value curve, business development 25–26, 25F, 26F

V

Vaccines, pharmaceutical companies 100
Valdecoxib, structure 658F, 681F

Valium, invention of, Sternbach, Leo 256–257
Valsartan, mechanism of action, GPCR targeting 773T
Value chain migration, outsourcing 212
Value drivers, pharmaceutical industry organization 106–108
Value per drug 124
Vancomycin, structure 46F
Vardenafil 927, 938–939
  as phosphodiesterase inhibitor 947T
  quinazoline derivatives 939
Varenicline
  nicotinic acetylcholine receptor binding 888
  structure 887F
Variable data quality, bioinformatics 317–318
Vascular endothelial growth factor receptor family kinases 969
  see also specific kinases
  angiogenesis 969
  in disease 963T
  inhibitors 969, 970F
Vascular tone regulator, phosphodiesterase-5 928
Vasoactive intestinal peptide (VIP), as drug
  N-terminus modification 583
  structural motif replacement 590
Vasopressin
  analogs 580
  as peptide/protein drugs 579T
  intranasal administration 593–594
Vasopressin 1B receptors, in human disease 775T
Vasopressin V1 receptor antagonists 618–619, 620F
Vasopressin V2 receptors, in human disease 775T
Vatalanib *see* PTK-787
Veblen, Thorstein, universities in commerce 93
Venlafaxine, structure 62F
Venture capital, biotechnology company funding 8
Verapamil
  characteristics 829T
  structure 75F, 829T
Vertex Pharmaceuticals 226T, 761T
Vertical structure, R&D management 249
Vesicular monoamine transporters (VMATs) 59, 60T
Veterinary products, pharmaceutical companies 100
VH1, three-dimensional structures 636T
Viagra *see* Sildenafil citrate
Vibrational circular dichroism, chiral drug analysis 720
Vinpocetine
  clinical trials 947

  phosphodiesterase-1 inhibition 931–932
Vinyl fluorides, carboxamide bioisosterism 668, 669F
Virtual companies *see* Pharmaceutical industry *vs.* biotech, drug discovery; Start-up companies
Visual impairments, calcium channels 853
Vitamin D receptor (VDR) 1007
  bone mineralization 1007
  calcium homeostasis 1007
  ligands 1018, 1018F
  see also specific drugs
  nomenclature 996T
  retinoid X receptor (RXR) heterodimerization 1007
Voltage-gated ion channels *see* Ion channels, voltage-gated

W

Water, production, mitochondria complex IV 1077
Water solubility, hydrophilic depoting 587
Water (solvent) suppression, NMR 514
Web-based programs, raw MS/MS data searching 346
Weight-loss drugs, oxidative phosphorylation uncoupling 1079
Well types, HTS data acquisition 451
White, Tom, management strategies 261
Winter, Jill, peptoid-based antibiotics, discovery of 259
Wittig reaction, solid support synthesis 468–469, 469F
Women's Health Initiative, Memory Study 53
Working climate
  effective leaders 242
  ineffective leaders 242
World Trade Organization (WTO), intellectual property protection 131
Written project reports, project management teams 148
Wyeth, R&D spending 101T

X

XL665, homogeneous time-resolved fluorescence 413, 414F
X-ray crystallography
  peptidomimetics 622–623
  protein kinases 964, 965T, 966F
  signal transduction 964

Src homology-2 domains 634, 637F, 638F
Src kinases 964, 966F
X-ray diffraction, chiral drug analysis 720
X-ray LIGA see Lithography/photolithography

Y

Yes kinase, in disease 963T

Z

Zactima see ZD-6474
Zaleplon
   scaffold hopping 682F
   structure 682F
Zap70 SH2, three-dimensional structures 636T
Zaprinast
   derivatives, phosphodiesterase-1 inhibitors 932
   phosphodiesterase-11 inhibitors 931
ZD-1839 see Gefitinib (ZD-1839)
ZD-6474 969–970
   characteristics 984T
   KIT inhibitor 972
   structure 970F
   VEGFR family kinase inhibitors 969
Zidovudine (azidothymidine: AZT), mitochondria, effects on 1082
Zifrosilone, structure 695F
Zinc finger proteins, expression profiling 312
Zinc-gated ion channel 896
   characteristics 881T
ZK-159222
   structure 1018F
   as vitamin D receptor ligand 1019
ZK-216348 1016–1017
   structure 1017F
Zoladex, controlled release formulations 592
Zolpidem
   scaffold hopping 682F
   structure 652F, 682F
Zonisamide 850
   characteristics 829T
   structure 829T
Zoom gels, protein expression mapping 385
Zopiclone
   scaffold hopping 682F
   structure 652F, 682F